Genomic and Personalized Medicine
Second Edition

Genomic and Personalized Medicine

Volume 2
Second Edition

Edited by

Geoffrey S. Ginsburg, M.D., Ph.D.
Director, Genomic Medicine
Duke Institute for Genome Sciences & Policy
Executive Director, Center for Personalized Medicine
Duke University Health System
Professor of Medicine
Duke University School of Medicine
Durham, North Carolina 27710

and

Huntington F. Willard Ph.D.
Institute Director
Duke Institute for Genome Sciences & Policy
Nanaline H. Duke Professor of Genome Sciences
Duke University
Durham, North Carolina 27708

AMSTERDAM • BOSTON • HEIDELBERG • LONDON • NEW YORK • OXFORD
PARIS • SAN DIEGO • SAN FRANCISCO • SYDNEY • TOKYO
Academic Press is an imprint of Elsevier

Academic Press is an imprint of Elsevier
32 Jamestown Road, London NW1 7BY, UK
225 Wyman Street, Waltham, MA 02451, USA
525 B Street, Suite 1800, San Diego, CA 92101-4495, USA

First edition 2009
Second edition 2013

Notice

No responsibility is assumed by the publisher for any injury and/or damage to persons
or property as a matter of products liability, negligence or otherwise, or from any use
or operation of any methods, products, instructions or ideas contained in the material
herein. Because of rapid advances in the medical sciences, in particular, independent
verification of diagnoses and drug dosages should be made

British Library Cataloguing-in-Publication Data
A catalogue record for this book is available from the British Library

Library of Congress Cataloging-in-Publication Data
A catalog record for this book is available from the Library of Congress

ISBN : 978-0-12-382227-7 (set)
ISBN : 978-0-12-415938-9 (vol. 1)
ISBN : 978-0-12-415937-2 (vol. 2)

For information on all Academic Press publications
visit our website at www.elsevierdirect.com

Typeset by MPS Limited, Chennai, India
www.adi-mps.com

Printed and bound in United States of America

12 13 14 15   10 9 8 7 6 5 4 3 2 1

Working together to grow
libraries in developing countries

www.elsevier.com | www.bookaid.org | www.sabre.org

ELSEVIER   BOOK AID
            International   Sabre Foundation

Table of Contents

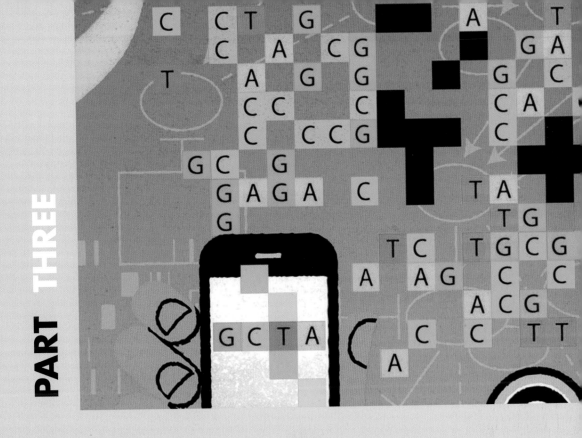

Disease-Based Genomic and Personalized Medicine: Genome Discoveries & Clinical Significance

Cardiovascular Genomic Medicine

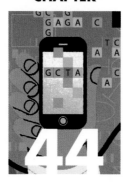

Hypertension

Patricia B. Munroe and Toby Johnson

INTRODUCTION

Elevated blood pressure (BP) or hypertension [$\geq$140 mmHg systolic blood pressure (SBP) and/or $\geq$90 mmHg diastolic blood pressure (DBP)] is highly prevalent, affecting 26.4% of people aged 20 or older worldwide (Kearney et al., 2005), and is the leading global cause of preventable death. The World Health Organization (WHO) estimated that in 2004 hypertension accounted for 13% of deaths worldwide; 16.8% in high-income countries and 7.5% in low-income countries, due mainly to coronary heart disease (CHD) and stroke (WHO, 2009). Other complications associated with hypertension include heart failure, peripheral vascular disease, renal impairment, retinal hemorrhage, and visual impairment (WHO, 2002). Prospective observational studies show that the risk of cardiovascular disease (CVD) increases in a roughly linear fashion across the normal population range of BP (Chobanian et al., 2003), and that in older age groups, the risk of cardiovascular disease doubles for each increment of 20 mmHg SBP and 10 mmHg DBP, starting as low as 115 mmHg SBP and 75 mmHg DBP.

Existing anti-hypertensive treatments are effective at the population level, reducing the risk of developing CHD events (fatal and nonfatal) by 25% and the risk of stroke by 33%, per 10 mmHg reduction in SBP and 5 mmHg reduction in DBP achieved, independent of pre-treatment BP level (Law et al., 2009). However, at the individual level, BP is often poorly controlled, and many patients do not achieve <140 mmHg SBP and <90 mmHg DBP targets. In the 2006 Health Survey for England, only 28% of patients achieved this target, even

though most were prescribed two or more anti-hypertensive drugs (Falaschetti et al., 2009). For many years, the first-choice therapies were beta-blockers and thiazide diuretics, which were developed more than 50 years ago (Borhani, 1959; Prichard, 1966). More recently angiotensin-converting enzyme inhibitors (ACEi), calcium channel blockers (CCBs), and angiotensin II receptor antagonists (ARBs) have become increasingly preferred, and current guidelines in the United Kingdom recommend ACEi or ARBs as first-line treatment for individuals <55 years, and CCBs prescribed for individuals >55 years and those of African or Caribbean ancestry (National Institute for Health and Clinical Excellence, 2011). The ACEi, ARBs, and thiazide diuretics all target the renin-angiotensin-aldosterone system (RAAS), which is a hormonal pathway that maintains normal BP and blood volume (Nguyen Dinh Cat and Touyz, 2011). There are very few new drugs in development for lowering BP in the general population, and only one new therapy has been granted regulatory approval in the past three years. Aliskirin is a first-in-class oral renin inhibitor that also targets the RAAS, and clinical trial data indicate effects on BP comparable to ACEi and ARBs (Musini et al., 2008).

The main causes of hypertension are well known. Lifestyle and genetic effects are both influential. The most important lifestyle risk factors are excess dietary sodium intake, body weight, alcohol consumption, stress, and lack of exercise (Chiong, 2008). Evidence for a genetic component comes from studies of families and twins, and suggests that the heritability (the fraction of BP variance contributed by genetic factors) for both SBP and DBP is between 30% and 50% (Havlik et al.,

Genomic and Personalized Medicine, 2nd edition
by Ginsburg & Willard. DOI: http://dx.doi.org/10.1016/B978-0-12-382227-7.00044-6

1979). However, heritability studies do not identify which genetic differences are important or by what mechanisms they exert their effects on BP. Recent advances in human genetics offer the opportunity to discover hitherto-unknown mechanisms and pathways affecting BP, which could, in principle, be targeted by novel therapeutic approaches and thus improve treatment of hypertension and prevention of CVD.

FINDING BLOOD PRESSURE GENES

Identification of the genetic basis of BP has been a longstanding and challenging research objective (Wallace et al., 2007). Mutations in specific genes causing rare, monogenic forms of hypertension have been successfully identified using linkage analysis and positional cloning. These mutations are in genes primarily expressed in the kidney, and affect salt/water homeostasis (Lifton et al., 2001). Although linkage analysis has had some success in mapping genes for other complex diseases, including type 1 diabetes and Crohn's disease (Brant and Shugart, 2004; Concannon et al., 2005), the use of linkage-based methods has not been successful for finding genes causing essential hypertension (high BP with no obvious medical cause), despite studies of sibling pairs and families with large sample sizes, and up to 400 microsatellite markers across the genome being analyzed (Caulfield et al., 2003; Cowley, 2006; Wu et al., 2006). Alongside genome-wide linkage studies, numerous candidate gene association studies for hypertension have also been carried out, albeit mostly with relatively small sample sizes, and with a general lack of consistency of the replicability of findings (Dominiczak et al., 2004).

However, over the past three years there has been substantial progress, and this success is largely attributable to the advent and rapid technological advances of genome-wide association studies (GWAS) (Hirschhorn and Daly, 2005). Modern GWAS use mass-produced DNA microarrays to genotype 300,000 or more single nucleotide polymorphisms (SNPs), distributed across the whole genome, in each individual. SNPs not directly genotyped can be imputed using correlation structures between multiple SNPs observed in reference panels densely genotyped by the HapMap (International HapMap Consortium, 2003) or 1000G studies (1000 Genomes Project Consortium, 2010). Typically a discovery analysis is conducted, in which hundreds or thousands of individuals are genotyped, and at each SNP (genotyped or imputed) the allele present is tested for association with hypertensive status and/or continuous BP phenotypes. The "BP phenotype" commonly used for GWAS studies is an untreated BP value, or an imputed BP value if the individual is taking anti-hypertensive medication (Tobin et al., 2005). Gender, age, age^2, and body mass index (BMI) are then included as covariates in an analysis, and a correction for population stratification is included if necessary. Associations that are suggestive or significant after multiple testing correction (genome-wide significant) are typically followed up by genotyping in further independent samples of

individuals to establish definitive genome-wide significance (McCarthy et al., 2008). Most GWAS have used unrelated individuals, for which tests of association require straightforward contingency table or linear regression analyses, although it is also possible to apply GWAS methodology in samples of related individuals (Scuteri et al., 2007).

The Wellcome Trust Case Control Consortium (WTCCC) published the first GWAS results for hypertension in 2007 (Wellcome Trust Case Control Consortium, 2007). This study tested for association by comparing 2000 unrelated hypertensive cases, versus 3000 "common controls." The WTCCC study design used the same "common controls" for seven different disease comparisons, which had a potentially important disadvantage because the controls were from general population samples and therefore were not all non-hypertensive. The WTCCC did not find any genome-wide significant SNPs for hypertension, suggesting common genetic variants affecting risk for hypertension were likely to have relatively small effect sizes and/or were not covered by the Affymetrix 500K genotyping arrays used.

A number of subsequent GWAS tested association with either hypertension or with SBP and DBP as continuous outcomes, and reported a significant association for SBP at SNPs on chromosome 2q24.3, in the STK39 (serine threonine kinase 39) gene (Wang et al., 2009), a significant association for DBP and hypertension at a SNP on chromosome 16q23.3, upstream of the CDH13 (adhesion glycoprotein T-cadherin) gene (Org et al., 2009), and a significant association for SBP and DBP at a SNP near the ATP2B1 (ATPase calcium transporting, plasma membrane 1) gene (ATP=adenosine triphosphate), reported by the Korean Association Resource (KARE) project (Cho et al., 2009). Only variants near ATP2B1 have been replicated in subsequent studies with much larger sample sizes, suggesting that the associations near STK39 and near CDH13 are either false positives or are population-specific effects.

At the same time, Newton-Cheh and colleagues reported results from a candidate gene association study for two BP biomarkers, atrial natriuretic peptide (ANP) and brain natriuretic peptide (BNP), which are peptides with known vasodilatory properties. They tested SNPs near the atrial natriuretic peptide precursor (NPPA) and brain natriuretic peptide precursor (NPPB) genes, and for two SNPs they found significant association with increased plasma ANP and BNP levels, and, at the same two SNPs, association with decreased SBP and DBP (Newton-Cheh et al., 2009b). This study highlights a more general idea that studying intermediate quantitative phenotypes may increase power to detect associations for the disease phenotype(s) that are ultimately of interest (Plomin et al., 2009).

The modest findings of individual GWAS for BP traits meant that a natural next step was to meta-analyze results from multiple GWAS, and thus achieve effectively much larger sample sizes. The greater power of such studies has led to the identification of 40 distinct loci, each harboring one or more genetic variants with robust and validated association with BP traits (Table 44.1).

TABLE 44.1	Summary of genetic variants robustly associated with blood pressure from peer-reviewed publications

SNP	Locus nickname	Chr	Base pair position (NCBI build 36)	MAF	Discovery cohort ethnicity	Reference
GWAS of SBP and DBP						
*rs17249754	ATP2B1	12	88,584,717	0.37	East Asian	Cho et al., 2009
Intermediate phenotype analysis						
*rs5068	NPPA/NPPB	1	11,828,511	0.06	European	Newton-Cheh et al., 2009b
Meta-analyses of GWAS of SBP, DBP, and hypertension						
Global BPgen Consortium						
rs17367504	MTHFR-NPPB	1	11,785,365	0.14	European	Newton-Cheh et al., 2009a
*rs11191548	CYP17A1-NT5C2	10	104,836,168	0.09	European	
*rs16998073	FGF5	4	81,541,520	0.21	European	
*rs1530440	c10orf107	10	63,194,597	0.19	European	
*rs653178	SH2B3	12	110,470,476	0.47	European	
*rs1378942	CYP1A1-CSK	15	72,864,420	0.35	European	
*rs16948048	ZNF652	17	44,795,465	0.39	European	
CHARGE Consortium						
rs1004467	CYP17A1	10	104,584,497	0.10	European	Levy et al., 2009
*rs381815	PLEKHA7	11	16,858,844	0.26	European	
rs2681492	ATP2B1	12	88,537,220	0.20	European	
rs3184504	SH2B3	12	110,368,991	0.47	European	
*rs9815354	ULK4	3	41,887,655	0.17	European	
*rs11014166	CACNB2	10	18,748,804	0.34	European	
*rs2384550	TBX3-TBX5	12	113,837,114	0.35	European	
rs6495122	CSK-ULK3	15	72,912,698	0.42	European	
*rs880315	CASZ1	1	10,719,453	0.36	East Asian	Takeuchi et al., 2010
AGEN-BP study						
rs17030613	ST7L-CAPZA1	1	112,971,190	0.47	East Asian	Kato et al., 2011
*rs16849225	FIGN-GRB14	2	164,615,066	0.40	East Asian	
*rs6825911	ENPEP	4	111,601,087	0.48	East Asian	
*rs1173766	NPR3	5	32,840,285	0.38	East Asian	
*rs11066280	RPL6-ALDH2	12	111,302,166	0.22	East Asian	
rs35444	TBX3	12	114,036,820	0.25	East Asian	

(continued)

TABLE 44.1	(Continued)						

SNP	Locus nickname	Chr	Base pair position (NCBI build 36)	MAF	Discovery cohort ethnicity	Reference
ICBP-GWAS for SBP and DBP						
*rs2932538	MOV10	1	113,018,066	0.25	European	Ehret et al., 2011
*rs13082711	SLC4A7	3	27,512,913	0.22	European	
*rs419076	MECOM	3	170,583,580	0.47	European	
*rs13107325	SLC39A8	4	103,407,732	0.05	European	
*rs13139571	GUCY1A3-GUCY1B3	4	156,864,963	0.24	European	
rs1173771	NPR3-C5orf23	5	32,840,285	0.40	European	
*rs11953630	EBF1	5	157,777,980	0.37	European	
*rs1799945	HFE	6	26,199,158	0.14	European	
*rs805303	BAT2-BAT5	6	31,724,345	0.41	European	
rs1813353	CACNB2(3')	10	18,747,454	0.45	European	
*rs932764	PLCE1	10	95,885,930	0.44	European	
*rs7129220	ADM	11	10,307,114	0.19	European	
*rs633185	FLJ32810-TMEM133	11	100,098,748	0.28	European	
*rs2521501	FES-FURIN	15	89,238,392	0.31	European	
*rs17608766	GOSR2	17	42,368,270	0.14	European	
*rs1327235	JAG1	20	10,917,030	0.46	European	
*rs6015450	GNAS-EDN3	20	57,184,512	0.12	European	
rs17367504	MTHFR-NPPB	1	11,785,365	0.15	European	
rs3774372	ULK4	3	41,852,418	0.17	European	
rs1458038	FGF5	4	81,383,747	0.29	European	
rs4373814	CACNB2 (5')	10	18,459,978	0.45	European	
rs4590817	C10orf107	10	63,137,559	0.16	European	
rs11191548	CYP17A1-NT5C2	10	104,836,168	0.09	European	
rs381815	PLEKHA7	11	16,858,844	0.26	European	
Rs17249754	ATP2B1	12	88,584,717	0.16	European	
rs3184504	SH2B3	12	110,368,991	0.47	European	
rs10850411	TBX5-TBX3	12	113,872,179	0.30	European	
rs1378942	CYP1A1-ULK3	15	72,864,420	0.35	European	
rs12940887	ZNF652	17	44,757,806	0.38	European	

(continued)

TABLE 44.1 (Continued)						
SNP	**Locus nickname**	**Chr**	**Base pair position (NCBI build 36)**	**MAF**	**Discovery cohort ethnicity**	**Reference**
ICBP-GWAS for PP and MAP						
rs13002573	FIGN	2	164,623,454	0.20	European	Wain et al., 2011
rs1446468	FIGN	2	164,671,732	0.47	European	
*rs319690	MAP4	3	47,902,488	0.49	European	
*rs871606	CHIC2	4	54,494,002	0.15	European	
*rs2071518	NOV	8	120,504,993	0.17	European	
*rs17477177	PIK3CG	7	106,199,094	0.28	European	
rs2782980	ADRB1	10	115,771,517	0.19	European	
*rs11222084	ADAMTS8	11	129,778,440	0.37	European	
Extreme case/control design						
*rs13333226	UMOD	16	20,273,155	0.17	European	Padmanabhan et al., 2010
Women's Genome Health Study						
*rs2898290	BLK-GATA4	8	11,471,318	0.47	European	Ho et al., 2011
Candidate genes – subset of GWAS						
*rs2004776	AGT	1	228,915,325	0.24	European	Johnson et al., 2011
*492	ADRB1	10	115,795,046	0.27	European	

This table lists all significantly associated SNPs with blood pressure per study. * indicates the study which reported the first association at this locus, the "40 independent blood pressure loci." Some studies found associations at the same locus, but the index SNP reported differed. In some cases the reported SNP was in high LD, for others it was an independent signal at a locus. The "locus nickname" indicates the genes nearest to the associated genetic variant as described in the original papers. Chr = chromosome, bp = base pair, MAF = minor allele frequency.

Meta-analyses of GWAS to Discover Blood Pressure Genes

The first large-scale meta-analyses of GWAS results for SBP, DBP, and case/control hypertension were published in May 2009 by two large international consortia, Global BPgen (GBPG) and the Cohorts for Heart and Aging Research in Genome Epidemiology (CHARGE) (Levy et al., 2009; Newton-Cheh et al., 2009a). Both consortia analyzed approximately 2.5 million SNPs (directly genotyped and imputed) in large numbers of individuals of European ancestry (GBPG N = 34,433 and CHARGE N = 29,136), and both followed up their top 10 independent signals from each scan by performing a simultaneous reciprocal exchange of association results, with the GBPG consortium also following up 12 significant signals by direct genotyping in a further 71,225 individuals of European ancestry.

Each consortium identified genome-wide significant ($P < 5 \times 10^{-8}$) SNP associations at eight distinct loci. Of these, three loci were simultaneously discovered by both consortia, and one association identified by CHARGE was the same as the previously discovered association at the ATP2B1 locus (Cho et

al., 2009); hence, in total, 12 novel associations were discovered (Table 44.1). The majority of the genome-wide significant SNPs were associated with both SBP and DBP and odds of hypertension, with the same direction of effect. The associated SNPs were common, with a minor allele frequency (MAF) >5%, and the effect sizes were ≤1.0 mmHg for SBP and ≤0.5 mmHg for DBP.

Since the GBPG and CHARGE meta-analyses, there has been a steady flow of new BP loci discoveries. Association at a SNP located near CASZ1, which had showed suggestive association in the CHARGE meta-analysis results, was replicated in samples of East Asian (Japanese) ancestry (Takeuchi et al., 2010). Combining the CHARGE meta-analysis results with data from a single, extremely large cohort of 23,019 individuals from the Women's Genome Health Study (WGHS) led to the discovery of an association near the BLK-GATA4 genes (Ho et al., 2011). This association is located on chromosome 8 in a large polymorphic inversion, in a linkage disequilibrium (LD) block spanning many genes, and allelic imbalance has also been reported at this locus (Nusbaum et al., 2006; Wagner et al., 2010). The Asian Genetic Epidemiology Network Blood Pressure (AGEN-BP) GWAS study meta-analyzed GWAS results from 19,608 individuals of East

Asian ancestry, followed up top hits in a further 31,000 individuals also of East Asian ancestry, and discovered six significant associations with BP, near *ST7L/CAPZA1*, *FIGN/GRB14*, *ENPEP*, *NPR3*, *ALDH2*, and *TBX3* (Kato et al., 2011). The association at the *TBX3* locus was independent ($r^2 = 0.001$ in Utah residents with Northern and Western European ancestry (CEU)) to the variant previously reported by CHARGE (Levy et al., 2009).

Padmanabhan and colleagues performed an "extreme case/control" GWAS comparing highly selected hypertensive cases (top 2% of the population) versus controls selected for low BP and low occurrence of cardiovascular events (bottom 9% of the population) (Padmanabhan et al., 2010). Despite genotyping a discovery sample of fewer than 4000 individuals, the stringency of the ascertainment scheme meant that this study discovered a novel SNP at *UMOD*, which was validated by follow-up in a further 36,386 individuals. The SNP allele associated with higher risk of hypertension had previously been observed to be also associated with impaired renal function (as measured by estimated glomerular filtration rate) and higher risk for chronic kidney disease (Kottgen et al., 2009). Johnson and colleagues selected 30 genes that were known targets for anti-hypertensive drugs and tested SNPs in this "candidate gene" set for association with BP and hypertension, using the CHARGE meta-analysis results. With follow-up using GBPG meta-analysis and WGHS GWAS results, genome-wide significant associations at two loci, angiotensinogen (*AGT)* and the beta-adrenergic receptor 1 (*ADRB1*) were discovered (Johnson et al., 2011). The association at *AGT* was the same as the one previously reported by Watkins and colleagues in a smaller candidate gene study (Watkins et al., 2010).

Although there is wide variation in study designs (e.g., studies in populations with different ethnic ancestries, cases/controls ascertained from the extremes of the population BP distribution, or testing only "candidate gene" subsets of GWAS data), it is unclear whether the new discoveries are being made as a direct consequence of the different study designs, or are merely a reflection of the fact that all genetic association studies for BP have been somewhat underpowered. Hence, each study (regardless of design) would detect a more-or-less random subset of the likely hundreds of genetic associations with individually small effect sizes (Park et al., 2010). The latter hypothesis is supported by the observation that GBPG and CHARGE, two large studies with very similar overall designs, had relatively little overlap in the loci that reached discovery significance thresholds (Munroe et al., 2009).

Many of the genetic associations just described have subsequently been replicated in samples of different ancestry from where they were initially discovered, including in East Asian (Hong et al., 2010a, b; Kato et al., 2011; Liu et al., 2011; Niu et al., 2010; Takeuchi et al., 2010), South Asian (Newton-Cheh et al., 2009a), and African ancestries (Fox et al., 2011; Zhu et al., 2011).

International Consortium for Blood Pressure Genome-wide Association Studies

In the autumn of 2008, GBPG and CHARGE joined forces, and with additional GWAS studies formed a new BP genetics consortium, the International Consortium for Blood Pressure Genome-wide Association Studies (ICBP-GWAS). This consortium evaluated associations between 2.5 million SNPs and SBP and DBP, and a second project focused on mean arterial pressure (MAP) and pulse pressure (PP). MAP and PP are calculated from SBP and DBP, but testing for association with MAP may increase power to detect genetic variants that influence both SBP and DBP with concordant effects, and testing for association with PP increases power to detect some genetic variants that influence SBP and DBP with discordant effects.

The meta-analysis of GWAS for SBP and DBP was performed in 69,395 individuals of European ancestry, and was followed by a three-staged validation study using 133,661 additional individuals of European ancestry (Ehret et al., 2011). A total of 29 SNPs at 28 loci were found to be significantly associated with SBP and DBP (all with $P < 5 \times 10^{-9}$), of which 16 of the associations were novel findings (Table 44.1), with one of the associations (at *NPR3*) discovered simultaneously by AGEN-BP (Kato et al., 2011). Analyses in this enlarged dataset did not support the association at the *PLCD3* locus that was reported previously by GBPG, illustrating that even large GWAS meta-analyses are susceptible to false positive results (which are controlled at 5% *per phenotype* per study by the conventional genome-wide significance threshold). The effect sizes of the new variants were similar to findings from other GWAS BP studies (mostly ≤1 mmHg for SBP and ≤0.5 mmHg for DBP), and observed directions of effect were concordant for SBP, DBP, and odds of hypertension.

A meta-analysis of GWAS for MAP and PP was performed in 74,064 individuals of European ancestry (Wain et al., 2011). Using "look-ups" in GWAS results from a further 48,607 individuals meant that a larger number of SNPs (99 in total) selected at a less stringent threshold ($P < 1 \times 10^{-5}$) could be followed up and potentially validated. This strategy revealed eight genome-wide significant associations; five for PP near *FIGN*, *CHIC2*, *PIK3CG*, *NOV*, and *ADAMTS8*, and three for MAP near *FIGN*, *ADRB1*, and *MAP4*. The SNPs near *FIGN* associated with PP and with MAP are two independent SNPs not in linkage disequilibrium (LD) in Europeans ($r^2 = 0.054$ in HapMap CEU), suggesting possible multiple causal variants. One of the SNPs near *FIGN* is the same as recently reported by Kato and colleagues (2011). The association at *ADRB1* is not in strong LD ($r^2 = 0.14$ in CEU) with the variant reported previously by Johnson and colleagues (2011).

Previous work had shown that some of the 13 significant associations discovered by GBPG and CHARGE in European ancestry samples also showed significant associations in samples of non-European ancestries (Hong et al., 2010a, b; Kato et al., 2011; Liu et al., 2011; Niu et al., 2010; Takeuchi et al., 2010; Zhu et al., 2011). To study this systematically in the largest sample sizes available, the ICBP-GWAS consortium tested the 29 SNPs for association with SBP and DBP in meta-analyses of results from non-European ancestries. Some of the individual SNPs showed significant associations in populations of East- or South Asian ancestry (or both) after correction for multiple testing (Ehret et al., 2011). The general lack of statistically significant associations likely reflects a lack of power due to the fact that the available sample sizes were small compared to the discovery sample size in Europeans. Arguably, the most meaningful analysis of ancestry-specific effects (or lack of effects) is to test whether each SNP is associated with different effect sizes (on a mmHg-per-allele scale) in different ancestries, which is equivalent to testing for a genotype-by-ancestry interaction effect and allows for allele frequency differences between populations (because, for example, SNPs monomorphic in a given population have infinitely wide confidence limits on a mmHg-per-allele scale). Using effect size estimates for European ancestry samples that are free of winners' curse bias, there is no evidence of such ancestry-specific effects (Figure 44.1A and B). This analysis addresses the most biologically informative question, whether there is evidence that an allele would be associated with a different effect size if it was segregating in multiple populations. Hence, on the data currently available, although genotype frequencies do differ between populations of different ancestries, there is no evidence that the expected phenotype given a particular genotype depends on population ancestry (i.e., the biological mechanisms that determine phenotype as a function of genotype are the same for all populations). If this is true more generally (beyond the 29 SNPs studied here by ICBP-GWAS for SBP and DBP), future studies could increase power by combining data from populations of multiple ancestries in a single meta-analysis, as has been done for other phenotypes [Chambers et al., 2010; Coronary Artery Disease (C4D) Genetics Consortium, 2011].

NEW INSIGHTS INTO BLOOD PRESSURE BIOLOGY

To date, large-scale meta-analyses of GWAS and other approaches have revealed 40 genetic loci for BP and hypertension. The genetic variants discovered are all common (MAF > 5%), which is expected because such variants are better covered by GWAS genotyping arrays and because association tests are more powerful for common variants. Of the 29 associations with SBP and DBP reported by ICBP-GWAS, for example, many of the associated SNPs are intergenic, with the nearest gene located several kb (kilobasepairs) away, whereas eight are within genes and are potentially functional as encoding amino acid changes in the protein sequence (non-synonymous SNPs; nsSNPs). Functional mechanisms involving gene regulation are suggested for at least 5/29 SNPs, which are cis-acting expression SNPs (eSNPs) associated with the expressed transcript levels for nearby genes (Ehret et al., 2011). Of all the robustly associated BP loci (Table 44.1), many have at least one biologically plausible gene in the associated interval (e.g., NPPA, NPPB, CYP17A1, GUCY1A3, GUC1B3, NPR3, ADM, GNAS, EDN3, ENPEP, GATA4, ADRB1, and AGT).

Some of the recently reported genetic associations are in or near genes previously suspected to affect BP on the basis of prior functional and physiological experiments. However, despite AGT being a known component of the RAAS and one of the most intensively studied candidate genes, ADRB1 being a known target for beta-blockers, and ADM encoding the adrenomedullin peptide with known vasodilatory properties, the small apparent effect sizes (for all genetic variants tested so far) have meant that very large sample sizes were needed to obtain robust confirmation of an association between BP and common genetic variation at these loci. The GWAS also identified genetic variants associated with BP that are in or near several genes that are part of the natriuretic peptide-guanylate cyclase-nitric oxide signaling pathway. Three distinct SNPs have now been reported near NPPA and NPPB (Newton-Cheh et al., 2009a, b; Tomaszewski et al., 2010), one SNP near the guanylate cyclase α and β subunits (GUCY1A3-GUCY1B3) (Ehret et al., 2011) and two SNPs near the C-type natriuretic peptide receptor (NPR3) (Ehret et al., 2011; Kato et al., 2011).

Many of the newly discovered BP loci do not contain genes that have previously been implicated in BP or cardiovascular disease. However, several of the genes have been studied in other contexts and have known functions that engender novel and potentially testable mechanistic hypotheses. One such example is the hemochromatosis (HFE) gene, where known mutations cause hereditary hemochromatosis, a disorder of iron overload that leads to hepatic cirrhosis and other complications (Pietrangelo, 2010). A second example is solute carrier family 39, member 8 (SLC39A8), which encodes a zinc transporter previously shown to play a role in cellular importation of zinc at the onset of inflammation. Its expression can be induced by tumor necrosis factor alpha (TNF-α) (Besecker et al., 2008), and this protein has also been implicated in cadmium and manganese transport (Himeno et al., 2009). The SNPs in these genes associated with BP are both nsSNPs: H63D in HFE is a low-penetrance allele for hereditary hemochromatosis, and A391T in SLC39A8 has also been associated with high-density lipoprotein levels (Teslovich et al., 2010) and body mass index (Speliotes et al., 2010). Functional studies of these genes and their effects on BP and CVD, using, for example, transgenic knockout mouse model systems, will be an interesting area for future research.

DEVELOPING NEW THERAPIES FOR CARDIOVASCULAR DISEASE

Although each genetic variant exhibits only a modest effect on BP (mostly <1 mmHg for SBP and <0.5 mmHg DBP per risk allele), this does not necessarily correlate with efficacy of a therapeutic agent at the level of the gene product. In this context, the examples of HMG CoA reductase (HMGCR) and cholesterol esterase transfer protein (CETP) are widely cited. Genetic variants in or near these genes are associated with small effects on low-density lipoprotein (LDL) cholesterol and

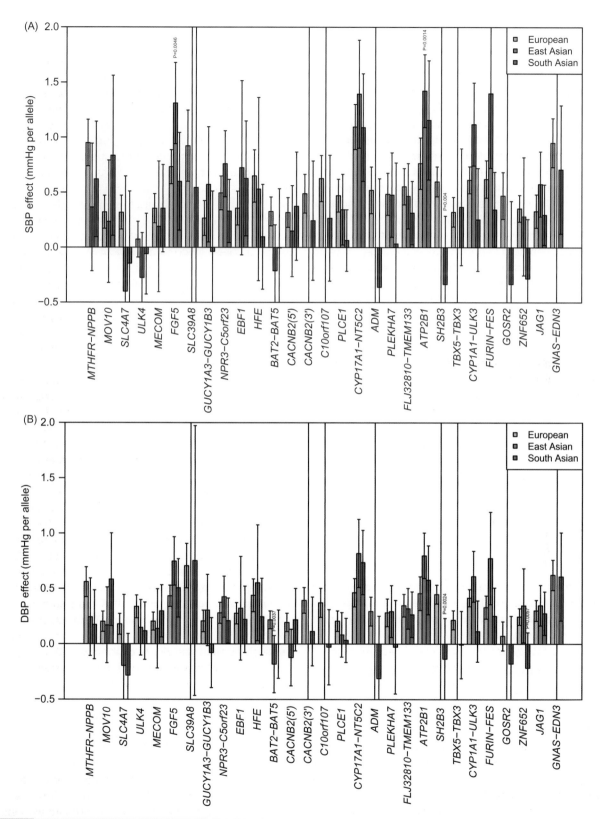

Figure 44.1 Comparison of ancestry-specific effect size estimates. For 29 SNPs associated with BP in European ancestry samples, per-allele effect size estimates are compared with those estimated in East Asian and South Asian ancestry samples. (A) SBP effect, (B) DBP effect. Each bar shows the per-allele effect size with 95% confidence interval (CI) (which can be very wide for SNPs with low MAF). All plotted data are results reported by ICBP-GWAS (Ehret et al., 2011). *P*-values for SNP × ancestry interaction effects are reported where suggestive (*P* < 1/116 for 116 tests: 29 SNPs × 2 phenotypes × 2 Asian ancestries versus European reference), but overall no interaction effects are significant after multiple testing correction (no *P* < 0.05/116 = 0.00043).

high-density lipoprotein (HDL) cholesterol levels respectively (Teslovich et al., 2010), but statins and CETP inhibitor therapies have more profound effects on the functions of these two genes, lowering LDL by 30–40% (Law et al., 2003) and raising HDL by approximately 30% (Fayad et al., 2011). Although it can be hoped that similar behavior may be found for some of the BP genes, at present there are few other experimentally validated examples, and future work will be needed to determine if this phenomenon extends more generally.

As summarized in Table 44.1, large-scale genetic studies have thus far yielded 40 BP loci, the likely candidate gene is known for some, and there is a wealth of existing functional data available. For the *NPPA*, *NPPB*, *NPR3*, *GUCY1A3*, and *GUCY1B3* genes, it may be worth revisiting existing therapies and those in development as potential new therapies for treating essential hypertension. Nesiritide is a recombinant form of human atrial natriuretic peptide B (encoded by *NPPB*) that was originally developed for treating decompensated congestive heart failure and its complications (O'Connor et al., 2011). This therapy is now in a Phase I clinical trial for stage 1 hypertension; the outcome of this will be interesting to follow. There are several small-molecule agonists targeting the C-type natriuretic peptide receptor (encoded by *NPR3*), and some pharmacological agents capable of directly stimulating soluble guanylate cyclases (encoded by *GUCY1A3* and *GUCY1B3* genes) are in development and in clinical trials. Riociguat, a soluble guanylate cyclase stimulator, is currently in Phase III clinical trials for pulmonary hypertension and chronic thromboembolic pulmonary hypertension. These and other molecules targeting the natriuretic peptide-guanylate cyclase-nitric oxide signaling pathway may provide leads to new anti-hypertensive medications, some of which may complement existing BP therapies in a similar fashion to what is already achieved by aliskirin.

Perhaps the greatest translational benefit will be new drug development based on the function and pathways of genes with no prior links to BP or cardiovascular disease. This is likely to be an area of increasing interest for both academics and the pharmaceutical industry.

UTILITY OF A GENETIC RISK SCORE

For disorders such as hypertension that have a complex genetic and environmental basis involving many genetic variants with small individual effects, it is clear that no single genetic variant will be usefully predictive for hypertension (although we note that the pharmacogenetic question of whether single genetic variants may indicate specific underlying causes or sub-types of hypertension, and thus may predict which individuals will benefit from particular therapies, remains largely unaddressed at this time). Many researchers have asked whether a "genetic risk score" (GRS) that combines information from multiple genetic variants, assuming a classical quantitative genetic or additive polygenic model (Falconer and Mackay, 1996), might

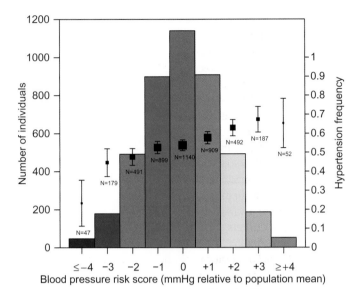

Figure 44.2 A genetic risk score (GRS) combining information from 29 SNPs associated with BP. This GRS predicts a genetic contribution to BP relative to the population mean (*x*-axis). In 2406 hypertensive cases and 1990 normotensive controls from the BRIGHT study (Caulfield et al., 2003), the distribution of the GRS is approximately normal (colored bars and left-hand *y*-axis). The frequency of hypertension increases according to the value of the GRS (squares with 95% CIs shown; right-hand *y*-axis). The difference in hypertension frequency between the bottom ~1% and top ~1% of individuals when stratified by the GRS is nonetheless extreme (OR > 6).

be usefully predictive for complex genetic disorders (Davies et al., 2010; Lin et al., 2009; Meigs et al., 2008).

The most extensive GRS analyses for BP and hypertension were reported by ICBP-GWAS, which constructed a GRS that combined the weighted effects of BP-increasing alleles at the 29 SNPs (Ehret et al., 2011). Because the effects of each SNP on continuous BP measures had been validated in multiple independent datasets, and because dichotomous hypertension status is defined using continuous BP measures, it is not surprising that this GRS was associated with dichotomous hypertension in the BRIGHT (British Genetics of Hypertension) case/control collection (see Figure 44.2). Given also that the 29 SNPs collectively explain approximately 3% of the heritability and thus approximately 1% of the total variation in continuous BP measures, it is not surprising that the GRS has little accuracy or utility for predicting hypertension status across the general population. However, the predictive accuracy can be much greater for the modest numbers of individuals at extremely low or high genetic risk of hypertension. For example, individuals in the bottom and top roughly 1% tail of the genetic risk score distribution (≤−4 and ≥+4 in Figure 44.2) show substantially lower and higher frequencies of hypertension compared to the BRIGHT sample as a whole. This illustrates an important point – that in assessing the predictive utility of findings from GWAS for complex traits in general, it is important to distinguish between predictive accuracy across the whole population [as measured, for example, by receiver operating characteristic (ROC) curves in many publications], and accuracy of screening for relatively small numbers of individuals at the extremes of risk. ICBP-GWAS also showed that the same GRS, predictive of DBP and SBP, was associated with occurrence of stroke and coronary artery disease (CAD) (Ehret et al., 2011). Under Mendelian randomization assumptions (Lawlor et al., 2008), these results can be interpreted as evidence of a causal effect of BP on stroke and CAD. These results are not surprising, since different classes of therapeutic agents that lower BP by a variety of distinct mechanisms (e.g., thiazide diuretics, beta-blockers, CCBs, and ACEi) have all been shown to lower risk of stroke and CAD in placebo-controlled, randomized clinical trials (RCTs). Because neither modern genetic association studies nor RCTs are susceptible to reverse causation or to confounding, we know that genetic variants tagged by the 29 SNPs must have causal effects on BP, and that in turn BP has causal effects on stroke and CAD. Hence the GRS results are interesting not because formally significant P-values were achieved, but

because the effect sizes were estimated from a range of general population samples (Figure 44.3). For obvious reasons, RCTs for anti-hypertensive therapies powered to detect effects on CVD endpoints have been conducted using hypertensive patients (only). The genetic results of ICBP-GWAS therefore complement prospective observational studies in suggesting that the beneficial effects of BP-lowering extend across the normal population range of BP levels. More surprisingly, in analyses using the same genetic risk score and in similar sample sizes, the ICBP-GWAS study observed no association with prevalent chronic kidney disease or prevalent microalbuminurea, or with continuous measures of kidney function. The latter results suggest that, across the normal population range, BP has a much weaker causal effect on kidney function and disease than on stroke or CAD. Specifically, the upper 95% confidence limit for odds ratio effect on kidney disease does not overlap the lower 95% confidence limit for odds or hazard ratio on stroke or CAD. Thus, the genetic results of ICBP-GWAS suggest that associations between BP and kidney function measures found in longitudinal observational studies (Perneger et al., 1993) may be driven by uncorrected confounding and/or causal effects of kidney function on BP (Lewis, 2010).

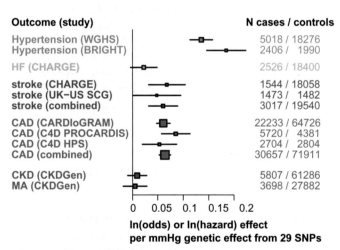

Outcome (study)	N cases / controls
Hypertension (WGHS)	5018 / 18276
Hypertension (BRIGHT)	2406 / 1990
HF (CHARGE)	2526 / 18400
stroke (CHARGE)	1544 / 18058
stroke (UK–US SCG)	1473 / 1482
stroke (combined)	3017 / 19540
CAD (CARDIoGRAM)	22233 / 64726
CAD (C4D PROCARDIS)	5720 / 4381
CAD (C4D HPS)	2704 / 2804
CAD (combined)	30657 / 71911
CKD (CKDGen)	5807 / 61286
MA (CKDGen)	3698 / 27882

ln(odds) or ln(hazard) effect
per mmHg genetic effect from 29 SNPs

Figure 44.3 Genetic risk score. Estimated effects on binary outcomes, per mmHg change in average BP due to genetic differences at 29 SNPs. HF=heart failure, CAD=coronary artery disease, CKD=chronic kidney disease, MA=microalbuminurea, WGHS=Womens' Genome Health Study, BRIGHT=British Genetics of Hypertension, CHARGE=Cohorts for Heart and Aging Research in Genomic Epidemiology, CARDIoGRAM=Coronary Artery DIsease Genome-wide Replication and Meta-analysis, C4D=Coronary Artery Disease Genetics consortium, CKDGen=Chronic Kidney Disease Genetics consortium, SCG=Stroke Collaborative Group, HPS=Heart Protection Study, PROCARDIS=precocious coronary artery disease consortium. All plotted data are results reported by ICBP-GWAS (Ehret et al., 2011). Placebo-controlled clinical trials for calcium channel blocker therapy on average achieve ~7mmHg lower average BP and ~0.7 hazard ratio for cardiovascular disease (Wright and Musini, 2009), which equates to a comparable ~0.05 ln(hazard) per mmHg change.

CONCLUSION

After decades of research on hypertension genetics in which the majority of published results subsequently failed to be replicated, the advent of modern GWAS and of meta-analyses that combine data from tens of thousands of individuals, genetic variants associated with BP that robustly replicate are finally being identified. The SNPs discovered so far are all common variants (MAF ≥ 5%), all have small effect sizes (mostly ≤1 mmHg for SBP and ≤0.5 mmHg for DBP), and they collectively explain only a small proportion (3–4%) of BP heritability. It could be argued that this is a modest return (thus far) given the cost and effort expended, but actual costs spent on BP genetics are hard to quantify, because the majority of the contributing individuals are from population studies, and so have been recruited and are being studied for many different inherited traits. The GWAS methodology has yielded new loci for many traits, and for BP there are now 40 loci that collectively harbor over one hundred new candidate genes. This represents a major advance, when three years ago almost no specific details were known about the genetic architecture of hypertension beyond the Mendelian disorders. The small fraction of BP heritability explained by the genetic variants cataloged so far inevitably triggers speculation about the genetic basis of the "missing" heritability. Estimates by ICBP-GWAS suggest there are probably more than one hundred more common genetic variants, all with small effects on BP (Ehret et al., 2011). Initial enthusiasm for the hypothesis that the missing heritability might be explained by structural genomic variation was tempered by two negative results. The WTCCC directly assayed copy number variants (CNVs) in 2000 hypertensives and 3000 common controls but did not find any significant associations (Craddock et al., 2010), and in a much larger sample size (N = 69,395) ICBP-GWAS did not find any significant associations at SNPs that tagged the CNVs assayed by the WTCCC (Ehret et al., 2011). At this time, we do not know if there are also rare variants associated with BP with large effects, and there has been no systematic study of whether there are multiple independently associated genetic variants at each locus. Fine-mapping studies of BP loci and sequence-based discovery of rare variants in extreme hypertensive cases and normotensive controls are ongoing, and the results of these studies will provide further insights into the underlying genetic causes of BP.

ACKNOWLEDGMENTS

T.J. is supported by the Wellcome Trust (grant number 093078/Z/10/Z). We acknowledge funding support from the British Heart Foundation and the Medical Research council, UK. This work forms part of the research themes contributing to the translational research portfolio for Barts and the London Cardiovascular Biomedical Research Unit, which is supported and funded by the National Institute for Health Research. P.B.M. would like to dedicate this chapter to the memory of Jonathan Munroe who died on 26th June 2011.

REFERENCES

1000 Genomes Project Consortium, 2010. A map of human genome variation from population-scale sequencing. Nature 467, 1061–1073.

Besecker, B., Bao, S., Bohacova, B., Papp, A., Sadee, W., Knoell, D.L., 2008. The human zinc transporter SLC39A8 (Zip8) is critical in zinc-mediated cytoprotection in lung epithelia. Am J Physiol Lung Cell Mol Physiol 294, L1127–L1136.

Borhani, N.O., 1959. The use of chlorothiazide and other thiazide derivatives in the treatment of hypertension. Int Rec Med 172, 509–516.

Brant, S.R., Shugart, Y.Y., 2004. Inflammatory bowel disease gene hunting by linkage analysis: Rationale, methodology, and present status of the field. Inflamm Bowel Dis 10, 300–311.

Caulfield, M., Munroe, P., Pembroke, J., et al., 2003. Genome-wide mapping of human loci for essential hypertension. Lancet 361, 2118–2123.

Chambers, J.C., Zhang, W., Lord, G.M., et al., 2010. Genetic loci influencing kidney function and chronic kidney disease. Nat Genet 42, 373–375.

Chiong, J.R., 2008. Controlling hypertension from a public health perspective. Int J Cardiol 127, 151–156.

Cho, Y.S., Go, M.J., Kim, Y.J., et al., 2009. A large-scale genome-wide association study of Asian populations uncovers genetic factors influencing eight quantitative traits. Nat Genet 41, 527–534.

Chobanian, A.V., Bakris, G.L., Black, H.R., et al., 2003. Seventh report of the Joint National Committee on Prevention, Detection, Evaluation, and Treatment of High Blood Pressure. Hypertension 42, 1206–1252.

Concannon, P., Erlich, H.A., Julier, C., et al., 2005. Type 1 diabetes: Evidence for susceptibility loci from four genome-wide linkage scans in 1435 multiplex families. Diabetes 54, 2995–3001.

Coronary Artery Disease (C4D) Genetics Consortium, 2011. A genome-wide association study in Europeans and South Asians identifies five new loci for coronary artery disease. Nat Genet 43, 339–344.

Cowley Jr., A.W., 2006. The genetic dissection of essential hypertension. Nat Rev Genet 7, 829–840.

Craddock, N., Hurles, M.E., Cardin, N., et al., 2010. Genome-wide association study of CNVs in 16,000 cases of eight common diseases and 3000 shared controls. Nature 464, 713–720.

Davies, R.W., Dandona, S., Stewart, A.F., et al., 2010. Improved prediction of cardiovascular disease based on a panel of single nucleotide polymorphisms identified through genome-wide association studies. Circ Cardiovasc Genet 3, 468–474.

Dominiczak, A., Brain, N., Charchar, F.J., McBride, M., Hanlon, N., Lee, W., 2004. Genetics of hypertension: Lessons learnt from Mendelian and polygenic syndromes. Clin Exp Hypertens 26, 611–620.

Ehret, G.B., Munroe, P.B., Rice, K.M., et al., 2011. Genetic variants in novel pathways influence blood pressure and cardiovascular disease risk. Nature 478, 103–109.

Falaschetti, E., Chaudhury, M., Mindell, J., Poulter, N., 2009. Continued improvement in hypertension management in England: Results from the Health Survey for England 2006. Hypertension 53, 480–486.

Falconer, D., Mackay, T., 1996. Introduction to Quantitative Genetics. Addison Wesley Longman, Harlow, UK.

Fayad, Z.A., Mani, V., Woodward, M., et al., 2011. Safety and efficacy of dalcetrapib on atherosclerotic disease using novel non-invasive multimodality imaging (dal-PLAQUE): A randomised clinical trial. Lancet 378, 1547–1559.

Fox, E.R., Young, J.H., Li, Y., et al., 2011. Association of genetic variation with systolic and diastolic blood pressure among African Americans: The Candidate Gene Association Resource study. Hum Mol Genet 20, 2273–2284.

Havlik, R.J., Garrison, R.J., Feinleib, M., Kannel, W.B., Castelli, W.P., McNamara, P.M., 1979. Blood pressure aggregation in families. Am J Epidemiol 110, 304–312.

Himeno, S., Yanagiya, T., Fujishiro, H., 2009. The role of zinc transporters in cadmium and manganese transport in mammalian cells. Biochimie 91, 1218–1222.

Hirschhorn, J.N., Daly, M.J., 2005. Genome-wide association studies for common diseases and complex traits. Nat Rev Genet 6, 95–108.

Ho, J.E., Levy, D., Rose, L., Johnson, A.D., Ridker, P.M., Chasman, D.I., 2011. Discovery and replication of novel blood pressure genetic loci in the Women's Genome Health Study. J Hypertens 29, 62–69.

Hong, K.W., Go, M.J., Jin, H.S., et al., 2010a. Genetic variations in ATP2B1, CSK, ARSG and CSMD1 loci are related to blood pressure and/or hypertension in two Korean cohorts. J Hum Hypertens 24, 367–372.

Hong, K.W., Jin, H.S., Lim, J.E., Kim, S., Go, M.J., Oh, B., 2010b. Recapitulation of two genomewide association studies on blood pressure and essential hypertension in the Korean population. J Hum Genet 55, 336–341.

International HapMap Consortium, 2003. The International HapMap Project. Nature 426, 789–796.

Johnson, A.D., Newton-Cheh, C., Chasman, D.I., et al., 2011. Association of hypertension drug target genes with blood pressure and hypertension in 86,588 individuals. Hypertension 57, 903–910.

Kato, N., Takeuchi, F., Tabara, Y., et al., 2011. Meta-analysis of genome-wide association studies identifies common variants associated with blood pressure variation in east Asians. Nat Genet 43, 531–538.

Kearney, P.M., Whelton, M., Reynolds, K., Muntner, P., Whelton, P.K., He, J., 2005. Global burden of hypertension: Analysis of worldwide data. Lancet 365, 217–223.

Kottgen, A., Glazer, N.L., Dehghan, A., et al., 2009. Multiple loci associated with indices of renal function and chronic kidney disease. Nat Genet 41, 712–717.

Law, M.R., Morris, J.K., Wald, N.J., 2009. Use of blood pressure lowering drugs in the prevention of cardiovascular disease: Meta-analysis of 147 randomised trials in the context of expectations from prospective epidemiological studies. BMJ 338, b1665.

Law, M.R., Wald, N.J., Rudnicka, A.R., 2003. Quantifying effect of statins on low density lipoprotein cholesterol, ischaemic heart disease, and stroke: Systematic review and meta-analysis. BMJ 326, 1423.

Lawlor, D.A., Harbord, R.M., Sterne, J.A., Timpson, N., Davey Smith, G., 2008. Mendelian randomization: Using genes as instruments for making causal inferences in epidemiology. Stat Med 27, 1133–1163.

Levy, D., Ehret, G.B., Rice, K., et al., 2009. Genome-wide association study of blood pressure and hypertension. Nat Genet 41, 677–687.

Lewis, J.B., 2010. Blood pressure control in chronic kidney disease: Is less really more? J Am Soc Nephrol 21, 1086–1092.

Lifton, R.P., Gharavi, A.G., Geller, D.S., 2001. Molecular mechanisms of human hypertension. Cell 104, 545–556.

Lin, X., Song, K., Lim, N., et al., 2009. Risk prediction of prevalent diabetes in a Swiss population using a weighted genetic score – the CoLaus Study. Diabetologia 52, 600–608.

Liu, C., Li, H., Qi, Q., et al., 2011. Common variants in or near FGF5, CYP17A1 and MTHFR genes are associated with blood pressure and hypertension in Chinese Hans. J Hypertens 29, 70–75.

McCarthy, M.I., Abecasis, G.R., Cardon, L.R., et al., 2008. Genomewide association studies for complex traits: Consensus, uncertainty and challenges. Nat Rev Genet 9, 356–369.

Meigs, J.B., Shrader, P., Sullivan, L.M., et al., 2008. Genotype score in addition to common risk factors for prediction of type 2 diabetes. N Engl J Med 359, 2208–2219.

Munroe, P.B., Johnson, T., Caulfield, M., 2009. The genetic architecture of blood pressure variation. Curr Cardiovasc Risk Rep 3, 418–425.

Musini, V.M., Fortin, P.M., Bassett, K., Wright, J.M., 2008. Blood pressure lowering efficacy of renin inhibitors for primary hypertension. Cochrane Database Syst Rev, 4. CD007066.

National Institute for Health and Clinical Excellence, 2011. Hypertension – Clinical management of primary hypertension in adults. NICE clinical guideline CG127. <http://guidance.nice.org.uk/CG127>.

Newton-Cheh, C., Johnson, T., Gateva, V., et al., 2009a. Genome-wide association study identifies eight loci associated with blood pressure. Nat Genet 41, 666–676.

Newton-Cheh, C., Larson, M.G., Vasan, R.S., et al., 2009b. Association of common variants in NPPA and NPPB with circulating natriuretic peptides and blood pressure. Nat Genet 41, 348–353.

Nguyen Dinh Cat, A., Touyz, R.M, 2011. A new look at the renin-angiotensin system: Focusing on the vascular system. Peptides 32, 2141–2150.

Niu, W., Zhang, Y., Ji, K., Gu, M., Gao, P., Zhu, D., 2010. Confirmation of top polymorphisms in hypertension genome-wide association study among Han Chinese. Clin Chim Acta 411, 1491–1495.

Nusbaum, C., Mikkelsen, T.S., Zody, M.C., et al., 2006. DNA sequence and analysis of human chromosome 8. Nature 439, 331–335.

O'Connor, C.M., Starling, R.C., Hernandez, A.F., et al., 2011. Effect of nesiritide in patients with acute decompensated heart failure. N Engl J Med 365, 32–43.

Org, E., Eyheramendy, S., Juhanson, P., et al., 2009. Genome-wide scan identifies CDH13 as a novel susceptibility locus contributing to blood pressure determination in two European populations. Hum Mol Genet 18, 2288–2296.

Padmanabhan, S., Melander, O., Johnson, T., et al., 2010. Genome-wide association study of blood pressure extremes identifies variant near UMOD associated with hypertension. PLoS Genet 6, e1001177.

Park, J.H., Wacholder, S., Gail, M.H., et al., 2010. Estimation of effect size distribution from genome-wide association studies and implications for future discoveries. Nat Genet 42, 570–575.

Perneger, T., Nieto, F., Whelton, P., Klag, M., Comstock, G., Szklo, M., 1993. A prospective study of blood pressure and serum creatinine. Results from the "Clue" Study and the ARIC Study. JAMA 269, 488–493.

Pietrangelo, A., 2010. Hereditary hemochromatosis: Pathogenesis, diagnosis, and treatment. Gastroenterology 139, 393–408. 408 e1-2.

Plomin, R., Haworth, C.M., Davis, O.S., 2009. Common disorders are quantitative traits. Nat Rev Genet 10, 872–878.

Prichard, B.N., 1966. The treatment of hypertension by beta adrenergic blocking drugs. Angiologica 3, 318–329.

Scuteri, A., Sanna, S., Chen, W.M., et al., 2007. Genome-wide association scan shows genetic variants in the FTO gene are associated with obesity-related traits. PLoS Genet 3, e115.

Speliotes, E.K., Willer, C.J., Berndt, S.I., et al., 2010. Association analyses of 249,796 individuals reveal 18 new loci associated with body mass index. Nat Genet 42, 937–948.

Takeuchi, F., Isono, M., Katsuya, T., et al., 2010. Blood pressure and hypertension are associated with 7 loci in the Japanese population. Circulation 121, 2302–2309.

Teslovich, T.M., Musunuru, K., Smith, A.V., et al., 2010. Biological, clinical and population relevance of 95 loci for blood lipids. Nature 466, 707–713.

Tobin, M.D., Sheehan, N.A., Scurrah, K.J., Burton, P.R., 2005. Adjusting for treatment effects in studies of quantitative traits: Antihypertensive therapy and systolic blood pressure. Stat Med 24, 2911–2935.

Tomaszewski, M., Debiec, R., Braund, P.S., et al., 2010. Genetic architecture of ambulatory blood pressure in the general population: Insights from cardiovascular gene-centric array. Hypertension 56, 1069–1076.

Wagner, J.R., Ge, B., Pokholok, D., Gunderson, K.L., Pastinen, T., Blanchette, M., 2010. Computational analysis of whole-genome differential allelic expression data in human. PLoS Comput Biol 6, e1000849.

Wain, L.V., Verwoert, G.C., O'Reilly, P.F., et al., 2011. Genome-wide association study identifies six new loci influencing pulse pressure and mean arterial pressure. Nat Genet 43, 1005–1011.

Wallace, C., Xue, M., Caulfield, M., Munroe, P., 2007. Genetics of Hypertension. In: Birkenhager, W., Reid, J. (Eds.), Handbook of Hypertension. Elsevier, Amsterdam.

Wang, Y., O'Connell, J.R., McArdle, P.F., et al., 2009. From the cover: Whole-genome association study identifies STK39 as a hypertension susceptibility gene. Proc Natl Acad Sci USA 106, 226–231.

Watkins, W.S., Hunt, S.C., Williams, G.H., et al., 2010. Genotype-phenotype analysis of angiotensinogen polymorphisms and essential hypertension: The importance of haplotypes. J Hypertens 28, 65–75.

Wellcome Trust Case Control Consortium, 2007. Genome-wide association study of 14,000 cases of seven common diseases and 3000 shared controls. Nature 447, 661–678.

WHO, 2002. The World Health Report 2002: Reducing risks, promoting healthy life. WHO Press, Geneva.

WHO, 2009. Global Health Risks: Mortality and burden of disease attributable to selected major risks. WHO Press, Geneva.

Wright, J., Musini, V.M., 2009. First line drugs for hypertension. Cochrane Database Syst Rev, 3. CD001841.

Wu, X., Kan, D., Province, M., et al., 2006. An updated meta-analysis of genome scans for hypertension and blood pressure in the NHLBI Family Blood Pressure Program (FBPP). Am J Hypertens 19, 122–127.

Zhu, X., Young, J.H., Fox, E., et al., 2011. Combined admixture mapping and association analysis identifies a novel blood pressure genetic locus on 5p13: Contributions from the CARe consortium. Hum Mol Genet 20, 2285–2295.

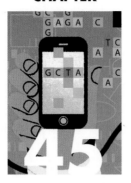

CHAPTER

45

Lipoprotein Disorders

Sekar Kathiresan and Daniel J. Rader

INTRODUCTION

Plasma lipoproteins are integral to energy and cholesterol metabolism, but disorders involving lipoprotein metabolism can predispose to atherosclerotic vascular disease (ASCVD). Genetic factors play an important role in influencing lipoprotein metabolism and therefore plasma levels of the major lipoproteins and cardiovascular risk. Molecular characterization of classic Mendelian monogenic lipoprotein disorders has provided major insight into the physiology and regulation of lipoprotein metabolism and new targets for therapeutic drug development. Recently, a large number of novel loci associated with plasma lipid traits have been identified through genome-wide association studies (GWAS). These discoveries are providing new insights into the genetic factors that influence the complex lipoprotein phenotypes that are much more common and important in influencing cardiovascular risk in the general population. Lipoprotein metabolism is a ripe area for the application of genomic medicine because of the frequency with which plasma lipids are measured in clinical practice, the quantitative importance of genetics in determining their levels, the large number of gene products involved in lipoprotein metabolism, and the broad clinical relevance of the field to the most important cause of morbidity and mortality in most of the world.

OVERVIEW OF LIPOPROTEIN METABOLISM

Lipoproteins are large macromolecular complexes that transport hydrophobic lipids (primarily triglycerides, cholesterol, and fat-soluble vitamins) through body fluids (plasma, interstitial fluid, and lymph) to and from tissues. Lipoproteins play an essential role in the absorption of dietary cholesterol, long-chain fatty acids, and fat-soluble vitamins; the transport of triglycerides, cholesterol, and fat-soluble vitamins from the liver to peripheral tissues; and the transport of cholesterol from peripheral tissues to the liver. Lipoproteins contain a core of hydrophobic lipids (triglycerides and cholesteryl esters) surrounded by hydrophilic lipids (phospholipids, unesterified cholesterol) and proteins that interact with body fluids. The plasma lipoproteins are divided into five major classes based on their relative density: chylomicrons, very-low-density lipoproteins (VLDL), intermediate-density lipoproteins (IDL), low-density lipoproteins (LDL), and high-density lipoproteins (HDL). Each lipoprotein class comprises a family of particles that vary slightly in density, size, migration during electrophoresis, and protein composition. The density of a lipoprotein is determined by the amount of lipid per particle. HDL is the smallest and most dense lipoprotein, whereas chylomicrons and VLDL are the largest and least dense lipoprotein particles. Most

Genomic and Personalized Medicine, 2nd edition
by Ginsburg & Willard. DOI: http://dx.doi.org/10.1016/B978-0-12-382227-7.00045-8

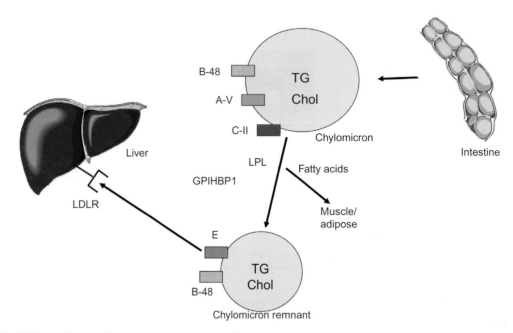

Figure 45.1   The exogenous pathway of lipoprotein metabolism. LDLR = LDL receptor, A-V = apoA-V, B-48 = apoB-48, C-II = apoC-II, E = apoE, TG = triglycerides.

plasma triglyceride is transported in chylomicrons or VLDL, and most plasma cholesterol is carried as cholesteryl esters in LDL and HDL.

The proteins associated with lipoproteins, called apolipoproteins, are required for the assembly, structure, and function of lipoproteins. Apolipoproteins activate enzymes important in lipoprotein metabolism, and act as ligands for cell surface receptors. ApoA-I, which is synthesized in the liver and intestine, is found on virtually all HDL particles. ApoA-II is the second most abundant HDL apolipoprotein, and is on approximately two-thirds of all HDL particles. ApoB is the major structural protein of chylomicrons, VLDL, IDL, and LDL; one molecule of apoB, either apoB-48 (chylomicrons) or apoB-100 (VLDL, IDL, or LDL), is present on each lipoprotein particle. The human liver synthesizes apoB-100 and the intestine makes apoB-48, which is derived from the same gene by mRNA editing. ApoE is present in multiple copies on chylomicrons, VLDL, and IDL, and plays a critical role in the metabolism and clearance of triglyceride-rich particles. ApoC-I, apoC-II, and apoC-III also participate in the metabolism of triglyceride-rich lipoproteins.

The exogenous pathway of lipoprotein metabolism involves the absorption and transport of dietary lipids to appropriate sites within the body (Figure 45.1). Dietary triglycerides are hydrolyzed by lipases within the intestinal lumen and emulsified with bile acids to form micelles. Dietary cholesterol, fatty acids, and fat-soluble vitamins are absorbed in the proximal small intestine. Cholesterol and retinol are esterified (by the addition of a fatty acid) in the enterocyte to form cholesteryl esters and retinyl esters, respectively. Longer-chain fatty acids (>12 carbons) are incorporated into triglycerides

and packaged with apoB-48, cholesteryl esters, retinyl esters, phospholipids, and cholesterol to form chylomicrons. Nascent chylomicrons are secreted into the intestinal lymph and are delivered via the thoracic duct directly to the systemic circulation, where they are extensively processed by peripheral tissues before reaching the liver. Lipoprotein lipase (LPL) is synthesized by adipocytes and cardiac and skeletal myocytes and is transported to the capillary endothelial surface, where it is anchored to GPI-anchored protein HDL binding protein (GPIHBP1; GPI = glycosylphosphatidylinsositol) and proteoglycans. The triglycerides of chylomicrons are hydrolyzed by LPL, and free fatty acids are released; apoC-II, which is transferred to circulating chylomicrons from HDL, acts as a cofactor for LPL in this reaction. The released free fatty acids are taken up by adjacent myocytes or adipocytes and are either oxidized to generate energy or re-esterified and stored as triglyceride. Some of the released free fatty acids bind albumin before entering cells and are transported to other tissues, especially the liver. The chylomicron particle progressively shrinks in size as the hydrophobic core is hydrolyzed and the hydrophilic lipids (cholesterol and phospholipids) and apolipoproteins on the particle surface are transferred to HDL, creating chylomicron remnants. Chylomicron remnants are rapidly removed from circulation by the liver through a process that requires apoE as a ligand for receptors in the liver.

The endogenous pathway of lipoprotein metabolism refers to the hepatic secretion of apoB-containing lipoproteins and their metabolism (Figure 45.2). VLDL particles resemble chylomicrons in protein composition, but contain apoB-100 rather than apoB-48 and have a higher ratio of cholesterol to triglyceride (~1 mg of cholesterol for every 5 mg of

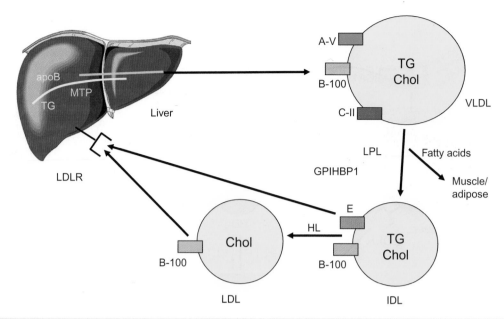

Figure 45.2 The endogenous pathway of lipoprotein metabolism. LDLR = LDL receptor, A-V = apoA-V, B-100 = apoB-100, C-II = apoC-II, Chol = cholesterol, E = apoE.

triglyceride). The triglycerides of VLDL are derived predominantly from the esterification of long-chain fatty acids in the liver. The packaging of hepatic triglycerides with the other major components of the nascent VLDL particle (apoB-100, cholesteryl esters, phospholipids, and vitamin E) requires the action of the enzyme microsomal triglyceride transfer protein (MTP). After secretion into the plasma, VLDL acquires multiple copies of apoE and apolipoproteins of the C series by transfer from HDL. As with chylomicrons, the triglycerides of VLDL are hydrolyzed by LPL, especially in muscle and adipose tissue. After the VLDL remnants dissociate from LPL, they are referred to as IDL, and contain roughly similar amounts of cholesterol and triglyceride. The liver removes approximately 40–60% of IDL by LDL receptor-mediated endocytosis via binding to apoE. The remainder of the IDL is remodeled by hepatic lipase (HL) to form LDL; during this process most of the triglyceride in the particle is hydrolyzed and all apolipoproteins except apoB-100 are transferred to other lipoproteins. The cholesterol in LDL accounts for more than half of the plasma cholesterol in most individuals. Approximately 70% of circulating LDL is cleared by LDL receptor-mediated endocytosis in the liver.

HDL metabolism is complex (Rader, 2006) (Figure 45.3). Nascent HDL particles are synthesized by the intestine and the liver. Newly secreted apoA-I rapidly acquires phospholipids and unesterified cholesterol from its site of synthesis (intestine or liver) via efflux promoted by the membrane protein ATP-binding cassette protein A1 (ABCA1; ATP = adenosine triphosphate). This process results in the formation of discoidal HDL particles, which then recruit additional unesterified cholesterol from the periphery. Within the HDL particle, the cholesterol is esterified by lecithin-cholesterol acyltransferase (LCAT), a plasma enzyme associated with HDL, and the more

hydrophobic cholesteryl ester moves to the core of the HDL particle. As HDL acquires more cholesteryl ester, it becomes spherical, and additional apolipoproteins and lipids are transferred to the particles from the surfaces of chylomicrons and VLDL during lipolysis. HDL cholesterol is transported to hepatocytes by both an indirect and a direct pathway. HDL cholesteryl esters can be transferred to apoB-containing lipoproteins in exchange for triglyceride by the cholesteryl ester transfer protein (CETP). The cholesteryl esters are then removed from the circulation by LDL receptor-mediated endocytosis. HDL cholesterol can also be taken up directly by hepatocytes via the scavenger receptor class BI (SR-BI), a cell surface receptor that mediates the selective transfer of lipids to cells. HDL particles undergo extensive remodeling within the plasma compartment by a variety of lipid transfer proteins and lipases. The phospholipid transfer protein (PLTP) has the net effect of transferring phospholipids from other lipoproteins to HDL. After CETP-mediated lipid exchange, the triglyceride-enriched HDL becomes a much better substrate for hepatic lipase (HL), which hydrolyzes the triglycerides and phospholipids to generate smaller HDL particles. A related enzyme called endothelial lipase (EL) hydrolyzes HDL phospholipids, generating smaller HDL particles that are catabolized faster. Remodeling of HDL influences the metabolism, function, and plasma concentrations of HDL.

INHERITED BASIS OF PLASMA LIPID TRAITS

Plasma total cholesterol was firmly established as an independent risk factor for cardiovascular disease (CVD) in 1961,

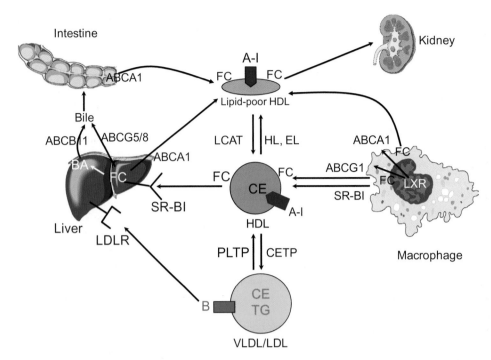

Figure 45.3 Pathways of HDL metabolism and reverse cholesterol transport. HDL metabolism. Liver and intestine, the two organs that synthesize apoA-I, are also primarily responsible for lipidating newly secreted lipid-poor apoA-I via ABCA1-mediated lipid efflux. Importantly, while liver and intestine ABCA1 may be the most critical for lipidating newly-synthesized lipid-free apoA-I, substantial additional cholesterol efflux to HDL occurs from other tissues. LDLR = LDL receptor, A-I = apoA-I, B = apoB, CE = cholesteryl ester, LXR = liver X receptor, BA = bile acid, FC = free cholesterol.

after researchers at the Framingham Heart Study demonstrated that participants with total cholesterol >245 mg/dL had a three-fold increased risk of future coronary heart disease compared with participants with a total cholesterol <210 mg/dL (Kannel et al., 1961). Subsequently, it was clarified that the two major lipoproteins carrying cholesterol, LDL and HDL, are associated with opposite influences on risk for cardiovascular disease, with LDL cholesterol associated with increased risk and HDL cholesterol with decreased risk. In observational studies, every 1 mg/dL increase in LDL cholesterol has been shown to be associated with a 2% increased risk for cardiovascular disease, whereas every 1 mg/dL increase in HDL cholesterol is associated with a 2–3% decreased risk (Gordon and Rifkind, 1989). For both LDL cholesterol and HDL cholesterol, there is a continuous, graded relationship between blood levels and subsequent risk for cardiovascular disease. Thus genetic factors that influence plasma levels of LDL cholesterol and HDL cholesterol are critically important to the risk of developing cardiovascular disease. The independent relationship of triglyceride levels and cardiovascular disease has been a topic of debate over the years, but recently a general consensus has developed that triglycerides are an independent predictor of cardiovascular risk. Thus genetic factors that influence triglycerides are also important determinants of cardiovascular risk. The fact that screening for

lipids is so widely performed clinically has resulted in frequent identification of extreme lipid phenotypes due to Mendelian syndromes, as well as widely available data for genetic association studies in populations.

Though lipid levels are affected by many non-genetic factors, inter-individual variability in lipids has been shown to have a strong inherited component. In the simplest terms, an important role for shared genes is suggested by the fact that the correlation between family members for LDL cholesterol, HDL cholesterol, or triglycerides is considerably greater than between unrelated individuals. Heritability, or the proportion of total phenotypic variance that is due to genetic variance, has been consistently estimated to be approximately 50% for the major blood lipid traits. For example, in the Framingham Heart Study, heritability for single time-point measurements of LDL cholesterol, HDL cholesterol, and triglycerides was 0.59, 0.52, and 0.48.

Broadly, traits that have an inherited basis may display a simple (or Mendelian) pattern of inheritance, where variation at a single genetic locus is both necessary and sufficient to cause a phenotype. Alternatively, traits may depend on multiple genetic loci (or multifactorial inheritance). For blood lipid disorders, both patterns of inheritance are operational. Mendelian syndromes where LDL cholesterol, HDL cholesterol, or triglyceride levels are extremely high or low have been successfully studied, and a number of specific genes have been isolated (see below). Knowledge derived from these genes has transformed our understanding of lipoprotein biology and treatment of cardiovascular disease. However, each of these syndromes is individually rare in the population and cannot explain the overall heritability of blood lipids.

Instead, blood lipid variation in the population depends on the additive effects of multiple loci. The additive effects of alleles at multiple loci can lead to a continuous trait that is normally distributed in the population – clearly the case for blood lipid levels. However, several aspects of the underlying genetic architecture for lipid levels are unknown. Some of the important unanswered questions are:

1. How many loci affect blood lipid variation?
2. At each locus, what are the specific variants that affect lipid levels?
3. What is the association of specific lipid-associated variants with coronary heart disease?
4. What are the mechanisms by which novel lipid genes exert their effects on plasma lipid levels?

MENDELIAN DISORDERS OF LIPOPROTEIN METABOLISM

Studies of the molecular basis of Mendelian disorders of lipoprotein metabolism have provided major insights into the regulation of lipoprotein metabolism, as well as novel targets for therapeutic development (Table 45.1). We first review the Mendelian disorders of high triglycerides, then high LDL-cholesterol (LDL-C), then low LDL-C, and finally low and high HDL-cholesterol (HDL-C).

Mendelian Disorders Primarily Causing Elevated Triglyceride Levels

Familial Chylomicronemia Syndrome (LPL Deficiency and ApoC-II Deficiency)

The familial hyperchylomicronemia syndrome (FCS) is caused by homozygosity for loss-of-function mutations in one of two genes encoding the proteins lipoprotein lipase (LPL) and apoC-II (Santamarina-Fojo, 1992). These conditions are virtual phenocopies and are therefore discussed together. FCS is characterized by extreme hypertriglyceridemia (>1000 mg/dL), usually presenting in childhood with acute pancreatitis, eruptive xanthomas, lipemia retinalis, and/or hepatosplenomegaly.

Chylomicron triglycerides are hydrolyzed in muscle and adipose capillary beds by lipoprotein lipase, with apoC-II acting as a required cofactor. Thus loss of function of either protein produces the phenotype of hyperchylomicronemia. Interestingly, despite markedly elevated triglyceride (and cholesterol) levels, premature atherosclerotic cardiovascular disease is not generally a feature of this disease. This observation contributed to our understanding of the importance of the nature of lipoprotein, not just the lipid, in determining

cardiovascular risk. Primary therapy is restriction of total dietary fat.

GPIHBP1 Deficiency

Lipoprotein lipase is synthesized by adipocytes and cardiac and skeletal myocytes, but requires active transport across the endothelial barrier to its site of action on the capillary luminal surface. The mouse knockout of the GPI-anchored protein HDL binding protein (GPIHBP1) unexpectedly revealed marked hypertriglyceridemia due to functional deficiency of LPL activity (Beigneux et al., 2007). Subsequently, mutations in *GPIHBP1* in humans were found in association with hyperchylomicronemia (Beigneux et al., 2009a; Wang and Hegele, 2007). Biochemical and physiological studies have demonstrated that GPIHBP1 serves as a chaperone for LPL from its site of synthesis to the endothelial surface, and helps to dock LPL to the endothelium (Beigneux et al., 2009b).

ApoA-V Deficiency

ApoA-V is an apolipoprotein associated with triglyceride-rich lipoproteins, that promotes LPL-mediated hydrolysis of lipoprotein triglycerides. Genetic deficiency of the apoA-V protein due to loss-of-function mutations in both alleles causes late-onset hyperchylomicronemia (Priore Oliva et al., 2005). Subjects with apoA-V deficiency have a severe lipolysis defect and markedly reduced VLDL catabolism. In addition, hetero-zygosity for mutations such as Q139X can lead to the same phenotype due to a dominant negative effect (Marcais et al., 2005). Subjects with severe hypertriglyceridemia have an excess of rare mutations in the *APOA5* gene (Johansen et al., 2010).

Familial Dysbetalipoproteinemia (Type III Hyperlipoproteinemia)

Familial dysbetalipoproteinemia (FD), or type III hyperlipoproteinemia, is caused by mutations in the gene for apolipoprotein E (apoE) (Mahley et al., 1999). ApoE on chylomicron and VLDL remnants normally mediate their catabolism by binding to receptors in the liver. FD is usually caused by homozygosity for a common variant called apoE2, which differs from the wild-type apoE3 form by a substitution of a cysteine for an arginine at position 158. ApoE2 has impaired binding to lipoprotein receptors such as the LDL receptor, resulting in defective removal of chylomicron and VLDL remnants. About 0.5% of individuals are homozygous for apoE2 but the prevalence of FD is only about 1 in 10,000, indicating that other genetic or environmental factors are required for expression of the phenotype. Because remnant lipoproteins are elevated and contain both triglycerides and cholesterol, plasma levels of both triglycerides and cholesterol are elevated. Palmar xanthomata and tuberoeruptive xanthomata on the elbows, knees, or buttocks are distinctive skin findings in this condition. Importantly, premature atherosclerotic CVD is common in this disorder, an observation that helped to clarify that remnant lipoproteins are highly atherogenic.

TABLE 45.1 Mendelian disorders of lipoprotein metabolism

Disorder	Gene	Protein	Lipid/lipoprotein changes	Genetic transmission
Elevated triglycerides				
Familial chylomicronemia syndrome	LPL	LPL	↑ Chylomicrons	AR
	APOC2	apoC-II	↑ Chylomicrons	AR
GPIHBP1 deficiency	GPIHBP1	GPIHBP1	↑ Chylomicrons	AR
ApoA-V deficiency	APOA5	apoA-V	↑ TG	AD
Familial dysbetalipoproteinemia	APOE	apoE	↑ TG, ↑chol, ↑remnants	AR/AD
Hepatic lipase deficiency	LIPC	HL	↑ TG, ↑chol, ↑remnants, ↑HDL	AR
Elevated LDL-C				
Familial hypercholesterolemia (FH)	LDLR	LDL receptor	↑ LDL-C	ACD
Familial defective apoB-100 (FDB)	APOB	apoB	↑ LDL-C	AD
Autosomal dominant hypercholesterolemia 3 (ADH3)	PCSK9 (GOF)	PCSK9	↑ LDL-C	ACD
Autosomal recessive hypercholesterolemia (ARH)	LDLRAP1	LDLRAP1	↑ LDL-C	AR
Sitosterolemia	ABCG5 or ABCG8	ABCG5 ABCG8	↑ LDL-C	AR
Reduced LDL-C				
Abetalipoproteinemia	MTP	MTP	↓ LDL, ↓TG	AR
Familial hypobetalipoproteinemia	APOB (truncations)	apoB	↓ LDL, ↓TG	ACD
PCSK9 deficiency	PCSK9 (LOF)	PCSK9	↓ LDL	ACD
Familial combined hypolipidemia	ANGPTL3	ANGPTL3	↓ LDL, ↓TG,	ACD
			↓HDL	AR
Reduced HDL-C				
ApoA-I deficiency	APOA1 (del, nonsense)	apoA-I	↓ HDL	ACD
ApoA-I structural mutations	APOA1 (missense)	apoA-I	↓ HDL	AD
Tangier disease	ABCA1	ABCA1	↓ HDL	ACD
LCAT deficiency/fish-eye disease	LCAT	LCAT	↓ HDL	AR
Elevated HDL-C				
CETP deficiency	CETP	CETP	↑ HDL, ↓LDL	AR

AD = autosomal dominant, ACD = autosomal co-dominant, AR = autosomal recessive, GOF = gain-of-function, LOF = loss-of-function, TG = triglycerides.

This disorder is one of the few genetic lipid disorders in which genotyping is clinically indicated for diagnosis; the finding of the apoE2/E2 genotype is diagnostic in the appropriate clinical setting.

Hepatic Lipase Deficiency

Hepatic lipase (HL) deficiency is caused by loss-of-function mutations in both alleles of the *LIPC* gene (Hegele et al., 1993). HL deficiency is characterized by elevated plasma levels of cholesterol and triglycerides due to the accumulation of circulating lipoprotein remnants as a result of lack of HL activity. Because HL also hydrolyzes HDL lipids, HDL-C is also elevated in this condition. HL deficiency is very rare; therefore it is difficult to determine its true relationship to atherosclerotic CVD.

Mendelian Disorders Primarily Causing Elevated LDL-C Levels

Familial Hypercholesterolemia

Familial hypercholesterolemia (FH) is caused by loss-of-function mutations in the LDL receptor (Hobbs et al., 1992). Homozygous FH, caused by mutations in both LDL receptor alleles, is a rare condition characterized by markedly elevated cholesterol (generally greater than 400 mg/dL), cutaneous and tendon xanthomas, and accelerated atherosclerosis developing in childhood. Because they work through up-regulation of the LDL receptor, statins, cholesterol absorption inhibitors, and bile acid sequestrants have only modest effects in reducing cholesterol. Liver transplantation is effective in decreasing LDL-C levels, and gene therapy has been attempted. LDL apheresis is the therapy of choice at this time. Heterozygous FH is one of the most common (approximately 1 in 500 persons) single-gene disorders. It is characterized by substantial elevations in LDL-C (usually 200–400 mg/dL), tendon xanthomas, and premature atherosclerotic cardiovascular disease. Treatment usually requires more than one drug, typically a statin plus a cholesterol absorption inhibitor and often a bile acid sequestrant and/or niacin. In some cases, drug therapy is inadequate and LDL apheresis is required.

Familial Defective Apolipoprotein B-100

Familial defective apoB-100 (FDB) (Tybjaerg-Hansen and Humphries, 1992) is caused by mutations in the receptor binding region of apoB-100 that impair its binding to the LDL receptor and delay clearance of LDL. The most common mutation causing FDB is a substitution of glutamine for arginine at position 3500 (R3500Q) in apoB-100; other mutations have also been reported that have a similar effect on apoB binding to the LDL receptor. FDB is generally recognized as an autosomal dominant condition, and occurs in approximately 1 in 700 persons of European descent. Due to a founder effect, the allele frequency of the R3500Q mutation in the Amish of Lancaster County, Pennsylvania, is ~12% (Shen et al., 2010). Like heterozygous FH, FDB is associated with elevated LDL-C and normal triglycerides, as well as with premature coronary disease (Tybjaerg-Hansen and Humphries, 1992). FDB is treated with statins and often additional drugs, similar to the treatment of heterozygous FH.

Autosomal Dominant Hypercholesterolemia 3

Linkage studies in some families with autosomal dominant hypercholesterolemia without causal mutations in the LDL receptor (LDLR) and apoB genes revealed what are now known to be gain-of-function mutations in the proprotein convertase subtilisin/kexin type 9 (PCSK9) gene (Abifadel et al., 2003). PCSK9 is secreted by hepatocytes and binds to cell-surface LDL receptors in the liver, promoting their lysosome degradation rather than recycling of the receptor (Horton et al., 2007). Gain-of-function mutations in PCSK9 enhance the ability of the PCSK9 protein to bind to the LDLR, which reduces the number of cell-surface LDL receptors and thus LDL clearance (Cunningham et al., 2007). This condition, now known as autosomal dominant hypercholesterolemia (ADH3), is also associated with premature coronary disease (Abifadel et al., 2003). Interestingly, loss-of-function mutations in this gene cause low LDL-C levels (see below).

Autosomal Recessive Hypercholesterolemia

Autosomal recessive hypercholesterolemia (ARH) is caused by mutations in LDLRAP1 (previously termed ARH) (Garcia et al., 2001), which produces a protein that regulates LDL receptor-mediated endocytosis in hepatocytes. Loss-of-function mutations in both alleles of LDLRAP1 prevent clathrin-dependent LDLR internalization, causing impaired LDL uptake and marked hypercholesterolemia (Mishra et al., 2002). ARH is very rare, and the clinical presentation resembles homozygous FH, though with presentation of cardiovascular disease generally being later (Arca et al., 2002). However, in contrast to FH, the condition is formally recessive, and obligate heterozygotes have normal cholesterol levels. Statins and other LDL-up-regulating therapies sometimes result in partial LDL-lowering response.

Sitosterolemia

Sitosterolemia is a recessive disorder caused by mutations in both alleles of one of two members of the adenosine triphosphate (ATP)-binding cassette (ABC) transporter family, ABCG5 and ABCG8 (Berge et al., 2000). These genes are expressed in the intestine and the liver, where they form a functional heterodimeric complex to limit intestinal absorption and promote biliary excretion of plant- and animal-derived neutral sterols. In sitosterolemia, the normally low level of intestinal absorption of plant sterols is markedly increased and biliary excretion of plant sterols is reduced, resulting in increased plasma levels of sitosterol and other plant sterols. Because the hepatic LDL receptor is down-regulated, LDL-C levels tend to be high in this condition. Patients with sitosterolemia often have tendon xanthomas and are at risk for premature cardiovascular disease (Salen et al., 1992). Treatment of sitosterolemia is focused on dietary counseling, cholesterol-absorption inhibitors, and bile acid sequestrants, rather than statins.

Mendelian Disorders Primarily Causing Reduced LDL-C Levels

Abetalipoproteinemia

Abetalipoproteinemia is caused by mutations in the gene encoding microsomal triglyceride transfer protein (MTP) (Sharp et al., 1993), a protein that transfers lipids to apoB in the endoplasmic reticulum (ER), forming nascent chylomicrons and VLDL in the intestine and liver, respectively. It is a very rare autosomal recessive disease characterized by extremely low plasma levels of cholesterol and no detectable apoB-containing lipoproteins in plasma (Rader and Brewer, 1993). Abetalipoproteinemia is characterized clinically by fat malabsorption, spinocerebellar degeneration, pigmented retinopathy, and acanthocytosis. Most clinical manifestations of abetalipoproteinemia result from defects in the absorption and transport of fat-soluble vitamins, especially alpha tocopherol (vitamin E), which is dependent on VLDL for efficient transport out of the liver. The discovery of MTP as the basis of abetalipoproteinemia led to the concept of MTP inhibition as a therapeutic target for lowering LDL-C levels.

Familial Hypobetalipoproteinemia

Familial hypobetalipoproteinemia generically refers to unusually low LDL-C levels that have a genetic basis. Historically it has been used to refer to low LDL-C due to mutations in apoB (Linton et al., 1993). There is a range of missense and nonsense mutations in apoB that have been shown to reduce secretion and/or accelerate catabolism of apoB. Individuals heterozygous for these mutations appear to be protected from the development of atherosclerotic vascular disease (Schonfeld et al., 2005). There are rare patients with homozygous hypobetalipoproteinemia who have mutations in both apoB alleles and have plasma lipids similar to those in abetalipoproteinemia. The discovery of apoB mutations as the basis of familial hypobetalipoproteinemia helped to support the idea of targeting apoB with oligonucleotide therapies as a therapeutic target for lowering LDL-C levels.

PCSK9 Deficiency

More recently, loss-of-function mutations in PCSK9 have also been shown to cause low LDL-C levels (Cohen et al., 2005; Kotowski et al., 2006). The mechanism involves reduced PCSK9-mediated targeting of the LDL receptor to degradation pathways, resulting in up-regulation of the hepatic LDL receptor and increased catabolism of LDL. This condition, which is more common in people of African descent, provided the opportunity to demonstrate that the effects of life-long low LDL-C levels are a substantial reduction in coronary heart disease (CHD) with no other adverse consequences (Cohen et al., 2006). This strongly supports the concept that aggressive LDL-C reduction is associated with long-term substantial reduction in cardiovascular risk. The discovery of PCSK9 gain-of-function mutations as the molecular basis of ADH3 and loss-of-function mutations as a cause of low LDL-C led to the identification of PCSK9 as a novel therapeutic target for reducing LDL-C.

ANGPTL3 Deficiency (Familial Combined Hypolipidemia)

Recently, loss-of-function mutations causing deficiency of angiopoetin-like protein 3 (*ANGPTL3*) were discovered to underlie another Mendelian syndrome of low LDL-C (Musunuru et al.,

2010a). A role for *ANGPTL3* in modulating human lipoprotein metabolism was initially suggested by resequencing efforts that identified multiple non-synonymous *ANGPTL3* variants in subjects drawn from the tails of the triglyceride distribution (Romeo et al., 2009). Functional analysis of the mutant proteins showed a decrease in protein secretion consistent with loss-of-function mutations in *ANGPTL3* resulting in reduced triglyceride (TG) levels. Independently, unbiased whole-exome sequencing was performed in a kindred with familial hypobetalipoproteinemia in which apoB mutations had been excluded (Musunuru et al., 2010a). Two siblings with the most extreme phenotype of combined hypolipidemia were found to be compound heterozygotes for two different nonsense mutations in *ANGPTL3*. The mutations co-segregated with low LDL-C (and low TG) in a co-dominant fashion in the extended pedigree; in addition, there are recessive effects of the mutations on reducing HDL-C. Biochemical studies indicated that ANGPTL3 has the ability to inhibit lipoprotein lipase, a plausible mechanism for its effects on triglycerides, and endothelial lipase, a potential mechanism for the HDL-C effects. However, the mechanism by which partial loss of ANGPTL3 function leads to reduction in LDL-C remains to be established. This example demonstrates how exome sequencing of patients with extreme lipid phenotypes has the potential to identify causal mutations and has established ANGPTL3 as another potential therapeutic target for reducing TG and LDL-C. Since the original observation in 2010, two reports have identified additional families where two loss-of-function ANGPTL3 alleles lead to low levels of TG, LDL-C, and HDL-C.

Mendelian Disorders Primarily Causing Reduced HDL-C Levels

ApoA-I Deficiency and Structural Mutations

A rare cause of extremely low HDL-C is complete deficiency of apoA-I either from apoA-I gene deletion or nonsense mutations, which result in virtually absent plasma HDL (Ng et al., 1994; Norum et al., 1982; Schaefer et al., 1982). Most of these cases are associated with premature CHD, consistent with the concept that apoA-I is atheroprotective and supportive of the concept that apoA-I elevation could be a therapeutic strategy. Another relatively rare cause of low HDL-C is missense or nonsense mutations that result in structurally abnormal or truncated apoA-I proteins. The best known of these mutations is apoA-I$_{Milano}$, where a substitution of cysteine for arginine at position 173 results in increased turnover of the mutant apoA-I$_{Milano}$ protein, as well of the wild-type apoA-I, and a substantial reduction in HDL-C (Chiesa and Sirtori, 2003). The low HDL-C levels, however, are not associated with an increased risk of atherosclerosis. Animal studies with intravenous infusion of recombinant apoA-I$_{Milano}$ show less atherosclerosis, and a small trial of the intravenous infusion of apoA-I$_{Milano}$-phospholipid complexes in humans demonstrated a reduction from baseline in coronary atheroma volume as measured by intravascular ultrasound (Nissen et al., 2003). There have been several other apoA-I structural mutations described that cause low HDL-C, but structural apoA-I mutations are rare, and in the general population apoA-I mutations are not a common source of variation in HDL-C levels.

Tangier Disease (ABCA1 Deficiency)

Tangier disease is caused by loss-of-function mutations in both alleles encoding the gene adenosine triphosphate-binding cassette protein A1 (*ABCA1*) (Bodzioch et al., 1999; Brooks-Wilson et al., 1999; Rust et al., 1999). It is characterized by cholesterol accumulation in the reticuloendothelial system, causing enlarged, orange tonsils, hepatosplenomegaly, intestinal mucosal abnormalities, and peripheral neuropathy, as well as markedly low HDL-C (<5 mg/dL) and apoA-I levels (Hobbs and Rader, 1999). The lack of ABCA1 results in markedly impaired efflux of cholesterol and phospholipids from cells to lipid-free apoA-I. The absence of intestinal and hepatic ABCA1 is probably largely responsible for the low HDL due to impaired lipidation of newly secreted apoA-I by these organs. Poorly lipidated apoA-I is then extremely rapidly catabolized. Impaired cholesterol efflux from other tissues, particularly macrophages, results in cholesterol accumulation, leading to many of the typical clinical characteristics of this disorder. However, Tangier disease patients do not develop rapidly accelerated atherosclerosis to the extent one might expect based on the cholesterol efflux defect and the extremely low HDL-C levels. Heterozygotes for *ABCA1* mutations have reduced HDL-C levels that are intermediate between Tangier disease and normal but have no evidence of cholesterol accumulation in tissues. Mutations in *ABCA1* have been found to cause low HDL-C levels in some families in which Tangier disease homozygotes are not found (Brooks-Wilson et al., 1999; Marcil et al., 1999). Rare private mutations in the *ABCA1* gene are also a cause of low HDL-C levels in the general population (Cohen et al., 2004). Whether individuals haploinsufficient for *ABCA1* are at increased risk of CHD is the topic of debate. As a result of the discovery of the molecular basis of Tangier disease, *ABCA1* is now a major target for the development of new therapies intended to up-regulate ABCA1 expression.

Lecithin-Cholesterol Acyltransferase Deficiency

Lecithin-cholesterol acyltransferase (LCAT) deficiency is caused by loss-of-function mutations in both alleles of the *LCAT* gene (Kuivenhoven et al., 1997). LCAT is the enzyme that esterifies the free cholesterol present on HDL to cholesteryl ester, creating a cholesteryl ester core and resulting in maturation of HDL. In the absence of functional LCAT and cholesterol esterification, mature HDL particles are not formed and nascent HDL particles containing apoA-I and apoA-II are rapidly catabolized (Rader et al., 1994). Two genetic forms of LCAT deficiency have been described: complete deficiency, known as classic LCAT deficiency, and partial deficiency, known as fish-eye disease (Kuivenhoven et al., 1997). In addition to extremely low HDL-C, both types of LCAT deficiency are characterized by corneal opacification, but only individuals with complete LCAT

deficiency have low-grade hemolytic anemia and progressive chronic kidney disease, leading to end-stage renal disease. Interestingly, neither form of LCAT deficiency is clearly associated with premature coronary disease, despite the markedly reduced HDL-C levels (Kuivenhoven et al., 1997), raising questions about the importance of LCAT in protecting against CHD. LCAT heterozygotes have slightly low to relatively normal HDL-C levels. Nevertheless, promotion of LCAT activity is of therapeutic interest as an HDL-raising approach.

Mendelian Conditions Primarily Causing Elevated HDL-C Levels

Cholesteryl Ester Transfer Protein (CETP) Deficiency

CETP deficiency is caused by loss-of-function mutations in both alleles of the *CETP* gene (Brown et al., 1989). CETP transfers cholesteryl esters from HDL to apoB-containing lipoproteins in exchange for triglycerides. Lack of functional CETP results in markedly elevated HDL-C levels due to lack of HDL remodeling, accumulation of cholesteryl esters in HDL, and slower turnover of apoA-I and apoA-II (Ikewaki et al., 1993). LDL-C levels are also low because of increased catabolism of LDL and apoB, with endogenous up-regulation of the LDL receptor (Ikewaki et al., 1995). CETP deficiency is extremely rare outside Japan; among the Japanese the most common mutations are a 5' donor splice site intron 14 G to A substitution and a missense mutation in exon 15 (D442G) (Inazu et al., 1990). Heterozygous individuals for CETP deficiency have 60–70% of normal CETP activity and only a modest increase in HDL-C levels, and otherwise normal LDL-C levels. Whether homozygous or heterozygous CETP deficiency is associated with increased, decreased, or unchanged cardiovascular risk remains to be resolved. Nevertheless, the identification of CETP deficiency led to the concept that CETP is a therapeutic target. Pharmacologic inhibition of CETP has been definitively established to raise HDL-C levels in humans. Although the first CETP inhibitor to enter full-scale clinical development, torcetrapib, failed (Barter et al., 2007), this outcome was likely due to off-target effects (Rader, 2007). Other CETP inhibitors continue in clinical development (Cannon et al., 2010; Stein et al., 2009), and this represents one of the most important clinical experiments in the field of lipoprotein metabolism and atherosclerosis.

Other Genes in Which Heterozygous Loss-of-Function Mutations Have Major Effects in Elevating HDL-C Levels

The line between Mendelian conditions and rare heterozygous mutations with non-trivial HDL-C elevating effects is becoming blurred. In the last few years, loss-of-function mutations in three HDL candidate genes have been reported in association with HDL-C elevation. Endothelial lipase (EL; gene name *LIPG*) has been shown in mice to be an important regulator of HDL-C levels (Ishida et al., 2003; Jaye et al., 1999; Jin et al., 2003). Deep candidate gene resequencing of the *LIPG* gene in subjects with extremes of HDL-C identified rare, non-synonymous loss-of-function variants in subjects with extremely high HDL-C levels not found in persons with low HDL-C (Edmondson et al., 2009). In addition, a low frequency (approximately 2% minor allele frequency) loss-of-function variant, N396S, was shown to be associated with significantly elevated HDL-C levels (Edmondson et al., 2009) and represents a "Goldilocks allele," useful for assessing the relationship of *LIPG* with CHD. The relationship of *LIPG* variants that reduce EL activity and raise HDL-C to CHD has not yet been definitively established. Apolipoprotein C-III (apoC-III; gene name *APOC3*) was previously known to influence TG and HDL metabolism in mice. A recent study in the Lancaster County Amish showed that a null mutation (R19X) in *APOC3* in the heterozygous state resulted in elevated HDL-C (and reduced TG and LDL-C) (Pollin et al., 2008). This variant, which occurs in about 5% of the Amish, also appears to be associated with reduced coronary atherosclerosis as assessed by coronary computed tomography (CT). Finally, the scavenger receptor class BI (SR-BI; gene name *SCARB1*) is a well-established candidate gene that mediates the selective uptake of cholesterol from HDL (Acton et al., 1996), and for which targeted deletion in mice causes markedly increased HDL-C levels as well as substantial increase in atherosclerosis (Braun et al., 2002). Recently, a family was reported in which heterozygosity for a loss-of-function mutation in SR-BI was associated with elevation in HDL-C levels but no obvious impact on atherosclerosis (Vergeer et al., 2011).

ASSOCIATION OF COMMON VARIANTS WITH PLASMA LIPID TRAITS AND RELATIONSHIP TO CARDIOVASCULAR DISEASE

Although the study of Mendelian lipoprotein disorders has been invaluable in helping to identify key proteins responsible for regulating lipoprotein metabolism, the vast majority of the genetic contribution to variation in plasma lipid traits in the general population is not explained by the rare mutations that cause Mendelian disorders. Through genotyping of hundreds of thousands of single nucleotide polymorphisms (SNPs) in tens of thousands of phenotyped subjects, genome-wide association studies (GWAS) have successfully identified common variants associated with lipid phenotypes and novel genes that contribute in a causal manner to lipid traits in the general population. The first GWAS for lipid traits, the Diabetes Genetics Initiative, genotyped approximately 3000 individuals of European descent, and identified SNPs near *APOE* and *APOB* as strongly associated with LDL-C levels, and SNPs near *CETP*, *LPL*, and *LIPC* as genome-wide significant determinants of HDL-C levels (Saxena et al., 2007). Subsequent GWAS studies added additional cohorts and involved almost 9000 individuals of European descent (Kathiresan et al., 2008b; Willer et al., 2008). In 2009, three additional lipid GWAS were reported. Kathiresan and colleagues (2009) utilized approximately 40,000 persons assembled from cohorts, case-control

studies, and clinical trials to identify 30 distinct loci associated with lipoprotein concentrations, including 11 loci that had not previously been identified as being genome-wide significant. The 11 newly defined loci were *ABCG8*, *MAFB*, *HNF1A*, and *TIMD4* associated with LDL cholesterol; *ANGPTL4*, *FADS1-FADS2-FADS3*, *HNF4A*, *LCAT*, *PLTP*, and *TTC39B* associated with HDL cholesterol; and *AMAC1L2*, *FADS1-FADS2-FADS3*, and *PLTP* associated with triglycerides. The use of an allelic dosage score suggested that the cumulative effect of multiple common variants contributes to polygenic dyslipidemia. Aulchenko and colleagues (2009) studied more than 20,000 persons from 16 different population-based cohorts, and identified 22 distinct loci associated with plasma lipids, including 6 new loci: *ABCG5*, *TMEM57*, *DNAH11*, and *FADS3-FADS2* associated with total and/or LDL cholesterol, and the *CTCF-PRMT8* region and *MADD-FOLH1* region associated with HDL. Sex-specific differences in effect size were demonstrated for *HMGCR*, *NCAN*, and *LPL*. Allelic dosage scores were associated with carotid intima media thickness and coronary heart disease incidence. The genetic risk score improves the screening of high-risk groups of dyslipidemia over classical risk factors. Sabatti and colleagues (2009) used the Northern Finland Birth Cohort 1966 for GWAS of lipid and other metabolic traits, and identified three new lipid loci: *AR* and *FADS1-FADS2* with LDL cholesterol and *NR1H3* (*LXRA*) with HDL cholesterol. A low-frequency variant in AR was found to be associated with substantially increased LDL-C in males – a mean LDL elevation of 28 mg/dL.

The largest-scale GWAS for lipid traits, the Global Lipids Genetics Consortium (GLGC), was published in 2010 (Teslovich et al., 2010). It involved a meta-analysis of more than 100,000 individuals of European descent, and identified 95 independent loci associated with at least one plasma lipid trait: 21 known lipid loci associated with an additional lipid phenotype for the first time, and 59 novel loci which had not been previously reported in association with lipid traits. Together, the 95 loci identified in the study account for 25–30% of the genetic variance in each lipid trait. The associated SNPs were shown to act additively to influence lipid phenotypes, and to be relevant not only to lipid traits in European populations, but also in South Asian, East Asian, and African-American populations. Common variants were found to be associated with plasma lipid traits in genes in which rare mutations cause Mendelian lipid disorders, including *LPL*, *APOA5*, *APOE*, and *LIPC* for TG; *LDLR*, *APOB*, *PCSK9*, *LDLRAP1*, *ABCG5/8*, and *ANGPTL3* for LDL cholesterol, and *ABCA1*, *LCAT*, *APOA1*, and *CETP* for HDL cholesterol. Known targets of LDL-lowering therapies, including *HMGCR* and *NPC1L1*, have also been identified as harboring common variants associated with LDL-C levels.

Testing association of "lipid SNPs" with coronary disease has the potential to provide insight into causality of the lipid–coronary artery disease (CAD) association – an approach known as Mendelian randomization. The GLGC tested the association between significant SNPs for lipid traits and coronary disease and found that 29 of the 95 loci identified were associated with CAD (Teslovich et al., 2010). Interestingly, most of the associated loci for CAD were in LDL-C genes, supporting the strong causal relationship between LDL-C and cardiovascular disease risk. The relationship of HDL-C loci with CAD was much more variable. GWAS focused directly on CAD phenotypes have been reported and in sum, they have identified ~25 loci for myocardial infarction (MI) or CHD (at genome-wide levels of significance) (Kathiresan et al., 2009b; Reilly et al., 2011; Samani et al., 2007; Schunkert et al., 2011). Of note, a substantial subset of the CAD loci defined in recent studies have been previously related to LDL-C; these include *LDLR*, *PCSK9*, *SORT1*, and *LPA*. These results further strengthen the now-unequivocal causal relationship between LDL cholesterol and CHD.

About two-thirds of the loci associated with plasma lipid traits by GWAS are novel, and the specific genes and mechanisms linking these loci to lipoprotein metabolism remain unknown. Indeed, functional validation of these novel loci is one of the most important challenges for the field. In an example of functional validation of a novel GWAS lipid locus, detailed studies were reported on the chromosome 1p13 locus strongly associated with both LDL-C and coronary heart disease (Musunuru et al., 2010b). The lead SNP at this locus has the strongest statistical association with LDL-C of any common SNP in the genome, with homozygotes for the major allele having an average of 16 mg/dL lower LDL-C (Kathiresan et al., 2008b). This same locus had been independently discovered by GWAS as being significantly associated with MI and cardiovascular disease (Kathiresan et al., 2009a; Samani et al., 2007). The minor allele at this locus is associated with substantially higher hepatic mRNA abundance of at least two genes at this locus, *SORT1* (encoding a protein known as sortilin) and *PSRC1*. The causal SNP at this locus, rs12740374, generates a CCAAT-enhancer binding protein (C/EBP) binding site, enabling increased transcription of these genes (Musunuru et al., 2010b). Overexpression of *SORT1*, but not *PSRC1*, in the livers of hypercholesterolemic mice resulted in reduced LDL-C, and knockdown of *SORT1* in mouse liver resulted in markedly increased LDL-C levels (Musunuru et al., 2010b). This is an example of "translation" of a novel GWAS locus into a previously unsuspected biological pathway clearly relevant to human physiology.

FUTURE DIRECTIONS

Medical resequencing – targeted, whole-exome, and whole-genome – will be required to identify the majority of the low-frequency variants and all of the rare variants (<0.5% allele frequency) that have a quantitatively large effect size on plasma lipids. One approach is the use of population-based samples and sequencing tails of the lipid traits or even whole samples. For example, resequencing of the coding region and proximal intronic regions of *ANGPTL3, 4, 5,* and *6*, in the multiethnic Dallas Heart Study, of 3551 individuals found an excess of non-synonymous variants in individuals with TG levels in the lowest quartile for ANGPTL3, 4, and 5 (Romeo et al., 2007; Romeo et al., 2009). One variant in ANGPTL4, E40K, was present in 3% of Caucasians and was associated with significantly lower plasma levels of TG and LDL-C and higher levels of HDL-C in two other large cohorts. As noted above, exome sequencing of a kindred with low TG, LDL-C, and HDL-C identified nonsense mutations in *ANGPTL3* that were causal for this novel Mendelian lipid disorder (Musunuru et al., 2010a).

Resequencing of individuals ascertained on the basis of extreme lipid traits, rather than those from the tails of a normal distribution, may provide a more efficient method of identifying rare mutations and low-frequency variants with large effect sizes on lipid phenotypes. For example, as noted above, resequencing of the *LIPG* gene in individuals with extremely high HDL-C levels and controls with relatively low HDL yielded a significant excess of rare variants in the high HDL group; analysis of variants created by site-directed mutagenesis proved them to be loss-of-function mutations (Edmondson et al., 2009). Resequencing of selected genes at GWAS TG loci in persons with extremely high TG yielded an excess of rare, nonsynonymous variants in the high TG individuals compared with normal TG controls (Johansen et al., 2010).

One of the critical questions regarding loci that influence plasma lipids is whether they also influence risk of CHD. This question can be addressed using Mendelian randomization. The key concept is that if a locus altering plasma lipid level is causal for CHD, then gene variants at that locus would also be associated with CHD in a manner consistent with the effect on lipids. As noted above, most of the LDL-C loci are also associated with CHD. In contrast, the association of HDL-C loci with CHD is much less consistent. Testing this hypothesis for the CETP locus is of particular interest as drug development programs targeting this protein are ongoing. Thompson and colleagues (2008) conducted a meta-analysis of studies reporting on the relationship between CETP gene variants, HDL cholesterol, and risk for CHD. They observed that CETP variants which increase HDL cholesterol modestly reduced the risk for CHD. Ridker and colleagues (2009) reported similar findings in a prospective cohort study. These studies are important, as they suggest that the HDL-raising effect of reduced CETP expression may reduce cardiovascular risk. Mendelian randomization studies that definitively evaluate the association of HDL "Goldilocks alleles" – those of low frequency but high effect size – with CHD will be of substantial importance to the field.

The explosion of GWAS for lipid traits and related phenotypes (such as response to lipid-lowering therapy) has the potential to impact on the evolution of personalized medicine. For example, genetic risk score based on the counting of the number of "adverse" alleles influencing lipids may enhance risk prediction compared with measurement of lipids alone (Kathiresan et al., 2008a). It is plausible that genotype – which represents a lifetime exposure to the lipid trait it influences – could be more predictive of CHD than a single measure of that lipid trait in plasma at one point in time. However, clinical adoption of genotyping genetic variants associated with lipid traits for CHD risk prediction will require unequivocal evidence that genotype predicts CHD even after adjusting for plasma lipids.

Genotype could also influence response to lipid-altering therapy in ways that could influence clinical management. However, no clinical trials have been reported in which lipid-associated genotype was used to stratify patients in order to formally assess pharmacogenetic responses and relationship to outcomes. Interestingly, safety and tolerability of statins is influenced by genotype in a manner that could be clinically relevant. For example, a GWAS of statin myopathy performed in the SEARCH study of two different doses of simvastatin revealed a highly significant association with a locus encoding the organic ion transporter SLCO1B1 responsible for transporting simvastatin into the liver (Link et al., 2008). Carriers of the rare allele were at significantly increased risk of statin myopathy and had a reduced response with regard to LDL lowering. In concept, genotyping for this allele could influence the choice of type or dose of statin to reduce the risk of myopathy.

The study of the genetics and genomics of lipoprotein metabolism has already led to several novel therapeutic targets and is likely to lead to more. Despite the success of statins, there remains a need for additional LDL-lowering therapies. Because mutations that impair the biosynthesis of apoB cause reduced LDL-C levels, the concept of targeting apoB therapeutically was developed. While small-molecule approaches to apoB inhibition are impractical, an approach using an antisense oligonucleotide (ASO) has been shown to be effective in animal models as well as in humans. The efficacy of an ASO targeted to apoB (mipomersen) in reducing LDL-C has been demonstrated in heterozygous FH (Akdim et al., 2010a) and homozygous FH (Raal et al., 2010) as well as in hypercholesterolemic patients on a stable dose of a statin (Akdim et al., 2010b). Small-interfering RNA (siRNA) molecules complementary to apoB mRNA also significantly reduced the plasma concentration of apoB-containing lipoproteins in non-human primates (Zimmermann et al., 2006). Inhibition of apoB expression using systemic administration of either an ASO or siRNA to apoB reduces LDL-C levels comparable to that seen in subjects heterozygous for familial hypobetalipoproteinemia.

The discovery that genetic deficiency of MTP causes the Mendelian disorder abetalipoproteinemia led to the concept that MTP inhibitors would disrupt VLDL assembly and secretion, with consequent reductions in circulating LDL (Burnett and Watts, 2007). An MTP inhibitor was shown to be effective in substantially reducing cholesterol in the Watanabe heritable hyperlipidemic (WHHL) rabbit, a model of homozygous familial hypercholesterolemia (Wetterau et al., 1998). The same compound was subsequently demonstrated to reduce LDL-C by approximately 50% in patients with homozygous FH, and kinetic studies demonstrated a marked reduction in LDL apoB production as the basis for the reduction in LDL-C and apoB (Cuchel et al., 2007). However, the trial demonstrated the mechanism-based increase in hepatic steatosis associated with MTP inhibition. MTP inhibition was also shown to be effective in reducing LDL-C in combination with ezetimibe (Samaha et al., 2008).

The story of PCSK9 provides one of the best examples in the lipoprotein field of how "unbiased" human genetics can identify novel therapeutic targets. As discussed above, through linkage studies, gain-of-function mutations in PCSK9 were found to cause autosomal dominant hypercholesterolemia (Abifadel et al., 2003). Subsequently, loss-of-function mutations were found to cause low LDL-C levels (Cohen et al., 2005) and reduce lifetime risk of CHD (Cohen et al., 2006). A compound heterozygote with loss-of-function mutations in both PCSK9 alleles has been reported to have a very low LDL-C but is in good health (Zhao et al., 2006). Thus, inhibition of PCSK9 is an attractive target based on the human genetics. Because PCSK9 is up-regulated by statin therapy, the addition of a PCSK9 inhibitor to a statin could result in additive reduction of LDL-C.

Investigation of the human genetics of HDL has also provided important information regarding potential therapeutic targets. The fact that LCAT deficiency is a cause of severely low HDL-C supports the concept that LCAT up-regulation or stimulation would be a strategy for raising HDL. The discovery of *ABCA1* as the mutated gene in Tangier disease created interest in *ABCA1* as a therapeutic target for up-regulation, to raise HDL-C levels and promote cholesterol efflux. Perhaps the best example of human genetics identifying a potential therapeutic target for HDL is CETP deficiency. As noted above, CETP-deficient subjects have markedly elevated HDL-C levels, directly identifying CETP as a target for inhibition to raise HDL; several CETP inhibitors are in clinical development. Other molecular causes of high HDL-C could provide additional new targets for therapeutic development.

In summary, Mendelian lipoprotein disorders and plasma lipid traits have proven to be fertile ground for the discovery of both rare and common genetic variants at more than one hundred independent genetic loci associated with plasma lipids. These discoveries have provided insight into normal human lipoprotein physiology and the relationship of plasma lipoprotein metabolism to CHD. They have begun to point the way toward the use of genotyping and sequencing of lipid genes to guide risk assessment and clinical management. Most importantly, they have identified several new therapeutic targets for the development of therapies to treat lipid disorders, and thereby prevent or treat CHD.

REFERENCES

Abifadel, M., Varret, M., Rabes, J.P., et al., 2003. Mutations in PCSK9 cause autosomal dominant hypercholesterolemia. Nat Genet 34, 154–156.

Acton, S., Rigotti, A., Landschultz, K.T., Xu, S., Hobbs, H.H., Krieger, M., 1996. Identification of scavenger receptor SR-BI as a high density lipoprotein receptor. Science 271, 460–461.

Akdim, F., Visser, M.E., Tribble, D.L., et al., 2010a. Effect of mipomersen, an apolipoprotein B synthesis inhibitor, on low-density lipoprotein cholesterol in patients with familial hypercholesterolemia. Am J Cardiol 105, 1413–1419.

Akdim, F., Stroes, E.S., Sijbrands, E.J., et al., 2010b. Efficacy and safety of mipomersen, an antisense inhibitor of apolipoprotein B, in hypercholesterolemic subjects receiving stable statin therapy. J Am Coll Cardiol 55, 1611–1618.

Arca, M., Zuliani, G., Wilund, K., et al., 2002. Autosomal recessive hypercholesterolaemia in Sardinia, Italy, and mutations in ARH: A clinical and molecular genetic analysis. Lancet 359, 841–847.

Aulchenko, Y.S., Ripatti, S., Lindqvist, I., et al., 2009. Loci influencing lipid levels and coronary heart disease risk in 16 European population cohorts. Nat Genet 41, 47–55.

Barter, P.J., Caulfield, M., Eriksson, M., et al., 2007. Effects of torcetrapib in patients at high risk for coronary events. N Engl J Med 357, 2109–2122.

Beigneux, A.P., Davies, B.S., Gin, P., et al., 2007. Glycosylphosphatidylinositol-anchored high-density lipoprotein-binding protein 1 plays a critical role in the lipolytic processing of chylomicrons. Cell Metab 5, 279–291.

Beigneux, A.P., Franssen, R., Bensadoun, A., et al., 2009a. Chylomicronemia with a mutant GPIHBP1 (Q115P) that cannot bind lipoprotein lipase. Arterioscler Thromb Vasc Biol 29, 956–962.

Beigneux, A.P., Weinstein, M.M., Davies, B.S., et al., 2009b. GPIHBP1 and lipolysis: An update. Curr Opin Lipidol 20, 211–216.

Berge, K.E., Tian, H., Graf, G.A., et al., 2000. Accumulation of dietary cholesterol in sitosterolemia caused by mutations in adjacent ABC transporters. Science 290, 1771–1775.

Bodzioch, M., Orso, E., Klucken, J., et al., 1999. The gene encoding ATP-binding cassette transporter 1 is mutated in Tangier disease. Nat Genet 22, 347–351.

Braun, A., Trigatti, B.L., Post, M.J., et al., 2002. Loss of SR-BI expression leads to the early onset of occlusive atherosclerotic coronary artery disease, spontaneous myocardial infarctions, severe cardiac dysfunction, and premature death in apolipoprotein E-deficient mice. Circ Res 90, 270–276.

Brooks-Wilson, A., Marcil, M., Clee, S.M., et al., 1999. Mutations in ABC1 in Tangier disease and familial high-density lipoprotein deficiency. Nat Genet 22, 336–345.

Brown, M.L., Inazu, A., Hesler, C.B., et al., 1989. Molecular basis of lipid transfer protein deficiency in a family with increased high-density lipoproteins. Nature 342, 448–451.

Burnett, J.R., Watts, G.F., 2007. MTP inhibition as a treatment for dyslipidaemias: Time to deliver or empty promises? Exp Opin Ther Targets 11, 181–189.

Cannon, C.P., Shah, S., Dansky, H.M., et al., 2010. Safety of anacetrapib in patients with or at high risk for coronary heart disease. N Engl J Med 363, 2406–2415.

Chiesa, G., Sirtori, C.R., 2003. Apolipoprotein A-I$_{Milano}$: Current perspectives. Curr Opin Lipidol 14, 159–163.

Cohen, J., Pertsemlidis, A., Kotowski, I.K., Graham, R., Garcia, C.K., Hobbs, H.H., 2005. Low LDL cholesterol in individuals of African descent resulting from frequent nonsense mutations in PCSK9. Nat Genet 37, 161–165.

Cohen, J.C., Boerwinkle, E., Mosley Jr., T.H., Hobbs, H.H., 2006. Sequence variations in PCSK9, low LDL, and protection against coronary heart disease. N Engl J Med 354, 1264–1272.

Cohen, J.C., Kiss, R.S., Pertsemlidis, A., Marcel, Y.L., McPherson, R., Hobbs, H.H., 2004. Multiple rare alleles contribute to low plasma levels of HDL cholesterol. Science 305, 869–872.

Cuchel, M., Bloedon, L.T., Szapary, P.O., et al., 2007. Inhibition of microsomal triglyceride transfer protein in familial hypercholesterolemia. N Engl J Med 356, 148–156.

Cunningham, D., Danley, D.E., Geoghegan, K.F., et al., 2007. Structural and biophysical studies of PCSK9 and its mutants linked to familial hypercholesterolemia. Nat Struct Mol Biol 14, 413–419.

Edmondson, A.C., Brown, R.J., Kathiresan, S., et al., 2009. Loss-of-function variants in endothelial lipase are a cause of elevated HDL cholesterol in humans. J Clin Invest 119, 1042–1050.

Garcia, C.K., Wilund, K., Arca, M., et al., 2001. Autosomal recessive hypercholesterolemia caused by mutations in a putative LDL receptor adaptor protein. Science 292, 1394–1398.

Gordon, D.J., Rifkind, B.M., 1989. High-density lipoproteins – the clinical implications of recent studies. N Engl J Med 321, 1311–1316.

Hegele, R., Little, J.A., Vezina, C., et al., 1993. Hepatic lipase deficiency: Clinical, biochemical, and molecular genetic characteristics. Arterioscler Thromb 13, 720–728.

Hobbs, H.H., Brown, M.S., Goldstein, J.L., 1992. Molecular genetics of the LDL receptor gene in familial hypercholesterolemia. Hum Mutat 1, 445–446.

Hobbs, H.H., Rader, D.J., 1999. ABC1: Connecting yellow tonsils, neuropathy, and very low HDL. J Clin Invest 104, 1015–1017.

Horton, J.D., Cohen, J.C., Hobbs, H.H., 2007. Molecular biology of PCSK9: Its role in LDL metabolism. Trends Biochem Sci 32, 71–77.

Ikewaki, K., Nishiwaki, M., Sakamoto, T., et al., 1995. Increased catabolic rate of low density lipoproteins in humans with cholesteryl ester transfer protein deficiency. J Clin Invest 96, 1573–1581.

Ikewaki, K., Rader, D.J., Sakamoto, T., et al., 1993. Delayed catabolism of high density lipoprotein apolipoproteins A-I and A-II in human cholesteryl ester transfer protein deficiency. J Clin Invest 92, 1650–1658.

Inazu, A., Brown, M.L., Hesler, C.B., et al., 1990. Increased high-density lipoprotein levels caused by a common cholesteryl-ester transfer protein gene mutation. N Engl J Med 323, 1234–1238.

Ishida, T., Choi, S., Kundu, R.K., et al., 2003. Endothelial lipase is a major determinant of HDL level. J Clin Invest 111, 347–355.

Jaye, M., Lynch, K.J., Krawiec, J., et al., 1999. A novel endothelial-derived lipase that modulates HDL metabolism. Nat Genet 21, 424–428.

Jin, W., Millar, J.S., Broedl, U., Glick, J.M., Rader, D.J., 2003. Inhibition of endothelial lipase causes increased HDL cholesterol levels in vivo. J Clin Invest 111, 357–362.

Johansen, C.T., Wang, J., Lanktree, M.B., et al., 2010. Excess of rare variants in genes identified by genome-wide association study of hypertriglyceridemia. Nat Genet 42, 684–687.

Kannel, W.B., Dawber, T.R., Kagan, A., Revotskie, N., Stokes III, J., 1961. Factors of risk in the development of coronary heart disease – six year follow-up experience. The Framingham Study. Ann Intern Med 55, 33–50.

Kathiresan, S., Melander, O., Anevski, D., et al., 2008a. Polymorphisms associated with cholesterol and risk of cardiovascular events. N Engl J Med 358, 1240–1249.

Kathiresan, S., Melander, O., Guiducci, C., et al., 2008b. Six new loci associated with blood low-density lipoprotein cholesterol, high-density lipoprotein cholesterol or triglycerides in humans. Nat Genet 40, 189–197.

Kathiresan, S., Voight, B.F., Purcell, S., et al., 2009a. Genome-wide association of early-onset myocardial infarction with single nucleotide polymorphisms and copy number variants. Nat Genet 41, 334–341.

Kathiresan, S., Willer, C.J., Peloso, G.M., et al., 2009b. Common variants at 30 loci contribute to polygenic dyslipidemia. Nat Genet 41, 56–65.

Kotowski, I.K., Pertsemlidis, A., Luke, A., et al., 2006. A spectrum of PCSK9 alleles contributes to plasma levels of low-density lipoprotein cholesterol. Am J Hum Genet 78, 410–422.

Kuivenhoven, J.A., Pritchard, H., Hill, J., Frohlich, J., Assmann, G., Kastelein, J., 1997. The molecular pathology of lecithin:cholesterol acyltransferase (LCAT) deficiency syndromes. J Lipid Res 38, 191–205.

Link, E., Parish, S., Armitage, J., et al., 2008. SLCO1B1 variants and statin-induced myopathy – a genomewide study. N Engl J Med 359, 789–799.

Linton, M.F., Farese Jr., R.V., Young, S.G., 1993. Familial hypobetalipoproteinemia. J Lipid Res 34, 521–541.

Mahley, R.W., Huang, Y., Rall, S.C.J., 1999. Pathogenesis of type III hyperlipoproteinemia (dysbetalipoproteinemia): Questions, quandaries, and paradoxes. J Lipid Res 40, 1933–1949.

Marcais, C., Verges, B., Charriere, S., et al., 2005. Apoa5 Q139X truncation predisposes to late-onset hyperchylomicronemia due to lipoprotein lipase impairment. J Clin Invest 115, 2862–2869.

Marcil, M., Brooks-Wilson, A., Clee, S.M., et al., 1999. Mutations in the ABC1 gene in familial HDL deficiency with defective cholesterol efflux [see comments]. Lancet 354, 1341–1346.

Mishra, S.K., Watkins, S.C., Traub, L.M., 2002. The autosomal recessive hypercholesterolemia (ARH) protein interfaces directly with the clathrin-coat machinery. Proc Natl Acad Sci USA 99, 16,099–16,104.

Musunuru, K., Pirruccello, J.P., Do, R., et al., 2010a. Exome sequencing, ANGPTL3 mutations, and familial combined hypolipidemia. N Engl J Med 363, 2220–2227.

Musunuru, K., Strong, A., Frank-Kamenetsky, M., et al., 2010b. From noncoding variant to phenotype via SORT1 at the 1p13 cholesterol locus. Nature 466, 714–719.

Ng, D., Leiter, L., Vezina, C., Connelly, P., Hegele, R., 1994. Apolipoprotein A-I Q[-2]X causing isolated apolipoprotein A-I deficiency in a family with analphalipoproteinemia. J Clin Invest 93, 223–229.

Nissen, S.E., Tsunoda, T., Tuzcu, E.M., et al., 2003. Effect of recombinant ApoA-I$_{Milano}$ on coronary atherosclerosis in patients with acute coronary syndromes: A randomized controlled trial. JAMA 290, 2292–2300.

Norum, R.A., Lakier, J.B., Goldstein, S., et al., 1982. Familial deficiency of apolipoproteins A-I and C-III and precocious coronary-artery disease. N Engl J Med 306, 1513–1519.

Pollin, T.I., Damcott, C.M., Shen, H., et al., 2008. A null mutation in human APOC3 confers a favorable plasma lipid profile and apparent cardioprotection. Science 322, 1702–1705.

Priore Oliva, C., Pisciotta, L., Li Volti, G., et al., 2005. Inherited apolipoprotein A-V deficiency in severe hypertriglyceridemia. Arterioscler Thromb Vasc Biol 25, 411–417.

Raal, F.J., Santos, R.D., Blom, D.J., et al., 2010. Mipomersen, an apolipoprotein B synthesis inhibitor, for lowering of LDL cholesterol concentrations in patients with homozygous familial hypercholesterolaemia: A randomised, double-blind, placebo-controlled trial. Lancet 375, 998–1006.

Rader, D.J., 2006. Molecular regulation of HDL metabolism and function: Implications for novel therapies. J Clin Invest 116, 3090–3100.

Rader, D.J., 2007. Illuminating HDL – is it still a viable therapeutic target? N Engl J Med 357, 2180–2183.

Rader, D.J., Brewer Jr., H.B., 1993. Abetalipoproteinemia: New insights into lipoprotein assembly and vitamin E metabolism from a rare genetic disease. JAMA 270, 865–869.

Rader, D.J., Ikewaki, K., Duverger, N., et al., 1994. Markedly accelerated catabolism of apolipoprotein A-II (ApoA-II) and high density lipoproteins containing ApoA-II in classic lecithin:cholesterol acyltransferase deficiency and fish-eye disease. J Clin Invest 93, 321–330.

Reilly, M.P., Li, M., He, J., et al., 2011. Identification of ADAMTS7 as a novel locus for coronary atherosclerosis and association of ABO with myocardial infarction in the presence of coronary atherosclerosis: Two genome-wide association studies. Lancet 377, 383–392.

Ridker, P.M., Pare, G., Parker, A.N., Zee, R.Y., Miletich, J.P., Chasman, D.I., 2009. Polymorphism in the CETP gene region, HDL cholesterol, and risk of future myocardial infarction: Genomewide analysis among 18,245 initially healthy women from the Women's Genome Health Study. Circ Cardiovasc Genet 2, 26–33.

Romeo, S., Pennacchio, L.A., Fu, Y., et al., 2007. Population-based resequencing of ANGPTL4 uncovers variations that reduce triglycerides and increase HDL. Nat Genet 39, 513–516.

Romeo, S., Yin, W., Kozlitina, J., et al., 2009. Rare loss-of-function mutations in ANGPTL family members contribute to plasma triglyceride levels in humans. J Clin Invest 119, 70–79.

Rust, S., Rosier, M., Funke, H., et al., 1999. Tangier disease is caused by mutations in the gene encoding ATP-binding cassette transporter 1. Nat Genet 22, 352–355.

Sabatti, C., Service, S.K., Hartikainen, A.-L., et al., 2009. Genome-wide association analysis of metabolic traits in a birth cohort from a founder population. Nat Genet 41, 35–46.

Salen, G., Shefer, S., Nguyen, L., Ness, G.C., Tint, G.S., Shore, V., 1992. Sitosterolemia. J Lipid Res 33, 945–955.

Samaha, F.F., McKenney, J., Bloedon, L.T., Sasiela, W.J., Rader, D.J., 2008. Inhibition of microsomal triglyceride transfer protein alone or with ezetimibe in patients with moderate hypercholesterolemia. Nat Clin Pract Cardiovasc Med 5, 497–505.

Samani, N.J., Erdmann, J., Hall, A.S., et al., 2007. Genomewide association analysis of coronary artery disease. N Engl J Med 357, 443–453.

Santamarina-Fojo, S., 1992. Genetic dyslipoproteinemias: Role of lipoprotein lipase and apolipoprotein C-II. Curr Opin Lipidol 3, 186–195.

Saxena, R., Voight, B.F., Lyssenko, V., et al., 2007. Genome-wide association analysis identifies loci for type 2 diabetes and triglyceride levels. Science 316, 1331–1336.

Schaefer, E.J., Heaton, W.H., Wetzel, M.G., Brewer Jr., H.B., 1982. Plasma apolipoprotein A-1 absence associated with a marked reduction of high density lipoproteins and premature coronary artery disease. Arteriosclerosis 2, 16–26.

Schonfeld, G., Lin, X., Yue, P., 2005. Familial hypobetalipoproteinemia: Genetics and metabolism. Cell Mol Life Sci 62, 1372–1378.

Schunkert, H., Konig, I.R., Kathiresan, S., et al., 2011. Large-scale association analysis identifies 13 new susceptibility loci for coronary artery disease. Nat Genet 43, 333–338.

Sharp, D., Blinderman, L., Combs, K.A., et al., 1993. Cloning and gene defects in microsomal triglyceride transfer protein associated with abetalipoproteinemia. Nature 365, 65–69.

Shen, H., Damcott, C.M., Rampersaud, E., et al., 2010. Familial defective apolipoprotein B-100 and increased low-density lipoprotein cholesterol and coronary artery calcification in the old order Amish. Arch Intern Med 170, 1850–1855.

Stein, E.A., Stroes, E.S., Steiner, G., et al., 2009. Safety and tolerability of dalcetrapib. Am J Cardiol 104, 82–91.

Teslovich, T.M., Musunuru, K., Smith, A.V., et al., 2010. Biological, clinical and population relevance of 95 loci for blood lipids. Nature 466, 707–713.

Thompson, A., Di Angelantonio, E., Sarwar, N., et al., 2008. Association of cholesteryl ester transfer protein genotypes with CETP mass and activity, lipid levels, and coronary risk. JAMA 299, 2777–2788.

Tybjaerg-Hansen, A., Humphries, S.E., 1992. Familial defective apolipoprotein B-100: A single mutation that causes hypercholesterolemia and premature coronary artery disease. Atherosclerosis 96, 91–107.

Vergeer, M., Korporaal, S.J., Franssen, R., et al., 2011. Genetic variant of the scavenger receptor BI in humans. N Engl J Med 364, 136–145.

Wang, J., Hegele, R.A., 2007. Homozygous missense mutation (G56R) in glycosylphosphatidylinositol-anchored high-density lipoprotein-binding protein 1 (GPI-HBP1) in two siblings with fasting chylomicronemia (MIM 144650). Lipids Health Dis 6, 23.

Wetterau, J.R., Gregg, R.E., Harrity, T.W., et al., 1998. An MTP inhibitor that normalizes atherogenic lipoprotein levels in WHHL rabbits. Science 282, 751–754.

Willer, C.J., Sanna, S., Jackson, A.U., et al., 2008. Newly identified loci that influence lipid concentrations and risk of coronary artery disease. Nat Genet 40, 161–169.

Zhao, Z., Tuakli-Wosornu, Y., Lagace, T.A., et al., 2006. Molecular characterization of loss-of-function mutations in PCSK9 and identification of a compound heterozygote. Am J Hum Genet 79, 514–523.

Zimmermann, T.S., Lee, A.C., Akinc, A., et al., 2006. RNAi-mediated gene silencing in non-human primates. Nature 441, 111–114.

Coronary Artery Disease and Myocardial Infarction

Samir B. Damani and Eric J. Topol

INTRODUCTION

Coronary artery disease (CAD) and myocardial infarction (MI) remain leading causes of death despite recent advances in diagnosis, prevention, and treatment (Lloyd-Jones et al., 2009). Further, the anticipated global aging of populations and worldwide increases in obesity and diabetes promise to propel ischemic heart disease (IHD) to epidemic levels in both developed and developing nations. However, recent advances in genomic technologies may help to avert this looming pandemic (Collins, 2010; Venter, 2010; A special report on the human genome June 19th 2010).

Recently, an unprecedented number of discoveries illuminating the genetic underpinnings of scores of complex traits have surfaced (Manolio, 2010). Many novel pathobiologic pathways have been identified, and previously recognized pathways in disease biology have been validated. The primary driver of these findings has been the genome-wide association study (GWAS), which can simultaneously assess hundreds of thousands of potentially disease-causing common single nucleotide polymorphisms (SNPs) in thousands of cases and controls. To date, nearly 1100 loci have been strongly linked to more than 150 complex "polygenic" traits using this technology (Lander, 2011; Manolio, 2010). Moreover, pharmacogenetic variants with substantial impact on drug efficacy and toxicity have emerged for a variety of commonly treated conditions (Damani

and Topol, 2011). Today, large-scale DNA sequencing programs spawned by striking reductions in sequencing costs are enabling the discovery of rare susceptibility variants (Johansen et al., 2010; Surolia et al., 2010; Venter, 2010). Together, these common and rare variants have exceptional relevance in the treatment of patients with CAD, and more importantly, promise to accelerate the development of more effective predictive algorithms and therapeutics that are based on one's unique biologic makeup and not a "shotgun" or "one size fits all" approach. In this chapter, we will review the most transformative genomic and pharmacogenomic findings that have surfaced to date as they relate to the prevention, diagnosis, and treatment of IHD and its related phenotypes.

SCANNING FOR SUSCEPTIBILITY SNPs

Prior to detailing recent developments in CAD genomics, it is vital that the reader understand the general schema behind how traits are inherited. Simple "monogenic" disorders typically follow classic Mendelian inheritance patterns (dominant, recessive, X-linked). Simple traits are relatively rare, with the most common Mendelian disorders affecting only one in several hundred individuals (Guttmacher and Collins, 2002). Examples from cardiovascular disease include familial hypercholesterolemia, hereditary hemochromatosis, and the long-QT syndromes. In contrast, complex "polygenic" traits

Genomic and Personalized Medicine, 2nd edition
by Ginsburg & Willard. DOI: http://dx.doi.org/10.1016/B978-0-12-382227-7.00046-X

stem from highly intricate gene–gene and gene–environment interactions (Hirschhorn and Daly, 2005). These complex disorders comprise the vast majority of commonly encountered diseases, and include CAD, diabetes, and various cancers, to name a few. Additionally, many important heritable intermediate phenotypes such as plasma lipids and hypertension also follow complex inheritance patterns.

Earlier approaches used to identify the genetic underpinnings of complex traits consisted of selectively investigating single or multiple candidate genes for disease association (Damani and Topol, 2007). Importantly, this hypothesis-driven approach precludes the discovery of gene variants with yet unknown but potentially substantial impact on disease susceptibility, progression, and even treatment response. Another traditional method – genome-wide linkage – involves assessing approximately 400 microsatellite markers evenly spaced (~10 centimorgans) across the genome for extensive allele sharing among family members affected by a disease, compared to unaffected members (Damani and Topol, 2007). Although less biased than candidate gene methods,

identifying the causative functional variant in a linkage analysis proved to be quite difficult given the large genomic region required to be surveyed after the general vicinity of a risk locus has been established (Figure 46.1). The required recruitment of families with affected and unaffected members was equally challenging. Overall, both candidate gene and linkage studies were largely unsuccessful in identifying and validating complex trait disease markers.

Importantly, these early failures have been overshadowed by the recent overwhelming success of GWAS. As noted earlier in this chapter, the genomic bases of more than 150 complex traits have been illuminated through GWAS (Manolio, 2010). These disorders have spanned the spectrum from common diseases and phenotypes such as CAD, hypertension, and diabetes, to more uncommon phenotypes such as exfoliative glaucoma (Table 46.1) (Altshuler et al., 2008). The main enabling aspect of GWAS has been the identification of the approximately 10 million SNPs that occur commonly, having a minor allele frequency (MAF) of at least 5% (Frazer et al., 2007; Frazer et al., 2009). These common SNPs are not inherited

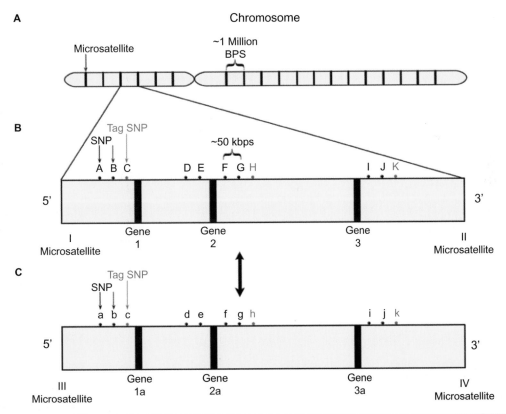

Figure 46.1   SNPs, tag SNPs, and microsatellites as genomic markers. (A) Autologous chromosome with evenly spaced microsatellites. (B) Segment of DNA between microsatellite markers. Single nucleotide polymorphisms are noted (A, B, C …) within the DNA segment. Tag SNPs (C, H, K) travel with other noted SNPs as blocks (haplotypes) and can serve as a surrogate for these haplotypes, and more importantly for disease-causing genes in close proximity. (C) DNA segment with alternative alleles and genomic markers of the same genes designated in part B of the figure. Note that the microsatellite markers are not as close in proximity to the genes as the noted SNPs. *Reproduced with permission from Damani and Topol, 2007.*

TABLE 46.1	Recent common and rare gene variants linked to CAD, MI, lipids, and ventricular fibrillation*				
Gene or locus	**Phenotype**	**Experimental method**	**Effect size (OR)[†] (single allele)**	**Effect size (OR) (multiple alleles)**	**References**
9p21.3	MI	GWAS	1.2–1.4	1.6–2.0	
(CDKN2A, CDKN2B)	AAA		1.31	1.74	Gretarsdottir et al., 2010;
	IA		1.29	1.72	Helgadottir et al., 2007;
	PAD		1.14	–	Helgadottir et al., 2008; McPherson et al., 2007; Samani et al., 2007; Wellcome Trust Case Control Consortium, 2007
DAB2IP	Early onset MI	GWAS	1.18	–	Gretarsdottir et al., 2010
	AAA		1.21	–	
	PE		1.20	–	
	PAD		1.14	–	
LPA	CAD	GWAS, candidate gene, resequencing	1.7–1.9	2.5–4.0	Clarke et al., 2009; Musunuru et al., 2010
APOE	CAD	GWAS, candidate gene, resequencing	1.1–1.4	1.2–1.6	
	LDL				Beekman et al., 2002; Kathiresan et al., 2009b
PCSK9	CAD	GWAS, resequencing	0.11–0.5; 1.13[‡]	–	
	LDL		–	–	Cohen et al., 2005; Jones et al., 2010; Kathiresan et al., 2009b
LDLR	CAD	GWAS	1.3	–	
	LDL		–	–	Kathiresan et al., 2009b
SORT1	CAD	GWAS	1.3	–	
	LDL		–	–	Kathiresan et al., 2009a; Yusuf et al., 2001
ANGPTL4	HDL	GWAS	–	–	Kathiresan et al., 2009b
CETP	HDL	GWAS, candidate gene	–	–	Kathiresan et al., 2009b
ABCA1	HDL	GWAS	–	–	Kathiresan et al., 2009b
ANGPTL3	TG	GWAS	–	–	Kathiresan et al., 2009b
APOA5	HTG	GWAS, resequencing	3.28	–	Kathiresan et al., 2009b; Teslovich et al., 2010
GCKR	HTG	GWAS, resequencing	1.75	–	Teslovich et al., 2010
LPL	HTG	GWAS, resequencing	0.32	–	Kathiresan et al., 2009b; Teslovich et al., 2010
APOB	HTG	GWAS, resequencing	1.67	–	Teslovich et al., 2010

(continued)

Gene or locus	Phenotype	Experimental method	Effect size (OR)† (single allele)	Effect size (OR) (multiple alleles)	References
21q21	VF	GWAS	1.5–1.8	–	
(CXADR)				–	Bezzina et al., 2010

OR – odds ratio; MI – myocardial infarction; AAA – abdominal aortic aneurysm; IA – intracranial aneurysm; PAD – peripheral arterial disease; PE – pulmonary embolism; CAD – coronary artery disease; LDL – low density lipoprotein; HDL – high density lipoprotein; HTG – hypertriglyceridemia; VF – ventricular fibrillation; GWAS – genome-wide association study; TG – Triglycerides.
*Newly discovered non-Mendelian common and rare susceptibility variants. All variants in genes listed have reached stringent statistical association criteria.
†Effect size for continuous variables not listed.
‡Rare loss-of-function variants in PCSK9 result in a significant protective effect, while common variants identified through GWAS confer heightened risk for CAD.

independently, but as "blocks" or "bins" called haplotypes. The genotype of one or a few SNPs can "tag" an entire haplotype. Thus, by assessing a few hundred thousand tag SNPs, GWAS are essentially assessing hundreds of thousands of haplotype blocks for disease and drug response associations. More importantly, the added statistical power provided by GWAS enables the use of population-based cases and controls, and eliminates the need for family-based cohorts in genomic investigations.

Another very recent advance that has great potential for transforming medicine was the discovery of rare DNA polymorphisms (MAF <5%) critically involved in disease biology (Johansen et al., 2010; Surolia et al., 2010). Through whole genome, exome, and targeted resequencing programs potentiated by advances in sequencing technologies and dramatic reductions in cost, investigators are now surveying for both common and rare disease susceptibility variants in the next generation of GWAS (Jones et al., 2010). Along those lines, Cohen and colleagues recently identified rare variants in PCSK9 (MAF 2% in African Americans) by resequencing the gene in 32 individuals with very low LDL levels from an original cohort of more than 2000 African Americans (Cohen et al., 2005). These loss-of-function variants mediated an impressive 40% reduction in plasma LDL levels, and in a subsequent study conferred a striking 50% reduction in risk for CAD, thereby underscoring the link between the lifelong reduction of LDL and reduced risk for IHD (Cohen et al., 2006).

MENDELIAN RANDOMIZATION

Several recent studies using principles of "Mendelian randomization" have provided meaningful insight into several critical intermediate phenotypes of CAD, including inflammation, plasma lipoproteins, and homocysteine (Kamstrup et al., 2009; Zacho et al., 2008). This approach leverages the fact that genes undergo recombination prior to gamete formation, and are subsequently distributed in random fashion from parent to offspring during conception (Clayton and McKeigue, 2001;

Nitsch et al., 2006). Thus, a Mendelian randomization study that seeks to establish a relationship between a gene product and intermediate phenotype is in essence tantamount to a randomized clinical trial. Potential confounders of such studies do exist, including inadequate accounting for population admixture from ancestral populations carrying different genotypes and risk for disease, as well as gene variants that are in linkage disequilibrium (LD) (Figure 46.2) with the variant being studied. However, both of these confounders can be easily accounted for with proper attention to study design.

The power of Mendelian randomization was elegantly demonstrated in a recent study that assessed the relationship between genetically elevated C-reactive protein (CRP) and the risk for IHD (Zacho et al., 2008). For decades, elevated CRP has been tied to a heightened risk for vascular events, but whether CRP was simply a marker or a contributor to disease remained unclear. Zacho and colleagues appropriately sought to address this question by assessing two independent populations of more than 50,000 cases and controls for the effect of four CRP polymorphisms on plasma CRP levels and on the risk for ischemic vascular disease. Notably, the polymorphisms resulted in as much as a 64% increase in plasma CRP levels, but did not lead to a heightened risk for CAD. In contrast, the link between CRP and vascular disease remained strong, with levels greater than 3 mg/dl culminating in a 60% increase in CAD rates and a 30% increase in stroke rates.

The importance of these seminal results was underscored with the publication of the highly publicized JUPITER (Justification for the Use of Statin in Prevention: An Intervention Trial Evaluating Rosuvastatin) trial (Ridker et al., 2008). In JUPITER, 17,802 otherwise healthy men with normal LDL (<130 mg/dl) and elevated CRP levels (>2 mg/dl) were randomized to rosuvastatin or placebo. Not surprisingly, individuals on rosuvastatin had an average 50% reduction in LDL cholesterol levels. Interestingly, they also had a 37% reduction in CRP levels. The trial was subsequently stopped early after two years when a corresponding 50% reduction in the risk for stroke and CAD was observed in the rosuvastatin group.

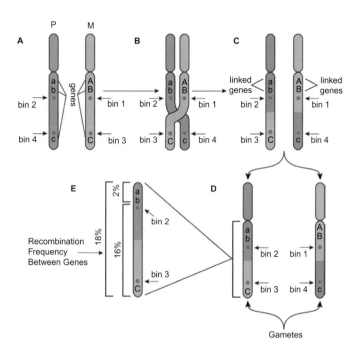

Figure 46.2 Schematic of genetic linkage and recombination. (A) Two homologous chromosomes: blue (paternal) and orange (maternal). Three genes with separate alleles and linkage disequilibrium "bins" noted. (A,a; B,b; C,c; bins 1–4). (B) Crossing over during meiosis. (C) Two alleles and their linked bins (C,c; bins 3 and 4) have switched locations via recombination. Four additional alleles and their associated bins (A,a; B,b; bins 1 and 2) have not switched and are considered linked. (D) Recombined haploid chromosomes segregate separately during meiosis as gametes prior to fertilization. (E) Sample recombination frequencies between genes demonstrating higher rates of recombination for genes further apart. *Reproduced with permission from Damani and Topol, 2007.*

Unfortunately, an additional control group of individuals with normal CRP and LDL levels was not included. Despite this fact, when integrating the JUPITER trial results with genetic data on CRP obtained by Zacho and colleagues, it is clear that the observed reduction in the primary endpoint was not mediated by a reduction in CRP. Instead, diminished inflammation as a result of statin therapy played a key role.

LIPID TRAITS

No discussion on personalized medicine and CAD would be complete without a discussion of lipids. Large twin studies have demonstrated that more than 50% of the variability in plasma lipid levels can commonly be attributed to genetic factors (Beekman et al., 2002). However, documenting the actual genes that contribute to this variability has been a much more

formidable task. Nevertheless, three recent seminal studies have shed significant light on this highly complex polygenic trait.

The first of these studies was a GWAS meta-analysis that assessed 2.6 million SNPs in more than 100,000 individuals from diverse ancestral populations (Teslovich et al., 2010). Remarkably, more than 95 SNPs were linked to plasma LDL, HDL, and triglycerides (TG) levels at a genome-wide significance level ($p < 5 \times 10^{-8}$). Of these 95 loci, 59 were novel, 39 were in genomic regions previously tied to lipid traits, and 18 were in genes involved in classic Mendelian lipid disorders. Moreover, a risk score developed based on these 95 loci demonstrated that individuals in the top quartile were 13 times more likely to have high LDL levels, four times as likely to have high HDL, and an impressive 44 times more likely to be hypertriglyceridemic when compared to individuals in the lowest risk quartile.

Subsequently, in an accompanying report, the investigators sought to further define the loci most significantly associated with LDL levels from the original GWAS meta-analysis ($p = 1 \times 10^{-170}$) (Musunuru et al., 2010). Interestingly, this locus (rs12740374) corresponds to a more favorable LDL profile, which includes a significantly lower LDL level (16 mg/dl), larger, more buoyant and less atherogenic LDL particles, and a 40% reduction in risk for ischemic heart disease in minor allele compared to major allele homozygotes. The protective allele resides in a non-coding genomic region between two genes, *CELSR2* and *PSRC1*, and near another gene, *SORT1*, that functions as a cell surface receptor-sorting protein. Importantly, before this study, none of these genes were known to participate in lipid metabolism.

The authors then designed a set of rigorous experiments to conclusively determine which of these genes actually regulate plasma LDL and how they exert their effects. First, they found that individuals who carried the protective SNP exhibited 12 times more liver *SORT1* expression. Notably, no elevation in *CELSR2 gene expression was detected*. Second, enhanced binding of the liver-specific transcription factor *CEBPA* to the promoter region of *SORT1 was observed in the hepatocytes of variant carriers*. Third, adenovirus transfection of mice with human *SORT1* resulted in substantially reduced concentrations of small dense and total LDL cholesterol, which mirrored the human phenotype. As final confirmation that *SORT1* was the culprit gene, knockdown of *SORT1* by means of small interfering RNA completely reversed the protective LDL phenotype. Together, these two studies incontrovertibly demonstrate that the most important common genetic contributor to plasma LDL concentrations acts through *SORT1* by creating a new binding site for the transcription factor *CEBPA*.

In the third and final study, a GWAS involving more than 2000 hypertriglyceridemia (HTG) cases and controls was followed by an extensive resequencing study (Johansen et al., 2010). Detailed analysis of the coding regions from the top four GWAS gene hits (*APOA5, GCKR, LPL, APOB*) revealed 47 rare functional variants in HTG cases versus only nine in the healthy controls ($p = 4.4 \times 10^{-5}$). Subsequently, logistic regression showed that common and rare sequence variants explained a greater proportion of risk for HTG than common clinical risk factors (22% vs 19%). This demonstrated for the first time that genotyping in common forms of dyslipidemia can lead to better predictive capabilities than using clinical risk factors alone.

Several additional progressive aspects of these studies require description. First, by using a meta-analytic approach, the investigators significantly enhanced the statistical power of their study for detecting important common genetic contributors to a highly intricate polygenic phenotype. This fact is bolstered by the several previous smaller GWAS that failed to detect the common variants found here. Second, by including individuals from diverse ancestries, the investigators have for the first time documented that key common lipid-related susceptibility variants exist across ethnic groups. Most importantly, the *SORT1* story serves as the new gold standard for how GWAS can illuminate previously unrecognized biologic pathways that serve as potential future therapeutic targets.

CORONARY ARTERY DISEASE AND MYOCARDIAL INFARCTION

Lipoprotein(a)

Of the 95 common lipid-altering SNPs discovered through GWAS, 14 confer statistically significant risk for CAD, albeit at modest levels (Teslovich et al., 2010). The vast majority of these 14 variants involve pathways of LDL metabolism. Although this finding supports the current emphasis that LDL reduction receives in primary and secondary prevention settings, it also indicates that a fair amount of the genetic contribution to CAD relates to factors outside of LDL cholesterol. Data now indicate that this missing CAD heritability is partially explained by common and rare susceptibility variants in the lipoprotein(a) gene, *LPA*. These recently discovered *LPA* sequence variants are some of the strongest genetic predictors of CAD and MI detected to date (Clarke et al., 2009; Tregouet et al., 2009). Interestingly, the lipoprotein(a) [Lp(a)] molecule is an LDL particle covalently linked to the apolipoprotein(a) molecule – a glycoprotein resembling the plasminogen protein (Figure 46.3). Similar to CRP, Lp(a) has been a putative risk factor for IHD for more than four decades, but the extent of this risk has been debatable until now.

The first study confirming the link between CAD and *LPA* was a three-stage GWAS that assessed 500,000 SNPs in more than 2000 CAD cases and controls (Tregouet et al., 2009). The most significantly associated SNPs ($p < 10^{-5}$) from stage one were on chromosome 6q26-27 and encompassed the *LPA* gene. The SNP associations were then validated in a second stage. Subsequently, four SNPs in two haplotype blocks on chromosome 6q26-27 were tested in a third stage. In a final adjusted analysis, both haplotypes were significantly

associated ($p = 1.0 \times 10^{-13}$, $p = 1.0 \times 10^{-15}$) with CAD, with odds ratios of 1.2 and 1.8 respectively.

Another seminal study assessed 48,000 common and rare SNPs from pertinent candidate genes for an association with CAD in more than 7000 cases and controls (Clarke et al., 2009). Remarkably, the same *LPA* locus defined in the GWAS above was also the strongest susceptibility locus in this study. In particular, two SNPs – one common (rs10455872; MAF 7%) and one rare (rs3798220; MAF 2%) variant – conferred a 70% and 92% increase in risk for CAD respectively. However, for individuals who carried both at-risk variants, their risk was an astounding 2.5-fold higher than wild-type (wt) allele carriers. These variants were also found to tag *LPA* alleles with a lower number of copy number polymorphisms (CNPs), which, as noted below, correlate with important qualitative features of the Apo(a) molecule and with a CAD risk that is independent of plasma Lp(a) levels.

In a final confirmatory study, Kamstrup and colleagues used Mendelian randomization principles to demonstrate that common kringle IV type II (KIV-2) CNPs result in elevated plasma Lp(a) levels (Kamstrup et al., 2009). They then illustrated that the same CNPs correlated to a 50% increase in risk for CAD in a prospective cohort of more than 9000 subjects followed for 16 years. Impressively, this heightened risk remained significant after adjustment for LP(a) levels, further underscoring the importance of *LPA* genotyping when trying to assess the patient's global risk for CAD.

Most recently, the European Atherosclerosis Society issued a statement recommending routine Lp(a) screening and subsequent treatment with niacin for individuals with levels greater

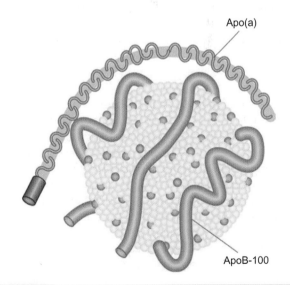

Apo(a)

ApoB-100

Figure 46.3 Lipoprotein(a) molecule. Lp(a) consists of an LDL particle and a glycoprotein molecule, apolipoprotein(a) [Apo(a)], attached to the ApoB-100 moiety of the LDL particle through a disulfide bond. Apo(a) size is determined by the number of kringle repeats. *Reproduced with permission from Danesh and Erqou, 2009.*

than 50 mg/dl (Beekman et al., 2002). Nevertheless, based on data covered here, a strong argument for *LPA* genotyping over Lp(a) phenotyping can be made on the following basis. First, current Lp(a) tests do not fully account for important qualitative features such as apolipoprotein size and CNPs (Erqou et al., 2010). Second, no consensus on the optimal Lp(a) assay exists. Third, currently available assays are highly variable with respect to their sensitivities and specificities for determining future risk for CAD. Finally, the effects of *LPA* variants appear to be independent of age and gender (Ariyo et al., 2003; Berglund and Ramakrishnan, 2004; Moliterno et al., 1995).

9p21

As alluded to earlier, the most exceptional aspect of GWAS is their ability to pinpoint previously unrecognized genomic regions and biologic pathways involved in disease predisposition. Recent GWAS findings tying 9p21.3 chromosomal variants to IHD exemplify this fact (Wellcome Trust Case Control Consortium, 2007). The 9p21 variants reside in a "desert" region of the genome, with the closest annotated genes being more than 100,000 base pairs away (*CDKN2A* and *CDKN2B*). (McPherson, 2010). With remarkable consistency in a compressed period of time, four independent GWAS published in 2007, which involved close to 60,000 cases and controls, found that European heterozygous (~50%) and homozygous variant carriers (~25%) harbored a 30% and 65% increase in risk for CAD and MI respectively (Helgadottir et al., 2007; McPherson et al., 2007; Samani et al., 2007; Wellcome Trust Case Control Consortium, 2007). The corresponding risk for early-onset MI cases was more than two-fold. Importantly, MI has been a much more accurately defined and clinically significant phenotype when compared to CAD (Damani and Topol, 2007). This makes the predictive nature of the 9p21 risk locus all the more significant.

It is noteworthy that 9p21 variants have since been widely replicated in independent populations involving more than 100,000 individuals, thereby making it one of the most validated markers of complex trait susceptibility discovered to date (Palomaki et al., 2010). Prominent gene dosage effects and more progressive atherosclerosis have also been documented (Dandona et al., 2010; Ye et al., 2008). Two large studies have now also linked 9p21 variants to the risk of two additional vascular phenotypes: abdominal aortic aneurysm and intracranial aneurysm (Gretarsdottir et al., 2010; Helgadottir et al., 2008). This surprising finding indicates a common thread between these previously distinct phenotypes.

Harismendy and colleagues demonstrated that the 9p21.3 locus is most likely exerting its effects through key inflammatory pathways triggered by local and long-distance gene interactions (Harismendy et al., 2010). In order to determine this, they first examined transcription factor binding and chromatin modification profiles in various human cell lines, which revealed that the 9p21.3 region contains over 30 enhancers – one of the highest densities for predicted enhancers in the human genome. Second, through computational modeling, they found that the

9p21.3 SNP rs10757278, which is the SNP most consistently associated with coronary artery disease in previous GWAS, resides within an enhancer site (ECAD9) and disrupts the binding of *STAT1* – a transcription factor critically involved in inflammation and interferon-γ (INF-γ) signaling.

Subsequently, the investigators showed that *STAT1* binding to the ECAD9 site in human umbilical vein endothelial cells increased in the presence of INF-γ, but not in cell lines homozygous for the coronary artery disease-associated SNP rs10757278. Moreover, INF-γ and *STAT1* binding inversely correlated with the expression of *CDKN2B*, which codes for a cell cycle regulator and is the closest gene to the 9p21.3 interval. Finally, using a technology known as chromatin conformation capture (3C), the investigators observed that INF-γ triggered genomic interactions between 9p21.3 enhancers and the gene *IFNA21* (interferon-α 21) – located 946,000 bases downstream on chromosome 9 – as well as other nearby and distant genes.

For years, epidemiologic data has suggested that inflammation is central to the progression and development of atherosclerotic coronary artery disease. Now, this study indelibly links inflammation to not only coronary artery disease, but to other 9p21.3-associated phenotypes, including intracranial and abdominal aortic aneurysm.

Ventricular Fibrillation

Important predictors of ventricular fibrillation (VF) are also emerging. A recent GWAS assessed 500,000 SNPs in 515 subjects with ventricular fibrillation and MI, and found a SNP (rs2824292, $p = 2.2 \times 10^{-10}$) present in 50% of the cases conferred a nearly two-fold increase in risk for ventricular fibrillation (Bezzina et al.). Interestingly, the at-risk variant is near *CXADR,* which encodes the coxsackie and adenovirus receptor protein and has been previously linked to both viral cardiomyopathies and sudden cardiac death. Further cohort and mechanistic studies will be required before the biological relevance and clinical significance of this variant can be fully understood. Nevertheless, in the future, individuals presenting with MI may be risk-stratified, with selective defibrillator therapy deployment based on this and other yet-to-be-discovered genomic sudden cardiac death predictors.

SNP Profiling Studies

The overwhelming number of CAD-related SNPs identified in recent years has spawned a number of studies investigating the value of these SNPs for risk prediction. Paynter and colleagues assessed the relationship of 101 SNPs to CAD in a cohort of 19,000 women followed for 12 years from the Women's Genome Health Study (WGHS) (Paynter et al., 2010). A "genetic risk score" based on these 101 SNPs revealed a significant relationship between higher genetic risk scores and the CAD phenotype, but failed to add incremental value to existing clinical models. Several critics of GWAS have used these null findings to support claims that genomic models of risk are inconsequential for common forms of CAD. Such claims are inaccurate and misleading for several reasons.

First, many current profiling studies fail to appropriately weigh the specific risk variants according to their effect size. For example, all 101 variants in the WGHS received equal statistical weighting in the study, despite clear evidence that variants such as those on chromosome 9p21.3 confer substantially greater risk for disease than other common susceptibility variants. Second, inconsistent phenotyping for CAD and the co-inclusion of MI in some but not all GWAS further confounds the extrapolation of GWAS results to populations that are different than the one originally studied. Third, many important genetic variants, such as those in *LPA,* have been omitted. Fourth, many SNP profiling studies, including the WGHS, have demonstrated that a family history continues to be predictive for CAD after adjusting for clinical variables. With much left to more fully explain heritability of coronary disease, only a relatively limited predictive capacity has been potentiated via common SNP variants. Recent resequencing studies have identified important rare CAD susceptibility variants and illustrate this fact. Finally, recent data indicate that a composite assessment of all relevant SNP data with appropriate weighting can proportionally increase the likelihood that an individual will manifest a heritable phenotype (Park et al., 2010). Hence, a proper genetic model should incorporate all common and rare susceptibility SNPs – including those yet to be discovered. Currently, such a model does not exist.

PHARMACOGENOMICS

Many of the common SNPs linked to diseases through GWAS confer modest increases in risk for disease. Several of the more transformative findings such as the relationship between 9p21 polymorphisms and MI will take years to fully decode and be translated to the clinical arena. In contrast, recently discovered pharmacogenetic variants have immediate ramifications with increased risk for adverse events approaching 50 in specific cases (Table 46.2) (Damani and Topol, 2011). In this section, we will briefly summarize the pertinent pharmacogenetic findings as they apply to the treatment of CAD and related phenotypes.

Antiplatelet Therapy

Clopidogrel

The benefits of clopidogrel therapy, including reduction of recurrent ischemia, stent thrombosis, and death in patients receiving coronary stents, are well established (Mehta et al., 2001; Steinhubl et al., 2002; Yusuf et al., 2001). In 2006, however, a variable antiplatelet effect was observed in healthy carriers of loss-of-function polymorphisms in the hepatic cytochrome 2C19 enzyme (*CYP2C19*) system (Hulot et al.,

2006). Several large cohort studies have since validated this effect in populations with acute coronary syndromes, and more importantly, have found an impressive two-fold and three-fold increase in risk for MI, death, and stent thrombosis respectively in variant carriers (Hulot et al., 2010). The *CYP2C19* system is a critical activator of clopidogrel – a prodrug – with loss-of-function variant carriers having a substantially reduced ability to convert the drug to its active form and hence a lower bioavailability of active clopidogrel. Now, gain-of-function variants (*17) have also been found to confer increased risk of bleeding (Sibbing et al., 2010; Wallentin et al., 2010). More than one-third of Europeans and nearly 50% of Asians and African Americans are carriers of the at-risk variants (Damani and Topol, 2010).

These data on the *CYP2C19* locus were initially discovered through hypothesis-driven candidate gene studies (Collet et al., 2009; Mega et al., 2009). Accordingly, Shuldiner and colleagues subsequently assessed more than 400,000 SNPs in a GWAS of platelet reactivity in 429 healthy individuals at baseline and after they received clopidogrel for seven days (Shuldiner et al., 2009). Not surprisingly, they found that *CYP2C19* SNPs were the most significant genetic predictor of platelet reactivity after seven days of clopidogrel, but not at baseline. They also found, in a separate cohort of individuals receiving stents, that carriers of the most common loss-of-function allele (*CYP2C19*2*) had a striking 350% increase in risk for MI, stent thrombosis, and death, thus incontrovertibly confirming earlier candidate gene study results.

Recent data on newer more potent antiplatelet agents such as prasugrel and ticagrelor appear to indicate that those agents may be superior to clopidogrel in preventing vascular events (Wallentin et al., 2010; Mega et al., 2010; Wallentin et al., 2009; Wiviott et al., 2007). However, a pharmacogenetic explanation for this benefit is clearly evident upon a closer look at the data. In the Platelet Inhibition and Patient Outcomes trial (PLATO), an initial analysis demonstrated superiority of ticagrelor versus clopidogrel for preventing hard clinical endpoints (8.6% vs 11.3%, $p = 0.0380$). When the analysis was restricted to *CYP2C19* extensive metabolizers however, the observed benefit no longer reached statistical significance (8.6% vs 10.0%, $p = 0.06$) (Wallentin et al., 2010). In Trial to Assess Improvement in Therapeutic Outcomes by Optimizing Platelet Inhibition Thrombolysis in Myocardial Infarction (TRITON TIMI-38), a prasugrel benefit over clopidogrel was also similarly abolished when examining those with wt *CYP2C19* alleles (Mega et al., 2010). Given that generic clopidogrel will be available in the next year at a fraction of the cost of newer agents, these pharmacogenetic data become highly relevant.

Curiously, in a recent highly publicized genetic substudy that analyzed the CURE (Clopidogrel in Unstable Angina to Prevent Recurrent Events) and ACTIVE A (Atrial Fibrillation Clopidogrel Trial with Irbesartan for Prevention of Vascular Events) trial populations, no excess cardiovascular risk in *CYP2C19* variant carriers was observed (Pare et al., 2010). On closer examination, however, for several reasons this study

TABLE 46.2	Recent pharmacogenomic breakthroughs			
Gene or locus	**Condition/ therapeutic use**	**Effect**	**Effect size (CI)**	**References**
LPA	CAD	Enhanced response to aspirin	2.2 (1.4–3.5)*	Chasman et al., 2009
SCLOB1	Lipid therapy	Myopathy	4.5 (2.6–7.7); 16.9 (4.7–61.1)[†]	Simon et al., 2009
		Musculoskeletal side effects	2.2 (1.4–3.6)	Voora et al., 2009
APOE	MI	Enhanced statin response in APOE4 variant carriers	1.8 (1.1–3.1)[‡]	Song et al., 2004
CYP2C19	Stent thrombosis (*2-*8)	Reduced conversion of clopidogrel to its active metabolite secondary to CYP2C19 LOF variants (*2-*8) results in excess risk for stent thrombosis, while GOF mutations result in increased bleeding events	3.5 (2.1–5.6); 4.7 (1.6–14.1)[†]	Hulot et al., 2010
	Bleeding (*17)		1.8 (1.03–3.14)	Sibbing et al., 2010; Wallentin et al., 2010
ABCB1	MI, death, stroke	Homozygote variant carriers of ABCB1 (3435 C→T) have reduced bioavailability of clopidogrel and an increased risk for adverse outcomes	1.7 (1.2–2.4)	Mega et al., 2010; Taubert et al., 2006

CI – 95% Confidence Interval; LOF – loss-of-function; GOF – gain-of-function.
*OR represents the increased risk for CAD in rs3798220 carriers. This enhanced risk was completely abrogated by aspirin therapy.
[†]Homozygous odds ratio.
[‡]OR represents heightened risk of future MI in APOE4 carriers, which was abolished with statin therapy.

does little to add to or refute the current evidence incriminating the CYP2C19 locus. The strongest evidence to date on CYP2C19 polymorphisms relates to a heightened risk for stent thrombosis. Thus the null findings from the study are not surprising, considering that fewer than 14% of CURE patients received stents. Second, the earlier studies incriminating CYP2C19 SNPs involved cohorts where a substantial portion of the population studied had had an acute MI (>80% in PLATO), which is in stark contrast to CURE where less than one-quarter of patients had had an MI (Mehta et al., 2001; Wallentin et al., 2009). Third, the evidence of benefit from clopidogrel in Atrial Fibrillation (AF) is tenuous at best. Ultimately, the cohorts studied in CURE and ACTIVE A are substantially different from the vast majority of patients who require clopidogrel today, making the results from the current study fairly meaningless. Therefore, based on the weight of the current evidence, we should continue to genotype for CYP2C19 alleles in all individuals who require coronary stenting in order to help prevent the lethal complication of stent thrombosis.

While CYP2C19 is an important contributor to clopidogrel nonresponsiveness, it only contributes to a relatively modest portion of the estimated 80% genetic contribution to clopidogrel response variability. Recent data now reveals where this missing heritability might lie (Bouman et al., 2011).

In elegant fashion, Bouman and colleagues first expressed enzymes – hepatic cytochromes and plasma esterases – thought to be involved in clopidogrel metabolism in a human embryonic kidney cell line. Interestingly, they found that clopidogrel bioactivation involves two critical sequential steps of oxidation and hydrolysis by plasma esterases. Notably, the second hydrolysis step by paraoxonases 1 and 3 (PON1 and PON3) was found to generate the pharmacologically active form of clopidogrel, which is a novel finding. They also found that a common coding polymorphism in PON1 (Q192R) correlated strongly with clopidogrel active metabolite levels, plasma paraoxonase activity, and the clopidogrel antiplatelet effect.

Moreover, in two independent cohorts involving over 9000 coronary artery disease patients, the homozygote carriers (QQ192) of the PON1 mutation had a striking 12-fold increase in risk for ST. A significant gene dose effect was also observed with heterozygotes (QR192) having a substantive 4-fold increase in risk for stent thrombosis (ST). Multivariate adjustment for known hepatic cytochrome variants did not attenuate the PON1 genotype effect, thus indicating a robust and independent effect. Questions regarding why PON1 variants have failed to show up in existing GWAS on clopidogrel response remain. Going forward, we can expect to hear much more about the biologic basis of *PON1*-associated clopidogrel resistance.

A final DNA sequence variant worth mentioning resides in the *ABCB1* gene, which encodes the intestinal efflux pump P-glycoprotein. This cell surface pump is responsible for the tranport of various molecules, including clopidogrel, across cell membranes (Taubert et al., 2006). Notably, homozygotes of *ABCB1* (3435 C → T) have reduced bioavailability of clopidogrel (Mega et al., 2010). More importantly, in TRITON TIMI 38, *ABCB1* homozygotes had a 70% increase in risk for stent thrombosis, MI, and death (Mega et al., 2010). An effect of similar magnitude was seen in another study involving 2000 French patients with acute MI (Mega et al., 2010; Simon et al., 2009). Oddly, the C allele of *ABCB1* (3435 T → C) in PLATO emerged as the risk locus. No clear explanation of these contrasting results is currently available. Nevertheless, the burden of evidence today points towards the T allele as the major risk locus. Going forward, integrating mechanistic studies with clinical trial data will hopefully clarify these paradoxical results.

Aspirin

A recent study reported that heterozygote carriers of a rare variant (rs3798220) in *LPA* (MAF ~3%) were found to have an enhanced response to aspirin therapy (Chasman et al., 2009). Strikingly, the same variant was also found to be the most significant genomic contributor to CAD in a recent limited GWAS of 7000 cases and controls (OR 1.8) (Clarke et al., 2009). The nonsynonymous variant results in an isoleucine for methionine substitution, and a corresponding eight-fold increase in plasma Lp(a) levels and two-fold increase in risk for MI (Chasman et al., 2009). Most importantly, the heightened risk for MI was completely abrogated with aspirin therapy. The mechanism behind the aspirin benefit in this select group of patients remains elusive, as does the mechanism of accelerated atherosclerosis with *LPA* variants. In the future, a systems biology approach using "omic" technologies will likely provide important insights into the biologic basis of *LPA*-mediated risk and the benefit from aspirin therapy. Nevertheless, a strong argument for initiating aspirin therapy in *LPA*-variant carriers could now be made based on these data.

In summary, the clopidogrel and aspirin stories represent the prototypical scenarios for the use of and benefit from pharmacogenetic algorithms in cardiovascular medicine (Damani and Topol, 2010). Clopidogrel is the third most highly prescribed drug in the world, with annual sales exceeding $6.5 billion, and is a staple for a procedure performed in over one million patients annually in the US alone. Clear alternatives exist for those genetically at risk for clopidogrel resistance, including the use of more potent antiplatelet agents such as prasugrel and ticagrelor, as well as higher dose clopidogrel with adjunctive platelet-function testing. Aspirin use is similarly widespread, with its benefit in the primary prevention setting being recently brought into question

given the substantial bleeding risk demonstrated in a recent well-conducted meta-analysis. Going forward, hopefully clinicians will be willing to incorporate compelling pharmacogenetic data in their clinical practice. To wait several years for further large-scale randomized clinical trial prospective validation that may never take place will place tens of thousands of individuals at risk for serious adverse events and deny patients state-of-the-art care.

Statins

Statins are the most widely used agents in the world for the treatment and prevention of CAD. Thus, it is important to emphasize several recent genomic findings on the safety and efficacy of these drugs. A nonsynonymous SNP in *SLCOB1* – encoding the organic anion-transporting polypeptide *OATP1B1* – conferred a four-fold increase in risk for statin-related myopathy in a recent GWAS involving just 160 cases (Link et al., 2008). For *SLCOB1* homozygotes, the risk for statin myopathy was a remarkable 1600% higher than wt carriers. In a more recent study, *SLCOB1* heterozygotes (MAF ~13%) were more likely to experience less severe but more common musculoskeletal side effects (Voora et al., 2009). Notably, those effects were confined to individuals on simvastatin only. Subsequent pharmacokinetic studies showed greater simvastatin accumulation and blood levels in variant compared to wt allele carriers, thereby providing a plausible biologic explanation for the findings. These effects and altered blood levels were not seen with atorvastatin or pravastatin. Given the different pharmacokinetic properties of these statins, this finding is not surprising. Nevertheless, lowering the dosage or changing therapy in patients with musculoskeletal side effects may be a useful alternative to discontinuing therapy.

In addition to genomic predictors of statin toxicity, genomic biomarkers indicating enhanced statin efficacy have been documented (Damani and Topol, 2011). *APOE4* has been linked to higher LDL cholesterol levels and CAD risk for decades (Song et al., 2004). More recently, in a five-year study of 966 MI survivors, *APOE4* carriers had a two-fold increase in risk for death that was conspicuously abolished by simvastatin therapy (Gerdes et al., 2000). Importantly, baseline characteristics were similar between E4 allele carriers and non-carriers. Also, the benefit derived from statin therapy was independent of LDL lowering, again exemplifying the salutary effects of statins that are independent of cholesterol reduction. Based on these data, an argument could be made for implementing statin therapy for primary prevention of CAD in *APOE4* heterozygotes with moderate risk factors. Such a strategy would also be suitable for prospective study if long-term follow-up and genetic data were available from an existing statin trial. Previous obstacles to *APOE4* testing have included psychosocial issues given the E4 allele's link to Alzheimer's. However, recent data indicate that this concern may be overblown (Green et al., 2009).

Accompanying the promises of genomic and personalized medicine are the potential pitfalls associated with prematurely adopting or promoting a pharmacogenetic test or algorithm. Along these lines, recent claims that *KIF6* enhanced statin response have been misleading. The *KIF6 719Arg* variant allele conferred a modest increase (OR 1.1–1.5) in risk for CAD in three observational studies (Iakoubova et al., 2008a, b; Shiffman et al., 2008). Further, in the PROVE IT-TIMI 22 (Pravastatin or Atorvastatin Evaluation and Infection Therapy in Myocardial Infarction) trial, statin therapy extinguished this excess risk (Iakoubova et al., 2008c). Consequently, more than 150,000 *KIF6* gene tests have been ordered since 2008 based on this tenuous data. This trend is quite troubling. Unlike *APOE4*, no link between *KIF6* and CAD or lipid traits has been established in more than ten GWAS conducted on this matter to date (Helgadottir et al., 2007; Johansen et al., 2010; Kathiresan et al., 2008; Kathiresan et al., 2009a, b; McPherson et al., 2007; Samani et al., 2007; Saxena et al., 2007; Teslovich et al., 2010; Willer et al., 2008; Wellcome Trust Case Control Consortium, 2007). Moreover, no convincing data on *KIF6* vascular expression has been reported, making a plausible biologic story for *KIF6* unlikely (Marian, 2008). Not surprisingly, recent meta-analysis findings in more than 17,000 individuals refute the claims that *KIF6* enhances statin efficacy and elevates CAD risk. This narrative should serve as a useful reminder for clinicians to stringently vet all pharmacogenetic data before implementing algorithms in clinical practice. Such vetting should include but not be limited to a literature search for biologic relevance and validation of SNPs through relevant GWAS.

NON-DNA APPROACHES TO CAD AND MI

Accompanying the transformative discoveries on genetic susceptibility variants are additional predictive CAD and MI markers that are emerging from related "omic" fields. Rosenberg and colleagues found that the gene expression signature of 23 genes obtained from the peripheral blood of non-diabetic patients undergoing coronary angiography for acute chest pain was effective in reclassifying ~20% patients when compared to traditional clinical models (Rosenberg et al., 2010). Importantly, the negative predictive value of 83% for the gene expression assay is comparable to that of typically used clinical tests such as myocardial perfusion imaging, but without the unnecessary and potentially harmful radiation exposure.

Another application being used with increased frequency for the detection of novel biomarkers is mass spectrometry and nuclear magnetic resonance for the characterization of the human metabolome in various cardiovascular phenotypes (Lewis et al., 2008). Notably, the metabolome is the most proximal reporter for all human disease pathways. Accordingly, perturbations in the genome, transcriptome, proteome, and the environment are frequently reflected by changes in the various metabolic pathways. Recently, metabolomic studies have been effective in detecting novel metabolites that predict statin response, exercise induced myocardial ischemia, and subsequent cardiovascular events in individuals with stable CAD, while also uncovering novel mechanisms and pathways of disease (Sabatine et al., 2005; Kaddurah-Daouk et al., 2010; Shah et al., 2010).

Like with the human genome, however, we are only beginning to build the comprehensive databases and functional tools necessary to help classify, characterize, validate, and subsequently implement the use of gene transcripts and metabolites in clinical cardiovascular settings.

FUTURE DIRECTIONS

Over the next several years, the genomic basis of most human diseases will be incrementally decoded through several parallel approaches. Rare susceptibility variants will rapidly be identified through targeted resequencing, exome, and whole-genome sequencing programs. Further advances in single molecule sequencing and more efficient approaches to *de novo* genome assembly will allow for better characterization of important structural variants such as gene copy number changes, as well as smaller base pair insertions and deletions. Detailed cataloging of tissue-specific epigenetic features including chromatin state, histone modifications, and DNA methylation will yield important insight on the intrinsic "omic" biology along with the environmental and lifestyle impacts on disease. A systems biology approach that includes relevant proteomic and metabolomic experiments will also be important for understanding the functional mechanisms that mediate the effects of newly discovered genomic and pharmacogenomic variants.

Many additional challenges not directly related to scientific inquiry will also need to be addressed in the years ahead. These challenges include the development of a robust infrastructure that enables the effortless sharing and integration of large genomic datasets and clinical information, so that appropriate bioinformatics analysis of the next generation of genome-wide association studies can be performed. Ethical issues surrounding dissemination of genomic information, consumer-driven genetic testing, and public concerns regarding genetic discrimination will also need to be fully addressed. Finally, physicians must be educated on the nuances of conducting and interpreting the results of genomic studies, and the proper methods of disseminating genomic data to other physicians, scientists, and patients. Only then can the full potential of a genomic and individualized medicine era be fully realized.

REFERENCES

Altshuler, D., Daly, M.J., Lander, E.S., 2008. Genetic mapping in human disease. Science 322, 881–888.

Ariyo, A.A., Thach, C., Tracy, R., 2003. Lp(a) lipoprotein, vascular disease, and mortality in the elderly. N Engl J Med 349, 2108–2115.

Beekman, M., Heijmans, B.T., Martin, N.G., et al., 2002. Heritabilities of apolipoprotein and lipid levels in three countries. Twin Res 5, 87–97.

A special report on the human genome. The Economist June 19th 2010.

Berglund, L., Ramakrishnan, R., 2004. Lipoprotein(a): An elusive cardiovascular risk factor. Arterioscler Thromb Vasc Biol 24, 2219–2226.

Bezzina C.R., Pazoki R., Bardai A., et al., 2010. Genome-wide association study identifies a susceptibility locus at 21q21 for ventricular fibrillation in acute myocardial infarction. Nat Genet 42, 688–691.

Bouman H.J., Schomig E., van Werkum J.W., et al., 2011. Paraoxonase-1 is a major determinant of clopidogrel efficacy. Nat Med 17, 110–116.

Chasman, D.I., Shiffman, D., Zee, R.Y., et al., 2009. Polymorphism in the apolipoprotein(a) gene, plasma lipoprotein(a), cardiovascular disease, and low-dose aspirin therapy. Atherosclerosis 203, 371–376.

Clarke, R., Peden, J.F., Hopewell, J.C., et al., 2009. Genetic variants associated with Lp(a) lipoprotein level and coronary disease. N Engl J Med 361, 2518–2528.

Clayton, D., McKeigue, P.M., 2001. Epidemiological methods for studying genes and environmental factors in complex diseases. Lancet 358, 1356–1360.

Cohen, J., Pertsemlidis, A., Kotowski, I.K., Graham, R., Garcia, C.K., Hobbs, H.H., 2005. Low LDL cholesterol in individuals of African descent resulting from frequent nonsense mutations in PCSK9. Nat Genet 37, 161–165.

Cohen, J.C., Boerwinkle, E., Mosley Jr., T.H., Hobbs, H.H., 2006. Sequence variations in PCSK9, low LDL, and protection against coronary heart disease. N Engl J Med 354, 1264–1272.

Collet, J.P., Hulot, J.S., Pena, A., et al., 2009. Cytochrome P450 2C19 polymorphism in young patients treated with clopidogrel after myocardial infarction: A cohort study. Lancet 373, 309–317.

Collins F., 2010. Has the revolution arrived? Nature 464, 674–675.

Damani, S.B., Topol, E.J., 2007. Future use of genomics in coronary artery disease. J Am Coll Cardiol 50, 479–486.

Damani S.B., Topol E.J., 2010. The case for routine genotyping in dual-antiplatelet therapy. J Am Coll Cardiol 56(2), 109–111.

Damani S.B., Topol E.J., 2011. Emerging clinical applications in cardiovascular pharmacogenomics. Wiley Interdiscip Rev Syst Biol Med 3, 206–215.

Dandona, S., Stewart, A., Chen, L., Williams, K., 2010. Gene dosage of the common variant 9p21 predicts severity of coronary artery disease. J Am Coll Cardiol 56, 479–486.

Danesh, J., Erqou, S., 2009. Lipoprotein(a) and coronary disease – moving closer to causality. Nat Rev Cardiol doi: 10.1038/nrcardio.2009.138.

Erqou S., Thompson A., Di Angelantonio E., et al., 2010. Apolipoprotein(a) isoforms and the risk of vascular disease: Systematic review of 40 studies involving 58,000 participants. J Am Coll Cardiol 55, 2160–2167.

Frazer, K.A., Ballinger, D.G., Cox, D.R., et al., 2007. A second generation human haplotype map of over 3.1 million SNPs. Nature 449, 851–861.

Frazer, K.A., Murray, S.S., Schork, N.J., Topol, E.J., 2009. Human genetic variation and its contribution to complex traits. Nat Rev 10, 241–251.

Gerdes, L.U., Gerdes, C., Kervinen, K., et al., 2000. The apolipoprotein epsilon4 allele determines prognosis and the effect on prognosis of simvastatin in survivors of myocardial infarction: A substudy of the Scandinavian simvastatin survival study. Circulation 101, 1366–1371.

Green, R.C., Roberts, J.S., Cupples, L.A., et al., 2009. Disclosure of APOE genotype for risk of Alzheimer's disease. N Engl J Med 361, 245–254.

Gretarsdottir S., Baas A.F., Thorleifsson G., et al., 2010. Genome-wide association study identifies a sequence variant within the DAB2IP gene conferring susceptibility to abdominal aortic aneurysm. Nat Genet 41, 692–697.

Guttmacher, A.E., Collins, F.S., 2002. Genomic medicine – a primer. N Engl J Med 347, 1512–1520.

Harismendy O., Notani D., Song X., et al., 2011. 9p21 DNA variants associated with coronary artery disease impair interferon-gamma signalling response. Nature 470, 264–268.

Helgadottir, A., Thorleifsson, G., Magnusson, K.P., et al., 2008. The same sequence variant on 9p21 associates with myocardial infarction, abdominal aortic aneurysm and intracranial aneurysm. Nat Genet 40, 217–224.

Helgadottir, A., Thorleifsson, G., Manolescu, A., et al., 2007. A common variant on chromosome 9p21 affects the risk of myocardial infarction. Science 316, 1491–1493.

Hirschhorn, J.N., Daly, M.J., 2005. Genome-wide association studies for common diseases and complex traits. Nat Rev 6, 95–108.

Hulot, J.S., Bura, A., Villard, E., et al., 2006. Cytochrome P450 2C19 loss-of-function polymorphism is a major determinant of clopidogrel responsiveness in healthy subjects. Blood 108, 2244–2247.

Hulot J.S., Collet J.P., Silvain J., et al., 2010. Cardiovascular risk in clopidogrel-treated patients according to cytochrome P450 2C19*2 loss-of-function allele or proton pump inhibitor coadministration: A systematic meta-analysis. J Am Coll Cardiol 56, 134–143.

Iakoubova, O., Shepherd, J., Sacks, F., 2008a. Association of the 719Arg variant of KIF6 with both increased risk of coronary events and with greater response to statin therapy. J Am Coll Cardiol 51, 2195, author reply 2196.

Iakoubova, O.A., Sabatine, M.S., Rowland, C.M., et al., 2008b. Polymorphism in KIF6 gene and benefit from statins after acute coronary syndromes: Results from the PROVE IT-TIMI 22 study. J Am Coll Cardiol 51, 449–455.

Iakoubova, O.A., Tong, C.H., Rowland, C.M., et al., 2008c. Association of the Trp719Arg polymorphism in kinesin-like protein 6 with myocardial infarction and coronary heart disease in 2 prospective trials: The CARE and WOSCOPS trials. J Am Coll Cardiol 51, 435–443.

Johansen C.T., Wang J., Lanktree M.B., et al., 2010. Excess of rare variants in genes identified by genome-wide association study of hypertriglyceridemia. Nat Genet 42, 684–687.

Jones S., Wang T.L., Shih I.M., et al., 2010. Frequent mutations of chromatin remodeling gene ARID1A in ovarian clear cell carcinoma. Science 330, 6001, 228–231.

Kaddurah-Daouk R., Baillie R.A., Zhu H., et al., 2010. Lipidomic analysis of variation in response to simvastatin in the Cholesterol and pharmacogenetics study. Metabolomics 6, 191–201.

Kamstrup, P.R., Tybjaerg-Hansen, A., Steffensen, R., Nordestgaard, B.G., 2009. Genetically elevated lipoprotein(a) and increased risk of myocardial infarction. JAMA 301, 2331–2339.

Kathiresan, S., Melander, O., Guiducci, C., et al., 2008. Six new loci associated with blood low-density lipoprotein cholesterol, high-density lipoprotein cholesterol or triglycerides in humans. Nat Genet 40, 189–197.

Kathiresan, S., Voight, B.F., Purcell, S., et al., 2009a. Genome-wide association of early-onset myocardial infarction with single nucleotide polymorphisms and copy number variants. Nat Genet 41, 334–341.

Kathiresan, S., Willer, C.J., Peloso, G.M., et al., 2009b. Common variants at 30 loci contribute to polygenic dyslipidemia. Nat Genet 41, 56–65.

Lander E.S., 2011. Initial impact of the sequencing of the human genome. Nature 470, 187–197.

Lewis, G.D., Asnani, A., Gerszten, R.E., 2008. Application of metabolomics to cardiovascular biomarker and pathway discovery. J Am Coll Cardiol 52, 117–123.

Link, E., Parish, S., Armitage, J., et al., 2008. *SLCO1B1* variants and statin-induced myopathy – a genomewide study. N Engl J Med 359, 789–799.

Lloyd-Jones, D., Adams, R., Carnethon, M., et al., 2009. Heart disease and stroke statistics – 2009 update: A report from the American Heart Association Statistics Committee and Stroke Statistics Subcommittee. Circulation 119, 480–486.

Manolio T.A., 2010. Genomewide association studies and assessment of the risk of disease. N Engl J Med 363, 166–176.

Marian, A.J., 2008. Surprises of the genome and "personalized" medicine. J Am Coll Cardiol 51, 456–458.

McPherson R., 2010. Chromosome 9p21 and coronary artery disease. N Engl J Med 362, 1736–1737.

McPherson, R., Pertsemlidis, A., Kavaslar, N., et al., 2007. A common allele on chromosome 9 associated with coronary heart disease. Science 316, 1488–1491.

Mega, J.L., Close, S.L., Wiviott, S.D., et al., 2009. Cytochrome p-450 polymorphisms and response to clopidogrel. N Engl J Med 360, 354–362.

Mega, J.L., Close, S.L., Wiviott, S.D., Shen, L., 2010. Genetic variants in *ABCB1* and *CYP2C19* and cardiovascular outcomes after treatment with clopidogrel and prasugrel in the TRITON-TIMI 38 trial: A pharmacogenetic analysis. Lancet 376(9749), 1312–1319.

Mehta, S.R., Yusuf, S., Peters, R.J., et al., 2001. Effects of pretreatment with clopidogrel and aspirin followed by long-term therapy in patients undergoing percutaneous coronary intervention: The PCI-CURE study. Lancet 358, 527–533.

Moliterno, D.J., Jokinen, E.V., Miserez, A.R., et al., 1995. No association between plasma lipoprotein(a) concentrations and the presence or absence of coronary atherosclerosis in African-Americans. Arterioscler Thromb Vasc Biol 15, 850–855.

Musunuru, K., Strong, A., Frank-Kamenetsky, M., et al., 2010. From noncoding variant to phenotype via *SORT1* at the 1p13 cholesterol locus. Nature 466, 714–719.

Nitsch, D., Molokhia, M., Smeeth, L., DeStavola, B.L., Whittaker, J.C., Leon, D.A., 2006. Limits to causal inference based on Mendelian randomization: A comparison with randomized controlled trials. Am J Epidemiol 163, 397–403.

Palomaki G.E., Melillo S., Bradley L.A., 2010. Association between 9p21 genomic markers and heart disease: A meta-analysis. JAMA 303, 648–656.

Pare, G., Mehta, S.R., Yusuf, S., 2010. Effects of *CYP2C19* genotype on outcomes of clopidogrel treatment. N Engl J Med 363, 1704–1714.

Park J.H., Wacholder S., Gail M.H., et al., 2010. Estimation of effect size distribution from genome-wide association studies and implications for future discoveries. Nat Genet 42, 570–575.

Paynter N.P., Chasman D.I., Pare G., et al., 2010. Association between a literature-based genetic risk score and cardiovascular events in women. JAMA 303, 631–637.

Ridker, P.M., Danielson, E., Fonseca, F.A., et al., 2008. Rosuvastatin to prevent vascular events in men and women with elevated C-reactive protein. N Engl J Med 359, 2195–2207.

Rosenberg S., Elashoff M.R., Beineke P., et al., 2010. Multicenter validation of the diagnostic accuracy of a blood-based gene expression test for assessing obstructive coronary artery disease in nondiabetic patients. Ann Intern Med 153, 425–434.

Sabatine, M.S., Liu, E., Morrow, D.A., et al., 2005. Metabolomic identification of novel biomarkers of myocardial ischemia. Circulation 112, 3868–3875.

Samani, N.J., Erdmann, J., Hall, A.S., et al., 2007. Genomewide association analysis of coronary artery disease. N Engl J Med 357, 443–453.

Saxena, R., Voight, B.F., Lyssenko, V., et al., 2007. Genome-wide association analysis identifies loci for type 2 diabetes and triglyceride levels. Science 316, 1331–1336.

Shah S.H., Bain J.R., Muehlbauer M.J., et al., 2010. Association of a peripheral blood metabolic profile with coronary artery disease and risk of subsequent cardiovascular events. Circ Cardiovasc Genet 3, 207–214.

Shiffman, D., Chasman, D.I., Zee, R.Y., et al., 2008. A kinesin family member 6 variant is associated with coronary heart disease in the Women's Health Study. J Am Coll Cardiol 51, 444–448.

Shuldiner, A.R., O'Connell, J.R., Bliden, K.P., et al., 2009. Association of cytochrome P450 2C19 genotype with the antiplatelet effect and clinical efficacy of clopidogrel therapy. JAMA 302, 849–857.

Sibbing D., Koch W., Gebhard D., et al., 2010. Cytochrome 2C19*17 allelic variant, platelet aggregation, bleeding events, and stent thrombosis in clopidogrel-treated patients with coronary stent placement. Circulation 121, 512–518.

Simon, T., Verstuyft, C., Mary-Krause, M., et al., 2009. Genetic determinants of response to clopidogrel and cardiovascular events. N Engl J Med 360, 363–375.

Song, Y., Stampfer, M.J., Liu, S., 2004. Meta-analysis: Apolipoprotein E genotypes and risk for coronary heart disease. Ann Intern Med 141, 137–147.

Steinhubl, S.R., Berger, P.B., Mann III, J.T., et al., 2002. Early and sustained dual oral antiplatelet therapy following percutaneous coronary intervention: A randomized controlled trial. JAMA 288, 2411–2420.

Surolia I., Pirnie S.P., Chellappa V., et al., 2010. Functionally defective germline variants of sialic acid acetylesterase in autoimmunity. Nature 466, 243–247.

Taubert, D., von Beckerath, N., Grimberg, G., et al., 2006. Impact of P-glycoprotein on clopidogrel absorption. Clin Pharmacol Ther 80, 486–501.

Teslovich, T.M., Musunuru, K., Smith, A.V., et al., 2010. Biological, clinical and population relevance of 95 loci for blood lipids. Nature 466, 707–713.

Tregouet, D.A., Konig, I.R., Erdmann, J., et al., 2009. Genome-wide haplotype association study identifies the *SLC22A3-LPAL2-LPA* gene cluster as a risk locus for coronary artery disease. Nat Genet 41, 283–285.

Venter J.C., 2010. Multiple personal genomes await. Nature 464, 676–677.

Voora, D., Shah, S.H., Spasojevic, I., et al., 2009. The SLCO1B1*5 genetic variant is associated with statin-induced side effects. J Am Coll Cardiol 54, 1609–1616.

Wallentin, L., Becker, R.C., Budaj, A., et al., 2009. Ticagrelor versus clopidogrel in patients with acute coronary syndromes. N Engl J Med 361, 1045–1057.

Wallentin, L., James, S., Storey, R.F., 2010. Effect of *CYP2C19* and *ABCB1* single nucleotide polymorphisms on outcomes of treatment with ticagrelor versus clopidogrel for acute coronary syndromes: A genetic substudy of the PLATO trial. Lancet 376(9749), 1320–1328.

Wellcome Trust Case Control Consortium, 2007. Genome-wide association study of 14,000 cases of seven common diseases and 3,000 shared controls. Nature 447, 661–678.

Willer, C.J., Sanna, S., Jackson, A.U., et al., 2008. Newly identified loci that influence lipid concentrations and risk of coronary artery disease. Nat Genet 40, 161–169.

Wiviott, S.D., Braunwald, E., McCabe, C.H., et al., 2007. Prasugrel versus clopidogrel in patients with acute coronary syndromes. N Engl J Med 357, 2001–2015.

Ye, S., Willeit, J., Kronenberg, F., Xu, Q., Kiechl, S., 2008. Association of genetic variation on chromosome 9p21 with susceptibility and progression of atherosclerosis: A population-based, prospective study. J Am Coll Cardiol 52, 378–384.

Yusuf, S., Zhao, F., Mehta, S.R., Chrolavicius, S., Tognoni, G., Fox, K.K., 2001. Effects of clopidogrel in addition to aspirin in patients with acute coronary syndromes without ST-segment elevation. N Engl J Med 345, 494–502.

Zacho, J., Tybjaerg-Hansen, A., Jensen, J.S., Grande, P., Sillesen, H., Nordestgaard, B.G., 2008. Genetically elevated C-reactive protein and ischemic vascular disease. N Engl J Med 359, 1897–1908.

CHAPTER

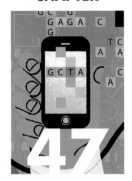

Atherosclerosis, Vulnerable Plaques, and Acute Coronary Syndromes

Jacob Fog Bentzon and Erling Falk

PLAQUE RUPTURE AND EROSION

Unstable angina and acute myocardial infarction (MI) are nearly always caused by a luminal thrombus superimposed on a coronary atherosclerotic plaque with or without concomitant vaso-spasm (Davies, 2000). In ST-segment elevation MI (STEMI), the thrombus is usually occlusive and sustained, whereas in unstable angina and non-STEMI, the thrombus is more often nonocclusive and dynamic. Acute or organized thrombus is found in the majority of victims of sudden coronary death; the rest die with severe coronary disease and myocardial scarring (old MI) in the absence of acute thrombosis (Davies, 2000; Virmani et al., 2000). Rare causes of acute coronary syndromes include thromboembolism, artery dissection, vasculitis, cocaine abuse, and trauma.

The most frequent cause of coronary thrombi is *plaque rupture*. In plaque rupture, a structural defect in the fibrous cap of the atherosclerotic plaque exposes the highly thrombogenic core to the blood (Figure 47.1) (Schaar et al., 2004). By definition, when no plaque rupture is identified despite a thorough search, the term plaque erosion is used (Figure 47.2). A large number of autopsy studies have been conducted in which fatal coronary thrombi were identified and the underlying plaques studied carefully to detect plaque rupture (Bentzon and Falk,

2010b). Overall, these studies show that plaque rupture is the major cause of coronary thrombosis, being responsible for approximately 75% of cases. Similar findings have been made *in vivo* in acute MI patients using optical coherence tomography imaging (Kubo et al., 2010a). Plaque rupture is a less frequent cause of thrombosis in women (60%) than in men (78%), and it is rare in a very small subgroup of patients, namely premenopausal women, who constitute less than 1% of heart attack victims (Burke et al., 1998). A few studies have reported that diabetes, smoking, and the level of hyperlipidemia are associated with the mechanism of thrombosis in acute coronary syndromes, but except for sex and menopause, no consistent relationships have been demonstrated (Falk et al., 2004).

PLAQUE VULNERABILITY

To predict which plaques are at risk of precipitating thrombosis and how it could be prevented, much focus is directed toward identifying so-called *vulnerable* plaques (or, synonymously, *thrombosis-prone* or *high-risk plaques*) – those plaques at high short-term risk of thrombosis (Schaar et al., 2004). The rationale is clear: if vulnerable plaques could be identified and rendered harmless before they would otherwise cause thrombosis, atherosclerosis would be a much less dangerous disease.

Genomic and Personalized Medicine, 2nd edition
by Ginsburg & Willard. DOI: http://dx.doi.org/10.1016/B978-0-12-382227-7.00047-1

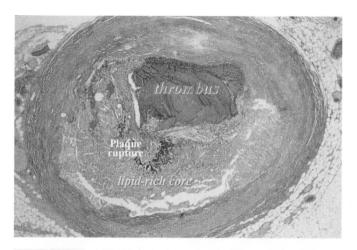

Figure 47.1 Thrombosis caused by plaque rupture. Cross-sectioned coronary artery showing a ruptured plaque with superimposed occlusive thrombosis. There is a structural defect in an extremely thin fibrous cap, with exposure of the highly thrombogenic, lipid-rich necrotic core to the blood. Trichrome, staining collagen blue and thrombus red (necrotic core colorless).

Since the mechanisms of superimposed thrombosis can be divided into two groups, plaque rupture and erosion, so can vulnerable plaques: *rupture-prone* and *erosion-prone plaques*.

Rupture-Prone Plaques

Table 47.1 outlines the features that characterize ruptured plaques. By inference, the same features, except thrombus, are assumed to characterize rupture-prone plaques. The prototypical rupture-prone plaque contains a large and soft lipid-rich necrotic core covered by a thin and inflamed fibrous cap (Figure 47.3) (Falk et al., 1995; Finn et al., 2010). Associated features include large plaque size, expansive remodeling mitigating luminal obstruction (mild stenosis by angiography), neovascularization, plaque hemorrhage, adventitial inflammation, and a "spotty" pattern of calcifications (Falk, 2006; Schaar et al., 2004). Many of these signs of vulnerability may be detected by imaging (Figure 47.4).

Necrotic Core

During atherogenesis, atherogenic lipoproteins are retained within intima, predominantly deeply in the abluminal part (Tabas, 2010). Some of these "pools" of lipids seem to attract

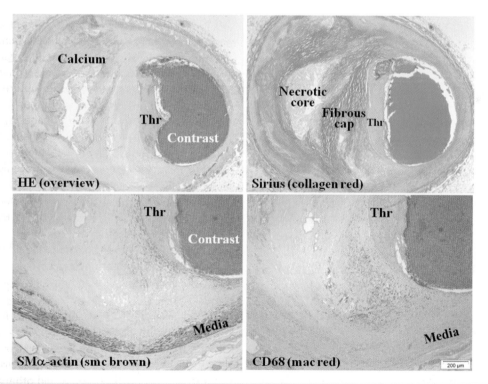

Figure 47.2 Thrombosis not caused by plaque rupture (erosion). Coronary artery containing a non-ruptured plaque with mural thrombosis (Thr), so-called *plaque erosion*. The endothelium between the plaque and the thrombus is missing, but no other morphologic features characterize this type of thrombosed plaques and their vulnerable precursors. The plaque is rich in smooth muscle cells (smc), but not macrophages (mac), just beneath the thrombus. Focal macrophage accumulation is seen at the shoulder region of the plaque but not located within a thin cap (i.e., this plaque morphology is not permissive for plaque rupture). HE, hematoxylin and eosin; SMα-actin, smooth muscle α-actin; CD, cluster of differentiation.

TABLE 47.1   Features of ruptured coronary plaques*
Thrombus
Weak fibrous cap
Thin (<65 µm)
Macrophage infiltration (inflammation)
Few smooth muscle cells (apoptosis)
Large lipid-rich core
Hemorrhage into the core
Other aspects
Big plaque
Expansive remodeling
Neovascularization
Spotty calcification
Adventitial inflammation

*By inference, the same features, except for thrombus, are assumed to characterize rupture-prone plaques.

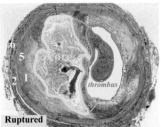

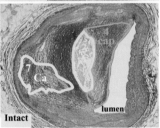

Unstable vs stable plaque

1. **Plaque size↑**
2. **Expansive remodeling**
3. **Necrotic core↑**
   - ~34% of plaque area
   - ~3.8 mm², ~9 mm long
4. **Fibrous cap**
   - thickness↓, ~23 mm (95% <65 mm)
   - macrophages (●)↑, ~26% of cap
   - smooth muscle cells↓ (apoptosis)
   - thrombus
5. **Angiogenesis↑**
   - intraplaque hemorrhage
6. **Perivascular inflammation**
7. **Calcification↓— spotty**

Figure 47.4   Targets for imaging of vulnerable (rupture-prone) plaques. For comparison, a ruptured coronary plaque with thrombus (top) and an intact and stable plaque (bottom) are depicted, and unstable (vulnerable) plaque features are listed to the right. By inference, vulnerable plaques of the rupture-prone type are presumed to look like plaque rupture except for an intact cap without thrombus. *Values obtained from Kolodgie et al., 2001.

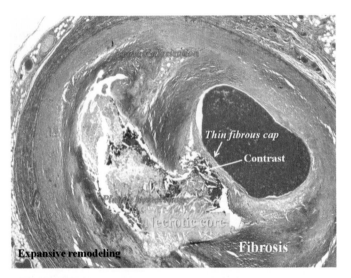

Figure 47.3   Thin-cap fibroatheroma (TCFA). A coronary TCFA containing a soft and destabilizing necrotic core devoid of supporting collagen and separated from the lumen only by a thin fibrous cap. Such a plaque is presumed to be rupture-prone and, if that is correct, constitutes the most common type of vulnerable plaques. In this case, postmortem contrast injection provided proof of vulnerability; the contrast has penetrated from the lumen into the necrotic core, indicating the presence of a disrupted fibrous cap. The contrast followed a path bordered by extravasated erythrocytes (plaque hemorrhage). Trichrome, staining collagen blue (necrotic core not stained).

macrophages that secrete proteolytic enzymes and engulf lipid until they die, leaving behind a soft and destabilizing lipid-rich cavity containing cholesterol crystals and devoid of supporting collagen and cells, the *necrotic core* (Tabas, 2010). Such a plaque is called an *atheroma* or *fibroatheroma* (Stary, 2000; Virmani et al., 2000).

Later on, plaque neovascularization (angiogenesis) supervenes (Sluimer et al., 2008). The new microvessels rarely originate from the lumen but usually from vasa vasorum in adventitia (Barger et al., 1984). They lack supporting cells and are fragile and leaky, giving rise to local extravasation of plasma proteins and erythrocytes (Sluimer et al., 2008). Such intraplaque bleedings are common and may expand the necrotic core, causing rapid progression of the lesion (Kolodgie et al., 2003). Another not uncommon source of plaque hemorrhage is extravasation of blood through a ruptured fibrous cap (Falk, 1983).

The presence of a necrotic core is the *sine qua non* of plaque rupture. If no necrotic core is present in the plaque, there is no overlying fibrous cap to rupture. However, a larger necrotic core also appears to confer greater risk than a small one (Gertz and Roberts, 1990; Virmani et al., 2000). The importance of necrotic core size for plaque stability is comprehensible, because the expansion of the core may erode the fibrous cap from below, and because the total lack of supporting collagen in the core confers greater tensile stress to the overlying fibrous cap. A large necrotic core may also promote thrombosis after plaque rupture and hence increase the risk of a clinical event because of its high amounts of prothrombotic apoptotic microparticles and tissue factor (Tedgui and Mallat, 2001).

Thin Fibrous Cap

Only very thin fibrous caps are at risk of rupturing. In the coronary arteries, the mean fibrous cap thickness in ruptured plaques at autopsy is 23 μm (the diameter of just *one* foam cell), and 95% of ruptured fibrous caps are below 65 μm (Burke et al., 1997). Based on these observations, Virmani and colleagues (2000) suggested the term *thin-cap fibroatheromas* (TCFAs) for coronary fibroatheromas with a fibrous cap thickness of less than 65 μm (Figure 47.3). This group of plaques encompasses the vast majority of rupture-prone plaques. On the other hand, probably not all TCFAs are destined to rupture (Finn et al., 2010). Indeed, ruptured plaques have larger necrotic cores, more cap inflammation and fewer cap SMCs than the average TCFA (Kolodgie et al., 2001).

Of note, the critical thickness of the fibrous cap seems to be artery-dependent. In the aorta and the carotid artery, plaque ruptures occur with slightly thicker fibrous caps; mean fibrous cap thickness of ruptured carotid plaques has been reported at 80 μm and that of ruptured aortic plaques at 130 μm (Felton et al., 1997; Trostdorf et al., 2005). This difference may reflect differences in vessel wall tension, being lowest in the coronary arteries, intermediate in carotid arteries, and highest in the aorta.

Thinning of the fibrous cap occurs by two concurrent and probably related mechanisms. One is the gradual loss of SMCs from the fibrous cap. Ruptured caps contain fewer SMCs and less collagen than intact caps (Davies et al., 1993; Kolodgie et al., 2001), and SMCs are usually absent at the actual site of rupture (Kolodgie et al., 2001; van der Wal et al., 1994). Since SMCs are the sole cell type responsible for synthesizing the extracellular fibers of the fibrous cap, their loss inevitably leads to cap weakening. At the same time, infiltrating macrophages degrade the collagen-rich cap matrix. Ruptured caps examined at autopsy are usually heavily infiltrated by macrophage foam cells (Falk, 1983; van der Wal et al., 1994), which secrete proteolytic enzymes such as plasminogen activators and a family of matrix metalloproteinases (MMPs: collagenases, gelatinases, and stromelysins) (Hansson, 2005). Although the macrophage density in ruptured caps is high, whole-plaque macrophage density rarely exceeds a few percent because ruptured caps are tiny (Figure 47.1) (Kolodgie et al., 2001).

Inflammation

Atherosclerosis is an inflammatory disease in which smoldering inflammatory activity is not confined to just a few atherosclerotic lesions but is present, more or less, in all such lesions throughout the body. In contrast, vulnerable plaques are relatively rare, and inflammation may play a causal role in plaque rupture only if located within a thin fibrous cap, i.e., the microstructure of the plaque needs to be permissive for the rupture. Thus, although plaque inflammation may be useful as a marker of disease activity, it is probably not useful as a stand-alone marker for plaque vulnerability.

Expansive Remodeling

During atherogenesis, the artery tends to remodel in such a way that the luminal obstruction is either attenuated (expansive remodeling) or accentuated (constrictive remodeling). Although vulnerable plaques of the rupture-prone type (TCFA) are usually large, they are often nonobstructive by angiography because of the expansive remodeling (Motoyama et al., 2009). This has been called the "remodeling paradox," because it is not good to preserve the lumen if it occurs at the expense of a higher risk of thrombosis and acute coronary syndrome (Pasterkamp et al., 1998). In contrast, plaques responsible for stable angina usually are smaller, but nevertheless are often associated with more severe luminal narrowing by angiography because of concomitant constrictive remodeling (Pasterkamp et al., 1998). The reasons for the different modes of remodeling remain to be elucidated, but recent clinical observations indicate that diabetes is accompanied by inadequate compensatory remodeling (Nicholls et al., 2008).

Other Features

A number of additional plaque features are more commonly found in ruptured plaques than in intact plaques, including increased neovascularization and adventitial inflammation (Kohchi et al., 1985; Virmani et al., 2005). Furthermore, culprit lesions responsible for acute coronary syndromes are generally less calcified than plaques responsible for stable angina, and the pattern of plaque calcification also differs (Ehara et al., 2004). These features are not independently associated with plaque rupture, however, and if they are causally involved in plaque rupture, their most likely mode of action is through modulation of the thickness or inflammation of the fibrous cap or the size of the lipid-rich core. Their special importance, however, lies in the fact that they may be targets for imaging.

Erosion-Prone Plaques

Vulnerable plaques of the erosion-prone type are heterogeneous and defined only by their fate (thrombosis, mostly mural) (Figure 47.2; Schaar et al., 2004; Virmani et al., 2005). The surface endothelium is missing, but whether it vanished before or after thrombosis remains unknown. No distinct morphological features have been identified (Virmani et al., 2005) but in general, eroded plaques with thrombosis are scarcely calcified, rarely associated with expansive remodeling, and only sparsely inflamed (Virmani et al., 2005). So, irrespective of plaque type, it is a misconception that vulnerable plaques are heavily inflamed. Advanced plaques, including those that appear vulnerable, are in general hypocellular and consist mainly of fibrous tissue, necrosis, and calcifications (Cheruvu et al., 2007; Falk, 2006). Thus, in contrast to vulnerable plaques of the rupture-prone type, detection of the erosion-prone type by imaging remains a challenge.

Location and Dynamics of Vulnerable Plaques

The number, distribution and natural history of vulnerable plaques are critical for choosing the most effective approach to identifying and treating them.

In two recent autopsy studies, the occurrence of TCFAs in the coronary tree, assumed to encompass the group of rupture-prone plaques, was examined. Cheruvu and colleagues (2007) analyzed 14 hearts with at least one plaque rupture (average 1.35 ruptures) and found a mean of 1.21 TCFAs at a second site. TCFAs were mostly located in the proximal part of the coronary tree, where also most coronary thrombi occur (Wang et al., 2004). Mauriello and colleagues, (2005) detected nearly two TCFAs (in addition to the culprit plaque) in acute MI patients. The true number of TCFAs in both studies was probably higher, however, because the applied sectioning of the coronary arteries was unlikely to reveal all TCFAs.

Whether critical thinning of the fibrous cap takes decades to evolve or is much more dynamic is not known. The fact that fibroatheromas are commonly seen from 30 years of age (Dalager et al., 2007), whereas acute coronary syndromes are exceedingly rare, seems to indicate that the development of rupture-prone plaques is usually a slow, smoldering process. On the other hand, a recent virtual histology intravascular ultrasound (VH-IVUS) study of non-culprit plaques found that during a 12 month period, most VH-TCFA (defined by the absence of recognizable fibrous cap by VH-IVUS) became more fibrotic without causing symptoms, while new VH-TCFA arose from less dangerous plaque types (Kubo et al., 2010b). The ability of the VH-IVUS technique to reliably assess changes in plaque components has recently been challenged, however (Thim et al., 2010b).

CHALLENGES FOR PREVENTION

Acute coronary syndromes and other complications of atherosclerosis are mainly problems of middle-aged and older people, but atherosclerosis is a life-long disease. Fatty streaks are present in the majority of people after puberty, intermediate lesions in most people from 20 to 30 years of age, and fibroatheromas are frequent from 30 years of age and beyond (Dalager et al., 2007; Stary et al., 1994, 1995). It is unknown whether the plaques that form first in life are also the first to cause symptoms, but the plaque that eventually causes an acute coronary syndrome probably has a history that spans several decades before the event. Thus, in principle there should be ample time to intervene against dangerous plaque development, but this is not achieved today. Less than 15% of non-diabetic patients hospitalized with a first atherosclerotic event are in preventive lipid-lowering treatment before admission (Sachdeva et al., 2009).

Clinical observations indicate that just modest life-long LDL cholesterol reductions can reduce the risk of acute coronary syndromes dramatically (Brown and Goldstein, 2006). Certain mutations in the *PCSK9* and *LDLR* genes (Cohen et al., 2006; Linsel-Nitschke et al., 2008) and near the *SORT1* gene (Musunuru et al., 2010) moderately lower LDL cholesterol, but the preventive effect against ischemic heart disease appears much greater than what can be achieved by larger reductions in LDL cholesterol late in life.

There should be substantial potential for prevention of atherosclerotic events if patients at risk could be identified earlier than they are now, and benefit from lifestyle changes, statins, and other risk factor-modifying drugs that are already available.

PERSONALIZED RISK PREDICTION

Atherothrombotic cardiovascular disease (CVD) is common, and is associated with increased morbidity, hospitalizations, and mortality (Lloyd-Jones et al., 2010). Despite, or rather because of, great advances in treating the disease, more patients are now surviving and living longer with chronic and costly disease (Fuster and Mearns, 2009; Lloyd-Jones et al., 2010). Today, CVD is three times more costly to treat than all cancers combined (Lloyd-Jones et al., 2010). Furthermore, sudden and unexpected death is still a common first manifestation of coronary atherosclerosis (Lloyd-Jones et al., 2010). The only effective approach to limit this undue loss of life, health, and resources is to prevent the disease from developing in the first place, i.e., primary prevention. The critical question is how to find the high-risk individuals who require personalized attention beyond that provided by a sensible public health (population) strategy. Despite an enormous effort, this goal appears elusive (Blankenberg et al., 2010; Shah et al., 2010). Although genome-wide association studies have substantially improved our understanding of the biology of atherothrombosis (Reilly et al., 2011), the impact of this new knowledge on personalized risk assessment is less obvious because of the small effect size of common genetic variants (Janssens and van Duijn, 2008; Kraft and Hunter, 2009; Manolio, 2010; Paynter et al., 2010). Unfortunately, testing for susceptibility genes, although potentially promising, has not yet proven useful for CVD risk stratification and cannot be recommended at this time (Greenland et al., 2010). In fact, the family history may provide more prognostic information than genetic testing (Kraft and Hunter, 2009; Paynter et al., 2010). A recent review article on genetic cardiovascular risk prediction concluded that:

> At this stage, clinicians should continue to inquire about family history for risk prediction, because this continues to represent a simple, cheap, and clinically useful risk factor for CVD that likely represents the net effect of hundreds of genetic risk variants that have yet to be discovered.
>
> Thanassoulis and Vasan, 2010

Risk Factors as Predictors

Causal risk factors for atherothrombotic CVD are known and constitute important therapeutic targets (Yusuf et al., 2004), but their usefulness as predictors for development of the disease is limited (Ware, 2006). Most heart attacks and strokes occur among people with an average risk factor level, misclassified as low- or intermediate-risk by traditional risk factor scoring (Ware, 2006). For example, in the Framingham Heart Study (Wilson et al., 2008), the Physicians' Health Study (Ridker et al., 2008), the Women's Health Study (Ridker et al., 2007), and the Northwick Park Heart Study (Shah et al., 2009), more than 75% of all hard coronary events occurred in people not classified as high-risk and consequently not offered optimal preventive therapy. In women, this figure may even pass 90% (Ridker et al., 2007).

Subclinical Atherosclerosis: Burden, Activity, and Vulnerability

An alternative and potentially much more rewarding risk assessment strategy seems ready for implementation. It is based not only on risk factors for developing atherosclerosis, but also incorporates direct and indirect evidence for already having it without knowing it (subclinical atherosclerosis), capturing the overall impact of all risk and susceptibility factors combined, known as well as unknown. Today, subclinical atherosclerosis can be detected directly by imaging, and its severity and extent (burden), progression rate (activity), and thrombosis-risk (vulnerability) can be assessed (Figure 47.4). The utility of such a personalized approach in the primary prevention of atherothrombotic CVD is being evaluated in several observational studies, including the population-based BioImage Study of the High-Risk Plaque (HRP) Initiative (Muntendam et al., 2010). The HRP Initiative focuses on near-term risk (Figure 47.5), and the predictive power of blood biomarkers (proteomics, lipidomics) is studied in collaboration with the Copenhagen Heart Study (Nordestgaard et al., 2010), and the Framingham Heart Study (http://www.hrpinitiative.com).

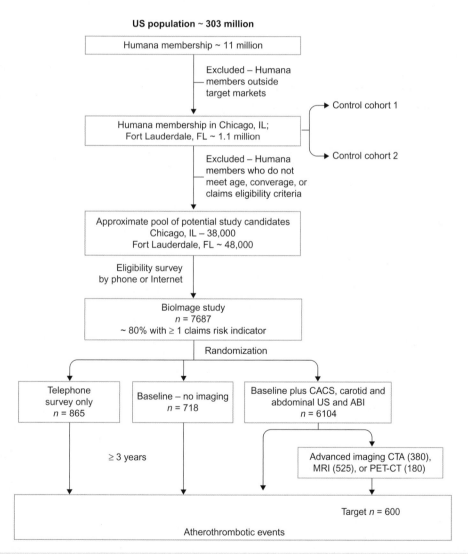

Figure 47.5 Flowchart for the BioImage Study. ABI, ankle-brachial index; CACS, coronary artery calcium score; CTA, computed tomography angiography; MRI, magnetic resonance imaging; PET-CT, positron emission tomography – computed tomography; US, ultrasound. *Reproduced with permission from Falk et al., 2011.*

Atherothrombotic disease is always preceded by a long preclinical phase in which subclinical atherosclerosis evolves and can be detected by non-invasive imaging, offering unique opportunities for timely and personalized preventive care. Coronary artery calcium (CAC) can be quantified by computed tomography (CT), and the CAC score correlates strongly with plaque burden and provides prognostic information beyond traditional risk factor scoring (Detrano et al., 2008; Erbel et al., 2010; Polonsky et al., 2010; Taylor et al., 2010). In particular, the CAC score may reclassify individuals of uncertain (intermediate) risk to lower or higher risk categories where the preventive approach is better defined (Erbel et al., 2010; Polonsky et al., 2010). CT coronary angiography may reveal plaque composition and vulnerability, but its clinical utility remains to be defined (Motoyama et al., 2009). In carotid arteries, the severity, progression and vulnerability of arterial disease may be assessed by ultrasonography (Nambi et al., 2010), and magnetic resonance imaging (MRI) (Underhill et al., 2010), and the incremental prognostic value provided by each of these modalities is under investigation.

Imaging inflammation in atherosclerosis serves two different purposes: detecting vulnerable patients (systemic activity) or vulnerable plaques of the rupture-prone type (focal activity). Because atherosclerosis is an innate inflammatory disease, inflammatory activity is not confined to just a few atherosclerotic lesions, but is present, more or less, in all such lesions throughout the body (Hansson, 2005). In contrast, vulnerable plaques are relatively rare (Waxman et al., 2006), and inflammation plays a causal role in plaque rupture only if the microstructure of the plaque is permissive for rupture (Figures 47.1, 47.3 and 47.4). With its high sensitivity, positron emission tomography (PET) imaging using fluorine-18-fluorodeoxyglucose has evolved as a promising method for the detection of arterial inflammation when combined with higher-resolution CT or MRI to help localize the signal to the arterial wall (Rudd et al., 2010).

CHALLENGES FOR TREATMENT OF ADVANCED ATHEROSCLEROSIS

Most of our understanding of the molecular mechanisms of atherosclerosis relates to the development of uncomplicated atherosclerotic plaques. When patients identify themselves at a late stage by suffering a clinical event, we know comparatively little about the relevant biological processes against which it may be possible to intervene. Our knowledge of what causes fibrous cap thinning and plaque rupture is incomplete, and the mechanisms leading to thrombosis on non-ruptured plaques are not known at all.

As the understanding of the morphology of vulnerable plaques becomes more clear, and as intravascular imaging techniques gain the resolution to identify them, it is imperative that we find out what to do about them. The potential to prevent new events by local treatment of co-existing asymptomatic lesions assumed to be vulnerable during PCI (percutaneous coronary intervention) for acute coronary syndromes is being explored. A target-lesion approach alone, however, will not eliminate the threat posed by all of the vulnerable plaques to come, and their overall risk determines the long-term prognosis.

To develop drugs that can protect vulnerable plaques against rupture and thrombosis, it is imperative to develop new, preferably small-animal, models in which the biology and dynamics of vulnerable plaques can be studied. Mouse models have proven themselves defective in that regard (Bentzon and Falk, 2010a), but the creation of knockout rats may provide new opportunities (Geurts et al., 2009). In addition, there is a need to increase the use and availability of large-animal atherosclerosis models for preclinical testing of emerging technologies in interventional cardiology and for the validation of imaging techniques, including especially those used for surrogate endpoints in clinical trials (Thim et al., 2010a, b). Several porcine models of advanced coronary atherosclerosis exist and may fulfill this need, and some of the most promising were recently reviewed (Granada et al., 2009).

REFERENCES

Barger, A.C., Beeuwkes III, R., Lainey, L.L., Silverman, K.J., 1984. Hypothesis: Vasa vasorum and neovascularization of human coronary arteries. A possible role in the pathophysiology of atherosclerosis. N Engl J Med 310, 175–177.

Bentzon, J.F., Falk, E., 2010a. Atherosclerotic lesions in mouse and man: Is it the same disease? Curr Opin Lipidol 21, 434–440.

Bentzon, J.F., Falk, E., 2010b. Pathogenesis of stable and acute coronary syndromes. In: Theroux, P. (Ed.), Acute Coronary Syndromes: A Companion to Braunwald's Heart Disease, second ed. Saunders, pp. 41–51.

Blankenberg, S., Zeller, T., Saarela, O., et al., 2010. Contribution of 30 biomarkers to 10-year cardiovascular risk estimation in 2 population cohorts: The MONICA, risk, genetics, archiving, and monograph (MORGAM) biomarker project. Circulation 121, 2388–2397.

Brown, M.S., Goldstein, J.L., 2006. Biomedicine: Lowering LDL – not only how low, but how long? Science 311, 1721–1723.

Burke, A.P., Farb, A., Malcom, G.T., Liang, Y.H., Smialek, J., Virmani, R., 1997. Coronary risk factors and plaque morphology in men with coronary disease who died suddenly. N Engl J Med 336, 1276–1282.

Burke, A.P., Farb, A., Malcom, G.T., Liang, Y., Smialek, J., Virmani, R., 1998. Effect of risk factors on the mechanism of acute thrombosis and sudden coronary death in women. Circulation 97, 2110–2116.

Cheruvu, P.K., Finn, A.V., Gardner, C., et al., 2007. Frequency and distribution of thin-cap fibroatheroma and ruptured plaques in human coronary arteries: A pathologic study. J Am Coll Cardiol 50, 940–949.

Cohen, J.C., Boerwinkle, E., Mosley Jr., T.H., Hobbs, H.H., 2006. Sequence variations in PCSK9, low LDL, and protection against coronary heart disease. N Engl J Med 354, 1264–1272.

Dalager, S., Paaske, W.P., Kristensen, I.B., Laurberg, J.M., Falk, E., 2007. Artery-related differences in atherosclerosis expression: Implications for atherogenesis and dynamics in intima-media thickness. Stroke 38, 2698–2705.

Davies, M.J., 2000. The pathophysiology of acute coronary syndromes. Heart 83, 361–366.

Davies, M.J., Richardson, P.D., Woolf, N., Katz, D.R., Mann, J., 1993. Risk of thrombosis in human atherosclerotic plaques: Role of extracellular lipid, macrophage, and smooth muscle cell content. Br Heart J 69, 377–381.

Detrano, R., Guerci, A.D., Carr, J.J., et al., 2008. Coronary calcium as a predictor of coronary events in four racial or ethnic groups. N Engl J Med 358, 1336–1345.

Ehara, S., Kobayashi, Y., Yoshiyama, M., et al., 2004. Spotty calcification typifies the culprit plaque in patients with acute myocardial infarction: An intravascular ultrasound study. Circulation 110, 3424–3429.

Erbel, R., Mohlenkamp, S., Moebus, S., et al., 2010. Coronary risk stratification, discrimination, and reclassification improvement based on quantification of subclinical coronary atherosclerosis: The Heinz Nixdorf Recall Study. J Am Coll Cardiol 56, 1397–1406.

Falk, E., 1983. Plaque rupture with severe pre-existing stenosis precipitating coronary thrombosis: Characteristics of coronary atherosclerotic plaques underlying fatal occlusive thrombi. Br Heart J 50, 127–134.

Falk, E., 2006. Pathogenesis of atherosclerosis. J Am Coll Cardiol 47, C7–C12.

Falk, E., Shah, P.K., Fuster, V., 1995. Coronary plaque disruption. Circulation 92, 657–671.

Falk, E., Shah, P.K., Fuster, V., 2004. Atherothrombosis and thrombosis-prone plaques. In: Fuster, V., Alexander, R.W., O'Rourke, R.A. (Eds.), Hurst's the Heart, eleventh ed. McGraw-Hill, pp. 1123–1139.

Falk, E., Sillesen, H., Muntendam, P., Fuster, V., 2011. The high-risk plaque initiative: Primary prevention of atherothrombotic events in the asymptomatic population. Curr Atheroscler Rep 13, 359–366.

Felton, C.V., Crook, D., Davies, M.J., Oliver, M.F., 1997. Relation of plaque lipid composition and morphology to the stability of human aortic plaques. Arterioscler Thromb Vasc Biol 17, 1337–1345.

Finn, A.V., Nakano, M., Narula, J., Kolodgie, F.D., Virmani, R., 2010. Concept of vulnerable/unstable plaque. Arterioscler Thromb Vasc Biol 30, 1282–1292.

Fuster, V., Mearns, B.M., 2009. The CVD paradox: Mortality vs prevalence. Nat Rev Cardiol 6, 669.

Gertz, S.D., Roberts, W.C., 1990. Hemodynamic shear force in rupture of coronary arterial atherosclerotic plaques. Am J Cardiol 66, 1368–1372.

Geurts, A.M., Cost, G.J., Freyvert, Y., et al., 2009. Knockout rats via embryo microinjection of zinc-finger nucleases. Science 325, 433.

Granada, J.F., Kaluza, G.L., Wilensky, R.L., Biedermann, B.C., Schwartz, R.S., Falk, E., 2009. Porcine models of coronary atherosclerosis and vulnerable plaque for imaging and interventional research. EuroIntervention 5, 140–148.

Greenland, P., Alpert, J.S., Beller, G.A., et al., 2010. ACCF/AHA guideline for assessment of cardiovascular risk in asymptomatic adults: A report of the American College of Cardiology Foundation/American Heart Association Task Force on Practice Guidelines. Circulation 122, e584–636.

Hansson, G.K., 2005. Inflammation, atherosclerosis, and coronary artery disease. N Engl J Med 352, 1685–1695.

Janssens, A.C., van Duijn, C.M., 2008. Genome-based prediction of common diseases: Advances and prospects. Hum Mol Genet 17, R166–R173.

Kohchi, K., Takebayashi, S., Hiroki, T., Nobuyoshi, M., 1985. Significance of adventitial inflammation of the coronary artery in patients with unstable angina: Results at autopsy. Circulation 71, 709–716.

Kolodgie, F.D., Burke, A.P., Farb, A., et al., 2001. The thin-cap fibroatheroma: A type of vulnerable plaque: The major precursor lesion to acute coronary syndromes. Curr Opin Cardiol 16, 285–292.

Kolodgie, F.D., Gold, H.K., Burke, A.P., et al., 2003. Intraplaque hemorrhage and progression of coronary atheroma. N Engl J Med 349, 2316–2325.

Kraft, P., Hunter, D.J., 2009. Genetic risk prediction: Are we there yet? N Engl J Med 360, 1701–1703.

Kubo, T., Imanishi, T., Kashiwagi, M., et al., 2010a. Multiple coronary lesion instability in patients with acute myocardial infarction as determined by optical coherence tomography. Am J Cardiol 105, 318–322.

Kubo, T., Maehara, A., Mintz, G.S., et al., 2010b. The dynamic nature of coronary artery lesion morphology assessed by serial virtual histology intravascular ultrasound tissue characterization. J Am Coll Cardiol 55, 1590–1597.

Linsel-Nitschke, P., Gotz, A., Erdmann, J., et al., 2008. Lifelong reduction of LDL-cholesterol related to a common variant in the LDL-receptor gene decreases the risk of coronary artery disease: A Mendelian Randomisation study. PLoS.One 3, e2986.

Lloyd-Jones, D., Adams, R.J., Brown, T.M., et al., 2010. Heart disease and stroke statistics – 2010 update: A report from the American Heart Association. Circulation 121, e46–e215.

Manolio, T.A., 2010. Genomewide association studies and assessment of the risk of disease. N Engl J Med 363, 166–176.

Mauriello, A., Sangiorgi, G., Fratoni, S., et al., 2005. Diffuse and active inflammation occurs in both vulnerable and stable plaques of the entire coronary tree: A histopathologic study of patients dying of acute myocardial infarction. J Am Coll Cardiol 45, 1585–1593.

Motoyama, S., Sarai, M., Harigaya, H., et al., 2009. Computed tomographic angiography characteristics of atherosclerotic plaques subsequently resulting in acute coronary syndrome. J Am Coll Cardiol 54, 49–57.

Muntendam, P., McCall, C., Sanz, J., Falk, E., Fuster, V., 2010. The BioImage Study: Novel approaches to risk assessment in the primary prevention of atherosclerotic cardiovascular disease – study design and objectives. Am Heart J 160, 49–57.

Musunuru, K., Strong, A., Frank-Kamenetsky, M., et al., 2010. From noncoding variant to phenotype via SORT1 at the 1p13 cholesterol locus. Nature 466, 714–719.

Nambi, V., Chambless, L., Folsom, A.R., et al., 2010. Carotid intima-media thickness and presence or absence of plaque improves prediction of coronary heart disease risk: The ARIC (Atherosclerosis Risk In Communities) study. J Am Coll Cardiol 55, 1600–1607.

Nicholls, S.J., Tuzcu, E.M., Kalidindi, S., et al., 2008. Effect of diabetes on progression of coronary atherosclerosis and arterial remodeling: A pooled analysis of 5 intravascular ultrasound trials. J Am Coll Cardiol 52, 255–262.

Nordestgaard, B.G., Adourian, A.S., Freiberg, J.J., Guo, Y., Muntendam, P., Falk, E., 2010. Risk factors for near-term myocardial infarction in apparently healthy men and women. Clin Chem 56, 559–567.

Pasterkamp, G., Schoneveld, A.H., Van der Wal, A.C., et al., 1998. Relation of arterial geometry to luminal narrowing and histologic markers for plaque vulnerability: The remodeling paradox. J Am Coll Cardiol 32, 655–662.

Paynter, N.P., Chasman, D.I., Pare, G., et al., 2010. Association between a literature-based genetic risk score and cardiovascular events in women. JAMA 303, 631–637.

Polonsky, T.S., McClelland, R.L., Jorgensen, N.W., et al., 2010. Coronary artery calcium score and risk classification for coronary heart disease prediction. JAMA 303, 1610–1616.

Reilly, M.P., Li, M., He, J., et al., 2011. Identification of ADAMTS7 as a novel locus for coronary atherosclerosis and association of ABO with myocardial infarction in the presence of coronary atherosclerosis: Two genome-wide association studies. Lancet 377, 383–392.

Ridker, P.M., Buring, J.E., Rifai, N., Cook, N.R., 2007. Development and validation of improved algorithms for the assessment of global cardiovascular risk in women: The Reynolds Risk Score. JAMA 297, 611–619.

Ridker, P.M., Paynter, N.P., Rifai, N., Gaziano, J.M., Cook, N.R., 2008. C-reactive protein and parental history improve global cardiovascular risk prediction: The Reynolds Risk Score for men. Circulation 118, 2243–2251.

Rudd, J.H., Narula, J., Strauss, H.W., et al., 2010. Imaging atherosclerotic plaque inflammation by fluorodeoxyglucose with positron emission tomography: Ready for prime time? J Am Coll Cardiol 55, 2527–2535.

Sachdeva, A., Cannon, C.P., Deedwania, P.C., et al., 2009. Lipid levels in patients hospitalized with coronary artery disease: An analysis of 136,905 hospitalizations in Get With The Guidelines. Am Heart J 157, 111–117.

Schaar, J.A., Muller, J.E., Falk, E., et al., 2004. Terminology for high-risk and vulnerable coronary artery plaques. Report of a meeting on the vulnerable plaque, June 17 and 18, 2003, Santorini, Greece. Eur Heart J 25, 1077–1082.

Shah, S.H., Granger, C.B., Hauser, E.R., et al., 2010. Reclassification of cardiovascular risk using integrated clinical and molecular biosignatures: Design of and rationale for the Measurement to Understand the Reclassification of Disease of Cabarrus and Kannapolis (MURDOCK) Horizon 1 Cardiovascular Disease Study. Am Heart J 160, 371–379.

Shah, T., Casas, J.P., Cooper, J.A., et al., 2009. Critical appraisal of CRP measurement for the prediction of coronary heart disease events: New data and systematic review of 31 prospective cohorts. Int J Epidemiol 38, 217–231.

Sluimer, J.C., Gasc, J.M., Van Wanroij, J.L., et al., 2008. Hypoxia, hypoxia-inducible transcription factor, and macrophages in human atherosclerotic plaques are correlated with intraplaque angiogenesis. J Am Coll Cardiol 51, 1258–1265.

Stary, H.C., 2000. Natural history and histological classification of atherosclerotic lesions: An update. Arterioscler Thromb Vasc Biol 20, 1177–1178.

Stary, H.C., Chandler, A.B., Dinsmore, R.E., et al., 1995. A definition of advanced types of atherosclerotic lesions and a histological classification of atherosclerosis. A report from the Committee on Vascular Lesions of the Council on Arteriosclerosis, American Heart Association. Arterioscler Thromb Vasc Biol 15, 1512–1531.

Stary, H.C., Chandler, A.B., Glagov, S., et al., 1994. A definition of initial, fatty streak, and intermediate lesions of atherosclerosis. A report from the Committee on Vascular Lesions of the Council on Arteriosclerosis, American Heart Association. Circulation 89, 2462–2478.

Tabas, I., 2010. Macrophage death and defective inflammation resolution in atherosclerosis. Nat Rev Immunol 10, 36–46.

Taylor, A.J., Cerqueira, M., Hodgson, J.M., et al., 2010. ACCF/SCCT/ACR/AHA/ASE/ASNC/SCAI/SCMR 2010 Appropriate use criteria for cardiac computed tomography. A report of the American College of Cardiology Foundation Appropriate Use Criteria Task Force, the Society of Cardiovascular Computed Tomography, the American College of Radiology, the American Heart Association, the American Society of Echocardiography, the American Society of Nuclear Cardiology, the Society for Cardiovascular Angiography and Interventions, and the Society for Cardiovascular Magnetic Resonance. Circulation 122, e525–555.

Tedgui, A., Mallat, Z., 2001. Apoptosis as a determinant of atherothrombosis. Thromb Haemost 86, 420–426.

Thanassoulis, G., Vasan, R.S., 2010. Genetic cardiovascular risk prediction: Will we get there? Circulation 122, 2323–2334.

Thim, T., Hagensen, M.K., Drouet, L., et al., 2010a. Familial hypercholesterolemic downsized pig with human-like atherosclerosis: A model for preclinical studies. EuroIntervention 6, 261–268.

Thim, T., Hagensen, M.K., Wallace-Bradley, D., et al., 2010b.Drouet, L., Unreliable assessment of necrotic core by virtual histology intravascular ultrasound in porcine coronary artery disease. Circ Cardiovasc Imaging 3, 384–391.

Trostdorf, F., Buchkremer, M., Harmjanz, A., et al., 2005. Fibrous cap thickness and smooth muscle cell apoptosis in high-grade carotid artery stenosis. Eur J Vasc Endovasc Surg 29, 528–535.

Underhill, H.R., Hatsukami, T.S., Fayad, Z.A., Fuster, V., Yuan, C., 2010. MRI of carotid atherosclerosis: Clinical implications and future directions. Nat Rev Cardiol 7, 165–173.

Van der Wal, A.C., Becker, A.E., Van der Loos, C.M., Das, P.K., 1994. Site of intimal rupture or erosion of thrombosed coronary atherosclerotic plaques is characterized by an inflammatory process irrespective of the dominant plaque morphology. Circulation 89, 36–44.

Virmani, R., Kolodgie, F.D., Burke, A.P., Farb, A., Schwartz, S.M., 2000. Lessons from sudden coronary death: A comprehensive morphological classification scheme for atherosclerotic lesions. Arterioscler. Thromb Vasc Biol 20, 1262–1275.

Virmani, R., Kolodgie, F.D., Burke, A.P., et al., 2005. Atherosclerotic plaque progression and vulnerability to rupture: Angiogenesis as a source of intraplaque hemorrhage. Arterioscler Thromb Vasc Biol 25, 2054–2061.

Wang, J.C., Normand, S.L., Mauri, L., Kuntz, R.E., 2004. Coronary artery spatial distribution of acute myocardial infarction occlusions. Circulation 110, 278–284.

Ware, J.H., 2006. The limitations of risk factors as prognostic tools. N Engl J Med 355, 2615–2617.

Waxman, S., Ishibashi, F., Muller, J.E., 2006. Detection and treatment of vulnerable plaques and vulnerable patients: Novel approaches to prevention of coronary events. Circulation 114, 2390–2411.

Wilson, P.W., Pencina, M., Jacques, P., Selhub, J., D'Agostino Sr., R., O'Donnell, C.J., 2008. C-reactive protein and reclassification of cardiovascular risk in the Framingham Heart Study. Circ Cardiovasc Qual Outcomes 1, 92–97.

Yusuf, S., Hawken, S., Ounpuu, S., et al., 2004. Effect of potentially modifiable risk factors associated with myocardial infarction in 52 countries (the INTERHEART study): Case-control study. Lancet 364, 937–952.

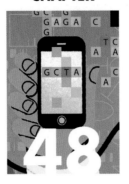

CHAPTER

Heart Failure

Ivor J. Benjamin

DEFINITIONS

Heart failure has been conveniently subdivided according to abnormalities in the cardiac cycle, namely, systolic heart failure (SHF) and diastolic heart failure (DHF). SHF is associated with decreased cardiac output and ventricular contractility, termed systolic dysfunction, and is attributed to a loss of ventricular muscle cells. Dilated cardiomyopathy (DCM) refers to the enlargement of one or both ventricular chambers, characterized by impaired systolic function and myocardial remodeling. Idiopathic dilated cardiomyopathy (IDCM) is primary myocardial disease in the absence of coronary, valvular, or systemic disease. The ventricular remodeling of DHF, however, is characterized by normal chamber size without impaired ventricular filling from abnormal myocardial stiffness during the relaxation phase. More recently, the clinical syndrome of heart failure with normal ejection fraction (HFNEF) – left ventricular ejection fraction >50% – has been recognized in several cross-sectional studies (Bhatia et al., 2006; Owan et al., 2006).

PREDISPOSITION – GENETIC AND NON-GENETIC

In Western societies, ischemic heart disease and hypertension are the most common causes of ventricular systolic dysfunction, but there is now irrefutable evidence for genetic defects whose onset and progression occur in adulthood (e.g., familial cardiomyopathy) (Benjamin and Schneider, 2005; Morita et al., 2005). Beginning in the late 1950s, distinct alterations in the size and geometry of the left ventricle, termed "ventricular remodeling," were being recognized, but the ensuing debate, which lasted for more than three decades, was primarily focused on morphological classifications. Hypertrophic cardiomyopathy (HCM) is characterized by predominant and marked thickening of the left circumferential ventricular wall (i.e., hypertrophy), small left ventricle (LV) cavity size, and hypercontractility. Such patients, including young athletes, were prone to sudden cardiac death attributed pathophysiologically to subaortic stenosis and cavity obliteration triggering inadequate cardiac output and lethal arrhythmias. In contrast, dilatation of left ventricular cavity and reduced systolic function are the hallmarks of DCM. In 1991, the Seidmans' laboratory at Harvard Medical School reported for the first time that mutations in the gene encoding the β-myosin heavy chain, a major structural and contractile protein, was the genetic basis for familial hypertrophic cardiomyopathy associated with sudden death, ushering in the present era of cardiovascular genomic medicine. This seminal discovery permanently shifted the paradigm from the morphological to the molecular, enabling the identification of disease-causing genes, the characterization of biochemical and metabolic pathways, and from such basic insights into disease pathogenesis to consider novel therapeutic interventions.

Genomic and Personalized Medicine, 2nd edition
by Ginsburg & Willard. DOI: http://dx.doi.org/10.1016/B978-0-12-382227-7.00048-3

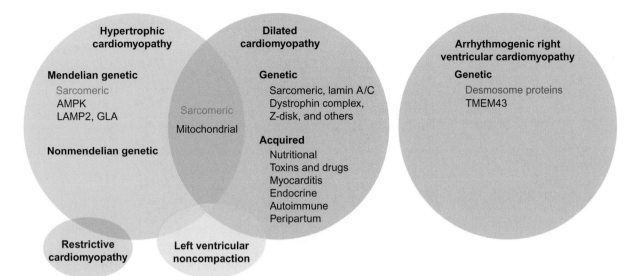

Figure 48.1 Clinical categories of inherited cardiomyopathies and their genetic basis. The clinical entities hypertrophic cardiomyopathy and dilated cardiomyopathy share some disease genes with each other, as well as with restrictive cardiomyopathy and left ventricular non-compaction, which are less common. Arrhythmogenic right ventricular cardiomyopathy appears to be a genetically distinct category, although its clinical phenotype cannot always be easily distinguished from that of dilated cardiomyopathy. AMPK denotes adenosine monophosphate-activated protein kinase, GLA α-galactosidase A, LAMP2 lysosomal-associated membrane protein 2, and TMEM43 transmembrane protein 43. Classes of genes shown in red are the overwhelmingly predominant cause of disease within the respective categories. *Reproduced with permission from Watkins et al., 2011.*

Many more single genetic defects are routinely being linked to familial heart failure (Watkins et al., 2011) (Figure 48.1), but an important future challenge is to establish how inherited and acquired or epigenetic factors conspire to drive the growing epidemic of heart failure. See Table 48.1 for a summary of the genes identified as associated with cardiomyopathies.

Severe occlusive coronary disease is the substrate for acute coronary syndromes, myocardial infarctions and subsequent pump failure, as shown in Figure 48.2. The high prevalence of heart failure in African-Americans with hypertension underscores potential gene–environment interactions in selected populations. Infectious etiologies (e.g., rheumatic heart disease) are declining, but valvular heart disease from iatrogenic causes (e.g., diet pills, toxins) remains an important risk factor (Figure 48.2). Viruses (e.g., Coxsackie's B3, parvovirus) are the major suspected culprits for IDCM, in which the post-viral sequelae of inflammation and apoptosis trigger ventricular remodeling and dilation (Liu and Mason, 2001). IDCM accounts for 30% of cases of DCM. Heart failure on presentation in the peripartum or post-partum period has a variable clinical course, from severe pump failure to complete recovery. The most common cause of right ventricular heart failure (RVHF) is left ventricular systolic dysfunction. In addition, RVHF is associated with congenital heart disease (e.g., tetralogy of Fallot), primary pulmonary hypertension, and arrhythmogenic right ventricular dysphasia and right ventricular infarction. Stress cardiomyopathy is a rare

reversible form of left ventricular dysfunction associated clinically with emotional stress, angiographically with "apical ballooning," and pathophysiologically with excess sympathetic activation (Wittstein et al., 2005). This entity remains a diagnosis of exclusion, which mimics ST-segment elevation MI (STEMI; MI=myocardial infarction) on presentation, but has a much more favorable clinical outcome than STEMI. Lastly, thyrotoxicosis, Paget's disease, and severe chronic anemia are rare causes of high-output heart failure. Individuals afflicted with heart failure with preserved ejection fraction are more commonly older age, female gender, and have a history of hypertension and atrial fibrillation.

SCREENING

The New York Heart Association (NYHA) functional classification scheme, an older but widely used screening tool, assesses the severity of functional limitations of individuals afflicted with heart failure. The four classes of the NYHA classification are linked to increasing severity of signs and symptoms and correlate well with prognosis. This classification scheme, however, has important limitations, since diverse pathophysiological processes leading to symptomatic heart failure are overlooked (Dunselman et al., 1988). Accordingly, the American College of Cardiology and American Heart Association (ACC/AHA) Classification of Chronic Heart Failure was developed to account for the multiple

TABLE 48.1	The genetic basis of cardiomyopathies		
Symbol	**Chromosome**	**Gene product**	**Cardiomyopathy type**
ACTC	15q11-14	Cardiac muscle α-actin	Hypertrophic and dilated
ABCC9	12p12.1	Member 9 of the superfamily C of ATP-binding cassette transporters	Dilated
CSRP3, MLP	11p15.1	Cysteine- and glycine-rich protein 3	Dilated
DES	2q35	Desmin	Dilated
DSP	6p24	Desmoplakin	Dilated
LMNA	1q21.2-21.3	Lamin A/C	Dilated
VCL	10q22.1-q23	Metavinculin	Dilated
MYBPC3	11p11.2	Cardiac myosin-binding protein C	Hypertrophic and dilated
MYH6	14q12	Cardiac muscle α-isoform of myosin heavy chain (heavy polypeptide 6)	Hypertrophic
MYH7	14q12	Cardiac muscle α-isoform of myosin heavy chain (heavy polypeptide 7)	Hypertrophic and dilated
MYL2	12q23-24.3	Myosin regulatory light chain associated with cardiac myosin-β (or slow) heavy chain	Hypertrophic
MYL3	3p21.2-21.3	Myosin light chain 3	Hypertrophic
PLN	6q22.1	Phospholamban	Dilated
PRKAG2	7q35-36	γ2 non-catalytic subunit of AMP-activated protein kinase	Hypertrophic
SGCB	4q12	β-sarcoglycan (43 kDa dystrophin-associated glycoprotein)	Dilated
SGCD	5q33-34	δ-sarcoglycan (35 kDa dystrophin-associated glycoprotein)	Dilated
TAZ, G4.5	Xq28	Tafazzin	Dilated
TTN	2q31	Titin	Hypertrophic and dilated
TCAP	17q12	Titin-cap	Dilated
TPM1	15q22.1	Tropomyosin 1 (α)	Hypertrophic
TNNI3	19q13.4	Troponin I, a subunit of the troponin complex of the thin filaments of striated muscle	Hypertrophic
TNNT2	1q32	Cardiac isoform of troponin T2, tropomyosin-binding subunit of the troponin complex	Hypertrophic and dilated

ATP = adenosine triphosphate, AMP = adenosine monophosphate.
Adapted with permission from Liew and Dzau, 2004.

stages and predisposing conditions associated with the clinical syndrome. Designed to encompass emerging scientific evidence, an expert panel periodically assembles these updates, which are the most widely used and authoritative sources on the evaluation, management, performance measures, and outcomes of heart failure (Bonow et al., 2005; Hunt et al., 2005; Radford et al., 2005). In turn, these guidelines incorporate preclinical stages, risk factors, pathophysiologic stages, and clinical recognition of heart failure, that are further subdivided into four stages:

Stage A patients are at high risk for developing heart failure, but have had neither symptoms nor evidence of structural cardiac abnormalities. Major risk factors include hypertension, diabetes mellitus, coronary artery disease, and family history of cardiomyopathy. In selected patients, the administration of an angiotensin-converting enzyme (ACE) inhibitor or an angiotensin receptor blocker (ARB) is recommended to prevent adverse ventricular remodeling.

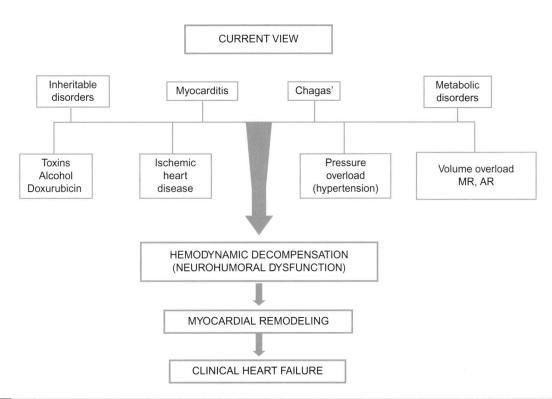

The different etiologies and multiple compensatory and adaptive pathways implicated in the clinical syndrome of heart failure. MR = mitral regurgitation, AR = aortic regurgitation.

Stage B patients have structural abnormalities from previous myocardial infarction, LV dysfunction or valvular heart disease, but have remained asymptomatic. Both ACE inhibitors and beta-blockers are recommended.

Stage C patients have evidence of structural abnormalities along with current or previous symptoms of dyspnea, fatigue, and impaired exercise tolerance. In addition to ACE inhibitors and beta-blockers, the optimal medical regimen may include diuretics, digoxin, and aldosterone antagonists.

Stage D patients have end-stage symptoms of heart failure that are refractory to standard maximal medical therapy. Such patients are candidates for left ventricular assist devices (LVADs) and other sophisticated maneuvers for hemodynamic resuscitation, myocardial salvage and/or end-of-life care.

PATHOPHYSIOLOGY

Neurohumoral Mechanisms

Low cardiac output and systemic hypoperfusion caused by LV dysfunction elicit a cascade of compensatory mechanisms, principally neurohumoral activation and the renin-angiotensin-aldosterone system (RAAS) (Figure 48.2). Sympathetic nervous system (SNS) release of catecholamines (i.e., norepinephrine, epinephrine) from the adrenal medulla increases heart rate and contractility and stimulates vasoconstriction and peripheral resistance, triggering increased afterload and myocardial oxygen consumption. Recruitment of RAAS, which is designed evolutionarily for fluid retention, sets up a vicious feedback loop whose deleterious consequences from volume overload can exacerbate decompensated heart failure. In contrast, natriuretic peptides released from specialized cells in the atria exert hormonal actions in distant vascular beds, stimulating vasodilation and diuresis. Afterload-reducing agents (i.e., ACE inhibitors, ARBs, and beta-adrenergic blockers) have significantly reduced morbidity and mortality while improving the survival of patients with heart failure. Likewise, antagonists of aldosterone, which promotes salt and water retention, have proven clinical benefits. Because β-blocker therapy has important implications for pharmacogenomics, a brief review of its clinical and translational development is warranted.

Early efforts to identify and characterize the subtypes of adrenergic receptors (e.g., α_1- and β_1-subtypes) have yielded important insights about receptor pharmacology, thus improving our understanding of the maladaptive regulatory mechanisms during heart failure. When Bristow and coworkers began their groundbreaking studies to identify whether elevations of catecholamines might down-regulate adrenergic receptors in heart failure (Bristow et al., 1986), there was initial skepticism within the cardiovascular community (Bristow, 2011). The notion that β-adrenergic blockade might mitigate the deleterious effects of excessive catecholamine caused by adrenergic stimulation seemed counterintuitive, since existing evidence suggested

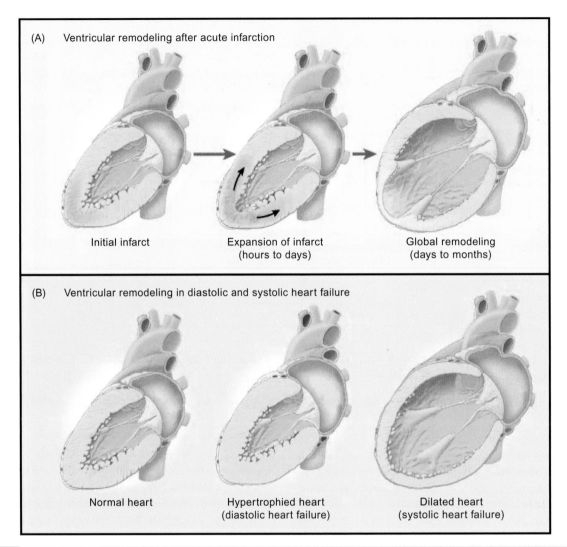

Figure 48.3   (A) Ventricular remodeling after infarction, and (B) in diastolic heart failure. *Reproduced from Jessup and Brozena, 2003.*

that decreased norepinephrine in the failing heart resulted from receptor denervation, leading to receptor hypersensitivity and catecholamine excess. Without doubt, the early clinical trials by the Göteborg Group, in which anti-adrenergic strategies were administered in patients with heart failure, defied and challenged the prevailing dogma (Waagstein et al., 1975).

Although the gold standard for efficacy in heart failure treatment was improvement in submaximal exercise tolerance and functional capacity, the sea change came from the PRECISE and MOCHA trials, which showed improvements with β-blockers (e.g., Carvedilol) in mortality and hospitalization rate (Bristow et al., 1996). These seminal observations paved the way for the comprehensively chronicled journey in which anti-adrenergic agents became mainstay therapy for congestive heart failure (CHF) (Bristow, 2011). Lastly, the storied history of β-blockers has important implications for pharmacogenomic targeting, as the collaborative studies on the $\beta_1$389Arg/Gly polymorphism by Dr Liggett have elegantly shown (Liggett et al., 2006).

Myocardial Remodeling

Left ventricular dysfunction and systolic heart failure secondary to myocardial infarction or ischemia are the prerequisites of low ejection fraction and elevated pulmonary pressures with congestion. Acquired or inherited conditions that either decrease cardiomyocyte viability and/or increase cell death will ultimately trigger pump failure and symptomatic heart failure. Given the heart's limited capacity for regeneration (Bergmann et al., 2009), terminally differentiated ventricular cardiomyocytes may undergo hypertrophy in response to increased metabolic and homodynamic demands. Activation of the "fetal gene program" orchestrates transcriptional up-regulation of genes encoding contractile and cytoskeletal proteins – the prerequisite for compensatory hypertrophy. Recruitment of such adaptive mechanisms provides a variable but stable and asymptomatic interval – perhaps lasting years – before cardiac decompensation. The ensuing ventricular dilatation is a pathologic form of adaptation called ventricular remodeling, affecting intrinsic cardiac mass, the extracellular matrix, collagen deposition, and fibrosis, as shown in Figures 48.2 and 48.3. Whereas low levels of reactive oxygen species (ROS) serve as stress signals in redox-dependent regulation, elevated levels of ROS caused by mitochondrial dysfunction may alter myocardial energetics and cardiac metabolism, and may trigger the release of cytochrome c, thereby activating cell survival/death pathways. Endothelial dysfunction gives rise to the aberrant release of nitric oxide, a potent vasodilator, and/or reactivity with reactive oxygen species to form peroxynitrite, which causes oxidative damage and cellular injury. Progressive remodeling, in attempts to maintain systolic function and homeostasis (Stage B), leads to valvular regurgitation from inadequate apposition of the mitral leaflets, increasing myocardial stress, and ultimately to decompensated heart failure (Stage C and Stage D).

Mechanisms of Cell Death in Heart Failure

Progressive loss of cardiomyocytes from either necrosis or apoptosis with diverse pathogenic states contributes to the pathogenesis of heart failure, as shown in Figure 48.3 (Liew and Dzau, 2004; Wencker et al., 2003). Both beneficial and deleterious effects of catecholamines ensue from SNS activation, such as increased renin secretion, cell death, fibrosis, and myocardial irritability, which are the underlying substrates and proximate substrates for lethal arrhythmias and sudden death. Apoptosis or programmed cell death is activated by signaling cascades via either the extrinsic or intrinsic cell survival/death pathways (Danial and Korsmeyer, 2004). Ligands such as TNF-α, which bind to cognate receptors at the plasma membrane, mediate cell death through the extrinsic pathway, whereas the Bcl-2 family – consisting of both pro- and anti-apoptotic proteins – regulates the intrinsic pathway. Mitochondria play a central role in cell survival/death, principally from the initiation of stress signals (e.g., ROS) and release of mitochondrial cytochrome c, which initiates complex formation and the activation of apoptotic proteases (e.g., caspase-9) (Danial and Korsmeyer, 2004). The role of apoptosis in chronic heart failure, which ranges between 80–250 myocytes per 100,000 nuclei in failing human hearts, was elegantly validated by Wencker and coworkers using transgenic mice harboring a fusion protein FK506 binding protein (FKBP) fused with a conditionally active caspase (Wencker et al., 2003). In contrast, low-level inhibition of apoptosis prevented DCM and death, suggesting possible therapeutic strategies for combating heart failure.

DIAGNOSIS

Patients newly diagnosed with heart failure most commonly seek medical attention for either gradual or abrupt onset of the classical signs and symptoms of pulmonary congestion. The clinical spectrum varies widely, but dyspnea on exertion, peripheral edema, orthopnea and paroxysmal nocturnal dyspnea are not uncommon. Exertional chest pain or angina at rest requires an immediate evaluation to determine if biochemical evidence of myocardial damage demands more aggressive management of acute coronary syndromes. Elevated jugular venous distension from right ventricular failure, ascites and cachexia are more ominous signs of low cardiac output and decompensation, requiring urgent attention, preferably from a provider who specializes in heart failure management.

Routine diagnostic studies include an electrocardiogram, chest radiograph, and B-type natriuretic peptide, the latter having the best predictive value for distinguishing between CHF and non-CHF patients (Maisel and McCullough, 2003). Noninvasive echocardiography is the most commonly used diagnostic tool for the assessment and follow-up of patients with heart failure with or without preserved ejection fraction. Coronary angiography should be performed to exclude reversible causes of left ventricular dysfunction or to guide prompt revascularization. If the coronary vessels are widely patent in the setting of global dysfunction, then endomyocardial biopsy should be considered to assess for reversible causes, including viral myocarditis (Liu and Mason, 2001). Equilibrium radionucleotide angiography (ERNA) is another noninvasive diagnostic study that assesses both left and right ventricular systolic function. Screening tools such as contrast-enhanced computed tomographic angiography and magnetic resonance imaging (MRI) are gaining attention as emerging technologies with equivalent sensitivity and specificity to the invasive angiogram for coronary arteriography. MRI may also uncover unsuspected infiltrative cardiomyopathy, arrhythmogenic right ventricular dysplasia, and allows the assessment of myocardial viability before revascularization. In the STICH (Surgical Treatment for Ischemic Heart Failure) trial, however, the assessment of myocardial viability using either single-photon emission computed tomography (SPECT) or dobutamine echocardiography did not provide a mortality benefit for patients with coronary artery disease and left ventricular dysfunction randomized to either CABG (coronary artery bypass and graft surgery) or optimal medical therapy (Bonow et al., 2011).

PROGNOSIS

More than 5.7 million Americans, or 1.5% of the US population, have chronic heart failure, and there is a similar number of Americans at risk of undiagnosed left ventricular dysfunction (Braunwald and Bristow, 2000). With more than 550,000 new cases of heart failure and 300,000 deaths each year, the disproportionate health and economic burden exceeds $34 billion annually (Heidenreich et al., 2011). Soon, heart failure will become the number one cause of death worldwide, eclipsing infectious disease (Bleumink et al., 2004). Although new pharmacologic management and revascularization techniques continue to improve survival after acute myocardial infarction, the prevalence of chronic heart failure appears to be increasing as the population ages (Braunwald and Bristow, 2000). Notwithstanding, heart failure accounts for 20% of all hospital admissions in patients older than 65, and the hospitalization rate has increased by 159% in the past decade

(Jessup and Brozena, 2003). Available treatments for heart failure have only modestly improved morbidity and mortality (Jessup and Brozena, 2003), and for patients with advanced heart failure, the prognosis still remains grim, with one-year mortality rates between 20% and 45%, overshadowing the worst forms of some cancers (Jessup and Brozena, 2003).

Pharmacogenomics

If the ultimate goal of pharmacogenomics is to apply an individual's genetic variation to predict either the best drug response or adverse outcome, then substantial hurdles must be overcome to improve the sensitivity and specificity of approaches in current practice (Arnett et al., 2007). Heart failure is a clinical syndrome arising from predisposing risk factors [e.g., coronary artery disease (CAD)]. Its phenotypic heterogeneity and the variability of responses among individuals taking pharmacologic agents has been attributed to common polymorphisms in the genome (Liggett, 2001; Roberts, 2008). A surmountable hurdle, however, is the robustness of the association between putative genetic markers and therapeutic response. Recent lessons from studies of human heart failure illustrate handsomely both the enormous potential and challenges for pharmacogenomics – a maturing discipline in which an individual's genetic determinants are used to predict drug response and outcomes (Evans and Relling, 1999; Liggett, 2001). Liggett and coworkers have demonstrated that non-synonymous single-nucleotide polymorphisms (SNPs) of the β_1-adrenergic receptor (β_1-AR), a member of the seven membrane-spanning receptor superfamily, alter therapeutic response to β-blockers during heart failure (Liggett et al., 2006). Stimulatory effects between β_1-AR and heterotrimeric G protein G_s mediate both beneficial and deleterious signal transduction pathways during the onset and progression of heart failure. Because β_1-AR is the major subtype in cardiac myocytes, increased catecholamines exert potent cardiomyopathic effects, cardiac remodeling and abrogation of gene expression, which are antagonized by β_1-AR blockers, resulting in improved outcomes (Lowes et al., 2002). A single-nucleotide variation at nucleotide 1165 in the gene encoding β_1-AR results in either Arg or Gly at position 389 residue (Liggett et al., 2006). In response to inotropic stimulation, human trabecular muscle with the β_1-Arg-389 residue from either non-failing or failing hearts exhibited significantly greater contractility than β_1-Gly-389 polymorphism.

As shown in the β-Blocker Evaluation of Survival Trial (BEST), which evaluated the β-blocker bucindolol for the treatment of Class III/IV heart failure, insights into the mechanisms for pharmacogenomic phenotypes involving the Arg/Gly polymorphism of the β_1-AR owe much credit to the DNA Study Group (Feldman et al., 2005, Liggett et al., 2006). The foresight of BEST investigators to recognize the power of genetic haplotyping underscores the importance for all future well-designed human trials to include contingencies for pharmacogenomics in this era of genomic medicine. Notwithstanding the success gleaned from a highly penetrant, single-gene trait such as Arg/Gly polymorphism of the β-AR, future advances will require undertaking the more formidable challenges related to multigene traits that influence drug metabolism and response for therapeutic individualization (Evans and McLeod, 2003).

The recent African-American Heart Failure Trial (A-HeFT) enrolled 1050 black patients with NYHA Class III or IV heart failure to receive a fixed dose of two well-established medications, isosorbide dinitrate and hydralazine, in a placebo-controlled, randomized, multicenter trial (Taylor et al., 2004). Combination nitrates and hydralazine, termed BiDil, when added to standard therapy, was efficacious, improving survival in blacks. But the implications of this high-profile study have drawn considerable scientific and ethical scrutiny owing to the marketing strategy of this therapy, which, under the proprietary label, was advanced as a novel approach for race-based management. Because physical and genetic traits are not interchangeable, A-HeFT *per se* might prove to be a poor surrogate for studies of pharmacogenetics, since neither BiDil's efficacy in other racial and ethnic groups nor genetic markers for predicting the response of blacks to BiDil were ever tested. For example, linkage studies of 354 subjects from A-HeFT indicate higher levels of aldosterone with the -344C allele of the aldosterone synthase promoter, suggesting that aldosterone receptor antagonists could be efficacious in reducing disease susceptibility (McNamara et al., 2006).

In contrast, extensive studies of patients with heart failure and polymorphisms in the angiotensin-converting enzyme (ACE) pathway have associated the ACE DD polymorphism with significantly higher deaths and need for transplants compared with II and ID genotypes (McNamara et al., 2001). With concurrent β-blocker treatment, patients with the ACE DD polymorphism showed improved survival but benefited from a higher ACE dosage (McNamara et al., 2004), supporting the clinical utility of genetic information in clinical management. For at-risk populations, the pace for moving bidirectional bench-to-bedside and bedside-to-community-based practices should accelerate, using evidence-based strategies emerging from disciplines such as health outcomes research.

MONITORING

Genomic Profiling

Heart failure encompasses dynamic events in which the activation or deactivation of distinct pathways and processes at different stages from onset and during disease progression suggest possible opportunities for intervention and even prevention before irreversible decompensation. A fundamental question, therefore, is how to develop improved diagnostic and prognostic indices that may guide improvements in diagnosis, treatment, outcomes, and perhaps prevention of heart failure. A major goal of microarray-based analyses is to identify genes whose similar patterns of expression accurately represent the disease state or biological process. Such information, however, is often insufficient to identify causal mechanisms, but provides a comprehensive picture of the underlying process, which can predict responses to therapy or disease stage.

Both unsupervised and supervised approaches are applied to determine if previously unrecognized or unexpected patterns of gene expression exist in the datasets. Hierarchical clustering, for example, is an unsupervised approach that may be used after gene expression profiling to identify interdependent pathways before the onset of overt heart failure. Identification and validation of genes or novel pathways that are activated earliest may improve early detection, and ultimately will be essential for designing therapies that prevent the natural history and progression of disease. If individual genes have different predictive power, then a "weighted voting scheme" based on levels of gene expression can be designed and tested before widespread application. Considerable caution in data interpretation is warranted, however, as comparisons from different laboratories may be skewed considerably by patient selection, different treatments and clinical stages.

Next-generation sequencing platforms are transforming the landscape of high-throughput DNA sequencing over conventional methods with the potential to obtain genome-wide-scale information on SNPs and copy number variations (CNV) at unparalleled speeds. While many technical details of massive parallel sequencing are beyond the scope of this chapter, considerable attention is being paid to standardization of data collection, normalization, and data reporting. We predict that meta-analyses of large datasets hold particular promise for finding normal human genome variation (Mardis, 2008), for predicting disease susceptibility, and for clinical staging from gene expression and genomic profiles.

The application of large-scale genome-wide studies (>2.4 million SNPs) to identify genomic loci associated with incident heart failure (HF) was recently undertaken among four community-based prospective populations comprising both European and African ancestries: the Atherosclerosis Risk in Communities Study, the Cardiovascular Health Study, the Framingham Heart Study, and the Rotterdam Study (Morrison et al., 2010; Smith et al., 2010). Smith and coworkers found two significant loci on chromosomes 15q22 and 12q14 (Smith et al., 2010), whereas Morrison and colleagues found one genome-wide significant locus on chromosome 3p22 when 2.3 million SNPs were analyzed for association with all-cause mortality (Morrison et al., 2010). Because there is an inverse relationship between the allele frequency and effect size, future challenges are not only directed at replicating such findings in other cohorts but also establishing the biological function of such loci to the pathogenesis of heart failure (Mestroni and Taylor, 2011).

Transcriptional Profiling of Heart Failure

Neurohumoral, hemodynamic, and environmental factors participate in remodeling the failing heart, but genetic, molecular, and cellular events are inscribed at the transcriptional level. Signaling pathways and biological processes implicated in the hypertrophic response of the heart are shown in Figure 48.4. Early

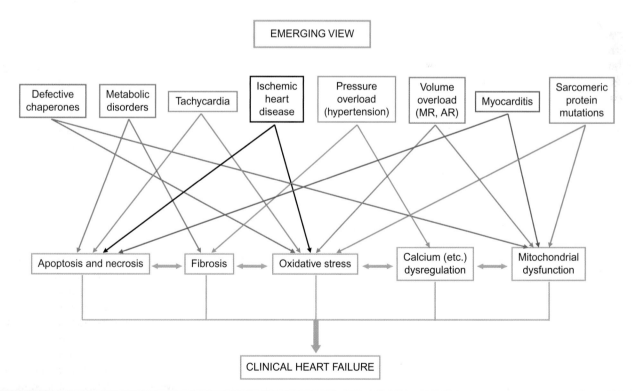

Figure 48.4   At the cellular and molecular levels, crosstalk among pathways related to oxidative stress and calcium dysregulation, for example, may contribute to secondary apoptosis and necrosis. In spite of considerable insights about mechanisms, current therapies focus on reversing neurohumoral imbalances but rarely on underlying mechanisms. MR = mitral regurgitation, AR = aortic regurgitation.

genetic markers of cardiac hypertrophy include transcriptional reprogramming of genes encoding contractile proteins, onco-genes, neurohumoral factors, and transcription factors. Proto-oncogenes *c-jun*, *c-fos*, *c-myc*, and skeletal α-actin and atrial natriuretic factor (*ANF*) are also activated in response to hyper-trophic stimuli (Izumo et al., 1988; Komuro et al., 1988; Mulvagh et al., 1987; Schwartz et al., 1986). In angiotensin II receptor type 1alpha knockout mice, cardiomyocytes were capable of evoking increased protein synthesis and microtubule-associated protein kinase (MAPK) activation when stretched, strengthening the primary role of mechanical stretch in maintaining the hyper-trophic phenotype. The mechanisms by which mechanical stress is converted into biological response are yet to be fully eluci-dated. High-density oligonucleotide arrays have also identified multiple genes representing diverse biological processes (e.g., myocardial structure, myocardial assembly and degradation, metabolism, protein synthesis, and stress response) that were differentially expressed in non-failing and failing human hearts (Yang et al., 2000). Other larger studies of human heart failure have confirmed the role of mitogen-activated protein kinases (MAPKs), mechanical stress, and neurohumoral pathways in heart failure (Kudoh et al., 1998). Likewise, genetic pathways identified during acute and chronic pressure overload reflect differential gene expression during distinct phases and may rep-resent potential targets for therapy. However, lack of reproduc-ibility and reliability in the sample sets owing to selection bias and differences in etiology, age, sex, mode of onset, treatment regimens, and clinical course remain substantial limitations.

End-stage heart failure is associated with increased activ-ity and alterations of multiple gene products, including those of the extracellular matrix/cytoskeleton (e.g., collagen types I and III, fibromodulin, fibronectin, and connexin 43) (Tan et al., 2002). When gene expression profiling was applied in a trans-genic model of tumor necrosis factor-α overexpression, the large number of immune response-related genes expressed, along with IgG deposition in the myocardium, supports activa-tion of the immune system and an inflammatory mechanism in the development and progression of heart failure (Feldman and McTiernan, 2004; Kubota et al., 1997).

Gene expression profiles of heart failure caused by alco-holic cardiomyopathy and familial cardiomyopathy suggest that onset and disease progression may involve different genetic determinants. A provocative study by Blaxall and coworkers reported that genomic profiling in a murine model of heart failure reverted to the normal phenotype after rescue by expression of β-adrenergic receptor kinase, and suggested that mild and advanced heart failure may be similar in mice and humans (Blaxall et al., 2003). As previously mentioned, the conclusions must be interpreted cautiously, owing to the complexities of heart failure related to the imprecision associ-ated with genetic, physiologic, and clinical phenotypes.

MicroRNAs in Heart Failure

MicroRNAs (miRNA) are small, ~22 nucleotide, single-stranded RNAs, which regulate gene expression by binding, through Watson–Crick base pairing, to the 3′-untranslated regions (3′-UTRs) of specific mRNAs of target genes. Such intricate forms of post-transcription regulation are likely related mechanisti-cally to profound effects evoked by subtle modulation of gene expression in the cardiovascular system (Small and Olson, 2011). Besides functions in development and tissue homeo-stasis, various pathophysiological stimuli are potent inducers of miRNAs with dynamic and rapid effects on diverse biologi-cal processes ranging from mitochondrial metabolism to car-diac hypertrophy to post-ischemic ventricular remodeling and heart failure (Matkovich et al., 2009; Roy et al., 2009; Thum et al., 2007; Van Rooij et al., 2006). In particular, the identifi-cation of ~1000 miRNAs in the genome has been combined with bioinformatics approaches to specify their key roles in "fine-tuning" gene expression of dozens of related biological targets. Oligonucleotide-based inhibitors, or miRNA mimet-ics, have been shown to manipulate the expression of cellular targets *in vivo*, propelling their delivery as drug targets of key biological pathways in disease states (Van Rooij et al., 2008). Lastly, investigators are seeking to identify and understand the biological roles that circulating miRNAs, whose remarkable stability might be suitable for distant cell-to-cell communica-tion, might play as biomarkers for heart failure and related disease states (Creemers et al., 2012).

A Case for Biologic Reclassification of Heart Failure

Gene expression profiling has significantly improved the diag-nostic classification of specific conditions (e.g., breast cancer, chronic myelogenous leukemia), but deciphering meaningful insights about the biological mechanisms underlying disease pathogenesis remains a formidable challenge (Quackenbush, 2006). Among inheritable forms of cardiovascular diseases, recent advances in research into single-gene disorders have fundamentally altered our understanding about the cellular processes, metabolic alterations, and transcriptional repro-gramming of the diseased heart (Seidman and Seidman, 2001). Beyond the availability of genetic tests for disease-causing mutations of cardiomyopathy (Morita et al., 2005), the development of genomic tools that are causally linked to disease pathogenesis, termed a "molecular signature," will likely accelerate progress for early detection, targeted ther-apy, and disease monitoring of inheritable heart failure (Bell, 2004). Tumor classification, based on genomic signatures, has been applied successfully in cancer therapeutics (Bell, 2004; Quackenbush, 2006). We suggest that opportunities exist for microarray-based profiling, proteomics, metabolomics, and next-generation sequencing technologies to propel the transi-tion from clinico-pathologic to clinico-genomic classifications for heart failure.

Different gene profiles for failing and non-failing hearts have already permitted differentiation among types of heart failure with different etiologies, as shown recently by Donahue and colleagues (2006). Considerable gaps exist, however, between our ability to identify predisposing factors for heart

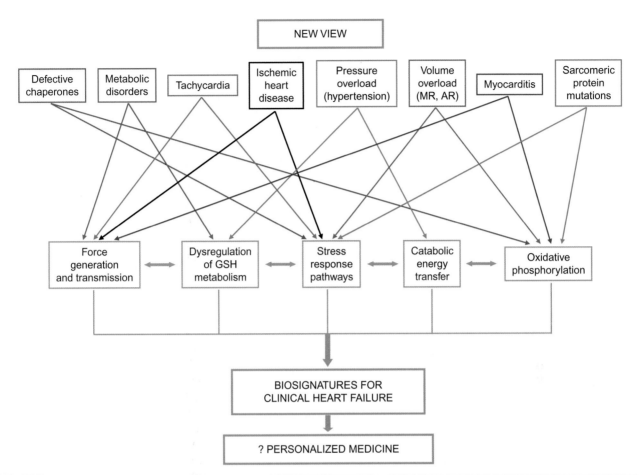

Figure 48.5 New diagnostic approaches based on information that integrates genes and molecular pathways at the onset, progression, and end stages are needed to improve heart failure classification. "Biosignatures" of heart failure, developed from microarray analysis technologies, proteomics, and genomic technologies, are proposed here to integrate the biological processes and molecular mechanisms for rational drug design and treatment in the post-genomic era of personalized medicine. MR = mitral regurgitation, AR = aortic regurgitation, GSH = glutathione.

failure using genomic profiling and translating such knowledge into clinical management.

Biomarkers vs Biosignatures for Heart Failure

The biological heterogeneity of heart failure demands more robust tools to guide clinical outcomes. Much recent attention has focused on biological markers, or biomarkers, which objectively measure and evaluate normal biological processes, pathologic processes, or pharmacological responses to therapeutic intervention (Vasan, 2006). Current enthusiasm for biomarker strategies, however, has also brought confusion and ambiguity for applications in clinical practice. Too often, highly fragmented information obtained from patients at different clinical stages precludes meaningful analysis and extrapolation to broader subclasses.

Accordingly, we propose that an integrative approach – one that encompasses our ability to predict the onset, rate of progression, and response to therapy and/or clinical outcome – requires the development of a molecular signature or "biosignature." In order to circumvent existing limitations of

biomarkers, proof-of-concept for such biosignatures, however, may require tissue sampling and serial analysis for identification and validation (Figure 48.5). Among eight individuals with IDCM but with similar clinical characteristics for chronic heart failure at baseline, Lowes and colleagues reported that serial sampling was superior to cross-sectional gene expression profiling, since there was less variance in the differences on gene chip analysis of endomyocardial biopsies from the same patient than among the different subjects with similar phenotypes (Lowes et al., 2006). Because these biological processes may precede the transition into heart failure and premature death, future work should address the intriguing possibility that distinct metabolic pathways might be linked to novel molecular signatures in disease pathogenesis (Figure 48.5).

Desmin-related myopathy (DRM) belongs to the group of myofibrillar diseases characterized by muscle weakness and ventricular remodeling, including DCM (Figure 48.6). Desmin, an intermediate filament protein, localizes to the Z-disk of myofibrils, where it is essential for muscle integrity and function. To maintain both contractile and functional properties,

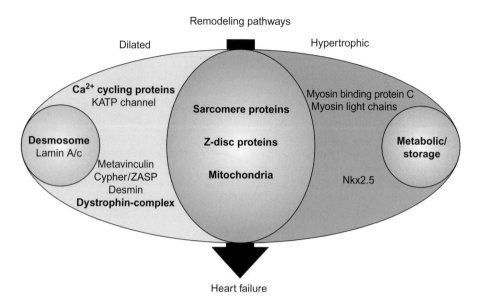

Figure 48.6 Human gene mutations can cause cardiac hypertrophy (blue), dilation (yellow), or both (green). In addition to these two patterns of remodeling, particular gene defects produce hypertrophic remodeling with glycogen accumulation (pink) or dilated remodeling with fibrofatty degeneration of the myocardium (orange). Sarcomere proteins denote β-myosin heavy chain, cardiac troponin T, cardiac troponin I, α-tropomyosin, cardiac actin, and titin. Metabolic/storage proteins denote AMP-activated protein kinase γ subunit, LAMP2, lysosomal acid α 1,4–glucosidase, and lysosomal hydrolase α-galactosidase A. Z-disc proteins denote MLP and telethonin. Dystrophin complex proteins denote δ-sarcoglycan, β-sarcoglycan, and dystrophin. Ca²⁺ cycling proteins denote PLN and RyR2. Desmosome proteins denote plakoglobin, desmin, desmoplakin, and plakophilin-2. *Reproduced with permission from Morita et al., 2005.*

molecular chaperones such as αB-crystallin (CryAB) are selectively expressed in striated muscle to facilitate proper folding and protein quality-control pathways. Because CryAB binds to its client protein desmin especially at Z-disk structures, mutations in genes encoding these proteins exhibit remarkably similar phenotypes in humans (Goldfarb and Dalakas, 2009).

Recent studies from our laboratory have demonstrated that the missense R120G mutation in the gene encoding the human small Heat Shock Protein (HSP) αB-crystallin (hR120GCryAB) recapitulates the cardiotoxicity and heart failure through a novel, toxic gain-of-function mechanism involving reductive stress in mice (Rajasekaran et al., 2007). Reductive stress refers to an abnormal increase in the amounts of reducing equivalents [e.g., glutathione, or the reduced form of nicotinamide adenine dinucleotide phosphate (NADPH)], which has been demonstrated in lower eukaryotes (Simons et al., 1995; Trotter and Grant, 2002) but has not been commonly shown in mammals and/or in disease states (Chance et al., 1979; Gores et al., 1989). Because the R120GCryAB cardiomyopathy and increased activity of glucose-6-phosphate dehydrogenase (G6PD), the rate-limiting enzyme of the pentose phosphate shunt, are rescued by G6PD deficiency, such genetic evidence established G6PD as a genetic modifier of such myofibrillar disease. We further hypothesize that certain interactions among metabolic and genetic pathways might represent potential biosignatures for disease states (Rajasekaran et al., 2008). It is conceivable that such metabolic

and biochemical changes might be detectable with both sensitivity and specificity before the onset of myopathic or pathologic alterations, providing the opportunity for interventions that correct such imbalances that contribute to cardiomyopathy and heart failure. Applications in redox proteomics and multiplex protein markers are presently being pursued to determine if glutathionylation, for example, of key components in mitochondrial and other metabolic pathways are causally linked to disease pathogenesis.

Molecular Diagnosis of Allograft Rejection

Although peripheral blood mononuclear cells (PBMCs) are abundant and highly accessible sources of genomic material, a potential for diagnostic inaccuracy and therapeutic failure exists if there is discordance between the information in PBMCs and an underlying condition in the diseased tissues. Significant progress has been made in patients after cardiac transplantation that could change existing paradigms for clinical decision-making and management of allograft rejection. Standard protocols after heart transplantation require patients to undergo serial endomyocardial biopsies (EMBs) as a means to monitor rejection and to guide immunosuppressive therapy. Such surveillance maneuvers are invasive, expensive, and carry considerable risks, such as perforation of the ventricular wall and subsequent life-threatening hemopericardium. Analysis of the histological data by expert pathologists is subject to inter-observer variability, and the diagnosis of acute rejection has been controversial

(Nielsen et al., 1993; Winters and McManus, 1996). Horwitz and colleagues were among the first to demonstrate that gene expression profiles of PBMCs might provide an alternative approach of the diagnosis of allograft rejection (Horwitz et al., 2004). Patients who subsequently developed acute rejection had a distinct genomic profile compared with patients without any rejection and, after treatment for rejection, the majority (98%) of differentially expressed genes returned to baseline.

The CARGO (Cardiac Allograft Rejection Gene Expression Observational) study prospectively investigated gene expression analysis from PBMCs as a diagnostic tool to predict transplant rejection (Mehra, 2005) (see Chapter 49). From the core group of 11 genes associated with immune response pathways, which were identified by quantitative real-time polymerase chain reaction (QT-PCR) and assigned weighted scores, CARGO investigators were able to predict rejection with a sensitivity and specificity of 80% and 60%, respectively (Deng et al., 2006). Owing to reduced sensitivity and specificity immediately after transplantation, the test may be unreliable for diagnosis of low/intermediate-grade rejection. Now commercially available (AlloMap®), this landmark study provides proof-of-concept that expression profiling of 11 genes in PBMCs was predictive for acute rejection pathways in cardiac transplant patients (see Chapter 49). One important implication is that genomic profiling of specific targets expressed in peripheral blood can be used as a sensitive marker to determine low, intermediate, and high groups for transplant rejection (Mehra et al., 2008), providing direct evidence that such monitoring could guide surveillance and therapeutic management.

NOVEL THERAPEUTICS AND FUTURE DIRECTIONS

Cardiac Stem and Progenitor Cells for Myocardial Regeneration

Stem cells are undifferentiated cells that give rise to more than 200 types of cells in the adult human body. Because of their capacity for self-renewal, embryonic stem cells (ESCs) have generated much interest for applications that might one day repair or replace damaged cells after a heart attack, restore regions of the brain after a stroke, and regenerate neurons severed after accidental injury of the spinal cord. Innovative approaches and treatments might, therefore, be on the horizon for certain neurodegenerative diseases (e.g., Alzheimer's disease, Parkinson's disease) and acquired diseases, including myocardial infarction.

Considerable public discussion and controversy, however, have focused on the sources of certain stem cell lines – the four- or five-day human embryos or fertilized cells discarded by fertility clinics. Opponents of stem cell research have viewed such sources as inhumane and have argued for an outright ban or severe restrictions on the availability of human ESCs in biomedical research. Yet, the promise of stem cell research for tissue regeneration has continuously intensified

as patients and their families and advocates have heightened expectations to find new cures for current intractable conditions.

In the era of genomic medicine, among the many approaches being investigated are efforts to exploit the resident population of cardiac stem and progenitor cells for regeneration or repair of damaged myocardium. Cardiac embryogenesis proceeds from the parallel contributions of two separate progenitor cell populations derived from a common progenitor at gastrulation, namely, the primary and secondary "heart fields" (Garry and Olson, 2006; Kelly et al., 2001). Genetic and morphogenic programs give rise to the mesoderm-derived primary heart field, the earliest population of cardiac progenitors, originating in the splanchnic mesoderm and subsequently migrating into the cardiac crescent. Cells from the cardiac crescent proceed to form the midline linear heart tube, consisting of the inner endocardial and outer myocardial layers. The four-chambered heart arises from rightward looping, differential growth, and from contributions of the secondary, or anterior, heart field (Buckingham et al., 2005). Whereas precursors of the primary heart field ultimately contribute to the developing left ventricle and atria; the right ventricle and outflow tract are derived from the secondary heart field. A landmark study by Laugwitz and colleagues has challenged existing dogma by demonstrating $Isl1^+$ cardiac progenitor cells with replicative capacity can be isolated from the adult mouse, rat, and human heart (Laugwitz et al., 2005). Fate mapping studies indicate $isl1^+$ progenitor cells migrate into the outflow region and contribute to development of the right atrium, right ventricle, and partial left ventricle.

Embryonic precursors of distinctive origins give rise to three cardiac lineages – cardiac, smooth muscle, and endothelial cell – that constitute the mature heart. Until recently, however, plasticity of the adult heart was widely disbelieved, due to the prevailing dogma that this post-mitotic, terminally differentiated organ was devoid of regenerative capacity. A new era of cardiac biology has emerged with the discovery of a resident subpopulation of myocardial cells with the capacity for replication, termed "cardiac progenitors," in both normal and pathological hearts.

The discovery of c-kit-positive, lineage-negative cardiac stem cells in the adult mammalian heart has intensified recent efforts to develop strategies for repairing or regenerating injured myocardial tissues after heart attacks (Beltrami et al., 2003). From studies of carbon-14 released from tests of nuclear bombs and integrated into the DNA of the human myocardium, Bergmann and colleagues have demonstrated renewal rates for cardiomyocytes of between 1% and 0.45% at 25 and 75 years, respectively (Bergmann et al., 2009). From the isolation of cardiac stem cells from patients undergoing selective cardiac procedures (e.g., endomyocardial biopsy), which exhibit the characteristics for clonogenicity, self-renewal, and multipotentiality, patients could conveniently receive autologous transplants by catheter-based approaches during surgeries for coronary artery bypass grafting (CABG), valvular replacement, or left ventricular assist device (LVAD) implantation.

Based on published reports in mice, the left ventricular apex and left atrium regions are the optimal sites for harvesting cardiac progenitors, but cardiac niches with progenitors have also been identified in the mid-region and basal regions (Urbanek et al., 2006). Currently unanswered questions include whether age, gender, risk factors, and other disease status might impact the plasticity, proliferation, or cellular functions of cardiac stem cells and progenitors.

Cardiac Cell Therapy

Existing therapies for heart failure are palliative and do not address the underlying root cause of the disease, supporting the quest for adjuvant strategies of cardiac cell therapy for myocardial regeneration (Orlic et al., 2001). As an emerging field, investigators first tackled whether administration of progenitor cells derived from bone marrow or circulating blood was feasible. Results of several human trials have either been equivocal or have shown modest improvement in selected endpoints but not in clinical outcomes. In the Reinfusion of Enriched Progenitor Cells and Infarct Remodeling in Acute Myocardial Infarction (REPAIR-AMI) trial, the largest study of cardiac stem cell therapy, patients who received an intracoronary infusion of progenitors derived from bone marrow (bone marrow cells, BMCs) after successful percutaneous intracoronary intervention for acute myocardial infarction showed an absolute improvement in left ventricular ejection fraction (LVEF) at four months compared with the placebo group (Schachinger et al., 2006). In contrast, results of the Bone Marrow Transfer to Enhance ST-Elevation Infarct Regeneration (BOOST) trial have dampened enthusiasm, since the relative improvement in LVEF at six months was abolished at 18 months (Meyer et al., 2006). An important question for clinical and translational scientists is whether cardiac cell therapy is efficacious in current victims of chronic heart failure. The TOPCARE-CHD trial (Transplantation of Progenitor Cells and Recovery of LV Function in patients with Chronic Ischemic Heart Disease) enrolled patients with LV dysfunction but healed scars an average of six years after myocardial infarction. Patients receiving BMCs showed greater improvement in global and regional LV function compared with a similar cohort receiving progenitors derived from circulating blood cells (Assmus et al., 2006). Notwithstanding, effective local regeneration by either BMCs or circulating progenitors appears limited, owing to follow-up animal studies indicating that the majority of infused cells (>97%) exit the myocardium (Hofmann et al., 2005). In the Stem Cell Infusion in Patients with Ischemic Cardiomyopathy (SCIPIO) Phase 1 trial, the investigators first harvested c-kit-positive, lineage-negative cardiac stem cells (CSCs) from patients with LV dysfunction (<40%) undergoing CABG before the subsequent infusion of one million cells delivered by intracoronary infusion. Among the 14 patients treated with CSC infusion, there was improvement in LVEF and reduced infarct size at both 4 and 12 months compared with control patients who received no

treatment (Bolli et al., 2011). While such proof-of-concept for cardiac cell therapy appears to be safe, an immediate challenge for the field is to provide evidence of improvement in clinical outcomes in patients with chronic heart failure. Without doubt, the chances of autologous infusion therapy becoming the widespread standard of care would be significantly curtailed by the prohibitive costs to achieve such individualized care.

DISEASE MODELING FOR PERSONALIZED HEALTHCARE

In 2006, Shinya Yamanaka and colleagues demonstrated for the first time that four factors (Oct3/4, Sox2, Kfl4, and c-Myc) fully recapitulated the reprogramming of murine and subsequently human somatic cells into induced pluripotent stem cells (iPSCs) (Takahashi et al., 2007; Takahashi and Yamanaka, 2006). In other words, such iPSCs, when reintroduced into mice, and later into human somatic cells, exhibited the capacity for germ-line transmission and repopulation into mesoderm, ectoderm, and endoderm lineages – the hallmark properties of ESCs. While c-Myc increases the efficiency of generating human iPSCs, this factor also carries a higher burden for producing tumors in iPSC-derived mice. When c-Myc has been replaced with L-Myc, however, this tumor potential was dramatically reduced, while retaining the efficiency of human iPSC formation.

Induced pluripotent stem cell technology affords numerous unprecedented opportunities for understanding basic biology and disease mechanisms, ranging from personalized *in vitro* models of disease to autologous cell therapy in patients. Without doubt, this platform for generating human iPSC-derived cells circumvents the limited capacity for generating human somatic cells from various organs. Beginning with immobilized peripheral blood cells, scientists have now successfully generated iPSCs that differentiate into cardiomyocytes, neurons, and other hematopoietic progenitor cellular types *in vitro*. These patient-derived iPSCs have the properties of human ES cells, such as the morphology and pluripotency markers, but retain the karyotype and epigenetic markers from the donor. However, specific questions remain about the sources of somatic cells for generating human iPSCs. What role does the cellular niche play in subsequent efficiency and capacity for differentiation? In other words, do somatic cells derived from the human myocardium for reprogramming in specific cell types exhibit similar properties to those obtained from peripheral blood or skin?

Notwithstanding, direct reprogramming of somatic cells into iPSCs is well established (Takahashi et al., 2007; Takahashi and Yamanaka, 2006), and several labs have successfully generated iPSC-derived cardiomyocytes (CMs), including atrial myocytes that recapitulate the long-QT syndrome (Itzhaki et al., 2011; Kamp, 2011; Moretti et al., 2010). Current studies are focusing on the generation of human iPSC-derived myocytes mimicking atrial fibrillation (AF) for disease modeling.

Progress in the field of cell therapy and iPSCs seeks to fundamentally impact two major facets of today's practice of cardiovascular medicine: diagnostics and advanced technologies. In cancer research, sophisticated molecular, cellular, pathological, and genomic profiles of available tissues have aided therapeutic decisions, such as the optimal responses to chemotherapeutic agents. Accordingly, iPSC technology is being actively investigated by cardiovascular biologists for personalized healthcare, through generation of human disease-specific CMs and other cell types.

CONCLUSIONS AND RECOMMENDATIONS

Current therapeutic interventions for heart failure primarily target end-stage manifestations (e.g., volume overload) without regard for the etiology, often with unpredictable consequences for the individual patient. If the goals of personalized medicine are to be realized soon, then significant breakthroughs in prevention, early detection, targeted therapies, and enhanced monitoring of disease states are essential to stem the looming heart failure epidemic in the genomic era. From the Human Genome Project and HapMap Consortium (Collins et al., 2003; International HapMap Consortium, 2005), computational analyses and molecular methods are beginning to influence diagnostics, guide therapies, and ultimately, inform us about preventative measures. Although specific therapies for many inheritable cardiac diseases have lagged substantially behind advances in other fields, new opportunities for improving their diagnosis, prognosis and clinical outcomes now seem within reach.

We would make the following recommendations:

1. Codify and foster stronger relationships among patients, providers, and institutional review boards (IRBs) as the "tripod" for genomic medicine.
2. Accelerate collection of tissue samples from patients with unexplained heart failure.
3. Focus more attention on discovery of basic disease mechanisms to identify "biosignatures" as the new basis for molecular classifications.
4. Encourage investments on small molecule screening to identify "high-quality" therapeutic targets for personalized medicine.
5. Enhance use of systems biology to refine and validate targeted therapies.

ACKNOWLEDGMENTS

Awards from the National Heart, Lung, and Blood Institute (ARRA Award 2 R01 HL063834-06 to IJB), 2009 NIH Director's Pioneer, VA Merit Review Award (IJB), NHLBI 5R01HL074370-03 (IJB), Leducq Transatlantic Network of Excellence, and Christi T. Smith Foundation provided support for this work. Jamila Roehrig provided excellent editorial assistance during preparation of this manuscript.

REFERENCES

Arnett, D.K., Baird, A.E., Barkley, R.A., et al., 2007. Relevance of genetics and genomics for prevention and treatment of cardiovascular disease: A scientific statement from the American Heart Association Council on Epidemiology and Prevention, the Stroke Council, and the Functional Genomics and Translational Biology Interdisciplinary Working Group. Circulation 115, 2878–2901.

Assmus, B., Honold, J., Schachinger, V., et al., 2006. Transcoronary transplantation of progenitor cells after myocardial infarction. N Engl J Med 355, 1222–1232.

Bell, J., 2004. Predicting disease using genomics. Nature 429, 453–456.

Beltrami, A.P., Barlucchi, L., Torella, D., et al., 2003. Adult cardiac stem cells are multipotent and support myocardial regeneration. Cell 114, 763–776.

Benjamin, I.J., Schneider, M.D., 2005. Learning from failure: Congestive heart failure in the postgenomic age. J Clin Invest 115, 495–499.

Bergmann, O., Bhardwaj, R.D., Bernard, S., et al., 2009. Evidence for cardiomyocyte renewal in humans. Science 324, 98–102.

Bhatia, R.S., Tu, J.V., Lee, D.S., et al., 2006. Outcome of heart failure with preserved ejection fraction in a population-based study. N Engl J Med 355, 260–269.

Blaxall, B.C., Spang, R., Rockman, H.A., Koch, W.J., 2003. Differential myocardial gene expression in the development and rescue of murine heart failure. Physiol Genomics 15, 105–114.

Bleumink, G.S., Knetsch, A.M., Sturkenboom, M.C., et al., 2004. Quantifying the heart failure epidemic: Prevalence, incidence rate, lifetime risk and prognosis of heart failure. The Rotterdam Study. Eur Heart J 25, 1614–1619.

Bolli, R., Chugh, A.R., D'Amario, D., et al., 2011. Cardiac stem cells in patients with ischaemic cardiomyopathy (SCIPIO): Initial results of a randomised phase 1 trial. Lancet 378, 1847–1857. doi: 10.1016/S0140-6736(11)61590-0.

Bonow, R.O., Bennett, S., Casey Jr., D.E., et al., 2005. ACC/AHA Clinical performance measures for adults with chronic heart failure: A report of the American College of Cardiology/American Heart Association Task Force on Performance Measures (Writing Committee to Develop Heart Failure Clinical Performance Measures) – endorsed by the Heart Failure Society of America. Circulation 112, 1853–1887.

Bonow, R.O., Maurer, G., Lee, K.L., et al., 2011. Myocardial viability and survival in ischemic left ventricular dysfunction. N Engl J Med 364, 1617–1625.

Braunwald, E., Bristow, M.R., 2000. Congestive heart failure: Fifty years of progress. Circulation 102, IV14–IV23.

Bristow, M.R., 2011. Treatment of chronic heart failure with beta-adrenergic receptor antagonists: A convergence of receptor pharmacology and clinical cardiology. Circ Res 109, 1176–1194.

Bristow, M.R., Gilbert, E.M., Abraham, W.T., et al., 1996. Carvedilol produces dose-related improvements in left ventricular function and survival in subjects with chronic heart failure. MOCHA Investigators. Circulation 94, 2807–2816.

Bristow, M.R., Ginsburg, R., Umans, V., et al., 1986. Beta 1- and beta 2-adrenergic-receptor subpopulations in nonfailing and failing human ventricular myocardium: Coupling of both receptor subtypes to muscle contraction and selective beta 1-receptor down-regulation in heart failure. Circ Res 59, 297–309.

Buckingham, M., Meilhac, S., Zaffran, S., 2005. Building the mammalian heart from two sources of myocardial cells. Nat Rev Genet 6, 826–835.

Chance, B., Sies, H., Boveris, A., 1979. Hydroperoxide metabolism in mammalian organs. Physiol Rev 59, 527–605.

Collins, F.S., Morgan, M., Patrinos, A., 2003. The Human Genome Project: Lessons from large-scale biology. Science 300, 286–290.

Creemers, E.E., Tijsen, A.J., Pinto, Y.M., 2012. Circulating microRNAs: Novel biomarkers and extracellular communicators in cardiovascular disease? Circ Res 110, 483–495.

Danial, N.N., Korsmeyer, S.J., 2004. Cell death: Critical control points. Cell 116, 205–219.

Deng, M.C., Eisen, H.J., Mehra, M.R., et al., 2006. Noninvasive discrimination of rejection in cardiac allograft recipients using gene expression profiling. Am J Transplant 6, 150–160.

Donahue, M.P., Marchuk, D.A., Rockman, H.A., 2006. Redefining heart failure: The utility of genomics. J Am Coll Cardiol 48, 1289–1298.

Dunselman, P.H., Kuntze, C.E., Van Bruggen, A., et al., 1988. Value of New York Heart Association classification, radionuclide ventriculography, and cardiopulmonary exercise tests for selection of patients for congestive heart failure studies. Am Heart J 116, 1475–1482.

Evans, W.E., McLeod, H.L., 2003. Pharmacogenomics – drug disposition, drug targets, and side effects. N Engl J Med 348, 538–549.

Evans, W.E., Relling, M.V., 1999. Pharmacogenomics: Translating functional genomics into rational therapeutics. Science 286, 487–491.

Feldman, A.M., McTiernan, C., 2004. Is there any future for tumor necrosis factor antagonists in chronic heart failure? Am J Cardiovasc Drugs 4, 11–19.

Feldman, D.S., Carnes, C.A., Abraham, W.T., Bristow, M.R., 2005. Mechanisms of disease: Beta-adrenergic receptors – alterations in signal transduction and pharmacogenomics in heart failure. Nat Clin Pract Cardiovasc Med 2, 475–483.

Garry, D.J., Olson, E.N., 2006. A common progenitor at the heart of development. Cell 127, 1101–1104.

Goldfarb, L.G., Dalakas, M.C., 2009. Tragedy in a heartbeat: Malfunctioning desmin causes skeletal and cardiac muscle disease. J Clin Invest 119, 1806–1813.

Gores, G.J., Flarsheim, C.E., Dawson, T.L., Nieminen, A.L., Herman, B., Lemasters, J.J., 1989. Swelling, reductive stress, and cell death during chemical hypoxia in hepatocytes. Am J Physiol 257, C347–C354.

Heidenreich, P.A., Trogdon, J.G., Khavjou, O.A., et al., 2011. Forecasting the future of cardiovascular disease in the United States: A policy statement from the American Heart Association. Circulation 123, 933–944.

Hofmann, M., Wollert, K.C., Meyer, G.P., et al., 2005. Monitoring of bone marrow cell homing into the infarcted human myocardium. Circulation 111, 2198–2202.

Horwitz, P.A., Tsai, E.J., Putt, M.E., et al., 2004. Detection of cardiac allograft rejection and response to immunosuppressive therapy with peripheral blood gene expression. Circulation 110, 3815–3821.

Hunt, S.A., Abraham, W.T., Chin, M.H., et al., 2005. ACC/AHA 2005 Guideline update for the diagnosis and management of chronic heart failure in the adult: A report of the American College of Cardiology/American Heart Association Task Force on Practice Guidelines (Writing Committee to Update the 2001 Guidelines for the Evaluation and Management of Heart Failure); developed in collaboration with the American College of Chest Physicians and the International Society for Heart and Lung Transplantation; endorsed by the Heart Rhythm Society. Circulation 112, e154–e235.

International HapMap Consortium, 2005. A haplotype map of the human genome. Nature 437, 1299–1320.

Itzhaki, I., Maizels, L., Huber, I., et al., 2011. Modelling the long-QT syndrome with induced pluripotent stem cells. Nature 471, 225–229.

Izumo, S., Nadal-Ginard, B., Mahdavi, V., 1988. Protooncogene induction and reprogramming of cardiac gene expression produced by pressure overload. Proc Natl Acad Sci USA 85, 339–343.

Jessup, M., Brozena, S., 2003. Heart failure. N Engl J Med 348, 2007–2018.

Kamp, T.J., 2011. An electrifying iPSC disease model: Long-QT syndrome type 2 and heart cells in a dish. Cell Stem Cell 8, 130–131.

Kelly, R.G., Brown, N.A., Buckingham, M.E., 2001. The arterial pole of the mouse heart forms from Fgf10-expressing cells in pharyngeal mesoderm. Dev Cell 1, 435–440.

Komuro, I., Kurabayashi, M., Takaku, F., Yazaki, Y., 1988. Expression of cellular oncogenes in the myocardium during the developmental stage and pressure-overloaded hypertrophy of the rat heart. Circ Res 62, 1075–1079.

Kubota, T., McTiernan, C.F., Frye, C.S., et al., 1997. Dilated cardiomyopathy in transgenic mice with cardiac-specific overexpression of tumor necrosis factor-alpha. Circ Res 81, 627–635.

Kudoh, S., Komuro, I., Hiroi, Y., et al., 1998. Mechanical stretch induces hypertrophic responses in cardiac myocytes of angiotensin II type 1a receptor knockout mice. J Biol Chem 273, 24,037–24,043.

Laugwitz, K.L., Moretti, A., Lam, J., et al., 2005. Postnatal isl1$^+$ cardioblasts enter fully differentiated cardiomyocyte lineages. Nature 433, 647–653.

Liew, C.C., Dzau, V.J., 2004. Molecular genetics and genomics of heart failure. Nat Rev Genet 5, 811–825.

Liggett, S.B., 2001. Pharmacogenetic applications of the Human Genome project. Nat Med 7, 281–283.

Liggett, S.B., Mialet-Perez, J., Thaneemit-Chen, S., et al., 2006. A polymorphism within a conserved beta(1)-adrenergic receptor motif alters cardiac function and beta-blocker response in human heart failure. Proc Natl Acad Sci USA 103, 11,288–11,293.

Liu, P.P., Mason, J.W., 2001. Advances in the understanding of myocarditis. Circulation 104, 1076–1082.

Lowes, B.D., Gilbert, E.M., Abraham, W.T., et al., 2002. Myocardial gene expression in dilated cardiomyopathy treated with beta-blocking agents. N Engl J Med 346, 1357–1365.

Lowes, B.D., Zolty, R., Minobe, W.A., et al., 2006. Serial gene expression profiling in the intact human heart. J Heart Lung Transplant 25, 579–588.

Maisel, A.S., McCullough, P.A., 2003. Cardiac natriuretic peptides: A proteomic window to cardiac function and clinical management. Rev Cardiovasc Med 4 (Suppl. 4), S3–S12.

Mardis, E.R., 2008. The impact of next-generation sequencing technology on genetics. Trends Genet 24, 133–141.

Matkovich, S.J., Van Booven, D.J., Youker, K.A., et al., 2009. Reciprocal regulation of myocardial microRNAs and messenger RNA in human cardiomyopathy and reversal of the microRNA signature by biomechanical support. Circulation 119, 1263–1271.

McNamara, D.M., Holubkov, R., Janosko, K., et al., 2001. Pharmacogenetic interactions between beta-blocker therapy and the angiotensin-converting enzyme deletion polymorphism in patients with congestive heart failure. Circulation 103, 1644–1648.

McNamara, D.M., Holubkov, R., Postava, L., et al., 2004. Pharmacogenetic interactions between angiotensin-converting enzyme inhibitor therapy and the angiotensin-converting enzyme deletion polymorphism in patients with congestive heart failure. J Am Coll Cardiol 44, 2019–2026.

McNamara, D.M., Tam, S.W., Sabolinski, M.L., et al., 2006. Aldosterone synthase promoter polymorphism predicts outcome in African Americans with heart failure: Results from the A-HeFT Trial. J Am Coll Cardiol 48, 1277–1282.

Mehra, M.R., 2005. The emergence of genomic and proteomic biomarkers in heart transplantation. J Heart Lung Transplant 24, S213–S218.

Mehra, M.R., Kobashigawa, J.A., Deng, M.C., et al., 2008. Clinical implications and longitudinal alteration of peripheral blood transcriptional signals indicative of future cardiac allograft rejection. J Heart Lung Transplant 27, 297–301.

Mestroni, L., Taylor, M.R., 2011. Pharmacogenomics, personalized medicine, and heart failure. Discov Med 11, 551–561.

Meyer, G.P., Wollert, K.C., Lotz, J., et al., 2006. Intracoronary bone marrow cell transfer after myocardial infarction: Eighteen months' follow-up data from the randomized, controlled BOOST (BOne marrOw transfer to enhance ST-elevation infarct regeneration) trial. Circulation 113, 1287–1294.

Moretti, A., Bellin, M., Welling, A., et al., 2010. Patient-specific induced pluripotent stem-cell models for long-QT syndrome. N Engl J Med 363, 1397–1409.

Morita, H., Seidman, J., Seidman, C.E., 2005. Genetic causes of human heart failure. J Clin Invest 115, 518–526.

Morrison, A.C., Felix, J.F., Cupples, L.A., et al., 2010. Genomic variation associated with mortality among adults of European and African ancestry with heart failure: The cohorts for heart and aging research in genomic epidemiology consortium. Circ Cardiovasc Genet 3, 248–255.

Mulvagh, S.L., Michael, L.H., Perryman, M.B., Roberts, R., Schneider, M.D., 1987. A hemodynamic load in vivo induces cardiac expression of the cellular oncogene, c-myc. Biochem Biophys Res Commun 147, 627–636.

Nielsen, H., Sorensen, F.B., Nielsen, B., Bagger, J.P., Thayssen, P., Baandrup, U., 1993. Reproducibility of the acute rejection diagnosis in human cardiac allografts. The Stanford Classification and the International Grading System. J Heart Lung Transplant 12, 239–243.

Orlic, D., Kajstura, J., Chimenti, S., et al., 2001. Bone marrow cells regenerate infarcted myocardium. Nature 410, 701–705.

Owan, T.E., Hodge, D.O., Herges, R.M., Jacobsen, S.J., Roger, V.L., Redfield, M.M., 2006. Trends in prevalence and outcome of heart failure with preserved ejection fraction. N Engl J Med 355, 251–259.

Quackenbush, J., 2006. Microarray analysis and tumor classification. N Engl J Med 354, 2463–2472.

Radford, M.J., Arnold, J.M., Bennett, S.J., et al., 2005. ACC/AHA key data elements and definitions for measuring the clinical management and outcomes of patients with chronic heart failure: A report of the American College of Cardiology/American Heart Association Task Force on Clinical Data Standards (Writing Committee to Develop Heart Failure Clinical Data Standards); developed in collaboration with the American College of Chest Physicians and the International Society for Heart and Lung Transplantation; endorsed by the Heart Failure Society of America. Circulation 112, 1888–1916.

Rajasekaran, N.S., Connell, P., Christians, E.S., et al., 2007. Human alphaB-crystallin causes oxido-reductive stress and protein aggregation cardiomyopathy in mice. Cell 130, 427–439.

Rajasekaran, N.S., Firpo, M.A., Milash, B.A., Weiss, R.B., Benjamin, I.J., 2008. Global expression profiling identifies a novel biosignature for protein aggregation R120GCryAB cardiomyopathy in mice. Physiol Genomics 35, 165–172.

Roberts, R., 2008. Personalized medicine: A reality within this decade. J Cardiovasc Transl Res 1, 11–16.

Roy, S., Khanna, S., Hussain, S.R., et al., 2009. MicroRNA expression in response to murine myocardial infarction: miR-21 regulates fibroblast metalloprotease-2 via phosphatase and tensin homologue. Cardiovasc Res 82, 21–29.

Schachinger, V., Erbs, S., Elsasser, A., et al., 2006. Intracoronary bone marrow-derived progenitor cells in acute myocardial infarction. N Engl J Med 355, 1210–1221.

Schwartz, K., De la Bastie, D., Bouveret, P., Oliviero, P., Alonso, S., Buckingham, M., 1986. Alpha-skeletal muscle actin mRNAs accumulate in hypertrophied adult rat hearts. Circ Res 59, 551–555.

Seidman, J.G., Seidman, C., 2001. The genetic basis for cardiomyopathy: From mutation identification to mechanistic paradigms. Cell 104, 557–567.

Simons, J.F., Ferro-Novick, S., Rose, M.D., Helenius, A., 1995. BiP/Kar2p serves as a molecular chaperone during carboxypeptidase Y folding in yeast. J Cell Biol 130, 41–49.

Small, E.M., Olson, E.N., 2011. Pervasive roles of microRNAs in cardiovascular biology. Nature 469, 336–342.

Smith, N.L., Felix, J.F., Morrison, A.C., et al., 2010. Association of genome-wide variation with the risk of incident heart failure in adults of European and African ancestry: A prospective meta-analysis from the cohorts for heart and aging research in genomic epidemiology (CHARGE) consortium. Circ Cardiovasc Genet 3, 256–266.

Takahashi, K., Tanabe, K., Ohnuki, M., et al., 2007. Induction of pluripotent stem cells from adult human fibroblasts by defined factors. Cell 131, 861–872.

Takahashi, K., Yamanaka, S., 2006. Induction of pluripotent stem cells from mouse embryonic and adult fibroblast cultures by defined factors. Cell 126, 663–676.

Tan, F.L., Moravec, C.S., Li, J., et al., 2002. The gene expression fingerprint of human heart failure. Proc Natl Acad Sci USA 99, 11,387–11,392.

Taylor, A.L., Ziesche, S., Yancy, C., et al., 2004. Combination of isosorbide dinitrate and hydralazine in blacks with heart failure. N Engl J Med 351, 2049–2057.

Thum, T., Galuppo, P., Wolf, C., et al., 2007. MicroRNAs in the human heart: A clue to fetal gene reprogramming in heart failure. Circulation 116, 258–267.

Trotter, E.W., Grant, C.M., 2002. Thioredoxins are required for protection against a reductive stress in the yeast Saccharomyces cerevisiae. Mol Microbiol 46, 869–878.

Urbanek, K., Cesselli, D., Rota, M., et al., 2006. Stem cell niches in the adult mouse heart. Proc Natl Acad Sci USA 103, 9226–9231.

Van Rooij, E., Marshall, W.S., Olson, E.N., 2008. Toward microRNA-based therapeutics for heart disease: The sense in antisense. Circ Res 103, 919–928.

Van Rooij, E., Sutherland, L.B., Liu, N., et al., 2006. A signature pattern of stress-responsive microRNAs that can evoke cardiac hypertrophy and heart failure. Proc Natl Acad Sci USA 103, 18,255–18,260.

Vasan, R.S., 2006. Biomarkers of cardiovascular disease: Molecular basis and practical considerations. Circulation 113, 2335–2362.

Waagstein, F., Hjalmarson, A., Varnauskas, E., Wallentin, I., 1975. Effect of chronic beta-adrenergic receptor blockade in congestive cardiomyopathy. Br Heart J 37, 1022–1036.

Watkins, H., Ashrafian, H., Redwood, C., 2011. Inherited cardiomyopathies. N Engl J Med 364, 1643–1656.

Wencker, D., Chandra, M., Nguyen, K., et al., 2003. A mechanistic role for cardiac myocyte apoptosis in heart failure. J Clin Invest 111, 1497–1504.

Winters, G.L., McManus, B.M., 1996. Consistencies and controversies in the application of the International Society for Heart and Lung Transplantation working formulation for heart transplant biopsy specimens. Rapamycin Cardiac Rejection Treatment Trial Pathologists. J Heart Lung Transplant 15, 728–735.

Wittstein, I.S., Thiemann, D.R., Lima, J.A., et al., 2005. Neurohumoral features of myocardial stunning due to sudden emotional stress. N Engl J Med 352, 539–548.

Yang, J., Moravec, C.S., Sussman, M.A., et al., 2000. Decreased SLIM1 expression and increased gelsolin expression in failing human hearts measured by high-density oligonucleotide arrays. Circulation 102, 3046–3052.

CHAPTER

Cardiac Transplant Rejection

Michael X. Pham and James Yee

INTRODUCTION

AlloMap® Molecular Expression Testing, based on peripheral blood gene expression profiling, has journeyed from a diagnostic test concept through candidate gene discovery, classifier development, regulatory approvals, and validation in a clinical comparative effectiveness study. The intended use of the non-invasive AlloMap® test, performed on RNA extracted from a peripheral venous blood sample, is to aid in the assessment of heart transplant recipients for acute cellular rejection. To provide clinical context, we summarize the types of allograft rejection that a recipient may experience, the current immunosuppressive regimens available to prevent rejection, and the rejection monitoring strategies employed. The development of the gene expression profiling test is then described along with its utility in clinical practice.

TYPES OF CARDIAC ALLOGRAFT REJECTION

Heart transplantation is the definitive therapy for end-stage heart failure, but can be complicated by recipient immune responses that may manifest as hyperacute, acute cellular, antibody-mediated, or chronic rejection of the allograft (Table 49.1).

Hyperacute rejection is very rare and manifests as severe graft failure within the first few minutes to hours after transplantation. This type of rejection is caused by circulating pre-formed antibodies to the ABO blood group (in cases of ABO blood group incompatibility), or to major histocompatability antigens, and/or to endothelial antigens in the cardiac allograft (Kemnitz et al., 1991).

Acute cellular rejection (ACR) (Billingham et al., 1990; Rodriguez, 2003; Stewart et al., 2005; Winters and McManus, 1996; Winters et al., 1998) occurs in 30–50% of heart transplant recipients in the first year following transplantation (Hershberger et al., 2005). ACR is primarily mediated by T-lymphocytes and is characterized by lymphocytic infiltration within the myocardium, direct myocyte damage, and subsequent graft dysfunction. The grading of ACR severity is standardized according to the International Society of Heart and Lung Transplantation (ISHLT) cardiac biopsy grading schema shown in Table 49.2, and reflects the distribution and extent of inflammation, and the presence or absence of myocyte damage. Although most episodes occur within the first 3–6 months, the risk of ACR persists several years post-transplantation (Gradek et al., 2001; Heimansohn et al., 1997; Hershberger et al., 2005; Stehlik et al., 2006) and is associated with cardiac allograft vasculopathy and graft loss (Brunner-La Rocca et al., 1998; Radovancevic et al., 2005; Raichlin et al., 2009).

Antibody-mediated rejection (AMR) occurs when a recipient's immune system generates antibodies to donor antigens. Binding of these antibodies to the lining of the graft vasculature promotes an inflammatory response involving innate immune components including complement and myeloid cells, which leads to endothelial dysfunction and graft failure. AMR can cause catastrophic graft loss in a small

Genomic and Personalized Medicine, 2nd edition
by Ginsburg & Willard. DOI: http://dx.doi.org/10.1016/B978-0-12-382227-7.00049-5

TABLE 49.1 | Types of allograft rejection

Type of rejection	Description
Hyperacute	Occurs within minutes to hours of the blood flow being reestablished and is caused by preformed antibodies to ABO blood group antigens, HLA, or endothelial antigens
Acute cellular rejection	May occur at any time after transplantation, but is most common within the first 3–6 months. Acute cellular rejection is a T-cell-mediated response with infiltration of lymphocytes and macrophages, resulting in myocytolysis; the diagnosis is made by endomyocardial biopsy
Acute antibody-mediated rejection	Occurs within days to weeks after transplantation and is initiated by alloantibodies directed against donor HLA or endothelial cell antigens. May be manifested in later time post-transplant, in isolation or recurrent pattern. Also referred to as humoral or vascular rejection
Cardiac allograft vasculopathy	Occurs months to years after transplantation; the mechanism is incompletely understood, but is known to occur as a result of the humoral and cellular consequences of allorecognition; also referred to as chronic rejection, which manifests as diffuse atherosclerosis with myointimal proliferation in the coronary arteries, resulting in ischemia and infarction

ACR = acute cellular rejection; AMR = acute antibody-mediated rejection; CAV = cardiac allograft vasculopathy, HLA = human leukocyte antigen.

TABLE 49.2 | ISHLT standardized cardiac biopsy grading

Grade	Description
Acute cellular rejection	
0R	No rejection
1R (mild)	Interstitial and/or perivascular mononuclear cell infiltrate with up to one focus of myocyte damage
2R (moderate)	Two or more foci of mononuclear cell infiltrate with associated myocyte damage
3R (severe)	Diffuse infiltrate with multifocal myocyte damage ± edema, ± hemorrhage ± vasculitis
Antibody-mediated rejection	
AMR 0	Negative for acute AMR
AMR 1	Positive immunofluorescence or immunoperoxidase for AMR (positive CD68, C4d)

ISHLT = International Society of Heart and Lung Transplantation; AMR = acute antibody mediated rejection. *Adapted from Stewart et al., 2005 with permission from Elsevier.*

to harmonize the nomenclature applied to the study of CAV, an ISHLT panel analyzed information pertaining to angiography, intravascular ultrasound imaging, microvascular function, cardiac allograft histology, circulating immune markers, non-invasive imaging tests, and gene-based and protein-based biomarkers (Mehra et al., 2010). Coronary angiography coupled with assessment of cardiac allograft function is considered to be the cornerstone of established methodology for the diagnosis and grading of CAV.

number of cases, tends to be more difficult to treat, and is more likely to be associated with hemodynamic compromise than ACR. AMR is a strong risk factor for the early development of chronic allograft vasculopathy (Book et al., 2003; Michaels et al., 2003; Querec et al., 2009). While progress has been made in understanding ACR, there are a number of controversies regarding AMR, summarized in Figure 49.1 (Kfoury and Hammond, 2010).

Cardiac allograft vasculopathy (CAV) has also been called chronic rejection. CAV manifests as progressive, often diffuse, myointimal cell proliferation in the coronary arteries of the allograft. The narrowing of the coronary arteries results in graft ischemia, heart failure and occasionally myocardial infarction. CAV is a major cause of graft loss after the first year post-transplantation. The pathogenesis of CAV involves innate and adaptive immune responses (Schmauss and Weis, 2008). In an effort

IMMUNOSUPPRESSION STRATEGIES TO PREVENT REJECTION

Most current immunosuppressive protocols employ a three-drug regimen consisting of a calcineurin inhibitor (CNI) such as tacrolimus or cyclosporine; an antiproliferative agent such as mycophenolate mofetil, mycophenolic acid, or azathioprine; and corticosteroids such as prednisone (Hunt and Haddad, 2008). The highest doses of immunosuppressives are administered in the initial months after transplantation and are tapered over the first year, with the objective of achieving the lowest maintenance levels of immune suppression sufficient to prevent graft rejection while minimizing drug toxicities. Although the use of modern immunosuppressive agent combinations has decreased the incidence of both rejection and life-threatening infections, these regimens are still associated with drug- and class-specific toxicities, including metabolic derangements

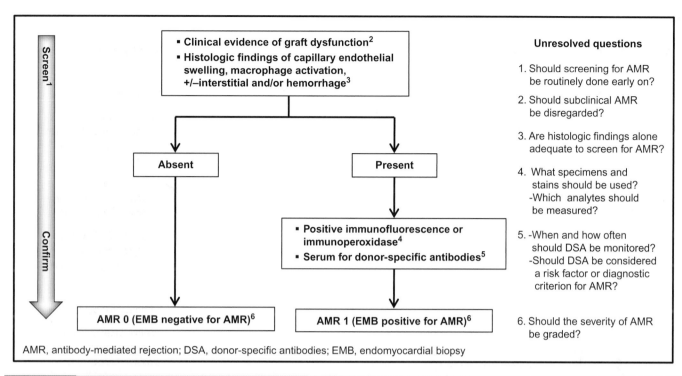

Controversies in the diagnostic algorithm for AMR based on current ISHLT guidelines. *Adapted from Kfoury and Hammond, 2010 with permission from Elsevier.*

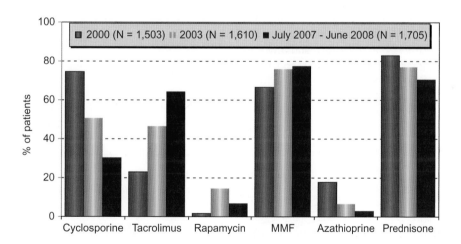

Maintenance immunosuppressive agents at one year post-transplant follow-ups in years 2000, 2003, and July 2007 through June 2008. MMF: mycophenolate mofetil. Note: different patients were analyzed in each time frame. Analysis is limited to the patients who were alive at time of follow-up. *Adapted from Taylor et al., 2009 with permission from Elsevier.*

(hypertension, dyslipidemia, diabetes mellitus) and renal dysfunction. There have been efforts to minimize CNI use in order to avoid the nephrotoxic effects of these cornerstone immunosuppressive agents (Zuckermann and Aliabadi, 2009).

The proliferation signal inhibitors (PSI) sirolimus (rapamycin) and its derivative everolimus have antiproliferative as well as immunosuppressive activity. The utility of the PSIs to reduce the doses of CNI or to inhibit the progression of CAV may be offset by adverse effects associated with PSIs (Zuckermann et al., 2008). According to a registry maintained by the ISHLT, the use of sirolimus peaked in 2003 at 14% of patients at one year post-transplant (2003), and was down to 7% in 2008. Everolimus use was 3% of patients at one year (2008). Mycophenolate mofetil (MMF) remains the predominant anti-proliferative agent, used in 77% of patients at one year. Prednisone is still used by 71% of patients at one year (Figure 49.2).

STRATEGIES FOR MONITORING TRANSPLANT RECIPIENTS

Advances in surgical techniques, control of infections, and immunosuppressive drug regimens have led to increases in recipient survival, most notably in the first two months after surgery (Figure 49.3) (Taylor et al., 2009). Regular monitoring of the patient's cardiac function, maintenance of lifelong immunosuppression, and vigilance for allograft rejection, drug toxicities, opportunistic infections, and neoplasms are considered essential to maximize patient survival.

Current practices include periodic clinic visits by the transplant recipient for surveillance testing that typically includes immunosuppressive drug level monitoring, histopathological evaluation of cardiac tissue obtained using endomyocardial biopsy (EMB), and assessment of graft function through measurement of intracardiac pressures at the time of EMB or echocardiographic imaging of cardiac function. The EMB (Figure 49.4) is performed via the right internal jugular vein or femoral vein by introducing a long forceps into the right ventricle and obtaining three to five pieces of endomyocardium, typically from the right ventricular septum. It has been the cornerstone of the diagnosis of ACR despite its known limitations, including significant interobserver variability, sampling error, and difficulty in differentiating benign nodular endocardial infiltrates (Quilty lesions) from the histologic changes observed during rejection (Marboe et al., 2005). The technical aspects of the biopsy procedure, and processing and evaluation by a pathologist under light microscope for evidence of inflammatory infiltrates and myocyte injury are described by Baumgartner et al., (2002), and Kirklin et al., (2002).

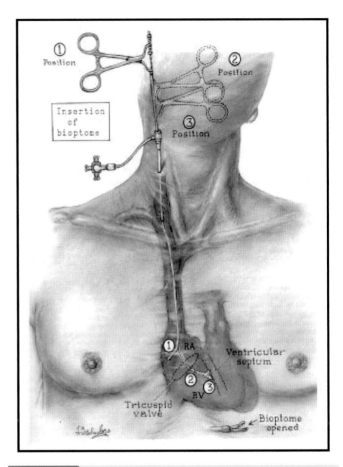

Figure 49.4 Diagram of endomyocardial biopsy procedure. *Reproduced by permission from Elsevier: Baumgartner et al., 2002.*

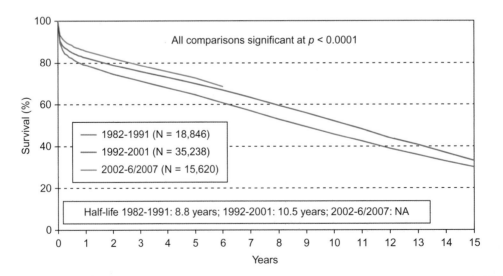

Figure 49.3 Kaplan–Meier survival by era for adult heart transplants performed from January 1982 through June 2007. The survival rates in the three time eras are each significantly different (p < 0.0001). *From Taylor et al., 2009. Reproduced with permission from Elsevier.*

Clinical monitoring practices have evolved empirically over the past 40 years. Typically, surveillance biopsies are performed weekly between weeks two and six following surgery, and then monthly through the first six months post-transplantation. The frequency of biopsies after the sixth month is decreased to every one to two months until the end of the first year, and then discontinued altogether or performed every three to six months, depending upon transplant center-specific protocols. The cumulative incidence of biopsy-proven ACR of ISHLT grade ≥2R at one year is between 24% and 36%, with the highest risk of rejection occurring during the first two months post-transplant (Hamour et al., 2008). The risk is much lower (<0.02 rejections/month) and constant by six months (Kirklin et al., 1992; Kubo et al., 1995; Stehlik et al., 2006). Because of the observed low incidence of rejections after six months, there has been disagreement regarding the utility of routine surveillance biopsies beyond the first year after transplantation. The risks of adverse events related to the biopsy include loss of jugular venous vascular access due to intravascular scarring from repetitive venous cannulation procedures, tricuspid regurgitation due to perforation or trauma to chordae tendineae, arrhythmias, perforation of the right ventricular free wall, pericardial effusions leading to cardiac tamponade, and formation of coronary artery fistulas. Although the risk of such adverse events is low with a single biopsy, the cumulative risk of at least one adverse event exceeds 20% when a patient has accumulated 12 or more procedures, which is common in the first year post-transplant (Hamour et al., 2008). There are no randomized, controlled clinical trials to establish the optimal frequency and duration of rejection surveillance with EMB with respect to patient outcomes (Costanzo et al., 2010).

The technical limitations of EMB interpretation, coupled with the uncertainty of the benefits of long-term continuation of surveillance biopsies described above, the relative inconvenience, and potential risks (Hamour et al., 2008) of this invasive procedure, have led investigators to evaluate an extensive list of possible non-invasive alternative approaches to the detection of cardiac allograft rejection (e.g., intramyocardial electrogram analysis, myocardial strain-rate imaging, anti-myosin scintigraphy, and lymphocyte-function assays) (Butler et al., 2009; Grasser et al., 2003; Hesse et al., 1995; Israeli et al., 2010; Marciniak et al., 2007). However, none of these non-invasive methods have been tested in a randomized, controlled clinical trial.

DEVELOPMENT OF A GENE EXPRESSION SIGNATURE FOR ACUTE CELLULAR REJECTION IN CARDIAC TRANSPLANT RECIPIENTS

The strength of gene expression profiling (GEP) is that a large number of genes can be individually measured, reflecting the influence of many biological pathways. Since the status of the host immune system can affect circulating blood cells, measurement of gene expression patterns in peripheral blood may provide useful diagnostic information for managing cardiac transplant patients. Blood is a complex tissue composed of numerous and dynamic cell types with significant individual variability. The complexity of a blood sample can be minimized by preparing peripheral blood mononuclear cells (PBMC), which removes most of the red cell component and mature granulocytes.

Studies in cardiac transplant patients using peripheral blood were conducted and reported by three independent investigative teams with similar aims of identifying genes and pathways associated with acute cellular rejection (Deng et al., 2006b; Horwitz et al., 2004; Schoels et al., 2004). Of these pioneering efforts, Deng and Schoels extracted RNA from PBMC preparations whereas the Horwitz group worked with RNA extracted from whole blood. Only the Deng gene-expression signature research sponsored by XDx, Inc (Brisbane, CA) continued through full development of a commercially available, FDA-cleared diagnostic test, named AlloMap® Molecular Expression Testing.

All of the AlloMap® test development and validation was based on RNA extracted from PBMC prepared from a venous blood specimen. The choice of blood sample preparation technique has substantial implications on the complexity and cost of specimen preparation, in addition to the characteristics of the isolated RNA. Whole blood collection is a simpler and lower-cost method for sample collection than a PBMC isolation approach, but has increased complexity for RNA extraction and purification. PBMC isolation, as elected for development of AlloMap®, begins with collection of a venous blood sample into a VACUTAINER® CPT™ (cell preparation tube), followed by centrifugation at the collection site. The PBMC fraction is then transferred to another tube for cell lysis. The RNA-containing lysate is frozen and shipped to a central laboratory for RNA purification. The CPT-tube PBMC isolation step was selected for the AlloMap® test to enhance the detection of small differences in expression levels of the informative genes compared to whole-blood collection, which contains RNA from reticulocytes and neutrophils, which are relatively abundant cells in comparison to PBMC alone. All stages of the AlloMap® test development used RNA collected from PBMC collected in CPT tubes.

The multi-center Cardiac Allograft Rejection Gene Expression Observational (CARGO) study enabled the development of a mathematical classifier that combines data from multiple genes expressed in peripheral blood in order to discriminate between the absence of rejection and the presence of moderate to severe rejection (ISHLT grade 2R or 3R) in cardiac transplant recipients (Deng et al., 2006b). The scheme for gene discovery, test development, and clinical validation from CARGO is shown in Figure 49.5.

Microarrays were used to identify changes in gene expression levels in PBMC samples ($n = 285$) from cardiac transplant recipients that are associated with moderate to severe acute cellular rejection. While an empiric discovery approach provided a comprehensive look at the genome, technical limitations of microarrays were known to include poor sensitivity for genes expressed at low levels or in small subsets of circulating cells, and non-specificity of array probes for individual transcripts, polymorphisms, or splice variants. Therefore, additional candidate genes were selected from existing literature, expert consultation, and bioinformatics databases. A set of 252 genes was selected for quantitative real-time polymerase chain reaction (qRT-PCR) assay development, based on results from the microarray data and from published studies in the field of allograft rejection. Of the 252 additional genes, 101 originated from microarray analysis.

The differential expression of 68 genes from the panel of 252 candidate genes was confirmed in 145 CARGO samples using qRT-PCR. Of the 68 genes confirmed by PCR, 28 were originally discovered in the microarray studies, and the remainder were included in the confirmation set based on existing evidence. qRT-PCR data from the 145 CARGO samples used for gene confirmation was further explored to identify the optimal set of genes for discrimination between samples from patients rejecting and samples from those not rejecting. Genes were entered singly or as subsets into multiple types of mathematical classification algorithms. From the 68 confirmed genes, an 11-gene classifier (5 from the microarrays) was developed using linear discriminant analysis to differentiate

between samples from the rejection and no-rejection groups (Figure 49.6). Nine additional genes were included in the gene expression profiling (GEP) test for normalization and quality control.

The classifier was validated by both analytical validation to characterize reproducibility of the GEP test and clinical validation to confirm diagnostic performance. The clinical validation that led to Food and Drug Administration (FDA) clearance of the test consisted of 300 independent samples that were not used in either the discovery or development phases described above. This method of excluding samples used in the validation study obtained from any patient who had already contributed samples to the discovery stage of gene selection was helpful in reducing the number of false positive candidate genes (Deng et al., 2006a).

Based on the CARGO study results, the GEP test was commercialized as AlloMap® Molecular Expression Testing by XDx, Inc. in Brisbane, CA. The test employs triplicate qRT-PCR assays for each of the 20 genes, and reports an AlloMap® score ranging from 0 to 40, with lower scores being associated with a very low likelihood of moderate to severe cardiac allograft acute cellular rejection. Figure 49.7 qualitatively illustrates how the test score number (0–40) value is generated from the levels of expression of the respective informative genes. The rationale for grouping the 11 informative genes into groups (steroid responsive, platelet activation, hematopoiesis, and others) is discussed in the next section.

Figure 49.8 illustrates a representative distribution of test scores from peripheral blood samples of transplant patients; because of the relatively low incidence of biopsy-proven ACR (2% of all biopsies in patients >6 months post-transplant), the positive predictive value (PPV) of test scores >34 remains relatively low. Scores between 35 and 39 have PPVs that range between 6% and 10% in the >2–6 month post-transplant time frame, and a PPV of up to 5.4% in samples from after 6 months post-transplant. Thus, the majority of patients who have test scores above a nominal threshold of 34 will not have rejection found on biopsy.

An excerpt of the test score performance characteristics between 28 and 34 is given in Table 49.3. The PPV of a test score result between 28 and 34, as shown in Table 49.3, tends to be slightly higher in the early post-transplant period (>2–6 months) than after 6 months, because of the higher incidence of ACR in the >2–6 month time frame. Nonetheless,

Stage 0. Precisely identify diagnostic test concept: a gene expression profile to rule out rejection in cardiac transplant patients.

Stage 1. *Discovery* and confirmation of candidate genes: 252 from microarrays and literature; 68 confirmed by qRT-PCR using 285 CARGO samples.

Stage 2. *Development* of diagnostic classifier: 20-gene assay panel and algorithm derived using 145 CARGO samples.

Stage 3. *Validation*: analytical and clinical using 300 CARGO samples from 154 independent patients.

Stage 4. Comparative effectiveness: IMAGE clinical outcomes randomized controlled trial with 602 patients.

Figure 49.5 Approach to the development of a gene expression profiling test for cardiac transplant rejection.

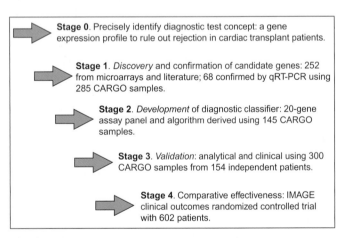

Gene candidates chosen from:	Gene candidates tested with qRT-PCR:	Genes confirmed by qRT-PCR:	Genes selected for final AlloMap® classifier:
Microarray results	101	28	5
Evidence-based from literature	151	40	6

Figure 49.6 Process that led to selection of the final 11 genes included in the AlloMap® test classifier.

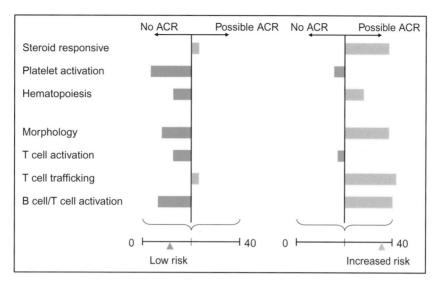

Figure 49.7 The AlloMap® test score: the relative expression levels of the informative genes are analyzed by an algorithm to render a numeric score between 0 and 40, with higher risk of rejection associated with higher scores. ACR = acute cellular rejection.

Interpreting Scores

Test developed to identify stable patients (no ACR)
- Scores below threshold – rarely risk of ACR
- Scores above threshold – increased risk of ACR

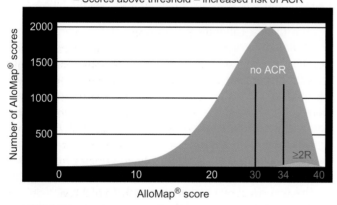

Figure 49.8 Illustration of a typical distribution of AlloMap® test scores from a general transplant population and relative proportion of associated "no ACR" and "ACR≥2R" that may be found in associated biopsy specimens. Note: the scale of the pink "≥2R" subpopulation should be about 2–5% relative to the 95% proportionate size of the green, "no ACR" majority. ACR = acute cellular rejection.

the primary utility of the test result is to help indicate a low probability of rejection [i.e., the high negative predicative value (>98 %) associated with a score below 34]. As described further below, many physicians have utilized the AlloMap® test scores, in conjunction with other clinical assessment, to identify patients with a low probability of ACR, eliminating the need for 75% of routine biopsies performed after the first 6 months post-transplantation.

PATHWAYS MONITORED BY THE ALLOMAP® TEST

The AlloMap® test combines the expression level of 11 genes that are best able to discriminate between the absence of ACR and the presence of moderate to severe ACR. The classifier algorithm contains seven gene groups – four groups based on single genes and three groups where the expression level of coordinately expressed genes is averaged (Table 49.4). Insight into the rejection-associated pathways underlying these differential patterns of gene expression can be gained by considering the source of the gene in PBMC, whether expression is generally increased or decreased in rejection, and information from the literature regarding function, expression patterns and rejection biology.

Two genes of the AlloMap® test are associated with T cell priming. In peripheral blood, *PDCD1* is expressed primarily by T cells. This gene encodes a cell surface protein that is expressed by antigen-specific T cells in circulation only during the course of an active immune response (Miller et al., 2008). Similarly, *ITGA4* encodes an adhesion molecule that is induced on T cells by activation (Pacheco et al., 1998) and is required for migration to sites of inflammation (Hammer et al., 2001). The expression levels of both *PDCD1* and *ITGA4* were higher in patients experiencing rejection, suggesting a state of increased T cell activation.

One group containing three genes – *IL1R2*, *FLT3*, and *ITGAM* – was observed to be coordinately expressed in the CARGO study, and the expression level correlated with the dose of corticosteroid. *IL1R2* expression is known to be transcriptionally regulated by corticosteroids (Re et al., 1994). On average, these genes are expressed at a lower level in ACR samples and may reflect a lower response to corticosteroids in patients with ACR.

TABLE 49.3	Excerpt of AlloMap® performance characteristics						
AlloMap® score	**>2–6 months (n =166 samples)**			**>6 months (n =134 samples)**			**AlloMap® score**
	% pts below	PPV ≥3A(2R)	NPV <3A(2R)	% pts below	PPV ≥3A(2R)	NPV <3A(2R)	
28	68.3%	3.3%	98.5%	39.1%	2.1%	98.9%	28
30	77.2%	4.6%	98.6%	50.6%	2.1%	98.7%	30
32	85.6%	2.9%	98.0%	63.1%	2.9%	99.0%	32
34	91.7%	5.0%	98.2%	79.1%	4.1%	98.9%	34

TABLE 49.4	Genes of the AlloMap® test	
Gene groups	**Expression in rejection**	**Putative biology**
Single gene groups		
PDCD1	Increased	T cell priming
ITGA4	Increased	T cell priming
SEMA7A	Increased	Unknown
RHOU	Increased	Unknown
Multi-gene groups		
PF4, C6orf25	Decreased	Platelet activation
MARCH8, WDR40A	Increased	Inflammation
IL1R2, FLT3, ITGAM	Decreased	Steroid response

Residual reticulocytes in the PBMC sample seem to be the primary source of WDR40A and MARCH8, which are more highly expressed in rejection samples. The expression of these genes correlates with other reticulocyte genes such as ALAS2 and HBB (Deng et al., 2006b), suggesting a connection between cardiac ACR and erythropoiesis. This may be related to the erythropoietic activity of inflammatory cytokines such as IL-6 (Sui et al., 1996) that are produced in the heart during rejection (Deng et al., 2002). Rejection was associated with reduced expression of PF4 and C6orf25, two genes predominantly expressed in platelets (McRedmond et al., 2004). ACR has been shown to be associated with an increase in the activation state of circulating platelets (Segal et al., 2001), probably as a consequence of the inflammatory response in the rejecting heart.

REGULATORY APPROVALS OF THE ALLOMAP® TEST

The AlloMap® test first became commercially available through the Clinical Laboratory Improvement Amendment (CLIA)-certified XDx reference laboratory in January, 2005.

The test was subsequently cleared by the FDA in 2008 as a "de novo" class II in vitro diagnostic multivariate index assay (IVDMIA) with the following indication for use:

> AlloMap® Testing is intended to aid in the identification of heart transplant recipients with stable allograft function who have a low probability of moderate/severe acute cellular rejection at the time of testing in conjunction with standard clinical assessment.

The AlloMap® test is one of only three IVDMIAs cleared by the FDA as of July, 2010. The regulatory approvals for the AlloMap® test are discussed below in the context of the evolving FDA regulatory policies for laboratory-developed tests (LDTs) (see Table 49.5).

The Clinical Laboratory Improvement Amendment (CLIA) of 1988 led to increased oversight of laboratory testing, including a certification process, accreditation requirements, periodic inspections, education, and training requirements. The focus of CLIA is on the quality of the laboratory and personnel performing the tests rather than on the validation and manufacture of the tests themselves. Under the LDT pathway regulated by CLIA, clinical validation (i.e., whether the test works as claimed) is not required. Clinicians are guided on the utility of an LDT based largely on published literature regarding the test. No independent premarket review of data and claims is required, so the LDT may be used to test patients while the clinical utility of the test may be undergoing further clinical research (see Figure 49.9).

The FDA defines in vitro devices (IVDs) to be a subset of medical devices which are "reagents, instruments, and systems intended for use in the diagnosis of disease or other conditions … (21 CFR 809.3)." The risk-based classification of IVDs has three levels: class I (common, low risk devices), class II (more complex, moderate risk), and class III (most complex, high risk, and novel intended uses). Recently, diagnostic tests such as the AlloMap® test have emerged, that combine the measurements of several analytes into a single result. This has prompted the FDA to create a new classification, the in vitro diagnostic multivariate index assay (IVDMIA), defined as a device that: (1) combines the values of multiple variables using an interpretation function to yield a single, patient-specific result (e.g., a "classification," "score," "index," etc.), that is intended for use in the diagnosis of disease or other conditions, and (2) provides a

TABLE 49.5	Regulatory pathways for laboratory developed tests	
	FDA	**CMS (CLIA)**
Registration/listing	Registration of establishment	Registration and certification of lab
	Publicly available listing of marketed tests	Lists of tests maintained by CMS (not currently publicly accessible)
Analytical validation	Premarket review of analytical data for class II and class III tests	Sampling after marketing during periodic laboratory inspections
Clinical validation	Premarket review of clinical claims for class II and class III tests; postmarket surveillance of clinical claims for class I tests	Not required
Quality system	GMPs, QS Regulations	Laboratory quality system
	Assessed by inspection	Assessed by inspection
Design controls	Required for class II and class III tests and all other devices with software	Not required Software not addressed by CLIA
Adverse event reporting	Yes	No
Postmarket surveillance	Yes	No
Recalls	Yes	No

CLIA = clinical laboratory improvement amendments, CMS = Center for Medicare and Medicaid Services; GMP = Good Manufacturing Practice; QS = Quality Systems.

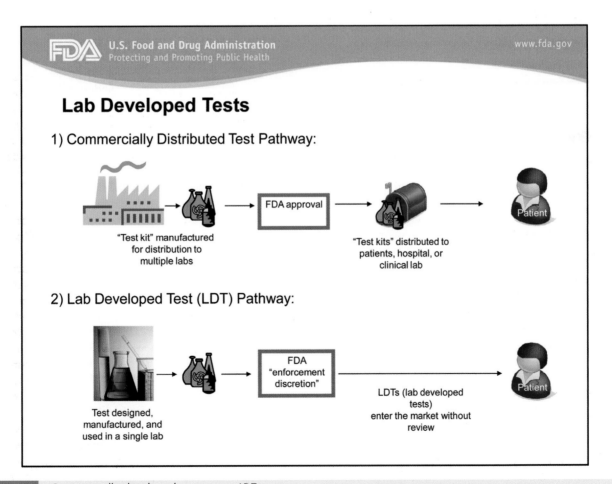

Figure 49.9   Commercially distributed tests versus LDTs.

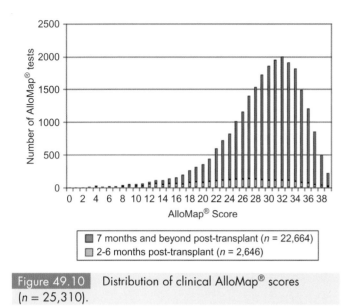

Figure 49.10 Distribution of clinical AlloMap® scores (n = 25,310).

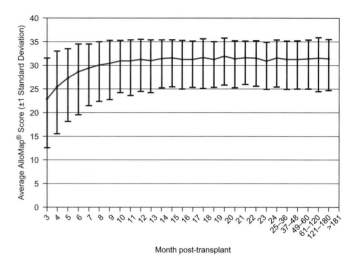

Figure 49.11 Mean AlloMap® test scores from patients at various times post-transplant.

result whose derivation is non-transparent and cannot be independently derived or verified by the end user. The IVDMIA Draft Guidance was first issued by the FDA in 2006 and discussed at a July 19, 2010 public workshop. The FDA observed that LDTs have evolved to be more like commercial IVDs, and stated that these devices should be subject to FDA regulation.

CLINICAL USE OF THE ALLOMAP® TEST

The data from early clinical use of GEP testing at three large US transplant centers confirmed a high negative predictive value (NPV) (100%): there were no events of biopsy grades ≥2R associated with AlloMap® test scores of ≤34 (Starling et al., 2006). The distribution of test score values from 25,310 AlloMap® tests from more than 7000 patients through July 2010 is shown in Figure 49.10. The relatively small proportion (12%) of tests ordered in the >2–6 month time frame reflects how the test has predominantly been used in patients who are >6 months post-transplant, perhaps indicating physician precaution in using the test in a population considered to be at a lower risk for ACR. The mean score for tests from patients at various times post-transplant is shown in Figure 49.11. The upward trend in mean score between month 2 and month 6 is attributed, in part, to reduction of maintenance corticosteroid doses, which are typically reduced from 10 to 5 mg of prednisone per day between month 2 and month 6, and further tapered thereafter. Prednisone doses of >20 mg per day may artificially decrease the AlloMap® score.

The clinical outcomes of patients who underwent rejection surveillance with either the AlloMap® GEP test or EMB were compared in the 13-center Invasive Monitoring Attenuation through Gene Expression (IMAGE) study (Pham et al., 2010). A total of 602 patients in the United States who

had undergone cardiac transplantation six months prior were randomly assigned to undergo rejection surveillance using either EMB or AlloMap® testing. Those patients assigned to the AlloMap® GEP test underwent a surveillance biopsy only if the non-invasive assay score was above a pre-specified threshold, or if there was evidence of graft dysfunction by history, physical examination, or echocardiography. The primary endpoint of the study was a composite of serious clinical outcomes – allograft dysfunction, death, or retransplantation. Over a median follow-up period of 19 months, patients who were managed with GEP and those who underwent routine EMBs had similar two-year cumulative primary event rates (14.5% and 15.3%, respectively; hazard ratio with gene expression profiling, 1.04; 95% confidence interval, 0.67 to 1.68) (Figure 49.12) (Pham et al., 2010).

The planned statistical non-inferiority analysis specified that the upper bound of the 95% confidence interval for the primary outcome measure could not be greater than 2.054 for non-inferiority to be demonstrated. Since the trial data showed a hazard ratio of 1.04 and a 95% confidence interval of 0.67 to 1.68, the criterion for non-inferiority was satisfied. In patients who underwent surveillance with the use of GEP, 0.5 cardiac biopsies were performed per patient-year of follow-up, as compared with 3.0 biopsies per patient-year in the EMB group (P < 0.001) (Figure 49.13) (Pham et al., 2010). The majority (87%) of patients in the GEP group had two or fewer biopsies per patient-year, and 50% did not require a biopsy during the study.

Patient satisfaction was assessed by scoring: "How satisfied are you with the current method of detecting rejection?" on an ordinal scale from 1 (very unhappy) to 10 (very happy). The patient satisfaction score significantly increased in the GEP group, from 6.9 at baseline to 8.2 after one year, and 8.7 after two years, whereas in the EMB group the score was

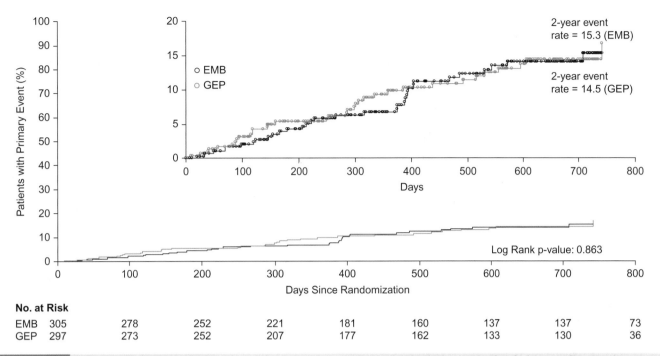

No. at Risk								
EMB	305	278	252	221	181	160	137	137	73
GEP	297	273	252	207	177	162	133	130	36

Figure 49.12 Two-year cumulative primary event rates in patients managed with GEP compared with those undergoing routine endomyocardial biopsies (EMBs). *From Pham et al., 2010. Reproduced by permission from Massachusetts Medical Society.*

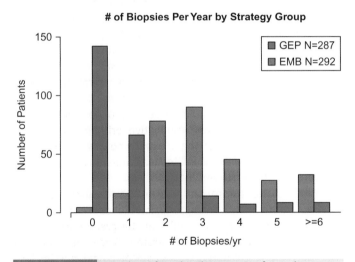

Figure 49.13 Number of cardiac biopsies performed per patient-year for the EMB and GEP strategy groups. *From Pham et al., 2010. Reproduced by permission from Massachusetts Medical Society.*

TABLE 49.6 Relative demographics of IMAGE patients compared to UNOS population

	IMAGE (n = 602)	UNOS/OPTN (n = 2207)
Age at transplant (%)		
12–17 years	0.8	4.8
18–34 years	11.1	10.6
35–49 years	20.8	19.9
50–64 years	52.5	43.5
65+ years	14.8	11.1
Male gender (%)	81.9	73.7
Race (%)		
White	77.7	67.5
Hispanic	6.5	9.4
African American	11.8	19.3
Asian/Pacific Is.	2.2	3.2
Other	1.8	0.7

OPTN = Organ Procurement and Transplantation Network.

unchanged from 6.7 at baseline to 6.6 at one year and 6.7 at two years.

Several limitations of the IMAGE study should be noted. The study population was restricted to patients who were at least six months post-transplantation, whereas the highest risk for rejection occurs in the first six months after surgery (Kirklin et al., 1992). In addition, only 20% of potentially eligible patients were enrolled, and therefore the cohort studied may not represent the complete population. However, the study population is representative of the demographics of all transplant patients in a recent United Network for Organ Sharing (UNOS) survey (see Table 49.6). These design limitations were necessary to ensure equipoise with respect to the two clinical strategies under comparison. The low incidence of the endpoint events (about 7%/year) and the limited number of available patients necessitated the choice of a wide (2.054) non-inferiority margin. The actual demonstrated 95% confidence interval (CI) margin was 0.67–1.68. In order to reduce the upper boundary of the confidence interval to 1.25, it is estimated the study would have needed 625 events and approximately 3000 patients. Such studies are difficult to perform in the field of heart transplantation due to the small number of patients (only about 2000 new transplants per year in the United States) and the reluctance of patients and physicians to volunteer for participation in investigational studies.

The IMAGE study demonstrated non-inferiority of clinical outcomes for patients monitored with the AlloMap® GEP test compared to patients monitored with the traditional biopsy method. The AlloMap® test is the only diagnostic test in the field of heart transplantation to have undergone rigorous validation in the context of a large, randomized, controlled clinical outcomes study. Additionally, it is recognized as a non-invasive method of excluding moderate to severe ACR in low-risk patients who are between six months and five years after heart transplantation in the ISHLT guidelines for the care of transplant recipients (class IIa recommendation, level of evidence B) (Costanzo et al., 2010). The AlloMap® test is among the first GEP-based tests to have completed the roadmap for development and validation of a therapeutically relevant genomic classifier (Simon, 2005).

FUTURE DIRECTIONS OF GENE EXPRESSION PROFILE TESTING

The proof-of-principle exhibited by the development of the AlloMap® test over the past ten years demonstrates the feasibility of applying GEP technologies to yield a non-invasive test that aids in the management of heart transplant recipients and reduces the need for invasive procedures such as endomyocardial biopsies. The challenges for the future are to further enhance the utility of the AlloMap® test for individual patients and to extend these techniques into other clinical applications to improve patient well-being and the cost-effectiveness of patient care.

Recognizing that the AlloMap® test score provides information about the relative probability of moderate or severe acute cellular rejection, some clinicians have applied the AlloMap® test as an aid to minimize or wean immunosuppressive agents, most often corticosteroids or a CNI (Scott et al., 2009). An ongoing randomized, controlled, single-center study compares AlloMap® testing to EMB starting two months post-transplant. In this protocol, these surveillance methods are compared for guiding corticosteroid weaning between months 6 and 12 post-transplant. The basic principle is that if a patient's AlloMap® test score remains low and stable following a reduction in corticosteroid dosage and other standard clinical assessments indicate that the patient is doing well, then a further step down in corticosteroid dosage is prescribed with

a follow-up AlloMap® test performed one month later. Ideally, a larger, multi-center randomized study should be conducted to confirm the safety and effectiveness of this approach for immunosuppressive weaning utilizing AlloMap® testing (Ashton-Chess et al., 2009). Ultimately, this could enable immunosuppression optimization on an individualized patient basis, instead of the current common practice of applying an empirical "one-regimen" schedule to all patients.

Another approach to improve the utility of the AlloMap® test is based on the observation that the longitudinal scores for individuals have significantly less variation over time than scores from different patients, supporting the hypothesis that individual stable cardiac allograft patients have a characteristic, reproducible longitudinal AlloMap® score range over time (Deng et al., 2009). This suggests that the established population-based reference scores could be augmented by individualized patient-specific score ranges to better "personalize" the use of the AlloMap® test score to interpret the probability of ACR.

Additional efforts are being made to better understand the meaning of the higher-end range of AlloMap® scores (34 or above). The positive predictive value of this higher-end range of AlloMap® scores for ACR is less than 20%, so in the vast majority of situations, these high AlloMap® scores are not expected to be associated with histopathology findings of moderate to severe ACR. Preliminary studies have suggested that AlloMap® scores greater than 34 may be associated with underlying conditions independent of existing ACR, such as future ACR rejection (Mehra et al., 2007, 2008), CAV (Yamani et al., 2007), Quilty lesions (Kabadi et al., 2009) or type of CNI (Patel et al., 2010). These studies have been limited by small sample sizes, and by single-center, retrospective, non-randomized study designs. Ongoing planned analyses on the large dataset from the 602 patient, multi-center, randomized IMAGE study will further explore these hypotheses as well as other potential uses of the AlloMap® test.

Each of these studies has evaluated the ability of the AlloMap® score to inform on the clinical condition. To improve on the information from AlloMap®, additional gene expression signatures may be required. Additionally, the coupling of the current AlloMap® test with proteomic markers of rejection may enable greater specificity for diagnosis of rejection. Investigators are actively seeking proteomic or gene signatures common across solid organ allografts that could add to or supersede the current test. Identification of gene expression or proteomic markers to distinguish acute cellular rejection from antibody-mediated rejection and/or cardiac allograft vasculopathy may be encouraged by the advances demonstrated with the development of AlloMap®. As better understanding and consensus are reached on the definitions of these conditions using existing methods, there may be an opportunity to apply genomic platforms such as GEP.

The AlloMap® test experience teaches us several lessons in undertaking the development of a genomics-based test. (1) Precise and accurate "gold standards" for defining the disease condition based on existing technologies are an essential foundation. The histology-based grading of rejection is a reference standard with limited accuracy due to sampling error, artifacts introduced during slide processing, or inter- or intra-reader variability of grading. New genomic tests for AMR or CAV will require well-defined reference standards. (2) Accumulation of sufficient numbers of well-curated specimens may be the most rate-limiting process. Biopsy-proven acute cellular rejection (with grade $\geq$2R) was ascertained by local pathology readings in 2 to 4% of 11,600 biopsies from heart transplant recipients participating in three large multi-center studies sponsored by XDx, Inc. (unpublished data). Up to half of these moderate to severe rejection events would be down-graded if re-read by independent pathologists (Marboe et al., 2005). Careful consideration must be given to estimating time and costs to accumulate the necessary number of the rarest specimens of interest in future projects. There may be strate-gies to enrich the yield of the rarest samples to enhance the cost-effectiveness of the research process. (3) The RNA col-lection method must be well-defined and robust. The overall cost of reagents, supplies and skilled labor to collect and pro-cess samples for the current 20 genes in triplicate for each AlloMap® test score is substantial. There is opportunity for improvements of simpler and faster methods to stabilize and extract RNA. (4) Definition of potentially confounding factors that may impact test performance is a challenge because of the complex and heterogeneous variety of concurrent medical conditions experienced by many transplant recipients. Many samples are needed in order to adequately study the impact of less common factors (e.g., recipients of dual kidney and heart allografts) on test performance. (5) Large randomized comparative effectiveness studies are needed to provide the highest level of clinical evidence of the utility of future tests. Such trials are costly and difficult to conduct. Physicians and patients may prefer to continue to use their past standard of care test methods rather than choosing to participate in clinical studies designed to establish clinical utility of the new method.

We are very grateful to the investigators and the patients who participated in the IMAGE study, which provides substan-tive evidence of the clinical utility of AlloMap® compared to endomyocardial biopsy for surveillance of heart transplant recipients for cellular rejection. The work also paves the way for consideration of more advanced studies of AlloMap® or derivative tests with greater confidence and equipoise. Finally, we observe that the development, introduction, and adoption of a new test such as AlloMap® requires substantial resources and long-term commitment. A broad assembly of genomic and laboratory sciences, clinical, statistical, bioinformatics, regu-latory, manufacturing, and business professionals from aca-demic, private practice, industry, and governmental agencies must work in a cohesive and complementary manner in order to achieve success.

REFERENCES

Ashton-Chess, J., Giral, M., Soulillou, J.P., Brouard, S., 2009. Can immune monitoring help to minimize immunosuppression in kidney transplantation? Transplant Int 22, 110–119.

Baumgartner, W.A., Reitz, B.A., Kasper, E.K., Theodore, J., 2002. Heart and Lung Transplantation. Elsevier Health Sciences.

Billingham, M.E., Cary, N.R., Hammond, M.E., et al., 1990. A work-ing formulation for the standardization of nomenclature in the diagnosis of heart and lung rejection: Heart Rejection Study Group. The International Society for Heart Transplantation. J Heart Transplant 9, 587–593.

Book, W.M., Kelley, L., Gravanis, M.B., 2003. Fulminant mixed humoral and cellular rejection in a cardiac transplant recipient: A review of the histologic findings and literature. J Heart Lung Transplant 22, 604–607.

Brunner-La Rocca, H.P., Schneider, J., Kunzli, A., Turina, M., Kiowski, W., 1998. Cardiac allograft rejection late after transplantation is a risk factor for graft coronary artery disease. Transplantation 65, 538–543.

Butler, C.R., Thompson, R., Haykowsky, M., Toma, M., Paterson, I., 2009. Cardiovascular magnetic resonance in the diagnosis of acute heart transplant rejection: A review. J Cardiovasc Magn Reson 11, 7.

Costanzo, M.R., Costanzo, M.R., Dipchand, A., et al., 2010. The International Society of Heart and Lung Transplantation Guidelines for the care of heart transplant recipients. J Heart Lung Transplant 29, 914–956.

Deng, M.C., Eisen, H.J., Mehra, M.R., 2006a. Methodological chal-lenges of genomic research – the CARGO study. Am J Transplant 6, 1086–1087.

Deng, M.C., Eisen, H.J., Mehra, M.R., et al., 2006b. Noninvasive dis-crimination of rejection in cardiac allograft recipients using gene expression profiling. Am J Transplant 6, 150–160.

Deng, M.C., Halpern, B., Wolters, H., et al., 2009. Patient-specific lon-gitudinal patterns of AlloMap test scores – path towards person-alized medicine? J Heart Lung Transplant 28, S229.

Deng, M.C., Plenz, G., Labarrere, C., et al., 2002. The role of IL6 cytokines in acute cardiac allograft rejection. Transplant Immunol 9, 115–120.

Gradek, W.Q., D'Amico, C., Smith, A.L., Vega, D., Book, W.M., 2001. Routine surveillance endomyocardial biopsy continues to detect significant rejection late after heart transplantation. J Heart Lung Transplant 20, 497–502.

Grasser, B., Iberer, F., Schreier, G., et al., 2003. Computerized heart allograft-recipient monitoring: A multicenter study. Transplant Int 16, 225–230.

Hammer, M.H., Zhai, Y., Katori, M., et al., 2001. Homing of in vitro-generated donor antigen-reactive CD4+ T lymphocytes to renal allografts is alpha 4 beta 1 but not alpha L beta 2 integrin dependent. J Immunol 166, 596–601.

Hamour, I.M., Burke, M.M., Bell, A.D., Panicker, M.G., Banerjee, R., Banner, N.R., 2008. Limited utility of endomyocardial biopsy in the first year after heart transplantation. Transplantation 85, 969–974.

Heimansohn, D.A., Robison, R.J., Paris III, J.M., Matheny, R.G., Bogdon, J., Shaar, C.J., 1997. Routine surveillance endomyocardial biopsy: Late rejection after heart transplantation. Ann Thorac Surg 64, 1231–1236.

Hershberger, R.E., Starling, R.C., Eisen, H.J., et al., 2005. Daclizumab to prevent rejection after cardiac transplantation. N Engl J Med 352, 2705–2713.

Hesse, B., Mortensen, S.A., Folke, M., Brodersen, A.K., Aldershvile, J., Pettersson, G., 1995. Ability of antimyosin scintigraphy monitoring to exclude acute rejection during the first year after heart transplantation. J Heart Lung Transplant 14, 23–31.

Horwitz, P.A., Tsai, E.J., Putt, M.E., et al., 2004. Detection of cardiac allograft rejection and response to immunosuppressive therapy with peripheral blood gene expression. Circulation 110, 3815–3821.

Hunt, S.A., Haddad, F., 2008. The changing face of heart transplantation. J Am Coll Cardiol 52, 587–598.

Israeli, M., Ben-Gal, T., Yaari, V., et al., 2010. Individualized immune monitoring of cardiac transplant recipients by noninvasive longitudinal cellular immunity tests. Transplantation 89, 968–976.

Kabadi, R.M., Ebert, B., Flannagan, C., Farber, J.L., Mather, P.J., 2009. The AlloMap test fails to distinguish between Quilty lesions and acute allograft rejection in heart transplant patients. J Heart Lung Transplant 27, S227.

Kemnitz, J., Cremer, J., Restrepo-Specht, I., et al., 1991. Hyperacute rejection in heart allografts. Case studies. Pathol Res Pract 187, 23–29.

Kfoury, A.G., Hammond, M.E., 2010. Controversies in defining cardiac antibody-mediated rejection: Need for updated criteria. J Heart Lung Transplant 29, 389–394.

Kirklin, J.K., Naftel, D.C., Bourge, R.C., White-Williams, C., Caulfield, J.B., Tarkka, M.R., et al., 1992. Rejection after cardiac transplantation: A time-related risk factor analysis. Circulation 86, II236–II241.

Kirklin, J.K.M., Young, J.B.M., McGiffin, D.C.M., 2002. Heart Transplantation. Elsevier Health Sciences.

Kubo, S.H., Naftel, D.C., Mills Jr., R.M., et al., 1995. Risk factors for late recurrent rejection after heart transplantation: A multiinstitutional, multivariable analysis. Cardiac Transplant Research Database Group. J Heart Lung Transplant 14, 409–418.

Marboe, C.C., Billingham, M., Eisen, H., et al., 2005. Nodular endocardial infiltrates (Quilty lesions) cause significant variability in diagnosis of ISHLT grade 2 and 3A rejection in cardiac allograft recipients. J Heart Lung Transplant 24, S219–S226.

Marciniak, A., Eroglu, E., Marciniak, M., et al., 2007. The potential clinical role of ultrasonic strain and strain rate imaging in diagnosing acute rejection after heart transplantation. Eur J Echocardiogr 8, 213–221.

McRedmond, J.P., Park, S.D., Reilly, D.F., et al., 2004. Integration of proteomics and genomics in platelets: A profile of platelet proteins and platelet-specific genes. Mol Cell Proteomics 3, 133–144.

Mehra, M.R., Crespo-Leiro, M.G., Dipchand, A., et al., 2010. International Society for Heart and Lung Transplantation working formulation of a standardized nomenclature for cardiac allograft vasculopathy – 2010. J Heart Lung Transplant 29, 717–727.

Mehra, M.R., Kobashigawa, J.A., Deng, M.C., et al., 2007. Transcriptional signals of T-cell and corticosteroid-sensitive genes are associated with future acute cellular rejection in cardiac allografts. J Heart Lung Transplant 26, 1255–1263.

Mehra, M.R., Kobashigawa, J.A., Deng, M.C., et al., 2008. Clinical implications and longitudinal alteration of peripheral blood transcriptional signals indicative of future cardiac allograft rejection. J Heart Lung Transplant 27, 297–301.

Michaels, P.J., Espejo, M.L., Kobashigawa, J., et al., 2003. Humoral rejection in cardiac transplantation: Risk factors, hemodynamic consequences and relationship to transplant coronary artery disease. J Heart Lung Transplant 22, 58–69.

Miller, J.D., Van Der Most, R.G., Akondy, R.S., et al., 2008. Human effector and memory CD8+ T cell responses to smallpox and yellow fever vaccines. Immunity 28, 710–722.

Pacheco, K.A., Tarkowski, M., Klemm, J., Rosenwasser, L.J., 1998. CD49d expression and function on allergen-stimulated T cells from blood and airway. Am J Respir Cell Mol Biol 18, 286–293.

Patel, J., Hicks, A.J., Rowe, T., Hankins, S.R., Eisen, H.J., 2010. Differential leukocyte gene expression in cardiac transplant recipients receiving tracrolimus compared to cyclosporine. J Heart Lung Transplant 29, S149.

Pham, M.X., Teuteberg, J.J., Kfoury, A.G., et al., 2010. Gene-expression profiling for rejection surveillance after cardiac transplantation. N Engl J Med 362, 1890–1900.

Querec, T.D., Akondy, R.S., Lee, E.K., et al., 2009. Systems biology approach predicts immunogenicity of the yellow fever vaccine in humans. Nat Immunol 10, 116–125.

Radovancevic, B., Konuralp, C., Vrtovec, B., et al., 2005. Factors predicting 10-year survival after heart transplantation. J Heart Lung Transplant 24, 156–159.

Raichlin, E., Edwards, B.S., Kremers, W.K., et al., 2009. Acute cellular rejection and the subsequent development of allograft vasculopathy after cardiac transplantation. J Heart Lung Transplant 28, 320–327.

Re, F., Muzio, M., De Rossi, M., et al., 1994. The type II "receptor" as a decoy target for interleukin 1 in polymorphonuclear leukocytes: Characterization of induction by dexamethasone and ligand binding properties of the released decoy receptor. J Exp Med 179, 739–743.

Rodriguez, E.R., 2003. The pathology of heart transplant biopsy specimens: Revisiting the 1990 ISHLT working formulation. J Heart Lung Transplant 22, 3–15.

Schmauss, D., Weis, M., 2008. Cardiac allograft vasculopathy: Recent developments. Circulation 117, 2131–2141.

Schoels, M., Dengler, T.J., Richter, R., Meuer, S.C., Giese, T., 2004. Detection of cardiac allograft rejection by real-time PCR analysis of circulating mononuclear cells. Clin Transplant 18, 513–517.

Scott, R.L., Kasper, D.L., Arabia, F.A., et al., 2009. Peripheral blood transcriptomic signature-based cortiosteroid minimization: A single center 24-month experience. J Heart Lung Transplant 28, S216.

Segal, J.B., Kasper, E.K., Rohde, C., et al., 2001. Coagulation markers predicting cardiac transplant rejection. Transplantation 72, 233–237.

Simon, R., 2005. Roadmap for developing and validating therapeutically relevant genomic classifiers. J Clin Oncol 23, 7332–7341.

Starling, R.C., Pham, M., Valantine, H., et al., 2006. Molecular testing in the management of cardiac transplant recipients: Initial clinical experience. J Heart Lung Transplant 25, 1389–1395.

Stehlik, J., Starling, R.C., Movsesian, M.A., et al., 2006. Utility of long-term surveillance endomyocardial biopsy: A multi-institutional analysis. J Heart Lung Transplant 25, 1402–1409.

Stewart, S., Winters, G.L., Fishbein, M.C., et al., 2005. Revision of the 1990 working formulation for the standardization of nomenclature in the diagnosis of heart rejection. J Heart Lung Transplant 24, 1710–1720.

Sui, X., Tsuji, K., Tajima, S., et al., 1996. Erythropoietin-independent erythrocyte production: Signals through gp130 and c-kit dramatically promote erythropoiesis from human CD34+ cells. J Exp Med 183, 837–845.

Taylor, D.O., Stehlik, J., Edwards, L.B., et al., 2009. Registry of the International Society for Heart and Lung Transplantation: Twenty-sixth Official Adult Heart Transplant Report – 2009. J Heart Lung Transplant 28, 1007–1022.

Winters, G.L., Marboe, C.C., Billingham, M.E., 1998. The International Society for Heart and Lung Transplantation grading system for heart transplant biopsy specimens: Clarification and commentary. J Heart Lung Transplant 17, 754–760.

Winters, G.L., McManus, B.M., 1996. Consistencies and controversies in the application of the International Society for Heart and Lung Transplantation working formulation for heart transplant biopsy specimens. Rapamycin Cardiac Rejection Treatment Trial Pathologists. J Heart Lung Transplant 15, 728–735.

Yamani, M.H., Taylor, D.O., Rodriguez, E.R., et al., 2007. Transplant vasculopathy is associated with increased AlloMap gene expression score. J Heart Lung Transplant 26, 403–406.

Zuckermann, A., Manito, N., Epailly, E., et al., 2008. Multidisciplinary insights on clinical guidance for the use of proliferation signal inhibitors in heart transplantation. J Heart Lung Transplant 27, 141–149.

Zuckermann, A.O., Aliabadi, A.Z., 2009. Calcineurin-inhibitor minimization protocols in heart transplantation. Transpl Int 22, 78–89.

Hypertrophic Cardiomyopathies

J. Martijn Bos, Steve R. Ommen, and Michael J. Ackerman

DEFINITIONS, CLINICAL PRESENTATION AND DIAGNOSIS

Introduction

Hypertrophic cardiomyopathy (HCM) is defined as unexplained left ventricular hypertrophy (LVH) in the absence of precipitating factors such as hypertension or aortic stenosis. HCM is a disease of enormous phenotypic and genotypic heterogeneity. Affecting an estimated 1 in 500 people, it is the most prevalent genetic cardiovascular disease, and more importantly, it is one of the most common causes of sudden cardiac death (SCD) before 40 years of age, especially in young athletes (Maron, 2002; Maron et al., 1995). HCM can manifest with negligible-to-extreme hypertrophy, minimal-to-extensive fibrosis and myocyte disarray, absent-to-severe left ventricular outflow tract obstruction, and distinct patterns of hypertrophy.

Nomenclature

HCM was described fully for the first time by Teare in 1958 as "asymmetrical hypertrophy of the heart in young adults" (Teare, 1958). Over the past half-century, HCM has since been known by a confusing array of names, reflecting its clinical heterogeneity and uncommon occurrence in daily practice. In 1968, the World Health Organization (WHO) defined cardiomyopathies as "diseases of different and often unknown etiology in which the dominant feature is cardiomegaly and

heart failure" (Abelmann, 1984). In 1980, cardiomyopathies were newly defined as "heart muscles diseases of unknown cause," thereby differentiating them from specific heart muscle diseases of known cause such as myocarditis. Throughout the years, names such as idiopathic hypertrophic subaortic stenosis (IHSS) (Braunwald et al., 1964), muscular subaortic stenosis (Pollick et al., 1982), and hypertrophic obstructive cardiomyopathy (HOCM) (Schoendube et al., 1995) have been used widely and interchangeably to define the same disease. In 1995, a WHO/International Society and Federation of Cardiology Task Force on the Definition and Classification of Cardiomyopathies classified the different cardiomyopathies by dominant pathophysiology, or if possible, by etiological/pathogenetic factors (Richardson et al., 1996). The four most important cardiomyopathies – dilated cardiomyopathy (DCM), restrictive cardiomyopathy (RCM), arrhythmogenic right ventricular cardiomyopathy (ARVC) and HCM – were recognized, next to a number of specific and mostly acquired cardiomyopathies such as ischemic and inflammatory variants (Richardson et al., 1996). Accordingly, HCM is described as left and/or right ventricular hypertrophy, usually asymmetric and involving the interventricular septum, with predominant autosomal dominant inheritance that can be caused by mutations in sarcomeric contractile proteins (Richardson et al., 1996). On a microscopic level, HCM is characterized by the classical triad of cardiomyocyte hypertrophy, interstitial fibrosis and myofibrillar disarray.

Genomic and Personalized Medicine, 2nd edition
by Ginsburg & Willard. DOI: http://dx.doi.org/10.1016/B978-0-12-382227-7.00050-1

Clinical Presentation and Diagnosis

The clinical presentation of HCM is underscored by extreme variability that can range from an asymptomatic disease course to that of severe heart failure, arrhythmias and SCD. Many patients remain asymptomatic or only mildly symptomatic throughout the course of life. HCM commonly manifests between the second and fourth decades of life, but can present at the extremes of age. The most common symptoms at presentation of disease are exertional dyspnea, chest pain, and syncope or presyncope (fainting; transient loss of consciousness and postural tone, characterized by rapid onset and spontaneous recovery, sometimes preceded by dizziness, loss of vision, hearing and/or pain and feeling, nausea; if these are not followed by loss of consciousness/postural tone, it is defined as pre-syncope). Infants and young children may present with severe hypertrophy leading to heart failure, and these patients have poor prognosis. More often, SCD can be the tragic sentinel event for HCM in children, adolescents, or young adults. Approximately 5% of patients with HCM progress to "end-stage" disease characterized by left ventricular dilatation and heart failure. In such cases, cardiac transplantation may be considered. Other serious, life-threatening complications include embolic stroke and cardiac arrhythmias.

Echocardiography

Conventional two-dimensional echocardiography is the initial diagnostic imaging modality of choice for the clinical diagnosis of HCM. On the echocardiogram, unexplained and usually asymmetric, diffuse or segmental hypertrophy associated with a non-dilated left ventricle (LV) independent of presence or absence of LV outflow obstruction (LVOTO) can be seen. A left ventricular wall thickness of ≤12 mm is typically regarded as normal, with measurements of 13–15 mm labeled as "borderline hypertrophy." A maximal LV end-diastolic wall thickness exceeding 15 mm represents the absolute dimension generally accepted for the clinical diagnosis of HCM in adults (in children, two or more standard deviations from the mean relative to body surface area) (Maron et al., 2003b). Echocardiography can also provide details of location and degree of hypertrophy. In general, while there are a large number morphologic appearances of the heart, four different morphological sub-types of HCM can be recognized in most cases (Table 50.1, Figure 50.1): sigmoid septum, reverse septal curvature, apical, and neutral contour variant.

Dynamic LVOTO is a common feature of HCM, but is not required for the diagnosis of HCM. The existence of LVOTO is diagnosed by demonstration of a resting or provocable Doppler gradient of >30 mm Hg. LVOTO is produced by the interaction of the hypertrophied septum and systolic anterior motion (SAM) of the mitral valve's anterior leaflet. The latter results from abnormal blood flow vectors across the valve and abnormal anterior positioning of the valve and its support structures. Variable in severity, posteriorly directed mitral regurgitation is a common finding.

Most patients at presentation manifest some degree of impaired diastolic function, ranging from abnormal relaxation to severe myocardial stiffness, elevated LV end-diastolic

TABLE 50.1	Septal morphologies in hypertrophic cardiomyopathy
Septal morphology	**Description**
Sigmoid septum	Septum concave to cavity with pronounced septal bulge Ovoid LV cavity
Reverse septal curvature	Predominant mid-septal convexity toward LV cavity Crescent-shaped cavity
Apical variant	Predominant apical distribution of hypertrophy
Neutral variant	Overall straight or variable convexity, neither predominantly convex or concave

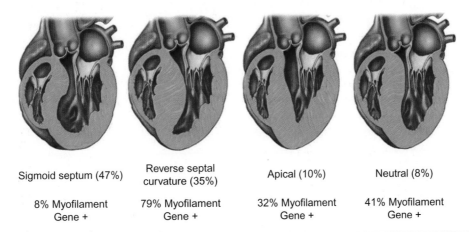

Sigmoid septum (47%)   Reverse septal curvature (35%)   Apical (10%)   Neutral (8%)

8% Myofilament Gene +   79% Myofilament Gene +   32% Myofilament Gene +   41% Myofilament Gene +

Figure 50.1   Septal morphologies in hypertrophic cardiomyopathy (HCM). Shown are the most common septal morphologies in HCM, their distribution in a large cohort of patients with HCM, and the yield of genetic testing for each morphological subgroup.

pressure, elevated atrial pressure, and pulmonary congestion, leading to exercise intolerance and fatigue. Systolic cardiac function, as measured by ejection fraction, is usually preserved. It is notable that more elegant measures of systolic performance, such as tissue velocity and strain imaging, suggest a decrease in systolic function in patients with HCM, which may even be present in HCM-genotype patients before hypertrophy can be detected. "End-stage disease" is characterized by LV dilatation, poor systolic function, and heart failure.

Magnetic Resonance Imaging (MRI)

Magnetic resonance imaging constitutes an important additional diagnostic tool, especially in patients with suboptimal echocardiography or unusual segmental involvement of the myocardium. Delayed gadolinium enhancement in MRI is an excellent tool to identify areas of scarring, replacement fibrosis, or irreversible myocardial injury. There are some recent reports of higher risk of SCD associated with delayed gadolinium enhancement. In a cohort of 42 patients, myocardial hyper-enhancement was found in 79%. Extent of hyper-enhancement was greater in patients with progressive disease (28.5% vs 8.7%, $p < 0.001$), in patients with two or more risk factors for SCD (15.7% vs 8.6%, $p = 0.02$) (Moon et al., 2003) as well as in genotype-positive HCM patients (75% vs 53%; $p < 0.001$) (Rubinshtein et al., 2010).

MOLECULAR GENETICS OF HYPERTROPHIC CARDIOMYOPATHY

Sarcomeric/Myofilament-HCM

Since the sentinel discovery of the first locus for familial HCM on chromosome 14 in 1989 and the first mutations involving the MYH7-encoded beta-myosin heavy chain in 1990 as the pathogenic basis for HCM (Geisterfer-Lowrance et al., 1990; Jarcho et al., 1989), several hundred mutations scattered in more than 20 genes encoding various sarcomeric proteins (myofilaments and Z-disc associated proteins) and calcium-handling proteins, as well in genes encoding proteins associated with phenocopies of the disease, have been identified (Tables 50.2 and 50.3) (Bos et al., 2009). The most common genetically mediated subtype of HCM is myofilament-HCM, with hundreds of disease-associated mutations in nine genes encoding proteins critical to the sarcomere's myofilament: myosin binding protein C (MYBPC3), beta-myosin heavy chain (MYH7), regulatory myosin light chain (MYL2), essential myosin light chain (MYL3), cardiac troponin T (TNNT2), cardiac troponin I (TNNI3), cardiac troponin C (TNNC1), α-tropomyosin (TPM1), and actin (ACTC) (Mogensen et al., 1999; Olson et al., 2002). Analysis of giant sarcomeric TTN-encoded titin, which extends throughout half of the sarcomere, has thus far revealed only few mutations (Arimura et al., 2009; Satoh et al., 1999). More recently, mutations have been described in the myofilament protein alpha-myosin heavy chain encoded by MYH6 (Carniel et al., 2005) as

TABLE 50.2	Summary of HCM-susceptibility genes		
Gene	**Locus**	**Protein**	**Frequency (%)**
Myofilament HCM			
MYBPC3	11p11.2	Cardiac myosin-binding protein C	15–25
MYH7	14q11.2-q12	β-myosin heavy chain	15–25
ACTC	15q14	α-cardiac actin	<1
MYH6	14q11.2-q12	α-myosin heavy chain	<1
MYL2	12q23-q24.3	Ventricular regulatory myosin light chain	<2
MYL3	3p21.2-p21.3	Ventricular essential myosin light chain	<1
TNNC1	3p21.3-p14.3	Cardiac troponin C	<1
TNNI3	19p13.4	Cardiac troponin I	<5
TNNT2	1q32	Cardiac troponin T	<5
TPM1	15q22.1	α-tropomyosin	<5
TTN	2q24.3	Titin	<1
Z-disc HCM			
ACTN2	1q42-q43	Alpha-actinin 2	<1
ANKRD1	10q23.33	Ankyrin repeat domain 1 (alias: CARP)	<1
CSRP3	11p15.1	Muscle LIM protein	<1
LBD3	10q22.2-q23.3	LIM binding domain 3 (alias: ZASP)	1–5
MYOZ2	4q26-q27	Myozenin 2	<1
TCAP	17q12-q21.1	Telethonin	<1
VCL	10q22.1-q23	Vinculin/ metavinculin	<1
Calcium-handling HCM			
JPH2	20q12	Junctophilin-2	<1
PLN	6q22.1	Phospholamban	<1

In bold: genes available as commercial genetic test for HCM.
Reproduced from Bos et al., 2009 with permission from Elsevier.

TABLE 50.3   HCM phenocopies

Gene	Locus	Protein	Syndrome
AGL	1p21	Amylo-1,6-glucosidase	Forbes disease
DTNA	18q12	α-dystrobrevin	Barth syndrome/LVNC
FXN	9q13	Frataxin	Friedrich's ataxia
GAA	17q25.2-q25.3	α-1,4-glucosidase deficiency	Pompe's disease
GLA	Xq22	α-galactosidase A	Fabry's disease
KRAS	12p12.1	v-Ki-ras2 Kirsten rat sarcoma viral oncogene homolog	Noonan's syndrome
LAMP2	Xq24	Lysosome-associated membrane protein 2	Danon's syndrome/WPW
PRKAG2	7q35-q36.36	AMP-activated protein kinase	WPW/HCM
PTPN11	12q24.1	Protein tyrosine phosphatase, non-receptor type 11, SHP-2	Noonan's syndrome, LEOPARD syndrome
RAF1	3p25	V-RAF-1 murine leukemia viral oncogene homolog 1	Noonan's syndrome, LEOPARD syndrome
SOS1	2p22-p21	Son of sevenless homolog 1	Noonan's syndrome
TAZ	Xq28	Tafazzin (G4.5)	Barth syndrome/LVNC

LVNC, left ventricular non-compaction; WPW, Wolff-Parkinson-White syndrome. In bold: genes available as commercial genetic test for HCM. *Reproduced from Bos et al., 2009 with permission from Elsevier.*

well as cardiac troponin C encoded by *TNNC1* (Landstrom et al., 2008). The prevalence of mutations in the eight most common myofilament-associated genes, in different international cohorts, ranges from 30–61%, leaving still a large number of patients with genetically unexplained disease (Van Driest et al., 2005b).

Z-disc and calcium-handling HCM

Over the last few years, the spectrum of HCM-associated genes has expanded beyond the various cardiac myofilaments

to encompass additional subgroups that could be classified as "Z-disc HCM" and "calcium-handling HCM" (Table 50.2). Due to the Z-disc's close proximity to the contractile apparatus of the myofilament, its structure–function relationship with regard to cyto-architecture, and its role in the stretch-sensor mechanism of the sarcomere, attention has focused on genes encoding the proteins that comprise the cardiac Z-disc as potential HCM-susceptibility genes. The Z-disc is an intricate assembly of proteins at the Z-line of the cardiomyocyte sarcomere. Proteins of the Z-disc are important in the structural and mechanical stability of the sarcomere as they appear to serve as a docking station for transcription factors, Ca^{2+} signaling proteins, kinases, and phosphatases (Frank et al., 2006; Pyle and Solaro, 2004). This assembly of proteins also seems to serve as a way station for proteins that regulate transcription, by aiding their controlled translocation between the nucleus and the Z-disc (Frank et al., 2006; Pyle and Solaro, 2004). Initial HCM-associated mutations were described in *CSRP3*-encoded muscle LIM protein (Geier et al., 2003) and *TCAP*-encoded telethonin (Hayashi et al., 2004), an observation replicated in a large cohort of unrelated patients with HCM (Bos et al., 2006). Subsequently, mutations in patients with HCM have been detected in *LDB3*-encoded LIM domain binding 3, *ACTN2*-encoded alpha-actinin 2 and *VCL*-encoded vinculin/metavinculin (Theis et al., 2006). Interestingly, although the first HCM-associated mutation in vinculin was found in the cardiac-specific insert of the gene, yielding the protein called metavinculin (Vasile et al., 2006b), the follow-up study also identified a mutation in the ubiquitously expressed protein vinculin (Vasile et al., 2006a). In 2007, a mutation in the Z-disc-associated *MYOZ2*-encoded myozenin 2 was reported as a novel gene for HCM (Osio et al., 2007), as was, more recently, *ANKRD1*-encoded ankyrin repeat domain-1 (also known as CARP, cardiac ankyrin repeat protein) (Arimura et al., 2009).

As the critical ion in the excitation-contraction coupling of the cardiomyocyte, calcium, and proteins involved in calcium-induced calcium release (CICR) have always been of high interest in the pathogenesis of HCM. Although with very low frequency, mutations have been described in the promoter and coding region of *PLN*-encoded phospholamban, an important inhibitor of cardiac muscle sarcoplasmic reticulum Ca(2+)-ATPase (SERCA) (Haghighi et al., 2006; Minamisawa et al., 2003), the *RYR2*-encoded cardiac ryanodine receptor (Fujino et al., 2006), and the *JPH2*-encoded type 2 junctophilin (Landstrom et al., 2008).

HCM Phenocopies

The last genetic subgroup is HCM phenocopies: patients with seemingly unexplained LVH mimicking the HCM phenotype. In these cases, LVH is usually the primary or even sole presenting feature of a much larger, complicated syndrome or disease. However, sometimes only genetics can elucidate the presence of this underlying disease and distinguish it from a primary cardiomyopathy. All currently discovered phenocopies capable of mimicking HCM are summarized in Table 50.3. In 2001, the first mutations in *PRKAG2*-encoded AMP-activated protein kinase gamma-2, a protein involved in the energy homeostasis of the heart, were described in two families with severe HCM and aberrant AV-conduction in some individuals (Blair et al., 2001). Further studies showed that patients with mutations in this gene lacked the microscopic features of HCM, but did show newly formed vacuoles filled with glycogen-associated granules. This glycogen storage disease, therefore, seemed to mimic HCM, distinguishing itself by electrophysiological abnormalities, particularly ventricular pre-excitation (Arad et al., 2002, 2005). In 2005, Arad and colleagues described mutations in lysosome-associated membrane protein-2 encoded by *LAMP2* (Danon's syndrome) and protein kinase gamma-2 encoded by *PRKAG2* in glycogen storage disease-associated genes found in patients presenting with the clinical phenotype of HCM (Arad et al., 2005). A 2006 community-based study showed that in 50 healthy individuals with idiopathic LVH, participants with either HCM-associated sarcomeric mutations or HCM secondary to perturbations in glycogen storage were indistinguishable clinically from those without mutations (Morita et al., 2006). In 2005, a mutation in *FXN*-encoded frataxin associated with Friedrich ataxia was described in a patient with HCM. Although this patient also harbored a myofilament mutation in *MYBPC3*-encoded myosin binding protein C, functional characterization showed significant influence of the FXN mutant on the phenotype, suggesting that the observed alterations in energetics may act in synergy with the present myofilament mutation (Van Driest et al., 2005a). Akin to PRKAG2 and LAMP2, Fabry's disease can express predominant cardiac features of seemingly unexplained LVH. Over the years, mutations in *GLA*-encoded alpha-galactosidose A have been found in patients with this multisystem disorder (Nakao et al., 1995; Sachdev et al., 2002; Sakuraba et al., 1990).

Role of Modifiers in HCM

The role of modifiers of the HCM phenotype, either by the presence of common polymorphisms or founder mutations, has been the subject of recent investigations. The most important subgroup of polymorphisms studied to date involves the major components of the renin-angiotensin-aldosterone system (RAAS). Polymorphisms in the RAAS pathway [angiotensinogen-I converting enzyme (ACE), angiotensin receptor 1 (AGTR1), chymase 1 (CMA), angiotensin I (AGT), and cytochrome P450, polypeptide 2 (CYP11B2): DD-*ACE*, CC-*AGTR1*, AA-*CMA*, T174M- and M235T-*AGT*, and CC-*CYP11B2*] appear to influence the HCM phenotype, in particular the severity of LVH (Ortlepp et al., 2002; Perkins et al., 2005). Among patients with the DD-ACE genotype, there was greater LVH than among those with an ID or II genotype (Lechin et al., 1995). Furthermore, a combined "pro-LVH" profile of five RAAS genes was associated with higher degree of LVH in one particular, founder MYPBC3-HCM pedigree (Ortlepp et al., 2002) and in a large cohort of myofilament-positive patients (Perkins et al., 2005).

In 2008, sex hormone polymorphisms were shown to modify the HCM phenotype (Lind et al., 2008). Fewer CAG repeats in *AR*-encoded androgen receptor were associated with thicker myocardial walls in males ($p = 0.008$) and male carriers of the A allele in the promoter of *ESR1*-encoded estrogen receptor 1 (SNP rs6915267) exhibited an 11% decrease in LV wall thickness ($p = 0.047$) compared to GG-homozygote males (Lind et al., 2008). HCM modifier polymorphisms such as these could contribute to the clinical differences observed between men and women with HCM (Bos et al., 2008; Olivotto et al., 2005). The release of the complete human genome sequence and the enormity of variation in individuals show a growing role for modifier genes and the search for effect by genome-wide studies. In 2007, Daw and colleagues performed the first study of this kind for HCM, and they identified multiple loci with suggestive linkage. Effect sizes on left ventricular mass in this cohort of 100 patients ranged from a minor shift (~8 grams) from one locus for the common allele to a major shift (~90 grams) in patients carrying the minor allele on another locus (Daw et al., 2007).

Genotype–Phenotype Relationships in HCM

For more than a decade, multiple studies have tried to identify phenotypic characteristics most indicative of myofilament-HCM to facilitate genetic counseling and strategically direct clinical genetic testing (Ackerman et al., 2002; Richard et al., 2003; Van Driest et al., 2002, 2003, 2004a, c; Woo et al., 2003). Until 2001, it was thought that specific mutations in these myofilament genes were inherently "benign" or "malignant" (Anan et al., 1994; Coviello et al., 1997; Elliott et al., 2000; Moolman et al., 1997; Niimura et al., 2002; Seidman and Seidman, 2001; Varnava et al., 1999; Watkins et al., 1992). However, these studies were based on highly penetrant, single families with HCM, and later genotype–phenotype studies involving a large cohort of unrelated patients have indicated that great caution must be exercised in assigning particular prognostic significance to any particular mutation (Ackerman et al., 2002; Van Driest et al., 2002, 2004b). Furthermore, these studies demonstrated that the two most common forms of genetically mediated HCM – MYH7-HCM and MYBPC3-HCM – were phenotypically similar (Van Driest et al., 2004c). While several phenotype–genotype relationships have emerged to enrich the yield of genetic testing, these patient profiles have not been particularly clinically informative on an individual level.

In 2006, an important discovery was made linking the echocardiographically determined septal morphology to the underlying genetic substrate (Binder et al., 2006). The first link to be drawn between septal morphologies was a result of an HCM study by Lever and colleagues in the 1980s, where septal contour (Table 50.1) was found to be age-dependent, with a predominance of sigmoidal-HCM noted in the elderly (Lever et al., 1989). Then, an early genotype–phenotype observation involving a small number of patients and family members noted that patients with MYH7-HCM generally had reversed curvature septal contours (reverse curve-HCM) (Solomon et al., 1993). Analysis of the echocardiograms of 382 previously genotyped and published patients (Van Driest et al., 2003, 2004a, c), revealed that sigmoidal-HCM (47% of cohort) and reverse curve-HCM (35% of cohort) were the two most prevalent anatomical subtypes of HCM, and that the septal contour was the strongest predictor for the presence of a myofilament mutation, regardless of age (odds ratio 21, $p < 0.001$) (Binder et al., 2006). The yield from the commercially equivalent HCM genetic research test for myofilament-HCM was 79% in reverse curve-HCM, but just 8% in patients with sigmoidal-HCM. These observations may facilitate echo-guided genetic testing with informed genetic counseling about the *a priori* probability of a positive genetic test based upon the patient's expressed anatomical phenotype of HCM (Figure 50.1).

Genotype–phenotype relationships for the complete sarcomere, especially focused on the cardiac Z-disc, have been even less informative. A main implication for the Z-disc is its involvement in the cardiomyocyte stretch sensing and response systems (Knoll et al., 2002), which has demonstrated that a lot of genes associated with HCM are also found to be implicated in DCM, with *ACTC*, *MYH7*, *TNNT2*, *TPM1*, *MYBPC3*, *TTN*, *ACTN2*, *LDB3*, *MLP*, *TCAP*, and *VCL* established as both HCM and DCM susceptibility genes (Daehmlow et al., 2002; Geier et al., 2003; Gerull et al., 2002; Hayashi et al., 2004; Kamisago et al., 2000; Mohapatra et al., 2003; Olson et al., 2000, 2001; Vasile et al., 2006b). Linking reverse-curve HCM to the presence of myofilament mutation, and recognizing that the Z-disc may transduce multiple signaling pathways during stress, translating into hypertrophic responses, cell growth, and remodeling (Frey et al., 2004), it was observed that Z-disc HCM, in contrast to myofilament HCM, is preferentially sigmoidal. In fact, 11 of 13 patients with Z-disc HCM had a sigmoidal septal contour, and no reverse septal curvatures were seen (Theis et al., 2006). It is speculated that Z-disc HCM leads to a hypertrophic response that is expressed in the areas of highest stress (i.e., LVOT), and therefore predisposes to a sigmoidal septal contour.

Recently, a longitudinal study in a large cohort of unrelated Italian patients was the first to describe long-term follow-up of genotyped patients with HCM. The study reported an increased risk of cardiovascular death, non-fatal stroke, or progression to New York Heart Association (NYHA) class III/IV among patients with a positive HCM genetic test involving any of the myofilament genes, compared to those patients with a negative genetic test (25% vs 7%, respectively; $p = 0.002$) (Figure 50.2A). Multivariate analysis showed genotype-positive HCM to be the strongest predictor of an adverse outcome [hazard ratio 4.27 (CI 1.43–12.48), $p = 0.008$] (Olivotto et al., 2008). Furthermore, patients with genotype-positive HCM had greater probability of developing severe LV systolic dysfunction ($p = 0.021$; Figure 50.2B) and restrictive LV filling ($p = 0.018$; Figure 50.2C).

Lastly, it has been observed that patients with multiple mutations (i.e., compound or double heterozygotes), detected in about 3–5% of genotype-positive patients, have a more severe phenotype and increased incidence of sudden death (Ho et al., 2000; Ingles et al., 2005; Van Driest et al.,

2004c), suggesting that a gene-dosage effect might contribute to disease severity. Interestingly, in the majority of cases of compound heterozygosity, one of the mutations usually involves *MYBPC3* (Van Driest et al., 2004c). In their longitudinal study, Olivotto and colleagues observed a similar trend, showing that patients with double mutations (of which one was usually *MYBPC3*) had greater disease severity than myofilament-negative patients or patients with a single *MYBPC3*, thick filament, or thin filament mutation combined ($p < 0.05$; Figure 50.2D). This dosage effect was also evident in a small subset of patients hosting 3 myofilament, as these patients, albeit rare, demonstrated increased risk of end-stage progression and ventricular arrhythmias (Girolami et al., 2010). In summary, although clinical prognostication must be rendered with great caution for specific gene domains or specific genetic mutations, a positive HCM genetic test in general portends a greater likelihood of disease progression, particularly as it pertains to systolic and diastolic dysfunction and propensity to develop symptoms. As such, clinical genetic testing may thereby aid in the prognostication of a patient's disease outcome.

Whole Genome Sequencing and HCM

Over the last few years, the field of genetic testing and DNA sequencing has evolved rapidly where now, albeit in research settings, a patient's complete genetic code ("whole genome" ~3 billion nucleotides) or the protein-translating part of their genome ("whole exome", 1–2% of whole genome, ~30,000 genes). Where previous gene discovery strategies focused on either a specific gene (candidate gene selection strategies), a region (via linkage analysis), or previously identified single nucleotide polymorphisms (SNPs) [genome-wide association studies (GWAS)], these next generation sequencing platforms offer all of a patient's genetic variants in one assay, opening the door for identification of novel susceptibility mutations and/or disease modifiers, and giving insight into possible novel pathways in disease pathogenesis. However, although the techniques exist to analyze one's genome, the results come with profound bioinformatics and interpretation challenges. For example, an average human sample, subjected to whole exome sequencing, hosts 3000–5000 genetic variants of which 500–1000 are novel, previously unseen variants. Therefore, appropriate filtering methods, such as excluding common variants and subtracting the whole exome profile of a phenotype-negative relative, can help in elucidating the possible genetic cause. With the rapid development of these techniques, it won't be long before they will be applied to HCM in a research setting and eventually, there will be a commercial equivalent. As discussed below, commercial genetic testing is available for the major HCM-susceptibility genes and such testing has major diagnostic, modest prognostic, and even mild therapeutic impact in clinical practice (Ackerman et al., 2011). The role and impact of whole genome/exome next generation sequencing on clinical practice has yet to be determined.

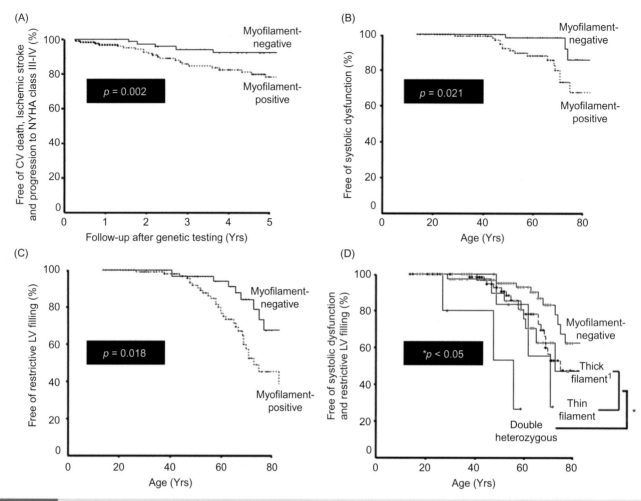

Figure 50.2 Relation of genetic test status to outcome in patients with HCM. Follow-up data shows that patients harboring HCM-associated myofilament mutation progress to CV death, ischemic stroke or NYHA-Class III-IV more rapidly than patients with a negative genetic test (A). Furthermore, patients with a myofilament mutation are more likely to develop systolic dysfunction (B) or a restrictive filling pattern (C), independent of the genotype involved (D). *Reproduced from Bos et al., 2009 with permission from Elsevier.*

New Insights and Approaches to Genomics of HCM

Not only in molecular genetics but also in other "-omics" fields, novel pathways underlying the pathophysiology of this heterogeneous disease have been identified using several new techniques to study large-scale transcriptional changes (Churchill, 2002; Holland, 2002; Velculescu et al., 1997). For example, a transcriptomic approach uses microarray analysis, a technique that gives a snapshot view of gene expression that, combined with complex analytic tools, can identify a transcriptional network of genes and pathways. Microarray chips can hold more than 30,000 genes, and can be utilized to compare expression levels in certain disease states with healthy controls. In 2002, Hwang and colleagues studied RNA from heart failure patients with either HCM or DCM, and found 192 genes to be up-regulated in both, as well as several genes differentially expressed between the two

diseases, providing information on different pathways and genes involved in the pathogenesis (Hwang et al., 2002). More recently, Rajan and colleagues performed microarray analysis on ventricular tissue of two previously developed transgenic HCM mice carrying mutations in alpha-tropomyosin (TPM1). Studying 22,600 genes, they discovered 754 expressed differentially between transgenic and non-transgenic mice, of which 266 were differentially regulated between the two different mutant hearts, showing most significant changes in genes belonging to the "secreted/extracellular matrix" (up-regulation) and "metabolic enzymes" (down-regulation) (Rajan et al., 2006).

The role of microRNAs (miRNAs) in cardiac development and (hypertrophic) heart disease is an emerging field. Extensively reviewed (Barringhaus and Zamore, 2009; Small et al., 2010), these fundamental cellular regulators were first described by Lee and colleagues in 1993 (Lee et al., 1993).

Consisting of non-coding RNA molecules 15–25 nucleotides in length, miRNAs generally function to silence genes through post-transcriptional regulation. Utilizing two mouse models of pathological hypertrophy – transverse aortic constriction (TAC) and calcineurin transgenic mice – six miRNAs were up-regulated, which in vitro were sufficient to induce hypertrophic growth of cardiomyocytes (Van Rooij et al., 2006). Furthermore, transgenic mouse models over-expressing miR-195 showed that a single miRNA could induce pathological hypertrophy and heart failure (Van Rooij et al., 2006). Multiple studies of cardiac hypertrophy have been published with miRNA expression profiles in different settings, in vivo and in vitro (Ikeda et al., 2007; Sayed et al., 2007; Tatsuguchi et al., 2007; Thum et al., 2007; Van Rooij et al., 2006, 2007). The sizes of miRNAs and the fact that one miRNA functions on multiple targets make them the ideal target for novel drugs targeted to prevent or slow down development of HCM. Also, novel techniques such as PMAGE, a messenger RNA (mRNA) profiling technology, which can detect mRNAs as rare as one transcript per three cells (Kim et al., 2007), or filter-based hybridization, which performs rapid deep-sequencing and can detect copy-number variation (Herman et al., 2009), will help over the coming years to expand the field and knowledge of biological pathways involved in pro-hypertrophic remodeling and epigenetic factors involved in disease pathogenesis. This in turn might lead to discovery of novel disease-causing genes, involved pathways and possible novel therapeutic targets.

SCREENING AND TREATMENT FOR HCM

HCM Genetic Testing in Clinical Practice

Currently, a number of commercial companies offer HCM genetic testing for the eight most common myofilament-associated genes; additional genes offered by some include those involved in the glycogen storage diseases or the recently discovered TNNC1-encoded troponin C-HCM. The HCM susceptibility genes available for commercial genetic testing are highlighted in bold in Tables 50.2 and 50.3. Although some of the new HCM susceptibility genes may surpass the prevalence of mutations found in some of the myofilament proteins, MYBPC3 and MYH7 remain by far the most common HCM-associated genes, with an estimated prevalence of 15–25% for each gene. Among the nine HCM-associated, myofilament-encoding genes, the prevalence of myofilament-HCM has ranged from 35–65% in several different international cohorts of unrelated patients who met the clinically accepted definition of HCM (Marian and Roberts, 2001; Van Driest et al., 2005b). As discussed previously, genetic screening for the proband thus far has limited implications for treatment and follow-up of the disease, however, it can provide insight for relatives of the proband and required screening on their part (Ackerman et al., 2011).

Family Screening in HCM

Clinical evaluation, including genetic testing of family members with HCM, plays an important role in the early diagnosis of the disease. Figure 50.3 shows a possible algorithm to follow after diagnosis of a proband with HCM. In general, all first-degree relatives and probably "athletic" second-degree relatives of an index case of HCM should be screened by ECG and echocardiogram. Annual screenings are recommended for young adults (aged 12–25 yrs) and athletes, and thereafter every 3–5 years. If an HCM-associated mutation is identified, confirmatory genetic testing in first-degree relatives should also be performed. Depending on the established familial versus sporadic pattern, confirmatory genetic testing should proceed in concentric circles of relatedness (Ackerman et al., 2011). If the putative HCM-associated mutation is not-found in a phenotype-negative family member, screening can be stopped. However, a decision to cease surveillance for HCM in a relative hinges critically on the certainty of the identified gene/mutation and its causative link, as well as the complete absence of any traditional evidence used to clinically diagnose HCM (i.e., an asymptomatic and normal echocardiogram) (Ackerman et al., 2011).

Follow-up and Sports Participation

Intense physical exertion can potentially trigger SCD in individuals with HCM. According to the 2005 Bethesda Conference recommendations, athletes with HCM should be excluded from participation in contact-sports as well as most organized competitive sports, with the possible exception of low-intensity sports classified as class IA sports (i.e., golf, bowling, cricket, billiards, and riflery) (Maron et al., 2005). Per the guidelines, the presence of an implantable cardioverter-defibrillator (ICD) does not alter these recommendations. However, these guideline-based recommendations are not uniformly embraced, and a national ICD sports registry, tracking athletes with ICDs because of various cardiac conditions including HCM, has been established (Estes et al., 2001; Lampert and Cannom, 2008).

Pharmacological Therapy for Obstructive HCM

The primary goal of pharmacologic therapy in obstructive HCM (also called HOCM) is to decrease symptoms. Understanding the pathophysiology of the LVOTO is crucial to devising appropriate medication recommendations. The LVOTO in HCM is dynamic and highly load-dependent. The obstruction is accentuated with decreases in preload (volume status) and afterload (vasodilation), and with increases in contractility. Simple physical exertion such as walking can cause all of these to occur simultaneously. As patients with HCM typically only have symptoms with effort, the goal of medical therapy is to decrease the effort-related augmentation of LVOTO. Due to negative inotropic and negative chronotropic effects, beta blockers are the traditional mainstay of HCM therapy. Beta blockers are used in symptomatic patients with or without obstruction to control heart failure and anginal chest

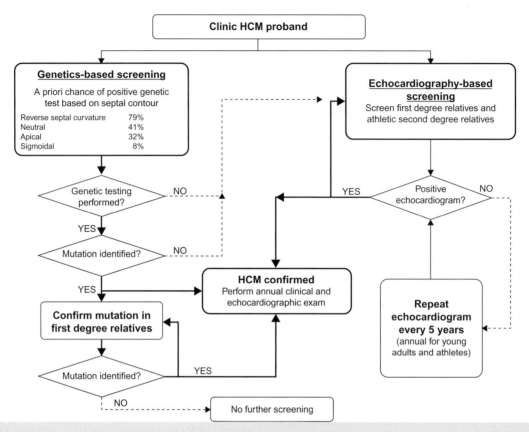

Figure 50.3 Genetic and echocardiographic-based screening in HCM. Flow chart showing a possible decision tree for genetic and echocardiography-based screening in HCM. Noted are the *a priori* chances for a positive genetic test result based on the echocardiographic-scored septal contour, as well as the steps to follow if a patient chooses not to pursue genetic testing. *Reproduced from Bos et al., 2009 with permission from Elsevier.*

pain. The dose–response relationship of these medications varies significantly from patient to patient. Commonly used beta blockers include propranolol, atenolol, metoprolol and nadolol (Elliott and McKenna, 2004; Spirito and Autore, 2006). Calcium channel blockers are used in HCM for their negative inotropic effect. They should be avoided in infants, and used with caution in patients with heart failure and/or very significant obstruction (Maron et al., 2003b). There are no data that beta blockers, calcium channel blockers, or disopyramide alter the risk of SCD (Maron et al., 2003b). Angiotensin-converting enzyme inhibitors (ACE inhibitors), angiotensin II blockers, nifedipine, and other pure afterload-reducing agents should be used with caution, as afterload reduction may worsen LVOTO (Roberts and Sigwart, 2005; Spirito and Autore, 2006). Beta-adrenergic agents such as dopamine, dobutamine, or epinephrine, that is, agents with increased inotropic activity, may worsen LVOTO (Maron et al., 2003b). Likewise, rapid or aggressive diuresis can decrease preload and worsen LVOTO.

Pharmacogenomics

Currently, no treatment has been identified that can stop or reverse the hypertrophic process in patients with or without

genetically proven HCM, although some animal models show promise. One of the first studies of its kind was performed in transgenic MYH7-R403Q mice models (designated $\alpha MHC^{403/+}$). In a randomized trial, $\alpha MHC^{403/+}$-mice treated with diltiazem, an L-type calcium inhibitor, showed significant improvement compared to mice treated with placebo in terms of cardiac systolic function as measured by increased end-diastolic and end-systolic volumes, decreased dP/dT_{max} values, and end-systolic elastance (Semsarian et al., 2002). Furthermore, diltiazem-treated mice showed significantly less hypertrophy at 30 and 39 weeks than age-matched $\alpha MHC^{403/+}$-untreated mice, as well as less fibrosis and myocyte disarray on microscopy (Semsarian et al., 2002). Recently, in a different transgenic mouse model (TNNT2–I79N), Westermann and colleagues showed that diltiazem improved diastolic function and prevented diastolic heart failure and sudden cardiac death compared to untreated mice (Westermann et al., 2006). RAAS polymorphisms modify the phenotype of HCM, particularly MYBPC3-HCM (Ortlepp et al., 2002; Perkins et al., 2005), and there is now growing evidence that ACE inhibitors, especially combined with low doses of aldosterone receptor blockers, may attenuate the progression of hypertrophy and fibrosis (Fraccarollo et al., 2003,

2005; Kalkman et al., 1999; Kambara et al., 2003; Monteiro de Resende et al., 2006). In early mouse models of transgenic cardiac troponin T (cTnT-Q^{92}) that exhibit myocyte disarray and fibrosis, a randomized, blinded trial comparing losartan (an angiotensin-II blocker) or placebo demonstrated that losartan significantly reversed fibrosis and expression of collagen 1α(I) and TGFβ-1 in the transgenic mice (Lim et al., 2001). In a similar study involving the same transgenic mice, losartan produced a 50% reduction in myocyte disarray compared to mice treated with placebo, as well as complete normalization of the collagen volume fraction (Tsybouleva et al., 2004). Therefore, with the identification of the patient's pathogenic substrate and polymorphism profile, and novel discoveries in the field of the HCM transcriptome and hypertrophy-associated miRNAs, one can imagine that specific therapies may someday emerge. In other cases, the proper and prompt recognition of an HCM phenocopy, such as cardiac Fabry's disease, can facilitate gene-specific pharmacotherapy such as enzyme replacement therapy. Albeit rare, such clinical sleuthing can enable early treatment and prevent the progression of the disease.

Septal Reduction Therapies for the Treatment of Obstructive HCM Refractory to Pharmacotherapy

Septal Myectomy Surgery

Ventricular septal myectomy remains the gold standard for treating drug-refractory, symptomatic, obstructive HCM; a procedure in which a piece of hypertrophied septum is removed in order to relieve the obstruction (Maron et al., 2004; Nishimura and Holmes, 2004; Spirito and Autore, 2006). Surgery is usually indicated in patients with peak instantaneous LVOT Doppler gradient of 50 mm Hg or higher under rest or provocation, and/or severely symptomatic patients (NYHA class III or IV) (Maron et al., 2003b, 2004; Spirito and Autore, 2006). This profile describes approximately 5% of patients with HCM (Elliott and McKenna, 2004). More extensive, extended septal myectomy involving the anterolateral papillary muscle and mitral valvuloplasty may be needed in patients with abnormal papillary muscle apparatus and mitral valve abnormalities (Ommen et al., 2005). Surgical mortality is <1% in most major HCM specialty centers (Maron et al., 2004; Poliac et al., 2006). Long-term survival after surgical myectomy is equal to that observed in the general population (Ommeni et al., 2005). Surgery provides long-term improvement in LVOT gradient and mitral valve regurgitation, and symptomatic improvement (Maron et al., 2004; Ommen et al., 2005; Poliac et al., 2006).

Septal Ablation

In an alcohol septal ablation technique, ethanol (95% alcohol 1–3 mL) is injected into specific septal branches of the left anterior descending artery to produce a controlled septal infarction, often providing dramatic symptomatic improvement in some patients (Faber et al., 1998; Gietzen et al., 1999; Kimmelstiel and Maron, 2004; Knight et al., 1997; Lakkis et al., 1998). The criteria for patient selection for alcohol septal ablation are similar to those for myectomy with the following caveat – the long-term impact of alcohol septal ablation on SCD risk is unknown. Scarring associated with alcohol septal ablation may create a permanent arrhythmogenic substrate (Maron et al., 2003b). Complications include complete atrioventricular blocks requiring permanent pacemakers (5–10% of patients), large myocardial infarction, acute mitral valve regurgitation, ventricular fibrillation and death (2–4%) (Nishimura and Holmes, 2004; Roberts and Sigwart, 2005; Spirito and Autore, 2006). Alcohol septal ablation is not suitable for patients with LVOTO secondary to abnormal mitral valve apparatus and unusual location of hypertrophy away from the area supplied by septal perforator. Given the unknown future risks of alcohol septal ablation, it is not recommended in children or young adults (Maron et al., 2003b). In contrast, alcohol septal ablation may be most appropriate for the elderly patient with unacceptable surgical co-morbidities (Sorajja et al., 2008).

Implantable Cardioverter-Defibrillator (ICD)

The implantable cardioverter-defibrillator (ICD) assumes an important role in primary and secondary prevention of SCD for patients with HCM. In a multi-center study of ICDs in patients with HCM, the device intervened appropriately, terminating ventricular tachycardia/fibrillation (VT/VF), at a rate of 5% per year for those patients implanted as primary prevention, and 11% per year for secondary prevention, over an average follow-up of three years (Maron et al., 2000). The current indications for implantation of an ICD are listed in Figure 50.4 and can be summarized as secondary prevention for all patients with a

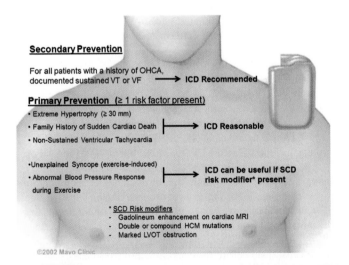

Figure 50.4 SCD risk stratification for ICD placement in patients with HCM. Shown are the risk factors for secondary and primary prevention of SCD in patients with HCM. An ICD should always be placed as secondary prevention or when one or more primary major risk factors (with or without additional SCD modifiers) are present.

history of out-of-hospital cardiac arrest (OHCA) or documented sustained VT or VF, or as primary prevention in patients with one or more established risk factors (Elliott et al., 2000, 2001; Gersh et al., 2011; Maron, 2002, 2003; Maron et al., 2003a, b; McKenna and Behr, 2002; Olivotto et al., 1999; Spirito et al., 2000, 2009). In contrast to other sudden death-predisposing diseases such as long-QT syndrome, where family history is not a personal risk factor (Kaufman et al., 2008), a family history of premature SCD is indeed one of the five *bona fide* personal risk factors in HCM, and may be equipotent to the other risk factors (Bos et al., 2010; Elliott et al., 2000; Maron et al., 2007).

CONCLUSIONS

HCM is a disease underscored by profound genotypic and phenotypic heterogeneity. Over the last two decades, numerous investigators and investigations have exposed some of the pathogenic mechanisms of the disease with the elucidation of many HCM susceptibility genes and an abundance of genotype–phenotype relationships. Hopefully, future research will establish whether these discoveries can be translated further into pathogenetic-based strategies for the treatment and prevention of this potentially life-threatening disease.

REFERENCES

Abelmann, W.H., 1984. Classification and natural history of primary myocardial disease. Prog Cardiovasc Dis 27 (2), 73–94.

Ackerman, M.J., Priori, S.G., Willems, S., et al., 2011. HRS/EHRA Expert Consensus Statement on the State of Genetic Testing for the Channelopathies and Cardiomyopathies. This document was developed as a partnership between the Heart Rhythm Society (HRS) and the European Heart Rhythm Association (EHRA). Heart Rhythm 8 (8), 1308–1339.

Ackerman, M.J., Van Driest, S.V., Ommen, S.R., et al., 2002. Prevalence and age-dependence of malignant mutations in the beta-myosin heavy chain and troponin T gene in hypertrophic cardiomyopathy: A comprehensive outpatient perspective. J Am Coll Cardiol 39 (12), 2042–2048.

Anan, R., Greve, G., Thierfelder, L., et al., 1994. Prognostic implications of novel beta cardiac myosin heavy chain gene mutations that cause familial hypertrophic cardiomyopathy. J Clin Invest 93 (1), 280–285.

Arad, M., Benson, D.W., Perez-Atayde, A.R., et al., 2002. Constitutively active AMP kinase mutations cause glycogen storage disease mimicking hypertrophic cardiomyopathy. J Clin Invest 109 (3), 357–362.

Arad, M., Maron, B.J., Gorham, J.M., et al., 2005. Glycogen storage diseases presenting as hypertrophic cardiomyopathy. N Engl J Med 352 (4), 362–372.

Arimura, T., Bos, J.M., Sato, A., et al., 2009. Cardiac ankyrin repeat protein gene (ANKRD1) mutations in hypertrophic cardiomyopathy. J Am Coll Cardiol 54 (4), 334–342.

Barringhaus, K.G., Zamore, P.D., 2009. MicroRNAs: Regulating a change of heart. Circulation 119 (16), 2217–2224.

Binder, J., Ommen, S.R., Gersh, B.J., et al., 2006. Echocardiography-guided genetic testing in hypertrophic cardiomyopathy: Septal morphological features predict the presence of myofilament mutations. Mayo Clin Proc 81 (4), 459–467.

Blair, E., Redwood, C., Ashrafian, H., et al., 2001. Mutations in the gamma(2) subunit of AMP-activated protein kinase cause familial hypertrophic cardiomyopathy: Evidence for the central role of energy compromise in disease pathogenesis. Hum Mol Genet 10 (11), 1215–1220.

Bos, J.M., Maron, B.J., Ackerman, M.J., et al., 2010. Role of family history of sudden death in risk stratification and prevention of sudden death with implantable defibrillators in hypertrophic cardiomyopathy. Am J Cardiol 106 (10), 1481–1486.

Bos, J.M., Poley, R.N., Ny, M., et al., 2006. Genotype–phenotype relationships involving hypertrophic cardiomyopathy-associated mutations in titin, muscle LIM protein, and telethonin. Mol Genet Metab 88 (1), 78–85.

Bos, J.M., Theis, J.L., Tajik, A.J., et al., 2008. Relationship between sex, shape, and substrate in hypertrophic cardiomyopathy. Am Heart J 155 (6), 1128–1134.

Bos, J.M., Towbin, J.A., Ackerman, M.J., 2009. Diagnostic, prognostic, and therapeutic implications of genetic testing for hypertrophic cardiomyopathy. J Am Coll Cardiol 54 (3), 201–211.

Braunwald, E., Lambrew, C.T., Rockoff, S.D., Ross Jr., J., Morrow, A.G., 1964. Idiopathic hypertrophic subaortic stenosis. I. A description of the disease based upon an analysis of 64 patients. Circulation 30 (Suppl. 4), 3–119.

Carniel, E., Taylor, M.R., Sinagra, G., et al., 2005. Alpha-myosin heavy chain: A sarcomeric gene associated with dilated and hypertrophic phenotypes of cardiomyopathy. Circulation 112 (1), 54–59.

Churchill, G.A., 2002. Fundamentals of experimental design for cDNA microarrays. Nat Genet 32 (Suppl.), 490–495.

Coviello, D.A., Maron, B.J., Spirito, P., et al., 1997. Clinical features of hypertrophic cardiomyopathy caused by mutation of a "hot spot" in the alpha-tropomyosin gene. J Am Coll Cardiol 29 (3), 635–640.

Daehmlow, S., Erdmann, J., Knueppel, T., et al., 2002. Novel mutations in sarcomeric protein genes in dilated cardiomyopathy. Biochem Biophys Res Commun 298 (1), 116–120.

Daw, E.W., Chen, S.N., Czernuszewicz, G., et al., 2007. Genome-wide mapping of modifier chromosomal loci for human hypertrophic cardiomyopathy. Hum Mol Genet 16 (20), 2463–2471.

Elliott, P., McKenna, W.J., 2004. Hypertrophic cardiomyopathy. Lancet 363 (9424), 1881–1891.

Elliott, P.M., Gimeno Blanes, J.R., Mahon, N.G., Poloniecki, J.D., McKenna, W.J., 2001. Relation between severity of left-ventricular hypertrophy and prognosis in patients with hypertrophic cardiomyopathy. Lancet 357 (9254), 420–424.

Elliott, P.M., Poloniecki, J., Dickie, S., et al., 2000. Sudden death in hypertrophic cardiomyopathy: Identification of high risk patients. J Am Coll Cardiol 36 (7), 2212–2218.

Estes 3rd, N.A., Link, M.S., Cannom, D., et al., 2001. Report of the NASPE policy conference on arrhythmias and the athlete. J Cardiovasc Electrophysiol 12 (10), 1208–1219.

Faber, L., Seggewiss, H., Gleichmann, U., 1998. Percutaneous transluminal septal myocardial ablation in hypertrophic obstructive cardiomyopathy: Results with respect to intraprocedural myocardial contrast echocardiography. Circulation 98 (22), 2415–2421.

Fraccarollo, D., Galuppo, P., Hildemann, S., et al., 2003. Additive improvement of left ventricular remodeling and neurohormonal activation by aldosterone receptor blockade with eplerenone and ACE inhibition in rats with myocardial infarction. J Am Coll Cardiol 42 (9), 1666–1673.

Fraccarollo, D., Galuppo, P., Schmidt, I., Ertl, G., Bauersachs, J., 2005. Additive amelioration of left ventricular remodeling and molecular alterations by combined aldosterone and angiotensin receptor blockade after myocardial infarction. Cardiovasc Res 67 (1), 97–105.

Frank, D., Kuhn, C., Katus, H.A., Frey, N., 2006. The sarcomeric Z-disc: A nodal point in signalling and disease. J Mol Med 84 (6), 1–23.

Frey, N., Katus, H.A., Olson, E.N., Hill, J.A., 2004. Hypertrophy of the heart: A new therapeutic target? Circulation 109 (13), 1580–1589.

Fujino, N., Ino, H., Hayashi, K., et al., 2006. A novel missense mutation in cardiac ryanodine receptor gene as a possible cause of hypertrophic cardiomyopathy: Evidence from familial analysis. Circulation 114 (18) II-165: 915.

Geier, C., Perrot, A., Ozcelik, C., et al., 2003. Mutations in the human muscle LIM protein gene in families with hypertrophic cardiomyopathy. Circulation 107 (10), 1390–1395.

Geisterfer-Lowrance, A.A., Kass, S., Tanigawa, G., et al., 1990. A molecular basis for familial hypertrophic cardiomyopathy: A beta cardiac myosin heavy chain gene missense mutation. Cell 62, 999–1006.

Gersh, B.J., Maron B.J., Bonow, R.O., et al., 2011. ACCF/AHA guideline for the diagnosis and treatment of hypertrophic cardiomyopathy: Executive summary: A report of the American College of cardiology foundation/American heart association task force on practice guidelines. 124 (24), e783–831.

Gerull, B., Gramlich, M., Atherton, J., et al., 2002. Mutations of TTN, encoding the giant muscle filament titin, cause familial dilated cardiomyopathy. Nat Genet 30 (2), 201–204.

Gietzen, F.H., Leuner, C.J., Raute-Kreinsen, U., et al., 1999. Acute and long-term results after transcoronary ablation of septal hypertrophy (TASH). Catheter interventional treatment for hypertrophic obstructive cardiomyopathy. Eur Heart J 20 (18), 1342–1354.

Girolami, F., Ho, C.Y., Semsarian, C., et al., 2010. Clinical features and outcome of hypertrophic cardiomyopathy associated with triple sarcomere protein gene mutations. J Am Coll Cardiol 55 (14), 1444–1453.

Haghighi, K., Kolokathis, F., Gramolini, A.O., et al., 2006. A mutation in the human phospholamban gene, deleting arginine 14, results in lethal, hereditary cardiomyopathy. Proc Natl Acad Sci USA 103 (5), 1388–1393.

Hayashi, T., Arimura, T., Itoh-Satoh, M., et al., 2004. Tcap gene mutations in hypertrophic cardiomyopathy and dilated cardiomyopathy. J Am Coll Cardiol 44 (11), 2192–2201.

Herman, D.S., Hovingh, G.K., Iartchouk, O., et al., 2009. Filter-based hybridization capture of subgenomes enables resequencing and copy-number detection. Nat Methods 6 (7), 507–510.

Ho, C.Y., Lever, H.M., DeSanctis, R., et al., 2000. Homozygous mutation in cardiac troponin T: Implications for hypertrophic cardiomyopathy. Circulation 102 (16), 1950–1955.

Holland, M.J., 2002. Transcript abundance in yeast varies over six orders of magnitude. J Biol Chem 277 (17), 14,363–14,366.

Hwang, J.J., Allen, P.D., Tseng, G.C., et al., 2002. Microarray gene expression profiles in dilated and hypertrophic cardiomyopathic end-stage heart failure. Physiol Genomics 10 (1), 31–44.

Ikeda, S., Kong, S.W., Lu, J., et al., 2007. Altered microRNA expression in human heart disease. Physiol Genomics 31 (3), 367–373.

Ingles, J., Doolan, A., Chiu, C., et al., 2005. Compound and double mutations in patients with hypertrophic cardiomyopathy: Implications for genetic testing and counselling. J Med Genet 42 (10), e59.

Jarcho, J.A., McKenna, W., Pare, J.A., et al., 1989. Mapping a gene for familial hypertrophic cardiomyopathy to chromosome 14q1. N Engl J Med 321 (20), 1372–1378.

Kalkman, E.A., van Haren, P., Saxena, P.R., Schoemaker, R.G., 1999. Early captopril prevents myocardial infarction-induced hypertrophy but not angiogenesis. Eur J Pharmacol 369 (3), 339–348.

Kambara, A., Holycross, B.J., Wung, P., et al., 2003. Combined effects of low-dose oral spironolactone and captopril therapy in a rat model of spontaneous hypertension and heart failure. J Cardiovasc Pharmacol 41 (6), 830–837.

Kamisago, M., Sharma, S.D., DePalma, S.R., et al., 2000. Mutations in sarcomere protein genes as a cause of dilated cardiomyopathy. N Engl J Med 343 (23), 1688–1696.

Kaufman, E.S., McNitt, S., Moss, A.J., et al., 2008. Risk of death in the long QT syndrome when a sibling has died. Heart Rhythm 5 (6), 831–836.

Kim, J.B., Porreca, G.J., Song, L., et al., 2007. Polony multiplex analysis of gene expression (PMAGE) in mouse hypertrophic cardiomyopathy. Science 316 (5830), 1481–1484.

Kimmelstiel, C.D., Maron, B.J., 2004. Role of percutaneous septal ablation in hypertrophic obstructive cardiomyopathy. Circulation 109 (4), 452–456.

Knight, C., Kurbaan, A.S., Seggewiss, H., et al., 1997. Nonsurgical septal reduction for hypertrophic obstructive cardiomyopathy: Outcome in the first series of patients. Circulation 95 (8), 2075–2081.

Knoll, R., Hoshijima, M., Hoffman, H.M., et al., 2002. The cardiac mechanical stretch sensor machinery involves a Z-disc complex that is defective in a subset of human dilated cardiomyopathy. Cell 111 (7), 943–955.

Lakkis, N.M., Nagueh, S.F., Kleiman, N.S., et al., 1998. Echocardiography-guided ethanol septal reduction for hypertrophic obstructive cardiomyopathy. Circulation 98 (17), 1750–1755.

Lampert, R., Cannom, D., 2008. Sports participation for athletes with implantable cardioverter-defibrillators should be an individualized risk-benefit decision. Heart Rhythm 5 (6), 861–863.

Landstrom, A.P., Parvatiyar, M.S., Pinto, J.R., et al., 2008. Molecular and functional characterization of novel hypertrophic cardiomyopathy susceptibility mutations in TNNC1-encoded troponin C. J Mol Cell Cardiol 45 (2), 281–288.

Lechin, M., Quinones, M.A., Omran, A., et al., 1995. Angiotensin-I converting enzyme genotypes and left ventricular hypertrophy in patients with hypertrophic cardiomyopathy. Circulation 92 (7), 1808–1812.

Lee, R.C., Feinbaum, R.L., Ambros, V., 1993. The C. elegans heterochronic gene lin-4 encodes small RNAs with antisense complementarity to lin-14. Cell 75 (5), 843–854.

Lever, H.M., Karam, R.F., Currie, P.J., Healy, B.P., 1989. Hypertrophic cardiomyopathy in the elderly. Distinctions from the young based on cardiac shape. Circulation 79 (3), 580–589.

Lim, D.S., Lutucuta, S., Bachireddy, P., et al., 2001. Angiotensin II blockade reverses myocardial fibrosis in a transgenic mouse model of human hypertrophic cardiomyopathy. Circulation 103 (6), 789–791.

Lind, J.M., Chiu, C., Ingles, J., et al., 2008. Sex hormone receptor gene variation associated with phenotype in male hypertrophic cardiomyopathy patients. J Mol Cell Cardiol 45 (2), 217–222.

Marian, A.J., Roberts, R., 2001. The molecular genetic basis for hypertrophic cardiomyopathy. J Mol Cell Cardiol 33 (4), 655–670.

Maron, B.J., 2002. Hypertrophic cardiomyopathy: A systematic review. JAMA 287 (10), 1308–1320.

Maron, B.J., 2003. Contemporary considerations for risk stratification, sudden death and prevention in hypertrophic cardiomyopathy. Heart 89 (9), 977–978.

Maron, B.J., Ackerman, M.J., Nishimura, R.A., et al., 2005. Task Force 4: HCM and other cardiomyopathies, mitral valve prolapse, myocarditis, and Marfan syndrome. J Am Coll Cardiol 45 (8), 1340–1345.

Maron, B.J., Dearani, J.A., Ommen, S.R., et al., 2004. The case for surgery in obstructive hypertrophic cardiomyopathy. J Am Coll Cardiol 44 (10), 2044–2053.

Maron, B.J., Estes, N.A.M., Maron, M.S., et al., 2003a. Primary prevention of sudden death as a novel treatment strategy in hypertrophic cardiomyopathy. Circulation 107, 2872–2875.

Maron, B.J., Gardin, J.M., Flack, J.M., et al., 1995. Prevalence of hypertrophic cardiomyopathy in a general population of young adults. Echocardiographic analysis of 4111 subjects in the CARDIA Study, Coronary Artery Risk Development in (Young) Adults. Circulation 92 (4), 785–789.

Maron, B.J., McKenna, W.J., Danielson, G.K., Kappenberger, L.J., Kuhn, H.J., et al., 2003b. American College of Cardiology/ European Society of Cardiology clinical expert consensus document on hypertrophic cardiomyopathy. A report of the American College of Cardiology Foundation Task Force on Clinical Expert Consensus Documents and the European Society of Cardiology Committee for Practice Guidelines. J Am Coll Cardiol 42 (9), 1687–1713.

Maron, B.J., Shen, W.K., Link, M.S., Epstein, A.E., Almquist, A.K., et al., 2000. Efficacy of implantable cardioverter-defibrillators for the prevention of sudden death in patients with hypertrophic cardiomyopathy [see comments]. N Engl J Med 342 (6), 365–373.

Maron, B.J., Spirito, P., Shen, W.K., Haas, T.S., Formisano, F., et al., 2007. Implantable cardioverter-defibrillators and prevention of sudden cardiac death in hypertrophic cardiomyopathy. JAMA 298 (4), 405–412.

McKenna, W.J., Behr, E.R., 2002. Hypertrophic cardiomyopathy: Management, risk stratification, and prevention of sudden death. Heart 87 (2), 169–176.

Minamisawa, S., Sato, Y., Tatsuguchi, Y., Fujino, T., Imamura, S., et al., 2003. Mutation of the phospholamban promoter associated with hypertrophic cardiomyopathy. Biochem Biophys Res Commun 304 (1), 1–4.

Mogensen, J., Klausen, I.C., Pedersen, A.K., Egeblad, H., Bross, P., et al., 1999. Alpha-cardiac actin is a novel disease gene in familial hypertrophic cardiomyopathy. J Clin Invest 103 (10), R39–R43.

Mohapatra, B., Jimenez, S., Lin, J.H., Bowles, K.R., Coveler, K.J., et al., 2003. Mutations in the muscle LIM protein and alpha-actinin-2 genes in dilated cardiomyopathy and endocardial fibroelastosis. Mol Genet Metab 80 (1–2), 207–215.

Monteiro de Resende, M., Kriegel, A.J., Greene, A.S., 2006. Combined effects of low-dose spironolactone and captopril therapy in a rat model of genetic hypertrophic cardiomyopathy. J Cardiovasc Pharmacol 48 (6), 265–273.

Moolman, J.C., Corfield, V.A., Posen, B., Ngumbela, K., Seidman, C., et al., 1997. Sudden death due to troponin T mutations. J Am Coll Cardiol 29 (3), 549–555.

Moon, J.C., McKenna, W.J., McCrohon, J.A., Elliott, P.M., Smith, G.C., et al., 2003. Toward clinical risk assessment in hypertrophic cardiomyopathy with gadolinium cardiovascular magnetic resonance. J Am Coll Cardiol 41 (9), 1561–1567.

Morita, H., Larson, M.G., Barr, S.C., Vasan, R.S., O'Donnell, C.J., et al., 2006. Single-gene mutations and increased left ventricular wall thickness in the community: The Framingham Heart Study. Circulation 113 (23), 2697–2705.

Nakao, S., Takenaka, T., Maeda, M., Kodama, C., Tanaka, A., et al., 1995. An atypical variant of Fabry's disease in men with left ventricular hypertrophy. N Engl J Med 333 (5), 288–293.

Niimura, H., Patton, K.K., McKenna, W.J., Soults, J., Maron, B.J., et al., 2002. Sarcomere protein gene mutations in hypertrophic cardiomyopathy of the elderly. Circulation 105 (4), 446–451.

Nishimura, R.A., Holmes Jr., D.R., 2004. Clinical practice. Hypertrophic obstructive cardiomyopathy. N Engl J Med 350 (13), 1320–1327.

Olivotto, I., Girolami, F., Ackerman, M.J., Nistri, S., Bos, J.M., et al., 2008. Myofilament protein gene mutation screening and outcome of patients with hypertrophic cardiomyopathy. Mayo Clin Proc 83 (6), 630–638.

Olivotto, I., Maron, B.J., Montereggi, A., Mazzuoli, F., Dolara, A., et al., 1999. Prognostic value of systemic blood pressure response during exercise in a community-based patient population with hypertrophic cardiomyopathy. J Am Coll Cardiol 33 (7), 2044–2051.

Olivotto, I., Maron, M.S., Adabag, A.S., Casey, S.A., Vargiu, D., et al., 2005. Gender-related differences in the clinical presentation and outcome of hypertrophic cardiomyopathy. J Am Coll Cardiol 46 (3), 480–487.

Olson, T.M., Doan, T.P., Kishimoto, N.Y., Whitby, F.G., Ackerman, M.J., et al., 2000. Inherited and de novo mutations in the cardiac actin gene cause hypertrophic cardiomyopathy. J Mol Cell Cardiol 32, 1687–1694.

Olson, T.M., Karst, M.L., Whitby, F.G., Driscoll, D.J., 2002. Myosin light chain mutation causes autosomal recessive cardiomyopathy with mid-cavitary hypertrophy and restrictive physiology. Circulation 105 (20), 2337–2340.

Olson, T.M., Kishimoto, N.Y., Whitby, F.G., Michels, V.V., 2001. Mutations that alter the surface charge of alpha-tropomyosin are associated with dilated cardiomyopathy. J Mol Cell Cardiol 33 (4), 723–732.

Ommen, S.R., Maron, B.J., Olivotto, I., et al., 2005. Long-term effects of surgical septal myectomy on survival in patients with obstructive hypertrophic cardiomyopathy. J Am Coll Cardiol 46 (3), 470–476.

Ortlepp, J.R., Vosberg, H.P., Reith, S., et al., 2002. Genetic polymorphisms in the renin-angiotensin-aldosterone system associated with expression of left ventricular hypertrophy in hypertrophic cardiomyopathy: A study of five polymorphic genes in a family with a disease causing mutation in the myosin binding protein C gene. Heart (Br Cardiac Soc) 87 (3), 270–275.

Osio, A., Tan, L., Chen, S.N., et al., 2007. Myozenin 2 is a novel gene for human hypertrophic cardiomyopathy. Circ Res 100 (6), 766–768.

Perkins, M.J., Van Driest, S.L., Ellsworth, E.G., et al., 2005. Gene-specific modifying effects of pro-LVH polymorphisms involving the renin-angiotensin-aldosterone system among 389 unrelated patients with hypertrophic cardiomyopathy. Eur Heart J 26 (22), 2457–2462.

Poliac, L.C., Barron, M.E., Maron, B.J., 2006. Hypertrophic cardiomyopathy. Anesthesiology 104 (1), 183–192.

Pollick, C., Morgan, C.D., Gilbert, B.W., Rakowski, H., Wigle, E.D., 1982. Muscular subaortic stenosis: The temporal relationship between systolic anterior motion of the anterior mitral leaflet and the pressure gradient. Circulation 66 (5), 1087–1094.

Pyle, W.G., Solaro, R.J., 2004. At the crossroads of myocardial signaling: The role of Z-discs in intracellular signaling and cardiac function. Circ Res 94 (3), 296–305.

Rajan, S., Williams, S.S., Jagatheesan, G., et al., 2006. Microarray analysis of gene expression during early stages of mild and severe cardiac hypertrophy. Physiol Genomics 27 (3), 309–317.

Richard, P., Charron, P., Carrier, L., et al., 2003. Hypertrophic cardiomyopathy: Distribution of disease genes, spectrum of mutations, and implications for a molecular diagnosis strategy. Circulation 107 (17), 2227–2232.

Richardson, P., McKenna, W., Bristow, M., et al., 1996. Report of the 1995 World Health Organization/International Society and Federation of Cardiology Task Force on the definition and classification of cardiomyopathies. Circulation 93 (5), 841–842.

Roberts, R., Sigwart, U., 2005. Current concepts of the pathogenesis and treatment of hypertrophic cardiomyopathy. Circulation 112 (2), 293–296.

Rubinshtein, R., Glockner, J.F., Ommen, S.R., et al., 2010. Characteristics and clinical significance of late gadolinium enhancement by contrast-enhanced magnetic resonance imaging in patients with hypertrophic cardiomyopathy. Circ Heart Fail 3 (1), 51–58.

Sachdev, B., Takenaka, T., Teraguchi, H., et al., 2002. Prevalence of Anderson-Fabry disease in male patients with late onset hypertrophic cardiomyopathy. Circulation 105 (12), 1407–1411.

Sakuraba, H., Oshima, A., Fukuhara, Y., et al., 1990. Identification of point mutations in the alpha-galactosidase A gene in classical and atypical hemizygotes with Fabry disease. Am J Hum Genet 47 (5), 784–789.

Satoh, M., Takahashi, M., Sakamoto, T., et al., 1999. Structural analysis of the titin gene in hypertrophic cardiomyopathy: Identification of a novel disease gene. Biochem Biophys Res Commun 262 (2), 411–417.

Sayed, D., Hong, C., Chen, I.Y., Lypowy, J., Abdellatif, M., 2007. MicroRNAs play an essential role in the development of cardiac hypertrophy. Circ Res 100 (3), 416–424.

Schoendube, F.A., Klues, H.G., Reith, S., et al., 1995. Long-term clinical and echocardiographic follow-up after surgical correction of hypertrophic obstructive cardiomyopathy with extended myectomy and reconstruction of the subvalvular mitral apparatus. Circulation 92 (Suppl. 9), 122–127.

Seidman, J.G., Seidman, C., 2001. The genetic basis for cardiomyopathy: From mutation identification to mechanistic paradigms. Cell 104 (4), 557–567.

Semsarian, C., Ahmad, I., Giewat, M., et al., 2002. The L-type calcium channel inhibitor diltiazem prevents cardiomyopathy in a mouse model. J Clin Invest 109 (8), 1013–1020.

Small, E.M., Frost, R.J., Olson, E.N., 2010. MicroRNAs add a new dimension to cardiovascular disease. Circulation 121 (8), 1022–1032.

Solomon, S.D., Wolff, S., Watkins, H., et al., 1993. Left ventricular hypertrophy and morphology in familial hypertrophic cardiomyopathy associated with mutations of the beta-myosin heavy chain gene. J Am Coll Cardiol 22 (2), 498–505.

Sorajja, P., Valeti, U., Nishimura, R.A., et al., 2008. Outcome of alcohol septal ablation for obstructive hypertrophic cardiomyopathy. Circulation 118 (2), 131–139.

Spirito, P., Autore, C., 2006. Management of hypertrophic cardiomyopathy. BMJ 332 (7552), 1251–1255.

Spirito, P., Autore, C., Rapezzi, C., et al., 2009. Syncope and risk of sudden death in hypertrophic cardiomyopathy. Circulation 119 (13), 1703–1710.

Spirito, P., Bellone, P., Harris, K.M., et al., 2000. Magnitude of left ventricular hypertrophy and risk of sudden death in hypertrophic cardiomyopathy. N Engl J Med 342 (24), 1778–1785.

Tatsuguchi, M., Seok, H.Y., Callis, T.E., et al., 2007. Expression of microRNAs is dynamically regulated during cardiomyocyte hypertrophy. J Mol Cell Cardiol 42 (6), 1137–1141.

Teare, D., 1958. Asymmetrical hypertrophy of the heart in young adults. Br Heart J 20 (1), 1–8.

Theis, J.L., Martijn Bos, J., Bartleson, V.B., et al., 2006. Echocardiographic-determined septal morphology in Z-disc hypertrophic cardiomyopathy. Biochem Biophys Res Commun 351 (4), 896–902.

Thum, T., Galuppo, P., Wolf, C., et al., 2007. MicroRNAs in the human heart: A clue to fetal gene reprogramming in heart failure. Circulation 116 (3), 258–267.

Tsybouleva, N., Zhang, L., Chen, S., et al., 2004. Aldosterone, through novel signaling proteins, is a fundamental molecular bridge between the genetic defect and the cardiac phenotype of hypertrophic cardiomyopathy. Circulation 109 (10), 1284–1291.

Van Driest, S.L., Ellsworth, E.G., Ommen, S.R., et al., 2003. Prevalence and spectrum of thin filament mutations in an outpatient referral population with hypertrophic cardiomyopathy. Circulation 108 (4), 445–451.

Van Driest, S.L., Jaeger, M.A., Ommen, S.R., et al., 2004a. Comprehensive analysis of the beta-myosin heavy chain gene in 389 unrelated patients with hypertrophic cardiomyopathy. J Am Coll Cardiol 44 (3), 602–610.

Van Driest, S.L., Maron, B.J., Ackerman, M.J., 2004b. From malignant mutations to malignant domains: The continuing search for prognostic significance in the mutant genes causing hypertrophic cardiomyopathy. [comment]. Heart (Br Cardiac Soc) 90 (1), 7–8.

Van Driest, S.L., Vasile, V.C., Ommen, S.R., et al., 2004c. Myosin binding protein C mutations and compound herterozygosity in hypertrophic cardiomyopathy. J Am Coll Cardiol 44 (9), 1903–1910.

Van Driest, S.L., Gakh, O., Ommen, S.R., Isaya, G., Ackerman, M.J., 2005a. Molecular and functional characterization of a human frataxin mutation found in hypertrophic cardiomyopathy. Mol Genet Metab 85 (4), 280–285.

Van Driest, S.L., Ommen, S.R., Tajik, A.J., Gersh, B.J., Ackerman, M.J., 2005b. Sarcomeric genotyping in hypertrophic cardiomyopathy. Mayo Clin Proc 80 (4), 463–469.

Van Driest, S.L., Ommen, S.R., Tajik, A.J., Gersh, B.J., Ackerman, M.J., 2005c. Yield of genetic testing in hypertrophic cardiomyopathy. Mayo Clin Proc 80 (6), 739–744.

Van Driest, S.V., Ackerman, M.J., Ommen, S.R., et al., 2002. Prevalence and severity of "benign" mutations in the beta myosin heavy chain, cardiac troponin-T, and alpha tropomyosin genes in hypertrophic cardiomyopathy. Circulation 106 (24), 3085–3090.

Van Rooij, E., Sutherland, L.B., Liu, N., et al., 2006. A signature pattern of stress-responsive microRNAs that can evoke cardiac hypertrophy and heart failure. Proc Natl Acad Sci USA 103 (48), 18,255–18,260.

Van Rooij, E., Sutherland, L.B., Qi, X., et al., 2007. Control of stress-dependent cardiac growth and gene expression by a microRNA. Science 316 (5824), 575–579.

Varnava, A., Baboonian, C., Davison, F., et al., 1999. A new mutation of the cardiac troponin T gene causing familial hypertrophic cardiomyopathy without left ventricular hypertrophy. Heart 82 (5), 621–624.

Vasile, V.C., Ommen, S.R., Edwards, W.D., Ackerman, M.J., 2006a. A missense mutation in a ubiquitously expressed protein, vinculin, confers susceptibility to hypertrophic cardiomyopathy. Biochem Biophys Res Commun 345 (3), 998–1003.

Vasile, V.C., Will, M.L., Ommen, S.R., et al., 2006b. Identification of a metavinculin missense mutation, R975W, associated with both hypertrophic and dilated cardiomyopathy. Mol Genet Metab 87 (2), 169–174.

Velculescu, V.E., Zhang, L., Zhou, W., et al., 1997. Characterization of the yeast transcriptome. Cell 88 (2), 243–251.

Watkins, H., Rosenzweig, A., Hwang, D.S., et al., 1992. Characteristics and prognostic implications of myosin missense mutations in familial hypertrophic cardiomyopathy. N Engl J Med 326 (17), 1108–1114.

Westermann, D., Knollmann, B.C., Steendijk, P., et al., 2006. Diltiazem treatment prevents diastolic heart failure in mice with familial hypertrophic cardiomyopathy. Eur J Heart Fail 8 (2), 115–121.

Woo, A., Rakowski, H., Liew, J.C., et al., 2003. Mutations of the beta myosin heavy chain gene in hypertrophic cardiomyopathy: Critical functional sites determine prognosis. Heart (Br Cardiac Soc) 89 (10), 1179–1185.

CHAPTER

51

Arrhythmias

Barry London

INTRODUCTION

Arrhythmias, or abnormal heart rhythms, remain a major cause of morbidity and mortality in the United States (Goldberger et al., 2011). Arrhythmias are categorized as slow (bradyarrhythmias) versus fast (tachyarrhythmias), and as originating in the atria or nodes (supraventricular) versus ventricles or lower conduction system (ventricular). Bradyarrhythmias, including sinus node disease and heart block, lead to more than a hundred thousand pacemaker implants in the United States and approximately half a million worldwide each year (Brunner et al., 2004). Atrial fibrillation, an irregularly irregular supraventricular tachyarrhythmia, affects over two million people in the United States, often requires systemic anticoagulation due to the increased risk of stroke, and usually requires pharmacological therapy or invasive procedures to either control the ventricular rate or maintain sinus rhythm (Kannel and Benjamin, 2008). Ventricular tachyarrhythmias, including ventricular tachycardia and ventricular fibrillation, lead to sudden cardiac death (SCD), defined as death from cardiac causes within an hour of symptom onset. Between 200,000 and 450,000 people annually die from SCD in the United States; this exceeds the mortality of all cancers combined (Goldberger et al., 2011; Smith and Cain, 2006). Pharmacological therapies do not effectively prevent SCD in those at high risk (Echt et al., 1991). Implantable cardioverter defibrillators (ICDs) are effective, but their use is associated with procedural complications, limited battery life, infections, premature device and lead failure, inappropriate shocks, limitations to quality of life, and cost (Tung et al., 2008). In addition, many ICDs must be placed to prevent one sudden death, and most sudden deaths occur in individuals not identified as being at high risk (Myerburg et al., 2009).

Mutations in the genes that control cardiac electrical activity cause inherited arrhythmia syndromes such as long-QT syndrome and Brugada syndrome, and are characterized by both atrial and ventricular arrhythmias (Priori, 2010; Priori and Napolitano, 2004). While rare, they have led to research that has greatly enhanced our understanding of arrhythmia mechanisms. Mutations in structural heart genes cause inherited cardiomyopathies such as arrhythmogenic right ventricular cardiomyopathy, hypertrophic cardiomyopathy, and familial dilated cardiomyopathy, and are also associated with arrhythmias and sudden death (Elliott et al., 2000). The vast majority of atrial and ventricular arrhythmias, however, occur in the setting of structural heart disease that results from coronary artery disease with myocardial ischemia or infarction, viral infections leading to cardiomyopathy, valve disease, hypertension, endocrine and metabolic disorders, and the use of substances toxic to the myocardium (Goldberger et al., 2011; Myerburg, 2002; Smith and Cain, 2006). While these do not have a simple genetic etiology, the prognosis and response of these disorders to therapy is likely influenced by genomic factors.

For patients with either inherited syndromes or structural heart disease, clinical predictors of the onset and/or severity of arrhythmias have been disappointing (Spooner, 2009). Similarly, the response to pharmacological and device therapy is variable and often unpredictable. In this chapter, we will

Genomic and Personalized Medicine, 2nd edition
by Ginsburg & Willard. DOI: http://dx.doi.org/10.1016/B978-0-12-382227-7.00051-3

discuss the development and use of personalized and genomic predictors to define the population at risk for arrhythmias and the efficacy of potential therapies. We will begin with inherited arrhythmia syndromes and proceed to the more complex arrhythmias associated with structural heart disease. Finally, we will summarize the current state of clinical genetic testing.

INHERITED ARRHYTHMIA SYNDROMES AND ARRHYTHMIA MECHANISMS

During the last 20 years, the genes and mutations responsible for the arrhythmia susceptibility in some individuals affected by inherited arrhythmia syndromes associated with sudden cardiac death have been determined by positional cloning and the candidate gene approach [~60% of long-QT syndrome (LQTS), ~30% of Brugada syndrome, >50% short-QT syndrome (SQTS), >50% catecholaminergic polymorphic ventricular tachycardia (CPVT); Table 51.1] (Katz et al., 2009; Priori, 2010; Schulze-Bahr, 2008; Zareba and Cygankiewicz, 2008). The majority of mutations that have been identified to date are in ion channels or ion-channel related genes. While it makes sense that ion channel genes should cause these syndromes, there is bias due to preferentially searching for mutations in ion channel genes and other genetic causes are probably underestimated.

Bradyarrhythmias result from functional and/or structural defects in the nodes and conduction system. Rare mutations in the pacemaker channel, HCN4, alter the I_f current in the sinoatrial (SA) node and result in an inherited form of bradycardia (Milanesi et al., 2006). Mutations in the cardiac Na^+ channel (SCN5A, $Na_v1.5$) cause inherited forms of sinus bradycardia and heart block via degeneration of the specialized conduction system, either as isolated syndromes or in conjunction with Brugada syndrome and long-QT syndrome type 3 (LQT3); the pathophysiological mechanisms that link Na^+ channel mutations to conduction disease have not been fully determined (Schott et al., 1999). In long-QT syndrome, marked prolongation of the action potential duration (APD) and effective refractory period (ERP) can lead to 2:1 second degree heart block when the sinus rate is rapid.

The mechanisms underlying tachyarrhythmias are usually grouped into the events that initiate the activity and those that maintain it. Extra beats initiating tachyarrhythmias may result from either abnormal automaticity or triggered activity, such as early or delayed afterdepolarizations (Aiba and Tomaselli, 2010 ; Shah et al., 2005). Early afterdepolarizations (EADs) are thought to result from reactivation of L-type Ca^{2+} channels, which is more likely in the setting of prolongation of APD, as is seen in long-QT syndrome. Delayed afterdepolarizations (DADs), on the other hand, result from abnormal Ca^{2+} release from the sarcoplasmic reticulum (SR) and subsequent depolarization mediated by the electrogenic Na-Ca exchanger (NCX). APD prolongation in long-QT syndrome could trigger DADs via Ca^{2+} overload, while abnormal Ca^{2+} release channels in the SR directly enhance DADs following exercise

and adrenergic stimulation in CPVT. Once initiated, ventricular tachycardia (VT) is maintained through re-entrant circuits in the ventricular myocardium, and can degenerate into ventricular fibrillation (VF); these re-entrant mechanisms require regions of slow conduction and conduction block. Since the inherited arrhythmia syndromes do not usually lead to anatomical abnormalities such as fibrosis or scars, functional abnormalities promote re-entry. In long-QT syndrome, marked increases in the dispersion of repolarization and refractoriness across the ventricular wall are responsible for the maintenance of re-entrant circuits (Antzelevitch, 2007; Baker et al., 2000). In Brugada syndrome, premature repolarization of the right ventricular epicardium leads to marked differences in refractoriness that can both initiate and maintain the tachyarrhythmias (Antzelevitch et al., 1999).

GENE- AND MUTATION-SPECIFIC FEATURES IN INHERITED ARRHYTHMIA SYNDROMES

Mutations in several different genes cause each of the inherited arrhythmia syndromes (Table 51.1), and the clinical phenotype within each syndrome differs to some extent depending on the mutated gene (Priori, 2010). For example, the electrocardiogram (ECG) differs between the common forms of long-QT syndrome: in LQT1 with KvLQT1 mutations, the T-wave is large and broad-based; in LQT2 with HERG mutations, the T-wave is low amplitude; in LQT3 with SCN5A mutations, the heart rate is slow and the T-wave is peaked (Moss et al., 1995). In Brugada syndrome, patients with SCN5A mutations have more conduction disease and a greater widening of the QRS interval on the surface ECG in response to Na^+ channel blockers, but less atrial fibrillation compared to patients without SCN5A mutations (Mahida et al., 2011; Smits et al., 2002).

Hundreds of different mutations in Na^+ and K^+ channels cause the common types of long-QT and Brugada syndromes. These include nonsense, frameshift, and splice site mutations that lead to truncated, nonfunctional proteins; missense mutations that alter channel biophysical properties and/or trafficking; and promoter and intronic mutations that alter channel expression or mRNA stability. Certain mutations, such as the A341V LQT1 mutation in *KCNQ1*, are known to produce a severe phenotype in unrelated families (Crotti et al., 2007). More generally, mutations in the highly conserved transmembrane domains of KvLQT1, those with a dominant negative effect on channel function *in vitro*, and those that cause a decreased rate of I_{Ks} current activation are associated with a higher risk of cardiac events (Jons et al., 2011; Moss et al., 2007).

The potential of gene-specific therapies has been explored in long-QT syndrome. Beta-adrenergic blocking agents (β-blockers) are the therapy of choice for LQT1 patients with mutations in the K^+ channel KvLQT1, the channel responsible for the I_{Ks} current that increases in response to beta-adrenergic stimulation (Moss et al., 2007). The efficacy of β-blockers for LQT2 patients may be lower, and there may be potential for

TABLE 51.1	Inherited arrhythmia genes and modifiers				
Locus	**Gene**	**Protein**	**Current**	**Locus**	**Modifiers**
Long-QT syndrome (Zareba and Cygankiewicz, 2008)					
LQT1	*KCNQ1*	KvLQT1	↓I_{Ks}	11p15.5	KCNE1 D85N (Lahtinen et al., 2011) NOS1AP (Crotti et al., 2009)
LQT2	*KCNH2*	HERG	↓I_{Kr}	7q35-36	KCNH2 K897T (Crotti et al., 2005)
LQT3	*SCN5A*	SCN5A, Na$_v$1.5	↑I_{Na}	3p21	SCN5A H558R (Marangoni et al., 2011; Ye et al., 2003) SCN5A S1103Y (Cheng et al., 2011)
LQT4	*ANK2*	Ankyrin-B	↑I_{Na} (?)	4q25-27	
LQT5	*KCNE1*	MinK	↓I_{Ks}	21q21-22	
LQT6	*KCNE2*	MiRP1	↓I_{Kr} (?)	21q21-22	
LQT7	*KCNJ2*	IRK1, K$_{ir}$2.1	↓I_{K1}	17q23.1	
LQT8	*CACNA1C*	Ca$_v$1.2	↑I_{Ca}	12p13.3	
LQT9	*CAV3*	Caveolin-3	↑I_{Na} (?)	3p25.3	
LQT10	*SCN4B*	SCN4B	↑I_{Na}	11q23.3	
LQT11	*AKAP9*	Yotiao	↓I_{Ks}	7q21-22	
LQT12	*SNTA1*	Alpha-1-syntrophin	↑I_{Na}	20q11.2	
LQT13	*KCNJ5*	GIRK4, K$_{ir}$3.4	↓I_{KACh}	11q24.5	
Short-QT syndrome (Zareba and Cygankiewicz, 2008)					
SQT1	*KCNH2*	HERG	↑I_{Kr}	7q35-36	
SQT2	*KCNQ1*	KvLQT1	↑I_{Ks}	11p15.5	
SQT3	*KCNJ2*	IRK1, K$_{ir}$2.1	↑I_{K1}	17q23.1	
Brugada syndrome (Ruan et al., 2009)					
BRUG1	*SCN5A*	SCN5A, Na$_v$1.5	↓I_{Na}	3p21	SCN5A H558R (Makielski et al., 2003; Marangoni et al., 2011; Poelzing et al., 2006; Ye et al., 2003)
BRUG2	*GPD1-L*	GPD1-L	↓I_{Na}	3p24	
BRUG3	*CACNA1C*	Ca$_v$1.2	↓I_{Ca}	12p13.3	
BRUG4	*CACNB2B*	Ca$_v$2b	↓I_{Ca}	10p12	
BRUG5	*KCNE3*	MiRP2	↑I_{to}	11q13.4	
BRUG6	*SCN1B*	SCN1B	↓I_{Na}	19q13	
Catecholaminergic polymorphic ventricular tachycardia (Katz et al., 2009)					
CPVT1	*RYR2*	Ryanodine Receptor	↑Ca$_i$	1q42.1	
CPVT2	*CASQ2*	Calsequestrin	↑Ca$_i$	1p13.3	
Bradycardia and heart block (Milanesi et al., 2006)					
BRADY1	*SCN5A*	SCN5A, Na$_v$1.5	↓I_{Na}	3p21	SCN5A H558R (Viswanathan et al., 2003)
BRADY2	*HCN4*	HCN4	↓I_f	15q24.1	

(Continued)

Locus	Gene	Protein	Current	Locus	Modifiers
Familial atrial fibrillation (Mahida et al., 2011)					
AFIB1	KCNQ1	KvLQT1	↑ I_{Ks}	11p15.5	
AFIB2	KCNE2	MiRP1	↑ I_{Ks}	21q21-22	
AFIB3	SCN5A	SCN5A, Na$_v$1.5	Δ I_{Na}	3p21	
AFIB4	KCNJ2	IRK1, K$_{ir}$2.1	↑ I_{K1}	17q23.1	
AFIB5	KCNA5	Kv1.5	↓ I_{Kur}	12p13.3	
AFIB6	GJA5	Connexin-40	↓ coupling	1q21.1	
AFIB7	KCNE5	KCNE1-like	↑ I_{Ks}	Xq22.3	
AFIB8	NUP155	Nucleoporin	?	5p13.1	
AFIB9	NPPA	Atrial Natiuretic Peptide	?	1p36.2	
AFIB10	SCN1B	SCN1B, Navβ1	Δ I_{Na}	19q13.1	
AFIB11	SCN2B	SCN2B, Navβ2	Δ I_{Na}	11q23	

Δ, changes properties of.

harm from β-blocker therapy in LQT3 patients with Na$^+$ channel mutations (Priori et al., 2004; Shimizu and Antzelevitch, 2000). The combination of oral K$^+$ supplementation and aldosterone antagonists has been shown to increase serum K$^+$ and decrease the corrected QT interval (QTc) in LQT2 patients with mutations in the HERG K$^+$ channel responsible for the I_{Kr} current that is suppressed in the setting of lowered extracellular K$^+$ (Etheridge et al., 2003). The Na$^+$ channel blockers mexilitine and flecainide block late inward currents from mutant Na$^+$ channels and decrease QTc in some LQT3 patients (Benhorin et al., 2000; Schwartz et al., 1995). In addition, the biophysical properties of the Na$^+$ channels may predict the patients who would be likely to respond to this therapy (Ruan et al., 2007). Unfortunately, although gene-specific therapies other than β-blockers have proven highly efficacious in individual patients, they may be harmful in some subsets. To date, no studies have proven that these pharmacological therapies provide a survival advantage, and ICDs remain the recommended therapy for high risk patients that fail β-blockade.

Quinidine and hydroquinidine, blockers of the I_{to} K$^+$ current, have been proposed as potential therapies for Brugada syndrome (Hermida et al., 2004). Two international trials are currently underway to assess the efficacy of this therapy in Brugada syndrome patients independent of genotype [US National Institutes of Health (NIH) clinical trials registry (http://www.clinicaltrials.gov) identifiers NCT00789165, NCT00927732]. Similarly, flecainide is being tested as a potential therapy for CPVT (NCT01117454) (Watanabe et al., 2009). Given the relatively small number of patients with each specific genotype and the large number of individual mutations, it is unclear whether gene-specific information on the therapies will result, and unlikely that mutation-specific information will become available.

GENETIC AND GENOMIC MODIFIERS IN INHERITED ARRHYTHMIA SYNDROMES

Most autosomal dominant inherited diseases are characterized by marked variation in disease severity and incomplete penetrance. This can, in part, be explained by other gene mutations and polymorphisms that modify the phenotype. Co-expression of the KCNH2 K897T polymorphism in the presence of the LQT2 A1116V mutation decreased I_{Kr} current *in vitro*, and the presence of the polymorphism was associated with a more malignant LQTS phenotype in the proband of a small family (Crotti et al., 2005). The KCNE1 D85N single nucleotide polymorphism (SNP) has been associated with QT prolongation in male but not female members of a large family with the KCNQ1 G589D LQT1 mutation (Lahtinen et al., 2011). The common SCN5A H558R polymorphism (allele frequency ~20%) decreases I_{Na} amplitude in the 2016 amino acid splice variant (Q1077) more than in the 2015 amino acid splice variant (Q1077del) *in vitro*. In addition, it rescues the kinetic changes in the T512I conduction system mutation, rescues trafficking deficiencies and/or modifies expressivity of abnormal currents in the M1766L and S216L Brugada/LQT3 syndrome mutations when expressed on the same channel/allele, and rescues trafficking of the R282H Brugada syndrome mutation whether on the same or on the adjacent channel/allele (Makielski et al., 2003; Marangoni et al., 2011; Poelzing et al., 2006; Viswanathan et al., 2003; Ye et al., 2003). The SCN5A S1103Y polymorphism present in ~13% of African Americans accentuates the late current caused by the SCN5A R680H mutation (Cheng et al., 2011).

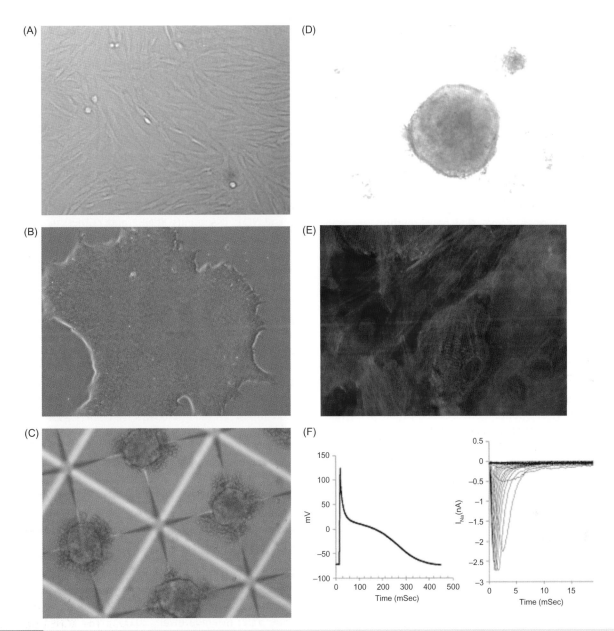

Figure 51.1 Using iPS cells to study inherited arrhythmias. (A) Fibroblasts isolated from a punch skin biopsy and grown to confluence. (B) iPS cells growing on matrigel following reprogramming of fibroblasts with the pluripotency genes *OCT4*, *KLF4*, *SOX2*, and *C-MYC*. (C) Day-3 embryoid bodies in an aggrewell plate being treated with the growth factors BMP4, bFGF, ActivinA, and DKK1 to induce cardiac-specific differentiation. (D) Day-25 embryoid body that exhibited spontaneous contraction. (E) Differentiated iPS-derived cardiac myocytes stained for F-actin (Phalloidin-Atto 633, Red), myosin (MF-20, green), and nuclei (DAPI, blue). Note the sarcomere striation patterns (green). (F) Action potential (left) and inward Na$^+$ current (right) from a single dissociated iPS-derived cardiac myocyte recorded using the whole-cell voltage-clamp technique.

Genome-wide association studies (GWAS) have identified polymorphisms in NOS1AP that are associated with QT interval prolongation and risk of sudden cardiac death in the general population (Arking et al., 2006; Kao et al., 2009). NOS1AP SNPs were associated with QT prolongation, symptoms, SCD, and aborted SCD in a large South African LQT1 family with the A341V mutation in *KCNQ1* (Crotti et al., 2009).

The identification and testing of modifier genes in inherited arrhythmia syndromes remains difficult, given the incomplete identification of potential modifiers and the fact that most mutations occur in small families. As a result, the potential clinical utility of the findings remains unclear. The recent engineering of induced pluripotent stem (iPS) cells from patient fibroblasts may circumvent many of these difficulties (Figure 51.1). iPS-derived cardiac myocytes from subjects with long-QT syndrome have prolonged action potentials (Moretti et al., 2010; Yazawa et al., 2011). It remains to be seen whether the action potential and ionic currents measured by whole-cell voltage clamp in these myocytes will correlate with clinical QT intervals and predict clinical disease severity both for specific mutations and within individual families. It is also unclear whether this type of analysis will prove useful for the study of the more common types of atrial and ventricular arrhythmias.

BRADYARRHYTHMIAS AND MECHANISMS OF HEART RATE CONTROL

Bradyarrhythmias, or abnormal, slow heart rhythms, are manifestations of diseases of the specialized conduction system (Lin and Callans, 2004; Schwartzman, 2004). Abnormalities in the sinus node can cause sick sinus syndrome with sinus bradycardia, sinus pauses, and/or sinus arrest. Abnormalities of the AV node and conduction system, on the other hand, cause second degree heart block (where some atrial beats conduct to the ventricle and some do not) and third degree or complete heart block (where conduction from atrium to ventricle fails, leading to A-V dissociation). The idiopathic forms of isolated heart block associated with fibrosis and/or calcification were called Lev-Lenegre syndrome.

Automaticity, or spontaneous depolarization in conduction system tissues leading to pacemaking, depends on the pacemaking current I_f and on an intrinsic Ca^{2+} clock (Lakatta and DiFrancesco, 2009). As noted above, mutations in cardiac ion channels and connexins can cause inherited forms of bradyarrhythmias. Bradycardia and heart block most often result from age-dependent degeneration of or damage to the conduction system, however.

GENETIC AND GENOMIC MODIFIERS OF HEART RATE

Resting heart rate is heritable, and a number of SNPs associated with heart rate have been identified by GWAS (Eijgelsheim et al., 2010). These include polymorphisms at 6q22 near GJA1 (connexin 43), 14q12 near MYH6 (α cardiac myosin), 14q12 near MYH7 (β cardiac myosin), 6q22 near PLN (Ca^{2+}-handling protein phospholamban), 1q32, 7q22, 11q12, and 12p12. The mechanism by which these variants alter heart rate remains unknown, and together they explain less than 2% of the heritable component of heart rate.

ATRIAL FIBRILLATION/FLUTTER AND ARRHYTHMIA MECHANISMS

Atrial fibrillation, an irregularly irregular supraventricular arrhythmia, is the most common arrhythmia, affecting over two million people in the United States and increasing in prevalence with age (to 9% by age 80–89) (Kannel and Benjamin,

2008). Atrial fibrillation increases the risk of heart failure, death, and stroke, and often requires systemic anticoagulation. In addition, it usually requires pharmacological therapy or invasive procedures to either control the ventricular rate or maintain sinus rhythm.

Atrial fibrillation is a chaotic rhythm that is maintained either by multiple re-entrant circuits (rotors) or by a single driving, or mother, rotor (often near the junction of the pulmonary veins and the left atrium) that degenerates with distance into multiple wavelets (Wakili et al., 2011). Atrial flutter is a more organized rhythm caused by a single re-entrant circuit. Mutations in the K^+ channels that encode I_{Ks} (α subunit KCNQ1, β subunits KCNE2 and KCNE5), I_{K1} (KCNJ2), and I_{Kur} (KCNA5); in the cardiac Na^+ channel (α subunit SCN5A, β subunits SCN1B and SCN2B); in the atrial connexin-40 gene (GJA5); in the atrial natriuretic peptide (NPPA); and in nucleoporin (NUP155) cause rare inherited forms of atrial fibrillation (Table 51.1) (Mahida et al., 2011). The primary mechanisms appear to be shortening of the APD and refractory period for the K^+ channel mutations, and altering intercellular conduction for connexins and Na^+ channel mutations. Most atrial fibrillation is acquired, however. Hypertension and valve disease increase the risk of the common forms of atrial fibrillation by increasing atrial size and fibrosis, thereby enhancing re-entry. In addition, electrophysiological changes that shorten APD and the atrial refractory period promote re-entry. In fact, atrial fibrillation begets atrial fibrillation because electrical remodeling following its onset leads to further shortening of the APD and refractory period, while structural remodeling leads to enhanced fibrosis (Wakili et al., 2011).

GENETIC AND GENOMIC MODIFIERS IN ATRIAL FIBRILLATION

Association studies have identified a number of genetic variants that potentially alter the likelihood of atrial fibrillation (Table 51.2) (Mahida et al., 2011). These include SNPs in ion channels similar to those identified in familial atrial fibrillation (K^+ channel α subunit KCNH2, K^+ channel β subunit KCNE1 and KCNE5, Na^+ channel α subunit SNC5A, connexin-40 GJA5, G-protein potentially modifying I_{K1} expression GNB3), along with polymorphisms in the renin-angiotensin system (ACE, AGT), atrial myocyte Ca^{2+} handling (SLN), inflammation (IL6, IL10) and extracellular matrix remodeling (MMP2). Most of the polymorphisms increase or decrease the risk of atrial fibrillation less than two-fold. Of note, many of these studies used relatively small populations and have not been adequately replicated.

Genome-wide association studies have identified loci at 4q25 (near the cardiac transcription factor PITX2), 1q21 (near the Ca^{2+}-activated K^+ channel KCNN3), and 16q22 (near the enhancer/transcription factor ZFHX3) that are strongly associated with atrial fibrillation, and modify the odds ratio for atrial fibrillation up to ~1.7-fold (Table 51.2) (Sinner et al., 2011). The mechanisms

TABLE 51.2	Polymorphisms that may affect arrhythmias				
Gene	**Protein**	**Locus**	**Variant (freq)**	**Disease**	**Mechanisms**
Ion channels, calcium handling, and related genes					
KCNQ1	KvLQT1	11p15.5	Tag SNPs	SCD (Arking et al., 2011)	Δ I_{Ks}
KCNH2	HERG	7q35-36	K897T (0.20 C)	AFib (Sinner et al., 2008)	Δ I_{Kr}
KCNE1	MinK	21q21-22	G38S (0.35 C)	AFib (Fatini et al., 2006; Lai et al., 2002)	Δ I_{Ks}
KCNE5	KCNE1L	Xq22.3	C97T (0.16 C)	AFib (Chen et al., 2007)	? Δ I_{Ks}
SCN5A	Na$_v$1.5	3p21	H558R (0.20 C) (0.27 AA) S1103Y (0.07 AA)	AFib (Chen et al., 2007) VT/VF, SCD, SIDS (Burke et al., 2005; Plant et al., 2006; Splawski et al., 2002; Sun et al., 2011)	Δ expression, trafficking Δ I_{Na} kinetics
CACNA1D	Ca$_v$1.3	3p14.3	Tag SNPs	SCD (Arking et al., 2011)	? Δ I_{Ca}
KCNN3	K$_{Ca}$2.3	1q21.3	Tag SNPs	AFib (Sinner et al., 2011)	? Δ atrial I_K
GJA5	Connexin 40	1q21.1	G-44A/A+71G (0.21 C)	AFib (Juang et al., 2007)	Δ cell coupling
CASQ2	Calsequestrin	1p13.3	Tag SNPs	SCD (Westaway et al., 2011)	? Δ Ca handling
PLN	Phospholamban	6q22.1	Tag SNPs	SCD (Arking et al., 2011)	? Δ Ca handling
SLN	Sarcolipin	11q22-23	G-65C (0.48 C)	AFib (Nyberg et al., 2007)	? Δ Ca handling
GPD1-L	GPD1-L	3p24	Tag SNPs	SCD (Westaway et al., 2011)	? Δ I_{Na}
NOS1AP	NOS1AP	1q23	Tag SNPs	SCD (Arking et al., 2011; Eijgelsheim et al., 2009; Kao et al., 2009; Westaway et al., 2011)	↑ APD
ADRB2	β2-AR	5q31-32	Q27E (0.45 C)	SCD (Gavin et al., 2011; Sotoodehnia et al., 2006)	? Δ downregulation
GNB3	G-Protein β$_3$	12p13	C825T (0.32 C) (0.82 AA)	AFib (Schreieck et al., 2004)	? Δ I_{K1}
Other genes					
GPC5	Glypican 5	13q32	Tag SNPs	SCD (Arking et al., 2010)	Unknown
PITX2	? Pitx2c	4q25	Tag SNPs	AFib (Sinner et al., 2011)	Unknown
ZFHX3	ZFHX3	16q22.3	Tag SNPs	AFib (Sinner et al., 2011)	Unknown
eNOS	NOS3	7q36	Multiple	AFib (Mahida et al., 2011)	Unknown
AGT	Angiotensinogen	1q42.2	Multiple	AFib (Mahida et al., 2011)	? Hyper, fibrosis
ACE	Angiotensin Converting Enzyme	17q23.3	Del/ins (0.48 C)	AFib (Bedi et al., 2006; Fatini et al., 2007)	? Hyper, fibrosis
MMP2	Matrix Metallo-peptidase 2	16q12.2	C1306T (0.08 A)	AFib (Kato et al., 2007)	? Fibrosis
IL10	Interleukin 10	1q32.1	A-592C (0.36 A)	AFib (Kato et al., 2007)	? Inflammation
IL6	Interleukin 6	7p15.3	G-174C (0.27 C)	AFib (Gaudino et al., 2003)	? Inflammation

Δ, changes properties of; SCD, sudden cardiac death; SIDS, sudden infant death syndrome; VT/VF, ventricular tachycardia and/or ventricular fibrillation; AFib, atrial fibrillation; A, Asian; C, Caucasian; AA, African American; APD, action potential duration; del, deletion; ins, insertion; hyper, hypertrophy.

by which variants in these genes may promote atrial fibrillation remain unclear. Other GWAS studies have focused on intermediate phenotypes potentially related to atrial fibrillation, such as prolongation of the PR interval on the surface ECG. The relevance of these loci for atrial fibrillation needs to be confirmed.

Given that atrial fibrillation is relatively common in the population, it is possible that a panel of SNPs could be used to predict the risk of atrial fibrillation in a clinically significant manner. For example, six independent SNPs at 4q25 identified a group of subjects (1% of total population) with a six-fold higher risk of atrial fibrillation (Lubitz et al., 2010). In addition, SNPs at 4q25 may predict the success of catheter ablation for the treatment of atrial fibrillation (Husser et al., 2010). While promising, a more complete catalogue of susceptibility loci, the mechanisms by which they act, and their interactions with each other and environmental causes remains years away.

VENTRICULAR TACHYARRHYTHMIAS AND MECHANISMS OF SUDDEN CARDIAC DEATH

Ventricular tachyarrhythmias are responsible for the majority of cases of sudden cardiac death (Schaffer and Cobb, 1975). While some cases of VT and VF result from inherited ion channel disorders as discussed above, the vast majority occur in the setting of structural heart disease (Goldberger et al., 2011). An arrhythmogenic myocardial substrate can result from ischemia or infarction due to coronary artery disease or coronary emboli, myocarditis, inherited cardiomyopathies (dilated or hypertrophic), infiltrative cardiomyopathies, valve disease, hypertension, endocrine and metabolic disorders, and the use of substances toxic to the myocardium.

Acute ischemia leads to alterations in APD, refractory period, resting membrane potential, conduction velocity, and Ca^{2+} handling. These changes can both promote afterdepolarizations and enhance re-entry by slowing conduction and increasing dispersion of repolarization. In addition, scar tissue from any cause creates anatomical barriers and islands of poorly coupled myocytes that promote re-entry.

Cardiomyopathy and heart failure lead to prolonged APD due to downregulation of the repolarizing K^+ currents I_{to} and I_{K1} or increased late Na^+ currents (Beuckelmann et al., 1993; Nabauer et al., 1993; Valdivia et al., 2005). The delayed repolarization can promote both EADs and DADs and contribute to arrhythmias and sudden death (El-Sherif et al., 1997; Frazier et al., 1989; Vermeulen et al., 1994). In addition, heterogeneities in ion channel distribution in the myopathic heart, increased dispersion of repolarization and refractoriness, anatomical abnormalities such as fibrosis and scars, and abnormalities in intercellular communication with slow intraventricular conduction can enhance both the formation of rotors and scroll waves responsible for re-entrant VT and their degeneration leading to the transition from VT to VF (Brahmajothi et al., 1997; Cabo and Boyden, 2006; London et al., 2003).

Heart failure also leads to changes in Ca^{2+} handling that can contribute to arrhythmias (Bers et al., 2006; Beuckelmann et al., 1992; London et al., 2003; O'Rourke et al., 1999). The expression of sarcoplasmic reticulum calcium ATPase 2a (SERCA2a; ATP = adenosine triphosphate) protein, which pumps Ca^{2+} into the SR, is decreased in heart failure. In addition, the ratio of SERCA2a to phospholamban (an inhibitor of SERCA2a) is decreased, further depressing SERCA2a activity. At the same time, the expression of the electrogenic Na-Ca exchanger (NCX) is increased. Together, these changes lead to smaller, but prolonged, Ca^{2+} transients, decreased myocyte contractile function, and an increased propensity to afterdepolarizations and re-entrant arrhythmias in both ischemic and nonischemic cardiomyopathies (Pu et al., 2000). Beta adrenergic receptor downregulation and changes in protein phosphorylation further decrease contractile reserve in heart failure and may contribute to arrhythmias (Anderson, 2007; Bristow et al., 1990; Houser, 2000).

GENETIC AND GENOMIC MODIFIERS OF SUDDEN CARDIAC DEATH

Sudden cardiac death has a significant heritable component (Spooner, 2009). In the Family History and Primary Cardiac Arrest study from Seattle, a family history of SCD in a parent or a first-degree relative [independent of a family history of myocardial infarction (MI)] was associated with a greater than twofold increased risk of cardiac arrest for subjects less than 65 years old and with no known history of coronary artery disease, congenital heart disease, or other severe illnesses (Friedlander et al., 1998). In the Paris Prospective Study, subjects with known ischemic heart disease who had a parent with a history of SCD had a 1.8-fold higher risk of SCD over five years of follow-up (Jouven et al., 1999). Similarly, two case-control studies have shown: (1) a 1.6-fold higher incidence of SCD in subjects with a family history of SCD compared to those with a family history of MI but no SCD, and a 2.2-fold higher risk of SCD in subjects with a family history of SCD compared to controls; and (2) a 2.7-fold higher incidence of familial sudden death in ST segment elevation MI patients with versus without primary VF (Dekker et al., 2006; Kaikkonen et al., 2006).

Rare mutations in ion channel genes similar to those that cause the long-QT and Brugada syndromes do not cause a significant fraction of the common forms of sudden cardiac death. Sequencing 67 SCD victims in the Oregon Sudden Unexpected Death Study (Oregon-SUDS) did not identify any causative mutations in SCN5A (Stecker et al., 2006), while sequencing 59 Australian SCD victims identified no mutations in *SCN5A* or *KvLQT1* (Doolan et al., 2008). Sequencing of *SCN5A*, *KCNQ1*, *KCNH2*, *KCNE1*, and *KCNE2* in 113 SCD victims from the Nurses' Health Study and the Health Professional Follow-up Study did identify five rare *SCN5A* variants of uncertain significance in six women, a frequency greater than in controls (Albert et al., 2008).

A number of relatively common SNPs that lead to amino acid changes are present in cardiac ion channel genes, along with many more variants in introns and non-coding regions that could affect RNA and protein expression. Given that rare mutations in these ion channel genes can cause inherited forms of sudden death in patients with structurally normal hearts (e.g., long-QT syndrome, Brugada syndrome), it seems plausible these common SNPs could predispose individuals with structural heart disease to arrhythmias and sudden death by causing relatively small changes in ion channel function or number. At the present time, however, there is limited data that common ion channel SNPs modulate the frequency of the common forms of life-threatening arrhythmias (Table 51.2). The strongest data relate to the S1103Y polymorphism in SCN5A, a variant that is present to a significant extent only in persons of African descent (allele frequency ~7%) and causes subtle increased late current in heterologous expression systems *in vitro*. The 1103Y variant was more common in African American adults with unexplained arrhythmias and sudden death, and in infants that died of sudden infant death syndrome (SIDS) (Burke et al., 2005; Plant et al., 2006; Splawski et al., 2002). In addition, in a cohort of African Americans with a reduced ejection fraction (EF) who received an ICD for primary prevention, those carrying the polymorphism were more likely to receive an appropriate ICD therapy (shock or antitachycardia pacing) than those without it (Sun et al., 2011). Thus, if validated, this variant may predict arrhythmic risk in heart failure patients of African descent.

Two common SNPs in the β_1-adrenergic receptor (S49G, R389G), three common SNPs in the β_2-adrenergic receptor (G16R, Q27E, T164I), and one deletion in the α_{2C}-adrenergic receptor that is common in African Americans and synergistic with β_1-AR R389G (Del322-325) have been heavily investigated due to the central role of the β-adrenergic system in heart failure and arrhythmias (Dorn and Liggett, 2009; McNamara et al., 2002). *In vitro*, these variants alter responsiveness and/or downregulation of the receptors in response to β-agonist treatment. Clinically, some of the variants have been associated with outcomes and/or altered pharmacogenetic responses to β-adrenergic blocking drugs in heart failure patients. In particular, studies have reported that homozygotes for the glutamine allele of the $\beta2$-AR Q27E polymorphism have a 1.1–1.6-fold increased risk of SCD, that was highly significant in a meta-analysis (Table 51.2) (Gavin et al., 2011; Sotoodehnia et al., 2006). It is unclear, however, whether the mechanism is related primarily to predisposing to arrhythmias versus promoting the development of coronary artery disease.

QTc prolongation on the ECG is associated with an increased risk of SCD and is a potential intermediate phenotype for sudden death (Algra et al., 1991). Genome-wide association studies based on QTc have identified a number of SNPs in the nitric oxide synthase 1 adaptor protein (NOS1AP) responsible for 3–5 ms of QTc prolongation per allele (Aarnoudse et al., 2007; Arking et al., 2006). Some SNPs in the same region of NOS1AP have been associated with

a 1.2–1.3-fold increased risk of SCD (Eijgelsheim et al., 2009; Kao et al., 2009). Interestingly, different SNPs appear to be most closely associated with QTc prolongation versus SCD. The mechanisms by which *NOS1AP* variants lead to QTc prolongation and sudden death are not currently known. In addition, the importance of a number of other loci associated with QTc prolongation in predisposing to arrhythmias and SCD is unclear (Pfeufer et al., 2009).

A recent study analyzed haplotype tagging SNPs for 18 candidate SCD genes in the Oregon-SUDS study and identified associations near the Ca^{2+}-binding protein gene calsequestrin (*CASQ2*), the Brugada gene glycerol-3-phosphate dehydrogenase 1 like (*GPD1-L*), and *NOS1AP* (Westaway et al., 2011). The findings for CASQ2 and GPD1-L need to be confirmed.

A GWAS of SCD victims from the Oregon-SUDS study using 424 cases and 226 controls failed to identify any loci with genome-wide significance, but did identify a SNP in the glypican 5 gene (*GPC5*), which encodes a heparin sulfate proteoglycan, that was associated with a lower risk of SCD and was reproduced using 521 SCD cases from the Atherosclerosis Risk in Communities Study (ARIC) and the Cardiovascular Health Study (CHS) (Arking et al., 2010). This finding awaits confirmation and the putative mechanism is unclear. More recently, a GWAS meta-analysis from five studies using 1283 SCD cases of European ancestry, with follow-up genotyping of 3119 SCD cases from eleven studies, identified a SNP at 2q24.2 in *BAZ2B* associated with SCD (Arking et al., 2011). The mechanism potentially linking this gene to SCD is unclear. Of note, the initial GWAS meta-analysis also identified two SNPs (chromosomes 16 and 17) that reached genome-wide significance but did not reproduce in the follow-up cohorts. In addition, the authors used the meta-analysis to study a number of SNPs related to ECG characteristics, and identified variants in/near the Ca^{2+} channel gene *CACNA1D*, the K^+ channel gene *KCNQ1*, the SR Ca^{2+}-handling gene *PLN*, and *NOS1AP* that were associated with SCD (although only the variant near *CACNA1D* was significant after correction for multiple comparisons; Table 51.2). Surprisingly, for the SNPs associated with *CACNA1D* and *KCNQ1*, QRS and QT shortening, respectively, were associated with SCD.

Many of the genes that cause ventricular tachyarrhythmias and SCD in the inherited arrhythmia syndromes are currently unknown. It is likely that variants in these genes could be associated with common forms of SCD, and it is not currently possible to search for SNPs in those genes in a directed manner. Potential candidate genes include those associated with inflammation, redox state, and omega-3 fatty acid metabolism (Hernesniemi et al., 2008; London et al., 2007). Additional linkage and candidate gene studies in families with inherited tachyarrhythmias may identify additional genes that can be studied by deep sequencing or by association using non-synonymous or tagging SNPs. GWAS using SCD as a potential endpoint clearly have utility, although GWAS may fail to identify low frequency SNPs with a moderate to large effect on SCD (Spooner, 2009).

GENETIC AND GENOMIC MODIFIERS OF ARRHYTHMIA THERAPIES

Acquired long-QT syndrome most commonly results from the use of pharmacological therapy (antiarrhythmics, antihistamines, antipsychotics) in susceptible individuals (Chiang, 2004; Roden and Viswanathan, 2005). Antiarrhythmic medications other than β-blockers for atrial and ventricular arrhythmias can increase the risk of sudden death (Echt et al., 1991; Singh et al., 1995). These proarrhythmic effects are often thought to be either due to increased susceptibility of ion channels to the medication, decreased clearance, or decreased repolarization reserve. While all of these are potentially genetically mediated, methods to identify those at risk are not currently available.

Medical therapy for heart disease including β-blockers, angiotensin-converting enzyme inhibitors (ACEIs), angiotensin receptor blockers (ARBs), and aldosterone inhibitors may decrease the rate of SCD (Anand et al., 2006; Jessup et al., 2009; Nessler et al., 2007). The efficacy of bucindolol, a nonselective β-blocking agent with intrinsic sympathomimetic activity used in heart failure, depends on the genotypes of the β_1-AR R389G and α_{2c}-AR Del 322-325 deletion polymorphisms (Bristow et al., 2010; Liggett et al., 2006). The potential relevance to arrhythmias is currently uncertain.

At least in part as a response to the limited efficacy of drugs, device therapy for heart failure has grown in popularity during the last few decades. ICDs were first used in the 1980s, with the primary indication being the treatment of malignant, refractory ventricular arrhythmias (Mirowski et al., 1980), and randomized controlled trials have subsequently shown that ICDs improve mortality both in survivors of aborted SCD (secondary prevention) and in subjects with ischemic or nonischemic cardiomyopathies (primary prevention) (AVID, 1997; Bardy et al., 2005; Moss et al., 1996, 2002). American College of Cardiology/American Heart Association/Heart Rhythm Society (HRS) guidelines recommend ICD placement for subjects with heart failure and an EF <35% due to a nonischemic cardiomyopathy or a prior MI at least 40 days previously, and for asymptomatic subjects with an EF <30% from an MI at least 40 days previously (Epstein et al., 2008). The majority of patients that receive an ICD for primary prevention for cardiomyopathy will never experience an appropriate shock, and complications associated with ICD implantation include infection and bleeding at the time of placement; device and lead failure at later times; inappropriate shocks for atrial tachycardias; and limitations to employment, insurability, quality of life, and leisure activities (Huikuri et al., 2001; Tung et al., 2008). Clinical predictors beyond EF have proven disappointing in their ability to identify high-risk patient subsets (Rosenbaum, 2008). Genomic predictors of SCD risk could potentially target ICD therapy to the highest risk subgroup of patients at risk for SCD. Two large prospective NIH-sponsored multicenter studies (Genetic Risk Assessment of Defibrillator Events, GRADE; Prospective Observational Study of the ICD in Sudden Cardiac Death Prevention, PROSE-ICD) are currently underway to test this hypothesis, both using appropriate ICD shocks as a surrogate for sudden cardiac death (Ellenbogen et al., 2006; Refaat et al., 2009).

CLINICAL APPLICATIONS OF GENETIC TESTING

Inherited arrhythmia syndromes and inherited forms of structural heart disease cause arrhythmias and SCD. Clinical investigations used for the diagnosis of these conditions include a history of syncope, palpitations, or aborted sudden death (all conditions); a family history of sudden death, pacemaker or defibrillator placement, heart failure, or miscarriage (all conditions); ECG [LQTS, SQTS, Brugada, arrhythmogenic right ventricular cardiomyopathy (ARVC)]; echocardiography, computed tomography (CT), or magnetic resonance imaging (MRI) [ARVC, hypertrophic cardiomyopathy (HCM), dilated cardiomyopathy (DCM), restrictive cardiomyopathy (RCM)]; stress testing (LQTS, CPVT); drug testing (Brugada, CPVT); and invasive electrophysiological testing with programmed stimulation (Brugada) (Priori, 2010). Genetic testing for these conditions should be performed on one clinically affected individual in a family, and is currently available through several commercially available sources including Correlagen [LQTS, Brugada, CPVT, familial atrial fibrillation (AFib), ARVC, HCM, DCM, RCM, left ventricular noncompaction (LVNC); http://www.correlagen.com], Familion/Transgenomic (LQTS, SQTS, Brugada, CPVT, familial AFib, ARVC, HCM, DCM, LVNC; http://www.familion.com), GeneDx (LQTS, SQTS, Brugada, CPVT, ARVC, HCM, DCM; http://www.genedx.com), and Partners (CPVT, ARVC, HCM, DCM, RCM, LVNC; http://pcpgm.partners.org). The genes screened do differ somewhat between companies, along with the potential out-of-pocket costs to patients. Fortunately, insurance carriers have been increasingly willing to cover the cost of the test, which ranges from several hundred dollars if a gene and mutation in a family are known, to several thousand dollars for a screening panel of genes for one of the inherited syndromes. A definitive genetic diagnosis of the condition can help to direct clinical recommendations, which can include periodic follow-up, avoiding medications that may exacerbate the condition, and avoiding strenuous activities such as competitive sports. In addition, specific genetic diagnoses may guide therapies such as the use of beta blockers in long-QT syndrome, ICD placement in SQTS, and experimental gene-based therapies as described above.

Current clinical practice guidelines recommend clinical screening of asymptomatic first-degree relatives and all potentially symptomatic relatives of patients with inherited arrhythmias. If a causative gene has been identified, genetic screening is also indicated, although insurance carriers will not, in general, reimburse for the genetic screening of asymptomatic patients. This is unfortunate, as they will reimburse for more expensive clinical testing that may need to be repeated every few years. In fact, the greatest utility in genetic testing at present lies in the ability to define, in a family with an inherited condition of known genetic etiology, those individuals who are not affected and therefore (1) require no further clinical follow-up and (2) cannot pass the condition to their children.

In young individuals who die suddenly with a normal autopsy, a mutation in an inherited arrhythmia gene can be identified up to one third of the time (Tester and Ackerman, 2007). This molecular autopsy has great potential utility to the family, who may be at risk for SCD and reluctant to undergo systematic clinical screening. Unfortunately, family members are seldom aware of this option, and they need to pay for genetic analysis themselves, as insurance coverage ends at the time of death.

A number of caveats regarding genetic testing need to be understood by both physicians and patients. First, only a fraction of the genes and mutations are known for each of these inherited conditions, and the failure to identify a mutation/gene in a proband and his/her family has minimal clinical utility as it does not exclude the diagnosis. Second, the optimal treatment of asymptomatic genetically affected individuals is often unknown, and the genetic diagnosis may place unnecessary limitations on lifestyle and/or lead to pharmacological or device therapies whose utility is not proven. Third, the genetic diagnosis may have implications regarding insurability and employment. Finally, we must remember that the genetic testing for arrhythmias is in its early stages. Variants of unknown significance are often identified and must not be assumed to be pathogenic. In addition, some variants currently called pathogenic mutations will eventually be shown to be benign.

CONCLUSION

The identification of genetic predictors of common arrhythmias is in its earliest stages. The diseases that lead to the common atrial and ventricular arrhythmia syndromes are diverse, as are the arrhythmia mechanisms that they support. It is likely that multiple genetic variants will interact to control risk, and that differential importance of the variants will coincide with the underlying disease states. In addition, interactions with race, gender, environment, and other cardiovascular conditions may further complicate the analysis.

Ultimately, a handful of SNPs that modulate the risk of each arrhythmic syndrome may be identified and used to develop an algorithm to predict the likelihood of developing the condition, its virulence, and its response to specific therapies based on genotype. Of course, any such personalized approach to arrhythmias will require prospective testing in large patient cohorts.

One potential question is whether genetic testing, as opposed to novel clinical predictors, can or should be used to restrict ICD placement. Currently, the beneficial effects of ICDs in preventing sudden death are counteracted by the risks, the reluctance of many physicians to strongly advise ICD placement in heart failure patients, and the reluctance of patients to accept ICD therapy. Better predictors of sudden death risk could identify the group of patients who would benefit most from ICD placement, along with patients with less severe left ventricular dysfunction at sufficient risk to warrant ICD placement. In addition, delineation of a group of patients at low risk for sudden death could shift the focus on that cohort from devices to optimal medical management. For the field of arrhythmias, this may represent the greatest promise of personalized medicine.

REFERENCES

Aarnoudse, A.J., Newton-Cheh, C., De Bakker, P.I., et al., 2007. Common NOS1AP variants are associated with a prolonged QTc interval in the Rotterdam Study. Circulation 116 (1), 10–16.

Aiba, T., Tomaselli, G.F., 2010. Electrical remodeling in the failing heart. Curr Opin Cardiol 25 (1), 29–36.

Albert, C.M., Nam, E.G., Rimm, E.B., et al., 2008. Cardiac sodium channel gene variants and sudden cardiac death in women. Circulation 117 (1), 16–23.

Algra, A., Tijssen, J.G., Roelandt, J.R., Pool, J., Lubsen, J., 1991. QTc prolongation measured by standard 12-lead electrocardiography is an independent risk factor for sudden death due to cardiac arrest. Circulation 83 (6), 1888–1894.

Anand, K., Mooss, A.N., Mohiuddin, S.M., 2006. Aldosterone inhibition reduces the risk of sudden cardiac death in patients with heart failure. J Renin Angiotensin Aldosterone Syst 7 (1), 15–19.

Anderson, M.E., 2007. Multiple downstream proarrhythmic targets for calmodulin kinase II: Moving beyond an ion channel-centric focus. Cardiovasc Res 73 (4), 657–666.

Antzelevitch, C., 2007. Role of spatial dispersion of repolarization in inherited and acquired sudden cardiac death syndromes. Am J Physiol Heart Circ Physiol 293 (4), H2024–H2038.

Antzelevitch, C., Yan, G.X., Shimizu, W., 1999. Transmural dispersion of repolarization and arrhythmogenicity: The Brugada syndrome versus the long QT syndrome. J Electrocardiol 32 (Suppl), 158–165.

Arking, D.E., Junttila, M.J., Goyette, P., et al., 2011. Identification of a sudden cardiac death susceptibility locus at 2q24.2 through genome-wide association in European ancestry individuals. PLoS Genet 7 (6), e1002158.

Arking, D.E., Pfeufer, A., Post, W., et al., 2006. A common genetic variant in the NOS1 regulator NOS1AP modulates cardiac repolarization. Nat Genet 38 (6), 644–651.

Arking, D.E., Reinier, K., Post, W., et al., 2010. Genome-wide association study identifies GPC5 as a novel genetic locus protective against sudden cardiac arrest. PLoS One 5 (3), e9879.

AVID (Antiarrhythmics versus Implantable Defibrillators) Investigators, 1997. A comparison of antiarrhythmic-drug therapy with implantable defibrillators in patients resuscitated from near-fatal ventricular arrhythmias. N Engl J Med 337 (22), 1576–1583.

Baker, L.C., London, B., Choi, B.R., Koren, G., Salama, G., 2000. Enhanced dispersion of repolarization and refractoriness in transgenic mouse hearts promotes reentrant ventricular tachycardia. Circ Res 86 (4), 396–407.

Bardy, G.H., Lee, K.L., Mark, D.B., et al., 2005. Amiodarone or an implantable cardioverter-defibrillator for congestive heart failure. N Engl J Med 352 (3), 225–237.

Bedi, M., McNamara, D., London, B., Schwartzman, D., 2006. Genetic susceptibility to atrial fibrillation in patients with congestive heart failure. Heart Rhythm 3 (7), 808–812.

Benhorin, J., Taub, R., Goldmit, M., et al., 2000. Effects of flecainide in patients with new SCN5A mutation: Mutation-specific therapy for long-QT syndrome? Circulation 101 (14), 1698–1706.

Bers, D.M., Despa, S., Bossuyt, J., 2006. Regulation of Ca^{2+} and Na^+ in normal and failing cardiac myocytes. Ann NY Acad Sci 1080, 165–177.

Beuckelmann, D.J., Nabauer, M., Erdmann, E., 1992. Intracellular calcium handling in isolated ventricular myocytes from patients with terminal heart failure. Circulation 85 (3), 1046–1055.

Beuckelmann, D.J., Nabauer, M., Erdmann, E., 1993. Alterations of K^+ currents in isolated human ventricular myocytes from patients with terminal heart failure. Circ Res 73 (2), 379–385.

Brahmajothi, M.V., Morales, M.J., Reimer, K.A., Strauss, H.C., 1997. Regional localization of ERG, the channel protein responsible for the rapid component of the delayed rectifier, K^+ current in the ferret heart. Circ Res 81 (1), 128–135.

Bristow, M.R., Hershberger, R.E., Port, J.D., et al., 1990. Beta-adrenergic pathways in nonfailing and failing human ventricular myocardium. Circulation 82 (2 Suppl), I12–I25.

Bristow, M.R., Murphy, G.A., Krause-Steinrauf, H., et al., 2010. An alpha2C-adrenergic receptor polymorphism alters the norepinephrine-lowering effects and therapeutic response of the beta-blocker bucindolol in chronic heart failure. Circ Heart Fail 3 (1), 21–28.

Brunner, M., Olschewski, M., Geibel, A., Bode, C., Zehender, M., 2004. Long-term survival after pacemaker implantation. Prognostic importance of gender and baseline patient characteristics. Eur Heart J 25 (1), 88–95.

Burke, A., Creighton, W., Mont, E., et al., 2005. Role of SCN5A Y1102 polymorphism in sudden cardiac death in blacks. Circulation 112 (6), 798–802.

Cabo, C., Boyden, P.A., 2006. Heterogeneous gap junction remodeling stabilizes reentrant circuits in the epicardial border zone of the healing canine infarct: A computational study. Am J Physiol Heart Circ Physiol 291 (6), H2606–H2616.

Chen, L.Y., Ballew, J.D., Herron, K.J., Rodeheffer, R.J., Olson, T.M., 2007. A common polymorphism in SCN5A is associated with lone atrial fibrillation. Clin Pharmacol Ther 81 (1), 35–41.

Cheng, J., Tester, D.J., Tan, B.H., et al., 2011. The common African American polymorphism SCN5A-S1103Y interacts with mutation SCN5A-R680H to increase late Na current. Physiol Genomics 43 (9), 461–466.

Chiang, C.E., 2004. Congenital and acquired long QT syndrome. Current concepts and management. Cardiol Rev 12 (4), 222–234.

Crotti, L., Lundquist, A.L., Insolia, R., et al., 2005. KCNH2-K897T is a genetic modifier of latent congenital long-QT syndrome. Circulation 112 (9), 1251–1258.

Crotti, L., Monti, M.C., Insolia, R., et al., 2009. NOS1AP is a genetic modifier of the long-QT syndrome. Circulation 120 (17), 1657–1663.

Crotti, L., Spazzolini, C., Schwartz, P.J., et al., 2007. The common long-QT syndrome mutation KCNQ1/A341V causes unusually severe clinical manifestations in patients with different ethnic backgrounds: Toward a mutation-specific risk stratification. Circulation 116 (21), 2366–2375.

Dekker, L.R., Bezzina, C.R., Henriques, J.P., et al., 2006. Familial sudden death is an important risk factor for primary ventricular fibrillation: A case-control study in acute myocardial infarction patients. Circulation 114 (11), 1140–1145.

Doolan, A., Langlois, N., Chiu, C., Ingles, J., Lind, J., Semsarian, C., 2008. Postmortem molecular analysis of KCNQ1 and SCN5A genes in sudden unexplained death in young Australians. Int J Cardiol 127 (1), 138–141.

Dorn, G.W., Liggett, S.B., 2009. Mechanisms of pharmacogenomic effects of genetic variation of the cardiac adrenergic network in heart failure. Mol Pharmacol 76 (3), 466–480.

Echt, D.S., Liebson, P.R., Mitchell, L.B., et al., 1991. Mortality and morbidity in patients receiving encainide, flecainide, or placebo. The Cardiac Arrhythmia Suppression Trial. N Engl J Med 324 (12), 781–788.

Eijgelsheim, M., Newton-Cheh, C., Aarnoudse, A.L., et al., 2009. Genetic variation in NOS1AP is associated with sudden cardiac death: Evidence from the Rotterdam Study. Hum Mol Genet 18 (21), 4213–4218.

Eijgelsheim, M., Newton-Cheh, C., Sotoodehnia, N., et al., 2010. Genome-wide association analysis identifies multiple loci related to resting heart rate. Hum Mol Genet 19 (19), 3885–3894.

Ellenbogen, K.A., Levine, J.H., Berger, R.D., et al., 2006. Are implantable cardioverter defibrillator shocks a surrogate for sudden cardiac death in patients with nonischemic cardiomyopathy? Circulation 113 (6), 776–782.

Elliott, P.M., Poloniecki, J., Dickie, S., et al., 2000. Sudden death in hypertrophic cardiomyopathy: Identification of high risk patients. J Am Coll Cardiol 36 (7), 2212–2218.

El-Sherif, N., Chinushi, M., Caref, E.B., Restivo, M., 1997. Electrophysiological mechanism of the characteristic electrocardiographic morphology of torsade de pointes tachyarrhythmias in the long-QT syndrome: Detailed analysis of ventricular tridimensional activation patterns. Circulation 96 (12), 4392–4399.

Epstein, A.E., DiMarco, J.P., Ellenbogen, K.A., et al., 2008. ACC/AHA/HRS 2008 Guidelines for Device-Based Therapy of Cardiac Rhythm Abnormalities: A report of the American College of Cardiology/American Heart Association Task Force on Practice Guidelines (Writing Committee to Revise the ACC/AHA/NASPE 2002 Guideline Update for Implantation of Cardiac Pacemakers and Antiarrhythmia Devices) developed in collaboration with the American Association for Thoracic Surgery and Society of Thoracic Surgeons. J Am Coll Cardiol 51 (21), e1–e62.

Etheridge, S.P., Compton, S.J., Tristani-Firouzi, M., Mason, J.W., 2003. A new oral therapy for long QT syndrome: Long-term oral potassium improves repolarization in patients with HERG mutations. J Am Coll Cardiol 42 (10), 1777–1782.

Fatini, C., Sticchi, E., Gensini, F., et al., 2007. Lone and secondary non-valvular atrial fibrillation: Role of a genetic susceptibility. Int J Cardiol 120 (1), 59–65.

Fatini, C., Sticchi, E., Genuardi, M., et al., 2006. Analysis of minK and eNOS genes as candidate loci for predisposition to non-valvular atrial fibrillation. Eur Heart J 27 (14), 1712–1718.

Frazier, D.W., Wolf, P.D., Wharton, J.M., Tang, A.S., Smith, W.M., Ideker, R.E., 1989. Stimulus-induced critical point. Mechanism for electrical initiation of reentry in normal canine myocardium. J Clin Invest 83 (3), 1039–1052.

Friedlander, Y., Siscovick, D.S., Weinmann, S., et al., 1998. Family history as a risk factor for primary cardiac arrest. Circulation 97 (2), 155–160.

Gaudino, M., Andreotti, F., Zamparelli, R., et al., 2003. The -174G/C interleukin-6 polymorphism influences postoperative interleukin-6 levels and postoperative atrial fibrillation. Is atrial fibrillation an inflammatory complication? Circulation 108 (Suppl. 1), II195–II199.

Gavin, M.C., Newton-Cheh, C., Gaziano, J.M., Cook, N.R., VanDenburgh, M., Albert, C.M., 2011. A common variant in the beta2-adrenergic receptor and risk of sudden cardiac death. Heart Rhythm 8 (5), 704–710.

Goldberger, J.J., Buxton, A.E., Cain, M., et al., 2011. Risk stratification for arrhythmic sudden cardiac death: Identifying the roadblocks. Circulation 123 (21), 2423–2430.

Hermida, J.S., Denjoy, I., Clerc, J., et al., 2004. Hydroquinidine therapy in Brugada syndrome. J Am Coll Cardiol 43 (10), 1853–1860.

Hernesniemi, J.A., Karhunen, P.J., Rontu, R., et al., 2008. Interleukin-18 promoter polymorphism associates with the occurrence of sudden cardiac death among Caucasian males: The Helsinki Sudden Death Study. Atherosclerosis 196 (2), 643–649.

Houser, S.R., 2000. When does spontaneous sarcoplasmic reticulum CA(2+) release cause a triggered arrythmia? Cellular versus tissue requirements. Circ Res 87 (9), 725–727.

Huikuri, H.V., Castellanos, A., Myerburg, R.J., 2001. Sudden death due to cardiac arrhythmias. N Engl J Med 345 (20), 1473–1482.

Husser, D., Adams, V., Piorkowski, C., Hindricks, G., Bollmann, A., 2010. Chromosome 4q25 variants and atrial fibrillation recurrence after catheter ablation. J Am Coll Cardiol 55 (8), 747–753.

Jessup, M., Abraham, W.T., Casey, D.E., et al., 2009. Focused update: ACCF/AHA Guidelines for the Diagnosis and Management of Heart Failure in Adults. A Report of the American College of Cardiology Foundation/American Heart Association Task Force on Practice Guidelines. Circulation 119 (14), 1977–2016.

Jons, C., O-Ochi, J., Moss, A.J., et al., 2011. Use of mutant-specific ion channel characteristics for risk stratification of long QT syndrome patients. Sci Transl Med 3 (76) 76ra28.

Jouven, X., Desnos, M., Guerot, C., Ducimetiere, P., 1999. Predicting sudden death in the population: The Paris Prospective Study I. Circulation 99 (15), 1978–1983.

Juang, J.M., Chern, Y.R., Tsai, C.T., et al., 2007. The association of human connexin 40 genetic polymorphisms with atrial fibrillation. Int J Cardiol 116 (1), 107–112.

Kaikkonen, K.S., Kortelainen, M.L., Linna, E., Huikuri, H.V., 2006. Family history and the risk of sudden cardiac death as a manifestation of an acute coronary event. Circulation 114 (14), 1462–1467.

Kannel, W.B., Benjamin, E.J., 2008. Status of the epidemiology of atrial fibrillation. Med Clin North Am 92 (1), 17–40. ix.

Kao, W.H., Arking, D.E., Post, W., et al., 2009. Genetic variations in nitric oxide synthase 1 adaptor protein are associated with sudden cardiac death in US white-community-based populations. Circulation 119 (7), 940–951.

Kato, K., Oguri, M., Hibino, T., et al., 2007. Genetic factors for lone atrial fibrillation. Int J Mol Med 19 (6), 933–939.

Katz, G., Arad, M., Eldar, M., 2009. Catecholaminergic polymorphic ventricular tachycardia from bedside to bench and beyond. Curr Probl Cardiol 34 (1), 9–43.

Lahtinen, A.M., Marjamaa, A., Swan, H., Kontula, K., 2011. KCNE1 D85N polymorphism – a sex-specific modifier in type 1 long QT syndrome? BMC Med Genet 12, 11.

Lai, L.P., Su, M.J., Yeh, H.M., et al., 2002. Association of the human minK gene 38G allele with atrial fibrillation: Evidence of possible genetic control on the pathogenesis of atrial fibrillation. Am Heart J 144 (3), 485–490.

Lakatta, E.G., DiFrancesco, D., 2009. What keeps us ticking: A funny current, a calcium clock, or both? J Mol Cell Cardiol 47 (2), 157–170.

Liggett, S.B., Mialet-Perez, J., Thaneemit-Chen, S., et al., 2006. A polymorphism within a conserved beta(1)-adrenergic receptor motif alters cardiac function and beta-blocker response in human heart failure. Proc Natl Acad Sci USA 103 (30), 11,288–11,293.

Lin, D., Callans, D.F., 2004. Sinus rhythm abnormalities. In: Zipes, D.P., Jalife, J. (Eds.), Cardiac Electrophysiology: From Cell to Bedside. Saunders, Philadelphia, pp. 479–484.

London, B., Albert, C., Anderson, M.E., et al., 2007. Omega-3 fatty acids and cardiac arrhythmias: Prior studies and recommendations for future research. A report from the National Heart, Lung, and Blood Institute and Office Of Dietary Supplements Omega-3 Fatty Acids and their Role in Cardiac Arrhythmogenesis Workshop. Circulation 116 (10), e320–e335.

London, B., Baker, L.C., Lee, J.S., et al., 2003. Calcium-dependent arrhythmias in transgenic mice with heart failure. Am J Physiol Heart Circ Physiol 284 (2), H431–H441.

Lubitz, S.A., Sinner, M.F., Lunetta, K.L., et al., 2010. Independent susceptibility markers for atrial fibrillation on chromosome 4q25. Circulation 122 (10), 976–984.

Mahida, S., Lubitz, S.A., Rienstra, M., Milan, D.J., Ellinor, P.T., 2011. Monogenic atrial fibrillation as pathophysiological paradigms. Cardiovasc Res 89 (4), 692–700.

Makielski, J.C., Ye, B., Valdivia, C.R., et al., 2003. A ubiquitous splice variant and a common polymorphism affect heterologous expression of recombinant human SCN5A heart sodium channels. Circ Res 93 (9), 821–828.

Marangoni, S., Di Resta, C., Rocchetti, M., et al., 2011. A Brugada syndrome mutation (p.S216L) and its modulation by p.H558R polymorphism: Standard and dynamic characterization. Cardiovasc Res 91 (4), 606–616.

McNamara, D.M., MacGowan, G.A., London, B., 2002. Clinical importance of beta-adrenoceptor polymorphisms in cardiovascular disease. Am J Pharmacogenomics 2 (2), 73–78.

Milanesi, R., Baruscotti, M., Gnecchi-Ruscone, T., DiFrancesco, D., 2006. Familial sinus bradycardia associated with a mutation in the cardiac pacemaker channel. N Engl J Med 354 (2), 151–157.

Mirowski, M., Mower, M.M., Reid, P.R., 1980. The automatic implantable defibrillator. Am Heart J 100 (6 Pt 2), 1089–1092.

Moretti, A., Bellin, M., Welling, A., et al., 2010. Patient-specific induced pluripotent stem-cell models for long-QT syndrome. N Engl J Med 363 (15), 1397–1409.

Moss, A.J., Hall, W.J., Cannom, D.S., et al., 1996. Improved survival with an implanted defibrillator in patients with coronary disease at high risk for ventricular arrhythmia. Multicenter Automatic Defibrillator Implantation Trial Investigators. N Engl J Med 335 (26), 1933–1940.

Moss, A.J., Shimizu, W., Wilde, A.A., et al., 2007. Clinical aspects of type-1 long-QT syndrome by location, coding type, and biophysical function of mutations involving the KCNQ1 gene. Circulation 115 (19), 2481–2489.

Moss, A.J., Zareba, W., Benhorin, J., et al., 1995. ECG T-wave patterns in genetically distinct forms of the hereditary long QT syndrome. Circulation 92 (10), 2929–2934.

Moss, A.J., Zareba, W., Hall, W.J., et al., 2002. Prophylactic implantation of a defibrillator in patients with myocardial infarction and reduced ejection fraction. N Engl J Med 346 (12), 877–883.

Myerburg, R.J., 2002. Scientific gaps in the prediction and prevention of sudden cardiac death. J Cardiovasc Electrophysiol 13 (7), 709–723.

Myerburg, R.J., Reddy, V., Castellanos, A., 2009. Indications for implantable cardioverter-defibrillators based on evidence and judgment. J Am Coll Cardiol 54 (9), 747–763.

Nabauer, M., Beuckelmann, D.J., Erdmann, E., 1993. Characteristics of transient outward current in human ventricular myocytes from patients with terminal heart failure. Circ Res 73 (2), 386–394.

Nessler, J., Nessler, B., Kitlinski, M., et al., 2007. Sudden cardiac death risk factors in patients with heart failure treated with carvedilol. Kardiol Pol 65 (12), 1417–1422, discussion 1423-1424.

Nyberg, M.T., Stoevring, B., Behr, E.R., Ravn, L.S., McKenna, W.J., Christiansen, M., 2007. The variation of the sarcolipin gene (SLN) in atrial fibrillation, long QT syndrome and sudden arrhythmic death syndrome. Clin Chim Acta 375 (1-2), 87–91.

O'Rourke, B., Kass, D.A., Tomaselli, G.F., Kaab, S., Tunin, R., Marban, E., 1999. Mechanisms of altered excitation-contraction coupling in canine tachycardia-induced heart failure, I: Experimental studies. Circ Res 84 (5), 562–570.

Pfeufer, A., Sanna, S., Arking, D.E., et al., 2009. Common variants at ten loci modulate the QT interval duration in the QTSCD Study. Nat Genet 41 (4), 407–414.

Plant, L.D., Bowers, P.N., Liu, Q., et al., 2006. A common cardiac sodium channel variant associated with sudden infant death in African Americans, SCN5A S1103Y. J Clin Invest 116 (2), 430–435.

Poelzing, S., Forleo, C., Samodell, M., et al., 2006. SCN5A polymorphism restores trafficking of a Brugada syndrome mutation on a separate gene. Circulation 114 (5), 368–376.

Priori, S.G., 2010. The fifteen years of discoveries that shaped molecular electrophysiology: Time for appraisal. Circ Res 107 (4), 451–456.

Priori, S.G., Napolitano, C., 2004. Genetics of cardiac arrhythmias and sudden cardiac death. Ann NY Acad Sci 1015, 96–110.

Priori, S.G., Napolitano, C., Schwartz, P.J., et al., 2004. Association of long QT syndrome loci and cardiac events among patients treated with beta-blockers. JAMA 292 (11), 1341–1344.

Pu, J., Robinson, R.B., Boyden, P.A., 2000. Abnormalities in Ca(i)-handling in myocytes that survive in the infarcted heart are not just due to alterations in repolarization. J Mol Cell Cardiol 32 (8), 1509–1523.

Refaat, M., Frangiskakis, J.M., Grimley, S., et al., 2009. The β2-adrenergic receptor Gln27 polymorphism is associated with increased ventricular arrhythmias in patients with severe heart failure. Heart Rhythm 6 (5), S456.

Roden, D.M., Viswanathan, P.C., 2005. Genetics of acquired long QT syndrome. J Clin Invest 115 (8), 2025–2032.

Rosenbaum, D.S., 2008. T-wave alternans in the sudden cardiac death in heart failure trial population: Signal or noise? Circulation 118 (20), 2015–2018.

Ruan, Y., Liu, N., Bloise, R., Napolitano, C., Priori, S.G., 2007. Gating properties of SCN5A mutations and the response to mexiletine in long-QT syndrome type 3 patients. Circulation 116 (10), 1137–1144.

Ruan, Y., Liu, N., Priori, S.G., 2009. Sodium channel mutations and arrhythmias. Nat Rev Cardiol 6 (5), 337–348.

Schaffer, W.A., Cobb, L.A., 1975. Recurrent ventricular fibrillation and modes of death in survivors of out-of-hospital ventricular fibrillation. N Engl J Med 293 (6), 259–262.

Schott, J.J., Alshinawi, C., Kyndt, F., et al., 1999. Cardiac conduction defects associate with mutations in SCN5A. Nat Genet 23 (1), 20–21.

Schreieck, J., Dostal, S., Von Beckerath, N., et al., 2004. C825T polymorphism of the G-protein beta3 subunit gene and atrial fibrillation: Association of the TT genotype with a reduced risk for atrial fibrillation. Am Heart J 148 (3), 545–550.

Schulze-Bahr, E., 2008. Susceptibility genes and modifiers for cardiac arrhythmias. Prog Biophys Mol Biol 98 (2-3), 289–300.

Schwartz, P.J., Priori, S.G., Locati, E.H., et al., 1995. Long QT syndrome patients with mutations of the SCN5A and HERG genes have differential responses to Na+ channel blockade and to increases in heart rate. Implications for gene-specific therapy. Circulation 92 (12), 3381–3386.

Schwartzman, D., 2004. Atrioventricular Block and Atrioventricular Dissociation. In: Zipes, D.P., Jalife, J. (Eds.), Cardiac Electrophysiology: From Cell to Bedside. Saunders, Philadelphia, pp. 485–489.

Shah, M., Akar, F.G., Tomaselli, G.F., 2005. Molecular basis of arrhythmias. Circulation 112 (16), 2517–2529.

Shimizu, W., Antzelevitch, C., 2000. Differential effects of beta-adrenergic agonists and antagonists in LQT1, LQT2 and LQT3 models of the long QT syndrome. J Am Coll Cardiol 35 (3), 778–786.

Singh, S.N., Fletcher, R.D., Fisher, S.G., et al., 1995. Amiodarone in patients with congestive heart failure and asymptomatic ventricular arrhythmia. Survival Trial of Antiarrhythmic Therapy in Congestive Heart Failure. N Engl J Med 333 (2), 77–82.

Sinner, M.F., Ellinor, P.T., Meitinger, T., Benjamin, E.J., Kaab, S., 2011. Genome-wide association studies of atrial fibrillation: Past, present, and future. Cardiovasc Res 89 (4), 701–709.

Sinner, M.F., Pfeufer, A., Akyol, M., et al., 2008. The non-synonymous coding IKr-channel variant KCNH2-K897T is associated with atrial fibrillation: Results from a systematic candidate gene-based analysis of KCNH2 (HERG). Eur Heart J 29 (7), 907–914.

Smith, T.W., Cain, M.E., 2006. Sudden cardiac death: Epidemiologic and financial worldwide perspective. J Interv Card Electrophysiol 17 (3), 199–203.

Smits, J.P., Eckardt, L., Probst, V., et al., 2002. Genotype–phenotype relationship in Brugada syndrome: Electrocardiographic features differentiate SCN5A-related patients from non-SCN5A-related patients. J Am Coll Cardiol 40 (2), 350–356.

Sotoodehnia, N., Siscovick, D.S., Vatta, M., et al., 2006. Beta2-adrenergic receptor genetic variants and risk of sudden cardiac death. Circulation 113 (15), 1842–1848.

Splawski, I., Timothy, K.W., Tateyama, M., et al., 2002. Variant of SCN5A sodium channel implicated in risk of cardiac arrhythmia. Science 297 (5585), 1333–1336.

Spooner, P.M., 2009. Sudden cardiac death: The larger problem ... the larger genome. J Cardiovasc Electrophysiol 20 (5), 585–596.

Stecker, E.C., Sono, M., Wallace, E., Gunson, K., Jui, J., Chugh, S.S., 2006. Allelic variants of SCN5A and risk of sudden cardiac arrest in patients with coronary artery disease. Heart Rhythm 3 (6), 697–700.

Sun, A.Y., Koontz, J.I., Shah, S.H., et al., 2011. The S1103Y cardiac sodium channel variant is associated with implantable cardioverter-defibrillator events in blacks with heart failure and reduced ejection fraction. Circ Cardiovasc Genet 4 (2), 163–168.

Tester, D.J., Ackerman, M.J., 2007. Postmortem long QT syndrome genetic testing for sudden unexplained death in the young. J Am Coll Cardiol 49 (2), 240–246.

Tung, R., Zimetbaum, P., Josephson, M.E., 2008. A critical appraisal of implantable cardioverter-defibrillator therapy for the prevention of sudden cardiac death. J Am Coll Cardiol 52 (14), 1111–1121.

Valdivia, C.R., Chu, W.W., Pu, J., et al., 2005. Increased late sodium current in myocytes from a canine heart failure model and from failing human heart. J Mol Cell Cardiol 38 (3), 475–483.

Vermeulen, J.T., McGuire, M.A., Opthof, T., et al., 1994. Triggered activity and automaticity in ventricular trabeculae of failing human and rabbit hearts. Cardiovasc Res 28 (10), 1547–1554.

Viswanathan, P.C., Benson, D.W., Balser, J.R., 2003. A common SCN5A polymorphism modulates the biophysical effects of an SCN5A mutation. J Clin Invest 111 (3), 341–346.

Wakili, R., Voigt, N., Kaab, S., Dobrev, D., Nattel, S., 2011. Recent advances in the molecular pathophysiology of atrial fibrillation. J Clin Invest 121 (8), 2955–2968.

Watanabe, H., Chopra, N., Laver, D., et al., 2009. Flecainide prevents catecholaminergic polymorphic ventricular tachycardia in mice and humans. Nat Med 15 (4), 380–383.

Westaway, S.K., Reinier, K., Huertas-Vazquez, A., et al., 2011. Common variants in CASQ2, GPD1L, and NOS1AP are significantly associated with risk of sudden death in patients with coronary artery disease. Circ Cardiovasc Genet 4 (4), 397–402.

Yazawa, M., Hsueh, B., Jia, X., et al., 2011. Using induced pluripotent stem cells to investigate cardiac phenotypes in Timothy syndrome. Nature 471 (7337), 230–234.

Ye, B., Valdivia, C.R., Ackerman, M.J., Makielski, J.C., 2003. A common human SCN5A polymorphism modifies expression of an arrhythmia causing mutation. Physiol Genomics 12 (3), 187–193.

Zareba, W., Cygankiewicz, I., 2008. Long QT syndrome and short QT syndrome. Prog Cardiovasc Dis 51 (3), 264–278.

Hemostasis and Thrombosis

Richard C. Becker, Deepak Voora, and Svati H. Shah

HEMOSTASIS AND BLOOD COAGULATION

Hemostasis, as a highly regulated process, and blood coagulation are cell-based, biochemical events designed to stem the loss of blood following vascular injury and to provide the necessary cellular and protein constituents for vascular growth and repair. Blood coagulation represents a complex yet well-coordinated series of events, which involve tissue factor-bearing cells and platelets. The initiation and propagation phases of coagulation are catalyzed by thrombin in small (nM) and large concentrations, respectively, with fibrin as the end product (Monroe and Hoffman, 2006).

The complex catalysts that participate in tissue factor-mediated thrombin generation each consist of a serine protease interacting with a receptor and/or cofactor protein anchored to a specific cellular surface. One must appreciate, however, that coagulation is balanced by stoichiometric and dynamic systems of inhibition. Specifically, the tissue factor concentration threshold for thrombin generation is steep, and the resulting product is dependent largely on the concentration of plasma proteins, surface inhibitors (including tissue factor pathway inhibitor, antithrombin III, protein C, protein S, protein Z, and activated protein C) and fibrin substrate regulatory proteases such as tissue plasminogen activator and urokinase-like plasminogen activator. Thus, the blood coagulation proteome can be studied in biochemically quantifiable terms (Mann et al., 2006).

EVOLUTIONARY GENOMICS AND COAGULATION

The complex framework of integrated biochemical events regulating mammalian coagulation comprises five proteases [factor II (fII), fVII, fIX, fX, and protein C] that interact with five cofactors (tissue factor, fVIII, fV, thrombomodulin, and membrane proteins) to generate fibrin (Davidson et al., 2003b). Although each component of the network has unique functional properties, data derived from gene organization, protein structure assessments, and sequence analysis suggest that coagulation regulatory proteins may have emerged more than 400 million years ago from duplication and diversification of only two gene structures: a vitamin K-dependent serine protease, composed of a γ-carboxylated glutamic acid-epidermal growth factor (EGF)-like domain structure (common to fVII, fIX, fX, and protein C), and the A1-A2-B-A3-C1-C2 domain structure common to fV and fVIII. Prothrombin (fII) is also a vitamin K-dependent serine protease, but it contains kringle domains rather than EGF domains, suggesting a replacement during gene duplication and exon shuffling. Analysis of active-site amino acid residues reveals distinguishing characteristics of thrombin from other serine proteases, supporting its position as the ancestral blood enzyme (Davidson et al., 2003a, b; Krem and Di Cera, 2002; McLysaght et al., 2002; Van Hylckama Vlieg et al., 2003).

Genomic and Personalized Medicine, 2nd edition
by Ginsburg & Willard. DOI: http://dx.doi.org/10.1016/B978-0-12-382227-7.00052-5

HERITABILITY OF COAGULATION

Given the evidence for evolutionary conservation of coagulation proteins, two fundamental questions arise: whether genetic factors determine variability in the levels of coagulation proteins, and whether they are related to the risk of clinical thrombosis. Heritability is defined as the proportion of the total variation in a given phenotype within a population that is attributable to genetic variance. Twin studies suggest that variability in the levels of coagulation-related proteins is heritable (Table 52.1) (Ariens et al., 2002; Bladbjerg et al., 2006; De Lange et al., 2001; Dunn et al., 2004; Nowak-Gottl et al., 2008; Souto et al., 2000). Other components of the thrombotic process are also heritable. O'Donnell and colleagues examined the heritability of platelet aggregation, which is known to play a pivotal role in arterial thrombosis leading to cerebrovascular disease (CVD) and stroke. In the Framingham Heart Study (O'Donnell et al., 2001) they found high heritability for the following platelet measures: 0.48, 0.44, and 0.62 for epinephrine-induced platelet aggregation, adenosine diphosphate (ADP)-induced platelet aggregation, and collagen lag time, respectively. Similarly, thrombotic risk as measured by an individual's response to medications may have a genetic component. Inadequate suppression of platelet function with aspirin therapy has been associated with risk of future myocardial infarction (MI), stroke, and death (Eikelboom et al., 2002; Frelinger et al., 2009;

Grundmann et al., 2003; Gum et al., 2003). A study of *ex vivo* platelet function after aspirin administration in 1880 asymptomatic subjects from families with premature CVD found that changes in measures of platelet function indirectly related to cyclooxygenase (COX)-1, such as aggregation of platelet-rich plasma with collagen, ADP, or epinephrine, were less pronounced and more variable than those directly related to COX-1 (Faraday et al., 2007). These measures were strongly heritable across races, with heritability ranging from 0.27–0.76 (Faraday et al., 2007). Venous function, which contributes to the pathogenesis of venous thrombosis, may itself also have a genetic component. Venous capacity and venous compliance as measured by impedance-plethysmography are highly heritable, suggesting that the genes related to these functions could contribute to risk for venous disease states (Brinsuk et al., 2004).

The search for genes underlying disease heritability has identified some of the genetic architecture of thrombotic and hemostatic processes. However, a large portion of the heritability for many proteins and biological processes remains only partially elucidated.

GENETICS OF PLATELET FUNCTION

Platelet function, as measured *ex vivo*, has been employed for decades to identify patients with suspected platelet dysfunction and as a screening tool for individuals or families with

TABLE 52.1	Heritability of coagulation				
Study/ investigator	**Reference**	**Subjects**	**Coagulation factor(s), endpoint**	**Heritability (% of variation)**	**Comments**
De Lange	De Lange et al., 2001	1474 twin pairs	fVII, fVIII, fXIII fibrinogen, vWF, PA	41–75 highest in monozygotes	1002 female twin pairs 149 monozygotic twin pairs 325 dizygotic twin pairs
Nowak-Gottl	Nowak-Gottl et al., 2008	28	Fibrinogen, PC, PZ, fII, fV, PS, TFPI	High Shared environmental effects	282 families with pediatric stroke
Bladbjerg	Bladbjerg et al., 2006	285 twin pairs	D-dimer TAFI	33 71	Aging Danish twin study 130 monozygotic twin pairs 155 dizygotic twin pairs
Ariens	Ariens et al., 2002	230 twin pairs	F1.2 TAT complex D-dimer	45 40 65	Markers of thrombin generation, thrombin activity, fibrinolysis
GAIT study	Souto et al., 2000	398 in 21 extended pedigrees	Thrombosis endpoint	60	Additive genetic heritability for thrombosis
Dunn	Dunn et al., 2004	137 twin pairs	Fibrin clot structure	High	137 twin pairs, heritable component for fiber size and pore size

F1.2 = prothrombin fragment 1.2; TAT = thrombin-antithrombin; GAIT = genetic analysis of idiopathic thrombosis; f = factor; vWF = von Willebrand Factor; PA = plasminogen activator; TAFI = thrombin activatable fibrinolytic inhibitor; PC = protein C; PS = protein S; PZ = protein Z; TFPI = tissue factor pathway inhibitor.

bleeding diatheses. Subsequent genetic analyses identified single nucleotide polymorphisms (SNPs) that led to the description of inherited platelet disorders following a Mendelian inheritance pattern (i.e., autosomal dominant or recessive) (Salles et al., 2008). More recently, platelet-specific genes have been the target of resequencing and genotyping to identify novel genetic variants that lead to a *heightened* platelet response to biochemical stimuli. The first follows the genetic understanding of Glanzmann's Thrombasthenia (GT) through sequencing of the *ITGB3* gene, which encodes the β3 integrin. A SNP responsible for a leucine-to-proline substitution at position 33, rs5918 (Newman et al., 1989), has received the most attention. This variant is in linkage disequilibrium with a variety of other variants in the *ITGB3* gene, which together comprise the PI^A or human platelet antigen 1 (HPA-1). The PI^{A1} antigen (marked by Leu33 and also known as HPA-1a) is the most common, and is carried by 85–88% of Caucasians and Africans, but is rare in Asians. Carriers of the minor allele for rs5918 (Pro33, or PI^{A2}, or HPA-1b antigen) display heightened platelet response to epinephrine (Feng et al., 1999) and ADP-induced aggregation (Feng et al., 1999; Michelson et al., 2000), enhanced thrombin generation (Undas et al., 2001), and shorter bleeding times (Szczeklik et al., 2000); however, several studies have failed to replicate these associations (Bennett et al., 2001; Cooke et al., 1998; Meiklejohn et al., 1999). The association of genetic variants at this locus with clinical events such as MI and stroke has also been inconsistent (Herrmann et al., 1997; Ridker et al., 1997), and although most studies have found increased risk in carriers of the risk allele/haplotype (Bojesen et al., 2003; Burr et al., 2003; Carter et al., 1998; Wagner et al., 1998; Weiss et al., 1996), a recent meta-analysis found no effect (Ye et al., 2006).

The *GP1BA* gene encodes the GP1ba transmembrane subunit of this complex and contains a pair of SNPs, rs2243093 (a T>C substitution five base pairs upstream of the initiation codon, also known as the "Kozak" polymorphism) and rs6065 (also known as the HPA-2 platelet antigen, which leads to a threonine to methionine substitution at position 145). Both SNPs are in strong linkage disequilibrium (D'=1.0) in Caucasians, Africans, and Asians, as well as with the number of *Variations in the Number of 13 amino acid Tandem Repeats* (VNTR; range 1–4). The presence of the Met145 allele varies by ethnicity and is present in 5% of Caucasians, 8% of Chinese, 16% of Japanese, and 27% of Africans. It is associated with an increased level of GPIb-IX-V on the platelet membrane surface (Afshar-Kharghan et al., 1999). Although some reports suggest that carriers of Met145 have an increased risk of MI (Bray et al., 2007a; Kenny et al., 2002; Meisel et al., 2001) and stroke (Maguire et al., 2008), a recent meta-analysis does not support a major role (Ye et al., 2006).

Finally, common genetic variations have been found in two additional platelet receptors *GP6* (Johnson et al., 2010) and *PEAR1* (Faraday et al., 2011; Johnson et al., 2010), that lead to their higher surface expression and heightened platelet function. None have yet been assessed for their influence on MI or stroke, however.

GENETIC–ENVIRONMENTAL INTERFACE

The structural homology among coagulation proteins suggests similarities in their biosynthesis (transcription, post-translational processing) and in the mechanisms that determine their kinetics. In a group of healthy subjects participating in the Leiden Thrombophilia Study (Van der Meer et al.), clustering was found among the plasma concentrations of vitamin K-dependent factors (fII, fVII, fIX, and fX) and those of fXI and fXII. Factors V and VIII clustered with fibrinogen and D-dimer. The anticoagulant factors (protein C, protein S, and antithrombin III) clustered together, whereas fXIII remained independent. The identification of several independent clusters within the group of coagulation proteins and regulatory proteins suggests that individual fluctuations in plasma levels may be influenced by factors at the transcriptional, post-transcriptional and epigenetic levels (Chavali et al., 2008).

RACE, CULTURE, AND ETHNICITY

Numerous studies have suggested race/ethnicity-related differences in the incidence of thrombotic diseases. When compared with Whites, African Americans have a higher risk for venous thromboembolism (VTE) (Dowling et al., 2003; Tsai et al., 2002) and a 40% higher incidence of VTE (White et al., 1998). Given the known association between genetic polymorphisms such as factor V Leiden and prothrombin G20210A with VTE, the question then arises, could genetic differences explain some of the variability in thrombotic disease risk seen in different races? It is well-established that linkage disequilibrium (LD) and the population frequency of genetic variants vary markedly by race. As such, genetic risk for common and rare diseases also has race/ethnicity-related components.

Studies have shown that factor V Leiden (FVL) and prothrombin G20210A are common in Whites (Margaglione et al., 1998; Svensson and Dahlback, 1994), but not in African American populations (Dilley et al., 1998; Hooper et al., 1996). When evaluating clinical factors that might explain racial differences in VTE, African Americans have a higher proportion of idiopathic VTE, with a significantly lower proportion with recent surgery, trauma or infection, family history of VTE, documented thrombophilia, or recent oral contraceptive or hormone therapy (Dowling et al., 2003; Heit et al., 2010). These data suggest that unidentified genetic factors may explain this disparity in the incidence of thrombotic disease in African Americans.

Plasma levels of coagulation proteins vary by race (Folsom et al., 1992; Lange et al., 2008; Lutsey et al., 2006), and, as detailed above, these coagulation proteins themselves have a strong heritable component. In the Atherosclerosis Risk in Communities (ARIC) study of over 12,000 participants, Blacks had higher mean values of fibrinogen, fVII, von Willebrand factor, and antithrombin III, and lower mean values of protein C (Folsom et al., 1992). Similarly, in the Multi-ethnic Study of Atherosclerosis (MESA), Blacks had the highest levels of von

Willebrand factor and the highest levels of fVIII and D-dimer (Lutsey et al., 2006). Similar results have been reported in other studies (Kadir et al., 1999; Kim et al., 2010). Accordingly, higher levels of fibrinogen, fVIII and D-dimer have been associated with increased risk of VTE among African Americans (Austin et al., 2000; Patel et al., 2003).

Racial differences may also exist for platelet function phenotypes. Studies have suggested that Blacks have higher levels of platelet activation (Folsom et al., 2009), higher maximum platelet-fibrin clot strength, and shorter time to platelet-fibrin clot formation (Gurbel et al., 2008). In one heritability study, both Whites and Blacks showed strong heritability for epinephrine- and ADP-induced platelet aggregation, but heritability for collagen-induced platelet aggregation in platelet-rich plasma was significant only in Blacks (Bray et al., 2007b).

Although there are few confirmatory data, a number of studies suggest that variation in the frequency of mutations for genes encoding components of the thrombotic and hemostatic pathways may explain some of the observed clinical heterogeneity of incidence of disease and response to medications in different races/ethnicities (Capodanno and Angiolillo, 2010; Man et al., 2010; Rieder et al., 2005; Small et al., 2010; Xie et al., 2002). Further studies in larger populations of racial/ethnic subgroups, and more use of "unbiased" molecular technologies to identify novel genes and/or novel mutations specific to these racial/ethnic subgroups are necessary to test this hypothesis fully.

PATIENT AND POPULATION SCREENING

A search for inherited thrombophilic conditions is justified among patients experiencing arterial thrombosis if at least one of the following characteristics is present: (1) recurrent thromboembolic event; (2) young age (≤50 years if male, ≤55 years if female); (3) lack of significant arterial stenosis at angiography; (4) age ≤55 years if male or ≤60 years if female and no apparent cause (i.e., lack of traditional cardiovascular risk factors, systemic illnesses, malignancies, offending drugs); or (5) age ≤55 years if male or ≤60 years if female and strong family history of thrombosis, particularly at a young age without clear risk factors (Table 52.2) (Andreotti and Becker, 2005). A VTE event should prompt further evaluation in the following individuals: age <50 years, thrombosis at unusual sites (cerebral, mesenteric, hepatic, portal veins), recurrent venous thrombosis, venous thrombosis and a strong family history of thrombotic disease, and women with recurring miscarriages and/or puerperal complications. The overall yield of diagnostic investigation is greatest in adult populations (Calhoon et al., 2010) . A targeted approach to testing strongly influences the likelihood of detecting an inherited thrombophilia with biological linkage (Table 52.3) (Seligsohn and Lubetsky, 2001). The importance of obtaining a family history, a simple but frequently underutilized tool available to all clinicians, has been stressed by the Centers for Disease Control and Prevention, Office of Genomics and Disease Prevention (Yoon et al., 2003). Selective testing for common inherited thrombophilias is more cost-effective than universal screening (Wu et al., 2006).

Genetic testing for family members of those with inherited thrombophilias is a subject of clinical relevance and ongoing debate. A systematic review (Segal et al., 2009) identified an increased risk of VTE in family members of adults with the FVL gene mutation. The risk was particularly high with homozygosity for the mutation [odds ratio 18.95%, confidence interval (CI) 7.8–40]. The evidence was insufficient for family members of those with the prothrombin G20210A gene mutation. A family history of VTE is also associated with a twofold increased risk of being an FVL gene mutation carrier among pregnant women. However, the sensitivity, specificity, and positive predictive value of a family history of VTE for identifying FVL carriers were 16.4%, 92.3%, and 5.6%, respectively (Horton et al., 2010). The relatively low sensitivity and predictive value for an isolated gene mutation supports a likelihood that multiple inherited factors contribute to the risk of VTE (Bezemer et al., 2009). The higher prevalence of transient risk factors and idiopathic VTE among Black Americans suggests that family history and heritability may be particularly important in that population (Heit et al., 2010a).

Family history of VTE may also contribute to a decision to perform thrombophilia screening among patients in high-risk settings, including women prescribed hormone replacement therapy and those undergoing major orthopedic surgery (Wu et al., 2006). The totality of evidence does not support routine thrombophilia screening in unselected populations as a guide for treatment decisions (Baglin et al., 2010). Testing should be considered in selected patients with VTE and in those who are from apparently risk-prone families as a means to guide duration of therapy and to assist with counseling.

TREATMENT CONSIDERATIONS IN THROMBOSIS AND PHARMACOGENETICS

The acute treatment of VTE should not differ in patients with a known, inherited thrombophila. Furthermore, for the primary prevention of venous thrombosis there are no definitive data or guidelines to support the use of genetic testing to guide primary prevention strategies among individuals at risk. Asymptomatic carriers of FVL mutation (Coppens et al., 2006) or family members of carriers with VTE (Middeldorp et al., 1998; Segal et al., 2009) are at an increased risk of a first VTE. Despite this increased risk, the absolute risk is low, making long-term anticoagulation for prophylaxis a difficult strategy to justify in terms of cost and the potential risk of bleeding. Even in clinical settings where the absolute risk of VTE is high (e.g., after major orthopedic surgery) and where carriers of FVL or prothrombin G20210A are known to be at increased risk for VTE (Baba-Ahmed et al., 2007; Wåhlander et al., 2002), the benefit of prophylaxis does

TABLE 52.2	Risk factors for thrombosis

Venous	**Arterial**
	Inherited
• Gene polymorphisms of the hemostatic system:	• Gene polymorphisms of the hemostatic system:
• Factor V Leiden (G1691A)	• Factor V Leiden (G1691A)
• G20210A prothrombin variant	• G20210A prothrombin variant
• Gain-of-function variants of factors VIII, IX, XI	• Gain-of-function variants of fibrinogen, factor VII, plasminogen activator inhibitor type-1
• Deficiencies of antithrombin, protein C, protein S	• Glycoprotein IIIa (Leu33Pro)
• Dysfibrinogenemia	• Family history of arterial thrombosis[†]
• Family history of venous thrombosis	• Homocystinurias
• Homocystinurias and MTHFR C677T variant	• Congenital dyslipidemias
• Varicose veins	
	Physiological
• Pregnancy, puerperium	• Male sex (before 65 years of age)
• Aging	• Aging
	Environmental
• Surgery, trauma, immobilization	• Smoking, cocaine use
• Oral contraceptives, hormone replacement therapy	• Oral contraceptives, hormone replacement therapy
• HIT	• HIT
• Antifibrinolytic agents, prothrombin complex concentrates	• Antifibrinolytic agents, prothrombin complex concentrates
• Endotoxemia	• Thienopyridine-related TTP
	Other
• Previous venous thromboembolism	• Previous arterial thrombosis
• Obesity	• Atherosclerosis, vasculitis
• Malignancies	• Hypercholesterolemia, metabolic syndrome and its components
• Antiphospholipid antibodies	• Congestive heart failure, renal failure
• Polycythemia vera, essential thrombocythemia	• Atrial fibrillation
• Congestive heart failure	• Antiphospholipid antibodies, SLE, rheumatoid arthritis
• Nephrotic syndrome	• Polycythemia vera, essential thrombocythemia
• Behçet's disease, other vasculitis	• Sickle cell anemia, macroglobulinemia
• Paroxysmal nocturnal hemoglobinuria	• Malignancies

TTP, thrombotic thrombocytopenic purpura; HIT, heparin-induced thrombocytopenia; SLE, systemic lupus erythematosus; MTHFR, methylenetetrahydrofolate reductase (NAD(P)H).

TABLE 52.3	Diagnostic approach to suspected inherited thrombophilia

- High index of suspicion
  - Positive family history of thrombosis
  - Clotting at unusual sites
  - Recurring episodes of thrombosis
  - Age <50 years
- Laboratory investigation
  - Activated protein C resistance screen: if positive, perform factor V Leiden gene mutation test
  - Prothrombin G20210A genotype
  - Antithrombin III activity, protein C activity and protein S activity
  - Factor VIII, IX, and XI activity
  - Lipoprotein A [Lp(a)] level
  - ? Platelet protein studies*

*Hyper-response to low concentration of epinephrine, ADP and collagen.

not differ (Baba-Ahmed et al., 2007), suggesting that testing is unlikely to alter management decisions. Carriers of FVL or prothrombin G20210A are at an increased risk of *recurrent* VTE compared to noncarriers (Segal et al., 2009). No prospective study using genetics to guide secondary prevention has been conducted and retrospective analyses do not support a genetic-guided strategy. For example, the intensity of long-term oral anticoagulation for secondary prevention (none vs low-dose warfarin, or low- vs standard-dose warfarin, or none vs ximelagatran – a direct thrombin inhibitor), is equally efficacious in those with and those without FVL or prothrombin G20210A (Kearon et al., 2008; Ridker et al., 2003; Wåhlander et al., 2006). Until differential treatment benefits are demonstrated in carriers vs noncarriers of genetically determined thrombophilia, it is unlikely that treatment decisions will be tailored to an individual's genetic profile.

In the setting of arterial thrombosis, there are neither definitive data nor guidelines to support the use of genetic testing to inform treatment decisions. However, for primary prevention there are retrospective data that support the earlier use of preventive agents such as aspirin for individuals at risk. A genetic variant at *LPA* appears to modify the protective effects of aspirin. An uncommon (found in <5% of Caucasians) variant, rs3798220, is associated with markedly higher concentrations of Lp(a) and a twofold increased risk of cardiovascular disease (Clarke et al., 2009). In a genetic substudy of the Women's Health Study in which more than 25,000 women were randomly assigned either placebo or low-dose aspirin, carriers of this variant had a more than twofold reduction in risk for cardiovascular disease with aspirin, whereas noncarriers (>95% of Caucasians) had none (Chasman et al., 2009). Therefore, this marker could be used to select patients for whom low-dose aspirin may be beneficial in the primary prevention of cardiovascular disease, although this hypothesis has not been tested.

PHARMACOGENETICS

Warfarin

Warfarin is metabolized primarily via oxidation in the liver by CYP2C9, and exerts its anticoagulant effect by inhibiting the protein VKORC1. Three single-nucleotide polymorphisms (SNPs), two in the *CYP2C9* gene (*2,*3) and one in the *VKORC1* gene, have been found to play key roles in determining the effect of warfarin therapy on coagulation. Although other variants in *CYP4F2*, *GGCX*, and *CALU* have also been associated with response to warfarin (Rieder et al., 2005), they will not be discussed here. The prevalence of these variants also varies by race, with 37% of Caucasians and 14% of African Americans carrying the A allele.

These three SNPs play key roles in determining: (1) the dose of warfarin required to produce a therapeutic International Normalized Ratio (INR) (typically 2.0 to 3.0), (2) the risk of bleeding or of producing supratherapeutic INR (>4), and (3) the time required to achieve a stable therapeutic dose.

Carriers of *CYP2C9*2* and *CYP2C9*3* require, on average, a 19% and 33% reduction per allele respectively, in warfarin dose vs those who carry the *1 allele. Carriers of the *VKORC1* A allele require, on average, a 28% reduction per allele in their warfarin dose compared to those who carry none (International Warfarin Pharmacogenetics Consortium, 2009; Rieder et al., 2005). Standard dosing algorithms (i.e., initiation with 5 mg/day) lead, on average, to a two- to threefold increased risk of serious or life threatening bleeding or an out-of-range INR (>4.0) in carriers of the *2 or *3 alleles of *CYP2C9* (Higashi et al., 2002). Similarly, carriers of the *VKORC1* A allele are also at a two- to threefold higher risk of an INR >4 during initiation of warfarin therapy when standard dosing algorithms are used (Schwarz et al., 2008). Finally, because of the sensitivity of these patients to warfarin and the additional dose adjustments required, the time required to achieve a "stable" INR between 2.0 and 3.0 is significantly delayed in carriers of all three SNPs (Higashi et al., 2002; Schwarz et al., 2008).

Prospective, genetically tailored warfarin therapy has been tested only in a limited fashion, often without adequate sample size or control groups (Anderson et al., 2007; Caraco et al., 2007; Epstein et al., 2010). To overcome these limitations, ongoing randomized controlled trials will prospectively test the hypothesis that genetically guided warfarin therapy will improve laboratory and, most importantly, thrombotic/hemorrhagic outcomes.

Clopidogrel

Variation in the laboratory response to clopidogrel can be assessed by ADP-induced platelet aggregation, and studies have shown that reduced inhibition is associated with an increased risk of future cardiovascular events (Gurbel et al., 2005). Because this variation is a highly heritable trait (Shuldiner et al., 2009), clopidogrel is a prime candidate for pharmacogenetic study. The vast majority of the evidence surrounding clopidogrel pharmacogenetics has centered around one gene, *CYP2C19*, though there is evidence for variation at additional loci (*ABCB1*, *CYP2C9*, *PON1*, and *P2RY12*) that will not be discussed here.

Clopidogrel (a second generation thienopyridine) is a prodrug that requires hepatic bioactivation via several cytochrome P450 (CYP) enzymes to generate an active metabolite, which irreversibly inhibits the platelet ADP receptor P2Y12 (Savi et al., 2000). CYP2C19 is one of the CYPs required in this process, and there are known genetic variants that eliminate the enzymatic activity of CYP2C19. The *2 variant ("star 2," rs4244285) is the most common reduced-function variant and results in complete loss of enzymatic activity (Desta et al., 2002). Carriers of *2 have reduced formation of clopidogrel's active metabolite and reduced clopidogrel-induced platelet inhibition (Brandt et al., 2006; Mega et al., 2009a). The prevalence of the *2 and *3 alleles varies by ethnicity. In Caucasians, African Americans, and Asians, the proportion who carry at least one copy of *2 is 25%, 30%, and 40–50% respectively, and for *3 is <1%, <1%, and 7%, respectively. Additional variants exist (e.g., *3-*8, and *17) but their influence on the clopidogrel response has not been completely characterized.

In patients who received percutaneous coronary intervention (PCI) after an acute coronary syndrome (ACS) and were treated with clopidogrel, carriers of at least one *2 allele experienced an increased risk of cardiovascular death, MI, or stroke compared to noncarriers (Mega et al., 2009a). In patients treated for ST-segment elevation MI, carriers of any two alleles (*2, *3, *4, or *5) who were treated with clopidogrel had a twofold increase in the risk of the same composite outcome during follow-up (Simon et al., 2009). The highest risk appears to be in young (age <45) patients with ST-segment elevation MI, with a threefold increased risk conferred by carrying at least one *2 allele. In addition to an increased risk of this composite endpoint, these and additional studies demonstrated that in patients treated with PCI, the incidence of stent thrombosis is increased threefold in carriers of at least one *2 allele, with highest risk (up to sixfold) in those who carry two

*2 alleles (Collet et al., 2009; Giusti et al., 2009; Mega et al., 2009a; Sibbing et al., 2009). These risks appear to be consistent across indications for PCI (elective vs ACS) and stent type (bare metal vs drug-eluting), and there is a graded, increased risk in those who carry zero, one, or two variant alleles (Hulot et al., 2010). Therefore, the associations with *CYP2C19* genetic variants and the laboratory response to clopidogrel are mirrored in the clinical response to clopidogrel.

Two potential alternative treatment strategies for carriers of *2 are either higher doses of clopidogrel or alternate P2Y12 inhibitors. Higher loading and maintenance doses (e.g., 1200 mg loading and 150 mg maintenance) appear, in part, to overcome the genetic deficiency of the *2 allele (Gladding et al., 2008; Hulot et al., 2010), although not completely (Jeong et al., 2010), and can require up to 600 mg/day individuals (Pena et al., 2009). Ticlopidine (first-generation thienopyridine) is also a prodrug, but it is unclear to what extent CYP2C19 is required for its bioactivation. Prasugrel (third-generation thienopyridine prodrug) and ticagrelor (a non-thienopyridine, non-prodrug), do not depend on CYP2C19 for their metabolism, such that carriers of the *2 allele produce equivalent concentrations of active metabolite and achieve similar degrees of platelet inhibition compared to noncarriers for both drugs (Brandt et al., 2007; Mega et al., 2009b; Tantry et al., 2010; Varenhorst et al., 2009). As was the case with the laboratory response, neither prasugrel nor ticagrelor are associated with an increased risk of cardiovascular death, MI, stroke, or stent thrombosis in carriers of the *2 allele (Mega et al., 2009b, 2010) thus providing the scientific rationale for prescribing these newer agents in patients with ACS receiving PCI who carry a *2 allele instead of clopidogrel, to mitigate their increased risk. As is the case with warfarin, a genetically guided theinopyridine strategy is currently being tested prospectively.

FUTURE DIRECTIONS IN GENETICS AND GENOMICS

The available evidence clearly shows a strong genetic component for both intermediate coagulation and platelet protein levels, as well as the end consequences of dysregulation of these processes, resulting in a wide variety of thrombotic disorders and thromboembolic diseases. While several genes have been identified and evaluated, much of the heritability underlying coagulation proteins and the familial clustering of thromboembolic disease remains unexplained. Emerging technologies, including genome-wide association studies (GWAS), whole exome/whole genome resequencing, microRNA profiling, and epigenetic analyses will likely play a major contributing role in better defining the genetic architecture of these complex processes.

REFERENCES

Afshar-Kharghan, V., Li, C.Q., Khoshnevis-Asl, M., Lopez, J.A., 1999. Kozak sequence polymorphism of the glycoprotein (GP) Ibalpha gene is a major determinant of the plasma membrane levels of the platelet GP Ib-IX-V complex. Blood 94, 186–191.

Anderson, J.L., Horne, B.D., Stevens, S.M., et al., 2007. Randomized trial of genotype-guided versus standard warfarin dosing in patients initiating oral anticoagulation. Circulation 116 (22), 2563–2570.

Andreotti, F., Becker, R.C., 2005. Atherothrombotic disorders: New insights from hematology. Circulation 111, 1855–1863.

Ariens, R.A., De Lange, M., Snieder, H., Boothby, M., Spector, T.D., Grant, P.J., 2002. Activation markers of coagulation and fibrinolysis in twins: Heritability of the prethrombotic state. Lancet 359, 667–671.

Austin, H., Hooper, W.C., Lally, C., et al., 2000. Venous thrombosis in relation to fibrinogen and factor VII genes among African-Americans. J Clin Epidemiol 53, 997–1001.

Baba-Ahmed, M., Le Gal, G., Couturaud, F., Lacut, K., Oger, E., Leroyer, C., 2007. High frequency of factor V Leiden in surgical patients with symptomatic venous thromboembolism despite prophylaxis. J Thromb Haemost 97, 171–175.

Baglin, T., Gray, E., Greaves, M., et al., 2010. Clinical guidelines for testing for heritable thrombophilia. Br J Haematol 149, 209–220.

Bennett, J.S., Catella-Lawson, F., Rut, A.R., et al., 2001. Effect of the PIA2 alloantigen on the function of {beta}3-integrins in platelets. Blood 97, 3093–3099.

Bezemer, I.D., Van der Meer, F.J., Eikenboom, J.C., Rosendaal, F.R., Doggen, C.J., 2009. The value of family history as a risk indicator for venous thrombosis. Arch Intern Med 169, 610–615.

Bladbjerg, E.M., De Maat, M.P., Christensen, K., Bathum, L., Jespersen, J., Hjelmborg, J., 2006. Genetic influence on thrombotic risk markers in the elderly – a Danish twin study. J Thromb Haemost 4, 599–607.

Bojesen, S.E., Juul, K., Schnohr, P., Tybjaerg-Hansen, A., Nordestgaard, B.G., 2003. Platelet glycoprotein IIb/IIIa PI(A2)/PI(A2) homozygosity associated with risk of ischemic cardiovascular disease and myocardial infarction in young men: The Copenhagen City Heart Study. J Am Coll Cardiol 42, 661–667.

Brandt, J.T., Close, S.L., Iturria, S.J., et al., 2007. Common polymorphisms of CYP2C19 and CYP2C9 affect the pharmacokinetic and pharmacodynamic response to clopidogrel but not prasugrel. J Thromb Haemost 5, 2429–2436.

Brandt, J.T., Kirkwood, S., Mukopadhay, N., 2006. CYP2C19*2 polymorphism contributes to a diminished pharmacodynamic response to clopidogrel. J Am Coll Cardiol 47, 380A.

Bray, P.F., Howard, T.D., Vittinghoff, E., Sane, D.C., Herrington, D.M., 2007a. Effect of genetic variations in platelet glycoproteins Ibalpha and VI on the risk for coronary heart disease events in postmenopausal women taking hormone therapy. Blood 109, 1862–1869.

Bray, P.F., Mathias, R.A., Faraday, N., et al., 2007b. Heritability of platelet function in families with premature coronary artery disease. J Thromb Haemost 5, 1617–1623.

Brinsuk, M., Tank, J., Luft, F.C., Busjahn, A., Jordan, J., 2004. Heritability of venous function in humans. Arterioscler Thromb Vasc Biol 24, 207–211.

Burr, D., Doss, H., Cooke, G.E., Goldschmidt-Clermont, P.J., 2003. A meta-analysis of studies on the association of the platelet PIA polymorphism of glycoprotein IIIa and risk of coronary heart disease. Stat Med 22, 1741–1760.

Calhoon, M.J., Ross, C.N., Pounder, E., Cassidy, D., Manco-Johnson, M.J., Goldenberg, N.A., 2010. High prevalence of thrombophilic traits in children with family histories of thromboembolism. *J Pediatr* 157 (3), 485–489.

Capodanno, D., Angiolillo, D.J., 2010. Impact of race and gender on antithrombotic therapy. J Thromb Haemost 104, 471–484.

Caraco, Y., Blotnick, S., Muszkat, M., 2007. CYP2C9 Genotype-guided warfarin prescribing enhances the efficacy and safety of anticoagulation: A prospective randomized controlled study. Clin Pharmacol Ther 83, 460–470.

Carter, A.M., Catto, A.J., Bamford, J.M., Grant, P.J., 1998. Platelet GP IIIa PlA and GP Ib variable number tandem repeat polymorphisms and markers of platelet activation in acute stroke. Arterioscler Thromb Vasc Biol 18, 1124–1131.

Chasman, D.I., Shiffman, D., Zee, R.Y., et al., 2009. Polymorphism in the apolipoprotein(a) gene, plasma lipoprotein(a), cardiovascular disease, and low-dose aspirin therapy. Atherosclerosis 203, 371–376.

Chavali, S., Sharma, A., Tabassum, R., Bharadwaj, D., 2008. Sequence and structural properties of identical mutations with varying phenotypes in human coagulation factor IX. Proteins 73, 63–71.

Clarke, R., Peden, J.F., Hopewell, J.C., et al., 2009. Genetic variants associated with Lp(a) lipoprotein level and coronary disease. N Engl J Med 361, 2518–2528.

Collet, J.-P., Hulot, J.-S., Pena, A., et al., 2009. Cytochrome P450 2C19 polymorphism in young patients treated with clopidogrel after myocardial infarction: A cohort study. Lancet 373, 309–317.

Cooke, G.E., Bray, P.F., Hamlington, J.D., Pham, D.M., Goldschmidt-Clermont, P.J., 1998. PlA2 polymorphism and efficacy of aspirin. Lancet 351, 1253.

Coppens, M., Van de Poel, M.H., Bank, I., et al., 2006. A prospective cohort study on the absolute incidence of venous thromboembolism and arterial cardiovascular disease in asymptomatic carriers of the prothrombin 20210A mutation. Blood 108, 2604–2607.

Davidson, C.J., Hirt, R.P., Lal, K., et al., 2003a. Molecular evolution of the vertebrate blood coagulation network. J Thromb Haemost 89, 420–428.

Davidson, C.J., Tuddenham, E.G., McVey, J.H., 2003b. 450 million years of hemostasis. J Thromb Haemost 1, 1487–1494.

De Lange, M., Snieder, H., Ariens, R.A., Spector, T.D., Grant, P.J., 2001. The genetics of haemostasis: A twin study. Lancet 357, 101–105.

Desta, Z., Zhao, X., Shin, J.G., Flockhart, D.A., 2002. Clinical significance of the cytochrome P450 2C19 genetic polymorphism. Clin Pharmacokinet 41, 913–958.

Dilley, A., Austin, H., Hooper, W.C., et al., 1998. Prevalence of the prothrombin 20210 G-to-A variant in blacks: Infants, patients with venous thrombosis, patients with myocardial infarction, and control subjects. J Lab Clin Med 132, 452–455.

Dowling, N.F., Austin, H., Dilley, A., Whitsett, C., Evatt, B.L., Hooper, W.C., 2003. The epidemiology of venous thromboembolism in Caucasians and African-Americans: The GATE study. J Thromb Haemost 1, 80–87.

Dunn, E.J., Ariens, R.A., De Lange, M., et al., 2004. Genetics of fibrin clot structure: A twin study. Blood 103, 1735–1740.

Eikelboom, J.W., Hirsh, J., Weitz, J.I., Johnston, M., Yi, Q., Yusuf, S., 2002. Aspirin-resistant thromboxane biosynthesis and the risk of myocardial infarction, stroke, or cardiovascular death in patients at high risk for cardiovascular events. Circulation 105, 1650–1655.

Epstein, R.S., Moyer, T.P., Aubert, R.E., et al., 2010. Warfarin genotyping reduces hospitalization rates results from the MM-WES (Medco-Mayo Warfarin Effectiveness Study). J Am Coll Cardiol 55 (25), 2804–2812.

Faraday, N., Yanek, L.R., Mathias, R., et al., 2007. Heritability of platelet responsiveness to aspirin in activation pathways directly and indirectly related to cyclooxygenase-1. Circulation 115, 2490–2496.

Faraday, N., Yanek, L.R., Yang, X.P., et al., 2011. Identification of a specific intronic PEAR1 gene variant associated with greater platelet aggregability and protein expression. Blood 118 (12), 3367–3375.

Feng, D., Lindpaintner, K., Larson, M.G., et al., 1999. Increased platelet aggregability associated with platelet GPIIIa PlA2 polymorphism: The Framingham offspring study. Arterioscler Thromb Vasc Biol 19, 1142–1147.

Folsom, A.R., Aleksic, N., Sanhueza, A., Boerwinkle, E., 2009. Risk factor correlates of platelet and leukocyte markers assessed by flow cytometry in a population-based sample. Atherosclerosis 205, 272–278.

Folsom, A.R., Wu, K.K., Conlan, M.G., et al., 1992. Distributions of hemostatic variables in blacks and whites: Population reference values from the Atherosclerosis Risk in Communities (ARIC) study. Ethn Dis 2, 35–46.

Frelinger III, A.L., Li, Y., Linden, M.D., et al., 2009. Association of cyclooxygenase-1-dependent and independent platelet function assays with adverse clinical outcomes in aspirin-treated patients presenting for cardiac catheterization. Circulation 120, 2586–2596.

Giusti, B., Gori, A.M., Marcucci, R., et al., 2009. Relation of cytochrome P450 2C19 loss-of-function polymorphism to occurrence of drug-eluting coronary stent thrombosis. Am J Cardiol 103, 806–811.

Gladding, P., Webster, M., Zeng, I., et al., 2008. The pharmacogenetics and pharmacodynamics of clopidogrel response: An analysis from the PRINC (Plavix Response in Coronary Intervention) trial. JACC: Cardiovasc Interv 1, 620–627.

Grundmann, K., Jaschonek, K., Kleine, B., Dichgans, J., Topka, H., 2003. Aspirin non-responder status in patients with recurrent cerebral ischemic attacks. J Neurol 250, 63–66.

Gum, P.A., Kottke-Marchant, K., Welsh, P.A., White, J., Topol, E.J., 2003. A prospective, blinded determination of the natural history of aspirin resistance among stable patients with cardiovascular disease. J Am Coll Cardiol 41, 961–965.

Gurbel, P.A., Bliden, K.P., Cohen, E., et al., 2008. Race and sex differences in thrombogenicity: Risk of ischemic events following coronary stenting. Blood Coagul Fibrinolysis 19, 268–275.

Gurbel, P.A., Bliden, K.P., Guyer, K., et al., 2005. Platelet reactivity in patients and recurrent events post-stenting: Results of the PREPARE post-stenting study. J Am Coll Cardiol 46, 1820–1826.

Heit, J.A., Beckman, M.G., Bockenstedt, P.L., et al., 2010. Comparison of characteristics from White- and Black-Americans with venous thromboembolism: A cross-sectional study. Am J Hematol 85, 467–471.

Herrmann, S.M., Poirier, O., Marques-Vidal, P., et al., 1997. The Leu33/Pro polymorphism (PlA1/PlA2) of the glycoprotein IIIa (GPIIIa) receptor is not related to myocardial infarction in the ECTIM study. Etude Cas-Temoins de l'Infarctus du Myocarde. J Thromb Haemost 77, 117911–117981.

Higashi, M.K., Veenstra, D.L., Kondo, L.M., et al., 2002. Association between CYP2C9 genetic variants and anticoagulation-related outcomes during warfarin therapy. JAMA 287, 1690–1698.

Hooper, W.C., Dilley, A., Ribeiro, M.J., et al., 1996. A racial difference in the prevalence of the Arg506→Gln mutation. Thromb Res 81, 577–581.

Horton, A.L., Momirova, V., Dizon-Townson, D., et al., 2010. Family history of venous thromboembolism and identifying factor V Leiden carriers during pregnancy. Obstet Gynecol 115, 521–525.

Hulot, J.-S., Collet, J.-P., Silvain, J., et al., 2010. Cardiovascular risk in clopidogrel-treated patients according to cytochrome P450 2C19*2 loss-of-function allele or proton pump inhibitor coadministration: A systematic meta-analysis. J Am Coll Cardiol 56, 134–143.

International Warfarin Pharmacogenetics Consortium, 2009. Estimation of the warfarin dose with clinical and pharmacogenetic data. N Engl J Med 360, 753–764.

Jeong, Y.H., Kim, I.S., Park, Y., et al., 2010. Carriage of cytochrome 2C19 polymorphism is associated with risk of high post-treatment platelet reactivity on high maintenance-dose clopidogrel of 150 mg/day results of the ACCEL-DOUBLE (Accelerated Platelet Inhibition by a Double Dose of Clopidogrel According to Gene Polymorphism) study. JACC Cardiovasc Interv, 3, 731–741.

Johnson, A.D., Yanek, L.R., Chen, M.-H., et al., 2010. Genome-wide meta-analysis identifies seven loci associated with platelet aggregation in response to agonists. Nat Genet 42, 608–613.

Kadir, R.A., Economides, D.L., Sabin, C.A., Owens, D., Lee, C.A., 1999. Variations in coagulation factors in women: Effects of age, ethnicity, menstrual cycle and combined oral contraceptive. J Thromb Haemost 82, 1456–1461.

Kearon, C., Julian, J.A., Kovacs, M.J., et al., 2008. Influence of thrombophilia on risk of recurrent venous thromboembolism while on warfarin: Results from a randomized trial. Blood 112, 4432–4436.

Kenny, D., Muckian, C., Fitzgerald, D.J., Cannon, C.P., Shields, D.C., 2002. Platelet glycoprotein Ib alpha receptor polymorphisms and recurrent ischaemic events in acute coronary syndrome patients. J Thromb Thrombolysis 13, 13–19.

Kim, C.X., Bailey, K.R., Klee, G.G., et al., 2010. Sex and ethnic differences in 47 candidate proteomic markers of cardiovascular disease: The Mayo Clinic proteomic markers of arteriosclerosis study. PLoS One 5, e9065.

Krem, M.M., Di Cera, E., 2002. Evolution of enzyme cascades from embryonic development to blood coagulation. Trends Biochem Sci 27, 67–74.

Lange, L.A., Reiner, A.P., Carty, C.L., Jenny, N.S., Cushman, M., Lange, E.M., 2008. Common genetic variants associated with plasma fibrin D-dimer concentration in older European- and African-American adults. J Thromb Haemost 6, 654–659.

Lutsey, P.L., Cushman, M., Steffen, L.M., et al., 2006. Plasma hemostatic factors and endothelial markers in four racial/ethnic groups: The MESA study. J Thromb Haemost 4, 2629–2635.

Maguire, J.M., Thakkinstian, A., Sturm, J., et al., 2008. Polymorphisms in platelet glycoprotein 1balpha and factor VII and risk of ischemic stroke: A meta-analysis. Stroke 39, 1710–1716.

Man, M., Farmen, M., Dumaual, C., et al., 2010. Genetic variation in metabolizing enzyme and transporter genes: Comprehensive assessment in 3 major East Asian subpopulations with comparison to Caucasians and Africans. J Clin Pharmacol 50, 929–940.

Mann, K.G., Brummel-Ziedins, K., Orfeo, T., Butenas, S., 2006. Models of blood coagulation. Blood Cells Mol Dis 36, 108–117.

Margaglione, M., Brancaccio, V., Giuliani, N., et al., 1998. Increased risk for venous thrombosis in carriers of the prothrombin G→A20210 gene variant. Ann Intern Med 129, 89–93.

McLysaght, A., Hokamp, K., Wolfe, K.H., 2002. Extensive genomic duplication during early chordate evolution. Nat Genet 31, 200–204.

Mega, J.L., Close, S.L., Wiviott, S.D., et al., 2009a. Cytochrome p-450 polymorphisms and response to clopidogrel. N Engl J Med 360, 354–362.

Mega, J.L., Close, S.L., Wiviott, S.D., et al., 2009b. Cytochrome P450 genetic polymorphisms and the response to prasugrel: Relationship to pharmacokinetic, pharmacodynamic, and clinical outcomes. Circulation doi: 10.1161/CIRCULATIONAHA.109.851949.

Mega, J.L., Close, S.L., Wiviott, S.D., et al., 2010. Genetic variants in ABCB1 and CYP2C19 and cardiovascular outcomes after treatment with clopidogrel and prasugrel in the TRITON-TIMI 38 trial: A pharmacogenetic analysis. Lancet 376, 1312–1319.

Meiklejohn, D.J., Urbaniak, S.J., Greaves, M., 1999. Platelet glycoprotein IIIa polymorphism HPA 1b (PlA2): No association with platelet fibrinogen binding. Br J Haematol 105, 664–666.

Meisel, C., Afshar-Kharghan, V., Cascorbi, I., et al., 2001. Role of Kozak sequence polymorphism of platelet glycoprotein Ibalpha as a risk factor for coronary artery disease and catheter interventions. J Am Coll Cardiol 38, 1023–1027.

Michelson, A.D., Furman, M.I., Goldschmidt-Clermont, P., et al., 2000. Platelet GP IIIa PI(A) polymorphisms display different sensitivities to agonists. Circulation 101, 1013–1018.

Middeldorp, S., Henkens, C.M., Koopman, M.M., et al., 1998. The incidence of venous thromboembolism in family members of patients with factor V Leiden mutation and venous thrombosis. Ann Intern Med 128, 15–20.

Monroe, D.M., Hoffman, M., 2006. What does it take to make the perfect clot? Arterioscler Thromb Vasc Biol 26, 41–48.

Newman, P.J., Derbes, R.S., Aster, R.H., 1989. The human platelet alloantigens, PlA1 and PlA2, are associated with a leucine33/proline33 amino acid polymorphism in membrane glycoprotein IIIa, and are distinguishable by DNA typing. J Clin Invest 83, 1778–1781.

Nowak-Gottl, U., Langer, C., Bergs, S., Thedieck, S., Strater, R., Stoll, M., 2008. Genetics of hemostasis: Differential effects of heritability and household components influencing lipid concentrations and clotting factor levels in 282 pediatric stroke families. Environ Health Perspect 116, 839–843.

O'Donnell, C.J., Larson, M.G., Feng, D., et al., 2001. Genetic and environmental contributions to platelet aggregation: The Framingham heart study. Circulation 103, 3051–3056.

Patel, R.K., Ford, E., Thumpston, J., Arya, R., 2003. Risk factors for venous thrombosis in the black population. J Thromb Haemost 90, 835–838.

Pena, A., Collet, J.-P., Hulot, J.-S., et al., 2009. Can we override clopidogrel resistance? Circulation 119, 2854–2857.

Ridker, P.M., Goldhaber, S.Z., Danielson, E., et al., 2003. Long-term, low-intensity warfarin therapy for the prevention of recurrent venous thromboembolism. N Engl J Med 348, 1425–1434.

Ridker, P.M., Hennekens, C.H., Schmitz, C., Stampfer, M.J., Lindpaintner, K., 1997. PIA1/A2 polymorphism of platelet glycoprotein IIIa and risks of myocardial infarction, stroke, and venous thrombosis. Lancet 349, 385–388.

Rieder, M.J., Reiner, A.P., Gage, B.F., et al., 2005. Effect of VKORC1 haplotypes on transcriptional regulation and warfarin dose. N Engl J Med 352, 2285–2293.

Salles, I.I., Feys, H.B., Iserbyt, B.F., De Meyer, S.F., Vanhoorelbeke, K., Deckmyn, H., 2008. Inherited traits affecting platelet function. Blood Rev 22, 155–172.

Savi, P., Pereillo, J.M., Uzabiaga, M.F., et al., 2000. Identification and biological activity of the active metabolite of clopidogrel. J Thromb Haemost 84, 891–896.

Schwarz, U.I., Ritchie, M.D., Bradford, Y., et al., 2008. Genetic determinants of response to warfarin during initial anticoagulation. N Engl J Med 358, 999–1008.

Segal, J.B., Brotman, D.J., Necochea, A.J., et al., 2009. Predictive value of factor V Leiden and prothrombin G20210A in adults with venous thromboembolism and in family members of those with a mutation: A systematic review. JAMA 301, 2472–2485.

Seligsohn, U., Lubetsky, A., 2001. Genetic susceptibility to venous thrombosis. N Engl J Med 344, 1222–1231.

Shuldiner, A.R., O'Connell, J.R., Bliden, K.P., et al., 2009. Association of cytochrome P450 2C19 genotype with the antiplatelet effect and clinical efficacy of clopidogrel therapy. JAMA 302, 849–857.

Sibbing, D., Stegherr, J., Latz, W., et al., 2009. Cytochrome P450 2C19 loss-of-function polymorphism and stent thrombosis following percutaneous coronary intervention. Eur Heart J 30, 916–922.

Simon, T., Verstuyft, C., Mary-Krause, M., et al., 2009. Genetic determinants of response to clopidogrel and cardiovascular events. N Engl J Med 360, 363–375.

Small, D.S., Kothare, P., Yuen, E., et al., 2010. The pharmacokinetics and pharmacodynamics of prasugrel in healthy Chinese, Japanese, and Korean subjects compared with healthy Caucasian subjects. Eur J Clin Pharmacol 66, 127–135.

Souto, J.C., Almasy, L., Borrell, M., et al., 2000. Genetic susceptibility to thrombosis and its relationship to physiological risk factors: The GAIT study. Genetic Analysis of Idiopathic Thrombophilia. Am J Hum Genet 67, 1452–1459.

Svensson, P.J., Dahlback, B., 1994. Resistance to activated protein C as a basis for venous thrombosis. N Engl J Med 330, 517–522.

Szczeklik, A., Undas, A., Sanak, M., Frolow, M., Wegrzyn, W., 2000. Relationship between bleeding time, aspirin and the PlA1/A2 polymorphism of platelet glycoprotein IIIa. Br J Haemost 110, 965–967.

Tantry, U.S., Bliden, K.P., Wei, C., et al., 2010. First analysis of the relation between CYP2C19 genotype and pharmacodynamics in patients treated with ticagrelor versus clopidogrel: The ONSET/ OFFSET and RESPOND genotype studies. Circ Cardiovasc Genet 3 (6), 556–566.

Tsai, A.W., Cushman, M., Rosamond, W.D., Heckbert, S.R., Polak, J.F., Folsom, A.R., 2002. Cardiovascular risk factors and venous thromboembolism incidence: The longitudinal investigation of thromboembolism etiology. Arch Intern Med 162, 1182–1189.

Undas, A., Brummel, K., Musial, J., Mann, K.G., Szczeklik, A., 2001. PlA2 polymorphism of {beta}3 integrins is associated with enhanced thrombin generation and impaired antithrombotic action of aspirin at the site of microvascular injury. Circulation 104, 2666–2672.

Van der Meer, F.J., Koster, T., Vandenbroucke, J.P., Briët, E., Rosendaal, F.R., 1997. The Leiden Thrombophilia Study (LETS). Thromb Haemost 78 (1), 631–635.

Van Hylckama Vlieg, A., Callas, P.W., Cushman, M., Bertina, R.M., Rosendaal, F.R., 2003. Inter-relation of coagulation factors and D-dimer levels in healthy individuals. J Thromb Haemost 1, 516–522.

Varenhorst, C., James, S., Erlinge, D., et al., 2009. Genetic variation of CYP2C19 affects both pharmacokinetic and pharmacodynamic responses to clopidogrel but not prasugrel in aspirin-treated patients with coronary artery disease. Eur Heart J 30, 1744–1752.

Wagner, K.R., Giles, W.H., Johnson, C.J., et al., 1998. Platelet glycoprotein receptor IIIa polymorphism PlA2 and ischemic stroke risk: The stroke prevention in young women study. Stroke 29, 581–585.

Wåhlander, K., Eriksson, H., Lundström, T., et al., 2006. Risk of recurrent venous thromboembolism or bleeding in relation to thrombophilic risk factors in patients receiving ximelagatran or placebo for long-term secondary prevention of venous thromboembolism. Br J Haematol 133, 68–77.

Wåhlander, K., Larson, G., Lindahl, T.L., et al., 2002. Factor V Leiden (G1691A) and prothrombin gene G20210A mutations as potential risk factors for venous thromboembolism after total hip or total knee replacement surgery. Thromb Haemost 87, 580–585.

Weiss, E.J., Bray, P.F., Tayback, M., et al., 1996. A polymorphism of a platelet glycoprotein receptor as an inherited risk factor for coronary thrombosis. N Engl J Med 334, 1090–1094.

White, R.H., Zhou, H., Romano, P.S., 1998. Incidence of idiopathic deep venous thrombosis and secondary thromboembolism among ethnic groups in California. Ann Intern Med 128, 737–740.

Wu, O., Robertson, L., Twaddle, S., et al., 2006. Screening for thrombophilia in high-risk situations: Systematic review and cost-effectiveness analysis. The Thrombosis: Risk and Economic Assessment of Thrombophilia Screening (TREATS) study. Health Tech Assess 10, 1–110.

Xie, H.G., Prasad, H.C., Kim, R.B., Stein, C.M., 2002. CYP2C9 allelic variants: Ethnic distribution and functional significance. Adv Drug Deliv Rev 54, 1257–1270.

Ye, Z., Liu, E.H., Higgins, J.P., et al., 2006. Seven haemostatic gene polymorphisms in coronary disease: Meta-analysis of 66,155 cases and 91,307 controls. Lancet 367, 651–658.

Yoon, P.W., Scheuner, M.T., Khoury, M.J., 2003. Research priorities for evaluating family history in the prevention of common chronic diseases. Am J Prev Med 24, 128–135.

Peripheral Arterial Disease

Arabindra B. Katwal, Ayotunde O. Dokun, and Brian H. Annex

INTRODUCTION

Advances in mouse and human genome research have markedly enhanced the opportunity to develop novel approaches to advance our understanding of human diseases. The knowledge and techniques emerging from the mouse and human genome projects have revolutionized the approach to mapping and identifying new disease-related genes. The use of genomic methodologies is providing insight leading not only to identification of novel disease-related genes, but also to understanding the effect of gene modulation in disease processes, with the hope that it will likely lead to improvements in the detection of diseases and guide novel strategies for therapeutic interventions, with the distant potential to personalize therapies.

While the use of genomic methodologies has led to great advances in knowledge in some fields, such as cancer research, their use is still in its infancy in others. In the field of vascular biology, and more specifically in studies of peripheral arterial disease (PAD), investigators are just beginning to take advantage of the unique opportunities these approaches provide. Here we describe epidemiology, risk factors, and clinical presentations of PAD, and some of the current and potential future avenues through which genomic methodologies might be used to further our understanding of the disease and its management.

EPIDEMIOLOGY AND RISK FACTORS FOR PERIPHERAL ARTERIAL DISEASE

Peripheral arterial disease is characterized by obstruction in arterial beds other than the coronary arteries and is caused by atherosclerosis in the vast majority of patients. The most common site is the lower extremity, where occlusive disease leads to impaired perfusion. The major established risk factors for the development of PAD are essentially the same as those recognized as important in generalized atherosclerosis, and include increasing age after 40 years, cigarette smoking, diabetes mellitus, hyperlipidemia and hypertension (Table 53.1) (Fowkes, 1990; Hiatt et al., 1995). Elevated levels of C-reactive protein and homocysteine may also be important risk factors (Darius et al., 2003; Gerhard et al., 1995; Graham et al., 1997).

Although previously under-recognized and under-diagnosed by the medical community and therefore viewed as less important than heart disease, PAD is now recognized to have a prevalence similar to that of ischemic heart disease (Gerhard et al., 1995; Kannel and McGee, 1985). It affects about 3–10% of adults in the world and 15–20% of those over 70 years, suggesting an increased prevalence with age, and with an aging population one would expect that the prevalence of PAD will continue to increase (Hirsch et al., 2001; Norgren et al., 2007).

Genomic and Personalized Medicine, 2nd edition

by Ginsburg & Willard. DOI: http://dx.doi.org/10.1016/B978-0-12-382227-7.00053-7

TABLE 53.1 Epidemiology, risk factors and clinical manifestations of peripheral arterial disease

	Intermittent claudication	Critical limb ischemia
Risk factors		
Diabetes	Known risk factor	Known risk factor
Smoking	Known risk factor	Known risk factor
Hypertension	Known risk factor	Known risk factor
Hyperlipidemia	Known risk factor	Known risk factor
C-reactive protein	Known risk factor	Known risk factor
Clinical characteristics		
Pain with ambulation	Usually present	Usually present
Pain at rest	Usually not present	Usually present
Ulceration or gangrene	Usually not present	May be present
ABI <0.9	Usually present	Sometimes associated with lower ABIs, but low ABI does not predict disease
Annual mortality	1–2%	20%
Amputation rate	1–2%	25–40%
Biochemical characteristics		
B2 microglobulin	Higher levels in PAD patients, but not shown to differentiate between IC and CLI*	
Angiogenic factors or receptors	Higher levels of sTie 2[†] in PAD patients, and also shown to differentiate between IC and CLI[‡]	

ABI, ankle-brachial index; IC, intermittent claudication; CLI, chronic limb ischemia.
*Wilson et al., 2007.
[†]The soluble form of the endothelial receptor for angiopoietins, tyrosine kinase with immunoglobulin-like and epidermal growth factor-like domains 2.
[‡]Findley et al., 2008.

CLINICAL MANIFESTATIONS OF PERIPHERAL ARTERIAL DISEASE

Most patients with PAD lack the classic symptoms of PAD and thus are often considered asymptomatic. Some estimates suggest that as many as 50% of PAD patients fall into this category (Hirsch et al., 2001). It is currently recommended that patients with multiple risk factors, especially smokers and diabetics with other risk factors, should undergo non-invasive testing, such as the ankle brachial index (ABI) detected by Doppler probe, to make the diagnosis. ABI measures the ratio of systolic blood pressure in the ankle to that of the brachial vessels. An ABI <0.9 is considered diagnostic for PAD (Hirsch et al., 2001; Newman et al., 1999; Norgren et al., 2007). Once PAD is diagnosed, further testing with duplex ultrasonography, segmental Doppler pressure or volume plethysmography, magnetic resonance angiography (MRA), computed tomographic angiography (CTA), or angiography may be utilized, depending on the clinical situation.

There are two major clinical manifestations of PAD: intermittent claudication (IC) and critical limb ischemia (CLI) (see Table 53.1 for comparison of IC and CLI). IC manifests as reduced blood flow to the extremities during exercise resulting in pain relieved only by rest, while CLI describes pain at rest that may be associated with non-healing leg ulcers or gangrene. Interestingly, while patients with IC have amputation and annual mortality rate of 1–2%, those with CLI have a 6-month amputation risk of 25–40% and an annual mortality of 20% (Ouriel, 2001).

The diagnosis of IC vs CLI is based upon time-tested clinical classification schemes, namely the Rutherford and the Fontaine classifications. In the Rutherford classification, IC encompasses categories 1–3 (mild, moderate, and severe claudication respectively), while CLI includes categories 4–6 (ischemic rest pain, minor tissue loss, and ulceration or gangrene). The Fontaine classification is more commonly used in Europe, with stages IIa and IIb describing IC, while stages III–IV are categories of CLI. There is no biomarker or hemodynamic measure that is pathognomonic for either IC or CLI.

THERAPEUTIC STRATEGIES FOR PERIPHERAL ARTERIAL DISEASE

Most treatment strategies for PAD are geared toward prevention of disease progression by addressing underlying risk factors and interventions to treat symptoms. Strategies for prevention of disease progression focus primarily on smoking cessation in smokers, and good glycemic control in diabetic patients. Treatment of hypertension and dyslipidemia, as well as use of anti-platelet therapy is also highly recommended. Interventions aimed at reducing symptoms and increasing walk-time in patients include medical therapy, with use of the phosphodiesterase inhibitor cilostazol (Hiatt, 2001). The methylxanthine derivative pentoxyphilline, which improves deformity of red blood cells, is also used, but has been shown to be less effective (Dawson et al., 2000). However, there are no pharmacological therapies available for PAD that have the ability to increase perfusion to the leg and thus correct the underlying problem of reduced blood flow.

Interestingly, one of the most effective therapies for PAD is actually not pharmacologically based. Numerous studies have shown exercise training to be effective in improving symptoms and increasing walk-time in patients with PAD (Christman et al., 2001; Leng et al., 2000). As a result, a supervised exercise program is now recommended as a key component of PAD management (Christman et al., 2001; Hirsch et al., 2006). However, the molecular mechanism by which exercise improves symptoms in PAD is poorly understood. One of the leading hypotheses is that exercise training leads to improved perfusion of the extremities in PAD patients through the growth of blood vessels, i.e., angiogenesis. The potential role of these mechanisms and factors that modulate them (e.g., growth factors) are currently being investigated.

In patients with CLI or with severe IC, revascularization with surgical bypass grafting or percutaneous peripheral intervention (PPI) is performed quite commonly, especially in the United States. The less invasive nature of PPI is making it a more common treatment option in these patient populations. However, not all patients are candidates for revascularization, for a variety of reasons, including small target vessels, diffuse PAD, and presence of other co-morbidities that preclude them from undergoing surgery or PPI. Such patients have limited therapeutic options aside from amputation. This has led to the development of novel therapeutic options currently in the experimental phase, such as therapeutic angiogenesis.

Therapeutic angiogenesis is an experimental approach in which vascular growth factors are delivered to ischemic tissues in an attempt to improve tissue perfusion. A variety of vascular growth factors are currently being investigated. They include vascular endothelial growth factor (VEGF), fibroblast-derived growth factor (FGF), hepatocyte growth factor (HGF), and angiopoietins (Jones and Annex, 2007). A Phase I/II clinical study of HGF-plasmid in CLI patients unsuitable for revascularization showed a significant increase from baseline in transcutaneous partial pressure of oxygen ($TcPO_2$, used as a measure of limb perfusion) in the high-dose group compared with placebo at 6 months. Larger trials will be needed to determine whether HGF-plasmid can improve wound healing, limb salvage or survival (Powell et al., 2008). NV1FGF is a new-generation DNA plasmid-based gene delivery system used for more efficient and sustained local expression of FGF-1. Previously, the double-blind, randomized, and placebo-controlled Phase II TALISMAN study, conducted in Europe, demonstrated a significant reduction in the risk of major amputation for CLI patients treated with Intramuscular (IM) NV1FGF (Nikol et al., 2008). However, data presented at the 2010 American Heart Association convention from the Phase III TAMARIS study, which was designed to evaluate the safety and efficacy of NV1FGF in 525 CLI patients with skin lesions (due to ulcer or gangrene) who were unsuitable for revascularization (http://www.ClinicalTrials.gov; identifier: NCT00566657), was disappointing. At one year, no significant difference was observed between the NV1FGF group versus the placebo arm in the primary end-point of time to major amputation or death (Hiatt et al., 2010).

INTERMITTENT CLAUDICATION AND CRITICAL LIMB ISCHEMIA ARE DISTINCT CLINICAL OUTCOMES OF PERIPHERAL ARTERIAL DISEASE

Major advances in the field of ischemic heart disease followed the recognition that ischemic heart disease is a continuum, but the two major clinical manifestations of PAD, IC and CLI, appear to be quite distinct. The observation that progressively lower ABIs are associated with worsening symptoms of claudication (McDermott et al., 2002) has led some to assume that IC and CLI are due to a continuum of reduced blood flow. In this line of thought, CLI is simply a worse form of IC. However, the ultimate symptomatic presentation of PAD is heterogeneous, with some individuals presenting with claudication while others present *de novo* with CLI. Moreover, many with IC never progress to CLI and many presenting with CLI do not report antecedent claudication (Boyd, 1962; Cronenwett et al., 1984; Imparato et al., 1975). Interestingly, even in patients with similar risk factors and atherosclerotic burden, and virtually the same peripheral hemodynamics, some will present with IC while others present with CLI. These observations suggest that factors other than the currently known risk factors for PAD may influence this clinical syndrome.

GENETIC SUSCEPTIBILITY TO PERIPHERAL ARTERIAL DISEASE

The well-established risk factors for PAD are nearly identical to those for atherosclerosis. There is, however, an emerging body of evidence that suggests that an individual's genetic background may be important in PAD pathogenesis. In one study, the prevalence of PAD in various ethnic backgrounds was evaluated. Results suggested higher prevalence in African Americans (AAs), even after adjusting for age and other traditional PAD risk factors (Kullo et al., 2003). Since individuals of the same ethnicity are likely to share certain ancestral genes, the higher prevalence of PAD in AAs suggests that gene polymorphisms may contribute to PAD. Further evidence supporting the possible role of genetic risk factors in PAD comes from both association and linkage studies. Estimates of the heritability of PAD using ABI as a surrogate have shown that the contribution of genetic factors to overall variation in ABI ranges from 21–48% (Carmelli et al., 2000; Kullo et al., 2006; Murabito et al., 2006). In the Multi-Ethnic Study of Atherosclerosis (MESA), self-reported race/ethnic group predicted the presence of PAD, independent of all other "established" and "novel" cardiovascular disease risk factors of atherosclerosis, with the lowest risk in Hispanics and Chinese and the highest in African Americans (Allison et al., 2006). While family studies suggest that PAD is heritable, the degree of genetic influence on the development, course, and outcome of PAD, irrespective of other risk factors, remains to be understood (Knowles et al., 2007).

TABLE 53.2 Gene polymorphisms associated with peripheral arterial disease

Thrombosis	Atherosclerosis	Unknown function
Factor II (*FII G20210A*)	*APO E*	*PAOD1*
P2Y12 (H2 allele)	*APO B*	
Fibrinogen (beta)	IL-6 (-174 G/C)	
	E-Selectin	
	ICAM-1 (469E/K)	
	MCP-1 (-2518 A/G)	
	MMP1 and *MMP3*	
	eNOS (-786C)	
	ACE D	
	CHRNA3	

Gene Polymorphisms Contributing to Peripheral Arterial Disease

Sequence variations in a variety of genes have shown statistically significant association with PAD. These genes can be classified into three different categories: pro-atherosclerotic, pro-atherothrombotic, and unknown, based on the function of the gene products (Table 53.2). Each category is discussed in greater detail below.

Gene Polymorphisms Contributing to Atherosclerosis and Peripheral Arterial Disease

PAD is a consequence of atherosclerosis in the lower extremities. Therefore it is not unexpected that polymorphisms of genes contributing to the development of atherosclerosis can be found to be associated with PAD. One of the major risk factors for atherosclerosis is hyperlipidemia. Monsalve and coworkers evaluated polymorphisms within the *APO B* gene among 205 patients with a diagnosis of PAD. Their data showed higher prevalence of an *R1* and *X1* allele among the PAD patients compared to non-PAD controls. However, since most of the individuals also had arterial disease other than PAD, they concluded that variations at the *apo B* locus contribute to predisposing individuals to the development of arterial disease, but they could not discriminate where in the arterial system the disease develops (Monsalve et al., 1991). Similarly, an association study from the Honolulu-Asia Aging Study, involving 3161 Japanese-American men aged 71–93, showed association of *APO E* polymorphism with PAD (Resnick et al., 2000). At this point, the specific mechanisms by which these gene polymorphisms contribute to the pathogenesis of PAD is poorly understood.

Inflammation is becoming increasingly recognized as an important factor in the pathogenesis of atherosclerosis, with its development and progression orchestrated by several molecules belonging to different families of inflammatory mediators, such as cytokines, chemokines, adhesion molecules, and proteolytic enzymes (Cesari et al., 2003; Flex et al., 2007). IL-6 is a pro-inflammatory, multifunctional cytokine produced by various cell types, including monocytes, adipocytes, and endothelial cells. Elevated levels of IL-6 have been found in patients with atherosclerotic disease (Biasucci et al., 1996; Milei et al., 1996). Sequence variations in the promoter region of the *IL-6* gene (*-174*) have been reported, with two different alleles identified resulting in three genotypes: GG, GC, and CC (Flex et al., 2002). These described polymorphisms influence the level of transcription of *IL-6* and IL-6 protein concentration in serum. Flex and coworkers, in their study of 84 patients with PAD, found that the GG genotype was more common in individuals with PAD, suggesting a role for this pro-inflammatory cytokine in the pathogenesis of PAD (Flex et al., 2002, 2007).

In a study of 157 PAD patients and 206 controls, Flex and coworkers analyzed gene polymorphisms including *IL-6* (*-174 G/C*), E-selectin (*E-sel*) *Ser128Arg*, intercellular adhesion molecule-1 (*ICAM-1*) *469 E/K*, monocyte chemoattractant protein-1 (*MCP-1*) *-2518 A/G* matrix metalloproteinase (*MMP*)-1 *-1607 1G/2G*, and *MMP-3 -1171 5A/6A*. The *IL-6*, *MCP-1*, and *MMP-3* polymorphisms influence the rate of transcription of the respective genes, substantially influencing protein plasma concentrations. MMP-1 plasma levels are also significantly affected by the polymorphism studied. The *ICAM-1* polymorphism changes the amino acid sequence of the immunoglobulin-like domain 5, which is important for ICAM-1 and leukocyte antigen interactions, and for B-cell adhesion. They found that *IL-6*, *E-sel*, *ICAM-1*, *MCP-1*, *MMP-1*, and *MMP-3* gene polymorphisms were significantly and independently associated with PAD. They also found that these pro-inflammatory polymorphisms act synergistically and determine genetic profiles that confer different levels of risk for PAD and CLI, depending on the number of high-risk genotypes concomitantly present in a given individual. This is an instance of how susceptibility to a disease results from functional interactions among a number of modifier genes (Flex et al., 2007).

Angiotensin-converting enzyme (ACE) processes the decapeptide angiotensin I to octapeptide angiotensin II, which is a strong vasoconstrictor. Several studies have addressed ACE insertion/deletion (I/D) polymorphism in the context of PAD. This polymorphism refers to the insertion/deletion of a 287-bp fragment in intron 16 of the *ACE* gene. The *ACE D* allele is associated with increased serum levels of the circulating enzyme (Rigat et al., 1990). Basar and colleagues and Li and coworkers found borderline significant association between *ACE I/D* polymorphism and PAD (Basar et al., 2007; Li et al., 2007a). Fatini and colleagues studied the *ACE I/D* polymorphism and the *-240 A>T* polymorphism in the promoter region of the *ACE* gene in the context of PAD, and showed that the *ACE D* allele and the *ACED/-240T* haplotype significantly and independently influenced predisposition to PAD (Fatini et al., 2009). ACE may affect the atherosclerotic process through bradykinin degradation and nitric oxide (NO) release reduction (Sticchi et al., 2010). NO is synthesized from L-arginine by at least three isoforms of NO synthase (NOS), one of which is endothelial (eNOS) (Moncada and Higgs, 1993). eNOS-derived NO acts as an antiatherogenic molecule, and reduced endogenous NO synthesis promotes the progression of atherosclerosis (Kawashima and Yokoyama, 2004). *eNOS* gene polymorphisms have been shown to cause decreased NO synthesis, thereby reducing NO availability and causing endothelial dysfunction (Hingorani, 2001). Of the several *eNOS* gene polymorphisms, the *-786T>C* and *eNOS 4a4b* polymorphisms are associated with variability in the NO plasma levels (Nakayama et al., 1999; Tsukada et al., 1998). The *-786C* allele significantly reduces eNOS promoter activity, and individuals carrying the *4a4a* genotype were found to have lower NO production (Dosenko et al., 2006). Fowkes and coworkers did not find an association between *eNOS 4a/4b* polymorphism and PAD (Fowkes et al., 2000). However, Sticchi and colleagues demonstrated an association between *eNOS* and *ACE* genes in increasing the susceptibility to PAD in smokers, providing evidence for a modifier gene–environment interaction. In smokers, but not in nonsmokers, the concomitant presence of the *eNOS -786C* and *4a* alleles was significantly associated with increased predisposition to PAD. Furthermore, the *eNOS-786/4a* haplotype also increased susceptibility to PAD in smokers (but not in nonsmokers) also carrying the *ACE D* allele (Sticchi et al., 2010).

Polymorphism of Pro-Atherothrombotic Genes and Peripheral Arterial Disease

It is fairly well established that thrombosis plays an important role in the pathogenesis of atherosclerosis (Yee et al., 2001). Thrombin is not only important in fibrin formation and platelet aggregation, but is also important in endothelial activation, and platelet and leukocyte recruitment (Coughlin, 2000). Consequently, various groups have hypothesized that polymorphisms in genes encoding hemostatic proteins may contribute to development of atherosclerosis. Indeed, studies looking at fibrinogen gene polymorphism found that certain genotypes were associated with an increased risk of peripheral atherosclerosis (Fowkes et al., 1992). Similarly, Reny and coworkers investigated an association of factor II and V polymorphisms in PAD (Reny et al., 2004). Although no association was found for factor V polymorphisms, a statistically significant association was shown between the *FII G20210A* allele of factor II and PAD. The *FII G20210A* polymorphism appears to increase local generation of thrombin, and thus contribute to the progression of atherothrombosis.

The importance of platelet aggregation in arterial thrombosis is also well known. The platelet adenosine diphosphate (ADP) receptor P2Y12 is a 7 transmembrane receptor which upon activation promotes platelet aggregation (Conley and Delaney, 2003; Gachet, 2001). Blockade of P2Y12 by thienopyridines has been shown to be beneficial in patients with cardiovascular disease. Polymorphism in the *P2Y12* gene has been described (Fontana et al., 2003a); one of the alleles (*H2*)

results in a gain of function haplotype on ADP-induced platelet aggregation. Hence it was hypothesized that this allele may be associated with increased risk of PAD (Fontana et al., 2003b). Indeed, an association was found for the presence of the *H2* allele and PAD even after adjustment for traditional PAD risk factors.

Thrombophilia can also be mediated through defects in the folate pathway (Zintzaras and Zdoukopoulos, 2009). Methylene tetrahydrofolate reductase is a critical folate-metabolizing enzyme involved in the folate/homocysteine pathway. Three studies found a significant positive association between *MTHFR 677 C/T* polymorphism and PAD (Sabino et al., 2009; Sofi et al., 2005; Todesco et al., 1999).

Other Gene Polymorphisms and Peripheral Arterial Disease

Recent studies have shown that loss of function mutation in the gene encoding the facilitative glucose transporter GLUT10 (*SLC2A10*) causes arterial tortuosity syndrome via upregulation of the transforming growth factor (TGF)-β. The *SLC2A10* gene is thought to be a candidate gene for vascular complications in type 2 diabetics (Brownlee, 2005; Coucke et al., 2006; Jiang et al., 2010). In a prospective cohort study of 372 diabetic patients, several common single nucleotide polymorphisms (SNPs) of the *SLC2A10* gene were significantly associated with PAD, with the strongest association shown by the T allele at *rs2179357*. They also identified a common haplotype (H4) conferring a strong risk of PAD in type 2 diabetic patients. Furthermore, carriers of the H4 haplotype were more likely to develop PAD during follow-up (Jiang et al., 2010).

Genetic Locus Conferring Susceptibility to Peripheral Arterial Disease

Human studies of the genetics of PAD are quite limited outside of genes contributing to atherosclerosis or thrombosis. Despite an extensive review, we were able to identify only a single family-based linkage study that has identified a genetic locus conferring susceptibility to PAD. This study of Icelandic families with multiple family members exhibiting PAD identified a locus termed *PAOD1*, which mapped to human chromosome 1p31 (Gudmundsson et al., 2002). Interestingly, other risk factors for PAD, such as hypertension, hyperlipidemia, and diabetes, did not contribute to the positive linkage. Despite these strong and convincing genetic data, the genes responsible for *PAOD1* have not been identified. This is not surprising in light of the difficulties involved in studying genetics of a complex disease such as PAD in humans.

Recently, a few genome-wide association studies (GWAS) have been published identifying genetic loci associated with susceptibility to PAD (Table 53.3). One study found a significant association between nicotine dependence and an SNP (*rs1051730*) on chromosome 15q2. This SNP is located within the *CHRNA3* gene, which is in a linkage disequilibrium block that contains genes encoding nicotinic acetylcholine receptors. Interestingly, this study also demonstrated a significant

TABLE 53.3	Gene polymorphisms associated with peripheral arterial disease identified by genome-wide association studies	
Gene polymorphisms	**Gene(s) involved or possibly involved**	**Putative functions of the gene(s)**
SNP rs1051730, chromosome 15q2	CHRNA3	Close to genes encoding nicotinic acetylcholine receptors
		SNP also associated with nicotine dependence
SNPs *rs1333049* or *rs10757278*, chromosome 9p21	Close to two genes – CDKN2A and CDKN2B	Involved in cell proliferation, aging, senescence, and apoptosis

association of this SNP with PAD and lung cancer, suggesting a gene–environment interaction (Thorgeirsson et al., 2008).

Guided by previous GWAS demonstrating a strong association of a common variant at chromosome 9p21 (tagged by the *rs1333049* or *rs10757278* SNP) with myocardial infarction, Helgadottir and colleagues showed that the same variant was associated with myocardial infarction and major arterial aneurysms, a form of peripheral arterial disease (Helgadottir et al., 2008). They also found an association with PAD in a sample younger than 75 years, which disappeared upon removal of those with a history of myocardial infarction (MI). However, data analyzed from three studies revealed that the C allele at *rs1333049* is associated with an increased prevalence of PAD and lower mean ABI in three older white populations, independent of the presence of previous MI and atherosclerotic risk factors (Cluett et al., 2009).

Identification of Novel Gene Polymorphisms Involved in Peripheral Arterial Disease

As described earlier, sequence variations in certain genes are associated with the development of PAD. However, these polymorphisms appear to contribute to the development of atherosclerotic and/or atherothrombotic disease in general, rather than specifically to PAD. Here we will focus on polymorphisms in genes involved in skeletal muscle function and adaptation to ischemic stress, which may be important in the pathogenesis of PAD. This hypothesis is supported by the observation that, in spite of similar atherosclerotic burden, some individuals present with IC while others present *de novo* with CLI. Furthermore, in a mouse preclinical model of PAD in which the mouse femoral artery is ligated and excised to introduce ischemic stress (hind limb ischemia, or HLI), our group and others observed that recovery is strain-dependent

(Fukino et al., 2003; Scholz et al., 2002), with C57BL/6 mice showing robust perfusion recovery and rare necrosis compared to the BALB/C or A/J mice after HLI. Since the ischemic stress in this model is independent of atherosclerosis and yet recovery is strain-dependent, this suggests a role for polymorphism in genes other than those contributing to atherosclerosis. Investigating the role of genetic polymorphisms involved in such novel mechanisms in PAD is quite challenging. For instance, it requires taking an unbiased approach in which one starts with a phenotype, and then through association or linkage studies identifies genes that are involved in the observed phenotype. Moreover, even when such an approach is taken, identifying the specific genes involved could still be quite challenging. This is perhaps best exemplified by the Gudmundsson study described earlier, in which *PAOD1* was identified. Although the investigators took an unbiased approach, starting with a phenotype leading to the identification of a genetic locus, the specific gene or genes conferring the *PAOD1* phenotype remain unknown. Nevertheless, current advances in genomic methodologies are allowing novel approaches to be taken to investigate polymorphisms that may contribute to PAD via mechanisms different from those involved in development of atherosclerosis or atherothrombosis.

Identification of a Quantitative Trait Locus (QTL) Involved in a Preclinical Model of Peripheral Arterial Disease

To identify novel polymorphisms that may be important modifiers of PAD, we took advantage of the strain-specific difference in recovery following HLI described above to map and identify candidate genes (Dokun et al., 2008). The F1 offspring of the cross of the C57BL/6 and BALB/C mice showed no difference in perfusion recovery or incidence of necrosis compared to the C57BL/6 mice, suggesting that a dominant allele(s) from C57BL/6 was responsible for the greater limb perfusion and less limb loss. To map the allele(s) in the C57BL/6 mice, N2 progeny was derived from one-way mating of F1 males with BALB/C females, and genome-wide linkage scans were done for the phenotypes of necrosis and perfusion ratio at day 21 after HLI surgery. Whereas previous approaches involved use of microsatellite markers to genotype and map involved genes, this approach was time-consuming and costly. An alternative approach involves use of commercially available mouse SNP linkage panels. This allows for rapid scanning of the entire mouse genome with subsequent identification of QTLs that are associated with the phenotype of interest. We identified a single QTL, limb salvage QTL1 (*LSq-1*) on chromosome 7, spanning ~31 Mb and centered at the SNP marker *rs13479513*, that showed significant linkage to the phenotypes of tissue necrosis and day 21 perfusion ratios, with peak logarithm of odds (LOD) scores of 7.96 and 3.71 respectively. This region contains 495 genes.

Refining an Identified QTL Using Haplotype Analysis

Once a QTL has been identified, it may be important to further refine the identified locus from a region containing hundreds of genes to one containing a handful of genes, making it easier to identify the specific genes responsible for the phenotype of interest. Traditionally, this would have involved making congenic mice strains generated by successive crosses between the F1 (generated from a cross between the two parental strains) and the parental strain with the phenotype of interest. While this approach has other benefits, including allowing for independently testing the role of the locus in the absence of other genetic components from the parent strain, it is, however, quite time-consuming and costly. An alternative to this time-consuming approach is to perform haplotype analysis. Most of the genetic variations between inbred mouse strains are found in ancestral haplotype blocks that are shared between strains. Although sequence variations unique to specific strains do exist, they are less frequent (Wade et al., 2002). Therefore, if the phenotypes of multiple mouse strains are known, then one can take advantage of ancestral haplotype sharing patterns within the inbred mouse lineage to quickly identify high-priority areas within a QTL interval that are likely to harbor the polymorphism involved in a given phenotype (Peters et al., 2007). This approach has been successfully used to quickly refine the boundaries of QTLs covering otherwise large genomic intervals into regions containing only a few genes (Sheehan et al., 2007; Wang et al., 2004). Taking advantage of ancestral haplotype blocks that are similar in strains with poor recovery but different in a strain with good recovery, we were able to refine the genetic region of interest to two regions containing a total of 37 genes (Dokun et al., 2008).

Identification of Candidate Genes

To identify the genes that may be involved in the differential recovery of mouse strains following experimental hind-limb ischemia, one approach has been to use gene expression profiling. In this approach, the differential expression of genes from post hind-limb ischemic tissue from different mouse strains is sought. Genes that are highly expressed in association with a given phenotype are then selected as candidate genes conferring a given phenotype. Unfortunately, as is the case with most studies using gene profiling, multiple differentially expressed genes may be identified. This creates the challenge of determining which genes are the most relevant in conferring the phenotype.

To avoid this predicament, gene expression profiling can be combined with QTL mapping and haplotype analysis (Figure 53.1). This allows one to focus only on genes within an identified QTL that are differentially expressed between strains. Gene expression profiling can be applied to tissue samples from patients with and without PAD. Candidate genes can then be identified among differentially expressed genes that localize within a QTL (e.g., *PAOD1*) known to be associated with PAD.

The overall approach used in our study has already led to the identification of genes associated with human disease. Taking advantage of differences in mouse strain susceptibility to the development of atherosclerosis, Paigen and coworkers found a locus on chromosome 1 (*Ath 1* gene), refined it

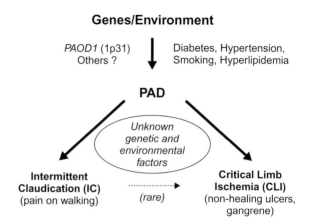

Genes/Environment

PAOD1 (1p31)
Others ?

Diabetes, Hypertension, Smoking, Hyperlipidemia

PAD

Unknown genetic and environmental factors

Intermittent Claudication (IC)
(pain on walking)

(rare)

Critical Limb Ischemia (CLI)
(non-healing ulcers, gangrene)

Figure 53.1 In PAD of the lower extremity, both genetic and environmental risk factors cause atherosclerotic occlusive disease (top arrow). *PAOD1* is an example of a genetic locus that is associated with the generation of peripheral atherosclerosis. We hypothesize that in the setting of the occlusion that causes atherosclerosis, genetic factors may influence the degree of the *clinical manifestations* of PAD. Differences in gene expression in the absence of or following ischemia may give clues as to the genetic factors.

to a narrower locus, and, using a combination of differential gene expression, gene targeting, and transgenic overexpression, successfully identified *Tnfsf4* as the responsible gene (Paigen et al., 1987; Phelan et al., 2002). They also identified the human orthologs and performed association studies in humans showing that *Tnfsf4* polymorphism was associated with increased risk of MI (Wang et al., 2005).

At present, the mechanisms by which *LSq-1* contributes to strain-dependent recovery after HLI is not completely understood. Differential angiogenic response after limb ischemia (Fukino et al., 2003) and differences in the extent of pre-existing collateral vessels and arteriogenesis (Scholz et al., 2002) have been thought to play a role in perfusion recovery after HLI. We showed a correlation between lower perfusion recovery and greater necrosis, suggesting that the locus effect may be via a mechanism contributing to restoration of perfusion after ischemia. Our data was consistent with a role for angiogenesis in strain-specific recovery, although the roles of arteriogenesis or other combinations of factors are not excluded. Faber and colleagues have previously proposed that differences in the extent (density and diameter) of the pre-existing collaterals, and/or their enlargement in response to ischemia (remodeling), underlie the different recovery phenotypes observed in C57BL/6 and BALB/C mice after HLI (Chalothorn et al., 2007; Chalothorn and Faber, 2010). They have shown that compared with C57Bl/6, BALB/C and A/J strains have fewer native collaterals in the hind limb, associated with greater reduction in perfusion immediately after femoral ligation, slower recovery of perfusion, and greater hind limb impairment. They also showed that a locus on chromosome 7 had a major protective effect on collateral extent and remodeling. Finally, highlighting the importance of the *LSq-1* locus in recovery after ischemia in general, Keum and coworkers recently demonstrated that a locus identical to *LSq-1* on mouse chromosome 7 determines the majority of the observed strain-dependent variation in the extent of ischemic tissue damage after distal middle cerebral artery occlusion in a mouse model of stroke (Keum and Marchuk, 2009).

BIOMARKERS OF PERIPHERAL ARTERIAL DISEASE

As mentioned above, despite the well-established high sensitivity and specificity of ankle-brachial index for diagnosing PAD, PAD remains under-recognized and under-diagnosed within the medical community, with tremendous implications for patients and the overall healthcare cost to the society. One of the factors contributing to the above is the need for busy physicians and healthcare providers to have specialized equipment and perform additional measures. One possible approach to improving this would be to develop a blood biomarker that changes the clinical index of suspicion for PAD and could identify patients who merit greater scrutiny. Furthermore, they can prognosticate the disease as well as improve our understanding of the mechanisms underlying the development of PAD.

Multiple studies have shown that levels of inflammatory biomarkers like C-reactive protein (CRP) are elevated in PAD patients (McDermott and Lloyd-Jones, 2009; Wildman et al., 2005). Increased levels of inflammatory biomarkers like CRP and soluble intercellular adhesion molecule (sICAM)-1 are also associated with increased risk of developing PAD (Pradhan et al., 2008; Ridker et al., 2001). Furthermore, higher baseline levels of some inflammatory biomarkers like interleukin-6, CRP, and sICAM-1 were found to be associated with greater declines in ABI over time (Tzoulaki et al., 2005). However, these biomarkers are not specific for PAD as they are elevated in other vascular diseases as well.

In an attempt to identify specific biomarkers of PAD, Wilson and colleagues used plasma proteomic profiling by mass spectroscopy. They found that 11 of 1619 protein peaks were stronger in 45 patients with PAD compared with 43 patients without PAD (Wilson et al., 2007). Six of these protein peaks represented beta-2 microglobulin (B2M, a 12-kDa protein). Higher levels of B2M in PAD compared with non-PAD patients were further validated in two confirmatory studies involving 40 and 237 patients respectively. Plasma B2M levels also correlated with ABI and functional capacity. A study of 540 patients showed that a biomarker panel score derived from B2M, cystatin C, high-sensitivity CRP, and glucose had an increased association with PAD, independent of the traditional risk factors (Fung et al., 2008). Our group used a candidate gene approach to determine if circulating levels of

proteins known to be involved in angiogenesis were differently expressed in control subjects versus those with PAD, and within the PAD group, those with intermittent claudication versus critical limb ischemia (Findley et al., 2008). Plasma levels of the soluble form of the endothelial receptor for angiopoietins, tyrosine kinase with immunoglobulin-like and epidermal growth factor-like domains 2 (sTie2), were elevated when compared to age- and sex-matched non-PAD patients. Perhaps more interestingly, these findings held even when compared to patients with intermittent claudication. From a clinical perspective, there is significant value in the possibility that one or more or a combination of these proteins could serve as biomarkers that are highly associated with a specific form of PAD. Further studies will be needed to determine whether B2M, sTie2 or other biomarkers are justified in being used as a screening tool.

FUTURE POTENTIAL USE OF GENOMIC METHODOLOGIES IN PERIPHERAL ARTERIAL DISEASE

As mentioned earlier, the use of genomic methodologies in studies of PAD is in its infancy. However, investigators in vascular biology are now beginning to take advantage of the novel approaches that these genomic methodologies offer. As described in detail above, genomic approaches are currently being used to identify novel genetic polymorphisms that may be important in understanding other mechanisms contributing to the pathogenesis of PAD in ways previously unappreciated. While gene association studies focus on the contribution of common variants to the common diseases, the relative role of rarer variants in determining genetic susceptibility will not be addressed by them. That will require whole-genome sequencing, which at present is prohibitively expensive for large-scale studies, but does have the ultimate potential to realize a completely "personalized" analysis of one's genome (Arnett et al., 2007).

There are numerous areas in studies of PAD where genomic methodologies may have significant impact. For instance, gene expression profiling may be applied to gain insight into candidate genes whose differential expression might be contributing to distinct clinical outcomes in patients with PAD. Novel insights into disease pathogenesis may be obtained by performing gene expression analysis on tissue samples from IC and CLI patients matched for anatomic distribution of disease, ABIs, and risk factors for PAD.

Another area where genomic strategies may be used in PAD studies is in investigation of how modulation of gene expression as a consequence of altered tissue environment may contribute to the pathogenesis of PAD.

A classic example of tissue environment alteration is the metabolic derangements seen in type 2 diabetes. As mentioned earlier, diabetes is a major risk factor for PAD and, along with smoking, accounts for about 80% of the risk associated with development of PAD. Interestingly, other than through its contribution to increased risk of atherosclerosis, how diabetes contributes to development of PAD is poorly understood. Previously our laboratory showed that there is alteration in the mRNA and protein expression of the potent angiogenic factor VEGF and its receptors in the skeletal muscle of mice with high fat diet-induced type 2 diabetes (Li et al., 2007b). This suggests that alterations in angiogenic factors may be important in the pathogenesis of PAD in diabetics. To identify additional angiogenic factors whose expression may be altered in diabetes, analysis of differential expression of all known angiogenic and vasculogenic factors can be sought in normal versus diabetic tissues via micro-array analysis. Once specific genes with altered expression are identified, the mechanism by which their altered expression contributes to PAD can then be studied.

The potential use of methodologies afforded by current genomics is not limited to understanding disease pathogenesis in PAD, but may be extended to providing solutions to some of the challenges faced with current therapeutic approaches.

Supervised exercise training has been shown to be an effective therapy for PAD patients (Gardner et al., 2000).

However, it can be quite expensive and patient compliance is typically a major challenge. Therefore, understanding the molecular mechanisms by which exercise training contributes to symptom improvement in PAD patients may lead to the development of less expensive and more practical therapeutic options. One possible approach is to investigate potential mechanisms involved by studying genes that may be differentially expressed in exercised versus non-exercised tissues. Of course, it is possible that exercise training leads to increased transcription of multiple genes, thus creating the challenge of identifying which genes are most relevant. Should this occur, priority can be given to those genes with known or presumed function in angiogenesis or tissue adaptation to ischemia. The specific role of these genes can then be tested in preclinical models of PAD by targeted gene silencing or over-expression. Ultimately, the role of the human orthologs of these genes can be assessed in humans with PAD.

In conclusion, although the use of genomic methodologies is still in its infancy in studies of vascular disease, especially in PAD, its use is growing exponentially and will likely continue to positively impact our understanding of the disease and its therapies.

REFERENCES

Allison, M.A., Criqui, M.H., McClelland, R.L., et al., 2006. The effect of novel cardiovascular risk factors on the ethnic-specific odds for peripheral arterial disease in the Multi-Ethnic Study of Atherosclerosis (MESA). J Am Coll Cardiol 48, 1190–1197.

Arnett, D.K., Baird, A.E., Barkley, R.A., et al., 2007. Relevance of genetics and genomics for prevention and treatment of cardiovascular disease: A scientific statement from the American Heart Association Council on Epidemiology and Prevention, the Stroke Council, and the Functional Genomics and Translational Biology Interdisciplinary Working Group. Circulation 115, 2878–2901.

Basar, Y., Salmayenli, N., Aksoy, M., Seckin, S., Aydin, M., Ozkok, E., 2007. ACE gene polymorphism in peripheral vascular disease. Horm Metab Res 39, 534–537.

Biasucci, L.M., Vitelli, A., Liuzzo, G., et al., 1996. Elevated levels of interleukin-6 in unstable angina. Circulation 94, 874–877.

Boyd, A.M., 1962. The natural course of arteriosclerosis of the lower extremities. Proc R Soc Med 55, 591–593.

Brownlee, M., 2005. The pathobiology of diabetic complications: A unifying mechanism. Diabetes 54, 1615–1625.

Carmelli, D., Fabsitz, R.R., Swan, G.E., Reed, T., Miller, B., Wolf, P.A., 2000. Contribution of genetic and environmental influences to ankle-brachial blood pressure index in the NHLBI Twin Study. Am J Epidemiol 151, 452–458.

Cesari, M., Penninx, B.W., Newman, A.B., et al., 2003. Inflammatory markers and onset of cardiovascular events: Results from the Health ABC study. Circulation 108, 2317–2322.

Chalothorn, D., Clayton, J.A., Zhang, H., Pomp, D., Faber, J.E., 2007. Collateral density, remodeling, and VEGF-A expression differ widely between mouse strains. Physiol Genomics 30, 179–191.

Chalothorn, D., Faber, J.E., 2010. Strain-dependent variation in collateral circulatory function in mouse hindlimb. Physiol Genomics 42, 469–479.

Christman, S.K., Ahijevych, K., Buckworth, J., 2001. Exercise training and smoking cessation as the cornerstones of managing claudication. J Cardiovasc Nurs 15, 64–77.

Cluett, C., McDermott, M.M., Guralnik, J., et al., 2009. The 9p21 myocardial infarction risk allele increases risk of peripheral artery disease in older people. Circ Cardiovasc Genet 2, 347–353.

Conley, P.B., Delaney, S.M., 2003. Scientific and therapeutic insights into the role of the platelet P2Y12 receptor in thrombosis. Curr Opin Hematol 10, 333–338.

Coucke, P.J., Willaert, A., Wessels, M.W., et al., 2006. Mutations in the facilitative glucose transporter GLUT10 alter angiogenesis and cause arterial tortuosity syndrome. Nat Genet 38, 452–457.

Coughlin, S.R., 2000. Thrombin signalling and protease-activated receptors. Nature 407, 258–264.

Cronenwett, J.L., Warner, K.G., Zelenock, G.B., et al., 1984. Intermittent claudication. Current results of nonoperative management. Arch Surg 119, 430–436.

Darius, H., Pittrow, D., Haberl, R., et al., 2003. Are elevated homocysteine plasma levels related to peripheral arterial disease? Results from a cross-sectional study of 6880 primary care patients. Eur J Clin Invest 33, 751–757.

Dawson, D.L., Cutler, B.S., Hiatt, W.R., et al., 2000. A comparison of cilostazol and pentoxifylline for treating intermittent claudication. Am J Med 109, 523–530.

Dokun, A.O., Keum, S., Hazarika, S., et al., 2008. A quantitative trait locus (LSq-1) on mouse chromosome 7 is linked to the absence of tissue loss after surgical hindlimb ischemia. Circulation 117, 1207–1215.

Dosenko, V.E., Zagoriy, V.Y., Haytovich, N.V., Gordok, O.A., Moibenko, A.A., 2006. Allelic polymorphism of endothelial NO-synthase gene and its functional manifestations. Acta Biochim Pol 53, 299–302.

Fatini, C., Sticchi, E., Sofi, F., et al., 2009. Multilocus analysis in candidate genes *ACE*, *AGT*, and *AGTR1* and predisposition to peripheral arterial disease: Role of ACE D/-240T haplotype. J Vasc Surg 50, 1399–1404.

Findley, C.M., Mitchell, R.G., Duscha, B.D., Annex, B.H., Kontos, C.D., 2008. Plasma levels of soluble Tie2 and vascular endothelial growth factor distinguish critical limb ischemia from intermittent claudication in patients with peripheral arterial disease. J Am Coll Cardiol 52, 387–393.

Flex, A., Gaetani, E., Angelini, F., et al., 2007. Pro-inflammatory genetic profiles in subjects with peripheral arterial occlusive disease and critical limb ischemia. J Intern Med 262, 124–130.

Flex, A., Gaetani, E., Pola, R., et al., 2002. The -174 G/C polymorphism of the interleukin-6 gene promoter is associated with peripheral artery occlusive disease. Eur J Vasc Endovasc Surg 24, 264–268.

Fontana, P., Dupont, A., Gandrille, S., et al., 2003a. Adenosine diphosphate-induced platelet aggregation is associated with P2Y12 gene sequence variations in healthy subjects. Circulation 108, 989–995.

Fontana, P., Gaussem, P., Aiach, M., et al., 2003b. P2Y12 H2 haplotype is associated with peripheral arterial disease: A case-control study. Circulation 108, 2971–2973.

Fowkes, F.G., 1990. Peripheral vascular disease: A public health perspective. J Public Health Med 12, 152–159.

Fowkes, F.G., Connor, J.M., Smith, F.B., Wood, J., Donnan, P.T., Lowe, G.D., 1992. Fibrinogen genotype and risk of peripheral atherosclerosis. Lancet 339, 693–696.

Fowkes, F.G., Lee, A.J., Hau, C.M., Cooke, A., Connor, J.M., Lowe, G.D., 2000. Methylene tetrahydrofolate reductase (MTHFR) and nitric oxide synthase (ecNOS) genes and risks of peripheral arterial disease and coronary heart disease: Edinburgh Artery Study. Atherosclerosis 150, 179–185.

Fukino, K., Sata, M., Seko, Y., Hirata, Y., Nagai, R., 2003. Genetic background influences therapeutic effectiveness of VEGF. Biochem Biophys Res Commun 310, 143–147.

Fung, E.T., Wilson, A.M., Zhang, F., et al., 2008. A biomarker panel for peripheral arterial disease. Vasc Med 13, 217–224.

Gachet, C., 2001. ADP receptors of platelets and their inhibition. Thromb Haemost 86, 222–232.

Gardner, A.W., Katzel, L.I., Sorkin, J.D., et al., 2000. Improved functional outcomes following exercise rehabilitation in patients with intermittent claudication. J Gerontol A Biol Sci Med Sci 55, M570–M577.

Gerhard, M., Baum, P., Raby, K.E., 1995. Peripheral arterial-vascular disease in women: Prevalence, prognosis, and treatment. Cardiology 86, 349–355.

Graham, I.M., Daly, L.E., Refsum, H.M., et al., 1997. Plasma homocysteine as a risk factor for vascular disease. The European Concerted Action Project. JAMA 277, 1775–1781.

Gudmundsson, G., Matthiasson, S.E., Arason, H., et al., 2002. Localization of a gene for peripheral arterial occlusive disease to chromosome 1p31. Am J Hum Genet 70, 586–592.

Helgadottir, A., Thorleifsson, G., Magnusson, K.P., et al., 2008. The same sequence variant on 9p21 associates with myocardial infarction, abdominal aortic aneurysm and intracranial aneurysm. Nat Genet 40, 217–224.

Hiatt, W.R., 2001. Medical treatment of peripheral arterial disease and claudication. N Engl J Med 344, 1608–1621.

Hiatt, W.R., Baumgartner, I., Nikol, S., et al., 2010. NV1FGF gene therapy on amputation-free survival in critical limb ischemia - Phase 3 randomized double-blind placebo-controlled trial. Circulation, 122.

Hiatt, W.R., Hoag, S., Hamman, R.F., 1995. Effect of diagnostic criteria on the prevalence of peripheral arterial disease. The San Luis Valley Diabetes Study. Circulation 91, 1472–1479.

Hingorani, A.D., 2001. Polymorphisms in endothelial nitric oxide synthase and atherogenesis: John French Lecture 2000. Atherosclerosis 154, 521–527.

Hirsch, A.T., Criqui, M.H., Treat-Jacobson, D., et al., 2001. Peripheral arterial disease detection, awareness, and treatment in primary care. JAMA 286, 1317–1324.

Hirsch, A.T., Haskal, Z.J., Hertzer, N.R., et al., 2006. ACC/AHA 2005 practice guidelines for the management of patients with peripheral arterial disease (lower extremity, renal, mesenteric, and abdominal aortic): A collaborative report from the American Association for Vascular Surgery/Society for Vascular Surgery, Society for Cardiovascular Angiography and Interventions, Society for Vascular Medicine and Biology, Society of Interventional Radiology, and the ACC/AHA Task Force on Practice Guidelines (Writing Committee to Develop Guidelines for the Management of Patients With Peripheral Arterial Disease): Endorsed by the American Association of Cardiovascular and Pulmonary Rehabilitation; National Heart, Lung, and Blood Institute; Society for Vascular Nursing; TransAtlantic Inter-Society Consensus; and Vascular Disease Foundation. Circulation 113, e463–e654.

Imparato, A.M., Kim, G.E., Davidson, T., Crowley, J.G., 1975. Intermittent claudication: Its natural course. Surgery 78, 795–799.

Jiang, Y.-D., Chang, Y.-C., Chiu, Y.-F., et al., 2010. SLC2A10 genetic polymorphism predicts development of peripheral arterial disease in patients with type 2 diabetes. BMC Med Genet 11, 126.

Jones, W.S., Annex, B.H., 2007. Growth factors for therapeutic angiogenesis in peripheral arterial disease. Curr Opin Cardiol 22, 458–463.

Kannel, W.B., McGee, D.L., 1985. Update on some epidemiologic features of intermittent claudication: The Framingham Study. J Am Geriatr Soc 33, 13–18.

Kawashima, S., Yokoyama, M., 2004. Dysfunction of endothelial nitric oxide synthase and atherosclerosis. Arterioscler Thromb Vasc Biol 24, 998–1005.

Keum, S., Marchuk, D.A., 2009. A locus mapping to mouse chromosome 7 determines infarct volume in a mouse model of ischemic stroke. Circ Cardiovasc Genet 2, 591–598.

Knowles, J.W., Assimes, T.L., Li, J., Quertermous, T., Cooke, J.P., 2007. Genetic susceptibility to peripheral arterial disease: A dark corner in vascular biology. Arterioscler Thromb Vasc Biol 27, 2068–2078.

Kullo, I.J., Bailey, K.R., Kardia, S.L., Mosley Jr., T.H., Boerwinkle, E., Turner, S.T., 2003. Ethnic differences in peripheral arterial disease in the NHLBI Genetic Epidemiology Network of Arteriopathy (GENOA) study. Vasc Med 8, 237–242.

Kullo, I.J., Turner, S.T., Kardia, S.L., Mosley Jr., T.H., Boerwinkle, E., De Andrade, M., 2006. A genome-wide linkage scan for ankle-brachial index in African American and non-Hispanic white subjects participating in the GENOA study. Atherosclerosis 187, 433–438.

Leng, G.C., Fowler, B., Ernst, E., 2000. Exercise for intermittent claudication. Cochrane Database Syst Rev CD000990.

Li, R., Nicklas, B., Pahor, M., et al., 2007a. Polymorphisms of angiotensinogen and angiotensin-converting enzyme associated with lower extremity arterial disease in the health, aging and body composition study. J Hum Hypertens 21, 673–682.

Li, Y., Hazarika, S., Xie, D., Pippen, A.M., Kontos, C.D., Annex, B.H., 2007b. In mice with type 2 diabetes, a vascular endothelial growth factor (VEGF)-activating transcription factor modulates VEGF signaling and induces therapeutic angiogenesis after hindlimb ischemia. Diabetes 56, 656–665.

McDermott, M.M., Greenland, P., Liu, K., et al., 2002. The ankle brachial index is associated with leg function and physical activity: The Walking and Leg Circulation Study. Ann Intern Med 136, 873–883.

McDermott, M.M., Lloyd-Jones, D.M., 2009. The role of biomarkers and genetics in peripheral arterial disease. J Am Coll Cardiol 54, 1228–1237.

Milei, J., Parodi, J.C., Fernandez Alonso, G., et al., 1996. Carotid atherosclerosis. Immunocytochemical analysis of the vascular and cellular composition in endarterectomies. Cardiologia 41, 535–542.

Moncada, S., Higgs, A., 1993. The L-arginine-nitric oxide pathway. N Engl J Med 329, 2002–2012.

Monsalve, M.V., Robinson, D., Woolcock, N.E., Powell, J.T., Greenhalgh, R.M., Humphries, S.E., 1991. Within-individual variation in serum cholesterol levels: Association with DNA polymorphisms at the apolipoprotein B and AI-CIII-AIV loci in patients with peripheral arterial disease. Clin Genet 39, 260–273.

Murabito, J.M., Guo, C.Y., Fox, C.S., D'Agostino, R.B., 2006. Heritability of the ankle-brachial index: The Framingham Offspring Study. Am J Epidemiol 164, 963–968.

Nakayama, M., Yasue, H., Yoshimura, M., et al., 1999. T-786→C mutation in the 5'-flanking region of the endothelial nitric oxide synthase gene is associated with coronary spasm. Circulation 99, 2864–2870.

Newman, A.B., Shemanski, L., Manolio, T.A., et al., 1999. Ankle-arm index as a predictor of cardiovascular disease and mortality in the cardiovascular health study. Arterioscler Thromb Vasc Biol 19, 538–545.

Nikol, S., Baumgartner, I., Van Belle, E., et al., 2008. Therapeutic angiogenesis with intramuscular NV1FGF improves amputation-free survival in patients with critical limb ischemia. Mol Ther 16, 972–978.

Norgren, L., Hiatt, W.R., Dormandy, J.A., et al., 2007. Inter-society consensus for the management of peripheral arterial disease (TASC II). Eur J Vasc Endovasc Surg 33 (Suppl. 1), S1–S75.

Ouriel, K., 2001. Peripheral arterial disease. Lancet 358, 1257–1264.

Paigen, B., Mitchell, D., Reue, K., Morrow, A., Lusis, A.J., Leboeuf, R.C., 1987. Ath-1, a gene determining atherosclerosis susceptibility and high density lipoprotein levels in mice. Proc Natl Acad Sci USA 84, 3763–3767.

Peters, L.L., Robledo, R.F., Bult, C.J., Churchill, G.A., Paigen, B.J., Svenson, K.L., 2007. The mouse as a model for human biology: A resource guide for complex trait analysis. Nat Rev Genet 8, 58–69.

Phelan, S.A., Beier, D.R., Higgins, D.C., Paigen, B., 2002. Confirmation and high resolution mapping of an atherosclerosis susceptibility gene in mice on chromosome 1. Mamm Genome 13, 548–553.

Powell, R.J., Simons, M., Mendelsohn, F.O., et al., 2008. Results of a double-blind, placebo-controlled study to assess the safety of intramuscular injection of hepatocyte growth factor plasmid to improve limb perfusion in patients with critical limb ischemia. Circulation 118, 58–65.

Pradhan, A.D., Shrivastava, S., Cook, N.R., Rifai, N., Creager, M.A., Ridker, P.M., 2008. Symptomatic peripheral arterial disease in women: Nontraditional biomarkers of elevated risk. Circulation 117, 823–831.

Reny, J.L., Alhenc-Gelas, M., Fontana, P., et al., 2004. The factor II G20210A gene polymorphism, but not factor V Arg506Gln, is associated with peripheral arterial disease: Results of a case-control study. J Thromb Haemost 2, 1334–1340.

Resnick, H.E., Rodriguez, B., Havlik, R., et al., 2000. Apo E genotype, diabetes, and peripheral arterial disease in older men: The Honolulu Asia-aging study. Genet Epidemiol 19, 52–63.

Ridker, P.M., Stampfer, M.J., Rifai, N., 2001. Novel risk factors for systemic atherosclerosis: A comparison of C-reactive protein, fibrinogen, homocysteine, lipoprotein(a), and standard cholesterol screening as predictors of peripheral arterial disease. JAMA 285, 2481–2485.

Rigat, B., Hubert, C., Alhenc-Gelas, F., Cambien, F., Corvol, P., Soubrier, F., 1990. An insertion/deletion polymorphism in the angiotensin I-converting enzyme gene accounting for half the variance of serum enzyme levels. J Clin Invest 86, 1343–1346.

Sabino, A., Fernandes, A.P., Lima, L.M., et al., 2009. Polymorphism in the methylene tetrahydrofolate reductase (C677T) gene and homocysteine levels: A comparison in Brazilian patients with coronary arterial disease, ischemic stroke and peripheral arterial obstructive disease. J Thromb Thrombolysis 27, 82–87.

Scholz, D., Ziegelhoeffer, T., Helisch, A., et al., 2002. Contribution of arteriogenesis and angiogenesis to postocclusive hindlimb perfusion in mice. J Mol Cell Cardiol 34, 775–787.

Sheehan, S., Tsaih, S.W., King, B.L., et al., 2007. Genetic analysis of albuminuria in a cross between C57BL/6J and DBA/2J mice. Am J Physiol Renal Physiol 293, F1649–F1656.

Sofi, F., Lari, B., Rogolino, A., et al., 2005. Thrombophilic risk factors for symptomatic peripheral arterial disease. J Vasc Surg 41, 255–260.

Sticchi, E., Sofi, F., Romagnuolo, I., et al., 2010. eNOS and ACE genes influence peripheral arterial disease predisposition in smokers. J Vasc Surg 52, 97–102 e1.

Thorgeirsson, T.E., Geller, F., Sulem, P., et al., 2008. A variant associated with nicotine dependence, lung cancer and peripheral arterial disease. Nature 452, 638–642.

Todesco, L., Angst, C., Litynski, P., Loehrer, F., Fowler, B., Haefeli, W.E., 1999. Methylene tetrahydrofolate reductase polymorphism, plasma homocysteine and age. Eur J Clin Invest 29, 1003–1009.

Tsukada, T., Yokoyama, K., Arai, T., et al., 1998. Evidence of association of the ecNOS gene polymorphism with plasma NO metabolite levels in humans. Biochem Biophys Res Commun 245, 190–193.

Tzoulaki, I., Murray, G.D., Lee, A.J., Rumley, A., Lowe, G.D., Fowkes, F.G., 2005. C-reactive protein, interleukin-6, and soluble adhesion molecules as predictors of progressive peripheral atherosclerosis in the general population: Edinburgh artery study. Circulation 112, 976–983.

Wade, C.M., Kulbokas 3rd, E.J., Kirby, A.W., et al., 2002. The mosaic structure of variation in the laboratory mouse genome. Nature 420, 574–578.

Wang, X., Korstanje, R., Higgins, D., Paigen, B., 2004. Haplotype analysis in multiple crosses to identify a QTL gene. Genome Res 14, 1767–1772.

Wang, X., Ria, M., Kelmenson, P.M., et al., 2005. Positional identification of TNFSF4, encoding OX40 ligand, as a gene that influences atherosclerosis susceptibility. Nat Genet 37, 365–372.

Wildman, R.P., Muntner, P., Chen, J., Sutton-Tyrrell, K., He, J., 2005. Relation of Inflammation to Peripheral Arterial Disease in the National Health and Nutrition Examination Survey, 1999–2002. Am J Cardiol 96, 1579–1583.

Wilson, A.M., Kimura, E., Harada, R.K., et al., 2007. Beta2-microglobulin as a biomarker in peripheral arterial disease: Proteomic profiling and clinical studies. Circulation 116, 1396–1403.

Yee, K.O., Ikari, Y., Schwartz, S.M., 2001. An update of the Grutzbalg hypothesis: The role of thrombosis and coagulation in atherosclerotic progression. Thromb Haemost 85, 207–217.

Zintzaras, E., Zdoukopoulos, N., 2009. A field synopsis and meta-analysis of genetic association studies in peripheral arterial disease: The CUMAGAS-PAD database. Am J Epidemiol 170, 1–11.

Congenital Heart Disease

Lisa J. Martin and D. Woodrow Benson

INTRODUCTION

Pediatric heart disease, or cardiovascular disease in the young, comprises varied phenotypes, including cardiomyopathies (see Chapter 50), cardiac arrhythmias (see Chapter 51), and congenital cardiac malformations or congenital heart defects (CHD). The genetic basis of CHD has been the subject of intense investigation and increasing understanding. CHD has long been recognized as a key feature of some cytogenetic syndromes, such as Down syndrome (Trisomy 21) and diGeorge syndrome (a.k.a. velocardiofacial syndrome, del22q11), and single gene defects were thought to be relatively uncommon. However, in the past two decades, linkage analysis and positional cloning in family-based studies have identified mutations in single genes as a cause of CHD. While these approaches have provided an explanation for a relatively small portion of CHD, the findings have supplied new tools for the cardiac developmental biologist. At the same time, this has raised additional questions regarding the genetic origins of CHD. The purpose of this chapter is to review accomplishments to date and appraise future directions in the post-genomic era.

WHAT IS CONGENITAL HEART DISEASE?

CHD refers to structural or functional abnormalities that are present at birth, even if discovered later in life (Hoffman, 1995a, b). CHD constitute a major portion of clinically significant birth defects with a definition-dependent estimated incidence of 4–50 per 1000 live births. A figure of 4–10 live-born infants per 1000 with a cardiac malformation (of which 40% are diagnosed in the first year of life) is often quoted. However, bicuspid aortic valve (BAV), the most common congenital cardiac malformation, is usually excluded from this estimate. BAV is associated with considerable morbidity and mortality in affected individuals, and by itself occurs in 10–20 per 1000 of the population (Siu and Silversides, 2010). When isolated aneurysm of the atrial septum and persistent left superior vena cava, which each occur in 5–10 per 1000 (Benson et al., 1998), are taken into account, the incidence of cardiac malformations approaches 50 per 1000 live births.

ORIGINS OF CONGENITAL HEART DISEASE

There has been considerable interest in understanding the origins of CHD. Early identification of environmental teratogens, such as rubella, thalidomide, and high altitude, focused investigations of CHD cause on environmental factors. However, the Baltimore–Washington Infant Study identified positive family history and maternal diabetes as common CHD risk factors (Ferencz et al., 1993). In addition, increased risk of CHD in family members of affected individuals has long been known (McKeown et al., 1953). Additional evidence supporting genetic origins came through the association of CHD and other birth defects with chromosomal abnormalities (Pierpont et al.,

Genomic and Personalized Medicine, 2nd edition
by Ginsburg & Willard. DOI: http://dx.doi.org/10.1016/B978-0-12-382227-7.00054-9

TABLE 54.1 Congenital heart disease genes identified by linkage

Gene	Inheritance	Phenotype	Citation
ELN	AD	Supravalvular aortic stenosis	Olson et al., 1993
JAG1	AD	Alagille syndrome	Hol et al., 1995
TBX5	AD	Holt-Oram syndrome	Basson et al., 1997; Li et al., 1997
ZIC3	X	X-linked hetertaxy syndrome	Gebbia et al., 1997
NKX2.5	AD	Familial CHD	Schott et al., 1998
TFAP2B	AD	Char syndrome	Satoda et al., 2000
EVC	AR	Ellis van Creveld syndrome	Ruiz-Perez et al., 2000
PTPN11	AD	Noonan syndrome	Tartaglia et al., 2001
EVC2 (LBN)	AR	Ellis van Creveld syndrome	Ruiz-Perez et al., 2003
GATA4	AD	Familial CHD	Garg et al., 2003
MYH6	AD	Familial atrial septal defect	Ching et al., 2005
NOTCH1	AD	LVOT disease	Garg et al., 2005

LVOT, left ventricular outflow tract

2007). On the basis of studies of recurrence and transmission risks, a hypothesis of multifactorial etiology was proposed (Nora, 1968). In this type of inheritance, an individual's genetic predisposition interacts with other genes and/or environmental factors to cause heart disease.

However, in the past two decades, Mendelian inheritance models have been used to exploit molecular genetic and cytogenetic observations in family-based studies (Table 54.1). From these investigations, it has been learned that the new genetic taxonomy of CHD does not precisely align with the clinical taxonomy used by anatomists, cardiologists, and surgeons. For example, there is not a tetralogy of Fallot gene or an atrial septal defect (ASD) gene. Further, family members harboring the same mutation may exhibit clinically distinct phenotypes that cross the clinical taxonomy boundaries. Indeed, genetic heterogeneity, reduced penetrance and variable expressivity have been a recurring theme in genetic studies of CHD.

CONGENITAL HEART DISEASE AND SINGLE GENE DISORDERS

In this section, we will review three examples of single gene disorders discovered through the use of linkage analysis and positional cloning (Table 54.1). Despite the universal presence of variable expressivity and/or reduced penetrance, this approach was successful in identifying disease-causing genetic variants.

Supravalvular aortic stenosis (SVAS; mendelian inheritance in man (MIM) #185500) describes an obstructive arteriopathy of varying severity most prominent at the aortic sinotubular junction. Phenotypic variation is manifested by stenoses of pulmonary arteries and both semilunar valves as well as other CHD, including aortic coarctation, patent ductus arteriosus, atrial and ventricular septal defects, and tetralogy of Fallot (Stamm et al., 2001). In 1993, studies linking familial SVAS to chromosome 7q11.2 (Ewart et al., 1993; Olson et al., 1993), and a translocation t(6;7)(p21.1;q11.23) that cosegregated with SVAS (Curran et al., 1993), led to the identification of mutations in the elastin (ELN) gene as the cause of SVAS. SVAS is a frequent feature of Williams–Beuren syndrome (WBS; MIM #194050), a contiguous gene deletion syndrome that includes hemizygous deletion of the ELN gene.

Linkage to a 5-cM region on chromosome 12q24.1 led to identification of PTPN11 as a gene causing Noonan syndrome (MIM #163950), which is characterized by cardiac and non-cardiac abnormalities (Tartaglia et al., 2001). The most common cardiac defects are pulmonary stenosis with a dysplastic valve, secundum atrial septal defects, atrioventricular septal defects, and/or hypertrophic cardiomyopathy. Non-cardiac defects include hypertelorism and short stature. Noonan syndrome is primarily inherited as an autosomal dominant condition, but inheritance may be sporadic. PTPN11 mutations account for approximately 50% of Noonan syndrome cases (Tartaglia et al., 2002). Studies of the Noonan syndrome-related syndromes – Noonan, Costello and cardiofaciocutaneous (CFC) syndromes – led to a new concept that these clinically related disorders are caused by dysregulation of members of the rat sarcoma (a family of genes) (RAS)/mitogen-activated protein kinase (MAPK) pathway (Tidyman and Rauen, 2009) (Figure 54.1).

Linkage analysis was also successful in patients with non-syndromic CHD and led to identification of mutations in the homeobox transcription factor NKX2.5 (MIM #600584) (Schott et al., 1998). Heterozygous NKX2.5 mutations are fully penetrant but exhibit variable expressivity (Benson, 2009; Benson et al., 1999). Atrial septal defects (~80%), and progressive atrioventricular block (>95%) are common. However, ventricular septal defects (~30%) including tetralogy of Fallot and double outlet right ventricles, and tricuspid valve abnormalities (~15%) including Ebstein anomaly, are also observed

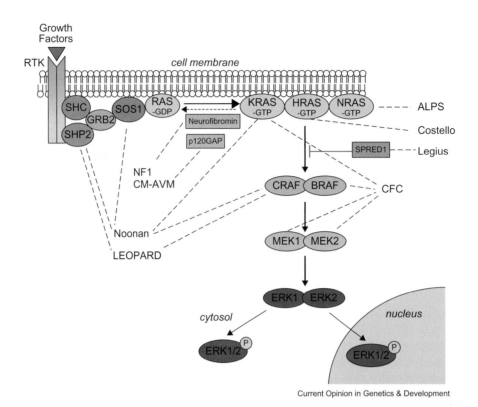

Figure 54.1 The RAS/mitogen-activated protein kinase (MAPK) signaling pathway and associated developmental syndromes (indicated by dashed lines). The MAPK signaling pathway of protein kinases is critically involved in cell proliferation, differentiation, motility, apoptosis, and senescence. The RAS/MAPK pathway proteins with germline mutations in their respective genes are associated with Noonan, LEOPARD, hereditary gingival fibromatosis 1, neurofibromatosis 1, capillary malformation–arteriovenous malformation, Costello, autoimmune lymphoproliferative (ALPS), cardiofaciocutaneous and Legius syndromes. *Reproduced with permission from Tidyman and Rauen, 2009.*

[reviewed in Benson and Martin (2010)]. *NKX2.5* mutations account for 3% of all CHD and 4% of tetralogy of Fallot (McElhinney et al., 2003).

In each of the three examples, linkage was used to identify a region that contained promising candidate genes. Successful linkage required broad definitions of phenotype to determine affected status. For example, the designation of the phenotype as Noonan syndrome rather than a specific CHD determined affected status, and in the case of *NKX2.5*, affected status was determined by the presence of any CHD or atrioventricular conduction block. At the chromosomal loci, mutation analyses of candidate genes identified rare genetic variants that segregated with disease in the kindred and were predicted to alter protein function, supporting causation. Subsequently, for all examples listed in Table 54.1, including the three examples above, genetic engineering in animal models produced other lines of biologic evidence supporting causation (Katsanis, 2009; Page et al., 2003). In retrospect, even though the associated traits exhibit characteristics of complex inheritance (e.g., variable expressivity, reduced penetrance, genetic heterogeneity), the assumption of simple inheritance worked adequately to determine linkage, possibly due to the

large size of the genetic effect. Taken together, these findings suggest that what appeared to be simple inheritance is actually complex inheritance, as genotype does not predict phenotype. This raises questions regarding the research approach in the future, with implications for utilization and interpretation of genetic testing.

ORIGINS OF PLEIOTROPIC HEART DEFECTS IN SINGLE GENE DISORDERS

To study genetic origins of pleiotropic heart defects with incomplete penetrance, the Jay laboratory analyzed the cardiac phenotype in mutant mice (*Nkx2.5$^{+/-}$*) from a C57Bl/6 background, and compared them with progeny of outcrosses (F1) to the strains FVB/N and A/J and to F1XF1 intercrosses or backcrosses to the parent strain (F2) (Winston et al., 2010). Detailed phenotyping identified cardiac phenotypes of atrial, ventricular, and atrioventricular septal defect and double outlet right ventricle. The analysis of more than 3000 *Nkx2.5$^{+/-}$* hearts from five F2 crosses demonstrated the profound influence of genetic modifiers on disease presentation. From the incidences and coincidences, they found that anatomically distinct cardiac malformations have shared and unique modifiers. All three strains carry susceptibility alleles at different loci for atrial and ventricular septal defects. Relative to C57Bl/6 and FVB/N, the A/J strain carries polymorphisms that confer greater susceptibility to atrial septal defect and atrioventricular septal defect, and C57Bl/6 to muscular ventricular septal defect. Segregation analyses revealed that two or more loci influence membranous ventricular septal defect susceptibility, whereas three or more loci and at least one epistatic interaction affect muscular ventricular septal defect and atrial septal defect. The authors conclude that alleles of modifier genes can either buffer perturbations on cardiac development or direct the manifestation of a defect, but in a genetically heterogeneous population the predominant effect of modifier genes is health.

This study's findings emphasize the importance of detailed cardiac phenotyping and provide insight into the challenges of genetic discovery in CHD. A model system involving perturbation of a single gene in a relatively simple (compared with human) genetic background would be expected to demonstrate simple Mendelian inheritance, but this was not the case. Segregation analysis revealed the presence of phenotype-specific genetic and epistatic modifiers. Although not powered to discover the modifiers, identifying their presence provides an explanation for the pleiotropic phenotypes observed in the mutant mice. Taken together, these findings further implicate CHD as a complex trait.

WHAT ARE COMPLEX TRAITS?

Complex traits exhibit evidence of an underlying genetic basis, for example they cluster in families, but they do not segregate in families in a Mendelian fashion. The reason they fail to exhibit Mendelian segregation is that complex traits may be affected by multiple genes as well as environmental factors. To further complicate the inheritance, the genetic and environmental factors may act in a non-additive fashion (e.g., gene–gene and gene–environment interactions). The presence of multiple genes, each of which can independently cause disease, is known as genetic heterogeneity. In the successful gene discovery examples shown in Table 54.1, the identified variants typically explain only a portion of the disease. This suggests that there are other genetic factors, such as genetic heterogeneity, responsible for the remaining disease. Further, genes may interact with other genes (gene–gene interaction), and these interactions may result in a single genetic variant exhibiting broad phenotypic variability in the presence of disease (penetrance) and in the manifestation of disease (variable expressivity).

Mouse models help to demonstrate the scope of this problem (Winston et al., 2010). Indeed, one of the problems in the CHD literature has been publication of pedigrees of large multiplex families; pedigree analysis suggests autosomal dominant inheritance with reduced penetrance. However, these studies typically do not include or discuss families with single affected individuals, a frequent occurrence in the CHD population. The presence of both single and multiplex affected families suggests there may be multiple modes (complex) of inheritance for CHD. This problem is well illustrated by bicuspid aortic valve (BAV), the most common form of CHD, where pedigree analysis of selected, large multiplex families was interpreted as indicating autosomal dominant inheritance with reduced penetrance (Cripe et al., 2004; Martin et al., 2007a). However, these studies did not include families with single affected individuals, which may account for as many as 50% of all BAV families.

ISSUES TO CONSIDER IN FUTURE STUDIES

While there are clear examples of genes causing CHD (see Table 54.1), discovery of genetic causes of CHD is not keeping pace with the amount of genetic information now available. Part of the challenge may be that the "low-hanging fruit" has been discovered. The first genetic discoveries had sufficient effect sizes to overcome the issues of variable expressivity, reduced penetrance, and genetic heterogeneity. However, for the next set of discoveries, it will be important to design studies that do not assume that these complexities will not override the genetic evidence. In the following sections we will review issues of importance to the design and execution of successful future genetic studies.

Study Design

There are two main questions in gene discovery: (1) are there genetic differences between affected and unaffected individuals? and (2) do genetic variants segregate with disease in families? The first question can be answered using family- or population-based data, but the second question can only be tested using family-based information.

Family sampling includes trios (proband and parents), nuclear families (first-degree relatives, i.e., parents and siblings), and extended families (multiple generations). Familial sampling is dependent on the prevalence (how common a condition is). For rare traits, a large number of family members must be sampled to find affected individuals. In these cases, either trios or sequential sampling is effective (Cripe et al., 2004). In sequential sampling, phenotypic information is collected on all first-degree relatives. If another affected individual is identified within the family, then all of that person's first-degree relatives are phenotyped. For rare traits, sequential sampling is superior to traditional extended-family sampling, because it focuses on informative arms of the family.

Population-based sampling includes the cohort and case-control designs. Because CHD is rare, case-control sampling is used. In this approach, the researcher recruits individuals with disease (cases) and unrelated individuals without disease (controls). The challenge is ensuring that the controls come from the same population as the cases and that controls are definitively unaffected. Differences between cases and controls can result in spurious findings.

Statistical Approaches to Gene Discovery

There are two main statistical approaches to gene discovery: *linkage* and *association* (See Figure 54.2). Linkage analysis searches for genetic markers, which segregate with disease and can be performed only using family-based samples. Linkage assumes that the genetic marker and the disease variant are in close proximity and are transmitted intact across generations (Bray and Ryan, 2000). Evidence of linkage is determined either by consistent transmission of both the disease and the marker from parent to offspring (parametric), or by an increased proportion of alleles shared between affected relative pairs (non-parametric). If the underlying model is unknown or complex, non-parametric methodology is appropriate. Linkage is suited to identify regions harboring rare variants with large effect. However, these regions may be large and may encompass hundreds of genes. Thus, linkage analysis is often coupled with association to identify individual variants at a locus. There have been several notable successes for linkage analysis, as described in Table 54.1.

Association analysis tests the hypothesis that a genetic variant occurs more or less frequently in individuals with a particular phenotype than in those without it, and can use population- or family-based samples. However, there are multiple reasons for association, including (1) causation, (2) correlation with causal variant (linkage disequilibrium), (3) confounding, and (4) random chance. The ultimate goal of gene discovery is to identify the causal variant (#1), but linkage disequilibrium (#2) can identify a narrow region that harbors the causal variant (Morton, 2008). Confounding (#3) occurs when the cases and controls are not adequately matched (Knowler et al., 1988). Random chance (#4) often occurs because researchers do not test a single variant but rather test many, which increases the likelihood of incorrectly identifying a statistically significant association. As such, gene discovery often employs more stringent p-values (Martin et al., 2007b). Unfortunately, it is impossible to know which of the multiple reasons are responsible for the association. The selection of the variants for association can be based on known biology (candidate gene) or using a genome-wide approach [genome-wide association studies (GWAS)]. While association is well positioned to find common variants for common traits with small effect, it is not well powered for rare variants that cause rare (low-prevalence) diseases (Visscher et al., 2008). As most CHD occurs at frequencies of 1% or less, common variants would have to have extremely low penetrance to be the causes of CHD. Thus the utility of GWAS for gene discovery in CHD is questionable.

Indeed, a GWAS has not been published for CHD, perhaps in part because of the rarity of CHD. Based on GWAS from other disciplines, rare traits will require multi-center consortia (Harley et al., 2008) in order to have sufficient power. Even then, the characteristics of previously identified variants and genes for CHD (locus heterogeneity, genetic heterogeneity, rare variants) raise the question of whether GWAS will be fruitful for CHD.

Identification of Genetic Variants

With completion of the Human Genome Project, it is clear that genetic variation occurs frequently in individuals and occurs on many different scales, ranging from gross karyotype alterations to single-nucleotide changes (Stankiewicz and Lupski, 2010; Udwadia et al., 1996). Currently, the most common markers for population- and family-based studies are point changes in nucleotides (single nucleotide polymorphisms, SNPs) (Evans and Cardon, 2004). There are more than three million SNPs in the genome (Pushkarev et al., 2009). Current SNP panels utilize only a fraction of those available (approximately one million). These panels contain variants with a broad range of minor allele frequencies (MAF), ranging from variants occurring in less than 1% of the population to those that occur in 50% of the population. Nonetheless, these high-throughput SNP chips include a disproportionate number of common variants (MAF >10%). Further, due to technical issues with genotype calling, rare variants (MAF <5%) are often excluded (Tabangin et al., 2009). As such, these SNP chips are positioned to identify the common, but not the rare variants.

Structural variation is being recognized as an important part of the variability in the human genome (Zhang et al., 2009). Indeed, chromosomal abnormalities are often associated with multiple congenital malformations including CHD (Lin et al., 2008), and account for approximately 10% of the known etiology of CHD (Botto and Correa, 2003; Ferencz et al., 1989). The most severe of these malformations influencing CHD are aneuploids, in which an entire chromosome is duplicated or deleted (Dulac et al., 2008; Macmahon et al., 1953). However, individuals with partial deletions of chromosomes may also be affected by CHD. The most commonly diagnosed micro-deletion, 22q11 Deletion syndrome (a.k.a. diGeorge syndrome and others) has a prevalence of 1 in 4000 live births (Lammer et al., 2009; Lin et al., 2008). While the variation affecting large chromosomal regions can result in large phenotypic perturbations, these tend to be rare in the population. Small/regional copy number variation (CNV) can be common and can have minimal-to-severe effects on a variety of phenotypes (Redon et al., 2006; Wain et al., 2009). The challenge for studies involving CNV variants is detection (Fanciulli et al., 2009), as copy number algorithms have poor agreement (Shtir et al., 2009). Thus, while CNV analysis offers promise, the technical and statistical assessment of CNV is evolving (Oldridge et al., 2010; Cooper et al., 2008).

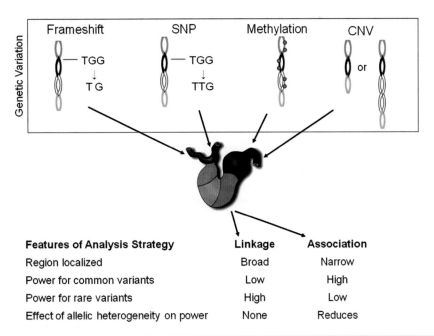

Features of Analysis Strategy	Linkage	Association
Region localized	Broad	Narrow
Power for common variants	Low	High
Power for rare variants	High	Low
Effect of allelic heterogeneity on power	None	Reduces

Figure 54.2 Strategy for continued congential heart disease genetic discovery. Various markers of genetic variation (frameshift, SNP, methylation, CNV) can be used to combine evidence from linkage and association analyses. The genetic and molecular elucidation of CHD will require complementary statistical approaches to realize discovery.

Candidate Gene Mutation Analysis/Sequencing

Neither linkage nor GWAS are sufficient to identify causal variants. As described above, linkage identifies regions and GWAS focuses on common variation. The common variants are unlikely to be the causal variants. Thus, these techniques will require follow-up analysis. As the cost of sequencing has dropped dramatically in recent years (Tucker et al., 2009), the next stage of query is often sequencing. For the sequencing approach, there are three main strategies for discovery: case-control, case-only, and family-based. In the case/control analysis, researchers will search for variants in cases that appear to be functionally significant (e.g., predicted to alter protein structure) and test whether these variants are present in the control group. If the variant is identified in at least one case but no controls and the variant has structural impact on protein, it is thought to be a causal variant (Dasgupta et al., 2001; Pizzuti et al., 2003; Zhao et al., 2010). There are two problems with such an approach. First, it is limited to only those variants that influence protein structure. There is a new field of genomics focusing on the regulatory genome (Alonso et al., 2009). This knowledge will be required to fully understand the genome and genetic variation. Second, in the case-control analysis, the lack of variants in a control population could be a function of the rarity of the variant rather than causality. Further, the presence of variants in the controls may be a reflection of inadequate phenotyping, a frequent problem for convenience controls. In the case-only analysis, the focus will again be those variants with putative functionality. Often the case-only analysis will look for evidence that multiple cases have functional variants in the same gene. As with the case-control analysis, this approach is limited by our knowledge of what function is and how it relates to the clinical phenotype of interest. Additionally, this approach assumes that a single gene will be responsible for most, if not all, of a specific disease. In the family-based analysis, the assumption is that the mutation will segregate with disease. While this could be considered the gold standard for sequence-level data, the challenge is with reduced penetrance, i.e., segregation may not be complete. Ultimately, demonstration of segregation with disease in family-based samples and much lower frequencies of variants in controls will provide the greatest protection against false positives.

This problem expands beyond candidate gene regions. With the promise of the $1000 genome project looming (Schadt et al., 2010), some may question whether whole genome sequencing is the next logical step in gene discovery. Whole genome sequencing has been used in case-only and family-based designs to identify variants for rare diseases (Arcelli et al., 2010; Gamazon et al., 2010, Lupski et al., 2010; Sobreira et al., 2010). While whole genome sequencing will delineate the genome of an individual, the challenge is how to interpret the three billion base pairs of data. To date, researchers have used assumed Mendelian models to narrow the number of variants considered. Researchers also tend to restrict the focus to those variants with putative functional effects. While this is a logical strategy to narrow the potential number of variants, it is limited by our current knowledge of genes and proteins. Thus it is likely that many causal variants will still be missed.

Epigenetics

While DNA sequence alterations, whether single nucleotide changes or gross karyotype alterations, have been linked to disease, DNA modifications without sequence change are increasingly being recognized as having important roles in disease. Epigenetic modification is an essential part of heart development, and thus could be an important source of variation for CHD. Epigenetic refers to heritable changes in phenotype or gene expression caused by mechanisms other than changes in the underlying DNA sequence. An important source of epigenetic modification is DNA methylation, the addition of a methyl group to the cytosine within a CG dinucleotide sequence. The presence of the methyl group can prevent transcription, thereby changing gene expression patterns (Bird and Wolffe, 1999). Detailed analysis of methylation across several chromosomes has demonstrated that the promoter regions of nearly 20% of genes, many of which influence transcription, are methylated (Eckhardt et al., 2006).

Mice lacking DNA methyltransferases, enzymes that catalyze the transfer of a methyl group to DNA, die *in utero* (Biniszkiewicz et al., 2002; Li et al., 1993). More specifically, methylation plays important roles in heart development (Jones et al., 2008; McGrew et al., 1996; Menheniott et al., 2008). In mice, alteration of methylation results in heart and valve defects, including ventricular septal defect and endocardial cushion defects (Li et al., 2005). Humans with CblC-type methylmalonic aciduria and homocystinuria (cblC) have abnormal DNA methylation during development, and 50% have CHD (Profitlich et al., 2009). Additionally, mutations in *TBX5*, a genetic cause of CHD (McDermott et al., 2005), alter demethylase and methyltransferase activity (Miller et al., 2008), supporting the importance of both genetic and epigenetic factors in heart development. Taken together, these data support the importance of methylation in heart development. Clearly, future research should explore this type of variation with respect to CHD.

Family History and Phenotype Determination: Recruiting Subjects for Genetic Studies

Family history can distinguish genetic conditions that are not usually inherited (e.g., Down syndrome or Trisomy 21) from genetic conditions that exhibit familial clustering (e.g., bicuspid aortic valve). The recognition of familial heart disease has been complicated by three genetic phenomena that obscure the familial nature: reduced penetrance, variable expressivity, and genetic heterogeneity. Further, while most patients believe family history is important, many are unfamiliar with important clinical detail. Too often, in the fast pace of a busy clinic, family history is asked about during the initial visit, recorded and never revisited (Acheson et al., 2000). A recording of precise family history may require revisiting the questions on more than one occasion and obtaining information from more than one family member. In addition, family history, like other elements of the medical history, is dynamic and subject to change with the passage of time (Hinton, 2008).

However, based on family history and clinical examination, the likelihood of a genetic etiology can be determined.

Clearly and precisely defining phenotype is essential. In addition to clinical examination, echocardiography has emerged as a gold standard (Cripe et al., 2004, Hinton et al., 2007). Phenotype definition that is too broad or too narrow may fail to identify an association with an existing genetic variant or identify a pathologic one. Thus, definition of the phenotype most aligned with the underlying genetic etiology is essential for successful identification of causative genetic variants. CHD has clearly defined clinical phenotypes, and individual defects are classified based on anatomy (e.g., atrial septal defect), physiology (e.g., left-to-right shunt defects), or embryology (e.g., endocardial cushion defects). While these approaches to classification have been useful in stratifying patients for clinical care, there is substantial phenotypic heterogeneity in clinical CHD definition. For instance, ventricular septal defect (VSD) is sub-classified into four types: conoseptal, conoventricular, muscular, and inlet (Hor et al., 2008). Evidence to date suggests that each VSD type has different developmental origins and possibly a distinct genetic basis. For example, some conoventricular VSDs are associated with 22q11 deletion syndrome, while inlet VSDs are often associated with Trisomy 21. Further, the second heart field and neural crest programs would be expected to contribute to conoseptal and conoventricular septation, but not inlet or muscular septation. In addition, in complex CHD like tetralogy of Fallot, defined by pulmonary stenosis, overriding aorta, and VSD, specific phenotypic details identify subtypes; tetralogy of Fallot phenotype heterogeneity may be sub-classified by variants involving aortic arch specification, pulmonary valve anatomy, and non-cardiac malformations.

Cardiac Development and the Genetics of Congenital Heart Disease

Recent advances in cardiac developmental biology have revealed new complexity in the genetic regulation of heart morphogenesis that may contribute to CHD phenotypic heterogeneity. The cardiac crescent, including the lateral plate mesoderm, is a major source of cardiac progenitor cells, but there are other, extra-cardiac, precursors derived from the second heart field, cardiac neural crest, and pro-epicardium that migrate into the heart and contribute to its development. Differentiation of the progenitor cells that contribute to the linear heart tube results in spatially and temporally organized lineages, including myocardium, conduction system, coronary vessels, and cardiac valves. The linear heart tube then loops and develops endocardial cushions that ultimately contribute to valve structures and chamber formation. Further, epigenetic factors such as blood flow perturbations that occur after primary cardiac morphogenesis but before birth may significantly impact and modify the ultimate phenotype. Thus, CHD is the result of complex interaction between multiple developmental factors, which may not be captured by the clinical CHD nomenclature.

Phenotypic heterogeneity may also affect the ability to detect genes and underlying CHD etiology. Thus, while there are clinical definitions of specific phenotypes essential for establishing a diagnosis and designing a clinical management strategy, these definitions do not directly address etiology and may not be sufficient for genetic discovery. The insights gained by human genetic and molecular biology studies must be reconciled with the established clinical taxonomy to provide the most precise classification scheme for CHD. Elucidation of the genetic basis of CHD should lead to a meaningful classification system that integrates the molecular basis of cardiac development with the clinical phenotype.

Relevance of CHD Genomics to Genetic Testing

Although still heavily research-focused, the genomics of CHD has made its way to the clinic, and advances in technology have had an impact on the availability of genetic testing. Genetests, a resource supported by the National Institutes of Health (NIH), keeps up-to-date information on genetic condition and clinical/research laboratory testing sites (see *Recommended Resources*). However, this same technology that performs mutation analysis of individual genes as well as large segments of the human genome in individual patients has led to the realization that each of us harbors thousands of rare, nonsynonymous DNA variants as well as noncoding variants of unknown significance. The implications of these findings for genetic testing indications and interpretation of results are slowly making their way to the clinic.

If completion of the personal medical history, family history, and clinical exam suggests a genetic etiology of heart disease, genetic testing may be considered. A step-wise process for genetic testing identification, counseling, and explanation of results begins with a decision about the type of test to order.

If there is a strong index of suspicion for a specific genetic or cytogenetic abnormality, then karyotyping, fluorescence *in situ* hybridization (FISH), or gene-specific mutation analysis are indicated based on that suspicion. If the characteristics are not typical for a known condition, karyotyping or comparative genomic hybridization (CGH) may be necessary to identify a rare or novel genetic change (see Figure 54.3).

Patients who decide to undergo genetic testing should be pre-counseled for the possible test results, which may include a positive test, negative result, or variant of unknown significance (VUS). Once a genetic test result is obtained, it can be used to make decisions about management, screening and prophylaxis. Additionally, patients with isolated CHD are at risk for secondary phenotypes that can be caused by their gene mutation. For example, patients with an *NKX2.5* mutation may have undergone successful surgery for a CHD, but they will continue to be at risk for atrioventricular block (Benson, 2009). They should receive regular electrocardiogram (ECG) screening to monitor that risk and encourage early treatment of any abnormal findings. Early detection of genetic status can improve screening and management, and as we come to understand the underlying pathogenesis, detection will be important for prophylactic treatment, such as the use of implantable cardiac defibrillators (ICDs) in patients with channelopathies.

Genetic information also has health management and psychosocial implications for extended family members. Individuals who have not previously had any symptoms or risks for CHD may become candidates for intensified screening due to the genetic diagnosis of a family member. Families with a negative clinical history of CHD may think they are not at risk for hereditary heart conditions, but because of reduced penetrance and variable expressivity, that conclusion may be

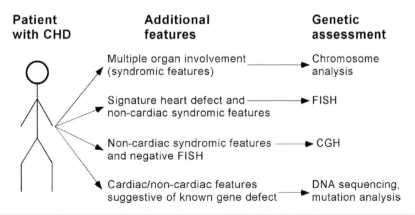

Figure 54.3 Diagram depicting a step-wise process for genetic testing. If there is a strong index of suspicion for a specific genetic or cytogenetic abnormality, then karyotyping, fluorescence *in situ* hybridization (FISH), or gene-specific mutation analysis are indicated. If not characteristic for a known condition, comparative genomic hybridization (CGH) may be necessary to identify a copy number variant.

erroneous. Once a genetic cause has been identified, it can be advantageous to rule out individuals who are not at risk for a condition. This can prevent unnecessary, expensive, and sometimes inconvenient screening practices. Importantly, information that has implications for family members must be managed cautiously, as some family members may not be interested in sharing genetic information, getting genetic testing, or carrying out prophylactic measures that are available to them.

Genetic testing can also be used for pre-implantation genetic diagnosis (PGD) in future pregnancies (McDermott et al., 2006). This procedure involves external fertilization of embryos, as used for *in vitro* fertilization (IVF), but adds a genetic screening step prior to reintroduction of the non-affected embryos to the uterus. Although PGD is infrequently used for non-life threatening conditions or adult-onset disease, its use may increase as the technology is improved and the cost decreases. PGD has already been used for Holt-Oram syndrome (He et al., 2004) and Marfan syndrome (Spits et al., 2006).

SUMMARY

CHD are a common and clinically important form of pediatric heart disease. Despite advances in diagnosis and therapy, CHD morbidity and mortality remain significant, and there is considerable interest in understanding the cause of these clinically important birth defects. While some progress has been made in identifying the genetic origins of CHD, in most cases the cause remains unknown. The frequent observation of genetic heterogeneity, reduced penetrance, and variable expressivity indicates that single genes do not fully explain inheritance, and the occurrence of sporadic cases suggests that inheritance is complex. Advances in methodology to determine individual human genetic variation provides an opportunity to revolutionize discovery in individuals with CHD. Continued genetic discovery of CHD-causing genetic variants poses challenging study design questions and suggests that complementary approaches may be required to realize discovery. Strategies to identify genetic causes of CHD will require precise phenotype assessment and an emphasis on statistical considerations and study design.

REFERENCES

Acheson, L.S., Wiesner, G.L., Zyzanski, S.J., Goodwin, M.A., Stange, K.C., 2000. Family history-taking in community family practice: Implications for genetic screening. Genet Med 2, 180–185.

Alonso, M.E., Pernaute, B., Crespo, M., Gomez-Skarmeta, J.L., Manzanares, M., 2009. Understanding the regulatory genome. Int J Dev Biol 53, 1367–1378.

Arcelli, D., Farina, A., Cappuzzello, C., et al., 2010. Identification of circulating placental mRNA in maternal blood of pregnancies affected with fetal congenital heart diseases at the second trimester of pregnancy: Implications for early molecular screening. Prenat Diagn 30, 229–234.

Basson, C.T., Bachinsky, D.R., Lin, R.C., et al., 1997. Mutations in human TBX5 [corrected] cause limb and cardiac malformation in Holt-Oram syndrome. Nat Genet 15, 30–35.

Benson, D.W., 2009. Genetic origins of pediatric heart disease. Pediatr Cardiol 31, 422–429.

Benson, D.W., Martin, L.J., 2010. Complex story of the genetic origins of pediatric heart disease. Circulation 121, 1277–1279.

Benson, D.W., Sharkey, A., Fatkin, D., et al., 1998. Reduced penetrance, variable expressivity, and genetic heterogeneity of familial atrial septal defects. Circulation 97, 2043–2048.

Benson, D.W., Silberbach, G.M., Kavanaugh-Mchugh, A., et al., 1999. Mutations in the cardiac transcription factor NKX2.5 affect diverse cardiac developmental pathways. J Clin Invest 104, 1567–1573.

Biniszkiewicz, D., Gribnau, J., Ramsahoye, B., et al., 2002. Dnmt1 overexpression causes genomic hypermethylation, loss of imprinting, and embryonic lethality. Mol Cell Biol 22, 2124–2135.

Bird, A.P., Wolffe, A.P., 1999. Methylation-induced repression – belts, braces, and chromatin. Cell 99, 451–454.

Botto, L.D., Correa, A., 2003. Decreasing the burden of congenital heart anomalies: An epidemiologic evaluation of risk factors and survival. Prog Pediatr Cardiol 18, 111–121.

Bray, G.A., Ryan, D.H., 2000. Clinical evaluation of the overweight patient. Endocrine 13, 167–186.

Ching, Y.H., Ghosh, T.K., Cross, S.J., et al., 2005. Mutation in myosin heavy chain 6 causes atrial septal defect. Nat Genet 37, 423–428.

Cooper, G.M., Zerr, T., Kidd, J.M., Eichler, E.E., Nickerson, D.A., 2008. Systematic assessment of copy number variant detection via genome-wide SNP genotyping. Nat Genet 40, 1199–1203.

Cripe, L., Andelfinger, G., Martin, L.J., Shooner, K., Benson, D.W., 2004. Bicuspid aortic valve is heritable. J Am Coll Cardiol 44, 138–143.

Curran, M.E., Atkinson, D.L., Ewart, A.K., et al., 1993. The elastin gene is disrupted by a translocation associated with supravalvular aortic stenosis. Cell 73, 159–168.

Dasgupta, C., Martinez, A.M., Zuppan, C.W., et al., 2001. Identification of connexin43 (alpha1) gap junction gene mutations in patients with hypoplastic left heart syndrome by denaturing gradient gel electrophoresis (DGGE). Mutat Res 479, 173–186.

Dulac, Y., Pienkowski, C., Abadir, S., Tauber, M., Acar, P., 2008. Cardiovascular abnormalities in Turner's syndrome: What prevention? Arch Cardiovasc Dis 101, 485–490.

Eckhardt, F., Lewin, J., Cortese, R., et al., 2006. DNA methylation profiling of human chromosomes 6, 20 and 22. Nat Genet 38, 1378–1385.

Evans, D.M., Cardon, L.R., 2004. Guidelines for genotyping in genomewide linkage studies: Single-nucleotide-polymorphism maps versus microsatellite maps. Am J Hum Genet 75, 687–692.

Ewart, A.K., Morris, C.A., Ensing, G.J., et al., 1993. A human vascular disorder, supravalvular aortic stenosis, maps to chromosome 7. Proc Natl Acad Sci USA 90, 3226–3230.

Fanciulli, M., Petretto, E., Aitman, T.J., 2009. Gene copy number variation and common human disease. Clin Genet 77 (3), 201–213.

Ferencz, C., Neill, C.A., Boughman, J.A., et al., 1989. Congenital cardiovascular malformations associated with chromosome abnormalities: An epidemiologic study. J Pediatr 114, 79–86.

Ferencz, C., Rubin, J.D., Loffredo, C.A., Magee, D.A., 1993. Epidemiology of congenital heart disease: The Baltimore-Washington Infant Study 1981–1989. In: Anderson, R.H. (Ed.), Perspectives in Pediatric Cardiology Futura Publishing, Mount Kisco, NY.

Gamazon, E.R., Zhang, W., Konkashbaev, A., et al., 2010. SCAN: SNP and copy number annotation. Bioinformatics 26, 259–262.

Garg, V., Kathiriya, I.S., Barnes, R., et al., 2003. GATA4 mutations cause human congenital heart defects and reveal an interaction with TBX5. Nature 424, 443–447.

Garg, V., Muth, A.N., Ransom, J.F., et al., 2005. Mutations in NOTCH1 cause aortic valve disease. Nature 437, 270–274.

Gebbia, M., Ferrero, G.B., Pilia, G., et al., 1997. X-linked situs abnormalities result from mutations in ZIC3. Nat Genet 17, 305–308.

Harley, J.B., Alarcon-Riquelme, M.E., Criswell, L.A., et al., 2008. Genome-wide association scan in women with systemic lupus erythematosus identifies susceptibility variants in ITGAM, PXK, KIAA1542 and other loci. Nat Genet 40, 204–210.

He, J., McDermott, D.A., Song, Y., et al., 2004. Preimplantation genetic diagnosis of human congenital heart malformation and Holt-Oram syndrome. Am J Med Genet A 126A, 93–98.

Hinton Jr., R.B., Martin, L.J., Tabangin, M.E., Mazwi, M., Cripe, L.H., Benson, D.W., 2007. Hypoplastic left heart syndrome is heritable. J Am Coll Cardiol 50, 1590–1595. PMID: 17936159.

Hinton Jr., R.B., 2008. The family history: Reemergence of an established tool. Crit Care Nurs Clin North Am 20, 149–158.

Hoffman, J.I., 1995a. Incidence of congenital heart disease: I. Postnatal incidence. Pediatr Cardiol 16, 103–113.

Hoffman, J.I., 1995b. Incidence of congenital heart disease: II. Prenatal incidence. Pediatr Cardiol 16, 155–165.

Hol, F.A., Hamel, B.C., Geurds, M.P., et al., 1995. Localization of Alagille syndrome to 20p11.2-p12 by linkage analysis of a three-generation family. Hum Genet 95, 687–690.

Hor, K.N., Border, W.L., Cripe, L.H., Benson, D.W., Hinton, R.B., 2008. The presence of bicuspid aortic valve does not predict ventricular septal defect type. Am J Med Genet A 146A, 3202–3205.

Jones, B., Su, H., Bhat, A., et al., 2008. The histone H3K79 methyltransferase Dot1L is essential for mammalian development and heterochromatin structure. PLoS Genet 4, e1000190.

Katsanis, N., 2009. From association to causality: The new frontier for complex traits. Genome Med 1, 23.

Knowler, W.C., Williams, R.C., Pettitt, D.J., Steinberg, A.G., 1988. Gm3;5,13,14 and type 2 diabetes mellitus: An association in American Indians with genetic admixture. Am J Hum Genet 43, 520–526.

Lammer, E.J., Chak, J.S., Iovannisci, D.M., et al., 2009. Chromosomal abnormalities among children born with conotruncal cardiac defects. Birth Defects Res A Clin Mol Teratol 85, 30–35.

Li, D., Pickell, L., Liu, Y., et al., 2005. Maternal methylenetetrahydrofolate reductase deficiency and low dietary folate lead to adverse reproductive outcomes and congenital heart defects in mice. Am J Clin Nutr 82 (1), 188–195.

Li, E., Beard, C., Forster, A.C., Bestor, T.H., Jaenisch, R., 1993. DNA methylation, genomic imprinting, and mammalian development. Cold Spring Harb Symp Quant Biol 58, 297–305.

Li, Q.Y., Newbury-Ecob, R.A., Terrett, J.A., et al., 1997. Holt-Oram syndrome is caused by mutations in TBX5, a member of the Brachyury (T) gene family. Nat Genet 15 (1), 21–29.

Lin, A.E., Basson, C.T., Goldmuntz, E., et al., 2008. Adults with genetic syndromes and cardiovascular abnormalities: Clinical history and management. Genet Med 10, 469–494.

Lupski, J.R., Reid, J.G., Gonzaga-Jauregui, C., et al., 2010. Whole-genome sequencing in a patient with Charcot-Marie-Tooth neuropathy. N Engl J Med 362, 1181–1191.

Macmahon, B., McKeown, T., Record, R.G., 1953. The incidence and life expectation of children with congenital heart disease. Br Heart J 15, 121–129.

Martin, L.J., Ramachandran, V., Cripe, L.H., et al., 2007a. Evidence in favor of linkage to human chromosomal regions 18q, 5q and 13q for bicuspid aortic valve and associated cardiovascular malformations. Hum Genet 121, 275–284.

Martin, L.J., Woo, J.G., Avery, C.L., Chen, H.-S., North, K.E., 2007b. Multiple testing in the genomics era: Findings from GAW15, Group 15. Genet Epidemiol 31 Suppl 1, S124–S131.

McDermott, D.A., Basson, C.T., Hatcher, C.J., 2006. Genetics of cardiac septation defects and their pre-implantation diagnosis. Methods Mol Med 126, 19–42.

McDermott, D.A., Bressan, M.C., He, J., et al., 2005. TBX5 genetic testing validates strict clinical criteria for Holt-Oram syndrome. Pediatr Res 58, 981–986.

McElhinney, D.B., Geiger, E., Blinder, J., Benson, D.W., Goldmuntz, E., 2003. NKX2.5 mutations in patients with congenital heart disease. J Am Coll Cardiol 42, 1650–1655.

McGrew, M.J., Bogdanova, N., Hasegawa, K., et al., 1996. Distinct gene expression patterns in skeletal and cardiac muscle are dependent on common regulatory sequences in the MLC1/3 locus. Mol Cell Biol 16, 4524–4534.

McKeown, T., Macmahon, B., Parsons, C.G., 1953. The familial incidence of congenital malformation of the heart. Br Heart J 15, 273–277.

Menheniott, T.R., Woodfine, K., Schulz, R., et al., 2008. Genomic imprinting of Dopa decarboxylase in heart and reciprocal allelic expression with neighboring Grb10. Mol Cell Biol 28, 386–396.

Miller, S.A., Huang, A.C., Miazgowicz, M.M., Brassil, M.M., Weinmann, A.S., 2008. Coordinated but physically separable interaction with H3K27-demethylase and H3K4-methyltransferase activities are required for T-box protein-mediated activation of developmental gene expression. Genes Dev 22, 2980–2993.

Morton, N.E., 2008. Into the post-HapMap era. Adv Genet 60, 727–742.

Nora, J.J., 1968. Multifactorial inheritance hypothesis for the etiology of congenital heart diseases. The genetic-environmental interaction. Circulation 38, 604–617.

Oldridge, D.A., Banerjee, S., Setlur, S.R., Sboner, A., Demichelis, F., 2010. Optimizingcopy number variatio analysis using genome-wide short sequence oligonucleotide arays. Nucleic Acids Res 38 (10), 3275–3286.

Olson, T.M., Michels, V.V., Lindor, N.M., et al., 1993. Autosomal dominant supravalvular aortic stenosis: Localization to chromosome 7. Hum Mol Genet 2, 869–873.

Page, G.P., George, V., Go, R.C., Page, P.Z., Allison, D.B., 2003. "Are we there yet?": Deciding when one has demonstrated specific genetic causation in complex diseases and quantitative traits. Am J Hum Genet 73, 711–719.

Pierpont, M.E., Basson, C.T., Benson, D.W., et al., 2007. Genetic basis for congenital heart defects: Current knowledge. A scientific statement from the American Heart Association Congenital Cardiac Defects Committee, Council on Cardiovascular Disease in the Young, endorsed by the American Academy of Pediatrics. Circulation 115, 3015–3038.

Pizzuti, A., Sarkozy, A., Newton, A.L., et al., 2003. Mutations of ZFPM2/FOG2 gene in sporadic cases of tetralogy of Fallot. Hum Mutat 22, 372–377.

Profitlich, L.E., Kirmse, B., Wasserstein, M.P., Diaz, G.A., Srivastava, S., 2009. High prevalence of structural heart disease in children with cblC-type methylmalonic aciduria and homocystinuria. Mol Genet Metab 98, 344–348.

Pushkarev, D., Neff, N.F., Quake, S.R., 2009. Single-molecule sequencing of an individual human genome. Nat Biotechnol 27 (9), 847–850.

Redon, R., Ishikawa, S., Fitch, K.R., et al., 2006. Global variation in copy number in the human genome. Nature 444, 444–454.

Ruiz-Perez, V.L., Ide, S.E., Strom, T.M., et al., 2000. Mutations in a new gene in Ellis-van Creveld syndrome and Weyers acrodental dysostosis. Nat Genet 24, 283–286.

Ruiz-Perez, V.L., Tompson, S.W., Blair, H.J., et al., 2003. Mutations in two nonhomologous genes in a head-to-head configuration cause Ellis-van Creveld syndrome. Am J Hum Genet 72, 728–732.

Satoda, M., Zhao, F., Diaz, G.A., et al., 2000. Mutations in TFAP2B cause Char syndrome, a familial form of patent ductus arteriosus. Nat Genet 25, 42–46.

Schadt, E.E., Turner, S., Kasarskis, A., 2010. A window into third-generation sequencing. Hum Mol Genet 15 19(R2), R227–R240.

Schott, J.J., Benson, D.W., Basson, C.T., et al., 1998. Congenital heart disease caused by mutations in the transcription factor NKX2-5. Science 281, 108–111.

Shtir, C., Pique-Regi, R., Siegmund, K., et al., 2009. Copy number variation in the Framingham Heart Study. BMC Proc 3 (Suppl. 7), S133.

Siu, S.C., Silversides, C.K., 2010. Bicuspid aortic valve disease. J Am Coll Cardiol 55, 2789–2800.

Sobreira, N.L., Cirulli, E.T., Avramopoulos, D., et al., 2010. Whole-genome sequencing of a single proband together with linkage analysis identifies a Mendelian disease gene. PLoS Genet 6, e1000991.

Spits, C., De Rycke, M., Verpoest, W., et al., 2006. Preimplantation genetic diagnosis for Marfan syndrome. Fertil Steril 86, 310–320.

Stamm, C., Friehs, I., Ho, S.Y., et al., 2001. Congenital supravalvar aortic stenosis: A simple lesion? Eur J Cardiothorac Surg 19, 195–202.

Stankiewicz, P., Lupski, J.R., 2010. Structural variation in the human genome and its role in disease. Annu Rev Med 61, 437–455.

Tabangin, M.E., Woo, J.G., Martin, L.J., 2009. The effect of minor allele frequency on the likelihood of obtaining false positives. BMC Proc 3 (Suppl. 7), S41.

Tartaglia, M., Kalidas, K., Shaw, A., et al., 2002. PTPN11 mutations in Noonan syndrome: Molecular spectrum, genotype-phenotype correlation, and phenotypic heterogeneity. Am J Hum Genet 70, 1555–1563.

Tartaglia, M., Mehler, E.L., Goldberg, R., et al., 2001. Mutations in PTPN11, encoding the protein tyrosine phosphatase SHP-2, cause Noonan syndrome. Nat Genet 29, 465–468.

Tidyman, W.E., Rauen, K.A., 2009. The RASopathies: Developmental syndromes of RAS/MAPK pathway dysregulation. Curr Opin Genet Dev 19, 230–236.

Tucker, T., Marra, M., Friedman, J.M., 2009. Massively parallel sequencing: The next big thing in genetic medicine. Am J Hum Genet 85, 142–154.

Udwadia, A.D., Khambadkone, S., Bharucha, B.A., Lokhandwala, Y., Irani, S.F., 1996. Familial congenital valvar pulmonary stenosis: Autosomal dominant inheritance. Pediatr Cardiol 17, 407–409.

Visscher, P.M., Andrew, T., Nyholt, D.R., 2008. Genome-wide association studies of quantitative traits with related individuals: Little (power) lost but much to be gained. Eur J Hum Genet 16, 387–390.

Wain, L.V., Armour, J.A., Tobin, M.D., 2009. Genomic copy number variation, human health, and disease. Lancet 374, 340–350.

Winston, J.B., Erlich, J.M., Green, C.A., et al., 2010. Heterogeneity of genetic modifiers ensures normal cardiac development. Circulation 121 (11), 1313–1321.

Zhang, F., Gu, W., Hurles, M.E., Lupski, J.R., 2009. Copy number variation in human health, disease, and evolution. Annu Rev Genomics Hum Genet 10, 451–481.

Zhao, T., Zhao, W., Chen, Y., Ahokas, R.A., Sun, Y., 2010. Vascular endothelial growth factor (VEGF)-A: Role on cardiac angiogenesis following myocardial infarction. Microvasc Res 80, 188–194.

RECOMMENDED RESOURCES

National Center for Biotechnology Information
    http://www.ncbi.nlm.nih.gov/
    Provides access to biomedical and genomic information
Online Mendelian Inheritance in Man (OMIM)
    http://www.ncbi.nlm.nih.gov/omim/
    Compendium of human genes and genetic phenotypes
Genetests
    http://www.genetests.org/
    Listing of US and international laboratories offering genetic testing

Hapmap
    http://www.hapmap.org
    Characterization of common variation across major racial groups
1000 Genomes Project
    http://www.1000genomes.org
    Sequence level genetic variation
Surgeon General's Family Health History Initiative
    http://www.hhs.gov/familyhistory/
    To help focus attention on family history importance

Perioperative Genomics

Mihai V. Podgoreanu

SCIENTIFIC RATIONALE FOR PERIOPERATIVE GENOMIC MEDICINE

More than 40 million patients annually undergo surgery in the US, at a cost of $450 billion. Although improvements in resuscitation, anesthesia, and critical care have made important contributions to survival following major surgery and trauma, approximately one million patients per year sustain medical complications after surgery, resulting in annual costs of $25 billion. The proportion of the US population older than 65 is predicted to double in the next two decades (to 20% of the overall population), leading to a 25% increase in the number of surgeries, a 50% increase in surgery-related costs, and a 100% increase in complications from surgery. This accelerated aging of the population, combined with increased reliance on surgery for the treatment of disease, has resulted in a significant surgical burden. Meanwhile, pre-surgical risk profiling remains inconsistent and by definition not personalized, so more robust prognostic markers are needed to improve the quality of surgical care (Mangano, 2004).

The perioperative period represents a unique and extreme example of gene–environment interaction. As we appreciate in our daily practice in the operating rooms and intensive care units, one hallmark of perioperative physiology is the striking variability in patient responses to the acute, robust, and systemic perturbations induced by surgical injury or trauma. Further complicating factors include the sometimes profound hemodynamic challenges, transient limb/organ ischemia, partial/total organ resection, a multitude of

organ system support technologies (vascular cannulation, extracorporeal circulation, intra-aortic balloon counterpulsation, mechanical ventilation), transfusions, anesthetic agents, and the pharmacopoeia used in the perioperative period. This translates into substantial interindividual variability in the incidence or severity of immediate perioperative adverse events, as well as long-term outcomes (Table 55.1). For decades we have attributed this variability to complexities such as age, nutritional state, comorbidities, or the heterogeneity of the surgical insult. But given equivalent magnitudes of surgery or trauma, adequate antibiotic prophylaxis, and appropriate clinical and surgical skill, why is it that some patients never develop complications (e.g., infection, organ dysfunction) but others do? Moreover, why is it that when such complications do occur, some patients recover while others progress inexorably to multiple organ failure and death? It is becoming increasingly recognized that perioperative and post-traumatic morbidity arises as a direct result of exposure to the acute environmental stress of surgery or trauma occurring on a landscape of susceptibility determined by an individual's clinical and genetic characteristics (constitutive factors). It may even occur in otherwise healthy individuals. Thus, overall genetic predisposition to perioperative complications stems not only from genetic contributions to the development of comorbid risk factors (such as coronary artery disease and reduced preoperative cardiopulmonary reserve) during the patient's lifetime, but also from genetic variability in specific biological pathways participating in the host response to surgical injury or trauma (Figure 55.1). Such adverse outcomes will develop

Genomic and Personalized Medicine, 2nd edition
by Ginsburg & Willard. DOI: http://dx.doi.org/10.1016/B978-0-12-382227-7.00055-0

TABLE 55.1	Categories of perioperative phenotypes

Immediate perioperative outcomes

In-hospital mortality

Perioperative myocardial infarction

Perioperative low cardiac output syndrome/acute decompensated heart failure/ventricular dysfunction

Perioperative vasoplegic syndrome

Perioperative dysrhythmias [atrial fibrillation, corrected QT interval (QTc) prolongation]

Postoperative bleeding

Perioperative venous thrombosis

Acute postoperative stroke

Postoperative delirium

Perioperative acute kidney injury

Acute perioperative lung injury/prolonged postoperative mechanical ventilation

Acute allograft dysfunction/rejection

Postoperative sepsis

Multiple organ dysfunction syndrome

Postoperative nausea and vomiting

Acute postoperative pain

Variability in response to anesthetics, analgesics, and other perioperative drugs

Intermediate phenotypes (plasma biomarker levels)

Long-term postoperative outcomes

Event-free survival/major adverse cardiac events

Progression of vein graft disease

Chronic allograft dysfunction/rejection

Postoperative cognitive dysfunction

Postoperative depression

Transition from acute to chronic pain

Cancer progression

Quality of life

only in patients whose combined burden of genetic and environmental risk factors exceeds a certain threshold, which may vary with age. In fact, physiologic stress associated with life-threatening injury exposes genetic anomalies that might otherwise go unnoticed. Identification of such genetic variations that not only contribute to disease causation and susceptibility, but also influence the individual patient's *responses* to disease and drug therapy, and incorporation of genetic risk

information in clinical decision-making, may lead to improved health outcomes and reduced costs. For instance, understanding the role of allotypic variation in pro-inflammatory and pro-thrombotic pathways, which remain the primary mechanisms underlying perioperative complications when they interact with metabolic failure, may contribute to the development of target-specific therapies, thereby limiting the incidence of adverse events in high-risk, susceptible patients.

With increasing evidence suggesting that genetic variation can significantly modulate risk of adverse perioperative events (Fox et al., 2004; Podgoreanu and Schwinn, 2005; Stuber and Hoeft, 2002; Ziegeler et al., 2003), the emerging field of *perioperative genomics* aims to apply functional genomic approaches to discover underlying biological mechanisms that explain why similar patients have such dramatically different outcomes after surgery. The field is justified by a unique combination of environmental insults and postoperative phenotypes that characterize surgical and critically ill patient populations.

This chapter serves as a primer in perioperative genomic medicine by highlighting the rapidly evolving current and future applications of molecular (including genomic) technologies for perioperative risk stratification, outcome prediction, and mechanistic understanding of surgical stress responses, as well as identification and validation of novel targets for perioperative organ protection. To increase clinical relevance for the practicing perioperative and critical care physician, we summarize below existing evidence by specific outcome, while highlighting candidate genes in relevant mechanistic pathways (Tables 55.2, 55.3, and 55.4).

PERIOPERATIVE CARDIAC ADVERSE EVENTS

Patients with underlying cardiovascular disease may be at increased risk for perioperative cardiac complications such as perioperative myocardial infarction (PMI) and ventricular dysfunction. The incidence of PMI following cardiovascular surgery remains between 7% and 19% (Domanski et al., 2011; Mangano, 1997), despite advances in surgical, cardioprotective, and anesthetic techniques, and is consistently associated with reduced short- and long-term survival in those patients. Over the last few decades, several multifactorial risk indices have been developed and validated for both non-cardiac (e.g., Lee's Revised Cardiac Risk Index; Lee et al., 1999) and cardiac surgical patients [e.g., Hannan score, EuroSCORE (Hannan et al., 1994; Nashef et al., 2002)], with the specific aim of stratifying risk for perioperative adverse events. However, because these multifactorial risk indices have only limited predictive value for identifying patients at the highest risk of PMI (Howell and Sear, 2004), it has been proposed that genomic approaches could aid in refining an individual's risk profile (Podgoreanu and Schwinn, 2005).

The pathophysiology of PMI after cardiac surgery involves systemic and local inflammation, "vulnerable" blood, and neuroendocrine stress (Podgoreanu and Schwinn, 2005). In non-cardiac

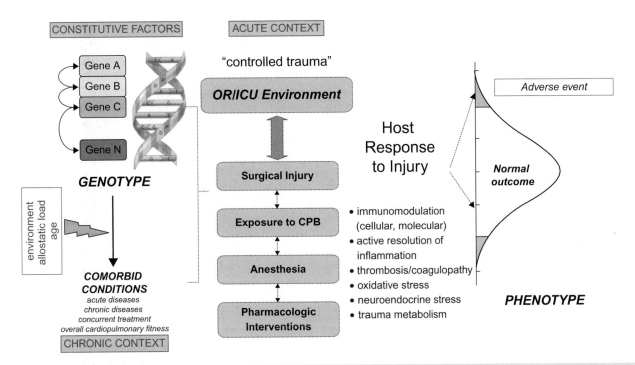

Perioperative adverse events are complex traits, characteristically involving an interaction between robust operative environmental perturbations (surgical trauma, hemodynamic challenges, exposure to extracorporeal circulation, drug administration) and multiple susceptibility genes. The observed variability in perioperative outcomes can in part be attributed to genetic variability modulating the host response to surgical injury. OR, operating room; CPB, cardiopulmonary bypass; ICU, intensive care unit.

surgery, PMI occurs as a result of two distinct mechanisms: (1) coronary plaque rupture and subsequent thrombosis triggered by a number of perioperative stressors (catecholamine surges, pro-inflammatory and pro-thrombotic states), and (2) myocardial oxygen supply–demand imbalance (Landesberg et al., 2009). Interindividual genetic variability in these mechanistic pathways is extensive, which may combine to modulate overall susceptibility to perioperative stress and ultimately the magnitude of myocardial injury. Nevertheless, until recently, only a few studies have explored the role of genetic factors in the development of PMI (DeLanghe et al., 1997; Ortlepp et al., 2001; Volzke et al., 2002), mainly conducted in patients undergoing coronary artery bypass grafting (CABG) surgery (Table 55.2).

Inflammatory Biomarkers and Perioperative Myocardial Outcomes

Although the role of inflammation in cardiovascular disease biology has long been established, we are just beginning to understand the relationship between genetically controlled variability in inflammatory responses to surgery and PMI pathogenesis. Recently, three inflammatory gene single nucleotide polymorphisms (SNPs) have been described to have an independent predictive value for incident PMI after cardiac surgery performed with cardiopulmonary bypass (CPB) (Podgoreanu et al., 2006). They include the pro-inflammatory cytokine interleukin-6 (*IL6*) and two adhesion

molecules – intercellular adhesion molecule-1 (*ICAM1*) and E-selectin (*SELE*). Furthermore, the predictive ability of a PMI model based only on traditional risk factors was improved by the addition of genotypic information. Similarly, a combined haplotype in the mannose-binding lectin gene (*MBL2* LYQA secretor haplotype), an important recognition molecule in the lectin complement pathway, has been independently associated with PMI in a cohort of Caucasian patients undergoing first-time CABG with CPB (Collard et al., 2007). Genetic variants in *IL6* and *TNFA* have also been associated with increased incidence of postoperative cardiovascular complications, including PMI after lung surgery for cancer (Shaw et al., 2005). In the setting of surgical or traumatic procedures, multiple SNPs present in pro-inflammatory signaling pathway genes lead to altered expression of relevant proteins. Examples include the promoter SNPs in *IL6* (Brull et al., 2001), which have been shown to prolong the length of hospital stay (Burzotta et al., 2001), the apolipoprotein E genotype (*ε4* allele) (Grocott et al., 2001), SNPs in the tumor necrosis factor genes (*TNFA, LTA*) (Roth-Isigkeit et al., 2001) associated with postoperative left ventricular dysfunction (Tomasdottir et al., 2003), and a functional SNP in the macrophage migration inhibitory factor (*MIF*) (Lehmann et al., 2006). Conversely, genetic variation regulating expression of the anti-inflammatory cytokine interleukin-10 (*IL10*) in response to CPB has been described, with high levels of IL10 being associated with

TABLE 55.2 Representative genetic polymorphisms associated with altered susceptibility to adverse perioperative cardiovascular events

Gene	Polymorphism	Type of surgery	OR	Reference(s)
Perioperative myocardial infarction, ventricular dysfunction, early vein graft failure				
IL6	-572G>C	Cardiac/CPB	2.47	Podgoreanu et al., 2006
	-174G>C	Thoracic	1.8	Shaw et al., 2005
ICAM-1	E469L	Cardiac/CPB	1.88	Podgoreanu et al., 2006
SELE	98G>T		0.16	Podgoreanu et al., 2006
MBL2	LYQA secretor haplotype	CABG/CPB	3.97	Collard et al., 2007
ITGB3	L33P	CABG/CPB	2.5[a]	Rinder et al., 2002
	(Pl^{A1}/Pl^{A2})	Major vascular	2.4	Faraday et al., 2004
GP1BA	T145M	Major vascular	3.4	Faraday et al., 2004
TNFA	-308G>A	Thoracic	2.5	Shaw et al., 2005
TNFB (LTA)	TNFB2	Cardiac/CPB	3.84	Tomasdottir et al., 2003
IL10	-1082G>A	Cardiac/CPB	n.r.	Galley et al., 2003
F5	R506Q(FVL)	CABG/CPB	3.29	Moor et al., 1998
CMA1	-1905A>G	CABG/CPB	n.r.	Ortlepp et al., 2001
ANRIL	rs10116277 G>T	CABG	1.79	Liu et al., 2010†
(chr 9p21 locus)	rs6475606 C>T		1.79	
	rs2383207 A>G		1.71	
NPR3	rs700923 A>G	CABG/CPB	4.28	Fox et al., 2009
	rs16890196 A>G		4.09	
	rs765199 C>T		4.27	
	rs700926 A>C		3.89	
NPPA_NPPB	rs632793 T>C	CABG/CPB	0.52	Fox et al., 2009
	rs6668352 G>A		0.44	
	rs549596 T>C		0.48	
	rs198388 C>T		0.51	
	rs198389 A>G		0.54	
PAI-1	4G/5G	CABG	n.r.	Rifon et al., 1997
Chr 3p22.3 locus	rs17691914	CABG	2.01	Fox et al., 2011*
Chr 3p14.2 locus	rs17061085		1.70	
Chr 11q23.2 locus	rs12279572		2.19	
Perioperative vasoplegia, vascular reactivity, coronary tone				
DDAH II	-449G>C	Cardiac/CPB	0.4	Ryan et al., 2006
NOS3	E298D		n.r.	Heusch et al., 2001; Philip et al., 1999
ACE	In/del		n.r.	Henrion et al., 1999; Lasocki et al., 2002

(continued)

TABLE 55.2	(Continued)			
Gene	**Polymorphism**	**Type of surgery**	**OR**	**Reference(s)**
ADRB2	Q27E	Tracheal intubation	11.7[b]	Kim et al., 2002
GNB3	825C>T	Response to α-AR agonists	n.r.	Heusch et al., 2001
PON1	Q192R	Resting coronary tone	n.r.	Heusch et al., 2001
TNFβ+250	-1082G>A	Hyperdynamic state		Iribarren et al., 2008
Postoperative arrhythmias: atrial fibrillation, QTc prolongation				
IL6	-174G>C	CABG/CPB	3.25	Gaudino et al., 2003a; Motsinger et al., 2006
		β-blocker failure	n.r.	Donahue and Roden, 2005
		Thoracic	1.8	Shaw et al., 2005
RANTES	-403G>A	β-blocker failure	n.r.	Donahue and Roden, 2005
TNFA	-308G>A	Thoracic	2.5	Shaw et al., 2005
ATFB5	rs2200733 C>T	Cardiac/CPB	1.97	Body et al., 2009*†
(chr 4q25 locus)	rs2220427 T>G		1.76	Virani et al., 2011*†
	rs10033464 G>T		1.47	
IL1B	-511T>C	Cardiac/CPB	1.44	Podgoreanu et al., 2007*
	5810G>A		0.66	
Postoperative death, MACE, late vein graft failure				
ADRB1	R389G	Non-cardiac with spinal block	1.87[c]	Zaugg et al., 2007
ANRIL (chr 9p21 locus)	rs10116277 G>T	CABG	1.7	Muehlschlegel et al., 2010†
ACE	In/del	CABG/CPB	3.1[d]	Volzke et al., 2002
ITGB3	L33P		4.7	Zotz et al., 2000
MTHFR	A222V	PTCA and CABG/CBP	2.8	Botto et al., 2004
ADRB2	R16G	Cardiac surgery/CPB	1.96	Podgoreanu et al., 2003
	Q27E		2.82	
HP	Hp1/Hp2	CABG	n.r.	Delanghe et al., 1997
CR1, KDR		CABG/CPB	n.r.	Ellis et al., 2007
MICA				
HLA-DPB1				
VTN				
LPL	HindIII		n.r.	Taylor et al., 2004
THBD	A455V	CABG/CPB	2.78	Lobato et al., 2011*
IL6	-174G>C	Non-cardiac vascular surgery	2.14	Stoica et al., 2010
	nt565 G>A		1.84	
IL10	-1082 G>A	Non-cardiac vascular surgery		Stoica et al., 2010
	-819 C>T		2.16	
	-592 C>A			
	ATA haplotype			

(continued)

TABLE 55.2	(Continued)			
Gene	**Polymorphism**	**Type of surgery**	**OR**	**Reference(s)**
Cardiac allograft rejection				
TNFA	-308G>A	Cardiac transplant	n.r.	Holweg et al., 2004
IL10	-1082G>A		n.r.	Holweg et al., 2004
ICAM1	K469E		n.r.	Borozdenkova et al., 2001
IL1RN	86-bp VNTR	Thoracic transplant	2.02	Vamvakopoulos et al., 2002
IL1B	3953C>T		20.5[e]	Vamvakopoulos et al., 2002
TGF-β	915G>C	Cardiac transplant	n.r.	Benza et al., 2009

CPB, cardiopulmonary bypass; in/del, insertion/deletion polymorphism; chr, chromosome; n.r., not reported; OR, odds ratio; VNTR, variable number tandem repeat; α-AR, alpha adrenergic receptors.
[a]Relative risk;
[b]F-value;
[c]hazard ratio;
[d]β-coefficient;
[e]in haplotype with *IL1RN* VNTR.
[*]Findings independently replicated.
[†]Replication of previous GWAS finding in the perioperative clinical setting.

TABLE 55.3	Representative genetic polymorphisms associated with altered susceptibility to adverse perioperative neurological events			
Gene	**Polymorphism**	**Type of surgery**	**OR**	**Reference(s)**
Perioperative stroke				
IL6	-174G>C	Cardiac/CPB	3.3	Grocott et al., 2005
CRP	1846C>T			
Perioperative cognitive dysfunction, neurodevelopmental dysfunction				
SELP	E298D	Cardiac/CPB	0.51	Mathew et al., 2007
CRP	1059G>C	Cardiac/CPB	0.37	Mathew et al., 2007
ITGB3	L33P (Pl^{A1}/Pl^{A2})	Cardiac/CPB	n.r.	Mathew et al., 2001
APOE	ε4	CABG/CPB (adults)	n.r.	Tardiff et al., 1997
	ε2	Cardiac/CPB (children)	7; 11	Gaynor et al., 2003; Zeltser et al., 2008
APOE	ε4	CABG/CPB	1.26	McDonagh et al., 2010
Postoperative delirium				
APOE	ε4	Major non-cardiac	3.64	Leung et al., 2007
		Critically ill	7.32	Ely et al., 2007

OR, odds ratio; n.r., not reported.

postoperative cardiovascular dysfunction (Galley et al., 2003). In patients undergoing elective surgical revascularization for peripheral vascular disease, several SNPs in *IL6* and *IL10* were associated with endothelial dysfunction and increased risk of a composite endpoint of acute postoperative cardiovascular events (Stoica et al., 2010). Overall, while genetic factors may

not be better predictors of outcomes than intermediate phenotypes (e.g., plasma cytokine levels), their greater ease of assessment, stability, and availability pre-procedure are significant advantages influencing potential future clinical utility.

C-reactive protein (CRP) is the prototypical acute-phase reactant and the most extensively studied inflammatory

TABLE 55.4	Representative genetic polymorphisms associated with other adverse perioperative outcomes				
Gene	**Polymorphism**	**Type of surgery**	**OR**	**Reference**	
Perioperative thrombotic events					
F5	FVL	Non-cardiac, cardiac	n.r.	Donahue, 2004	
Perioperative bleeding					
F5	R506Q(FVL)	Cardiac/CPB	−1.25[a]	Donahue et al., 2003	
PAI-1	4G/5G		10[b]	Duggan et al., 2007	
ITGA2	-52C>T, 807C>T	CABG/CPB	−0.15[a]	Welsby et al., 2005	
GP1BA	T145M		−0.22[a]	Welsby et al., 2005	
TF	-603A>G		−0.03[a]	Welsby et al., 2005	
TFPI	-399C>T		−0.05[a]	Welsby et al., 2005	
F2	20210G>A		0.38[a]	Welsby et al., 2005	
ACE	In/del		0.15[a]	Welsby et al., 2005	
ITGB3	L33P (Pl[A1]/Pl[A2])		n.r.	Morawski et al., 2005	
PAI-1	4G/5G	Cardiac/CPB	10b	Duggan et al., 2007	
TNFA	-238G>A	Brain AVM treatment	3.5[c]	Achrol et al., 2007	
APOE	ε2		10.9[c]	Achrol et al., 2007	
ELAM-1	98 G/T	CABG/CPB	n.r.	Welsby et al., 2010	
	561 A/C				
Perioperative acute kidney injury					
IL6	-572G>C	CABG/CPB	20.04[d]	Stafford-Smith et al., 2005	
AGT	M235T		32.19[d]	Stafford-Smith et al., 2005	
NOS3	E298D		4.29[d]	Stafford-Smith et al., 2005	
APOE	ε4		−0.13[a]	MacKensen et al., 2004; Stafford-Smith et al., 2005	
Perioperative severe sepsis					
APOE	ε3		0.28[e]	Moretti et al., 2005	

n.r., not reported; OR, odds ratio; AVM, arterio-venous malformations.
[a]β-coefficient;
[b]odds ratio;
[c]hazard ratio;
[d]F-value;
[e]relative risk.

marker in clinical studies, and high-sensitivity CRP (hs-CRP) has emerged as a robust predictor of cardiovascular risk at all stages, from healthy subjects to patients with acute coronary syndromes and acute decompensated heart failure (Willerson and Ridker, 2004). Whether CRP is merely a marker or is also a mediator of inflammatory processes is yet unclear, but several lines of evidence support the latter theory. In perioperative medicine, elevated preoperative CRP levels have been associated with increased short- and long-term morbidity and mortality in patients undergoing primary elective CABG (cutoff >3 mg/l) (Perry et al., 2010) as well as in higher-acuity CABG patients (cutoff >10 mg/l) (Kangasniemi et al., 2006). Interestingly, in a retrospective analysis of patients with elevated baseline hs-CRP levels undergoing off-pump CABG surgery, preoperative statin therapy was associated with reduced postoperative myocardial injury and need for dialysis (Song et al., 2010). In elective major non-cardiac surgery patients, preoperative CRP levels (cutoff >3.4 mg/l) independently predicted perioperative major cardiovascular events [composite of myocardial infarction (MI), pulmonary edema, cardiovascular death] and significantly improved the predictive power of revised cardiac risk index (RCRI) in receiver operating characteristic analysis (Choi et al., 2010). In addition to the already established heritability of elevated baseline plasma CRP levels, recent reports indicate that the acute phase rise in postoperative plasma CRP levels is also genetically determined. The CRP1059G>C polymorphism was associated with lower peak postoperative serum CRP following both elective CABG with CPB (Perry et al., 2009) and esophagectomy for thoracic esophageal cancer (Motoyama et al., 2009). Furthermore, CRP-717C>T polymorphism was associated with stress hyperglycemia in patients undergoing esophagectomy for cancer leading to increased postoperative infectious complications and length of intensive care unit (ICU) stay (Motoyama et al., 2010).

Thrombosis Biomarkers and Perioperative Myocardial Outcomes

The host response to surgery is also characterized by alterations in the coagulation system, manifested as increased fibrinogen concentration, platelet adhesiveness, and plasminogen activator inhibitor-1 (PAI-1) production. Those changes can be more pronounced after cardiac surgery, where the complex and multifactorial effects of hypothermia, hemodilution, and CPB-induced activation of coagulation, fibrinolytic, and inflammatory pathways are combined. Dysfunction of the coagulation system following cardiac surgery can manifest on a continuum ranging from increased thrombotic complications such as coronary graft thrombosis, PMI, stroke, and pulmonary embolism at one end of the spectrum, to excessive bleeding at the other extreme. The balance between normal hemostasis, bleeding, and thrombosis is markedly influenced by the rate of thrombin formation and platelet activation, with genetic variability known to modulate each of those mechanistic pathways (Voetsch and Loscalzo, 2004), suggesting significant heritability of the prothrombotic state (see Table 55.4 for an overview of genetic variants associated with postoperative bleeding). Several genotypes in hemostatic genes have been associated with increased risk of coronary graft thrombosis and myocardial injury following CABG. A genetic variant in the promoter of the PAI-1 gene, consisting of an insertion (5G)/deletion (4G) polymorphism at position -675 has been associated with changes in the plasma levels of PAI-1. Since PAI-1 is an important negative regulator of fibrinolytic activity, its polymorphism has been associated with increased risk of early graft thrombosis after CABG (Rifon et al., 1997), and, in a meta-analysis, with increased incidence of MI (Iacoviello et al., 1998). Similarly, a polymorphism in the platelet glycoprotein IIIa gene (ITGB3) resulting in increased platelet aggregation (PI[A2] polymorphism) has been associated with higher levels of postoperative troponin I release following CABG (Rinder et al., 2002), and with increased risk of thrombotic coronary graft occlusion, myocardial infarction,

and death one year following CABG (Zotz et al., 2000). In the setting of non-cardiac surgery, two polymorphisms in platelet glycoprotein receptors (ITGB3 and GP1BA) have been shown to be independent risk predictors of PMI in patients undergoing major vascular surgery, and resulted in improved discrimination in an ischemia risk assessment tool when added to historic and procedural risk factors (Faraday et al., 2004). Finally, a point mutation in coagulation factor V resulting in resistance to activated protein C (factor V Leiden) was also associated with various postoperative thrombotic complications following non-cardiac surgery (Donahue, 2004). Conversely, in patients undergoing cardiac surgery, factor V Leiden was associated with significant reductions in postoperative blood loss and overall risk of transfusion (Donahue et al., 2003). Nevertheless, in a prospective study of CABG patients with routine three-month postoperative angiographic follow-up, carriers of factor V Leiden had a higher incidence of graft occlusion (Moor et al., 1998).

Natriuretic Peptides and Perioperative Myocardial Outcomes

Circulating B-type natriuretic peptide (BNP) is a powerful biomarker of cardiovascular outcomes in many circumstances. Produced mainly in the ventricular myocardium, BNP is formed by cleavage of its prohormone by the enzyme corin into the biologically active C-terminal fragment (BNP) and an inactive N-terminal fragment (NT-proBNP). Known stimuli of BNP activation are myocardial mechanical stretch (from volume or pressure overload), acute ischemic injury, and a variety of other pro-inflammatory and neurohormonal stimuli inducing myocardial stress. Although secreted in 1:1 ratio, circulating levels of BNP and NT-proBNP differ considerably due to different clearance characteristics.

A large number of studies have reported consistent associations of baseline plasma BNP or NT-proBNP levels with a variety of postoperative short- and long-term morbidity and mortality endpoints, independent of the traditional risk factors. For non-cardiac surgery, these have been summarized in two meta-analyses that overall indicate an approximately 20-fold increase in risk of adverse perioperative cardiovascular outcomes (Karthikeyan et al., 2009; Ryding et al., 2009). Similarly, for cardiac surgery patients, preoperative BNP was a strong independent predictor of in-hospital postoperative ventricular dysfunction, length of hospital stay and five-year mortality following primary CABG (Fox et al., 2008), performing better than peak postoperative BNP (Fox et al., 2010). The current guidelines for preoperative cardiac risk assessment in non-cardiac surgery list BNP and NT-proBNP measurements as class IIa/level B indications (Poldermans et al., 2009). However, despite the large number of studies conducted in both cardiac and non-cardiac surgery, precise cut-off levels for BNP still need to be determined and adjusted for age, gender, and renal function. Similarly, no BNP-based, goal-directed therapies have been reported in the perioperative period.

However, a role for BNP assays in monitoring aortic valve disease for optimal timing of surgery has been described (Shaw et al., 2008).

Furthermore, a recent study identified genetic variation in natriuretic peptide precursor genes (*NPPA/NPPB*) to be independently associated with decreased risk of postoperative ventricular dysfunction following primary CABG, whereas variants in natriuretic peptide receptor *NPR3* were associated with an increased risk (Table 55.2) (Fox et al., 2009), offering additional clues into the molecular mechanisms underlying postoperative ventricular dysfunction.

The Role of Genetic Variability in Perioperative Vascular Reactivity

The perioperative period is characterized by robust activation of the sympathetic nervous system, which plays an important role in the pathophysiology of PMI. Thus, patients with coronary artery disease (CAD) who carry specific polymorphisms in adrenergic receptor (AR) genes can be at high risk for catecholamine toxicity and cardiovascular complications. Several functional variants modulating the AR pathways have been described (Zaugg and Schaub, 2005). The Arg389Gly polymorphism in the β_1-AR gene (*ADRB1*) has been associated with increased risk of composite cardiovascular morbidity at one year after non-cardiac surgery under spinal anesthesia (Zaugg et al., 2007), whereas, surprisingly, perioperative beta-blockade had no effect. These findings prompted the investigators to suggest that stratification by AR genotype in future trials may help identify patients likely to benefit from perioperative beta-blocker therapy. Significantly increased vascular responsiveness to alpha-adrenergic stimulation (phenylephrine) has been observed in carriers of the endothelial nitric oxide synthase (*NOS3*) 894>T polymorphism (Philip et al., 1999), and angiotensin-converting enzyme (ACE) insertion/deletion (indel) polymorphism (Henrion et al., 1999; Lasocki et al., 2002) undergoing cardiac surgery with CPB. Differences in perioperative vascular reactivity in relation to genetic variants of the β_2-AR gene (*ADRB2*) have also been noted in patients undergoing non-cardiac surgery. In patients with a common functional *ADRB2* SNP (Glu27), increased blood pressure responses to endotracheal intubation were observed in one study (Kim et al., 2002). In a different study of obstetric patients who had spinal anesthesia for cesarean delivery, the incidence and severity of maternal hypotension and response to treatment was affected by *ADRB2* genotype (Gly16 and/or Glu27 led to lower vasopressor use for the treatment of hypotension) (Smiley et al., 2006). In patients undergoing cardiac surgery, the frequently observed vasoplegic syndrome and vasopressor requirements have been associated with a common polymorphism in the dimethylarginine dimethylaminohydrolase II (*DDAH II*) gene, an important regulator of nitric oxide synthase activity (Ryan et al., 2006).

Two recent studies have reported the association of a common SNP at the 9p21 locus with both perioperative myocardial injury independent of CAD severity (Liu et al., 2010), and all-cause mortality after primary CABG, independent of the extent of myocardial injury (Muehlschlegel et al., 2010). This SNP has previously been correlated with a wide array of vascular phenotypes in ambulatory populations (including CAD, MI, carotid atherosclerosis, abdominal aortic aneurysms, intracranial aneurysms) in replicated genome-wide association study (GWAS) analyses. The mechanism of action of this variant in the development of PMI and mortality is not understood, but it involves altered regulation of cell proliferation, senescence, and apoptosis. It seems that cardiac surgery with CPB may trigger the effects of the 9p21 gene variant, leading to accumulation of senescent cells or cells that show evidence of necrotic death with cellular edema and lysis.

PERIOPERATIVE ATRIAL FIBRILLATION

Perioperative atrial fibrillation (PoAF) remains a significant clinical problem after cardiac and non-cardiac thoracic procedures. With an incidence of 27%–40%, PoAF is associated with increased morbidity, length of hospital stay, rehospitalization, and healthcare costs, and reduced survival. The high incidence of PoAF has prompted several investigators to develop comprehensive risk indices for the prediction of PoAF based on demographic, clinical, electrocardiographic, and procedural risk factors. Nevertheless, the predictive accuracy of these risk indices remains limited (Mathew et al., 2004), suggesting that genetic variation may play a significant role in the occurrence of PoAF. Heritable forms of atrial fibrillation (AF) have been described in the ambulatory non-surgical population, and it appears that both monogenic forms like "lone" AF and polygenic predisposition to more common, acquired forms such as PoAF do exist (Brugada, 2005). A recent GWAS for AF found two polymorphisms on chromosome 4q25 to be significantly associated with AF (Gudbjartsson et al., 2007); findings replicated in other patient groups from Sweden, the United States, and Hong Kong, and subsequently also associated with increased risk of ischemic strokes and AF recurrence after catheter ablation. Recently, this locus was also associated with new-onset PoAF after cardiac surgery with CPB (CABG with or without concurrent valve surgery) (Body et al., 2009). The results were validated in an independent study that also identified associations with increased risk of postoperative long-term AF and mortality, but not with long-term stroke, and, interestingly, suggested differential therapeutic responses in carriers of those SNPs (increased risk with beta-blockers, reduced risk with statins) (Virani et al., 2011). The mechanism of action of the genetic locus identified by these two noncoding SNPs remains unknown, but it lies close to several genes involved in the development of the pulmonary myocardium, or the sleeve of cardiomyocytes extending from the left atrium into the initial portion of the pulmonary veins. Clinical studies have demonstrated that ectopic foci of electric activity arising from within the pulmonary veins and posterior left atrium play a substantial role in initiating and maintaining AF.

Other candidate susceptibility genes for PoAF include those that determine the duration of action potential (voltage-gated ion channels, ion transporters), responses to extracellular factors (adrenergic and other hormone receptors, heat shock proteins), remodeling processes, and magnitude of inflammatory and oxidative stress. It has been described that inflammation, reflected by elevated baseline CRP or IL-6 levels and exaggerated postoperative leukocytosis, predicts the occurrence of PoAF. A link between inflammation and the development of PoAF is also supported by evidence that postoperative administration of non-steroidal anti-inflammatory drugs may reduce the incidence of PoAF. Several recent studies have found that a functional SNP in the *IL6* promoter is associated with higher perioperative plasma IL-6 levels and several adverse outcomes after CABG, including PoAF (Gaudino et al., 2003a, b; Motsinger et al., 2006). In non-cardiac surgery, polymorphisms in the *IL6* and *TNFA* genes have been shown to be associated with an increased risk of postoperative morbidity, including new-onset arrhythmias (Shaw et al., 2005). There is, however, a contradictory lack of association between C-reactive protein levels (strongly regulated by IL-6) and PoAF in women undergoing cardiac surgery (Hogue et al., 2006), which may reflect gender-related differences. On the other hand, a recent study reported that both pre- and postoperative PAI-1 levels were independently associated with development of PoAF following cardiac surgery (Pretorius et al., 2007).

POSTOPERATIVE EVENT-FREE SURVIVAL

Several large randomized clinical trials examining the benefits of CABG surgery and percutaneous coronary interventions relative to medical therapy and/or to one another have refined our knowledge of early and long-term survival after CABG. While these studies have helped define the subgroups of patients who benefit from surgical revascularization, they also demonstrated substantial variability in long-term survival after CABG, that was altered by important demographic and environmental risk factors. Increasing evidence suggests that the *ACE* gene indel polymorphism may influence post-CABG complications, with carriers of the *D* allele having higher mortality and restenosis rates after CABG surgery compared with those with the allele (Volzke et al., 2002). As discussed above, a prothrombotic amino acid alteration in the β_3-integrin chain of the glycoprotein IIb/IIIa platelet receptor (the PI^{A2} polymorphism) is associated with increased risk for major adverse cardiac events (a composite of myocardial infarction, coronary bypass graft occlusion, or death) following CABG surgery (Table 55.2) (Zotz et al., 2000). We found preliminary evidence for association of two functional SNPs modulating β_2-adrenergic receptor activity (Arg16Gly and Gln27Glu) with incidence of death or major adverse cardiac events following cardiac surgery (Podgoreanu et al., 2003), and

recently identified in replicated analyses a functional polymorphism in thrombomodulin (*THBD* Ala455Val) to be independently associated with increased five-year mortality after CABG and to improve the predictive accuracy of EuroSCORE (Lobato et al., 2011).

POSTOPERATIVE STROKE AND NEUROCOGNITIVE DYSFUNCTION

Despite advances in surgical and anesthetic techniques, significant neurological morbidity continues to occur following cardiac surgery, ranging in severity from coma and focal stroke (incidence 1%–3%) to more subtle cognitive deficits (incidence up to 69%), with a substantial impact on the risk of perioperative death, quality of life, and resource utilization. Variability in the reported incidence of both early and late neurological deficits remains poorly explained by procedural risk factors, suggesting that environmental (operative) and genetic factors may interact to determine disease onset, progression, and recovery. The pathophysiology of perioperative neurological injury is thought to involve complex interactions between primary pathways associated with atherosclerosis and thrombosis, and secondary response pathways such as inflammation, vascular reactivity, and direct cellular injury. Many functional genetic variants have been reported in each of these mechanistic pathways involved in modulating the magnitude and the response to neurological injury, which may have implications in chronic as well as acute perioperative neurocognitive outcomes. For example, Grocott and colleagues examined 26 SNPs in relationship to the incidence of acute postoperative ischemic stroke in 1635 patients undergoing cardiac surgery, and found that the interaction of minor alleles of the C-reactive protein (1846C>T) and IL-6 promoter SNP -174G>C significantly increases the risk of acute stroke (Grocott et al., 2005). Similarly, a recent study suggests that the P-selectin and CRP genes both contribute to modulating susceptibility to postoperative cognitive decline (POCD) following cardiac surgery (Mathew et al., 2007). Specifically, the loss-of-function minor alleles of *CRP* 1059G>C and *SELP* 1087G>A are independently associated with a reduction in the observed incidence of POCD after adjustment for known clinical and demographic covariates (Table 55.3).

Our group has previously demonstrated a significant association between the apolipoprotein E (*APOE*) E4 genotype and adverse cerebral outcomes in cardiac surgery patients (Newman et al., 2001a; Tardiff et al., 1997). This is consistent with the role of the *APOE* genotype in recovery from acute brain injury, such as intracranial hemorrhage (Alberts et al., 1995), closed head injury (Teasdale et al., 1997), and stroke (Slooter et al., 1997), as well as experimental models of cerebral ischemia-reperfusion injury (Sheng et al., 1998). Two subsequent studies in CABG patients, however, have not replicated those initial findings. Furthermore, the incidence of postoperative delirium following major non-cardiac surgery in the elderly (Leung et al., 2007) and in critically ill patients (Ely et al., 2007) is increased in carriers of the *APOE* ε4 allele. Unlike adult cardiac surgery patients, infants possessing the *APOE* ε2 allele are at increased risk for developing adverse neurodevelopmental sequelae following cardiac surgery (Gaynor et al., 2003; Zeltser et al., 2008). The mechanisms by which the *APOE* genotypes might influence neurological outcomes have yet to be determined, but do not seem to be related to alterations in global cerebral blood flow of oxygen metabolism during CPB (Ti et al., 2001). However, genotypic effects in modulating the inflammatory response (Grocott et al., 2001), extent of aortic atheroma burden (Ti et al., 2003), and risk for premature coronary atherosclerosis (Newman et al., 2001b) may play a role.

Recent studies have suggested a role for platelet activation in the pathophysiology of adverse neurological sequelae. Genetic variants in surface platelet membrane glycoproteins, important mediators of platelet adhesion and platelet–platelet interactions, have been shown to increase susceptibility to prothrombotic events. Among those, the PI^{A2} polymorphism in glycoprotein IIb/IIIa has been related to various adverse thrombotic outcomes, including acute coronary thrombosis (Weiss et al., 1996) and atherothrombotic stroke (Carter et al., 1998). We found the PI^{A2} allele to be associated with more severe neurocognitive decline after CPB (Mathew et al., 2001), which could represent exacerbation of platelet-dependent thrombotic processes associated with plaque embolism.

In non-cardiac surgery, a study conducted in patients undergoing carotid endarterectomy has demonstrated that preoperative plasma levels of fibrinogen and high-sensitivity CRP (hs-CRP) were independently associated with new periprocedural cerebral ischemic lesions caused by microembolic events, as determined by diffusion-weighted magnetic resonance imaging (MRI) (Heider et al., 2007).

PERIOPERATIVE ACUTE KIDNEY INJURY

Acute kidney injury (AKI) is a common, serious complication of cardiac surgery; about 8%–15% of patients develop moderate renal injury (>1.0 mg/dl peak creatinine rise), and up to 5% of them develop renal failure requiring dialysis (Mangano et al., 1998). AKI is independently associated with in-hospital mortality rates exceeding 60% in patients requiring dialysis (Mangano et al., 1998). Several studies have demonstrated that inheritance of genetic polymorphisms in the *APOE* gene (ε4 allele) (MacKensen et al., 2004) and in the promoter region of the *IL6* gene (Gaudino et al., 2003b) are associated with AKI following CABG surgery (Table 55.4). Stafford-Smith and colleagues have reported that major differences in peak postoperative serum creatinine rise after CABG are predicted by possession of combinations of polymorphisms that, interestingly, differ by race: the angiotensinogen (*AGT*) 842T>C and *IL6* -572G>C

variants in Caucasians, and the endothelial nitric oxide synthase (*NOS3*) 894G>T and angiotensin-converting enzyme (*ACE*) indel in African-Americans, are associated with more than 50% reduction in postoperative glomerular filtration rate (Stafford-Smith et al., 2005). Recently, pre-operative BNP levels have been shown to be associated with AKI following cardiac surgery, and modestly improved risk prediction compared to clinical parameters alone (Patel et al., 2012). Further identification of biomarkers predictive of adverse perioperative renal outcomes could facilitate individually tailored therapy, risk stratification of patients for interventional trials targeting the gene product itself, and aid in medical decision-making (e.g., selecting medical over surgical management).

In summary, the above sections illustrate that variation in genes encoding important pathways involved in the host response to injury seem to confer a constitutional predisposition to adverse perioperative and long-term postoperative outcomes following surgery or trauma. A next crucial step in understanding the complexity of adverse perioperative outcomes is to assess the contribution of variations in many genes simultaneously and their interaction with traditional risk factors to the longitudinal prediction of outcomes in individual patients. The use of such outcome predictive models that incorporate genetic information may help stratify mortality and morbidity in surgical patients, improve prognostication, direct medical decision-making both intra-operatively and during postoperative follow-up, and may even suggest novel targets for therapeutic intervention in the perioperative period.

DYNAMIC GENOMIC MARKERS OF PERIOPERATIVE OUTCOMES

The "static" view of constitutive DNA sequence variants potentially involved in the pathophysiology of perioperative complications can be complemented by a "dynamic" view that integrates their functionality. There is increasing evidence that variability in gene expression levels underlies complex disease and is determined by regulatory DNA polymorphisms affecting transcription, splicing, and translation efficiency in a tissue- and stimulus-specific manner (Stranger et al., 2007). Thus, analysis of large-scale variability in the pattern of RNA and protein expression both at baseline and in response to the multidimensional perioperative stimuli (*dynamic genomics*) using microarray and proteomic approaches provides a much-needed understanding of the overall regulatory networks involved in the pathophysiology of adverse postoperative outcomes. This information is complementary to the assessment of genetic variability at the DNA sequence level using various genotyping techniques, as described in previous sections (*static genomics*). Such dynamic genomic markers can be incorporated in genomic classifiers and used clinically to improve perioperative risk stratification or to monitor postoperative recovery (Hopf,

2003). This concept of *molecular classification* involves the description of informational features in a training dataset, using changes in relative RNA and protein abundance in the context of genetic predisposition, and applying them to a test dataset to recognize a defined "fingerprint" characteristic of a particular perioperative phenotype (Table 55.5). For example, Feezor and colleagues used a combined genomic and proteomic approach to identify expression patterns of 138 genes from peripheral blood leukocytes and the concentrations of seven circulating plasma proteins, a fingerprint that discriminated patients who developed multiple organ dysfunction syndrome (MODS) after thoracoabdominal aortic aneurysm repair from those who did not. Importantly, these patterns of genome-wide gene expression and plasma protein concentration were observed *before* surgical trauma and visceral ischemia-reperfusion injury, suggesting that patients who developed MODS differed in either their genetic predisposition or their pre-existing inflammatory state (Feezor et al., 2004).

Alternatively, dynamic genomic markers can be used to improve mechanistic understanding of perioperative stress and to evaluate and catalog organ-specific responses to surgical stress and severe systemic stimuli such as CPB and endotoxemia, which can be subsequently used to identify and validate novel targets for organ protective strategies (Hughes et al., 2000). Similarly, integrated transcriptomic and proteomic analyses have characterized the peripheral blood molecular response signatures to cardiac surgery with and without CPB, a robust trigger of systemic inflammation (Tomic et al., 2005). The study demonstrated that rather than being the primary source of serum cytokines, peripheral blood leukocytes only assume a "primed" phenotype upon contact with the extracorporeal circuits that facilitate their trapping and subsequent tissue-associated inflammatory response. Interestingly, many inflammatory mediators achieved similar systemic levels following off-pump surgery, but with delayed kinetics, offering novel insights into the concepts of contact activation and compartmentalization of inflammatory responses to major surgery.

Several studies have profiled myocardial gene expression in the ischemic heart, demonstrating alterations in the expression of immediate-early genes (*c-fos*, *junB*) as well as genes coding for calcium-handling proteins (calsequestrin, phospholamban), extracellular matrix, and cytoskeletal proteins (Sehl et al., 2000). Upregulation of transcripts mechanistically involved in cytoprotection (heat shock proteins), resistance to apoptosis, and cell growth has been found in stunned myocardium (Depre et al., 2001). Moreover, cardiac gene expression profiling after cardiopulmonary bypass and cardioplegic arrest has identified the upregulation of inflammatory and transcription activators, apoptotic genes, and stress genes (Ruel et al., 2003), which appear to be age-related (Konstantinov et al., 2004). Microarray technology has also been used in the quest for novel cardioprotective genes, with the ultimate goal of designing strategies

to activate those genes and prevent myocardial injury. Preconditioning is one such well-studied model of cardioprotection. It can be induced by various triggers, including intermittent ischemia, osmotic or redox stress, heat shock, toxins, and inhaled anesthetics. The main functional categories of genes identified as potentially involved in cardioprotective pathways include a host of transcription factors, heat shock proteins, antioxidant genes (heme-oxygenase, glutathione peroxidase), and growth factors, but different gene programs appear to be activated in ischemic versus anesthetic preconditioning, resulting in two distinct cardioprotective phenotypes (Sergeev et al., 2004). More recently, a transcriptional response pattern consistent with late preconditioning has been reported in peripheral blood leukocytes following sevoflurane administration in healthy volunteers. This pattern is characterized by reduced expression of L-selectin, as well as downregulation of genes involved in fatty acid oxidation and the PGC-1α (peroxisome proliferators-activated receptor γ coactivator-1α) pathway (Lucchinetti et al., 2007a), which mirrors changes observed in the myocardium from patients undergoing off-pump CABG (Table 55.5) (Lucchinetti et al., 2007b). Deregulation of these novel survival pathways thus appears to generalize across tissues, making them important targets for cardioprotection, but further studies are needed to correlate perioperative gene expression response patterns in end organs, such as the myocardium, to those in readily available surrogate tissues, such as peripheral blood leukocytes.

Investigations of the transcriptional responses to AF in human atrial appendage myocardium collected at the time of cardiac surgery or in preclinical models (Table 55.5) have identified a ventricular-like genomic signature in fibrillating atria, with increased ratios of ventricular to atrial isoforms, suggesting dedifferentiation (Barth et al., 2005). It remains unclear whether this "ventricularization" of atrial gene expression reflects a cause or an effect of AF, but it likely represents an adaptive energy-saving process to the high metabolic demand of fibrillating atrial myocardium, akin to chronic hibernation. Recently, a different mechanism has been proposed to be involved in PoAF. It has been found that patients who exhibit PoAF after cardiac surgery display a differential genomic response to CPB in their peripheral blood leukocytes, characterized by upregulation of oxidative stress genes, which correlated with a significantly larger increase in oxidant stress both systemically (as measured by total peroxide levels) as well as at the myocardial level (as measured in the right atrium) (Ramlawi et al., 2007a).

Cardiac surgical patients who develop postoperative cognitive dysfunction (POCD) demonstrate inherently different genetic responses to cardiopulmonary bypass from those without POCD, as evidenced by acute deregulation in peripheral blood leukocytes of gene expression pathways involving inflammation, antigen presentation, and cellular adhesion (Ramlawi et al., 2007b). Those findings corroborate with proteomic changes, in which patients with POCD similarly

TABLE 55.5		Summary of gene expression studies with implications for perioperative cardiovascular outcomes	
Tissue (species)	**Stimulus/method**	**Genomic signature: number/types of genes**	**Reference(s)**
Myocardium (rat)	Ischemia/μA	14 (wound-healing, Ca-handling)	Sehl et al., 2000
Myocardium (human)	CPB/circulatory arrest/μA	58 (inflammation, transcription activators, apoptosis, stress response) – adults	Konstantinov et al., 2004; Ruel et al., 2003
		50 (cardioprotective, antiproliferative, antihypertrophic) – neonates	
Myocardium (rat)	IPC vs APC/μA	566 differentially regulated/56 jointly regulated (cell defense)	Sergeev et al., 2004
Myocardium (rat)	APC vs APostC/μA	Opposing genomic profiles, 8 gene clusters, <2% jointly regulated genes	Lucchinetti et al., 2005
Myocardium (human)	APC, OPCAB,	319 upregulated and 281 downregulated gene sets in response to OPCAB; deregulation of fatty acid oxidation, DNA-damage signaling and G-CSF	Lucchinetti et al., 2007b
	postoperative LV function/μA	Survival (perioperative) and PGC-1α (constitutive) pathways predict improved LV function in sevoflurane treated patients	
PBMC (human)	APC, sevoflurane/μA	Deregulation of late preconditioning, PGC-1α, fatty acid oxidation, and L-selectin pathways	Lucchinetti et al., 2007a
Atrial myocardium (pig)	Pacing-induced AF/μA+P	81 (MCL-2 ventricular/atrial isoform shift)	Lai et al., 2004
Atrial myocardium (human)	AF/μA	1434 (ventricular-like genomic signature)	Barth et al., 2005
PBMC (human)	Cardiac surgery, PoAF/μA	1303 genes uniquely deregulated in PoAF/401 upregulated (oxidative stress), 902 downregulated	Ramlawi et al., 2007a
PBMC (human)	Cardiac surgery, POCD/μA	1201 genes uniquely deregulated in POCD/531 upregulated, 670 downregulated (inflammation, antigen presentation, cell adhesion, and apoptosis)	Ramlawi et al., 2007b
PBMC (human)	Heart transplant/μA	30 (profile correlated with biopsy-proven rejection; persistent immune activation in response to treatment)	Horwitz et al., 2004
PBMC (human)	Heart transplant /RT-PCR	11 (AlloMap, AlloMap score)	Pham et al., 2010
Myocardium (human)	Heart transplant/P	2 (increased αB-crystallin and tropomyosin serum levels)	Borozdenkova et al., 2004
PBMC, plasma (human)	TAAA/μA+P	138 genes and 7 plasma proteins predicted MODS	Feezor et al., 2004
PBMC (human)	Obstructive CAD in non-diabetic patients/RT-PCR	23-gene expression signature	Rosenberg et al., 2010
Ventricular myocardium (human)	End-stage cardiomyopathy on LVAD/μA	Combined signature of 28 microRNAs and 29 mRNAs had superior performance to classify status and predict recovery	Matkovich et al., 2009

AF, atrial fibrillation; APC, anesthetic preconditioning, APostC, anesthetic postconditioning; G-CSF granulocyte colony stimulating factor; IPC, ischemic preconditioning; LV, left ventricle; μA, microarray; MCL-2, myosin light chain 2; MODS, multiple organ dysfunction syndrome; OPCAB, off-pump coronary artery bypass; P, proteomics; PBMC, peripheral blood mononuclear cells; PGC-1α, peroxisome proliferators-activated receptor γ cofactor-1α; PoAF, postoperative atrial fibrillation; POCD, postoperative cognitive decline; RT-PCR, real time polymerase chain reaction; TAAA, thoracoabdominal aortic aneurysm repair; LVAD, left ventricular assist device.

have significantly higher serologic inflammatory indices compared with patients without POCD (Ramlawi et al., 2006a, b). This adds to the increasing evidence that CPB does not cause an indiscriminate variation in gene expression, but rather distinct patterns in specific pathways that are highly associated with the development of postoperative complications such as POCD. Implications for perioperative medicine include identifying populations at risk who might benefit not only from improved informed consent, stratification, and resource allocation, but also from targeted anti-inflammatory strategies.

Because alternative splicing, a wide variety of post-translational modifications, and protein–protein interactions responsible for biological function would remain undetected by gene expression profiling, proteomic studies specifically examine dynamic protein products, with the goal of identifying proteins that undergo changes in abundance, modification, or localization in response to a particular disease state, trauma, stress, or therapeutic intervention (for a review, see Atkins and Johansson, 2006). Several preclinical proteomic studies relevant to perioperative medicine have

characterized the temporal changes in brain protein expression in response to various inhaled anesthetics (Futterer et al., 2004; Kalenka et al., 2007), or following cardiac surgery with hypothermic circulatory arrest (Sheikh et al., 2006). This may focus further studies aimed at identifying new anesthetic binding sites, and the development of neuroprotective strategies. Furthermore, detailed knowledge of the plasma proteome has profound implications in perioperative transfusion medicine (Queloz et al., 2006), in particular related to peptide and protein changes that occur during storage of blood products. The development of protein arrays and real-time proteomic analysis technologies has the potential to allow the use of these versatile and rigorous high-throughput methods for clinical applications, and is the object of intense investigation.

Emerging metabolomic tools have created the opportunity to establish metabolic signatures of myocardial injury. In a population of patients undergoing alcohol septal ablation for hypertrophic obstructive cardiomyopathy, a human model of planned (albeit chemical) myocardial infarction that recapitulates spontaneous myocardial infarction, targeted

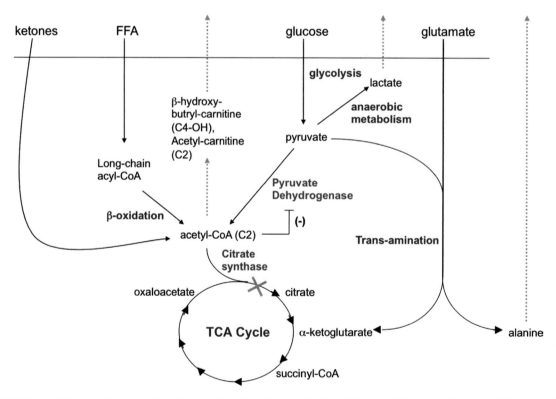

Figure 55.2   Proposed model of myocardial metabolism after surgical ischemia-reperfusion: myocardial uptake of substrates is impaired, particularly in the case of LV dysfunction; fatty acids and ketones are bound to carnitine and transported into the inner mitochondrial matrix where they are oxidized to acetyl-CoA; oxidation of acetyl-CoA (and 3-hydroxybutyryl-CoA) is impaired, and as it accumulates, it is released into plasma in the form of acetylcarnitine; the increased acetyl-CoA levels inhibit pyruvate dehydrogenase, promoting anaerobic metabolism of pyruvate into lactate; additionally, the increased alanine elution postreperfusion suggests that pyruvate and glutamate undergo enhanced transamination to alanine and α-ketoglutarate to restore tricarboxylic acid (TCA) intermediates; FFA, free fatty acids.

mass-spectrometry-based metabolite profiling identified changes in circulating levels of metabolites participating in pyrimidine metabolism, the tricarboxylic acid (TCA) cycle, and the pentose phosphate pathway as early as 10 minutes after MI in an initial derivation group, and these changes were validated in a second, independent group. Coronary sinus sampling distinguished cardiac-derived from peripheral metabolic changes. To assess generalizability, the planned-MI-derived metabolic signature (consisting of aconitic acid, hypoxanthine, trimethylamine N-oxide, and threonine) differentiated with high accuracy patients with spontaneous MI (Lewis et al., 2008). We applied a similar approach to cardiac surgical patients undergoing planned global myocardial ischemia-reperfusion (I/R), and identified clear differences in metabolic fuel uptake based on the pre-existing ventricular state (left ventricular dysfunction, coronary artery disease, or neither) as well as altered metabolic signatures predictive of postoperative hemodynamic course and perioperative myocardial infarction (Turer et al., 2009) (Figure 55.2). While simultaneous assessment of coronary sinus effluent in addition to the peripheral blood improves cardiac specificity of the observed signatures, direct measurements of metabolites in myocardial tissue allows marked enrichment and easier detection of potential biomarkers compared to plasma, as well as an assessment of how metabolic substrates are utilized in the tissue of interest. Such studies are possible in cardiac surgical patients where atrial tissues are routinely removed. For example, one study using high-resolution ^{1}H-nuclear magnetic resonance (NMR) spectroscopy identified alterations in myocardial ketone metabolism associated with persistent atrial fibrillation, and the ratio of glycolytic end products to end products of lipid metabolism correlated positively with time of onset of postoperative atrial fibrillation (Mayr et al., 2008). Furthermore, rapid metabolic phenotyping approaches are emerging for both real-time intraoperative diagnostic (e.g., intelligent surgical knives for tissue/tumor diagnosis) (Balog et al., 2010) and for perioperative prognostic stratification, monitoring, and optimization of surgical (Kinross et al., 2011) and critically injured patients (Cohen et al., 2010).

CONCLUSIONS

Genomic medicine has ushered in a new era of mechanistic understanding of many medical diseases. Recognizing the role of genetic variation in modulating host responses to acute surgical and traumatic injury has also increased understanding of post-procedure adverse events and outcomes in critically ill patients. Most studies published so far in the field of periprocedural genetic risk prediction have been limited in size, mostly unreplicated, and not focused on the development of clinico-genomic predictive scores, and few have reported the predictive accuracy of their models. Therefore, like all other areas of genomic medicine, in order to take full advantage of the unique opportunities offered by the genomic revolution, the cycle of innovation in perioperative medicine must include a comprehensive and standardized definition of the phenotypes of interest (including short-term and long-term adverse outcomes such as organ injury/dysfunction, adverse drug responses, transition to chronic pain), followed by identification of the underlying genes, characterization of the mechanism from DNA to phenotype, and rigorous development and validation of risk prediction and diagnostic or therapeutic applications, using high quality methodology and reporting. The broader potential future impact of such findings stems from the ability to inform therapeutic decisions based on a "personalized" susceptibility profile, for instance in optimizing care for coronary artery disease patients between medical therapy vs percutaneous revascularization vs surgical revascularization using genetically stratified population cohorts in clinical trials, or by testing personalized cardioprotective strategies in patients susceptible to myocardial ischemia-reperfusion injury following elective cardiovascular interventions (e.g., adding targeted anti-inflammatory agents to cardioplegia regimens).

ACKNOWLEDGMENTS

Supported in part by NIH grants HL075273 and HL092071 (to MVP).

REFERENCES

Achrol, A.S., Kim, H., Pawlikowska, L., et al., 2007. Association of tumor necrosis factor-alpha-238G>A and apolipoprotein E2 polymorphisms with intracranial hemorrhage after brain arteriovenous malformation treatment. Neurosurgery 61, 731–739, discussion 740.

Alberts, M.J., Graffagnino, C., McClenny, C., et al., 1995. ApoE genotype and survival from intracerebral haemorrhage. Lancet 346, 575.

Atkins, J.H., Johansson, J.S., 2006. Technologies to shape the future: Proteomics applications in anesthesiology and critical care medicine. Anesth Analg 102, 1207–1216.

Balog, J., Szaniszlo, T., Schaefer, K.C., et al., 2010. Identification of biological tissues by rapid evaporative ionization mass spectrometry. Anal Chem 82, 7343–7350.

Barth, A.S., Merk, S., Arnoldi, E., et al., 2005. Reprogramming of the human atrial transcriptome in permanent atrial fibrillation: Expression of a ventricular-like genomic signature. Circ Res 96, 1022–1029.

Benza, R.L., Coffey, C.S., Pekarek, D.M., et al., 2009. Transforming growth factor-beta polymorphisms and cardiac allograft rejection. J Heart Lung Transpl 28, 1057–1062.

Body, S.C., Collard, C.D., Shernan, S.K., et al., 2009. Variation in the 4q25 chromosomal locus predicts atrial fibrillation after coronary artery bypass graft surgery. Circ Cardiovasc Genet 2, 499–506.

Borozdenkova, S., Smith, J., Marshall, S., Yacoub, M., Rose, M., 2001. Identification of ICAM-1 polymorphism that is associated with protection from transplant associated vasculopathy after cardiac transplantation. Hum Immunol 62, 247–255.

Borozdenkova, S., Westbrook, J.A., Patel, V., et al., 2004. Use of proteomics to discover novel markers of cardiac allograft rejection. J Proteome Res 3, 282–288.

Botto, N., Andreassi, M.G., Rizza, A., et al., 2004. C677T polymorphism of the methylenetetrahydrofolate reductase gene is a risk factor of adverse events after coronary revascularization. Int J Cardiol 96, 341–345.

Brugada, R., 2005. Is atrial fibrillation a genetic disease? J Cardiovasc Electrophysiol 16, 553–556.

Brull, D.J., Montgomery, H.E., Sanders, J., et al., 2001. Interleukin-6 gene -174g>c and -572g>c promoter polymorphisms are strong predictors of plasma interleukin-6 levels after coronary artery bypass surgery. Arterioscler Thromb Vasc Biol 21, 1458–1463.

Burzotta, F., Iacoviello, L., Di Castelnuovo, A., et al., 2001. Relation of the -174 G/C polymorphism of interleukin-6 to interleukin-6 plasma levels and to length of hospitalization after surgical coronary revascularization. Am J Cardiol 88, 1125–1128.

Carter, A.M., Catto, A.J., Bamford, J.M., Grant, P.J., 1998. Platelet GP IIIa PIA and GP Ib variable number tandem repeat polymorphisms and markers of platelet activation in acute stroke. Arterioscler Thromb Vasc Biol 18, 1124–1131.

Choi, J.H., Cho, D.K., Song, Y.B., et al., 2010. Preoperative NT-proBNP and CRP predict perioperative major cardiovascular events in non-cardiac surgery. Heart 96, 56–62.

Cohen, M.J., Grossman, A.D., Morabito, D., Knudson, M.M., Butte, A.J., Manley, G.T., 2010. Identification of complex metabolic states in critically injured patients using bioinformatic cluster analysis. Critical Care 14, R10.

Collard, C.D., Shernan, S.K., Fox, A.A., et al., 2007. The MBL2 'LYQA secretor' haplotype is an independent predictor of postoperative myocardial infarction in whites undergoing coronary artery bypass graft surgery. Circulation 116, I106–I112.

DeLanghe, J., Cambier, B., Langlois, M., et al., 1997. Haptoglobin polymorphism, a genetic risk factor in coronary artery bypass surgery. Atherosclerosis 132, 215–219.

Depre, C., Tomlinson, J.E., Kudej, R.K., et al., 2001. Gene program for cardiac cell survival induced by transient ischemia in conscious pigs. Proc Natl Acad Sci USA 98, 9336–9341.

Domanski, M.J., Mahaffey, K., Hasselblad, V., et al., 2011. Association of myocardial enzyme elevation and survival following coronary artery bypass graft surgery. JAMA 305, 585–591.

Donahue, B.S., 2004. Factor V Leiden and perioperative risk. Anesth Analg 98, 1623–1634.

Donahue, B.S., Gailani, D., Higgins, M.S., Drinkwater, D.C., George Jr., A.L., 2003. Factor V Leiden protects against blood loss and transfusion after cardiac surgery. Circulation 107, 1003–1008.

Donahue, B.S., Roden, D., 2005. Inflammatory cytokine polymorphisms are associated with beta-blocker failure in preventing postoperative atrial fibrillation. Anesth Analg 100 SCA30 (abstract).

Duggan, E., O'Dwyer, M.J., Caraher, E., et al., 2007. Coagulopathy after cardiac surgery may be influenced by a functional plasminogen activator inhibitor polymorphism. Anesth Analg 104, 1343–1347.

Ellis, S.G., Chen, M.S., Jia, G., Luke, M., Cassano, J., Lytle, B., 2007. Relation of polymorphisms in five genes to long-term aortocoronary saphenous vein graft patency. Am J Cardiol 99, 1087–1089.

Ely, E.W., Girard, T.D., Shintani, A.K., et al., 2007. Apolipoprotein E4 polymorphism as a genetic predisposition to delirium in critically ill patients. Crit Care Med 35, 112–117.

Faraday, N., Martinez, E.A., Scharpf, R.B., et al., 2004. Platelet gene polymorphisms and cardiac risk assessment in vascular surgical patients. Anesthesiology 101, 1291–1297.

Feezor, R.J., Baker, H.V., Xiao, W., et al., 2004. Genomic and proteomic determinants of outcome in patients undergoing thoracoabdominal aortic aneurysm repair. J Immunol 172, 7103–7109.

Fox, A.A., Collard, C.D., Shernan, S.K., et al., 2009. Natriuretic peptide system gene variants are associated with ventricular dysfunction after coronary artery bypass grafting. Anesthesiology 110, 738–747.

Fox, A.A., Muehlschlegel, J.D., Body, S.C., et al., 2010. Comparison of the utility of preoperative versus postoperative B-type natriuretic peptide for predicting hospital length of stay and mortality after primary coronary artery bypass grafting. Anesthesiology 112, 842–851.

Fox, A.A., Pretorius, M., Liu, K.Y., et al., 2011. Genome-wide assessment for genetic variants associated with ventricular dysfunction after primary coronary artery bypass graft surgery. PloS ONE 6, e24593.

Fox, A.A., Shernan, S.K., Body, S.C., 2004. Predictive genomics of adverse events after cardiac surgery. Semin Cardiothorac Vasc Anesth 8, 297–315.

Fox, A.A., Shernan, S.K., Collard, C.D., et al., 2008. Preoperative B-type natriuretic peptide is an independent predictor of ventricular dysfunction and mortality after primary coronary artery bypass grafting. J Thorac Cardiovasc Surg 136, 452–461.

Futterer, C.D., Maurer, M.H., Schmitt, A., Feldmann Jr., R.E., Kuschinsky, W., Waschke, K.F., 2004. Alterations in rat brain proteins after desflurane anesthesia. Anesthesiology 100, 302–308.

Galley, H.F., Lowe, P.R., Carmichael, R.L., Webster, N.R., 2003. Genotype and interleukin-10 responses after cardiopulmonary bypass. Br J Anaesth 91, 424–426.

Gaudino, M., Andreotti, F., Zamparelli, R., et al., 2003a. The -174G/C interleukin-6 polymorphism influences postoperative interleukin-6 levels and postoperative atrial fibrillation. Is atrial fibrillation an inflammatory complication? Circulation 108 (Suppl. 1), II195–II199.

Gaudino, M., Di Castelnuovo, A., Zamparelli, R., et al., 2003b. Genetic control of postoperative systemic inflammatory reaction and pulmonary and renal complications after coronary artery surgery. J Thorac Cardiovasc Surg 126, 1107–1112.

Gaynor, J.W., Gerdes, M., Zackai, E.H., et al., 2003. Apolipoprotein E genotype and neurodevelopmental sequelae of infant cardiac surgery. J Thorac Cardiovasc Surg 126, 1736–1745.

Grocott, H.P., Newman, M.F., El-Moalem, H., Bainbridge, D., Butler, A., Laskowitz, D.T., 2001. Apolipoprotein E genotype differentially influences the proinflammatory and anti-inflammatory response to cardiopulmonary bypass. J Thorac Cardiovasc Surg 122, 622–623.

Grocott, H.P., White, W.D., Morris, R.W., et al., 2005. Genetic polymorphisms and the risk of stroke after cardiac surgery. Stroke 36, 1854–1858.

Gudbjartsson, D.F., Arnar, D.O., Helgadottir, A., et al., 2007. Variants conferring risk of atrial fibrillation on chromosome 4q25. Nature 448, 353–357.

Hannan, E.L., Kilburn Jr., H., Racz, M., Shields, E., Chassin, M.R., 1994. Improving the outcomes of coronary artery bypass surgery in New York State. JAMA 271, 761–766.

Heider, P., Poppert, H., Wolf, O., et al., 2007. Fibrinogen and high-sensitive C-reactive protein as serologic predictors for perioperative cerebral microembolic lesions after carotid endarterectomy. J Vasc Surg 46, 449–454.

Henrion, D., Benessiano, J., Philip, I., et al., 1999. The deletion genotype of the angiotensin I-converting enzyme is associated with an increased vascular reactivity in vivo and in vitro. J Am Coll Cardiol 34, 830–836.

Heusch, G., Erbel, R., Siffert, W., 2001. Genetic determinants of coronary vasomotor tone in humans. Am J Physiol Heart Circ Physiol 281, H1465–H1468.

Hogue Jr., C.W., Palin, C.A., Kailasam, R., et al., 2006. C-reactive protein levels and atrial fibrillation after cardiac surgery in women. Ann Thorac Surg 82, 97–102.

Holweg, C.T., Weimar, W., Uiterlinden, A.G., Baan, C.C., 2004. Clinical impact of cytokine gene polymorphisms in heart and lung transplantation. J Heart Lung Transpl 23, 1017–1026.

Hopf, H.W., 2003. Molecular diagnostics of injury and repair responses in critical illness: What is the future of "monitoring" in the intensive care unit? Crit Care Med 31, S518–S523.

Horwitz, P.A., Tsai, E.J., Putt, M.E., et al., 2004. Detection of cardiac allograft rejection and response to immunosuppressive therapy with peripheral blood gene expression. Circulation 110, 3815–3821.

Howell, S.J., Sear, J.W., 2004. Perioperative myocardial injury: Individual and population implications. Br J Anaesth 93, 3–8.

Hughes, T.R., Marton, M.J., Jones, A.R., et al., 2000. Functional discovery via a compendium of expression profiles. Cell 102, 109–126.

Iacoviello, L., Burzotta, F., Di Castelnuovo, A., Zito, F., Marchioli, R., Donati, M.B., 1998. The 4G/5G polymorphism of PAI-1 promoter gene and the risk of myocardial infarction: A meta-analysis. Thromb Haemost 80, 1029–1030.

Iribarren, J.L., Sagasti, F.M., Jimenez, J.J., et al., 2008. TNFbeta + 250 polymorphism and hyperdynamic state in cardiac surgery with extracorporeal circulation. Interact Cardiovasc Thorac Surg 7, 1071–1074.

Kalenka, A., Hinkelbein, J., Feldmann Jr., R.E., Kuschinsky, W., Waschke, K.F., Maurer, M.H., 2007. The effects of sevoflurane anesthesia on rat brain proteins: A proteomic time-course analysis. Anesth Analg 104, 1129–1135.

Kangasniemi, O.P., Biancari, F., Luukkonen, J., et al., 2006. Preoperative C-reactive protein is predictive of long-term outcome after coronary artery bypass surgery. Eur J Cardiothorac Surg 29, 983–985.

Karthikeyan, G., Moncur, R.A., Levine, O., et al., 2009. Is a pre-operative brain natriuretic peptide or N-terminal pro-B-type natriuretic peptide measurement an independent predictor of adverse cardiovascular outcomes within 30 days of noncardiac surgery? A systematic review and meta-analysis of observational studies. J Am Coll Cardiol 54, 1599–1606.

Kim, N.S., Lee, I.O., Lee, M.K., Lim, S.H., Choi, Y.S., Kong, M.H., 2002. The effects of beta2 adrenoceptor gene polymorphisms on pressor response during laryngoscopy and tracheal intubation. Anaesthesia 57, 227–232.

Kinross, J.M., Holmes, E., Darzi, A.W., Nicholson, J.K., 2011. Metabolic phenotyping for monitoring surgical patients. Lancet 377, 1817–1819.

Konstantinov, I.E., Coles, J.G., Boscarino, C., et al., 2004. Gene expression profiles in children undergoing cardiac surgery for right heart obstructive lesions. J Thorac Cardiovasc Surg 127, 746–754.

Lai, L.P., Lin, J.L., Lin, C.S., et al., 2004. Functional genomic study on atrial fibrillation using cDNA microarray and two-dimensional protein electrophoresis techniques and identification of the myosin regulatory light chain isoform reprogramming in atrial fibrillation. J Cardiovasc Electrophysiol 15, 214–223.

Landesberg, G., Beattie, W.S., Mosseri, M., Jaffe, A.S., Alpert, J.S., 2009. Perioperative myocardial infarction. Circulation 119, 2936–2944.

Lasocki, S., Iglarz, M., Seince, P.F., et al., 2002. Involvement of renin-angiotensin system in pressure-flow relationship: Role of angiotensin-converting enzyme gene polymorphism. Anesthesiology 96, 271–275.

Lee, T.H., Marcantonio, E.R., Mangione, C.M., et al., 1999. Derivation and prospective validation of a simple index for prediction of cardiac risk of major noncardiac surgery. Circulation 100, 1043–1049.

Lehmann, L.E., Schroeder, S., Hartmann, W., et al., 2006. A single nucleotide polymorphism of macrophage migration inhibitory factor is related to inflammatory response in coronary bypass surgery using cardiopulmonary bypass. Eur J Cardiothorac Surg 30, 59–63.

Leung, J.M., Sands, L.P., Wang, Y., et al., 2007. Apolipoprotein E e4 allele increases the risk of early postoperative delirium in older patients undergoing noncardiac surgery. Anesthesiology 107, 406–411.

Lewis, G.D., Wei, R., Liu, E., et al., 2008. Metabolite profiling of blood from individuals undergoing planned myocardial infarction reveals early markers of myocardial injury. J Clin Invest 118, 3503–3512.

Liu, K.Y., Muehlschlegel, J.D., Perry, T.E., et al., 2010. Common genetic variants on chromosome 9p21 predict perioperative myocardial injury after coronary artery bypass graft surgery. J Thorac Cardiovasc Surg 139, 483–488.

Lobato, R.L., White, W.D., Mathew, J.P., et al., 2011. Thrombomodulin gene variants are associated with increased mortality after coronary artery bypass surgery in replicated analyses. Circulation 124, S143–S148.

Lucchinetti, E., Aguirre, J., Feng, J., et al., 2007a. Molecular evidence of late preconditioning after sevoflurane inhalation in healthy volunteers. Anesth Analg 105, 629–640.

Lucchinetti, E., Da Silva, R., Pasch, T., Schaub, M.C., Zaugg, M., 2005. Anaesthetic preconditioning but not postconditioning prevents early activation of the deleterious cardiac remodelling programme: Evidence of opposing genomic responses in cardioprotection by pre- and postconditioning. Br J Anaesth 95, 140–152.

Lucchinetti, E., Hofer, C., Bestmann, L., et al., 2007b. Gene regulatory control of myocardial energy metabolism predicts postoperative cardiac function in patients undergoing off-pump coronary artery bypass graft surgery: Inhalational versus intravenous anesthetics. Anesthesiology 106, 444–457.

MacKensen, G.B., Swaminathan, M., Ti, L.K., et al., 2004. Preliminary report on the interaction of apolipoprotein E polymorphism with aortic atherosclerosis and acute nephropathy after CABG. Ann Thorac Surg 78, 520–526.

Mangano, C.M., Diamondstone, L.S., Ramsay, J.G., Aggarwal, A., Herskowitz, A., Mangano, D.T., 1998. Renal dysfunction after myocardial revascularization: Risk factors, adverse outcomes, and hospital resource utilization. The Multicenter Study of Perioperative Ischemia Research Group. Ann Intern Med 128, 194–203.

Mangano, D.T., 1997. Effects of acadesine on myocardial infarction, stroke, and death following surgery. A meta-analysis of the 5 international randomized trials. The Multicenter Study of Perioperative Ischemia (McSPI) Research Group. JAMA 277, 325–332.

Mangano, D.T., 2004. Perioperative medicine: NHLBI working group deliberations and recommendations. J Cardiothorac Vasc Anesth 18, 1–6.

Mathew, J.P., Fontes, M.L., Tudor, I.C., et al., 2004. A multicenter risk index for atrial fibrillation after cardiac surgery. JAMA 291, 1720–1729.

Mathew, J.P., Podgoreanu, M.V., Grocott, H.P., et al., 2007. Genetic variants in P-selectin and C-reactive protein influence susceptibility to cognitive decline after cardiac surgery. J Am Coll Cardiol 49, 1934–1942.

Mathew, J.P., Rinder, C.S., Howe, J.G., et al., 2001. Platelet PlA2 polymorphism enhances risk of neurocognitive decline after cardiopulmonary bypass. Multicenter Study of Perioperative Ischemia (McSPI) Research Group. Ann Thorac Surg 71, 663–666.

Matkovich, S.J., Van Booven, D.J., Youker, K.A., et al., 2009. Reciprocal regulation of myocardial microRNAs and messenger RNA in human cardiomyopathy and reversal of the microRNA signature by biomechanical support. Circulation 119, 1263–1271.

Mayr, M., Yusuf, S., Weir, G., et al., 2008. Combined metabolomic and proteomic analysis of human atrial fibrillation. J Am Coll Cardiol 51, 585–594.

McDonagh, D.L., Mathew, J.P., White, W.D., et al., 2010. Cognitive function after major noncardiac surgery, apolipoprotein E4 genotype, and biomarkers of brain injury. Anesthesiology 112, 852–859.

Moor, E., Silveira, A., Van't Hooft, F., et al., 1998. Coagulation factor V (Arg506>Gln) mutation and early saphenous vein graft occlusion after coronary artery bypass grafting. Thromb Haemost 80, 220–224.

Morawski, W., Sanak, M., Cisowski, M., et al., 2005. Prediction of the excessive perioperative bleeding in patients undergoing coronary artery bypass grafting: Role of aspirin and platelet glycoprotein IIIa polymorphism. J Thorac Cardiovasc Surg 130, 791–796.

Moretti, E.W., Morris, R.W., Podgoreanu, M., et al., 2005. APOE polymorphism is associated with risk of severe sepsis in surgical patients. Crit Care Med 33, 2521–2526.

Motoyama, S., Miura, M., Hinai, Y., et al., 2009. C-reactive protein 1059G>C genetic polymorphism influences serum C-reactive protein levels after esophagectomy in patients with thoracic esophageal cancer. J Am Coll Surg 209, 477–483.

Motoyama, S., Miura, M., Hinai, Y., Maruyama, K., Murata, K., Ogawa, J., 2010. C-reactive protein -717C>T genetic polymorphism associates with esophagectomy-induced stress hyperglycemia. World J Surg 34, 1001–1007.

Motsinger, A.A., Donahue, B.S., Brown, N.J., Roden, D.M., Ritchie, M.D., 2006. Risk factor interactions and genetic effects associated with post-operative atrial fibrillation. Pac Symp Biocomput, 584–595.

Muehlschlegel, J.D., Liu, K.Y., Perry, T.E., et al., 2010. Chromosome 9p21 variant predicts mortality after coronary artery bypass graft surgery. Circulation 122, S60–S65.

Nashef, S.A., Roques, F., Hammill, B.G., et al., 2002. Validation of European System for Cardiac Operative Risk Evaluation

(EuroSCORE) in North American cardiac surgery. Eur J Cardiothorac Surg 22, 101–105.

Newman, M.F., Booth, J.V., Laskowitz, D.T., Schwinn, D.A., Grocott, H.P., Mathew, J.P., 2001a. Genetic predictors of perioperative neurological and cognitive injury and recovery. Best Pract Res Clin Anesth 15, 247–276.

Newman, M.F., Laskowitz, D.T., White, W.D., et al., 2001b. Apolipoprotein E polymorphisms and age at first coronary artery bypass graft. Anesth Analg 92, 824–829.

Ortlepp, J.R., Janssens, U., Bleckmann, F., et al., 2001. A chymase gene variant is associated with atherosclerosis in venous coronary artery bypass grafts. Coron Artery Dis 12, 493–497.

Patel, U.D., Garg, A.X., Krumholz, H.M., et al., 2012. Pre-operative serum brain natriuretic peptide and risk of acute kidney injury after cardiac surgery. Circulation doi: 10.1161/CIRCULATIONAHA.111.029686.

Perry, T.E., Muehlschlegel, J.D., Liu, K.Y., et al., 2009. C-Reactive protein gene variants are associated with postoperative C-reactive protein levels after coronary artery bypass surgery. BMC Med Genet 10, 38.

Perry, T.E., Muehlschlegel, J.D., Liu, K.Y., et al., 2010. Preoperative C-reactive protein predicts long-term mortality and hospital length of stay after primary, nonemergent coronary artery bypass grafting. Anesthesiology 112, 607–613.

Pham, M.X., Teuteberg, J.J., Kfoury, A.G., et al., 2010. Gene-expression profiling for rejection surveillance after cardiac transplantation. N Engl J Med 362, 1890–1900.

Philip, I., Plantefeve, G., Vuillaumier-Barrot, S., et al., 1999. G894T polymorphism in the endothelial nitric oxide synthase gene is associated with an enhanced vascular responsiveness to phenylephrine. Circulation 99, 3096–3098.

Podgoreanu, M.V., Booth, J.V., White, W.D., et al., 2003. Beta adrenergic receptor polymorphisms and risk of adverse events following cardiac surgery. Circulation 108, IV434.

Podgoreanu, M.V., Morris, R., Zhang, Q., Mathew, J.P., Schwinn, D.A., 2007. Gene variants in interleukin 1-beta are associated with early QTc prolongation after cardiac surgery. Anesthesiology 107, A1287 (abstract).

Podgoreanu, M.V., Schwinn, D.A., 2005. New paradigms in cardiovascular medicine: Emerging technologies and practices – perioperative genomics. J Am Coll Cardiol 46, 1965–1977.

Podgoreanu, M.V., White, W.D., Morris, R.W., et al., 2006. Inflammatory gene polymorphisms and risk of postoperative myocardial infarction after cardiac surgery. Circulation 114, I275–I281.

Poldermans, D., Bax, J.J., Boersma, E., et al., 2009. Guidelines for pre-operative cardiac risk assessment and perioperative cardiac management in non-cardiac surgery. Eur Heart J 30, 2769–2812.

Pretorius, M., Donahue, B.S., Yu, C., Greelish, J.P., Roden, D.M., Brown, N.J., 2007. Plasminogen activator inhibitor-1 as a predictor of postoperative atrial fibrillation after cardiopulmonary bypass. Circulation 116, I1–I7.

Queloz, P.A., Thadikkaran, L., Crettaz, D., Rossier, J.S., Barelli, S., Tissot, J.D., 2006. Proteomics and transfusion medicine: Future perspectives. Proteomics 6, 5605–5614.

Ramlawi, B., Otu, H., Mieno, S., et al., 2007a. Oxidative stress and atrial fibrillation after cardiac surgery: A case-control study. Ann Thorac Surg 84, 1166–1172, discussion 1172–1173.

Ramlawi, B., Otu, H., Rudolph, J.L., et al., 2007b. Genomic expression pathways associated with brain injury after cardiopulmonary bypass. J Thorac Cardiovasc Surg 134, 996–1005.

Ramlawi, B., Rudolph, J.L., Mieno, S., et al., 2006a. C-Reactive protein and inflammatory response associated to neurocognitive decline following cardiac surgery. Surgery 140, 221–226.

Ramlawi, B., Rudolph, J.L., Mieno, S., et al., 2006b. Serologic markers of brain injury and cognitive function after cardiopulmonary bypass. Ann Surg 244, 593–601.

Rifon, J., Paramo, J.A., Panizo, C., Montes, R., Rocha, E., 1997. The increase of plasminogen activator inhibitor activity is associated with graft occlusion in patients undergoing aorto-coronary bypass surgery. Br J Haematol 99, 262–267.

Rinder, C.S., Mathew, J.P., Rinder, H.M., et al., 2002. Platelet PlA2 polymorphism and platelet activation are associated with increased troponin I release after cardiopulmonary bypass. Anesthesiology 97, 1118–1122.

Rosenberg, S., Elashoff, M.R., Beineke, P., et al., 2010. Multicenter validation of the diagnostic accuracy of a blood-based gene expression test for assessing obstructive coronary artery disease in nondiabetic patients. Ann Intern Med 153, 425–434.

Roth-Isigkeit, A., Hasselbach, L., Ocklitz, E., et al., 2001. Interindividual differences in cytokine release in patients undergoing cardiac surgery with cardiopulmonary bypass. Clin Exp Immunol 125, 80–88.

Ruel, M., Bianchi, C., Khan, T.A., et al., 2003. Gene expression profile after cardiopulmonary bypass and cardioplegic arrest. J Thorac Cardiovasc Surg 126, 1521–1530.

Ryan, R., Thornton, J., Duggan, E., et al., 2006. Gene polymorphism and requirement for vasopressor infusion after cardiac surgery. Ann Thorac Surg 82, 895–901.

Ryding, A.D., Kumar, S., Worthington, A.M., Burgess, D., 2009. Prognostic value of brain natriuretic peptide in noncardiac surgery: A meta-analysis. Anesthesiology 111, 311–319.

Sehl, P.D., Tai, J.T., Hillan, K.J., et al., 2000. Application of cDNA microarrays in determining molecular phenotype in cardiac growth, development, and response to injury. Circulation 101, 1990–1999.

Sergeev, P., Da Silva, R., Lucchinetti, E., et al., 2004. Trigger-dependent gene expression profiles in cardiac preconditioning: Evidence for distinct genetic programs in ischemic and anesthetic preconditioning. Anesthesiology 100, 474–488.

Shaw, A.D., Vaporciyan, A.A., Wu, X., et al., 2005. Inflammatory gene polymorphisms influence risk of postoperative morbidity after lung resection. Ann Thorac Surg 79, 1704–1710.

Shaw, S.M., Lewis, N.T., Williams, S.G., Tan, L.B., 2008. A role for BNP assays in monitoring aortic valve disease for optimal timing of surgery. Int J Cardiol 127, 328–330.

Sheikh, A.M., Barrett, C., Villamizar, N., et al., 2006. Proteomics of cerebral injury in a neonatal model of cardiopulmonary bypass with deep hypothermic circulatory arrest. J Thorac Cardiovasc Surg 132, 820–828.

Sheng, H., Laskowitz, D.T., Bennett, E., et al., 1998. Apolipoprotein E isoform-specific differences in outcome from focal ischemia in transgenic mice. J Cereb Blood Flow Metab 18, 361–366.

Slooter, A.J., Tang, M.X., Van Duijn, C.M., et al., 1997. Apolipoprotein E epsilon4 and the risk of dementia with stroke. A population-based investigation. JAMA 277, 818–821.

Smiley, R.M., Blouin, J.L., Negron, M., Landau, R., 2006. beta2-adrenoceptor genotype affects vasopressor requirements during spinal anesthesia for cesarean delivery. Anesthesiology 104, 644–650.

Song, Y., Kwak, Y.L., Choi, Y.S., Kim, J.C., Heo, S.B., Shim, J.K., 2010. Effect of preoperative statin therapy on myocardial protection and morbidity endpoints following off-pump coronary bypass surgery in patients with elevated C-reactive protein level. Korean J Anesthesiol 58, 136–141.

Stafford-Smith, M., Podgoreanu, M., Swaminathan, M., et al., 2005. Association of genetic polymorphisms with risk of renal injury after coronary bypass graft surgery. Am J Kidney Dis 45, 519–530.

Stoica, A.L., Stoica, E., Constantinescu, I., Uscatescu, V., Ginghina, C., 2010. Interleukin-6 and interleukin-10 gene polymorphism, endothelial dysfunction, and postoperative prognosis in patients with peripheral arterial disease. J Vasc Surg 52, 103–109.

Stranger, B.E., Nica, A.C., Forrest, M.S., et al., 2007. Population genomics of human gene expression. Nat Genet 39, 1217–1224.

Stuber, F., Hoeft, A., 2002. The influence of genomics on outcome after cardiovascular surgery. Curr Opin Anaesthesiol 15, 3–8.

Tardiff, B.E., Newman, M.F., Saunders, A.M., et al., 1997. Preliminary report of a genetic basis for cognitive decline after cardiac operations. The Neurologic Outcome Research Group of the Duke Heart Center. Ann Thorac Surg 64, 715–720.

Taylor, K.D., Scheuner, M.T., Yang, H., et al., 2004. Lipoprotein lipase locus and progression of atherosclerosis in coronary-artery bypass grafts. Genet Med 6, 481–486.

Teasdale, G.M., Nicoll, J.A., Murray, G., Fiddes, M., 1997. Association of apolipoprotein E polymorphism with outcome after head injury. Lancet 350, 1069–1071.

Ti, L.K., MacKensen, G.B., Grocott, H.P., et al., 2003. Apolipoprotein E4 increases aortic atheroma burden in cardiac surgical patients. J Thorac Cardiovasc Surg 125, 211–213.

Ti, L.K., Mathew, J.P., MacKensen, G.B., et al., 2001. Effect of apolipoprotein E genotype on cerebral autoregulation during cardiopulmonary bypass. Stroke 32, 1514–1519.

Tomasdottir, H., Hjartarson, H., Ricksten, A., Wasslavik, C., Bengtsson, A., Ricksten, S.E., 2003. Tumor necrosis factor gene polymorphism is associated with enhanced systemic inflammatory response and increased cardiopulmonary morbidity after cardiac surgery. Anesth Analg 97, 944–949.

Tomic, V., Russwurm, S., Moller, E., et al., 2005. Transcriptomic and proteomic patterns of systemic inflammation in on-pump and off-pump coronary artery bypass grafting. Circulation 112, 2912–2920.

Turer, A.T., Stevens, R.D., Bain, J.R., et al., 2009. Metabolomic profiling reveals distinct patterns of myocardial substrate use in humans with coronary artery disease or left ventricular dysfunction during surgical ischemia/reperfusion. Circulation 119, 1736–1746.

Vamvakopoulos, J.E., Taylor, C.J., Green, C., et al., 2002. Interleukin 1 and chronic rejection: Possible genetic links in human heart allografts. Am J Transplant 2, 76–83.

Virani, S.S., Brautbar, A., Lee, V.V., et al., 2011. Usefulness of single nucleotide polymorphism in chromosome 4q25 to predict in-hospital and long-term development of atrial fibrillation and survival in patients undergoing coronary artery bypass grafting. Am J Cardiol 107, 1504–1509.

Voetsch, B., Loscalzo, J., 2004. Genetic determinants of arterial thrombosis. Arterioscler Thromb Vasc Biol 24, 216–229.

Volzke, H., Engel, J., Kleine, V., et al., 2002. Angiotensin I-converting enzyme insertion/deletion polymorphism and cardiac mortality and morbidity after coronary artery bypass graft surgery. Chest 122, 31–36.

Weiss, E.J., Bray, P.F., Tayback, M., et al., 1996. A polymorphism of a platelet glycoprotein receptor as an inherited risk factor for coronary thrombosis. N Engl J Med 334, 1090–1094.

Welsby, I.J., Podgoreanu, M.V., Phillips-Bute, B., et al., 2005. Genetic factors contribute to bleeding after cardiac surgery. J Thromb Haemost 3, 1206–1212.

Welsby, I.J., Podgoreanu, M.V., Phillips-Bute, B., et al., 2010. Association of the 98T ELAM-1 polymorphism with increased bleeding after cardiac surgery. J Cardiothorac Vasc Anesth 24, 427–433.

Willerson, J.T., Ridker, P.M., 2004. Inflammation as a cardiovascular risk factor. Circulation 109, II2–II10.

Zaugg, M., Bestmann, L., Wacker, J., et al., 2007. Adrenergic receptor genotype but not perioperative bisoprolol therapy may determine cardiovascular outcome in at-risk patients undergoing surgery with spinal block: the Swiss Beta Blocker in Spinal Anesthesia (BBSA) study – a double-blinded, placebo-controlled, multicenter trial with 1-year follow-up. Anesthesiology 107, 33–44.

Zaugg, M., Schaub, M.C., 2005. Genetic modulation of adrenergic activity in the heart and vasculature: Implications for perioperative medicine. Anesthesiology 102, 429–446.

Zeltser, I., Jarvik, G.P., Bernbaum, J., et al., 2008. Genetic factors are important determinants of neurodevelopmental outcome after repair of tetralogy of Fallot. J Thorac Cardiovasc Surg 135, 91–97.

Ziegeler, S., Tsusaki, B.E., Collard, C.D., 2003. Influence of genotype on perioperative risk and outcome. Anesthesiology 99, 212–219.

Zotz, R.B., Klein, Dauben, H.P., Moser, C., Gams, E., Scharf, R.E., 2000. Prospective analysis after coronary-artery bypass grafting: Platelet GP IIIa polymorphism (HPA-1b/PIA2) is a risk factor for bypass occlusion, myocardial infarction, and death. Thromb Haemost 83, 404–407.

Stroke

Matthew B. Lanktree, Tisha R. Joy, and Robert A. Hegele

INTRODUCTION

Stroke is a major cause of adult neurological disability and mortality. Stroke is an umbrella term for a rapid loss of neurologic function in a particular vascular territory, with symptoms lasting longer than 24 hours. Clinically, it is divided into subtypes based upon the responsible mechanism, namely hemorrhagic versus ischemic. Traditional cardiovascular risk factors, such as age, sex, smoking, hypertension, and dyslipidemia contribute to stroke risk, but some individuals with many risk factors remain unaffected, while others with no risk factors suffer a stroke. As stroke is the clinical culmination of many complex processes and interacting pathways, none of which is individually necessary nor sufficient to cause the outcome, the analysis of etiological factors is complicated (Dichgans, 2007). A broad range of divergent pathophysiologies contribute to stroke risk, including atherosclerosis, dyslipidemia, cardiac arrhythmia and dysfunction, platelet dysfunction and disordered coagulation, deficits affecting structure and function of the arterial wall, and inflammatory and immune mechanisms. Discovery and replication of association between genetic variation and stroke will provide researchers with new insights into the molecular disease processes, and will give providers the opportunity to identify high-risk patients for risk reduction strategies.

Exciting advances in accurate and high-throughput genotyping technologies have revolutionized the scope and nature of genomic data that can be obtained to understand a complex disease such as stroke. In particular, the genome-wide association study (GWAS), which evaluates association between genetic variation and carefully defined stroke and related phenotypes, is an agnostic experimental framework to uncover novel contributors to pathogenic mechanisms. In this chapter we provide a summary of the evidence for a heritable component of stroke, a discussion of monogenic forms of stroke, novel discoveries in the genetics of stroke, and future directions in the genomic analysis of stroke.

DISSECTING THE STROKE PHENOTYPE

The primary mechanisms for a stroke are either blockage of blood flow, causing ischemia, or rupture of a blood vessel, resulting in hemorrhage into the surrounding tissue (Figure 56.1) (Runchey and McGee, 2010). Blockage of blood flow may be caused either by a thrombus, the development of a blood clot at the site of blockage, or through the creation of an embolus upstream of the eventual blockage. Embolus formation occurs either within the heart, due to hemostasis within the atria because of atrial fibrillation or a patent foramen ovale, or within the arteries, due to atherosclerosis or a primary or secondary hypercoagulability. Hemorrhagic strokes, on the other hand, are subdivided according to the location of bleeding as either intracerebral or within the subarachnoid space. Intracerebral hemorrhage is often the result of hypertension or a cerebral amyloid angiopathy, whereas subarachnoid hemorrhage is often the result of vascular malformation or the rupture of a berry aneurysm. Thus, the relative contribution of different pathological

Genomic and Personalized Medicine, 2nd edition
by Ginsburg & Willard. DOI: http://dx.doi.org/10.1016/B978-0-12-382227-7.00056-2

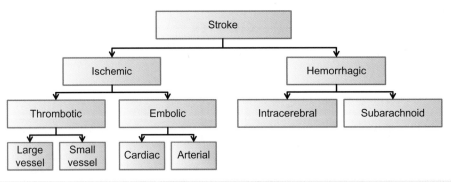

Figure 56.1   Stroke classified by etiology.

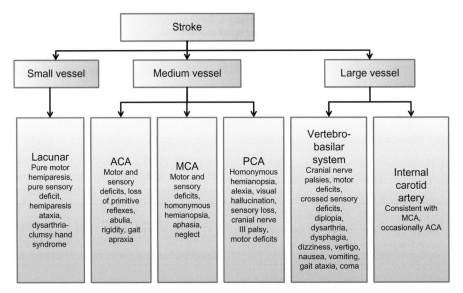

Figure 56.2   Stroke classified by affected arterial territory. Commonly observed clinical symptoms are included in each box. Anterior cerebral artery, ACA; middle cerebral artery, MCA; posterior cerebral artery, PCA.

mechanisms can vary greatly depending upon the etiology of the stroke.

The clinical signs and symptoms of the stroke are characteristic of the vascular bed affected (Figure 56.2). Through careful medical examination and knowledge of the function of the specific regions of the brain, physicians can specifically narrow the location of the stroke. Appropriate neuroimaging techniques, such as computed tomography (CT) scan, magnetic resonance imaging (MRI), and angiography of the cerebral vasculature can confirm the exact location of the blockage or hemorrhage. It remains unknown if the genetic contribution to stroke varies depending on the location of the stroke.

Several phenomena may present similarly to stroke, such as transient ischemic attacks (TIAs), migraines, head trauma, brain tumors, and metabolic and toxic insults (hypoglycemia, hypothyroidism, renal or liver failure, certain drug intoxications). The only difference between TIA and stroke is the duration of symptoms, as the neurologic deficits secondary to TIAs

resolve within 24 hours. Within research studies, classification systems are typically employed to identify stroke subtypes and provide a means of clinical comparison between studies. The Trial of ORG 10172 in Acute Stroke Treatment (TOAST) classification system, originally created for a randomized controlled drug trial of a low molecular weight heparinoid drug, and the Oxfordshire Community Stroke Project (OCSP), based upon efforts to identify the natural history of stroke, are two such classification systems that are commonly employed (Dichgans and Markus, 2005). Classification systems that attempt to group cases by specific pathophysiology, as opposed to predicted prognosis, are the best suited to research. However, any categorization system is dependent upon the quality of clinical information and experience of the raters (Dichgans and Markus, 2005). Ensuring a homogenous stroke phenotype between cases is of the utmost importance to maximize study power within all forms of genetic and genomic analysis of stroke.

HERITABILITY OF STROKE

Evidence for the heritability of stroke has been drawn from studies of twins and the family history of stroke cases. A meta-analysis of three twin studies found that monozygotic twins were 1.65 times more likely to be concordant for stroke than dizygotic twins [odds ratio (OR) = 1.65; 95% confidence interval (CI) = 1.2–2.3] (Flossmann et al., 2004). This concordance is substantially lower than for many other complex diseases; collection of concordant twin pairs for stroke is difficult due to the late-age of onset of stroke. A positive family history was a significant risk factor for stroke in both case-control (OR = 1.76; 95% CI = 1.7–1.9) and population cohort studies (OR = 1.3; 95% CI = 1.2–1.5) (Flossmann et al., 2004). There also appears to be a greater genetic liability for stroke at a younger age (Flossmann et al., 2004). It should be noted that there is a substantial heritable component to traditional cardiovascular risk factors such as blood pressure and plasma lipid concentration (Knuiman et al., 1996). Based upon this data suggesting a heritable component to stroke risk, investigations have been undertaken to identify the genetic determinants of stroke.

MONOGENIC FORMS OF STROKE

Only a small proportion of strokes in the population are due to a single genetic cause, but there is much to learn from these patients. In a monogenic disease, carriage of the responsible genetic variation (or carriage of two copies of the genetic variation in the case of a recessive monogenic disease) is sufficient to cause the disease, irrespective of environmental factors. Identification of a gene in which the genetic variation lies uncovers an important player in a required pathway in stroke prevention. Screening the population for the variation would be ineffective due to its low prevalence, but testing of individuals with strong family history for stroke may be useful. With the advancement of whole genome sequencing technologies, genetic variants may be discovered more readily in asymptomatic individuals. Identifying the genetic cause of a number of monogenic forms of stroke adds to our knowledge base on the pathophysiology of stroke.

Mutations that dramatically increase the risk of stroke have been identified in more than 15 different genes (Table 56.1). In cerebral autosomal dominant arteriopathy with subcortical infarcts and leukoencephalopathy (CADASIL) and its recessive form (CARASIL), the small arteries are primarily affected, and patients suffer from recurrent ischemic strokes beginning in the fourth decade of life followed by progressive cognitive decline and psychiatric disturbance (Dichgans, 2002). Bilateral involvement of the temporal white matter is required for the diagnosis (Dichgans, 2002). The responsible mutations have been discovered in NOTCH3 and HTRA1, respectively. Despite similar phenotypes, the connection between the molecular functions of the two genes remains unknown, as NOTCH3 is a transmembrane

TABLE 56.1	Monogenic forms of stroke	
Phenotype	**OMIM**	**Gene(s)**
Small vessel vasculopathy		
CARASIL	600142	HTRA1
CADASIL	125310	NOTCH3
Vasculopathy, retinal with cerebral leukodystrophy	192315	TREX1
Large vessel vasculopathy		
Moyamoya disease	252350	MYMY1, MYMY2, MYMY3
Neurofibromatosis, type I	162200	NF1
Small and large vessel vasculopathy		
Fabry disease	301500	GLA
Sickle cell disease	603903	HBB
Homocystinuria	236200	CBS
Pseudoxanthoma elasticum	177850	ABCC6
Smooth muscle alpha-actin mutations	102620	ACTA2
Connective tissue disorders		
Ehlers-Danlos syndrome, type VI	225400	PLOD1
Marfan syndrome	154700	FBN1
Cerebrovascular malformations		
Cerebral cavernous malformation	116860	CCM1/KRIT1, CCM2, CCM3/PDCD10

OMIM, Online Inheritance in Man identification number.

receptor that affects transcriptional programming (Joutel et al., 2004), while loss of HTRA1 appears to interfere with transforming growth factor-β (TGF-β) signaling (Hara et al., 2009). Also preferentially affecting the small vessels is a vasculopathy with retinal involvement and cerebral leukodystrophy, previously known as hereditary endotheliopathy with retinopathy, nephropathy, and stroke (HERNS). The responsible mutation has been identified in the single-exon gene encoding a 3′ DNA exonuclease repair protein (TREX1) (Richards et al., 2007). In moyamoya disease, caused by mutations in one of three genes (MYMY1, MYMY2, MYMY3), progressive bilateral occlusion of the carotid arteries occurs at the base of the circle of Willis, accompanied by the development of telangiectatic vessels in the region of the basal ganglia. Patients with moyamoya disease often have ischemic strokes in childhood, and rupture of the telangiectatic vessels can result in hemorrhage (Ikezaki et al., 1997).

Monogenic conditions exist that disrupt metabolism and cause stroke as a secondary component, such as Fabry's disease (alpha-galactosidase deficiency), sickle cell anemia, and homocystinuria (Majersik and Skalabrin, 2006). In monogenic connective tissue disorders such as Ehlers-Danlos and Marfan syndromes, affected individuals are at elevated risk for aortic dissection or aneurysm, and may thus be at elevated risk for TIA, ischemic stroke, and subdural hematoma (Dichgans, 2007). Finally, disorders that affect key pathogenic pathways in stroke have been identified, such as monogenic cardiac dysrhythmias (e.g., atrial fibrillation), cardiomyopathies (e.g., dilated cardiomyopathy), coagulopathies (e.g., anti-thrombin III deficiency), dyslipidemias (e.g., familial combined hyperlipidemia), and vasculopathies (e.g., Kawasaki disease).

GENOMICS OF COMMON COMPLEX STROKE

Candidate Gene Association Studies

Extensive effort has been expended to identify association between single nucleotide polymorphisms (SNPs) in functional candidate genes and stroke. Based on our growing understanding of the genetic architecture of complex traits, genetic association studies require samples of at least 1000 cases and controls to have suitable power (Matarin et al., 2010). Several genes in multiple implicated pathways, such as hemostasis and coagulation, lipid metabolism, inflammation, and the renin–angiotensin–aldosterone system, have been examined, and meta-analyses have evaluated the overall evidence primarily with the ischemic form of stroke (Ariyaratnam et al., 2007; Bersano et al., 2008; Dichgans, 2007; Matarin et al., 2010). The genes with the greatest evidence for association from candidate gene studies include: factor V Leiden (*F5*), prothrombin (*F2*), plasminogen activator inhibitor-1 (*PAI-1*), Gp1b/IX/ complex (*GP1BA*), angiotensin converting enzyme (*ACE*), methyl tetrahydrofolate reductase (*MTHFR*), and apolipoprotein E (*APOE*) (Bersano et al., 2008; Dichgans, 2007). Moreover, many of these associations appear to hold true in persons of both European and non-European descent (Ariyaratnam et al., 2007).

Genome-wide Association Studies

To date, GWAS of stroke phenotypes has been performed in seven cohorts, resulting in nine publications, and reporting ten novel loci associated with stroke (Table 56.2) (Bilguvar et al., 2008; Debette et al., 2010; Gretarsdottir et al., 2008; Gudbjartsson et al., 2009; Hata et al., 2007; Ikram et al., 2009; Kubo et al., 2007; Matarin et al., 2007; Yamada et al., 2009). Novel genes observed to be associated with stroke in GWAS may uncover new biology, as none of them have known roles in stroke. For instance, sex determining region Y-box 17 (*SOX17*) is a transcription factor required for formation and

maintenance of endothelial vascular endothelium (Bilguvar et al., 2008). Cadherin epidermal growth factor laminin A seven-pass G-type receptor 1 (*CELSR1*) is a cell surface receptor of relatively unknown function (Yamada et al., 2009). A homolog of *CELSR1*, *CELSR2* has been repeatedly associated with plasma low density lipoprotein (LDL) cholesterol concentrations (Lanktree et al., 2009a; Teslovich et al., 2010). Genetic variation in both paired-like homeodomain transcription factor 2 (*PITX2*) and zinc finger homeobox 3 (*ZFHX3*) were observed to be associated with ischemic stroke and atrial fibrillation, and pathway analysis suggests that both variants may play a role in cardiac development (Gretarsdottir et al., 2008; Gudbjartsson et al., 2009). However, no single locus has yet been identified in two independent stroke GWAS at a genome-wide level of significance ($P < 5 \times 10^{-7}$). Additionally, these GWAS have failed to confirm an association between stroke and a locus with previous evidence from candidate gene studies of stroke risk, nor an association between stroke and a gene causative of a monogenic form of stroke.

Of the novel GWAS hits reported in Table 56.2, only ninjurin-2 (*NINJ2*) has been thoroughly examined in a replication sample in a published report. Unfortunately, in a sample of 8637 cases and 8733 controls of European ancestry, and a population-based cohort of 22,054 participants with 278 ischemic stroke events, no association was observed for the reported variants in *NINJ2* ($P > 0.40$) (International Stroke Genetics Consortium and Welcome Trust Case-Control Consortium 2, 2010). The authors of the replication study suggest that a false positive result is the most likely reason for the non-replication. All of the novel genes identified require replication, but they create multiple new leads for mechanisms underlying stroke.

Chromosome 9p21

Association of an intergenic region at chromosome 9p21 with coronary artery disease and myocardial infarction was first reported in 2007 (McPherson et al., 2007; Samani et al., 2007; Welcome Trust Case Control Consortium, 2007). The same locus was also reported to be associated with a range of vascular phenotypes including both abdominal aortic and intracranial aneurysms (Helgadottir et al., 2008). In the first GWAS of intracranial aneurysm, significant association was also observed with chromosome 9p21 (Bilguvar et al., 2008). Since these reports, many studies have tested the 9p21 locus for association with ischemic stroke (Table 56.3). While no studies of ischemic stroke have reached a standard GWAS significance threshold despite large sample sizes, a consistent marginal association ($0.05 < P < 9 \times 10^{-4}$) has been observed.

Mitochondrial DNA

A two-fold excess of maternal versus paternal stroke has been observed in female stroke probands (Touze and Rothwell, 2008). Since mitochondrial DNA (mtDNA) is inherited

TABLE 56.2	Genome-wide association studies of stroke				
Reference	Phenotype	Discovery sample size	Discovery sample ancestry	Associated regions[†] (nearest gene)
Hata et al., 2007; Kubo et al., 2007	Ischemic stroke	1112 cases 1112 controls*	Japanese	PRKCH AGTRL1
Matarin et al., 2007	Ischemic stroke	249 cases 268 controls	White	None
Gretarsdottir et al., 2008; Gudbjartsson et al., 2009	Ischemic stroke	1661 cases 10815 controls	Icelandic	PITX2 ZFHX3
Bilguvar et al., 2008	Intracranial aneurysm	2100 cases 8000 controls	Finnish, Dutch, Japanese	PLCL1 SOX17 CDNK2A-CDNK2B-ANRIL
Ikram et al., 2009	Ischemic stroke	cohort of 19,602, 1164 events	White	NINJ2
Yamada et al., 2009	Ischemic stroke	992 cases 5349 controls	Japanese	CELSR1
Debette et al., 2010	MRI-defined brain infarcts	9889 individuals (population-based)	White	MACROD2

*Tested 52,608 SNPs in a targeted gene array, remainder of studies included >300,000 SNPs.
[†]with $P < 5 \times 10^{-7}$.

completely from mothers, it has been suggested that genetic variation in mtDNA could be partially responsible for this phenomenon (Chinnery et al., 2010). Large-effect mutations of mtDNA have been identified as responsible for a rare maternally inherited mitochondrial encephalomyopathy with lactic acidosis and stroke-like episodes (MELAS) (Sproule and Kaufmann, 2008). A recent study of 950 cases and 2939 controls identified association between the common mitochondrial haplogroup K, composed of multiple common genetic variants marking a mitochondrial lineage, and susceptibility to TIA and stroke ($P < 1.0 \times 10^{-5}$) (Chinnery et al., 2010). Further studies will be required to replicate and validate this association.

Intermediate Phenotypes

Intermediate phenotypes, also referred to as biomarkers, subphenotypes, endophenotypes, subclinical traits, or disease attributes, are targeted measures of disease progression that may be more directly related to a specific disease etiology than the cumulative disease endpoint. For instance, carotid intima-media thickness (IMT), which is the most commonly used non-invasive measure of sub clinical atherosclerosis, is strongly predictive of subsequent cardiovascular events and may represent an intermediate phenotype more closely linked to certain specific pathophysiologic processes underlying the pathogenesis of stroke (Manolio et al., 2004). In the first

gene-centric microarray-based association study of IMT, two novel loci were associated with IMT: histone deacetylase 4 (HDAC4) and natriuretic peptide receptor a/guanylate cylcase A (NPR1) (Lanktree et al., 2009b). Many candidate gene association studies of IMT have been reported, and in the largest meta-analysis to date, which included more than 5000 individuals, three genes were consistently associated with IMT: APOE, ACE, and MTHFR (Paternoster et al., 2010). All of these genes have been reliably associated with stroke in candidate gene studies, but none have been identified as associated with stroke in GWAS. Large-scale GWAS of IMT may identify novel mechanisms of atherosclerosis.

Mendelian Randomization Studies

If a genotype is associated with elevated levels of a biomarker and the biomarker is in turn associated with a clinical endpoint, then the size and strength of the direct association between genotype and endpoint can be predicted and tested (Sheehan et al., 2008). This procedure is known as Mendelian randomization, and it is considered to be one approach to provide evidence for a causative role for the biomarker in the pathogenesis of the endpoint (Sheehan et al., 2008). If the association between the biomarker and the endpoint is due to association with a confounding variable or reverse causation (e.g., the endpoint causes elevation of the biomarker), then the effect of the genotype on the endpoint

TABLE 56.3	Summary of evidence for association between chromosome 9p21 and stroke				
Reference	**Phenotype**	**Lead SNP**	**Sample size**	**Ethnicity**	**P-value**
Bilguvar et al., 2008	Intracranial aneurysm	rs133040	2100 cases 8000 controls	White, Japanese	1.4×10^{-10}
Helgadottir et al., 2008	Intracranial aneurysm	rs10757278	1134 cases 15,481 controls	White	2.5×10^{-6}
Matarin et al., 2008	Ischemic stroke	rs1333040	249 cases 268 controls	White	0.104
Wahlstrand et al., 2009	All stroke	rs10757278	163 cases 5099 controls	White	0.006
Lemmens et al., 2009	Ischemic stroke	rs10757278	648 cases 828 controls	White	0.60
Gschwendtner et al., 2009	Atherosclerotic stroke	rs1537378	961 cases 4202 controls	White	0.0005
Smith et al., 2009	Ischemic stroke	rs2383207	2725 cases 1840 controls	White	0.002
Ding et al., 2009	Ischemic stroke	rs10757278	1000 cases 1059 controls	Chinese	0.003
Karvanen et al., 2009	Cerebral infarction	rs1333049	571 cases 32,711 controls	White	0.11
Hashikata et al., 2010	Intracranial aneurysm	rs1333040	419 cases 408 controls	Japanese	0.02
Nakaoka et al., 2010	Intracranial aneurysm	rs1333040	981 cases 699 controls	Japanese	1.5×10^{-6}
Meta-analysis					
Anderson et al., 2010	Ischemic stroke	rs10757278	9632 cases 30,716 controls	White, Chinese, African-American	0.001

should be insignificant. For instance, the effect of methyl tetrahydrofolate reductase (*MTHFR*) genotypes on both homocysteine and stroke risk was examined using the Mendelian randomization approach (Casas et al., 2005). The difference in stroke risk between *MTHFR* genotype groups was similar to that expected given the association between *MTHFR* genotypes and homocysteine concentrations, and the association between homocysteine concentrations and stroke risk, indicating that homocysteine concentrations were likely to play a direct role in mediating stroke risk (Casas et al., 2005). However, the Mendelian randomization framework needs to be interpreted in the context of assumptions regarding the absence of pleiotropy and interactions.

Non-DNA based Genomic Approaches to Stroke Diagnosis and Prognosis

Profiling of mRNA to detect global patterns of aberrant expression from diseased tissues has been applied with some success in the field of oncology. Some investigators have suggested that profiling of mRNA from peripheral blood leukocytes can similarly be used to investigate stroke and its subtypes (Baird, 2010). Circulating monocytes and lymphocytes are involved in atherogenesis, which underlies many forms of cardiovascular disease, including stroke. Also particular to stroke, peripheral blood mononuclear cells can play a role in the evolution of brain infarction, reperfusion injury, tissue repair and remodeling (Baird, 2010). Typically, genome-wide screening of expression profiles from peripheral blood is accomplished using microarray technology. Once candidate mRNAs are identified that are either significantly over- or under-expressed in diseased states, independent validation can be accomplished using real-time polymerase chain reaction experiments. This overall strategy presents yet another potential approach that could help to develop a rapid clinical test for stroke – for instance, to quickly diagnose a thrombotic or hemorrhagic stroke profile, and potentially to guide and predict response to treatment (Baird, 2010).

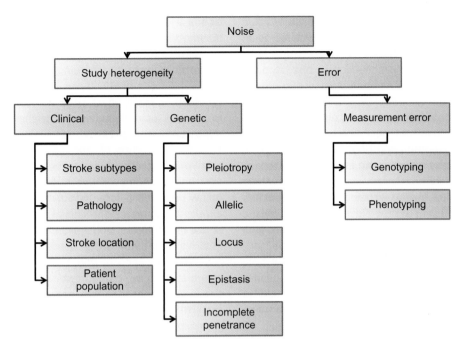

Figure 56.3 Obstacles to the identification of replicable genetic associations with stroke. Both clinical and genetic heterogeneity contribute to overall study heterogeneity. Clinical heterogeneity consists of differences in stroke subtypes as depicted in Figure 56.1, differences in the pathological mechanisms that lead to those subtypes, differences in anatomical location of the stroke as depicted in Figure 56.2, and differences in the patient population with respect to age, ascertainment, and ethnicity. Genetic heterogeneity consists of: pleiotropy, where one gene affects multiple pathologies; epistasis, where multiple genes interact to affect one pathology; allelic heterogeneity, where multiple alleles within the same locus have different effects; locus heterogeneity, where alleles in multiple loci have similar effects; and incomplete penetrance, where the same allele has different effects in different individuals.

PHARMACOGENETICS

Through the promise of individualized pharmacological stroke prevention, pharmacogenetics has the potential to reduce the time to achieve the therapeutic window and minimize the risk of adverse events (Meschia, 2009). European White individuals display a greater than 20-fold inter-individual variation in required anticoagulant dose as a preventive therapy for reducing thromboembolic events (Wadelius et al., 2009). GWAS of the required warfarin dose to obtain therapeutic efficacy has identified the genes encoding enzyme cytochrome P450 2C9 (*CYP2C9*), cytochrome P450 4F2 (*CYP4F2*) and vitamin K epoxide reductase complex subunit 1 (*VKORC1*) (Takeuchi et al., 2009). Clinical trial evidence shows that a pharmacogenetic dosing algorithm is better at predicting dose than an algorithm consisting of clinical variables (Lenzini et al., 2010). Similarly, the potential value of pharmacogenetic testing to predict response to anti-platelet agents, such as clopidogrel, while a topic of some interest in the literature, needs to be further evaluated before it can be applied clinically in stroke (Wang et al., 2011).

OVERCOMING HETEROGENEITY IN THE STROKE PHENOTYPE

Replicable genetic associations with stroke have proven difficult to identify, within both candidate gene studies and GWAS. Replication of a genetic association finding is the gold standard for validation, especially completely independent replication by different investigators in different samples (Hegele, 2002). Early studies suffered from inadequate power due to smaller sample sizes and unrealistic expectations of genetic effect size. Four reasons exist for lack of reproducibility of a genetic association with stroke: (1) the original report was a false-positive, (2) the validation study is a false-negative, (3) the discovery and validation studies differ in methodology, and (4) the discovery and validation study populations differ in either clinical ascertainment or ethnicity. Systematic analysis of replication studies, in the form of meta-analysis, can more precisely estimate effect sizes, but homogeneity of phenotype remains of prime importance.

Perhaps the largest obstacles to successful identification of genetic variants reliably associated with stroke are study heterogeneity and measurement error (Figure 56.3). Sources of study heterogeneity can broadly be classified as either clinical or genetic. As discussed, a "stroke" represents the endpoint of a heterogeneous collection of disease pathways. In stroke, confounding genetic features such as pleiotropy, incomplete penetrance, epistasis, allelic heterogeneity, and locus heterogeneity are likely the rule, not the exception. Monogenic forms of stroke may be exacerbated by, but are likely independent of, common contributors to "garden variety" stroke risk, and hence may obscure association with common risk variants. Such genetic heterogeneity can overwhelm statistical approaches to find association signals, especially with inappropriate sample sizes or clinically heterogeneous samples. As for measurement error, new genotyping technologies produce exceedingly low error rates, with miscall rates <0.5% (Teo et al., 2007). Careful categorization of the responsible pathological mechanism of each stroke patient using predefined phenotyping protocols that include quantitative measures will improve the likelihood of identifying meaningful, reproducible results (Dichgans and Markus, 2005).

"Phenomics" is an important concept that entails both the collection of a wide breadth of fine resolution phenotypes and the careful analysis of phenotype data to extract the most information possible (Freimer and Sabatti, 2003). Due to the large degree of clinical heterogeneity in stroke patients, the genomic analysis of stroke may particularly benefit from phenomics. Clustering analysis may be used to identify subgroups of patients with a more homogeneous disease pathophysiology and more specific associations with a limited number of genetic variants (Lanktree et al., 2010). Text mining has been proposed for creating deep phenotypes from electronic medical records, and pathway analysis strategies have been proposed for mining relational databases of gene–disease associations, further refining deep phenotypes and uncovering new associations (Lussier and Liu, 2007).

FUTURE DIRECTIONS

Since GWAS have become available, is there a future for candidate gene studies of stroke (Sharma and Bentley, 2010)? The largest limitation of the candidate gene approach is the lack of ancestral informative markers to correct for population stratification. Though, as fewer statistical tests are performed, candidate gene studies need not deal with the complicated issue of multiple testing. However, candidate gene studies are unable to compare the strength of association of the tested variant to the distribution of results obtained from the plethora of unassociated variants in GWAS, as is done in quantile–quantile plots and the calculation of a genomic control inflation factor (Devlin et al., 2004). Acquiring the enormous sample sizes required to obtain genome-wide significance for small effect variants is no small feat. Future candidate gene association studies will require physiological evidence of a functional effect in support of the tested hypothesis, with replication

of the association in additional samples remaining the gold standard for validation (Hegele, 2002).

Efforts to combine large population-based cohorts into mega-meta-analysis, including more than 100,000 participants, hold the possibility to identify common variants with exceedingly small effect sizes on stroke and its risk factors. However, such Herculean efforts may provide diminishing returns, as the effect sizes discovered may be too small to affect the care of an individual patient. Nevertheless, small-effect genetic variants may reveal interesting biological insights. Moreover, due to the high prevalence of stroke, follow-up of such large population cohorts may provide an excellent opportunity for prospective genomic analysis of stroke in the future, and small effect sizes may become amplified in a cohort with the progression of time.

While the agnostic nature of GWAS is clearly a strength, it also leads to the complicated issue of multiple testing. Given the large number of tests performed, SNPs with apparently excellent P-values representing extraordinary association would be expected simply by chance. The current standard methodology for multiple testing correction, the Bonferroni correction, involves the correction of the P-value for the number of tests performed to maintain an appropriate number of false positives. The order or rank of the association results remains unchanged, potentially overlooking true associations with disproportionate significance levels due to either small effect size or higher background noise. Bayesian methods, which include prior probability of association, either due to previously reported association for a SNP or given the function of the locus or pathway in which the SNP resides (Torkamani and Schork, 2009), removes the unbiased nature of GWAS but can change the rank of the SNPs (Wei et al., 2010). As additional insight into the genetics of stroke is obtained, the exclusion of previous knowledge in favor of an agnostic approach may become unwise.

Finally, next generation sequencing technologies promise to interrogate the genome at the resolution of a single base pair. Deep resequencing, in which a candidate region is sequenced in samples of cases and controls, has successfully identified genes that appear to contain an excess, or "accumulation," of rare sequence changes in a range of complex phenotypes (Ahituv et al., 2007; Cohen et al., 2004; Johansen et al., 2010). Technological progression from deep resequencing to whole genome sequencing, akin to the relationship between candidate gene association studies and GWAS, lies in the not-so-distant future, but will present enormous hurdles for data storage, analysis, and interpretation (Wheeler et al., 2008).

CLINICAL IMPLICATIONS OF GENOMIC DISCOVERIES IN STROKE

There is currently no evidence to support the testing of common genetic variants associated with stroke risk identified from candidate gene studies or GWAS. However, the identification of rare mutations would help the diagnosis of patients suspected of being affected with a monogenic cause of stroke. An individual with a strong family history of stroke or knowledge of a family member with a monogenic form of stroke may benefit from genetic testing. Genotyping variants associated with elevated cardiovascular risk factors may prove useful, but additional evidence is required (Johansen and Hegele, 2009). Currently, the type of genetic testing with the greatest potential for clinical relevance for stroke is pharmacogenetics, specifically for assisting with anticoagulant dosing (Lenzini et al., 2010).

CONCLUSION

Advancements in the genomic analysis of stroke have yielded several exciting and novel developments. The genetic cause of many monogenic forms of stroke has been discovered, but consistent, robust associations with common, complex stroke have yet to be identified. Efforts to remove study heterogeneity and reduce measurement error in genomic studies of stroke will help to improve the reproducibility of novel discoveries. Additional studies of both intermediate phenotypes and clinical endpoints, including candidate gene, GWAS, and Mendelian randomization approaches, are needed. The genomic analysis of stroke requires intelligent, appropriately powered, multidisciplinary studies, incorporating knowledge from clinical medicine, epidemiology, genomics, and molecular biology.

ACKNOWLEDGMENTS

Dr Lanktree is supported by the Canadian Institutes of Health Research (CIHR) MD/PhD Studentship Award, the University of Western Ontario MD/PhD Program, and is a CIHR Fellow in Vascular Research. Dr Hegele is a Career Investigator of the Heart and Stroke Foundation of Ontario, holds the Edith Schulich Vinet Canada Research Chair (Tier I) in Human Genetics, the Martha G. Blackburn Chair in Cardiovascular Research, and the Jacob J. Wolfe Distinguished Medical Research Chair at the University of Western Ontario. This work was supported by CIHR (MOP-13430, MOP-79523, CTP-79853), the Heart and Stroke Foundation of Ontario (NA-6059, T-6018, PRG-4854), Genome Canada through Ontario Genomics Institute, and the Pfizer Jean Davignon Distinguished Cardiovascular and Metabolic Research Award.

REFERENCES

Ahituv, N., Kavaslar, N., Schackwitz, W., et al., 2007. Medical sequencing at the extremes of human body mass. Am J Hum Genet 80, 779–791.

Anderson, C.D., Biffi, A., Rost, N.S., et al., 2010. Chromosome 9p21 in ischemic stroke: Population structure and meta-analysis. Stroke 41, 1123–1131.

Ariyaratnam, R., Casas, J.P., Whittaker, J., et al., 2007. Genetics of ischaemic stroke among persons of non-European descent: A meta-analysis of eight genes involving approximately 32,500 individuals. PLoS Med 4, e131.

Baird, A,E., 2010. Genetics and genomics of stroke: Novel approaches. J Am Coll Cardiol 56, 245–253.

Bersano, A., Ballabio, E., Bresolin, N., Candelise, L., 2008. Genetic polymorphisms for the study of multifactorial stroke. Hum Mutat 29, 776–795.

Bilguvar, K., Yasuno, K., Niemela, M., et al., 2008. Susceptibility loci for intracranial aneurysm in European and Japanese populations. Nat Genet 40, 1472–1477.

Casas, J.P., Bautista, L.E., Smeeth, L., et al., 2005. Homocysteine and stroke: Evidence on a causal link from Mendelian randomisation. Lancet 365, 224–232.

Chinnery, P.F., Elliott, H.R., Syed, A., Rothwell, P.M., 2010. Mitochondrial DNA haplogroups and risk of transient ischaemic attack and ischaemic stroke: A genetic association study. Lancet Neurol 9, 498–503.

Cohen, J.C., Kiss, R.S., Pertsemlidis, A., et al., 2004. Multiple rare alleles contribute to low plasma levels of HDL cholesterol. Science 305, 869–872.

Debette, S., Bis, J.C., Fornage, M., et al., 2010. Genome-wide association studies of MRI-defined brain infarcts: Meta-analysis from the charge consortium. Stroke 41, 210–217.

Devlin, B., Bacanu, S.A., Roeder, K., 2004. Genomic Control to the Extreme. Nat Genet 36, 1129–1130, author reply 1131.

Dichgans, M., 2002. Cerebral autosomal dominant arteriopathy with subcortical infarcts and leukoencephalopathy: Phenotypic and mutational spectrum. J Neurol Sci 203–204, 77–80.

Dichgans, M., 2007. Genetics of ischaemic stroke. Lancet Neurol 6, 149–161.

Dichgans, M., Markus, H.S., 2005. Genetic association studies in stroke: Methodological issues and proposed standard criteria. Stroke 36, 2027–2031.

Ding, H., Xu, Y., Wang, X., et al., 2009. 9p21 is a shared susceptibility locus strongly for coronary artery disease and weakly for ischemic stroke in Chinese Han population. Circ Cardiovasc Genet 2, 338–346.

Flossmann, E., Schulz, U.G., Rothwell, P.M., 2004. Systematic review of methods and results of studies of the genetic epidemiology of ischemic stroke. Stroke 35, 212–227.

Freimer, N., Sabatti, C., 2003. The human phenome project. Nat Genet 34, 15–21.

Gretarsdottir, S., Thorleifsson, G., Manolescu, A., et al., 2008. Risk variants for atrial fibrillation on chromosome 4q25 associate with ischemic stroke. Ann Neurol 64, 402–409.

Gschwendtner, A., Bevan, S., Cole, J.W., et al., 2009. Sequence variants on chromosome 9p21.3 confer risk for atherosclerotic stroke. Ann Neurol 65, 531–539.

Gudbjartsson, D.F., Holm, H., Gretarsdottir, S., et al., 2009. A sequence variant in ZFHX3 on 16q22 associates with atrial fibrillation and ischemic stroke. Nat Genet 41, 876–878.

Hara, K., Shiga, A., Fukutake, T., et al., 2009. Association of HTRA1 mutations and familial ischemic cerebral small-vessel disease. N Engl J Med 360, 1729–1739.

Hashikata, H., Liu, W., Inoue, K., et al., 2010. Confirmation of an association of single-nucleotide polymorphism rs1333040 on 9p21 with familial and sporadic intracranial aneurysms in Japanese patients. Stroke 41, 1138–1144.

Hata, J., Matsuda, K., Ninomiya, T., et al., 2007. Functional SNP in an Sp1-binding site of AGTRL1 gene is associated with susceptibility to brain infarction. Hum Mol Genet 16, 630–639.

Hegele, R.A., 2002. SNP judgments and freedom of association. Arterioscler Thromb Vasc Biol 22, 1058–1061.

Helgadottir, A., Thorleifsson, G., Magnusson, K.P., et al., 2008. The same sequence variant on 9p21 associates with myocardial infarction, abdominal aortic aneurysm and intracranial aneurysm. Nat Genet 40, 217–224.

Ikezaki, K., Han, D.H., Kawano, T., et al., 1997. A clinical comparison of definite moyamoya disease between South Korea and Japan. Stroke 28, 2513–2517.

Ikram, M.A., Seshadri, S., Bis, J.C., et al., 2009. Genomewide association studies of stroke. N Engl J Med 360, 1718–1728.

International Stroke Genetics Consortium and Welcome Trust Case-Control Consortium 2, 2010. Failure to validate association between 12p13 variants and ischemic stroke. N Engl J Med 362, 1547–1550.

Johansen, C.T., Hegele, R.A., 2009. Predictive genetic testing for coronary artery disease. Crit Rev Clin Lab Sci 46, 343–360.

Johansen, C.T., Wang, J., Lanktree, M.B., et al., 2010. Excess of rare variants in genes identified by genome-wide association study of hypertriglyceridemia. Nat Genet 42, 684–687.

Joutel, A., Monet, M., Domenga, V., et al., 2004. Pathogenic mutations associated with cerebral autosomal dominant arteriopathy with subcortical infarcts and leukoencephalopathy differently affect Jagged1 binding and Notch3 activity via the RBP/JK signaling pathway. Am J Hum Genet 74, 338–347.

Karvanen, J., Silander, K., Kee, F., et al., 2009. The impact of newly identified loci on coronary heart disease, stroke and total mortality in the MORGAM prospective cohorts. Genet Epidemiol 33, 237–246.

Knuiman, M.W., Divitini, M.L., Welborn, T.A., Bartholomew, H.C., 1996. Familial correlations, cohabitation effects, and heritability for cardiovascular risk factors. Ann Epidemiol 6, 188–194.

Kubo, M., Hata, J., Ninomiya, T., et al., 2007. A nonsynonymous SNP in PRKCH (protein kinase C eta) increases the risk of cerebral infarction. Nat Genet 39, 212–217.

Lanktree, M.B., Anand, S.S., Yusuf, S., Hegele, R.A., 2009a. Replication of genetic associations with plasma lipoprotein traits in a multi-ethnic sample. J Lipid Res 50, 1487–1496.

Lanktree, M.B., Hassell, R.G., Lahiry, P., Hegele, R.A., 2010. Phenomics: Expanding the role of clinical evaluation in genomic studies. J Investig Med 58, 700–706.

Lanktree, M.B., Hegele, R.A., Yusuf, S., Anand, S.S., 2009b. Multi-ethnic genetic association study of carotid intima-media thickness using a targeted cardiovascular SNP microarray. Stroke 40 (10), 3173–3179.

Lemmens, R., Abboud, S., Robberecht, W., et al., 2009. Variant on 9p21 strongly associates with coronary heart disease, but lacks association with common stroke. Eur J Hum Genet 17 (10), 1287–1293.

Lenzini, P., Wadelius, M., Kimmel, S., et al., 2010. Integration of genetic, clinical, and INR data to refine warfarin dosing. Clin Pharmacol Ther 87, 572–578.

Lussier, Y.A., Liu, Y., 2007. Computational approaches to phenotyping: High-throughput phenomics. Proc Am Thorac Soc 4, 18–25.

Majersik, J.J., Skalabrin, E.J., 2006. Single-gene stroke disorders. Semin Neurol 26, 33–48.

Manolio, T.A., Boerwinkle, E., O'Donnell, C.J., Wilson, A.F., 2004. Genetics of ultrasonographic carotid atherosclerosis. Arterioscler Thromb Vasc Biol 24, 1567–1577.

Matarin, M., Brown, W.M., Scholz, S., et al., 2007. A genome-wide genotyping study in patients with ischaemic stroke: Initial analysis and data release. Lancet Neurol 6, 414–420.

Matarin, M., Brown, W.M., Singleton, A., et al., 2008. Whole genome analyses suggest ischemic stroke and heart disease share an association with polymorphisms on chromosome 9p21. Stroke 39, 1586–1589.

Matarin, M., Singleton, A., Hardy, J., Meschia, J., 2010. The genetics of ischaemic stroke. J Intern Med 267, 139–155.

McPherson, R., Pertsemlidis, A., Kavaslar, N., et al., 2007. A common allele on chromosome 9 associated with coronary heart disease. Science 316, 1488–1491.

Meschia, J.F., 2009. Pharmacogenetics and stroke. Stroke 40, 3641–3645.

Nakaoka, H., Takahashi, T., Akiyama, K., et al., 2010. Differential effects of chromosome 9p21 variation on subphenotypes of intracranial aneurysm: Site distribution. Stroke 41, 1593–1598.

Paternoster, L., Martinez-Gonzalez, N.A., Charleton, R., et al., 2010. Genetic effects on carotid intima-media thickness: Systematic assessment and meta-analyses of candidate gene polymorphisms studied in more than 5000 subjects. Circ Cardiovasc Genet 3, 15–21.

Richards, A., van den Maagdenberg, A.M., Jen, J.C., et al., 2007. C-terminal truncations in human 3′–5′ DNA exonuclease TREX1 cause autosomal dominant retinal vasculopathy with cerebral leukodystrophy. Nat Genet 39, 1068–1070.

Runchey, S., McGee, S., 2010. Does this patient have a hemorrhagic stroke?: Clinical findings distinguishing hemorrhagic stroke from ischemic stroke. JAM 303, 2280–2286.

Samani, N.J., Erdmann, J., Hall, A.S., et al., 2007. Genomewide association analysis of coronary artery disease. N Engl J Med 357, 443–453.

Sharma, P., Bentley, P., 2010. Down but not out: Candidate gene-based studies still have value in a world dominated by whole genome approaches. Circ Res 106, 1019–1021.

Sheehan, N.A., Didelez, V., Burton, P.R., Tobin, M.D., 2008. Mendelian randomisation and causal inference in observational epidemiology. PLoS Med 5, e177.

Smith, J.G., Melander, O., Lovkvist, H., et al., 2009. Common genetic variants on chromosome 9p21 confer risk of ischemic stroke: A large-scale genetic association study. Circ Cardiovasc Genet 2, 159–164.

Sproule, D.M., Kaufmann, P., 2008. Mitochondrial encephalopathy, lactic acidosis, and strokelike episodes: Basic concepts, clinical phenotype, and therapeutic management of MELAS syndrome. Ann N Y Acad Sci 1142, 133–158.

Takeuchi, F., McGinnis, R., Bourgeois, S., et al., 2009. A genome-wide association study confirms VKORC1, CYP2C9, and CYP4F2 as principal genetic determinants of warfarin dose. PLoS Genet 5, e1000433.

Teo, Y.Y., Inouye, M., Small, K.S., et al., 2007. A genotype calling algorithm for the Illumina BeadArray platform. Bioinformatics 23, 2741–2746.

Teslovich, T.M., Musunuru, K., Smith, A.V., et al., 2010. Biological, clinical and population relevance of 95 loci for blood lipids. Nature 466, 707–713.

Torkamani, A., Schork, N.J., 2009. Pathway and network analysis with high-density allelic association data. Methods Mol Biol 563, 289–301.

Touze, E., Rothwell, P.M., 2008. Sex differences in heritability of ischemic stroke: A systematic review and meta-analysis. Stroke 39, 16–23.

Wadelius, M., Chen, L.Y., Lindh, J.D., et al., 2009. The largest prospective warfarin-treated cohort supports genetic forecasting. Blood 113, 784–792.

Wahlstrand, B., Orho-Melander, M., Delling, L., et al., 2009. The myocardial infarction associated *CDKN2A/CDKN2B* locus on chromosome 9p21 is associated with stroke independently of coronary events in patients with hypertension. J Hypertens 27, 769–773.

Wang, L., McLeod, H.L., Weinshilboum, R.M, 2011. Genomics and drug response. N Engl J Med 364, 1144–1153.

Wei, Y.C., Wen, S.H., Chen, P.C., et al., 2010. A simple Bayesian mixture model with a hybrid procedure for genome-wide association studies. Eur J Hum Genet 18, 942–947.

Welcome Trust Case Control Consortium, 2007. Genome-wide association study of 14,000 cases of seven common diseases and 3000 shared controls. Nature 447, 661–678.

Wheeler, D.A., Srinivasan, M., Egholm, M., et al., 2008. The complete genome of an individual by massively parallel DNA sequencing. Nature 452, 872–876.

Yamada, Y., Fuku, N., Tanaka, M., et al., 2009. Identification of CELSR1 as a susceptibility gene for ischemic stroke in Japanese individuals by a genome-wide association study. Atherosclerosis 207 (1), 144–149.

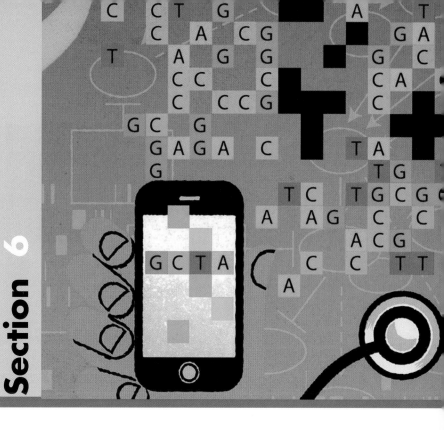

Oncology Genomic Medicine

Section 6

Lymphomas

Sandeep Dave and Katherine Walsh

INTRODUCTION

Lymphomas are a heterogeneous group of malignancies that arise from lymphocytes. Since the first report of lymphoma by Thomas Hodgkin in 1832, lymphomas have been grouped into either Hodgkin's lymphoma or non-Hodgkin's lymphoma (NHL). In 2009, there were 601,180 patients with lymphoma (148,460 Hodgkin's lymphoma and 452,720 NHL) in the United States alone. Advances in technology have revealed multiple additional subtypes of NHL. Of these subtypes, diffuse large B-cell lymphoma (DLBCL) accounts for the largest percentage at 40% of cases, and in the United States alone there are 23,000 new diagnoses per year. Lymphomas are generally treatable with chemotherapy. Even so, the majority of patients with DLBCL and a number of other lymphoma subtypes will eventually succumb to their disease. This chapter will begin with an overview of lymphoma and discuss the influence of genomics on diagnosis, prognosis, and treatment, with emphasis on DLBCL, mantle-cell lymphoma, Burkitt's lymphoma, and Hodgkin's lymphoma.

CLINICAL PRESENTATION AND STAGING

There is no effective screen for lymphoma. Clinical presentation of the disease includes constitutional symptoms (fever, fatigue, weight loss, night sweats), itching, swollen lymph nodes, or pain. Initial evaluation includes physical examination and blood work with complete blood count (CBC) with differential and lactate dehydrogenase (LDH). The diagnosis relies on obtaining a biopsy with an assessment of histology and of expression of surface proteins using flow cytometry and immunohistochemistry, as well as chromosomal analysis. Extent of disease is usually evaluated with computed tomography (CT) scans [or positron emission tomography (PET)/CT scan] and bone marrow biopsy. Lumbar puncture is included if central nervous system involvement is suspected.

Lymphoma staging is primarily based on the Ann Arbor Staging System, World Health Organization (WHO) classification, and the International Prognostic Index. The Ann Arbor system is anatomically based, with Stage I, single lymph node region; Stage II, two separate regions on the same side of the diaphragm; Stage III, lymph node regions on both sides of the diaphragm; and Stage IV, diffuse disease to at least one extranodal organ. Hodgkin's lymphoma is divided into nodular lymphocyte-predominant Hodgkin's (CD20 positive, usually lack Reed Sternberg cells) and classical Hodgkin's lymphoma (composed of four entities: nodular sclerosing, lymphocyte rich, lymphocyte depleted, and mixed cellularity subtypes). NHL is classified based primarily on the cell of origin, with 27 distinct B-cell lymphomas and 20 T-cell or natural killer (NK)-cell lymphoma types. The International Prognostic Index is a prognostic score assigning one point each for age >60, elevated LDH, poor performance status, Stage III–IV, and one or more extranodal sites for NHL. For Hodgkin's disease, the prognosis is based on seven criteria: serum albumin less than 4 g/dl, hemoglobin below 10.5, male gender, Stage IV, leukocytosis

Genomic and Personalized Medicine, 2nd edition
by Ginsburg & Willard. DOI: http://dx.doi.org/10.1016/B978-0-12-382227-7.00057-4

[white blood cell count (WBC) 15,000 or higher], lymphocytopenia (either below 600/mm³ or <8% of WBC).

Standard therapy can range from watchful waiting for indolent lymphomas to radiation and combination chemotherapy for more aggressive diseases. Chemotherapy is adriamycin-based and for NHL typically consists of CHOP (cyclophosphamide, hydroxydaunorubicin, oncovin, and prednisone) with rituximab if the molecular target CD20 is expressed. Hodgkin's disease is typically treated with ABVD (adriamycin, bleomycin, vinblastine, dacarbazine). For those patients whose tumors relapse, bone marrow transplantation remains the only curative option. Current research in genomics is working to improve accuracy of diagnosis and to identify predictors of disease aggressiveness and response to therapy. The following section will provide an overview of genomics.

THE ROLE OF KEY GENES IN LYMPHOMA

Genetic studies in lymphoma have identified a role for a number of key genes that are recurrently deregulated, including

BCL6, *PRDM1* (BLIMP1), and *XBP1* (Staudt and Dave, 2005). These four genes are important in the germinal center reaction of lymph nodes, where B cells are responding to antigen presentation with increased proliferation, differentiation, and class-switching to mount an antibody response. During this process, naïve cells undergo the germinal center (GC) reaction. GC B cells divide rapidly as they differentiate into memory B cells or plasma cells. The stage of B cells appears to be a critical determinant of the kind of lymphoma that can develop when this process of normal differentiation is deregulated. Thus, lymphomas frequently maintain the features of, including the expression of a number of markers related to, their normal cell counterparts (Figure 57.1).

The *BCL6* (B-cell chronic lymphocytic leukemia/lymphoma 6) gene encodes a transcription factor that has a repressor function and works to promote B-cell maturation. It is a critical regulator of GC differentiation. Approximately a quarter of DLBCL cases have a chromosomal translocation that causes a loss of regulation of BCL6, leading to reduced differentiation of these cells. The high expression of BCL6 in lymphomas derived from GC cells indicates that these malignant cells retain the biology of their cell of origin and express genes typical of the germinal center.

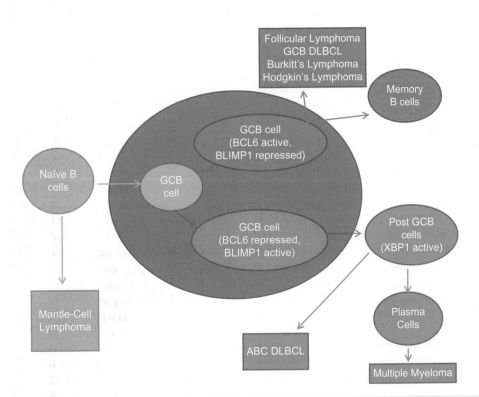

Figure 57.1 Schematic representing B-cell differentiation through the germinal center. In normal differentiation, naïve B cells enter the germinal center and develop into memory B cells if BCL6 is active, or develop into plasma cells if BLIMP1 is active. The type of lymphoma that develops clinically depends on where in normal development the malignant transformation takes place. Naïve B cells give rise to mantle-cell lymphoma. Germinal center B cells are the cell of origin for follicular lymphoma, germinal center B-cell-like (GCB) DLBCL, Burkitt's lymphoma, and Hodgkin's lymphoma. Post germinal center B cells give rise to ABC (activated B-cell-like) DLBCL. Plasma cells are the cells of origin for multiple myeloma.

PRDM1 (positive regulatory domain containing 1, with zinc finger domain) encodes BLIMP1 (B-lymphocyte-induced maturation protein 1), which is a transcription factor that promotes plasma-cell differentiation. Normally, BCL6 represses BLIMP1 and the two proteins exist in a state of equilibrium. Inhibition of BLIMP1 favors memory B-cell formation. In the setting of BCL6 inhibition, BLIMP1 de-repression stimulates plasma cell development through a coordinated process of increasing CD38 expression and down-regulating *MYC* and a number of B-cell markers.

A number of genes that have normal functions in B cells also have roles that can be subverted by B-cell malignancies. For instance, the *MYC* gene has high association with poor outcome (Rosenwald et al., 2002) and functions as a transcription factor that regulates cellular metabolism and growth, including determination of cell size. *XBP1* (X-box binding protein 1) is a key gene in plasma-cell differentiation. It is responsible for production of the endoplasmic reticulum (ER) and golgi. Both the ER and golgi apparatus are involved in protein production and secretion. This contributes to the typical features of plasma cells, including high rates of secretion of immunoglobulins and large cell size. These features are well preserved in their malignant counterparts such as multiple myeloma.

The vast majority of NHL arises from these mature B cells (85%), and from the multiple identified subtypes of NHL, the largest percentages develop from GC B cells or post-GC B cells, pointing to the germinal center reaction as a likely mechanism for the acquisition of genetic alterations.

MicroRNA

MicroRNAs (miRNAs, or miRs) are defined as chains of 18–22 nucleotide RNA molecules, with their function being to monitor and control gene expression, including regulating each stage of B-cell differentiation. This includes regulation of the transcription factors LMO2 and BLIMP1. The *LMO2* gene encodes a transcription factor that is vital for angiogenesis and is highly expressed in DLBCLs that arise from the germinal center. A number of these miRNAs appear to be important in sustaining the clinical features of normal B cells and lymphomas. Consistent with this notion, miRNAs were found to correctly predict B-cell lineage in over 95% of lymphoma patients (Zhang et al., 2009).

Experimental evidence points to a role for miR223, miR17-92, and several others in lymphomas. For instance, miR223 has been found to regulate *LMO2*, with increased expression of miR223 leading to increased expression of *LMO2*. MiR223 was found to be highly expressed at two distinct transition points during B-cell differentiation: the switch from naïve B cells to GC B cells and the transformation from GC B cells to memory cells. The miR17-92 family was highly expressed in GC cells and deregulation of the miRNA leads to lymphoma in transgenic mouse models.

GENE EXPRESSION SIGNATURES

Beyond the role of individual genes, gene expression signatures provide a robust estimate of underlying biological function of different genes acting in concerted fashion. Gene expression signatures comprise genes that have similar expression patterns based on a shared feature and encode gene products that have similar biologic effects, such as on cell type (for example, T-cell signature including genes for the T-cell receptor and markers expressed on the cell surface), differentiation state (e.g., germinal center B-cell signature, discussed further under DLBCL section), or signaling response [e.g., nuclear factor kappa B (NFκB)] (Shaffer et al., 2001). Differential gene expression can be due to characteristics of the tumor cells, or due to the reaction of the non-malignant cells in the adjacent tissue. These signatures are developed using DNA microarrays that allow quantitative evaluation of thousands of genes at the same time (Schena et al., 1995). Individual gene expression signatures with clinical associations in lymphoma (see Table 57.1) include those associated with good prognosis [major histocompatibility (MHC) class II], worse prognosis (proliferation), and host response (lymph node) and are discussed below.

The MHC class II (major histocompatibility complex II) signature includes genes associated with the antigen-presenting function of macrophages, dendritic cells, and B cells, and encodes the components of the MHC class II molecules. Patient samples with increased expression of MHC II had better survival in DLBCL (Rosenwald et al., 2002). The hypothesis is that when gene expression of MHC class II molecules is reduced, the ability to monitor for and remove malignant cells is lost. This is consistent with studies showing that DLBCLs without MHC class II expression have diminished CD8+ T cells in the tumor samples. Differences in the characteristics of the lymphoma cells were confirmed to be the cause of differential expression of MHC class II, rather than this reflecting host

TABLE 57.1 Summary of the gene expression signatures discussed in the text, function of the genes included, and their relevance to clinical prognosis

Signature	Characteristics	Clinical relevance
GCB	Associated with germinal center B-cell differentiation	Good prognosis
MHC class II	Associated with antigen-presenting function of the immune system	Good prognosis
Proliferation	Associated with cell-cycle events and cell growth	Poor prognosis
Lymph node	Associated with immune response in the microenvironment	Good prognosis

GCB, germinal center B-cell-like.

response to the tumor. MHC class II reflects the patient's ability to detect and mount an antibody response to a tumor, and has prognostic importance in lymphoma.

The proliferation signature is one of the largest signatures and includes genes involved in multiple cell-cycle events (cell growth, cell-cycle progression, DNA synthesis and repair) (Alizadeh et al., 2000; Shaffer et al., 2001). Within this large group of genes, there are individual genes highly associated with poor outcome (e.g., *c-myc*) and subgroups of genes associated with particular aspects of cell division that are closely co-regulated (Rosenwald et al., 2003b). The components of the signature associated with poor survival are genes involved in cell growth and metabolism, primarily functioning in the nucleolus, and are targets of c-myc, including nucleostemin and nucleophosmin-3/NPM3. The highest *MYC* expression was present in the lymphoma cells derived from the post-germinal center B cells (Staudt and Dave, 2005).

In addition to reflecting tumor characteristics, gene expression signatures can also identify immune response to the malignant cells in the surrounding microenvironment. One example of this is the lymph node (LN) signature, which was named based on the finding that the genes were more associated with the non-malignant lymph node components (Alizadeh et al., 2000). The genes in this signature correspond to monocytes, macrophages, connective tissue growth factors (CTGFs), and those causing changes in the extracellular matrix as a result of tumor infiltration. Patients with high expression of this signature typically have a better prognosis. Two theories were proposed for the improved survival seen with expression of this signature: (1) the LN reflects patients who are able to develop an antibody response to the tumor, or (2) the malignant cells become dependent on the support of the microenvironment to survive, and therefore cannot spread distally. An example of this process is seen in multiple myeloma, where the upregulation of c-maf leads to increased integrins that bind E-cadherin and result in production of vascular endothelial growth factor (VEGF) in the bone marrow, providing increased angiogenesis as well as causing the myeloma cells to bind more tightly to stromal cells (Hurt et al., 2004).

Another approach has been to combine known gene expression signatures with individual predictive genes to generate predictors. One example of this is a 17-gene model created with 16 genes from known signatures (3 germinal center B cell, 4 MHC class II, 6 LN, and 3 proliferation genes) that predicted survival after chemotherapy for DLBCL with high certainty (Rosenwald et al., 2002). Gene expression signatures have been shown to predict survival independently from known clinical indices. Thus, gene expression signatures may have additional uses such as improving diagnostic accuracy and identifying distinct lymphoma subtypes, which will be discussed in the next section.

GENOMICS BY LYMPHOMA TYPE

Non-Hodgkin's lymphoma (NHL) contains multiple subtypes based on the most recent WHO classifications, with large variation in prognosis and survival. Genomics and gene expression profiling have identified separate subtypes that are grouped together under the current classification, which may explain some of the variation in response within the WHO-defined subtypes. The influence of genomics on diagnostics, prognosis, and prediction of response to chemotherapy will be discussed for different NHLs (follicular, DLBCL, mantle cell, and Burkitt's lymphoma) and Hodgkin's lymphoma, with a summary provided in Table 57.2.

Follicular Lymphoma

Follicular lymphoma ranks as the second most frequent NHL, and is generally an indolent disease, with a median survival of 10 years. The malignant cells of this disease arise from GC B cells. Patients display a wide range of survival, from under 1 year to more than 20 years. Two gene expression signatures have been identified that together strongly predicted survival in a synergistic manner: immune response 1 (good prognosis) and 2 (poor prognosis) (Dave et al., 2004). The immune response 1 gene expression signature included genes encoding T-cell markers (e.g., *CD8* and *STAT4*). Immune response 2 included genes expressed in macrophages and/or dendritic cells (e.g., *CD68*, *TLR5*, and *C3AR1*). The survival of 191 follicular lymphoma patients was found to be significantly different when patients were grouped into quartiles based on their expression of these two gene expression signatures: 13.6 years, 11.1 years, 10.8 years, and 3.9 years. Further evaluation after sorting by flow cytometry confirmed that both immune response signatures were expressed predominantly in the non-malignant, CD19-negative cells. Further, compared to the gene expression profiles of normal hematopoietic cells, genes from immune responses 1 and 2 were over-expressed in T cells and monocytes, but not in B cells. Taken together, these data suggest that the prognostically important immune response gene expression signatures are derived from non-malignant cells, and highlights the interaction of the patient's immune system and the malignant cells as an important factor in tumor progression (Dave et al., 2004).

Diffuse Large B-Cell Lymphoma

DLBCL is a more aggressive lymphoma compared to follicular lymphoma. The notion of DLBCL as a single disease was first challenged when gene expression profiling demonstrated that DLBCL comprised at least two distinct subgroups with different gene expression and clinical behavior. These subgroups were termed germinal-center-like B cells (GCBs) and activated B-cell-like B cells (ABCs), with GCBs arising from normal GC B cells and ABCs arising from a normal post-GC B-cell stage. There was a striking difference in survival between these subgroups of patients, with overall survival at five years of 76% for GCB and 16% for ABC DLBCL patients (Alizadeh et al., 2000). These distinctions have been confirmed in several additional studies (Rosenwald et al., 2002; Wright et al., 2003). These entities share a number of other features such as the MHC class II gene expression.

TABLE 57.2	Summary of lymphoma subtypes including their point of origin during B-cell differentiation, key oncogenic events, and associated gene expression signatures		
Subtype	**Origin**	**Oncogenic events**	**Signatures**
ABC DLBCL	Post germinal center	Trisomy 3, *INK4alpha*, *ARF* deletion, BLIMP1, *XBP1*, NFκB pathway	ABC
GCB DLBCL	Germinal center	t(14;18), *bcl-2*, *c-rel*, *LMO2*, *BCL6*	Good prognosis: GCB, LN, MHC-2, stromal 1 Bad prognosis: proliferation, stromal 2
PMBL DLBCL	Germinal center	*PDL2* on chromosome 9p, NFκB pathway	30% overlap with Hodgkin's lymphoma
Follicular lymphoma	Germinal center	t(14;18), *bcl-2*, *BCL6*	Good prognosis: immune response 1 Poor prognosis: immune response 2
Burkitt's lymphoma	Germinal center	*c-myc*	mBL signature; composite of myc, subset GCB, and low level NFκB and MHC class 1
Hodgkin's lymphoma	Germinal center	*PDL2* on chromosome 9p, NFκB	30% overlap with PMBL
MCL	Pre-germinal center	t(11;14); cyclin D1; deletion of *INK4alpha/ARF*	MCL proliferation

ABC DLBCL, activated B-cell-like DLBCL; GCB (DLBCL), germinal-center B-cell-like (DLBCL); PMBL, primary mediastinal B-cell lymphoma; mBL, molecular Burkitt's lymphoma; MCL, mantle cell lymphoma.

The GCB DLBCL subgroup showed a number of previously identified oncogenic events, including the t(14;18) translocation affecting the *BCL2* gene, overexpression of *C-REL* from chromosome 2p, as well as upregulation of *LMO2*. Those events were not present in the ABC subgroup (Rosenwald et al., 2002). The GCB gene expression signature (Alizadeh et al., 2000) suggests that GCB DLBCL cells originate from normal GC B cells (Dave et al., 2004,) and exhibit many of the biological features of normal GC B cells. For instance, these tumors are active in proliferation with several G2/M phase mediators and low levels of metabolic genes such as *MYC* (Shaffer et al., 2001). From the GCB signature, three genes stand out as the most powerfully linked with survival (*BCL6*, *SERPINA9*, and *GCET2*) (Rosenwald et al., 2002; Staudt and Dave, 2005). At the genetic level, copy number gains involving miR17-92 and allelic losses involving *PTEN* were specifically associated with the GCB subtype (Lenz et al., 2008).

The malignant cells of the ABC subgroup of DLBCLs arise from post-germinal-center B cells that are poised to differentiate into plasma cells. The genes in the ABC DLBCL-defining gene expression signatures are associated with plasma-cell differentiation, with high secretion of proteins and immunoglobulins. Genetic alterations associated with the ABC subtype include deletion of the known tumor suppressor *INK4a/ARF*, and trisomy 3 (Lenz et al., 2008). Unlike GCB DLBCLs, which express *BCL6*, the ABC subgroup of DLBCLs overexpress *IRF4*, which normally functions to increase cell division after the presentation of an antigen, but becomes permanently active in the development of DLBCL (Alizadeh et al., 2000). The genes *PRKCB1* and *PDE4B* were more highly expressed in ABC than GCB DLBCL (Staudt and Dave, 2005).

A third distinct subtype of DLBCL has since emerged, called primary mediastinal B-cell lymphoma (PMBL) which overlaps with the other subtypes with regard to expression of the LN signature and increased activity of NFκB, and its signature had one-third of its genes in common with a gene expression signature reflecting Hodgkin's lymphoma. There was also a clinical overlap with Hodgkin's disease, with most cases occurring in young women as mediastinal disease, with better five-year overall survival compared to DLBCL (64 vs 46%). The gene that was best able to distinguish PMBL from the other DLBCL subtypes was *PDL2* (programmed death ligand 2), which is located on chromosome 9p and is also characteristic of Hodgkin's lymphoma (Rosenwald et al., 2003a).

The three defined subtypes of DLBCL differ in their survival outcomes and responses to therapy. Five-year overall survival for GCB, PMBL, and ABC has been reported to be 59%, 64%, and 31% respectively (Alizadeh et al., 2000; Rosenwald et al., 2002; Wright et al., 2003). These survival differences were independent of the microarray type used to determine the gene expression profiles (Wright et al., 2003). In general, GCB DLBCL shows a better prognosis than ABC DLBCL, but considerable variability remains in the outcome, with 30% mortality for GCB patients at two years from diagnosis and around one-quarter of ABC DLBCL patients still alive more than ten years from diagnosis. Prognostic gene expression signatures have been developed in DLBCL that reflect GC B cells, LN, proliferation, and MHC class II, which are highly efficacious in prognosticating survival in these patients (Rimsza et al., 2008).

Similar to follicular lymphoma, the microenvironment and immune response are important in DLBCL (Dave et al., 2004). In addition to the previously described signatures, two gene expression signatures, identified as stromal 1 and stromal 2, were predictive of response to therapy with CHOP or R-CHOP (rituximab in combination with CHOP chemotherapy)

(Lenz et al., 2008). The stromal 1 gene expression signature comprised genes of the extracellular matrix responsible for fibronectin, osteonectin, and collagen. In patient samples, the products of these genes were found in the fibrous strands between tumor cells as well as histiocytic cells in the DLBCL biopsies. This signature is believed to represent a monocytic immune response to the lymphoma cells. The stromal 2 signature includes genes for endothelial cells and blood vessel formation. This correlated clinically with high expression of the signature producing increased blood vessel density in patients.

These data indicate that gene expression signatures in DLBCL are robust measures of the underlying biology of these tumors, and have shed new light with regard to the observed clinical heterogeneity in these patients.

Mantle-cell Lymphoma

Mantle-cell lymphoma (MCL) accounts for 6% of lymphomas and has a disproportionately higher mortality rate. It arises from naïve B cells that have not yet undergone the germinal center reaction. Characteristic genetic events include t(11;14), *INK4a/ARF* deletion, p53 deletion and *ATM* deletion, in addition to increased expression of *BMI-1*.

In MCL, a gene expression signature indicating proliferation was associated with poorer survival, with a survival difference of 6.7 years (vs 0.8 years) based on the level of expression. Although cyclin D1 over-expression is a hallmark of the tumor, an additional subset of cyclin-D1-negative MCL patients were identified using gene expression profiling. These patients are believed to over-express cyclin D2 or D3 in the place of cyclin D1 (Rosenwald et al., 2003b) with similar downstream consequences.

Burkitt's Lymphoma

Burkitt's lymphoma (BL) is an aggressive lymphoma with high cure rates after treatment with methotrexate/cytarabine (90% children, 70% adults) and high mortality without treatment. The need for accurate diagnosis, particularly differentiating Burkitt's and DLBCL, is crucial given the intense nature of BL chemotherapy compared to the standard CHOP-based regimens for DLBCL. Diagnostic accuracy for BLs remains low, with an estimated inter-observer agreement of approximately 50% among experts diagnosing the disease.

Gene expression profiling has demonstrated an improvement to diagnostic accuracy. Hummel and colleagues defined biologic Burkitt's lymphoma as cases that showed increased expression of their molecular Burkitt's lymphoma (mBL) signature, containing 58 genes. Cases expressing the mBL signature had an overall good outcome, with five-year overall survival of 75% (Hummel et al., 2006). A parallel study (Dave et al., 2006) defined a classifier with three components: high *MYC* target gene expression, increased expression of a subset of GCB genes, and low expression of both MHC class I and NFκB target genes. Both these studies indicate the feasibility and accuracy of gene expression profiling for rendering this clinically important distinction, which is difficult using existing methods in pathology.

SUMMARY

Lymphomas are a heterogeneous group of malignancies with respect to both clinical behavior and genetic alterations. The application of genomics has shed new light on these complex tumors and enabled a better understanding of the molecular heterogeneity that underlies the clinical heterogeneity. Advances include the identification of subgroups with distinct genomics and responses to therapy within the previously broad category of lymphoma, as well as the identification of prognostic and diagnostic gene expression signatures. The clinical translation of this work will improve multiple aspects of care, including diagnosis, counseling of patients regarding prognosis, identification of patients most likely to benefit from therapy, and the design of clinical trials to further develop the current knowledge base.

REFERENCES

Alizadeh, A.A., Eisen, M.B., Davis, R.E., et al., 2000. Distinct types of diffuse large B-cell lymphoma identified by gene expression profiling. Nature 403, 503–511.

Dave, S.S., Fu, K., Wright, G.W., et al., 2006. Molecular diagnosis of Burkitt's lymphoma. N Engl J Med 354, 2431–2442.

Dave, S.S., Wright, G., Tan, B., et al., 2004. Prediction of survival in Follicular Lymphoma based on molecular features of tumor-infiltrating immune cells. N Engl J Med 351, 2159–2169.

Hummel, M., Bentink, S., Berger, H., et al., 2006. A biologic definition of Burkitt's Lymphoma from transcriptional and genomic profiling. N Engl J Med 354, 2419–2430.

Hurt, E.M., Wiestner, A., Rosenwald, A., et al., 2004. Overexpression of c-maf is a frequent oncogenic event in multiple myeloma that promotes proliferation and pathologic interactions with bone marrow stroma. Cancer Cell 5, 191–199.

Lenz, G., Wright, G.W., Emre, N.C.T., et al., 2008. Molecular subtypes of diffuse large B-cell lymphoma arise by distinct genetic pathways. Proc Natl Acad Sci USA 105, 13,520–13,525.

Rimsza, L.M., LeBlanc, M.L., Unger, J.M., et al., 2008. Gene expression predicts overall survival in paraffin-embedded tissues of diffuse large B-cell lymphoma treated with R-CHOP. Blood 112, 3425–3433.

Rosenwald, A., Wright, G., Chan, W.C., et al., 2002. The use of molecular profiling to predict survival after chemotherapy for diffuse large B-cell lymphoma. N Engl J Med 346, 1937–1947.

Rosenwald, A., Wright, G., Leroy, K., et al., 2003a. Molecular diagnosis of primary mediastinal B cell lymphoma identifies a clinically favorable subgroup of diffuse large B cell lymphoma related to Hodgkin lymphoma. J Exp Med 198, 851–862.

Rosenwald, A., Wright, G., Wiestner, A., et al., 2003b. The proliferation gene expression signature is a quantitative integrator of oncogenic events that predicts survival in mantle cell lymphoma. Cancer Cell 3, 185–197.

Schena, M., Shalon, D., Davis, R.W., Brown, P.O., 1995. Quantitative monitoring of gene expression patterns with a complementary DNA microarray. Science 270, 467–470.

Shaffer, A.L., Rosenwald, A., Hurt, E.M, et al., 2001. Signatures of the immune response. Immunity 15, 375–385.

Staudt, L.M., Dave, S.S., 2005. The biology of human lymphoid malignancies revealed by gene expression profiling. In: Alt, F. (Ed.), Advanced Immunology, Vol 87. Academic Press, pp. 163–208.

Wright, G., Tan, B., Rosenwald, A., et al., 2003. A gene expression-based method to diagnose clinically distinct subgroups of diffuse large B cell lymphoma. Proc Natl Acad Sci USA 100, 9991–9996.

Zhang, J., Jima, D.D., Jacobs, C., et al., 2009. Patterns of microRNA expression characterize stages of human B-cell differentiation. Blood 13, 4586–4594.

CHAPTER

58

Leukemias

Lars Bullinger and Stefan Fröhling

INTRODUCTION

Recently, our understanding of the hematopoietic system has expanded dramatically, revealing that leukemias exhibit an extraordinary biologic and clinical heterogeneity. While the broad classification of leukemias is still based on the cell phenotype (myeloid versus lymphoid) and the rapidity of the clinical course (acute versus chronic), during the last decade, efforts to characterize additional biologically and clinically relevant leukemia entities were successful. Today, the World Health Organization (WHO) classification of tumors of hematopoietic and lymphoid tissues summarizes a consensus (Swerdlow et al., 2008) that divides acute myeloid leukemia (AML) into several subclasses, which can be further subdivided into distinct AML subtypes (Table 58.1). For many subclasses, however, the underlying pathogenic events are still unknown, and well-defined leukemia subgroups, such as AML cases with translocation t(8;21) or inversion inv(16), display considerable clinical heterogeneity (Bullinger and Döhner, 2010). Thus, an improved understanding of the molecular mechanisms leading to leukemia is needed.

Today, insights into the molecular anatomy of leukemias have increased tremendously, based on the advancement of technologies allowing genome-wide analyses of transcriptomic, genomic, and epigenomic changes (Figure 58.1) (Bullinger and Armstrong, 2010; Maciejewski and Mufti, 2008; Wouterset al., 2009a). Improved understanding of molecular aberrations has started to translate into daily routine. For instance, the revised WHO classification now includes

molecularly defined AML subgroups (such as AML with mutated *NPM1*), AML guidelines support genotype-specific treatment approaches, and new targeted therapies are investigated within clinical trials (Döhner et al., 2010; Rowe, 2009).

LEUKEMIA CYTOGENETICS AND MOLECULAR GENETICS

Leukemias have a long tradition of being "first." For example, chronic myelogenous leukemia (CML) was the first malignancy associated with a recurring chromosomal abnormality, the Philadelphia chromosome, which was later found to result from a balanced translocation (Rowley, 1973). The fusion gene resulting from this translocation t(9;22), *BCR-ABL1*, was then shown to be responsible for the myeloproliferation observed in CML (Daley et al., 1990; Konopka et al., 1985; Shtivelman et al., 1985). Today, chromosome banding analysis is still the gold standard to screen for genomic aberrations in leukemias, and well over 100 chromosome translocations as well as numerous recurring chromosomal gains and losses have been identified in leukemic cells (see the Mitelman Database of Chromosome Aberrations and Gene Fusions in Cancer, accessible through http://cgap.nci.nih.gov/Chromosomes/Mitelman).

Further characterization at the molecular level showed that many chromosomal translocations result in deregulated expression of oncogenes, such as *MYC* in acute lymphoblastic leukemia (ALL). Alternatively, translocations can result in the creation of chimeric fusion proteins, many of which alter

Genomic and Personalized Medicine, 2nd edition
by Ginsburg & Willard. DOI: http://dx.doi.org/10.1016/B978-0-12-382227-7.00058-6

TABLE 58.1	WHO classification of tumors of hematopoietic and lymphoid tissues – leukemias

"Acute" clinical course	"Chronic" clinical course
Myeloid neoplasm	
AML with recurrent genetic abnormalities	**Myeloproliferative neoplasms (MPN)**
AML with t(8;21)(q22;q22); *RUNX1-RUNX1T1*	Chronic myelogenous leukemia, *BCR-ABL1* positive
AML with inv(16)(p13.1q22); *CBFB-MYH11*	Chronic neutrophilic leukemia
APL with t(15;17)(q22;q12); *PML-RARA*	Polycythemia vera
AML with t(9;11)(p22;q23); *MLLT3-MLL*	Primary myelofibrosis
AML with t(6;9)(p23;q34); *DEK-NUP214*	Essential thrombocythemia
AML with inv(3)(q21q26.2); *RPN1-EVI1*	Chronic eosinophilic leukemia
AML with t(1;22)(p13;q13); *RBM15-MKL1*	Mastocytosis
Provisional entity: AML with mutated *NPM1*	Myeloproliferative neoplasms, unclassifiable
Provisional entity: AML with mutated *CEBPA*	
AML with myelodysplasia-related changes	**Myelodysplastic/myeloproliferative neoplasms**
Therapy-related myeloid neoplasms	Chronic myelomonocytic leukemia
AML, not otherwise specified	Atypical CML, *BCR-ABL1* negative
Myeloid sarcoma	Juvenile myelomonocytic leukemia
Myeloid prolif. related to Down Syndrome	Myelodysplastic/Myeloproliferative neoplasm, unclassifiable
Blastic plasmacytoid dendritic cell neoplasm	**Myelodysplastic syndrome (MDS)**
Acute leukemias of ambiguous lineage	**Myeloid and lymphoid neoplasms associated with eosinophilia and abnormalities of *PDGFRA*, *PDGFRB*, or *FGFR1***
Lymphoid neoplasm	
B lymphoblastic leukemia/lymphoma	**Mature B-cell neoplasms**
B lymphoblastic leukemia/lymphoma, NOS	CLL
B lymphoblastic leukemia/lymphoma with recurrent genetic abnormalities:	B-cell prolymphocytic leukemia
with t(9;22)(q34;q11.2); *BCR-ABL 1*	Hairy cell leukemia
with t(v;11q23); *MLL* rearranged	
with t(12;21)(p13;q22) *TEL-AML1* (*ETV6-RUNX1*)	**Mature T-cell neoplasms**
	T-cell prolymphocytic leukemia
T lymphoblastic leukemia/lymphoma	T-cell granular lymphocytic leukemia

APL, acute promyelocytic leukemia; CLL, chronic lymphocytic leukemia; CML, chronic myelogenous leukemia.
Modified from Swerdlow et al., 2008.

transcriptional programs and block cell differentiation. For example, deregulation of the core binding factor (CBF) transcription factor complex activity by t(8;21) or inv(16) (resulting in a *RUNX1-RUNX1T1* and *CBFB-MYH11* fusion, respectively) in AML leads to altered gene expression and disrupts cell differentiation. However, it has been shown that most fusion genes are not sufficient to cause a leukemic phenotype (Döhner, 2007;

Fröhling et al., 2005). For malignant transformation, additional cooperating events that can induce myeloproliferation are necessary, such as constitutive activation of FLT3 (fms-related tyrosine kinase 3) or RAS [RAS viral (v-ras) oncogene homolog]. Thus, a combination of differentiation-blocking and proliferation-inducing mechanisms is likely to be involved in leukemogenesis (Döhner and Döhner, 2008; Fröhling et al., 2005). Today,

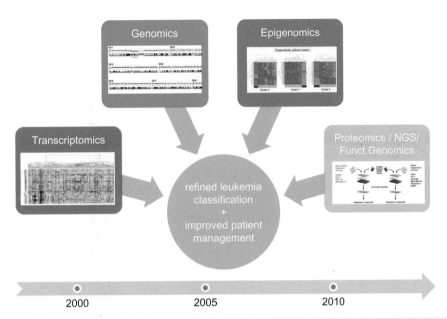

Figure 58.1 A decade of "omics" research in AML. At the beginning of this decade, microarray-based gene expression profiling studies began providing new insight into leukemia pathogenesis. Since then, novel single nucleotide polymorphism (SNP) microarray-based genomics studies have unraveled novel findings, and now genome-wide epigenomics promises to provide a deeper understanding of leukemia. Proteomics, next-generation sequencing (NGS) and functional genomics approaches are also underway and will provide additional insights. Future efforts will aim toward integrated data analyses, leading to a more comprehensive view of the biology of the disease – a prerequisite for refined leukemia classification and improved patient management. *Reproduced with permission from Bullinger and Armstrong, 2010.*

several lines of evidence implicate a multistep leukemogenesis, and based on advances in molecular genetics, many pathogenetically relevant mutations have been identified in both myeloid (Table 58.2 provides an overview of recent advances in the detection of gene mutations in AML) (Bullinger and Döhner, 2010; Döhner and Döhner, 2008) and lymphoid leukemias (Armstrong and Look, 2005; Cools, 2010).

Targeted Treatment of Leukemogenic Genomic Aberrations

Consistent with the tradition of leukemias being "first," acute promyelocytic leukemia (APL) was one of the first malignancies to be successfully treated with a molecularly targeted therapy, all-trans retinoic acid (ATRA). ATRA specifically targets the transforming potential of the fusion gene product PML-RARA, which results from a t(15;17), which is commonly detected in APL (Huang et al., 1988). Similarly, CML was the first disorder in which a small molecule inhibitor was designed to specifically target the disease-causing underlying molecular defect, the BCR-ABL1 fusion protein (Druker, 2009).

Recently, novel targeted therapies have been shown to be effective in leukemias, including numerous tyrosine kinase inhibitors (TKIs, e.g., FLT3 inhibitors such as midostaurin), farnesyltransferase inhibitors (e.g., tipifarnib), demethylating agents (e.g., decitabine), histone deacetylase inhibitors (e.g., valproic acid), immunomodulatory drugs (e.g., lenalidomide), and monoclonal antibodies (e.g., the anti-CD33

antibody gemtuzumab ozogamicin) (Haferlach, 2008), which are currently investigated within clinical trials in combination with conventional chemotherapy. While steady progress in the development of effective treatments has led to a cure rate of more than 80% in children with ALL (Pui et al., 2008), it is hoped that these novel treatment regimens will boost the generally low cure rates in adults and subgroups of children with high-risk leukemia. Hopefully, advances in our understanding of the pathobiology of leukemia, fuelled by emerging molecular technologies, will lead to the development of drugs that specifically target the genetic defects of leukemic cells and will ultimately change management of the disease.

A DECADE OF TRANSCRIPTOMICS IN LEUKEMIA

Insights from Gene Expression Profiling

Offering the ability to capture the molecular variation underlying the biological and clinical heterogeneity of leukemia, gene expression profiling (GEP) has been successfully applied to improve the molecular classification of AML (Bullinger, 2006; Wouters et al., 2009a). Again, it was leukemias in which the feasibility of genomics-based prediction of known tumor classes (*class prediction*) was demonstrated for the first time. Based on distinct gene expression profiles, Golub and colleagues could accurately predict AML and ALL without

TABLE 58.2 Biological and clinical features associated with gene mutations in acute myeloid leukemia (AML)

Gene	Biological features	Clinical features	Targeted therapy
NPM1	Phosphoprotein with pleiotropic functions; mutations found in 25–35% of adult AML; associated with FLT3-ITD and FLT3 TKD mutations	Genotype "mutant NPM1 without FLT3-ITD" associated with favorable RFS (relapse-free survival) and OS (overall survival)	ATRA (all trans retinoic acid)
CEBPA	Master regulatory transcription factor in hematopoiesis; mutations found in 10–18% of cytogenetically normal AML (CN-AML)	Mutations associated with higher CR (complete remission) rate and better RFS and OS	–
FLT3	Class III receptor tyrosine kinase; plays an important role in proliferation, survival and differentiation of hematopoietic cells		Tyrosine kinase inhibitors (TKI; e.g., midostaurin, sunitinib)
ITD	Internal tandem duplications (ITD) found in about 20% of all AML	Association with inferior outcome in CN-AML	
TKD	Point mutations of the tyrosine kinase domain (TKD) found in 5–10%	Prognostic significance controversial	
MLL	DNA binding protein which regulates gene expression in hematopoiesis; MLL-PTD found in 5–11% of CN-AML	Associated with shorter CR duration and inferior RFS and EFS	Potentially DOT1L inhibitors
NRAS	Membrane-associated protein regulating mechanisms of proliferation, differentiation and apoptosis, mutations found in 9–14% of CN-AML	Mutant NRAS may confer sensitivity to cytarabine	–
KRAS	See also NRAS; mutations found in 5–17% of CBF AML	So far, no prognostic significance found	Potentially STK33 inhibitors
KIT	Class III receptor tyrosine kinase; key role in survival, proliferation, differentiation, and functional activation of hematopoietic progenitor cells; mutations found in about 30% of CBF AML	Mutations, in particular in exon 17, associated with inferior outcome in many but not all studies	TKI (e.g., dasatinib)
WT1	Transcription factor implicated in regulation of apoptosis, proliferation, and differentiation; mutations found in about 10% of CN-AML	Initial studies suggest an association with induction failure, prognostic impact currently under investigation	–
RUNX1	Transcription factor involved in normal hematopoietic differentiation; mutations found in 10–13% of CN-AML	Associated with lower CR rate and shorter RFS and OS; independent poor prognostic factor for OS	–
IDH1/ IDH2	Isocitrate dehydrogenase, found in ~16% of CN-AML; mutations lead to an excess accumulation R(2)-2-hydroxyglutarate (2HG)	Remains to be determined	–
TET2	Belongs to the tet family, may catalyze the conversion of methyl-cytosine (5mc) to 5-hydroxymethylcytosine (hmc), mutations in 12–17% of all AML cases	Probable negligible prognostic impact in NPM1 mutated cases	–
ASXL1	Putative polycomb group protein; protein is thought to disrupt chromatin in localized areas	Remains to be determined	–
JAK2	Protein tyrosine kinase; mutations common in myeloproliferative disease, but rare events in AML	So far, no prognostic significance found	TKI (e.g., ruxolitinib)
CBL	Adaptor protein for receptor protein-tyrosine kinases, found in ~1–5% of AML cases, association with CBF AML	Remains to be determined	–
TP53	Tumor protein p53, which responds to diverse cellular stresses; incidence of mutations in AML with complex karyotypic 56–78%	Associated with inferior outcome	–

Table adapted from Döhner and Döhner, 2008.

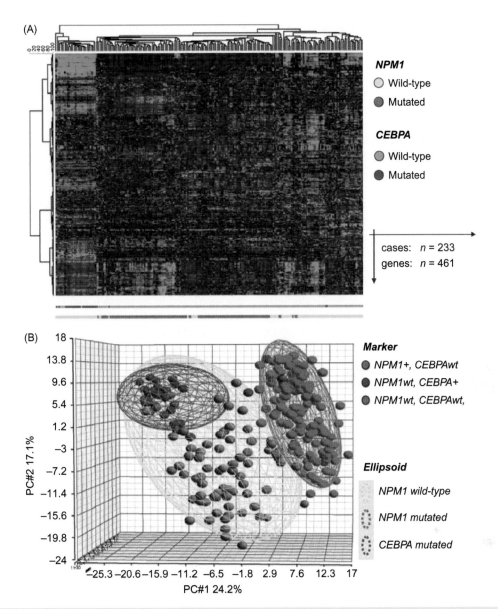

(A)

NPM1
○ Wild-type
● Mutated

CEBPA
◐ Wild-type
● Mutated

cases: *n* = 233
genes: *n* = 461

(B)

PC#2 17.1%

PC#1 24.2%

Marker
● NPM1+, CEBPAwt
● NPM1wt, CEBPA+
● NPM1wt, CEBPAwt,

Ellipsoid
NPM1 wild-type
NPM1 mutated
CEBPA mutated

Figure 58.2   Integrative signature for *NPM1* and *CEBPA* mutations in CN-AML. (A) Visualization of the hierarchical clustering analysis results from a multicenter signature for CN-AML (*n* = 233 cases, rows) and 461 differentially expressed genes (columns) that were calculated from two-group comparisons according to both *NPM1* and *CEBPA* mutation status. Normalized gene expression values are color coded [standard deviation (s.d.) from mean]: red indicates high and green low expression. (B) The integrated signature for *NPM1* and *CEBPA* was further explored by principal component analysis. For all 233 CN-AML cases (colored according to *NPM1* and *CEBPA* mutation status), each sphere represents a single gene expression profile based on the 461 differentially expressed probe sets. Ellipsoids with two-fold s.d. are drawn for wild-type *NPM1*, mutated *NPM1*, and mutated *CEBPA* cases (for more details, see Kohlmann et al., 2010).

previous knowledge of the respective leukemia classes (Golub et al., 1999). Since then, DNA microarray technology has contributed significantly to the field of leukemia research, leading, for example, to the identification of a general gene dosage effect for the expression of genes located in areas of genomic gain or loss in both myeloid and lymphoid leukemias such as chronic lymphocytic leukemia (CLL) (Haslinger et al., 2004). First demonstrated in childhood ALL (Yeoh et al., 2002), today many well-defined cytogenetic and molecular genetic leukemia

subgroups can be predicted with high accuracy (Figure 58.2) (Haferlach et al., 2010; Kohlmann et al., 2010; Verhaak et al., 2009), thereby not only providing a novel diagnostic tool, but also additional molecular insights, as exemplified by the identification of the role of deregulated *FLT3* expression in t(11q23)-rearranged leukemia (Armstrong et al., 2002).

Furthermore, in subgroups where class prediction does not perform well, it often turned out that GEP seems to better capture disease biology. For example, in AML the GEP-based prediction of the *FLT3* mutation class can outperform the molecular marker itself with regard to correlation with outcome (Bullinger et al., 2008), and in *CEBPA*-mutated cases only *CEBPA* double mutants exhibit a characteristic gene expression pattern (Wouters et al., 2009b). GEP has also been a very powerful means for the discovery of novel, yet undefined, leukemia subclasses (*class discovery*). Screening large AML patient cohorts has identified novel, biologically and clinically meaningful subgroups (Bullinger et al., 2004; Valk et al., 2004), and gene expression patterns reflecting the clinical heterogeneity of the disease could be identified (Bullinger et al., 2007).

GEP has also been successfully used for *surrogate marker discovery* and *outcome prediction* in leukemia. For example, *ZAP70* (encoding a tyrosine kinase essential for T cell signaling) expression was identified as a surrogate for prognostically unfavorable unmutated V_H genes in patients with CLL (Rosenwald et al., 2001). The clinical utility of this novel marker, which can be readily measured at the protein level by flow cytometry, could be independently validated, and the marker has now been clinically implemented (Orchard et al., 2004). Similarly, GEP-based prediction of response to chemotherapeutic agents can provide significant information regarding treatment outcome in both ALL (Holleman et al., 2004; Lugthart et al., 2005) and AML (Heuser et al., 2005), suggesting novel, individualized therapeutic strategies. GEP-based outcome prediction has also been demonstrated in AML (Bullinger et al., 2004), with findings being validated and refined by independent groups (Metzeler et al., 2008; Radmacher et al., 2006).

In addition to these observational studies, leukemia was again one of the first diseases in which DNA microarray technology was applied to study the effects of specific drugs, such as ATRA in APL (Tamayo et al., 1999), imatinib mesylate in CML (Tipping et al., 2003), fludarabine in CLL (Rosenwald et al., 2004), and L-asparaginase in ALL (Fine et al., 2005). Furthermore, drug response signatures for both conventional chemotherapy (see above) and novel drugs, such as immunomodulatory substances, could be defined (Ebert et al., 2008a; Giannopoulos et al., 2009). In addition to the identification and prioritization of potential therapeutic targets, GEP plays an important role in drug discovery and has been a powerful tool in pharmacogenomics (Walgren et al., 2005). For example, GEP-based high-throughput screening approaches for chemical compounds with differentiation-inducing activity in leukemia identified an inhibitor of epidermal growth factor receptor (EGFR) kinase activity (Stegmaier et al., 2004) and it was demonstrated that the clinically-approved EGFR inhibitor gefitinib promoted the differentiation of human AML cells (Stegmaier et al., 2005). Notably, the analyzed AML cells did not express EGFR, indicating an EGFR-independent mechanism of gefitinib-induced differentiation in AML. By an integrated proteomic and RNA interference (RNAi)-based strategy to identify the "off-target" anti-AML mechanism, the tyrosine kinase SYK was identified as a target (Hahn et al., 2009). That demonstrated the power of integrating diverse chemical, proteomic, and genomic screening approaches to identify therapeutic strategies for cancer.

In summary, since its invention, GEP has contributed an important new facet to the exploration of hematologic malignancies. In the future, new technologies such as exon microarrays (Kwan et al., 2008) and next-generation sequencing (NGS) will further increase our knowledge, by providing additional insights into, for example, alternative splice variants and 3′ mRNA modifications that can lead to altered mRNA stability and miRNA processing (Mayr and Bartel, 2009).

miRNA Profiling in Leukemia

Recently, deregulation of microRNAs (miRNAs), which are involved in the post-transcriptional regulation of gene expression, has been shown to play an important role in leukemogenesis, as miRNAs can function as oncogenes and tumor suppressor genes in leukemia (Calin and Croce, 2006; Chen et al., 2010). Initial microarray-based genomics approaches have successfully demonstrated that miRNAs can be reliably measured in leukemia (Calin et al., 2005; Lu et al., 2005). Subsequent miRNA profiling revealed expression patterns characteristic of cytogenetic subgroups (Jongen-Lavrencic et al., 2008; Li et al., 2008) and molecular leukemia subtypes, such as AML with *CEBPA* and *NPM1* mutations (Jongen-Lavrencic et al., 2008; Marcucci et al., 2008). Being correlated with cytogenetic and molecular genetic aberrations as well as altered gene expression, miRNA expression signatures also have been shown to confer prognostic information (Calin et al., 2005; Garzon et al., 2008; Marcucci et al., 2008). With regard to novel biological insights, the list of miRNA–target gene interactions likely involved in tumorigenesis is rapidly growing (Chen et al., 2010). For example, miRNA profiling of CLL cases revealed the miR-15a/miR-16-1 cluster to be not only of prognostic relevance (Calin et al., 2005), but also showed that its control of B-cell proliferation and its deletion is involved in CLL pathogenesis (Klein et al., 2010). While these first studies have already increased our knowledge, future in-depth studies of the miRNA transcriptome using NGS in combination with mRNA analyses will provide additional insights by being less error-prone than hybridization-based technologies (Kuchenbauer et al., 2008).

GENOMICS HAS OPENED NEW AVENUES IN LEUKEMIA RESEARCH
Microarray-based Detection of Genomic Aberrations

With the advent of microarray-based comparative genomic hybridization (arrayCGH) and single-nucleotide polymorphism (SNP) microarrays, which provide both copy number and allele-specific information, it has now become possible to scan the genome for subtle copy number alterations (CNAs) and regions with loss of heterozygosity (LOH) at high resolution (Maciejewski and Mufti, 2008; Pollack and Iyer, 2002). Genome-wide SNP analyses revealed leukemia-specific acquired homozygosity in the form of segmental uniparental disomy (UPD), a partial duplication of the maternal or the paternal chromosome caused by chromosomal non-disjunction or homologous recombination during mitosis (Fitzgibbon et al., 2005). Parallel analysis of GEP and array-CGH-defined CNAs has helped to delineate candidate genes in the respective genomic regions, as demonstrated in AML with complex aberrations (Rücker et al., 2006).

SNP microarray profiling has revealed the most interesting insights in ALL in recent years. Here, careful simultaneous analysis of both tumor and corresponding germline DNA (most often derived from blood samples taken in complete remission or skin biopsies) led to the discovery that both B-progenitor and T-lineage ALL harbor recurring CNAs that target genes regulating lymphoid development, tumor suppression, cell cycle, apoptosis, lymphoid signaling, and drug responsiveness (Mulligan and Downing, 2009). These alterations are often focal (less than 1 Mb in size), not detectable by conventional cytogenetic analysis, and very common. For instance, in B-progenitor ALL, alterations in genes regulating B-lymphoid development can be found in 60% of cases. These comprise CNAs, gene mutations, and translocations of lymphoid transcription factors such as *PAX5*, *IKZF1*, and *EBF1* (Mulligan et al., 2007). Loss of function in these genes accelerates leukemogenesis in experimental models, suggesting

that these genetic alterations contribute to leukemogenesis. One prominent example is the deletion of *IKZF1* (encoding the early lymphoid transcription factor IKAROS), which was shown to be a near-obligate lesion in *BCR-ABL1* ALL (Mulligan et al., 2008). *IKZF1* deletions were usually hemizygous and affected either the entire *IKZF1* locus or a subset of exons, resulting in the expression of an internally truncated transcript that might act as a dominant-negative isoform. While the presence of an *IKZF1* deletion at diagnosis was also associated with poor outcome in ALL, comparison of CNAs at diagnosis and relapse identified a marked evolution of genetic alterations. Furthermore, a common clonal origin of diagnosis and relapse leukemic cells in most cases suggests selection for specific genetic changes during treatment that confer resistance to therapy (Mulligan and Downing, 2009).

In line with findings in ALL, recent progress in high-resolution profiling of genomic aberrations in AML showed that 40% of cases with an abnormal karyotype had additional CNAs, although the average number of aberrations was much lower than in ALL (Walter et al., 2009). A recent study in AML cases with cytogenetically normal AML (CN-AML) demonstrated CNAs in 49% of cases, including submicroscopic losses (for example, in band 9q21; see Figure 58.3) as well as cryptic

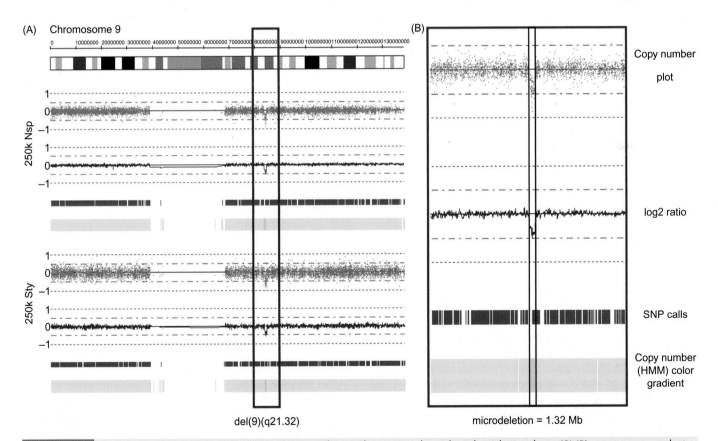

Figure 58.3 Acquired copy number alterations (CNAs) detected using single-nucleotide polymorphism (SNP) microarray analysis. (A) SNP microarray profiling (500k platform, 250k Nsp and Sty data shown separately) demonstrates a distinct region of genomic loss on chromosome 9 in a cytogenetically normal acute myeloid leukemia. The micro-deletion is highlighted by a blue box and depicted in an enlarged view (B), showing that the aberration was detected by hidden Markov model (HMM).

translocations [t(6;11) (Bullinger et al., 2010b)], and UPD in 12–19% of CN-AML cases (Bullinger et al., 2010b; Gupta et al., 2008; Walter et al., 2009). Especially in this AML subgroup, UPD may play a crucial role in the multistep process of leukemogenesis by causing homozygosity of gene mutations located in the respective genomic regions, or by duplication of a chromosomal segment harboring imprinted genes. For example, UPDs on chromosome arms 11p and 13q have been shown to be associated with homozygous *WT1* (Wilms tumor 1 gene) mutations and *FLT3* internal tandem duplications (ITD), respectively (Bullinger et al., 2010, Fitzgibbon et al., 2005, Gupta et al., 2008).

SNP Microarray-based Genome-wide Association Studies

More recently, genome-wide association studies (GWAS) have also proven that SNP microarrays are a powerful tool for identification of common, low-penetrance polymorphic loci contributing to cancer susceptibility, irrespective of prior knowledge of location or function of the respective alleles. GWAS have been reported for many cancer subtypes, including hematologic malignancies, and have already revealed more than 90 new cancer susceptibility loci (Fletcher and Houlston, 2010). Identification of these variants provides fresh insights into cancer biology, as few of the genes implicated by the GWAS scans have previously been evaluated in targeted association studies, due to limited knowledge of the underlying tumor biology. As a consequence, new insights into tumorigenesis are emerging, as exemplified by a GWAS that showed that common low-penetrance susceptibility alleles in *IKZF1*, *CEBPE*, and *ARID5B*, genes involved in transcriptional regulation and differentiation of B-cell progenitors, contribute to the risk of developing childhood ALL (Papaemmanuil et al., 2009). Here, the 10q21.2 (*ARID5B*) ALL association represents a striking genotype–phenotype relationship, as it has been shown to be highly selective for the subset of B-cell precursor ALL with hyperdiploidy, suggesting that germline variants can affect susceptibility to and characteristics of specific ALL subtypes (Papaemmanuil et al., 2009; Trevino et al., 2009).

Similar to ALL, two of the susceptibility loci associated with CLL, 6p25.3 (*IRF4*) and 16q24.1 (*IRF8*), were found to affect genes involved in B-cell lineage specification, immunoglobulin rearrangement, and regulation of germinal center reaction (Crowther-Swanepoel et al., 2010; Di Bernardo et al., 2008). However, many loci identified by SNP microarray analysis also map to non-coding regions of the genome. For example, the 8q24.21 region, to which the cancer associations map, is bereft of genes and predicted transcripts (Crowther-Swanepoel et al., 2010). It is possible that the effect of these other cancer risk loci might be mediated through long-range *cis*-acting mechanisms.

Despite the success of GWAS analyses in hematologic cancers, the currently identified loci explain only a small portion of the overall familial risk. As the underlying genetic architecture of leukemia susceptibility may be embodied in

a multitude of common susceptibility alleles, each of which accounts for a small portion of the inherited risk, larger GWAS, as well as combined studies across multiple analyses, should identify additional susceptibility loci (Fletcher and Houlston, 2010). Structural variations such as genomic copy number variations (CNVs) are also likely to contribute to cancer predisposition, and as these aberrations have not yet been thoroughly studied in most GWAS, further investigations are warranted.

Next-Generation Sequencing: A New Era in Leukemia Research

In the very near future, the rapid evolution of next-generation sequencing (NGS) technology is likely to usher in a new era of genomics in leukemia research. It will allow genome-wide detection of CNV, sequence variation, and structural rearrangements of leukemia genomes, as well as analysis of transcriptomic and epigenomic alterations. While the costs of analyzing large numbers of leukemia genomes continue to be high, and data analysis pipelines are still in development, the first whole-genome sequencing effort in a typical case of AML has clearly demonstrated the potential of this technology (Ley et al., 2008). Using NGS, ten genes with acquired mutations were discovered in that case. Of the ten, eight represented new mutations of as yet unknown functional significance, thereby further supporting the concept of a multistep leukemogenesis, although not all of the newly identified gene mutations necessarily need to be "driver" mutations (Fröhling et al., 2007).

Recently, the massively parallel DNA sequencing of another *de novo* CN-AML genome and a matched germline DNA sample derived from normal skin identified 12 acquired mutations within the coding sequences of genes (Mardis et al., 2009). While mutations in *NRAS* and *NPM1* had been identified previously in patients with AML, two other yet unidentified mutations appeared to be recurrent, with one of them, *IDH1* gene mutation, being present in 15 of 187 additional AML genomes tested. However, in accordance with the first AML genome sequenced, this case contained a large number of point mutations (~750), of which only a small fraction are likely to be relevant to leukemogenesis. Thus, large cohorts will need to be studied in order to get a representative picture of leukemia heterogeneity, a task that will be performed within The Cancer Genome Atlas (http://cancergenome.nih.gov/) and the International Cancer Genome Consortium (ICGC; http://www.icgc.org/), which was launched to coordinate large-scale cancer genome studies in tumors from 50 different cancer types including hematologic tumors (ICGC, 2010).

In addition to the opportunity for genome-wide studies, in the future, targeted resequencing using NGS technology will offer the ability to comprehensively profile tumor samples for known molecular markers at low cost – an essential prerequisite for improved patient management that is still not available to all leukemia patients. The feasibility of such an approach has recently been demonstrated in a study analyzing cases of chronic myelomonocytic leukemia (CMML) for mutations in the *CBL, JAK2, MPL, NRAS, KRAS, RUNX1,* and *TET2* genes by NGS technology (Kohlmann et al., 2010). As the number of molecular markers used to categorize lymphoid and myeloid neoplasms is constantly increasing, NGS screening will be a tool for comprehensive characterization of leukemia, with patterns of molecular mutations translating into different biologically and prognostically relevant leukemia classes.

LEUKEMIA EPIGENOMICS
Altered DNA Methylation Patterns in Leukemia

Epigenetic modifications, including aberrant promoter cytosine methylation in CpG islands and the resulting gene silencing, have been shown to be involved in leukemia development. As epigenetic changes are potentially reversible, improved insights become especially valuable, since genomic aberrations underlying cancer are largely immutable. In AML, numerous factors, such as fusion genes resulting from chromosome translocations, can influence/deregulate the epigenetic profile, thereby contributing to leukemogenesis (Chen et al., 2010). For example, aberrant CpG island methylation in the promoter regions of *CDKN2B* and *CDKN2A* (encoding cyclin-dependent kinase inhibitors 2B and 2A) has been a well-described phenomenon, but the development of novel technologies for genome-wide screens based on microarray and NGS technology now allow us to further decipher the methylome in AML (Estecio and Issa, 2009).

Recently, a first comprehensive, genome-wide promoter DNA methylation profiling study in AML demonstrated that cytogenetic and molecular genetic leukemia subgroups have distinct methylation patterns, while on the other hand most AML cases seem to share a "common" DNA methylation pattern (Figueroa et al., 2010). Furthermore, recent evidence that methylation-based outcome prediction in AML is feasible (Bullinger et al., 2010a) was confirmed, as a robust DNA methylation-based outcome predictor could be built based on the promoter microarray data (Figueroa et al., 2010). Similarly, first genome-wide studies in ALL and CLL suggest leukemia subtype-specific DNA methylation patterns associated with disease biology (Kanduri et al., 2010; Schafer et al., 2010). While there are numerous additional studies providing evidence for the role of aberrant DNA methylation in leukemia (Plass et al., 2008), future efforts to extend epigenetic analysis will still be necessary. They will also have to go beyond CpG island promoter methylation analysis, as first NGS-based studies demonstrated that CpG island methylation sites represent dynamic epigenetic marks undergoing extensive changes during cellular differentiation (Meissner et al., 2008) and that DNA methylation in non-CpG contexts may play an important role (Lister et al., 2009).

Histone Modifications and Leukemogenesis

Recent technological advances also allow other chromatin modifications that impact leukemia development significantly to be taken into account, such as covalent histone modifications. Here, chromatin immunoprecipitation (ChIP), followed by either microarray or NGS-based analysis, will also provide novel insights. For example, a first analysis showed that DNA-bound PML-RARA recruits HDAC1 (histone deacetylase 1) and leads to loss of histone H3 acetylation, increased tri-methylation of histone H3 lysine 9, and, unexpectedly, increased tri-methylation of histone H3 lysine 4, thereby regulating key cancer-related genes and pathways by inducing a repressive chromatin state in its direct target genes (Hoemme et al., 2008). Similarly, ectopic histone methylation of lysine 79 in histone H3 has been shown to be important for the maintenance of fusion gene-driven gene expression in MLL-AFF1 (also known as MLL-AF4) leukemia (Krivtsov et al., 2008). It has been shown that several fusion partners of MLL (such as AF4) bind and potentially recruit DOT1L (DOT1-like, histone H3 methyltransferase), a histone methyltransferase that methylates histone 3 at lysine 79 (H3K79), and first ChIP sequencing results in primary t(9;11) leukemias suggest similar histone methylation patterns (Figure 58.4).

While first studies in leukemia have provided promising insights, for a better interpretation of aberrant DNA methylation and histone modifications in leukemia, malignant cells will have to be compared with their non-malignant counterparts. For hematopoietic tumors, especially acute leukemias, there is still a large debate about what the ideal control might be, as the cells of origin are still not known. Ongoing efforts within, for example, the National Institutes of Health (NIH) Roadmap Epigenomics Mapping Consortium (http://www.roadmapepigenomics.org/) will generate a public resource of human epigenomic data that will further catalyze disease-oriented research, including epigenomics in leukemia.

PROTEOMICS AND FUNCTIONAL GENOMICS – THE FUTURE HAS BEGUN
Proteomics

Proteomic approaches are being developed and, analogous to those using GEP, have been used to address several essential questions in normal and malignant hematopoiesis, including studies on drug response and prediction of AML patient response and survival (Kornblau et al., 2009). While, as discussed above, GEP studies have made a major contribution to our current understanding of leukemia, they also have limitations, as mRNA expression does not always correlate with protein expression levels. There are several causes for this lack of correlation, including mRNA instability, post-transcriptional regulation, translation initiation, and protein stability, as well as mRNA splice variants and post-translational modifications. All of these factors represent critical parameters controlling the resulting protein function, including protein localization, activity, interaction with other molecules (e.g., other proteins, nucleic acids, lipids), and turnover. As most cell phenotypes can be ascribed to protein expression levels

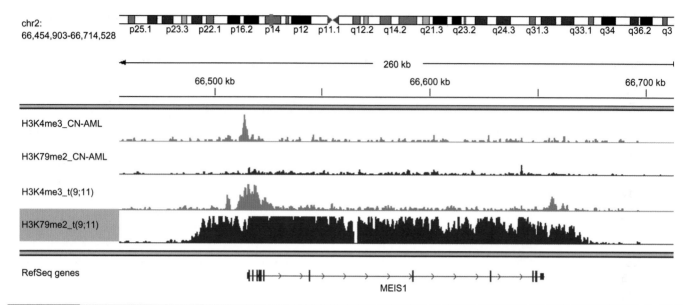

Figure 58.4 ChIP-Seq (chromatin immunoprecipitation followed by NGS) in AML. ChIP-Seq profiles of two diagnostic AML samples, one CN-AML and one t(9;11) case. Chromatin immunoprecipitation was performed using antibodies against tri-methylated histone 3 at lysine 4 (H3K4me3), and bi-methylated histone at lysine 79 (H3K79me2). The individual ChIP-Seq data have been visualized using the integrated genomics viewer (IGV). Displayed is a region on chromosome 2 (chr2:66,454, 903–66,714,528), harboring the *MEIS1* (Meis homeobox 1) gene commonly expressed in t(9;11) AML, which displays a prominent activating H3K79me2 mark. *(Data kindly provided by Drs S. Vempati and A.L. Kung, Dana-Farber Cancer Institute, Boston, MA, USA.)*

and activities/interactions, combined characterization of cell types by GEP and proteomic analysis is clearly requisite for a full understanding of normal and disease-specific biological processes as well as how the proteome is reflected by the transcriptome. Future efforts including novel technologies will shed further light on leukemia biology.

Functional Genomics

As described above, recent progress in genomics technology holds great promise for providing novel insights into leukemia biology and for identifying abnormalities that can be used as prognostic parameters or may serve as targets for therapy. However, cytogenetics, analysis of DNA copy number changes, DNA sequencing, gene expression and miRNA profiling, and epigenomic analyses are primarily descriptive approaches that require subsequent functional studies to assess the pathogenic relevance of the identified alterations.

The challenge of delineating functionally relevant genes or pathways based on genomic screens alone is best illustrated by recent large-scale mutational screens in cancer. Among the most striking findings of the first genome-wide DNA resequencing studies in epithelial cancers and AML was the realization that cancer genomes may harbor dozens of somatically acquired non-synonymous sequence variants (Greenman et al., 2007; Ley et al., 2008; Sjöblom et al., 2006; Wood et al., 2007). This indicated the need to distinguish "driver" mutations – which are causally implicated in

malignant transformation and therefore have been positively selected during cancer development – from "passenger" alterations, innocent bystanders that have occurred by chance in the immediate progenitor cell of the malignant clone, but confer no clonal growth advantage and therefore are not subject to selection. To identify drivers, these studies used statistical models that incorporate mutation type, mutation frequency, and sequence context, and compare the observed frequency of non-synonymous mutations with that expected by chance (Greenman et al., 2007; Sjöblom et al., 2006). Any excess of non-synonymous mutations was interpreted as positive selection pressure, which was thought to be indicative of the presence of driver mutations.

While this approach has resulted in the identification of novel disease alleles, such as mutant *IDH* family members in AML (Mardis et al., 2009), other studies have highlighted some of the potential limitations of statistical methods for assessing the functional relevance of DNA sequence variants. In particular, it has been shown that methods that rely on mutational frequency for the identification of driver mutations might fail to detect driver mutations occurring at frequencies that would not allow them to be distinguished from unselected passenger changes. This is exemplified by several mutations in the *FLT3* gene that were present in less than 1% of patients with AML but were nevertheless found to be gain-of-function alleles that result in increased kinase activity (Fröhling et al., 2007). It has also been demonstrated that mutations in conserved and

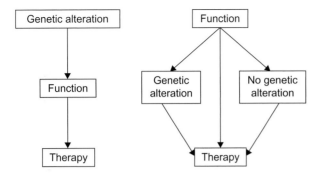

- Gene function as primary readout
- Immediate establishment of genotype–phenotype correlations
- Identification of "normal" genes with pathogenic relevance

Figure 58.5 Identifying therapeutic targets through functional genomics. The classic strategy to identify novel therapeutic targets is based on structural genetic characterization of primary tumor samples and requires subsequent functional studies to assess the pathogenic relevance of the identified alterations. Functional genetic studies are based on the experimental perturbation of gene activity, followed by direct assessment of the resulting consequences in phenotypic assays. Such an approach allows the upfront establishment of causal relationships between genotypes and phenotypes and has potential for discovering functionally relevant genes that are not mutated or affected by changes in DNA copy number or mRNA expression.

functionally relevant domains need not necessarily be implicated in cancer pathogenesis (Fröhling et al., 2007).

The challenges associated with structural genetic analyses are also illustrated by the observation that most genomic gains or losses identified by chromosome banding or high-resolution CGH and SNP genotyping analyses affect large genomic regions (Fröhling and Döhner, 2008). Since such aberrations typically involve multiple genes, the identification of their functionally relevant targets has proven to be difficult, even though "filtering" the genes within regions of CNAs to identify those that are also altered at the RNA or protein level has been used to nominate new candidate cancer/leukemia genes (Santarius et al., 2010).

As an alternative to structural genetic characterization coupled with statistical analysis for the discovery of cancer genes, it is also possible to perform functional genetic studies based on the experimental perturbation of gene activity, followed by direct assessment of the resulting consequences in phenotypic assays. Functional genetic analyses have two major advantages: they allow the upfront establishment of causal relationships between genotypes and phenotypes, and they have great potential for discovering functionally relevant genes that are not mutated or affected by changes in DNA copy number or mRNA expression (Figure 58.5). One particularly promising functional genetic approach is to suppress gene function using RNAi technology. The scale of RNAi studies can range

from individual experiments, in which the functional relevance of single candidate genes is assessed, to comprehensive RNAi screens. The latter can be targeted screens, which are often guided by structural genetic information and are particularly valuable to study regions of genomic imbalances that include multiple genes, or large-scale screens, in which entire gene families or even complete cancer/leukemia genomes can be examined. Positive RNAi screens are based on the ability of an interfering RNA to rescue a cell from a cytotoxic or cytostatic influence, and are therefore ideal for identifying tumor suppressor genes. Negative screens are based on the ability of an interfering RNA to produce a cytostatic or cytotoxic phenotype, and are therefore suited to identifying genes that promote the proliferation and survival of tumor cells.

Thus far, most RNAi screens have been conducted in epithelial cancer cells, perhaps reflecting the difficulty of delivering short interfering RNA (siRNA) or short hairpin (shRNA) molecules to hematopoietic cells. However, one prime example of a targeted RNAi screen was a recent study by Ebert and colleagues (2008b) in which graded down-regulation of multiple candidate genes by RNAi was used to identify *RPS14* as a haploinsufficient tumor suppressor gene in myelodysplastic syndrome (MDS) with isolated del(5q), a subtype of MDS characterized by a 1.5-Mb commonly deleted region on chromosome band 5q32 (Boultwood et al., 2002). It is anticipated that similar studies will be performed to evaluate other genomic imbalances, which are increasingly being discovered in AML using modern SNP array technology but remain uncharacterized with regard to their functionally relevant targets (Bullinger et al., 2010b; Radtke et al., 2009; Walter et al., 2009).

In addition to targeted screens guided by prior knowledge of genetic alterations associated with specific cancer types, it is now possible to perform unbiased screens, as several laboratories have generated genome-wide RNAi libraries to facilitate large-scale functional genomic studies (Chang et al., 2006; Echeverri et al., 2006; Root et al., 2006). Again, these reagents have primarily been used in epithelial malignancies, but the feasibility of such an approach for the identification of genetic vulnerabilities in myeloid leukemia cells has recently been demonstrated (Figure 58.6) (Scholl et al., 2009). Importantly, the results of several large-scale RNAi screens support the emerging concept that cancer cells may develop specific dependencies on genes that are not activated by structural alterations or overexpressed to an extent that directly promotes tumorigenesis (Barbie et al., 2009; Schlabach et al., 2008; Scholl et al., 2009; Shaffer et al., 2008; Silva et al., 2008). This phenomenon has been termed "non-oncogene addiction" (Luo et al., 2009).

A particular challenge associated with functional genetic approaches occurs in the analysis of primary human cancer/leukemia cells. They are often scarce and difficult to maintain in culture, unlike the immortalized cell lines that have been used in most RNAi screens reported to date. Remarkably, a recent study demonstrated that it was possible to screen a set of approximately 90 genes (all known tyrosine kinase genes, *NRAS*, and *KRAS*) in freshly isolated leukemic cells from 30

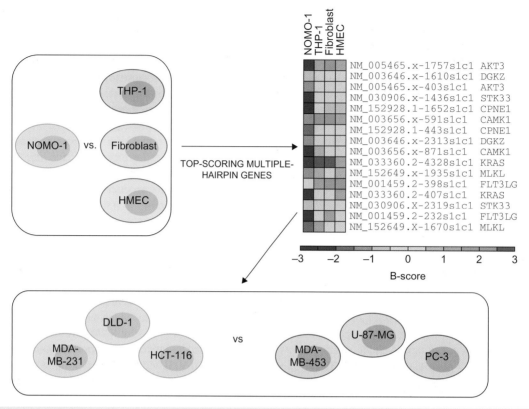

Figure 58.6 Experimental strategy of large-scale RNAi screens for genes required by human leukemia cell lines harboring mutant *KRAS*. Screening data obtained in mutant *KRAS*-expressing NOMO-1 cells were compared with data from screens conducted in *KRAS* wildtype THP-1 cells, and untransformed human fibroblasts and mammary epithelial cells (HMEC). Genes selectively required by NOMO-1 cells were identified based on supervised analysis of shRNA performance, and shRNAs targeting the top-ranking candidate genes were subsequently evaluated in additional cancer cell lines included in the primary RNAi screens (cell lines harboring mutant or wildtype *KRAS* are indicated in red and blue, respectively; untransformed cell types are indicated in green; for more details see Scholl et al., 2009). *Reproduced with permission from Scholl et al., 2009.*

patients with ALL or various myeloid malignancies, to identify known and previously unrecognized driver mutations as well as dependencies on genes that were not structurally altered (Tyner et al., 2009).

In addition to the identification of driver genes that are implicated in tumorigenesis and thus may represent direct targets for treatment, RNAi can also be used to pinpoint genes and pathways whose inhibition sensitizes cancer cells to specific therapeutic agents. For example, such a "synthetic lethality" screen revealed that suppression of components of the Wnt/Ca^{2+}/NFAT pathway enhanced killing of human CML cells by the BCR-ABL1 inhibitors imatinib and dasatinib (Gregory et al., 2010).

FUTURE CHALLENGES OF GENOMICS IN LEUKEMIA

Although genomics has led to a tremendous growth in our understanding of leukemia biology over the last decade, many challenges remain, including the management of ever-growing data sets. Success in the life sciences in general and in leukemia research in particular will depend on the ability to properly interpret the large-scale, high-dimensional data sets generated by the technologies discussed above. This will require advances in informatics that can address the complex interplay of the different hallmarks of cancer, as all of the alterations mentioned above are likely to be interconnected.

Furthermore, most of the studies reported to date have been performed with "bulk" tumor material; a better understanding of normal hematopoiesis and sophisticated analyses of tumor subpopulations will shed additional light on the molecular basis of the disease, especially with regard to the cell of origin (Chan and Huntly, 2008). Here, it will be also important to expand the integrative exploration of findings in murine models and primary human leukemias to rigorously defined cancer/leukemia-initiating cells. Finally, future efforts should also take into account the influence of the tumor environment, the so-called "stem cell niche," which despite being incompletely characterized seems to be of great importance

(Lane et al., 2009). We are still far from understanding the complex nature underlying all leukemia subtypes. However, within the near future, genomics technologies will enter leukemia diagnostics, and systematic studies of cancer genomes at the genomic, epigenomic, and transcriptomic levels will reveal the repertoire of oncogenic mutations, will define clinically relevant subtypes for prognosis and therapeutic management, and will contribute to the development of new cancer therapies.

ACKNOWLEDGMENTS

The authors have no conflicts of interest and have nothing to disclose.

REFERENCES

Armstrong, S.A., Look, A.T., 2005. Molecular genetics of acute lymphoblastic leukemia. J Clin Oncol 23, 6306–6315.

Armstrong, S.A., Staunton, J.E., Silverman, L.B., et al., 2002. MLL translocations specify a distinct gene expression profile that distinguishes a unique leukemia. Nat Genet 30, 41–47.

Barbie, D.A., Tamayo, P., Boehm, J.S., et al., 2009. Systematic RNA interference reveals that oncogenic KRAS-driven cancers require TBK1. Nature 462, 108–112.

Boultwood, J., Fidler, C., Strickson, A.J., et al., 2002. Narrowing and genomic annotation of the commonly deleted region of the 5q-syndrome. Blood 99, 4638–4641.

Bullinger, L., 2006. Gene expression profiling in acute myeloid leukemia. Haematologica 91, 733–738.

Bullinger, L., Armstrong, S.A., 2010. HELP for AML: Methylation profiling opens new avenues. Cancer Cell 17, 1–3.

Bullinger, L., Döhner, K., 2010. The molecular basis of adult AML. Hematol Educ 4, 19–30.

Bullinger, L., Döhner, K., Bair, E., et al., 2004. Use of gene-expression profiling to identify prognostic subclasses in adult acute myeloid leukemia. N Engl J Med 350, 1605–1616.

Bullinger, L., Döhner, K., Kranz, R., et al., 2008. An FLT3 gene-expression signature predicts clinical outcome in normal karyotype AML. Blood 111, 4490–4495.

Bullinger, L., Ehrich, M., Döhner, K., et al., 2010a. Quantitative DNA methylation predicts survival in adult acute myeloid leukemia. Blood 115, 636–642.

Bullinger, L., Krönke, J., Schön, C., et al., 2010b. Identification of acquired copy number alterations and uniparental disomies in cytogenetically normal acute myeloid leukemia using high-resolution single-nucleotide polymorphism analysis. Leukemia 24, 438–449.

Bullinger, L., Rücker, F.G., Kurz, S., et al., 2007. Gene-expression profiling identifies distinct subclasses of core binding factor acute myeloid leukemia. Blood 110, 1291–1300.

Calin, G.A., Croce, C.M., 2006. MicroRNA signatures in human cancers. Nat Rev Cancer 6, 857–866.

Calin, G.A., Ferracin, M., Cimmino, A., et al., 2005. A microRNA signature associated with prognosis and progression in chronic lymphocytic leukemia. N Engl J Med 353, 1793–1801.

Chan, W.I., Huntly, B.J., 2008. Leukemia stem cells in acute myeloid leukemia. Semin Oncol 35, 326–335.

Chang, K., Elledge, S.J., Hannon, G.J., 2006. Lessons from Nature: microRNA-based shRNA libraries. Nat Methods 3, 707–714.

Chen, J., Odenike, O., Rowley, J.D., 2010. Leukaemogenesis: More than mutant genes. Nat Rev Cancer 10, 23–36.

Cools, J., 2010. Role of oncogenic mutations and fusion genes in acute lymphoblastic leukemia. Hematol Educ 4, 1–5.

Crowther-Swanepoel, D., Broderick, P., Di Bernardo, M.C., et al., 2010. Common variants at 2q37.3, 8q24.21, 15q21.3 and 16q24.1 influence chronic lymphocytic leukemia risk. Nat Genet 42, 132–136.

Daley, G.Q., Van Etten, R.A., Baltimore, D., 1990. Induction of chronic myelogenous leukemia in mice by the P210bcr/abl gene of the Philadelphia chromosome. Science 247, 824–830.

Di Bernardo, M.C., Crowther-Swanepoel, D., Broderick, P., et al., 2008. A genome-wide association study identifies six susceptibility loci for chronic lymphocytic leukemia. Nat Genet 40, 1204–1210.

Döhner, H., 2007. Implication of the molecular characterization of acute myeloid leukemia. Am Soc Hematol Educ Prog, 412–419.

Döhner, H., Estey, E.H., Amadori, S., et al., 2010. Diagnosis and management of acute myeloid leukemia in adults: Recommendations from an international expert panel, on behalf of the European LeukemiaNet. Blood 115, 453–474.

Döhner, K., Döhner, H., 2008. Molecular characterization of acute myeloid leukemia. Haematologica 93, 976–982.

Druker, B.J., 2009. Perspectives on the development of imatinib and the future of cancer research. Nat Med 15, 1149–1152.

Ebert, B.L., Galili, N., Tamayo, P., et al., 2008a. An erythroid differentiation signature predicts response to lenalidomide in myelodysplastic syndrome. PLoS Med 5, e35.

Ebert, B.L., Pretz, J., Bosco, J., et al., 2008b. Identification of RPS14 as a 5q- syndrome gene by RNA interference screen. Nature 451, 335–339.

Echeverri, C.J., Beachy, P.A., Baum, B., et al., 2006. Minimizing the risk of reporting false positives in large-scale RNAi screens. Nat Methods 3, 777–779.

Estecio, M.R., Issa, J.P., 2009. Tackling the methylome: Recent methodological advances in genome-wide methylation profiling. Genome Med 1, 106.

Figueroa, M.E., Lugthart, S., Li, Y., et al., 2010. DNA methylation signatures identify biologically distinct subtypes in acute myeloid leukemia. Cancer Cell 17, 13–27.

Fine, B.M., Kaspers, G.J., Ho, M., Loonen, A.H., Boxer, L.M., 2005. A genome-wide view of the in vitro response to l-asparaginase in acute lymphoblastic leukemia. Cancer Res 65, 291–299.

Fitzgibbon, J., Smith, L.L., Raghavan, M., et al., 2005. Association between acquired uniparental disomy and homozygous gene mutation in acute myeloid leukemias. Cancer Res 65, 9152–9154.

Fletcher, O., Houlston, R.S., 2010. Architecture of inherited susceptibility to common cancer. Nat Rev Cancer 10, 353–361.

Fröhling, S., Döhner, H., 2008. Chromosomal abnormalities in cancer. N Engl J Med 359, 722–734.

Fröhling, S., Scholl, C., Gilliland, D.G., Levine, R.L., 2005. Genetics of myeloid malignancies: Pathogenetic and clinical implications. J Clin Oncol 23, 6285–6295.

Fröhling, S., Scholl, C., Levine, R.L., et al., 2007. Identification of driver and passenger mutations of FLT3 by high-throughput DNA sequence analysis and functional assessment of candidate alleles. Cancer Cell 12, 501–513.

Garzon, R., Volinia, S., Liu, C.G., et al., 2008. MicroRNA signatures associated with cytogenetics and prognosis in acute myeloid leukemia. Blood 111, 3183–3189.

Giannopoulos, K., Dmoszynska, A., Kowal, M., et al., 2009. Thalidomide exerts distinct molecular antileukemic effects and combined thalidomide/fludarabine therapy is clinically effective in high-risk chronic lymphocytic leukemia. Leukemia 23, 1771–1778.

Golub, T.R., Slonim, D.K., Tamayo, P., et al., 1999. Molecular classification of cancer: Class discovery and class prediction by gene expression monitoring. Science 286, 531–537.

Greenman, C., Stephens, P., Smith, R., et al., 2007. Patterns of somatic mutation in human cancer genomes. Nature 446, 153–158.

Gregory, M.A., Phang, T.L., Neviani, P., et al., 2010. Wnt/Ca2+/NFAT signaling maintains survival of Ph+ leukemia cells upon inhibition of Bcr-Abl. Cancer Cell 18, 74–87.

Gupta, M., Raghavan, M., Gale, R.E., et al., 2008. Novel regions of acquired uniparental disomy discovered in acute myeloid leukemia. Genes Chromosomes Cancer 47, 729–739.

Haferlach, T., 2008. Molecular genetic pathways as therapeutic targets in acute myeloid leukemia. Am Soc Hematol Educ Prog, 400–411.

Haferlach, T., Kohlmann, A., Wieczorek, L., et al., 2010. Clinical utility of microarray-based gene expression profiling in the diagnosis and subclassification of leukemia: Report from the International Microarray Innovations in Leukemia Study Group. J Clin Oncol 28, 2529–2537.

Hahn, C.K., Berchuck, J.E., Ross, K.N., et al., 2009. Proteomic and genetic approaches identify Syk as an AML target. Cancer Cell 16, 281–294.

Haslinger, C., Schweifer, N., Stilgenbauer, S., et al., 2004. Microarray gene expression profiling of B-cell chronic lymphocytic leukemia subgroups defined by genomic aberrations and VH mutation status. J Clin Oncol 22, 3937–3949.

Heuser, M., Wingen, L.U., Steinemann, D., et al., 2005. Gene-expression profiles and their association with drug resistance in adult acute myeloid leukemia. Haematologica 90, 1484–1492.

Hoemme, C., Peerzada, A., Behre, G., et al., 2008. Chromatin modifications induced by PML-RARalpha repress critical targets in leukemogenesis as analyzed by ChIP-Chip. Blood 111, 2887–2895.

Holleman, A., Cheok, M.H., Den Boer, M.L., et al., 2004. Gene-expression patterns in drug-resistant acute lymphoblastic leukemia cells and response to treatment. N Engl J Med 351, 533–542.

Huang, M.E., Ye, Y.C., Chen, S.R., et al., 1988. Use of all-trans retinoic acid in the treatment of acute promyelocytic leukemia. Blood 72, 567–572.

ICGC, 2010. International network of cancer genome projects. Nature 464, 993–998.

Jongen-Lavrencic, M., Sun, S.M., Dijkstra, M.K., Valk, P.J., Löwenberg, B., 2008. MicroRNA expression profiling in relation to the genetic heterogeneity of acute myeloid leukemia. Blood 111, 5078–5085.

Kanduri, M., Cahill, N., Goransson, H., et al., 2010. Differential genome-wide array-based methylation profiles in prognostic subsets of chronic lymphocytic leukemia. Blood 115, 296–305.

Klein, U., Lia, M., Crespo, M., et al., 2010. The DLEU2/miR-15a/16-1 cluster controls B cell proliferation and its deletion leads to chronic lymphocytic leukemia. Cancer Cell 17, 28–40.

Kohlmann, A., Bullinger, L., Thiede, C., et al., 2010. Gene expression profiling in AML with normal karyotype can predict mutations for molecular markers and allows novel insights into perturbed biological pathways. Leukemia 24, 1216–1220.

Konopka, J.B., Watanabe, S.M., Singer, J.W., Collins, S.J., Witte, O.N., 1985. Cell lines and clinical isolates derived from Ph1-positive chronic myelogenous leukemia patients express c-abl proteins with a common structural alteration. Proc Natl Acad Sci USA 82, 1810–1814.

Kornblau, S.M., Tibes, R., Qiu, Y.H., et al., 2009. Functional proteomic profiling of AML predicts response and survival. Blood 113, 154–164.

Krivtsov, A.V., Feng, Z., Lemieux, M.E., et al., 2008. H3K79 methylation profiles define murine and human MLL-AF4 leukemias. Cancer Cell 14, 355–368.

Kuchenbauer, F., Morin, R.D., Argiropoulos, B., et al., 2008. In-depth characterization of the microRNA transcriptome in a leukemia progression model. Genome Res 18, 1787–1797.

Kwan, T., Benovoy, D., Dias, C., et al., 2008. Genome-wide analysis of transcript isoform variation in humans. Nat Genet 40, 225–231.

Lane, S.W., Scadden, D.T., Gilliland, D.G., 2009. The leukemic stem cell niche: Current concepts and therapeutic opportunities. Blood 114, 1150–1157.

Ley, T.J., Mardis, E.R., Ding, L., et al., 2008. DNA sequencing of a cytogenetically normal acute myeloid leukaemia genome. Nature 456, 66–72.

Li, Z., Lu, J., Sun, M., et al., 2008. Distinct microRNA expression profiles in acute myeloid leukemia with common translocations. Proc Natl Acad Sci USA 105, 15,535–15,540.

Lister, R., Pelizzola, M., Dowen, R.H., et al., 2009. Human DNA methylomes at base resolution show widespread epigenomic differences. Nature 462, 315–322.

Lu, J., Getz, G., Miska, E.A., et al., 2005. MicroRNA expression profiles classify human cancers. Nature 435, 834–838.

Lugthart, S., Cheok, M.H., Den Boer, M.L., et al., 2005. Identification of genes associated with chemotherapy cross-resistance and treatment response in childhood acute lymphoblastic leukemia. Cancer Cell 7, 375–386.

Luo, J., Emanuele, M.J., Li, D., et al., 2009. A genome-wide RNAi screen identifies multiple synthetic lethal interactions with the Ras oncogene. Cell 137, 835–848.

Maciejewski, J.P., Mufti, G.J., 2008. Whole genome scanning as a cytogenetic tool in hematologic malignancies. Blood 112, 965–974.

Marcucci, G., Radmacher, M.D., Maharry, K., et al., 2008. MicroRNA expression in cytogenetically normal acute myeloid leukemia. N Engl J Med 358, 1919–1928.

Mardis, E.R., Ding, L., Dooling, D.J., et al., 2009. Recurring mutations found by sequencing an acute myeloid leukemia genome. N Engl J Med 361, 1058–1066.

Mayr, C., Bartel, D.P., 2009. Widespread shortening of 3′UTRs by alternative cleavage and polyadenylation activates oncogenes in cancer cells. Cell 138, 673–684.

Meissner, A., Mikkelsen, T.S., Gu, H., et al., 2008. Genome-scale DNA methylation maps of pluripotent and differentiated cells. Nature 454, 766–770.

Metzeler, K.H., Hummel, M., Bloomfield, C.D., et al., 2008. An 86-probe-set gene-expression signature predicts survival in cytogenetically normal acute myeloid leukemia. Blood 112, 4193–4201.

Mulligan, C.G., Downing, J.R., 2009. Genome-wide profiling of genetic alterations in acute lymphoblastic leukemia: Recent insights and future directions. Leukemia 23, 1209–1218.

Mulligan, C.G., Goorha, S., Radtke, I., et al., 2007. Genome-wide analysis of genetic alterations in acute lymphoblastic leukaemia. Nature 446, 758–764.

Mulligan, C.G., Miller, C.B., Radtke, I., et al., 2008. BCR-ABL1 lymphoblastic leukaemia is characterized by the deletion of Ikaros. Nature 453, 110–114.

Orchard, J.A., Ibbotson, R.E., Davis, Z., et al., 2004. ZAP-70 expression and prognosis in chronic lymphocytic leukaemia. Lancet 363, 105–111.

Papaemmanuil, E., Hosking, F.J., Vijayakrishnan, J., et al., 2009. Loci on 7p12.2, 10q21.2 and 14q11.2 are associated with risk of childhood acute lymphoblastic leukemia. Nat Genet 41, 1006–1010.

Plass, C., Oakes, C., Blum, W., Marcucci, G., 2008. Epigenetics in acute myeloid leukemia. Semin Oncol 35, 378–387.

Pollack, J.R., Iyer, V.R., 2002. Characterizing the physical genome. Nat Genet 32 (Suppl), 515–521.

Pui, C.H., Robison, L.L., Look, A.T., 2008. Acute lymphoblastic leukaemia. Lancet 371, 1030–1043.

Radmacher, M.D., Marcucci, G., Ruppert, A.S., et al., 2006. Independent confirmation of a prognostic gene-expression signature in adult acute myeloid leukemia with a normal karyotype: A Cancer and Leukemia Group B study. Blood 108, 1677–1683.

Radtke, I., Mulligan, C.G., Ishii, M., et al., 2009. Genomic analysis reveals few genetic alterations in pediatric acute myeloid leukemia. Proc Natl Acad Sci USA 106, 12,944–12,949.

Root, D.E., Hacohen, N., Hahn, W.C., Lander, E.S., Sabatini, D.M., 2006. Genome-scale loss-of-function screening with a lentiviral RNAi library. Nat Methods 3, 715–719.

Rosenwald, A., Alizadeh, A.A., Widhopf, G., et al., 2001. Relation of gene expression phenotype to immunoglobulin mutation genotype in B cell chronic lymphocytic leukemia. J Exp Med 194, 1639–1647.

Rosenwald, A., Chuang, E.Y., Davis, R.E., et al., 2004. Fludarabine treatment of patients with chronic lymphocytic leukemia induces a p53-dependent gene expression response. Blood 104, 1428–1434.

Rowe, J.M., 2009. Optimal induction and post-remission therapy for AML in first remission. Hematology Am Soc Hematol Educ Program, 396–405. doi: 10.1182/asheducation-2009.1.396.

Rowley, J.D., 1973. Identificaton of a translocation with quinacrine fluorescence in a patient with acute leukemia. Ann Genet 16, 109–112.

Rücker, F.G., Bullinger, L., Schwaenen, C., et al., 2006. Disclosure of candidate genes in acute myeloid leukemia with complex karyotypes using microarray-based molecular characterization. J Clin Oncol 24, 3887–3894.

Santarius, T., Shipley, J., Brewer, D., Stratton, M.R., Cooper, C.S., 2010. A census of amplified and overexpressed human cancer genes. Nat Rev Cancer 10, 59–64.

Schafer, E., Irizarry, R., Negi, S., et al., 2010. Promoter hypermethylation in MLL-r infant acute lymphoblastic leukemia: Biology and therapeutic targeting. Blood 115, 4798–4809.

Schlabach, M.R., Luo, J., Solimini, N.L., et al., 2008. Cancer proliferation gene discovery through functional genomics. Science 319, 620–624.

Scholl, C., Fröhling, S., Dunn, I.F., et al., 2009. Synthetic lethal interaction between oncogenic *KRAS* dependency and STK33 suppression in human cancer cells. Cell 137, 821–834.

Shaffer, A.L., Emre, N.C., Lamy, L., et al., 2008. IRF4 addiction in multiple myeloma. Nature 454, 226–231.

Shtivelman, E., Lifshitz, B., Gale, R.P., Canaani, E., 1985. Fused transcript of *abl* and *bcr* genes in chronic myelogenous leukaemia. Nature 315, 550–554.

Silva, J.M., Marran, K., Parker, J.S., et al., 2008. Profiling essential genes in human mammary cells by multiplex RNAi screening. Science 319, 617–620.

Sjöblom, T., Jones, S., Wood, L.D., et al., 2006. The consensus coding sequences of human breast and colorectal cancers. Science 314, 268–274.

Stegmaier, K., Corsello, S.M., Ross, K.N., 2005. Gefitinib induces myeloid differentiation of acute myeloid leukemia. Blood 106, 2841–2848.

Stegmaier, K., Ross, K.N., Colavito, S.A., 2004. Gene expression-based high-throughput screening (GE-HTS) and application to leukemia differentiation. Nat Genet 36, 257–263.

Swerdlow, S.H., Campo, E., Harris, N.L., et al., 2008. WHO Classification of Tumours of Haematopoietic and Lymphoid Tissues. IARC Press, Lyon, France, Chapter 6, 109–147.

Tamayo, P., Slonim, D., Mesirov, J., et al., 1999. Interpreting patterns of gene expression with self-organizing maps: Methods and application to hematopoietic differentiation. Proc Natl Acad Sci USA 96, 2907–2912.

Tipping, A.J., Deininger, M.W., Goldman, J.M., Melo, J.V., 2003. Comparative gene expression profile of chronic myeloid leukemia cells innately resistant to imatinib mesylate. Exp Hematol 31, 1073–1080.

Trevino, L.R., Yang, W., French, D., et al., 2009. Germline genomic variants associated with childhood acute lymphoblastic leukemia. Nat Genet 41, 1001–1005.

Tyner, J.W., Deininger, M.W., Loriaux, M.M., et al., 2009. RNAi screen for rapid therapeutic target identification in leukemia patients. Proc Natl Acad Sci USA 106, 8695–8700.

Valk, P.J., Verhaak, R.G., Beijen, M.A., et al., 2004. Prognostically useful gene-expression profiles in acute myeloid leukemia. N Engl J Med 350, 1617–1628.

Verhaak, R.G., Wouters, B.J., Erpelinck, C.A., et al., 2009. Prediction of molecular subtypes in acute myeloid leukemia based on gene expression profiling. Haematologica 94, 131–134.

Walgren, R.A., Meucci, M.A., McLeod, H.L., 2005. Pharmacogenomic discovery approaches: Will the real genes please stand up? J Clin Oncol 23, 7342–7349.

Walter, M.J., Payton, J.E., Ries, R.E., et al., 2009. Acquired copy number alterations in adult acute myeloid leukemia genomes. Proc Natl Acad Sci USA 106, 12,950–12,955.

Wood, L.D., Parsons, D.W., Jones, S., et al., 2007. The genomic landscapes of human breast and colorectal cancers. Science 318, 1108–1113.

Wouters, B.J., Löwenberg, B., Delwel, R., 2009a. A decade of genome-wide gene expression profiling in acute myeloid leukemia: Flashback and prospects. Blood 113, 291–298.

Wouters, B.J., Löwenberg, B., Erpelinck-Verschueren, C.A., 2009b. Double CEBPA mutations, but not single CEBPA mutations, define a subgroup of acute myeloid leukemia with a distinctive gene expression profile that is uniquely associated with a favorable outcome. Blood 113, 3088–3091.

Yeoh, E.J., Ross, M.E., Shurtleff, S.A., et al., 2002. Classification, subtype discovery, and prediction of outcome in pediatric acute lymphoblastic leukemia by gene expression profiling. Cancer Cell 1, 133–143.

RECOMMENDED RESOURCES

LEUKEMIA

American Society of Hematology
  http://www.hematology.org/
European Hematology Association
  http://www.ehaweb.org/
National Cancer Institute (NCI)
  http://www.cancer.gov

GENOMICS IN LEUKEMIA

Gene expression omnibus
  http://www.ncbi.nlm.nih.gov/geo/

Cancer Genome Anatomy Project
  http://cgap.nci.nih.gov/cgap.html
Mitelman Database of Chromosome Aberrations and Gene Fusions in Cancer
  http://cgap.nci.nih.gov/Chromosomes/Mitelman
International Cancer Genome Consortium (ICGC)
  http://www.icgc.org/
NIH Roadmap Epigenomics Mapping Consortium
  http://www.roadmapepigenomics.org/
The Cancer Genome Atlas
  http://cancergenome.nih.gov

CHAPTER

Lung Cancer

Hector Marquez, Praveen Govender, Jerome S. Brody, and Hasmeena Kathuria

INTRODUCTION

In the early 1900s, lung cancer was a rare disease. Today it is the most common cause of cancer death in the United States and in the world. Worldwide in 2010, there were more than 1.3 million lung cancer deaths. In the United States alone, there were 226,160 new cases of lung cancer in 2012, with 160,340 deaths, accounting for 28% of all cancer deaths in men and in women (Siegel et al., 2012). The incidence of lung cancer tracks, with a 30–40-year lag, the frequency of cigarette smoking in a population. Smoking incidence and lung cancer rates are similar in the United States and Western Europe, but are 30% higher in Eastern Europe and in China, where 70% of men smoke and an epidemic of lung cancer is predicted in the mid-21st century.

The fact that only 10–15% of smokers develop lung cancer suggests that there are genetic factors that influence individual susceptibility to the carcinogenic effects of cigarette smoke. However, the late age at which lung cancer usually develops limits traditional genetic studies, since parents and siblings are not often available to study. Unfortunately, survival rates for newly diagnosed lung cancer remain at approximately 15%. This has remained virtually unchanged for the past four to five decades, largely because the disease is most often diagnosed at an advanced stage and there is presently no way to determine which of the 10–15% of current and former smokers are at highest risk for developing lung cancer.

This review summarizes the progress that has been made in understanding the molecular mechanisms and pathogenesis of lung cancer, and attempts to describe how gene expression profiling of human lung cancer, airway, and peripheral blood specimens is leading to new approaches in lung cancer screening, classification, diagnosis, prognosis, and treatment. The review focuses mainly on genomic studies of lung cancer, including a brief discussion on recent advances in the role of epigenetics in lung tumorigenesis. Studies using single nucleotide polymorphism (SNP) and protein arrays are beyond the scope of this review, as are studies of *in vitro* cell lines and mouse models. Although we have attempted to be as inclusive as possible, our list of studies reviewed is likely to be incomplete.

EARLY DIAGNOSIS/SCREENING OF LUNG CANCER
Smoking, Lung Cancer, and Genetics

In the early 20th century, lung cancer was a rare disease. It was first suspected that cigarette smoking caused lung cancer in the late 1920s. By 1964, there was enough evidence for the US Surgeon General to support the conclusion that cigarette smoking caused lung cancer. Since then, smoking rates in men have decreased to 20–25%, with a subsequent leveling off and then fall in lung cancer incidence in the 1990s. Smoking rates in women, however, began to rise in the 1950s, and lung cancer became the leading cause of death from cancer in women in the 1980s (reviewed by Spiro and Silvestri, 2005). By 2030, smoking incidence is projected to plateau at 20% of the adult United States population in both men and women, ensuring a

Genomic and Personalized Medicine, 2nd edition
by Ginsburg & Willard. DOI: http://dx.doi.org/10.1016/B978-0-12-382227-7.00059-8

constant high rate of lung cancer throughout the first half of the 21st century.

Recently, there have been increasing reports of lung cancer occurring in never-smokers (individuals who have had a lifetime exposure of fewer than 100 cigarettes). A review by Scagliotti and colleagues showed lung cancer in never-smokers ranging from 14.4–20.8 per 100,000 person-years in women and 4.8–13.7 per 100,000 person-years in men (Scagliotti et al., 2009). Higher incidence of lung cancer occurred in women from Asiatic countries compared to Western countries. The authors showed a higher risk of lung cancer in never-smokers whose spouses were smokers, and noted increases in those exposed to domestic coal use for heating, high-dose vitamin E supplementation, indoor air pollution from solid fuel use, cannabis smoking, and cooking fumes (Scagliotti et al., 2009). Studies suggest that polymorphisms in DNA repair genes increase the risk of lung cancer in Chinese non-smoking women exposed to cooking oil fumes (Luo et al., 1996; Yin et al., 2009; Zhou et al., 2000). These observations raise the question of whether lung cancer in other never-smokers might result from a variety of unrecognized toxic exposures, especially in genetically susceptible individuals.

We now know that although 80–85% of patients with lung cancer have a history of smoking, only 10–15% of patients who smoke actually develop lung cancer, suggesting that hereditary factors may explain susceptibility to the carcinogenic effects of smoking. It is well established that smokers who are first-degree relatives in families with a history of lung cancer have a two- to three-fold increased risk of developing lung cancer (Thun et al., 2002). A systemic review of cohort and case-control studies showed a significant increase in lung cancer risk in cases where lung cancer was diagnosed at a younger age in the relative and those with multiple family members being affected (Matakidou et al., 2005). Hereditary occurrence is also suggested, since both smokers and non-smokers with a positive family history are at higher risk for developing cancer.

Epidemiological studies show that Native-American and African-American smokers are more susceptible to lung cancer than whites, while Latinos and Japanese Americans are less susceptible (Haiman et al., 2006). These differences are accentuated in those who have smoked fewer than 10 cigarettes per day, and tend to disappear in those who have smoked more than 30 cigarettes per day, suggesting a genetically determined carcinogenic susceptibility that is masked at high levels of smoke exposure. The reasons for the observed racial/ethnic differences in response to the carcinogenic effects of cigarette smoke are not yet known, but it is likely that both genetic and environmental differences are involved.

A large number of studies have attempted to define heritable causes of lung cancer susceptibility. Polymorphic variants in almost 50 genes have been claimed to be associated with either reduced or elevated risk of lung cancer (reviewed by Cooper, 2005). A number of studies have focused on polymorphisms in genes involved in the metabolism of the toxic components of cigarette smoke. Most of the compounds in cigarettes are activated by phase I drug-metabolizing enzymes (DMEs), such as cytochrome p450, to become active and carcinogenic. These phase I genes code for substances that convert smoke constituents to highly reactive intermediates that can bind to and mutate DNA. Of the phase I enzymes, two specific *CYP1A1* polymorphisms (Msp1 and exon 7) have been associated with higher risks of lung cancer (Le Marchand et al., 2003; Vineis et al., 2004). Phase II DMEs, such as N-acetyltransferases, sulfotransferases, and glutathianone S-transferases, detoxify carcinogenic products by conjugating those reactive species to less reactive, excretable products. While some studies support the importance of polymorphisms in both phase 1 and 2 genes, others have found no relation between polymorphisms in these genes and the incidence of lung cancer (Perera et al., 2006; Raunio et al., 1999; Taioli et al., 2003; Wikman et al., 2001).

It is likely that there are a number of genes – some involved in interacting pathways – that account for individual susceptibility to the carcinogenic effects of smoking. Studies have described polymorphisms in the cell cycle checkpoint genes in African-American smokers with lung cancer, and in DNA repair genes in smokers both with and without lung cancer (David-Beabes and London, 2001; Wenzlaff et al., 2005). A promoter polymorphism that increases promoter activity in caspase-9, a cysteine protease involved in the caspase-mediated apoptotic pathway, was recently found to contribute to genetic susceptibility in lung cancer (Park et al., 2006). In addition, a number of studies have demonstrated linkage to lung cancer on chromosome 6qa23 (Bailey-Wilson et al., 2004) and on chromosome 12q (Sy et al., 2004). Mouse modeling studies have identified similar linkage between lung cancer and chronic obstructive pulmonary disease (COPD), and many of the genes in the area of linkage are involved in inflammation, a process likely to be central in the pathogenesis of both diseases (Bauer et al., 2004).

A relatively new field, "genetical genomics," which attempts to link gene expression in affected tissues with genetic polymorphisms, promises to provide insights into heritable factors in lung cancer (Li and Burmeister, 2005). This field is derived from recent technological and computational advances that allow one to measure levels of gene expression in the affected tissue and relate them to high-density SNP discovery in genomic DNA (Pastinen et al., 2006).

Genome-wide association studies have identified lung cancer susceptibility loci at 15q25, 6p21, and 5p15.33. A confirmatory study by Truong and colleagues in more than 26,000 lung cancer cases and control subjects validated an association between lung cancer risk and 15q25 and 5p15, but not 6p21. Interestingly, associations between 15q25 and the risk of lung cancer were replicated in white smokers, but not in never-smokers or in Asians (Truong et al., 2010). There is evidence that the 15q25 locus is associated with smoking status and nicotine dependence. The association region contains several genes, including three that encode nicotinic acetylcholine receptor subunits (*CHRNA5*, *CHRNA3*, and *CHRNB4*).

These subunits are expressed in neurons, alveolar epithelial cells, and pulmonary neuroendocrine cells, and they bind to N'-nitrosonornicotine and potential lung carcinogens, reinforcing interest in nicotinic acetylcholine receptors as potential chemopreventative targets (Hung et al., 2008).

Epidemiological data combined with genetic data may help identify high-risk individuals for lung cancer. Patterns of gene expression in smoke-exposed tissue and linkage to SNPs in the affected genes are likely to lead to new insights into heritable causes of lung cancer.

Identifying Smokers at Risk

The risk of developing lung cancer increases with accumulated exposure to cigarette smoke, most often expressed as pack-years (calculated by multiplying the number of packs of cigarettes smoked per day by the number of years the person has smoked). As noted earlier, even in high-risk populations of smokers, the incidence of lung cancer is only about 15% over a lifetime. The incidence increases slightly in those over the age of 60 with greater than 30 pack-years, previous lung cancer, history of COPD, and/or atypia of bronchial cells in the sputum.

Because of the lack of effective diagnostic biomarkers that identify which current and former smokers are at the greatest risk for developing cancer, lung cancer is most often diagnosed at a late stage, after it has spread. In contrast to most cancers, five-year survival rates for lung cancer (about 15%) have not changed appreciably over the past four to five decades. Previous screening trials with frequent chest X-rays and sputum cytology have not demonstrated an effect on lung cancer mortality (reviewed by Jett and Midthun, 2004). Spiral computerized tomography (CT) scan screening can detect lung tumors at an early stage. The I-ELCAP study, a systematic case-control observational study that included more than 31,000 subjects who were at risk for lung cancer, found that screening by CT resulted in a diagnosis of lung cancer in 484 participants, of which 412 (85%) had clinical stage I lung cancer (Henschke et al., 2006). Spiral CTs, however, while highly sensitive, can be non-specific, and many newly detected small lesions have proven on resection to be non-malignant rather than early lung cancers (Jett and Midthun, 2004). Thus, it is not yet known if this approach will alter lung cancer mortality or justify the relatively high cost of large-scale screening.

Recently, the National Lung Screening Trial (NLST), a randomized trial involving greater than 53,000 current and former heavy smokers, compared low-dose helical CT and standard chest X-ray on lung cancer mortality. They noted 20% fewer lung cancer deaths among participants screened with low-dose helical CT (Aberle et al., 2011). A number of other trials are ongoing, including the DANTE and ITALUNG trials with results expected before 2016 (Infante et al., 2009; Lopes et al., 2009).

Developing biomarkers that are highly sensitive, specific, and that target individuals with early stages of cancer is clearly necessary to improve lung cancer mortality. A study by Rahman and colleagues examined proteomic profiles in fresh-frozen bronchial epithelial and lung tissue samples using matrix-assisted laser desorption/ionization time-of-flight mass spectrometry (MALDI-TOF). This profile predicted the tumor status of the entire blinded set of samples from the 53 patients, and of a previously published dataset, with a high degree of accuracy (Rahman et al., 2005). Using bronchial tissue from this 2005 patient cohort, this group recently derived and validated a signature from the proteomic analysis of bronchial lesions that could predict the diagnosis of lung cancer. The performance of their prediction model in identifying lung cancer was tested in an independent cohort of bronchial specimens from 60 patients, and was validated by Western blotting and immunohistochemistry. These results show that proteomic analysis of endobronchial lesions may facilitate the diagnosis of lung cancer (Rahman et al., 2011).

Several groups have used genomic profiling to identify the subset of smokers who are at higher risk of developing lung cancer. Powell and colleagues compared the gene expression profiles of tumor and matched normal tissues from smokers and non-smokers (Powell et al., 2003). Although hierarchical clustering did not separate tumors from smokers versus non-smokers, it did separate tumor and non-tumor tissue. Four times more genes were altered between tumor and lung in non-smokers than in smokers. The findings demonstrate that underlying normal tissues from smokers and non-smokers differ, consistent with the concept of "field cancerization" in smokers, where genes are altered in the entire respiratory epithelium.

To begin to understand the mechanisms by which some individuals protect themselves from the carcinogenic effects of smoking, Spira and colleagues used high-throughput genomic and bioinformatic tools to define the genome-wide impact of smoking and smoking cessation on bronchial airway epithelium (Spira et al., 2004). They demonstrated changes in both antioxidant and drug-metabolizing genes in airway epithelial cells, as well as increases in putative oncogenes and decreases in tumor-suppressor genes. They also noted that the expression level of smoking-induced genes among former smokers began to resemble that of never-smokers after two years of smoking cessation. Genes that reverted to normal within two years of cessation tended to serve metabolizing and antioxidant functions. The authors also found that several genes, including putative oncogenes and tumor-suppressor genes, failed to revert to never-smoker levels years after smoking cessation, perhaps explaining the continued risk for developing lung cancer many years after individuals have stopped smoking (Spira et al., 2004).

Using histologically normal large-airway epithelial cells obtained at bronchoscopy, Spira and colleagues described an 80-gene expression signature that distinguished smokers with and without lung cancer. The biomarker had an accuracy of 83% (80% sensitivity, 84% specificity) in an independent test set and was subsequently validated on an independent prospective series (Spira et al., 2007). The biomarker's overall accuracy improved when used in combination with clinical features, providing a paradigm for a clinicogenomic approach to comprehensive lung cancer diagnosis and treatment (Beane et al., 2008). More recently, this group used airway transcriptome sequencing to elucidate the mechanisms of response to tobacco smoke and to identify additional biomarkers of lung cancer risk and novel targets for chemoprevention. These authors found a significant correlation between the RNA sequencing gene expression data and Affymetrix microarray data generated from the same samples (P < 0.001) (Beane et al., 2011).

Detection of mutations or aberrant methylation in sputum or serum is another promising approach to the early diagnosis of lung cancer. Detection of promoter hypermethylation in the *CDKN2A*, *DAPK1*, *RASSF1*, *CDH1*, *GSTP1*, *RAB*, *CDH13*, *APC*, *MLH1*, *MSH2*, and *MGMT* genes has been demonstrated, using sputum or bronchial lavage (reviewed by Cooper, 2005). Recent studies have also demonstrated that gene promoter hypermethylation in sputum could identify people at high risk for lung cancer. One study demonstrated *FHIT* methylation in pre-neoplastic lesions from smoking-damaged bronchial epithelium (Zochbauer-Muller et al., 2001). Another study demonstrated hypermethylation of the *CDKN2A* gene promoter prior to clinical evidence of lung carcinoma (Kersting et al., 2000). Lastly, Palmisano and colleagues showed that *p16* and/or *MGMT* could be detected in DNA from sputum up to three years before clinical diagnosis of squamous cell cancer (Palmisano et al., 2000).

Reports are also surfacing documenting the clinical potential of detecting tumor-related genes, mutations, loss of heterozygosity, polymorphisms, and promoter hypermethylation of circulating tumor DNA in serum. High-throughput methods for analyzing the methylation status of hundreds of genes simultaneously are being applied to the discovery of methylation signatures that distinguish normal from cancerous tissue samples (Bibikova et al., 2006; Wilson et al., 2006). However, it is important to note that none of these studies have tested the diagnostic potential in prospective multi-center trials.

Future Directions

Lung cancer is one of the few cancers for which screening is not recommended, even in high-risk individuals. Although spiral computerized tomography (CT) may increase the detection of early lesions, these procedures are costly and/or relatively invasive. Also, they have not been shown to be specific for the presence of lung cancer, and it is not yet known if screening will alter mortality. Since molecular, genetic, and epigenetic abnormalities precede morphological changes in bronchi and alveoli, biomarkers may help select a group of high-risk patients who would benefit from spiral CT and/or fluorescent bronchoscopy.

To predict those at risk for developing cancer and/or detect lung cancer at an earlier stage, screening profiles must be capable of identifying the few abnormal cells among many normal cells, and ideally samples should be obtained in a relatively non-invasive fashion. Recently, Sridhar and colleagues studied gene expression profiles in epithelial cells that line the intrathoracic (bronchial) and extrathoracic (buccal and nasal) airways in healthy and current smokers. The authors identified a common set of genes induced by tobacco smoke in buccal, nasal, and bronchial epithelium, supporting the concept that smoking induces a common field of injury throughout the airway (Sridhar et al., 2008). Similarly, a recent study by Boyle and colleagues using whole-genome gene expression profiling of punch biopsies from the buccal mucosa of 40 healthy smokers and 40 healthy non-smokers showed a strong relationship between the gene expression responses to smoking in buccal and bronchial epithelium (Boyle et al., 2010). These studies suggest that easily collected buccal and nasal epithelium may provide a biomarker that identifies smokers at high risk for developing lung cancer.

The availability of a number of agents that might be effective in reversing the pre-malignant changes in airway epithelial cells (see below) has driven the search for biomarkers that identify individuals at highest risk for developing lung cancer. Genetic abnormalities can be detected from bronchial biopsies, respiratory cells from sputum, and circulating DNA, and gene expression profiles generated from those specimens offer a wide area of investigation for biomarker development. To be applied in a screening program, the biomarkers must be specific and cost-effective, with high efficiency. The focus has been on intermediate markers that indicate risk but may be reversible with appropriate treatment.

Chemoprevention in high-risk current or former smokers represents one of the most important areas of current research in the prevention of lung cancer. It has been a topic of great interest for more than 20 years. Despite numerous trials, beginning with high doses of retinoids in the 1980s, no studies to date have demonstrated a positive outcome in terms of mortality from lung cancer, and few studies have demonstrated an effect on intermediate end points such as sputum atypia or genomic markers of epithelial cell damage. Several publications review the results of previous randomized trials and the rationale for use of new chemopreventive agents (Hirsch and Lippman, 2005; Kelloff et al., 2006; Khuri and Cohen, 2004). Chemoprevention (chemoprophylaxis) has assumed increasing importance, as former smokers now account for 50% of new lung cancer cases in the United States. Since there are approximately 45 million former smokers in the United States, identifying which former smokers are at highest risk for developing lung cancer in the future is the most important first step in any chemoprevention program.

Molecular alterations in the field of injury that persist after smoking cessation may help explain persistent lung-cancer risk in former smokers, and risk-related enzymes that metabolize carcinogens in cigarette smoke suggest possible targets for lung cancer chemoprevention (reviewed by Steiling et al., 2008). A clinical study conducted in 10 patients to assess the potential chemopreventive effect of myo-inositol in smokers with bronchial dysplasia showed a significant increase in the rate of regression of pre-existing dysplastic lesions (Lam et al., 2006). Gustafson and colleagues recently observed a significant increase in a genomic signature of phosphatidylinositol 3-kinase (PI3K) pathway activation in cytologically normal bronchial airway of smokers with lung cancer and smokers with dysplastic lesions. The authors reported decreased PI3K activity in the airways of high-risk smokers who had significant regression of dysplasia after treatment with myo-inositol, an inhibitor of the PI3K pathway *in vitro* (Gustafson et al., 2010).

Several groups are beginning to integrate clinical markers of lung cancer risk with alterations in gene expression or DNA repair capacity in constructing comprehensive models for lung cancer diagnosis. One such genomic approach is leveraging the Connectivity Map to discover novel chemopreventive agents (http://www.broadinstitute.org/genome_bio/connectivitymap.html). The Connectivity Map is a collection of genome-wide transcriptional expression data from cultured human cells treated with bioactive small molecules, and simple pattern-matching algorithms, that together enable the discovery of functional connections between drugs, genes, and diseases. Using this approach, Boyle and colleagues, discovered heat-shock protein 90 inhibitor geldanamycin as a lung cancer chemopreventive agent that could potentially reverse epithelial gene expression changes associated with tobacco smoke exposure (Boyle et al., 2010). The potential impact of chemoprevention is large, but the field awaits the emergence of intermediate markers of cancer risk that must be validated in prospective studies.

CLASSIFICATION AND PROGNOSIS
Histological Classification of Lung Tumors and TNM Staging

Lung tumors are classified as small-cell lung carcinomas (SCLC) or non-small-cell carcinomas (NSCLC). Small-cell lung cancers have neuroendocrine features that are identified by immunohistochemistry and histology. Non-small-cell carcinomas are sub-categorized as adenocarcinomas (the most common), squamous-cell carcinomas, and large-cell carcinomas, and are clinically distinct from SCLC. The pathological distinction between SCLC and NSCLC is very important since the tumor types are treated differently (reviewed by Spira and Ettinger, 2004).

Pre-malignant lesions termed low- and high-grade dysplasia, carcinoma *in situ* (CIS), and atypical adenomatous hyperplasia (AAH) are associated with increased risk for developing lung cancer (reviewed by Kerr, 2001). When the airways are exposed to carcinogens, cellular and molecular changes occur over broad mucosal surfaces, a concept called "field cancerization." Accumulation of mutations and epigenetic alterations ensues, eventually leading to invasive lung cancer, a process called "multistep carcinogenesis" (reviewed by Cooper, 2005).

Histological analysis is currently used to identify and classify cancers, but definitive diagnoses are often difficult to make. Furthermore, tumors with the same histological classification behave differently, and morphological classification has not been effective in predicting the aggressiveness of a cancer or how the cancer will respond to therapeutic agents.

In NSCLC, the anatomic extent or stage of the tumor as defined by Tumor-Node-Metastasis (TNM) descriptors is the most important prognostic variable for predicting survival. In late 2009, the International Association for the Study of Lung Cancer (IASLC) published the seventh revision of the TNM staging system (summarized by Detterbeck et al., 2009). Based on clinical and pathological data from 67,725 cases, significant changes were made to improve the alignment of the TNM stage with prognosis. The TNM staging system takes into account the degree of spread of the primary tumor (T), the extent of regional lymph node involvement (N), and the presence or absence of metastases (M). Briefly, the following are new in the seventh revision: (1) T1 and T2 have "a" and "b" subcategories based on tumor size, and tumors >7 cm are upstaged to T3; (2) separate tumor nodules in the same lobe, in a different ipsilateral lobe, and in a contralateral lobe are classified as T3, T4, and M1a respectively; (3) M1 has "a" and "b" subcategories, with the M1a category including malignant pleural and pericardial effusions. With these alterations, T2bN0M0 and T2aN1M0 cases are classified as stage IIA, and T4N(0 or 1)M0 cases are now reclassified as IIIA (see Figure 59.1 for details of the seventh revision). Despite these changes and improvements in diagnosis, including both non-invasive methods [CT, positron emission tomography (PET)] and invasive methods (endoscopic ultrasound (EUS), endobronchial ultrasound (EBUS)), patients who are diagnosed with similar stages and are treated using similar protocols often respond quite differently and have varying survival rates.

Several combined clinical, histological, and laboratory variables such as age, stage or grade of tumor, and serum protein levels can be used to assess a patient's prognosis with variable accuracy. But these criteria are not able to provide important information about the prognostic diversity within each stage, such as how aggressive a particular subtype will be or how a patient will respond to therapy. Some tumors, even when found at an early stage, will rapidly progress to metastatic disease, reflecting the limitation of relying solely on anatomic extent to define prognosis. By combining clinical variables and histopathology with gene expression profiling, predicting a patient's prognosis could theoretically be improved.

Molecular Classification and Prognostic Value of Lung Cancer Genomics

Since lung cancer is genetically heterogeneous, morphological tumor classification does not always accurately predict the patient's clinical behavior. For example, patients with stage IA lung cancer resected for cure still have 30% mortality from local recurrence, distant metastases, and/or new occurrence. Within each clinical stage, there is variability in the presence of specific mutations, deletions of tumor-suppressor genes, amplifications of oncogenes, and chromosomal abnormalities. Genomics is a powerful tool for classifying tumor subtypes. Lung cancer patients with biomarkers that predict a poor outcome could be selected for adjuvant chemotherapy, while those that predict a good prognosis may be able to avoid the toxicity and cost of unnecessary chemotherapy.

(A)

Descriptor	Definitions
T	**Primary tumor**
T0	No primary tumor
T1	Tumor <3 cm, surrounded by lung or visceral pleura, not more proximal than the lobar bronchus
T1a	Tumor <2 cm
T1b	Tumor >2 but ≤3 cm
T2	Tumor >3 but ≤7 cm, or tumor with any of the following: invades visceral pleura, involves main bronchus <2 cm distal to the carina, atelectasis/obstructive pneumonia extending to hilum but not involving the entire lung
T2a	Tumor >3 but ≤5 cm
T2b	Tumor >5 but ≤7 cm
T3	Tumor >7 cm or directly invading chest wall, diaphragm, phrenic nerve, mediastinal pleura, or parietal pericardium or tumor in the main bronchus <2 cm distal to the carina or atelectasis/obstructive pneumonitis of entire lung or separate tumor nodules in the same lobe
T4	Tumor of any size with invasion of heart, great vessels, trachea, recurrent laryngeal nerve, esophagus, vertebral body, or carina; or separate tumor nodules in a different ipsilateral lobe
N	**Regional lymph nodes**
N0	No regional node metastasis
N1	Metastasis in ipsilateral peribronchial and/or perihilar lymph nodes and intrapulmonary nodes, including involvement by direct extension
N2	Metastasis in ipsilateral mediastinal and/or subcarinal lymph nodes
N3	Metastasis in contralateral mediastinal, contralateral hilar, ipsilateral or contralateral scalene, or supraclavicular lymph nodes
M	**Distant metastasis**
M0	No distant metastasis
M1a	Separate tumor nodules in a contralateral lobe, or tumor with pleural nodules or malignant pleural dissemination
M1b	Distant metastasis
Special situations	
TX, NX, MX	T, N, or M status not able to be assessed
Tis	Focus of *in situ* cancer
T1	Superficial spreading tumor of any size but confined to the wall of the trachea or main stem bronchus

(B) UICC/AJCC 7th edition of TNM lung cancer staging system

M1a, M1b = IV 2/13%	T1a	T1b	T2a	T2b	T3	T4
N0	IA	50/73%	IB 43/58%	IIA	IIB	IIIA
N1	IIA	36/46%	IIA	IIB25/36%		
N2	IIIA	19/24%	IIIA		IIIA	IIIB
N3	IIIB	7/9%	IIIB		IIIB	

(C) UICC/AJCC 6th edition of TNM lung cancer staging system

M1=IV	T1	T2	T3	T4
N0	IA	IB	IIB	IIIB
N1	IIA	IIB	IIIA	IIIB
N2	IIIA	IIIA	IIIA	IIIB
N3	IIIB	IIIB	IIIB	IIIB

Figure 59.1 (A) Definitions of the TNM descriptors; (B) 7th edition of TNM lung cancer staging system: overall survival at five years is provided for each stage as assessed by (clinical stage/pathological stage); (C) 6th edition of TNM lung cancer staging system. The juxtaposition of the 7th and 6th editions illustrates the changes in the TNM groupings. *Part (B) modified from Detterbeck et al., 2009. UICC/AJCC- Union Internationale Contre le Cancer/ American Joint Cancer Committee.*

Recently, microarray studies have identified and validated specific genes whose expression differs between normal and tumor tissue. Tomida and colleagues developed a set of classifier genes from analyzing 8644 genes in NSCLC cases. The classifier was used to predict an independent set of six NSCLC patients; in four of the six cases, the outcome classifier was correct (Tomida et al., 2004). Using complementary DNA (cDNA) arrays, Wikman and colleagues compared gene

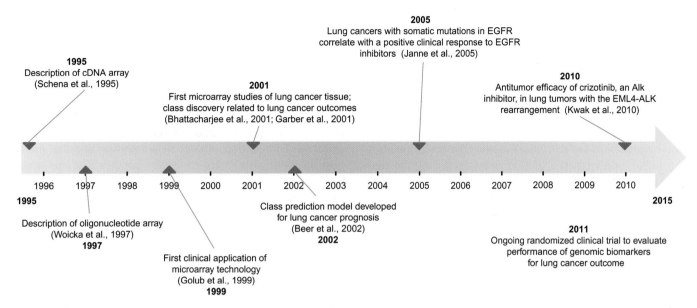

1995
Description of cDNA array
(Schena et al., 1995)

2005
Lung cancers with somatic mutations in EGFR
correlate with a positive clinical response to EGFR
inhibitors (Janne et al., 2005)

2001
First microarray studies of lung cancer tissue;
class discovery related to lung cancer outcomes
(Bhattacharjee et al., 2001; Garber et al., 2001)

2010
Antitumor efficacy of crizotinib, an Alk
inhibitor, in lung tumors with the EML4-ALK
rearrangement (Kwak et al., 2010)

1996 1997 1998 1999 2000 2001 2002 2003 2004 2005 2006 2007 2008 2009 2010

1995

2015

Description of oligonucleotide array
(Woicka et al., 1997)
1997

Class prediction model developed
for lung cancer prognosis
(Beer et al., 2002)
2002

2011
Ongoing randomized clinical trial to evaluate
performance of genomic biomarkers
for lung cancer outcome

First clinical application of
microarray technology
(Golub et al., 1999)
1999

Figure 59.2 Timeline for the development of DNA microarray technology and its application to lung cancer. The key advances in the field of microarray technology and their clinical applications to lung cancer tissue for prognostic purposes are highlighted.

expression profiles of 14 pulmonary adenocarcinoma patients with normal lung tissue (Wikman et al., 2002). The authors demonstrated marked differences in gene expression levels between normal lung and adenocarcinomas.

Another study by Yamagata and associates showed that a profile based on gene expression could be used on blinded samples to differentiate primary NSCLC from normal lung and lung metastases (Yamagata et al., 2003). His group identified 62 genes that had a significantly different expression level between the metastatic lesion and the lung primary tumor, and the genes were able to identify the correct origin in 12 of 13 tumor samples chosen as a test sample. Furthermore, groups of genes were identified that were able to identify known histological subgroups of NSCLCs.

Genomic high-throughput technologies have been used in many studies to identify gene expression signatures that predict patient survival and/or relapse rates. A considerable number of expression profiling studies have been performed on clinical lung cancer specimens (summarized by Cooper, 2005; Granville and Dennis, 2005; Kopper and Timar, 2005; Santos et al., 2009). Not only can gene expression profiles group tumor samples consistently with classical histology, but they can also identify subgroups within histologic subclasses. In an attempt to discover previously unrecognized subtypes of lung cancer, two microarray studies aimed at class discovery by hierarchical clustering using primary lung cancer specimens were published in 2001 (see Figure 59.2). Garber and colleagues identified gene subsets that are characteristic of each of the known morphological subtypes in NSCLC (Garber et al., 2001). The authors also found that the tumors could be further divided into subgroups with significant differences in patient survival.

By gene expression analysis of 186 lung carcinomas, a study by Bhattacharjee and coworkers showed that biologically

distinct subclasses of lung carcinomas exist (Bhattacharjee et al., 2001). In the study, a subclass of adenocarcinomas defined by having neuroendocrine gene expression had a less favorable outcome, whereas a subset of patients with predominantly type II pneumocyte expression had a more favorable outcome. In another study, Hayes and colleagues were able to identify reproducible adenocarcinoma subtypes in different cohorts of patients. These subtypes significantly differed in clinically important behaviors, including age-specific survival (Hayes et al., 2006). More recently, Guo and colleagues identified a 37-gene signature predicting prognosis in a cohort of 86 adenocarcinomas (Guo et al., 2008). When applied to an independent cohort of 84 adenocarcinomas, the signature separated patients into three prognostic groups – good (mean survival 66.9 months), moderate (mean survival 27.6 months), and poor (mean survival 22.4 months), with 96% accuracy.

In a study designed to investigate whether gene expression patterns could predict survival, Beer and colleagues demonstrated that expression profiles based on microarray analysis could be used to predict disease progression and clinical outcome in early-stage lung adenocarcinomas (Beer et al., 2002). Subdivision of the lung tumors based on gene expression patterns matched the morphological classification of tumors into squamous, large-cell, adenocarcinoma, and small-cell lung cancer. Adenocarcinomas could be further subclassified based on their gene expression profiles, that correlated with the degree of tumor differentiation and patient survival. They demonstrated by gene profiling that a list of 50 genes were most effective at dividing patients with stage I lung adenocarcinoma into high- and low-risk groups for mortality, thus potentially identifying a subset of stage I NSCLC cancer patients who would benefit from adjuvant therapy. This group has extended their observations to the squamous-cell carcinoma subset of

NSCLC and have generated a similar risk-of-progression stratifier which, combined with the adenocarcinoma tool, will cover approximately 80% of all NSCLC (Raponi et al., 2006).

New attempts to compare results from several different studies are being undertaken. Parmigiani and colleagues evaluated to what extent three such studies (Beer et al., 2002; Bhattacharjee et al., 2001; Garber et al., 2001) agree, and whether the results could be integrated. Correlation for diagnosis of squamous-cell carcinoma versus adenocarcinoma was high (0.85), and increased (0.925) when using only the most consistent genes in all three studies. In their review of the data, 14 genes were identified as significant predictors of survival in all three studies, and led to a combined statistical significance level of 0.00675 (Parmigiani et al., 2004).

Future Directions

Currently, lung cancer patients within a given clinical stage and tumor type receive the same treatment despite the genomic heterogeneity that exists between patients. Although expression profiling of lung cancer has identified prognostic genes and subtypes of adenocarcinomas, until recently those profiles have not been ready to be incorporated into clinical practice. Comparison of array studies has been difficult because the array platforms, sample preparation, and technical factors have been different (see Chapters 12 and 19). In addition, many molecular classification studies do not match the classification based on tumor histology, perhaps because use of whole lungs for microarray studies may not accurately reflect gene expression in the cancer cells due to contamination and differences in abundance of stromal and surrounding normal cells. So far, little has been published on the ability of prognostic methods for lung cancer to perform in larger datasets or with independent validation samples.

To address these issues, a multi-institutional collaborative study was conducted to generate gene expression profiles from a large number of lung cancer specimens, the results of which will hopefully lead to a gene expression signature that will improve diagnostic options. A pooled analysis of 442 lung adenocarcinomas from multiple institutions established the performance of gene expression signatures across different patient populations and different laboratories. Investigators analyzed whether gene expression data alone or combined with clinical data could be used to predict overall survival in patients with lung cancer. This study identified predictors of survival based on clinical and gene expression microarray data, with better accuracy to predict survival with combined clinical and molecular data (Shedden et al., 2008).

Redefining tumor classification from strictly morphology-based schemes to molecular-based classifications promises to provide clinically important information on tumor subsets within morphological classes. Prognostic profiles and, eventually, a prognostic-specific gene chip may help guide clinical decision-making by identifying high-risk lung cancer patients who would benefit from improved diagnostic and treatment options.

PATHOGENESIS AND TREATMENT OF LUNG CANCER

Molecular Alterations in Lung Cancer

Proto-oncogenes

Proto-oncogenes such as *Ras* and *Myc* were first discovered in the 1970s and 80s. It was only in the 1980s and 90s that these genes were integrated into signaling pathways and that mutations in them were linked to human cancers. The *Ras* genes (*Hras, Kras*, and *Nras*) encode GTPase proteins that help transduce survival- and growth-promoting signals. When oncogenic mutations occur, the normal abrogation of RAS signaling by hydrolysis of bound guanosine triphosphate (GTP) to guanosine diphosphate (GDP) is impaired, resulting in persistent signaling (reviewed by Singhal et al., 2005). Point mutations (found most frequently in codons 12, 13, and 61) are detected in 20–30% of lung adenocarcinomas (Slebos et al., 1990), and 90% of these mutations are found in *Kras* (see Table 59.1). Mutations in *Kras* are markers for poor prognosis in NSCLCs (Graziano et al., 1999), as mutations rarely occur in SCLC. *Myc* genes *(MYCL, MYCN*, and *CMYC)*, which encode transcription factors that regulate genes involved in cell cycle regulation, DNA synthesis, and RNA metabolism, become activated by loss of transcriptional control or by gene amplification, resulting in MYC protein overexpression. C-Myc amplification occurs in 5–10% of NSCLCs (Richardson and Johnson, 1993).

The anaplastic lymphoma kinase (*ALK*) fusion gene was first described in lymphoma in 1994. In 2007, Soda and colleagues described an *EML4-ALK* fusion oncogene that generates aberrant signaling in NSCLC (Soda et al., 2007). The fusion protein arises from an inversion on the short arm of chromosome 2 that joins exons 1–13 of *EML4* to exons 20–29 of *ALK*. Lung tumors that contain the *EML4-ALK* fusion oncogene or its variants are associated with specific clinical features, including never- or light-smoking history, younger age, and adenocarcinoma. *ALK* gene arrangements are largely mutually exclusive with EGFR or KRAS mutations and occur in 2–7% of NSCLC patients (Takahashi et al., 2010). A recent trial demonstrated the antitumor efficacy of crizotinib, an Alk inhibitor, in lung tumors with the *ALK* rearrangement, resulting in tumor shrinkage or stable disease in most patients (Kwak et al., 2010). Two secondary mutations within the kinase domain of *EML4-ALK* in tumor cells have been demonstrated in patients who relapsed during Alk inhibitor treatment (Choi et al., 2010).

Growth Factors and Their Receptors

Lung tumors often express growth factors and their receptors, and the resulting regulatory loops can stimulate tumor growth. The epidermal growth factor receptor (EGFR), also known as ERBB-1, is highly expressed in many epithelial tumors, including a subset of lung adenocarcinomas (Reissman et al., 1999). When ligand [EGF or transforming growth factor (TGF)-α] binds to EGFR, there is receptor tyrosine kinase activation and a series of downstream signaling events, including the mitogen-activated

TABLE 59.1	Common molecular alterations in lung cancer	
	Type of mutation	**Frequency**
Proto-oncogenes		
k-Ras	Point mutation	NSCLC 20–30% (mostly adeno)
MYC	Translocation	SCLC 30–40%
	Amplification	NSCLC 5–10%
Bcl-2	Translocation	NSCLC (30%)
		Squamous 25%; adeno 10%
EML4-ALK	Translocation	NSCLC 2–7% (mostly adeno)
Growth factors		
EGFR (ERBB-1)	Deletion/mutation	SCLC 0%
	Amplification	NSCLC 10% (BAC 25%)
HER2 (erbB2)	Translocation	NSCLC 30–40%
	Amplification	
Tumor-suppressors		
p53	Deletion/LOH	SCLC 75%
		NSCLC 50%
Rb	Point mutation	SCLC 90%
		NSCLC 15–30%
P16 (CDKN2)	Deletion /LOH	SCLC >80%
		NSCLC 30–50%
FHIT	Deletion/LOH	SCLC 100%
		NSCLC 60%
Lkb1	Mutation	NSCLC 30%

BAC, Bronchioloalveolar carcinoma; LOH, loss of heterozygosity.

protein kinase (MAPK), PI3/AKT, and JAK/STAT pathways (see Figure 59.3). These signaling events can result in cellular proliferation, increased cell motility, tumor invasion, anti-apoptosis, and resistance to chemotherapy (summarized by Baselga, 2006). Therefore, EGFR receptor tyrosine kinase was proposed as a target for cancer therapy 20 years ago.

Anti-EGFR drugs approved for cancer treatment include monoclonal antibodies directed against the extracellular domain of the receptor (anti-EGFR Mabs) and small molecule inhibitors of EGFR's tyrosine kinase activity (TKIs). Examples of anti-EGFR Mabs include cetuximab (erbitux), panitumumab, and pertuzumab. Examples of TKIs include gefitinib (iressa) and erlotinib (tarceva). In NSCLCs, two trials have demonstrated that in unselected patient populations, anti-EGFR TKIs have modest levels of antitumor activity. In 10% of patients, however, tumor response is often dramatic, but the responses tend to be transient. From multivariate analysis, non-smoking status with Asian ancestry appears to be an independent factor for prolonged survival, rather than being a specific predictor of benefit with EGFR TKIs (reviewed by Perez-Soler, 2009).

Subsequent analyses have demonstrated that lung cancers with somatic mutations in EGFR correlate with a positive clinical response to EGFR inhibitors (reviewed by Janne et al., 2005; Thomas et al., 2006). In four separate studies of EGFR TKIs in patients with metastatic NSCLC, a significantly longer overall survival was observed in patients harboring EGFR mutations. In the INTEREST study, EGFR mutation status was identified as the only variable predictor of progression-survival benefit with gefitinib (Douillard et al., 2008). Mok and colleagues recently reported results from the I-PASS trial, which showed that non-smoking Asian patients with EGFR mutations had a longer progression-free survival with first-line gefitinib than with first-line chemotherapy (Mok et al., 2009). Unfortunately, initial responders often develop resistance to these drugs. A second EGFR somatic mutation, T790M, occurs in some cases of NSCLC that recur after an initial response to TKIs (Kobayashi et al., 2005). Germline T790M mutations have been described in a family with multiple cases of lung cancer (Bell et al., 2005).

Clinical trials using alternate EGFR inhibitors, such as irreversible EGFR inhibitors, are underway in NSCLC patients. Some centers have begun to sequence EGFR in all resected tumors in order to tailor drug therapy to specific mutations, although further research is needed to more completely define the appropriate population for EGFR testing (reviewed by Sequist et al., 2006). The complexity of EGFR mutations is highlighted by a report identifying different EGFR mutations in the primary tumor versus lesions metastatic from the primary site (Italiano et al., 2006). Furthermore, Engelman and colleagues recently reported that allelic dilution of a biologically significant resistance mutation in EGFR-amplified lung cancer may be undetected by direct sequencing (Engelman and Cantley, 2006).

Tumor-suppressor Genes

The role of the tumor-suppressor gene p53 is to help maintain genomic integrity after DNA damage. When cells undergo stress, such as from carcinogen exposure, ultraviolet radiation, and/or hypoxia, p53 becomes up-regulated. It then acts as a transcription factor to increase genes such as p21 (which in turn controls G1/2 cell cycle transition) and induces apoptosis by activating genes such as BAX, PERP, and others (reviewed by Fong et al., 2003; Singhal et al., 2005). In lung cancer, missense mutations causing loss of p53 function occur in more than 75% of SCLCs and 50% of NSCLCs (Toyooka et al., 2003).

The loss of the Lkb1 tumor-suppressor gene (by point mutations or deletions) occurs in at least 15–35% of NSCLC, and concurrent Kras and Lkb1 mutation is observed in 4–10% of NSCLC (Carretero et al., 2004). The Lkb1 gene encodes a

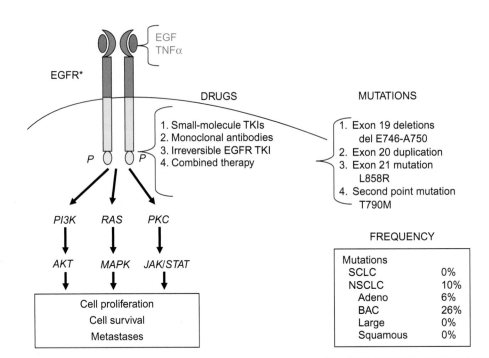

Figure 59.3 Downstream events following ligand binding to EGFR. Receptor expression can be increased by gene duplication (*). Most of the tyrosine kinase domain and all activating EGFR mutations thus far are in exons 18–21. Deletions in exon 19 and substitution mutations in exon 21 account for the majority of mutations. Mutations occur most frequently in bronchoalveolar carcinoma (BAC), but not in squamous-cell carcinoma or SCLC (Marchetti et al., 2005; Shigematsu et al., 2005).

serine/threonine kinase that phosphorylates several substrates, many of which are also kinases. Decreased expression of p16 by promoter hypermethylation, mutations, or allelic loss occurs in 30–50% of NSCLCs. The $p16^{INK4A}$-cyclin D1-CDK4-RB pathway controls G1/S cell cycle transition, and loss of p16 releases the tumor cells from RB-mediated cell cycle arrest (reviewed by Fong et al., 2003). Alternatively, the *Rb* gene can be inactivated directly by deletions, point mutations, or alternative splicing, which is more commonly found in SCLCs (>90%), than NSCLCs (15–30%) (see Table 59.1).

Oncogenic Pathway Signatures

The malignancy process is complex, involving multiple mutations leading to deregulation of signaling pathways (see above). The role of microRNAs (miRNAs) in this is emerging. miRNAs are very small, non-coding RNA products that play a key role in regulatory networks by regulating the translation and stability of mRNAs (reviewed by Calin and Croce, 2006; see Chapter 12). Recent studies have demonstrated altered miRNA gene expression in lung cancers. Navarro and colleagues examined expression of 18 miRNAs in 33 paired lung-cancer/normal-lung samples and found that the expression of all members of the *let-7* miRNA family except *let-7e*, and *miR-222* and *-155* were down-regulated, whereas miRNAs of the miR-17-92 cluster, *miR-221*, *-7e*, and *–98*, were up-regulated (Navarro et al., 2009). In a study by Yanaihara and colleagues analyzing 104 pairs of NSCLC and corresponding normal lung tissues, an expression profile was identified with 43 miRNAs differentially expressed.

In this study, high *miR-155* and low *let-7a-2* correlated with poor survival in lung adenocarcinomas (Yanaihara et al., 2006). Reduced expression of *let-7* miRNA family members has been found to correlate with shorter post-operative survival in potentially curative lung cancer resection (Takamizawa et al., 2004). Yu and colleagues recently identified a five-microRNA signature (which included *let-7a*) that predicted overall survival in stage I–III NSCLC patients (Yu et al., 2008). *Let-7* miRNA family members have been shown to directly regulate *Ras* genes (Johnson et al., 2005). These combined studies suggest that *let-7* may be a promising therapeutic agent to treat lung cancers caused by activating mutations in *Ras* genes (reviewed by Esquela-Kerscher and Slack, 2006).

Epigenetic events are central in tumor progression (see Chapter 2). Genome-wide scans for aberrant promoter methylation have identified novel targets for methylation. Restriction landmark genomic scanning (RLGS) is an assay that evaluates the DNA methylation status of thousands of NotI or AscI restriction sites that are preferentially located in CpG island sequences. This assay was used in primary NSCLC tumor samples and matched normal controls, and determined that about 4.8% of all CpG island promoters in a lung cancer genome are targeted for aberrant DNA methylation (Risch and Plass, 2008).

DNA microarray-based gene expression signatures have been developed that have the ability to define oncogenic pathways. The ability to predict the dysregulation of various oncogenic pathways (both the mutated gene product itself and its downstream targets) offers an opportunity for new therapeutic drugs that are pathway-specific. Diederichs and colleagues identified gene expression changes in pathways involved in metastasis (Diederichs et al., 2004). Their array data provided evidence that expression of S100 proteins and trypsinogens is associated with metastasis.

Another study, by Bild and colleagues, demonstrated that DNA microarray-based gene expression signatures can not only differentiate cells expressing oncogenic activity from control cells, but can also predict deregulation of various oncogenic pathways in specific tumor types derived from mouse cancer models and human cancer specimens (Bild et al., 2006). The authors used multiple experimental models, including functional cell-based assays, mouse lung cancer models, and human cancer specimens, to demonstrate that Ras pathway status clearly correlates with lung adenocarcinoma relative to squamous-cell carcinoma subtypes. Furthermore, independent of tumor histology, patients displaying deregulation of multiple pathways (Ras with Src, Myc, β-catenin) had poor survival.

Molecularly Targeted Therapy to Reduce Lung Cancer Mortality

Biomarkers provide an opportunity to identify subpopulations of patients who are most likely to respond to a given therapy and to identify new targets for drug development. A study by Olaussen and colleagues, for example, showed that in lung tumor specimens, the absence of ERCC1 (an enzyme that participates in the repair of DNA damage caused by cisplatin) was associated with a survival benefit from cisplatin-based adjuvant chemotherapy, whereas patients whose tumor expressed the enzyme failed to benefit from chemotherapy (Olaussen et al., 2006). In a study using oligonucleotide microarrays to analyze 16 NSCLC cancer specimens, lysosomal protease inhibitors serpin B3 and cystatin C predicted clinical response to platinum-based therapy with 72% accuracy (Petty et al., 2006). Kikuchi and colleagues compared gene expression microarray data from 37 laser-capture microdissected primary lung tumor cells with measurements of the sensitivity of six anti-cancer drugs. The authors found significant associations between expression of several genes and chemosensitivity of NSCLCs (Kikuchi et al., 2003).

Several clinical studies have now shown the therapeutic efficacy and safety of tyrosine kinase inhibitors for specific tumor subtypes. Many of the 90 proteins with tyrosine kinase domains are aberrantly activated in human tumors. Intriguingly, in some tumor subtypes, after treatment with tyrosine kinase inhibitors there is marked reduction not only in tumor growth, but also in tumor viability. This concept of dependence of the tumor-mutated oncogenes on EGFR-mediated survival signals is referred to as *oncogene dependence* or *oncogene addiction* (Weinstein, 2002; Varmus, 2006). Oncogene dependence is also demonstrated in mouse models overexpressing constitutively active K-ras, where the mutant K-ras is then silenced. In this model, lung tumors rapidly regress as a result of apoptosis when K-ras is silenced, even in the absence of important tumor suppressor genes (Fisher et al., 2001). An alternative hypothesis to explain oncogene

addiction is through differential signal attenuation of multiple pro-apoptotic and pro-survival signals, a term coined *oncogenic shock* (Sharma et al., 2006).

We now know that no more than 10% of the general population of lung cancer patients will respond to TKIs. However, when only patients who carry gene mutations are treated, the response rate to these agents can be as high as 70%. Improved survival in NSCLC patients taking EGFR TKIs does not seem to be limited only to patients with EGFR mutations. Markers such as *EGFR* gene amplification, ErbB3 levels, and *Her2* mutations may also be predictors of responsiveness (reviewed by Engelman and Cantley, 2006). In addition, the *Kras* gene is mutated in about 20–30% of lung adenocarcinomas, and the presence of *Kras* mutations may predict a poor response to TKIs.

Studies have shown that a high-*EGFR* gene copy number recognized by fluorescence *in situ* hybridization (FISH positivity) is predictive of a clinical benefit with EGFR TKIs (reviewed by Hirsch et al, 2009). In the ISEL study, FISH positivity, which was reported in 114 of 370 patients, was associated with improved survival benefit (Thatcher et al., 2005). Ongoing studies are needed to investigate EGFR TKIs as first-line therapy in selected patients with mutations in exons 18–21, increased *EGFR* copy number or protein expression, and with clinical characteristics associated with response (reviewed by Johnson, 2006). Although the majority of lung tumor cells acquire resistance mutations during therapy with EGFR TKIs, these cells seem to be to be sensitive to a new group of TKIs that covalently cross-link the receptor (Carter et al., 2005). In addition, a recent study using microarray gene expression profiling demonstrated a pattern of gene expression associated with sensitivity to EGFR (Coldren et al., 2006). Mouse lung models with inducible expression of mutations in *EGFR* have been generated, and may help further understanding of the pathogenesis of human lung cancer and aid in the validation of cancer therapeutics (reviewed by Dutt and Wong, 2006). Similar efforts to identify predictive markers for EGFR inhibition have been undertaken in the proteomics area.

Given the complexity of NSCLC, it is likely that these tumors are dependent on more than one oncogenic signaling pathway. Recently, Shaw and colleagues reported that patients with the *EML4/ALK* fusion gene have a distinct subtype of NSCLC with identifiable clinical characteristics such as young age and being non-smokers, and have a high likelihood of responding to ALK-targeted agents (Shaw et al., 2009).

Alterations in cancer-specific copy number and loss of heterozygosity (LOH) are important changes found in cancer cells, and SNP array analyses may reveal pathways disrupted in tumorigenesis. SNP arrays have been used in genomic studies to detect LOH in lung cancer, and have detected previously unknown regions of copy-number change as well as known regions of both amplification and homozygous deletion (see Chapters 8 and 10). Using cell lines derived from tumors representing all major subtypes of NSCLC tumors, Sos and colleagues analyzed gene copy-number alterations using high-resolution SNP arrays. They compared the profile of amplifications, deletions, and mutations in the cell line collection with that of a set of 371 primary lung adenocarcinomas, revealing a striking similarity between the two datasets. Furthermore, using cell-based compound screening coupled with diverse computational approaches, they identified genomic predictors of therapeutic response to clinically relevant compounds (Sos et al., 2009).

It is likely that combination-targeted therapy directed at multiple oncogenic pathways may not only prove more effective than single agents alone, but may also prevent or delay secondary resistance (see Chapter 69). Therefore, it will be crucial to establish tumor banks to study the molecular phenotype of patients who are sensitive and resistant to therapy in order to identify the appropriate patients to treat. Recently, gene expression signatures from tumor biopsy specimens have been developed that can predict sensitivity to individual chemotherapeutics. Furthermore, these chemotherapy response signatures are being integrated with signatures of oncogenic pathway deregulation to potentially identify new therapeutic strategies (Bild et al., 2006). Results from a phase II clinical trial program, BATTLE (Biomarker-integrated Approaches of Targeted Therapy for Lung Cancer Elimination), suggest that patients prescribed existing drugs based on their tumor biomarkers benefit more than patients whose treatments are not based on their tumor biomarkers. This trial underscores the need to characterize cancers not only by histopathology, but also by molecular targets and pathways (reviewed by Mahapatra, 2010). The considerable potential of using gene expression profiling of tumors to define causal molecular pathways and potential therapeutic targets for individuals is summarized in a review from Nevins' group (Bild et al., 2006).

It is anticipated that during the next decade, most common forms of human cancer will be systematically surveyed to identify somatic changes in gene copy number, sequence, and expression. Next-generation sequencing can simultaneously measure nucleotide sequence and copy number, detect somatic mutations in a subset of tumor cells, and identify structural alterations including insertions, deletions, gene duplications, and rearrangements (see Chapter 10). Next-generation sequencing is now feasible, and the hope is that it will improve lung cancer diagnosis, prognosis, and therapy (reviewed by Metzker, 2010).

CONCLUSION

To date, significant advances in lung cancer treatment have come from the addition of adjuvant therapy in early-stage disease and from trials combining chemotherapy and radiation in stage IIIA disease. The mortality rate from lung cancer, however, is higher than the next three major cancers combined. A review by Shepard stressed the importance of a multi-targeted approach, focusing on prevention, early detection, and molecularly targeted therapy to optimize lung cancer diagnosis and treatment (Shepard, 2005). Tobacco use, particularly in underdeveloped countries where smoking is still on the rise, must be the first target, since prevention is the most effective way to

reduce lung cancer mortality. Early detection must be the second target, using state-of-the-art computer tomography (CT) scanning and positron emission tomography (PET) combined with molecular profiling to identify therapeutically exploitable differences between normal, precancerous, and cancer cells. These approaches must be combined with genomic and genetic biomarkers that identify current and former smokers at highest risk for developing lung cancer. High-risk former smokers may well benefit from one of the many chemopreventive medications now being tested in clinical trials. The final target must be better treatment. Already, based upon discoveries in the research lab, molecular targets have been developed, leading to agents that interfere with the EGFR pathway and those that block endothelial growth factors. Patients' clinical presentations, along with molecular characteristics of their lung cancers, will have a significant effect on response and survival.

Although they are promising, few of the molecular advances have yet reached clinical application. Better understanding of cancer heterogeneity, including variability in prognosis and patient response to therapy, requires stronger working relationships and collaborations between bench scientists and their clinical counterparts, including oncologists, pathologists, and thoracic surgeons, both in the community and in academia (see Chapter 43). Large databases will need to be constructed with standardized methods and study design for data collection and analysis. Clinical trials

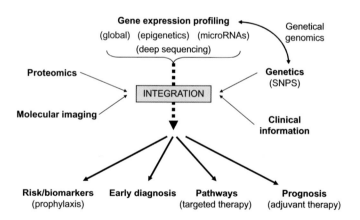

Figure 59.4 Diagram depicting contributions of gene expression profiling which, combined with genetics, clinical information, proteomics, and imaging studies, can be applied to developing (1) risk-assessment tools to identify current and former smokers at highest risk for developing lung cancer and who might benefit from chemopreventive therapy; (2) tools for early diagnosis of current and former smokers with stage I, potentially resectable lung cancer; (3) tools to define molecular pathways that have led to individual lung cancers, and to predict what pharmacological approaches might be used to define best approaches to treatment of individual lung cancers; and (4) prognosis of resected cancers, defining which patients have high probability of recurrence or metastasis, and should therefore receive adjuvant therapy.

(several are in phase II trials) need to be designed to verify results before being incorporated into routine clinical practice. Combinations of datasets from lung cancer specimens, including gene expression array data, SNP array data, and sequencing data, are being collected by some researchers so that key molecular changes can be more readily identified. Complete genome sequencing has already provided insights into the mutation spectra of a number of cancer types, including lung cancer. The contributions of modern genomic technologies, particularly those that provide measures of global gene expression, are depicted in Figure 59.4. Gene expression profiling provides a new approach to assessing risk of developing lung cancer, potential tools for early diagnosis, and new approaches to determining prognosis and oncogenic pathways that will lead to individually targeted therapies. Combined with genome-wide SNP screens that hold the promise of determining heritable predispositions to developing lung cancer and defining pathway-based pharmacogenomics, there is hope that lung cancer may eventually no longer be the number one cause of cancer death in the world.

REFERENCES

Aberle, D.R., Adams, A.M., Berg, C.D., et al., 2011. Reduced lung-cancer mortality with low-dose computed tomographic screening. N Engl J Med 365, 395–409.

Bailey-Wilson, J.E., Amos, C.I., Pinney, S.M., et al., 2004. A major lung cancer susceptibility locus maps to chromosome 6q23-25. Am J Hum Genet 75, 460–474.

Baselga, J., 2006. Targeting tyrosine kinases in cancer: The second wave. Science 312, 1175–1178.

Bauer, A.K., Malkinson, A.M., Kleeberger, S.R., 2004. Susceptibility to neoplastic and non-neoplastic pulmonary diseases in mice: Genetic similarities. Am J Physiol Lung Cell Mol Physiol 287, L685–L703.

Beane, J., Sebastiani, P., Whitfield, T.H., et al., 2008. A prediction model for lung cancer diagnosis that integrates genomic and clinical features. Cancer Prev Res (Phila) 1, 56–64.

Beane, J., Vick, J., Schembri, F., et al., 2011. Characterizing the impact of smoking and lung cancer on the airway transcriptome using RNA-Seq. Cancer Prev Res (Phila) 4, 803–817.

Beer, D.G., Kardia, S.L., Huang, C.C., et al., 2002. Gene-expression profiles predict survival of patients with lung adenocarcinoma. Nat Med 8, 816–824.

Bell, D.W., Gore, I., Okimoto, R.A., et al., 2005. Inherited susceptibility to lung cancer may be associated with the T790M drug resistance mutation in EGF. Nat Genet 37, 1315–1316.

Bhattacharjee, A., Richards, W.G., Staunton, J., et al., 2001. Classification of human lung carcinomas by mRNA expression profiling reveals distinct adenocarcinoma subclasses. Proc Natl Acad Sci USA 98, 13,790–13,795.

Bibikova, M., Lin, Z., Zhou, L., et al., 2006. High-throughput DNA methylation profiling using universal bead arrays. Genome Res 16, 383–393.

Bild, A.H., Potti, A., Nevins, J.R., 2006. Linking oncogenic pathways with therapeutic opportunities. Nat Rev Cancer 6, 735–741.

Boyle, J.O., Gumus, Z.H., Kacker, A., et al., 2010. Effects of cigarette smoke on the human oral mucosal transcriptome. Cancer Prev Res (Phila) 3, 266–278.

Calin, G.A., Croce, C.M., 2006. MicroRNA-cancer connection: The beginning of a new tale. Cancer Res 66, 7390–7394.

Carretero, J., Medina, P.P., Pio, R., Montuenga, L.M., Sanchez-Cespedes, M., 2004. Novel and natural knockout lung cancer cell lines for the LKB1/STK11 tumor suppressor gene. Oncogene 23, 4037–4040.

Carter, T.A., Wodicka, L.M., Shah, N.P., et al., 2005. Inhibition of drug-resistant mutants of ABL, KIT, and EGF receptor kinases. Proc Natl Acad Sci USA 102, 11,011–11,016.

Choi, Y.L., Soda, M., Yamashita, Y., et al., 2010. EML4-ALK mutations in lung cancer that confer resistance to ALK inhibitors. N Engl J Med 363, 1734–1739.

Coldren, C.D., Helfrich, B.A., Witta, S.E., et al., 2006. Baseline gene expression predicts sensitivity to gefitinib in non-small-cell lung cancer cell lines. Mol Cancer Res 4, 521–528.

Cooper, D.N., 2005. The Molecular Genetics of Lung Cancer. Springer, Heidelberg, Germany.

David-Beabes, G.L., London, S.J., 2001. Genetic polymorphism of XRCC1 and lung cancer risk among African-Americans and Caucasians. Lung Cancer 34, 333–339.

Detterbeck, F.C., Boffa, D.J., Tanoue, L.T., 2009. The new lung cancer staging system. Chest 136, 260–271.

Diederichs, S., Baumer, N., Ji, P., et al., 2004. Identification of interaction partners and substrates of the cyclin A1-CDK2 complex. J Biol Chem 279, 33,727–33,741.

Douillard, J.Y., Shepherd, F.A., Hirsh, V., et al., 2008. Molecular predictors of outcome with gefitinib and docetaxel in previously treated non-small-cell lung cancer: Data from the randomized phase III INTEREST trial. J Clin Oncol 28, 744–752.

Dutt, A., Wong, K.K., 2006. Mouse models of lung cancer. Clin Cancer Res 12, 4396s–4402s.

Dziadziuszko, R., Hirsch, F.R., Varella-Garcia, M., Bunn Jr., P.A., 2006. Selecting lung cancer patients for treatment with epidermal growth factor receptor tyrosine kinase inhibitors by immunohistochemistry and fluorescence in situ hybridization – why, when, and how? Clin Cancer Res 12 (14 Pt 2), 4409s–4415s.

Engelman, J.A., Cantley, L.C., 2006. The role of the ErbB family members in non-small-cell lung cancers sensitive to epidermal growth factor receptor kinase inhibitors. Clin Cancer Res 12, 4372s–4376s.

Engelman, J.A., Mukohara, T., Zejnullahu, K., et al., 2006. Allelic dilution obscures detection of a biologically significant resistance mutation in EGFR-amplified lung cancer. J Clin Invest 116 (10), 2695–2706.

Esquela-Kerscher, A., Slack, F.J., 2006. Oncomirs – microRNAs with a role in cancer. Nat Rev Cancer 6, 259–269.

Fisher, G.H., Wellen, S.L., Klimstra, D., et al., 2001. Induction and apoptotic regression of lung adenocarcinomas by regulation of a K-Ras transgene in the presence and absence of tumor suppressor genes. Genes Dev 15, 3249–3262.

Fong, K.M., Sekido, Y., Gazdar, A.F., Minna, J.D., 2003. Lung cancer, 9. Molecular biology of lung cancer: Clinical implications. Thorax 58, 892–900.

Garber, M.E., Troyanskaya, O.G., Schluens, K., et al., 2001. Diversity of gene expression in adenocarcinoma of the lung. Proc Natl Acad Sci USA 98, 13,784–13,789.

Golub, T.R., Slonim, D.K., Tamayo, P., et al., 1999. Molecular classification of cancer: class discovery and class prediction by gene expression monitoring. Science 286 (5439), 531–537.

Granville, C.A., Dennis, P.A., 2005. An overview of lung cancer genomics and proteomics. Am J Respir Cell Mol Biol 32, 169–176.

Graziano, S.L., Gamble, G.P., Newman, N.B., et al., 1999. Prognostic significance of K-ras codon 12 mutations in patients with resected stage I and II non-small-cell lung cancer. J Clin Oncol 17, 668–675.

Guo, N.L., Wan, Y.W., Tosun, K., et al., 2008. Confirmation of gene expression-based prediction of survival in non-small cell lung cancer. Clin Cancer Res 14, 8213–8220.

Gustafson, A.M., Soldi, R., Anderlind, C., et al., 2010. Airway PI3K pathway activation is an early and reversible event in lung cancer development. Sci Transl Med 2 26ra25.

Haiman, C.A., Stram, D.O., Wilkens, L.R., et al., 2006. Ethnic and racial differences in the smoking-related risk of lung cancer. N Engl J Med 354, 333–342.

Hayes, D.N., Monti, S., Parmigiani, G., et al., 2006. Gene expression profiling reveals reproducible human lung adenocarcinoma subtypes in multiple independent patient cohorts. J Clin Oncol 24, 5079–5090.

Henschke, C.I., Yankelevitz, D.F., Libby, D.M., Pasmantier, M.W., Smith, J.P., Miettinen, O.S., 2006. Survival of patients with stage I lung cancer detected on CT screening. N Engl J Med 355, 1763–1771.

Hirsch, F.R., Lippman, S.M., 2005. Advances in the biology of lung cancer chemoprevention. J Clin Oncol 23, 3186–3197.

Hirsch, F.R., Varella-Garcia, M., Cappuzzo, F., 2009. Predictive value of EGFR and HER2 overexpression in advanced non-small-cell lung cancer. Oncogene 28 (Suppl 1), S32–S37.

Hung, R.J., McKay, J.D., Gaborieau, V., et al., 2008. A susceptibility locus for lung cancer maps to nicotinic acetylcholine receptor subunit genes on 15q25. Nature 452, 633–637.

Infante, M., Cavuto, S., Lutman, F.R., et al., 2009. A randomized study of lung cancer screening with spiral computed tomography: Three-year results from the DANTE trial. Am J Respir Crit Care Med 180, 445–453.

Italiano, A., Vandenbos, F.B., Otto, J., et al., 2006. Comparison of the epidermal growth factor receptor gene and protein in primary non-small-cell-lung cancer and metastatic sites: Implications for treatment with EGFR-inhibitors. Ann Oncol 17, 981–985.

Janne, P.A., Engelman, J.A., Johnson, B.E., 2005. Epidermal growth factor receptor mutations in non-small-cell lung cancer: Implications for treatment and tumor biology. J Clin Oncol 23, 3227–3234.

Jemal, A., Siegel, R., Xu, J., Ward, E., 2010. Cancer statistics, 2010. CA Cancer J Clin 60, 277–300.

Jett, J.R., Midthun, D.E., 2004. Screening for lung cancer: Current status and future directions. Thomas A. Neff lecture. Chest 125, 158S–162S.

Johnson, D.H., 2006. Targeted therapies in combination with chemotherapy in non-small cell lung cancer. Clin Cancer Res 12, 4451s–4457s.

Johnson, S.M., Grosshans, H., Shingara, J., et al., 2005. RAS is regulated by the let-7 microRNA family. Cell 120, 635–647.

Kelloff, G.J., Lippman, S.M., Dannenberg, A.J., et al., 2006. Progress in chemoprevention drug development: The promise of molecular biomarkers for prevention of intraepithelial neoplasia and cancer – a plan to move forward. Clin Cancer Res 12, 3661–3697.

Kerr, K.M., 2001. Pulmonary preinvasive neoplasia. J Clin Pathol 54, 257–271.

Kersting, M., Friedl, C., Kraus, A., Behn, M., Pankow, W., Schuermann, M., 2000. Differential frequencies of p16(INK4a) promoter hypermethylation, p53 mutation, and K-ras mutation in exfoliative material mark the development of lung cancer in symptomatic chronic smokers. J Clin Oncol 18, 3221–3229.

Khuri, F.R., Cohen, V., 2004. Molecularly targeted approaches to the chemoprevention of lung cancer. Clin Cancer Res 10, 4249s–4253s.

Kikuchi, T., Daigo, Y., Katagiri, T., et al., 2003. Expression profiles of non-small-cell lung cancers on cDNA microarrays: Identification of genes for prediction of lymph-node metastasis and sensitivity to anti-cancer drugs. Oncogene 22, 2192–2205.

Kobayashi, S., Boggon, T.J., Dayaram, T., et al., 2005. EGFR mutation and resistance of non-small-cell lung cancer to gefitinib. N Engl J Med 352, 786–792.

Kopper, L., Timar, J., 2005. Genomics of lung cancer may change diagnosis, prognosis and therapy. Pathol Oncol Res 11, 5–10.

Kwak, E.L., Bang, Y.J., Camidge, D.R., et al., 2010. Anaplastic lymphoma kinase inhibition in non-small-cell lung cancer. N Engl J Med 363, 1693–1703.

Lam, S., McWilliams, A., LeRiche, J., MacAulay, C., Wattenberg, L., Szabo, E., 2006. A phase I study of myo-inositol for lung cancer chemoprevention. Cancer Epidemiol Biomarkers Prev 15, 1526–1531.

Le Marchand, L., Guo, C., Benhamou, S., et al., 2003. Pooled analysis of the CYP1A1 exon 7 polymorphism and lung cancer (United States). Cancer Causes Control 14, 339–346.

Li, J., Burmeister, M., 2005. Genetical genomics: Combining genetics with gene expression analysis. Hum Mol Genet 14 (Spec No. 2), R163–R169.

Lopes Pegna, A., Picozzi, G., Mascalchi, M., et al., 2009. Design, recruitment and baseline results of the ITALUNG trial for lung cancer screening with low-dose CT. Lung Cancer 64, 34–40.

Luo, R.X., Wu, B., Yi, Y.N., Huang, Z.W., Lin, R.T., 1996. Indoor-burning coal air pollution and lung cancer – a case-control study in Fuzhou, China. Lung Cancer 14, S113–S119.

Mahapatra, A., 2010. Lung cancer – genomics and personalized medicine. ACS Chem Biol 5, 529–531.

Marchetti, A., Martella, C., Felicioni, L., et al., 2005. EGFR mutations in non–small-cell lung cancer: analysis of a large series of cases and development of a rapid and sensitive method for diagnostic screening with potential implications on pharmacologic treatment. J Clin Oncol 23 (4), 857–865.

Matakidou, A., Eisen, T., Houlston, R.S., 2005. Systematic review of the relationship between family history and lung cancer risk. Br J Cancer 93, 825–833.

Metzker, M.L., 2010. Sequencing technologies – the next generation. Nat Rev Genet 11, 31–46.

Mok, T., Wu, Y.L., Zhang, L., 2009. A small step towards personalized medicine for non-small-cell lung cancer. Discov Med 8, 227–231.

Navarro, A., Marrades, R.M., Vinolas, N., et al., 2009. MicroRNAs expressed during lung cancer development are expressed in human pseudoglandular lung embryogenesis. Oncology 76, 162–169.

Olaussen, K.A., Dunant, A., Fouret, P., et al., 2006. DNA repair by ERCC1 in non-small-cell lung cancer and cisplatin-based adjuvant chemotherapy. N Engl J Med 355, 983–991.

Palmisano, W.A., Divine, K.K., Saccomanno, G., et al., 2000. Predicting lung cancer by detecting aberrant promoter methylation in sputum. Cancer Res 60, 5954–5958.

Park, J.Y., Park, J.M., Jang, J.S., et al., 2006. Caspase 9 promoter polymorphisms and risk of primary lung cancer. Hum Mol Genet 15, 1963–1971.

Parmigiani, G., Garrett, E., Anbazhagan, R., Gabrielson, E., 2004. Molecular classification of lung cancer: A cross-platform comparison of gene expression data sets. Chest 125, 103S.

Pastinen, T., Ge, B., Hudson, T.J., 2006. Influence of human genome polymorphism on gene expression. Hum Mol Genet 15 (Spec No 1), R9–R16.

Perera, F.P., Tang, D., Brandt-Rauf, P., et al., 2006. Lack of associations among cancer and albumin adducts, ras p21 oncoprotein levels, and CYP1A1, CYP2D6, NAT1, and NAT2 in a nested case-control study of lung cancer within the physicians' health study. Cancer Epidemiol Biomarkers Prev 15, 1417–1419.

Perez-Soler, R., 2009. EGFR TKIs for advanced NSCLC: Practical questions. Oncology (Williston Park) 24 (400-401), 409.

Petty, R.D., Kerr, K.M., Murray, G.I., et al., 2006. Tumor transcriptome reveals the predictive and prognostic impact of lysosomal protease inhibitors in non-small-cell lung cancer. J Clin Oncol 24, 1729–1744.

Powell, C.A., Spira, A., Derti, A., et al., 2003. Gene expression in lung adenocarcinomas of smokers and nonsmokers. Am J Respir Cell Mol Biol 29, 157–162.

Rahman, S.M., Gonzalez, A.L., Li, M., et al., 2011. Lung cancer diagnosis from proteomic analysis of preinvasive lesions. Cancer Res 71, 3009–3017.

Rahman, S.M., Shyr, Y., Yildiz, P.B., et al., 2005. Proteomic patterns of preinvasive bronchial lesions. Am J Respir Crit Care Med 172, 1556–1562.

Raponi, M., Zhang, Y., Yu, J., et al., 2006. Gene expression signatures for predicting prognosis of squamous cell and adenocarcinomas of the lung. Cancer Res 66, 7466–7472.

Raunio, H., Hakkola, J., Hukkanen, J., et al., 1999. Expression of xenobiotic-metabolizing CYPs in human pulmonary tissue. Exp Toxicol Pathol 51, 412–417.

Reissmann, P.T., Koga, H., Figlin, R.A., Holmes, E.C., Slamon, D.J., 1999. Amplification and overexpression of the cyclin D1 and epidermal growth factor receptor genes in non-small-cell lung cancer. Lung Cancer Study Group. J Cancer Res Clin Oncol 125, 61–70.

Richardson, G.E., Johnson, B.E., 1993. The biology of lung cancer. Semin Oncol 20, 105–127.

Risch, A., Plass, C., 2008. Lung cancer epigenetics and genetics. Int J Cancer 123, 1–7.

Santos, E.S., Blaya, M., Raez, L.E., 2009. Gene expression profiling and non-small-cell lung cancer: Where are we now? Clin Lung Cancer 10, 168–173.

Scagliotti, G.V., Longo, M., Novello, S., 2009. Non-small-cell lung cancer in never smokers. Curr Opin Oncol 21, 99–104.

Schena, M., Shalon, D., Davis, R.W., Brown, P.O., 1995. Quantitative monitoring of gene expression patterns with a complementary DNA microarray. Science 270, 467–470.

Sequist, L.V., Joshi, V.A., Janne, P.A., et al., 2006. Epidermal growth factor receptor mutation testing in the care of lung cancer patients. Clin Cancer Res 12, 4403s–4408s.

Sharma, S.V., Fischbach, M.A., Haber, D.A., Settleman, J., 2006. "Oncogenic shock": Explaining oncogene addiction through differential signal attenuation. Clin Cancer Res 12, 4392s–4395s.

Shaw, A.T., Yeap, B.Y., Mino-Kenudson, M., et al., 2009. Clinical features and outcome of patients with non-small-cell lung cancer who harbor EML4-ALK. J Clin Oncol 27, 4247–4253.

Shedden, K., Taylor, J.M., Enkemann, S.A., et al., 2008. Gene expression-based survival prediction in lung adenocarcinoma: A multisite, blinded validation study. Nat Med 14, 822–827.

Shepherd, F.A., 2005. A targeted approach to reducing lung cancer mortality. J Clin Oncol 23, 3173–3174.

Shigematsu, H., Lin, L., Takahashi, T., et al., 2005. Clinical and biological features associated with epidermal growth factor receptor gene mutations in lung cancers. J Natl Cancer Inst 97 (5), 339–346.

Siegel, R., Naishadham, D., Jemal, A., 2012. Cancer statistics. CA: Cancer J. Clin. 62, 10–29.

Singhal, S., Vachani, A., Antin-Ozerkis, D., Kaiser, L.R., Albelda, S.M., 2005. Prognostic implications of cell cycle, apoptosis, and angiogenesis biomarkers in non-small cell lung cancer: A review. Clin Cancer Res 11, 3974–3986.

Slebos, R.J., Kibbelaar, R.E., Dalesio, O., et al., 1990. K-ras oncogene activation as a prognostic marker in adenocarcinoma of the lung. N Engl J Med 323, 561–565.

Soda, M., Choi, Y.L., Enomoto, M., et al., 2007. Identification of the transforming EML4-ALK fusion gene in non-small-cell lung cancer. Nature 448, 561–566.

Sos, M.L., Michel, K., Zander, T., et al., 2009. Predicting drug susceptibility of non-small cell lung cancers based on genetic lesions. J Clin Invest 119, 1727–1740.

Spira, A., Beane, J., Shah, V., et al., 2004. Effects of cigarette smoke on the human airway epithelial cell transcriptome. Proc Natl Acad Sci USA 101, 10,143–10,148.

Spira, A., Beane, J.E., Shah, V., et al., 2007. Airway epithelial gene expression in the diagnostic evaluation of smokers with suspect lung cancer. Nat Med 13, 361–366.

Spira, A., Ettinger, D.S., 2004. Multidisciplinary management of lung cancer. N Engl J Med 350, 379–392.

Spiro, S.G., Silvestri, G.A., 2005. One hundred years of lung cancer. Am J Respir Crit Care Med 172, 523–529.

Sridhar, S., Schembri, F., Zeskind, J., et al., 2008. Smoking-induced gene expression changes in the bronchial airway are reflected in nasal and buccal epithelium. BMC Genomics 9, 259.

Steiling, K., Ryan, J., Brody, J.S., Spira, A., 2008. The field of tissue injury in the lung and airway. Cancer Prev Res (Phila) 1, 396–403.

Sy, S.M., Fan, B., Lee, T.W., et al., 2004. Spectral karyotyping indicates complex rearrangements in lung adenocarcinoma of nonsmokers. Cancer Genet Cytogenet 153, 57–59.

Taioli, E., Gaspari, L., Benhamou, S., et al., 2003. Polymorphisms in CYP1A1, GSTM1, GSTT1 and lung cancer below the age of 45 years. Int J Epidemiol 32, 60–63.

Takahashi, T., Sonobe, M., Kobayashi, M., et al., 2010. Clinicopathologic features of non-small-cell lung cancer with EML4-ALK fusion gene. Ann Surg Oncol 17, 889–897.

Takamizawa, J., Konishi, H., Yanagisawa, K., et al., 2004. Reduced expression of the let-7 microRNAs in human lung cancers in association with shortened postoperative survival. Cancer Res 64, 3753–3756.

Takeuchi, T., Tomida, S., Yatabe, Y., et al., 2006. Expression profile-defined classification of lung adenocarcinoma shows close relationship with underlying major genetic changes and clinicopathologic behaviors. J Clin Oncol 24, 1679–1688.

Thatcher, N., Chang, A., Parikh, P., et al., 2005. Gefitinib plus best supportive care in previously treated patients with refractory advanced non-small-cell lung cancer: Results from a randomised, placebo-controlled, multicentre study (Iressa Survival Evaluation in Lung Cancer). Lancet 366, 1527–1537.

Thomas, R.K., Weir, B., Meyerson, M., 2006. Genomic approaches to lung cancer. Clin Cancer Res 12, 4384s–4391s.

Thun, M.J., Henley, S.J., Calle, E.E., 2002. Tobacco use and cancer: An epidemiologic perspective for geneticists. Oncogene 21, 7307–7325.

Tomida, S., Koshikawa, K., Yatabe, Y., et al., 2004. Gene expression-based, individualized outcome prediction for surgically treated lung cancer patients. Oncogene 23, 5360–5370.

Toyooka, S., Tsuda, T., Gazdar, A.F., 2003. The TP53 gene, tobacco exposure, and lung cancer. Hum Mutat 21, 229–239.

Truong, T., Sauter, W., McKay, J.D., et al., 2010. International Lung Cancer Consortium: Coordinated association study of 10 potential lung cancer susceptibility variants. Carcinogenesis 31, 625–633.

Varmus, H., 2006. The new era in cancer research. Science 312, 1162–1165.

Vineis, P., Veglia, F., Anttila, S., et al., 2004. CYP1A1, GSTM1 and GSTT1 polymorphisms and lung cancer: A pooled analysis of gene–gene interactions. Biomarkers 9, 298–305.

Weinstein, I.B., 2002. Cancer. Addiction to oncogenes – the Achilles heal of cancer. Science 297, 63–64.

Wenzlaff, A.S., Cote, M.L., Bock, C.H., et al., 2005. CYP1A1 and CYP1B1 polymorphisms and risk of lung cancer among never smokers: A population-based study. Carcinogenesis 26, 2207–2212.

Wikman, H., Kettunen, E., Seppanen, J.K., et al., 2002. Identification of differentially expressed genes in pulmonary adenocarcinoma by using cDNA array. Oncogene 21, 5804–5813.

Wikman, H., Thiel, S., Jager, B., et al., 2001. Relevance of N-acetyltransferase 1 and 2 (NAT1, NAT2) genetic polymorphisms in non-small-cell lung cancer susceptibility. Pharmacogenetics 11, 157–168.

Wilson, I.M., Davies, J.J., Weber, M., et al., 2006. Epigenomics: Mapping the methylome. Cell Cycle 5, 155–158.

Wodicka, L., Dong, H.L., Mittmann, M., Ho, M.H., Lockhart, D.J., 1997. Genome wide expression monitoring in saccharomyces cerevisiae. Nat. Biotechnol. 15 (13), 1359–1367.

Yamagata, N., Shyr, Y., Yanagisawa, K., et al., 2003. A training-testing approach to the molecular classification of resected non-small cell lung cancer. Clin Cancer Res 9, 4695–4704.

Yanaihara, N., Caplen, N., Bowman, E., et al., 2006. Unique microRNA molecular profiles in lung cancer diagnosis and prognosis. Cancer Cell 9, 189–198.

Yin, Z., Su, M., Li, X., et al., 2009. ERCC2, ERCC1 polymorphisms and haplotypes, cooking oil fume and lung adenocarcinoma risk in Chinese non-smoking females. J Exp Clin Cancer Res 28, 153.

Yu, S.L., Chen, H.Y., Chang, G.C., et al., 2008. MicroRNA signature predicts survival and relapse in lung cancer. Cancer Cell 13, 48–57.

Zhou, B.S., Wang, T.J., Guan, P., Wu, J.M., 2000. Indoor air pollution and pulmonary adenocarcinoma among females: A case-control study in Shenyang, China. Oncol Rep 7, 1253–1259.

Zochbauer-Muller, S., Fong, K.M., Maitra, A., et al., 2001. 5′ CpG island methylation of the FHIT gene is correlated with loss of gene expression in lung and breast cancer. Cancer Res 61, 3581–3585.

CHAPTER

Breast Cancer

Philip S. Bernard

INTRODUCTION

Breast cancer is the most prevalent cancer in the world (~4.4 million) with a high yearly incidence (~1.15 million newly diagnosed) and relatively lower death rate (~465,000) (Parkin et al., 2005). Predicting who will develop breast cancer and who will survive breast cancer continues to be challenging for modern medicine. Even in an era in which we know the sequence of all genes (coding and non-coding), and can decipher many of the protein interactions and regulatory mechanisms, there are many unknown factors that can lead to the development, resistance, and progression of the disease.

RISK FACTORS

The average American woman has an approximately one in eight chance of developing breast cancer in her life. A woman's relative risk of developing breast cancer is at least doubled for those with a major risk factor, such as: (1) having a germline mutation in *BRCA1/BRCA2*, (2) having a previous diagnosis of atypical hyperplasia or pre-invasive breast cancer, (3) having received chest radiation before age 30, or (4) having a first-degree relative diagnosed with breast cancer before age 60 (Schwartz et al., 2008). Genetic testing for hereditary cancer predisposition syndromes has become integrated into the practice of medical oncology and is leading to better strategies of surveillance and prevention (Garber and Offit, 2005; Robson et al., 2010). Most of the high risk familial breast cancer genes have already been discovered and are inherited as autosomal dominant mutations, including the hereditary breast cancer syndromes (*BRCA1/BRCA2*), Li-Fraumeni syndrome (*TP53*), Cowden syndrome (*PTEN*), and Peutz-Jeghers syndrome (*STK11*). In addition, concurrent mutations in other genes can provide additional risk (i.e., modifier genes). For example, *RAD51* can modify the risk of breast cancer in *BRCA2* carriers but has little effect on risk in non-carriers (Antoniou et al., 2007). While modest and intermediate (e.g., *CHEK2* or *ATM*) risk genes and modifier genes are known to exist, they are not routinely tested because there is no actionable clinical plan if detected. The American Society of Clinical Oncology has set forth guidelines for genetic and genomic testing for cancer susceptibility (Robson et al., 2010). These guidelines recommend that any variants/mutations detected must be adequately interpreted and result in a change of medical/surgical management. Depending on the level of risk, intervention may range from increased vigilance [e.g., semi-annual breast exams or magnetic resonance imaging (MRI)], to non-surgical therapy (e.g., chemoprevention with tamoxifen), to surgical intervention (e.g., bilateral mastectomy).

PROGNOSTIC FACTORS

The meaning of risk is different between predisposition to disease and the actual diagnosis of breast cancer. After a diagnosis of invasive breast cancer, the most important prognostic indicators routinely assessed are anatomic staging

Genomic and Personalized Medicine, 2nd edition
by Ginsburg & Willard. DOI: http://dx.doi.org/10.1016/B978-0-12-382227-7.00060-4

(tumor size, node status, and metastasis), histological grade, and molecular status of estrogen receptor (ER), progesterone receptor (PR), and *HER2*. Depending on stage and biomarker status, there is then more specialized testing that can be performed. Many breast cancer signatures have been validated for prognosis in early stage disease, including the 21-gene recurrence score (Onco*type* Dx®; Paik et al., 2004), the Netherland Cancer Institute 70-gene signature (MammaPrint®; Van de Vijver et al., 2002), the 14-gene distant metastasis signature (BreastOncPx™; Tutt et al., 2008), the PAM50 "intrinsic" subtyping test (Parker et al., 2009), the 76-gene Rotterdam predictor (Desmedt et al., 2007; Wang et al., 2005), and the index of sensitivity to endocrine therapy (SET index; Symmans et al., 2010). All these signatures perform similarly for prognosis in node-negative, ER+ disease, and could be used to identify high-risk patients requiring the addition of chemotherapy to endocrine blockade therapy. Alternatively, none can satisfactorily identify low-risk node-positive, ER+ patients, as defined by having less than 10% recurrence at 10 years given only adjuvant endocrine blockade therapy. Although the prognostic signatures for early

stage, ER+ breast cancer appear to be tracking similar biology, there is remarkably little overlap in the gene sets (Fan et al., 2006). A comparison of the gene overlap for four commercially available clinical assays shows that only 18 out of a total of 118 genes are used in two or more tests, and none are in common across all tests (Table 60.1). The lack of overlap in genomic signatures, even when used for the same clinical purpose, is due to several factors including: (1) redundancy in biology (i.e., genes can serve as surrogates for other genes), (2) different patient populations and technology platforms used for signature development, and (3) different methods of bioinformatics.

Genes involved in cell cycle regulation (i.e., proliferation) are repeatedly identified as important prognostic biomarkers for determining relapse-free survival in early stage, ER+ breast cancer (Dai et al., 2005; Nielsen et al., 2010; Perreard et al., 2006; Sotiriou et al., 2006; Tutt et al., 2008). Studies have shown that incorporating Ki67 immunohistochemistry (IHC) staining can improve outcome predictions (Cheang et al., 2009; Dowsett et al., 2007; Ring et al., 2006). Tumors that have a decrease in the Ki-67 index after two weeks of

TABLE 60.1	Genes in common across multiple prognostic breast cancer signatures				
Gene symbol	Gene name	PAM50	Onco*type* Dx	BreastOncPx	MammaPrint
BAG1	BCL2-associated athanogene	×	×	–	–
BCL2	B-cell CLL/lymphoma 2	×	×	–	–
BIRC5	Baculoviral IAP repeat containing 5	×	×	–	–
C16orf61	Chromosome 16 open reading frame 61	–	–	×	×
CCNB1	Cyclin B1	×	×	×	–
CENPA	Centromere protein A	–	–	×	×
DIAPH3	Diaphanous homolog 3	–	–	×	×
ERBB2	v-erb-b2 erythroblastic leukemia viral oncogene homolog 2	×	×	–	–
ESR1	Estrogen receptor 1	×	×	–	–
GRB7	Growth factor receptor-bound protein 7	×	×	–	–
MELK	Maternal embryonic leucine zipper kinase	×	–	×	×
MKI67	Antigen identified by monoclonal antibody Ki-67	×	×	–	–
MMP11	Matrix metallopeptidase 11	×	×	–	–
MYBL2	v-myb myeloblastosis viral oncogene homolog-like 2	×	×	×	–
ORC6	Origin recognition complex, subunit 6	×	–	×	×
PGR	Progesterone receptor	×	×	–	–
RFC4	Replication factor C (activator 1) 4, 37 kDa	–	–	×	×
SCUBE2	Signal peptide, CUB domain, EGF-like 2	–	×	–	×

endocrine blockade therapy in the neo-adjuvant setting have a lower risk of relapse (Dowsett et al., 2007). Using more quantitative methods of analysis than IHC [e.g., quantitative real-time polymerase chain reaction (RT-qPCR)] and using a panel of proliferation markers rather than just Ki67, provide additional prognostic information (Nielsen et al., 2010). As shown in Figure 60.1, the PAM50 and Onco*type* Dx® have considerable overlap in proliferation genes.

Onco*type* Dx uses a recurrence score (RS) to convey risk of relapse in tamoxifen-treated ER+ breast cancer patients (Paik et al., 2004, 2006). Conversely, the PAM50 gives a categorical subtype call of Luminal A, Luminal B, HER2-enriched or Basal-like (Nielsen et al., 2010; Parker et al., 2009). Patients with Luminal A tumors are ER+ and have the most favorable prognosis in early-stage disease without systemic treatment. Luminal B tumors are also ER+, but patients have a worse prognosis than Luminal A cancers. These more aggressive ER+ tumors can have genomic instability and harbor mutations in *TP53* (Bergamaschi et al., 2006). Luminal B tumors are sometimes clinically *HER2+* and often endocrine blockade resistant (Nielsen et al., 2010). A risk of relapse (ROR) score has been developed for the PAM50 from a Cox proportional hazards model that uses the distance of a new sample to each subtype centroid and considers the prognosis associated with each subtype (Parker et al., 2009). The ROR provides a "genomic summary" that accounts for the continuous nature of breast cancer biology. ER+ tumors that classify in the center of the Luminal A centroid have the lowest risk, and risk increases in tumors classified more towards the Luminal B centroid. Basal-like tumors have low expression of ER and ER-regulated genes; however, as many as 30% of HER2-enriched tumors may be clinically called ER+ (Nielsen et al., 2010; Parker et al., 2009). The genomic subtype classification of HER2-enriched has proven to be more accurate than clinical ER status, since ER+/ HER2-enriched tumors have poor prognosis with endocrine blockade therapy alone (Nielsen et al., 2010).

PREDICTING THERAPEUTIC RESPONSE
Endocrine Blockade Therapy and Resistance

Resistance to endocrine blockade may be due to the genetic make-up of the tumor (i.e., HER2-enriched or basal-like subtype) or the physiology of the patient. Estrogen causes increased growth and proliferation of ER+ tumors. The most widely prescribed anti-estrogen is tamoxifen. In order for tamoxifen to be effective it must be metabolized in the liver by the cytochrome P450 system (CYP) (Desta et al., 2004).

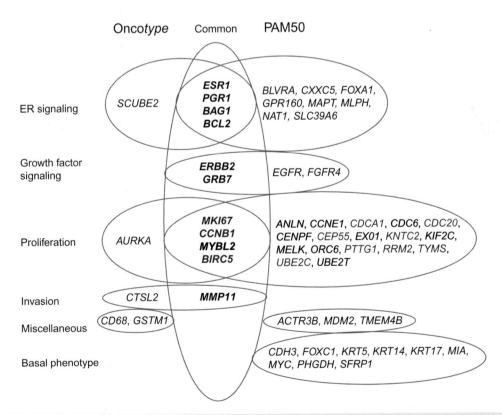

Figure 60.1   Shared genes between Onco*type* Dx and PAM50. Genes in bold (center) are the same across the tests. Genes in blue are used together (meta-genes) in the PAM50 assay to generate a "proliferation" score (cell cycle regulation). Since the PAM50 is also relevant for estrogen receptor-negative disease, there are additional genes important for characterizing these tumors, specifically biomarkers expressed in HER2-enriched and basal-like tumors.

This reaction is best catalyzed by the enzyme CYP2D6 to form endoxifen, the most active metabolite of tamoxifen for ER inhibition (Wu et al., 2009). There are at least 75 known variants in *CYP2D6* that determine the efficacy of tamoxifen treatment (Ingelman-Sundberg, 2005). Wild-type *CYP2D6* alleles (1*, 2*, 33*, 35*) are considered to have normal activity, while other alleles have decreased or no function, and duplication of some wild-type alleles leads to increased function (Gaedigk et al., 2007, 2008; Ingelman-Sundberg, 2005). Although some reports suggest that patients with inherited non-functional alleles will be resistant to tamoxifen therapy (Ingelman-Sundberg, 2005), this topic remains heavily debated (Lash et al., 2009). The menopausal status of women with breast cancer should also be considered. Tamoxifen is most often prescribed to pre-menopausal women, and so the *CYP2D6* genotype becomes a larger issue in this group of patients. In post-menopausal women, blockade of extragonadal estrogens by an aromatase inhibitor has a safe toxicity profile, and may be more effective than tamoxifen (Thurlimann et al., 2005).

Therapy in High-Risk Disease

Multivariable analyses show that both genomics and standard clinical-pathologic information are independent predictors of survival (Prat and Perou, 2011). Patients considered high risk by either genomics and/or current clinical-pathologic standards (ER−, and/or >2 cm tumor, and/or node positive) will benefit most from personalized treatment plans. These may be in directed biologic therapies, such as the use of herceptin that targets HER2-expressing tumors (Slamon et al., 1987, 2001), and/or tailored chemotherapy regimens. For instance, anthracyclines are known to improve outcomes in high-risk patients (Ejlertsen et al., 2007; Gluck, 2005; Levine et al., 2005), but have significant toxicities including neutropenia and cardiac toxicity. Meta-analyses of multiple studies have revealed that most of the additional benefit occurs in women with HER2+ tumors (Gennari et al., 2008). Unfortunately, cardiac toxicities are compounded when anthracyclines are used in combination with herceptin therapy for HER2+ breast cancer (Slamon and Pegram, 2001; Slamon et al., 2001). A promising option for HER2+ breast cancer patients is replacing anthracyclines with platinum and taxane-based chemotherapies (Coudert et al., 2007; Hayes et al., 2007). A trend away from anthracycline therapy for all ER− breast cancer is supported by a population-based study in British Columbia finding that triple-negative disease does not benefit from anthracycline-containing regimens (Cheang et al., 2008).

Basal-like and triple-negative (ER−, PR−, and HER2−) tumors are often used synonymously, however, basal-like is a genomic definition (Hu et al., 2006; Perou et al., 2000; Sorlie et al., 2001, 2003) and is not well characterized using IHC panels (Parker et al., 2009). Although there is an overlap in genes expressed in basal-like and triple-negative breast cancers (e.g., *EGFR* and *KRTNs* 5, 14, 17) (Nielsen et al., 2004), an accurate diagnosis requires many more biomarkers and more quantitative analyses. Parker and colleagues found that only 65% of triple-negative tumors are classified as basal-like by the PAM50 (Parker et al., 2009). Furthermore, tumors that are called triple-negative by IHC and are not basal-like receive a much lower complete pathologic response to neo-adjuvant chemotherapy than basal-like tumors that are not triple negative, suggesting there are significant biologic differences between these tumors that are important for treatment decisions (Parker et al., 2009). Basal-like breast cancers occur more frequently in patients of African descent, and may have inherited mutations in the *BRCA1* gene (Carey et al., 2006; Comen et al., 2011; Sorlie et al., 2003). Both Basal-like tumors and tumors with *BRCA1* mutations commonly have defective DNA repair mechanisms, which may provide higher sensitivity to platinum-based chemotherapy (Koshy et al., 2010; Silver et al., 2010) and inhibitors of poly(ADP)-ribose polymerases (PARPs; ADP = adenosine diphosphate) (Comen and Robson, 2010). Figure 60.2 outlines a diagnostic and therapeutic strategy integrating several types of data for making treatment decisions.

DISCOVERY OF NEW BIOMARKERS

The state of the art for breast cancer medicine incorporates traditional and new methods of risk prediction. The discovery of non-coding RNAs (ncRNAs) and microRNAs (miRNAs), in particular, provide promising new analytes for cancer testing (Mattick and Makunin, 2006). Patterns of miRNA expression regulate many cancer processes (e.g., proliferation, angiogenesis, metastasis, etc.) and correlate with histological diagnosis, stage, and other clinical variables (Lee and Dutta, 2008; Lu et al., 2005). In breast cancer, miRNAs have been shown to have differential expression based on tumor subtype (Blenkiron et al., 2007). Perhaps the greatest potential for using miRNAs in breast cancer medicine will be for monitoring or early detection of the disease from blood (Chen et al., 2008). The field of cancer genomics is advancing at a rapid pace with next-generation sequencing, which provides hundreds of millions of sequence reads per sample and can detect single nucleotide substitutions, small insertions/deletions, large structural changes, and copy number variances (Ding et al., 2010b). The ability to identify molecular markers of metastatic potential and chemoresistance in tumors prior to treatment will allow physicians to develop personalized treatment plans up front, and tailor diagnostics and therapies during the course of disease. Resequencing of matching primary and metastatic breast cancers from the same patient is revealing the evolution of the disease (Ding et al., 2010a; Shah et al., 2009). For a given individual, the mutation spectrum can vary between the primary and metastatic cancers and may depend on the biology of the tumor, treatment decisions, and time to recurrence. For instance, it was found that there were very few differences in the somatic mutation profile between a basal-like cancer that was deep sequenced

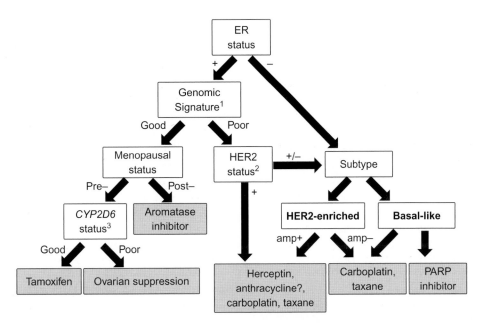

Figure 60.2 Diagnostic and treatment strategy for early stage breast cancer treatment. Treatment options in gray boxes and genomic subtypes in bold. [1]Several options for assessing risk by genomics in early stage, ER+ disease: PAM50 (Parker et al., 2009), MammaPrint® (Van de Vijver et al., 2002), BreastOncPx™ (Tutt et al., 2008), and Oncotype Dx® (Paik et al., 2004). [2]Clinical HER2 status assessed by immunohistochemistry and/or FISH/CISH. Equivocal cases referred to quantitative gene expression methods. [3]Genotypes resistant to tamoxifen would seek alternative methods of ovarian suppression.

from the primary tumor, the tumor graft in the mouse, and the brain metastasis (Ding et al., 2010a). Interestingly, the engraftment of the patient's tumor in the mouse showed a similar clonal enrichment pattern to the metastasis, suggesting this is a suitable pre-clinical model for studying the disease *in vivo*. In contrast, the deep sequencing analyses of a lobular cancer, which are commonly characterized as indolent ER+ tumors, showed that only 5 out of 32 somatic non-synonymous coding mutations detected in the metastatic pleural effusion could be found in the primary (Shah et al., 2009). By gene expression analyses, lobular tumors are almost always luminal subtype and have much greater molecular heterogeneity than basal-like tumors (Parker et al., 2009). Other possibilities are that the larger heterogeneity between the primary and metastasis in the lobular cancer may have been due to radiation treatment in the pleural effusion tested and/or the longer time to recurrence (nine years in the lobular vs less than one year in the Basal-like tumor).

INDIVIDUALIZING TREATMENT

Clinical trials are beginning to be designed to prospectively enroll patients and apply these powerful technologies to distinguish low-risk patients that do not need extensive therapy intervention from those who are at high risk of recurrence and would greatly benefit from a personalized plan. These include patients with germline variants causing altered drug metabolism and either a reduction in efficacy (e.g., tamoxifen) or increased toxicity (e.g., taxane), and patients with tumors that could be biologically targeted with drugs against angiogenesis (VEGF inhibitors) (Robert et al., 2011), growth factor interactions (EGFR-HER2 dimerization) (Cameron et al., 2008), and other signaling pathways (MAPK/ERK and/or PI3K/AKT inhibition) (Dave et al., 2011). As we understand more about inherited and acquired mutations that confer sensitivity or resistance to commonly used systemic therapies, we will become more effective in individualizing breast cancer treatment.

REFERENCES

Antoniou, A.C., Sinilnikova, O.M., Simard, J., et al., 2007. RAD51 135G–>C modifies breast cancer risk among BRCA2 mutation carriers: Results from a combined analysis of 19 studies. Am J Hum Genet 81, 1186–1200.

Bergamaschi, A., Kim, Y.H., Wang, P., et al., 2006. Distinct patterns of DNA copy number alteration are associated with different clinicopathological features and gene-expression subtypes of breast cancer. Genes Chromosomes Cancer 45 (11), 1033–1040.

Blenkiron, C., Goldstein, L.D., Thorne, N.P., et al., 2007. MicroRNA expression profiling of human breast cancer identifies new markers of tumor subtype. Genome Biol 8, R214.

Cameron, D., Casey, M., Press, M., et al., 2008. A phase III randomized comparison of lapatinib plus capecitabine versus capecitabine alone in women with advanced breast cancer that has progressed on trastuzumab: Updated efficacy and biomarker analyses. Breast Cancer Res Treat 112, 533–543.

Carey, L.A., Perou, C.M., Livasy, C.A., et al., 2006. Race, breast cancer subtypes, and survival in the Carolina breast cancer study. JAMA 295, 2492–2502.

Cheang, M.C., Chia, S.K., Voduc, D., et al., 2009. Ki67 index, HER2 status, and prognosis of patients with luminal B breast cancer. J Natl Cancer Inst 101 (10), 736–750.

Cheang, M.C., Voduc, D., Bajdik, C., et al., 2008. Basal-like breast cancer defined by five biomarkers has superior prognostic value than triple-negative phenotype. Clin Cancer Res 14, 1368–1376.

Chen, X., Ba, Y., Ma, L., et al., 2008. Characterization of microRNAs in serum: A novel class of biomarkers for diagnosis of cancer and other diseases. Cell Res 18, 997–1006.

Comen, E., Davids, M., Kirchhoff, T., Hudis, C., Offit, K., Robson, M., 2011. Relative contributions of BRCA1 and BRCA2 mutations to "triple-negative" breast cancer in Ashkenazi Women. Breast Cancer Res Treat 129 (1), 185–190.

Comen, E.A., Robson, M., 2010. Inhibition of poly(ADP)-ribose polymerase as a therapeutic strategy for breast cancer. Oncology (Williston Park) 24, 55–62.

Coudert, B.P., Largillier, R., Arnould, L., et al., 2007. Multicenter phase II trial of neoadjuvant therapy with trastuzumab, docetaxel, and carboplatin for human epidermal growth factor receptor-2-over-expressing stage II or III breast cancer: Results of the GETN(A)-1 trial. J Clin Oncol 25, 2678–2684.

Dai, H., Van't Veer, L., Lamb, J., et al., 2005. A cell proliferation signature is a marker of extremely poor outcome in a subpopulation of breast cancer patients. Cancer Res 65, 4059–4066.

Dave, B., Migliaccio, I., Gutierrez, M.C., et al., 2011. Loss of phosphatase and tensin homolog or phosphoinositol-3 kinase activation and response to trastuzumab or lapatinib in human epidermal growth factor receptor 2-overexpressing locally advanced breast cancers. J Clin Oncol 29, 166–173.

Desmedt, C., Piette, F., Loi, S., et al., 2007. Strong time dependence of the 76-gene prognostic signature for node-negative breast cancer patients in the TRANSBIG multicenter independent validation series. Clin Cancer Res 13, 3207–3214.

Desta, Z., Ward, B.A., Soukhova, N.V., Flockhart, D.A., 2004. Comprehensive evaluation of tamoxifen sequential biotransformation by the human cytochrome P450 system in vitro: Prominent roles for CYP3A and CYP2D6. J Pharmacol Exp Ther 310, 1062–1075.

Ding, L., Ellis, M.J., Li, S., et al., 2010a. Genome remodelling in a basal-like breast cancer metastasis and xenograft. Nature 464, 999–1005.

Ding, L., Wendl, M.C., Koboldt, D.C., Mardis, E.R., 2010b. Analysis of next-generation genomic data in cancer: Accomplishments and challenges. Hum Mol Genet 19, R188–R196.

Dowsett, M., Smith, I.E., Ebbs, S.R., et al., 2007. Prognostic value of Ki67 expression after short-term presurgical endocrine therapy for primary breast cancer. J Natl Cancer Inst 99, 167–170.

Ejlertsen, B., Mouridsen, H.T., Jensen, M.B., et al., 2007. Improved outcome from substituting methotrexate with epirubicin: Results from a randomised comparison of CMF versus CEF in patients with primary breast cancer. Eur J Cancer 43, 877–884.

Fan, C., Oh, D.S., Wessels, L., et al., 2006. Concordance among gene-expression-based predictors for breast cancer. N Engl J Med 355, 560–569.

Gaedigk, A., Ndjountche, L., Divakaran, K., et al., 2007. Cytochrome P4502D6 (CYP2D6) gene locus heterogeneity: Characterization of gene duplication events. Clin Pharmacol Ther 81, 242–251.

Gaedigk, A., Simon, S.D., Pearce, R.E., Bradford, L.D., Kennedy, M.J., Leeder, J.S., 2008. The CYP2D6 activity score: Translating genotype information into a qualitative measure of phenotype. Clin Pharmacol Ther 83, 234–242.

Garber, J.E., Offit, K., 2005. Hereditary cancer predisposition syndromes. J Clin Oncol 23, 276–292.

Gennari, A., Sormani, M.P., Pronzato, P., et al., 2008. HER2 status and efficacy of adjuvant anthracyclines in early breast cancer: A pooled analysis of randomized trials. J Natl Cancer Inst 100, 14–20.

Gluck, S., 2005. Adjuvant chemotherapy for early breast cancer: Optimal use of epirubicin. Oncologist 10, 780–791.

Hayes, D.F., Thor, A.D., Dressler, L.G., et al., 2007. HER2 and response to paclitaxel in node-positive breast cancer. N Engl J Med 357, 1496–1506.

Hu, Z., Fan, C., Oh, D.S., et al., 2006. The molecular portraits of breast tumors are conserved across microarray platforms. BMC Genomics 7, 96.

Ingelman-Sundberg, M., 2005. Genetic polymorphisms of cytochrome P450 2D6 (CYP2D6): Clinical consequences, evolutionary aspects and functional diversity. Pharmacogenomics J 5, 6–13.

Koshy, N., Quispe, D., Shi, R., Mansour, R., Burton, G.V., 2010. Cisplatin-gemcitabine therapy in metastatic breast cancer: Improved outcome in triple negative breast cancer patients compared to non-triple negative patients. Breast 19, 246–248.

Lash, T.L., Lien, E.A., Sorensen, H.T., Hamilton-Dutoit, S., 2009. Genotype-guided tamoxifen therapy: Time to pause for reflection? Lancet Oncol 10, 825–833.

Lee, Y.S., Dutta, A., 2008. MicroRNAs in cancer. Annu Rev Pathol 4, 199–227.

Levine, M.N., Pritchard, K.I., Bramwell, V.H., Shepherd, L.E., Tu, D., Paul, N., 2005. Randomized trial comparing cyclophosphamide, epirubicin, and fluorouracil with cyclophosphamide, methotrexate, and fluorouracil in premenopausal women with node-positive breast cancer: Update of National Cancer Institute of Canada Clinical Trials Group Trial MA5. J Clin Oncol 23, 5166–5170.

Lu, J., Getz, G., Miska, E.A., et al., 2005. MicroRNA expression profiles classify human cancers. Nature 435, 834–838.

Mattick, J.S., Makunin, I.V., 2006. Non-coding RNA. Hum Mol Genet 15 Spec No 1, R17–R29.

Nielsen, T.O., Hsu, F.D., Jensen, K., et al., 2004. Immunohistochemical and clinical characterization of the basal-like subtype of invasive breast carcinoma. Clin Cancer Res 10, 5367–5374.

Nielsen, T.O., Parker, J.S., Leung, S., et al., 2010. A comparison of PAM50 intrinsic subtyping with immunohistochemistry and clinical prognostic factors in tamoxifen-treated estrogen receptor-positive breast cancer. Clin Cancer Res 16, 5222–5232.

Paik, S., Shak, S., Tang, G., et al., 2004. A multigene assay to predict recurrence of tamoxifen-treated, node-negative breast cancer. N Engl J Med 351, 2817–2826.

Paik, S., Tang, G., Shak, S., et al., 2006. Gene expression and benefit of chemotherapy in women with node-negative, estrogen receptor-positive breast cancer. J Clin Oncol 24, 3726–3734.

Parker, J.S., Mullins, M., Cheang, M.C., et al., 2009. Supervised risk predictor of breast cancer based on intrinsic subtypes. J Clin Oncol 27, 1160–1167.

Parkin, D.M., Bray, F., Ferlay, J., Pisani, P., 2005. Global cancer statistics, 2002. CA Cancer J Clin 55, 74–108.

Perou, C.M., Sorlie, T., Eisen, M.B., et al., 2000. Molecular portraits of human breast tumours. Nature 406, 747–752.

Perreard, L., Fan, C., Quackenbush, J.F., et al., 2006. Classification and risk stratification of invasive breast carcinomas using a real-time quantitative RT-PCR assay. Breast Cancer Res 8, R23.

Prat, A., Perou, C.M., 2011. Deconstructing the molecular portraits of breast cancer. Mol Oncol 5, 5–23.

Ring, B.Z., Seitz, R.S., Beck, R., et al., 2006. Novel prognostic immunohistochemical biomarker panel for estrogen receptor-positive breast cancer. J Clin Oncol 24, 3039–3047.

Robert, N.J., Dieras, V., Glaspy, J., et al., 2011. RIBBON-1: Randomized, double-blind, placebo-controlled, phase III trial of chemotherapy with or without bevacizumab for first-line treatment of human epidermal growth factor receptor 2-negative, locally recurrent or metastatic breast cancer. J Clin Oncol 29, 1252–1260.

Robson, M.E., Storm, C.D., Weitzel, J., Wollins, D.S., Offit, K., 2010. American Society of Clinical Oncology policy statement update: Genetic and genomic testing for cancer susceptibility. J Clin Oncol 28, 893–901.

Schwartz, G.F., Hughes, K.S., Lynch, H.T., et al., 2008. Proceedings of the international consensus conference on breast cancer risk, genetics, and risk management, April 2007. Cancer 113, 2627–2637.

Shah, S.P., Morin, R.D., Khattra, J., et al., 2009. Mutational evolution in a lobular breast tumour profiled at single nucleotide resolution. Nature 461, 809–813.

Silver, D.P., Richardson, A.L., Eklund, A.C., et al., 2010. Efficacy of neoadjuvant cisplatin in triple-negative breast cancer. J Clin Oncol 28, 1145–1153.

Slamon, D., Pegram, M., 2001. Rationale for trastuzumab (herceptin) in adjuvant breast cancer trials. Semin Oncol 28, 13–19.

Slamon, D.J., Clark, G.M., Wong, S.G., Levin, W.J., Ullrich, A., McGuire, W.L., 1987. Human breast cancer: Correlation of relapse and survival with amplification of the HER-2/neu oncogene. Science 235, 177–182.

Slamon, D.J., Leyland-Jones, B., Shak, S., et al., 2001. Use of chemotherapy plus a monoclonal antibody against HER2 for metastatic breast cancer that overexpresses HER2. N Engl J Med 344, 783–792.

Sorlie, T., Perou, C.M., Tibshirani, R., et al., 2001. Gene expression patterns of breast carcinomas distinguish tumor subclasses with clinical implications. Proc Natl Acad Sci USA 98, 10,869–10,874.

Sorlie, T., Tibshirani, R., Parker, J., et al., 2003. Repeated observation of breast tumor subtypes in independent gene expression data sets. Proc Natl Acad Sci USA 100, 8418–8423.

Sotiriou, C., Wirapati, P., Loi, S., et al., 2006. Gene expression profiling in breast cancer: Understanding the molecular basis of histologic grade to improve prognosis. J Natl Cancer Inst 98, 262–272.

Symmans, W.F., Hatzis, C., Sotiriou, C., et al., 2010. Genomic index of sensitivity to endocrine therapy for breast cancer. J Clin Oncol 28, 4111–4119.

Thurlimann, B., Keshaviah, A., Coates, A.S., et al., 2005. A comparison of letrozole and tamoxifen in postmenopausal women with early breast cancer. N Engl J Med 353, 2747–2757.

Tutt, A., Wang, A., Rowland, C., et al., 2008. Risk estimation of distant metastasis in node-negative, estrogen receptor-positive breast cancer patients using an RT-PCR based prognostic expression signature. BMC Cancer 8, 339.

Van De Vijver, M.J., He, Y.D., Van't Veer, L.J., et al., 2002. A gene-expression signature as a predictor of survival in breast cancer. N Engl J Med 347, 1999–2009.

Wang, Y., Klijn, J.G., Zhang, Y., et al., 2005. Gene-expression profiles to predict distant metastasis of lymph-node-negative primary breast cancer. Lancet 365, 671–679.

Wu, X., Hawse, J.R., Subramaniam, M., Goetz, M.P., Ingle, J.N., Spelsberg, T.C., 2009. The tamoxifen metabolite, endoxifen, is a potent antiestrogen that targets estrogen receptor alpha for degradation in breast cancer cells. Cancer Res 69, 1722–1727.

Ovarian Cancer

Tanja Pejovic, Matthew L. Anderson, and Kunle Odunsi

ORIGINS OF EPITHELIAL OVARIAN CANCER

Epithelial ovarian cancer (EOC) arises primarily from the ovarian surface epithelium (OSE), with a subset possibly originating in the adjacent fimbria (Levanon et al., 2008). The OSE forms a monolayer of cells surrounding the ovary. Developmentally it derives from the celomic epithelium that also gives rise to the peritoneal mesothelium and oviductal epithelium (Auersperg et al., 2001). The OSE appears generally stable, uniform and quiescent, though it is competent to undergo proliferation *in vivo* (Wright et al., 2008). No physiological role for the primate OSE has been established (Wright et al., 2010), and the lack of any obvious function could contribute to the asymptomatic nature of early-stage EOC.

In other organs, such as the colon, distinct premalignant lesions have been identified and found to accumulate genetic defects that ultimately result in malignancy. However, the search to identify similar epithelial precursors in the human ovary has proven only partially fruitful, in large part because normal ovaries are rarely biopsied or examined. A number of investigators have described histologic findings consistent with a preinvasive lesion for ovarian cancer in ovaries removed from women who eventually developed peritoneal carcinomas, ovaries from high-risk women undergoing prophylactic oophorectomy, and in areas of ovarian epithelium adjacent to early-stage ovarian cancers that demonstrate a transition from normal to malignant cells (Schlosshauer et al., 2003). The

hypothesis that these lesions are premalignant is strengthened by observations that regions of epithelial irregularity express levels of TP53 and Ki-67 intermediate between those found in normal ovarian epithelium and ovarian cancers.

Each of these observations is consistent with the hypothesis that, similar to cancers originating in other organs, ovarian cancer evolves from an intraepithelial precursor. If so, improved means to detect and/or eradicate these lesions may prove fruitful for preventing ovarian cancer.

INHERITED SYNDROMES OF OVARIAN CANCERS

Linkage analysis of familial breast and ovarian cancers has provided some of the first insights into the molecular basis of ovarian cancer. These efforts ultimately identified two genes, *BRCA1* and *BRCA2*, each of which were clearly associated with an increased incidence of ovarian cancer. Although only a minority (8–10%) of diagnosed ovarian cancers are familial, most familial ovarian cancers (76–92%) are associated with mutations at the *BRCA1* locus, which is located on 17q21. Hundreds of mutations in *BRCA1* have now been identified, most commonly as loss-of-function, nonsense, or frameshift mutations. Two specific mutations, 185delAG and 5382insC, are found in 1% and 0.1% respectively of Ashkenazi Jewish women.

Functionally, *BRCA1* regulates TP53, an oncogene frequently implicated in ovarian cancer. Thus, loss of *BRCA1*

Genomic and Personalized Medicine, 2nd edition
by Ginsburg & Willard. DOI: http://dx.doi.org/10.1016/B978-0-12-382227-7.00061-6

allows DNA damage to accumulate via a loss of its activation of TP53. However, mutations in *BRCA1* also likely contribute to ovarian cancer by mechanisms other than its interactions with *p53*. These include its ability to specifically regulate X chromosome gene expression mediated by an association of *Xist* with the inactive X chromosome (Ganesan et al., 2005). Consistent with this observation, site-specific dysregulation of X-linked gene expression in *BRCA1*-associated epithelial ovarian malignancies have been described.

Although mutations in *BRCA1/2* are rarely observed in sporadic, nonfamilial ovarian cancers, it is possible that the mutations in pathways by which BRCA1 regulates X-chromosome gene expression do, in fact, contribute to this disease. Characterization of genome-wide patterns of gene expression in sporadic breast cancers has allowed investigators to classify these tumors as either *BRCA1*-like or *BRCA2*-like in the patterns of their gene expression. This observation implicates the contribution of alterations in other components of the *BRCA1/2*-regulated pathways to sporadic breast cancers and, possibly, to ovarian cancer. Any understanding of the role of *BRCA1* in ovarian cancer is further complicated by reports of women with high-risk mutations in *BRCA1* who fail to develop ovarian cancer. These observations speak clearly to the role of genetic modifiers in determining whether *BRCA1/2* mutations ultimately lead to malignancy.

Genomic Instability

Genomic instability – manifested as a cell's ability to tolerate DNA damage – is a hallmark of all cancer, including epithelial ovarian cancers. Tolerance to DNA damage can be achieved by alterations in any of the six major DNA repair pathways: base excision repair, mismatch repair, nucleotide excision repair, homologous recombination, nonhomologous recombination, and translesion DNA synthesis (Figure 61.1). The specific DNA pathway affected often predicts the specific type of mutations observed in particular cancers, its sensitivity to drugs, as well as clinical outcome of affected patients.

Fanconi Anemia DNA Repair Pathway

Studies of the pathogenesis of rare inherited DNA repair disorders such as Fanconi anemia (FA) have helped define the molecular basis of defective DNA damage responses linked to cancer risk. FA is a rare genetic disorder characterized by skeletal anomalies, progressive bone marrow failure, cancer susceptibility, and cellular hypersensitivity to DNA cross-linking agents. To date, 13 FA genes have been cloned: *FANCA, -B, -C, -D1, -D2, -E, -F, -G, -J, -L, -M, -N,* and *-I*. Of these, *FANCA, FANCB, FANCC, FANCE, FANCF, FANCG, FANCL,* and *FANCM* form a nuclear core complex. Although the functional scope of this complex has not been fully defined, it is clear that it must be completely intact to facilitate monoubiquitination of the downstream FANCD2 and FANCI proteins, a change that permits FANCD2 and FANCI to colocalize with BRCA1, BRCA2, (and presumably *FANCJ* and *FANCN*) and RAD51 in damage-induced nuclear foci (Table 61.1) (Garcia-Higuera et al., 2001).

Four lines of evidence link the FA pathway with ovarian carcinogenesis. First, *BRCA2* has been identified as the FA gene *FANCD1*. As a result, heterozygotes for *BRCA2* mutations have a high risk of tissue-specific epithelial cancers, while homozygotes develop FA. Second, an increased prevalence of epithelial cancers, including ovarian malignancies, has been observed in *FANCD2*-nullizygous mice. Functionally

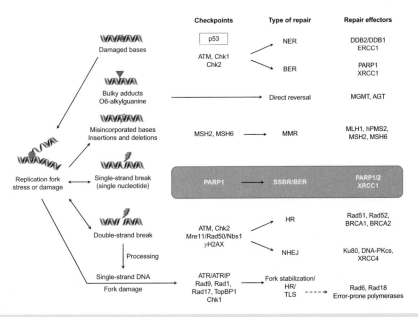

Figure 61.1 DNA Repair Mechanisms.
In response to stress/DNA damage cells activate one of the 6 major DNA repair mechanisms – Nucleotide excision repair (NER), Base excision repair (BER), mismatch repair (MMR), non-homologous end joining (NHER), homologous recombination (HR), and Translesion synthesis (TLS).

TABLE 61.1	Incidence of Fanconi anemia gene mutations in epithelial ovarian cancer		
Gene	**Unique mutations (n = 706) (%)**	**Total mutations (n = 2002) (%)**	**Chromosomal location**
FANCA	60.9	57.0	16q24.3
FANCB	2.0	0.9	Xq22.31
FANCC	7.6	14.4	9q22.3
FANCD1 (BRCA2)	5.0	2.8	13q12–13
FANCD2	4.7	3.9	3p25.3
FANCE	3.5	2.3	6p21–22
FANCF	2.0	2.0	11p15
FANCG	6.9	11.2	9p13
FANCI	2.8	1.7	15q25–26
FANCJ (BACH1/Brip1)	1.7	2.4	17q22–24
FANCL	0.4	0.2	2p16.1
FANCM (Hef)	0.3	0.2	14q21.3
FANCN (PALB2)	2.1	0.8	16p12
*FANCO (Rad51C)	–	–	17q25.1
FAAP100	–	–	17q25.1
FAAP24	–	–	19q13.11
FAN1	–	–	15q13.2–q13.3

*RAD51C biallelic mutations found to be associated with FA-like syndrome.
Adapted from Fanconi Anemia Mutation Database at Rockefeller University (http://www.rockefeller.edu/fanconi/mutate/).

significant silencing of *FANCF* in ovarian cancer through promoter hypermethylation has also been described. Lastly, low levels of FANCD2 protein are found in ovarian surface epithelia from women at risk for ovarian cancer who tested negative for BRCA mutations. Taken together, these observations suggest that the FA pathway is important in defining predisposition to ovarian (and breast) cancer, and that aberrations of FA genes may account for some familial ovarian cancer cases not accounted for by *BRCA1* and *BRCA2* mutations.

In sporadic ovarian cancers, the epigenetic silencing of the FA pathway through methylation of the FA gene promoter region is one of the frequent mechanisms of inactivation. One study found that 4 of 19 primary ovarian carcinomas had *FANCF* methylation, although a larger study of 106 ovarian tumors did not identify loss of *FANCF* expression. Loss of *BRCA2* mRNA and protein has been reported in 13% of ovarian carcinomas, and in contrast to other FA genes, methylation is not a cause of the protein loss. Epigenetic silencing of *BRCA1* through methylation was found in 23% of advanced ovarian carcinomas. Interestingly, tumors with inactivated *BRCA2* are responsive to cisplatin. However, due to their low accuracy of DNA repair, these cells accumulate secondary genetic modifications that can lead to reversal of *BRCA2*

mutation, allowing them to acquire resistance to cross-linking agents (Sakai et al., 2008).

Other DNA Repair Pathways

Similar to the FA/BRCA pathway, disruptions of other DNA repair pathways have been observed in ovarian cancer. These disruptions account, at least in part, for the specific drug sensitivity of the tumors. Recent studies indicate that translesion DNA synthesis defect in ovarian cancer is a consequence of elevation in activity of POLB, an error-prone polymerase. Inhibition of POLB in these cells results in resensitization to cisplatin (Boudsocq et al., 2005). Overall, it is believed that although inactivation of one DNA repair pathway may confer advantage to tumors, cancer cells may rely more on other repair pathways. Therefore, inactivation of the second pathway would be deleterious for the cells, causing synthetic lethality. An RNA interference screen identified the ATM pathway to be synthetically lethal with FA (Kennedy et al., 2007). Similarly, a strategy for synthetic lethality using base excision repair PARP1 inhibitors for the treatment of homologous recombination-deficient ovarian cancer is under investigation (Bryant et al., 2005; Farmer et al., 2005). PARP [poly (ADP-ribose) polymerase] inhibition has been shown to be up to one thousand times selectively more toxic to cancer cells than to wild-type cells. PARP inhibitors act by exploiting a tumor cell's defect in homologous recombination (HR), a type of DNA repair. This is because following PARP inhibition, cells require HR to repair common types of DNA damage. While normal cells can use HR for repair of this damage and survive, certain types of tumors (e.g., those with BRCA1 or BRCA2 defects) have lost the ability to repair by HR and will die.

GENOME-WIDE ASSOCIATION STUDIES

Identification of common ovarian cancer susceptibility variants may have clinical implications in the future for identifying patients at greatest risk of the disease. In this regard, several genome-wide association studies (GWAS) have been performed in ovarian cancer. The most striking of these was a recent study by the Ovarian Cancer Association Consortium designed to identify common ovarian cancer susceptibility alleles (Song et al., 2009). A total of 507,094 single nucleotide polymorphisms (SNPs) were genotyped in 1817 cases and 2353 controls from the UK; 22,790 top-ranked SNPs were also genotyped in 4274 cases and 4809 controls of European ancestry from Europe, the United States, and Australia. Twelve SNPs associated with disease risk were identified at 9p22 ($P < 10^{-8}$). The most significant SNP (rs3814113; $P = 2.5 \times 10^{-17}$) was genotyped in a further 2670 ovarian cancer cases and 4668 controls, confirming its association [combined data odds ratio (OR) = 0.82, 95% confidence interval (CI) 0.79–0.86, $P_{trend} = 5.1 \times 10^{-19}$]. The association was strongest for serous ovarian cancers (OR = 0.77, 95% CI 0.73–0.81, $P_{trend} = 4.1 \times 10^{-21}$).

TRANSCRIPTIONAL PROFILING OF OVARIAN CANCER HISTOLOGIC SUBTYPES

Several gene expression studies using cDNA microarrays have been performed in ovarian cancer. Additionally, several studies have focused on alterations in DNA copy number (Mayr et al., 2006; Meinhold-Heerlein et al., 2005; Nakayama et al., 2007). Array-based technology has shown that the different histological subtypes of ovarian carcinoma are distinguishable based on their overall genetic expression profiles. A common finding among several studies is the ability to distinguish between low-grade serous ovarian carcinoma and high-grade carcinoma based on gene expression profiles (Bonome et al., 2005; Gilks et al., 2005; Hough et al., 2001; Hough et al., 2000; Meinhold-Heerlein et al., 2005; Schwartz et al., 2002). A number of genes shown to be differentially expressed in EOC are known to be involved in many important cellular mechanisms, including cell cycle regulation, apoptosis, tumor invasion and control of local immunity (Berchuck et al., 2004; Gilks et al., 2005; Landen et al., 2008).

Increased mutagenic signaling by receptor tyrosine kinases plays a major role in ovarian carcinogenesis. Overexpression of EGFR (ERBB1), ERBB2/HER-2/neu, and c-FMS has been reported. One of the major downstream mediators of signaling initiated by these receptors is the phosphatidylinositol 3-kinase (PIK3K)-AKT pathway. Aberrations in this pathway, including increased AKT1 kinase activity, AKT2 and PIK3 amplification, and PIK3R1 mutations, may provide opportunities for therapeutic intervention.

It has been reported that more than 75% of ovarian carcinomas are resistant to transforming growth factor-β (TGF-β) (Hu et al., 2000), and the loss of TGF-β responsiveness may play an important role in the pathogenesis and/or progression of ovarian cancer. It has also been shown that TGF-β_1, the TGF-β receptors (TβR-II and TβR-I), and the TGF-β-signaling component Smad2 are altered in ovarian cancer. Alterations in TβR-II have been identified in 25% of ovarian carcinomas, whereas mutations in TβR-I were reported in 33% of such cancers (Chen et al., 2001). Proto-oncogene transformation might lead either to an overexpression of mitogenic molecules or an inactivation of those with inhibitory action, thus contributing to neoplastic transformation and development. The most important proto-oncogenes of the first group are undoubtedly c-FMS and HER-2/neu. The first one encodes a transmembrane tyrosine kinase receptor that binds MCSF. It is possible that c-FMS stimulates epithelial cell proliferation and induces a chemical attraction for macrophages, which in turn can produce mitogenic-stimulating factor. Elevated plasma concentrations of macrophage colony stimulating factor (MCSF) are present in the sera of 70% of patients with ovarian cancer (Van Haaften-Day et al., 2001). The second proto-oncogene, HER-2/neu, encodes another tyrosine kinase that is similar to the epidermal growth factor receptor (EGFR). Its action may consist of amplification of mitogenic action in target cells. This oncogene is overexpressed in 30–35% of ovarian cancer and is associated with a poor prognosis (Coronado Martin et al., 2007).

METASTASIS OF OVARIAN CANCERS

Metastasis involves the invasion of transformed epithelial cells across their basement membrane, through the underlying stroma, and into blood vessels and lymphatic channels, which subsequently disseminate them to distant sites. Only a tiny fraction of cells released into the circulation by a tumor ever result in metastasis. A wide variety of genes and gene products implicated in the metastasis of other cancers have also been implicated in the metastasis of ovarian cancer. These include growth factor receptors such as EGFR, insulin-like growth factor receptors (IGFRs) and kinases such as jak/stat, focal adhesion kinase, PI-3 kinase and c-met. Comparisons of primary and metastatic ovarian cancers by transcriptional profiling have failed to reveal significant differences in the expression of gene products likely related to the metastatic process.

Particular attention has recently been focused on the role of lysophosphatidic acid (LPA) in promoting the metastasis of ovarian cancers. LPA is constitutively produced by mesothelial cells lining the peritoneal cavity. Its levels are increased in the ascites of women with both early- and late-stage ovarian cancers (Ren et al., 2006). *In vitro*, LPA promotes both the migration of these cells in a manner dependent on Ras MEK kinase-1, and their invasion across artificial barriers analogous to a basement membrane. At the molecular level, exogenous LPA enhances ovarian cancer invasiveness both by activating matrix metalloproteinase-2 via membrane-type-1-matrix metalloproteinase (MT1-MMP) and by down-regulating the expression of specific tissue inhibitors of metalloproteinases (TIMP-2 and -3) (Sengupta et al., 2007). Other observations are consistent with the idea that LPA promotes dissemination of ovarian cancer by loss of cell adhesion (Do et al., 2007). However, LPA has been shown to promote the invasiveness of ovarian cancers by additional mechanisms dependent on interleukin-8. The G12/13-RhoA and cyclooxygenase pathways have also been implicated in the LPA-induced migration of ovarian cancers. These mechanisms appear to be independent of the ability of LPA to induce changes in *MMP2* expression.

Until recently, ovarian cancer metastasis has been almost exclusively studied as a process involving individual cells. However, multicellular clusters of self-adherent cells known as *spheroids* can be isolated from the ascitic fluid of women with ovarian cancer. Spheroids readily adhere to both extracellular matrix proteins such collagen IV and mesothelial cells in monolayer culture using beta-1 integrins. Once adherent, the cells contained in spheroids disaggregate, allowing them to invade underlying mesothelial cells and create invasive foci (Burleson et al., 2004). These observations are consistent with the hypothesis that ovarian cancer spheroids play an important role in the metastatic potential of ovarian cancer. Recent evidence has shown that a loss of circulating gonadotropins results in a dose-dependent decrease in the expression of *VEGF* in the outer, proliferating cells of ovarian cancer spheroids (Schiffenbauer et al., 1997), indicating that these cell clusters remain responsive to signals in their microenvironment that may promote metastasis.

The presence of spheroids in ascites may also help to explain the frequent persistence and frequent recurrence of ovarian cancer after treatment. Spheroids express high levels of cyclin-dependent kinase inhibitor 1B (CDKN1B) and P-glycoprotein that contribute, at least in part, to their relative resistance to the cytotoxic effects of paclitaxel when compared with ovarian cancer cells in monolayer culture. Ovarian cancer spheroids have also been shown to be relatively resistant to the cytotoxic effects of radiation (Griffon et al., 1995). These observations are consistent with *in vitro* studies that demonstrate that the signals generated by adhesion to specific components of the extracellular matrix, such as collagen IV, can modify the sensitivity of ovarian cancers to chemotherapy. However, the mechanisms by which the aggregation of malignant cells promotes or enhances cell survival remain unclear.

It is also unclear how the aggregation of these malignant cells might promote or enhance the migration, attachment, or invasion of ovarian cancer cells. However, experimental evidence has recently demonstrated a number of observations that identify novel mechanisms by which ovarian cancer spheroids confer a chemoresistant phenotype, such as overexpression of gene products involved in epigenetic regulation. These include EZH2 and other members of the PRC2 complex. Data indicate that increased expression of EZH2 can be associated with chemotherapy resistance. Notably, the PRC2 complex has been implicated in regulating the pleuripotency of stem cells. Thus, formation of spheroids in ovarian cancer ascites may enable the survival of ovarian cancer cells in women undergoing front line treatment by creating a unique niche, in which expression of gene products typically associated with pleuripotency not only enables metastasis but also the ability of these cells to persist through treatment. The molecular nature of this niche is not currently known.

INTEGRATED GENOMIC ANALYSES OF OVARIAN CARCINOMA

Cancer Genome Atlas Research Network

Recently, the National Institutes of Health (NIH)-sponsored The Cancer Genome Atlas (TCGA) consortium has comprehensively examined a large series of epithelial ovarian cancer specimens. This represents the most comprehensive approach yet undertaken to decipher the molecular events involved in ovarian cancer. Specifically, TCGA investigators analyzed 489 high-grade serous ovarian cancers [International Federation of Gynecology and Obstetrics (FIGO) stage II–IV] using a series of approaches that included targeted sequencing of the ovarian cancer exome, mutational analysis of protein-coding genes, mRNA and microRNA (miRNA) expression profiling, examination of DNA copy number variation, and chip-based testing for patterns of CpG methylation. This project examined more than 180,000 exons from ~18,500 known protein-coding genes. Data generated by the TCGA is publicly available at http://tcga-data.nci.nih.gov/dox/publications/ov_2011. Key findings include the observation that *TP53* was mutated in 303 of the 216 samples tested. Also, similar to published reports, mutations in DNA damage-repair proteins BRCA1 and BRCA2 were observed in 9% and 8% of sporadic ovarian cancer cases tested. Six other gene products were found to be consistently mutated at levels that exceeded statistical expectations. These included NF1, FAT3, CDMD3, GABRA6, and CDK12. Integrated analyses identified 168 genes where increased methylation is thought to lead to decreased expression in ovarian cancer. Collectively, integration of the TCGA dataset was used to define four subsets of ovarian cancer patients defined by their transcriptional profiles: immunoreactive, differentiated, proliferative, and mesenchymal. Each of these subsets is distinguished from the others based on the expression of key gene products associated with predominant features of that group. For example, immunoreactive ovarian cancers are particularly noted for their expression of T-cell chemokine ligand CXCL11 and CXCL10, as well as the receptor CXCR3. In addition, integration of survival analyses was used to identify a set of 193 genes whose patterns of transcription predicted ovarian survival (108 correlated with poor survival and 85 associated with good survival). These investigators were able to confirm the validity of this profile using a previously reported database detailing patterns of gene expression profiled in an independent set of ovarian cancers. TCGA observations provide not only unique insight into the nature of this disease but also allow investigators to better integrate their research into hypothesis-driven pictures, that can be tested and used to develop novel molecular-based treatments.

ANGIOGENESIS

Angiogenesis is tightly regulated by a balance of pro- and anti-angiogenic factors. These include growth factors such as TGF-β, vascular endothelial growth factor (VEGF), and platelet-derived growth factor; prostaglandins such as prostaglandin E2; cytokines such as interleukin-8; and other factors, such as the angiopoietins (Ang-1, Ang-2), and hypoxia-inducible factor-1alpha (HIF-1α). Many of these angiogenic factors have been implicated in ovarian cancer. For example, VEGF is a family of secreted polypeptides with critical roles in both normal development and human disease. Many cancers, including ovarian carcinomas, release VEGF in response to the hypoxic or acidic conditions typical in solid tumors. Variable levels of VEGF expression have been reported in ovarian cancers, in which higher levels correlate with advanced disease and poor clinical prognosis (Kassim et al., 2004). Circulating levels of VEGF have also been reported to be higher in the serum of women with ovarian cancers when compared with those with benign tumors. Expression of hypoxia-inducible factor 1 (HIF-1) correlates well with microvessel density in ovarian cancers and has been proposed to upregulate VEGF expression (Jiang and Feng 2006). Culturing ovarian cancer cell lines under hypoxic conditions stimulates the expression of both HIF-1α and VEGF in ovarian cancer cell lines; addition of prostaglandin E2 potentiates the ability of hypoxia to induce the expression of both proangiogenic factors (Zhu et al., 2004).

Ironically, many of the molecules implicated in regulating angiogenesis in cancer, such as c-met, also regulate other processes critical for cancer metastasis, such as cell migration and invasiveness. Inhibition of PI-3 kinase decreases transcription of VEGF in ovarian cancer cells, an effect that is reversed by the forced expression of *AKT*. Such observations are consistent with reports that hypoxia not only induces angiogenesis, but also increases the invasiveness of ovarian cancer cells (Imai et al., 2003). Likewise, an acidic environment induces increased interleukin-8 expression in ovarian cancer in a manner dependent on transcription factors AP-1 and nuclear

factor-kappa B-like factor, suggesting that feedback between these pathways may also determine how tumors interact with their external environment. Undoubtedly, better insight into these interactions will help to define the suitability of these molecules as therapeutic targets.

EPIGENETICS

It has become increasingly apparent that epigenetic events can lead to cancer as frequently as loss of gene function due to mutations or loss of heterozygosity. The overall level of genomic methylation is reduced in cancer (global hypomethylation), but hypermethylation of promoter regions of specific genes is a common event often associated with transcriptional inactivation of specific genes (Baylin and Ohm, 2006). This is critical because the silenced genes are often tumor suppressor genes. Epigenetic gene silencing is a complex series of events that includes DNA hypermethylation of CpG islands within gene-promoter regions, histone deacetylation, methylation or phosphorylation, or histone demethylation. Global hypermethylation of CpG islands appears to be prevalent but highly variable in ovarian cancer tissue (Wei et al., 2002). Multiple genes are abnormally methylated in ovarian cancer compared with normal ovarian tissue, including the genes for p16, RARbeta, H-cadherin, leukotriene B4 receptor, progesterone receptor, and estrogen receptor-α, *GSTP1*, *MGMT*, *RASSF1A*, *MTHFR*, *CDH1*, *IGSF4*, *BRCA1*, *TMS1*, the putative tumor suppressor *km23* (TGF-β component), and others (Balch et al., 2004). The degree of DNA methylation and the demethylation activity of chemotherapeutic drugs, and the sensitive relations of histone acetylation and the specificity of demethylation of select genes, are important to ensure the success of treatment and prevent disease recurrence.

PHARMACOGENOMICS AND OVARIAN CANCER

Ovarian cancer shows considerable variability in its chemoresponse. This is largely attributed to acquired or inherited genetic, as well as epigenetic, variations between individuals. Several candidate loci have been identified in ovarian cancer, including *ERCC1*, *ABCB1* and *TP53* variants (Paige and Brown, 2008). Additionally, pharmacoepigenomic modulators of key genes and pathways, such as promoter methylation (*MLH1* and *BRCA1* genes) and miRNA regulation (PTEN/AKT and NF-κB pathways), have been implicated in ovarian cancer chemoresponse.

Most of the studies have been focused on the variability in clinical paclitaxel sensitivity. One of the most comprehensive taxane-platinum pharmacogenetic studies in ovarian cancer assessed 27 polymorphisms from 16 key genes involved in taxane and platinum pathways in 914 patients from the

SCOTROC1 phase III trial (Marsh et al., 2007). However, no clear candidates for taxane-platinum pharmacogenetic markers were found, underlying the increasing need for validation of potential markers in a very large sample set using genome-wide association studies.

SUMMARY

Recent years have witnessed tremendeous advancements in identifying molecular determinants of ovarian cancer predisposition, as well as biochemical determinants of ovarian cancer development. These successes have been facilitated by development of high-throughput technologies, as well as large-scale genomic association studies in clinically well described populations. Implementing these basic science results into clinical treatment has been started by emergence of the first generation of poly (ADP-ribose) polymerase (PARP) inhibitors for treatment of BRCA mutation carriers with ovarian cancer, and holds promise for application in a large group of patients with sporadic ovarian cancer. Integration of all available genetic data and their critical analysis allows us to consider complex questions about disease pathways, identify novel molecular targets in ovarian cancer, stratify this heterogenous disease based on molecular profiles, and apply this knowledge into clinical practice.

REFERENCES

Auersperg, N., Wong, A.S., Choi, K.C., et al., 2001. Ovarian surface epithelium: Biology, endocrinology, and pathology. Endocr Rev 22, 255–288.

Balch, C., Huang, T.H., Brown, R., Nephew, K.P., 2004. The epigenetics of ovarian cancer drug resistance and resensitization. Am J Obstet Gynecol 191, 1552.

Baylin, S.B., Ohm, J.E., 2006. Epigenetic gene silencing in cancer – a mechanism for early oncogenic pathway addiction? Nat Rev Cancer 6, 107.

Berchuck, A., Iversen, E.S., Lancaster, J.M., et al., 2004. Prediction of optimal versus suboptimal cytoreduction of advanced-stage serous ovarian cancer with the use of microarrays. Am J Obstet Gynecol 190, 910–925.

Bonome, T., Lee, J.Y., Park, D.C., et al., 2005. Expression profiling of serous low malignant potential, low-grade, and high-grade tumors of the ovary. Cancer Res 65, 10,602–10,612.

Boudsocq, F., Benaim, P., Canitrot, Y., et al., 2005. Modulation of cellular response to cisplatin by a novel inhibitor of DNA polymerase beta. Mol Pharmacol 67, 1485.

Bryant, H.E., Schultz, N., Thomas, H.D., et al., 2005. Specific killing of BRCA2-deficient tumours with inhibitors of poly(ADP-ribose) polymerase. Nature 434, 913–917.

Burleson, K.M., Hansen, L.K., Skubitz, A.P., 2004. Ovarian carcinoma spheroids disaggregate on type I collagen and invade live human mesothelial cell monolayers. Clin Exp Metastasis 21, 685.

Chen, T., Triplett, J., Dehner, B., et al., 2001. Transforming growth factor-beta receptor type I gene is frequently mutated in ovarian carcinomas. Cancer Res 61, 4679.

Coronado Martin, P.J., Fasero Laiz, M., Garcia Santos, J., Ramirez Mena, M., Vidart Aragon, J.A., 2007. Overexpression and prognostic value of p53 and HER2/neu proteins in benign ovarian tissue and in ovarian cancer. Med Clin (Barc) 128, 1.

Do, T.V., Symowicz, J.C., Berman, D.M., et al., 2007. Lysophosphatidic acid down-regulates stress fibers and up-regulates pro-matrix metalloproteinase-2 activation in ovarian cancer cells. Mol Cancer Res 5, 121.

Farmer, H., McCabe, N., Lord, C.J., et al., 2005. Targeting the DNA repair defect in BRCA mutant cells as a therapeutic strategy. Nature 434, 917–921.

Ganesan, S., Richardson, A.L., Wang, Z.C., et al., 2005. Abnormalities of the inactive X chromosome are a common feature of BRCA1 mutant and sporadic basal-like breast cancer. Cold Spring Harb Symp Quant Biol 70, 93.

Garcia-Higuera, I., Taniguchi, T., Ganesan, S., et al., 2001. Interaction of the Fanconi anemia proteins and BRCA1 in a common pathway. Mol Cell 7, 249.

Gilks, C.B., Vanderhyden, B.C., Zhu, S., Van de Rijn, M., Longacre, T.A., 2005. Distinction between serous tumors of low malignant potential and serous carcinomas based on global mRNA expression profiling. Gynecol Oncol 96, 684–694.

Griffon, G., Marchal, C., Merlin, J.L., et al., 1995. Radiosensitivity of multicellular tumour spheroids obtained from human ovarian cancers. Eur J Cancer 31A, 85.

Hough, C.D., Cho, K.R., Zonderman, A.B., Schwartz, D.R., Morin, P.J., 2001. Coordinately up-regulated genes in ovarian cancer. Cancer Res 61, 3869–3876.

Hough, C.D., Sherman-Baust, C.A., Pizer, E.S., et al., 2000. Large-scale serial analysis of gene expression reveals genes differentially expressed in ovarian cancer. Cancer Res 60, 6281–6287.

Hu, W., Wu, W., Nash, M.A., et al., 2000. Anomalies of the TGF-beta postreceptor signaling pathway in ovarian cancer cell lines. Anticancer Res 20, 729.

Imai, T., Horiuchi, A., Wang, C., et al., 2003. Hypoxia attenuates the expression of E-cadherin via up-regulation of SNAIL in ovarian carcinoma cells. Am J Pathol 163, 1437.

Jiang, H., Feng, Y., 2006. Hypoxia-inducible factor 1alpha (HIF-1alpha) correlated with tumor growth and apoptosis in ovarian cancer. Int J Gynecol Cancer 16 (Suppl. 1), 405.

Kassim, S.K., El-Salahy, E.M., Fayed, S.T., et al., 2004. Vascular endothelial growth factor and interleukin-8 are associated with poor prognosis in epithelial ovarian cancer patients. Clin Biochem 37, 363.

Kennedy, R.D., Chen, C.C., Stuckert, P., et al., 2007. Fanconi anemia pathway-deficient tumor cells are hypersensitive to inhibition of ataxia telangiectasia mutated. J Clin Invest 117, 1440–1449.

Landen Jr., C.N., Birrer, M.J., Sood, A.K., 2008. Early events in the pathogenesis of epithelial ovarian cancer. J Clin Oncol 26, 995–1005.

Levanon, K., Crum, C., Drapkin, R., 2008. New insights into the pathogenesis of serous ovarian cancer and its clinical impact. J Clin Oncol 26, 5284–5293.

Marsh, S., Paul, J., King, C.R., et al., 2007. JCO Pharmacogenetic assessment of toxicity and outcome after platinum plus taxane chemotherapy in ovarian cancer: The Scottish randomized trial in ovarian cancer. J Clin Oncol 25, 4528–4535.

Mayr, D., Kanitz, V., Anderegg, B., et al., 2006. Analysis of gene amplification and prognostic markers in ovarian cancer using comparative genomic hybridization for microarrays and immunohistochemical analysis for tissue microarrays. Am J Clin Pathol 126, 101–109.

Meinhold-Heerlein, I., Bauerschlag, D., Hilpert, F., et al., 2005. Molecular and prognostic distinction between serous ovarian carcinomas of varying grade and malignant potential. Oncogene 24, 1053–1065.

Nakayama, K., Nakayama, N., Jinawath, N., et al., 2007. Amplicon profiles in ovarian serous carcinomas. Int J Cancer 120, 2613–2617.

Paige, A.J., Brown, R., 2008. Pharmaco (Epi)genomics in ovarian cancer. Pharamacogenomics 9, 1825–1834.

Ren, J., Xiao, Y.J., Singh, L.S., et al., 2006. Lysophosphatidic acid is constitutively produced by human peritoneal mesothelial cells and enhances adhesion, migration, and invasion of ovarian cancer cells. Cancer Res 66, 3006.

Sakai, W., Swisher, E.M., Karlan, B.Y., et al., 2008. Secondary mutations as a mechanism of cisplatin resistance in BRCA2-mutated cancers. Nature 451, 1116–1120.

Schiffenbauer, Y.S., Abramovitch, R., Meir, G., et al., 1997. Loss of ovarian function promotes angiogenesis in human ovarian carcinoma. Proc Natl Acad Sci USA 94, 13,203.

Schlosshauer, P.W., Cohen, C.J., Penault-Llorca, F., et al., 2003. Prophylactic oophorectomy: A morphologic and immunohistochemical study. Cancer 98, 2599.

Schwartz, D.R., Kardia, S.L., Shedden, K.A., et al., 2002. Gene expression in ovarian cancer reflects both morphology and biological behavior, distinguishing clear cell from other poor-prognosis ovarian carcinomas. Cancer Res 62, 4722–4729.

Sengupta, S., Kim, K.S., Berk, M.P., et al., 2007. Lysophosphatidic acid down-regulates tissue inhibitor of metalloproteinases, which are negatively involved in lysophosphatidic acid-induced cell invasion. Oncogene 26, 2894–2901.

Song, H., Ramus, S.J., Tyrer, J., et al., 2009. A genome-wide association study identifies a new ovarian cancer susceptibility locus on 9p22.2. Nat Genet 41, 996–1000.

Van Haaften-Day, C., Shen, Y., Xu, F., et al., 2001. OVX1, macrophage-colony stimulating factor, and CA-125-II as tumor markers for epithelial ovarian carcinoma: A critical appraisal. Cancer 92, 2837.

Wei, S.H., Chen, C.M., Strathdee, G., et al., 2002. Methylation microarray analysis of late-stage ovarian carcinomas distinguishes progression-free survival in patients and identifies candidate epigenetic markers. Clin Cancer Res 8, 2246.

Wright, J.W., Pejovic, T., Fanton, J., Stouffer, R.L., 2008. Induction of proliferation in the primate ovarian surface epithelium in vivo. Hum Reprod 23, 129–138.

Wright, J.W., Pejovic, T., Lawson, M., et al., 2010. Ovulation in the absence of the ovarian surface epithelium in the primate. Biol Reprod 82, 599–605.

Zhu, G., Saed, G.M., Deppe, G., Diamond, M.P., Munkarah, A.R., 2004. Hypoxia up-regulates the effects of prostaglandin E2 on tumor angiogenesis in ovarian cancer cells. Gynecol Oncol 94, 422.

Colorectal Cancer

Ad Geurts van Kessel, Ramprasath Venkatachalam, and Roland P. Kuiper

INTRODUCTION

Colorectal cancer represents a multistep process involving various molecular events that underlie its initiation and progression (Attolini et al., 2010; Foulds, 1958; Vogelstein and Kinzler, 1993; Vogelstein et al., 1988). The sequence of events from aberrant crypt proliferation or hyperplasia to benign adenoma, then to carcinoma *in situ* and finally to metastatic carcinoma is well defined. The tumor morphology and the likelihood of tumor progression are predominantly dependent on the temporal accumulation of genetic or genomic changes, accompanied by epigenetic alterations.

GENETICS OF COLORECTAL CANCER

In the earliest stages of the adenoma-carcinoma sequence, known as aberrant crypt foci (ACF), mutations in the *APC* gene are associated with the degree of dysplasia of these small lesions (Jen et al., 1994). An additional oncogenic *KRAS* mutation is required for adenoma growth, and the synergistic action of mutated *APC* and *KRAS* genes underlies clonal expansion and dysplasia in the nascent colorectal tumor (Fodde et al., 2001; Smith et al., 1994). Subsequent malignant transformation is driven by additional mutations and allelic losses, presumably centered around the tumor suppressor genes

SMAD4 and *TP53* (Vogelstein et al., 1988). Chromosome 17p losses, including the *TP53* gene, are found in 75% of colorectal carcinomas but infrequently in benign lesions, which indicates that loss of p53 function is involved in late tumor progression rather than initiation (Rodrigues et al., 1990). Overall, it has been predicted that at least seven genetic hits are required for full-blown colorectal cancer development (Fodde et al., 2001). Indeed, whole genome profiling efforts have yielded a handful of commonly mutated genes and an even larger number of infrequently mutated genes (Wood et al., 2007).

EPIGENETICS OF COLORECTAL CANCER

While genetic contributions to colorectal cancer are clear, they tell only part of the story, which includes a clear role for epigenetic modifications in both the initiation and progression of cancer. In addition to the above-mentioned gene mutations, global DNA methylation patterns in colorectal cancer cells have been found to differ considerably from those in their normal counterparts (Esteller, 2008; Feinberg and Vogelstein, 1983; Gargiulo and Minucci, 2009). Hypomethylation phenomena may convey diverse effects upon living cells, including an increase in genome instability, over-expression of a variety of genes, and loss of imprinting of particular genes, such as *IGF2*, the last of which has indeed been implicated in

Genomic and Personalized Medicine, 2nd edition
by Ginsburg & Willard. DOI: http://dx.doi.org/10.1016/B978-0-12-382227-7.00062-8

the pathogenesis of colorectal cancer (Cui et al., 2003). Besides global hypomethylation, discrete hypermethylation events that target promoter regions of specific genes have also frequently been observed in colorectal cancer cells (Esteller, 2008). In addition, genes that are mutated or hypermethylated in sporadic tumors may similarly be affected in inherited cases, as exemplified by the *MLH1* gene (see below) (Cui et al., 1998; Esteller et al., 2001).

At least two molecular pathways have been identified that may lead to colorectal cancer development and reflect a combination of genomic and epigenetic changes: the chromosomal instability (CIN) pathway and the CpG island methylator phenotype (CIMP) pathway (Grady and Carethers, 2008; Issa, 2008; Shen et al., 2007). CIN tumors account for approximately 80% of colorectal cancers in which genetic instability drives the adenoma-carcinoma sequence (Vogelstein et al., 1988). Tumors that develop through this pathway exhibit a microsatellite-stable but aneuploid phenotype. In contrast to CIN tumors, CIMP tumors show less severe aneuploidy. Instead, *BRAF*, or occasionally *KRAS*, mutation-induced activation of the RAS-RAF-MEK-ERK pathway is an important characteristic of these tumors (Jass et al., 2002; Kambara et al., 2004; Shen et al., 2007; Weisenberger et al., 2006). Recently, it was shown that DNA methylation patterns within normal mucosa vary with age and region and, as such, may also be associated with CIMP or CIN pathway-specific predispositions to colorectal cancer development (Worthley et al., 2010).

GENETICS OF COLORECTAL CANCER-ASSOCIATED SYNDROMES

Adding to the emerging picture of the genetics of colorectal cancer, during the past decades we have also witnessed the elucidation of the genetic basis of several familial cancer syndromes, many of which include colorectal cancer (De la Chapelle, 2004; Garber and Offit, 2005; Kuiper et al., 2010; McDermott et al., 2011). The identification of genes involved in these syndromes, to be summarized in this chapter, has resulted in immediate benefits for genetic testing and counseling, as well as pre-symptomatic screening and prevention, and has provided important insights into the molecular mechanisms underlying cancer susceptibility (Stoffel and Chittenden, 2010).

Several colorectal cancer syndromes with Mendelian dominant inheritance patterns have been identified, which can be subdivided into two major groups: presenting with non-polyposis or polyposis, respectively (De la Chapelle, 2004; Michils et al., 2005).

Non-Polyposis Colorectal Cancer

Lynch syndrome [also known as hereditary non-polyposis colorectal cancer (HNPCC)] represents the major non-polyposis colorectal cancer syndrome, and is caused by defects in DNA mismatch repair (MMR) genes, *MLH1*, *MSH2*, *PMS2*, or *MSH6*. These defects lead to, among other things, microsatellite instability in the tumors. Carriers of mutations in either one of these genes have a high risk of developing colorectal cancer (60–90%) and/or endometrial cancer (20–60%) and, to a lesser extent, cancer of the small bowel, stomach, ovary, urinary tract, and/or hepatobiliary tract (Lynch and De la Chapelle, 2003; Van der Post et al., 2010). It has been reported that a disease-causing germline mutation in one of the MMR genes can be identified in 78% of families suspected of Lynch syndrome in whom tumors are characterized by microsatellite instability and the absence of hypermethylation of the *MLH1* promoter (Overbeek et al., 2007). Subtypes or variants of Lynch syndrome include Muir-Torre syndrome (Kruse et al., 1998), and Turcot syndrome (De Vos et al., 2004; Hamilton et al., 1995).

Polyposis Colorectal Cancer

A well-defined hereditary form of polyposis colorectal cancer is caused by germline mutations in the *APC* gene. Carriers of such mutations will inevitably develop familial adenomatous polyposis (FAP). Once the role of *APC* in FAP was established, mutations in this gene were determined to be one of the most frequent early events in sporadic colorectal cancers as well (Kinzler and Vogelstein, 1996). In addition to FAP, an attenuated form (AFAP) of colorectal cancer exists, which exhibits fewer polyps and which is associated with mutations at the extreme 5′ and 3′ ends of the *APC* gene (Spirio et al., 1993). Other, less frequent, polyposis-associated colorectal cancer syndromes have also been described, and include Gardner syndrome, which has been diagnosed in patients with extracolonic features of FAP. At present, there is consensus that this latter syndrome is synonymous to FAP (Galiatsatos and Foulkes, 2006). Furthermore, Turcot syndrome is characterized by adenomatous polyposis together with early-onset central nervous system tumors and can be caused by either *APC* gene mutations or (bi-allelic) MMR gene mutations (see above; Hamilton et al., 1995; De Vos et al., 2004).

In contrast to all above-mentioned syndromes, *MUTYH*-associated polyposis (MAP) syndrome exhibits an autosomal recessive inheritance pattern and results from bi-allelic mutations in the *MUTYH* gene (Al-Tassan et al., 2002; Balaguer et al., 2007). In addition, germline mutations in the *LKB1/STK11* gene have been shown to cause Peutz-Jeghers syndrome (Hemminki et al., 1998; Jenne et al., 1998), and mutations in the *SMAD4* and *BMPR1A* genes have been shown to underlie juvenile polyposis syndrome (Houlston et al., 1998; Howe et al., 1998).

MISSING HERITABILITY IN COLORECTAL CANCER

Together, the above-mentioned syndromes are responsible for ~5% of all colorectal cancers, whereas twin studies have suggested that yet another 20–30% of colorectal cancers may be

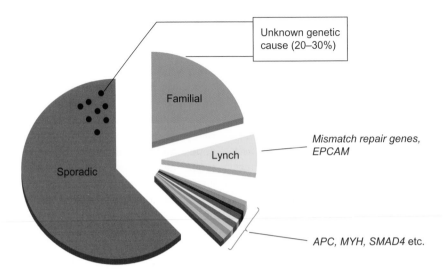

Figure 62.1 Colorectal cancers can be divied into sporadic, familial, and hereditary subtypes. In total, 35% of colorectal cancers are believed to result from a genetic defect but, as yet, only ~5% of them have been associated with a highly penetrant dominant or recessive inherited syndrome, such as Lynch syndrome. In 20–30% of familial and *de novo* early-onset colorectal cancer patients (the latter usually classified as sporadic; black dots), the genetic defect remains to be resolved. *Adapted from Lynch and Lynch, 2004.*

attributed to an underlying genetic susceptibility (Figure 62.1) (De la Chapelle, 2004; Lichtenstein et al., 2000; Lynch and De la Chapelle, 2003). A majority of this risk is believed to be due to low to moderate susceptibility genetic variants, according to two alternative hypotheses: the "common disease, common variant" (CDCV) hypothesis and the "common disease, rare variant" (CDRV) hypothesis (Bodmer and Bonilla, 2008).

The CDCV hypothesis states that multiple common variants (population frequency >5%), each attributing to only a minor increase in susceptibility, substantially influence the overall risk for a common disease. Recent developments in genotyping technologies have enabled the identification of such genetic variants in a more-or-less unbiased manner, by analyzing thousands of affected individuals and matched controls for hundreds of thousands of single nucleotide polymorphisms (SNPs) covering the entire human genome. Such so-called genome-wide association studies (GWAS) have so far revealed at least 14 different loci associated with an increased risk for colorectal cancer (Houlston et al., 2010; Tenesa and Dunlop, 2009; Tomlinson et al., 2010). In two studies it was shown that a common variant on chromosome 8q24 can affect Wnt signaling and *MYC* expression (Pomerantz et al., 2009; Tuupanen et al., 2009). In addition, it has been shown that a common variant on chromosome 8q23.3 can act as a *cis*-acting regulator of the eukaryotic translation initiation factor 3 subunit H gene *EIF3H*, whose increased expression results in colorectal cancer growth and invasiveness (Pittman et al., 2010). These studies have provided the first functional clues on how common genetic variants can influence colorectal cancer susceptibility. So far, however, these variants still may explain only a small percentage of its heritability.

According to the competing CDRV hypothesis, at least part of the so-far missing heritability might be explained by a cumulative effect of multiple rare variants (population frequency <1%). These variants might act as independent dominant susceptibility factors, each conveying a moderate but significant increase in cancer risk (Bodmer and Bonilla, 2008). Such rare variants may be identified by hypothesis-driven strategies based on the assumption that they affect genes involved in, for instance, Wnt signaling or DNA mismatch repair in colorectal cancer (Fearnhead et al., 2004; Frayling et al., 1998; Lipkin et al., 2004). These variants, however, will not be detected even by very extensive GWAS, because of their low frequency, specificity to an individual, and limited contributions to the overall risk of tumor development. Thus, alternative genomic strategies will be required for their identification.

In the past few years, constitutional copy number variants (CNVs), usually deletions, have been encountered in cancer-prone families and have been shown to be of critical importance for the identification of novel predisposing genes. It was also found that CNVs occur in about 4–15% of such families, which has turned DNA dosage analysis into an essential component of genetic screening for cancer predisposition (Bunyan et al., 2004; Michils et al., 2005; Wagner et al., 2003; Wijnen et al., 1998). Van der Klift and colleagues (2005) performed a systematic search for genomic variants in a cohort of over 400 Lynch syndrome families, and identified 48 rearrangements in 68 unrelated kindreds, of which six left the coding region of the presumptive disease gene intact. Interestingly, these rearrangements comprised four deletions upstream of the *MSH2* gene, suggesting that they might interfere – most likely through epigenetic mechanisms – with *MSH2* gene function.

"EPIMUTATIONS" AND HEREDITARY COLORECTAL CANCER

About one-third of all cases that meet the criteria for Lynch syndrome do not exhibit pathogenic mutations in one of the known MMR genes. For at least two genes – *MLH1* and *MSH2* – the causal mechanism of gene dysfunction appears to involve as yet incompletely understood heritable epigenetic changes.

Gazzoli and colleagues (2002) first identified a person with early-onset colorectal cancer who lacked MMR gene mutations but, instead, showed constitutional hemi-allelic hypermethylation of the *MLH1* gene. The tumor of this patient showed microsatellite instability and a complete loss of *MLH1* expression due to somatic deletion of the second allele. The subsequent identification of similar familial cases allowed a further delineation of what have been termed germline "epimutations" at the *MLH1* gene.

Hypermethylation of CpG sites in different patients is restricted to a single allele at the *MLH1* gene, as shown by employing a heterozygous SNP within the gene promoter (Hitchins et al., 2005; Miyakura et al., 2004; Suter et al., 2004). Transcriptional silencing of the affected allele was also confirmed at the mRNA level in normal cells by exploiting heterozygous SNPs located within coding regions (exons) of the *MLH1* gene. Notably, it was found that *MLH1* epimutation carriers exhibited mono- or hemi-allelic *MLH1* hypermethylation patterns even throughout their normal somatic tissues, suggestive of a germline origin (Hitchins et al., 2007).

Such germline *MLH1* epimutations are distinct from somatic epigenetic *MLH1* silencing events, which can be observed in sporadic microsatellite-unstable colorectal cancers of the elderly, where bi-allelic hypermethylation of *MLH1* and its flanking genes is essentially confined to the tumors (Hitchins and Ward, 2009). It is thought that *de novo* hypermethylation of the core promoter of this gene is seeded from flanking methylated sites, which may also explain the apparent stochastic process of *de novo* hypermethylation events observed in these tumors (Clark and Melki, 2002). Additionally, the penetrance of germline *MLH1* promoter hypermethylation may not be complete, as was exemplified by a family in which a male Lynch syndrome proband was found to have inherited the epimutation from his unaffected mother (Morak et al., 2008).

Hitchins and colleagues (2007) were also able to show that germline *MLH1* epimutations may follow a non-Mendelian inheritance pattern and, in addition, may be reversible. Together, these observations suggest that in different patients germline epimutations may be caused by different mechanisms, associated with different inheritance and transmission patterns. Whereas the exact nature of the mechanism(s) underlying these epimutations still remains to be established, it has been suggested that a *trans*-acting factor (such as a modifier, see below) may be involved (Hesson et al., 2010), which could explain the epimutation reversibility between generations and the non-Mendelian inheritance patterns observed.

Importantly, such epimutations are not limited to *MLH1*. Chan and colleagues (2006) reported a germline *MSH2* epimutation in a three-generation family presenting with Lynch-associated tumors and varying levels of *MSH2* methylation in normal blood and rectal mucosa-derived cells. At that time, the mechanism underlying that inherited epimutation also remained elusive.

GENE SILENCING FROM TRANSCRIPTIONAL READ-THROUGH

The above findings have raised the intriguing possibility that colorectal cancer susceptibility in a significant proportion of families may have an epigenetic basis. Based on their results and building on the well-known "two hit" hypothesis of cancer initiation, Chan and colleagues (2006) hypothesized that the observed germline *MSH2* epimutation could provide a first hit towards colorectal cancer development in the family studied. Through high-resolution DNA copy number analysis, Ligtenberg and colleagues (2009) encountered microdeletions in the 3′ region of the *EPCAM* gene, located upstream of *MSH2*, in four Dutch MMR-deficient families. As a result of these microdeletions, the transcription-terminating polyadenylation signals in the *EPCAM* gene were lost, leading to extended transcription (read-through) beyond the normal termination site and into the adjacent *MSH2* gene. Subsequently, it was found that all patients with such a microdeletion showed *cis*-limited hypermethylation of the *MSH2* promoter in the microsatellite-positive tumors as well as in their normal epithelial tissues, showing a strong correlation with *EPCAM* expression levels.

These data indicate that *EPCAM* 3′ deletions represent a novel type of epigenetic germline mutation that leads to tissue-specific inactivation of the *MSH2* gene. Strikingly, a very similar deletion and concomitant promoter-silencing mechanism was observed in the family originally described by Chan and colleagues, thus corroborating the results obtained (Chan et al., 2006; Ligtenberg et al., 2009). Thus far, 19 different *EPCAM* deletions have been described in 45 Lynch syndrome families from multiple geographical origins (Kuiper et al., 2011). These deletions, though variable in size and location, invariably encompass the last two exons of the *EPCAM* gene, including its polyadenylation signals. Carriers of these deletions have a high risk of developing colorectal cancer, very similar to that noted earlier in carriers of *MLH1* or *MSH2* gene mutations. In contrast to the high colorectal cancer risk, however, the *EPCAM* deletion carriers were found to exhibit a strikingly low risk of developing endometrial cancer. This contrasts with patients with deletions extending close to the *MSH2* promoter, who appeared to have an unchanged risk of developing endometrial cancer (Kempers et al., 2011).

Although the basis for this remarkable clinical difference remains to be established, the identification and characterization of *EPCAM* deletions by copy number analysis now suggests an optimized protocol for the recognition and targeted prevention of cancer in *EPCAM* deletion carriers. Hence, implementation of *EPCAM* deletion mapping in routine clinical diagnostics should be considered for suspected Lynch syndrome families (Kempers et al., 2011; Kuiper et al., 2011). Specifically, clinical screening could be focused on colorectal cancer and, by doing so, some women might be saved the necessity of surveillance and prophylactic hysterectomy.

Despite the fact that the exact mechanism underlying transcription-mediated epigenetic silencing in colorectal cancer remains to be established, there are examples in the literature describing similar phenomena. For example, aberrant constitutional hypermethylation and silencing of the α-globin (*HBA2*) gene has been reported in an individual with α-thalassemia (Tufarelli et al., 2003). In this case, a 3′ truncating deletion of the *LUC7L* gene caused antisense transcriptional read-through and hypermethylation of the adjacent *HBA2* gene promoter and a concomitant *cis*-limited silencing of the gene, leaving the gene itself intact. Similarly, Yu and colleagues (2008) found that antisense *p15* RNA expression can induce epigenetic chromatin changes resulting in *cis* hypermethylation of the *p15* gene promoter and silencing of the gene. It has also been shown that maternal imprinting of the *Gnas* locus in mouse oocytes depends on transcription across the entire locus from the upstream *Nesp* gene promoter. Notably, maternal microdeletions of the orthologous *NESP55* gene in humans can cause pseudohypoparathyroidism type 1b (Bastepe et al., 2005; Chotalia et al., 2009), clearly pointing to a correlation between transcription and DNA methylation. One possible explanation for this phenomenon could be RNA–DNA duplex formation in the promoter region that may, either directly or indirectly, induce recruitment of the DNA methylation machinery, resulting in epigenetic remodeling of the gene promoter (Hawkins et al., 2009). Whether a similar mechanism underlies *MSH2* methylation in *EPCAM* deletion carriers, however, remains to be established.

NOVEL COLORECTAL CANCER PREDISPOSING GENES

Based on the above observations, it has been hypothesized that additional rare CNVs leading to the disruption and/or silencing of critical genes could result in the identification of novel genetic variants involved in moderate- to high-penetrance forms of cancer predisposition. In order to test this hypothesis, a discovery cohort of colorectal cancer patients with unexplained microsatellite stability was assembled based on stringent criteria (Venkatachalam et al., 2011). Subsequent high-resolution genomic profiling of these patients resulted in the identification of novel variants affecting a number of genes, including the bone morphogenetic protein antagonist family gene *GREM1*, the breakpoint cluster region gene *BCR*, the protein tyrosine phosphatase gene *PTPRJ*, the cadherin gene *CDH18* and the predicted protein-coding gene *KIAA1797* (Venkatachalam et al., 2011). In addition, two genomic deletions were encountered encompassing microRNA genes, *miR-491* located within the *KIAA1797* gene and *miR-646*. In keeping with the CDRV hypothesis introduced earlier, these results indicate that novel, rare germline CNVs may indeed play a role in colorectal cancer susceptibility and could account for at least part of the currently unaccounted-for heritability (Kuiper et al., 2010).

Several of the newly identified colorectal cancer-susceptibility candidate genes could be functionally linked to known colorectal cancer-associated cellular pathways, but as yet the strongest support seems to be available for *PTPRJ*. Initially, mouse *Ptprj* was identified as the target of a colorectal cancer-susceptibility locus, *Scc1* (Ruivenkamp et al., 2002). Subsequently, loss of heterozygosity at the human *PTPRJ* gene was found in several tumor types, including colon cancer (Iuliano et al., 2004; Ruivenkamp et al., 2003). *PTPRJ* loss of heterozygosity was also frequently encountered in aberrant crypt foci, the earliest neoplastic lesions of the colon, suggesting a major role for this gene in early colon neoplasia (Luo et al., 2006). The *PTPRJ* gene codes for the density-enhanced phosphatase DEP-1 (also designated CD148), which is a widely expressed candidate tumor suppressor, involved in the control of cellular growth and transformation by antagonizing receptor tyrosine kinase activities (Balavenkatraman et al., 2006; Iuliano et al., 2004; Keane et al., 1996; Kovalenko et al., 2000; Massa et al., 2004; Palka et al., 2003). Additional loss-of-function analyses have revealed that DEP-1 may inhibit the motility and, presumably, invasion of tumor cells (Petermann et al., 2011). Interestingly, it has also been found that a novel germline partial duplication in *PTPRJ* (Venkatachalam et al., 2010) results in transcriptional read-through and concomitant allele-specific hypermethylation and silencing of, in this case, the wild-type *PTPRJ* locus. Based on this latter observation, it is tempting to speculate that a variety of genomic alterations, including deletions and duplications, but also inversions and translocations, may induce aberrant promoter methylation through transcriptional read-through (Kuiper et al., 2010). Since most tumor suppressor genes associated with Mendelian cancer syndromes contain CpG island promoters, which can be affected by transcriptional read-through, targeted analysis of their neighboring genes seems warranted (Figure 62.2).

FUNCTIONAL GROUPING OF VARIANTS IN PATHWAYS

Despite the fact that the identified copy number variants are rare, many of the affected genes can be functionally linked to pathways associated with colorectal cancer development. Previously, DEP-1 has been found to play an important role in

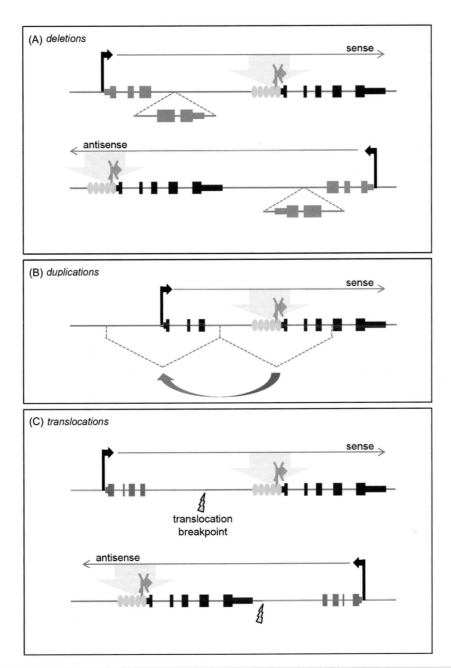

Figure 62.2 Epigenetic gene silencing (indicated by orange ovals) induced by sense or antisense read-through of truncated transcriptional units into a CpG island promoter. Transcriptional read-through can occur through (A) deletion of the 3′-region of an upstream gene, (B) a partial gene duplication, or (C) translocation of an active promoter element into another genomic region. Germline epigenetic silencing may be mosaic due to a tissue-specific activity of the promoter driving the read-through transcript. *From Kuiper et al., 2010.*

the regulation of cell–cell adhesion through dephosphorylation of the cadherin-catenin complex at adherence junctions (Grazia Lampugnani et al., 2003; Jandt et al., 2003; Kellie et al., 2004; Lilien and Balsamo, 2005; Östman et al., 2006; Palka et al., 2003). It has also been reported that deregulation of cadherin-catenin complexes may contribute to tumor development by affecting the adhesion of epithelial cells in the colon (Goss and Groden, 2000). Furthermore, it has been found that

this deregulation may affect Wnt signaling by (de)regulating the stability and sub-cellular localization of β-catenin (Cavallaro et al., 2004; Fodde et al., 2001). E-cadherins may have profound effects on the morphogenesis of tissues mediated by cell surface adhesion, cell–cell recognition, and regulation of the actin cytoskeleton (Tepass et al., 2000). Additionally, β-catenin has been implicated, next to Wnt signaling, in the rearrangement of mitotic spindles (Cabello et al., 2010). Classical cadherins affect tissue organization by acting as cues that orient the mitotic spindle during symmetric cell divisions in mammalian epithelia (Den Elzen et al., 2009). Thus, cell adherens junction components, including E-cadherin and β-catenin, may also be responsible for spindle orientation during mitosis (Cabello et al., 2010; Inaba et al., 2010; Tepass, 2002).

Interestingly, it has been proposed that the colorectal cancer-predisposing gene *APC* acts as a multifunctional tumor suppressor that, besides suppressing canonical Wnt signaling, also promotes tumorigenesis through loss of cell adhesion (Aoki and Taketo, 2007; Bienz and Hamada, 2004; Birchmeier et al., 1995). Indeed, a mutation in the *Apc* gene in mouse intestinal epithelial cells was found to decrease the level of E-cadherin at the cell membrane (Carothers et al., 2001). Correspondingly, human *APC* has been found to regulate microtubule networks and to play a role in microtubule-mediated processes such as cell migration and mitotic spindle formation (Fodde et al., 2001; Nathke, 2006). So, the various gene discovery and functional annotation strategies mentioned have not only been successful in the identification of colorectal cancer-susceptibility genes, but also in exposing important functional pathways that may play a crucial mechanistic role in colorectal cancer susceptibility. Thus, grouping of variants based on gene function in single biochemical pathways and networks may be considered a strategy to search for novel, rare genetic colorectal cancer-susceptibility variants (Bodmer and Tomlinson, 2010).

MODERATE RISKS AND MODIFIER GENES

Modifier genes are defined as genes that affect the phenotypic and/or molecular expression of other genes. Genetic modifiers can affect penetrance, dominance, expressivity, and pleiotropy (Nadeau, 2001). A classical example of a modifier gene comes from the *APC^{min/+}* mouse model, a murine counterpart of human FAP (Moser et al., 1990). These mice exhibit wide phenotypic variation, for example, in the number of polyps, depending on the genetic background. By linkage analysis, a modifier gene originally called *Mom1* for "modifier of Min" in the *APC^{min/+}* mouse model was mapped to the distal part of chromosome 4 (Dietrich et al., 1993). This modifier was able to explain 50% of the genetic variance in polyp number (Nadeau, 2001).

Such a scenario may apply to some of the above-mentioned newly identified colorectal cancer candidate genes, including *PTPRJ*. In fact, mouse *Ptprj* has been suggested to function as a genetic modifier affecting penetrance (Lesueur et al., 2005; Ruivenkamp et al., 2002). Also, in colorectal adenomas, loss of heterozygosity at the *PTPRJ* locus was found to occur as an early event, followed by the loss of chromosome 18q12-21, which includes genes like *SMAD4* and *SMAD2* (Ruivenkamp et al., 2003). Since there is ample evidence indicating that independent genetic factors can modify cancer

risks of mutation carriers in Mendelian cancer syndromes (Antoniou and Chenevix-Trench, 2010), it is conceivable that the newly identified candidate genes may act equally as modifiers.

So far, however, candidate gene studies have not been very successful in identifying such putative modifiers and most attempts appear to have been underpowered. Only GWAS studies have been successful in the identification of variants that may modify cancer risks (Antoniou and Chenevix-Trench, 2010). Based on current data, we conclude that as an adjunct to GWAS studies, high-resolution copy number profiling may be employed as an approach to identify genetic modifiers in cancer syndromes.

PERSPECTIVES

Predictive clinical genetic testing is currently available for various Mendelian cancer syndromes, including FAP (*APC*, *MUTYH*) and Lynch syndrome (*MSH2*, *EPCAM*, *MLH1*, *MSH6*, *PMS2*), and patients carrying mutations in any one of these highly penetrant genes may immediately be subjected to surveillance. However, genetic testing and counseling still remain a challenge in families with an obvious strong suspicion of a genetic predisposition, but lacking overt pathogenic mutations in any of the known highly penetrant genes. Hence, for clinical practice, the validation and application of additional susceptibility variants, including those with lower penetrance, will be of paramount importance.

Unfortunately, the various low-penetrance colorectal cancer-susceptibility loci that have been identified via GWAS studies in the last five years have an as-yet limited clinical relevance, since they each contribute little to an individual's risk. Hence, an ideal strategy for better understanding colorectal cancer susceptibility would be to integrate information from SNP, CNV and whole-genome next-generation sequencing studies. This wealth of information will, without any doubt, augment novel genetic testing strategies and genetic counseling. In addition, even with a modest genotype–phenotype association, the identified candidate genes may offer considerable new translational prospects through the identification of modifiable cellular pathways, leading to novel options for therapeutic intervention and/or the generation of preventive strategies (McCarthy et al., 2008).

It has become clear that rare CNVs can impact genes that act in cancer-related pathways, thereby resulting in an increased cancer risk. Since these CNVs can easily be interrogated in genome-wide screens, genomic profiling strategies can efficiently be deployed in the search for novel cancer-predisposing genes. In the near future, the detection of new CNVs, including microdeletions and duplications, by novel strategies such as high-throughput paired-end mapping using large scale genomic (re-)sequencing (Korbel et al., 2007) is anticipated to further revolutionize this field of research. With these technical possibilities within reach, new concepts of gene inactivation may be uncovered, as has been shown, for example, for the deletion of the 3′ end of the *EPCAM* gene that leads to Lynch syndrome through epigenetic silencing of the downstream *MSH2* gene. The observed transcriptional read-through and concomitant allele-specific epigenetic silencing of the *PTPRJ* locus indicates that gene silencing by transcriptional read-through may represent a general mutational mechanism in colorectal cancer.

It is anticipated that the application of new comprehensive genome profiling strategies will lead to the identification of novel colorectal cancer-susceptibility genes and mechanisms at an increasingly rapid pace. This, in turn, is expected to provide further insight into the etiology of colorectal cancer and to open up new avenues for improved diagnosis, prognosis, and (personalized) therapy of both hereditary and non-hereditary forms of colorectal cancer.

ACKNOWLEDGMENTS

The authors thank M. Ligtenberg and N. Hoogerbrugge for their contributions. This work was supported by the Dutch Cancer Society (NKB-KWF) and the Netherlands Organization for Health Research and Development (ZonMW).

REFERENCES

Al-Tassan, N., Chmiel, N.H., Maynard, J., et al., 2002. Inherited variants of MYH associated with somatic G:C→T:A mutations in colorectal tumors. Nat Genet 30, 227–232.

Antoniou, A.C., Chenevix-Trench, G., 2010. Common genetic variants and cancer risk in Mendelian cancer syndromes. Curr Opin Genet Dev 20, 299–307.

Aoki, K., Taketo, M.M., 2007. Adenomatous polyposis coli (APC): A multi-functional tumor suppressor gene. J Cell Sci 120, 3327–3335.

Attolini, C.S., Cheng, Y.K., Beroukhim, R., et al., 2010. A mathematical framework to determine the temporal sequence of somatic gene events in cancer. Proc Natl Acad Sci USA 107, 17,604–17,609.

Balaguer, F., Castellví-Bel, S., Castells, A., et al., 2007. Identification of MYH mutation carriers in colorectal cancer: A multicenter, case-control, population-based study. Clin Gastroenterol Hepatol 5, 379–387.

Balavenkatraman, K.K., Jandt, E., Friedrich, K., et al., 2006. DEP-1 protein tyrosine phosphatase inhibits proliferation and migration of colon carcinoma cells and is upregulated by protective nutrients. Oncogene 25, 6319–6324.

Bastepe, M., Fröhlich, L.F., Linglart, A., et al., 2005. Deletion of the NESP55 differentially methylated region causes loss of maternal GNAS imprints and pseudohypoparathyroidism type Ib. Nat Genet 37, 25–27.

Bienz, M., Hamada, F., 2004. Adenomatous polyposis coli proteins and cell adhesion. Curr Opin Cell Biol 16, 528–535.

Birchmeier, W., Hulsken, J., Behrens, J., 1995. Adherens junction proteins in tumour progression. Cancer Surv 24, 129–140.

Bodmer, W., Bonilla, C., 2008. Common and rare variants in multifactorial susceptibility to common diseases. Nat Genet 40, 695–701.

Bodmer, W., Tomlinson, I., 2010. Rare genetic variants and the risk of cancer. Curr Opin Genet Dev 20, 262–267.

Bunyan, D.J., Eccles, D.M., Sillibourne, J., et al., 2004. Dosage analysis of cancer predisposition genes by multiplex ligation-dependent probe amplification. Br J Cancer 91, 1155–1159.

Cabello, J., Neukomm, L.J., Günesdogan, U., et al., 2010. The Wnt pathway controls cell death engulfment, spindle orientation, and migration through CED-10/Rac. PLoS Biol 8, e1000297.

Carothers, A.M., Melstrom Jr, K.A., Mueller, J.D., Weyant, M.J., Bertagnolli, M.M., 2001. Progressive changes in adherens junction structure during intestinal adenoma formation in Apc mutant mice. J Biol Chem 276, 39,094–39,102.

Cavallaro, U., Christofori, G., 2004. Cell adhesion and signalling by cadherins and Ig-CAMs in cancer. Nat Rev Cancer 4, 118–132.

Chan, T.L., Yuen, S.T., Kong, C.K., et al., 2006. Heritable germline epimutation of MSH2 in a family with hereditary nonpolyposis colorectal cancer. Nat Genet 38, 1178–1183.

Chotalia, M., Smallwood, S.A., Ruf, N., et al., 2009. Transcription is required for establishment of germline methylation marks at imprinted genes. Genes Dev 23, 105–117.

Clark, S.J., Melki, J., 2002. DNA methylation and gene silencing in cancer: Which is the guilty party? Oncogene 21, 5380–5387.

Cui, H., Cruz-Correa, M., Giardiello, F.M., et al., 2003. Loss of IGF2 imprinting: A potential marker of colorectal cancer risk. Science 299, 1753–1755.

Cui, H., Horon, I.L., Ohlsson, R., Hamilton, S.R., Feinberg, A.P., 1998. Loss of imprinting in normal tissue of colorectal cancer patients with microsatellite instability. Nat Med 4, 1276–1280.

De la Chapelle, A., 2004. Genetic predisposition to colorectal cancer. Nat Rev Cancer 4, 769–780.

Den Elzen, N., Buttery, C.V., Maddugoda, M.P., Ren, G., Yap, A.S., 2009. Cadherin adhesion receptors orient the mitotic spindle during symmetric cell division in mammalian epithelia. Mol Biol Cell 20, 3740–3750.

De Vos, M., Hayward, B.E., Picton, S., Sheridan, E., Bonthron, D.T., 2004. Novel PMS2 pseudogenes can conceal recessive mutations causing a distinctive childhood cancer syndrome. Am J Hum Genet 74, 954–964.

Dietrich, W.F., Lander, E.S., Smith, J.S., et al., 1993. Genetic identification of Mom-1, a major modifier locus affecting Min-induced intestinal neoplasia in the mouse. Cell 75, 631–639.

Esteller, M., 2008. Epigenetics in cancer. N Engl J Med 358, 1148–1159.

Esteller, M., Fraga, M.F., Guo, M., et al., 2001. DNA methylation patterns in hereditary human cancers mimic sporadic tumorigenesis. Hum Mol Genet 10, 3001–3007.

Fearnhead, N.S., Wilding, J.L., Winney, B., et al., 2004. Multiple rare variants in different genes account for multifactorial inherited susceptibility to colorectal adenomas. Proc Natl Acad Sci USA 101, 15,992–15,997.

Feinberg, A.P., Vogelstein, B., 1983. Hypomethylation distinguishes genes of some human cancers from their normal counterparts. Nature 301, 89–92.

Fodde, R., Smits, R., Clevers, H., 2001. APC, signal transduction and genetic instability in colorectal cancer. Nat Rev Cancer 1, 55–67.

Foulds, L., 1958. The natural history of cancer. J Chronic Dis 8, 2–37.

Frayling, I.M., Beck, N.E., Ilyas, M., et al., 1998. The APC variants I1307K and E1317Q are associated with colorectal tumors, but not always with a family history. Proc Natl Acad Sci USA 95, 10,722–10,727.

Galiatsatos, P., Foulkes, W.D., 2006. Familial adenomatous polyposis. Am J Gastroenterol 101, 385–398.

Garber, J.E., Offit, K., 2005. Hereditary cancer predisposition syndromes. J Clin Oncol 23, 276–292.

Gargiulo, G., Minucci, S., 2009. Epigenomic profiling of cancer cells. Int J Biochem Cell Biol 41, 127–135.

Gazzoli, I., Loda, M., Garber, J., Syngal, S., Kolodner, R.D., 2002. A hereditary nonpolyposis colorectal carcinoma case associated with hypermethylation of the MLH1 gene in normal tissue and loss of heterozygosity of the unmethylated allele in the resulting microsatellite instability-high tumor. Cancer Res 62, 3925–3928.

Goss, K.H., Groden, J., 2000. Biology of the adenomatous polyposis coli tumour suppressor. J Clin Oncol 18, 1967–1979.

Grady, W.M., Carethers, J.M., 2008. Genomic and epigenetic instability in colorectal cancer pathogenesis. Gastroenterology 135, 1079–1099.

Grazia Lampugnani, M., Zanetti, A., Corada, M., et al., 2003. Contact inhibition of VEGF-induced proliferation requires vascular endothelial cadherin, beta-catenin, and the phosphatase DEP-1/CD148. J Cell Biol 161, 793–804.

Hamilton, S.R., Liu, B., Parsons, R.E., et al., 1995. The molecular basis of Turcot's syndrome. N Engl J Med 332, 839–847.

Hawkins, P.G., Santoso, S., Adams, C., Anest, V., Morris, K.V., 2009. Promoter targeted small RNAs induce long-term transcriptional gene silencing in human cells. Nucleic Acids Res 37, 2984–2995.

Hemminki, A., Markie, D., Tomlinson, I., et al., 1998. A serine/threonine kinase gene defective in Peutz-Jeghers syndrome. Nature 391, 184–187.

Hesson, L.B., Hitchins, M.P., Ward, R.L., 2010. Epimutations and cancer predisposition: Importance and mechanisms. Curr Opin Genet Dev 20, 290–298.

Hitchins, M.P., Ward, R.L., 2009. Constitutional (germline) MLH1 epimutation as an aetiological mechanism for hereditary non-polyposis colorectal cancer. J Med Genet 46, 793–802.

Hitchins, M.P., Williams, R., Cheong, K., et al., 2005. MLH1 germline epimutations as a factor in hereditary nonpolyposis colorectal cancer. Gastroenterology 129, 1392–1399.

Hitchins, M.P., Wong, J.J., Suthers, G., et al., 2007. Inheritance of a cancer-associated MLH1 germ-line epimutation. N Engl J Med 356, 697–705.

Houlston, R.S., Bevan, S., Williams, A., et al., 1998. Mutations in DPC4 (SMAD4) cause juvenile polyposis syndrome, but only account for a minority of cases. Hum Mol Genet 7, 1907–1912.

Houlston, R.S., Cheadle, J., Dobbins, S.E., et al., 2010. Meta-analysis of three genome-wide association studies identifies susceptibility loci for colorectal cancer at 1q41, 3q26.2, 12q13.13 and 20q13.33. Nat Genet 42, 973–977.

Howe, J.R., Roth, S., Ringold, J.C., et al., 1998. Mutations in the SMAD4/DPC4 gene in juvenile polyposis. Science 280, 1086–1088.

Inaba, M., Yuan, H., Salzmann, V., Fuller, M.T., Yamashita, Y.M., 2010. E-cadherin is required for centrosome and spindle orientation in *Drosophila* male germline stem cells. PLoS One 5, e12473.

Issa, J.P., 2008. Colon cancer: It's CIN or CIMP. Clin Cancer Res 14, 5939–5940.

Iuliano, R., Le Pera, I., Cristofaro, C., et al., 2004. The tyrosine phosphatase PTPRJ/DEP-1 genotype affects thyroid carcinogenesis. Oncogene 23, 8432–8438.

Jandt, E., Denner, K., Kovalenko, M., Ostman, A., Böhmer, F.D., 2003. The protein-tyrosine phosphatase DEP-1 modulates growth factor-stimulated cell migration and cell-matrix adhesion. Oncogene 22, 4175–4185.

Jass, J.R., Whitehall, V.L., Young, J., et al., 2002. Emerging concepts in colorectal neoplasia. Gastroenterology 123, 862–876.

Jen, J., Powell, S.M., Papadopoulos, N., et al., 1994. Molecular determinants of dysplasia in colorectal lesions. Cancer Res 54, 5523–5526.

Jenne, D.E., Reimann, H., Nezu, J., et al., 1998. Peutz-Jeghers syndrome is caused by mutations in a novel serine threonine kinase. Nat Genet 18, 38–43.

Kambara, T., Simms, L.A., Whitehall, V.L., et al., 2004. BRAF mutation is associated with DNA methylation in serrated polyps and cancers of the colorectum. Gut 53, 1137–1144.

Keane, M.M., Ettenberg, S.A., Lowrey, G.A., Russell, E.K., Lipkowitz, S., 1996. The protein tyrosine phosphatase DEP-1 is induced during differentiation and inhibits growth of breast cancer cells. Cancer Res 56, 4236–4243.

Kellie, S., Craggs, G., Bird, I.N., Jones, G.E., 2004. The tyrosine phosphatase DEP-1 induces cytoskeletal rearrangements, aberrant cell-substratum interactions and a reduction in cell proliferation. J Cell Sci 117, 609–618.

Kempers, M.J., Kuiper, R.P., Ockeloen, C.W., et al., 2011. Risk of colorectal and endometrial cancers in EPCAM deletion-positive Lynch syndrome: A cohort study. Lancet Oncol 12, 49–55.

Kinzler, K.W., Vogelstein, B., 1996. Lessons from hereditary colorectal cancer. Cell 87, 159–170.

Korbel, J.O., Urban, A.E., Affourtit, J.P., et al., 2007. Paired-end mapping reveals extensive structural variation in the human genome. Science 318, 420–426.

Kovalenko, M., Denner, K., Sandström, J., et al., 2000. Site-selective dephosphorylation of the platelet-derived growth factor beta-receptor by the receptor-like protein-tyrosine phosphatase DEP-1. J Biol Chem 275, 16,219–16,226.

Kruse, R., Rütten, A., Lamberti, C., et al., 1998. Muir-Torre phenotype has a frequency of DNA mismatch-repair-gene mutations similar to that in hereditary nonpolyposis colorectal cancer families defined by the Amsterdam criteria. Am J Hum Genet 63, 63–70.

Kuiper, R.P., Ligtenberg, M.J., Hoogerbrugge, N., Geurts Van Kessel, A., 2010. Germline copy number variation and cancer risk. Curr Opin Genet Dev 20, 282–289.

Kuiper, R.P., Vissers, L.E., Venkatachalam, R., et al., 2011. Recurrence and variability of germline EPCAM deletions in Lynch syndrome. Hum Mutat 32, 407–414.

Lesueur, F., Pharoah, P.D., Laing, S., et al., 2005. Allelic association of the human homologue of the mouse modifier Ptprj with breast cancer. Hum Mol Genet 14, 2349–2356.

Lichtenstein, P., Holm, N.V., Verkasalo, P.K., et al., 2000. Environmental and heritable factors in the causation of cancer – analyses of cohorts of twins from Sweden, Denmark, and Finland. N Engl J Med 343, 78–85.

Ligtenberg, M.J.L., Kuiper, R.P., Chan, T.L., et al., 2009. Heritable somatic methylation and inactivation of MSH2 in families with Lynch syndrome due to deletion of the 3′ exons of TACSTD1. Nat Genet 41, 112–117.

Lilien, J., Balsamo, J., 2005. The regulation of cadherin-mediated adhesion by tyrosine phosphorylation/dephosphorylation of β-catenin. Curr Opin Cell Biol 17, 459–465.

Lipkin, S.M., Rozek, L.S., Rennert, G., et al., 2004. The MLH1 D132H variant is associated with susceptibility to sporadic colorectal cancer. Nat Genet 36, 694–699.

Luo, L., Shen, G.Q., Stiffler, K.A., et al., 2006. Loss of heterozygosity in human aberrant crypt foci (ACF), a putative precursor of colon cancer. Carcinogenesis 27, 1153–1159.

Lynch, H.T., De la Chapelle, A., 2003. Hereditary colorectal cancer. N Engl J Med 348, 919–932.

Lynch, H.T., Lynch, J.F., 2004. Lynch syndrome: History and current status. Dis Markers 20, 181–198.

Massa, A., Barbieri, F., Aiello, C., et al., 2004. The expression of the phosphotyrosine phosphatase DEP-1/PTPeta dictates the responsivity of glioma cells to somatostatin inhibition of cell proliferation. J Biol Chem 279, 29,004–29,012.

McCarthy, M.I., Abecasis, G.R., Cardon, L.R., et al., 2008. Genome-wide association studies for complex traits: Consensus, uncertainty and challenges. Nat Rev Genet 9, 356–369.

McDermott, U., Downing, J.R., Stratton, M.R., 2011. Genomics and the continuum of cancer care. N Engl J Med 364, 340–350.

Michils, G., Tejpar, S., Thoelen, R., et al., 2005. Large deletions of the APC gene in 15% of mutation-negative patients with classical polyposis (FAP): A Belgian study. Hum Mut 25, 125–134.

Miyakura, Y., Sugano, K., Akasu, T., et al., 2004. Extensive but hemi-allelic methylation of the hMLH1 promoter region in early-onset sporadic colon cancers with microsatellite instability. Clin Gastroenterol Hepatol 2, 147–156.

Morak, M., Schackert, H.K., Rahner, N., et al., 2008. Further evidence for heritability of an epimutation in one of 12 cases with MLH1 promoter methylation in blood cells clinically displaying HNPCC. Eur J Hum Genet 16, 804–811.

Moser, A.R., Pitot, H., Dove, W.F., 1990. A dominant mutation that predisposes to multiple intestinal neoplasia in the mouse. Science 247, 322–324.

Nadeau, J.H., 2001. Modifier genes in mice and humans. Nat Rev Genet 2, 165–174.

Nathke, I., 2006. Cytoskeleton out of the cupboard: Colon cancer and cytoskeletal changes induced by loss of APC. Nat Rev Cancer 6, 967–974.

Östman, A., Hellberg, C., Böhmer, F.D., 2006. Protein-tyrosine phosphatases and cancer. Nat Rev Cancer 6, 307–320.

Overbeek, L.I., Kets, C.M., Hebeda, K.M., et al., 2007. Patients with an unexplained microsatellite unstable tumour have a low risk of familial cancer. Br J Cancer 96, 1605–1612.

Palka, H.L., Park, M., Tonks, N.K., 2003. Hepatocyte growth factor receptor tyrosine kinase Met is a substrate of the receptor protein-tyrosine phosphatase DEP-1. J Biol Chem 278, 5728–5735.

Petermann, A., Haase, D., Wetzel, A., et al., 2011. Loss of the protein-tyrosine phosphatase DEP-1/PTPRJ drives meningioma cell motility. Brain Pathol 21, 405–418.

Pittman, A.M., Naranjo, S., Jalava, S.E., et al., 2010. Allelic variation at the 8q23.3 colorectal cancer risk locus functions as a cis-acting regulator of EIF3H. PLoS Genet 6.pii, e1001126.

Pomerantz, M.M., Ahmadiyeh, N., Jia, L., et al., 2009. The 8q24 cancer risk variant rs6983267 shows long-range interaction with MYC in colorectal cancer. Nat Genet 41, 882–884.

Rodrigues, N.R., Rowan, A., Smith, M.E., et al., 1990. p53 mutations in colorectal cancer. Proc Natl Acad Sci USA 87, 7555–7559.

Ruivenkamp, C.A., Hermsen, M., Postma, C., et al., 2003. LOH of PTPRJ occurs early in colorectal cancer and is associated with chromosomal loss of 18q12-21. Oncogene 22, 3472–3474.

Ruivenkamp, C.A., Van Wezel, T., Zanon, C., et al., 2002. Ptprj is a candidate for the mouse colon-cancer susceptibility locus Scc1 and is frequently deleted in human cancers. Nat Genet 31, 295–300.

Shen, L., Toyota, M., Kondo, Y., et al., 2007. Integrated genetic and epigenetic analysis identifies three different subclasses of colon cancer. Proc Natl Acad Sci USA 104, 18,654–18,659.

Smith, A.J., Stern, H.S., Penner, M., et al., 1994. Somatic APC and K-ras codon 12 mutations in aberrant crypt foci from human colons. Cancer Res 54, 5527–5530.

Spirio, L., Olschwang, S., Groden, J., et al., 1993. Alleles of the APC gene: An attenuated form of familial polyposis. Cell 75, 951–957.

Stoffel, E.M., Chittenden, A., 2010. Genetic testing for hereditary colorectal cancer: Challenges in identifying, counseling, and managing high-risk patients. Gastroenterology 139, 1436–1441.

Suter, C.M., Martin, D.I., Ward, R.L., 2004. Germline epimutation of MLH1 in individuals with multiple cancers. Nat Genet 36, 497–501.

Tenesa, A., Dunlop, M.G., 2009. New insights into the aetiology of colorectal cancer from genome-wide association studies. Nat Rev Genet 10, 353–358.

Tepass, U., 2002. Adherens junctions: New insight into assembly, modulation and function. Bioessays 24, 690–695.

Tepass, U., Truong, K., Godt, D., Ikura, M., Peifer, M., 2000. Cadherins in embryonic and neural morphogenesis. Nat Rev Mol Cell Biol 1, 91–100.

Tomlinson, I.P., Dunlop, M., Campbell, H., et al., 2010. COGENT (COlorectal cancer GENeTics): An international consortium to study the role of polymorphic variation on the risk of colorectal cancer. Br J Cancer 102, 447–454.

Tufarelli, C., Stanley, J.A., Garrick, D., et al., 2003. Transcription of antisense RNA leading to gene silencing and methylation as a novel cause of human genetic disease. Nat Genet 34, 157–165.

Tuupanen, S., Turunen, M., Lehtonen, R., et al., 2009. The common colorectal cancer predisposition SNP rs6983267 at chromosome 8q24 confers potential to enhanced Wnt signaling. Nat Genet 41, 885–890.

Van der Klift, H., Wijnen, J., Wagner, A., et al., 2005. Molecular characterization of the spectrum of genomic deletions in the mismatch repair genes MSH2, MLH1, MSH6, and PMS2 responsible for hereditary nonpolyposis colorectal cancer (HNPCC). Genes Chromosomes Cancer 44, 123–138.

Van der Post, R.S., Kiemeney, L.A., Ligtenberg, M.J., et al., 2010. Risk of urothelial bladder cancer in Lynch syndrome is increased, in particular among MSH2 mutation carriers. J Med Genet 47, 464–470.

Venkatachalam, R., Ligtenberg, M.J., Hoogerbrugge, N., et al., 2010. Germline epigenetic silencing of the tumor suppressor gene PTPRJ in early-onset familial colorectal cancer. Gastroenterology 139, 2221–2224.

Venkatachalam, R., Verwiel, E.T., Kamping, E.J., et al., 2011. Identification of candidate predisposing copy number variants in familial and early-onset colorectal cancer patients. Int J Cancer 129, 1635–1642.

Vogelstein, B., Fearon, E.R., Hamilton, S.R., et al., 1988. Genetic alterations during colorectal-tumor development. N Engl J Med 319, 525–532.

Vogelstein, B., Kinzler, K.W., 1993. The multistep nature of cancer. Trends Genet 9, 138–141.

Wagner, A., Barrows, A., Wijnen, J.T., et al., 2003. Molecular analysis of hereditary nonpolyposis colorectal cancer in the United States: High mutation detection rate among clinically selected families and characterization of an American founder genomic deletion of the MSH2 gene. Am J Hum Genet 72, 1088–1100.

Weisenberger, D.J., Siegmund, K.D., Campan, M., et al., 2006. CpG island methylator phenotype underlies sporadic microsatellite instability and is tightly associated with BRAF mutation in colorectal cancer. Nat Genet 38, 787–793.

Wijnen, J., Van der Klift, H., Vasen, H., et al., 1998. MSH2 genomic deletions are a frequent cause of HNPC. Nat Genet 20, 326–328.

Wood, L.D., Parsons, D.W., Jones, S., et al., 2007. The genomic landscapes of human breast and colorectal cancers. Science 318, 1108–1113.

Worthley, D.L., Whitehall, V.L., Buttenshaw, R.L., et al., 2010. DNA methylation within the normal colorectal mucosa is associated with pathway-specific predisposition to cancer. Oncogene 29, 1653–1662.

Yu, W., Gius, D., Onyango, P., et al., 2008. Epigenetic silencing of tumour suppressor gene p15 by its antisense RNA. Nature 451, 202–206.

CHAPTER

63

Prostate Cancer

Aubrey R. Turner, Junjie Feng, Wennuan Liu, Jin Woo Kim, and Jianfeng Xu

INTRODUCTION

In personalized medicine (PM), the aim is to provide individual risk assessment for medical conditions or to predict the efficacy of measures intended to monitor, prevent, or treat these conditions (Personalized Medicine Coalition, 2009). The approaches of PM could be important in addressing the clinical and public health issues involved in a variety of diseases, including cancers that are detected via population-level screening. This could be particularly relevant to prostate cancer (PCa), where concerns have been raised regarding prostate-specific antigen (PSA) screening, subsequent over-diagnosis of low-grade diseases, and ultimately overtreatment of many indolent cancers that for the most part are not life-threatening. These interrelated issues have prompted a significant effort to identify markers that can effectively differentiate individuals who have different risks for PCa onset or progression. Improved risk estimation may help to address this major public health problem, as the prostate is the most common site of cancer diagnosis, accounting for approximately 30% of all new cancer diagnoses and 11% of cancer deaths in US men. This translates to an estimated 220,900 prostate cancer diagnoses and 28,900 deaths in the US each year (American Cancer Society, 2011).

Findings in several genetics/genomics-based fields have provided new insights regarding the development and progression of PCa, and have offered great potential to be translated into clinically useful biomarkers. There is now a great opportunity to utilize these findings in order to address some of the most vexing issues in the healthcare of men affected by PCa. In this chapter, we will discuss promising discoveries from germline genetics, somatic genetics, and epigenetics, as well as their potential implementation as clinical tools. We will conclude by outlining key research areas expected to shape the future of the field.

GERMLINE GENETICS OF PROSTATE CANCER

A variety of risk factors have been proposed as underlying PCa risk and mortality, but the only confirmed risk factors to date are age, race, and family history. It is arguable that genetic inheritance plays a significant role in the pathogenesis of PCa, as demonstrated by the observation that the risk of PCa increases with additional affected first-degree relatives (Johns and Houlston, 2003), and by the findings from a large twin study determining that the heritability of PCa is 42%, being notably the greatest among all common cancers (Lichtenstein et al., 2000). These findings have shown the need to identify and study the role of inherited (germline) genetic variants in association with altered risks for PCa onset or progression.

Approaches for Studying Germline Genetics of Prostate Cancer

Genome-wide Linkage Studies

Genome-wide linkage studies were previously used to identify chromosomal regions harboring susceptibility genes in

Genomic and Personalized Medicine, 2nd edition
by Ginsburg & Willard. DOI: http://dx.doi.org/10.1016/B978-0-12-382227-7.00063-X

hereditary prostate cancer (HPC) families, typically defined as three or more first-degree affected relatives. Linkage results have led to the identification of several loci potentially involved in HPC, including *HPC1* (Tavtigian et al., 2001), *RNASEL* (Carpten et al., 2002), and *MSR1* (Xu et al., 2002). However, these are not major risk genes on a par with *BRCA1* and *BRCA2* for hereditary breast cancer or *MSH2* and *MLH1* for hereditary nonpolyposis colorectal cancer (HNPCC). Accordingly, it appears unlikely that the results of linkage studies to date will lead to molecular testing for patients in HPC families that are at significantly increased risk for PCa.

Candidate Gene-based Association Studies

In parallel to the HPC linkage studies, genetic association studies have been conducted using case/control designs to examine variants in candidate genes, and later in candidate pathways. For the most part, these candidate gene-based association studies have failed to consistently demonstrate that specific genetic variants are associated with PCa risk or prognosis, findings that are thus unlikely to lead to clinical applications.

Genome-wide Association Studies

In contrast, the recently emerging genome-wide association studies (GWAS) have made significant progress toward the goal of clinical applications in PCa. GWAS have provided comprehensive and unbiased assessment of single nucleotide polymorphisms (SNPs) across the genome for their associations with disease phenotypes. So far, GWAS have achieved great successes in identifying novel SNP associations with a variety of complex diseases, such as cancer, diabetes, and cardiovascular disease. For PCa, more than 30 SNPs have been identified by GWAS that are associated with PCa risk (summarized in Table 63.1). These associations all exceed the genome-wide statistical significance ($P \leq 10^{-7}$), greatly minimizing the potential of chance associations due to multiple comparisons. Notably, these results have been validated in independent study populations, suggesting that they represent true associations with PCa risk. This consistent replication is an important distinction of the GWAS findings versus previous association studies based on gene/pathway approaches.

Translational and Clinical Research

Many of the PCa-associated SNPs identified by GWAS are in introns or intergenic regions (i.e., "gene-desert" regions) of the genome; thus the molecular mechanisms underlying these associations remain largely unknown. Nonetheless, the consistent replications of these results have led to an intensive effort to evaluate these risk variants for clinical applications, particularly in prediction of disease risk for unaffected men. Although the risk-associated alleles of these SNPs are common (usually >10%) in the general population, each has a small individual effect on disease risk, with odds ratios (ORs) generally in the range of 1.1–1.3 (Foulkes, 2008). However, larger ORs are observed when multiple SNPs are combined in a risk prediction model to assess cumulative lifetime risk. For example, when the first five PCa risk-associated SNPs identified from GWAS and family history were examined, men who had five or more of these risk factors had an OR of 9.46 for PCa, as compared with men without any of the risk factors (Zheng et al., 2008). In a later study that examined the first 14 PCa risk-associated SNPs identified in this way, it was found that 55-year-old men with a family history and 14 risk alleles had a 52% risk of being diagnosed with PCa in the next 20 years. In comparison, without knowledge of genotype and family history, these men had an average population absolute risk of 13% (Xu et al., 2009). The cumulative level of PCa risk as predicted by associated risk alleles of SNPs is comparable to current population risk screening methods for various other types of cancers, such as screening for lung cancer based on smoking status (Spitz et al., 2008), or screening for breast cancer based on mammography (Barlow et al., 2006).

For perspective, it can be helpful to compare the new PCa risk-associated SNPs with variables that are already used clinically for risk prediction that may ultimately lead many men to have prostate biopsy. Unfortunately, existing clinical variables such as age, family history, and PSA have limited ability to predict outcomes of prostate biopsy. Typically, only 25–30% patients who are recommended for prostate biopsy based on existing clinical models have positive biopsy results. This leads to a phenomenon of over-biopsy in many western countries. In a Swedish cohort of men who underwent a prostate biopsy during 2005–2007, it has been shown that a prediction model that included PCa risk-associated SNPs and existing clinical variables (age, PSA, free-to-total PSA, and family history) performed significantly better than the clinical model only. The genetic model group required significantly fewer biopsies (22.7%) than the non-genetic model group, at a cost of missing a PCa diagnosis in 3% of patients characterized as having an aggressive disease (Aly et al., 2011). These results suggest the genetic markers can be used to improve risk prediction, and may ultimately help reduce the number of unnecessary biopsies.

Future Directions of Germline Genetic Studies of Prostate Cancer

The prediction of disease risk for unaffected men represents the most straightforward clinical application of the risk-associated SNPs discovered by GWAS. Such risk prediction could be used independently, but most likely will be combined with existing predictors such as PSA in order to guide biopsy decision-making. Physicians could use refined risk information as a basis to recommend initial biopsy among men at increased risk. Another application could be guiding how aggressively to pursue a schedule of repeat PSA tests or repeat biopsy among men whose prior biopsy results were negative.

An important area requiring additional research is the identification of germline variants associated with PCa aggressiveness or survival. Because most prostate tumors are indolent, treatment can cause significant morbidity among men

TABLE 63.1	Summary of SNPs reproducibly associated with prostate cancer, based on GWAS before December 2010					
Chr	SNP	Region	Position	Known genes	m/M allele	Risk allele
2	rs1465618	2p21	43,407,453	THADA	A/G	A
2	rs721048	2p15	62,985,235	EHBP1	A/G	A
2	rs12621278	2q31.1	173,019,799	ITGA6	G/A	A
3	rs2660753	3p12	87,193,364	–	T/C	T
3	rs10934853	3q21.3	129,521,063	EEFSEC	A/C	A
4	rs17021918	4q22.3	95,781,900	PDLIM5	T/C	C
4	rs7679673	4q24	106,280,983	TET2	A/C	C
6	rs9364554	6q25	160,753,654	SLC22A3	T/C	T
7	rs10486567	7p15	27,943,088	JAZF1	A/G	G
7	rs6465657	7q21	97,654,263	LMTK2	T/C	C
8	rs2928679	8p21.2	23,494,920	SLC25A37	A/G	A
8	rs1512268	8p21.2	23,582,408	NKX3.1	T/C	T
8	rs10086908	8q24 (5)	128,081,119	–	C/T	T
8	rs16901979	8q24 (2)	128,194,098	–	A/C	A
8	rs16902094	8q24.21	128,389,528	–	N/A	G
8	rs620861	8q24 (4)	128,404,855	–	A/G	G
8	rs6983267	8q24 (3)	128,482,487	–	G/T	G
8	rs1447295	8q24 (1)	128,554,220	–	A/C	A
9	rs1571801	9q33	123,467,194	DAB21C	G/A	A
10	rs10993994	10q11	51,219,502	MSMB	T/C	T
10	rs4962416	10q26	126,686,862	CTBP2	C/T	C
11	rs7127900	11p15.5	2,190,150	IGF2, IGF2AS, INS, TH	G/A	A
11	rs12418451	11q13 (2)	68,691,995	–	A/G	A
11	rs10896449	11q13 (1)	68,751,243	MYEOV	A/G	G
17	rs11649743	17q12 (2)	33,149,092	HNF1B	A/G	G
17	rs4430796	17q12 (1)	33,172,153	HNF1B	A/G	A
17	rs1859962	17q24.3	66,620,348	–	G/T	G
19	rs8102476	19q13.2	43,427,153	PPP1R14A	T/C	C
19	rs887391	19q13	46,677,464	–	C/T	T
19	rs2735839	19q13	56,056,435	KLK3	A/G	G
22	rs9623117	New 22q13	38,782,065	TNRC6B	C/T	C
22	rs5759167	New 22q13.2	41,830,156	TTLL1, BIK, MCAT, PACSIN2	T/G	G
23	rs5945619	Xp11	51,258,412	NUDT10, NUDT11, LOC340602	C/T	C

Chr, chromosome; m/M, minor allele/major allele.
Table adapted from Lu et al., 2011.

who would never have gone on to develop clinical symptoms or die from PCa. Most of the identified GWAS SNPs are not associated with aggressiveness or survival, which is to be expected because the original studies primarily used early-stage cases for association discovery and validation. However, recent reports have identified SNPs (rs4054823 at 17p12, rs6497287 at 15q13, rs2735839 at 19q13, and rs7679673 at 4q24) specifically associated with PCa progression or survival (FitzGerald et al., 2011; Gallagher et al., 2010; Pomerantz et al., 2011; Xu et al., 2010). Efforts to confirm these initial associations have been inconsistent and were likely limited by different definitions of progression, varying study designs, and small numbers of advanced PCa cases. Yet the initial findings offer important leads and design guidance for future studies in this critical area.

It should be mentioned that the current generation of GWAS target mostly common genetic variants, with minor allele frequency (MAF) ≥0.05. As the ongoing 1000 Genomes project (http://www.1000genomes.org/) has revealed, a tremendous number of rare genetic variants (MAF < 0.05) have been and are being discovered. These rare variants are usually not included in conventional GWAS, but nonetheless bear the potential to influence complex diseases such as PCa in a non-trivial way. With the decreasing costs associated with next-generation sequencing technologies and exome SNP arrays, it is expected that additional rare genetic polymorphisms will be identified and studied for their association with PCa. Findings from such studies may have a clinical impact on PCa, particularly if the discovered variants have statistical and functional significance.

SOMATIC GENETICS OF PROSTATE CANCER

As in many other human cancers, massive genomic changes, including base substitutions (mutations), insertion/deletions, translocations, inversions, gene fusions, and copy number alterations (CNAs) have been identified in prostate tumors. Some of these somatic-level genetic alterations are found to be frequent across different prostatic tumor samples, and are thus speculated to play a crucial role in the tumorigenesis and progression of PCa. Notably, PCa is a heterogeneous disease – various combinations of somatic-level genomic alterations have been identified among different tumors, or even from different cancer cells within the same tumor tissue (Berger et al., 2011). This genetic heterogeneity is considered the primary obstacle for distinguishing aggressive from indolent forms of PCa and for treating the disease effectively. In this section, we will highlight several significant discoveries in this field and discuss the potential implications of these findings for the clinical management of PCa.

Major Types of Somatic Genetic Alterations
Copy Number Alterations

CNAs refer to changes in the number of copies of DNA in specific regions of the genome that range in size from 1 kb

(kilobasepair) to a complete chromosome arm, and mainly involve copy number losses (deletions) or gains (amplifications). Current genome-wide analyses of PCa have revealed that CNAs are a major component of the landscape of the PCa tumor genome. The most frequently detected CNAs in primary PCa tumors include amplifications of genomic regions containing oncogenic *MYC* (8q24.21; 20–35%), and deletions of DNA sequences containing tumor-suppressor genes such as *NKX3-1* (8p21.2; 35–70%), *PTEN* (10q23.31; 10–40%), *CDKN1B* (12p13.1; 5–40%), *RB1* (13q14.2; 25–45%), and *TP53* (17p13.1; 20–30%) (Figure 63.1). Another frequently amplified gene, found primarily in hormone-refractory cancers (Visakorpi et al., 1995) and metastases (Liu et al., 2009), is *AR* on chromosome Xq12 (30–60%). The molecular mechanisms by which these well-known oncogenes and tumor-suppressor genes influence cell proliferation, apoptosis, senescence, angiogenesis, and metastasis in tumorigenesis have been well established and will not be the subject of this chapter. It is notable that the frequencies of these recurrent CNAs usually vary greatly among different tumor samples, depending on the composition of cohorts, tumor grade, and degree of tumor cell heterogeneity, reflecting the complex roles of these CNAs in PCa. For example, high-grade, metastatic prostatic tumors typically contain more frequent CNAs than clinically localized PCa (Sun et al., 2007; Taylor et al., 2010). The majority of the PCa-specific CNAs resulting in deletions of tumor-suppressor genes involve loss of only one copy of the genes. It is speculated that either these genes are haploinsufficient or these hemizygous deletions are complemented by additional genetic and epigenetic alterations on the other copy of the gene. For example, *NKX3-1* has been reported to be affected by both deletion and methylation (Asatiani et al., 2005). In addition, it has been reported that each patient with a metastatic PCa typically harbors a unique signature of CNAs that is shared among tumors from different metastatic sites, and that each of these tumors also accumulate a pattern of CNAs that is distinctive from each other, suggesting a clonally evolving nature of metastatic prostate tumors (Liu et al., 2009).

Gene Fusions

Unlike most other solid tumors, but resembling some types of hematological cancers such as leukemias and lymphomas, PCa contains a high frequency of fusion genes that typically result from chromosomal rearrangements such as translocations, interstitial deletions, and chromosomal inversions. For example, the archetypal *TMPRSS2-ERG* fusion gene, formed as a consequence of translocation or deletion, has been found in 40–70% of PCa tumors (Tomlins et al., 2005) (Figure 63.2). To date, more than 10 such fusion genes have been identified in PCa tumor genomes. One interesting common feature of these fusion genes is that they often involve combinations of 5′ promoter/untranslated regions of androgen-regulated genes such as *TMPRSS2*, *SLC45A3*, *KLK2*, *ACSL2*, etc., and 3′ coding regions of E-twenty-six (ETS) transcription factors such as *ERG*, *ETV1*, *ETV4*, and *ETV5* (Kumar-Sinha et al., 2008). Whereas these

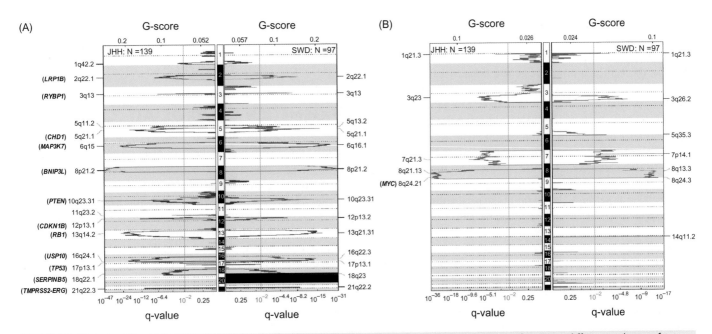

Significant genomic deletions (blue) and amplifications (red) in the tumor genome from two different cohorts of patients with prostate cancer. JHH = Johns Hopkins Hospital, SWD = Sweden. G-score is a measure of significance of a copy number alteration event, which considers the amplitude of the aberration as well as the frequency of its occurrence across samples. q-value is a measure of False Discovery Rate for the aberrant regions.

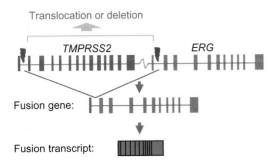

Figure 63.2 *TMPRSS2-ERG* gene fusion caused by DNA rearrangements.

potentially oncogenic ETS transcription factors remain silent in normal prostatic epithelial cells, gene fusions cause androgen-induced overexpression of the ETS proteins. Interestingly, it has been demonstrated that androgen receptors (AR) play a critical role in the formation of at least some of these fusion genes, such as *TMPRSS2-ERG/ETV1* (Haffner et al., 2010; Lin et al., 2009). While the functions of these fusion genes are yet to be sufficiently defined, current evidence supports the notion that they may represent an early, crucial genetic lesion, which, in collaboration with other genetic/molecular alterations, drives the whole oncogenic process of PCa.

Point Mutations

Point mutations refer to changes in the sequence of DNA bases, and include substitutions, insertions, and deletions of one or more bases. Although PCa has the lowest point

mutation rate [~0.33/Mb (megabase)] among the major human cancers (Kan et al., 2010), relatively high frequencies of point mutations at cancer-related genes (oncogenes or tumor-suppressor genes) *TP53* (24%), *PTEN* (15%), *RB1* (9%), *EGFR* (8%), *KRAS* (7%), *CTNNB1* (7%), *BRAF* (6%), and *CDKN2A* (3%) have been documented in prostate adenocarcinoma [see the Catalogue of Somatic Mutations in Cancer (COSMIC), at http://www.sanger.ac.uk/genetics/CGP/cosmic]. Usually the frequency of mutation at each of the genes varies greatly among different PCa tumor samples, depending on a number of factors including cohorts and detection methods used, which possibly reflects the genetic heterogeneity of PCa. Although point mutation frequencies are relatively low compared with other cancer types, the mutations at these genes may still contribute significantly to the pathogenesis of PCa, possibly by complementing the alterations at other genes in their corresponding pathways to favor the oncogenesis of PCa.

Methods for Detecting Somatic Alterations

Several approaches have been applied for the identification of somatic genomic alterations in PCa, including fluorescence *in situ* hybridization (FISH) and comparative genomic hybridization (CGH). FISH is the major method for detection of genetic abnormality in clinical settings. In addition to detection of loss and gain at the genes of interest, it has been instrumental for identification and confirmation of *TMPRSS2-ETS* fusions, especially for balanced translocation, which cannot be detected by CGH. Multicolored FISH has been used to detect alterations in at least six genes (Anderson et al., 2011). CGH, such as metaphase (or conventional) and bacterial artificial chromosome

(BAC)/oligo(nucleotide)/complementary DNA (cDNA)/SNP (array-based) hybridization with different resolutions, has been used to identify a number of recurrent losses and gains affecting the tumor-suppressor genes, oncogenes, and fusion genes described above in the PCa tumor genome. In addition to copy number data, the SNP array can generate genotype information that allows allele-specific analysis, which has the advantage of minimizing the contamination of DNA from nonmalignant cells in tumor tissues, because the other allele can be used as an internal control for hemizygous deletion or amplification. However, this approach has not yet been widely used in clinical settings. In addition, several other advanced technologies such as Sanger sequencing and next-gen (next-generation) sequencing are becoming more robust and affordable, and are expected to play an increasingly important role in clinical settings for genetic tests.

Translational and Clinical Research

Promising progress has been achieved in the translational research of these PCa-associated somatic genomic alterations, which investigates whether they can be used as biomarkers, alone or in combination with already-established clinicopathologic parameters, to improve screening, diagnosis, staging, and risk prediction of PCa. For example, loss of *PTEN* has been shown to be a late genetic event and is predictive of a shorter time to biochemical recurrence of PCa (Yoshimoto et al., 2007). The relative copy number gain of *MYC* is reportedly associated with worse PCa-specific survival (Ribeiro et al., 2007). By combining the cancer-specific *TMPRSS2-ERG* score with levels of PSA and PCA3 (the product of prostate cancer gene 3), Tomlins and colleagues have recently demonstrated a more accurate stratification of men at high risk for developing clinically significant PCa (Tomlins et al., 2011). While promising, application in the clinical setting is still years away.

Determination of the profiles of somatic genetics in individual PCa patients can be used to direct anticancer therapy for the disease. There are currently 32 US Food and Drug Administration (FDA)-approved drugs for use in oncology that have pharmacogenomic information in their labels; a strong majority of these target somatic genetic alterations (US FDA, 2012). For example, dysregulation of the PI3K-PTEN-AKT pathway, usually caused by deletion/mutation of *PTEN*, amplification/mutation-induced upregulation of *PIK3CA* and AKT genes, etc., has been closely implicated in prostate carcinogenesis and progression, and thus is the target of several pharmaceutical drugs that have been developed or are under development. Given the general inter-patient genetic heterogeneity, the clinical success of these therapeutic drugs can be maximized by prospectively identifying patients harboring molecular abnormalities in this pathway, who may have a higher likelihood of responding. Furthermore, because cancer cells are able to evolve genetically, especially under pressure of therapeutic drugs, genome-wide analysis of further acquired alterations of DNA in response to cancer treatments will shed light on prognosis, new drug targets, and selection of strategies for the effective management of PCa for personalized medicine.

EPIGENETICS OF PROSTATE CANCER

The US National Institutes of Health (NIH) has defined epigenetics as "an emerging field of science that studies heritable changes caused by the activation and deactivation of genes without any change in the underlying DNA sequence of the organism" (http://www.genome.gov/Glossary/index.cfm?id=528). Epigenetics consists of a variety of molecular mechanisms, including DNA methylation, histone modifications, and noncoding RNAs. A common feature of these mechanisms is the potential to modify gene expression and alter the growth state of cells, thus contributing to carcinogenesis and tumor progression. Because regulation of genes by epigenetic mechanisms is considered reversible, it may be possible to treat prostate cancer using molecular therapies that selectively correct epigenetic processes that are going awry during tumorigenesis.

Epigenetic Mechanisms

DNA Methylation

DNA methylation is a well-characterized epigenetic modification that plays a key role in gene regulation, with alterations contributing to tumorigenesis in PCa. This involves the covalent addition of a methyl group at the 5-position carbon of a cytosine ring in mammals. Methylation usually occurs at CG dinucleotides, where a cytosine is joined by a phosphate to a guanine, and is frequently abbreviated as CpG. When a cluster of CpGs occur within a DNA sequence, they may be referred to as a CpG island, and thus a likely location for methylation. The majority of DNA methylation in normal human cells is observed in noncoding regions of the genome, with 34% of intragenic CpG islands methylated, as opposed to only 2% of CpG islands in promoters (Maunakea et al., 2010). Throughout the process of PCa initiation and progression, global hypomethylation is observed, particularly in intragenic regions (Feinberg and Vogelstein, 1983; Kim et al., 2011). Nonetheless, promoter hypermethylation has been reported for more than 60 genes in PCa, and may be independent of global hypomethylation status in PCa (Ongenaert et al., 2008). For example, promoter hypermethylation of *GSTP1*, which is the most frequently studied DNA methylated gene in PCa, is observed in >90% of prostate tumors and results in reduced expression of this detoxification enzyme, presumably creating an environment that is more permissive to DNA damage and ultimately tumorigenesis and progression (Brooks et al., 1998). DNA methylation patterns are established and maintained by DNA methyltransferases (DNMTs). The activity of these enzymes can be modified by several small-molecule chemicals such as vidaza (5-azacytidine) and decitabine (5-aza-2′-deoxycytidine), whose chemotherapeutic potentials for PCa are being explored.

Histone Modifications

A variety of chemical modifications have been identified on histone proteins, including methylation, acetylation, phosphorylation, ubiquitination, sumoylation, and adenosine diphosphate (ADP)-ribosylation. These modifications may change the structure and dynamics of the chromatin where the modified histones reside, and could potentially influence the expression of genes located nearby. Among these, the reversible methylation and acetylation of histones, that is, the attachment or removal of methyl and acetyl groups on histone proteins via the actions of histone methyltransferases (HMTs), acetyl transferases (HATs), deacetylases (HDACs) and their coregulators, have been most studied to date, and have an established role in regulation of gene expression in mammalian cells. In general, transcription of genes within a chromosomal region is reduced when specific lysine residues of histones are deacetylated and/or methylated, but, on the other hand, becomes more permissive or enhanced when such sites are acetylated and/or demethylated. Due to these regulatory roles in controlling gene expression, dysregulation of histone methylation and acetylation has been widely implicated in various human cancers, including PCa. For example, methylation changes on histone H3 at lysines 4 (K4) and 27 (K27) have been associated with extensive gene expression reprogramming of primary prostate cells during epithelial-to-mesenchymal transition (EMT) and early transformation steps, suggesting a crucial role of these histone methylations in PCa carcinogenesis (Ke et al., 2010). In addition, compared with normal tissues, prostate tumors have been observed to harbor histone acetylation and demethylation in specific patterns that are permissive to transcription, and these modifications are linked to the risk of prostate cancer recurrence and metastasis (Seligson et al., 2005; Yu et al., 2007).

Noncoding RNAs

A significant proportion of genomic DNA is transcribed into RNAs that are not translated into proteins and are thus termed noncoding RNAs. The functions of this type of RNA were not well understood until recent years, when growing evidence revealed that they may be involved in regulating diverse cellular processes such as splicing, transcription, translation, genomic imprinting, telomere maintenance, etc., and are implicated in many human diseases, including cancers. The best understood of these noncoding RNAs are microRNAs, consisting of approximately 22 nucleotides that can base-pair with complementary sequences of mRNA, causing degradation of the mRNA or preventing its translation into a protein. MicroRNAs have been suggested to be important players in the tumorigenesis and progression of PCa. For example, microRNA-205, which prevents cell growth and migration via apoptosis and cell cycle arrest, has been observed to be down-regulated in prostate tumors compared to normal tissues (Ambs et al., 2008; Gandellini et al., 2009; Porkka et al., 2007; Tong et al., 2009). Expression of certain microRNAs has been shown to be also associated with metastatic PCa (Carlsson et al., 2011; Watahiki et al., 2011). It should be pointed out, however, that while the area is promising, overall results have remained inconsistent. Various technical hurdles remain, including study design, sample handling, and assay design. As standards are developed, perhaps clearer patterns will emerge.

Translational and Clinical Research

A better understanding of DNA methylation, histone modifications, and noncoding RNAs may allow for the development of clinical applications that target elements of these molecular mechanisms. Disease screening based on DNA methylation status is promising. Many studies have examined the measurement of GSTP1 promoter methylation as a predictive test. A recent meta-analysis observed high specificity (0.89) for GSTP1 promoter methylation in a variety of body fluids, suggesting that this measure may be useful as complement to PSA screening as it can help address the relatively low specificity of PSA, which is a major shortcoming of this conventional screening approach (Wu et al., 2011). Disease staging represents another potential application of epigenetic assays. For example, patterns of histone alteration could be measured in specific regions and across the genome, and used to predict clinical staging and outcomes, and thus could also be useful in guiding treatment decisions.

Molecular guided therapies represent another promising direction. DNMT inhibitors (vidaza and decitabine) are already FDA-approved for treating myelodysplastic syndrome. A phase I/II clinical trial is underway to study the side effects and best doses of vidaza and docetaxel when given together with prednisone, and to see how well they work in treating patients with metastatic prostate cancer that is not responsive to hormone therapy (identifier NCT00503984 at the ClinicalTrials.gov registry, see http://clinicaltrials.gov/ct2/show/NCT00503984). There is a body of research evaluating HDAC inhibitors, acting via mechanisms such as cell cycle arrest and triggering of host immune responses, as a means to induce apoptosis in tumors (Bolden et al., 2006; Dokmanovic and Marks, 2005). Two phase II clinical trials are currently underway – one is evaluating the HDAC inhibitor SB939 in patients with recurrent or metastatic prostate cancer (NCT01075308), the other is evaluating the HDAC inhibitor SAHA (virinostat) in patients with progressive metastatic prostate cancer (NCT00330161). Another exciting area in the future will be epigenetic approaches to cancer prevention, such as a trial of compounds found in cruciferous vegetables that may prevent prostate cancer development by inhibiting histone deacetylase (HDAC) activity (NCT01265953).

Future Directions of Epigenetic Studies of Prostate Cancer

There are many opportunities for future research in the epigenetics of prostate cancer. On the technical side, it is expected that genome-wide epigenetic profiling will be more extensively employed in PCa epigenetic studies as operating costs continue to decrease. The first genome-wide DNA methylation study using next-generation sequencing was recently published (Kim et al., 2011), and comprehensive studies of quantitative differences in DNA methylation as well as intercellular heterogeneity will be possible using newer approaches such as pyrosequencing and MALDI-TOF (matrix-assisted laser desorption/ionization with time-of-flight) mass spectrometry (EpiTYPER). Each epigenetic mechanism can be studied by coupling next-generation sequencing with other techniques; these include bisulfate modification for DNA methylation (MethylC-seq) and chromatin immunoprecipitation for histone modification (ChIP-seq). These approaches are expected to provide a more precise epigenetic landscape of the human genome and reveal undiscovered prostate cancer-specific differentiation that is a result of epigenetic modifications. On the translational/clinical side, it is expected that more studies will be exploring the potential of translating these emerging epigenetic findings into clinical biomarkers for PCa. For example, a growing number of studies have been exploring the association between DNA methylation in PCa and clinicopathological parameters such as Gleason grade, Tumor-Node-Metastasis (TNM) stage, PSA level, recurrence, and survival time.

CONCLUSIONS

The burgeoning fields of germline genetics, somatic genetics, and epigenetics have dramatically reshaped studies of PCa, one of the major public healthcare problems. The various germline and somatic, genetic and epigenetic variations or alterations identified to be associated with increased risk for PCa onset and progression have not only widened our view and deepened our understanding of the complex mechanisms underlying the pathogenesis of PCa, but also offer great potential to be translated into clinically useful biomarkers or novel drug targets for personalized medicine of this disease.

REFERENCES

Aly, M., Wiklund, F., Xu, J., et al., 2011. Polygenic risk score improves prostate cancer risk prediction: Results from the Stockholm-1 cohort study. Eur Urol 60, 21–28.

Ambs, S., Prueitt, R.L., Yi, M., et al., 2008. Genomic profiling of micro-RNA and messenger RNA reveals deregulated microRNA expression in prostate cancer. Cancer Res 68, 6162–6170.

American Cancer Society, <http://www.cancer.org/Cancer/Prostate Cancer/DetailedGuide/prostate-cancer-key-statistics>.

Anderson, K., Lutz, C., Van Delft, F.W., et al., 2011. Genetic variegation of clonal architecture and propagating cells in leukaemia. Nature 469, 356–361.

Asatiani, E., Huang, W.X., Wang, A., et al., 2005. Deletion, methylation, and expression of the NKX3.1 suppressor gene in primary human prostate cancer. Cancer Res 65, 1164–1173.

Barlow, W.E., White, E., Ballard-Barbash, R., et al., 2006. Prospective breast cancer risk prediction model for women undergoing screening mammography. J Natl Cancer Inst 98, 1204–1214.

Berger, M.F., Lawrence, M.S., Demichelis, F., et al., 2011. The genomic complexity of primary human prostate cancer. Nature 470, 214–220.

Bolden, J.E., Peart, M.J., Johnstone, R.W., 2006. Anticancer activities of histone deacetylase inhibitors. Nat Rev Drug Discov 5, 769–784.

Brooks, J.D., Weinstein, M., Lin, X., et al., 1998. CG island methylation changes near the GSTP1 gene in prostatic intraepithelial neoplasia. Cancer Epidemiol Biomarkers Prev 7, 531–536.

Carlsson, J., Davidsson, S., Helenius, G., et al., 2011. A miRNA expression signature that separates between normal and malignant prostate tissues. Cancer Cell Int 11, 14.

Carpten, J., Nupponen, N., Isaacs, S., et al., 2002. Germline mutations in the ribonuclease L gene in families showing linkage with HPC1. Nat Genet 30, 181–184.

Dokmanovic, M., Marks, P.A., 2005. Prospects: Histone deacetylase inhibitors. J Cell Biochem 96, 293–304.

Feinberg, A.P., Vogelstein, B., 1983. Hypomethylation distinguishes genes of some human cancers from their normal counterparts. Nature 301, 89–92.

FitzGerald, L.M., Kwon, E.M., Conomos, M.P., et al., 2011. Genome-wide association study identifies a genetic variant associated with risk for more aggressive prostate cancer. Cancer Epidemiol Biomarkers Prev 20, 1196.

Foulkes, W.D., 2008. Inherited susceptibility to common cancers. N Engl J Med 359, 2143–2153.

Gallagher, D.J., Vijai, J., Cronin, A.M., et al., 2010. Susceptibility loci associated with prostate cancer progression and mortality. Clin Cancer Res 16, 2819.

Gandellini, P., Folini, M., Longoni, N., et al., 2009. miR-205 exerts tumor-suppressive functions in human prostate through down-regulation of protein kinase C-epsilon. Cancer Res 69, 2287–2295.

Haffner, M.C., Aryee, M.J., Toubaji, A., et al., 2010. Androgen-induced TOP2B-mediated double-strand breaks and prostate cancer gene rearrangements. Nat Genet 42, 668–675.

Johns, L.E., Houlston, R.S., 2003. A systematic review and meta-analysis of familial prostate cancer risk. BJU Int 91, 789–794.

Kan, Z., Jaiswal, B.S., Stinson, J., et al., 2010. Diverse somatic mutation patterns and pathway alterations in human cancers. Nature 466, 869–873.

Ke, X.S., Qu, Y., Cheng, Y., et al., 2010. Global profiling of histone and DNA methylation reveals epigenetic-based regulation of gene expression during epithelial to mesenchymal transition in prostate cells. BMC Genomics 11, 669.

Kim, J.H., Dhanasekaran, S.M., Prensner, J.R., et al., 2011. Deep sequencing reveals distinct patterns of DNA methylation in prostate cancer. Genome Res 21, 1028–1041.

Kumar-Sinha, C., Tomlins, S.A., Chinnaiyan, A.M., 2008. Recurrent gene fusions in prostate cancer. Nat Rev Cancer 8, 497–511.

Lichtenstein, P., Holm, N.V., Verkasalo, P.K., et al., 2000. Environmental and heritable factors in the causation of cancer – analyses of cohorts of twins from Sweden, Denmark, and Finland. N Engl J Med 343, 78–85.

Lin, C., Yang, L., Tanasa, B., et al., 2009. Nuclear receptor-induced chromosomal proximity and DNA breaks underlie specific translocations in cancer. Cell 139, 1069–1083.

Liu, W., Laitinen, S., Khan, S., et al., 2009. Copy number analysis indicates monoclonal origin of lethal metastatic prostate cancer. Nat Med 15, 559–565.

Lu, Y., Zhang, Z., Yu, H., et al., 2011. Functional annotation of risk loci identified through genome-wide association studies for prostate cancer. Prostate 71, 955–963.

Maunakea, A.K., Nagarajan, R.P., Bilenky, M., et al., 2010. Conserved role of intragenic DNA methylation in regulating alternative promoters. Nature 466, 253–257.

Ongenaert, M., Van Neste, L., De Meyer, T., Menschaert, G., Bekaert, S., Van Criekinge, W., 2008. PubMeth: A cancer methylation database combining text-mining and expert annotation. Nucleic Acids Res 36, D842–D846.

Personalized Medicine Coalition, 2009. The Case for Personalized Medicine. <http://www.personalizedmedicinecoalition.org/communications/TheCaseforPersonalizedMedicine_5_5_09.pdf>.

Pomerantz, M.M., Werner, L., Xie, W., et al., 2011. Association of prostate cancer risk loci with disease aggressiveness and prostate cancer-specific mortality. Cancer Prev Res 4, 719.

Porkka, K.P., Pfeiffer, M.J., Waltering, K.K., Vessella, R.L., Tammela, T.L., Visakorpi, T., 2007. MicroRNA expression profiling in prostate cancer. Cancer Res 67, 6130–6135.

Ribeiro, F.R., Henrique, R., Martins, A.T., Jeronimo, C., Teixeira, M.R., 2007. Relative copy number gain of MYC in diagnostic needle biopsies is an independent prognostic factor for prostate cancer patients. Eur Urol 52, 116–125.

Seligson, D.B., Horvath, S., Shi, T., et al., 2005. Global histone modification patterns predict risk of prostate cancer recurrence. Nature 435, 1262–1266.

Spitz, M.R., Etzel, C.J., Dong, Q., et al., 2008. An expanded risk prediction model for lung cancer. Cancer Prev Res (Phila) 1, 250–254.

Sun, J., Liu, W., Adams, T.S., et al., 2007. DNA copy number alterations in prostate cancers: A combined analysis of published CGH studies. Prostate 67, 692–700.

Tavtigian, S.V., Simard, J., Teng, D.H., et al., 2001. A candidate prostate cancer susceptibility gene at chromosome 17p. Nat Genet 27, 172–180.

Taylor, B.S., Schultz, N., Hieronymus, H., et al., 2010. Integrative genomic profiling of human prostate cancer. Cancer Cell 18, 11–22.

Tomlins, S.A., Aubin, S.M., Siddiqui, J., et al., 2011. Urine TMPRSS2:ERG fusion transcript stratifies prostate cancer risk in men with elevated serum PSA. Sci Transl Med 3, 94ra72.

Tomlins, S.A., Rhodes, D.R., Perner, S., et al., 2005. Recurrent fusion of TMPRSS2 and ETS transcription factor genes in prostate cancer. Science 310, 644–648.

Tong, A.W., Fulgham, P., Jay, C., et al., 2009. MicroRNA profile analysis of human prostate cancers. Cancer Gene Ther 16, 206–216.

US Federal Drug Adminstration, 2012. Table of Pharmacogenomic Biomarkers in Drug Labels. <http://www.fda.gov/drugs/scienceresearch/researchareas/pharmacogenetics/ucm083378.htm>.

Visakorpi, T., Hyytinen, E., Koivisto, P., et al., 1995. In vivo amplification of the androgen receptor gene and progression of human prostate cancer. Nat Genet 9, 401–406.

Watahiki, A., Wang, Y., Morris, J., et al., 2011. MicroRNAs associated with metastatic prostate cancer. PLoS One 6, e24950.

Wu, T., Giovannucci, E., Welge, J., et al., 2011. Measurement of GSTP1 promoter methylation in body fluids may complement PSA screening: A meta-analysis. Br J Cancer 105, 65–73.

Xu, J., Sun, J., Kader, A.K., et al., 2009. Estimation of absolute risk for prostate cancer using genetic markers and family history. Prostate 69, 1565–1572.

Xu, J., Zheng, S.L., Isaacs, S.D., et al., 2010. Inherited genetic variant predisposes to aggressive but not indolent prostate cancer. Proc Natl Acad Sci USA 107, 2136–2140.

Xu, J., Zheng, S.L., Komiya, A., et al., 2002. Germline mutations and sequence variants of the macrophage scavenger receptor 1 gene are associated with prostate cancer risk. Nat Genet 32, 321–325.

Yoshimoto, M., Cunha, I.W., Coudry, R.A., et al., 2007. FISH analysis of 107 prostate cancers shows that PTEN genomic deletion is associated with poor clinical outcome. Br J Cancer 97, 678–685.

Yu, J., Yu, J., Rhodes, D.R., et al., 2007. A polycomb repression signature in metastatic prostate cancer predicts cancer outcome. Cancer Res 67, 10,657–10,663.

Zheng, S.L., Sun, J., Wiklund, F., et al., 2008. Cumulative association of five genetic variants with prostate cancer. N Engl J Med 358, 910–919.

CHAPTER

64

Head and Neck Cancer

Giovana R. Thomas and Gina Jefferson

HEAD AND NECK SQUAMOUS CELL CARCINOMA

Predisposition

The predominant environmental risk factors for developing head and neck squamous cell carcinoma (HNSCC) identified to date are the use of alcohol and tobacco, immunosuppression, and exposure to high-risk human papilloma virus (HPV). However, not all smokers and drinkers develop cancers. In fact, only 10–15% of smokers develop lung cancer, and an even a smaller proportion is diagnosed with HNSCC (Ho et al., 2007). Therefore, genetic predisposition and other host factors may play an equally important role in tumorigenesis.

Processed tobacco contains at least 30 known carcinogens, and cigarette smoke contains approximately 50 known carcinogens and procarcinogens. Nearly all carcinogens and procarcinogens require activation via metabolizing enzymes. Likewise, detoxifying enzymes act to deactivate carcinogens and their intermediate byproducts. The collective metabolizing enzymes and detoxifying enzymes are known as xenobiotic-metabolizing enzymes, and are found in the liver and upper aerodigestive tract (UADT) mucosa. Genetic polymorphisms of these enzymes modify an individual's response to carcinogen exposure, and consequently the potential of the exposure to incite the development of HNSCC (Ho et al., 2007).

The most significant of the procarcinogens contained in cigarette smoke are the aromatic hydrocarbons such as benzo(a)pyrene, tobacco-specific nitrosamines, and aromatic amines. Benzo(a)pyrenediol epoxide (BPDE) is a toxic metabolite of benzo(a)pyrene that exerts its mutagenic effect mostly by irreversibly binding to DNA, forming BPDE-DNA adducts, or by DNA oxidation. Nitrosamines produce free radical compounds known to cause DNA fragmentation. The individual capacity for nucleotide excision and repair of normal DNA after BPDE-induced damage was shown to correlate with HNSCC development independent of age, sex, ethnicity, smoking status, or alcohol use in a case-control study (Cheng et al., 1998).

The genetic determination of DNA repair capacity is also associated with an individual's susceptibility to the development of HNSCC. DNA repair capacity for removing BPDE-induced DNA adducts is considered an independent biomarker for the risk of developing smoking-related HNSCC (Wang et al., 2010). Mutagen hypersensitivity or an individual's capacity for DNA repair against free radical damage and its relevance to the development of HNSCC was examined by Schantz and coworkers (1997). The relationship between the risk of HNSCC, nutrition, and mutagen hypersensitivity was studied in HNSCC patients and controls. Dietary intake of various antioxidants such as vitamins C and E and carotenoids, cigarette smoking, alcohol use, and body mass index (BMI) were also examined. The study found that mutagen hypersensitivity was strongly associated with increased risk of HNSCC. Also, high intakes of vitamins C and E and some carotenoids were independently related to a decreased risk of HNSCC.

The role of genomic instability in the development of HNSCC was further elucidated in a study of patients with Fanconi anemia, a rare autosomal recessive disorder characterized by a high degree of spontaneous chromosomal

Genomic and Personalized Medicine, 2nd edition
by Ginsburg & Willard. DOI: http://dx.doi.org/10.1016/B978-0-12-382227-7.00064-1

Copyright © 2013, Elsevier Inc.
All rights reserved.

742

aberrations (Kutler et al., 2003). The overall incidence of HNSCC in Fanconi anemia patients was 3%, as compared with an incidence of 0.038% in the general population, and their median age was 31. The cumulative rate of disease recurrence after treatment by the age of 40 was 50% in this patient population. This was attributed to defects in DNA damage repair causing increased genomic instability in epithelial cells of the head and neck region.

Although the incidence of laryngeal, oral cavity, and hypopharyngeal cancers has continued to decline since the 1980s, coincident with the decline of active smokers, the incidence of oropharyngeal carcinomas continues to rise, particularly among individuals in the 40–60 age group. HPV type 16 is a causative agent in up to 70% of oropharyngeal carcinomas (Pai and Westra, 2009). Gillison and colleagues demonstrated that these cancers, as compared to non-HPV-associated oropharyngeal HNSCC, tended to occur in non-drinkers and non-smokers, have basaloid features on histology, and have no association with p53 mutation, and the patients have a 59% reduction in risk of dying from cancer, after adjustment for stage of disease, morbidity, and other confounding factors (Gillison et al., 2008). The association between HPV16 infection and HNSCC in specific sites suggests that the strongest and most consistent association is with tonsil cancer, and the magnitude of this association is consistent with an infectious etiology (Hobbs et al., 2006). Recently, marijuana use was identified as an independent risk factor for HPV-positive HNSCC, with the strength of this association increasing with intensity, duration, and cumulative years of marijuana smoking. Cannabinoid binding to the CB2 receptor can suppress immune responses, decrease host response to viral pathogens, and attenuate tumor activity (Pai and Westra, 2009).

Screening

A multistep carcinogenesis process results in epigenetic and metabolic changes that give rise to histologically distinct precursor phenotypes that harbor specific genetic alterations. Premalignant lesions of the UADT can present as leukoplakia, a white patch, and erythroplakia, a red patch. Histologically, leukoplakia may demonstrate benign hyperkeratosis of the surface epithelium, epithelial hyperplasia or dysplasia, or invasive carcinoma. Up to 18% of leukoplastic lesions may progress to invasive squamous cell carcinoma (SCC), whereas erythroplakia has a much greater potential for malignancy. Approximately 90% of erythroplastic lesions may demonstrate severe dysplasia, carcinoma *in situ*, or invasive SCC. The transformation rate of dysplasia to cancer has been reported to be as high as 36.4% among erythroplastic lesions.

Improvements in overall survival in patients with HNSCC rest on early identification of premalignant lesions and intervention in patients at risk prior to the development of advanced-stage disease. Clinical trials have shown the utility of toluidine blue (TB) in identifying asymptomatic oral SCC as well as premalignant lesions at risk of progressing to HNSCC that may otherwise remain undetected until the lesion becomes more advanced. One such study, which demonstrated a greater than six-fold increase in cancer risk for TB-positive lesions, showed that TB retention predicts progression to cancer of even oral premalignant lesions with benign histology or low-grade dysplasia, and likewise exhibited a four-fold increase in cancer progression (Epstein et al., 2007).

Previous reports of individual serologic inflammatory, angiogenesis, and tumor growth factors have shown correlation to HNSCC disease status. Simultaneous testing of various cytokines, growth factors and tumor antigens, or a serum multiplex panel of biomarkers, is being investigated for correlation with HNSCC disease status. The multimarker panel demonstrated 84.5% sensitivity with 98% specificity, classifying 92% of patients correctly. This panel comprised 25 biomarkers, including epidermal growth factor (EGF), epidermal growth factor receptor (EGFR), interleukin 8 (IL-8), granulocyte colony-stimulating factor (G-CSF), alpha-fetoprotein (AFP), matrix metalloproteinase-2 (MMP-2), interferon gamma (IFN-γ), and soluble vascular cell adhesion molecule (sVCAM) among others (Linkov et al., 2007). Thus, testing of serum for multiplex biomarkers appears more promising than testing for any single class in determination of HNSCC phase of disease. However, further study remains necessary.

Most recently, oral cancer screening visualization aids have been developed to assist dental professionals in the early identification of precancerous and cancerous lesions. The VELscope® (Visually Enhanced Lesion scope) and ViziLite® take advantage of the changes in light absorption between high-risk lesions and normal mucosa. In the ViziLite® system, a diffuse bluish-white chemiluminescent light is used. Normal cells absorb the light and have a bluish color, whereas the light is reflected by abnormal cells, which appear white with more distinct borders. The reported sensitivity was 100% and specificity ranged from 0–14%. The VELscope® system detects the loss of fluorescence in visible and non-visible high-risk oral lesions by applying direct fluorescence. Normal mucosa emits a green auto-fluorescence, whereas abnormal mucosa absorbs fluorescent light and appears dark. The reported sensitivity was 97–98% and specificity 94–100%. However, several studies have reported that the use of ViziLite® or VELscope® was not beneficial in identifying dysplasia or cancer. More randomized trials are needed to confirm their usefulness (Huber, 2009; Mehrotra et al., 2010; Patton et al., 2008; Trullenque-Eriksson et al., 2009).

Diagnosis

Because the probability of curing patients with HNSCC is related to the stage of disease, the importance of diagnosing HNSCC at an early stage cannot be overemphasized. Visual inspection is the most common method of detecting oral and oropharyngeal squamous carcinoma. The diagnosis of HNSCC, however, is frequently delayed, because symptoms for which patients will seek medical attention such as pain, dysphagia, and shortness of breath occur late in the stage of disease. Laryngeal squamous carcinoma remains one of the few subsites that is likely to have disease detection at earlier stages due to the presenting symptom of persistent hoarseness, which can occur with mild alterations of the vibratory surfaces of the true vocal cords.

Radiologic imaging modalities such as computed tomography (CT), magnetic resonance imaging (MRI), positron-emission tomography (PET), combined PET/CT, ultrasound, and lymphoscintigraphy are critical tools that provide information on extent of tissue invasion, involvement of regional lymph nodes, and presence of distant metastatic disease. The information provided by imaging modalities is critical for staging and subsequent treatment planning. However, there remains significant variability in outcomes for patients within the same tumor-node-metastasis (TNM) stage classification. Therefore, attempting to identify high-risk patients through molecular markers is an active area of investigation.

Exposure to carcinogens such as tobacco and alcohol may result in premalignant epithelial changes over a wide surface area of epithelium within the aerodigestive tract. This clinical phenomenon is referred to as "field cancerization," and may lead to frequent occurrence of multiple primary tumors in epithelial areas affected by widespread premalignant disease, and possibly distant related primary tumors in the UADT. For this reason, it is imperative to evaluate mucosa of the UADT, including the esophagus and trachea, with directed biopsies during endoscopic examination.

By identifying patients at risk for cervical lymph node metastasis and extracapsular spread without the need for surgical node dissection, tissue or serum biomarkers could play a vital role in clinical decision-making. Molecular profiling of primary tumors from HNSCC remains an ongoing investigation for developing a "genomic fingerprint signature" with the potential for predicting the presence of lymph node metastasis at the time of diagnosis. DNA microarray gene expression of primary tumors of the oral cavity and oropharynx found that signature or predictor gene sets can detect local lymph node metastases using material from primary HNSCC with better performance than current clinical diagnosis (Belbin et al., 2005; O'Donnell et al., 2005; Roepman et al., 2005). Other investigators have found that the gene expression profile of 53 genes with roles in cell differentiation, adhesion, signal transduction, and transcription regulation are associated with depth of invasion in patients with oral SCC (Toruner et al., 2004).

A large number of genes found in tumors from patients with oral SCC are abnormally expressed in comparison to control groups. A large percentage of these were upregulated proteins, including metalloproteinases (MMPs), proteins involved in regulation of cell adhesion, and cell-cycle-related proteins. Additional abnormal genes are present as downregulated tumor suppressors. Together, these findings suggest that oral SCCs show aberrant expression of genes involved in proliferation, apoptosis, extracellular matrix degradation, and other cellular regulation pathways (Kornberg et al., 2005).

The literature contains more than 20 studies utilizing microarray analysis to show genetic changes in oral SCC, which have not been validated in an independent cohort. However, each study uses different gene expression arrays and platforms, rendering direct comparison of data impossible. Also, no study tests the ability of the utilized genetic array(s) to predict the progression of a premalignant lesion to oral SCC against an independent validation dataset. Incorporation of "omics" technologies remains an active field of investigation (Nagaraj, 2009). Biomarker discovery, validation, and integration into clinical practice may eventuate early-stage disease detection and lead to improved outcomes.

Prognosis

TNM staging currently remains the most important factor correlating with prognosis of HNSCC. Large primary tumor size, positive margins after surgical excision, perineural invasion, and presence of lymph node metastasis are reliable indicators of poor clinical outcome in patients with HNSCC. However, the single most important factor that determines survival is the metastatic status of the cervical lymph nodes at the time of diagnosis. Particularly, the presence of extracapsular spread in cervical lymph node metastasis remains the most significant clinical prognostic indicator of survival, local-regional recurrence, and distant metastasis in patients with HNSCC. Although these clinical prognostic parameters provide the best possible criteria for deciding adjuvant treatment modalities, they are limited in discerning the future behavior of aggressive HNSCC.

The search for novel molecular prognostic markers with potentially significant predictive value for biological aggressiveness of HNSCC has exploded in recent years. Better prediction of the risk of developing distant metastases would help introduce a more selective treatment approach, according to the biological aggressiveness of the tumor. Several molecular mediators of tumor progression, invasion, and metastasis that function in growth factor signaling, metastasis, and suppressor genes have been well investigated in HNSCC, and are described below.

TP53 is a tumor-suppressor gene located on 17p13. It consists of 11 exons that encode a protein, p53, and functions in carcinogenesis by initiating G1 arrest in response to certain DNA damage and apoptosis. The prevalence of *TP53* mutations is reportedly greater than 50% in HNSCC. A number of studies have shown that *TP53* gene mutations are associated with increased risk for locoregional recurrence and poor outcome (Erber et al., 1998; Mineta et al., 1998). Mutated p53 protein overexpression is also associated with reduced therapeutic responsiveness, tumor recurrence, poor overall survival rate, increased rates of locoregional failure, and decreased disease-free survival in HNSCC. These mutations confer varying degrees of dysregulation based upon where the mutation occurs at the chromosomal level, resulting in a variegation of clinical impact on the p53 protein structure, stability, and DNA-binding properties (Pai and Westra, 2009).

Epidermal growth factor receptor (EGFR) is a transmembrane tyrosine kinase capable of promoting neoplastic transformation. The downstream signaling events upon ligand binding include activation of tyrosine kinase and activation of intracellular Ras, Raf, and mitogen-activated protein kinase (MAPK) cascades. EGFR expression has been extensively studied in HNSCC, and its overexpression has been reported in more than 90% of HNSCCs (Pai and Westra, 2009). This marker is significantly associated with short disease-free survival and overall survival and poor prognosis in patients with HNSCC. However, only a small subset of these HNSCCs overexpressing EGFR actually demonstrate amplified copy numbers or mutational activation of the EGFR gene. Autocrine and paracrine loops take effect via high expression of the EGFR with various ligands, and binding of these ligands results in autophosphorylation of the intracellular kinase domain and subsequent activation of multiple oncogenic pathways. This scenario explains the only modest success of EGFR blockade as monotherapy in the treatment of patients with HNSCC (Pai and Westra, 2009).

HPV, particularly type 16, is found in more than 70% of HNSCCs, and this form of HNSCC behaves distinctly differently than its smoking- and alcohol-related counterpart. HPV-positive HNSCCs express the viral oncoproteins E6 and E7, overexpress the p16 gene, and infrequently harbor TP53 gene mutations. These HPV-positive tumors are associated with increased radiosensitivity and improved prognosis. This behavior is postulated to occur as a result of immune surveillance for viral-specific tumor antigens, an intact apoptotic response to radiation, and absence of the widespread genetic alterations associated with smoking (Pai and Westra, 2009). In 2006, the HPV vaccine gardasil was approved by the US Food and Drug Administration (FDA) as an effective means of preventing cervical cancer and precancerous lesions due to HPV types 6, 11, 16, and 18. Clinical trials are currently underway with the vaccine in HNSCC (Pai and Westra, 2009).

In larger studies of patients with HNSCC, DNA microarray analysis has been used to identify distinct gene expression signatures associated with clinical outcome in HNSCC. Distinct subtypes of HNSCC based upon gene expression patterns obtained from tumor samples from patients with HNSCC were described. These subtypes had significant differences in clinical outcomes, including recurrence-free survival and overall survival, and patterns of expression were identified that could predict the presence of lymph node metastases in HNSCC tumors (Chung et al., 2004). Biomarker investigation remains an active area of ongoing research in the realm of screening, diagnosis, and prognosis.

Pharmacogenomics

Organ preservation therapy using chemotherapy with cisplatin-based agents, taxanes, and biotherapeutic agents combined with radiotherapy has been used effectively in patients with advanced HNSCC. Newer studies have shown that combining cisplatin and 5-FU (fluorouracil) with a taxane (TPF regimen) as induction chemotherapy increases organ preservation rate and overall response. Disease free survival was also improved when this regimen was used before concurrent chemoradiotherapy. Despite remarkable improvements in radiation techniques and novel chemotherapeutic agents, however, overall survival is still poor. Since there is significant inter- and intra-individual variability in clinical outcome, improvements in survival rates will likely require the identification of patients, prior to treatment, who are most likely to have chemo-radiotherapeutic benefit and patients with the highest risk of suffering genotoxic side effects. Inter-individual variability in treatment response frequently leads to treatment failure or treatment-related death.

The response to chemotherapy and radiation and their side effects in patients with HNSCC is dependent on several factors, such as site of primary tumor, disease stage, patient characteristics, and comorbidities. The interplay among these factors is still poorly understood. Genetic co-variables may influence toxicity and tumor responses to chemotherapy and/or radiotherapy. Knowledge of inter-individual pharmacokinetic variability and genetic profiling provides a novel scientific basis for an improved and individualized therapeutic approach.

Multidrug resistance (MDR) to chemotherapeutic agents and radiotherapy occurs in many types of tumors as well as in HNSCC, presents a major obstacle to the effectiveness of chemoradiotherapy, and subsequently leads to treatment failure. MDR is a process in which cells acquire simultaneous resistance to a group of drugs that appear to be unrelated structurally and functionally. The main mechanism that gives rise to the MDR phenotype in cancer is the overexpression of drug efflux transporters in the plasma membrane. ATP-binding cassette (ABC; ATP = adenosine triphosphate) transporters contribute to drug resistance via ATP-dependent drug efflux, extruding anticancer agents or their metabolites from cells. P-glycoprotein (Pgp), which is encoded by the MDR1 gene, confers resistance to certain anticancer agents. Very little information is known about the importance of MDR and Pgp expression in HNSCC. Although Pgp levels and mRNA have been noted in recurrent oral SCC and in oral mucosa with increasing severity of dysplasia, insignificant Pgp levels have been found in oral SCC cell lines (Pérez-Sayáns et al., 2010). Treatment of different cell lines with vincristine shows that Pgp can be induced by genetic induction of the MDR1 gene. Theile and colleagues (2010) studied HNSCC cell lines for drug transporter expression and susceptibility to cisplatin, paclitaxel and 5-FU, and found that cisplatin and paclitaxel resistances were inversely correlated. However, none of the cell lines expressed the well-established Pgp/ABCB1 drug transporter. Other ABC transporters not linked to MDR were induced. These findings questioned the significance Pgp in MDR in HNSCC.

Other drug-transporter-independent mechanisms involved in the process of MDR have been described. They include increased detoxification of drugs by glutathione S-transferase, downregulation of pharmacological targets such as DNA topoisomerase II, increased catabolism of drugs by multidrug-resistance-associated proteins (MRP), inadequate anabolism of prodrugs, and altered regulation of acid pH in the tumor microenvironment [via vacuolar ATPase (V-ATPase)]. MRP-1 mRNA has been found in human and murine oral SCC lines treated with vincristine. Hirata and colleagues (2000) examined the expression levels of mRNA for the MDR1, MRP, human canalicular multispecific organic anion transporter (cMOAT), lung resistance-related protein (LRP), topoisomerase II alpha and beta (Topo II alpha and beta), and topoisomerase I (Topo I) genes in human HNSCC and mucosa specimens. No significant differences were observed in the expression level of the six genes between samples exposed to platinum drugs and those not exposed to platinum drugs. The changes in cytosolic pH play an important role in chemotherapy drug resistance. The more acidic extracellular pH in solid tumors interferes with the absorption of basic chemotherapy drugs, reducing their effect on tumor cells. Recent evidence suggests that V-ATPases may play a role in acidification of the tumor microenvironment by secreting protons through the plasma membrane. Pretreatment with proton pump inhibitors has been found to sensitize tumor cell lines to the effects of different chemotherapy drugs (Pérez-Sayáns et al., 2010). Although it has been shown that pH regulation in oral SCC lines is mediated by vacuolar proton-pump ATPases, and that this pH regulation is involved in cell transformation, none of these inhibitors were proven to be useful in Oral Squamous cell carcinomas (OSCC). Nevertheless, whether the above mechanisms play a significant role in MDR in HNSCC remains uncertain.

Clinical studies have shown that EGFR targeting has synergistic effects with chemotherapy in HNSCC and reverses chemoresistance of epithelial tumors. However, increasing evidence suggests that a majority of patients do not respond to these therapies, and those who show initial response ultimately become refractory to treatment. This suggests the development of acquired resistance. Potential mechanisms of resistance to EGFR-targeted therapies involve EGFR and Ras mutations, epithelial-mesenchymal transition, and activation of alternative and downstream pathways.

Resistance to cisplatin chemotherapy has been shown to be significantly correlated to expression of mutant p53 (Bradford et al., 2003; Cabelguenne et al., 2000) and overexpression of anti-apoptotic proteins Bcl-2 and Bcl-x_L (Bauer et al., 2005). Overexpression of mutant-type p53 in HNSCC was also shown to be associated with increased sensitivity to ionizing radiation (Ogawa et al., 1998; Servomaa et al., 1996). Low expression of Bcl-x_L in tumor specimens from patients with HNSCC has been shown to be correlated with response to induction chemotherapy (Bradford et al., 2003). Induction of mutant p53 in HNSCC lines resulted in decreased expression of Bcl-2 and increased susceptibility to cisplatin-induced apoptosis, and implicates Bcl-2 in the deregulation of p53-induced apoptosis (Andrews et al., 2004). Increased tumor resistance to cytotoxic agents, including radiotherapy, has been associated with EGFR overexpression in HNSCC (Milas et al., 2003), in addition to its association with more aggressive tumor behavior.

In the future, individualization of treatment for patients with advanced HNSCC by genetic profiling will require prospective, properly powered, randomized clinical trials. However, research for effective ways of overcoming MDR in HNSCC is still in its infancy.

Monitoring

Patients treated for HNSCC are followed clinically for evidence of recurrent disease, development of second primary lesions or distant metastasis. The chance of developing a second primary tumor has been estimated at 2–3% per year in patients with HNSCC. In addition, 20–30% of patients treated for HNSCC will develop recurrent disease at the primary site, and such recurrence is the most common cause of treatment failure. Because prognosis of late-stage recurrent disease is dismal, early detection is imperative. Distant metastatic disease occurs in 11–15% of patients treated for HNSCC, and at this stage, treatment is palliative. Identifying molecular markers in primary tumors that are associated with locoregional relapse may allow for early identification of patients needing additional surveillance and treatment, and may have the potential to decrease the probability of distant disease.

There was an initial anticipation that EGFR expression would serve as predictive biomarker for likelihood of response to cetuximab therapy. However, studies have shown that immunohistochemistry (IHC)-based assays measuring EGFR expression are not predictors for response to cetuximab therapy. In addition, no mutations in EGFR have been identified to date that are reliable predictors for antibody-based EGFR therapies. EGFR overexpression has been shown to be an independent prognostic factor for neck node relapse in primary specimens of patients with laryngeal SCC undergoing primary resection (Almadori et al., 1999). In addition, when surgical margins of primary HNSCC are examined for mutational changes, there is an increased risk of local recurrence when positive margins demonstrating clonal alterations in *TP53* are observed (Van Houten et al., 2004).

Gene expression signatures identified using DNA microarray technology have potential utility as biomarkers to predict patients at risk for locoregional recurrence. Several gene expression signatures from HNSCC tumors from various anatomical sites in the head and neck have been described.

Novel and Emerging Therapeutics

Surgery was the primary treatment modality for malignant neoplasia of the head and neck until the Veterans Affairs larynx trial. Since then, the combination of radiation therapy and chemotherapy, administered concurrently, has allowed organ preservation and treatment of locally advanced HNSCC with improved outcomes (Forastiere et al., 2003). For tumors in the oropharynx and supraglottis, FDA-approved transoral robotic-assisted surgery (TORS) decreases the potential side effects of primary concurrent chemoradiotherapy and may potentially improve quality of life in patients expected to have a prolonged survival.

CONCLUSION

Current knowledge of head and neck molecular pathogenesis favors the existence of deregulated and redundant pathways that may be targets for disease diagnosis, monitoring, and treatment. Proper validation of new biomarkers is of paramount importance for disease diagnosis, monitoring, and evaluating potential treatment responses; however, standardization of scientific techniques is still lacking.

Many of the genetic alterations involved in the development and progression of HNSCC, such as in TP53 and EGFR, have been well characterized. Nevertheless, in recent years, the field of head and neck cancer therapy has witnessed the emergence of novel targeted therapies against EGFR and VEGFR (vascular endothelial growth factor receptor) that inhibit specific pathways and key molecules in growth and progression of squamous carcinoma of the head and neck. Clinical trials using EGFR in advanced HNSCC are an area of active and ongoing investigation. However, correlation between EGFR expression in tumors and treatment response has not been demonstrated thus far. Perhaps specific mutations in the EGFR genes may offer powerful new opportunities to predict those patients more likely to benefit from the use of targeted treatments.

Although we have seen recent success in the targeted therapy of head and neck cancer, several major gaps in our knowledge remain regarding the predictive potential of many other biomarkers in HNSCC. Understanding of specific molecular and cellular mechanisms whereby biomarkers in HNSCC contribute to carcinogenesis or progression of HNSCC is critical. New genomic technologies such as DNA and tissue microarray coupled with advanced bioinformatics tools may make it feasible to study biomarkers that are able to reliably and accurately predict outcome during cancer management and treatment.

Future work in this field should include studies that incorporate complementary advanced imaging technology with molecular surveillance in the monitoring of patients with HNSCC. Methods to identify mutations or molecular footprints to predict those patients most likely to respond favorably to targeted therapy in combination with traditional treatments are needed.

REFERENCES

Almadori, G., Cadoni, G., Galli, J., et al., 1999. Epidermal growth factor receptor expression in primary laryngeal cancer: An independent prognostic factor of neck node relapse. Int J Cancer 84 (2), 188–191.

Andrews, G.A., Xi, S., Pomerantz, R.G., et al., 2004. Mutation of p53 in head and neck squamous cell carcinoma correlates with Bcl-2 expression and increased susceptibility to cisplatin-induced apoptosis. Head Neck 26 (10), 870–877.

Bauer, J.A., Trask, D.K., Kumar, B., et al., 2005. Reversal of cisplatin resistance with a BH3 mimetic,(-)-gossypol, in head and neck cancer cells: Role of wild-type p53 and Bcl-xl. Mol Cancer Ther 4 (7), 1096–1104. <http://www.ncbi.nlm.nih.gov/pubmed/16020667/>.

Belbin, T.J., Singh, B., Smith, R.V., et al., 2005. Molecular profiling of tumor progression in head and neck cancer. Arch Otolaryngol 131 (1), 10–18.

Bradford, C.R., Zhu, S., Ogawa, H., et al., 2003. P53 mutation correlates with cisplatin sensitivity in head and neck squamous cell carcinoma lines. Head Neck 25 (8), 654–661.

Cabelguenne, A., Blons, H., De Waziers, I., et al., 2000. p53 alterations predict tumor response to neoadjuvant chemotherapy in head and neck squamous cell carcinoma: A prospective series. J Clin Oncol 18 (7), 1465–1473.

Cheng, L., Eicher, S.A., Guo, Z., Hong, W.K., Spitz, M.R., Wei, Q., 1998. Reduced DNA repair capacity in head and neck cancer patients. Cancer Epidemiol Biomarkers Prev 7 (6), 465–468.

Chung, C.H., Parker, J.S., Karaca, G., et al., 2004. Molecular classification of head and neck squamous cell carcinomas using patterns of gene expression. Cancer Cell 5 (5), 489–500.

Epstein, J.B., Sciubba, J., Silverman Jr, S., Sroussi, H.Y., 2007. Utility of toluidine blue in oral premalignant lesions and squamous cell carcinoma: Continuing research and implications for clinical practice. Head Neck 29 (10), 948–958.

Erber, R., Conradt, C., Homann, N., et al., 1998. TP53 DNA contact mutations are selectively associated with allelic loss and have a strong clinical impact in head and neck cancer. Oncogene 16 (13), 1671–1679.

Forastiere, H., Goepfert, M., Maor, M., 2003. Concurrent chemotherapy and radiotherapy for organ preservation in advanced laryngeal cancer. N Engl J Med 349 (22), 2091–2098.

Gillison, M.L., D'Souza, G., Westra, W., et al., 2008. Distinct risk factor profiles for human papillomavirus type 16-positive and human papillomavirus type 16-negative head and neck cancers. J Natl Cancer Inst 100 (6), 407–420.

Hirata, S., Katoh, O., Oguri, T., Watanabe, H., Yajin, K., 2000. Expression of drug resistance-related genes in head and neck squamous cell carcinomas and normal mucosa. Jpn J Cancer Res 91 (1), 84–90.

Ho, T., Wei, Q., Sturgis, E.M., 2007. Epidemiology of carcinogen metabolism genes and risk of squamous cell carcinoma of the head and neck. Head Neck 29 (7), 682–699.

Hobbs, C.G., Sterne, J.A., Bailey, M., Heyderman, R.S., Birchall, M.A., Thomas, S.J., 2006. Human papillomavirus and head and neck cancer: A systematic review and meta-analysis. Clin Otolaryngol 31 (4), 259–266.

Huber, M.A., 2009. Assessment of the VELscope as an adjunctive examination tool. Tex Dent J 126 (6), 528–535.

Kornberg, L.J., Villaret, D., Popp, M., et al., 2005. Gene expression profiling in squamous cell carcinoma of the oral cavity shows abnormalities in several signaling pathways. Laryngoscope 115 (4), 690–698.

Kutler, D.I., Wreesmann, V.B., Goberdhan, A., et al., 2003. Human papillomavirus DNA and p53 polymorphisms in squamous cell carcinomas from Fanconi anemia patients. J Natl Cancer Inst 95 (22), 1718–1721.

Linkov, F., Lisovich, A., Yurkovetsky, Z., et al., 2007. Early detection of head and neck cancer: Development of a novel screening tool using multiplexed immunobead-based biomarker profiling. Cancer Epidemiol Biomarkers Prev 16 (1), 102–107.

Mehrotra, R., Singh, M., Thomas, S., et al., 2010. Cross-sectional study evaluating chemiluminescence and autofluorescence in the detection of clinically innocuous precancerous and cancerous oral lesions. J Am Dent Assoc 141 (2), 151–156.

Milas, L., Mason, K.A., Ang, K.K., 2003. Epidermal growth factor receptor and its inhibition in radiotherapy: In vivo findings. Int J Radiat Biol 79 (7), 539–545.

Mineta, H., Borg, A., Dictor, M., Wahlberg, P., Akervall, J., Wennerberg, J., 1998. p53 mutation, but not p53 overexpression, correlates with survival in head and neck squamous cell carcinoma. Br J Cancer 78 (8), 1084–1090.

Nagaraj, N.S., 2009. Evolving 'omics' technologies for diagnostics of head and neck cancer. Brief Funct Genomic Proteomic 8 (1), 49–59. <http://www.ncbi.nlm.nih.gov/pubmed/19273537/>.

O'Donnell, R.K., Kupferman, M., Wei, S.J., et al., 2005. Gene expression signature predicts lymphatic metastasis in squamous cell carcinoma of the oral cavity. Oncogene 24 (7), 1244–1251.

Ogawa, Y., Nishioka, A., Hamada, N., et al., 1998. Changes of mutant-type p53 expression in squamous cell carcinoma of the head and neck during radiation therapy and its clinical significance: Comparison of an immunohistochemical method and PCR-SSCP assay. Oncol Rep 5 (5), 1053–1059.

Pai, S.I., Westra, W.H., 2009. Molecular pathology of head and neck cancer: Implications for diagnosis, prognosis, and treatment. Annu Rev Pathol 4, 49–70.

Patton, L.L., Epstein, J.B., Kerr, A.R., 2008. Adjunctive techniques for oral cancer examination and lesion diagnosis: A systematic review of the literature. J Am Dent Assoc 139 (7), 896–905.

Pérez-Sayáns, M., Somoza-Martín, J.M., Barros-Angueira, F., Diz, P.G., Rey, J.M., García-García, A., 2010. Multidrug resistance in oral squamous cell carcinoma: The role of vacuolar ATPases. Cancer Lett 295 (2), 135–143.

Roepman, P., Wessels, L.F., Kettelarij, N., et al., 2005. An expression profile for diagnosis of lymph node metastases from primary head and neck squamous cell carcinomas. Nat Genet 37 (2), 182–186.

Schantz, S.P., Zhang, Z.F., Spitz, M.S., Sun, M., Hsu, T.C., 1997. Genetic susceptibility to head and neck cancer: Interaction between nutrition and mutagen sensitivity. Laryngoscope 107 (6), 765–781.

Servomaa, K., Kiuru, A., Grénman, R., Pekkola-Heino, K., Pulkkinen, J.O., Rytömaa, T., 1996. p53 mutations associated with increased sensitivity to ionizing radiation in human head and neck cancer cell lines. Cell Prolif 29 (5), 219–230.

Theile, D., Ketabi-Kiyanvash, N., Herold-Mende, C., et al., 2010. Evaluation of drug transporters' significance for multidrug resistance in head and neck squamous cell carcinoma. Head Neck 33 (7), 959–968. doi: 10.1002/hed.21559 [Epub.].

Toruner, G.A., Ulger, C., Alkan, M., et al., 2004. Association between gene expression profile and tumor invasion in oral squamous cell carcinoma. Cancer Genet Cytogenet 154 (1), 27–35.

Trullenque-Eriksson, A., Muñoz-Corcuera, M., Campo-Trapero, J., Cano-Sánchez, J., Bascones-Martínez, A., 2009. Analysis of new diagnostic methods in suspicious lesions of the oral mucosa. Med Oral Patol Oral Cir Bucal 14 (5), E210–E216.

Van Houten, V.M., Leemans, C.R., Kummer, J.A., et al., 2004. Molecular diagnosis of surgical margins and local recurrence in head and neck cancer patients: A prospective study. Clin Cancer Res 10 (11), 3614–3620.

Wang, L.E., Hu, Z., Sturgis, E.M., et al., 2010. Reduced DNA repair capacity for removing tobacco carcinogen-induced DNA adducts contributes to risk of head and neck cancer but not tumor characteristics. Clin Cancer Res 16 (2), 764–774.

Brain Tumors and Gliomas

Sean E. Lawler and E. Antonio Chiocca

INTRODUCTION

Approximately 20,000 new patients with primary brain tumors are diagnosed in the United States each year (De Angelis, 2001). These constitute the most common solid tumors in children, and rank first among all cancer types in average years lost. They rarely metastasize outside the central nervous system (CNS), however more than 100,000 patients per year die with symptomatic intracranial metastases due to systemic primary cancer.

Tumor type is currently determined by histopathologic analysis of biopsy samples, and graded on a scale of I (benign) to IV (highly malignant) based on a range of histological tumor features (frequency of mitotic figures, necrosis, nuclear atypia, and vascularity) according to the World Health Organization (WHO) classification of nervous system tumors (Kleihues and Cavenee, 2000; Louis et al., 2007). The various brain tumor types are similar in clinical presentation, diagnosis, and treatment. Most are treated by aggressive surgical resection when possible, followed by chemo- and/or radiotherapy. Brain tumors are especially challenging, because they are often resistant to therapies, progress rapidly, and infiltrate normal brain tissue. Even a benign brain tumor may seriously compromise normal brain function, and surgical tumor excision must be carried out without damaging vital brain structures. Delivery of drugs to the central nervous system and therapy-induced neurotoxicity also present critical barriers to effective treatment.

This chapter will focus mainly on gliomas, the best-studied and most common primary brain tumor type, accounting for approximately half of all cases (reviewed by Furnari et al., 2007; Wen and Kesari, 2008). Gliomas are usually classed as either astrocytic or oligodendroglial (summarized in Table 65.1). Low-grade gliomas such as pilocytic astrocytomas are benign and have a good prognosis. These do not typically progress, and are genetically distinct. The survival range for grade II and III astrocytomas is broad, and these usually progress to higher-grade tumors. The most common and most aggressive astrocytic tumor is the grade IV glioblastoma multiforme. It is among the deadliest of human cancers, with a median survival of around 14 months, even with aggressive treatment. Oligodendrogliomas and the mixed oligoastrocytomas are less common than pure astrocytomas, and typically have longer survival times. Most low-grade oligodendrogliomas progress to higher-grade tumors. With early diagnosis by magnetic resonance imaging (MRI), and chemotherapy, mean survival is 16 years.

There are several well-established molecular alterations commonly seen in gliomas, which are also observed in many other cancer types (reviewed in Louis, 2006; Schwartzbaum et al., 2006). "Traditional" molecular analyses performed in the 1980s and 1990s showed that gliomas are typically characterized by increased tyrosine kinase receptor activation. This is reflected in amplified levels of receptors and/or ligands, such as the commonly observed epidermal growth factor receptor

Genomic and Personalized Medicine, 2nd edition
by Ginsburg & Willard. DOI: http://dx.doi.org/10.1016/B978-0-12-382227-7.00065-3

TABLE 65.1	Classification of gliomas		
Tumor type	**Characteristics**	**Peak incidence**	**Survival**
Astrocytic			
Pilocytic astrocytoma (I)	Slow growing, often cystic	Children	Infrequently fatal
Diffuse astrocytoma (II)	Slow growing, invasive, tendency to progress	Young adults	Mean 6–8 years, highly variable
Anaplastic astrocytoma (III)	Tendency to progress	40–60 years	Mean TTP 2 years
Glioblastoma (IV)	*De novo* or secondary mitotic figures, anaplasia, necrosis, vascularity	45–70 years	12 months
Oligodendroglial			
Oligodendroglioma (II)	Grow diffusely in cortex and white matter	50–60 years	3–15 years
Anaplastic oligodendroglioma (III)			
Mixed			
Oligoastrocytoma (II)	Oligodendrocytic and astrocytic characteristics	35–50 years	4–7 years
Anaplastic oligoastrocytoma (III)			

(EGFR) amplification (Libermann et al., 1985) or the presence of mutated constitutively active receptors such as EGFRvIII (Sugawa et al., 1990). This leads to activation of multiple downstream intracellular signaling pathways, with the PI3K/Akt pathway being considered extremely important by promoting cell survival. In high-grade gliomas, PI3K activation is further potentiated by loss or mutation of the tumor suppressor *PTEN* (Li et al., 1997). The cell cycle is deregulated through disruption of *p53* (Chung et al., 1991) and RB pathway alteration. The *CDKN2A* tumor suppressor locus that encodes both p16 and p14ARF is commonly deleted in glioblastoma (Simon et al., 1999). These alterations constitute the core pathways known to be disrupted in gliomas. This view of glioma genetics has been supported by recent high-throughput, chip-based global molecular profiling techniques such as array-CGH (comparative genomic hybridization) and gene expression profiling on large sets of patient tumor samples. The most significant recent development in brain tumor genomics is the development of The Cancer Genome Atlas (TCGA) (Cancer Genome Atlas Research Network, 2008). This initiative was started by the National Cancer Institute and the National Human Genome Research Institute due to the recalcitrant nature of glioblastoma and a strong need to understand its genetics. This freely available resource contains DNA copy number, gene expression, microRNA, and methylation data on more than 250 pathologically diagnosed glioblastomas (http://cancergenome.nih.gov/). The exons of more than 600 candidate genes have also been sequenced, in order to identify tumor-associated mutations. In another important study, next-generation sequencing was performed on 20,661 genes, as well as profiling in 22 glioblastomas (Parsons et al., 2008). These studies have identified novel mutations and confirmed previously known alterations in core pathways, and mark the beginning of a new era in glioblastoma

research and discovery, as described in later sections. These studies have identified many alterations which will be of value in stratifying patients in future trials, give valuable prognostic information, and identify potential novel mechanisms, for example *IDH1* mutations as described below.

Based on age of onset and pathology, two different kinds of glioblastoma exist, with characteristic genetic alterations, although they are clinically similar (Kleihues and Cavenee, 2000). Glioblastoma in older patients shows no sign of previous low-grade tumors, and is known as either *de novo* or primary glioblastoma (95%). Typically these tumors show EGFR amplification, *p16* deletions and PTEN deletions. Secondary glioblastoma occurs in younger patients (5%), and occurs due to progression from a grade II or III glioma. These are characterized by *p53* mutation, amplification of PDGF signaling, and the presence of mutations in *IDH1* (Parsons et al., 2008). In contrast to astrocytomas, oligodendrogliomas are characterized by allelic loss of 1p and 19q. However, tumor progression is associated with changes similar to those seen in astrocytomas (see Figure 65.1).

PREDISPOSITION

Relatively few genetic and environmental factors that influence brain tumor development have been clearly identified (reviewed in Schwartzbaum et al., 2006), and only a small minority of tumors can be attributed to inherited predisposition. For example, a shared susceptibility to breast cancer, brain tumors, and Fanconi anemia was reported in four families with germline *BRCA2* mutations (Offit et al., 2003). In addition, genetic predisposition is associated with various familial cancer syndromes (Turcot's, NF1, NF2, and Li-Fraumeni), accounting for 1–2% of all brain tumors.

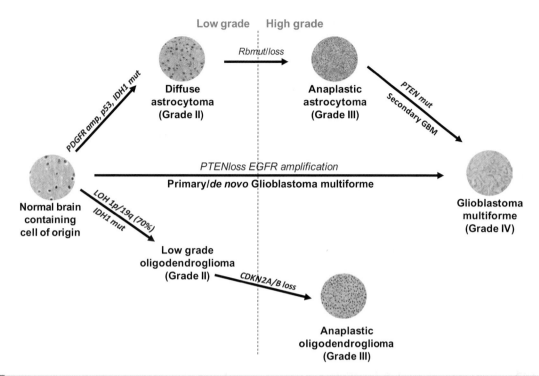

Figure 65.1 Pathways to glioma formation. Three pathways of genetic alteration are currently known that define oligodendroglioma, primary glioblastoma, and secondary glioblastoma. The major genetic alterations associated with tumor progression are shown, with their approximate frequency. Mut = mutation, LOH = loss of heterozygosity, amp = amplification, GBM = glioblastoma multiforme.

Epidemiological studies have shown that glioblastoma has higher incidence in the white population and is more common in men than women (ratio 3:2), for unknown reasons. Many environmental risk factors have been investigated, including the use of cell phones, electromagnetic fields, viral infection, diet, tobacco, and previous head trauma or injury (Schwartzbaum et al., 2006). However, only gamma radiation is a widely accepted risk factor for gliomas. Indeed, genetic sensitivity to radiation or environmental carcinogens may play a role in brain tumor pathogenesis (Bondy et al., 2001).

Further evidence of an inherited component in glioblastoma susceptibility comes from studies that show the risk of glioma is elevated two-fold in first-degree relatives of victims (Hemminki and Li, 2003). Inherited polymorphisms that may influence brain tumor formation involve oxidation, DNA repair, and immune function. For example, the DNA repair gene *XRCC7* G7621T variant, leads to a 1.8-fold increased risk of glioblastoma (Wang et al., 2004), and the *hTERT* MNS16A allele results in a two-fold increase in survival, due to higher expression levels of this allele (Wang et al., 2006). A consistent inverse association between self-reported allergic conditions and glioma has been observed. Asthma-associated polymorphisms in IL-4RA and IL-13 were inversely associated with glioblastoma incidence. This is a provocative observation, because these factors have also been shown to inhibit glioma growth, and it suggests a link between glioblastoma and immune function (Schwartzbaum et al., 2005). This is supported by the demonstration that polymorphisms in the IL-4 receptor gene are

significantly associated with improved survival (Scheurer et al., 2008a). Effects of combined single nucleotide polymorphisms (SNPs) in multiple DNA repair genes increase glioma risk (Liu et al., 2008). Evidence is beginning to accumulate suggesting a role of cytomegalovirus (CMV) in glioma progression, a finding that awaits further investigation (Cobbs et al., 2008; Prins et al., 2008; Scheurer et al., 2008b). The role of CMV in glioma biology is a matter of current debate.

Two recent large-scale genome-wide association studies (GWAS) identified several SNPs strongly associated with glioma risk. In 2009, Shete and colleagues conducted a meta-analysis of two GWAS from the US and UK with genotyping data on 550 K tagging SNPs, in a total of 1878 glioma cases and 3670 controls, as well as validation in three independent large studies. This identified five risk loci for glioma in genes encoding *TERT* (OR 1.27), *CCDC26* (OR 1.36), *CDKN2A/CDKN2B* (OR 1.24), *RTEL1* (1.28), and *PHLDB1* (OR 1.18) (Shete et al., 2009). In the second study, Wrensch and colleagues (2009) examined 692 high-grade glioma cases and 3992 controls, and identified an association with SNPs related to *CDKN2B* and *RTEL1*. *TERT* was also identified in this study, but was below the level of statistical significance. *CDKN2A* and *CDKN2B* are frequently inactivated in glioma tumor samples and would normally block cell cycle progression by acting as inhibitors of CDK4 and CDK6. *RTEL1* is a DNA helicase critical for regulation of telomere length in mice. Its loss is associated with shortened telomere length, chromosome breaks, and translocations. The data from these studies clearly establish genetic risk factors for glioma and suggest that glioma risk is enhanced by common, multiple, low-penetrance susceptibility alleles. Shete and colleagues suggest that these alleles account for approximately 7–14% of the excess familial risk of glioma. However, despite increased knowledge of the molecular features of glioblastoma, the majority of cases cannot be explained by either environmental or genetic causes.

SCREENING

Many brain tumors are comparatively rare, tissue is not easily accessible, predisposition is poorly understood, and reliable serum markers have not been established – therefore screening is not routinely carried out. In the future, however, improved imaging techniques may lead to earlier detection of presymptomatic lesions, providing the opportunity for effective surgical intervention. Screening for brain tumors would be greatly facilitated by the identification of robust genetic links or consistent robust serum markers. Through efforts in gene expression profiling and proteomics, candidate markers are emerging, as described in later sections. A novel approach to identifying brain tumor-specific markers has emerged through studies aimed at identifying circulating exosomes in glioblastoma patients (Skog et al., 2008). Exosomes are small vesicles secreted by many cell types, including cancer cells. This study showed that the tumor-specific EGFRvIII mutant could be detected in RNA-isolated exosomes from the serum of glioblastoma patients, thus identifying a new method for the identification of brain tumor-specific biomarkers.

DIAGNOSIS AND PROGNOSIS

Brain tumor diagnosis is currently based largely on histologic analysis of tumor biopsies according to WHO guidelines, as described in the introductory section. Major prognostic factors in gliomas are tumor type and grade, patient age, symptom duration, degree of surgery, and neurological deficit. The few long-term survivors of glioblastoma are young, in otherwise good health, and able to undergo gross total resection followed by chemo- and radiotherapy. Although increasing molecular information is now available, age remains the best predictor of survival, with 5-year survival at 13% for patients between 15 and 45 years of age, and only 1% for patients 75 and over (CBTRUS, 2012). The current WHO tumor classification system suffers from well-recognized problems due to interobserver variability, tumor heterogeneity, and ambiguous tumor types (Louis et al., 2001). Very few further molecular markers are in common use at present, although co-deletion of 1p and 19q is useful in predicting clinical responsiveness in oligodendroglioma, as described below. The incorporation of molecular information would allow less subjective diagnosis, therapeutic guidance, and more precise prognostic information.

The most commonly used molecular diagnostic tool based on chromosomal alterations at present is in oligodendroglioma. Oligodendrogliomas frequently show a remarkable sensitivity to PCV (procarbazine, CCNU and vincristine) chemotherapy, with a response rate of approximately 75%. Thus, within a histologically indistinguishable entity, there exist subgroups that show different biological behavior. Molecular genetic studies revealed that loss of heterozygosity (LOH) of chromosome 1p, which is found in 60–80% of oligodendrogliomas and often accompanied by LOH of 19q (Cairncross et al., 1998), was the underlying reason for the increase in sensitivity. This occurs in both high- and low-grade oligodendrogliomas, and also applies to radiotherapy and temozolomide (Hoang-Xuan et al., 2004), the current chemotherapeutic regimen of choice in glioma. Combined loss of 1p and 19q are strong correlates of longer survival in oligodendroglioma (Cairncross et al., 1998), and has been associated with increased survival in glioblastoma patients (Ino et al., 2000). Therefore, genetic analysis for 1p and 19q is now commonly performed for oligodendroglioma, and can be used to determine the most suitable therapeutic strategy. The prognostic significance of 1p/19q has been confirmed in recent large trials (Van den Bent et al., 2006).

Identification of the key genes involved at 1p and 19q may lead to novel treatment strategies broadly applicable to gliomas in general. It has recently been shown that 1p/19q co-deletion is a result of an imbalanced translocation. This finding may lead to a better understanding of the functional importance of 1p/19q co-deletion (Jenkins et al., 2006). Studies also show that oligodendrogliomas and oligoastrocytomas may be further subgrouped on the basis of distinct genetic changes in addition to 1p and 19q alterations. CGH revealed a hemizygous deletion in the 500 kb region in 11q13 and the 300 kb region in 13q12 in virtually all low-grade oligodendrogliomas, regardless of 1p/19q status (Rossi et al., 2005), suggesting that further markers exist.

Other molecular diagnostic tools include analysis of EGFR amplification, which is seen commonly in gliomas. Glioblastomas with amplified EGFR often occur in older patients and can be difficult to distinguish from anaplastic oligodendroglioma due to their small cell appearance. Assessment of EGFR along with 1p and 19q can therefore be helpful in this situation (Burger et al., 2001). In summary, at present, to distinguish between different glioma groups the following changes are the most typical, and some are used diagnostically. Loss of chromosomes 1p/19q is typical of oligodendrogliomas, whereas gains of chromosome 7 in the setting of intact 1p/19q are more typical of astrocytomas. The detection of amplified EGFR favors the diagnosis of high-grade astrocytomas over anaplastic oligodendroglioma, which is especially relevant for small cell astrocytomas.

Chromosomal Alterations in Brain Tumors

Glioblastoma has a highly rearranged genome, and is known for its extreme genomic instability. High-resolution genomic analysis has uncovered a complex range of somatic alterations on the genomic level, which may harbor tumor-related oncogenes and tumor suppressors. Several CGH studies along with the TCGA project and linked transcriptome profiling studies have provided a more complete view of the landscape of genetic alterations and their linked pathways.

A range of characteristic genetic changes has been observed, which in high-grade gliomas are variable and complex. Benign tumor types are characterized by fewer changes, the most prominent being alteration of chromosome 22. Cytogenetic analysis often reveals a normal karyotype for grade I astrocytomas, whereas grade II gliomas often have mutations in the *TP53* gene, and overexpression of *PDGFRA*. The most commonly observed genetic alteration in glioblastoma is loss of chromosome 10 (60–80%), resulting in reduced PTEN expression, although it is possible that other genes on chromosome 10 are important. Gain of chromosome 7 (EGFR), and loss of 9p (p16, ARF), are also consistently observed (shown in Figure 65.1). Other consistently observed chromosomal alterations are shown in Table 65.2.

Recent detailed, microarray-based high-resolution genomic mapping studies have confirmed these studies, revealing the presence of numerous recurrent regional copy number alterations (CNAs) (Ichimura et al., 2006; Kotliarov et al., 2006; Maher et al., 2006; Mulholland et al., 2006; Nigro et al., 2005; Rossi et al., 2005; Wiedemeyer et al., 2008). The large numbers of CNAs detected by these studies and the long lists of resident genes associated with them make the distinction of genuine glioma-associated genes from bystanders a complicated task.

Array-CGH can readily distinguish primary from secondary glioblastoma, and also showed that secondary glioblastoma falls into two distinct groups (Maher et al., 2006). Kotliarov and colleagues (2006) reported the use of Affymetrix high density 100 K SNP arrays to analyze genomic alterations in 178 glioma samples at an unprecedented resolution of 25 Kb. Each tumor group had many more unique events than shared, and therefore may benefit from quite different therapeutic approaches. Complementary DNA (cDNA) microarray-based CGH was able to predict tumor type successfully based on common genetic alterations (Bredel et al., 2005). All of these studies revealed large numbers of previously unidentified alterations. Other studies have reported deletions in chromosomes 6, 21, and 22 (Lassman et al., 2005). Two regions in 6q26 were found to be commonly deleted, containing three novel genes that may have tumor suppressor functions (Ichimura et al., 2006). Using TCGA and other available datasets, a network model of a cooperative genetic landscape was developed in order to better understand genetic changes in glioblastoma (Bredel et al., 2009). This approach led to the identification of a seven gene set – POLD2, CYCS, MYC, AKR1C3, YME1L1, ANXA7, and PDCD4 – whose alteration leads to an unfavorable prognosis.

Transcriptional Alterations in Brain Tumors

Global transcriptional profiling using microarrays has revealed many novel changes in brain tumors. It is now well established that expression profiles can readily distinguish different histologically classified tumor types, and also may predict survival and treatment responses. For example, microarray data could separate glioblastoma from oligodendroglioma on the basis of a 170- (Shai et al., 2003) or a 70-gene signature (Nutt et al., 2003). EGFR-amplified glioblastoma was distinguished from non-amplified EGFR tumors using a 90-gene signature (Mischel et al., 2003), grade II and grade III oligodendrogliomas by a 200-gene subset (Watson et al., 2002) and also oligodendrogliomas with and without 1p loss (Mukasa et al., 2002). Gene expression analysis has also identified tumor subgroups that are indistinguishable histologically. For example, clustering analysis separated oligodendrogliomas into two survival groups based on gene expression (Huang et al., 2004). Two microarray studies of glioblastoma identified gene signatures corresponding to different survival groups, and suggest that increased expression of genes that promote infiltration can lead to a poor prognosis (Liang et al., 2005; Rich et al., 2005). Furthermore, microarrays were better predictors of survival than histological grading for oligodendrogliomas (Nutt et al., 2003). Gene expression profiling therefore may provide useful diagnostic and prognostic information.

In a comprehensive study, gene expression profiles were obtained from 76 resected astrocytomas with known survival data (Phillips et al., 2006). Tumors were separated into either short or relatively long survival groups, and cluster analysis then segregated tumors into three distinct, discrete sample sets based on 35 signature genes. These were defined by the predominant class of gene expressed, as either proneural (PN), proliferative (Prolif), or mesenchymal (Mes). The PN subset was associated with substantially longer survival, was more

TABLE 65.2	Chromosomal alterations found in gliomas	
Region	**Alterations**	**Candidate glioma genes**
1p (34-pter) (various)	Gains and deletions	Unknown
1q32	Gains	*RIPK5, MDM4, PIK3C2B*
4q	Deletions	*NEK1, NIMA*
7p11.2–p12	Amplification/gain	*EGFR*
9p21–p24	Deletions	*CDKN2*
10q23	Deletions	*PTEN*
10q25–q26	Deletions	*MGMT*
12q13.3–q15	Amplification	*MDM2, CDK4*
13p11–p13	Loss	*RB*
19q13	Loss	*GLTSCR1, GLTSCR2, LIG1, PSCD2*
22q11.2–q12.2	Loss	28 genes, including *INI1*
22q13.1–13.3	Loss	Not known

prominent in younger patients, and showed activated Notch signaling as seen by elevated DLL3 expression. Poor prognosis was associated with up-regulation of proliferative markers such as PCNA, TOP2A, and angiogenesis markers such as VEGF and its receptors, and the endothelial cell marker PECAM1. Markers of neural stem cells were also associated with poor prognosis. All WHO-classified grade II tumors were *PN*, whereas for grade IV tumors 31% were *PN*, 20% were *Prolif*, and 49% were *Mes*, with expression profiles shifting toward the *Mes* class after recurrence. This suggests that the identified subtypes represent stages of tumorigenesis rather than different tumors. This study also correlated gene expression profiles with genetic abnormalities, demonstrating that losses of chromosome 10 and gains on chromosome 7 are associated with *Prolif* and *Mes* phenotypes (around 80%), compared with 20% for the *PN* class. The authors therefore propose a model in which *PN* phenotype progresses to a *Prolif* phenotype, and ultimately the *Mes* phenotype, with the poorest prognosis. This kind of study demonstrates that useful information could be obtained from a small subset of signature genes. Based on this study, Carro and colleagues (2010) examined transcription factor binding sites in genes involved in the *Mes* signature, and identified STAT3 and C/EBPβ as important mediators of the mesenchymal phenotype. This demonstrates that secondary studies of microarray-based datasets can identify further targets of interest.

A similar study to that of Phillips and colleagues identified a 44-gene expression signature to classify gliomas into previously unrecognized biological and prognostic groups which outperformed histology-based classification in survival prediction (Freije et al., 2004). In agreement with Phillips, key genes identified include *DLL3* (good prognosis), *TOP2A* (poor) *LIF* (poor), *S100A4* (poor), and *VEGF* (poor). Efforts to classify gliomas based on transcriptional profiling have continued (Petalidis et al., 2008). As described in earlier studies, this data reveals three distinct molecular classes of glioma, and suggests that grading based on molecular features can outperform traditional histologic methods. In order to develop a clinically useful tool to distinguish treatment-sensitive tumors from treatment-refractory tumors, an analysis of gene expression was performed from four different datasets (Colman et al., 2010). This led to the identification of a consensus 38-gene survival set, with worse outcome associated with genes involved in mesenchymal differentiation and angiogenesis. In order to render the list useful in practical terms, nine of these genes were further identified as suitable for polymerase chain reaction (PCR) from formalin-fixed paraffin-embedded (FFPE) patient tumor samples.

Analysis of TCGA transcriptional profiling data has led to the identification of a gene expression-based molecular classification of glioblastoma (GBM) into proneural, neural, classical, and mesenchymal phenotypes. These phenotypes show associations with known chromosomal alterations, and response to radiation and temozolomide differs by subtype, with the greatest benefit in the classical subtype, and none in the proneural subtype (Verhaak et al., 2010). This study reaches very similar conclusions to Phillips and colleagues', but there are some

differences, potentially because this study is limited to glioblastomas, whereas Phillips also looked at astrocytomas in their samples. Thus, there are clearly distinct molecular subtypes of glioblastoma (GBM) in which chromosomal abnormalities and gene expression patterns are associated. They have prognostic value, and may lead to subtype-specific therapeutic approaches in the coming years.

The most common alterations seen in gene expression profiling studies of gliomas are in genes involved in immune system regulation, hypoxia, cell proliferation, angiogenesis, neurogenesis, and cell motility. Ten of the most commonly reported up-regulated genes in glioblastoma are summarized in Table 65.3. Many of these changes were not predicted by analysis of chromosomal alterations. Of these highly up-regulated genes, *GPNMB* (Kuan et al., 2006) and *CHI3L1* (*YKL-40*) (Hormigo et al., 2006) have been proposed as prognostic markers. *YKL-40* has emerged from gene screening studies as one of the most robust and consistently observed markers in glioblastoma (Freije et al., 2004; Phillips et al., 2006). *YKL-40* mRNA was on average 82-fold higher in glioblastoma than anaplastic oligodendrogliomas (Nutt et al., 2003). A further study showed that that *YKL-40* may be a better marker than *GFAP*, which is currently the standard in distinguishing diagnostically challenging gliomas, and suggests that combining *YKL-40* and *GFAP* staining is the best approach at present (Nutt et al., 2005). *YKL-40* expression is also associated with radioresistance (Pelloski et al., 2005).

TABLE 65.3	Genes up-regulated in glioblastoma	
Gene	**Symbol**	**Function**
Fibronectin 1	FN1	ECM – angiogenesis, invasion
Insulin-like growth factor binding protein 2	IGFBP2	Promotes invasion
Collagens (IV and VI)	COL6A/ COL4A	ECM, promotes invasion
Topoisomerase 2A	TOP2A	DNA replication and transcription
Biglycan	BGN	ECM proteoglycan. Role in glioma unknown
Chitinase 3-like-1	CHI3LI	ECM, invasion, angiogenesis, survival
Vascular endothelial growth factor	VEGF	Angiogenesis, invasion
Vimentin	VIM	Intermediate filament, invasion
Epidermal growthfactor receptor	EGFR	Survival, growth, invasion
Transforming growth factor β1	TGFB1	Proliferation, differentiation, invasion

ECM, extracellular matrix.

A potential drawback of transcriptional profiling is that mRNA levels in a given sample may not be representative of the whole tumor, and also that they do not provide an accurate indication of the translation into protein. mRNA must be bound to polysomes in order to be translated, and a study of polysomal RNA in glioma cell lines revealed marked differences with global RNA in the same sample (Rajasekhar et al., 2003). Other genetic factors, including methylation and microRNAs, have potential diagnostic predictive and therapeutic applications, as described below.

Mutations in Brain Tumors

Two studies have greatly improved our knowledge of mutational events in protein coding genes in glioblastoma. In addition to methylome, transcriptome, and CGH data of in excess of 250 pathologically diagnosed glioblastomas, along with treatment and survival data, the TCGA dataset contains sequence data from the coding sequences of more than 600 genes. This has revealed that the core pathways of tyrosine kinase/PI3K, p53 and RB signaling, as identified previously, are the most altered pathways in glioblastoma (Cancer Genome Atlas Research Network, 2008) (Figure 65.2). Because of the level of detail of this study, accurate frequencies of alterations in each component could be ascertained. In addition to well-known alterations, the project has identified homozygous deletions of NF1 and PARK2, and amplification of AKT3. Sequence analysis of 601 candidate genes uncovered 452 validated, non-silent somatic mutations in 223 unique genes. Statistically,

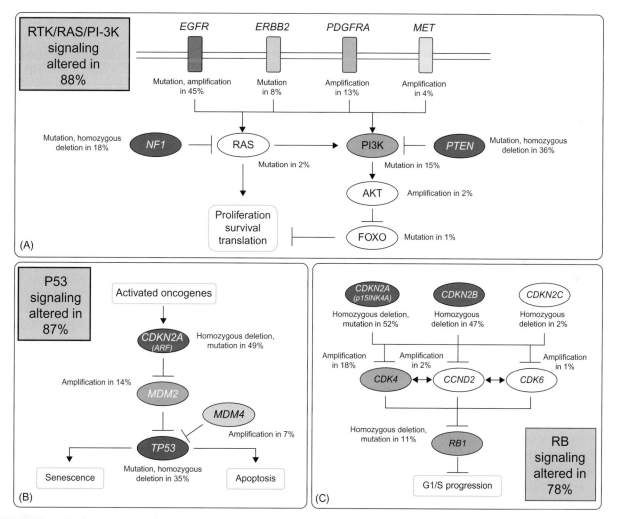

Frequent genetic alterations in three critical signaling pathways. Primary sequence alterations and significant copy number changes for components of the (A) RTK/RAS/PI-3K, (B) p53, and (C) RB signaling pathways are shown. Red indicates activating genetic alterations, with frequently altered genes showing deeper shades of red. Conversely, blue indicates inactivating alterations, with darker shades corresponding to a higher percentage of alteration. For each altered component of a particular pathway, the nature of the alteration and the percentage of tumors affected are indicated. Blue boxes contain the final percentages of glioblastomas with alterations in at least one known component gene of the designated pathway. *Reproduced with permission from the Cancer Genome Atlas Research Network, 2008. Copyright 2008 Nature Publishing.*

eight genes were identified as significantly mutated, including *p53*. Another important discovery was the identification of *NF1* as a glioblastoma suppressor gene, with 23% of patients carrying homozygous deletions or somatic mutations. Large numbers of mutations in EGFR were identified, as well as some in the related receptor *ERBB2*. Downstream mutations were detected in the PI3K enzyme complex, some of which were novel.

In contrast to TCGA, in which selected genes were sequenced, Parsons and colleagues (2008) sequenced 20,661 protein coding genes in 22 glioblastoma patients using next generation sequencing. This identified similar mutations to TCGA, and identified recurrent mutations not previously identified in glioblastoma in the *IDH1* gene. These mutations were seen in 12% of patients, who were largely young and had been diagnosed with secondary glioblastoma. This has been verified in independent studies including low-grade diffuse gliomas (Hartmann et al., 2009; Yan et al., 2009), and is associated with improved survival. The mutation changes Arg 172 to His in the active site of *IDH1*. A similar mutation was detected in the homologous *IDH2* enzyme in the absence of *IDH1* mutations. *IDH1* is involved in the metabolism of isocitrate to succinate, and the mutation prevents this reaction and causes the enzyme to essentially work in reverse and catalyze the production of hydroxyglutarate from succinate. The role of this altered biochemistry in tumor formation is not yet known, but is a fascinating, rapidly developing story (Dang et al., 2009). *IDH1* mutations also were proposed to lead to increased levels of HIF-1α, which increases tumor growth (Zhao et al., 2009). *IDH1* mutations therefore have important functional consequences, and are very common in gliomas, with the exception of primary glioblastomas, in which they are rarely detected.

The identification of *IDH1* or *IDH2* mutations in small biopsies could be useful to confirm diagnosis, and could find an application in distinguishing grade II astrocytoma from grade I pilocytic astrocytoma, which does not have this mutation. This could be done by mutation-specific PCR or with an antibody specific for the R172H mutation (Camelo-Piragua, 2010).

Methylation Alterations in Brain Tumors

Some promising prognostic information has emerged from analysis of TCGA data, which has given a picture of the first integrated view of the glioblastoma (GBM) methylome. Much evidence has accumulated that epigenetic alterations, including DNA methylation, play important roles in cancer by silencing tumor suppressor genes (Jones and Baylin, 2007). A comprehensive study of the glioblastoma methylome focused on data gathered as part of the TCGA project identified some potentially interesting prognostic markers (Noushmehr et al., 2010). The TCGA study used Illumina Golden Gate and Infinium platforms to measure global methylation patterns across 272 TCGA samples. In the same way as the transcriptome-based studies described above, it was possible to divide the glioblastoma patient cohort into three subgroups. One of these groups was reminiscent of the CpG island methylation promoter phenotype (CIMP) seen in colorectal cancer (Toyota et al., 1999), which was characterized

by hypermethylation of a subset of genes. A total of 90% of glioma-CIMP (G-CIMP) tumors appear to represent approximately one-third of the proneural subgroup identified by Phillips and colleagues (2006). G-CIMP-positive proneural tumors arose in younger patients and had a better prognosis than other proneural tumors. G-CIMP-positive tumors also had a significant association with *IDH1* mutations, as well as several genes known to be associated with a poor outcome, including *FABP5*, *PDPN*, *CHI3L1*, and *LGALS3*. The G-CIMP methylome signature was distilled into an eight-gene panel. It will be of great interest to see how this study and others like it may translate into improved clinical practice.

MicroRNA Alterations in Brain Tumors

MicroRNAs can have a profound impact on gene expression, either by causing degradation of specific transcripts or by preventing translation of the mRNA. MicroRNA expression is altered in many cancers (Volinia et al., 2006), including brain tumors, with miR-21 being particularly over-represented in glioblastoma (Chan et al., 2005; Ciafre et al., 2005). This also has functional importance, because knockdown of miR-21 leads to apoptosis in glioma cells (Chan et al., 2005).

An increasing number of microRNAs have been characterized in glioblastoma. MiR-21 antagonism and re-expression of weakly expressed microRNAs, such as miR-7, miR-124, and miR-128, all have effects on glioma cell proliferation and/or invasion, and target important genes, such as *EGFR* (miR-7) (Kefas et al., 2008), *CDK6* (miR-124) (Silber et al., 2008), and *Bmi1* (miR-128) (Godlewski et al., 2008). Also, miR-296 was identified in endothelial cells as a pro-angiogenic microRNA (Würdinger et al., 2008). These microRNAs have important functional effects (reviewed in Lawler and Chiocca, 2009) and may also be prognostic (Godlewski et al., 2010). A recent study suggested levels of miR-181b and miR-181c predict response to temozolomide in glioblastoma patients (Slaby et al., 2010). Undoubtedly, microRNAs will play an important part in improving our understanding of gliomas and in the development of new therapies and biomarkers, as microRNAs were detected in circulating exosomes of glioblastoma patients (Skog et al., 2008).

SNPs in Gliomas

Two studies previously mentioned identified SNPs involved in glioma predisposition. This approach has been extended (Liu et al., 2010) to the identification of SNPs that could predict survival. The top 100 glioma-susceptibility SNPs were examined in a panel of 590 glioblastoma patients with known survival information. Polymorphisms in *LIG4* and *BTBD2* were predictors of short-term survival (<12 months), and polymorphisms in *CCDC26*, *HMGA2*, and *RTEL1* predicted long-term survival (>36 months). Indeed, a combination of *RTEL1* and *HMGA2* polymorphisms gave 7.8-year median survival. These genes are all involved in double-strand-break DNA repair, and illustrate its importance in glioblastoma. Further studies in this area will likely identify further polymorphisms and investigate their functional consequences on cancer cell behavior.

Proteomic Studies in Brain Tumors

Proteomic studies on brain tumors have typically used the two-dimensional (2d) gel electrophoresis and mass spectrometry approach. Comparison of glioblastoma with non-tumor tissue identified 11 up-regulated and 4 down-regulated proteins, including FABP7 – also identified as a prognostic marker in microarrays (Hiratsuka et al., 2003, Liang et al., 2005). The study closely examined SIRT2, a tubulin deacetylase, down-regulated in 12 out of 17 gliomas, and suggested it may act as a tumor suppressor. Another study used 2d gel-based proteomics to molecularly classify and predict survival in 85 samples that included various tumors and normal brain. Expression patterns were identified that could distinguish tumor types and predict survival (Iwadate et al., 2004). Proteomics may find utility in the identification of serum biomarkers that could lead to much simpler diagnosis and monitoring. Few studies have yet been performed in this area, although cathepsin D was found to be elevated in the serum of glioblastoma patients and associated with poor prognosis (Fukuda et al., 2005). MMP-9 and YKL-40 have also been detected in the serum of glioblastoma patients (Hormigo et al., 2006). However, like other studies in this area, independent analysis on further patient groups is not yet available.

Direct mass spectrometry of tissue samples is emerging as a powerful tool, and may become increasingly influential in the diagnostic arena (Schwamborn and Caprioli, 2010). Analysis of 162 patient biopsies from glioma patients using this approach, known as direct tissue matrix-assisted laser desorption ionization mass spectroscopy (MALDI-MS), generated a novel classification scheme based on protein expression profiles that segregated long-term and short-term survivors (Schwartz et al., 2005). The authors were able to distinguish normal brain from glioma biopsies and distinguish various grades of tumor by protein expression in these samples. Because MALDI-MS technology can analyze numerous samples, and analysis time is only five minutes per sample, this technology is amenable to high-throughput tissue screening in a clinical setting. Significant developments can be expected in this area in the near future.

PHARMACOGENOMICS

At present, many brain tumors are treated in a similar way regardless of their classification, but this situation is changing as patient responsiveness to various therapies is correlated with genetic changes in the tumor. For example, patients with combined 1p and 19q loss may opt for chemotherapy but withhold radiation until recurrence, therefore avoiding the problematic effects of extensive radiation to the brain in long-surviving patients. On the other hand, a patient with intact chromosome 1p and EGFR amplification may opt to forego PCV chemotherapy and have radiation and a novel therapy, thereby avoiding the myelotoxic effects of PCV (Louis, 2006).

The benefit of chemotherapy in glioblastoma and anaplastic astrocytoma is small, with response plus stabilization in 20–50% of patients, because these tumors have intrinsic chemoresistance. Esteller and colleagues (2000) first described the mechanism by which some gliomas are resistant to nitrosourea alkylating agents such as carmustine (BCNU). These agents kill by alkylation of the O6 position of guanine, thereby crosslinking adjacent DNA strands. These crosslinks can be repaired by the DNA repair enzyme O6-methylguanine-DNA methyltransferase (MGMT), which rapidly reverses alkylation. Around half of glioblastomas have reduced levels of MGMT, and this correlates closely with chemosensitivity. Methylation of the promoter region of the gene is thought to account for this variation. Out of 19 high-grade glioma patients with methylated MGMT promoters, 12 had a partial or complete response to BCNU, whereas only 1 of 28 patients with an unmethylated promoter had a response (Esteller et al., 2000). MGMT promoter methylation also is linked to responsiveness to temozolomide in glioblastoma (Hegi et al., 2005), pediatric glioblastomas (Pollack et al., 2006), and low-grade gliomas (Everhard et al., 2006). In a large trial studying concomitant temozolomide and radiation therapy, Hegi and colleagues showed that 46% of patients with methylated MGMT were alive at two years versus only 23% with unmethylated MGMT. Because some patients with unmethylated MGMT showed benefit from treatment, it is unclear whether a test for MGMT would be used to impact treatment in any way. Accumulating recent evidence suggests that MGMT methylation leads to a better prognosis independent of alkylating therapy. Thus MGMT may predict a prognostically favorable molecular phenotype (reviewed in Jansen et al., 2010; Rivera et al., 2010; Weller et al., 2010). At present, MGMT testing is not thought to be necessary, but due to its high profile it is recommended that MGMT analysis be incorporated into clinical trials (Jansen et al., 2010).

Recently, several groups have reported that high levels of the stem cell marker CD133 are associated with poor prognosis in gliomas (Beier et al., 2008; Rebetz et al., 2008; Zeppernick et al., 2008). Resistance to temozolomide chemotherapy also appears to be linked to a stem cell signature (Murat et al., 2008).

A second example of pharmacogenomics is in the use of EGFR kinase inhibitors for glioma treatment. Molecularly targeted therapeutics such as EGFR inhibitors are an attractive option for glioma because of the frequently observed amplification of EGFR in these tumors. However, trials so far have shown only a 20% response rate (Mellinghoff et al., 2005). Molecular determinants of responsiveness were analyzed, and it was found that combined EGFRvIII mutation and the presence of PTEN sensitizes tumors, suggesting that secondarily targeting PI3K signaling may improve this kind of therapy, and that patients with the appropriate profile should be enlisted in further trials of these drugs. As glioma treatment becomes increasingly sophisticated, further details and examples of chemoresistance mechanisms, which can affect therapeutic decision-making, are sure to emerge. It was also shown that EGFR inhibitors may be ineffective due to the activation of multiple tyrosine kinase receptors (Stommel et al., 2007). Ultimately, the use of targeted inhibitors will be dependent on and guided by the molecular make-up of the specific tumor.

The increasingly detailed genetic information on glioblastoma is allowing the development of preclinical therapeutics based on alterations seen in subsets of tumors. These provide support for the development of future patient-tailored therapies. For example, targeting of CDK4 and CDK6 with a small molecule inhibitor PD0332991 was strongly dependent on the co-deletion of CDKN2A and CDKN2C, in glioma cell lines (Wiedemeyer et al., 2010). Thus, at least in cell lines, genome-informed drug sensitivity studies identified a subset of glioblastomas that would be likely to respond to CDK4/6 inhibition. Indeed, CDK4/6 inhibitors are relatively well tolerated by patients.

MONITORING

Currently, the determination of treatment response is monitored using imaging techniques such as MRI or computed tomography. Patients' cognitive abilities and other neurological symptoms are also examined. For low-grade tumors, monitoring is carried out once or twice yearly by MRI scanning. Often extensive tumor progression occurs in the interval between different imaging studies and/or goes undetected due to lack of contrast enhancement. Alternative markers of tumor burden could potentially permit early detection of treatment failure and allow for more rapid changes in therapeutic strategy. Effective screening, monitoring and diagnosis may be enhanced by the identification of serum markers, which can be readily detected in the presence of a brain tumor. For example, YKL-40 is over-expressed in glioblastoma and can be detected in the serum of patients, and has been proposed as a potentially useful surrogate marker of tumor burden, response to treatment, or relapse (Hormigo et al., 2006). Other markers have been proposed, such as low-molecular weight caldesmon (Zheng et al., 2005), and cathepsin D (Fukuda et al., 2005). The utility of these markers for monitoring and prognostic purposes needs to be established in prospective studies with large patient numbers.

NOVEL AND EMERGING THERAPEUTICS

The fact that brain tumors can be clearly separated into molecular subgroups with different prognosis or drug response suggests that, ultimately, specific therapies may be necessary, tailored according to the molecular alterations seen in an individual tumor. An effective therapeutic strategy must eliminate tumor cells with minimal neurological damage, cross the blood–brain barrier (unless directly delivered to the tumor), and overcome elevated interstitial tumor pressure and active resistance mechanisms. The existence of various cell types within a tumor suggests that a cocktail of drugs

may be required. A wide range of novel approaches are being examined for brain tumor treatment, in both clinical trials and in experimental animal models, generally using glioma cells intracranially implanted into the brains of mice or rats. The important current strategies are described below and can be summarized as follows:

- Optimizing and enhancing existing strategies
- Improved drug delivery
- Targeted therapies in which molecules and pathways and key processes involved in gliomagenesis are targeted with either specific small molecule inhibitors, immunotoxins, or gene therapy agents
- Vaccines and immunostimulatory methods to enhance immune recognition and destruction of the tumor
- Oncolytic viruses engineered to specifically replicate in and destroy tumor cells
- Combinations of therapeutic methods designed to target multiple processes.

The current standard of care for glioblastoma employs aggressive resection followed by radiotherapy and chemotherapy with temozolomide. This treatment doubles the two-year survival rate compared with radiotherapy alone (Stupp et al., 2005). The effectiveness of temozolomide and other chemotherapeutic agents is limited by chemoresistance, due at least partly to MGMT expression. Therefore, inhibition of DNA repair pathways may be effective in conjunction with alkylator chemotherapy in patients expressing MGMT. For example, the MGMT inhibitor O6-benzyl guanine enhances the effectiveness of alkylator therapy in glioma cells (Liu and Gerson, 2006). In addition, other approaches that can build on the conventional approaches or combine effectively would seem sensible. For example, enhancement of temozolomide with perifosine (an Akt inhibitor) showed additive effects in a mouse glioma model (Momota et al., 2005).

Some problems, such as systemic toxicity and the presence of the blood–brain barrier, may be overcome by local drug delivery directly to the resected tumor, thereby achieving high local concentration of the therapeutic agent. This can be done by implanting chemotherapeutic-infused wafers (e.g., gliadel) into the resected tumor cavity. Another approach being tested in patients is to deliver the therapeutic agent from an external reservoir by a method known as convection-enhanced delivery (CED), in which a drug is slowly infused into the tumor area over a period of days. Both conventional and novel therapeutics such as gene therapy agents and immunotoxins may be suited to this method (reviewed by Chiocca et al., 2004).

Due to the limited progress made with chemotherapy and radiotherapy over the years, researchers have begun to investigate the development of more specific, targeted treatment modalities that exploit the molecular pathogenesis of cancers in the brain. The identification of molecular alterations by traditional molecular techniques, and now by high-throughput methods, is leading to the identification of many potential targets. The development of small molecule inhibitors and monoclonal antibodies designed to target key enzymes on which cancer cells depend raises the possibility that rational approaches can be used. Targets such as EGFR, PDGFR and VEGFR tyrosine kinase domains have been considered promising for the development of small molecule inhibitors, such as erlotinib and gefitinib (Mellinghoff et al., 2005). Monoclonal antibodies such as bevacizumab (humanized VEGF monoclonal antibody) and trastazumab (Her-2/neu monoclonal antibody) have been examined. These inhibitors are being used in a wide range of combinations (for example, with traditional therapies or other pathway inhibitors such as rapamycin) to improve their efficacy. Recent reports indicate that bevacizumab in combination with chemotherapy leads to a very high response rate in glioblastoma (Pope et al., 2006). Anti-angiogenic approaches show promise in early trials. These include the VEGFR inhibitor cediranib (AZD2171) (Batchelor et al., 2007) and the integrin inhibitor cilengitide (Reardon et al., 2008). The anti-VEGF monoclonal antibody avastin showed excellent radiographic responses in trials for recurrent glioblastoma in combination with CPT11 (Vredenburgh et al., 2007). It is not yet clear whether these impressive responses will translate into improved survival (Norden et al., 2009). Avastin will now be investigated in newly diagnosed glioblastoma in combination with temozolomide and radiation.

To improve the prospects for these therapies, it will be important to profile genetic and transcriptional alterations in patients involved in trials. It is becoming clear that these agents are highly dependent upon the profile of genetic alterations in a given tumor. For example, inhibitors of Notch or Akt could be improved significantly by enrolling patients with a *PN* or *Prolif/ Mes* signature respectively (Phillips et al., 2006).

Tumor-targeted immunotoxins consist of a tumor cell receptor ligand conjugated to a highly potent toxin, and have been associated with encouraging long-term survivals (Rainov and Heidecke, 2004; Sampson et al., 2005). Antibodies to cell surface markers highly expressed on gliomas, such as IL-13, EGFR, and IL-4, have been examined, with toxin-conjugated IL-13 involved in a large-scale multi-institutional clinical trial for recurrent glioblastoma at the time of writing.

Strategies that harness the power of the immune system hold some promise in glioma treatment (reviewed by Hussain and Heimberger, 2005). Immunosuppression is a major feature of gliomas; it prevents normal immune function and may facilitate tumor growth. In a particularly interesting study, Steiner and colleagues (2004) explored the use of immunotherapy as an adjuvant to standard radiotherapy in patients with glioblastoma. A vaccine prepared from patient tumor cells, infected with Newcastle disease virus (an avian paramyxovirus) as a non-specific immunostimulant, was administered to patients every 3–4 weeks, and doubled median overall survival, although this needs verification in randomized trial. Recently published data indicates potential efficacy of an immunotherapy approach based on the use of EGFRvIII-based vaccine, CDX-110 (Heimberger and Sampson, 2009).

Viral and gene-based therapies have been explored for many brain tumors, and have been in clinical trials for approximately 15 years – they have shown safety but a lack of efficacy. Research in this area is ongoing, and improved approaches are slowly coming to the clinic. A recent trial using non-replicating adenovirus to deliver the thymidine kinase gene showed improved patient survival, indicating that progress is still being made in this area (Immonen et al., 2004). Viruses with oncolytic properties have also been exploited (Chiocca, 2002). These agents either naturally target tumor cells or can be engineered to do so. They can also be armed with therapeutic transgenes or used in combination with other therapeutic strategies, leading to increased anti-tumor efficacy. Recently it was reported that temozolomide can synergize with oncolytic herpes virus therapy in a mouse glioma model, indicating that these two separate approaches may be compatible (Aghi et al., 2006).

A final area of great interest is in the targeting of so-called brain tumor stem cells, which may constitute the root of the tumor, and whose elimination could be required for successful therapy. Until recently, brain tumors were thought to arise from glial cells residing in the brain parenchyma. But recent evidence in human and animal studies suggests that neural stem cells are an alternative cellular origin. In experimental animal models, both astrocytes and neural progenitor cells can give rise to neoplasms that recapitulate the histopathological hallmarks of human gliomas (discussed in Louis, 2006). The adult forebrain contains abundant neural stem cells, and human glioblastomas contain tumorigenic neural stem-like cells (Singh et al., 2004). Moreover, stem-like cells are much more tumorigenic than non-stem-like cells from the same tumor. It would seem that unless the stem-like component of gliomas can be effectively targeted, treatment failure is inevitable. One of the reasons for the failure of traditional therapies may be due to the fact that they are designed to target dividing cells, which only represent a small proportion of the actual tumor cells, and stem cells are thought to divide slowly. It has been reported that stem-like cells in brain tumors are radioresistant, indicating that special measures may be needed to take care of this population therapeutically (Bao et al., 2006). CD133-positive cells also express higher MGMT and are more resistant to chemotherapy (Liu et al., 2006). These data suggest that the stem-like component of gliomas is a major reason for the difficulty in treatment, and that novel approaches may be required for their elimination. There has been a large amount of work in the glioma stem cell area recently. Importantly, some of the genetic changes in glioblastoma can be correlated with increased stem cell-like behavior. Zheng and colleagues (2008) identified higher numbers of p53 mutations in primary glioblastoma than expected. They then found that inactivation of PTEN and p53 in murine neural stem cells promoted an undifferentiated state with high self-renewal potential and an increase in myc levels, and an associated gene-expression signature. The importance of myc was demonstrated in further functional studies.

CONCLUSIONS

The prognosis for many brain tumors remains dismal. However, the wealth of data currently being generated by microarray-based studies in particular should have a profound effect on brain tumor management. Headway is currently being made in four main areas:

1. *Identification of robust molecular markers and signatures that define tumor types.* Recent studies have revealed that molecular signatures readily distinguish tumor types, and can also further sub-divide histologically similar tumors into previously unknown groups. This opens up the possibility that molecular information will be increasingly useful in brain tumor diagnosis.
2. *Correlation of molecular signatures with patient prognosis.* Molecular information may provide the ability to predict patient survival more reliably than at present. Recent studies show that molecular classification can improve prediction of patient outcome compared with histology in some cases.
3. *Correlation of molecular data with therapeutic response.* Molecular alterations play a role in drug responsiveness. The identification of these alterations is beginning to impact clinical decision-making; 1p/19q is now widely used, and MGMT and EGFR status may be used in the near future. These represent the first steps toward the development of patient-tailored therapies.
4. *Identification of novel therapeutic targets.* Improved understanding of brain tumor biology is allowing the development of new therapeutic strategies based on novel molecular targets revealed by global genetic analyses.

The most promising approaches so far may involve tumor classification by gene expression profiling using a panel of relevant genes. The identification of the *PN*, *Mes*, and *Prolif* tumor subgroups (Phillips et al., 2006) provides a strong foundation for such an approach. Microarray studies have identified many candidates that could be of use in prognosis and monitoring. The most promising of these so far may be YKL-40, which can differentiate astrocytic from oligodendroglial tumors, and is associated with poor prognosis. The association between MGMT promoter methylation and improved treatment response in glioblastoma patients has provoked much interest in this area. The incorporation of MGMT expression analysis may therefore soon emerge as a useful predictive clinical tool.

The list of molecular alterations is growing, but is still far from complete. In addition, the role of epigenetic changes has barely been studied as yet in gliomas. The major challenge at this point is to translate this increasingly large amount of data into clinically useful tools. New multi-dimensional datasets from efforts such as TCGA provide a valuable resource for future studies. Recent studies also show that the integration of genomic, functional, and pharmacologic data can be exploited to inform the development of targeted therapies directed against specific cancer pathways.

Next-generation whole genome sequencing is likely to be the next major technical breakthrough in cancer. Indeed, the whole genome of the U87 glioma cell line was recently sequenced, at a cost of $35,000 (Clark et al., 2010). This should lead to the development of a new genomic approach to glioma treatment, using molecular signatures to provide accurate diagnosis and to stratify patients for the most effective therapy, or for targeted therapies. These advances should finally lead to a better outlook for these cancers.

REFERENCES

Aghi, M., Rabkin, S., Martuza, R.L., 2006. Effect of chemotherapy-induced DNA repair on oncolytic herpes simplex viral replication. J Natl Cancer Inst 98, 38–50.

Bao, S., Wu, Q., McLendon, R.E., et al., 2006. Glioma stem cells promote radioresistance by preferential activation of the DNA damage response. Nature 444, 756–760.

Batchelor, T.T., Sorensen, A.G., di Tomaso, E., et al., 2007. AZD2171, a pan-VEGF receptor tyrosine kinase inhibitor, normalizes tumor vasculature and alleviates edema in glioblastoma patients. Cancer Cell 11, 83–95.

Beier, D., Wischhusen, J., Dietmaier, W., et al., 2008. CD133 expression and cancer stem cells predict prognosis in high-grade oligodendroglial tumors. Brain Pathol 18, 370–377.

Bondy, M.L., Wang, L.E., El-Zein, R., et al., 2001. Gamma-radiation sensitivity and risk of glioma. J Natl Cancer Inst 93, 1553–1557.

Bredel, M., Bredel, C., Juric, D., et al., 2005. Functional network analysis reveals extended gliomagenesis pathway maps and three novel MYC-interacting genes in human gliomas. Cancer Res 65, 8679–8689.

Bredel, M., Scholtens, D.M., Harsh, G.R., et al., 2009. A network model of a cooperative genetic landscape in brain tumors. JAMA 15, 261–275.

Burger, P.C., Pearl, D.K., Aldape, K., et al., 2001. Small cell architecture: A histological equivalent of EGFR amplification in glioblastoma multiforme? J Neuropathol Exp Neurol 60, 1099–1104.

Cairncross, J.G., Ueki, K., Zlatescu, M.C., et al., 1998. Specific genetic predictors of chemotherapeutic response and survival in patients with anaplastic oligodendrogliomas. J Natl Cancer Inst 90, 1473–1479.

Camelo-Piragua, S., Jansen, M., Ganguly, A., Kim, J.C., Louis., D.N., Nutt, C.L., 2010. Mutant IDH1-specific immunohistochemistry distinguishes diffuse astrocytoma from astrocytosis. Acta Neuropathol 119, 509–511.

Cancer Genome Atlas Research Network, 2008. Comprehensive genomic characterization defines human glioblastoma genes and core pathways. Nature 455, 1061–1068.

Carro, M.S., Lim, W.K., Alvarez, M.J., et al., 2010. The transcriptional network for mesenchymal transformation of brain tumours. Nature 463, 318–325.

CBTRUS, 2012. http://www.cbtrus.org/2012-NPCR-SEER/CBTRUS_Report_2004-2008_3-23-2012.pdf

Chan, J.A., Krichevsky, A.M., Kosik, K.S., 2005. MicroRNA-21 is an antiapoptotic factor in human glioblastoma cells. Cancer Res 65, 6029–6033.

Chiocca, E.A., 2002. Oncolytic viruses. Nat Rev Cancer 2, 938–950.

Chiocca, E.A., Broaddus, W.C., Gillies, G.T., Visted, T., Lamfers, M.L., 2004. Neurosurgical delivery of chemotherapeutics, targeted toxins, genetic and viral therapies in neuro-oncology. J Neurooncol 69, 101–117.

Chung, R., Whaley, J., Kley, N., et al., 1991. TP53 gene mutations and 17p deletions in human astrocytomas. Genes Chromosomes Cancer 3, 323–331.

Ciafre, S.A., Galardi, S., Mangiola, A., et al., 2005. Extensive modulation of a set of microRNAs in primary glioblastoma. Biochem Biophys Res Commun 334, 1351–1358.

Clark, M.J., Homer, N., O'Connor, B.D., et al., 2010. U87MG decoded: The genomic sequence of a cytogenetically aberrant human cancer cell line. PLoS Genet 6, e1000832.

Cobbs, C.S., Soroceanu, L., Denham, S., Zhang, W., Kraus, M.H., 2008. Modulation of oncogenic phenotype in human glioma cells by cytomegalovirus IE1-mediated mitogenicity. Cancer Res 68, 724–730.

Colman, H., Zhang, L., Sulman, E.P., et al., 2010. A multigene predictor of outcome in glioblastoma. Neuro Oncol 12, 49–57.

Dang, L., White, D.W., Gross, S., et al., 2009. Cancer-associated IDH1 mutations produce 2-hydroxyglutarate. Nature 465, 966–972.

De Angelis, L.M., 2001. Brain tumors. N Engl J Med 344, 114–123.

Esteller, M., Garcia-Foncillas, J., Andion, E., et al., 2000. Inactivation of the DNA-repair gene MGMT and the clinical response of gliomas to alkylating agents. N Engl J Med 343, 1350–1354.

Everhard, S., Kaloshi, G., Crinière, E., et al., 2006. MGMT methylation: A marker of response to temozolomide in low-grade gliomas. Ann Neurol 60, 740–743.

Freije, W.A., Castro-Vargas, F.E., Fang, Z., et al., 2004. Gene expression profiling of gliomas strongly predicts survival. Cancer Res 64, 6503–6510.

Fukuda, M.E., Iwadate, Y., Machida, T., et al., 2005. Cathepsin D is a potential serum marker for poor prognosis in glioma patients. Cancer Res 65, 5190–5194.

Furnari, F.B., Fenton, T., Bachoo, R.M., et al., 2007. Malignant astrocytic glioma: Genetics, biology, and paths to treatment. Genes Dev 21, 2683–2710.

Godlewski, J., Nowicki, M.O., Bronisz, A., et al., 2008. Targeting of the Bmi-1 oncogene/stem cell renewal factor by microRNA-128 inhibits glioma proliferation and self-renewal. Cancer Res 68, 9125–9130.

Godlewski, J., Nowicki, M.O., Bronisz, A., et al., 2010. MicroRNA-451 regulates LKB1/AMPK signaling and allows adaptation to metabolic stress in glioma cells. Mol Cell 37, 620–632.

Hartmann, C., Meyer, J., Balss, J., et al., 2009. Type and frequency of IDH1 and IDH2 mutations are related to astrocytic and oligodendroglial differentiation and age: A study of 1010 diffuse gliomas. Acta Neuropathol 118, 469–474.

Hegi, M.E., Diserens, A.C., Gorlia, T., et al., 2005. MGMT gene silencing and benefit from temozolomide in glioblastoma. N Engl J Med 352, 997–1003.

Heimberger, A.B., Sampson, J.H., 2009. The PEPvIII-KLH (CDX-110) vaccine in glioblastoma multiforme patients. Expert Opin Biol Ther 9, 1087–1098.

Hemminki, K., Li, X., 2003. Familial risk in nervous system tumors. Cancer Epidemiol Biomarkers Prev 12, 1137–1142.

Hiratsuka, M., Inoue, T., Toda, T., et al., 2003. Proteomics-based identification of differentially expressed genes in human gliomas: Down-regulation of SIRT2 gene. Biochem Biophys Res Commun 309, 558–566.

Hoang-Xuan, K., Capelle, L., Kujas, M., et al., 2004. Temozolomide as initial treatment for adults with low-grade oligodendrogliomas or oligoastrocytomas and correlation with chromosome 1p deletions. J Clin Oncol 22, 133–138.

Hormigo, A., Gu, B., Karimi, S., et al., 2006. YKL-40 and matrix metalloprotease-9 as potential serum biomarkers for patients with high-grade gliomas. Clin Cancer Res 12, 5698–5704.

Huang, H., Okamoto, Y., Yokoo, H., et al., 2004. Gene expression profiling and subgroup identification of oligodendrogliomas. Oncogene 23, 6012–6022.

Hussain, S.F., Heimberger, A.B., 2005. Immunotherapy for human glioma: Innovative approaches and recent results. Expert Rev Anticancer Ther 5, 777–790.

Ichimura, K., Mungall, A.J., Fiegler, H., et al., 2006. Small regions of overlapping deletions on 6q26 in human astrocytic tumours identified using chromosome 6 tile path array-CGH. Oncogene 25, 1261–1271.

Immonen, A., Vapalahti, M., Tyynela, K., et al., 2004. AdvHSV-tk gene therapy with intravenous ganciclovir improves survival in human malignant glioma: A randomised, controlled study. Mol Ther 10, 967–972.

Ino, Y., Zlatescu, M.C., Sasaki, H., et al., 2000. Long survival and therapeutic responses in patients with histologically disparate high-grade gliomas demonstrating chromosome 1p loss. J Neurosurg 92, 983–990.

Iwadate, Y., Sakaida, T., Hiwasa, T., et al., 2004. Molecular classification and survival prediction in human gliomas based on proteome analysis. Cancer Res 64, 2496–2501.

Jenkins, R.B., Blair, H., Ballman, K.V., et al., 2006. A t(1;19)(q10;p10) mediates the combined deletions of 1p and 19q and predicts a better prognosis of patients with oligodendroglioma. Cancer Res 66, 9852–9861.

Jones, P.A., Baylin, S.B., 2007. The epigenomics of cancer. Cell 128, 683–692.

Kefas, B., Godlewski, J., Comeau, L., et al., 2008. MicroRNA-7 inhibits the epidermal growth factor receptor and the Akt pathway and is down-regulated in glioblastoma. Cancer Res 68, 3566–3572.

Kleihues, P., Cavenee, W.K., 2000. World Health Organization Classification of Tumours: Pathology and Genetics: Tumours of the Nervous System. IARC Press.

Kotliarov, Y., Steed, M.E., Christopher, N., et al., 2006. High-resolution global genomic survey of 178 gliomas reveals novel regions of copy number alterations and allelic imbalances. Cancer Res. 66, 9428–9436.

Kuan, C.T., Wakiya, K., Dowell, J.M., et al., 2006. Glycoprotein non-metastatic melanoma protein B, a potential molecular therapeutic target in patients with glioblastoma multiforme. Clin Cancer Res 12, 1970–1982.

Lassman, A.B., Rossi, M.R., Raizer, J.J., et al., 2005. Molecular study of malignant gliomas treated with epidermal growth factor receptor inhibitors: Tissue analysis from North American Brain Tumor Consortium trials 01–03 and 00–01. Clin Cancer Res 11, 7841–7850.

Lawler, S., Chiocca, E.A., 2009. Emerging functions of microRNAs in glioblastoma. J Neurooncol 92, 297–306.

Li, J., Yen, C., Liaw, D., et al., 1997. PTEN, a putative protein tyrosine phosphatase gene mutated in human brain, breast, and prostate cancer. Science 275, 1943–1947.

Liang, Y., Diehn, M., Watson, N., et al., 2005. Gene expression profiling reveals molecularly and clinically distinct subtypes of glioblastoma multiforme. Proc Natl Acad Sci USA 102, 5814–5819.

Libermann, T.A., Nusbaum, H.R., Razon, N., et al., 1985. Amplification, enhanced expression and possible rearrangement of EGF receptor gene in primary human brain tumours of glial origin. Nature 313, 144–147.

Liu, G., Yuan, X., Zeng, Z., et al., 2006. Analysis of gene expression and chemoresistance of CD133+ cancer stem cells in glioblastoma. Mol Cancer 5, 67.

Liu, L., Gerson, S.L., 2006. Targeted modulation of MGMT: Clinical implications. Clin Cancer Res 12, 328–331.

Liu, Y., Zhou, K., Zhang, H., et al., 2008. Polymorphisms of LIG4 and XRCC4 involved in the NHEJ pathway interact to modify risk of glioma. Hum Mutat 29, 381–389.

Liu, Y., Shete, S., Etzel, C.J., et al., 2010. Polymorphisms of LIG4,BTBD2, HMGA2, and RTEL1 genes involved in the double-strand break repair pathway predict glioblastoma survival. J Clin Oncol 28, 2467–2474.

Louis, D.N., 2006. Molecular pathology of malignant gliomas. Annu Rev Pathol Mech Dis 1, 97–117.

Louis, D.N., Holland, E.C., Cairncross, J.G., 2001. Glioma classification: A molecular reappraisal. Am J Pathol 159, 779–786.

Louis, D.N., Ohgaki, H., Wiestler, O.D., et al., 2007. The 2007 WHO classification of tumors of the central nervous system. Acta Neuropathol (Berl) 114, 97–109.

Maher, E.A., Brennan, C., Wen, P.Y., et al., 2006. Marked genomic differences characterize primary and secondary glioblastoma subtypes and identify two distinct molecular and clinical secondary glioblastoma entities. Cancer Res 66, 11,502–11,513.

Mellinghoff, I.K., Wang, M.Y., Vivanco, I., et al., 2005. Molecular determinants of the response of glioblastomas to EGFR kinase inhibitors. N Engl J Med 353, 2012–2024.

Mischel, P.S., Shai, R., Shi, T., et al., 2003. Identification of molecular subtypes of glioblastoma by gene expression profiling. Oncogene 22, 2361–2373.

Momota, H., Nerio, E., Holland, E.C., 2005. Perifosine inhibits multiple signaling pathways in glial progenitors and cooperates with temozolomide to arrest cell proliferation in gliomas in vivo. Cancer Res 65, 7429–7435.

Mukasa, A., Ueki, K., Matsumoto, S., et al., 2002. Distinction in gene expression profiles of oligodendrogliomas with and without allelic loss of 1p. Oncogene 21, 3961–3968.

Mulholland, P.J., Fiegler, H., Mazzanti, C., et al., 2006. Genomic profiling identifies discrete deletions associated with translocations in glioblastoma multiforme. Cell Cycle 5, 783–791.

Murat, A., Migliavacca, E., Gorlia, T., et al., 2008. Stem cell-related "self-renewal" signature and high epidermal growth factor receptor expression associated with resistance to concomitant chemoradiotherapy in glioblastoma. J Clin Oncol 26, 3015–3024.

Nigro, J.M., Misra, A., Zhang, L., et al., 2005. Integrated array-comparative genomic hybridization and expression array profiles identify clinically relevant molecular subtypes of glioblastoma. Cancer Res 65, 1678–1686.

Norden, A.D., Drappatz, J., Muzikansky, A., et al., 2009. An exploratory survival analysis of anti-angiogenic therapy for recurrent malignant glioma. J Neurooncol 92, 149–155.

Noushmehr, H., Weisenberger, D.J., Diefes, K., et al., 2010. Identification of a CpG island methylator phenotype that defines a distinct subgroup of glioma. Cancer Cell 17, 510–522.

Nutt, C.L., Betensky, R.A., Brower, M.A., Batchelor, T.T., Louis, D.N., Stemmer-Rachamimov, A.O., 2005. YKL-40 is a differential diagnostic marker for histologic subtypes of high-grade gliomas. Clin Cancer Res 11, 2258–2264.

Nutt, C.L., Mani, D.R., Betensky, R.A., et al., 2003. Gene expression-based classification of malignant gliomas correlates better with survival than histological classification. Cancer Res 63, 1602–1607.

Offit, K., Levran, O., Mullaney, B., et al., 2003. Shared genetic susceptibility to breast cancer, brain tumors, and Fanconi anemia. J Natl Cancer Inst 95, 1548–1551.

Parsons, D.W., Jones, S., Zhang, X., et al., 2008. An integrated genomic analysis of human glioblastoma multiforme. Science 321, 1807–1812.

Pelloski, C.E., Mahajan, A., Maor, M., et al., 2005. YKL-40 expression is associated with poorer response to radiation and shorter overall survival in glioblastoma. Clin Cancer Res 11, 3326–3334.

Petalidis, L.P., Oulas, A., Backlund, M., et al., 2008. Improved grading and survival prediction of human astrocytic brain tumors by artificial neural network analysis of gene expression microarray data. Mol Cancer Ther 7, 1013–1024.

Phillips, H.S., Kharbanda, S., Chen, R., et al., 2006. Molecular subclasses of high-grade glioma predict prognosis, delineate a pattern of disease progression, and resemble stages in neurogenesis. Cancer Cell 9, 157–173.

Pollack, I.F., Hamilton, R.L., Sobol, R.W., et al., 2006. O6-methylguanine-DNA methyltransferase expression strongly correlates with outcome in childhood malignant gliomas: Results from the CCG-945 cohort. J Clin Oncol 24 (21), 3431–3437.

Pope, W.B., Lai, A., Nghiemphu, P., Mischel, P., Cloughesy, T.F., 2006. MRI in patients with high-grade gliomas treated with bevacizumab and chemotherapy. Neurology 66, 1258–1260.

Prins, R.M., Cloughesy, T.F., Liau, L.M., 2008. Cytomegalovirus immunity after vaccination with autologous glioblastoma lysate. N Engl J Med 359, 539–541.

Rainov, N.G., Heidecke, V., 2004. Long term survival in a patient with recurrent malignant glioma treated with intratumoral infusion of an IL4-targeted toxin (NBI-3001). J Neurooncol 66, 197–201.

Rajasekhar, V.K., Viale, A., Socci, N.D., Wiedmann, M., Hu, X., Holland, E.C., 2003. Oncogenic Ras and Akt signaling contribute to glioblastoma formation by differential recruitment of existing mRNAs to polysomes. Mol Cell 12, 889–901.

Reardon, D.A., Nabors, L.B., Stupp, R., Mikkelsen, T., 2008. Cilengitide: An integrin-targeting arginine-glycine-aspartic acid peptide with promising activity for glioblastoma multiforme. Expert Opin Investig Drugs 17, 1225–1235.

Rebetz, J., Tian, D., Persson, A., et al., 2008. Glial progenitor-like phenotype in low-grade glioma and enhanced CD133-expression and neuronal lineage differentiation potential in high-grade glioma. PLoS ONE 3, e1936.

Rich, J.N., Hans, C., Jones, B., et al., 2005. Gene expression profiling and genetic markers in glioblastoma survival. Cancer Res 65, 4051–4058.

Rossi, M.R., Gaile, D., Laduca, J., et al., 2005. Identification of consistent novel submegabase deletions in low-grade oligodendrogliomas using array-based comparative genomic hybridization. Genes Chromosomes Cancer 44, 85–96.

Sampson, J.H., Reardon, D.A., Friedman, A.H., et al., 2005. Sustained radiographic and clinical response in patient with bifrontal recurrent glioblastoma multiforme with intracerebral infusion of the recombinant targeted toxin TP-38: Case study. Neuro-oncol 7, 90–96.

Scheurer, M.E., Amirian, E., Cao, Y., et al., 2008a. Polymorphisms in the interleukin-4 receptor gene are associated with better survival in patients with glioblastoma. Clin Cancer Res 14, 6640–6646.

Scheurer, M.E., Bondy, M.L., Aldape, K.D., Albrecht, T., El-Zein, R., 2008b. Detection of human cytomegalovirus in different histological types of gliomas. Acta Neuropathol 116, 79–86.

Schwamborn, K., Caprioli, R.M., 2010. Molecular imaging by mass spectrometry – looking beyond classical histology. Nat Rev Cancer 10, 639–646.

Schwartz, S.A., Weil, R.J., Thompson, R.C., et al., 2005. Proteomic-based prognosis of brain tumor patients using direct-tissue matrix-assisted laser desorption ionization mass spectrometry. Cancer Res 65, 7674–7681.

Schwartzbaum, J., Ahlbom, A., Malmer, B., et al., 2005. Polymorphisms associated with asthma are inversely related to glioblastoma multiforme. Cancer Res 65, 6459–6465.

Schwartzbaum, J.A., Fisher, J.L., Aldape, K.D., Wrensch, M., 2006. Epidemiology and molecular pathology of glioma. Nat Clin Pract Neurol 2, 494–503.

Shai, R., Shi, T., Kremen, T.J., et al., 2003. Gene expression profiling identifies molecular subtypes of gliomas. Oncogene 22, 4918–4923.

Shete, S., Hosking, F.J., Robertson, L.B., et al., 2009. Genome-wide association study identifies five susceptibility loci for glioma. Nat Genet 41, 899–904.

Silber, J., Lim, D.A., Petritsch, C., et al., 2008. miR-124 and miR-137 inhibit proliferation of glioblastoma multiforme cells and induce differentiation of brain tumor stem cells. BMC Med 6, 14.

Simon, M., Köster, G., Menon, A.G., Schramm, J., 1999. Functional evidence for a role of combined CDKN2A (p16-p14(ARF))/CDKN2B (p15) gene inactivation in malignant gliomas. Acta Neuropathol (Berl) 98, 444–452.

Singh, S.K., Hawkins, C., Clarke, I.D., et al., 2004. Identification of human brain tumour initiating cells. Nature 432, 396–401.

Skog, J., Würdinger, T., van Rijn, S., et al., 2008. Glioblastoma microvesicles transport RNA and proteins that promote tumour growth and provide diagnostic biomarkers. Nat Cell Biol 10, 1470–1476.

Slaby, O., Lakomy, R., Fadrus, P., et al., 2010. MicroRNA-181 family predicts response to concomitant chemoradiotherapy with temozolomide in glioblastoma patients. Neoplasma 57, 264–269.

Steiner, H.H., Bonsanto, M.M., Beckhove, P., et al., 2004. Antitumor vaccination of patients with glioblastoma multiforme: A pilot study to assess feasibility, safety, and clinical benefit. J Clin Oncol 22, 4272–4281.

Stommel, J.M., Kimmelman, A.C., Ying, H., et al., 2007. Coactivation of receptor tyrosine kinases affects the response of tumor cells to targeted therapies. Science 318, 287–290.

Stupp, R., Mason, W.P., Van den Bent, M.J., et al., 2005. Radiotherapy plus concomitant and adjuvant temozolomide for glioblastoma. N Engl J Med 352, 987–996.

Sugawa, N., Ekstrand, A.J., James, C.D., Collins, V.P., 1990. Amplified and rearranged epidermal growth factor receptor genes in human glioblastomas reveal deletions of sequences encoding portions of the N- and/or C-terminal tails. Proc Natl Acad Sci USA 87, 8602–8606.

Toyota, M., Ahuja, N., Suzuki, H., et al., 1999. Aberrant methylation in gastric cancer associated with the CpG island methylator phenotype. Cancer Res 59, 5438–5442.

Van den Bent, M.J., Carpentier, A.F., Brandes, A.A., et al., 2006. Adjuvant procarbazine, lomustine, and vincristine improves progression-free survival but not overall survival in newly diagnosed anaplastic oligodendrogliomas and oligoastrocytomas: A randomized European organisation for research and treatment of cancer phase III trial. J Clin Oncol 24, 2715–2722.

Verhaak, R.G., Hoadley, K.A., Purdom, E., et al., 2010. Integrated genomic analysis identifies clinically relevant subtypes of glioblastoma characterized by abnormalities in PDGFRA, IDH1, EGFR, and NF1. Cancer Cell 17, 98–110.

Volinia, S., Calin, G.A., Liu, C.G., et al., 2006. A microRNA expression signature of human solid tumors defines cancer gene targets. Proc Natl Acad Sci USA 103, 2257–2261.

Vredenburgh, J.J., Desjardins, A., Herndon II, J.E., et al., 2007. Bevacizumab plus irinotecan in recurrent glioblastoma multiforme. J Clin Oncol 25, 4722–4729.

Wang, L., Wei, Q., Wang, L.E., et al., 2006. Survival prediction in patients with glioblastoma multiforme by human telomerase genetic variation. J Clin Oncol 24, 1627–1632.

Wang, L.E., Bondy, M.L., Shen, H., et al., 2004. Polymorphisms of DNA repair genes and risk of glioma. Cancer Res 64, 5560–5563.

Watson, M.A., Perry, A., Budhraja, V., Hicks, C., Shannon, W.D., Rich, K.M., 2002. Gene expression profiling with oligonucleotide microarrays distinguishes World Health Organization grade of oligodendrogliomas. Cancer Res 61, 1825–1829.

Weller, M., Felsberg, J., Hartmann, C., et al., 2009. Molecular predictors of progression-free and overall survival in patients with newly diagnosed glioblastoma: A prospective translational study of the German Glioma Network. J Clin Oncol 27, 5743–5750.

Wen, P.Y., Kesari, S., 2008. Malignant gliomas in adults. N Engl J Med 359, 492–507.

Wiedemeyer, R., Brennan, C., Heffernan, T.P., et al., 2008. Feedback circuit among INK4 tumor suppressors constrains human glioblastoma development. Cancer Cell 13, 355–364.

Wiedemeyer, W.R., Dunn, I.F., Quayle, S.N., et al., 2010. Pattern of retinoblastoma pathway inactivation dictates response to CDK4/6 inhibition in GBM. Proc Natl Acad Sci USA 107, 11,501–11,506.

Wrensch, M., Jenkins, R.B., Chang, J.S., et al., 2009. Variants in the CDKN2B and RTEL1 regions are associated with high-grade glioma susceptibility. Nat Genet 41, 905–910.

Würdinger, T., Tannous, B.A., Saydam, O., et al., 2008. miR-296 regulates growth factor receptor overexpression in angiogenic endothelial cells. Cancer Cell 14, 382–393.

Yan, H., Parsons, D.W., Jin, G., et al., 2009. IDH1 and IDH2 mutations in gliomas. N Engl J Med 360, 765–773.

Zeppernick, F., Ahmadi, R., Campos, B., et al., 2008. Stem cell marker CD133 affects clinical outcome in glioma patients. Clin Cancer Res 14, 123–129.

Zhao, S., Lin, Y., Xu, W., et al., 2009. Glioma-derived mutations in IDH1 dominantly inhibit IDH1 catalytic activity and induce HIF-1alpha. Science 324, 261–265.

Zheng, H., Ying, H., Yan, H., et al., 2008. p53 and Pten control neural and glioma stem/progenitor cell renewal and differentiation. Nature 455, 1129–1133.

Zheng, P.P., Hop, W.C., Sillevis Smitt, P.A., et al., 2005. Low-molecular weight caldesmon as a potential serum marker for glioma. Clin Cancer Res 11, 4388–4392.

Melanoma

Christina K. Augustine, Jennifer A. Freedman,
Georgia M. Beasley, and Douglas S. Tyler

INTRODUCTION

The incidence of melanoma is increasing at a rate faster than any other cancer, with an estimated 68,720 new cases in 2009 (Jemal et al., 2009). Even more critically, however, mortality rates for melanoma continue to increase – a marked contrast to the decreasing mortality rates for all other cancer cases combined (Jemal et al., 2009). This underscores the urgent need to better understand the biology driving melanomagenesis and to identify more effective treatment strategies. In this chapter, we will summarize current knowledge regarding melanoma biology, the use of high-throughput genomic profiling technology to increase our understanding of the biology underlying each melanoma, and the clinical utility of genomic profiling for more personalized treatment of melanoma patients.

MELANOMA PROGRESSION

Melanoma arises from mutated melanocytes that have escaped normal growth control. Melanocytes derive from neural crest cells and migrate predominantly to the skin and hair, where the melanin produced by these cells determines the pigmentation and acts to absorb ultraviolet (UV) radiation, providing protection to the skin from sun-induced damage (Ibrahim and Haluska, 2009). Melanocytes also migrate to other sites such as the uveal tract of the eye and ectodermal

mucosa. Melanoma can be grouped into three families based on the primary site of the tumor: cutaneous, mucosal, and ocular. The dominant site of melanoma occurrence is the skin, with cutaneous melanoma accounting for 91.2% of all melanomas and the remainder occurring at ocular (5.8%), mucosal (1.3%), or unknown sites. Although sun exposure is considered the predominant environmental risk factor for melanoma, the relationship between sun exposure and melanoma is in fact complex, with melanoma occurring not only on chronic and intermittent sun-exposed sites, but also at sites with little to no sun exposure, such as acral (palm and sole) and mucosal melanoma. Classification of cutaneous melanoma has historically involved four histopathological descriptions – superficial spreading melanoma (SSM; occurring predominantly on intermittently sun-exposed sites such as the trunk and extremities), lentigo maligna melanoma (LMM; occurring predominantly on chronic sun-exposed sites such as the face and neck), nodular malignant melanoma (NMM; occurring on intermittently sun-exposed sites), and acral lentiginous melanoma (ALM; occurring on sites not exposed to the sun) (Duncan, 2009; Palmieri et al., 2009). More recent attempts to classify melanoma based on patterns of genetic alterations (such as gains or losses of DNA) suggest very distinct pathways in the development of melanoma based on the level of sun exposure (chronic, non-chronic/intermittent, or no sun exposure) (Curtin et al., 2005).

Progression from melanocyte to melanoma generally begins with a benign nevus – a clonal population of melanocytes

Genomic and Personalized Medicine, 2nd edition
by Ginsburg & Willard. DOI: http://dx.doi.org/10.1016/B978-0-12-382227-7.00066-5

that have proliferated into a hyperplastic lesion, but which are in a state of cellular senescence and hence do not progress. There are several factors, both inherited genetic risk factors and environmental risk factors such as sun exposure, which likely contribute to this aberrant proliferative state. In response to appropriate stimuli, these hyperplastic lesions exit senescence and begin to grow as dysplastic nevi and then progress to a radial growth phase (RGP), where the lesions spread superficially in an area confined to the epidermis, with little invasive potential. Eventually these RGP lesions progress to a vertical growth phase (VGP), at which point they begin to invade the dermis and eventually metastasize. Metastatic spread of cutaneous melanoma usually occurs as regional lymph node metastases, loco-regional satellite or in-transit metastases, or distant metastases (in 50, 20, and 30% of patients with recurrent melanoma, respectively), with lymph node, satellite, and in-transit metastases frequently progressing to distant metastatic sites (Leiter et al., 2004). The time of recurrence depends on the metastatic pathway, such that distant metastases usually appear after 24 months, and lymph node, satellite, and in-transit metastases appear after 16 to 19 months. Five year survival rates for patients with melanoma that has metastasized to distant sites is a grim 16% (compared to 99% for primary melanoma that is locally confined (Jemal et al., 2009)) underscoring both the importance of early detection as well as the tremendous need to identify better treatments for those patients whose disease has spread beyond the primary site.

GENETICS OF MELANOMA

Inherited Genetic Lesions

Although familial, or hereditary, melanoma accounts for only a small percentage of all cases of melanoma (<7%; Olsen et al., 2010) it can provide a valuable insight into genetic factors contributing to the aberrant growth of a melanocyte. The cyclin-dependent kinase inhibitor 2A (*CDKN2A*) locus on chromosome 9 is the most clearly linked, high-risk gene associated with melanoma, with estimates of 25 to 50% of familial melanoma patients harboring mutations at this locus (Nelson and Tsao, 2009). Two tumor suppressor proteins critical in the regulation of the cell cycle are encoded by the *CDKN2A* locus, p16^{INK4a} and p14^{ARF}, and genetic mutations at this locus can lead to deregulated signaling in both the retinoblastoma (Rb) and p53 pathways. Under normal conditions, p16^{INK4a} binds Cdk4 (cyclin-dependent kinase 4), inhibiting its protein kinase activity and ultimately inhibiting transcription of S phase genes by the transcription factor E2F1. Loss of p16^{INK4a} activity leads to hyperphosphorylation of the tumor suppressor protein Rb, induction of the synthesis of S phase genes, and an increase in cell cycling allowing the melanocyte to leave G1 arrest and enter the G1 to S transition. Mutation of p16^{INK4a} (9% of melanomas) or loss of p16^{INK4a} (frequently due to hypermethylation; 50% of melanomas) can thus lead to increased cell cycle progression. p14^{ARF} binds human homolog

of murine Mdm2 (HDM2), sequestering it in the nucleolus and preventing HDM2 from interacting with and destabilizing the tumor suppressor protein p53. Decreased p14^{ARF} activity due to loss or mutation leads to increased interaction of HDM2 with p53, the subsequent ubiquitination and degradation of p53 and ultimately to genomic instability, as the cell is unable to detect genetic damage, signal for DNA repair, or activate apoptotic pathways when DNA damage is too extensive for repair. Germline mutations in *CDK4*, although rare, are likewise high-risk for melanoma. Mutations in *CDK4*, which are mutually exclusive with p16^{INK4a} mutations likely because of redundancy (Soufir et al., 1998), block binding with p16^{INK4a} leading to an increase in cell cycle progression, as described above. Another gene associated with melanoma is the melanocortin-1 receptor (*MC1R*) gene, which is important in the regulation of skin color and when activated by α-melanocyte stimulating hormone (α-MSH) triggers melanocytes to switch production from pheomelanin (red/yellow melanin) to eumelanin (brown/black melanin). Variants in *MC1R* that lead to a loss of this switch in melanin production are associated with not only red hair and fair skin, but also an increased risk for melanoma. These variants in *MC1R* represent a low-penetrance genetic risk factor for melanoma and individuals harboring both p16^{INK4a} mutation and one or more variant of the *MC1R* gene are prone to presenting with melanoma at an earlier age (Ibrahim and Haluska, 2009). A high incidence of inactivating mutations in *TP53* occur in many cancer types, and germline mutations in *TP53* occur in melanoma, but only rarely (~9%) (Sekulic et al., 2008). Altered regulation of p53 in melanoma generally comes as a consequence of altered signaling at an early point in cell cycle progression, as in the case of mutations in p14^{ARF} described above.

Acquired Genetic Lesions

For most melanomas (~90%) the genetic lesions contributing to progression of the disease are acquired after birth, rather than inherited, and are termed sporadic melanomas. Several somatic mutations causing deregulation of key intracellular signaling pathways important in proliferation and apoptosis have been identified to date in melanoma. One of the first somatic mutations noted to occur predominantly in melanoma was an activating mutation in the BRaf serine/threonine protein kinase (Davies et al., 2002). Up to 70% of melanomas have been shown to harbor a single base pair change leading to a glutamate for valine substitution at codon 600 in the kinase domain. This mutation leads to constitutive activation of BRaf kinase, with the resultant unchecked stimulation of the mitogen-activated protein kinase/extracellular signal-regulated kinase (MEK) and extracellular signal-regulated kinase (ERK) pathway leading to increased activity in several critical pathways involved in proliferation and survival. Although common in melanoma, the timing of this mutation is controversial, with some reports suggesting that it is an early occurrence (Pollock et al., 2003) and others suggesting that it occurs later as melanoma progresses from RGP to VGP (Greene et al., 2009).

BRAF mutation alone is not considered a transforming event, but rather works in concert with other mutations (Ibrahim and Haluska, 2009) such as loss of function mutations in p53 (Patton et al., 2005). The relationship between sun exposure and *BRAF* mutation is complex, as the mutation itself is not a typical ultraviolet B (UVB) signature mutation. *BRAF* mutations have been shown to occur more frequently on intermittently sun-exposed sites (Palmieri et al., 2009), while in individuals with chronic sun exposure or in sites with no sun exposure, *BRAF* mutation frequency is much less (Curtin et al., 2005; Hacker et al., 2010). Although less common, activating mutations in *NRAS* also occur in ~15 to 30% of melanomas, leading to similar deregulation in pathways such as MEK and ERK that are important in proliferation and survival (Haluska et al., 2006). Likely because of redundancy in downstream signaling cascades, *NRAS* and *BRAF* mutations are mutually exclusive in melanoma.

Phosphatase and tensin homolog (*PTEN*) was identified as a frequently deleted gene in melanoma more than 20 years ago (Parmiter et al., 1988). *PTEN* is located on chromosome 10q23-24, a region that is frequently lost at an early stage in melanoma (Fountain et al., 1990). PTEN functions as a tumor suppressor protein regulating the levels of phosphatidylinositol phosphate (PIP3). PIP3 activates the Akt protein kinase signaling pathway leading to reduced apoptosis. In this way PTEN functions to promote apoptosis by blocking the anti-apoptotic activity of the Akt signaling pathway. *PTEN* loss has been reported in 37% of melanomas, while nevi, both benign and dysplastic, have been shown to retain *PTEN* expression (Tsao et al., 2003). The high frequency of *PTEN* loss in melanoma, compared to the relatively low incidence of *PTEN* mutation, suggests that epigenetic silencing of *PTEN* is important in melanomagenesis, and the multiple methylation sites identified in the *PTEN* promoter region lend support to this theory (Palmieri et al., 2009). *PTEN* and *BRAF* mutations are frequent concurrent events in melanoma, with this dual mutation acting to deregulate both the mitogen-activated protein kinase (MAPK) and the Akt signaling pathways (Haluska et al., 2006). In contrast, *PTEN* and *NRAS* mutations are mutually exclusive, which is likely a result of the redundant effects both PTEN loss of function and NRAS gain of function would have on the Akt pathway signaling (Haluska et al., 2006).

Micropthalmia-associated transcription factor (MITF) and calcium selective transient receptor potential channel (TRPM1) are two proteins important in melanocyte biology. MITF is a transcription factor that regulates the expression of genes involved in melanocyte differentiation, melanin production, and cell survival (Steingrimsson et al., 2004). Although less is known about the functional role of TRPM1, a recent study showed a correlation between melanin content and TRPM1 expression and suggested a role for TRPM1 in normal melanocyte pigmentation (Oancea et al., 2009). The role that *MITF* plays in the progression from melanocyte to melanoma is complex, as it has been shown to be an amplified oncogene important for survival in some tumors, but to induce

the expression of tumor suppressor proteins (Garraway et al., 2005; Levy et al., 2006; Loercher et al., 2005; Ugurel et al., 2007). More recently, *MITF* mutations have been identified in primary metastatic melanomas; however, the effects on cell proliferation and gene regulation during melanomagenesis is unclear (Cronin et al., 2009). TRPM1 (melastatin) has been shown to negatively correlate with the aggressiveness of melanoma and to show reduced expression in metastatic melanomas (Duncan et al., 1998). MITF can bind to the promoter region and increase expression of *TRPM1* (Miller et al., 2004). Reports have also shown that oncogenic BRaf-induced increases in ERK activity can lead to decreased *MITF* expression (Wellbrock and Marais, 2005).

The cadherin family of cell adhesion proteins plays an important role in both normal melanocyte function and malignant transformation. Melanocytes are characterized by a largely E-cadherin-expressing phenotype, which provides for high-affinity anchoring to neighboring E-cadherin-expressing keratinocytes and serves as a tumor suppression mechanism (Hsu et al., 2000). With malignant transformation, N-cadherin is up-regulated as a dominant phenotype, generally at the vertical growth phase. As melanoma progresses from a pre-invasive to a metastatic form, this switch in cadherin expression from E-cadherin to N-cadherin leads to acquisition of invasive, migratory capabilities and an ability to transit the endothelial vascular barrier and undergo epithelial-to-mesenchymal transition (Alonso et al., 2007; Huber et al., 2005; Qi et al., 2005). Src tyrosine kinase, which has been shown to be activated during transendothelial migration of melanoma cells, and other intracellular signaling pathways may be altered by this switch in cadherin expression leading to downstream consequences such as increased proliferation, survival, angiogenesis, and decreased apoptosis (Qi et al., 2006).

c-Kit is a tyrosine kinase receptor that is important in melanocyte migration from the neural crest to the dermis during development (Masson and Ronnstrand, 2009). Activation by binding of stem cell factor (SCF) to the extracellular domain leads to dimerization of the receptor, autophosphorylation, and activation of the tyrosine kinase (Alexeev and Yoon, 2006). Downstream events triggered by the activated receptor are numerous and the specific role played in malignant transformation and melanoma progression is complex. Early reports demonstrated a loss of c-Kit expression during progression from benign nevi to metastatic melanoma (Montone et al., 1997) and a more recent report suggests that, while activated c-Kit is important for melanocyte migration, it is not essential for malignant transformation (Alexeev and Yoon, 2006). More recently, however, there is evidence to suggest that *CKIT* is an important oncogene, but only in the setting of non-sun-exposed (mucosal or acral) melanomas and chronic sun-exposed melanomas – notably melanoma types that do not commonly harbor *BRAF* mutations (Curtin et al., 2006). Multiple pathways have been shown to be downstream of activated c-Kit, many of which can lead to enhanced cell proliferation, including MAPK, phosphoinositide 3 (PI3)-kinase, and Src (Masson and Ronnstrand,

TABLE 66.1	Genetics of melanoma				
	Protein	**Normal function**	**Effect of lesion**	**Affected cellular process**	**Oncogenic signaling pathway**
Inherited genetic lesions					
CDKN2A	p16/INK4α	Tumor suppressor	Copy number loss (9p21.3), inactivating mutation	Cell cycle progression	E2F1
	p14/ARF	Tumor suppressor	Copy number loss (9p21.3), inactivating mutation	Cell cycle progression	p53
CDK4	Cdk4	Oncogene	Activating mutation	Cell cycle progression	E2F1
MC1R	MC1 receptor	Melanogenesis	Variants	Pigmentation	
TP53	p53	Tumor suppressor	Inactivating mutation	Cell cycle progression	p53
Acquired genetic lesions					
BRAF	BRaf	Oncogene	Activating mutation, copy number gain (7q34)	Proliferation; survival	Ras
NRAS	Nras	Oncogene	Activating mutation	Proliferation; survival	Ras, PI3K
PTEN	PTEN	Tumor suppressor	Copy number loss (10q23.31)	Cell cycle progression; survival	PI3K, Akt
MITF	MITF	Transcription factor	Mutation; copy number gain (3p13)	Pigmentation	
TRPM1	TRPM1	Ion channel	Loss of expression	Pigmentation	
CDH1	E-cadherin	Adhesion to keratinocytes	Loss of expression	Cellular adhesion	
CDH2	N-cadherin		Gain of expression	Cellular adhesion	
CKIT	c-Kit	Oncogene	Activating mutation	Proliferation; differentiation	Src, PI3K

2009), suggesting potential transforming capabilities of aberrant c-Kit signaling. See Table 66.1 for a summary of the genetics of melanoma.

GENOMIC STRATEGIES TO CHARACTERIZE MELANOMA

DNA Copy Number and Oncogenic Mutations

In general, all cancers exhibit a large degree of genetic heterogeneity, with multiple mutations and genetic alterations combining to generate the cancer phenotype. As can be inferred from the number of genetic lesions and somatic mutations identified to date, melanoma is no exception to this paradigm. Multiple studies documenting chromosomal copy number alterations, loss of heterozygosity, and mutations in oncogenes support the existence of substantial genetic heterogeneity across melanoma patients (see also Chapter 10). Utilizing an algorithm designed to interrogate comparative genomic hybridization (CGH) data and single-nucleotide polymorphism (SNP) array data (Genomic Identification of Somatic Targets In Cancer, or GISTIC), Lin and colleagues (2008) identified 14 regions of copy number gains and 13 regions of copy number losses significantly present in a large collection of cultured melanoma cells. Significant copy number gains were identified at chromosome 7q34, which harbors the BRAF gene, and 3p13, harboring the MITF gene, while significant loses were detected at 9p21.3 (CDKN2A) and 10q23.31 (PTEN). Similar results were obtained using data derived from primary melanomas. Unsupervised hierarchical clustering analysis classified the melanoma cell line data into six molecularly distinct groups that were characterized by copy number alterations and oncogene mutation status. CGH analysis of benign melanocytic nevi and melanomas showed significantly fewer aberrations in benign nevi (7%) compared to melanomas (96% showing some aberration), and subgroups of melanomas were identified, based on DNA copy number alterations, that correlated with anatomical site and sun-exposure pattern (Bastian et al., 2003). BRaf mutation status has likewise been shown to correlate with sun exposure, with non-chronic sun-exposed melanomas showing a much higher incidence compared to chronic sun-exposed, acral, and mucosal lesions (Maldonado et al., 2003). Similarly, an analysis of DNA copy number and oncogene mutation status across 126 melanoma lesions revealed unique patterns based on the degree of ultraviolet light exposure, with non-chronic sun-exposed lesions showing the highest incidence of BRaf or NRas mutation, lesions with wildtype BRaf and NRas showing higher incidence of copy number gain in *CDK4* and cyclin *D1* genes, and mucosal and acral melanomas showing the highest degree overall of chromosomal aberration (Curtin et al., 2005). Unsupervised hierarchical clustering analysis of DNA copy number data classified sun-exposed melanomas into two groups – chronic and non-chronic sun exposed.

RNA Expression Patterns – Gene Expression Profiling

Patterns of gene expression, identified using microarray-based platforms, also support the existence of substantial genetic heterogeneity among melanomas. In an early example of the use of gene expression data to classify melanoma samples, Bittner and colleagues (2000) identified a distinct subgroup that showed unique patterns of expression across several genes that are differentially regulated in melanomas capable of forming primitive tubular networks in an *in vitro* setting. This gene expression-based classification of melanoma has the potential for clinical implications in terms of prognosis because formation of these tubular networks is a characteristic of more aggressive melanomas (Maniotis et al., 1999). There are several examples of the use of gene expression profiling to characterize melanoma progression on the basis of underlying molecular mechanisms. Haqq and colleagues (2005) performed unsupervised hierarchical clustering on gene expression data derived from nine benign nevi, six primary, and 19 metastatic melanomas. Two distinct subgroups of metastatic melanomas were identified that showed differential patterns of gene expression. Notably, one of the metastatic melanoma subtypes showed patterns of gene expression similar to that observed in radial growth phase melanomas, which have traditionally been considered to have less metastatic potential. Furthermore, distinct patterns of gene expression gains and losses were identified in the transition from nevus to primary melanoma (Haqq et al., 2005). Other examples of gene expression analysis of melanoma progression have demonstrated distinct alterations in gene expression and biological processes that are associated with progression from early-stage to advanced-stage melanomas (Smith et al., 2005) as well as differential gene expression patterns associated with common nevi, dysplastic nevi, RGP, VGP, and metastatic melanomas (Scatolini et al., 2010). Jaeger and colleagues

(2007) identified 308 genes that were differentially expressed between primary and metastatic melanomas. Several biological processes were represented in these genes, including those associated with cell cycle regulation and cell adhesion. Furthermore, subgroups of primary melanoma based on tumor thickness revealed differential patterns of gene expression (Jaeger et al., 2007). In addition to classifying melanomas based on progression, gene expression analysis of melanoma has been used to identify patterns associated with clinical outcome. In one such study, expression across 254 genes was associated with four-year distant metastasis-free survival of 58 primary melanoma patients (Winnepenninckx et al., 2006). More recently, analysis of gene expression data from 57 stage IV melanomas revealed four subgroups with characteristic patterns of expression of genes associated with distinct biological processes: high-immune response, proliferative, pigmentation, and normal-like (Jonsson et al., 2010). Notable expression patterns included consistently higher *MITF* expression in the pigmentation subgroup and frequent *CDKN2A* deletions as well as high frequencies of BRaf and NRas mutation in the proliferative subgroup. Clinical relevance of these subgroups is supported by the correlation between the proliferative subgroup and poorer survival. An evaluation of 30 immune response-related genes revealed lower expression of these genes in the proliferative subgroup compared to the high immune-response and pigmentation subgroups (Jonsson et al., 2010). Importantly, significantly poorer survival was observed across patients with tumors with low expression of these immune-response genes suggesting a potential application in predicting response to immunotherapy.

Oncogenic Signaling Pathways

Another approach to characterizing both the genetic heterogeneity that is present in melanoma and the distinct stages in melanoma progression is to examine the oncogenic signaling pathways that are altered in melanomas. Many studies have implicated particular signaling networks in melanomagenesis based on the underlying genetic lesions (reviewed in Hocker et al., 2008). For example, the Raf arm of the Ras pathway is thought to be important for melanocytic proliferation, while the PI3k-Akt arm of the Ras pathway is thought to be important for melanoma progression. The Cdkn2a/Cdk4 network has been implicated in melanocytic senescence, while the Bcl-2/p53 network has been implicated in the resistance of melanoma cells to apoptosis and chemotherapeutic reagents. Alterations in signaling pathways involved in melanocyte development, such as c-Kit and Mitf, have also been implicated in melanomagenesis.

There are several advantages to defining subgroups of melanomas based on the oncogenic signaling pathways that are deregulated. First, although characteristics of morphology as well as biochemical assays for the mutational status of individual genes can reveal subgroups of tumors with distinct underlying biology, classifying melanomas with respect to alterations in oncogenic signaling pathways that have contributed to disease progression allows tumors to be stratified into more homogenous subgroups. Second, in spite of the contribution that a single mutation in a gene makes toward transformation, melanomagenesis is driven by multiple accumulated genetic alterations and mutations. Third, a further understanding of intratumoral heterogeneity, with respect to deregulation of unique oncogenic signaling pathways contributing to progression in particular melanomas, not only increases knowledge of the biology underlying an individual's disease, but also has the potential to guide therapeutic strategies leading to treatment options that might be more effective against each individual melanoma.

Activation of an oncogenic signaling pathway in a cell generates a number of unique gene expression changes reflective of the biological state. The unique changes in gene expression observed in cells in which a particular oncogenic signaling pathway is turned "off" or "on" (gene expression signature) can then be used to guide the analysis of samples in which oncogenic signaling pathway activation is not known. Previous work has established methods to generate gene expression signatures reflecting the activity of a number of oncogenic signaling pathways (Bild et al., 2006; Gatza et al., 2010). In brief, human mammary epithelial cells were infected with a recombinant adenovirus containing either a control DNA insert expressing green fluorescent protein, the pathway "off" state, or an insert expressing an activated oncogene, the pathway "on" state, and the RNA expression measured using DNA microarray technology. Supervised methods of analysis identified sets of genes that exhibited a consistent pattern of expression with respect to the activation of a particular oncogenic pathway (Bild et al., 2006). Gene expression signatures reflecting the activity of several oncogenic signaling pathways have been used to successfully predict the status of the respective pathways in murine models of cancer as well as human tumors (Bild et al., 2006; Gatza et al., 2010; Mori et al., 2008).

In melanoma, these gene expression signatures of oncogenic pathway activation were recently applied to gene expression data from 17 patients with in-transit metastatic disease, to evaluate the degree to which multi-nodal lesions are concordant with respect to predictions derived from an analysis of gene expression data (Augustine et al., 2010a). Predicted probabilities of activated signaling in the Ras, PI3-kinase, Myc, Src, E2F3, and β-catenin pathways were strongly concordant within a given patient, with the range of predictions not exceeding 0.3 (on a scale of 0 to 1.0) in 70% or more of the patients, demonstrating the potential clinical utility of these gene expression signatures in melanoma patients presenting with multiple lesions. Distinct patterns of oncogenic pathway activation can be seen across in-transit melanomas as shown in Figure 66.1. Notably, Src, Myc, and β-catenin show high probabilities of pathway activation compared to Ras, E2F3 and PI3-kinase, while multiple lesions showed high probabilities of activation in both Src and β-catenin pathways.

Gene signatures of oncogenic pathway activation, or altered biological processes such as proliferation and

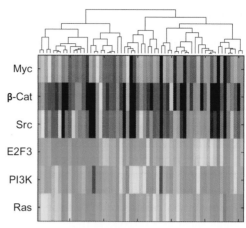

In-transit metastatic melanoma samples

Figure 66.1 The status of oncogenic signaling pathway activities analyzed in in-transit metastatic melanoma samples. MAS5.0 gene-expression data was log transformed and normalized using Bayesian Factor Regression Modeling (BFRM). Within the heatmap, columns represent samples and rows represent the oncogenic signaling pathways analyzed, as noted. Pathway activities are color-coded on a low (blue) to high (red) continuum.

immune-response, clearly have clinical potential as a means to couple therapeutic options with distinct subgroups of melanomas that share a common underlying biology. In addition, identification of multiple deregulated oncogenic pathways and biological processes in a tumor can help to guide combination drug regimens. As such, the classification of melanoma on the basis of patterns of gene expression provides an important and novel approach to the development of potentially more effective targeted therapeutic options for patients.

CLINICAL APPLICATIONS OF GENOMICS IN MELANOMA

Tumors have traditionally been classified by descriptive characteristics, such as tissue of origin, histology, aggressiveness, and extent of spread. Genetic studies of tumors using microarray and other high-throughput genomic profiling technologies provide tools to classify tumors based on underlying biology, and have been utilized to characterize melanoma progression and predict clinical outcomes (Bittner et al., 2000; Haqq et al., 2005; Jonsson et al., 2010) (see also Chapters 7, 8, 10, 12, and 13). An evolving use of genetic studies in cancer is the classification of tumors based on predicted response to a drug (Lee et al., 2007; Staunton et al., 2001; Williams et al., 2009). By utilizing genetic markers in this way, treatment strategies can be customized to each patient while at the same time optimized so that toxicity and adverse response to treatment are minimized (See Table 66.2).

TABLE 66.2	Clinical applications of Genomics in Melanoma
Targeted therapies	
Imatinib	*c-Kit* mutation
PLX4032	*BRaf* mutation
Sorafenib	*BRaf* mutation
RAF-265	*BRaf* mutation
GSK2118436	*BRaf* mutation
Chemotherapies	
DTIC, TMZ	MGMT
	Gene expression signature (Augustine et al., 2010a)
Immunotherapies	
Interferon α-2b	HLA-Cw∗06 positive
Prognosis	
Relapse-free survival >5 years	cytokines: IL-1α and β, IL6, TNFα; chemokines: MIP-1α and β
Aggressive phenotype	Gene expression signature (Bittner et al., 2000)
Distant metastasis-free survival	Gene expression signature (Winnepenninckx et al., 2006)
Poorer survival	Gene expression signature (Jonsson et al., 2010)

MGMT, O-6-methylguanine-DNA methyltransferase;
DTIC, dacarbazine; TMZ, Temozolomide.

Targeted Therapeutics

One of the most extensively targeted proteins for the treatment of melanoma is BRaf kinase because of the frequent occurrence of activating mutations. Several targeted therapeutics directed toward BRaf kinase have been developed, including PLX4032 (Plexxikon/Roche), that targets the V600E mutation specifically in BRaf kinase; Sorafenib (Bayer) and RAF-265 (Novartis), that target several tyrosine kinases in addition to BRaf, including CRaf kinase, vascular endothelial growth factor (VEGF) receptor, and platelet-derived growth factor (PDGF) receptor; and GSK2118436 (GlaxoSmithKline), that targets the three Raf kinases, ARaf, CRaf and BRaf (for review see Shepherd et al., 2010). Early results from a phase I study of PLX4032 dosed orally at ≥240 mg twice daily noted 10 of 16 V600E BRaf mutant patients had partial response, while one showed a complete response (Flaherty et al., 2009). Dosed at the maximum tolerated dose of 960 mg twice daily has shown, to date, a RECIST (response evaluation criteria in solid tumors) objective response in 25 of 32 V600E BRaf mutant patients (Shepherd et al., 2010). Furthermore, a recent clinical report

suggests that PLX4032 may also target the V600K BRaf mutation, a less common melanoma lesion that, nevertheless, can occur in up to 30% of melanomas (Rubinstein et al., 2010). Although encouraging, these results must be taken with caution as acquired resistance with subsequent relapse has been shown to occur (Shepherd et al., 2010). Strategies to overcome resistance to BRaf inhibition with drugs such as PLX4032 are important. One study suggests that resistance to BRaf inhibition may arise as a result of a rebounding of MAPK signaling due to recovery of phosphorylated ERK, and that combination therapy with MEK and BRaf inhibition could be a rational approach to overcoming resistance (Paraiso et al., 2010). Although in preclinical studies Sorafenib has been shown to enhance the response of melanoma to regional chemotherapy with melphalan (Augustine et al., 2010b), clinical trials to date have not shown any improvement in response either with Sorafenib alone or when given in combination with systemic chemotherapy in patients with advanced melanoma (Eisen et al., 2006; Hauschild et al., 2009).

Imatinib is a tyrosine kinase inhibitor of Abelson murine leukemia viral oncogene homolog (Abl), PDGF receptor, and c-Kit (CD117). Although infrequent, gain-of-function mutations or copy number increases in *CKIT* do occur in up to 39% of mucosal melanomas, 36% of acral melanomas, and 28% of chronically sun-damaged skin (Curtin et al., 2006). Critically, several melanomas have been identified that harbor a specific *CKIT* mutation at the juxtamembrane domain in exon 11 that predicts sensitivity to inhibition with imatinib (Beadling et al., 2008; Corless et al., 2004). A recent clinical trial demonstrated a marked response to imatinib therapy in a patient with primary anal melanoma harboring an activating mutation in exon 11 of the c-Kit gene (Hodi et al., 2008). Furthermore, preclinical studies showed that response to imatinib therapy in mucosal melanoma cell lines correlated with c-Kit mutational status suggesting that targeting imatinib therapy to the patients with tumors harboring the appropriate genetic lesion has the potential to yield promising clinical outcomes (Jiang et al., 2008).

Predicting Response to Chemotherapy

Metastatic melanoma is notable for its exceptional resistance to chemotherapy, with overall response rates of only 10–15% seen with dacarbazine (DTIC) or its imidotetrazine derivative temozolomide (TMZ), the most active cytotoxic agents for treatment of patients with advanced systemic melanoma (Middleton et al., 2000). Other systemic treatments for melanoma include immunotherapy, alone and in combination with chemotherapy, notably interleukin 2 (IL-2) and interferon-α (IFN-α) (Keilholz and Eggermont, 2000; Kirkwood et al., 2004). Evidence suggests, however, that a similarly limited number of patients will respond to immunotherapy alone and the combination of immunotherapy and traditional chemotherapy can lead to severe toxicity, limiting the clinical utility for this treatment. Furthermore, the inherent chemoresistance and molecular complexity of melanoma make traditional single candidate markers of inflammation or tissue damage, such as c-reactive protein and lactate dehydrogenase, of limited use in predicting response (Ekmekcioglu et al., 2003; Tartour et al., 1996). In light of these limitations of current chemotherapy treatments for melanoma, identification of predictive genomic markers that could be used to prospectively select patients most likely to benefit from a given chemotherapy would be beneficial. The comprehensive nature of gene expression profiling has the potential to capture the molecular complexity of melanoma into clinically useful predictors of therapeutic response.

In-transit metastatic melanoma provides a unique setting to evaluate the utility of genetic markers to predict chemosensitivity. Traditional therapy for in-transit melanoma confined to the extremity involves hyperthermic isolated limb perfusion (HILP) or isolated limb infusion (ILI) whereby chemotherapy is delivered regionally to the isolated extremity at a dose several orders of magnitude higher than used for systemic administration (Aloia et al., 2005; Beasley et al., 2008). Microarray-based gene expression profiling studies of in-transit melanoma lesions have aimed to identify gene expression signatures that are predictive of response to regional chemotherapeutic agents currently in use – melphalan, the most commonly used chemotherapeutic for regional therapy, and temozolomide (TMZ), used systemically and currently in clinical trial for use in regional chemotherapy. Initial studies to develop a genomic predictor of response to melphalan relied on the use of the National Cancer Institute panel of 60 cancer cell lines (NCI60), for which published *in vitro* drug response data and microarray gene expression profiles are available, and an independent panel of melanoma cell lines to generate a gene signature predictive of response to melphalan. The gene expression signatures, however, were insufficiently robust to accurately predict response across an independent panel of melanoma cell lines (unpublished data) suggesting that the complexity of the tumor environment – particularly the vasculature – is both crucial in determining tumor response to melphalan and inadequately captured in the setting of *in vitro* cell culture. Ongoing work in our laboratory consists of utilizing patient tumor samples with known outcomes to melphalan ILI to develop a gene expression signature that shows a more robust capacity to predict response to melphalan in an independent patient population. In a similar way, *in vitro* drug response data and microarray gene expression profiles were utilized to identify a gene signature predictive of response to temozolomide-based treatment regimens. In contrast to melphalan, the gene signature derived from *in vitro* studies of melanoma cell lines showed a robust capacity to predict response to TMZ across an independent panel of melanoma cell lines (Augustine et al., 2010a). Notably, however, TMZ response was equally well predicted by expression of the DNA repair enzyme O-6-methylguanine-DNA methyltransferase (MGMT) alone (Augustine et al., 2009). An ongoing phase I trial of isolated limb infusion utilizing intra-arterial TMZ in patients with advanced extremity melanoma will be used to validate the cell-line-derived gene expression signature of TMZ sensitivity and the predictive utility of MGMT expression in the clinical setting.

Predicting Response to Immunotherapy

Recent studies have undertaken to identify genomic markers that are predictive of response to immunotherapy. As an example, genomic DNA obtained from 284 high-risk melanoma patients enrolled in a randomized phase III study of one month versus one year of adjuvant high-dose interferon α-2b was molecularly typed for human leukocyte antigen (HLA) class I and II, and compared to 246 healthy patients as controls (Gogas et al., 2010; Pectasides et al., 2009). This study found that HLA-Cw∗06-positive patients had better relapse-free (RFS) and overall (OS) survival (Gogas et al., 2010). Utilizing a high-throughput profiling technology (xMAP multiplex immunobead assay from Luminex Corp.), 29 cytokines, chemokines, angiogenic and growth factors, and soluble receptors were simultaneously measured in sera obtained from 179 high-risk melanoma patients and 378 healthy age- and gender-matched controls. Melanoma patients were randomly chosen for this study based on disease status, and 93 received GMK vaccine (ganglioside G_{M2} conjugated to keyhole limpet hemocyanin mixed with QS-21 adjuvant) while 86 received high-dose IFN-α2b therapy. The results of this study revealed a relationship between pretreatment levels of four pro-inflammatory cytokines (IL-1β, IL-1α, IL-6, TNF-α) and two chemokines (MIP-1α and MIP-1β) and relapse-free survival, such that patients with relapse-free survival of greater than five years showed higher sera levels of these cyto- and chemokines compared to patients with shorter relapse-free survival (Tarhini and Kirkwood, 2009). Although still exploratory, these studies demonstrate the potential use of biomarkers to help guide systemic treatment with immunotherapy reagents in patients with metastatic melanoma. See Table 66.2 for a summary of clinical applications of genomics in melanoma.

CONCLUSION

Melanoma is an increasing health problem with dismal prognosis for stage IV patients who have metastatic disease (Ross, 2006). The increasing incidence of melanoma and the traditionally poor response rates to chemotherapy, due in large part to the inherently chemoresistant nature that characterizes melanoma, have combined to yield discouraging statistics for melanoma. These statistics are particularly troubling in an era where so many advances are being made in our understanding of and ability to treat not just cancer but many other human diseases. The substantial genetic heterogeneity observed in melanoma has profound implications for the treatment of this disease and makes clear that distinct mechanisms, and thus disease biology, are associated with each melanoma patient. This underscores the importance of developing personalized treatment strategies for melanoma, and points to the need to dissect and better understand the distinct subgroups of melanomas. While targeted therapies are appealing and have potential for more personalized treatment options, they are only useful to the extent to which they can be matched in a rational way with the characteristics of the patient's tumor. The effectiveness of existing targeted therapies and the identification of targets for development of novel therapeutics will depend on the ability to identify and select subgroups of later stage melanoma patients who share a common underlying biology. The ability to prospectively predict patient response to treatment using genetic markers that reflect the underlying biology of a patient's disease will facilitate the design and implementation of more effective and less toxic treatment strategies for melanoma patients.

REFERENCES

Alexeev, V., Yoon, K., 2006. Distinctive role of the cKit receptor tyrosine kinase signaling in mammalian melanocytes. J Invest Dermatol 126, 1102–1110.

Aloia, T.A., Grubbs, E., Onaitis, M., et al., 2005. Predictors of outcome after hyperthermic isolated limb perfusion: Role of tumor response. Arch Surg 140, 1115–1120.

Alonso, S.R., Tracey, L., Ortiz, P., et al., 2007. A high-throughput study in melanoma identifies epithelial-mesenchymal transition as a major determinant of metastasis. Cancer Res 67, 3450–3460.

Augustine, C.K., Jung, S.H., Sohn, I., et al., 2010a. Gene expression signatures as a guide to treatment strategies for in-transit metastatic melanoma. Mol Cancer Ther 9, 779–790.

Augustine, C.K., Toshimitsu, H., Jung, S.H., et al., 2010b. Sorafenib, a multikinase inhibitor, enhances the response of melanoma to regional chemotherapy. Mol Cancer Ther 9, 2090–2101.

Augustine, C.K., Yoo, J.S., Potti, A., et al., 2009. Genomic and molecular profiling predicts response to temozolomide in melanoma. Clin Cancer Res 15, 502–510.

Bastian, B.C., Olshen, A.B., Leboit, P.E., Pinkel, D., 2003. Classifying melanocytic tumors based on DNA copy number changes. Am J Pathol 163, 1765–1770.

Beadling, C., Jacobson-Dunlop, E., Hodi, F.S., et al., 2008. *KIT* gene mutations and copy number in melanoma subtypes. Clin Cancer Res 14, 6821–6828.

Beasley, G.M., Petersen, R.P., Yoo, J., et al., 2008. Isolated limb infusion for in-transit malignant melanoma of the extremity: A well-tolerated but less effective alternative to hyperthermic isolated limb perfusion. Ann Surg Oncol 15, 2195–2205.

Bild, A.H., Yao, G., Chang, J.T., et al., 2006. Oncogenic pathway signatures in human cancers as a guide to targeted therapies. Nature 439, 353–357.

Bittner, M., Meltzer, P., Chen, Y., et al., 2000. Molecular classification of cutaneous malignant melanoma by gene expression profiling. Nature 406, 536–540.

Corless, C.L., Fletcher, J.A., Heinrich, M.C., 2004. Biology of gastrointestinal stromal tumors. J Clin Oncol 22, 3813–3825.

Cronin, J.C., Wunderlich, J., Loftus, S.K., et al., 2009. Frequent mutations in the MITF pathway in melanoma. Pigment Cell Melanoma Res 22, 435–444.

Curtin, J.A., Busam, K., Pinkel, D., Bastian, B.C., 2006. Somatic activation of *KIT* in distinct subtypes of melanoma. J Clin Oncol 24, 4340–4346.

Curtin, J.A., Fridlyand, J., Kageshita, T., et al., 2005. Distinct sets of genetic alterations in melanoma. N Engl J Med 353, 2135–2147.

Davies, H., Bignell, G.R., Cox, C., et al., 2002. Mutations of the *BRAF* gene in human cancer. Nature 417, 949–954.

Duncan, L.M., 2009. The classification of cutaneous melanoma. Hematol Oncol Clin North Am 23, 501–513, ix.

Duncan, L.M., Deeds, J., Hunter, J., et al., 1998. Down-regulation of the novel gene melastatin correlates with potential for melanoma metastasis. Cancer Res 58, 1515–1520.

Eisen, T., Ahmad, T., Flaherty, K.T., et al., 2006. Sorafenib in advanced melanoma: A phase II randomised discontinuation trial analysis. Br J Cancer 95, 581–586.

Ekmekcioglu, S., Ellerhorst, J.A., Mumm, J.B., et al., 2003. Negative association of melanoma differentiation-associated gene (mda-7) and inducible nitric oxide synthase (iNOS) in human melanoma: *MDA-7* regulates iNOS expression in melanoma cells. Mol Cancer Ther 2, 9–17.

Flaherty, K., Puzanov, I., Sosman, J., et al., 2009. Phase I study of PLX4032: Proof of concept for V600E *BRAF* mutation as a therapeutic target in human cancer. J Clin Oncol 27, 9000.

Fountain, J.W., Bale, S.J., Housman, D.E., Dracopoli, N.C., 1990. Genetics of melanoma. Cancer Surv 9, 645–671.

Garraway, L.A., Widlund, H.R., Rubin, M.A., et al., 2005. Integrative genomic analyses identify *MITF* as a lineage survival oncogene amplified in malignant melanoma. Nature 436, 117–122.

Gatza, M.L., Lucas, J.E., Barry, W.T., et al., 2010. A pathway-based classification of human breast cancer. Proc Natl Acad Sci USA 107, 6994–6999.

Gogas, H., Kirkwood, J.M., Falk, C.S., et al., 2010. Correlation of molecular human leukocyte antigen typing and outcome in high-risk melanoma patients receiving adjuvant interferon. Cancer 116, 4326–4333.

Greene, V.R., Johnson, M.M., Grimm, E.A., Ellerhorst, J.A., 2009. Frequencies of *NRAS* and *BRAF* mutations increase from the radial to the vertical growth phase in cutaneous melanoma. J Invest Dermatol 129, 1483–1488.

Guo, J., Si, L., Kong, Y., et al., 2011. Phase II, Open-Label, Single-Arm trial of imatinib mesylate in patients with metastatic melanoma harboring c-Kit mutation or amplification. J Clin Oncol 29 (21), 2904–2909.

Hacker, E., Hayward, N.K., Dumenil, T., James, M.R., Whiteman, D.C., 2010. The association between *MC1R* genotype and *BRAF* mutation status in cutaneous melanoma: Findings from an Australian population. J Invest Dermatol 130, 241–248.

Haluska, F.G., Tsao, H., Wu, H., Haluska, F.S., Lazar, A., Goel, V., 2006. Genetic alterations in signaling pathways in melanoma. Clin Cancer Res 12, 2301s–2307s.

Haqq, C., Nosrati, M., Sudilovsky, D., et al., 2005. The gene expression signatures of melanoma progression. Proc Natl Acad Sci USA 102, 6092–6097.

Hauschild, A., Agarwala, S.S., Trefzer, U., et al., 2009. Results of a phase III, randomized, placebo-controlled study of sorafenib in combination with carboplatin and paclitaxel as second-line treatment in patients with unresectable stage III or stage IV melanoma. J Clin Oncol 27, 2823–2830.

Hocker, T.L., Singh, M.K., Tsao, H., 2008. Melanoma genetics and therapeutic approaches in the 21st century: Moving from the benchside to the bedside. J Invest Dermatol 128, 2575–2595.

Hodi, F.S., Friedlander, P., Corless, C.L., et al., 2008. Major response to imatinib mesylate in KIT-mutated melanoma. J Clin Oncol 26, 2046–2051.

Hsu, M.Y., Meier, F.E., Nesbit, M., et al., 2000. E-cadherin expression in melanoma cells restores keratinocyte-mediated growth control and down-regulates expression of invasion-related adhesion receptors. Am J Pathol 156, 1515–1525.

Huber, M.A., Kraut, N., Beug, H., 2005. Molecular requirements for epithelial-mesenchymal transition during tumor progression. Curr Opin Cell Biol 17, 548–558.

Ibrahim, N., Haluska, F.G., 2009. Molecular pathogenesis of cutaneous melanocytic neoplasms. Annu Rev Pathol 4, 551–579.

Jaeger, J., Koczan, D., Thiesen, H.J., et al., 2007. Gene expression signatures for tumor progression, tumor subtype, and tumor thickness in laser-microdissected melanoma tissues. Clin Cancer Res 13, 806–815.

Jemal, A., Siegel, R., Ward, E., Hao, Y., Xu, J., Thun, M.J., 2009. Cancer statistics, 2009. CA Cancer J Clin 59, 225–249.

Jiang, X., Zhou, J., Yuen, N.K., et al., 2008. Imatinib targeting of KIT-mutant oncoprotein in melanoma. Clin Cancer Res 14, 7726–7732.

Jonsson, G., Busch, C., Knappskog, S., et al., 2010. Gene expression profiling-based identification of molecular subtypes in stage IV melanomas with different clinical outcome. Clin Cancer Res 16, 3356–3367.

Keilholz, U., Eggermont, A.M., 2000. The role of interleukin-2 in the management of stage IV melanoma: The EORTC melanoma cooperative group program. Cancer J Sci Am 6 (Suppl. 1), S99–S103.

Kirkwood, J.M., Manola, J., Ibrahim, J., Sondak, V., Ernstoff, M.S., Rao, U., 2004. A pooled analysis of eastern cooperative oncology group and intergroup trials of adjuvant high-dose interferon for melanoma. Clin Cancer Res 10, 1670–1677.

Lee, J.K., Havaleshko, D.M., Cho, H., et al., 2007. A strategy for predicting the chemosensitivity of human cancers and its application to drug discovery. Proc Natl Acad Sci USA 104, 13,086–13,091.

Leiter, U., Meier, F., Schittek, B., Garbe, C., 2004. The natural course of cutaneous melanoma. J Surg Oncol 86, 172–178.

Levy, C., Khaled, M., Fisher, D.E., 2006. MITF: Master regulator of melanocyte development and melanoma oncogene. Trends Mol Med 12, 406–414.

Lin, W.M., Baker, A.C., Beroukhim, R., et al., 2008. Modeling genomic diversity and tumor dependency in malignant melanoma. Cancer Res 68, 664–673.

Loercher, A.E., Tank, E.M., Delston, R.B., Harbour, J.W., 2005. MITF links differentiation with cell cycle arrest in melanocytes by transcriptional activation of INK4A. J Cell Biol 168, 35–40.

Maldonado, J.L., Fridlyand, J., Patel, H., et al., 2003. Determinants of BRAF mutations in primary melanomas. J Natl Cancer Inst 95, 1878–1890.

Maniotis, A.J., Folberg, R., Hess, A., et al., 1999. Vascular channel formation by human melanoma cells in vivo and in vitro: Vasculogenic mimicry. Am J Pathol 155, 739–752.

Masson, K., Ronnstrand, L., 2009. Oncogenic signaling from the hematopoietic growth factor receptors c-Kit and Flt3. Cell Signal 21, 1717–1726.

Middleton, M.R., Grob, J.J., Aaronson, N., et al., 2000. Randomized phase III study of temozolomide versus dacarbazine in the treatment of patients with advanced metastatic malignant melanoma. J Clin Oncol 18, 158–166.

Miller, A.J., Du, J., Rowan, S., Hershey, C.L., Widlund, H.R., Fisher, D.E., 2004. Transcriptional regulation of the melanoma prognostic marker melastatin (TRPM1) by MITF in melanocytes and melanoma. Cancer Res 64, 509–516.

Montone, K.T., Van Belle, P., Elenitsas, R., Elder, D.E., 1997. Proto-oncogene c-kit expression in malignant melanoma: Protein loss with tumor progression. Mod Pathol 10, 939–944.

Mori, S., Rempel, R.E., Chang, J.T., et al., 2008. Utilization of pathway signatures to reveal distinct types of B lymphoma in the Emicromyc model and human diffuse large B-cell lymphoma. Cancer Res 68, 8525–8534.

Nelson, A.A., Tsao, H., 2009. Melanoma and genetics. Clin Dermatol 27, 46–52.

Oancea, E., Vriens, J., Brauchi, S., Jun, J., Splawski, I., Clapham, D.E., 2009. TRPM1 forms ion channels associated with melanin content in melanocytes. Sci Signal 2 (ra21).

Olsen, C.M., Carroll, H.J., Whiteman, D.C., 2010. Familial melanoma: A meta-analysis and estimates of attributable fraction. Cancer Epidemiol Biomarkers Prev 19, 65–73.

Palmieri, G., Capone, M., Ascierto, M.L., et al., 2009. Main roads to melanoma. J Transl Med 7, 86.

Paraiso, K.H., Fedorenko, I.V., Cantini, L.P., et al., 2010. Recovery of phospho-ERK activity allows melanoma cells to escape from BRAF inhibitor therapy. Br J Cancer 102, 1724–1730.

Parmiter, A.H., Balaban, G., Clark Jr., W.H., Nowell, P.C., 1988. Possible involvement of the chromosome region 10q24-q26 in early stages of melanocytic neoplasia. Cancer Genet Cytogenet 30, 313–317.

Patton, E.E., Widlund, H.R., Kutok, J.L., et al., 2005. BRAF mutations are sufficient to promote nevi formation and cooperate with p53 in the genesis of melanoma. Curr Biol 15, 249–254.

Pectasides, D., Dafni, U., Bafaloukos, D., et al., 2009. Randomized phase III study of 1 month versus 1 year of adjuvant high-dose interferon alfa-2b in patients with resected high-risk melanoma. J Clin Oncol 27, 939–944.

Pollock, P.M., Harper, U.L., Hansen, K.S., et al., 2003. High frequency of BRAF mutations in nevi. Nat Genet 33, 19–20.

Qi, J., Chen, N., Wang, J., Siu, C.H., 2005. Transendothelial migration of melanoma cells involves N-cadherin-mediated adhesion and activation of the beta-catenin signaling pathway. Mol Biol Cell 16, 4386–4397.

Qi, J., Wang, J., Romanyuk, O., Siu, C.H., 2006. Involvement of Src family kinases in N-cadherin phosphorylation and beta-catenin dissociation during transendothelial migration of melanoma cells. Mol Biol Cell 17, 1261–1272.

Rose, A.E., Poliseno, L., Wang, J., et al., 2011. Integrative genomics identifies molecular alterations that challenge the linear model of melonoma progression. Cancer Res 71 (7), 2561–2571.

Ross, M.I., 2006. Early-stage melanoma: Staging criteria and prognostic modeling. Clin Cancer Res 12, 2312s–2319s.

Rubinstein, J.C., Sznol, M., Pavlick, A.C., et al., 2010. Incidence of the V600K mutation among melanoma patients with BRAF mutations, and potential therapeutic response to the specific BRAF inhibitor PLX4032. J Transl Med 8, 67.

Scatolini, M., Grand, M.M., Grosso, E., et al., 2010. Altered molecular pathways in melanocytic lesions. Int J Cancer 126, 1869–1881.

Sekulic, A., Haluska Jr., P., Miller, A.J., et al., 2008. Malignant melanoma in the 21st century: The emerging molecular landscape. Mayo Clin Proc 83, 825–846.

Shepherd, C., Puzanov, I., Sosman, J.A., 2010. B-RAF inhibitors: An evolving role in the therapy of malignant melanoma. Curr Oncol Rep 12, 146–152.

Smith, A.P., Hoek, K., Becker, D., 2005. Whole-genome expression profiling of the melanoma progression pathway reveals marked molecular differences between nevi/melanoma in situ and advanced-stage melanomas. Cancer Biol Ther 4, 1018–1029.

Sosman, J.A., Kim, K.B., Schuchter, L., et al., 2012. Survival in BRAF V600-Mutant advanced melonoma treated with vemurafenib. N Engl J Med 366 (8), 707–714.

Soufir, N., Avril, M.F., Chompret, A., et al., 1998. Prevalence of p16 and CDK4 germline mutations in 48 melanoma-prone families in France. The French Familial Melanoma Study Group. Hum Mol Genet 7, 209–216.

Staunton, J.E., Slonim, D.K., Coller, H.A., et al., 2001. Chemosensitivity prediction by transcriptional profiling. Proc Natl Acad Sci USA 98, 10,787–10,792.

Steingrimsson, E., Copeland, N.G., Jenkins, N.A., 2004. Melanocytes and the microphthalmia transcription factor network. Annu Rev Genet 38, 365–411.

Tarhini, A.A., Kirkwood, J.M., 2009. Clinical and immunologic basis of interferon therapy in melanoma. Ann NY Acad Sci 1182, 47–57.

Tartour, E., Blay, J.Y., Dorval, T., et al., 1996. Predictors of clinical response to interleukin-2-based immunotherapy in melanoma patients: A French multi-institutional study. J Clin Oncol 14, 1697–1703.

Tsao, H., Mihm Jr., M.C., Sheehan, C., 2003. PTEN expression in normal skin, acquired melanocytic nevi, and cutaneous melanoma. J Am Acad Dermatol 49, 865–872.

Ugurel, S., Houben, R., Schrama, D., et al., 2007. Microphthalmia-associated transcription factor gene amplification in metastatic melanoma is a prognostic marker for patient survival, but not a predictive marker for chemosensitivity and chemotherapy response. Clin Cancer Res 13, 6344–6350.

Walia, V., Mu, E.W., Lin, J.C., Samuels, Y., 2012. Delving into somatic variation in sporadic melanoma. Pigment Cell Melanoma Res 25 (2), 155–170.

Wellbrock, C., Marais, R., 2005. Elevated expression of MITF counteracts B-RAF-stimulated melanocyte and melanoma cell proliferation. J Cell Biol 170, 703–708.

Williams, P.D., Cheon, S., Havaleshko, D.M., et al., 2009. Concordant gene expression signatures predict clinical outcomes of cancer patients undergoing systemic therapy. Cancer Res 69, 8302–8309.

Winnepenninckx, V., Lazar, V., Michiels, S., et al., 2006. Gene expression profiling of primary cutaneous melanoma and clinical outcome. J Natl Cancer Inst 98, 472–482.

RECOMMENDED RESOURCES

http://www.stat.duke.edu/research/software/west/bFRm/
http://www.ncbi.nlm.nih.gov/geo/

Metastatic Cancer

Jude Alsarraj and Kent W. Hunter

INTRODUCTION

Metastasis is an extremely complex process that remains a major problem in the clinical management of cancer, because most cancer deaths are attributed to a disseminated disease rather than the primary tumor. It has been observed that patients who have no evidence of tumor dissemination at diagnosis of the primary tumor are at risk for metastatic disease after years or even decades from diagnosis. Approximately one-third of women who are sentinel lymph node negative at the time of surgical resection of the primary breast tumor will develop clinically detectable secondary lesions (Heimann et al., 2000). Therefore, the ability to control the spread, or effectively treat patients with or at risk of metastatic disease would significantly improve the overall outcome of the disease.

For metastasis to occur, tumor cells must go through a series of steps to successfully form clinically detectable lesions at a distant site (Figure 67.1), and failure to complete any of these steps results in a complete loss of the metastatic behavior of cancer cells. These steps include detachment from the primary tumor and invasion of the surrounding tissues and the basement membranes, where tumor cells will invade small blood vessels, lymphatic channels or body cavities and arrest in a distant target organ either by receptor-mediated adhesion or physical trapping. The cells will then adapt to the foreign microenvironment either to initiate growth within the blood or lymphatic vessel or to extravasate into the surrounding tissues, followed by proliferation and induction of angiogenesis. Invasion through the basement membrane is the hallmark

that defines malignancy, but not all neoplasms are invasive (e.g., ductal carcinoma *in situ* of the breast and prostatic intra-epithelial neoplasia), although they can progress towards malignancy. Similarly, the ability to metastasize is not an inherent property of all tumor cells. Some tumors are highly aggressive, forming secondary lesions with high frequency (e.g., small cell carcinoma of the lung, melanoma, and pancreatic carcinoma), while others rarely metastasize to distant sites despite being locally invasive (e.g., basal cell carcinomas of the skin and glioblastoma multiforme) (reviewed in Eccles and Welch, 2007). Some tumor cells might acquire the ability to metastasize by conversion from polarized epithelial cells to motile mesenchymal cells, a process called *epithelial-to-mesenchymal transition* (EMT) (Thompson et al., 2005; Xu et al., 2006). Cells that undergo EMT not only exhibit enhanced motility, but also are resistant to apoptosis. Those are key requirements for successful metastasis (Robson et al., 2006).

In addition to cell-intrinsic factors affecting metastasis, recent results have demonstrated an important role of non-tumor cells in tumor progression (Guy et al., 1992; Lin et al., 2001; Wyckoff et al., 2007). Millions of tumor cells can be shed daily into the bloodstream or the lymphatic system, but very few clinically relevant metastases are formed in most individuals. This could be explained by the inefficiency of the metastatic process. Most studies show that most of the cells entering the vasculature fail to form macroscopic foci at distant sites, despite the fact that tumor cells in the blood stream arrest in the capillary beds, extravasate with high efficiency, and reside dormant in the secondary sites for long periods of time – sometimes for years (Luzzi et al., 1998; Riethmuller and

Genomic and Personalized Medicine, 2nd edition
by Ginsburg & Willard. DOI: http://dx.doi.org/10.1016/B978-0-12-382227-7.00067-7

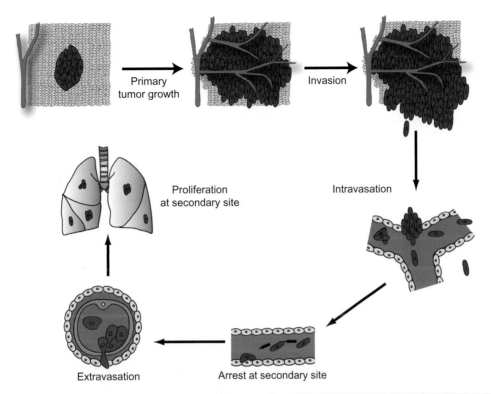

Figure 67.1 The metastatic process. Metastasis is a complex process that enables a neoplastic cell to leave a primary tumor, disseminate from its origin either by lymphatic or blood vessels, and form a viable lesion in a distant organ. The initial steps of metastasis require growth of the primary tumor, which causes the tumor to produce angiogenic factors, which facilitate growth of new blood vessels, or vascularization. The tumor will then invade through adjacent tissues and basement membranes. This process continues until the tumor invades blood vessels, lymphatic channels, or body cavities, when individual tumor cells, either in aggregates or in isolation, detach from the primary tumor mass, enter the bloodstream, and are carried to a distant target organ. Subsequently, tumor cells arrest in small vessels within the distant organ, extravasate into the surrounding tissue, and proliferate at the secondary site.

Klein, 2001). The lack of correlation between clinical metastases and the capability of large numbers of tumor cells to enter and exit the vasculature suggests that metastatic capacity is not solely dependent upon the properties of the tumor cell itself. How tumor cells interact with host stroma (see e.g., Kaplan et al., 2005), and differential functionality of both the tumor cell and stroma as a consequence of hereditary variation (ex. Park et al., 2005), also appear to play important roles in metastasis formation.

Application of new technologies over the past few years has resulted in significant advances in our understanding of metastatic disease, although much is yet to be learned. While space constraints do not permit a comprehensive presentation of all of the novel advances, some of the recent highlights of genomic investigation into metastasis are presented here.

THE GENETICS OF METASTASIS
Mouse Models of Cancer

Mouse models have provided insights into the biological functions of genes and signaling pathways involved in cancer, and

have allowed the generation of advanced concepts of metastasis. Although the mouse model systems do not perfectly match the human disease in genetic background (Radiloff et al., 2008; Rivera and Tessarollo, 2008; Thyagarajan et al., 2003), tumor growth and metastasis in mouse models still replicate many aspects of human cancer progression. Therefore, to some extent, these mouse models allow us to understand the relationship between tumorigenesis or metastasis and the effect of a single gene in each step of the metastatic process, thereby facilitating understanding of how the disease develops, progresses, and how it can be prevented and treated.

Mouse models of human cancer have proven particularly useful in defining genomic regions harboring *quantitative trait loci* (QTLs), or susceptibility loci that encompass polymorphic gene(s) that modify the risk of cancer. QTLs are defined by correlating a measurable trait (e.g., tumor size, metastasis frequency) with allelic variation in linked polymorphic genetic markers (e.g., microsatellite alleles) in a given population. This type of study frequently utilizes animals generated by back- or intercrosses, since these models have been shown to be a robust means of generating low-resolution localization of genes that modulate phenotype (Hunter and Williams, 2002).

Following identification of a QTL, the aim is then to determine the identity of the gene(s) within the QTL that are responsible for the observed linkage. This task is frequently laborious, primarily because it necessitates the use of positional cloning within a locus that is often many millions of base pairs in length. Given that genomic regions of this size typically contain many thousands of genes, the reader can appreciate the magnitude of the task facing researchers, although the difficulty of such approaches has been partially alleviated by the publication of the complete genome sequence. It is therefore likely that in the future many more cancer susceptibility genes will be identified in this manner.

A number of transgenic mice have been found to metastasize. Since these animals develop metastatic disease in a heritable and highly penetrant manner, they offer the potential to utilize the power of mouse genetics to identify and characterize the modifier/suppressor loci known to be present in the mouse genome. Transgenic mice expressing the polyoma virus middle T antigen (PyMT) under the transcriptional control of the mouse mammary tumor virus (MMTV) promoter initiate tumors in the mammary glands, resulting in the appearance of multifocal mammary adenocarcinomas involving all mammary glands, with rapid onset (Guy et al., 1992). Furthermore, 85–95% of these mice develop pulmonary metastases by 100 days of age. This mouse model has been used by a variety of investigators, because in addition to its rapid induction it shares many characteristics with human breast tumors, such as gradual loss of estrogen and progesterone receptors and the multistage progression from hyperplasia to full malignancy. Also, metastatic potential appears to be independent of hormonal fluctuations, with a reproducible progression rate (Fluck and Haslam, 1996; Guy et al., 1992).

The highly metastatic MMTV/PyMT transgenic mice have therefore been a valuable tool in the study of the genetics of metastasis. To investigate the role of germline variations in metastasis development, PyMT mice were crossed with different inbred strains and the metastatic capacity of the progeny was determined (Lifsted et al., 1998). Some of the F$_1$ progeny animals developed ~10-fold fewer pulmonary metastases compared to the parental PyMT strain, while others displayed a ~2–3-fold increase in pulmonary lesions. The fact that all of these tumors were induced by the same event – the activation of the transgene – suggests that inherited polymorphisms influence the process of metastasis in addition to the somatic events that occur during tumor evolution.

To further study and characterize this observation, quantitative trait genetic mapping was performed. A probable metastasis efficiency locus called *Mtes1* was mapped on mouse chromosome 19 (Hunter et al., 2001). A subsequent study showed that further loci are present on chromosomes 7, 9, and 17 (Lancaster et al., 2005). A *multiple cross mapping* (MCM) *strategy* was then used to identify potential candidate genes within the *Mtes1* interval. MCM strategy is a technique that exploits shared haplotypes among different inbred strains of mice used in genetic mapping studies to reduce the number of potential candidate genes in a given candidate interval (Hitzemann et al., 2002; Wiltshire et al., 2003). This strategy narrowed the list of identified genes to one gene, "signal-induced proliferation-associated gene 1" (*Sipa1*, also known as *Spa1*) (Park et al., 2005). *Sipa1* is a mitogen-inducible gene that encodes a GTPase-activating protein (GAP) specific for the Ras-related proteins Rap1 and Rap2 GTPases, and negatively regulates Rap through its Rap-GAP activity, which catalyzes the hydrolysis of active GTP-Rap to inactive GDP-Rap. Further analysis of *Sipa1* identified an amino acid polymorphism in the PDZ-protein interaction domain, which reduced the Rap-GAP activity of the molecule. Furthermore, polymorphisms within the human *Sipa1* gene were found to be associated with markers of poor outcome and with breast cancer incidence in Caucasian population-based breast cancer patient cohorts (Crawford et al., 2006; Hsieh et al., 2009). *Sipa1* expression has also been linked to metastasis in human cervical cancers (Brooks et al., 2010). These findings suggest that *Sipa1*, and by extension other polymorphic genes, may play an important role in establishing metastatic susceptibility in humans as well as in mice.

THE GENOMICS OF METASTASIS
Microarray Gene Expression Analysis

In recent years, microarray analysis of gene expression in both primary and secondary tumors has come to play a pivotal role in research into the mechanistic aspects of metastatic progression. However, microarray analysis appears to be incompatible with the conventional theory of metastasis, the *somatic progression model* originally proposed by Nowell (1976) (Figure 67.2). That model suggests that the ability to metastasize is an acquired characteristic of tumor cells, generated by the sequential accumulation of somatic alterations as a tumor evolves. Some of these changes occur in the genome,

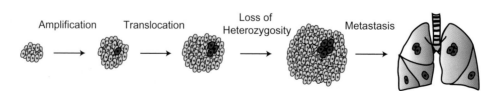

Amplification　　Translocation　　Loss of Heterozygosity　　Metastasis

Figure 67.2 The progression model of metastasis. In the progression model, a series of random mutational events occurs within a tumor, resulting in a small subpopulation of the tumor cells acquiring full metastatic potential.

which leads to the silencing of metastasis suppressors and the up-regulation of metastasis promoters (Fidler, 1970; Nowell, 1976). *Metastasis suppressors* are genes that have the ability to suppress the development of metastasis without affecting the primary tumor, which distinguishes them from tumor suppressors. Metastasis suppressor genes are not always mutated in cancer cells, however, down-regulation of the expression of these genes occurs frequently due to epigenetic changes such as gene methylation, or through post-transcriptional/translational modifications (reviewed in Kauffman et al., 2003). The existence of these metastasis suppressor genes makes a strong argument for the occurrence of mutational alterations during the metastatic process.

Two studies involving microarray gene expression signatures of primary tumors and their matched metastases, however, showed that the metastatic potential of human tumors is encoded in the bulk of the primary tumor (Ramaswamy et al., 2003; van't Veer et al., 2002), challenging the conventional progression model. In those studies it was suggested that the ability to metastasize depends on the genetic state of the primary tumor rather than the emergence of rare cells with the metastatic phenotype, and that the metastatic capacity is likely to be encoded early in tumorigenesis by the particular collections of oncogenic events that initiate the tumor. Therefore, a pro-metastatic gene signature expressed within a small subpopulation of cells within a primary tumor (as would be predicted by the progression hypothesis) would be masked by the larger bulk of the tumor. This observation was also supported by the finding that the tendency to metastasize is largely determined by the mutant alleles acquired by tumor cells relatively early during tumorigenesis (Bernards and Weinberg, 2002).

These observations could be explained by tumor heterogeneity. It is possible that the metastasis signature is represented by the bulk population, but not by every cell within that population. Also, some subpopulations within the tumor could express one of the metastasis signature genes, whereas others could express larger proportions (or all) of the signature genes. Since metastases are usually clonal, successful cells must individually express all required properties, or be able to depend upon host support for those they lack (reviewed in Eccles and Welch, 2007). This possibility is supported by the observation that several of the genes identified in metastasis signatures are stromal in origin (Allinen et al., 2004; Kurose et al., 2001), suggesting the contribution of the host in metastasis (Kopfstein and Christofori, 2006). Gene expression profiles from mouse models further sustained the contribution of the host in the metastatic process. Gene expression profiles of tumors of high- or low-metastatic genotype, and also from normal tissues from these two metastatic genotypes, identified a number of differentially regulated genes associated with metastasis, including genes encoding the stromally derived basement membrane components (Lukes et al., 2009). The gene expression difference from both high- versus low-metastatic genotype mouse tumors and normal tissues was found to predict human breast cancer outcome. These data

suggest that the metastasis-predictive gene signatures are likely due to a combination of preexisting signatures established by inherited factors present in all tissues (Figure 67.3), as well as somatic mutations within the tumor epithelium, and also suggest that host genetics and epigenetic effects are important components of metastatic progression.

Copy Number Analysis

As discussed above, it was suggested that metastasis-predictive gene expression signatures are likely due to a combination of preexisting signatures established by inherited factors present in all tissues, as well as somatic mutations within the tumor. One of the technologies widely employed to access somatic mutations within cancer genomes is *comparative genomic hybridization* (CGH). CGH technology produces a map of DNA sequence copy number as a function of chromosomal location throughout the entire genome (Kallioniemi et al., 1992). Differentially labeled test DNA and normal reference DNA are hybridized simultaneously to normal chromosome spreads. The hybridization is detected with two different fluorescent dyes. Regions of gain or loss of DNA sequences such as deletions, duplications, or amplifications are detected as changes in the ratio of the intensities of the two fluorescent dyes along the target chromosomes. The principal disadvantage of CGH is that its resolution is limited and only has the ability to detect chromosomal aberrations at intervals of ~10–20 Mb (Albertson, 2003), which is primarily a reflection of its dependence on the use of metaphase chromosomes to map aberrations. This approach is gradually being superseded by microarray-based formats, or *array CGH* (aCGH), which has a number of advantages over the use of chromosomes. Array CGH experimentation uses microarrays that possess representations of the genome spotted on the array surface. The spots are typically one of a number of commonly utilized formats, including bacterial artificial chromosomes, complementary DNA (cDNA) clones, and oligonucleotides (Figure 67.4). In many aspects, the experimental procedures for CGH and aCGH are very similar and typically involve differential labeling of test and reference genomes followed by hybridization and visualization.

Array CGH has been widely used to compare the genomes of primary tumors and their metastases. For example, the genomes of primary breast invasive duct carcinoma, its sentinel and more distal lymph node metastases, and also primary breast invasive duct carcinoma without nodal metastasis, were compared using this technique (Wang et al., 2009). There was a high degree of similarity between primary tumors and their nodal metastases, as well as between metastases to the sentinel and distal lymph nodes. Examples of tumor types where novel chromosome aberrations have been successfully characterized in metastatic lesions using aCGH include metastatic colorectal (Buffart et al., 2005; Tanami et al., 2005), endocervical (Hirai et al., 2004), and nasopharyngeal cancers (Yan et al., 2005).

Genome-wide single nucleotide polymorphism (SNP) *analysis* is another method used to study alterations in the

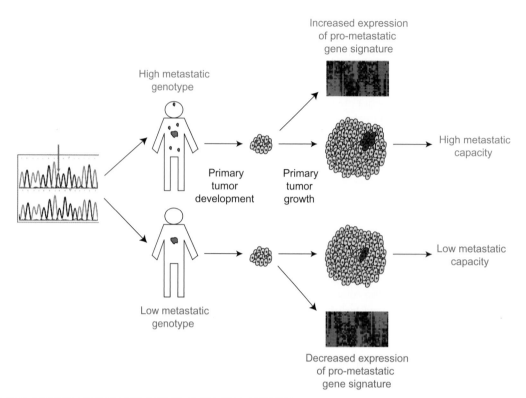

Figure 67.3 Inherited germline modification of metastasis. A growing body of evidence implicates genetic background as a factor involved in modulation of metastatic propensity within a primary tumor. That is, an individual will be more or less susceptible to tumor dissemination as a consequence of constitutional polymorphisms, and thus it may be possible to classify individuals as possessing either a "high" or "low" metastatic genotype. Such germline variations influence all aspects of the metastatic cascade, including the expression of pro-metastatic gene expression signatures within the primary tumor.

number of copies of genomic DNA during metastatic cancer. It is a high-throughput, high-resolution, oligonucleotide-based SNP array technology to analyze the copy number alteration for >100,000 SNP loci in different cancer genomes. For example, in lymph node negative breast cancer patients, copy number alterations that correlated with the time in developing distant metastases were identified. First, chromosome regions with prognostic copy number alterations were identified. An 81-gene copy number alteration signature was then generated by correlating these copy number data with already-published gene expression signatures (Zhang et al., 2009). This study suggested that combining DNA copy number analysis and gene expression analysis provides an additional and better means of risk assessment in breast cancer patients. Genome-wide SNP analysis was also used to compare copy number alteration across the genome of a poorly metastatic breast cancer cell line, and a cell line variant that is highly tumorigenic and also maintains the propensity to metastasize in a mouse xenograft model (Andrews et al., 2010). A number of large regions of chromosome copy number difference were identified between the two cell lines. DNA methylation profiles were also generated and compared with the copy number difference between the two cell lines. Many of the relative losses in copy number correlated with an apparent gain

in hypermethylation, while increases in copy number tended to correlate with losses in DNA methylation. This approach enhances opportunities, particularly in the context of patient tumor material, to identify therapeutic targets for breast cancer treatment that are epigenetically regulated by alterations in DNA methylation.

Next-generation DNA Sequencing Technology

Next-generation DNA sequencing (NGS) technology is a new tool that is used to characterize all somatic coding mutations, including point mutations, rearrangements, and copy number changes, that occur during the development of cancer. The technology facilitates the comparison of multiple samples taken from the same patient from the primary tumor at diagnosis and later from metastatic lesions, to address the genetic basis of tumor progression and metastasis. NGS technology has three major applications:

1. *Full-genome sequencing,* or more targeted discovery of mutations or polymorphisms, as well as copy number variations and rearrangements
2. *RNA-Seq* or *high-throughput transcriptome sequencing,* which is analogous to expressed sequence tags or serial analysis of gene expression, where shotgun libraries

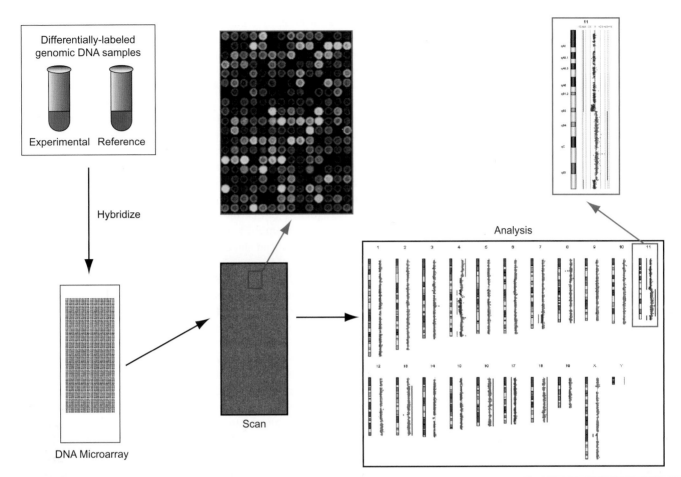

Figure 67.4 Array comparative genomic hybridization (aCGH) is a powerful means to assess differences in genomic copy number between samples. The format of microarrays varies, but all involve differential labeling of two genomic DNA samples (e.g., genomic DNAs derived from a primary tumor and a metastatic lesion). Following labeling, the two samples are combined, hybridized to a DNA microarray, and the fluorescent intensity of each probe feature on the microarray is determined by scanning. Probes are printed on the microarrays in a manner that allows scanned feature intensity to be correlated with probe genomic location. Subsequent analysis of the relative fluorescence of the reference and experimental dye intensities for each microarray feature allows an estimation of probe copy number differences between the two samples. Software packages allow a genome-wide assessment of probe copy number, which in turn can be used to assess gross chromosomal aberrations. For example, the output shown here (CGH Analytics©, Agilent Technologies, Palo Alto, CA) shows a number of statistically significant chromosomal abnormalities between the two mouse experimental and reference genomic DNA samples, the location of which are denoted by solid blue lines adjacent to the relevant chromosomes.

derived from mRNA or small RNAs are deeply sequenced; the counts corresponding to individual species can be used for quantification over a broad dynamic range, and the sequences themselves can be used for annotation

3. *ChIP-Seq* or *genome-wide mapping of DNA–protein interactions* by deep sequencing of DNA fragments pulled down by chromatin immunoprecipitation (reviewed in Shendure and Ji, 2008).

NGS offers many advantages over microarray-based assays, in that it has relatively fewer artifacts such as noise in the form of cross-hybridization generated by the hybridization step on microarrays. Furthermore, genome coverage is not limited in NGS technology by the repertoire of probe sequences fixed on the array. This is particularly important for the analysis of repetitive regions of the genome, which are typically masked out on microarrays (reviewed in Huang et al., 2010). The RNA-Seq approach avoids the need for bacterial cloning of the cDNA input, and the resulting sequence reads are individually mapped to the source genome and counted to obtain the number and density of reads corresponding to RNA from each known exon, splice event, or new candidate gene (Mortazavi et al., 2008).

NGS has been used to investigate metastatic progression of breast cancer in several studies. For example, it was performed to identify somatic non-synonymous coding mutations in metastatic lobular breast cancer patients (Shah et al., 2009). Lobular breast cancer is an estrogen receptor positive subtype of breast cancer that is of low-intermediate histological grade and can recur many years after diagnosis. *Genome paired-end sequencing* and RNA-Seq approaches were performed in the study. In genome paired-end sequencing, short reads are generated from both ends to millions of DNA fragments, to systematically detect and characterize somatic mutations (Campbell et al., 2008). In the Shah study, application of the two techniques led to the observation that there was significant heterogeneity in tumor somatic mutation content in the primary tumors and metastases. A total of 32 somatic non-synonymous mutations were present in the metastases, and the frequency of those somatic mutations in DNA from primary tumors from the same patients (which arose nine years earlier) was measured. Of the 32 mutations, five were prevalent in the DNA of the primary tumors, six were present at lower frequencies, 19 were not detected in the primary tumor, and two were not determined. The authors concluded that single nucleotide mutational heterogeneity can be a property of low- or intermediate-grade primary breast cancers, and that significant evolution can occur with disease progression, which is consistent with the somatic evolution model of breast cancer.

In another study of a case of basal-like breast cancer, a paired-end sequencing strategy was applied to determine the difference between the primary tumor and the resulting metastases. Basal-like breast cancer is characterized by the absence of estrogen receptor expression, the lack of *ERBB2* gene amplification, and a high mitotic index (Ding et al., 2010). The authors compared metastasis and xenograft derived from the primary tumor with the primary tumors, and found that there were differential mutational frequencies and structural pattern in metastasis when compared to primary tumors, with the metastasis having higher allele frequencies. Two *de novo* mutations were discovered in the metastatic tumor but were not detected in the primary or xenograft tumor genomes. The metastasis and the xenograft tumor genomes also exhibited elevated copy number alterations compared to the primary tumor genome. Taken together, these observations suggest that metastasis may arise from rare cells within the primary tumor, consistent with the conventional progression model proposed by Nowell (1976). However, the authors demonstrate that although additional somatic mutations, copy number alterations, and structural variations do occur during the clinical course of the disease, most of the original mutations and structural variants present in the primary tumor are propagated.

Although the NGS technology is still evolving, this technology is likely to enable us to identify genetic variations within individual human subjects that help explain the genetic predisposition to cancer. The NGS technology could also be very useful in the clinical development of genomic diagnostics. Sequencing-based cancer assays such as expression profiles and tumor mutations, for example, can be developed, which will help in tailoring therapy to the genetic background of an individual. The integration of the data generated from this technology in the clinical setting will lead to a more refined diagnosis and selection of effective treatment.

MicroRNA (miRNA) Expression Analysis

miRNAs are a class of 19–24 nucleotide-long, non-coding RNAs that act by silencing the expression of their target genes, leading to RNA degradation or translational inhibition (Bartel, 2004). In recent years, growing evidence has suggested that miRNAs play a crucial role in cancer progression, invasion, and metastasis. For example, it has been shown that miR-10b is highly expressed in breast cancer cell lines, and positively regulates tumor invasion and metastasis (Ma et al., 2007). Moreover, miR-21 overexpression was shown to be associated with advanced clinical stage, lymph node metastasis, and poor prognosis in human breast cancer (Yan et al., 2008).

Profiling of miRNAs in cancer has demonstrated that their expression varies between malignant and normal tissues. Several miRNAs are often up-regulated in specific cancers and have been shown to possess oncogenic properties (He et al., 2005; Si et al., 2007; Volinia et al., 2006). However, the majority of miRNAs are down-regulated as tumors become less differentiated, consistent with a role for miRNAs in determining cellular proliferation (Lu et al., 2005). The miRNA signature profile of metastasizing cancer has been investigated for different types of cancers, such as hepatocellular carcinoma, uveal melanoma, and other solid tumors such as breast, lung, colon, and bladder (Baffa et al., 2009; Budhu et al., 2008; Worley et al., 2008). When primary solid tumors and their matched lymph node metastases were compared by miRNA gene signature analysis, a number of miRNAs were found to be differentially expressed in primary versus metastatic cancer (Baffa et al., 2009). This suggests that specific miRNAs may be directly involved in cancer metastasis and may represent a novel diagnostic tool for the characterization of metastatic cancer gene targets. Moreover, in the same study the authors found that there was differential expression of several miRNAs between the primary tumors and the matched metastases specific for tissue type of origin. This is a very interesting observation, because miRNA signature could be then used to help identify the organ of origin in patients of unknown primary cancer metastatic disease (Rosenfeld et al., 2008). These patients, who constitute approximately 5% of solid tumor-related cases, typically present with disseminated disease but have no clinically detectable primary tumor or only a small, well-differentiated lesion that is found at autopsy (Riethmuller and Klein, 2001).

miRNA expression patterns were also found to be associated with prognosis and survival of colon adenocarcinoma. It was shown that the expression patterns of miRNAs are systematically altered compared to normal colon mucosa as colon adenomas progress to adenocarcinomas (Schetter et al., 2008). Overexpression of miR-21 was shown to be associated with poor survival and poor response to adjuvant chemotherapy in colon adenocarcinomas. High miR-21 expression in tumors from patients receiving chemotherapy predicted a shorter overall survival, which gives preliminary support to the hypothesis that high miR-21 expression is associated with poor therapeutic outcome. Other studies have found that specific miRNA profiles are associated with poor prognosis in other cancers, such as ovarian cancer (Nam et al., 2008), esophageal adenocarcinoma (Mathe et al., 2009), and melanoma (Satzger et al., 2010). Taken together, these observations suggest that miRNA profiles could be of potential importance as diagnostic and prognostic tools as well as predictors of response to therapy. A list of several recently reported miRNAs that have pro- and anti-metastatic effects in regulating critical steps of metastasis is shown in Table 67.1.

The combination of cancer miRNA profiles with gene expression profiles is another potentially useful application that has not yet been fully explored. It has been demonstrated

that 35% of translationally repressed miRNA target genes have reduced levels of mRNA (Baek et al., 2008). One of the studies that combined miRNA and mRNA datasets was performed on a set of microsatellite stable and unstable colorectal cancer (Lanza et al., 2007). About 15% of colorectal carcinomas develop through the microsatellite instability pathway, where the tumors show stable karyotype, low frequencies of allelic losses, and diploid nuclear DNA content. The authors identified a molecular signature that could distinguish between these two types of colorectal cancer patients. The authors also concluded that the combination of mRNA/miRNA expression signatures may represent a general approach for improving the tumor subtype classification of human cancers.

It was recently suggested that miRNAs could be classified as cancer predisposing genes, as miRNAs may be targets of germline or somatic mutations linked to cancer development and progression. SNPs found in primary, precursor, or mature miRNAs may alter their maturation, expression, and binding to their target mRNA, and therefore may contribute to cancer prognosis and survival. In breast cancer patients, for example, SNPs in predicted miRNA-binding sites in the integrin genes were shown to be associated with aggressive tumor characteristics as well as poor survival (Brendle et al., 2008). Integrins control cell attachment to the extracellular matrix and play an important role in mediating cell proliferation, migration, and survival. In non-small cell lung cancer (NSCLC) patients, it was demonstrated that polymorphisms in a specific pre-miRNA were associated with survival (Hu et al., 2008). These data suggest that SNPs in specific miRNAs could be prognostic markers for different human cancers, and further characterization of miRNA SNPs may open new avenues for therapeutic intervention.

TABLE 67.1	Pro-metastatic and anti-metastatic miRNAs regulating tumor metastasis	
miRNA	**Influence on metastasis**	**Step(s) of the invasion-metastasis cascade affected**
miR-10b	Pro-metastatic	EMT, migration, and invasion
miR-21	Pro-metastatic	Migration and invasion
miR-27a/b	Pro-metastatic	Angiogenesis
miR-31	Anti-metastatic	Migration and invasion, colonization
miR-34b/c	Anti-metastatic	Migration and invasion
miR-143	Pro-metastatic	Migration and invasion
miR-146a/b	Anti-metastatic	Migration and invasion
miR-148a	Anti-metastatic	Migration and invasion
miR-182	Pro-metastatic	Migration and invasion
miR-200a/b/c	Anti-metastatic	EMT
miR-206	Anti-metastatic	Migration and invasion
miR-335	Anti-metastatic	Migration and invasion
miR-373	Pro-metastatic	Migration and invasion
miR-520c	Pro-metastatic	Migration and invasion

NEW TREATMENT POSSIBILITIES FOR METASTATIC CANCER

Recent advances aside, cancer relapse and death rates remain unacceptably high. There is a clear need for clinicians to have access to investigative technologies that will allow both the degree of treatment required and overall prognosis to be determined with improved accuracy. Despite the existence of microarray technology and the recent emergence of NGS technology, multiple challenges remain for the achievement of personalized medicine in cancer and metastatic disease. NGS technology will help to complete the catalog of driver somatic mutations, which will open the way for new molecular innovations, both predictive markers and therapies, progressing toward personalized treatment of cancer. Germline variation in the form of SNPs could also be a tool to determine the likelihood of metastatic disease development. It should be noted, however, that it is exceedingly unlikely that any single SNP or combination of SNPs within an individual gene will possess sufficient prognostic power to be used as a stand-alone predictive assay. More likely, a clinically viable prognostic assay will require the development of a panel of SNPs within a collection of genes, with each gene represented shown independently to be a modifier of metastatic efficiency.

Accumulating evidence in a wide range of malignancies suggests that a subpopulation of tumor cells with distinct stem cell properties is responsible for tumor initiation. In human breast cancer, for example, a subpopulation of cells with dramatically enhanced tumorigenic capacity was isolated when transplanted into immunodeficient mice (Al-Hajj et al., 2003). Two groups succeeded in repopulating an entire mouse mammary gland from a single cell, identifying the putative stem cell with a series of surface markers (Shackleton et al., 2006; Stingl et al., 2006). These so called *cancer stem cells* (CSC), or tumor-initiating cells, are able to self-renew and generate diverse cells that comprise the tumor (Bonnet and Dick, 1997; Clarke et al., 2006; Reya et al., 2001). The important question here is whether metastases can directly arise from CSCs. Recent reports have supported the concept of metastatic CSCs in cancers such as breast, pancreas, and liver (Hermann et al., 2007; Liu et al., 2007; Shipitsin et al., 2007; Yang et al., 2008). These observations reflect the generation of a distinct metastatic CSC within the tumor, or the evolution of a second CSC with a different immunophenotype from the first CSC (reviewed in Visvader and Lindeman, 2008).

CSCs share a number of properties with normal stem cells, such as relative quiescence, an active DNA repair capacity, resistance to apoptosis, and resistance to drugs and toxins through the expression of several ATP-binding cassette (ABC) transporters (reviewed in Dean et al., 2005). These intrinsic similarities between normal stem cells and CSCs make the development of targeted therapy very challenging. In hematopoietic malignancies, different agents that selectively target CSCs but not normal stem cells were used in mice (Guzman et al., 2005; Jin et al., 2006; Krause et al., 2006; Yilmaz et al., 2006). These agents need to be studied to carefully establish whether they will attack the normal stem cells in cancer patients. From a clinical prospective, all CSCs need to be eradicated to achieve long-term, disease-free survival. Quiescent CSCs, however, are thought to be more resistant to chemotherapy and targeted therapy (Ishikawa et al., 2007; Ito et al., 2008). In the clinic, it would potentially be important to track residual CSCs following treatment in order to monitor the efficacy of potential anti-CSC-targeted therapies.

CLINICAL DIAGNOSTICS
Circulating Tumor Cells

It has been recently proposed that *circulating tumor cells* (CTCs) provide an important diagnostic source in patients with metastatic disease. The CTCs are present in the peripheral blood of cancer patients and have been shown to be detectable months to years after removal of the primary tumors (Pantel et al., 2008). Immunocytochemistry assays using antibodies against proteins that are expressed on epithelial cells have been used to quantify the number of CTCs in whole blood of metastatic prostate (Danila et al., 2007), colorectal (Cohen et al., 2006), and breast cancer (Cristofanilli et al., 2004). However, the detection of the mRNA levels of these epithelial markers seems to have higher diagnostic sensitivity (Ring et al., 2005). For example, the detection of CD-19 mRNA-positive CTCs in early breast cancer patients before and after adjuvant therapy was found to be an independent factor associated with reduced disease-free survival and overall survival (Xenidis et al., 2009). Recently, comparison of gene expression profiles in CTC-enriched blood samples of metastatic breast cancer patients and their corresponding primary tumor showed that most CTC samples clustered well with the corresponding primary tumor tissues (Sieuwerts et al., 2011). In this study, the authors determined the expression of 65 epithelial tumor cell-specific miRNA and mRNA expression levels in CTCs. Based on the expression of the CTC-specific 65-gene set, healthy blood donors and breast cancer patients without detectable CTCs clustered closely together and could be clearly separated from breast cancer patients with detectable CTCs. The authors also found some discrepancies in molecular characteristics between primary tumor tissues and CTCs, stressing the importance of further studies on molecular characterization of CTCs (Sieuwerts et al., 2011).

The ability to detect and analyze CTCs at the molecular level may therefore provide valuable clinical information including early diagnosis, risk assessment for recurrence, and improved stratification of patients, as well as monitoring the effectiveness of therapeutic regimens and, in the future, this could lead to the development of improved individualized therapy.

Cancer of Unknown Primary

Carcinoma of unknown primary (CUP), also called tumor of unknown origin (TUO), represents approximately 3–5% of all diagnosed cancer (Pavlidis and Fizazi, 2009; Varadhachary et al., 2004). Patients diagnosed with CUP present with metastatic disease for which the primary site cannot be identified. As therapy and clinical management relies on the primary tumor entity and tumor stage, the decision of specific treatment regimen is problematic. Nonetheless, some CUP patients have been defined in subcategories that can be treated according to the most likely primary. For other CUP patients who do not belong to a subcategory, standardized chemotherapy has been proposed.

The identification of the primary tumor in CUP remains challenging. Different tests have been developed in an attempt to identify the primary tumor, such as serum tumor markers and advanced imaging techniques. Immunohistochemistry approaches, with a growing panel of antibodies specific for certain tumors, have also been used to improve the rate of primary identification (Briasoulis et al., 2005; Kaufmann et al., 2002; Varadhachary et al., 2004). Recently, gene expression assays to classify tumors by tissue of origin have been developed. These assays employ microarray methods to quantify mRNA transcripts (Bhattacharjee et al., 2001; Bloom et al., 2004; Ma et al., 2006; Monzon et al., 2009; Ramaswamy et al., 2001; Su et al., 2001; Talantov et al., 2006; Tothill et al., 2005) or miRNA (Ferracin et al., 2011; Rosenfeld et al., 2008). Currently only one molecular test for tissue of origin determination is cleared by the Food and Drug Administration (FDA): the Pathwork Tissue of Origin Test developed by Pathwork Diagnostics, which is a 1550-gene microarray-based test used on frozen tissues. This assay was developed on a gene expression microarray that showed adequate reproducibility in laboratory comparison studies, suggesting that this platform could be used for clinical application (Dumur et al., 2008; Monzon et al., 2009).

There are also other tests available in the United States that are currently in clinical trials. Most of these assays have shown promising results in the internal validations, and have been translated to quantitative reverse transcriptase polymerase chain reaction (qRT-PCR) or microarray platform. However, external validation is needed to determine whether these assays will perform adequately with samples other than those used for their development (reviewed in Monzon and Koen, 2010). One of the criticisms regarding these assays is that they used samples of known origin to evaluate the assay, and this may not reflect the biology of CUP samples. The comparison of molecular signatures between CUP oriented to a specific primary site by molecular criteria and primary tumors of the relevant sites may disclose important differences (Pentheroudakis et al., 2007; Pentheroudakis and Pavlidis, 2009).

In the coming years, adjustments of existing tests and the development of new assays are likely to continue. An important consideration in the development of these tests is not only the molecular classification of CUP followed by the primary site-specific therapy, but also whether these tests will improve the patient's outcome and morbidity, which should be validated by prospective clinical trials.

CONCLUSIONS

Owing to the eventual outcome of metastatic disease in many forms of solid cancer, the quest to gain fresh insights into the molecular mechanisms of the metastasis process remains at the forefront of cancer research. Animal models have yielded important findings about the mechanisms of metastasis, but they must be tested in clinical settings. Development of targeted treatments that specifically target metastases is urgently needed, and clinical trials that target metastasis are urgently required, but the design of such trials is challenging. The emergence of the latest high-throughput technologies should lead to the collection of new genomic and genetic data that will open new roads in the conceptual design of translational and clinical research, as well as new therapeutic targets.

REFERENCES

Albertson, D.G., 2003. Profiling breast cancer by array CGH. Breast Cancer Res Treat 78 (3), 289–298.

Al-Hajj, M., Wicha, M.S., Benito-Hernandez, A., Morrison, S.J., Clarke, M.F., 2003. Prospective identification of tumorigenic breast cancer cells. Proc Natl Acad Sci USA 100 (7), 3983–3988.

Allinen, M., Beroukhim, R., Cai, L., et al., 2004. Molecular characterization of the tumor microenvironment in breast cancer. Cancer Cell 6 (1), 17–32.

Andrews, J., Kennette, W., Pilon, J., et al., 2010. Multi-platform whole-genome microarray analyses refine the epigenetic signature of breast cancer metastasis with gene expression and copy number. PLoS One 5 (1), e8665.

Baek, D., Villen, J., Shin, C., et al., 2008. The impact of microRNAs on protein output. Nature 455 (7209), 64–71.

Baffa, R., Fassan, M., Volinia, S., et al., 2009. MicroRNA expression profiling of human metastatic cancers identifies cancer gene targets. J Pathol 219 (2), 214–221.

Bartel, D.P., 2004. MicroRNAs: Genomics, biogenesis, mechanism, and function. Cell 116 (2), 281–297.

Bernards, R., Weinberg, R.A., 2002. A progression puzzle. Nature 418 (6900), 823.

Bhattacharjee, A., Richards, W.G., Staunton, J., et al., 2001. Classification of human lung carcinomas by mRNA expression profiling reveals distinct adenocarcinoma subclasses. Proc Natl Acad Sci USA 98 (24), 13,790–13,795.

Bloom, G., Yang, I.V., Boulware, D., et al., 2004. Multi-platform, multi-site, microarray-based human tumor classification. Am J Pathol 164 (1), 9–16.

Bonnet, D., Dick, J.E., 1997. Human acute myeloid leukemia is organized as a hierarchy that originates from a primitive hematopoietic cell. Nat Med 3 (7), 730–737.

Brendle, A., Lei, H., Brandt, A., et al., 2008. Polymorphisms in predicted microRNA-binding sites in integrin genes and breast cancer: ITGB4 as prognostic marker. Carcinogenesis 29 (7), 1394–1399.

Briasoulis, E., Tolis, C., Bergh, J., Pavlidis, N., 2005. ESMO Minimum Clinical Recommendations for diagnosis, treatment and follow-up of cancers of unknown primary site (CUP). Ann Oncol 16 (Suppl 1), i75–i76.

Brooks, R., Kizer, N., Nguyen, L., et al., 2010. Polymorphisms in MMP9 and SIPA1 are associated with increased risk of nodal metastases in early-stage cervical cancer. Gynecol Oncol 116 (3), 539–543.

Budhu, A., Jia, H.L., Forgues, M., et al., 2008. Identification of metastasis-related microRNAs in hepatocellular carcinoma. Hepatology 47 (3), 897–907.

Buffart, T.E., Coffa, J., Hermsen, M.A., et al., 2005. DNA copy number changes at 8q11-24 in metastasized colorectal cancer. Cell Oncol 27 (1), 57–65.

Campbell, P.J., Stephens, P.J., Pleasance, E.D., et al., 2008. Identification of somatically acquired rearrangements in cancer using genome-wide massively parallel paired-end sequencing. Nat Genet 40 (6), 722–729.

Clarke, M.F., Dick, J.E., Dirks, P.B., et al., 2006. Cancer stem cells – perspectives on current status and future directions: AACR Workshop on cancer stem cells. Cancer Res 66 (19), 9339–9344.

Cohen, S.J., Alpaugh, R.K., Gross, S., et al., 2006. Isolation and characterization of circulating tumor cells in patients with metastatic colorectal cancer. Clin Colorectal Cancer 6 (2), 125–132.

Crawford, N.P., Ziogas, A., Peel, D.J., et al., 2006. Germline polymorphisms in SIPA1 are associated with metastasis and other indicators of poor prognosis in breast cancer. Breast Cancer Res 8 (2), R16.

Cristofanilli, M., Budd, G.T., Ellis, M.J., et al., 2004. Circulating tumor cells, disease progression, and survival in metastatic breast cancer. N Engl J Med 351 (8), 781–791.

Danila, D.C., Heller, G., Gignac, G.A., et al., 2007. Circulating tumor cell number and prognosis in progressive castration-resistant prostate cancer. Clin Cancer Res 13 (23), 7053–7058.

Dean, M., Fojo, T., Bates, S., 2005. Tumour stem cells and drug resistance. Nat Rev Cancer 5 (4), 275–284.

Ding, L., Ellis, M.J., Li, S., et al., 2010. Genome remodelling in a basal-like breast cancer metastasis and xenograft. Nature 464 (7291), 999–1005.

Dumur, C.I., Lyons-Weiler, M., Sciulli, C., et al., 2008. Interlaboratory performance of a microarray-based gene expression test to determine tissue of origin in poorly differentiated and undifferentiated cancers. J Mol Diagn 10 (1), 67–77.

Eccles, S.A., Welch, D.R., 2007. Metastasis: Recent discoveries and novel treatment strategies. Lancet 369 (9574), 1742–1757.

Ferracin, M., Pedriali, M., Veronese, A., et al., 2011. MicroRNA profiling for the identification of cancers with unknown primary tissue-of-origin. J Pathol 225 (1), 43–53.

Fidler, I.J., 1970. Metastasis: Quantitative analysis of distribution and fate of tumor emboli labeled with 125 I-5-iodo-2′-deoxyuridine. J Natl Cancer Inst 45 (4), 773–782.

Fluck, M.M., Haslam, S.Z., 1996. Mammary tumors induced by polyomavirus. Breast Cancer Res Treat 39 (1), 45–56.

Guy, C.T., Cardiff, R.D., Muller, W.J., 1992. Induction of mammary tumors by expression of polyomavirus middle T oncogene: A transgenic mouse model for metastatic disease. Mol Cell Biol 12 (3), 954–961.

Guzman, M.L., Rossi, R.M., Karnischky, L., et al., 2005. The sesquiterpene lactone parthenolide induces apoptosis of human acute myelogenous leukemia stem and progenitor cells. Blood 105 (11), 4163–4169.

He, L., Thomson, J.M., Hemann, M.T., et al., 2005. A microRNA polycistron as a potential human oncogene. Nature 435 (7043), 828–833.

Heimann, R., Lan, F., McBride, R., Hellman, S., 2000. Separating favorable from unfavorable prognostic markers in breast cancer: The role of E-cadherin. Cancer Res 60 (2), 298–304.

Hermann, P.C., Huber, S.L., Herrler, T., et al., 2007. Distinct populations of cancer stem cells determine tumor growth and metastatic activity in human pancreatic cancer. Cell Stem Cell 1 (23), 313–323.

Hirai, Y., Utsugi, K., Takeshima, N., et al., 2004. Putative gene loci associated with carcinogenesis and metastasis of endocervical adenocarcinomas of uterus determined by conventional and array-based CGH. Am J Obstet Gynecol 191 (4), 1173–1182.

Hitzemann, R., Malmanger, B., Cooper, S., et al., 2002. Multiple cross mapping (MCM) markedly improves the localization of a QTL for ethanol-induced activation. Genes Brain Behav 1 (4), 214–222.

Hsieh, S.M., Smith, R.A., Lintell, N.A., Hunter, K.W., Griffiths, L.R., 2009. Polymorphisms of the SIPA1 gene and sporadic breast cancer susceptibility. BMC Cancer 9, 331.

Hu, Z., Chen, J., Tian, T., et al., 2008. Genetic variants of miRNA sequences and non-small cell lung cancer survival. J Clin Invest 118 (7), 2600–2608.

Huang, Y.W., Huang, T.H., Wang, L.S., 2010. Profiling DNA methylomes from microarray to genome-scale sequencing. Technol Cancer Res Treat 9 (2), 139–147.

Hunter, K.W., Broman, K.W., Voyer, T.L., et al., 2001. Predisposition to efficient mammary tumor metastatic progression is linked to the breast cancer metastasis suppressor gene. Brms1 Cancer Res 61 (24), 8866–8872.

Hunter, K.W., Williams, R.W., 2002. Complexities of cancer research: Mouse genetic models. ILAR J 43 (2), 80–88.

Ishikawa, F., Yoshida, S., Saito, Y., et al., 2007. Chemotherapy-resistant human AML stem cells home to and engraft within the bone-marrow endosteal region. Nat Biotechnol 25 (11), 1315–1321.

Ito, K., Bernardi, R., Morotti, A., et al., 2008. PML targeting eradicates quiescent leukaemia-initiating cells. Nature 453 (7198), 1072–1078.

Jin, L., Hope, K.J., Zhai, Q., Smadja-Joffe, F., Dick, J.E., 2006. Targeting of CD44 eradicates human acute myeloid leukemic stem cell. Nat Med 12 (10), 1167–1174.

Kallioniemi, A., Kallioniemi, O.P., Sudar, D., et al., 1992. Comparative genomic hybridization for molecular cytogenetic analysis of solid tumors. Science 258 (5083), 818–821.

Kaplan, R.N., Riba, R.D., Zacharoulis, S., et al., 2005. VEGFR1-positive haematopoietic bone marrow progenitors initiate the pre-metastatic niche. Nature 438 (7069), 820–827.

Kauffman, E.C., Robinson, V.L., Stadler, W.M., Sokoloff, M.H., Rinker-Schaeffer, C.W., 2003. Metastasis suppression: The evolving role of metastasis suppressor genes for regulating cancer cell growth at the secondary site. J Urol 169 (3), 1122–1133.

Kaufmann, O., Fietze, E., Dietel, M., 2002. Immunohistochemical diagnosis in cancer metastasis of unknown primary tumor. Pathologe 23 (3), 183–197.

Kopfstein, L., Christofori, G., 2006. Metastasis: Cell-autonomous mechanisms versus contributions by the tumor microenvironment. Cell Mol Life Sci 63 (4), 449–468.

Krause, D.S., Lazarides, K., von Andrian, U.H., Van Etten, R.A., 2006. Requirement for CD44 in homing and engraftment of BCR-ABL-expressing leukemic stem cells. Nat Med 12 (10), 1175–1180.

Kurose, K., Hoshaw-Woodard, S., Adeyinka, A., et al., 2001. Genetic model of multi-step breast carcinogenesis involving the epithelium and stroma: Clues to tumour-microenvironment interactions. Hum Mol Genet 10 (18), 1907–1913.

Lancaster, M., Rouse, J., Hunter, K.W., 2005. Modifiers of mammary tumor progression and metastasis on mouse chromosomes 7, 9, and 17. Mamm Genome 16 (2), 120–126.

Lanza, G., Ferracin, M., Gafa, R., et al., 2007. mRNA/microRNA gene expression profile in microsatellite unstable colorectal cancer. Mol Cancer 6, 54.

Lifsted, T., Le, V.T., Williams, M., et al., 1998. Identification of inbred mouse strains harboring genetic modifiers of mammary tumor age of onset and metastatic progression. Int J Cancer 77 (4), 640–644.

Lin, E.Y., Nguyen, A.V., Russell, R.G., Pollard, J.W., 2001. Colony-stimulating factor 1 promotes progression of mammary tumors to malignancy. J Exp Med 193 (6), 727–740.

Liu, R., Wang, X., Chen, G.Y., et al., 2007. The prognostic role of a gene signature from tumorigenic breast-cancer cells. N Engl J Med 356 (3), 217–226.

Lu, J., Getz, G., Miska, E.A., et al., 2005. MicroRNA expression profiles classify human cancers. Nature 435 (7043), 834–838.

Lukes, L., Crawford, N.P., Walker, R., Hunter, K.W., 2009. The origins of breast cancer prognostic gene expression profiles. Cancer Res 69 (1), 310–318.

Luzzi, K.J., MacDonald, I.C., Schmidt, E.E., et al., 1998. Multistep nature of metastatic inefficiency: Dormancy of solitary cells after successful extravasation and limited survival of early micrometastases. Am J Pathol 153 (3), 865–873.

Ma, L., Teruya-Feldstein, J., Weinberg, R.A., 2007. Tumour invasion and metastasis initiated by microRNA-10b in breast cancer. Nature 449 (7163), 682–688.

Ma, X.J., Patel, R., Wang, X., et al., 2006. Molecular classification of human cancers using a 92-gene real-time quantitative polymerase chain reaction assay. Arch Pathol Lab Med 130 (4), 465–473.

Mathe, E.A., Nguyen, G.H., Bowman, E.D., et al., 2009. MicroRNA expression in squamous cell carcinoma and adenocarcinoma of the esophagus: Associations with survival. Clin Cancer Res 15 (19), 6192–6200.

Monzon, F.A., Koen, T.J., 2010. Diagnosis of metastatic neoplasms: Molecular approaches for identification of tissue of origin. Arch Pathol Lab Med 134 (2), 216–224.

Monzon, F.A., Lyons-Weiler, M., Buturovic, L.J., et al., 2009. Multicenter validation of a 1550-gene expression profile for identification of tumor tissue of origin. J Clin Oncol 27 (15), 2503–2508.

Mortazavi, A., Williams, B.A., McCue, K., Schaeffer, L., Wold, B., 2008. Mapping and quantifying mammalian transcriptomes by RNA-Seq. Nat Methods 5 (7), 621–628.

Nam, E.J., Yoon, H., Kim, S.W., et al., 2008. MicroRNA expression profiles in serous ovarian carcinoma. Clin Cancer Res 14 (9), 2690–2695.

Nowell, P.C., 1976. The clonal evolution of tumor cell populations. Science 194 (4260), 23–28.

Pantel, K., Brakenhoff, R.H., Brandt, B., 2008. Detection, clinical relevance and specific biological properties of disseminating tumour cells. Nat Rev Cancer 8 (5), 329–340.

Park, Y.G., Zhao, X., Lesueur, F., et al., 2005. Sipa1 is a candidate for underlying the metastasis efficiency modifier locus Mtes1. Nat Genet 37 (10), 1055–1062.

Pavlidis, N., Fizazi, K., 2009. Carcinoma of unknown primary (CUP). Crit Rev Oncol Hematol 69 (3), 271–278.

Pentheroudakis, G., Briasoulis, E., Pavlidis, N., 2007. Cancer of unknown primary site: Missing primary or missing biology? Oncologist 12 (4), 418–425.

Pentheroudakis, G., Pavlidis, N., 2009. Cancer of unknown primary: What kind of chemotherapy? What kind of disease? Onkologie 32 (4), 159–160.

Radiloff, D.R., Rinella, E.S., Threadgill, D.W., 2008. Modeling cancer patient populations in mice: Complex genetic and environmental factors. Drug Discov Today Dis Models 4 (2), 83–88.

Ramaswamy, S., Ross, K.N., Lander, E.S., Golub, T.R., 2003. A molecular signature of metastasis in primary solid tumors. Nat Genet 33 (1), 49–54.

Ramaswamy, S., Tamayo, P., Rifkin, R., et al., 2001. Multiclass cancer diagnosis using tumor gene expression signatures. Proc Natl Acad Sci USA 98 (26), 15,149–15,154.

Reya, T., Morrison, S.J., Clarke, M.F., Weissman, I.L., 2001. Stem cells, cancer, and cancer stem cells. Nature 414 (6859), 105–111.

Riethmuller, G., Klein, C.A., 2001. Early cancer cell dissemination and late metastatic relapse: Clinical reflections and biological approaches to the dormancy problem in patients. Semin Cancer Biol 11 (4), 307–311.

Ring, A.E., Zabaglo, L., Ormerod, M.G., Smith, I.E., Dowsett, M., 2005. Detection of circulating epithelial cells in the blood of patients with breast cancer: Comparison of three techniques. Br J Cancer 92 (5), 906–912.

Rivera, J., Tessarollo, L., 2008. Genetic background and the dilemma of translating mouse studies to humans. Immunity 28 (1), 1–4.

Robson, E.J., Khaled, W.T., Abell, K., Watson, C.J., 2006. Epithelial-to-mesenchymal transition confers resistance to apoptosis in three murine mammary epithelial cell lines. Differentiation 74 (5), 254–264.

Rosenfeld, N., Aharonov, R., Meiri, E., et al., 2008. MicroRNAs accurately identify cancer tissue origin. Nat Biotechnol 26 (4), 462–469.

Satzger, I., Mattern, A., Kuettler, U., et al., 2010. MicroRNA-15b represents an independent prognostic parameter and is correlated with tumor cell proliferation and apoptosis in malignant melanoma. Int J Cancer 126 (11), 2553–2562.

Schetter, A.J., Leung, S.Y., Sohn, J.J., et al., 2008. MicroRNA expression profiles associated with prognosis and therapeutic outcome in colon adenocarcinoma. JAMA 299 (4), 425–436.

Shackleton, M., Vaillant, F., Simpson, K.J., et al., 2006. Generation of a functional mammary gland from a single stem cell. Nature 439 (7072), 84–88.

Shah, S.P., Morin, R.D., Khattra, J., et al., 2009. Mutational evolution in a lobular breast tumour profiled at single nucleotide resolution. Nature 461 (7265), 809–813.

Shendure, J., Ji, H., 2008. Next-generation DNA sequencing. Nat Biotechnol 26 (10), 1135–1145.

Shipitsin, M., Campbell, L.L., Argani, P., et al., 2007. Molecular definition of breast tumor heterogeneity. Cancer Cell 11 (3), 259–273.

Si, M.L., Zhu, S., Wu, H., et al., 2007. miR-21-mediated tumor growth. Oncogene 26 (19), 2799–2803.

Sieuwerts, A.M., Mostert, B., Bolt-de, V.J., et al., 2011. mRNA and microRNA expression profiles in circulating tumor cells and primary tumors of metastatic breast cancer patients. Clin Cancer Res 17 (11), 3600–3618.

Stingl, J., Eirew, P., Ricketson, I., et al., 2006. Purification and unique properties of mammary epithelial stem cells. Nature 439 (7079), 993–997.

Su, A.I., Welsh, J.B., Sapinoso, L.M., et al., 2001. Molecular classification of human carcinomas by use of gene expression signatures. Cancer Res 61 (20), 7388–7393.

Talantov, D., Baden, J., Jatkoe, T., et al., 2006. A quantitative reverse transcriptase-polymerase chain reaction assay to identify metastatic carcinoma tissue of origin. J Mol Diagn 8 (3), 320–329.

Tanami, H., Tsuda, H., Okabe, S., et al., 2005. Involvement of cyclin D3 in liver metastasis of colorectal cancer, revealed by genome-wide copy-number analysis. Lab Invest 85 (9), 1118–1129.

Thompson, E.W., Newgreen, D.F., Tarin, D., 2005. Carcinoma invasion and metastasis: A role for epithelial-mesenchymal transition? Cancer Res 65 (14), 5991–5995.

Thyagarajan, T., Totey, S., Danton, M.J., Kulkarni, A.B., 2003. Genetically altered mouse models: The good, the bad, and the ugly. Crit Rev Oral Biol Med 14 (3), 154–174.

Tothill, R.W., Kowalczyk, A., Rischin, D., et al., 2005. An expression-based site of origin diagnostic method designed for clinical application to cancer of unknown origin. Cancer Res 65 (10), 4031–4040.

van't Veer, L.J., Dai, H., van de Vijver, M.J., et al., 2002. Gene expression profiling predicts clinical outcome of breast cancer. Nature 415 (6871), 530–536.

Varadhachary, G.R., Abbruzzese, J.L., Lenzi, R., 2004. Diagnostic strategies for unknown primary cancer. Cancer 100 (9), 1776–1785.

Visvader, J.E., Lindeman, G.J., 2008. Cancer stem cells in solid tumours: Accumulating evidence and unresolved questions. Nat Rev Cancer 8 (10), 755–768.

Volinia, S., Calin, G.A., Liu, C.G., et al., 2006. A microRNA expression signature of human solid tumors defines cancer gene targets. Proc Natl Acad Sci USA 103 (7), 2257–2261.

Wang, C., Iakovlev, V.V., Wong, V., et al., 2009. Genomic alterations in primary breast cancers compared with their sentinel and more distal lymph node metastases: An aCGH study. Genes Chromosomes Cancer 48 (12), 1091–1101.

Wiltshire, T., Pletcher, M.T., Batalov, S., et al., 2003. Genome-wide single-nucleotide polymorphism analysis defines haplotype patterns in mouse. Proc Natl Acad Sci USA 100 (6), 3380–3385.

Worley, L.A., Long, M.D., Onken, M.D., Harbour, J.W., 2008. MicroRNAs associated with metastasis in uveal melanoma identified by multiplexed microarray profiling. Melanoma Res 18 (3), 184–190.

Wyckoff, J.B., Wang, Y., Lin, E.Y., et al., 2007. Direct visualization of macrophage-assisted tumor cell intravasation in mammary tumors. Cancer Res 67 (6), 2649–2656.

Xenidis, N., Ignatiadis, M., Apostolaki, S., et al., 2009. Cytokeratin-19 mRNA-positive circulating tumor cells after adjuvant chemotherapy in patients with early breast cancer. J Clin Oncol 27 (13), 2177–2184.

Xu, J., Wang, R., Xie, Z.H., et al., 2006. Prostate cancer metastasis: Role of the host microenvironment in promoting epithelial to mesenchymal transition and increased bone and adrenal gland metastasis. Prostate 66 (15), 1664–1673.

Yan, L.X., Huang, X.F., Shao, Q., et al., 2008. MicroRNA miR-21 overexpression in human breast cancer is associated with advanced clinical stage, lymph node metastasis and patient poor prognosis. RNA 14 (11), 2348–2360.

Yan, W., Song, L., Wei, W., et al., 2005. Chromosomal abnormalities associated with neck nodal metastasis in nasopharyngeal carcinoma. Tumour Biol 26 (6), 306–312.

Yang, Z.F., Ho, D.W., Ng, M.N., et al., 2008. Significance of CD90+ cancer stem cells in human liver cancer. Cancer Cell 13 (2), 153–166.

Yilmaz, O.H., Valdez, R., Theisen, B.K., et al., 2006. Pten dependence distinguishes haematopoietic stem cells from leukaemia-initiating cells. Nature 441 (7092), 475–482.

Zhang, Y., Martens, J.W., Yu, J.X., et al., 2009. Copy number alterations that predict metastatic capability of human breast cancer. Cancer Res 69 (9), 3795–3801.

RECOMMENDED RESOURCES

Bild, A.H., Parker, J.S., Gustafson, A.M., et al., 2009. An integration of complementary strategies for gene-expression analysis to reveal novel therapeutic opportunities for breast cancer. Breast Cancer Res 11 (4), R55.

Desmarais, V., Yamaguchi, H., Oser, M., et al., 2009. N-WASP and cortactin are involved in invadopodium-dependent chemotaxis to EGF in breast tumor cells. Cell Motil Cytoskeleton 66 (6), 303–316.

Gatza, M.L., Lucas, J.E., Barry, W.T., et al., 2010. A pathway-based classification of human breast cancer. Proc Natl Acad Sci USA 107 (15), 6994–6999.

Patsialou, A., Wyckoff, J., Wang, Y., et al., 2009. Invasion of human breast cancer cells *in vivo* requires both paracrine and autocrine loops involving the colony-stimulating factor-1 receptor. Cancer Res 69 (24), 9498–9506.

CHAPTER

68

Bioinformatics in Personalized Cancer Care

Kenneth H. Buetow

INTRODUCTION

The biomedical informatics necessary to support cancer research is at the vanguard of personalized medicine, much as cancer itself is on the leading edge of personalized care. Cancer has commonly been the proving ground for the concepts and applications of personalized medicine. For example, approximately 50 years ago, there were only two forms of blood cancer described, leukemia and lymphoma. Today, as a consequence of cytogenetic analysis – the first-generation whole-genome scanning technology – there are 38 known subtypes of leukemia and 51 known subtypes of lymphoma. The identification of these different molecular subtypes has been key to understanding the biology underpinning the disease. Moreover, the subtypes are used to direct therapeutic intervention.

The elucidation of the mechanisms underpinning cytogenetically classified subtypes has required the coupling of state-of-the-art molecular characterization technology with information technology (IT). Informatics has been essential to managing the volume and complexity of the data. Early cytogenetic findings in cancer foreshadowed a tremendous complexity and heterogeneity in underlying molecular etiology. As more cancer classification is translated into molecular mechanisms, it is likely that all cancers will come to be considered "orphan" (rare diseases found only in very small fractions of the population) diseases. The incredible volume of cancer subtypes and the diverse armamentarium of interventions will strain the fundamental infrastructure necessary to generate evidence of therapeutic efficacy.

BIOMEDICAL INFORMATICS CAPABILITIES DRIVEN BY THE STUDY OF CANCER

Many of the biomedical informatics capabilities required to drive personalized medicine in cancer are not unique to the field and are covered elsewhere in this volume. However, given the leading-edge nature of adoption of molecular characterization technology within cancer, there are unique capabilities first generated within the cancer community. These capabilities provide insights that will likely be useful outside of cancer and will become part of the larger biomedical community's informatics fabric.

Systems analysis of cancer has established that, at a molecular level, each cancer has a complex and diverse profile. This complexity includes tremendous heterogeneity at a nucleotide level, both inherited and somatically generated. It also includes heterogeneity in the molecular mechanism changing the "state" of genes. Different cancers in different individuals may exhibit nucleotide substitutions, insertions, deletions, rearrangements, translocations, copy-number alterations, epigenetic alterations, microRNA changes, gene expression changes, proteomic alterations, and/or glycomic changes. No one dimension is sufficient to capture the underlying molecular etiology. As such, to understand the biology of a given tumor, it is necessary to examine multidimensional molecular characterizations. Because of the dimensionality of the problem, it is also necessary to have large sample sizes in order to

Genomic and Personalized Medicine, 2nd edition
by Ginsburg & Willard. DOI: http://dx.doi.org/10.1016/B978-0-12-382227-7.00068-9
2013 Published by Elsevier Inc.

distinguish signal from noise. The cancer community has created the informatics capabilities to share these large-scale, multidimensional datasets.

Multidimensional Data Access Portals

It is not practical to catalog all individual datasets generated and shared by the cancer community. Many of these datasets reside on individual institution websites or are referenced in supplemental material sections indicating availability on request. However, several large, canonical datasets are available that serve as references for the entire community.

The largest dataset available is that associated with The Cancer Genome Atlas (TCGA). TCGA is generating multidimensional molecular characterizations of more than 20

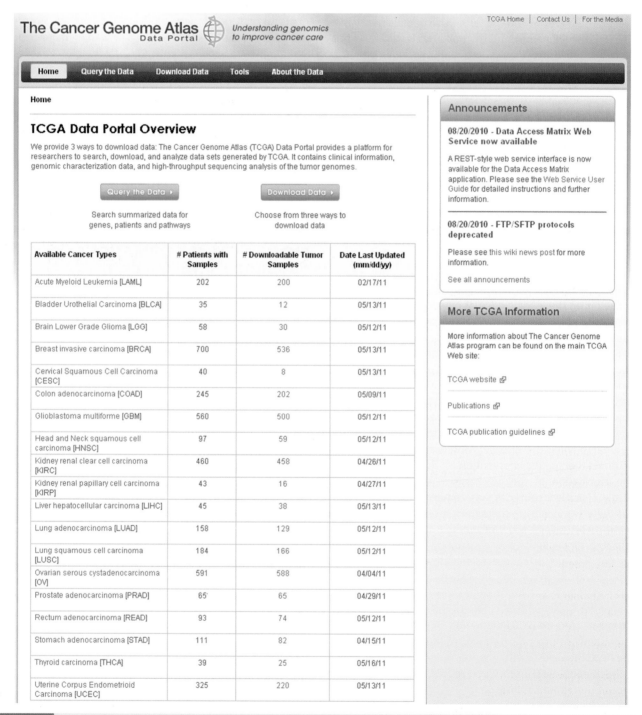

Figure 68.1 TCGA Data Portal. Home page of the Data Portal, which provides navigation to data browsing tools ("Query the Data") and multiple approaches to downloading the entire dataset or selected subsets.

tumor types (Figure 68.1). It is projected that 500 cancer and 500 non-cancer samples with rich clinical annotation will ultimately be available for each type. The material is characterized for nucleotide substitutions, insertions, deletions, rearrangements, translocations, copy number, epigenetic state, microRNA expression, and gene expression.

TCGA data are available for download via The Cancer Genome Atlas Data Portal (Figure 68.1). The data portal permits users to search the data, download the raw data via multiple mechanisms, and perform simple analysis of the data. It contains clinical information (with access restricted to authorized users for human subjects' protection), genomic characterization data, and sequence analysis information. Data download sets can be selected through the portal's data access matrix, bulk download of dataset files, HTTP directories, and through a REST-style web service interface. A query interface provides simple summary information for genes, individuals within the dataset, and pathways.

Complementing TCGA is the Therapeutically Applicable Research to Generate Effective Targets (TARGET) project, which is performing comparable analysis in pediatric cancer. A total of five cancers are under study: acute lymphoblastic leukemia, myeloid leukemia, neuroblastoma, osteosarcoma, and high-risk Wilm's tumor. Data generated from TARGET are available in two tiers. An open-access tier contains information that is deemed to present minimal risk of participant re-identification. The majority of TARGET data are included in the open-access tier. A controlled-access tier contains broader demographic, clinical, and genotypic information. Open-access tier data are available for download through the TARGET Data Matrix.

Similar to TCGA and TARGET, an International Cancer Genome Consortium (ICGC) has been initiated to coordinate a large number of international research projects that are comprehensively assessing genomic changes in cancers that contribute to worldwide cancer burden. The ICGC is generating comprehensive catalogs of genomic abnormalities (somatic mutations, abnormal expression of genes, epigenetic modifications) in tumors in up to 50 different cancer types. Genomic analyses of tumors of the pancreas, liver, breast, lung, and skin are currently available. ICGC data are available for download, browsing, and analysis via the ICGC BioMart Portal. The ICGC BioMart Portal supports programmatic interfaces using REST/SOAP, SPARQL, and Java, as well as allowing bulk downloads.

The Integrated Cancer Biology Program (ICBP) provides access to several large-scale cancer genomic datasets generated by its member academic institutions. An index to all of the data available through the ICBP website is obtained by clicking on the Resources tab. The website links to more than 100 datasets from the Broad Institute Cancer Program. The Broad datasets contain gene expression, single nucleotide polymorphism (SNP) microarray assays, and RNA interference (RNAi) screens of cancer types including breast, colorectal, hepatocellular, leukemia, lung, lymphoma, medulloblastoma, melanoma, nasopharyngeal, neuroblastoma, prostate, and sarcoma. Similarly, links to data from the Memorial Sloan–Kettering Cancer Center's Prostate Cancer analysis are listed on the ICBP website. The site also links to the Sage Bionetworks Commons site that contains "globally coherent" cancer datasets.

Multidimensional Data Browsers

Bulk download provides access for users who have significant computer infrastructure and comparable analytic capabilities. For a much larger fraction of the community, access to data browsers that have already assembled, integrated, and processed the data is more scientifically meaningful. Several cancer-specific data browsers have emerged to fill this niche.

One of the first efforts to provide integrated access to multidimensional cancer data was the Glioblastoma Molecular Diagnostics Initiative's Repository of Molecular Brain Neoplasia Data (REMBRANDT). REMBRANDT is a cancer clinical genomics database and a web-based data mining and analysis platform that contains genome variation, gene expression, and clinical data for glioblastoma (Madhavan et al., 2009). In addition to simple browsing of the data, users can readily query gene expression or copy-number data and graph changes in survival rate (Figure 68.2). REMBRANDT also supports simple access to applications that perform higher-order gene expression analyses [such as class comparison, clustering, and Principal Component Analysis (PCA)]. For more sophisticated, *ab initio* analysis, REMBRANDT provides access to the Broad Institute's GenePattern analytic engine (Reich et al., 2006), which combines a scientific workflow platform with more than 90 computational and visualization tools for analysis of genomic data.

One of the earliest and most comprehensive, true multidimensional cancer data browsers is the Cancer Genome Workbench (CGWB). The CGWB (Zhang et al., 2007) is a visualization tool that integrates somatic mutation data with copy-number alteration, gene expression, methylation, *in vivo* imaging and microRNA expression data. The CGWB is a cancer-specific extension of the University of California Santa Cruz's Genome Browser (Figure 68.3). The CGWB has integrated data from TCGA, TARGET, the Sanger Center's COSMIC (Catalogue of Somatic Mutations in Cancer) initiative, the National Human Genome Research Institute (NHGRI)'s Tumor Sequencing Project (TSP), whole-genome somatic mutation data from the Vogelstein laboratory at Johns Hopkins University and GlaxoSmithKline Cancer Cell Line Genomic Profiling Data, thereby facilitating the integration and interpretation of diverse, high-quality raw data. In order for data from multiple independent sources to be visualized in an integrated manner, it is necessary for the data to be accumulated and processed consistently. As such, primary somatic copy-number data from different sources are recomputed using matching tumor/normal samples (where available), so that somatic alterations can be distinguished from germline copy-number variations. Putative sequence mutations are also recomputed from raw sequence chromatograms and are available for inspection so that a researcher can evaluate the quality of predicted mutations without implementing a complex analytical process. All

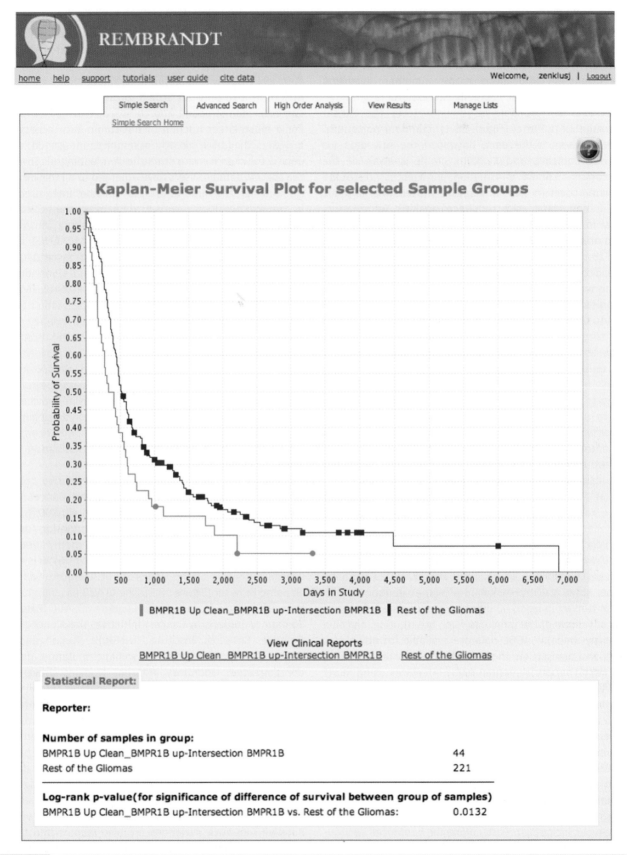

Figure 68.2 REMBRANDT screenshot showing a Kaplan–Meier (K–M) survival plot contrasting glioma samples upregulated for BMRP1 with those without upregulation. K–M estimates are based on the last follow-up time. The Kaplan–Meier estimates are then plotted against the survival time. Significance of the difference in survival between any two groups of samples is based on a log-rank *p*-value.

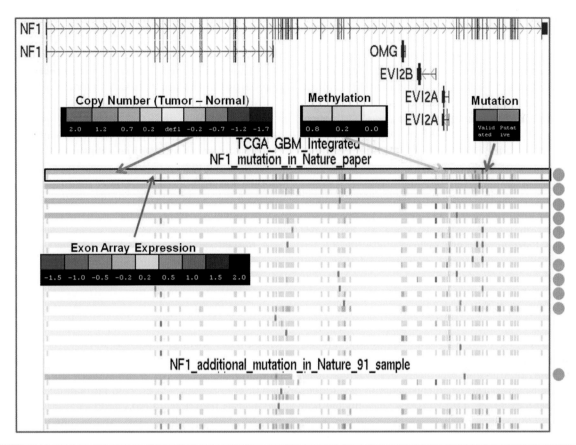

Figure 68.3 CGWB integrated viewer. Integrated genomic view of TCGA data at the *NF1* locus in Glioblastoma Multiforme (GBM) (from CGWB Tutorial). Each sample has two rows of data. The top row shows somatic copy-number alterations (CNAs) (blue for deletion, red for amplification, white for copy-number neutral), methylation (white, light yellow, and dark yellow for no methylation, medium methylation ($0.2 < \beta < 0.8$) and high methylation ($\beta > 0.8$), respectively), and somatic sequence mutation (dark green for validated mutation and light green for putative mutation). The bottom row shows exon expression measured by Affymetrix Human Exon 1.0 ST. Medium expression is shown in white. Dark shades of gray indicate higher expression; dark shades of purple indicate lower expression. Samples have *NF1* somatic mutations identified in the 91 samples used for the Cancer Genome Atlas Network publication [Cancer Genome Atlas Research Network, 2008]. Mutations identified after the first publication are in the group "NF1_additional_mutation_in_Nature_91_sample."

mutations (validated and predicted) are annotated with amino acid changes and potentially deleterious changes in evolutionarily conserved protein domains, as well as mapped onto three-dimensional protein structures.

The University of California Santa Cruz (UCSC) Genome Browser has been extended to support the presentation and analysis of cancer data. The UCSC Cancer Genomics Browser represents a suite of tools that visualize, integrate, and analyze multidimensional cancer genomic and clinical data. The UCSC Cancer Genomics Browser provides access to histopathologic images of tumors, when available. A unique feature of the UCSC Cancer Genome Browser is its set of "stock" genome signatures constructed from published scientific papers.

Biologic Process Network Analysis

Biologic phenomena emerge as consequences of the action of genes and their products in pathways. Diseases arise through alteration of these complex networks (Altomare and Testa, 2005;

Bianco et al., 2006; Hanahan and Weinberg, 2000; Matoba et al., 2006; Parsons et al., 2005). It is increasingly clear that the complex multidimensional data observed in cancer obtain coherence when projected onto biologic networks (Cancer Genome Atlas Research Network, 2008). Cancer has been at the leading edge in defining methods to examine complex data using networks. Bioinformatics methods that examine biologic processes can be classified in three categories: *ab initio* network generation, statistical tests for oversampling of canonical networks, and methods that leverage interactions in networks.

Ab initio methods generate the networks that they analyze, either with the data being evaluated or from independent data. As the network interactions of most molecules are still unknown, this represents a powerful approach to defining new relationships. A drawback to this approach is that it is difficult to compare the products of different independent analyses. It also does not leverage the vast literature of known biologic relationships. One of the earliest applications of this approach was Segal and colleagues (2004), which defined biologic modules that are activated and de-activated in various cancers. Similarly, Rhodes and coworkers (2005) generated sub-networks by their association with sets of genes identified through their over- or under-expression in each biologic phenotype. One of the most successful set of applications (Margolin et al., 2006) predicts interactions through reverse engineering algorithms, such as ARACNE (Algorithm for the Reconstruction of Accurate Cellular Networks). Modulation of interactions is predicted by MINDY, an algorithm for the prediction of modulators of transcriptional interactions.

Analytic approaches that look for over-representation of genes selected in gene signatures in predefined pathways represent the most common form of network analysis. Early work of this type was performed by Bild and colleagues (2006) and Glinski and coworkers (2005), who demonstrated that gene signatures determined by a small set of preselected canonical pathways can distinguish tumor characteristics. Gene set enrichment analysis (GSEA) (Subramanian et al., 2005) represents the dominant approach in this category. GSEA allows a user to choose a set of genes that distinguishes a given phenotype, and to determine the relative statistical significance in the collection of genes that defines a pathway. Pathway membership is measured to assess combined contributions. GSEA does not make use of the structure of the network. Since the method starts with the discrimination of single genes, it can only build on existing statistical inference, and leverages differences that come from the interdependency of multiple gene interactions.

Early work using network interaction analysis in cancer was performed by Efroni and colleagues (2007). Using the tool PathOlogist, a user can systematically evaluate the interaction structure across predefined canonical networks. In measuring the state of the interaction, it combines information from gene state and network structure and allows each interaction to be scored as "active" or "consistent" depending upon states of genes in the canonical network. Efroni provides examples of the coherence that emerges from the analysis of biologic network interactions.

TRANSLATABLE INFORMATICS

The multidimensional molecular nature of cancer and the essential role of biologic networks have implications for the generation of evidence required for translation of cancer biology into clinical practice. Once identified, molecularly defined subgroups should permit much smaller sample sizes for generating significant evaluations. However, identifying such subgroups requires testing very large numbers of individuals with extensive clinical evaluations. More specifically, finding groups that have a specific array of molecular characteristics may require the screening of hundreds of thousands of individuals in order to identify the hundreds with the requisite signature in which the personalized intervention will be tested. Moreover, the current model of randomized clinical trials

will be overwhelmed with the number of independent trials that will be needed to contrast all possible targeted interventions with each molecular subgroup. The cancer community has experimented with novel infrastructure and clinical trial approaches to address these challenges.

Novel Cancer Research Infrastructure

As indicated above, successfully implementing personalized medicine in cancer requires integrating unprecedented amounts of information from a diverse ecosystem of individuals, organizations, and institutions. To achieve the necessary sample sizes required for statistically significant results, one must assemble information from diverse communities of researchers. Collecting, integrating, and analyzing multidimensional biologic and clinical data represents a challenge that traditional research practices, organizational structures, and analytical tools have not been designed to address. To date, most biomedical research has been conducted by individuals or small teams in institutions in close geographic proximity. Personalized medicine requires collaboration by multidisciplinary teams working in dispersed locations and sharing limited resources.

This multidisciplinary research model is only possible if data from multiple, dispersed sources are readily accessible and understandable – a process facilitated by interoperable IT capabilities to represent data and link researchers. This infrastructure can facilitate data sharing between researchers within a laboratory or an organization, or between institutions.

Increasingly, the technical capacity to generate the vast quantities of molecular data required to enable personalized medicine is not the rate-limiting step in clinical research. The larger challenge comes from joining molecular data with clinical data. Unfortunately, the vast majority of clinical data are often collected in proprietary data formats from commercial, in-house or legacy systems that seriously restrict the ability to use them for downstream, integrated analysis. As multisite trials become commonplace, the need to easily collect and compare data generated by different systems at diverse locations becomes critical. The National Cancer Institute (NCI) of the National Institutes of Health, through its Cancer Biomedical Informatics Grid (caBIG®) community, has developed a prototype solution to this challenge.

Foundational to this solution is the adoption of widely recognized data standards where they exist, and close collaboration with industry associations and assorted international organizations to develop vocabularies, and define common data elements (CDEs) and data models where such standards do not exist. The caBIG® community has worked in partnership with key industry standards organizations such as the Clinical Data Interchange Standards Consortium (CDISC) and Health Level 7 (HL7), and is developing tools that adhere to these standards. The Biomedical Research Integrated Domain Group (BRIDG) Model group represents an example of such an effort where stakeholders from the cancer community, CDISC and the HL7 Regulated Clinical Research Information Management

Technical Committee are creating a common information model for the representation of clinical care and research data.

Interoperable electronic health records (EHRs) represent an additional link in obtaining clinical data that can be joined with molecular information. The caBIG® community has worked with standards developers to use BRIDG in the generation of EHRs. The caBIG® community and the American Society of Clinical Oncology (ASCO) have collaborated to specify an oncology EHR that will utilize common standards for interoperability to integrate clinical outcomes data with genomic profiling information, helping to advance personalized medicine.

The cancer community has developed tools that leverage international standards and facilitate electronic access. These tools, collectively referred to as the cancer Common Ontologic Representation Environment (caCORE), generate standards-based resources with open interfaces written in common electronic "dialects" (e.g., RESTful, SOAP). To further facilitate access, the electronic resources are inventoried in public repositories which describe the resource's information models and standards used in their construction. For resources not created using standards, adapters have been created by the cancer community that translate data from its original format to a standards-based representation (caAdapter and Clinical Integration Hub). An open-source effort that is broader than the cancer community has created an adapter that supports the translation of many different clinical information representations (Mirth Connect).

The cancer community has also developed a prototype framework to enable connectivity among people, organizations, data, and analysis tools – caGrid. Built on the infrastructure that supports the Internet and the World Wide Web, caGrid is largely invisible to the end user. caGrid is a model-driven, service-oriented architecture that provides core "services," tools, and interfaces so that the community can connect applications and share data and analyses. Because these services are standardized, new applications created from them automatically share identical methods for handling data and can pass data back and forth easily. A real-time list of organizations, data, and services currently available on caGrid can be found at the caGrid portal. Data from projects such as TCGA are available through grid services.

An increasing array of tools and resources are available that support the above standards and interoperability specifications. A variety of community-generated capabilities and tools are accessible through the Cancer eScientist Portal. The Cancer eScientist Portal directs users to a wide spectrum of community-based capabilities such as applications, databases, and research platforms, for those whose interests include: *in silico* research, clinical trials management, biospecimens, imaging, genome annotation, proteomics, microarrays, data analysis and statistical tools, data sharing, infrastructure, vocabularies, translational research, and other biomedical research activities.

Cancer biologists can manage large collections of molecular data using caArray and perform analysis using GenePattern. They can examine genomic information such as SNPs or copy-number variation using geWorkbench, can perform integrated analysis with clinical outcomes data using caIntegrator, and then can share their data and results with collaborators in other labs or other institutions.

Clinical researchers can simplify the execution of clinical trials by using interoperable tools such as the Cancer Central Clinical Patient Registry (C3PR) to register patients, the Patient Study Calendar (PSC) to manage participant schedules, and the cancer Adverse Events Reporting System (caAERS) to handle adverse event reporting. Researchers can collect and manage clinical data produced during the trial using numerous electronic data capture platforms, including OpenClinica and Oracle Clinical-based C3D.

Biobank managers can inventory, annotate, and track samples and their derivatives, and even manage multiple dispersed physical inventories from a single easy-to-use interface in caTissue Suite.

Radiologists can collect and manage DICOM (the Digital Imaging and Communications in Medicine standard) images from a variety of imaging modalities in a central repository and annotate those images with standardized, sharable information using the National Biomedical Imaging Archive (NBIA) and the Annotation Image Markup tools (AIM).

Next-generation Adaptive Clinical Trials Infrastructure

Evidence generation to support personalized medicine requires novel study designs supported by informatics platforms that join all of the components described above. An example of such an integrated approach is the Investigation of Serial Studies to Predict Your Therapeutic Response with Imaging and Molecular Analysis 2 (I-SPY 2 TRIAL). The I-SPY 2 TRIAL is a multisite, adaptive clinical trial for women with breast cancer, designed to determine whether adding investigational drugs to standard chemotherapy pre-surgery is superior to standard chemotherapy alone for women with high-risk biology. The trial uses the information from each participant who completes the study treatment to help determine treatment for women who join the trial in future, creating a "learning trial" and helping the study researchers to understand rapidly which investigational drugs will be most beneficial for women with certain tumor characteristics. The I-SPY 2 TRIAL tests tailoring of treatment based on molecular markers to identify which patients should be treated with investigational drugs.

TRANSCEND (Translational Informatics System to Coordinate Emerging Biomarkers, Novel Agents, and Clinical Data) is an informatics platform built to facilitate next-generation, adaptive phase II trials such as the I-SPY 2 TRIAL. TRANSCEND integrates an open-source EHR, a randomization engine, and several caBIG® community tools, providing a range of functionalities and capabilities for clinicians and researchers. The objective of the TRANSCEND project is to develop and deploy a nationally scalable, open-source informatics infrastructure that is built upon a robust architecture capable of meeting the needs of the cancer research community. TRANSCEND automates the integration of the adaptive trials infrastructure into the care delivery environment through EHRs. It supports multiple standardized methods and varying levels of sophistication for data exchange between clinical research and care delivery systems to ensure that as many as possible can receive and share data. The platform integrates the adaptive trials infrastructure with clinical trial, laboratory, adverse event, and regulatory systems using NCI Enterprise Services to ensure data consistency, reliability, and availability. It leverages the caBIG® community's Imaging Workstation and National Biomedical Imaging Archive (NBIA) to enhance visualization capabilities, improve image management, and facilitate sharing of images across sites, which currently is not possible. Finally, TRANSCEND integrates patient-reported symptoms and quality-of-life information into the clinical research environment.

The TRANSCEND platform provides an infrastructure that can be adopted wholesale, as an architectural foundation, or as a set of components that can integrate with other tools as necessary, providing the utility and scalability needed for all types of adaptive trials. As new standards emerge, they will be incorporated into TRANSCEND, so that the platform can continue to evolve over time and can continue to meet the needs of the research community.

TOWARD AN IT-ENABLED ECOSYSTEM

Twenty-first century medicine will be predictive, personalized, preemptive, and participatory. To achieve this will require unprecedented "data liquidity" among a vast collection of stakeholders across the ecosystem described above. The cancer community has provided proof-of-concept that it is technically possible to connect its various members on an international scale (Figure 68.4). To move beyond this demonstration will require large-scale deployment of biomedical informatics resources that leverage the increasing capabilities of new information technology while supporting individual organizations' historic investments. It will be equally important to address the cultural, legal, and ethical barriers that are often confounded with the technical challenges. As in the past, cancer has the potential to lead a new, molecularly based, data-driven biomedical ecosystem that transforms medicine.

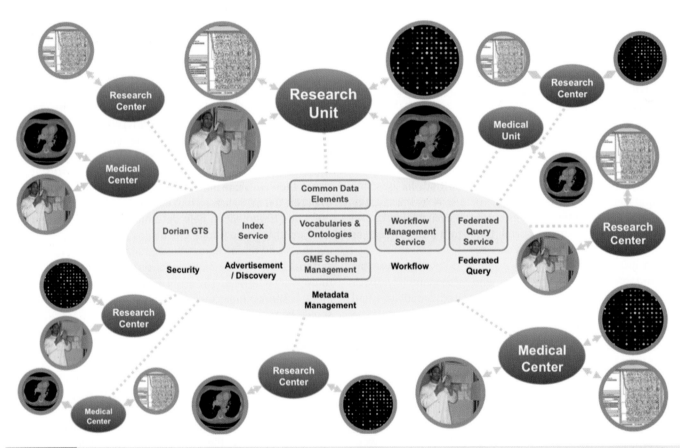

Figure 68.4 The IT-enabled ecosystem. First-generation, proof-of-concept infrastructure created by the cancer community through the caBIG® program that connects disparate data types within organizations and provides secure, controlled access to data and analytic resources on a global scale.

REFERENCES

Altomare, D.A., Testa, J.R., 2005. Perturbations of the AKT signaling pathway in human cancer. Oncogene 24, 7455–7464.

Bianco, R., Melisi, D., Ciardiello, F., Tortora, G., 2006. Key cancer cell signal transduction pathways as therapeutic targets. Eur J Cancer 42, 290–294.

Bild, A.H., Yao, G., Chang, J.T., et al., 2006. Oncogenic pathway signatures in human cancers as a guide to targeted therapies. Nature 439, 353–357.

Cancer Genome Atlas Research Network, 2008. Comprehensive genomic characterization defines human glioblastoma genes and core pathways. Nature 455, 1061–1068.

Efroni, S., Schaefer, C.F., Buetow, K.H., 2007. Identification of key processes underlying cancer phenotypes using biologic pathway analysis. PLoS ONE 2, e425.

Glinsky, G.V., Berezovska, O., Glinskii, A.B., 2005. Microarray analysis identifies a death-from-cancer signature predicting therapy failure in patients with multiple types of cancer. J Clin Invest 115, 1503–1521.

Hanahan, D., Weinberg, R.A., 2000. The hallmarks of cancer. Cell 100, 57–70.

Madhavan, S., Zenklusen, J.C., Kotliarov, Y., Sahni, H., Fine, H.A., Buetow, K.H., 2009. Rembrandt: Helping personalized medicine become a reality through integrative translational research. Mol Cancer Res 7, 157–167.

Margolin, A.A., Nemenman, I., Basso, K., et al., 2006. ARACNE: An algorithm for the reconstruction of gene regulatory networks in a mammalian cellular context. BMC Bioinform 7 (Suppl. 1), S7.

Matoba, S., Kang, J.G., Patino, W.D., et al., 2006. p53 regulates mitochondrial respiration. Science 312, 1650–1653.

Parsons, D.W., Wang, T.L., Samuels, Y., et al., 2005. Colorectal cancer: Mutations in a signalling pathway. Nature 436, 792.

Reich, M., Liefeld, T., Gould, J., Lerner, J., Tamayo, P., Mesirov, J.P., 2006. GenePattern 2.0. Nat Genet 38, 500–501.

Rhodes, D.R., Tomlins, S.A., Varambally, S., et al., 2005. Probabilistic model of the human protein–protein interaction network. Nat Biotechnol 23, 951–959.

Segal, E., Friedman, N., Koller, D., Regev, A., 2004. A module map showing conditional activity of expression modules in cancer. Nat Genet 36, 1090–1098.

Subramanian, A., Tamayo, P., Mootha, V.K., et al., 2005. Gene set enrichment analysis: A knowledge-based approach for interpreting genome-wide expression profiles. Proc Natl Acad Sci USA 102, 15,545–15,550.

Zhang, J., Finney, R.P., Rowe, W., et al., 2007. Systematic analysis of genetic alterations in tumors using Cancer Genome WorkBench (CGWB). Genome Res 17, 1111–1117.

RECOMMENDED RESOURCES

TCGA Data Portal
http://tcga-data.nci.nih.gov/tcga

TARGET Data Portal
http://target.nci.nih.gov/dataMatrix/TARGET_DataMatrix.html

ICGC Data Portal
http://dcc.icgc.org

ICBP Databases
http://icbp.nci.nih.gov/resources/icbp_software/databases

Broad Institute Cancer Datasets
http://www.broadinstitute.org/cgi-bin/cancer/datasets.cgi

Sage Bionetwork Datasets
http://www.sagebase.org/commons

REMBRANDT Data Browser
https://caintegrator.nci.nih.gov/rembrandt/

Cancer Genome Workbench
http://cgwb.nci.nih.gov

UCSC Cancer Genomics Browser
http://genome-cancer.soe.ucsc.edu

ARACNE, MINDY downloads
http://wiki.c2b2.columbia.edu/califanolab/index.php/Software

GSEA download
http://www.broadinstitute.org/gsea

PathOlogist download
ftp://ftp1.nci.nih.gov/pub/pathologist/

BRIDG Model
http://www.bridgmodel.org

caCORE
http://cabig.nci.nih.gov/community/concepts/caCORE_overview

caAdpater
http://cabig.nci.nih.gov/community/tools/caAdapter

Clinical Integration Hub
http://cabig.nci.nih.gov/community/tools/caBIGIntegrationHub

Mirth Connect
http://www.mirthcorp.com/products/mirth-connect

caGrid
http://cagrid-portal.nci.nih.gov

Cancer eScientist Portal
http://cabig.cancer.gov/escientist

CHAPTER

69

Diagnostic-Therapeutic Combinations

Jeffrey S. Ross

INTRODUCTION

The molecular diagnostic industry continues to grow at a double-digit pace to meet increasing demand for the integration of diagnostic tests with the selection of therapy and the development of truly personalized medicine. A wide variety of drugs for the treatment of cancer in late preclinical and early clinical development continue to be targeted to tumor-specific gene and protein signatures that may ultimately require co-approval of diagnostic and therapeutic products by the regulatory agencies. Members of the increasingly educated public, using the Internet and other resources, are demanding more and more information about their specific forms of cancer and how they might be arrested or cured with new therapies custom-designed for their individual clinical status. To respond to this demand, major pharmaceutical companies continue to partner with diagnostics companies as well as developing their own in-house capabilities, enabling them to efficiently produce more effective and less toxic integrated personalized medicine *"drug and test"* products. For diagnostic laboratories, functional imagers, and oncologists, this integration of diagnostics and therapeutics represents a major new opportunity to further advance cancer care as a paradigm of the new medicine based on the use of test results for selection, dosage, route of administration, and multi-drug

combinations. In the past several years, targeted therapies featuring combinations of drugs and diagnostics tests have become the standard of care in several approved indications. The resulting high level of uptake, coupled with premium prices, has made targeted therapies the leading therapy class in the oncology market.

The regulatory approvals in the United States and Europe of trastuzumab (herceptin®) for the treatment of HER2-overexpressing metastatic breast cancer (Figure 69.1), and imatinib mesylate (Gleevec®) for the treatment of patients with bcr/abl translocation-positive chronic myelogenous leukemia (CML) (Figure 69.2) and gastrointestinal stromal tumors featuring an activating c-*kit* growth factor receptor mutation, have created enthusiasm for anticancer targeted therapy in both the scientific and public communities (Mauro and Druker, 2001; O'Dwyer and Druker, 2001). Recent major news magazines and other public media have highlighted interest in new anticancer drugs that exploit disease-specific genetic defects as the target of their mechanism of action (Brown et al., 2001; Lemonick and Park, 2001). It is now widely held that the continued integration of molecular oncology and molecular diagnostics will further revolutionize oncology drug discovery and development, will customize the selection, dosing, and route of administration of both previously approved traditional agents and new therapeutics in clinical trials, and

Genomic and Personalized Medicine, 2nd edition
by Ginsburg & Willard. DOI: http://dx.doi.org/10.1016/B978-0-12-382227-7.00069-0

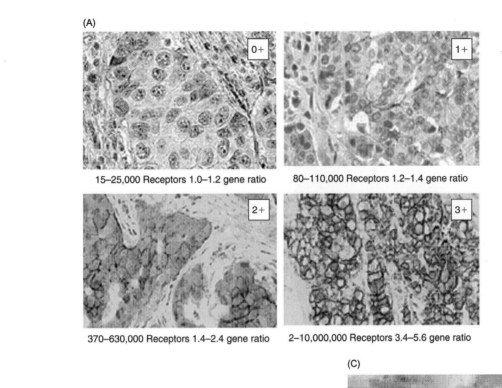

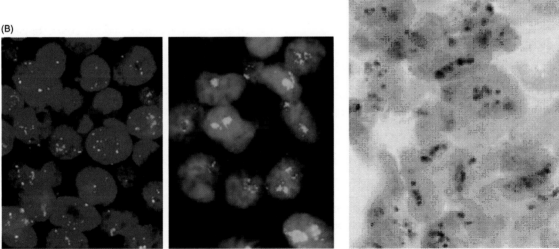

Figure 69.1 (A) HER-2/neu protein expression in infiltrating ductal breast cancer measured by immunohistochemistry using the Herceptest™ slide scoring system. Upper Left: 0+ (negative) staining for HER-2/neu protein. This level of staining is typically associated with 15,000–25,000 surface receptor molecules per cell and HER-2/neu gene copy to chromosome 17 copy ratios measured by fluorescence *in situ* hybridization (FISH) of 1.0 to 1.2. Upper Right: 1+ staining, associated with 80,000 to 110,000 receptors and gene ratio of 1.2 to 1.4. Lower Left: 2+ staining with membranous distribution, but no total cell encirclement, associated with 370,000 to 630,000 receptors and gene ratio of 1.4 to 2.4. Lower Right: 3+ staining with diffuse positive membranous distribution, total cell encirclement and "chicken wire" appearance, associated with 2,000,000 to 10,000,000 receptors and gene ratio of 3.4 to 5.6 (peroxidase – anti-peroxidase with Herceptest™ antibody × 200). Receptor count and FISH gene ratio data provided by Dr Kenneth Bloom, USLabs, Inc., Irvine, CA. (B) HER-2/neu gene amplification in infiltrating ductal breast cancer detected by FISH. Left: HER-2/neu gene amplification demonstrated by the Abbott-Vysis Pathvysion™ method, showing significant increase in HER-2/neu gene signals (red) compared to chromosome 17 signals (green), with a HER-2/neu gene ratio of 3.9. Right: HER-2/neu gene amplification using the Ventana Inform™ method, showing another breast cancer specimen with an absolute (raw) HER-2/neu gene copy number of 24. (C) HER-2/neu gene amplification in infiltrating breast cancer detected by chromogenic *in situ* hybridization (CISH) using anti- HER-2/neu probe and immunohistochemistry (IHC) with diaminobenzidine chromagen (SpotLight™ HER-2/neu probe, Zymed Corp., South San Francisco, CA). *Reproduced with permission from Ross and Hortobagyi, 2004.*

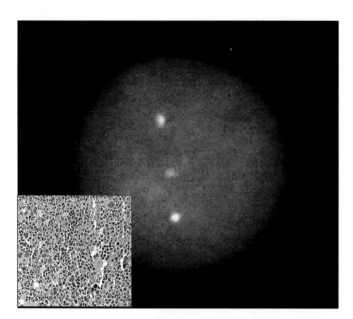

Figure 69.2 Chronic myelogenous leukemia and imatinib (Gleevec®) therapy. Photomicrograph demonstrates a *bcr/abl* translocation detected by fluorescence *in situ* hybridization (FISH) in a patient with a packed bone marrow biopsy diagnostic of chronic myelogenous leukemia (inset). Note the yellow "fusion" gene product, indicating the apposition of one green (chromosome 22) and one red (chromosome 9) resulting from the translocation. This patient was treated with single agent imatinib (Gleevec®) and achieved complete remission of bone marrow histology and absence of *bcr/abl* by routine cytogenetics and FISH assessment. *Reproduced with permission from Ross and Hortobagyi, 2004.*

will individualize medical care for the cancer patient (Amos and Paitnaik, 2002; Bottles, 2001; Evans and McLeod, 2003; Weinshilboum, 2003).

TARGETED THERAPIES FOR CANCER

From the regulatory perspective, targeted therapy has been defined as a drug with a specific reference in the approval label to a simultaneously or previously approved diagnostic test that must be performed before the patient can be considered eligible to receive that specific drug. This has enabled a new field known as "companion diagnostics." The co-approvals of the anti-breast cancer antibody trastuzumab (Herceptin®) and the required tissue-based tests for patient eligibility (Herceptest®, Pathway®, InSite®, and Pathvysion®) exemplify this strict definition of targeted therapy. However, for many

scientists and oncologists, anti-cancer drugs are considered to be "targeted" when they feature a focused mechanism that specifically acts on a well-defined target or biologic pathway that, when inactivated, causes regression or destruction of the malignant process. Examples of this less-rigorous definition of targeted therapy include hormonal-based therapies for breast cancer, small-molecule inhibitors of the epidermal growth factor receptor (EGFR), inhibitors of a mutated *BRAF* gene, inhibitors of a translocated *ALK* gene, blockers of invasion and metastasis-enabling proteins and enzymes, anti-angiogenesis agents, pro-apoptotic drugs, and proteasome inhibitors. Another definition of targeted therapy involves anticancer antibody therapeutics that seek out and kill malignant cells bearing the target antigen.

THE IDEAL TARGET

The ideal cancer target (Box 69.1) can be defined as a macromolecule that is crucial to the malignant phenotype and is not significantly expressed in vital organs and tissues; that has biologic relevance that can be reproducibly measured in readily obtained clinical samples; that is definably correlated with clinical outcome; and that interruption, interference, or inhibition of such a macromolecule yields a clinical response in a significant proportion of patients whose tumors express the target, with minimal to absent responses in patients whose tumors do not express the target. For antibody therapeutics, additional important criteria include the use of cell surface targets that, when complexed with the therapeutic naked or conjugated antibody, internalize the antigen–antibody complex by reverse pinocytosis, thus facilitating tumor cell killing.

Box 69.1 Features of the ideal anticancer target

- Crucial to the malignant phenotype
- Not significantly expressed in vital organs and tissues
- A biologically-relevant molecular feature
- Reproducibly measurable in readily obtained clinical samples
- Correlated with clinical outcome
- When interrupted, interfered with or inhibited, the result is a clinical response in a significant proportion of patients whose tumors express the target
- Responses in patients whose tumors do not express the target are minimal.

Reproduced with permission from Ross and Hortobagyi, 2004.

THE FIRST DIAGNOSTIC-THERAPEUTIC COMBINATION

Targeted therapy for cancer began in the early 1970s with the introduction of the estrogen receptor (ER) biochemical assay to select patients with painful metastatic breast cancer for surgical ablation of estrogen-producing organs (ovaries, adrenals) (Figure 69.3) (Osborne, 1998). The ER assay was followed by a similar, dextran-coated charcoal biochemical assay for the progesterone receptor (PR), and was subsequently converted to an immunohistochemistry (IHC) platform when the decreased size of primary tumors associated with self-examination and mammography-based screening programs prevented the use of the biochemical

test (Wilbur et al., 1992). The drug tamoxifen (Nolvadex®), which has both hormonal and non-hormonal mechanisms of action, has been the most widely prescribed anti-estrogen for the treatment of metastatic breast cancer and chemoprevention of the disease in high-risk women (Ciocca and Elledge, 2000; Jordan, 2003a). Although ER and progesterone receptor testing is the front line for predicting tamoxifen response, additional biomarkers including HER-2/neu (HER-2) and cathepsin D testing have been used to further refine therapy selection (Locker, 1998). The introductions of specific estrogen response modulators and aromatase inhibitors such as anastrozole (Arimidex®), letrozole (Femara®), and the combination chemotherapeutic estramustine (Emcyt®) (Buzdar et al., 2002; Ibrahim and Hortobagyi, 1999; Jordan, 2003b;

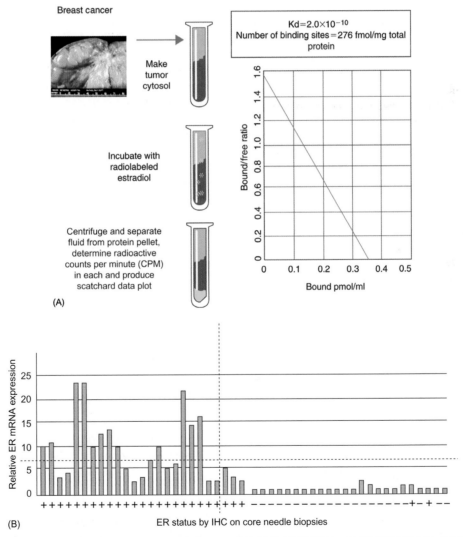

Figure 69.3 ER status determination. (A) Biochemical competitive binding assay for ER status determination. (B) Comparison of ER messenger RNA expression detected by microarray profiling, and corresponding ER protein expression measured by IHC. The concordance between ER levels determined by IHC and ER levels determined by gene expression profiling was about 95%. *Reprinted from Ross and Hortobagyi, 2004 with permission from the publisher.*

Miller et al., 2002b), have added new strategies for evaluating tumors for hormonal therapy. Recently, the College of American Pathologists (CAP) and the American Society of Clinical Oncology (ASCO) have issued joint guidelines covering the performance of ER and PR testing in clinical breast cancer specimens (Hammond et al., 2010).

Most recently, the Oncotype Dx® (Genomic Health, Redwood City, CA) multigene reverse transcriptase polymerase chain reaction (RT-PCR) multiplex assay, using a 21-gene probe set and mRNA extracted from paraffin blocks of stored breast cancer tissues, was introduced as a new guide to the use of tamoxifen in ER-positive, node-negative breast cancer patients (Paik et al., 2004). The assay features 16 cancer-related genes and five reference genes that were selected based on a series of transcriptional profiling experiments. The cancer-related genes include markers of proliferation such as Ki-67, markers of apoptosis such as survivin, invasion-associated protease genes such as *MMP11* and cathepsin L2, ER and HER2/*neu* gene family members, the glutathione S transferase genotype M1, CD68, a lysosomal monocyte/macrophage marker, and BAG1, a co-chaperone glucocorticoid receptor associated with bcl-2 and apoptosis. Using a cohort of 688 lymph node-negative, ER+ tumors obtained from patients enrolled in the NSABP B-14 clinical trial treated with tamoxifen alone, the 21-gene assay produced three prognosis scores of low, intermediate, and high risk. The recurrence rates for these patients at 10 years follow-up was 7% for the low-risk group, 14% for the intermediate-risk group and 31% for the high-risk group. The difference in relapse rates between the low-risk and high-risk patients was highly significant ($p < 0.001$). On multivariate analysis this assay predicted adverse outcome independent of tumor size and also predicted overall survival (Paik et al., 2004). Although not currently approved by the Food and Drug Administration (FDA), the interest in this new assay has been intense and it has become commercially available in a centralized format for new patients. Data presented at the 2005 ASCO Meeting showed that Oncotype Dx® is also capable of performing as a stand-alone prognostic test based on the test results in an untreated patient population (Paik et al., 2005). Detailed evaluation of the gene set in the Oncotype Dx® assay indicates that the mRNA levels of ER appear to be the most significant predictors in the node-negative ER(IHC)-positive population. Further studies are needed to validate the assay, and learn its best uses and limitations, given the evolving approach to hormonal therapy with non-tamoxifen drugs, the wide use of cytotoxic agents in the adjuvant setting for node-negative patients, and the availability of both RT-PCR-based and non-RT-PCR approaches to predicting breast cancer response to anti-estrogen and other anti-neoplastic agents used for treatment of the disease (Bast and Hortobagyi, 2004). A number of additional multigene predictors have been commercially developed for planning therapy for breast cancer patients. These assays use both fresh and paraffin-embedded material and a variety of test platforms, including IHC, FISH, RT-PCR and microarray-based gene expression profiling (Ross et al., 2008).

DIAGNOSTIC-THERAPEUTIC COMBINATIONS FOR LEUKEMIA AND LYMPHOMA

The introduction of immunophenotyping for leukemia and lymphoma was followed by the first applications of DNA-based assays, the polymerase chain reaction, and RNA-based molecular technologies in these diseases, that complemented continuing advances in tumor cytogenetics (Gleissner and Thiel, 2001; Rubnitz and Pui, 1999). In addition to imatinib (Gleevec®), a targeted therapy for chronic myelogenous leukemia, molecular targeted therapy in hematologic malignancies includes the use of all-*trans*-retinoic acid (ATRA) for the treatment of acute promyelocytic leukemia (Grimwade and Lo Coco, 2002), anti-CD20 antibody therapeutics such as rituximab (Rituxan®) (Kiyoi and Naoe, 2002) that target non-Hodgkin's lymphomas, and the emerging Flt-3 target for a subset of acute myelogenous leukemia patients (see below) (Ross et al., 2003a).

HER-2-Positive Breast Cancer and Trastuzumab (Herceptin®)

After the introduction of hormone receptor testing, some 30 years elapsed before the next major targeted cancer chemotherapy program for a solid tumor was developed. In the mid-1980s, the discovery of the *HER-2* (c-*erb*B2) gene and protein, and their subsequent association with an adverse outcome in breast cancer, provided clinicians with a new biomarker that could be used to guide adjuvant chemotherapy (Slamon et al., 2001). The development of trastuzumab (Herceptin®) – a humanized monoclonal antibody designed to treat advanced metastatic breast cancer that had failed first- and second-line chemotherapy – caused a rapid, wide adoption of HER-2 testing of patients' primary tumors (Schnitt and Jacobs, 2001). The importance of accurate HER-2 testing has recently been emphasized by the dramatic results of the use of trastuzumab in the adjuvant setting (Piccart-Gebhart et al., 2005; Romond et al., 2005). However, soon after its approval, widespread confusion concerning the most appropriate diagnostic test to determine HER-2 status in formalin-fixed paraffin-embedded breast cancer tissues substantially impacted trastuzumab use (Hayes and Thor, 2002; Hortobagyi, 2001; Masood and Bui, 2002; Paik et al., 2002; Tanner et al., 2000; Wang et al., 2000, 2001; Zhao et al., 2002). Since its launch in 1998, trastuzumab has become an important therapeutic option for patients with HER-2-positive breast cancer (Bast et al., 2001; Ligibel and Winer, 2002; McKeage and Perry, 2002; Shawver et al., 2002).

In general, when specimens have been carefully fixed, processed, and embedded, there has been excellent correlation between *HER-2* gene copy status determined by FISH and HER-2 protein expression levels determined by IHC (Slamon et al., 2001). The main use of either method in current clinical practice is focused on the negative prediction of response to trastuzumab. Currently, both ASCO and CAP consider HER-2 testing to be part of the standard work-up and management of breast cancer (Hammond et al., 2000; Pawlowski et al., 2000; Ross et al., 2009). Recently, the chromogenic (non-fluorescent) *in situ* hybridization (CISH) technique has been used to determine *HER-2* gene amplification status with promising results (Figure 69.1) (Ross et al., 2009). In a recent study, the CISH technique achieved a 98% concordance with the FDA-approved FISH technique (Hanna and Kwok, 2006). Non-morphologic approaches for determining HER-2 status have also been developed. The RT-PCR technique, which has predominantly been used to detect HER-2 mRNA in peripheral blood and bone marrow samples, has correlated more with gene amplification status than IHC levels of primary tumors, but has been less successful as a predictor of survival (Bieche et al., 1999; Dressman et al., 2003; Tubbs et al., 2001). With the advent of laser capture microscopy and the acceptance of RT-PCR as a routine and reproducible laboratory technique, the use of RT-PCR for the determination of HER-2 status may increase in the future. The complementary DNA (cDNA) microarray-based method of detecting HER-2 mRNA expression levels has recently received interest as an alternative method for measuring HER-2/neu status in breast cancer (Fornier et al., 2005; Pusztai et al., 2003). Finally, the FDA-approved serum HER-2/enzyme-linked immunosorbent assay (ELISA), test measuring circulating HER-2 (p185neu) protein, has seen increased clinical use as a method for monitoring response to trastuzumab (Ross et al., 2009). A summary of HER-2 testing methods in breast cancer is shown in Table 69.1.

TABLE 69.1	Methods of detection of HER-2/neu status in breast cancer		
Method	**Target**	**FDA-approved**	**Slide-based**
IHC	Protein	Yes*	Yes
FISH	Gene	Yes*	Yes
CISH	Gene	No	Yes
Southern blot	Gene	No	No
RT-PCR	mRNA	No	No
Microarray TP	mRNA	No	No
Tumor ELISA	Protein	No	No
Serum ELISA	Protein	Yes†	No

IHC = immunohistochemistry, FISH = fluorescence *in situ* hybridization, CISH = chromogenic *in situ* hybridization, TP = transcriptional profiling, ELISA = enzyme-linked immunosorbent assay.
*For prognosis and prediction of response and eligibility to receive trastuzumab therapy.
†For monitoring response of breast cancer to treatment.
Table reprinted from Ross and Hortobagyi, 2004 with permission from the publisher.

Other Targeted Anticancer Therapies Using Antibodies

An unprecedented number and variety of targeted small molecule and antibody-based therapeutics are currently in early development and clinical trials for the treatment of cancer. Therapeutic antibodies have become a major strategy in clinical oncology because of their ability to specifically bind to primary and metastatic cancer cells with high affinity, and create antitumor effects by complement-mediated cytolysis and antibody-dependent cell-mediated cytotoxicity (naked antibodies), or by the focused delivery of radiation or cellular toxins (conjugated antibodies) (Goldenberg, 2002; Hemminki, 2002; Milenic, 2002; Reichert, 2002; Ross et al., 2003b; Ross et al., 2009). Currently, there are eight anticancer therapeutic antibodies approved by the US Food and Drug Administration (FDA) for sale in the United States (Table 69.2). Therapeutic monoclonal antibodies are typically of the IgG class, containing two heavy and two light chains. The heavy chains form a fused "Y" structure, with two light chains running in parallel to the open portion of the heavy chain. The tips of the heavy–light chain pairs form the antigen binding sites, with the primary antigen-recognition regions known as the complementarity-determining regions.

The early promise of mouse monoclonal antibodies for the treatment of human cancers was not realized, because (1) unfocused target selection led to the identification of target antigens that were not critical for cancer cell survival and progression, (2) there was a low overall potency of naked mouse antibodies as anticancer drugs, (3) antibodies penetrated tumor cells poorly, (4) there was limited success in producing radioisotope and toxin conjugates, and (5) the development of human anti-mouse antibodies (HAMA) prevented the use of multiple dosing schedules (Reilly et al., 1995).

The next advance in antibody therapeutics began in the early 1980s when recombinant DNA technology was applied to antibody design, to reduce the antigenicity of murine and other rodent-derived monoclonal antibodies. Chimeric antibodies were developed where the constant domains of the human IgG molecule were combined with the murine variable regions by transgenic fusion of the immunoglobulin genes; the chimeric monoclonal antibodies were produced from engineered hybridomas and Chinese hamster ovary (CHO) cells (Merluzzi et al., 2000; Winter and Harris, 1993). The use of chimeric antibodies significantly reduced the HAMA responses, but did not completely eliminate them (Kuus-Reichel et al., 1994; Merluzzi et al., 2000). Although several chimeric antibodies achieved regulatory approval, certain targets required humanized antibodies to achieve appropriate dosing. Partially humanized antibodies were then developed where the six complementarity-determining regions of the heavy and light chains, and a limited number of structural amino acids of the murine monoclonal antibody, were grafted by recombinant technology to the complementarity-determining region-depleted human IgG scaffold (Milenic, 2002). Although this process further reduced or eliminated the HAMA responses,

TABLE 69.2	Antibody therapeutics for cancer					
Name	**FDA approval**	**Source** *partners*	**Type**	**Target**	**Indication(s) (both approved and investigational)**	**Companion diagnostic**
Alemtuzumab *Campath®*	May–01	BTG *ILEX Oncology Schering AG*	Monoclonal antibody, humanized Anticancer, immunological Multiple sclerosis treatment Immunosuppressant	CD52	Cancer, leukemia, chronic lymphocytic Cancer, leukemia, chronic myelogenous Multiple sclerosis, chronic progressive	No (leukemia immuno-phenotyping)
Daclizumab *Zenapax®*	Mar–02	Protein Design Labs *Hoffmann-La Roche*	Monoclonal IgG$_1$ Chimeric Immunosuppressant Antipsoriasis Antidiabetic Multiple sclerosis treatment		Transplant rejection, general Transplant rejection, bone marrow Uveitis Multiple sclerosis, relapsing-remitting Multiple sclerosis, chronic progressive Cancer, leukemia, general Psoriasis Diabetes, Type I Asthma Colitis, ulcerative	No
Rituximab *Rituxan®*	Nov–97	IDEC *Genentech Hoffmann-La Roche Zenyaku Kogyo*	Monoclonal IgG$_1$ Chimeric Anticancer, immunological Antiarthritic, immunological Immunosuppressant	CD20	Cancer, lymphoma, non-Hodgkin's Cancer, lymphoma, B-cell Arthritis, rheumatoid Cancer, leukemia, chronic lymphocytic Thrombocytopenic purpura	No (lymphoma immunophenotyping)
Trastuzumab *Herceptin®*	Sep–98	Genentech *Hoffmann-La Roche ImmunoGen*	Monoclonal IgG$_1$, humanized Anticancer, immunological	p185neu	Cancer, breast Cancer, lung, non-small cell Cancer, pancreatic	Yes
Gemtuzumab *Mylotarg®*	May–00	Wyeth/AHP	Monoclonal IgG$_4$ Humanized	CD33/ coleacheamycin	Cancer, leukemia, AML (patients older than 60 years)	No (leukemia typing)
Ibritumomab *Zevalin®*	Feb–02	IDEC	Monoclonal IgG$_1$ Murine Anticancer	CD20/^{90}Yttrium	Cancer, lymphoma, low grade, follicular, transformed non-Hodgkin's (relapsed or refractory)	No (lymphoma immunophenotyping)
Tositomumab *Bexxar®*	June–03	Corixa	Anti-CD 20 murine monoclonal antibody with ^{131}I conjugation	CD20	Cancer, lymphoma, non-Hodgkin's	No (lymphoma immunophenotyping)

(continued)

TABLE 69.2	(Continued)					
Name	**FDA approval**	**Source _partners_**	**Type**	**Target**	**Indication(s) (both approved and investigational)**	**Companion diagnostic**
Cetuximab _Erbitux®_	Feb–04	Imclone Bristol Myers Squibb	Anti-EGFR monoclonal antibody	EGFR	Approved for third line treatment of metastatic colorectal cancer that has failed primary chemotherapy	Yes Originally EGFR expression test by IHC. Now _KRAS_ mutation test by gene sequencing.
Panitumumab _Vectibix®_	Sep–06	Amgen	Anti-EGFR monoclonal antibody	EGFR	Approved for third line treatment of metastatic colorectal cancer that has failed primary chemotherapy	Yes Originally EGFR expression test by IHC. Now _KRAS_ mutation test by gene sequencing.
Bevacizumab _Avastin®_	Feb–04	Genentech	Anti-VEGF (ligand)	VEGF	Avastin is approved for use in combination with intravenous 5-fluorouracil-based chemotherapy as a treatment for patients with first-line, or previously untreated, metastatic colorectal cancer.	No
Edrecolomab _Panorex™_	Jan–95 (Europe only, not FDA-approved)	Glaxo-Smith-Kline	Monoclonal IgG$_{2A}$ Murine Anticancer	Epithelial cell adhesion molecule (EpCAM)	Cancer, colorectal	No

Table adapted from Ross and Hortobagyi, 2004 with permission from the publisher.

in many cases significant further antibody design procedures were needed to reestablish the required specificity and affinity of the original murine antibody (Isaacs, 2001; Jones et al., 1986; Purim, 1994; Watkins and Ouwehand, 2000).

A second approach to reducing the immunogenicity of monoclonal antibodies has been to replace immunogenic epitopes in the murine variable domains with benign amino acid sequences, resulting in a de-immunized variable domain. The de-immunized variable domains are genetically linked to human IgG constant domains, to yield a de-immunized antibody (Biovation, Aberdeen, Scotland). Additionally, primatized antibodies were subsequently developed that featured a chimeric antibody structure of human and monkey that, as a nearly exact copy of a human antibody, further reduced immunogenicity and enabled continuous repeat dosing and chronic therapy (Reff et al., 2002). Finally, fully human antibodies have now been developed using murine sources and transgenic techniques (Reff et al., 2002).

Using modern antibody design and de-immunization technologies, scientists and clinicians have attempted to improve the efficacy and reduce the toxicity of anticancer antibody therapeutics (Chester and Hawkins, 1995; Nielsen and Marks, 2000; Reff and Heard, 2001; Reilly et al., 1995; Ross et al., 2009). The bacteriophage antibody design system has facilitated the development of high-affinity antibodies by increasing antigen binding rates and reducing corresponding detachment rates (Nielsen and Marks, 2000). Increased antigen binding is also achieved in bivalent antibodies with multiple attachment sites, a feature known as avidity. Modern antibody design has endeavored to create small antibodies that can penetrate to cancerous sites but maintain their affinity and avidity. A variety of approaches have been used to increase antibody efficacy (Ross et al., 2009). Clinical trials have recently combined anticancer antibodies with conventional cytotoxic drugs, yielding promising results (Goldenberg, 2002; Hemminki, 2002; Milenic, 2002; Reichert, 2002; Ross et al., 2003b; Ross et al., 2009). The applications of radioisotope, small molecule cytotoxic drug, and protein toxin conjugation have yielded promising results in clinical trials and achieved regulatory approval for several drugs now on the market (see below). Antibodies have also been designed to increase their enhancement of effector functions of antibody-dependent cellular cytotoxicity. Another cause of toxicity of conjugated antibodies has been the limitations of the conjugation technology, which can restrict the ratio of the number of toxin molecules per antibody molecule (Goldenberg, 2002; Ross et al., 2009; Watkins and Ouwehand, 2000). Methods designed to overcome the toxicity of conjugated antibodies include the use of antibody-targeted, liposomal, small molecule drug conjugates, and the use of antibody conjugates with drugs in nanoparticle formats to enhance bonding strength that enable controlled release of the cytotoxic agent. Another technique that uses site-selective prodrug activation to reduce bystander tissue toxicity is the antibody-directed enzyme prodrug therapy (ADEPT). An antibody-bound enzyme is targeted to tumor cells. This allows for selective activation of a nontoxic prodrug to a cytotoxic agent at the tumor site for cancer therapy.

A variety of factors can reduce antibody efficacy: (1) limited penetration of the antibody into a large solid tumor or into vital regions such as the brain, (2) reduced extravasations of antibodies into target sites due to decreased vascular permeability, (3) cross-reactivity and nonspecific binding of antibody to normal tissues reduces targeting effect, (4) heterogeneous tumor uptake results in untreated zones, (5) increased metabolism of injected antibodies reduces therapeutic effects, and (6) HAMA and human anti-human antibodies form rapidly and inactivate the therapeutic antibody.

Toxicity has been a major obstacle in the development of therapeutic antibodies for cancer (Goldenberg, 2002; Ross et al., 2003b; Ross et al., 2009; Watkins and Ouwehand, 2000). Cross-reactivity with normal tissues can cause significant side effects for unconjugated (naked) antibodies, which can be enhanced when the antibodies are conjugated with toxins or radioisotopes. Immune-mediated complications can include dyspnea from pulmonary toxicity, occasional central and peripheral nervous system complications, and decreased liver and renal function. On occasion, unexpected toxic complications can be seen, such as the cardiotoxicity associated with the HER-2 targeting antibody trastuzumab. Radioimmunotherapy with isotopic-conjugated antibodies can also cause bone marrow suppression (see below).

Unconjugated or naked antibodies include a variety of targeting molecules both on the market and in early and late clinical development. A variety of mechanisms have been cited to explain the therapeutic benefit of these drugs, including enhanced immune effector functions and direct inactivation of the targeted pathways, as seen in the antibodies directed at surface receptors, such as HER-1 (EGFR) and HER-2 (Amos and Patnaik, 2002; Brown et al., 2001; Lemonick and Park,

2001; Mauro and Druker, 2001). Surface receptor targeting can reduce intracellular signaling, resulting in decreased cell growth and increased apoptosis (Reff et al., 2002).

As seen in Table 69.2, of the nine anticancer antibodies on the market in the US, two are conjugated with a radioisotope – ^{90}Y-ibritumomab tiuxetan (Zevalin®) and ^{131}I-tositomumab (Bexxar®) – and one is conjugated to a complex natural product toxin – gemtuzumab ozogamicin (Mylotarg®). Conjugation procedures have been designed to improve the efficacy of antibody therapy and have used a variety of methods to complex the isotope, toxin, or cytotoxic agent to the antibody (Goldenberg, 2002; Ross et al., 2009). Cytotoxic small molecule drug conjugates have been widely tested, but enthusiasm for this approach has been limited by the relatively low potency of the compounds (Ross et al., 2009). Fungal-derived, potent toxins have yielded greater success. The calicheamicin-conjugated anti-CD33 antibody gemtuzumab ozogamicin has been approved for the treatment of acute myelogenous leukemia, and a variety of antibodies conjugated with the fungal toxin maytansanoid (DM-1) are in preclinical development and early clinical trials. The interest in radioimmunotherapy increased significantly in 2001 with the FDA approvals of the ^{90}Y-conjugated anti-CD20 antibody ^{90}Y-ibritumomab tiuxetan and the ^{131}I-conjugated anti-CD20 antibody ^{131}I-tositumomab. A variety of isotopes are under investigation in addition to ^{90}Y as potential conjugates for anticancer antibodies (Goldenberg, 2002). Radioimmunotherapy features the phenomenon of the bystander effect, in which if antigen expression is heterogeneous, extensive tumor cell killing can still take place, even on non-expressing cells, but it can also lead to significant toxicity when the neighboring cells are vital non-neoplastic tissues such as the bone marrow and liver.

Antibody Therapeutics for Hematologic Malignancies

The earliest and most successful clinical use of antibodies in oncology has been for the treatment of hematologic malignancies (Burke et al., 2002; Goldenberg, 2002; Linenberger et al., 2002; Reff et al., 2002; Ross et al., 2003b; Ross et al., 2009; Stevenson et al., 2002; Watkins and Ouwehand, 2000; Wiseman et al., 2002).

Rituximab (Rituxan®)

Approved in 1997, rituximab (Rituxan®) is arguably the most commercially successful anticancer drug of any type since the introduction of taxanes. Rituximab sales exceeded $700 million in the United States in 2001 (Reichert, 2002). Targeting the CD20 surface receptor common to many B cell non-Hodgkin's lymphoma subtypes, rituximab is a chimeric monoclonal IgG$_1$ antibody that induces apoptosis, antibody-dependent cell cytotoxicity and complement-mediated cytotoxicity (Reff et al., 2002). It has achieved significantly improved disease-free survival rates compared with patients receiving cytotoxic agents alone (Coiffier, 2002; Dillman, 2001; Grillo-Lopez, 2002; Grillo-Lopez et al., 2002).

^{90}Y-ibritumomab Tiuxetan (Zevalin®)

^{90}Y-ibritumomab tiuxetan (Zevalin®) consists of the murine version of the anti-CD20 chimeric monoclonal antibody rituximab, which has been covalently linked to the metal chelator MD-DTPA, permitting stable binding of ^{111}In when used for radionucleotide tumor imaging and ^{90}Y when used to produce enhanced targeted cytotoxicity (Dillman, 2002; Gordon et al., 2002; Krasner and Joyce, 2001; Wagner et al., 2002). In early 2002, ^{90}Y-ibritumomab tiuxetan became the first radioconjugated antibody therapeutic for cancer approved by the FDA. Since its approval, numerous patients who have received ^{90}Y-ibritumomab tiuxetan after becoming refractory to a rituximab-based regimen have achieved significant responses (Dillman, 2002; Gordon et al., 2002).

Gemtuzumab Ozogamicin (Mylotarg®)

FDA approval of gemtuzumab ozogamicin (Mylotarg®) in 2000 marked the first introduction of a plant toxin-conjugated antibody therapeutic (Bross et al., 2001; Larson et al., 2002; Nabhan and Tallman, 2002; Sievers and Linenberger, 2001; Stadtmauer, 2002).

Alemtuzumab (Campath®)

Alemtuzumab (Campath®), a humanized monoclonal antibody, was approved in mid-2001 for the treatment of B-cell chronic lymphocytic leukemia in patients who have been treated with alkylating agents and who have failed fludarabine therapy (Dumont, 2002; Pangalis et al., 2001).

Daclizumab (Zenapax®)

Daclizumab (Zenapax®) is a chimeric monoclonal antibody that targets the interleukin-2 receptor.

^{131}I-tositumomab (Bexxar®)

^{131}I-tositumomab (Bexxar®) is a radiolabeled anti-CD20 murine monoclonal antibody approved in 2003 for the treatment of relapsed and refractory follicular/low-grade and transformed non-Hodgkin's lymphoma (Cheson, 2002; Zelenetz, 2003).

Antibody Therapeutics for Solid Tumors

Interest in the development of antibody therapeutics for solid tumors among many commercial organizations and universities has been significantly impacted by the technologic advances in antibody engineering and the approval and recent clinical and commercial success of trastuzumab, the only therapeutic antibody approved by the FDA for the treatment of solid tumors (edrecolomab is approved in Germany, but not in the United States).

Trastuzumab (Herceptin®)

Trastuzumab (Herceptin®) has been described above. During the six years since the FDA approved trastuzumab, two additional antibodies have been approved for the treatment of solid tumors (cetuximab and bevacizumab). Recently, a conjugated version of trastuzumab, trastuzumab-DM1, has achieved significant benefit in patients with advanced HER-2-positive breast cancer refractory to treatment with trastuzumab naked antibody (Krop et al., 2010). More recently, trastuzumab has been approved in Europe and is under consideration at the FDA for the treatment of HER-2-overexpressing gastric and gastroesophageal junction adenocarcinomas (Bang et al., 2010b). The addition of the small molecule HER-1/HER-2 inhibitor lapatinib (see below) has shown promise for the treatment of HER-2-positive breast cancer metastatic to the brain (Yip et al., 2010). Progress is continuing in this field, and a number of both late-stage and early-stage products in development show substantial promise.

Cetuximab (Erbitux®) and Panitumumab (Vectabix®)

The epidermal growth factor receptor (EGFR), also known as HER-1, is the target of two FDA-approved small molecule drugs (see below) and one FDA-approved antibody (Mendelsohn and Baselga, 2000). Cetuximab (Erbitux®) is a chimeric monoclonal antibody that binds to the EGFR with high affinity, blocking growth factor binding, receptor activation, and subsequent signal transduction events, and leading to cell proliferation (Baselga, 2001). Cetuximab enhanced the antitumor effects of chemotherapy and radiotherapy in preclinical models by inhibiting cell proliferation, angiogenesis, and metastasis, and by promoting apoptosis (Baselga, 2001). Cetuximab has been evaluated both alone and in combination with radiotherapy and various cytotoxic chemotherapeutic agents in a series of phase II/III studies that primarily treated patients with either head and neck or colorectal cancer (Baselga, 2001; Herbst and Langer, 2002). Breast cancer trials are also underway (Leonard et al., 2002). Although the FDA approval process for cetuximab was initially slowed because of concerns over clinical trial design and outcome data management (Reynolds, 2002), in February 2004 the antibody was approved for use in the treatment of advanced metastatic colorectal cancer (CRC). Similar to trastuzumab, the development of cetuximab also included an immunohistochemical test for determining EGFR overexpression, to define patient eligibility to receive the antibody (Wong, 2005). However, there have been conflicting reports suggesting that the use of a pharmacodiagnostic test (EGFR immunostaining) is unnecessary for the selection of cetuximab in colorectal cancer therapy (Saltz, 2005).

During clinical development, it was noted that cetuximab showed evidence of benefit in only 10–20% of patients with metastatic CRC. Thus, to reduce costs and improve potential patient outcomes, investigators sought novel biomarkers that could predict resistance to anti-EGFR therapy for CRC. The *KRAS* oncogene is a major component of the ras/raf pathway known to regulate the cell cycle in normal cells and CRC cells. Studies have indicated that *KRAS* mutation is found in between 27% and 53% of CRC cases (Ross et al., 2010). The vast majority of *KRAS* mutations are located in the second and third exons in codons 12, 13, and 61. Mutated *KRAS* CRC, with few exceptions, appears to be completely resistant to benefit from treatment by cetuximab and panitumumab in the metastatic setting. Data in support of this association appear fully confirmed for mutations in codons 12 and 13 (Ross et al., 2010). For codons 61 and 146, the evidence for conferring resistance to cetuximab and panitumumab is not as robust as for codons 12 and 13 (Van Cutsem et al., 2009). In mid 2009, ASCO recommended that all patients with metastatic CRC who are candidates for anti-EGFR antibody therapy have their tumors tested for *KRAS* mutations in a Clinical Laboratory Improvement Amendments-accredited laboratory. If *KRAS* mutations are identified in codons 12 or 13, ASCO recommends that patients with metastatic CRC should not receive anti-EGFR antibody therapy as part of their treatment. More recently, the FDA has incorporated this ASCO provisional opinion, and essentially all patients with newly diagnosed CRC considered for anti-EGFR antibody therapeutics are first tested for *KRAS* mutation, and only patients with wild-type *KRAS* gene tumors are considered eligible to receive these drugs.

Recent clinical trials have found significant efficacy for cetuximab in the treatment of head and neck squamous cell cancers, often in combination with radiation treatment (Harari and Huang, 2006).

Bevacizumab (Avastin®)

Bevacizumab (rhuMAb-VEGF) is a humanized murine monoclonal antibody targeting the vascular endothelial growth factor ligand (VEGF). It was approved by the FDA in 2004 for the front-line or first-line treatment of metastatic colorectal cancer, in combination with chemotherapy. VEGF regulates both vascular proliferation and permeability, and functions as an anti-apoptotic factor for newly formed blood vessels (Chen et al., 2000; Ferrara, 2005; Rosen, 2002). In addition to its approved indication in colorectal cancer, bevacizumab has shown promising efficacy in combination with cytotoxic drugs for the treatment of non-small cell lung cancer (Wakelee and Belani, 2005), renal cell carcinoma (Stadler, 2005), pancreatic cancer (Bruckner et al., 2005), breast cancer (De Gramont and Van Cutsem, 2005), and prostate cancer (Berry and Eisenberger, 2005). Unlike cetuximab, the development of bevacizumab has not included a diagnostic eligibility test. Neither direct measurement of VEGF expression in tumor, circulating VEGF levels in serum or urine, or assessment of tumor microvessel density have been incorporated into the clinical trials or linked to the response rates to the antibody. To date, a number of theories have been proposed as to the actual mechanism of action of bevacizumab and the relative contributions of direct anti-angiogenesis and other tumor vasculature stabilization, and cytotoxic chemotherapy potentiation effects of the antibody (Blagosklonny, 2005;

Hurwitz and Kabbinavar, 2005). In summary, currently used without an integrated diagnostic eligibility test, bevacizumab cannot be considered a true targeted therapy, and further development of this agent for use in prostatic, breast, lung, renal, and other cancers may well be inhibited by the inability to individually select patients who will be more likely to benefit from its use, either alone or in combination with other traditional cytotoxic drugs, antibodies, and novel drugs.

Edrecolomab (Panorex®)

Edrecolomab is a murine IgG$_{2A}$ monoclonal antibody that targets the human tumor-associated antigen Ep-CAM (17-1A).

SELECTED TARGETED ANTICANCER THERAPIES USING SMALL MOLECULES

Table 69.3 lists selected small molecule drugs designed to target specific genetic events and biologic pathways critical to cancer growth, invasion, and metastasis.

Targeted Small Molecule Drugs for Hematologic Malignancies

ATRA

Arguably the first truly targeted therapy after the development of hormonal therapy for breast cancer was the development of ATRA for the treatment of acute promyelocytic leukemia, a subset of acute nonlymphocytic leukemia featuring a disease-defining retinoic acid receptor activating t(15:17) reciprocal translocation (Fang et al., 2002; Parmar and Tallman, 2003). For these selected patients, direct targeting of the retinoic acid receptor with ATRA has resulted in very high response rates, delay in disease progression, and long-term cures (Fang et al., 2002; Parmar and Tallman, 2003).

Imatinib (Gleevec®)

The development in 2001 of imatinib for patients with CML ushered in new excitement in both the scientific and public communities for targeted anticancer therapy. Imatinib received fast-track approval by the FDA as an ATP-competitive selective inhibitor of *bcr-abl*. It has unprecedented efficacy for the treatment of early-stage CML, typically achieving durable, complete hematologic and complete cytogenetic remissions, with minimal toxicity (Druker, 2003; Goldman and Melo, 2003; O'Brien et al., 2003). Imatinib is a true targeted therapy for leukemia in that a test for the *bcr/abl* translocation must be performed before a patient will be considered eligible to receive the drug. The prediction of resistance to imatinib in early-phase CML has been the subject of numerous studies (Lange et al., 2005; O'Hare et al., 2006). The current goal is to predict resistance emergence with gene mutation testing, and employ novel tyrosine kinase inhibitors to attempt to overcome blast cells that have lost the ability to bind

TABLE 69.3	Selected small molecule drugs designed to target specific genetic events and biologic pathways critical to cancer growth and progression				
Target	**Drug**	**Source**	**Clinical development status**	**Comment**	**Companion diagnostic**
PML-RARA	ATRA	Promega	Approved	First true targeted therapy since the introduction of ER testing and hormonal therapy for breast cancer	No (leukemia typing)
Bcr/abl in CML	Imatinib	Novartis	Approved	Has emerged as standard of care for early stage CML	Yes (bcr/abl translocation detection by FISH or RT-PCR)
Bcr/abl in CML	Dasatinib	Bristol Myers	Approved	Originally approved for imatinib-resistant CML. Recently has been used in front-line CML treatment. In clinical trials as a src kinase inhibitor in solid tumors	Yes (bcr/abl translocation detection by FISH or RT-PCR)
c-Kit in GIST PDGF-R	Imatinib	Novartis	Approved	Responses in relapsed/metastatic GIST can be predicted by the location of the activating c-kit mutation	Yes (cKit expression by IHC)
Flt-3 in AML	SU5416 PKC412 MLN-518	Pfizer Novartis Millennium	Early stage clinical trials	Small molecule drugs that target the flt-3 internal tandem duplication seen in 30% of AML	Pending
EGFR in NSCLC	Gefitinib	Astra Zeneca	Approved/ withdrawn	No survival benefit in NSCLC phase III trial. Has returned to the market using EGFR mutation status as an entry screen for treatment selection	Yes (EGFR mutation status)
EGFR in NSCLC and pancreatic cancer	Erlotinib	Genentech/ OSI	Approved	Survival benefit demonstrated in NSCLC and pancreatic cancer. EGFR mutation status currently used as favored entry diagnostic for treatment selection	Yes (EGFR mutation status)
Anti-angiogenesis in renal cell carcinoma	BAY 43-9006	Bayer	Approved	Raf kinase inhibitor also targets PDGFR and VEGFR	No
Anti-angiogenesis in myelodysplastic syndrome	Lenolidamide	Celgene	Approved	Also in clinical trials for the treatment of multiple myeloma	No
Other anti-angiogenesis	Thalidomide Sunitinib	Celgene Pfizer/Sugen	Approved Approved	Multiple Myeloma Gastrointestinal tumor	No
Proteasome in Multiple Myeloma	Bortezomib	Millennium	Approved	Proteasome inhibition effective in hematologic malignancies, but of uncertain potential for the treatment of solid tumors	No (studies pending)
HER-2 in breast cancer	Lapatinib	Glaxo-Smith Kline	Approved	HER-1/HER-2 tyrosine kinase inhibition in HER-2-positive breast cancer	Yes (HER-2 overexpression and amplification by IHC and FISH)

(continued)

Target	Drug	Source	Clinical development status	Comment	Companion diagnostic
TABLE 69.3 (Continued)					
BRAF (V600E mutation)	Vemurafenib	Plexxikon/ Roche/ Daiichi-Sankyo	Approved	Significant efficacy in phase I-II trial for the treatment of metastatic melanoma and thyroid cancer	Yes (BRAF mutation testing by Roche Cobas Genotyping)
EML4-ALK translocation	Critozinib	Pfizer	Approved	Significant efficacy in phase II trial of NSCLC (adenocarcinoma) featuring EML4-ALK translocation (5–8% of NSCLC).	Yes (ELM4-ALK translocation testing by FISH)

AML = acute myeloid leukemia, GIST = gastrointestinal stromal tumor, NSCLC = non-small cell lung cancer, PDGF = platelet-derived growth factor. Table adapted from Ross and Hortobagyi, 2004 with permission from the publisher.

imatinib to the ATP-binding pocket of the fusion gene (Lange et al., 2005; O'Hare et al., 2006). Recently, the development of the tyrosine kinase inhibitor dasatinib has shown significant promise for treatment of patients who have developed resistance or cannot tolerate continuing doses for imatinib (Talpaz et al., 2006).

Imatinib has also achieved regulatory approval for the treatment of relapsed and metastatic gastrointestinal stromal tumors (GISTs), which characteristically feature an activating point mutation in the c-*kit* receptor tyrosine kinase gene (Von Mehren, 2003). For GISTs, the response to imatinib treatment appears to be predictable based on the location of the c-*kit* mutation (Verweij et al., 2003). The use of imatinib in GIST is also an example of targeted therapy, as a measurement of c-*kit* expression, usually performed by IHC, is required to confirm the diagnosis and render the patient eligible for treatment. Interestingly, most commercially available antibodies for c-*kit* recognize the total c-*kit* and do not distinguish the activated or phosphorylated version, which is the actual target of imatinib. Currently, the high treatment failure rate is directly linked to the test used to characterize the patients. It is anticipated that either the use of specific antibodies designed to identify the activated c-*kit* gene or directed sequencing of the c-*kit* gene may be required before imatinib is prescribed for patients with recurrent or metastatic GIST. An alternative to c-*kit* mutation testing for the prediction of resistance to imatinib, functional imaging after initial dosing of the drugs, has been employed for patients with metastatic GIST (Heinicke et al., 2005). In early 2006, the anti-angiogenesis agent sunitinib was approved by the FDA for the treatment of advanced GIST that has become resistant to imatinib therapy.

Dasatinib and Nerlotinib

Since the approval of imatinib, two additional bcr-abl tyrosine kinase inhibitors, dasatinib (Sprycel®) and nerlotinib (Tasigna®), have been approved for the treatment of CML (Gora-Tybor, 2012). These drugs were originally developed for

the treatment of imatinib-resistant CML, but have found their way into front-line treatment in certain clinical trials and individual patient treatments. Dasatinib is also an src kinase inhibitor and is in early stage clinical trials for the treatment of solid tumors, such as triple-negative breast cancer, that overexpress src kinase.

Flt-3 Targeted Therapy

In approximately 30% of cases of acute myelogenous leukemia, and less frequently in other forms of leukemia, a *flt-3* gene mutation creates an internal tandem duplication that creates an abnormal FLT3 receptor that promotes the growth and survival of the leukemic cells (Advani, 2005; Gilliland and Griffin, 2002; Kelly et al., 2002; Sawyers, 2002).

Targeted Small Molecule Drugs for Solid Tumors
Gefitinib (Iressa®)

Gefitinib was originally approved by the FDA in 2003 as a monotherapy for the treatment of patients with locally advanced or metastatic non-small cell lung cancer after failure of both platinum-based and docetaxel chemotherapies (Ranson, 2002; Schiller, 2003). Gefitinib is a small molecule drug that targets the EGFR. In contrast with the approval of trastuzumab, the approval of gefitinib did not include an eligibility requirement reference to a specific tumor diagnostic test designed to select patients who were more likely to respond to the drug. Overexpression of EGFR, typically identified by IHC, is extremely common in both lung and breast cancers (Campiglio et al., 2004; Ranson, 2002; Schiller, 2003), but in contrast with HER-2 overexpression, which is virtually limited to cases with gene amplification, multiple mechanisms of dysregulation of EGFR and associated activation of signaling pathways have been described for both of these tumors (Campiglio et al., 2004; Ranson, 2002; Schiller, 2003). Thus, it has been difficult to develop this drug for expanded indications or combination therapies in the absence of a well-defined efficacy test. More recently, however, two independent groups reported their similar discovery of a specific activating mutation in the tyrosine kinase domain of the EGFR receptor that was associated with a high response of patients with non-small cell lung cancer to gefitinib (Lynch et al., 2004; Paez et al., 2004). Of interest have been the consistent observations that both a bronchioloalveolar histology and a persistent skin rash have been the best clinical signals of gefitinib response in lung cancer (Dudek et al., 2006). In addition, although specific activating mutations in the EGFR gene have been reproduced in a number of studies (Chan et al., 2006), some studies have failed to demonstrate this association, and other biomarkers, including EGFR gene amplification, have also been found to be predictive of tumor response (Carbone, 2004; Kobayashi et al., 2005). Most recently, follow-on studies of gefitinib in lung cancer revealed that the increased response rates that led to the approval of the drug were not accompanied by a clinical survival advantage (Twombly, 2005). However, with the introduction of EGFR clinical mutation testing (see below), the efficacy of gefitinib for the treatment of NSCLC has significantly increased.

Erlotinib (Tarceva®)

Erlotinib is another targeted small molecule inhibitor of EGFR. It was approved by the FDA in 2005 for the treatment of non-small cell lung cancer and pancreatic cancer (Moore, 2005; Smith, 2005). To date, similar to gefitinib, the clinical trials and FDA approval of erlotinib have not included an assessment of EGFR status or other diagnostic tests for eligibility to receive the drug. In lung cancer, the predictors of tumor response such as skin rash and bronchioloalveolar histology have also applied to erlotinib, as have the somewhat conflicting associations of both activating EGFR mutations and EGFR gene amplification as predictors of drug response (Chan et al., 2006; Silvestri and Rivera, 2005). Although clinical trials demonstrated that erlotinib does add a survival benefit to the treatment of both lung and pancreatic cancers, the drug was originally used without a companion diagnostic eligibility test. Subsequently, a number of approaches have been used to develop a clinically useful diagnostic for erlotinib, including gene mutation assays, gene copy number assays using FISH and CISH, and IHC overexpression assays for EGFR and other proteins (Ross, 2010). Currently, it is widely held that EGFR gene mutation testing is the most predictive assay of erlotinib/gefitinib benefit in NSCLC, and this approach has become standard. Tumors are tested either when high-risk cases are resected, or at the time of first recurrence for the EGFR mutation status. Both wild-type tumors and tumors with specific sequence-based tyrosine kinase resistance-associated mutations are given standard front-line treatment, but patients with EGFR mutations associated with drug sensitivity are given either gefitinib or erlotinib in a front-line regimen (Chirieac and Dacic, 2010; Ross, 2010).

BAY 43-9006 (Sorafenib®)

BAY 43-9006 is a RAF kinase inhibitor that also inhibits the PDGFR (platelet-derived growth factor receptor) and VEGFR growth factor receptors. It is thus considered to be an anti-angiogenesis drug. This oral agent was approved in late 2005 by the FDA for the treatment of metastatic renal cell carcinoma (Staehler et al., 2005). Sorafenib is now widely believed to function primarily as an anti-angiogenesis agent (Gollob et al., 2006). Currently, there are no diagnostic tests associated with the selection of this agent and clinical trials for other types of cancer are ongoing.

SU-11228 (Sunitinib)

Sunitinib is a multi-kinase small molecule inhibitor approved by the FDA in 2006 for the treatment of advanced renal cell carcinoma and GISTs that have become resistant to imatinib (Reddy, 2006). To date, no companion diagnostic has been developed for selection of this agent.

Other Small Molecule Anti-angiogenesis Agents

A variety of small molecule drugs that target the establishment and growth of tumor blood vessels are currently in clinical trials for the treatment of solid tumors (Khalil et al., 2003; Mendel et al., 2000; Thomas and Kantarjian, 2000; Zogakis and Libutti, 2001). Additional compounds that target matrix metalloproteases, such as the drug marimastat, are also considered to be angiogenesis inhibitors (Brown, 2000; Dell'Eva et al., 2002; Miller et al., 2002a). The anti-angiogenesis drug, lenalidomide (Revlimid®) was approved by the FDA in late 2005 for the treatment of myelodysplastic syndrome, and more recently for multiple myeloma (List, 2005). To date, none of these compounds has linked a diagnostic test, such as tumor microvessel density or the expression of an angiogenesis-promoting gene or protein, in their clinical development plans.

Bortezomib (Velcade®)

Recently, drugs targeting the proteasome have been developed that are designed to impact downstream pathways regulating angiogenesis, tumor growth, adhesion, and resistance to apoptosis (Adams, 2002; Elliott and Ross, 2001). One of these agents, bortezomib (PS-341), was recently approved for the treatment of advanced refractory multiple myeloma (Richardson et al., 2003). Bortezomib has shown both preclinical activity in animal studies and biologic activity in early clinical trials involving patients with a variety of solid tumors, but to date no trials using this agent alone or in combination with other drugs have progressed to phase III. Although pharmacogenomic studies of bortezomib use in multiple myeloma have been conducted, to date no specific pattern of gene expression or other specific test has emerged that could be a guide to the selection of patients for treatment. Bortezomib is currently widely used for the treatment of multiple myeloma and mantle cell lymphoma (Genin et al., 2010), but has not yet achieved an approved indication for the treatment of solid tumors, despite continuing clinical investigations (Genin et al., 2010).

Lapatinib (Tykerb®)

Lapatinib is an orally available small molecule dual inhibitor of the EGFR and HER-2 tyrosine kinases (Geyer et al., 2006; Medina and Goodin, 2008). Lapatinib was approved by the FDA in 2007 for use in combination with capecitabine for the treatment of HER-2-positive metastatic breast cancer that has progressed with standard treatment (Burstein et al., 2008). Lapatinib has been used in combination both with cytotoxic agents and with trastuzumab. There is continued interest in the potential of lapatinib to treat HER-2-positive breast cancer that has spread to the brain, given that trastuzumab will not cross the blood–brain barrier. Similar to trastuzumab, lapatinib has shown significant efficacy in both the adjuvant and neo-adjuvant clinical treatment settings (Moy and Goss, 2006).

Vemuraenib PLX4032

PLX4032 is a small molecule inhibitor of the activated *BRAF* gene associated with the specific V600E mutation (Flaherty et al., 2010). In a phase I clinical trial of 55 patients with the *BRAF* V600E-mutated metastatic malignant melanoma, 10 patients developed a partial response and one had a complete response. Among the 32 patients in the trial extension cohort, 24 had a partial response and two had a complete response. The estimated median progression-free survival among all patients was more than seven months, and was considered to be a major breakthrough in the treatment of this challenging malignancy. The *BRAF* V600E mutation in melanoma occurs in approximately 45% of cases with metastatic disease. PLX4032 has also been used to treat *BRAF*-mutated metastatic thyroid cancer. The regulatory approval of this drug will require a pre-treatment test (gene sequencing) that confirms that the tumor harbors the V600E *BRAF* mutation.

Critozinib (PF-02341066)

In 4–7% of non-small cell lung cancers, the *EML4* gene is translocated to the site of the anaplastic lymphoma kinase gene (*ALK*), causing activation of alk-mediated signaling. Critozinib is a small molecule inhibitor of *ALK* that has shown dramatic efficacy in patients with recurrent NSCLC that features EML4-ALK translocation (Bang et al., 2010a; Kwak et al., 2009). Current development has featured the FISH technique for detecting the EML4-ALK translocation.

PHARMACOGENOMICS

Targeted therapy in oncology has been a major stimulus for the evolving field of pharmacogenomics. In its broadest definition, pharmacogenomics can encompass both germline and somatic (disease) gene and protein measurements used to predict the likelihood that a patient will respond to a specific single or multi-agent chemotherapy regimen, and to predict the risk of toxic side effects (Ross et al., 2004; Weinstein, 2000). In breast cancer, whole genome transcriptional profiling has been

used as a technique for classification and prognosis (Bertucci et al., 2000; Sorlie et al., 2001; Van de Vijver et al., 2002; Van't Veer et al., 2002). Gene expression profiles can define cellular functions, biochemical pathways, cell proliferation activity, and regulatory mechanisms. The hierarchical clustering technique of data analysis from transcriptional profiling of clinical samples known to have responded or to have been resistant to a single agent or combination of anticancer drugs (Figure 69.4) has recently been employed as a guide to anticancer drug therapy in cancers of the breast and other organs (Ntzani and Ioannidis, 2003). Using transcriptional profiling, the microarray technique has been able to generate 81% accuracy for predicting the presence or absence of pathologic complete response after preoperative chemotherapy with sequential weekly paclitaxel and 5-FU, doxorubicin, and cyclophosphamide (FAC) in breast cancer (Ayers et al., 2004). Interestingly, the highest-rated single gene predictor in this study has also predicted paclitaxel response in an on-slide immunohistochemistry format (Rouzier et al., 2005). Currently, there is great interest in

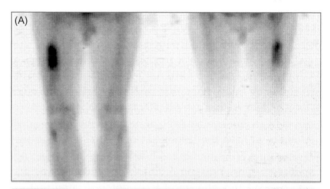

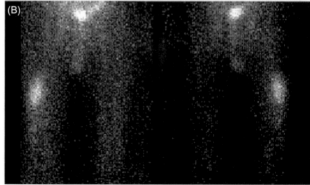

Figure 69.4 Prostate specific membrane antigen (PSMA) expression in non-prostate cancer. (A) Traditional bone scan demonstrating bilateral activity in the femur, indirectly indicating the presence of metastatic renal cell carcinoma. (B) ^{111}I-huJ591$_{EXT}$ diagnostic immunoscintiscan of the same patient showing direct localization of the anti-PSMA antibody conjugate to the sites of metastatic renal cell carcinoma that feature PSMA expression in the tumor neovasculature. *Reproduced with permission from Ross and Hortobagyi, 2004.*

both the scientific and commercial communities in learning whether the high-density genomic microarrays will ultimately be used as diagnostic assays themselves, or will yield to more familiar technologies testing small subsets of the discovered markers on platforms already entrenched in the clinical laboratory (Ross, 2005). Transcriptional profiling has also successfully been used both to classify (Sorlie et al., 2001) and grade (Sotiriou et al., 2006) breast cancer. Given the general agreement of gene profiling results across multiple platforms and data analysis approaches (Fan et al., 2006), it is anticipated that this technique will continue to be employed by investigators of an increasing number of tumors. What is not known at this time is whether these dense multi-thousand gene array platforms will ultimately be developed as stand-alone, integrated diagnostic tests or, in the end, will yield to more familiar platforms such as RT-PCR and IHC.

PHARMACOGENETICS

Testing the germline gene sequences of patients for certain drug metabolism, transport, and excretion genes (genotyping) is frequently referred to as pharmacogenetics. This approach has been used to predict altered metabolism of a variety of anticancer drugs.

UGT1A1. Polymorphisms in the *UGT1A1* gene have been strongly linked to toxicity of irinotecan-based treatment for CRC (Palomaki et al., 2009). One particular polymorphism, homozygosity for the 7-repeat allele (also known as UGT1A1*28), is associated with severe diarrhea when irinotecan is administered (Palomaki et al., 2009). Although *UGT1A1* genotyping is being performed on a regular basis in some facilities, widespread use has not occurred due both to the presence of conflicting negative data and the lack of current endorsement for testing by specialty societies (ASCO, CAP) or regulatory agencies.

CYP2D6. The CYP2D6 polymorphism is associated with reduced metabolic activation of a variety of drugs, and can also be determined by blood-based germline genotyping. It has been used to predict the efficacy of tamoxifen for the treatment and prevention of estrogen receptor-positive breast cancer. A number of studies indicate that women with impaired CYP2D6 metabolism have lower endoxifen concentrations and greater risk of breast cancer recurrence. However, the use of CYP2D6 genotyping remains controversial, with some oncologists using the test widely to select hormonal therapy and others using only classic clinicopathologic parameters to make this decision (Sideras et al., 2010).

WHOLE GENOME ANALYSIS BY NEXT-GENERATION DNA SEQUENCING

As seen above, traditional single gene "hotspot" approaches to sequence analysis are now widely used to guide therapy for patients diagnosed with lung and colorectal cancer, as well as for melanoma, sarcomas (e.g., GIST), and subtypes of leukemia and lymphoma (Ross and Cronin, 2011). The next-generation sequencing (NGS) approach holds a number of potential advantages over the traditional methods, including the ability to fully sequence large numbers of genes (hundreds to thousands) all in a single test, and simultaneously detect deletions, insertions, copy number alterations, translocations, and exome-wide base substitutions in all known cancer-related genes. Adoption of clinical NGS testing will place significant demands on laboratory infrastructure, will require extensive computational expertise, and require a deep knowledge of cancer medicine and biology needed to generate truly useful "clinically actionable" reports. It is anticipated that continuing advances in NGS technology will lower the overall cost, speed up the turn-around time, increase the breadth of genome sequencing, and detect epigenetic markers and other important genomic parameters, while becoming applicable to smaller and smaller specimens including circulating tumor cells and circulating free DNA in plasma.

DEVELOPING DIAGNOSTIC-THERAPEUTIC COMBINATIONS

The FDA has established regulations and requirements for the submission of specific data to Investigational New Drug (IND), New Drug Application (NDA), or Biologic License Application (BLA) submissions, and for already-approved NDAs (or BLAs) (Pietrusko, 2004). The FDA has provided definitions for biomarkers in their approach to defining what biomarker data need to be submitted to the agency and at what stage of development, regardless of whether or not they are pharmacogenomic or pharmacogenetic in nature. The FDA has defined a biomarker as a characteristic that is objectively measured and evaluated as an indicator of normal biologic processes, pathogenic processes, or pharmacologic responses to a therapeutic intervention.

A valid biomarker is defined as a biomarker that is measured in an analytical test system with well-established performance characteristics, and for which there is an established scientific framework or body of evidence that elucidates the physiologic, toxicologic, pharmacologic, or clinical significance of the test results (Pietrusko, 2004). A known valid biomarker is a biomarker that is measured in an analytical test system with well-established performance characteristics, and for which there is widespread agreement in the medical or scientific community about the physiologic, toxicologic, pharmacologic, or clinical significance of the results (Pietrusko, 2004). Validation of a biomarker is context-specific, and the criteria for validation will vary with its intended use. A probable valid biomarker is a biomarker that is measured in an analytical test system with well-established performance characteristics, and for which there is a scientific framework or body of evidence that appears to elucidate the physiologic, toxicologic, pharmacologic, or clinical significance of the test results. According to Pietrusko, a probable valid biomarker may not have reached the status of a known valid biomarker because of any of the following reasons: (1) data elucidating its significance may have been generated within a single company and may not be available for public scientific scrutiny; (2) data elucidating its significance, although highly suggestive, may not be conclusive; and (3) independent verification of the results may not have occurred (Pietrusko, 2004). If sponsors possess human data that suggest that a particular biomarker is a probable biomarker for evaluating the safety of the drug being evaluated, the data must be submitted to the IND, because it could potentially aid in evaluation of the safety of the investigations per the regulations. The designation for pharmacogenomic data submitted voluntarily to the FDA is "voluntary genomic data submission." Data in this category would include results from pharmacogenomic tests that are not known or probable valid biomarkers. General exploratory or research data, such as gene expression screening, collection of sera or tissue samples, or pharmacogenomic tests that are not known or probable valid biomarkers, do not need to be submitted to an IND or an approved NDA or BLA. However, the FDA recommends inclusion of such data in new NDA, BLA, or supplemental submissions as synopses (Pietrusko, 2004).

Combination products within this definition can be considered "virtual" combination products that, in concept and labeling, must be used together to achieve the intended use, indication, or effect, but are not packaged together. An example would be if therapy with a targeted product for prostate cancer required a molecular test to define the subpopulation that would respond. Under FDA regulations, a biomarker test will need to be approved if mentioned specifically in the label of the drug product or if use of the product requires an established laboratory service capable of achieving the specific results (Pietrusko, 2004). This has been applicable initially to breast cancer treatments. It has been extended to other oncology products as therapy becomes more targeted and likely to be effective only in tumors with certain features of overexpression or mutations. The approval of Erbitux® (cetuximab) used in combination with irinotecan for the treatment of epidermal growth factor (EGFR, HER-1)-expressing metastatic colorectal carcinoma in patients who are refractory for irinotecan-based chemotherapy took place in 2004. Patients enrolled in the clinical studies were required to have immunohistochemical evidence of positive EGFR expression. Similar to the same-day approval of trastuzumab and the Herceptest IHC assay in 1998, the 2004 approval of cetuximab accompanied by the approval of a PMA application for EGFR pharm Dx kit (by DakoCytomation, Gosstrup, Denmark). The premarket approval (PMA) for the EGFR pharm Dx kit was submitted to the FDA in parallel with the cetuximab submission.

A new Office of Combination Products has been formed within the Commissioner's Office of the FDA, to facilitate the premarket review and approval process among the review divisions in the different centers of the FDA. However, further clarity is needed to define how the primary center for review and approval, such as the Oncology Drug Products Division, works with the Office of *In Vitro* Diagnostic Device Evaluation and Safety within the Center for Devices and Radiological Health to have the product, as well as the required test, approved and available simultaneously. The degree of complexity may be different if one company develops both the drug product and device, two companies collaborate to develop both, or if two companies act independently to develop the drug and device separately (Pietrusko, 2004). A major problem could occur if the drug is ready for approval but the device is not ready for clearance at the same time. Finally, a process is now needed to ensure that labeling for both the drug product and the diagnostic test is updated and remains up-to-date, accurate, and mutually conforming.

CONCLUSION

During the next several years, the field of oncology drug development and cancer medicine will see numerous products developed with integrated diagnostic tests. These diagnostic-therapeutic combinations will enter the market designed to "personalize" their use, dosage, route of administration, and length of treatment for each patient, one at a time. Only time will tell whether this new approach to anticancer pharmaceuticals will yield breakthrough results, reducing morbidity and mortality, and improving outcomes for the new cancer patients of the personalized medicine era.

REFERENCES

Adams, J., 2002. Development of the proteasome inhibitor PS-341. Oncologist 7, 9–16.

Advani, A.S., 2005. FLT3 and acute myelogenous leukemia: Biology, clinical significance and therapeutic applications. Curr Pharm Des 11, 3449–3457.

Amos, J., Patnaik, M., 2002. Commercial molecular diagnostics in the US: The Human Genome Project to the clinical laboratory. Hum Mutat 19, 324–333.

Ayers, M., Symmans, W.F., Stec, J., et al., 2004. Gene expression profiles predict complete pathologic response to neoadjuvant paclitaxel and fluorouracil, doxorubicin, and cyclophosphamide chemotherapy in breast cancer. J Clin Oncol 22, 2284–2293.

Bang, Y., Kwak, E.L., Shaw, A.T., et al., 2010a. Clinical activity of the oral ALK inhibitor PF-02341066 in ALK-positive patients with non-small cell lung cancer (NSCLC). J Clin Oncol 28 (suppl; abstr 3), 18s.

Bang, Y.J., Van Cutsem, E., Feyereislova, A., et al., 2010b. Trastuzumab in combination with chemotherapy versus chemotherapy alone for treatment of HER2-positive advanced gastric or gastro-oesophageal junction cancer (ToGA): A phase 3, open-label, randomised controlled trial. Lancet 376, 687–697.

Baselga, J., 2001. The EGFR as a target for anticancer therapy – focus on cetuximab. Eur J Cancer 37 (suppl 4), S16–S22.

Bast Jr, R.C., Hortobagyi, G.N., 2004. Individualized care for patients with cancer – a work in progress. N Engl J Med 351, 2865–2867.

Bast Jr, R.C., Ravdin, P., Hayes, D.F., et al., 2001. 2000 update of recommendations for the use of tumor markers in breast and colorectal cancer: Clinical practice guidelines of the American Society of Clinical Oncology. J Clin Oncol 19, 1865–1878.

Berry, W., Eisenberger, M., 2005. Achieving treatment goals for hormone-refractory prostate cancer with chemotherapy. Oncologist 10 (Suppl 3), 30–39.

Bertucci, F., Houlgatte, R., Benziane, A., et al., 2000. Gene expression profiling of primary breast carcinomas using arrays of candidate genes. Hum Mol Genet 9, 2981–2991.

Bieche, I., Onody, P., Laurendeau, I., et al., 1999. Real-time reverse transcription-PCR assay for future management of ERBB2-based clinical applications. Clin Chem 45, 1148–1156.

Blagosklonny, M.V., 2005. How Avastin potentiates chemotherapeutic drugs: Action and reaction in antiangiogenic therapy. Cancer Biol Ther 4, 1307–1310.

Bottles, K., 2001. A revolution in genetics: Changing medicine, changing lives. Physician Exec 27, 58–63.

Bross, P.F., Beitz, J., Chen, G., et al., 2001. Approval summary: Gemtuzumab ozogamicin in relapsed acute myeloid leukemia. Clin Cancer Res 7, 1490–1496.

Brown, E., Lewis, P.H., Nocera, J., 2001. In search of the silver bullet. Fortune 143, 166–170.

Brown, P.D., 2000. Ongoing trials with matrix metalloproteinase inhibitors. Expert Opin Investig Drugs 9, 2167–2177.

Bruckner, H.W., Hrehorovich, V.R., Sawhney, H.S., 2005. Bevacizumab as treatment for chemotherapy-resistant pancreatic cancer. Anticancer Res 25, 3637–3639.

Burke, J.M., Jurcic, J.G., Scheinberg, D.A., 2002. Radioimmunotherapy for acute leukemia. Cancer Control 9, 106–113.

Burstein, H.J., Storniolo, A.M., Franco, S., et al., 2008. A phase II study of lapatinib monotherapy in chemotherapy-refractory HER2-positive and HER2-negative advanced or metastatic breast cancer. Ann Oncol 19, 1068–1074.

Buzdar, A.U., Robertson, J.F., Eiermann, W., et al., 2002. An overview of the pharmacology and pharmacokinetics of the newer generation aromatase inhibitors anastrozole, letrozole, and exemestane. Cancer 95, 2006–2016.

Campiglio, M., Locatelli, A., Olgiati, C., et al., 2004. Inhibition of proliferation and induction of apoptosis in breast cancer cells by the epidermal growth factor receptor (EGFR) tyrosine kinase inhibitor ZD1839 ("Iressa") is independent of EGFR expression level. J Cell Physiol 198, 259–268.

Carbone, D.P., 2004. Biomarkers of response to gefitinib in non-small-cell lung cancer. Nat Clin Pract Oncol 1, 66–67.

Chan, S.K., Gullick, W.J., Hill, M.E., 2006. Mutations of the epidermal growth factor receptor in non-small cell lung cancer: Search and destroy. Eur J Cancer 42, 17–23.

Chen, H.X., Gore-Langton, R.E., Cheson, B.D., 2000. Clinical trials referral resource: Current clinical trials of the anti-VEGF monoclonal antibody bevacizumab. Oncology 15, 1023–1026.

Cheson, B.D., 2002. Bexxar (Corixa/GlaxoSmithKline). Curr Opin Investig Drugs 3, 165–170.

Chester, K.A., Hawkins, R.E., 1995. Clinical issues in antibody design. Trends Biotechnol 13, 294–300.

Chirieac, L.R., Dacic, S., 2010. Targeted therapies in lung cancer. Surg Pathol Clin 3, 71–82.

Ciocca, D.R., Elledge, R., 2000. Molecular markers for predicting response to tamoxifen in breast cancer patients. Endocrine 13, 1–10.

Coiffier, B., 2002. Rituximab in the treatment of diffuse large B-cell lymphomas. Semin Oncol 29 (1 suppl 2), 30–35.

De Gramont, A., Van Cutsem, E., 2005. Investigating the potential of bevacizumab in other indications: Metastatic renal cell, non-small cell lung, pancreatic and breast cancer. Oncology 69 (Suppl 3), 46–56.

Dell'Eva, R., Pfeffer, U., Indraccolo, S., et al., 2002. Inhibition of tumor angiogenesis by angiostatin: From recombinant protein to gene therapy. Endothelium 9, 3–10.

Dillman, R.O., 2001. Monoclonal antibody therapy for lymphoma: An update. Cancer Pract 9, 71–80.

Dillman, R.O., 2002. Radiolabeled anti-CD20 monoclonal antibodies for the treatment of B-cell lymphoma. J Clin Oncol 20, 3545–3557.

Dressman, M.A., Baras, A., Malinowski, R., et al., 2003. Gene expression profiling detects gene amplification and differentiates tumor types in breast cancer. Cancer Res 63, 2194–2199.

Druker, B.J., 2003. Imatinib alone and in combination for chronic myeloid leukemia. Semin Hematol 40, 50–58.

Dudek, A.Z., Kmak, K.L., Koopmeiners, J., et al., 2006. Skin rash and bronchoalveolar histology correlates with clinical benefit in patients treated with gefitinib as a therapy for previously treated advanced or metastatic non-small cell lung cancer. Lung Cancer 51, 89–96.

Dumont, F.J., 2002. CAMPATH (alemtuzumab) for the treatment of chronic lymphocytic leukemia and beyond. Expert Rev Anticancer Ther 2, 23–35.

Elliott, P.J., Ross, J.S., 2001. The proteasome: A new target for novel drug therapies. Am J Clin Pathol 116, 637–646.

Evans, W.E., McLeod, H.L., 2003. Pharmacogenomics – drug disposition, drug targets, and side effects. N Engl J Med 348, 538–549.

Fan, C., Oh, D.S., Wessels, L., et al., 2006. Concordance among gene-expression-based predictors for breast cancer. N Engl J Med 355, 560–569.

Fang, J., Chen, S.J., Tong, J.H., et al., 2002. Treatment of acute promyelocytic leukemia with ATRA and As2O3: A model of molecular target-based cancer therapy. Cancer Biol Ther 1, 614–620.

Ferrara, N., 2005. VEGF as a therapeutic target in cancer. Oncology 69 (Suppl 3), 11–16.

Flaherty, K.T., Puzanov, I., Kim, K.B., et al., 2010. Inhibition of mutated, activated BRAF in metastatic melanoma. N Engl J Med 363, 809–819.

Fornier, M.N., Seidman, A.D., Schwartz, M.K., et al., 2005. Serum HER2 extracellular domain in metastatic breast cancer patients treated with weekly trastuzumab and paclitaxel: Association with HER2 status immunohistochemistry and fluorescence in situ hybridization and with response rate. Ann Oncol 16, 234–239.

Genin, E., Reboud-Ravaux, M., Vidal, J., 2010. Proteasome inhibitors: Recent advances and new perspectives in medicinal chemistry. Curr Top Med Chem 10, 232–256.

Geyer, C.E., Forster, J., Lindquist, D., et al., 2006. Lapatinib plus capecitabine for HER2-positive advanced breast cancer. N Engl J Med 355, 2733–2743.

Gilliland, D.G., Griffin, J.D., 2002. The roles of FLT3 in hematopoiesis and leukemia. Blood 100, 1532–1542.

Gleissner, B., Thiel, E., 2001. Detection of immunoglobulin heavy chain gene rearrangements in hematologic malignancies. Expert Rev Mol Diagn 1, 191–200.

Goldenberg, D.M., 2002. Targeted therapy of cancer with radiolabeled antibodies. J Nucl Med 43, 693–713.

Goldman, J.M., Melo, J.V., 2003. Chronic myeloid leukemia – advances in biology and new approaches to treatment. N Engl J Med 349, 1451–1464.

Gollob, J.A., Wilhelm, S., Carter, C., et al., 2006. Role of Raf kinase in cancer: Therapeutic potential of targeting the Raf/MEK/ERK signal transduction pathway. Semin Oncol 33, 392–406.

Gora-Tybor, J., 2012. Emerging therapies in chronic myeloid leukemia. Curr Cancer Drug Targets 12, 458–470.

Gordon, L.I., Witzig, T.E., Wiseman, G.A., et al., 2002. Yttrium 90 ibritumomab tiuxetan radioimmunotherapy for relapsed or refractory low-grade non-Hodgkin's lymphoma. Semin Oncol 29 (1 suppl 2), 87–92.

Grillo-Lopez, A.J., 2002. AntiCD20 mAbs: Modifying therapeutic strategies and outcomes in the treatment of lymphoma patients. Expert Rev Anticancer Ther 2, 323–329.

Grillo-Lopez, A.J., Hedrick, E., Rashford, M., et al., 2002. Rituximab: Ongoing and future clinical development. Semin Oncol 29 (1 suppl 2), 105–112.

Grimwade, D., Lo Coco, F., 2002. Acute promyelocytic leukemia: A model for the role of molecular diagnosis and residual disease monitoring in directing treatment approach in acute myeloid leukemia. Leukemia 16, 1959–1973.

Hammond, M.E., Fitzgibbons, P.L., Compton, C.C., et al., 2000. College of American Pathologists Conference XXXV: Solid tumor prognostic factors – which, how and so what? Summary document and recommendations for implementation. Cancer Committee and Conference Participants. Arch Pathol Lab Med 124, 958–965.

Hammond, M.E., Hayes, D.F., Dowsett, M., Allred, D.C., et al., 2010. American Society of Clinical Oncology/College of American Pathologists guideline recommendations for immunohistochemical testing of estrogen and progesterone receptors in breast cancer (unabridged version). Arch Pathol Lab Med 134, e48–e72.

Hanna, W.M., Kwok, K., 2006. Chromogenic *in-situ* hybridization: A viable alternative to fluorescence *in-situ* hybridization in the HER2 testing algorithm. Mod Pathol 19, 481–487.

Harari, P.M., Huang, S., 2006. Radiation combined with EGFR signal inhibitors: Head and neck cancer focus. Semin Radiat Oncol 16, 38–44.

Hayes, D.F., Thor, A.D., 2002. c-erbB-2 in breast cancer: Development of a clinically useful marker. Semin Oncol 29, 231–245.

Heinicke, T., Wardelmann, E., Sauerbruch, T., et al., 2005. Very early detection of response to imatinib mesylate therapy of gastrointestinal stromal tumours using 18fluoro deoxyglucose-positron emission tomography. Anticancer Res 25, 4591–4594.

Hemminki, A., 2002. From molecular changes to customised therapy. Eur J Cancer 38, 333–338.

Herbst, R.S., Langer, C.J., 2002. Epidermal growth factor receptors as a target for cancer treatment: The emerging role of IMC-C225 in the treatment of lung and head and neck cancers. Semin Oncol 29 (1 suppl 4), 27–36.

Hortobagyi, G.N., 2001. Overview of treatment results with trastuzumab (Herceptin) in metastatic breast cancer. Semin Oncol 28, 43–47.

Hurwitz, H., Kabbinavar, F., 2005. Bevacizumab combined with standard fluoropyrimidine-based chemotherapy regimens to treat colorectal cancer. Oncology 69 (Suppl 3), 17–24.

Ibrahim, N.K., Hortobagyi, G.N., 1999. The evolving role of specific estrogen receptor modulators (SERMs). Surg Oncol 8, 103–123.

Isaacs, J.D., 2001. From bench to bedside: Discovering rules for antibody design, and improving serotherapy with monoclonal antibodies. Rheumatology 40, 724–738.

Jones, P.T., Dear, P.H., Foote, J., et al., 1986. Replacing the complementarity-determining regions in a human antibody with those from a mouse. Nature 321, 522–525.

Jordan, V.C., 2003a. Antiestrogens and selective estrogen receptor modulators as multifunctional medicines. 1. Receptor interactions. J Med Chem 46, 883–908.

Jordan, V.C., 2003b. Antiestrogens and selective estrogen receptor modulators as multifunctional medicines. 2. Clinical considerations and new agents. J Med Chem 46, 1081–1111.

Kelly, L.M., Yu, J.C., Boulton, C.L., et al., 2002. CT53518, a novel selective FLT3 antagonist for the treatment of acute myelogenous leukemia (AML). Cancer Cell 1, 421–432.

Khalil, M.Y., Grandis, J.R., Shin, D.M., 2003. Targeting epidermal growth factor receptor: Novel therapeutics in the management of cancer. Expert Rev Anticancer Ther 3, 367–380.

Kiyoi, H., Naoe, T., 2002. FLT3 in human hematologic malignancies. Leukemia Lymphoma 43, 1541–1547.

Kobayashi, S., Boggon, T.J., Dayaram, T., et al., 2005. EGFR mutation and resistance of non-small-cell lung cancer to gefitinib. N Engl J Med 352, 786–792.

Krasner, C., Joyce, R.M., 2001. Zevalin: 90yttrium labeled anti-CD20 (ibritumomab tiuxetan), a new treatment for non-Hodgkin's lymphoma. Curr Pharm Biotechnol 2, 341–349.

Krop, I.E., Beeram, M., Modi, S., et al., 2010. Phase I study of trastuzumab-DM1, an HER2 antibody-drug conjugate, given every 3 weeks to patients with HER2-positive metastatic breast cancer. J Clin Oncol 28, 2698–2704.

Kuus-Reichel, K., Grauer, L.S., Karavodin, L.M., et al., 1994. Will immunogenicity limit the use, efficacy, and future development of therapeutic monoclonal antibodies? Clin Diagn Lab Immunol 1, 365–372.

Kwak, E.L., Camidge, D.R., Clark, J., et al., 2009. Clinical activity observed in a phase I dose escalation trial of an oral c-met and ALK inhibitor, PF-02341066. J Clin Oncol 27, (suppl; abstr 3509), 15s.

Lange, T., Park, B., Willis, S.G., et al., 2005. BCR-ABL kinase domain mutations in chronic myeloid leukemia: Not quite enough to cause resistance to imatinib therapy? Cell Cycle 4, 1761–1766.

Larson, R.A., Boogaerts, M., Estey, E., et al., 2002. Antibody-targeted chemotherapy of older patients with acute myeloid leukemia in first relapse using Mylotarg (gemtuzumab ozogamicin). Leukemia 16, 1627–1636.

Lemonick, M.D., Park, A., 2001. New hope for cancer. Time 157, 62–69.

Leonard, D.S., Hill, A.D., Kelly, L., et al., 2002. Anti-human epidermal growth factor receptor 2 monoclonal antibody therapy for breast cancer. Br J Surg 89, 262–271.

Ligibel, J.A., Winer, E.P., 2002. Trastuzumab/chemotherapy combinations in metastatic breast cancer. Semin Oncol 29, 38–43.

Linenberger, M.L., Maloney, D.G., Bernstein, I.D., 2002. Antibody-directed therapies for hematological malignancies. Trends Mol Med 8, 69–76.

List, A.F., 2005. Emerging data on IMiDs in the treatment of myelo-dysplastic syndromes (MDS). Semin Oncol 32, S31–S35.

Locker, G.Y., 1998. Hormonal therapy of breast cancer. Cancer Treat Rev 24, 221–240.

Lynch, T.J., Bell, D.W., Sordella, R., et al., 2004. Activating mutations in the epidermal growth factor receptor underlying responsive-ness of non-small-cell lung cancer to Gefitinib. N Engl J Med 350, 2129–2139.

Masood, S., Bui, M.M., 2002. Prognostic and predictive value of HER2/neu oncogene in breast cancer. Microsc Res Tech 59, 102–108.

Mauro, M.J., Druker, B.J., 2001. STI571: Targeting BCR-ABL as therapy for CML. Oncologist 6, 233–238.

McKeage, K., Perry, C.M., 2002. Trastuzumab: A review of its use in the treatment of metastatic breast cancer overexpressing HER2. Drugs 62, 209–243.

Medina, P.J., Goodin, S., 2008. Lapatinib: A dual inhibitor of human epidermal growth factor receptor tyrosine kinases. Clin Ther 30, 1426–1447.

Mendel, D.B., Laird, A.D., Smolich, B.D., et al., 2000. Development of SU5416, a selective small molecule inhibitor of VEGF recep-tor tyrosine kinase activity, as an anti-angiogenesis agent. Anticancer Drug Des 15, 29–41.

Mendelsohn, J., Baselga, J., 2000. The EGF receptor family as targets for cancer therapy. Oncogene 19, 6550–6565.

Merluzzi, S., Figini, M., Colombatti, A., et al., 2000. Humanized antibod-ies as potential drugs for therapeutic use. Adv Clin Path 4, 77–85.

Milenic, D.E., 2002. Monoclonal antibody-based therapy strategies: Providing options for the cancer patient. Curr Pharm Des 8, 1749–1764.

Miller, K.D., Gradishar, W., Schuchter, L., et al., 2002a. A randomized phase II pilot trial of adjuvant marimastat in patients with early-stage breast cancer. Ann Oncol 13, 1220–1224.

Miller, W.R., Anderson, T.J., Dixon, J.M., 2002b. Anti-tumor effects of letrozole. Cancer Invest 20, 15–21.

Moore, M.J., 2005. Brief communication: A new combination in the treatment of advanced pancreatic cancer. Semin Oncol 32, 5–6.

Moy, B., Goss, P.E., 2006. Lapatinib: Current status and future direc-tions in breast cancer. Oncologist 11, 1047–1057.

Nabhan, C., Tallman, M.S., 2002. Early phase I/II trials with gemtu-zumab ozogamicin (Mylotarg®) in acute myeloid leukemia. Clin Lymph 2 (suppl 1), S19–S23.

Nielsen, U.B., Marks, J.D., 2000. Internalizing antibodies and targeted cancer therapy: Direct selection from phage display libraries. PSTT 3, 282–291.

Ntzani, E.E., Ioannidis, J.P., 2003. Predictive ability of DNA microarrays for cancer outcomes and correlates: An empirical assessment. Lancet 362, 1439–1444.

O'Brien, S.G., Guilhot, F., Larson, R.A., et al., 2003. Imatinib compared with interferon and low-dose cytarabine for newly diagnosed chronic-phase chronic myeloid leukemia. N Engl J Med 348, 994–1004.

O'Dwyer, M.E., Druker, B.J., 2001. Chronic myelogenous leukaemia – new therapeutic principles. J Intern Med 250, 3–9.

O'Hare, T., Corbin, A.S., Druker, B.J., 2006. Targeted CML therapy: Controlling drug resistance, seeking cure. Curr Opin Genet Dev 16, 92–99.

Osborne, C.K., 1998. Steroid hormone receptors in breast cancer management. Breast Cancer Res Treat 51, 227–238.

Paez, J.G., Janne, P.A., Lee, J.C., et al., 2004. EGFR mutations in lung cancer: Correlation with clinical response to gefitinib therapy. Science 304, 1497–1500.

Paik, S., Shak, S., Tang, G., et al., 2004. A multigene assay to predict recurrence of tamoxifen-treated, node-negative breast cancer. N Engl J Med 351, 2817–2826.

Paik, S., Shak, S., Tang, G., et al., 2005. Expression of the 21 genes in the Recurrence Score assay and tamoxifen clinical benefit in the NSABP study B-14 of node negative, estrogen receptor positive breast cancer. Proc ASCO 24, 510.

Paik, S., Tan-chui, E., Bryan, J., et al., 2002. Successful quality assur-ance program for HER-2 testing in the NSAPB trial for Herceptin. Breast Cancer Res Treat 76 (suppl 1), S31.

Palomaki, G.E., Bradley, L.A., Douglas, M.P., et al., 2009. Can UGT1A1 genotyping reduce morbidity and mortality in patients with met-astatic colorectal cancer treated with irinotecan? An evidence-based review. Genet Med 11, 21–34.

Pangalis, G.A., Dimopoulou, M.N., Angelopoulou, M.K., et al., 2001. Campath-1H (anti-CD52) monoclonal antibody therapy in lym-phoproliferative disorders. Med Oncol 18, 99–107.

Parmar, S., Tallman, M.S., 2003. Acute promyelocytic leukaemia: A review. Expert Opin Pharmacother 4, 1379–1392.

Pawlowski, V., Revillion, F., Hornez, L., et al., 2000. A real-time one-step reverse transcriptase-polymerase chain reaction method to quantify c-erbB-2 expression in human breast cancer. Cancer Detect Prev 24, 212–223.

Piccart-Gebhart, M.J., Procter, M., Leyland-Jones, B., et al., 2005. Trastuzumab after adjuvant chemotherapy in HER2-positive breast cancer. N Engl J Med 353, 1659–1672.

Pietrusko, R., 2004. Regulatory Aspects and Implications for Molecular Testing in the United States. In: Ross, J.S., Hortobagyi, G.N. (Eds.), Molecular Oncology of Breast Cancer. Jones and Bartlett, Sudbury MA, pp. 478–486.

Pusztai, L., Ayers, M., Stec, J., et al., 2003. Gene expression profiles obtained from fine-needle aspirations of breast cancer reli-ably identify routine prognostic markers and reveal large-scale molecular differences between estrogen-negative and estrogen-positive tumors. Clin Cancer Res 9, 2406–2415.

Ranson, M., 2002. ZD1839 (Iressa): For more than just non-small cell lung cancer. Oncologist 7 (suppl 4), 16–24.

Reddy, K., 2006. Phase III study of sunitinib malate (SU11248) versus interferon-alpha as first-line treatment in patients with meta-static renal cell carcinoma. Clin Genitourin Cancer 5, 23–25.

Reff, M.E., Hariharan, K., Braslawsky, G., 2002. Future of monoclonal antibodies in the treatment of hematologic malignancies. Cancer Control 9, 152–166.

Reff, M.E., Heard, C., 2001. A review of modifications to recombinant antibodies: Attempt to increase efficacy in oncology applica-tions. Crit Rev Oncol Hematol 40, 25–35.

Reichert, J.M., 2002. Therapeutic monoclonal antibodies: Trends in development and approval in the US. Curr Opin Mol Ther 4, 110–118.

Reilly, R.M., Sandhu, J., Alvarez-Diez, T.M., et al., 1995. Problems of delivery of monoclonal antibodies: Pharmaceutical and pharma-cokinetic solutions. Clin Pharmacokinet 28, 126–142.

Reynolds, T., 2002. Biotech firm faces challenges from FDA, falling stock prices. J Natl Cancer Inst 94, 326–328.

Richardson, P.G., Barlogie, B., Berenson, J., et al., 2003. A phase 2 study of bortezomib in relapsed, refractory myeloma. N Engl J Med 348, 2609–2617.

Romond, E.H., Perez, E.A., Bryant, J., et al., 2005. Trastuzumab plus adjuvant chemotherapy for operable HER2-positive breast cancer. N Engl J Med 353, 1673–1684.

Rosen, L.S., 2002. Clinical experience with angiogenesis signaling inhibitors: Focus on vascular endothelial growth factor (VEGF) blockers. Cancer Control 9 (2 suppl), 36–44.

Ross, J.S., 2005. Genomic microarrays in cancer molecular diagnostics: Just biomarker discovery tools or future bedside clinical assays? Expert Rev Mol Diagn 5, 837–838.

Ross, J.S., 2010. Biomarker update for breast, colorectal and non-small cell lung cancer. Drug News Perspect 23, 82–88.

Ross, J.S., Cronin, M., 2011. Whole cancer genome sequencing by next-generation methods. Am J Clin Pathol 136, 527–539.

Ross, J.S., Fletcher, J.A., Linette, G.P., et al., 2003a. The Her-2/neu gene and protein in breast cancer 2003: Biomarker and target of therapy. Oncologist 8, 307–325.

Ross, J.S., Gray, K., Gray, G.S., et al., 2003b. Anticancer antibodies. Am J Clin Pathol 119, 472–485.

Ross, J.S., Hatzis, C., Symmans, W.F., et al., 2008. Commercialized multigene predictors of clinical outcome for breast cancer. Oncologist 13, 477–493.

Ross, J.S., Hortobagyi, G.H. (eds) 2004. The Molecular Oncology of Breast Cancer. Sudbury, MA: Jones and Bartlett, Inc.

Ross, J.S., Schenkein, D.P., Kashala, O., et al., 2004. Pharmacogenomics. Adv Anat Pathol 11, 211–220.

Ross, J.S., Slodkowska, E.A., Symmans, W.F., Pusztai, L., Ravdin, P.M., Hortobagyi, G.N., 2009. The HER-2 receptor and breast cancer: Ten years of targeted anti-HER-2 therapy and personalized medicine. Oncologist 14, 320–368.

Ross, J.S., Torres-Mora, J., Wagle, N., Jennings, T.A., Jones, D.M., 2010. Biomarker-based prediction of response to therapy for colorectal cancer: Current perspective. Am J Clin Pathol 134, 478–490.

Rouzier, R., Rajan, R., Wagner, P., et al., 2005. Microtubule-associated protein tau: A marker of paclitaxel sensitivity in breast cancer. Proc Natl Acad Sci USA 102, 8315–8320.

Rubnitz, J.E., Pui, C.H., 1999. Molecular diagnostics in the treatment of leukemia. Curr Opin Hematol 6, 229–235.

Saltz, L., 2005. Epidermal growth factor receptor-negative colorectal cancer: Is there truly such an entity? Clin Colorectal Cancer 5 (Suppl 2), S98–S100.

Sawyers, C.L., 2002. Finding the next Gleevec: FLT3 targeted kinase inhibitor therapy for acute myeloid leukemia. Cancer Cell 1, 413–415.

Schiller, J.H., 2003. New directions for ZD1839 in the treatment of solid tumors. Semin Oncol 30 (1 suppl 1), 49–55.

Schnitt, S.J., Jacobs, T.W., 2001. Current status of HER2 testing: Caught between a rock and a hard place. Am J Clin Pathol 116, 806–810.

Shawver, L.K., Slamon, D., Ullrich, A., 2002. Smart drugs: Tyrosine kinase inhibitors in cancer therapy. Cancer Cell 1, 117–123.

Sideras, K., Ingle, J.N., Ames, M.M., et al., 2010. Coprescription of tamoxifen and medications that inhibit CYP2D6. J Clin Oncol 28, 2768–2776.

Sievers, E.L., Linenberger, M., 2001. Mylotarg: Antibody-targeted chemotherapy comes of age. Curr Opin Oncol 13, 522–527.

Silvestri, G.A., Rivera, M.P., 2005. Targeted therapy for the treatment of advanced non-small cell lung cancer: A review of the epidermal growth factor receptor antagonists. Chest 128, 3975–3984.

Slamon, D.J., Leyland-Jones, B., Shak, S., et al., 2001. Use of chemotherapy plus a monoclonal antibody against HER2 for metastatic breast cancer that overexpresses HER2. N Engl J Med 344, 783–792.

Smith, J., 2005. Erlotinib: Small-molecule targeted therapy in the treatment of non-small-cell lung cancer. Clin Ther 27, 1513–1534.

Sorlie, T., Perou, C.M., Tibshirani, R., et al., 2001. Gene expression patterns of breast carcinomas distinguish tumor subclasses with clinical implications. Proc Natl Acad Sci USA 98, 10,869–10,874.

Sotiriou, C., Wirapati, P., Loi, S., et al., 2006. Gene expression profiling in breast cancer: Understanding the molecular basis of histologic grade to improve prognosis. J Natl Cancer Inst 98, 262–272.

Stadler, W.M., 2005. Targeted agents for the treatment of advanced renal cell carcinoma. Cancer 104, 2323–2333.

Stadtmauer, E.A., 2002. Trials with gemtuzumab ozogamicin (Mylotarg®) combined with chemotherapy regimens in acute myeloid leukemia. Clin Lymph 2 (suppl 1), S24–S28.

Staehler, M., Rohrmann, K., Haseke, N., et al., 2005. Targeted agents for the treatment of advanced renal cell carcinoma. Curr Drug Targets 6, 835–846.

Stevenson, G.T., Anderson, V.A., Leong, W.S., 2002. Engineered antibody for treating lymphoma. Recent Results Cancer Res 159, 104–112.

Talpaz, M., Shah, N.P., Kantarjian, H., et al., 2006. Dasatinib in imatinib-resistant Philadelphia chromosome-positive leukemias. N Engl J Med 354, 2531–2541.

Tanner, M., Gancberg, D., Di Leo, A., et al., 2000. Chromogenic in situ hybridization: A practical alternative for fluorescence in situ hybridization to detect HER-2/neu oncogene amplification in archival breast cancer samples. Am J Pathol 157, 1467–1472.

Thomas, D.A., Kantarjian, H.M., 2000. Current role of thalidomide in cancer treatment. Curr Opin Oncol 12, 564–573.

Tubbs, R.R., Pettay, J.D., Roche, P.C., et al., 2001. Discrepancies in clinical laboratory testing of eligibility for trastuzumab therapy: Apparent immunohistochemical false-positives do not get the message. J Clin Oncol 19, 2714–2721.

Twombly, R., 2005. Failing survival advantage in crucial trial, future of Iressa is in jeopardy. J Natl Cancer Inst 97, 249–250.

Van Cutsem, E., Köhne, C.H., Hitre, E., et al., 2009. Cetuximab and chemotherapy as initial treatment for metastatic colorectal cancer. N Engl J Med 360, 1408–1417.

Van de Vijver, M.J., He, Y.D., Van't Veer, L.J., et al., 2002. A gene-expression signature as a predictor of survival in breast cancer. N Engl J Med 347, 1999–2009.

Van't Veer, L.J., Dai, H., Van de Vijver, M.J., et al., 2002. Gene expression profiling predicts clinical outcome of breast cancer. Nature 415, 530–536.

Verweij, J., Van Oosterom, A., Blay, J.Y., et al., 2003. Imatinib mesylate (STI-571 Glivec, Gleevec) is an active agent for gastrointestinal stromal tumours, but does not yield responses in other soft-tissue sarcomas that are unselected for a molecular target. Results from an EORTC Soft Tissue and Bone Sarcoma Group phase II study. Eur J Cancer 39, 2006–2011.

Von Mehren, M., 2003. Gastrointestinal stromal tumors: A paradigm for molecularly targeted therapy. Cancer Invest 21, 553–563.

Wagner Jr, H.N., Wiseman, G.A., Marcus, C.S., et al., 2002. Administration guidelines for radioimmunotherapy of non-Hodgkin's lymphoma with ^{90}Y-labeled anti CD20 monoclonal antibody. J Nucl Med 43, 267–272.

Wakelee, H., Belani, C.P., 2005. Optimizing first-line treatment options for patients with advanced NSCLC. Oncologist 10 (Suppl 3), 1–10.

Wang, S., Saboorian, M.H., Frenkel, E., et al., 2000. Laboratory assessment of the status of Her-2/neu protein and oncogene in breast cancer specimens: Comparison of immunohistochemistry assay with fluorescence *in situ* hybridization assays. J Clin Pathol 53, 374–381.

Wang, S., Saboorian, M.H., Frenkel, E.P., et al., 2001. Assessment of HER-2/neu status in breast cancer. Automated Cellular Imaging System (ACIS)-assisted quantitation of immunohistochemical assay achieves high accuracy in comparison with fluorescence *in situ* hybridization assay as the standard. Am J Clin Pathol 116, 495–503.

Watkins, N.A., Ouwehand, W.H., 2000. Introduction to antibody engineering and phage display. Vox Sang 78, 72–79.

Weinshilboum, R., 2003. Inheritance and drug response. N Engl J Med 348, 529–537.

Weinstein, J.N., 2000. Pharmacogenomics – teaching old drugs new tricks. N Engl J Med 343, 1408–1409.

Wilbur, D.C., Willis, J., Mooney, R.A., et al., 1992. Estrogen and progesterone detection in archival formalin-fixed paraffin embedded tissue from breast carcinoma: A comparison of immunocytochemistry with dextran coated charcoal assay. Mod Pathol 5, 79–84.

Winter, G., Harris, W.J., 1993. Humanized antibodies. Immunol Today 14, 243–246.

Wiseman, G.A., Gordon, L.I., Multani, P.S., et al., 2002. Ibritumomab tiuxetan radioimmunotherapy for patients with relapsed or refractory non-Hodgkin lymphoma and mild thrombocytopenia: A phase II multicenter trial. Blood 99, 4336–4342.

Wong, S.F., 2005. Cetuximab: An epidermal growth factor receptor monoclonal antibody for the treatment of colorectal cancer. Clin Ther 27, 684–694.

Yip, A.Y., Tse, L.A., Ong, E.Y., Chow, L.W., 2010. Survival benefits from lapatinib therapy in women with HER2-overexpressing breast cancer: A systematic review. Anticancer Drugs 21, 487–493.

Zelenetz, A.D., 2003. A clinical and scientific overview of tositumomab and iodine I 131 tositumomab. Semin Oncol 30 (2 suppl 4), 22–30.

Zhao, J., Wu, R., Au, A., et al., 2002. Determination of HER2 gene amplification by chromogenic *in situ* hybridization (CISH) in archival breast carcinoma. Mod Pathol 15, 657–665.

Zogakis, T.G., Libutti, S.K., 2001. General aspects of anti-angiogenesis and cancer therapy. Expert Opin Biol Ther 1, 253–275.

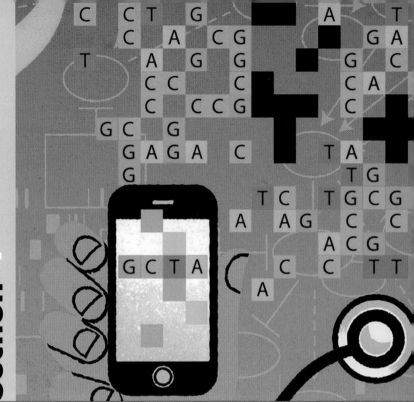

Inflammatory and Metabolic Disease Genomic Medicine

Section 7

CHAPTER

Autoimmune Disorders

Sergio E. Baranzini

INTRODUCTION

Autoimmune disorders arise when physiological tolerance to "self" antigens is lost. Although several mechanisms may be involved in this pathogenic process, dysregulation of T-cell and B-cell activation and of pathways leading to inflammation are logical candidates. Susceptibility to autoimmune diseases has been associated with multiple factors including genetics, epigenetics, and the environment. While the modest concordance rate in monozygotic twins suggests that environmental factors are major players in most autoimmune diseases, increased heritability within families and the decrease in risk with degree of relatedness all argue in favor of genetic factors. With the advent of high-throughput genomic and other *'omic* technologies, massive amounts of genetic data are being produced and reported almost on a monthly basis. Although considerable insight has been gained from each of these individual studies, a detailed comparative analysis will likely identify both unique and common pathways operating in auto-immunity. This kind of analysis may set the basis for more targeted and rational therapeutic approaches.

Certain autoimmune disorders co-occur significantly in a single individual or within nuclear families more often than expected, suggesting the presence of genetic variants that predispose to or protect against autoimmunity generally [Barcellos et al., 2006; International Multiple Sclerosis

Genetics Consortium (IMSGC), 2009; Lin et al., 1998; Tait et al., 2004]. In a recent analysis, Rzhetzky and colleagues reviewed 1.5 million medical records involving 161 diseases, and computed pairwise correlations of disease co-occurrences (Rzhetsky et al., 2007). Indeed, several autoimmune disorders co-occurred in the same individuals more often than expected by chance. Type 1 diabetes (T1D) most often correlated with the presence of type 2 diabetes (T2D), but also with rheumatoid arthritis (RA) and psoriasis. Similarly, multiple sclerosis (MS) correlated with systemic lupus erythematosus (SLE), T1D, T2D and psoriasis, while RA strongly correlated with SLE, ankylosing spondylitis, T1D, T2D, Sjogren's syndrome, and psoriasis. Although these data were derived from medical records and not from genetic analyses, the results strongly suggest that common genetic and pathologic mechanisms may be at play in different autoimmune diseases (AID).

Genetic polymorphisms are heritable sequence variants in the genome that contribute to phenotypic variability and can modulate the expression and/or function of genes, thus affecting the behavior of biological pathways, potentially underlying the magnitude of one's susceptibility to diseases. Modern genomics developments have made available miniaturization and automation of genotyping platforms, and more than 200 genome-wide association studies (GWAS) have been performed in different diseases to date [Hindorff et al., 2009; National Human Genome Research Institute (NHGRI), 2009]

Genomic and Personalized Medicine, 2nd edition
by Ginsburg & Willard. DOI: http://dx.doi.org/10.1016/B978-0-12-382227-7.00070-7

including 31 studies in seven common AID that are the focus of this chapter [SLE, T1D, MS, RA, psoriasis, Crohn's disease (CD) and celiac disease (CeD)].

In this chapter, I will summarize the findings of these studies, elaborate hypotheses about the possible pathogenic mechanisms implicated in each disorder, and provide a global view of shared and specific genes that characterize them.

HLA AND AUTOIMMUNITY

The human leukocyte antigen (HLA) region has been associated with hundreds of human diseases, including most autoimmune diseases (e.g., the B27 allele with ankylosing spondylitis; DR3 with Graves' disease, myasthenia gravis, SLE, and T1D; DR15 with MS) (for a review, see Shiina et al., 2004). For most of these diseases, however, it has not yet been possible to show the molecular mechanisms underlying disease association with a particular HLA molecule(s). One possible explanation for this shortcoming is that it has often been difficult to unequivocally ascertain the primary disease-risk HLA genes due to the remarkably strong linkage disequilibrium (LD) across the major histocompatibility region (MHC) (Miretti et al., 2005). There is still debate, for example, as to whether the *HLA-DRB1*1501* association explains the entire MHC–class II genetic linkage signal in MS (Fernandez et al., 2004; Fukazawa et al., 2000; Khare et al., 2005; Prat et al., 2005; Sospedra et al., 2006), and whether susceptibility genes also exist within the class III region (De Jong et al., 2002) and/or are telomeric to the class I region (Marrosu et al., 2001, Rubio et al., 2002, Yeo et al., 2007). In addition, HLA-associated diseases can be the result of the combination of different HLA molecules expressed at various loci (class-I and/or class-II) rather than the result of one HLA variant only. The situation has been complicated by the fact that nearly all HLA-associated diseases are multifactorial polygenic diseases in which particular HLA alleles, in combination with other genetic variants and environmental factors, are involved in disease susceptibility. Moreover, disease-relevant autoantigens are largely unknown, which prevents a thorough three-dimensional analysis of HLA–peptide interactions. Last, but not least, human autoimmune diseases are phenotypically very heterogeneous in terms of clinical presentation, age at onset, association with other autoimmune disorders, and severity or rapidity of evolution, and it is likely that different alleles, or allelic combinations at different loci, will predispose to different forms of the disease (Caillat-Zucman, 2009).

GENE DISCOVERY IN AUTOIMMUNITY: THE EXAMPLE OF MULTIPLE SCLEROSIS

In this section, I illustrate the path of gene discovery in various AID, citing MS as the major example (see also Chapter 72). The aim of GWAS is to characterize the genetic architecture of complex genetic traits through the identification of disease variants against the background of random variation seen in a population as a whole. In a typical study, hundreds of thousands of markers covering a significant portion of the common variation in the population are tested simultaneously in cases and controls, and the allelic frequencies of each marker are compared between the two groups. The large number of common genetic variants in the human population means that prior odds that a randomly chosen marker is relevant for a given disease are extremely low [for example, estimated at 10^{-5} for MS (Oksenberg et al., 2008)]. For this reason, and in the absence of other prior knowledge, only highly significant markers are taken into consideration.

Linkage Versus Association Studies

The establishment of genetic linkage requires the collection of family pedigrees with more than one affected member to track the inheritance and identify discrete chromosomal segments in which variants deviate from independent segregation and co-segregate with the trait of interest. The dramatic success of the linkage approach for monogenic disorders, and the limited – but tantalizing – achievements in complex disorders including apolipoprotein E (*APOE*) in late-onset Alzheimer disease (Corder et al., 1993) among other examples (Horikawa et al., 2000; Hugot et al., 2001; Ogura et al., 2001; Stefansson et al., 2002), combined with improved maps of highly polymorphic markers for all chromosomes, fueled the wide application of this approach in autoimmune disorders, including MS. Linkage screens with different levels of resolution and genome coverage have been completed in more than 30 datasets of familial MS cases (Fernald et al., 2005). Each of these studies suggested multiple chromosomal regions with potential involvement in disease susceptibility, consistent with the long-held view that MS is a polygenic disorder. However, only the *HLA-class II* locus within the MHC region on chromosome 6p21.3 has ever exceeded the threshold for formal statistical significance, and no other region of statistically confident linkage has ever been observed. Using the values of *HLA* allele sharing by descent in sibships, it has been estimated that the *HLA* locus accounts for 20–60% of the genetic susceptibility in MS (Haines et al., 1998). Even at the higher end of this estimate, much of the genetic component of MS remains to be explained.

The IMSGC reported in late 2005 the results of a linkage screen in 730 multi-case MS families using more than 4500 single nucleotide polymorphisms (SNPs) (Sawcer et al., 2005). The peak logarithm of the odds (LOD) score of 11.7 (a LOD > 3 is considered conventional proof of linkage) found in the *HLA* region illustrates the substantially greater power achieved by this high-density screen compared with earlier efforts. Strikingly, however, although numerous regions on multiple chromosomes revealed possible linkage signals of interest, no other locus reached genome-wide significance. Given the magnitude of the study, these data indicate that any susceptibility allele common in the population and outside the *HLA* region is likely to increase MS risk by less than a factor of two. Similar findings have been reported for other AID.

Early success in identifying the relevance of *HLA*, where nominally significant association was detected with just 32 cases (Compston et al., 1976) and a highly significant association was shown with fewer than 200 (Olerup and Hillert, 1991), created a false sense of confidence in the research community, which assumed that other genes of relevance would be identified with equal ease by testing only 100–200 cases. Unfortunately, but not surprisingly, candidate gene studies using this approach have met with only modest success in identifying MS-causing genes, because of the difficulty in selecting from among the many plausible candidates and because of the limited statistical power of the relatively small tested datasets.

Genome-wide Association Studies

Compared to linkage, association studies have greater statistical power to detect common genetic variants that confer a modest risk for a disease (Risch and Merikangas, 1996). Association analysis determines whether specific genetic variants predispose to disease at the population level by comparing the frequency of the alleles at a polymorphic site in well-matched groups of affected and unaffected individuals. Although population stratification and the introduction of phenocopies due to the nature of phenotypic ascertainment of study participants are important confounding factors for association studies, inadequate sample size is undoubtedly the main reason why most published claims of association are suspected type I errors (Hirschhorn et al., 2002). Hence, the identification of genes influencing the development of MS needs to rely primarily on association-based methods and must involve very large patient cohorts.

The results of six GWAS in MS have been reported to date, including an early study that genotyped only non-synonymous coding SNPs and a recent scan of a high-risk isolate of Finland (Baranzini et al., 2009b; Burton et al., 2007; Comabella et al., 2008; Hafler et al., 2007; IMSGC, 2009; Jakkula et al., 2010). The classic *HLA-DRB1* risk locus stood out in all studies with remarkably strong statistical significance (e.g., $P < 1 \times 10^{-32}$ in Baranzini et al., 2009b). These were followed by extensive replication efforts of the top hits (Ban et al., 2009; De Jager et al., 2009b; D'Netto et al., 2009; Hafler et al., 2009; Hoppenbrouwers et al., 2008; Rubio et al., 2008) that, together with a comprehensive replicated meta-analysis (De Jager et al., 2009b), provided robust evidence for approximately a dozen novel loci affecting disease susceptibility (Table 70.1). It is important to note that these markers may not represent necessarily the causal disease variant themselves, explaining in part the very modest independent odds values for each allele. Additional follow-up experiments refined some of the association signals and revealed early mechanistic insights into the functional role of the identified genes, most notable a change in the soluble vs membrane-bound ratio for the interleukin (IL)2 and IL7 receptors (De Jager et al., 2009b; Gregory et al., 2007; Maier et al., 2009) and diminished expression of CD58 mRNA (De Jager et al., 2009a).

It is noteworthy that some of the MS-associated allelic variants have also been proposed to be involved in other autoimmune diseases, suggesting common mechanisms underlying different autoimmune conditions (Baranzini, 2009; Hafler et al., 2009; IMSGC, 2009; Sirota et al., 2009). For example, effects on disease susceptibility mediated by variants at the *IL2RA* locus are shared among MS, T1D, Graves' disease, and RA (Maier et al., 2009). Interestingly, the direction of the association is not consistent across diseases, as the *IL2RA* allelic variant associated with increased susceptibility to MS appears to confer resistance to T1D. A second allele confers susceptibility to both diseases, whereas yet a third allele is associated with susceptibility to T1D only (Dendrou et al., 2009; Maier et al., 2009). Similarly, while the minor allele of the R620W polymorphism in *PTPN22* has been associated with susceptibility to T1D, RA [Gregersen et al., 2009; Plenge et al., 2007; Raychaudhuri et al., 2008; Wellcome Trust Case Control Consortium (WTCCC), 2007], and SLE (Harley et al., 2008, Hom et al., 2008), it appears to confer protection to CD (Barrett et al., 2008). Functional studies aimed at detecting tissue- or cell-specific variation in the expression or function of the target gene may contribute to elucidate the pathogenic mechanisms operating in each AID (Dendrou et al., 2009).

This modularity of human disease is not limited to autoimmunity (Goh et al., 2007; Oti and Brunner, 2007; Oti et al., 2009). By merging conventional reductionism (i.e., correlations between distinct experimental endpoints and clinical phenotypes) and the non-reductionist approach of systems biomedicine (i.e., incorporating network biology principles), a new disease classification based on coherent genotypes and phenotypes might be possible (Baranzini, 2009; Loscalzo et al., 2007). A detailed description of such an approach is provided in the following section.

Altogether, the GWAS data seem to support the long-held view that MS susceptibility rests on the action of common sequence variants (i.e., those with a risk allele frequency of >5%) in multiple genes (IMSGC, 2010). Even with this expanding roster of risk loci, the understanding of MS genetics remains incomplete. For example, the sorting and classification of duplications or deletions of genomic segments generating copy number variants (CNVs) lag behind. CNVs are a major source of human genetic diversity and have been shown to influence rare genomic disorders as well as complex traits and diseases (Marshall et al., 2008; Wain et al., 2009; Zhang et al., 2009b). Recent studies have reported associations between CNVs and autoimmune diseases in humans such as SLE, psoriasis, CD, RA and T1D (Schaschl et al., 2009). A 10,000-case GWAS with a high-SNP/CNV-density platform, which is adequately powered to identify common risk alleles with an odds ratio of 1.2 or more, was proposed to be a reasonable next step in the genetic analysis of complex diseases in general and MS in particular (Altshuler et al., 2008; Oksenberg et al., 2008). Such a screen including MS, ankylosing spondylitis, psoriasis, and ulcerative colitis (UC) is now near completion by an international mega-group consortium in collaboration with the Wellcome Trust (https://www.wtccc.org.uk/ccc2).

TABLE 70.1	Top associated genes in seven common autoimmune diseases	
Disease	**Gene**	**Reference**
Celiac	IL21	Hunt et al., 2008; Van Heel et al., 2007
	RGS1	Hunt et al., 2008
	HLA-DQA1	Van Heel et al., 2007
MS	HLA-DRB1	ANZgene, 2009; Comabella et al., 2008; De Jager et al., 2009b; Hafler et al., 2007
	METTL1, CYP27B1	ANZgene, 2009
	CD58	ANZgene, 2009; De Jager et al., 2009a
	HLA-B	De Jager et al., 2009b
	TNFRSF1A	De Jager et al., 2009b
	IL2RA	De Jager et al., 2009b; Hafler et al., 2007
Psoriasis	HLA-C	Capon et al., 2008; Liu et al., 2008; Nair et al., 2009
	IL12B	Nair et al., 2009; Zhang et al., 2009a
	TNIP1	Nair et al., 2009
	IL13	Nair et al., 2009
	TNFAIP3	Nair et al., 2009
	LCE3D, LCE3A	Zhang et al., 2009a
Crohn's	IL23R	Barrett et al., 2008; Libioulle et al., 2007; Raelson et al., 2007; Rioux et al., 2007; WTCCC*, 2007
	ATG16L1	Barrett et al., 2008; Rioux et al., 2007; WTCCC, 2007
	PTGER4	Barrett et al., 2008
	NOD2	Barrett et al., 2008; Raelson et al., 2007; WTCCC, 2007
	ZNF365	Barrett et al., 2008
	PTPN2	Barrett et al., 2008; Parkes et al., 2007; WTCCC, 2007
	NKX2-3	Barrett et al., 2008; Parkes et al., 2007
	IRGM	Barrett et al., 2008; Parkes et al., 2007
	IL12B	Barrett et al., 2008
	MST1	Barrett et al., 2008; Parkes et al., 2007
	CCR6	Barrett et al., 2008
	STAT3	Barrett et al., 2008
	LRRK2, MUC19	Barrett et al., 2008
	TNFSF15	Barrett et al., 2008
	CDKAL1	Barrett et al., 2008
	BSN, MST1	WTCCC, 2007
	CARD15	Rioux et al., 2007
RA	PTPN22	Gregersen et al., 2009; Plenge et al., 2007; Raychaudhuri et al., 2008; WTCCC, 2007
	REL	Gregersen et al., 2009
	OLIG3, TNFIP3	Plenge et al., 2007; Raychaudhuri et al., 2008
	HLA-DRB1	Plenge et al., 2007; Raychaudhuri et al., 2008; WTCCC, 2007

(continued)

TABLE 70.1	(Continued)	
Disease	**Gene**	**Reference**
	HLA-DQA1, HLA-DQA2	Julia et al., 2008; WTCCC, 2007
	TRAF1-C5	Plenge et al., 2007
SLE	TNFAIP3	Graham et al., 2008
	STAT4	Graham et al., 2008; Hom et al., 2008
	HLA-DQA1	
	IRF5, TNPO3	Hom et al., 2008
	ITGAM, ITGAX	Hom et al., 2008
	C8orf13, BLK	Hom et al., 2008
	BANK1	Kozyrev et al., 2008
T1D	MHC	Barrett et al., 2008; Cooper et al., 2008; Hakonarson et al., 2007; WTCCC, 2007
	PTPN22	Barrett et al., 2008; Cooper et al., 2008; Hakonarson et al., 2007; Todd et al., 2007; WTCCC, 2007
	INS	Barrett et al., 2008; Hakonarson et al., 2007; Todd et al., 2007
	C10orf59	Barrett et al., 2008
	SH2B3	Barrett et al., 2008; WTCCC, 2007
	ERBB3	Barrett et al., 2008; Cooper et al., 2008; Hakonarson et al., 2008; Todd et al., 2007; WTCCC, 2007
	CLEC16A	Barrett et al., 2008; Cooper et al., 2008
	CTLA4	Barrett et al., 2008; Cooper et al., 2008
	PTPN2	Barrett et al., 2008; Cooper et al., 2008; Todd et al., 2007
	IL2RA	Barrett et al., 2008
	IL27	Barrett et al., 2008
	C6orf173	Barrett et al., 2008
	IL2	Barrett et al., 2008
	ORMDL3	Barrett et al., 2008
	GLIS3	Barrett et al., 2008
	CD69	Barrett et al., 2008
	UBASH3A	Barrett et al., 2008; Grant et al., 2009
	IFIH1	Barrett et al., 2008; Todd et al., 2007
	BACH2	Barrett et al., 2008; Cooper et al., 2008
	CTSH	Barrett et al., 2008; Cooper et al., 2008
	PRKCQ	Barrett et al., 2008; Cooper et al., 2008
	C1QTNF6	Barrett et al., 2008
	C12orf30	Cooper et al., 2008
	C1QTNF6	Cooper et al., 2008
	KIAA0350	Hakonarson et al., 2007; Todd et al., 2007; WTCCC, 2007
	C12orf30	Todd et al., 2007

*WTCCC, Wellcome Trust Case Control Consortium.

SYSTEMATIC ASSESSMENT OF COMMONALITIES AND DIFFERENCES AMONG AUTOIMMUNE DISEASES

In most GWAS, the number of markers in which the evidence for association exceeds the genome-wide significance threshold is small, and markers that do not exceed this threshold are generally neglected. A straightforward solution to the problem of low-powered datasets is to conduct larger studies. This is the rationale behind the second phase of the Wellcome Trust Case Control Consortium (WTCCC2), a massive collaborative project that is genotyping 120,000 samples in 13 diseases (including, among the AID, ankylosing spondylitis, MS, and psoriasis) and two quantitative phenotypes (WTCCC2, 2008).

An alternative strategy to increase the prior odds of finding a true significant marker is to incorporate prior biological knowledge into the GWAS data in the form of gene ontologies or pathways (Lesnick et al., 2007; Marchini et al., 2005; Torkamani et al., 2008; Wang et al., 2007). The advantage of these methods is that even if markers in individual genes do not reach genome-wide significance, several modest associations in genes from the same biological pathway may highlight collectively its involvement in a disease process. Building on this rationale, we merged statistical evidence from GWAS analyses with experimental evidence of protein interaction from yeast two-hybrid or chromatin immunoprecipitation (ChIP) studies to discover sub-networks (or modules) of interacting proteins associated with MS susceptibility (Baranzini et al., 2009a). Through this approach we were able to identify novel susceptibility pathways in MS, such as axon guidance and glutamate metabolism (Figure 70.1). We also uncovered genetic overlaps between MS and Alzheimer's disease and between MS and bipolar disorder. In addition, the presence of common variants in the MHC region between MS, RA, and T1D, but not T2D or CD with MHC alleles, was highlighted (Figure 70.2).

A survey of all studies reported in the GWAS Catalog (NHGRI, 2009) (as of July 2009) in seven common AID (CeD, CD, MS, psoriasis, RA, SLE, and T1D), as well as T2D, revealed that variants in 45 genes (seven of which are MHC-related) are associated with disease susceptibility (Table 70.1). Only markers with highly significant associations ($p < 10^{-10}$) or identified at $p < 10^{-7}$ in two or more studies are shown in this list. However, several truly associated variants may never reach this significance due to the limited power of most studies. As exemplified previously, exploratory analyses using lower significance thresholds may uncover important associations, particularly if they occur in candidate genes or pathways. Although by simply tabulating data, genes associated with more than one disease can be easily identified (e.g., *PTPN22* with RA and T1D, *IL2RA* with MS and T1D, *IL12B* in psoriasis and CD), this task becomes more difficult at a lower significance cut-off as the number of genes increases considerably. Unfortunately, the GWAS Catalog only lists associations at 10^{-6} or lower, thus preventing any analysis using a more liberal significance threshold.

Analytic and computational approaches that integrate results from multiple GWAS datasets represent an alternative strategy that may strengthen previous conclusions, suggest novel loci or pathways, and refine the localization of association signals (Chen and Witte, 2007; Forabosco et al., 2009). For example, a recent meta-analysis of three MS studies identified *CD6*, *TNFRSF1A*, and *IRF8*, three non-MHC genes not found in any of the previous GWAS in this disease (De Jager et al., 2009b). In a larger study, Johnson and O'Donnell collected and cataloged results from 118 GWAS published through March 1, 2008, all of which tested trait associations with >50,000 markers (Johnson and O'Donnell, 2009). This study listed all the p-values as provided by the authors in the original publications, in some cases as relaxed as $p < 0.05$. Although several autoimmune diseases were included in that set, comparing genes across traits was out of the scope of that work. We then extracted all moderately significant ($p < 10^{-4}$) associations from each study (plus those in T2D) to analyze and compare the genetic contribution to these autoimmune disorders (Table 70.2). When multiple studies for the same disorder reported on the same gene, p-values were combined using the Fisher's method (Fisher, 1948). Altogether, 1201 genes with modest evidence for association in at least one of these autoimmune disorders were identified.

Recently, Goh and colleagues integrated all available genetic data from the Online Mendelian Inheritance in Man (OMIM) database using a bipartite network-based visualization approach (Goh et al., 2007). Since OMIM focuses primarily on Mendelian disorders, genetic data on complex disorders was derived from literature mining, and thus, less comprehensively represented in this analysis. Nevertheless, this strategy identified groups of diseases that shared susceptibility genes and grouped them together, thus creating a disease landscape based on genetic similarity. To address whether genes involved in one AID also confer susceptibility to another, we carried out a similar approach to that used by Goh, using evidence from GWAS (Figure 70.3). When the network is visualized with only those genes exceeding the genome-wide significance level of $p < 10^{-7}$, a large connected core of genes and diseases can be observed (Figure 70.3A). In this visualization, a strong cluster of MHC-associated genes is readily identified, in particular for RA and T1D, but also for MS and SLE. This is evidenced by the prominent circles of red diamonds in the center of the figure.

All of the MHC-related genes associated with SLE and CeD were also shared by RA, T1D, or MS. However, this observation may be a consequence of the strong linkage disequilibrium operating in that region of the genome. The only two diseases showing no MHC associations were CD and T2D. This finding is likely a consequence of the absence of GWAS signals in chromosome 6 reported in several studies in CD and the fact that T2D is primarily a metabolic disease, not an AID, and is included in this analysis only for comparison (Kumar et al., 2005). Despite this observation, CD is still connected to the main core by sharing *MCTP1* with RA; *IL23* and *IL12B* with psoriasis; *ORMDL3*, *PTPN2*, and *PTPN22* with T1D; and *CDKAL1* with T2D.

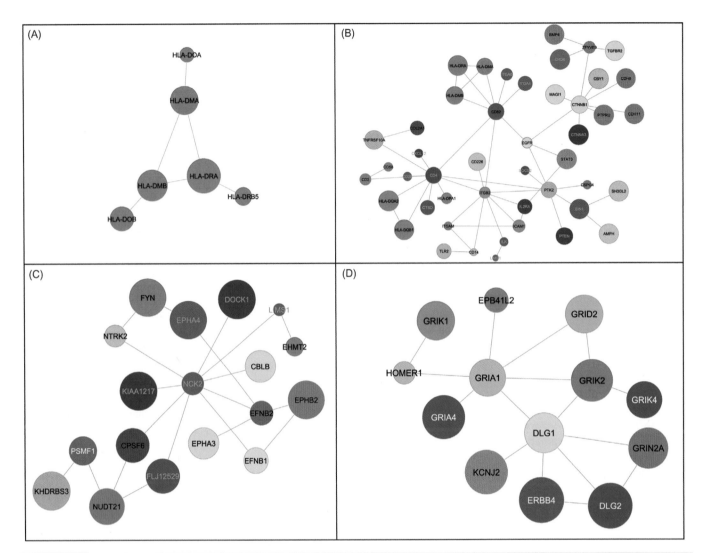

Figure 70.1 Integration of GWAS and interactome data for MS. Nodes represent proteins and connections represent physical interactions as determined by the human protein interaction network. The size of each node is proportional to the −log(10) p-value of association. Nodes are colored by chromosome. Each of the four panels depicts one significant sub-network (module) (HLA, extended immune module, axonal guidance module, and glutamate receptors module). (A) HLA module. This is the highest scoring module in MS, possibly due to the high significance of HLA-DRA and its interaction with other linked genes in the HLA region. (B) Extended immune module. In addition to HLA genes, this module contains other immune-related genes with more modest p-values of association. The significance of the entire module is possibly the result of the many interactions between these genes. (C) MS neural module 1. Seven genes encoding axon guidance molecules are part of this small module. (D) MS neural module 2. Seven glutamate receptors (gene symbols starting with GR) and two glutamate-related genes (*HOMER1* and *DLG1*) are included in this module.

While the connection among diseases through MHC-associated genes is illuminating, the extensive linkage disequilibrium in this region obscures the identification of additional shared genes. If the MHC locus is ignored, 16 genes are still associated with more than one disease at this significance level (displayed towards the center of Figure 70.3A). This select list of potentially "general" autoimmunity genes includes *PTPN22*, a tyrosine phosphatase strongly associated with T1D (aggregate $p < 10^{-226}$), RA (aggregate $p < 10^{-90}$), and to a lesser extent with CD (aggregate $p < 10^{-8}$). The risk allele

of this non-synonymous SNP (R620W), disrupts the P1 proline-rich motif that is important for interaction with cytoplasmic tyrosine kinase, potentially altering the protein's normal function as a negative regulator of T-cell activation. *PTPN22* has also been associated with other autoimmune diseases, including Addison's disease (Husebye and Lovas, 2009) and Graves' thyroiditis (Smyth et al., 2004). *TNFAIP3* is also highly associated with RA (aggregate $p < 10^{-20}$), SLE (aggregate $p < 10^{-11}$), and psoriasis (aggregate $p < 10^{-11}$), and moderately associated with CD (aggregate $p < 10^{-5}$). This TNFα-induced gene

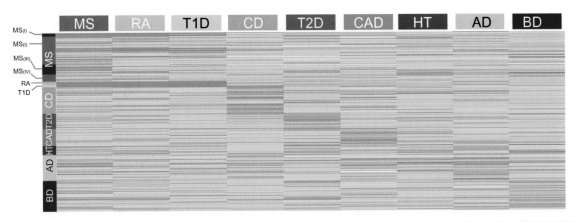

Figure 70.2 Module specificity. The p-values of genes from the modules identified in several complex diseases are displayed as a heatmap. Each row corresponds to a single gene. Genes are organized by their membership of modules. Genes corresponding to the four modules described for MS (Figure 70.1A) are at the top, followed by genes corresponding to modules from all other diseases. Because modules from different diseases may share one or more genes (e.g., HLA in autoimmune diseases), these may be represented more than once in the figure. Color-coded bars next to each module mark the genes that the module comprises. The same color code in the column headers indicates the disease for which the p-values are represented below. In general, genes from modules identified in one disease show the highest p-values for that disease, and less significant p-values for most other diseases. A notable exception concerns the HLA genes, which show overlap between MS, RA, and T1D, all of which are autoimmune diseases. Interestingly, some of the genes from the AD and BD modules show significant p-values also in MS.

is essential for limiting inflammation by terminating NF-κB responses.

If a more relaxed threshold ($p < 10^{-4}$) is used to visualize the reported associations, 71 non-MHC genes are identified as shared by at least two diseases (Figure 70.3B), seven by three diseases, and only two by four diseases (Table 70.3). In addition to *PTPN22* and *TNFAIP3* described above, *IL23R* and the *KIAA1109* locus appear as additional general autoimmunity genes. IL23R is also a key component of the T-cell activation pathway and plays a critical role in differentiation, expansion, and stabilization of proinflammatory TH17 cells (Abraham and Cho, 2009). KIAA1109 is located within a region of high linkage disequilibrium on chromosome 4 that also encompasses the genes for ADAD1, IL2, and IL21. The role of IL2 in T- and B-cell proliferation and its potential implications in autoimmunity have been extensively documented (Malek and Bayer, 2004; Nelson, 2004). Meanwhile, IL-21 acts as a co-stimulator of proliferation, enhances memory response, and modulates homeostasis. Within the innate immune system, IL-21 has a role in the terminal differentiation of Natural killer (NK) cells, enhancing cytotoxic function while also decreasing cellular viability. These immune maturation and stimulating functions have resulted in IL-21 being tested in a variety of models of immunity (Ettinger et al., 2008a, b; Jones et al., 2009). Finally, another gene with seemingly general autoimmune properties is *CTLA4*, although in this analysis it only reached genome-wide significance (aggregate $p < 10^{-7}$) in T1D and RA. When engaged by its receptors CD80 or CD86, CTLA4 initiates signals resulting in inhibition of T-cell activation. Several additional reports relating *CTLA4* with

multiple autoimmune diseases exist (e.g., Scalapino and Daikh, 2008), but most of these followed a candidate gene approach, and thus were not included in the analysis here.

Altogether, the data presented here suggest that genes involved in activation, proliferation, and homeostasis of cells involved in adaptive immune responses are more likely to represent general autoimmunity genes. This is further supported by the observation that a large proportion of these genes physically interact among each other (S. Baranzini, unpublished observation), thus possibly taking part in the same or highly overlapping biological pathways. A corollary to this observation indicates that associations unique to each disease would be responsible for attracting immune responses to specific tissues, although more functional studies in animal models will be needed to confirm this.

While, at $p < 10^{-4}$, several of these discoveries may represent false positives, potential associations with autoimmunity genes may be discovered if genes are shared by two or more diseases at a higher proportion than would be expected by chance alone. To this end, and as shown in Table 70.4, we counted all shared association between all pairs of diseases and assessed enrichment by computing a chi-squared test (Baranzini, 2009). We found that RA shared significantly more genes than expected with all other diseases studied. For example, it shared 12 genes with MS (1.49 expected, $p < 10^{-18}$), 12 with T1D (2.13 expected, $p < 10^{-12}$), and 18 with T2D (3.6 expected, $p < 10^{-14}$). On the other hand, genes shared between T2D and most other diseases (except RA), barely exceeded the expected number (Table 70.4).

TABLE 70.2	GWAS studies used for network analysis				
Phenotype	**Cases**	**Controls**	**Analyzed SNPs**	**#SNPs reported (criteria)**	**Reference**
CeD	778	1422	310,605	50 ($p < 10^{-2}$)	Van Heel et al., 2007
CD	382 (trios)	-	164,279	62 ($p < 10^{-3}$)	Raelson et al., 2007
CD	393	399	92,387	139 ($p < 10^{-2}$)	Franke et al., 2007
CD	547	548	308,332	6 (top)	Duerr et al., 2006
CD	547	928	302,451	1 (top)	Libioulle et al., 2007
CD	946	977	304,413	23 ($p < 10^{-4}$)	Rioux et al., 2007
CD	94	752	72,738	4 ($p < 10^{-3}$)	Yamazaki et al., 2005
CD	2000	3000	469,557	502 ($p < 10^{-3}$)	WTCCC, 2007
MS	931	2431	334,923	114 ($p < 10^{-3}$)	Hafler et al., 2007
MS	978	883	551,642	44 ($p < 10^{-3}$)	Baranzini et al., 2009b
Psoriasis	318	288	313,830	3 ($p < 10^{-4}$)	Capon et al., 2007
RA	1522	1850	297,086	193 ($p < 10^{-2}$)	Plenge et al., 2007
RA	625	558	203,269	14 ($p < 10^{-2}$)	Steer et al., 2007
RA (CCP+)	397	1211	79,853	205 ($p < 10^{-3}$)	Plenge et al., 2007
RA	2000	3000	469,557	380 ($p < 10^{-3}$)	WTCCC, 2007
SLE	94	538	52,608	1 (top)	Kamatani et al., 2008
SLE	51	54	262,264	5 (top)	Cervino et al., 2007
SLE	720	2337	265,648	35 ($p < 10^{-2}$)	Harley et al., 2008
T1D	1028	1143	534,071	88 (top)	Hakonarson et al., 2007
T1D	2000	3000	469,557	102 ($p < 10^{-2}$)	WTCCC, 2007
T2D	1464	1467	386,371	102 ($p < 10^{-2}$)	Saxena et al., 2007
T2D	105	102	115,352	72 ($p < 10^{-2}$)	Hanson et al., 2007
T2D	640	674	80,044	89 (top)	Hayes et al., 2007
T2D	124	295	82,485	125 (top)	Rampersaud et al., 2007
T2D	500	497	315,917	7 (top)	Salonen et al., 2007
T2D	1161	1174	315,635	97 ($p < 10^{-2}$)	Scott et al., 2007
T2D	661	614	392,935	50 ($p < 10^{-2}$)	Sladek et al., 2007
T2D	1399	5275	313,179	48 (top)	Steinthorsdottir et al., 2007
T2D	3757	5346	393,453	65 ($p < 10^{-3}$)	Zeggini et al., 2007
T2D	307 (trios)	-	66,543	45 ($p < 10^{-3}$)	Florez et al., 2007
T2D	91	1083	70,987	5 (top)	Meigs et al., 2007

THE NEXT STEPS IN COMPLEX GENETICS RESEARCH

Despite the success of GWAS in identifying novel susceptibility loci that withstood the challenge of independent replication, many questions remain concerning the genetic architecture of autoimmune diseases (Oksenberg and Baranzini, 2010). To illustrate the complexity and great challenges ahead of us, some observers draw on the example of human height, a complex trait with high (approximately 80%) estimated heritability; at least 40 loci have been associated with height, yet they only explain about 5% of phenotypic variance (Goldstein, 2009).

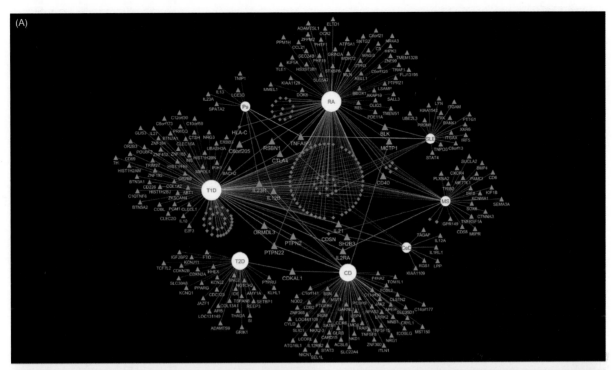

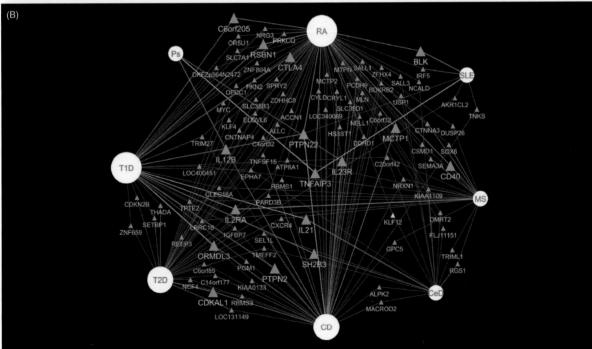

Figure 70.3 Autoimmune disease–gene network. (A) Top genetic associations in seven autoimmune diseases and T2D. The most significant SNP per gene was selected. Only associations with significance of at least $p < 10^{-7}$ are visualized. If a given gene was identified in more than one disease, multiple lines connecting it with each disease were drawn. Lines are colored using a "heat" scheme according to the evidence for association. Thus "hot" edges (e.g., red, orange) represent more significant associations than "cold" edges (e.g., purple, blue). Diseases are depicted by circles of size proportional to the number of associated genes. Non-MHC genes are shown by gray triangles, and genes in the MHC region are shown as red diamonds. (B) Similar to A, but with the threshold of significance lowered to $p < 10^{-4}$. To aid visualization, only genes shared by at least two diseases are shown. Abbreviations are as in the text; in addition, Ps = psoriasis.

TABLE 70.3 Genes shared by at least three diseases at (aggregate) $p < 10^{-4}$

Gene	Chr	Description	Crohn's	RA	T1D	Celiac	MS	SLE	Psoriasis
PTPN22	1	Protein tyrosine phosphatase, non-receptor type	10^{-8}	10^{-90}	10^{-226}			10^{-5}	
IL23R	1	Interleukin 23 receptor	10^{-102}	10^{-4}					10^{-7}
NRXN1	2	Neurexin 1 isoform beta precursor	10^{-4}	10^{-4}			10^{-4}		
KIAA1109	4	Hypothetical protein LOC84162		10^{-5}		10^{-12}	10^{-5}		
EPHA7	6	Ephrin receptor EphA7		10^{-4}	10^{-4}	10^{-5}			
TRIM27	6	Tripartite motif-containing 27	10^{-5}	10^{-6}					
TNFAIP3	6	Tumor necrosis factor alpha-induced protein 3	10^{-5}	10^{-20}	10^{-11}			10^{-11}	10^{-11}
TNKS	8	Tankyrase	10^{-4}				10^{-4}	10^{-6}	
C20orf42	20	Fermitin family homolog 1			10^{-5}	10^{-4}	10^{-4}		

TABLE 70.4 Number of non-MHC genes shared across diseases

	CD			MS			Psoriasis			RA			SLE			T1D			T2D		
	O	E	p	O	E	p	O	E	p	O	E	p	O	E	p	O	E	p	O	E	p
CeD	4	0.48	**4.0E-07**	3	0.2	**8.0E-10**	0	0.01	0.89	7	0.7	**1.7E-14**	0	0.06	0.8	3	0.3	**6.7E-7**	2	0.5	**0.03**
CD				3	1.05	0.06	3	0.09	**1.2E-22**	4	0.44	**5.2E-8**	3	0.3	**1E-6**	6	1.5	**2.3E-4**	10	2.54	**2.3E-6**
MS							0	0.04	0.84	12	1.49	**3.4E-18**	7	0.13	**0.02**	7	0.64	**1.75E-15**	5	1.08	**1.5E-4**
Psoriasis										2	0.13	**1.1E-7**	0	0.01	**7.4E-21**	0	0.05	0.81	0	0.09	0.76
RA													4	0.43	**5.2E-8**	12	2.13	**9E-12**	18	3.6	**1.4E-14**
SLE																1	0.2	0.06	1	0.32	0.22
T1D																			4	1.56	**0.05**

p-values of 0.05 or less are in bold
O = Observed, E = expected, p = chi-squared p value.

This may imply the futility of using this type of genetic information to generate reliable diagnostic signatures.

In other words, given that none of the high-frequency associated alleles are sufficient to cause disease and none is obligatory for the development of disease, most individuals in a population carry similar levels of genetically determined risk and very few can have their disease status accurately predicted from genetic testing (Sawcer et al., 2010). It is simple to attend to this rationale; the likelihood of unaccountable epistatic interactions, etiological heterogeneity, epigenetic and random events, suggest that even if the model includes all variant risks, there would still be considerable uncertainties when predicting case-control status. It is not unconceivable, however, that predictive modeling can be substantially improved when more penetrant variants, transcriptional signatures, accurate environmental exposures, and other variables are fitted into the model (Corvol et al., 2008; De Jager et al., 2009b). This translational potential extends to the possibility to redefine disease classification, prognosticate progression, and predict response and/or side effects to therapies.

It is noteworthy that all of the reported associations are derived from microarray-based genotyping studies, where only common variants (typically with a minor allele frequency >10%) are interrogated in GWAS. Thus, the almost certain influence of evolutionarily younger (rare) variants has not been adequately evaluated. With the advent of next-generation sequencing, more data are expected to be gathered in the near future to address this important question. A recent report found that rare variants in the interferon-inducible gene *IFIH1* are protective against T1D (Nejentsev et al., 2009). While this study resequenced exons and splice junctions of a few candidate genes, more rare variants are likely to be found when sequencing of entire genomes becomes mainstream. The 1000 Genomes Project (http://www.1000genomes.org), an international research consortium formed to create the most detailed and medically useful picture to date of human genetic variation, is currently producing whole-genome high-quality sequences of ~1000 individuals of different ethnic backgrounds and health conditions. Data emerging from this effort will undoubtedly uncover a multitude of private variants giving rise to a new catalog of human variation, an invaluable resource in the search for novel disease associations.

In a modest but important first step to assess this hypothesis, we recently sequenced the entire genomes of a female MS-discordant monozygotic twin pair, generating over one billion high-quality, shotgun whole-genome reads, corresponding to approximately 22-fold coverage of each genome (Baranzini et al., 2010). Remarkably, among ~3.6 million SNPs, ~200,000 insertion-deletion polymorphisms (indels), 27 CNVs, and 1.1 million methyl-CpG dinucleotides detected, no SNPs, indels, or CNVs, or methylation of two cytosine residues differed between the twins. We also deciphered the full mRNA transcriptome and epigenome sequences of CD4+ lymphocytes from three pairs of monozygotic, MS-discordant twins. Approximately 19,000 genes were expressed in each of the CD4+ T-cell preparations, but no reproducible transcriptional differences were identified between MS-affected and unaffected twins. Surprisingly, only 2 to 176 differences in methylation of ~2 million CpG dinucleotides were detected between siblings in three twin pairs, in contrast to ~800 methylation differences between T cells of unrelated individuals and several thousand differences between tissues or between normal and cancerous tissues. The sequences of many more discordant twin pairs, patients, and control genomes will be necessary to maximize the power of this approach.

This study coincided with a report presenting full-genome sequencing of a person affected with Charcot–Marie–Tooth disease (Lupski et al., 2010) and the whole-genome sequencing of a family of four, consisting of two siblings and their parents. One of the offspring of this family had primary ciliary diskinesia, and the other Miller syndrome, two rare genetic diseases. These two studies inaugurated the era of whole genome sequencing in human disease.

SYSTEMS BIOLOGY

Traditionally, researchers have coped with the enormous complexity of the immune system in a purely reductionist way: by subdividing it into specific cell populations and characterizing each of them in isolation. Although that approach has enjoyed remarkable success, the next challenge is to translate such valuable information into a better understanding of how several higher-order properties of the immune system, such as the development of tolerance or autoimmunity, emerge from such a complex interplay. Under the reductionist paradigm, a positive correlation between a single biological parameter and the occurrence of a disease is often considered a major success, even though the complete pathogenic mechanism may remain largely unknown. Notably, this approach has been followed by the pharmaceutical industry, despite a remarkably low rate of success (only 11% of new therapeutic targets reach the market as new drugs) (Kola and Landis, 2004).

In a conceptually broader approach, systems biology focuses on understanding not only the components of a given system (thus complementing the reductionist approach), but also the effect of interactions among them and the interaction of the system with its environment. Like physiology, systems biology is deeply rooted in the principle that the whole is more than the sum of its parts (Baranzini, 2006) (see also Chapter 5).

Biological systems offer multiple examples of collective properties, those in which the behavior of the whole cannot be predicted from the detailed study of individual components. In systems biology, as in the science of complexity, those intrinsic properties are referred to as *emergent* (Germain, 2001). Emergent properties are present in all physiological systems and include the maintenance of blood volume, blood pressure, tissue pH, and body temperature. Analyses based on single-cell neurobiology, immunology, or molecular biology are very useful for characterizing the individual behavior of each component. However, these properties are the result of nonrandom interactions of highly specialized cells assembled in large networks, and they can only be understood by not disrupting these structures. Based on this and other examples of systems composed of interconnected elements (e.g., acquaintances among people, the Internet, airports), inductive models have been created that display surprisingly generalizable properties (Barabasi and Oltvai, 2004; Watts and Strogatz, 1998).

It is imperative that any scientific research, especially if it requires an interdisciplinary approach, be grounded in the philosophy of science and furnished with the logical tools that permit the translation of empirical data into useful knowledge and worthwhile means of advancement (Fogel and Eylan, 1963). The knowledge generated in the last few decades on the pathogenesis of autoimmune diseases, as well as the development of high-throughput methods of analysis, biotechnology, and computational biology, has provided unprecedented opportunities for developing new disease-modifying therapies. I argue that the integration of this knowledge into the theoretical framework and the tools provided by systems biology will be an invaluable help in this process (Villoslada et al., 2009). However, therapeutic decisions made by physicians as they treat patients will always be based on their expertise and on evidence-based medical practice. The genomic portrait of an individual might allow a predictive and personalized approach to therapy. In the near future, this process will be aided by powerful computers, with information technologies that will manage the available information from a number of sources, including the patient's tests, the patient's medical history and genetic history, and a series of ever-evolving and expanding scientific databases. Although systems biology offers a new framework for the study of human disease, better characterization of biological processes, quantitative data, and dynamic information are still required to successfully translate this paradigm from the bench/chip to the bedside.

CONCLUSIONS

The genetic bases of autoimmunity are just starting to be uncovered. While several *bona fide* susceptibility genes have been identified in most common autoimmune diseases, technological advances in high-throughput genotyping platforms and more affordable next-generation sequencing methods will contribute to significantly expand these lists. In addition, structural variants (insertion/deletion polymorphisms, copy number variations, etc.) are also likely to play a significant role in determining susceptibility to AID. Taking into account the known susceptibility loci for each trait, disease-specific custom genotyping chips will be designed so as to cover a wide spectrum of variants in larger cohorts of individuals. At the same time, deep sequencing of candidate regions will be carried out to identify rare (private) mutations and structural variants that affect only a few individuals. Together, these approaches will eventually discover most if not all of the genetic contribution to these diseases and allow for the systematic search for similarities and differences among them.

While the identification of the precise pathways involved in susceptibility to AID will clearly require additional time and effort, integration of data from multiple diseases represents the logical next step in discovering similarities and differences among them. While similarities will shed light onto the general mechanism behind autoimmunity, genetic features private to a given disease will help pinpoint the basis for tissue specificity. One obvious benefit of this new knowledge would be the cross-utilization of drugs for diseases with a similar genetic fingerprint. Ultimately, this high-resolution genetic disease landscape will contribute to more accurate models of pathogenesis setting the bases for the development of more rational therapeutic approaches.

REFERENCES

Abraham, C., Cho, J.H., 2009. IL-23 and autoimmunity: New insights into the pathogenesis of inflammatory bowel disease. Annu Rev Med 60, 97–110.

Altshuler, D., Daly, M.J., Lander, E.S., 2008. Genetic mapping in human disease. Science 322, 881–888.

ANZgene, 2009. Genome-wide association study identifies new multiple sclerosis susceptibility loci on chromosomes 12 and 20. Nat Genet 41, 824–828.

Ban, M., Goris, A., Lorentzen, 2009. Replication analysis identifies TYK2 as a multiple sclerosis susceptibility factor. Eur J Hum Genet 17, 1309–1313.

Barabasi, A.L., Oltvai, Z.N., 2004. Network biology: Understanding the cell's functional organization. Nat Rev Genet 5, 101–113.

Baranzini, S.E., 2006. Systems-based medicine approaches to understand and treat complex diseases. The example of multiple sclerosis. Autoimmunity 39, 651–662.

Baranzini, S.E., 2009. The genetics of autoimmune diseases: A networked perspective. Curr Opin Immunol 21, 596–605.

Baranzini, S.E., Galwey, N.W., Wang, J., et al., 2009a. Pathway and network-based analysis of genome-wide association studies in multiple sclerosis. Hum Mol Genet 18, 2078–2090.

Baranzini, S.E., Mudge, J., Van Velkinburgh, J.C., et al., 2010. Genome, epigenome and RNA sequences of monozygotic twins discordant for multiple sclerosis. Nature 464, 1351–1356.

Baranzini, S.E., Wang, J., Gibson, R.A., et al., 2009b. Genome-wide association analysis of susceptibility and clinical phenotype in multiple sclerosis. Hum Mol Genet 18, 767–778.

Barcellos, L.F., Kamdar, B.B., Ramsay, P.P., et al., 2006. Clustering of autoimmune diseases in families with a high risk for multiple sclerosis: A descriptive study. Lancet Neurol 5, 924–931.

Barrett, J.C., Hansoul, S., Nicolae, D.L., et al., 2008. Genome-wide association defines more than 30 distinct susceptibility loci for Crohn's disease. Nat Genet 40, 955–962.

Burton, P.R., Clayton, D.G., Cardon, L.R., et al., 2007. Association scan of 14,500 nonsynonymous SNPs in four diseases identifies autoimmunity variants. Nat Genet 39, 1329–1337.

Caillat-Zucman, S., 2009. Molecular mechanisms of HLA association with autoimmune diseases. Tissue Antigens 73, 1–8.

Capon, F., Bijlmakers, M.J., Wolf, N., et al., 2008. Identification of ZNF313/RNF114 as a novel psoriasis susceptibility gene. Hum Mol Genet 17, 1938–1945.

Capon, F., Di Meglio, P., Szaub, J., et al., 2007. Sequence variants in the genes for the interleukin-23 receptor (IL23R) and its ligand (IL12B) confer protection against psoriasis. Hum Genet 122, 201–206.

Cervino, A.C., Tsinoremas, N.F., Hoffman, R.W., 2007. A genome-wide study of lupus: Preliminary analysis and data release. Ann NY Acad Sci 1110, 131–139.

Chen, G.K., Witte, J.S., 2007. Enriching the analysis of genomewide association studies with hierarchical modeling. Am J Hum Genet 81, 397–404.

Comabella, M., Craig, D.W., Camina-Tato, M., et al., 2008. Identification of a novel risk locus for multiple sclerosis at 13q31.3 by a pooled genome-wide scan of 500,000 single nucleotide polymorphisms. PLoS ONE 3, e3490.

Compston, D.A., Batchelor, J.R., McDonald, W.I., 1976. B-lymphocyte alloantigens associated with multiple sclerosis. Lancet 2, 1261–1265.

Cooper, J.D., Smyth, D.J., Smiles, A.M., et al., 2008. Meta-analysis of genome-wide association study data identifies additional type 1 diabetes risk loci. Nat Genet 40, 1399–1401.

Corder, E.H., Saunders, A.M., Strittmatter, W.J., et al., 1993. Gene dose of apolipoprotein E type 4 allele and the risk of Alzheimer's disease in late onset families. Science 261, 921–923.

Corvol, J.-C., Pelletier, D., Henry, R.G., et al., 2008. Abrogation of T cell quiescence characterizes patients at high risk for multiple sclerosis after the initial neurological event. Proc Natl Acad Sci USA 105, 11,839–11,844.

D'Netto, M.J., Ward, H., Morrison, K.M., et al., 2009. Risk alleles for multiple sclerosis in multiplex families. Neurology 72, 1984–1988.

De Jager, P.L., Baecher-Allan, C., Maier, L.M., et al., 2009a. The role of the CD58 locus in multiple sclerosis. Proc Natl Acad Sci USA 106, 5264–5269.

De Jager, P.L., Jia, X., Wang, J., et al., 2009b. Meta-analysis of genome scans and replication identify CD6, IRF8 and TNFRSF1A as new multiple sclerosis susceptibility loci. Nat Genet 41, 776–782.

De Jong, B.A., Huizinga, T.W.J., Zanelli, E., et al., 2002. Evidence for additional genetic risk indicators of relapse-onset MS within the HLA region. Neurology 59, 549–555.

Dendrou, C.A., Plagnol, V., Fung, E., et al., 2009. Cell-specific protein phenotypes for the autoimmune locus IL2RA using a genotype-selectable human bioresource. Nat Genet 41, 1011–1015.

Duerr, R.H., Taylor, K.D., Brant, S.R., et al., 2006. A genome-wide association study identifies IL23R as an inflammatory bowel disease gene. Science 314, 1461–1463.

Ettinger, R., Kuchen, S., Lipsky, P.E., 2008a. Interleukin 21 as a target of intervention in autoimmune disease. Ann Rheum Dis 67 (Suppl 3), iii83–iii86.

Ettinger, R., Kuchen, S., Lipsky, P.E., 2008b. The role of IL-21 in regulating B-cell function in health and disease. Immunol Rev 223, 60–86.

Fernald, G.H., Yeh, R.F., Hauser, S.L., Oksenberg, J.R., Baranzini, S.E., 2005. Mapping gene activity in complex disorders: Integration of expression and genomic scans for multiple sclerosis. J Neuroimmunol 167, 157–169.

Fernandez, O., Fernandez, V., Alonso, A., et al., 2004. DQB1*0602 allele shows a strong association with multiple sclerosis in patients in Malaga, Spain. J Neurol 251, 440–444.

Fisher, R.A., 1948. Combining independent tests of significance. Am Stat 2, 30.

Florez, J.C., Manning, A.K., Dupuis, J., et al., 2007. A 100K genome-wide association scan for diabetes and related traits in the Framingham Heart Study: Replication and integration with other genome-wide datasets. Diabetes 56, 3063–3074.

Fogel, A., Eylan, E., 1963. Preventive medicine and the physician. WHO Chron 17, 350–358.

Forabosco, P., Bouzigon, E., Ng, M.Y., et al., 2009. Meta-analysis of genome-wide linkage studies across autoimmune diseases. Eur J Hum Genet 17, 236–243.

Franke, A., Hampe, J., Rosenstiel, P., et al., 2007. Systematic association mapping identifies NELL1 as a novel IBD disease gene. PLoS ONE 2, e691.

Fukazawa, T., Sasaki, H., Kikuchi, S., Hamada, T., Tashiro, K., 2000. Genetics of multiple sclerosis. Biomed Pharmacother 54, 103–106.

Germain, R.N., 2001. The art of the probable: System control in the adaptive immune system. Science 293, 240–245.

Goh, K.I., Cusick, M.E., Valle, D., Childs, B., Vidal, M., Barabasi, A.L., 2007. The human disease network. Proc Natl Acad Sci USA 104, 8685–8690.

Goldstein, D.B., 2009. Common genetic variation and human traits. N Engl J Med 360, 1696–1698.

Graham, R.R., Cotsapas, C., Davies, L., et al., 2008. Genetic variants near TNFAIP3 on 6q23 are associated with systemic lupus erythematosus. Nat Genet 40, 1059–1061.

Grant, S.F., Qu, H.Q., Bradfield, J.P., et al., 2009. Follow-up analysis of genome-wide association data identifies novel loci for type 1 diabetes. Diabetes 58, 290–295.

Gregersen, P.K., Amos, C.I., Lee, A.T., et al., 2009. REL, encoding a member of the NF-kappaB family of transcription factors, is a newly defined risk locus for rheumatoid arthritis. Nat Genet 41, 820–823.

Gregory, S.G., Schmidt, S., Seth, P., et al., 2007. Interleukin 7 receptor alpha chain (IL7R) shows allelic and functional association with multiple sclerosis. Nat Genet 39, 1083–1091.

Hafler, D.A., Compston, A., Sawcer, S., et al., 2007. Risk alleles for multiple sclerosis identified by a genomewide study. N Engl J Med 357, 851–862.

Hafler, J.P., Maier, L.M., Cooper, J.D., et al., 2009. CD226 Gly307Ser association with multiple autoimmune diseases. Genes Immun 10, 5–10.

Haines, J.L., Terwedow, H.A., Burgess, K., et al., 1998. Linkage of the MHC to familial multiple sclerosis suggests genetic heterogeneity. The Multiple Sclerosis Genetics Group. Hum Mol Genet 7, 1229–1234.

Hakonarson, H., Grant, S.F., Bradfield, J.P., et al., 2007. A genome-wide association study identifies KIAA0350 as a type 1 diabetes gene. Nature 448, 591–594.

Hakonarson, H., Qu, H.Q., Bradfield, J.P., et al., 2008. A novel susceptibility locus for type 1 diabetes on Chr12q13 identified by a genome-wide association study. Diabetes 57, 1143–1146.

Hanson, R.L., Bogardus, C., Duggan, D., et al., 2007. A search for variants associated with young-onset type 2 diabetes in American Indians in a 100K genotyping array. Diabetes 56, 3045–3052.

Harley, J.B., Alarcon-Riquelme, M.E., Criswell, L.A., et al., 2008. Genome-wide association scan in women with systemic lupus erythematosus identifies susceptibility variants in ITGAM, PXK, KIAA1542 and other loci. Nat Genet 40, 204–210.

Hayes, M.G., Pluzhnikov, A., Miyake, K., et al., 2007. Identification of type 2 diabetes genes in Mexican Americans through genome-wide association studies. Diabetes 56, 3033–3044.

Hindorff, L.A., Sethupathy, P., Junkins, H.A., et al., 2009. Potential etiologic and functional implications of genome-wide association loci for human diseases and traits. Proc Natl Acad Sci USA 106, 9362–9367.

Hirschhorn, J.N., Lohmueller, K., Byrne, E., Hirschhorn, K., 2002. A comprehensive review of genetic association studies. Genet Med 4, 45–61.

Hom, G., Graham, R.R., Modrek, B., et al., 2008. Association of systemic lupus erythematosus with C8orf13-BLK and ITGAM-ITGAX. N Engl J Med 358, 900–909.

Hoppenbrouwers, I.A., Aulchenko, Y.S., Ebers, G.C., et al., 2008. EVI5 is a risk gene for multiple sclerosis. Genes Immun 9, 334–337.

Horikawa, Y., Oda, N., Cox, N.J., et al., 2000. Genetic variation in the gene encoding calpain-10 is associated with type 2 diabetes mellitus. Nat Genet 26, 163–175.

Hugot, J.P., Chamaillard, M., Zouali, H., et al., 2001. Association of NOD2 leucine-rich repeat variants with susceptibility to Crohn's disease. Nature 411, 599–603.

Hunt, K.A., Zhernakova, A., Turner, G., et al., 2008. Newly identified genetic risk variants for celiac disease related to the immune response. Nat Genet 40, 395–402.

Husebye, E., Lovas, K., 2009. Pathogenesis of primary adrenal insufficiency. Best Pract Res Clin Endocrinol Metab 23, 147–157.

IMSGC, 2009. The expanding genetic overlap between multiple sclerosis and type I diabetes. Genes Immun 10, 11–14.

IMSGC, 2010. Comprehensive follow-up of the first genome-wide association study of multiple sclerosis identifies KIF21B and TMEM39A as susceptibility loci. Hum Mol Genet 19, 953–962.

Jakkula, E., Leppa, V., Sulonen, A.M., et al., 2010. Genome-wide association study in a high-risk isolate for multiple sclerosis reveals associated variants in STAT3 gene. Am J Hum Genet 86, 285–291.

Johnson, A.D., O'Donnell, C.J., 2009. An open access database of genome-wide association results. BMC Med Genet 10, 6.

Jones, J.L., Phuah, C.L., Cox, A.L., et al., 2009. IL-21 drives secondary autoimmunity in patients with multiple sclerosis, following therapeutic lymphocyte depletion with alemtuzumab (Campath-1H). J Clin Invest 119, 2052–2061.

Julia, A., Ballina, J., Canete, J.D., et al., 2008. Genome-wide association study of rheumatoid arthritis in the Spanish population: KLF12 as a risk locus for rheumatoid arthritis susceptibility. Arthritis Rheum 58, 2275–2286.

Kamatani, Y., Matsuda, K., Ohishi, T., et al., 2008. Identification of a significant association of a single nucleotide polymorphism in TNXB with systemic lupus erythematosus in a Japanese population. J Hum Genet 53, 64–73.

Khare, M., Mangalam, A., Rodriguez, M., David, C.S., 2005. HLA DR and DQ interaction in myelin oligodendrocyte glycoprotein-induced experimental autoimmune encephalomyelitis in HLA class II transgenic mice. J Neuroimmunol 169, 1–12.

Kola, I., Landis, J., 2004. Can the pharmaceutical industry reduce attrition rates? Nat Rev Drug Discov 3, 711–715.

Kozyrev, S.V., Abelson, A.K., Wojcik, J., et al., 2008. Functional variants in the B-cell gene BANK1 are associated with systemic lupus erythematosus. Nat Genet 40, 211–216.

Kumar, V., Abbas, A.K., Fausto, N., Robbins, S.L., Cotran, R.S., 2005. Robbins and Cotran Pathologic Basis of Disease. Elsevier Saunders, Philadelphia, PA.

Lesnick, T.G., Papapetropoulos, S., Mash, D.C., et al., 2007. A genomic pathway approach to a complex disease: Axon guidance and Parkinson disease. PLoS Genet 3, e98.

Libioulle, C., Louis, E., Hansoul, S., et al., 2007. Novel Crohn disease locus identified by genome-wide association maps to a gene desert on 5p13.1 and modulates expression of PTGER4. PLoS Genet 3, e58.

Lin, J.P., Cash, J.M., Doyle, S.Z., et al., 1998. Familial clustering of rheumatoid arthritis with other autoimmune diseases. Hum Genet 103, 475–482.

Liu, Y., Helms, C., Liao, W., et al., 2008. A genome-wide association study of psoriasis and psoriatic arthritis identifies new disease loci. PLoS Genet 4, e1000041.

Loscalzo, J., Kohane, I., Barabasi, A.L., 2007. Human disease classification in the postgenomic era: A complex systems approach to human pathobiology. Mol Syst Biol 3, 124.

Lupski, J.R., Reid, J.G., Gonzaga-Jauregui, C., et al., 2010. Whole-genome sequencing in a patient with Charcot–Marie–Tooth neuropathy. N Engl J Med 362, 1181–1191.

Maier, L.M., Lowe, C.E., Cooper, J., et al., 2009. IL2RA genetic heterogeneity in multiple sclerosis and type 1 diabetes susceptibility and soluble interleukin-2 receptor production. PLoS Genet 5, e1000322.

Malek, T.R., Bayer, A.L., 2004. Tolerance, not immunity, crucially depends on IL-2. Nat Rev Immunol 4, 665–674.

Marchini, J., Donnelly, P., Cardon, L.R., 2005. Genome-wide strategies for detecting multiple loci that influence complex diseases. Nat Genet 37, 413–417.

Marrosu, M.G., Murru, R., Murru, M.R., et al., 2001. Dissection of the HLA association with multiple sclerosis in the founder isolated population of Sardinia. Hum Mol Genet 10, 2907–2916.

Marshall, C.R., Noor, A., Vincent, J.B., et al., 2008. Structural variation of chromosomes in autism spectrum disorder. Am J Hum Genet 82, 477–488.

Meigs, J.B., Manning, A.K., Fox, C.S., et al., 2007. Genome-wide association with diabetes-related traits in the Framingham Heart Study. BMC Med Genet 8 (Suppl 1), S16.

Miretti, M.M., Walsh, E.C., Ke, X., et al., 2005. A high-resolution linkage-disequilibrium map of the human major histocompatibility complex and first generation of tag single-nucleotide polymorphisms. Am J Hum Genet 76, 634–646.

Nair, R.P., Duffin, K.C., Helms, C., et al., 2009. Genome-wide scan reveals association of psoriasis with IL-23 and NF-kappaB pathways. Nat Genet 41, 199–204.

Nejentsev, S., Walker, N., Riches, D., Egholm, M., Todd, J.A., 2009. Rare variants of IFIH1, a gene implicated in antiviral responses, protect against type 1 diabetes. Science 324, 387–389.

Nelson, B.H., 2004. IL-2, regulatory T cells, and tolerance. J Immunol 172, 3983–3988.

NHGRI, 2009. A Catalog of Published Genome-Wide Association Studies. Bethesda, MD: NHGRI. <http://www.genome.gov/26525384>.

Ogura, Y., Inohara, N., Benito, A., Chen, F.F., Yamaoka, S., Nunez, G., 2001. Nod2, a Nod1/Apaf-1 family member that is restricted to monocytes and activates NF-kappaB. J Biol Chem 276, 4812–4818.

Oksenberg, J.R., Baranzini, S.E., 2010. Multiple sclerosis genetics – is the glass half full, or half empty? Nat Rev Neurol 6, 429–437.

Oksenberg, J.R., Baranzini, S.E., Sawcer, S., Hauser, S.L., 2008. The genetics of multiple sclerosis: SNPs to pathways to pathogenesis. Nat Rev Genet 9, 516–526.

Olerup, O., Hillert, J., 1991. HLA class II-associated genetic susceptibility in multiple sclerosis: A critical evaluation. Tissue Antigens 38, 1–15.

Oti, M., Brunner, H.G., 2007. The modular nature of genetic diseases. Clin Genet 71, 1–11.

Oti, M., Huynen, M.A., Brunner, H.G., 2009. The biological coherence of human phenome databases. Am J Hum Genet 85, 801–808.

Parkes, M., Barrett, J.C., Prescott, N.J., et al., 2007. Sequence variants in the autophagy gene IRGM and multiple other replicating loci contribute to Crohn's disease susceptibility. Nat Genet 39, 830–832.

Plenge, R.M., Seielstad, M., Padyukov, L., et al., 2007. TRAF1-C5 as a risk locus for rheumatoid arthritis – a genomewide study. N Engl J Med 357, 1199–1209.

Prat, E., Tomaru, U., Sabater, L., et al., 2005. HLA-DRB5*0101 and -DRB1*1501 expression in the multiple sclerosis-associated HLA-DR15 haplotype. J Neuroimmunol 167, 108–119.

Raelson, J.V., Little, R.D., Ruether, A., et al., 2007. Genome-wide association study for Crohn's disease in the Quebec Founder Population identifies multiple validated disease loci. Proc Natl Acad Sci USA 104, 14,747–14,752.

Rampersaud, E., Damcott, C.M., Fu, M., et al., 2007. Identification of novel candidate genes for type 2 diabetes from a genome-wide association scan in the Old Order Amish: Evidence for replication from diabetes-related quantitative traits and from independent populations. Diabetes 56, 3053–3062.

Raychaudhuri, S., Remmers, E.F., Lee, A.T., et al., 2008. Common variants at CD40 and other loci confer risk of rheumatoid arthritis. Nat Genet 40, 1216–1223.

Rioux, J.D., Xavier, R.J., Taylor, K.D., et al., 2007. Genome-wide association study identifies new susceptibility loci for Crohn disease and implicates autophagy in disease pathogenesis. Nat Genet 39, 596–604.

Risch, N., Merikangas, K., 1996. The future of genetic studies of complex human diseases. Science 273, 1516–1517.

Roach, J.C., Glusman, G., Smit, A.F., et al., 2010. Analysis of genetic inheritance in a family quartet by whole-genome sequencing. Science 328, 636–639.

Rubio, J.P., Bahlo, M., Butzkueven, H., et al., 2002. Genetic dissection of the human leukocyte antigen region by use of haplotypes of Tasmanians with multiple sclerosis. Am J Hum Genet 70, 1125–1137.

Rubio, J.P., Stankovich, J., Field, J., et al., 2008. Replication of KIAA0350, IL2RA, RPL5 and CD58 as multiple sclerosis susceptibility genes in Australians. Genes Immun 9, 624–630.

Rzhetsky, A., Wajngurt, D., Park, N., Zheng, T., 2007. Probing genetic overlap among complex human phenotypes. Proc Natl Acad Sci USA 104, 11,694–11,699.

Salonen, J.T., Uimari, P., Aalto, J.M., et al., 2007. Type 2 diabetes whole-genome association study in four populations: The DiaGen consortium. Am J Hum Genet 81, 338–345.

Sawcer, S., Ban, M., Maranian, M., et al., 2005. A high-density screen for linkage in multiple sclerosis. Am J Hum Genet 77, 454–467.

Sawcer, S., Ban, M., Wason, J., Dudbridge, F., 2010. What role for genetics in the prediction of multiple sclerosis? Ann Neurol 67, 3–10.

Saxena, R., Voight, B.F., Lyssenko, V., et al., 2007. Genome-wide association analysis identifies loci for type 2 diabetes and triglyceride levels. Science 316, 1331–1336.

Scalapino, K.J., Daikh, D.I., 2008. CTLA-4: A key regulatory point in the control of autoimmune disease. Immunol Rev 223, 143–155.

Schaschl, H., Aitman, T.J., Vyse, T.J., 2009. Copy number variation in the human genome and its implication in autoimmunity. Clin Exp Immunol 156, 12–16.

Scott, L.J., Mohlke, K.L., Bonnycastle, L.L., et al., 2007. A genome-wide association study of type 2 diabetes in Finns detects multiple susceptibility variants. Science 316, 1341–1345.

Shiina, T., Inoko, H., Kulski, J.K., 2004. An update of the HLA genomic region, locus information and disease associations: 2004. Tissue Antigens 64, 631–649.

Sirota, M., Schaub, M.A., Batzoglou, S., Robinson, W.H., Butte, A.J., 2009. Autoimmune disease classification by inverse association with SNP alleles. PLoS Genet 5, e1000792.

Sladek, R., Rocheleau, G., Rung, J., et al., 2007. A genome-wide association study identifies novel risk loci for type 2 diabetes. Nature 445, 881–885.

Smyth, D., Cooper, J.D., Collins, J.E., et al., 2004. Replication of an association between the lymphoid tyrosine phosphatase locus (LYP/PTPN22) with type 1 diabetes, and evidence for its role as a general autoimmunity locus. Diabetes 53, 3020–3023.

Sospedra, M., Muraro, P.A., Stefanova, I., et al., 2006. Redundancy in antigen-presenting function of the HLA-DR and -DQ molecules in the multiple sclerosis-associated HLA-DR2 haplotype. J Immunol 176, 1951–1961.

Steer, S., Abkevich, V., Gutin, A., et al., 2007. Genomic DNA pooling for whole-genome association scans in complex disease: Empirical demonstration of efficacy in rheumatoid arthritis. Genes Immun 8, 57–68.

Stefansson, H., Sigurdsson, E., Steinthorsdottir, V., et al., 2002. Neuregulin 1 and susceptibility to schizophrenia. Am J Hum Genet 71, 877–892.

Steinthorsdottir, V., Thorleifsson, G., Reynisdottir, I., et al., 2007. A variant in CDKAL1 influences insulin response and risk of type 2 diabetes. Nat Genet 39, 770–775.

Tait, K.F., Marshall, T., Berman, J., et al., 2004. Clustering of autoimmune disease in parents of siblings from the Type 1 diabetes Warren repository. Diabet Med 21, 358–362.

Todd, J.A., Walker, N.M., Cooper, J.D., et al., 2007. Robust associations of four new chromosome regions from genome-wide analyses of type 1 diabetes. Nat Genet 39, 857–864.

Torkamani, A., Topol, E.J., Schork, N.J., 2008. Pathway analysis of seven common diseases assessed by genome-wide association. Genomics 92, 265–272.

Van Heel, D.A., Franke, L., Hunt, K.A., et al., 2007. A genome-wide association study for celiac disease identifies risk variants in the region harboring IL2 and IL21. Nat Genet 39, 827–829.

Villoslada, P., Steinman, L., Baranzini, S.E., 2009. Systems biology and its application to the understanding of neurological diseases. Ann Neurol 65, 124–139.

Wain, L.V., Armour, J.A., Tobin, M.D., 2009. Genomic copy number variation, human health, and disease. Lancet 374, 340–350.

Wang, K., Li, M., Bucan, M., 2007. Pathway-based approaches for analysis of genomewide association studies. Am J Hum Genet 81, 1278–1283.

Watts, D.J., Strogatz, S.H., 1998. Collective dynamics of "small-world" networks. Nature 393, 440–442.

WTCCC, 2007. Genome-wide association study of 14,000 cases of seven common diseases and 3000 shared controls. Nature 447, 661–678.

WTCCC2, 2008. Wellcome Trust Case Control Consortium 2. <https://www.wtccc.org.uk/ccc2/index.shtml>.

Yamazaki, K., McGovern, D., Ragoussis, J., et al., 2005. Single nucleotide polymorphisms in TNFSF15 confer susceptibility to Crohn's disease. Hum Mol Genet 14, 3499–3506.

Yeo, T.W., De Jager, P.L., Gregory, S.G., et al., 2007. A second major histocompatibility complex susceptibility locus for multiple sclerosis. Ann Neurol 61, 228–236.

Zeggini, E., Weedon, M.N., Lindgren, C.M., et al., 2007. Replication of genome-wide association signals in UK samples reveals risk loci for type 2 diabetes. Science 316, 1336–1341.

Zhang, X.J., Huang, W., Yang, S., et al., 2009a. Psoriasis genome-wide association study identifies susceptibility variants within LCE gene cluster at 1q21. Nat Genet 41, 205–210.

Zhang, Y., Martens, J.W., Yu, J.X., et al., 2009b. Copy number alterations that predict metastatic capability of human breast cancer. Cancer Res 69, 3795–3801.

Rheumatoid Arthritis

Robert M. Plenge

INTRODUCTION

Rheumatoid arthritis (RA) is a systemic, chronic inflammatory disorder whose root cause is unclear. The clinical hallmark of RA is an inflammatory arthritis with a predilection for specific diarthrodial (freely movable) joints. It is the most common form of inflammatory arthritis, with an estimated prevalence of up to ~1% in the adult population. Females are at greater risk than males for developing the disease, with a female:male ratio of ~2.5:1. While the disease can occur at any age, the peak age on onset is in the 40s, with an increasing incidence with age (Silman and Pearson, 2002).

As with many complex diseases – those influenced by multiple genes and environmental exposures – there is substantial clinical heterogeneity in RA (Klareskog et al., 2009). The clinical features of new-onset RA are highlighted in Table 71.1. With longstanding disease, articular erosions and joint deformities occur. Autoantibodies [rheumatoid factor (RF) and anti-cyclic citrullinated proteins (CCP) antibodies] have important diagnostic and prognostic features and have proven very useful in the clinical management of RA. Most efforts aimed at understanding the molecular basis of RA have focused on genetic studies of disease susceptibility. To date, common alleles from >30 gene loci have been convincingly demonstrated to influence risk of RA (Stahl et al., 2010). There are no reproducible associations with response to therapy. Other genomic technologies such as large-scale expression and proteomic profiling in RA are less mature, but offer promise in understanding disease etiology.

Genomic medicine – translating genetic information into prediction of disease susceptibility, characterization of gene–environment interactions, identification of new therapeutic targets, and development of novel gene-based diagnostics – has had very little impact to date on the clinical management of patients with RA. For example, the association of susceptibility to *HLA-DRB1* alleles in the major histocompatibility complex (MHC) region has been known since the 1970s, yet has not translated into novel diagnostics or therapeutic treatments. Genome-wide association studies (GWAS) have identified >30 risk alleles outside of the MHC, but these have not yet had an impact on clinical care of patients. Nonetheless, there is reason to be optimistic, as many of these genetic associations have only recently been discovered. An intriguing observation offering hope for improved therapy is that several risk loci discovered by GWAS harbor genes known to be the targets of biological disease-modifying anti-rheumatic drugs (DMARDs).

CLINICAL FEATURES

The range of presenting clinical symptoms in RA is quite variable, but the *sine qua non* of RA is an inflammatory arthritis,

Genomic and Personalized Medicine, 2nd edition
by Ginsburg & Willard. DOI: http://dx.doi.org/10.1016/B978-0-12-382227-7.00071-9

TABLE 71.1 Common clinical features of new-onset rheumatoid arthritis

Symptoms

Joint swelling

Pain/stiffness (commonly in morning and lasting >1 hour)

Fatigue

Malaise

Articular characteristics

Palpation tenderness

Synovial thickening

Effusion

Erythema

Distribution

Symmetrical

Distal (e.g., hands and feet) more commonly than proximal (e.g., spine)

PIP, MCP/MTP, wrist/ankle more commonly than elbow/knee, shoulder/hip

PIP = proximal interphalangeal joint, MCP = metacarpophalangeal joint, MTP = metatarsophalangeal joint.

manifested by symmetrical joint pain, stiffness, and swelling of diarthrodial joints. In most cases, onset of symptoms is insidious, and several months of symptoms are required to establish a diagnosis. Other systemic diseases may mimic RA (including arthritis secondary to inflammatory bowel disease, Lyme disease, or psoriasis), although the clinical presentation, joint distribution, and radiographic changes are often sufficient to eliminate these disorders from the differential diagnosis. The American College of Rheumatology (ACR) has established criteria for the diagnosis of RA (Arnett et al., 1988). These criteria have ~90% sensitivity and specificity for RA when compared with non-RA rheumatic disease control subjects.

The clinical course of RA is extremely variable: some patients suffer mild, self-limiting arthritis while others develop progressive multi-system inflammation with profound morbidity and mortality (Pincus, 1995). Approximately 15% of patients who meet the ACR criteria for RA become free of symptoms at one year's time. For the majority of patients, however, chronic disease ensues; if left untreated, joint destruction from synovitis occurs. Radiographic evidence of RA is present on standard X-rays in more than 70% within two years. More sensitive techniques such as magnetic resonance imaging (MRI) can identify changes such as synovial hypertrophy, bone edema, and erosive changes as early as four months after disease onset (McGonagle et al., 1999; McQueen et al., 1998). Ultimately, irreversible articular damage leads to

physical deformity and functional disability, which carries an economic burden to society.

For those patients with chronic disease, RA can be divided generally into those who are "seropositive" and those who are "seronegative," based upon the presence of circulating autoantibodies. Classically, rheumatoid factor (RF) is the autoantibody that establishes whether a patient is seropositive or negative. More recently, antibodies directed against citrullinated proteins (anti-CCP antibodies) have been identified. These autoantibodies demonstrate higher sensitivity and specificity for RA when compared to RF. In general, seropositive patients progress more rapidly and have more severe disease than seronegative patients. (See below for a more detailed discussion on autoantibodies.)

These patient characteristics are important when interpreting genetic and epidemiological studies in RA. Conclusions from a collection of new-onset, early arthritis cohort may be quite different than a collection of patients with long-standing disease. Furthermore, the extent of clinical heterogeneity underscores the importance of identifying genomic risk factors that not only predict risk of disease, but subset RA patients into meaningful clinical categories.

PREDISPOSITION

Despite decades of research, the root cause of RA is unclear (Firestein, 2003). Genes and environment together contribute to development of RA (Klareskog et al., 2009). Anti-CCP antibodies have emerged as a specific marker for RA. That anti-CCP antibodies predate the diagnosis of RA by years suggests that these autoantibodies are pathogenic rather than simply a marker of chronic inflammation. Together, these risk factors suggest a hypothetical model that an environment trigger (e.g., smoking) invokes a generalized inflammatory response in genetically susceptible hosts, which leads to the formation of autoantibodies and eventually RA (Figure 71.1) (Klareskog et al., 2006).

Genetic Basis of RA

The genetic contribution to RA susceptibility in humans has been demonstrated through twin studies (MacGregor et al., 2000), family studies (Bali et al., 1999), and genome-wide linkage scans (Amos et al., 2006; Cornelis et al., 1998; Etzel et al., 2006; Jawaheer et al., 2001, 2003; MacKay et al., 2002; Shiozawa et al., 1998). Heritability refers to the amount of phenotypic variation due to additive genetic factors, rather than common environmental factors, stochastic variation, gene–environment interactions, and gene–gene interactions. One such study demonstrated that approximately 60% of disease variability is inherited (MacGregor et al., 2000). A small study of twins indicated that the heritability in CCP+ disease is similar to CCP− disease (Van der Woude et al., 2009). Another measure of genetic contribution to disease activity is to compare prevalence of disease in family members compared to

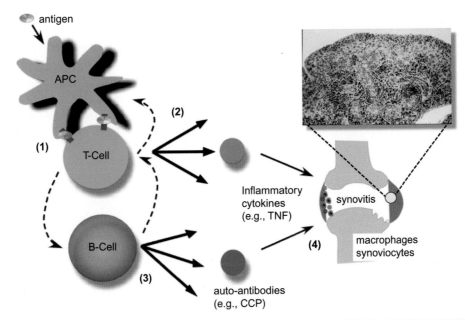

Figure 71.1   Model of RA pathogenesis. There are four conceptual steps. (1) Aberrant systemic selection and/or activation of T-cells is driven via antigens presented by MHC class II molecules on APCs; (2) CD4+ T-cells respond to antigenic stimuli, leading to activation and proliferation; (3) T-cells communicate with B-cells, which secrete autoantibodies. A growing list of autoantibodies is associated with RA, including serologic reactivity to citrullinated proteins (anti-CCP antibodies). The advent of a B-lymphocyte directed biotherapeutic (anti-CD20, rituximab) provides further evidence of a role for B-cells in RA pathogenesis. (4) Effector molecules of synovial inflammation include components of both the innate and adaptive immune system, including pro-inflammatory cytokines (e.g., TNF, IL1, and IL6). Therapeutic inhibition of several pro-inflammatory cytokines (e.g.,TNF) is clinically effective in patients with active RA. APC = antigen presenting cell; TNF = tumor necrosis factor; IL = interleukin.

the general population. Whereas the overall population risk of RA is ~1%, the monozygotic (MZ) twin of a patient with RA has a risk of ~15% (Aho et al., 1986; Jarvinen and Aho, 1994; Silman et al., 1993). Moreover, the relative risk to the sibling of a proband with RA (known as λ_s) is ~5 (Deighton et al., 1989; Hasstedt et al., 1994; Wolfe et al., 1988), although the number varies depending on the population studied (Jawaheer et al., 2001).

The MHC-region and HLA-DRB1 Susceptibility Alleles

The MHC region spans ~3.6 megabases (Mb) on the short arm of human chromosome 6 and contains hundreds of genes, including many involved in immune function (Horton et al., 2004; Stewart et al., 2004). It has been estimated that the MHC region of the human genome accounts for approximately one-third of the overall genetic component of RA risk (Deighton et al., 1989; Rigby et al., 1991). Genome-wide linkage scans using both microsatellite (Cornelis et al., 1998; Jawaheer et al., 2001, 2003; MacKay et al., 2002; Shiozawa et al., 1998) and single nucleotide polymorphism (SNP) markers (Amos et al., 2006) have consistently identified this region as important in RA pathogenesis. These genome-wide scans have demonstrated that the MHC region has the largest genetic contribution in RA, and the relative contribution of

MHC genes (λ_{MHC}) was found to be ~1.75 (Cornelis et al., 1998; Jawaheer et al., 2001).

Much, but probably not all, of the risk attributable to the MHC-region is associated with alleles within the *HLA-DRB1* gene. An association between RA and the class II HLA (human leukocyte antigen) proteins was first noted in the 1970s, when the mixed lymphocyte culture (MLC) type Dw4 (related to the serological subtype DR4) was observed to be more common among patients with RA compared to controls (Stastny, 1978; Stastny and Fink, 1977). Subsequently, investigation of the molecular diversity of class II proteins (subunits of HLA-DR, -DQ and -DP) localized the serological Dw4 subtype to the *HLA-DRB1* gene (Gregersen et al., 1986a, b). When the susceptible DR subtypes were considered as a group, Gregersen and colleagues noted a shared amino acid (a.a.) sequence at positions 70–74 of the HLA-DRB1 protein (Gregersen et al., 1987). These residues are important in peptide binding, and thus it was hypothesized that RA-associated alleles bind specific peptides, which in turn facilitates the development of autoreactive T-cells. These alleles are now known collectively as "shared epitope" (SE) alleles due to the related sequence composition in the third hypervariable region (Table 71.2): the susceptibility alleles result in missense a.a. changes, where the shared susceptibility a.a. motif is ^{70}Q/R-K/R-R-A-A^{74}.

TABLE 71.2	HLA-DRB1 "shared epitope" (SE) alleles					
DRB1 alleles	**Low-resolution**	**a.a (location)**				
		70	71	72	73	74
		Q	R	R	A	A
DRB1*0101	DR1	Q	R	R	A	A
DRB1*0102	DR1	–	–	–	–	–
DRB1*0103	DR1	D	E	–	–	–
DRB1*03	DR3	–	K	–	G	R
DRB1*0401	DR4	–	K	–	–	–
DRB1*0402	DR4	D	E	–	–	–
DRB1*0403	DR4	–	–	–	–	E
DRB1*0404	DR4	–	–	–	–	–
DRB1*0405	DR4	–	–	–	–	–
DRB1*0407	DR4	–	–	–	–	E
DRB1*0408	DR4	–	–	–	–	–
DRB1*0411	DR4	–	–	–	–	E
DRB1*07	DR7	D	–	–	G	Q
DRB1*08	DR8	D	–	–	–	L
DRB1*0901	DR9	R	–	–	–	E
DRB1*1001	DR10	R	–	–	–	–
DRB1*1101	DR11	D	–	–	–	–
DRB1*1102	DR11	D	E	–	–	–
DRB1*1103	DR11	D	E	–	–	–
DRB1*1104	DR11	D	–	–	–	–
DRB1*12	DR12	D	–	–	–	–
DRB1*1301	DR13	D	E	–	–	–
DRB1*1302	DR13	D	E	–	–	–
DRB1*1303	DR13	D	K	–	–	–
DRB1*1323	DR13	D	E	–	–	–
DRB1*1401	DR14	R	–	–	–	–
DRB1*1402	DR14	–	–	–	–	–
DRB1*1404	DR14	R	–	–	–	E
DRB1*15	DR2	–	A	–	–	–
DRB1*16	DR16	D	–	–	–	–

HLA-DRB1 SE alleles are classified by amino acid (a.a.) sequence at positions 70–74. The consensus a.a. sequence (QRRAA) is shown at the top; the identical a.a. is indicated by a dash (–) and variable a.a. indicated by appropriate nomenclature. Alleles in bold are associated with RA susceptibility.

The HLA-DRB1 gene encodes for a protein that is part of MHC class II molecules. These molecules, heterodimers of alpha and beta proteins, are found on professional antigen-presenting cells and display peptides derived from extracellular proteins to CD4+ T-cells (T-"helper" cells). The genes that encode the class II molecules are found in three subregions (DR, DQ, and DP) spanning ~1 Mb within the MHC region. Within the DR subregion, one alpha- and three beta-chain genes have been described; the alpha-chain gene and two beta-chain genes, *DRB1* and *DRB2*, are clearly expressed. The DQ subregion contains two sets of alpha- and beta-chain genes, DX- and DQ-alpha and beta. With the exception of DR-alpha, all of the expressed genes display considerable allelic diversity. Classification of HLA-DRB1 molecules includes serological nomenclature (e.g., DR1, DR4 and DR10), MCL nomenclature (e.g., Dw4) and DNA-sequence based nomenclature (see below).

Since the initial observation, a large number of population studies have confirmed the association between RA and allelic variants at *HLA-DRB1* (Ollier and Thomson, 1992). At the level of DNA, the most common (>5% population frequency) *HLA-DRB1* shared epitope susceptibility alleles include *0101, *0401, and *0404 in individuals of European ancestry, and *0405 and *0901 in individuals of Asian ancestry; less common shared epitope alleles include *0102, *0104, *0408, *0413, *0416, *1001, and *1402. Of note, the *0901 allele observed among Asian populations does not strictly conform to the SE a.a. sequence motif (^{70}R-R-R-A-E^{74}, see Table 71.2), and the classic SE alleles may not contribute to risk in African-American and Hispanic-American RA populations (McDaniel et al., 1995; Teller et al., 1996). Thus, additional exploration of the molecular basis of *HLA-DRB1* susceptibility alleles is needed in the future.

While *HLA-DRB1* susceptibility alleles are often considered as a group, the strength of the genetic association to RA susceptibility differs across the *DRB1* alleles. There are at least two classes of *HLA-DRB1* risk alleles (high and moderate). In general, DRB1*0401 and *0405 alleles exhibit a high level of risk, with a relative risk (RR) of approximately 3. The DRB1*0101, *0404, *0901, and *1001 alleles exhibit a more moderate relative risk in the range of 1.5.

It is becoming increasingly clear that *HLA-DRB1* shared epitope alleles only influence the development of seropositive RA, and more specifically anti-CCP+ RA (see below for discussion on autoantibodies) (Huizinga et al., 2005; Irigoyen et al., 2005). Collectively, the shared epitope alleles have an odds ratio of over 5 if CCP+ RA patients are compared to matched healthy controls. Because these alleles are quite common in the general population (collectively, allele frequency ~40% in individuals of European ancestry), the attributable risk for SE alleles is quite high.

Several investigators have proposed a refined classification of shared epitope alleles, as this hypothesis alone cannot explain all of the genetic risk attributable to the *HLA-DRB1* locus (De Vries et al., 2002; Gao et al., 1991; Michou et al., 2006; Zanelli et al., 1998). No consensus has emerged, however. Some of these studies suggest that a protective allele may be in linkage disequilibrium with the *HLA-DRB1* alleles. It has been hypothesized that the presence of an asparagine a.a. at position 70 of the HLA-DRB1 protein (D70) may be associated with protection from the development of RA (once the effect of the shared epitope alleles has been taken into consideration) (Mattey et al., 2001a; Ruiz-Morales et al., 2004).

Numerous studies have shown that *HLA-DRB1* susceptibility alleles influence disease severity in longstanding disease,

particularly the development of bony erosions (for example: Chen et al., 2002; Gorman et al., 2004; Moxley and Cohen, 2002). More recently, however, it has been suggested that this association is primarily due to the presence of CCP autoantibodies (Huizinga et al., 2005). It remains to be determined whether *HLA-DRB1* alleles contribute additional risk of developing erosive disease independent of CCP autoantibodies (Van der Helm-Van Mil et al., 2006). This more recent observation may be an important explanation for why some studies have demonstrated that SE alleles predict erosive changes, but only in RF⁻ patients (El-Gabalawy et al., 1999; Mattey et al., 2001b). One hypothesis to explain this observation is that RF− patients in these older studies are actually CCP+ [and it is known that SE alleles have a stronger association with CCP+ than RF+ RA (Irigoyen et al., 2005)]. In the future, it will be important to assess the relationship between *HLA-DRB1* alleles and clinical outcome, controlling for the effect of CCP as well as other important clinical variables.

Despite decades of research, it is not fully known how the *HLA-DRB1* alleles cause risk of RA, and direct functional proof has been elusive (Goronzy and Weyand, 1993; Nepom, 2001). Hypotheses include that the SE alleles influence (1) thresholds for T-cell activation (based on avidity between the T-cell receptor, MHC, and peptide, especially in the context of post-translational modification events important in RA pathogenesis) (Hill et al., 2003); (2) thymic selection of high-affinity self-reactive T-cells (based on the T-cell synovial repertoire) (Yang et al., 1999); and (3) molecular mimicry of microbial antigens (Albani et al., 1992). It is worthwhile noting that the third hypervariable region of the protein (location of SE allelic variants) contains a peptide-binding groove that serves to present peptides to CD4+ T-cells (Seyfried et al., 1988) and that citrullination of certain peptides triggers a strong immune response to citrullinated peptides in *HLA-DRB1* *0401 transgenic mice (Hill et al., 2003).

Other MHC-region Genes

Several studies suggest that, once the effect of *HLA-DRB1* has been taken into consideration, additional genes within the MHC likely contribute to disease susceptibility (Jawaheer et al., 2002; Kochi et al., 2004; Mulcahy et al., 1996; Singal et al., 1999; Zanelli et al., 2001). For example, an extended haplotype that includes *HLA-DRB1* DR3 alleles may be associated with RA (Jawaheer et al., 2002). The associated haplotype spans ~500 kb and contains class III MHC genes, including the TNF-alpha region implicated in other studies (Mulcahy et al., 1996; Ota et al., 2001; Waldron-Lynch et al., 2001). One study suggests that this association is restricted to CCP− patients (Irigoyen et al., 2005). Application of large-scale SNP genotyping across the MHC provided continued support for additional alleles, including alleles that are found on the conserved A1-B8-DR3 (8.1) haplotype and alleles near the *HLA-DPB1* gene (Ding et al., 2008; Lee et al., 2008; Vignal et al., 2008).

Non-MHC Genes

The most dramatic improvement in our understanding of the genetic basis of RA over recent years comes from

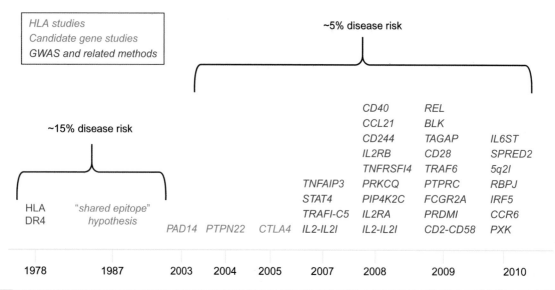

Figure 71.2 RA risk gene loci discovered over time.

genome-wide association studies (GWAS) of common DNA variants. These studies have identified >30 regions outside of the MHC that are associated with risk of RA at a stringent level of statistical significance (Barton et al., 2008; Chang et al., 2008; Gregersen et al., 2009; Kochi et al., 2010; Kurreeman et al., 2007; Plenge et al., 2007a, b; Raychaudhuri et al., 2008; Raychaudhuri et al., 2009b; Remmers et al., 2007; Stahl et al., 2010; Suzuki et al., 2008; Thomson et al., 2007; Wellcome Trust Case Control Consortium, 2007; Zhernakova et al., 2007). In contrast to the MHC, the common alleles identified by GWAS and related methodologies have much smaller effects on risk, with odds ratios of approximately 1.20 or less per copy of the risk allele. While the MHC alleles account for ~15% of the overall disease burden, the non-MHC alleles discovered to date explain approximately 5% of disease risk (Raychaudhuri, 2010; Stahl et al., 2010).

The first convincing non-HLA gene associated with RA risk was *PTPN22* (Begovich et al., 2004), a finding that has been replicated across multiple independent studies among cases of European ancestry (for example: Dieude et al., 2005; Harrison et al., 2006; Hinks et al., 2005; Lee et al., 2005; Orozco et al., 2005; Plenge et al., 2005; Steer et al., 2005; Viken et al., 2005; Zhernakova et al., 2005). The susceptibility allele is a missense variant that changes an arginine to tryptophan a.a. (R620W), resulting in alteration of T-cell activation (Begovich et al., 2004; Vang et al., 2005). The magnitude of the genetic effect, as measured by the odds ratio (OR), is substantially less than for *HLA-DRB1**0401 but similar to other SE alleles (*PTPN22* has OR ~1.75). Interestingly, this allele is absent for East Asians, and thus not associated with susceptibility in Japanese populations (Ikari et al., 2006).

Extensive genotype–phenotype correlations have not been conducted for *PTPN22* (as they have for *HLA-DRB1*).

Like *HLA-DRB1* alleles, *PTPN22* only associates with seropositive RA. There is some evidence that *PTPN22* influences age of onset (Plenge et al., 2005), and may have a more significant effect in males compared to females (Pierer et al., 2006; Plenge et al., 2005), but no evidence that it influences disease activity or radiographic erosions (Harrison et al., 2006; Wesoly et al., 2005).

A list of the other gene loci associated with risk of RA is shown in Figure 71.2. Several themes are beginning to emerge:

1. **Shared autoimmune risk loci.** The success in mapping RA risk loci has been paralleled by the success in other autoimmune and inflammatory diseases, including type 1 diabetes (T1D), celiac disease, systemic lupus erythematosus (SLE), inflammatory bowel disease, and multiple sclerosis. Many of these loci predispose to more than one autoimmune disease (Behrens et al., 2008; Fung et al., 2009; International Multiple Sclerosis Genetics Consortium, 2008; Plenge, 2008; Raychaudhuri et al., 2008; Remmers et al., 2007; Smyth et al., 2008; Zhernakova et al., 2007). Up to half of the known non-MHC risk loci in RA also appear to confer risk to at least one additional autoimmune disease. In most examples, it is the same allele that is associated with risk in both diseases [e.g., a common allele at the *STAT4* gene locus, and risk of both RA and SLE (Remmers et al., 2007)]. There are gene loci, however, that appear to have multiple autoimmune alleles [e.g., the *6q23/ TNFAIP3* locus and risk of RA, SLE, and T1D (Fung et al., 2009; Graham et al., 2008; Musone et al., 2008; Plenge et al., 2007a; Thomson et al., 2007)]. Finally, there are loci, such as *PTPN22*, in which an allele is *predisposed* to one autoimmune disease [e.g., RA (Begovich et al., 2004)],

but the same allele is *protective* in another autoimmune disease [e.g., Crohn's disease (Barrett et al., 2008)].

2. **Subsets of disease: CCP-positive vs CCP-negative.** All large-scale genetic studies to date have been biased towards autoantibody-positive RA patients, and CCP+ patients in particular. Despite this bias, it is clear that many validated RA risk alleles are more strongly, if not exclusively, associated with CCP+ disease (Van der Helm-Van Mil and Huizinga, 2008). The two best examples are the *HLA-DRB1* risk alleles (Ding et al., 2008; Huizinga et al., 2005; Irigoyen et al., 2005) and *PTPN22* (Lee et al., 2005). As yet, there are no validated risk alleles that are associated exclusively with autoantibody-negative disease. The observation that RA risk alleles are able to subset disease into clinically meaningful categories – in this case, the presence or absence of CCP autoantibodies – raises the possibility that RA risk alleles may identify additional clinical subsets of disease (e.g., patients who respond to treatment or who develop specific co-morbid conditions).

3. **Ethnic differences in RA risk alleles.** Most of the large-scale genetic studies have been conducted among individuals of self-described "white" European ancestry (Barton et al., 2008; Gregersen et al., 2009; Plenge et al., 2007a, b; Raychaudhuri et al., 2008, 2009b; Remmers et al., 2007; Stahl et al., 2010; Thomson et al., 2007), with only a single GWAS among individuals of Asian ancestry (Kochi et al., 2010). As such, the literature is very biased towards alleles that are associated with risk of RA among patients of European ancestry. Nonetheless, several large candidate-gene based studies have been conducted among individuals of East Asian ancestry, leading to the identification of the *PADI4* and *CCR6* risk loci (among others) (Kochi et al., 2010; Suzuki et al., 2003, 2008). A meta-analysis of the four studies (3713 cases and 2485 controls) provides unequivocal support for an association of a common variant of *PADI4* (combined $P = 4 \times 10^{-13}$; OR = 1.31 per copy of the risk allele) (Takata et al., 2008). In contrast, six studies in individuals of European ancestry – a total of 5530 cases and 13,176 controls – fail to show any strong evidence of a *PADI4* association with risk of RA [OR = 1.03 (0.98–1.09)] (Barton et al., 2004; Martinez et al., 2005; Plenge et al., 2005; Poor et al., 2007; Raychaudhuri et al., 2008; Wellcome Trust Case Control Consortium, 2007). The frequency of the *PTPN22* risk allele varies among patients of European ancestry – it is present in ~10% of controls from northern Europe and ~3% of controls from southern Europe. In patients of East Asian ancestry, however, the allele is absent (Ikari et al., 2006). As such, there is no contribution to risk of disease in patients of Asian ancestry. Very few large-scale genetic studies have been conducted among individuals of African, Hispanic, or other ancestries, and this remains important work for the future.

4. **Pathways that lead to RA.** When considering pathways implicated by genetic studies, there are two important caveats to the way in which risk alleles are often described. First, the risk locus, typically ascertained by a single or a few clustered SNPs, is often assigned a gene name (e.g., *TNFAIP3*), even though there may be no direct evidence that the gene itself has been disrupted by the risk allele. Generally, the most compelling biological candidate gene near the SNP or SNPs in question is chosen – in the case of *TNFAIP3*, a knock-out mouse model demonstrates severe inflammation, consistent with the phenotype of RA (Lee et al., 2000) – although in some cases more objective criteria are used (Raychaudhuri et al., 2009a, b). And second, the reported risk allele most likely represents a highly correlated proxy for the as-yet unidentified causal allele (i.e., the causal allele has not yet been identified). In other words, for most RA risk loci, the exact causal mutation and the exact causal gene have not yet been determined.

Despite these caveats, recent genetic discoveries have identified pathways that are important in RA pathogenesis (Figure 71.1). Moreover, objective computational methods clearly indicate that there are biological relationships among genes across RA risk loci (Raychaudhuri et al., 2009a, b). One important pathway is the NF-κB (nuclear factor kappa-light-chain-enhancer of activated B-cells) signaling pathway. NF-κB, a protein complex that acts as a transcription factor, is found in almost all cell types and is involved in a variety of cellular responses to stimuli. Both TNF and CD40 signal through NF-κB to exert their effects on the immune system. Several of the RA risk loci contain genes that are involved in NF-κB signaling, including *CD40*, *TRAF1*, *TNFAIP3*, *PRKCQ*, and *TNFRSF14*.

Another important pathway involves antigen presentation via MHC class II molecules to the T-cell receptor-CD3 complex. The *CD247* risk gene, which encodes for the zeta chain of the T-cell receptor-CD3 complex, directly implicates the T-cell receptor (TCR) and downstream signaling as a critical mechanism of RA pathogenesis. Other RA risk genes known to modulate T-cell activation/differentiation include (but are not limited to): *CTLA4* and *CD28*, which are cell-surface proteins on T-cells that serve as co-stimulatory molecules following T-cell activation through the TCR (Noel et al., 1996); *PTPN22*, which encodes an intracellular phosphatase, and the missense variant is a potent negative regulator of T-lymphocyte activation (Vang et al., 2005); *TAGAP*, which is up-regulated upon T-cell activation (Mao et al., 2004); *RBPJ* (encoding recombination site binding protein J and also known as CSL), which encodes a transcription factor within the Notch signaling pathway, and *RBPJ*-deficient mice have no T-cell development (Rothenberg et al., 2008); *CCR6*, which encodes a cell surface protein that distinguishes Th17 cells from other CD4+ helper T-cells (Hirota et al., 2007); and *STAT4*, which encodes a transcription factor important in differentiation of T-helper lymphocyte cells (Watford et al., 2004).

Environmental Factors

Sex Bias and Hormonal Factors

Perhaps the strongest non-genetic risk factor for the development of RA is female sex; females are more than twice as likely to develop RA compared to males, and this disparity is even greater at a younger age (Linos et al., 1980; Symmons et al., 1994). In females, the risk of developing RA increases with age, and peaks around the time of menopause (Doran et al., 2002; Goemaere et al., 1990; Karlson et al., 2004). These observations have led to considerable effort in examining the role of hormonal and pregnancy factors in disease occurrence.

Smoking

Smoking is the environmental risk factor with the strongest demonstrated risk of developing RA. Over the past 20 years, many studies have convincingly shown that smoking increases risk of developing RA in both males and females by a factor of ~2.0 (Gorman, 2006; Stolt et al., 2003). Most of these studies demonstrate that the increased risk is in developing autoantibody-positive RA and that the risk is greatest for heavy, current smokers; the risk remains, however, for >10 years following smoking cessation. Because the link between smoking and autoantibody-positive RA is reminiscent of the genetic association between *HLA-DRB1* SE alleles and autoantibody-positive RA, Padyukov and colleagues investigated the risk of developing disease in carriers of these two established risk factors: a 16-fold risk of developing RF + RA in smokers who carry two copies of the SE alleles was observed (Padyukov et al., 2004).

Other putative environmental risk factors that have not been as widely replicated as smoking include blood transfusions, obesity, occupational silica and mineral oil exposure, and socio-economic class (Silman and Pearson, 2002; Symmons, 2003). It is interesting that several of these putative environmental factors are inhaled through the lungs. Silica dust and mineral oil exposure, similar to exposure to cigarette smoke, was a risk factor only for seropositive RA. Mineral oil can also act as an adjuvant capable of inducing experimental arthritis in rodent models.

Autoantibodies

RF and CCP Autoantibodies

Autoantibodies have proven useful in the diagnosis and prognosis of RA patients. Autoantibodies are detected in about two-thirds of patients with RA and predict severe disease. The two major types of autoantibodies used clinically to create RA subsets, as introduced briefly earlier, are those against rheumatoid factor (RF), which is an immunoglobulin specific to the Fc region of IgG, and anti-cyclic citrullinated peptide (CCP) antibodies, which are antibodies directed against peptides that have arginine post-translationally modified to citrulline (Schellekens et al., 2000). These autoantibodies are strongly correlated with each other, but may represent distinct clinical subsets of RA.

RF autoantibodies are part of the diagnostic criteria for RA (Arnett et al., 1988). The RF assay, however, remains suboptimal as a diagnostic test, as it lacks sensitivity (50–90%) and specificity (50–90%) (Shmerling and Delbanco, 1991). Furthermore, it is present in many other disease states (including those that mimic RA) and patients that smoke, and its incidence increases with age. In contrast, anti-CCP antibodies have moderate sensitivity (60–80%) but increased specificity for RA (>90%), and predict functional status and radiographic erosions in patients with early-onset RA (Kroot et al., 2000; Van Jaarsveld et al., 1999; Van Zeben et al., 1992).

There has been much debate about whether these autoantibodies are causal or whether they represent a non-specific response to systemic inflammation. Several lines of evidence suggest that these autoantibodies – and CCP in particular – are pathogenic: (1) anti-CCP antibodies appear before disease onset (Rantapaa-Dahlqvist et al., 2003; Schellekens et al., 2000); (2) the presence of CCP is very specific to RA; (3) anti-CCP antibodies enhance tissue injury in a murine model of arthritis (collagen-induced arthritis) (Kuhn et al., 2006); and (4) genetic variation in an enzyme, PADI4, involved in the citrullination pathway, appears associated with RA susceptibility (Suzuki et al., 2003) (see *Non-MHC Genes*, above).

Insights into Predisposition from Mouse Models of Inflammatory Arthritis

There are several mouse models of inflammatory arthritis that provide insight into the predisposition of RA. Collagen-induced arthritis (CIA), which is among the most widely used, occurs following immunization of susceptible mice by type II collagen (Holmdahl et al., 2002). CIA is genetically controlled by class II MHC molecules, involves the activation of antigen-specific CD4+ T-cells, and is dependent on both T-cells and B-cells. Another mouse model is the K/BxN serum transfer model (Monach et al., 2007). Mice expressing the KRN T-cell receptor transgene and a specific MHC class II molecule (K/BxN mice) develop severe inflammatory arthritis. Serum from these mice causes similar arthritis in a wide range of mouse strains, owing to pathogenic autoantibodies to glucose-6-phosphate isomerase. Finally, a spontaneous point mutation of the *ZAP-70* gene, a key signal transduction molecule in T-cells, causes chronic autoimmune arthritis in mice (Sakaguchi et al., 2003). Altered signal transduction from TCR through the aberrant ZAP-70 changes the thresholds of T-cells to thymic selection, leading to the positive selection of otherwise negatively selected autoimmune T-cells.

Hypothetical Model of RA Predisposition

Based upon the above information, it is tempting to speculate about a hypothetical model of RA pathogenesis. The following is one interesting model that accounts for (1) *HLA-DRB1* allelic associations, (2) genes implicated by GWAS outside of the MHC region (especially those involved in TCR signaling), (3) smoking as an environmental risk factor, (4) pathogenic autoantibodies against citrullinated peptides, and (5) mouse models of inflammatory arthritis. In this model, as suggested by Klareskog and colleagues (2006), disease predisposition starts early in life, during the neonatal and pre-adolescent periods when the thymus removes autoreactive T-cells from the body. In genetically predisposed hosts, autoreactive T-cells escape the thymus, as TCR-mediated signaling (which is required to select out autoantigens) is attenuated by polymorphisms in genes such as *PTPN22*, *CD247*, *CTLA4*, and *CD28*. However, the pre-clinical phase of disease does not begin until later in life when autoreactive T-cells encounter citrullinated proteins that act as autoantigens (Klareskog et al., 2009). The association of smoking as a risk factor for RA suggests that citrullination of peptides occurs in the lungs by inhaled toxins (Klareskog et al., 2006). MHC class II molecules present citrullinated peptides, which act as disease-specific autoantigens, to autoreactive T-cells. This initiates a cascade of events that includes the formation of CCP autoantibodies from B-cells and the release of inflammatory cytokines in the joint.

While aspects of this model are almost certainly correct, it will not explain RA pathogenesis in all patients. Continued integration of clinical and basic science research is necessary to revise, or refine, models of RA pathogenesis. The more that is learned about the functional consequences of the RA risk alleles, the more we will understand about RA pathogenesis.

SCREENING

No reliable methods to screen a population prior to disease onset exist for RA. It may be that the presence of CCP autoantibodies together with genetic susceptibility variants will increase risk substantially, to the point where screening the healthy population becomes cost-effective (Berglin et al., 2004; Johansson et al., 2005). However, an aggregate genetic risk score based on the *HLA-DRB1* and non-MHC risk alleles is only able to stratify risk 10-fold per individual (Karlson et al., 2010). Under this scenario, it remains to be determined whether effective intervention to prevent RA will emerge; currently, RA therapies are directed at symptoms rather than a cure. Overall, no clear paradigm has emerged as to how screening the population would influence clinical decision-making, and additional attention should be paid to this area in the future.

DIAGNOSIS, PROGNOSIS, AND MONITORING

The diagnosis of RA is based on established clinical criteria (Arnett et al., 1988). The clinical course for any individual patient is highly variable. Extent of synovial inflammation and presence of autoantibodies at the time of initial diagnosis portend a poor prognosis, but cannot alone predict prognosis in any individual patient. Similarly, the presence of *HLA-DRB1* SE alleles predicts more severe disease, but does not routinely enter into decision-making in clinical patient care. Because treatment strategies have improved dramatically over the last

decade with the advent of TNF-alpha inhibitors, prognosis is more influenced by response to treatment rather than aggressiveness of the underlying disease in an untreated patient.

A major goal of therapy is early institution of disease-modifying anti-rheumatic drugs (DMARDs). The most popular disease modifying therapy is low-dose weekly methotrexate. This drug is used as a monotherapy or in combination with other synthetic molecules (anti-malarials, sulfasalazine, leflunomide) or with biologic response modifiers. Biologics that neutralize the pro-inflammatory cytokine TNF-alpha have been a major advance in the therapy of RA. Monoclonal antibodies (infliximab and adalimumab) or the p75 TNF-alpha soluble receptor fusion protein (etanercept) are effective as a monotherapy or in combination with methotrexate in reducing the signs and symptoms of the disease, as well as improving quality of life and slowing the rate of radiographic progression. These therapies are very effective when combined with methotrexate. Newer biologics recently approved for use in rheumatoid arthritis include abatacept (blocks the CTLA4 co-stimulation pathway) and rituximab (a B-cell depleting strategy). Early intervention with methotrexate and combination therapy (including anti-TNF-alpha therapy) have had a major impact on this disease.

PHARMACOGENOMICS

In contrast to literature on RA susceptibility, the literature on pharmacogenomics is less mature. No single gene is clearly associated with response to therapy, although variants in the *MTHFR* gene may influence methotrexate response and side-effect profile (Kumagai et al., 2003; Urano et al., 2002; Van Ede et al., 2001; Weisman et al., 2006). Variants within the MHC region (Criswell et al., 2004; Kang et al., 2005; Marotte et al., 2006; Martinez et al., 2004; Padyukov et al., 2003; Seitz et al., 2007) and at the *PTPRC*/CD45 locus (Cui et al., 2010) may influence response to anti-TNF-α agents, although additional studies are required.

CONCLUSIONS

While the concepts of genomic medicine have yet to have a large impact on the clinical management of RA patients, the field is still in its infancy. The advent of large genome-wide association studies has increased the number of confirmed RA risk alleles, but in total, these risk alleles (including MHC risk alleles) explain only 20% of disease burden. Once a gene is identified, the next immediate challenge will be to identify the causal allele(s), incorporate genetic knowledge into evolving models of RA pathogenesis, and correlate genotypes to sub-clinical phenotypes such as radiographic erosions, disease severity, and treatment response. Ultimately, the true measure of success of genomic medicine will be whether genetic information improves patient care, through novel diagnostic or therapeutic interventions. The goal of the field should be nothing short of a cure for rheumatoid arthritis.

REFERENCES

Aho, K., Koskenvuo, M., Tuominen, J., Kaprio, J., 1986. Occurrence of rheumatoid arthritis in a nationwide series of twins. J Rheumatol 13, 899–902.

Albani, S., Carson, D.A., Roudier, J., 1992. Genetic and environmental factors in the immune pathogenesis of rheumatoid arthritis. Rheum Dis Clin North Am 18, 729–740.

Amos, C.I., Chen, W.V., Lee, A., et al., 2006. High-density SNP analysis of 642 Caucasian families with rheumatoid arthritis identifies two new linkage regions on 11p12 and 2q33. Genes Immun 7, 277–286.

Arnett, F.C., Edworthy, S.M., Bloch, D.A., et al., 1988. The American Rheumatism Association 1987 revised criteria for the classification of rheumatoid arthritis. Arthritis Rheum 31, 315–324.

Bali, D., Gourley, S., Kostyu, D.D., et al., 1999. Genetic analysis of multiplex rheumatoid arthritis families. Genes Immun 1, 28–36.

Barrett, J.C., Hansoul, S., Nicolae, D.L., et al., 2008. Genome-wide association defines more than 30 distinct susceptibility loci for Crohn's disease. Nat Genet 40, 955–962.

Barton, A., Bowes, J., Eyre, S., et al., 2004. A functional haplotype of the *PADI4* gene associated with rheumatoid arthritis in a Japanese population is not associated in a United Kingdom population. Arthritis Rheum 50, 1117–1121.

Barton, A., Thomson, W., Ke, X., et al., 2008. Re-evaluation of putative rheumatoid arthritis susceptibility genes in the post-genome wide association study era and hypothesis of a key pathway underlying susceptibility. Hum Mol Genet 17, 2274–2279.

Begovich, A.B., Carlton, V.E., Honigberg, L.A., et al., 2004. A missense single-nucleotide polymorphism in a gene encoding a protein tyrosine phosphatase (PTPN22) is associated with rheumatoid arthritis. Am J Hum Genet 75, 330–337.

Behrens, E.M., Finkel, T.H., Bradfield, J.P., et al., 2008. Association of the *TRAF1-C5* locus on chromosome 9 with juvenile idiopathic arthritis. Arthritis Rheum 58, 2206–2207.

Berglin, E., Padyukov, L., Sundin, U., et al., 2004. A combination of autoantibodies to cyclic citrullinated peptide (CCP) and *HLA-DRB1* locus antigens is strongly associated with future onset of rheumatoid arthritis. Arthritis Res Ther 6, R303–R308.

Chang, M., Rowland, C.M., Garcia, V.E., et al., 2008. A large-scale rheumatoid arthritis genetic study identifies association at chromosome 9q33.2. PLoS Genet 4, e1000107.

Chen, J.J., Mu, H., Jiang, Y., King, M.C., Thomson, G., Criswell, L.A., 2002. Clinical usefulness of genetic information for predicting radiographic damage in rheumatoid arthritis. J Rheumatol 29, 2068–2073.

Cornelis, F., Faure, S., Martinez, M., et al., 1998. New susceptibility locus for rheumatoid arthritis suggested by a genome-wide linkage study. Proc Natl Acad Sci USA 95, 10,746–10,750.

Criswell, L.A., Lum, R.F., Turner, K.N., et al., 2004. The influence of genetic variation in the *HLA-DRB1* and *LTA-TNF* regions on the response to treatment of early rheumatoid arthritis with methotrexate or etanercept. Arthritis Rheum 50, 2750–2756.

Cui, J., Saevarsdottir, S., Thomson, B., et al., 2010. Rheumatoid arthritis risk allele *PTPRC* is also associated with response to anti-tumor necrosis factor alpha therapy. Arthritis Rheum 62, 1849–1861.

Deighton, C.M., Walker, D.J., Griffiths, I.D., Roberts, D.F., 1989. The contribution of HLA to rheumatoid arthritis. Clin Genet 36, 178–182.

De Vries, N., Tijssen, H., Van Riel, P.L., Van de Putte, L.B., 2002. Reshaping the shared epitope hypothesis: HLA-associated risk for rheumatoid arthritis is encoded by amino acid substitutions at positions 67–74 of the HLA-DRB1 molecule. Arthritis Rheum 46, 921–928.

Dieude, P., Garnier, S., Michou, L., et al., 2005. Rheumatoid arthritis seropositive for the rheumatoid factor is linked to the protein tyrosine phosphatase nonreceptor 22-620W allele. Arthritis Res Ther 7, R1200–R1207.

Ding, B., Padyukov, L., Lundstrom, E., et al., 2008. Different patterns of associations with anti-citrullinated protein antibody-positive and anti-citrullinated protein antibody-negative rheumatoid arthritis in the extended major histocompatibility complex region. Arthritis Rheum 60, 30–38.

Doran, M.F., Pond, G.R., Crowson, C.S., O'Fallon, W.M., Gabriel, S.E., 2002. Trends in incidence and mortality in rheumatoid arthritis in Rochester, Minnesota, over a forty-year period. Arthritis Rheum 46, 625–631.

El-Gabalawy, H.S., Goldbach-Mansky, R., Smith 2nd, D., et al., 1999. Association of HLA alleles and clinical features in patients with synovitis of recent onset. Arthritis Rheum 42, 1696–1705.

Etzel, C.J., Chen, W.V., Shepard, N., et al., 2006. Genome-wide meta-analysis for rheumatoid arthritis. Hum Genet 119, 634–641.

Firestein, G.S., 2003. Evolving concepts of rheumatoid arthritis. Nature 423, 356–361.

Fung, E.Y., Smyth, D.J., Howson, J.M., et al., 2009. Analysis of 17 autoimmune disease-associated variants in type 1 diabetes identifies 6q23/TNFAIP3 as a susceptibility locus. Genes Immun 10, 188–191.

Gao, X., Gazit, E., Livneh, A., Stastny, P., 1991. Rheumatoid arthritis in Israeli Jews: Shared sequences in the third hypervariable region of *DRB1* alleles are associated with susceptibility. J Rheumatol 18, 801–803.

Goemaere, S., Ackerman, C., Goethals, K., et al., 1990. Onset of symptoms of rheumatoid arthritis in relation to age, sex and menopausal transition. J Rheumatol 17, 1620–1622.

Gorman, J.D., 2006. Smoking and rheumatoid arthritis: Another reason to just say no. Arthritis Rheum 54, 10–13.

Gorman, J.D., Lum, R.F., Chen, J.J., Suarez-Almazor, M.E., Thomson, G., Criswell, L.A., 2004. Impact of shared epitope genotype and ethnicity on erosive disease: A meta-analysis of 3240 rheumatoid arthritis patients. Arthritis Rheum 50, 400–412.

Goronzy, J.J., Weyand, C.M., 1993. Interplay of T lymphocytes and HLA-DR molecules in rheumatoid arthritis. Curr Opin Rheumatol 5, 169–177.

Graham, R.R., Cotsapas, C., Davies, L., et al., 2008. Genetic variants near *TNFAIP3* on 6q23 are associated with systemic lupus erythematosus. Nat Genet 40, 1059–1061.

Gregersen, P.K., Amos, C.I., Lee, A.T., et al., 2009. *REL*, encoding a member of the NF-kappaB family of transcription factors, is a newly defined risk locus for rheumatoid arthritis. Nat Genet 41, 820–823.

Gregersen, P.K., Moriuchi, T., Karr, R.W., et al., 1986a. Polymorphism of HLA-DR beta chains in DR4, -7, and -9 haplotypes: Implications for the mechanisms of allelic variation. Proc Natl Acad Sci USA 83, 9149–9153.

Gregersen, P.K., Shen, M., Song, Q.L., et al., 1986b. Molecular diversity of *HLA-DR4* haplotypes. Proc Natl Acad Sci USA 83, 2642–2646.

Gregersen, P.K., Silver, J., Winchester, R.J., 1987. The shared epitope hypothesis. An approach to understanding the molecular genetics of susceptibility to rheumatoid arthritis. Arthritis Rheum 30, 1205–1213.

Harrison, P., Pointon, J.J., Farrar, C., Brown, M.A., Wordsworth, B.P., 2006. Effects of PTPN22 C1858T polymorphism on susceptibility and clinical characteristics of British Caucasian rheumatoid arthritis patients. Rheumatology (Oxford) 45, 1009–1011.

Hasstedt, S.J., Clegg, D.O., Ingles, L., Ward, R.H., 1994. HLA-linked rheumatoid arthritis. Am J Hum Genet 55, 738–746.

Hill, J.A., Southwood, S., Sette, A., Jevnikar, A.M., Bell, D.A., Cairns, E., 2003. Cutting edge: The conversion of arginine to citrulline allows for a high-affinity peptide interaction with the rheumatoid arthritis-associated HLA-DRB1*0401 MHC class II molecule. J Immunol 171, 538–541.

Hinks, A., Barton, A., John, S., et al., 2005. Association between the *PTPN22* gene and rheumatoid arthritis and juvenile idiopathic arthritis in a UK population: Further support that *PTPN22* is an autoimmunity gene. Arthritis Rheum 52, 1694–1699.

Hirota, K., Yoshitomi, H., Hashimoto, M., et al., 2007. Preferential recruitment of CCR6-expressing Th17 cells to inflamed joints via CCL20 in rheumatoid arthritis and its animal model. J Exp Med 204, 2803–2812.

Holmdahl, R., Bockermann, R., Backlund, J., Yamada, H., 2002. The molecular pathogenesis of collagen-induced arthritis in mice – a model for rheumatoid arthritis. Ageing Res Rev 1, 135–147.

Horton, R., Wilming, L., Rand, V., et al., 2004. Gene map of the extended human MHC. Nat Rev Genet 5, 889–899.

Huizinga, T.W., Amos, C.I., Van der Helm-Van Mil, A.H., et al., 2005. Refining the complex rheumatoid arthritis phenotype based on specificity of the HLA-DRB1 shared epitope for antibodies to citrullinated proteins. Arthritis Rheum 52, 3433–3438.

Ikari, K., Momohara, S., Inoue, E., et al., 2006. Haplotype analysis revealed no association between the *PTPN22* gene and RA in a Japanese population. Rheumatology (Oxford) 45, 1345–1348.

International Multiple Sclerosis Genetics Consortium, 2008. The expanding genetic overlap between multiple sclerosis and type I diabetes. Genes Immun 10. doi: 10.1038/gene.2008.83.

Irigoyen, P., Lee, A.T., Wener, M.H., et al., 2005. Regulation of anti-cyclic citrullinated peptide antibodies in rheumatoid arthritis: Contrasting effects of HLA-DR3 and the shared epitope alleles. Arthritis Rheum 52, 3813–3818.

Jarvinen, P., Aho, K., 1994. Twin studies in rheumatic diseases. Semin Arthritis Rheum 24, 19–28.

Jawaheer, D., Li, W., Graham, R.R., et al., 2002. Dissecting the genetic complexity of the association between human leukocyte antigens and rheumatoid arthritis. Am J Hum Genet 71, 585–594.

Jawaheer, D., Seldin, M.F., Amos, C.I., et al., 2001. A genomewide screen in multiplex rheumatoid arthritis families suggests genetic overlap with other autoimmune diseases. Am J Hum Genet 68, 927–936.

Jawaheer, D., Seldin, M.F., Amos, C.I., et al., 2003. Screening the genome for rheumatoid arthritis susceptibility genes: A replication study and combined analysis of 512 multicase families. Arthritis Rheum 48, 906–916.

Johansson, M., Arlestig, L., Hallmans, G., Rantapaa-Dahlqvist, S., 2005. PTPN22 polymorphism and anti-cyclic citrullinated peptide antibodies in combination strongly predicts future onset of rheumatoid arthritis and has a specificity of 100% for the disease. Arthritis Res Ther 8, R19.

Kang, C.P., Lee, K.W., Yoo, D.H., Kang, C., Bae, S.C., 2005. The influence of a polymorphism at position -857 of the tumour necrosis factor alpha gene on clinical response to etanercept therapy in rheumatoid arthritis. Rheumatology (Oxford) 44, 547–552.

Karlson, E.W., Chibnik, L.B., Kraft, P., et al., 2010. Cumulative association of 22 genetic variants with seropositive rheumatoid arthritis risk. Ann Rheum Dis 69, 1077–1085.

Karlson, E.W., Mandl, L.A., Hankinson, S.E., Grodstein, F., 2004. Do breast-feeding and other reproductive factors influence future risk of rheumatoid arthritis? Results from the Nurses' Health Study. Arthritis Rheum 50, 3458–3467.

Klareskog, L., Catrina, A.I., Paget, S., 2009. Rheumatoid arthritis. Lancet 373, 659–762.

Klareskog, L., Stolt, P., Lundberg, K., et al., 2006. A new model for an etiology of rheumatoid arthritis: Smoking may trigger HLA-DR (shared epitope)-restricted immune reactions to autoantigens modified by citrullination. Arthritis Rheum 54, 38–46.

Kochi, Y., Okada, Y., Suzuki, A., et al., 2010. A regulatory variant in CCR6 is associated with rheumatoid arthritis susceptibility. Nat Genet 42, 515–519.

Kochi, Y., Yamada, R., Kobayashi, K., et al., 2004. Analysis of single-nucleotide polymorphisms in Japanese rheumatoid arthritis patients shows additional susceptibility markers besides the classic shared epitope susceptibility sequences. Arthritis Rheum 50, 63–71.

Kroot, E.J., De Jong, B.A., Van Leeuwen, M.A., et al., 2000. The prognostic value of anti-cyclic citrullinated peptide antibody in patients with recent-onset rheumatoid arthritis. Arthritis Rheum 43, 1831–1835.

Kuhn, K.A., Kulik, L., Tomooka, B., et al., 2006. Antibodies against citrullinated proteins enhance tissue injury in experimental autoimmune arthritis. J Clin Invest 116, 961–973.

Kumagai, K., Hiyama, K., Oyama, T., Maeda, H., Kohno, N., 2003. Polymorphisms in the thymidylate synthase and methylenetetrahydrofolate reductase genes and sensitivity to the low-dose methotrexate therapy in patients with rheumatoid arthritis. Int J Mol Med 11, 593–600.

Kurreeman, F.A., Padyukov, L., Marques, R.B., et al., 2007. A candidate gene approach identifies the TRAF1/C5 region as a risk factor for rheumatoid arthritis. PLoS Med 4, e278.

Lee, A.T., Li, W., Liew, A., et al., 2005. The PTPN22 R620W polymorphism associates with RF positive rheumatoid arthritis in a dose-dependent manner but not with HLA-SE status. Genes Immun 6, 129–133.

Lee, E.G., Boone, D.L., Chai, S., et al., 2000. Failure to regulate TNF-induced NF-kappaB and cell death responses in A20-deficient mice. Science 289, 2350–2354.

Lee, H.S., Lee, A.T., Criswell, L.A., et al., 2008. Several regions in the major histocompatibility complex confer risk for anti-CCP-antibody positive rheumatoid arthritis, independent of the *DRB1* locus. Mol Med 14, 293–300.

Linos, A., Worthington, J.W., O'Fallon, W.M., Kurland, L.T., 1980. The epidemiology of rheumatoid arthritis in Rochester, Minnesota: A study of incidence, prevalence, and mortality. Am J Epidemiol 111, 87–98.

MacGregor, A.J., Snieder, H., Rigby, A.S., et al., 2000. Characterizing the quantitative genetic contribution to rheumatoid arthritis using data from twins. Arthritis Rheum 43, 30–37.

MacKay, K., Eyre, S., Myerscough, A., et al., 2002. Whole-genome linkage analysis of rheumatoid arthritis susceptibility loci in 252 affected sibling pairs in the United Kingdom. Arthritis Rheum 46, 632–639.

Mao, M., Biery, M.C., Kobayashi, S.V., et al., 2004. T lymphocyte activation gene identification by coregulated expression on DNA microarrays. Genomics 83, 989–999.

Marotte, H., Pallot-Prades, B., Grange, L., et al., 2006. The shared epitope is a marker of severity associated with selection for, but not with response to, infliximab in a large rheumatoid arthritis population. Ann Rheum Dis 65, 342–347.

Martinez, A., Salido, M., Bonilla, G., et al., 2004. Association of the major histocompatibility complex with response to infliximab therapy in rheumatoid arthritis patients. Arthritis Rheum 50, 1077–1082.

Martinez, A., Valdivia, A., Pascual-Salcedo, D., et al., 2005. PADI4 polymorphisms are not associated with rheumatoid arthritis in the Spanish population. Rheumatology (Oxford) 44, 1263–1266.

Mattey, D.L., Dawes, P.T., Gonzalez-Gay, M.A., et al., 2001a. *HLA-DRB1* alleles encoding an aspartic acid at position 70 protect against development of rheumatoid arthritis. J Rheumatol 28, 232–239.

Mattey, D.L., Hassell, A.B., Dawes, P.T., et al., 2001b. Independent association of rheumatoid factor and the *HLA-DRB1* shared epitope with radiographic outcome in rheumatoid arthritis. Arthritis Rheum 44, 1529–1533.

McDaniel, D.O., Alarcon, G.S., Pratt, P.W., Reveille, J.D., 1995. Most African-American patients with rheumatoid arthritis do not have the rheumatoid antigenic determinant (epitope). Ann Intern Med 123, 181–187.

McGonagle, D., Conaghan, P.G., O'Connor, P., et al., 1999. The relationship between synovitis and bone changes in early untreated rheumatoid arthritis: A controlled magnetic resonance imaging study. Arthritis Rheum 42, 1706–1711.

McQueen, F.M., Stewart, N., Crabbe, J., et al., 1998. Magnetic resonance imaging of the wrist in early rheumatoid arthritis reveals a high prevalence of erosions at four months after symptom onset. Ann Rheum Dis 57, 350–356.

Michou, L., Croiseau, P., Petit-Teixeira, E., et al., 2006. Validation of the reshaped shared epitope *HLA-DRB1* classification in rheumatoid arthritis. Arthritis Res Ther 8, R79.

Monach, P., Hattori, K., Huang, H., et al., 2007. The K/BxN mouse model of inflammatory arthritis: Theory and practice. Meth Mol Med 136, 269–282.

Moxley, G., Cohen, H.J., 2002. Genetic studies, clinical heterogeneity, and disease outcome studies in rheumatoid arthritis. Rheum Dis Clin North Am 28, 39–58.

Mulcahy, B., Waldron-Lynch, F., McDermott, M.F., et al., 1996. Genetic variability in the tumor necrosis factor-lymphotoxin region influences susceptibility to rheumatoid arthritis. Am J Hum Genet 59, 676–683.

Musone, S.L., Taylor, K.E., Lu, T.T., et al., 2008. Multiple polymorphisms in the *TNFAIP3* region are independently associated with systemic lupus erythematosus. Nat Genet 40, 1062–1064.

Nepom, G.T., 2001. The role of the DR4 shared epitope in selection and commitment of autoreactive T cells in rheumatoid arthritis. Rheum Dis Clin North Am 27, 305–315.

Noel, P.J., Boise, L.H., Thompson, C.B., 1996. Regulation of T cell activation by CD28 and CTLA4. Adv Exp Med Biol 406, 209–217.

Ollier, W., Thomson, W., 1992. Population genetics of rheumatoid arthritis. Rheum Dis Clin North Am 18, 741–759.

Orozco, G., Sanchez, E., Gonzalez-Gay, M.A., et al., 2005. Association of a functional single-nucleotide polymorphism of *PTPN22*, encoding lymphoid protein phosphatase, with rheumatoid arthritis and systemic lupus erythematosus. Arthritis Rheum 52, 219–224.

Ota, M., Katsuyama, Y., Kimura, A., et al., 2001. A second susceptibility gene for developing rheumatoid arthritis in the human MHC is localized within a 70-kb interval telomeric of the TNF genes in the HLA class III region. Genomics 71, 263–270.

Padyukov, L., Lampa, J., Heimburger, M., et al., 2003. Genetic markers for the efficacy of tumour necrosis factor blocking therapy in rheumatoid arthritis. Ann Rheum Dis 62, 526–529.

Padyukov, L., Silva, C., Stolt, P., Alfredsson, L., Klareskog, L., 2004. A gene–environment interaction between smoking and shared epitope genes in HLA-DR provides a high risk of seropositive rheumatoid arthritis. Arthritis Rheum 50, 3085–3092.

Pierer, M., Kaltenhauser, S., Arnold, S., et al., 2006. Association of *PTPN22* 1858 single-nucleotide polymorphism with rheumatoid arthritis in a German cohort: Higher frequency of the risk allele in male compared to female patients. Arthritis Res Ther 8, R75.

Pincus, T., 1995. Assessment of long-term outcomes of rheumatoid arthritis. How choices of measures and study designs may lead to apparently different conclusions. Rheum Dis Clin North Am 21, 619–654.

Plenge, R.M., 2008. Shared genetic risk factors for type 1 diabetes and celiac disease. N Engl J Med 359, 2837–2838.

Plenge, R.M., Cotsapas, C., Davies, L., et al., 2007a. Two independent alleles at 6q23 associated with risk of rheumatoid arthritis. Nat Genet 39, 1477–1482.

Plenge, R.M., Padyukov, L., Remmers, E.F., et al., 2005. Replication of putative candidate-gene associations with rheumatoid arthritis in >4000 samples from North America and Sweden: Association of susceptibility with PTPN22, CTLA4, and PADI4. Am J Hum Genet 77, 1044–1060.

Plenge, R.M., Seielstad, M., Padyukov, L., et al., 2007b. *TRAF1-C5* as a risk locus for rheumatoid arthritis – a genomewide study. N Engl J Med 357, 1199–1209.

Poor, G., Nagy, Z.B., Schmidt, Z., Brozik, M., Meretey, K., Gergely Jr., P., 2007. Genetic background of anti-cyclic citrullinated peptide autoantibody production in Hungarian patients with rheumatoid arthritis. Ann NY Acad Sci 1110, 23–32.

Rantapaa-Dahlqvist, S., De Jong, B.A., Berglin, E., et al., 2003. Antibodies against cyclic citrullinated peptide and IgA rheumatoid factor predict the development of rheumatoid arthritis. Arthritis Rheum 48, 2741–2749.

Raychaudhuri, S., 2010. Recent advances in the genetics of rheumatoid arthritis. Curr Opin Rheumatol 22, 109–118.

Raychaudhuri, S., Plenge, R.M., Rossin, E.J., et al., 2009a. Identifying relationships among genomic disease regions: Predicting genes at pathogenic SNP associations and rare deletions. PLoS Genet 5, e1000534.

Raychaudhuri, S., Remmers, E.F., Lee, A.T., et al., 2008. Common variants at *CD40* and other loci confer risk of rheumatoid arthritis. Nat Genet 40, 1216–1223.

Raychaudhuri, S., Thomson, B.P., Remmers, E.F., et al., 2009b. Genetic variants at *CD28*, *PRDM1* and *CD2/CD58* are associated with rheumatoid arthritis risk. Nat Genet 41, 1313–1318.

Remmers, E.F., Plenge, R.M., Lee, A.T., et al., 2007. STAT4 and the risk of rheumatoid arthritis and systemic lupus erythematosus. N Engl J Med 357, 977–986.

Rigby, A.S., Silman, A.J., Voelm, L., et al., 1991. Investigating the HLA component in rheumatoid arthritis: An additive (dominant) mode of inheritance is rejected, a recessive mode is preferred. Genet Epidemiol 8, 153–175.

Rothenberg, E.V., Moore, J.E., Yui, M.A., 2008. Launching the T-cell-lineage developmental programme. Nature reviews 8, 9–21.

Ruiz-Morales, J.A., Vargas-Alarcon, G., Flores-Villanueva, P.O., et al., 2004. *HLA-DRB1* alleles encoding the "shared epitope" are associated with susceptibility to developing rheumatoid arthritis whereas *HLA-DRB1* alleles encoding an aspartic acid at position 70 of the beta-chain are protective in Mexican Mestizos. Hum Immunol 65, 262–269.

Sakaguchi, N., Takahashi, T., Hata, H., et al., 2003. Altered thymic T-cell selection due to a mutation of the *ZAP-70* gene causes autoimmune arthritis in mice. Nature 426, 454–460.

Schellekens, G.A., Visser, H., De Jong, B.A., et al., 2000. The diagnostic properties of rheumatoid arthritis antibodies recognizing a cyclic citrullinated peptide. Arthritis Rheum 43, 155–163.

Seitz, M., Wirthmuller, U., Moller, B., Villiger, P.M., 2007. The -308 tumour necrosis factor-alpha gene polymorphism predicts therapeutic response to TNFalpha-blockers in rheumatoid arthritis and spondyloarthritis patients. Rheumatology (Oxford) 46, 93–96.

Seyfried, C.E., Mickelson, E., Hansen, J.A., Nepom, G.T., 1988. A specific nucleotide sequence defines a functional T-cell recognition epitope shared by diverse *HLA-DR* specificities. Hum Immunol 21, 289–299.

Shiozawa, S., Hayashi, S., Tsukamoto, Y., et al., 1998. Identification of the gene loci that predispose to rheumatoid arthritis. Int Immunol 10, 1891–1895.

Shmerling, R.H., Delbanco, T.L., 1991. The rheumatoid factor: An analysis of clinical utility. Am J Med 91, 528–534.

Silman, A.J., MacGregor, A.J., Thomson, W., et al., 1993. Twin concordance rates for rheumatoid arthritis: Results from a nationwide study. Br J Rheumatol 32, 903–907.

Silman, A.J., Pearson, J.E., 2002. Epidemiology and genetics of rheumatoid arthritis. Arthritis Res 4 (Suppl 3), S265–S272.

Singal, D.P., Li, J., Lei, K., 1999. Genetics of rheumatoid arthritis (RA): Two separate regions in the major histocompatibility complex contribute to susceptibility to RA. Immunol Lett 69, 301–306.

Smyth, D.J., Plagnol, V., Walker, N.M., et al., 2008. Shared and distinct genetic variants in type 1 diabetes and celiac disease. N Engl J Med 359, 2767–2777.

Stahl, E.A., Raychaudhuri, S., Remmers, E.F., et al., 2010. Genome-wide association study meta-analysis identifies seven new rheumatoid arthritis risk loci. Nat Genet 42, 508–514.

Stastny, P., 1978. Association of the B-cell alloantigen DRw4 with rheumatoid arthritis. N Engl J Med 298, 869–871.

Stastny, P., Fink, C.W., 1977. HLA-Dw4 in adult and juvenile rheumatoid arthritis. Transplant Proc 9, 1863–1866.

Steer, S., Lad, B., Grumley, J.A., Kingsley, G.H., Fisher, S.A., 2005. Association of R602W in a protein tyrosine phosphatase gene with a high risk of rheumatoid arthritis in a British population: Evidence for an early onset/disease severity effect. Arthritis Rheum 52, 358–360.

Stewart, C.A., Horton, R., Allcock, R.J., et al., 2004. Complete MHC haplotype sequencing for common disease gene mapping. Genome Res 14, 1176–1187.

Stolt, P., Bengtsson, C., Nordmark, B., et al., 2003. Quantification of the influence of cigarette smoking on rheumatoid arthritis: Results from a population-based case-control study, using incident cases. Ann Rheum Dis 62, 835–841.

Suzuki, A., Yamada, R., Chang, X., et al., 2003. Functional haplotypes of *PADI4*, encoding citrullinating enzyme peptidylarginine deiminase 4, are associated with rheumatoid arthritis. Nat Genet 34, 395–402.

Suzuki, A., Yamada, R., Kochi, Y., et al., 2008. Functional SNPs in *CD244* increase the risk of rheumatoid arthritis in a Japanese population. Nat Genet 40, 1224–1229.

Symmons, D.P., 2003. Environmental factors and the outcome of rheumatoid arthritis. Best Pract Res Clin Rheumatol 17, 717–727.

Symmons, D.P., Barrett, E.M., Bankhead, C.R., Scott, D.G., Silman, A.J., 1994. The incidence of rheumatoid arthritis in the United Kingdom: Results from the Norfolk Arthritis Register. Br J Rheumatol 33, 735–739.

Takata, Y., Inoue, H., Sato, A., et al., 2008. Replication of reported genetic associations of *PADI4*, *FCRL3*, *SLC22A4* and *RUNX1* genes with rheumatoid arthritis: Results of an independent Japanese population and evidence from meta-analysis of East Asian studies. J Hum Genet 53, 163–173.

Teller, K., Budhai, L., Zhang, M., Haramati, N., Keiser, H.D., Davidson, A., 1996. *HLA-DRB1* and *DQB* typing of Hispanic American patients with rheumatoid arthritis: The "shared epitope" hypothesis may not apply. J Rheumatol 23, 1363–1368.

Thomson, W., Barton, A., Ke, X., et al., 2007. Rheumatoid arthritis association at 6q23. Nat Genet 39, 1431–1433.

Urano, W., Taniguchi, A., Yamanaka, H., et al., 2002. Polymorphisms in the methylenetetrahydrofolate reductase gene were associated with both the efficacy and the toxicity of methotrexate used for the treatment of rheumatoid arthritis, as evidenced by single locus and haplotype analyses. Pharmacogenetics 12, 183–190.

Van der Helm-Van Mil, A.H., Huizinga, T.W., 2008. Advances in the genetics of rheumatoid arthritis point to subclassification into distinct disease subsets. Arthritis Res Ther 10, 205.

Van der Helm-Van Mil, A.H., Verpoort, K.N., Breedveld, F.C., Huizinga, T.W., Toes, R.E., De Vries, R.R., 2006. The *HLA-DRB1* shared epitope alleles are primarily a risk factor for anti-cyclic citrullinated peptide antibodies and are not an independent risk factor for development of rheumatoid arthritis. Arthritis Rheum 54, 1117–1121.

Van der Woude, D., Houwing-Duistermaat, J.J., Toes, R.E., et al., 2009. Quantitative heritability of anti-citrullinated protein antibody-positive and anti-citrullinated protein antibody-negative rheumatoid arthritis. Arthritis Rheum 60, 916–923.

Van Ede, A.E., Laan, R.F., Blom, H.J., et al., 2001. The C677T mutation in the methylenetetrahydrofolate reductase gene: A genetic risk factor for methotrexate-related elevation of liver enzymes in rheumatoid arthritis patients. Arthritis Rheum 44, 2525–2530.

Vang, T., Congia, M., Macis, M.D., et al., 2005. Autoimmune-associated lymphoid tyrosine phosphatase is a gain-of-function variant. Nat Genet 37, 1317–1319.

Van Jaarsveld, C.H., Ter Borg, E.J., Jacobs, J.W., et al., 1999. The prognostic value of the antiperinuclear factor, anti-citrullinated peptide antibodies and rheumatoid factor in early rheumatoid arthritis. Clin Exp Rheumatol 17, 689–697.

Van Zeben, D., Hazes, J.M., Zwinderman, A.H., Cats, A., Van der Voort, E.A., Breedveld, F.C., 1992. Clinical significance of rheumatoid factors in early rheumatoid arthritis: Results of a follow-up study. Ann Rheum Dis 51, 1029–1035.

Vignal, C., Bansal, A.T., Balding, D.J., et al., 2008. Genetic association of the major histocompatibility complex with rheumatoid arthritis implicates two non-*DRB1* loci. Arthritis Rheum 60, 53–62.

Viken, M.K., Amundsen, S.S., Kvien, T.K., et al., 2005. Association analysis of the 1858C>T polymorphism in the *PTPN22* gene in juvenile idiopathic arthritis and other autoimmune diseases.. Genes Immun 6, 271–273.

Waldron-Lynch, F., Adams, C., Amos, C., et al., 2001. Tumour necrosis factor 5′ promoter single nucleotide polymorphisms influence susceptibility to rheumatoid arthritis (RA) in immunogenetically defined multiplex RA families. Genes Immun 2, 82–87.

Watford, W.T., Hissong, B.D., Bream, J.H., Kanno, Y., Muul, L., O'Shea, J.J., 2004. Signaling by IL-12 and IL-23 and the immunoregulatory roles of STAT4. Immunol Rev 202, 139–156.

Weisman, M.H., Furst, D.E., Park, G.S., et al., 2006. Risk genotypes in folate-dependent enzymes and their association with methotrexate-related side effects in rheumatoid arthritis. Arthritis Rheum 54, 607–612.

Wellcome Trust Case Control Consortium, 2007. Genome-wide association study of 14,000 cases of seven common diseases and 3000 shared controls. Nature 447, 661–678.

Wesoly, J., Van der Helm-Van Mil, A.H., Toes, R.E., et al., 2005. Association of the *PTPN22* C1858T single-nucleotide polymorphism with rheumatoid arthritis phenotypes in an inception cohort. Arthritis Rheum 52, 2948–2950.

Wolfe, F., Kleinheksel, S.M., Khan, M.A., 1988. Prevalence of familial occurrence in patients with rheumatoid arthritis. Br J Rheumatol 27 (Suppl 2), 150–152.

Yang, H., Rittner, H., Weyand, C.M., Goronzy, J.J., 1999. Aberrations in the primary T-cell receptor repertoire as a predisposition for synovial inflammation in rheumatoid arthritis. J Investig Med 47, 236–245.

Zanelli, E., Huizinga, T.W., Guerne, P.A., et al., 1998. An extended *HLA-DQ-DR* haplotype rather than *DRB1* alone contributes to RA predisposition. Immunogenetics 48, 394–401.

Zanelli, E., Jones, G., Pascual, M., et al., 2001. The telomeric part of the HLA region predisposes to rheumatoid arthritis independently of the class II loci. Hum Immunol 62, 75–84.

Zhernakova, A., Alizadeh, B.Z., Bevova, M., et al., 2007. Novel association in chromosome 4q27 region with rheumatoid arthritis and confirmation of type 1 diabetes point to a general risk locus for autoimmune diseases. Am J Hum Genet 81, 1284–1288.

Zhernakova, A., Eerligh, P., Wijmenga, C., Barrera, P., Roep, B.O., Koeleman, B.P., 2005. Differential association of the *PTPN22* coding variant with autoimmune diseases in a Dutch population. Genes Immun 6, 459–461.

Multiple Sclerosis

Francisco J. Quintana and Howard L. Weiner

INTRODUCTION

Multiple sclerosis (MS) is an autoimmune disorder in which the central nervous system (CNS) is targeted by the dysregulated activity of the immune system, resulting in focal lesions and progressive neurological dysfunction. MS is heterogeneous in its clinical symptoms, rate of progression, and response to therapy, reflecting the existence of several pathogenic mechanisms that make different contributions to the disease. The etiology of MS is still unknown, but it is thought to result from a combination of genetic and environmental factors. In this chapter, we analyze the contributions of genomics, transcriptomics, immunomics, and proteomics in delineating these factors, as well as their utility for monitoring disease progression and response to therapy, identifying new targets for therapeutic intervention, and detecting individuals at risk of developing the disease later on in life.

GENOMICS

The first observation suggesting a genetic contribution to MS susceptibility was the identification of familial aggregation: first-, second-, and third-degree relatives of MS patients have an increased risk of developing the disease (Mackay, 1950). A sibling of an MS patient, for example, has a 20-times-greater lifetime risk of developing MS than an individual from the general population (Sawcer et al., 2005). Further studies have shown that familial aggregation in MS results from sharing predisposing genetic elements, and not from exposure to environmental factors (Dyment et al., 2004).

MS is considered to be a complex genetic disease in which many polymorphic genes have small or at most moderate effects on the overall MS risk, disease severity, rate of progression, and age of onset, among several clinical outcomes. To date, the major histocompatibility complex (MHC) locus on chromosome 6p21 remains the strongest and most convincing chromosomal region linked to MS. The human leukocyte antigen (HLA) class II histocompatibility antigen DRB1*15 (DRB1*15) and the DRB1*17 alleles increase the risk for MS, while the DRB1*14 has a disease-protective effect (Baranzini et al., 2004; GAMES and the Transatlantic Multiple Sclerosis Genetics Cooperative, 2003; Jersild et al., 1972). Several non-MHC candidate loci have also been linked to MS (Baranzini et al., 2004), but it has proven difficult to validate their association in independent studies. The difficulty in identification of non-MHC genes associated with MS may result from the genetic heterogeneity existing among MS patients, meaning that different combinations of genes may lead to the same end phenotype. In this scenario, methods such as linkage analysis may not be sensitive enough for the detection of genes bearing only modest effects on MS susceptibility; thus

Genomic and Personalized Medicine, 2nd edition
by Ginsburg & Willard. DOI: http://dx.doi.org/10.1016/B978-0-12-382227-7.00072-0

association studies in large cohorts of patients and controls may be needed.

Briefly, two approaches are used for identification of genes linked to MS pathogenesis and progression: linkage and association mapping.

Linkage Mapping

Linkage mapping is based on the study of the co-inheritance of genetic markers and phenotypes in families over several generations. Linkage mapping is successful in finding genes for rare Mendelian monogenic diseases inherited in a dominant fashion. However, in diseases such as MS, where several loci make a small contribution to the phenotype under study, linkage studies only identify those loci that have the strongest influence. Also, one of the premises implicit in linkage studies is that all the families studied have their susceptibility determined by the same genes, an assumption at odds with mounting evidence suggesting that susceptibility to MS is genetically heterogeneous.

A study of 2692 individuals in 730 families of northern European descent analyzed 4506 genetic markers, looking for co-inheritance of genetic markers and MS (Sawcer et al., 2005). Multipoint non-parametric linkage analysis could only find one significant linkage, which unsurprisingly pointed to the MHC locus on chromosome 6p21. This study therefore confirms the identification of the MHC locus as a genetic determinant of susceptibility to MS, but it also highlights two limitations of linkage mapping. First, linkage mapping tends to identify large chromosomal fragments because of the small number of recombination events analyzed in familial pedigrees; the MHC locus, for example, contains more than 200 genes. Second, the authors also found suggestive linkage on chromosomes 17q23, 5q33, and 19p13, but the data were inconclusive about these loci even when a high density of genetic markers was used. Other mapping strategies are needed to identify loci with modest effects on MS.

Association Mapping

Association mapping looks for genetic markers with higher frequencies in MS patients than controls, suggesting an association between a disease phenotype and allelic variation (Hafler and De Jager, 2005).

Although genetic susceptibility to MS has been linked to the MHC locus, the particular gene or genes underlying susceptibility to MS within this locus are a matter of discussion, particularly because of the punctuated pattern of linkage disequilibrium observed for this chromosomal location (Jeffreys et al., 2001). As a result, other genes within the MHC locus besides *HLA-DRB1* might be associated with MS; *tnf* (tumor necrosis factor) and other loci in HLA (human leukocyte antigen) class III and HLA class I are logical candidates. Lincoln and coworkers genotyped 4203 individuals from Finland and Canada with a high-density single nucleotide polymorphism (SNP) panel to identify the gene or genes in the MHC locus responsible for the increased susceptibility to MS (Lincoln

et al., 2005). Their results identified the DRB1 allele in the HLA class II region as the single major susceptibility locus in the MHC region, although there is still a small chance that a closely adjacent locus contains the true susceptibility variant. Thus, association mapping confined the MS susceptibility to the approximately 100 kb of sequence flanking *HLA-DRB1*. On the one hand, this work illustrates the increased power of association over linkage mapping studies. On the other hand, it highlights the dependency of association mapping on linkage disequilibrium – the uneven distribution of recombination events in this region leads to the identification of a locus for MS susceptibility big enough to contain several genes.

Admixture Mapping

Admixture mapping is an association mapping strategy based on a difference in the prevalence of a disease between two ethnic groups and the existence of a third, admixed population (Smith and O'Brien, 2005). Northern Europeans and Africans have different susceptibilities to many autoimmune, circulatory, and metabolic disorders, among them MS (Smith and O'Brien, 2005). This differential susceptibility is also reflected in North America, where MS is more prevalent in European Americans than African Americans. On average, there have been only six generations since African and European populations came into contact in North America, resulting in little recombination between chromosomes of African and European ancestry in the history of African-American populations. Thus, the chromosomal segments of one or the other ancestry are long in the admixed population, and few genetic markers can be used to classify the genome of an African American into sections with African or European origins. It is therefore possible to study African Americans to identify genomic regions where individuals with MS tend to have an unusually high proportion of ancestry from either Europeans or Africans, indicative of the presence of an MS variant that differs in frequency between the two ancestral populations. This mapping strategy was initially suggested several years ago (Chakraborty and Weiss, 1988), but it could only be fully implemented with the advent of genetic markers that identify human populations of different ancestry (Sachidanandam et al., 2001).

Admixture mapping led to the identification of a new locus associated with MS risk around the chromosome 1 centromere, as a result of the analysis of 1166 genetic markers in 1648 samples (605 MS patients and 1043 controls) (Reich et al., 2005). Notably, this linkage could not be replicated using an independent set of 143 African Caribbeans with MS, from Martinique and the UK, suggesting that either the locus in chromosome 1 does not have a role in African-Caribbean populations or that the African-Caribbean cohort was not large enough. Future studies should attempt to clone the susceptibility gene on chromosome 1.

Genome Wide Association Studies

Genome scans aimed at identifying non-MHC genes linked to MS usually fail when initial promising candidates cannot be validated in independent populations. New experimental techniques and analytical methods, however, are seeking to change this trend. The meta-analysis of the results of whole-genome screens for linkage or association in 18 populations has identified several genomic regions that were then verified in a different set of samples (Abdeen et al., 2006), leading to the identification of several non-MHC candidate genes that modify the risk for MS. The list of candidates includes genes involved in CNS development and regeneration (*NTN1*, *NCAM1*, *ADAM22*, and *ADAMTS10*), as well as genes directly linked to inflammation (*TGFA*, *TGFBR*, *IL18*, and *IL10RA*) (Abdeen et al., 2006). Although these new gene candidates are still awaiting independent validation, the meta-analysis used for their identification may constitute a new method for short-listing candidates to be further analyzed in independent replication studies.

The identification of polymorphisms in the α-chain of the interleukin-7 (IL-7) receptor (IL-7Ra) was a significant step in the study of non-MHC genetic variants linked to MS (Gregory et al., 2007; Hafler et al., 2007; Lundmark et al., 2007). Although these polymorphisms make only a small contribution to genetic susceptibility to MS, they are a significant step toward the identification of genetic determinants for MS outside the MHC locus. The *IL-7Ra* allele associated with MS favors a relative decrease in the membrane-bound IL-7α (Gregory et al., 2007). IL-7 is produced by stromal cells in lymphoid tissues. Its availability is controlled through its uptake by the membrane-bound IL-7R on T cells (Mazzucchelli and Durum, 2007). Thus, considering the positive effects that IL-7 has on lymphocyte survival and proliferation (Mazzucchelli and Durum, 2007), the decrease in membrane IL-7R may result in increased levels of IL-7 available to fuel the inflammatory T cell response in MS.

The α-chain IL-2 receptor (*IL-2Ra*) gene has also been recently linked to MS (Hafler et al., 2007). *IL-2Ra* allelic variation has been previously associated with other autoimmune diseases such as type I diabetes, but at a different genomic position (Lowe et al., 2007). IL-2 is required for the development of regulatory T cells (T_{reg}) (Malek, 2008; Suzuki et al., 1995), and indeed, deficits in T_{reg} activity characterize relapsing-remitting MS (Viglietta et al., 2004). Thus *IL-2Ra* polymorphisms may be related to the immune dysregulation observed in MS. Notably, *IL-2Ra*-specific antibodies have shown promising beneficial effects for the treatment of MS in phase II clinical trials (Bielekova et al., 2004; Rose et al., 2004). Although the link between *IL-2Ra* polymorphisms and MS is still awaiting further validation, the association of *IL-7Ra* and *IL-2Ra* variants with MS supports the use of genome-wide studies to delineate pathways contributing to disease pathogenesis.

TRANSCRIPTOMICS

Large-scale studies of mRNA expression in MS have been directed at characterizing the two main components of the disease: the lesion and the immune response.

Characterization of the Lesion

The hallmark of MS is the presence of demyelinated plaques in the white matter area of the CNS (Lassmann, 1998). These lesions are heterogeneous, reflecting the contribution of diverse mechanisms to disease progression (Lassmann et al., 2001). The study of the transcriptional activity in the lesions is therefore of importance for the characterization of the processes that drive MS, and for the identification of new targets of immune intervention.

Lindberg and coworkers studied the transcriptional activity of lesions and normal-appearing white matter (NAWM) in samples taken from six secondary progressive MS (SPMS) patients and 12 matched controls (Lindberg et al., 2004). Within the lesions, only 21% of the up-regulated genes were associated with the immune response; 77% of those immune-related transcripts corresponded to the cellular response. However, the four most significant immune up-regulated genes were linked to humoral immunity (immunoglobulins). Notably, tertiary lymph nodes that support the maturation of antibody-secreting plasma cells within the CNS have been identified in SPMS brains (Aloisi and Pujol-Borrell, 2006; Corcione et al., 2005). In addition, the genes that showed a decreased expression mainly belonged to one of two categories: genes with well-known anti-inflammatory activities (pointing to a deficit in immunoregulatory mechanisms), and genes involved in neural homeostasis. Interestingly, the NAWM showed a significant up-regulation in immune-related genes, mainly involved in signaling and effector functions such as blood–brain barrier (BBB) disruption and lymphocyte activation. These findings provide a molecular insight into the pathological mechanisms driving MS, and confirm the presence of physiological abnormalities in NAWM.

The study of the transcriptional profile of MS lesions can lead to the identification of new targets for immune intervention. Lock and coworkers studied the expression profile in acute MS lesions (with signs of inflammation) and in silent lesions (without inflammation, but showing clear signs of demyelination and scarring) (Lock et al., 2002). Both types of lesions showed an up-regulated expression of genes associated with MHC class II antigen presentation, immunoglobulin synthesis, complement, and pro-inflammatory cytokines. Neuron-associated genes and those associated with myelin production were under-expressed.

Different expression profiles were found in active and silent lesions. α-integrin was found to be elevated in chronic silent MS lesions; notably, antibodies to α4-integrin reverse and reduce the rate of relapse in relapsing-remitting experimental autoimmune encephalomyelitis (EAE) (Yednock et al., 1992), and a humanized version of this antibody showed promising effects in treatment of human MS (Polman et al., 2006).

Expression of IgE and IgG Fc receptors was up-regulated in chronic silent MS lesions (Lock et al., 2002). Accordingly, mice harboring impaired IgE and IgG Fc receptors develop a milder EAE than their wild-type counterparts (Lock et al., 2002). These effects were stronger from day 20 onward following the induction of the disease, in accordance with microarray data showing that Fc receptor transcripts are elevated in chronic but not in acute MS lesions. Granulocyte colony-stimulating factor (GCSF) was found to be up-regulated in acute active lesions; its administration to mice before challenge with an encephalopathogenic peptide also led to an amelioration of the disease, suggesting that endogenous GCSF may participate in the natural regulation of acute attacks (Lock et al., 2002).

In a separate study, the large-scale sequencing of non-normalized complementary DNA (cDNA) libraries derived from plaques dissected from brains of patients with MS showed an increased frequency of transcripts coding for osteopontin (OPN), a Th1 cytokine involved in the immune response to infectious diseases (Chabas et al., 2001). OPN transcripts were detected exclusively in the MS mRNA population, but not in control brain mRNA. The up-regulated expression of OPN was confirmed by immunohistochemistry in human MS plaques, which showed OPN expression in microvascular endothelial cells and macrophages, and in the white matter areas adjacent to plaques. Similar patterns of expression were seen in mice and rats representing a model of relapsing-remitting and monophasic EAE, respectively. To test the relevance of OPN in MS, OPN-deficient mice were generated. They showed a reduced severity of experimental autoimmune encephalomyelitis (EAE). Neutralization of OPN with neutralizing antibodies also led to an amelioration of EAE (Blom et al., 2003), validating the use of cDNA microarrays in the search for new therapeutic targets for MS. The up-regulation of OPN levels in MS plaques (Tajouri et al., 2005) and in the circulation (Comabella et al., 2005; Vogt et al., 2004) of MS patients was replicated in independent studies, prompting the search for polymorphisms in the *opn* gene associated with MS. Although some controversy remains (Caillier et al., 2003), polymorphisms in the *opn* locus have been associated with increased levels of circulating OPN and disease course (Chiocchetti et al., 2005). OPN is therefore an example of how results obtained in transcriptomics studies may lead to the identification of genetic polymorphisms linked to MS.

Characterization of the Immune Response

The analysis of the transcriptional profile could also be applied to study the immune response in MS, to monitor progression of the disease and response to therapy. Two points, however, should be kept in mind when considering the use of cDNA arrays for the analysis of the immune response in MS patients. First, these studies assume that changes in the periphery somehow reflect the ongoing situation within the CNS. Second, these studies are limited by the normal "noise" that exists in basal gene expression, originating from diverse factors such as the relative proportion of different blood cell subsets, gender, age, and the time of day at which the sample was taken (Whitney et al., 2003).

Conversion to Definite MS

Transcriptional profiling of CD4+ T cells in MS has been recently used to predict the conversion of clinically isolated syndrome (CIS) patients to definite MS (Corvol et al., 2008). Baranzini and coworkers identified a set of genes whose expression is linked to the conversion to definite MS in less than a year. One of those genes is *TOB1*, a critical regulator of cell proliferation and a potential player in MS pathogenesis. Decreased *TOB1* expression at the RNA and protein levels was found in MS patients, and in EAE. Moreover, a genetic association was observed between *TOB1* variation and MS progression. All in all, these results link alterations in the regulation of CD4+ T cell quiescence with progression of MS (Corvol et al., 2008).

Follow-up of Disease Activity

Achiron and coworkers followed the transcriptional activity of peripheral blood mononuclear cells (PBMC) prepared from relapsing-remitting MS (RRMS) patients during the relapses and remission of MS (Achiron et al., 2004). The authors identified a transcriptional signature associated with the relapse that included genes involved in the recruitment of immune cells, epitope spreading, and escape from immune-regulation. Although encouraging, these results should be validated using an independent set of samples and in longitudinal studies to evaluate their predictive value.

Response to Therapy

Gene expression profiling can also be used to classify patients according to different clinical criteria, such as responders or non-responders to therapy. β-interferon (IFNβ) is widely used for the treatment of MS (Kappos and Hartung, 2005), but its precise mechanism of action is not known, nor are biomarkers that would allow the identification of patients that will benefit from it. Weinstock-Guttman and colleagues used cDNA microarrays to study the effects of IFNβ therapy on the transcriptional activity of monocyte-depleted PBMCs (Weinstock-Guttman et al., 2003). This pharmacodynamic study found that upon one hour of IFNβ administration, significant changes are detected in the expression of genes involved in anti-viral response and IFNβ signaling, and markers of lymphocyte activation. These studies provided a molecular description of the effects of IFNβ on RRMS patients, and were later extended to identify transcriptional signatures associated with a favorable response to treatment with IFNβ (Sturzebecher et al., 2003). They also demonstrated that MS patients showing a positive response to treatment with IFNβ, as assessed by longitudinal gadolinium-enhanced magnetic resonance imaging (MRI) scans and clinical disease activity, are characterized by specific patterns of gene expression (Sturzebecher et al., 2003). Based on these observations, Oksenberg and coworkers identified groups of genes whose expression has a prognostic value for the identification of MS patients likely to respond to treatment with IFNβ (Baranzini et al., 2004). The work of Oksenberg and colleagues is remarkable for two reasons. First, it demonstrates that gene expression profiling can be used in the management of MS to select a therapeutic regime suited to the patient's metabolism. Second, it used a methodology [reverse transcription polymerase chain reaction (RT-PCR)] accessible to clinical laboratories, facilitating the translation of their results into daily medical practice.

The combination of data generated in transcriptomics and genomics studies can be an invaluable source of information and new hypotheses. Aune and colleagues compared the genes differentially expressed by lymphocytes in rheumatoid arthritis (RA), systemic lupus erythematosus (SLE), insulin-dependent diabetes mellitus (IDDM) and multiple sclerosis (MS), concluding that they are clustered within chromosomal domains in the genome (Aune et al., 2004). Strikingly, they found that the chromosomal domains containing the genes differentially expressed in autoimmune disorders could be mapped to disease susceptibility loci associated to those diseases by genetic linkage studies (Aune et al., 2004). These results suggest that the expression of disease-associated genes is co-regulated as a result of shared genetic regulatory elements or local patterns of chromatin condensation. Recently, Baranzini and coworkers studied the genetic concordance between gene expression and genetic linkage in MS. They first compiled the data on gene expression available for MS and EAE, and then superimposed it with all of the known susceptibility loci identified in MS and EAE. They identified the MS susceptibility genes located in the MHC locus as overlapping with clusters of differentially expressed genes in MS and murine EAE. However, they also identified an interesting region on chromosome X that may contribute to the sexual dimorphism observed in MS. The integration of the data generated by different platforms such as transcriptomics, genomics, and proteomics is therefore likely to deepen our understanding of the mechanisms driving MS.

IMMUNOMICS

The autoimmune nature of MS suggests that the study of the immune response should be useful for early diagnosis, prognosis, and monitoring of MS patients. With this aim, new tools have been developed that allow a high-throughput analysis of the T cell and antibody-mediated immune response.

T Cell Response

MHC arrays have recently been developed based on the immobilization of peptide/MHC tetramers used to activate peptide-specific T cells and follow cytokine secretion or adhesion (Chen et al., 2005; Soen et al., 2003). Their use for the study of human immunology is limited by the high degree of polymorphism existing at the MHC locus. Nevertheless, these arrays have recently been used to characterize cellular response in tumor-vaccinated humans (Chen et al., 2005).

Reverse phase arrays consist of arrays of spotted lysates prepared from primary cells, which are interrogated with antibodies specific for phosphorylated or dephosphorylated proteins involved in signaling transduction pathways of interest (Chan et al., 2004). By combining different antibodies and

inhibitors of specific signal transduction pathways, detailed signaling maps can be constructed describing the molecular events that lead to the activation of a specific cell population. This approach has recently been used to study $CD4^+CD25^+$ regulatory T cells (T_{reg}) (Chan et al., 2004). Since defects in T_{reg} function have been described in MS patients (Viglietta et al., 2004), these arrays could be useful in the identification of signaling defects in the immune system of MS patients, and how these pathways are modified in immunomodulatory regimes. We have recently used reverse protein arrays to characterize signaling pathways activated in cells of the innate immune system in the CNS during the progressive phases of MS or its experimental models. Using this approach, we identified a signaling axis initiated by the activation of toll-like receptor 2 that culminated with the activation of poly(ADP-ribose) polymerase-1 (PARP-1; ADP = adenosine diphosphate), which controls inflammatory and neurotoxic activities in macrophages, microglia, and astrocytes (Farez et al., 2009). This axis is therefore a potential therapeutic target to modulate the chronic innate immune activation thought to promote neurodegeneration and disease progression in the progressive stages of MS.

Multidimensional FACS (fluorescence-activated cell sorting) is based on the simultaneous computation of the signals produced by more than 16 colors, and their analysis using advanced mathematical algorithms (Perfetto et al., 2004). This technique could be useful to interrogate the immune response to specific antigens using a combination of tetramers, intracellular cytokine staining, and surface markers (Chattopadhyay et al., 2005).

Cell migration arrays (Kuschel et al., 2006). To reach the CNS and cause tissue destruction, autoimmune T cells must express surface molecules that allow them to cross the blood–brain barrier (BBB) and initiate inflammation (Luster et al., 2005). The importance of this process for MS progression is shown by the beneficial effects that natalizumab, a humanized antibody to α4 integrin that interferes with T cell extravasation to the CNS and has showed promising preliminary results in the treatment of MS (Polman et al., 2006). Thus, the characterization of the adhesion properties exhibited by the different cell populations in MS patients may be of help to predict and monitor response to treatment and to predict and prevent the development of relapses.

B Cell Response

Antibodies are of interest in MS because they may have a pathological role in MS (Genain et al., 1995), and more importantly, because antibody responses are thought to reflect the activity of the T cell compartment (Robinson et al., 2002b). These antibodies can be produced by B cells in the periphery and make their way to the inflamed CNS via the disrupted BBB, but they are also produced by intrathecal germinal centers that seem to form tertiary lymph nodes (Aloisi and Pujol-Borrell, 2006; Corcione et al., 2005). It is easier to assay reproducible antibody reactivity than antigen-specific T cell responses; thus efforts have been invested in the development of new technologies for monitoring humoral response in MS patients and autoimmunity (Quintana et al., 2004; Robinson et al., 2002a).

Antigen arrays can be used to detect changes in the repertoire of antibodies, reflecting the antigen spreading that accompanies EAE progression (Robinson et al., 2003). The information obtained about the degree of antigen spreading observed in each mouse was used to design tailored immunomodulatory vaccines to control EAE (Robinson et al., 2003). In humans, it has been used to identify new lipid targets of antibodies present in the cerebrospinal fluid (CSF) of MS patients; some of the new lipid targets found were also validated on EAE (Kanter et al., 2006). Future experiments should study the antibody response in the serum of MS patients, searching for patterns of antibody reactivity that predict progression of MS or response to therapy, as has been shown for other autoimmune disorders, such as RA (Hueber et al., 2005), autoimmune diabetes (Quintana et al., 2004), and SLE (Li et al., 2005). Antigen arrays have also been shown to identify the individual mice that will develop autoimmune diabetes later in life by studying an experimental model that shows incomplete penetrance.

Using antigen arrays, we characterized patterns of antibody reactivity in MS serum against a panel of CNS protein and lipid autoantigens and heat shock proteins (Quintana et al., 2008b). We found unique autoantibody patterns that distinguished RRMS, SPMS, and primary progressive MS (PPMS) patients from both healthy controls and from other neurologic or autoimmune-driven diseases, including Alzheimer's disease, adrenoleukodystrophy, and lupus erythematosus. RRMS was characterized by autoantibodies to heat shock proteins that were not observed in PPMS or SPMS. Also, RRMS, SPMS, and PPMS were characterized by unique patterns of reactivity to CNS antigens (Quintana et al., 2008b). Furthermore, we examined sera from patients with different immunopathological patterns of MS as determined by brain biopsy, and we identified unique patterns of antibodies to lipids and CNS-derived peptides that were linked to each type of pathology (Quintana et al., 2008b). Further characterization of the role of these lipids in EAE identified them as direct effectors in MS pathogenesis by the activation of the innate immune system in the CNS (Quintana et al., 2008b). Demonstration of unique serum immune signatures linked to different stages and pathological processes in MS provides an avenue to monitor MS and to characterize immunopathogenic mechanisms and therapeutic targets in the disease.

Thus, antigen arrays may be useful for identification of patients at risk of developing MS before the overt onset of the symptoms (Quintana et al., 2004). In addition, antigen arrays may be used to interrogate the antibody repertoire in search of new targets of the MS autoimmune attack, and therefore new targets for immunomodulation.

PROTEOMICS

Proteomic studies in MS can identify new targets of the autoimmune process, complementing the information provided by antigen arrays; they can also identify new processes contributing to disease pathology, and biomarkers for early diagnosis and monitoring of MS patients.

Identification of New Targets of Autoimmunity

Using two-dimensional gels on brain extracts, Almeras and coworkers identified 14 new targets recognized by antibodies present in CSF (Almeras et al., 2004). The targets included heat shock proteins, structural proteins, and enzymes involved in glucose metabolism (Almeras et al., 2004). Although these results have been partially validated in the experimental model of MS EAE (Zephir et al., 2006), they are still awaiting validation in an independent patient cohort.

Identification of New Pathogenic Processes

Proteomic studies have also identified new mechanisms that contribute to MS pathology. The existence of a link between Epstein–Barr virus infection (EBV) and MS has recently been strengthened by the work of Cepok and coworkers, who used protein expression arrays to characterize the reactivity of antibodies in the CSF of MS patients. Most of those antibodies recognized EBV epitopes (Cepok et al., 2005). The results suggest that EBV reactivation may elicit an abnormal immune response in susceptible individuals that contributes to MS (Sundstrom et al., 2004).

Brinkmeier and coworkers analyzed the CSF of MS and Guillain–Barré syndrome (GBS) patients to characterize molecules that might affect ion-channel function (Brinkmeier et al., 2000). They identified an endogenous pentapeptide (QYNAD) that works as a reversible Na^+ channel blocker (Brinkmeier et al., 2000). This pentapeptide is present in the CSF of healthy individuals, but its levels are up-regulated 3–14-fold in MS and GBS patients. This blocking peptide may be involved in the fast exacerbations and relapses commonly seen in demyelinating autoimmune diseases. It could become a valuable marker of disease activity and the target of future therapeutic interventions. These observations, however, could not be replicated by independent researchers (Cummins et al., 2003), highlighting the need for independent validation of potential proteomic biomarkers.

A large proteomic study of MS lesions by Steinman and coworkers has recently identified the dysregulation of the clotting cascade as a new player in MS pathogenesis (Han et al., 2008). In vivo administration of hirudin or recombinant activated protein C reduced disease severity in experimental autoimmune encephalomyelitis, and suppressed T_H1 and T_H17 cytokines in astrocytes and immune cells (Han et al., 2008). Administration of mutant forms of recombinant activated protein C showed that both its anticoagulant and its signaling functions were essential for optimal amelioration of EAE (Han et al., 2008). Thus, the coagulation cascade may constitute a new therapeutic target for MS.

Identification of Biomarkers

Proteomics can identify biomarkers useful in monitoring the different processes that contribute to MS pathology. In this direction, several studies have suggested that cytokines, chemokines, complement, and adhesion molecules can be used as indicators of the inflammatory process in MS, while the levels of actin, tubulin, neurofilaments, tau, GFAP (glial-fibrillary acidic protein) and S-100 proteins can be taken as indicators of axonal loss and gliosis (reviewed in Bielekova and Martin, 2004; Miller, 2004; Teunissen et al., 2005). However, for these markers to be of widespread clinical use, they should be easy to collect and provide reproducible results. This motivation has boosted efforts aimed at finding markers of inflammation and neurodegeneration detectable in blood (Bielekova and Martin, 2004), urine (Bielekova and Martin, 2004), and tears (Devos et al., 2001). Simultaneously, it has fostered the development of new technologies aimed at detecting minute amounts of proteins in body fluids (Moriguchi et al., 2006).

PLATFORMS FOR THE IDENTIFICATION OF NEW THERAPEUTIC TARGETS

Screenings aimed at identifying genes or drugs controlling immune response cannot be easily undertaken in mice because they are based on crossing, maintaining, and screening large numbers of animals, an expensive, time-, space-, and labor-intensive task. New experimental models are needed. The zebra fish (*Danio rerio*) harbors an adaptive immune system that resembles the mammalian immune system (Langenau and Zon, 2005; Trede et al., 2004) and offers several experimental advantages for the study of pathways controlling vertebrate processes of interest. As part of our work on the zebra fish to identify pathways controlling immunity, we have identified the ligand-activated transcription factor aryl hydrocarbon receptor (AHR) as a regulator of the generation of T_{reg} and T_H17 cells (Quintana et al., 2008a). AHR activation by its ligand 2,3,7,8-tetrachlorodibenzo-p-dioxin (TCDD) results in the generation of functional T_{reg} that inhibit the development of EAE by a TGFβ1-dependent mechanism. Surprisingly, AHR activation by an alternative ligand, 6-formylindolo[3,2-b]carbazole, interferes with T_{reg} differentiation, boosts T_H17 differentiation, and worsens EAE. Recently, we have demonstrated that AHR also plays an important role during the differentiation of IL-10-producing type 1 regulatory T cells (Tr1 cells) (Apetoh et al., 2010). We have also shown that small chemicals that activate AHR can promote the differentiation of functional human regulatory T cells (Gandhi et al., 2010). Thus, AHR is a unique therapeutic target for treatment of autoimmune diseases. Our findings also suggest that the experimental advantages offered by the zebra fish can be exploited to characterize metabolic pathways controlling immunity in vertebrates, and to identify new targets for therapeutic intervention.

CONCLUSION

How can we apply the information provided by genomics, transcriptomics, immunomics, and proteomics to the early diagnosis, prevention, monitoring, and therapy of MS? A first step is adaptation of the technologies mentioned in this chapter to a clinical setup, in terms of costs and practicality. The Personal Genome Project, for example, aims to develop, within the next 10 years, cheap, fast, and reliable methods to sequence the whole genome of an individual for $1000 or less (Shendure et al., 2004). This would facilitate the early detection of individuals at risk of developing MS, while identifying genetic markers known to influence disease progression or response to therapy. Similar projects are aimed at facilitating the proteomic analysis of clinical samples (Newcombe et al., 2005). However, note that each one of these technologies is focused on only one of the biochemical processes or layers that describe the organism in a particular physiological state. Thus, if we want to predict the behavior of such a complex biological system as the human body (its response to environmental stimuli, therapy, etc.), then it is imperative to identify not only all of the relevant cellular and molecular layers that make it up, but also to describe qualitatively and quantitatively how those layers interact. In other words, we need to know how the genomics, transcriptomics, immunomics, and proteomics of an individual influence each other. How are the genomics and transcriptomics of MS-linked loci related (Aune et al., 2004; Fernald et al., 2005)? How is the transcriptional activity reflected in the immune response? How is that immune response related to the proteomic profile of the individual? How does the genomic makeup of an MS patient influence her/his response to therapy (Danesi et al., 2000)? Addressing these points will require the development of computational tools for integration of networks and pathways into accurate quantitative models (Bauch and Superti-Furga, 2006; Hwang et al., 2005), and will rely on graphical tools that facilitate the visualization of these models (Efroni et al., 2003).

MS results from a complex dialogue between a susceptible individual and a fostering environment. This dialogue is likely to be unique to each individual, and many of its words are currently unknown. Genomics, transcriptomics, immunomics, and proteomics will allow us to construct accurate models to identify those missing words and prevent, diagnose, and cure MS.

REFERENCES

Abdeen, H., Heggarty, S., Hawkins, S.A., Hutchinson, M., McDonnell, G.V., Graham, C.A., 2006. Mapping candidate non-MHC susceptibility regions to multiple sclerosis. Genes Immun 7, 494–502.

Achiron, A., Gurevich, M., Friedman, N., Kaminski, N., Mandel, M., 2004. Blood transcriptional signatures of multiple sclerosis: Unique gene expression of disease activity. Ann Neurol 55, 410–417.

Almeras, L., Lefranc, D., Drobecq, H., et al., 2004. New antigenic candidates in multiple sclerosis: Identification by serological proteome analysis. Proteomics 4, 2184–2194.

Aloisi, F., Pujol-Borrell, R., 2006. Lymphoid neogenesis in chronic inflammatory diseases. Nat Rev Immunol 6, 205–217.

Apetoh, L., Quintana, F.J., Pot, C., et al., 2010. The aryl hydrocarbon receptor interacts with c-Maf to promote the differentiation of type 1 regulatory T cells induced by IL-27. Nat Immunol 11, 854–861.

Aune, T.M., Parker, J.S., Maas, K., Liu, Z., Olsen, N.J., Moore, J.H., 2004. Co-localization of differentially expressed genes and shared susceptibility loci in human autoimmunity. Genet Epidemiol 27, 162–172.

Baranzini, S.E., Mousavi, P., Rio, J., et al., 2004. Transcription-based prediction of response to IFNbeta using supervised computational methods. PLoS Biol 3, e2.

Bauch, A., Superti-Furga, G., 2006. Charting protein complexes, signaling pathways, and networks in the immune system. Immunol Rev 210, 187–207.

Bielekova, B., Martin, R., 2004. Development of biomarkers in multiple sclerosis. Brain 127, 1463–1478.

Bielekova, B., Richert, N., Howard, T., et al., 2004. Humanized anti-CD25 (daclizumab) inhibits disease activity in multiple sclerosis patients failing to respond to interferon beta. Proc Natl Acad Sci USA 101, 8705–8708. (Epub.).

Blom, T., Franzen, A., Heinegard, D., Holmdahl, R., 2003. Comment on "The influence of the proinflammatory cytokine, osteopontin, on autoimmune demyelinating disease". Science 299, 1845.

Brinkmeier, H., Aulkemeyer, P., Wollinsky, K.H., Rudel, R., 2000. An endogenous pentapeptide acting as a sodium channel blocker in inflammatory autoimmune disorders of the central nervous system. Nat Med 6, 808–811.

Caillier, S., Barcellos, L.F., Baranzini, S.E., et al., 2003. Osteopontin polymorphisms and disease course in multiple sclerosis. Genes Immun 4, 312–315.

Cepok, S., Zhou, D., Srivastava, R., et al., 2005. Identification of Epstein-Barr virus proteins as putative targets of the immune response in multiple sclerosis. J Clin Invest 115, 1352–1360.

Chabas, D., Baranzini, S.E., Mitchell, D., et al., 2001. The influence of the proinflammatory cytokine, osteopontin, on autoimmune demyelinating disease. Science 294, 1731–1735.

Chakraborty, R., Weiss, K.M., 1988. Admixture as a tool for finding linked genes and detecting that difference from allelic association between loci. Proc Natl Acad Sci USA 85, 9119–9123.

Chan, S.M., Ermann, J., Su, L., Fathman, C.G., Utz, P.J., 2004. Protein microarrays for multiplex analysis of signal transduction pathways. Nat Med 10, 1390–1396.

Chattopadhyay, P.K., Yu, J., Roederer, M., 2005. A live-cell assay to detect antigen-specific CD4+ T cells with diverse cytokine profiles. Nat Med 11, 1113–1117. (Epub.).

Chen, D.S., Soen, Y., Stuge, T.B., et al., 2005. Marked differences in human melanoma antigen-specific T cell responsiveness after vaccination using a functional microarray. PLoS Med 2, e265.

Chiocchetti, A., Comi, C., Indelicato, M., et al., 2005. Osteopontin gene haplotypes correlate with multiple sclerosis development and progression. J Neuroimmunol 163, 172–178.

Comabella, M., Pericot, I., Goertsches, R., et al., 2005. Plasma osteopontin levels in multiple sclerosis. J Neuroimmunol 158, 231–239.

Corcione, A., Aloisi, F., Serafini, B., et al., 2005. B-cell differentiation in the CNS of patients with multiple sclerosis. Autoimmun Rev 4, 549–554.

Corvol, J.C., Pelletier, D., Henry, R.G., et al., 2008. Abrogation of T cell quiescence characterizes patients at high risk for multiple sclerosis after the initial neurological event. Proc Natl Acad Sci USA 105, 11,839–11,844.

Cummins, T.R., Renganathan, M., Stys, P.K., et al., 2003. The pentapeptide QYNAD does not block voltage-gated sodium channels. Neurology 60, 224–229.

Danesi, R., Mosca, M., Boggi, U., Mosca, F., Del Tacca, M., 2000. Genetics of drug response to immunosuppressive treatment and prospects for personalized therapy. Mol Med Today 6, 475–482.

Devos, D., Forzy, G., De Seze, J., et al., 2001. Silver stained isoelectrophoresis of tears and cerebrospinal fluid in multiple sclerosis. J Neurol 248, 672–675.

Dyment, D.A., Ebers, G.C., Sadovnick, A.D., 2004. Genetics of multiple sclerosis. Lancet Neurol 3, 104–110.

Efroni, S., Harel, D., Cohen, I.R., 2003. Toward rigorous comprehension of biological complexity: Modeling, execution, and visualization of thymic T-cell maturation. Genome Res 13, 2485–2497.

Farez, M.F., Quintana, F.J., Gandhi, R., Izquierdo, G., Lucas, M., Weiner, H.L., 2009. Toll-like receptor 2 and poly(ADP-ribose) polymerase 1 promote central nervous system neuroinflammation in progressive EAE. Nat Immunol 10, 958–964.

Fernald, G.H., Yeh, R.F., Hauser, S.L., Oksenberg, J.R., Baranzini, S.E., 2005. Mapping gene activity in complex disorders: Integration of expression and genomic scans for multiple sclerosis. J Neuroimmunol 167, 157–169.

Gandhi, R., Kumar, D., Burns, E.J., et al., 2010. Activation of the aryl hydrocarbon receptor induces human type 1 regulatory T cell-like and Foxp3(+) regulatory T cells. Nat Immunol 11 (9), 846–853. doi:10.1038/ni.1915.

Genain, C.P., Nguyen, M.H., Letvin, N.L., et al., 1995. Antibody facilitation of multiple sclerosis-like lesions in a nonhuman primate. J Clin Invest 96, 2966–2974.

Genetic Analysis of Multiple Sclerosis in EuropeanS (GAMES) and the Transatlantic Multiple Sclerosis Genetics Cooperative, 2003. A meta-analysis of whole genome linkage screens in multiple sclerosis. J Neuroimmunol 143, 39–46.

Gregory, S.G., Schmidt, S., Seth, P., et al., 2007. Interleukin 7 receptor alpha chain (IL7R) shows allelic and functional association with multiple sclerosis. Nat Genet 39, 1083–1091.

Hafler, D.A., Compston, A., Sawcer, S., et al., 2007. Risk alleles for multiple sclerosis identified by a genomewide study. N Engl J Med 357, 851–862.

Hafler, D.A., De Jager, P.L., 2005. Applying a new generation of genetic maps to understand human inflammatory disease. Nat Rev Immunol 5, 83–91.

Han, M.H., Hwang, S.I., Roy, D.B., et al., 2008. Proteomic analysis of active multiple sclerosis lesions reveals therapeutic targets. Nature 451, 1076–1081. (Epub.).

Hueber, W., Kidd, B.A., Tomooka, B.H., et al., 2005. Antigen microarray profiling of autoantibodies in rheumatoid arthritis. Arthritis Rheum 52, 2645–2655.

Hwang, D., Rust, A.G., Ramsey, S., et al., 2005. A data integration methodology for systems biology. Proc Natl Acad Sci USA 102, 17,296–17,301.

Jeffreys, A.J., Kauppi, L., Neumann, R., 2001. Intensely punctate meiotic recombination in the class II region of the major histocompatibility complex. Nat Genet 29, 217–222.

Jersild, C., Svejgaard, A., Fog, T., 1972. HL-A antigens and multiple sclerosis. Lancet 1, 1240–1241.

Kanter, J.L., Narayana, S., Ho, P.P., et al., 2006. Lipid microarrays identify key mediators of autoimmune brain inflammation. Nat Med 12, 138–143.

Kappos, L., Hartung, H.P., 2005. 10 years of interferon beta-1b (beta feron therapy). J Neurol 252 (Suppl. 3), iii1–iii2.

Kuschel, C., Steuer, H., Maurer, A.N., Kanzok, B., Stoop, R., Angres, B., 2006. Cell adhesion profiling using extracellular matrix protein microarrays. Biotechniques 40, 523–531.

Langenau, D.M., Zon, L.I., 2005. The zebrafish: A new model of T-cell and thymic development. Nat Rev Immunol 5, 307–317.

Lassmann, H., 1998. Neuropathology in multiple sclerosis: New concepts. Mult Scler 4, 93–98.

Lassmann, H., Bruck, W., Lucchinetti, C., 2001. Heterogeneity of multiple sclerosis pathogenesis: Implications for diagnosis and therapy. Trends Mol Med 7, 115–121.

Li, Q.Z., Xie, C., Wu, T., et al., 2005. Identification of autoantibody clusters that best predict lupus disease activity using glomerular proteome arrays. J Clin Invest 115, 3428–3439.

Lincoln, M.R., Montpetit, A., Cader, M.Z., et al., 2005. A predominant role for the HLA class II region in the association of the MHC region with multiple sclerosis. Nat Genet 37, 1108–1112. (Epub.).

Lindberg, R.L., De Groot, C.J., Certa, U., et al., 2004. Multiple sclerosis as a generalized CNS disease – comparative microarray analysis of normal appearing white matter and lesions in secondary progressive MS. J Neuroimmunol 152, 154–167.

Lock, C., Hermans, G., Pedotti, R., et al., 2002. Gene-microarray analysis of multiple sclerosis lesions yields new targets validated in autoimmune encephalomyelitis. Nat Med 8, 500–508.

Lowe, C.E., Cooper, J.D., Brusko, T., et al., 2007. Large-scale genetic fine mapping and genotype-phenotype associations implicate polymorphism in the IL2RA region in type 1 diabetes. Nat Genet 39, 1074–1082. (Epub.).

Lundmark, F., Duvefelt, K., Iacobaeus, E., et al., 2007. Variation in interleukin 7 receptor alpha chain (IL7R) influences risk of multiple sclerosis. Nat Genet 39, 1108–1113.

Luster, A.D., Alon, R., Von Andrian, U.H., 2005. Immune cell migration in inflammation: Present and future therapeutic targets. Nat Immunol 6, 1182–1190.

Mackay, R.P., 1950. The familial occurrence of multiple sclerosis and its implications. Ann Intern Med 33, 298–320.

Malek, T.R., 2008. The biology of interleukin-2. Annu Rev Immunol 26, 453–479.

Mazzucchelli, R., Durum, S.K., 2007. Interleukin-7 receptor expression: Intelligent design. Nat Rev Immunol 7, 144–154.

Miller, D.H., 2004. Biomarkers and surrogate outcomes in neurodegenerative disease: Lessons from multiple sclerosis. NeuroRx 1, 284–294.

Moriguchi, T., Hamada, M., Morito, N., et al., 2006. MafB is essential for renal development and F4/80 expression in macrophages. Mol Cell Biol 26, 5715.

Newcombe, J., Eriksson, B., Ottervald, J., Yang, Y., Franzen, B., 2005. Extraction and proteomic analysis of proteins from normal and multiple sclerosis postmortem brain. J Chromatogr B Analyt Technol Biomed Life Sci 815, 191–202.

Perfetto, S.P., Chattopadhyay, P.K., Roederer, M., 2004. Seventeen-colour flow cytometry: Unravelling the immune system. Nat Rev Immunol 4, 648–655.

Polman, C.H., O'Connor, P.W., Havrdova, E., et al., 2006. A randomized, placebo-controlled trial of natalizumab for relapsing multiple sclerosis. N Engl J Med 354, 899–910.

Quintana, F.J., Basso, A.S., Iglesias, A.H., et al., 2008a. Control of T(reg) and T(H)17 cell differentiation by the aryl hydrocarbon receptor. Nature 453, 65–71.

Quintana, F.J., Farez, M.F., Viglietta, V., et al., 2008b. Antigen microarrays identify unique serum autoantibody signatures in clinical and pathologic subtypes of multiple sclerosis. Proc Natl Acad Sci USA 105, 18,889–18,894.

Quintana, F.J., Hagedorn, P.H., Elizur, G., Merbl, Y., Domany, E., Cohen, I.R., 2004. Functional immunomics: Microarray analysis of IgG autoantibody repertoires predicts the future response of mice to induced diabetes. Proc Natl Acad Sci USA 101 (Suppl. 2), 14,615–14,621.

Reich, D., Patterson, N., De Jager, P.L., et al., 2005. A whole-genome admixture scan finds a candidate locus for multiple sclerosis susceptibility. Nat Genet 37, 1113–1118. (Epub.).

Robinson, W.H., Digennaro, C., Hueber, W., et al., 2002a. Autoantigen microarrays for multiplex characterization of autoantibody responses. Nat Med 8, 295–301.

Robinson, W.H., Fontoura, P., Lee, B.J., et al., 2003. Protein microarrays guide tolerizing DNA vaccine treatment of autoimmune encephalomyelitis. Nat Biotechnol 21, 1033–1039.

Robinson, W.H., Garren, H., Utz, P.J., Steinman, L., 2002b. Millennium Award. Proteomics for the development of DNA tolerizing vaccines to treat autoimmune disease. Clin Immunol 103, 7–12.

Rose, J.W., Watt, H.E., White, A.T., Carlson, N.G., 2004. Treatment of multiple sclerosis with an anti-interleukin-2 receptor monoclonal antibody. Ann Neurol 56, 864–867.

Sachidanandam, R., Weissman, D., Schmidt, S.C., et al., 2001. A map of human genome sequence variation containing 1.42 million single nucleotide polymorphisms. Nature 409, 928–933.

Sawcer, S., Ban, M., Maranian, M., et al., 2005. A high-density screen for linkage in multiple sclerosis. Am J Hum Genet 77, 454–467. (Epub.).

Shendure, J., Mitra, R.D., Varma, C., Church, G.M., 2004. Advanced sequencing technologies: Methods and goals. Nat Rev Genet 5, 335–344.

Smith, M.W., O'Brien, S.J., 2005. Mapping by admixture linkage disequilibrium: Advances, limitations and guidelines. Nat Rev Genet 6, 623–632.

Soen, Y., Chen, D.S., Kraft, D.L., Davis, M.M., Brown, P.O., 2003. Detection and characterization of cellular immune responses using peptide-MHC microarrays. PLoS Biol 1, E65.

Sturzebecher, S., Wandinger, K.P., Rosenwald, A., et al., 2003. Expression profiling identifies responder and non-responder phenotypes to interferon-beta in multiple sclerosis. Brain 126, 1419–1429.

Sundstrom, P., Juto, P., Wadell, G., et al., 2004. An altered immune response to Epstein-Barr virus in multiple sclerosis: A prospective study. Neurology 62, 2277–2282.

Suzuki, H., Kundig, T.M., Furlonger, C., et al., 1995. Deregulated T cell activation and autoimmunity in mice lacking interleukin-2 receptor beta. Science 268, 1472–1476.

Tajouri, L., Mellick, A.S., Tourtellotte, A., Nagra, R.M., Griffiths, L.R., 2005. An examination of MS candidate genes identified as differentially regulated in multiple sclerosis plaque tissue, using absolute and comparative real-time Q-PCR analysis. Brain Res Brain Res Protoc 15, 79–91.

Teunissen, C.E., Dijkstra, C., Polman, C., 2005. Biological markers in CSF and blood for axonal degeneration in multiple sclerosis. Lancet Neurol 4, 32–41.

Trede, N.S., Langenau, D.M., Traver, D., Look, A.T., Zon, L.I., 2004. The use of zebrafish to understand immunity. Immunity 20, 367–379.

Viglietta, V., Baecher-Allan, C., Weiner, H.L., Hafler, D.A., 2004. Loss of functional suppression by CD4+ CD25+ regulatory T cells in patients with multiple sclerosis. J Exp Med 199, 971–979.

Vogt, M.H., Floris, S., Killestein, J., et al., 2004. Osteopontin levels and increased disease activity in relapsing-remitting multiple sclerosis patients. J Neuroimmunol 155, 155–160.

Weinstock-Guttman, B., Badgett, D., Patrick, K., et al., 2003. Genomic effects of IFN-beta in multiple sclerosis patients. J Immunol 171, 2694–2702.

Whitney, A.R., Diehn, M., Popper, S.J., et al., 2003. Individuality and variation in gene expression patterns in human blood. Proc Natl Acad Sci USA 100, 1896–1901.

Yednock, T.A., Cannon, C., Fritz, L.C., Sanchez-Madrid, F., Steinman, L., Karin, N., 1992. Prevention of experimental autoimmune encephalomyelitis by antibodies against alpha 4 beta 1 integrin. Nature 356, 63–66.

Zephir, H., Almeras, L., El Behi, M., et al., 2006. Diversified serum IgG response involving non-myelin CNS proteins during experimental autoimmune encephalomyelitis. J Neuroimmunol 179, 53–64.

CHAPTER

Inflammatory Bowel Disease

Eleanora Anna Margaretha Festen, Cisca Wijmenga, and Rinse K Weersma

INTRODUCTION

The inflammatory bowel diseases (IBDs) are relapsing inflammatory diseases of the gastrointestinal tract, consisting mainly of Crohn's disease (CD) and ulcerative colitis (UC). A small part of the IBDs consists of microscopic colitis (collagenic and lymphocytic). These forms of IBD have hardly been considered in genetic research and therefore fall outside the scope of this chapter.

EPIDEMIOLOGY

The prevalence of IBD varies greatly among populations, with the highest prevalence seen in the western world and a much lower rate in developing countries and Asia (Logan and Bowlus, 2010). UC is more common than CD, with a prevalence of approximately 235/100,000 in North America, Western Europe, and the UK, and a prevalence of approximately 20/100,000 in developing countries. The prevalence of CD is approximately 200/100,000 in North America and 10/100,000 in developing countries (Logan and Bowlus, 2010). Whereas the incidence and prevalence of CD and UC in the western world is stabilizing, the rates for both diseases are still rising in southern Europe, in developing countries, and in Asia. For several ethnic groups it has been shown that after migration to a different geographical area, the prevalence of IBD in the ethnic group becomes equal to the local level of prevalence within one to two generations. This phenomenon, along with the rising level of IBD in developing countries, implies that factors in the western lifestyle and environment have an effect on IBD risk (Logan and Bowlus, 2010).

DISEASE PHENOTYPES

Crohn's Disease

CD is typically diagnosed in the second or third decade of life, while an earlier age of onset is correlated with a more severe disease phenotype (Sauer and Kugathasan, 2009). Inflammatory pathways involved in CD are the T-helper 1 (Th1) pathway, effectuating cell-mediated immunity through tumor necrosis factor-beta (TNF-β) and interferon-gamma (IFN-γ), and the T-helper 17 (Th17) pathway, recruiting neutrophils to the site of inflammation (Baumgart and Carding, 2007). Inflammation can occur anywhere in the digestive tract, from the oral cavity to the anus, but it mainly affects the terminal ileum. Stretches of inflamed mucosa are typically sharply demarcated and alternate with areas of uninvolved mucosa (Baumgart and Sandborn, 2007). Initially lesions resemble small aphthous ulcers, with neutrophil infiltration. Subsequent inflammation in CD is often transmural, at times causing abscesses, strictures, and fistulas from the bowel to skin, bladder, vagina, or other bowel segments. The presentation of CD can range from abdominal pain and weight loss to fever and bloody diarrhea, and largely depends on disease localization and severity (Baumgart and Sandborn, 2007). The preferential therapy for CD is drug therapy, and initially the disease is

Genomic and Personalized Medicine, 2nd edition
by Ginsburg & Willard. DOI: http://dx.doi.org/10.1016/B978-0-12-382227-7.00073-2

treated with corticosteroids. If corticosteroid therapy is insufficient, immunosuppressive therapy with thiopurines should be considered. CD that is resistant to therapy with both corticosteroids and thiopurines can be treated with anti-TNF-alpha (TNF-α) drugs, which neutralize TNF-α by preventing it from binding to the TNF-α receptor (Baumgart and Sandborn, 2007; Bernstein et al., 2010; Colombel et al., 2010). In the majority of CD patients, the disease eventually progresses from a purely inflammatory disease, which can be treated with drug therapy, to a more complicated disease with strictures, abscesses or fistulas, which require surgical intervention (Baumgart and Sandborn, 2007; Bernstein et al., 2010). Due to the extensive nature of CD, surgical intervention can never cure the disease and the principal treatment will always be drug therapy (Bernstein et al., 2010).

Ulcerative Colitis

The characteristic age of onset of UC is between the second and third decade of life, but the condition may arise in both younger, or considerably older individuals. The main immune pathway underlying inflammation in UC is the T-helper 2 (Th2) pathway and, to a lesser extent, the Th17 pathway (Baumgart and Carding, 2007). UC is confined to the colon, with occasionally limited involvement of the terminal ileum called "backwash ileitis" in patients with severe pancolitis (Baumgart and Sandborn, 2007). Inflammation is continuous, extending from the rectum, and typically only involves the mucosa and submucosa. Affected mucosa may show slight reddening and easy bleeding in mild disease to extensive ulceration with pseudopolyps of regenerating mucosa in severe disease. Most UC patients present with bloody mucoid diarrhea and lower abdominal pain. Rarely, the initial attack of UC leads to acute cessation of bowel function and toxic dilatation and necrosis of the colon (toxic megacolon) (Baumgart and Sandborn, 2007). An important complication of UC is malignant dysplasia of the colon mucosa as a result of long-term inflammation (Baumgart and Sandborn, 2007). It is estimated that 10% of patients with pancolitis will progress to having a carcinoma (Bernstein et al., 2001). Currently UC patients are regularly checked for malignant dysplasia by colonoscopy, as yet there are no methods (such as gene-expression tests) to predict malignant transformation. UC confined to the rectum can be managed by local therapy with 5-aminosalicylic acid suppositories, if necessary in combination with local corticosteroids (Baumgart and Sandborn, 2007; Bernstein et al., 2010). For patients with inflammation extending further proximally, enemas should be substituted for the suppositories. In left-sided UC or pancolitis, or when local therapy is insufficient in distal disease, oral therapy with 5-aminosalicylic acid should be started. If this yields insufficient results, oral corticosteroids can be added to the regime. Thiopurines can be added to prevent relapses (Baumgart and Sandborn, 2007; Bernstein et al., 2010). Fulminant relapsing disease, disease resistance to drug therapy and the occurrence of dysplasia are indications for surgical intervention in UC. Such intervention is usually a two-step procedure, starting with a proctocolectomy and ileostomy, followed by reestablishing the continuity of the gastrointestinal tract by anastomosing and creating an ileal pouch which is connected to the anus (Caprilli et al., 2007; Vella et al., 2007). Since UC is limited to the colon, a complete proctocolectomy will cure the disease.

Unclassified Inflammatory Bowel Disease

In patients with disease limited to the colon, it is not always possible to make a definite diagnosis of either CD or UC. When a patient with chronic colitis does not show clear clinical, endoscopic or histological features of CD or UC, a diagnosis of unclassified inflammatory bowel disease is made. Gene-expression studies in both CD and UC have been performed showing expression of many candidate genes and a large range of genes in inflammatory pathways (Noble et al., 2008; Noble et al., 2010). However, gene-expression tests are as yet not specific enough to differentiate between CD and UC in intermediate lesions. In many IBD-U patients, progression of the disease over time leads to a more definite diagnosis of their disease. However, in some patients the diagnosis remains unclear, which can significantly complicate treatment.

PATHOGENESIS AND RISK FACTORS

It is hypothesized that inflammation in IBD arises as an aberrant reaction to the normal commensal microflora in genetically predisposed individuals. In the normal state, the intestinal immune system is tolerant to its commensal flora (Horwitz, 2007; Nell et al., 2010). In both CD and UC, the innate and adaptive immune response and the epithelial barrier play an important role in the loss of this tolerance (McGuckin et al., 2009; Su et al., 2009).

CD and UC are complex diseases, meaning that both environmental and genetic factors play a role in disease risk. The most well-known environmental factors influencing IBD risk are smoking and prior appendectomy (Danese et al., 2004; Molodecky and Kaplan, 2010). Prior appendectomy has a protective effect against UC. Smoking has a remarkably paradoxical effect on CD and UC – it is a strong risk factor for the development of CD and has a strong negative influence on the course of the disease. In contrast, stopping smoking is a risk factor for the development of UC (Danese et al., 2004; Molodecky and Kaplan, 2010).

As the inflammation in IBD is thought to be a reaction to the microflora of the bowel, analysis of the microflora of IBD patients and comparison to healthy controls is important for gaining insight into the pathomechanism of the disease. It is known that the gut microflora can influence the homeostasis of the gut mucosa by, for example, influencing adhesion to the epithelium or localized pH. Sequencing technology will be instrumental in analysis of the role of the microbiota living in the gut in promoting and protecting against IBD. Reference gene sets for the human gut microbiome have been generated by collaborative networks and are already being used to identify associations between individual bacterial species and diseases, including IBD (Ballal et al., 2011; Renz et al., 2011).

The importance of genetic risk factors in IBD is reflected in a higher concordance of disease in monozygotic twins than in dizygotic twins and in the increased risk for IBD in first-degree relatives of IBD patients (Brant, 2010; Loftus, 2004). In CD, the concordance of disease for monozygotic twins is 56% and for dizygotic twins 4%. The twin concordance for UC is 18% for monozygotic twins and 5% for dizygotic twins (Brant, 2010; Loftus, 2004; Orholm et al., 2000). These data suggest that the influence of genetic susceptibility is more prominent in CD than in UC. In contrast to diseases with a Mendelian inheritance (in which a mutation in a single risk gene causes disease), the genetic risk in IBD is conveyed by multiple genes, each of which contributes a modest effect to the individual's risk of disease (Brant, 2010).

WHAT WAS KNOWN BEFORE GENOME-WIDE ASSOCIATION STUDIES

It is widely recognized that understanding the genetic background of IBD will lead to a better understanding of the diseases' pathogenesis, and will thereby contribute to the development of new therapies for IBD. The first studies to uncover part of the genetic background of IBD were genome-wide linkage studies performed in families with a high prevalence of the disease (Hugot et al., 2001; Ogura et al., 2001). Such studies identified nine genetic loci, called *IBD1–IBD9*, that segregate with IBD in high-incidence families, and these may contain genetic variants that convey a substantial risk of disease (Schreiber et al., 2005). *IBD1* is the most consistently replicated locus, and it was found to contain the strongest CD risk gene: *NOD2* (also known as *CARD15*) (Abreu et al., 2002; Akolkar et al., 2001; Hampe et al., 2001; Hugot et al., 2001; Lesage et al., 2002; Ogura et al., 2001; Vermeire et al., 2002). NOD2 is an intracellular pattern recognition receptor that, after activation by bacterial peptidoglycan or muramyl dipeptide (MDP), activates the NF-κB pathway and triggers the innate immune system (McGovern et al., 2001; Van Heel et al., 2001). Three *NOD2* variants (Arg702Trp, Gly908Arg, and leu1007fsinsC), all located in or near the leucine-rich repeat sequence in the *NOD2* gene, were identified as conferring risk of CD (Hugot et al., 2001; Ogura et al., 2001). Heterozygotes for any of these *NOD2* variants have an odds ratio (OR) of 2.4 for CD, whereas homozygotes and compound heterozygotes for any of the three variants have an OR of 17.1. Despite the high ORs for the *NOD2* risk variants, these variants are neither sufficient in themselves nor necessary to cause CD (Hugot et al., 2001; Ogura et al., 2001).

Conclusive risk genes were not identified from any of the other IBD susceptibility loci. *IBD5*, a risk locus that was also replicated in case-control studies of unrelated individuals, contains several possible candidate genes: *SLC22A4* (also known

as *OCTN1*) and *SLC22A5* (or *OCTN2*), both encoding organic cation transporters (Rioux et al., 2001). Studies trying to identify the causal variants in this locus have not yet been successful (Fisher et al., 2006; Silverberg et al., 2007). *IBD3* is located on 6p21 and encompasses the human leukocyte antigen (HLA)-region (Duerr et al., 2000). The association of CD with certain HLA subtypes seems to be highly variable between populations (Kawasaki et al., 2000; Lappalainen et al., 2008; Silverberg et al., 2003; Trachtenberg et al., 2000). Such associations are more consistent for UC: the *HLA-DRB1*0103* haplotype seems to be the strongest HLA risk factor for UC in Caucasians (Aizawa et al., 2009; Fisher et al., 2008; Franke et al., 2008; Lappalainen et al., 2008; Matsumura et al., 2008; Silverberg et al., 2003; Silverberg et al., 2009).

Results from genome-wide linkage studies combined with knowledge about the pathogenesis of IBD prompted researchers to select candidate genes to study for genetic association with IBD (Hirschhorn and Daly, 2005). A large number of these studies have been performed with functional candidate genes involved in the mucosal barrier function, innate immune pathways and adaptive immune pathways. However, consistent replication of the results of candidate gene studies has often proved to be a problem. Inconsistent association results have been obtained for variants in *TLR-4*, *TLR-5*, *TLR-9*, *NOD1*, *ICAM-1*, *MYOIXB*, *MDR1*, and *DLG5* with IBD (Arnott et al., 2004; Bodegraven et al., 2006; Browning et al., 2008; Cario and Podolsky, 2000; Croucher et al., 2003; Hong et al., 2007; Stoll et al., 2004; Torok et al., 2004; Van Schwab et al., 2003; Yang et al., 1995; Zouali et al., 2003). More robust associations with IBD have been reported for *IRF5*, *CARD9*, and *IL18RAP*, all of which are players in the innate immune system (Dideberg et al., 2007; Wang et al., 2010; Zhernakova et al., 2008; Zhernakova et al., 2008). IRF5 is a transcription factor that influences the expression of type I interferon genes and induces pro-inflammatory cytokines through the toll-like-receptor pathway (Krausgruber et al., 2010). CARD9 is part of a protein complex that forms a crucial connection in the process of stimulating innate immune signaling by intracellular and extracellular pathogens (Colonna, 2007). IL18RAP encodes the beta subunit of the IL-18 receptor. IL-18 is mainly produced by antigen-presenting cells, and it stimulates the production of interferon-gamma and several pro-inflammatory cytokines (Cheung et al., 2005).

GENOME-WIDE ASSOCIATION STUDY FINDINGS

Despite extensive research efforts made in genome-wide linkage studies and candidate gene association studies, they have often failed to deliver definitive results. In genome-wide association studies (GWAS), large numbers of cases and controls are genotyped and compared for thousands of single nucleotide polymorphisms (SNPs) distributed over the whole genome. They were developed to overcome the issues that restrict the effectiveness of both candidate gene and linkage studies (Hirschhorn and Daly, 2005). The fact that GWAS are hypothesis-free means that, unlike candidate gene association studies, the results are not limited to what is known about the pathophysiology and mechanisms of the disease. GWAS are especially suitable for finding common risk variants with a small or modest effect that are likely to influence complex diseases. This provides a major advantage over linkage studies, which are more suitable for finding rare risk variants with a large effect (Hirschhorn and Daly, 2005).

CD is one of the first diseases in which GWAS have been performed, and with the most success (Craddock et al., 2010; Duerr et al., 2006; Franke et al., 2007; Franke et al., 2008; Hampe et al., 2007; Libioulle et al., 2007; Parkes et al., 2007; Rioux et al., 2007; The Wellcome Trust Case Control Consortium, 2007; Yamazaki et al., 2005) (see Table 73.1). Because of the relatively large influence of genetic factors on disease risk, CD is a good candidate for testing the GWAS method. The identification of the *NOD2* and *IBD5* loci using GWAS provided an important proof-of-principle for the application of GWAS in complex diseases (Franke et al., 2007; The Wellcome Trust Case Control Consortium, 2007). To date, nine GWAS have been performed for CD, and have identified many novel CD risk loci (Craddock et al., 2010; Duerr et al., 2006; Franke et al., 2007; Franke et al., 2008; Hampe et al., 2007; Libioulle et al., 2007; Parkes et al., 2007; Rioux et al., 2007; The Wellcome Trust Case Control Consortium, 2007; Yamazaki et al., 2005). To increase the power of GWAS analysis, several CD GWAS datasets were combined in a meta-analysis of several thousand CD cases and controls (Barrett et al., 2008). While 10 years of linkage and candidate gene studies only yielded three consistently replicated CD risk loci, three years of GWAS increased the number of replicated CD risk loci to more than 30, and current studies are expected to increase that number to more than 70. All of these risk loci are of small effect, with ORs usually <1.5 (with the exception of *NOD2*). Together, these risk factors explain approximately 20% of the genetic risk for CD (Barrett et al., 2008). The identification of two CD risk loci containing *IRGM* and *ATG16L1*, both genes involved in cellular autophagy, pointed for the first time to a role for autophagy in CD pathogenesis (Hampe et al., 2007; Parkes et al., 2007). Furthermore, GWAS showed the importance of the intestinal microflora and the IL23 pathway for CD pathogenesis (Duerr et al., 2006). Other new CD risk loci, like *PTPN2*, indicate shared pathophysiology and disease mechanisms between CD and other immune-related diseases (Barrett et al., 2008; The Wellcome Trust Case Control Consortium, 2007; Todd et al., 2007).

After the first CD GWAS, it took more than a year until the first UC GWAS was performed (Fisher et al., 2008). In the meantime, several groups identified shared UC and CD risk genes by testing the newly found CD risk loci in UC (Fisher et al., 2008; Franke et al., 2008). The first GWAS in UC was a low-coverage study including only non-synonymous SNPs; it identified one risk locus, *ECM1*, that had not been identified in CD GWAS (Fisher et al., 2008). The four subsequent UC GWAS showed that there is a large genetic overlap between UC and CD (Franke et al., 2008; Franke et al., 2010; McGovern et al., 2010; Silverberg et al., 2009) (see Table 73.2). Loci such as *IL23R*, *NKX2-3*, *13q14*, *JAK2*, *MST1*, and *KIF21B*, which were first discovered as risk loci for CD in GWAS, were also found to be associated with UC. In addition to the many similarities, the GWAS also revealed marked differences between the two diseases. The strongest known CD-associated locus, *NOD2*, is not associated with UC, nor are the loci containing the autophagy genes *ATG16L1* and *IRGM*. And while no consistent signal in the major histocompatibility complex (MHC), region has been observed for CD, the association signal in this region is consistent between UC GWAS, probably as a result of association with the *HLA-DRB1* subclass (Franke et al., 2008; Franke et al., 2008; Franke et al., 2010; McGovern et al., 2010; Silverberg et al., 2009). The four UC GWAS have identified several additional UC risk loci that point to UC-specific pathobiology. For example, three loci containing the cell–cell adhesion genes *CDH1*, *HNF4α*, and *LAMB1* suggest that epithelial integrity is a major factor in UC pathogenesis (Franke et al., 2010; McGovern et al., 2010). The *IL10* gene, the main gene in the UC-associated locus on chromosome 1q32, encodes a protein with anti-inflammatory properties (Franke et al., 2008). This protein, interleukin 10, is mainly expressed by monocytes and Th2 cells, and can inhibit the synthesis of pro-inflammatory cytokines by macrophages and Th1 cells. The association of this locus with UC suggests that defects in the anti-inflammatory pathway play a role in its pathogenesis (Louis et al., 2009).

GWAS have revealed genetic overlap between UC and CD, probably consistent with an overlap in their pathogenesis (see Table 73.3). Even with 15 GWAS now published in IBD, the number of shared risk loci keeps rising (Imielinski et al., 2009; Kugathasan et al., 2008). Many risk loci discovered for one of the diseases have a relatively small effect on the risk for the other disease. Now that GWAS are increasing in number and size, the power to discover risk loci with a relatively small effect is also increasing. Interestingly, GWAS in other immune-related diseases such as celiac disease, rheumatoid arthritis, and type 1 diabetes have also discovered genes that are associated with CD and/or UC (Barrett et al., 2009; Dubois et al., 2010; Hunt et al., 2008; Lettre and Rioux, 2008; Smyth et al., 2008; The Wellcome Trust Case Control Consortium, 2007; Wang et al., 2010; Zhernakova et al., 2007; Zhernakova et al., 2009;). This pleiotropy suggests that there is a broad sharing of pathobiology between different immune-related diseases (Lettre and Rioux, 2008; Zhernakova et al., 2009). In fact, functional studies of the pathology underlying the different immune-related diseases had previously shown such sharing, and these findings are being recognized, since a range of immune-related diseases are now being treated with only a small number of anti-inflammatory drugs. Our increasing knowledge of the disease-specific and pleiotropic pathobiology should contribute to the development of both disease-specific and new, broadly applicable therapies (Melgar and Shanahan, 2010).

| TABLE 73.1 | Risk loci for Crohn's disease |

Number	Position (chr)	Gene	Gene function	Odds ratio	Reference
1	10q21.2	Intergenic		0.39–0.81	The Wellcome Trust Case Control Consortium, 2007; Rioux et al., 2007; Barrett et al., 2008
2	16q12.1	NOD2	Antigen recognition and activation of innate immune system	1.29–1.71	The Wellcome Trust Case Control Consortium, 2007; Duerr et al., 2006; Libioulle et al., 2007; Rioux et al., 2007; Franke et al., 2007; Barrett et al., 2008; Kugathasan et al., 2008; Raelson et al., 2007
3	18p11.21	PTPN2	T cell protein tyrosine phosphatase, a key negative regulator of inflammation	1.15–1.35	The Wellcome Trust Case Control Consortium, 2007; Parkes et al., 2007; Barrett et al., 2008
4	1q23.3	ITLN1	Epithelial integrity of enetrocyte brush border	1.14–1.59	Barrett et al., 2008
5	1q24.3	Intergenic		1.14–1.19	Parkes et al., 2007; Barrett et al., 2008
6	21q22.2	Intergenic		1.15–1.41	Parkes et al., 2007; Kugathasan et al., 2008
7	2q37.1	ATG16L1	Antigen autophagy and innate immune activation	1.19–1.45	The Wellcome Trust Case Control Consortium, 2007; Rioux et al., 2007; Barrett et al., 2008
8	5p13.1	PTGER4	Activation of T cell signaling	1.32–1.54	The Wellcome Trust Case Control Consortium, 2007; Libioulle et al., 2007; Franke et al., 2007; Barrett et al., 2008
9	5q31.1	Intergenic		1.23–1.55	The Wellcome Trust Case Control Consortium, 2007; Barrett et al., 2008
10	5q33.1	IRGM	Antigen autophagy and innate immune activation	1.33–1.54	The Wellcome Trust Case Control Consortium, 2007; Parkes et al., 2007; Barrett et al., 2008
11	9q32	TNFSF15	T cell effector stimulation by the innate immune system	1.21–1.22	Barrett et al., 2008; Kugathasan et al., 2008
12	21q22.3	ICOSLG	Co-stimulatory signal for T and B cell proliferation and differentiation	1.13	Barrett et al., 2008
13	11q13.5	C11orf30	–	1.16	Barrett et al., 2008
14	6q27	CCR6	Chemokine receptor for memory T cells	1.21–1.37	Barrett et al., 2008
15	1q31.2	Intergenic		1.47	Parkes et al., 2007
16	7p12.2	Intergenic		1.2	Barrett et al., 2008
17	8q24.13	Intergenic		1.08	Barrett et al., 2008
18	12q12	**MUC19**, LRRK2	Mucus glycoprotein, protective mucous layer of bowel	1.54–1.74	Barrett et al., 2008
19	11p15.1	NELL1	Chondro- and osteogenesis	1.30	Franke et al., 2007
20	10p15.1	Intergenic		1.11–1.16	The Wellcome Trust Case Control Consortium, 2007

(continued)

Number	Position (chr)	Gene	Gene function	Odds ratio	Reference
TABLE 73.1 (Continued)					
21	7q36.1	Intergenic		1.38	The Wellcome Trust Case Control Consortium, 2007
22	17q12	ORMDL3	Protein folding and unfolded protein response	1.2	Barrett et al., 2008
23	1p13.2	PTPN22	T cell receptor signaling pathway	1.31	Barrett et al., 2008
24	1q32.1	Intergenic		1.12	Barrett et al., 2008

This table contains all risk loci for Crohn's disease. The function of the main candidate gene (indicated in bold) is shown.

Number	Position (chr)	Gene	Gene function	Odds ratio	Reference
TABLE 73.2 Risk loci for ulcerative colitis					
1	12q15	**IL26**, IFNG, IL22	Increase of IL10 secretion and subsequent anti-inflammatory effect	1.16–1.54	Silverberg et al., 2009
2	1p36.13	OTUD3, PLA2G2E	Unknown	1.17–1.41	Franke et al., 2008; Silverberg et al., 2009; Barrett et al., 2009
3	1q32.1	IL10	Anti-inflammatory from Th2 on Th1 and monocytes	1.19–1.46	Franke et al., 2008; Franke et al., 2010; Barrett et al., 2009
4	20q13.12	HNF4A	Epithelial integrity	1.20	Barrett et al., 2009
5	6p21.32	**HLA-DRA**, BTNL2	Antigen recognition and presentation	1.40–1.92	Franke et al., 2008; Silverberg et al., 2009; Franke et al., 2010; Kugathasan et al., 2008; Barrett et al., 2009; Asano et al., 2009
6	7q31.1	**LAMB1**, SLC26A3, DLD	Epithelial integrity	1.09–1.32	Silverberg et al., 2009; Barrett et al., 2009; Asano et al., 2009
7	1p36.12	Intergenic		1.10	Barrett et al., 2009
8	16q22.1	CDH1	Epithelial integrity	1.14	Barrett et al., 2009
9	22q13.33	IL17REL	Th17 pro-inflammatory pathway	1.11	Franke et al., 2010
10	9q21.32	Intergenic		1.20	Silverberg et al., 2009
11	7q22.1	**SMURF1**, KPNA7	Smad ubiquitination regulatory factor 1	1.56	Franke et al., 2010
12	13q12.13	USP12	Deubiquitinating enzyme	1.35	Asano et al., 2009
13	1p36.13	RNF186	Unknown	1.17–1.41	Silverberg et al., 2009
14	1q23.3	FCGR2A	Uptake and phagocytosis of IgG-coated antigens	1.59	Asano et al., 2009
15	1q32.1	KIF21B	Transport of cellular components along axonal and dendritic microtubules	1.23–1.46	Barrett et al., 2009

This table contains all risk loci for ulcerative colitis. The function of the main candidate gene (indicated in bold) is shown.

TABLE 73.3	Shared risk loci for Crohn's disease and ulcerative colitis				
Number	**Position (chr)**	**Gene**	**Gene function**	**Odds ratio**	**Reference**
1	10q24.2	*NKX2-3*	T and B cell proliferation	1.18–1.22	The Wellcome Trust Case Control Consortium, 2007; Parkes et al., 2007; Barrett et al., 2008; Franke et al., 2010; Barrett et al., 2009
2	13q14.11	Intergenic		1.10–1.25	Barrett et al., 2008; Barrett et al., 2009
3	1p31.3	*IL23R*	Th1 and Th17 pro-inflammatory pathways	1.27–2.92	Silverberg et al., 2007; The Wellcome Trust Case Control Consortium, 2007; Duerr et al., 2006; Libioulle et al., 2007; Rioux et al., 2007; Barrett et al., 2008; Barrett et al., 2008; Kugathasan et al., 2008; Barrett et al., 2009; Raelson et al., 2007
4	3p21.31	**MST1**, BSN	Anti-inflammatory pathway macrophages	1.09–1.20	The Wellcome Trust Case Control Consortium, 2007; Parkes et al., 2007; Barrett et al., 2008; Barrett et al., 2009
5	6p21.33	*HLA*	Antigen recognition and presentation	1.40–1.76	The Wellcome Trust Case Control Consortium, 2007; Barrett et al., 2008; Asano et al., 2009
6	9p24.1	*JAK2*	Th1 and Th17 pro-inflammatory pathways	1.12–1.34	Barrett et al., 2008; Asano et al., 2009
7	5q33.3	*IL12B*	Th1 and Th17 pro-inflammatory pathways	1.11–1.45	Parkes et al., 2007; Barrett et al., 2008; Anderson et al., 2009
8	17q21.2	*STAT3*	Th1 and Th17 pro-inflammatory pathways	1.18	Barrett et al., 2008; Anderson et al., 2009
9	6p22.3	*CDKAL1*	unknown	1.21	Barrett et al., 2008; Anderson et al., 2009
10	10p11.21	Intergenic		1.16	Barrett et al., 2008
11	6p25	*LYRM4*	Mitochondrial iron-sulfur protein biosynthesis	1.10	Barrett et al., 2008; Anderson et al., 2009
12	2q11	*IL18RAP*	Stimulation of innate and adaptive immune system by antigen-presenting cells	1.09	Zhernakova et al., 2008
13	9q34.3	*CARD9*	Antigen recognition and activation of innate immune system	1.12–1.18	Zhernakova et al., 2008

This table contains all risk loci identified as shared risk loci for Crohn's disease and ulcerative colitis. The function of the main candidate gene (indicated in bold) is shown.

IMPLICATIONS OF DISEASE-ASSOCIATED LOCI TO PATHOGENESIS

The recently identified CD-specific, UC-specific, or shared IBD risk loci provide new insight into the biology of these diseases (Figure 73.1). Four observations support the hypothesis that the intestinal microflora incite the inflammatory process leading to the CD and UC phenotypes: (Logan and Bowlus, 2010) the presence of antibodies against intestinal bacteria in the blood of CD patients, (Sauer and Kugathasan, 2009) inflammation predominantly occurring in areas of the gastrointestinal tract with high numbers of bacteria, (Baumgart and

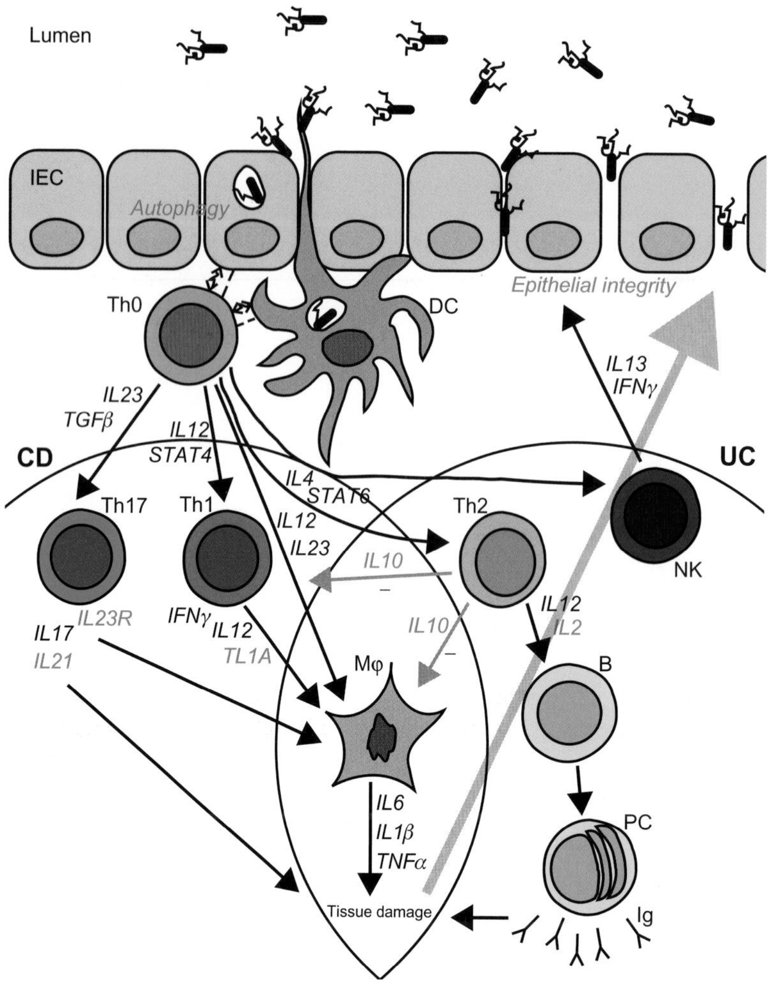

Figure 73.1 Inflammatory bowel disease pathology. This figure schematically shows the pathomechanisms for Crohn's disease (left) and ulcerative colitis (right) from the initial stimulation by the intestinal microflora (top) to the tissue damage (bottom). Elements in the pathogenesis for which genetic risk loci have been identified are indicated in red. The intestinal microflora is sampled by dendritic cells and invaded intestinal epithelial cells that, after autophagy, present antigen to undifferentiated T helper cells. These T helper cells proliferate and differentiate to T-helper 17 or T-helper 1 in Crohn's disease. Through direct effector functions of these pathways and through stimulation of pro-inflammatory activity of macrophages, these pathways lead to tissue damage. In ulcerative colitis, T-helper 2 cells stimulate B cell differentiation and natural killer cells proliferate that lead to direct (effector functions of the individual pathways) and indirect (through stimulation of macrophages) tissue damage. Lumen: the bowel lumen, IEC: intestinal epithelial cell, Tho: undifferentiated T-helper cell, DC: dendritic cell, Th#: T-helper # cell, NK: natural killer cell, Mφ: macrophage, B: B cell, PC: plasma cell, IL#: interleukin #, TGFβ: tumor growth factor beta, STAT#: signal transducer and activator of transcription #, IL23: interleukin 23 receptor, TL1A: TNF-like ligand 1A, TNFα: tumor necrosis factor alpha, Ig: immunoglobulin.

Carding, 2007) the anti-inflammatory effect of antibiotic treatment in some patients, and (Baumgart and Sandborn, 2007) the difference in the composition of the commensal microflora between IBD patients and healthy controls (Elliott et al., 2005; Fava and Danese, 2010; Israeli et al., 2005; Juste et al., 2008; Nell et al., 2010; Porter et al., 2008; Wong and Bressler, 2008). Normally a range of processes prevent the gastrointestinal immune system from mounting an inflammatory reaction to the commensal microflora or normal food ingredients, for example, a mechanism called oral tolerance (Baumgart and Carding, 2007; Shale and Ghosh, 2009). In IBD patients, the first breach in oral tolerance seems to be caused by a defect of the barrier between the microflora in the gut lumen and the immune cells in the mucosa and submucosa (McGuckin et al., 2009; Su et al., 2009). The first line of defense in this barrier is the mucus layer covering the bowel epithelium and preventing contact between microbes and the mucosa. It has been shown that this mucus layer has a different composition and is often thinner in IBD patients than in normal controls (Gersemann et al., 2009; Moehle et al., 2006). These findings, and the association between CD and a risk locus containing the *MUC19* gene, which encodes a mucus glycoprotein, seem to indicate that a defect in this mucus layer plays a role in CD pathogenesis (Phillips et al., 2010).

In UC, several genetic risk loci now point to the presence of a fundamental defect in epithelial integrity. *LAMB1*, a strong candidate gene from a newly identified UC risk locus, encodes a main collagen component of the basal membrane of the mucosa (Barrett et al., 2009). Laminin-1, the protein encoded by this gene, contributes to cell–cell adhesions and tissue survival (Barrett et al., 2009). A second UC risk locus contains *CDH1*, a gene encoding E-cadherin; a vital component of cell–cell adhesions (Barrett et al., 2009). E-cadherin is exploited by different bacteria that down-regulate the protein to mediate adhesion to epithelial cells and to disrupt epithelial integrity. Studies have shown that loss or mislocalization of E-cadherin can catalyze inflammation of the gut mucosa in mice through disruption of cell–cell contacts and increased permeability (Berx et al., 1995; Berx et al., 1996; Machado et al., 2001; Perl et al., 1998; Strumane et al., 2004; Takeno et al., 2004). A third candidate gene in UC with a role in the mucosal barrier is *HNF4A*. The protein encoded by *HNF4A* is a nuclear transcription factor important for epithelium homeostasis, cell function, and cell architecture in the liver and the gastrointestinal tract (Barrett et al., 2009). Interestingly, E-cadherin and HNF4α interact in the Wnt/β-catenin signaling pathway. Loss of HNF4α induces mislocalization of E-cadherin and subsequently an increased intestinal permeability (Cattin et al., 2009). Changes in the integrity of the mucosal barrier enable pathogens to reach and penetrate the intestinal epithelium. It has indeed been shown that IBD patients have more adherent invasive microbes than healthy controls (Darfeuille-Michaud et al., 2004; Rolhion and Darfeuille-Michaud, 2007).

In healthy controls, microbes invading the intestinal mucosa are processed by the intestinal epithelial cells and are presented to cells of the innate immune system. As an extra line of defense, cells of the innate immune system patrol the epithelium, processing and presenting microbes that come near to or cross the mucosal barrier to the rest of the immune system. For this system of monitoring and presentation, it is vital for intestinal epithelial cells and the cells of the innate immune system to be able to process microbes for presentation. This mechanism is termed autophagy, and it appears to be impaired in patients with CD. A number of CD risk genes, e.g., *IRGM*, *ATG16L1*, and *NOD2*, are involved in the mechanism of autophagy and the subsequent activation of the immune system (Kuballa et al., 2008; McCarroll et al., 2008; Rosenstiel et al., 2007). It has been observed that the CD-associated *IRGM* allele leads to a lower expression of the gene. Moreover, a lower expression of *IRGM* reduces the cells' ability for autophagic processing of *Salmonella typhimurium in vitro* (McCarroll et al., 2008). *ATG16L1* is a member of the ATG family of autophagy genes. A non-synonymous SNP (Ala281Thr) is associated with CD, with the minor threonine allele conferring some protection from disease (Hampe et al., 2007).

ATG16L1 has a role in the intestinal epithelium of mice and CD patients by selectively influencing the cell biology and regulatory properties of Paneth cells, which are cells at the base of the ileal crypt involved in mucosal immunity (Cadwell et al., 2008). Furthermore, some insight into the complex interaction between genes and environmental factors has recently been gained since it was shown that a specific interaction between ATG16L1 and a murine norovirus induces multiple features in mice resembling Crohn's disease (Cadwell et al., 2010).

In the group of CD-associated innate immune genes, *NOD2* is a special case. It is an intracellular sensor of bacterial peptidoglycan, and it activates innate inflammation through the NFκB pathway. *Nod2⁻ᐟ⁻* mice exhibit increased systemic translocation of orally administered *Listeria monocytogenes* (Kobayashi et al., 2005). After systemic administration of the pathogen, *Nod⁻ᐟ⁻* mice show similar colonization to wild-type mice, demonstrating the role of NOD2 in immune responses after oral exposure to pathogens (Kobayashi et al., 2005). The CD-associated *NOD2* mutations could contribute to disease risk both by impaired induction of tolerance to repeated stimulation by luminal microflora, and by a diminished ability to mount an inflammatory reaction to invading pathogens (Cho and Abraham, 2007). Despite 10 years of extensive research on the subject, this question still has not been answered.

While autophagy genes only seem to be associated with CD, *CARD9*, a gene with a related function in the innate immune system, is associated with both CD and UC (Zhernakova et al., 2008). The *CARD9* gene represents another major link between sensing of pathogens by intestinal epithelial cells and activation of the immune system. Like *NOD2*, the *CARD9* gene is a member of the family of pattern recognition receptors. In reaction to intracellular pathogens, *CARD9* interacts with *NOD2* in a signaling pathway aimed at activating the innate immune system (Colonna, 2007). A gene involved in diminishing the inflammatory activity of the innate immune system, *MST1*, was found to be associated with both CD and UC (Fisher et al., 2008; Goyette et al., 2008). MSP, the protein encoded by *MST1*, binds to a receptor on macrophages and limits their inflammatory activity in reaction to bacterial lipopolysaccharide (LPS) (Wang et al., 1994).

After the sensing of pathogens by antigen-presenting cells from the innate immune system, activation of the adaptive immune system is necessary for an effective immune response. Several genes that are directly involved in the communication between innate and adaptive immune cells have been found to be associated with both CD and UC. The UC-associated *FCGR2A* gene encodes an Fc receptor found on phagocytic cells of the innate immune system, ensuring the uptake and phagocytosis of IgG-coated antigens (McGovern et al., 2010). *CCR6*, a CD-associated gene, encodes a chemokine receptor expressed by certain antigen-presenting cells of the innate immune system's memory T cells (Wang et al., 2009). This interaction through the *CCR6*-encoded receptor stimulates B cell differentiation and T cell attraction in reaction to the macrophage inflammatory protein 3α (MIP3α) (Liao et al., 1999). Another gene at the border between the innate and the adaptive immune systems is *TNFSF15*, which is associated with CD and multiple other immune-related diseases (Wang et al., 2010; Yamazaki et al., 2005; Zhang et al., 2009; Zinovieva et al., 2009). The protein encoded by *TNFSF15*, TL1A, stimulates T cell effector function in reaction to sensing of pathogens by the innate immune system. Co-expression of TL1A with different co-stimulatory cytokines determines which T cell subclass is stimulated (Prehn et al., 2004; Takedatsu et al., 2008).

Several genes whose main role is in the adaptive immune response have been found to be associated with IBD. One of the IBD-associated genes involved in the adaptive immune response is *NKX2-3* (Fisher et al., 2008). The protein encoded by *NKX2-3* is involved in the localization of T and B cells in the spleen during their maturation (Tarlinton et al., 2003). Another adaptive immune gene, *ICOSLG*, is only associated with CD (Barrett et al., 2008). This gene encodes a protein that provides a co-stimulatory signal for T and B cell proliferation and differentiation. Finally, one of the first and strongest IBD risk genes identified by GWAS, *IL23R*, is essentially an adaptive immune gene. *IL23R* is expressed by T helper cells, cytotoxic T cells, and natural killer cells (Duerr et al., 2006). In particular, Th17 cells, a subclass of T helper cells, express high levels of *IL23R* (Kobayashi et al., 2008). In other subclasses of T helper cells, *IL23R* expression is down-regulated by subset-specific cytokines, whereas in Th17 cells IL23R stimulation ensures expansion and survival, making *IL23R* expression a hallmark for Th17 cells (Sutton et al., 2009). High numbers of Th17 cells are present in the blood and inflamed mucosa of CD patients (Kobayashi et al., 2008; Park et al., 2005). *IL23R* is part of a pathway that seems to have a high affinity with CD and UC, since several of its players are associated with both diseases. IL23, the substrate for IL23R, consists of two subunits, one of which is encoded by *IL12B*, a gene that is also associated with both CD and UC (Cho, 2008). *JAK2* and *STAT3*, both genes encoding proteins that effectuate signal transduction after stimulation of IL23R by IL23, have also been found to be associated with CD and UC. This underlines the importance of the IL23R pathway for IBD pathogenesis (Cho, 2008).

The fact that components of both the innate and adaptive immune system play multiple roles in the immune response means that extensive research is of vital importance before these genetic findings can be used as a basis for therapy and modifying immune pathways at any level.

RISK PREDICTION

Risk prediction in IBD could have three possible purposes: predicting an individual's chance of developing the disease, predicting the course of the disease (i.e., its progression) in an existing IBD patient, and distinguishing between CD and UC in patients with IBD-U.

Even though the use of risk prediction for a disease that cannot be prevented might be considered debatable, it was widely hoped that the recently identified IBD risk genes would provide the basis for a genetic risk test for IBD. However, since the currently known CD risk genes only account for 20% of the genetic risk for CD, and many of the risk genes for IBD are shared with other immune-related diseases, a predictive test based on these genes will have neither a high sensitivity nor specificity (Barrett et al., 2008). A more specific genetic risk prediction test can be constructed by not only using the most significantly associated risk genes, but also by taking into account the mass of markers that only show a slight association with the disease (Purcell et al., 2009). Among these borderline associated markers there will be many false-positive hits, but the unique combination of true-positive hits and their share in genetic disease risk greatly increases the sensitivity and specificity of a predictive test (Purcell et al., 2009). It has been suggested that high dietary intake of fat and meat increase the risk for IBD and that high fiber and fruit intake might decrease the risk, but no real preventative measures are known for IBD (Hou et al., 2011). For this reason, and because genetic risk prediction does not yet reach sufficient sensitivity and specificity, screening risk genes does not yet contribute to diagnostic or treatment strategies for (potential) patients.

For the second possible purpose of genetic risk prediction – predicting the presentation or progression of disease – it is necessary to know what risk variants predispose an individual to a certain disease course. Unfortunately, only a few studies have reported risk variants that influence disease presentation or progression. Such influences have been identified and replicated for the strongest CD risk locus: the *NOD2* risk variants for CD have been found to be associated with disease of the terminal ileum and a more severe disease course (Cuthbert et al., 2002; Hampe et al., 2002; Lesage et al., 2002). Detecting phenotype-specific genetic associations requires highly powered genetic studies with detailed phenotype information. It is likely that the genes associated with disease occurrence are different from those modifying disease. At the moment, it is not yet feasible to predict disease course with genetic markers.

Determining whether a patient with indeterminate colitis has either CD or UC is very important for choosing the most appropriate therapy. For CD, several disease-specific risk genes have been identified. However, only relatively few CD patients actually carry these risk variants. In addition, a considerable part of the general population without disease also carries these variants, thus making the use of a differentiating genetic test very limited. Possibly biomarkers such as serum perinuclear Anti-Neutrophil Cytoplasmic Antibodies (pANCAs), that are mainly associated with UC, and Anti-*Saccharomyces cerevisiae* antibodies (ASCA), that have particularly been associated with CD, can improve the differentiating ability of a genetic test (Oudkerk Pool et al., 1993; Oudkerk Pool et al., 1995).

Although these markers can be used in combination for some degree of differentiation between UC and CD, the presence of these types of markers has not been associated with a particular course of the disease, nor do they have enough sensitivity and specificity to help delineate non-affected people at risk of developing IBD (Yang et al., 1995). We do expect, however, that comprehensive collaborative efforts with uniform phenotyping data will contribute to the number of known genotype-phenotype associations, which would increase the predictive value of a genetic test. These new genetic markers, combined with factors such as age, gender, family history, known biomarkers and environmental risk factors, might even contribute to a composite test to predict the prognosis of newly diagnosed IBD patients.

PHARMACOGENOMICS

Compared to the current knowledge of the genetic background of IBD, current knowledge on pharmacogenetics is limited. GWAS on the subject will be necessary to bring the knowledge on pharmacogenetics to the same level. A predictive test for the effectiveness of an IBD therapy would significantly improve patient care since IBD pharmacotherapeutics are generally expensive and their side-effects can be severe.

One well-established polymorphism influencing drug response is the thiopurine methyl transferase gene (*TPMT*) in relation to the outcome of therapy with thiopurines, of which azathioprine and 6-mercaptopurine are frequently used in IBD patients. The metabolism of azathioprine or 6-mercaptopurine is rather complicated, involving various enzymatic steps with inter-individual differences in genotype and phenotype. Essentially, the therapeutic activity of thiopurines is related to the concentration of the metabolite 6-thioguanine nucleosides (6-TGN), probably mediated by the phosphorylated forms of this metabolite (Neurath et al., 2005; Poppe et al., 2006). The pivotal enzyme, TPMT, methylates 6-mercaptopurine and its metabolite 6-thioinosine-monophosphate (6-TIMP) into 6-methylmercaptopurine (6-MMP) and 6-methyl-thioinosine-monophosphate (6-MTIMP), respectively. The activity of TPMT is genetically regulated (Krynetski et al., 1996; Yates et al., 1997). Approximately 1 in 9 individuals is heterozygous for the common polymorphisms, and 1 in 300 is homozygous. Patients with one mutant (dysfunctional) *TPMT* allele have diminished TPMT activity, and patients with two non-functional mutant alleles have no TPMT activity. Low TPMT activity will result in an increased amount of azathioprine or 6-mercaptopurine being metabolized by hypozanthine-guanine phospho-ribosyl transferase (HPRT), resulting in high levels of 6-TGN. Hence, TPMT deficiency leads to a potentially life-threatening myelosuppression as 6-TGN accumulates. TPMT with higher-than-average activity leads to generation of high levels of 6-MMP, which has been associated with hepatotoxicity in one study in a specific group of children suffering from CD, and 6-MTIMP, and low levels of 6-TGN (Dubinsky et al., 2000). It is also associated with therapeutic inefficacy. Taken together, determination of the TPMT alleles may be helpful in predicting response to thiopurine therapy and in assessment of risk of myelotoxicity or hepatotoxicity. Their determination prior to treatment with thiopurines has therefore been

advocated. Nevertheless, other factors contribute to adverse events associated with thiopurines, as has been shown by retrospective analysis of more than 40 IBD patients in whom thiopurines had to be withdrawn due to myelotoxicity. The majority of them had wild-type *TPMT* alleles, although most patients with myelodepression in the first weeks of treatment were homozygous for mutant alleles (Colombel et al., 2000).

The response to glucocorticosteroids also seems to be partly genetically determined. Polymorphisms in multi-drug resistance gene-1 (*MDR-1*) are excellent candidates to affect the absorption and concentrations of *MDR-1* substrates (Farrell et al., 2000; Hoffmeyer et al., 2000). In a study of Hungarian patients, it has been suggested that the *DLG5* 113A allele, which was shown not to be associated with disease susceptibility, may confer resistance to steroids (Lakatos et al., 2006).

IBD patients homozygous for the methylenetetrahydrofolate reductase (*MTHFR*) 1298C allele are more likely to experience side effects than patients homozygous for the wild-type A allele (21.0% versus 6.3%, $P < 0.05$) when treated with the immunosuppressor methotrexate (MTX) (Herrlinger et al., 2005).

With advanced understanding of human genetic variation, it is now possible to perform GWAS in large cohorts of responders and non-responders to identify genetic risk and efficacy factors related to the effects of such drugs; this is similar to what is being done to identify disease susceptibility genes. Such research may have broad implications, since it is likely that the response to a particular drug will not be restricted to the IBD phenotype. Hence, in future, these genetic profiles may help tailor disease treatment and patient care by increasing efficacy and preventing adverse drug reactions (Egan et al., 2006).

As mentioned previously, broader understanding of IBD disease pathology will deliver novel drug targets for disease intervention. It is anticipated that genomics and proteomics technologies will assist in compound identification in drug discovery. Infliximab, a chimeric monoclonal antibody against TNF-α, is such a target, that neutralizes one of the critical inflammatory mediators in IBD. Nevertheless, the drug is only effective in 40–66% of patients after two years of treatment (Hanauer et al., 2002; Present et al., 1999; Sands et al., 2004). It is clear that drug response, in general, is partly genetically determined. Carriage of *NOD2* seems to play no role in response to infliximab (Mascheretti et al., 2002; Vermeire et al., 2002), whereas another study on CD patients who had received infliximab showed that patients homozygous for the V allele of the FcgammaRIIIa-158 polymorphism had a better biological and possibly better clinical response to infliximab (Louis et al., 2004). In addition, there has been a suggestion that the *IBD5* locus is involved in response to infliximab (Urcelay et al., 2005). Recently there has also been a report suggesting that the lack of response to infliximab can be attributed in part to certain genetic polymorphisms in two apoptosis genes, Fas ligand and caspase-9 (Hlavaty et al.,

2005). Additional studies are needed to corroborate these findings since none have been independently confirmed.

CONCLUSION

The recent surge of GWAS has, in a short period of time, greatly improved our knowledge of the genetics of CD and UC. In both CD and UC, defects in the innate immunity and T cell differentiation (in particular Th1 and Th17 responses) are involved in pathogenesis. Furthermore, in both diseases the commensal microflora also play an important role in disease pathogenesis. The identification of more than 30 genetic risk loci for CD has taught us that the process of autophagy plays an important role in CD pathogenesis. GWAS in UC have identified more than 20 UC risk loci, highlighting the role of the epithelial barrier function. The number of confirmed IBD genes is expected to rise to 100 in 2010, with large scale meta-analyses underway.

Before using these genetic findings as starting points for diagnosis, prognosis, and development of new therapies, we should use them to learn about the actual pathogenesis of CD and UC. By the time we understand how the disease works, we will be able to predict, treat, and perhaps even prevent it. The scientific community will now have to aim its efforts at translating these genetic findings into functional knowledge and clinical applications, such as risk prediction and new therapies.

REFERENCES

Abreu, M.T., Taylor, K.D., Lin, Y.C., et al., 2002. Mutations in *NOD2* are associated with fibrostenosing disease in patients with Crohn's disease. Gastroenterology 123 (3), 679–688.

Aizawa, H., Kinouchi, Y., Negoro, K., et al., 2009. HLA-B is the best candidate of susceptibility genes in HLA for Japanese ulcerative colitis. Tissue Antigens 73 (6), 569–574.

Akolkar, P.N., Gulwani-Akolkar, B., Lin, X.Y., et al., 2001. The IBD1 locus for susceptibility to Crohn's disease has a greater impact in Ashkenazi Jews with early onset disease. Am J Gastroenterol 96 (4), 1127–1132.

Anderson, C.A., Massey, D.C., Barrett, J.C., et al., 2009. Investigation of Crohn's disease risk loci in ulcerative colitis further defines their molecular relationship. Gastroenterology 136 (2), 523–529.

Arnott, I.D., Nimmo, E.R., Drummond, H.E., et al., 2004. *NOD2/CARD15*, *TLR4* and *CD14* mutations in Scottish and Irish Crohn's disease patients: Evidence for genetic heterogeneity within Europe? Genes Immun 5 (5), 417–425.

Asano, K., Matsushita, T., Umeno, J., et al., 2009. A genome-wide association study identifies three new susceptibility loci for ulcerative colitis in the Japanese population. Nat Genet 41 (12), 1325–1329.

Ballal, S.A., Gallini, C.A., Segata, N., Huttenhower, C., Garrett, W.S., 2011. Host and gut microbiota symbiotic factors: Lessons from inflammatory bowel disease and successful symbionts. Cell Microbiol 13 (4), 508–517.

Barrett, J.C., Clayton, D.G., Concannon, P., et al., 2009. Genome-wide association study and meta-analysis find that over 40 loci affect risk of type 1 diabetes. Nat Genet 41 (6), 703–707.

Barrett, J.C., Hansoul, S., Nicolae, D.L., et al., 2008. Genome-wide association defines more than 30 distinct susceptibility loci for Crohn's disease. Nat Genet 40 (8), 955–962.

Barrett, J.C., Lee, J.C., Lees, C.W., et al., 2009. Genome-wide association study of ulcerative colitis identifies three new susceptibility loci, including the HNF4A region. Nat Genet 41 (12), 1330–1334.

Baumgart, D.C., Carding, S.R., 2007. Inflammatory bowel disease: Cause and immunobiology. Lancet 369 (9573), 1627–1640.

Baumgart, D.C., Sandborn, W.J., 2007. Inflammatory bowel disease: Clinical aspects and established and evolving therapies. Lancet 369 (9573), 1641–1657.

Bernstein, C.N., Blanchard, J.F., Kliewer, E., Wajda, A., 2001. Cancer risk in patients with inflammatory bowel disease: A population-based study. Cancer 91 (4), 854–862.

Bernstein, C.N., Fried, M., Krabshuis, J.H., et al., 2010. World Gastroenterology Organization Practice Guidelines for the diagnosis and management of IBD in 2010. Inflamm Bowel Dis 16 (1), 112–124.

Berx, G., Cleton-Jansen, A.M., Nollet, F., et al., 1995. E-cadherin is a tumour/invasion suppressor gene mutated in human lobular breast cancers. EMBO J 14 (24), 6107–6115.

Berx, G., Cleton-Jansen, A.M., Strumane, K., et al., 1996. E-cadherin is inactivated in a majority of invasive human lobular breast cancers by truncation mutations throughout its extracellular domain. Oncogene 13 (9), 1919–1925.

Brant, S.R., 2010. Update on the heritability of inflammatory bowel disease: The importance of twin studies. Inflamm Bowel Dis.

Browning, B.L., Annese, V., Barclay, M.L., et al., 2008. Gender-stratified analysis of DLG5 R30Q in 4707 patients with Crohn disease and 4973 controls from 12 Caucasian cohorts. J Med Genet 45 (1), 36–42.

Cadwell, K., Liu, J.Y., Brown, S.L., et al., 2008. A key role for autophagy and the autophagy gene *Atg16l1* in mouse and human intestinal Paneth cells. Nature 456 (7219), 259–263.

Cadwell, K., Patel, K.K., Maloney, N.S., et al., 2010. Virus-plus-susceptibility gene interaction determines Crohn's disease gene *Atg16L1* phenotypes in intestine. Cell 141 (7), 1135–1145.

Caprilli, R., Viscido, A., Latella, G., 2007. Current management of severe ulcerative colitis. Nat Clin Pract Gastroenterol Hepatol 4 (2), 92–101.

Cario, E., Podolsky, D.K., 2000. Differential alteration in intestinal epithelial cell expression of toll-like receptor 3 (TLR3) and TLR4 in inflammatory bowel disease. Infect Immun 68 (12), 7010–7017.

Cattin, A.L., Le Beyec, J., Barreau, F., et al., 2009. Hepatocyte nuclear factor 4alpha, a key factor for homeostasis, cell architecture, and barrier function of the adult intestinal epithelium. Mol Cell Biol 29 (23), 6294–6308.

Cheung, H., Chen, N.J., Cao, Z., Ono, N., Ohashi, P.S., Yeh, W.C., 2005. Accessory protein-like is essential for IL-18-mediated signaling. J Immunol 174 (9), 5351–5357.

Cho, J.H., 2008 June. The genetics and immunopathogenesis of inflammatory bowel disease. Nat Rev Immunol 8 (6), 458–466.

Cho, J.H., Abraham, C., 2007. Inflammatory bowel disease genetics: Nod2. Annu Rev Med 58, 401–416.

Colombel, J.F., Ferrari, N., Debuysere, H., et al., 2000. Genotypic analysis of thiopurine S-methyltransferase in patients with Crohn's disease and severe myelosuppression during azathioprine therapy. Gastroenterology 118 (6), 1025–1030.

Colombel, J.F., Sandborn, W.J., Reinisch, W., et al., 2010. Infliximab, azathioprine, or combination therapy for Crohn's disease. N Engl J Med 362 (15), 1383–1395.

Colonna, M., 2007. All roads lead to CARD9. Nat Immunol 8 (6), 554–555.

Craddock, N., Hurles, M.E., Cardin, N., et al., 2010. Genome-wide association study of CNVs in 16,000 cases of eight common diseases and 3000 shared controls. Nature 464 (7289), 713–720.

Croucher, P.J., Mascheretti, S., Foelsch, U.R., Hampe, J., Schreiber, S., 2003. Lack of association between the C3435T MDR1 gene polymorphism and inflammatory bowel disease in two independent Northern European populations. Gastroenterology 125 (6), 1919–1920.

Cuthbert, A.P., Fisher, S.A., Mirza, M.M., et al., 2002. The contribution of *NOD2* gene mutations to the risk and site of disease in inflammatory bowel disease. Gastroenterology 122 (4), 867–874.

Danese, S., Sans, M., Fiocchi, C., 2004. Inflammatory bowel disease: The role of environmental factors. Autoimmun Rev 3 (5), 394–400.

Darfeuille-Michaud, A., Boudeau, J., et al., 2004. High prevalence of adherent-invasive *Escherichia coli* associated with ileal mucosa in Crohn's disease. Gastroenterology 127 (2), 412–421.

Dideberg, V., Kristjansdottir, G., Milani, L., et al., 2007. An insertion-deletion polymorphism in the interferon regulatory factor 5 (IRF5) gene confers risk of inflammatory bowel diseases. Hum Mol Genet 16 (24), 3008–3016.

Dubinsky, M.C., Lamothe, S., Yang, H.Y., et al., 2000. Pharmacogenomics and metabolite measurement for 6-mercaptopurine therapy in inflammatory bowel disease. Gastroenterology 118 (4), 705–713.

Dubois, P.C., Trynka, G., Franke, L., et al., 2010. Multiple common variants for celiac disease influencing immune gene expression. Nat Genet. 42 (4), 295–302.

Duerr, R.H., Barmada, M.M., Zhang, L., Pfutzer, R., Weeks, D.E., 2000. High-density genome scan in Crohn disease shows confirmed linkage to chromosome 14q11-12. Am J Hum Genet 66 (6), 1857–1862.

Duerr, R.H., Taylor, K.D., Brant, S.R., et al., 2006. A genome-wide association study identifies *IL23R* as an inflammatory bowel disease gene. Science 314 (5804), 1461–1463.

Egan, L.J., Derijks, L.J., Hommes, D.W., 2006. Pharmacogenomics in inflammatory bowel disease. Clin Gastroenterol Hepatol 4 (1), 21–28.

Elliott, P.R., Moore, G.T., Bell, S.J., Connell, W.R., 2005. Severe recurrent Crohn's disease of the ileocolonic anastomosis disappearing completely with antibacterial therapy. Gut 54 (12), 1818–1819.

Farrell, R.J., Murphy, A., Long, A., et al., 2000. High multidrug resistance (P-glycoprotein 170) expression in inflammatory bowel disease patients who fail medical therapy. Gastroenterology 118 (2), 279–288.

Fava, F., Danese, S., 2010. Crohn's disease: Bacterial clearance in Crohn's disease pathogenesis. Nat Rev Gastroenterol Hepatol 7 (3), 126–128.

Fisher, S.A., Hampe, J., Onnie, C.M., et al., 2006. Direct or indirect association in a complex disease: The role of SLC22A4 and SLC22A5 functional variants in Crohn disease. Hum Mutat 27 (8), 778–785.

Fisher, S.A., Tremelling, M., Anderson, C.A., et al., 2008. Genetic determinants of ulcerative colitis include the ECM1 locus and five loci implicated in Crohn's disease. Nat Genet 40 (6), 710–712.

Franke, A., Balschun, T., Karlsen, T.H., et al., 2008. Replication of signals from recent studies of Crohn's disease identifies previously unknown disease loci for ulcerative colitis. Nat Genet 40 (6), 713–715.

Franke, A., Balschun, T., Karlsen, T.H., et al., 2008. Sequence variants in IL10, ARPC2 and multiple other loci contribute to ulcerative colitis susceptibility. Nat Genet. 40 (11), 1319–1323.

Franke, A., Balschun, T., Sina, C., et al., 2010. Genome-wide association study for ulcerative colitis identifies risk loci at 7q22 and 22q13 (IL17REL). Nat Genet 42 (4), 292–294.

Franke, A., Fischer, A., Nothnagel, M., et al., 2008. Genome-wide association analysis in sarcoidosis and Crohn's disease unravels a common susceptibility locus on 10p12.2. Gastroenterology 135 (4), 1207–1215.

Franke, A., Hampe, J., Rosenstiel, P., et al., 2007. Systematic association mapping identifies *NELL1* as a novel IBD disease gene. PLoS One 2 (1), e691.

Gersemann, M., Becker, S., Kubler, I., et al., 2009. Differences in goblet cell differentiation between Crohn's disease and ulcerative colitis. Differentiation 77 (1), 84–94.

Goyette, P., Lefebvre, C., Ng, A., et al., 2008. Gene-centric association mapping of chromosome 3p implicates MST1 in IBD pathogenesis. Mucosal Immunol 1 (2), 131–138.

Hampe, J., Cuthbert, A., Croucher, P.J., et al., 2001. Association between insertion mutation in *NOD2* gene and Crohn's disease in German and British populations. Lancet 357 (9272), 1925–1928.

Hampe, J., Franke, A., Rosenstiel, P., et al., 2007. A genome-wide association scan of nonsynonymous SNPs identifies a susceptibility variant for Crohn disease in ATG16L1. Nat Genet 39 (2), 207–211.

Hampe, J., Grebe, J., Nikolaus, S., et al., 2002. Association of *NOD2* (*CARD15*) genotype with clinical course of Crohn's disease: A cohort study. Lancet 359 (9318), 1661–1665.

Hanauer, S.B., Feagan, B.G., Lichtenstein, G.R., et al., 2002. Maintenance infliximab for Crohn's disease: The ACCENT I randomised trial. Lancet 359 (9317), 1541–1549.

Herrlinger, K.R., Cummings, J.R., Barnardo, M.C., Schwab, M., Ahmad, T., Jewell, D.P., 2005. The pharmacogenetics of methotrexate in inflammatory bowel disease. Pharmacogenet Genomics 15 (10), 705–711.

Hirschhorn, J.N., Daly, M.J., 2005. Genome-wide association studies for common diseases and complex traits. Nat Rev Genet 6 (2), 95–108.

Hlavaty, T., Pierik, M., Henckaerts, L., et al., 2005. Polymorphisms in apoptosis genes predict response to infliximab therapy in luminal and fistulizing Crohn's disease. Aliment Pharmacol Ther 22 (7), 613–626.

Hoffmeyer, S., Burk, O., Von Richter, O., et al., 2000. Functional polymorphisms of the human multidrug-resistance gene: Multiple sequence variations and correlation of one allele with P-glycoprotein expression and activity *in vivo*. Proc Natl Acad Sci USA 97 (7), 3473–3478.

Hong, J., Leung, E., Fraser, A.G., Merriman, T.R., Vishnu, P., Krissansen, G.W., 2007. Polymorphisms in *NFKBIA* and *ICAM-1* genes in New Zealand Caucasian Crohn's disease patients. J Gastroenterol Hepatol 22 (10), 1666–1670.

Horwitz, B.H., 2007. The straw that stirs the drink: Insight into the pathogenesis of inflammatory bowel disease revealed through the study of microflora-induced inflammation in genetically modified mice. Inflamm Bowel Dis 13 (4), 490–500.

Hou, J.K., Abraham, B., El-Serag, H., 2011. Dietary intake and risk of developing inflammatory bowel disease: A systematic review of the literature. Am J Gastroenterol 106 (4), 563–573.

Hugot, J.P., Chamaillard, M., Zouali, H., et al., 2001. Association of NOD2 leucine-rich repeat variants with susceptibility to Crohn's disease. Nature 411 (6837), 599–603.

Hunt, K.A., Zhernakova, A., Turner, G., et al., 2008. Newly identified genetic risk variants for celiac disease related to the immune response. Nat Genet 40 (4), 395–402.

Imielinski, M., Baldassano, R.N., Griffiths, A., et al., 2009. Common variants at five new loci associated with early-onset inflammatory bowel disease. Nat Genet 41 (12), 1335–1340.

Israeli, E., Grotto, I., Gilburd, B., et al., 2005. Anti-*Saccharomyces cerevisiae* and antineutrophil cytoplasmic antibodies as predictors of inflammatory bowel disease. Gut 54 (9), 1232–1236.

Juste, R.A., Elguezabal, N., Garrido, J.M., et al., 2008. On the prevalence of *M. avium* subspecies paratuberculosis DNA in the blood of healthy individuals and patients with inflammatory bowel disease. PLoS One 3 (7), e2537.

Kawasaki, A., Tsuchiya, N., Hagiwara, K., Takazoe, M., Tokunaga, K., 2000. Independent contribution of HLA-DRB1 and TNF alpha promoter polymorphisms to the susceptibility to Crohn's disease. Genes Immun 1 (6), 351–357.

Kobayashi, K.S., Chamaillard, M., Ogura, Y., et al., 2005. Nod2-dependent regulation of innate and adaptive immunity in the intestinal tract. Science 307 (5710), 731–734.

Kobayashi, T., Okamoto, S., Hisamatsu, T., et al., 2008. IL23 differentially regulates the Th1/Th17 balance in ulcerative colitis and Crohn's disease. Gut 57 (12), 1682–1689.

Krausgruber, T., Saliba, D., Ryzhakov, G., Lanfrancotti, A., Blazek, K., Udalova, I.A., 2010. IRF5 is required for late-phase TNF secretion by human dendritic cells. Blood 115 (22), 4421–4430.

Krynetski, E.Y., Tai, H.L., Yates, C.R., et al., 1996. Genetic polymorphism of thiopurine S-methyltransferase: Clinical importance and molecular mechanisms. Pharmacogenetics 6 (4), 279–290.

Kuballa, P., Huett, A., Rioux, J.D., Daly, M.J., Xavier, R.J., 2008. Impaired autophagy of an intracellular pathogen induced by a Crohn's disease-associated ATG16L1 variant. PLoS One 3 (10), e3391.

Kugathasan, S., Baldassano, R.N., Bradfield, J.P., et al., 2008. Loci on 20q13 and 21q22 are associated with pediatric-onset inflammatory bowel disease. Nat Genet 40 (10), 1211–1215.

Lakatos, P.L., Fischer, S., Claes, K., et al., 2006. DLG5 R30Q is not associated with IBD in Hungarian IBD patients but predicts clinical response to steroids in Crohn's disease. Inflamm Bowel Dis 12 (5), 362–368.

Lappalainen, M., Halme, L., Turunen, U., et al., 2008. Association of IL23R, TNFRSF1A, and HLA-DRB1*0103 allele variants with inflammatory bowel disease phenotypes in the Finnish population. Inflamm Bowel Dis 14 (8), 1118–1124.

Lesage, S., Zouali, H., Cezard, J.P., et al., 2002. *CARD15/NOD2* mutational analysis and genotype-phenotype correlation in 612 patients with inflammatory bowel disease. Am J Hum Genet 70 (4), 845–857.

Lettre, G., Rioux, J.D., 2008. Autoimmune diseases: Insights from genome-wide association studies. Hum Mol Genet 17 (R2), R116–R121.

Liao, F., Rabin, R.L., Smith, C.S., Sharma, G., Nutman, T.B., Farber, J.M., 1999. CC-chemokine receptor 6 is expressed on diverse memory subsets of T cells and determines responsiveness to macrophage inflammatory protein 3 alpha. J Immunol 162 (1), 186–194.

Libioulle, C., Louis, E., Hansoul, S., et al., 2007. Novel Crohn disease locus identified by genome-wide association maps to a gene desert on 5p13.1 and modulates expression of PTGER4. PLoS Genet 3 (4), e58.

Loftus Jr., E.V., 2004. Clinical epidemiology of inflammatory bowel disease: Incidence, prevalence, and environmental influences. Gastroenterology 126 (6), 1504–1517.

Logan, I., Bowlus, C.L., 2010. The geoepidemiology of autoimmune intestinal diseases. Autoimmun Rev 9 (5), A372–A378.

Louis, E., El Ghoul, Z., Vermeire, S., et al., 2004. Association between polymorphism in IgG Fc receptor IIIa coding gene and biological response to infliximab in Crohn's disease. Aliment Pharmacol Ther 19 (5), 511–519.

Louis, E., Libioulle, C., Reenaers, C., Belaiche, J., Georges, M., 2009. Genetics of ulcerative colitis: The come-back of interleukin 10. Gut 58 (9), 1173–1176.

Machado, J.C., Oliveira, C., Carvalho, R., et al., 2001. E-cadherin gene (*CDH1*) promoter methylation as the second hit in sporadic diffuse gastric carcinoma. Oncogene 20 (12), 1525–1528.

Mascheretti, S., Hampe, J., Croucher, P.J., et al., 2002. Response to infliximab treatment in Crohn's disease is not associated with mutations in the *CARD15* (*NOD2*) gene: An analysis in 534 patients from two multicenter, prospective GCP-level trials. Pharmacogenetics 12 (7), 509–515.

Matsumura, Y., Kinouchi, Y., Nomura, E., et al., 2008. HLA-DRB1 alleles influence clinical phenotypes in Japanese patients with ulcerative colitis. Tissue Antigens 71 (5), 447–452.

McCarroll, S.A., Huett, A., Kuballa, P., et al., 2008. Deletion polymorphism upstream of IRGM associated with altered IRGM expression and Crohn's disease. Nat Genet. 40 (9), 1107–1112.

McGovern, D.P., Gardet, A., Torkvist, L., et al., 2010. Genome-wide association identifies multiple ulcerative colitis susceptibility loci. Nat Genet 42 (4), 332–337.

McGovern, D.P., Van Heel, D.A., Ahmad, T., Jewell, D.P., 2001. *NOD2* (*CARD15*), the first susceptibility gene for Crohn's disease. Gut 49 (6), 752–754.

McGuckin, M.A., Eri, R., Simms, L.A., Florin, T.H., Radford-Smith, G., 2009. Intestinal barrier dysfunction in inflammatory bowel diseases. Inflamm Bowel Dis 15 (1), 100–113.

Melgar, S., Shanahan, F., 2010. Inflammatory bowel disease: From mechanisms to treatment strategies. Autoimmunity.

Moehle, C., Ackermann, N., Langmann, T., et al., 2006. Aberrant intestinal expression and allelic variants of mucin genes associated with inflammatory bowel disease. J Mol Med 84 (12), 1055–1066.

Molodecky, N.A., Kaplan, G.G., 2010. Environmental risk factors for inflammatory bowel disease. Gastroenterol Hepatol (NY) 6 (5), 339–346.

Nell, S., Suerbaum, S., Josenhans, C., 2010. The impact of the microbiota on the pathogenesis of IBD: Lessons from mouse infection models. Nat Rev Microbiol 8 (8), 564–577.

Neurath, M.F., Kiesslich, R., Teichgraber, U., et al., 2005. 6-thioguanosine diphosphate and triphosphate levels in red blood cells and response to azathioprine therapy in Crohn's disease. Clin Gastroenterol Hepatol 3 (10), 1007–1014.

Noble, C.L., Abbas, A.R., Cornelius, J., et al., 2008. Regional variation in gene expression in the healthy colon is dysregulated in ulcerative colitis. Gut 57 (10), 1398–1405.

Noble, C.L., Abbas, A.R., Lees, C.W., et al., 2010. Characterization of intestinal gene expression profiles in Crohn's disease by genome-wide microarray analysis. Inflamm Bowel Dis 16 (10), 1717–1728.

Ogura, Y., Bonen, D.K., Inohara, N., et al., 2001. A frameshift mutation in NOD2 associated with susceptibility to Crohn's disease. Nature 411 (6837), 603–606.

Orholm, M., Binder, V., Sorensen, T.I., Rasmussen, L.P., Kyvik, K.O., 2000. Concordance of inflammatory bowel disease among Danish twins. Results of a nationwide study. Scand J Gastroenterol 35 (10), 1075–1081.

Oudkerk Pool, M., Bouma, G., Meuwissen, S.G., et al., 1995. Serological markers to differentiate between ulcerative colitis and Crohn's disease. J Clin Pathol 48 (4), 346–350.

Oudkerk Pool, M., Ellerbroek, P.M., Ridwan, B.U., et al., 1993. Serum antineutrophil cytoplasmic autoantibodies in inflammatory bowel disease are mainly associated with ulcerative colitis. A correlation study between perinuclear antineutrophil cytoplasmic autoantibodies and clinical parameters, medical, and surgical treatment. Gut 34 (1), 46–50.

Park, H., Li, Z., Yang, X.O., et al., 2005. A distinct lineage of CD4 T cells regulates tissue inflammation by producing interleukin 17. Nat Immunol 6 (11), 1133–1141.

Parkes, M., Barrett, J.C., Prescott, N.J., et al., 2007. Sequence variants in the autophagy gene IRGM and multiple other replicating loci contribute to Crohn's disease susceptibility. Nat Genet 39 (7), 830–832.

Perl, A.K., Wilgenbus, P., Dahl, U., Semb, H., Christofori, G., 1998. A causal role for E-cadherin in the transition from adenoma to carcinoma. Nature 392 (6672), 190–193.

Phillips, A.M., Nimmo, E.R., Van Limbergen, J., Drummond, H.E., Smith, L., Satsangi, J., 2010. Detailed haplotype-tagging study of germline variation of MUC19 in inflammatory bowel disease. Inflamm Bowel Dis 16 (4), 557–558.

Poppe, D., Tiede, I., Fritz, G., et al., 2006. Azathioprine suppresses ezrin-radixin-moesin-dependent T cell-APC conjugation through inhibition of Vav guanosine exchange activity on Rac proteins. J Immunol 176 (1), 640–651.

Porter, C.K., Tribble, D.R., Aliaga, P.A., Halvorson, H.A., Riddle, M.S., 2008. Infectious gastroenteritis and risk of developing inflammatory bowel disease. Gastroenterology 135 (3), 781–786.

Prehn, J.L., Mehdizadeh, S., Landers, C.J., et al., 2004. Potential role for TL1A, the new TNF-family member and potent costimulator of IFN-gamma, in mucosal inflammation. Clin Immunol 112 (1), 66–77.

Present, D.H., Rutgeerts, P., Targan, S., et al., 1999. Infliximab for the treatment of fistulas in patients with Crohn's disease. N Engl J Med 340 (18), 1398–1405.

Purcell, S.M., Wray, N.R., Stone, J.L., et al., 2009. Common polygenic variation contributes to risk of schizophrenia and bipolar disorder. Nature 460 (7256), 748–752.

Raelson, J.V., Little, R.D., Ruether, A., et al., 2007. Genome-wide association study for Crohn's disease in the Quebec Founder Population identifies multiple validated disease loci. Proc Natl Acad Sci USA 104 (37), 14,747–14,752.

Renz, H., Von Mutius, E., Brandtzaeg, P., Cookson, W.O., Autenrieth, I.B., Haller, D., 2011. Gene-environment interactions in chronic inflammatory disease. Nat Immunol 12 (4), 273–277.

Rioux, J.D., Daly, M.J., Silverberg, M.S., et al., 2001. Genetic variation in the 5q31 cytokine gene cluster confers susceptibility to Crohn disease. Nat Genet 29 (2), 223–228.

Rioux, J.D., Xavier, R.J., Taylor, K.D., et al., 2007. Genome-wide association study identifies new susceptibility loci for Crohn disease and implicates autophagy in disease pathogenesis. Nat Genet 39 (5), 596–604.

Rolhion, N., Darfeuille-Michaud, A., 2007. Adherent-invasive Escherichia coli in inflammatory bowel disease. Inflamm Bowel Dis 13 (10), 1277–1283.

Rosenstiel, P., Sina, C., End, C., et al., 2007. Regulation of DMBT1 via NOD2 and TLR4 in intestinal epithelial cells modulates bacterial recognition and invasion. J Immunol 178 (12), 8203–8211.

Sands, B.E., Blank, M.A., Patel, K., Van Deventer, S.J., 2004. Long-term treatment of rectovaginal fistulas in Crohn's disease: Response to infliximab in the ACCENT II Study. Clin Gastroenterol Hepatol 2 (10), 912–920.

Sauer, C.G., Kugathasan, S., 2009. Pediatric inflammatory bowel disease: Highlighting pediatric differences in IBD. Gastroenterol Clin North Am 38 (4), 611–628.

Schreiber, S., Rosenstiel, P., Albrecht, M., Hampe, J., Krawczak, M., 2005. Genetics of Crohn disease, an archetypal inflammatory barrier disease. Nat Rev Genet 6 (5), 376–388.

Schwab, M., Schaeffeler, E., Marx, C., et al., 2003. Association between the C3435T MDR1 gene polymorphism and susceptibility for ulcerative colitis. Gastroenterology 124 (1), 26–33.

Shale, M., Ghosh, S., 2009. How intestinal epithelial cells tolerise dendritic cells and its relevance to inflammatory bowel disease. Gut 58 (9), 1291–1299.

Silverberg, M.S., Cho, J.H., Rioux, J.D., et al., 2009. Ulcerative colitis-risk loci on chromosomes 1p36 and 12q15 found by genome-wide association study. Nat Genet 41 (2), 216–220.

Silverberg, M.S., Duerr, R.H., Brant, S.R., et al., 2007. Refined genomic localization and ethnic differences observed for the IBD5 association with Crohn's disease. Eur J Hum Genet 15 (3), 328–335.

Silverberg, M.S., Mirea, L., Bull, S.B., et al., 2003. A population- and family-based study of Canadian families reveals association of HLA DRB1*0103 with colonic involvement in inflammatory bowel disease. Inflamm Bowel Dis 9 (1), 1–9.

Smyth, D.J., Plagnol, V., Walker, N.M., et al., 2008. Shared and distinct genetic variants in type 1 diabetes and celiac disease. N Engl J Med 359 (26), 2767–2777.

Stoll, M., Corneliussen, B., Costello, C.M., et al., 2004. Genetic variation in DLG5 is associated with inflammatory bowel disease. Nat Genet 36 (5), 476–480.

Strumane, K., Berx, G., Van Roy, F., 2004. Cadherins in cancer. Handb Exp Pharmacol 2004 (165), 69–103.

Su, L., Shen, L., Clayburgh, D.R., et al., 2009. Targeted epithelial tight junction dysfunction causes immune activation and contributes to development of experimental colitis. Gastroenterology 136 (2), 551–563.

Sutton, C.E., Lalor, S.J., Sweeney, C.M., Brereton, C.F., Lavelle, E.C., Mills, K.H., 2009. Interleukin-1 and IL-23 induce innate IL-17 production from gammadelta T cells, amplifying Th17 responses and autoimmunity. Immunity 31 (2), 331–341.

Takedatsu, H., Michelsen, K.S., Wei, B., et al., 2008. TL1A (TNFSF15) regulates the development of chronic colitis by modulating both T-helper 1 and T-helper 17 activation. Gastroenterology 135 (2), 552–567.

Takeno, S., Noguchi, T., Fumoto, S., Kimura, Y., Shibata, T., Kawahara, K., 2004. E-cadherin expression in patients with esophageal squamous cell carcinoma: Promoter hypermethylation, Snail overexpression, and clinicopathologic implications. Am J Clin Pathol 122 (1), 78–84.

Tarlinton, D., Light, A., Metcalf, D., Harvey, R.P., Robb, L., 2003. Architectural defects in the spleens of Nkx2-3-deficient mice are intrinsic and associated with defects in both B cell maturation and T cell-dependent immune responses. J Immunol 170 (8), 4002–4010.

Todd, J.A., Walker, N.M., Cooper, J.D., et al., 2007. Robust associations of four new chromosome regions from genome-wide analyses of type 1 diabetes. Nat Genet 39 (7), 857–864.

Torok, H.P., Glas, J., Tonenchi, L., Bruennler, G., Folwaczny, M., Folwaczny, C., 2004. Crohn's disease is associated with a toll-like receptor-9 polymorphism. Gastroenterology 127 (1), 365–366.

Trachtenberg, E.A., Yang, H., Hayes, E., et al., 2000. HLA class II haplotype associations with inflammatory bowel disease in Jewish (Ashkenazi) and non-Jewish caucasian populations. Hum Immunol 61 (3), 326–333.

Urcelay, E., Mendoza, J.L., Martinez, A., et al., 2005. IBD5 polymorphisms in inflammatory bowel disease: Association with response to infliximab. World J Gastroenterol 11 (8), 1187–1192.

Van Bodegraven, A.A., Curley, C.R., Hunt, K.A., et al., 2006. Genetic variation in myosin IXB is associated with ulcerative colitis. Gastroenterology 131 (6), 1768–1774.

Van Heel, D.A., McGovern, D.P., Jewell, D.P., 2001. Crohn's disease: Genetic susceptibility, bacteria, and innate immunity. Lancet 357 (9272), 1902–1904.

Vella, M., Masood, M.R., Hendry, W.S., 2007. Surgery for ulcerative colitis. Surgeon 5 (6), 356–362.

Vermeire, S., Louis, E., Rutgeerts, P., et al., 2002. *NOD2/CARD15* does not influence response to infliximab in Crohn's disease. Gastroenterology 123 (1), 106–111.

Vermeire, S., Wild, G., Kocher, K., et al., 2002. *CARD15* genetic variation in a Quebec population: Prevalence, genotype–phenotype relationship, and haplotype structure. Am J Hum Genet 71 (1), 74–83.

Wang, K., Baldassano, R., Zhang, H., et al., 2010. Comparative genetic analysis of inflammatory bowel disease and type 1 diabetes implicates multiple loci with opposite effects. Hum Mol Genet 19 (10), 2059–2067.

Wang, K., Zhang, H., Kugathasan, S., et al., 2009. Diverse genome-wide association studies associate the IL12/IL23 pathway with Crohn Disease. Am J Hum Genet 84 (3), 399–405.

Wang, M.H., Cox, G.W., Yoshimura, T., Sheffler, L.A., Skeel, A., Leonard, E.J., 1994. Macrophage-stimulating protein inhibits induction of nitric oxide production by endotoxin- or cytokine-stimulated mouse macrophages. J Biol Chem 269 (19), 14,027–14,031.

The Wellcome Trust Case Control Consortium, 2007. Genome-wide association study of 14,000 cases of seven common diseases and 3000 shared controls. Nature 447 (7145), 661–678.

Wong, K., Bressler, B., 2008. Mild to moderate Crohn's disease: An evidence-based treatment algorithm. Drugs 68 (17), 2419–2425.

Yamazaki, K., McGovern, D., Ragoussis, J., et al., 2005. Single nucleotide polymorphisms in *TNFSF15* confer susceptibility to Crohn's disease. Hum Mol Genet 14 (22), 3499–3506.

Yang, H., Vora, D.K., Targan, S.R., Toyoda, H., Beaudet, A.L., Rotter, J.I., 1995. Intercellular adhesion molecule 1 gene associations with immunologic subsets of inflammatory bowel disease. Gastroenterology 109 (2), 440–448.

Yang, P., Jarnerot, G., Danielsson, D., Tysk, C., Lindberg, E., 1995. P-ANCA in monozygotic twins with inflammatory bowel disease. Gut 36 (6), 887–890.

Yates, C.R., Krynetski, E.Y., Loennechen, T., et al., 1997. Molecular diagnosis of thiopurine S-methyltransferase deficiency: Genetic basis for azathioprine and mercaptopurine intolerance. Ann Intern Med 126 (8), 608–614.

Zhang, J., Wang, X., Fahmi, H., et al., 2009. Role of TL1A in the pathogenesis of rheumatoid arthritis. J Immunol 183 (8), 5350–5357.

Zhernakova, A., Alizadeh, B.Z., Bevova, M., et al., 2007. Novel association in chromosome 4q27 region with rheumatoid arthritis and confirmation of type 1 diabetes point to a general risk locus for autoimmune diseases. Am J Hum Genet 81 (6), 1284–1288.

Zhernakova, A., Festen, E.M., Franke, L., et al., 2008. Genetic analysis of innate immunity in Crohn's disease and ulcerative colitis identifies two susceptibility loci harboring CARD9 and IL18RAP. Eur J Hum Genet 16 p06.070, 304. (Abstract).

Zhernakova, A., Festen, E.M., Franke, L., et al., 2008. Genetic analysis of innate immunity in Crohn's disease and ulcerative colitis identifies two susceptibility loci harboring CARD9 and IL18RAP. Am J Hum Genet 82 (5), 1202–1210.

Zhernakova, A., Van Diemen, C.C., Wijmenga, C., 2009. Detecting shared pathogenesis from the shared genetics of immune-related diseases. Nat Rev Genet 10 (1), 43–55.

Zinovieva, E., Bourgain, C., Kadi, A., et al., 2009. Comprehensive linkage and association analyses identify haplotype, near to the *TNFSF15* gene, significantly associated with spondyloarthritis. PLoS Genet 5 (6), e1000528.

Zouali, H., Lesage, S., Merlin, F., et al., 2003. *CARD4/NOD1* is not involved in inflammatory bowel disease. Gut 52 (1), 71–74.

ONGOING RESEARCH

Anderson, C.A., Boucher, G., Lees, C.W., et al., 2011. Meta-analysis identifies 29 additional ulcerative colitis risk loci, increasing the number of confirmed associations to 47. Nat Genet 43 (3), 246–252.

Franke, A., McGovern, D.P., Barrett, J.C., et al., 2010. Genome-widemeta-analysis increases to 71 the number of confirmed Crohn's disease susceptibility loci. Nat Genet 42 (12), 1118–1125.

Fransen, K., Mitrovic, M., van Diemen, C.C., Weersma, R.K., 2011. The quest for genetic risk factors for Crohn's disease in the post-GWAS era. Genome Med 3 (2), 13.

Rivas, M.A., Beaudoin, M., Gardet, A., et al., 2011. Deep resequencing of GWAS loci identifies independent rare variants associated with inflammatory bowel disease. Nat Genet 43 (11), 1066–1073.

RECOMMENDED RESOURCES

http://www.ibdgenetics.org/

CHAPTER 74

Asthma

Scott T. Weiss

INTRODUCTION

Since the initial completion of the Human Genome Project, there has been an explosion of knowledge about complex traits generally, and asthma in particular. Most of this new knowledge has been identification of genes associated with the asthma phenotype, and this effort has led to one translational initiative to try to prevent the disease, an application of genomics that has fundamentally altered the concept of asthma as a disease. This chapter will review the genes that have been identified for asthma and the methods used to find those genes, and then will focus on one discovery, based on a positionally cloned gene, the vitamin D receptor, that has led to a novel theory of asthma prevention currently being evaluated in clinical trials.

To form a basis for examining the current state of the field, it is necessary to review the fundamentals of a genetic association study and the major threats to validity of such an experiment. In a genetic association study, one relates single nucleotide polymorphisms (SNPs) to a defined phenotype, such as asthma. There are two basic study designs for this type of experiment: a case-control study, where allele frequencies of a SNP are compared in asthma cases to controls without asthma; or a family-based study, where the number of alleles transmitted from heterozygous parents to an affected offspring are compared to the untransmitted alleles from that parent. In studies such as these, there are several potential sources of error: (1) there can be inadequate coverage of the gene with too few SNPs, or SNPs that don't accurately reflect the evolutionary history of the gene (its linkage disequilibrium); (2) there may be a small sample size leading to reduced statistical power; (3) by mixing different allele frequencies from different ethnic groups in the cases and controls, population stratification can occur, leading to spurious result; and (4) error in genotyping or error in phenotyping of the subjects can lead to misclassification. These four sources of error are not shared equally by the two types of study designs, with the family-based design being more susceptible to bias due to genotyping error and the case-control design being more susceptible to population stratification.

CANDIDATE GENE STUDIES

Many hundreds of candidate gene studies have been performed for asthma and a large number of genes have been identified. They are summarized in Figure 74.1, where the genes are plotted on the *x*-axis and the number of positive studies on the *y*-axis. The major criticism of these studies is that, while successful, they do not identify novel biology, but merely reinforce what is already known about asthma as a disease and hence do not advance the field. For example, it was well known that *IL4*, *IL13*, and *TNF alpha*, three of the most-replicated asthma candidate genes in Figure 74.1, were associated with a TH2 type of inflammatory response, and thus the new knowledge that resulted from these candidate gene studies was at the level of molecular mechanisms that might be related to disease pathobiology.

Genomic and Personalized Medicine, 2nd edition
by Ginsburg & Willard. DOI: http://dx.doi.org/10.1016/B978-0-12-382227-7.00074-4

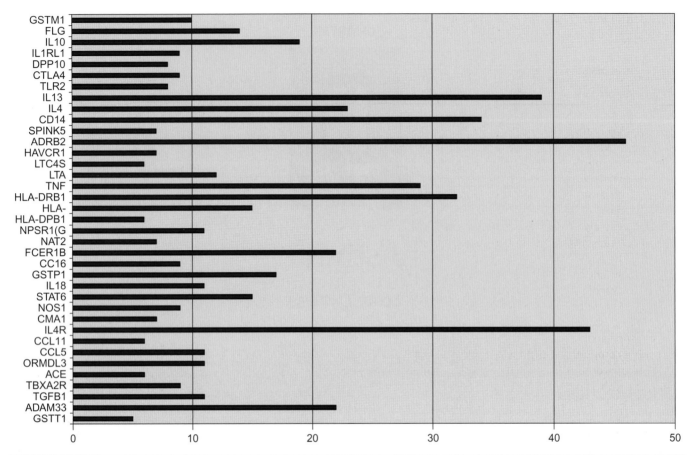

Figure 74.1   Candidate gene studies in asthma. Genes are shown on the x-axis and the number of positive studies on the y-axis. *Reproduced with permission from Postma et al., 2011.*

While it is fair to say that these studies did not change the concept of asthma as a disease, they gave researchers valuable experience with these types of experiments and how they were conducted. To date, 37 well-replicated asthma genes have been identified in multiple candidate gene genetic association studies, with 20 of those genes being significant in 10 or more studies (Figure 74.1) (Postma et al., 2011). At the present time, none of these candidate genes have been put together in a clinically useful test to predict disease occurrence.

POSITIONALLY CLONED GENES

The second type of study that was performed utilized a combination of linkage and association. In a linkage study, polymorphic microsatellite markers (di-, tri-, and tetra-nucleotide repeats) across the genome are associated with a phenotype of interest using a statistical technique similar to logistic regression. This approach identifies a broad region of a chromosome that is harboring one or more significant loci, and then by using the approach of fine mapping, (i.e., genotyping many

SNP markers in the chromosomal region), additional associational analyses are performed, and the chromosomal region of interest is thus narrowed. The major validity threats to this experimental approach are similar to those of a candidate gene study, but with one additional caveat – it may be impossible to find the actual gene or variant leading to the association because of linkage disequilibrium. Linkage disequilibrium is the nonrandom association of alleles in the genome. Two SNPs will be in complete linkage disequilibrium if the r2 between them is 1. They will be closely correlated. The degree of linkage disequilibrium varies throughout the genome and is dependent on evolutionary history, genetic distance (i.e., how far apart the SNPs are from each other), SNP allele frequency, and other factors. By 2003, several novel genes for asthma were identified using this approach, such as ADAM33, the Vitamin D receptor (VDR), DPP10, PHF11, HLA G, and GPR154 (NPSR1) (Weiss et al., 2009). Functional variants were identified for Adam33, VDR, and GPR154, and a potential functional role for these variants in asthma pathobiology has begun to be defined. Although the linkage and positional cloning approach was able to identify novel genes, this approach was not assessing the whole genome in an unbiased way, so investigators moved on to Genome Wide

Association Studies. Again no real effort has been made to develop a clinically useful test to predict disease occurrence.

WHOLE-GENOME ASSOCIATION STUDIES

The next major advance in asthma genetics came in 2007 with the first whole-genome association studies. The rationale behind those studies was that by using linkage disequilibrium across the whole genome, one could type as few as 500,000 SNPs and, because of the linkage disequilibrium, get information on most of the 13 million SNPs across the genome. Because of the large number of statistical comparisons, a very high level of statistical stringency was set on the results (usually $p < 10^{-7}$). The early genome-wide association studies (GWAS) had two deficiencies – the SNP chips did not give complete genome coverage, and the studies were of smaller sample size, due to the cost of the chips. However, the rationale was that these studies would be adequately powered to discover common genetic variation (e.g., allele frequencies >20%) associated with a complex trait such as asthma. The initial GWAS in asthma also incorporated a unique approach, combining gene expression with genetic association to find a novel locus on chromosome 17q21 (Moffatt et al., 2007). The initial paper that identified ORMDL3 as the most significant asthma locus in children demonstrated the power of the central dogma: using SNPs to predict gene expression level, and then using these expression SNPs (eSNPs) to direct the search for asthma genes in a region of the genome, overcoming the problem of linkage disequilibrium not giving a precise signal in a genomic region. Other small GWAS were published, identifying several novel genes for asthma, including *PDE4D, TLE4, ADRA1B, PRNP, DPP10, RAD50-IL13,* and *DENND1B* (Hancock et al., 2009; Li et al., 2010; Mathias et al., 2010; Sleiman et al., 2009). Some of these were only identified in African Americans: *ADRA1B, PRNP,* and *DPP10* (Galanter et al., 2008; Himes et al., 2009; Sleiman et al., 2008). Of the initial wave of published GWAS, the only locus that has been found consistently across all studies is the ORMDL3 locus. This is because this locus has a large effect on asthma susceptibility in children and the effect of any other locus is much smaller in comparison. Recent functional data would suggest that this locus is actually a three-gene locus that functions with an insulator and a repressor regulating ZPBP2/GSDMB/ORMDL3 (Verlaan et al., 2009).

A major concern with these early studies was limited sample size. Given that most of the loci for asthma were of small effect, larger studies were needed to demonstrate statistical significance and find more genes. To accomplish that goal, two large consortia were formed, one in Europe (GABRIEL, Moffatt et al., 2010), and one in the US (EVE).

GABRIEL is a European consortium that included Caucasian cases from all over Europe. They had 10,365 asthma cases, both children and adults, and 16,110 controls (Moffatt et al., 2010).

TABLE 74.1	Asthma risk loci in EVE and GABRIEL		
Gene (region)	**EVE**	**GABRIEL**	**Ethnicity**
ORMDL3/GSDML (17q21)	1.2×10^{-14}	$6.4 \times 10^{-23*}$	All
ILRL1/L18R1 (chr 2)	1.4×10^{-8}	3.4×10^{-9}	All
TSLP (chr 5)	7.3×10^{-10}	1.0×10^{-7}	All
IL33 (chr 9)	2.5×10^{-7}	9.2×10^{-10}	All
PYHIN1 (chr 1)	3.6×10^{-7}	–	African ancestry
HLA-DQ (chr 6)	1.4×10^{-2}	7.0×10^{-14}	All
SMAD3 (chr 15)	1.9×10^{-4}	3.9×10^{-9}	European/ European American
IL2RB (chr 22)	ns	1.1×10^{-8}	European
SLC22A5 (chr 5)	ns	2.2×10^{-7}	European
IL13 (chr 5)	ns	1.4×10^{-7}	European
RORA (chr 15)	2.1×10^{-3}	1.1×10^{-7}	European/ European American

Data courtesy of Carole Ober and EVE consortium investigators.
EVE Consortium – http://eve.uchicago.edu/, GABRIEL Consortium – http://www.gabriel-fp6.org/
chr, chromosome; ns, not sufficient.
*Childhood asthma.

EVE used a mixed study design that included 3246 asthma cases and 3385 controls, 1702 asthma trios, 355 family-based cases, and 468 family-based controls from three ethnic groups – European Americans, African Americans/Afro-Caribbeans, and Hispanics (Torgerson et al., 2011). Table 74.1 summarizes the results of these two large consortium studies. The results are mostly confirmatory and mutually reinforcing. Both groups confirmed that the 17q21 locus was the most significant finding. For GABRIEL, the next most significant finding was *HLA-DQ*. While this was confirmed in EVE, it was not at the same level of significance as in GABRIEL. The other three major novel loci confirmed in both studies were *IL1RL/IL18R1, TSLP,* and *IL33*.

Ten Years of Asthma Genetic Association Studies

Several features emerge from the now-extensive literature on asthma genetics. First, as is true for all autoimmune diseases, HLA (human leukocyte antigen) is an important genetic locus for asthma. Several different HLA types have been implicated in asthma, and this links asthma to other autoimmune diseases in which HLA has been deemed an important locus, such as Crohn's disease, type 1 diabetes mellitus, multiple sclerosis, rheumatoid arthritis, and celiac disease. The second feature is that many of the novel loci relate to immune system functioning – TSLP, IL33, and the IL1 complex clearly implicate immune processes as being central to the asthma phenotype. With the exception of the 17q21 locus, it is difficult to implicate any one gene, or gene locus, as being a major factor in asthma causation, since the effect estimates for any of the identified genes is in the range of 1.1–1.3, with only the 17q21 locus having an effect size close to 1.5. Thus, for most of these studies, the results are driven by the large sample size rather than the overwhelming effect of any one individual gene or genes. Prediction is still a problem, since the amount of explained variability of the asthma phenotype by significant GWAS results is less than 5%. No one has begun to put all the candidate genes, positionally cloned genes, and GWAS genes into a multivariate predictive model. To date, the clinical relevance of asthma genetics is still of minimal impact.

Asthma Pharmacogenomics

Asthma pharmacogenomics has focused on three common drug response pathways: beta agonist, corticosteroid, and leukotrienes. Three phenotypes have been the primary focus: change in lung function either after a short-acting beta agonist or after eight weeks of inhaled corticosteroid; exacerbations, defined as hospitalizations, emergency room visits, or prednisone bursts; and airway responsiveness to methacholine. Prior work had identified *Arg1* as a gene predicting short acting response to beta agonist (Litonjua et al., 2008). Recent work has localized the effect to a specific regulatory haplotype of the *Arg1* gene (Duan et al., 2011). In addition, a novel gene, the thyroid hormone receptor, has been localized as a predictor of short-acting beta agonist response based on a study of selected transcription factors active in airway epithelium and airway smooth muscle cells (Duan et al., 2012). After assessing 98 transcription factors, a SNP in the *THRβ* gene determined bronchodilator response in several clinical trials populations (Duan et al., 2012). The low affinity IgE receptor gene *FCER2* has previously been associated with asthma exacerbations on inhaled corticosteroids (Tantisira et al., 2007). Recent work has confirmed this association with the T2206C SNP in two independent populations (Koster et al., 2011). Finally, a report appeared relating a promoter SNP in GLCCI1 to change in lung function after eight weeks of inhaled corticosteroid in several different asthma clinical trials of both adults and children (Tantisira et al., 2011). A single SNP in the promoter of the gene was associated with decrease in *GLCCI1* gene expression and decreased luciferase reporter activity (Tantisira et al., 2011). The SNP accounted for 6.6% of the variability in inhaled glucocorticoid treatment response (Tantisira et al., 2011). Grundberg and coworkers looked for *cis*-acting expressed quantitative trait loci (eQTLs) in a set of primary human bone cells using 18 pharmacologic stimulants, including dexamethasone (Grundberg et al., 2011). They identified a number of novel eQTLs but these were not treatment specific. Use of pharmacologic stimuli does allow dissection of the specific interactions between the cellular environment and the *cis*-acting variants (Grundberg et al., 2011).

Asthma Genomics

Aside from the GWAS papers in asthma, referred to above, the field of asthma genomics has been relatively quiet over the past year. The eMERGE investigators found that the chromosome 17q locus was a significant determinant of peripheral blood leukocyte count in a large population of 13,923 subjects (Crosslin et al., 2011). Recent interest has been seen in the possible importance of microRNAs in asthma. MicroRNAs control transcription and modulate gene expression. SNPs can influence microRNAs in a number of ways, by inhibiting their transcription or by controlling their expression. Recently the Let-7 microRNA has been shown to inhibit Il-13 expression and hence decrease allergic inflammation in asthma (Kumar et al., 2011).

VITAMIN D AND ASTHMA

By 2003, the VDR had been recognized as one of the genes on chromosome 12q31 to be associated with asthma, and these results had been replicated in other populations (Raby et al., 2004; Poon et al., 2004). There were also negative studies (Saadi et al., 2009), but the best studies methodologically demonstrated a positive association. This led to an inquiry as to what might be the mechanism for a relationship between vitamin D and asthma.

Several lines of reasoning suggested that there might be such a relationship. First, vitamin D was known to influence immune function by influencing both regulatory T cells (Treg) and dendritic cell function, as indicated in Figure 74.2 (Lange et al., 2009). Second, vitamin D was known to influence the development of the fetal lung *in utero*, particularly the process of alveolarization in the third trimester. Third, changes in vitamin D intake in US populations were substantial over the same time period as the increase in asthma and allergy prevalence. Finally, the vast majority of children and adults, in the US and worldwide, were deficient or insufficient in their vitamin D levels. Since asthma is a disease of early childhood usually occurring before the age of six years, the easiest way to test whether there was a relationship between vitamin D and asthma was to do a birth cohort study (see Table 74.2).

Utilizing a birth cohort in a large health maintenance organization (HMO), Camargo and colleagues examined association of maternal intake of vitamin D in quartiles during pregnancy to asthma risk in the offspring. There were 186 cases of recurrent wheeze in the children, and there was a 59% reduction in the risk of recurrent wheezing in the highest quartile of maternal vitamin D intake compared to the lowest quartile (Figure 74.3) (Camargo et al., 2007). This effect persisted after adjustment for a large number of potential confounders.

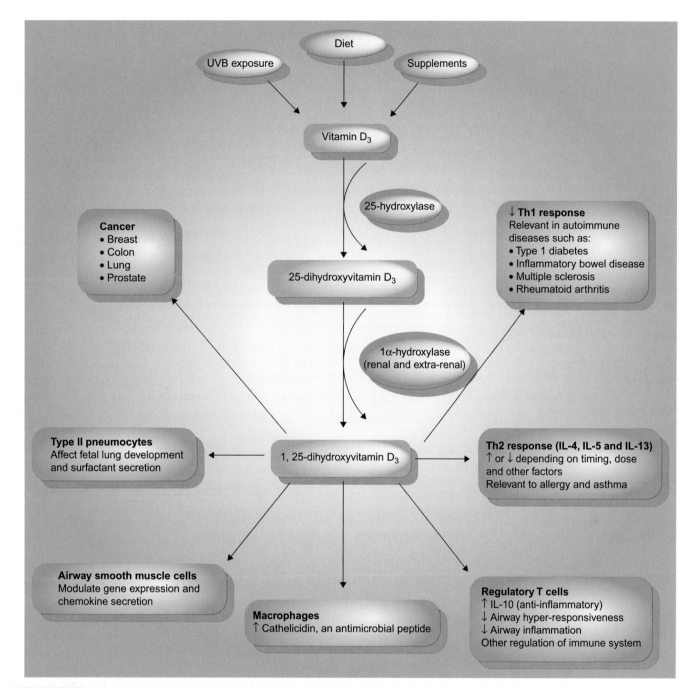

Diet
UVB exposure
Supplements

Figure 74.2 Extraskeletal effects of vitamin D. *Reproduced with permission from Lange et al., 2009.*

Similar results were obtained in a second birth cohort study in Aberdeen, Scotland, where the intake of vitamin D was much lower than it was in the Boston HMO (Table 74.3), with the highest intake in Aberdeen being equivalent to the lowest quintile of intake in Boston (Devereux et al., 2007).

Nevertheless, the results were very similar, with the highest quintile of vitamin D intake during pregnancy having the lowest asthma risk in the offspring (Table 74.3) (Devereux et al., 2007). In this study, the effects were strong for all classes of wheezing,

but the results for doctor-diagnosis of asthma at age five were not significant. An additional observational birth cohort study in Finland yielded similar results (Erkkola et al., 2009).

Several significant points could be made about these studies. First, none of the observational birth cohort studies looked at maternal vitamin D levels during pregnancy. This is probably a good thing, since even at the range of intakes we observed, the levels of maternal vitamin D achieved would have been in the insufficient or deficient range.

TABLE 74.2	Project Viva: maternal vitamin D intake and asthma				
	Quartile of vitamin D intake by mother during pregnancy				
	1	2	3	4	P for trend
Median intake (IU)	356	513	603	724	
Range	60–445	446–562	563–658	659–1145	
n	298	299	299	298	
Recurrent wheeze, n = 186 cases					
Unadjusted	1	0.55 (0.36, 0.84)	0.55 (0.36, 0.83)	0.39 (0.25, 0.62)	<0.001
Multivariable model*	1	0.49 (0.30, 0.80)	0.56 (0.34, 0.92)	0.41 (0.24, 0.70)	0.001

*Multivariable model adjusts for sex, birth weight, income, maternal age, maternal pre-pregnancy body mass index, passive smoking exposure, breastfeeding duration at one year, number of children less than 12 years of age in household, maternal history of asthma, and paternal history of asthma. IU, international units.
Adapted with permission from Camargo et al., 2007.

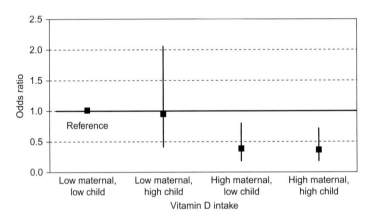

Figure 74.3 Relationship between maternal and child vitamin D intake and asthma risk in the offspring. *Reproduced with permission from Camargo et al., 2007.*

TABLE 74.3	Maternal vitamin D intake and early childhood wheezing					
	Quintile of vitamin D intake by mother during pregnancy					
	1 (lowest)	2	3	4	5 (highest)	P for trend
Median energy-adjusted intake (IU/day)	77	104	128	157	275	
5th–95th centile	46–92	94–115	117–139	142–182	189–751	
Recurrent wheeze at five years of age						
Unadjusted	1	0.86 (0.45–1.66)	0.81 (0.42–1.56)	0.63 (0.32–1.24)	0.39 (0.18–0.84)	0.01
Multivariable model*	1	0.89 (0.44–1.82)	0.71 (0.34–1.48)	0.47 (0.21–1.08)	0.34 (0.14–0.83)	0.005

*Multivariable model is adjusted for maternal atopy, maternal age, maternal smoking, maternal age of leaving full-time education, paternal social class, deprivation index based on area of residence, breast feeding, infant gender, infant antibiotic use in first year, birth weight, birth order, season of last menstrual period (LMP).
Table adapted from Devereux et al., 2007.

Second, although the studies adjust for potential confounding variables and looked at a wide range of vitamin D intakes, they did not look at a dose of vitamin D that would result in "sufficient" levels of vitamin D in the mother, which would be >30 ng/ml. Levels <20 ng/ml are deficient and 20–30 ng/ml is insufficient. While it is not known what the optimal level is for immune function, most vitamin D experts would suggest levels above 30 ng/ml. To achieve this level, the intake would have needed to be at least 4000 IU/day (Hollis and Wagner, 2004; Johnson et al., 2011).

The argument for an effect of vitamin D in asthma pathogenesis does not simply rest on these epidemiologic studies, as there is substantial data suggesting that there were structural effects of vitamin D on lung development and also on immune function. It also rests on the very large number of vitamin D-insufficient and -deficient pregnant women (roughly 75% of all pregnancies). Vitamin D also fits the model for autoimmune diseases generally, where it is postulated that there is genetic susceptibility (HLA being the most important), and that usually an infection (or other antigen) in conjunction with this genetic susceptibility leads to disease. While HLA is the most important locus for autoimmune disease, other genes such as VDR are involved and play a secondary role. In particular, VDR polymorphisms and vitamin D levels help to determine how the genetically susceptible host handles viral illness. In the case of asthma, respiratory syncytial virus (RSV) and other viruses lead to severe respiratory infections, hospitalizations, and recurrent wheezing. In the case of type 1, or juvenile, diabetes mellitus, the viral illness is adenovirus, which then permanently damages the islet cells of the pancreas, causing reduced insulin levels.

While doing more genetics to define the biochemical effects of the vitamin D pathway genes that were conferring risk and doing genomic studies to find additional vitamin D genes involved in the process, the observational epidemiology suggested that the definitive study to determine if vitamin D deficiency caused asthma would be a randomized controlled clinical trial in pregnant women. This is particularly true since deficiency is so widespread – observational studies would have a significant null bias due to lack of variation. In 2009,

such a trial was funded by the National Heart, Lung, and Blood Institute (UO1 HL091528; clinical trials.gov # NCT00920621). A total of 870 pregnant women with a family history of asthma and allergies were randomized to 4400 IU of vitamin D/day, or just to multivitamins (400 IU/day). They are currently being followed through their pregnancy and their offspring will be followed to age three for wheezing illnesses and asthma. At the present time, it is unclear if the trial will be successful or not. If successful, the power of the trial will derive from the vast number of insufficient and deficient women, and hence, the large population-attributable risk, if these women were properly repleted with vitamin D.

LESSONS LEARNED

What lessons can be learned from these data in attempting to translate genetic and genomic discoveries into clinical practice for asthma?

1. There are no loci of large effect for asthma. The locus with the greatest effect, chromosome 17q21, while significant, is, in virtually every study, only important in childhood disease.

2. Like in all autoimmune diseases, the HLA locus is important in asthma; next to 17q21, this locus has the greatest effect in most studies.

3. Translating genetic effects into clinical practice for asthma is likely to both be difficult and take a long time, since individual genes have small effects, and those effects are linked to the functioning of other genes and pathways. A systems genetics approach might be the best way forward.

4. The vitamin D story provides an unusual example of a nonstandard way that genomics might translate into clinical practice. This is an example of using discovery genetics to find novel biology, but especially a novel exposure (vitamin D) that had been completely overlooked.

5. We need to consider novel ways to functionally assess genes found to be associated with asthma and to accelerate translation into clinical practice.

REFERENCES

Camargo Jr., C.A., Rifas-Shiman, S.L., Litonjua, A.A., et al., 2007. Maternal intake of vitamin D during pregnancy and risk of recurrent wheeze in children at 3 yrs of age. Am J Clin Nutr 85, 788–795.

Crosslin, D.R., McDavid, A., Weston, N., et al., 2011. Genetic variants associated with the white blood cell count in 13,923 subjects in the eMERGE network. Hum Genet, 131. doi: 10.1007/s00439-011-1103-9 [Epub.].

Devereux, G., Litonjua, A.A., Turner, S.W., et al., 2007. Maternal vitamin D intake during pregnancy and early childhood wheezing. Am J Clin Nutr 85, 853–859.

Duan, Q.L., Du, R., Lasky-Su, J., et al., 2012. A polymorphism in the thyroid hormone receptor gene is associated with bronchodilator response in asthmatics. Pharmacogenomics doi: 10.1038/tpj.2011.56 [Epub.].

Duan, Q.L., Gaume, B.R., Hawkins, G.A., et al., 2011. Regulatory haplotypes in ARG1 are associated with altered bronchodilator response. Am J Respir Crit Care Med 183, 449–454.

Erkkola, M., Kaila, M., Nwaru, B.I., et al., 2009. Maternal vitamin D intake during pregnancy is inversely associated with asthma and allergic rhinitis in 5-year old children. Clin Exp Allergy 39, 875–882.

Galanter, J., Choudhry, S., Eng, C., et al., 2008. Ormdl3 gene is associated with asthma in three ethnically diverse populations. Am J Respir Crit Care Med 177, 1194–1200.

Grundberg, E., Adoue, V., Kwan, T., et al., 2011. Global analysis of the impact of environmental perturbation on cis-regulation of gene expression. PLoS Genet 7, e1001279.

Hancock, D.B., Romieu, I., Shi, M., et al., 2009. Genome-wide association study implicates chromosome 9q21.31 as a susceptibility locus for asthma in Mexican children. PLoS Genet 5, e1000623.

Himes, B.E., Hunninghake, G.M., Baurley, J.W., et al., 2009. Genome-wide association analysis identifies pde4d as an asthma-susceptibility gene. Am J Hum Genet 84, 581–593.

Hollis, B.W., Wagner, C.L., 2004. Assessment of dietary vitamin D requirements during pregnancy and lactation. Am J Clin Nutr 79, 717–726.

Johnson, D.D., Wagner, C.L., Hulsey, T.C., et al., 2011. Vitamin D deficiency and insufficiency is common during pregnancy. Am J Perinatol 28, 7–12.

Koster, E.S., Maitland-van der Zee, A.H., Tavendale, R., et al., 2011. FCER2 T2206C variant associated with chronic symptoms and exacerbations in steroid-treated asthmatic children. Allergy 66, 1546–1552.

Kumar, M., Ahmad, T., Sharma, A., et al., 2011. Let-7 microRNA-mediated regulation of IL-13 and allergic airway inflammation. J Allergy Clin Immunol 128, 1077–1085.

Lange, N.E., Litonjua, A., Hawrylowicz, C.M., Weiss, S., 2009. Vitamin D, the immune system and asthma. Expert Rev Clin Immunol 5, 693–702.

Li, X., Howard, T.D., Zheng, S.L., et al., 2010. Genome-wide association study of asthma identifies rad50-il13 and hla-dr/dq regions. J Allergy Clin Immunol 125, 328–335. e311.

Litonjua, A.A., Lasky-Su, J., Schneiter, K., et al., 2008. ARG1 is a novel bronchodilator response gene: Screening and replication in four asthma cohorts. Am J Respir Crit Care Med 178, 688–694.

Mathias R.A., Grant A.V., Rafaels N., et al., 2010. A genome-wide association study on African-ancestry populations for asthma. J Allergy Clin Immunol 125, 336–346, e334.

Moffatt, M.F., Gut, I.G., Demenais, F., et al., 2010. GABRIEL Consortium. A large-scale, consortium-based genome-wide association study of asthma. N Engl J Med 363, 1211–1221.

Moffatt, M.F., Kabesch, M., Liang, L., et al., 2007. Genetic variants regulating ormdl3 expression contribute to the risk of childhood asthma. Nature 448, 470–473.

Poon, A.H., Laprise, C., Lemire, M., et al., 2004. Association of vitamin D receptor genetic variants with susceptibility to asthma and atopy. Am J Respir Crit Care Med 170, 967–973.

Postma, D.S., Kerkhof, M., Boezen, H.M., Koppelman, G.H., 2011. Asthma and COPD: Common genes, common environments? Am J Respir Crit Care Med, 183. doi: 10.1164/rccm.201011-1796PP.

Raby, B.A., Lazarus, R., Silverman, E.K., et al., 2004. Association of vitamin D receptor gene polymorphisms with childhood and adult asthma. Am J Respir Crit Care Med 170, 1057–1065.

Saadi, A., Gao, G., Li, H., et al., 2009. Association study between vitamin D receptor gene polymorphisms and asthma in the Chinese Han population: A case-control study. BMC Med Genet 10, 71.

Sleiman, P.M., Annaiah, K., Imielinski, M., et al., 2008. Ormdl3 variants associated with asthma susceptibility in North Americans of European ancestry. J Allergy Clin Immunol 122, 1225–1227.

Sleiman, P.M., Flory, J., Imielinski, M., et al., 2009. Variants of dennd1b associated with asthma in children. N Engl J Med 362, 36–44.

Tantisira, K.G., Lasky-Su, J., Harada, M., et al., 2011. Genome-wide association between GLCCI1 and response to glucocorticoid therapy in asthma. N Engl J Med 365, 1173–1183.

Tantisira, K.G., Silverman, E.S., Mariani, T.J., et al., 2007. FCER2: A pharmacogenetic basis for severe exacerbations in children with asthma. J Allergy Clin Immunol 120, 1285–1291.

Torgerson, D.G., Ampleford, E.J., Chiu, G.Y., et al., 2011. Meta-analysis of genome-wide association studies of asthma in ethnically diverse North American populations. Nat Gen 43, 887–892.

Verlaan, D.J., Berlivet, S., Hunninghake, G.M., et al., 2009. Allele-specific chromatin remodeling in the ZPBP2/GSDMB/ORMDL3 locus associated with the risk of asthma and autoimmune disease. Am J Hum Genet 85, 377–393.

Weiss, S.T., Raby, B.A., Rogers, A., 2009. Asthma genetics and genomics. Curr Opin Genet Dev 19, 279–282.

Chronic Obstructive Pulmonary Disease

Peter J. Barnes

SUMMARY

Chronic obstructive pulmonary disease (COPD) is characterized by progressive and largely irreversible airflow limitation due to narrowing and fibrosis of small airways and loss of airway alveolar attachments as a result of emphysema. Cigarette smoking is the most important risk factor, but other noxious gases are important in developing countries. Genes likely determine which smokers are susceptible to the development of airflow obstruction, but these genes have not yet been consistently identified. Several single nucleotide polymorphisms of candidate genes have now been associated with COPD susceptibility, but most have not been replicated. More recent genome-wide association studies (GWAS) have shown few consistent findings apart from polymorphisms of the nicotine receptor (which may reflect nicotine addiction) and abnormalities in the hedgehog signaling pathways involved in lung development. There is a specific pattern of inflammation characterized by increased numbers of macrophages, neutrophils, and T-lymphocytes (particularly $CD8^+$ cells), which is an amplification of the inflammation seen in normal cigarette smokers and increases with disease progression. This amplification may be due to reduced histone deacetylase (HDAC) activity, which could be genetically determined. Proteinases, particularly MMP-9, result in degradation of alveolar wall elastin, resulting in emphysema. There is an abnormal pattern of inflammatory protein expression in COPD, and most of these proteins are regulated at a transcriptional level. Gene microarray analysis of peripheral lung and specific cell types shows up-regulation of many genes in inflammatory and immune signaling pathways. Proteomic analysis also shows increased inflammatory proteins in the plasma of COPD patients. Management of COPD involves smoking cessation (which may be influenced by genetic factors). Bronchodilator therapy reduces hyperinflation, and long-acting inhaled β_2-agonists and anticholinergics are the preferred therapy. Inhaled corticosteroids do not reduce inflammation because of reduced HDAC activity. Pharmacogenomics has not yet been applied to COPD, but polymorphisms of the β_2-adrenoceptor have little influence on responses to β_2-agonists. Future therapies may be identified by targeting new genes (or pathways) that regulate disease activity.

INTRODUCTION

COPD is characterized by progressive development of airflow limitation that is not fully reversible (Barnes, 2000). The term COPD encompasses chronic obstructive bronchiolitis with obstruction of small airways and emphysema with enlargement of airspaces and destruction of lung parenchyma, loss of lung elasticity, and closure of small airways (Hogg, 2004). Chronic bronchitis, by contrast, is defined by a productive

Genomic and Personalized Medicine, 2nd edition
by Ginsburg & Willard. DOI: http://dx.doi.org/10.1016/B978-0-12-382227-7.00075-6

cough of more than three months duration for more than two successive years; this reflects mucous hypersecretion and is not necessarily associated with airflow limitation.

COPD is common and is increasing globally. It is now the fourth-leading cause of death in the US, and the only common cause of death that is increasing. This is probably an underestimate as COPD is likely to be contributory to other common causes of death. It is predicted to become the third most common cause of death and the fifth most common cause of chronic disability worldwide in the next few years (Mannino and Buist, 2007). It currently affects more than 5% of the adult population and is grossly underdiagnosed in the community.

There is persuasive evidence that susceptibility to develop COPD is genetically determined, although the exact genes have not yet been identified. Genomics and proteomics are currently in use to characterize the abnormal protein expression in this disease as a way of understanding its complex pathophysiology. There are clearly many different phenotypes of COPD that need to be better defined, and this will be increasingly important in genetic studies. Finally, pharmacogenomics may have an impact on COPD therapy in the future (Hersh, 2010). This chapter gives an overview of COPD, and highlights where genomic medicine is relevant to future understanding and management of this common disease.

PREDISPOSITION

Several environmental and endogenous factors, including genes, increase the risk of developing COPD (see Table 75.1).

TABLE 75.1 Risk factors for COPD	
Environmental factors	**Endogenous (host) factors**
Cigarette smoking:	α_1-antitrypsin deficiency
Active	Other genetic factors
Passive	Ethnic factors
Maternal	
Air pollution:	Airway hyper-responsiveness?
Outdoor	
Indoor: biomass fuels	
Occupational exposure	Low birth weight
Dietary factors:	
High salt	
Low antioxidant vitamins	
Low unsaturated fatty acids	
Infections	

Environmental Factors

In industrialized countries, cigarette smoking accounts for most cases of COPD, but in developing countries other environmental pollutants, such as particulates associated with cooking in confined spaces, are important causes (Salvi and Barnes, 2009). It is likely that there are important interactions between environmental factors and a genetic predisposition to the disease. Air pollution (particularly sulfur dioxide and particulates), exposure to certain occupational chemicals such as cadmium, and passive smoking may also be risk factors. The roles of airway hyper-responsiveness and allergy as risk factors for COPD is still uncertain. Atopy, serum IgE and blood eosinophilia are not important risk factors. However, this is not necessarily the same type of abnormal airway responsiveness that is seen in asthma. Low birth weight is also a risk factor for COPD, probably because poor nutrition in fetal life results in small lungs, so that decline in lung function with age starts from a lower peak value.

Genetic Factors

Longitudinal monitoring of lung function in cigarette smokers reveals that only a minority (15–40%, depending on definition) develop significant airflow obstruction due to an accelerated decline in lung function [two- to five-fold higher than the normal decline of 15 to 30 mL forced expiratory volume in 1 second (FEV_1)/year compared to the normal population and the remainder of smokers who have consumed an equivalent number of cigarettes] (Figure 75.1). The rate of decline in lung function in the general population (Framingham Study) has a heritability of about 50%. This strongly suggests that genetic factors may determine which smokers are susceptible and develop airflow limitation. Familial clustering of patients with early-onset COPD and the differences in COPD prevalence between different ethnic groups (Wan and Silverman, 2009) provide further evidence that genetic factors are important. Patients with α_1-antitrypsin deficiency [proteinase inhibitor (Pi)ZZ phenotype with α_1-antitrypsin levels <10% of normal values] develop early emphysema that is exacerbated by smoking, indicating a clear genetic predisposition to COPD (Kelly et al., 2010). However, α_1-antitrypsin deficiency accounts for <1% of patients with COPD, and many other genetic variants of α_1-antitrypsin that are associated with lower-than-normal serum levels of this proteinase inhibitor have not been clearly associated with an increased risk of COPD, although analysis of several studies indicates a small risk, with an odds ratio from COPD in PiMZ versus PiMM individuals of approximately two.

This has led to a search for associations between COPD and single nucleotide polymorphisms (SNPs) of other candidate genes that may be involved in its pathophysiology (Castaldi et al., 2010). Various SNPs have been associated with COPD, as defined by a reduced FEV_1, but there is emerging evidence that different aspects of COPD may relate to different genotypes. For example, a reduction in gas diffusion is

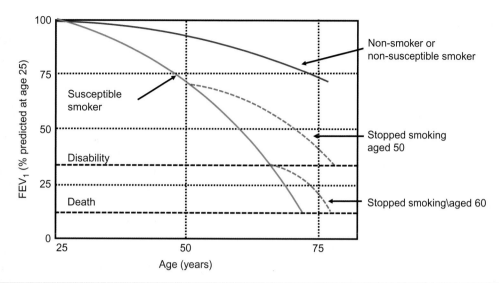

Figure 75.1 Natural history of COPD. Annual decline in airway function showing accelerated decline in susceptible smokers and effects of smoking cessation. Patients with COPD usually show an accelerated annual decline in forced expiratory volume in 1 second (FEV₁), often greater than 50 mL/year, compared to the normal decline of approximately 20 mL/year, although this is variable between patients. The classic studies of Fletcher and Peto established that 10–20% of cigarette smokers are susceptible to this rapid decline. However, with longer follow-up, more smokers may develop COPD. The propensity to develop COPD among smokers is only weakly related to the amount of cigarettes smoked; this suggests that other factors play an important role in determining susceptibility. Most evidence points toward genetic factors, although the genes determining susceptibility have not yet been determined.

correlated in SNPs of the microsomal epoxide hydrolase gene, reduced exercise performance is correlated with SNPs of the latent transforming growth factor-beta binding protein-4 (*LTBP4*), whereas dyspnea was linked to three SNPs in transforming growth factor β1 (TGF-β1). This suggests that more careful phenotyping will be required in the future to sort out susceptibility genes. A 10-fold increased risk of COPD in individuals who have a polymorphism in the promoter region of the gene for tumor necrosis factor-α (TNF-α) that is associated with increased TNF-α production has been reported in a Taiwanese population, but not confirmed in Caucasian populations. So far, few significant associations have been detected between SNPs and disease, and even those reported have not usually been replicated in other studies or have no obvious functional effects. Several other genes have been implicated in COPD, but very few have been replicated in different populations (Castaldi et al., 2010). One problem is that the number of patients in most studies is too low to detect genetic effects odds ratios (OR) of 1.2–1.5. In a meta-analysis of adequately powered studies, only four genes were significantly associated with COPD susceptibility: *GSTM1* (OR 1.45), *TGFβ1* (OR 0.73), *TNF-α* (OR 1.19) and *SOD3* (OR 1.97) (see Table 75.2). In addition to candidate genes that have been linked to the progressive lung disease, genetic factors are also important in addictive behavior, such as the nicotine addiction of smokers, which has a high degree of heritability. For example, SNPs of the dopamine transporter gene *SLC6A3* and the dopamine

Table 75.2 Some of the genes associated with COPD susceptibility

Candidate genes	Risk
α₁-antitrypsin	ZZ genotype high risk
	MZ, SZ genotypes small risk
α₁-chymptrypsin	Associated in some populations
Matrix metalloproteinase-1, -2, -9, -12	Associated in some studies
Microsomal epoxide hydrolase	Increased risk
Glutathione S-transferase	Increased risk
Heme oxygenase-1	Small risk but consistent
Interleukin-13	Small risk
Vitamin D binding protein	Inconsistent
TNF-α promoter	Inconsistent
TGF-β promoter	Inconsistent
α-nicotinic receptor	From GWAS: marker of nicotine addiction?
Hedgehog-interacting protein	From GWAS: involved in lung development

TNF, tumor necrosis factor, TGF, transforming growth factor.

receptor 2 gene have been associated with nicotine addiction and inability to quit (Erblich et al., 2005).

One of the problems in the genetics of COPD is the clinical heterogeneity of the disease. There is now an effort to link different aspects of the disease to specific genetic associations, in order to classify disease subtypes that might be useful in selecting patients for different types of therapy in the future. Several SNPs were reported to be associated with severe COPD in almost 400 patients who participated in the National Emphysema Treatment Trial (NETT). Two of those genes, microsomal epoxide hydrolase (*EPHX1*) and *SERPINE2*, were found to be significantly associated with hypoxema, and in a separate population of early-onset COPD patients, the same SNPs were associated with requirement of domiciliary oxygen (Castaldi et al., 2010). Another SNP in the surfactant protein B (*SFTPB*) gene was associated with pulmonary hypertension. Bronchodilator responsiveness to the inhaled β_2-agonist albuterol has been linked to SNPs in the *SERPINE2* and *EPHX1* genes, as well as in the β_2-adrenergic receptor (*ADRB2*) gene (Kim et al., 2009). Extracellular superoxide dismutase (SOD3) is an important antioxidant enzyme in the lungs, and two novel SNPs of *SOD3* have been linked to reduced FEV_1 values in a Danish population study; these polymorphisms were associated with hospitalization and mortality from COPD (Dahl et al., 2008). A SNP in the promoter region of *MMP12* has been shown to be protective against the development of COPD in smokers (OR 0.65) (Hunninghake et al., 2009).

Several genome-wide association studies (GWAS) have been reported and are currently underway, such as COPDGene. The first reported GWAS identified three SNPs on chromosome 5 corresponding to the nicotinic acetylcholine receptor (CHRNA3, CHRNA5) (Pillai et al., 2009). A similar association has previously been made with lung cancer and almost certainly represents a marker of nicotine addiction. A weak association was also found with hedgehog interacting protein (HHIP), which is involved in lung development and is located on chromosome 4. A GWAS based on the Framingham Heart Study found an SNP in the region of HHIP (Wilk et al., 2009), and this has been confirmed in another study (Van Durme et al., 2010). A further GWAS has shown that the *BICD1* gene is linked to emphysema (Kong et al., 2010). This gene encodes a protein that is associated with telomere length and therefore may be a marker of susceptibility to aging.

PATHOPHYSIOLOGY

COPD includes chronic obstructive bronchiolitis with fibrosis and obstruction of small airways, and emphysema with enlargement of airspaces and destruction of lung parenchyma, loss of lung elasticity, and closure of small airways. Chronic bronchitis, by contrast, is defined by a productive cough of more than three months duration for more than two successive years; this reflects mucous hypersecretion and is not necessarily associated with airflow limitation. Most patients with COPD have all three pathological mechanisms (chronic obstructive bronchitis, emphysema, and mucus plugging), as all are induced by smoking, but may differ in the proportion of emphysema and obstructive bronchitis (Figure 75.2).

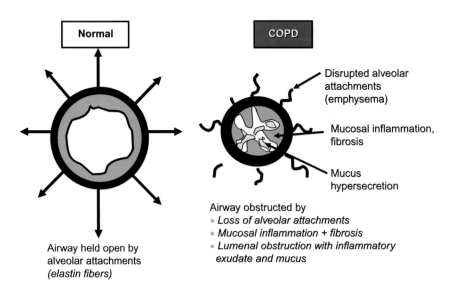

Figure 75.2   Mechanisms of airflow limitation in COPD. The airway in normal subjects is distended by alveolar attachments during expiration, allowing alveolar emptying and lung deflation. In COPD, these attachments are disrupted because of emphysema, contributing to airway closure during expiration, trapping gas in the alveoli and resulting in hyperinflation (Hogg, 2001). Peripheral airways are also obstructed and distorted by airway inflammation and fibrosis (chronic obstructive bronchiolitis) and by occlusion of the airway lumen by mucous secretions that may be trapped in the airways because of poor mucociliary clearance.

Small Airways

There has been debate about the predominant mechanism of progressive airflow limitation – recent pathological studies suggest that is closely related to the degree of inflammation, narrowing, and fibrosis in small airways (Barnes, 2004; Hogg et al., 2004). Emphysema may contribute to airway narrowing in the more advanced stages of the disease, with disruption of alveolar attachments facilitating small airway closure and gas trapping. This combined effect of small airway disease and early closure on expiration results in lung hyperinflation, which leads to progressive exertional dyspnea, the predominant symptom of COPD.

Emphysema

Emphysema describes loss of alveolar walls due to destruction of matrix proteins (predominantly elastin) and loss of type 1 pneumocytes as a result of apoptosis. Several patterns of emphysema are recognized: centriacinar emphysema radiates from the terminal bronchiole, panacinar emphysema involves more widespread destruction, and bullae are large airspaces. Emphysema results in airway obstruction by loss of elastic recoil so that intrapulmonary airways close more readily during expiration. Emphysema with loss of gas-exchanging surface also leads to progressive hypoxia and eventually to respiratory failure.

Pulmonary Hypertension

Chronic hypoxia may lead to hypoxic vasoconstriction, with structural changes in pulmonary vessels that eventually lead to secondary pulmonary hypertension. Inflammatory changes similar to those seen in small airways are also seen in pulmonary arterioles. Only a small proportion of COPD patients develop pulmonary hypertension, and it is likely that genetic susceptibility plays a role.

Systemic Features

Patients with severe COPD also develop systemic features, which may have an adverse effect on prognosis (Barnes and Celli, 2009). The most common systemic feature is weight loss due to loss of skeletal muscle bulk. This may contribute to muscle weakness as a result of impaired mobility due to dyspnea. Osteoporosis and depression are other systemic features and may be due to overspill of inflammatory mediators from the lung into the systemic circulation. There appears to be a difference in susceptibility to systemic features among patients as they do not always correlate with disease severity, and this may be genetically determined. Patients with COPD also have comorbidities, particularly cardiovascular diseases. Smoking is a common risk factor for ischemic heart disease and COPD, but there may be shared genetic susceptibilities. There may be common genetic predispositions to COPD and cardiovascular diseases. COPD patients with coexisting cardiovascular disease have increased circulating levels of

C-reactive protein (CRP). There is evidence for genetic determinants of plasma CRP concentrations in COPD patients (Hersh et al., 2006), and genetic factors may determine the likelihood of developing systemic features and the prevalence of comorbidities.

Exacerbations

Exacerbations are an important feature of COPD, with worsening of dyspnea and an increase in sputum production. They may lead to hospitalization, and account for a high proportion of the costs of COPD. Exacerbations are usually caused by infections, either due to bacteria (especially *Haemophilus influenzae* or *Streptococcus pneumoniae*) or upper respiratory tract virus infections (especially rhinovirus or respiratory syncytial virus) (Celli and Barnes, 2007). COPD exacerbations tend to increase as the disease progresses, but some patients appear to have more frequent exacerbations than others, which may suggest predisposing genetic factors.

CELLULAR AND MOLECULAR MECHANISMS

Inflammation

In early stages of the disease, chronic inflammation occurs, predominantly in small airways and the lung parenchyma, with an increase in numbers of macrophages and neutrophils, indicating an enhanced innate immune response. In more advanced stages of the disease, there is an increase in lymphocytes (particularly cytotoxic $CD8^+$ T cells), including lymphoid follicles that contain B- and T-lymphocytes, indicating acquired immunity (Barnes, 2008b; Hogg et al., 2004) (Figure 75.3). Alveolar macrophages play a critical role in the orchestration of this pulmonary inflammation since they are activated by inhaled irritants such as cigarette smoke, and release chemokines that attract inflammatory cells such as monocytes, neutrophils, and T cells into the lungs (Barnes et al., 2003).

Inflammatory Mediators

Prominent mediators are those that amplify inflammation, such as tumor necrosis factor-α (TNF-α), interleukin-1β (IL-1β) and IL-6, and chemokines, which attract inflammatory cells such as CXCL8, CXCL1, CXCL10, CCL1, and CCL5 (Barnes, 2009b). Elastolytic enzymes account for the tissue destruction caused by emphysema and include neutrophil elastase and matrix metalloproteinase-9 (MMP-9). There is an imbalance between increased production of elastases and a deficiency of endogenous antiproteases, such as α_1-antritrypsin, secretory leukoprotease inhibitor, and tissue inhibitors of MMPs. MMP-9 may be the predominant elastolytic enzyme causing emphysema, and also activates transforming growth factor-β, a cytokine expressed particularly in small airways that may result in the characteristic peribronchiolar fibrosis.

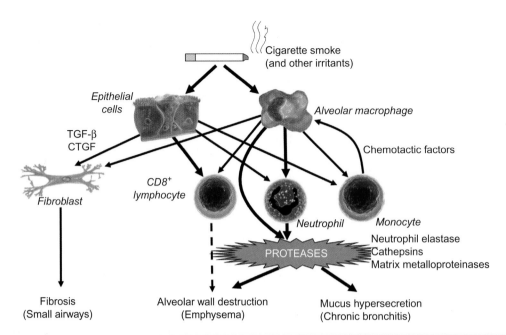

Figure 75.3 Inflammatory mechanisms in COPD. Cigarette smoke and other irritants activate macrophages in the respiratory tract that release neutrophil chemotactic factors, including interleukin-8 (IL-8) and leukotriene B_4 (LTB$_4$). These cells then release proteases that break down connective tissue in the lung parenchyma, resulting in emphysema, and also stimulate mucus hypersecretion. These enzymes are normally counteracted by protease inhibitors, including α_1-antitrypsin, secretory leukoprotease inhibitor (SLPI) and tissue inhibitor of matrix metalloproteinases (TIMP). Cytotoxic T cells (CD8$^+$) may also be recruited and may be involved in alveolar wall destruction. CTGF, connective tissue growth factor.

Oxidative stress is a prominent feature of COPD and is due to exogenous oxidants in cigarette smoke and endogenous oxidants released from activated inflammatory cells such as neutrophils and macrophages (Bowler et al., 2004). Endogenous antioxidants may also be defective. Oxidative stress enhances inflammation and may lead to corticosteroid resistance.

Differences with Asthma

Although both COPD and asthma involve chronic inflammation of the respiratory tract, there are marked differences in the inflammatory processes of the diseases, as summarized in Table 75.3 (Barnes, 2008b). While there are striking differences between the inflammation in mild asthma and COPD, patients with severe asthma become much more similar to patients with COPD, with involvement of neutrophils, macrophages, TNF-α, CXCL8, oxidative stress, and a poor response to corticosteroids. This suggests that there may be similarities in genetic predisposition in smokers with COPD and in patients with severe asthma. Several novel susceptibility genes that have been identified in severe asthma, including *DPP10*, *GPRA*, *PHF11* and *ADAM33*, have now also been seen in COPD patients (Van Diemen et al., 2005).

Gene Regulation

Proinflammatory mediators such as TNF-α and IL-1β activate the transcription factor nuclear factor-κB (NF-κB), which is activated in the airways and lung parenchyma of COPD patients,

particularly in epithelial cells and macrophages. There is further activation of NF-κB during exacerbations. NF-κB switches on many of the inflammatory genes that are activated in COPD lungs, including chemokines, adhesion molecules such as ICAM-1 and E-selectin, inflammatory enzymes such as cyclooxygenase-2 and inducible nitric oxide synthase, elastolytic enzymes such as MMP-9, and proinflammatory mediators such as TNF-α and IL-1β, which themselves activate NF-κB. NF-κB-activated genes result in acetylation of core histones (particularly histone-4), which is necessary for activation of inflammatory genes; this is reversed by histone deacetylase-2 (HDAC2). There is a marked reduction in HDAC2 activity and expression in lung parenchyma, airways and alveolar macrophages of COPD patients (Barnes, 2009a). This mechanism may account for the amplified pulmonary inflammation seen in COPD compared to smokers with normal lung function, and also explains why COPD patients are not responsive to corticosteroids, since HDAC2 is the mechanism whereby corticosteroids switch off activated inflammatory genes (Barnes and Adcock, 2009) (Figure 75.4). The reduction in HDAC2 expression in COPD appears to be secondary to increased oxidative stress, which leads to tyrosine nitration and phosphorylation of HDAC2.

Genomics

Better knowledge of the complex pathophysiology underlying the disease has led to the identification of many candidate genes that appear to be involved in the molecular

TABLE 75.3	Differences between inflammation in COPD and asthma	
Inflammation	**COPD**	**Asthma**
Inflammatory cells	Neutrophils	Eosinophils
	CD8$^+$ T cells +++	Mast cells
	CD4$^+$ T cells +	CD4$^+$ T cells
	Macrophages +++	Macrophages +
Inflammatory mediators	LTB$_4$	LTD$_4$, histamine
	TNF-α	IL-4, IL-5, IL-13
	CXCL1, CXCL8	CCL11
	Oxidative stress +++	Oxidative stress +
Inflammatory effects	Epithelial metaplasia	Epithelial shedding
	Fibrosis ++	Fibrosis +
	Mucus secretion +++	Mucus secretion +
	AHR ±	AHR +++
Location	Peripheral airways predominantly	All airways
	Parenchymal destruction	No parenchymal effects
Response to corticosteroids	±	+++

LT, leukotriene; TNF, tumor necrosis factor; IL, interleukin; AHR, airway hyper-responsiveness.

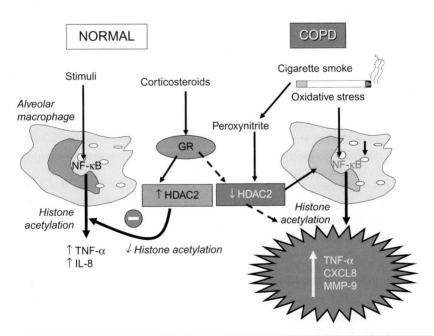

Figure 75.4 Proposed mechanism of corticosteroid resistance in COPD patients. Stimulation of normal alveolar macrophages activates nuclear factor-κB (NF-κB) and other transcription factors to switch on histone acetyltransferase, leading to histone acetylation and subsequently to transcription of genes encoding inflammatory proteins, such as tumor necrosis factor-α (TNF-α) and interleukin-8 (IL-8). Corticosteroids reverse this by binding to glucocorticoid receptors (GR) and recruiting histone deacetylase-2 (HDAC2). This reverses the histone acetylation induced by NF-κB and switches off the activated inflammatory genes. In COPD patients, cigarette smoke activates macrophages, as in normal subjects, but oxidative stress (acting through the formation of peroxynitrite) impairs the activity of HDAC2. This amplifies the inflammatory response to NF-κB activation, but also reduces the anti-inflammatory effect of corticosteroids, as HDAC2 is now unable to reverse histone acetylation.

mechanisms of COPD. Polymorphisms of several of them have been investigated for association with COPD. With gene microarrays it is now possible to study the global expression of all human genes in cells and tissues (transcriptomics). Few studies of global gene expression have been conducted in COPD, but several laboratories are now looking at the differences in gene expression between COPD patients and smokers exposed to similar numbers of cigarettes who have normal lung function. Those studies are using alveolar macrophages, circulating monocytes, airway epithelial cells, and peripheral lung. Serial analysis of gene expression and microarray analysis in peripheral lung from mild to moderate COPD patients versus normal smokers has demonstrated that a total of 261 genes showed differential expression (Ning et al., 2004). Peripheral lung of COPD patients shows up-regulation in genes involved in extracellular matrix, tissue repair, and apoptosis, whereas genes involved in suppressing the inflammatory response were down-regulated (Wang et al., 2008). Gene expression profiling of alveolar macrophages has demonstrated 40 genes up-regulated and 35 down-regulated in smokers compared to nonsmokers. Most of these genes belong to the functional categories of immune/inflammatory response, cell adhesion and extracellular matrix, proteases/antiproteases, antioxidants, signal transduction, and regulation of transcription (Heguy et al., 2006). The amplification of inflammation in COPD patients compared to equivalent smokers with normal lung function may be genetically determined, through amplification of inflammatory genes or defective expression of anti-inflammatory or protective mechanisms. However, there is a high degree of variability in gene array patterns between different subjects, which means that large numbers of samples are likely to be needed to characterize the COPD transcriptome (Zeskind et al., 2008).

Studies looking at susceptibility to the effects of cigarette smoke in different strains of mice may also be informative. For example, in two murine strains susceptible to cigarette smoke (DBA/2 and C57BL/6J) there is an increase in gene clusters involved in acute responses, adhesion molecule expression, and emphysema, whereas a smoke-insensitive strain (ICR) does not show any increase in those genes after tobacco exposure (Cavarra et al., 2009). Molecular genomics may identify markers of risk, but may also reveal novel molecular targets for the development of treatments in the future.

Proteomics

While gene microarray analysis gives considerable information about proteins that are transcriptionally regulated, it is now recognized that many genes involved in inflammation are regulated post-transcriptionally and that it is necessary to measure protein expression using proteomic approaches (Bowler et al., 2006). A proteomic approach can be used to investigate the plasma concentrations of multiple inflammatory and immune proteins during COPD exacerbations compared to baseline measurements (Hurst et al., 2006). CRP was found to be the most discriminate marker, and was linked to various proteins associated with monocytic and lymphocytic inflammatory pathways. In another study, 143 plasma proteins were assayed in COPD patients compared to control subjects; 25 were found to show some correlation with clinical parameters in the COPD patients (Pinto-Plata et al., 2006). Proteomic analysis may also be applied to cells, lung tissue or sputum (Barnes et al., 2006).

Another area where gene array analysis may be useful is metagenomic profiling using 16S rRNA to identify DNA of multiple microbes, including bacteria, viruses, or fungi, in the respiratory tract. This has revealed an enormous variety of organisms in normal and COPD lungs, although the pathological relevance of this is not yet established (Hilty et al., 2010; Huang et al., 2010). The relationship between this airway microbiome and disease severity, progression, and acute exacerbations is now under investigation.

DIAGNOSIS AND SCREENING
Diagnosis

COPD diagnosis is commonly made from the history of progressive dyspnea in a chronic smoker and is confirmed by spirometry that shows an FEV_1/vital capacity (VC) ratio of <70% and FEV_1 <80% predicted. Staging of severity is made on the basis of FEV_1, but exercise capacity and the presence of systemic features may be more important determinants of clinical outcome (Barnes and Celli, 2009). Measurement of lung volumes by body plethysmography shows an increase in total lung capacity, residual volume and functional residual capacity, with consequent reduction in inspiratory capacity, representing hyperinflation as a result of small airway closure. This results in dyspnea, which can be measured by dyspnea scales, and reduced exercise tolerance, which can be measured by a six-minute or shuttle walking test. Carbon monoxide diffusion is reduced in proportion to the extent of emphysema (Figure 75.5).

A chest X-ray is rarely useful, but may show hyperinflation of the lungs and the presence of bullae. High resolution computerized tomography demonstrates emphysema but is not used as a routine diagnostic test. Blood tests are rarely useful; a normocytic normochromic anemia is more commonly seen in patients with severe disease than polycythemia due to chronic hypoxia. Arterial blood gases demonstrate hypoxia and, in some patients, hypercapnia.

Screening

As symptoms present late in the progression of disease, COPD is grossly underdiagnosed (Mannino and Buist, 2007). Population screening with measurement of FEV_1 and FEV_1/VC ratio would pick up early COPD in smokers over 40 years old. However, this would not identify COPD due to non-smoking causes, which constitute 10–20% of the total. This would require screening spirometry of all individuals over 40 in addition to the 20–40% of the population who smoke. It is hoped

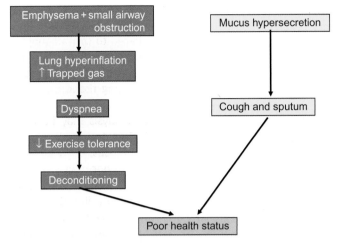

Emphysema + small airway obstruction

Lung hyperinflation ↑ Trapped gas

Dyspnea

↓ Exercise tolerance

Deconditioning

Mucus hypersecretion

Cough and sputum

Poor health status

Figure 75.5 Symptoms of COPD. The most prominent symptom of COPD is dyspnea, which is largely caused by hyperinflation of the lungs, a result of small airway collapse due to emphysema and narrowing due to fibrosis, so that the alveoli are not able to empty. Hyperinflation induces an uncomfortable sensation and reduces exercise tolerance. This leads to immobility and deconditioning, and results in poor health status. Other common symptoms of COPD are cough and sputum production as a result of mucus hypersecretion, but not all patients have these symptoms and many smokers with these symptoms do not have airflow obstruction (simple chronic bronchitis).

that gene expression studies might indicate a biomarker that is correlated with COPD susceptibility, but this is unlikely as so many genes are probably involved (Barnes et al., 2006).

PROGNOSIS

COPD is slowly progressive, with an accelerated decline in FEV_1 leading to slowly increasing symptoms, fall in lung function, and eventually to respiratory failure (Figure 75.1). The only strategy to reduce disease progression is smoking cessation, although this is relatively ineffective once FEV_1 has fallen below 50% predicted and the patient is symptomatic. Patients with more severe exacerbations develop acute exacerbations, which have a prolonged effect on quality of life for many months. There is still debate about the role of acute exacerbations on disease progression, but the decline in lung function may be accelerated further following an acute exacerbation (Celli and Barnes, 2007). Factors other than FEV_1 are important in the prognosis of the disease, including exercise performance and muscle weakness, which signal a greater mortality. Patients who develop right heart failure (*cor pulmonale*) also have poor survival, although this may be improved by long-term oxygen therapy. It is hoped that in the future, genetic approaches will help to improve the prediction of prognosis and identify the most effective management strategies.

MANAGEMENT

COPD is managed according to the severity of the disease, with progressive escalation of therapy as the disease progresses (Barnes and Stockley, 2005).

Anti-smoking Measures

Smoking cessation is the only measure shown so far to slow the progression of COPD, but in advanced disease, stopping smoking has little effect and the chronic inflammation persists. Nicotine replacement therapy (gum, transdermal patch, inhaler) helps in quitting smoking, but varenecline, a partial nicotinic agonist, is currently the most effective drug for treatment of nicotine addiction. As discussed above, there may be genetic determinants of smoking cessation that make it very difficult for some patients to quit.

Bronchodilators

Bronchodilators are the mainstay of current COPD drug therapy. Bronchodilator response measured by an increase in FEV_1 is limited in COPD, but bronchodilators may improve symptoms by reducing hyperinflation and therefore dyspnea, and may improve exercise tolerance, despite the fact that there is little improvement in spirometric measurements. Short-acting bronchodilators, including β_2-agonists and anticholinergics, have now largely been replaced by long-acting bronchodilators, including inhaled long-acting β_2-agonists (long-acting beta-adrenergic agonist (LABA), salmeterol and formoterol) and the once-daily inhaled anticholinergic tiotropium bromide. In patients with more severe disease, these therapies appear to be additive. Theophylline is also used as an add-on bronchodilator in patients with very severe disease, but systemic side effects limit its value.

Antibiotics

Acute exacerbations of COPD are commonly assumed to be due to bacterial infections since they may be associated with increased volume and purulence of the sputum. But it is increasingly recognized that exacerbations may also be due to upper respiratory tract viral infections or may be non-infective, which leads to questioning the place of antibiotic treatment in many patients. Controlled trials of antibiotics in COPD show a relatively minor benefit in terms of clinical outcomes and lung function. Although antibiotics are still widely used in exacerbations of COPD, methods that can reliably diagnose bacterial infection in the respiratory tract are needed so that antibiotics are not used inappropriately. There is no evidence that prophylactic antibiotics prevent acute exacerbations.

Oxygen

Home oxygen accounts for a large proportion (over 30% in the USA) of healthcare spending on COPD. Long-term oxygen therapy reduces mortality and improves quality of life in patients with severe COPD and chronic hypoxemia ($PaO_2 < 55$ mm Hg).

Corticosteroids

Inhaled corticosteroids are now the mainstay of chronic asthma therapy, and the recognition that chronic inflammation is also present in COPD provided a rationale for their use in COPD. Indeed, inhaled corticosteroids are now widely prescribed in the treatment of COPD. However, the inflammation in COPD shows little or no suppression even by high doses of inhaled or oral corticosteroids (Barnes, 2010). This may reflect the fact that neutrophilic inflammation is not suppressible by corticosteroids, as neutrophil survival is prolonged by steroids. There is also evidence of an active cellular resistance to corticosteroids, with no evidence that even high doses of corticosteroids suppress the synthesis of inflammatory mediators or enzymes. This is related to decreased activity and expression of HDAC2 (Barnes, 2009a). Approximately 10% of patients with stable COPD show some symptomatic and objective improvement with oral corticosteroids. It is likely that those patients have concomitant asthma, as both diseases are very common. Furthermore, those patients have elevated sputum eosinophils and exhaled nitric oxide (NO), which are features of asthmatic inflammation. Long-term treatment with high doses of inhaled corticosteroids fails to reduce disease progression, even at the early stages of the disease. However, there is a small protective effect against acute exacerbations (approximately 20% reduction) in patients with severe disease. In view of the risk of systemic side effects in this susceptible population, inhaled corticosteroids are only recommended in patients with $FEV_1 < 50\%$ predicted who have two or more severe exacerbations a year. There is a small beneficial effect of systemic corticosteroids in treating acute exacerbations of COPD, with improved clinical outcome and reduced length of hospital admission.

Other Drug Therapies

Systematic reviews show that mucolytic therapies reduce exacerbation by about 20%, but most of the benefit appears to derive from N-acetyl cysteine, which is also an antioxidant. A controlled trial, however, did not show any overall benefit in reducing exacerbations or slowing disease progression (Decramer et al., 2005).

Pulmonary Rehabilitation

Pulmonary rehabilitation consists of a structured program of education, exercises, and physiotherapy. In controlled trials, it has been shown to improve exercise capacity and quality of life of patients with severe COPD, with a reduction in healthcare utilization (Casaburi and ZuWallack, 2009). Pulmonary rehabilitation is now an important part of the management plan in patients with severe COPD. Most of the benefits appear to relate to exercise, so that modified, simplified programs are now often used.

Lung Volume Reduction

Surgical removal of emphysematous lung improves ventilatory function in carefully selected patients. Reduction in hyperinflation improves the mechanical efficiency of the inspiratory muscles. Careful patient selection after a period of pulmonary rehabilitation is essential. Patients with localized upper lobe emphysema with poor exercise capacity do best, but there is a relatively high operative mortality, particularly with patients who have a low diffusing capacity. Significant functional improvements include increased FEV_1, reduced total lung capacity and functional residual capacity, improved function of respiratory muscles, improved exercise capacity and improved quality of life. Benefits persist for at least a year in most patients, but careful long-term follow-up is needed to evaluate the long-term benefits of this therapy. More recently, non-surgical bronchoscopic lung volume reduction has been achieved by insertion of one-way valves by fiber optic bronchoscopy. This results in significant improvement in some patients and appears to be safe, but collateral ventilation reduces the efficacy of this treatment so that significant deflation of the affected lung may not be achieved.

Management of Acute Exacerbations

Acute exacerbations of COPD should be managed by supplementary oxygen therapy, initially 24% oxygen, checking that there is no depression of ventilation. Antibiotics should be given if the sputum is purulent or if there are other signs of bacterial infection. High doses of corticosteroids reduce hospital stays and are routinely given. Chest physiotherapy is usually given, but there is little objective evidence of benefit. Non-invasive ventilation is indicated for incipient respiratory failure and reduces the need for intubation.

Pharmacogenomics

There is little or no information about the impact of pharmacogenomics on COPD therapy (Hersh, 2010). Polymorphisms of the β_2-adrenergic receptor (ADRB2) have little clinical effect on bronchodilator response to short- and long-acting β_2-agonists in asthma (Bleecker et al., 2007). It is possible that polymorphisms in corticosteroid signaling pathways may contribute to corticosteroid resistance in COPD, but this has not yet been investigated. Using genomics to better define clinical phenotypes may lead to a more rational use of specific therapies in the future.

NEW TREATMENTS

New Bronchodilators

New long-acting inhaled bronchodilators include once-daily β_2-agonists such as indacaterol and vilanterol and new once-daily anticholinergics such as glycopyrrolate and aclidinium. Combination inhalers with once-daily β_2-agonists and anticholinergics are also in clinical development.

New Anti-inflammatory Treatments

Apart from quitting smoking, no other treatments, including corticosteroids, slow the progression of COPD. Yet this disease is associated with an active inflammatory process and progressive proteolytic injury of lung tissues even in its most advanced stages, suggesting that pharmacological intervention may be possible. A better understanding of the cellular and molecular mechanisms involved in COPD provides new molecular targets for the development of new drugs, and several classes of new drugs are now in development (Barnes, 2008a; Hansel and Barnes, 2009). Blocking individual inflammatory mediators such as TNF-α, CXCL8 and LTB_4 has so far been unsuccessful, probably reflecting the fact that so many different mediators are involved. It is possible that there may be individuals who respond better to specific antagonists and that they may in the future be identified by pharmacogenomic profiling. Protease inhibitors are also in development, particularly targeting neutrophil elastase and MMP-9, but so far there are no clinical studies. Non-steroidal anti-inflammatory treatments are also in development. The most advanced are phosphodiesterase (PDE)4 inhibitors, which have an inhibitory effect on key inflammatory cells involved in COPD, including macrophages, neutrophils, and T lymphocytes. Roflumilast, a PDE4 inhibitor, has recently been approved for severe COPD in some countries, and provides some improvement in lung function and reduction in exacerbations, but the oral dose is limited by side effects such as nausea and headaches (Calverley et al., 2009). Other novel anti-inflammatory approaches in development include inhibitors of NF-κB and inhibitors of p38 mitogen-activated protein kinase, which are in clinical trials.

Impacts of Genomics on Therapy

As indicated above, molecular genetics and genomics may identify novel targets for the development of new therapies. Pharmacogenomics could have an impact on choice of therapy in the future. It is unlikely that gene-based therapies will be useful, but silencing of inflammatory genes by inhaled small interfering RNAs and antisense oligonucleotides may be a feasible approach in the future (Ulanova et al., 2006). Micro-RNAs (miRNAs) are small regulatory RNAs that inhibit the translation of several activated genes, and may have application for the suppression of multiple inflammatory genes in the future (Ying et al., 2006).

CONCLUSIONS

COPD is a major, increasing global disease. Although the major causes are cigarette smoking and biomass fuels, relatively little is understood about its underlying inflammatory and immune mechanisms. The accelerated decline in lung function in the 10–20% of smokers who develop COPD is likely to be genetically determined, but there is no agreement on which genes are involved. More studies – particularly high-density, genome-wide approaches – are needed to identify susceptibility genes and to link them to different patient phenotypes. Identification of susceptibility genes may facilitate screening and early identification of disease. Microarray analysis is now revealing up-regulation of various inflammatory, immune, and repair genes and down-regulation of other genes that may be protective in peripheral lung, macrophages, and epithelial cells. More studies are needed in relevant cells from carefully phenotyped patients. At the moment it is not possible to give any guidance to family members of COPD patients (apart from those with PiZ α_1-antitrypsin deficiency) about genetic risks, since the genes determining risk have not yet been identified. Pharmacogenetics needs to be applied to COPD therapies in order to select patients who would benefit more from one treatment compared to another. Current therapy for COPD is unsatisfactory, and there are no effective anti-inflammatory treatments. Identification of novel genes involved in COPD may also lead to the discovery of new drug targets.

REFERENCES

Barnes, P.J., 2000. Chronic obstructive pulmonary disease. N Engl J Med 343, 269–280.

Barnes, P.J., 2004. Small airways in COPD. N Engl J Med 350, 2635–2637.

Barnes, P.J., 2008a. Frontrunners in novel pharmacotherapy of COPD. Curr Opin Pharmacol 8, 300–307.

Barnes, P.J., 2008b. Immunology of asthma and chronic obstructive pulmonary disease. Nat Immunol Rev 8, 183–192.

Barnes, P.J., 2009a. Role of HDAC2 in the pathophysiology of COPD. Ann Rev Physiol 71, 451–464.

Barnes, P.J., 2009b. The cytokine network in COPD. Am J Respir Cell Mol Biol 41, 631–638.

Barnes, P.J., 2010. Inhaled corticosteroids in COPD: A controversy. Respiration 80, 89–95.

Barnes, P.J., Adcock, I.M., 2009. Glucocorticoid resistance in inflammatory diseases. Lancet 342, 1905–1917.

Barnes, P.J., Celli, B.R., 2009. Systemic manifestations and comorbidities of COPD. Eur Respir J 33, 1165–1185.

Barnes, P.J., Chowdhury, B., Kharitonov, S.A., et al., 2006. Pulmonary biomarkers in chronic obstructive pulmonary disease. Am J Respir Crit Care Med 174, 6–14.

Barnes, P.J., Shapiro, S.D., Pauwels, R.A., 2003. Chronic obstructive pulmonary disease: Molecular and cellular mechanisms. Eur Respir J 22, 672–688.

Barnes, P.J., Stockley, R.A., 2005. COPD: Current therapeutic interventions and future approaches. Eur Respir J 25, 1084–1106.

Bleecker, E.R., Postma, D.S., Lawrance, R.M., Meyers, D.A., Ambrose, H.J., Goldman, M., 2007. Effect of ADRB2 polymorphisms on response to longacting beta2-agonist therapy: A pharmacogenetic analysis of two randomised studies. Lancet 370, 2118–2125.

Bowler, R.P., Barnes, P.J., Crapo, J.D., 2004. The role of oxidative stress in chronic obstructive pulmonary disease. J COPD 2, 255–277.

Bowler, R.P., Ellison, M.C., Reisdorph, N., 2006. Proteomics in pulmonary medicine. Chest 130, 567–574.

Calverley, P.M., Rabe, K.F., Goehring, U.M., Kristiansen, S., Fabbri, L.M., Martinez, F.J., 2009. Roflumilast in symptomatic chronic obstructive pulmonary disease: Two randomised clinical trials. Lancet 374, 685–694.

Casaburi, R., ZuWallack, R., 2009. Pulmonary rehabilitation for management of chronic obstructive pulmonary disease. N Engl J Med 360, 1329–1335.

Castaldi, P.J., Cho, M.H., Cohn, M., Langerman, F., et al., 2010. The COPD genetic association compendium: A comprehensive online database of COPD genetic associations. Hum Mol Genet 19, 526–534.

Cavarra, E., Fardin, P., Fineschi, S., et al., 2009. Early response of gene clusters is associated with mouse lung resistance or sensitivity to cigarette smoke. Am J Physiol Lung Cell Mol Physiol 296, L418–L429.

Celli, B.R., Barnes, P.J., 2007. Exacerbations of chronic obstructive pulmonary disease. Eur Respir J 29, 1224–1238.

Dahl, M., Bowler, R.P., Juul, K., Crapo, J.D., Levy, S., Nordestgaard, B.G., 2008. Superoxide dismutase 3 polymorphism associated with reduced lung function in two large populations. Am J Respir Crit Care Med 178, 906–912.

Decramer, M., Rutten-van Molken, M., Dekhuijzen, P.N., et al., 2005. Effects of N-acetylcysteine on outcomes in chronic obstructive pulmonary disease (Bronchitis Randomized on NAC Cost-Utility Study, BRONCUS): A randomised placebo-controlled trial. Lancet 365, 1552–1560.

Erblich, J., Lerman, C., Self, D.W., Diaz, G.A., Bovbjerg, D.H., 2005. Effects of dopamine D2 receptor (DRD2) and transporter (SLC6A3) polymorphisms on smoking cue-induced cigarette craving among African-American smokers. Mol Psychiatry 10, 407–414.

Hansel, T.T., Barnes, P.J., 2009. New drugs for COPD exacerbations based on modulation of innate immune responses. Lancet 374, 744–755.

Heguy, A., O'Connor, T.P., Luettich, K., et al., 2006. Gene expression profiling of human alveolar macrophages of phenotypically normal smokers and nonsmokers reveals a previously unrecognized subset of genes modulated by cigarette smoking. J Mol Med 84, 318–328.

Hersh, C.P., 2010. Pharmacogenetics of chronic obstructive pulmonary disease: Challenges and opportunities. Pharmacogenomics 11, 237–247.

Hersh, C.P., Miller, D.T., Kwiatkowski, D.J., Silverman, E.K., 2006. Genetic determinants of C-reactive protein in chronic obstructive pulmonary disease. Eur Resp J 28, 1156–1162.

Hilty, M., Burke, C., Pedro, H., et al., 2010. Disordered microbial communities in asthmatic airways. PLoS One 5, e8578.

Hogg, J.C., 2001. Chronic obstructive pulmonary disease: An overview of pathology and pathogenesis. Novartis Found Symp 234, 4–19.

Hogg, J.C., 2004. Pathophysiology of airflow limitation in chronic obstructive pulmonary disease. Lancet 364, 709–721.

Hogg, J.C., Chu, F., Utokaparch, S., et al., 2004. The nature of small-airway obstruction in chronic obstructive pulmonary disease. N Engl J Med 350, 2645–2653.

Huang, Y.J., Kim, E., Cox, M.J., et al., 2010. A persistent and diverse airway microbiota present during chronic obstructive pulmonary disease exacerbations. OMICS 14, 9–59.

Hunninghake, G.M., Cho, M.H., Tesfaigzi, Y., et al., 2009. MMP12, lung function, and COPD in high-risk populations. N Engl J Med 361, 2599–2608.

Hurst, J.R., Donaldson, G.C., Perea, W.R., et al., 2006. Utility of plasma biomarkers at exacerbation of chronic obstructive pulmonary disease. Am J Respir Crit Care Med 174, 867–874.

Kelly, E., Greene, C.M., Carroll, T.P., McElvaney, N.G., O'Neill, S.J., 2010. Alpha-1 antitrypsin deficiency. Respir Med 104, 763–772.

Kim, W.J., Hersh, C.P., Demeo, D.L., Reilly, J.J., Silverman, E.K., 2009. Genetic association analysis of COPD candidate genes with bronchodilator responsiveness. Respir Med 103, 552–557.

Kong, X., Cho, M.H., Anderson, W., et al., 2010. Genome-wide association study identifies BICD1 as a susceptibility gene for emphysema. Am J Respir Crit Care Med 183, 43–49.

Mannino, D.M., Buist, A.S., 2007. Global burden of COPD: Risk factors, prevalence, and future trends. Lancet 370, 765–773.

Ning, W., Li, C.J., Kaminski, N., Feghali-Bostwick, C.A., et al., 2004. Comprehensive gene expression profiles reveal pathways related to the pathogenesis of chronic obstructive pulmonary disease. Proc Natl Acad Sci USA 101, 14,895–14,900.

Pillai, S.G., Ge, D., Zhu, G., et al., 2009. A genome-wide association study in chronic obstructive pulmonary disease (COPD): Identification of two major susceptibility loci. PLoS Genet 5, e1000421.

Pinto-Plata, V., Toso, J., Lee, K., et al., 2006. Use of proteomic patterns of serum biomarkers in patients with chronic obstructive pulmonary disease: Correlation with clinical parameters. Proc Am Thorac Soc 3, 465–466.

Salvi, S.S., Barnes, P.J., 2009. Chronic obstructive pulmonary disease in non-smokers. Lancet 374, 733–743.

Ulanova, M., Schreiber, A.D., Befus, A.D., 2006. The future of antisense oligonucleotides in the treatment of respiratory diseases. BioDrugs 20, 1–11.

Van Diemen, C.C., Postma, D.S., Vonk, J.M., Bruinenberg, M., Schouten, J.P., Boezen, H.M., 2005. A disintegrin and metalloprotease 33 polymorphisms and lung function decline in the general population. Am J Respir Crit Care Med 172, 329–333.

Van Durme, Y.M., Eijgelsheim, M., Joos, G.F., et al., 2010. Hedgehog-interacting protein is a COPD susceptibility gene: The Rotterdam Study. Eur Respir J 36, 89–95.

Wan, E.S., Silverman, E.K., 2009. Genetics of COPD and emphysema. Chest 136, 859–866.

Wang, I.M., Stepaniants, S., Boie, Y., et al., 2008. Gene expression profiling in patients with chronic obstructive pulmonary disease and lung cancer. Am J Respir Crit Care Med 177, 402–411.

Wilk, J.B., Chen, T.H., Gottlieb, D.J., et al., 2009. A genome-wide association study of pulmonary function measures in the Framingham Heart Study. PLoS Genet 5, e1000429.

Ying, S.Y., Chang, D.C., Miller, J.D., Lin, S.L., 2006. The microRNA: Overview of the RNA gene that modulates gene functions. Methods Mol Biol 342, 1–18.

Zeskind, J.E., Lenburg, M.E., Spira, A., 2008. Translating the COPD transcriptome: Insights into pathogenesis and tools for clinical management. Proc Am Thorac Soc 5, 834–841.

Interstitial Lung Disease

Mark P. Steele, Eric B. Meltzer, and Paul W. Noble

INTRODUCTION

Interstitial lung diseases (ILDs), also referred to as diffuse parenchymal lung diseases (DPLDs), are a diverse group of lung diseases, that can be classified according to clinical, radiologic, physiologic, or pathologic criteria, that result in fibrosis of the alveolar interstitium and adjacent structures, which impairs gas exchange. The term DPLD more accurately describes these entities, since the interstitium of the lung is not the only structure of the lung involved in these diseases. For example, among the ILDs there is variable involvement of terminal and respiratory bronchioles, and lymphatics along the bronchovascular bundle and interlobular septae. A general classification scheme of ILD/DPLD categorizes DPLDs as related to connective tissue diseases, drug-induced diseases, occupational and environmental exposures, granulomatous diseases, inherited conditions such as familial interstitial pneumonia (FIP) and Hermansky–Pudlak syndrome, unique conditions such as eosinophilic granuloma and amyloidosis, and the idiopathic interstitial pneumonias (IIPs). While the pathogenic mechanisms are known or inferred in some of the DPLDs, such as those related to environmental exposure, drug exposure, or autoimmune mechanisms, the pathogenesis of most of these entities is poorly understood. Furthermore, it is well recognized that there is considerable individual variation in the natural history of DPLD, particularly regarding the susceptibility to

agents known to cause pulmonary fibrosis, such as radiation or asbestos. Since the lung responds to these injuries in a limited fashion, when using standard radiologic and pathologic evaluation there is significant overlap and diagnostic uncertainty with DPLD. Consequently, there is considerable interest in the study of the genetics, genomics, and proteomics of ILD/DPLD to better understand the pathogenesis of these diseases and individual disease susceptibility, to discover biomarkers for disease diagnosis and prognosis, and to develop effective treatment interventions.

Four lines of evidence suggest that the development of pulmonary fibrosis is at least partially determined by genetic factors. First, clustering of pulmonary fibrosis, an uncommon disease, has been reported in monozygotic twins raised in different environments (Bonanni et al., 1965; Javaheri et al., 1980; Solliday et al., 1973), in genetically related members of several families (Bitterman et al., 1986; Bonanni et al., 1965; Hughes, 1964; Swaye et al., 1969), in consecutive generations of the same families (Bonanni et al., 1965; Hodgson et al., 2002; Lee et al., 2005), and in family members separated at an early age (Swaye et al., 1969). While a single report suggests that FIP is inherited as an autosomal recessive trait (Tsukahara and Kajii, 1983), other pedigrees demonstrate an autosomal dominant pattern of inheritance (Adelman et al., 1966; McKusick and Fisher, 1958; Swaye et al., 1969), perhaps with reduced penetrance (Adelman et al., 1966; Bitterman et al., 1986; Hughes,

Genomic and Personalized Medicine, 2nd edition
by Ginsburg & Willard. DOI: http://dx.doi.org/10.1016/B978-0-12-382227-7.00076-8

1964; Javaheri et al., 1980; Marshall et al., 2000; Musk et al., 1986; Solliday et al., 1973; Swaye et al., 1969). Second, pulmonary fibrosis is observed in genetic disorders with pleiotropic presentation, including Hermansky–Pudlak syndrome (Depinho and Kaplan, 1985), neurofibromatosis (Riccardi, 1981), tuberous sclerosis (Harris et al., 1969; Makle et al., 1970), Neimann–Pick disease (Terry et al., 1954), Gaucher disease (Schneider et al., 1977), familial hypocalciuric hypercalcemia (Auwerx et al., 1985), familial surfactant protein C mutation (Thomas et al., 2002), and dyskeratosis congenita (Dokal and Vulliamy, 2000). Third, there is considerable variability in the development of pulmonary fibrosis among workers exposed to similar concentrations of fibrogenic dusts or organic antigens. For instance, following exposure to asbestos, similarly exposed individuals may experience very different outcomes (Polakoff et al., 1979; Selikoff et al., 1979). Fourth, inbred strains of mice differ in their susceptibility to fibrogenic agents. In comparison to BALB/c or 129 mice, C57BL/6 mice develop more lung fibrosis when challenged with either bleomycin (Ortiz et al., 1998; Rossi et al., 1987) or asbestos (Corsini et al., 1994; Warshamana et al., 2002).

This review will focus on the genetic, genomic, and proteomic approaches to identification of disease susceptibility genes that predispose to DPLD and to improve the diagnostic classification of DPLD. To better understand the pathogenesis of pulmonary sarcoidosis, both candidate gene and genome-wide linkage strategies have been utilized, as well as proteomic approaches to identification of pathogenic antigens causing granulomatous inflammation. Surfactant proteins have been linked to familial cases of pediatric and adult interstitial pneumonia, and represent the most successful candidate gene approach to studying genetic susceptibility to DPLD. Dyskeratosis congenita, a rare inherited disorder sometimes associated with DPLD, has been associated with variants of the telomerase complex, and mutations in telomerase proteins are associated with FIP. A single nucleotide polymorphism (SNP) in the promoter region of the mucin 5B gene (*MUC5B*) was identified by genome-wide linkage analysis to predispose to FIP and sporadic idiopathic pulmonary fibrosis, and is present in the majority of patients. Microarray-based expression profiling is being successfully employed to aid classification of DPLDs and to identify novel disease susceptibility genes. Recombinant inbred and congenic mouse strains have been successfully used to map susceptibility loci related to radiation, bleomycin-, and asbestos-induced pulmonary fibrosis.

GENETIC DETERMINANTS OF DIFFUSE PARENCHYMAL LUNG DISEASE IN MOUSE STRAINS

Investigation of inbred mouse strains that differ in their susceptibility to pulmonary fibrosis can be used to identify pulmonary fibrosis susceptibility genes. No mouse strain has been identified that spontaneously develops pulmonary fibrosis that resembles any IIP, but several mouse strains have been identified as being susceptible or resistant to pulmonary fibrosis following various stresses to the lung. The most commonly studied model is pulmonary fibrosis following intratracheal administration of bleomycin in the susceptible C57BL6 strain compared to the more resistant C2Hf/Kam and C3H/HeJ strains. One limitation of the bleomycin mouse model is that the lung injury induced by bleomycin tends to resolve over time with relatively little pulmonary fibrosis compared to the permanent damage seen in humans. Differences in susceptibility to radiation-induced (Franko et al., 1996; Haston and Travis, 1997; Haston et al., 2002b; Sharplin and Franko, 1989), bleomycin-induced, and paraquat-induced pulmonary fibrosis in different inbred mouse strains have been used to identify loci linked to pulmonary fibrosis (Haston et al., 1996, 2002a, 2005; Lemay and Haston, 2005). In the bleomycin-induced

TABLE 76.1 Mouse models of diffuse parenchymal lung disease

Animal model	Mouse strain	Locus	Result	Reference
Bleomycin	C57BL6	Chr 17	LOD score 2.8, marker D17Mit198/D17Mit16 localized to 2.7 cM region of MHC accounting for 40% of genetic risk	Haston et al., 2002a
		Chr 11	LOD score 3.3, D11Mit272/D11Mit310 with evidence for interactions between Chr 17 and 11	Haston et al., 2002a
Bleomycin	C57BL6 and A/J congenic strain	Chr 9	LOD 4.9 at D9Mit236, 246 differentially expressed genes mapped to the interval	Lemay and Haston, 2005
Bleomycin	C3H-H2 reduced congenic	MHC susceptible	Reduced levels of bleomycin hydrolase activity in the fibrosis-prone strains	Haston et al., 2002a
Radiation	C57BL/6JChr 17	Chr 17	LOD 4.2 at D17Mit16 within the bleomycin linked region	Haston et al., 2002b
		Chr 1	LOD 4.5 at D1Mit206	Haston et al., 2002b
		Chr 6	LOD 4.6 at D6Mit254	

LOD, logarithmic odds.

pulmonary fibrosis model, the bleomycin hydrolase gene appears to be at least one of the candidate genes. In those studies, lung fibrosis is histologically scored as a quantitative trait (QTL) for QTL analysis. Given that linked QTL intervals typically span large regions (up to 10–50 Mb), microarray analysis can be targeted to linked regions to identify differentially expressed genes that are potential candidate genes (Haston et al., 2005; Katsuma et al., 2001). Results of these studies are summarized in Table 76.1. Potential candidate genes located in intervals identified from mouse studies have not been followed up in any human studies.

GENETIC DETERMINANTS OF SARCOIDOSIS

Evidence for Genetic Basis

Sarcoidosis is a complex disease, with racial and ethnic differences in prevalence, an association with environmental exposures, and a genetic predisposition. A genetic basis for pulmonary sarcoidosis is suggested by familial clustering of sarcoidosis (Buck and McKusick, 1961; Harrington et al., 1994; Headings et al., 1976; Moura et al., 1990; Wiman, 1972) and racial differences in disease prevalence (Rybicki et al., 2001a, b). A large, multicenter, case-control study investigating the familial aggregation of sarcoidosis in the United States demonstrated a 5.8-fold increase in the relative risk of developing sarcoidosis among first-degree relatives (Rybicki et al., 2001a, 2005a).

Candidate Gene Studies

Sarcoidosis exhibits characteristic granulomatous inflammation, with morphologic features of the granuloma having a concentration of CD4-positive T cells at the central core, CD8-positive T cells at the periphery, and a host of associated immunologic abnormalities (American Thoracic Society, 1999). Immunologic abnormalities observed in patients with sarcoidosis include expansion of T cells bearing restricted T cell receptor. This suggests oligoclonality, increased expression of members of the TNF-ligand and TNF-receptor superfamilies by T cells, B cell hyperactivity and spontaneous *in situ* production of immunoglobulin, and accumulation of monocytes/macrophages with antigen-presenting capacity associated with increased release of macrophage-derived cytokines [IL-1, IL-6, IL-8, IL-15, TNF-α, IFN-χ, granulocyte macrophage colony stimulating factor (GM-CSF)], chemokines [regulated and normal T cell expressed and secreted (RANTES), MIP-1α, IL-16], and fibrogenic cytokines (TGF-β, PDGF). Genes involved in these pathways are plausible biologic candidate genes.

Given the known racial and ethnic differences in disease prevalence and the characteristic immunologic features of the disease, attention has been directed to human leukocyte antigen (HLA) region genes that predispose to the development of sarcoidosis. HLA class II alleles have been most frequently reported to be associated with the risk of developing sarcoidosis. Studies performed in different populations have produced conflicting results, demonstrating either increased risk or protection from various HLA alleles. There is a consistent association with HLA-DR3 haplotypes and a more favorable prognosis in Czech, German, Italian, Japanese, Polish, and Scandinavian populations (Berlin et al., 1997; Bogunia-Kubik et al., 2001; Foley et al., 2001; Gardner et al., 1984; Ina et al., 1989; Ishihara et al., 1994; Martinetti et al., 1995; Swider et al., 1999). The HLA-DRB1 and HLA-DQB1 alleles have been associated with milder forms of the disease (erythema nodosum, Löfgren's syndrome, stage 0/1 CXR findings) in patients from Scandinavia, the United Kingdom, and the Netherlands (Berlin et al., 1997; Sato et al., 2002). Of note, HLA-DQB1 is in linkage disequilibrium (LD) with HLA-DRB1, which is in close proximity to non-HLA-related genes, such as TNF, that may be similarly influencing outcomes in sarcoidosis (Mrazek et al., 2005). The favorable outcome associated with increased TNF-α production based on the A2 promoter allele may be related to a common haplotype shared by HLA-DR3 (Seitzer et al., 2002). Other HLA loci that confer disease susceptibility include HLA-A1, -B8, -B22, -B13, -DR15, and -DR16, whereas protection from disease or milder forms of the disease have been associated with HLA-DR17 and -DRw52 (Berlin et al., 1997; Grunewald et al., 2004; Maliarik et al., 1998a; Martinetti et al., 1995; Rossman et al., 2003).

A number of non-HLA genes have been investigated for an association with sarcoidosis. The most heavily investigated is an insertion/deletion located in intron 16 of the gene for angiotensin-converting enzyme (ACE), which is known to affect serum ACE levels, although there appears to be no relationship to disease susceptibility (Alia et al., 2005; Maliarik et al., 1998b; McGrath et al., 2001). Other non-HLA candidate genes studied in sarcoidosis and their associations are summarized in Table 76.2. When interpreting these data, it is important to look closely at the method of choosing control populations to avoid spurious associations due to population stratification, as well as statistical genetic methods such as verification that genetic markers are in Hardy–Weinberg equilibrium, and correction for multiple comparisons. It is also important to look for evidence of LD flanking putative disease susceptibility single nucleotide polymorphisms (SNPs), since unknown mutations in LD with the putative disease susceptibility SNP may be responsible for the apparent association. In general, replication studies in multiple populations are necessary before a candidate gene SNP is unequivocally linked as a sarcoidosis susceptibility gene. Currently, utilizing mutation screening in candidate genes, only HLA alleles have been validated unequivocally as susceptibility genes for sarcoidosis.

Genome-wide Association Studies

There have been two genome-wide association studies (GWAS) in sarcoidosis (Hofmann et al., 2008, 2011). Both GWAS have been done by the same investigators using similar sarcoidosis patients from Germany utilizing different methodologic approaches to genotyping. One study used the

TABLE 76.2	Non-HLA candidate gene polymorphisms in sarcoidosis		
Candidate gene	**Polymorphism**	**Result**	**Reference**
ACE	Intron 16 in/del	Population specific	Berlin et al., 1997; McGrath et al., 2001
Vitamin D receptor	BsmI RFLP	Increased risk	Niimi et al., 2000b
IL-1 cluster	IL-a-889	Increase risk 2×	Niimi et al., 2000a
TGF b3	4785A	Fibrotic sarcoid	Kruit et al., 2006
HSP-70 hom	2763, 2437	Löfgren's syndrome	Bogunia-Kubik et al., 2006
BTNL2	10 intron/exon 5 SNP	Increased risk	Rybicki et al., 2005b
TLR4	A299G, T399I	Increased in chronic sarcoid	Pabst et al., 2006
Nod2/Card15			
TNFa	A2 promoter allele	Favorable prognosis	Sabounchi-Schutt et al., 2003
CCR5	HHC haplotype	Persistent lung disease	Spagnolo et al., 2005
BTNL2	3 locus haplotype	Increased risk	Akahoshi et al., 2004
HSP70-hom	C2437T	Lofgren's syndrome	Zorzetto et al., 2002
TGFβ	4875A	Fibrotic sarcoid	Takada et al., 2001

Affymetrix GeneChip Human Mapping 100k set, while the other utilized the Affymetrix 5.0 Array set. Since these two chips share only a small proportion of their SNPs, different results were obtained in the two studies. In one of the studies, the strongest signal was on chromosome 10q22.3 in the region *annexin 11* gene (1). The SNP with the strongest association after validation in an independent population was rs278679 [$P = 10^{-11}$, odds ratio (OR) 0.6 (0.52,0.69)] located in the 3′ untranslated region (UTR) of the annexin gene. The SNP rs278679 was in strong LD with five surrounding SNPs (rs1049550, rs1953600, rs2573346, rs2784773, rs7071579). Haplotype analysis of these five SNPs demonstrated five common haplotypes, and the most common haplotype, AGCATT, was less frequent in cases (32% in cases compared to 40% in controls). *Annexin 11* was shown to be highly expressed in normal and sarcoidosis lung, particularly in lung epithelium and mononuclear inflammatory cells. The exact function and consequences of these SNP variants in *annexin 11* remain unknown. In the second study, using the Affymetrix 100k chip, the SNP rs10484410 on chromosome 6p12.1 was significantly associated with cases of sarcoidosis, and remained significant after validation in a second sarcoid population ($P = 10^{-4}$). Fine mapping of this region narrowed the interval to a region containing five genes, *BAG2*, *C6orf65*, *KIAA1586*, *ZNF451*, and *RAB23*. Only *RAB23* and *ZNF451* are expressed in lung. Logistic regression using nine SNPs in the regions that remained significant after validation and fine mapping indicated several variants in *RAB23* to be the SNPs principally associated with cases. The SNP rs1040461 (G > A) changes the RAB23 protein sequence by substitution of glycine for serine. Further studies are needed to determine the importance and function of this variant of RAB23 in sarcoidosis. The GWAS studies show

that several genes increase the risk of sarcoidosis. The overall increased risk of developing sarcoidosis from both annexin 11 and RAB23 variants requires further testing before these variants can be clinically used to predict the risk of developing sarcoidosis.

Genetic Epidemiology

Evidence that sarcoidosis aggregates in some families and the ability to accurately identify and phenotype those families are critical to the successful completion of family-based linkage studies. The most comprehensive study of the familial aggregation of sarcoidosis comes from A Case Control Etiologic Study of Sarcoidosis (ACCESS) (Rybicki et al., 2001a). The study population was drawn from 10,862 first-degree and 17,047 second-degree relatives identified by 706 sarcoidosis case-control pairs. Controls were matched to cases by race, sex, age, and three-digit phone numbers. The familial relative risk of sarcoidosis was 5.8 [OR 5.8 (2.1, 15.9)] for sibs, and 3.8 [OR 3.8 (1.2, 11.3)] for parents.

Linkage Analysis

The first published genome-wide linkage study in sarcoidosis was from 63 German families consisting mostly of affected sibling pairs, having 138 affected siblings and 95 first-degree relatives (Schurmann et al., 2001). This genome-wide scan used 225 microsatellite markers, and was analyzed using the NPL score (Genehunter 2.0), a nonparametric, model-free approach where modes of inheritance and penetrance values are not required. The highest NPL score was 2.99, $P < 0.001$, obtained with marker D6S1666, which resides in the MHC class II gene. The genome-wide significant NPL score is approximately 3.6. Therefore, the nonparametric lod score (NPL) of 2.99 is very suggestive but not conclusive evidence for linkage to the MHC class II gene. Other chromosome regions showing minor peaks ($P < 0.05$) include NPL scores of 2.39 on chromosome 3p21, 1.87 on chromosome 1p22, 1.82 for chromosome 9q33, 1.64 on chromosome X, and 1.92 on chromosomes 7q22 and 7q36. The second linkage study performed in sarcoidosis was from the Sarcoidosis Genetic Analysis Consortium (SAGA). It reported linkage analysis of sibling pairs from 229 African-American families using 380 microsatellite markers. The investigators reported *P*-values based on Haseman-Elston regression. The strongest signal was at marker D5S407 ($P = 0.005$) on 5q11.2. Fine mapping of the chromosome 5 interval also identified a protective variant on 5p15.2 (Gray-McGuire et al., 2006). Interestingly, despite using a higher density of markers in the MHC class II region, the SAGA investigators did not find evidence for linkage in the MHC class II region. The different linkage results in the German and African-American populations would be consistent with ethnicity-related locus heterogeneity. Also, the sample sizes in these two studies are relatively small for a complex disease, and that may account for the different results. Nevertheless, these studies indicate that genome-wide linkage studies can be employed to identify disease susceptibility genes in sarcoidosis.

Based on the initial linkage analysis of the 63 German families demonstrating linkage to chromosome 6p21, SNP-based fine mapping of the region was performed using extended families and trios to conduct transmission disequilibrium tests (TDTs) and case-control association analysis (Valentonyte et al., 2005). The results demonstrated an association with SNP rs2076530 ($P_{TDT} = 3 \times 10^{-6}$, $P_{\text{case-control}} = 1.1 \times 10^{-8}$) located in the butyrophilin-like2 gene (*BTNL2*) located adjacent to HLA-DR1. Additional analysis demonstrated G to A transition in *BTNL2* that leads to the use of a cryptic slice site resulting in a 4 bp deletion that generates a premature stop codon, and the resulting protein lacks the C-terminal immunoglobulin-like constant domain and transmembrane domain. The authors demonstrate that the disease allele encodes a protein that is located in the cytoplasm rather than on the plasma membrane. The OR for developing sarcoidosis when heterozygous for the susceptibility allele is 1.6, and 2.75 in homozygotes. The authors did not comment on the overall frequency of the susceptibility allele in their population. Given the complexity of sarcoidosis, and the modest OR in the range of 1.6–2.75, one would expect that *BTNL2* is one of the several sarcoid susceptibility genes, and further investigations of *BTNL2* in other, non-German, populations will be important. Additional studies of the role of the *BTNL2* gene in sarcoidosis may identify novel mechanisms predisposing individuals to sarcoidosis or other granulomatous inflammation.

Gene Expression Studies

Little is known about the genetic regulation of granulomatous inflammation at the level of tissue transcription. However, two recent studies help to shed light on these processes. First, Crouser and colleagues were able to analyze gene expression patterns from whole lung tissues obtained from sarcoidosis patients versus normal control tissues (Crouser et al., 2009). They found that upregulated genes in sarcoidosis were predominantly associated with helper T cell activity; notably, STAT-1 and CCL5 transcripts were highly expressed. It was also reported that IL-7, MMP-12, and ADAMDEC1 were upregulated in sarcoidosis; this association was previously unknown. A second study, by Lockstone and colleagues, described the use of gene expression profiles to differentiate between clinical subtypes of sarcoidosis (Lockstone et al., 2010). The objective of the study was to determine if gene expression profiles and associated molecular pathways could distinguish self-limited sarcoidosis from the progressive/fibrotic form of the disease. Using gene set enrichment analysis (GSEA) to evaluate whether sets of genes grouped together by biologic processes, as defined by gene ontology, show enrichment among genes expressed differentially between self-limited and progressive, fibrotic sarcoidosis, the authors found gene set enrichment was limited to the progressive, fibrotic group. This study shows that genes related to immune activity and to host defense functions at the tissue transcription level are increased in patients with progressive, fibrotic sarcoidosis.

GENOMIC MEDICINE AND SARCOIDOSIS

The identification of *BTNL2* mutations, or mutations in other genes that lead to sarcoidosis susceptibility, will likely aid several aspects of the clinical approach to patients with sarcoidosis. GWAS studies have identified several candidate genes that increase the risk of developing sarcoidosis. Most likely there are many genes that increase the risk of sarcoidosis. Additional studies are needed to determine how combinations of these risk alleles increase the risk of developing sarcoidosis or alter its clinical course. A very clinically relevant application of sarcoid susceptibility alleles would be the development of a diagnostic test that distinguishes chronic progressive forms of sarcoidosis from spontaneous remitting disease. Another useful risk allele would be one that identifies risk for extra-pulmonary sarcoidosis, such as cardiac or central nervous system involvement, since involvement of those organs results in substantial morbidity and mortality. The identification of risk alleles that distinguish progressive forms of the disease or forms of the disease associated with higher morbidity would identify a group of patients needing aggressive anti-inflammatory therapy, while sparing those with milder, self-limited disease the toxicity of those therapies. The study by Lockstone and colleagues (2010) implies future application of gene expression technology as a tool for predicting prognosis in the setting of sarcoidosis, a disease with variable outcomes. In addition, the identification of sarcoid susceptibility genes might provide insight into development of new pharmacologic therapies.

PROTEOMICS IN SARCOIDOSIS

Sarcoidosis is characterized by granulomatous inflammation with expansion of T cells bearing restricted T cell receptor, suggesting oligoclonality and polyclonal antibody production. The intradermal injection of the Kveim reagent – homogenates of diseased sarcoid tissue obtained from spleen or lymph nodes – induces characteristic granulomatous inflammation known as the Kveim reaction (Siltzbach, 1961). The biochemical features of the Kveim reagent include neutral detergent insolubility; heat, acid, and protease resistance; and sensitivity to potent denaturants; suggesting a poorly soluble protein or protein aggregate (Chase and Siltzbach, 1967; Lyons et al., 1992). Proteomic approaches have been applied to better characterize the inflammatory response of sarcoidosis and to identify potential antigenic proteins present in the Kveim reagent. The Kveim reaction can also be obtained by intradermal injection of bronchoalveolar lavage fluid (BALF) concentrates. Hence, proteomic analysis of BALF, a less complex mixture than lymph node or spleen homogenates, has been conducted to attempt to identify potential antigenic proteins in BALF. Several studies have analyzed sarcoid BALF protein by two-dimensional gel electrophoresis and mass spectroscopy. These studies demonstrate increased concentrations of many plasma proteins in sarcoid BALF (Magi et al., 2002; Rottoli et al., 2005; Sabounchi-Schutt et al., 2003). However, specific proteins or antigens in BALF responsible for the Kveim reaction have not been identified.

Epidemiologic studies suggest an association between infective agents and sarcoidosis based on seasonal variations and case clustering (Baughman et al., 2003). Several lines of evidence suggest mycobacteria as a candidate organism. First, while mycobacteria in general cannot be isolated from sarcoid tissue using conventional microbiologic techniques, cell wall-deficient L forms persisting as an intracellular organism can be identified. Second, several polymerase chain reaction (PCR)-based studies have identified mycobacterium DNA in sarcoid tissue homogenates (Song et al., 2005). Utilizing biochemical techniques similar to that used to isolate the Kveim reagent, and immunoblotting using immunoglobulin obtained from the serum of sarcoid patients, investigators identified immuno-reactive proteins unique to sarcoid tissue homogenates. These antigenic bands were excised from gels, subjected to trypsin digestion and analyzed by matrix-assisted laser desorption/ionization with time-of-flight (MALDI-TOF) mass spectroscopy. The peptide fingerprints were identified as mycobacterium tuberculosis catalase–peroxidase. The authors hypothesize that insoluble aggregates of mycobacteria-derived catalase–peroxidase drive the sarcoid immune response in genetically predisposed individuals. It is quite likely that infectious agents such as mycobacteria, acid-fast cell-wall-deficient forms of bacteria, and potentially other infectious as well as non-infectious agents, are antigens that drive an abnormal immune response in a genetically predisposed individual leading to a common phenotype clinically recognized as sarcoidosis.

SURFACTANT PROTEINS AND DIFFUSE PARENCHYMAL LUNG DISEASE

Pulmonary Surfactant

Pulmonary surfactant is a complex mixture of phospholipids and proteins that functions to reduce surface tension at the alveolar air interface preventing atelectasis. Deficiency of pulmonary surfactant is the principal cause of respiratory distress syndrome in premature infants (Whitsett and Weaver, 2002). Four surfactant-associated proteins, surfactant proteins A, B, C, and D, have been described, and deficiency of surfactant proteins B and C has been associated with respiratory distress in newborns. Surfactant protein C (SP-C) is a highly hydrophobic protein that enhances the surface tension-lowering properties of pulmonary surfactant. Familial cases of neonatal respiratory distress have been associated with surfactant protein B deficiency, but respiratory distress of neonates is not considered to be a form of DPLD/ILD (Nogee et al., 2000). Genetic variants of SP-A and SP-C have been associated with increased risk of idiopathic pulmonary fibrosis (IPF) (Lawson et al., 2004).

Family Studies with SP-C

Nogee and colleagues (2001) described a full-term baby girl born to a woman who had desquamative interstitial pneumonia, a type of IIP, at one year of age. The infant's maternal grandfather had died of an unknown lung disease. The infant developed respiratory distress at the age of six weeks, and surgical lung biopsy demonstrated non-specific interstitial pneumonia (NSIP). Both the infant and the mother had minimal SP-C by either immunohistochemical staining or immunoblotting of lung tissue. DNA sequence analysis of the SP-C gene demonstrated a heterozygous substitution of A to G at the first base of intron 4 that abolished the normal donor splice site, resulting in a truncated mRNA. Subsequently, there have been several other similar families with SP-C mutations and interstitial pneumonia (Chibbar et al., 2004; Thomas et al., 2002). In the largest kindred, a heterozygous T-to-A substitution was identified in exon 5, causing glutamine for threonine substitution. In this pedigree, there was both adult-onset interstitial pneumonia of the usual interstitial pneumonia (UIP) histology and children with the cellular NSIP (Thomas et al., 2002). Immunohistochemical analysis of these patients demonstrated intracellular aggregates of SP-C, and in vitro expression studies demonstrated abnormal intracellular processing of SP-C in alveolar type II cells. SP-C variants causing ILD/DPLD in adults have been described by Van Moorsel and colleagues (2011) in a Dutch population. The authors identified SP-C mutations in five unrelated individuals from 20 families with familial pulmonary fibrosis. The mutations identified in the five patients were methionine to valine at amino acid 71, isoleucine to serine at amino acid 412, and three patients with isoleucine to threonine at amino acid 73. These variants were not identified in 121 sporadic cases of IPF. In contrast, less than 5% of families with familial pulmonary fibrosis from the US have coding variants in SP-C (Lawson and Loyd, 2010).

SP-C Genetic Variants Associated with IPF

In a single study by Selman and colleagues, the SP-A1 6A^4 haplotype is associated with a substitution of three amino acids at positions 19, 50, and 219, with the 219 variant being a tryptophan-for-arginine substitution (Lawson et al., 2004; Selman et al., 2003). The amino acid 219 variant is associated with IPF in smokers and nonsmokers [OR 3.67 (1.34, 10.07), $p = 0.01$]. It appears that both the frequency and type of variant SP-C causing ILD/DPLD in adults or children is quite variable, and differs across ethnicity. Hence, approaches to personalized medicine to predict risk of DPLD/ILD due to SP-C mutations will require further studies to determine the prevalence of high-risk variants across different populations.

Surfactant Protein A2

Surfactant protein A (SP-A) is the most abundant surfactant protein, and is structurally homologous to a family of innate immune proteins known as collectins. Collectins contain a collagen-like N-terminal domain that is involved in trimerization of proteins, and they contain a C-terminal carbohydrate-recognition lectin domain. Surfactant protein A assembles into an 18-subunit complex arranged into a "flower bouquet" pattern, and mice that lack SP-A are susceptible to infection (Wright, 2005). There are two SP-A genes, A1 and A2, which are >98% homologous in their coding regions. Two rare variants of SP-A2 (glycine to valine at codon 231, phenylalanine to serine at codon 198) have been identified in two families with lung cancer and pulmonary fibrosis, and these variants were not found in 1000 healthy controls. These two families are characterized by a young age of onset of pulmonary fibrosis (<50 years) and highly penetrant, autosomal-dominant inheritance (Wang et al., 2009).

ATP-Binding Cassette A3 Gene (ABCA3)

ABCA3 is expressed on the limiting membrane of lamellar bodies, the secretory vesicle for surfactant in the alveolar type II epithelial cell, and is involved in surfactant lipid transport into the lamellar body (Ban et al., 2007). Mutations in *ABCA3* have been associated with neonatal respiratory distress and surfactant deficiency (Shulenin et al., 2004), children with DPLD (Nogee, 2006), and adolescents with IPF (Young et al., 2008).

Genomic Medicine and Surfactant Proteins

Disruption of the surfactant proteins or surface tension properties of surfactant also disrupts homeostasis of the alveolar surface, and predisposes to either neonatal respiratory distress or the development of pulmonary fibrosis later in life. Pulmonary fibrosis from surfactant protein variants appears to result in onset of the disease at an early age (before age 50), which is considerably earlier than idiopathic pulmonary fibrosis and other types of idiopathic interstitial pneumonia. Neonates, children, and young adults with DPLD can be considered for genetic screening to identify variants of the surfactant proteins and surfactant-associated proteins. Genetic screening for mutations in surfactant protein B and C genes can be considered in those with a family history of neonatal respiratory distress syndrome or ILD/DPLD with age of onset in childhood or young adulthood. Given the large number of different mutations that have been reported to date, and the lack of large population based studies to determine the precise frequency of specific mutations, DNA sequencing of the coding regions of these genes is the current recommended approach to screening.

GENETIC DETERMINANTS OF PULMONARY FIBROSIS IDENTIFIED IN RARE INHERITED DISORDERS

Pulmonary fibrosis is observed in genetic disorders with pleiotropic presentation, including Hermansky–Pudlak syndrome, neurofibromatosis, tuberous sclerosis, Neimann–Pick disease, Gaucher disease, familial hypocalciuric hypercalcemia, familial SP-C mutation, and most recently in dyskeratosis congenita. Recently, mutations associated with dyskeratosis congenita syndrome have been seen in a small percentage of FIP cases (Armanios et al., 2007). Specifically, 8% of 73 families more than one case of IIP were found to have heterozygous mutations in telomerase reverse transcriptase, resulting in shortening of telomeres. The authors suggest that telomere shortening may cause apoptosis of the alveolar epithelium. Approximately 2% of familial cases of pulmonary fibrosis have mutations in the gene encoding the RNA component of telomerase (*TERC*), and 15% will have mutations in the gene for the reverse transcriptase component of telomerase (*TERT*). Genetic screening of family members for *TERC* and *TERT* mutations should be considered in familial pulmonary fibrosis (Schwartz et al., 2010). Given the large number of different mutations described, DNA sequencing of the coding region of these genes is the recommended approach to genetic screening.

Telomerase Variants and Adult-Onset FIP and Sporadic IPF

Two groups independently reported an association of mutations in *TERT* and familial and sporadic IPF (Armanios et al., 2007; Tsakiri et al., 2007). Variants in *TERC* have also been identified. These studies were from a relatively small number of cases in the United States. Recently, abnormalities in *TERT* have been associated with sporadic IPF from a genome-wide association study. In a larger cohort of patients from Japan, investigators conducted a GWAS by genotyping 159 patients with IPF and 934 controls for 214,508 tagged SNPs. SNP rs2736100, located in intron 2 of *TERT*, had a P-value of 2.8×10^{-6}, and in a replication set of 83 IPF patients and 535 controls, a P-value of 3.6×10^{-3} (Mushiroda et al., 2008). This *TERT* intron 2 SNP is located within an LD block (D' and $r^2 > 0.8$) within *TERT*, suggesting that rs2736100 may be associated with some other unidentified variation within *TERT* in this population of IPF patients from Japan.

Armanios and colleagues identified missense mutations in *TERT* in patients from the US with dyskeratosis congenita who had pulmonary fibrosis. They also sequenced *TERT* in 73 kindreds with familial IPF. They identified 8% of probands heterozygous for *TERT* mutations and one proband heterozygous for a mutation in *TERC*. Telomere length in the affected individuals was less than the 10th percentile in lymphocytes. Taskiri and colleagues identified two large families with pulmonary fibrosis, performed whole-genome linkage and found evidence for linkage to chromosome 5 in a region containing *TERT*. Variants in *TERT* were found in the families.

Recently, the possible role of telomere shortening has been generalized to sporadic and familial IPF patients without identified mutations in *TERT/TERC* (Cronkhite et al., 2008). These investigators used a modified Southern blot technique restriction fragment length method and a quantitative polymerase chain reaction method to measure telomere length of genomic DNA from peripheral blood. A significantly higher proportion of probands with familial pulmonary fibrosis (24%), and sporadic cases (23%) in which no coding mutations in *TERT* or *TERC* were found, had telomere length less than the 10th percentile ($P = 2.6 \times 10^{-8}$). Pulmonary affection status was significantly associated with telomerase restriction fragment lengths even after controlling for age, sex, and ethnicity ($P = 6.1 \times 10^{-11}$). Overall, 25% of sporadic IPF and 37% of familial IPF had telomere lengths less than the 10th percentile. These studies were performed in circulating lymphocytes, and it will be important to replicate these results in future studies in lung cells, particularly the alveolar epithelium.

Emerging Concepts from Genomic Studies of Associated DPLD

An interesting feature of these studies is the variable histopathologic features among family members sharing the identical SP-C mutation, suggesting modification of disease phenotype by other unknown factors. Mutations in SP-C and telomerase reverse transcriptase also suggest that abnormalities of the alveolar epithelium, particularly alveolar type II epithelial cells (the major source of pulmonary surfactant and also the progenitor cell for alveolar type I epithelial cell), are critical for the development of pulmonary fibrosis and interstitial pneumonia. In the instance of SP-C mutations, intracellular aggregates of abnormally processed SP-C may induce functional abnormalities of alveolar epithelial cells. Telomerase shortening may predispose to alveolar epithelial cell senescence, or the inability to re-epithelialize the alveolar surface after injury. These studies have resulted in a paradigm shift in understanding of the mechanism and pathogenesis of IIP, away from an inflammatory hypothesis and toward one of abnormal injury and repair of the alveolar epithelium and the fibroproliferative response. These data support lack of efficacy of glucocorticoid and cytotoxic treatments of IPF.

GENETIC DETERMINANTS OF FAMILIAL INTERSTITIAL PNEUMONIA

Clinical Features

The interstitial pneumonias can be classified as sporadic or idiopathic (IIP), in which no positive family history can be identified, or familial (FIP). The term familial interstitial pneumonia is preferred over the term familial idiopathic pulmonary fibrosis or familial pulmonary fibrosis since FIP accounts for the observation that several histopathologic types of interstitial pneumonia can be seen in a single family. There are no data on the relative proportion of interstitial pneumonias that are familial, but estimates are in the range of 5–10%. Familial aggregation has been reported in a variety of studies in twins, siblings raised apart, and multigenerational families. Steele and colleagues (2005) reported the largest collection of families, identifying 111 families from the United States and 20 multigenerational pedigrees that were consistent with autosomal dominant inheritance. A total of 45% of the families demonstrated phenotypic heterogeneity, with some families having bronchiolitis obliterans or NSIP and UIP within the same pedigree. Cigarette smoking was associated with affection status among siblings [OR 3.6 (1.3, 9.8), $P = 0.01$]. Histopathologic heterogeneity in these families has been confirmed in a subsequent study with multiple independent pathologists. UIP is found in 40% of the families, but the predominant histopathologic pattern (60%) for the FIP cohort was difficult to classify (Steele et al., 2007). The most common histopathologic pattern was non-classifiable advanced fibrosis with a high incidence of microscopic honeycombing, smooth muscle proliferation and fibrosis, and variable diffuse alveolar septal and airway-centered scarring. Similar to the findings with SP-C mutations, these data suggest that histologically distinct forms of pulmonary fibrosis may have common pathogenic mechanisms, and that cigarette smoking may contribute to the development of pulmonary fibrosis in individuals who are genetically prone to the disease. Idiopathic pulmonary fibrosis (IPF), the most common of the idiopathic interstitial pneumonias (IIP), most often occurs as sporadic cases and continues to be a devastating and poorly understood disease. Familial forms represent about 5–10% of IPF cases. Genetic studies of both familial and sporadic IPF are providing new insights into potential disease mechanisms.

Candidate Genes Identified by Microarray Analysis in Familial Pulmonary Fibrosis

Yang and colleagues performed microarray analysis of 16 cases of sporadic IIP (14 UIP, 2 NSIP), and 10 cases of familial IIP (6 UIP, 4 NSIP), compared to 9 normal lung controls. RNA was extracted from diseased lung tissue (specimens from surgical lung biopsy, autopsy, or explanted lung at the time of lung transplant), or normal controls (9 lung samples taken from donor lung at the time of lung transplant). Complementary RNA (cRNA) was synthesized and hybridized onto whole human genome arrays modified with an additional 657 probes for genes/expressed sequence tags (ESTs) that would be potentially informative based on preliminary linkage data (Yang et al., 2005). Expression profiling was performed using standard protocols (co-hybridized with human universal cell line reference RNA, replicates with the two dyes swapped, lowess-normalized intensities, analysis using the TGR MIDAS and MeV software), and differentially expressed genes were identified using significance analysis of microarrays (SAM) with 100 permutations. A total of 558 transcripts expressed differentially in cases compared with controls were identified, with 135 genes being up- or down-regulated greater than 1.8-fold. When hierarchical clustering was applied to the set of 135 genes, all but two samples clustered according to disease versus no disease, and

familial disease segregated from sporadic disease (Figure 76.1). A total of 69 differentially expressed genes were identified that distinguish sporadic and familial IP, and these are broadly grouped into functional classes (Figure 76.2), with a wide variety of chemokines, extracellular matrix, and growth-related genes that are differentially expressed. These data indicate that familial and sporadic IIP are transcriptionally distinct and also suggest many similarities between the histologic subtypes of UIP and NSIP. This study indicates that additional candidate genes that may be important in the pathogenesis of IIP, or that might be useful as biomarkers of disease activity, can be identified using whole-genome microarray analysis. Genes that are differentially expressed in lung tissue and are extracellular proteins are potential genes that could be explored as diagnostic biomarkers in blood to facilitate determining the type of IIP without surgical lung biopsy.

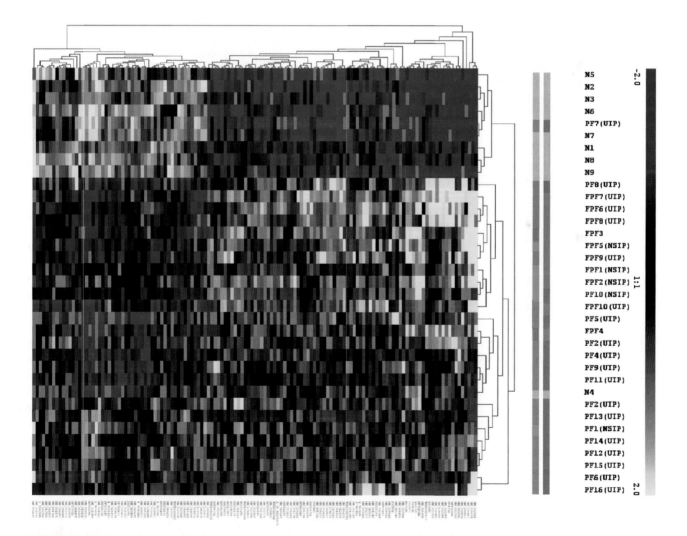

Figure 76.1 Hierarchical clustering of 35 cases and 16 controls. Hierarchical clustering of 35 samples based on 135 transcripts that best differentiate patients with pulmonary fibrosis from normal controls. Samples are color-coded according to the disease inheritance status (orange – normal, blue – sporadic IIP, magenta – familial IIP) or histological features (orange – normal, green – NSIP, red – UIP, gray – no histological evaluation). Average linkage clustering with Euclidian distance metric was used.

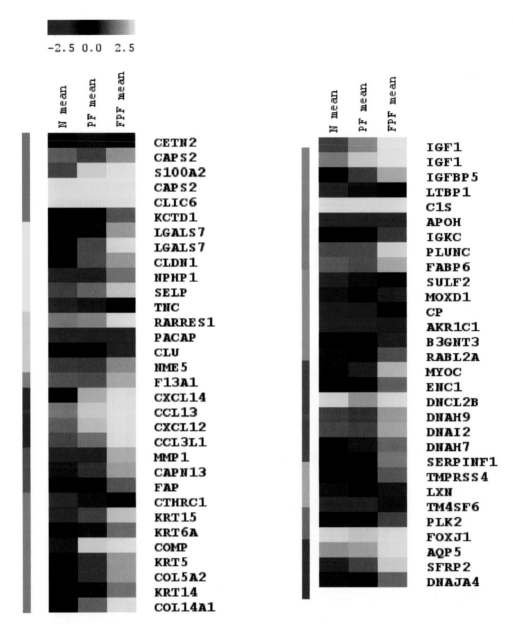

Figure 76.2 Mean expression ratios in normal, sporadic IIP, and familial IIP. Groups of 69 genes with known function that best differentiate familial IIP from sporadic IIP. Genes are arranged and color-coded according to their function: dark green – calcium/potassium ion binding and transport; yellow – cell adhesion; orange – cell proliferation and death; red – coagulation cascade; blue – cytokines, chemokines, and their receptors; purple – extracellular matrix (ECM) degradation; gray – ECM structural component; bright green – growth factors and their receptors; magenta – immune response; turquoise – metabolism; navy – motor activity; lavender – proteases and inhibitors; brown – regulation of I-kappaB/NF-kappaB cascade; and black – other.

Relevance of Genomics and Microarray Studies to DPLD/ILD

A number of troublesome clinical challenges in DPLD may be addressed by proper application of microarray-based genomic technology. One such challenge is to improve diagnostic accuracy. Due to the overlap of clinical, radiological, and pathological findings in DPLD, a definitive consensus diagnosis can only be achieved in 85% of cases. However, gene expression profiling may have the capacity to improve diagnostic accuracy. This capacity is limited by the absence of a true "gold standard" for the diagnosis of DPLD. Nevertheless, Selman and colleagues demonstrated a distinct expression profile in *hypersensitivity pneumonitis* when compared with *idiopathic pulmonary fibrosis* (Selman et al., 2006). A similar challenge has been predicting rates of disease progression in patients with pulmonary fibrosis; this elusive goal is essential to effectively managing this illness. Boon and colleagues recently identified unique patterns of gene expression that distinguish patients with relatively stable pulmonary fibrosis from those with progressive disease (Boon et al., 2009). Meanwhile, the pathogenesis of pulmonary fibrosis remains poorly understood. This creates a barrier to developing new effective treatments for this devastating disease. However, several microarray-based studies have recently identified genes in pulmonary fibrosis that were either previously unexpected or unreported. Such genes include *MMP-7* (Rosas et al., 2008; Zuo et al., 2002), *osteopontin* (Pardo et al., 2005), *caveolin-1* (Wang et al., 2006), *HIF-1alpha* (Tzouvelekis et al., 2007), *CCNA2*, and alpha-defensins (Konishi et al., 2009). Prognostic biomarkers are needed to better stratify patients in clinical treatment trials of IPF. For example, pirfendione is approved for use in IPF in Japan and Europe, but was denied approval in the US by the Food and Drug Administration (FDA) due to conflicting results from clinical trials performed in the US. In the US trials, pirfendione was effective in one study but ineffective in a second study (Noble et al., 2011). Detailed analysis of these trials shows different survival rates in the control populations indicating substantial heterogeneity of disease with respect to rate of disease progression. Gene-expression studies are a potential tool to identify novel prognostic biomarkers for IPF.

Linkage Studies in Pulmonary Fibrosis

The only published study performing genome-wide linkage analysis in FIP comes from Finland (Hodgson et al., 2006). Using six pedigrees, NPL scores of 1.7 on chromosome 3 (marker D3S1278) and chromosome 4 (marker D4S424), and an NPL score of 1.6 on chromosome 13 (D13S265) were obtained. On chromosome 4, a shared haplotype was identified among 8 of 24 multiplex families. A candidate gene located in the region of interest, ELMOD2, was further investigated by resequencing of exons and exon/intron boundaries. No mutations in ELMOD2 in these locations were identified. Reverse transcriptase (RT)-PCR and *in situ* hybridization demonstrated decreased levels of ELMOD2 mRNA in six cases of sporadic IPF compared to controls. The largest linkage study was performed in 82 families (Seibold et al., 2011), and demonstrated linkage to a 3.4 Mb region of chromosome 11p15 having a LOD score of 3.3. Fine mapping of this region using 306 tagging SNPs in a case-control association analysis in two independent validation populations, a familial cohort ($n = 85$), a sporadic IPF cohort ($n = 494$), compared to controls ($n = 331$), demonstrated several significant associations with SNPs in lung-expressed, gel-forming mucin genes. The SNP rs35705950 (G to T), 3 kb upstream of the *MUC5B* transcription start site, was the SNP with the most significant association with both sporadic and familial disease. The SNP rs3570590 was strongly associated with both FIP ($P = 6.3 \times 10^{-12}$) and IPF ($P = 6.6 \times 10^{-31}$). The relative risks of disease for subjects heterozygous and homozygous for the rarer allele of this SNP were 6.4 [95% confidence interval (CI) 3.7–11.1] and 20.2 (95% CI 3.7–110.5) for FIP, and 8.5 (95%

CI 5.9–12.3) and 22.9 (95% CI 5.3–98.2) for IPF. Furthermore, this SNP is associated with a 37-fold increase in *MUC5B* mRNA in lung. Importantly, approximately 55–60% of patients with sporadic or familial disease carry either one or two copies of the risk allele, and to date, this variant is the most common susceptibility allele for pulmonary fibrosis. The mechanism whereby *MUC5B* overexpression predisposes to pulmonary fibrosis is not clear. If screening for genetic variants predisposing to pulmonary fibrosis is considered in either familial or sporadic pulmonary fibrosis, genotyping for rs3570590 should be performed.

CONCLUSION

The ILDs/DPLDs are a heterogeneous group of diseases with complex pathogenesis, diverse histopathology, and variable natural history. It is becoming increasingly clear that these entities occur in genetically susceptible individuals when combined with other triggers such as environmental and drug exposures. Currently, obtaining a thorough family history and environmental history will identify at-risk individuals. In the future, at-risk individuals will likely be identified by genotyping for high-risk alleles. High-risk alleles for the development of sarcoidosis have recently been identified, and the application of these risk alleles to the diagnosis, prognosis, and treatment of sarcoidosis will likely evolve rapidly over the next several years. Also, proteomic approaches have demonstrated success in identification of mycobacteria species as a potential environmental trigger to the development of sarcoidosis, and similar approaches are likely to identify environmental triggers in other DPLDs. Linkage, GWAS, and candidate genes studies will continue to identify additional susceptibility genes for the development of pulmonary fibrosis.

Microarray expression profiling and proteomic profiling may play an increasing role in the diagnosis of ILDs/DPLDs. It is likely that in the future, the diagnosis of a specific type of ILD/DPLD will include the use of a panel of disease-specific antibodies to stain lung tissue in a manner similar to the histopathologic diagnosis of carcinoma or lymphoma, or the use of other biomarkers identified using genomic approaches. Secondly, substantial variation in rates of disease progression exists for most of the IIPs. Some patients experience stable disease lasting several years, and then experience acute exacerbations characterized by rapid disease progression resulting in precipitous drop of lung function and death. This variable natural history confounds several relevant clinical problems such as the appropriate timing of referral for lung transplantation, an operation that has a clear survival advantage in IPF and end-stage pulmonary sarcoidosis, but is not uniformly successful in all patients. The variable natural history of these diseases also confounds the appropriate design of clinical treatment trials. The identification of biomarkers using genomic and proteomic profiling that accurately predict prognosis or treatment responsiveness would be of great clinical value. Proteomic profiling of lung tissue or bronchoalveolar lavage fluid may clarify environmental exposure and assist in the diagnosis of hypersensitivity pneumonitis, since the triggering antigen is sometimes not identified by the environmental history. Currently, we have little insight into the mechanism of acute exacerbations of IPF or the pathogenesis of IIPs in general. Through the use of genomic medicine, it is likely that better insight into disease mechanisms will be obtained, leading to better treatment strategies in these disorders.

Clinical diagnostic evaluation of patients with ILD/DPLD remains at this time primarily a radiologic and histopathologic diagnosis. High-resolution chest CT combined with transbronchial biopsy remains the method of choice for the diagnosis of patients with pulmonary sarcoidosis. Patients with pulmonary sarcoidosis with dyspnea and abnormal lung function warrant a trial of glucocorticoid therapy to determine a therapeutic response. One goal of personalized medicine in sarcoidosis is to identify genetic or protein markers that will aid in diagnosis and predict steroid-responsive lung disease. Similarly, the diagnosis of IIP is based on clinical history, high-resolution chest CT, and, when performed, histopathologic confirmation by surgical lung biopsy. One goal of personalized medicine in the management of IIP is to determine biomarkers that distinguish the IIPs, to distinguish slowly and rapidly progressive IPF, and predict disease responsive to anti-inflammatory therapy. Currently,

there is sufficient validation of susceptibility mutations in *SP-C*, telomerase genes (*TERT/TERC*), and *MUC5B* to suggest a role for screening family members with FIP for mutations in these genes to assess individual risk of disease (Table 76.3). In the near future, it is likely that susceptibility variants for sarcoidosis and sporadic IIP will be available for clinical testing.

TABLE 76.3 Molecular testing in familial interstitial pneumonia

Gene	% of FIP	Method	Mutations*	Test available[†]
TERT	10–15	Sequence analysis	Sequence variants	Yes
TERC	2	Sequence analysis	Sequence variants	Yes
SP-C	1	Sequence analysis	Sequence variants	Yes
MUC5B	60	SNP assay	G to T	No

*Sequence variants such as small insertion/deletions, missense, nonsense, or splice mutations.
[†]US Clinical Laboratory Improvement Amendment-licensed laboratory.

REFERENCES

Adelman, A.G., Chertkow, G., Hayton, R.C., 1966. Familial fibrocystic pulmonary dysplasia: A detailed family study. Can Med Assoc J 95, 603–610.

Akahoshi, M., Ishihara, M., Remus, N., et al., 2004. Association between IFNA genotype and the risk of sarcoidosis. Hum Genet 114, 503–509.

Alia, P., Mana, J., Capdevila, O., Alvarez, A., Navarro, M.A., 2005. Association between ACE gene I/D polymorphism and clinical presentation and prognosis of sarcoidosis. Scand J Clin Lab Invest 65, 691–697.

American Thoracic Society, 1999. Statement on sarcoidosis. Joint Statement of the American Thoracic Society (ATS), the European Respiratory Society (ERS) and the World Association of Sarcoidosis and Other Granulomatous Disorders (WASOG) adopted by the ATS Board of Directors and by the ERS Executive Committee, February 1999. Am J Respir Crit Care Med 160, 736–755.

Armaniaos, M.Y., 2007. Telomerase mutations in families with idiopathic pulmonary fibrosis. N Engl J Med 356, 1317–1326.

Auwerx, J., Boogaerts, M., Ceuppens, J.L., Demedts, M., 1985. Defective host defence mechanisms in a family with hypocalciuric hypercalcaemia and coexisting interstitial lung disease. Clin Exp Immunol 62, 57–64.

Ban, N., Matsumura, Y., Sakai, H., et al., 2007. ABCA3 as a lipid transporter in pulmonary surfactant biogenesis. J Biol Chem 282, 9628–9634.

Baughman, R.P., Lower Elyse, E., Du Bois, R.M., 2003. Sarcoidosis. Lancet 361, 1111–1118.

Berlin, M., Fogdell-Hahn, A., Olerup, O., Eklund, A., Grunewald, J., 1997. HLA-DR predicts the prognosis in Scandinavian patients with pulmonary sarcoidosis. Am J Respir Crit Care Med 156, 1601–1605.

Bitterman, P.B., Rennard, S.I., Keogh, B.A., et al., 1986. Familial idiopathic pulmonary fibrosis: Evidence of lung inflammation in unaffected members. N Engl J Med 314, 1343–1347.

Bogunia-Kubik, K., Koscinska, K., Suchnicki, K., Lange, A., 2006. HSP70-hom gene single nucleotide (12763 G/A and 12437 C/T) polymorphisms in sarcoidosis. Int J Immunogenet 33, 135–140.

Bogunia-Kubik, K., Tomeczko, J., Suchnicki, K., Lange, A., 2001. HLA-DRB1*03, DRB1*11 or DRB1*12 and their respective DRB3 specificities in clinical variants of sarcoidosis. Tissue Antigens 57, 87–90.

Bonanni, P.P., Frymoyer, J.W., Jacox, R.F., 1965. A family study of idiopathic pulmonary fibrosis. A possible dysproteinemic and genetically determined disease. Am J Med 39, 411–421.

Boon, K., Bailey, N.W., Yang, J., et al., 2009. Molecular phenotypes distinguished patients with relatively stable from progressive idiopathic pulmonary fibrosis (IPF). PlosOne 4, e5134.

Buck, A.A., McKusick, V.A., 1961. Epidemiologic investigations of sarcoidosis. III. Serum proteins; syphilis; association with tuberculosis: Familial aggregation. Am J Hyg 74, 174–188.

Chase, M., Siltzbach, L.E., 1967. Concentration of the active principle responsible for the Kveim reaction. La Sarcoidose 196, 150–160.

Chibbar, R., Shih, F., Baga, M., et al., 2004. Nonspecific interstitial pneumonia and usual interstitial pneumonia with mutation in surfactant protein C in familial pulmonary fibrosis. Mod Pathol 17, 973–980.

Corsini, E., Luster, M.I., Mahler, J., et al., 1994. A protective role for T lymphocytes in asbestos-induced pulmonary inflammation and collagen deposition. Am J Respir Cell Mol Biol 11, 531–539.

Cronkhite, J.T., Xing, C., Raghu, G., et al., 2008. Telomere shortening in familial and sporadic pulmonary fibrosis. Am J Respir Crit Care Med 178, 729–737.

Crouser, E.D., Culver, D.A., Knox, K.S., et al., 2009. Gene expression profiling identifies MMP-12 and ADAMDEC1 as potential pathogenic mediators of pulmonary sarcoidosis. Am J Respir Crit Care Med 179, 929–938.

Depinho, R.A., Kaplan, K.L., 1985. The Hermansky–Pudlak syndrome: Report of three cases and review of pathophysiology and management considerations. Medicine 64, 192–202.

Dokal, I., Vulliamy, T., 2000. Dyskeratosis congenital in all its forms. Br J Haematol 110, 768–769.

Foley, P.J., McGrath, D.S., Puscinska, E., et al., 2001. Human leukocyte antigen-DRB1 position 11 residues are a common protective marker for sarcoidosis. Am J Respir Cell Mol Biol 25, 272–277.

Franko, A.J., Sharplin, J., Ward, W.F., Taylor, J.M., 1996. Evidence for two patterns of inheritance of sensitivity to induction of lung fibrosis in mice by radiation, one of which involves two genes. Radiat Res 146, 68–74.

Gardner, J., Kennedy, H.G., Hamblin, A., Jones, E., 1984. HLA associations in sarcoidosis: A study of two ethnic groups. Thorax 39, 19–22.

Gray-McGuire, C., Sinha, R., Iyengar, S., et al., 2006. Genetic characterization and fine mapping of susceptibility loci for sarcoidosis in African Americans on chromosome 5. Hum Genet 120, 420–430.

Grunewald, J., Eklund, A., Olerup, O., 2004. Human leukocyte antigen class I alleles and the disease course in sarcoidosis patients. Am J Respir Crit Care Med 169, 696–702.

Harrington, D., Major, M., Rybicki, B., et al., 1994. Familial analysis of 91 families. Sarcoidosis 11, 240–243.

Harris, J., Waltuck, B., Swenson, E., 1969. The pathophysiology of the lungs in tuberous sclerosis. A case report and literature review. Am Rev Respir Dis 100, 379–387.

Haston, C.K., Amos, C.I., King, T.M., Travis, E.L., 1996. Inheritance of susceptibility to bleomycin-induced pulmonary fibrosis in the mouse. Cancer Res 56, 2596–2601.

Haston, C.K., Tomko, T.G., Godin, N., Kerckhoff, L., Hallett, M.T., 2005. Murine candidate bleomycin induced pulmonary fibrosis susceptibility genes identified by gene expression and sequence analysis of linkage regions. J Med Genet 42, 464–473.

Haston, C.K., Travis, E.L., 1997. Murine susceptibility to radiation-induced pulmonary fibrosis is influenced by a genetic factor implicated in susceptibility to bleomycin-induced pulmonary fibrosis. Cancer Res 57, 5286–5291.

Haston, C.K., Wang, M., Dejournett, R.E., et al., 2002a. Bleomycin hydrolase and a genetic locus within the MHC affect risk for pulmonary fibrosis in mice. Hum Mol Genet 11, 1855–1863.

Haston, C.K., Zhou, X., Gumbiner-Russo, L., et al., 2002b. Universal and radiation-specific loci influence murine susceptibility to radiation-induced pulmonary fibrosis. Cancer Res 62, 3782–3788.

Headings, V.E., Weston, D., Young Jr., R.C., Hackney Jr., R.L., 1976. Familial sarcoidosis with multiple occurrences in eleven families: A possible mechanism of inheritance. Ann NY Acad Sci 278, 377–385.

Hodgson, U., Laitinen, T., Tukiainen, P., 2002. Nationwide prevalence of sporadic and familial idiopathic pulmonary fibrosis: Evidence of founder effect among multiplex families in Finland. Thorax 57, 338–342.

Hodgson, U., Pulkkinen, V., Dixon, M., et al., 2006. ELMOD2 is a candidate gene for familial idiopathic pulmonary fibrosis. Am J Hum Genet 79, 149–154.

Hofmann, S., Fischer, A., Till, A., et al., 2011. A genome-wide association study reveals evidence of association with sarcoidosis at 6p12.1. ERJ Express 38., 1127–1135.

Hofmann, S., Franke, A., Fischer, A., et al., 2008. Genome-wide association study identifies *ANXA11* as a new susceptibility locus for sarcoidosis. Nature Genetics 40, 1103–1106.

Hughes, E.W., 1964. Familial interstitial pulmonary fibrosis. Thorax 19, 515–525.

Ina, Y., Takada, K., Yamamoto, M., et al., 1989. HLA and sarcoidosis in the Japanese. Chest 95, 1257–1261.

Ishihara, M., Ohno, S., Ishida, T., et al., 1994. Molecular genetic studies of HLA class II alleles in sarcoidosis. Tissue Antigens 43, 238–241.

Javaheri, S., Lederer, D.H., Pella, J.A., Mark, G.J., Levine, B.W., 1980. Idiopathic pulmonary fibrosis in monozygotic twins. The importance of genetic predisposition. Chest 78, 591–594.

Katsuma, S., Nishi, K., Tanigawara, K., et al., 2001. Molecular monitoring of bleomycin-induced pulmonary fibrosis by cDNA microarray-based gene expression profiling. Biochem Biophys Res Commun 288, 747–751.

Konishi, K., Gibson, K.F., Lindell, K.O., et al., 2009. Gene expression profiles of acute exacerbations of idiopathic pulmonary fibrosis. Am J Respir Crit Care Med 180, 167–175.

Kruit, A., Grutters, J.C., Ruven, H.J., et al., 2006. Transforming growth factor-beta gene polymorphisms in sarcoidosis patients with and without fibrosis. Chest 129, 1584–1591.

Lawson, W.E., Grant, S.W., Ambrosini, V., et al., 2004. Genetic mutations in surfactant protein C are a rare cause of sporadic cases of IPF. Thorax 59, 977–980.

Lawson, W.E., Loyd, J.E., 2010. Will the genes responsible for familial pulmonary fibrosis provide clues to the pathogenesis of IPF? Am J Respir Crit Care Med 182, 1342–1343.

Lee, H.L., Ryu, J.H., Wittmer, M.H., et al., 2005. Familial idiopathic pulmonary fibrosis: Clinical features and outcome. Chest 127, 2034–2041.

Lemay, A.M., Haston, C.K., 2005. Bleomycin-induced pulmonary fibrosis susceptibility genes in AcB/BcA recombinant congenic mice. Physiol Genomics 23, 54–61.

Lockstone, H.E., Sanderson, S., Kulakova, N., et al., 2010. Gene set analysis of lung samples provides insight into pathogenesis of progressive, fibrotic pulmonary sarcoidosis. Am J Respir Crit Care Med 181, 1367–1375.

Lyons, D.J., Donald, S., Mitchell, D.N., Asherson, G.L., 1992. Chemical inactivation of the Kveim reagent. Respiration 59, 22–26.

Magi, B., Bini, L., Perari, M.G., et al., 2002. Bronchoalveolar lavage fluid protein composition in patients with sarcoidosis and idiopathic pulmonary fibrosis: A two-dimensional electrophoretic study. Electrophoresis 23, 3434–3444.

Makle, S.K., Pardee, N., Martin, C.J., 1970. Involvement of the lung in tuberous sclerosis. Chest 58, 538–540.

Maliarik, M., Chen, K., Major, M., et al., 1998a. Analysis of HLA-DPB1 polymorphisms in African-Americans with sarcoidosis. Am J Respir Crit Care Med 158, 111–114.

Maliarik, M.J., Rybicki, B.A., Malvitz, E., et al., 1998b. Angiotensin-converting enzyme gene polymorphism and risk of sarcoidosis. Am J Respir Crit Care Med 158, 1566–1570.

Marshall, R., Puddicombe, A., Cookson, W., Laurent, G., 2000. Adult familial cryptogenic fibrosing alveolitis in the United Kingdom. Thorax 55, 143–146.

Martinetti, M., Tinelli, C., Kolek, V., et al., 1995. The sarcoidosis map: A joint survey of clinical and immunogenetic findings in two European countries. Am J Respir Crit Care Med 152, 557–564.

McGrath, D.S., Foley, P.J., Petrek, M., et al., 2001. Ace gene I/D polymorphism and sarcoidosis pulmonary disease severity. Am J Respir Crit Care Med 164, 197–201.

McKusick, V.A., Fisher, A.M., 1958. Congenial cystic disease of the lung with progressive pulmonary fibrosis and carcinomatosis. Ann Inter Med 48, 774–790.

Moura, M., Carre, P., Larios-Ramos, L., Didier, A., Leophonte, P., 1990. Sarcoidosis and heredity. 3 familial cases. Rev Pneumol Clin 46, 28–30.

Mrazek, F., Holla, L.I., Hutyrova, B., et al., 2005. Association of tumour necrosis factor-alpha, lymphotoxin-alpha and HLA-DRB1 gene polymorphisms with Lofgren's syndrome in Czech patients with sarcoidosis. Tissue Antigens 65, 163–171.

Mushiroda, T., Wattanapokayakit, S., Takahasi, A., et al., 2008. A genome-wide association study identifies an association of a common variant in TERT with susceptibility to idiopathic pulmonary fibrosis. J Med Genet 45, 654–656.

Musk, A., Zilko, P., Manners, P., Kay, P., Kamboh, M., 1986. Genetic studies in familial fibrosing alveolitis. Possible linkage with immunoglobulin allotypes (gm). Chest 89, 206–210.

Niimi, T., Sato, S., Tomita, H., et al., 2000a. Lack of association with interleukin 1 receptor antagonist and interleukin-1beta gene polymorphisms in sarcoidosis patients. Respir Med 94, 1038–1042.

Niimi, T., Tomita, H., Sato, S., et al., 2000b. Vitamin D receptor gene polymorphism and calcium metabolism in sarcoidosis patients. Sarcoidosis Vasc Diffuse Lung Dis 17, 266–269.

Noble, P.W., Albera, C., Bradford, W.Z., et al., 2011. Pirfenidone in patients with idiopathic pulmonary fibrosis (CAPACITY): Two randomized trials. Lancet 377, 1760–1769.

Nogee, L.M., 2006. Genetics of pediatric interstitial lung disease. Curr Opin Pediatr 18 (3), 287–292.

Nogee, L.M., Dunbar 3rd, A.E., Wert, S.E., et al., 2001. A mutation in the surfactant protein C gene associated with familial interstitial lung disease. N Engl J Med 344, 573–579.

Nogee, L.M., Wert, S.E., Proffit, S.A., Hull, W.M., Whitsett, J.A., 2000. Allelic heterogeneity in hereditary surfactant protein B (SP-B) deficiency. Am J Respir Crit Care Med 161, 973–981.

Ortiz, L.A., Lasky, J., Hamilton Jr., R.F., et al., 1998. Expression of TNF and the necessity of TNF receptors in bleomycin-induced lung injury in mice. Exp Lung Res 24, 721–743.

Pabst, S., Baumgarten, G., Stremmel, A., et al., 2006. Toll-like receptor (TLR) 4 polymorphisms are associated with a chronic course of sarcoidosis. Clin Exp Immunol 143, 420–426.

Pardo, A., Gibson, K., Cisneros, J., et al., 2005. Up-regulation and profibrotic role of osteopontin in human idiopathic pulmonary fibrosis. PLoS Med 2, e251.

Polakoff, P.L., Horn, B.R., Scherer, O.R., 1979. Prevalence of radiographic abnormalities among Northern California shipyard workers. Ann NY Acad Sci 33, 333–339.

Riccardi, V., 1981. Von Recklinghausen neurofibromatosis. N Engl J Med 305, 1617–1627.

Rosas, I.O., Richards, T.J., Konishi, K., et al., 2008. MMP1 and MMP7 as potential peripheral blood biomarkers in idiopathic pulmonary fibrosis. PLoS Med 5 (4), e93.

Rossi, G., Szapiel, S., Ferrans, V., Crystal, R., 1987. Susceptibility to experimental interstitial lung disease is modified by immune- and non-immune-related genes. Am Rev Respir Dis 135, 448–455.

Rossman, M.D., Thompson, B., Frederick, M., et al., 2003. HLA-DRB1*1101: A significant risk factor for sarcoidosis in blacks and whites. Am J Hum Genet 73, 720–735.

Rottoli, P., Magi, B., Perari, M.G., et al., 2005. Cytokine profile and proteome analysis in bronchoalveolar lavage of patients with sarcoidosis, pulmonary fibrosis associated with systemic sclerosis and idiopathic pulmonary fibrosis. Proteomics 5, 1423–1430.

Rybicki, B.A., Hirst, K., Iyengar, S.K., et al., 2005a. A sarcoidosis genetic linkage consortium: The sarcoidosis genetic analysis (SAGA) study. Sarcoidosis Vasc Diffuse Lung Dis 22, 115–122.

Rybicki, B.A., Iannuzzi, M.C., Frederick, M.M., et al., 2001a. Familial aggregation of sarcoidosis. A case-control etiologic study of sarcoidosis (ACCESS). Am J Respir Crit Care Med 164, 2085–2091.

Rybicki, B.A., Kirkey, K.L., Major, M., et al., 2001b. Familial risk ratio of sarcoidosis in African-American sibs and parents. Am J Epidemiol 153, 188–193.

Rybicki, B.A., Walewski, J.L., Maliarik, M.J., Kian, H., Iannuzzi, M.C., 2005b. The BTNL2 gene and sarcoidosis susceptibility in African Americans and Whites. Am J Hum Genet 77, 491–499.

Sabounchi-Schutt, F., Astrom, J., Hellman, U., Eklund, A., Grunewald, J., 2003. Changes in bronchoalveolar lavage fluid proteins in sarcoidosis: A proteomics approach. Eur Respir J 21, 414–420.

Sato, H., Grutters, J.C., Pantelidis, P., et al., 2002. HLA-DQB1*0201: A marker for good prognosis in British and Dutch patients with sarcoidosis. Am J Respir Cell Mol Biol 27, 406–412.

Schneider, E., Epstein, C., Kaback, M., Brandes, D., 1977. Severe pulmonary involvement in adult Gaucher's disease: Report of three cases and review of the literature. Am J Med 63, 475–480.

Schurmann, M., Reichel, P., Muller-Myhsok, B., et al., 2001. Results from a genome-wide search for predisposing genes in sarcoidosis. Am J Respir Crit Care Med 164, 840–846.

Seibold, M.A., Wise, A.L., Speer, M.C., et al., 2011. A polymorphism in MUC5B promoter plays a role in the development of familial interstitial pneumonia (FIP) and idiopathic pulmonary fibrosis (IPF). N Engl J Med 364, 1503–1512.

Seitzer, U., Gerdes, J., Muller-Quernheim, J., 2002. Genotyping in the MHC locus: Potential for defining predictive markers in sarcoidosis. Respir Res 3, 6.

Selikoff, I.J., Lilis, R., Nicholson, W.J., 1979. Asbestos disease in United States shipyards. Ann NY Acad Sci 330, 293–311.

Selman, M., et al., 2006. Gene expression profiles distinguish idiopathic pulmonary fibrosis from hypersensitivity pneumonitis. Am J Respir Crit Care Med 173, 188–198.

Selman, M., Lin, H.M., Montano, M., et al., 2003. Surfactant protein A and B genetic variants predispose to idiopathic pulmonary fibrosis. Hum Genet 113, 542–550.

Sharplin, J., Franko, A., 1989. Quantitative histological study of strain-dependent differences in the effects of irradiation on mouse lung during the early phase. Radiation Res 119, 1–14.

Shulenin, S., Nogee, L.M., Annilo, T., et al., 2004. ABCA3 gene mutations in newborns with fatal surfactant deficiency. N Engl J Med 350 (13), 1296–1303.

Siltzbach, L.E., 1961. The Kveim test in sarcoidosis. A study of 750 patients. JAMA 178, 476–482.

Solliday, N.H., Williams, J.A., Gaensler, E.A., Coutu, R.E., Carrington, C., 1973. Familial chronic interstitial pneumonia. Am Rev Respir Dis 108, 193–204.

Song, Z., Marzilli, L., Greenlee, B.M., et al., 2005. Mycobacterial catalase-peroxidase is a tissue antigen and target of the adaptive immune response in systemic sarcoidosis. J Exp Med 201, 755–767.

Spagnolo, P., Renzoni, E.A., Wells, A.U., et al., 2005. C-C chemokine receptor 5 gene variants in relation to lung disease in sarcoidosis. Am J Respir Crit Care Med 172, 721–728.

Steele, M.P., Brown, K., Wahidi, M., 2007. The contrasting histopathologic and radiographic features of familial interstitial pneumonia and idiopathic pulmonary fibrosis. Am J Respir Crit Care Med 3, A242.

Steele, M.P., Speer, M.C., Loyd, J.E., et al., 2005. Clinical and pathologic features of familial interstitial pneumonia. Am J Respir Crit Care Med 172, 1146–1152.

Swaye, P., van Ordstrand, H.S., McCormack, L.J., Wolpaw, S.E., 1969. Familial Hamman-Rich syndrome. Chest 55, 7–12.

Swider, C., Schnittger, L., Bogunia-Kubik, K., et al., 1999. TNF-alpha and HLA-DR genotyping as potential prognostic markers in pulmonary sarcoidosis. Eur Cytokine Netw 10, 143–146.

Takada, T., Suzuki, E., Ishida, T., et al., 2001. Polymorphism in RANTES chemokine promoter affects extent of sarcoidosis in a Japanese population. Tissue Antigens 58, 293–298.

Terry, R., Sperry, W., Brodoff, B., 1954. Adult lipoidosis resembling Neimann–Pick disease. Am J Pathol 30, 263–286.

Thomas, A.Q., Lane, K., Phillips III, J., et al., 2002. Heterozygosity for a surfactant protein C gene mutation associated with usual interstitial pneumonitis and cellular nonspecific interstitial pneumonitis in one kindred. Am J Respir Crit Care Med 165, 1322–1328.

Tsakiri, K.D., Cronkhite, J.T., Kuan, P.J., et al., 2007. Adult-onset pulmonary fibrosis caused by mutations in telomerase. Proc Natl Acad Sci USA 104, 7552–7557.

Tsukahara, M., Kajii, T., 1983. Interstitial pulmonary fibrosis in two sisters. Possible autosomal recessive inheritance. Jinrui Idengaku Zasshi 28, 263–267.

Tzouvelekis, A., Harokopos, V., Paparountas, T., et al., 2007. Comparative expression profiling in pulmonary fibrosis suggests a role of hypoxia-inducible factor-1alpha in disease pathogenesis. Am J Respir Crit Care Med 176, 1108–1119.

Valentonyte, R., Hampe, J., Huse, K., et al., 2005. Sarcoidosis is associated with a truncating splice site mutation in BTNL2. Nat Genet 37, 357–364.

Van Moorsel, C.H., Van Oosterhout, M.F., Barlo, N.P., et al., 2011. Surfactant protein C mutations are the basis of a significant portion of adult familial pulmonary fibrosis in a Dutch cohort. Am J Respir Crit Care Med 182, 1419–1425.

Wang, X.M., Zhang, Y., Kim, H.P., et al., 2006. Caveolin-1: A critical regulator of lung fibrosis in idiopathic pulmonary fibrosis. J Exp Med 203, 2895–2906.

Wang, Y., Kuan, P.J., Xing, C., et al., 2009. Genetic defects in surfactant protein A2 are associated with pulmonary fibrosis and lung cancer. Am J Hum Genet 84, 52–59.

Warshamana, G.S., Pociask, D.A., Sime, P., Schwartz, D.A., Brody, A.R., 2002. Susceptibility to asbestos-induced and transforming growth factor-beta1-induced fibroproliferative lung disease in two strains of mice. Am J Respir Cell Mol Biol 27, 705–713.

Whitsett, J.A., Weaver, T.E., 2002. Hydrophobic surfactant proteins in lung function and disease. N Engl J Med 347, 2141–2148.

Wiman, L.G., 1972. Familial occurrence of sarcoidosis. Scand J Respir Dis Suppl 80, 115–119.

Wright, J.R., 2005. Immunoregulatory functions of surfactant proteins. Nat Rev Immunolog 5, 58–68.

Yang, I.V., et al., 2005. Gene expression profiling distinguishes familial and non-familial forms of pulmonary fibrosis. Proc of the Am Thorac Soc 2, A242.

Young, L.R., Nogee, L.M., Barnett, B., et al., 2008. Usual interstitial pneumonia in an adolescent with ABCA3 mutations. Chest 134 (1), 192–195.

Zorzetto, M., Bombieri, C., Ferrarotti, I., et al., 2002. Complement receptor 1 gene polymorphisms in sarcoidosis. Am J Respir Cell Mol Biol 27, 17–23.

Zuo, F., Kaminski, N., Eugui, E., et al., 2002. Gene expression analysis reveals matrilysin as a key regulator of pulmonary fibrosis in mice and humans. Proc Natl Acad Sci USA 99, 6292–6297.

RECOMMENDED RESOURCES

http://pulmonary.medicine.duke.edu/research/
   clinical-research/interstitial-lung-disease-clinical-trials

http://www.nationaljewish.org/healthinfo/conditions/
   pulmonary-fibrosis/index.aspx

http://www.ncbi.nlm.nih.gov/books/NBK1230/
   ?&log$=disease5_name

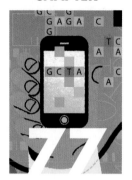

Peptic Ulcer Disease

John Holton

INTRODUCTION

Ulceration of the stomach can be caused by a number of conditions including Crohn's disease, ischemia, hypersecretory states [e.g., Zollinger–Ellison syndrome (gastrinoma)], antral G-cell hyperplasia, mastocytosis and multiple endocrine neoplasms (MEN-1), acute stress [such as burns, trauma, admission to an intensive care unit (ICU)], other medical conditions such as cirrhosis, chronic pulmonary disease, renal failure, the use of non-steroidal anti-inflammatory agents (NSAIDs), and *Helicobacter pylori*. The latter is the principal cause of peptic ulceration, followed closely by NSAID use [although recently in the UK, NSAID use has overtaken *H. pylori* as the main cause of peptic ulcer disease (PUD)]. This chapter will concentrate upon those two causes.

HELICOBACTER PYLORI

Bacteriology

Host–pathogen interactions leading to the development of disease inevitably involve aspects of both the host genome and the microbial genome from the perspective of risk factors and virulence factors respectively. In order to understand this complex interaction and how genomic and proteomic analysis can assist in understanding the pathophysiology of disease as well as shaping diagnosis and management, a brief introduction to the organism is necessary.

The *H. pylori* genome is about 1.60 Mbp (megabase pairs) long, and consists of about 1600 predicted genes. The organism has a large number of outer membrane proteins and putative adhesion molecules, testifying to the importance of adhesion in its biology. It also encodes for about 25 restriction endonucleases and 27 methylases, underpinning the importance of genetic exchange to *H. pylori*. Additionally, there are large numbers of homo-polymorphic tracts that are in part the basis for the known genetic diversity of the organism. The genome has several regions [pathogenicity islands (PAI)] that it has acquired from other organisms by lateral transfer of DNA and that usually carry virulence genes. Important areas of the *H. pylori* genome include the locus for the vacuolating cytotoxin A, the cagA pathogenicity island (*cag*PAI), and the plasticity zone (PZ). The first locus encodes for a cytotoxin that induces vacuolation in exposed cells, the second locus encodes for a Type IV secretion system (T4SS), and CagA (cytotoxin-associated gene) a protein that is transferred to the host cell and interferes with host cell signaling, leading to altered cell morphology, cell division, or apoptosis. The third locus is an area of hyper-mutation.

Epidemiology

It is now recognized that *H. pylori* colonizes more than 50% of the population of the world (Hunt et al., 2010). In developed countries, about 30% of the population is colonized and this percentage is gradually decreasing, but in developing countries as many as 70–90% are colonized.

Pathophysiology

Helicobacter pylori is an important gastroduodenal pathogen and is recorded as a class I carcinogen, making it not

Genomic and Personalized Medicine, 2nd edition
by Ginsburg & Willard. DOI: http://dx.doi.org/10.1016/B978-0-12-382227-7.00077-X

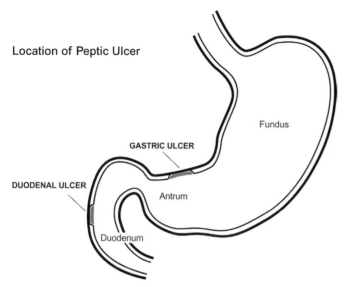

Location of Peptic Ulcer

This diagram shows the usual location of gastric and duodenal ulcers. Duodenal ulcers are found in the antrum of the stomach or the first part of the duodenum. Gastric ulcers are located at the junction of the antrum and fundus, or body of the stomach.

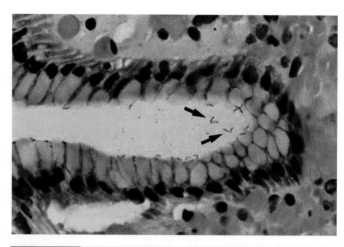

Figure 77.2 A silver-stained section of stomach showing *Helicobacter pylori* located within a gastric gland.

only a significant global pathogen but also of great scientific interest as the only bacterium thus far to be linked to carcinogenesis. The main adhesion protein of the organism is BabA (coded for by *BabA2*), which binds the Lewis b blood group antigen expressed on gastric epithelial cells. A second adhesion protein is SabA, which binds Lewis x expressed on inflamed mucosa. There are two patterns of binding that affect the clinical presentation: binding to the antrum of the stomach (antral gastritis) leads to a high acid output and duodenal ulceration (DU); binding throughout the stomach (pan-gastritis) leads to a low acid state with gastric ulcer (GU) or the development of adenocarcinoma of the stomach [gastric cancer (GCA)] (Figure 77.1).

There are four general pathophysiological mechanisms whereby *H. pylori* causes gastroduodenal disease: the production of surface-damaging agents (the ammonium ion, phospholipase, vacuolating cytotoxin); interference with normal acid secretion mechanisms leading to increased acid production; interference with cell signaling and cell dynamics (e.g., the CagA protein); and, finally, inflammatory (*cag*PAI) and immunopathological mechanisms causing bystander damage to epithelial cells (due to activation of granulocytes and the release of free radicals), and autoimmunity directed at parietal cells leading to hypochlorhydria.

Helicobacter pylori is a good example of a "slow" infection, as acquisition occurs in childhood but related diseases occur in adulthood. Although most persons colonized by *H. pylori* will remain symptom free, about 20% will go on to develop peptic ulcer disease and about 1% gastric cancer (adenocarcinoma) and mucosa-associated lymphoid tissue (MALT) lymphoma (see sections "*The Helicobacter Genome*

and *Virulence Markers*" and "*Host Polymorphisms*" below). The role of *H. pylori* in non-ulcer or functional dyspepsia (NUD, FD) is not clear.

Diagnosis

Diagnosis of infection with *H. pylori* is made by endoscopic or non-endoscopic means; the former involves taking biopsies which are investigated by culture, histology (Figure 77.2), detection of bacterial urease activity, and polymerase chain reaction (PCR) for detection of the organism. Non-endoscopic methods include serology; the stool antigen test (Ricci et al., 2007) – an immunoassay on the feces to detect the organism – and the urea breath test (Ricci et al., 2007), which detects the presence of the organism by measuring the evolution of CO_2 expelled in the breath after taking a test meal containing urea.

Treatment

The standard first-line therapy for *H. pylori* eradication is a proton pump inhibitor (PPI) combined with two antibiotics: clarithromycin (500 mg BD) and amoxicillin (1000 mg BD), or metronidazole (500 mg BD) and amoxicillin (1000 mg BD), all given for 7–10 days. Although, more recently, sequential therapy has been suggested as first-line therapy, where amoxicillin is given for five days prior to the introduction of clarithromycin and concomitant with ten days of a PPI. As indicated later (*Pharmacogenomics* section), tailoring the management regimen to the individual genetic profile of the host is of increasing importance.

Non-steroidal Anti-inflammatory Agents (NSAIDs)

There are more than 25 different NSAIDs. Pharmacologically they fall into several categories: derivatives of salicylic acid (e.g., aspirin), proprionic acid (e.g., ibuprofen), acetic acid (e.g., indomethacin), enolic acid (e.g., meloxicam), fenamic acid (e.g., mefenamic acid), and the selective cyclooxygenase-2 inhibitors (e.g., celecoxib). The common mode of action is to inhibit the activity of cyclooxygenase-1 and -2 (COX-1, COX-2), which are responsible for the production of prostaglandins and thromboxanes from arachidonic acid. Prostaglandins are important in maintaining an adequate gastric mucosal defense by decreasing acid production and maintaining mucus secretion; inhibition of prostaglandins leads to gastric ulceration. Some of the NSAIDs also have a direct toxic effect on gastric epithelial cells.

Ulcer Formation

The pathophysiology of ulcer development in the stomach is not clearly identified although, in general terms, two main factors may be involved: the disruption of the protective mucus layer and the exposure of the underlying epithelium to cytotoxic compounds. The lumen of the stomach contains many proteases (i.e., the digestive enzymes such as trypsin) in an acid solution (because of the secretion of hydrochloric acid from the parietal cells) at a pH of about 1.5–2.5. The reduction in protection and the subsequent exposure of the epithelium to damaging agents has been likened to a leaking roof, allowing these agents to directly affect the gastric epithelium and induce ulceration (Figures 77.3, 77.4).

Metagenomics of the Gastric Bacterial Flora

See also Chapter 14. Studies using culture demonstrate that relatively few bacteria are resident in the stomach, and those that are found are thought to be simply passing through. However, three studies have used a metagenomic approach to investigate the gastric microflora in individuals (Andersson et al., 2008; Bik et al., 2006; Monstein et al., 2000). The results of these studies illustrate that the stomach has abundant and diverse microflora with high inter-subject variation. The main phylotypes identified belong to the Firmicutes, Actinobacteria, Bacteroidetes, Proteobacteria, and Fusobacteria with the major component being the Firmicutes.

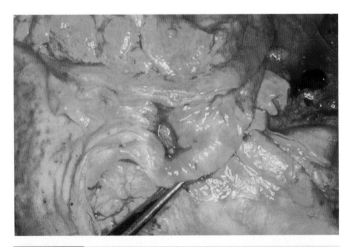

Figure 77.3   The macroscopic appearance of a duodenal ulcer in the antrum of the stomach.

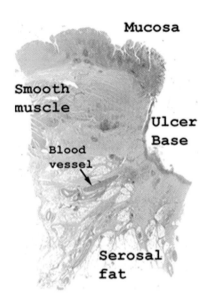

Figure 77.4   Photomicrograph of a hematoxylin- and eosin-stained section of stomach illustrating an ulcer which has eroded the epithelium, exposing the underlying tissues.

More recently, two studies have investigated the gastric microflora in patients with either gastric cancer (Dicksved et al., 2009) or gastritis in the absence of *H. pylori* (Li et al., 2009). In the former study, the microflora from ten patients with gastric cancer were compared to five patients with dyspepsia using terminal restriction fragment length polymorphism (T-RFLP) and 16S rRNA cloning and sequencing. Surprisingly, there was little difference between the patients with cancer and the dyspeptic controls. A total of 102 phylotypes were identified, representing five phyla: the Firmicutes (60%), Bacteroidetes (11%), Proteobacteria (6%), Actinobacteria (7%), and Fusobacteria (3%), confirming the results of the previous studies. *H. pylori* was detected in low numbers, with the predominant genera belonging to the oral bacteria: *Streptococcus*, *Prevotella*, *Lactobacillus*, and *Veillonella*. In the second study, which used 16S rRNA cloning and sequencing, 133 phylotypes representing eight phyla were identified from five normal individuals and five with antral gastritis. Paired samples were taken from corpus and antral mucosa. The identity and distribution of the phyla were similar to the other studies. The five normal individuals showed no difference between the antral and corpus specimens, but in those with antral gastritis (negative for *H. pylori* and not taking NSAIDs) there was an increase in the Firmicutes from the specimens compared to the normal mucosa in the corpus (44% vs 22%) and a relative decrease in the Proteobacteria (20% vs 37%).

In summary, these preliminary results clearly show the complex nature of the gastric flora and that the different studies detect the same principal phyla, also showing that the phylum Firmicutes is the predominant one. These results naturally raise the question of whether the microflora of the stomach may modulate the clinical presentation of *H. pylori* infection.

More generally, the use of metagenomic techniques to quantify and identify the microflora in different body compartments is likely to be used more frequently and to yield exciting new associations with physiology and pathophysiology.

THE *HELICOBACTER* GENOME AND VIRULENCE MARKERS

Since the first isolation of *H. pylori* and its subsequent association with peptic ulcer disease and gastric carcinoma/lymphoma, there has understandably been a search for virulence markers to explain these outcomes, particularly as, compared to gastric ulcer and carcinoma, the development of duodenal ulcer has different pathophysiological processes. On the one hand, antral gastritis has a high acid output leading to duodenal ulcer, but on the other hand, a pangastritis has a diminished acid output leading to gastric ulcer and carcinoma. However, the majority of those infected remain asymptomatic despite having gastritis. The main question this raises is how one organism can lead to different clinical outcomes.

Genomics

Helicobacter pylori has one of the highest levels of allelic diversity, with the most evolutionarily dynamic genomes of any bacterium (Baltrus et al., 2009). Much of the diversity is localized within the PAI or PZ. A strong Darwinian selective pressure has been shown to act upon certain genes of *H. pylori* following colonization, which leads to genetic variability and therefore may be an explanation of different genes acting as markers for different clinical outcomes (Torres-Morquecho et al., 2010).

In summary, there are a number of markers of virulence associated with more severe disease that are principally outer membrane proteins linked to adhesion and the immediate pro-inflammatory effect induced by bacterial adhesion. The main virulence marker identified to date is the presence of the cagA PAI, although it is not unreasonable to suppose that further virulence markers are likely to be identified in the near future.

Colonization and BabA

As the ability to colonize the stomach by penetrating the mucus, surviving the gastric acid, and adhering to the gastric epithelial cells are the first steps in the pathogenesis of disease, motility, the spiral shape, urease activity, and the adhesion proteins BabA/SabA are important initial virulence characteristics (Fischer et al., 2009; Ilver et al., 1998; Younson et al., 2009). The Bab locus consists of *BabA1* (silent), *BabA2*, and *BabB*. Recombination between *BabA1* and *BabB* will also allow binding to Lewis b blood group antigen (Leb), but at a lower level compared to *BabA2* (Backstrom et al., 2004). *BabA* strains are significantly associated with both DU and GCA (Gerhard et al., 1999) but are not routinely looked for in a clinical setting.

CagA and VacA

Helicobacter pylori can be divided into Type I and Type II strains based upon the presence of CagA and the secretion of VacA. Type I strains are CagA positive and secrete VacA; Type II stains are CagA negative and do not secrete VacA (Xiang et al., 1995). However, the totality of the cag pathogenicity island is important, because if the T4SS is not functional due to deletion in the *cag*PAI, then CagA will not be injected into the host cell, and if *cag-E* is deleted, interleukin-8 (IL-8) induction is diminished (Figure 77.5). Thus the ability to induce IL-8 production in cells is correlated with the presence of the *cag*-PAI and the presence of an intact *cag*PAI is correlated with the development of more severe pathology.

Nearly all strains carry the *vac-A* locus, which secretes a cytotoxin whose activity is determined by host polymorphisms within the gene (Figure 77.6). There is allelic diversity within the leader sequence and mid-region of the *vac-A* gene. There are four variations of the leader sequence, designated s1a, s1b, s1c, and s2, and several variations in the mid-region, designated m1a, m1b, m1c, m1t, msa, m2b, and chimeric types such as m1b-m2 (Atherton et al., 1995; Pan et al., 1998; Strobel et al., 1998; Van Doorn et al., 1998a, 1999). Two further polymorphic regions have been identified in the *vac-A* gene: the intermediate region (i-region) (Rhead et al., 2007) and the d-region (Ogiwara et al., 2009), either without an 81-bp deletion (d1-367-379 bp) or with the 81-bp deletion (d2-298bp). Strains with an s1 leader sequence are more often associated with PUD than s2-containing strains. The mid-region is also an independent marker for cytotoxin activity, and strains with s1/m1 alleles are more cytotoxic than s2/m2 strains in *in vitro* assays. Strains that are s1/m1 appear to produce high levels of cytotoxin, and are more frequently found in *cag-A* positive strains (Atherton, 1997). Additionally, there may be a modulation of toxin production based upon levels of transcription of the *vac-A* gene (Forsyth et al., 1998), which could affect the phenotype of the strain and clinical outcomes.

Induced by Contact with Epithelium Gene (ice-A)

Genes induced in *H. pylori* by contact with gastric epithelium were identified as *ice-A1* and *ice-A2* loci. *Ice-A1* strains were significantly associated with IL-8 induction and clinical presentation of DU (Peek et al., 1998).

Oip-A

Sequence analysis of *H. pylori* demonstrated the existence of a large number of outer membrane proteins (which include BabA2). Transposon mutation analysis demonstrated that IL-8 induction was significantly reduced in mutants of HP0638 (outer inflammatory protein −*oip-A*). *Oip-A* can be in the "on" or "off" position depending on its expression, which is related to the presence of several dinucleotide repeats resulting in truncation of the gene (Yamaoka et al., 2000).

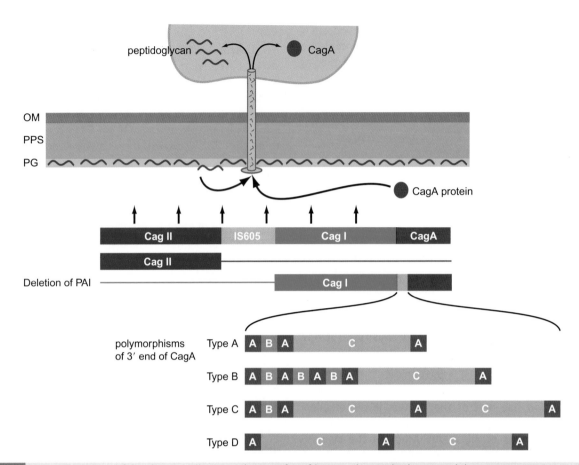

Figure 77.5 The upper part of the diagram illustrates the transfer of bacterial peptidoglycan and the CagA protein into the host cell by a Type IV secretion mechanism. The middle part of the diagram illustrates the structure of the cagPAI, having Cag I and Cag II regions separated by an insertion sequence, IS605. The lower part of the diagram illustrates polymorphisms found in the 3′ region adjacent to the CagA protein. OM = outer membrane, PPS = periplasmic space, PG = peptidoglycan, IS = insertion sequence.

Dup-A

A study of 500 clinical isolates originating from East Asia and South America in patients with gastritis, DU, GU, and GCA demonstrated one gene (*jhp0917-jhp0918*), called a duodenal-ulcer-promoting gene (*dup-A*), was significantly associated with DU [odds ratio (OR) 3.1] and less frequent in GCA (OR 0.42) (Lu et al., 2005b).

Hom-B

Hom-B is a further outer membrane protein that has been identified as a virulence marker. Knockout mutants had a reduced induction of IL-8 and a reduced ability to bind gastric tissue, suggesting a dual role in pathogenesis. The role of *hom-B* has been investigated in 190 *H. pylori* isolates from both children and adults, and has been associated with PUD and correlated with the presence of *cag-A*, *bab-A2*, *vac-A* s1, *hop-Q1*, and *oip-A* ("on" phenotype) (Oleastro et al., 2008).

HopQ

Two families of the outer membrane proteins HopQ are present in *Helicobacter pylori*: HopQ1 and 2. HopQ1 is significantly associated with Type I strains (*cag-A* positive *vac-A*

s1) from patients with PUD compared to Type II strains from patients colonized by *cag-A* negative *vac-A* s2 strains (Cao and Cover, 2002). HopQ appears to modulate binding of *H. pylori*, as some mutants result in increased binding with subsequent higher levels of phosphorylated CagA (Loh et al., 2008).

Jhp0562 – Glycosyl Transferase

The product of gene *jhp0562* is a glycosyl transferase involved in synthesis of lipopolysaccharide and has been associated with the development of PUD in children (Oleastro et al., 2006). This gene, and a similar gene (*jhp0563*) that codes for a galactosyl-transferase, were further investigated in children, and *jhp0562* was again strongly associated with PUD and found more frequently in *cag*PAI, *vac-A* s1, *bab-A2*, *hom-B*, *oip-A* "on" and *hop-Q1* positive strains. The best discriminator between PUD and NUD in children was the *cag*PAI+ *jhp0562*+ *hom-B*+ genotype (Oleastro et al., 2010).

Ethno-geographic Variation in Helicobacter Polymorphisms

Virulence markers for clinical outcome generally were initially identified in strains isolated from Western countries, but as more data has accumulated from other countries it has

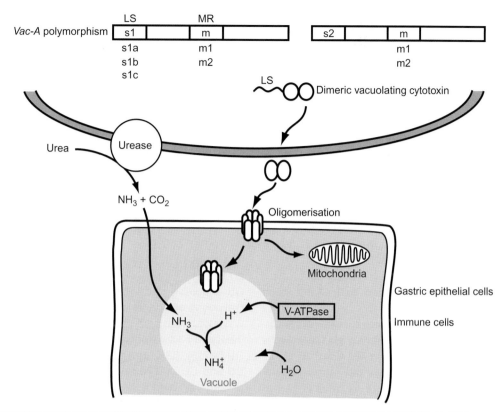

Figure 77.6 The upper part of the diagram illustrates the polymorphisms of the vacA cytotoxin. The rest of the diagram illustrates the secretion by a Type V secretion mechanism, oligomerization, and binding of the toxin to cells. It can form pores in the cytoplasmic membrane leading to leakage of nutrients from the cell; be taken up by cells and target the mitochondria, which ultimately induces apoptosis; be taken up by endocytosis and bind to the late endosome membranes forming anion-selective channels and inducing vacuolation. An acid environment is produced within the vesicles by V-ATPase, which is present on the vesicle membrane. As NH_3, produced by the *Helicobacter* urease, passively diffuses into the vesicles, it becomes protonated and the vesicle accumulates water due to increased osmotic pressure, thereby enlarging. LS=leader sequence, MR=mid-region, V-ATPase=vacuolar ATPase (ATP=adenosine triphosphate).

become apparent that there are ethno-geographic differences in the polymorphisms. Thus possession of the *cag*PAI is unequally distributed in strains isolated from different areas of the world. In East Asian countries, more than 90% of isolates possess the *cag*PAI, versus 60% of strains isolated from Western populations (Covacci et al., 1999).

In summary, the various recognized virulence markers, first identified from Western isolates, have been sought in isolates from other ethno-geographic locations around the globe.

Europe

In European countries, *cag-A vac-A* s1 strains are more frequently linked to severe gastroduodenal disease. In Lithuania, no correlation between PUD and *ice-A1* was found, although *cag-A* and *vac-A* s1 was more common in DU than GCA (Miciuleviciene et al., 2008). In Bulgaria, *ice-A1* positive strains were *cag-A* positive, *vac-A* s1a. *Bab-A2* positive strains were *cag-A* positive, *ice-A1*, and *vac-A* s1a. Strains that were *cag-A* positive and *vac-A* s1a were associated with patients with DU, although singly neither *ice-A1* nor *bab-A2* was more common

in DU patients. However, *iceA1* was a marker for clarithromycin susceptibility (Boyanova et al., 2010). A study of Sicilian patients illustrated the following frequencies of virulence markers: *cag-A* (51%), *vac-A* s1 (80%), *vac-A* m1 (35%), *bab-A2* (47%), *hop-Q1* (67%), and they were present in GCA irrespective of whether DU was present or not (Chiarini et al., 2009). A study of isolates from four European countries demonstrated an association between *vac-A* s1/*cag-A*/*bab-A2* and PUD in Germany, Portugal, and Finland, but not in Sweden, compared to isolates lacking *bab-A2* (Olfat et al., 2005). In one Italian study, the incidence of polymorphisms within the *vac-A* biomarker was studied over a 15-year period. The *vac-A* s1/m1 genotype had increased while the *vac-A* s2/m2 had decreased. In particular, *vac-A* s1/m1 and *vac-A* s1/m2 had increased while *vac-A* s2/m2 had decreased in patients with dyspepsia, but only *vac-A* s1/m1 had increased in patients with PUD. This finding questions whether *vac-A* s1/m1 is a valid biomarker for serious gastroduodenal disease as previously thought, as the incidence of PUD in Italy is actually falling despite the increase in *vac-A* s1/m1 (De Francesco et al., 2009). A study of

203 patients in Italy confirmed the association of *cag-A* and i1 *vac-A* with PUD, particularly duodenal ulcer for i1 *vac-A* (Basso et al., 2008). A study of *dup-A* in various ethnic groups from Europe, Australia, and the Far East showed that it was present in FD, DU, and GCA in Sweden at comparable prevalence, but was more frequent in DU/GCA in Chinese compared to FD (62/54% vs 28% respectively) (Schmidt et al., 2009).

Middle East

In a study in Turkey, *cag-A* was detected in more than 90% of strains with the predominant *vac-A* genotype, s1/m2, associated with DU (68%). The *ice-A1* allele was found in 68% of DU, with the *ice-A2* allele in 22% of gastritis. Both *ice-A* alleles were significantly linked to *cag-A* and *vac-A* s1a/m2 (Caner et al., 2007). In Turkey, a study of the *cag-A* 3′ region demonstrated that two or more EPIYA-C motifs (an amino acid sequence, using the single-letter amino acid code) were strongly associated with PUD compared to gastritis (Salih et al., 2010).

In Iran, a study using PCR and RFLP could not find any correlation between *cag-A* status and GU, although there was a correlation between the *vac-A* gene and GU (Farshad et al., 2007). A separate study in Saudi Arabia did find a correlation between PUD and either *cag-A* or *ice-A1* (Momenah and Tayeb, 2007). A study of the genotype of *H. pylori* in relation to different ethnic groups in Iran demonstrated that *cag-A* was common in all groups, but *cag-E* was uncommon in Turks and Persians (30–47%) while it occurred at 77% in other ethnic groups (Kurds, Lurs, Afghanis, Arabs). The common *vac-A* genotype was s1/m1 but there was no correlation with disease outcome (Dabiri et al., 2009). A study by the same group comparing the differences in genotypes between Iranian and immigrant Afghanis showed that *cag-A* was more common in Afghanis compared to Iranians (67% vs 60%), with *cag-E* present in comparable frequencies (53% vs 51%). *Vac-A* s1 was found in 80% vs 67%, with *vac-A* s1/m1 being the commonest in Afghanis (53%) and *vac-A* s1/m2 commonest in Iranians (47%). No correlation between any of the markers and clinical presentation was noted (Dabiri et al., 2010; Jafari et al., 2008). However, in a study of 50 clinical isolates, *vac-A* s1/m1 was commonly found in patients with DU (68%) and GU (56%) (Salari et al., 2009). No association was found between the *cag-A* status and the degree of inflammation in 166 patients nor between the *vac-A* genotypes, except for *vac-A* s1, which was present in patients with more severe inflammation (Molaei et al., 2010).

In a study of 59 patients from Iran and 49 from Iraq, there was no difference in the *cag-A* status; there was an association with DU in Iraq but not in Iran. In Iran, *vac-A* polymorphisms were not correlated with PUD, although in Iraq *vac-A* i1 was associated with GU. The duodenal-ulcer-promoting gene, *dup-A*, was found in similar numbers in both Iran and Iraq, and was associated with PUD in Iraq but not Iran (Hussein et al., 2008). A further study in Iran investigating the prevalence of *cag*PAI deletions demonstrated that 62% of isolates (n = 231) had a partial PAI deletion, but there was no correlation with clinical outcome (Baghaei et al., 2009). A similar study from Iran did find an association between the presence of *cag-A* and DU (80%) and GU (77%) compared to gastritis (46%), but no correlation between disease outcome and 3′ *cag* region subtypes (Salehi et al., 2009). An additional study from Iran demonstrated an association between *cag-A* and PUD/GCA compared to NUD (90–100% vs 64%), and *cag-E* and PUD compared to GCA/NUD (95–96% vs 70–73%) (Ghasemi et al., 2008). In Egypt, a serological study (Said Essa et al., 2008) demonstrated anti-CagA antibodies associated with PUD and gastric cancer compared to non-ulcer dyspepsia (NUD).

India and Pakistan

In a study of 115 isolates from Pakistan, very few strains had a complete *cag*PAI. A complete *cag*PAI was associated with GU and GCA, particularly the *cag-A* promoter region for GU (Yakoob et al., 2009b). In a further study of 224 isolates, *cag-A* was associated with chronic gastritis (CG), DU and GU in 73%, 72%, and 63% respectively, and *vac-A* s1a/m1 in GU, GCA and DU in 72%, 58%, and 53%. VacA s1b/m1 was associated with DU and GU in 37% and 28% (Yakoob et al., 2009a). In India, a study comparing asymptomatic patients and those with DU could not find any difference in the histological characteristics nor the distribution of *H. pylori* genotypes in either population, despite the difference in clinical outcome (Saha et al., 2009). Diversity in the *cag-A* gene and the PAI has been investigated in India, showing large differences between north and south India, with partial deletion of the PAI more frequent in north India (65% vs 53%) and variation in both the 5′ and 3′ of the *cag-A* gene, the majority of the isolates harboring a western-specific sequence (WSS; see next section) (Kumar et al., 2010).

Far East

In a study from Thailand, *cag-A*, *vac-A*, and *ice-A1* were found in the majority of specimens, with the predominant *vac* alleles being s1a and s1a/s1c. A correlation between the presence of *cag-A* and mixed s1a/s1c strains and DU was noted (Linpisarn et al., 2007). However, a separate study showed that *vac-A* s1/m1, *cag-A*, *cag-E*, *ice-A1*, and *bab-A2* were the predominant alleles, but singly or combined did not correlate with clinical outcome (Chomvarin et al., 2008). In Korea, *vac-A* s1/i1/m1 was the predominant genotype reported in one study, and associated with gastritis, PUD, and GCA in decreasing frequency (Jang et al., 2010). In Japan, a study of 220 isolates confirmed the presence of Eastern and Western strains, the former represented by the East-Asia-specific sequence (EASS) and *vac-A* s1c/m1b strains, with no correlation to PUD, and the latter by the WSS and *vac-A* s1a/m1a or s2, with a correlation to PUD (Yamazaki et al., 2005). The locus *ice-A* has been identified as a marker for peptic ulcer in Western populations (Van Doorn et al., 1998b), but this was not found in a study of 140 Japanese isolates where the *ice-A1* allele bore no relationship to clinical outcome (Ito et al., 2000). A similar result was found in Taiwan, where the *ice-A1* allele was associated more with chronic gastritis than with PUD and, surprisingly for Asia, the *vac-A* s1c allele was underrepresented compared to surrounding countries, but the *vac-A* s1a allele was a marker for PUD (Wu et al., 2005). However, in a study comparing isolates from Bogotá, Houston, Seoul, and Kyoto, *cag-A* positive, *ice-A1* and *vac-A* s1c/m1 were predominant in Japan and Korea; *cag-A* positive, *ice-A2*, *vac-A* s1b/m1 in North America; and *cag-A* positive, *iceA2*, *vac-A* s1a/m1 in South America. None of the markers were associated with clinical outcome in any of the countries (Yamaoka et al., 1999). There have been several studies examining the 3′ region in different ethnic groups and different clinical presentations. Using PCR on paraffin sections of stomach from 71 young Japanese patients, all samples possessed the EASS in both PUD and GCA (Ueda et al., 2006). Analysis of *cag-A* subtypes in Malaysia demonstrated that type A was more common in Chinese compared to Indians or Malaysians and type B was more common in Malaysians, although there was no correlation with PUD. However, PUD is more common in Chinese, and *cag-A* type A may represent a marker for more severe disease in Chinese (Ramelah et al., 2005). *H. pylori* from 78 patients with different clinical conditions in Japan, Hong Kong, India, and the US were studied, but no correlation could be found between the 3′ region and either geographical location or disease (Park et al., 2010). An analysis of the d-region found that d1 (no deletion) was associated with gastric cancer but not PUD (Ogiwara et al., 2009). In a Chinese population, the *dup-A* locus was positively associated with DU and inversely with GU and gastric cancer (Zhang et al., 2008). A Japanese study of *dup-A* did not show any association with clinical outcome (Nguyen et al., 2010).

Americas

In a study from Cuba, *cag-A* was present in 73% of isolates (n = 130), 82% were *bab-A2* positive and the predominant *vac-A* types were s1/m1. Strains with the genotypes *cag-A*, *vac-A* s1, and *cag-A*, *vac-A* s1, *bab-A2* were significantly associated with DU (Torres et al., 2009). In Argentina, a study found a strong correlation between *cag-A* positive, *vac-A* s1/m1, *ice-A1* strains and PUD and clarithromycin resistance (Vega et al., 2010). In the US, *cag*PAI and *vac-A* s1 were significantly linked to DU but *vac-A* m1 was not, and in Colombia there was no association between *cag*PAI or the vac alleles. The only allele to have a low significant association was *bab-A2* (Yamaoka et al., 2002). A study of 51 children from the US demonstrated that *in situ cag-A* expression was associated with presence of peptic ulcers (Rick et al., 2010). A meta-analysis of 2358 patients demonstrated that *dup-A* was present in 48% and was significantly associated with DU (OR 1.4). The geographic prevalence was variable, with the highest rates in South America, and a significant association was noted between *dup-A* negative isolates and GU/GCA, although in China there was a correlation between *dup-A* positive isolates and GU/GCA (Hussein,

2010). A study of restriction fragment length polymorphisms in the *ure-C* and *ure-B* loci in *H. pylori* from patients in Brazil could demonstrate no differences in the patterns between PUD or gastritis (Roesler et al., 2008). In a study of North and South American strains, *hom-B* was more a marker associated with GCA than with DU (Jung et al., 2009). In a study of isolates from East Asian and Western countries, *hom-B*, *cag-A*, and *vac-A* s1 were not related to clinical outcome in the former, but related to PUD in the latter. *Hom-A* was more common in NUD and found in *cag-A* negative isolates, whereas *hom-B* was also correlated with *cag-A* positive, *vac-A* s1 strains (Oleastro et al., 2009).

Analysis of three strains, one isolated in the US from a case of DU, one isolated in Japan from a case of GCA, and one isolated in the US from a case of GU, demonstrated that the *cag*PAI was present in all three, and the following *vac-A* alleles were present in each: s1b/m1 (DU), s1c/m1 (GCA), and s1a/m2 (GU). The DU strain has 23 strain-specific genes, the GCA 22, and the GU 51 (McClain et al., 2009).

This has led to an appreciation that the same virulence marker is detected at different rates in different countries, and even at different rates within the same country in separate ethnic populations (Table 77.1). In some countries and populations there is an association with disease outcome, in other countries or populations there is no disease association – even for the same allele. This gives rise to a rather confusing picture: in part due to small numbers of individuals in some reported studies, in part due to evolutionary relationships with different ethnic groups, and in part due to population migrations. It also signifies that none of the virulence markers individually nor in combination can accurately predict clinical outcome and that one has to recognize the importance of the bacteria–host interaction in the final outcome.

Genome-wide Studies of Helicobacter pylori Using Sequencing and Micro-array Technology

A study of paired isolates of *H. pylori* from the same 13 individuals over a period of 7–10 years using PCR, amplified fragment length polymorphism (AFLP), and sequencing has illustrated the development of genomic diversity between the initial and final isolates. This *in vivo* genetic drift was also accompanied by phenotypic differences in Lewis antigen expression (Kuipers et al., 2000). A similar study used whole-genome sequencing of five sets of isolates from 454 Columbian patients with separation intervals of 3–16 years. In the chronically infected patients, there were 27–232 different single nucleotide polymorphisms detected between the first and last isolate, and evidence of 16–441 imported lengths of DNA as a result of recombination. The mutation rate was calculated at 2.5×10^{-5}/year/site and the recombination rate at 5.5×10^{-5}/year/site (Kennemann et al., 2011). An analysis of recently infected Mexican children did not show any genetic divergence over a time period of 90 days, suggesting that recent acquisition does not trigger genomic changes (Salama et al., 2007). These various results suggest long-term colonization, particularly with multiple

TABLE 77.1	Ethno-geographic distribution of *H. pylori* polymorphisms and relation to disease outcome			
Location	**Frequency of allele in isolates (%)**	**Frequent allele combinations in strains**	**Association with PUD**	**Doubtful association with PUD**
Europe	*iceA1* (69)	*cag-A–ice-A–vac-A* s1a	*cag-A*	*ice-A*
	iceA2 (30)	*cag-A–vac-A* s1a–*bab-A2–ice-A1*	*vac-A* s1	*vac-A* s1/m1
	cag-A (51–82)		*vac-A* i1	
	vac-A s1 (80–89)		*cag-A-vac-A* s1a	
	vac-A s2 (10)		*bab-A2*	
	vac-A m1 (35–39)			
	vac-A m2 (60)			
	bab-A (47–48)			
	hop-Q (67)			
Middle East	*cag-A* (93)	*cag-A–vac-A* s1/m2–*ice-A*	*ice-A1*	*cag-A-vac-A* s1/m2
	cag-A (60–67)		*vac-A*	*ice-A2*
	cag-E (30–47/77)		*cag-A*	*cag-A*
	va-cA s1m2 (63–68)		*vac-A* s1/m1	*cag-E*
	vac-A s1m1 (53)			*vac-A*
	vac-A s1m2 (47)			*vac-A* s1/m1
	ice-A2 (66–68)			
India and Pakistan	*cagPAI* (28–45)	Not recorded	*cagPAI*	*vac-A* s1b/m1
	cag-A (94)		*cag-A*	
	cag-E (56–94)		*vac-A* s1a/m1	
	vac-A s1 (36–47)			
	vac-A s2 (1)			
	vac-A m1 (62–72)			
	vac-A m2 (19–34)			
	ice-A1 (41–58)			
	ice-A2 (36–38)			
Far East	*cag-A* (88–98)	*cag-A–vac-A* s1a/s1c	*cag-A*	Multiple strains per person
	cag-E (88)	*cag-A–vac-A* s1a/m1a	*vac-A* s1a/s1c	*cag-A*
	vac-A s1 (98–100)	*cag-A–cag-E*	*vac-A* s1a/m1a	*vac-A*
	vac-A s1a (40)	*vac-A* s1/i1/m1	*dup-A*	*cag-E*
	vac-A s1c (16)	*vac-A* s1c/m1b		*ice-A*
	vac-A s1a/s1c (41)	*cag-A–ice-A1–vacA* s1c/m1		*bab-A2*
	vac-A s2 (0)			*vac-A* s1c/m1b
	ice-A (89)			*dup-A*
	ice-A1 (45–69)			

(continued)

TABLE 77.1	(Continued)			
Location	**Frequency of allele in isolates (%)**	**Frequent allele combinations in strains**	**Association with PUD**	**Doubtful association with PUD**
	ice-A2 (10–33)			
	bab-A2 (92)			
Americas	*cag-A* (73–100)	*cag-A–ice-A2–vac-A* s1b/m1 (NA)	*cag*PAI	*cag-A*
	vac-A s1 (73)		*vac-A* s1	*vac-A*
	vac-A s2 (26)	*cag-A–ice-A2–vac-A* s1b/m1 (SA)	*vac-A* s1b/m1	*ice-A*
	vac-A m1 (57)	*cag-A–vac-A* s1/m1	*vac-A* s1a/m2	*vac-A* m1
	vac-A m2 (40)		*cag-A–vac-A* s1	*hom-B*
	vac-A s1a/m1 (10–40)	*vac-A* s1/m1–*bab-A2*	*cag-A–vac-A* s1–*bab-A2*	
		cag-A–bab-A2		
	vac-A s1b/m1 (30–50)		*cag-A–vac-A* s1/m1–*ice-A1*	
	vac-A s1c/m1 (0–10)		*dup-A*	
	vac-A s1a/m2 (10–15)			
	vac-A s1b/m2 (10)			
	vac-A s1c/m2 (0)			
	vac-A s2 (10–30)			
	ice-A1 (15–30)			
	ice-A2 (44–78)			
	bab-A2 (82)			

isolates over a period of years, leads to the development of multiple strains by mutation and recombination which may have evolutionary significance in adaptation to different areas of the stomach and may also affect disease outcome (Salama et al., 2000). A whole-genome micro-array study of 42 isolates showed that 341 (25%) were polymorphic, one third of the polymorphism being in the plasticity region. The number of genes always associated with DU and less frequently found in FD or GCA was 17, and an additional 30 genes were either more or less frequently associated with specific disease, three less frequently found in DU (*omp2*, *mod*, *vap-D*) and four more frequently found, including *JHP0951* (integrase), *JHP0950*, *HP1405* (low pH response), and *HP1381* (Romo-Gonzalez et al., 2009).

One approach to identifying disease-associated markers is genome-wide studies of isolates from infected patients as described above. An alternative approach is to experimentally infect an animal model with different isolates as described below.

In order to investigate the clinical outcome of infection a GU-promoting isolate (B128 – originally a human

isolate of *Helicobacter pylori* from a patient with GU) and a DU-promoting isolate (G1.1 – originally a human isolate of *H. pylori* from a patient with DU) were compared in the gerbil model, and strain-specific markers identified by microarray. Both isolates induced antral gastritis, and 93% of gerbils infected with B128 developed corpus gastritis within one month, compared to only 5% for G1.1. Inflammatory scores, antral erosions, and disrupted gastric glands were higher in B128-infected animals and 17% developed a GU, whereas none did in G1.1-infected animals. In B128-infected animals, 83% showed parietal cell loss, but no effect on parietal cells was seen in G1.1-infected animals, and there was increased apoptosis and mucosal proliferation in both antrum and corpus with B128. The significant genomic difference between B128 and G1.1 was a partial deletion of the *cag*PAI including *cag-E* in G1.1, which affected IL-8 induction (Israel et al., 2001). Studies of *H. mustelae*, which lacks many of the virulence factors of *H. pylori* yet causes ulceration in naturally infected ferrets, may also reveal PUD biomarkers (O'Toole et al., 2010).

Proteomics

The Bacterial Proteome

The proteome of various strain types of *H. pylori* have been studied and identified, with many of the proteins undergoing post-translational modification (Backert et al., 2005b; Cho et al., 2002; Lock et al., 2001; Zhang et al., 2009). The first protein–protein interaction map for *H. pylori*, which obtained 46% coverage of the proteome, was reported in *Nature*. A graphic interface, PIMRider (http://pim.hybrigenics.com) was developed to analyze the data, and revealed novel interactions previously not recognized (Rein et al., 2001). A study specifically targeting DU was undertaken by two-dimensional electrophoresis of a strain isolated from a patient with DU followed by immunoblotting with sera from 10 patients with DU and, as a comparator, 10 patients with GCA. A total of 11 proteins were identified by the DU-sera, of which elongation factor EF-G, catalase, and the A-subunit of urease were most prominent in patients with DU. The discrimination between DU and GCA increased when all three proteins were used as biomarkers compared to the proteins singly (OR 1.85 vs 5.71 respectively) (Lin et al., 2007). A proteomic study using Surface enhanced laser desorption ionization – Time of Flight Mass Spectrometry (SELDI-TOF-MS) of 129 *H. pylori* strains from Korea and Colombia in patients with either DU or GCA revealed 18 biomarkers that distinguished the two conditions, of which three were identified: neutrophil-activating protein (NapA) and an RNA-binding protein (HPAG1-0813), which were more common in GCA, and a histone-like protein (jhp0228), which was more common in DU. None of the nine proteins from each country distinguishing the two conditions were identical, testifying to the strain and strain/ethnic interaction diversity (Khoder et al., 2009).

Proteomic Studies of the Host

Proteins from gastric juice were analyzed by two-dimensional Polyacrylamide gel electrophoresis (PAGE-MS), and it was demonstrated that GU had higher levels of protein compared to DU and that in GU (and GCA but not DU) alpha1-antitrypsin was a significant marker compared to controls (42% vs 6%) (Hsu et al., 2007). However, in another study, GU had higher levels of alkyl hydroperoxide reductase (AhpC) compared to DU but antibody levels of elongation factor G (FusA), catalase (KatA), and the urease subunit UreA were more common in DU compared to GCA (70% vs 45%, 50% vs 41%, and 44% vs 27%, respectively). The presence of gamma glutamyl transpeptidase has also been identified as a biomarker for PUD (Gong et al., 2010; Wu et al., 2008). A serological study using Western blots showed that VacA was more strongly linked to the clinical outcome of DU, and neither CagA nor several outer membrane proteins (p33, p30, p19) were associated with histological parameters or PUD (Filipec Kanizaj et al., 2009).

Also, these studies again confirm ethno-geographic differences between the bacteria isolated from the same clinical condition, but do show within each study there are differences which correlate with the clinical outcome. Further studies

TABLE 77.2	Proteomics of *H. pylori* related to disease outcome		
DU	**GU**	**GCA**
Bacterial proteins		
Elongation factor-G (FusA)		RNA-binding protein (HPAG1-0813)
Catalase (KatA)		
UreaseA (UreA)		
Neutrophil-activating protein (NapA)		
Histone-like protein (Jhp0228)		
Host proteins		
Anti-FusA	Alpha1-antitrypsin	Alpha1-antitrypsin
Anti-KatA	AhpC	
Anti-UreA		
Anti-VacA		

DU, duodenal ulceration; GU, gastric ulcer; GCA, gastric cancer.

interrogating the proteome in clinically diverse outcomes of infection with *H. pylori* is likely to lead to a more detailed understanding of the pathophysiology of disease as well as perhaps identifying general biomarkers for specific outcomes. As matrix-assisted laser desorption/ionization with time-of-flight mass spectrometry (MALDI-TOF MS) is increasingly used in routine clinical practice it is likely that proteomic information from both the patient and the isolate will lead to a clearer understanding of the host–pathogen interaction in patients with severe gastroduodenal disease, including PUD.

In summary, these proteomic studies have identified specific bacterial proteins associated with the development of DU and confirmed that the host responses to these proteins are also associated with DU (Table 77.2).

The Human Transcriptome Response to Helicobacter pylori

An alternative approach to investigate the host/pathogen interaction is to analyze the transcriptome of host gastric epithelial cells exposed to *H. pylori*. In 43 patients with *Helicobacter*-gastritis, there was modulation of mRNA transcripts associated with biologically appropriate pathways consistent with the known pathophysiology of infection, of pathogen recognition (e.g., TLR4), intercellular signaling (e.g., MAPK8), chemokines (e.g., CCL3), and complement activation. The response in the antrum was more vigorous than that noted in the fundus, and correlated better with the genotype of the infecting strain (Hofman et al., 2007). After co-culture of *H. pylori* with a gastric cell line, 434 genes were downregulated and 284 genes upregulated at 48 hours post inoculation using a micro-array analysis (Sepulveda et al., 2002). A combination of a transcriptome and proteome investigation of co-culture of *H. pylori* with a human stomach adenocarcinoma cell line (AGS) cells demonstrated an upregulation of 179 proteins and downregulation of 11, several of which were dependent on the T4SS. After co-culture, 81 and 59 genes putatively linked to cancer development were either up- or downregulated respectively. This study demonstrated a rapid, dynamic, complex, multiphasic response of the host to the organism (Backert et al., 2005a). A study using laser capture microdissection and epithelial tissue from patients with PUD before and after *H. pylori* eradication illustrated 98 differentially regulated genes. Those showing the highest value for a decrease in expression were immune-related; those showing an increase were metabolic- and cell-adhesion-related (Resnick et al., 2006). A micro-array analysis of non-coding microRNA (miRNA) has also been used to investigate host–pathogen interaction. It demonstrated that some miRNAs correlate with infection, but not with any specific clinical outcome (Matsushima et al., 2010).

The Host (Rat) Transcriptome Response Associated with Ulcer Formation

An *in vivo* experiment in rats using cysteamine as the ulcer-inducing agent and real-time PCR and reverse transcription PCR (RT-PCR) demonstrated that 40 genes had a significant change in expression in gastric epithelial cells. They included genes linked to transcription factors, cell-surface antigens, DNA-binding proteins, transport proteins, receptor proteins, and ion channels, specifically: endothelin receptor B (ETRB); endothelin-1; caspase 3; transcription factors erg-1, sp-1, and STAT-3; vascular endothelial growth factor (VEGF); platelet-derived growth factor (PDGF); potassium channel drk-1; and genes of unknown function-expressed sequence tag (EST)1,2. All of the above showed increased expression. Some genes showed decreased expression: NFkB, EST 3-6, and p19 of IL-23. Early changes (0.5–2 hrs) involved genes affecting vascular permeability and late changes (>12 hrs) involved genes affecting synthesis of growth factors, apoptosis, and necrosis (Deng et al., 2008).

Transcriptomics

In summary, the modulation of the transcriptome, by *H. pylori* infection in humans or chemically induced ulcer formation in animals, is consistent with recognized, biologically reasonable pathophysiological mechanism. None of the changes identified to date, however, specifically identify a pathway to PUD. Nor is there an unambiguous relationship with the genotype of the infecting strain. Further studies in different ethnic groups infected and with different genotypes of *H. pylori* are likely to highlight possible associations. For the time being, transcriptomics is likely to remain a research tool rather than a diagnostic or to guide therapy.

HOST POLYMORPHISMS
Helicobacter-induced Ulcers

Engagement of the T4SS with the gastric epithelial cell upregulates NFkB and ultimately leads to secretion of IL-8 and other pro-inflammatory cytokines from the epithelial cell with consequent recruitment of granulocytes. Simultaneously the CagA protein and peptidoglycan are injected into the epithelial cells, inducing an altered cell morphology (the "humming bird" phenotype) and disturbance of the cell cycle, leading to an increase in cell proliferation of gastric epithelial cells and an alteration in epithelial barrier function. Host polymorphisms in the different pathways involved in the inflammatory response to the presence of the organism have been sought and correlated with clinical outcome. IL-1 and TNF are two important pro-inflammatory cytokines that have been investigated.

Polymorphisms in the IL-1 gene and receptor antagonist [*IL-1B-511T/-31C* and *IL-1RN*2* – the penta-allelic variable number tandem repeat (VNTR) in intron 2 of *IL-1RN*] have been linked to serious gastroduodenal disease and act synergistically with bacterial markers of virulence, notably *cag-A/vac-A* s1, to increase the risk of severe inflammation and atrophy from OR = 1.7 to 24 and 9.5 for lymphocytic and granulocytic infiltration respectively (El Omar et al., 2000; Rad et al., 2003). In a study of polymorphisms of *IL-1B+3954* and *IL-1RN* in 179 Spanish patients with PUD and a matched control population of 99 healthy individuals, there was no difference in allele frequency of *IL-1B* and *IL-1RN* in relation to clinical outcome or control population. However, in PUD, there was a strong association between alleles of *IL-1B*2+3954* and *IL-1RN*2*, which was specific for DU rather than GU. Non-carriers of *IL-1B+3954*2* were more often carriers of *IL-1RN*2* and vice versa. However, carriage of *IL-1B+3954*2* and *IL-1RN*2* was associated with a reduced risk of DU (OR 0.37) (Garcia-Gonzalez et al., 2001). Simultaneous carriage of *IL-1RN*2, IL-1B-511*C, IL-1B-31*T*, and *IL-1B+3954*C* was a risk factor for developing DU (OR 3.22) (Garcia-Gonzalez et al., 2003). A further study including polymorphisms in *IL-1B, IL-1RN, TNFA*, and *TNFB* in 118 Spanish patients compared to 99 healthy controls, demonstrated that the frequency of polymorphisms was equally distributed in patients and controls but in aggregate *IL-1B-31T/IL-1B-511C/IL-1B+3954C/IL-1RN*2/TNF*-haplotype E was slightly more common in *H. pylori* DU patients (OR 1.85) (Garcia-Gonzalez et al., 2005b). Polymorphisms in the promoter regions of *TNFA* (-238, -308) and the first intron of *TNFB* (lymphotoxin alpha (LTA) – RFLP-defined) which occur as five haplotypes (*TNF-C, E, H, I*, and *P*) were investigated in 130 Spanish patients with DU, 50 with GU and 102 healthy controls. Patients with GU were less likely to carry *TNFA-308*2* and more likely to carry a *TNFBNcol2.2* genotype and *TNF-I* haplotype compared to patients with DU, whereas DU patients were more likely to carry a *TNF-E* haplotype and carriage of the *TNF-I* haplotype was a marker for development of PUD (Lanas et al., 2001). Investigation of polymorphisms (*1031T/C*, *863C/A*, *857C/T*, *806C/T*, and *308G/A*) in the *TNF* promoter region demonstrated that *1031T/C* and *863C/A* were independent risk factors for PUD in patients colonized by *H. pylori*. The odds ratio increased from 2.46 to 6.06 if both polymorphisms were present (Lu et al., 2005a). These studies are not inconsistent with those of El Omar or Rad, as the histological pathway in those latter studies would more likely have resulted in GU rather than DU. Indeed, in a study in a Chinese population, the *IL-1B+3954T/T* polymorphism was also shown to be protective for ulcer development (Yang et al., 2008). However, in a Taiwanese study of 147 patients with DU and 168 control subjects, *IL-RN*2* was demonstrated to be an independent risk factor for the development of DU – the OR was 2.7 but increased to 22 when colonization by *H. pylori* was taken into account, and 27 if also blood group O (Hsu et al., 2004). A study in children reported that in patients colonized by *ice-A* positive strains, the *TNFA* polymorphism *G-238A* was associated with PUD (Wilschanski et al., 2007).

Polymorphisms in other cytokines, including IL-6, IL-8, IL-10, and IL-12, and the intracellular molecule NOD-1, have also been investigated. An association between *IL-8-251A/T* and *NOD-1 A796A* with DU was noted compared to controls in a Hungarian population (65% vs 36% and 20% vs 6%, respectively) although no correlations between *IL-6*, or *TNFA TLR4* or *CD14* promoter polymorphisms and disease were noted (Gyulai et al., 2004; Hofner et al., 2007). Similarly, no correlation could be found between *IL-6-174 G/C* polymorphisms and ulcer disease in a Brazilian population (Lobo Gatti et al., 2005). A relationship between *IL-8-251A/T* and gastric ulceration has also been noted in a Japanese population (Ohyauchi et al., 2005). The *IL-10* polymorphisms *-1082(A/G)*, *-819(T/C)*, and *-592(A/C)* were also investigated in Japanese patients, showing that *-592C* and *-819C* were not linked to PUD although they were linked to GCA, and that the *-1082* polymorphism was not linked to either (Sugimoto et al., 2007). Genetic polymorphisms in *IL-6*, *IL-8*, and *IL-10* were studied in a Korean population, showing that *IL-10-1082G* was more common in GCA and GU irrespective of the presence of *H. pylori*. The *IL-8-251A/A* genotype was more frequently found in patients colonized with *H. pylori* who had GCA or GU, and the frequency of *IL-6-572G/G* was less in *H. pylori* positive patients with DU (Kang et al., 2009). An association between *IL-12* and PUD could not be demonstrated (Garcia-Gonzalez et al., 2005a).

Apart from the cytokine network, other loci that may be related to ulcer development have also been investigated. CD11 is a component of the B-integrins which mediate neutrophil attachment. A study investigating the CD11 complex in 315 *H. pylori* positive erosion/ulcer patients demonstrated that the G/A haplotype of *CD11c* exon 15 and intron 31 was associated with ulceration (OR 2.4) (Hellmig et al., 2005). Myloperoxidase is secreted from neutrophils, and a study has demonstrated that the polymorphism *MP-468A/A* is strongly linked to DU (OR 8.7) when colonized by *H. pylori* (Hsu et al., 2005). Heat shock proteins are upregulated during cellular

stress and a study of 447 Mexican patients, of which 58 had DU, demonstrated that significant association between *HSP70-1C/G* and DU existed (OR 16.19) (Partida-Rodriguez et al., 2010).

Matrix metalloproteases (MMPs) are involved in degrading extracellular matrix proteins and *H. pylori* increases their secretion. A study of polymorphisms in *MMP1, 3, 7*, and *9* from 599 infected patients showed a strong association between promoter alleles of *MMP7* (*-181G*), the *GAAGG* haplotype in exon 6 of *MMP9*, and GU. However, the haplotypes *GGAGG* and *TGGAA* in *MMP* were protective of ulceration (Hellmig et al., 2006).

Transforming growth factor B (TGFB) is associated with tissue regeneration, and a study of polymorphisms (*MspA1/T+869C* and *Sau961 G+915C*) demonstrated that *C+869C* was linked to a reduced risk of developing ulceration (OR 0.32) but the *T+869C* allele was linked to an increased risk (Garcia-Gonzalez et al., 2006). Polymorphism in the transcription factor Nrf2 (a regulator of cellular redox homeostasis) has been investigated for its relation to gastritis and ulceration (Arisawa et al., 2007a). The *C-650C* allele was related to the grade of gastritis, and the alleles *-686A* and *-650A* were decreased in patients with ulceration. Histamine is the stimulus for acid production and is metabolized by histamine N-methyltransferase (NMT). The *C314T* polymorphism in the histamine NMT gene showed no association with DU in a Chinese population (Hailong et al., 2008).

Vascular endothelial growth factor (VEGF) stimulates the growth of new blood vessels and is involved in tissue regeneration. As most ulcers develop on the lesser curvature of the stomach, which has a poor sub-mucosal plexus, polymorphisms of *VEGF* may indicate an abnormal angiogenic response and thus explain the site of ulcer development. A study investigating the role of polymorphism of *VEGF* illustrated that subjects with the promoter polymorphisms *-1780T/C* and *-1780CC* compared to *-1780TT* had a poor angiogenic response, and carriage of these polymorphisms may predict ulcer development (Kim et al., 2008). However, in a separate study, the genotypes *G1612A* and *C936T* present in the 3' untranslated region of *VEGF* were not markers for PUD, although *G1612A* was a marker for GCA susceptibility (Tahara et al., 2009).

As indicated above, several human polymorphisms are associated with binding of *H. pylori* or ulcer development, but one group of polymorphisms thus far not investigated are the genes involved in glycosylation of the mucin covering the stomach, which may also be a factor in binding and subsequent events as the gastric mucin is a primary defense against potential damaging agents (Niv, 2010).

Thus far, there have been no genome-wide association studies related to disease outcome. However, one such study has demonstrated evolution of the organism from the same individual over a period of a decade.

In summary, as one might anticipate, polymorphisms in genes involved in the whole pathophysiological pathway have

TABLE 77.3	Host polymorphisms associated with PUD	
Increased risk of PUD	**Decreased risk of PUD**	
IL-1B-511T/-31C	IL-1B*2+3954 and IL-1RN*2 (DU)	
IL-1B*2+3954 (DU)		
IL-1RN*2 (DU)	IL-1B+3954T/T	
IL-1RN*2, IL-1B-511*C, IL-1B-31*T, IL-1B+3954*C (DU)	tgfb-1+869C/C	
LTANcol2.2 (GU)		
TNF-I haplotype (GU)		
TNF-E haplotype (DU)		
TNF1031T/C		
TNF863C/A		
IL-8-251A/T		
nod-1 796A/A		
CD11c haplotype G/A		
mmp-468A/A (DU)		
mmp-7-181G+ mmp9 haplotype (GU)		
hsp70-1C/C (DU)		
tgfb-1+869T/C		
vegf-1780T/C		
vegf-1780C/C		
vegf-1780T/T		
NSAID-Induced PUD		
cox-1 A842G/C50T		
cox-1-T1676C (GU)		
IL-1B 511T polymorphisms		
CYP2C9*3		

an effect on the clinical outcome, from cytokines (e.g., IL-1B, IL-8), host defense factors (e.g., myloperoxidase and integrins), growth factors (e.g., TGF, VEGF), to tissue-damaging factors (e.g., matrix metalloproteases) (Table 77.3). When an individual's polymorphism is combined with strain polymorphism, the prediction of a specific clinical outcome is higher than if either is taken alone. It is likely that by combining these databases with data from proteomics in future studies, a more accurate prediction of an individual's chance of developing an ulcer or cancer will be obtained.

NSAID-Induced Ulcers

NSAIDs act by inhibition of cyclooxygenase (COX). Polymorphisms in cox may be related to the risk of developing ulcers when given NSAIDs. One study that found subjects heterozygous for the A-842G/C50T haplotype had greater inhibition of prostaglandin synthesis by aspirin compared to the homozygous (Halushka et al., 2003), but were less likely to be associated with ulcer bleeding (OR 0.75) compared to the wild-type (Van Oijen et al., 2006).

A Japanese study of the effects of the polymorphisms A842G/C50T and T1676C on gastric ulceration illustrated that T1676C was significantly associated with GU but not DU in the absence of NSAID use. In the presence of NSAID use, T1676C was associated with PUD although male gender but the presence of H. pylori was not. The A842G/C50T polymorphism was rare in Japan (Arisawa et al., 2007b). The polymorphism G765C in cox-2 was significantly associated with GCA irrespective of the presence of H. pylori, but significantly associated with PUD only if H. pylori was also present (Saxena et al., 2008).

Polymorphisms in cytochrome P450 may also be associated with NSAID-induced ulceration. In a study of 109 patients takings NSAIDs, 17% had ulcers but there was no association with allelic variants of CYP2C9 (Ma et al., 2008). On the other hand, in France CYP2C9*3 was strongly linked to non-aspirin NSAID use and acute GI bleeding, but not with aspirin use and bleeding (Carbonell et al., 2010). In animals HSP70 expression was shown to be gastro-protective, and in humans induction of HSP70 by geranylgeranylacetone was also gastro-protective of NSAID-induced ulceration (Suemasu et al., 2009; Yanaka et al., 2007). In patients taking aspirin for cardiovascular prophylaxis, ulceration was more common if the patient was polymorphic for IL-1B511/31 alleles (Shiotani et al., 2009).

In summary, polymorphisms in the genes involved in metabolism of NSAIDs and in the mechanism of action of NSAIDs are linked to a greater or lesser extent to developing ulceration. In the future, data such as this may be very relevant to tailoring a specific NSAID to an individual, thus decreasing unwanted side effects.

PHARMACOGENOMICS

See also Chapter 31. Successful eradication of H. pylori is dependent upon antibiotic resistance, compliance, and efficacy of acid suppression during therapy. The latter is affected by polymorphisms of the cytochrome P450 class of enzymes (e.g., CYP2C19), which affect the degree of metabolism of many drugs, including proton pump inhibitors (PPIs) and antibiotics, drug transporters (e.g., MDR-1), and the levels of inflammatory cytokines, particularly IL-1B.

The serum levels of PPI are related to polymorphisms in CYP2C19 (Furuta et al., 2005; Sakai et al., 2001; Sapone et al., 2003). A recently discovered variant of CYP2C19 identified as CYP2C19*17, which was initially believed to have increased activity, has been shown to have no effect on the eradication rate of *H. pylori* (Kurzawski et al., 2006; Li-Wan-Po et al., 2010). The effect on eradication rates of *H. pylori* by polymorphisms in *CYP2C19* (*2, *3) and *MDR1* (C3435T) was investigated. Of two regimens used (omeprazole, amoxicillin, and clarithromycin, and pantoprazole, amoxicillin, and metronidazole) the latter regimen was more successful in first-round eradication, and there was a strong correlation between heterozygous *CYP2C19*1/*2* and homozygous *T3435T* and successful first-round eradication (34% vs 8% and 38% vs 13%, respectively) (Gawronska-Szklarz et al., 2005). However, there is conflicting evidence about the relevance of metabolizer status and eradication of *H. pylori*. A meta-analysis of 17 articles indicated there was little difference between poor metabolizers (PM) and heterogeneous extensive metabolizers (HetEM) in eradication rate, but differences were noted depending on the type of PPI used in respect of homozygous extensive metabolizers (HomEM). *CYP2C19* polymorphisms appeared to be relevant if omeprazole was used but not if lansoprazole or rabeprazole were used (Padol et al., 2006). However, a subsequent meta-analysis of 20 articles demonstrated significant differences in the eradication rate between PM and HetEM, PM and HomEM, and HetEM and HomEM. Both omeprazole and lansoprazole gave a higher eradication rate in PM compared to HomEM, and in HetEM compared to HomEM. The eradication rate with rabeprazole was unaffected by metabolizer status (Zhao et al., 2008). However, a pharmacodynamic analysis in rabeprazole therapy demonstrated that drug clearance was lowest in PM compared to EM, with a consequent higher *Helicobacter* eradication rate (Yang and Lin, 2010; Yang et al., 2009). A study investigating the relevance of *CYP2C19* polymorphisms to regimens containing pantoprazole and esomeprazole showed overall comparable eradication rates (82–88%) but when analyzed for metabolizer status, PM were better than EM (97% vs 83%) with no difference for the type of PPI (Kang et al., 2008). On the other hand, a similar study comparing metabolizer status and eradication rate with esomeprazole or pantoprazole showed greater acid inhibition with esomeprazole compared to pantoprazole, with the latter affected by *CYP2C19* metabolizer genotype (Hunfeld et al., 2010). A study of esomeprazole- or rabeprazole-based quadruple therapy after initial failure of standard triple therapy showed that rabeprazole had a higher eradication rate compared to esomeprazole, and that a *CYP2C19* HomEM status was a predictor of failure (Kuo et al., 2010). A further study of metabolizer status comparing lansoprazole and rabeprazole showed eradication rates for lansoprazole of 73% (HomEM), 80% (HetEM), and 85% (PM), and for rabeprazole 68% (HomEM), 73% (HetEM), and 71% (PM) (Lee et al., 2010). A study of the effect of *CYP2C19* in northern Indians, who were genotyped prior to being given 20mg omeprazole, 750mg amoxicillin, and 500mg tinidazole for seven days to eradicate *H. pylori*, showed that the eradication rates in EM and PM were 37% and 92% respectively. Extensive metabolizers where *H. pylori* had not been eradicated were then given 40mg omeprazole latin

notation bis in die which means twice daily (BD) and 500mg amoxicillin latin notation quater die sumendus which means four times daily (QDS) for 14 days, and achieved 90% eradication (Chaudhry et al., 2009). On the other hand, a study of the effect of *CYP2C19* and *MDR1* on *Helicobacter* eradication in a Korean population did not find any significant effect (Oh et al., 2009).

Liver disease can affect the levels of PPI. In a study using pantoprazole, irrespective of whether the subject was EM or PM, the levels of the PPI were higher than in the control group without liver disease (Shao et al., 2009). A study of 95 patients with cirrhosis and PUD given rabeprazole as the PPI as part of standard triple therapy showed *Helicobacter* eradication rates of 80% (HomEM), 89% (HetEM), and 100% (PM), which was similar to the ulcer cure rates (Lay and Lin, 2010).

The effects of polymorphisms in *MRD1* and *CYP2C19* and bacterial susceptibility on the eradication rate of *H. pylori* were investigated in 313 Japanese patients using lansoprazole as part of the triple therapy. The eradication rate in *MRD1* 3435C/C, C/T, and T/T were 82%, 81%, and 67% respectively. For *CYP2C19*, the eradication rate was 66% (HomEM), 79% (HetEM), and 89% (PM), and for clarithromycin resistance 89% (sensitive) and 42% (resistant) (Furuta et al., 2007). A similar study, in 139 Polish subjects, of the polymorphisms *CYP2C19* (*2,*3,*17), *MDR1* C3435T, and *IL-1B* C3954T treated with pantoprazole-containing triple therapy only showed a difference in *CYP2C19* between the eradicated and non-eradicated groups (Gawronska-Szklarz et al., 2010).

In summary, these studies have shown that genes related to the metabolism and transport of PPIs have an effect upon the eradication rate of *H. pylori*. The polymorphisms in these genes are listed in Table 77.4. The principle genes are *CYP2C19* and *MDR-1*, and they affect all the PPIs to some degree although different studies do give slightly different results. The rate of metabolism of the PPI and therefore success of eradication depends upon whether the person is heterozygous or homozygous for the polymorphism. Again, ethnic differences have been noted. Currently it is in the area of pharmacogenomics that knowledge of host polymorphisms is of greater benefit to patients in relation to management, compared to diagnosis of ulcer development.

PROTEOMICS AND GENOMICS OF ULCER DETECTION

Other than endoscopy, there is no specific test to identify the presence of a peptic ulcer. A number of molecular techniques can be used to detect, type, and assess the eradication of *H. pylori* (Megraud and Lehours, 2007; Van Doorn, 2001) and proteomics offers hope of identifying specific biomarkers of *Helicobacter* infection (Albala, 2001). Detection of markers of virulence and antibiotic resistance has been achieved largely by polymerase chain reaction (PCR) with appropriate primers (Tables 77.1, 77.2). The detection of

TABLE 77.4	Polymorphisms affecting ulcer treatment				
Gene	Ranking and eradication	Eradication affected	Eradication not affected	Specific comparisons	
CYP2C19*1,*2,*17	PM>HetEM>HomEM	O, L, R, P, ES	R, P	P > O	
		Liver disease		P > ES	
				P = ES	
				L > R	
				L/R > O	
				R > ES	
MDR-1 C3436T	C/C>C/T>T/T				

P=pantoprazole, O=omeprazole, ES=esomeprazole, L=lansoprazole, R=rabeprazole, PM=poor metabolizer, HetEM=heterozygous extensive metabolizer, HomEM=homozygous extensive metabolizer.

a Type I isolate does not indicate the presence of an ulcer, and the detection of a Type II strain does not indicate the absence of an ulcer in every case.

Specimens used for the detection and genotyping of *Helicobacter* have largely been gastric biopsies, although stool specimens have also been used. A number of different methods of typing *H. pylori* (Ge and Taylor, 1998) in both coding and non-coding areas (Bereswell et al., 2000) have been used.

Quantifying the levels of pepsinogen I (PG I) and II (PG II) and gastrin, and the presence of antibodies to *H. pylori* can be used to indicate the presence of *H. pylori*, but do not provide conclusive evidence of ulceration. Elevation of both PG I (e.g., 70ng/ml compared to 50ng/ml) and PG II (e.g., 25ng/ml compared to 10ng/ml) with a reduction of the PG I/PG II ratio (e.g., 3.5 compared to 6.2) is found in *H. pylori*-associated gastritis compared to *H. pylori* negative individuals. Some studies have shown that PG I levels are even more elevated in peptic ulcer disease compared to gastritis.

GENOMICS AND TREATMENT OF PEPTIC ULCERS

An exciting development in relation to the treatment of peptic ulcers has been demonstrated by Deng and colleagues (2004), who showed that rats who received intra-duodenal administration of an adenoviral vector carrying VEGF or PDGF, or naked VEGF/PDGF DNA, had smaller ulcers compared to controls. Further studies are required to confirm these findings.

CONCLUSIONS

To date, a number of virulence markers (the *cag*PAI, *cag-A*, *vac-A*, *bab-A*, *oip-A*, *du-pA*, *ice-A*, *hom-B*, *hop-Q*, and *Jhp0562*) have been identified and linked to serious gastroduodenal disease compared to asymptomatic gastritis. However, currently there is no single virulence marker that has been unambiguously identified as an allele identifying the risk of developing peptic ulcer disease, leading to the conclusion that either ulceration may represent the long-term effects of inflammation caused by persistent colonization with no specific identifiable marker, or that several alleles from both the host and the bacterium interact in a way that contributes to the eventual clinical outcome, with the most important polymorphisms or interactions perhaps yet to be identified. The increasing use of micro-array and proteome studies of host/pathogen interaction in relation to clinical outcome may lead to specific biomarkers to identify the risk of developing a specific illness, although because of the co-evolution of *H. pylori* with *Homo sapiens*, it is likely these markers will be geo-ethnically specific. The most useful role for genomic information in relation to both *H. pylori* and NSAID use is in pharmacogenomics, by identifying metabolism of appropriate drugs and tailoring the treatment to the specific host/pathogen genotype.

ACKNOWLEDGMENTS

We wish to thank Professor Sampurna Roy of Calcutta University, India, for permission to use Figures 77.3 and 77.4.

REFERENCES

Albala, J.S., 2001. Beyond our genome: The application of proteomics in molecular diagnostics. Exp Rev Mol Diagn 1, 243–244.

Andersson, A.F., Lindberg, M., Jakobsson, H., et al., 2008. Comparative analysis of human gut microbiota by barcode pyrosequencing. PLoS ONE 3, e2836.

Arisawa, T., Tahara, T., Shibata, T., et al., 2007a. The relationship between *Helicobacter pylori* infection and promoter polymorphism of the Nrf2 gene in chronic gastritis. Int J Mol Med 19, 43–148.

Arisawa, T., Tahara, T., Shibata, T., et al., 2007b. Association between genetic polymorphism in the cyclooxygenase-1 gene promoter and peptic ulcer disease in Japan. Int J Mol Med 20, 373–378.

Atherton, J.C., 1997. The clinical relevance of strain types of *Helicobacter pylori*. Gut 40, 701–703.

Atherton, J.C., Cao, P., Peek, R.M., et al., 1995. Mosaicism in vacuolating cytotoxin alleles of *Helicobacter pylori*. J Biol Chem 270, 17,771–17,777.

Backert, S., Gressmann, H., Kwok, T., et al., 2005a. Gene expression and protein profiling of AGS gastric epithelial cells upon infection with *Helicobacter pylori*. Proteomics 5, 3902–3918.

Backert, S., Kwok, T., Schmid, M., et al., 2005b. Subproteomes of soluble and structure bound *Helicobacter pylori* proteins analyzed by 2-dimensional gel electrophoresis and mass spectrometry. Proteomics 5, 1331–1345.

Backstrom, A., Lundberg, C., Kersulyte, D., et al., 2004. Metastability of *Helicobacter pylori bab* adhesin genes and dynamics of Lewis b antigen binding. Proc Natl Acad Sci USA 101, 16,923–16,928.

Baghaei, K., Shokrzadeh, L., Jafari, F., et al., 2009. Determination of *Helicobacter pylori* virulence by analysis of the cag pathogenicity island isolated from Iranian patients. Dig Liver dis 4, 634–638.

Baltrus, D.A., Blaser, M.J., Guillemin, K., 2009. *Helicobacter pylori* genome plasticity. Genome Dyn 6, 75–90.

Basso, D., Zambon, C.F., Letley, D.P., et al., 2008. Clinical relevance of *Helicobacter pylori cagA* and *vacA* gene polymorphisms. Gastroenterology 135, 91–99.

Bereswell, S., Schonenberger, R., Thies, C., et al., 2000. New approaches for genotyping of *Helicobacter pylori* based on amplification of polymorphisms in intergenic DNA regions and at the insertion site of the cag pathogenicity island. Med Microbiol Immunol 189, 105–113.

Bik, E.M., Eckburg, P.B., Gil, S.R., et al., 2006. Molecular analysis of the bacterial microbiota in the human stomach. Proc Natl Acad Sci USA 103, 732–737.

Boyanova, L., Yordanov, D., Gergova, G., Markovska, R., Mitov, I., 2010. Association of *iceA* and *babA* genotypes in *Helicobacter pylori* strains with patient and strain characteristics. Antonie Van Leeuwenhoek 98, 343–350.

Caner, V., Yilmaz, M., Yonetci, N., et al., 2007. *H. pylori iceA* alleles are disease specific virulence factors. World J Gastroenterol 13, 2581–2585.

Cao, P., Cover, T.L., 2002. Two different families of *hopQ* alleles in *Helicobacter pylori*. J Clin Microbiol 40, 4504–4511.

Carbonell, N., Verstuyft, C., Massard, J., et al., 2010. *CYP2C9*3* loss of function allele is associated with acute upper gastrointestinal bleeding related to the use of NSAIDs other than aspirin. Clin Pharmacol Ther 87, 693–698.

Chaudhry, A.S., Kochhar, R., Kohli, K.K., 2009. Importance of CYP2C19 genetic polymorphism in the eradication of *Helicobacter pylori* in North Indians. Indian J Med Res 130, 437–443.

Chiarini, A., Cala, C., Bonura, C., et al., 2009. Prevalence of virulence associated genotypes of *Helicobacter pylori* and correlation with severity of gastric physiology in patients from Western Sicily, Italy. Eur J Microbiol Infect Dis 28, 437–446.

Cho, M.J., Jeon, B.S., Park, J.W., et al., 2002. Identifying the major proteome components of *Helicobacter pylori* strain 26695. Electrophoresis 23, 1161–1173.

Chomvarin, C., Namwat, W., Chaicumpar, K., et al., 2008. Prevalence of *Helicobacter pylori vacA, cagA, cagE, iceA* and *babA2* genotypes in Thai dyspeptic patients. Int J Infect Dis 12, 30–36.

Covacci, A., Telford, J.L., Giudice, G.D., Parsonnet, J., Rappuoli, R., 1999. *Helicobacter pylori* virulence and genetic geography. Science 384, 1328–1333.

Dabiri, H., Bolfion, M., Mirsalehian, A., et al., 2010. Analysis of *Helicobacter pylori* genotypes in Afghani and Iranian isolates. Pol J Microbiol 59, 61–66.

Dabiri, H., Maleknejad, P., Yamaoka, Y., et al., 2009. Distribution of *Helicobacter pylori cagA, cagE, oipA* and *vacA* in different major ethnic groups in Tehran, Iran. J Gastroenterol Hepatol 24, 1380–1386.

De Francesco, V., Margiotta, M., Zullo, A., et al., 2009. *Helicobacter pylori vacA* arrangement and related diseases: A retrospective study over a period of 15 years. Dig Dis Sci 54, 97–102.

Deng, X., Szabo, S., Khomenko, T., et al., 2008. Detection of duodenal ulcer associated genes in rats. Dig Dis Sci 53, 375–384.

Deng, X., Szabo, S., Khomenko, T., Jadus, M.R., Yoshida, M., 2004. Gene therapy with adenoviral plasmids or naked DNA of vascular endothelial growth factor and platelet derived growth factor accelerates healing of duodenal ulcers in rats. J Pharmacol Exp Ther 311, 982–988.

Dicksved, J., Lindberg, M., Rosenquist, M., et al., 2009. Molecular characterization of the stomach microbiota in patients with gastric cancer and in controls. J Mol Microbiol 58, 509–516.

El-Omar, E.M., Carrington, M., Chow, W.H., et al., 2000. Interleukin-1 polymorphisms associated with increased risk of gastric cancer. Nature 404, 398–402.

Farshad, S., Japoni, A., Alborzi, A., et al., 2007. Restriction fragment length polymorphism of virulence genes *cagA, vacA* and *ureAB* of *Helicobacter pylori* strains isolated from Iranian patients with gastric ulcer and non-ulcer disease. Saudi Med J 28, 529–534.

Filipec Kanizaj, T., Katicic, M., Presecki, V., et al., 2009. Serum antibodies positivity to 12 *Helicobacter pylori* virulence antigens in patients with benign or malignant gastroduodenal disease: Cross sectional study. Croat Med J 50, 124–132.

Fischer, W., Prassi, S., Haas, R., 2009. Virulence mechanisms and persistence strategies of the human gastric pathogen *Helicobacter pylori*. Curr Top Microbiol Immunol 337, 129–171.

Forsyth, M.H., Atherton, J.C., Blaser, M.J., Cover, T.L., 1998. Heterogeneity of levels of vacuolating cytotoxin gene (*vacA*) transcription among *Helicobacter pylori* strains. Infect Immun 66, 3088–3094.

Furuta, T., Shirai, N., Sugimoto, M., et al., 2005. Influence of *CYP2C19* pharmacogenetic polymorphism on proton pump inhibitor-based therapies. Clin Pharmacokinet 44, 441–466.

Furuta, T., Sugimoto, M., Shirai, N., et al., 2007. Effects of *MDR1* C3435T polymorphism on cure rates of *Helicobacter pylori* by triple therapy with lansoprazole, amoxicillin and clarithromycin in relation to *CYP2C19* genotypes and *23 rRNA* genotypes of *H. pylori*. Aliment Pharmacol Ther 26, 693–703.

Garcia-Gonzalez, M.A., Lanas, A., Santolaria, S., et al., 2001. The polymorphic *IL-1B* and *IL-1RN* genes in the aetiopathogenesis of peptic ulcer. Clin Exp Immunol 125, 368–375.

Garcia-Gonzalez, M.A., Lanas, A., Savelkoul, P.H., et al., 2003. Association of interleukin 1 gene family polymorphisms with duodenal ulcer disease. Clin Exp Immunol 134, 525–531.

Garcia-Gonzalez, M.A., Lanas, A., Wu, J., et al., 2005a. Lack of association of *IL-12p40* gene polymorphism with peptic ulcer disease. Hum Immunol 66, 72–76.

Garcia-Gonzalez, M.A., Savelkoul, P.H., Benito, R., et al., 2005b. No allelic variant associations of the *IL-1* and *TNF* gene polymorphisms in the susceptibility to duodenal ulcer disease. Int J Immunogenet 32, 299–306.

Garcia-Gonzalez, M.A., Strunk, M., Piazuelo, E., et al., 2006. *TGF1* gene polymorphisms: Their relevance in susceptibility to *Helicobacter pylori*-related disease. Genes Immun 7, 640–646.

Gawronska-Szklarz, B., Siuda, A., Kurzawski, M., et al., 2010. Effects of CYP2C19, MDR1 and interleukin 1-B gene variants on the eradication rate of Helicobacter pylori infection by triple therapy with pantoprazole, amoxicillin and metronidazole. Eur J Clin Pharmacol 66, 681–687.

Gawronska-Szklarz, B., Wrzesniewska, J., Starzynska, T., et al., 2005. Effect of CYP2C19 and MRD1 polymorphisms on cure rate in patients with acid related disorders with Helicobacter pylori. Eur J Clin Pharmacol 61, 375–379.

Ge, Z., Taylor, D.E., 1998. Helicobacter pylori: Molecular genetics and diagnostic typing. Br Med Bull 54, 31–38.

Gerhard, M., Lehn, N., Neumayer, N., et al., 1999. Clinical relevance of the Helicobacter gene for blood group antigen binding adhesion. Proc Natl Acad Sci USA 96, 12,778–12,783.

Ghasemi, A., Shiraz, M.H., Ranjbar, R., et al., 2008. The prevalence of cagA and cagE genes in Helicobacter pylori strains isolated from different patient groups by PCR. Pak J Biol Sci 15, 2579–2583.

Gong, M., Ling, S.S., Lui, S.Y., Yeoh, K.G., Ho, B., 2010. Helicobacter pylori gamma-glutamyl transpeptidase is a pathogenic factor in the development of peptic ulcer disease. Gastroenterology 139, 564–573.

Gyulai, Z., Klausz, G., Tiszai, A., et al., 2004. Genetic polymorphism of interleukin 8 (IL-8) is associated with Helicobacter pylori induced duodenal ulcer. Eur Cytokine Netw 15, 353–358.

Hailong, C., Mei, Q., Zhang, L., et al., 2008. C314T polymorphism in histamine N methyltransferase gene and susceptibility to duodenal ulcer in Chinese population. Clin Chim Acta 389, 51–54.

Halushka, M.K., Walker, L.P., Halushka, P.V., 2003. Genetic variation in cyclooxygenase-1: Effects on response to aspirin. Clin. Pharmacol Ther 73, 122–130.

Hellmig, S., Mascheretti, S., Renz, J., et al., 2005. Haplotype analysis of the CD11 gene cluster in patients with chronic Helicobacter pylori infection and gastric ulcer disease. Tissue Antigens 65, 271–274.

Hellmig, S., Ott, S., Rosenstiel, P., et al., 2006. Genetic variants in matrix metalloproteinase genes are associated with development of gastric ulcer in Helicobacter pylori infection. Am J Gastroenterol 101, 29–35.

Hofman, V.J., Morilhon, C., Brest, P.D., et al., 2007. Gene expression profiling in human gastric mucosa infected with Helicobacter pylori. Mod Pathol 20, 974–989.

Hofner, P., Gyulai, Z., Kiss, Z.F., et al., 2007. Genetic polymorphism of nod1 and IL-* but not polymorphism of tlr4 genes are associated with Helicobacter pylori-induced duodenal ulcer and gastritis. Helicobacter 12, 124–131.

Hsu, P.-I., Chen, C.-H., Hsieh, C.-S., et al., 2007. Alpha1-antitrypsin precursor in gastric juice is a novel biomarker for gastric cancer and ulcer. Clin Cancer Res 13, 876–883.

Hsu, P.I., Jwo, J.J., Tseng, H.H., et al., 2005. Association of the myloperoxidase-468G: A polymorphism with gastric inflammation and duodenal ulcer risk. World J Gastroenterol 11, 2796–2801.

Hsu, P.I., Li, C.N., Tseng, H.H., et al., 2004. The interleukin-1 RN polymorphism and Helicobacter pylori infection in the development of duodenal ulcer. Helicobacter 9, 605–613.

Hunfeld, N.G., Touw, D.J., Mathot, R.A., et al., 2010. A comparison of the acid inhibitory effect of esomeprazole and pantoprazole in relation to pharmacokinetics and CYP2C19 polymorphism. Aliment Pharmacol Ther 31, 150–159.

Hunt, R.H., Xiao, S.D., Megraud, F., et al., 2010. World gastroenterology organization global guideline: Helicobacter pylori in developing countries. <http://www.worldgastroenterology.org/helicobacter-pylori-in-developing-countries.html>.

Hussein, N.R., 2010. The association of dupA and Helicobacter pylori related gastroduodenal diseases. Eur J Clin Microbiol Infect Dis 29, 817–821.

Hussein, N.R., Mohammadi, M., Talebkhan, Y., et al., 2008. Differences in virulence markers between Helicobacter pylori strains from Iraq and those from Iran: Potential importance of regional differences in H. pylori-associated disease. J Clin Microbiol 46, 1774–1779.

Ilver, D., Arnqvist, A., Ogren, J., et al., 1998. Helicobacter pylori adhesin binding fucosylated histo-blood group antigens revealed by retagging. Science 279, 373–377.

Israel, D.A., Salama, N., Arnold, C.N., et al., 2001. Helicobacter strain differences in genetic content identified by microarrays influence host inflammatory responses. J Clin Invest 107, 611–620.

Ito, Y., Azuma, T., Ito, S., et al., 2000. Sequence analysis and clinical significance of the iceA gene from Helicobacter pylori strains in Japan. J Clin Microbiol 38, 483–488.

Jafari, F., Shokrzadeh, L., Dabiri, H., et al., 2008. VacA genotypes of Helicobacter pylori in relation to cagA status and clinical outcome in Iranian populations. Jpn J Infect Dis 61, 290–293.

Jang, S., Jones, K.R., Olsen, C.H., et al., 2010. Epidemiological link between gastric disease and polymorphisms in vacA and cagA. J Clin Microbiol 48, 559–567.

Jung, S.W., Sugimoto, M., Graham, D.Y., et al., 2009. HomB status of Helicobacter pylori as a novel marker to distinguish gastric cancer from duodenal ulcer. J Clin Microbiol 47, 3241–3245.

Kang, J.M., Kim, N., Lee, D.H., et al., 2008. Effect of the CYP2C19 polymorphism on the eradication rate of Helicobacter pylori infection by 7-day triple therapy with regular proton pump inhibitor dosage. J Gastroenterol Hepatol 23, 1287–1291.

Kang, J.M., Kim, N., Lee, D.H., et al., 2009. The effects of genetic polymorphisms of IL-6, IL-8 and IL-10 on Helicobacter induced gastroduodenal disease in Korea. J Clin Gastroenterol Hepatol 43, 420–428.

Kennemann, L., Didelot, X., Aebischer, T., et al., 2011. Helicobacter pylori genome evolution during infection. Proc Natl Acad Sci USA 108, 5033–5038.

Khoder, G., Yamaoka, Y., Fauchere, J.-L., Burucoa, C., Atanassov, C., 2009. Proteomic Helicobacter pylori biomarkers discriminating between duodenal ulcer and gastric cancer. J Chromatogr B Analyt Technol Biomed Life Sci 877, 1193–1199.

Kim, Y.S., Park, S.W., Kim, M.H., et al., 2008. Novel single nucleotide polymorphism of the VEGF gene as a risk predictor for gastroduodenal ulcers. J Gastroenterol Hepatol 23 (Suppl 2), S131–S139.

Kuipers, E.J., Israel, D.A., Kusters, J.G., et al., 2000. Quasispecies development of Helicobacter pylori observed in paired isolates obtained years apart from the same host. J Infect Dis 181, 273–282.

Kumar, S., Kumar, A., Dixit, V.K., 2010. Diversity in the cag pathogenicity island of Helicobacter pylori isolates in populations from North and South India. J Med Microbiol 59, 32–40.

Kuo, C.H., Wang, S.S., Hsu, W.H., et al., 2010. Rabeprazole can overcome the impact of CYP2C19 polymorphism on quadruple therapy. Helicobacter 15, 265–272.

Kurzawski, M., Gawronska-Szklarz, B., Wrzesniewska, J., et al., 2006. Effect of CYP2C19*17 gene variant on Helicobacter pylori eradication in peptic ulcer patients. Eur J Clin Pharmacol 62, 877–880.

Lanas, A., Garcia-Gonzales, M.A., Santolaria, S., et al., 2001. TNF and LTA gene polymorphisms reveal different risk in gastric and duodenal ulcer patients. Genes Immun 2, 415–421.

Lay, C.S., Lin, J.R., 2010. Correlation of *CYP2C19* genetic polymorphisms with *Helicobacter pylori* eradication in patients with cirrhosis and peptic ulcer. J Chin Med Assoc 73, 188–193.

Lee, J.H., Jung, H.Y., Choi, K.D., et al., 2010. The influence of *CYP2C19* polymorphism on eradication of *Helicobacter pylori*: A prospective randomized study of lansoprazole and rabeprazole. Gut Liver 4, 201–206.

Li, X.-X., Wong, G.L.-H., To, K.-F., et al., 2009. Bacterial microbiota profiling in gastritis without *Helicobacter pylori* infection or non-steroidal anti-inflammatory drug use. PLoS ONE 4, e7985. doi:10.137/journal.pone.0007985.

Lin, Y.-F., Chen, C.-Y., Tsai, M.-H., et al., 2007. Duodenal ulcer related antigens from *Helicobacter pylori*: Immunoproteome and protein microarray approaches. Mol Cell Proteomics 6, 1018–1026.

Linpisarn, S., Suwan, W., Lertprasertsuk, N., et al., 2007. *Helicobacter pylori cagA, vacA* and *iceA* genotypes in northern Thai patients with gastric disease. Southeast Asian J Trop Med Public health 38, 356–362.

Li-Wan-Po, A., Girard, T., Farndon, P., Cooley, C., Lithgow, J., 2010. Pharmacogenetics of CYP2C19: Functional and clinical implications of a new variant *CYP2C19*17*. Br J Clin Pharmacol 69, 222–230.

Lobo Gatti, L., Zanbaldi Tunes, M., De Labio, R.W., et al., 2005. Interleukin 6 polymorphism and *Helicobacter pylori* infection in Brazilian adult patients with chronic gastritis. Clin Exp Med 5, 112–116.

Lock, R.A., Cordwell, S.J., Coombs, G.W., Walsh, B.J., Forbes, G.M., 2001. Proteome analysis of *Helicobacter pylori*: Major proteins of Type strain 11637. Pathology 33, 365–374.

Loh, J.T., Torres, V.J., Algood, H.M., McClain, M.S., Cover, T.L., 2008. *Helicobacter pylori* HopQ outer membrane proteins attenuates bacterial adherence to gastric epithelial cells. FEMS Microbiol Lett 289, 53–58.

Lu, C.C., Sheu, B.S., Chen, T.W., et al., 2005a. Host *TNF-alpha*-1031 and 863 promoter single nucleotide polymorphisms determine the risk of benign ulceration after *H. pylori* infection. Am J Gastroenterol 100, 1274–1282.

Lu, H., Hsu, P.-I., Graham, D.Y., et al., 2005b. Duodenal ulcer promoting gene of *Helicobacter pylori*. Gastroenterology 128, 833–848.

Ma, J., Yang, X.Y., Qiao, L., Liang, L.Q., Chen, M.H., 2008. *CYP2C9* polymorphism in NSAID induced gastropathy. J Dig Dis 9, 79–83.

Matsushima, K., Isomoto, H., Inoue, N., et al., 2010. MicroRNA signatures in *Helicobacter pylori* infected gastric mucosa. Int J Cancer 128, 361–370.

McClain, M.S., Shaffer, C.L., Israel, D.A., Peek, R.M., Cover, T.L., 2009. Genome sequence analysis of *Helicobacter pylori* strains associated with gastric ulceration and gastric cancer. BMC Genomics 10, 3. doi: 101186/1471-2164-10-3.

Megraud, F., Lehours, P., 2007. *Helicobacter pylori* detection and antimicrobial susceptibility testing. Clin Microbiol Rev 20, 280–322.

Miciuleviciene, J., Calkauskas, H., Jonaitis, L., et al., 2008. *Helicobacter pylori* genotypes in Lithuanian patients with chronic gastritis and duodenal ulcer. Medicina (Kaunas) 44, 449–454.

Molaei, M., Foroughi, F., Mashayekhi, R., et al., 2010. *Cag A* status and *vacA* subtypes of *Helicobacter pylori* in relation to histopathological findings in Iranian population. Indian J Pathol Microbiol 53, 24–27.

Momenah, A.M., Tayeb, M.T., 2007. *Helicobacter pylori cagA* and *iceA* genotypes status and risk of peptic ulcer in Saudi patients. Saudi Med J 28, 382–385.

Monstein, H.J., Tiveljung, A., Kraft, C.H., Borch, K., Jonasson, J., 2000. Profiling of bacterial flora in gastric biopsies from patients with *Helicobacter* associated gastritis and histologically normal control individuals by temperature gradient gel electrophoresis and 16SrDNA sequence analysis. J Med Microbiol 49, 817–822.

Nguyen, L.T., Uchida, T., Tsukamoto, Y., et al., 2010. *Helicobacter pylori dupA* gene is not associated with clinical outcome in the Japanese population. Clin Microbiol Infect 16, 1264–1269.

Niv, Y., 2010. *H. pylori*/NSAID-negative peptic ulcer – the mucin theory. Med Hypotheses 75, 435–533.

Ogiwara, H., Sugimoto, M., Ohon, T., et al., 2009. Role of deletion located between the intermediate and middle regions of the *Helicobacter pylori vacA* gene in cases of gastroduodenal disease. J Clin Microbiol 47, 3493–3500.

Oh, J.H., Dong, M.S., Choi, M.G., et al., 2009. Effects of *CYP2C19* and *MDR1* genotype on the eradication rate of *Helicobacter pylori* infection by triple therapy with pantoprazole, amoxicillin and clarithromycin. J Gastroenterol Hepatol 24, 294–298.

Ohyauchi, M., Imatani, A., Yonechi, M., et al., 2005. The polymorphism interleukin 8 -251A/T influences the susceptibility of *Helicobacter pylori* related gastric disease in the Japanese population. Gut 54, 330–335.

Oleastro, M., Cordeiro, R., Manard, A., et al., 2008. Evaluation of the clinical significance of HomB, a novel candidate marker of *Helicobacter pylori* strains associated with peptic ulcer disease. J Infect Dis 198, 1379–1387.

Oleastro, M., Cordeiro, R., Yamaoka, Y., et al., 2009. Disease association with two *Helicobacter pylori* duplicate outer membrane protein genes, *homB* and *homA*. Gut Pathog 1, 12.

Oleastro, M., Monterion, L., Lehours, P., Megraud, F., Menard, A., 2006. Identification of markers for *Helicobacter pylori* strains isolated from children with peptic ulcer disease by suppressive subtractive hybridization. Infect Immun 74, 4064–4074.

Oleastro, M., Santos, A., Cordeiro, R., et al., 2010. Clinical relevance and diversity of two homologous genes encoding glycosyltransferases in *Helicobacter pylori*. J Clin Microbiol 48, 2885–2891.

Olfat, F.O., Zheng, Q., Oleastro, M., et al., 2005. Correlation of the *Helicobacter pylori* adherence factor BabA with duodenal ulcer disease in four European countries. FEMS Immunol Med Microbiol 44, 151–156.

O'Toole, P.W., Snelling, W.J., Canchaya, C., et al., 2010. Comparative genomics and proteomics of *Helicobacter mustalae*, an ulcerogenic and carcinogenic gastric pathogen. BMC Genomics 11, 164.

Padol, S., Yuan, Y., Thabane, M., Padol, I.T., Hunt, R.H., 2006. The effect of *CYP2C19* polymorphisms on *Helicobacter pylori* eradication rate in dual and triple first line PPI therapies: A meta analysis. Am J Gastroenterol 101, 1467–1475.

Pan, Z.-J., Berg, D.E., Van der Hulst, R.W.M., et al., 1998. Prevalence of vacuolating cytotoxin production and distribution of distinct *vacA* alleles in *Helicobacter pylori* from China. J Infect Dis 178, 220–226.

Park, S.Y., Lee, Y.D., Kim, S.K., 2010. Examination of geographical, clinical and intra-host variation in the 3′ repeat region of the *cagA* gene in *Helicobacter pylori*. J Korean Med Sci 25, 61–66.

Partida-Rodriguez, O., Torres, J., Flores-Luna, L., et al., 2010. Polymorphisms in *tnf* and *hsp-70* show a significant association with gastric cancer and duodenal ulcer. Int J Cancer 126, 1861–1868.

Peek, R.M., Thompson, S.A., Donahue, J.P., et al., 1998. Adherence to gastric epithelial cells induces expression of a *Helicobacter pylori* gene, *iceA*, that is associated with clinical outcome. Proc Assoc Am Physicians 110, 531–544.

Rad, R., Prinz, C., Neu, B., et al., 2003. Synergistic effect of *Helicobacter pylori* virulence factors and interleukin-1 polymorphisms for the development of severe histological changes in the gastric mucosa. J Infect Dis 188, 272–281.

Rain, J.-C., Selig, L., De Reuse, H., et al., 2001. The protein–protein interaction map of *Helicobacter pylori*. Nature 409, 211–215.

Ramelah, M., Aminuddin, A., Alfizah, H., et al., 2005. *CagA* gene variants in Malaysian *Helicobacter pylori* strains isolated from patients of different ethnic groups. FEMS Immunol Med Microbiol 44, 239–242.

Resnick, M.B., Sabo, E., Meitner, M.P., et al., 2006. Global analysis of the human gastric epithelial transcriptome altered by *Helicobacter pylori* eradication *in vivo*. Gut 55, 1717–1724.

Rhead, J.L., Letley, D.P., Mohammadi, M., et al., 2007. A new *Helicobacter pylori* vacuolating cytotoxin determinant, the intermediate region, is associated with gastric cancer. Gastroenterology 133, 926–936.

Ricci, C., Holton, J., Vaira, D., 2007. Diagnosis of *Helicobacter pylori*: Invasive and non-invasive tests. Best Pract Res Clin Gastroenterol 21, 299–313.

Rick, J.R., Goldman, M., Semino-Mora, C., et al., 2010. *In-situ* expression of CagA and risk of gastroduodenal disease in *Helicobacter pylori*-infected children. J Pediatr Gastroenterol Nutr 50, 167–172.

Roesler, B.M., De Oliveira, T.B., Bonon, S.H., et al., 2008. Restriction fragment length polymorphisms of urease C and urease B genes of *Helicobacter pylori* strains isolated from Brazilian patients with peptic ulcer and chronic gastritis. Dig Dis Sci 54, 1487–1493.

Romo-Gonzalez, C., Salama, N.R., Burgeno-Ferreira, J., et al., 2009. Differences in genome content among *Helicobacter pylori* isolates from patients with gastritis, duodenal ulcer or gastric cancer reveal novel disease associated genes. Infect Immun 77, 2201–2211.

Saha, D.R., Datta, S., Chattopadhyay, S., et al., 2009. Indistinguishable cellular changes in gastric mucosa between *Helicobacter pylori* infected asymptomatic tribal and duodenal ulcer patients. World J Gastroenterol 15, 1105–1112.

Said Essa, A., Alaa Eldeen Nouh, M., Mohammed Ghaniam, N., Graham, D.Y., Said Sabry, H., 2008. Prevalence of CagA in relation to clinical presentation of *Helicobacter pylori* infection in Egypt. Scand J Infect Dis 40, 730–733.

Sakai, T., Aoyama, N., Kita, T., et al., 2001. *CYP2C19* genotype and pharmacokinetics of three proton pump inhibitors in healthy subjects. Pharm Res 18, 721–727.

Salama, NR., Gonzales-Valencia, G., Deatherage, B., et al., 2007. Genetic analysis of Helicobacter pylori strain populations colonizing the stomach at different times post infection. J Bacteriol 189, 3834–3845.

Salama, N., Guillemin, K., McDaniel, T.K., et al., 2000. A whole-genome microarray reveals genetic diversity among *Helicobacter pylori* strains. Proc Natl Acad Sci USA 97, 14,668–14,673.

Salari, M.H., Shirazi, M.H., Hadaiti, M.A., et al., 2009. Frequency of *Helicobacter pylori* vacA genotypes in Iranian patients with gastric and duodenal ulcers. J Infect Public health 2, 204–208.

Salehi, Z., Jelodar, M.H., Rassa, M., et al., 2009. *Helicobacter pylori* CagA status and peptic ulcer disease in Iran. Dig Dis Sci 54, 608–613.

Salih, B.A., Bolek, B.K., Arikan, S., 2010. DNA sequence analysis of *cagA* 3′ motifs of *Helicobacter pylori* strains from patients with peptic ulcer disease. J Med Microbiol 59, 144–148.

Sapone, A., Vaira, D., Trespidi, S., et al., 2003. The clinical role of cytochrome P450 genotypes in *Helicobacter pylori* management. Am J Gastroenterol 98, 1010–1015.

Saxena, A., Prasad, K.N., Ghoshal, U.C., et al., 2008. Polymorphism of 765G>C *cox2* is a risk factor for gastric adenocarcinoma and peptic ulcer disease in addition to *H. pylori* infection: A study from northern India. World J Gastroenterol 14, 1498–1503.

Schmidt, H.-M.A., Andres, S., Kaakoush, N.O., et al., 2009. The prevalence of the duodenal ulcer promoting gene (*dupA*) in *Helicobacter pylori* isolates varies by ethnic group and is not universally associated with disease development: A case control study. Gut Pathog 1, 5.

Sepulveda, A.R., Tao, H., Carloni, E., et al., 2002. Screening gene expression profiles in gastric epithelial cells induced by *Helicobacter pylori* using microarray analysis. Aliment Pharmacol Ther 16 (Suppl 2), 145–157.

Shao, J.G., Jiang, W., Li, K.Q., Lu, J.R., Sun, Y.Y., 2009. Blood concentrations of pantoprazole sodium is significantly high in hepatogenic ulcer patients, especially those with a poor CYP2C19 metabolism. J Dig Dis 10, 55–60.

Shiotani, A., Sakakibara, T., Yamanaka, Y., et al., 2009. The preventive factors for aspirin induced peptic ulcer: Aspirin ulcer and corpus atrophy. J Gastroenterol 44, 717–725.

Strobel, S., Bereswill, S., Balig, P., et al., 1998. Identification and analysis of new *vacA* genotypes of *Helicobacter pylori* in different patient groups in Germany. J Clin Microbiol 36, 1285–1289.

Suemasu, A., Tanaka, K., Namba, T., et al., 2009. A role for HSP70 in protecting against indomethocin-induced gastric lesions. J Biol Chem 284, 19,705–19,715.

Sugimoto, M., Furuta, T., Shirai, N., et al., 2007. Effects of interleukin-10 gene polymorphism on the development of gastric cancer and peptic ulcer in Japanese subjects. J Gastroenterol Hepatol 22, 1443–1449.

Tahara, T., Shibata, T., Nakamura, M., et al., 2009. Effects of polymorphisms in the 3′ untranslated region (3′UTR) of vascular endothelial factor gene on gastric cancer and peptic ulcer disease in Japan. Mol Carcinog 48, 1030–1037.

Torres, L.E., Melian, K., Moreno, A., et al., 2009. Prevalence of *vacA*, *cagA* and *babA2* genes in Cuban *Helicobacter pylori* isolates. World J Gastroenterol 15, 204–210.

Torres-Morquecho, A., Giono-Cerezo, S., Camorlinga-Ponce, M., Vargas-Mendoza, C.F., Torres, J., 2010. Evolution of bacterial genes: Evidence of positive Darwinian selection and fixation of base substitutions in virulence genes of *Helicobacter pylori*. Infect Genet Evol 10, 764–776.

Ueda, H., Ito, M., Eguci, H., et al., 2006. Development of a novel method to detect *Helicobacter pylori cagA* genotype from paraffin-embedded materials: Comparison between patients with duodenal ulcer and gastric cancer in young Japanese. Digestion 73, 47–53.

Van Doorn, L.-J., 2001. Detection of *Helicobacter pylori* virulence-associated genes. Exp Rev Mol Diagn 1, 290–298.

Van Doorn, L.-J., Figueiredo, C., Megraud, F., et al., 1999. Geographic distribution of *vacA* allelic types of *Helicobacter pylori*. Gastroenterology 116, 823–830.

Van Doorn, L.-J., Figueiredo, C., Sanna, R., et al., 1998a. Expanding allelic diversity of *Helicobacter pylori vacA*. J Clin Microbiol 36, 2597–2603.

Van Doorn, L.J., Figueiredo, C., Sanna, R., et al., 1998b. Clinical relevance of the *cagA*, *vacA* and *iceA* status of *Helicobacter pylori*. Gastroenterology 115, 58–66.

Van Oijen, M.G., Laheij, M., De Kleine, E., et al., 2006. Effect of specific cyclooxygenase gene polymorphism (A842G/C50T) on the occurrence of peptic ulcer haemorrhage. Dig Dis Sci 51, 2348–2352.

Vega, A.E., Cortinas, T.I., Puig, O.N., et al., 2010. Molecular characterization and susceptibility testing of *Helicobacter pylori* strains isolated in western Argentina. Int J Infect Dis 14 (Suppl 3), e85–e92.

Wilschanski, M., Schlesinger, Y., Faber, J., et al., 2007. Combination of *Helicobacter pylori* strain and tumour necrosis factor-alpha polymorphism of the host increases the risk of peptic ulcer disease in children. J Pediatr Gastroenterol Nutr 45, 199–203.

Wu, C.-C., Chou, P.-Y., Hu, C.-T., et al., 2005. Clinical relevance of the *vacA*, *iceA*, *cagA* and *flaA* genes of *Helicobacter pylori* strains isolated in eastern Taiwan. J Clin Microbiol 43, 2913–2915.

Wu, M.-S., Chow, L.-P., Lin, J.-Y., et al., 2008. Proteomic identification of biomarkers related to *Helicobacter pylori*-associated gastroduodenal disease: Challenges and opportunities. J Gastroenterol Hepatol 23, 1657–1661.

Xiang, Z., Censini, S., Bayeli, P.F., et al., 1995. Analysis of expression of CagA and VacA virulence factors in 43 strains of *Helicobacter pylori* reveals that clinical isolates can be divided into two major types and that cagA is not necessary for expression of the vaculating cytotoxin. Infect Immun 63, 94–98.

Yakoob, J., Abid, S., Abbas, Z., et al., 2009a. Distribution of *Helicobacter pylori* virulence markers in patients with gastroduodenal disease in Pakistan. BMC Gastroenterol 9, 87.

Yakoob, J., Jafri, W., Abbas, Z., et al., 2009b. Low prevalence of the intact *cag* pathogenicity island in clinical isolates of *Helicobacter pylori* in Karachi, Pakistan. Br J Biomed Sci 66, 137–142.

Yamaoka, Y., Kodama, T., Gutierrez, O., et al., 1999. Relationship between *Helicobacter pylori iceA*, *cagA*, *vacA* status and clinical outcome in four different countries. J Clin Microbiol 37, 2274–2279.

Yamaoka, Y., Kwon, D.H., Graham, D.Y., 2000. A M(r) 34,000 pro-inflammatory outer membrane protein (OipA) of *Helicobacter pylori*. Proc Natl Acad Sci USA 97, 7533–7538.

Yamaoka, Y., Souchek, J., Odenbreit, S., et al., 2002. Discrimination between cases of duodenal ulcer and gastritis on the basis of putative virulence factors of *Helicobacter pylori*. J Clin Microbiol 40, 2244–2246.

Yamazaki, S., Yamakawa, A., Okuda, T., et al., 2005. Distinct diversity of *vacA*, *cagA* and *cagE* genes of *Helicobacter pylori* associated with peptic ulcer in Japan. J Clin Microbiol 43, 3906–3916.

Yanaka, A., Zhang, S., Sato, D., et al., 2007. Geranylgeranylacetone protects the human gastric mucosa from diclofenac-induced injury via induction of heat shock protein 70. Digestion 75, 148–155.

Yang, G.F., Luo, J., Wu, H.W., et al., 2008. Interleukin-1 gene family polymorphisms from paraffin-embedded archival tissue of Chinese patients with gastric ulcer and gastric cancer. Hepatogastroenterology 55, 1878–1881.

Yang, J.C., Lin, C.J., 2010. *CYP2C19* genotype in the pharmacokinetics/pharmacodynamics of proton pump inhibitor based therapy of *Helicobacter pylori* infection. Exp Opin Drug Metab Toxicol 6, 29–41.

Yang, J.C., Yang, Y.F., Uang, Y.S., Lin, C.J., Wang, T.H., 2009. Pharmacokinetic–pharmacodynamic analysis of the role of *CYP2C19* genotypes in short term rabeprazole based triple therapy against *Helicobacter pylori*. Br J Clin Pharmacol 67, 503–510.

Younson, J., O'Mahony, R., Liu, H., et al., 2009. A human domain antibody and Lewis b glycoconjugates that inhibit binding of *Helicobacter pylori* to Lewis b receptor and adhesion to human gastric epithelium. J Clin Microbiol 200, 1574–1582.

Zhang, M.J., Zaho, F., Xiao, D., et al., 2009. Comparative proteomic analysis of passaged *Helicobacter pylori*. J Basic Microbiol 49, 482–490.

Zhang, Z., Zheng, Q., Chen, X., et al., 2008. The *Helicobacter pylori* duodenal ulcer promoting gene *dupA* in China. BMC Gastroenterol 8, 49.

Zhao, F., Wang, J., Yang, Y., et al., 2008. Effect of *CYP2C19* genetic polymorphism on the efficacy of proton pump inhibitor based triple therapy for *Helicobacter pylori* eradication: A meta analysis. Helicobacter 13, 532–541.

ONGOING RESEARCH

Devi, S.H., Taylor, T.D., Avasthi, T.S., et al., 2010. Genome of *Helicobacter pylori* strain 908. J Bacteriol 192, 6488–6489.

Viroriano, U., Saraiva-Pava, K.D., Rocha-Goncalves, A., 2011. Ulcerogenic *Helicobacter pylori* strains isolated from children: A contribution to get insight into the virulence of the bacteria. PLoS One 6, e26265.

Jung, S.W., Sugimoto, M., Shiota, S., Graham, D.Y., Yamaoka, Y., 2012. The intact dupA cluster is more a reliable *Helicobacter pylori* virulence marker than dupA alone. Infect Immun 80, 381–387.

Aviles-Jimenez, F., Reyes-Leon, A., Nieto-Patlan, E., et al., 2012. In vivo expression of *Helicobacter pylori* virulence genes in patients with gastritis, ulcer and gastric cancer. Infect Immun 80, 594–601.

Martinez-Becerra, F., Castillo-Rojas, G., de Leon, S.P., Lopez-Vidal, Y., 2012. IgG subclass against *Helicobacter pylori* isolates: An important tool for disease characterization. Scand J Immunol 76, 26–32.

Abadi, A.T., Taghvaei, T., Wolfram, L., Kusters, J.G., 2012. Infection with *Helicobacter pylori* strains lacking dupA is associated with an increased risk of gastric ulcer and gastric cancer development. J Med Microbiol 61, 23–30.

Salagacka, A., Bartczak, M., Zebrowska, M., et al., 2011. C3435T polymorphism of the *ABCB1* gene: Impact on genetic susceptibility to peptic ulcers. Pharmacological Reports 63, 992–998.

Zebrowska, M., Jazdzyk, M., Salagacka, A., et al., 2012. Investigation of ABCB1 1236 and 2677 SNP's in patients with peptic ulcer. Scand J Gastroenterol 47, 22–27.

Niv, Y., Boltin, D., Halpern, 2012. Membrane bound mucins and mucin terminal glycans expression in idiopathic or Helicobacter pylori, NSAID associated peptic ulcers. Dig Dis Sci 57, 2535–2544.

Ohyama, K., Shiokawa, A., Ito, K., 2012. Toxicoproteomic analysis of a mouse model of nonsteroidal anti-inflammatory drug-induced gastric ulcers. BBRC 420, 210–215.

CHAPTER

Cirrhosis

Nicholas A. Shackel, Keyur Patel, and John McHutchison

INTRODUCTION

The liver has been called the "the custodian of the milieu intérieur." Consistent with its many varied metabolic functions, the liver has a complex transcriptome and proteome. The normal liver has many diverse functions, including synthesis of vitamins and proteins, bile production, and immune defense, as well as metabolism of carbohydrates, lipids, and toxins. Despite being only 2.5% of body weight, the liver receives 25% of cardiac output, which is essential for maintaining its metabolic and synthetic functions. The functional unit of the liver is the hepatic lobule, which is arranged in an organized, repeating fashion around a central venule to form the intact organ (Figure 78.1). Liver injury is characterized by progressive fibrosis tissue deposition within the lobule, leading to the eventual disruption of normal lobular architecture that is characteristic of cirrhosis (Figure 78.2) (Friedman, 2000). In the genomic era, the molecular classification of fibrosis has been limited, with cirrhosis development predominantly characterized by morphological changes.

Cirrhosis is the pathogenic hallmark of advanced liver injury. Cirrhosis is morphologically defined by distortion of hepatic architecture by dense bands of fibrosis "scar," leading to "islands" or nodules of hepatocytes (Friedman, 2000). The causes of cirrhosis are varied, with many unique aspects based on etiology (Table 78.1). The teleological role of this response is the confinement of injury while maintaining hepatic function. Cirrhosis is a premalignant condition, with virtually all cases of liver cancer [also known as hepatocellular carcinoma (HCC)] developing in individuals with cirrhosis. Various intrahepatic cell populations are essential in understanding the development of cirrhosis. The pivotal cell involved in the fibrosis leading to cirrhosis is the hepatic stellate cell. However, the functional unit of the liver is the hepatocyte, the cell type from which HCC develops. The mechanisms of fibrosis leading to cirrhosis will be discussed in the context of understanding the pathogenic mechanisms by which this process evolves. Genomic medicine promises new insights into the pathogenesis of fibrosis, and brings the promise of individualized, predictive medicine aiming to avoid cirrhosis and the sequelae of liver failure and HCC.

LIVER STRUCTURE

Disruption of normal hepatic structure is a hallmark of liver injury. The liver is a complex organ made up of heterogeneous cell types. The functional cell of the liver is the 20–30-μm-diameter hepatocyte that is arranged in interconnecting plates within the structural unit of the liver, the hepatic lobule (Figure 78.1). There are 50,000–100,000 lobules in the adult liver. A vascular space known as the hepatic sinusoid is lined by fenestrated, sinusoidal endothelial cells (SECs) and Kupffer cells surrounding the hepatocytes (Figure 78.1). The SECs perform a central role in the recruitment and retention of inflammatory cells within the liver. Kupffer cells

Genomic and Personalized Medicine, 2nd edition
by Ginsburg & Willard. DOI: http://dx.doi.org/10.1016/B978-0-12-382227-7.00078-1

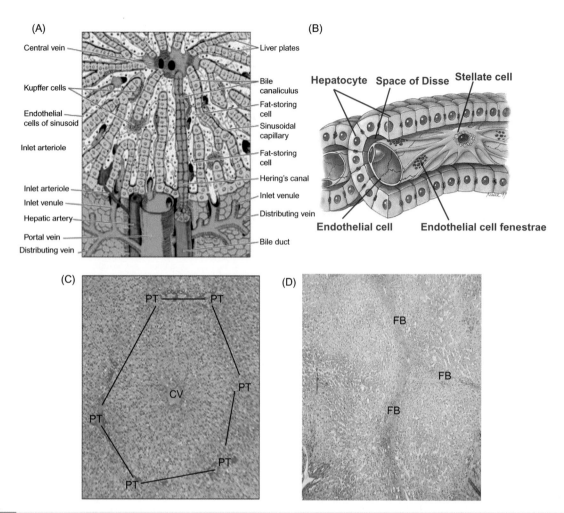

Figure 78.1 Hepatic structure. (A) The liver is composed of plates of hepatocytes surround by sinusoids (vascular spaces) through which blood flows from either the portal vein or hepatic artery located in the portal triad to a central vein. (B) Sinusoids consist of hepatocyte plates surrounded by a fenestrated endothelial layer. The space between the endothelial layer and the hepatocyte is the space of Disse, in which the hepatic stellate cell is found. (C) A normal liver consists of multiple hepatic lobules with portal triads (PT) at the periphery and a central vein (CV) within the middle of the lobule. (D) Within a cirrhotic liver, there is disruption of normal lobular architecture by dense connecting bands of fibrosis (FB) [contrast with normal liver in (C)].

are intrahepatic macrophages attached to the sinusoidal walls. These cells are preferentially located around portal tracts and are essential cells involved in modulating intrahepatic immune responses. The hepatic stellate cell (HSC) is the main mediator of intrahepatic fibrogenesis. It is located between the SECs and the hepatocytes in the space of Disse. Bile from hepatocytes is secreted into bile canaliculi formed between hepatocytes, which drain into epithelium-lined bile ducts. The cellular interactions within the liver determine the normal homeostatic functions and response to injury.

The space of Disse between the endothelium and hepatocytes, in which the HSCs are located, is the initial site of intrahepatic fibrogenesis. In normal liver, the space of Disse is composed of non-fibril-forming collagens, particularly types IV, VI, and XIV, proteoglycans (principally perlecan),

and glycoproteins (principally fibronectin, laminin, and tenascin). In contrast, during liver injury the space of Disse is enlarged as the extracellular matrix (ECM) is remodeled, with an initial increase in fibronectin and tenascin. Subsequently, fibrogenesis proceeds with type III then I collagens, elastin, and laminin deposition in the space of Disse. The molecular pathogenesis of abnormal matrix production is poorly understood.

FIBROSIS AND CIRRHOSIS

Despite the development of genomic medicine approaches in recent years, the morphological changes of intrahepatic fibrosis and cirrhosis are the hallmarks of liver injury.

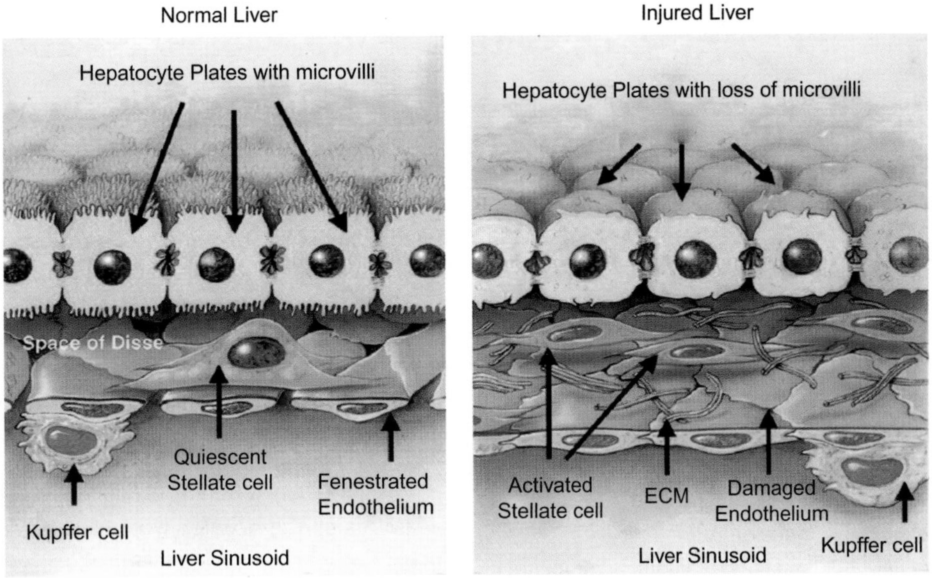

Figure 78.2 Liver fibrogenesis. In liver fibrosis development, the quiescent stellate cell is transformed into an activated phenotype with extracellular matrix changes characterized by the deposition of a scar-like matrix within the space of Disse. This change is accompanied by a loss of hepatocyte microvilli and endothelial fenestration. ECM = extracellular matrix.

In end-stage fibrosis, there is "encapsulation" of islands of hepatocytes by bands of fibrosis, a condition known as cirrhosis (Figures 78.1, 78.2). To date, no gene or protein signature defines the risk of hepatic failure or likelihood of developing hepatocellular carcinoma as well as the morphological change of cirrhosis. The hepatic fibrotic response represents a wound-healing response with morphological features of matrix remodeling, contraction, and "scarring" (Eng and Friedman, 2000). The hepatic stellate cell is central to this fibrogenic process (Figure 78.2). The fibrogenic process can be regarded in stages based on morphology or on molecular pathogenic events (Friedman, 2000). Fibrosis of the liver has been regarded classically as an irreversible disease process. However, it is now clear that advanced fibrosis, possibly even cirrhosis, can significantly improve and possibly resolve completely in some cases (Benyon and Iredale, 2000; Corbett et al., 1993; Dufour et al., 1998; Tsushima et al., 1999). Molecular studies of intrahepatic fibrogenesis in progressive injury indicated that this process is dynamic and characterized by distinct events associated with initiation, perpetuation, and regression (Bataller and Brenner, 2005). The remodeling of the ECM in fibrogenesis is clearly a dynamic process, and improved understanding of the molecular pathways involved in this process will help in developing future targeted therapeutic agents.

Liver injury characterized by fibrosis is extremely rare in the absence of intrahepatic inflammatory changes. The evolution of both innate and adaptive immune response in chronic liver injury is an essential factor perpetuating and driving intrahepatic injury resulting in liver fibrogenesis. Therefore, the separate study of fibrogenic pathways characterized by ECM remodeling and intrahepatic inflammatory response is often completely arbitrary.

Therefore, in liver disease pathogenesis there are many liver cell subpopulations to be considered – hepatocytes (the parenchymal cells of the liver), hepatic stellate cells, endothelial cells and immune cells. A major limitation in studying liver disease pathogenesis has been apparent for decades, but has been highlighted by genomics studies – the lack of membrane markers that uniquely define liver cell subpopulations. This has hampered the classical reductionist approach to studies of liver disease pathogenesis, as isolation of pure liver cell subpopulations is difficult. Also, such approaches are hampered by isolation procedure artifacts, which remove the intracellular interactions that often define disease. It is clear from genomics studies that the liver has a highly diverse genome and transcriptome that changes specifically with disease. Although there are gene and protein signature changes associated with the development of fibrosis, cirrhosis, and HCC, it is now apparent that these defined phenotypes are the result of many potential pathways that in varied combinations may give rise to disease.

DIAGNOSIS OF CIRRHOSIS

An accurate assessment of cirrhosis and the preceding evolution of intrahepatic fibrosis is a frequent diagnostic challenge. Routine biochemical analysis and clinical signs are at

TABLE 78.1	Causes of cirrhosis and/or chronic liver disease

Infectious diseases

Brucellosis

Capillariasis

Echinococcosis

Schistosomiasis

Toxoplasmosis

Viral hepatitis (hepatitis B, C, D; cytomegalovirus; Epstein–Barr virus)

Inherited and metabolic disorders

α1-antitrypsin deficiency

Alagille syndrome

Biliary atresia

Familial intrahepatic cholestasis (FIC) types 1–3

Fanconi's syndrome

Galactosemia

Gaucher's disease

Glycogen storage disease

Hemochromatosis

Hereditary fructose intolerance

Hereditary tyrosinemia

Wilson's disease

Drugs and toxins

Alcohol

Amioradone

Arsenicals

Oral contraceptives (Budd–Chiari)

Pyrrolizidine alkaloids (veno-occlusive disease)

Other causes

Biliary obstruction (chronic)

Cystic fibrosis

Graft-versus-host disease

Jejunoileal bypass

Non-alcoholic steatohepatitis

Primary biliary cirrhosis

Primary sclerosing cholangitis

Sarcoidosis

Causes of non-cirrhotic hepatic fibrosis

Idiopathic portal hypertension (non-cirrhotic portal fibrosis, Banti's syndrome); three variants: intrahepatic phlebosclerosis and fibrosis, portal and splenic vein sclerosis, and portal and splenic vein thrombosis

Schistosomiasis (*"pipe-stem"* fibrosis with pre-sinusoidal portal hypertension)

Congenital hepatic fibrosis (may be associated with polycystic disease of liver and kidneys)

best suggestive but not diagnostic of the grade of fibrosis or the development of cirrhosis. To date, liver biopsy has been the principal approach to fibrosis assessment. However, liver biopsy is an invasive procedure prone to sampling error and significant observer variability, and it may be associated with morbidity and mortality (Siegel et al., 2005). This has led to the development of non-invasive measures of fibrosis, including the appraisal of panels of profibrogenic serum markers and assessment of intrahepatic compliance changes using modified ultrasonography (elastometry). Other techniques of imaging the liver, such as ultrasound, computed tomography, or magnetic resonance imaging, can provide indicators of cirrhosis, but interpretation is often operator-dependent and lacks sensitivity for accurate staging of advanced disease. However, imaging and biochemistry may be useful in providing indicators of worsening hepatic function from cirrhosis, such as ascites or the development of HCC. To date, there are no molecular markers of fibrosis progression or cirrhosis used in clinical practice.

TREATMENT OF CIRRHOSIS

Treatment options in cirrhosis are limited, and in all cases the best treatment option is the minimization or avoidance of the causative agent. Significant advances have been made in the treatment of liver injury with the advent of antiviral therapy that can both control and eradicate viral hepatitis B and C (see Chapter 99). However, antiviral therapy is frequently ineffective. Additionally, the pathobiology of many causes of liver disease is poorly understood. As a result, research has intensified to develop agents that may lessen the progression of fibrosis, with a view to delaying or avoiding the development of cirrhosis. Approaches to control intrahepatic damage are frequently shown to be effective but are often limited by a lack of specificity for liver injury.

Minimization of intrahepatic inflammation is a promising approach to the control of liver injury. The use of interleukin (IL)-10 (a Th2 cytokine) opposes the pro-inflammatory Th1 cytokine profile frequently observed in intrahepatic inflammation, driving profibrotic liver injury. IL-10 has been successfully used in a number of animal models, as well as in individuals with chronic hepatitis C virus (HCV) infection, and it has been shown to decrease the development of fibrosis (Boyer and Marcellin, 2000). Further, anti-inflammatory agents that target the Kupffer cell, such as inhibitors of the 5-lipoxygenase pathway, show promise in animal models of fibrosis. However, there is presently little clinical data supporting the use of anti-inflammatory agents to treat intrahepatic fibrosis.

The prevention of liver injury through modulation of apoptosis is another approach to control liver injury (Kaplowitz, 2000; Liu et al., 2004; Sancho-Bru et al., 2005). This concept is supported by the development and progression of caspase inhibitors into phase II clinical trials. In theory, while the avoidance of hepatocyte apoptosis may be a reasonable approach to the management of liver injury, the absence of cell-specific delivery means that individuals treated this way may be exposed to a plethora of potential adverse effects, including an increased risk of developing malignancies. Non-specific approaches, including the use of antioxidants such as alpha-tocopherol, the flavonoid silymarin, and the Japanese herb sho-saiko-to, show promise in *in vitro* and animal studies, but the results to date in clinical studies have been disappointing (Rockey, 2005).

Measures designed to stop or eliminate activated hepatic stellate cells represent an approach to liver injury treatment receiving a lot of attention. The most promising opposes the effects of the renin angiotensin system, which activates HSCs (Warner et al., 2007). With abundant angiotensin (AT) receptors on hepatic stellate cells that are important in HSC activation, the use of angiotensin converting enzyme (ACE) and ATII receptor inhibitors appears to be a logical and promising approach. Initial studies have suggested that both ACE and ATIIR inhibitors are effective in decreasing cirrhosis. Further, as these agents have already been extensively used in other diseases, their safety profile is well known, and they are likely to be readily tolerated by patients with liver disease. Other approaches designed to target the HSCs are at best promising but clinically unproven concepts. They include the use of selective endothelin receptor antagonists, leptin antagonists, adiponectin agonists, and cannabinoids to treat intrahepatic fibrosis (Lotersztajn et al., 2007; Mallat and Lotersztajn, 2006; Rockey, 2005). The use of PPAR (peroxisome proliferator-activated receptor) α/γ agonists also shows promise, with thiazolidinediones being shown in animal models to be effective in reducing liver fibrosis (Rockey, 2005). These drugs have been extensively used in diabetes mellitus, have known safety profiles, and are currently being actively investigated in clinical trials as possible liver antifibrotics in non-alcoholic steatohepatitis.

Immune modulators such as interferon-α and -γ have been shown to have an antifibrotic effect *in vitro* (Rockey, 2005). However, lack of specificity is a major limitation to the use of these agents solely as antifibrotics. Similarly, strategies such as transforming growth factor beta (TGF-β) antagonism, while promising, also lack specificity and are likely to be complicated by significant adverse side effects.

GENETICS OF CIRRHOSIS

Cirrhosis pathogenesis is characterized by many conserved as well as divergent pathways involved in the development of progressive fibrosis. Clearly, a number of important host genetic factors influence the development of fibrosis, rate of progression, and the subsequent development of complications (Bataller et al., 2003). Older individuals and males are known to have significantly greater rates of liver fibrogenesis and subsequent development of cirrhosis (Poynard et al., 2005). Multiple candidate genes thought to be central to cirrhosis development have been identified using rodent models, and many human gene

polymorphisms are likewise associated with fibrosis development (Table 78.2) (Bataller et al., 2003). Genetic polymorphisms have been implicated at all stages of fibrosis development, including disease susceptibility (e.g., HCV and haptoglobin), injury (e.g., LPS (lipopolysaccharide) interaction with CD14), immune responses [e.g., human leukocyte antigen II (HLA-II) alleles], activation of HSCs (e.g., angiotensinogen), and increase in fibrogenic ECM (e.g., plasminogen and TGF-β) (Bataller et al., 2003). The single most comprehensive analysis of genetic polymorphisms in fibrosis analyzed 24,832 putative functional single nucleotide polymorphisms (SNPs) in 916 individuals with HCV infection. The study identified a missense mutation in the DEAD box polypeptide 5 (*DDX5*) gene, associated with two *POLG2* SNPs, that is linked with increased risk of advanced fibrosis (Huang et al., 2006). The same study identified another missense SNP in carnitine palmitoyltransferase 1A (*CPT1A*) associated with decreased risk of fibrosis. The fibrosis association of these polymorphisms was validated in a separate cohort of 483 individuals (Huang et al., 2006). However, it is unclear what the function of these genes is in HCV liver injury. Further, it is unclear if the identified polymorphisms are important only in HCV liver injury, or more generally in other types of liver injury.

The identification of candidate genes and polymorphisms has seen a number of novel molecular approaches adopted in an attempt to treat liver disease (Prosser et al., 2006). These have included the use of caspase inhibitors, TGF-β blockade, tumor necrosis factor (TNF) blockade, platelet-derived growth factor (PDGF) blockade, and inhibition of Kupffer cell activity (Prosser et al., 2006). Further approaches have attempted to change ECM composition in fibrosis via the administration of matrix metalloproteinases and plasminogen activator (Prosser et al., 2006). Although modern molecular approaches have identified multiple genes involved in pathogenesis of cirrhosis, most of these findings have not yet led to therapeutic agents in the clinical setting.

The identification of genetic loci associated with the development of progressive liver injury has focused on the underlying disease rather than the development of fibrosis and eventual cirrhosis. Immunogenetics of autoimmune liver disease and viral hepatitis is well characterized, with a number of genetic loci being associated with disease severity and/or progression (Donaldson, 2004). However, the immunogenetics of other forms of liver injury is poorly characterized. The precise association of many of these markers with liver disease severity remains to be determined. In the future, however, determination of the genetic haplotypes conferring susceptibility to fibrosis or predicting the likely progression of liver disease is likely to be incorporated into routine clinical practice.

THE LIVER TRANSCRIPTOME

Evolving genomic medicine approaches to liver disease require an understanding of the complexity of genome expression

TABLE 78.2 Candidate genes and polymorphisms involved in cirrhosis

Gene	Candidate gene	Gene polymorphism	Effect of protein on fibrosis
ADH		√	Unknown
ALDH		√	Unknown
Angiotensinogen		√	Increases
ApoE		√	Increases
CD14		√	Increases
CPT1A	√		Decreases
CTLA-4		√	Increases
CYP2E1		√	Unknown
DDX5		√	Increases
Fas	√		Increases
HFE		√	Unknown
HLA-II haplotypes		√	Variable
IFN-γ	√		Decreases
IL-1R		√	Increases
IL-10	√	√	Decreases
IL-13	√		Unknown
IL-1β		√	Increases
IL-6	√		Unknown
IL-12 associated	√	√	Unknown
IL-28B		√	Unknown
Leptin	√		Increases
MnSOD		√	Increases
NOS-2	√		Decreases
OB-R	√		Increases
Plasminogen	√		Decreases
PNPLA3	√	√	Increases
SMAD-3	√		Increases
TAP2		√	Unknown
Telomerase	√		Decreases
TGF-β1	√	√	Increases
TIMP-1	√		Increases
TNF	√		Unknown

within normal and diseased liver. There are estimated to be approximately 20,000–25,000 protein-encoding genes in the human genome. Further, there may be more than 150,000 functionally significant, alternately spliced transcripts arising from those genes. How many of those mRNA transcripts are expressed in the liver is unknown. Methods of identifying and comparing organ transcriptomes are now available. In the past, inferring complexity could be done by examining GenBank human UniGene clusters of non-redundant gene sets (Yuan et al., 2001). These UniGene clusters are compiled from annotated and uncharacterized mRNA sequences, and as a group represent a species' transcriptome (Yuan et al., 2001). The human UniGene assembly of clusters has approximately seven million sequences, representing 122,755 non-redundant transcripts. Parsing keyword searches (keywords in UniGene Homo sapiens Build #229), approximately 16% of transcripts were identified in liver tissue; this compares to brain (34%), lung (26%), kidney (21%), colon/gut (19%), and heart (12%). Coulouarn and colleagues used a similar approach, and identified 12,638 non-redundant clusters from liver tissue (Coulouarn et al., 2004). An alternate approach, in which serial analysis of gene expression (SAGE) libraries were examined, can also provide insights into the complexity of the liver transcriptome (Yamashita et al., 2000, 2004a). Two normal human liver SAGE libraries identified 15,496 and 18,081 unique transcripts from a total of 66,308 and 125,700 tags, respectively (Yamashita et al., 2000, 2004a). However, in a SAGE comparison of multiple organs, 32,131 unique tags were identified (from a total of 455,325 tags); of these 56% were expressed in the liver compared to brain (75%), breast (81%), and colon (91%) (Yamashita et al., 2004b). Therefore, it is clear that the normal liver has a complex transcriptome expressing many thousands of mRNA transcripts. The answer to the question of how complex the human liver transcriptome is will be answered in the next decade, with the use of next-generation sequencing of multiple individual liver RNA pools.

Normal liver transcriptome expression varies to account for phenotypic differences such as sex and age variation (Cao et al., 2001; Tadic et al., 2002; Yang et al., 2006). Microarray analysis of normal human liver by Yano and colleagues highlights the variability of the non-diseased liver transcriptome (Yano et al., 2001). A total of 2418 genes were examined in five normal patients, with only half of those transcripts being detected in four out of five patients. Further, only 27% of genes had coordinate expression in these non-diseased, apparently normal patients. Therefore, in addition to the liver having a complex transcriptome, there appears to be significant individual variability in transcript expression. This is further highlighted by the observation of Enard and colleagues that duplicate liver samples from the same individual differed by 12% (technical variation), but that intraspecies variation was as pronounced as interspecies variation in hepatic mRNA transcript expression, comparing chimpanzees and humans (Enard et al., 2002). Focused specialized arrays, such as the Liverpool nylon array, targeting the liver transcriptome have now been synthesized and include in excess of 10,000 target genes (Coulouarn et al., 2004). However, such approaches fail to detect differential gene expression for transcripts not expressed in normal liver that are subsequently expressed with the development of liver pathobiology. This is a particularly important consideration in genomic medicine, as transcriptomes in disease can markedly increase in complexity, especially in the presence of neoplastic transformation and inflammation (Feezor et al., 2005; Scriver, 2004).

THE LIVER PROTEOME

The human proteome is complex and variable depending on the organ or cell population being studied. Proteins can be subject to more than 100 different types of post-translational modifications (Cantin and Yates, 2004). Therefore, the estimated 25,000 genes in the human genome may give rise to more than one million distinct proteins (Neverova and Van Eyk, 2005; Righetti et al., 2005). By parsing GenBank with keywords, we can gain insight into human organ proteome complexity (keywords in UniGene Homo sapiens Build #229). Proteomes from human brain, liver, lung, kidney, bowel/colon, heart and serum express 40%, 13%, 10%, 18%, 6%, 7%, and 6%, respectively, of the proteins found in these organs. This highlights the complexity of the proteome within the liver and other solid organs.

Contained within a cell, approximately 90% of the cellular protein mass is due to the 100 most abundant proteins, and a further 1200 proteins account for another 7% of the protein mass (Lefkovits et al., 2000, 2001). However, the remaining 3% of the protein mass includes 2800 proteins (more than 50% of the different protein species), and frequently is below the detection limit of most proteomic detection methods (Lefkovits et al., 2000, 2001). Therefore, it is important to consider the frequency of protein expression within a homogenous cellular sample (e.g., a cell line) compared to a heterogeneous cellular sample (e.g., an organ). This is an especially important consideration in genomic medicine, as the non-parenchymal cell subpopulation abundance in biopsy specimens is low, and sample representation of the whole organ is often incorrect, a problem known to occur in more than 15% of liver biopsy samples (Ratziu et al., 2005; Regev et al., 2002).

Genomic medicine approaches often assume with transcriptome analysis that changes in mRNA expression reflect changes in corresponding protein expression. However, there are multiple examples where protein expression or function is not controlled by mRNA expression. Indeed, in intact, non-diseased liver tissue, approximately 25% of the changes in the mRNA transcript expression are not accompanied by changes in protein expression (Anderson and Seilhamer, 1997). Studies comparing mRNA and protein expression are rare in all organs, including the liver. Anderson and colleagues showed a poor correlation of the liver tissue abundance of 19 proteins and corresponding mRNA transcripts (correlation

coefficient of only 0.48) (Anderson and Seilhamer, 1997). Additionally, they isolated 50 abundant mRNA transcripts, of which 29 encoded secreted proteins (Anderson and Seilhamer, 1997). However, this result contrasted with the 50 most abundant proteins they isolated, as none were secreted (Anderson and Seilhamer, 1997). Overall, within mRNA transcriptomes, compared to the corresponding proteomes there is a bias in expression toward an over-representation of mRNA transcripts encoding secreted proteins, and high abundance of mRNA transcripts such as G3PDH has been repeatedly demonstrated (Anderson and Seilhamer, 1997; Jansen and Gerstein, 2000; Miklos and Maleszka, 2001a, 2001b; Ter Kuile and Westerhoff, 2001). However, it is salient to remember that protein expression in every cell is controlled by the transcriptome, although the relationship between individual gene transcripts and the corresponding protein expression may not at first be apparent.

DEVELOPMENT OF LIVER FIBROSIS

Liver fibrosis is characterized by activation of the HSC. Invariably, there is associated inflammation and an intrahepatic immune response driving the perpetuation of the pro-fibrogenic phenotype of the HSC. Therefore, fibrosis development can be examined by focusing on the pro-fibrogenic HSC, or alternately by studying the various types of liver injury, associated inflammatory response, and the pathways of HSC activation.

Hepatic Stellate Cells

The hepatic stellate cell (HSC) is the principal cell type mediating the matrix remodeling and degradation of fibrosis (Figure 78.3). The HSC normally constitutes approximately 5% of the intrahepatic cell numbers in the non-diseased liver, and increases to 10–15% in the diseased liver (Friedman, 2000; Mehal et al., 2001). In contrast to the endodermal origin of hepatic parenchymal cells, the HSC is likely to have evolved from a neural crest origin, as suggested by the presence of markers such as glial fibrillary acidic protein, nestin, and N-CAM (Levy et al., 1999; Nakatani et al., 1996; Niki et al., 1999; Vogel et al., 2000). The use of gene array analysis has led to the identification of a number of additional neural markers in cirrhosis, including brain-derived neurotrophic factor (BDNF), glial-cell-line-derived neurotrophic factor (GDNF), and neuromodulin, which are now thought to be previously unrecognized markers of HSCs (Shackel et al., 2002). However, heterogeneity in the intrahepatic HSC population suggests that the intrahepatic stellate cells may not derive from a single embryonic source (Levy et al., 1999; Nakatani et al., 1996; Sell, 2001). The normal stellate cell exists in a quiescent state, and with injury it is transformed into a proliferative, fibrogenic, and contractile myofibroblast, a response known as stellate cell activation (Friedman, 2000). The activation of the HSC involves both initiation and perpetuation of the activated cell phenotype. Additionally, the activated phenotype can resolve

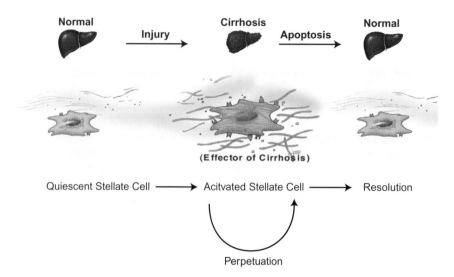

Normal   Injury →   Cirrhosis   Apoptosis →   Normal

(Effector of Cirrhosis)

Quiescent Stellate Cell → Acitvated Stellate Cell → Resolution

Perpetuation

Figure 78.3 Hepatic stellate cell activation. Intrahepatic injury is accompanied by activation of the hepatic stellate cell. This transformation sees the quiescent stellate cell, which stores retinoid, transformed into a profibrogenic cell that actively remodels the extracellular matrix (ECM), leading to cirrhosis. Activation of the stellate cell leads to a state in which the activated cell phenotype is perpetuated by both the initial injurious insult and also by autocrine cell activation. Resolution of fibrosis is due to removal of the activated stellate cell through the process of apoptosis.

in the liver with cessation of injury. It is now clear that distinct molecular pathways are involved in the processes of initiation, perpetuation, and resolution of the activated HSC (Friedman, 2000).

Overall, fibrosis is characterized by a three- to six-fold increase in collagen and non-collagenous components of the extracellular matrix (ECM) (Gressner, 1998). Activated stellate cells produce hyaluronan, fibronectin, entactin, tenescin, undulin, elastin, and laminin (Gressner, 1998). The quiescent stellate cell produces predominantly type IV collagen, whereas the activated myofibroblast produces predominantly collagen type I, as well as collagen type IV and type III (Friedman, 2000; Gressner, 1998). Additionally, myofibroblasts produce proteoglycan, with chondroitin sulfate being the predominant proteoglycan, as well as LTBP (latent transforming growth factor beta binding protein) sulfate and heparin sulfate (Gressner, 1998).

The phenotypic changes associated with HSC activation include matrix remodeling, proliferation, and contractility. Upon activation, the morphology of the HSC changes, with a loss of intracellular vitamin A stores, development of more pronounced cytoplasmic processes, and flattening of the cell (Friedman, 2000; Gressner, 1998). This is accompanied by increased HSC contractility, especially due to the production of NO and endothelin-1, which is an important determinant of the increased portal venous pressure during liver injury (Rockey, 1997). Proliferation of the activated HSC occurs in response to a number of growth factors, most of which signal through receptor tyrosine kinases (Ankoma Sey et al., 1998; Friedman, 2000). Growth factors such as PDGF, TGF-β, and EGF are initially

secreted from adjacent Kupffer cells, hepatocytes, and cells of the inflammatory infiltrate mitogen (Callahan et al., 1985; Marra et al., 1994; Pinzani et al., 1996). Subsequently, an autocrine loop that produces these **growth factors is established by the activated HSC, especially PDGF and TGF-β (Friedman, 2000; Marra et al., 1994; Pinzani et al., 1996).

The dynamic nature of the fibrotic response is apparent given the documented resolution seen in many animal models of fibrosis, and the improvement seen with treatment of some cases of human liver disease. The resolution of the activated HSC is unlikely to involve "retrodifferentiation," or transition back to the quiescent phenotype, but appears to involve elimination by apoptosis of activated HSCs (Benyon and Iredale, 2000). In animal models of fibrosis, there is a significant increase in the rate of activated HSC apoptosis. Activated HSCs are more sensitive to Fas-ligand-induced apoptosis compared to quiescent HSCs. In response to liver injury, intrahepatic natural killer (NK) cells up-regulate Fas-ligand (FasL) expression (Benyon and Iredale, 2000; Moroda et al., 1997; Tsutsui et al., 1996). Persistence of the activated HSCs despite increased Fas-ligand and Fas-ligand sensitivity seems to be due to a concurrent increase in proliferative factors, especially PDGF, that are anti-apoptotic (Benyon and Iredale, 2000; Simakajornboon et al., 2001; Staiger and Loffler, 1998). Similarly, other autocrine factors may act to enhance activated HSC apoptosis, including expression of IL-10 and members of the Bcl-2 family (Galle, 1997; Kanzler and Galle, 2000; Ockner, 2001). This pro- and anti-apoptotic response seen with HSC activation may be a means of autocrine regulation limiting "scar" formation.

Genomic Studies of Hepatic Stellate Cells

Genomic studies of HSCs have led to the development of *ex vivo* cell culture systems and a better understanding of *in vivo* injury attributable to HSCs. Microarray analysis has been utilized to identify differential gene expression in activated HSCs from *in vitro* culture models (Lee et al., 2004; Liu et al., 2004). In one study, a number of novel and previously recognized genes were identified, including mitogen-activated protein kinase (MAPK) pathway genes, osteopontin and *ERK-1* (Lee et al., 2004; Qiang et al., 2006). *ERK-1* was subsequently shown in RNA interference experiments to be necessary for HSC proliferation, as well as being anti-apoptotic, with expression maintaining the activated cell phenotype (Qiang et al., 2006). Differentially expressed genes in murine activated HSCs included those involved in protein synthesis (*RP16*), cell cycle regulation (*Cdc7*), apoptosis (*Nip3*), and DNA damage response (*MAT1*) (Liu et al., 2004). Further, genomic studies addressing the perpetuation of the activated HSC phenotype identified expression of the telomerase catalytic subunit (human telomerase reverse transcriptase, *hTERT*) in human activated HSCs, which immortalizes these cells and maintains an activated HSC phenotype (Schnabl et al., 2002). Senescent HSCs expressed reduced levels of extracellular matrix proteins, including collagens, tenascin, and fibronectin. Furthermore, maintenance of telomere length represents an important survival factor for activated human HSCs (Schnabl et al., 2003). Therefore, an immortalized human HSC line has been generated by infecting primary human HSCs with a retrovirus overexpressing hTERT (Schnabl et al., 2002). The hTERT-positive HSCs do not undergo oncogenic transformation, and exhibit morphologic and functional characteristics of *in vivo* activated HSCs. Subsequent GeneChip™ and reverse transcription polymerase chain reaction (RT-PCR) analysis showed that mRNA expression patterns in telomerase-positive HSCs are similar to those in primary *in vivo* activated human HSCs (Schnabl et al., 2002). Similarly, another two immortalized HSC lines, LX-1 and LX-2, were characterized by microarray analysis and determined to have a gene expression profile similar to that of *in vivo* activated human primary HSCs (Xu et al., 2005). These newly developed cell lines are proving to be valuable tools to study the biology of human HSCs. Importantly, although microarray and GeneChip™ studies have demonstrated that these *ex vivo* HSCs are similar to *in vivo* activated HSCs, with a greater than 70% similarity in transcript expression, a number of differences have been identified. Therefore, *ex vivo* cell culture of stable HSCs may not be completely representative of *in vivo* injury, highlighting the importance of intracellular interactions within the liver in determining transcriptome expression.

TRANSCRIPTOME ANALYSIS OF LIVER DISEASE

Presently there are hundreds of published gene array studies of human liver disease or studies that utilize human liver tissue. Most of these studies attempt to understand the liver by examining mRNA transcript expression. There are few publications in human liver disease where gene expression is correlated with clinical outcome. Two of the most common causes of liver injury globally, viral hepatitis B and C, are discussed in more detail in Chapter 99. Unfortunately, many forms of liver injury cannot be studied in commonly used laboratory animal models, and most animal studies focus on the pathways of fibrosis development.

Hepatitis B Virus Infection

Functional genomics studies of acute hepatitis B virus (HBV) infection in chimpanzees has led to unique insights into how the virus evades immune response and causes injury (Wieland et al., 2004). Initially, there is no apparent significant immune/inflammation-associated differential gene expression during the early phase of HBV infection and viral replication. Therefore, HBV infection acts in the initial phase as a "stealth virus," failing to induce a significant innate immune response (Wieland et al., 2004; Wieland and Chisari, 2005). Intrahepatic gene induction is first seen during the phase of attempted viral clearance. Gene expression during the early phase of infection is associated with T cell receptor and antigen presentation. Following this, T cell effector function (granzymes), T cell recruitment (chemokines), and monocyte activation-associated gene expression has been demonstrated. A later phase of clearance is associated with the expression of B cell-related genes. Chronic HBV infection is characterized by gene expression profiles consistent with active inflammatory response (increased IRF-6 and CCL16), cell proliferation (increased cyclin-H and p53) and cellular apoptosis (14-3-3 interacting gene increased) (Honda et al., 2001, 2006).

Hepatitis C Virus Infection

Acute and chronic hepatitis C virus (HCV) infections have been studied using functional genomics techniques. These experiments have examined acute HCV infection in primates, as well as the sequelae of chronic infection. The results from microarray studies of acute HCV infection in the chimpanzee are intriguing. Acute HCV infection is characterized by a rapid (within two weeks) as well as a delayed (up to six weeks) induction of genes that are involved in the innate immune response (Bigger et al., 2001). Most of these genes are associated with interferon gene expression, and are known as interferon response genes (IRGs, including *ISG15*, *ISG16*, *CXCL9*, *CXCL10*, *Mx-1*, *stat-1*, 2'5'-oligoadenylate synthetase, and *p27*). Viral clearance appears to be associated with rapid induction of these IRGs. Overall, HCV persistence appears to be associated with less induction of IRGs compared to viral clearance. Further, chronic HCV-related liver injury appears to be characterized by an IRG-associated Th1 immune response, which is insufficient to clear the virus but is chronic and responsible for ongoing liver injury. The situation with interferon treatment of HCV-infected individuals, which is aimed at viral eradication, is similar to acute infection, as an immune response characterized by a significant increase in IRG expression following treatment is associated with greater likelihood of a sustained long-term therapy response. The recent identification of SNPs in the *IL-28B* gene that predict the likelihood of response to HCV treatment with interferon and spontaneous clearance of the virus has highlighted the importance of the IRGs in determining HCV outcomes. Clearly, the immune response to HCV drives fibrogenesis, as interferon administration is associated with a reduction in intrahepatic inflammation and fibrosis even in the absence of long-term virological clearance following treatment. The newly identified *IL-28B* polymorphisms are the first specific example of a genomic medicine approach that will lead to individualized treatment approaches in liver disease.

Steatosis is a hallmark of genotype 3 HCV liver injury. Studies directly comparing genotype-specific injuries are rare, although genotype differences have been studied in the setting of definite hepatic steatosis and non-alcoholic steatohepatitis (Younossi et al., 2009). Gene expression profiling has identified genotype 3-specific genes *SOCS1* and *IFITM1*, and genotype 1-specific genes *CCL3*, *CCL4*, *IFNAR*, and *PRKRIR* (Younossi et al., 2009). Importantly, fibrotic injury did not have a genotype-specific signature (Younossi et al., 2009). Unique HCV-associated fibrosis signatures have been identified and have led to the correlation of the corresponding proteins of three of these genes; inter-α inhibitor H1, serpin peptidase inhibitor clade F member 2, and transthyretin being correlated with stages of fibrosis development (Caillot et al., 2009). Clearly, such proteins may well be useful biomarkers of HCV fibrosis development in the future.

Alcoholic Liver Disease

Globally, alcohol is a leading cause of progressive liver fibrosis leading to cirrhosis. Intrahepatic gene profiling using microarrays in ethanol-fed baboons has identified increased expression of multiple annexin-related genes (including A1 and A2) that were not previously implicated in the progression of fibrosis in alcoholic liver disease (ALD) (Seth et al., 2003). Further, the intrahepatic transcriptome profile in alcohol-associated liver injury is significantly different from other forms of liver disease (Deaciuc et al., 2004; Seth et al., 2003). Transcriptome profiling has allowed differentiation of alcoholic hepatitis from alcoholic steatosis. Genes known to be involved in alcohol injury, such as alcohol dehydrogenases, acetaldehyde dehydrogenases, interleukin-8, S-adenosyl methionine synthetase, phosphatidylethanolamine N-transferase, and several solute carriers, have been shown to be differentially expressed in alcoholic hepatitis versus alcoholic steatosis. In alcoholic hepatitis, many novel, differentially expressed genes have been identified, including claudins, osteopontin, *CD209*, selenoprotein, annexin A2, and genes related to bile duct proliferation (Seth et al., 2003). Differentially expressed genes involved cell adhesion, extracellular matrix proteins, oxidative stress, and coagulation common to alcoholic hepatitis and end-stage ALD. Importantly, genes associated with fibrosis, cell adhesion, and ECM remodeling are increased in human advanced ALD, consistent with the fibrotic nature of ALD. However, many of these genes are not specific to alcohol-induced liver injury, and have been reported in other forms of liver cirrhosis such as primary biliary cirrhosis (Shackel et al., 2001, 2002).

Non-Alcoholic Fatty Liver Disease

Non-alcoholic fatty liver disease (NAFLD) is the "Western" liver disease related to obesity, and forms part of the metabolic syndrome characterized by increased body mass index (BMI), hypertension and insulin resistance. Non-alcoholic steatohepatitis (NASH) is the clinicopathological syndrome in NAFLD, in which lipid deposition within the liver is accompanied by inflammation. It is being widely studied using gene array analysis of mRNA transcript expression. The inflammation in NAFLD that results in NASH is of considerable importance, as NASH (not NAFLD) is characterized by progressive injury and eventual cirrhosis. Studies have identified differentially expressed genes in end-stage NASH cirrhosis compared to other disease states (Sreekumar et al., 2003; Younossi et al., 2005a, 2005b). Decreased expression of genes associated with mitochondrial function and increased expression of genes associated with the acute phase response are observed (Sreekumar et al., 2003). The latter increases were speculated to be associated with insulin resistance, a feature of NAFLD (Sreekumar et al., 2003). Further differential expression of genes involved in lipid metabolism, ECM remodeling, regeneration, apoptosis, and detoxification have all been observed in NASH following microarray analysis (Younossi et al., 2005a).

Autoimmune Hepatitis

Autoimmune hepatitis (AIH) is an uncommon autoimmune disease affecting the hepatic lobule, with a number of morphological similarities to the injury caused by HBV and HCV. The only available data on AIH are a comparison between HCV- and AIH-associated cirrhosis (Shackel et al., 2002). One of the key findings in that study was the observation that three inhibitors of apoptosis (IAP) genes were selectively differentially expressed in AIH. This is an intriguing finding. If this gene expression was identified to be in the intrahepatic lymphocyte population, then lack of apoptosis of such cells may be an important pathogenic pathway in the perpetuation of AIH. In this comparison to HCV, AIH was associated with an inflammatory gene pathway that consisted of a mix of Th1- and Th2-associated genes. Therefore, an intrahepatic Th1 immune response appears to be more fibrogenic, as is evident from this work, HCV liver injury, and animal models of liver injury (Shi et al., 1997).

Biliary Liver Injury

Viral hepatitis and alcoholic liver injury typically affect the hepatic lobule within the liver, and are referred to classically as "hepatitic" or "lobular" liver injury. However, a number of other types of liver injury are characterized by insults primarily to the bile ducts, and are commonly referred to as "biliary" diseases. Classically, biliary diseases include autoimmune-mediated bile duct injury of primary biliary cirrhosis (PBC) and the poorly understood condition of bile ducts known as primary sclerosing cholangitis (PSC), in which inflammation, infection, and fibrosis initially affect the bile ducts and then the liver parenchyma.

One of the major findings in a gene array examination of PBC end-stage liver disease was the identification of a subset of genes associated with the Wnt pathway (Shackel et al., 2001). In particular, *Wnt13*, *Wnt5A*, and *Wnt12* were shown to be differentially expressed. Other genes particularly up-regulated in PBC included transcription initiation factor 250 kDa subunit (TAFII-250), PAX3/forkhead transcription factor, and patched homolog (PTC). A consistent feature of the gene array analysis of PBC was the repeated identification of *Drosophila* gene homologs that were differentially expressed (Wnt genes, hedgehog pathway, notch pathway) (Shackel et al., 2001). Further studies have compared the gene expression signature of ursodeoxycholic acid-treated PBC with non-treated PBC (Chen et al., 2008). In this study, a 221-gene expression profile identified a striking down-regulation of genes involved in protein biosynthetic pathways in the livers of individuals treated with ursodeoxycholic acid (Chen et al., 2008).

The only available data on PSC come in a comparison with PBC (Shackel et al., 2001). Compared with PBC, a far greater number of genes showed differential expression in PSC versus non-diseased liver (compared with PBC and non-diseased liver). This included genes associated with epithelial biology (amphiregulin, bullous pemphigoid antigen), inflammation (T cell secreted protein I-309, CTLA4), apoptosis-related genes (Bcl-2 interacting killer, Bcl-x, death-associated protein 3), and intracellular kinases such as CDK7 and JAK1.

Biliary atresia (BA), a disease in which bile ducts are absent, has been studied, comparing gene expression in embryonic versus perinatal forms of the disease (Zhang et al., 2004). Gene profiling clearly separated these two conditions. The most remarkable difference was in the expression of so-called regulatory genes. In embryonic BA, 45% of differentially expressed genes were in this category, versus 15% in the perinatal form. These genes included imprinting genes and genes associated with RNA processing and cell cycle regulation that were not present in the perinatal form of BA. Overall, the results from genomics studies of biliary liver disease are consistent with distinct patterns of injury, each associated with a unique gene signature. A further study in BA has now identified a 150-gene molecular signature distinguishing an inflammatory from a fibrotic BA subtype (Moyer et al., 2010). Importantly, these subtypes of BA could not be distinguished in any other way, and the fibrosis subtype had a decreased transplant-free survival (Moyer et al., 2010). Although not validated in a different cohort, this has direct implications in BA clinical management.

The first clinically adopted genomic medicine approach in liver disease arose from the development of a gene array resequencing chip capable of diagnosing a number of rare, inherited inborn errors of metabolism resulting in neonatal hyperbilirubinemia (Liu et al., 2007; Watchko et al., 2002).

Hepatocellular Carcinoma

Liver injury characterized by cirrhosis is a premalignant condition. Despite the many diverse causes of liver injury, the development of HCC invariably occurs in the setting of cirrhosis. Indeed, the development of HCC without cirrhosis is extremely rare. Functional genomics methodologies are advancing understanding of how fibrosis progression to cirrhosis is required for malignancy development. However, this sequence of events is still poorly understood.

Neoplastic proliferation in HCV-associated HCCs has been studied by microarray analysis. A plethora of potential novel tumor markers have been identified. These include the serine/threonine kinase 15 (STK15) and phospholipase A2 (PLA2G13 and PLA2G7), which were found to be increased in more than half of the tumors identified (Smith et al., 2003). However, different studies identify different gene groups in HCV-associated HCC, such as cytoplasmic dynein light chain, hepatoma-derived growth factor, ribosomal protein L6, TR3 orphan receptor, and c-myc (Shirota et al., 2001). The clustering analysis in this study showed that the expression of 22 genes in HCC related to differentiation of the malignancy, with more than half of the genes being transcription factors or related to cell development or differentiation (Shirota et al., 2001). Although many of these genes can be implicated in HCC development, they are often identified in gene sets obtained from end-stage diseased tissue. Therefore, whether these genes represent cause or effect is unknown. HBV-associated HCC has been studied by several groups (Kim et al., 2004a, b; Iizuka et al., 2002, 2004). Genes associated with cell proliferation, cell cycle, apoptosis, and angiogenesis were dysregulated in HCC tissues. Increased expression of cyclin-dependent kinases was seen, while several cell cycle negative regulators decreased. Metastatic development has also been studied using gene arrays (Pan et al., 2003; Qin and Tang, 2004; Tang et al., 2004; Ye et al., 2003). Genes identified with metastatic development include osteopontin, and *in vivo* neutralizing antibodies to osteopontin block tumor invasion (Ye et al., 2003). A study of unsupervised gene profiling of HCC patients revealed a set of genes associated with decreased survival, including *RhoC* (Wang et al., 2004). These genes include a subset of pro-proliferative, anti-apoptotic genes, as well as genes involved in ubiquitination and histone modification. Gene profiles in HCC have also demonstrated patterns of gene expression associated with tumor differentiation, vascular invasion, and recurrence after surgery (Tang et al., 2004).

Gene Arrays in Animal Models

Animal models studied have included acute liver regeneration, drug toxicity, liver fibrosis, fatty liver, biliary obstruction, liver transplantation, and carcinogenesis. Drug toxicity studies are numerous and include effects induced by clofibrate, PPAR alpha agonists, carbon tetrachloride, amiodarone, arsenic, and methotrexate (Bulera et al., 2001; Cunningham et al., 2000; Huang et al., 2004; Jiang et al., 2004; Jolly et al., 2005; Jung et al., 2004; Minami et al., 2005; Shankar et al., 2003; Ulrich et al., 2004; Waring et al., 2001a, b). In one study, a novel complementary DNA (cDNA) library highly enriched for genes expressed under a variety of hepatotoxic conditions was created, and used to develop a custom oligonucleotide library (Waring et al., 2003).

An expression signature for rat liver fibrosis was identified using a 14,814 cDNA gene microarray (Utsunomiya et al., 2004). The "genetic fibrosis index" identified consisted of 95 genes (87 up-regulated, 8 down-regulated). They included genes associated with cytoskeletal proteins, cell proliferation, and protein synthesis. Bile obstruction in the mouse identified three sequential main biological processes. At day 1, enzymes involved in steroid metabolism were over-expressed. This was followed by an increase in cell cycle/proliferation-associated genes at day 7, occurring at a time of maximum cholangiocyte proliferation. During days 14–21, genes associated with the inflammatory response and matrix remodeling were identified. Similar temporal gene expression was identified in the model of acute liver regeneration. Steroid and lipid metabolism genes were down-regulated as early as 2 hours post hepatectomy, while genes associated with cytoskeletal assembly and DNA synthesis became up-regulated by 16 hours, and remained elevated at the 40 hour time point at the peak of S phase.

Alcoholic liver disease has been studied in the mouse chronic enterogastric ethanol infusion model (Deaciuc et al., 2004). A total of 12,422 genes were analyzed, and several cytochrome P450 genes were shown to be up-regulated, while several genes involved in fatty acid metabolism and fatty acid synthase were down-regulated. In contrast, genes associated with glutathione-s-transferase were markedly up-regulated. Interestingly, a novel intestinal factor was 50-fold down-regulated. Therefore, it is postulated that chronic alcohol ingestion may affect healthy intestinal epithelium and down-regulate this gene, causing intestinal permeability changes.

PROTEOMIC STUDIES OF LIVER DISEASE

Proteomic studies of liver disease have fallen into the following four groups: (1) discovery of previously unrecognized proteins in a cell population or disease state (Eng and Friedman, 2000; Kawada et al., 2001; Kristensen et al., 2000), (2) biomarker discovery (Schwegler et al., 2005; Seow et al., 2001), (3) hepatic toxicological prediction/profiling, (Fella et al., 2005; Fountoulakis and Suter, 2002; Gao et al., 2004; Guzey and Spigset, 2002; Kaplowitz, 2004; Low et al., 2004; Meneses-Lorente et al., 2004; Merrick and Bruno, 2004; Nordvarg et al., 2004; Roelofsen et al., 2004), and (4) studies of

known proteins or classes of proteins (Greenbaum et al., 2002; Joyce et al., 2004) (see Tables 78.1 and 78.2). Proteomics has successfully been used in biodiscovery of proteins in hepatocytes, HSCs (Eng and Friedman, 2000; Kawada et al., 2001; Kristensen et al., 2000), HCC (Seow et al., 2001), and viral hepatitis (Comunale et al., 2004; Garry and Dash, 2003; Kim et al., 2003; Lu et al., 2004; Rosenberg, 2001; Scholle et al., 2004). In these studies, tens to hundreds of proteins were identified. However, the studies are limited, as they sample rather than profile the proteome. Further, observed changes in protein expression may reflect weak associations rather than a direct role in the development of pathobiology. The biodiscovery approach has been used successfully in toxicological models, and has been used to develop characteristic profiles of protein expression that may predict intrahepatic toxicological responses (Fella et al., 2005; Fountoulakis and Suter, 2002; Gao et al., 2004; Guzey and Spigset, 2002; Kaplowitz, 2004; Low et al., 2004; Meneses-Lorente et al., 2004; Merrick and Bruno, 2004; Nordvarg et al., 2004; Roelofsen et al., 2004). This is an area of intense research focus for pharmaceutical companies as they strive to reduce development costs and aim to predict drug toxicity earlier in the drug development cycle. Biomarker discovery is another area receiving attention using proteomic methods. However, for years there has not been a new US Food and Drug Administration (FDA)-approved serum marker, as this is an immense research challenge. Biological sample protein concentrations vary by 12–15 orders of magnitude, and specific serum markers are likely to be expressed at nanomolar or lower concentrations. One approach to overcome these limitations is to use a combination of potential markers that are easier to detect, but with each protein marker alone having a lower specificity but high sensitivity. This is an approach currently used in serum tests of hepatic fibrosis, and proteomic methods are being used to try to identify new serum markers of hepatic fibrogenesis (Henkel et al., 2005; Poon et al., 2005; Xu et al., 2004). One of the most promising approaches is the use of accurate mass tags (AMT) or suicide substrates that selectively and reproducibly target a sub-proteome (Bogdanov and Smith, 2005; Greenbaum et al., 2002; Joyce et al., 2004; Pasa-Tolic et al., 2004). This has the added advantage of aiding pre-fractionation and increasing resolution of proteins, as the tag can be captured on an affinity surface (Greenbaum et al., 2002; Joyce et al., 2004).

Hepatitis B Virus Infection

In contrast to gene array experiments, a number of studies have used proteomics on sera to examine different stages of chronic HBV infection. In one study, altered proteomic profiles were identified for haptoglobin beta and alpha 2 chain, apolipoprotein A-1 and A-1V, alpha-1 antitrypsin, transthyretin, and DNA topoisomerase 11 beta (He et al., 2003). Some of these proteins are among the most abundant serum proteins secreted by the liver and are generally associated with acute-phase inflammatory responses. What was apparent in this study was that different isoforms of some of the proteins showed distinct changes in HBV infection itself and differed at times between patients with low inflammatory scores versus high inflammatory scores. Examples include a decrease in cleaved haptoglobin beta peptides and apoA-1 fragments in patients with higher inflammatory scores. In comparison, some alpha-1 antitrypsin fragments were increased in patients with higher inflammatory scores. An alternate approach studied serum protein profiles and correlated this with disease severity using a surface-enhanced laser desorption/ionization (SELDI) protein chip analysis and artificial neural network models (Poon et al., 2005). The researchers found six fragments with a positive prediction and 24 with a negative prediction of fibrosis stage, and subsequently developed a fibrosis index with excellent, precise values for significant fibrosis and cirrhosis based on the Ishak fibrosis score. The Ishak fibrosis score is a widely accepted scoring system for assessment of fibrosis and necroinflammation. It takes into account the degree of interface hepatitis, lobular necrosis, portal infiltrate and confluent necrosis to determine the grade of injury. The inclusion of clinical biochemical parameters such as ALT (alanine transaminase), bilirubin, total protein, hemoglobin, and international normalized ratio (INR) strengthened the accuracy of their predictive model.

Hepatitis C Virus Infection

Proteomic methodologies have been applied to a number of aspects of HCV-related liver injury. They include the study of HCV-related HCC development in which overexpression of alpha enolase was identified and correlated with poorly differentiated HCC (Kuramitsu and Nakamura, 2005; Takashima et al., 2005). The response of hepatocyte cell lines to interferon (IFN) gamma treatment has uncovered more than 54 IFN response genes, including many novel targets, an approach that may pave the way for novel therapies. Examination of protein extracts that bind to the HCV IRES (internal ribosome entry site) has identified a number of novel protein targets such as Ewing Sarcoma breakpoint 1 region protein (EWS) and TRAF-3. The final aspect of HCV liver injury receiving attention is the study of potential biomarkers such as heat shock protein HSP-70, associated with HCV infection progression to HCC (Takashima et al., 2003).

AIH-, PBC-, and PSC-Associated Liver Disease

Few published proteomic studies address the pathophysiology of these diseases. Cholangiocarcinoma associated with PSC has been studied using proteomics techniques (Koopmann et al., 2004). Using tandem mass spectroscopy, Koopman and colleagues identified Mac-2-binding protein (Mac-2BP) as a diagnostic marker in biliary carcinoma. The diagnostic accuracy of serum Mac-2BP expression in biliary carcinoma was superior to the established marker CA19-9. This study highlights the progression of proteomic research in liver disease – a focus initially on malignancy and biomarker discovery followed by studies of pathophysiology.

Alcoholic Liver Disease and Non-Alcoholic Fatty Liver Disease

As with gene array studies, proteomic studies of alcoholism have been an eclectic mix of research, examining HCC development associated with alcohol, hepatocyte alcohol-related biology, and neural aspects of alcohol addiction. Studies of intrahepatic toxic effects using proteomics have helped outline toxicology profiles that can be used for screening as well as trying to understand alcohol-associated liver injury. Mitochondrial ethanol hepatotoxicity is thought to involve modification of protein thiol redox state. Using two-dimensional gel proteomic studies, Venkatraman and colleagues were able to demonstrate a decrease in the reduced thiols on aldehyde dehydrogenase and glucose-regulated protein 78 (Venkatraman et al., 2004). The change in aldehyde dehydrogenase-reduced thiols was accompanied by a reduction in specific activity of the enzyme. The term "alcoholomics" has been coined to refer to the study of those proteins (i.e., the subproteome) that are directly or indirectly affected by alcohol.

There has been only a single study to date evaluating proteomic profiling in NAFLD (Younossi et al., 2005a). The study used SELDI-TOF MS (SELDI time-of-flight mass spectrometry) to profile serum samples from 91 patients with NAFLD and seven obese controls. Twelve unique protein peaks were identified that associated with NAFLD (four associated with steatosis, four with steatosis with non-specific inflammation, and four with NASH). Unfortunately, although peak mass is shown, SELDI-TOF MS lacks the accuracy required to give mass determination enabling equivocal protein identification.

Hepatocellular Carcinoma

To date, biomarker profiling has predominantly focused on studies of malignant tissue (Liu et al., 2005; Wiesner, 2004; Wong et al., 2004). In one study, HCC development in chronic HBV infection was characterized by a significant decrease in a fragment of complement-3 and an isoform of apolipoprotein A1 (Steel et al., 2003). In tissue studies, expression of variants of aldehyde dehydrogenase and tissue ferritin light chain has been identified in HCC, but not in surrounding tissues. Similar studies have also identified fructose-bisphosphatase, argininosuccinate synthetase, and cathepsin B pre-protein as being down-regulated in HCC tissues. In an extensive study using laser capture microscopy, Li and colleagues identified 261 proteins differentially expressed between HCC and non-HCC hepatocytes (Xu et al., 2004). Kinases in the Eph family were identified, along with the Ras-like family of Rho proteins. In addition, a DEAD box polypeptide was down-regulated, while three members of the spliceosome and heterogeneous nuclear ribonucleoprotein K were up-regulated. Also, SELDI-TOF MS was used to examine the sera of 82 patients with cirrhosis (38 without and 44 with HCC). An algorithm including the six highest scoring peaks allows the prediction of HCC in more than 90% of cases (Paradis et al., 2005). The highest discriminating peak was a C-terminal peptide of vitronectin.

"Next-Generation" Liver Function Tests?

Several studies have evaluated proteomic analysis of serum protein as a diagnostic test to assess the severity of liver disease, and in particular for non-invasive assessment of liver fibrosis. These studies are in their infancy, with basic methodologies still unresolved. However, early studies show that the technique has promise. In a pilot study of 46 patients with chronic hepatitis B, an artificial neural network (ANN) model was derived from the proteomic fingerprint and used to derive a fibrosis index (Poon et al., 2005). The ANN fibrosis index strongly correlated with Ishak scores and stages of fibrosis. The areas under the receiver operating characteristic (ROC) curve for significant fibrosis (Ishak score >2) and cirrhosis (Ishak score >4) were both >0.90. Inclusion of the international normalized ratio (INR), total protein, bilirubin, alanine aminotransferase, and hemoglobin in the ANN model improved its predictive power, yielding accuracies >90% for prediction of fibrosis and cirrhosis. Another study found that pretreatment of serum proteins to remove N-glycosylation enhanced the resolution of serum polypeptide profiles (Comunale et al., 2004). This technique has the potential to improve diagnostic serum proteomics.

Chen and colleagues developed a method of glycoproteomic analysis to discover serum markers that can assist in the early detection of HBV-induced liver cancer (Chen et al., 2004). The authors showed that woodchucks diagnosed with HCC have dramatically higher levels of serum-associated core alpha-1,6-linked fucose. One glycoprotein, Golgi protein 73 (GP73), was found to be elevated and hyper-fucosylated in the serum of animals and humans with a diagnosis of HCC. Serum profiling was used to distinguish HCC from earlier stages of HCV-related liver disease (Schwegler et al., 2005). The proteomic model distinguished chronic HCV from HCV-HCC with moderate sensitivity and specificity. Inclusion of known serum markers alpha fetoprotein and des-gamma carboxyprothrombin, along with GP73, significantly improved the diagnostic accuracy of HCC detection.

Proteomics in Other Liver Disease

The metalloproteome is defined as the set of proteins that have metal-binding capacity by being metalloproteins or having metal-binding sites. The Cu and Zn metalloproteomes were defined in human hepatoma lines (She et al., 2003). Although the gene for Wilson disease has been identified, the mechanisms by which excess copper leads to oxidative stress, acute liver failure, or cirrhosis are not fully understood. Using an *in vitro* model of copper loading, novel copper-binding proteins were isolated using proteomic techniques (Roelofsen et al., 2004).

Although there has been limited proteomic analysis of liver tissue in models of disease, liver-associated pathobiological processes have been examined. In one study of the liver aging process, 85 differentially expressed proteins comprising anti-oxidation, glucose/amino acid metabolism, signal transduction, and cell cycle systems were identified using proteomics tools (Cho et al., 2003). In aging, the anti-oxidation system showed a large increase in glutathione peroxidase and a decrease in glutathione-s-transferase. Similarly, levels of t-glycolytic enzymes were decreased in the aging animal. Furthermore, levels of proteins associated with signal transduction/apoptosis, for example cathepsin B, were decreased in the aging process. However, it is unclear whether the identified genes explain the increased rate of fibrosis progression seen with increasing age.

Proteomics has also been used to identify genes associated with LPS-induced liver injury. Proteins such as TRAIL receptor 2 were down-regulated in the liver of LPS-treated mice, while TNFAIP1 was significantly up-regulated. Four different proteins were novel in the fatty liver proteome (aconitase, succinate dehydrogenase, propanol Co-A carboxylase alpha chain, and 3-hydroxyanthrilate 3-4-dioxygenase). LPS is thought to be an important mediator of injury in a number of conditions, such as alcohol-related liver injury.

GENETIC MARKERS AND PHARMACOGENOMICS OF LIVER FIBROSIS

Genetic markers of liver fibrosis are typically studied in a disease-specific context, and studies are frequently designed to detect susceptibility to disease progression to cirrhosis rather than the earlier stages of progressive fibrosis. The greatest advance in understanding the genetics of fibrotic injury has been the identification of susceptibility loci to the various causes of liver diseases (Karlsen et al., 2010a; Krawczyk et al., 2010). Classical pharmacogenomics studies in liver fibrosis and cirrhosis development are rare due to the absence of generalized anti-fibrotic treatments.

Viral Hepatitis

The strongest association of a polymorphism with a treatment outcome in managing human disease has been the identification of polymorphisms in the *IL-28B* gene encoding IFN-λ3. This is discussed in detail in Chapter 99, as it is strongly associated with interferon treatment outcomes as well as spontaneous clearance of hepatitis (Ge et al., 2009; Suppiah et al., 2009; Thomas et al., 2009). It is not established whether *IL-28B* polymorphisms confer an increased risk of fibrosis progression in individuals infected with HCV.

In the context of HCV-associated liver injury, a seven-gene expression profile was used to develop a cirrhosis risk score (Marcolongo et al., 2009). The genes *AZIN1*, *TLR4*, *TRPM5*, and *AQP2* had seven SNPs that were predictive of both fibrosis progression to even mild-stage fibrosis and cirrhosis (Marcolongo et al., 2009). Unfortunately, the seven-SNP signature was not validated in an independent cohort, and it was not described how common the signature is in a normal population. However, it is clear that host genetics do define the risk of fibrosis progression, at least in HCV-associated cirrhosis.

PBC- and PSC-Associated Liver Disease

In PBC, genome-wide association studies (GWAS) in a European population have identified IL12-related pathways, *SPIB*, *IRF5-TNPO3*, *STAT4*, *DENND1B*, *CD80*, *IL7R*, *CXCR5*, *TNFRSF1A*, *CLEC16A*, *NFKB1*, and 17q12-21 locus genes *GSDMB*, *ZPBP2*, and *IKZF3* associated with susceptibility to disease (Mells et al., 2011). However, in an independent cohort from Japan, only 17q12-21 locus genes *GSDMB*, *ZPBP2*, and *IKZF3* were associated with PBC susceptibility (Tanaka et al., 2011). Therefore, while there are common susceptibility loci, there are clearly different susceptibility genes, or possible modifiers, within different ethnic groups.

GWAS have been used to study primary sclerosing cholangitis susceptibility. In a comparison of 285 PSC patients with 298 healthy controls, a number of susceptibility loci were identified (Juran and Lazaridis, 2010; Karlsen et al., 2010b). Candidate genes associated with disease included coupled bile acid receptor 1 and macrophage-stimulating 1 (Karlsen et al., 2010b). Interestingly, the association of PSC with inflammatory bowel disease (IBD) was evident in this GWAS, as established IBD gene susceptibility loci were identified on 2q35 and 3p21 in PSC (Karlsen et al., 2010b).

Non-Alcoholic Fatty Liver Disease and Alcoholic Liver Disease

In NAFLD, a number of susceptibility loci have been identified, but the most consistent finding is the identification of a non-synonymous coding SNP L148M in the patatin-like phospholipase domain-containing protein 3 (*PNPLA3*) (also known as adiponutrin). This *PNPLA3* SNP is strongly associated with both hepatic fat and inflammation (Romeo et al., 2008). In further validation cohorts, SNPs in the *PNPLA3* gene have been associated with biopsy-proven NAFLD, liver fat disposition, inflammation, and fibrosis. Importantly, this was independent of age, sex, diabetes, and alcohol consumption (Romeo et al., 2008; Rotman et al., 2010). A further three SNPs were variably associated with the development of NAFLD (Rotman et al., 2010). Although this *PNPLA3* SNP confers NAFLD susceptibility, it is unclear if screening for it will change treatment regimes for the underlying disease. It could be argued that most current treatments in NAFLD, such as diabetic control and weight loss, would be necessary irrespective of the SNP result. However, selective or earlier therapy targeting those individuals with the unfavorable SNP is likely, and is an example of evolving personalized medicine. The efficacy of this management approach will be tested in the next 5–10 years.

Two aspects of ALD have been examined in GWAS studies: the propensity for alcohol dependence and the development of liver injury. Dependency in confirmed alcoholism has been strongly linked to the expression of the serotonin receptor gene *HTR7* on chromosome 10q23 (Zlojutro et al., 2011). However, in population-based studies, no clear genetic susceptibility has been identified with the risk of alcoholism (Heath et al., 2011). The selected population used to study an association in confirmed alcoholics compared to screening populations partially explains the apparent discrepancy. In isolation, the identified SNPs are insufficient to identify susceptibility, although it remains unclear whether combining the SNPs with simple clinical variables would significantly improve the performance of any potential screening test. Interestingly, alcoholic liver injury is morphologically similar to the injury in NAFLD, and it is therefore not surprising that a *PNPLA3* SNP identified in NAFLD is also associated with liver injury due to alcohol (Tian et al., 2010).

FUTURE IMPACT OF GENOMICS

Although genomics studies have already made significant contributions to our understanding of liver injury, many unanswered questions remain. In particular, the canonical molecular pathways mediating progressive fibrosis progression and the development of cirrhosis are poorly understood. In the next decade, genomics studies will help to predict susceptibility to fibrosis and the likelihood of developing sequelae such as liver failure and HCC. Long-term proteomics promises to develop non-invasive markers of liver injury, and will help to screen individuals for HCC. The goal in the future is to develop newer therapeutic approaches based on our understanding of the molecular events involved in the development of cirrhosis. Also, the identification of genomic susceptibility markers will help to further individualize risk assessment.

CONCLUSIONS

Cirrhosis is the pathogenic consequence of a remarkably conserved response to injury within the liver, characterized by progressive fibrosis tissue deposition and eventual disruption of normal hepatic architecture. The hallmark of liver injury is activation of the hepatic stellate cell, which mediates the development of fibrous tissue and change in the composition of the extracellular matrix. Inflammation and immune responses are the driving forces behind the transition of the quiescent hepatic stellate cell to an activated phenotype. Additionally, liver response to injury includes changes in cellular proliferation and apoptosis that may explain the premalignant potential of cirrhosis. Although remarkably consistent in its development, cirrhosis can arise from many diverse causes, is often difficult to diagnose, and once established is often difficult to treat. Unfortunately, it is clear that better pharmaceutical agents are needed to alter the natural history of fibrosis and subsequent development of cirrhosis. Importantly, the promise of genomics approaches is now being realized, as these technologies are being used to better understand liver disease pathogenesis, and determine response to treatment and susceptibility to progression, as well enabling the development of novel therapies. Consequently, individualized patient assessment and tailored therapy will be possible in liver disease due to genomics approaches.

REFERENCES

Anderson, L., Seilhamer, J., 1997. A comparison of selected mRNA and protein abundances in human liver. Electrophoresis 18 (3–4), 533–537.

Ankoma Sey, V., 1998. Coordinated induction of VEGF receptors in mesenchymal cell types during rat hepatic wound healing. Oncogene 17 (1), 115–121.

Bataller, R., Brenner, D.A., 2005. Liver fibrosis. J Clin Invest 115 (2), 209–218.

Bataller, R., North, K.E., Brenner, D.A., 2003. Genetic polymorphisms and the progression of liver fibrosis: A critical appraisal. Hepatology 37 (3), 493–503.

Benyon, R.C., Iredale, J.P., 2000. Is liver fibrosis reversible? Gut 46 (4), 443–446.

Bigger, C.B., Brasky, K.M., Lanford, R.E., 2001. DNA microarray analysis of chimpanzee liver during acute resolving hepatitis C virus infection. J Virol 75 (15), 7059–7066.

Bogdanov, B., Smith, R.D., 2005. Proteomics by FTICR mass spectrometry: Top down and bottom up. Mass Spectrom Rev 24 (2), 168–200.

Boyer, N., Marcellin, P., 2000. Pathogenesis, diagnosis and management of hepatitis C. J Hepatol 32 (1 Suppl.), 98–112.

Bulera, S.J., 2001. RNA expression in the early characterization of hepatotoxicants in Wistar rats by high-density DNA micro-arrays. Hepatology 33 (5), 1239–1258.

Caillot, F., 2009. Novel serum markers of fibrosis progression for the follow-up of hepatitis C virus-infected patients. Am J Pathol 175 (1), 46–53.

Callahan, M., Cochran, B.H., Stiles, C.D., 1985. The PDGF-inducible "competence genes": Intracellular mediators of the mitogenic response. Ciba Found Symp 116, 87–97.

Cantin, G.T., Yates III, J.R., 2004. Strategies for shotgun identification of post-translational modifications by mass spectrometry. J Chromatogr A 1053 (1–2), 7–14.

Cao, S.X., 2001. Genomic profiling of short- and long-term caloric restriction effects in the liver of aging mice. Proc Natl Acad Sci USA 98 (19), 10,630–10,635.

Chen, J., 2004. Proteome analysis of gastric cancer metastasis by two-dimensional gel electrophoresis and matrix assisted laser desorption/ionization-mass spectrometry for identification of metastasis-related proteins. J Proteome Res 3 (5), 1009–1016.

Chen, L., 2008. Gene expression profiling of early primary biliary cirrhosis: Possible insights into the mechanism of action of ursodeoxycholic acid. Liver Int 28 (7), 997–1010.

Cho, Y.M., 2003. Differential expression of the liver proteome in senescence accelerated mice. Proteomics 3 (10), 1883–1894.

Comunale, M.A., 2004. Comparative proteomic analysis of de-N-glycosylated serum from hepatitis B carriers reveals polypeptides that correlate with disease status. Proteomics 4 (3), 826–838.

Corbett, C.E., Duarte, M.I., Bustamante, S.E., 1993. Regression of diffuse intralobular liver fibrosis associated with visceral leishmaniasis. Am J Trop Med Hyg 49 (5), 616–624.

Coulouarn, C., 2004. Altered gene expression in acute systemic inflammation detected by complete coverage of the human liver transcriptome. Hepatology 39 (2), 353–364.

Cunningham, M.J., 2000. Gene expression microarray data analysis for toxicology profiling. Ann NY Acad Sci 919 (1), 52–67.

Deaciuc, I.V., 2004. Large-scale gene profiling of the liver in a mouse model of chronic, intragastric ethanol infusion. J Hepatol 40 (2), 219–227.

Donaldson, P.T., 2004. Genetics of liver disease: Immunogenetics and disease pathogenesis. Gut 53 (4), 599–608.

Dufour, J.F., DeLellis, R., Kaplan, M.M., 1998. Regression of hepatic fibrosis in hepatitis C with long-term interferon treatment. Dig Dis Sci 43 (12), 2573–2576.

Enard, W., 2002. Intra- and interspecific variation in primate gene expression patterns. Science 296 (5566), 340–343.

Eng, F.J., Friedman, S.L., 2000. Fibrogenesis I. New insights into hepatic stellate cell activation: The simple becomes complex. Am J Physiol Gastrointest Liver Physiol 279 (1), G7–G11.

Feezor, R.J., 2005. Functional genomics and gene expression profiling in sepsis: Beyond class prediction. Clin Infect Dis 41 (Suppl. 7), S427–S435.

Fella, K., 2005. Use of two-dimensional gel electrophoresis in predictive toxicology: Identification of potential early protein biomarkers in chemically induced hepatocarcinogenesis. Proteomics 5 (7), 1914–1927.

Fountoulakis, M., Suter, L., 2002. Proteomic analysis of the rat liver. J Chromatogr B Analyt Technol Biomed Life Sci 782 (1–2), 197–218.

Friedman, S.L., 2000. Molecular regulation of hepatic fibrosis, an integrated cellular response to tissue injury. J Biol Chem 275 (4), 2247–2250.

Galle, P.R., 1997. Apoptosis in liver disease. J Hepatol 27 (2), 405–412.

Gao, J., 2004. Identification of in vitro protein biomarkers of idiosyncratic liver toxicity. Toxicol In Vitro 18 (4), 533–541.

Garry, R.F., Dash, S., 2003. Proteomics computational analyses suggest that hepatitis C virus E1 and pestivirus E2 envelope glycoproteins are truncated class II fusion proteins. Virology 307 (2), 255–265.

Ge, D., 2009. Genetic variation in IL28B predicts hepatitis C treatment-induced viral clearance. Nature 461 (7262), 399–401.

Greenbaum, D., 2002. Chemical approaches for functionally probing the proteome. Mol Cell Proteomics 1 (1), 60–68.

Gressner, A.M., 1998. The cell biology of liver fibrogenesis – an imbalance of proliferation, growth arrest and apoptosis of myofibroblasts. Cell Tissue Res 292 (3), 447–452.

Guzey, C., Spigset, O., 2002. Genotyping of drug targets: A method to predict adverse drug reactions? Drug Saf 25 (8), 553–560.

He, Q.Y., 2003. Serum biomarkers of hepatitis B virus infected liver inflammation: A proteomic study. Proteomics 3 (5), 666–674.

Heath, A.C., 2011. A quantitative-trait genome-wide association study of alcoholism risk in the community: Findings and implications. Biol Psychiatry 70 (6), 513–518.

Henkel, C., 2005. Identification of fibrosis-relevant proteins using DIGE (difference in gel electrophoresis) in different models of hepatic fibrosis. Z Gastroenterol 43 (1), 23–29.

Honda, M., 2001. Differential gene expression between chronic hepatitis B and C hepatic lesion. Gastroenterology 120 (4), 955–966.

Honda, M., et al., 2006. Different signaling pathways in the livers of patients with chronic hepatitis B or chronic hepatitis C. Hepatology 44 (5), 1122–1138.

Huang, H., 2006. Identification of two gene variants associated with risk of advanced fibrosis in patients with chronic hepatitis C. Gastroenterology 130 (6), 1679–1687.

Huang, Q., 2004. Gene expression profiling reveals multiple toxicity endpoints induced by hepatotoxicants. Mutat Res 549 (1–2), 147–167.

Iizuka, N., 2002. Comparison of gene expression profiles between hepatitis B virus- and hepatitis C virus-infected hepatocellular carcinoma by oligonucleotide microarray data on the basis of a supervised learning method. Cancer Res 62 (14), 3939–3944.

Iizuka, N., 2004. Molecular signature in three types of hepatocellular carcinoma with different viral origin by oligonucleotide microarray. Int J Oncol 24 (3), 565–574.

Jansen, R., Gerstein, M., 2000. Analysis of the yeast transcriptome with structural and functional categories: Characterizing highly expressed proteins. Nucleic Acids Res 28 (6), 1481–1488.

Jiang, Y., 2004. Changes in the gene expression associated with carbon tetrachloride-induced liver fibrosis persist after cessation of dosing in mice. Toxicol Sci 79 (2), 404–410.

Jolly, R.A., 2005. Pooling samples within microarray studies: A comparative analysis of rat liver transcription response to prototypical toxicants. Physiol Genomics 22 (3), 346–355.

Joyce, J.A., 2004. Cathepsin cysteine proteases are effectors of invasive growth and angiogenesis during multistage tumorigenesis. Cancer Cell 5 (5), 443–453.

Jung, J.W., 2004. Gene expression analysis of peroxisome proliferators- and phenytoin-induced hepatotoxicity using cDNA microarray. J Vet Med Sci 66 (11), 1329–1333.

Juran, B.D., Lazaridis, K.N., 2010. Update on the genetics and genomics of PBC. J Autoimmun 35 (3), 181–187.

Kanzler, S., Galle, P.R., 2000. Apoptosis and the liver. Semin Cancer Biol 10 (3), 173–184.

Kaplowitz, N., 2000. Mechanisms of liver cell injury. J Hepatol 32 (1 Suppl.), 39–47.

Kaplowitz, N., 2004. Drug-induced liver injury. Clin Infect Dis 38 (Suppl. 2), S44–S48.

Karlsen, T.H., Melum, E., Franke, A., 2010a. The utility of genome-wide association studies in hepatology. Hepatology 51 (5), 1833–1842.

Karlsen, T.H., 2010b. Genome-wide association analysis in primary sclerosing cholangitis. Gastroenterology 138 (3), 1102–1111.

Kawada, N., 2001. Characterization of a stellate cell activation-associated protein (STAP) with peroxidase activity found in rat hepatic stellate cells. J Biol Chem 276 (27), 25,318–25,323.

Kim, B.Y., 2004a. Feature genes of hepatitis B virus-positive hepatocellular carcinoma, established by its molecular discrimination approach using prediction analysis of microarray. Biochim Biophys Acta 1739 (1), 50–61.

Kim, J.W., 2004b. Cancer-associated molecular signature in the tissue samples of patients with cirrhosis. Hepatology 39 (2), 518–527.

Kim, W., 2003. Comparison of proteome between hepatitis B virus- and hepatitis C virus-associated hepatocellular carcinoma. Clin Cancer Res 9 (15), 5493–5500.

Koopmann, J., 2004. Mac-2-binding protein is a diagnostic marker for biliary tract carcinoma. Cancer 101 (7), 1609–1615.

Krawczyk, M., 2010. Genome-wide association studies and genetic risk assessment of liver diseases. Nat Rev Gastroenterol Hepatol 7 (12), 669–681.

Kristensen, D.B., 2000. Proteome analysis of rat hepatic stellate cells. Hepatology 32 (2), 268–277.

Kuramitsu, Y., Nakamura, K., 2005. Current progress in proteomic study of hepatitis C virus-related human hepatocellular carcinoma. Expert Rev Proteomics 2 (4), 589–601.

Lee, S.H., 2004. Effects and regulation of osteopontin in rat hepatic stellate cells. Biochem Pharmacol 68 (12), 2367–2378.

Lefkovits, I., Kettman, J.R., Frey, J.R., 2000. Global analysis of gene expression in cells of the immune system I. Analytical limitations in obtaining sequence information on polypeptides in two-dimensional gel spots. Electrophoresis 21 (13), 2688–2693.

Lefkovits, I., Kettman, J.R., Frey, J.R., 2001. Proteomic analysis of rare molecular species of translated polypeptides from a mouse fetal thymus cDNA library. Proteomics 1 (4), 560–573.

Levy, M.T., 1999. Fibroblast activation protein: A cell surface dipeptidyl peptidase and gelatinase expressed by stellate cells at the tissue remodelling interface in human cirrhosis. Hepatology 29 (6), 1768–1778.

Liu, A.Y., 2005. Analysis of prostate cancer by proteomics using tissue specimens. J Urol 173 (1), 73–78.

Liu, C., 2007. Novel resequencing chip customized to diagnose mutations in patients with inherited syndromes of intrahepatic cholestasis. Gastroenterology 132 (1), 119–126.

Liu, X.J., 2004. Association of differentially expressed genes with activation of mouse hepatic stellate cells by high-density cDNA microarray. World J Gastroenterol 10 (11), 1600–1607.

Lotersztajn, S., 2007. CB2 receptors as new therapeutic targets for liver diseases. Br J Pharmacol 153 (2), 286–289.

Low, T.Y., 2004. A proteomic analysis of thioacetamide-induced hepatotoxicity and cirrhosis in rat livers. Proteomics 4 (12), 3960–3974.

Lu, H., 2004. Riboproteomics of the hepatitis C virus internal ribosomal entry site. J Proteome Res 3 (5), 949–957.

Mallat, A., Lotersztajn, S., 2006. Endocannabinoids as novel mediators of liver diseases. J Endocrinol Invest 29 (3 Suppl.), 58–65.

Marcolongo, M., 2009. A seven-gene signature (cirrhosis risk score) predicts liver fibrosis progression in patients with initially mild chronic hepatitis C. Hepatology 50 (4), 1038–1044.

Marra, F., 1994. Regulation of platelet-derived growth factor secretion and gene expression in human liver fat-storing cells. Gastroenterology 107 (4), 1110–1117.

Mehal, W.Z., Azzaroli, F., Crispe, I.N., 2001. Immunology of the healthy liver: Old questions and new insights. Gastroenterology 120 (1), 250–260.

Mells, G.F., 2011. Genome-wide association study identifies 12 new susceptibility loci for primary biliary cirrhosis. Nat Genet 43 (4), 329–332.

Meneses-Lorente, G., 2004. A proteomic investigation of drug-induced steatosis in rat liver. Chem Res Toxicol 17 (5), 605–612.

Merrick, B.A., Bruno, M.E., 2004. Genomic and proteomic profiling for biomarkers and signature profiles of toxicity. Curr Opin Mol Ther 6 (6), 600–607.

Miklos, G.L., Maleszka, R., 2001a. Integrating molecular medicine with functional proteomics: Realities and expectations. Proteomics 1 (1), 30–41.

Miklos, G.L., Maleszka, R., 2001b. Protein functions and biological contexts. Proteomics 1 (2), 169–178.

Minami, K., 2005. Relationship between hepatic gene expression profiles and hepatotoxicity in five typical hepatotoxicant-administered rats. Toxicol Sci 87 (1), 296–305.

Moroda, T., 1997. Autologous killing by a population of intermediate T-cell receptor cells and its NK1.1+ and NK1.1− subsets, using Fas ligand/Fas molecules. Immunology 91 (2), 219–226.

Moyer, K., 2010. Staging of biliary atresia at diagnosis by molecular profiling of the liver. Genome Med 2 (5), 33.

Nakatani, K., 1996. Expression of neural cell adhesion molecule (N-CAM) in perisinusoidal stellate cells of the human liver. Cell Tissue Res 283 (1), 159–165.

Neverova, I., Van Eyk, J.E., 2005. Role of chromatographic techniques in proteomic analysis. J Chromatogr B Analyt Technol Biomed Life Sci 815 (1–2), 51–63.

Niki, T., 1999. Class VI intermediate filament protein nestin is induced during activation of rat hepatic stellate cells. Hepatology 29 (2), 520–527.

Nordvarg, H., 2004. A proteomics approach to the study of absorption, distribution, metabolism, excretion, and toxicity. J Biomol Tech 15 (4), 265–275.

Ockner, R.K., 2001. Apoptosis and liver diseases: Recent concepts of mechanism and significance. J Gastroenterol Hepatol 16 (3), 248–260.

Pan, H.W., 2003. Overexpression of osteopontin is associated with intrahepatic metastasis, early recurrence, and poorer prognosis of surgically resected hepatocellular carcinoma. Cancer 98 (1), 119–127.

Paradis, V., 2005. Identification of a new marker of hepatocellular carcinoma by serum protein profiling of patients with chronic liver diseases. Hepatology 41 (1), 40–47.

Pasa-Tolic, L., 2004. Proteomic analyses using an accurate mass and time tag strategy. Biotechniques 37 (4), 621–624, 626–633.

Pinzani, M., 1996. Expression of platelet-derived growth factor and its receptors in normal human liver and during active hepatic fibrogenesis. Am J Pathol 148 (3), 785–800.

Poon, T.C., 2005. Prediction of liver fibrosis and cirrhosis in chronic hepatitis B infection by serum proteomic fingerprinting: A pilot study. Clin Chem 51 (2), 328–335.

Poynard, T., 2005. Age and gender will survive to competing risks as fibrosis factors. Gastroenterology 128 (2), 519–520, author reply 520–521.

Prosser, C.C., Yen, R.D., Wu, J., 2006. Molecular therapy for hepatic injury and fibrosis: Where are we? World J Gastroenterol 12 (4), 509–515.

Qiang, H., 2006. Differential expression genes analyzed by cDNA array in the regulation of rat hepatic fibrogenesis. Liver Int 26 (9), 1126–1137.

Qin, L.X., Tang, Z.Y., 2004. Recent progress in predictive biomarkers for metastatic recurrence of human hepatocellular carcinoma: A review of the literature. J Cancer Res Clin Oncol 130 (9), 497–513.

Ratziu, V., 2005. Sampling variability of liver biopsy in nonalcoholic fatty liver disease. Gastroenterology 128 (7), 1898–1906.

Regev, A., 2002. Sampling error and intraobserver variation in liver biopsy in patients with chronic HCV infection. Am J Gastroenterol 97 (10), 2614–2618.

Righetti, P.G., 2005. Prefractionation techniques in proteome analysis: The mining tools of the third millennium. Electrophoresis 26 (2), 297–319.

Rockey, D., 1997. The cellular pathogenesis of portal hypertension: Stellate cell contractility, endothelin, and nitric oxide. Hepatology 25 (1), 2–5.

Rockey, D.C., 2005. Antifibrotic therapy in chronic liver disease. Clin Gastroenterol Hepatol 3 (2), 95–107.

Roelofsen, H., Balgobind, R., Vonk, R.J., 2004. Proteomic analyses of copper metabolism in an in vitro model of Wilson disease using surface enhanced laser desorption/ionization-time of flight-mass spectrometry. J Cell Biochem 93 (4), 732–740.

Romeo, S., 2008. Genetic variation in PNPLA3 confers susceptibility to nonalcoholic fatty liver disease. Nat Genet 40 (12), 1461–1465.

Rosenberg, S., 2001. Recent advances in the molecular biology of hepatitis C virus. J Mol Biol 313 (3), 451–464.

Rotman, Y., 2010. The association of genetic variability in patatin-like phospholipase domain-containing protein 3 (PNPLA3) with histological severity of nonalcoholic fatty liver disease. Hepatology 52 (3), 894–903.

Sancho-Bru, P., 2005. Genomic and functional characterization of stellate cells isolated from human cirrhotic livers. J Hepatol 43 (2), 272–282.

Schnabl, B., 2002. Immortal activated human hepatic stellate cells generated by ectopic telomerase expression. Lab Invest 82 (3), 323–333.

Schnabl, B., 2003. Replicative senescence of activated human hepatic stellate cells is accompanied by a pronounced inflammatory but less fibrogenic phenotype. Hepatology 37 (3), 653–664.

Scholle, F., 2004. Virus-host cell interactions during hepatitis C virus RNA replication: Impact of polyprotein expression on the cellular transcriptome and cell cycle association with viral RNA synthesis. J Virol 78 (3), 1513–1524.

Schwegler, E.E., 2005. SELDI-TOF MS profiling of serum for detection of the progression of chronic hepatitis C to hepatocellular carcinoma. Hepatology 41 (3), 634–642.

Scriver, C.R., 2004. After the genome – the phenome? J Inherit Metab Dis 27 (3), 305–317.

Sell, S., 2001. Heterogeneity and plasticity of hepatocyte lineage cells. Hepatology 33 (3), 738–750.

Seow, T.K., 2001. Hepatocellular carcinoma: From bedside to proteomics. Proteomics 1 (10), 1249–1263.

Seth, D., 2003. Gene expression profiling of alcoholic liver disease in the baboon (Papio hamadryas) and human liver. Am J Pathol 163 (6), 2303–2317.

Shackel, N.A., 2001. Identification of novel molecules and pathogenic pathways in primary biliary cirrhosis: cDNA array analysis of intrahepatic differential gene expression. Gut 49 (4), 565–576.

Shackel, N.A., 2002. Insights into the pathobiology of hepatitis C virus-associated cirrhosis: Analysis of intrahepatic differential gene expression. Am J Pathol 160 (2), 641–654.

Shankar, K., 2003. Activation of PPAR-alpha in streptozotocin-induced diabetes is essential for resistance against acetaminophen toxicity. Faseb J 17 (12), 1748–1750.

She, Y.M., 2003. Identification of metal-binding proteins in human hepatoma lines by immobilized metal affinity chromatography and mass spectrometry. Mol Cell Proteomics 2 (12), 1306–1318.

Shi, Z., Wakil, A.E., Rockey, D.C., 1997. Strain-specific differences in mouse hepatic wound healing are mediated by divergent T helper cytokine responses. Proc Natl Acad Sci USA 94 (20), 10,663–10,668.

Shirota, Y., 2001. Identification of differentially expressed genes in hepatocellular carcinoma with cDNA microarrays. Hepatology 33 (4), 832–840.

Siegel, C.A., 2005. Liver biopsy 2005: When and how? Cleve Clin J Med 72 (3), 199–201.

Simakajornboon, N., 2001. In vivo PDGF beta receptor activation in the dorsocaudal brainstem of the rat prevents hypoxia-induced apoptosis via activation of Akt and BAD. Brain Res 895 (1–2), 111–118.

Smith, M.W., 2003. Hepatitis C virus and liver disease: Global transcriptional profiling and identification of potential markers. Hepatology 38 (6), 1458–1467.

Sreekumar, R., 2003. Hepatic gene expression in histologically progressive nonalcoholic steatohepatitis. Hepatology 38 (1), 244–251.

Staiger, H., Loffler, G., 1998. The role of PDGF-dependent suppression of apoptosis in differentiating 3T3-L1 preadipocytes. Eur J Cell Biol 77 (3), 220–227.

Steel, L.F., 2003. A strategy for the comparative analysis of serum proteomes for the discovery of biomarkers for hepatocellular carcinoma. Proteomics 3 (5), 601–609.

Suppiah, V., 2009. IL28B is associated with response to chronic hepatitis C interferon-alpha and ribavirin therapy. Nat Genet 41 (10), 1100–1104.

Tadic, S.D., 2002. Sex differences in hepatic gene expression in a rat model of ethanol-induced liver injury. J Appl Physiol 93 (3), 1057–1068.

Takashima, M., 2003. Proteomic profiling of heat shock protein 70 family members as biomarkers for hepatitis C virus-related hepatocellular carcinoma. Proteomics 3 (12), 2487–2493.

Takashima, M., 2005. Overexpression of alpha enolase in hepatitis C virus-related hepatocellular carcinoma: Association with tumor progression as determined by proteomic analysis. Proteomics 5 (6), 1686–1692.

Tanaka, A., 2011. Replicated association of 17q12-21 with susceptibility of primary biliary cirrhosis in a Japanese cohort. Tissue Antigens 78 (1), 65–68.

Tang, Z.Y., 2004. A decade's studies on metastasis of hepatocellular carcinoma. J Cancer Res Clin Oncol 130 (4), 187–196.

Ter Kuile, B.H., Westerhoff, H.V., 2001. Transcriptome meets metabolome: hierarchical and metabolic regulation of the glycolytic pathway. FEBS Lett 500 (3), 169–171.

Thomas, D.L., 2009. Genetic variation in IL28B and spontaneous clearance of hepatitis C virus. Nature 461 (7265), 798–801.

Tian, C., 2010. Variant in PNPLA3 is associated with alcoholic liver disease. Nat Genet 42 (1), 21–23.

Tsushima, H., Kawata, S., Tamura, S., et al., 1999. Reduced plasma transforming growth factor-beta1 levels in patients with chronic hepatitis C after interferon-alpha therapy: Association with regression of hepatic fibrosis. J Hepatol 30 (1), 1–7.

Tsutsui, H., 1996. IFN-gamma-inducing factor up-regulates Fas ligand-mediated cytotoxic activity of murine natural killer cell clones. J Immunol 157 (9), 3967–3973.

Ulrich, R.G., 2004. Overview of an interlaboratory collaboration on evaluating the effects of model hepatotoxicants on hepatic gene expression. Environ Health Perspect 112 (4), 423–427.

Utsunomiya, T., 2004. A gene-expression signature can quantify the degree of hepatic fibrosis in the rat. J Hepatol 41 (3), 399–406.

Venkatraman, A., 2004. Oxidative modification of hepatic mitochondria protein thiols: Effect of chronic alcohol consumption. Am J Physiol Gastrointest Liver Physiol 286 (4), G521–G527.

Vogel, S., 2000. An immortalized rat liver stellate cell line (HSC-T6): A new cell model for the study of retinoid metabolism in vitro. J Lipid Res 41 (6), 882–893.

Wang, W., 2004. Genomic analysis reveals RhoC as a potential marker in hepatocellular carcinoma with poor prognosis. Br J Cancer 90 (12), 2349–2355.

Waring, J.F., 2001a. Microarray analysis of hepatotoxins in vitro reveals a correlation between gene expression profiles and mechanisms of toxicity. Toxicol Lett 120 (1–3), 359–368.

Waring, J.F., 2001b. Clustering of hepatotoxins based on mechanism of toxicity using gene expression profiles. Toxicol Appl Pharmacol 175 (1), 28–42.

Waring, J.F., 2003. Development of a DNA microarray for toxicology based on hepatotoxin-regulated sequences. EHP Toxicogenomics 111 (1T), 53–60.

Warner, F.J., 2007. Liver fibrosis: A balance of ACEs? Clin Sci (Lond) 113 (3), 109–118.

Watchko, J.F., Daood, M.J., Biniwale, M., 2002. Understanding neonatal hyperbilirubinaemia in the era of genomics. Semin Neonatol 7 (2), 143–152.

Wieland, S., 2004. Genomic analysis of the host response to hepatitis B virus infection. Proc Natl Acad Sci USA 101 (17), 6669–6674.

Wieland, S.F., Chisari, F.V., 2005. Stealth and cunning: Hepatitis B and hepatitis C viruses. J Virol 79 (15), 9369–9380.

Wiesner, A., 2004. Detection of tumor markers with ProteinChip technology. Curr Pharm Biotechnol 5 (1), 45–67.

Wong, Y.F., 2004. Protein profiling of cervical cancer by protein-biochips: Proteomic scoring to discriminate cervical cancer from normal cervix. Cancer Lett 211 (2), 227–234.

Xu, C., 2004. Identification and characterization of 177 unreported genes associated with liver regeneration. Genomics Proteomics Bioinform 2 (2), 109–118.

Xu, L., 2005. Human hepatic stellate cell lines, LX-1 and LX-2: New tools for analysis of hepatic fibrosis. Gut 54 (1), 142–151.

Yamashita, T., et al., 2000. Comprehensive gene expression profile of a normal human liver. Biochem Biophys Res Commun 269 (1), 110–116.

Yamashita, T., 2004a. Genome-wide transcriptome mapping analysis identifies organ-specific gene expression patterns along human chromosomes. Genomics 84 (5), 867–875.

Yamashita, Y., 2004b. cDNA microarray analysis in hepatocyte differentiation in Huh 7 cells. Cell Transplant 13 (7–8), 793–799.

Yang, X., 2006. Tissue-specific expression and regulation of sexually dimorphic genes in mice. Genome Res 16 (8), 995–1004.

Yano, N., 2001. Profiling the adult human liver transcriptome: Analysis by cDNA array hybridization. J Hepatol 35 (2), 178–186.

Ye, Q.H., 2003. Predicting hepatitis B virus-positive metastatic hepatocellular carcinomas using gene expression profiling and supervised machine learning. Nat Med 9 (4), 416–423.

Younossi, Z.M., 2005a. A genomic and proteomic study of the spectrum of nonalcoholic fatty liver disease. Hepatology 42 (3), 665–674.

Younossi, Z.M., 2005b. Hepatic gene expression in patients with obesity-related non-alcoholic steatohepatitis. Liver Int 25 (4), 760–771.

Younossi, Z.M., 2009. Gene expression profile associated with superimposed non-alcoholic fatty liver disease and hepatic fibrosis in patients with chronic hepatitis C. Liver Int 29 (9), 1403–1412.

Yuan, J., 2001. Genome analysis with gene-indexing databases. Pharmacol Ther 91 (2), 115–132.

Zhang, D.Y., 2004. Coordinate expression of regulatory genes differentiates embryonic and perinatal forms of biliary atresia. Hepatology 39 (4), 954–962.

Zlojutro, M., 2011. Genome-wide association study of theta band event-related oscillations identifies serotonin receptor gene HTR7 influencing risk of alcohol dependence. Am J Med Genet B Neuropsychiatr Genet 156B (1), 44–58.

Systemic Sclerosis

Sevdalina Lambova and Ulf Müller-Ladner

DEFINITION

Systemic sclerosis (SSc) is a chronic, multisystem connective tissue disease characterized by microangiopathy, fibrosis of the skin and internal organs, and activation of humoral as well as cellular immune responses.

SUBSETS OF SSC

SSc is divided into two major subsets defined by the extension of skin involvement: limited SSc, characterized by skin involvement at the face, neck, and the skin distal to the elbows and knees, and diffuse SSc, with skin thickening distal and proximal to the knees and elbows, and thickening involving the trunk. CREST syndrome is a form of limited SSc.

EPIDEMIOLOGY

The prevalence and incidence of SSc vary in different populations, suggesting the involvement of genetic and/or environmental factors. The prevalence of SSc is 276 cases per million adults in the United States (Mayes et al., 2003) and 8 cases per 100,000 adults in Europe (Allcock et al., 2004). The annual incidence of SSc is 18–23 cases per million. SSc usually begins between 30 and 50 years of age and is three times more common in women than in men (Matucci-Cerinic et al., 2009; Mayes et al., 2003).

ETIOLOGY AND PATHOGENESIS

The Role of Causative Agents – Environmental and Infectious Agents

Environmental factors that have been proposed to be SSc causative agents include organic solvents, vinyl chloride, silica, metal dust, certain pesticides, hair dyes, and toxins (Matucci-Cerinic et al., 2009; Mayes, 1999). Several infectious agents (e.g., herpes viruses, retroviruses, and human cytomegalovirus) triggering molecular mimicry have also been hypothesized to be causative agents in SSc. This hypothesis has been supported by the presence of IgG anti-human cytomegalovirus antibodies and sequence homologies between retroviral proteins and the target of anti-Scl-70 antibody – a topoisomerase I antigen (Matucci-Cerinic et al., 2009; Namboodiri et al., 2004).

Microchimerism

Microchimeric cells, which can be transferred from one person to another during pregnancy, blood transfusion, and bone marrow and solid-organ transplantation, have been detected in peripheral blood and tissue of SSc patients, and may support a graft-versus-host disease triggering SSc (Matucci-Cerinic et al., 2009).

Pathogenic Pathways

The pathogenesis of SSc is complex, and is based on (a) microvascular damage, (b) activation of the immune system, and (c) progressive fibrosis.

Genomic and Personalized Medicine, 2nd edition
by Ginsburg & Willard. DOI: http://dx.doi.org/10.1016/B978-0-12-382227-7.00079-3

Vascular Damage

Vascular dysfunction can be found very early in SSc and consists of apoptosis of microvascular endothelial cells induced by anti-endothelial antibodies, entailing the subsequent inflammatory and finally fibrotic complications (Sgonc et al., 1996). Increased levels of free oxygen radicals, anti-endothelial antibodies, and cytokines produced by activated lymphocytes contribute to this activation (Herrick and Matucci-Cerinic, 2001; Kahaleh et al., 2003; Matucci-Cerinic et al., 2009). The endothelial injury in SSc results in decreased production of vasodilators such as nitric oxide (NO) and prostacyclin (which also inhibits platelet aggregation), and increased levels of vasoconstrictors such as endothelin-1 (ET). The exposition of the subendothelium to the bloodstream may induce platelet adhesion and intravascular thrombus formation (Block and Sequeira, 2001; Ho and Belch, 1998; Konttinen et al., 2003). Pathomorphologic alterations of the blood vessels in SSc include intimal proliferation, narrowing of the vessel lumen, and reduced blood flow (Matucci-Cerinic et al., 2009; Silver et al., 2005).

Immune Activation

Mononuclear cell infiltrates are found in affected skin and visceral organs in SSc. CD4-positive T cells are the predominant cell type (Matucci-Cerinic et al., 2009). The activation of humoral immunity and the role of B cells is demonstrated by the numerous autoantibodies detected in the serum of SSc patients (Sato et al., 2004). Antinuclear autoantibodies (ANA) are present in more than 90% of SSc patients. The predominant antibodies are directed to the centromere (anti-centromere antibodies – ACA) in limited SSc and to DNA topoisomerase I (topo I) in diffuse SSc (Ho and Reveille, 2003; Kuwana et al., 1994; Meyer, 2006). Additional serum ANA, including anti-U 1 RNP, anti-PM-Scl, anti-Ku, anti-Th RNP, anti-U3 RNP (fibrillarin), and anti-RNA polymerases have also been detected in SSc sera. Novel autoantibodies include anti-endothelial cell antibodies, antibodies against matrix metalloproteinases, and antibodies against platelet-derived growth factor receptor (PDGFR) (Kraaij and Van Laar, 2008; Lunardi et al., 2006).

Fibrosis

In SSc, transforming growth factor-β (TGF-β) is one of the key cytokines involved in tissue fibrosis (Kraaij and Van Laar, 2008; Sgonc and Wick, 2008; Varga and Abraham, 2007). Human B lymphocytes, among other cells, can be a source of TGF-β and express receptors for this cytokine (Kehrl et al., 1986; Kraaij and Van Laar, 2008). TGF-β promotes collagen deposition, and inhibits collagen degradation by decreasing matrix-metalloproteinases and increasing the expression of tissue inhibitors of metalloproteinases (Sgonc and Wick, 2008; Verrecchia et al., 2006). In addition, a cross-talk between TGF-β and platelet-derived growth factor (PDGF) has been found with positive up-regulation of PDGF α receptors after stimulation of scleroderma fibroblasts by TGF-β (Trojanowska, 2008). Connective tissue growth factor (CTGF), which is produced by fibroblasts, vascular smooth muscle cells, and endothelial cells in response to TGF-β stimulation, is another potentially important mediator of fibrosis (Matucci-Cerinic et al., 2009).

GENETIC MARKERS

It has been shown that risk of SSc is increased among first-degree relatives of patients compared to the general population. In a study of 703 families in the US, including 11 multiplex SSc families, the relative risk for first-degree relatives to develop SSc was approximately 13 with a 1.6% recurrence rate, compared to 0.026% in the general population (Arnett et al., 2001). The risk ratio for siblings was approximately 15 (ranging from 10–27 across cohorts). However, twin studies revealed that fewer than 5% of mono- and dizygotic twins are concordant for SSc. On the other hand, when gene expression was compared by high-throughput microarray techniques, the pattern of dermal fibroblasts in unaffected monozygotic twins was not significantly different from that of SSc patients. In this study, additional functional experiments revealed that healthy fibroblasts that were incubated with serum from an SSc-affected patient or with serum of the unaffected monozygotic twin developed a typical SSc pattern with an increased expression of collagen1 A2, SPARC (secreted protein, acidic and rich in cysteine; osteonectin) and CTGF (Zhou, 2005).

At present, no whole genome or proteome analysis has revealed a unique pattern in SSc patients, tissues, or cells (Ahmed and Tan, 2003; Feghali-Bostwick, 2005), although several approaches have found distinct polymorphisms and disease associations (Allanore et al., 2010, 2011; Arnett et al., 2010).

Knowledge from genetics addressing target genes and the respective protein groups, such as the topoisomerase I complex, will likely provide novel ideas in this field (Czubaty, 2005). In addition, complementary DNA (cDNA) array techniques were successfully used to identify the gene expression of disease-related cell types such as endothelial cells. When compared to normal skin endothelial cells, it has been demonstrated that microvascular endothelial cells of patients with SSc show abnormalities in a variety of genes that are able to account for defective angiogenesis (Giusti, 2006). In another approach, cDNA arrays were used to compare the gene expression profiles of peripheral blood mononuclear cells of patients with early SSc, which revealed a distinct up-regulation of 18 interferon-inducible genes, selectins, and integrins, supporting the idea of an infectious trigger in the early phases of the disease (Tan, 2005). CD8-positive lung T cells, on the other hand, resulted in two distinct gene cluster groups, with one showing a type II T cell activation in combination with profibrotic factors and matrix metalloproteinases (MMPs) (Luzina, 2003).

With regard to genetic markers, a limited but increasing number of studies have examined the presence and pattern of single nucleotide polymorphisms (SNPs) in SSc (Assassi, 2005). Also, as SSc is not inherited in a Mendelian manner,

TABLE 79.1	Loci of interest of genetic abnormalities of different molecules involved in SSc pathophysiology
Gene	**Loci**
TGF-β	Codon 10, −1133bp in promoter
Fibrillin-1	SNP in 5′-untranslated region, CT insertion in exon A
TNF-α	TNF-α 13 microsatellite, −863A allele
TNIP1	SNP at the 5q33 locus
IL-1	CTG/CTG diplotype SNP, −889 allele
	Polymorphism
CCL2	SNP
ACE	Insertions/deletions in chromosome 17
eNOS	Polymorphisms

experimental and clinical research has focused on genetic alterations in numerous genes known to be operative in SSc path physiology. This has revealed interesting aspects, especially with regard to growth factors, matrix-related molecules, and inflammation markers such as TGF, fibrillin-1, tumor necrosis factor-α (TNF-α), TNF-α receptor type II, TNF-β, and interleukin-1 (IL-1) (Table 79.1).

The initial analysis of the TGF gene revealed no strong genetic abnormalities, nor were any found for PDGF (Zhou, 2000). However, a detailed analysis at codon 10 showed that SSc patients are prone to high TGF synthesis, irrespective of limited or diffuse disease (Crilly, 2002). Interestingly, adenoviral gene transfer of TGF receptor type I (TGF-RI) into fibroblasts in combination with cDNA array revealed a distinct TGF-RI-induced profibrotic phenotype with up-regulation of collagen type I and CTGF (Pannu, 2006). CTGF is another promising candidate for detailed gene and polymorphism analysis (Zhu, 2004).

Further research on matrix metabolism-regulating genes showed an association of the stromelysin promoter with SSc (Marasini, 2001), but no association of the MMP-1 promoter with the disease (Johnson, 2001).

Tumor necrosis factor (TNF) is one of the driving pro-inflammatory molecules in several autoimmune diseases. Based on knowledge of the inflamed initial stages in SSc pathophysiology, numerous groups have examined the presence of TNF gene polymorphisms in SSc patients. In the first intron at locus 252, the two homozygous genotypes of TNF were significantly associated in Japanese SSc patients. In the TNF gene itself, however, statistical power was not sufficient to prove a similar association (Pandey and Takeuchi, 1999). Notably, the rare GG genotype in exon 6 of the TNF receptor

type II gene was found to be more frequent in another European diffuse SSc cohort (Tolusso, 2005).

The three strongly associated SNPs at the 5q33 locus are located within the TNFAIP3 interacting protein 1 (*TNIP1*) gene. *TNIP1* is a very interesting new candidate gene for SSc. The protein encoded by this gene exerts a negative regulation of NF-kappaB (NF-κB). The reduced inhibition of NF-κB favors inflammatory/immune responses and potentially contributes to the overproduction of extra-cellular matrix. Subsequently, in a recent large genome-wide association study of SSc, two new SSc-risk loci were identified, psoriasis susceptibility 1 candidate 1 (PSORS1C1) and TNIP1 (Allanore et al., 2011).

Genomic evaluation of the interleukin-1 (*IL-1*) gene revealed distinct genetic aberrations in Japanese SSc patients, and SNP analyses showed that a distinct CTG/CTG diplotype associated strongly with the development of interstitial lung disease in those patients (Kawaguchi, 2003). A study in Czech patients revealed also a polymorphism in the *IL-1A* gene at position 889 (Hutyrova, 2004).

A SNP of the gene encoding monocyte chemoattractant protein-1 (*CCL2*) has previously been suggested to be involved in susceptibility to SSc, but in a recent study that gene was not implicated in susceptibility nor in the SSc phenotype (Radstake et al., 2009).

With regard to cellular immunology, CD19-positive B cells appear to bear a 499G-T polymorphism in the CD19 coding region in SSc patients, which was also associated with susceptibility to the disease (Tsuchiya, 2004). In Korean patients, transporters associated with antigen processing-1 and -2 polymorphisms were found to be independent of other human leukocyte antigen DR (HLADR) associations (Takeuchi, 1994), suggesting different roles of genomic alterations in antigen-presenting cells in SSc (Song, 2005). However, distinct human leukocyte antigen (HLA) alleles appear to be linked directly to SSc subtypes, as it could be shown that in male SSc patients, HLA class II allele DQA1*0501 was associated with diffuse but not limited disease (Lambert, 2000a). In that population, maternal HLA compatibility was not a significant risk factor for development of the disease (Lambert, 2000b). A recent study found that DRB1*0407 and *1304 are independent risk factors for the development of scleroderma renal crisis (Nguyen et al., 2011).

Another study has confirmed the presence of an association between the *TNFSF4* gene promoter polymorphism and SSc genetic susceptibility, especially in patients with the limited form of the disease who are also positive for ACA. The *TNFSF4* gene encodes OX40L, which is expressed on activated antigen-presenting cells and endothelial cells in acute inflammation. Furthermore, it enhances B cell proliferation and differentiation and promotes proliferation and survival of T cells (Bossini-Castilo et al., 2011).

Alterations in genes regulating microvasculature development, intravascular thrombosis, dysregulated fibrinolysis and perivascular fibrosis have also been addressed by genomic analyses. A study in an Italian population showed that patients with SSc appear to have a higher prevalence of angiotensin-converting enzyme (ACE) insertion/deletions on chromosome 17 and polymorphisms within the endothelial NO synthase gene (Fatini, 2004). In contrast, other groups showed that in a French Caucasian population, *eNOS* polymorphisms neither influenced the course of SSc nor did they enhance susceptibility (Allanore, 2004; Tikly, 2005). A recent investigation performed in Korea also could not find a difference in the frequencies of *ACE* insertion/deletion genotypes between patients and controls, nor between diffuse and limited SSc patients (Joung, 2006).

CLINICAL PRESENTATION

Initial Symptoms

Raynaud's phenomenon (RP) is the initial symptom in a majority of patients with the limited form of the disease, while in those with the diffuse form, the classic initial complaints are swelling of the hands, skin thickening, and sometimes arthritis. Occasionally, the earliest symptom is visceral involvement, such as gastroesophageal involvement (dysphagia, heartburn) or dyspnea (Silver et al., 2005).

Raynaud's Phenomenon

RP is one of the most frequent SSc symptoms. It occurs in the course of SSc in more than 95% of patients. Usually it is the initial symptom, which can precede the other features of the disease by years. RP affects hands, feet, and, less frequently, the tip of the nose, the earlobes, and the tongue. RP in SSc is severe, and often presents with digital ulcers (Seibold and Steen, 1994; Silver et al., 2005). Small areas of ischemic necrosis or ulceration of the fingertip are a frequent finding, often leaving pitting scars. Digital necrosis of the terminal portions also develops in SSc patients (Figure 79.1). The capillaroscopic pattern in SSc is the most specific among rheumatic diseases, and is characterized by the presence of dilated and giant capillaries, hemorrhages, avascular areas, and neoangiogenesis. There are three phases of capillaroscopic changes in SSc: an early phase (appearance of few dilated and/or giant capillaries and few hemorrhages), an active phase (a high number of giant capillaries and hemorrhages, moderate loss of capillaries, slight derangement and diffuse pericapillary edema), and a late phase (extensive avascular areas, appearance of ramified and bushy capillaries) (Bollinger and Fagrell, 1990; Cutolo et al., 2000, 2003, 2005; Maricq et al., 1980, 1983) (Figure 79.2).

Skin

The earliest skin changes in SSc are tight, puffy fingers, especially in the morning (the edematous phase). Edema may last indefinitely (in patients with limited SSc) or may be replaced gradually by thickening and tightening of the skin (the indurative phase) after several weeks or months. According to

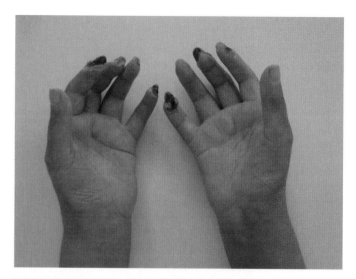

Figure 79.1 Female patient with limited SSc and digital necrosis.

the extent of skin changes, SSc is divided into two distinct subsets – limited and diffuse. Facial skin changes may result in the characteristic expressionless appearance, with thin, tightly pursed lips, vertical folds around the mouth and reduced oral aperture (Figure 79.3). After several years, the dermis tends to soften, and in many cases it reverts to normal and becomes thinner than normal (the atrophic phase) (Silver et al., 2005). The modified Rodnan skin score is a simple, inexpensive, reliable, and reproducible method for the assessment of skin thickening in SSc. It includes four grades: 0 – normal, 1 – thickened skin, 2 – thickened and unable to pinch, 3 – thickened and unable to move. The maximum skin score is 51 (Clements et al., 1995; Kahaleh et al., 1986). Skin thickness score correlates closely with skin biopsy thickness (Clements et al., 1995; Verrecchia et al., 2007). High-frequency ultrasound and a plicometer test have also been reported to be feasible methods for assessment of skin thickness in SSc, with very low interobserver variability (Moore et al., 2003; Nives Parodi et al., 1997).

Calcinosis

Patients with limited SSc or late-stage diffuse disease commonly develop intra- or subcutaneous calcifications composed of hydroxyapatite. They are located mainly in the digital pads and periarticular tissues, along the extensor surface of the forearms, in the olecranon bursae, prepatellar areas, and buttocks. Calcinosis may be complicated by ulceration of the overlying skin and secondary bacterial infection (Scheja and Akkeson, 1997; Silver et al., 2005).

Lung Involvement

Pulmonary involvement occurs in more than 70% of patients and is the leading cause of morbidity and mortality in patients with SSc. Alveolitis, membrane thickening, and/or modification of microvascular function are the characteristic features of lung

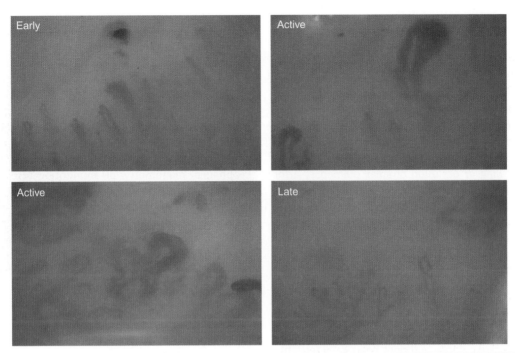

Figure 79.2   "Scleroderma"-type capillaroscopic pattern – early, active, and late phase.

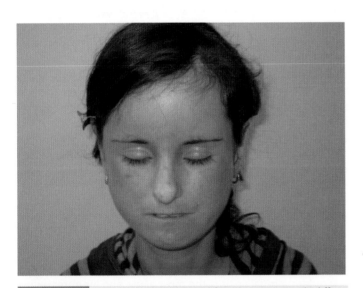

Figure 79.3   Face of a 24-year-old female patient with diffuse SSc with long duration of the disease, since childhood. Note skin thickening and microstomia.

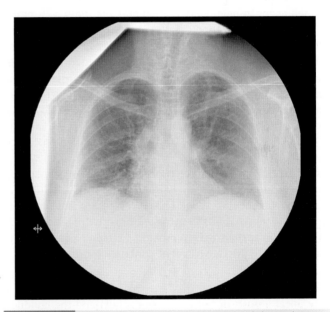

Figure 79.4   X-ray of the lungs in a patient with limited SSc, demonstrating bilateral pulmonary fibrosis. Note interstitial thickening in a reticular pattern.

involvement (Kim et al., 2002; Matucci-Cerinic et al., 2009). During progression of the disease, these changes may lead to interstitial lung disease and to pulmonary arterial hypertension (PAH) (McLaughlin et al., 2009) and pulmonary fibrosis. Chest radiography has low sensitivity for detection of early lung disease, and may be negative in more than 10% of symptomatic patients in the early stage of disease (Peters-Golden et al., 1984; Wells et al., 1997) (Figure 79.4). At present, chest high-resolution computed tomography is the preferred non-invasive

standard technique for diagnosis of interstitial lung disease in SSc (Remy-Jardin et al., 1993a, b). The usefulness of broncho-alveolar lavage (BAL) to define alveolitis in SSc has been questioned (Witt et al., 1999). Neutrophilia has been found to be associated with early mortality (Goh et al., 2007).

Pulmonary Arterial Hypertension

The prevalence of PAH in SSc is 10–12% (Condiffe et al., 2009), varying in different reports between 4.9% and 26.7%

(Proudman et al., 2007). Isolated PAH, in the absence of lung fibrosis, was found to be more frequent in the limited form of SSc (45%) than in the diffuse form of the disease (26%) (Walker et al., 2007). In SSc, two pathogenic pathways for development of the syndrome are recognized: PAH associated with pulmonary fibrosis and PAH without pulmonary fibrosis. Pulmonary fibrosis can be found in more than one-third of SSc patients with either the diffuse or limited form of the disease. PAH in SSc patients without pulmonary fibrosis is a severe complication due to narrowing or occlusion of small pulmonary arteries caused by smooth muscle hypertrophy, intimal hyperplasia, inflammation, thrombosis in situ. The rate of progression of dyspnea from normal exercise tolerance to oxygen dependency in this form of PAH in SSc is about 6–12 months with a subsequent mean duration of survival of two years. In contrast, SSc patients with PAH in the context of interstitial lung disease develop a similar degree of disability, but progress more slowly for a period of more than two and up to 10 years (Humbert et al., 2004; Jeffery and Morrell, 2002; McLaughlin et al., 2009; Silver et al., 2005). If PAH is suspected, a transthoracic Doppler echocardiography and right heart catheterization (RHC) must be performed (McLaughlin et al., 2009; Proudman et al., 2007). According to echocardiographic findings, PAH is defined as mean pulmonary arterial pressure (PAP) >25 mmHg at rest, >30 mmHg during exercise, or systolic pulmonary pressure >40 mmHg. Echocardiographic findings suggestive of the complication are elevated tricuspid regurgitation velocity jet above 2.8 m/s, dilated right ventricle or dilated atrium. As the mean PAP cannot easily be determined by echocardiography, an estimated systolic PAP >35 mmHg and/or an increased tricuspid velocity are used as indicators of probable PAH (McLaughlin et al., 2009; Proudman et al., 2007). Pulmonary functional tests, particularly carbon monoxide diffusing capacity (DLCO) values, are of crucial value for prediction of PAH. Diagnostic criteria for PAH include both a mean PAP >25 mmHg and a pulmonary vascular resistance >3 Wood units. With regard to biomarkers for PAH, N-terminal fragment of pro-brain natriuretic peptide (NT-pro-BNP) appears to be a promising screening parameter in SSc-related PAH (McLaughlin et al., 2009; Meyer, 2006).

Cardiac Involvement

Subclinical fibrosis of both the myocardium and the conducting system with "patchy" distribution has been described in SSc (Bulkley et al., 1976; Kahan et al., 1985). Malignant ventricular arrhythmias are dreadful complications, frequently requiring a pacemaker.

Gastrointestinal Involvement

Gastrointestinal abnormalities in SSc involve motility dysfunction and mucosal damage in different segments of the alimentary tract. Myogenic and neurogenic factors are involved in the pathogenesis of these abnormalities. Although any part of the gastrointestinal tract can be involved, esophageal disease occurs in nearly all patients with SSc. Common esophageal manifestations in SSc include esophageal dysmotility,

which occurs in 75–90% of SSc patients, gastroesophageal reflux, Barrett's esophagus, adenocarcinoma, and infectious esophagitis (Wielosz et al., 2010). Stomach involvement occurs in 10–75% of SSc patients, small bowel involvement in 30–70%, colon involvement in 10–50% and ano-rectal involvement in 50–70% (Ntoumazios et al., 2006; Wielosz et al., 2010). Watermelon stomach is a rare finding, which presents with gastric antral venous ectasia. It may result in persistent bleeding in patients with SSc (Elkayam et al., 2000). Esophageal manometry, pH-monitoring tests and endoscopy are the technical methods for evaluation of the upper gastrointestinal tract disorders (those of the esophagus and the stomach). The upper intestine can be evaluated with deep duodenal biopsy, transit time, H_2-breath tests, D-xylose tests, lactose tolerance tests, beta-carotin, and endomysium and transglutaminase autoantibodies. The lower intestine can be evaluated with colonoscopy, computed tomography scans, and rectoscopy (Davidson et al., 1985; Goldblatt et al., 2002; Greydanus and Camilleri, 1989; Segel et al., 1985; Weber et al., 2000; Wegener et al., 1994).

Musculoskeletal Involvement

Joint symptoms have been noted in 12–66% of patients at the time of diagnosis and in 24–97% of patients at a given time during the course of their illness. Histological evidence of synovitis has been found in up to 66% of synovial biopsies from patients with SSc, but, clinically, arthralgias have been considered to occur more frequently than true arthritis. Radiographic abnormalities, which are detected in 46% of SSc cases, are juxta-articular demineralization, terminal phalangeal tuft resorption, articular calcification, joint erosions, and joint space narrowing (Allali et al., 2007; Baron et al., 1982) (Figure 79.5). About 20% of patients have a primary myopathy. Typically, this is a subtle process with weakness, noted by the examining physician, mild or no serum muscle enzyme elevation, perimysial and endomysial fibrosis on histologic examination. A minority of patients develop more pronounced proximal muscle weakness classified as SSc overlap with polymyositis (Clements et al., 1978; Silver et al., 2005).

Kidney Involvement

Scleroderma renal crisis is the most acute and life-threatening manifestation of SSc. It occurs most often in patients with SSc with diffuse cutaneous involvement. Until the availability of ACE inhibitors in 1979, it was almost always associated with malignant hypertension, rapidly progressive renal failure and early mortality (Steen et al., 1990).

DEVELOPMENT OF THE CLASSIFICATION CRITERIA FOR SSC

American College of Rheumatology (ACR) criteria from 1980 are used, at present, for the diagnosis of SSc (Masi et al., 1980) (Table 79.2). Presence of the major criterion or two

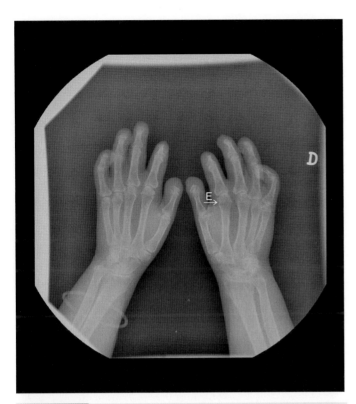

Figure 79.5 X-ray of the hands of a female SSc patient with limited SSc, demonstrating acroosteolysis of all of the fingers bilaterally and erosion (E) of the distal surface of the second right metacarpal bone.

TABLE 79.3	Valentini disease activity index for SSc	
Parameter		**Score**
1. Modified Rodnan skin score >14		1
2. Scleredema		0.5
3. Change in skin symptoms in the last month*		2
4. Digital necrosis		0.5
5. Change in vascular symptoms in the last month*		0.5
6. Arthritis		0.5
7. DLCO less than 80% of the predicted values		0.5
8. Change in cardiopulmonary symptoms*		2
9. Erythrocyte sedimentation rate (ESR) >30 mm for the first hour, Westergren method		1.5
10. Hypocomplementemia (C3 and/or C4)		1
Total disease activity index score		**10**

*Assessed by the patient.
Adapted from Valentini et al. (2003).

TABLE 79.2	ACR criteria for the classification of SSc
I. Major criterion	Skin thickening, proximal of metacarpophalangeal joints
II. Minor criteria	1. Sclerodactyly
	2. Fingertip pitting scars
	3. Bibasilar pulmonary fibrosis

Adapted from Masi et al. (1980).

minor criteria are required for the diagnosis. Nagy and Czirjac (2004) found, in 447 patients with connective tissue diseases, a high specificity of capillaroscopy for early diagnosis of SSc. Abnormal "scleroderma-type" capillaroscopic pattern was found in patients who did not fulfill the ACR criteria – those patients presented with sclerodactyly, teleangiectasia, subcutaneous calcinosis, esophageal dysmotility, and other symptoms. It has been found that the sensitivity of the ACR criteria could be improved from 33.6% to 82.9% in limited SSc by inclusion of abnormal capillaroscopic pattern as a criterion, and adding the presence of ACA increased sensitivity to 91.5% (Hudson et al., 2007b, Lonzetti et al., 2001; Walker, 2007;

Ziegler et al., 2003). To improve early diagnosis of SSc, Le Roy and Medsger proposed that patients with RP and abnormal nailfold capillaroscopic changes and/or testing positive for SSc-specific autoantibodies (anti-Scl-70 and ACA) should be diagnosed as "pre-scleroderma" or limited SSc, even in the absence of other manifestations of the disease (Le Roy and Medsger, 2001).

PROGNOSTIC MARKERS AND SYSTEMIC SCLEROSIS ACTIVITY SCORE

The difficulty of measuring disease activity is an important barrier in the study of SSc. The *Valentini Disease Activity Index* appears to be simple and easy to use, and is now increasingly being used in research settings (Table 79.3). The disease is considered active if the sum of the scores of detected items is ≥3 (Hudson et al., 2007a; Valentini et al., 2003).

TREATMENT
Immunosuppressive and Antifibrotic Therapies

Although used frequently, owing to the unfavorable outcome of controlled trials (Clements et al., 1999, 2004; Steen et al., 1982), D-penicillamine is not recommended for treatment of SSc by the European League Against Rheumatism (EULAR) (Kowal-Bielecka et al., 2009). Treatment with methotrexate in patients with early (less than three years) diffuse SSc for one year led to significantly improved modified Rodnan skin scores compared with placebo (Pope et al., 2001) and can be used in early active disease.

Although IL-2-dependent pathways are active in SSc, the nephrotoxicity of cyclosporin limits its use in SSc patients (Clements et al., 1993; Denton and Black, 2005). The production of fibrogenic cytokines, such as IL-4, IL-6, IL-17, and TGF-β1, was attenuated by rapamycin, but convincing clinical studies are lacking (Yoshizaki et al., 2010). A randomized controlled trial (RCT) comparing azathioprine and placebo after cyclophosphamide induction therapy found only marginal benefits (Hoyles et al., 2006; Ramos-Casals et al., 2010).

Mycophenolate mofetil is being tested, and further evaluation of its effect in interstitial lung disease is ongoing (Gerbino et al., 2008; Ramos Casals et al., 2010).

Specific Management of the Different Disease Manifestations

Raynaud's Phenomenon

Patient education in avoiding exposure to cold and emotional stress, in wearing warm clothes, and in smoking cessation is very important (Block and Sequeira, 2001; Herrick, 2003; Wigley, 2002). A number of pharmacologic agents have shown therapeutic efficacy in double-blind studies, including nifedipine, prazosin, pentoxiphyllin, prostaglandin E1, prostacyclin, and losartan (Silver et al., 2005). Calcium-channel blockers (CCBs) such as nifedipine (30–60 mg daily), felodipine (2.5–10 mg bid), amlodipine, nicardipine, or isradipine are usually the first choice in RP (Thompson et al., 2001). Diltiazem (30–120 mg tid) also shows a good therapeutic effect in RP (Leighton, 2001). ACE inhibitors are potent vasodilators, but the potential benefit for renin-angiotensin blockade for vascular disease in organs other than the kidney is unknown (Dziadzio et al., 1999; Herrick, 2003). The direct vasodilator prazosin and the oral serotonin antagonist ketanserin have also demonstrated therapeutic efficacy (Leighton, 2001).

Intravenous infusion with prostacyclin is a treatment of choice in patients with severe secondary RP with digital ulcers. The starting dose of epoprostenol is 1–2 ng/kg/min, which is increased to 20–40 ng/kg/min (Badesch, 2000). Treprostinil is given as a continuous subcutaneous infusion at the dose of 10–20 ng/kg/min (McLaughlin et al., 2003). Iloprost is administered as intravenous infusion for 6–8 hours, at the dose of 0.5–3 ng/kg/min. The mean duration of a therapeutic course is 5–10 days. These effects may be maintained by a one-day infusion at intervals of several weeks according to the therapeutic effect (Della Bella et al., 2001; Mittag and Beckheinrich, 2001; Scorza et al., 2001).

NO is a potent vasodilator and an inhibitor of platelet activation and vascular smooth muscle proliferation. Synthesis of NO is regulated by the family of NO synthases, and its effect is mediated via cyclic guanosine monophosphate (cGMP). The intracellular concentration of cGMP is regulated by phosphodiesterases (PDEs), which rapidly degrade cGMP *in vivo*. Pentoxiphylline is a widely used drug from this group (dose 400 mg tid), but it has not proven to be effective in severe forms of RP (Leighton, 2001; Wigley, 2002). Sildenafil

and tadalafil are specific inhibitors of the PDE-5 isoform. Sildenafil, at a dose of 50 mg three or four times per day, leads to improved blood flow in patients with severe secondary RP (Galie et al., 2005; Hoeper and Welte, 2006; Kamata et al., 2005). Tadalafil is currently used in patients who fail to improve after treatment with sildenafil (Kamata et al., 2005; Kumana et al., 2004; Rosenkranz et al., 2003).

Improvement of RP, with significant reductions in the Raynaud's condition score and patient assessment by visual analog scale, has been observed during atorvastatin treatment of SSc patients at the dose 10 mg daily for 24 months. Statins may thus be beneficial in treating vascular manifestations of SSc (Kuwana et al., 2009).

The endothelin-receptor antagonist (ERA) bosentan has no proven efficacy in the treatment of active digital ulcers in SSc patients. However, it has proven efficacy in two high quality RCTs (the RAPIDS-1 and -2 studies) to prevent digital ulcers in diffuse SSc patients, particularly in those with multiple digital ulcers (Korn et al., 2004). CCBs, prostanoids and bosentan are included in the current EULAR recommendations for the management of RP in SSc, which consist of 14 statements referring to the specific disease manifestations (Table 79.4) (Kowal-Bielecka et al., 2009).

Lung

In patients with evidence of inflammatory alveolitis, cyclophosphamide is the gold standard, with doses of 1.5–2 mg/kg per day for a period of up to one year (Akesson et al., 1994; Kowal-Bielecka et al., 2009).

Pulmonary Arterial Hypertension

It has been accepted that the therapeutic approach in severe PAH in the context of autoimmune disease should follow the same treatment strategy as in primary PAH patients. As a general recommendation, low-level aerobic exercise, such as walking, is recommended for PAH patients. Restriction of sodium consumption to 2400 mg per day in patients with right ventricular failure is also a common measure in these patients. Routine immunizations against influenza and pneumococcal pneumonia are advised to prevent severe infections. The 2009 American College of Cardiology (ACC) guidelines recommend oxygen therapy in PAH patients with pulse oximetry saturation of less than 90% at rest or with exercise (McLaughlin et al., 2009). Abnormalities of the activated clotting system, impaired fibrinolysis, abnormal platelet function, and histological evidence of microvascular thrombosis provide the rationale for anticoagulation therapy in these patients (Johnson et al., 2006; Mathai and Hassoun, 2009; McLaughlin et al., 2009). The 2009 ACC treatment guidelines recommend warfarin anticoagulation in patients with idiopathic PAH, titrated to an international normalized ratio of 1.5:2.5. Treatment of right heart failure with digoxin, diuretics, and oxygen therapy is thought to offer little more than palliation (Proudman et al., 2007). Epoprostenol at the above-mentioned doses improves exercise capacity in SSc-related PAH

TABLE 79.4 The European League Against Rheumatism (EULAR) recommendations for the treatment of SSc

I. SSc-related digital vasculopathy (RP, digital ulcers)

1. A meta-analysis on dihydropiridine-type CCBs and a meta-analysis on prostanoids indicate that nifedipine and intravenous (IV) iloprost reduce the frequency and severity of SSc-related RP attacks. Dihydropiridine-type CCBs, usually oral nifedipine, should be considered for first-line therapy for SSc-related RP, and intravenous iloprost, or other available IV prostanoids, should be considered for severe SSc-related RP.

2. Two RCTs indicate that intravenous prostanoids (particularly iloprost) are efficacious in healing digital ulcers in patients with SSc. IV prostanoids (in particular iloprost) should be considered for treatment of active digital ulcers in patients with SSc.

3. Bosentan has no proven efficacy in the treatment of active digital ulcers in SSc patients. In two high-quality RCTs, bosentan has proven efficacy to prevent digital ulcers in diffuse SSc patients, particularly those with multiple digital ulcers. Bosentan should be considered in diffuse SSc with multiple digital ulcers, after failure of CCBs and, usually, prostanoid therapy.

II. SSc-related PAH

4. Two high-quality RCTs indicate that bosentan improves exercise capacity, functional class, and some hemodynamic measures in PAH. Bosentan should be strongly considered to treat SSc-related PAH.

5. Two high-quality RCTs indicate that sitaxsentan improves exercise capacity, functional class, and some hemodynamic measures in PAH. At present, sitaxsentan may also be considered to treat SSc-PAH.

6. One high-quality RCT indicates that sildenafil improves exercise capacity, functional class and some hemodynamic parameters in PAH. Sildenafil may be considered to treat SSc-PAH.

7. One high-quality RCT indicates that continuous intravenous epoprostenol improves exercise capacity, functional class, and hemodynamic measures in SSc-PAH. Sudden drug withdrawal could be life-threatening. Intravenous epoprostenol should be considered for the treatment of patients with severe SSc-PAH.

III. SSc-related skin involvement

8. Two RCTs have shown that methotrexate improves skin score in early diffuse SSc. Positive effects on other organ manifestations have not been established. Methotrexate may be considered for treatment of skin manifestations of early diffuse SSc.

IV. Scleroderma interstitial lung disease

9. In view of the results from two high-quality RCTs and despite its known toxicity, cyclophosphamide should be considered for treatment of SSc-related interstitial lung disease.

V. Scleroderma renal crisis

10. Despite the lack of RCTs, experts believe that ACE inhibitors should be used in the treatment of scleroderma renal crisis.

11. Four retrospective studies suggest that steroids are associated with a higher risk of scleroderma renal crisis. Patients on steroids should be carefully monitored for blood pressure and renal function.

VI. SSc-related gastrointestinal disease

12. Despite the lack of specific RCTs, experts believe that proton pump inhibitors should be used for prevention of SSc-related gastroesophageal reflux disease (GERD), esophageal ulcers and strictures.

13. Despite the lack of specific RCTs, experts believe that prokinetic drugs should be used for the management of SSc-related symptomatic motility disturbances (dysphagia, GERD, early satiety response, bloating, pseudo-obstruction, etc.)

14. Despite the lack of specific RCTs, experts believe that, when malabsorption is due to bacterial overgrowth, rotating antibiotics may be useful in SSc.

Adapted from Kowal-Bielecka et al. (2009).

and improves survival (Badesch, 2000; McLaughlin et al., 2009). Both intravenous and subcutaneous treprostinil are possible, and produce similar hemodynamic effects to those of epoprostenol in patients with PAH. Prostaglandins for inhalation are also used in PAH, and inhaled iloprost has been found to be effective therapy for patients with severe PAH. When administered via inhalation, its pulmonary vasodilative potency was similar to that of prostacyclin, but its effects lasted for 30–90 minutes, as compared with 15 minutes for prostracyclin (Hoeper et al.,

2002). Beraprost is the first oral prostacyclin analog with vasodilative and antiplatelet action, and a half-life of approximately one hour (Galie et al., 2002; Saji et al., 1996).

Bosentan is approved for treatment of PAH associated with systemic rheumatic diseases. It is used at a dose of 62.5 mg twice daily for four weeks before titration up to 125–250 mg twice daily (Korn et al., 2004; McLaughlin et al., 2009; Rubin et al., 2002). In 2008, bosentan was also approved in the European Union for the treatment of mild PAH (WHO class

II) in SSc (EARLY study) (Galie et al., 2008). Side effects are headache and elevation of liver enzymes, anemia, and edema (Denton et al., 2006; Galie et al., 2008; Korn et al., 2004; Rubin et al., 2002). The Food and Drug Administration (FDA) requires monthly liver function tests and the hematocrit to be checked every three months (McLaughlin et al., 2009). Sitaxsentan is a new oral, once-daily, highly selective ERA. It is used at the dose of 50–100 mg per day. The observed side effects are similar to those during therapy with bosentan (Waxman, 2007). Ambrisentan is a potent and selective inhibitor of ET-A receptors (ET-A:ET-B − 4000:1). The half-life of the medication is 9–15 hours, which allows once-daily dosing (2.5–10 mg per day). Effectiveness of ambrisentan was evaluated in two Phase III RCTs (ARIES-1, -2) in patients with PAH. Ambrisentan improves exercise capacity, with a dose-dependent effect.

The isoform 5 of the enzyme PDE is present in large amounts in the lung (Giaid and Saleh, 1995; Kamata et al., 2005). All three PDE-5 inhibitors (sildenafil, 50 mg daily dose; tadalafil, 20, 40, and 60 mg daily dose; vardenafil, 10 and 20 mg daily dose) have been found to cause significant pulmonary vasorelaxation. Significant improvement in arterial oxygenation (equally to NO inhalation), however, was only noted with sildenafil.

The combination therapy of sildenafil with intravenous epoprostenol has led to improvement in patients with idiopathic PAH, while in SSc-related PAH, efficacy remains to be determined. The combination of sildenafil and bosentan improved the functional class in idiopathic PAH, while patients with PAH in SSc did not experience significant improvement (Mathai and Hassoun, 2009).

Imatinib is a small molecule tyrosine kinase inhibitor that binds competitively to the ATP-binding pocket of Abelsonkinase (c-Abl) and thereby efficiently blocks its tyrosine kinase activity. C-Abl is an important downstream signaling molecule of TGF-β and PDGF. Thus, imatinib targets simultaneously and selectively two major pro-fibrotic pathways activated in SSc (Distler and Distler, 2008; Hassoun, 2009). Based on evidence that PDGF signaling is an important process in the pathophysiology of PAH, imatinib has been tested and shown to be effective in experimental models of PAH.

Of the available surgical options, atrial septostomy creates a right-to-left inter-atrial shunt, decreasing right heart loading pressures and improving right heart function and left heart load in patients with worsened right heart function in severe PAH. Single and double lung transplantation and combined heart and lung transplantation are ultimate therapeutic options in patients with end-stage disease (McLaughlin et al., 2009).

Gastrointestinal Involvement

An upright posture and avoidance of exertion after eating are common measures in cases with esophageal disease to minimize the possibility of heartburn (Matucci-Cerinic et al., 2009). For symptoms of heartburn and dyspepsia, proton pump inhibitors should be used. In refractory cases, doubling of the dose could be considered as well as the addition of a H2 blocker at bedtime, because of their different mechanisms of action

to prevent nocturnal breakthrough (Baron et al., 2010; Khan et al., 2007; Mainie et al., 2008; Williams and McColl, 2006). In cases of gastroparesis, pro-motility agents such as erythromycin (100–150 mg qid), azithromycin (400 mg daily), metoclopramide (10–15 mg qid), and domperidon (10–20 mg qid), may be helpful (Baron et al., 2010; Ntoumazios et al., 2006; Ramirez-Mata et al., 1977). Watermelon stomach is treated by several transendoscopic procedures: laser photocoagulation, bipolar electrocoagulation, heater probe coagulation, and injection sclerotherapy. Surgical intervention (partial gastrectomy, antrectomy) is indicated in some cases (Elkayam et al., 2000).

Malabsorption due to suspected bacterial overgrowth irrespective of the results of breath testing is treated with rotating broad-spectrum antibiotics. A 10-day course of a selective antibiotic could be tried, with subsequent treatment with a selective antibiotic for the first 10 days of four consecutive months (Baron et al., 2010; Marie et al., 2009). Therapy with octreotide may be considered in cases with refractory symptoms from small bowel involvement (Baron et al., 2010; Perlemuter et al., 1999). Pseudo-obstruction of the small and large intestines occurs in patients with either the limited or diffuse form of SSc, and is frequently lethal regardless of the therapeutic efforts as outlined above (Silver et al., 2005).

Musculoskeletal Involvement

Articular complaints can be treated with non-steroidal antiinflammatory drugs. Corticosteroids are rarely necessary, but patients with true synovitis may benefit from prednisone, 5–7.5 mg daily. Patients with diffuse cutaneous involvement often develop digital contractures, with deformity and severe impairment of hand function. A vigorous exercise program should be recommended with use of analgesics before each exercise session to relieve the pain and maximize the results. In cases of severe finger contractures, the hand is vulnerable to trauma. Repeated skin breakdown and infection, including septic arthritis, are potential complications. In these cases, joint fusion is a treatment option, while joint replacement will not be effective. Active polymyositis should be treated with moderate doses of corticosteroids (prednisone, 15–20 mg/ day). In cases of incomplete response, methotrexate, azathioprine, or another immunosuppressive agent can be added (Silver et al., 2005).

Kidney Involvement

All patients with SSc who have new-onset hypertension, with or without evidence of renal involvement, should be treated and maintained on a maximum tolerable dose of an ACE inhibitor, even if renal failure develops (Steen et al., 1990). Patients with scleroderma renal crisis who received ACE inhibitors had an impressive one-year survival of 76% and a five-year survival of 65%, compared with 15% one-year survival and 10% five-year survival of patients not receiving ACE inhibitors, despite other aggressive antihypertensive treatment (Steen et al., 1990). In a small group of patients with "normotensive" scleroderma renal crisis, ACE inhibitors also have been shown to improve survival (Helfrich et al., 1989). High-dose corticosteroids are associated with a higher risk of scleroderma renal crisis, and it is recommended that patients on steroids should be carefully monitored for blood pressure and renal function (Kowal-Bielecka et al., 2009) (Table 79.4).

New Therapies

Biologic Agents in SSc

Rituximab has been examined in SSc patients. It has shown to be well tolerated and may have efficacy for skin disease, but clinical trials in larger patient groups are warranted (Smith et al., 2010, Van Laar, 2010). Trials using TNF inhibitors have not been optimistic for the use of TNF as a specific therapeutic target in SSc (Allanore et al., 2006; Denton et al. 2009; Ostor et al., 2004; Ramos-Casals et al., 2010).

Stem Cell Transplantation

Autologous hematopoietic stem cell transplantation (autologous HSCT) has emerged as a new therapeutic procedure for patients affected by severe SSc refractory to conventional treatments. In a long-term study of 11 SSc patients, Tsukamoto and colleagues evaluated the efficacy of HSCT. Clinical and immunological analyses were conducted for four years. Skin sclerosis markedly improved within six months and was maintained for 48 months in all patients. Progression of interstitial pneumonia was prevented for 48 months (Tsukamoto et al., 2006). In 26 SSc patients, Vonk and colleagues observed that autologous HSCT in selected patients with severe diffuse cutaneous SSc resulted in sustained improvement of skin thickening and stabilization of organ function (e.g., pulmonary, cardiac, and renal) up to seven years after transplantation (Vonk et al., 2008). Multicenter, randomized controlled trials of autologous HSCT for SSc, such as the ASTIS and SCOT trials, are currently ongoing in Europe and the US. The results are urgently needed to determine the true role of HSCT in SSc (Ramos-Casals et al., 2010).

A successful treatment of digital ulcers with local stem cell transplantation has also been reported (Nevskaya et al., 2009).

PROGNOSIS

Factors negatively affecting survival include male sex and older age at diagnosis. Survival of patients with scleroderma renal crisis after initiation of treatment with ACE inhibitors is improved. Thus, currently PAH is one of the complications of SSc that determines poor prognosis. Historically, the one-year survival rate in SSc-associated PAH was 45% (Condiffe et al., 2009; Koh et al., 1996). A recent six-year follow-up (2001–2006) of 315 SSc patients with PAH who have been documented in the UK national registry has revealed one-, two- and three-year survival rates of 78, 58, and 47%, respectively. Although survival in patients with SSc-associated PAH is better in the modern treatment era than in historical cohorts, it remains unacceptably poor (Condiffe et al., 2009).

REFERENCES

Ahmed, S.S., Tan, F.K., 2003. Identification of novel targets in scleroderma, update on population studies, cDNA arrays, SNP analysis, and mutations. Curr Opin Rheumatol 15, 766–771.

Akesson, A., Schema, A., Lundin, A., Wollheim, F.A., 1994. Improved pulmonary function in systemic sclerosis after treatment with cyclophosphamide. Arthritis Rheum 37, 729–735.

Allali, F., Tahiri, L., Senjari, A., et al., 2007. Erosive arthropathy in systemic sclerosis. BMC Public Health 7, 260.

Allanore, Y., 2004. Lack of association of eNOS (G894T) and p22hox NAPDH oxidase subunit (C242T) polymorphism with systemic sclerosis in a cohort of French Caucasian patients. Clin Chim Acta 350, 51–55.

Allanore, Y., Devos-François, G., Caramella, C., et al., 2006. Fatal exacerbation of fibrosing alveolitis associated with systemic sclerosis in a patient treated with adalimumab. Ann Rheum Dis 65, 834–835.

Allanore, Y., Dieude, P., Boileau, C., 2010. Updating the genetics of systemic sclerosis. Curr Opin Rheumatol 22, 665–670.

Allanore, Y., Saad, M., Dieude, P., et al., 2011. Genome-wide scan identifies TNIP1, PSORS1C1, and RHOB as novel risk loci for systemic sclerosis. PLoS Genetics 7(7), e1002091 doi:10.1371/journal.pgen.100209.

Allcock, R.J., Forrest, I., Corris, P.A., et al., 2004. A study of the prevalence of systemic scerosis in northeast England. Rheumatology (Oxford) 43, 596–602.

Arnett, F.C., Cho, M., Chatterjee, S., et al., 2001. Familial occurrence frequencies and relative risks for systemic sclerosis (scleroderma) in three United States cohorts. Arthritis Rheum 44, 1359–1362.

Arnett, F.C., Gourh, P., Shete, S., et al., 2010. Major histocompatibility complex (MHC) class II alleles, haplotypes, and epitopes which confer susceptibility or protection in the fibrosing autoimmune disease systemic sclerosis: Analyses in 1300 Caucasian, African-American and Hispanic cases and 1000 controls. Ann Rheum Dis 69, 822–827.

Assassi, S., 2005. Polymorphisms of endothelial nitric oxide synthase and angiotensin-converting enzyme in systemic sclerosis. Am J Med 118, 907–911.

Badesch, D.B., 2000. Continuous intravenous epoprostenol for pulmonary hypertension due to the scleroderma spectrum disease. Ann Intern Med 132, 434–452.

Baron, M., Bernier, P., Cote, L.F., et al., 2010. Screening and management for malnutrition and related gastro-intestinal disorders in systemic sclerosis: Recommendations of a North American expert panel. Clin Exp Rheumatol 28 (Suppl 58), S42–S46.

Baron, M., Lee, P., Keystone, E.C., 1982. The articular manifestations of progressive systemic sclerosis (scleroderma). Ann Rheum Dis 41, 147–152.

Block, J.A., Sequeira, W., 2001. Raynaud's phenomenon. Lancet 357, 2042–2048.

Bollinger, A., Fagrell, B., 1990. Clinical capillaroscopy: A guide to its use in clinical research and practice. Hogrete & Huber Publishers, Toronto, 1–158.

Bossini-Castillo, L., Broen, J.C., Simeon, C.P., et al., 2011. A replication study confirms the association of TNFSF4 (OX40L) polymorphisms with systemic sclerosis in a large European cohort. Ann Rheum Dis 70, 638–641.

Bulkley, B.H., Ridolfi, R.L., Salyer, W.R., et al., 1976. Myocardial lesions of progressive systemic sclerosis: A cause of cardiac dysfunction. Circulation 53, 483–490.

Clements, P., Lachenbruch, P., Seibold, J., et al., 1995. Inter- and intraobserver variability of total skin thickness score (modified Rodnan TSS) in systemic sclerosis. J Rheumatol 22, 1281–1285.

Clements, P.J., Furst, D.E., Campion, D.S., et al., 1978. Muscle disease in progressive systemic sclerosis: Diagnostic and therapeutic considerations. Arthritis Rheum 21, 62–71.

Clements, P.J., Furst, D.E., Wong, W.K., et al., 1999. High-dose versus low-dose D-penicillamine in early diffuse systemic sclerosis: Analysis of a two-year, double-blind, randomized, controlled clinical trial. Arthritis Rheum 42, 1194–1203.

Clements, P.J., Lachenbruch, P.A., Sterz, M., et al., 1993. Cyclosporine in systemic sclerosis – results of a forty-eight-week open safety study in ten patients. Arthritis Rheum 36, 75–83.

Clements, P.J., Seibold, J.R., Furst, D.E., et al., 2004. High-dose versus low-dose D-penicillamine in early diffuse systemic sclerosis trial: Lessons learned. Semin Arthritis Rheum 33, 249–263.

Condiffe, R., Kiely, D.G., Peacock, A.J., et al., 2009. Connective tissue disease-associated pulmonary arterial hypertension in the modern treatment era. Am J Respir Crit Care Med 179, 151–167.

Crilly, A., 2002. Analysis of transforming growth factor beta1 gene polymorphisms in patients with systemic sclerosis. Ann Rheum Dis 61, 678–681.

Cutolo, M., Grassi, W., Matucci-Cerinic, M., 2003. Raynaud's phenomenon and the role of capillaroscopy. Arthritis Rheum 48, 3023–3030.

Cutolo, M., Pizzorni, C., Sulli, A., 2005. Capillaroscopy. Best Pract Res Clin Rheumatol 19, 437–452.

Cutolo, M., Sulli, A., Pizzorni, C., Accardo, S., 2000. Nailfold videocapillaroscopy assessment of microvascular damage in systemic sclerosis. J Rheumatol 27, 155–160.

Czubaty, A., 2005. Proteomic analysis of complexes formed by human topoisomerase I. Biochim Biophys Acta 1749, 133–141.

Davidson, A., Russell, C., Littlejohn, G.O., 1985. Assessment of esophageal abnormalities in progressive systemic sclerosis using radionuclide transit. J Rheumatol 12, 472–477.

Della Bella, S., Molteni, M., Mocellin, C., et al., 2001. Novel mode of action of iloprost: In vitro downregulation of endothelial cell adhesion molecules. Prostaglandins Other Lipid Mediat 65, 73–83.

Denton, C.P., Black, C.M., 2005. Targeted therapy comes in age of scleroderma. Trends Immunol 26, 596–602.

Denton, C.P., Engelhart, M., Tvede, N., et al., 2009. An open-label pilot study of infliximab therapy in diffuse cutaneous systemic sclerosis. Ann Rheum Dis 68, 1433–1439.

Denton, C.P., Humbert, M., Rubin, L., Black, C.M., 2006. Bosentan treatment for pulmonary arterial hypertension related to connective tissue disease: A subgroup analysis of the pivotal clinical trials and their open-label extensions. Ann Rheum Dis 65, 1336–1340.

Distler, J.H.W., Distler, O., 2008. Intracellular tyrosine kinase as a novel target for anti-fibrotic therapy in systemic sclerosis. Rheumatology 47, v10–v11.

Dziadzio, M., Denton, C.P., Smith, R., et al., 1999. Losartan therapy for Raynaud's phenomenon and scleroderma. Arthritis Rheum 42, 2646–2655.

Elkayam, O., Oumanski, M., Yaron, M., Caspi, D., 2000. Watermelon stomach following and preceding systemic sclerosis. Semin Arthritis Rheum 30, 127–131.

Fatini, C., 2004. Vascular injury in systemic sclerosis, angiotensin-converting enzyme insertion/deletion polymorphism. Curr Rheumatol Rep 6, 149–155.

Feghali-Bostwick, C.A., 2005. Genetics and proteomics in scleroderma. Curr Rheumatol Rep 7, 129–134.

Galie, N., Ghofrani, H.A., Torbicki, A., et al., 2005. Sildenafil citrate therapy for pulmonary arterial hypertension. N Engl J Med 353, 2148–2157.

Galie, N., Humbert, M., Vachiery, J.L., et al., 2002. Effects of beraprost sodium, an oral prostacyclin analogue, in patients with pulmonary arterial hypertension: A randomized, double-blind, placebo-controlled trial. J Am Coll Cardiol 39, 1496–1502.

Galie, N., Rubin, L.J., Koeper, M.M., et al., 2008. Treatment of patients with mildly symptomatic pulmonary arterial hypertension with bosentan (EARLY study): A double-blind, randomized controlled trial. Lancet 371, 2093–2100.

Gerbino, A.J., Goss, C.H., Molitor, J.A., 2008. Effect of mycophenolate mofetil on pulmonary function in scleroderma-associated interstitial lung disease. Chest 133, 455–460.

Giaid, A., Saleh, D., 1995. Reduced expression of endothelial nitric oxide synthase in the lungs of patients with pulmonary hypertension. N Engl J Med 333, 214–221.

Giusti, B., 2006. A model of anti-angiogenesis, differential transcriptosome profiling of microvascular endothelial cells from diffuse systemic sclerosis patients. Arthritis Res Ther 8, R115.

Goh, N.S., Veeraraghavan, S., Desai, S.R., et al., 2007. Bronchoalveolar lavage cellular profiles in patients with systemic sclerosis-associated interstitial lung disease are not predictive of disease progression. Arthritis Rheum 56, 2005–2012.

Goldblatt, F., Gordon, T.P., Waterman, S.A., 2002. Antibody-mediated gastrointestinal dysmotility in scleroderma. Gastroenterology 123, 1144–1150.

Greydanus, M.P., Camilleri, M., 1989. Abnormal postcibal antral and small bowel motility due to neuropathy or myopathy in systemic sclerosis. Gastroenterology 96, 110–115.

Hassoun, P.M., 2009. Therapies for scleroderma-related pulmonary arterial hypertension. Exp Rev Resp Med 3, 187–196.

Helfrich, D.J., Banner, B., Steen, V.D., Medsger Jr., T.A., 1989. Normotensive renal failure in systemic sclerosis. Arthritis Rheum 32, 1128–1134.

Herrick, A.L., 2003. Treatment of Raynaud's phenomenon: New insights and developments. Curr Rheumatol Rep 5, 168–174.

Herrick, A.L., Matucci-Cerinic, M., 2001. The emerging problem of oxidative stress and the role of antioxidants in systemic sclerosis. Clin Exp Rheumatol 19, 4–8.

Ho, K.T., Reveille, J.D., 2003. The clinical relevance of autoantibodies in scleroderma. Arthritis Res Ther 5, 80–93.

Ho, M., Belch, J.J.F., 1998. Raynaud's phenomenon: State of the art 1998. Scand J Rheumatol 27, 319–322.

Hoeper, M.M., Spiekerkoetter, E., Westerkamp, V., Gatzke, R., Fabel, H., 2002. Intravenous iloprost for treatment failure of aerosolised iloprost in pulmonary arterial hypertension. Eur Respir J 20, 339–343.

Hoeper, M.M., Welte, T., 2006. Sildenafil citrate therapy for pulmonary arterial hypertension. N Engl J Med 354, 1091–1093.

Hoyles, R.K., Ellis, R.W., Wellsbury, J., et al., 2006. A multicenter, prospective, randomized, double-blind, placebo-controlled trial of corticosteroids and intravenous cyclophosphamide followed by oral azathioprine for the treatment of pulmonary fibrosis in scleroderma. Arthritis Rheum 54, 3962–3970.

Hudson, M., Steele, R., Baron, M., 2007a. Update on indices of disease activity in systemic sclerosis. Semin Arthritis Rheum 37, 93–98.

Hudson, M., Taillefer, S., Steele, R., et al., 2007b. Improving the sensitivity of the American College of Rheumatology classification criteria for systemic sclerosis. Clin Exp Rheumatol 25, 754–757.

Humbert, M., Morrell, N.W., Archer, A.L., et al., 2004. Cellular and molecular pathobiology of pulmonary arterial hypertension. J Am Coll Cardiol 43 (12), 13S–24S.

Hutyrova, B., 2004. Interleukin 1alpha single-nucleotide polymorphism associated with systemic sclerosis. J Rheumatol 31, 81–84.

Jeffery, T.K., Morrell, N.W., 2002. Molecular and cellular basis of pulmonary vascular remodeling in pulmonary hypertension. Prog Cardiovasc Dis 45, 173–202.

Johnson, K.L., 2001. Fetal cell microchimerism in tissue from multiple sites in women with systemic sclerosis. Arthritis Rheum 44, 1848–1854.

Johnson, S.R., Mehta, S., Granton, J.T., 2006. Anticoagulation in pulmonary arterial hypertension: A qualitative systematic review. Eur Respir J 28, 999–1004.

Joung, C.I., 2006. Angiotensin-converting enzyme gene insertion/deletion polymorphism in Korean patients with systemic sclerosis. J Korean Med Sci 21, 329–332.

Kahaleh, B., Meyer, O., Scorza, R., 2003. Assessment of vascular involvement. Clin Exp Rhematol 21 (Suppl. 29), S9–S14.

Kahaleh, M.B., Sultany, G.L., Smith, E.A., et al., 1986. A modified scleroderma skin scoring method. Clin Exp Rheumatol 4, 367–369.

Kahan, A., Nitenberg, A., Foult, J.M., et al., 1985. Decreased coronary reserve in primary scleroderma myocardial disease. Arthritis Rheum 28, 637–646.

Kamata, Y., Kamimura, T., Iwamoto, M., Minota, S., 2005. Comparable effects of sildenafil citrate and alprostadil on severe Raynaud's phenomenon in a patient with systemic sclerosis. Clin Exp Dermatol 30, 451.

Kawaguchi, Y., 2003. Association of IL1A gene polymorphisms with susceptibility to and severity of systemic sclerosis in the Japanese population. Arthritis Rheum 48, 186–192.

Kehrl, J.H., Roberts, A.B., Wakefield, L.M., et al., 1986. Transforming growth factor beta is an important immunomodulatory protein for human B lymphocytes. J Immunol 137, 3855–3860.

Khan, M., Santana, J., Donnelan, C., et al., 2007. Medical treatments in the short term management of reflux esophagitis. Cochrane Database Syst Rev, 18.

Kim, D.S., Yoo, B., Lee, J.S., 2002. The major histopathologic pattern of pulmonary fibrosis in scleroderma is non specific interstitial pneumonia. Sarcoidosis Vasc Diffuse Lung Dis 19, 121–127.

Koh, E.T., Lee, P., Gladman, D., Abu-Shakram, M., 1996. Pulmonary hypertension in systemic sclerosis: An analysis of 17 patients. Br J Rheumatol 35, 989–993.

Konttinen, Y.T., Mackiewicz, Z., Ruuttila, P., et al., 2003. Vascular damage and lack of angiogenesis in systemic sclerosis skin. Clin Rheumatol 22, 196–202.

Korn, J.H., Mayes, M., Matucci-Cerinic, M., et al., 2004. Digital ulcers in systemic sclerosis: Prevention by treatment with bosentan, an oral endothelin receptor antagonist. Arthritis Rheum 50, 3985–3993.

Kowal-Bielecka, O., Landewé, R., Avouac, J., et al., 2009. EULAR recommendations for the treatment of systemic sclerosis: A report from the EULAR Scleroderma Trials and Research group (EUSTAR). Ann Rheum Dis 68, 620–628.

Kraaij, M.D., Van Laar, J.M., 2008. The role of B cells in systemic sclerosis. Biologics 2, 389–395.

Kumana, C.R., Cheung, G.T.Y., Lau, C.S., 2004. Severe digital ischaemia treated with phosphodiesterase inhibitors. Ann Rheum Dis 63, 1522–1524.

Kuwana, M., Okano, Y., Kaburaki, J., Tojo, T., Medsger Jr., T.A., 1994. Racial differences in the distribution of systemic sclerosis – related antinuclear antibodies. Arthritis Rheum 37, 902–906.

Kuwana, M., Okazaki, Y., Kaburaki, J., 2009. Long-term beneficial effects of statins on vascular manifestations in patients with systemic sclerosis. Mod Rheumatol 19, 530–535.

Lambert, N.C., 2000a. Cutting edge, persistent fetal microchimerism in T lymphocytes is associated with HLA-DQA1*0501, implications in autoimmunity. J Immunol 164, 5545–5548.

Lambert, N.C., 2000b. HLA-DQA1*0501 is associated with diffuse systemic sclerosis in Caucasian men. Arthritis Rheum 43, 2005–2010.

Le Roy, E.C., Medsger Jr., T.A., 2001. Criteria for the classification of early systemic sclerosis. J Rheumatol 28, 1573–1576.

Leighton, C., 2001. Drug treatment in scleroderma. Drugs 61, 419–427.

Lonzetti, L.S., Joyal, F., Raynauld, J.P., et al., 2001. Updating the American College of Rheumatology preliminary classification criteria for systemic sclerosis: Addition of severe nailfold capillaroscopy abnormalities markedly increases the sensitivity for limited scleroderma. Arthritis Rheum 44, 735–736.

Lunardi, C., Dolcino, M., Peterlana, D., et al., 2006. Antibodies against human cytomegalovirus in the pathogenesis of systemic sclerosis: A gene array approach. PLoS Med 3, e2.

Luzina, I.G., 2003. Occurrence of an activated, profibrotic pattern of gene expression in lung CD8 T cells from scleroderma patients. Arthritis Rheum 48, 2262–2274.

Mainie, I., Tutuian, R., Castell, D.O., 2008. Addition of a H2 receptor antagonist to PPI improves acid control and decreases nocturnal acid breakthrough. J Clin Gastroenterol 42, 676–679.

Marasini, B., 2001. Stromelysin promoter polymorphism is associated with systemic sclerosis. Rheumatology 40, 475–476.

Maricq, H.R., Harper, F.E., Khan, M.M., et al., 1983. Microvascular abnormalities as possible predictors of disease subsets in Raynaud phenomenon and early connective tissue disease. Clin Exp Rheumatol 1, 195–205.

Maricq, H.R., LeRoy, E.C., D'Angelo, W.A., et al., 1980. Diagnostic potential of in vivo capillary microscopy in scleroderma and related disorders. Arthritis Rheum 23, 183–189.

Marie, I., Ducrotte, P., Denis, P., Menard, J.F., Levesque, H., 2009. Small intestine bacterial overgrowth in systemic sclerosis. Rheumatology (Oxford) 48, 1314–1319.

Masi, A., Rodnan, G.P., Medsger Jr., T.A., et al., 1980. Preliminary criteria for the classification of systemic sclerosis (scleroderma). Subcommittee for Scleroderma criteria of the American Rheumatism Association Diagnostic and Therapeutic Criteria Committee. Arthritis Rheum 23, 581–590.

Mathai, S.C., Hassoun, P.M., 2009. Therapy for pulmonary arterial hypertension associated with systemic sclerosis. Curr Opin Rheumatol 21, 642–648.

Matucci-Cerinic, M., Miniati, I., Denton, C.P., 2009. Systemic sclerosis. In: Bijlsma, J.W.J. (Ed.), EULAR Compendium on Rheumatic Diseases, 1st ed BMJ Publishing Group, London, pp. 290–296.

Mayes, M.D., 1999. Epidemiologic studies of environmental agents and systemic autoimmune diseases. Environ Health Perspect 107 (Suppl 5), 743–748.

Mayes, M.D., Lacey Jr., J.V., Beebe-Dimmer, J., et al., 2003. Prevalence, incidence, survival, and disease characteristics of systemic sclerosis in a large US population. Arthritis Rheum 48, 2246–2255.

McLaughlin, V.V., Archer, S.L., Badesch, D.B., et al., 2009. ACCF/AHA 2009 Expert Consensus Document on Pulmonary Hypertension: A Report of the American College of Cardiology Foundation Task Force on Expert Consensus Documents and the American Heart Association. J Am Coll Cardiol 53, 1573–1619.

McLaughlin, V.V., Gaine, S.P., Barst, R.J., et al., 2003. Efficacy and safety of treprostinil: An epoprostenol analog for primary pulmonary hypertension. J Cardiovasc Pharmacol 41, 293–299.

Meyer, O., 2006. Prognostic markers for systemic sclerosis. Joint Bone Spine 73, 490–494.

Mittag, M.P., Beckheinrich, U.F., 2001. Systemic sclerosis-related Raynaud's phenomenon: Effects of iloprost infusion therapy on serum cytokine, growth factor and soluble adhesion molecule levels. Acta Derm Venereol 81, 294–297.

Moore, T.L., Lunt, M., McManus, B., Anderson, M.E., Herrick, A.L., 2003. Seventeen-point dermal ultrasound scoring system – a reliable measure of skin thickness in patients with systemic sclerosis. Rheumatology (Oxford) 42, 1559–1563.

Nagy, Z., Czirjac, L., 2004. Nailfold digital capillaroscopy in 447 patients with connective tissue disease and Raynaud's disease. J Eur Acad Dermatol Venerol 18, 62–68.

Namboodiri, A.M., Rocca, K.M., Pandey, J.P., 2004. IgG antibodies to human cytomegalovirus late protein UL94 in patients with systemic sclerosis. Autoimmunity 37, 241–244.

Nevskaya, T., Ananieva, L., Bykovskaia, S., et al., 2009. Autologous progenitor cell implantation as a novel therapeutic intervention for ischaemic digits in systemic sclerosis. Rheumatology (Oxford) 48, 61–64.

Nguyen, B., Mayes, M.D., Arnett, F.C., 2011. HLA-DRB1*0407 and *1304 are risk factors for scleroderma renal crisis. Arthritis Rheum 63, 530–534.

Nives Parodi, M., Castagneto, C., Filaci, G., et al., 1997. Plicometer skin test: A new technique for the evaluation of cutaneous involvement in systemic sclerosis. Br J Rheumatol 36, 244–250.

Ntoumazios, S.K., Voulgari, P.V., Potsis, K., et al., 2006. Esophageal involvement in scleroderma: Gastroesophageal reflux, the common problem. Semin Arthritis Rheum 36, 173–181.

Ostor, A.J., Crisp, A.J., Somerville, M.F., Scott, D.G., 2004. Fatal exacerbation of rheumatoid arthritis associated fibrosing alveolitis in patients given infliximab. BMJ 329, 1266.

Pandey, J.P., Takeuchi, F., 1999. TNF-alpha and TNF-beta gene polymorphisms in systemic sclerosis. Hum Immunol 60, 1128–1130.

Pannu, J., 2006. Increased levels of transforming growth factor beta receptor type I and up-regulation of matrix gene program: A model of scleroderma. Arthritis Rheum 54, 3011–3021.

Perlemuter, G., Cacoub, P., Chaussade, S., et al., 1999. Octreotide treatment of chronic intestinal pseudoobstruction secondary to connective tissued diseases. Arthritis Rheum 42, 1545–1549.

Peters-Golden, M., Wise, R.A., Hochberg, M.C., et al., 1984. Carbon monoxide diffusing capacity as predictor of outcome in systemic sclerosis. Am J Med 77, 1027–1034.

Pope, J.E., Bellamy, N., Seibold, J.R., et al., 2001. A randomized, controlled trial of methotrexate versus placebo in early diffuse scleroderma. Arthritis Rheum 44, 1351–1358.

Proudman, S.M., Stevens, W.M., Sahhar, J., Celermajer, D., 2007. Pulmonary arterial hypertension in systemic sclerosis: The need for early detection and treatment. Int Med J 37, 485–494.

Radstake, T.R., Vonk, M.C., Dekkers, M., 2009. The -2518A>G promoter polymorphism in the CCL2 gene is not associated with systemic sclerosis susceptibility or phenotype: Results from a multicenter study of European Caucasian patients. Hum Immunol 70, 130–133.

Ramirez-Mata, M., Ibañez, G., Alarcon-Segovia, D., 1977. Stimulatory effect of metoclopramide on the esophagus and lower esophageal sphincter of patients of patients with PSS. Arthritis Rheum 20, 30–34.

Ramos-Casals, M., Fonollosa-Pla, V., Brito-Zerón, P., Sisó-Almirall, A., 2010. Targeted therapy for systemic sclerosis: How close are we? Nat Rev Rheumatol 6, 269–278.

Remy-Jardin, M., Giraud, F., Remy, J., Copin, M.C., Gosselin, B., Duhamel, A., 1993a. Importance of ground-glass attenuation in chronic diffuse infiltrative lung disease: Pathologic-CT correlation. Radiology 189, 693–698.

Remy-Jardin, M., Remy, J., Wallaert, B., 1993b. Pulmonary involvement in progressive systemic sclerosis: Sequential evaluation with CT, pulmonary tests and broncho-alveolar lavage. Radiology 188, 499–506.

Rosenkranz, S., Diet, F., Karasch, T., et al., 2003. Sildenafil improved pulmonary hypertension and peripheral blood flow in a patient with scleroderma-associated lung fibrosis and Raynaud's phenomenon. Ann Int Med 139, 871–873.

Rubin, L.J., Badesch, D.B., Barst, R.J., et al., 2002. Bosentan therapy for pulmonary hypertension. N Engl J Med 346, 896–903.

Saji, T., Ozawa, Y., Ishikita, T., et al., 1996. Short term hemodynamic effect of a new oral PgI2 analogue, beraprost, in pulmonary and secondary pulmonary hypertension. Am J Cardiol 78, 244–247.

Sato, S., Fujimoto, M., Hasegawa, M., et al., 2004. Altered blood B lymphocyte homeostasis in systemic sclerosis: Expanded naive B cells and diminished but activated memory B cells. Arthritis Rheum 50, 1918–1927.

Scheja, A., Akkeson, A., 1997. Comparison of high frequency (20 MHz) ultrasound and palpation for the assessment of skin involvement in systemic sclerosis (scleroderma). Clin Exp Rheumatol 15, 283–288.

Scorza, R., Caronni, M., Mascagni, B., et al., 2001. Effects of long-term cyclic iloprost therapy in systemic sclerosis with Raynaud's phenomenon. A randomized controlled study. Clin Exp Rheumatol 19, 503–508.

Segel, M.C., Campbell, W.L., Medsger Jr., T.A., Roumm, A.D., 1985. Systemic sclerosis (scleroderma) and esophageal adenocarcinoma: Is increased patient screening necessary? Gastroenterology 89, 485–488.

Seibold, J.R., Steen, V.D., 1994. Systemic sclerosis. In: Klippel, J.H., Dieppe, P.A. (Eds.), Rheumatology, 1st ed Mosby, London, pp. 6.8–6.11.

Sgonc, R., Gruschwitz, M., Dietrich, H., et al., 1996. Endothelial cell apoptosis is a primary pathogenetic event underlying skin lesions in avian and human scleroderma. J Clin Invest 98, 785–792.

Sgonc, R., Wick, G., 2008. Pro- and anti-fibrotic effects of TGF-β in scleroderma. Rheumatology (Oxford) 47, v5–v7.

Silver, R.M., Medsger Jr., T.A., Bolster, M.B., 2005. Systemic sclerosis and scleroderma variants: Clinical aspects. In: Koopman, W.J.,

Moreland, L.W. (Eds.), Arthritis and Allied Conditions: A textbook in rheumatology, 15th ed Lippincot, Williams & Wilkins, Philadelphia, pp. 1633–1680.

Smith, V., Van Praet, J.T., Vandooren, B., et al., 2010. Rituximab in diffuse cutaneous systemic sclerosis: An open-label clinical and histopathological study. Ann Rheum Dis 69, 193–197.

Song, Y.W., 2005. Association of TAP1 and TAP2 gene polymorphisms with systemic sclerosis in Korean patients. Hum Immunol 66, 810–817.

Steen, V.D., Costantino, J.P., Shapiro, A.P., Medsger Jr., T.A., 1990. Outcome of renal crisis in systemic sclerosis: Relation to availability of angiotensin-converting enzyme (ACE) inhibitors. Ann Int Med 113, 352–357.

Steen, V.D., Medsger Jr., T.A., Rodnan, G.P., 1982. D-penicillamine therapy in progressive systemic sclerosis (scleroderma). A retrospective analysis. Ann Intern Med 97, 652–659.

Takeuchi, F., 1994. Association of HLA-DR with progressive systemic sclerosis in Japanese. J Rheumatol 21, 857–863.

Tan, F.K., 2005. Classification analysis of the transcriptosome of nonlesional cultured dermal fibroblasts from systemic sclerosis patients with early disease. Arthritis Rheum 52, 856–876.

Thompson, A.E., Shea, B., Welch, V., et al., 2001. Calcium-channel blockers for Raynaud's phenomenon in systemic sclerosis. Arthritis Rheum 44, 1841–1847.

Tikly, M., 2005. Lack of association of eNOS(G849T) and p22hox NADPH oxidase submit (C242T) polymorphisms with systemic sclerosis in a cohort of French Caucasian patients. Clin Chim Acta 358, 196–197.

Tolusso, B., 2005. 238 and 489 TNF-alpha along with TNFRII gene polymorphisms associate with the diffuse phenotype in patients with systemic sclerosis. Immunol Lett 96, 103–108.

Trojanowska, M., 2008. Role of PDGF in fibrotic disease and systemic sclerosis. Rheumatology (Oxford) 47, v2–v4.

Tsuchiya, N., 2004. Association of a functional CD19 polymorphism with susceptibility to systemic sclerosis. Arthritis Rheum 50, 4002–4007.

Tsukamoto, H., Nagafuji, K., Horiuchi, T., et al., 2006. A phase I-II trial of autologous peripheral blood stem cell transplantation in the treatment of refractory autoimmune disease. Ann Rheum Dis 65, 508–514.

Valentini, G., Silman, A.J., Veale, D., 2003. Assessment of disease activity. Clin Exp Rheumatol 21 (3 Suppl 29), S39–S41.

Van Laar, J.M., 2010. B-cell depletion with rituximab: A promising treatment for diffuse cutaneous systemic sclerosis. Arthritis Res Ther 12, 112.

Varga, J., Abraham, D., 2007. Systemic sclerosis: A prototypic multisystem fibrotic disorder. J Clin Invest 117, 557–567.

Verrecchia, F., Laboureau, J., Verola, O., et al., 2007. Skin involvement in scleroderma – where histological and clinical scores meet. Rheumatology (Oxford) 46, 833–841.

Verrecchia, F., Mauviel, A., Farge, D., 2006. Transforming growth factor-beta signaling through the Smad proteins: Role in systemic sclerosis. Autoimmun Rev 5, 563–569.

Vonk, M.C., Marjanovic, Z., van den Hoogen, F.H., et al., 2008. Long-term follow-up results after autologous haematopoietic stem cell transplantation for severe systemic sclerosis. Ann Rheum Dis 67, 98–104.

Walker, J.G., 2007. The development of systemic sclerosis classification criteria. Clin Rheumatol 26, 1401–1409.

Walker, U.A., Tyndall, A., Czirják, L.O., et al., 2007. Clinical risk assessment of organ manifestations in systemic sclerosis: A report from the EULAR Scleroderma Trials and Research group database. Ann Rheum Dis 66, 754–763.

Waxman, A.B., 2007. A review of sitaxsentan sodium in patients with pulmonary arterial hypertension. Vasc Health Risk Manag 3, 151–157.

Weber, P., Ganser, G., Frosch, M., et al., 2000. Twenty-four hour intraesophageal pH monitoring in children and adolescents with scleroderma and mixed connective tissue disease. J Rheumatol 27, 2692–2695.

Wegener, M., Adamek, R.J., Wedmann, J., et al., 1994. Gastrointestinal transit through oesophagus, stomach, small and large intestine in patients with progressive systemic sclerosis. Dig Dis Sci 39, 2209–2215.

Wells, A.U., Hansell, D.M., Rubens, M.B., et al., 1997. Fibrosing alveolitis in systemic sclerosis: Indices of lung function in relation to extent of disease in computed tomography. Arthritis Rheum 40, 1229–1236.

Wielosz, E., Borys, O., Żychowska, I., Majdan, M., 2010. Gastrointestinal involvement in patients with systemic sclerosis. Pol Arch Med Wewn 120, 132–135.

Wigley, F.M., 2002. Raynaud's phenomenon. N Engl J Med 347, 1001–1007.

Williams, C., McColl, K.E., 2006. Review article: Proton pump inhibitors and bacterial overgrowth. Aliment Pharmacol Ther 23, 3–10.

Witt, C., Borges, A.C., John, M., et al., 1999. Pulmonary involvement in diffuse cutaneous systemic sclerosis: Broncheoalveolar fluid granulocytosis predicts progression of fibrosing alveolitis. Ann Rheum Dis 58, 635–640.

Yoshizaki, A., Yanaba, K., Yoshizaki, A., et al., 2010. Treatment with rapamycin prevents fibrosis in tight-skin and bleomycin-induced mouse models of systemic sclerosis. Arthritis Rheum 62, 2476–2487.

Zhou, X., 2000. Microsatellites and intragenic polymorphisms of transforming growth factor beta and platelet-derived growth factor and their receptor genes in Native Americans with systemic sclerosis (scleroderma), a preliminary analysis showing no genetic associations. Arthritis Rheum 43, 1068–1073.

Zhou, X., 2005. Monozygotic twins clinically discordant for scleroderma show concordance from fibroblast gene expression profiles. Arthritis Rheum 52, 3305–3314.

Zhu, H., 2004. Polymorphisms of the TGF-beta1 promoter in tight skin (TSK) mice. Autoimmunity 37, 51–55.

Ziegler, S., Brunner, M., Eigenbauer, E., Minar, E., 2003. Long-term outcome of primary Raynaud's phenomenon and its conversion to connective tissue disease: A 12-year prospective patient analysis. Scand J Rheumatol 32, 343–347.

Systemic Lupus Erythematosus

Benjamin Rhodes and Timothy J. Vyse

CLASSIFICATION AND CLINICAL FEATURES

Systemic lupus erythematosus (SLE or lupus) is a potentially fatal systemic autoimmune disease that predominantly affects women of childbearing age (Bertoli and Alarcon, 2007). Although the pathogenesis of SLE remains poorly understood, there is clearly widespread immune dysfunction encompassing both the innate and adaptive immune pathways. The immunological hallmark of SLE is the production of high-affinity autoantibodies directed against ubiquitous intracellular antigens, particularly double-stranded DNA and nucleoprotein complexes (Rahman and Isenberg, 2008). The clinical features of lupus are notoriously heterogeneous, with symptoms related to immune-mediated dysfunction of the kidney, skin, central nervous system, blood, and other organs. Accelerated cardiovascular disease is an important late-stage complication (Elliott et al., 2007; Urowitz et al., 1976).

This diversity of clinical presentation raises issues of how to define SLE cases for the purposes of enrollment into research studies. At present, the 1982 American College of Rheumatology (ACR) criteria are almost universally used (Table 80.1) (Tan et al., 1982). Later, we will discuss how use of the ACR criteria impacts upon the interpretation and utility of genetic studies. The key issue is that lupus is so clinically heterogeneous that patients fulfilling the ACR criteria may be very different in

terms of their disease manifestations; indeed it is theoretically possible for lupus patients to share no clinical features at all. The implication of this is that SLE may also be genetically heterogeneous, with some genes associated with disease in general and others associated with specific clinical sub-phenotypes.

Although considerable progress has been made in the management of lupus over the last 10 years (recent estimates suggest an 80–90% 10-year survival rate), it remains an incurable disease (Ramsey-Goldman and Gladman, 2007). The current therapeutic options of corticosteroids, small-molecule immunosuppressants, and even targeted biological agents all really fail to address fundamental disease mechanisms in SLE. As a consequence, treatment regimens remain suboptimal both in terms of therapeutic efficacy and unwanted side effects.

Perhaps the key motivation for studying lupus genetics is that it has the potential to highlight fundamental disease pathways that are part of the causative pathogenic process – pathways that will then be amenable to therapeutic intervention. However, we are also aware that demonstrable immunological dysfunction precedes the development of clinical lupus, often by many years, and it is during this window of opportunity that we perhaps have the best chances of intervening with targeted therapeutics to prevent the onset of overt disease (Arbuckle et al., 2003). The identification of genetic variants contributing to disease susceptibility may also eventually allow us to identify individuals who are at high risk of developing the

Genomic and Personalized Medicine, 2nd edition
by Ginsburg & Willard. DOI: http://dx.doi.org/10.1016/B978-0-12-382227-7.00080-X

TABLE 80.1 The 1982 revised ACR criteria for the diagnosis of lupus

Classification criterion	Details
Malar rash	The "butterfly rash" of SLE
Discoid rash	Raised erythematosus plaques. May scar
Photosensitivity	Skin rash as unusual response to sunlight
Oral ulcers	Usually painless
Arthritis	Non-erosive in >2 peripheral joints
Serositis	Pleuritis or pericarditis
Renal disorder	Persistent proteinuria or cellular casts
Neurological disorder	Seizures or psychosis
Hematological disorder	Hemolytic anemia or other cytopenia
Anti-nuclear antibodies	In the absence of "drug-induced" lupus
Other immunological disorder	e.g., anti-dsDNA or anti-Sm antibodies

SLE diagnosed for the purposes of clinical study if >4 criteria present. dsDNA, double stranded DNA; Sm, smooth muscle; ACR, American College of Rheumatology. *Adapted from Tan et al., 1982.*

disease and who can therefore be targeted for preventative, or pre-clinical treatment strategies. Whether we can utilize our genetic knowledge further, to predict disease manifestations or to guide therapeutic decisions, is more speculative and dependent on further research (Rhodes and Vyse, 2010).

EPIDEMIOLOGY AND ETIOLOGY

Family studies provide good evidence that genetics is an important determinant of disease susceptibility in lupus. Twin studies have shown that the 2–5% disease concordance in dizygotic twins is dramatically different from the 24–57% concordance in monozygotic twins (Block et al., 1975; Deapen et al., 1992). Similarly the sibling risk ratio (λ_s) has been reported to be around 29 (Alarcon-Segovia et al., 2005; Hochberg, 1987). The heritability of lupus has been estimated at 66 ± 11% (Lawrence et al., 1987). A striking feature of the epidemiology of SLE is the difference in prevalence among ethnic groups. The lowest documented prevalence (~20/100,000) is found in white northern European populations, while the highest prevalence (~150–200/100,000) is found in African-Americans or African-Caribbean populations living in the US or UK (Borchers et al., 2010; Chakravarty et al., 2007; Hopkinson et al., 1995). Hispanics, Southeast Asians and South Asians all also appear at greater risk than northern European populations (Borchers et al., 2010). The prevalence among populations of African ancestry still living in Africa remains puzzling. While reported to be a rare disease in these populations, it remains possible that this simply reflects a difficulty in

recording cases, or alternatively a modification of genetic risk by common environmental factors such as endemic infection (Bae et al., 1998). Epidemiological observations therefore demonstrate that the genetic basis of this disease is strong enough to make the hunt for susceptibility genes worthwhile, and that we may expect to observe differences in lupus genetics between populations of different ancestry.

MENDELIAN SLE – THE ROLE OF RARE VARIANTS

A number of rare but highly penetrant variants have been identified that confer a high risk of lupus (or lupus-like) syndromes in a small number of families. The first to be identified was complete deficiency of complement component C1q, which causes a lupus-like syndrome (characterized by rash, glomerulonephritis, and anti-nuclear antibodies) in about 90% of individuals with the inherited deficiency (Pickering and Walport, 2000; Walport et al., 1998). Since 10% of patients deficient in C1q appear not to develop lupus-like disease, gene × gene or gene × environment interactions must play a role in determining overall phenotype. The nature of these interactions is currently unknown. Deficiencies of C1r and C1s are even rarer but also confer a similar risk of disease (Loos and Heinz, 1986; Pickering and Walport, 2000). Approximately 75% of individuals with homozygous C4 deficiency and 33% of individuals with homozygous C2 deficiency will also develop a lupus-like disease of moderate severity (Pickering and Walport, 2000). Both C2 and C4 are encoded by genes in the *MHC*, a region of the genome that presents particular difficulties for study. C4 deficiency will be discussed further in the next section. Another rare but penetrant variant was recently identified by re-sequencing the DNA exonuclease gene *TREX1* (Lee-Kirsch et al., 2007). Unlike the complement component mutations, this was not associated with complete gene deletion, but resulted in an altered intracellular distribution of the *TREX1* protein product.

The number of rare variants contributing to lupus may be higher than we currently think, but identifying them will require the large-scale screening of SLE cohorts by extensive re-sequencing to detect them. Assuming that many of these rare functional variants are likely to be due to non-synonymous coding changes, particular interest is likely to focus on exome sequencing or RNA sequencing techniques. There are currently no published large-scale sequencing projects in SLE.

SLE AND THE MHC

Variation at the major histocompatibility complex (*MHC*) was the first genetic susceptibility effect to be identified in lupus, and consistently shows the most significant association with disease (Grumet et al., 1971; Waters et al., 1971). Although the disease risk associated with *MHC* variants is high, the genetic architecture of this region makes it particularly challenging to study. The classical *MHC* consists of a 3.6 Mb region of chromosome 6 that contains at least 250 genes. The class I and class II *MHC* primarily contain the genes which encode the antigen-presenting *HLA* proteins, while the class III *MHC*, which lies between the class I and class II regions, contains 58 genes, many of which are of particular immunological relevance (e.g., *TNF*, *C4A*, *C4B*, and *C2*).

The most consistent *MHC* associations are with the class II alleles *DRB1*0301* (*DR3*) and the less common *DRB1*1501* (*DR2*) in white European populations, with rather inconsistent *HLA* associations in other ethnicities (Fernando et al., 2008). A notable feature of the *MHC* is that linkage disequilibrium between variants is strong and extends over a considerable genomic interval (up to 2 Mb) (Graham et al., 2002). In white Europeans, *DRB1*0301* is carried as part of an extended ancestral haplotype (AH8.1) incorporating multiple genetic variants that have been associated with lupus, including the *TNF -308A* allele and the *C4A* complement null allele. Due to this extended linkage disequilibrium, until recently it has been impossible to determine whether there is more than one functional effect arising from the *MHC*, or whether multiple indirect associations reflecting the same functional variant are being reported (Graham et al., 2002). The use of dense *MHC* single nucleotide polymorphism (SNP) maps in large SLE cohorts with analysis conditioned on the *HLADR* genotype has begun to answer this question. Independent class II signals have been reported arising from the *C6orf10* and *HLADQB2* regions and the *TAP2* gene (Barcellos et al., 2009; Ramos et al., 2009). Class III signals have been reported from the region of the *SKIV2L*, *MICB*, and *CREBL1* genes, while from the class I region signals have been seen from the region of the *TRIM27* and *OR2H2* genes (Barcellos et al., 2009; Fernando et al., 2007; Harley et al., 2008). The *TNF -308A* variant can be excluded as a functional effect in SLE on the basis of these studies (Fernando et al., 2007). It remains too early to consider any of these as definitive SLE susceptibility genes. The important message is that there are likely to be multiple independent effects arising from the *MHC*, which thus forms the genomic interval with both the most numerous and the strongest genetic susceptibility effects in SLE.

We have already noted the association between complement C4 deficiency and SLE. C4 is encoded by two class II *MHC* genes, *C4A* and *C4B*, which have minor sequence differences. The *C4* genes are inherited in a discrete "RCCX module" containing one *C4* gene (either *C4A* or *C4B*) and three neighboring genes. Each individual carries between two and eight copies of this module, and hence between two and eight *C4* genes, which may be either *C4A* or *C4B* (Yang et al., 2007). Carrying less than two copies of *C4A* is a risk factor for SLE (Yang et al., 2007). The AH8.1 haplotype contains a monomodular RCCX unit with a single *C4B* gene and no *C4A*, raising the possibility that the observed *C4A* association is present simply due to linkage disequilibrium between the gene copy number and functional variants in other *MHC* genes (Stewart et al., 2004). Some of the technical obstacles in assaying *C4* gene copy number have been overcome, and we should shortly have an answer to this question (Fernando et al., 2010).

GENOME WIDE ASSOCIATION STUDIES

By the beginning of the 21st century we had recognized the association between SLE and variants at the *MHC*, the Fc-Gamma receptor locus, and the genes encoding the early components of the complement cascade. In the last five years we have begun to develop a clear picture of some of the other important variants mediating susceptibility to sporadic lupus. The key to this progress has been the collection of large cohorts of SLE patients and the use of large-scale genetic association studies. Initially, genes were selected for study on the basis of their biological plausibility for a role in SLE, or because of an association with another autoimmune diseases. More recently the genome-wide association (GWA) methodology has been used with great effect (Table 80.2 and Figure 80.1) and further genes have been identified by meta-analysis and replication of the signals identified in these studies (Gateva et al., 2009; Graham et al., 2009).

TABLE 80.2	Lupus genome-wide association studies (published up to 2010)			
Study	**Ancestry of primary cohort**	**Genotyping platform**	**Numbers in primary cohort**	**Replication**
Harley et al., 2008; (SLEGEN)	White European	Illumina Infinium HumanHap 300	720 cases 2337 controls	1846 cases 1825 controls
Hom et al., 2008	White European	Illumina HumanHap 550	1311 cases 1783 controls	793 cases 857 controls
Graham et al., 2008b	White European	Affymetrix human SNP array 5.0	431 cases 2155 controls	470 SLE trios
Han et al., 2009a	Mainland Chinese Han	Illumina 610-Quad BeadChip	1047 cases 1205 controls	3152 cases 7050 controls
Yang et al., 2010	Hong Kong Chinese	Illumina 610-Quad BeadChip	320 cases 1500 controls	3300 cases 4200 controls

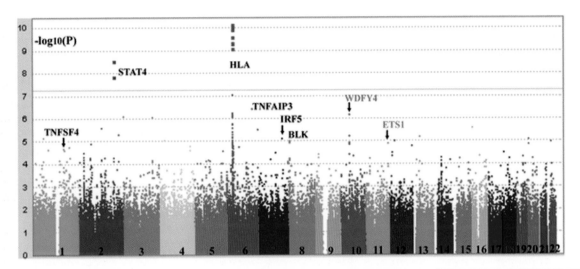

Figure 80.1 Genome-wide association signals from the Hong Kong Chinese study. The y-axis represents the level of association attributed to that individual SNP, positioned by chromosomal location (on the x-axis). The SNPs which have been annotated by gene name are those which have been additionally replicated either as part of this study or elsewhere. *Reproduced with permission from Yang et al., 2010.*

Table 80.3 lists the key non-*MHC* genes for which we have strongest evidence for association with SLE, and Figure 80.2 highlights the pathogenic pathways in SLE in which these genes may function. It should be noted that a number of other genes may play a role in lupus susceptibility, but the level of evidence is slightly lower due to either a failure to replicate in large studies or inconsistency over the associated SNPs or the risk alleles. This group includes genes such as *LYN*, *TYK2*, *PDCD1*, and *CRP*. There is also good evidence, backed up by functional data, to support an association between SLE and non-synonymous variants at *FCGR2B*, *FCGR3A*, and *FCGR3B* (discussed below), however the high level of sequence similarity and evidence of copy-number variation in this region makes these genes less amenable to study by current high-throughput techniques (Brown et al., 2007).

For some of these genes, we have identified putative functional variants on the basis of genetics alone (e.g., the *ITGAM R77H* variant). For others, there is data from functional experiments to back up the genetic data (e.g., *IRF5* splice and polyadenylation variants) (Cunninghame et al., 2007; Graham et al., 2006; Han et al., 2009b; Nath et al., 2008). For most genes, however, study is at an early stage and, at best, we can say that a functional genetic effect arises somewhere in the region of the highlighted gene. Fine mapping may reveal the true functional variant (or variants) and may lead to the effect estimate being revised upwards.

It can be seen from Table 80.3 that the majority of existing SLE genetic data comes from studies of white European patients – actually the population in which disease prevalence is lowest. We have also had some good Chinese data over the last two years, but we are clearly lacking data from African/African-American populations, where lupus is most prevalent.

It is notable that some of the strongest genetic effects in the European patients are absent from the African-American population due to absence, or low frequency, of the risk allele. This includes strong effects such as *HLADRB*0103*, *STAT4*, and *PTPN22*. Clearly, in this population, strong effects will be found elsewhere in the genome, and their identification will represent a considerable advance in our understanding of this disease.

SLE AND COPY NUMBER VARIATION

So far, we have focused on genetic variation due to single base pair substitutions, a field that has advanced rapidly due to the relative ease of SNP discovery and genotyping. We have long been aware of particular situations in which whole genes may be deleted or duplicated as an inherited genetic variant (for example the α-globin gene), but the publication of two genetic analyses in 2004 raised the possibility that this may be a widespread genomic phenomenon (Iafrate et al., 2004; Orkin 1978; Sebat et al., 2004). It is now realized that copy number variation (CNV) is a major source of human genetic variation, observed in up to 12% of the genome (Conrad et al., 2010; Redon et al., 2006). Indeed, CNV and smaller insertion/deletion variants account for a much greater proportion of human genomic variation than do SNPs (Pang et al., 2010). SLE was one of the first complex genetic diseases in which a CNV region was identified as underlying disease susceptibility: *C4* copy number variation (discussed earlier) (Yang et al., 2007). Recently, interest has focused on another important region of complex CNV: the low affinity immunoglobulin receptor gene cluster on chromosome 1.

| TABLE 80.3 | Established non-*MHC* SNP associations in lupus |

Gene	Odds ratio for the key associated SNP			References
	White European	**African-American**	**Chinese**	
TNFAIP3	1.9	1.6	1.7	Cai et al., 2010; Gateva et al., 2009; Graham et al., 2008b; Han et al., 2009a; Lodolce et al., 2010; Musone et al., 2008; Yang et al., 2010
IRF5	1.7	1.4	1.5	Gateva et al., 2009; Han et al., 2009a; Harley et al., 2008; Hom et al., 2008; Kelly et al., 2008; Sigurdsson et al., 2005; Yang et al., 2010
STAT4	1.6	(Weak association)	1.6	Han et al., 2009a; Harley et al., 2008; Hom et al., 2008; Namjou et al., 2009; Remmers et al., 2007; Yang et al., 2009a; Yang et al., 2010
ITGAM	1.6	1.6	2.2 (Low frequency risk allele)	Gateva et al., 2009; Harley et al., 2008; Hom et al., 2008; Nath et al., 2008; Yang et al., 2009b
PTPN22	1.4	(Low frequency risk allele)	(Low frequency risk allele)	Gateva et al., 2009; Harley et al., 2008; Kyogoku et al., 2004; Lee et al., 2007
TNFSF4	1.4		1.4	Chang et al., 2009; Gateva et al., 2009; Graham et al., 2008a; Graham et al., 2009; Han et al., 2009a; Yang et al., 2010
BANK1	1.4		1.3	Chang et al., 2009; Gateva et al., 2009; Kozyrev et al., 2008; Yang et al., 2010
IRAK1/MECP2	1.4	1.1		Gateva et al., 2009; Jacob et al., 2007; Sawalha et al., 2008
BLK	1.3		1.4	Gateva et al., 2009; Han et al., 2009a; Harley et al., 2008; Hom et al., 2008; Yang et al., 2009a; Yang et al., 2010
FCGR2A	1.3	2.0	Not associated (limited data)	Brown et al., 2007; Gateva et al., 2009; Harley et al., 2008; Karassa et al., 2002; Salmon et al., 1996
TNIP1	1.3		1.2	Gateva et al., 2009; Han et al., 2009a
ATG5	1.2		1.3	Gateva et al., 2009; Graham et al., 2009; Han et al., 2009a; Harley et al., 2008
UBE2L3	1.2		1.3	Gateva et al., 2009; Graham et al., 2009; Han et al., 2009a; Harley et al., 2008
PXK	1.2		Not associated	Gateva et al., 2009; Graham et al., 2009; Harley et al., 2008; Yang et al., 2009a
PHRF1	1.2			Gateva et al., 2009; Graham et al., 2009; Harley et al., 2008
PRDM1	1.2			Gateva et al., 2009
JAZF1	1.2			Gateva et al., 2009
UHRF1BP1	1.2			Gateva et al., 2009
IL10	1.2			Gateva et al., 2009
PTTG1	1.2			Gateva et al., 2009; Graham et al., 2009; Harley et al., 2008
IKZF1			1.4	Han et al., 2009a
SLC15A4			1.3	Han et al., 2009a
RASGRP3			1.4	Han et al., 2009a
ETS1			1.3	Han et al., 2009a; Yang et al., 2010
WDFY4			1.2	Han et al., 2009a; Yang et al., 2010

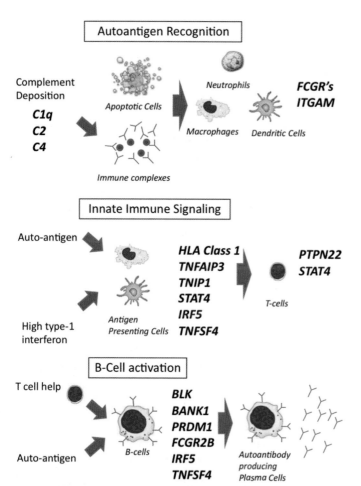

Autoantigen Recognition

Complement
Deposition

C1q
C2
C4

Apoptotic Cells

Immune complexes

Neutrophils

Macrophages   *Dendritic Cells*

FCGR's
ITGAM

Innate Immune Signaling

Auto-antigen

High type-1
interferon

*Antigen
Presenting Cells*

HLA Class 1
TNFAIP3
TNIP1
STAT4
IRF5
TNFSF4

T-cells

PTPN22
STAT4

B-Cell activation

T cell help

Auto-antigen

B-cells

BLK
BANK1
PRDM1
FCGR2B
IRF5
TNFSF4

*Autoantibody
producing
Plasma Cells*

Figure 80.2 Schematic diagram of the immunopathogenesis of lupus, highlighting key susceptibility genes. The pathogenesis of SLE is thought to involve humoral and cellular components of both the innate and adaptive immune system. The possible site of action of key SLE susceptibility genes is highlighted in bold. Other genes (such as *UBE2L3* and *ATG5*) have been identified and may be important, but our current understanding of their function is insufficient to place them in this scheme.

The immunoglobulin receptor FcγRIIIb exists as three allotypes termed HNA-1a, HNA-1b and HNA-1c (Bertrand et al., 2004). Allotypes of FcγRIIIb have been associated with SLE, although rather inconsistently, initially suggesting a minor or population-specific effect (Brown et al., 2007; Niederer et al., 2010). However, interest in the *FCGR3B* gene and SLE was renewed by the identification that *FCGR3B* lies within a region of CNV involving a large segment of the Fcγ receptor gene cluster, including *FCGR2C*. The initial report showed an association between low *FCGR3B* copy number and immune-mediated glomerulonephritis in humans and rats (Aitman et al., 2006). Subsequently it has been established that each individual carries between zero and six copies of *FCGR3B*, and that low copy number (particularly <2) is associated with SLE (Fanciulli et al., 2007; Willcocks et al., 2008). Finally, to draw

together the allotype and CNV data, it has been shown that the risk associated with low *FCGR3B* copy number is contingent on the FcγRIIIb allotype, with deletion of HNA-1a variants conferring a greater risk than deletion of HNA-1b variants (Morris et al., 2010).

EPIGENETICS

A number of lines of evidence suggest that epigenetic modifications are important in the pathogenesis of SLE. This is currently an area of active research and has been summarized in recent reviews (Pan and Sawalha, 2009; Sekigawa et al., 2006). The key lines of evidence include (Deng et al., 2003; Jacob et al., 2007; Javierre et al., 2010; Krapf et al., 1989; Luo et al., 2008; Richardson et al., 1990, 1992; Sawalha, 2009; Sawalha et al., 2008; Yung et al., 1995):

- The global hypomethylation of T-lymphocytes from SLE patients, with reduced expression of the DNA methyltransferases
- Hypomethylated CpG DNA motifs can be found in the serum of SLE patients, and these motifs can induce SLE in mouse models
- Demethylating agents may induce lupus-like phenotypes in *in vitro* and *in vivo* models
- Some of the monozygotic twin discordance for SLE can be explained by differences in methylation pattern
- Variation in the methylation pathway gene *MECP2* has been associated with human lupus (although there is some debate as to whether this is a truly functional effect).

Therefore, there is interesting circumstantial evidence from *in vitro* and animal models to implicate epigenetic phenomena in the pathogenesis of SLE, and direct demonstration of differences in DNA methylation in individuals with and without lupus in a relatively small-scale twin study. Clearly, further confirmation of methylation changes in the context of a case-control study will be required before this can be considered a definitive disease mechanism; this will require technological advances before it is feasible on a large scale.

Both DNA-encoded genetic change and epigenetic events may influence gene expression. There has been considerable interest in whether changes in the profile of expressed genes can be used to explore potential pathogenic mechanisms, or even to help diagnose or classify disease. In SLE, this approach was first highlighted in 2003 with the demonstration that numerous genes involved in interferon signaling pathways were up-regulated in peripheral blood mononuclear cells (PBMCs) from patients with lupus compared with healthy controls (Baechler et al., 2003; Bennett et al., 2003). Measuring gene expression in a mixed PBMC population (consisting of T- and B-lymphocytes, monocytes, and dendritic cells), when the proportion of cell types varies from patient to patient and between patients and controls, is clearly a suboptimal approach. In addition, these early studies emphasized the dramatic effect that immunosuppressive medication (especially corticosteroids) could have on the observed gene expression profiles. Recent studies have addressed some of these issues by studying gene expression in purified PBMC subsets, and demonstrate firstly that gene expression profiles do indeed differ between cell subsets and, secondly, that this approach allows greater discrimination between SLE-derived leucocytes and those from other autoimmune rheumatic diseases such as anti-neutrophil cytoplasmic antibody (ANCA)-associated vasculitis and rheumatoid arthritis (Juang et al., 2010; Lyons et al., 2010). Although these studies suggest gene expression profiling (be that through array-based analysis or RNA sequencing) may be a useful approach, currently only small-scale studies have been performed. While this technique seems able to distinguish SLE gene signatures from those of healthy controls, it remains to be seen how reproducibly the SLE signature can be distinguished from those of other system autoimmune diseases or, indeed, chronic viral infections, which are known to be potent inducers of interferon-regulated gene expression (Genin et al., 2009). Technological advances will allow the cost-effective evaluation of gene expression from larger populations, and undoubtedly some of these questions will be answered in the near future.

THE CURRENT GENETIC MODEL AND ITS UNCERTAINTIES

The picture that emerges from the current data is that, in a few rare cases, genetic susceptibility to lupus is mediated largely by a single highly penetrant mutation. In the majority of cases, however, SLE susceptibility behaves as a complex genetic trait mediated by multiple variants, each of which is common in the population and each of which contributes modestly to disease risk. The data therefore fits with the so-called "common disease, common variant" hypothesis (Becker, 2004). It is notable, however, that in comparison with other well-studied complex genetic diseases, the magnitude of effect from the leading SNPs is actually quite strong. For example, in cardiovascular disease even the most-associated variant has an odds ratio (OR) <1.75, with the majority below 1.2 (Humphries et al., 2010). As we discuss below, this strength of genetic effect impacts very favorably on our ability to develop a predictive genetic test for lupus.

There remain many unanswered questions. We have already alluded to some of these: the uncertainty over the contribution of rare variants to sporadic lupus, the uncertainty over the impact of gene copy number variation on disease susceptibility, and the uncertainty over the role of epigenetics. It also remains uncertain whether disease susceptibility increases in a simple additive or a multiplicative manner, depending on the number of risk alleles at different loci, or

whether there is significant epistasis between susceptibility loci. Although the current data suggest a simple model is appropriate, studies have not really been designed to look for epistasis, and it is possible that it is more important than we have thought.

Also unknown are the environmental triggers that act upon this genetic background to trigger disease. The nature of the environmental triggers remains largely unknown, although there has been some interest in the role of the Epstein-Barr virus (James et al., 2006). Also unknown is whether there are significant gene–environment interactions.

GENOMICS AND THE PREDICTION OF DISEASE

As we have discussed above, each individual genetic variant that contributes to the overall susceptibility in most patients with SLE exerts only a modest effect, certainly not large enough to make any meaningful prediction of the risk of disease in that individual. In order to develop a test that will predict individuals at very high risk of developing disease, information on a whole profile of informative variants will be required. Typically, the predictive ability of a test such as this is measured by the area under the receiver-operating characteristic curve (AUC). The AUC for a panel of SLE susceptibility variants was calculated as part of the analysis of the SLEGEN (International Consortium on the Genetics of Systemic Lupus Erythematosus) whole-genome analysis (Harley et al., 2008). Using a risk-weighted count of the seven most significant variants, the authors calculated an AUC of 0.67; in other words, these variants are moderately able to predict disease. However, for an effective screening test to highlight patients at high risk of disease we would really need a test with an AUC >0.80, while for a pre-symptomatic diagnostic test an AUC ≥0.99 would be required. We now have considerably more robust genetic data than was identified in the SLEGEN analysis, and we argue that it may well be feasible to develop a screening test, or even a diagnostic test, based on a profile of genetic variants.

The key factors that determine the ability of a profile of genetic variants to predict the disease they are associated with are: the prevalence of the disease, the heritability or genetic contribution to disease susceptibility, the overall number of genetic susceptibility variants, and the risk of disease (odds ratio) associated with each of them. The interplay among these factors and how they impact on the AUC has been outlined in a series of articles by Janssens, Van Duijn and colleagues (Janssens et al., 2006, 2007). Diagnostic tests perform better in rare diseases where the prior probability of any individual having the disease is low. Clearly a genetic test will perform better where the genetic component of disease susceptibility is high. The low prevalence of SLE (20–200/100,000) and its high heritability (66%) mean that, theoretically at least, an AUC in excess of 0.98 is possible for a predictive genetic test in lupus (Janssens et al., 2006, 2007).

A simulation based on the number and magnitude of SNP effects also gives a very optimistic picture for a predictive genetic test based on genetics. Figure 80.3 demonstrates the expected AUC for three simulated profiles of risk variants. It can be seen that for model 1, a good predictive profile (AUC >0.9) could be achieved with approximately 50 variants, and with 100 variants a very good predictive test (AUC >0.95) can be achieved. Even with the currently available genetic data, we have good reason to believe that the profile of SLE risk variants would be even more informative than simulated model 1. First, we have at least four genetic effects with an odds ratio >2.0 (*HLADR*, *MHC* class III, *FCRG3B* CNV, *TNFAIP3*). Critically, however, the frequency of these risk alleles is considerably higher than those modeled in the simulations.

For example, *DRB*0301* (OR 2.3) has an allele frequency of about 0.13 in a white UK population, while the *SKIV2L* risk variant (OR 2.0) has a much greater minor allele frequency of about 0.37 (Fernando et al., 2007). Although we await a formal SLE analysis, these simulations suggest that, providing we can identify more than about 50 susceptibility genes, a predictive test will become feasible.

We need to recognize that the examples described above are just simulations and assume a fairly simplistic model of

genetic risk effect. We have already discussed uncertainties about the true SLE genetic model (epistasis, rare variants, gene–environment interactions, etc.), and this, of course, introduces uncertainty in our simulated predictive test. It could also be argued that the true disease risk associated with any SNP may be overestimated in a case-control study, because the disease cases enrolled into studies are not representative of general disease cases in the population. Ironically, the relatively low prevalence of SLE makes this problem less likely, as investigators are happy to recruit any lupus patient to make up numbers, rather than selectively picking individuals with particularly severe disease or early onset. Nonetheless, it is critical that simulation is replaced with experimental data before we can be sure that disease prediction is a realistic possibility. Ideally, this experimental data should take the form of a population prospective-based study to provide a robust estimate of the size of risk effect associated with the genetic variants, although we recognize that for a relatively rare disease this may not yet be feasible.

GENOMICS AND THE PREDICTION OF CLINICAL MANIFESTATIONS

We have seen that the use of genetic data to provide a personalized estimate of the risk of developing lupus is certainly a

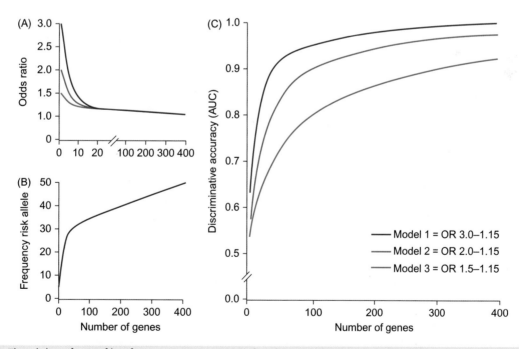

Figure 80.3 The ability of a profile of genetic variants to predict disease. The three models differ by the disease risk attributed to the 20 variants imparting highest risk: model 1, odds ratios 3.0–1.15, model 2, 2.0–1.15 and model 3, 1.5–1.15 (panel A). The odds ratios for the remaining 380 genes range from 1.15–1.05 in all models. The allele frequencies of these variants range from 0.05 for the strongest effect to 0.50 for the weakest effect (panel B). The discriminative ability of these three models is demonstrated in panel C. *Reproduced with permission from Janssens et al., 2006.*

theoretical possibility. We also now need to consider whether more sophisticated personalized disease assessments can be made using genetic profiles. For example, can we look at patients with newly diagnosed lupus and predict how severe their disease will be or what clinical features they will develop. The assumption this makes is that sub-phenotype or severity is, indeed, determined by genetics and not by environment. There is some evidence to support this assumption, as the disease sub-phenotypes observed in North Americans of white European ancestry are correlated with their ancestry as defined by a panel of ancestry-informative genetic markers – those with northern European ancestry tend to develop mucocutaneous disease, while those with southern European ancestry tend to develop more renal disease and immunological features (Chung et al., 2009; Richman et al., 2009).

It should be noted that current SLE association studies are not primarily designed to address the question of sub-phenotype association. We discussed in the introduction that patients are usually enrolled into SLE genetic studies if they fulfill standard ACR diagnostic criteria, which permit a very broad range of clinical subgroups to be used in the same analysis. This may be an advantage when considering the question of disease prediction, as these individuals are fairly representative of the range of lupus seen in the clinic, but it raises issues when using these data to interpret sub-phenotype associations. While a number of studies provide secondary sub-phenotype analyses, the most common approach is to perform a case-control analysis, with the cases limited to those with lupus and a particular sub-phenotype. This approach makes it difficult to directly assess the contribution of a genetic variant to particular clinical manifestations over and above its contribution to overall disease susceptibility. The alternative approach is to analyze lupus cases only, with cases (lupus and sub-phenotype of interest) compared with controls (lupus but not the sub-phenotype). This approach directly answers the question we are interested in, but at the cost of a dramatic reduction in the available sample size and hence power to demonstrate associations. Only a few studies adopt this approach. There also remain unresolved issues of how to correct for the testing of genetic variants against multiple clinical manifestations that are not completely independent of each other, presenting a difficulty in defining significance thresholds.

In summary, current literature suggests that the SLE susceptibility variants of *STAT4* appear associated with more severe disease in younger patients, with nephritis, immunological features, and accelerated ischemic vascular disease: *TNFAIP3* appears associated with nephritis, hematological manifestations, and immunological features, *ITGAM* and the *FCGR* variants appear particularly associated with nephritis, while *IRF5* tends to be associated with cutaneous manifestations (Bates et al., 2009; Brown et al., 2007; Cai et al., 2010; Jarvinen et al., 2010; Kim-Howard et al., 2009; Svenungsson et al., 2010; Taylor et al., 2008; Yang et al.,

2009b). Given the methodological issues discussed above, and the practical problem (for a relatively rare disease) of collecting cohorts that are large enough (and hence well powered) to reliably detect sub-phenotype association, we would appear to be a long way from translating these hints of genetic effects into a predictive algorithm that could be used to counsel patients and to guide treatment and monitoring strategies.

GENOMICS AND INDIVIDUALIZED PRESCRIBING

To some extent, pharmacogenetics (the influence of genetic variation on the individual response to therapeutic agents), is already utilized in the lupus clinic. Azathioprine is a commonly used immunosuppressive in patients with moderate and severe disease. It is metabolized by the enzyme thiopurine S-methyltransferase. The gene encoding this enzyme (*TPMT*) contains a functioning genetic variant that influences the metabolism of azathioprine, with individuals homozygous for the low or non-functional variant being at high risk of potentially fatal toxicity (Yates et al., 1997). Screening patients for *TPMT* variants before commencing azathioprine treatment is now routinely performed in many clinics. It is likely that pharmacogenetics will also be applied to other treatments in the future. For example, the rate of metabolism of mycophenolate mofetil, another immunosuppressant increasingly used in the treatment of SLE, is influenced by genetic variation at *UGT1A9* and *UGT1A8*, the genes encoding the uridine diphosphate-glucuronyltransferase enzymes, and this in turn may influence drug side effects (Betonico et al., 2008a, b).

An alternative question to whether we can use genetics to predict side-effects to particular therapeutic agents is whether we can use genetics to guide treatment choices or predict response to treatment. As the number of identified susceptibility genes increases, a pattern appears to emerge in which the genes cluster into particular biological pathways: antigen recognition (*ITGAM*, *FCGR2A*, *FCGR3B*), innate immune signaling (*IRF5*, *STAT4*, *TNFAIP3*, *IL10*), B-cell signaling (*BLK*, *LYN*, *BANK1*), B-cell/T-cell cross talk and T-cell activation (*PTPN22*, *TNFSF4*, *STAT4*, *HLA-DR*), and ubiquitination (*UBE2L3*, *TNFAIP3*, *TNIP1*) (Graham et al., 2009; Moser et al., 2009; Rhodes and Vyse, 2008). An understanding of how these pathways are genetically dysregulated is going to be fundamental to elucidating the pathogenesis of lupus. It is tempting to speculate that it may eventually allow target prescribing. For example, will treatment with BLyS inhibitors (targeting B-cells) be most effective in individuals with susceptibility variants clustering in B-cell gene pathways (Favas and Isenberg, 2009)? At present, these questions remain highly speculative, and until target drugs are available for SLE, remain of theoretical interest only.

REFERENCES

Aitman, T.J., Dong, R., Vyse, T.J., et al., 2006. Copy number polymorphism in Fcgr3 predisposes to glomerulonephritis in rats and humans. Nature 439, 851–855.

Alarcon-Segovia, D., Alarcon-Riquelme, M.E., Cardiel, M.H., et al., 2005. Familial aggregation of systemic lupus erythematosus, rheumatoid arthritis, and other autoimmune diseases in 1,177 lupus patients from the GLADEL cohort. Arthritis Rheum 52, 1138–1147.

Arbuckle, M.R., McClain, M.T., Rubertone, M.V., et al., 2003. Development of autoantibodies before the clinical onset of systemic lupus erythematosus. N Engl J Med 349, 1526–1533.

Bae, S.C., Fraser, P., Liang, M.H., 1998. The epidemiology of systemic lupus erythematosus in populations of African ancestry: A critical review of the "prevalence gradient hypothesis". Arthritis Rheum 41, 2091–2099.

Baechler, E.C., Batliwalla, F.M., Karypis, G., et al., 2003. Interferon-inducible gene expression signature in peripheral blood cells of patients with severe lupus. Proc Natl Acad Sci USA 100, 2610–2615.

Barcellos, L.F., May, S.L., Ramsay, P.P., et al., 2009. High-density SNP screening of the major histocompatibility complex in systemic lupus erythematosus demonstrates strong evidence for independent susceptibility regions. PLoS Genet 5, e1000696.

Bates, J.S., Lessard, C.J., Leon, J.M., et al., 2009. Meta-analysis and imputation identifies a 109 kb risk haplotype spanning TNFAIP3 associated with lupus nephritis and hematologic manifestations. Genes Immun 10, 470–477.

Becker, K.G., 2004. The common variants/multiple disease hypothesis of common complex genetic disorders. Med Hypotheses 62, 309–317.

Bennett, L., Palucka, A.K., Arce, E., et al., 2003. Interferon and granulopoiesis signatures in systemic lupus erythematosus blood. J Exp Med 197, 711–723.

Bertoli, A.M., Alarcon, G.S., 2007. Epidemiology of systemic lupus erythematosus. In: Tsokos, G.C., Gordon, C., Smolen, J.S. (Eds.), Systemic Lupus Erythematosus, first ed. Mosby, Philadelphia, pp. 1–18.

Bertrand, G., Duprat, E., Lefranc, M.P., Marti, J., Coste, J., 2004. Characterization of human FCGR3B*02 (HNA-1b, NA2) cDNAs and IMGT standardized description of FCGR3B alleles. Tissue Antigens 64, 119–131.

Betonico, G.N., Abbud-Filho, M., Goloni-Bertollo, E.M., et al., 2008a. Influence of UDP-glucuronosyltransferase polymorphisms on mycophenolate mofetil-induced side effects in kidney transplant patients. Transplant Proc 40, 708–710.

Betonico, G.N., Abudd-Filho, M., Goloni-Bertollo, E.M., Pavarino-Bertelli, E., 2008b. Pharmacogenetics of mycophenolate mofetil: A promising different approach to tailoring immunosuppression? J Nephrol 21, 503–509.

Block, S.R., Winfield, J.B., Lockshin, M.D., D'Angelo, W.A., Christian, C.L., 1975. Studies of twins with systemic lupus erythematosus. A review of the literature and presentation of 12 additional sets. Am J Med 59, 533–552.

Borchers, A.T., Naguwa, S.M., Shoenfeld, Y., Gershwin, M.E., 2010. The geoepidemiology of systemic lupus erythematosus. Autoimmun Rev 9, A277–A287.

Brown, E.E., Edberg, J.C., Kimberly, R.P., 2007. Fc receptor genes and the systemic lupus erythematosus diathesis. Autoimmunity 40, 567–581.

Cai, L.Q., Wang, Z.X., Lu, W.S., et al., 2010. A single-nucleotide polymorphism of the TNFAIP3 gene is associated with systemic lupus erythematosus in Chinese Han population. Mol Biol Rep 37, 389–394.

Chakravarty, E.F., Bush, T.M., Manzi, S., Clarke, A.E., Ward, M.M., 2007. Prevalence of adult systemic lupus erythematosus in California and Pennsylvania in 2000: Estimates obtained using hospitalization data. Arthritis Rheum 56, 2092–2094.

Chang, Y.K., Yang, W., Zhao, M., et al., 2009. Association of BANK1 and TNFSF4 with systemic lupus erythematosus in Hong Kong Chinese. Genes Immun 10, 414–420.

Chung, S.A., Tian, C., Taylor, K.E., et al., 2009. European population substructure is associated with mucocutaneous manifestations and autoantibody production in systemic lupus erythematosus. Arthritis Rheum 60, 2448–2456.

Conrad, D.F., Pinto, D., Redon, R., et al., 2010. Origins and functional impact of copy number variation in the human genome. Nature 464, 704–712.

Cunninghame Graham, D.S., Manku, H., Wagner, S., et al., 2007. Association of IRF5 in UK SLE families identifies a variant involved in polyadenylation. Hum Mol Genet 16, 579–591.

Deapen, D., Escalante, A., Weinrib, L., et al., 1992. A revised estimate of twin concordance in systemic lupus erythematosus. Arthritis Rheum 35, 311–318.

Deng, C., Lu, Q., Zhang, Z., et al., 2003. Hydralazine may induce autoimmunity by inhibiting extracellular signal-regulated kinase pathway signaling. Arthritis Rheum 48, 746–756.

Elliott, J.R., Kao, A.H., Manzi, S., 2007. The heart and systemic lupus erythematosus. In: Tsokos, G.C., Gordon, C., Smolen, J.S. (Eds.), Systemic Lupus Erythematosus, first ed. Mosby, Philadelphia, pp. 361–373.

Fanciulli, M., Norsworthy, P.J., Petretto, E., et al., 2007. FCGR3B copy number variation is associated with susceptibility to systemic, but not organ-specific, autoimmunity. Nat Genet 39, 721–723.

Favas, C., Isenberg, D.A., 2009. B-cell-depletion therapy in SLE – what are the current prospects for its acceptance? Nat Rev Rheumatol 5, 711–716.

Fernando, M.M., Boteva, L., Morris, D.L., et al., 2010. Assessment of complement C4 gene copy number using the paralog ratio test. Hum Mutat 31, 866–874.

Fernando, M.M., Stevens, C.R., Sabeti, P.C., et al., 2007. Identification of two independent risk factors for lupus within the MHC in United Kingdom families. PLoS Genet 3, e192.

Fernando, M.M.A., Stevens, C.R., Walsh, E.C., et al., 2008. Defining the role of the MHC in autoimmunity: A review and pooled analysis. PLoS Genet 4, e1000024.

Feuk, L., Carson, A.R., Scherer, S.W., 2006. Structural variation in the human genome. Nat Rev Genet 7, 85–97.

Gateva, V., Sandling, J.K., Hom, G., et al., 2009. A large-scale replication study identifies TNIP1, PRDM1, JAZF1, UHRF1BP1 and IL10 as risk loci for systemic lupus erythematosus. Nat Genet 41, 1228–1233.

Genin, P., Vaccaro, A., Civas, A., 2009. The role of differential expression of human interferon-a genes in antiviral immunity. Cytokine Growth Factor Rev 20, 283–295.

Graham, D.S., Graham, R.R., Manku, H., et al., 2008a. Polymorphism at the TNF superfamily gene TNFSF4 confers susceptibility to systemic lupus erythematosus. Nat Genet 40, 83–89.

Graham, R.R., Cotsapas, C., Davies, L., et al., 2008b. Genetic variants near TNFAIP3 on 6q23 are associated with systemic lupus erythematosus. Nat Genet 40, 1059–1061.

Graham, R.R., Hom, G., Ortmann, W., Behrens, T.W., 2009. Review of recent genome-wide association scans in lupus. J Intern Med 265, 680–688.

Graham, R.R., Kozyrev, S.V., Baechler, E.C., et al., 2006. A common haplotype of interferon regulatory factor 5 (IRF5) regulates splicing and expression and is associated with increased risk of systemic lupus erythematosus. Nat Genet 38, 550–555.

Graham, R.R., Ortmann, W.A., Langefeld, C.D., et al., 2002. Visualizing human leukocyte antigen class II risk haplotypes in human systemic lupus erythematosus. Am J Hum Genet 71, 543–553.

Grumet, F.C., Coukell, A., Bodmer, J.G., Bodmer, W.F., McDevitt, H.O., 1971. Histocompatibility (HL-A) antigens associated with systemic lupus erythematosus. A possible genetic predisposition to disease. N Engl J Med 285, 193–196.

Han, J.W., Zheng, H.F., Cui, Y., et al., 2009a. Genome-wide association study in a Chinese Han population identifies nine new susceptibility loci for systemic lupus erythematosus. Nat Genet 41, 1234–1237.

Han, S., Kim-Howard, X., Deshmukh, H., et al., 2009b. Evaluation of imputation-based association in and around the integrin-alpha-M (ITGAM) gene and replication of robust association between a non-synonymous functional variant within ITGAM and systemic lupus erythematosus (SLE). Hum Mol Genet 18, 1171–1180.

Harley, J.B., Alarcon-Riquelme, M.E., Criswell, L.A., et al., 2008. Genome-wide association scan in women with systemic lupus erythematosus identifies susceptibility variants in ITGAM, PXK, KIAA1542 and other loci. Nat Genet 40, 204–210.

Hochberg, M.C., 1987. The application of genetic epidemiology to systemic lupus erythematosus. J Rheumatol 14, 867–869.

Hom, G., Graham, R.R., Modrek, B., et al., 2008. Association of systemic lupus erythematosus with C8orf13-BLK and ITGAM-ITGAX. N Engl J Med 358, 900–909.

Hopkinson, N.D., Muir, K.R., Oliver, M.A., Doherty, M., Powell, R.J., 1995. Distribution of cases of systemic lupus erythematosus at time of first symptom in an urban area. Ann Rheum Dis 54, 891–895.

Humphries, S.E., Drenos, F., Ken-Dror, G., Talmud, P.J., 2010. Coronary heart disease risk prediction in the era of genome-wide association studies: Current status and what the future holds. Circulation 121, 2235–2248.

Iafrate, A.J., Feuk, L., Rivera, M.N., et al., 2004. Detection of large-scale variation in the human genome. Nat Genet 36, 949–951.

Jacob, C.O., Reiff, A., Armstrong, D.L., et al., 2007. Identification of novel susceptibility genes in childhood-onset systemic lupus erythematosus using a uniquely designed candidate gene pathway platform. Arthritis Rheum 56, 4164–4173.

James, J.A., Harley, J.B., Scofield, R.H., 2006. Epstein-Barr virus and systemic lupus erythematosus. Curr Opin Rheumatol 18, 462–467.

Janssens, A.C., Aulchenko, Y.S., Elefante, S., et al., 2006. Predictive testing for complex diseases using multiple genes: Fact or fiction? Genet Med 8, 395–400.

Janssens, A.C., Moonesinghe, R., Yang, Q., et al., 2007. The impact of genotype frequencies on the clinical validity of genomic profiling for predicting common chronic diseases. Genet Med 9, 528–535.

Jarvinen, T.M., Hellquist, A., Koskenmies, S., et al., 2010. Tyrosine kinase 2 and interferon regulatory factor 5 polymorphisms are associated with discoid and subacute cutaneous lupus erythematosus. Exp Dermatol 19, 123–131.

Javierre, B.M., Fernandez, A.F., Richter, J., et al., 2010. Changes in the pattern of DNA methylation associate with twin discordance in systemic lupus erythematosus. Genome Res 20, 170–179.

Juang, Y.T., Peoples, C., Kafri, R., et al., 2010. A systemic lupus erythematosus gene expression array in disease diagnosis and classification: A preliminary report. Lupus 20, 243–249.

Karassa, F.B., Trikalinos, T.A., Ioannidis, J.P., 2002. Role of the Fcgamma receptor IIa polymorphism in susceptibility to systemic lupus erythematosus and lupus nephritis: A meta-analysis. Arthritis Rheum 46, 1563–1571.

Kelly, J.A., Kelley, J.M., Kaufman, K.M., et al., 2008. Interferon regulatory factor-5 is genetically associated with systemic lupus erythematosus in African Americans. Genes Immun 9, 189–194.

Kim-Howard, X., Maiti, A.K., Anaya, J.M., et al., 2009. ITGAM coding variant (rs1143679) influences the risk of renal disease, discoid rash, and immunologic manifestations in lupus patients with European ancestry. Ann Rheum Dis 69, 1329–1332.

Kozyrev, S.V., Abelson, A.K., Wojcik, J., et al., 2008. Functional variants in the B-cell gene BANK1 are associated with systemic lupus erythematosus. Nat Genet 40, 211–216.

Krapf, F., Herrmann, M., Leitmann, W., Kalden, J.R., 1989. Antibody binding of macromolecular DNA and RNA in the plasma of SLE patients. Clin Exp Immunol 75, 336–342.

Kyogoku, C., Langefeld, C.D., Ortmann, W.A., et al., 2004. Genetic association of the R620W polymorphism of protein tyrosine phosphatase PTPN22 with human SLE. Am J Hum Genet 75, 504–507.

Lawrence, J.S., Martins, C.L., Drake, G.L., 1987. A family survey of lupus erythematosus. 1. Heritability. J Rheumatol 14, 913–921.

Lee, Y.H., Rho, Y.H., Choi, S.J., et al., 2007. The PTPN22 C1858T functional polymorphism and autoimmune diseases – a meta-analysis. Rheumatology (Oxford) 46, 49–56.

Lee-Kirsch, M.A., Gong, M., Chowdhury, D., et al., 2007. Mutations in the gene encoding the 3'-5' DNA exonuclease TREX1 are associated with systemic lupus erythematosus. Nat Genet 39, 1065–1067.

Lodolce, J.P., Kolodziej, L.E., Rhee, L., et al., 2010. African-derived genetic polymorphisms in TNFAIP3 mediate risk for autoimmunity. J Immunol 184, 7001–7009.

Loos, M., Heinz, H.P., 1986. Component deficiencies. 1. The first component: C1q, C1r, C1s. Prog Allergy 39, 212–231.

Luo, Y., Li, Y., Su, Y., et al., 2008. Abnormal DNA methylation in T cells from patients with subacute cutaneous lupus erythematosus. Br J Dermatol 159, 827–833.

Lyons, P.A., McKinney, E.F., Rayner, T.F., et al., 2010. Novel expression signatures identified by transcriptional analysis of separated leucocyte subsets in systemic lupus erythematosus and vasculitis. Ann Rheum Dis 69, 1208–1213.

Morris, D.L., Roberts, A.L., Witherden, A.S., et al., 2010. Evidence for both copy number and allelic (NA1/NA2) risk at the FCGR3B locus in systemic lupus erythematosus. Eur J Hum Genet 18, 1027–1031.

Moser, K.L., Kelly, J.A., Lessard, C.J., Harley, J.B., 2009. Recent insights into the genetic basis of systemic lupus erythematosus. Genes Immun 10, 373–379.

Musone, S.L., Taylor, K.E., Lu, T.T., et al., 2008. Multiple polymorphisms in the TNFAIP3 region are independently associated with systemic lupus erythematosus. Nat Genet 40, 1062–1064.

Namjou, B., Sestak, A.L., Armstrong, D.L., et al., 2009. High-density genotyping of STAT4 reveals multiple haplotypic associations with systemic lupus erythematosus in different racial groups. Arthritis Rheum 60, 1085–1095.

Nath, S.K., Han, S., Kim-Howard, X., et al., 2008. A nonsynonymous functional variant in integrin-alpha(M) (encoded by ITGAM) is associated with systemic lupus erythematosus. Nat Genet 40, 152–154.

Niederer, H.A., Clatworthy, M.R., Willcocks, L.C., Smith, K.G., 2010. FcgammaRIIB, FcgammaRIIIB, and systemic lupus erythematosus. Ann NY Acad Sci 1183, 69–88.

Orkin, S.H., 1978. The duplicated human alpha globin genes lie close together in cellular DNA. Proc Natl Acad Sci USA 75, 5950–5954.

Pan, Y., Sawalha, A.H., 2009. Epigenetic regulation and the pathogenesis of systemic lupus erythematosus. Transl Res 153, 4–10.

Pang, A.W., MacDonald, J.R., Pinto, D., et al., 2010. Towards a comprehensive structural variation map of an individual human genome. Genome Biol 11, R52.

Pickering, M.C., Walport, M.J., 2000. Links between complement abnormalities and systemic lupus erythematosus. Rheumatology (Oxford) 39, 133–141.

Rahman, A., Isenberg, D.A., 2008. Systemic lupus erythematosus. N Engl J Med 358, 929–939.

Ramos, P.S., Langefeld, C.D., Bera, L.A., et al., 2009. Variation in the ATP-binding cassette transporter 2 gene is a separate risk factor for systemic lupus erythematosus within the MHC. Genes Immun 10, 350–355.

Ramsey-Goldman, R., Gladman, D., 2007. Disease development and outcome. In: Tsokos, G.C., Gordon, C., Smolen, J.S. (Eds.), Systemic Lupus Erythematosus. Mosby, Philadelphia, pp. 24–31.

Redon, R., Ishikawa, S., Fitch, K.R., et al., 2006. Global variation in copy number in the human genome. Nature 444, 444–454.

Remmers, E.F., Plenge, R.M., Lee, A.T., et al., 2007. STAT4 and the risk of rheumatoid arthritis and systemic lupus erythematosus. N Engl J Med 357, 977–986.

Rhodes, B., Vyse, T.J., 2008. The genetics of SLE: An update in the light of genome-wide association studies. Rheumatology (Oxford) 47, 1603–1611.

Rhodes, B., Vyse, T.J., 2010. Using genetics to deliver personalized SLE therapy – A realistic prospect? Nat Rev Rheumatol 6, 373–377.

Richardson, B., Scheinbart, L., Strahler, J., et al., 1990. Evidence for impaired T cell DNA methylation in systemic lupus erythematosus and rheumatoid arthritis. Arthritis Rheum 33, 1665–1673.

Richardson, B.C., Strahler, J.R., Pivirotto, T.S., et al., 1992. Phenotypic and functional similarities between 5-azacytidine-treated T cells and a T cell subset in patients with active systemic lupus erythematosus. Arthritis Rheum 35, 647–662.

Richman, I.B., Chung, S.A., Taylor, K.E., et al., 2009. European population substructure correlates with systemic lupus erythematosus endophenotypes in North Americans of European descent. Genes Immun 11, 515–521.

Salmon, J.E., Millard, S., Schachter, L.A., et al., 1996. Fc gamma RIIA alleles are heritable risk factors for lupus nephritis in African Americans. J Clin Invest 97, 1348–1354.

Sawalha, A.H., 2009. Xq28 and lupus: IRAK1 or MECP2? Proc Natl Acad Sci USA 106, E62.

Sawalha, A.H., Webb, R., Han, S., et al., 2008. Common variants within MECP2 confer risk of systemic lupus erythematosus. PLoS One 3, e1727.

Sebat, J., Lakshmi, B., Troge, J., et al., 2004. Large-scale copy number polymorphism in the human genome. Science 305, 525–528.

Sekigawa, I., Kawasaki, M., Ogasawara, H., et al., 2006. DNA methylation: Its contribution to systemic lupus erythematosus. Clin Exp Med 6, 99–106.

Sigurdsson, S., Nordmark, G., Goring, H.H., et al., 2005. Polymorphisms in the tyrosine kinase 2 and interferon regulatory factor 5 genes are associated with systemic lupus erythematosus. Am J Hum Genet 76, 528–537.

Stewart, C.A., Horton, R., Allcock, R.J., et al., 2004. Complete MHC haplotype sequencing for common disease gene mapping. Genome Res 14, 1176–1187.

Svenungsson, E., Gustafsson, J., Leonard, D., et al., 2010. A STAT4 risk allele is associated with ischaemic cerebrovascular events and anti-phospholipid antibodies in systemic lupus erythematosus. Ann Rheum Dis 69, 834–840.

Tan, E.M., Cohen, A.S., Fries, J.F., et al., 1982. The 1982 revised criteria for the classification of systemic lupus erythematosus. Arthritis Rheum 25, 1271–1277.

Taylor, K.E., Remmers, E.F., Lee, A.T., et al., 2008. Specificity of the STAT4 genetic association for severe disease manifestations of systemic lupus erythematosus. PLoS Genet 4, e1000084.

Urowitz, M.B., Bookman, A.A., Koehler, B.E., et al., 1976. The bimodal mortality pattern of systemic lupus erythematosus. Am J Med 60, 221–225.

Walport, M.J., Davies, K.A., Botto, M., 1998. C1q and systemic lupus erythematosus. Immunobiology 199, 265–285.

Waters, H., Konrad, P., Walford, R.L., 1971. The distribution of HL-A histocompatibility factors and genes in patients with systemic lupus erythematosus. Tissue Antigens 1, 68–73.

Willcocks, L.C., Lyons, P.A., Clatworthy, M.R., et al., 2008. Copy number of FCGR3B, which is associated with systemic lupus erythematosus, correlates with protein expression and immune complex uptake. J Exp Med 205, 1573–1582.

Yang, W., Ng, P., Zhao, M., et al., 2009a. Population differences in SLE susceptibility genes: STAT4 and BLK, but not PXK, are associated with systemic lupus erythematosus in Hong Kong Chinese. Genes Immun 10, 219–226.

Yang, W., Shen, N., Ye, D.Q., et al., 2010. Genome-wide association study in asian populations identifies variants in ETS1 and WDFY4 associated with systemic lupus erythematosus. PLoS Genet 6, e1000841.

Yang, W., Zhao, M., Hirankarn, N., et al., 2009b. ITGAM is associated with disease susceptibility and renal nephritis of systemic lupus erythematosus in Hong Kong Chinese and Thai. Hum Mol Genet 18, 2063–2070.

Yang, Y., Chung, E.K., Wu, Y.L., et al., 2007. Gene copy-number variation and associated polymorphisms of complement component C4 in human systemic lupus erythematosus (SLE): Low copy number is a risk factor for and high copy number is a protective factor against SLE susceptibility in European Americans. Am J Hum Genet 80, 1037–1054.

Yates, C.R., Krynetski, E.Y., Loennechen, T., et al., 1997. Molecular diagnosis of thiopurine S-methyltransferase deficiency: Genetic basis for azathioprine and mercaptopurine intolerance. Ann Intern Med 126, 608–614.

Yung, R.L., Quddus, J., Chrisp, C.E., Johnson, K.J., Richardson, B.C., 1995. Mechanism of drug-induced lupus. I. Cloned Th2 cells modified with DNA methylation inhibitors in vitro cause autoimmunity in vivo. J Immunol 154, 3025–3035.

RECOMMENDED RESOURCES

Books

Lahita, R.G., Tsokos, G., Buyon, J., Koike, T. (Eds.), 2011. Systemic Lupus Erythematosus, Fifth Ed Elsevier

Tsokos, G.C., Gordon, C., Smolen, J.S. (Eds.), 2007. Systemic Lupus Erythematosus. Mosby, Philadelphia.

Review Articles

Deng, Y., Tsao, B.P., 2010. Genetic susceptibility to systemic lupus erythematosus in the genomic era. Nat Rev Rheumatol 6, 683–692.

Moser, K.L., Kelly, J.A., Lessard, C.J., Harley, J.B., 2009. Recent insights into the genetic basis of systemic lupus erythematosus. Genes Immun 10, 373–379.

Rhodes, B., Vyse, T.J., 2008. The genetics of SLE: An update in the light of genome-wide association studies. Rheumatology 47, 1603–1822.

Rhodes, B., Vyse, T.J., 2010. Using genetics to deliver personalized SLE therapy – A realistic prospect? Nat Rev Rheumatol 6, 373–377.

Osteoarthritis

Virginia Byers Kraus

PERSONALIZED MEDICINE – CHOICE VERSUS COST

In formulating an individual care plan, traditionally one considers a patient's signs and symptoms, family history, social circumstances, environment, risk factors, and behaviors. Using a personalized medicine approach, one seeks to optimize and tailor the care of the individual patient through reliance on traditional medical information complemented by some form of companion diagnostic to more fully characterize the condition and disease or health status of the individual patient. For treatment of osteoarthritis (OA), a musculoskeletal disease, the companion diagnostic can take the form of an *in vitro* or an *in vivo* test. *In vitro* diagnostics could be based in genetics, genomics (see Chapter 6), proteomics (see Chapter 13), metabolomics, or any other molecular profiling mechanism (Garnero, 2006). *In vivo* diagnostics for OA take the form of specialized imaging methodologies such as magnetic resonance imaging (MRI), including sodium MRI (Madelin et al., 2010) and T2 mapping, T1rho mapping, and delayed gadolinium-enhanced magnetic resonance imaging of cartilage (dGEMRIC) (Burstein et al., 2009), positron emission tomography and scintigraphy (Omoumi et al., 2009), and analysis of X-ray bone trabecular integrity by fractal signature analysis (Kraus et al., 2009), to name a few.

Although diagnostic advances make aspects of personalized medicine a reality, invariably, economic (Chapter 39) considerations play an increasing role in shaping healthcare choices. Moreover, identification of those who cannot safely stay on "standard care" does not necessarily imply that there will be third-party reimbursement (Chapter 41) for individualized non-standard care. The current practice of evidence-based medicine is guided by cost-effectiveness (Chapter 39) research. The cost-effectiveness of a therapeutic or preventive intervention is the ratio of the cost of the intervention to a relevant measure of its effect. To reach some theoretical threshold of cost-effectiveness, it would generally be necessary for an intervention with only incremental benefit to cost very little, or for a costly intervention to improve the health of many or result in a dramatic health benefit for a few. Cost-effectiveness analyses are designed to generally address which "one size" fits all and provides benefit for the most individuals at the least cost.

So long as the focus continues to be on late-stage manifestations of OA, these two approaches to medical therapy, personalization versus cost-effectiveness, can be at odds. This is because late-stage OA, diagnosed by X-ray, is a prevalent, recalcitrant, chronic disease process. Personalized medicine targeting end-stage subsets of OA likely would entail a proliferation of expensive salvage procedures for each joint type and site in the body. In contrast, a focus on development of pharmacologic agents targeting specific etiologic pathways of early molecular and pre-radiographic disease would likely constitute a more cost-effective strategy and have the added prospective cost and health benefits accruing from the prevention of disability.

Genomic and Personalized Medicine, 2nd edition
by Ginsburg & Willard. DOI: http://dx.doi.org/10.1016/B978-0-12-382227-7.00081-1

OSTEOARTHRITIS – THE MAGNITUDE OF THE PROBLEM

OA is the most common joint disorder in the US and worldwide; worldwide estimates are that 9.6% of men and 18.0% of women over 60 years old have symptomatic OA (World Health Organization, 2012). OA increases with age, and is estimated to affect 40% of individuals over 70 years of age (Dieppe and Lohmander, 2005; Valdes, 2010). OA affects an estimated 27 million Americans and is the most common cause of disability in elderly people in the developed world (Croft and Hay, 2006; Hogue and Mersfelder, 2002; Lawrence et al., 2008). The World Health Organization (WHO) estimates that 80% of those with OA will have limitations in movement, and that 25% cannot perform the major daily activities of their lives (WHO, 2012). Annual medical expenditures in the US attributable to OA are estimated to be as high as $185.5 billion, or 19% of the aggregate medical expenditures for the US adult population (Kotlarz et al., 2009; Murphy et al., 2010). Thus the potential cost implications of personalized medicine for such a prevalent and disabling condition as OA are quite great. For aspects of personalized medicine to be adopted as mainstream medical practice, there will need to be a reconciliation of these two important approaches and aspects of medical care – namely a reconciliation of the focus on the needs of the individual patient and the often conflicting focus on cost-effectiveness.

Where personalized medicine and cost-effective therapy are likely to synergize is in the early identification and prevention of a chronic disease process. The cost of identifying and treating the disease in its early stages would likely be a fraction of the cost of treating the established late stages, and have the added benefit of greater likelihood of disease modification and prevention of disability. Further cost savings could likely be realized with a strategy for identifying and screening individuals more susceptible to OA. Thus, development of an early identification strategy is of fundamental importance as it would promote cost-effective prevention and early-stage treatment strategies. The cost-effective means of identifying early OA and monitoring treatment of the molecular and pre-radiographic stages of the disease clearly lies in biomarker discovery and validation. So, in this chapter, I consider the application of biomarkers in a personalized medicine strategy for OA based on the targeting of pre-radiographic OA and the early molecular stage of OA.

PARADIGMS FOR STUDYING EARLY OSTEOARTHRITIS EVENTS

OA follows a consistent pattern of pathological progression (Goldring and Goldring, 2010) to a common end-stage radiographic manifestation. Radiographic OA prevalence increases in conjunction with estrogen deficiency and aging (Herrero-Beaumont et al., 2009). Based on current knowledge,

susceptibility to osteoarthritis involves multiple pathways including signaling cascades that regulate skeletal morphogenesis, extracellular matrix proteins, and chondrocyte phenotype, apoptosis pathways, as well as inflammatory molecules related to cytokine production, prostaglandin and arachidonic acid metabolism, and proteases (Bos et al., 2008; Valdes, 2010; Valdes and Spector, 2010). To date, three loci demonstrate compelling association with OA across a broad range of ethnic groups: the 7q22 locus, the *GDF5* gene encoding growth/differentiation factor 5, which has a role in joint development, and the *DIO2* gene encoding type II iodothyronine deiodinase, responsible for conversion of thyroid hormone to its active form (Loughlin, 2011). There is also an interaction between hereditary and environmental risk factors for OA (Englund et al., 2004) suggesting that genetic profiling early in life could contribute to preventive strategies to minimize risk of OA. Recent synovial fluid analyses suggest that OA may be a disease involving uric acid, inflammasome activation, and innate immunity (Denoble et al., 2011).

OA may also occur as a sequela of major joint insult such as infection or joint injury (Lohmander et al., 2007; Roos, 2005). Unlike primary idiopathic OA, post-traumatic OA has a known time of onset, making it much more tractable as a method of monitoring the early events of the OA disease process. Acute trauma to the anterior cruciate ligament (ACL) or meniscus has been demonstrated to be a major risk factor for the development of knee OA, with a 50% chance of a patient developing symptomatic OA 10–20 years post-injury (Lohmander et al., 2007). Post-traumatic OA is believed to account for up to 5.6 million cases per year, or 12% of the total US cases of symptomatic OA, with an estimated cost in 2006 of $3.06 billion per year or 0.15% of total US direct healthcare costs (Brown et al., 2006; Buckwalter et al., 2004). The predisposition to post-traumatic OA is also likely influenced by genetic determinants, suggested by the increased prevalence of post-traumatic OA after meniscectomy in individuals with hand OA (Doherty et al., 1983; Englund et al., 2004), a form of OA with a strong genetic component (Michou, 2011).

Although there are some additional factors that predict risk of progression for established disease, such as subchondral bone trabecular integrity and knee malalignment (Kraus et al., 2009), age, generalized OA, and vitamin D deficiency (Cheung et al., 2010) to name a few, to date only a few factors, such as joint injury and obesity (Cheung et al., 2010), have been identified as predictors of incident disease, and there are no biomarkers yet qualified as indicators of the pre-radiographic molecular stage of the disease. To identify the molecular phases of the disease, it will be necessary to rely on molecular profiling provided by *in vitro* and *in vivo* biomarkers.

IDENTIFYING THE MOLECULAR STAGE OF OSTEOARTHRITIS

Based on studies of established OA, biomarkers with particular utility for monitoring OA onset and disease process are macromolecules originating from joint structures whose levels in serum, urine, and synovial fluid reflect processes taking place locally in the joint (recent comprehensive review by Van Spil et al., 2010). Direct evidence for a molecular stage of OA comes from a recent study demonstrating that four serum proteins [matrix metalloproteinase (MMP)-7, interleukin (IL)-15, plasminogen activator inhibitor (PAI)-1 and soluble vascular adhesion protein (sVAP)-1] were already different in samples obtained 10 years before the appearance of incident radiographic knee or hand OA (Ling et al., 2009); of note, IL-15, MMP-7 and VAP-1 increased and PAI-I decreased in incident OA cases relative to controls. Another study demonstrated that systemic concentrations of joint tissue components, cartilage oliogmeric matrix protein (COMP) and hyaluronan (HA), can predict, up to seven years later, the occurrence of incident knee joint space narrowing (by both COMP and HA) and bony spurs termed osteophytes (by COMP) (Golightly et al., 2010). Several studies suggest that serum COMP predicts development of radiographic hip OA 6–8 years later (Chaganti et al., 2008, Kelman et al., 2006). COMP has also been shown to be increased in the absence of signs of radiographic hip OA in patients with symptoms of hip abnormality (Dragomir et al., 2002). Although these biomarkers could be considered prognostic of incident radiographic OA, alternatively they could be considered diagnostic of the molecular stages of the disease. Other studies using MRI have confirmed the existence of a symptomatic pre-radiographic stage of the disease (Cibere et al., 2009). Taken together, these studies demonstrate that degenerative changes in the articular cartilage occur long before radiological changes are observed. They provide data to support a new concept of the natural history of the OA disease process (Figure 81.1) (based on Kraus, 2010; Kraus et al., 2011).

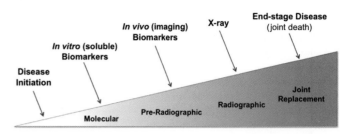

Figure 81.1 The natural history of osteoarthritis (OA) – a new paradigm. The triangle denotes the progression of disease from a clinically undetectable but serologically detectable molecular stage to the clinically detectable pre-radiographic and radiographic stages. The pre-radiographic stage refers to sensitive detection of disease features by magnetic resonance imaging, ultrasound, or other sensitive imaging modality. The radiographic stage refers to the detection of disease by the traditional method of X-ray imaging. The uphill slant of the triangle depicts the increasing difficulty, as disease progresses, to repair or modify the accrued structural damage. This graphic portrays the increased likelihood of successfully instituting cost-effective personalized medicine approaches with the earliest possible diagnosis and intervention.

JOINT INJURY AS A PARADIGM FOR EARLY OA IDENTIFICATION AND TREATMENT

As mentioned above, the joint injury paradigm provides a potential gateway to understanding the early stages of OA. Unlike primary idiopathic OA, OA arising from joint injury has a known date of onset, so the early events can be unequivocally dated and monitored. Biomarkers after joint injury can be monitored in the serum, urine, and, most proximally to the joint, through the joint or synovial fluid. Longitudinal studies of the aftermath of severe joint injury have incontrovertibly established the existence of a prolonged molecular phase of the disease characterized by biomarker abnormalities (Larsson et al., 2009; Lohmander, 1995; Lohmander et al., 1989, 1993a, b, c, 1994). After knee injury, cartilage degradation is favored over repair, with increased collagen cleavage (Aurich et al., 2005; Catterall et al., 2010; Lohmander et al., 2003). Within the first month after joint injury in humans, elevations have been documented in synovial fluid

concentrations of cartilage proteoglycan fragments (Catterall et al., 2010; Lohmander et al., 1989), metalloproteinases (MMP-3/stromelysin-1) (Lohmander et al., 1993a, 1994), and collagen fragments (Catterall et al., 2010). The aggrecanase cleavage at the 392Glu-393Ala bond in the interglobular domain (IGD) of aggrecan that releases N-terminal 393ARGS fragments has been shown to be one of the early key events in arthritis and joint injuries (Larsson et al., 2009). The challenge will be to extend these discoveries from synovial fluid to the serum. In this regard, a recent study showed significant correlation of serum and synovial fluid concentrations of several biomarkers in the setting of acute human knee injury (Catterall et al., 2010), including serum CTxI, NTx, osteocalcin, and MMP-3.

We noted that the pattern of biomarker alterations observed after joint injury matches the pattern of cartilage components released from cartilage stimulated *in vitro* with pro-inflammatory cytokines (Catterall et al., 2010) (Figure 81.2). These early events are characterized by a loss of proteoglycan followed by loss of collagen, considered a critical and possibly irreversible step in a course of inexorable joint degeneration and functional decline. This is supported by the fact that elevations of cartilage components in the serum can persist over decades after joint injury (Lohmander, 1995; Lohmander et al., 1993a, b, c, 1994). The sustained increased release of cartilage macromolecular fragments after joint trauma is thought to be responsible for the frequent development of post-traumatic OA in patients with injuries. These biomarker observations provide great hope that disease-modifying therapies are within reach since there are already

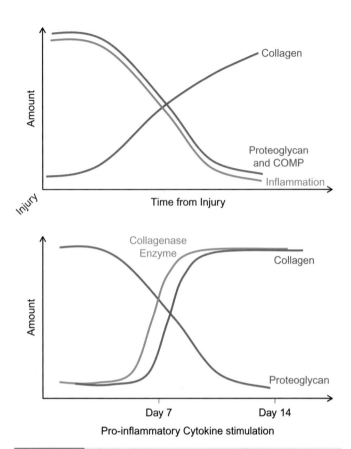

Figure 81.2 The pattern of synovial fluid biomarkers after severe joint injury. During the first 4–6 weeks after a severe sports injury, the inflamed joint contains molecular evidence of the breakdown and turnover of the cartilage. The same pattern of cartilage breakdown and turnover can be achieved by adding pro-inflammatory cytokines to a cartilage specimen *in vitro*. There is an initial phase of loss of proteoglycan fragments followed by a phase of loss of collagen fragments, considered a probable irreversible event. *Graphic kindly provided by Dr Jonathan Catterall and based on data reported by Catterall and colleagues* (2010).

many known pharmacologic agents with chondroprotective effects in the cartilage explant model and acute injury animal models of OA (Catterall et al., 2010; Lotz and Kraus, 2010).

It is also likely that early OA biomarkers identified through the study of joint injury will inform strategies for identifying early spontaneous idiopathic OA. This is illustrated by a recent article evaluating guinea pig knee OA, comparing and contrasting biomarkers in joint fluid from spontaneous OA versus injury-induced (anterior cruciate ligament transection) cartilage damage; there was no qualitative difference in the biomarkers, just a quantitative difference (higher levels post-injury than in spontaneous OA). This suggests that the two processes are, in some respects, qualitatively similar (Wei et al., 2010). Biomarker abnormalities common to both injury-related and spontaneous OA included higher synovial fluid stromal-cell-derived factor, collagen fragment C2C, proteoglycan fragments,

matrix metalloproteinase-13, and interleukin-1beta, and lower lubricin in synovial fluid. In another study comparing lipid peroxidation and antioxidant biomarkers in joint fluid from primary end-stage knee OA and post-injury cartilage damage, a number of biomarkers (thiobarbituric acid reactive substances, glutathione, glutathione peroxidase, and superoxide dismutase) were qualitatively and quantitatively similar in the two conditions (Sutipornpalangkul et al., 2009).

ADVANTAGES OF EARLY ARTHRITIS IDENTIFICATION ILLUSTRATED BY RHEUMATOID ARTHRITIS

Interestingly, a precedent now exists in rheumatoid arthritis (RA) for earlier diagnosis. New criteria (a revision of the 1987 criteria) for an RA diagnosis have recently been published. They were motivated by a desire to increase the likelihood of earlier diagnosis so that prompt disease-modifying treatment could be initiated (Aletaha et al., 2010) because it is now well recognized that delays in initiating therapy have a deleterious effect on long-term outcomes (Dale and Porter, 2010). The presence of RA joint erosions on radiographs was omitted from the criteria because of the new emphasis on early diagnosis. This shift of focus from diagnosing and alleviating the consequences of established joint damage to early diagnosis and intervention presents a compelling paradigm that should inform the approach to the diagnosis and treatment of OA.

In the case of RA, it is postulated that the early period of immune dysregulation is potentially reversible and that the subsequent development of pannus (the inflammatory synovial tissue found in RA joints) and erosive joint damage represents a stage of auto-immune dysfunction that is less amenable to immunomodulatory therapy (Dale and Porter, 2010). In RA, the persistence of synovitis and cumulative exposure to inflammation correlate with disease progression (summarized in Dale and Porter, 2010). We recently identified inflammasome activation and innate immunity as an indicator of severity and risk of progression of OA (Denoble et al., 2011) suggesting that in OA, as in RA, cumulative exposure to inflammation may be a key factor in severity and progression and represent a key target for intervention.

In addition to genetic predisposition, RA, as hypothesized for OA, also has an asymptomatic stage characterized by serological abnormalities; in RA this is characterized by autoantibody production and high C-reactive protein prior to onset of clinical manifestations of RA (Deane et al., 2010). The appropriate pre-clinical markers have been posited to be a potential trigger for pre-emptive treatments of "at risk" individuals (Dale and Porter, 2010). In early RA, high risk for progression has been associated with high baseline serum COMP, Glc-Gal-PyD, CTX-I and CTX-II, MMP-3, first year time-averaged erythrocyte sedimentation rate (ESR), the serum osteoprotegerin:RANKL ratio, positive rheumatoid factor, anti-cyclic citrullinated peptide antibody, and the HLA-DRB1 shared allele (reviewed in Landewe, 2007).

Moreover, efforts are ongoing to identify predictors of response to particular therapies (Marotte and Miossec, 2010). This enables identification of risk factors and prompt treatment at first clinical signs due to unequivocal identification earlier. In the case of RA, this is anticipated to minimize joint destruction through early intervention, minimize adverse events through minimizing exposure of patients to ineffective treatments, and ensure more appropriate use of healthcare funds. For instance, low serum COMP, a measure of joint tissue degradation, predicts a more rapid and sustained clinical response to treatment in RA with an anti-TNF (Tumor Necrosis Factor alpha) agent compared with high baseline serum COMP (Morozzi et al., 2007a, b), suggesting that serum COMP may reflect some irreversible aspects of joint deterioration in RA. Taken together, these results suggest that biomarkers may facilitate rapid, rational, and aggressive response in the individuals at high risk for progressive and severe disease.

OUTLOOK FOR THE FUTURE

A combination of biomarkers and sensitive imaging modalities improves the ability to diagnose and prognose established OA (Dam et al., 2009; Garnero, 2006), so it is also likely that combinatorial markers will provide the best discriminatory power for pre-radiographic disease. As noted by a working group on OA prevention and risk reduction (Jordan et al., 2011), it will be necessary to discern when a specific value of the biomarker(s) truly represents a "disease-free" state, and to establish an understanding of the rapidity of the biomarker change (if treated as a continuous variable) or conversion (if treated as a discrete variable) in relation to the development of disease. Ideally, the biomarker will have been previously validated against a clinically relevant endpoint for its use as a surrogate measure (Bauer et al., 2006).

Although the majority (95%) of drugs used are for chronic or complex diseases, there are currently no Food and Drug Administration (FDA)- or European Medicines Agency (EMA)-approved pharmacological interventions to modify the OA disease process. This is in large part due to the dearth of sensitive *in vitro* or *in vivo* biomarkers suitable for monitoring efficacy of interventions. However, there are many OA-related biomarkers in various stages of qualification and many ways in

which biomarkers may contribute to the development of drugs for OA (Kraus et al., 2011). An Osteoarthritis Research Society International (OARSI) working group on OA biomarkers has suggested that the biomarkers likely to have the earliest beneficial impact on clinical trials include those that would identify subjects likely to either respond and/or progress within a short time frame, and those that would provide early feedback for pre-clinical decision-making that a drug was having the desired biochemical effect (Kraus et al., 2011).

The ultimate in personalized medicine is to make new organs from a patient's own cells. In contrast to the current dearth of pharmacological therapy to impact joint degeneration, an FDA-approved autologous tissue repair procedure for OA (http://www.fda.gov/BiologicsBloodVaccines/CellularGeneTherapyProducts/ApprovedProducts/ucm134025.htm) was introduced in 1997. Engineered tissue offers great promise for cartilage regeneration. Allogeneic chondrocytes derived from human juvenile articular cartilage offer an alternative and may represent an improved source of cells for cartilage repair (Adkisson et al., 2010). This rapidly-evolving area may ultimately form the basis of a larger armamentarium of OA therapeutics (Evans et al., 2005; Nixon et al., 2007). It is expected that both *in vivo* and *in vitro* biomarkers will be central to the monitoring of these personalized medicine approaches to the treatment of OA.

In summary (Box 81.1), biomarkers are central to personalized medicine as they establish the foundation for identifying early disease stages and the molecular pathophysiology to inform individualized therapeutic recommendations and decision-making. Early disease identification with biomarkers has great potential for stimulating cost-effective personalized medicine approaches for the treatment of OA with the hope of preventing the long-term disabling consequences. Perhaps one day, a "joint health panel" of blood tests will become part of an annual health screening, particularly for high-risk individuals with known genetic susceptibility and/or joint trauma history. As for other "silent" diseases such as hypertension, osteoporosis, and early heart disease (Kraus, 2010), we may establish a means of detecting the disease prior to the stroke, fracture, or heart attack equivalent, namely, prior to the appearance of joint damage on an X-ray. We might thereby institute healthy lifestyle modifications and pharmacologic agents to slow or halt the disease process prior to loss of organ function.

Box 81.1   Summary of role of biomarkers for early osteoarthritis identification

- The most cost-effective application of personalized medicine in the chronic disease of OA is likely to be related to the identification, monitoring, and treatment of early OA, i.e., in the pre-radiographic and molecular stages.
- Rapid advances in the genetic knowledge of molecular pathways involved in OA susceptibility will inform and facilitate the development of biomarkers for early disease detection and monitoring.
- Biomarkers indicative of joint tissue metabolism in the first several years after severe acute joint injury could provide

insights into biomarkers that could possibly detect the early molecular stages of spontaneous OA.
- A combinatorial biomarker will likely provide greatest discriminatory power for early OA detection.
- The list of biomarkers for OA needs to expand beyond established radiographic OA to encompass pre-radiographic OA and molecular-stage OA, incidence and progression.

REFERENCES

Adkisson, H.D.T., Martin, J.A., Amendola, R.L., et al., 2010. The potential of human allogeneic juvenile chondrocytes for restoration of articular cartilage. Am J Sports Med 38, 1324–1333.

Aletaha, D., Neogi, T., Silman, A.J., et al., 2010. 2010 rheumatoid arthritis classification criteria: An American College of Rheumatology/European League Against Rheumatism collaborative initiative. Ann Rheum Dis 69, 1580–1588.

Aurich, M., Squires, G.R., Reiner, A., et al., 2005. Differential matrix degradation and turnover in early cartilage lesions of human knee and ankle joints. Arthritis Rheum 52, 112–119.

Bauer, D.C., Hunter, D.J., Abramson, S.B., et al., 2006. Classification of osteoarthritis biomarkers: A proposed approach. Osteoarthritis Cartilage 14, 723–727.

Bos, S.D., Slagboom, P.E., Meulenbelt, I., 2008. New insights into osteoarthritis: Early developmental features of an ageing-related disease. Curr Opin Rheumatol 20, 553–559.

Brown, T.D., Johnston, R.C., Saltzman, C.L., Marsh, J.L., Buckwalter, J.A., 2006. Posttraumatic osteoarthritis: A first estimate of incidence, prevalence, and burden of disease. J Orthop Trauma 20, 739–744.

Buckwalter, J.A., Saltzman, C., Brown, T., 2004. The impact of osteoarthritis: Implications for research. Clin Orthop Relat Res, S6–S15.

Burstein, D., Gray, M., Mosher, T., Dardzinski, B., 2009. Measures of molecular composition and structure in osteoarthritis. Radiol Clin North Am 47, 675–686.

Catterall, J.B., Stabler, T.V., Flannery, C.R., Kraus, V.B., 2010. Changes in serum and synovial fluid biomarkers after acute injury (NCT00332254). Arthritis Res Ther 12, R229.

Chaganti, R.K., Kelman, A., Lui, L., et al., 2008. Change in serum measurements of cartilage oligomeric matrix protein and association with the development and worsening of radiographic hip osteoarthritis. Osteoarthritis Cartilage 16, 566–571.

Cheung, P.P., Gossec, L., Dougados, M., 2010. What are the best markers for disease progression in osteoarthritis (OA)? Best Pract Res Clin Rheumatol 24, 81–92.

Cibere, J., Zhang, H., Garnero, P., et al., 2009. Association of biomarkers with pre-radiographically defined and radiographically defined knee osteoarthritis in a population-based study. Arthritis Rheum 60, 1372–1380.

Croft, P., Hay, E., 2006. Osteoarthritis in primary care. BMJ 333, 867–868.

Dale, J., Porter, D., 2010. Pharmacotherapy: Concepts of pathogenesis and emerging treatments. Optimising the strategy of care in early rheumatoid arthritis. Best Pract Res Clin Rheumatol 24, 443–455.

Dam, E.B., Loog, M., Christiansen, C., et al., 2009. Identification of progressors in osteoarthritis by combining biochemical and MRI-based markers. Arthritis Res Ther 11, R115.

Deane, K.D., Norris, J.M., Holers, V.M., 2010. Preclinical rheumatoid arthritis: Identification, evaluation, and future directions for investigation. Rheum Dis Clin North Am 36, 213–241.

Denoble, A.E., Huffman, K.M., Stabler, T.V., et al., 2011. Uric acid is a danger signal of increasing risk for osteoarthritis through inflammasome activation. Proc Natl Acad Sci USA 108, 2088–2093.

Dieppe, P.A., Lohmander, L.S., 2005. Pathogenesis and management of pain in osteoarthritis. Lancet 365, 965–973.

Doherty, M., Watt, I., Dieppe, P., 1983. Influence of primary generalised osteoarthritis on development of secondary osteoarthritis. Lancet 2, 8–11.

Dragomir, A.D., Kraus, V.B., Renner, J.B., et al., 2002. Serum cartilage oligomeric matrix protein and clinical signs and symptoms of potential pre-radiographic hip and knee pathology. Osteoarthritis Cartilage 10, 687–691.

Englund, M., Paradowski, P.T., Lohmander, L.S., 2004. Association of radiographic hand osteoarthritis with radiographic knee osteoarthritis after meniscectomy. Arthritis Rheum 50, 469–475.

Evans, C.H., Ghivizzani, S.C., Herndon, J.H., Robbins, P.D., 2005. Gene therapy for the treatment of musculoskeletal diseases. J Am Acad Orthop Surg 13, 230–242.

Garnero, P., 2006. Use of biochemical markers to study and follow patients with osteoarthritis. Curr Rheumatol Rep 8, 37–44.

Goldring, M.B., Goldring, S.R., 2010. Articular cartilage and subchondral bone in the pathogenesis of osteoarthritis. Ann NY Acad Sci 1192, 230–237.

Golightly, Y., Marshall, S., Kraus, V., et al., 2010. Serum cartilage oligomeric matrix protein, hyaluronan, high-sensitivity C-reactive protein, and keratan sulfate as predictors of incident radiographic knee osteoarthritis: Differences by chronic knee symptoms. Osteoarthritis Cartilage 18, S62–S63.

Herrero-Beaumont, G., Roman-Blas, J.A., Castaneda, S., Jimenez, S.A., 2009. Primary osteoarthritis no longer primary: Three subsets with distinct etiological, clinical, and therapeutic characteristics. Semin Arthritis Rheum 39, 71–80.

Hogue, J.H., Mersfelder, T.L., 2002. Pathophysiology and first-line treatment of osteoarthritis. Ann Pharmacother 36, 679–686.

Jordan, J.M., Sowers, M.F., Messier, S.P., et al., 2011. Methodologic issues in clinical trials for prevention or risk reduction in osteoarthritis. Osteoarthritis Cartilage 19, 500–508.

Kelman, A., Lui, L., Yao, W., Krumme, A., Nevitt, M., Lane, N.E., 2006. Association of higher levels of serum cartilage oligomeric matrix protein and N-telopeptide crosslinks with the development of radiographic hip osteoarthritis in elderly women. Arthritis Rheum 54, 236–243.

Kotlarz, H., Gunnarsson, C.L., Fang, H., Rizzo, J.A., 2009. Insurer and out-of-pocket costs of osteoarthritis in the US: Evidence from national survey data. Arthritis Rheum 60, 3546–3553.

Kraus, V., 2010. Waiting for action on the osteoarthritis front. Curr Drug Targets 11, 1–2.

Kraus, V., Burnett, B., Coindreau, J., et al., 2011. Application of biomarkers in the development of drugs intended for the treatment of osteoarthritis. Osteoarthritis Cartilage 19, 515–542.

Kraus, V.B., Feng, S., Wang, S., et al., 2009. Trabecular morphometry by fractal signature analysis is a novel marker of osteoarthritis progression. Arthritis Rheum 60, 3711–3722.

Landewe, R., 2007. Predictive markers in rapidly progressing rheumatoid arthritis. J Rheumatol Suppl 80, 8–15.

Larsson, S., Lohmander, L.S., Struglics, A., 2009. Synovial fluid level of aggrecan ARGS fragments is a more sensitive marker of joint disease than glycosaminoglycan or aggrecan levels: A cross-sectional study. Arthritis Res Ther 11, R92.

Lawrence, R.C., Felson, D.T., Helmick, C.G., et al., 2008. Estimates of the prevalence of arthritis and other rheumatic conditions in the United States: Part II. Arthritis Rheum 58, 26–35.

Ling, S.M., Patel, D.D., Garnero, P., et al., 2009. Serum protein signatures detect early radiographic osteoarthritis. Osteoarthritis Cartilage 17, 43–48.

Lohmander, L.S., 1995. The release of aggrecan fragments into synovial fluid after joint injury and in osteoarthritis. J Rheumatol Suppl 43, 75–77.

Lohmander, L.S., Atley, L.M., Pietka, T.A., Eyre, D.R., 2003. The release of crosslinked peptides from type II collagen into human synovial fluid is increased soon after joint injury and in osteoarthritis. Arthritis Rheum 48, 3130–3139.

Lohmander, L.S., Dahlberg, L., Ryd, L., Heinegard, D., 1989. Increased levels of proteoglycan fragments in knee joint fluid after injury. Arthritis Rheum 32, 1434–1442.

Lohmander, L.S., Englund, P.M., Dahl, L.L., Roos, E.M., 2007. The long-term consequence of anterior cruciate ligament and meniscus injuries: Osteoarthritis. Am J Sports Med 35, 1756–1769.

Lohmander, L.S., Hoerrner, L.A., Dahlberg, L., Roos, H., Bjornsson, S., Lark, M.W., 1993a. Stromelysin, tissue inhibitor of metalloproteinases and proteoglycan fragments in human knee joint fluid after injury. J Rheumatol 20, 1362–1368.

Lohmander, L.S., Hoerrner, L.A., Lark, M.W., 1993b. Metalloproteinases, tissue inhibitor, and proteoglycan fragments in knee synovial fluid in human osteoarthritis. Arthritis Rheum 36, 181–189.

Lohmander, L.S., Neame, P.J., Sandy, J.D., 1993c. The structure of aggrecan fragments in human synovial fluid. Evidence that aggrecanase mediates cartilage degradation in inflammatory joint disease, joint injury, and osteoarthritis. Arthritis Rheum 36, 1214–1222.

Lohmander, L.S., Roos, H., Dahlberg, L., Hoerrner, L.A., Lark, M.W., 1994. Temporal patterns of stromelysin-1, tissue inhibitor, and proteoglycan fragments in human knee joint fluid after injury to the cruciate ligament or meniscus. J Orthop Res 12, 21–28.

Lotz, M.K., Kraus, V.B., 2010. New developments in osteoarthritis. Posttraumatic osteoarthritis: Pathogenesis and pharmacological treatment options. Arthritis Res Ther 12, 211.

Loughlin, J., 2011. Osteoarthritis year 2010 in review: Genetics. Osteoarthritis Cartilage 19, 342–345.

Madelin, G., Lee, J.-S., Inati, S., Jerschow, A., Regatte, R., 2010. Sodium inversion recovery MRI of the knee joint in vivo at 7T. J Magn Reson doi: 10.1016/j.jmr.2010.08.003.

Marotte, H., Miossec, P., 2010. Biomarkers for prediction of TNFalpha blockers response in rheumatoid arthritis. Joint Bone Spine 77, 297–305.

Michou, L., 2011. Genetics of digital osteoarthritis. Joint Bone Spine 78, 347–351.

Morozzi, G., Fabbroni, M., Bellisai, F., Cucini, S., Simpatico, A., Galeazzi, M., 2007a. Low serum level of COMP, a cartilage turnover marker, predicts rapid and high ACR70 response to adalimumab therapy in rheumatoid arthritis. Clin Rheumatol 26, 1335–1338.

Morozzi, G., Fabbroni, M., Bellisai, F., Pucci, G., Galeazzi, M., 2007b. Cartilage oligomeric matrix protein level in rheumatic diseases: Potential use as a marker for measuring articular cartilage damage and/or the therapeutic efficacy of treatments. Ann NY Acad Sci 1108, 398–407.

Murphy, L.B., Helmick, C.G., Cisternas, M.G., Yelin, E.H., 2010. Estimating medical costs attributable to osteoarthritis in the US population: Comment on the article by Kotlarz et al. Arthritis Rheum 62, 2566–2567, author reply 2567-2568.

Nixon, A.J., Goodrich, L.R., Scimeca, M.S., et al., 2007. Gene therapy in musculoskeletal repair. Ann NY Acad Sci 1117, 310–327.

Omoumi, P., Mercier, G.A., Lecouvet, F., Simoni, P., Vande Berg, B.C., 2009. CT arthrography, MR arthrography, PET, and scintigraphy in osteoarthritis. Radiol Clin North Am 47, 595–615.

Roos, E.M., 2005. Joint injury causes knee osteoarthritis in young adults. Curr Opin Rheumatol 17, 195–200.

Sutipornpalangkul, W., Morales, N.P., Charoencholvanich, K., Harnroongroj, T., 2009. Lipid peroxidation, glutathione, vitamin E, and antioxidant enzymes in synovial fluid from patients with osteoarthritis. Int J Rheum Dis 12, 324–328.

Valdes, A., 2010. Molecular pathogenesis and genetics of osteoarthritis: Implications for personalized medicine. Personalized Med 7, 49–63.

Valdes, A.M., Spector, T.D., 2010. The clinical relevance of genetic susceptibility to osteoarthritis. Best Pract Res Clin Rheumatol 24, 3–14.

Van Spil, W.E., Degroot, J., Lems, W.F., Oostveen, J.C., Lafeber, F.P., 2010. Serum and urinary biochemical markers for knee and hip-osteoarthritis: A systematic review applying the consensus BIPED criteria. Osteoarthritis Cartilage 18, 605–612.

Wei, L., Fleming, B.C., Sun, X., et al., 2010. Comparison of differential biomarkers of osteoarthritis with and without posttraumatic injury in the Hartley guinea pig model. J Orthop Res 28, 900–906.

World Health Organization, 2012, Chronic diseases and health promotion, <http://www.who.int/chp/topics/rheumatic/en/index.html>.

Diabetes

Jose C. Florez

INTRODUCTION

Diabetes mellitus is defined by the occurrence of hyperglycemia in the basal or post-prandial state. Hyperglycemia results when the pancreatic beta cell is unable to secrete enough insulin to maintain normal glucose levels given the metabolic context of the organism. Therefore, diabetes can be caused by a variety of mechanisms, but the final common pathway results from relative insulin insufficiency. Clinical complications shared by all forms of diabetes are thought to result from long-standing hyperglycemia, but are compounded to a greater or lesser extent by co-morbid conditions.

The best characterized forms of diabetes include autoimmune diabetes (type 1 diabetes and latent autoimmune diabetes in the adult), which result from autoimmune destruction of pancreatic beta cells, and rarer monogenic forms such as maturity-onset diabetes of the young (MODY) or neonatal diabetes, which arise from specific single-gene mutations. However, the most common form of diabetes, responsible for ~90% of all diabetes cases worldwide, is type 2 diabetes (T2D), caused by a protean conglomerate of pathophysiological processes that lead to insulin resistance and β-cell failure in varying degrees, and are often accompanied and/or influenced by obesity, hypertension, and dyslipidemia. Therefore, for the purposes of this chapter, we will focus solely on T2D. Readers interested in the fascinating and largely parallel exploration of genetic determinants in type 1 diabetes are referred to excellent and timely reviews on this subject (Bluestone et al.,

2010; Concannon et al., 2009; Grant and Hakonarson, 2009; Naik et al., 2009; Pociot et al., 2010; Todd, 2010). Similarly, monogenic or syndromic forms of diabetes are reviewed elsewhere (Ellard et al., 2008; Greeley et al., 2010; Hattersley et al., 2009; Vaxillaire et al., 2009; Vaxillaire and Froguel, 2008).

EPIDEMIOLOGY AND GENETICS
Epidemiology of Type 2 Diabetes and its Complications

Type 2 diabetes is a leading cause of poor health worldwide (Bonow and Gheorghiade, 2004). In the United States, diabetes is the leading cause among adults of new blindness, end-stage renal disease and non-traumatic lower limb amputation (Engelgau et al., 2004; National Institute of Diabetes and Digestive and Kidney Diseases (NIDDK), 2005). The NHANES III study estimated that diabetes accounted for 8.6% of hospitalizations, 12.3% of nursing home admissions, and 10.3% of deaths among US adults during 1988–94 (Russell et al., 2005). As a sedentary lifestyle and excess caloric intake spread worldwide, diabetes is reaching pandemic proportions in developing nations as well (Chan et al., 2009; Mbanya et al., 2010; Yoon et al., 2006). The human toll, opportunity costs, and financial burdens imposed by growing numbers of people with diabetes threaten to severely damage the US and world economy (American Diabetes Association, 2008; Hamer and El Nahas, 2006; Ritz et al., 1999).

Genomic and Personalized Medicine, 2nd edition
by Ginsburg & Willard. DOI: http://dx.doi.org/10.1016/B978-0-12-382227-7.00082-3

Heritability of Type 2 Diabetes

The quick rise in T2D prevalence over the past century (during which the human genome has obviously remained largely unchanged) suggests that development of the disease is powerfully influenced by the environment and related behaviors, including access to food, dietary intake, and physical activity. However, several lines of evidence suggest that individuals vary in their susceptibility to these exposures, at least in part, as a function of genetics (Elbein, 2002; Florez et al., 2003; Groop, 1997). First, the risk of diabetes varies depending on the populations studied, even when these share a similar environment. For example, Mexican- or Native Americans have a higher risk of diabetes than their white compatriots (Carter et al., 1996). Second, family history is a key independent diabetes risk factor: in contrast to the population risk of 5–10%, a first-degree relative of someone with T2D has a 40% lifetime risk of developing diabetes, and this risk nearly doubles if he or she has two first-degree diabetic relatives (Meigs et al., 2000; Weijnen et al., 2002). Third, the effect of genes can be distinguished from that of a shared environment early in life through the use of twin studies. Rates of concordance are much higher for monozygotic twins (who share identical copies of the genome sequence) when compared to dizygotic twins, whose rates of concordance are similar to those of other first-degree relatives (parent–offspring or sibling pairs). For example, Barnett and colleagues found that 48 of 53 identical twin pairs were concordant for T2D if followed for long enough (Barnett et al., 1981), while Poulsen and colleagues found a concordance rate of 63% in Danish monozygotic twins vs 43% in dizygotic twins (Poulsen et al., 1999). Finally, as mentioned above, mutations in specific genes transmitted in a Mendelian fashion also have been found to confer a strong risk for diabetes (Greeley et al., 2010; Vaxillaire et al., 2009); while these mutations are rare, they provide a proof-of-concept that genetic variation underlies the risk for diabetes. All of the above provides compelling evidence that many forms of diabetes have a strong genetic component.

THE SEARCH FOR GENETIC DETERMINANTS OF TYPE 2 DIABETES

Early Studies

Linkage Analysis

Throughout the last 30 years, linkage mapping and positional cloning have allowed for scanning of the human genome prior to the availability of its full sequence, and enabled the discovery of hundreds of genetic mutations responsible for relatively rare human diseases. Linkage analysis is predicated on the principle of identity by descent, by which genomic segments containing the causal variant are inherited from the same ancestor by affected members of a pedigree, while unaffected relatives inherit the same segment from someone else who lacks the variant. It is particularly powerful in the setting of rare variants (i.e., more likely to segregate via a single uniparental line) that have a strong effect on phenotype, where there is close to a 1:1 relationship between mutation and disease. In these cases, the influence of other modifier genes or environmental factors is quite small when compared to that of the index mutation. In linkage, single recombination events in meiosis can significantly help narrow down the relevant DNA segment; by the same token, a single genotyping artifact can introduce substantial error.

In complex diseases, however, the phenotype arises through the interplay of many genetic variants, some or all of which interact with the environment. Individually, each of those variants has a modest effect on risk, and many of them may be quite common in the overall population (because their modest effect size, the late age of disease onset, and/or the kind of trait acquired in a different environmental context have not subjected them to strong purifying selection). Linkage analysis met with quick success when extended to type 1 diabetes, where as much as 50% of the genetic contribution to the phenotype can be explained by the human leukocyte antigen (HLA) region (Davies et al., 1994). This early finding, now viewed as exceptional, falsely reassured investigators that linkage could be deployed to other complex diseases with relative ease. The large number of inconclusive studies that followed, coupled with statistical simulations, showed that sample sizes were inadequate to generate linkage signals of sufficient magnitude to merit their subsequent exploration (Risch, 2000; Risch and Merikangas, 1996).

Association Analysis

Association analysis entails a fundamentally different approach. It simply asks whether a genetic variant is overrepresented in disease versus health. It asks about presence, rather than inheritance: the relationship with phenotype is probabilistic and not deterministic and therefore more applicable to unraveling the architecture of polygenic diseases. It can be carried out in family or unrelated datasets; while it also requires large sample sizes to detect modest effects, it is quite robust to genotyping error. The otherwise unexplained overrepresentation of a genetic variant in cases when compared to controls indicates that the genomic segment tagged by such a variant must be causal in disease, since (as opposed to other epidemiological associations) it precedes the phenotype and is unaffected by it.

The First T2D Genes

Unlike linkage, where anonymous genetic markers anchoring the entire genome facilitated genome-wide analyses, association studies require previous knowledge of the genetic variant to be queried. Thus, prior to the completion of the reference human genome sequence, only candidate gene association studies were possible. Investigators typically chose a particular gene based on its implication in T2D pathophysiology, sequenced it to discover common polymorphisms, and genotyped these in appropriate samples to establish whether a specific allele deviated from the expected frequency distribution under the null hypothesis of no association. From the many candidate gene studies performed in T2D, two associations have stood the test of time: the P12A polymorphism in the peroxisome proliferator-activated receptor γ2 gene (*PPARG*) (Altshuler et al., 2000), and the highly correlated E23K and A1369S variants in the two subunits that comprise the islet ATP-dependent potassium channel, encoded by the adjacent genes *KCNJ11* and *ABCC8* respectively (Florez et al., 2004; Gloyn et al., 2003). Interestingly, both of these loci involve missense changes that alter protein function, and both encode targets of anti-diabetic medications (thiazolidinediones activate PPARγ, and sulfonylureas bind to the sulfonylurea receptor/potassium channel complex).

Immediately before the advent of genome-wide association studies (GWAS), a combination of approaches led to the identification of the common genetic variant that has the strongest effect on T2D risk documented to date. A linkage study reported a number of suggestive peaks (Reynisdottir et al., 2003), and saturation fine-mapping of the various peaks uncovered a strong association in chromosome 10, which was replicated in two independent populations and did not explain the original linkage signal (Grant et al., 2006). The association mapped to the transcription factor 7-like 2 (*TCF7L2*) gene, and seemed to stem from an intronic single nucleotide polymorphism (rs7903146). It has since been replicated in most ethnic groups (Cauchi et al., 2007; Florez, 2007); the risk allele is present at a ~30% frequency in populations of European or African origin and confers a ~40% excess risk per allele carried.

Genome-wide Association Studies

The First Generation

GWAS became possible through several major advances in the field. The complete sequencing of the human genome (Lander et al., 2001; Venter et al., 2001), the ensuing cataloging of common human genetic variation (Reich et al., 2003; Sachidanandam et al., 2001) and the empiric realization that the existing correlation among genetic variants leads to reduced haplotypic diversity (Daly et al., 2001; Gabriel et al., 2002; Jorde, 2000) all gave rise to the International HapMap project (International HapMap Consortium, 2003), which by 2005 had produced a comprehensive view of the statistical relationships between common single nucleotide polymorphisms (SNPs) around the genome in four major world populations (International HapMap Consortium, 2005). Technology development kept pace, with several companies using this information to manufacture arrays that allowed for the generation of accurate genotype calls at hundreds of thousands of genomic sites. And the rising consciousness among scientists that genetic variants of modest effects could only be detected through very large sample sizes (particularly given the high number of tests performed) facilitated international collaborations and the sharing of data, analytical tools, and methodologies. By the midpoint of the first decade in the millennium, the field was ripe for genetic association testing to be deployed genome-wide across many human traits.

Type 2 diabetes was one of the first diseases to undergo high-density genome-wide association scanning. In a much-anticipated landmark paper, in early 2007 a French and French-Canadian collaboration reported genetic associations with T2D at *TCF7L2*, the *SLC30A8* gene encoding the zinc transporter ZnT8, a region around genes encoding a pancreatic transcription factor (*HHEX*), the insulin-degrading enzyme (*IDE*), and two other loci that later proved to be irreproducible (Sladek et al., 2007). The convincing replication of the *TCF7L2* signal at the top of the list validated both the specific genetic association as well as the genome-wide approach, and suggested that few to no other common variants surpassed its impact on T2D risk.

Shortly thereafter, several other GWAS came to light. A collaboration between the Diabetes Genetics Initiative (DGI) (Diabetes Genetics Initiative of Broad Institute of Harvard and MIT et al., 2007), the Wellcome-Trust Case Control Consortium (WTCCC) (Zeggini et al., 2007) and the FUSION group (Scott et al., 2007) jointly confirmed the *TCF7L2*, *SLC30A8*, and *HHEX* associations and identified *CDKAL1*, *CDKN2A/B*, and *IGF2BP2* as additional T2D susceptibility loci. Independently, deCODE investigators also reported the *CDKAL1* association (Steinthorsdottir et al., 2007). In the span of a few months, the number of *bona fide* T2D loci had grown from three to eight, an approach had been validated, and a fast-track path had been opened for investigators to begin populating the lists of genetic variants that increase T2D risk.

Meta-analyses

While well-powered candidate gene studies continued to yield associated loci, such as common SNPs in the Wolfram syndrome gene *WFS1* (Franks et al., 2008; Sandhu et al., 2007) or the MODY gene *HNF1B* (Gudmundsson et al., 2007; Winckler et al., 2007), the emphasis had shifted to the integration of genome-wide datasets via meta-analysis. This strategy makes sense, in that many true associations remain hidden below the stringent statistical threshold used to declare genome-wide significance, empirically established at $P < 5 \times 10^{-8}$ for common variants in populations of non-African descent (Pe'er et al., 2008). Larger sample sizes should raise statistical confidence that these false negative results represent real signals. In this manner, the DGI, WTCCC, and FUSION cohorts were combined into the Diabetes Genetics Replication and Meta-analysis (DIAGRAM) consortium, which included additional *in silico* or *de novo* replication resources. This effort, involving analysis of many tens of thousands of samples, generated six new T2D loci (*JAZF1*, *CDC123-CAMK1D*, *TSPAN8-LGR5*, *THADA*, *ADAMTS9*, and *NOTCH2-ADAM30*) (Zeggini et al., 2008). In parallel, follow-up of the French/French-Canadian GWAS identified *IRS1* as a biologically plausible T2D gene, through both genetic and functional means (Rung et al., 2009). A more comprehensive meta-analysis of GWAS datasets informative for T2D, termed DIAGRAM+, has recently reported 12 additional loci (*BCL11A*, *ZBED3*, *KLF14*, *TP53INP1*, *TLE4*, *KCNQ1*, *ARAP1*, *HMGA2*, *HNF1A*, *ZFAND6*, *PRC1*, and *DUSP9*) (Voight et al., 2010).

All of these GWAS have been conducted in populations of European descent. Study of other ethnic groups may detect novel signals, as statistical power may be increased for variants that are more common in these populations or have a stronger effect size via ethnic-specific gene–gene or gene–environment interactions. In an illustration of this concept, two Japanese groups independently reported that variants in *KCNQ1*, which are much more frequent in Asians than in Europeans, also increase T2D risk (Unoki et al., 2008; Yasuda et al., 2008). These variants (which are distinct from those later reported by DIAGRAM+) do produce a nominally significant association signal in Europeans, but due to their rare frequencies they had not reached the arbitrary level required for subsequent replication and had thus escaped detection in earlier analyses.

All T2D loci reported to date are listed in Table 82.1, and the year of discovery with approximate effect size are displayed in Figure 82.1.

GWAS for Quantitative Traits

A complementary strategy in the quest for T2D genes involves the study of related quantitative traits. Because T2D is defined by hyperglycemia and often results from insulin resistance, a search for genetic determinants of serum glucose or insulin levels may identify other loci implicated in T2D pathogenesis. This approach has the advantage of greater numbers afforded by population cohorts when compared to case/control collections, because nearly all individuals will be informative for such quantitative studies. Diabetic participants must be excluded, however, because diabetic treatment and/or the disease process itself may affect the very parameters one intends to measure. Restricting the study of these traits to the sub-diabetic range allows for the additional distinction between genetic variants that cause progressive pathology from those that simply regulate homeostatic glycemia.

Glucokinase is the rate-limiting enzyme regulating glucose-induced insulin secretion, thus functioning as the glucose sensor in the pancreatic β cell. Early candidate gene studies had suggested that its gene GCK not only harbors loss-of-function mutations that cause MODY but also polymorphisms of subtler effect that raise fasting glucose and over time can contribute to T2D (Weedon et al., 2006). In the same pathway, the gene encoding the glucose-6-phosphate catalytic subunit (*G6PC2*) was also

TABLE 82.1	Genetic variants associated with type 2 diabetes at genome-wide levels of statistical significance					
Marker	**Chr**	**Nearest gene**	**Type of mutation**	**Allele (effect/other)**	**Odds ratio (95% CI)**	**Discovery cohort**
rs10923931	1	NOTCH2	Intronic	T/G	1.13 (1.08–1.17)	Zeggini et al., 2008
rs340874	1	PROX1*	2 kb upstream	C/T	1.07 (1.05–1.09)	Dupuis et al., 2010
rs243021	2	BCL11A	99 kb downstream	A/G	1.08 (1.06–1.10)	Voight et al., 2010
rs780094	2	GCKR*	Intronic	C/T	1.06 (1.04–1.08)	Dupuis et al., 2010
rs2943641	2	IRS1	502 kb upstream	C/T	1.19 (1.13–1.25)	Rung et al., 2009
rs7578597	2	THADA	Missense: Thr1187Ala	T/C	1.15 (1.10–1.20)	Zeggini et al., 2008
rs4607103	3	ADAMSTS9	38 kb upstream	C/T	1.09 (1.06–1.12)	Zeggini et al., 2008
rs11708067	3	ADCY5*	Intronic	A/G	1.12 (1.09–1.15)	Dupuis et al., 2010
rs4402960	3	IGF2BP2	Intronic	T/G	1.14 (1.11–1.18)	Diabetes Genetics Initiative of Broad Institute of Harvard and MIT et al., 2007; Scott et al., 2007; Zeggini et al., 2007
rs1801282	3	PPARG	Missense: Pro12Ala	C/G	1.18 (1.09–1.29)†	Altshuler et al., 2000
rs10010131	4	WFS1	Intron-exon junction	G/A	1.11 (1.08–1.16)	Sandhu et al., 2007
rs4457053	5	ZBED3	41 kb upstream	G/A	1.08 (1.06–1.11)	Voight et al., 2010
rs7754840	6	CDKAL1	Intronic	C/G	1.12 (1.08–1.16)	Diabetes Genetics Initiative of Broad Institute of Harvard and MIT et al., 2007; Scott et al., 2007; Steinthorsdottir et al., 2007; Zeggini et al., 2007

(continued)

TABLE 82.1 (Continued)						
Marker	**Chr**	**Nearest gene**	**Type of mutation**	**Allele (effect/other)**	**Odds ratio (95% CI)**	**Discovery cohort**
rs2191349	7	*DGKB/ TMEM195**	Intergenic region	T/G	1.06 (1.04–1.08)	Dupuis et al., 2010
rs4607517	7	*GCK**	36 kb upstream	A/G	1.07 (1.05–1.10)	Dupuis et al., 2010
rs972283	7	*KLF14*	47 kb upstream	G/A	1.07 (1.05–1.10)	Voight et al., 2010
rs864745	7	*JAZF1*	Intronic	T/C	1.10 (1.07–1.13)	Zeggini et al., 2008
rs13266634	8	*SLC30A8*	Missense: Arg325Trp	C/T	1.12 (1.07–1.16)†	Sladek et al., 2007
rs896854	8	*TP53INP1*	Intronic	T/C	1.06 (1.04–1.09)	Voight et al., 2010
rs1081161	9	*CDKN2A/B*	125 kb upstream	T/C	1.20 (1.14–1.25)	Diabetes Genetics Initiative of Broad Institute of Harvard and MIT et al., 2007; Scott et al., 2007; Zeggini et al., 2007
rs13292136	9	*TLE4* (formerly *CHCHD9*)	234 kb upstream	C/T	1.11 (1.07–1.15)	Voight et al., 2010
rs12779790	10	*CDC123/ CAMK1D*	Intergenic region	G/A	1.11 (1.07–1.14)	Zeggini et al., 2008
rs1111875	10	*HHEX*	7.7 kb downstream	C/T	1.13 (1.08–1.17)†	Sladek et al., 2007
rs7903146	10	*TCF7L2*	Intronic	T/C	1.37 (1.28–1.47)†	Grant et al., 2006
rs1552224	11	*ARAP1* (formerly *CENTD2*)	5′ UTR	A/C	1.14 (1.11–1.17)	Voight et al., 2010
rs5219/ rs757110	11	*KCNJ11/ ABCC8*	Missense: Glu23Lys/ Ala1369Ser	T/C	1.14 (1.10–1.19)	Gloyn et al., 2003
rs2237892	11	*KCNQ1*	Intronic	C/T	1.40 (1.34–1.47)	Unoki et al., 2008; Yasuda et al., 2008
rs231362	11	*KCNQ1*	Intronic	G/A	1.08 (1.06–1.10)	Voight et al., 2010
rs10830963	11	*MTNR1B**	Intronic	G/C	1.09 (1.05–1.12)	Prokopenko et al., 2009
rs1531343	12	*HMGA2*	43 kb upstream	C/G	1.10 (1.07–1.14)	Voight et al., 2010
rs7957197	12	*HNF1A*	20 kb downstream	T/A	1.07 (1.05–1.10)	Voight et al., 2010
rs7961581	12	*TSPAN8/LGR5*	Intronic	C/T	1.09 (1.06–1.12)	Zeggini et al., 2008
rs8042680	15	*PRC1*	Intronic	A/C	1.07 (1.05–1.09)	Voight et al., 2010
rs11634397	15	*ZFAND6*	1.5 kb downstream	G/A	1.06 (1.04–1.08)	Voight et al., 2010
rs8050136	16	*FTO*	Intronic	A/C	1.23 (1.18–1.32)	Frayling et al., 2007
rs757210	17	*HNF1B*	Intronic	A/G	1.10 (1.04–1.16)†	Gudmundsson et al., 2007; Winckler et al., 2007
rs5945326	X	*DUSP9*	8 kb upstream	G/A	1.27 (1.18–1.37)	Voight et al., 2010

Loci are arranged alphabetically by chromosome number. Chr = chromosome, CI = confidence interval.
*Discovery of T2D association followed detection in GWAS for quantitative glycemic traits (see Table 82.2).
†Odds ratio from DIAGRAM (Zeggini et al., 2008).

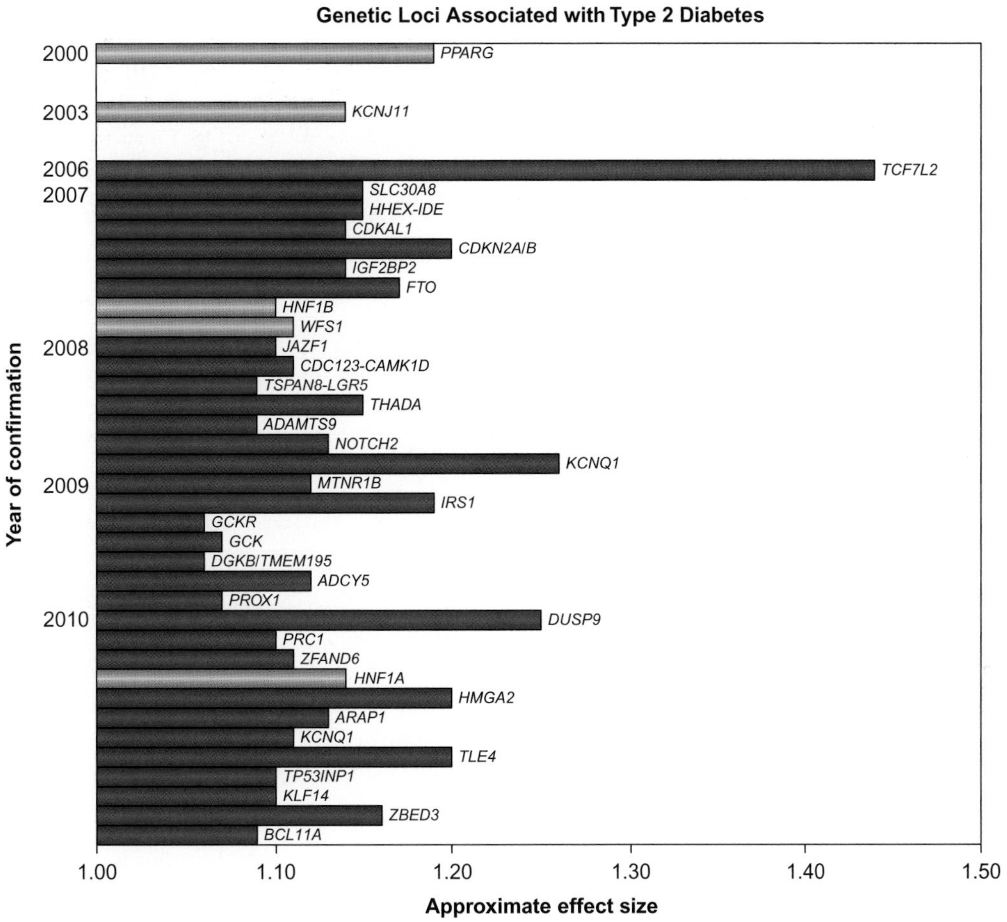

Genetic Loci Associated with Type 2 Diabetes

Figure 82.1 Chronological listing of T2D-associated genes, plotted by year of definitive publication and approximate effect size (in terms of relative risk of T2D per allele, compared to the population average, 1.0). Genes initially studied because they were candidate genes are shown in yellow, while genes discovered via agnostic genome-wide association approaches are shown in blue. *TCF7L2* (shown in green) was discovered by dense fine-mapping under a linkage signal and subsequently replicated by numerous genome-wide studies. Approximate allelic effect sizes are derived as in Table 82.1.

found to regulate fasting glucose (Bouatia-Naji et al., 2008; Chen et al., 2008). Interestingly, the glucose-raising SNP does not contribute to T2D susceptibility, and in fact leads to improved glucose-stimulated insulin secretion (Dupuis et al., 2010; Ingelsson et al., 2010). Recent elegant fine-mapping functional work has shown that the implicated SNP alters *G6PC2* promoter activity (Bouatia-Naji et al., 2010).

In a complementary approach to the GWAS conducted in case/control collections, multiple groups applied the same techniques in samples with quantitative glycemic phenotypes. Many cohorts coalesced to form the Meta-Analysis of Glucose and Insulin-related traits Consortium (MAGIC). An initial exchange of top findings among participating groups revealed that SNPs in the gene that encodes the melatonin receptor 2 (*MTNR1B*) influence fasting glucose and increase

T2D risk (Prokopenko et al., 2009); this was corroborated in independent work (Bouatia-Naji et al., 2009) and was ascribed to an impairment in β-cell function (Lyssenko et al., 2009). A more formal meta-analysis of all participating samples led to the discovery and/or confirmation of multiple loci associated with fasting glucose (besides the previously known variants at *TCF7L2, SLC30A8, GCK, G6PC2, MTNR1B, DGKB,* and *GCKR*, also novel associations in or near *ADCY5, MADD, ADRA2A, CRY2, FADS1, GLIS3, SLC2A2, PROX1,* and *C2CD4B*), fasting insulin (*GCKR* and *IGF1*) (Dupuis et al., 2010) and two-hour glucose (*ADCY5, GIPR,* and *VPS13C,* besides *TCF7L2* and *GCKR*) (Saxena et al., 2010). Of these, *GCK, MTNR1B, DGKB, GCKR, PROX1,* and *ADCY5* were also associated with T2D (Dupuis et al., 2010). Loci discovered by association with quantitative glycemic traits are listed in Table 82.2.

Genetic variants associated with quantitative glycemic traits at genome-wide levels of statistical significance

Marker	Chr	Nearest gene	Type of mutation	Allele (effect/other)	Trait	Beta (SE)	Discovery cohort
rs340874	1	PROX1*	2 kb upstream	C/T	FG	0.013 (0.003)	Dupuis et al., 2010
rs560887	2	G6PC2	Intronic	C/T	FG	0.075 (0.004)	Dupuis et al., 2010
					HOMA-B	−0.042 (0.004)	
rs780094	2	GCKR*	Intronic	C/T	FG	0.029 (0.003)	Dupuis et al., 2010
					FI	0.032 (0.004)	
					HOMA-IR	0.035 (0.004)	
rs1260326	2	GCKR	Missense: Leu446Pro	T/C	2-hr G	0.10 (0.01)	Saxena et al., 2010
rs11708067	3	ADCY5*	Intronic	A/G	FG	0.027 (0.003)	Dupuis et al., 2010
					HOMA-B	−0.023 (0.004)	
rs2877716	3	ADCY5*	Intronic ($r^2 = 0.82$ with rs11708067)	C/T	2-hr G	0.07 (0.01)	Saxena et al., 2010
rs11920090	3	SLC2A2	Intronic	T/A	FG	0.02 (0.004)	Dupuis et al., 2010
rs2191349	7	DGKB/TMEM195*	Intergenic region	T/G	FG	0.03 (0.003)	Dupuis et al., 2010
rs4607517	7	GCK*	36 kb upstream	A/G	FG	0.062 (0.004)	Dupuis et al., 2010
rs13266634	8	SLC30A8*	Missense: Arg325Trp	C/T	FG	0.027 (0.004)	Dupuis et al., 2010
rs7034200	9	GLIS3	Intronic	A/C	FG	0.018 (0.003)	Dupuis et al., 2010
					HOMA-B	−0.020 (0.004)	
rs10885122	10	ADRA2A	210 kb downstream	G/T	FG	0.022 (0.004)	Dupuis et al., 2010
rs7903146	10	TCF7L2*	Intronic	T/C	FG	0.023 (0.004)	Dupuis et al., 2010
rs12243326	10	TCF7L2*	Intronic ($r^2 = 0.79$ with rs7903146)	C/T	2-hr G	0.07 (0.01)	Saxena et al., 2010
rs11605924	11	CRY2	Intronic	A/C	FG	0.015 (0.003)	Dupuis et al., 2010
rs174550	11	FADS1	Intronic	T/C	FG	0.017 (0.003)	Dupuis et al., 2010
					HOMA-B	−0.020 (0.003)	
rs10830963	11	MTNR1B*	Intronic	G/C	FG	0.067 (0.003)	Dupuis et al., 2010
					HOMA-B	−0.034 (0.004)	
rs7944584	11	MADD	Intronic	A/T	FG	0.021 (0.003)	Dupuis et al., 2010
rs35767	12	IGF1	1.2 kb upstream	G/A	FI	0.01 (0.006)	Dupuis et al., 2010
					HOMA-IR	0.013 (0.006)	
rs11071657	15	C2CD4B	21 kb downstream	A/G	FG	0.008 (0.003)	Dupuis et al., 2010
rs17271305	15	VPS13C	Intronic	G/A	2-hr G	0.07 (0.01)	Saxena et al., 2010
rs10423928	19	GIPR	Intronic	A/T	2-hr G	0.11 (0.01)	Saxena et al., 2010

Loci are arranged alphabetically by chromosome number. Chr = chromosome, SE = standard error, FG = fasting glucose (mmol/L), FI = fasting insulin (pmol/L), 2-hr G = 2-hour glucose (FG adjusted, mmol/L), HOMA-B and HOMA-IR = β-cell function and insulin resistance by homeostasis model assessment respectively.
* Also associated with type 2 diabetes.

INSIGHTS GAINED FROM GENETIC STUDIES IN TYPE 2 DIABETES
From Genetic Association to Function

Despite robust signals of association at multiple loci, in most cases the specific genomic sequences that cause the molecular phenotype have not been identified. Indeed, the SNPs identified thus far merely signal genomic regions – at times far away from known genes – where an association has been found, but do not necessarily represent the causal variants; so fine-mapping and functional studies must be carried out before the true contribution of these loci to T2D can be meaningfully assessed. Thus, while GWAS can rapidly and systematically uncover new associations, they do not circumvent the process of finding the precise "causal" DNA sequences. Indeed, variants may exert their molecular effects at remote sites even when relatively close to other uninvolved genes. Thus, loci identified by GWAS require in-depth sequencing and functional studies. Given the very recent discovery of many of these associations, such functional evidence has only been generated for a handful of selected loci, reviewed below. The presumed mechanisms for some of the better characterized loci are shown in Figure 82.2.

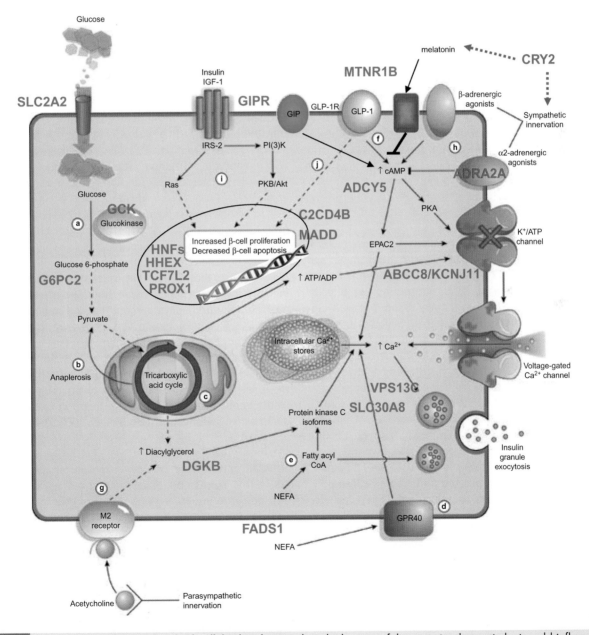

Figure 82.2 Proposed mechanisms and subcellular localization by which some of the examined genetic loci could influence glycemic regulation in the β cell and thereby increase risk for T2D. *Modified from Ingelsson et al., 2010.*

KCNJ11/ABCC8

The E23K polymorphism in *KCNJ11* (minor allele frequency ~30% in Europeans) was one of the first T2D associations proven beyond reasonable doubt (Gloyn et al., 2003). Its presence in a gene clearly involved in the transduction of glucose signaling in the β cell and the non-conservative nature of the mutation both made this variant a likely causal candidate. Initial functional studies seemed to support this notion (Schwanstecher et al., 2002). However, this SNP is highly correlated with another missense variant in a nearby and functionally related gene, *ABCC8*; each of them encodes a separate subunit of the islet ATP-sensitive potassium channel, and because both genes are next to each other and little recombination has occurred between the two SNPs, carriers of the risk K allele at *KCNJ11* E23K almost always carry the risk A allele at *ABCC8* A1369S in most ethnic groups (Florez et al., 2004). Thus, the association with T2D could not be distinguished on genetic grounds alone, and functional experiments required the design of constructs whereby the paired relationship could be experimentally truncated. This has been accomplished, and, at least with regard to response to the sulfonylurea gliclazide, the causal mutation appears to be the A1369S polymorphism (Hamming et al., 2009). Interestingly, risk allele carriers at this locus seem to respond better to gliclazide therapy than their wild-type counterparts (Feng et al., 2008).

GCKR

GCKR encodes the glucokinase regulatory protein (GKRP), which modulates the activity of hepatic hexokinase and thereby gates the entry of glucose into the glycolytic and glycogen synthesis pathways. It is not present in β cells. Interestingly, the index SNP was initially associated with elevated circulating triglycerides (Diabetes Genetics Initiative of Broad Institute of Harvard and MIT et al., 2007); follow-up of this association showed that the triglyceride-raising allele was associated with lower fasting glucose and a lower level of hepatic insulin resistance, as indicated by the homeostasis model assessment (HOMA-IR) (Orho-Melander et al., 2008; Sparso et al., 2008). Its association with lower HOMA-IR was confirmed at the genome-wide significant level by MAGIC; the alternate (glucose-raising but triglyceride-lowering) allele was associated with T2D (Dupuis et al., 2010). Fine-mapping in people of African descent implicated a nearby missense polymorphism (P446L) as the likely causal variant (Orho-Melander et al., 2008). At first sight, the association of one allele with higher triglyceride levels but lower glucose and diminished T2D risk appears counterintuitive, as hypertriglyceridemia and hyperglycemia tend to cluster in metabolic dysregulation. Elegant functional experiments have demonstrated that the presence of the 446L allele makes GKRP less responsive to binding fructose-6-phosphate, a known inducer of GKRP/GCK binding; diminished interaction between the two proteins leads to higher GCK activity and promotes the flux of glucose into glycolysis, thereby augmenting the availability of

malonyl-CoA for *de novo* lipogenesis and providing a unifying mechanism for the apparent epidemiological discrepancy (Beer et al., 2009).

Discovery of Novel Pathways

The unbiased genome-wide search for genetic determinants of T2D and the implementation of rigorous statistical criteria before declaring a true association have enabled the identification of loci previously unsuspected to play a role in T2D pathophysiology. While any one of a number of genes under an association signal could give rise to the phenotype, in a few instances fine-mapping and functional work has implicated a specific gene (and even a variant within it) as causal. This type of follow-up validation is more advanced for loci discovered in the early phases of this effort; two such examples are offered here.

TCF7L2

As mentioned above, *TCF7L2* harbors the common genetic variant that confers the highest risk of T2D reported to date (Florez, 2007). Early fine-mapping efforts, aided by the different haplotype structure in people of African origin, concluded that the intronic SNP rs7903146 was the source of the association signal (Grant et al., 2006; Helgason et al., 2007). *In vivo* phenotyping in humans soon documented that risk allele carriers had impaired β-cell function (Florez et al., 2006; Lyssenko et al., 2007). This is accompanied by higher proinsulin levels relative to fasting insulin (Gonzalez-Sanchez et al., 2008; Kirchhoff et al., 2008; Loos et al., 2007; Stolerman et al., 2009), as well as a diminished incretin response (Lyssenko et al., 2007; Pilgaard et al., 2009; Schafer et al., 2007; Villareal et al., 2010). Functional work suggests that *TCF7L2* may cause these effects by interfering with incretin receptor expression (Shu et al., 2009), β-cell proliferation during development (Liu and Habener, 2008; Shu et al., 2008), and/or insulin vesicle transport and fusion (Da Silva Xavier et al., 2009). Elegant fine-mapping work has shown that the diabetes risk T allele at rs7903146 affects TCF7L2 expression, pinpointing the exact nucleotide that gives rise to the functional defect (Gaulton et al., 2010). The precise nature of the genes targeted by this transcription factor is currently under active investigation (Hatzis et al., 2008; Zhao et al., 2010).

SLC30A8

SLC30A8 encodes the ZnT8 zinc transporter (Chimienti et al., 2004). It is expressed in β cells where it is responsible for zinc transport into insulin granules, necessary for insulin packaging and secretion (Chimienti et al., 2006). Although this gene had been implicated in β-cell biology, it was not known that alterations in this pathway could lead to T2D until GWAS associated the missense variant R325W with T2D (Sladek et al., 2007). Subsequent animal investigation showed that mice null for this gene exhibit subtle abnormalities in insulin secretion when challenged by glucose (Nicolson et al., 2009; Pound et al., 2009) or a high-fat diet (Lemaire et al., 2009). These phenotypes are obvious even in β-cell-specific knock-out mice (Wijesekara et al., 2010) and can be ascribed to loss of function induced by the risk

R325 allele. Review of the protein's crystal structure, however, indicates that the implicated amino acid substitution may not affect function substantially (Weijers, 2010).

Focus on β-cell Function

The majority of loci associated with T2D seem to impair β-cell function (Florez, 2008). While this observation could be partially explained by the ascertainment of lean T2D cases in two early GWAS (Diabetes Genetics Initiative of Broad Institute of Harvard and MIT et al., 2007; Sladek et al., 2007) (thus minimizing the relative contribution of obesity-related insulin resistance), the phenomenon still holds for more comprehensive, recent meta-analyses (Voight et al., 2010). When an unbiased genomic search for homeostasis estimates of beta-cell function (HOMA-B) or insulin resistance (HOMA-IR) is conducted in a diabetes-free population of equal size for both traits, the preponderance of the genome-wide significant signals are associated with HOMA-B rather than HOMA-IR (Dupuis et al., 2010). Examination of the *P*-value distributions for each trait shows that many more signals are to be found for HOMA-B than for HOMA-IR, despite the same sample size, the same set of SNPs, and the same biochemical measurements of fasting glucose and insulin. The dearth of "insulin resistance" signals is evident even when fasting insulin is chosen as the surrogate measure, as a way to obviate the need for a mathematical formula to derive the trait. This suggests that there exists a different genetic architecture underlying both traits, and is consistent with the higher heritability estimates typically ascribed to insulin secretion when compared with insulin resistance (Dupuis et al., 2010).

Put another way, the contribution of the environment seems to be stronger for insulin resistance than for β-cell function. This is consistent with the notion that the recent diabetes pandemic is the result of an environmentally driven increase in insulin resistance occurring on a background of genetic predisposition to β-cell failure. In order to discover genetic determinants of insulin resistance, therefore, integration of environmental variables (e.g., those causing obesity) into the statistical analyses may bring out genetic effects more likely to manifest in an obesogenic context. It is possible that a more refined measure of insulin resistance, such as those derived from oral glucose tolerance tests or hyperinsulinemic euglycemic clamps, may provide greater power to detect an association (although they are available in a much lower number of individuals). Be that as it may, some insulin resistance signals have emerged from GWAS, including *PPARG*, *FTO*, *IRS1*, *IGF1*, *GCKR*, and *KLF14* (Dupuis et al., 2010; Frayling et al., 2007; Rung et al., 2009; Voight et al., 2010). Integration of body mass index into genome-wide searches has recently led to the discovery of additional insulin resistance signals (cite Manning et al. PMID 22581228).

Hyperglycemia and T2D are not Equivalent

Another insight gained from these early genetic studies concerns the lack of a perfect overlap between variants that increase T2D risk and those that raise fasting glucose. While genetic causes of T2D must contribute to hyperglycemia (by definition), the converse is not necessarily true: there may be genetic variants that simply raise the homeostatic point of glycemia slightly, but do so in a stable manner and without reaching thresholds that are injurious to the β cell. On the other hand, other variants that increase glycemia do so by directly impacting β-cell mass or function, eventually leading to β-cell deterioration and ensuing T2D.

This point is most clearly illustrated by *G6PC2*, which has a substantial impact on fasting glycemia, and yet displays no association with T2D whatsoever. In contrast, *TCF7L2*, which robustly increases T2D risk, has a much lower effect on fasting glucose. Similarly, *MADD* and *ADCY5* have comparable allele frequencies and the same effect on fasting glucose, while the former has no evidence of T2D association and the latter does so at genome-wide significance levels (Dupuis et al., 2010). Therefore, as summed up by the MAGIC investigators:

> variation in fasting glucose in healthy individuals is not necessarily an endophenotype for T2D, which posits the hypothesis that the mechanism by which glucose is raised, rather than a mere elevation in fasting glucose levels, is a key contributor to disease progression
>
> (Dupuis et al., 2010).

Support for Prior Epidemiological Observations

Epidemiologic studies have long supported the clustering of metabolically deleterious factors around the so-called "metabolic syndrome" (Meigs et al., 1997; Reaven, 1988). The suggestion has been made that a "common soil" gave rise to conditions such as insulin resistance, obesity, hypertension, and dyslipidemia. Nevertheless, there has been little molecular evidence linking these various processes. As GWAS have been conducted around several of these traits, there was hope that some genetic markers would have pleiotropic effects across these various phenotypes. Such has been the case for *FADS1* and *HNF1A*, which influence both circulating lipid and glucose levels (Dupuis et al., 2010; Kathiresan et al., 2009; Voight et al., 2010).

A link between derangements in circadian rhythmicity and glycemic dysregulation is also emerging. Observational (Knutson et al., 2006; Nilsson et al., 2004) and interventional (Scheer et al., 2009; Spiegel et al., 1999) studies have suggested that alterations in sleep and/or circadian oscillations can increase T2D risk. Human data have been supported by phenotyping studies in mice that lack certain circadian genes and also exhibit metabolic abnormalities, including defects in pancreatic insulin secretion (Marcheva et al., 2010; Turek et al., 2005). The recent identification of two classical circadian genes, *MTNR1B* (encoding the melatonin receptor expressed in β cells) and *CRY2* (encoding a cryptochrome isoform) as harboring variants that elevate fasting glucose, with the former also increasing T2D risk, is another genetic building block that connects these two areas of physiology.

As a third illustration, it has long been known that low birth weight predisposes to insulin resistance and T2D in adulthood

(Barker et al., 1993). The relative contributions of inheritance and the uterine environment to such a phenotype is not completely elucidated, but gestational traits are undergoing the same intense and fruitful scrutiny afforded by modern genetic techniques. The fetal insulin hypothesis maintains that a relative impairment in insulin secretion or action during early development (when insulin also functions as a growth factor) leads to diminished fetal growth, and, if permanent, also to the organism's inability to compensate for insulin resistance later in life. Whether the phenotype is caused by the maternal or fetal risk allele can be distinguished in studies that have access to both DNA samples and incorporate mother/fetus pairs that are discordant for the risk genotype (Freathy et al., 2009; Weedon et al., 2006). In a targeted study that examined established T2D loci, the T2D-causing alleles at *CDKAL1* and *HHEX* also caused decreased birth weight (Freathy et al., 2009). It is particularly intriguing that an initial well-powered meta-analysis of GWAS for birth weight has independently uncovered the fasting glucose and T2D locus *ADCY5* (encoding adenylate cyclase type 5) as also influencing birth weight (Freathy et al., 2010). These findings have been confirmed in separate populations (Andersson et al., 2010).

In summary, the coalescing of cohorts that have undergone extensive phenotyping, together with dense genotyping platforms and widespread collaboration, has begun to provide potential genetic and molecular bases for heretofore unexplained epidemiological relationships.

Genetic Prediction is Not Yet Practical

The presence of genetic information at birth and its essential immutability during the lifetime of an individual make genetic markers potentially attractive variables for use in prediction models. However, our ability to distinguish modest genetic effects in very large population samples does not imply that such effects will be of use at the individual level. When distributions of a quantitative trait (e.g., fasting glucose or risk-allele count) overlap substantially between two groups, a significant difference between the population means does not necessarily translate into the unequivocal placement of an individual in one of the two groups. Furthermore, clinical variables obtained in the course of routine medical care already perform quite well at predicting outcomes (Wilson et al., 2007), and therefore genetic data may not add much new information.

This has been tested in multiple populations through the use of the receiver operator characteristics curves, whose area-under-the-curve (the c-statistic) essentially estimates the proportion of times such predictions will be correct. It is best applied in large prospective population cohorts and should contain the most up-to-date genetic results available. Two recent studies conducted in people of European descent (one in Europe and one in North America) reached comparable conclusions (Lyssenko et al., 2008; Meigs et al., 2008). A genotype score composed of ~18 T2D-associated variants (the list has grown since then) adds little to existing clinical information, particularly when fasting glucose is incorporated into prediction models and these variables are obtained shortly before the development of diabetes (in

which case the c-statistic can be as high as 90%). This may be so because these early genotype scores explain only a small proportion of T2D heritability, and/or because clinical variables such as fasting glucose or body mass index already contain a fair amount of the information furnished by genetic data. On the other hand, genetic markers perform better when tested in younger people or in those in whom diabetes developed much later, highlighting that clinical variables are less informative in those settings and suggesting that genetic results might be used earlier in life as a way to deploy appropriate prevention strategies.

Another way of testing prediction involves the generation of a discrimination index, i.e., assessing how genetic information modifies the placement of an individual into low-, moderate- or high-risk categories. It is best used when treatment decisions change depending on the stratum of risk. Again, discrimination indices based on genetic data work better in younger individuals (Meigs et al., 2008), although their clinical relevance is presently unclear. It is likely that genetic risk scores will capture more information as the number of discovered markers increases, although whether they will substantially alter existing prediction strategies will need to be determined.

Pharmacogenetics is in its Infancy

A general observation in the field of complex diseases is that most common genetic variants only confer a modest effect on risk. This stands to reason, in that strongly deleterious mutations would have been subject to purifying selection and kept at very low frequencies in the population. It is therefore hoped that extending the bounds of allele frequencies to uncommon variants, through next-generation sequencing techniques applied to extreme cases or varied populations, will lead to the identification of rarer variants that have stronger effects. However, while selection pressure may have been acting for a long time (in evolutionary terms) on certain phenotypes, modern pharmacology is relatively recent. Thus, there may not have been enough time for genetic determinants of drug response to undergo purifying selection, and therefore genetic effects on drug response may be stronger than those seen for disease pathogenesis (Link et al., 2008; Mega et al., 2008; Shuldiner et al., 2009).

In T2D, pharmacogenetic investigation remains at a very early stage. In retrospective studies conducted in the GoDARTS cohort, for which clinical outcomes are electronically available and DNA samples have been collected, carriers of the risk variant at *TCF7L2* showed diminished response to sulfonylurea therapy (Pearson et al., 2007) and were more likely to require insulin therapy (Kimber et al., 2007). This finding suggests that the impairment on insulin secretion conferred by *TCF7L2* cannot be overcome with an insulin secretagogue, and stands in contrast to that seen for the β-cell sulfonylurea receptor/potassium channel, where carriers of the risk alleles at *KCNJ11/ABCC8* displayed improved response to the specific sulfonylurea drug (gliclazide) that has shown an *in vitro* allelic effect (Feng et al., 2008; Hamming et al., 2009).

An early report that a missense polymorphism in the OCT1 metformin transporter (which is responsible for the absorption

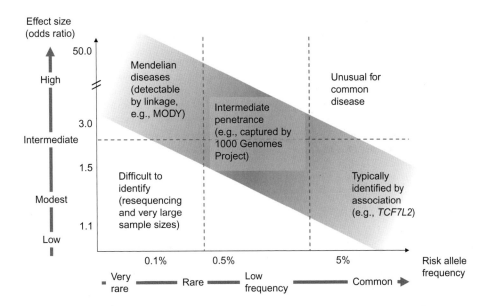

Figure 82.3 A portion of the heritability that is missing from present-day GWAS in T2D may be explained by low-frequency variants with intermediate penetrance. Current GWAS arrays may not have captured alleles with low frequencies. Next-generation high-throughput sequencing technology and novel genotyping arrays may help detect these low-frequency variants; larger sample sizes may be needed to identify significant signals.

of the drug into hepatocytes) alters response to metformin (Shu et al., 2007) has not been substantiated in the GoDARTS study (Zhou et al., 2009). On the other hand, a polymorphism in a different metformin transporter (MATE1), which catalyzes the disposition of drug into bile and urine, seemed to affect metformin response in a different small retrospective study (Becker et al., 2009); this has been confirmed in the Diabetes Prevention Program, where ~1000 participants at high risk of diabetes who received metformin for approximately three years for diabetes prevention showed little benefit from metformin treatment only if they were homozygous for the risk allele (Jablonski et al., 2010).

It is hoped that the integration of clinical datasets in which pharmacotherapeutic information is available will allow for the deployment of the same genome-wide methods that have enabled the discovery of disease genes, but this time in a search for genetic determinants of drug response. Such an undertaking would move the field from the current "trial and error" or "one size fits all" mentality in T2D therapeutics to a setting where drug choices will be predicated on individual characteristics. A GWAS for metformin response has been recently completed (cite PMID 21186350).

produce the observed phenotypic changes at the molecular level, is currently ongoing. Once the precise mechanism is identified, cellular and animal models can be created for in-depth characterization of the defect involved, and assessment of potential therapeutic strategies to overcome it. The emergence of next-generation sequencing techniques that produce enormous amounts of sequence data at a fraction of the cost, and their deployment in appropriate cohorts (e.g., extreme cases with early and severe disease, "hypercontrols" resistant to the development of disease despite multiple risk factors, or those at both extremes of a quantitative trait distribution) will extend the universe of known and assayable genetic variation into lower realms of allele frequencies, where effects are expected to be larger (Figure 82.3). Similarly, as these techniques are expanded into non-European populations through initiatives like the 1000 Genomes Project (http://www.1000genomes.org), variants that are unique or more common in these populations will also be identified and tested in case/control cohorts. Ultimately, whether this newfound genetic knowledge will lead to better patient outcomes will need to be tested in appropriately designed clinical trials.

CONCLUSIONS AND FUTURE DIRECTIONS

In conclusion, contemporary genomic approaches have yielded a rich trove of novel findings that illuminate new pathways of T2D pathogenesis. The laborious fine-mapping of these association signals, in a search for the actual causal variants that

ACKNOWLEDGMENTS

The author wishes to thank Dr Liana Billings for assistance with the preparation of the Tables and Figure 82.3, Drs David Altshuler and David Nathan for their mentorship, and his colleagues in the diabetes genetics community for continued intellectual stimulation through the years.

REFERENCES

Altshuler, D., Hirschhorn, J.N., Klannemark, M., et al., 2000. The common PPARγ Pro12Ala polymorphism is associated with decreased risk of type 2 diabetes. Nat Genet 26, 76–80.

American Diabetes Association, 2008. Economic costs of diabetes in the U.S. in 2007. Diabetes Care 31, 596–615.

Andersson, E.A., Pilgaard, K., Pisinger, C., et al., 2010. Type 2 diabetes risk alleles near *ADCY5*, *CDKAL1* and *HHEX-IDE* are associated with reduced birthweight. Diabetologia 53, 1908–1916.

Barker, D.J., Hales, C.N., Fall, C.H., et al., 1993. Type 2 (non-insulin-dependent) diabetes mellitus, hypertension and hyperlipidaemia (syndrome X): Relation to reduced fetal growth. Diabetologia 36, 62–67.

Barnett, A.H., Eff, C., Leslie, R.D., Pyke, D.A., 1981. Diabetes in identical twins. A study of 200 pairs. Diabetologia 20, 87–93.

Becker, M.L., Visser, L.E., Van Schaik, R.H.N., et al., 2009. Genetic variation in the multidrug and toxin extrusion 1 transporter protein influences the glucose-lowering effect of metformin in patients with diabetes: A preliminary study. Diabetes 58, 745–749.

Beer, N.L., Tribble, N.D., McCulloch, L.J., et al., 2009. The P446L variant in *GCKR* associated with fasting plasma glucose and triglyceride levels exerts its effect through increased glucokinase activity in liver. Hum Mol Genet 18, 4081–4088.

Bluestone, J.A., Herold, K., Eisenbarth, G., 2010. Genetics, pathogenesis and clinical interventions in type 1 diabetes. Nature 464, 1293–1300.

Bonow, R.O., Gheorghiade, M., 2004. The diabetes epidemic: A national and global crisis. Am J Med 116 (Suppl. 5A), 2S–10S.

Bouatia-Naji, N., Bonnefond, A., Baerenwald, D.A., et al., 2010. Genetic and functional assessment of the role of the rs13431652-A and rs573225-A alleles in the *G6PC2* promoter that strongly associate with elevated fasting glucose levels. Diabetes 59, 2662–2671.

Bouatia-Naji, N., Bonnefond, A., Cavalcanti-Proenca, C., et al., 2009. A variant near *MTNR1B* is associated with increased fasting plasma glucose levels and type 2 diabetes risk. Nat Genet 41, 89–94.

Bouatia-Naji, N., Rocheleau, G., Van Lommel, L., et al., 2008. A polymorphism within the *G6PC2* gene is associated with fasting plasma glucose levels. Science 320, 1085–1088.

Carter, J.S., Pugh, J.A., Monterrosa, A., 1996. Non-insulin-dependent diabetes mellitus in minorities in the United States. Ann Intern Med 125, 221–232.

Cauchi, S., El Achhab, Y., Choquet, H., et al., 2007. *TCF7L2* is reproducibly associated with type 2 diabetes in various ethnic groups: A global meta-analysis. J Mol Med 85, 777–782.

Chan, J.C., Malik, V., Jia, W., et al., 2009. Diabetes in Asia: Epidemiology, risk factors, and pathophysiology. JAMA 301, 2129–2140.

Chen, W.-M., Erdos, M.R., Jackson, A.U., et al., 2008. Association studies in caucasians identify variants in the *G6PC2/ABCB11* region regulating fasting glucose levels. J Clin Invest 118, 2620–2628.

Chimienti, F., Devergnas, S., Favier, A., Seve, M., 2004. Identification and cloning of a β-cell-specific zinc transporter, ZnT-8, localized into insulin secretory granules. Diabetes 53, 2330–2337.

Chimienti, F., Devergnas, S., Pattou, F., et al., 2006. In vivo expression and functional characterization of the zinc transporter ZnT8 in glucose-induced insulin secretion. J Cell Sci 119, 4199–4206.

Concannon, P., Rich, S.S., Nepom, G.T., 2009. Genetics of type 1A diabetes. N Engl J Med 360, 1646–1654.

Daly, M.J., Rioux, J.D., Schaffner, S.F., Hudson, T.J., Lander, E.S., 2001. High-resolution haplotype structure in the human genome. Nat Genet 29, 229–232.

Da Silva Xavier, G., Loder, M.K., McDonald, A., et al., 2009. TCF7L2 regulates late events in insulin secretion from pancreatic islet beta-cells. Diabetes 58, 894–905.

Davies, J.L., Kawaguchi, Y., Bennett, S.T., et al., 1994. A genome-wide search for human type 1 diabetes susceptibility genes. Nature 371, 130–136.

Diabetes Genetics Initiative of Broad Institute of Harvard and MIT, Lund University and Novartis Institutes for BioMedical Research, 2007. Genome-wide association analysis identifies loci for type 2 diabetes and triglyceride levels. Science 316, 1331–1336.

Dupuis, J., Langenberg, C., Prokopenko, I., et al., 2010. New genetic loci implicated in fasting glucose homeostasis and their impact on type 2 diabetes risk. Nat Genet 42, 105–116.

Elbein, S.C., 2002. The search for genes for type 2 diabetes in the post-genomic era. Endocrinology 143, 2012–2018.

Ellard, S., Bellanne-Chantelot, C., Hattersley, A.T., 2008. Best practice guidelines for the molecular genetic diagnosis of maturity-onset diabetes of the young. Diabetologia 51, 546–553.

Engelgau, M.M., Geiss, L.S., Saaddine, J.B., et al., 2004. The evolving diabetes burden in the United States. Ann Intern Med 140, 945–950.

Feng, Y., Mao, G., Ren, X., et al., 2008. Ser1369Ala variant in sulfonylurea receptor gene *ABCC8* is associated with antidiabetic efficacy of gliclazide in Chinese type 2 diabetic patients. Diabetes Care 31, 1939–1944.

Florez, J.C., 2007. The new type 2 diabetes gene *TCF7L2*. Curr Opin Clin Nutr Metab Care 10, 391–396.

Florez, J.C., 2008. Newly identified loci highlight beta cell dysfunction as a key cause of type 2 diabetes: Where are the insulin resistance genes? Diabetologia 51, 1100–1110.

Florez, J.C., Burtt, N., de Bakker, P.I.W., et al., 2004. Haplotype structure and genotype-phenotype correlations of the sulfonylurea receptor and the islet ATP-sensitive potassium channel gene region. Diabetes 53, 1360–1368.

Florez, J.C., Hirschhorn, J.N., Altshuler, D., 2003. The inherited basis of diabetes mellitus: Implications for the genetic analysis of complex traits. Annu Rev Genomics Hum Genet 4, 257–291.

Florez, J.C., Jablonski, K.A., Bayley, N., et al., 2006. *TCF7L2* polymorphisms and progression to diabetes in the Diabetes Prevention Program. N Engl J Med 355, 241–250.

Franks, P.W., Rolandsson, O., Debenham, S.L., et al., 2008. Replication of the association between variants in *WFS1* and risk of type 2 diabetes in European populations. Diabetologia 51, 458–463.

Frayling, T.M., Timpson, N.J., Weedon, M.N., et al., 2007. A common variant in the *FTO* gene is associated with body mass index and predisposes to childhood and adult obesity. Science 316, 889–894.

Freathy, R.M., Bennett, A.J., Ring, S.M., et al., 2009. Type 2 diabetes risk alleles are associated with reduced size at birth. Diabetes db08–1739.

Freathy, R.M., Mook-Kanamori, D.O., Sovio, U., et al., 2010. Variants in *ADCY5* and near *CCNL1* are associated with fetal growth and birth weight. Nat Genet 42, 430–435.

Gabriel, S.B., Schaffner, S.F., Nguyen, H., et al., 2002. The structure of haplotype blocks in the human genome. Science 296, 2225–2229.

Gaulton, K.J., Nammo, T., Pasquali, L., et al., 2010. A map of open chromatin in human pancreatic islets. Nat Genet 42, 255–259.

Gloyn, A.L., Weedon, M.N., Owen, K.R., et al., 2003. Large-scale association studies of variants in genes encoding the pancreatic β-cell KATP channel subunits Kir6.2 (*KCNJ11*) and SUR1 (*ABCC8*) confirm that the *KCNJ11* E23K variant is associated with type 2 diabetes. Diabetes 52, 568–572.

Gonzalez-Sanchez, J.L., Martinez-Larrad, M.T., Zabena, C., Perez-Barba, M., Serrano-Rios, M., 2008. Association of variants of the *TCF7L2* gene with increases in the risk of type 2 diabetes and the proinsulin:insulin ratio in the Spanish population. Diabetologia 51, 1993–1997.

Grant, S.F., Hakonarson, H., 2009. Genome-wide association studies in type 1 diabetes. Curr Diab Rep 9, 157–163.

Grant, S.F.A., Thorleifsson, G., Reynisdottir, I., et al., 2006. Variant of transcription factor 7-like 2 (*TCF7L2*) gene confers risk of type 2 diabetes. Nat Genet 38, 320–323.

Greeley, S.A., Tucker, S.E., Worrell, H.I., et al., 2010. Update in neonatal diabetes. Curr Opin Endocrinol Diabetes Obes 17, 13–19.

Groop, L.C., 1997. The molecular genetics of non-insulin-dependent diabetes mellitus. J Intern Med 241, 95–101.

Gudmundsson, J., Sulem, P., Steinthorsdottir, V., et al., 2007. Two variants on chromosome 17 confer prostate cancer risk, and the one in *TCF2* protects against type 2 diabetes. Nat Genet 39, 977–983.

Hamer, R.A., El Nahas, A.M., 2006. The burden of chronic kidney disease is rising rapidly worldwide. BMJ 332, 563–564.

Hamming, K.S., Soliman, D., Matemisz, L.C., et al., 2009. Coexpression of the type 2 diabetes susceptibility gene variants KCNJ11 E23K and ABCC8 S1369A alter the ATP and sulfonylurea sensitivities of the ATP-sensitive K(+) channel. Diabetes 58, 2419–2424.

Hattersley, A., Bruining, J., Shield, J., Njolstad, P., Donaghue, K.C., 2009. The diagnosis and management of monogenic diabetes in children and adolescents. Pediatr Diabetes 10 (Suppl. 12), 33–42.

Hatzis, P., Van der Flier, L.G., Van Driel, M.A., et al., 2008. Genome-wide pattern of TCF7L2/TCF4 chromatin occupancy in colorectal cancer cells. Mol Cell Biol 28, 2732–2744.

Helgason, A., Palsson, S., Thorleifsson, G., et al., 2007. Refining the impact of *TCF7L2* gene variants on type 2 diabetes and adaptive evolution. Nat Genet 39, 218–225.

Ingelsson, E., Langenberg, C., Hivert, M.F., et al., 2010. Detailed physiologic characterization reveals diverse mechanisms for novel genetic loci regulating glucose and insulin metabolism in humans. Diabetes 59, 1266–1275.

International HapMap Consortium, 2003. The International HapMap Project. Nature 426, 789–796.

International HapMap Consortium, 2005. A haplotype map of the human genome. Nature 437, 1266–1275.

Jablonski, K.A., McAteer, J.B., de Bakker, P.I., et al., 2010. Common variants in 40 genes assessed for diabetes incidence and response to metformin and lifestyle interventions in the Diabetes Prevention Program. Diabetes 59, 2672–2681.

Jorde, L.B., 2000. Linkage disequilibrium and the search for complex disease genes. Genome Res 10, 1435–1444.

Kathiresan, S., Willer, C.J., Peloso, G.M., et al., 2009. Common variants at 30 loci contribute to polygenic dyslipidemia. Nat Genet 41, 56–65.

Kimber, C.H., Doney, A.S., Pearson, E.R., et al., 2007. *TCF7L2* in the Go-DARTS study: Evidence for a gene dose effect on both diabetes susceptibility and control of glucose levels. Diabetologia 50, 1186–1191.

Kirchhoff, K., Machicao, F., Haupt, A., et al., 2008. Polymorphisms in the *TCF7L2*, *CDKAL1* and *SLC30A8* genes are associated with impaired proinsulin conversion. Diabetologia 51, 597–601.

Knutson, K.L., Ryden, A.M., Mander, B.A., Van Cauter, E., 2006. Role of sleep duration and quality in the risk and severity of type 2 diabetes mellitus. Arch Intern Med 166, 1768–1774.

Lander, E.S., Linton, L.M., Birren, B., et al., 2001. Initial sequencing and analysis of the human genome. Nature 409, 860–921.

Lemaire, K., Ravier, M.A., Schraenen, A., et al., 2009. Insulin crystallization depends on zinc transporter ZnT8 expression, but is not required for normal glucose homeostasis in mice. Proc Natl Acad Sci USA 106, 14,872–14,877.

Link, E., Parish, S., Armitage, J., et al., 2008. *SLCO1B1* variants and statin-induced myopathy – a genome-wide study. N Engl J Med 359, 789–799.

Liu, Z., Habener, J.F., 2008. Glucagon-like peptide-1 activation of TCF7L2-dependent Wnt signaling enhances pancreatic beta cell proliferation. J Biol Chem 283, 8723–8735.

Loos, R.J.F., Franks, P.W., Francis, R.W., et al., 2007. *TCF7L2* polymorphisms modulate proinsulin levels and β-cell function in a British Europid population. Diabetes 56, 1943–1947.

Lyssenko, V., Jonsson, A., Almgren, P., et al., 2008. Clinical risk factors, DNA variants, and the development of type 2 diabetes. N Engl J Med 359, 2220–2232.

Lyssenko, V., Lupi, R., Marchetti, P., et al., 2007. Mechanisms by which common variants in the *TCF7L2* gene increase risk of type 2 diabetes. J Clin Invest 117, 2155–2163.

Lyssenko, V., Nagorny, C.L., Erdos, M.R., et al., 2009. Common variant in *MTNR1B* associated with increased risk of type 2 diabetes and impaired early insulin secretion. Nat Genet 41, 82–88.

Marcheva, B., Ramsey, K.M., Buhr, E.D., et al., 2010. Disruption of the clock components CLOCK and BMAL1 leads to hypoinsulinaemia and diabetes. Nature 466, 627–631.

Mbanya, J.C., Motala, A.A., Sobngwi, E., Assah, F.K., Enoru, S.T., 2010. Diabetes in sub-Saharan Africa. Lancet 375, 2254–2266.

Mega, J.L., Close, S.L., Wiviott, S.D., et al., 2008. Cytochrome P-450 polymorphisms and response to clopidogrel. N Engl J Med NEJMoa0809171.

Meigs, J.B., Cupples, L.A., Wilson, P.W.F., 2000. Parental transmission of type 2 diabetes mellitus: The Framingham Offspring Study. Diabetes 49, 2201–2207.

Meigs, J.B., D'Agostino Sr., R.B., Wilson, P.W., et al., 1997. Risk variable clustering in the insulin resistance syndrome. The Framingham Offspring Study. Diabetes 46, 1594–1600.

Meigs, J.B., Shrader, P., Sullivan, L.M., et al., 2008. Genotype score in addition to common risk factors for prediction of type 2 diabetes. N Engl J Med 359, 2208–2219.

Naik, R.G., Brooks-Worrell, B.M., Palmer, J.P., 2009. Latent autoimmune diabetes in adults. J Clin Endocrinol Metab 94, 4635–4644.

Nicolson, T.J., Bellomo, E.A., Wijesekara, N., et al., 2009. Insulin storage and glucose homeostasis in mice null for the granule zinc transporter ZnT8 and studies of the type 2 diabetes-associated variants. Diabetes 58, 2070–2083.

NIDDK, 2005. National Diabetes Statistics fact sheet: General information and national estimates on diabetes in the United States, 2005.

MD, US Department of Health and Human Services, National Institute of Health, Bethesda.

Nilsson, P.M., Roost, M., Engstrom, G., Hedblad, B., Berglund, G., 2004. Incidence of diabetes in middle-aged men is related to sleep disturbances. Diabetes Care 27, 2464–2469.

Orho-Melander, M., Melander, O., Guiducci, C., et al., 2008. Common missense variant in the glucokinase regulatory protein gene is associated with increased plasma triglyceride and C-reactive protein but lower fasting glucose concentrations. Diabetes 57, 3112–3121.

Pearson, E.R., Donnelly, L.A., Kimber, C., et al., 2007. Variation in TCF7L2 influences therapeutic response to sulfonylureas: A GoDARTs study. Diabetes 56, 2178–2182.

Pe'er, I., Yelensky, R., Altshuler, D., Daly, M.J., 2008. Estimation of the multiple testing burden for genome-wide association studies of nearly all common variants. Genet Epidemiol 32, 381–385.

Pilgaard, K., Jensen, C.B., Schou, J.H., et al., 2009. The T allele of rs7903146 TCF7L2 is associated with impaired insulinotropic action of incretin hormones, reduced 24 h profiles of plasma insulin and glucagon, and increased hepatic glucose production in young healthy men. Diabetologia 52, 1298–1307.

Pociot, F., Akolkar, B., Concannon, P., et al., 2010. Genetics of type 1 diabetes: What's next? Diabetes 59, 1561–1571.

Poulsen, P., Kyvik, K.O., Vaag, A., Beck-Nielsen, H., 1999. Heritability of type II (non-insulin-dependent) diabetes mellitus and abnormal glucose tolerance – a population-based twin study. Diabetologia 42, 139–145.

Pound, L.D., Sarkar, S.A., Benninger, R.K., et al., 2009. Deletion of the mouse Slc30a8 gene encoding zinc transporter-8 results in impaired insulin secretion. Biochem J 421, 371–376.

Prokopenko, I., Langenberg, C., Florez, J.C., et al., 2009. Variants in MTNR1B influence fasting glucose levels. Nat Genet 41, 77–81.

Reaven, G., 1988. Banting lecture 1988. Role of insulin resistance in human disease. Diabetes 37, 1595–1607.

Reich, D.E., Gabriel, S.B., Altshuler, D., 2003. Quality and completeness of SNP databases. Nat Genet 33, 457–458.

Reynisdottir, I., Thorleifsson, G., Benediktsson, R., et al., 2003. Localization of a susceptibility gene for type 2 diabetes to chromosome 5q34-q35.2. Am J Hum Genet 73, 323–335.

Risch, N., Merikangas, K., 1996. The future of genetic studies of complex human diseases. Science 273, 1516–1517.

Risch, N.J., 2000. Searching for genetic determinants in the new millennium. Nature 405, 847–856.

Ritz, E., Rychlik, I., Locatelli, F., Halimi, S., 1999. End-stage renal failure in type 2 diabetes: A medical catastrophe of worldwide dimensions. Am J Kidney Dis 34, 795–808.

Rung, J., Cauchi, S., Albrechtsen, A., et al., 2009. Genetic variant near IRS1 is associated with type 2 diabetes, insulin resistance and hyperinsulinemia. Nat Genet 41, 1110–1115.

Russell, L.B., Valiyeva, E., Roman, S.H., et al., 2005. Hospitalizations, nursing home admissions, and deaths attributable to diabetes. Diabetes Care 28, 1611–1617.

Sachidanandam, R., Weissman, D., Schmidt, S.C., et al., 2001. A map of human genome sequence variation containing 1.42 million single nucleotide polymorphisms. Nature 409, 928–933.

Sandhu, M.S., Weedon, M.N., Fawcett, K.A., et al., 2007. Common variants in WFS1 confer risk of type 2 diabetes. Nat Genet 39, 951–953.

Saxena, R., Hivert, M.F., Langenberg, C., et al., 2010. Genetic variation in GIPR influences the glucose and insulin responses to an oral glucose challenge. Nat Genet 42, 142–148.

Schafer, S.A., Tschritter, O., Machicao, F., et al., 2007. Impaired glucagon-like peptide-1-induced insulin secretion in carriers of transcription factor 7-like 2 (TCF7L2) gene polymorphisms. Diabetologia 50, 2443–2450.

Scheer, F.A., Hilton, M.F., Mantzoros, C.S., Shea, S.A., 2009. Adverse metabolic and cardiovascular consequences of circadian misalignment. Proc Natl Acad Sci USA 106, 4453–4458.

Schwanstecher, C., Meyer, U., Schwanstecher, M., 2002. KIR6.2 polymorphism predisposes to type 2 diabetes by inducing overactivity of pancreatic β-cell ATP-sensitive K+ channels. Diabetes 51, 875–879.

Scott, L.J., Mohlke, K.L., Bonnycastle, L.L., et al., 2007. A genome-wide association study of type 2 diabetes in Finns detects multiple susceptibility variants. Science 316, 1341–1345.

Shu, L., Matveyenko, A.V., Kerr-Conte, J., et al., 2009. Decreased TCF7L2 protein levels in type 2 diabetes mellitus correlate with down-regulation of GIP- and GLP-1 receptors and impaired beta-cell function. Hum Mol Genet 18, 2388–2399.

Shu, L., Sauter, N.S., Schulthess, F.T., et al., 2008. Transcription factor 7-like 2 regulates β-cell survival and function in human pancreatic islets. Diabetes 57, 645–653.

Shu, Y., Sheardown, S.A., Brown, C., et al., 2007. Effect of genetic variation in the organic cation transporter 1 (OCT1) on metformin action. J Clin Invest 117, 1422–1431.

Shuldiner, A.R., O'Connell, J.R., Bliden, K.P., et al., 2009. Association of cytochrome P450 2C19 genotype with the antiplatelet effect and clinical efficacy of clopidogrel therapy. JAMA 302, 849–857.

Sladek, R., Rocheleau, G., Rung, J., et al., 2007. A genome-wide association study identifies novel risk loci for type 2 diabetes. Nature 445, 828–830.

Sparso, T., Andersen, G., Nielsen, T., et al., 2008. The GCKR rs780094 polymorphism is associated with elevated fasting serum triacylglycerol, reduced fasting and OGTT-related insulinaemia, and reduced risk of type 2 diabetes. Diabetologia 51, 70–75.

Spiegel, K., Leproult, R., Van Cauter, E., 1999. Impact of sleep debt on metabolic and endocrine function. Lancet 354, 1435–1439.

Steinthorsdottir, V., Thorleifsson, G., Reynisdottir, I., et al., 2007. A variant in CDKAL1 influences insulin response and risk of type 2 diabetes. Nat Genet 39, 770–775.

Stolerman, E.S., Manning, A.K., McAteer, J.B., et al., 2009. TCF7L2 variants are associated with increased proinsulin/insulin ratios but not obesity traits in the Framingham Heart Study. Diabetologia 52, 614–620.

Todd, J.A., 2010. Etiology of type 1 diabetes. Immunity 32, 457–467.

Turek, F.W., Joshu, C., Kohsaka, A., et al., 2005. Obesity and metabolic syndrome in circadian Clock mutant mice. Science 308, 1043–1045.

Unoki, H., Takahashi, A., Kawaguchi, T., et al., 2008. SNPs in KCNQ1 are associated with susceptibility to type 2 diabetes in East Asian and European populations. Nat Genet 40, 1098–1102.

Vaxillaire, M., Froguel, P., 2008. Monogenic diabetes in the young, pharmacogenetics and relevance to multifactorial forms of type 2 diabetes. Endocr Rev 29, 254–264.

Vaxillaire, M., Bonnefond, A., Froguel, P., 2009. Breakthroughs in monogenic diabetes genetics: From pediatric forms to young adulthood diabetes. Pediatr Endocrinol Rev 6, 405–417.

Venter, J.C., Adams, M.D., Myers, E.W., et al., 2001. The sequence of the human genome. Science 291, 1304–1351.

Villareal, D.T., Robertson, H., Bell, G.I., et al., 2010. TCF7L2 variant rs7903146 affects the risk of type 2 diabetes by modulating incretin action. Diabetes 59, 479–485.

Voight, B.F., Scott, L.J., Steinthorsdottir, V., et al., 2010. Twelve type 2 diabetes susceptibility loci identified through large-scale association analysis. Nat Genet 42, 579–589.

Weedon, M.N., Clark, V.J., Qian, Y., et al., 2006. A common haplotype of the glucokinase gene alters fasting glucose and birth weight: Association in six studies and population-genetics analyses. Am J Hum Genet 79, 991–1001.

Weijers, R.N., 2010. Three-dimensional structure of beta-cell-specific zinc transporter, ZnT-8, predicted from the type 2 diabetes-associated gene variant SLC30A8 R325W. Diabetol Metab Syndr 2, 33.

Weijnen, C.F., Rich, S.S., Meigs, J.B., Krolewski, A.S., Warram, J.H., 2002. Risk of diabetes in siblings of index cases with Type 2 diabetes: Implications for genetic studies. Diabetic Med 19, 41–50.

Wijesekara, N., Dai, F.F., Hardy, A.B., et al., 2010. Beta cell-specific Znt8 deletion in mice causes marked defects in insulin processing, crystallisation and secretion. Diabetologia 53, 1656–1668.

Wilson, P.W., Meigs, J.B., Sullivan, L., et al., 2007. Prediction of incident diabetes mellitus in middle-aged adults: The Framingham Offspring Study. Arch Intern Med 167, 1068–1074.

Winckler, W., Weedon, M.N., Graham, R.R., et al., 2007. Evaluation of common variants in the six known Maturity-Onset Diabetes of the Young (MODY) genes for association with type 2 diabetes. Diabetes 56, 685–693.

Yasuda, K., Miyake, K., Horikawa, Y., et al., 2008. Variants in *KCNQ1* are associated with susceptibility to type 2 diabetes mellitus. Nat Genet 40, 1092–1097.

Yoon, K.H., Lee, J.H., Kim, J.W., et al., 2006. Epidemic obesity and type 2 diabetes in Asia. Lancet 368, 1681–1688.

Zeggini, E., Scott, L.J., Saxena, R., et al., 2008. Meta-analysis of genome-wide association data and large-scale replication identifies additional susceptibility loci for type 2 diabetes. Nat Genet 40, 638–645.

Zeggini, E., Weedon, M.N., Lindgren, C.M., et al., 2007. Replication of genome-wide association signals in UK samples reveals risk loci for type 2 diabetes. Science 316, 1336–1341.

Zhao, J., Schug, J., Li, M., Kaestner, K.H., Grant, S.F., 2010. Disease-associated loci are significantly over-represented among genes bound by transcription factor 7-like 2 (TCF7L2) in vivo. Diabetologia 53, 2340–2346.

Zhou, K., Donnelly, L.A., Kimber, C.H., et al., 2009. Reduced function *SLC22A1* polymorphisms encoding Organic Cation Transporter 1 (OCT1) and glycaemic response to metformin: A Go-DARTS study. Diabetes 58, 1434–1439.

CHAPTER

83

The Metabolic Syndrome

Matthew B. Lanktree, Tisha R. Joy, and Robert A. Hegele

INTRODUCTION

The clustering of several metabolic abnormalities, including dyslipidemia [elevated serum triglycerides and depressed high-density lipoprotein (HDL) cholesterol], dysglycemia, hypertension, and central obesity, has been termed the metabolic syndrome (MetS) (Reaven, 1988). The term syndrome, originating from Greek and literally meaning "running together," refers to a set of signs or symptoms occurring together where the underlying pathophysiology leading to the concurrence is unknown (Jablonski, 1991). Several lines of evidence suggest that a MetS definition is important both clinically and as a research tool, though some controversy exists regarding the importance of a MetS diagnosis for prediction of the development of diabetes and atherosclerotic cardiovascular disease, compared to the sum of the risk factors independently (Eckel et al., 2010; Ford et al., 2008; Gami et al., 2007). Nonetheless, it is clinically apparent that these risk factors occur together more often than one would expect if they were independent processes, and a five-fold increased risk of diabetes and two-fold increased risk for cardiovascular disease is observed with a diagnosis of MetS (Alberti et al., 2009). Additionally, in patients with an extreme perturbation of one of the components of MetS, as observed, for example, in lipodystrophy, disruption of the other components almost inevitably follows (Hegele, 2003). The common form of MetS is a classic complex genetic trait involving the interaction of a multitude of genetic and environmental factors, and genetic and genomic investigations into MetS may yield insights into the responsible mechanisms.

Defining Metabolic Syndrome

At least six different organizations have published criteria for MetS diagnosis (Alberti et al., 2005; Balkau and Charles, 1999; Einhorn et al., 2003; Grundy et al., 2005; NCEP, 2001; WHO, 2007). A recent consensus statement of stakeholders has unified the definition for clinical use, as well as for epidemiological and basic research studies (Alberti et al., 2009). The debate over the criteria for metabolic syndrome has largely revolved around whether elevated abdominal obesity should be a mandatory component, and what the threshold for continuous variables should be. The revised definition includes five criteria, three of which must be met for a diagnosis (Table 83.1). The criteria include elevated waist circumference, triglycerides, blood pressure, and fasting glucose, and depressed HDL cholesterol. Differences in baseline waist circumference observed between the sexes and ethnicities have also been a concern, and sex- and ethnicity-specific guidelines have been specified (Table 83.2).

Regardless of the definition used, MetS is a common diagnosis (Grundy, 2008). Data from the Third National Health and Nutrition Examination Survey (NHANES III), which used the National Cholesterol Education Program Adult Treatment Panel III (NCEP-ATPIII) MetS criteria, found the prevalence of MetS to be 24% in the United States (Ford et al., 2002). Using the International Diabetes Foundation MetS definition, in a large study including >26,000 participants from 52 countries, the average MetS prevalence was 16.8% (Mente et al., 2010). Due to its high prevalence, an improvement to our understanding of MetS could yield large benefits to public health.

Genomic and Personalized Medicine, 2nd edition
by Ginsburg & Willard. DOI: http://dx.doi.org/10.1016/B978-0-12-382227-7.00083-5

TABLE 83.1	Criteria for metabolic syndrome diagnosis
Metabolic syndrome component	**Threshold for criteria**
Waist circumference	Population- and sex-specific definitions*
HDL cholesterol	<1.0 mmol/L
Triglycerides	≥1.7 mmol/L or drug therapy for high triglycerides†
Blood pressure	SBP ≥ 130 mmHg or DBP ≥ 85 mmHg or drug therapy for hypertension
Fasting glucose	≥5.5 mmol/L or drug therapy for elevated glucose

HDL, high density lipoprotein; SBP, systolic blood pressure;
DBP, diastolic blood pressure.
*See Table 83.2 for waist circumference definitions.
†Fibrates and nicotinic acid are the most commonly used triglyceride lowering therapies.
Data taken from stakeholder consensus on metabolic syndrome definition (Alberti et al., 2009).

TABLE 83.2	Sex- and ethnicity-specific waist circumference thresholds for metabolic syndrome definition as given by the International Diabetes Federation	
Ethnicity	**Waist circumference threshold for metabolic syndrome criteria**	
	Male	**Female**
Europid	≥94 cm	≥80 cm
Asian	≥90 cm	≥80 cm
Middle Eastern	≥94 cm	≥80 cm
Mediterranean	≥94 cm	≥80 cm
Sub-Saharan African	≥94 cm	≥80 cm
Ethnic Central and South American	≥90 cm	≥80 cm

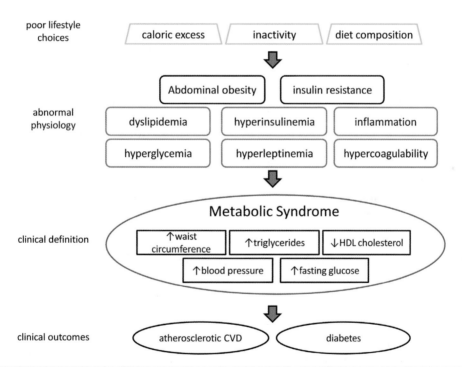

Figure 83.1 Poor lifestyle choices lead to the development of inciting factors for metabolic syndrome (black boxes) and abnormal physiology. Disrupted metabolism is clinically measured by the components of the metabolic syndrome. Metabolic syndrome subsequently increases risk for development of atherosclerotic cardiovascular disease (CVD) and diabetes.

Pathophysiology

The inciting factors in the development of MetS are abdominal obesity and insulin resistance (Reaven, 1988) (Figure 83.1). The accumulation of visceral fat, typically caused by over-nutrition and physical inactivity, results in the release of free fatty acids, leading to lipotoxicity and insulin resistance (Samuel et al., 2010), and eventually hyperinsulinemia and hyperglycemia

(Pollex and Hegele, 2006). Insulin has numerous molecular effects beyond glucose homeostasis: upregulation of amino acid uptake and protein synthesis, activation of lipoprotein lipase, and inhibition of very low-density lipoprotein (VLDL) secretion (Cornier et al., 2008). Abundance of fatty acids and diacylglycerol within skeletal muscle inhibits insulin signaling and reduces its ability to transport and utilize glucose (Samuel et al., 2010). Visceral fat accumulation creates a dysregulation of adipokine secretion, specifically hyposecretion of adiponectin and hypersecretion of both leptin and pro-inflammatory cytokines (such as tumor necrosis factor alpha and interleukin 6), each of which may contribute to the MetS pathophysiology (Kadowaki et al., 2006; Shoelson et al., 2006). In response to hyperinsulinemia and hyperglycemia, the liver secretes C-reactive protein and pro-thrombotic molecules such as fibrinogen and plasminogen activator inhibitor-1 (PAI-1) (Cornier et al., 2008). Clearly, MetS is a complicated phenotype involving a complex network of causative and associated biochemical players; disentangling their relationships to further our understanding and develop novel therapeutics is thus a difficult task.

HERITABILITY OF METABOLIC SYNDROME

Strong evidence exists for the heritability of both MetS and its components, arising from studies of both twins and families. In a study of 2508 pairs of American male twins, the MetS concordance between monozygotic pairs was 31.6% compared to 6.3% for dizygotic twins (Carmelli et al., 1994). In a study of 109 American female twin pairs, MetS and its components were also significantly better correlated in monozygotic than dizygotic twins (Edwards et al., 1997). Among 803 individuals from 89 Caribbean-Hispanic families, the heritability of a MetS diagnosis was 24%, with significant heritability for lipid/glucose/obesity (44%) and hypertension (20%) components (Lin et al., 2005). In a study of 1277 Omani Arab individuals in five large consanguineous families, a large degree of heritability was observed for MetS (38%), while the heritability of the individual MetS components ranged from 38% to 63% (Bayoumi et al., 2007). In an examination of 1942 Korean twin pairs and their families, significant heritability of both MetS (51–60%) and all of its components (46–77%) was observed (Sung et al., 2009). Variation in the reported heritability of MetS and its components is likely partially due to differences in ethnicity, variation in environmental exposures between family members, and the statistical techniques employed. Based upon the demonstrated heritability of MetS and its components, investigations to identify responsible genetic variants have been undertaken using both linkage analysis and association mapping strategies, the results of which are discussed later in this chapter.

LESSONS FROM MONOGENIC MODELS OF METABOLIC SYNDROME

Monogenic diseases, also termed Mendelian diseases, are the result of a single genetic mutation and thus have high penetrance, follow Mendelian inheritance patterns, and include small or non-existent environmental components (Moore et al., 2005). In contrast, in complex diseases such as MetS, there is no apparent inheritance pattern, many genetic loci

are involved, and there are large environmental components (Moore et al., 2005). In the last 30 years, the genetic basis of several monogenic diseases that include multiple components of MetS have been discovered, garnering insight into potential mechanisms of MetS, despite being responsible for a very small proportion of MetS in the population.

Since monogenic disorders segregate through families, linkage analysis of severely affected families followed by sequencing of genes within the identified region has discovered the responsible genetic variant for many monogenic models of MetS. Whole-exome studies, in which the transcribed portion of the genome is sequenced, have already been successful identifying variants responsible for Mendelian disorders (e.g., Ng et al., 2010). With the increasing availability of next-generation sequencing technologies, our ability to identify rare genetic variants responsible for monogenic models of MetS will only improve.

Lipodystrophy

Lipodystrophy is a heterogeneous group of disorders characterized by selective or generalized atrophy of anatomical adipose tissue stores (Table 83.3) (Garg, 2004). Loss of the ability to retain excess lipids in "classical" adipose tissue stores leads to the over-development of ectopic fat stores, such as within and around skeletal muscle, heart, liver, pancreas, and kidneys and within the arterial wall, presenting as atherosclerosis (Dulloo et al., 2004). In patients with congenital generalized lipodystrophy (CGL), an absence of adipose tissue is noted in early infancy (Garg, 2004). In familial partial lipodystrophy (FPLD), patients have normal fat distribution during childhood, but during or shortly after puberty there is a progressive and gradual loss of subcutaneous adipose tissue of the extremities, a triggering event for the development of the other components of MetS including extreme insulin resistance, dyslipidemia and hypertension (Garg, 2004).

A total of 11 genes have been identified that contain mutations causative of at least one form of lipodystrophy (Lanktree et al., 2010) (Table 83.3). The two most commonly mutated genes causing lipodystrophy are those encoding nuclear lamin A/C (*LMNA*) and peroxisome proliferator-activated receptor γ (*PPARG*). Currently, over 200 mutations have been identified in *LMNA*, causing 13 different diseases, together termed laminopathies: FPLD, Emery–Dreifuss muscular dystrophy, limb-girdle muscular dystrophy type 1B, dilated cardiomyopathy type 1A, Charcot–Marie–Tooth, Hutchinson–Gilford progeria syndrome, atypical Werner syndrome, and a range of overlapping syndromes (Stenson et al., 2009). *LMNA* mutations have also been identified in patients with a less severe form of lipodystrophy, closely resembling common MetS, which was named "metabolic laminopathy" (Decaudain et al., 2007). *LMNA* encodes an intermediate filament protein vital for the structural integrity of the nuclear envelope, transcriptional regulation, nuclear pore functioning, and heterochromatin organization. However, it remains unknown how the mutations observed in *LMNA* specifically cause the observed phenotypes.

TABLE 83.3 Monogenic lipodystrophy observed in humans

Disease	OMIM #
Generalized	
Congenital generalized lipodystrophy (CGL) (Berardinelli–Seip)	
CGL1 – caused by mutations in AGPAT2	608594
CGL2 – caused by mutations in BSCL2	269700
CGL3 – caused by mutations in CAV1, PTRF	612526
Partial	
Familial partial lipodystrophy (FPLD)	
FPLD2 (Dunnigan) – caused by mutations in LMNA	151660
FPLD3 – caused by mutations in PPARG	604367
FPLD caused by newly-discovered mutations – CAV1, AKT2	
Acquired partial lipodystrophy (APL) (Barraquer–Simmons)	
APL – some cases associated with LMNB2 mutations	608709
Syndromes that include lipodystrophy as a component	
Mandibuloacral dysplasia (MAD)	
MADA – caused by mutations in LMNA	248370
MADB – caused by mutations in ZMPSTE24	608612
SHORT syndrome – potentially caused by mutations in PITX2	269880
Hutchinson–Gilford progeria syndrome (HGPS) – caused by mutations in LMNA	176670
Werner syndrome (WRN) – caused by mutations in RECQL2, LMNA	277700

TABLE 83.4 Monogenic diseases of obesity and insulin resistance observed in humans

Disease	OMIM #
Obesity	
Alstrom syndrome – ALMS1	203800
Bardet–Biedl syndrome – BBS1-14	209900
Cohen syndrome – COH1	216550
Leptin deficiency – LEP	164160
Leptin receptor deficiency – LEPR	601007
Melanocortin-4 receptor deficiency – MC4R	155541
NTRK2 mutation	600456
Prader–Willi syndrome – SNRPN	176720
Prohormone convertase-1 deficiency – PCSK1	600955
Proopiomelanocortin deficiency – POM1	609734
Pseudohypoparathyroidism – GNAS	103580
SIM1 deletion	603128
WAGR syndrome – BDNF	612469
Insulin resistance	
Donohue syndrome – INSR	246200
Disrupted insulin signaling – AKT2	164731

PPARG was selected as a candidate for sequencing in FPLD patients due to its important role as a ligand-inducible transcription factor regulating adipogenesis, and the re-partitioning of fat stores observed in patients taking the thiazolidinedione (TZD) class of drugs, which are agonists for PPARG. Sequencing of *PPARG* resulted in the discovery of causative mutations (Agarwal and Garg, 2002; Hegele et al., 2002). Through functional studies, different *PPARG* mutations were observed to work through both a dominant negative mechanism, in which the mutant receptor is able to inhibit the action of the wild-type receptor (Southam et al., 2009), and a haploinsufficiency mechanism, in which the 50% reduction in wild-type expression was sufficient to create the phenotype (Hegele et al., 2002).

Monogenic Diseases of Obesity and Insulin Resistance

Over 20 genes have been implicated in rare monogenic diseases that include extreme obesity and/or insulin resistance, that could provide insight into more common forms of MetS (O'Rahilly, 2009) (Table 83.4). The most common monogenic cause of extreme obesity is mutation in melanocortin receptor 4 (*MC4R*) [Online Mendelian Inheritance in Man (OMIM) number: 155541], accounting for approximately 4% of extremely obese individuals (Tan et al., 2009). *MC4R* is primarily expressed in the brain and is thought to impact obesity through central effects on appetite and satiety (Farooqi et al., 2003). Proopiomelanocortin (POMC) (OMIM: 176830) is a precursor protein for multiple biologically active peptide hormones, including but not limited to adrenocorticotropic hormone (ACTH), β-liptropin, β-endorphin, and α-, β- and γ-melanocyte-stimulating hormone, which bind with varying affinity to five homologous melanocortin receptors, including MC4R (Krude et al., 2003). *Pomc -/-* mice develop obesity and abnormal pigmentation (Yaswen et al., 1999), and complete loss-of-function mutations in *POMC* were first described in patients with hypocortisolism, red hair, and early-onset extreme obesity (Krude et al., 1998). Melanocortin signaling has also been implicated in modulating both blood pressure and lipid metabolism, independent of weight and insulin signaling, indicating a potentially greater role for *MC4R* and *POMC* in MetS (Greenfield et al., 2009; Nogueiras et al., 2007).

Bardet–Biedl syndrome (BBS) is a group of at least six molecularly distinct but clinically similar diseases that include obesity as a central component (OMIM: 209900). As well as abdominal obesity, patients with BBS have diabetes, hypertension, mental retardation, dysmorphic extremities, retinal dystrophy or pigmentary retinopathy, hypogonadism and abnormal kidney structure and/or function (Moore et al., 2005). Knockout mice for three of the responsible genes have been developed: *Bbs2 -/-*, *Bbs4 -/-*, and *Bbs6 -/-* (Rahmouni et al., 2008). All three mice strains were hyperphagic, had low locomotor activity, and elevated circulating leptin concentrations (Rahmouni et al., 2008). Further examination of the knockout mice, as well as cellular work, suggests that the proteins mutated in BBS are required for leptin receptor signaling, and the impaired leptin signaling was associated with decreased *Pomc* expression (Seo et al., 2009). Rare mutations in the genes encoding both leptin (*LEP*) (OMIM: 164160) and the leptin receptor (*LEPR*) (OMIM: 601007) also cause obesity, hyperinsulinemia, and insulin resistance, and are knocked out in the commonly studied ob^-/ob^- and db^-/db^- mice, respectively.

Alstrom syndrome is an autosomal recessive disorder characterized by childhood obesity, insulin resistance, hyperglycemia, hyperlipidemia, and neurosensory defects, caused by mutations in a gene at chromosomal locus 2p13 subsequently named *ALMS1* (Collin et al., 2002) (OMIM: 203800). Dilated cardiomyopathy, hepatic dysfunction, and hyperthyroidism are also variably present. Mice with *ALMS1* knocked out recapitulate the findings observed in humans, providing evidence that loss of *ALMS1* alone is sufficient to cause the phenotype (Collin et al., 2005).

Mutations in the insulin receptor (*INSR*) can directly cause insulin resistance, creating several syndromes with variable insulin receptor dysfunction: leprechaunism (Donahue syndrome), Rabson–Mendenhall syndrome, and type A insulin resistance (OMIM: 147670). Patients with the most severe syndrome, Donahue syndrome, have marked hyperinsulinemia, fasting hypoglycemia, post-prandial hyperglycemia, growth restriction, and premature mortality (Longo et al., 2002).

GENETICS OF COMMON METABOLIC SYNDROME

Linkage Analysis

Studies to identify genes contributing to common MetS began by searching for chromosomal regions that were transmitted between affected individuals in large families using linkage analysis. While at least 38 regions have been reported to be linked with one or more MetS components (Teran-Garcia and Bouchard, 2007), few conclusions can be drawn, for three reasons: (1) linkage analysis is poorly powered in the context of a

genetic locus that explains only a small percentage of variation in the trait, (2) resolution is very poor and identified regions often include 100 genes, and (3) there has been little concordance between regions identified.

Association Studies

Genetic association studies test if alleles at a single nucleotide polymorphism (SNP) are found more often in cases than in controls, or in individuals with elevated levels of a quantitative trait. To find genetic variations of small effect, association studies have greater resolution and are better powered than linkage analysis. The first genetic association studies were candidate gene studies, with >20 studies focused upon investigations of genes with known roles in MetS (Pollex and Hegele, 2006). Genes that have been replicated in multiple studies include those in: lipid metabolism pathways, such as apolipoprotein A5 (*APOA5*) (Grallert et al., 2007; Yamada et al., 2007, 2008), and apolipoprotein C3 (*APOC3*) (Guettier et al., 2005; Pollex et al., 2007); inflammation, such as interleukin-6 (*IL6*) (Hamid et al., 2005; Stephens et al., 2007); and adipose tissue partitioning, such as *LMNA* (Hegele et al., 2000; Steinle et al., 2004) and *PPARG* (Frederiksen et al., 2002; Meirhaeghe et al., 2005).

Since the publication of the reference human genome sequence, advancements in high-throughput genotyping technologies and databases of common genetic variation have enabled genome-wide association studies (GWAS) to test over a million SNPs in a single experiment. To date, no GWAS of MetS has been reported in the literature. Genome-wide investigations into the individual components of MetS using GWAS, however, have been enormously successful. As reviewed in other chapters in this book, GWAS of the components of MetS including blood lipid and lipoprotein concentrations, blood pressure, body mass index, and fasting blood glucose have identified >100 genes involved in metabolic pathways relevant to MetS (Table 83.5). Many of the identified genes have previously known functional roles or are mutated in monogenic disease and represent positive controls for the GWAS approach. Additionally, many novel genes with no previously known roles in metabolic pathways have been robustly associated and their biological functions are under study.

Pleiotropy describes the situation where a gene impacts multiple phenotypic traits. In the interacting metabolic pathways involved in MetS pathophysiology, pleiotropy is to be expected. While no genes have been identified to be associated with all five components of the metabolic syndrome, many of the identified genes are associated with more than one component. Due to the high correlation between plasma HDL cholesterol and triglyceride concentration, it is unsurprising that 14 genes are associated with both traits (Table 83.5). With respect to genes associated with more than one of the other MetS components, glucokinase (hexokinase 4) regulatory protein (*GCKR*) is associated with both fasting triglyceride and glucose concentrations (Teslovich et al., 2010; Zeggini et al., 2008). The fat mass- and obesity-associated gene (*FTO*)

TABLE 83.5	Genes associated with components of metabolic syndrome in genome-wide association studies			
HDL	**Triglycerides**	**Blood pressure**	**Obesity***	**Fasting glucose and insulin-related†**
PABPC4	ANGPTL3	CASZ1	NEGR1	PROX1
ZNF648	**GALNT2**	MTHFR	SEC16B	NOTCH2
GALNT2	**APOB**	ULK4	SDCCAG8	THADA
APOB	**GCKR**	ITGA9	TMEM18	**IRS1**
COBLL1	**COBLL1**	FGF5	ETV5	BCL11A
IRS1	**IRS1**	CACNB2	GNPDA2	**GCKR**
SLC39A8	MSL2L1	c10orf170	PCSK1	CALPN10
ARL15	KLHL8	CYP17A1	NCR3-BAT2	G6PC2
C6orf106	MAP3K1	PLEKHA7	PTER	ADAMTS9
CITED2	TIMD4	ATP2B1	TNKS-MSRA	ADCY5
LPA	HLA	SH2B3	MTCH2	SLC2A2
MLXIPL	TYW1	TBX3-5	BDNF	PPARG
KLF14	**MLXIPL**	CSK-ULK3	FAIM2	IGF2BP2
PPP1R3B	PINX1	CYP1A2	MAF	WFS1
LPL	NAT2	ZNF652	NRXN3	CDKAL1
TRPS1	**LPL**	PLCD3	SH2B1	VEGFA
TRIB1	**TRIB1**		**FTO**	GCK
TTC39B	JMJD1C		NPC1	TMEM195
ABCA1	CYP26A1		MC4R	GLIS3
AMPD3	**FADS1-2-3**		KCTD15	JAZF1
LRP4	**APOA1**			SLC30A8
FADS1-2-3	LRP1			CDKN2B
APOA1	ZNF664			CDC123
UBASH3B	CAPN3			TCF7L2
PDE3A	FRMD5			ADRA2A
LRP1	**LIPC**			HHEX
MVK	**CETP**			KCNJ11
SBNO1	CTF1			MTNR1B
ZNF664	CILP2			MADD
SCARB1	**APOE**			CRY2
LIPC	**PLTP**			FADS1
LACTB	PLA2G6			**FTO**
CETP				HNF1A-TCF2
LCAT				TSPAN8
CMIP				C2CD4B

(continued)

TABLE 83.5 (Continued)				
HDL	**Triglycerides**	**Blood pressure**	**Obesity***	**Fasting glucose and insulin-related†**
STARD3				
ABCA8				
PGS1				
LIPG				
MC4R				
ANGPTL4				
LOC55908				
APOE				
LILRA3				
HNF4A				
PLTP				
UBE2L3				

Genes in bold type are associated with multiple components of metabolic syndrome.
*Genes for obesity taken from studies of waist circumference, body mass index and extreme obesity.
†Genes for fasting glucose, fasting insulin and type 2 diabetes.
Data from Dupuis et al., 2010; Levy et al., 2009; Newton-Cheh et al., 2009; Scherag et al., 2010; Takeuchi et al., 2010; Teslovich et al., 2010; Zeggini et al., 2008.

is associated with both adiposity and measures of insulin sensitivity (Do et al., 2008). Variants in insulin receptor substrate 1 (IRS1) are associated with type 2 diabetes risk, insulin resistance, HDL cholesterol, and triglycerides (Rung et al., 2009; Teslovich et al., 2010). Contrarily, it is interesting that some genes that are associated with one of the MetS components with large effect are not associated with other components. For example, common variants in endothelial lipase (LIPG) are robustly associated with HDL concentration, but have not been identified to be associated with plasma triglycerides (Teslovich et al., 2010). As meta-analyses of GWAS are performed, creating sufficient power to identify genetic variants of smaller effect, the pleiotropic effect of more genes is likely to be uncovered.

THE THRIFTY GENE HYPOTHESIS

The "thrifty gene" hypothesis has been proposed to explain the high prevalence of obesity, diabetes, and subsequent MetS in modern times. In this hypothesis, genetic variants that lead to the accumulation of adipose stores for preservation of nutrition until times when it would be required are under positive selection, and thus become more common in the population (Neel, 1962). Thus, in current times of caloric excess, previously beneficial alleles have become deleterious. Contradicting this hypothesis are the facts that in times of starvation, death is primarily caused by infection, not loss of adipose tissue, and that obese individuals may be at higher risk of succumbing to predation (Speakman, 2007). Furthermore, as many common genetic variants have now been identified that contribute to variation in obesity and diabetes, it has become possible to test the hypothesis by examining the obesity-, diabetes- and metabolic-risk alleles. Work by Southam and colleagues examined the characteristics of risk alleles as either minor or major alleles, ancestral or derived alleles, as well as the population differentiation statistics [fixation statistic (F_{st})]. Ultimately, they were able to uncover little evidence in support of the thrifty gene hypothesis (Southam et al., 2009).

FINDING THE MISSING HERITABILITY

Despite the success of GWAS for identifying genes involved in individual MetS components, the genetic variants identified as associated with such components typically explain <10% of variation in the traits across the population, despite heritability measurements of >50% in family studies. This contrast reflects what has been termed the "missing heritability," and efforts to identify genetic variants that explain additional trait variation is the focus of a tremendous research effort (Manolio et al., 2009). Examples of potential sources of missing heritability are rare variants, copy number variation (CNV), and micro RNA genes (*miR*), as well as gene–gene and gene–environment interactions.

Sequencing of genes identified by GWAS reveals rare reduced- or loss-of-function mutations, especially in individuals in the extremes of the distribution of the trait (Johansen et al., 2010). CNV involves the duplication and deletion of genomic DNA, typically defined as >1000 base pairs in size. CNVs are common in the population (Conrad et al., 2010) and have been associated with extreme obesity (Bochukova et al., 2010; Walters et al., 2010). Micro RNAs are short transcribed RNA segments that bind complementary messenger RNA (mRNA), causing translational repression and transcript degradation (Ghildiyal and Zamore, 2009). Two such *miR* genes, *miR-33b* and *miR-33a*, lie within the introns of sterol regulatory element-binding protein-1 (*SREBP1*) and -2 (*SREBP2*), genes encoding two transcription factors that control important glucose and fatty acid regulatory programs, respectively (Najafi-Shoushtari et al., 2010; Rayner et al., 2010). These micro RNAs target genes with opposing actions to *SREBP1* and *SREBP2*, most notably ATP binding cassette transporter A1 (*ABCA1*) (Najafi-Shoushtari et al., 2010; Rayner et al., 2010). Increased *SREBP1* and *miR-33b* expression leads to increased fatty acid synthesis and decreased cholesterol efflux to HDL, two of the hallmarks of MetS (Brown et al., 2010).

Since a relatively small percentage of variation in MetS components and MetS risk has been explained through variants identified in genetic studies, and since environment undoubtedly plays a role in MetS etiology, additional research into potential gene–environment interactions is required. In order to ensure valid assessment of gene–environment interactions in studies of MetS, suitable sample sizes and appropriate statistical methodologies must be employed (Thomas, 2010). Measurement of plasma fatty acid concentration may be less biased than reported dietary fat measurement (Phillips et al., 2010), and two recent studies have attempted to examine gene–environment interactions in the context of plasma fatty acid concentration and MetS. In a study of 1754 European participants, a significant association between leptin receptor (*LEPR*) polymorphisms and MetS was modulated by plasma fatty acid concentrations, which was replicated in a separate cohort (Phillips et al., 2010). Similarly, a gene–environment interaction has been reported between plasma fatty acid concentration and polymorphisms in the adiponectin gene (*ADIPOQ*) and its receptors (*ADIPOR1* and *ADIPOR2*) (Ferguson et al., 2010). These reports require additional validation but provide an interesting basis for further research into gene–environment interaction in MetS.

In systems biology, computational techniques attempt to find patterns in large networks of data derived from high-throughput technologies such as genomic variation, gene expression, proteomics, and metabolomics (Lusis et al., 2008). In this manner, MetS can be evaluated using a broad perspective, incorporating information from multiple sources and multiple physiological pathways, including gene–gene and gene–environment interactions. Using this approach in rodents, evidence for the involvement of lipoprotein lipase (*Lpl*), lactamase β (*Lactb*), and protein phosphate 1-like (*Ppm1L*) in metabolic syndrome and obesity was obtained (Chen et al., 2008). Similar approaches in human systems would also likely yield novel insights into MetS biology.

CLINICAL IMPLICATIONS TO GENETIC FINDINGS IN METABOLIC SYNDROME

Genetic testing of common variants associated with small changes in MetS components, such as lipid concentration or fasting glucose, or associated with risk of MetS complications, such as coronary artery disease or type 2 diabetes, is of little clinical utility at this point (Humphries et al., 2010; Johansen and Hegele, 2009; Talmud et al., 2010). Identification of large-effect mutations may assist with diagnosis in individuals with monogenic forms of MetS, such as lipodystrophy. However, it is premature to currently recommend genetic testing in the context of general MetS diagnosis or treatment.

CONCLUSION

MetS is a complex, heterogeneous diagnosis with numerous causative and associated biochemical players. Much can be learned about MetS pathophysiology by the identification and characterization of individuals with extreme disturbance of one of the MetS components. GWAS of the MetS components have been very successful, verifying known genes and uncovering novel genes, but a GWAS of MetS diagnosis is yet to be reported. As our understanding of the components of MetS improves, we need to begin to understand how these pathways interact and combine to form MetS, and the subsequent risk for atherosclerosis and diabetes.

ACKNOWLEDGMENTS

Dr Lanktree is supported by the Canadian Institutes of Health Research (CIHR) MD/PhD Studentship Award, the University of Western Ontario MD/PhD Program, and is a CIHR Fellow in Vascular Research. Dr Hegele is a Career Investigator of the Heart and Stroke Foundation of Ontario, holds the Edith Schulich Vinet Canada Research Chair (Tier I) in Human Genetics, the Martha G. Blackburn Chair in Cardiovascular Research, and the Jacob J. Wolfe Distinguished Medical Research Chair at the University of Western Ontario. This work was supported by CIHR (MOP-13430, MOP-79523, CTP-79853), the Heart and Stroke Foundation of Ontario (NA-6059, T-6018, PRG-4854), Genome Canada through Ontario Genomics Institute, and the Pfizer Jean Davignon Distinguished Cardiovascular and Metabolic Research Award.

REFERENCES

Agarwal, A.K., Garg, A., 2002. A novel heterozygous mutation in peroxisome proliferator-activated receptor-gamma gene in a patient with familial partial lipodystrophy. J Clin Endocrinol Metab 87, 408–411.

Alberti, K.G., Eckel, R.H., Grundy, S.M., et al., 2009. Harmonizing the metabolic syndrome: A joint interim statement of the International Diabetes Federation Task Force on Epidemiology and Prevention; National Heart, Lung, and Blood Institute; American Heart Association; World Heart Federation; International Atherosclerosis Society; and International Association for the Study of Obesity. Circulation 120, 1640–1645.

Alberti, K.G., Zimmet, P., Shaw, J., 2005. The metabolic syndrome – a new worldwide definition. Lancet 366, 1059–1062.

Balkau, B., Charles, M.A., 1999. Comment on the provisional report from the WHO consultation. European Group for the study of Insulin Resistance (EGIR). Diabet Med 16, 442–443.

Bayoumi, R.A., Al-Yahyaee, S.A., Albarwani, S.A., et al., 2007. Heritability of determinants of the metabolic syndrome among healthy Arabs of the Oman family study. Obesity (Silver Spring) 15, 551–556.

Bochukova, E.G., Huang, N., Keogh, J., et al., 2010. Large, rare chromosomal deletions associated with severe early-onset obesity. Nature 463, 666–670.

Brown, M.S., Ye, J., Goldstein, J.L., 2010. Medicine. HDL miR-ed down by SREBP introns. Science 328, 1495–1496.

Carmelli, D., Cardon, L.R., Fabsitz, R., 1994. Clustering of hypertension, diabetes, and obesity in adult male twins: Same genes or same environments? Am J Hum Genet 55, 566–573.

Chen, Y., Zhu, J., Lum, P.Y., et al., 2008. Variations in DNA elucidate molecular networks that cause disease. Nature 452, 429–435.

Collin, G.B., Cyr, E., Bronson, R., et al., 2005. Alms1-disrupted mice recapitulate human Alstrom syndrome. Hum Mol Genet 14, 2323–2333.

Collin, G.B., Marshall, J.D., Ikeda, A., et al., 2002. Mutations in ALMS1 cause obesity, type 2 diabetes and neurosensory degeneration in Alstrom syndrome. Nat Genet 31, 74–78.

Conrad, D.F., Pinto, D., Redon, R., et al., 2010. Origins and functional impact of copy number variation in the human genome. Nature 464, 704–712.

Cornier, M.A., Dabelea, D., Hernandez, T.L., et al., 2008. The metabolic syndrome. Endocr Rev 29, 777–822.

Decaudain, A., Vantyghem, M.C., Guerci, B., et al., 2007. New metabolic phenotypes in laminopathies: LMNA mutations in patients with severe metabolic syndrome. J Clin Endocrinol Metab 92, 4835–4844.

Do, R., Bailey, S.D., Desbiens, K., et al., 2008. Genetic variants of FTO influence adiposity, insulin sensitivity, leptin levels, and resting metabolic rate in the Quebec family study. Diabetes 57, 1147–1150.

Dulloo, A.G., Antic, V., Montani, J.P., 2004. Ectopic fat stores: Housekeepers that can overspill into weapons of lean body mass destruction. Int J Obes Relat Metab Disord 28 (Suppl 4), S1–S2.

Dupuis, J., Langenberg, C., Prokopenko, I., et al., 2010. New genetic loci implicated in fasting glucose homeostasis and their impact on type 2 diabetes risk. Nat Genet 42, 105–116.

Eckel, R.H., Alberti, K.G., Grundy, S.M., Zimmet, P.Z., 2010. The metabolic syndrome. Lancet 375, 181–183.

Edwards, K.L., Newman, B., Mayer, E., et al., 1997. Heritability of factors of the insulin resistance syndrome in women twins. Genet Epidemiol 14, 241–253.

Einhorn, D., Reaven, G.M., Cobin, R.H., et al., 2003. American College of Endocrinology position statement on the insulin resistance syndrome. Endocr Pract 9, 237–252.

Farooqi, I.S., Keogh, J.M., Yeo, G.S., et al., 2003. Clinical spectrum of obesity and mutations in the melanocortin 4 receptor gene. N Engl J Med 348, 1085–1095.

Ferguson, J.F., Phillips, C.M., Tierney, A.C., et al., 2010. Gene–nutrient interactions in the metabolic syndrome: Single nucleotide polymorphisms in ADIPOQ and ADIPOR1 interact with plasma saturated fatty acids to modulate insulin resistance. Am J Clin Nutr 91, 794–801.

Ford, E.S., Giles, W.H., Dietz, W.H., 2002. Prevalence of the metabolic syndrome among US adults: Findings from the third National Health and Nutrition Examination Survey. JAMA 287, 356–359.

Ford, E.S., Li, C., Sattar, N., 2008. Metabolic syndrome and incident diabetes: Current state of the evidence. Diabetes Care 31, 1898–1904.

Frederiksen, L., Brodbaek, K., Fenger, M., et al., 2002. Comment: Studies of the Pro12Ala polymorphism of the PPAR-gamma gene in the Danish MONICA cohort: Homozygosity of the Ala allele confers a decreased risk of the insulin resistance syndrome. J Clin Endocrinol Metab 87, 3989–3992.

Gami, A.S., Witt, B.J., Howard, D.E., et al., 2007. Metabolic syndrome and risk of incident cardiovascular events and death: A systematic review and meta-analysis of longitudinal studies. J Am Coll Cardiol 49, 403–414.

Garg, A., 2004. Acquired and inherited lipodystrophies. N Engl J Med 350, 1220–1234.

Ghildiyal, M., Zamore, P.D., 2009. Small silencing RNAs: An expanding universe. Nat Rev Genet 10, 94–108.

Grallert, H., Sedlmeier, E.M., Huth, C., et al., 2007. APOA5 variants and metabolic syndrome in Caucasians. J Lipid Res 48, 2614–2621.

Greenfield, J.R., Miller, J.W., Keogh, J.M., et al., 2009. Modulation of blood pressure by central melanocortinergic pathways. N Engl J Med 360, 44–52.

Grundy, S.M., 2008. Metabolic syndrome pandemic. Arterioscler Thromb Vasc Biol 28, 629–636.

Grundy, S.M., Cleeman, J.I., Daniels, S.R., et al., 2005. Diagnosis and management of the metabolic syndrome: An American Heart Association/National Heart, Lung, and Blood Institute Scientific Statement. Circulation 112, 2735–2752.

Guettier, J.M., Georgopoulos, A., Tsai, M.Y., et al., 2005. Polymorphisms in the fatty acid-binding protein 2 and apolipoprotein C-III genes are associated with the metabolic syndrome and dyslipidemia in a South Indian population. J Clin Endocrinol Metab 90, 1705–1711.

Hamid, Y.H., Rose, C.S., Urhammer, S.A., et al., 2005. Variations of the interleukin-6 promoter are associated with features of the metabolic syndrome in Caucasian Danes. Diabetologia 48, 251–260.

Hegele, R.A., 2003. Monogenic forms of insulin resistance: Apertures that expose the common metabolic syndrome. Trends Endocrinol Metab 14, 371–377.

Hegele, R.A., Cao, H., Frankowski, C., et al., 2002. PPARG F388L, a transactivation-deficient mutant, in familial partial lipodystrophy. Diabetes 51, 3586–3590.

Hegele, R.A., Cao, H., Harris, S.B., et al., 2000. Genetic variation in LMNA modulates plasma leptin and indices of obesity in aboriginal Canadians. Physiol Genomics 3, 39–44.

Humphries, S.E., Drenos, F., Ken-Dror, G., Talmud, P.J., 2010. Coronary heart disease risk prediction in the era of genome-wide association studies: Current status and what the future holds. Circulation 121, 2235–2248.

Jablonski, S., 1991. Syndrome: Le mot de jour. Am J Med Genet 39, 342–346.

Johansen, C.T., Hegele, R.A., 2009. Predictive genetic testing for coronary artery disease. Crit Rev Clin Lab Sci 46, 343–360.

Johansen, C.T., Wang, J., Lanktree, M.B., et al., 2010. Excess of rare variants in genes identified by genome-wide association study of hypertriglyceridemia. Nat Genet 42, 684–687.

Kadowaki, T., Yamauchi, T., Kubota, N., et al., 2006. Adiponectin and adiponectin receptors in insulin resistance, diabetes, and the metabolic syndrome. J Clin Invest 116, 1784–1792.

Krude, H., Biebermann, H., Luck, W., et al., 1998. Severe early-onset obesity, adrenal insufficiency and red hair pigmentation caused by POMC mutations in humans. Nat Genet 19, 155–157.

Krude, H., Biebermann, H., Schnabel, D., et al., 2003. Obesity due to proopiomelanocortin deficiency: Three new cases and treatment trials with thyroid hormone and ACTH4-10. J Clin Endocrinol Metab 88, 4633–4640.

Lanktree, M.B., Johansen, C.T., Joy, T.R., Hegele, R.A., 2010. A translational view of the genetics of lipodystrophy and ectopic fat deposition. Prog Mol Bio Transl Sci 94, 159–196.

Levy, D., Ehret, G.B., Rice, K., et al., 2009. Genome-wide association study of blood pressure and hypertension. Nat Genet 41, 677–687.

Lin, H.F., Boden-Albala, B., Juo, S.H., et al., 2005. Heritabilities of the metabolic syndrome and its components in the Northern Manhattan Family Study. Diabetologia 48, 2006–2012.

Longo, N., Wang, Y., Smith, S.A., et al., 2002. Genotype–phenotype correlation in inherited severe insulin resistance. Hum Mol Genet 11, 1465–1475.

Lusis, A.J., Attie, A.D., Reue, K., 2008. Metabolic syndrome: From epidemiology to systems biology. Nat Rev Genet 9, 819–830.

Manolio, T.A., Collins, F.S., Cox, N.J., et al., 2009. Finding the missing heritability of complex diseases. Nature 461, 747–753.

Meirhaeghe, A., Cottel, D., Amouyel, P., Dallongeville, J., 2005. Association between peroxisome proliferator-activated receptor gamma haplotypes and the metabolic syndrome in French men and women. Diabetes 54, 3043–3048.

Mente, A., Yusuf, S., Islam, S., et al., 2010. Metabolic syndrome and risk of acute myocardial infarction: A case-control study of 26,903 subjects from 52 countries. J Am Coll Cardiol 55, 2390–2398.

Moore, S.J., Green, J.S., Fan, Y., et al., 2005. Clinical and genetic epidemiology of Bardet-Biedl syndrome in Newfoundland: A 22-year prospective, population-based, cohort study. Am J Med Genet A 132, 352–360.

Najafi-Shoushtari, S.H., Kristo, F., Li, Y., et al., 2010. MicroRNA-33 and the SREBP host genes cooperate to control cholesterol homeostasis. Science 328, 1566–1569.

NCEP, 2001. Executive summary of the third report of the National Cholesterol Education Program (NCEP) expert panel on detection, evaluation, and treatment of high blood cholesterol in adults (Adult Treatment Panel III). JAMA 285, 2486–2497.

Neel, J.V., 1962. Diabetes mellitus: A "thrifty" genotype rendered detrimental by "progress"? Am J Hum Genet 14, 353–362.

Newton-Cheh, C., Johnson, T., Gateva, V., et al., 2009. Genome-wide association study identifies eight loci associated with blood pressure. Nat Genet 41, 66–76.

Ng, S.B., Buckingham, K.J., Lee, C., et al., 2010. Exome sequencing identifies the cause of a Mendelian disorder. Nat Genet 42, 30–35.

Nogueiras, R., Wiedmer, P., Perez-Tilve, D., et al., 2007. The central melanocortin system directly controls peripheral lipid metabolism. J Clin Invest 117, 3475–3488.

O'Rahilly, S., 2009. Human genetics illuminates the paths to metabolic disease. Nature 462, 307–314.

Phillips, C.M., Goumidi, L., Bertrais, S., et al., 2010. Leptin receptor polymorphisms interact with polyunsaturated fatty acids to augment risk of insulin resistance and metabolic syndrome in adults. J Nutr 140, 238–244.

Pollex, R.L., Ban, M.R., Young, T.K., et al., 2007. Association between the -455T>C promoter polymorphism of the APOC3 gene and the metabolic syndrome in a multi-ethnic sample. BMC Med Genet 8, 80.

Pollex, R.L., Hegele, R.A., 2006. Genetic determinants of the metabolic syndrome. Nat Clin Pract Cardiovasc Med 3, 482–489.

Rahmouni, K., Fath, M.A., Seo, S., et al., 2008. Leptin resistance contributes to obesity and hypertension in mouse models of Bardet–Biedl syndrome. J Clin Invest 118, 1458–1467.

Rayner, K.J., Suarez, Y., Davalos, A., et al., 2010. MiR-33 contributes to the regulation of cholesterol homeostasis. Science 328, 1570–1573.

Reaven, G.M., 1988. Banting lecture 1988: Role of insulin resistance in human disease. Diabetes 37, 1595–1607.

Rung, J., Cauchi, S., Albrechtsen, A., et al., 2009. Genetic variant near IRS1 is associated with type 2 diabetes, insulin resistance and hyperinsulinemia. Nat Genet 41, 1110–1115.

Samuel, V.T., Petersen, K.F., Shulman, G.I., 2010. Lipid-induced insulin resistance: Unravelling the mechanism. Lancet 375, 2267–2277.

Scherag, A., Dina, C., Hinney, A., et al., 2010. Two new loci for body-weight regulation identified in a joint analysis of genome-wide association studies for early-onset extreme obesity in French and German study groups. PLoS Genet 6, e1000916.

Seo, S., Guo, D.F., Bugge, K., et al., 2009. Requirement of Bardet–Biedl syndrome proteins for leptin receptor signaling. Hum Mol Genet 18, 1323–1331.

Shoelson, S.E., Lee, J., Goldfine, A.B., 2006. Inflammation and insulin resistance. J Clin Invest 116, 1793–1801.

Southam, L., Soranzo, N., Montgomery, S.B., et al., 2009. Is the thrifty genotype hypothesis supported by evidence based on confirmed type 2 diabetes- and obesity-susceptibility variants? Diabetologia 52, 1846–1851.

Speakman, J.R., 2007. A nonadaptive scenario explaining the genetic predisposition to obesity: The "predation release" hypothesis. Cell Metab 6, 5–12.

Steinle, N.I., Kazlauskaite, R., Imumorin, I.G., et al., 2004. Variation in the lamin A/C gene: Associations with metabolic syndrome. Arterioscler Thromb Vasc Biol 24, 1708–1713.

Stenson, P.D., Mort, M., Ball, E.V., et al., 2009. The human gene mutation database: 2008 update. Genome Med 1, 13.

Stephens, J.W., Hurel, S.J., Lowe, G.D., et al., 2007. Association between plasma IL-6, the IL6 -174G>C gene variant, and the metabolic syndrome in type 2 diabetes mellitus. Mol Genet Metab 90, 422–428.

Sung, J., Lee, K., Song, Y.M., 2009. Heritabilities of the metabolic syndrome phenotypes and related factors in Korean twins. J Clin Endocrinol Metab 94, 4946–4952.

Takeuchi, F., Isono, M., Katsuya, T., et al., 2010. Blood pressure and hypertension are associated with 7 loci in the Japanese population. Circulation 121, 2302–2309.

Talmud, P.J., Hingorani, A.D., Cooper, J.A., et al., 2010. Utility of genetic and non-genetic risk factors in prediction of type 2 diabetes: Whitehall II prospective cohort study. BMJ 340, b4838.

Tan, K., Pogozheva, I.D., Yeo, G.S., et al., 2009. Functional characterization and structural modeling of obesity-associated mutations in the melanocortin 4 receptor. Endocrinology 150, 114–125.

Teran-Garcia, M., Bouchard, C., 2007. Genetics of the metabolic syndrome. Appl Physiol Nutr Metab 32, 89–114.

Teslovich, T.M., Musunuru, K., Smith, A.V., et al., 2010. Biological, clinical and population relevance of 95 loci for blood lipids. Nature 466, 707–713.

Thomas, D., 2010. Gene–environment-wide association studies: Emerging approaches. Nat Rev Genet 11, 259–272.

Walters, R.G., Jacquemont, S., Valsesia, A., et al., 2010. A new highly penetrant form of obesity due to deletions on chromosome 16p11.2. Nature 463, 671–675.

WHO (World Health Organization), 2007. Definition, diagnosis, and classification of diabetes mellitus and its complications. Report of a WHO Consultation. http://whqlibdoc.who.int/hq/1999/WHO_NCD_NCS_99.2.pdf

Yamada, Y., Ichihara, S., Kato, K., et al., 2008. Genetic risk for metabolic syndrome: Examination of candidate gene polymorphisms related to lipid metabolism in Japanese people. J Med Genet 45, 22–28.

Yamada, Y., Kato, K., Hibino, T., et al., 2007. Prediction of genetic risk for metabolic syndrome. Atherosclerosis 191, 298–304.

Yaswen, L., Diehl, N., Brennan, M.B., Hochgeschwender, U., 1999. Obesity in the mouse model of pro-opiomelanocortin deficiency responds to peripheral melanocortin. Nat Med 5, 1066–1070.

Zeggini, E., Scott, L.J., Saxena, R., et al., 2008. Meta-analysis of genome-wide association data and large-scale replication identifies additional susceptibility loci for type 2 diabetes. Nat Genet 40, 638–645.

Neuropsychiatric Disease Genomic Medicine

Section 8

CHAPTER

84

Neuroscience and the Genomic Revolution: An Overview

Neelroop N. Parikshak and Daniel H. Geschwind

INTRODUCTION

The application of genomics methods to neuroscience is relatively new despite the striking similarities among the scientific and clinical questions the two fields are now addressing. Both neuroscience and genomics have experienced an impressive surge in the amount of information that can be acquired, resulting in a revision of earlier – and, in hindsight, simplistic – mechanisms in favor of more sophisticated models, where the nature versus nurture boundary has been blurred and interactions between different levels of biological regulation and the environment are beginning to be appreciated. On a spatial scale, the range and resolution of data acquisition has become important in both fields to study large-scale entities, such as neural circuits and transcriptomes. On a temporal scale, dynamic variations in what were previously considered static processes have ascribed a greater role in neuroscience to experience-dependent neural plasticity and regeneration; at the same time, in genetics, the epigenetic modulation of the genome as a mechanism of environmental influence has emerged in a similar role.

In neuroscience, the arrival of sophisticated imaging modalities including magnetic resonance imaging (MRI) and functional MRI (fMRI), as well as improved electrophysiological recording and manipulation methods, has enabled a shift from the study of single-lesion deficits to circuit- and brain-wide changes. In genetics, single nucleotide polymorphism (SNP) chips and RNA microarrays, as well as second-generation whole-genome

and transcriptome sequencing, are now allowing inquiry into genome-scale phenomena. These parallel advances have resulted in the same major challenge for both fields: turning the massive amount of complex information into knowledge. The goal for personalized and genomic medicine is to take a defined set of neural phenotypes – perhaps including a clinical work-up, imaging, and behavioral tests – and combine them with genomic information to refine the data into disease- or disorder-specific risks, diagnoses, prognoses, and therapies.

The aptly named field of neurogenetics combines the approaches from genetics and neurosciences and is now adopting large-scale, high-throughput methods to elucidate the genetic etiology of neuropsychiatric disease. It has modeled some of its approaches after the field of cancer biology, which was the first to transition from single-gene, single-pathway analysis to genomic methods, and has also been among the first to translate these methods from the research lab to the clinic. Neurogenetics has been a decade or so slower on the uptake of these methods, partly because they were initially perceived as too difficult to apply to the spatial heterogeneity of the brain or were seen as irrelevant to the behavioral phenotypes in neuropsychiatry. However, the genomics approach has already proven valuable in neuroscience, for high-throughput, large-scale approaches allow the dissection of spatial and temporal complexity and will improve our understanding of the molecular underpinnings of neurological and psychiatric disorders and diseases (Geschwind and Konopka, 2009).

Genomic and Personalized Medicine, 2nd edition
by Ginsburg & Willard. DOI: http://dx.doi.org/10.1016/B978-0-12-382227-7.00084-7

In this chapter, we provide a broad introduction to the unique challenges facing the field of neurogenetics. These include biological and practical issues surrounding the brain and its disorders, the genetic and genomic architecture of neuropsychiatric disease, the interplay of environment and genes in the brain, and future directions for bringing both bench and computational neurogenetics to the clinic. Based on information presented in earlier chapters in this volume, we will assume the reader has a good grasp of current concepts and technologies in genetics, but will also briefly review some historical developments to give a perspective of how far we have come and where the field is going next.

CHALLENGES FOR GENETICS IN NEUROSCIENCE

The central theme in neurogenetics that encompasses both its challenges and promises is the spatial and temporal complexity of the brain. There is an enormous and still incompletely characterized range of cell types with differing properties in the brain that can change in regional composition and connectivity by interacting with the environment through not just the canonical periods of prenatal and postnatal development, but throughout life. One way to categorize the challenges facing neurogenetics is to consider a few broad themes:

- The structural and functional heterogeneity of the brain
- The difficulty of acquiring quality tissue specimens and developing relevant model organisms
- The way in which neuropsychiatric phenotypes are defined and observed.

Since behavior is based on the interaction between perception (via sensory inputs), internal representations (knowledge), and motor output, models for the genetic control of neural phenotypes must take into account the entire spatial and temporal range of information storage and transfer in the brain (Figure 84.1). An individual neuron can respond to a stimulus on the millisecond timescale, while simultaneously storing information within the cell and in networks of cells for hours, days, and even decades. At the ultrastructural level, neurons localize RNA transcripts (for instance, by shuttling them to synapses), and RNAs can interact with proteins and non-coding RNAs (ncRNAs) to effect environment-induced changes such as learning and memory. At the gross level, structures and their functions in the brain vary by region, and connections between these regions can change though the gain and loss of synapses over time. In addition, glia and immune cells play a role in circuit development, neuronal function, and plasticity, but this is a relatively unexplored area that will only further complicate the development of comprehensive mechanistic models of disease.

Brain Regions, Circuits, and Networks

The functional regionalization of the human brain was appreciated early on by neuroscientists such as Broca, Wernicke,

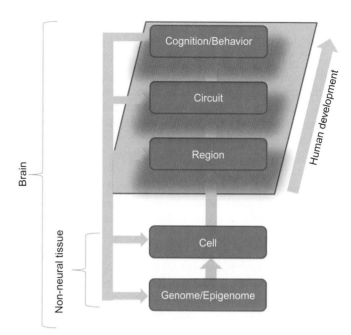

Figure 84.1 Properties of the brain that complicate the definition of genetic etiology and accurate phenotyping. The brain has regional, circuit, and cognitive/behavioral properties that, in addition to various internal feedback mechanisms (e.g., circuit function changing cellular architecture), have superimposed the dimension of human development and therefore change throughout life. Additionally, cognition and behavior can also regulate more microscopic processes to modify their underlying biology, further complicating the dissection of cause from effect. Most other non-central nervous system (CNS) organs, such as the kidney, heart, and lungs, manifest disease at the genomic, epigenomic, and cellular levels and, although complex, do not have to be considered in terms of these CNS levels of complexity.

Brodmann, and Penfield, and was eventually attributed to the different biological characteristics of each brain region, including cell-type composition and projections within and between regions. From a neurogenetics standpoint, the challenge is to explain how this variation occurs, given that these cells mostly contain the same foundational DNA sequence (the genome). Since regionalization has its origin in evolution on the species scale and in development on the individual scale, comparative cross-species studies as well as neurodevelopmental studies are essential in mapping the genetic diversity of the brain.

An interesting example of the power of the comparative approach relates to the *FOXP2* gene. The discovery of a Mendelian form of speech and language dysfunction associated with this gene in a human family led to it being dubbed a "language gene." Further investigation revealed a two-amino-acid change in the sequence of the *FOXP2* gene that likely occurred in the past 200,000 years of human evolution. Intriguingly, this is also around the time when tool-usage and language emerged in humans. This small change in sequence over a short evolutionary time period was put forward as evidence of accelerated evolution (Enard et al., 2002; Lai et al., 2001) and subsequently has been shown to result in significant changes in the encoded protein's functions, one of which is to serve as a transcription factor. For instance, mutating the two amino acids in mice results in different vocalizations and appears to affect cortico-basal ganglia circuitry (Enard et al., 2009). This same amino-acid change also affects gene expression in human differentially chimpanzees, suggesting that structural changes in transcription factors like *FOXP2* could result in pervasive changes that may play a key role in the development of higher cognition and complex behavior (Konopka et al., 2009). This demonstrates the potential of using evolutionary genetics to find important molecules involved in higher-level behaviors, and suggests that small – yet important – sequence changes could culminate in behavior-specific changes in neural function. The next step in these studies is to connect these molecular changes mechanistically with the observed changes at the level of brain circuitry, so as to bridge gene, brain, and behavior.

Developmental studies have revealed that there are critical windows for higher-level cognitive functions such as language acquisition and the ability to perceive complex stimuli. They have shown that structures of the brain have normal temporal patterns of growth and shrinkage throughout development, which are perturbed in several disorders (Courchesne et al., 2001; Giedd et al., 1999; Gogtay et al., 2004; Schumann et al., 2010). An interesting example of this has been appreciated in fMRI studies, where functional connectivity between brain regions increases throughout normal aging while connectivity within these regions decreases. These methods have been used to predict age in normal individuals by assessing the changes in inter- and intra-region connectivity, allowing for behavior-independent assessment of developmental age (Dosenbach et al., 2010). Such an approach emphasizes the notion of developmental trajectory as an important phenotype on its own, and other functional approaches to study active neurological function may serve as valuable quantifiable phenotypes in future investigations and clinical applications.

In both neurodegenerative disease and neurodevelopmental disorders, the very notion of pathophysiology related to dysfunction of a lone, specific brain region is being challenged. As suggested as early as the studies of Braak and Braak (1991), the idea that neurodegeneration affects large-scale neural networks in the brain has become more prevalent. An important recent study showed different, brain network-specific changes in regional network function based on structural and functional MRI metrics in several neurodegenerative syndromes including Alzheimer's disease (AD), behavioral variant frontotemporal dementia, semantic dementia, progressive non-fluent aphasia, and corticobasal syndrome (Seeley et al., 2009). Functional brain-wide changes have also been observed in neurodevelopmental diseases, where, for instance, a specific mutation associated with autism has been associated with altered frontal lobe network connectivity (Scott-Van Zeeland et al., 2010). Brain circuit and network findings of this sort emphasize the value of looking beyond single regions to the brain at large. Integration of brain connectivity networks with gene networks and multiple

levels of phenotypes will be necessary to provide a truly personalized view of the neuroscience-relevant genome in the individual (Geschwind and Konopka, 2009).

Tissue is the Issue

A practical challenge for neurogenetics is accessing the brain. Biopsies of brain tissue are only available in rare cases, so human brain tissue must be acquired postmortem. Such tissue, even if promptly frozen and preserved, is prone to artifacts from various antemortem and postmortem factors, which have to be considered. In early-onset disorders, tissue is rarely available. For late-onset disorders, where tissue is more readily available, there are still innumerable environmental variables that must be controlled for, as well as the possibility of other insults (hypoxia, inflammation) in addition to the neurological disease of interest. Cerebrospinal fluid provides an alternative avenue for biochemical characterization and has been valuable for studies of neurodegenerative disease (Bateman et al., 2007; Bian et al., 2008), and immunological and metabolite changes in blood can be useful in some disorders. Cell culture models have also been useful in the investigation of some disorders, and induced pluripotent stem cells (see Chapter 33) may one day provide an avenue to study disease and follow prognosis in a personalized manner (Lee et al., 2009; Marchetto et al., 2010; Paşca et al., 2011). However, none of these indirect methods allows assessment of the brain at the level of local or regional circuits. These difficulties necessitate organized projects for collecting specimens from human brains to acquire enough quality data for reliable analysis, and highlight the importance of developing consortia to make progress in studying these diseases.

For functional genetic investigation, neuroscience has relied largely on animal models to dissect the genetic underpinnings of behavior. Although advances in neuroimaging are now permitting *in vivo* investigation of human brain structure and function, exploration of mechanism at microcircuit and cellular levels requires animal models that are often studied without serious regard to the vast evolutionary differences between man and model organism (Miller et al., 2010; Preuss, 2000; Smulders, 2009). The most commonly used genetic model organism that strikes a balance between evolutionary similarity and laboratory practicality in neuroscience is the mouse. However, mouse disease models may not display the full range of human phenotypes for neurodevelopmental or neurodegenerative diseases. Additionally, mice often require genetic changes that are more drastic than those found in humans to induce a disease phenotype. How certain aspects of mouse behavior are related to complex human behavior – for example in the realm of vocal communication or social cognition – is largely unknown. Rather than overlooking these shortcomings as is often the case, the challenge is to use animal models judiciously, so as to fully exploit their power while understanding their potential limitations.

The Need for Better Phenotypes

Perhaps the most high-yield and logistically most difficult change to implement is the clinical characterization of neurological and psychiatric diseases in humans (Congdon et al., 2010; Sabb et al., 2009). From a biological perspective, a similar behavioral or cognitive phenotype can emerge in different ways in different individuals. However, many neuropsychiatric disorders are categorized as if they share a single common etiology. For instance, the common usage of the term "autism" encompasses many heterogeneous neurodevelopmental disorders from an etiological standpoint, whose cohesive overarching features include difficulty in socialization, repetitive behavior or interests, and (usually) aberrant language development. The same is true of schizophrenia (SZ), attention deficit hyperactivity disorder (ADHD), and even pathologically defined neurodegenerative disorders such as AD. It will ultimately benefit many neuropsychiatric diseases to be further categorized and staged based on their underlying biology and etiology, an area where genetics is now playing a major role.

One approach to overcome issues related to clinical disease classification is for neuropsychiatric studies to study more brain-based intermediate phenotypes or endophenotypes, instead of categorically defined disease labels (Gottesman and Gould, 2003). Endophenotypes reflect some biologically relevant trait such as functional connectivity, structural size, performance on a psychological test, or any other potentially heritable trait that is related to the disease. Endophenotypes enable stratification within disease by making the studied phenotype more biologically homogenous. However, they are not necessarily simple or a panacea, as they can increase the problem of multiple comparisons and may encounter trouble when faced with pleiotropy, where one gene affects multiple endophenotypes.

Clearly the ultimate solution is to find a balance between rigorous scientific criteria and convenient clinical nosology. This will likely include a move toward more thorough criteria for documenting disease, as is already done, for example, in the context of cancer grading and staging, which stratifies diagnosis and treatment, but retains ease in clinical discussion and management. Another step in this direction will be to incorporate more quantitative phenotypes, such as imaging, cerebrospinal fluid (CSF) or blood biomarkers, electrophysiology, and well-defined cognitive measures into disease classification, as well as temporal phenotypes based on change or trajectory. Accepting the phenotypic complexity of these diseases will enable advances in understanding basic science and ultimately provide better, more personalized management. Indeed, the upcoming diagnostics and statistical manual for mental disorders, 5th edition (DSM-5), the standard for clinical diagnoses in neuropsychiatric disorders, will see a shift from categorical definitions continuous descriptions similar to the endophenotype concept (see link at end of chapter).

THE GENETIC ARCHITECTURE OF NEUROPSYCHIATRIC DISORDERS

The history of genetic interrogation of neuropsychiatric disease gives an idea of the interplay among the discovery of biological mechanisms of heritability, the development of high-throughput technology, and the advances in computational statistics that continue to drive the field today. The field began with the hope that most disease etiology would be explained by high-frequency, high-risk alleles. Had this been true, a few genes would have accounted for most of the genetic variability that led to disease. After realizing such alleles were the exception and not the rule, combinations of lower-frequency, lower-risk alleles have been investigated as suitable technologies have become available, and now interactions between those alleles and the environment are being investigated. In this section we discuss the historical progress and future challenges facing the fields of neurodegeneration and neurodevelopment, but similar narratives apply to other major neuropsychiatric disorders.

From Monogenic Cause to Multifactorial Risk

Early genetic studies of the brain had some major successes in identifying monogenic causes and high-risk, high-penetrance alleles for disease. These initial studies were based on the premise that a few high frequency mutations with large effect sizes could account for the heritable component of most neuropsychiatric diseases. They involved looking at the inheritance of genetic markers across pedigrees with clear Mendelian patterns of disease inheritance. Linkage analysis and positional cloning eventually led to the identification of several key genes, which have now also been functionally modeled in animals. Two areas of considerable success were Alzheimer's Disease (AD) and Parkinson's Disease (PD). These neurodegenerative disorders illustrate how investigators have gone about searching for monogenic causes to multifactorial risks for disease.

Linkage analysis and positional cloning led to the discoveries of familial mutations in *APP* (Goate et al., 1991), *PSEN1*, and *PSEN2* (Rogaev et al., 1995; Sherrington et al., 1995) for AD, while Mendelian forms of PD identified *SNCA* (Polymeropoulos et al., 1997; Singleton et al., 2003), *LRRK2* (Gilks et al., 2005; Healy et al., 2008; Lesage et al., 2006), *PINK1* (Valente et al., 2004), *PARK2*, and *PARK7* (Kitada et al., 1998) (see also Chapters 85 and 86). However, the monogenic forms of disease only accounted for 1–2% and 5–10% of the total disease prevalance in AD and PD, respectively. The next step was to elucidate the etiology of the non-Mendelian yet still heritable forms of these diseases.

Until the completion of the Human Genome Project, candidate gene studies were the major design used to find additional causative genes for the sporadic forms of AD and PD. In AD, these studies were not very productive, with over 700 candidate genes yielding only a few potential loci of interest, significantly among them the *APOE* risk allele (Bertram et al., 2010). In PD, these studies showed that the *SNCA* and *LRRK2* genes also contribute to sporadic PD risk, implying some common pathology between the familial and late-onset forms of disease (Hardy, 2010). Overall, many candidate gene studies were lackluster and not replicable, and the inability to find many verifiable disease-susceptibility variants from these approaches is now attributed partially to small sample sizes (underpowered studies) and flawed study design (population stratification, technical artifacts). However, the major reason these studies were ineffectual was that there is a low prior probability that the right candidate genes would be chosen from known biology if there are multiple pathways contributing to a disorder (Congdon et al., 2010).

This led to the appreciation that many neuropsychiatric diseases may have a more complex etiology than initially surmised, and two overarching hypotheses have been used to approach this. The common disease, common variant (CDCV) hypothesis suggests that high-frequency variants with low penetrance are the cause of complex diseases, while the rare variant hypothesis suggests that rare, high-penetrance variants converge on common disease pathology.

From Common to Rare Variants

The CDCV hypothesis has been tested by the genome-wide association study (GWAS) approach, which has essentially eclipsed the candidate gene method due to its more unbiased and near complete coverage of common variation the genome. GWAS have provided many insights into the genetic architecture of neuropsychiatric disease and potential common pathways that can underlie similar diseases. In AD, for example, the monogenic mutations mentioned above are directly related to one of the pathological hallmarks of the disease, protein aggregates of extracellular β-amyloid. GWAS in AD, on the other hand, have implicated the *PICALM*, *CLU*, and *CR1* genes (Harold et al., 2009), which are *not* directly related to amyloid plaque formation. The *CR1* gene codes for complement receptor 1, the main receptor for the C3b protein in the complement pathway. The C3 protein is cleaved by C3 convertase to C3a and C3b, the latter of which may have a protective role in disease since the C3b fragment can bind to amyloid β to promote its clearance (Rogers et al., 2006; Webster et al., 1997). This implies a role for innate immunity in AD pathogenesis, which demonstrates the power of an agnostic, genome-wide approach (Gandhi and Wood, 2010). Intriguingly, GWAS of AD have not uncovered a role for additional proteins directly in the amyloid pathway in sporadic AD, suggesting that our current models of AD pathogenesis are far from complete.

GWAS can also find unexpected common links between seemingly heterogenous diseases, as in the case of the *MAPT* H1 haplotype (Pankratz et al., 2009) in PD. The *MAPT* gene codes for tau protein, which is involved in several other tauopathies including AD (Pittman et al., 2006), where a pathological hallmark is hyperphosphorylated intracellular tau. Here, genetic findings bring together diseases that are clinically considered to be distinct and suggests the potential for a common neurodegenerative pathway. The same is true with behaviorally defined neurodevelopmental disorders such as SZ, ADHD, and autism, where copy number variations (CNVs) common across disorders have been identified (Cantor and Geschwind, 2008; Pinto et al., 2010; Williams et al., 2010).

However, GWAS have shown that common variants as assayed thus far do not contribute the majority of the genetic risk to common disease. It may be that the contribution of any one gene to disease risk is too low in neuropsychiatric diseases for GWAS to detect, or that the causative variant is rare,

occuring in <5% of cases, in which case GWAS would be underpowered to detect effects. In the former case, GWAS would need an astronomically higher number of subjects in a study to declare a genome variant significant (Spencer et al., 2009). In the latter case, a combination of second-generation, high-throughput sequencing of subjects (such as that by the 1000 Genomes Project), combined with statistical methods such as imputation, can detect these rare variants with greater sensitivity. There is already evidence that these approaches, as well as the resequencing of suspect regions, will reveal many new, potentially causative, disease-related mutations (Tsuji, 2010).

Recently, whole exome studies, which assess rare variants in protein-coding DNA and their flanking regions, have begun to help unravel the full genetic architecture of neuropsychiatric disorders such as ASD (Devlin et al., 2012). These studies usually involve sequencing affected children and unaffected parents (trios), though other designs are also proving fruitful. The design of these studies is key to their interpretation, as there may be some baseline level of rare *de novo* mutation in all individuals, and particularly in those families where disorders are prevalent. Thus, extended pedigree structures, at least including a matched sibling, are useful for determining causality and large sample sizes are essential. Whole exome sequencing has also suggested a role for *de novo* mutations in schizophrenia and amyotropic lateral sclerosis (Johnson et al., 2010; Xu et al., 2011). The development of appropriate study designs and analytical methods for future whole exome, and eventually whole genome, studies will be essential to elucidating the genetic variants underlying brain disorders and diseases.

Beyond Single Nucleotide Changes: Copy Number Variation

The field of autism and the autism spectrum disorders (ASD) is instructive when considering the need to assess rare genetic variation (see also Chapter 90). The etiology of ASD has proven to be extremely heterogenous, as expected from the broad definition of ASDs (Abrahams and Geschwind, 2008). CNVs are believed to account for 5–20% of the total cases in ASD (Walsh and Engel, 2010). Hetereozygous CNVs associated with ASD can be found in the same loci as some Mendelian forms such as *MECP2* and the 15q region, but also in novel loci including 16p11.2, 22q11.2, and 15q13 (State, 2010; Weiss et al., 2008). There are also homozygous deletion CNVs that can be found in what appear to be non-coding regions of the genome. A complication that arises in neuropsychiatric diseases such as ASD that lack a clear biological phenotype is the pleiotropic nature of many of these CNVs. For instance, the 22q11.2 deletion and the 16p11.2 duplication are also implicated in SZ (Bassett et al., 1998; McCarthy et al., 2009), implying a mechanistic relationship between the two disorders at some level.

There are also several promising diagnostic and therapeutic tools that have come from genetic investigation of neurodevelopmental disorders. The discovery of CNVs as a major contributor to neuropsychiatric disease has led to a call for microarray-based screening (which has a higher sensitivity since it includes structural changes <5 Mb) in newborns for

ASD to supplant G-banded karyotyping as a first-tier genetic test (Liang et al., 2008; Thuresson et al., 2007). The elucidation of the pathways in two disorders in the autism spectrum, fragile X and Rett syndromes, has led to potential druggable targets, and therefore opened the way to new therapies (Dölen et al., 2007; Egger et al., 2004). This exemplifies how a mutation causing early cytoarchitectural abnormalities and brain malformations, may be partially or wholly ameliorated by pharmacological intervention at a later time.

Addressing Epistatic Interactions Between Genes

The above approaches can miss potential epistatic effects where multiple genes interact with each other and the environment to affect a phenotype in a non-additive manner (Phillips and Belknap, 2002). Examples of epistasis exist in both model organisms and in human neuropsychiatric phenotypes (Ni et al., 2009; Pezawas et al., 2008; Saraceno et al., 2009). However, epistasis in humans is difficult to study in a systematic manner and still remains relatively unexplored. Progress has been made by applying multivariate regression analyses as well as methods for grouping genes by pathway, but it is debated whether this approach will find the remaining large fraction of the unexplained heritability for most diseases (Eichler et al., 2010; Heng et al., 2010). This has led to the search for disease cause and risk in heritable genetic changes that have their basis in different biological mechanisms.

ENVIRONMENTAL AND EPIGENETIC INTERACTIONS IN NEUROGENETICS

Variations in DNA structure and sequence cannot by themselves account for the persistent changes seen in the brain. Regulatory sequences interact with transcription factors to modulate gene expression, while alternative splicing aids in the generation of transcript diversity. The epigenome, comprising biochemical and structural modifications of DNA bases and chromatin structure, influences DNA access and, in turn, RNA transcript expression. The product of all of these mechanisms of regulation, the transcriptome, can interact with these mechanisms themselves and the proteome (Horsthemke and Wagstaff, 2008). Thus, a seemingly anarchical system emerges and disease mechanisms become all the more difficult to elucidate (Figure 84.2). Personalized genomics in neuropsychiatric disease will likely not have direct access to all of these levels of regulation in human brain, but understanding them and relating them to sequence variation and epigenetic changes in surrogate tissue can aid in the development of therapies.

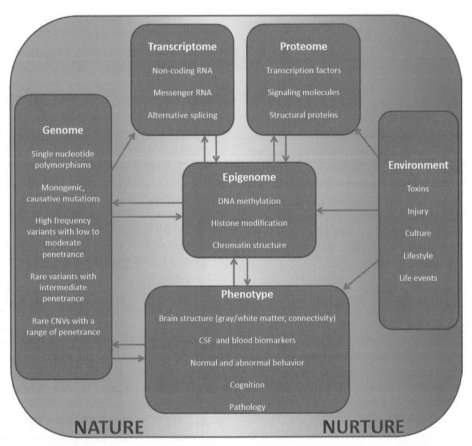

Figure 84.2 The nature/nurture gradient in neurogenetics. This is one representation of the levels of regulation due to genetic variation in the human brain, and interactions with the environment. Alternative conceptualizations are also valid, with the major point being that genetic and environmental interactions converge via multiple mechanisms to yield phenotype, and phenotypes (behavior) can modify the genome and epigenome.

Between the Genotype and Phenotype: The Epigenome and Transcriptome

Epigenetic mechanisms were initially thought to be restricted to replication-dependent, heritable chemical modifications of DNA structure, including histone acetylation and DNA methylation that affected transcription in a static manner by setting an on/off state in the germline. However, the role of epigenetic modifications in neural plasticity has proved to be important despite the low mitotic rate in neruons. This has broadened the definition of epigenetics to encompass any change in the structure of chromosomes that can serve an adaptive role and result in the modification of transcriptional levels (Meaney and Ferguson-Smith, 2010).

In neuroscience, this is of particular significance, since such mechanisms can ultimately change neuronal and glial function and underlie processes such as cell differentiation, neurodegeneration, and experienced-related phenomena such as drug addiction and chronic stress (Tsankova et al., 2007). These epigenetic mechanisms further emphasize the challenges of brain complexity faced by neurogenetics, as epigenetic changes can vary from cell to cell and region to region in the brain.

Another dynamic entity within the brain is the transcriptome. Extrinsic events, such as growth factor gradients and environmental interactions during development, have led to regionally unique functional transcriptional profiles through differential transcription factor binding of regulatory sequences, epigenetic modulation of the genome, and the alternative splicing of pre-mRNA (Li et al., 2007). Such transcriptional diversity allows for region-specific tagging of cell types for functional characterization of brain regions (Oldham et al., 2008), enables region-specific immunological staining, and will be essential to harness if targeted drugs and cell-specific gene therapy are to be developed. These expression-level changes were initially studied with *in situ* hybridization and Northern blotting, but now oligonucleotide microarrays, complementary DNA (cDNA) microarrays and, more recently, RNA-seq (see Chapter 12) are allowing quantitative high-throughput analysis. Disease- and development-specific changes in the transcriptome can also be observed, making it a more relevant endophenotype for quantifying biological function in neuropsychiatric disease (Johnson et al., 2009; Miller et al., 2008).

A Caveat to Consider: Heritable Environmental Effects

A well-documented exception to the claim that genome sequence alone is the source of heritability is transgenerational epigenetic inheritance (Morgan and Whitelaw, 2008). This phenomenon has the potential to confound the heritability metric, as the epigenetics of an individual can be affected in the womb of the mother, and as early as when the mother was in the grandmother's womb (since the egg making the offspring of interest has already developed in the fetus that will be its future mother). Stressful events during pregnancy can result in hormonal and metabolic changes which have the potential to modify brain function (Mandyam et al., 2008; Weaver et al., 2004). There is also evidence that emotionally stressful events for mothers before pregnancy may result in an increased incidence of SZ in offspring (Khashan et al., 2008).

THE OUTLOOK FOR PERSONALIZED AND GENOMIC MEDICINE IN NEUROSCIENCE

Our understanding of the genetic architecture of neurodevelopmental and neurodegenerative disease has changed over time as technologies have enabled us to probe the genome deeper and with greater dynamic range. The personalized genome will likely include a combination of common and rare CNVs, along with both common and rare single-nucleotide variants. The advent of cost-efficient, second-generation sequencing (see Chapter 7) will alleviate the challenges of reading the genome, but it will leave the major tasks of properly designing studies and accurately phenotyping patients. A similar principle will apply for milder cognitive phenotypes and normal development, although they have not been directly addressed here. Even after overcoming these technological hurdles and optimizing analysis, the whole genetic etiology of many diseases will still likely remain unsolved. There are certain factors that cannot be assayed, and stochastic biology combined with hidden environmental variables are fundamental limitations.

Making the Most of Our Information

From a practical stance, such incomplete knowledge, while imperfect, will still be sufficient to contribute to and vastly improve patient care. Our current genetic knowledge has already been applied to testing for rare disorders, and it is now routinely applied in many clinical settings to pick the ideal treatment and to inform management for a patient. Currently, the first level of assessment is done by the clinician, and the second level is performed by a lab that will test for a sequence variant. Examples include polymorphisms affecting warfarin dose response (Wadelius et al., 2005) and polymorphisms involved in responses to various psychiatric drugs (Lohoff and Ferraro, 2010). The goal in neuropsychiatric pharmacogenomics is to both improve existing drug therapies and expedite future clinical trials (De Leon, 2008).

Disease- and disorder-relevant biomarkers will be necessary to accelerate the translation of scientific findings to clinical tools. Other than the behavioral and neuroimaging endophenotypes discussed above, many neurological diseases have pervasive effects that might allow clinicians to assay tissue other than the brain to predict and follow pathology. A key goal of this work is the development of multimodal disease classifiers and risk calculators that could quantify changes in disease status to improve clinical management and research.

Consortia such as the Alzheimer's Disease Neuroimaging Initiative have longitudinally compiled MRI and positron emission tomography (PET) scans from normal aging, mild-cognitive impairment, and AD subjects to assess the feasibility of SNP, blood, CSF, and neuroimaging biomarkers (Mueller et al., 2005), with the goal of developing multimodal metrics to enable earlier diagnosis of AD. The methods of standardizing protocols, integrating data, and communicating between many sites in this consortium make it a prime example of how clinicians and researchers might go about designing technically sound studies with enough subjects. Some sites are attempting to combine existing GWAS data for better statistical analysis, such as the Psychiatric GWAS Consortium (Sullivan, 2010). Other groups, such as Autism Genetics Resource Exchange (Lajonchere, 2010) and the Simons Foundation consortium (Fischbach and Lord, 2010), are attempting to gather integrated data from multiple sites to power studies of heterogeneous diseases such as autism (see relevant web-pages in the *Recommended Resources* section at the end of this chapter). But, given that sample size is always a weakness, much more progress will need to be made in gathering study participants and developing large collaborations.

From Genetic Change to Therapy: An Emerging Paradigm

The inability to assay changes beyond genomic sequence and structural change in the individual does not preclude treatment. The elucidation of disease mechanisms can lead to the identification of druggable targets. For instance, downregulating mGluR signaling has now been identified as a potential therapeutic target in fragile X syndrome (Krueger and Bear, 2010). Understanding epigenetic disease mechanisms has opened avenues for treatment once a mutation is found. In Rett syndrome, aberrant epigenetic modfications can be reversed by the use of a histone deacetylase (HDAC) inhibitor. HDAC inhibitors may be useful in several other central nervous system disorders involving aberrant DNA methylation, including Rubinstein–Taybi syndrome, fragile X syndrome, motor neuron and polyglutamine diseases, addiction, and some psychiatric and mood disorders (Kazantsev and Thompson, 2008). A combination of the ability to clinically define etiology based on an individual's genome, coupled with the knowledge of disease mechanisms from the population at large, will enable new, genetically tailored therapies.

CONCLUSION

As neurogenetics looks forward, there will be inevitable advances driven by the increased spatial and temporal resolution of genomic technologies. There will also be unexpected advances in understanding disease mechanisms and environmental interactions. From a clinical standpoint, it will serve the field well to better standardize criteria for disease evaluation and establish databases so that classification systems can be improved and large studies may be efficiently designed. In the near future, as the cost of a sequenced genome drops below that of an MRI of the brain, whole-genome sequencing may become a routine part of the clinical neuropsychiatric work-up. But, without better phenotyping and book-keeping, the interpretation of these data will be muddied.

After taking all of this into consideration, one might well wonder: have we thrown out Occam's razor for what seem to be overly complicated, subdivided, and clinically intractable approaches? This is clearly not the case, as the emerging genetic picture reveals that even neurologic diseases that were formerly thought to be well understood may not have a simple, unique genetic solution (Lupski, 2007). It is likely that we are still underestimating the complexity of what lies ahead for neuropsychiatric disease. What is certain is that we must continue to put our resources into transforming information into knowledge so that we can use personalized genomics to better treat neuropsychiatric disorders and diseases.

REFERENCES

Abrahams, B.S., Geschwind, D.H., 2008. Advances in autism genetics: On the threshold of a new neurobiology. Nat Rev Genet 9 (5), 341–355.

Bassett, A.S., Hodgkinson, K., Chow, E.W., Correia, S., Scutt, L.E., Weksberg, R., 1998. 22q11 deletion syndrome in adults with schizophrenia. Am J Med Genet 81 (4), 328–337.

Bateman, R.J., Wen, G., Morris, J.C., Holtzman, D.M., 2007. Fluctuations of CSF amyloid-β levels. Neurology 68 (9), 666–669.

Bertram, L., Lill, C.M., Tanzi, R.E., 2010. The genetics of Alzheimer disease: Back to the future. Neuron 68 (2), 270–281.

Bian, H., Van Swieten, J.C., Leight, S., et al., 2008. CSF biomarkers in frontotemporal lobar degeneration with known pathology. Neurology 70 (19 Pt 2), 1827–1835.

Braak, H., Braak, E., 1991. Neuropathological stageing of Alzheimer-related changes. Acta Neuropathol 82, 239–259.

Cantor, R.M., Geschwind, D.H., 2008. Schizophrenia: Genome, interrupted. Neuron 58 (2), 165–167.

Congdon, E., Poldrack, R.A., Freimer, N.B., 2010. Neurocognitive phenotypes and genetic dissection of disorders of brain and behavior. Neuron 68 (2), 218–230.

Courchesne, E., Karns, C.M., Davis, H.R., et al., 2001. Unusual brain growth patterns in early life in patients with autistic disorder. Neurology 57 (2), 245–254.

de Leon, J., 2009. Pharmacogenomics: The promise of personalized medicine for CNS disorders. Neuropsychopharmacology 34 (1), 159–172.

Devlin, B., Scherer, S.W., 2012. Genetic architecture in autism spectrum disorder. Curr Op Genet Dev 22, 229–237.

Dosenbach, N.U.F., Nardos, B., Cohen, A.L., et al., 2010. Prediction of individual brain maturity using fMRI. Science 329, 1358–1361.

Dölen, G., Osterweil, E., Rao, B.S.S., et al., 2007. Correction of fragile X syndrome in mice. Neuron 56 (6), 955–962.

Egger, G., Liang, G., Aparicio, A., Jones, P.A., et al., 2004. Epigenetics in human disease and prospects for epigenetic therapy. Nature 429 (6990), 457–463.

Eichler, E.E., Flint, J., Gibson, et al., 2010. Missing heritability and strategies for finding the underlying causes of complex disease. Nat Rev Genet 11 (6), 446–450.

Enard, W., Gehre, S., Hammerschmidt, K., et al., 2009. A humanized version of Foxp2 affects cortico-basal ganglia circuits in mice. Cell 137 (5), 961–971.

Enard, W., Przeworski, M., Fisher, S.E., et al., 2002. Molecular evolution of FOXP2, a gene involved in speech and language. Nature 418 (6900), 869–872.

Fischbach, G.D., Lord, C., 2010. The simons simplex collection: A resource for identification of autism genetic risk factors. Neuron 68 (2), 192–195.

Gandhi, S., Wood, N.W., 2010. Genome-wide association studies: The key to unlocking neurodegeneration? Nat Neurosci 13 (7), 789–794.

Geschwind, D.H., Konopka, G., 2009. Neuroscience in the era of functional genomics and systems biology. Nature 461 (7266), 908–915.

Giedd, J.N., Blumenthal, J., Jeffries, N.O., et al., 1999. Brain development during childhood and adolescence: A longitudinal MRI study. Nat Neurosci 2 (10), 861–863.

Gilks, W.P., Abou-Sleiman, P.M., Gandhi, S., et al., 2005. A common LRRK2 mutation in idiopathic Parkinson's disease. Lancet 365 (9457), 415–416.

Goate, A., Chartier-Harlin, M.C., Mullan, M., et al., 1991. Segregation of a missense mutation in the amyloid precursor protein gene with familial Alzheimer's disease. Nature 349 (6311), 704–706.

Gogtay, N., 2004. From the Cover: Dynamic mapping of human cortical development during childhood through early adulthood. Proc Natl Acad Sci USA 101 (21), 8174–8179.

Gottesman, I.I., 2003. The endophenotype concept in psychiatry: Etymology and strategic intentions. Am J Psychiatry 160 (4), 636–645.

Hardy, J., 2010. Genetic analysis of pathways to Parkinson disease. Neuron 68 (2), 201–206.

Harold, D., Abraham, R., Hollingworth, P., et al., 2009. Genome-wide association study identifies variants at CLU and PICALM associated with Alzheimer's disease. Nat Genet 41 (10), 1088–1093.

Healy, D.G., Falchi, M., O'Sullivan, S.S., et al., 2008. Phenotype, genotype, and worldwide genetic penetrance of LRRK2-associated Parkinson's disease: A case-control study. Lancet Neurol 7 (7), 583–590.

Heng, H.H.Q., 2010. Missing heritability and stochastic genome alterations. Nat Rev Genet 11 (11), 812.

Horsthemke, B., Wagstaff, J., 2008. Mechanisms of imprinting of the Prader-Willi/Angelman region. Am J Med Genet A 146A (16), 2041–2052.

Iossifov, I., Ronemus, M., Levy, D., et al., 2012. De novo gene disruptions in children on the autistic spectrum. Neuron 74, 285–299.

Johnson, J.O., Mandrioli, J., Benatar, M., et al., 2010. Exome sequencing reveals VCP mutations as a cause of familial ALS. Neuron 68, 857–864.

Johnson, M.B., Kawasawa, Y.I., Mason, C.E., et al., 2009. Functional and evolutionary insights into human brain development through global transcriptome analysis. Neuron 62 (4), 494–509.

Kazantsev, A.G., Thompson, L.M., 2008. Therapeutic application of histone deacetylase inhibitors for central nervous system disorders. Nat Rev Drug Discov 7 (10), 854–868.

Khashan, A.S., Abel, K.M., McNamee, R., et al., 2008. Higher risk of offspring schizophrenia following antenatal maternal exposure to severe adverse life events. Arch Gen Psychiatry 65 (2), 146–152.

Kitada, T., Asakawa, S., Hattori, N., et al., 1998. Mutations in the parkin gene cause autosomal recessive juvenile parkinsonism. Nature 392 (6676), 605–608.

Konopka, G., Bomar, J.M., Winden, K., et al., 2009. Human-specific transcriptional regulation of CNS development genes by FOXP2. Nature 462 (7270), 213–217.

Krueger, D.D., Bear, M.F., 2011. Toward fulfilling the promise of molecular medicine in fragile X syndrome. Annu Rev Med 62, 411–429.

Ku, C.S., Polychronakos, C., Tan, E.K., et al., 2012. A new paradigm emerges from the study of de novo mutations in the context of neurodevelopmental disease. Mol. Psychiatry.

Lai, C.S.L., Fisher, S.E., Hurst, J.A., Vargha-Khadem, F., Monaco, A.P., 2001. A forkhead-domain gene is mutated in a severe speech and language disorder. Nature 413 (6855), 519–523.

Lajonchere, C.M., 2010. Changing the landscape of autism research: The autism genetic resource exchange. Neuron 68 (2), 187–191.

Lee, G., Papapetrou, E.P., Kim, H., et al., 2009. Modelling pathogenesis and treatment of familial dysautonomia using patient-specific iPSCs. Nature 461 (7262), 402–406.

Lesage, S., Dürr, A., Tazir, M., et al., 2006. LRRK2 G2019S as a cause of Parkinson's disease in North African Arabs. N Engl J Med 354 (4), 422–423.

Li, Q., Lee, J., Black, D.L., 2007. Neuronal regulation of alternative pre-mRNA splicing. Nat Rev Neurosci 8 (11), 819–831.

Liang, J., Shimojima, K., Yamamoto, T., 2008. Application of array-based comparative genome hybridization in children with developmental delay or mental retardation. Pediatr Neonatol 49 (6), 213–217.

Lohoff, F.W., Ferraro, T.N., 2010. Pharmacogenetic considerations in the treatment of psychiatric disorders. Expert Opin Pharmacother 11 (3), 423–439.

Lupski, J.R., 2007. Structural variation in the human genome. N Engl J Med 356 (11), 1169–1171.

Mandyam, C.D., Crawford, E.F., Eisch, A.J., Rivier, C.L., Richardson, H.N., 2008. Stress experienced in utero reduces sexual dichotomies in neurogenesis, microenvironment, and cell death in the adult rat hippocampus. Dev Neurobiol 68 (5), 575–589.

Marchetto, M.C.N., Carromeu, C., Acab, A., et al., 2010. A model for neural development and treatment of Rett syndrome using human induced pluripotent stem cells. Cell 143 (4), 527–539.

McCarthy, S.E., Makarov, V., Kirov, G., et al., 2009. Microduplications of 16p11.2 are associated with schizophrenia. Nat Genet 41 (11), 1223–1227.

Meaney, M.J., Ferguson-Smith, A.C., 2010. Epigenetic regulation of the neural transcriptome: The meaning of the marks. Nat Neurosci 13 (11), 1313–1318.

Miller, J.A., Horvath, S., Geschwind, D.H., 2010. Divergence of human and mouse brain transcriptome highlights Alzheimer disease pathways. Proc Natl Acad Sci USA 107 (28), 12,698–12,703.

Miller, J.A., Oldham, M.C., Geschwind, D.H., 2008. A systems level analysis of transcriptional changes in Alzheimer's disease and normal aging. J Neurosci 28 (6), 1410–1420.

Morgan, D.K., Whitelaw, E., 2008. The case for transgenerational epigenetic inheritance in humans. Mamm Genome 19 (6), 394–397.

Mueller, S.G., Weiner, M.W., Thal, L.J., et al., 2005. The Alzheimer's disease neuroimaging initiative. Neuroimaging Clin N Am 15 (4), 869–877. xi-xii.

Neale, B.M., Kou, Y., Liu, L., et al., 2012. Patterns and rates of exonic de novo mutations in autism spectrum disorders. Nature, 1–5.

Ni, X., Chan, D., Chan, K., McMain, S., Kennedy, J.L., 2009. Serotonin genes and gene–gene interactions in borderline personality disorder in a matched case-control study. Prog Neuro-Psychopharmacol Biol Psychiatry 33 (1), 128–133.

O'Roak, B.J., Vives, L., Girirajan, S., et al., 2012. Sporadic autism exomes reveal a highly interconnected protein network of de novo mutations. Nature, 1–7.

Oldham, M.C., Konopka, G., Iwamoto, K., et al., 2008. Functional organization of the transcriptome in human brain. Nat Neurosci 11 (11), 1271–1282.

Pankratz, N., Wilk, J.B., Latourelle, J.C., et al., 2009. Genomewide association study for susceptibility genes contributing to familial Parkinson disease. Hum Genet 124 (6), 593–605.

Paşca, S.P., Portmann, T., Voineagu, I., et al., 2011. Using iPSC-derived neurons to uncover cellular phenotypes associated with Timothy syndrome. Nat. Med. 17, 1657–1662.

Pezawas, L., Meyer-Lindenberg, A., Goldman, A.L., et al., 2008. Evidence of biologic epistasis between BDNF and SLC6A4 and implications for depression. Mol Psychiatry 13 (7), 709–716.

Phillips, T.J., Belknap, J.K., 2002. Complex-trait genetics: Emergence of multivariate strategies. Nat Rev Neurosci 3 (6), 478–485.

Pinto, D., Pagnamenta, A.T., Klei, L., et al., 2010. Functional impact of global rare copy number variation in autism spectrum disorders. Nature 466 (7304), 368–372.

Pittman, A.M., 2006. Untangling the tau gene association with neurodegenerative disorders. Hum Mol Genet 15 (Spec no 2), R188–R195.

Polymeropoulos, M.H., Lavedan, C., Leroy, E., et al., 1997. Mutation in the alpha-synuclein gene identified in families with Parkinson's disease. Science 276 (5321), 2045–2047.

Preuss, T.M., 2000. Taking the measure of diversity: Comparative alternatives to the model-animal paradigm in cortical neuroscience. Brain Behav Evol 55 (6), 287–299.

Rogaev, E.I., Sherrington, R., Rogaeva, E.A., et al., 1995. Familial Alzheimer's disease in kindreds with missense mutations in a gene on chromosome 1 related to the Alzheimer's disease type 3 gene. Nature 376 (6543), 775–778.

Rogers, J., Li, R., Mastroeni, D., et al., 2006. Peripheral clearance of amyloid beta peptide by complement C3-dependent adherence to erythrocytes. Neurobiol Aging 27 (12), 1733–1739.

Sabb, F.W., Burggren, A.C., Higier, R.G., et al., 2009. Challenges in phenotype definition in the whole-genome era: Multivariate models of memory and intelligence. Neuroscience 164 (1), 88–107.Sanders, S.J., Murtha, M.T., Gupta, A.R., et al., 2012. De novo mutations revealed by whole-exome sequencing are strongly associated with autism. Nature 485, 237–241.

Saraceno, L., Munafó, M., Heron, J., Craddock, N., van den Bree, M.B.M., 2009. Genetic and non-genetic influences on the development of co-occurring alcohol problem use and internalizing symptomatology in adolescence: A review. Addiction 104 (7), 1100–1121.

Schumann, C.M., Bloss, C.S., Barnes, C.C., et al., 2010. Longitudinal magnetic resonance imaging study of cortical development through early childhood in autism. J Neurosci 30 (12), 4419–4427.

Scott-Van Zeeland, A.A., Abrahams, B.S., Alvarez Retuerto, A.I., et al., 2010. Altered functional connectivity in frontal lobe circuits is associated with variation in the autism risk gene CNTNAP2. Sci Transl Med 2 (56), 56ra80.

Seeley, W.W., Crawford, R.K., Zhou, J., Miller, B.L., Greicius, M.D., 2009. Neurodegenerative diseases target large-scale human brain networks. Neuron 62 (1), 42–52.

Sherrington, R., Rogaev, E.I., Liang, Y., et al., 1995. Cloning of a gene bearing missense mutations in early-onset familial Alzheimer's disease. Nature 375 (6534), 754–760.

Singleton, A.B., 2003. Alpha-synuclein locus triplication causes Parkinson's disease. Science 302 (5646), 841.

Smulders, T.V., 2008. The relevance of brain evolution for the biomedical sciences. Biol Lett 5 (1), 138–140.

Spencer, C.C.A., Su, Z., Donnelly, P., Marchini, J., 2009. Designing genome-wide association studies: Sample size, power, imputation, and the choice of genotyping chip. PLoS Genet. 5, e1000477.

State, M.W., 2010. The genetics of child psychiatric disorders: Focus on autism and Tourette syndrome. Neuron 68 (2), 254–269.

Sullivan, P.F., 2010. The psychiatric GWAS consortium: Big science comes to psychiatry. Neuron 68 (2), 182–186.

Thuresson, A.C., Bondeson, M.L., Edeby, C., et al., 2007. Whole-genome array-CGH for detection of submicroscopic chromosomal imbalances in children with mental retardation. Cytogenet Genome Res 118 (1), 1–7.

Tsankova, N., Renthal, W., Kumar, A., Nestler, E.J., 2007. Epigenetic regulation in psychiatric disorders. Nat Rev Neurosci 8 (5), 355–367.

Tsuji, S., 2010. Genetics of neurodegenerative diseases: Insights from high-throughput resequencing. Hum Mol Genet. <http://hmg.oxfordjournals.org/content/early/2010/04/22/hmg.ddq162.abstract>.

Valente, E.M., 2004. Hereditary early-onset Parkinson's disease caused by mutations in PINK1. Science 304 (5674), 1158–1160.

Wadelius, M., Chen, L.Y., Downes, K., et al., 2005. Common VKORC1 and GGCX polymorphisms associated with warfarin dose. Pharmacogenomics J 5 (4), 262–270.

Walsh, C.A., Engle, E.C., 2010. Allelic diversity in human developmental neurogenetics: Insights into biology and disease. Neuron 68 (2), 245–253.

Weaver, I.C.G., Cervoni, N., Champagne, F.A., et al., 2004. Epigenetic programming by maternal behavior. Nat Neurosci 7 (8), 847–854.

Webster, S., Bradt, B., Rogers, J., Cooper, N., 1997. Aggregation state-dependent activation of the classical complement pathway by the amyloid beta peptide. J Neurochem 69 (1), 388–398.

Weiss, L.A., Arking, D.E., Gene Discovery Project of Johns Hopkins & the Autism Consortium, Daly, M.J., Chakravarti, A., 2009. A genome-wide linkage and association scan reveals novel loci for autism. Nature 461, 802–808.

Williams, N.M., Zaharieva, I., Martin, A., et al., 2010. Rare chromosomal deletions and duplications in attention-deficit hyperactivity disorder: A genome-wide analysis. Lancet 376 (9750), 1401–1408.

Xu, B., Roos, J.L., Dexheimer, P., et al., 2011. Exome sequencing supports a de novo mutational paradigm for schizophrenia. Nat Genet 43, 864–868.

RECOMMENDED RESOURCES

Gene Expression Atlases

Allen Brain Institute atlases
http://www.brain-map.org/
Gene Expression Nervous System Atlas
http://www.gensat.org

Examples of Integrative Data Approaches

Alzheimer's Disease Neuro-imaging Initiative
http://adni.loni.ucla.edu/
Enhancing Neuro-imaging Genetics through Meta-analysis
http://enigma.loni.ucla.edu/

Autism Genetic Resource Exchange
http://research.agre.org/
Simons Foundation Autism Research Initiative
https://sfari.org
Psychiatric GWAS Consortium
https://pgc.unc.edu/

Clinical Classification of Neuropsychiatric Disorders

Diagnostics and Statistical Manual of Mental Disorders, 5th edition
http://www.dsm5.org/

Alzheimer's Disease

Lars Bertram

INTRODUCTION

The fact that many neurodegenerative disorders – including Alzheimer's disease (AD) – often show extensive familial aggregation was recognized as a feature of these diseases decades before any of the underlying molecular genetic and biochemical properties were known. As a matter of fact, it was often only the identification of specific, disease-segregating mutations in previously unknown genes that directed the attention of molecular biologists to certain proteins and pathways that are now considered be crucial to the development and pathogenesis of the various diseases. These include the discovery of mutations in the β-amyloid precursor protein (gene: *APP*) causing Alzheimer's disease, mutations in α-synuclein (*SNCA*) causing Parkinson's disease (see Chapter 86), and mutations in microtubule-associated protein tau (*MAPT*) causing frontotemporal dementia with Parkinsonism (reviewed in Lill and Bertram, 2011). Another feature observed in AD and in many other common neurodegenerative syndromes is the co-existence of rare and familial forms (that follow Mendelian inheritance) with more common and seemingly non-familial forms (that do not follow Mendelian inheritance). Historically, the latter were also frequently described as "sporadic," although this terminology has proved to be oversimplistic since a large proportion of these cases are likely also substantially influenced by inherited genetic factors.

The identification of genuine risk factors for AD, the most common neurodegenerative disease in the elderly, is hindered by several circumstances. First, while diagnostic criteria have been proposed and revised, a "definite" diagnosis of AD can only be made after neuropathological examination at autopsy. A diagnosis based on purely clinical grounds without autopsy – which is how the vast majority of AD cases are defined in typical genetic studies – can at best represent "probable" AD. Thus, such a sample may actually be a conglomerate of predominantly Alzheimer's, but also other forms of dementia (i.e., "phenocopies"). A second and related issue is that most cases of AD manifest in old age, i.e., beyond 60 or 70 years. This makes the assessment of reliable familial histories – a prerequisite in many types of genetic analyses – very difficult because relatives may not have lived through the typical onset age, or may suffer from other conditions that can mask or mimic the phenotype of interest. Third, and much like many other common, adult-onset diseases, AD displays a large degree of genetic and phenotypic heterogeneity. This means that the same phenotype can be caused or modified by a number of different genetic loci and alleles, but also that mutations or polymorphisms in the same gene may lead to clinically different syndromes.

Collectively, these and other issues have led to the accumulation of a large number of proposed AD susceptibility genes and environmental risk factors (Bertram et al., 2007). Until recently, however, these have traditionally only received inconsistent support from subsequent and independent studies, apart from a few notable exceptions. This situation has changed to some degree since the advent of massively parallel genotyping (and more recently, sequencing) techniques. Currently, the most popular approach is based on genome-wide association screening where up to around one million

Genomic and Personalized Medicine, 2nd edition
by Ginsburg & Willard. DOI: http://dx.doi.org/10.1016/B978-0-12-382227-7.00085-9

genetic markers are simultaneously genotyped and assessed for potential correlations with disease risk and other phenotypic variables (e.g., disease onset, progression, survival). Since 2005, the genetics community has seen a deluge of genome-wide association studies (GWAS), including over a dozen reported in AD. While the success rate still varies from study to study, a number of well-replicated AD loci have already emerged from these projects, and more are likely to be discovered over the coming years (Bertram, 2011).

EARLY-ONSET FAMILIAL AD WITH MENDELIAN TRANSMISSION

Mendelian Forms of AD are Rare

Less than 5% of all AD cases are represented by early-onset familial AD (EOFAD) transmitted following Mendelian inheritance. Despite its rarity, genetic studies of this form of AD are facilitated by the availability of large, multigenerational pedigrees. These have allowed genetic linkage analysis and subsequent positional cloning, which is usually not possible in late-onset AD (LOAD) families, where fewer relatives survive the family-specific onset age and genetic information from parents is almost always lacking (see below). The search for causative mutations beyond those already identified is expected to be vastly facilitated by means of massively parallel ("next-generation") sequencing (NGS) technologies. The application of NGS to familial forms of other neurodegenerative diseases [Parkinson's disease (PD) (see Chapter 86), amyotrophic lateral sclerosis (ALS), and frontotemporal dementia (FTD)] has very recently led to a number of breakthrough findings (reviewed in Lill and Bertram, 2011).

Mutations in APP, PSEN1, and PSEN2 Can Cause AD

In 1991, the first pathogenic missense mutation for EOFAD was identified, located in the gene encoding β-amyloid precursor protein (APP). Since then, over 30 additional AD-causing mutations have been reported to reside in APP (Table 85.1; for an up-to-date overview of AD mutations visit the AD and FTD Mutation Database at http://www.molgen.ua.ac.be/ADMutations/) (Cruts and Van Broeckhoven, 1998). Most of the APP mutations occur near the putative γ-secretase site between residues 714 and 717, suggesting that especially the enzymatic γ-cleavage event of APP and/or its (dys)regulation are critical for the development of AD (Figure 85.1). In addition to these "classic" pathogenic mutations, unfolding their disease-causing effects via a change in the amino acid sequence, other APP variants exist that can also lead to EOFAD with high penetrance. For instance, this includes locus duplications [a special form of a copy number variant (CNV)] of the APP-containing chromosomal region on chromosome 21q (Rovelet-Lecrux et al., 2006). This discovery is in line with the decades-old observation that AD neuropathology almost invariably develops in patients with trisomy 21 (Down's syndrome), in which the extra copy of APP

leads to increases in the expression of APP and deposition of amyloid-β protein (Aβ).

Only one year after the discovery of the first APP mutation, a second AD linkage region – on chromosome 14q24 – was reported almost simultaneously by four independent laboratories (reviewed in Tanzi and Bertram, 2005). It took three more years to clone the responsible gene (PSEN1) and identify the first AD-causing mutations (see AD and FTD Mutation Database). It is now known that PSEN1 encodes a highly conserved polytopic membrane protein, presenilin 1 (PS1), that plays an essential role in mediating intramembranous, γ-secretase processing of APP to generate Aβ from APP (Figure 85.1). Even more than a decade after the original description of PSEN1, there are several new AD-causing mutations reported in this gene every year, currently reaching a total of over 180 (Table 85.1). Soon after the discovery of PSEN1 as an AD gene, a second member of the presenilin family of proteins was identified by searching the then available databases. It displayed significant homology to PSEN1 at the genomic as well as at the protein level (see AD and FTD Mutation Database), and, therefore, this gene was named PSEN2 (protein: PS2). It maps to the long arm of chromosome 1, and mutations in this gene account for the smallest fraction of all EOFAD cases (Table 85.1). On average, mutations in PSEN2 also display a later age of onset and slower disease progression than APP or PSEN1 mutations.

In conclusion, while the currently known AD-causing mutations occur in three different genes located on three different chromosomes, they all share a common biochemical pathway, i.e., the altered production of Aβ leading to a relative overabundance of the Aβ42 species, which eventually results in neuronal cell death and dementia. Collectively, these discoveries provided the essential connection between the long-known familial aggregation of EOFAD and the increase in Aβ production observed in the brains of autopsied AD patients, which originally gave rise to the "amyloid hypothesis of AD" (reviewed in Tanzi and Bertram, 2005).

Other Potential EOFAD Genes

Although no additional EOFAD gene has been identified since the discovery of PSEN2 in 1995, several lines of evidence suggest that further genetic factors remain to be identified for this form of AD. There exist early-onset families that do not show mutations in APP, PSEN1, or PSEN2 despite extensive sequencing efforts, and there are several additional key proteins involved in aggregation and deposition of Aβ (e.g., nicastrin, APH1, PEN2, BACE1; Figure 85.1). It can be expected that the application of NGS (e.g., via whole-exome or whole-genome sequencing) will allow the identification of novel AD-causing mutations in additional genes, should such mutations indeed exist. As described elsewhere in this volume (see Chapters 7 and 8) NGS will allow one to assess individual samples or small families that are otherwise not suitable for conventional genetic analyses. The application of whole-exome or whole-genome sequencing has already been successfully

Gene	Protein	Location	Inheritance	P-value	Possible molecular effect(s)*
EOFAD					
APP	β-amyloid precursor protein	21q21	dominant	N/A	increase in Aβ production or Aβ$_{42}$/Aβ$_{40}$ ratio
PSEN1	Presenilin 1	14q24	dominant	N/A	increase in Aβ$_{42}$/Aβ$_{40}$ ratio
PSEN2	Presenilin 2	1q31	dominant	N/A	increase in Aβ$_{42}$/Aβ$_{40}$ ratio
LOAD					
ABCA7	ATP-binding cassette, sub-family A, member 7	19p13.3	OR ~1.23	7.8E−23	lipid homeostasis/transport; Aβ production
APOE	Apolipoprotein E	19q13	OR ~3.69	1.0E−118	aggregation and clearance of Aβ; cholesterol metabolism
BIN1	Bridging integrator 1	2q14	OR ~1.17	3.3E−28	production and clearance of Aβ
CD2AP	CD2-associated protein	6p12	OR ~1.12	1.7E−10	cell membrane remodeling/trafficking
CD33	CD33 molecule (siglec 3)	19q13.3	OR ~1.12	2.0E−10	Cell–cell signaling; immune response
CLU	Clusterin	8p21.1	OR ~1.14	3.4E−37	aggregation and clearance of Aβ; inflammation
CR1	Complement component (3b/4b) receptor 1	1q32	OR ~1.17	4.7E−21	inflammation and immune activation
MS4A4E	Membrane-spanning 4-domains, subfamily A, member 4E	11q12.2	OR ~1.08	9.5E−10	unknown; integrity/homeostasis of cell membrane (?)
MS4A6A	Membrane-spanning 4-domains, subfamily A, member 6A	11q12.1	OR ~1.10	2.3E−8	unknown; integrity/homeostasis of cell membrane (?)
PICALM	Phosphatidylinositol-binding clathrin assembly protein	11q14	OR ~1.14	2.9E−20	production and clearance of Aβ; synaptic transmission

TABLE 85.1   Currently known AD loci and their possible pathogenic relevance

EOFAD = early-onset familial AD; LOAD = late-onset AD; dominant = autosomal dominant transmission (Mendelian EOFAD genes); OR = odds ratio per risk allele using additive transmission models (non-Mendelian LOAD genes); N/A = not applicable; ORs and P-values based on information in the AlzGene database (current as of November 15, 2011).

*Selection of proposed pathophysiological effects as described in the literature; note that the evidence for these loci is often scarce (see Lill and Bertram, 2011, for more details).

applied to a number of other neurodegenerative diseases, including ALS, FTD, PD, and Charcot–Marie–Tooth (CMT) disease (reviewed in Lill and Bertram, 2011).

LATE-ONSET AD WITHOUT MENDELIAN TRANSMISSION

In contrast to EOFAD, risk for late-onset Alzheimer's disease (LOAD; typically showing an onset age ≥65 years) is governed by a considerably more complex pattern of genetic factors that not only interact with one another but also with non-genetic factors. Thus, LOAD can be considered a typical genetically complex disease. Adding to this complexity are methodological difficulties inherent to common diseases in general, and late-onset diseases like AD in particular (see

above). Notwithstanding these difficulties, a recent twin study from Sweden estimated that genetic factors may account for up to 80% of the heritability of LOAD (Gatz et al., 2006). These figures indicate that gene-finding efforts in LOAD are likely worthwhile, possibly more so than trying to elucidate non-genetic factors, or those related to epigenetic mechanisms (see below).

As is the case for a large number of other genetically complex diseases (including many neurodegenerative disorders), the advent and completion of genome-wide association studies (GWAS) have revealed over three dozen potential new LOAD-susceptibility genes during recent years. While only a subset of these GWAS signals could be confirmed in subsequent independent replication efforts to date (Table 85.1), this success rate is still much higher than in the pre-GWAS era (Bertram, 2011).

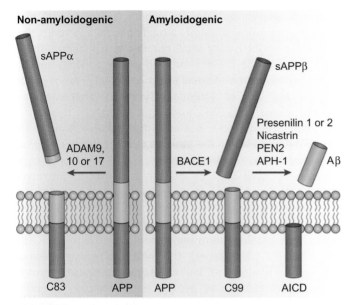

Non-amyloidogenic | Amyloidogenic

Figure 85.1 APP proteolysis and Aβ generation.

The amyloid-β (Aβ) peptide is derived via proteolysis from a larger, precursor molecule called the amyloid precursor protein (APP), a type 1 transmembrane protein consisting of 695–770 amino acids. APP can undergo proteolytic processing by one of two pathways. Most is processed through the non-amyloidogenic pathway, which precludes Aβ formation. The first enzymatic cleavage is mediated by α-secretase, of which three putative candidates belonging to the family of a disintegrin and metalloprotease (ADAM) have been identified: ADAM9, ADAM10, and ADAM17. Cleavage by α-secretase occurs within the Aβ domain, thereby preventing the generation and release of the Aβ peptide. Two fragments are released, the larger ectodomain (sAPPα) and the smaller carboxy-terminal fragment (C83). Furthermore, C83 can undergo an additional cleavage mediated by γ-secretase to generate P3 (not shown). APP molecules that are not cleaved by the non-amyloidogenic pathway become a substrate for β-secretase (β-site APP-cleaving enzyme 1; BACE1), releasing an ectodomain (sAPPβ), and retaining the last 99 amino acids of APP (known as C99) within the membrane. The first amino acid of C99 is the first amino acid of Aβ. C99 is subsequently cleaved 38–43 amino acids from the amino terminus to release Aβ, by the α-secretase complex, which is made up of presenilin 1 or 2, nicastrin, anterior pharynx defective and presenilin enhancer 2. This cleavage predominantly produces Aβ1–40 and the more amyloidogenic Aβ1–42 at a ratio of 10:1. AICD, APP intracellular domain; APH-1, anterior pharynx defective; PEN2, presenilin enhancer 2. *Reproduced with permission from Macmillan Publishers Ltd:* Nature Reviews Neuroscience *(LaFerla et al., 2007), copyright (2007). The original figure legend has been abbreviated for this article; see primary publication for more details.*

Apolipoprotein E (APOE) is the Most Important Genetic Risk Factor in LOAD

Several years prior to the availability of GWAS technology, one gene was already established as a LOAD risk factor, namely the ε4 allele of the apolipoprotein E gene (*APOE*) on chromosome 19q (Strittmatter et al., 1993). In contrast to most other association-based findings from the pre-GWAS era in AD, the risk effect of *APOE*-ε4 had been consistently replicated in a large number of studies across many ethnic groups. Even after the completion of nearly 20 GWAS in AD (see below), *APOE*-ε4 (or genetic markers highly correlated with it) remains by a margin the single most important genetic risk factor for AD, both in terms of effect size and statistical significance (Table 85.1). The three major alleles of the *APOE* locus, for historical reasons termed "ε2," "ε3," and "ε4," correspond to combinations of two amino acid changes at residues 112 and 158 (ε2: Cys_{112}/Cys_{158}; ε3: Cys_{112}/Arg_{158}; ε4: Arg_{112}/Arg_{158}). The two amino acid changes themselves are invoked by two separate single nucleotide polymorphisms (SNPs), rs429358 (residue 112) and rs7412 (residue 158). In most populations studied to date, odds ratios (ORs) vary between ~4 for heterozygous to ~15 for homozygous carriers of the ε4 allele, as compared to subjects carrying the "wildtype" ε3/ε3 genotype (Bertram et al., 2007). In addition to the increased risk exerted by the ε4-allele, several studies have also reported a weak, albeit significant, protective effect for the minor ε2-allele. Unlike the above-mentioned mutations in the known EOFAD genes, the *APOE*-ε4 allele is neither necessary nor sufficient to cause AD, but instead operates as a genetic risk-modifier by decreasing the age of onset in a dose-dependent manner.

Despite its long known and well-established genetic association, the biochemical consequences of *APOE*-ε4 in AD pathogenesis are not yet fully understood. Current hypotheses revolve around the observation that Aβ-accumulation is clearly enhanced in the brains of carriers as well as in transgenic mice expressing the human ε4 allele and mutant APP (reviewed in Vance and Hayashi, 2010). Further, apolipoprotein E normally plays a role in cholesterol transport and lipid metabolism, and *APOE*-ε4 predisposes to vascular disease as a result of its association with increased plasma cholesterol levels. High plasma cholesterol, in turn, has been correlated with increased β-amyloid deposition in the brain. Interestingly, cholesterol has also been shown to both increase Aβ production and to stabilize the peptide in the brains of transgenic AD mice. Thus, it is possible that *APOE*-ε4 confers risk for AD via a mechanism that is shared in common with its effect on vascular disease by increasing a carrier's risk for hypercholesterolemia, as this would also elevate accumulation of Aβ.

Genome-wide Association Studies (GWAS) in LOAD

As outlined above, GWAS have substantially reshaped the landscape of genetics research of common diseases, including AD, during the course of only a few years. Currently, the most promising GWAS findings in AD in addition to *APOE* relate to the identification of variants in or near *ABCA7*, *BIN1*, *CD2AP*, *CD33*, *CLU*, *CR1*, *MS4A4E*, *MS4A6A*, and *PICALM* (Bertram et al., 2008; Harold et al., 2009; Hollingworth et al., 2011; Lambert et al., 2009; Naj et al., 2011). As can be seen in Table 85.1, these loci all share the characteristic of showing association with AD risk at genome-wide significance, i.e., P-values at or below 5×10^{-8}, when all currently available data are taken into account [for up-to-date summaries of these and all other genetic association findings in AD, please visit the AlzGene database (http://www.alzgene.org)] (Bertram et al., 2007). Despite the strong statistical support for these associations, it needs to be emphasized that ORs exerted by the associated alleles of these genes are collectively small, i.e., ranging between ~1.1 and ~1.2 per copy of the respective risk alleles. The same small effect sizes resulting from comprehensive and well-powered GWAS are also observed in most other genetically complex diseases and highlight a number of issues.

First, it is worth remembering that it constitutes the very nature of genetically complex diseases that their underlying genetic architecture is characterized by a great deal of locus and allelic heterogeneity. A subject's liability to disease is, therefore, the result of the combined action of dozens to hundreds of different susceptibility alleles of small effect. Extreme examples can be found in schizophrenia (Purcell et al., 2009) or multiple sclerosis (Bush et al., 2010), which display an extensive polygenic basis judging from recent GWAS data (see Chapters 88 and 72, respectively). Second, it is unlikely that the polymorphisms currently showing the strongest association with disease risk actually represent the functionally responsible DNA sequence changes. They more likely represent proxies for the true functional variants, which only rarely are captured by current GWAS microarrays. Thus, in order to decipher the precise biochemical effects underlying the statistical associations, further fine-mapping of the associated regions is needed. This can be accomplished *in silico*, for instance, by utilizing high-resolution whole-genome resequencing data [e.g., as made available by the 1000 Genomes Project (1000G)], or *in vitro* by performing direct resequencing of risk-allele carriers, e.g., by using next-generation technologies. Third, a sizeable fraction of the heritability not explained by GWAS results (also sometimes referred to as the "dark matter" of genomics) may be due to the action of relatively uncommon sequence variants exerting large effects, e.g., by being located in functionally active gene regions. Current GWAS microarrays are poorly suited to directly measure such rare variants. Rather, in most scenarios, their detection will require large-scale resequencing of both affected and unaffected individuals.

Molecular Pathways Highlighted by Recent GWAS Signals

Determining how susceptibility alleles might biochemically impact the pathogenesis of complex phenotypes has often

been the rate-limiting step in genetic association studies. Indeed, unequivocal proof of causality has thus far only been demonstrated for a very small number of risk genes. This is in part due to the reasons outlined in the previous section, i.e., the combination of small effect sizes and – more importantly – the lack of potentially functional trait-associated polymorphisms, e.g., SNPs invoking a change in the amino-acid sequence, or those located in functional regions of the gene. However, even in the absence of an allele-specific proof of pathogenicity, researchers often assess *in silico* whether disease-associated SNPs or loci can be clustered into overarching "themes," e.g., into distinct molecular pathways that reveal meaning about likely pathogenesis. When this is done for the 13 currently known AD genes shown in Table 85.1 (using gene ontology term enrichments for the three EOFAD genes and 10 LOAD genes), the most compelling biological processes to emerge are "immune system response," "protein processing/maturation by peptide bond cleavage," "endocytosis," "response to stress," "lipid transport," and "apoptosis" (LB, unpublished observations). Interestingly, overlapping sets of pathways have also been previously implicated in AD and other neurodegenerative diseases, albeit mostly referring to cases carrying disease-causing mutations. The novel AD GWAS signals outlined above now provide further evidence to suggest that these molecular pathways may also contribute to the risk for non-Mendelian forms of AD and other disorders.

AD GENETICS BEYOND GWAS

Clearly, GWAS do not represent the only field of research in AD genetics, just as DNA sequence variants are not the only genetic features of potential relevance with respect to disease predisposition. The three major fields of research not covered in this review are (1) transcriptomics/expression quantitative trait locus (eQTL) mapping, (2) epigenetics, and (3) the application of NGS technologies. Compared to the recent knowledge gain afforded by GWAS, however, the currently available output of these three fields with respect to AD must be considered meager at best (for recent reviews see Sutherland et al., 2011; Urdinguio et al., 2009). In part, this is due to methodological issues (transcriptomics and epigenetics) or to the current lack of a sufficient number of studies (NGS). However, the application of NGS to uncover rare sequence variants, in particular, has recently led to spectacular findings in other neurodegenerative diseases (reviewed in Lill and Bertram, 2011). Despite the current lack of similar progress in AD, this situation is likely to change quite substantially over the coming years owing to an increasingly widespread application of NGS, including transcriptomics and epigenetics.

CONCLUSION

In only three years, non-Mendelian AD genetics has seen more progress than during the three decades before. This is owing to the increasing application of microarray technology that now allows genome-wide screenings in the quest for novel disease genes. To date, GWAS in AD have uncovered unequivocal evidence implicating nine loci in addition to *APOE*, which had already been reported during the candidate gene era. An even more pronounced boost to our understanding of AD genetics (both Mendelian and non-Mendelian) will result from the widespread application of NGS technology. But even NGS-derived, rare sequence variants invoking large risk effects will require solid support from functional assessments before any of this knowledge can be translated into the clinical setting.

REFERENCES

Bertram, L., 2011. Alzheimer's genetics in the GWAS era: A continuing story of "replications and refutations." Curr Neurol Neurosci Rep 11, 246–253.

Bertram, L., Lange, C., Mullin, K., et al., 2008. Genome-wide association analysis reveals putative Alzheimer's disease susceptibility loci in addition to APOE. Am J Hum Genet 83, 623–632.

Bertram, L., McQueen, M.B., Mullin, K., Blacker, D., Tanzi, R.E., 2007. Systematic meta-analyses of Alzheimer disease genetic association studies: The AlzGene database. Nat Genet 39, 17–23.

Bush, W.S., Sawcer, S.J., De Jager, P.L., et al., 2010. Evidence for polygenic susceptibility to multiple sclerosis – the shape of things to come. Am J Hum Genet 86, 621–625.

Cruts, M., Van Broeckhoven, C., 1998. Molecular genetics of Alzheimer's disease. Ann Med 30, 560–565.

Gatz, M., Reynolds, C.A., Fratiglioni, L., et al., 2006. Role of genes and environments for explaining Alzheimer disease. Arch Gen Psychiatry 63, 168–174.

Harold, D., Abraham, R., Hollingworth, P., et al., 2009. Genome-wide association study identifies variants at CLU and PICALM associated with Alzheimer's disease. Nat Genet 41, 1088–1093.

Hollingworth, P., Harold, D., Sims, R., et al., 2011. Common variants at ABCA7, MS4A6A/MS4A4E, EPHA1, CD33 and CD2AP are associated with Alzheimer's disease. Nat Genet 43, 429–435.

LaFerla, F.M., Green, K.N., Oddo, S., 2007. Intracellular amyloid-beta in Alzheimer's disease. Nat Rev Neurosci 8, 499–509.

Lambert, J.-C., Heath, S., Even, G., et al., 2009. Genome-wide association study identifies variants at CLU and CR1 associated with Alzheimer's disease. Nat Genet 41, 1094–1099.

Lill, C.M., Bertram, L., 2011. Towards unveiling the genetics of neurodegenerative diseases. Semin Neurol 41 (5), 531–541.

Naj, A.C., Jun, G., Beecham, G.W., et al., 2011. Common variants at MS4A4/MS4A6E, CD2AP, CD33 and EPHA1 are associated with late-onset Alzheimer's disease. Nat Genet 43, 436–441.

Purcell, S.M., Wray, N.R., Stone, J.L., et al., 2009. Common polygenic variation contributes to risk of schizophrenia and bipolar disorder. Nature 460, 748–752.

Rovelet-Lecrux, A., Hannequin, D., Raux, G., et al., 2006. APP locus duplication causes autosomal dominant early-onset Alzheimer disease with cerebral amyloid angiopathy. Nat Genet 38, 24–26.

Strittmatter, W.J., Saunders, A.M., Schmechel, D., et al., 1993. Apolipoprotein E: High-avidity binding to beta-amyloid and increased frequency of type 4 allele in late-onset familial Alzheimer disease. Proc Natl Acad Sci USA 90, 1977–1981.

Sutherland, G.T., Janitz, M., Kril, J.J., 2011. Understanding the pathogenesis of Alzheimer's disease: Will RNA-seq realize the promise of transcriptomics? J Neurochem 116, 937–946.

Tanzi, R.E., Bertram, L., 2005. Twenty years of the Alzheimer's disease amyloid hypothesis: A genetic perspective. Cell 120, 545–555.

Urdinguio, R.G., Sanchez-Mut, J.V., Esteller, M., 2009. Epigenetic mechanisms in neurological diseases: Genes, syndromes, and therapies. Lancet Neurol 8, 1056–1072.

Vance, J.E., Hayashi, H., 2010. Formation and function of apolipoprotein E-containing lipoproteins in the nervous system. Biochim Biophys Acta 1801, 806–818.

RECOMMENDED RESOURCES

1000 Genomes Project
http://www.1000genomes.org/

AD & FTD Mutation database
http://www.molgen.ua.ac.be/admutations/

ALSGene database
http://www.alsgene.org

AlzGene database
http://www.alzgene.org

MSGene database
http://www.msgene.org

PD Mutation database
http://www.molgen.ua.ac.be/PDmutDB/

PDGene database
http://www.pdgene.org

SZGene database
http://www.szgene.org

CHAPTER 86

Parkinson's Disease

Jing Zhang and Tessandra Stewart

INTRODUCTION

Parkinson's disease (PD) is one of the most common neurological disorders, affecting around 1% of the population over the age of 65. It is characterized by motor dysfunction resulting from the progressive loss of nigrostriatal input to the basal ganglia, as well as an array of non-motor symptoms, which likely result from deficits outside the striatum and/or in non-dopaminergic systems (Lang and Obeso, 2004). Familial transmission occurs in a minority of cases, and several gene mutations causing inherited PD have been identified (Bekris et al., 2010). These include mutations in *SNCA*, which codes for α-synuclein (a small protein found in Lewy bodies, which characterize brain pathology of PD); leucine-rich repeat kinase (*LRRK2*), which encodes an apparent tyrosine-kinase like protein; *PARK2*, encoding the E3 ubiquitin ligase parkin; *PINK1*, which encodes a mitochondria-localized kinase believed to function upstream of parkin; and *DJ-1*, which codes for a cytosolic protein that can translocate to mitochondria and appears to have antioxidant activity. However, like many other complex neurological disorders of adult life (see Chapter 84), most PD cases are idiopathic and are believed to result from complex interactions of genetic and environmental risk factors (Horowitz and Greenamyre, 2010; Vance et al., 2010). Although precise mechanisms responsible for PD development remain undefined, the common underlying pathways in sporadic and genetic PD are believed to relate to mitochondrial dysfunction, oxidative stress, and mishandling of protein degradation by an impaired ubiquitin-proteosome system (UPS).

THE NEED FOR RELIABLE PARKINSON'S DISEASE BIOMARKERS

The development of reliable biomarkers for PD is vital for understanding its pathogenesis and for improving diagnostic and predictive criteria. Currently, diagnosis of PD relies on assessment of clinical symptoms, which can overlap significantly with other disorders, including multiple system atrophy (MSA) and progressive supra-nuclear palsy (PSP). Additionally, identification of markers to monitor disease outcome would facilitate personalization of treatment. Of particular interest are measures allowing prediction of disease progression, as this would aid physicians in assessing prognosis and managing the disease. However, there are currently no established biomarkers capable of diagnosing PD patients, predicting or categorizing PD progression, or identifying non-symptomatic patients at risk of developing sporadic PD. Therefore, the development of objective criteria for these goals is a major area of study.

The advent of "omic" technologies provides an ideal opportunity for the development of biomarkers. These

Genomic and Personalized Medicine, 2nd edition
by Ginsburg & Willard. DOI: http://dx.doi.org/10.1016/B978-0-12-382227-7.00086-0

high-throughput techniques have been utilized for investigating genetic variation, gene expression, metabolic profiles, and complement of proteins associated with PD. The next few sections will summarize several important technologies used in biomarker discovery and validation, with an emphasis on the primary functional units of biological processes: proteins.

Genetic Susceptibility

Several genome-wide association studies (GWAS) have made use of high-throughput screening strategies to identify variants in the human genome that signal potential susceptibility factors for PD. Although this type of study was originally somewhat limited in utility due to low reproducibility, loci previously associated with neurodegenerative disease, including *SNCA* (encoding α-synuclein) and *MAPT* (encoding tau), have been identified in patients with a familial history of PD (Pankratz et al., 2008). Remarkably, this result was replicated by later studies (Edwards et al., 2010; Simón-Sánchez et al., 2009), unequivocally implicating variations in *SNCA* and *MAPT* as susceptibility markers for sporadic PD. As newer technologies become available, several groups have started including deeper sequencing, exon profiling, as well as investigations of epigenetics of genes associated with PD. Such studies not only suggest putative etiologic mechanisms, but also reveal novel targets/pathways that can be explored as biomarkers.

Expression Profiling

Profiling studies focusing on changes in the level of mRNA transcripts (see Chapter 12) have implicated biological processes likely to be involved in PD (Altar et al., 2008; Greene, 2010). Studies of transcripts isolated from the Substantia nigra (SN) have found changes in functional groups related to mitochondria/electron transport, protein degradation, synaptic transmission, and cell signaling/transduction (Bossers et al., 2009; Duke et al., 2007; Grünblatt et al., 2004; Mandel et al., 2005; Miller et al., 2006; Moran et al., 2006). Genes associated with dopamine signaling pathways were also found to be altered in some datasets, but studies controlling for the loss of dopaminergic neurons did not demonstrate such findings (Cantuti-Castelvetri et al., 2007; Simunovic et al., 2009), suggesting that this change may simply reflect the specific loss of the neuronal population in which these processes occur. Alterations in pathways important in synaptic transmission, signal transduction, and mitochondrial function, however, were detected, further implicating these pathways. One study (Scherzer et al., 2007) was able to use a transcriptional profile of whole blood to establish a risk marker based on eight genes, and particularly noted the ability of the expression of the heat-shock protein 70-interacting protein ST13 to differentiate between PD and controls. However, a subsequent study could not replicate this result (Shadrina et al., 2010).

Metabolomics

The metabolome is the complement of small molecules such as glucose, cholesterol, adenosine triphosphate (ATP), amine neurotransmitters, lipid signaling molecules, and other molecules present within a system. As their concentrations are dependent on the interactions of other processes (transcription, translation, cellular signaling, etc.), they have been investigated as reporters for metabolism in PD. One study used an unbiased design to find analytes in blood capable of differentiating between PD patients (medicated or unmedicated) and controls (Bogdanov et al., 2008). Among the differing analytes were two related to oxidative stress: levels of glutathione increased, while those of uric acid decreased. Another study comparing plasma metabolomes of controls, patients with idiopathic PD, and symptomatic and non-symptomatic carriers of the *LRRK2* mutation, found that groups could be distinguished from each other, including between idiopathic and *LRRK2* PD patients (Johansen et al., 2009). Interestingly, this study also found decreased uric acid in both PD cohorts compared to controls, with a lesser decrease in asymptomatic *LRRK2* carriers. Although this difference was not significant, it may indicate one early marker of PD.

Proteomics

Changes in mRNA levels reflect differential regulation of transcription, but the number of transcripts correlates only poorly with the cellular level of the corresponding protein (Nelson and Keller, 2007); in contrast, direct measurement of proteins more closely reflects biological function. Protein levels also capture not only disease-causing changes, but compensatory changes, as well as general effects of neurodegeneration, and can therefore potentially provide markers for both disease state and progression. Further, functionality depends not only on the levels of transcription and translation, but also on the posttranslational modifications (PTM), such as phosphorylation or glycosylation, that mature proteins often undergo. These modifications can dramatically alter the activity of proteins, even in the absence of changes in their overall levels, and cannot be detected by expression profiling. Additionally, proteins are more stable than mRNA and metabolites, a major advantage in adapting a biomarker to clinical settings. Therefore, proteomic technologies (see Chapter 13) have been applied to PD for biomarker discovery as well as for mechanistic analysis.

TECHNOLOGY FOR PROTEIN IDENTIFICATION AND VALIDATION

Identification and measurement of proteins in biological samples requires separating proteins, as by two-dimensional polyacrylamide gel electrophoresis (2D-PAGE) or liquid chromatography (LC). 2D-PAGE separates proteins first by their electric charge, then by size, and can be modified by the addition of dye labels to quantify differences between samples (Friedman et al., 2004, 2007). Both isoforms and unrelated proteins are separated, and each spot can be identified and quantified by mass spectrometry (MS) techniques. However, 2D-PAGE is laborious and of limited utility for proteins that migrate together and cannot be well resolved, as found especially in complicated human samples; it is also not useful for very small or highly hydrophobic peptides (Haynes and Yates, 2000; Rabilloud et al., 2010). LC strategies such as multidimensional protein identification technology (MudPIT) separate complex mixtures of peptides without requiring purification of each individually (Link et al., 1999; Washburn et al., 2001). This results in greater dynamic range, since the presence of low abundance proteins will be unmasked by separation from more abundant proteins, thus increasing the number of proteins that can be identified per experiment. However, it is less effective in detecting PTMs and quantitative analysis is not straightforward (see also Chapter 13).

Following separation, protein identification can be carried out using a variety of MS strategies (Aebersold and Mann, 2003). The interpretation of the resulting data relies on bioinformatics tools, making database development and search algorithms for peptide identification a major interest for proteomic fields. Because peptide ionization of individual samples is highly variable between experiments, quantitative comparisons of separately measured samples are challenging. Novel techniques based on the use of stable isotope labeling to tag peptides have been developed to address this difficulty. Samples are labeled and mixed, allowing them to be analyzed together, eliminating uncontrollable variables in the experimental process, and then distinguished based on the ratio of the heavy and light forms of the peptide. Several mechanisms of labeling are currently available, including isobaric tag for relative and absolute quantitation (iTRAQ), isotope-coded affinity tag (ICAT), and ^{18}O labeling. More recently, label-free techniques have also been investigated (Aebersold and Mann, 2003; Anderson, 2005; Pan et al., 2008), although reports of their utilization in neurodegenerative diseases are rather few.

Following identification by unbiased proteomic techniques, the target protein must be validated as a clinically useful marker. This is most commonly achieved using enzyme-linked immunosorbent assays (ELISA) or, more recently, bead-based multi-analyte profiling assays such as Luminex (Olsson et al., 2005). These technologies are notably limited by their dependence on the availability of antibodies directed against the specified proteins. Several groups are therefore developing MS-based targeted proteomic techniques, which depend on isotope labeling followed by addition of a known isotope-labeled synthetic peptide, which serves as a signature marker for identification and quantification of the target peptide (Aebersold and Mann, 2003; Anderson, 2005). These strategies improve identification of target peptides in complex samples, particularly those in low abundance, making them a promising potential source of future clinical applications.

UNBIASED PROFILING USING CEREBROSPINAL FLUID

Discovery of relevant protein biomarkers for clinical use presents a number of technical challenges. Protein alterations will be most apparent at the site of disease pathology, which, for central nervous system (CNS) disease, is in the brain itself. Studies using tissues that can be easily collected at multiple time-points provide greater power and flexibility by allowing longitudinal observation of progression in patients diagnosed at early disease stages. However, studies making use of peripheral fluids (blood, urine, saliva) have met with only limited success, due to the complexity of the fluids and the confounding influence of other organ systems on their composition. Further, the most relevant proteins likely cross the blood–brain barrier in only limited quantities, making them a very small fraction of the total protein content of peripheral fluids; thus, disease-related changes are likely to be masked by the presence of much more abundant proteins. In contrast, cerebrospinal fluid (CSF) is proximal to the site of disease pathology, being in contact with the extracellular space of the brain, and can be collected with minimal risk by lumbar puncture. As such, it is both more likely to contain relevant proteins than other fluids, and obtainable for repeated measurements. Thus, the optimal strategy for effective development of biomarkers for CNS disease may be to characterize disease-related changes in CSF protein levels in small cohorts in a research setting, followed by quantitative analysis of the same markers in peripheral fluids for wider diagnostic application.

By this rationale, several studies have explored disease-related changes in the CSF proteome. In a study using iTRAQ labeling followed by MudPIT and identification by tandem MS (MS/MS), CSF proteins were simultaneously measured in samples from control, PD, Alzheimer's disease (AD), and dementia with Lewy bodies (DLB) patients (Abdi et al., 2006). This strategy yielded not only the isolation of over 1500 proteins present in human CSF, but also identified dozens of quantitative changes specific to each condition, eight of which were confirmed by Western blot. Two of the confirmed proteins, apolipoprotein H/beta-2-glycoprotein 1 and ceruloplasmin, could distinguish between PD and non-PD with high sensitivity and specificity. A later validation study demonstrated the ability of a panel of eight of these CSF proteins to distinguish PD patients from control and AD patients using a bead-based multiplex assay, which is highly adaptable to a clinical setting (Zhang et al., 2008). Other studies of CSF have suggested autotaxin, Pigment epithelium derived factor (PEDF), C4α (Guo et al., 2009), tetranectin (Wang et al., 2010), and neuronal pentraxin receptor (Yin et al., 2009) as potential CSF markers for differentiating between PD and AD.

Similar unbiased techniques can make use of proteomic technology to identify alterations in isoforms of proteins present in CSF. For example, glycosylation is the most common and complex type of PTM, but information on the role of altered glycosylation in PD remains sparse. In order to better understand the role of glycosylation in neurodegenerative disease, efforts at unbiased profiling of the glycoproteome have identified many glycoproteins in the CSF (Hwang et al., 2010; Pan et al., 2006). Further development of this strategy should reveal alterations in this complex type of regulation in neurodegenerative disease.

Despite the increasing number of proteins identified in human CSF, several caveats must be observed in examining data from such studies. It cannot be overemphasized that, among the numerous candidate proteins related to PD and AD in human CSF, only a minority of the CSF proteins identified by proteomic profiling are related to the structure or function of the CNS. Furthermore, as with genomics and metabolomics, the reproducibility of candidate proteins by proteomic profiling is low between groups. There are many factors contributing to this phenomenon, including heterogeneity of human diseases, variations in collection techniques and technology used to identify proteins, and limitations associated with mass spectrometry (variation in ionization and sampling ionized peptides). To address some of these issues, the field has adopted a new strategy: unbiased profiling in brain tissue, with consideration of candidate biomarkers identified by other technologies, followed by targeted analysis of candidate proteins in CSF.

UNBIASED PROFILING USING BRAIN TISSUE

Tissue-based studies have also attempted to characterize protein expression in PD. In a study of substantia nigra pars compacta (SNpc), 2D-PAGE was used to separate proteins from control and PD tissue, identifying 44 proteins, of which nine differed between conditions (Basso et al., 2004). Another 2D-PAGE-based study identified 16 proteins that were differentially expressed, including proteins related to neuronal cell death (ApoD, annexin V), iron metabolism (ferritin L and H), redox [glutathione-S-transferase (GST) Mu3, Pi1, and omega 1], and glial activation (Werner et al., 2008).

Another study used iTRAQ or ICAT to label midbrain samples before identification by MS (Jin et al., 2006; Kitsou et al., 2008). These studies fractionated the samples into subcellular components, to allow deeper analysis of the tissue and detection of changes in relative proportion of the protein in various cellular locations. A total of 1263 proteins were identified by this method, many of which showed quantitative changes in control vs PD tissue. Of particular note, mortalin was further examined in an *in vitro* model of PD, which not only validated this change as a marker of PD, but implicated functional relevance to the disease condition (Jin et al., 2006).

A study comparing frontal cortices of patients at varying stages of disease progression revealed a large number of proteins with altered expression between PD progression groups (Shi et al., 2008). Intriguingly, this included 15 proteins that consistently, and in some cases progressively, decreased in the PD groups, including mortalin, glutathione-S-transferase Mu 3, and excitatory amino acid transporter 2, implying the potential of this technique to reveal not only biomarkers for diagnosis of the disease, but for objectively quantifying its progression. Similarly, when the cortical tissue sample enriched for synaptosomal proteins was examined, several proteins were found to either progressively increase (GSTP1, SH3GL2, ubiquitin

carboxyl-terminal hydrolase isozyme L1) or decrease (CNPase) with greater disease pathology (Shi et al., 2009).

While tissue-based profiling has identified a large number of potential CNS-specific candidate biomarkers, a number of caveats must also be noted. Because human protein databases are incomplete, as with CSF proteomics, biomarkers need to be validated, either by independent investigations or by alternative techniques, before they can be used in clinical settings. Additionally, tissue-based techniques suffer from challenges in sample quality and availability: artificial changes may occur postmortem, and this tissue is typically obtained only at advanced stages of disease. In other words, it remains important to improve proteomics technologies, allowing deeper analysis of the proteome of CSF, which can be sampled repeatedly in living patients.

TARGETED VALIDATION OF CANDIDATE BIOMARKERS

Given the shortcomings associated with "omics" of human diseases specific to the CNS, a complementary strategy for biomarker validation is necessary. The rationale behind targeted validation is to explore the levels of proteins with suspected relevance, based on unbiased discovery or other prior experiments, in relevant tissue or fluid samples. Fortunately, this line of research can benefit tremendously from studies focused on gene products known to be critical to PD pathogenesis.

Alpha-synuclein, the major component of Lewy bodies, has been identified in CSF and plasma (Borghi et al., 2000; El-Agnaf et al., 2003; Li et al., 2007; Tokuda et al., 2006). Several studies have found reduced levels of α-synuclein in CSF from PD patients compared with controls (Hong et al., 2010; Mollenhauer et al., 2008, 2011; Tokuda et al., 2006), including one study in which the reduction in α-synuclein correlated with the severity of parkinsonism (Tokuda et al., 2006). However, other studies have not detected a decrease in CSF α-synuclein (Ohrfelt et al., 2009; Spies et al., 2009). A number of potential factors may explain this discrepancy, including the possibility that different antibody pairs might detect different subsets of the α-synuclein species within a tissue. Variations in CSF contamination by blood may also play a major role; 98% of blood α-synuclein is contained in red blood cells (Shi et al., 2010), a source of the protein which not only dramatically affects apparent levels in blood-contaminated CSF, but also likely explains highly variable results when plasma/serum α-synuclein levels are measured (Lee et al., 2006; Li et al., 2007).

A recent study has also investigated α-synuclein in saliva, where red blood cells are not typically present (Devic et al., 2011), finding that α-synuclein was detectable in saliva and trended toward reduction in PD, but this trend did not reach statistical significance. Another strategy has been to examine plasma and CSF for α-synuclein oligomers. Elevated levels of oligomeric α-synuclein were detected in both plasma and CSF of PD patients (El-Agnaf et al., 2006) and could distinguish between PD and controls (Tokuda et al., 2010). Using the ratio of oligomeric α-synuclein/total α-synuclein improved this measure to a specificity of 89.3% and sensitivity of 90.6%.

Alpha-synuclein also undergoes a number of PTMs, which might be altered in disease conditions. Several PTMs seem to affect the tendency of α-synuclein to aggregate (Oueslati et al., 2010). The primary form of α-synuclein contained in Lewy bodies is phosphorylated at serine-129 (Anderson et al., 2006; Fujiwara et al., 2002), a modification that alters the affinity of α-synuclein for other protein-binding partners (McFarland et al., 2008). In addition, *in vitro* studies suggest aggregation of α-synuclein alters phosphorylation and glycosylation of other proteins (Kulathingal et al., 2009), and wild-type, but not mutant, parkin can ubiquitinate a glycosylated form of α-synuclein (Shimura et al., 2001), highlighting the complexity of its influence on cellular function. If similar effects can be identified in peripheral fluids available for diagnostic procedures, they may prove a valuable indicator of disease status.

Another protein implicated in familial and sporadic PD, DJ-1, has also been examined in CSF and plasma. As with α-synuclein, study results have been inconsistent. One group found that increased levels of DJ-1 in plasma correlated with disease severity (Waragai et al., 2007), and that a similar increase in CSF was more apparent at early disease stages (Waragai et al., 2006). However, other groups found contradictory results, with no differences observed between control and PD serum (Maita et al., 2008), and a decrease in DJ-1 in CSF (Hong et al., 2010). As with α-synuclein, the issues contributing to these discrepancies likely include variations in antibodies used, blood contamination of CSF [95% of blood DJ-1 is contained in red blood cells (Shi et al., 2010)], and heterogeneity of human diseases.

Non-proteomic techniques have also provided putative targets for proteomic validation. Genetic association studies implicated the *MAPT* gene in sporadic PD (Edwards et al., 2010; Pankratz et al., 2008; Simón-Sánchez et al., 2009). Interestingly, tau has been used to differentiate AD from DLB and/or Parkinson's disease with dementia (PDD), as total tau protein is reliably elevated in AD compared to other dementias (Alves et al., 2010; Bibl et al., 2010; Mollenhauer et al., 2011; Parnetti et al., 2008). However, while total tau and phosphorylated tau levels can distinguish between AD and DLB in groups, variability complicates their use in diagnosis of individuals (Mollenhauer et al., 2005; Vanderstichele et al., 2006). Elevated tau and p-tau have been reported in PDD compared to PD without dementia (PD-ND) or normal controls (Compta et al., 2009; Mollenhauer et al., 2005), although another study found only a non-statistically significant difference in tau levels between PDD and PD-ND (Parnetti et al., 2008). Additionally, a number of studies, especially those including larger numbers of control or disease cases, have reported decreased levels of tau species in PD compared to controls (Abdo et al., 2007; Montine et al., 2010; Shi et al., 2011).

Another protein typically associated with AD (see Chapter 85), amyloid-β (Aβ), has also been examined in PD. Levels of Aβ peptide are decreased in CSF of AD patients, and, together with the increased levels of tau, can differentiate between AD and other dementias (Hansson et al., 2006; Shaw et al., 2007). Aβ40 and Aβ42 were reported as reduced in PD compared to controls (Alves et al., 2010), and comparison of controls to PD with and without dementia revealed the highest levels in controls, intermediate levels in PD-ND, and lowest levels in PDD (Compta et al., 2009). Aβ1-42 has been reported as higher in patients with DLB (Gomez-Tortosa et al., 2003), but CSF levels of Aβ1-40 and Aβ1-42 do not discriminate between DLB and AD (Vanderstichele et al., 2006). An oxidized form of Aβ1-40 was increased in DLB (Bibl et al., 2006, 2007). In addition to their potential to serve as clinical biomarkers for PD, these results suggest the fascinating possibility of a role for Aβ and tau in PD pathogenesis. The observation that neurofibrillary tangles and amyloid plaques are not generally found in PD brains has led to the hypothesis that lowered levels of these proteins occur in PD CNS when soluble tau/p-tau and Aβ oligomers, which are often considered more toxic (Haass and Selkoe, 2007), deposit in the brain. This process may in turn affect the tendency of α-synuclein to oligomerize and deposit in Lewy bodies. Therefore, tau and Aβ may play a previously overlooked role in the pathogenesis of PD and should be investigated further.

Inflammatory cytokines are another major category of candidates for targeted validation, on the basis of several lines of evidence suggesting their potential involvement in PD. First, recent GWAS data suggest that human leukocyte antigen (HLA) genes are important in PD (Hamza et al., 2010; Pankratz et al., 2008; Simón-Sánchez et al., 2009). Second, there is repeated confirmation of involvement of inflammatory pathways by various "omics" approaches (Abdi et al., 2006; Duke et al., 2007; Elstner et al., 2011; Grünblatt et al., 2004). Finally, multiple investigations have consistently identified alterations in the levels of many candidate cytokines/chemokines in PD tissue (Boka et al., 1994; Mogi et al., 1994a, b, 1995, 1996a, b) and various body fluids (Blum-Degen et al., 1995; Brodacki et al., 2008; Dobbs et al., 1999; Mogi et al., 1995, 1996a; Rentzos et al., 2009).

A few studies have attempted to determine the diagnostic utility of cytokines as biomarkers. One such study found that the level of flt3 ligand in CSF could differentiate between MSA and PD, while the ratio of fractalkine/Aβ1-42 correlated well with severity of PD symptoms (Shi et al., 2011). Interleukin-6 (IL-6) levels were elevated in CSF of PD patients, and correlated with the severity of symptoms (Müller et al., 1998). Epidermal growth factor (EGF) levels in plasma correlated with cognitive performance, and predicted which patients would convert to PDD (Chen-Plotkin et al., 2010). In a prospective study, elevated IL-6 levels were associated with future development of PD, suggesting possible chronic inflammation preceding PD diagnosis (Chen et al., 2008), although some conflicting studies indicate this association may not be straightforward (Blum-Degen et al., 1995; Hofmann et al., 2009). A panel of 18 cytokines was able to distinguish AD from other neurological conditions, including PD (Ray et al., 2007), lending credence to the possibility of future diagnostic tests based on simple tests of blood cytokines.

FROM DISCOVERY TO VALIDATION

As discussed above, a common problem of all unbiased profiling techniques, including those examining genes, mRNA, proteins, and metabolites, is that individual genes or their products are not usually consistently detected between studies. The candidates with the highest potential for validation as biomarkers are those that have been confirmed, either by alternative techniques or independent detection by other investigators, but these comprise only a small minority of the proteins identified by unbiased studies undertaken to date (Table 86.1). However, these techniques do frequently identify related systems if one considers not individual molecules, but their functional groups. Using this type of analysis, unbiased "omics" strategies consistently implicate systems related to oxidation, inflammation, and protein degradation in PD. Therefore, it makes sense to think of these experiments as a way of determining likely sources of potential targets. Additionally, MS technology suffers from a distinct bias for detection of abundant proteins, making rigorous confirmation of putative biomarkers even more important in driving the transition to useful biomarkers.

Our strategy has been to first identify potential biomarkers in tissue, considering candidate proteins identified in CSF by proteomics as well as other technologies, then apply criteria such as previous implication in the disease mechanism, known expression or role in the CNS, or membership of a functional group of particular interest, to identify high-priority candidates (Figure 86.1). The presence of these proteins, and the disease-related changes suggested by unbiased profiling, must be confirmed in the same tissue by alternative techniques (Western blot, immunohistochemistry) to ensure their biological relevance and correct identification. Next, experimental models can be used to analyze the mechanistic role of the protein in the disease, which also leads to the identification of interacting proteins that may themselves be useful as biomarkers. These candidates can also be pursued in peripheral fluids using targeted approaches, which are not impaired by the problem of their low concentration compared to other components. Finally, relevant changes in disease-related proteins in fluids can be translated to clinical settings, in which they can serve as objective biomarkers.

Our recent studies demonstrate the utility of this strategy. We have performed tissue-based proteomic profiling to identify a large number of proteins differentially expressed between control and PD brains (Jin et al., 2006; Kitsou et al., 2008), or between PD brains with different degrees of disease progression (Pan et al., 2007; Shi et al., 2008, 2009). Expression levels of high-priority candidates including GST-P (Shi et al., 2009) and mortalin (Shi et al., 2008) were confirmed by Western blot to change according to disease condition. Further, the functional significance of both proteins to PD has been demonstrated

TABLE 86.1	High-potential candidates identified in unbiased profiling studies		
Protein	**Sample type**	**Techniques**	**Reference**
ALDH1A1	SN	MS and 2DGE	Jin et al., 2006; Werner et al., 2008
Sorcin A	SN	MS and 2DGE	Jin et al., 2006; Werner et al., 2008
Huntingtin-interacting protein 2	Frontal cortex/blood	MS and microarray	Scherzer et al., 2007; Shi et al., 2008
MAPT, SNCA	–	GWAS	Edwards et al., 2010; Pankrantz et al., 2008; Simón-Sánchez et al., 2009
Mortalin	SNpc	MS and Western blot	Jin et al., 2006
GSTP1, SH3GL2, CNPase	Frontal cortex	MS and Western blot	Shi et al., 2009
ApoC1, ApoH, ceruloplasmin, chromogranin B, VitD binding protein	CSF	MS and Western blot	Abdi et al., 2006

Proteins/gene loci considered high-potential candidates for targeted analysis, based on (1) identification by multiple independent investigations using unbiased study designs, (2) validation by alternative techniques after unbiased discovery, or (3) indication of disease-related changes based on studies from two or more "omic" technologies.

in *in vitro* models following proteomic identification (Jin et al., 2006; Shi et al., 2009). Evaluation of these confirmed candidates as biomarkers in body fluids, such as blood or CSF, is now being undertaken, with confidence as to their relevance.

CONCLUDING REMARKS

Identification of novel biomarkers should improve patient care, accelerate elucidation of the mechanisms of pathogenesis, and aid discovery of therapeutic targets in PD. Targeted and unbiased techniques have revealed a variety of promising potential markers for diagnosing disease and tracking its progression. However, results frequently do not align well between studies. While some of this variability likely results from heterogeneity

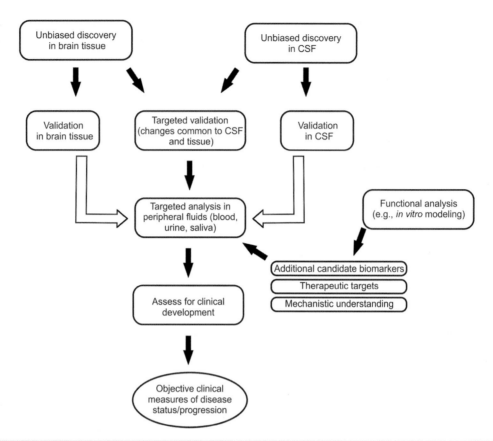

Figure 86.1 Suggested work flow for discovery and validation of protein biomarkers. Unbiased discovery strategies lead to the identification of large numbers of candidate biomarkers. These can be validated by alternative techniques within the same tissue, or by independent identification of the same disease-related changes in protein levels. Changes that occur in both brain tissue and CSF are of particular interest. Validated candidates can then be examined in peripheral fluids, which are easily obtained in a clinical setting. Functional analysis of the roles of proteins (identified by proteomic or other techniques) in disease not only identifies potential therapeutic targets and improves understanding of the mechanistic etiology of the disease, but also identifies interacting proteins that may themselves be useful as biomarkers. These can also be explored by targeted analysis in peripheral fluids. Changes validated in peripheral fluids should then be assessed for clinical utility as biomarkers (i.e., sensitivity and specificity in identifying the disease or predicting progression), leading to the identification of objective clinical biomarkers.

of the disease itself, standardization of sample collection and techniques is of critical importance. As proteomic technologies improve, allowing increasing depth and breadth of protein characterization at decreasing cost, better methods for identifying the most stable and useful markers will be vital. Such advances, along with collaborative experimental efforts involving large samples of patients, should be pursued to manage neurodegenerative diseases in an aging population.

ACKNOWLEDGMENTS

The efforts of Drs Stewart and Zhang are supported by grants from the Institute of National Health (ES004696, NS057567, NS060252, NS062684, AG025327, and AG033398) as well as Shaw Endowment. We also deeply appreciate those who donate their samples for our research.

REFERENCES

Abdi, F., Quinn, J.F., Jankovic, J., et al., 2006. Detection of biomarkers with a multiplex quantitative proteomic platform in cerebrospinal fluid of patients with neurodegenerative disorders. J Alzheimer's Dis 9, 293–348.

Abdo, W.F., Bloem, B.R., Van Geel, W.J., Esselink, R.A.J., Verbeek, M.M., 2007. CSF neurofilament light chain and tau differentiate multiple system atrophy from Parkinson's disease. Neurobiol Aging 28, 742–747.

Aebersold, R., Mann, M., 2003. Mass spectrometry-based proteomics. Nature 422, 198–207.

Altar, C.A., Vawter, M.P., Ginsberg, S.D., 2008. Target Identification for CNS diseases by transcriptional profiling. Neuropsychopharmacology 34, 18–54.

Alves, G., Brønnick, K., Aarsland, D., et al., 2010. CSF amyloid-beta and tau proteins, and cognitive performance, in early and untreated Parkinson's disease: The Norwegian ParkWest study. J Neurol Neurosurg Psychiatry 81, 1080–1086.

Anderson, J.P., Walker, D.E., Goldstein, J.M., et al., 2006. Phosphorylation of Ser-129 is the dominant pathological modification of alpha-synuclein in familial and sporadic Lewy body disease. J Biol Chem 281, 29,739–29,752.

Anderson, L., 2005. Candidate-based proteomics in the search for biomarkers of cardiovascular disease. J Physiol 563, 23–60.

Basso, M., Giraudo, S., Corpillo, D., et al., 2004. Proteome analysis of human substantia nigra in Parkinson's disease. Proteomics 4, 3943–3952.

Bekris, L.M., Mata, I.F., Zabetian, C.P., 2010. The genetics of Parkinson disease. J Geriatr Psychiatry Neurol 23, 228–242.

Bibl, M., Esselmann, H., Lewczuk, P., et al., 2010. Combined analysis of CSF tau, Aβ42, Aβ1-42% and Aβ1-40% in Alzheimer's disease, dementia with Lewy bodies and Parkinson's disease dementia. Int J Alzheimer's Dis.

Bibl, M., Mollenhauer, B., Esselmann, H., et al., 2006. CSF amyloid-beta-peptides in Alzheimer's disease, dementia with Lewy bodies and Parkinson's disease dementia. Brain 129, 1177–1187.

Bibl, M., Mollenhauer, B., Lewczuk, P., et al., 2007. Validation of amyloid-β peptides in CSF diagnosis of neurodegenerative dementias. Mol Psychiatry 12, 671–680.

Blum-Degen, D., Müller, T., Kuhn, W., et al., 1995. Interleukin-1 beta and interleukin-6 are elevated in the cerebrospinal fluid of Alzheimer's and de novo Parkinson's disease patients. Neurosci Lett 202, 17–20.

Bogdanov, M., Matson, W.R., Wang, L., et al., 2008. Metabolomic profiling to develop blood biomarkers for Parkinson's disease. Brain 131, 389–396.

Boka, G., Anglade, P., Wallach, D., et al., 1994. Immunocytochemical analysis of tumor necrosis factor and its receptors in Parkinson's disease. Neurosci Lett 172, 151–154.

Borghi, R., Marchese, R., Negro, A., et al., 2000. Full length alpha-synuclein is present in cerebrospinal fluid from Parkinson's disease and normal subjects. Neurosci Lett 287, 65–67.

Bossers, K., Meerhoff, G., Balesar, R., et al., 2009. Analysis of gene expression in Parkinson's disease: Possible involvement of neurotrophic support and axon guidance in dopaminergic cell death. Brain Pathol 19, 91–107.

Brodacki, B., Staszewski, J., Toczylowska, B., et al., 2008. Serum interleukin (IL-2, IL-10, IL-6, IL-4), TNFα, and INFγ concentrations are elevated in patients with atypical and idiopathic parkinsonism. Neurosci Lett 441, 158–162.

Cantuti-Castelvetri, I., Keller-McGandy, C., Bouzou, B., et al., 2007. Effects of gender on nigral gene expression and Parkinson disease. Neurobiol Dis 26, 606–614.

Chen, H., O'Reilly, E.J., Schwarzschild, M.A., Ascherio, A., 2008. Peripheral inflammatory biomarkers and risk of Parkinson's disease. Am J Epidemiol 167, 90–95.

Chen-Plotkin, A.S., Hu, W.T., Siderowf, A., et al., 2010. Plasma epidermal growth factor levels predict cognitive decline in Parkinson disease. Ann Neurol 69, doi: 10.1002/ana.22271.

Compta, Y., Martí, M.J., Ibarretxe-Bilbao, N., et al., 2009. Cerebrospinal tau, phospho-tau, and beta-amyloid and neuropsychological functions in Parkinson's disease. Mov Disord 24, 2203–2210.

Devic, I., Hwang, H., Edgar, J.S., et al., 2011. Salivary α-synuclein and DJ-1: Potential biomarkers for Parkinson's disease. Brain 134, doi: 10.1093/brain/awr015.

Dobbs, R.J., Charlett, A., Purkiss, A.G., et al., 1999. Association of circulating TNF-alpha and IL-6 with ageing and parkinsonism. Acta Neurol Scand 100, 34–41.

Duke, D.C., Moran, L.B., Pearce, R.K.B., Graeber, M.B., 2007. The medial and lateral substantia nigra in Parkinson's disease: mRNA profiles associated with higher brain tissue vulnerability. Neurogenetics 8, 83–94.

Edwards, T.L., Scott, W.K., Almonte, C., et al., 2010. Genome-wide association study confirms SNPs in SNCA and the MAPT region as common risk factors for Parkinson disease. Ann Hum Genet 74, 97–109.

El-Agnaf, O.M.A., Salem, S.A., Paleologou, K.E., et al., 2003. Alpha-synuclein implicated in Parkinson's disease is present in extracellular biological fluids, including human plasma. FASEB J 17, 1945–1947.

El-Agnaf, O.M.A., Salem, S.A., Paleologou, K.E., et al., 2006. Detection of oligomeric forms of alpha-synuclein protein in human plasma as a potential biomarker for Parkinson's disease. FASEB J 20, 419–425.

Elstner, M., Morris, C.M., Heim, K., et al., 2011. Expression analysis of dopaminergic neurons in Parkinson's disease and aging links transcriptional dysregulation of energy metabolism to cell death. Acta Neuropathol 122, doi: 10.1007/s00401-011-0828-9.

Friedman, D.B., Hill, S., Keller, J.W., et al., 2004. Proteome analysis of human colon cancer by two-dimensional difference gel electrophoresis and mass spectrometry. Proteomics 4, 793–811.

Friedman, D.B., Wang, S.E., Whitwell, C.W., Caprioli, R.M., Arteaga, C.L., 2007. Multivariable difference gel electrophoresis and mass spectrometry. Mol Cell Proteomics 6, 150–169.

Fujiwara, H., Hasegawa, M., Dohmae, N., et al., 2002. α-synuclein is phosphorylated in synucleinopathy lesions. Nat Cell Biol 4, 160–164.

Gomez-Tortosa, E., Gonzalo, I., Fanjul, S., et al., 2003. Cerebrospinal fluid markers in dementia with Lewy bodies compared with Alzheimer disease. Arch Neurol 60, 1218–1222.

Greene, J.G., 2010. Current status and future directions of gene expression profiling in Parkinson's disease. Neurobiol Dis 45, doi: 10.1016/j.nbd.2010.10.022.

Grünblatt, E., Mandel, S., Jacob-Hirsch, J., et al., 2004. Gene expression profiling of parkinsonian substantia nigra pars compacta; alterations in ubiquitin-proteasome, heat shock protein, iron and oxidative stress regulated proteins, cell adhesion/cellular matrix and vesicle trafficking genes. J Neural Transm 111, 1543–1573.

Guo, J., Sun, Z., Xiao, S., et al., 2009. Proteomic analysis of the cerebrospinal fluid of Parkinson's disease patients. Cell Res 19, 1401–1403.

Haass, C., Selkoe, D.J., 2007. Soluble protein oligomers in neurodegeneration: Lessons from the Alzheimer's amyloid beta-peptide. Nat Rev Mol Cell Biol 8, 101–112.

Hamza, T.H., Zabetian, C.P., Tenesa, A., et al., 2010. Common genetic variation in the HLA region is associated with late-onset sporadic Parkinson's disease. Nat Genet 42, 781–785.

Hansson, O., Zetterberg, H., Buchhave, P., et al., 2006. Association between CSF biomarkers and incipient Alzheimer's disease in patients with mild cognitive impairment: A follow-up study. Lancet Neurol 5, 228–234.

Haynes, P.A., Yates III, J.R., 2000. Proteome profiling – pitfalls and progress. Yeast 17, 81–87.

Hofmann, K.W., Schuh, A.F.S., Saute, J., et al., 2009. Interleukin-6 serum levels in patients with Parkinson's disease. Neurochem Res 34, 1401–1404.

Hong, Z., Shi, M., Chung, K.A., et al., 2010. DJ-1 and alpha-synuclein in human cerebrospinal fluid as biomarkers of Parkinson's disease. Brain 133, 713–726.

Horowitz, M.P., Greenamyre, J.T., 2010. Gene–environment interactions in Parkinson's disease: The importance of animal modeling. Clin Pharmacol Ther 88, 467–474.

Hwang, H., Zhang, J., Chung, K.A., et al., 2010. Glycoproteomics in neurodegenerative diseases. Mass Spectrom Rev 29, 79–125.

Jin, J., Hulette, C., Wang, Y., et al., 2006. Proteomic identification of a stress protein, mortalin/mthsp70/GRP75: Relevance to Parkinson disease. Mol Cell Proteomics 5, 1193–1204.

Johansen, K.K., Wang, L., Aasly, J.O., et al., 2009. Metabolomic profiling in LRRK2-related Parkinson's disease. PLoS One 4, e7551.

Kitsou, E., Pan, S., Zhang, J., et al., 2008. Identification of proteins in human substantia nigra. Proteomics Clin Appl 2, 776–782.

Kulathingal, J., Ko, L.-W., Cusack, B., Yen, S.-H., 2009. Proteomic profiling of phosphoproteins and glycoproteins responsive to wild-type alpha-synuclein accumulation and aggregation. Biochim Biophys Acta 1794, 211–224.

Lang, A.E., Obeso, J.A., 2004. Challenges in Parkinson's disease: Restoration of the nigrostriatal dopamine system is not enough. Lancet Neurol 3, 309–316.

Lee, P.H., Lee, G., Park, H.J., Bang, O.Y., Joo, I.S., Huh, K., 2006. The plasma alpha-synuclein levels in patients with Parkinson's disease and multiple system atrophy. J Neural Transm 113, 1435–1439.

Li, Q.-X., Mok, S.S., Laughton, K.M., et al., 2007. Plasma alpha-synuclein is decreased in subjects with Parkinson's disease. Exp Neurol 204, 583–588.

Link, A.J., Eng, J., Schieltz, D.M., et al., 1999. Direct analysis of protein complexes using mass spectrometry. Nat Biotechnol 17, 676–682.

Maita, C., Tsuji, S., Yabe, I., et al., 2008. Secretion of DJ-1 into the serum of patients with Parkinson's disease. Neurosci Lett 431, 86–89.

Mandel, S., Grunblatt, E., Riederer, P., et al., 2005. Gene expression profiling of sporadic Parkinson's disease substantia nigra pars compacta reveals impairment of ubiquitin-proteasome subunits, SKP1A, aldehyde dehydrogenase, and chaperone HSC-70. Ann NY Acad Sci 1053, 356–375.

McFarland, M.A., Ellis, C.E., Markey, S.P., Nussbaum, R.L., 2008. Proteomics analysis identifies phosphorylation-dependent alpha-synuclein protein interactions. Mol Cell Proteomics 7, 2123–2137.

Miller, R.M., Kiser, G.L., Kaysser-Kranich, T.M., et al., 2006. Robust dysregulation of gene expression in substantia nigra and striatum in Parkinson's disease. Neurobiol Dis 21, 305–313.

Mogi, M., Harada, M., Kondo, T., et al., 1994a. Interleukin-1 beta, interleukin-6, epidermal growth factor and transforming growth factor-alpha are elevated in the brain from parkinsonian patients. Neurosci Lett 180, 147–150.

Mogi, M., Harada, M., Kondo, T., et al., 1995. Transforming growth factor-beta 1 levels are elevated in the striatum and in ventricular cerebrospinal fluid in Parkinson's disease. Neurosci Lett 193, 129–132.

Mogi, M., Harada, M., Kondo, T., Riederer, P., Nagatsu, T., 1996a. Interleukin-2 but not basic fibroblast growth factor is elevated in parkinsonian brain. J Neural Transm 103, 1077–1081.

Mogi, M., Harada, M., Narabayashi, H., et al., 1996b. Interleukin (IL)-1 beta, IL-2, IL-4, IL-6 and transforming growth factor-alpha levels are elevated in ventricular cerebrospinal fluid in juvenile parkinsonism and Parkinson's disease. Neurosci Lett 211, 13–16.

Mogi, M., Harada, M., Riederer, P., et al., 1994b. Tumor necrosis factor-alpha (TNF-alpha) increases both in the brain and in the cerebrospinal fluid from parkinsonian patients. Neurosci Lett 165, 208–210.

Mollenhauer, B., Cepek, L., Bibl, M., et al., 2005. Tau protein, Abeta42 and S-100B protein in cerebrospinal fluid of patients with dementia with Lewy bodies. Dement Geriatr Cogn Disord 19, 164–170.

Mollenhauer, B., Cullen, V., Kahn, I., et al., 2008. Direct quantification of CSF alpha-synuclein by ELISA and first cross-sectional study in patients with neurodegeneration. Exp Neurol 213, 315–325.

Mollenhauer, B., Locascio, J.J., Schulz-Schaeffer, W., et al., 2011. α-Synuclein and tau concentrations in cerebrospinal fluid of patients presenting with parkinsonism: A cohort study. Lancet Neurol 10, 230–240.

Montine, T.J., Shi, M., Quinn, J.F., et al., 2010. CSF Aβ(42) and tau in Parkinson's disease with cognitive impairment. Mov Disord 25, 2682–2685.

Moran, L.B., Duke, D.C., Deprez, M., et al., 2006. Whole genome expression profiling of the medial and lateral substantia nigra in Parkinson's disease. Neurogenetics 7, 1–11.

Müller, T., Blum-Degen, D., Przuntek, H., Kuhn, W., 1998. Interleukin-6 levels in cerebrospinal fluid inversely correlate to severity of Parkinson's disease. Acta Neurol Scand 98, 142–144.

Nelson, P.T., Keller, J.N., 2007. RNA in brain disease. J Neuropathol Exp Neurol 66, 461–468.

Ohrfelt, A., Grognet, P., Andreasen, N., et al., 2009. Cerebrospinal fluid alpha-synuclein in neurodegenerative disorders – a marker of synapse loss? Neurosci Lett 450, 332–335.

Olsson, A., Vanderstichele, H., Andreasen, N., et al., 2005. Simultaneous measurement of β-amyloid(1-42), total tau, and phosphorylated tau (Thr181) in cerebrospinal fluid by the xMAP technology. Clin Chem 51, 336–345.

Oueslati, A., Fournier, M., Lashuel, H.A., 2010. Role of post-translational modifications in modulating the structure, function and toxicity of alpha-synuclein: Implications for Parkinson's disease pathogenesis and therapies. Prog Brain Res 183, 115–145.

Pan, S., Rush, J., Peskind, E.R., et al., 2008. Application of targeted quantitative proteomics analysis in human cerebrospinal fluid using a liquid chromatography matrix-assisted laser desorption/ionization time-of-flight tandem mass spectrometer (LC MALDI TOF/TOF) platform. J Proteome Res 7, 720–730.

Pan, S., Wang, Y., Quinn, J.F., et al., 2006. Identification of glycoproteins in human cerebrospinal fluid with a complementary proteomic approach. J Proteome Res 5, 2769–2779.

Pankratz, N., Wilk, J.B., Latourelle, J.C., et al., 2008. The PSG-PROGENI and GenePD investigators, coordinators and molecular genetic laboratories. Genomewide association study for susceptibility genes contributing to familial Parkinson disease. Hum Genet 124, 593–605.

Parnetti, L., Tiraboschi, P., Lanari, A., et al., 2008. Cerebrospinal fluid biomarkers in Parkinson's disease with dementia and dementia with Lewy bodies. Biol Psychiatry 64, 850–855.

Rabilloud, T., Chevallet, M., Luche, S., Lelong, C., 2010. Two-dimensional gel electrophoresis in proteomics: Past, present and future. J Proteomics 73, 2064–2077.

Ray, S., Britschgi, M., Herbert, C., et al., 2007. Classification and prediction of clinical Alzheimer's diagnosis based on plasma signaling proteins. Nat Med 13, 1359–1362.

Rentzos, M., Nikolaou, C., Andreadou, E., et al., 2009. Circulating interleukin-10 and interleukin-12 in Parkinson's disease. Acta Neurol Scand 119, 332–337.

Scherzer, C.R., Eklund, A.C., Morse, L.J., et al., 2007. Molecular markers of early Parkinson's disease based on gene expression in blood. Proc Natl Acad Sci USA 104, 955–960.

Shadrina, M.I., Filatova, E.V., Karabanov, A.V., et al., 2010. Expression analysis of suppression of tumorigenicity 13 gene in patients with Parkinson's disease. Neurosci Lett 473, 257–259.

Shaw, L.M., Korecka, M., Clark, C.M., Lee, V.M.-Y., Trojanowski, J.Q., 2007. Biomarkers of neurodegeneration for diagnosis and monitoring therapeutics. Nat Rev Drug Discov 6, 295–303.

Shi, M., Bradner, J., Bammler, T.K., et al., 2009. Identification of glutathione S-transferase pi as a protein involved in Parkinson disease progression. Am J Pathol 175, 54–65.

Shi, M., Bradner, J., Hancock, A.M., et al., 2011. Cerebrospinal fluid biomarkers for Parkinson disease diagnosis and progression. Ann Neurol 69, 570–580.

Shi, M., Jin, J., Wang, Y., et al., 2008. Mortalin: A protein associated with progression of Parkinson disease? J Neuropathol Exp Neurol 67, 117–124.

Shi, M., Zabetian, C.P., Hancock, A.M., et al., 2010. Significance and confounders of peripheral DJ-1 and alpha-synuclein in Parkinson's disease. Neurosci Lett 480, 78–82.

Shimura, H., Schlossmacher, M.G., Hattori, N., et al., 2001. Ubiquitination of a new form of α-synuclein by parkin from human brain: Implications for Parkinson's disease. Science 293, 263–269.

Simón-Sánchez, J., Schulte, C., Bras, J.M., et al., 2009. Genome-wide association study reveals genetic risk underlying Parkinson's disease. Nat Genet 41, 1308–1312.

Simunovic, F., Yi, M., Wang, Y., et al., 2009. Gene expression profiling of substantia nigra dopamine neurons: Further insights into Parkinson's disease pathology. Brain 132, 1795–1809.

Spies, P.E., Melis, R.J.F., Sjögren, M.J.C., Rikkert, M.G.M.O., Verbeek, M.M., 2009. Cerebrospinal fluid alpha-synuclein does not discriminate between dementia disorders. J Alzheimer's Dis 16, 363–369.

Tokuda, T., Qureshi, M.M., Ardah, M.T., et al., 2010. Detection of elevated levels of α-synuclein oligomers in CSF from patients with Parkinson disease. Neurology 75, 1766–1772.

Tokuda, T., Salem, S.A., Allsop, D., et al., 2006. Decreased alpha-synuclein in cerebrospinal fluid of aged individuals and subjects with Parkinson's disease. Biochem Biophys Res Commun 349, 162–166.

Vance, J.M., Ali, S., Bradley, W.G., Singer, C., Di Monte, D.A., 2010. Gene–environment interactions in Parkinson's disease and other forms of parkinsonism. Neurotoxicology 31, 598–602.

Vanderstichele, H., De Vreese, K., Blennow, K., et al., 2006. Analytical performance and clinical utility of the INNOTEST PHOSPHO-TAU181P assay for discrimination between Alzheimer's disease and dementia with Lewy bodies. Clin Chem Lab Med 44, 1472–1480.

Wang, E.-S., Sun, Y., Guo, J.-G., et al., 2010. Tetranectin and apolipoprotein A-I in cerebrospinal fluid as potential biomarkers for Parkinson's disease. Acta Neurol Scand 122, 350–359.

Waragai, M., Nakai, M., Wei, J., et al., 2007. Plasma levels of DJ-1 as a possible marker for progression of sporadic Parkinson's disease. Neurosci Lett 425, 18–22.

Waragai, M., Wei, J., Fujita, M., et al., 2006. Increased level of DJ-1 in the cerebrospinal fluids of sporadic Parkinson's disease. Biochem Biophys Res Commun 345, 967–972.

Washburn, M.P., Wolters, D., Yates III, J.R., 2001. Large-scale analysis of the yeast proteome by multidimensional protein identification technology. Nat Biotechnol 19, 242–247.

Werner, C.J., Heyny-von Haussen, R., Mall, G., Wolf, S., 2008. Proteome analysis of human substantia nigra in Parkinson's disease. Proteome Sci 6, 8.

Yin, G.N., Lee, H.W., Cho, J.-Y., Suk, K., 2009. Neuronal pentraxin receptor in cerebrospinal fluid as a potential biomarker for neurodegenerative diseases. Brain Res 1265, 158–170.

Zhang, J., Sokal, I., Peskind, E.R., et al., 2008. CSF multianalyte profile distinguishes Alzheimer and Parkinson diseases. Am J Clin Pathol 129, 526–529.

CHAPTER

Epilepsy

Sanjay M. Sisodiya

INTRODUCTION

Epilepsy: What Is It and Why Is It Important?

One in twenty people will have a single epileptic seizure at some time in their life (Duncan et al., 2006). Epilepsy is the tendency to have recurrent spontaneous seizures. It is estimated that, around the world, some 50 million people have epilepsy, making it amongst the most common serious chronic neurological diseases. Epilepsy can be a burdensome condition. For a given individual with epilepsy, its consequences may pervade every domain, from the anxiety of unpredictable seizures, to the injurious physical consequences of individual seizures, the cumulative cerebral damage from recurrent seizures, the neurological and somatic results of long-term antiepileptic drug exposure, the risk of altered fertility and teratogenicity, and reduced lifespan. Loss of driving privileges, employment options, and earnings may ensue. Rates of psychiatric illness and suicide are increased (Gaitatzis et al., 2004); somatic comorbidity is probably underestimated. Socioeconomic deprivation is associated with increased incidence of epilepsy, even in developed economies (Heaney et al., 2002).

About 60% of people with epilepsy live in resource-poor countries, with little access to healthcare and higher rates of associated conditions such as neurocysticercosis and cerebral malaria. Even in the burgeoning economy of China, the standardized mortality ratio for young people in the rural majority exceeds 23 (Ding et al., 2006). The societal burden of epilepsy is large. The cost to the European economy has been estimated at over 15 billion euros a year (Pugliatti et al., 2007). In the US, a recent study suggested that the total cost for employees with epilepsy was, on average, over $8000 per annum more than the cost of employees without epilepsy (Ivanova et al., 2010).

Additional impacts are becoming understood. Epilepsy is associated with an increased risk of drowning, especially when epilepsy is uncontrolled (Bell et al., 2008). There is an increased risk of sudden unexplained death, which may rise to a risk of 1 in 75 per annum in the most vulnerable people with epilepsy (Bell et al., 2010). There is also an increased risk of certain cancers (Singh et al., 2009). Most of these associations are unexplained.

While the availability of antiepileptic drugs (AEDs) may be limited in countries with low per capita GDP, even for cheaper, older agents such as phenobarbitone, many AEDs are now available in developed economies. In fact, most patients (about two-thirds) respond well to the first or second AED tried (Kwan and Brodie, 2000). Control of seizures is the best means of countering the burdens associated with epilepsy. But even for those who do respond, use-limiting side effects may occur, significant adverse reactions may emerge, and it can take time to find the right dose of the right drug, during which interval the dangers associated with seizures may still manifest (Kasperaviciute and Sisodiya, 2009). For children, there are added concerns about cognitive, behavioral, and psychosocial development (Camfield and Camfield, 2007). For women of child-bearing age, teratogenicity is a major concern (Hill et al., 2010). Surrounding all these personal difficulties is the stigma that epilepsy still attracts (Jacoby et al., 2009).

Genomic and Personalized Medicine, 2nd edition
by Ginsburg & Willard. DOI: http://dx.doi.org/10.1016/B978-0-12-382227-7.00087-2

For all these reasons, epilepsy is an important condition that still needs much research. We lack understanding to the extent that we cannot explain in any single individual why a seizure occurs at a given precise time, nor can we explain why one seizure may sometimes evolve to secondary generalization while another in the same patient does not; we also cannot explain why an AED may work in one patient with a given epilepsy but not in another with apparently the same epilepsy. We do not understand why one in three people with epilepsy continue to have seizures whichever AED is prescribed; these are the people at greatest risk of the negative impact of epilepsy. There is therefore much to be learned still about causation, manifestations, consequences, and management of epilepsy. Clean, effective new treatments are required, and prevention seems further off still. Current approaches, especially to drug development, are proving somewhat sterile, largely generating agents with the same apparent target profiles. New perspectives are urgently required.

The human brain can only react to disease in a number of ways – loss of function, coma, and seizures are three examples. From a basic science perspective, understanding epilepsy can also reveal much about normal human brain function, including cognition, memory, degeneration, and post-natal human neurogenesis. These are additional powerful reasons for the study of epilepsy.

WHY GENOMIC AND PERSONALIZED MEDICINE IN EPILEPSY?

Epilepsy is not a single condition. There are many reasons for the brain dysfunction that causes seizures. There are many different types of seizure and many different epilepsies. In fact, "the epilepsies" is often a better term than "epilepsy," reflecting this inherent heterogeneity. Our comparatively limited understanding of the epilepsies is reflected in ongoing attempts at its organization (Berg et al., 2010), and the often spirited debate any classification effort provokes. Syndromes (or "constellations") (Berg et al., 2010) are clearly recognized: some are phenotypically more homogeneous (e.g., Dravet syndrome, see later section), but even these may be genetically heterogeneous. Other syndromes are less homogeneous, but an argument can readily be made, especially in the clinical context, for definition of such syndromes. Mesial temporal lobe epilepsy with hippocampal sclerosis is one such example (Wieser, 2004). How unified a condition this is has yet to be completely established; neither its heritability nor its genetic landscape has been defined.

Heterogeneity extends to many aspects of the epilepsies. Causation, susceptibility, clinical manifestation, natural history, cognitive consequences, long-term outcome, treatment response, and adverse effects of treatment are all variable. Two patients with mesial temporal lobe epilepsy with hippocampal sclerosis may show widely differing outcomes with respect to effectiveness of AEDs and loss of memory, for example.

In routine clinical practice, variation in treatment response is the most commonly encountered unpredictability. If trivial causes of intra- or inter-individual variation in response, such as non-compliance, co-medication, or alcohol consumption, are excluded, the mechanisms underlying variable response are largely unknown for most AEDs. Clinical trial regulatory structure prevents formal studies of new AEDs against placebo alone in patients with new onset epilepsy; studies are almost always undertaken comparing a new AED against placebo in patients with chronic epilepsy on other AEDs. Response profiles to the new AED alone can only be approximated from attempts to decrease treatment to monotherapy, but this automatically selects good responders. Thus data on the true variation in response to a single AED across a population of patients are often not available. On the whole, published studies show that for any new AED, the rate of achievement of seizure freedom – the gold standard – is small, typically a few percent (Bialer et al., 2004). There is inevitable clinical heterogeneity in patients studied, but for greater external validity, this is desirable (Sisodiya, 2007). If one considers any AED and patients being newly treated for their epilepsy, ~50% will respond well to the first AED and a further ~16% to the second AED (Kwan and Brodie, 2000). These observations are all unexplained. Heritability studies of AED responsiveness have not been undertaken, but it must be a possibility that individual genomic variation contributes to phenotypic drug response variation.

The Example of Dravet Syndrome

Dravet syndrome (DS) is a distinctive epileptic syndrome, characterized by acute onset, often with *status epilepticus*, in the first 6–18 months of life. It is characterized by multiple seizure types that are resistant to drug treatment, in association with cognitive impairment that is often marked (Mullen and Scheffer, 2009). It is considered to be an epileptic encephalopathy, with seizures themselves being largely blamed for cognitive, and other, dysfunction. Premature mortality is common, while survivors are often entirely or heavily dependent on others for all daily activities. Seizure-freedom is rare, and some AEDs may actually aggravate the seizures (Guerrini et al., 1998).

The finding that DS is usually (70–80% of cases) caused by heterozygous mutation in the gene *SCN1A* has proven to be a landmark discovery (see review by Meisler et al., 2010). Other genes involved in DS include *SCN1B* (Patino et al., 2009), *GABRG2* (Harkin et al., 2002), and *SCN2A* (Kamiya et al., 2004; Shi et al., 2009). *PCDH19* mutations (Depienne et al., 2009) and deletions involving the chromosome 2q *SCN* cluster (Davidsson et al., 2008; Lossin, 2009; Meisler et al., 2010; Pereira et al., 2004) have also been reported in DS-like syndromes. DS therefore illustrates well the genetic heterogeneity that can be seen even in a rare distinctive epilepsy.

It is, however, the *SCN1A* discovery and the research that has followed that, in epilepsy, best illustrate the power of personalized medicine and the contribution of genetics and genomics to that end. Characterization of the syndrome allowed genetic discovery, which in turn enabled the generation of animal models. Mice heterozygous for either null (Yu et al., 2006) or truncating (Ogiwara et al., 2007) *Scn1a* alleles have severe phenotypes including seizures, ataxia, and premature death; reduced sodium currents in hippocampal and cortical GABAergic interneurons, but not pyramidal neurons, lead to altered firing patterns and hyperexcitability (Martin et al., 2010; Tang et al., 2009; Yu et al., 2006). Abnormal sodium flux in cerebellar Purkinje cells may underlie ataxia (Yu et al., 2006). Further parallels with DS include sensitivity to body temperature elevation, causing seizures and interictal epileptiform discharges (Oakley et al., 2009). The imbalance between inhibitory interneuronal hypofunction and preserved pyramidal neuronal function could, in theory, be exacerbated by sodium-channel acting (blocking) AEDs. Indeed, this is directly observed in clinical practice (Guerrini et al., 1998).

Conversely, agents that promote inhibitory activity should improve seizure control; the agents in clinical practice that are most effective are valproate, benzodiazepines, topiramate, levetiracetam, and stiripentol, which all have some pro-GABAergic activity (Fisher, 2009). In clinical practice, this understanding makes a tangible difference. For example, even after decades of uncontrolled seizures, withdrawal of a sodium channel-blocking AED and institution of levetiracetam led to a significant improvement in seizure control and cognitive function. This finding was greatly facilitated by recognition of the syndromic diagnosis and its molecular confirmation, which was valuable in view of the patient's advanced age in comparison with most patients with DS (Catarino et al., 2011). Syndrome-specific treatment can thus be based on more rational grounds than has often been possible, DS being one of the best understood epilepsies.

Even more personalized therapy may become possible with increasing understanding of the consequences of particular *SCN1A* mutations. Anecdotally, individual clinicians report particular patients with DS who appear to require phenytoin to have any reasonable degree of seizure control, a finding that appears counter-intuitive given the known molecular pathophysiology of DS above. Rusconi and colleagues (2007) showed, however, that in an *in vitro* model of an *SCN1A* mutation capable of causing DS (Annesi et al., 2003), phenytoin was capable of rescuing the trafficking-defective mutant protein. For some personal *SCN1A* mutations on an individual genetic background, therefore, phenytoin may conceivably be effective, despite its sodium channel-blocking effect. This idea has not been proven, but the methodology now exists to test this hypothesis. DS, a rare genetic epilepsy, has thus considerably enhanced our knowledge of epileptogenic mechanisms and led to direct treatment benefits for patients. It is widely anticipated that an imminent spate of genetic discoveries in the epilepsies will have similar consequences for many individuals with epilepsy, although this is yet to come. The rest of this chapter focuses on existing genomic and pharmacogenomic knowledge with implications for individuals with epilepsy.

GENOMICS IN EPILEPSY: WHAT DO WE KNOW?

Most progress in the epilepsies has been in specific epilepsies, where the effect of change in a single gene is marked (Reid et al., 2009). These epilepsies include a number of familial epilepsies (Baulac and Baulac, 2009; Helbig et al., 2008), those associated with brain malformations (Guerrini and Parrini, 2010), and several progressive myoclonic epilepsies (Ramachandran et al., 2009). There are also several hundred chromosomal rearrangements, additions, and deletions that produce syndromes with mental retardation and/or multiple congenital anomalies with epilepsy as an additional manifestation. These conditions have been well documented and thoroughly reviewed.

In all these cases, establishing a clinical diagnosis depends in the first place on clinical acumen. A definitive genetic diagnosis is critical and:

- Reveals a cause for an individual's problems, which is often a source of relief in itself
- Prevents further unnecessary, wasteful, and potentially unpleasant investigations
- May instigate counseling
- May point to specific additional screening tests
- May influence treatment choices and inform discussion about prognosis.

For each individual, a genetic diagnosis must be integrated into a holistic view not only of the epilepsy [as magnetic resonance imaging (MRI), electroencephalogram (EEG) data are integrated], but also of the whole person. The physician's perspective and that of the person with epilepsy may not coincide.

Causation

Few people with epilepsy have more than one affected relative (12–15% depending on the type of epilepsy). Very few people have a strong family history with what could be considered a Mendelian pattern of inheritance (Ottman et al., 1997). In some cases, thoughtful consideration of variable expressivity may prove important. But for the majority of individuals, epilepsy appears sporadic, or there are only distantly related affected individuals, without a clear pattern of inheritance. Epilepsy has thus been considered a "complex" trait. There are a few examples of oligogenic inheritance, or evidence to think it may occur (Mefford et al., 2010; Thomas et al., 2009), or modifier variants (Singh et al., 2009). One animal model demonstrates a balanced effect between a pro-epileptogenic variant and an antiepileptogenic variant (Meisler et al., 2010), a phenomenon that may greatly complicate analyses in humans if it can be shown.

In "complex" traits, recent efforts have been focused on the common variant(s)–common disease hypothesis, with remarkable success in many common conditions (for example, see Chapters 44, 45, 72, 73, 82, 92, 101). The common epilepsies are a heterogeneous group of conditions. If one considers only the partial epilepsies, which account, as a group, for the bulk of the epilepsies, the range of events thought to "cause" such epilepsies is very broad indeed. Epilepsies are attributed to developmental brain disorders, birth injury, head injury, substance abuse, stroke, hypoxic-ischemic injury, cerebral infection, neurodegeneration, toxins, and many other conditions. In most cases, the identification of a structural finding on brain MRI often leads either to cessation of the search for a cause, or investigation for a specific cause, though that is rarely mutation screening in a specific gene. For example, MRI can clearly identify hippocampal sclerosis as the possible underlying "cause" of drug-resistant mesial temporal lobe epilepsy, the most common single category of drug-resistant partial epilepsy in adults. In most cases, the detection of hippocampal sclerosis will be considered revelation of the cause of that individual's epilepsy; there may be a history of febrile seizures in infancy, and this will be considered in keeping with the overall syndrome (the febrile seizure may be itself considered the cause of the hippocampal sclerosis, or demonstration of some pre-febrile seizure vulnerability to this pathology). In some studies of the genetics of the epilepsies, cases with such structural "causation" will be excluded from analysis. But, in most cases, the reasons that an individual has such a structural "cause" and the reason that one individual with such a cause should develop epilepsy while another with the same cause does not, are not known. Typically, for example, even after long-term follow-up, 13–43% of people who suffer a potential "cause" of epilepsy (such as head injuries of a serious degree of severity) actually develop epilepsy; the figure is not 100%, and the reasons for this are obscure (Kharatishvili and Pitkänen, 2010; Raymont et al., 2010).

Family and twin studies prove complex heritability in the partial epilepsies. For example, all published twin studies show higher concordance rates in monozygotic (MZ) compared to dizygotic twins (DZ) (Ottman et al., 1996). Population-based twin studies suggest heritability in the epilepsies of up to 70% (Kjeldsen et al., 2001), including evidence of significant genetic contribution to partial epilepsies (Hemminki et al., 2006; Winawer and Shinnar, 2005). Specifically, concordance rates for partial epilepsies are significantly higher in monozygotic than in dizygotic twins, for example, in Berkovic's detailed referral center-based study (MZ = 0.36; DZ = 0.05) (Berkovic et al., 1998), and in Miller's population-based survey (MZ = 0.67; DZ = 0.11) (Miller et al., 1998). In the largest unselected population-based multinational survey, comprising 47,626 twin pairs, concordance rates for *all seizure* types were also significantly higher in MZ than DZ twins (Kjeldsen et al., 2005).

Published twin studies in the partial epilepsies are hardly ever subdivided by syndrome, but ultimately all partial seizures, by definition, require neuronal discharges and thus the involvement of a set of proteins. Moreover, there are striking, shared biological features in the partial epilepsies across syndromes, suggesting universal, syndrome-independent, mechanisms. Thus, partial epilepsies share "final common pathways" as supported by:

- *Clinical observations* – seizure semiology is limited in range and independent of etiology, and secondary generalization of partial seizures may be seen in any partial epilepsy
- *Investigations* – objective biological phenomena such as EEG abnormalities are shared across partial epilepsies, differing in localization rather than form
- *Treatment responses* – the same medications can be effective in different partial epilepsies.

Many of these shared phenomena are undoubtedly influenced by "host" genetic variation, as we and others have shown (Chorlian et al., 2007; Degen et al., 1993; Tate et al., 2005, 2006). For example, up to 5% of the variance in dosage of a particular medication across syndromes can be attributed to a single common variant (Tate et al., 2005, 2006). These data support the hypothesis that there are shared genetic variants predisposing to the characteristics common to all partial epilepsies. For the idiopathic generalized epilepsies, heritability estimates are higher still. It is therefore appropriate to consider how one might identify common genetic variants predisposing to seizures in either group of epilepsies, irrespective of syndrome. Such an aim is worthwhile in terms of understanding disease biology, and might generate powerful new therapeutic strategies effective across the partial epilepsies, irrespective of syndrome. Given that most strategies currently in use for the identification of new antiepileptic drugs are shared by those seeking to discover new treatments and have been in use for many years, it is perhaps unsurprising that most available AEDs share common mechanisms of action. It is therefore, perhaps, also unsurprising that the availability of newer AEDs has had little effect on the proportion of people with epilepsy with drug-resistant seizures.

Genome-wide Studies: Common and Rare Variants

These observations have motivated an unbiased genome-wide approach to search for common variants that might cause or contribute to the "common" epilepsies, both generalized and partial. One such study has now been published for the partial epilepsies, and others are underway for the generalized epilepsies. The study of the partial epilepsies was a classic genome-wide single nucleotide polymorphism (SNP)-based case-control association study (Kasperaviciute et al., 2010). Over 3500 people with epilepsy and almost 7000 control subjects were examined. The study had good power to detect common variants with odds ratio of >1.3, with results applicable to populations of Western European origin. Despite the size of the study, no genome-wide significant common variants associated with the disease phenotype (partial epilepsy) were detected (Kasperaviciute et al., 2010). Certain classes of genes did appear to have an over-representation amongst the lowest p-values that were observed, including genes encoding ion channels. Data on the generalized epilepsies are not yet available.

For the "common" epilepsies, these results have two main implications. Firstly, at least for populations of Western European extraction, common associated variants are unlikely to have anything more than a modest effect size, and their detection will require considerably larger cohorts. Secondly, the results encourage a reconsideration of our approach to detection of causal variants, suggesting a focusing of attention on rare variants.

Such a realignment of thinking towards a rare variant–common disease hypothesis is further supported by other findings. For both generalized and partial epilepsies, copy number change and, in particular, microdeletions have recently been reported (De Kovel et al., 2010; Dibbens et al., 2009; Heinzen et al., 2010; Helbig et al., 2009). These are not only microdeletions causing multiple congenital anomaly or mental retardation syndromes in association with epilepsy, but also "common" or non-syndromic epilepsies, both generalized and partial. Overall, microdeletions that are either large (e.g., >2 Mb) or recurrent (e.g., at chromosome 16p13.11) are significantly over-represented in the epilepsies, and account for about 1% of common epilepsies, thus representing one of the most common known classes of genetic variant in the epilepsies. A major effort is now required to search for rare variants in well-characterized epilepsies. Existing data suggest that no class of epilepsy ought to be excluded in such studies, although some may represent a greater chance of detection of causal variants.

Pharmacogenomics

There is great variation in clinical response to antiepileptic drugs, occurrence of adverse effects, and teratogenesis associated with AEDs. The consequences for patients and healthcare systems are sizeable (Ivanova et al., 2010; Pugliatti et al., 2007). There are over 20 licensed AEDs. Guidelines, with variable degrees of underlying evidence base, exist but do not provide rational individualized treatment strategies. In fact, there are very few clinically useful pointers to choice of the best drug for any particular individual. As illustrated elsewhere in this volume, personalized medicine is increasingly becoming a reality in many conditions. In epilepsy, however, there are very few replicated pharmacogenomic findings that have achieved clinical acceptance. Such findings have emerged from candidate gene studies, as, to date, there have been no published full genome-wide studies. The resources, methods, and knowledge permitting such studies are in place, and formal analysis should soon emerge.

There are two accepted pharmacogenetic associations for AED usage. Firstly, the HLA-B*1502 variant is strongly associated with Stevens–Johnson syndrome, or toxic epidermal necrolysis, in people of Han Chinese, Thai, Malay, and Indian, but not Japanese, extraction exposed to carbamazepine, oxcarbazepine, and phenytoin (Chung et al., 2010). Testing for this human leukocyte antigen (HLA) allele in prone populations is now recommended prior to use of these drugs. This is undoubtedly an important discovery, and the first major achievement in epilepsy pharmacogenomics. But, in practice, the finding may be of limited value, even in at-risk populations, where reasonable alternative agents, including those without an aromatic ring, are readily available. In addition, this allele appears not to be present in many Western European populations, in whom no genetic at-risk variants have yet been uncovered (Alfirevic et al., 2006).

The second association is for phenytoin metabolism and rare variants in the gene encoding the major enzyme, CYP2C9, responsible for its degradation. The *2 and *3 alleles result in slowed metabolism and the potential for clinical phenytoin neurotoxicity. This association has been known for some years, with strong additional recent support (Löscher et al., 2009). However, testing has not become adopted routine clinical practice: in many countries where phenytoin may represent one of the few treatment options, resources and laboratories for pre-prescription testing are rarely available; in other countries where such facilities might be available, it is still common practice to prescribe, evaluate, and ramp-up dosage according to clinical response and serum phenytoin levels. Lastly, until point-of-care testing becomes available, any delay incurred in initiating treatment with an AED is likely to be deemed unacceptable.

The reality of adoption of pharmacogenomic findings to regular clinical practice is likely to prove challenging and is considered further below.

Other Areas

As emphasized throughout this chapter, many aspects of the epilepsies display unexplained heterogeneity. Heritability estimates are rarely available for anything but causation. Nevertheless, as genome-wide cohort studies are undertaken, the impact upon variables such as cognitive outcome, risk of sudden unexplained death (SUDEP), psychiatric and somatic co-morbidities of individual genetic variation will come under scrutiny. Though animal models for some phenomena, including SUDEP, have been described, human susceptibility variants currently remain largely elusive, though ion channel genes remain prime candidates (Aurlien et al., 2009; Glasscock et al., 2010; Goldman et al., 2009; Le Gal et al., 2010; Nashef et al., 2007).

THE FUTURE: CHALLENGES AND POTENTIAL

Most of the genetic discovery in the epilepsies is yet to come. Most people with epilepsy do not have an illuminating family history. Genome-wide association studies (GWAS), to date, have not explained heritability, for which there may be a number of explanations. The "missing heritability" may yet be shown to reside in common variants, or may perhaps be found in rare, or complex combinations of rare, variants. Large-cohort GWAS and exome or whole-genome sequencing in well-characterized cohorts of patients will be required to uncover causal variants. Epigenomics and metabolomics have yet even to be considered. Tissue-specific studies are difficult in epilepsy, with little knowledge of the representativeness of extra-neural tissues for brain. But judging from rapid progress in other fields, the effort is likely to prove worthwhile, especially for pharmacogenomics (Ashley et al., 2010). In all such efforts, it will be crucial to engage people with epilepsy in such endeavors. After the discovery of putatively causal variants, particular attention will need to be paid to tissue- and cell-specific consequences, underlying molecular mechanisms, and detailed evaluation of phenotypic results. For pharmacogenomic research, meaningful description of clinically useful phenotypes will be critical – the aim simply to discover causal variants irrespective of phenotype should not be an end in itself. In this pursuit, multidisciplinary collaboration is likely to prove most fruitful. The potential benefit to people with epilepsy is great – genomics is likely to prove as valuable an advance as MRI has done over the last 20 years. The next 20 promise to be truly exciting.

REFERENCES

Alfirevic, A., Jorgensen, A.L., Williamson, P.R., et al., 2006. HLA-B locus in Caucasian patients with carbamazepine hypersensitivity. Pharmacogenomics 7, 813–818.

Annesi, G., Gambardella, A., Carrideo, S., et al., 2003. Two novel SCN1A missense mutations in generalized epilepsy with febrile seizures plus. Epilepsia 44, 1257–1258.

Ashley, E.A., Butte, A.J., Wheeler, M.T., et al., 2010. Clinical assessment incorporating a personal genome. Lancet 375, 1525–1535.

Aurlien, D., Leren, T.P., Tauboll, E., Gjerstad, L., 2009. New SCN5A mutation in a SUDEP victim with idiopathic epilepsy. Seizure 18, 158–160.

Baulac, S., Baulac, M., 2009. Advances on the genetics of Mendelian idiopathic epilepsies. Neurol Clin 27, 1041–1061.

Bell, G.S., Gaitatzis, A., Bell, C.L., Johnson, A.L., Sander, J.W., 2008. Drowning in people with epilepsy: How great is the risk? Neurology 71, 578–582.

Bell, G.S., Sinha, S., Tisi, J., et al., 2010. Premature mortality in refractory partial epilepsy: Does surgical treatment make a difference? J Neurol Neurosurg Psychiatry 81, 716–718.

Berg, A.T., Berkovic, S.F., Brodie, M.J., et al., 2010. Revised terminology and concepts for organization of seizures and epilepsies: Report of the ILAE Commission on Classification and Terminology, 2005–2009. Epilepsia 51, 676–685.

Berkovic, S.F., Howell, R.A., Hay, D.A., Hopper, J.L., 1998. Epilepsies in twins: Genetics of the major epilepsy syndromes. Ann Neurol 43, 435–445.

Bialer, M., Johannessen, S.I., Kupferberg, H.J., et al., 2004. Progress report on new antiepileptic drugs: A summary of the Seventh Eilat Conference (EILAT VII). Epilepsy Res 61, 1–48.

Camfield, C.S., Camfield, P.R., 2007. Long-term social outcomes for children with epilepsy. Epilepsia 48 (Suppl 9), 3–5.

Catarino, C.B., Liu, J.Y., Liagkouras, I., et al., 2011. Dravet syndrome as epileptic encephalopathy: Evidence from long-term course and neuropathology. Brain 134 (Pt 10), 2982–3010.

Chorlian, D.B., Tang, Y., Rangaswamy, M., et al., 2007. Heritability of EEG coherence in a large sib-pair population. Biol Psychol 75, 260–266.

Chung, W.H., Hung, S.I., Chen, Y.T., 2010. Genetic predisposition of life-threatening antiepileptic-induced skin reactions. Expert Opin Drug Saf 9, 15–21.

Davidsson, J., Collin, A., Olsson, M.E., Lundgren, J., Soller, M., 2008. Deletion of the SCN gene cluster on 2q24.4 is associated with severe epilepsy: An array-based genotype–phenotype correlation and a comprehensive review of previously published cases. Epilepsy Res 81, 69–79.

Degen, R., Degen, H.E., Koneke, B., 1993. On the genetics of complex partial seizures: Waking and sleep EEGs in siblings. J Neurol 240, 151–155.

De Kovel, C.G., Trucks, H., Helbig, I., et al., 2010. Recurrent microdeletions at 15q11.2 and 16p13.11 predispose to idiopathic generalized epilepsies. Brain 133, 23–32.

Depienne, C., Trouillard, O., Saint-Martin, C., et al., 2009. Spectrum of SCN1A gene mutations associated with Dravet syndrome: Analysis of 333 patients. J Med Genet 46, 183–191.

Dibbens, L.M., Mullen, S., Helbig, I., et al., 2009. Familial and sporadic 15q13.3 microdeletions in idiopathic generalized epilepsy: Precedent for disorders with complex inheritance. Hum Mol Genet 18, 3626–3631.

Ding, D., Wang, W., Wu, J., et al., 2006. Premature mortality in people with epilepsy in rural China: A prospective study. Lancet Neurol 5, 823–827.

Duncan, J.S., Sander, J.W., Sisodiya, S.M., Walker, M.C., 2006. Adult epilepsy. Lancet 367, 1087–1100.

Fisher, J.L., 2009. The anti-convulsant stiripentol acts directly on the GABA(A) receptor as a positive allosteric modulator. Neuropharmacology 56, 190–197.

Gaitatzis, A., Trimble, M.R., Sander, J.W., 2004. The psychiatric comorbidity of epilepsy. Acta Neurol Scand 110, 207–220.

Glasscock, E., Yoo, J.W., Chen, T.T., Klassen, T.L., Noebels, J.L., 2010. Kv1.1 potassium channel deficiency reveals brain-driven cardiac dysfunction as a candidate mechanism for sudden unexplained death in epilepsy. J Neurosci 30, 5167–5175.

Goldman, A.M., Glasscock, E., Yoo, J., et al., 2009. Arrhythmia in heart and brain: KCNQ1 mutations link epilepsy and sudden unexplained death. Sci Transl Med 1, 2ra6.

Guerrini, R., Dravet, C., Genton, P., et al., 1998. Lamotrigine and seizure aggravation in severe myoclonic epilepsy. Epilepsia 39, 508–512.

Guerrini, R., Parrini, E., 2010. Neuronal migration disorders. Neurobiol Dis 38, 154–166.

Harkin, L.A., Bowser, D.N., Dibbens, L.M., et al., 2002. Truncation of the GABA(A)-receptor gamma2 subunit in a family with generalized epilepsy with febrile seizures plus. Am J Hum Genet 70, 530–536.

Heaney, D.C., MacDonald, B.K., Everitt, A., et al., 2002. Socioeconomic variation in incidence of epilepsy: Prospective community based study in south-east England. BMJ 325, 1013–1016.

Heinzen, E.L., Radtke, R.A., Urban, T.J., et al., 2010. Rare deletions at 16p13.11 predispose to a diverse spectrum of sporadic epilepsy syndromes. Am J Hum Genet 86, 707–718.

Helbig, I., Mefford, H.C., Sharp, A.J., et al., 2009. 15q13.3 microdeletions increase risk of idiopathic generalized epilepsy. Nat Genet 41, 160–162.

Helbig, I., Scheffer, I.E., Mulley, J.C., Berkovic, S.F., 2008. Navigating the channels and beyond: Unravelling the genetics of the epilepsies. Lancet Neurol 7, 231–245.

Hemminki, K., Li, X., Johansson, S.E., Sundquist, K., Sundquist, J., 2006. Familial risks for epilepsy among siblings based on hospitalizations in Sweden. Neuroepidemiology 27, 67–73.

Hill, D.S., Wlodarczyk, B.J., Palacios, A.M., Finnell, R.H., 2010. Teratogenic effects of antiepileptic drugs. Expert Rev Neurother 10, 943–959.

Ivanova, J.I., Birnbaum, H.G., Kidolezi, Y., et al., 2010. Economic burden of epilepsy among the privately insured in the US. Pharmacoeconomics 28, 675–685.

Jacoby, A., Snape, D., Baker, G.A., 2009. Determinants of quality of life in people with epilepsy. Neurol Clin 27, 843–863.

Kamiya, K., Kaneda, M., Sugawara, T., et al., 2004. A nonsense mutation of the sodium channel gene SCN2A in a patient with intractable epilepsy and mental decline. J Neurosci 24, 2690–2698.

Kasperaviciute, D., Catarino, C.B., Heinzen, E.L., et al., 2010. Common genetic variation and susceptibility to partial epilepsies: A genome-wide association study. Brain 133, 2136–2147.

Kasperaviciute, D., Sisodiya, S.M., 2009. Epilepsy pharmacogenetics. Pharmacogenomics 10, 817–836.

Kharatishvili, I., Pitkänen, A., 2010. Post-traumatic epilepsy. Curr Opin Neurol 23, 183–188.

Kjeldsen, M.J., Kyvik, K.O., Christensen, K., Friis, M.L., 2001. Genetic and environmental factors in epilepsy: A population-based study of 11900 Danish twin pairs. Epilepsy Res 44, 167–178.

Kjeldsen, M.J., Corey, L.A., Solaas, M.H., et al., 2005. Genetic factors in seizures: A population-based study of 47,626 US, Norwegian and Danish twin pairs. Twin Res Hum Genet 8, 138–147.

Kwan, P., Brodie, M.J., 2000. Early identification of refractory epilepsy. N Engl J Med 342, 314–319.

Le Gal, F., Korff, C.M., Monso-Hinard, C., et al., 2010. A case of SUDEP in a patient with Dravet syndrome with SCN1A mutation. Epilepsia 51 (9), 1915–1918.

Löscher, W., Klotz, U., Zimprich, F., Schmidt, D., 2009. The clinical impact of pharmacogenetics on the treatment of epilepsy. Epilepsia 50, 1–23.

Lossin, C., 2009. A catalog of SCN1A variants. Brain Dev 31, 114–130.

Martin, M.S., Dutt, K., Papale, L.A., et al., 2010. Altered function of the SCN1A voltage-gated sodium channel leads to gamma-aminobutyric acid-ergic (GABAergic) interneuron abnormalities. J Biol Chem 285, 9823–9834.

Mefford, H.C., Muhle, H., Ostertag, P., et al., 2010. Genome-wide copy number variation in epilepsy: Novel susceptibility loci in idiopathic generalized and focal epilepsies. PLoS Genet 6, e1000962.

Meisler, M.H., O'Brien, J.E., Sharkey, L.M., 2010. Sodium channel gene family: Epilepsy mutations, gene interactions and modifier effects. J Physiol 588, 1841–1848.

Miller, L.L., Pellock, J.M., DeLorenzo, R.J., Meyer, J.M., Corey, L.A., 1998. Univariate genetic analyses of epilepsy and seizures in a population-based twin study: The Virginia Twin Registry. Genet Epidemiol 15, 33–49.

Mullen, S.A., Scheffer, I.E., 2009. Translational research in epilepsy genetics: Sodium channels in man to interneuronopathy in mouse. Arch Neurol 66, 21–26.

Nashef, L., Hindocha, N., Makoff, A., 2007. Risk factors in sudden death in epilepsy (SUDEP): The quest for mechanisms. Epilepsia 48, 859–871.

Oakley, J.C., Kalume, F., Yu, F.H., Scheuer, T., Catterall, W.A., 2009. Temperature- and age-dependent seizures in a mouse model of severe myoclonic epilepsy in infancy. Proc Natl Acad Sci USA 106, 3994–3999.

Ogiwara, I., Miyamoto, H., Morita, N., et al., 2007. Na(v)1.1 localizes to axons of parvalbumin-positive inhibitory interneurons: A circuit basis for epileptic seizures in mice carrying an Scn1a gene mutation. J Neurosci 27, 5903–5914.

Ottman, R., Hauser, W.A., Barker-Cummings, C., et al., 1997. Segregation analysis of cryptogenic epilepsy and an empirical test of the validity of the results. Am J Hum Genet 60, 667–675.

Ottman, R., Lee, J.H., Risch, N., Hauser, W.A., Susser, M., 1996. Clinical indicators of genetic susceptibility to epilepsy. Epilepsia 37, 353–361.

Patino, G.A., Claes, L.R., Lopez-Santiago, L.F., et al., 2009. A functional null mutation of SCN1B in a patient with Dravet syndrome. J Neurosci 29, 10,764–10,778.

Pereira, S., Vieira, J.P., Barroca, F., et al., 2004. Severe epilepsy, retardation, and dysmorphic features with a 2q deletion including SCN1A and SCN2A. Neurology 63, 191–192.

Pugliatti, M., Beghi, E., Forsgren, L., Ekman, M., Sobocki, P., 2007. Estimating the cost of epilepsy in Europe: A review with economic modeling. Epilepsia 48, 2224–2233.

Ramachandran, N., Girard, J.M., Turnbull, J., Minassian, B.A., 2009. The autosomal recessively inherited progressive myoclonus epilepsies and their genes. Epilepsia 50 (Suppl 5), 29–36.

Raymont, V., Salazar, A.M., Lipsky, R., et al., 2010. Correlates of posttraumatic epilepsy 35 years following combat brain injury. Neurology 75, 224–229.

Reid, C.A., Berkovic, S.F., Petrou, S., 2009. Mechanisms of human inherited epilepsies. Prog Neurobiol 87, 41–57.

Rusconi, R., Scalmani, P., Cassulini, R.R., et al., 2007. Modulatory proteins can rescue a trafficking defective epileptogenic Nav1.1 Na+ channel mutant. J Neurosci 27, 11,037–11,046.

Shi, X., Yasumoto, S., Nakagawa, E., et al., 2009. Missense mutation of the sodium channel gene SCN2A causes Dravet syndrome. Brain Development 31, 758–762.

Singh, G., Fletcher, O., Bell, G.S., McLean, A.E., Sander, J.W., 2009. Cancer mortality amongst people with epilepsy: A study of two cohorts with severe and presumed milder epilepsy. Epilepsy Res 83, 190–197.

Singh, N.A., Pappas, C., Dahle, E.J., et al., 2009. A role of SCN9A in human epilepsies, as a cause of febrile seizures and as a potential modifier of Dravet syndrome. PLoS Genet 5, e1000649.

Sisodiya, S.M., 2007. Choice of agent in treatment of epilepsy. In: Rothwell, P.M. (Ed.), The Lancet: Treating Individuals: From Randomised Trials to Personalised Medicine. Elsevier, Edinburgh, pp. 297–305.

Tang, B., Dutt, K., Papale, L., et al., 2009. A BAC transgenic mouse model reveals neuron subtype-specific effects of a generalized epilepsy with febrile seizures plus (GEFS+) mutation. Neurobiol Dis 35, 91–102.

Tate, S.K., Depondt, C., Sisodiya, S.M., et al., 2005. Genetic predictors of the maximum doses patients receive during clinical use of the anti-epileptic drugs carbamazepine and phenytoin. Proc Natl Acad Sci USA 102, 5507–5512.

Tate, S.K., Singh, R., Hung, C.C., et al., 2006. A common polymorphism in the SCN1A gene associates with phenytoin serum levels at maintenance dose. Pharmacogenet Genomics 16, 721–726.

Thomas, E.A., Reid, C.A., Berkovic, S.F., Petrou, S., 2009. Prediction by modeling that epilepsy may be caused by very small functional changes in ion channels. Arch Neurol 66, 1225–1232.

Wieser, H.G., 2004. ILAE commission report. Mesial temporal lobe epilepsy with hippocampal sclerosis. Epilepsia 45, 695–714.

Winawer, M.R., Shinnar, S., 2005. Genetic epidemiology of epilepsy or what do we tell families? Epilepsia 46 (Suppl 10), 24–30.

Yu, F.H., Mantegazza, M., Westenbroek, R.E., et al., 2006. Reduced sodium current in GABAergic interneurons in a mouse model of severe myoclonic epilepsy in infancy. Nat Neurosci 9, 1142–1149.

Schizophrenia and Bipolar Disorder

Hywel J. Williams, Michael C. O'Donovan, Nicholas Craddock, and Michael J. Owen

INTRODUCTION

At the end of the nineteenth century, Emil Kraepelin suggested that the complex mixture of severe mood and psychotic illnesses commonly encountered in adults with severe psychiatric illness could be split into two broad categories. He called these "dementia praecox," a chronic, disabling, often deteriorating illness with prominent delusions, hallucinations, and cognitive dysfunction, and "manic-depressive insanity," an episodic illness with prominent mood disturbances and good inter-episode functioning. This dichotomy, reflected in the modern categories of "schizophrenia" and "bipolar disorder" (BD), remains central to the classification of severe mental illness in the versions of the *Diagnostic and Statistical Manual of Mental Disorders* and the *International Classification of Diseases* used today. The shortcomings of such a simple formulation have been apparent for some time (Owen et al., 2007). However, scientific progress during the twentieth century was insufficient to overturn it, despite the fact that most new treatments did not show specificity to one or the other category (for example, "antipsychotics" are effective for some of the symptoms of both schizophrenia and BD). For this reason, the majority of genomic studies have examined the two "disorders" separately, although, as we shall see, as positive results have accrued, evidence for both genetic overlap and differences has been obtained.

Both schizophrenia and bipolar disorder are common, with lifetime risks of about 1% (Gottesman et al., 2010), and, like other common disorders, they show familial aggregation [sibling recurrence risk ratio (λ_s) of 9.0 and 7.9 for schizophrenia and BD, respectively (Lichtenstein et al., 2009)] but complex, non-Mendelian patterns of inheritance. Unlike many common disorders, however, these are syndromic diagnoses, and the categories encompass widespread clinical as well as etiological heterogeneity. There are no clear phenotypic boundaries between them, nor are there clear boundaries between them and other psychiatric abnormalities and normality. The genetic epidemiology of both disorders has been well researched (Cardno and Gottesman, 2000; Craddock and Jones, 1999; Gottesman et al., 2010; McGuffin et al., 2003), and there is extensive evidence that both disorders are highly heritable [heritability (h^2) of 0.64 and 0.59 for schizophrenia and BD, respectively (Lichtenstein et al., 2009)], and this has encouraged genomic studies. Until recently, the consensus of opinion was that the two disorders tended to "breed true," suggesting distinct genetic risk. However, most studies were likely underpowered to detect cross-disorder risk, and a recent study based on >2 million nuclear families identified from the Swedish population and hospital discharge registers showed that there are increased risks of *both* schizophrenia and BD in first-degree relatives of probands with either disorder.

Genomic and Personalized Medicine, 2nd edition
by Ginsburg & Willard. DOI: http://dx.doi.org/10.1016/B978-0-12-382227-7.00088-4

Moreover, there is evidence from the study of half-siblings and adopted-away relatives that this is due substantially to genetic factors (Lichtenstein et al., 2009).

Early Genetic Studies

The complexity and relative inaccessibility of the human brain, together with difficulties inherent in developing valid model systems, have provided major obstacles to understanding disease biology. Given the evidence for significant heritability, geneticists have long harbored the hope that identification of specific risk genes might allow crucial insights into disease pathogenesis to be obtained.

Work in psychiatric molecular genetics began with linkage studies of small numbers of high-density families, aimed at detecting high-penetrance alleles and conducting studies of functional candidate genes. It then progressed to larger scale linkage studies and the analysis of positional candidate genes as well as studies of microscopic chromosomal abnormalities. Thousands of functional candidate gene studies have been reported, but no strongly replicated findings have emerged (http://www.szgene.org/). This no doubt reflects in part our limited understanding of pathophysiology, but also the fact that the great majority of studies have been underpowered to detect either common alleles of small effect or rare alleles of large effect.

Positional studies appear to have been more successful in that a small number of promising susceptibility genes have been implicated in multiple genetic studies (Ross et al., 2006). Those for which the evidence is strongest are *NRG1*, *DTNBP1*, *DAOA*, and *DISC1*. Data consistent with the involvement of these genes in schizophrenia have additionally come from studies of extended and intermediate phenotypes, principally neurocognitive and neuroimaging in nature, and from analyses of gene expression and other aspects of neurobiology. However, in spite of this impressive body of evidence, in no case have specific risk alleles yet been unambiguously implicated as causal, and in no case do the strength and consistency (that is, the same alleles or haplotypes across studies) of the genetic evidence equal those for genes now known to be involved in other complex disorders (Manolio et al., 2009).

Thus, progress has until recently been largely disappointing, but in the last three years the application of novel genomic approaches has begun to yield a number of important new insights. Recent advances have made possible a new wave of genetic studies of complex diseases by allowing genome-wide association studies (GWAS) and genome-wide studies of copy number variants (CNVs). These have met with more success and are the main subject of this chapter.

GWAS FOR SCHIZOPHRENIA AND BIPOLAR DISORDER

To date, over 150 GWAS of various phenotypes have been published, and over 500 cases of genome-wide significant association to one or more single nucleotide polymorphisms (SNPs)

(Dudbridge and Gusnanto, 2008) have been reported [see the National Center for Biotechnology Informatics (NCBI) database of Genotypes and Phenotypes (dbGaP), http://www.ncbi.nlm.nih.gov/gap]. This amply demonstrates the feasibility of using this approach to search for common variants for complex diseases, such as schizophrenia and BD. Recent studies in both disorders have yielded evidence for genome-wide significant associations. Specifically, for schizophrenia, the associated markers implicated in the disorder at this level of support map within or in the vicinity of genes for zinc-finger binding protein 804A (*ZNF804A*) (O'Donovan et al., 2008a), transcription factor 4 (*TCF4*) (Stefansson et al., 2009), neurogranin (*NRGN*) (Stefansson et al., 2009), and an extended region spanning the *MHC* locus on chromosome 6 (Purcell et al., 2009; Shi et al., 2009; Stefansson et al., 2009) (see Table 88.1). For BD, the evidence implicates the genes for calcium channel, voltage-dependent, L type, alpha 1C subunit (*CACNA1C*) (Ferreira et al., 2008); ankyrin 3, node of Ranvier (*ANK3*) (Ferreira et al., 2008); and polybromo-1 (*PBRM1*) (McMahon et al., 2010) (see Table 88.2).

An interesting and highly significant finding from the International Schizophrenia Consortium (ISC) study was the observation that by summing the effects of alleles from loci showing very weak trends for association in one sample into quantitative (polygenic) scores and relating these to disease state in independent samples, it is possible to demonstrate a substantial polygenic component to schizophrenia that is likely to involve thousands of risk alleles of very small effect (Purcell et al., 2009). Simulation studies suggest that many of these alleles are common and that the polygenic component accounts for at least 30% of the total variation in liability.

Another observation from the above studies is that there is support for a substantial genetic overlap between schizophrenia and BD, as suggested by family studies. This comes from analyses of specific risk loci, including *ZNF804A* (Williams et al., 2010a) and *CACNA1C* (Green et al., 2010). Evidence for overlap in the identity of genes shows nominally significant gene-wide association signals in GWAS of both schizophrenia and BD (Moskvina et al., 2009); further, and most compellingly, there is strong evidence that polygenic scores derived from alleles that are more common in schizophrenia samples are greater in bipolar subjects than in controls (Purcell et al., 2009).

A number of implications follow from these findings. Most notably, we can now refute the hypothesis that schizophrenia is a disease caused entirely by rare, highly-penetrant variants, that occur in the region of 1 in a 1000 of the population or are specific to a single family (McClellan et al., 2007). This hypothesis had been based on the theory that the reduced fecundity observed in schizophrenia would be incompatible with the existence of common risk alleles, as selection would drive these out of the population. There are a number of lines of evidence against such an extreme rare variant explanation of schizophrenia (Craddock et al., 2007), but the GWAS data now give us empirical evidence that a number of common risk variants exist that escape purification. While it is not known entirely why they are not removed from the population, this is in part, presumably,

TABLE 88.1	Schizophrenia genome-wide associated loci					
Study	**Locus**	**SNP**	**P-value**	**OR**	**95% CI**
Stefansson et al., 2009	MHC	rs6932590	1.4×10^{-12}	1.15	1.05–1.26
Stefansson et al., 2009	MHC	rs3131296	2.3×10^{-10}	1.21	1.08–1.36
Stefansson et al., 2009	TCF4	rs9960767	4.1×10^{-9}	1.30	1.11–1.51
Stefansson et al., 2009	NRGN	rs12807809	2.4×10^{-9}	1.15	1.10–1.20
Williams et al., 2010a	ZNF804A	rs1344706	2.5×10^{-11}	1.10	1.07–1.14

P-value from Armitage trend test; OR=odds ratio, CI=confidence interval.

TABLE 88.2	Bipolar disorder genome-wide associated loci					
Study	**Locus**	**SNP**	**P-value**	**OR**	**95% CI**
McMahon et al., 2010	PBRM1	rs2251219	1.7×10^{-9}	0.88	0.85–0.92
Ferreira et al., 2008*	ANK3	rs1938526	1.3×10^{-8}	1.40	–
Ferreira et al., 2008*	CACNA1C	rs1006737	7.0×10^{-8}	1.18	–

P-value from Armitage trend test; OR=odds ratio, CI=confidence interval.
*No details of 95% CI given in original manuscript.

due to their individually small effect sizes and/or because they confer a small but balancing advantage to unaffected carriers.

While disease risk is therefore not due *entirely* to rare variants, what the GWAS findings cannot do is rule out *some* role for rare variants in the overall risk of schizophrenia. In fact, a small part of the polygenic contribution observed by the ISC may well be derived from such variants. However, it is likely that the genetic basis of schizophrenia and BD, like that of other common diseases, is going to consist of a mix of both common alleles of small effect and rare alleles, some of which confer a relatively large effect on disease risk. At the present time, the proportion of each is unknown and more data are required to resolve this issue. There is a compelling argument that the best way to achieve this is through the collection of more samples, as has been done in other complex diseases. For example, in complex diseases such as type 2 diabetes, sample sizes in the range of 42,500 cases and 99,000 controls have been accrued through meta-analysis, which has increased the number of confirmed disease loci to 38 (Voight et al., 2010). This contrasts to the situation in schizophrenia, where the largest amassed sample size to date consisted of approximately 19,000 schizophrenic cases and 39,000 controls (Williams et al., 2010a); although large for a psychiatric phenotype, this is still less than half that of the type 2 diabetes study.

STUDIES OF COPY NUMBER VARIATION

A welcome by-product of genome-wide arrays (SNP chips), beyond their ability to detect SNP associations, is their ability to detect CNVs on a genome-wide scale. This work, like that of GWAS, is still in its infancy and therefore strict quality controls have been used to ensure that the CNVs identified are valid given the resolution of the current technology. For that reason, researchers have focused on large (typically >100 kb) CNVs that are present at a frequency of about 1% or less in the population, as these are the most robust in terms of detection and replication. These variants are an important source of inter-individual variation as they are found ubiquitously in the population (McCarroll et al., 2008). They can disrupt gene function in many ways, including altering gene dosage, perturbing gene expression, disrupting gene structure, and possibly in a number of yet to be identified ways (Henrichsen et al., 2009). Although CNVs can be detected by various platforms, the use of genome arrays has advantages, as it is the large sample sizes commonly used in GWAS that has helped to determine the full spectrum of CNVs in the population and has led to evidence that, in addition to their role in a number of uncommon syndromes, they contribute to the etiology of more common disorders. Neurodevelopmental disorders, in particular autism and mental disability, have been the most widely studied, with compelling evidence that CNVs in the size range of 100 kb (which is the approximate threshold for accurate calling on SNP platforms) or larger contribute to at least 10% of cases in both (Alkan et al., 2011; Cooper et al., 2011).

The role of CNVs in the pathogenesis of schizophrenia has long been suspected. It is over 15 years (Pulver et al., 1994) since it was noted that deletions of 22q11.2 that give rise to velo-cardio-facial syndrome are associated with an increased risk of schizophrenia. Later, larger and more systematic studies have estimated the increased risk at around 20-fold

(Murphy et al., 1999). More recently, a number of large, rare but recurrent CNVs have been identified that are individually associated with schizophrenia susceptibility (Table 88.3), and evidence has been obtained that the cumulative burden of rarer events is increased in cases relative to controls. The scope of this chapter is not to review in detail all these findings, and so we will highlight just the main ones. However, there are a number of excellent reviews that the interested reader is encouraged to pursue (Kirov, 2009a; O'Donovan et al., 2008b; Sebat et al., 2004; St Clair, 2009).

ROLE OF CNVs IN SCHIZOPHRENIA AND BIPOLAR DISORDER

A general finding is that the burden of rare (<1%), large (>100 kb) CNVs is increased in subjects with schizophrenia when compared to healthy controls. Estimation of the magnitude of this excess varies between studies, possibly reflecting methodological or sample differences or both, and highlighting the requirement of very large sample sizes to unambiguously demonstrate association to rare events (the focus of most of CNV studies to date). In the largest published schizophrenia study (Purcell et al., 2009), the excess burden of CNVs was statistically highly significant but the overall burden itself was only modest, a 1.15-fold excess. Of note, this increased to 1.6 when only large (>500 kb) deletions were compared. While the evidence for an excess burden of CNVs in schizophrenia is now convincing, these analyses do not yet identify individual loci or allow the effect sizes conferred by each to be estimated.

The analysis of large datasets from GWAS has allowed a number of individual CNV loci to be identified as conferring risk with high degrees of statistical confidence. The loci identified are deletions of chromosomes 1q21.1, 15q13.3, and 3q29, and duplications at 16p11.2 and 16p13.1 (Table 88.3). These are rare (control frequencies <0.1%), involve less than 1% of cases, and confer fairly large effects on disease risk, with odds ratios (ORs) recently estimated to be ~10 (Kirov et al., 2009a). However, the CNVs are far from fully penetrant, with only around 10% of carriers having schizophrenia (based upon the data in the Kirov study). It is important to stress that the samples used to date are only powered to detect relatively rare events with very large effect sizes, and the expectation must be that there will be CNVs with larger effect sizes that will be rarer and others with smaller effect sizes that may be more common. Compatible with this, a more common deletion at 15q11.2 (Stefansson et al., 2008) is associated with schizophrenia (P = 2.8×10^{-8}) (Kirov et al., 2009a), but confers a smaller effect size (OR < 3) and has a very modest penetrance of about 2%. A number of other specific CNVs have been implicated in schizophrenia, and these are reviewed elsewhere (O'Donovan et al., 2009).

While these associations are strong, the CNVs implicated span multiple genes. Resolving which of these are relevant to pathophysiology is likely to prove challenging, as illustrated by the difficulty implicating a single specific

schizophrenia-susceptibility gene in the 22q11.2 deletion region. A notable exception is the accumulating evidence that rare deletions interrupting only the *NRXN1* gene are associated with increased risk of schizophrenia. *NRXN1* encodes the neuronal cell adhesion molecule neurexin 1 (Kirov et al., 2009b). Neurexins are found pre-synaptically, and are believed to interact with post-synaptic neuroligins in excitatory and inhibitory synapses in the brain, and together these proteins play a pivotal role in the development of normal synaptic function (Sudhof, 2008). These findings are therefore of considerable interest since they point towards the importance of abnormalities of synaptic development and/or function in the pathogenesis of schizophrenia.

NRXN1 deletions have also been found in autism, and other schizophrenia-associated CNVs have also been found in association with a range of neurodevelopmental phenotypes including autism, mental retardation, attention-deficit hyperactivity disorder (ADHD), and idiopathic generalized epilepsy (Sebat, et al., 2009; Williams et al., 2010b). However, each has also been observed in individuals who lack an apparent neurodevelopmental phenotype. A question that has been raised in the context of 22q11.2 deletions is whether the CNV-related psychoses represent "true" forms of schizophrenia. Aside from this begging the question of whether there is such a thing as a true form of schizophrenia, two points should be noted. First, although there is probably an enrichment of comorbidities such as epilepsy and cognitive impairment, there is no evidence that the schizophrenia phenotype is distinctive in cases with deletions. This has not been studied systematically for the recently identified deletions, but cases with 22q11.2 deletions show schizophrenia symptoms that closely resemble those seen in cases without 22q11.2 deletions (Bassett et al., 2003). Second, it is clear that schizophrenia in those individuals with CNVs is not simply a nonspecific manifestation of another "primary" disorder (e.g., mental retardation or epilepsy), since schizophrenia occurs in some CNV carriers in the absence of any other disorder. Rather than implying the existence of an atypical subtype, the CNV findings suggest that schizophrenia might be viewed as one type of phenotypic expression that can arise from underlying developmental abnormality, with others including mental retardation, epilepsy, autism, and ADHD.

Currently, those CNVs for which there is strong support for association with schizophrenia occur in only about 2–3% of cases, although there is weaker evidence for the potential involvement of others (Sebat et al., 2009). It seems reasonable to suppose this figure will rise with the use of better resolution platforms and larger samples. Despite the strong evidence implicating specific CNVs, it is clear that none is sufficient to cause disease. Instead, it is likely that expression of disorder, and the form of neurodevelopmental phenotype taken, depends upon an unknown combination of genetic (rare, common, and polygenic) and environmental factors, and possibly chance. These uncertainties aside, the relatively high individual risk associated with specific CNVs suggests that these findings will be more easily applicable to cellular and animal

TABLE 88.3	Schizophrenia-associated copy number variants							
Deletions					**Duplications**			
Locus	**Case (%)**	**Control (%)**	**OR**		**Locus**	**Case (%)**	**Control (%)**	**OR**
1q21.1*	0.2	0.02	9.1		16p11.2*	0.3	0.03	8.4
3q29†	0.08	0.003	17.0		16p13.1§	0.3	0.09	3.3
15q11.2*	0.6	0.3	2.8					
15q13.3*	0.2	0.008	11.4					
22q11.2‡	0.3	0	21.6					

OR = odds ratio.
*Kirov, 2010.
†Mulle, et al., 2010.
‡ISC, 2008.
§Ingason et al., 2009.

models than common risk alleles, which individually confer much smaller individual risks and thereby have the potential to pinpoint novel pathogenic mechanisms. There is thus a strong case for further work in this area to complement SNP-based analyses.

There were early reports of specific CNVs in individuals with BD (Lachman et al., 2007; Wilson et al., 2006) but these did not address the question of whether the genomic burden of CNVs is increased in BD. The first systematic study did not find any increase in overall CNV load in BD, although there was a nominally significant increase of "singleton" CNVs in cases compared with controls (Zhang et al., 2009). A more recent study (Grozeva et al., 2010) also found that the burden of CNVs in BD was not increased compared with controls, and was significantly less than in schizophrenia cases. Moreover, CNVs previously implicated in the etiology of schizophrenia were not more common in cases. These findings suggest that schizophrenia and BD differ with respect to CNV burden in general and association with specific CNVs in particular. A most recent paper (Malhotra et al., 2011) examined *de novo* CNVs in ~180 cases each of schizophrenia and BD; while the confidence intervals were wide, the data support the conclusion of increased frequency of *de novo* CNVs in both BD and schizophrenia.

While more studies are clearly needed, the data to date are consistent with the possibility that possession of large, rare deletions may modify the phenotype in those at risk of psychosis: those possessing such events are more likely to be diagnosed with schizophrenia; those without such events are more likely to be diagnosed with BD.

CONCLUSIONS AND FUTURE DIRECTIONS

Genomic studies of schizophrenia and BD have evolved in parallel with technology and, with the advent of GWAS and related approaches, are bearing fruit. There are strongly supported associations to common alleles for both disorders, and, in the case of schizophrenia and perhaps BD, clear evidence for involvement of CNVs. There is evidence for increased burden of common risk alleles in both disorders and for an increased burden of large, rare CNVs in schizophrenia. However, the robustly associated loci are still few in number, and cumulatively account for a small amount of the population variance in risk. Nevertheless, they represent a step change in that they clearly (and finally) indicate the tractability of these disorders to genomic approaches. Moreover, the strong evidence for overlap between schizophrenia and BD with respect to SNP associations, and for overlap between schizophrenia and other neurodevelopmental disorders with respect to CNVs, indicates important directions for future research and serves further to remind us of the limitations of current descriptive approaches to psychiatric classification and diagnosis.

It seems safe to conclude that more highly powered GWAS are likely to identify further specific risk loci. However, it would also appear to be the case that, given the highly polygenic nature of schizophrenia and bipolar disorder, it will not be possible to identify all, or even the majority of, common risk alleles even with samples of 30,000 or so cases. This begs the question of whether, or for how long, we should continue to undertake GWAS, or whether we should focus instead on other approaches, such as searches for rare risk variants using whole-exome and/or whole-genome sequencing (see Chapter 1).

Sequencing-based approaches are often presented as alternatives to the collection of large samples required by GWAS. For example, it might be possible to search for rare highly penetrant mutations in large multiply affected pedigrees that show evidence for linkage, and which might therefore plausibly be enriched for such mutations (McClellan et al., 2007). However, large multiplex pedigrees with these disorders are uncommon, and those in which linkage points unequivocally to specific regions of the genome almost nonexistent. In the absence of clues to their presence in specific individuals and to their location from unequivocal linkage signals, the majority of rare, disease-related mutations will also have to be identified through genome-wide approaches. Given the low prior probabilities for any given mutation and their low frequency, as for CNVs, very large sample sizes will be required to provide credible support (Altshuler et al., 2008). Thus all genetic approaches will require collection of data from large samples, but as genetic technology improves and becomes cheaper it will be possible to use these samples to test an increasing number of genetic and genomic hypotheses. It seems to us that attention should now be give to developing the most economical approaches to ascertaining and phenotyping such samples and to making the genetic and phenotypic data available to as many researchers as possible.

It is also argued that rare risk alleles, by virtue of having larger effect sizes, will give greater biological insights than common risk alleles of small effect. Studies in other common diseases have shown that this view of the utility of genetic data is not justified and that there are ample grounds for believing that the identification of common risk alleles will also bring valuable insights to disease mechanisms (Abraham and Cho, 2009; Vimaleswaran and Loos, 2010). One potential advantage of identifying higher penetrance alleles is that their effects might be easier to model in cellular and whole animal systems. While this is likely to be true, this must be balanced against the possibility that identification of pathways based upon common alleles might offer greater insights into more common mechanisms and thereby to novel drug targets with wider applicability than those derived from ultra-rare variants. Powerful methods for identifying disease pathways from GWAS data have been developed (Holmans et al., 2009) and successfully applied to Alzheimer's disease (Jones et al., 2010). Application of these approaches to schizophrenia and BD is likely to be limited by the current lack of fine-grained annotation of the protein complexes and pathways involved in neuronal function, a challenge that is currently being addressed by computational biologists (Pocklington et al., 2006).

It is impossible to predict at what point the law of diminishing returns will render the accumulation of additional data about loci of small effect uninformative, since we cannot predict either the rate at which crucial biological insights will accrue or how steeply the costs of genomic analysis will decline. However, given that the same samples can be used for GWAS and for sequencing, arguments about the cost of sample collection for GWAS by those who favor whole-exome/genome sequencing as a primary approach are misguided (McClellan et al., 2007). We certainly both expect and welcome whole-exome/genome sequencing as sequencing costs come down, but, in the meantime, the fact that GWAS technology is less than a small fraction of the cost of sequencing argues powerfully for collecting increasingly large samples and subjecting them to GWAS in the first instance, not for abandoning the approach just as clear successes are being made.

The shortcomings of current diagnostic approaches have been pointed out (e.g., Craddock et al., 2009b). Schizophrenia is highly heterogeneous, and it is self-evident that if we could sub-divide it into more homogeneous groups, or define dimensional phenotypes that more closely index pathophysiology, we would increase the power of genetic studies. It is therefore highly encouraging that, despite heterogeneity, risk loci can be robustly identified by GWAS using samples defined by current diagnostic criteria. Moreover, armed with sufficient GWAS data it should be possible to identify categorical and dimensional phenotypes that relate to specific sets of risk alleles, and recent findings suggest that such approaches should not be constrained by the current diagnostic approaches (Craddock and Owen, 2007; Craddock et al., 2009a).

In conclusion, while only small parts of the heritability of schizophrenia and BD have been accounted for to date, large-scale genomic approaches have yielded substantive advances, and developments in genetic technology and phenotyping will continue to improve the prospects. These studies will require large, well-phenotyped samples, and both funding agencies and researchers must ensure that these are put in place if psychiatry is to continue to benefit from developments in genetics.

REFERENCES

Abraham, C., Cho, J.H., 2009. Inflammatory bowel disease. N Engl J Med 361, 2066–2078.

Alkan, C., Coe, B.P., Eichler, E.E., 2011. Genome structural variation discovery and genotyping. Nat Rev Genet 12, 363–376.

Altshuler, D., Daly, M.J., Lander, E.S., 2008. Genetic mapping in human disease. Science 322, 881–888.

Bassett, A.S., Chow, E.W., AbdelMalik, P., et al., 2003. The schizophrenia phenotype in 22q11 deletion syndrome. Am J Psychiatry 160, 1580–1586.

Cardno, A.G., Gottesman, I.I., 2000. Twin studies of schizophrenia: From bow-and-arrow concordances to star wars Mx and functional genomics (review). Am J Med Genet 97, 12–17.

Cooper, G.M., Coe, B.P., Girirajan, S., et al., 2011. A copy number variation morbidity map of developmental delay. Nat Genet 43, 838–846.

Craddock, N., Jones, I., 1999. Genetics of bipolar disorder. J Med Genet 36, 585–594.

Craddock, N., Kendler, K., Neale, M., et al., 2009a. Dissecting the phenotype in genome-wide association studies of psychiatric illness. Br J Psychiatry 195, 97–99.

Craddock, N., O'Donovan, M.C., Owen, M.J., 2007. Phenotypic and genetic complexity of psychosis. Invited commentary on … Schizophrenia: A common disease caused by multiple rare alleles. Br J Psychiatry 190, 200–203.

Craddock, N., O'Donovan, M.C., Owen, M.J., 2009b. Psychosis genetics: Modeling the relationship between schizophrenia, bipolar disorder, and mixed (or "schizoaffective") psychoses. Schizophr Bull 35, 482–490.

Craddock, N., Owen, M.J., 2007. Rethinking psychosis: The disadvantages of a dichotomous classification now outweigh the advantages. World Psychiatry 6, 20–27.

Dudbridge, F., Gusnanto, A., 2008. Estimation of significance thresholds for genomewide association scans. Genet Epidemiol 32, 227–234.

Ferreira, M.A., O'Donovan, M.C., Meng, Y.A., et al., 2008. Collaborative genome-wide association analysis supports a role for ANK3 and CACNA1C in bipolar disorder. Nat Genet 40, 1056–1058.

Gotteman, I.I., Laursen, T.M., Bertelsen, A., Mortensen, P.B., 2010. Severe mental disorders in offspring with two psychiatrically ill parents. Arch Gen Psychiatry 67, 252–257.

Green, E.K., Grozeva, D., Jones, I., et al., 2010. The bipolar disorder risk allele at CACNA1C also confers risk of recurrent major depression and of schizophrenia. Mol Psychiatry 15, 1016–1022.

Grozeva, D., Kirov, G., Ivanov, D., et al., 2010. Rare copy number variants: A point of rarity in genetic risk for bipolar disorder and schizophrenia. Arch Gen Psychiatry 67, 318–327.

Henrichsen, C.N., Chaignat, E., Reymond, A., 2009. Copy number variants, diseases and gene expression. Hum Mol Genet 18, R1–R8.

Holmans, P., Green, E.K., Pahwa, J.S., et al., 2009. Gene ontology analysis of GWA study data sets provides insights into the biology of bipolar disorder. Am J Hum Genet 85, 13–24.

Ingason, A., Rujescu, D., Cichon, S., et al., 2009. Copy number variations of chromosome 16p13.1 region associated with schizophrenia. Mol Psychiatry 16. doi: 10.1038/mp.2009.101.

ISC, 2008. Rare chromosomal deletions and duplications increase risk of schizophrenia. Nature 455, 237–241.

Jones, L., Holmans, P.A., Hamshere, M.L., et al., 2010. Genetic evidence implicates the immune system and cholesterol metabolism in the aetiology of Alzheimer's disease. PLoS1 5, e13950. doi: 10.1371/journal.pone.0013950.

Kirov, G., 2010. The role of copy number variation in schizophrenia. Exp Rev Neurother 10, 25–32.

Kirov, G., Grozeva, D., Norton, N., et al., 2009a. Support for the involvement of large CNVs in the pathogenesis of schizophrenia. Hum Mol Genet 18, 1497–1503.

Kirov, G., Rujescu, D., Ingason, A., et al., 2009b. Neurexin 1 (NRXN1) deletions in schizophrenia. Schizophr Bull 35, 851–854.

Lachman, H.M., Pedrosa, E., Petruolo, O.A., et al., 2007. Increase in GSK3beta gene copy number variation in bipolar disorder. Am J Med Genet B Neuropsychiatr Genet 144, 259–265.

Lichtenstein, P., Yip, B.H., Bjork, C., et al., 2009. Common genetic determinants of schizophrenia and bipolar disorder in Swedish families: A population-based study. Lancet 373, 234–239.

Malhotra, D., McCarthy, S., Michaelson, J.J., et al., 2011. High frequencies of de novo CNVs in bipolar disorder and schizophrenia. Neuron 72, 951–963.

Manolio, T.A., Collins, F.S., Cox, N.J., et al., 2009. Finding the missing heritability of complex diseases. Nature 461, 747–753.

McCarroll, S.A., Kuruvilla, F.G., Korn, J.M., et al., 2008. Integrated detection and population-genetic analysis of SNPs and copy number variation. Nat Genet 40, 1166–1174.

McClellan, J.M., Susser, E., King, M.C., 2007. Schizophrenia: A common disease caused by multiple rare alleles. Br J Psychiatry 190, 194–199.

McGuffin, P., Rijsdijk, F., Andrew, M., et al., 2003. The heritability of bipolar affective disorder and the genetic relationship to unipolar depression. Arch Gen Psychiatry 60, 497–502.

McMahon, F.J., Akula, N., Schulze, T.G., et al., 2010. Meta-analysis of genome-wide association data identifies a risk locus for major mood disorders on 3p21.1. Nat Genet 42, 128–131.

Moskvina, V., Craddock, N., Holmans, P., et al., 2009. Gene-wide analyses of genome-wide association data sets: Evidence for multiple common risk alleles for schizophrenia and bipolar disorder and for overlap in genetic risk. Mol Psychiatry 14, 252–260.

Mulle, J.G., Dodd, A.F., McGrath, J.A., et al., 2010. Microdeletions of 3q29 confer high risk for schizophrenia. Am J Hum Genet 87, 229–236.

Murphy, K.C., Jones, L.A., Owen, M.J., 1999. High rates of schizophrenia in adults with velo-cardio-facial syndrome. Arch Gen Psychiatry 56, 940–945.

O'Donovan, M.C., Craddock, N., Norton, N., et al., 2008a. Identification of loci associated with schizophrenia by genome-wide association and follow-up. Nat Genet 40, 1053–1055.

O'Donovan, M.C., Craddock, N.J., Owen, M.J., 2009. Genetics of psychosis; insights from views across the genome. Hum Genet 126, 3–12.

O'Donovan, M.C., Kirov, G., Owen, M.J., 2008b. Phenotypic variations on the theme of CNVs. Nat Genet 40, 1392–1393.

Owen, M.J., Craddock, N., Jablensky, A., 2007. The genetic deconstruction of psychosis. Schizophrenia Bull 33, 905–911.

Pocklington, A.J., Cumiskey, M., Armstrong, J.D., Grant, S.G., 2006. The proteomes of neurotransmitter receptor complexes form modular networks with distributed functionality underlying plasticity and behaviour. Mol Syst Biol 2, 2006.0023 doi: 10.1038/msb4100041.

Pulver, A.E., Nestadt, G., Goldberg, R., et al., 1994. Psychotic illness in patients diagnosed with velo-cardio-facial syndrome and their relatives. J Nerv Ment Dis 182, 476–478.

Purcell, S.M., Wray, N.R., Stone, J.L., et al., 2009. Common polygenic variation contributes to risk of schizophrenia and bipolar disorder. Nature 460, 748–752.

Ross, C.A., Margolis, R.L., Reading, S.A., et al., 2006. Neurobiology of schizophrenia. Neuron 52, 139–153.

Sebat, J., Lakshmi, B., Troge, J., et al., 2004. Large-scale copy number polymorphism in the human genome. Science 305, 525–528.

Sebat, J., Levy, D.L., McCarthy, S.E., 2009. Rare structural variants in schizophrenia: One disorder, multiple mutations; one mutation, multiple disorders. Trends Genet 25, 528–535.

Shi, J., Levinson, D.F., Duan, J., et al., 2009. Common variants on chromosome 6p22.1 are associated with schizophrenia. Nature 460, 753–757.

St Clair, D., 2009. Copy number variation and schizophrenia. Schizophr Bull 35, 9–12.

Stefansson, H., Ophoff, R.A., Steinberg, S., et al., 2009. Common variants conferring risk of schizophrenia. Nature 460, 744–747.

Stefansson, H., Rujescu, D., Cichon, S., et al., 2008. Large recurrent microdeletions associated with schizophrenia. Nature 455, 232–236.

Sudhof, T.C., 2008. Neuroligins and neurexins link synaptic function to cognitive disease. Nature 455, 903–911.

Vimaleswaran, K.S., Loos, R.J., 2010. Progress in the genetics of common obesity and type 2 diabetes. Exp Rev Mol Med 12, e7.

Voight, B.F., Scott, L.J., Steinthorsdottir, V., et al., 2010. Twelve type 2 diabetes susceptibility loci identified through large-scale association analysis. Nat Genet 42, 579–589.

Williams, H.J., Norton, N., Dwyer, S., et al., 2010a. Fine mapping of ZNF804A and genome-wide significant evidence for its involvement in schizophrenia and bipolar disorder. Mol Psychiatry 16. doi: 10.1038/mp.2010.36.

Williams, N.M., Zaharieva, I., Martin, A., et al., 2010b. Rare chromosomal deletions and duplications in attention-deficit hyperactivity disorder: A genome-wide analysis. Lancet 376. doi: 10.1016/S0140-6736(10)61109-9.

Wilson, G.M., Flibotte, S., Chopra, V., et al., 2006. DNA copy-number analysis in bipolar disorder and schizophrenia reveals aberrations in genes involved in glutamate signaling. Hum Mol Genet 15, 743–749.

Zhang, D., Cheng, L., Qian, Y., et al., 2009. Singleton deletions throughout the genome increase risk of bipolar disorder. Mol Psychiatry 14, 376–380.

CHAPTER

89

Depression

Brigitta Bondy

INTRODUCTION

Depression is a highly prevalent, potentially life-threatening condition that affects hundreds of millions of people all over the world. It can occur at any age from childhood to later life, and exerts a tremendous cost upon society. There is a triad of core symptoms: low or depressed mood, anhedonia (reduced ability to experience pleasure from natural rewards), and low energy or fatigue. Other symptoms are also often present, such as irritability, difficulty in concentrating, sleep and psychomotor disturbances, feelings of guilt, low self-esteem, and suicidal tendencies, as well as autonomous and gastrointestinal disturbances.

Depression is not a homogeneous disorder but rather is a complex phenomenon with many subtypes. The symptomatology ranges from mild to severe symptoms with or without psychotic features. Interactions with other psychiatric and somatic disorders are quite common. In addition to mortality associated with suicidality, depressed patients are more likely to develop coronary artery disease with worsened prognosis after myocardial infarction (Frasure-Smith and Lesperance, 2006).

The etiology of depression is multifactorial, including nongenetic and genetic factors. As with most human complex disorders, the relationship between genes and the disorder is not straightforward but reflects complicated interactions between multiple genes and the environment. Similarly, the response to antidepressant medication varies among patients, and is also at least partly influenced by genetic variants.

DIAGNOSIS, PREVALENCE, AND COURSE OF DEPRESSION

The clinical course of major depression (MD; formerly unipolar depression) is characterized by one or more major depressive episodes without a history of manic, mixed, or hypomanic episodes. According to the diagnostic criteria of the *Diagnostic and Statistical Manual of Mental Disorders* (DSM-IV) (American Psychiatric Association, 1994), five of the following symptoms have to be present for a minimum of two weeks: depressed mood, loss of interest or pleasure, significant alteration in weight and appetite, insomnia or hyposomnia, disturbances in psychomotor activity with either agitation or retardation, fatigue or loss of energy, feelings of worthlessness, diminished ability to think or concentrate, and, last but not least, recurrent thoughts of death and suicidal ideation or acts. Recurrent episodes of major depressive disorders may differ in symptomatology and thus show pleomorphic manifestations within one individual (Oquendo et al., 2004).

The lifetime prevalence of depression is between 10 and 20% in the general population worldwide, with a female to male ratio of about 5:2. Typically, the course of the disease is recurrent, and most patients recover from major depressive

Genomic and Personalized Medicine, 2nd edition
by Ginsburg & Willard. DOI: http://dx.doi.org/10.1016/B978-0-12-382227-7.00089-6

episodes. However, a substantial proportion of patients become chronic, and after 5 or 10 years of prospective follow-up, 12% and 7%, respectively, are still depressed (Keller et al., 1997). But patients who recover also have a high rate of recurrence, as approximately 75% of patients experience more than one episode of major depression within 10 years. There is a high level of comorbidity between anxiety and depressive disorders, with co-occurrence rates up to 60% (Gorman, 1996). This suggests that comorbid anxiety and depression are the rule rather than the exception. Furthermore, there is high co-occurrence of neuroticism, which is characterized by dysphoria, tension, and emotional reactivity, and is often a premorbid personality structure and a robust predictor for future onset of depression. It is estimated that both anxiety and neuroticism share about 50% of the genetic factors with depression.

Depressive disorders are complex phenomena in terms of symptomatology and are multifactorial in etiopathogenesis. Although the specific causes have not been elucidated in detail, there is overwhelming evidence that depression is caused by a multiple interaction of genetic risk factors and environmental and neurobiological factors (Mann and Currier, 2006). Cross-influences can be seen between all these pathways, and genes contribute to all of them.

MOLECULAR NEUROBIOLOGY OF DEPRESSION

Major depression is a psychiatric syndrome that appears to be associated with subtle cellular and molecular alterations in a complex neural network. It is widely accepted that depression is caused by neurochemical imbalances in regions of the brain that are known to control mood, anxiety, cognition, and fear. These regions include the hippocampus, prefrontal and cingulate cortices, nucleus accumbens, and amygdala, which display dynamic neuroplastic adaptations to endocrine and immunologic stimuli, arising from within and outside the central nervous system. The discovery that depression is characterized by structural alterations of the brain, which result from atrophy and loss of neurons and glia in specific limbic regions and circuits, has contributed a fundamental change in our understanding of the illness. These structural changes are accompanied by dysregulation of neuroprotective mechanisms and neurotransmitter signaling, and alteration in immune function (Duman, 2009).

Dysfunction within highly interconnected limbic regions, which regulate emotion, reward, and executive functions, has been implicated in depression and antidepressant action. Both postmortem and neuroimaging studies have revealed reductions in grey matter volume and glial density in the prefrontal cortex and the hippocampus, regions thought to mediate the cognitive aspect of depression, such as feelings of worthlessness and guilt. However, findings are often complicated by comorbid disorders, such as anxiety, and the findings are not

consistent. In studies assessing brain function, such as functional magnetic resonance imaging (fMRI), it can be demonstrated that the activity within the amygdala and cingulate cortex is strongly correlated with dysphoric emotions. Indices of neuronal activity within these regions are increased by transient sadness in healthy volunteers and are chronically increased in depressed individuals, reverting to normal after successful treatment (Krishnan and Nestler, 2008).

The Role of Monoamines

For almost 50 years, the assumption of decreased brain monoaminergic function formed the core of pathophysiological hypothesis in major depression. Particularly from the mechanisms of action of antidepressant drugs, but also from biological studies, there is evidence that transmission of the monoamines serotonin (5-HT), norepinephrine (NE) and also dopamine (DA), which modulates many behavioral symptoms, is decreased in depression. The consequences of this reduced monoaminergic function are alterations in mood, vigilance, motivation, and fatigue, and psychomotor agitation or retardation, all major symptoms of the depressive state.

Abnormal function may arise from altered synthesis, storage, or release of the neurotransmitters, as well as from disturbed sensitivity of their respective receptors or sub-cellular messenger functions within the synapse. Transport proteins play a crucial role in neural transmission, as they reduce the availability of the neurotransmitters in the synaptic cleft and thus terminate their effect on pre- and postsynaptic receptors. In this respect, it is noteworthy that most of the available antidepressants increase monoaminergic transmission either by inhibiting neuronal reuptake (for example, selective 5-HT or NE reuptake inhibitors), by inhibiting degradation (for example, monoamine-oxidase inhibitors), or by coupling and thus affecting monoamine receptors on the pre- or postsynaptic membranes.

Although there is no doubt that this alteration in central monoaminergic function may contribute to vulnerability to depression, the cause of depression is far from being a simple deficiency in central monoamines (Krishnan and Nestler, 2008). Mere increase or decrease in monoaminergic concentrations, as reached with drugs or experimental monoamine depletion, may have a minimal effect in depressed patients, but do not alter mood in healthy controls (Ruhe et al., 2007). There is now consensus that alterations in the amount of synaptic monoamines are one part of the alteration in neuroplasticity, a fundamental mechanism of neural adaptation which is disrupted in mood disorders (Pittenger and Duman, 2008).

Stress, Depression, and Neuroplasticity

Stress is usually defined as a state of disturbed homeostasis, inducing somatic and mental adaptive reactions, globally termed "stress response." All these reactions aim to reconstitute the initial homeostasis or to create a new level of homeostasis after successful adaptation. The major neuroendocrine response mediating stress adaptation is activation of the hypothalamic–pituitary–adrenal (HPA) axis, with stimulation of corticotropin-releasing hormone (CRH) and vasopressin (VP) from the hypothalamus, leading to stimulation of NE secretion in the brain and adrenocorticotropin (ACTH) secretion from the pituitary, which finally increases glucocorticoid secretion from the adrenal cortex. Both basal production and transient increases of glucocorticoids and their hypothalamic regulators during acute stress are essential for neuronal plasticity, normal brain function, and survival. However, chronic exposure to stress hormones can predispose to psychological, metabolic, and immune alterations. Thus, prompt termination of the stress response via intact feedback mechanism is essential to prevent the negative effects of inappropriate levels of CRH and glucocorticoids (Aguilera, 2010).

There is wide consensus and support from preclinical and clinical data that stress exposure conceivably plays a causal role in the etiology of MD and depression-like disorders (Kendler et al., 2006). The influence of chronic stress and adverse life events (acute stress) has been the subject of numerous investigations, and most findings show an excess of severely threatening events prior to the onset of a depressive episode (Paykel, 2001). As such events do not always trigger depression, it has been suggested that certain people have a predisposition, probably on a genetic basis, towards adverse reactions. Furthermore, there are gender differences in the reaction to stress, with women being more susceptible to interpersonal stress and men to legal- or work-related stressful life events (Kendler et al., 2001). A very important factor is early life stress, including childhood neglect or sexual abuse, both of which are major risk factors for the onset of depression in later life (Caspi et al., 2010). An explanation for this delayed reaction could be an impact of stress on the formation of enduring effects on biological and psychological development (Mann and Currier, 2006).

Among the consistent findings in psychiatry is that a significant proportion of depressed patients hypersecrete CRH, ACTH, and finally stress hormones, such as cortisol or norepinephrine, during the acute state of the disease (Holsboer, 2000). This chronic increase in stress hormones results not only in inadequate glucocorticoid feedback mechanisms in depression with sustained CRH release, but also in disturbances of the serotonergic and immune systems on several levels. It is known that sustained CRH or cortisol overdrive down-regulates the serotonergic system in terms of turnover, activity of the neurons, and alterations of the receptor basis (Van Praag, 2005).

Although specific mechanisms have not yet been fully elucidated, growing evidence indicates that several neurotransmitters and neuropeptides, as well as immune and inflammatory mediators, are likely intermediate links between stress exposure, depressive symptoms, and major depression (Bartolomucci and Leopardi, 2009). Chronic stress causes alteration to the number and shape of neurons and glia in brain regions implicated in mood disorders, leading to neural damage, atrophy, and cell loss (Duman, 2009), which can be demonstrated in brain imaging studies of depressed patients (Campbell and MacQueen, 2006). Preclinical and postmortem studies of signal transduction pathways and target genes have demonstrated dysregulation of neurotrophic factors and neuroprotective mechanisms in response to stress in depressed patients at a molecular level. Conversely, administration of antidepressants blocks the effects of stress or leads to induction of neurotrophic or neuroprotective pathways, and thus counteracts the effects of stress on neural function (Duman, 2009; Manji et al., 2003).

A number of different mechanisms induce dentritic atrophy of the neural cells, decreased neurogenesis in the hippocampus and prefrontal cortex, and decreased proliferation of glia cells. These include alterations of neurotrophic factors, such as brain-derived neurotrophic factor (BDNF), nerve growth factors (NGF), and proinflammatory cytokines, and metabolic alterations. BDNF and NGF not only influence proliferation and differentiation in the developing brain but also play a critical role in the mature brain. Exposure to stress and the increased concentration of glucocorticoids result in a dramatic decrease of BDNF and decreased neurogenesis, especially in the hippocampus and prefrontal cortex, brain areas relevant for emotional regulation. Also, increase in proinflammatory cytokines, such as interleukin 1β (IL-1β), contributes to the decrease of growth factors (Barrientos et al., 2003; Smith et al., 1995). Abnormal levels of glutamate, the main excitatory neurotransmitter in the central nervous system, may contribute to cell damage and death, mainly via induction of oxidative stress, membrane damage, and DNA degradation (Duman, 2009).

The influence of these factors on cell function and survival could occur rapidly after a single major event, or gradually over time, with the accumulation of more insults. The effects of these cellular stressors are also influenced by genetic factors that can either increase susceptibility to cellular damage or conversely decrease susceptibility and increase resilience and neuroprotection (Duman, 2009). Thus, a relationship between BDNF, morphology, and behavior is supported by genetic studies of BDNF. A functional polymorphism in the BDNF gene, Val66Met, has been repeatedly investigated, and the Met allele was associated with reduced hippocampal size and decreased memory and executive function in humans (Frodl et al., 2007). The Met allele was also associated with patients who have an increased incidence of depression when exposed to stress and trauma (Kaufman et al., 2006; Kim et al., 2007) (see *Gene–Environment Interplay* section, below).

Altogether, there is no doubt that chronic stress and depression induce structural changes within the central nervous system, accompanied by dysregulation of neuroprotective signaling mechanisms. Conversely, behavioral and therapeutic interventions can reverse these structural alterations by stimulation of neuroprotective and neurotrophic pathways, and by blocking the effects of chronic stress (Pittenger and Duman, 2008).

GENETIC BASIS OF MAJOR DEPRESSION

Twin and family studies have documented that genetic factors play an important role in MD, as first-degree relatives of mood disorder patients have an approximately three-fold increased risk of developing depression (Malhi et al., 2000). This number is further increased up to eight-fold in recurrent, early-onset depression. Moreover, relatives of depressed patients are more susceptible to anxiety, substance abuse, or social impairment compared to the offspring of non-depressed parents (Weissman et al., 2006). While less heritable than schizophrenia and bipolar disorder (see Chapter 88), heritability of major depression can now be estimated at being about 40%, and is thus close to that of type 2 diabetes mellitus (see Chapter 82) and other complex disorders for which several genetic risk factors have recently been identified (Krishnan and Nestler, 2008).

The few adoption studies point to an important impact of parental depression, but also to a significant environmental influence of maternal depression in mediating disease among adopted adolescents (Tully et al., 2008). This clearly strengthens the theory that familial loading could be the result of shared environmental factors, thus suggesting that vulnerability to depression could be due to nurture rather than nature. Furthermore, there is no doubt that more than one single gene is responsible for the increased vulnerability, and that considerable gene–gene and gene–environment interactions complicate the identification of relevant susceptibility genes (Levinson, 2006).

Early Molecular Genetic Analyses: Linkage and Association Studies

In the late 1980s and early 1990s, there was enormous optimism among psychiatrists that major genes could be identified as susceptibility factors for depression and thus contribute to the knowledge of pathophysiological mechanisms. Principally there are two main approaches to determine genes conferring risk to a disorder, *linkage studies* and *association studies*.

Linkage Studies

Linkage studies find genomic regions in families, which are shared among affected individuals, but not among unaffected relatives. This method has the advantage that prior knowledge of disease etiology is not necessary and that susceptibility genes may be identified with their approximate chromosomal localization. Linkage analyses are ideal for detecting causative disease genes or genes of large effect; however, they fail in identifying genes of small effect (Sanders and Gill, 2007).

The enthusiasm of the initial linkage studies quickly faded when the studies were confronted with the many problems of psychiatric genetic studies, which are as follows: (1) no single gene appears to be necessary and sufficient for MD, (2) each susceptibility gene contributes only a small fraction of total risk, and (3) complex genetic heterogeneity is evident, meaning that multiple partially overlapping sets of susceptibility genes (which interact with the environment) can predispose individuals to similar syndromes that are indistinguishable on clinical grounds (Lohoff, 2010). Further critical points are the size of the study samples, the number of affected individuals in each family, incomplete penetrance, and last but not least, the problem of defining the phenotype (Burmeister et al., 2008). Phenotype definition might be a challenging issue, especially in MD patients with high rates of comorbidity.

Despite these adversities, some selected regions on chromosomes 3, 4, 6, 12, 15, and 18, especially in combination with depression-related personality traits such as neuroticism and harm avoidance, have received support from more than one study over more than two decades of research (reviewed by Levinson, 2006). However, although some linkage studies in MD have suggested regions in the genome that might harbor risk alleles, findings have been inconsistent and, thus far, no established universal genetic risk factor or causative gene for depression has been identified and widely accepted as a susceptibility gene for depression (Burmeister et al., 2008; O'Donovan et al., 2009).

Association Studies

In contrast to linkage studies, genetic association studies are more powerful in complex disorders, when the genetic variants have only small individual effects. Variations in DNA of selected genes (candidate genes) are examined in a sample of cases and compared to controls. Statistical differences in frequencies among samples would then indicate that the DNA variant under study, or another in close proximity, confers risk to susceptibility for a given disease. Typically, association studies have been used to investigate genes coding for proteins which are thought to be implicated in the etiology of depression, mainly in the serotonergic, noradrenergic, and dopaminergic systems. Other genes or genomic regions of interest are, of course, those particular areas where linkage analyses have already observed positive results (Sanders and Gill, 2007).

Due to the prominent importance of the serotonergic system, association studies have so far focused on polymorphisms in genes encoding the rate-limiting synthesizing enzyme tryptophan hydroxylase (TPH), the serotonin transporter (5-HTT), the different 5-HT receptors, and the degrading enzyme monoamine oxidase-A (MAOA). The majority of studies were carried out with the serotonin transporter gene (*SLC6A4*), which codes for a protein crucial within the synapse as it cleaves 5-HT from the synaptic cleft, and is the target protein of the selective 5-HT reuptake inhibitors and other antidepressants. A polymorphism within the promoter region of the *5-HTT* gene, comprising a 44-bp insertion/deletion resulting in short (S) and long (L) alleles, was shown to affect transcription and thus the function of the gene (Heils et al., 1995). Many studies followed the first findings that the S allele is associated with anxiety (Lesch et al., 1996), but the results were discrepant. Methodological issues might be one explanation for the failure to find such an association, because in most studies the samples were small and clinically heterogeneous.

Overall, similarly to linkage studies, association studies have mostly yielded results that could not reliably be replicated. In light of the multiple positive and negative reports for various candidate genes, several recent meta-analyses have tried to dissect the genetic factors for depression. The hope was that by combining several underpowered, small studies into one large study, the meta-analysis would have sufficient power to detect risk alleles.

One of the most comprehensive meta-analyses in depression included 183 articles, covering almost 400 single nucleotide polymorphisms (SNPs) in 102 genes (Lopez-Leon et al., 2008). This analysis calculated that the S allele of the *5-HTT* gene conferred an increased risk for MD, although with a very small effect [odds ratio (OR) of 1.11]. Calculations also revealed other genes with potential roles as susceptibility factors: apolipoprotein E2 (*APOE*), guanine nucleotide-binding protein β-3 (*GNB3*), the dopamine D2 type receptor (*DRD2*), the dopamine transporter (*SLCC6A3*), and methylene tetrahydrofolate reductase (*MTHFR*) (Lopez-Leon et al., 2008). Several further positive findings, such as the immune-related gene *P2RX7*, await further replication (Goltser-Dubner et al., 2010).

Considering the immensity of efforts over 20 years of research, the overall results are disappointing. Although the methodology and power of association studies has improved over time, they still experience a lot of criticism because of small samples used and because of insufficient statistical handling of multiple testing. With all these methodological pitfalls, it is not surprising that in all cases the putative effect sizes are small, with OR < 2, and that negative association studies have constantly followed the positive findings. Also, several meta-analyses, compiling different single studies, have retrieved inconsistent results thus underlining the early skepticism about the use of molecular genetic methods in psychiatric disorders (Penrose, 1971).

Genome-wide Association Studies (GWAS)

Genomics, in particular the recently demonstrated ability of genome-wide association studies to identify common risk variants, is transforming medicine. The obvious success of GWAS in somatic complex disorders has raised hope among psychiatrists that this would be the ideal methodology to identify risk genes for psychiatric conditions. GWAS involve the use of arrays that simultaneously genotype several hundred thousand SNPs per individual at reasonable cost. This enables a hypothesis-free search of every gene of the genome, in samples of unrelated patients and controls. In this respect, GWAS resemble genome-wide linkage studies (genome scans), but with several major advantages: they are not dependent on the recruitment of families (which is quite difficult with psychiatric patients), and they have a better resolution since they detect linkage disequilibrium with susceptibility variants, which usually extend over smaller genomic regions (in the range of a few tens of thousands of base pairs). This means that the majority of all genetic

variation is captured, although only a proportion of all possible SNPs are genotyped. Finally, GWAS have greater power to detect small genetic effects (Barrett and Cardon, 2006).

Despite these advantages, it has turned out that there are several serious difficulties in evaluating the results of GWAS. The extensive multiple testing, being inherent to this approach, requires large samples (preferable thousands or tens of thousands of patients and controls) to provide sufficient power to detect variants with modest effects. A large number of SNPs may be tested for association within the same analysis, and this may generate false positive findings. It is therefore necessary to correct for multiple testing, and this increases the threshold for genome-wide significance, which is currently a P-value of 5×10^{-8}. This high threshold could, on the one side, reduce the risk for false positive findings, but might on the other side also hamper the detection of genes with small disease risk (Nothen et al., 2010).

Thus far, five large GWAS have been published for major depression, including those based on different European and American samples. Although several regions of interest could be identified in the genome, including the piccolo gene (*PCLO*), cyclin-D2 (*CCND2*), and some intronic markers in the genes *ATP6V1B2*, *GRM7*, and *SP4*, the results of these studies were essentially negative, as none of them reached the high level of genome-wide significance (for a review, see Lohoff, 2010). Recently, data from the Genetic Association Information Network GWAS were used to explore previously reported candidate genes' association for depression (Bosker et al., 2010). The authors found support for the significant involvement of four genes (*C5orf20*, *NPY*, *TNF*, and *SLC6A2*), but pointed to the fact that, due to the large number of SNPs tested, these susceptibility genes might turn out to be false positive findings. In particular, the findings of the comprehensive recent meta-analysis that proposed several susceptibility genes for major depression (Lopez-Leon et al., 2008) could not be replicated in this analysis. The complex phenotype of MD, the interaction with environmental factors, and the impact of lifestyle might all be factors that decrease the efficiency of genetic approaches, even in GWAS.

Gene–Environment Interplay

It has long been discussed that gene–environment interactions might be an explanation for the failure of molecular genetic investigations and the diversity and inconsistency of results (see Chapter 4). Twin studies have been consistent in showing that environmental influences account for a substantial proportion of the population variance (Plomin et al., 1994). This means that subtle differences in genetic factors might cause people to respond differently to the same environmental agents. In particular, early stressful life events or early environmental features, such as sexual or physical abuse, but also more immediate negative life events, carry a long-term threat with negative effects (Rutter, 2010). However, the fact that even with unusually severe or prolonged deprivation there is considerable heterogeneity in response, raises the question of whether at least part of the heterogeneity derives from genetic influences.

Emerging evidence from human and animal studies indicates that a relative loss in *5-HTT* gene function not only increases anxiety, but also exerts a negative influence on the capacity to cope with stress, which further increases the risk for mood disorders. This apparent relation between the response to stress (environment) and genetic makeup was underlined by a study by Caspi and colleagues, who investigated a representative cohort in a prospective, longitudinal study (Caspi et al., 2003). They showed that individuals with one or two copies of the S allele at *5-HTT* exhibited more depressive symptoms, more diagnosable depression, and more suicidality in relation to earlier stressful life events than individuals who were homozygous for the L allele. This finding suggested that genetic variants may act to promote one's resistance to environmental pathogens, and that the S allele moderates the depressogenic influence of stressful life events.

Many studies followed this first report on gene–environment interaction. Although the majority of the studies could replicate the initial findings, others observed an interaction between gene and environment only in females or only with certain types of adversity, and some studies did not find an interaction at all. In an updated review of these studies, it turned out that all the studies that used objective measures or structured interviews were able to replicate the initial findings (Uher and McGuffin, 2010). However, other authors had serious doubts about the association between *5-HTT* variants, stressful life events, and depression, as a meta-analysis received several points of criticism (Risch et al., 2009). It could thus be concluded that variation in sampling and measurement account for most of the variation in these findings. But it may still be premature to conclude that a true understanding of the variation in findings has been achieved (Rutter, 2010).

Although the main focus of studies has so far been on the association between the *5-HTT* gene and early life stress, other genes of interest are being investigated in relation to gene–environment interaction. Among other results, reduced BDNF levels were reported in carriers of the *BDNF* Met allele who experienced childhood abuse (Elzinga et al., 2010). Other studies will follow, as the initial findings demonstrated a strong likelihood of a valid gene–environment effect in the genesis of depression (Rutter, 2010). On the other hand, even in the case of negative findings, epigenetic mechanisms operating in early life might be responsible for the biological changes, without any relation to genetic variants (Uher and McGuffin, 2010). While it remains too early for direct clinical implications, these findings provide leads on the possible ways in which adverse environments have enduring effects that persist beyond the experience in relation to the genes.

Epigenetic Mechanisms

Not all of the many non-replicable results of molecular genetics can be attributed to methodological issues of the studies. During the past decade, convincing evidence has accumulated that variations in DNA sequence alone are not likely to fully explain the genetic contribution in mental disorders. Through a process known as epigenetics, environmental agents are able to modify gene expression, independently from (and in concert with) the primary DNA sequence of the genome. Several biochemical mechanisms, among them methylation, acetylation, and phosphorylation, regulate chromatin structure in the cell nucleus, thus altering access to transcription factors in the promoter regions of genes (Bird, 2002) (see Chapter 1). As a consequence of these biochemical reactions, a gene may be turned on (expressed) or off (silenced) (Stahl, 2010). Epigenetic modifications may be *de novo* and stochastic, or may be mitotically heritable changes of the DNA structure (i.e., not the sequence), which are then linked to long-lasting gene-expression changes.

These mechanisms are very important during development, as the initial epigenetic pattern of a neuron is set during neurodevelopment to give each neuron its own lifelong "personality" (Nestler, 2009). However, under various circumstances, such as childhood abuse, chronic stress, dietary deficiencies, or other environmental factors, epigenetics may change the chromatin also in mature, differentiated cells, leading to favorable or non-favorable effects in the character of neurons (Stahl, 2010). Increasing evidence now accumulates that the dynamic regulation of epigenetic mechanisms is part of the normal gene–environment interface, and thus contributes to psychiatric and behavioral phenotypes.

Mental illness is defined solely on the basis of behavioral abnormalities which typically develop over a relatively long period of time, but can persist over life once formed. The reversal of these abnormalities in response to treatment (either medication or psychotherapeutic intervention) also occurs slowly, with gradual and progressive improvements in case of successful treatment. This suggests involvement of slowly developing, but stable and long-lasting, changes in the brain. Regulation of gene expression via epigenetic modifications could represent one such mechanism (Nestler, 2009). Epigenetics now promises the possibility to determine the state of activation or inactivation of genes in localized brain regions, and to characterize the detailed mechanisms responsible for gene activity.

Epigenetics is a new but increasing approach in psychiatric genomics, and so far only a few studies have been carried out. However, first results point to epigenetic alterations in the glucocorticoid receptor (Turner et al., 2010), or to the distinction between individuals with or without lifetime depression on the basis of genome-wide DNA methylation profiles (Uddin et al., 2010). Another study showed higher methylation of the *5-HTT* gene promoter region, which could be associated with increased risk of unresolved responses to loss or trauma in carriers of the usually protective L allele (Van IJzendoorn et al., 2010). This finding indicates that epigenetic methylation may serve as the interface between adverse environment and the developing organism. The future will show whether we can await reliable results or similar discrepancies as with the findings of genetic analyses (Mill et al., 2008).

CONCLUSIONS AND FUTURE ASPECTS OF DEPRESSION GENOMICS

Despite extensive research over more than 20 years, the identification of genetic risk factors for major depression remains a rather challenging task. Considering the initial goal of "simply" unraveling the underlying genetic susceptibility, we have to acknowledge that the promises have not been met today. Locus and allelic heterogeneity, involvement of unidentified environmental factors, heterogeneous phenotypes, the probable contribution of a multitude of rare variants, multiple genes with small effects interacting with each other and with environmental factors, are all complicating the reliable detection of susceptibility genes.

In the investigation of complex disorders, it has become apparent that testing one or a few polymorphisms within a gene for association with a disease, which was the "publication standard" for almost two decades, is insufficient. The increasing number of known gene variants has made clear that one gene contains many SNPs that could influence its function. Further, both the analytical and the statistical methods are not yet optimal to cope with this complexity. Although sample sizes have increased and methodological designs have improved, there is considerable debate as to when plausibly reported association findings should be accepted and when they still need to be confirmed. Both false positive and false negative results may arise, due to ethnic stratification, insufficient correction for multiple testing, and due to publication bias, favoring positive findings (Burmeister et al., 2008).

A primary challenge in psychiatric genetics is the lack of completely validated systems of classification. However, without a well-defined phenotype, the establishment of a relationship between gene and a disorder is difficult, since heterogeneity of the sample may dilute any existing effects. Thus patients being inadequately classified for a broad disease such as depression would reduce the estimated odds ratio against any particular genetic cause for a (not yet clinically definable) subset of depression. It is therefore not surprising that relatively few findings have been replicated.

The interaction with environmental and lifestyle factors is frequently quoted as an important criterion in the complexity of MD. The increasing knowledge about the effects of stress on the developing and mature brain is changing our pathophysiological concepts of depression. Structural changes of the brain, together with alterations in neuroplasticity after childhood or lifetime exposure to stressors, and together with their interaction with genetic variants, will focus future research on this interaction. So far, few studies have included environment in genetic analyses, but this field is now expanding and may advance basic science to generate broader approaches for studying complex diseases and traits.

Gene–environment interactions are closely associated with epigenetic mechanisms, which can alter gene transmission and function. Increased knowledge about the impact of environment in the disease process could open a new window towards unraveling pathophysiological mechanisms of disorders. Classical genetics deals with the sequence of DNA that is inherited, but epigenetics is confronting psychiatry with the hypothesis that both "normal" genes as well as risk genes can contribute to a mental disorder. If normal genes make normal gene products but at the wrong time, either being epigenetically expressed in neurons when they should be silenced or epigenetically silenced in neurons when they should be expressed, this can contribute to the development of psychiatric symptoms or disorders (Stahl, 2010). Epigenetics appears to offer a potentially powerful tool to investigate the transcriptional mechanisms in acute major depression and during treatment.

REFERENCES

Aguilera, G., 2010. HPA axis responsiveness to stress: Implications for healthy aging. Exp Gerontol 46, 90–95.

American Psychiatric Association, 1994. Diagnostic and Statistical Manual of Mental Disorders, Washington DC.

Barrett, J.C., Cardon, L.R., 2006. Evaluating coverage of genome-wide association studies. Nat Genet 38, 659–662.

Barrientos, R.M., Sprunger, D.B., Campeau, S., et al., 2003. Brain-derived neurotrophic factor mRNA downregulation produced by social isolation is blocked by intrahippocampal interleukin-1 receptor antagonist. Neuroscience 121, 847–853.

Bartolomucci, A., Leopardi, R., 2009. Stress and depression: Preclinical research and clinical implications. PLoS One 4, e4265.

Bird, A., 2002. DNA methylation patterns and epigenetic memory. Genes Dev 16, 6–21.

Bosker, F.J., Hartman, C.A., Nolte, I.M., et al., 2010. Poor replication of candidate genes for major depressive disorder using genome-wide association data. Mol Psychiatry 16, 516–532.

Burmeister, M., McInnis, M., Zöllner, S., 2008. Psychiatric genetics: Progress amid controversy. Nat Rev Genet 9, 527–540.

Campbell, S., MacQueen, G., 2006. An update on regional brain volume differences associated with mood disorders. Curr Opin Psychiatry 19, 25–33.

Caspi, A., Hariri, A.R., Holmes, A., Uher, R., Moffitt, T.E., 2010. Genetic sensitivity to the environment: The case of the serotonin transporter gene and its implications for studying complex diseases and traits. Am J Psychiatry 167, 509–527.

Caspi, A., Sugden, K., Moffitt, T.E., et al., 2003. Influence of life stress on depression: Moderation by a polymorphism in the 5-HTT gene. Science 301, 386–389.

Duman, R.S., 2009. Neuronal damage and protection in the pathophysiology and treatment of psychiatric illness: Stress and depression. Dialogues Clin Neurosci 11, 239–255.

Elzinga, B.M., Molendijk, M.L., Oude Voshaar, R.C., et al., 2010. The impact of childhood abuse and recent stress on serum brain-derived neurotrophic factor and the moderating role of BDNF Val(66)Met. Psychopharmacology (Berl) 214, 319–328.

Frasure-Smith, N., Lesperance, F., 2006. Depression and coronary artery disease. Herz 31 (Suppl. 3), 64–68.

Frodl, T., Schule, C., Schmitt, G., et al., 2007. Association of the brain-derived neurotrophic factor Val66Met polymorphism with reduced hippocampal volumes in major depression. Arch Gen Psychiatry 64, 410–416.

Goltser-Dubner, T., Galili-Weisstub, E., Segman, R.H., 2010. Genetics of unipolar major depressive disorder. Isr J Psychiatry Relat Sci 47, 72–82.

Gorman, J.M., 1996. Comorbid depression and anxiety spectrum disorders. Depress Anxiety 4, 160–168.

Heils, A., Teufel, A., Petri, S., et al., 1995. Functional promoter and polyadenylation site mapping of the human serotonin (5-HT) transporter gene. J Neural Transm Gen Sect 102, 247–254.

Holsboer, F., 2000. The corticosteroid receptor hypothesis of depression. Neuropsychopharmacology 23, 477–501.

Kaufman, J., Yang, B.Z., Douglas-Palumberi, H., et al., 2006. Brain-derived neurotrophic factor-5-HTTLPR gene interactions and environmental modifiers of depression in children. Biol Psychiatry 59, 673–680.

Keller, M.B., Hirschfeld, R.M., Hanks, D., 1997. Double depression: A distinctive subtype of unipolar depression. J Affect Disord 45, 65–73.

Kendler, K.S., Gardner, C.O., Prescott, C.A., 2006. Toward a comprehensive developmental model for major depression in men. Am J Psychiatry 163, 115–124.

Kendler, K.S., Thornton, L.M., Gardner, C.O., 2001. Genetic risk, number of previous depressive episodes, and stressful life events in predicting onset of major depression. Am J Psychiatry 158, 582–586.

Kim, J.M., Stewart, R., Kim, S.W., et al., 2007. Interactions between life stressors and susceptibility genes (5-HTTLPR and BDNF) on depression in Korean elders. Biol Psychiatry 62, 423–428.

Krishnan, V., Nestler, E.J., 2008. The molecular neurobiology of depression. Nature 455, 894–902.

Lesch, K.P., Bengel, D., Heils, A., et al., 1996. Association of anxiety-related traits with a polymorphism in the serotonin transporter gene regulatory region. Science 274, 1527–1531.

Levinson, D.F., 2006. The genetics of depression: A review. Biol Psychiatry 60, 84–92.

Lohoff, F.W., 2010. Overview of the genetics of major depressive disorder. Curr Psychiatry Rep 12, 539–546.

Lopez-Leon, S., Janssens, A.C., Gonzalez-Zuloeta Ladd, A.M., et al., 2008. Meta-analyses of genetic studies on major depressive disorder. Mol Psychiatry 13, 772–785.

Malhi, G.S., Moore, J., McGuffin, P., 2000. The genetics of major depressive disorder. Curr Psychiatry Rep 2, 165–169.

Manji, H.K., Quiroz, J.A., Sporn, J., et al., 2003. Enhancing neuronal plasticity and cellular resilience to develop novel, improved therapeutics for difficult-to-treat depression. Biol Psychiatry 53, 707–742.

Mann, J.J., Currier, D., 2006. Effects of genes and stress on the neurobiology of depression. Int Rev Neurobiol 73, 153–189.

Mill, J., Tang, T., Kaminsky, Z., et al., 2008. Epigenomic profiling reveals DNA-methylation changes associated with major psychosis. Am J Hum Genet 82, 696–711.

Nestler, E.J., 2009. Epigenetic mechanisms in psychiatry. Biol Psychiatry 65, 189–190.

Nothen, M.M., Nieratschker, V., Cichon, S., Rietschel, M., 2010. New findings in the genetics of major psychoses. Dialogues Clin Neurosci 12, 85–93.

O'Donovan, M.C., Craddock, N.J., Owen, M.J., 2009. Genetics of psychosis; insights from views across the genome. Hum Genet 126, 3–12.

Oquendo, M.A., Barrera, A., Ellis, S.P., et al., 2004. Instability of symptoms in recurrent major depression: A prospective study. Am J Psychiatry 161, 255–261.

Paykel, E.S., 2001. The evolution of life events research in psychiatry. J Affective Disord 62, 141–149.

Penrose, L.S., 1971. Psychiatric genetics. Psychol Med 1, 265–266.

Pittenger, C., Duman, R.S., 2008. Stress, depression, and neuroplasticity: A convergence of mechanisms. Neuropsychopharmacology 33, 88–109.

Plomin, R., Owen, M.J., McGuffin, P., 1994. The genetic basis of complex human behaviors. Science 264, 1733–1739.

Risch, N., Herrell, R., Lehner, T., et al., 2009. Interaction between the serotonin transporter gene (5-HTTLPR), stressful life events, and risk of depression: A meta-analysis. JAMA 301, 2462–2471.

Ruhe, H.G., Mason, N.S., Schene, A.H., 2007. Mood is indirectly related to serotonin, norepinephrine and dopamine levels in humans: A meta-analysis of monoamine depletion studies. Mol Psychiatry 12, 331–359.

Rutter, M., 2010. Gene–environment interplay. Depress Anxiety 27, 1–4.

Sanders, J., Gill, M., 2007. Unravelling the genome: A review of molecular genetic research in schizophrenia. Ir J Med Sci 176, 5–9.

Smith, M.A., Makino, S., Kvetnansky, R., Post, R.M., 1995. Effects of stress on neurotrophic factor expression in the rat brain. Ann NY Acad Sci 771, 234–239.

Stahl, S.M., 2010. Methylated spirits: Epigenetic hypotheses of psychiatric disorders. CNS Spectr 15, 220–230.

Tully, E.C., Iacono, W.G., McGue, M., 2008. An adoption study of parental depression as an environmental liability for adolescent depression and childhood disruptive disorders. Am J Psychiatry 165, 1148–1154.

Turner, J.D., Alt, S.R., Cao, L., et al., 2010. Transcriptional control of the glucocorticoid receptor: CpG islands, epigenetics and more. Biochem Pharmacol 80, 1860–1868.

Uddin, M., Koenen, K.C., Aiello, A.E., Wildman, D.E., De Los, S.R., Galea, S., 2010. Epigenetic and inflammatory marker profiles associated with depression in a community-based epidemiologic sample. Psychol Med, 1–11.

Uher, R., McGuffin, P., 2010. The moderation by the serotonin transporter gene of environmental adversity in the etiology of depression: 2009 update. Mol Psychiatry 15, 18–22.

Van IJzendoorn, M.H., Caspers, K., Bakermans-Kranenburg, M.J., Beach, S.R., Philibert, R., 2010. Methylation matters: Interaction between methylation density and serotonin transporter genotype predicts unresolved loss or trauma. Biol Psychiatry 68, 405–407.

Van Praag, H.M., 2005. Can stress cause depression? World J Biol Psychiatry 6 (Suppl. 2), 5–22.

Weissman, M.M., Wickramaratne, P., Nomura, Y., Warner, V., Pilowsky, D., Verdeli, H., 2006. Offspring of depressed parents: 20 years later. Am J Psychiatry 163, 1001–1008.

Autism Spectrum Disorders

Timothy W. Yu, Michael Coulter, Maria Chahrour, and Christopher A. Walsh

INTRODUCTION

Views on genetics of autism spectrum disorders (ASD) have changed dramatically in recent years. Although twin studies have supported a strong role for genetics for some time (Bailey et al., 1995; Folstein and Piven, 1991; Hallmayer et al., 2011), the paucity of known specific genetic causes has sustained persistent skepticism about the roles of genes (as opposed to, for example, environmental agents such as vaccines) and controversy as to the sorts of genetic risk factors that might be prevalent in ASD. The last few years have seen an explosion in the identification and study of highly penetrant, quasi-Mendelian ASD genes. The identification of these ASD genes reflects continuous developments in genomic technology, including the ability to determine copy number systematically at all sites in the genome and, more recently, the ability to sequence in a rapid and high-throughput fashion the entire genome or, in other cases, the entire portion of the genome that encodes proteins, known as the exome (Gillis and Rouleau, 2011). Currently, we are in the midst of an explosion of understanding of the genetic architecture of the disease due to the widespread application (as yet only beginning) of these powerful new sequencing methods.

Three major concepts have emerged recently about the genetics of ASD. First and foremost, ASD are extremely heterogeneous in every way (Abrahams and Geschwind, 2008;

Geschwind, 2008). The behavioral phenotype of autism itself is heterogeneous, in the sense that the term encompasses children with a very wide range of clinical presentations. Moreover, some children diagnosed on the broad autism spectrum can move into or out of the diagnosis of autism over the years, so the diagnosis is not always static. However, beyond even that, even if one focuses on children who look very similar clinically or phenotypically, children with indistinguishable conditions can reflect mutations in a very wide range of different genes, to the point that we presently do not have clinical tools capable of predicting a specific genetic etiology.

A second important concept is the effect of evolutionary selection on ASD-associated mutations. Since ASD patients, like those with intellectual disability or other severely disabling childhood illnesses, show severely reduced fertility, they have a greatly reduced likelihood of passing their disease-associated mutations to progeny, by comparison, for example, to typically developing children and/or those with adult-onset diseases. Therefore, mutations that confer a large relative risk in the heterozygous state tend to be very rarely transmitted. Hence, the presence of these highly penetrant heterozygous mutations in the population typically represents *de novo* mutations, i.e., those present in the affected child but not found in either parent (Abrahams and Geschwind, 2008; Gauthier et al., 2009; Kumar et al., 2008; Moessner et al., 2007; Sebat et al., 2007; Weiss et al., 2008).

Genomic and Personalized Medicine, 2nd edition
by Ginsburg & Willard. DOI: http://dx.doi.org/10.1016/B978-0-12-382227-7.00090-2

Such *de novo* mutations are a much more common cause of severe childhood illness than is generally recognized, though ASD do not so far appear to show a higher *de novo* mutation rate than other diseases with similarly severely reduced rates of transmission.

A third important concept is that no gene identified to date causes only ASD and nothing else; rather, the genes found to be mutated in ASD can be mutated in other developmental brain disorders that do not have ASD as a clinical feature. This observation was known from the earliest stages of analysis for Mendelian disorders such as tuberous sclerosis and fragile X, for which a certain proportion of patients present with typical ASD, whereas other affected patients can show intellectual disability or other features without prominent social defects. However, more recently, the spectrum of behavioral manifestations of virtually any genetic mutation that greatly increases the risk of ASD has been broadened, so that, for example, two well-documented copy number variants (CNVs), deletions at 22q11.2 or 16p11.2, can be associated with ASD, intellectual disabilities, or other neuropsychiatric disorders such as schizophrenia (Guilmatre et al., 2009; Lionel et al., 2011; Stone et al., 2008). No well-documented genetic cause of ASD so far causes "just autism," and this is true of so many genetic mutations that it has become highly likely to be a general rule. Hence, notwithstanding the clinical and diagnostic challenges, it should not be surprising that children with a variety of phenotypes can have similar underlying genetic pathology.

DEFINING AUTISM IN THE CLINIC AND IN THE LABORATORY

Autism is defined behaviorally by a core triad of social defects: poor language development, poor social behavior, and repetitive or stereotyped behaviors. Although these symptoms may be immediately evident upon clinical assessment, quantitative and standardized behavioral measures are often used to confirm the diagnosis and to quantify specific features in some cases. In the present *Diagnostic and Statistical Manual* version IV (DSM-IV-TR), ASD are synonymous with pervasive developmental disorders (PDD) and comprise five categories: autistic disorders, PDD not otherwise specified (PDD-NOS), Asperger syndrome, Rett syndrome, and childhood disintegrative disorders, though these categories are certain to be reorganized with the impending publication of DSM-V.

The term "autism" itself is a difficult one to apply robustly in a genetic research setting, since the diagnosis does not correlate with pathogenesis, level of function, genetic diagnosis, or prognosis. This becomes problematic in a research context since patients ascertained in different settings (e.g., school versus hospital clinic) can have very different levels of function associated with their ASD and hence probably have different sorts of underlying causation. This can be compounded by the fact that some clinicians tend to exclude from ASD those children with clinical features of ASD, but who have a known genetic condition (e.g., fragile X or tuberous sclerosis) even though the results of the revised autism diagnostic interview (ADI-R) or autism diagnostic observation schedule (ADOS) would classify them as autistic. As more and more specific genetic causes of ASD are identified, this creates a challenge for systematic classification. Fortunately, ASD are increasingly regarded as falling on a continuum with other developmental brain disorders, showing considerable overlap with many of them at a genetic, pathogenetic, syndromic, and behavioral level.

"SYNDROMIC" ASSOCIATIONS WITH ASD

Many of the earliest known genes associated with ASD were identified based on the fact that they also caused broader, multi-organ syndromes, and several such syndromes have ASD as a frequent manifestation. Tuberous sclerosis syndrome (TSC) is an autosomal dominant disorder, associated with a high spontaneous mutation rate, that manifests with cardiac tumors, skin lesions, and hamartomas in many organs, including the brain. Up to 50% of affected children show autistic symptoms, most in the setting of intellectual disability (Wiznitzer, 2004). So far, there is no clear understanding of the determinants of ASD symptoms in patients with TSC. Recently, animal models have shown that there are widespread abnormalities in myelination (Meikle et al., 2007), axonal connectivity, and other developmental events in mice heterozygous for TSC mutations, suggesting that humans with TSC mutations may also have widespread neuronal abnormalities beyond the focal lesions that are only evident with newer, sophisticated magnetic resonance imaging (MRI) methods (Choi et al., 2008).

Fragile X syndrome, associated often with mild dysmorphic features of the ears and head, is associated with ASD in about 50% of children and can rarely present with ASD in the absence of severe intellectual disability. Fragile X is probably the most common single-gene disorder causing ASD (Belmonte and Bourgeron, 2006; Bolton, 2009; McLennan et al., 2011).

The progressive neurodevelopmental disorder Rett syndrome (RTT; Mendelian Inheritance in Man (MIM) 312750) is caused by mutations in the X-linked gene *MECP2* (Amir et al., 1999). RTT is characterized by normal development in the first 6–18 months of life, followed by loss of any acquired speech and the replacement of purposeful hand use with stereotypic movements. *MECP2* mutations, as well as increased gene dosage, can result in a range of neurobehavioral abnormalities, including autism, mild learning disabilities, X-linked mental retardation, and infantile encephalopathy (Chahrour and Zoghbi, 2007). RTT phenotypes overlap with nonsyndromic autism, and RTT is included in DSM-IV. *MECP2* mutations have been reported in ~1% of children diagnosed with autism (Moretti and Zoghbi, 2006), and males with *MECP2* duplications often present with autism (Ramocki and Zoghbi, 2008; Ramocki et al., 2009). The MeCP2 protein regulates the expression of its target genes, and a better understanding of its role in maintaining neuronal function will have implications for autism.

A plethora of other Mendelian syndromes are associated with autistic symptoms as well, but less commonly or universally. As many as 300 different genetic syndromes have been reported at some level to be associated with ASD (Betancur, 2011), although the proportion of patients with the genetic syndrome that manifest ASD is highly variable, so that the strength of these associations is variable. It is important to note that, given that up to 1% of all children suffer from ASD, ASD will of course at some point be reported in children with any possible condition as a coincidence, and larger series will eventually be needed to prioritize which genetic syndromes are truly most commonly associated with ASD.

CHROMOSOMAL DISORDERS AND COPY NUMBER VARIANTS

A growing number of chromosomal rearrangements and duplications or deletions have been associated with ASD. Two of the earliest identified recurrent chromosomal disorders are deletions of 22q11.2 and duplications of 15q11-14. Deletions of 22q11.2, which are associated with the velo-cardio-facial syndrome, or duplications of the same interval, have also been associated with intellectual disability, schizophrenia, and ASD (Eliez, 2007; Fine et al., 2005; Lo-Castro et al., 2009; Ramelli et al., 2008). Inverted duplications of 15q11-14 were found in about 1% of children with ASD and were found to be exclusively maternally inherited, suggesting that an imprinted locus, likely on proximal 15q, is responsible for the ASD symptoms (Cook et al., 1997). These recurrent genomic syndromes served as a model for genetic causation of ASD, but determining the prevalence of such genomic rearrangements, or the potential role of smaller deletions and duplications in ASD, required new technology to scan genomes at high resolution.

A major breakthrough in autism genetics was the discovery of spontaneous (i.e., *de novo*) CNVs as a fairly common cause of ASD (Table 90.1) (Christian et al., 2008; Gilman et al., 2011; Guilmatre et al., 2009; Kusenda and Sebat, 2008; Lee and Lupski, 2006; Marshall et al., 2008; Morrow, 2010; Pinto et al., 2010; Sanders et al., 2011; Schaaf and Zoghbi, 2011; Sebat et al., 2007; Sykes et al., 2009). The availability of higher resolution microarrays has allowed systematic analysis of even small regions of the genome and led to the recognition of spontaneous CNVs in as many as 10% of ASD patients. The most recent and most rigorous studies, using the most modern genome arrays, suggest a consistent excess of CNVs in ~5% of ASD patients, but probably not much higher (Gilman et al., 2011; Levy et al., 2011; Pinto et al., 2010; Sanders et al., 2011; Shen et al., 2010).

Overall, CNVs illustrate several critical aspects of ASD genetics: the high degree of genetic heterogeneity, the important role for *de novo* mutation due to the severe negative evolutionary selection against the condition, and the seemingly indirect mapping of behavior onto genetics, with similar mutations causing a range of phenotypes.

Diagnostic Testing with Chromosomal Microarrays

The identification of spontaneous CNVs as a cause of ASD emphasized that the genetic landscape of ASD is likely to involve a major role of rare variation, including spontaneous mutation, and has hastened the pursuit of other rare variants (e.g., *de novo* point mutations and rare recessive mutations) as well. These chromosome microarray (CMA) studies have led to the recognition of many new regions of the genome that are recurrently affected in unrelated patients, including in 16p11.2, 7q11.23, and 15q13-14. Nonetheless, more than half of the regions that are deleted or duplicated in ASD-related CNVs have very low recurrence and are highly heterogeneous, sometimes almost unique to a given patient. *De novo* CNVs also occur in unaffected individuals (Sebat et al., 2004), making it surprisingly difficult to ascertain exactly which CNVs are likely to be causative and which are benign. Nonetheless, the

TABLE 90.1	Summary of recurrent *de novo* CNVs in ASD			
Genomic region	**Frequency in ASD**	**Deletion/duplication**	**Evidence level**	**Other disease associations**
16p11.2	0.5%	Deletion	++++	ID, ADHD, schizophrenia, obesity
		Duplication	++	ID, epilepsy, schizophrenia, microcephaly
15q11.2-13.1	0.2%	Duplication	+++	ID, epilepsy, schizophrenia
NRXN1 locus	0.13%	Deletion	+	Schizophrenia
22q11.2	0.13%	Deletion	++	VCFS, ID, schizophrenia
		Duplication	+	ID, ADHD
7q11.23	0.1%	Duplication	+	

The table summarizes some of the most common CNVs associated with ASD, and indicates the approximate level of evidence supporting their association. The right column also indicates additional neuropsychiatric or other disorders that have been associated with these same CNVs, to illustrate the variety of phenotypes associated with a given mutation. ID = intellectual disability, ADHD = attention deficit hyperactivity disorder, VCFS = velo-cardio-facial syndrome.
Adapted from Sanders et al., 2011.

widespread use of diagnostic CMA has revolutionized much of the culture of ASD clinical testing, since CMA is the single diagnostic test with the highest yield, leading to a consensus that CMA should be performed in all patients who are diagnosed clinically as being on the autism spectrum (Miller et al., 2011).

The 16p11.2 Deletion/Duplication Syndrome

The 16p11.2 deletion/duplication syndrome, noted previously (Kumar et al., 2008; Sebat et al., 2007) but most clearly delineated by Weiss and colleagues (2008), is now arguably one of the best-characterized ASD genetic syndromes. The 16p11.2 deletion syndrome illustrates themes that recur with other genetic forms of ASD. The 16p11.2 deletions are frequently *de novo*, but can be inherited in perhaps 50% of families, from parents who are either mildly affected or clinically normal, emphasizing the importance of *de novo* variation (Weiss et al., 2008). Deletions of 16p11.2 are not by any means invariably associated with ASD. Overall, about one third of deletion patients meet strict research criteria for autism, being indistinguishable in any syndromic or psychological way from "idiopathic" autism (Hanson et al., 2010; Miller et al., 1993; Sanders et al., 2011), even when phenotyped "blind" to genotype. On the other hand, another third or so of patients fulfill some criteria for ASD but not others; and the last third of patients with 16p11.2 deletion do not have ASD at all. Instead, they may be clinically normal or may have other neurodevelopmental conditions such as intellectual disability, schizophrenia, obesity, ADHD, epilepsy, or multiple disorders (Fernandez et al., 2010; Hanson et al., 2010; Miller et al., 1993; Yu et al., 2011). This heterogeneity even holds when the same deletion is inherited within a family (Shen et al., 2011), and in some families in which a deletion segregates, affected patients may lack a deletion present in other family members, or *vice versa* (Sanders et al., 2011). Duplication of the identical region on 16p11.2 also greatly increases the risk of ASD (Weiss et al., 2008), but 16p11.2 duplications are even more heterogeneous, being associated with brain malformations in some cases (agenesis of the corpus callosum, microcephaly), a high rate of epilepsy (Bedoyan et al., 2010), schizophrenia (McCarthy et al., 2009), and often more severe intellectual disability (Shinawi et al., 2010).

The Simons Simplex Collection

The Simons Simplex Collection (SSC) was initiated by the Simons Foundation in 2007 as a prospective registry and DNA sample collection that combined detailed phenotyping with intensive genetic study (Fischbach and Lord, 2010). It was conceived based on the suggestion that spontaneous mutation in ASD might be more common in "simplex" families, in which only a single individual is affected with ASD (Sebat et al., 2007), but the remarkable structure of the pedigrees has also facilitated studies of other genetic mechanisms in ASD. Recent study of the SSC has highlighted several important points. The study confirms that CNVs are a consistently reproducible

and highly significant cause of ASD, present and likely causative in ~5% of patients in the SSC. The SSC has confirmed the high statistical association of 16p11.2 CNV with ASD and has highlighted other newly described CNV regions (Gilman et al., 2011; Levy et al., 2011; Sanders et al., 2011). The SSC studies also show that patients with CNVs as a cause of ASD are not typically distinguishable from patients with "idiopathic" autism by any observable feature (e.g., IQ, social scales, etc.).

One region discovered to be associated with ASD in the SSC studies is the Williams syndrome region on 7q11.23. The Williams syndrome region is intriguing, because here a deletion is associated with a condition characterized by abnormal positive sociability, whereas the duplication is associated with poor socialization, suggesting a "dose-response" relationship to behavior. Hints of similar "dose effects" are seen for 16p11.2 deletion/duplication in head size (large and small, respectively), and body mass (obese and thin, respectively) (Hanson et al., 2010; Tannour-Louet et al., 2010; Yu et al., 2011), and have also been claimed for duplication/deletion of the 22q11.2 regions (Eliez, 2007; Lo-Castro et al., 2009; Niklasson et al., 2009). The physiological bases of these apparent dose effects are not clear.

"NONSYNDROMIC" GENES FOR ASD

A major step forward in our understanding of autism genetics came with identification of *NLGN3* and *NLGN4X* as the first clear, "nonsyndromic" causes of ASD, meaning that patients with these mutations can be indistinguishable from ASD patients without a clear genetic diagnosis (unlike, for example, TSC or fragile X syndrome, in which patients have additional somatic signs that can be diagnostic) (Jamain et al., 2003). *NLGN4X* was first identified as an X-linked candidate gene because it was deleted in a patient with ASD (Jamain et al., 2003); subsequent sequence analysis revealed additional point mutations in other males with ASD and identified a large family with an inherited mutation in *NLGN4X* present in 19 affected males, four of whom showed ASD, with the remaining affected males showing intellectual disability, emphasizing the overlap of ASD and intellectual disabilities (Laumonnier et al., 2004).

SHANK3 was also identified as a candidate gene based on the discovery of small deletions (Durand et al., 2007), and subsequent resequencing of *SHANK3* in patients with ASD has shown point mutations in up to 0.75% of ASD patients (Gauthier et al., 2009; Moessner et al., 2007). A chromosome rearrangement, as well as a rare recessive disorder associated with autistic symptoms (Strauss et al., 2006), also led to the identification of *CNTNAP2* as an ASD gene (Alarcon et al., 2008; Arking et al., 2008; Bakkaloglu et al., 2008; Stephan, 2008).

Although deletions were instrumental in implicating *SHANK3* and *NLGN4X* as ASD candidate regions, resequencing of genes contained in other recurrent CNV intervals has not always been successful in implicating a single gene as causative for the ASD. For example, in the 22q11.2 region, the 16p11.2 region (Konyukh et al., 2011), and the 7q11.23 region, the specific ASD gene is not yet known, suggesting the possibility that ASD in these cases reflects the combinatorial action of more than one gene in the interval. Suggestive evidence has also recently been presented for "oligogenic heterozygosity," implicating the coexistence of heterozygous mutations at more than one candidate gene as a potential causative mechanism of ASD (Schaaf et al., 2011).

RECESSIVE MUTATIONS IN ASD

The observation that all known ASD mutations appear to act by a loss-of-function mechanism suggests that other ASD mutations may also be loss-of-function; wherever this is the case, recessively acting or inherited mutations are likely to be important. For example, for human developmental brain malformations, one or two of the most common genetic mutations are dominant or X-linked, but the vast bulk of the heterogeneity of the condition arises from many rare autosomal recessive syndromes (Manzini and Walsh, 2011). Similarly, intellectual disability (whether associated with autistic features or not) is known to be caused by >70 X-linked recessive genes (Ropers, 2006), but there are already estimates of hundreds of autosomal recessive causes (Ropers, 2006, 2008), of which only dozens have been identified so far (Basel-Vanagaite et al., 2006; Mochida et al., 2009; Najmabadi et al., 2011).

Identification of recessive mutations for rare diseases has been aided recently by the advent of whole-exome sequencing (Bilguvar et al., 2010; Choi et al., 2009; Ng et al., 2010; Shendure and Ji, 2008). Typically, this still requires the ascertainment of large families, often with parental consanguinity (Mochida et al., 2009; Najmabadi et al., 2011; Yu et al., 2010), because the tremendous diversity/heterogeneity of the condition makes it virtually impossible to confidently identify recurrent mutations in unrelated families. Thus, while recessive mutations have been barely explored as a cause of ASD, one would expect that, by analogy to other brain disorders, recessive mutations may contribute greatly to the genetic heterogeneity of ASD. In support of this model, many syndromes associated with ASD also act in a recessive fashion (Betancur, 2011).

In the sole study of consanguineous families with ASD reported to date, several lines of evidence support a contribution of recessive genes to ASD. First, multiplex families (i.e., with more than one affected family) in which the parents are also related, show a lower male/female ratio of affected children (<3M/1F) than offspring of unrelated parents (>4M/1F); a more equal male/female ratio would be consistent with a higher contribution of autosomal mutations. Moreover, multiplex offspring of consanguineous parents show a lower frequency of de novo CNVs that segregate with disease, again suggesting other mechanisms at work. Finally, CNV analysis in a cohort of 78 probands identified five probands with homozygous deletions that appear to be tolerated in the heterozygous carrier state, but are causative

in the homozygous state. At least one of these deletions implicated a gene (*SLC9A9/NHE9*) that showed a significant excess of mutations in unrelated cases of ASD as well (Morrow et al., 2008). Some analysis of non-consanguineous families also suggests potential roles of recessive mutations (Casey et al., 2011; Chahrour et al., 2012). These data suggest that appropriate study designs to identify recessive mutations in ASD (e.g., further study of consanguineous families) may be very valuable.

WHAT ARE THE ROLES OF COMMON VARIANTS IN ASD?

A number of genetic association studies have been performed on large cohorts of ASD patients and have also suggested potential roles for "common" alleles in ASD, although these studies remain somewhat underpowered, and further work in the upcoming years is likely to be more illuminating (Levitt and Campbell, 2009). One interesting association has been with the c-Met gene (Campbell et al., 2006). A number of genome-wide association studies (GWAS) have recently been published, though the results have not been entirely clear. Two loci on chromosome 5 have been implicated (Glessner et al., 2009; Wang et al., 2009; Weiss et al., 2009), although one of the studies has been criticized on technical grounds (McClellan and King, 2010), and a subsequent meta-analysis did not confirm these two regions (Anney et al., 2010). Given the great heterogeneity in autism and the confirmed role for rare variants, larger studies may be needed, perhaps even studies in which patients with obvious high-risk CNVs or other high-risk alleles are removed, in order to detect a stronger signal of common variation.

EARLY RESULTS OF WHOLE-EXOME AND WHOLE-GENOME SEQUENCING IN ASD

Given that ASD mutations are strongly selected against evolutionarily, several studies have attempted to identify risk mutations directly by increasingly large-scale sequencing of candidate genes to identify mutations not present in parents but present in patients. One large study performed systematic resequencing of genes encoding proteins of the synapse and spine apparatus of neurons, testing for *de novo* mutations, and identified such mutations in *SHANK3* in several patients with ASD (Gauthier et al., 2009). Another study pioneered the use of whole-exome sequencing in ASD and suggested that a significant proportion of ASD patients may have *de novo* mutations (Ng et al., 2009; O'Roak et al., 2011). If further confirmed, these results would greatly expand the proportion of autism that can be explained genetically and would begin to argue strongly that whole-exome sequencing, along with a sensitive analysis of CNVs, would be central to the diagnostic evaluation of children on the autism spectrum in the not-too-distant future.

MECHANISTIC INSIGHTS INTO AUTISM FROM GENETIC STUDIES

Since the ultimate goal of genetic analysis in disease is the development and improvement of treatment, it is appropriate to ask what we have learned thus far from the genetic analysis of ASD. Space does not permit an in-depth analysis of this topic, which has been reviewed extensively elsewhere (Kelleher and Bear, 2008; Ramocki and Zoghbi, 2008; Schaaf and Zoghbi, 2011; Walsh et al., 2008), but a few comments may provide some general perspective.

A great deal of evidence supports the general interpretation that most, if not all, genes associated with ASD encode proteins with roles in the regulation of synapses. This association with synapses includes very direct roles, such as *NRXN1* and *NLGN3/4* genes, which encode structural adhesion molecules that constitute important components of the synaptic specialization itself. Other ASD genes encode intracellular scaffolding proteins that function in synaptic spines, such as *SHANK3* (Schaaf and Zoghbi, 2011). Additional ASD-related genes regulate levels of local protein translation in the dendritic spine, such as the FMRP protein, or the TSC proteins. Proteomic studies have identified up to 1000 proteins in synapses and spines, which would conveniently account for the expected tremendous genetic heterogeneity of ASD. Finally, still other ASD-related proteins function "downstream," in the nucleus, to regulate mRNA synthesis that is tightly regulated by neuronal activity (Guy et al., 2011). In some sense then, ASD-related genes are increasingly defining in some depth a pathway that appears to take us from neuronal activity to the plastic changes that underlie learning and memory.

The increasing focus of ASD genes on synaptic plasticity is tremendously exciting therapeutically, because it suggests a potential that drugs affecting many steps in this pathway may nonetheless hold some promise for children with diverse genetic mutations (Abrahams and Geschwind, 2008; Geschwind, 2008; Kelleher and Bear, 2008; Walsh et al., 2008). A number of animal studies have demonstrated the remarkable degree to which several mutant mice strains can improve dramatically by gene replacement (Giacometti et al., 2007; Guy et al., 2007), or pharmacological manipulation (Kelleher and Bear, 2008; Tropea et al., 2009), even after much of early brain development is complete, creating a surge of interest in the pharmacological treatment of ASD. While it is extremely early days in this challenging area, there is reason for some optimism.

ACKNOWLEDGMENTS

Research was supported by grants from the National Institute of Mental Health (RO1 MH083565; 1RC2MH089952) to CAW, the Dubai Harvard Foundation for Medical Research, the Nancy Lurie Marks Foundation, the Simons Foundation, the Autism Consortium, and the Manton Center for Orphan Disease Research. CAW is an Investigator of the Howard Hughes Medical Institute.

REFERENCES

Abrahams, B.S., Geschwind, D.H., 2008. Advances in autism genetics: On the threshold of a new neurobiology. Nat Rev Genet 9, 341–355.

Alarcon, M., Abrahams, B.S., Stone, J.L., et al., 2008. Linkage, association, and gene-expression analyses identify CNTNAP2 as an autism-susceptibility gene. Am J Hum Genet 82, 150–159.

Amir, R.E., Van den Veyver, I.B., Wan, M., et al., 1999. Rett syndrome is caused by mutations in X-linked MECP2, encoding methyl-CpG-binding protein 2. Nat Genet 23, 185–188.

Anney, R., Klei, L., Pinto, D., et al., 2010. A genome-wide scan for common alleles affecting risk for autism. Hum Mol Genet 19, 4072–4082.

Arking, D.E., Cutler, D.J., Brune, C.W., et al., 2008. A common genetic variant in the neurexin superfamily member CNTNAP2 increases familial risk of autism. Am J Hum Genet 82, 160–164.

Bailey, A., Le Couteur, A., Gottesman, I., et al., 1995. Autism as a strongly genetic disorder: Evidence from a British twin study. Psychol Med 25, 63–77.

Bakkaloglu, B., O'Roak, B.J., Louvi, A., et al., 2008. Molecular cytogenetic analysis and resequencing of contactin associated protein-like 2 in autism spectrum disorders. Am J Hum Genet 82, 165–173.

Basel-Vanagaite, L., Attia, R., Yahav, M., et al., 2006. The CC2D1A, a member of a new gene family with C2 domains, is involved in autosomal recessive non-syndromic mental retardation. J Med Genet 43, 203–210.

Bedoyan, J.K., Kumar, R.A., Sudi, J., et al., 2010. Duplication 16p11.2 in a child with infantile seizure disorder. Am J Med Genet 152A, 1567–1574.

Belmonte, M.K., Bourgeron, T., 2006. Fragile X syndrome and autism at the intersection of genetic and neural networks. Nat Neurosci 9, 1221–1225.

Betancur, C., 2011. Etiological heterogeneity in autism spectrum disorders: More than 100 genetic and genomic disorders and still counting. Brain Res 1380, 42–77.

Bilguvar, K., Ozturk, A.K., Louvi, A., et al., 2010. Whole-exome sequencing identifies recessive WDR62 mutations in severe brain malformations. Nature 467, 207–210.

Bolton, P.F., 2009. Medical conditions in autism spectrum disorders. J Neurodev Disord 1, 102–113.

Campbell, D.B., Sutcliffe, J.S., Ebert, P.J., et al., 2006. A genetic variant that disrupts MET transcription is associated with autism. Proc Natl Acad Sci USA 103, 16,834–16,839.

Casey, J.P., Magalhaes, T., Conroy, J.M., et al., 2011. A novel approach of homozygous haplotype sharing identifies candidate genes in autism spectrum disorder. Hum Genet doi: 10.1007/s00439-011-1094-6.

Chahrour, M., Zoghbi, H.Y., 2007. The story of Rett syndrome: From clinic to neurobiology. Neuron 56, 422–437.

Chahrour, MH., Yu, T.W., Lim, E.T., et al., 2012. Whole exome sequencing and homozygosity analysis implicate depolarization-regulated neuronal genes in autism. PLoS Genet 8 (4), e1002635.

Choi, M., Scholl, U.I., Ji, W., et al., 2009. Genetic diagnosis by whole exome capture and massively parallel DNA sequencing. Proc Natl Acad Sci USA 106, 19,096–19,101.

Choi, Y.J., Di Nardo, A., Kramvis, I., et al., 2008. Tuberous sclerosis complex proteins control axon formation. Genes Dev 22, 2485–2495.

Christian, S.L., Brune, C.W., Sudi, J., et al., 2008. Novel submicroscopic chromosomal abnormalities detected in autism spectrum disorder. Biol Psychiatry 63, 1111–1117.

Cook Jr., E.H., Lindgren, V., Leventhal, B.L., et al., 1997. Autism or atypical autism in maternally but not paternally derived proximal 15q duplication. Am J Hum Genet 60, 928–934.

Durand, C.M., Betancur, C., Boeckers, T.M., et al., 2007. Mutations in the gene encoding the synaptic scaffolding protein SHANK3 are associated with autism spectrum disorders. Nat Genet 39, 25–27.

Eliez, S., 2007. Autism in children with 22q11.2 deletion syndrome. J Am Acad Child Adolesc Psychiatry 46, 433–434, author reply 434.

Fernandez, B.A., Roberts, W., Chung, B., et al., 2010. Phenotypic spectrum associated with de novo and inherited deletions and duplications at 16p11.2 in individuals ascertained for diagnosis of autism spectrum disorder. J Med Genet 47, 195–203.

Fine, S.E., Weissman, A., Gerdes, M., et al., 2005. Autism spectrum disorders and symptoms in children with molecularly confirmed 22q11.2 deletion syndrome. J Autism Dev Disord 35, 461–470.

Fischbach, G.D., Lord, C., 2010. The Simons Simplex Collection: A resource for identification of autism genetic risk factors. Neuron 68, 192–195.

Folstein, S.E., Piven, J., 1991. Etiology of autism: Genetic influences. Pediatrics 87, 767–773.

Gauthier, J., Spiegelman, D., Piton, A., et al., 2009. Novel de novo SHANK3 mutation in autistic patients. Am J Med Genet 150B, 421–424.

Geschwind, D.H., 2008. Autism: Many genes, common pathways? Cell 135, 391–395.

Giacometti, E., Luikenhuis, S., Beard, C., Jaenisch, R., 2007. Partial rescue of MeCP2 deficiency by postnatal activation of MeCP2. Proc Natl Acad Sci USA 104, 1931–1936.

Gillis, R.F., Rouleau, G.A., 2011. The ongoing dissection of the genetic architecture of autistic spectrum disorder. Mol Autism 2, 12.

Gilman, S.R., Iossifov, I., Levy, D., et al., 2011. Rare de novo variants associated with autism implicate a large functional network of genes involved in formation and function of synapses. Neuron 70, 898–907.

Glessner, J.T., Wang, K., Cai, G., et al., 2009. Autism genome-wide copy number variation reveals ubiquitin and neuronal genes. Nature 459, 569–573.

Guilmatre, A., Dubourg, C., Mosca, A.L., et al., 2009. Recurrent rearrangements in synaptic and neurodevelopmental genes and shared biologic pathways in schizophrenia, autism, and mental retardation. Arch Gen Psychiatry 66, 947–956.

Guy, J., Cheval, H., Selfridge, J., Bird, A., 2011. The role of MeCP2 in the brain. Annu Rev Cell Dev Biol 27, 631–652.

Guy, J., Gan, J., Selfridge, J., Cobb, S., Bird, A., 2007. Reversal of neurological defects in a mouse model of Rett syndrome. Science 315, 1143–1147.

Hallmayer, J., Cleveland, S., Torres, A., et al., 2011. Genetic heritability and shared environmental factors among twin pairs with autism. Arch Gen Psychiatry 68, 1095–1102.

Hanson, E., Nasir, R.H., Fong, A., et al., 2010. Cognitive and behavioral characterization of 16p11.2 deletion syndrome. J Dev Behav Pediatr 31, 649–657.

Jamain, S., Quach, H., Betancur, C., et al., 2003. Mutations of the X-linked genes encoding neuroligins NLGN3 and NLGN4X are associated with autism. Nat Genet 34, 27–29.

Kelleher 3rd, R.J., Bear, M.F., 2008. The autistic neuron: Troubled translation? Cell 135, 401–406.

Konyukh, M., Delorme, R., Chaste, P., et al., 2011. Variations of the candidate SEZ6L2 gene on Chromosome 16p11.2 in patients with autism spectrum disorders and in human populations. PLoS ONE 6, e17289.

Kumar, R.A., KaraMohamed, S., Sudi, J., et al., 2008. Recurrent 16p11.2 microdeletions in autism. Hum Mol Genet 17, 628–638.

Kusenda, M., Sebat, J., 2008. The role of rare structural variants in the genetics of autism spectrum disorders. Cytogenet Genome Res 123, 36–43.

Laumonnier, F., Bonnet-Brilhault, F., Gomot, M., et al., 2004. X-linked mental retardation and autism are associated with a mutation in the NLGN4X gene, a member of the neuroligin family. Am J Hum Genet 74, 552–557.

Lee, J.A., Lupski, J.R., 2006. Genomic rearrangements and gene copy-number alterations as a cause of nervous system disorders. Neuron 52, 103–121.

Levitt, P., Campbell, D.B., 2009. The genetic and neurobiologic compass points toward common signaling dysfunctions in autism spectrum disorders. J Clin Invest 119, 747–754.

Levy, D., Ronemus, M., Yamrom, B., et al., 2011. Rare de novo and transmitted copy-number variation in autistic spectrum disorders. Neuron 70, 886–897.

Lionel, A.C., Crosbie, J., Barbosa, N., et al., 2011. Rare copy number variation discovery and cross-disorder comparisons identify risk genes for ADHD. Sci Transl Med 3 95ra75.

Lo-Castro, A., Galasso, C., Cerminara, C., et al., 2009. Association of syndromic mental retardation and autism with 22q11.2 duplication. Neuropediatrics 40, 137–140.

Manzini, M.C., Walsh, C.A., 2011. What disorders of cortical development tell us about the cortex: One plus one does not always make two. Curr Opin Genet Dev 21, 333–339.

Marshall, C.R., Noor, A., Vincent, J.B., et al., 2008. Structural variation of chromosomes in autism spectrum disorder. Am J Hum Genet 82, 477–488.

McCarthy, S.E., Makarov, V., Kirov, G., et al., 2009. Microduplications of 16p11.2 are associated with schizophrenia. Nat Genet 41, 1223–1227.

McClellan, J., King, M.C., 2010. Genetic heterogeneity in human disease. Cell 141, 210–217.

McLennan, Y., Polussa, J., Tassone, F., Hagerman, R., 2011. Fragile X syndrome. Curr Genomics 12, 216–224.

Meikle, L., Talos, D.M., Onda, H., et al., 2007. A mouse model of tuberous sclerosis: Neuronal loss of Tsc1 causes dysplastic and ectopic neurons, reduced myelination, seizure activity, and limited survival. J Neurosci 27, 5546–5558.

Miller, D.T., Adam, M.P., Aradhya, S., et al., 2011. Consensus statement: Chromosomal microarray is a first-tier clinical diagnostic test for individuals with developmental disabilities or congenital anomalies. Am J Hum Genet 86, 749–764.

Miller, D.T., Nasir, R., Sobeih, M.M., et al., 1993. 16p11.2 Microdeletion. In: Pagon R.A., Bird T.D., Dolan C.R., Stephens K. (Eds.) GeneReviews [Internet]. Seattle (WA): University of Washington.

Mochida, G.H., Mahajnah, M., Hill, A.D., et al., 2009. A truncating mutation of TRAPPC9 is associated with autosomal-recessive intellectual disability and postnatal microcephaly. Am J Hum Genet 85, 897–902.

Moessner, R., Marshall, C.R., Sutcliffe, J.S., et al., 2007. Contribution of SHANK3 mutations to autism spectrum disorder. Am J Hum Genet 81, 1289–1297.

Moretti, P., Zoghbi, H.Y., 2006. MeCP2 dysfunction in Rett syndrome and related disorders. Curr Opin Genet Dev 16, 276–281.

Morrow, E.M., 2010. Genomic copy number variation in disorders of cognitive development. J Am Acad Child Adolesc Psychiatry 49, 1091–1104.

Morrow, E.M., Yoo, S.Y., Flavell, S.W., et al., 2008. Identifying autism loci and genes by tracing recent shared ancestry. Science 321, 218–223.

Najmabadi, H., Hu, H., Garshasbi, M., et al., 2011. Deep sequencing reveals 50 novel genes for recessive cognitive disorders. Nature 478, 57–63.

Ng, S.B., Buckingham, K.J., Lee, C., et al., 2010. Exome sequencing identifies the cause of a Mendelian disorder. Nat Genet 42, 30–35.

Ng, S.B., Turner, E.H., Robertson, P.D., et al., 2009. Targeted capture and massively parallel sequencing of 12 human exomes. Nature 461, 272–276.

Niklasson, L., Rasmussen, P., Oskarsdottir, S., Gillberg, C., 2009. Autism, ADHD, mental retardation and behavior problems in 100 individuals with 22q11 deletion syndrome. Res Dev Disabil 30, 763–773.

O'Roak, B.J., Deriziotis, P., Lee, C., et al., 2011. Exome sequencing in sporadic autism spectrum disorders identifies severe de novo mutations. Nat Genet 43, 585–589.

Pinto, D., Pagnamenta, A.T., Klei, L., et al., 2010. Functional impact of global rare copy number variation in autism spectrum disorders. Nature 466, 368–372.

Ramelli, G.P., Silacci, C., Ferrarini, A., et al., 2008. Microduplication 22q11.2 in a child with autism spectrum disorder: Clinical and genetic study. Dev Med Child Neurol 50, 953–955.

Ramocki, M.B., Tavyev, Y.J., Peters, S.U., 2009. The MECP2 duplication syndrome. Am J Med Genet 152A, 1079–1088.

Ramocki, M.B., Zoghbi, H.Y., 2008. Failure of neuronal homeostasis results in common neuropsychiatric phenotypes. Nature 455, 912–918.

Ropers, H.H., 2006. X-linked mental retardation: Many genes for a complex disorder. Curr Opin Genet Dev 16, 260–269.

Ropers, H.H., 2008. Genetics of intellectual disability. Curr Opin Genet Dev 18, 241–250.

Sanders, S.J., Ercan-Sencicek, A.G., Hus, V., et al., 2011. Multiple recurrent de novo CNVs, including duplications of the 7q11.23 Williams syndrome region, are strongly associated with autism. Neuron 70, 863–885.

Schaaf, C.P., Sabo, A., Sakai, Y., et al., 2011. Oligogenic heterozygosity in individuals with high-functioning autism spectrum disorders. Hum Mol Genet 20, 3366–3375.

Schaaf, C.P., Zoghbi, H.Y., 2011. Solving the autism puzzle a few pieces at a time. Neuron 70, 806–808.

Sebat, J., Lakshmi, B., Malhotra, D., et al., 2007. Strong association of de novo copy number mutations with autism. Science 316, 445–449.

Sebat, J., Lakshmi, B., Troge, J., et al., 2004. Large-scale copy number polymorphism in the human genome. Science 305, 525–528.

Shen, Y., Chen, X., Wang, L., et al., 2011. Intra-family phenotypic heterogeneity of 16p11.2 deletion carriers in a three-generation Chinese family. Am J Med Genet B Neuropsychiatr Genet 156, 225–232.

Shen, Y., Dies, K.A., Holm, I.A., et al., 2010. Clinical genetic testing for patients with autism spectrum disorders. Pediatrics 125, e727–e735.

Shendure, J., Ji, H., 2008. Next-generation DNA sequencing. Nat Biotechnol 26, 1135–1145.

Shinawi, M., Liu, P., Kang, S.H., et al., 2010. Recurrent reciprocal 16p11.2 rearrangements associated with global developmental delay, behavioural problems, dysmorphism, epilepsy, and abnormal head size. J Med Genet 47, 332–341.

Stephan, D.A., 2008. Unraveling autism. Am J Hum Genet 82, 7–9.

Stone, J.L., O'Donovan, M.C., Gurling, H., et al., 2008. Rare chromosomal deletions and duplications increase risk of schizophrenia. Nature 455. doi: 10.1038/nature07239.

Strauss, K.A., Puffenberger, E.G., Huentelman, M.J., et al., 2006. Recessive symptomatic focal epilepsy and mutant contactin-associated protein-like 2. N Engl J Med 354, 1370–1377.

Sykes, N.H., Toma, C., Wilson, N., et al., 2009. Copy number variation and association analysis of SHANK3 as a candidate gene for autism in the IMGSAC collection. Eur J Hum Genet 17, 1347–1353.

Tannour-Louet, M., Han, S., Corbett, S.T., et al., 2010. Identification of de novo copy number variants associated with human disorders of sexual development. PLoS ONE 5, e15392.

Tropea, D., Giacometti, E., Wilson, N.R., et al., 2009. Partial reversal of Rett Syndrome-like symptoms in MeCP2 mutant mice. Proc Natl Acad Sci USA 106, 2029–2034.

Walsh, C.A., Morrow, E.M., Rubenstein, J.L., 2008. Autism and brain development. Cell 135, 396–400.

Wang, K., Zhang, H., Ma, D., et al., 2009. Common genetic variants on 5p14.1 associate with autism spectrum disorders. Nature 459, 528–533.

Weiss, L.A., Arking, D.E., Daly, M.J., Chakravarti, A., 2009. A genome-wide linkage and association scan reveals novel loci for autism. Nature 461, 802–808.

Weiss, L.A., Shen, Y., Korn, J.M., et al., 2008. Association between microdeletion and microduplication at 16p11.2 and autism. N Engl J Med 358, 667–675.

Wiznitzer, M., 2004. Autism and tuberous sclerosis. J Child Neurol 19, 675–679.

Yu, T.W., Mochida, G.H., Tischfield, D.J., et al., 2010. Mutations in WDR62, encoding a centrosome-associated protein, cause microcephaly with simplified gyri and abnormal cortical architecture. Nat Genet 42, 1015–1020.

Yu, Y., Zhu, H., Miller, D.T., et al., 2011. Age- and gender-dependent obesity in individuals with 16p11.2 deletion. J Genet Genomics 38, 403–409.

ONGOING RESEARCH

Kong, A., Frigge, M.L., Masson, G., et al., 2012. Rate of de novo mutations and the importance of father's age to disease risk. Nature 488, 471–475.

Iossifov, I., Ronemus, M., Levy, D., et al., 2012. De novo gene disruptions in children on the autistic spectrum. Neuron 74, 285–299.

Neale, B.M., Kou, Y., Liu, L., et al., 2012. Patterns and rates of exonic de novo mutations in autism spectrum disorders. Nature 485, 242–245.

O'Roak, B.J., Vives, L., Girirajan, S., et al., 2012. Sporadic autism exomes reveal a highly interconnected protein network of de novo mutations. Nature 485, 246–250.

Sanders, S.J., Murtha, M.T., Gupta, A.R., et al., 2012. De novo mutations revealed by whole-exome sequencing are strongly associated with autism. Nature 485, 237–241.

Eye Diseases

Janey L. Wiggs

INTRODUCTION

Overview of Eye Structure and Function

The eye functions to transduce light into an electrical signal that is transmitted to the brain. A variety of tissues and specialized cells carry out these complex processes. The ocular globe is divided into two fluid-filled compartments, called the anterior and posterior chambers (Figure 91.1). The anterior chamber is filled with an aqueous fluid called aqueous humor, and the posterior chamber is filled with a viscous substance called the vitreous humor. The globe is supported by the sclera, a tough outer shell that also supports the optic nerve as it exits the eye.

The cornea is a transparent tissue located on the anterior ocular surface that allows light to enter the eye and also helps focus the light on the retina. Inside the eye are a number of structures, including the iris and pupil (to regulate the amount of light entering the eye), the lens (to focus light on the retina), the ciliary body (to make aqueous humor), and a trabecular meshwork (to drain aqueous humor). Under normal circumstances, the rate of production of aqueous humor equals the rate of removal.

Light traveling through the cornea, pupil, and lens is focused on the retina, which carries out the phototransduction of light to produce an electrical signal that is transmitted through the optic nerve to the brain. The retina is a complex tissue made up of 10 distinct layers (Figure 91.2). The most external cell layer is the retinal pigment epithelium, which provides metabolic support and is attached to a basement membrane (Bruch's membrane). Next to the retinal pigment epithelium are the rod and cone photoreceptors, which are the cells where phototransduction occurs. Connected to the photoreceptors are the amacrine, bipolar, and horizontal cells that modulate the signal output from the rods (dim light) and cones (bright light). The signal from the photoreceptor goes through the bipolar cells to the ganglion cells. The axons of the ganglion cells form the optic nerve, and send the signal to their first synapse at the lateral geniculate body. The signal from the retina eventually forms an image in the occipital lobe of the brain.

Eye movements are controlled by six muscles located on the outside of the globe (extraocular muscles), which are innervated by cranial nerves III, IV, and VI.

Blinding diseases result from the interruption of the normal function of the ocular structures. Inherited early-onset disorders typically have Mendelian inheritance, while common adult-onset disorders are inherited as complex traits. The advances in genetics and genomics over the past decade have provided insight into the molecular processes underlying many ophthalmic disorders. In this chapter, recent discoveries using genomic approaches will be reviewed.

GENOME-WIDE ASSOCIATION STUDIES

Genome-wide association studies (GWAS) have successfully identified genetic factors contributing to a number of common ocular disorders with complex inheritance (Table 91.1). The

Genomic and Personalized Medicine, 2nd edition
by Ginsburg & Willard. DOI: http://dx.doi.org/10.1016/B978-0-12-382227-7.00091-4

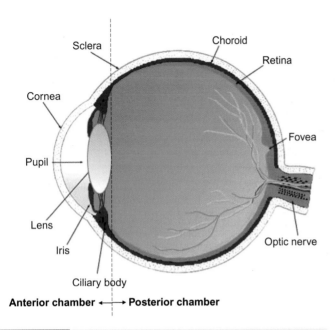

Sclera

Choroid

Retina

Cornea

Fovea

Pupil

Lens

Optic nerve

Iris

Ciliary body

Anterior chamber ◄——► **Posterior chamber**

Figure 91.1 Vertical sagittal section of the adult human eye.

2005 discovery of the association between the complement factor H gene (*CFH*) and **age-related macular degeneration** (ARMD) was one of the first successes of the GWAS approach (Edwards et al., 2005; Haines et al., 2005; Klein et al., 2005). ARMD causes destruction of the retina in the macular region that is responsible for best visual acuity. Subsequent to the original GWAS findings, a major locus on chromosome 10 has been identified (Dewan et al., 2006), as well as a number of other loci, including those that encode other members of the complement family (Chen et al., 2010; Kopplin et al., 2010; Neale et al., 2010). Together, these findings indicate that inflammation is a major process influencing susceptibility to this common blinding disease (reviewed in Telander, 2011).

Glaucoma is a term describing a complex group of disorders that have in common deterioration of the optic nerve (Fan and Wiggs, 2010). Recent GWAS for primary open angle glaucoma (POAG), the most common glaucoma subtype in the western world, have identified disease-associated single nucleotide polymorphisms (SNPs) in the chromosome regions containing *CAV1/CAV2* genes (Thorliefsson et al., 2010; Wiggs et al., 2011), *TMC01* (Burdon et al., 2011), and *CDKN2BAS* (Burdon et al., 2011; Wiggs et al., 2012). The *CDKN2BAS* association is particularly robust, and secondary analyses suggest that this chromosome region strongly influences degeneration of the optic nerve in glaucoma (Wiggs et al., 2012). *LOXL1*, a gene contributing to another common form of glaucoma, exfoliation syndrome, has also been identified using a GWAS approach (Thorliefsson et al., 2007). GWAS have also successfully identified susceptibility alleles for **myopia** (near-sightedness) (Hysi et al., 2010; Li et al., 2011; Saluki et al., 2010; Shi et al., 2011b), and **age-related corneal degeneration** (Fuch's endothelial dystrophy) (Baratz et al., 2010). A GWAS

for **cataract** from the eMERGE network is currently underway (McCarty et al., 2011). The formation of multiple consortia and collaborations has been crucial for success of the GWAS approach by increasing sample sizes, thereby increasing statistical power, enabling replication of findings from individual studies and establishing common methods of analysis (Manolio et al., 2009).

Quantitative Ocular Traits

Many ocular traits contributing to disease states are quantitative, and genetic factors influencing these traits have been discovered using genome-wide continuous trait analyses of large cohorts of normal individuals. SNPs near the *ATOH7* gene influence optic disc size, a risk factor for glaucoma (Khorr et al., 2011; Macgregor et al., 2010; Ramdas et al., 2010), and SNPs in the *CDKN2BAS* gene region, now known to influence POAG risk (Burdon et al., 2011; Fan et al., 2011; Ramdas et al., 2011; Wiggs et al., 2012) were originally discovered as SNPs associated with the quantitative optic nerve parameter vertical cup-to-disc-ratio (vCDR) (Ramdas et al., 2010). A number of other SNPs have been associated with these quantitative optic nerve parameters, reflecting their complex inheritance (Khorr et al., 2011; Macgregor et al., 2010; Ramdas et al., 2010). The thickness of the central cornea (CCT) is a highly heritable quantitative trait that is also a risk factor for glaucoma (Dueker et al., 2007), and several genomic studies have yielded a number of genes influencing this trait (Lu et al., 2010; Vitart et al., 2010; Vithana et al., 2011).

FUNCTIONAL GENOMICS

Investigations using genomic arrays to identify changes in gene expression patterns in ocular disease states have been completed for ARMD (Strunnikova et al., 2010), cataract (Dawes et al., 2007), and cornea (Joyce et al., 2011). An interesting approach using microdissection laser capture of retinal ganglion cells and a rat model of glaucoma has identified several genes with altered gene expression at various stages of ganglion cell death in glaucoma (Guo et al., 2011; Wang et al., 2010).

PHARMACOGENOMICS

Risk factors influencing response to ophthalmic medications are emerging. It is well known that glaucoma patients can have a variable response to beta-adrenergic antagonists, a class of medications commonly used to lower intraocular pressure, a risk factor for glaucoma-related optic nerve disease. This differential response to beta-adrenergic antagonists may be related to variation in the gene coding for the beta1-adrenergic receptor (Schwartz et al., 2005). Vascular endothelial growth factor (VEGF) is a critical factor influencing neovascularization in ARMD, a devastating development that is a harbinger of

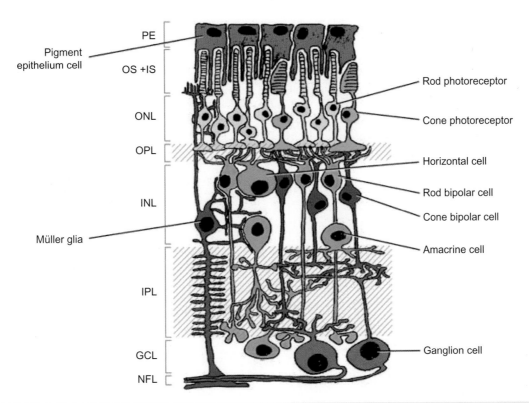

PE

Pigment
epithelium cell

OS +IS

ONL

OPL

INL

Müller glia

IPL

GCL

NFL

Rod photoreceptor

Cone photoreceptor

Horizontal cell

Rod bipolar cell

Cone bipolar cell

Amacrine cell

Ganglion cell

Figure 91.2   Schematic diagram of the human retina. PE, pigment epithelium; OS, outer segments; IS, inner segment; ONL, outer nuclear layer; OPL, outer plexiform layer; INL, inner nuclear layer; IPL, inner plexiform layer; GCL, ganglion cell layer; NFL, nerve fiber layer.

ensuing blindness (Bressler et al., 2011). Ocular injections of humanized VEGF antibodies, ranibizumab (Lucentis) and beva-cizumab (Avastin), have had a dramatic effect overall on the vision loss caused by this devastating complication (Bressler et al., 2011). However, carriers of some ARMD risk alleles, including *VEGF* polymorphism, show variation in response to anti-VEGF therapies (Kloeckener-Gruissem et al., 2011; Nakata et al., 2011), potentially influencing the individual success of this therapy. Ongoing genomic studies will also help define genetic factors that predispose to complications of ocular therapies, such as elevation of intraocular pressure caused by corticosteroid use (Jones and Rhee, 2006) and acute bilateral angle-closure glaucoma caused by systemic use of topiramate and other related drugs (Van Issum et al., 2011).

WHOLE-EXOME AND WHOLE-GENOME SEQUENCING

Over the past two years, there have been a number of ocular disease genes discovered using whole-exome and whole-genome approaches (Sergouniotis et al., 2011; Shi et al., 2011a). So far, these have been primarily aimed at genes responsible for ocular disorders with Mendelian inheritance patterns. Based on this recent success, it is very likely that these impressive technologies will reveal additional gene

mutations contributing to both Mendelian and complex forms of ocular disease, as well as those influencing response to ocular therapies and other disease outcomes.

GENE-BASED THERAPIES FOR OCULAR DISEASE

The accessibility of ocular tissue makes the eye an ideal target for gene-specific therapies. Mutations in eight genes can cause **Leber's congenital amaurosis** (LCA), a devastating disease leading to blindness at birth or in infancy (Stone, 2007). Using gene-replacement strategies for one of these gene defects, loss of RPE65 function, successful vision has been restored in an animal model of this condition (Acland et al., 2005), and clinical trials in humans have demonstrated promising results (Ashtari et al., 2011). As additional gene-replacement strategies are developed, diagnostic genotyping to identify causative mutations will become an important first step to choosing the appropriate treatment.

EPIGENETICS

Epigenetic gene regulation in ocular development and disease is an important area of current investigation. Cell type-specific

TABLE 91.1 Recent genome-wide association studies identifying significant SNPs in association with common ocular traits and disorders

Disease or trait	Associated SNP(s)*	Nearest gene(s)	Reference
Macular degeneration	Y402H, P = 5.0×10^{-10} (OR 2-4)	CFH	Edwards et al., 2005; Haines et al., 2005; Klein et al., 2005
	rs10490924, P = 4.1×10^{-12} (OR 1.7)	HTRA1/ARMS2	Dewan et al., 2006
	rs2230199, P = 1.0×10^{-10} (OR 1.74)	C3	Chen et al., 2010
	rs9621532, P = 1.1×10^{-11} (OR 1.41)	TIMP3	Chen et al., 2010; Neale et al., 2010
	rs641153, P = 6.43×10^{-9} (OR 0.32)	CFB, C2	Gold et al., 2006
	rs10033900, P = 2.4×10^{-11} (OR 1.31)	CFI	Yu et al., 2011
	rs1999930, P = 1.1×10^{-8} (OR 0.87)	FRK/COL10A1	
	rs4711751, P = 8.7×10^{-9} (OR 1.15)	VEGFA	
	rs10468017, P = 9×10^{-9} (OR 0.82)	LIPC	Neale et al., 2010
	rs429608, P = 1.1×10^{-11} (OR 0.54)	SKIV2L, BF	Kopplin et al., 2010
Glaucoma-POAG (primary open angle glaucoma)	rs4236601, P = 5.0×10^{-10} (OR 1.36)	CAV1/CAV2	Thorliefsson et al., 2010
	rs4656461, P = 6.1×10^{-10} (OR 1.68)	TMCO1	Burdon et al., 2011
	rs4977756, P = 4.7×10^{-9} (OR 1.50)	CDKN2BAS	Burdon et al., 2011; Wiggs et al., 2012
Glaucoma-NTG (normal tension)	rs2157719, P = 1.2×10^{-12} (OR 0.58)	CDKN2BAS	Wiggs et al., 2012
	rs284489, P = 8.88×10^{-10} (OR 0.62)	LRP12, ZFPM2	
Glaucoma-ES (exfoliation syndrome)	rs2165241, P = 1.0×10^{-27} (OR 3.62)	LOXL1	Thorliefsson et al., 2007
Myopia	rs634990, P = 2.2×10^{-14} (OR 1.4)	GJD2, ACTC1	Saluki et al., 2010
	rs8027411, P = 2.1×10^{-9}	RASGRF1	Hysi et al., 2010
	P = 8.0×10^{-13}	MYP11	Li et al., 2011
	P = 2.0×10^{-16}	MIPEP	Shi et al., 2011b
Central corneal thickness[†] (CCT)	rs96067, P = 5.40×10^{-13}	COL8A1	Vithana et al., 2011
	rs12447690, P = 4.4×10^{-9}	ZNF469	Lu et al., 2010; Vitart et al., 2010
	rs2721051, P = 5.0×10^{-10}	FOXO1	Lu et al., 2010
	rs1536482, P = 5.1×10^{-8}	COL5A1	Vitart et al., 2011
	rs1034200, P = 3.5×10^{-9}	AVGR8	
	rs6496932, P = 1.4×10^{-8}	AKAP13	
Optic nerve size[†]	rs1900004, P = 2.0×10^{-10}	ATOH7/PBLD	Khorr et al., 2011; Macgregor et al., 2010; Ramdas et al., 2010

(continued)

TABLE 91.1	(Continued)			
Disease or trait	**Associated SNP(s)***		**Nearest gene(s)**	**Reference**
	rs1192415, P= 3.0×10^{-28}		CDC7, TGFBR3	Khorr et al., 2011; Ramdas et al., 2010
	rs1362756, P = 5.0×10^{-9}		SALL1	Ramdas et al., 2010
	rs9607469, P = 3.0×10^{-12}		CARD10	Khorr et al., 2011
Optic nerve vertical cup-to-disc ratio[†]	rs1063192, P = 4.0×10^{-15}		CDKN2B	Ramdas et al., 2010
	rs10483727, P = 1.0×10^{-11}		SIX1, SIX6	
	rs17146964, P = 4.0×10^{-9}		SCYL1	
	rs1926320, P = 1.0×10^{-8}		DCLK1	
	rs1547014, P = 1.0×10^{-8}		CHEK2	
	rs1900004, P = 2.0×10^{-8}		ATOH7	
	rs8068952, P = 3.0×10^{-8}		BCAS3	
Fuch's corneal dystrophy	rs613872, P = 1.0×10^{-18} (OR 4.22)		TCF4	Baratz et al., 2010

*For conditions with multiple SNPs located near/in the same gene, the SNP with the best P value for association is listed. Only SNPs with P values for association $<5 \times 10^{-8}$ are included. The P values listed are those found in the initial discovery study. OR=odds ratio.
†The results for these quantitative traits were derived from a continuous trait analysis and ORs are not available.

DNA methylation patterns have been identified in photo-receptor-specific genes and during ocular development (Boatright et al., 2000; Pelzel et al., 2010; Zhong and Kowluru, 2011). Histone modifications have been identified in damaged retinal ganglion cells (Biermann et al., 2010), and epigenetic phenomena leading to inactivation of mitochondria may affect the development of diabetic retinopathy (Zhong and Kowluru, 2011). As epigenetics may explain at least part of the heritability of complex traits (Petronis, 2010), ongoing studies investigating how epigenetic regulation coupled with environmental factors could contribute to common complex ocular disorders such as glaucoma and ARMD will be of great interest.

OCULAR GENETIC AND GENOMIC RESOURCES

Several genetic and genomic databases contain information especially useful for ocular genomic medicine and research. The NEIbank (Wistow et al., 2008; http://neibank.nei.nih.gov/index.shtml), an integrated genomics resource, includes gene expression data for many eye tissues, as well as annotation of known human eye disease genes and candidate disease gene loci. eyeGENE (Brooks et al., 2008; http://www.nei.nih.gov/resources/eyegene.asp) is a network of Clinical Laboratory Improvement Amendment (CLIA)-certified laboratories offering genetic testing for patients affected by inherited ocular disease, that is coupled to a registry of clinical information. eyeGENE is also creating a database to be used by investigations exploring the relationships between genotypes and phenotypes for heritable ophthalmic disorders. A number of disease-based ocular databases are also available, including RETNET [http://www.sph.uth.tmc.edu/retnet/ (describes genes contributing to retinal diseases and degenerations)], MYOC [http://myocilin.com/ (lists mutations in MYOC known to cause Mendelian forms of juvenile and adult-onset glaucoma)], and eOPA1 [http://lbbma.univ-angers.fr/lbbma.php?id=9 (lists mutations in the OPA1 gene known to cause autosomal dominant optic atrophy)].

OUTLOOK FOR THE FUTURE

The number of genes known to contribute to ocular disease has risen dramatically over the past decade, making it possible to envision a personalized approach to ophthalmic care (Wiggs, 2008). Currently, a molecular diagnosis can be provided for many patients affected with disorders caused by known genes, informing genetic counseling and therapeutic decisions. As more genes associated with common forms of ocular disease are found, effective screening tests can be developed to identify individuals at risk, and also to predict outcomes and prognosis. Identifying gene–environment interactions could reveal modifiable risk factors. Understanding the molecular mechanisms of disease processes could lead to disease biomarkers and more effective treatments, and gene-directed therapies (both gene-replacement and biological therapies) could prevent and/or preempt blindness related to heritable ocular disease.

REFERENCES

Acland, G.M., Aguirre, G.D., Bennett, J., et al., 2005. Long-term restoration of rod and cone vision by single dose rAAV-mediated gene transfer to the retina in a canine model of childhood blindness. Mol Ther 6, 1072–1082.

Ashtari, M., Cyckowski, L.L., Monroe, J.F., et al., 2011. The human visual cortex responds to gene therapy-mediated recovery of retinal function. J Clin Invest 121, 2160–2168.

Baratz, K.H., Tosakulwong, N., Ryu, E., et al., 2010. E2-2 protein and Fuchs's corneal dystrophy. N Engl J Med 363, 1016–1024.

Biermann, J., Grieshaber, P., Goebel, U., et al., 2010. Valproic acid-mediated neuroprotection and regeneration in injured retinal ganglion cells. Invest Ophthalmol Vis Sci 51, 526–534.

Boatright, J.H., Nickerson, J.M., Borst, D.E., 2000. Site-specific DNA hypomethylation permits expression of the IRBP gene. Brain Res 887, 211–221.

Bressler, N.M., Doan, Q.V., Varma, R., et al., 2011. Estimated cases of legal blindness and visual impairment avoided using ranibizumab for choroidal neovascularization: Non-Hispanic white population in the United States with age-related macular degeneration. Arch Ophthalmol 129, 709–717.

Brooks, B.P., Macdonald, I.M., Tumminia, S.J., et al., 2008. Genomics in the era of molecular ophthalmology: Reflections on the National ophthalmic disease genotyping network (eyeGENE). Arch Ophthalmol 126, 424–435.

Chen, W., Stambolian, D., Edwards, A.O., et al., 2010. Genetic variants near TIMP3 and high-density lipoprotein-associated loci influence susceptibility to age-related macular degeneration. Proc Natl Acad Sci USA 107, 7401–7406.

Dawes, L.J., Elliott, R.M., Reddan, J.R., Wormstone, Y.M., Wormstone, I.M., 2007. Oligonucleotide microarray analysis of human lens epithelial cells: TGFbeta regulated gene expression. Mol Vis 13, 1181–1197.

Dewan, A., Liu, M., Hartman, S., et al., 2006. HTRA1 promoter polymorphism in wet age-related macular degeneration. Science 314, 989–992.

Dueker, D.K., Singh, K., Lin, S.C., et al., 2007. Corneal thickness measurement in the management of primary open-angle glaucoma: A report by the American Academy of Ophthalmology. Ophthalmology 114, 1779–1787.

Edwards, A.O., Ritter III, R., Abel, K.J., et al., 2005. Complement factor H polymorphism and age-related macular degeneration. Science 308, 421–424.

Fan, B.J., Wiggs, J.L., 2010. Glaucoma: Genes, phenotypes, and new directions for therapy. J Clin Invest 120, 3064–3072.

Fan, B.J., 2011. Genetic variants associated with optic nerve vertical cup-to-disc ratio are risk factors for primary open angle glaucoma in a US Caucasian population. Invest Ophthalmol Vis Sci 52, 1788–1792.

Gold, B., Merriam, J.E., Zernant, J., et al., 2006. Variation in factor B (BF) and complement component 2 (C2) genes is associated with age-related macular degeneration. Nat Genet 38, 458–462.

Guo, Y., Johnson, E.C., Cepurna, W.O., Dyck, J.A., Doser, T., Morrison, J.C., 2011. Early gene expression changes in the retinal ganglion cell layer of a rat glaucoma model. Invest Ophthalmol Vis Sci 52, 1460–1473.

Haines, J.L., Hauser, M.A., Schmidt, S., et al., 2005. Complement factor H variant increases the risk of age-related macular degeneration. Science 308, 419–421.

Hysi, P.G., Young, T.L., Mackey, D.A., et al., 2010. A genome-wide association study for myopia and refractive error identifies a susceptibility locus at 15q25. Nat Genet 42, 902–905.

Jones III, R., Rhee, D.J., 2006. Corticosteroid-induced ocular hypertension and glaucoma: A brief review and update of the literature. Curr Opin Ophthalmol 17, 163–167.

Joyce, N.C., Harris, D.L., Zhu, C.C., 2011. Age-related gene response of human corneal endothelium to oxidative stress and DNA damage. Invest Ophthalmol Vis Sci 52, 641–649.

Khor, C.C., Ramdas, W.D., Vithana, E.N., et al., 2011. Genome-wide association studies in Asians confirm the involvement of ATOH7 and TGFBR3, and further identify CARD10 as a novel locus influencing optic disc area. Hum Mol Genet 20, 1864–1872.

Klein, R.J., Zeiss, C., Chew, E.Y., et al., 2005. Complement factor H polymorphism in age-related macular degeneration. Science 308, 385–389.

Kloeckener-Gruissem, B., Barthelmes, D., Labs, S., et al., 2011. Genetic association with response to intravitreal ranibizumab in patients with neovascular AMD. Invest Ophthalmol Vis Sci 52, 4694–4702.

Kopplin, L.J., Igo Jr., R.P., Wang, Y., et al., 2010. Genome-wide association identifies SKIV2L and MYRIP as protective factors for age-related macular degeneration. Genes Immun 8, 609–621.

Li, Z., Qu, J., Xu, X., et al., 2011. A genome-wide association study reveals association between common variants in an intergenic region of 4q25 and high-grade myopia in the Chinese Han population. Hum Mol Genet 20, 2861–2868.

Lu, Y., Dimasi, D.P., Hysi, P.G., et al., 2010. Common genetic variants near the Brittle cornea syndrome locus ZNF469 influence the blinding disease risk factor central corneal thickness. PLoS Genet 6, e1000947.

Macgregor, S., Hewitt, A.W., Hysi, P.G., et al., 2010. Genome-wide association identifies ATOH7 as a major gene determining human optic disc size. Hum Mol Genet 19, 2716–2724.

Manolio, T.A., Collins, F.S., Cox, N.J., et al., 2009. Finding the missing heritability of complex diseases. Nature 461, 747–753.

McCarty, C.A., Chisholm, R.L., Chute, C.G., et al., 2011. The eMERGE Network: A consortium of biorepositories linked to electronic medical records data for conducting genomic studies. BMC Med Genomics 4, 13.

Nakata, I., Yamashiro, K., Nakanishi, H., et al., 2011. VEGF gene polymorphism and response to intravitreal bevacizumab and triple therapy in age-related macular degeneration. Jpn J Ophthalmol 55, 435–443.

Neale, B.M., Fagerness, J., Reynolds, R., et al., 2010. Genome-wide association study of advanced age-related macular degeneration identifies a role of the hepatic lipase gene (LIPC). Proc Natl Acad Sci USA 107, 7395–7400.

Pelzel, H.R., Schlamp, C.L., Nickells, R.W., 2010. Histone H4 deacetylation plays a critical role in early gene silencing during neuronal apoptosis. BMC Neurosci 11, 62.

Petronis, A., 2010. Epigenetics as a unifying principle in the aetiology of complex traits and diseases. Nature 465, 721–727.

Ramdas, W.D., Van Koolwijk, L.M., Ikram, M.K., et al., 2010. A genome-wide association study of optic disc parameters. PLoS Gene 6, e1000978.

Ramdas, W.D., 2011. Common genetic variants associated with open-angle glaucoma. Hum Mol Genet 20, 2464–2471.

Saluki, A.M., Verhoeven, V.J., van Duijn, C.M., et al., 2010. A genome-wide association study identifies a susceptibility locus for refractive errors and myopia at 15q14. Nat Genet 42, 897–901.

Schwartz, S.G., Puckett, B.J., Allen, R.C., Castillo, I.G., Leffler, C.T., 2005. Beta1-adrenergic receptor polymorphisms and clinical efficacy of betaxolol hydrochloride in normal volunteers. Ophthalmology 112, 2131–2136.

Sergouniotis, P.I., Davidson, A.E., Mackay, D.S., et al., 2011. Recessive mutations in KCNJ13, encoding an inwardly rectifying potassium channel subunit, cause Leber congenital amaurosis. Am J Hum Genet 89, 183–190.

Shi, Y., Li, Y., Zhang, D., et al., 2011a. Exome sequencing identifies ZNF644 mutations in high myopia. PLoS Genet 7, e1002084.

Shi, Y., Qu, J., Zhang, D., et al., 2011b. Genetic variants at 13q12.12 are associated with high myopia in the Han Chinese population. Am J Hum Genet 88, 805–813.

Stone, E.M., 2007. Leber congenital amaurosis: A model for efficient genetic testing of heterogeneous disorders: LXIV Edward Jackson memorial lecture. Am J Ophthalmol 144, 791–811.

Strunnikova, N.V., Maminishkis, A., Barb, J.J., et al., 2010. Transcriptome analysis and molecular signature of human retinal pigment epithelium. Hum Mol Genet 19, 2468–2486.

Telander, D.G., 2011. Inflammation and age-related macular degeneration (AMD). Semin Ophthalmol 3, 192–197.

Thorleifsson, G., Magnusson, K.P., Sulem, P., et al., 2007. Common sequence variants in the LOXL1 gene confer susceptibility to exfoliation glaucoma. Science 317, 1397–1400.

Thorleifsson, G., Walters, G.B., Hewitt, A.W., et al., 2010. Common variants near CAV1 and CAV2 are associated with primary open-angle glaucoma. Nat Genet 42, 906–909.

Van Issum, C., Mavrakanas, N., Schutz, J.S., Shaarawy, T., 2011. Topiramate-induced acute bilateral angle closure and myopia: Pathophysiology and treatment controversies. Eur J Ophthalmol 21, 404–409.

Vitart, V., Bencić, G., Hayward, C., et al., 2010. New loci associated with central cornea thickness include COL5A1, AKAP13 and AVGR8. Hum Mol Genet 19, 4304–4311.

Vithana, E.N., Aung, T., Khor, C.C., et al., 2011. Collagen-related genes influence the glaucoma risk factor, central corneal thickness. Hum Mol Genet 20, 649–658.

Wang, D.Y., Ray, A., Rodgers, K., et al., 2010. Global gene expression changes in rat retinal ganglion cells in experimental glaucoma. Invest Ophthalmol Vis Sci 51, 4084–4095.

Wiggs, J.L., 2008. Genomic promise: Personalized medicine for ophthalmology. Arch Ophthalmol 126, 422–423.

Wiggs, J.L., Kang, J.H., Yaspan, B.L., et al., 2011. Common variants near CAV1 and CAV2 are associated with primary open-angle glaucoma in caucasians from the United States. Hum Mol Genet, 2011. doi: 10.1093/hmg/ddr382.

Wiggs, J.L., Yaspan, B.L., Hauser, M.A., Kang, J.H., Allingham, R.R., Olson, L.M., 2012. Common variants at 9p21 and 8q22 are associated with increased susceptibility to optic nerve degeneration in glaucoma. PLoS Genet. 8 (4), e1002654.

Wistow, G., Peterson, K., Gao, J., et al., 2008. NEIBank: Genomics and bioinformatics resources for vision research. Mol Vis 14, 1327–1337.

Yu, Y., Bhangale, T.R., Fagerness, J., et al., 2011. Common variants near FRK/COL10A1 and VEGFA are associated with advanced age-related macular degeneration. Hum Mol Genet 20, 3699–3709.

Zhong, Q., Kowluru, R.A., 2011. Epigenetic changes in mitochondrial superoxide dismutase in the retina and the development of diabetic retinopathy. Diabetes 60, 1304–1313.

Glaucoma

Yutao Liu and R. Rand Allingham

INTRODUCTION

Glaucoma is a heterogeneous group of disorders and is the most common cause of irreversible blindness worldwide (Quigley and Broman, 2006). Glaucoma is defined by the progressive loss of retinal ganglion cells and is associated with a characteristic optic neuropathy and visual field loss. Most forms of glaucoma are felt to have significant genetic susceptibility. A number of genes and chromosomal loci have been identified through genetic linkage or association analysis that are associated with or, in rare cases, causative for different types of glaucoma. Genome-wide approaches are improving our understanding of the genetic basis of glaucoma. What follows is a comprehensive discussion of the role of genetics in those types of glaucoma in which the genetic contribution to the disease is better understood; this includes primary open-angle glaucoma (POAG), exfoliation glaucoma (XFG), primary congenital glaucoma (PCG), and developmental glaucoma.

PRIMARY OPEN-ANGLE GLAUCOMA

POAG is the most common type of glaucoma in the world. It is characterized by the presence of glaucomatous optic neuropathy without an identifiable secondary cause. Well-recognized risk factors include elevated intraocular pressure (IOP), positive family history, refractive error, and Hispanic or African-American ancestry (Allingham et al., 2009; Kwon et al., 2009;

Libby et al., 2005). First-degree relatives of POAG-affected individuals have a seven- to ten-fold higher risk of developing the disease than the general population (Wolfs et al., 1998). This, as well as other similar studies, indicates that there is a strong hereditary component to POAG (Libby et al., 2005).

Genetic linkage analysis has identified at least 16 chromosomal loci for POAG or POAG-related phenotypes, such as elevated IOP, which are shown in Table 92.1 (Allingham et al., 2009). Several genes associated with POAG have been identified within these loci, including those for myocilin (GLC1A), optineurin (GLC1E), WD repeat domain 36 (GLC1G), and neurotrophin (GLC1O) (Chen et al., 2010; Monemi et al., 2005; Pasutto et al., 2009; Rezaie et al., 2002; Stone et al., 1997). It is estimated that variants or mutations in these genes contribute to approximately 5% of POAG cases (Allingham et al., 2009; Kwon et al., 2009). The contribution of these genes to glaucoma will be discussed below.

Myocilin (MYOC)

Myocilin was identified by Stone and colleagues in GLC1A locus, the first reported locus for POAG located on chromosome 1 (Sheffield et al., 1993; Stone et al., 1997). Myocilin is also known as trabecular meshwork inducible-glucocorticoid response (TIGR) (Polansky et al., 1997). The myocilin gene contains three coding exons, and myocilin protein contains two major homology domains, an N-terminal myosin-like domain and a C-terminal olfactomedin-like domain. More than 70 different mutations have been reported, most of which are located in the third exon, which codes for the olfactomedin-like

Genomic and Personalized Medicine, 2nd edition
by Ginsburg & Willard. DOI: http://dx.doi.org/10.1016/B978-0-12-382227-7.00092-6

Table 92.1	Currently reported POAG chromosomal loci			
Chromosomal location	**POAG phenotype**	**Locus name**	**Candidate gene**	**Reference**
1q23-q24	JOAG*, adult-onset	GLC1A	MYOC	Sheffield et al., 1993; Stone et al., 1997
2cen-q13	Adult-onset	GLC1B		Akiyama et al., 2008; Charlesworth et al., 2006; Stoilova et al., 1996
2p12	Elevated IOP			Duggal et al., 2007
2p16.3-p15	JOAG, adult-onset	GLC1H		Lin et al., 2008
3p22-p21	Adult-onset	GLC1L		Baird et al., 2005
3q21-q24	Adult-onset	GLC1C		Kitsos et al., 2001; Wirtz et al., 1997
5q22.1	Adult-onset	GLC1G	WDR36	Monemi et al., 2005
5q22.1-q32	JOAG	GLC1M		Pang et al., 2006
7q35-q36	Adult-onset	GLC1F		Wirtz et al., 1999
8q23	Adult-onset	GLC1D		Trifan et al., 1998
9q22	JOAG	GLC1J		Wiggs et al., 2004
10p13	Adult-onset, NTG	GLC1E	OPTN	Rezaie et al., 2002; Sarfarazi et al., 1998
15q11-q13	Adult-onset	GLC1I		Allingham et al., 2005; Woodroffe et al., 2006
15q22-q24	JOAG	GLC1N		Wang et al., 2006
19p13.2	Elevated IOP			Duggal et al., 2007
19q13.3	Adult-onset	GLC1O	NTF4	Liu et al., 2010; Pasutto et al., 2009
20p12	JOAG	GLC1K		Sud et al., 2008; Wiggs et al., 2004

*JOAG = juvenile open-angle glaucoma.

domain (http://www.myocilin.com) (Hewitt et al., 2008a). Myocilin-mutation-associated POAG is inherited as an autosomal dominant trait. POAG patients with myocilin mutations generally have juvenile or early-adult-onset form of glaucoma (Alward et al., 1998; Shimizu et al., 2000). The glaucoma is clinically severe, and frequently there is highly elevated IOP, which responds poorly to medication and often requires surgical intervention. Myocilin mutations account for 3–5% of adult POAG patients in most populations around the world (Allingham et al., 2009; Fan et al., 2006; Libby et al., 2005), and this is the single most common form of inherited glaucoma.

Although widely expressed in most ocular tissues and other tissues throughout the body (Kwon et al., 2009), the only known clinical consequence of a myocilin mutation is glaucoma. The molecular mechanism that leads to the glaucoma phenotype remains unclear. Missense mutations are most common, and these are associated with the more severe phenotype; it is interesting that the most common mutation is Q368X, a nonsense mutation, which is associated with later adult-onset form of glaucoma (Alward et al., 1998). Myocilin

mutations do not cause glaucoma via haploinsufficiency or overexpression (Kwon et al., 2009; Libby et al., 2005). Decreased expression or overexpression or knockout of myocilin in mice does not cause glaucoma or elevated IOP (Kwon et al., 2009). These data suggest that disease-causing myocilin mutations alter protein function, leading to elevated IOP and eventually loss of retinal ganglion cells (Gould et al., 2006; Senatorov et al., 2006; Zhou et al., 2008).

Current studies have provided a partial understanding of the protein function of myocilin. Wild-type myocilin appears to be secreted via an N-terminal signal sequence and is found in the aqueous humor or the eye; however glaucoma-associated myocilin is poorly secreted into the aqueous humor (Jacobson et al., 2001). Recent studies suggest that myocilin protein is secreted into the aqueous humor via the shedding of small vesicles called exosomes (Hoffman et al., 2009). Exosomes are known to contain ligands that participate in autocrine and paracrine signaling, suggesting a potential role in trabecular meshwork homeostasis. Myocilin may play a role in exosome shedding, which may contribute to trabecular meshwork dysfunction and the elevated intraocular pressure seen in most cases of POAG with myocilin mutations. Others have suggested that myocilin mutations may interfere with protein trafficking and result in the intracellular accumulation of misfolded protein (Kwon et al., 2009). More recent studies indicate that mutated myocilin may also sensitize the retinal ganglion cells to oxidative stress and that different myocilin mutations may confer varying sensitivity to oxidative stress (Joe and Tomarev, 2010).

Optineurin (OPTN)

Optineurin (*OPTN*) was the second gene identified as associated with POAG (Rezaie et al., 2002). *OPTN* is located on chromosome 10 and consists of 16 exons, encoding a protein with a molecular weight of 66 kD. In contrast to the high IOP found in myocilin-associated glaucoma, *OPTN*-associated glaucoma is characterized by normal or only moderately elevated IOP, a form of POAG referred to as normal tension glaucoma (NTG). NTG is very common, comprising as much as 40% of all cases of POAG. However, *OPTN*-related POAG is very rare, found in no more than 1% of NTG cases. Among all *OPTN* mutations, E50K has the strongest evidence for a causal role for POAG (Allingham et al., 2009). POAG patients with E50K mutations tend to have a more severe form of glaucoma compared to patients without E50K mutations. These patients have an earlier onset, develop more advanced optic nerve cupping, and require surgical intervention more frequently than their non-*OPTN* NTG counterparts (Aung et al., 2005).

The OPTN protein interacts with Ras-associated protein RAB8, myosin VI, and transferrin receptor (Sahlender et al., 2005). *OPTN* may play a neuroprotective role by reducing retinal ganglion cell susceptibility to apoptosis, in which *OPTN* negatively regulates *TNFα*-induced *NF-κB* activation that affects the apoptotic threshold (Zhu et al., 2007). The *OPTN* E50K mutation increases binding to *TBK1* (TANK-binding kinase 1), which forms a complex that regulates *TNFα* and its pro-apoptotic effects (Morton et al., 2008). In another study, De Marco and colleagues found that overexpression of *OPTN* blocks cytochrome c release from mitochondria and protects cells from hydrogen-peroxide-induced cell death (De Marco et al., 2006). Overexpression of *OPTN* with E50K mutation in retinal ganglion cells inhibits the translocation of *OPTN* to the nucleus and compromises mitochondrial membrane integrity, leading to apoptosis from external stressors. More recently, Park and colleagues discovered that overexpression of E50K *OPTN*, compared to the wild-type, results in a more pronounced impairment of transferrin uptake in human retinal pigment epithelial cells and RGC5 cell line, suggesting that the E50K mutation induces defective protein trafficking (Park et al., 2010). Chi and coworkers reported that E50K mutant mice showed massive apoptosis and degeneration of the entire retina, leading to approximately 28% reduction of retina thickness (Chi et al., 2010a). The mutation disrupts the interaction between OPTN and Rab8, suggesting the involvement of *OPTN* with retinal degeneration.

Recently it was reported that mutations in *OPTN* cause familial amyotrophic lateral sclerosis (ALS) in the Japanese population (Maruyama et al., 2010). ALS-specific *OPTN* mutations abolished the inhibition of activation of NF-κB. Wild-type and E50K *OPTN* did not have this effect on NF-κB activation. It appears that *OPTN* may exert its primary effect on retinal ganglion cells, increasing their susceptibility to premature cell death. This is in contrast to the effect of myocilin mutations, which increase IOP and secondarily cause retinal ganglion cell death.

WD Domain Repeat 36 (WDR36)

The third gene identified for POAG is *WDR36*, located on chromosome 5 with 23 exons encoding a protein of 105 kD molecular weight (Monemi et al., 2005). *WDR36* is expressed ubiquitously in various tissues of the body and throughout the structures of the eye. Although the original report estimated the prevalence of *WDR36* sequence variations in POAG patients to be 1.6–17%, subsequent studies have failed to identify genetic variants in *WDR36* as the causal agent (Allingham et al., 2009). Hauser and colleagues reported that POAG patients with *WDR36* sequence variants are associated with a more severe disease phenotype than those without, suggesting that sequence variants in *WDR36* may play a role in disease susceptibility or severity rather than causation (Hauser et al., 2006). *WDR36* may be involved in ribosomal RNA processing and interact with the p53-mediated stress response pathway (Footz et al., 2009). It was recently reported that mutations in *WDR36* directly affect axon growth of the retinal ganglion cell, which leads to progressive retinal degeneration in mice (Chi et al., 2010b).

Neurotrophin (NTF4)

A recent report found that *NTF4* was associated with POAG. *NTF4* contains two exons, and the protein is about 22 kD in size. Mutations in *NTF4* were identified in 1.7% of 892 POAG cases in the European population (Pasutto et al., 2009). *NTF4* variants were predicted to affect either NTF4 dimer stability or the interaction between NTF4 dimer and its receptor TrkB. Although a recent study in the Chinese population indicates that *NTF4* mutations are a rare cause (0.6%) of POAG in that country (Vithana et al., 2010), other reports in Caucasian and Indian populations have failed to detect an association between these *NTF4* variants and POAG (Liu et al., 2010; Rao et al., 2010).

Genome-wide Association Studies in POAG

Genome-wide association studies (GWAS) are a powerful approach that enables researchers to test the association of hundreds of thousands of single nucleotide polymorphisms (SNPs) with a disease in hundreds or thousands of individuals (Manolio, 2010). Study designs vary for GWAS, ranging from case-control studies or cohort studies to clinical trials. GWAS have revolutionized the search for genetic factors in various complex human diseases and traits, and have been used to identify a large number of disease-associated genetic variants that can be viewed in the GWAS catalog (http://www.genome.

gov/gwastudies). Among the disorders that GWAS technology has contributed to are major eye diseases such as age-related macular degeneration, exfoliation glaucoma, and POAG.

Several GWAS have been done with POAG in different populations. The first GWAS for POAG was performed in the Japanese population (Nakano et al., 2009). The dataset consisted of 827 POAG cases and 748 controls. Although none of the SNPs reached genome-wide significance, this study identified three potential loci for POAG on chromosomes 1, 10, and 12. The second GWAS, also performed in the Japanese population, included 305 NTG cases and 355 controls (Meguro et al., 2010). Only one SNP, rs3213787, reached genome-wide significance. This SNP is located in the intron of the *SRBD1* gene (S1 RNA binding domain 1) on chromosome 2. Another SNP, rs735860, was not significant genome-wide. However, it lies in the 3′-untranslated region of *ELOVL5* (ELOVL family member 5, elongation of long chain fatty acids) with p-value of 4×10^{-6}. Both genes are reported to be involved in the induction of cell growth inhibition or apoptosis and might play role in the risk of glaucoma development. Findings from these studies have not been replicated to date (Rao et al., 2009).

The most recent landmark GWAS in POAG was reported in the Icelandic population (Thorleifsson et al., 2010). Based on 1263 POAG cases and 34,877 controls, this study identified a locus on chromosome 7q31 (rs4236601) associated with POAG (p-value 5×10^{-10}). The association was successfully replicated in the Caucasian (2175 cases/2064 controls) and Chinese populations (299 cases/580 controls). This finding has also been replicated by our group in Caucasian and African-American populations (authors' unpublished data). The risk variant is located in an intergenic region between the caveolin 1 (*CAV1*) and caveolin 2 (*CAV2*) genes. These two genes are expressed in the trabecular meshwork, the aqueous humor outflow tissue, and retinal ganglion cells. CAV1 and CAV2 are proteins involved in the formation of caveolae. How this intergenic association is related to POAG susceptibility is unknown.

Two large GWAS based in the US, GENEVA (GLAUGEN) (Genes and Environment Initiative in Glaucoma) and NEIGHBOR (NEI Glaucoma Human Genetics Collaboration), are currently underway (http://www.nei.nih.gov/funding/gen_resources.asp and http://www.nei.nih.gov/strategicplanning/genetics1.asp) (Cornelis et al., 2010). More than 3600 POAG cases and 3600 controls will be included in the study (personal communications with Janey L. Wiggs, M.D., Ph.D., and Louis R. Pasquale, M.D.). The resulting genotype data and clinical phenotype, as well as epidemiologic and environmental exposure data, will be made available to the research community through the National Center for Biotechnology Informatics (NCBI) database of Genotypes and Phenotypes (dbGAP). These GWAS studies (Wiggs et al., 2012) identified significant associations between two loci (CDKN2BAS region on 9p21 and the SIX1/SIX6 region on 14q23). In a subgroup analysis, two loci were significantly with normal pressure glaucoma: 9p21 containing the CDKN2BAS gene (CDKN2B antisense RNA) and a probable regulatory region on 8q22. An earlier GWAS study (Burdon et al., 2011) identified two glaucoma

susceptibility loci for POAG at transmembrane and coiled-coil domains 1 (TMCO1) and CDKN2BAS.

POAG-related Phenotype Associations

Another way to study POAG is to investigate the phenotypes that are associated with glaucoma, such as central corneal thickness, cup-to-disc ratio, and optic disc area. A number of GWAS have been reported for these ocular phenotypes. Two studies (Lu et al., 2010; Vitart et al., 2010) have identified sequence variants near or in the genes *ZNF469*, *COL5A1*, *AKAP13*, and *AVGR8*, that are associated with central corneal thickness. GWAS with cup-to-disc ratio have identified variants in the *ATOH7*, *CDKN2B*, and *SIX1* genes (Macgregor et al., 2010; Ramdas et al., 2010). Variants in the genes *ATOH7* and *CDC7* were associated with optic disc area. How these associations relate to the POAG disease phenotype in various populations will be of great interest.

EXFOLIATION GLAUCOMA

Exfoliation (also known as pseudoexfoliation) glaucoma (XFG) is the most common identifiable form of open-angle glaucoma. XFG is associated with exfoliation syndrome (XFS), which affects an estimated 60–70 million people worldwide. XFS was initially described by Lindberg in 1917 (Ritch and Schlotzer-Schrehardt, 2001). XFS is a systemic disorder of the extracellular matrix characterized by the accumulation of an abnormal fibrillary material on the lens surface (Figure 92.1) and in various ocular and non-ocular tissues in the body. In addition to its effect on the eye, XFS may also be associated with cardiovascular and cerebrovascular diseases (Schlotzer-Schrehardt and Naumann, 2006). The prevalence of XFS increases dramatically after age 60. The prevalence of XFS varies significantly, with none reported in the indigenous populations of Greenland and as high as 30% in Scandinavian populations of Iceland and Finland (Ritch and Schlotzer-Schrehardt, 2001). XFG has been reported to account for approximately 25% of all open-angle glaucoma cases worldwide, making this the most common secondary form of open-angle glaucoma (Ritch and Schlotzer-Schrehardt, 2001). In certain populations, such as on the Arabian Peninsula, XFG accounts for up to 77% of all cases of open-angle glaucoma (Bialasiewicz et al., 2005). XFG affects virtually all populations globally and has a more severe prognosis, presents with more advanced vision loss, and is more likely to lead to blindness than POAG (Ritch, 2008; Ritch and Schlotzer-Schrehardt, 2001).

Genetic factors have been shown to play important roles in the pathogenesis of XFS/XFG. Several chromosomal regions have been associated with XFS and XFG. Lemmela and colleagues identified genomic regions that showed strong linkage to XFS (Lemmela et al., 2007). Recently in a GWAS of XFG/XFS subjects in the Icelandic and Swedish populations, three SNPs in the lysyl oxidase-like 1 gene (*LOXL1*) were found in strong association with both XFS and XFG

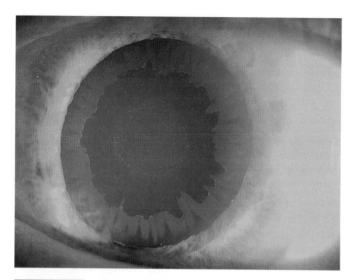

Figure 92.1 Exfoliation of the lens. Note the "target" distribution of white, flaky exfoliation material on the lens surface. This material is found throughout the body, and when produced within the aqueous drainage system is believed to be directly involved in blocking normal fluid flow from the eye. This results in elevation of eye pressure and glaucomatous optic nerve damage. *Courtesy of Dr Joseph Halabis, Durham VA Medical Center.*

(Thorleifsson et al., 2007). Two of these sequence variants, rs1048661 and rs3825942, are located in exon 1 of the gene and code for amino acid changes R141L and G153D, respectively. It has been considered that either or both of these coding variants may be functionally important in the pathobiology of XFS, although no evidence beyond genetic association has been reported. The association between *LOXL1* and XFS/XFG has been widely replicated (Chen et al., 2010). Variants of *LOXL1* are not associated with POAG (Liu et al., 2008), confirming that XFG and POAG are genetically distinct. Recent studies of *LOXL1* have found that rs1048661 is not associated with XFS/XFG in several populations (Chen et al., 2010). Furthermore, although the G allele of rs1048661 is associated with increased risk of XFS/XFG in the Caucasian populations, the opposite (T) allele is associated with XFS/XFG in the Chinese and Japanese populations (Chen et al., 2010), as summarized in Table 92.2. The second coding SNP, rs3825942, has been associated with XFS/XFG in all non-African populations studied to date. In non-African populations, the G allele is associated with disease. A recent report from our group examining *LOXL1* variants in the South African black population confirmed that both SNPs rs1048661 and rs3825942 are highly associated with XFG (Williams et al., 2010). However, in sharp contrast to the pattern reported in all non-African populations, in the South African black population the opposite (A) allele of rs3825942 was found to be the risk allele. There are no other coding variants in *LOXL1* that are associated with XFG in the black South African population. Interestingly, XFG is rare in persons of West African descent

(Herndon et al., 2002). Why XFG is so rare in people of West African descent is puzzling, especially in light of a report confirming the presence of major *LOXL1* risk variants in these populations (Liu et al., 2008).

The preceding discussion indicates that despite the extensive genetic study of *LOXL1* in XFG, the major functional risk variants in *LOXL1* have not been identified and do not appear to be secondary to coding variants of the protein. *LOXL1* mRNA and protein expression appears to be up-regulated in early stages of exfoliation only to be down-regulated in patients with XFG (Khan et al., 2010; Schlotzer-Schrehardt et al., 2008). Recently, Ferrell and colleagues reported that SNP rs16958477 in the promoter region of *LOXL1* affects promoter activity *in vitro*. In these studies, the C allele had a significantly greater effect than the A allele on promoter activity (Ferrell et al., 2009). However, this specific variant was not associated with XFG in the black South African or Caucasian populations (Williams et al., 2010). The search for causal variants in the *LOXL1* gene is still in progress and will help to reveal the functional role of *LOXL1* in glaucoma.

Since the discovery of *LOXL1* in XFS/XFG, sequence variants in the *CNTNAP2* gene (contactin-associated protein-like 2) were found to be significantly associated with XFS/XFG in a German population of 160 cases and 80 controls (Krumbiegel et al., 2011). Although it was not confirmed in an Italian cohort of 249 cases and 190 controls, this association was replicated in an independent German cohort of 610 cases and 364 controls. The lack of association in the Italian cohort may be due to the small sample size. The DNA samples were pooled in this study, instead of genotyped separately. The pooling strategy, which has been used widely, has significantly reduced the cost and effort for this GWAS. *CNTNAP2* was ubiquitously expressed in human ocular tissues, including retina. This gene has been associated with various neuropsychiatric disorders, including autism, mental retardation, schizophrenia, and epilepsy. However, the function of the gene still remains largely unknown.

PRIMARY CONGENITAL GLAUCOMA

Primary congenital glaucoma (PCG) is the single most common childhood glaucoma, accounting for about 25% of all pediatric glaucoma cases. In PCG, a developmental abnormality of the anterior chamber angle leads to obstruction of aqueous outflow. When the onset of glaucoma and elevated IOP occurs within the first few years of life, it produces enlargement of the globe of the eye termed buphthalmos ("ox eyed" in Greek). In addition to being physically enlarged, eyes may have cloudy corneas and the patients frequently suffer from photophobia and epiphora (tearing) (Figure 92.2). Significant progress has been made in identifying the genetic causes. Currently, depending on the population being studied, from 10 to over 80% of inherited pediatric glaucoma can be attributed to known chromosomal loci (Vasiliou and Gonzalez, 2008). At

TABLE 92.2	Genetic association of coding variants in *LOXL1* gene with XFS/XFG in different populations						
Studied population	**rs1048661 G allele**		**Significant association**	**rs3825942 G allele**		**Significant association**	**References**
	Case	**Control**		**Case**	**Control**		
Icelandic	0.781	0.651	Yes	0.984	0.847	Yes	Thorleifsson et al., 2007
Swedish	0.834	0.682	Yes	0.995	0.879	Yes	Thorleifsson et al., 2007
American	0.819	0.600	Yes	0.986	0.880	Yes	Fingert et al., 2007
Australian	0.78	0.660	Yes	0.95	0.84	Yes	Hewitt et al., 2008b
American	0.787	0.665	Yes	0.939	0.844	Yes	Challa et al., 2008
American	N/A	N/A	N/A	1.000	0.856	Yes	Yang et al., 2008
American	0.843	0.703	Yes	0.959	0.798	Yes	Aragon-Martin et al., 2008
Austrian	0.841	0.671	Yes	0.994	0.817	Yes	Mossbock et al., 2008
German	0.818	0.644	Yes	0.951	0.857	Yes	Pasutto et al., 2008
Italian	0.825	0.693	Yes	1.000	0.821	Yes	Pasutto et al., 2008
Finnish	0.825	0.683	Yes	0.968	0.823	Yes	Lemmela et al., 2009
German	0.844	0.660	Yes	0.992	0.856	Yes	Wolf et al., 2010
Chinese	0.110	0.480	Yes	1.000	0.900	Yes	Chen et al., 2009
Japanese	0.036	0.493	Yes	1.000	0.877	Yes	Fuse et al., 2008
Japanese	0.008	0.460	Yes	1.000	0.857	Yes	Hayashi et al., 2008
Japanese	0.006	0.450	Yes	0.994	0.853	Yes	Mabuchi et al., 2008
Japanese	0.005	0.474	Yes	0.995	0.850	Yes	Mori et al., 2008
Japanese	0.005	0.497	Yes	0.986	0.863	Yes	Ozaki et al., 2008
Japanese	0.005	0.554	Yes	0.993	0.806	Yes	Tanito et al., 2008
Indian	0.721	0.634	No	0.923	0.742	Yes	Ramprasad et al., 2008
Chinese	0.542	0.444	No	0.992	0.918	Yes	Lee et al., 2009
American	0.829	0.719	No	0.988	0.795	Yes	Fan et al., 2008
African-American	N/A	N/A	N/A	N/A	0.599	N/A	Liu et al., 2008
Ghanaian	N/A	N/A	N/A	N/A	0.570	N/A	Liu et al., 2008
South African	0.990	0.810	Yes	0.130	0.620	Yes	Williams et al., 2010

N/A = not applicable.

least 12 genetic loci have been linked to childhood glaucoma (Table 92.3). Three genetic loci (*GLC3A*, *GLC3B*, and *GLC3C*) are linked to PCG. Variants in the gene for cytochrome P450 1B1 (*CYP1B1*) were identified as causative for PCG within the *GLC3A* region on chromosome 2 (Stoilov et al., 1997).

Most commonly, PCG is inherited as an autosomal recessive disorder where both parents contribute a disease-associated variant of the gene. More than 80 such PCG-associated mutations have been described to date (Vasiliou and Gonzalez, 2008). These include missense and nonsense mutations, deletions,

insertions, and duplications. *CYP1B1* mutations have been found in patients with Peter's anomaly, Axenfeld–Rieger syndrome, and adult forms of POAG. This suggests that *CYP1B1* mutations are associated with a broad range of clinical phenotypes. *CYP1B1* has also been described as a modifier gene for myocilin-associated glaucoma, while mutations in the myocilin gene are associated with both juvenile and adult-onset primary open-angle glaucoma (Vasiliou and Gonzalez, 2008; Vincent et al., 2002). Interestingly, the additional presence of *CYP1B1* mutations correlates with an earlier onset of glaucoma in these

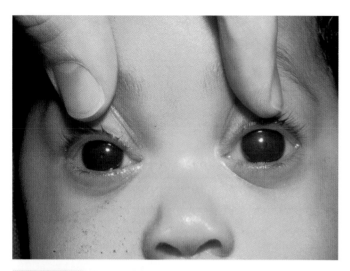

Figure 92.2 Congenital glaucoma in an infant. Note the enlargement of the eyes, a condition called buphthalmos. This enlargement of the eye is caused by elevated intraocular pressure. The mechanical stretching seen in the eyes of infants and young children does not occur in older children and adults because their eyes are much less compliant. *Courtesy of Dr David Wallace, Duke Eye Center.*

individuals. In a knockout mouse model, disease-associated genetic variants of *CYP1B1* generally cause loss of gene function. The prevalence of *CYP1B1* mutations in patients with PCG varies widely between populations. For example, the proportion of PCP patients with *CYP1B1* mutations is 100% in Saudi Arabian and Slovakian Rom, 50% in Brazilian, 30% in Indonesian, 20% in Japanese, and about 20% in Caucasian patients (Chakrabarti et al., 2006).

CYP1B1 is highly expressed in the iris and ciliary body, with lower expression in the cornea, retina, and trabecular meshwork. It functions in metabolism of 17-beta-estradiol and reduces estrogenic activity. It is suggested that *CYP1B1* may play a role in early differentiation of ocular tissues. Incomplete penetrance has also been described for *CYP1B1* mutations (Bejjani et al., 2000). This means that while some patients carrying *CYP1B1* mutations develop PCG, others do not. This suggests that additional, unknown genetic and environmental factors play a role in the development of PCG. The other two loci specific to PCG are *GLC3B* at chromosome 1p36.2-p36.1 and *GLC3C* at chromosome 14q24.3-q31.1 (Table 92.4) (Akarsu et al., 1996; Sivadorai et al., 2008; Stoilov and Sarfarazi, 2002). However, genes related to the development of PCG have not been identified in these loci. Mutations in the *LTBP2* gene (latent transforming growth factor beta binding protein 2) have been reported to cause PCG (Ali et al., 2009; Azmanov et al., 2011; Narooie-Nejad et al., 2009). *LTBP2* is located at chromosome 14q24.3, but is around 1.3 Mb proximal to the documented *GLC3C* region. Thus, it remains to be determined whether the gene in *GLC3C* is *LTBP2* or a second, adjacent gene.

Recent technological developments will help to sequence all the coding regions (the exome) in the human genome (see also Chapter 7). This could be done by sequence capturing and high-throughput sequencing with next-generation sequencing technology (Teer and Mullikin, 2010). By selecting different individuals from the same family and unrelated families with the same phenotype, it will be possible to identify the specific coding mutations in different diseases, including PCG. This innovative approach has and will significantly shorten the time to identify genetic mutations for rare Mendelian disorders (Bilguvar et al., 2010; Ng et al., 2010a, b, c).

DEVELOPMENTAL GLAUCOMA

Developmental glaucoma results from alterations in the development of the anterior segment of the eye. The anterior segment includes the cornea, iris, and anterior chamber angle, tissues that are essential for normal aqueous humor drainage. Therefore, abnormalities of these tissues can often lead to impairment of aqueous humor outflow and elevated IOP. The ocular abnormalities seen in patients with developmental glaucoma include anterior segment dysgenesis, Axenfeld–Rieger syndrome (Figure 92.3), Peter's anomaly (Figure 92.4), iridogoniodysgenesis, iris hypoplasia, and aniridia. A number of different terms have been used to describe developmental glaucoma, including anterior chamber cleavage syndrome, Axenfeld anomaly, Rieger syndrome, and mesodermal dysgenesis of the cornea and iris. Approximately half of patients with these developmental anomalies will develop glaucoma. Onset of glaucoma can be at birth or later in adulthood.

Due to the availability of extended, often multigenerational, families, and an earlier age of onset with clearly identifiable ocular and non-ocular phenotypes, developmental glaucoma is more amenable to traditional forms of investigation, such as genetic linkage analysis. Mendelian inheritance of this group of disorders can be either autosomal recessive or dominant. A number of genes have been identified as causing developmental glaucoma (Table 92.3). They include *PITX2*, *PITX3*, *FOXC1*, *FOXE3*, *PAX6*, *LMX1B*, and *MAF* (Gould and John, 2002; Gould et al., 2004), most of which code for transcription factors. Mutations in these transcription factors interfere with cellular and extracellular matrix signaling during development. Two additional loci – at chromosome 13q14 and 16q24 – have been linked with Axenfeld–Rieger syndrome, but the responsible genes have not been identified. As mentioned previously for PCG, exome sequencing approaches should prove useful to identify causative genes in these loci.

GLAUCOMA GENETICS AND PERSONALIZED MEDICINE

Significant progress has been made in understanding the genetic architecture of complex human disease with the

TABLE 92.3 Genetic loci of developmental glaucoma

Locus	Chromosomal location	Gene	Phenotype	Reference
RIEG1	4q25	*PITX2*	ARS, IGD, IH, PA	Semina et al., 1996
RIEG2	13q14		ARS and sensorineural hearing loss	Phillips et al., 1996; Stathacopoulos et al., 1987
RIEG3	6p25	*FOXC1*	ARS, PA, PCG, IGD	Gould et al., 1997; Nishimura et al., 2001
	11p13	*PAX6*	Aniridia, ARS, PA	Hanson et al., 1994; Jordan et al., 1992
	10q24-25	*PITX3*	ASD and cataracts	Semina et al., 1997, 1998
	1p32	*FOXE3*	ASD and cataracts, congenital aphakia	Blixt et al., 2000; Semina et al., 2001; Valleix et al., 2006
	9q34.1	*LMX1B*	Nail-patella syndrome and open-angle glaucoma	Dreyer et al., 1998; McIntosh et al., 1998, Vollrath et al., 1998
	16q22-23	*MAF*	ASD and congenital cataracts	Jamieson et al., 2002; Vanita et al., 2006
	16q24	?	ARS	Pal et al., 2004

PA = Peter's anomaly, ARS = Axenfeld–Rieger's syndrome, PCG = primary congenital glaucoma, IGD = iridogoniodysgenesis, IH = iris hypoplasia, ASD = anterior segment dysgenesis.

TABLE 92.4 Genetic loci of primary congenital glaucoma

Locus	Chromosomal location	Gene	Phenotype	Reference
GLC3A	2p22-p21	*CYP1B1*	PCG, PA, ARS	Plasilova et al., 1998; Sarfarazi et al., 1995; Stoilov et al., 1997
GLC3B	1p36.2-p36.1		PCG	Akarsu et al., 1996
GLC3C	14q24.3-q31.1	*LTBP2?*	PCG	Stoilov and Sarfarazi, 2002

PCG = primary congenital glaucoma, PA = Peter's anomaly, ARS = Axenfeld–Rieger's syndrome.

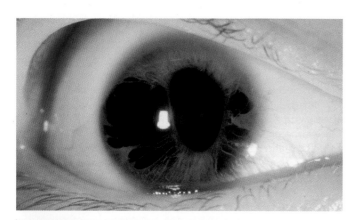

Figure 92.3 Axenfeld–Rieger syndrome with characteristic alterations of the iris. Approximately 50% of these patients will develop glaucoma. *Courtesy Dr David Wallace, Duke Eye Center.*

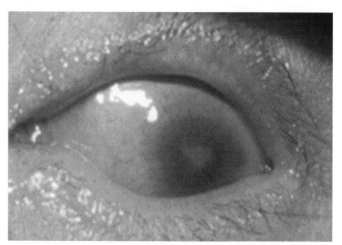

Figure 92.4 Peter's anomaly with central corneal clouding. The anterior segment of the eye develops abnormally in these cases. Glaucoma occurs in about 50% of affected children. *Courtesy Dr David Wallace, Duke Eye Center.*

powerful new methods for genetic discovery, logarithmic growth in genotyping technology, and the assembly of increasingly large, robust clinical cohorts. This genetic progress will ultimately lead to a vastly improved understanding of the molecular mechanisms of complex diseases like glaucoma. Ultimately, combined with other clinical information including family history, genetic screening with DNA markers will enable physicians to predict the risk of developing disease as well as disease severity. As importantly, expanding comprehension of the molecular pathways that produce disease will guide the development of more effective treatment options for individuals as well as their families. For diseases like POAG, where end organ damage is untreatable, the ability to accurately predict disease prior to symptomatic vision loss is essential.

Currently, identification and diagnosis of new glaucoma cases is achieved either by routine screening or examinations prompted by perceived risk. Traditional vision screening for disorders like POAG consumes limited and costly resources. Furthermore, the resulting examination provides, at best, a "snapshot" in the lives of those screened, where a negative result applies only to a single time-point. So screening becomes a continuous, inefficient process that consumes enormous resources. For this reason, it is not surprising that there is considerable debate regarding the utility of glaucoma screening.

Interestingly, genetic and genomic screening, a technology that is becoming more widespread and increasingly affordable, could be a powerful screening approach for developed countries as well as areas of the world that are medically underserved. Once risk assessment reaches a sufficient threshold of accuracy, genetic screening would provide focused delivery of medical resources on much smaller at-risk populations. Of course, genetic screening would apply to multiple disorders where the genetic architecture of the disease is known.

In the case of the glaucomas, genetic analysis could lead to the identification of causative mutations for PCG and other inherited forms of childhood glaucoma that are a major source of blindness in the pediatric population, particularly in developing nations. This information would be of great utility in counseling parents or other family members about a child's prognosis after diagnosis, and possible risk to those who are related. Genetic analysis can be used to counsel prospective parents concerned about the risk of having affected children. For example, Hollander and coworkers have found that specific mutations and combinations of mutations for *CYP1B1* can be correlated with ocular pathology and the severity of glaucoma (Hollander et al., 2006). This type of information could guide treatment, for example, where a specific mutation causes agenesis of Schlemm's canal in congenital glaucoma, where traditional goniotomy would likely fail and other treatment options might be preferable.

Even when a specific genetic condition cannot be identified with certainty, genetic counseling can prove useful to parents who are trying to assess risk of transmission to their offspring, or who may wish to be connected to ongoing research efforts. A convenient resource for families who wish to pursue genetic counseling is http://www.nsgc.org, the website of the National Society of Genetic Counselors. The site allows searching for qualified genetic counselors by the patient's postal code, as well as the counselor's area of specialization and institution. See also the *Recommended Resources* section at the end of the chapter.

REFERENCES

Akarsu, A.N., Turacli, M.E., Aktan, S.G., et al., 1996. A second locus (GLC3B) for primary congenital glaucoma (buphthalmos) maps to the 1p36 region. Hum Mol Genet 5, 1199–1203.

Akiyama, M., Yatsu, K., Ota, M., et al., 2008. Microsatellite analysis of the GLC1B locus on chromosome 2 points to NCK2 as a new candidate gene for normal tension glaucoma. Br J Ophthalmol 92, 1293–1296.

Ali, M., Mckibbin, M., Booth, A., et al., 2009. Null mutations in LTBP2 cause primary congenital glaucoma. Am J Hum Genet 84, 664–671.

Allingham, R.R., Liu, Y., Rhee, D.J., 2009. The genetics of primary open-angle glaucoma: A review. Exp Eye Res 88, 837–844.

Allingham, R.R., Wiggs, J.L., Hauser, E.R., et al., 2005. Early adult-onset POAG linked to 15q11-13 using ordered subset analysis. Invest Ophthalmol Vis Sci 46, 2002–2005.

Alward, W.L., Fingert, J.H., Coote, M.A., et al., 1998. Clinical features associated with mutations in the chromosome 1 open-angle glaucoma gene (GLC1A). N Engl J Med 338, 1022–1027.

Aragon-Martin, J.A., Ritch, R., Liebmann, J., et al., 2008. Evaluation of LOXL1 gene polymorphisms in exfoliation syndrome and exfoliation glaucoma. Mol Vis 14, 533–541.

Aung, T., Rezaie, T., Okada, K., et al., 2005. Clinical features and course of patients with glaucoma with the E50K mutation in the optineurin gene. Invest Ophthalmol Vis Sci 46, 2816–2822.

Azmanov, D.N., Dimitrova, S., Florez, L., et al., 2011. LTBP2 and CYP1B1 mutations and associated ocular phenotypes in the Roma/Gypsy founder population. Eur J Hum Genet 19, 326–333.

Baird, P.N., Foote, S.J., Mackey, D.A., Craig, J., Speed, T.P., Bureau, A., 2005. Evidence for a novel glaucoma locus at chromosome 3p21-22. Hum Genet 117, 249–257.

Bejjani, B.A., Stockton, D.W., Lewis, R.A., et al., 2000. Multiple CYP1B1 mutations and incomplete penetrance in an inbred population segregating primary congenital glaucoma suggest frequent de novo events and a dominant modifier locus. Hum Mol Genet 9, 367–374.

Bialasiewicz, A.A., Wali, U., Shenoy, R., Al-Saeidi, R., 2005. Patients with secondary open-angle glaucoma in pseudoexfoliation (PEX) syndrome among a population with high prevalence of PEX. Clinical findings and morphological and surgical characteristics. Ophthalmologe 102, 1064–1068.

Bilguvar, K., Ozturk, A.K., Louvi, A., et al., 2010. Whole-exome sequencing identifies recessive WDR62 mutations in severe brain malformations. Nature 467, 207–210.

Blixt, A., Mahlapuu, M., Aitola, M., Pelto-Huikko, M., Enerback, S., Carlsson, P., 2000. A forkhead gene, FoxE3, is essential for lens epithelial proliferation and closure of the lens vesicle. Genes Dev 14, 245–254.

Chakrabarti, S., Kaur, K., Kaur, I., et al., 2006. Globally, CYP1B1 mutations in primary congenital glaucoma are strongly structured by geographic and haplotype backgrounds. Invest Ophthalmol Vis Sci 47, 43–47.

Challa, P., Schmidt, S., Liu, Y., et al., 2008. Analysis of LOXL1 polymorphisms in a United States population with pseudoexfoliation glaucoma. Mol Vis 14, 146–149.

Charlesworth, J.C., Stankovich, J.M., Mackey, D.A., et al., 2006. Confirmation of the adult-onset primary open angle glaucoma locus GLC1B at 2cen-q13 in an Australian family. Ophthalmologica 220, 23–30.

Chen, H., Chen, L.J., Zhang, M., et al., 2010. Ethnicity-based subgroup meta-analysis of the association of LOXL1 polymorphisms with glaucoma. Mol Vis 16, 167–177.

Chen, L., Jia, L., Wang, N., et al., 2009. Evaluation of LOXL1 polymorphisms in exfoliation syndrome in a Chinese population. Mol Vis 15, 2349–2357.

Chi, Z.L., Akahori, M., Obazawa, M., et al., 2010a. Overexpression of optineurin E50K disrupts Rab8 interaction and leads to a progressive retinal degeneration in mice. Hum Mol Genet 19, 2606–2615.

Chi, Z.L., Yasumoto, F., Sergeev, Y., et al., 2010b. Mutant WDR36 directly affects axon growth of retinal ganglion cells leading to progressive retinal degeneration in mice. Hum Mol Genet 19, 3806–3815.

Cornelis, M.C., Agrawal, A., Cole, J.W., et al., 2010. The Gene, Environment Association Studies consortium (GENEVA): Maximizing the knowledge obtained from GWAS by collaboration across studies of multiple conditions. Genet Epidemiol 34, 364–372.

De Marco, N., Buono, M., Troise, F., Diez-Roux, G., 2006. Optineurin increases cell survival and translocates to the nucleus in a Rab8-dependent manner upon an apoptotic stimulus. J Biol Chem 281, 16,147–16,156.

Dreyer, S.D., Zhou, G., Baldini, A., et al., 1998. Mutations in LMX1B cause abnormal skeletal patterning and renal dysplasia in nail patella syndrome. Nat Genet 19, 47–50.

Duggal, P., Klein, A.P., Lee, K.E., Klein, R., Klein, B.E., Bailey-Wilson, J.E., 2007. Identification of novel genetic loci for intraocular pressure: A genomewide scan of the Beaver Dam Eye Study. Arch Ophthalmol 125, 74–79.

Fan, B.J., Pasquale, L., Grosskreutz, C.L., et al., 2008. DNA sequence variants in the LOXL1 gene are associated with pseudoexfoliation glaucoma in a US clinic-based population with broad ethnic diversity. BMC Med Genet 9, 5.

Fan, B.J., Wang, D.Y., Lam, D.S., Pang, C.P., 2006. Gene mapping for primary open angle glaucoma. Clin Biochem 39, 249–258.

Ferrell, G., Lu, M., Stoddard, P., et al., 2009. A single nucleotide polymorphism in the promoter of the LOXL1 gene and its relationship to pelvic organ prolapse and preterm premature rupture of membranes. Reprod Sci 16, 438–446.

Fingert, J.H., Alward, W.L., Kwon, Y.H., et al., 2007. LOXL1 mutations are associated with exfoliation syndrome in patients from the midwestern United States. Am J Ophthalmol 144, 974–975.

Footz, T.K., Johnson, J.L., Dubois, S., Boivin, N., Raymond, V., Walter, M.A., 2009. Glaucoma-associated WDR36 variants encode functional defects in a yeast model system. Hum Mol Genet 18, 1276–1287.

Fuse, N., Miyazawa, A., Nakazawa, T., Mengkegale, M., Otomo, T., Nishida, K., 2008. Evaluation of LOXL1 polymorphisms in eyes with exfoliation glaucoma in Japanese. Mol Vis 14, 1338–1343.

Gould, D.B., John, S.W., 2002. Anterior segment dysgenesis and the developmental glaucomas are complex traits. Hum Mol Genet 11, 1185–1193.

Gould, D.B., Mears, A.J., Pearce, W.G., Walter, M.A., 1997. Autosomal dominant Axenfeld–Rieger anomaly maps to 6p25. Am J Hum Genet 61, 765–768.

Gould, D.B., Reedy, M., Wilson, L.A., Smith, R.S., Johnson, R.L., John, S.W., 2006. Mutant myocilin nonsecretion in vivo is not sufficient to cause glaucoma. Mol Cell Biol 26, 8427–8436.

Gould, D.B., Smith, R.S., John, S.W., 2004. Anterior segment development relevant to glaucoma. Int J Dev Biol 48, 1015–1029.

Hanson, I.M., Fletcher, J.M., Jordan, T., et al., 1994. Mutations at the PAX6 locus are found in heterogeneous anterior segment malformations including Peters' anomaly. Nat Genet 6, 168–173.

Hauser, M.A., Allingham, R.R., Linkroum, K., et al., 2006. Distribution of WDR36 DNA sequence variants in patients with primary open-angle glaucoma. Invest Ophthalmol Vis Sci 47, 2542–2546.

Hayashi, H., Gotoh, N., Ueda, Y., Nakanishi, H., Yoshimura, N., 2008. Lysyl oxidase-like 1 polymorphisms and exfoliation syndrome in the Japanese population. Am J Ophthalmol 145, 582–585.

Herndon, L.W., Challa, P., Ababio-Danso, B., et al., 2002. Survey of glaucoma in an eye clinic in Ghana, West Africa. J Glaucoma 11, 421–425.

Hewitt, A.W., Mackey, D.A., Craig, J.E., 2008a. Myocilin allele-specific glaucoma phenotype database. Hum Mutat 29, 207–211.

Hewitt, A.W., Sharma, S., Burdon, K.P., et al., 2008b. Ancestral LOXL1 variants are associated with pseudoexfoliation in Caucasian Australians but with markedly lower penetrance than in Nordic people. Hum Mol Genet 17, 710–716.

Hoffman, E.A., Perkumas, K.M., Highstrom, L.M., Stamer, W.D., 2009. Regulation of myocilin-associated exosome release from human trabecular meshwork cells. Invest Ophthalmol Vis Sci 50, 1313–1318.

Hollander, D.A., Sarfarzai, M., Stoilov, I., Wood, I.S., Fredrick, D.R., Alvarado, J.A., 2006. Genotype and phenotype correlations in congenital glaucoma: CYP1B1 mutations, goniodysgenesis, and clinical characteristics. Am J Ophthalmol 142, 993–1004.

Jacobson, N., Andrews, M., Shepard, A.R., et al., 2001. Non-secretion of mutant proteins of the glaucoma gene myocilin in cultured trabecular meshwork cells and in aqueous humor. Hum Mol Genet 10, 117–125.

Jamieson, R.V., Perveen, R., Kerr, B., et al., 2002. Domain disruption and mutation of the bZIP transcription factor, MAF, associated with cataract, ocular anterior segment dysgenesis and coloboma. Hum Mol Genet 11, 33–42.

Joe, M.K., Tomarev, S.I., 2010. Expression of myocilin mutants sensitizes cells to oxidative stress-induced apoptosis: Implication for glaucoma pathogenesis. Am J Pathol 176, 2880–2890.

Jordan, T., Hanson, I., Zaletayev, D., et al., 1992. The human PAX6 gene is mutated in two patients with aniridia. Nat Genet 1, 328–332.

Khan, T.T., Li, G., Navarro, I.D., et al., 2010. LOXL1 expression in lens capsule tissue specimens from individuals with pseudoexfoliation syndrome and glaucoma. Mol Vis 16, 2236–2241.

Kitsos, G., Eiberg, H., Economou-Petersen, E., et al., 2001. Genetic linkage of autosomal dominant primary open angle glaucoma to chromosome 3q in a Greek pedigree. Eur J Hum Genet 9, 452–457.

Krumbiegel, M., Pasutto, F., Schlotzer-Schrehardt, U., et al., 2011. Genome-wide association study with DNA pooling identifies variants at CNTNAP2 associated with pseudoexfoliation syndrome. Eur J Hum Genet 19, 186–193.

Kwon, Y.H., Fingert, J.H., Kuehn, M.H., Alward, W.L., 2009. Primary open-angle glaucoma. N Engl J Med 360, 1113–1124.

Lee, K.Y., Ho, S.L., Thalamuthu, A., et al., 2009. Association of LOXL1 polymorphisms with pseudoexfoliation in the Chinese. Mol Vis 15, 1120–1126.

Lemmela, S., Forsman, E., Onkamo, P., et al., 2009. Association of LOXL1 gene with finnish exfoliation syndrome patients. J Hum Genet 54, 289–297.

Lemmela, S., Forsman, E., Sistonen, P., Eriksson, A., Forsius, H., Jarvela, I., 2007. Genome-wide scan of exfoliation syndrome. Invest Ophthalmol Vis Sci 48, 4136–4142.

Libby, R.T., Gould, D.B., Anderson, M.G., John, S.W., 2005. Complex genetics of glaucoma susceptibility. Annu Rev Genomics Hum Genet 6, 15–44.

Lin, Y., Liu, T., Li, J., et al., 2008. A genome-wide scan maps a novel autosomal dominant juvenile-onset open-angle glaucoma locus to 2p15-16. Mol Vis 14, 739–744.

Liu, Y., Liu, W., Crooks, K., Schmidt, S., Allingham, R.R., Hauser, M.A., 2010. No evidence of association of heterozygous NTF4 mutations in patients with primary open-angle glaucoma. Am J Hum Genet 86, 498–499, author reply 500.

Liu, Y., Schmidt, S., Qin, X., et al., 2008. Lack of association between LOXL1 variants and primary open-angle glaucoma in three different populations. Invest Ophthalmol Vis Sci 49, 3465–3468.

Lu, Y., Dimasi, D.P., Hysi, P.G., et al., 2010. Common genetic variants near the Brittle Cornea Syndrome locus ZNF469 influence the blinding disease risk factor central corneal thickness. PLoS Genet 6, e1000947.

Mabuchi, F., Sakurada, Y., Kashiwagi, K., Yamagata, Z., Iijima, H., Tsukahara, S., 2008. Lysyl oxidase-like 1 gene polymorphisms in Japanese patients with primary open angle glaucoma and exfoliation syndrome. Mol Vis 14, 1303–1308.

Macgregor, S., Hewitt, A.W., Hysi, P.G., et al., 2010. Genome-wide association identifies ATOH7 as a major gene determining human optic disc size. Hum Mol Genet 19, 2716–2724.

Manolio, T.A., 2010. Genomewide association studies and assessment of the risk of disease. N Engl J Med 363, 166–176.

Maruyama, H., Morino, H., Ito, H., et al., 2010. Mutations of optineurin in amyotrophic lateral sclerosis. Nature 465, 223–226.

McIntosh, I., Dreyer, S.D., Clough, M.V., et al., 1998. Mutation analysis of LMX1B gene in nail-patella syndrome patients. Am J Hum Genet 63, 1651–1658.

Meguro, A., Inoko, H., Ota, M., Mizuki, N., Bahram, S., 2010. Genome-wide association study of normal tension glaucoma: Common variants in SRBD1 and ELOVL5 contribute to disease susceptibility. Ophthalmology 117, 1331–1338. e5.

Monemi, S., Spaeth, G., Dasilva, A., et al., 2005. Identification of a novel adult-onset primary open-angle glaucoma (POAG) gene on 5q22.1. Hum Mol Genet 14, 725–733.

Mori, K., Imai, K., Matsuda, A., et al., 2008. LOXL1 genetic polymorphisms are associated with exfoliation glaucoma in the Japanese population. Mol Vis 14, 1037–1040.

Morton, S., Hesson, L., Peggie, M., Cohen, P., 2008. Enhanced binding of TBK1 by an optineurin mutant that causes a familial form of primary open angle glaucoma. FEBS Lett 582, 997–1002.

Mossbock, G., Renner, W., Faschinger, C., Schmut, O., Wedrich, A., Weger, M., 2008. Lysyl oxidase-like protein 1 (LOXL1) gene polymorphisms and exfoliation glaucoma in a Central European population. Mol Vis 14, 857–861.

Nakano, M., Ikeda, Y., Taniguchi, T., et al., 2009. Three susceptible loci associated with primary open-angle glaucoma identified by genome-wide association study in a Japanese population. Proc Natl Acad Sci USA 106, 12,838–12,842.

Narooie-Nejad, M., Paylakhi, S.H., Shojaee, S., et al., 2009. Loss of function mutations in the gene encoding latent transforming growth factor beta binding protein 2, LTBP2, cause primary congenital glaucoma. Hum Mol Genet 18, 3969–3977.

Ng, S.B., Bigham, A.W., Buckingham, K.J., et al., 2010a. Exome sequencing identifies MLL2 mutations as a cause of Kabuki syndrome. Nat Genet 42, 790–793.

Ng, S.B., Buckingham, K.J., Lee, C., et al., 2010b. Exome sequencing identifies the cause of a Mendelian disorder. Nat Genet 42, 30–35.

Ng, S.B., Nickerson, D.A., Bamshad, M.J., Shendure, J., 2010c. Massively parallel sequencing and rare disease. Hum Mol Genet 19, R119–R124.

Nishimura, D.Y., Searby, C.C., Alward, W.L., et al., 2001. A spectrum of FOXC1 mutations suggests gene dosage as a mechanism for developmental defects of the anterior chamber of the eye. Am J Hum Genet 68, 364–372.

Ozaki, M., Lee, K.Y., Vithana, E.N., et al., 2008. Association of LOXL1 gene polymorphisms with pseudoexfoliation in the Japanese. Invest Ophthalmol Vis Sci 49, 3976–3980.

Pal, B., Mohamed, M.D., Keen, T.J., et al., 2004. A new phenotype of recessively inherited foveal hypoplasia and anterior segment dysgenesis maps to a locus on chromosome 16q23.2-24.2. J Med Genet 41, 772–777.

Pang, C.P., Fan, B.J., Canlas, O., et al., 2006. A genome-wide scan maps a novel juvenile-onset primary open angle glaucoma locus to chromosome 5q. Mol Vis 12, 85–92.

Park, B., Ying, H., Shen, X., et al., 2010. Impairment of protein trafficking upon overexpression and mutation of optineurin. PLoS One 5, e11547.

Pasutto, F., Krumbiegel, M., Mardin, C.Y., et al., 2008. Association of LOXL1 common sequence variants in German and Italian patients with pseudoexfoliation syndrome and pseudoexfoliation glaucoma. Invest Ophthalmol Vis Sci 49, 1459–1463.

Pasutto, F., Matsumoto, T., Mardin, C.Y., et al., 2009. Heterozygous NTF4 mutations impairing neurotrophin-4 signaling in patients with primary open-angle glaucoma. Am J Hum Genet 85, 447–456.

Phillips, J.C., Del Bono, E.A., Haines, J.L., et al., 1996. A second locus for Rieger syndrome maps to chromosome 13q14. Am J Hum Genet 59, 613–619.

Plasilova, M., Ferakova, E., Kadasi, L., et al., 1998. Linkage of autosomal recessive primary congenital glaucoma to the GLC3A locus in Roms (Gypsies) from Slovakia. Hum Hered 48, 30–33.

Polansky, J.R., Fauss, D.J., Chen, P., et al., 1997. Cellular pharmacology and molecular biology of the trabecular meshwork inducible glucocorticoid response gene product. Ophthalmologica 211, 126–139.

Quigley, H.A., Broman, A.T., 2006. The number of people with glaucoma worldwide in 2010 and 2020. Br J Ophthalmol 90, 262–267.

Ramdas, W.D., Van Koolwijk, L.M., Ikram, M.K., et al., 2010. A genome-wide association study of optic disc parameters. PLoS Genet 6, e1000978.

Ramprasad, V.L., George, R., Soumittra, N., et al., 2008. Association of non-synonymous single nucleotide polymorphisms in the LOXL1 gene with pseudoexfoliation syndrome in India. Mol Vis 14, 318–322.

Rao, K.N., Kaur, I., Chakrabarti, S., 2009. Lack of association of three primary open-angle glaucoma-susceptible loci with primary glaucomas in an Indian population. Proc Natl Acad Sci USA 106, E125–E126, author reply E127.

Rao, K.N., Kaur, I., Parikh, R.S., et al., 2010. Variations in NTF4, VAV2, and VAV3 genes are not involved with primary open-angle and primary angle-closure glaucomas in an Indian population. Invest Ophthalmol Vis Sci 51, 4937–4941.

Rezaie, T., Child, A., Hitchings, R., et al., 2002. Adult-onset primary open-angle glaucoma caused by mutations in optineurin. Science 295, 1077–1079.

Ritch, R., 2008. The management of exfoliative glaucoma. Prog Brain Res 173, 211–224.

Ritch, R., Schlotzer-Schrehardt, U., 2001. Exfoliation (pseudoexfoliation) syndrome: Toward a new understanding. Proceedings of the first international think tank. Acta Ophthalmol Scand 79, 213–217.

Sahlender, D.A., Roberts, R.C., Arden, S.D., et al., 2005. Optineurin links myosin VI to the Golgi complex and is involved in Golgi organization and exocytosis. J Cell Biol 169, 285–295.

Sarfarazi, M., Akarsu, A.N., Hossain, A., et al., 1995. Assignment of a locus (GLC3A) for primary congenital glaucoma (buphthalmos) to 2p21 and evidence for genetic heterogeneity. Genomics 30, 171–177.

Sarfarazi, M., Child, A., Stoilova, D., et al., 1998. Localization of the fourth locus (GLC1E) for adult-onset primary open-angle glaucoma to the 10p15-p14 region. Am J Hum Genet 62, 641–652.

Schlotzer-Schrehardt, U., Naumann, G.O., 2006. Ocular and systemic pseudoexfoliation syndrome. Am J Ophthalmol 141, 921–937.

Schlotzer-Schrehardt, U., Pasutto, F., Sommer, P., et al., 2008. Genotype-correlated expression of lysyl oxidase-like 1 in ocular tissues of patients with pseudoexfoliation syndrome/glaucoma and normal patients. Am J Pathol 173, 1724–1735.

Semina, E.V., Brownell, I., Mintz-Hittner, H.A., Murray, J.C., Jamrich, M., 2001. Mutations in the human forkhead transcription factor FOXE3 associated with anterior segment ocular dysgenesis and cataracts. Hum Mol Genet 10, 231–236.

Semina, E.V., Ferrell, R.E., Mintz-Hittner, H.A., et al., 1998. A novel homeobox gene PITX3 is mutated in families with autosomal-dominant cataracts and ASMD. Nat Genet 19, 167–170.

Semina, E.V., Reiter, R., Leysens, N.J., et al., 1996. Cloning and characterization of a novel bicoid-related homeobox transcription factor gene, RIEG, involved in Rieger syndrome. Nat Genet 14, 392–399.

Semina, E.V., Reiter, R.S., Murray, J.C., 1997. Isolation of a new homeobox gene belonging to the Pitx/Rieg family: Expression during lens development and mapping to the aphakia region on mouse chromosome 19. Hum Mol Genet 6, 2109–2116.

Senatorov, V., Malyukova, I., Fariss, R., et al., 2006. Expression of mutated mouse myocilin induces open-angle glaucoma in transgenic mice. J Neurosci 26, 11,903–11,914.

Sheffield, V.C., Stone, E.M., Alward, W.L., et al., 1993. Genetic linkage of familial open angle glaucoma to chromosome 1q21-q31. Nat Genet 4, 47–50.

Shimizu, S., Lichter, P.R., Johnson, A.T., et al., 2000. Age-dependent prevalence of mutations at the GLC1A locus in primary open-angle glaucoma. Am J Ophthalmol 130, 165–177.

Sivadorai, P., Cherninkova, S., Bouwer, S., et al., 2008. Genetic heterogeneity and minor CYP1B1 involvement in the molecular basis of primary congenital glaucoma in Gypsies. Clin Genet 74, 82–87.

Stathacopoulos, R.A., Bateman, J.B., Sparkes, R.S., Hepler, R.S., 1987. The Rieger syndrome and a chromosome 13 deletion. J Pediatr Ophthalmol Strabismus 24, 198–203.

Stoilov, I., Akarsu, A.N., Sarfarazi, M., 1997. Identification of three different truncating mutations in cytochrome P4501B1 (CYP1B1) as the principal cause of primary congenital glaucoma (buphthalmos) in families linked to the GLC3A locus on chromosome 2p21. Hum Mol Genet 6, 641–647.

Stoilov, I.R., Sarfarazi, M., 2002. The third genetic locus (GLC3C) for primary congenital glaucoma (PCG) maps to chromosome 14q24.3. Invest Ophthalmol Vis Sci 43. E-Abstract 3015.

Stoilova, D., Child, A., Trifan, O.C., Crick, R.P., Coakes, R.L., Sarfarazi, M., 1996. Localization of a locus (GLC1B) for adult-onset primary open angle glaucoma to the 2cen-q13 region. Genomics 36, 142–150.

Stone, E.M., Fingert, J.H., Alward, W.L., et al., 1997. Identification of a gene that causes primary open angle glaucoma. Science 275, 668–670.

Sud, A., Del Bono, E.A., Haines, J.L., Wiggs, J.L., 2008. Fine mapping of the GLC1K juvenile primary open-angle glaucoma locus and exclusion of candidate genes. Mol Vis 14, 1319–1326.

Tanito, M., Minami, M., Akahori, M., et al., 2008. LOXL1 variants in elderly Japanese patients with exfoliation syndrome/glaucoma, primary open-angle glaucoma, normal tension glaucoma, and cataract. Mol Vis 14, 1898–1905.

Teer, J.K., Mullikin, J.C., 2010. Exome sequencing: The sweet spot before whole genomes. Hum Mol Genet 19, R145–R151.

Thorleifsson, G., Magnusson, K.P., Sulem, P., et al., 2007. Common sequence variants in the LOXL1 gene confer susceptibility to exfoliation glaucoma. Science 317, 1397–1400.

Thorleifsson, G., Walters, G.B., Hewitt, A.W., et al., 2010. Common variants near CAV1 and CAV2 are associated with primary open-angle glaucoma. Nat Genet 42, 906–909.

Trifan, O.C., Traboulsi, E.I., Stoilova, D., et al., 1998. A third locus (GLC1D) for adult-onset primary open-angle glaucoma maps to the 8q23 region. Am J Ophthalmol 126, 17–28.

Valleix, S., Niel, F., Nedelec, B., et al., 2006. Homozygous nonsense mutation in the FOXE3 gene as a cause of congenital primary aphakia in humans. Am J Hum Genet 79, 358–364.

Vanita, V., Singh, D., Robinson, P.N., Sperling, K., Singh, J.R., 2006. A novel mutation in the DNA-binding domain of MAF at 16q23.1 associated with autosomal dominant "cerulean cataract" in an Indian family. Am J Med Genet A 140, 558–566.

Vasiliou, V., Gonzalez, F.J., 2008. Role of CYP1B1 in glaucoma. Annu Rev Pharmacol Toxicol 48, 333–358.

Vincent, A.L., Billingsley, G., Buys, Y., et al., 2002. Digenic inheritance of early-onset glaucoma: CYP1B1, a potential modifier gene. Am J Hum Genet 70, 448–460.

Vitart, V., Bencic, G., Hayward, C., et al., 2010. New loci associated with central cornea thickness include COL5A1, AKAP13 and AVGR8. Hum Mol Genet 19, 4304–4311.

Vithana, E.N., Nongpiur, M.E., Venkataraman, D., Chan, S.H., Mavinahalli, J., Aung, T., 2010. Identification of a novel mutation in the NTF4 gene that causes primary open-angle glaucoma in a Chinese population. Mol Vis 16, 1640–1645.

Vollrath, D., Jaramillo-Babb, V.L., Clough, M.V., et al., 1998. Loss-of-function mutations in the LIM-homeodomain gene, LMX1B, in nail–patella syndrome. Hum Mol Genet 7, 1091–1098.

Wang, D.Y., Fan, B.J., Chua, J.K., et al., 2006. A genome-wide scan maps a novel juvenile-onset primary open-angle glaucoma locus to 15q. Invest Ophthalmol Vis Sci 47, 5315–5321.

Wiggs, J.L., Lynch, S., Ynagi, G., et al., 2004. A genomewide scan identifies novel early-onset primary open-angle glaucoma loci on 9q22 and 20p12. Am J Hum Genet 74, 1314–1320.

Williams, S.E., Whigham, B.T., Liu, Y., et al., 2010. Major LOXL1 risk allele is reversed in exfoliation glaucoma in a black South African population. Mol Vis 16, 705–712.

Wirtz, M.K., Samples, J.R., Kramer, P.L., et al., 1997. Mapping a gene for adult-onset primary open-angle glaucoma to chromosome 3q. Am J Hum Genet 60, 296–304.

Wirtz, M.K., Samples, J.R., Rust, K., et al., 1999. GLC1F, a new primary open-angle glaucoma locus, maps to 7q35-q36. Arch Ophthalmol 117, 237–241.

Wolf, C., Gramer, E., Muller-Myhsok, B., et al., 2010. Lysyl oxidase-like 1 gene polymorphisms in German patients with normal tension glaucoma, pigmentary glaucoma and exfoliation glaucoma. J Glaucoma 19, 136–141.

Wolfs, R.C., Klaver, C.C., Ramrattan, R.S., et al., 1998. Genetic risk of primary open-angle glaucoma. Population-based familial aggregation study. Arch Ophthalmol 116, 1640–1645.

Woodroffe, A., Krafchak, C.M., Fuse, N., et al., 2006. Ordered subset analysis supports a glaucoma locus at GLC1I on chromosome 15 in families with earlier adult age at diagnosis. Exp Eye Res 82, 1068–1074.

Yang, X., Zabriskie, N.A., Hau, V.S., et al., 2008. Genetic association of LOXL1 gene variants and exfoliation glaucoma in a Utah cohort. Cell Cycle 7, 521–524.

Zhou, Y., Grinchuk, O., Tomarev, S.I., 2008. Transgenic mice expressing the Tyr437His mutant of human myocilin protein develop glaucoma. Invest Ophthalmol Vis Sci 49, 1932–1939.

Zhu, G., Wu, C.J., Zhao, Y., Ashwell, J.D., 2007. Optineurin negatively regulates TNFalpha- induced NF-kappaB activation by competing with NEMO for ubiquitinated RIP. Curr Biol 17, 1438–1443.

RECOMMENDED RESOURCES

Children's Glaucoma Foundation
http://www.childrensglaucoma.com
Glaucoma Research Foundation
http://www.glaucoma.org
International Glaucoma Association
http://www.glaucoma-association.com

The Glaucoma Foundation
http://www.glaucomafoundation.org
National Glaucoma Research
http://www.ahaf.org/glaucoma

Infectious Disease Genomic Medicine

Section 9

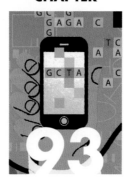

Diagnosis and Classification of Pathogens

Octavio Ramilo and Asunción Mejías

INTRODUCTION

Despite major advances in immunizations and antimicrobial therapies, infectious diseases continue to represent a major cause of morbidity and mortality around the world (World Health Organization, 2008). There is often a perception that the health impact of infectious diseases is limited to developing countries, and especially to the pediatric population [Fontaine et al., 2009; Rudan et al., 2007; UNAIDS (Joint United Nations Programme on HIV/AIDS), 2008]. Unfortunately, this is an oversimplification of the current situation. Infectious diseases represent a major public health challenge in the developed world as well. Recent examples such as the 2009 H1N1 influenza pandemic (Centres for Disease Control and Prevention, 2009; Libster et al., 2010), the ongoing epidemic of methicillin-resistant *Staphylococcus aureus* (MRSA) infections (Klevens et al., 2007; Zetola et al., 2005), as well as the increased frequency of hospital-acquired infections caused by multiple-resistant gram-negative bacilli (Kunz and Brook, 2010) and highly virulent strains of *Clostridium difficile* (Ananthakrishnan, 2011) highlight the challenges we continue to encounter in managing patients with infectious diseases.

In the clinical setting, physicians are constantly challenged by the uncertainty of the bacterial or viral diagnosis in patients with an acute febrile illness. The need to promptly start appropriate antimicrobial therapy in order to control a mild infection before it can progress to a more severe form must be balanced with the need for prudent use of antibiotics, especially in the current situation where outbreaks of emergent and re-emergent pathogens are linked to increased resistance to our current antimicrobial armamentarium. Within this context, it is obvious that there is a need for improved diagnostic tools that can advance our ability to diagnose and classify patients with infectious diseases more precisely, which in turn should allow the implementation of more targeted therapies.

HYPOTHESIS AND *IN VITRO* STUDIES

Our ability to identify infectious agents to establish an appropriate diagnosis remains inadequate, particularly if the organism is not present in the blood or another easily accessible site. These diagnostic obstacles can delay initiation of appropriate therapy, which can result in unnecessary morbidity and even death (Relman, 2002a). Traditional microbiologic diagnostic tests have relied on laboratory identification of the pathogen causing the infection. Microbial pathogens are detected in clinically relevant specimens using a variety of assays, including cultures, rapid antigen detection tests, and, more recently, polymerase chain reaction (PCR) assays. To date, growth of the specific pathogen (bacterium, virus, or

Genomic and Personalized Medicine, 2nd edition
by Ginsburg & Willard. DOI: http://dx.doi.org/10.1016/B978-0-12-382227-7.00093-8

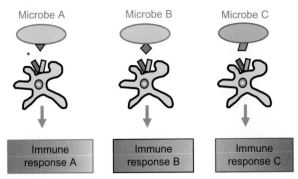

* Pattern recognition receptors

Figure 93.1 Different classes of pathogens trigger specific pattern-recognition receptors (PRRs) differentially expressed on immune cells, and elicit distinct immune responses.

fungus) remains the ultimate reference gold standard for identification. However, many pathogens grow slowly or require complex media (Cummings and Relman, 2002), and a significant number of clinically important microbial pathogens remain unrecognized as they are resistant to cultivation in the laboratory, limiting the physician's clinical decision-making (Cummings and Relman, 2002; Relman, 2002b). The introduction of more sensitive molecular diagnostic assays has dramatically improved our ability to diagnose viral infections (Balada-Llasat et al., 2011). Unfortunately, this has not been the case for bacterial pathogens. Moreover, in the clinical scenario it is not uncommon to encounter situations in which the sole identification of a pathogen is not sufficient to establish causality, for example, the detection of respiratory viruses in patients with pneumonia, which are often found in patients with possible bacterial coinfections.

An alternative approach to the traditional pathogen-detection strategy is based on a comprehensive analysis of the host response to the infection caused by different microbial pathogens (Ramilo and Mejias, 2009; Ramilo et al., 2007). Different classes of pathogens trigger specific pattern-recognition receptors (PRRs) differentially expressed on leukocytes (Medzhitov and Janeway, 1997; Medzhitov and Janeway, 2000). Leukocytes are components of the innate immune system (granulocytes, natural killer cells), the adaptive immune system (T and B lymphocytes), or both (monocytes and dendritic cells) (Figure 93.1).

Blood represents both a reservoir and a migration compartment for these immune cells, that become educated and implement their function by circulating between central and peripheral lymphoid organs and migrating to and from the site of infection via blood. Therefore, blood leukocytes constitute an accessible source of clinically relevant information, and a comprehensive molecular phenotype of these cells can be obtained using gene expression microarrays (Chaussabel et al., 2010). Because they provide a comprehensive assessment of the immune-related cells and pathways, genomic studies have been shown to be well suited to study host–pathogen

interaction. In fact, studies have shown that different classes of pathogens induce distinct gene expression profiles that can be identified by analysis of blood leukocytes (Figure 93.2) (Berry et al., 2010; Chaussabel et al., 2005; Chaussabel et al., 2008; Pankla et al., 2009; Ramilo et al., 2007).

IN VITRO STUDIES

The initial evidence supporting the hypothesis that pathogen-specific gene expression profiles can be measured in immune cells was derived from *in vitro* studies. A number of experimental studies demonstrated that different transcriptional programs could be triggered upon exposure of immune cells to various pathogens *in vitro* (Boldrick et al., 2002; Chaussabel et al., 2003; Huang et al., 2001; Nau et al., 2002). Comparative analysis of a compendium of host–pathogen microarray datasets identified both a common host transcriptional response to infection and pathogen-specific signatures (Jenner and Young, 2005). For instance, broad similarities exist in the context of fungal, bacterial, or viral infections, with dynamic cascades of cytokines and chemokines involved in the activation and recruitment of immune cells (Blumenthal et al., 2005; Cortez et al., 2006; Kim et al., 2005; Mysorekar et al., 2002; Piqueras et al., 2006). However, two factors seem to contribute to the specificity of transcriptional responses to infections: (1) the diversity of the molecular mechanisms involved in pathogen recognition, and (2) alterations of the immune response mounted by the host against a pathogen. Upon activation, Toll-like receptor (TLR) family members trigger signaling pathways that share common components while retaining unique characteristics, accounting in part for this specificity of transcriptional responses (Schmitz et al., 2004). Hence, qualitative and quantitative differences in the responses to gram-positive and gram-negative bacteria, recognized by TLR2 and TLR4 respectively, have been observed (Boldrick et al., 2002; Nau et al., 2002). Furthermore, responses measured in dendritic cells exposed to influenza virus (through TLR3), *E. coli* (through TLR4), and Candida (through TLR2/TLR4) were also found to be markedly different (Huang et al., 2001). Reprogramming of host cells by pathogens also contributes significantly to the diversification of transcriptional responses to infection. As measured by gene expression analysis, mycobacterial products are able to inhibit interferon (IFN)-gamma-induced gene regulation in macrophages (Pai et al., 2004). Similarly, microarray studies have demonstrated the ability of herpes virus, pseudorabies virus, hepatitis C, varicella-zoster virus, and rhinovirus to limit the ability of the host to develop effective antiviral responses, by a variety of mechanisms (Brukman and Enquist, 2006; Jhaveri et al., 2005; Jones and Arvin, 2006; Mossman et al., 2001; Peng et al., 2006). The vast body of *in vitro* experimental data accumulated over the past years suggests that hosts can mount pathogen-specific transcriptional responses to infections.

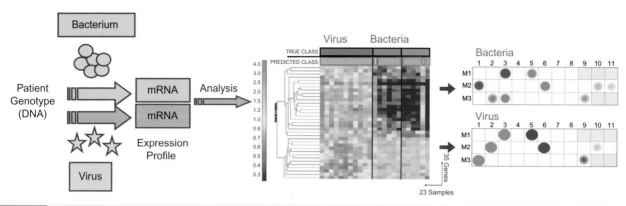

Figure 93.2 Using microarray technology, the differences in gene expression patterns present in blood immune cells can be measured as induced by various types of infectious agents.

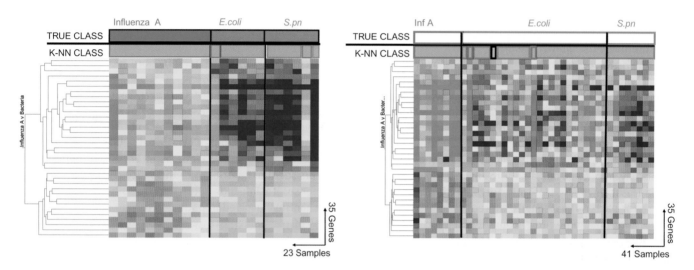

Figure 93.3 Gene expression patterns discriminate viral vs bacterial infections. Left: a set of 35 genes discriminates patients with acute Influenza A infection (green) and bacterial infections (red). The discriminative pattern is shown by the heatmap (red = over-expression, blue = under-expression). Right: the diagnostic signature was tested in an independent set of patients.

INITIAL HUMAN STUDIES: PROOF OF CONCEPT

Global changes in transcriptional responses have been measured in the blood of patients with different infectious diseases. Studies tested the hypothesis that leukocytes isolated from peripheral blood of patients with acute infections carry unique transcriptional signatures, which would, in turn, permit pathogen discrimination and thus patient classification (Ramilo et al., 2007). In those initial studies, gene expression patterns in peripheral blood mononuclear cells (PBMCs) from 95 pediatric patients with acute infections caused by four common human pathogens were analyzed: (1) influenza A (an RNA virus); (2) *Staphylococcus aureus* and (3) *Streptococcus pneumoniae* (two gram-positive bacteria); and (4) *Escherichia coli* (a gram-negative bacterium).

Pair-wise comparisons and class prediction analysis (K-NN algorithm) (Alizadeh et al., 2000; Golub et al., 1999) identified

35 genes that discriminated patients with influenza A viral infection from patients with bacterial infection caused by either *E. coli* or *S. pneumoniae* with 95% accuracy (Figure 93.3) (Ramilo et al., 2007).

Further analyses in this cohort of patients allowed us to identify 137 classifier genes, which were applied to a population of 27 pediatric patients with pneumonia and seven healthy children to determine whether we could differentiate patients presenting with similar symptoms according to the different etiologic pathogens (Figure 93.4).

Hierarchical clustering of genes and samples identified four prototypical expression profiles: healthy controls, influenza A infection, which showed increased expression of interferon-inducible genes and was clearly different from the third profile, which characterized bacterial infections caused by *S. aureus* and *S. pneumoniae*, and showed over-expression of neutrophil-associated genes. Three samples belonging to the influenza A group and one from the *S. aureus* group were

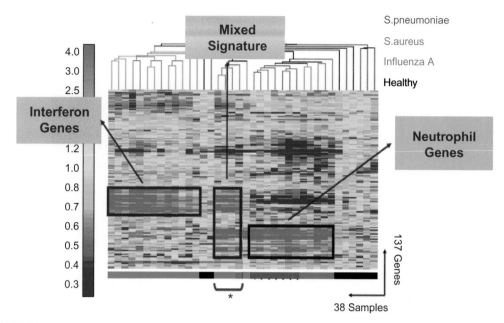

Figure 93.4 Blood diagnostic biosignatures discriminate patients with pneumonia caused by viral and bacterial pathogens. Hierarchical clustering of 137 discriminative genes in 34 individuals (27 patients with pneumonia and seven age-matched healthy controls). Each column represents an individual patient (n=38); each row represents one gene (n=137). Values were normalized to the median expression of each gene across all samples. Red indicated genes expressed at higher levels and blue lower levels than the median expression value.

characterized by a fourth profile, which combined elements of the previous ones, suggesting the possibility of a coinfection caused by both a viral and a bacterial pathogen (Ramilo et al., 2007). These initial studies demonstrated that blood leukocytes' gene expression patterns can be used to distinguish patients with acute infections caused by four different pathogens: influenza A virus, *E. coli*, *S. aureus*, and *S. pneumoniae*, which are among the most common infections leading to hospitalization in children.

To evaluate studies involving gene expression analysis in infectious diseases, it is important to consider aspects that might account for differences in gene expression levels observed in blood leukocytes, namely: (1) changes in transcriptional activity, (2) altered cellular composition, (3) the anatomical site of infection, and (4) the dynamic changes of the infectious process. We will review these aspects separately.

Changes in Transcriptional Activity

Changes in transcriptional activity such as the over-expression of interferon-inducible genes, have been described in patients with acute influenza infection and autoimmune disorders such as systemic lupus erythematosus (SLE) (Bennett et al., 2003; Chaussabel et al., 2008; Pascual et al., 2008; Ramilo et al., 2007).

Altered Cellular Composition of Blood Samples

It is well established in clinical practice that routine white blood cell and differential counts cannot distinguish between

viral and bacterial infections, and even less between infections caused by gram-positive and gram-negative bacteria. In the initial studies in children, review of the routine white cell counts did not show major differences in the cellular composition of blood samples obtained from patients infected by different classes of pathogens (Ramilo et al., 2007). To further understand how cell composition affected transcriptional profiles and whether the changes observed in gene expression patterns in patients with acute infections simply reflected alterations in immune cell numbers, a comprehensive study in patients with staphylococcal infections was conducted (Ardura et al., 2009). Patients with *S. aureus* infections demonstrated a distinct and robust gene expression profile that revealed marked over-expression of genes related to the innate immune response and significant under-expression of genes related to adaptive immunity, and, more specifically, a significant under-expression of T- and B-cell-related genes. However, flow cytometry studies performed in parallel blood samples from the same cohort revealed no differences in total numbers of B and T cells between infected patients and healthy controls, and only decreased numbers of central memory CD4+ T cells and CD8+ T cells (Figure 93.5) (Ardura et al., 2009).

Monocyte numbers were significantly increased and correlated with the increased expression of innate immune-related genes. These findings indicate that expression profiles do not simply reflect changes in blood immune cell populations, but reveal the complex effect of the pathogen on the

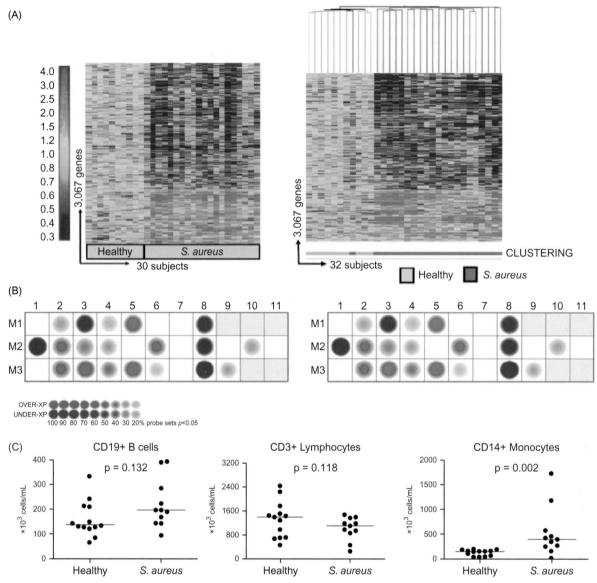

Figure 93.5 (A) *S. aureus* gene biosignature. Statistical group comparisons between 10 healthy individuals and 20 patients with acute *S. aureus* infections yielded 3067 genes differently expressed (Mann–Whitney p < 0.01 and 1.25-fold change) in the training set of patients (left panel). The same 3067-gene list was used to perform a cluster analysis on 22 new subjects (test set) with *S. aureus* infections, and correctly grouped 21 of the 22 patients solely by gene expression (right panel). Patients and significant genes were organized using hierarchical clustering, where each row represents a single gene and each column represents an individual. (B) Module mapping of gene expression profile changes identifies a *S. aureus*-specific fingerprint. Gene expression levels were compared between patients with *S. aureus* infections and healthy controls on a module-by-module basis. Colored spots represent the percentage of significantly (Mann–Whitney p < 0.05) over-expressed (red) or under-expressed (blue) transcripts within a module in patients with *S. aureus* infections; spot intensity represents the magnitude of the gene expression change, blank modules demonstrate no significant differences between groups (p > 0.05). Information is displayed on a grid, with the coordinates corresponding to one of 28 modules with the lower key representing functional interpretation of modules. A total of 19 modules are shown to be significantly different between healthy subjects and patients with *S. aureus* infection in the training and test set of patients. (C) Peripheral blood monocytes are significantly expanded in patients with invasive *S. aureus* infections. PBMCs obtained from age-matched healthy donors (n = 13) and patients with *S. aureus* infection (n = 11) were analyzed by flow cytometry for the expression of CD19 (left panel), CD3 (middle), and CD14 (right) markers. Results are expressed as absolute number of cells per ml of blood. Bars represent median values. Mann–Whitney test was applied for statistical analysis.

immune system, reflected by changes in the numbers of the different immune cells and in their transcriptional programs.

Anatomical Site of Disease Involvement

The anatomical site involved in the disease may reflect the predilection of certain pathogen species to infect certain organs. Predictably, gene expression changes in the blood of patients with systemic infections (*S. aureus* or *S. pneumoniae* bacteremia) were clearly evident. However, marked and distinctive expression signatures were also found in patients with localized infections (influenza A respiratory tract infection or *E. coli* urinary tract infection). Therefore, measuring changes in host transcriptional profiles in the blood may prove to be of diagnostic value, even in situations where the causative pathogen may not be present in the sample (Figure 93.4).

Dynamic Changes of the Infectious Disease Process

The majority of clinical illnesses in infectious diseases appear as acute events. In most studies, pathogen-specific biosignatures have been derived from a single time-point after the initial exposure, as shown in patients with malaria, dengue, *Burkholderia pseudomallei*, *S. aureus*, respiratory viruses, or tuberculosis (Ardura et al., 2009; Griffiths et al., 2005; Nascimento et al., 2009; Pankla et al., 2009; Popper et al., 2009; Ramilo et al., 2007). To further understand the disease process, it is crucial to capture the dynamic changes or the lack of variation in gene expression profiles that occurs during the course of the infection. Thompson and colleagues studied patients infected with *Salmonella typhi*, and demonstrated a strong and distinct peripheral blood biosignature that remained unchanged even months after the acute infection (Thompson et al., 2009). To a certain point, it is remarkable that even though in the majority of studies patients' samples were collected at different time-points after exposure to the pathogens and disease onset, a robust and pathogen-specific biosignature was derived and validated in independent cohorts of patients in a complete and controlled setting (Ardura et al., 2009; Ramilo et al., 2007; Zaas et al., 2009). Additional studies will be necessary to evaluate the merits of this approach in other relevant clinical settings, such as the emergency room or the physician's office, and not only in hospitalized patients. Furthermore, it will be important to determine whether transcriptional analysis of blood leukocytes can provide information that would permit clinicians to follow the progression of the disease and assess risks of complications.

NEW AREAS FOR IMPROVING DIAGNOSIS

Acute respiratory infections are responsible for significant morbidity and mortality worldwide. Acute upper respiratory tract infections are the most common reason for outpatient visits in the United States across all age groups. Lower respiratory tract infections and pneumonia represent one of the most frequent reasons for hospitalization and are the leading cause of death in children under five years of age (Bryce et al., 2005; Johnstone et al., 2008). Traditionally, most efforts have been focused on diagnosis and treatment of bacterial infections, but increasing evidence indicates that viral respiratory infections are associated with substantial morbidity and mortality as well, especially among children (Hall et al., 2009). Despite the impact of these infections, in current clinical practice, establishing a precise etiologic diagnosis of respiratory infections, or even simply discriminating viral from bacterial respiratory infections, remains challenging. This is true when evaluating outpatients in the clinic or the emergency room as well as hospitalized individuals. Unfortunately, the difficulties in differentiating viral from bacterial infections and the pressure to achieve a rapid resolution of symptoms commonly lead medical practitioners to take an overcautious approach and to treat many patients unnecessarily with antibiotics (Wachter et al., 2008). Obviously, this is a flawed approach which, aside from driving up healthcare costs, facilitates the development of antimicrobial resistance. In recent years, the introduction of molecular PCR-based assays has dramatically improved our ability to accurately diagnose viral respiratory infections caused by well-known and recently discovered viruses (Mahony, 2008). Despite these advances in the diagnosis of viral infections, molecular assays have been less successful for the diagnosis of bacterial respiratory infections, especially those caused by pneumococcus and other invasive bacteria (Anevlavis et al., 2009; Nolte, 2008). To complicate matters further, often the detection of a specific microbe in clinical samples is not sufficient to make a specific diagnosis. Hence, there is an urgent need for new methodologies to improve the diagnosis of acute respiratory infections.

A recent study by Zaas and colleagues (2009) provides an initial proof-of-concept of how application of blood gene expression profiles could represent an alternative approach for the diagnosis of respiratory viral infections. In their study, blood gene expression profiles from healthy volunteers experimentally infected with human rhinovirus (HRV), respiratory syncytial virus (RSV), or influenza A virus were analyzed. They identified an "acute respiratory viral signature" that clearly distinguished symptomatic, infected individuals from asymptomatic and uninfected volunteers using factor analysis. In this acute respiratory viral signature, the investigators identified genes (e.g., *RSAD2*, *IFI44L*, and *LAMP3*) that were expressed in individuals infected with any of the three different viruses. However, they also identified genes that were more specific for each viral infection, such as *FCRGR1A*, *GBP1*, and *LAP3* in RSV infection; *OAS2*, *CXCL10*, and *SOCS1* in individuals with HRV infection; and *TNFAIP1*, *SEPT4*, and *IFI27* in cases of influenza. These observations provide evidence of the potential value of this approach to define diagnostic signatures that are common for a group of pathogens and those which are pathogen-specific. To validate their findings, they applied the newly identified viral signature to an independent, previously

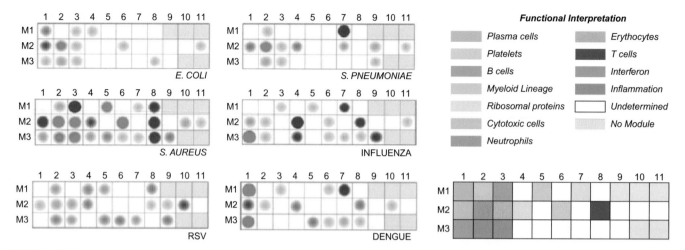

Figure 93.6   Mapping transcriptional changes at the module-level identifies disease-specific biosignatures in patients with infectious diseases. Expression levels were compared between patients and appropriately matched healthy controls on a module-by-module basis. The spots represent the percentage of significantly over-expressed (red) or under-expressed (blue) transcripts within a module (i.e., set of coordinately expressed genes). Blank spots indicate that there are no differences in the genes included in that module between patients and healthy controls. This information is displayed on a grid, with the coordinates corresponding to one of 28 module IDs (e.g., module M3.1 is at the intersection of the third row and first column). Each pathogen induces a disease-specific biosignature that is easy identifiable.

published dataset of patients with community-acquired respiratory infections (Ramilo et al., 2007). Despite the technical challenges involved in such analysis and the differences in the patient cohorts analyzed (children with naturally acquired infection vs adults with experimental infection), their experimentally identified acute respiratory viral signature classified pediatric patients naturally infected with influenza A from healthy, age-matched controls with 100% accuracy (Zaas et al., 2009).

These observations confirmed the potential value of blood gene signatures for clinical application in infectious diseases. The data clearly support the hypothesis that different pathogens elicit distinct host immune responses, and that the information is easily accessible from blood immune cells. It is remarkable that blood signatures can achieve such accuracy for diagnosis of respiratory viral pathogens, thought to be confined to the respiratory tract. From the practical perspective, in most clinical situations obtaining a blood sample is more feasible than obtaining infected tissue.

New Analytical Tools: Modular Analysis

One of the challenges for the practical application of gene profiling assays in the clinical setting is having appropriate tools that facilitate the analysis of the large datasets generated from these studies. Recently, Chaussabel and colleagues (Chaussabel et al., 2008; Wang and Marincola, 2008) developed a novel data-mining algorithm that radically simplifies these analyses. They extracted blood transcriptional modules from a set of 239 peripheral blood mononuclear cells (PBMC) samples obtained from patients with eight different diseases,

all immune-related (Chaussabel et al., 2008). The analysis initially identified 28 transcriptional modules comprising 4742 individual transcripts. These transcriptional modules are sets of genes that follow similar expression patterns across a large number of samples in multiple studies (identified through co-expression meta-analysis) and have a similar function. Once modules have been defined for a given biological system, statistical comparisons can be performed between study groups, on a module-by-module basis, which greatly decreases the background noise generated in this type of analysis, by reducing the number of variables to be studied. Although this strategy was initially designed to analyze patients with auto-immune diseases, module-level changes observed in patients' PBMC transcriptional profiles can be represented for different types of conditions. We have recently applied this tool to analyze gene expression profiles of patients presenting with a variety of infectious diseases (Figure 93.6).

When comparing the module maps or "disease fingerprints" in different infectious disease processes, we found that diseases are characterized by a unique modular combination (Figure 93.3). In fact, the expression of some modules alone is sufficient to distinguish different types of conditions. In studies currently underway, preliminary work using modular analysis identified distinct gene profiles in patients infected by different pathogens. Figure 93.3 displays unique disease fingerprints in patients with a variety of bacterial and viral infections. Certain specific modules demonstrate similar expression patterns in two different diseases. However, each disease displays a unique combination of modules that generates a characteristic disease fingerprint. Thus, disease fingerprinting is made possible by modular analysis derived from patient blood leukocyte gene expression profiles. This modular analysis will facilitate the application of gene expression profiles as a novel tool for diagnosis of infectious diseases.

Diagnostic Biosignatures and Disease Severity

The application of gene signatures to accurately classify patients according to the infectious etiologic agent has been shown to be noteworthy, but the comprehensive assessment of the host response also provides an opportunity to use them in a broader clinical context. A major advantage of this approach is the ability to correlate clinical markers of disease severity with changes in host gene expression by developing a genomic severity score that can help classify patients according to clinical characteristics (Chaussabel et al., 2008). Developing an objective genomic score of disease severity in patients with infectious diseases will be extremely helpful to monitor disease status, response to therapy, and eventually to predict clinical outcomes. The tools necessary to implement this approach in the clinical setting are currently available (Chaussabel et al., 2008; Ramilo and Mejias, 2009). Indeed, a recent study conducted by a multinational group of investigators led by Dr A. O'Garra has confirmed the value of gene expression profiles and modular analysis as a powerful diagnostic tool in patients with pulmonary tuberculosis (Berry et al., 2010). Initial analysis conducted in the United Kingdom identified a distinct 393-transcript profile in individuals with bacteriologically confirmed pulmonary tuberculosis that was different than in patients with latent tuberculosis. This profile was validated in an independent patient cohort of patients recruited in South Africa. The investigators also applied the molecular distance to health (MDTH) score (Pankla et al., 2009) to determine the value of gene expression profiles to assess clinical disease severity. MDTH scores significantly correlated with the radiologic extent of the disease, as patients with more severe pulmonary disease had higher MDTH scores. More importantly, as patients received appropriate treatment, MDTH scores progressively decreased and normalized at around 12 months.

This study highlights the strength and value of applying gene expression profiling to improve diagnosis and classification of patients with infectious diseases. The challenge now is to design relevant clinical studies to identify and validate the diagnostic and prognostic signatures induced by relevant pathogens. Studies should combine state-of-the-art molecular and traditional microbiologic methods with host gene signatures. Combining the detection of the pathogen with a comprehensive assessment of the host immune response will provide a broad new understanding of the correlations between specific etiologic agents, the corresponding host response and the clinical manifestations of the disease. This combined approach will help to develop a large databank of diagnostic signatures and biomarkers of disease severity. It will also improve our understanding of disease pathogenesis, and will help to determine the role of a pathogen detected in a clinical sample – whether it implies that the microbe is causing the disease, or that it simply reflects colonization and/or asymptomatic shedding. These assays will also be extremely valuable to study the significance of dual or triple viral coinfections as well as viral–bacterial coinfections.

REFERENCES

Alizadeh, A.A., Eisen, M.B., Davis, R.E., et al., 2000. Distinct types of diffuse large B-cell lymphoma identified by gene expression profiling. Nature 403, 503–511.

Ananthakrishnan, A.N., 2011. *Clostridium difficile* infection: Epidemiology, risk factors and management. Nat Rev Gastroenterol Hepatol 8, 17–26.

Anevlavis, S., Petroglou, N., Tzavaras, A., et al., 2009. A prospective study of the diagnostic utility of sputum Gram stain in pneumonia. J Infect 59, 83–89.

Ardura, M.I., Banchereau, R., Mejias, A., et al., 2009. Enhanced monocyte response and decreased central memory T cells in children with invasive *Staphylococcus aureus* infections. PLoS ONE 4, e5446.

Balada-Llasat, J.M., LaRue, H., Kelly, C., Rigali, L., Pancholi, P., 2011. Evaluation of commercial ResPlex II v2.0, MultiCode-PLx, and xTAG respiratory viral panels for the diagnosis of respiratory viral infections in adults. J Clin Virol 50, 42–45.

Bennett, L., Palucka, A.K., Arce, E., et al., 2003. Interferon and granulopoiesis signatures in systemic lupus erythematosus blood. J Exp Med 197, 711–723.

Berry, M.P., Graham, C.M., McNab, F.W., et al., 2010. An interferon-inducible neutrophil-driven blood transcriptional signature in human tuberculosis. Nature 466, 973–977.

Blumenthal, A., Lauber, J., Hoffmann, R., et al., 2005. Common and unique gene expression signatures of human macrophages in response to four strains of *Mycobacterium avium* that differ in their growth and persistence characteristics. Infect Immun 73, 3330–3341.

Boldrick, J.C., Alizadeh, A.A., Diehn, M., et al., 2002. Stereotyped and specific gene expression programs in human innate immune responses to bacteria. Proc Natl Acad Sci USA 99, 972–977.

Brukman, A., Enquist, L.W., 2006. Suppression of the interferon-mediated innate immune response by pseudorabies virus. J Virol 80, 6345–6356.

Bryce, J., Boschi-Pinto, C., Shibuya, K., Black, R.E., 2005. WHO estimates of the causes of death in children. Lancet 365, 1147–1152.

Centres for Disease Control and Prevention, 2009. H1N1 Flu. <http://www.cdc.gov/h1n1flu/>.

Chaussabel, D., Allman, W., Mejias, A., et al., 2005. Analysis of significance patterns identifies ubiquitous and disease-specific gene-expression signatures in patient peripheral blood leukocytes. Ann NY Acad Sci 1062, 146–154.

Chaussabel, D., Pascual, V., Banchereau, J., 2010. Assessing the human immune system through blood transcriptomics. BMC Biol 8, 84.

Chaussabel, D., Quinn, C., Shen, J., et al., 2008. A modular analysis framework for blood genomics studies: Application to systemic lupus erythematosus. Immunity 29, 150–164.

Chaussabel, D., Semnani, R.T., McDowell, M.A., Sacks, D., Sher, A., Nutman, T.B., 2003. Unique gene expression profiles of human macrophages and dendritic cells to phylogenetically distinct parasites. Blood 102, 672–681.

Cortez, K.J., Lyman, C.A., Kottilil, S., et al., 2006. Functional genomics of innate host defense molecules in normal human monocytes in response to Aspergillus fumigatus. Infect Immun 74, 2353–2365.

Cummings, C.A., Relman, D.A., 2002. Genomics and microbiology. Microbial forensics – "cross-examining pathogens". Science 296, 1976–1979.

Fontaine, O., Kosek, M., Bhatnagar, S., et al., 2009. Setting research priorities to reduce global mortality from childhood diarrhoea by 2015. PLoS Med 6, e41.

Golub, T.R., Slonim, D.K., Tamayo, P., et al., 1999. Molecular classification of cancer: Class discovery and class prediction by gene expression monitoring. Science 286, 531–537.

Griffiths, M.J., Shafi, M.J., Popper, S.J., et al., 2005. Genomewide analysis of the host response to malaria in Kenyan children. J Infect Dis 191, 1599–1611.

Hall, C.B., Weinberg, G.A., Iwane, M.K., et al., 2009. The burden of respiratory syncytial virus infection in young children. N Engl J Med 360, 588–598.

Huang, Q., Liu, D., Majewski, P., et al., 2001. The plasticity of dendritic cell responses to pathogens and their components. Science 294, 870–875.

Jenner, R.G., Young, R.A., 2005. Insights into host responses against pathogens from transcriptional profiling. Nat Rev Microbiol 3, 281–294.

Jhaveri, R., Kundu, P., Shapiro, A.M., Venkatesan, A., Dasgupta, A., 2005. Effect of heptitis C virus core protein on cellular gene expression: Specific inhibition of cyclooxygenase 2. J Infect Dis 191, 1498–1506.

Johnstone, J., Majumdar, S.R., Fox, J.D., Marrie, T.J., 2008. Viral infection in adults hospitalized with community-acquired pneumonia: Prevalence, pathogens, and presentation. Chest 134, 1141–1148.

Jones, J.O., Arvin, A.M., 2006. Inhibition of the NF-kappaB pathway by varicella-zoster virus in vitro and in human epidermal cells in vivo. J Virol 80, 5113–5124.

Kim, H.S., Choi, E.H., Khan, J., et al., 2005. Expression of genes encoding innate host defense molecules in normal human monocytes in response to Candida albicans. Infect Immun 73, 3714–3724.

Klevens, R.M., Morrison, M.A., Nadle, J., et al., 2007. Invasive methicillin-resistant Staphylococcus aureus infections in the United States. JAMA 298, 1763–1771.

Kunz, A.N., Brook, I., 2010. Emerging resistant gram-negative aerobic bacilli in hospital-acquired infections. Chemotherapy 56, 492–500.

Libster, R., Bugna, J., Coviello, S., et al., 2010. Pediatric hospitalizations associated with 2009 pandemic influenza A (H1N1) in Argentina. N Engl J Med 362, 45–55.

Mahony, J.B., 2008. Detection of respiratory viruses by molecular methods. Clin Microbiol Rev 21, 716–747.

Medzhitov, R., Janeway Jr., C., 2000. Innate immune recognition: Mechanisms and pathways. Immunol Rev 173, 89–97.

Medzhitov, R., Janeway Jr., C.A., 1997. Innate immunity: The virtues of a nonclonal system of recognition. Cell 91, 295–298.

Mossman, K.L., Macgregor, P.F., Rozmus, J.J., Goryachev, A.B., Edwards, A.M., Smiley, J.R., 2001. Herpes simplex virus triggers and then disarms a host antiviral response. J Virol 75, 750–758.

Mysorekar, I.U., Mulvey, M.A., Hultgren, S.J., Gordon, J.I., 2002. Molecular regulation of urothelial renewal and host defenses during infection with uropathogenic Escherichia coli. J Biol Chem 277, 7412–7419.

Nascimento, E.J., Braga-Neto, U., Calzavara-Silva, C.E., et al., 2009. Gene expression profiling during early acute febrile stage of dengue infection can predict the disease outcome. PLoS ONE 4, e7892.

Nau, G.J., Richmond, J.F., Schlesinger, A., Jennings, E.G., Lander, E.S., Young, R.A., 2002. Human macrophage activation programs induced by bacterial pathogens. Proc Natl Acad Sci USA 99, 1503–1508.

Nolte, F.S., 2008. Molecular diagnostics for detection of bacterial and viral pathogens in community-acquired pneumonia. Clin Infect Dis 47 (Suppl. 3), S123–S126.

Pai, R.K., Pennini, M.E., Tobian, A.A., Canaday, D.H., Boom, W.H., Harding, C.V., 2004. Prolonged toll-like receptor signaling by Mycobacterium tuberculosis and its 19-kilodalton lipoprotein inhibits gamma interferon-induced regulation of selected genes in macrophages. Infect Immun 72, 6603–6614.

Pankla, R., Buddhisa, S., Berry, M., et al., 2009. Genomic transcriptional profiling identifies a candidate blood biomarker signature for the diagnosis of septicemic melioidosis. Genome Biol 10, R127.

Pascual, V., Allantaz, F., Patel, P., Palucka, A.K., Chaussabel, D., Banchereau, J., 2008. How the study of children with rheumatic diseases identified interferon-alpha and interleukin-1 as novel therapeutic targets. Immunol Rev 223, 39–59.

Peng, T., Kotla, S., Bumgarner, R.E., Gustin, K.E., 2006. Human rhinovirus attenuates the type I interferon response by disrupting activation of interferon regulatory factor 3. J Virol 80, 5021–5031.

Piqueras, B., Connolly, J., Freitas, H., Palucka, A.K., Banchereau, J., 2006. Upon viral exposure, myeloid and plasmacytoid dendritic cells produce 3 waves of distinct chemokines to recruit immune effectors. Blood 107, 2613–2618.

Popper, S.J., Watson, V.E., Shimizu, C., Kanegaye, J.T., Burns, J.C., Relman, D.A., 2009. Gene transcript abundance profiles distinguish Kawasaki disease from adenovirus infection. J Infect Dis 200, 657–666.

Ramilo, O., Allman, W., Chung, W., et al., 2007. Gene expression patterns in blood leukocytes discriminate patients with acute infections. Blood 109, 2066–2077.

Ramilo, O., Mejias, A., 2009. Shifting the paradigm: Host gene signatures for diagnosis of infectious diseases. Cell Host Microbe 6, 199–200.

Relman, D.A., 2002a. New technologies, human-microbe interactions, and the search for previously unrecognized pathogens. J Infect Dis 186 (Suppl. 2), S254–S258.

Relman, D.A., 2002b. The human body as microbial observatory. Nat Genet 30, 131–133.

Rudan, I., El Arifeen, S., Black, R.E., Campbell, H., 2007. Childhood pneumonia and diarrhoea: Setting our priorities right. Lancet Infect Dis 7, 56–61.

Schmitz, F., Mages, J., Heit, A., Lang, R., Wagner, H., 2004. Transcriptional activation induced in macrophages by Toll-like receptor (TLR) ligands: From expression profiling to a model of TLR signaling. Eur J Immunol 34, 2863–2873.

Thompson, L.J., Dunstan, S.J., Dolecek, C., et al., 2009. Transcriptional response in the peripheral blood of patients infected with Salmonella enterica serovar Typhi. Proc Natl Acad Sci USA 106, 22,433–22,438.

UNAIDS (Joint United Nations Programme on HIV/AIDS), 2008. Report on the Global AIDS Epidemic. <http://www.unaids.org/en/dataanalysis/epidemiology/2008reportontheglobalaidsepidemic/>.

Wachter, R.M., Flanders, S.A., Fee, C., Pronovost, P.J., 2008. Public reporting of antibiotic timing in patients with pneumonia: Lessons from a flawed performance measure. Ann Intern Med 149, 29–32.

Wang, E., Marincola, F.M., 2008. Bottom up: A modular view of immunology. Immunity 29, 9–11.

World Health Organization, 2008. WHO global burden of disease: 2004 update. <http://www.who.int/healthinfo/global_burden_disease/2004_report_update/en/index.html>.

Zaas, A.K., Chen, M., Varkey, J., et al., 2009. Gene expression signatures diagnose influenza and other symptomatic respiratory viral infections in humans. Cell Host Microbe 6, 207–217.

Zetola, N., Francis, J.S., Nuermberger, E.L., Bishai, W.R., 2005. Community-acquired methicillin-resistant *Staphylococcus aureus*: An emerging threat. Lancet Infect Dis 5, 275–286.

Host–Pathogen Interactions

Scott D. Kobayashi and Frank R. DeLeo

INTRODUCTION

The innate immune system is essential for host protection against invading microorganisms and is indispensable for overall maintenance of human health. Polymorphonuclear leukocytes (PMNs or neutrophils) are key components of the innate immune system and are the most abundant cellular component of the host immune system. The importance of PMNs in protection against pathogenic microorganisms is exemplified by a heightened susceptibility to infection in individuals with neutrophil deficiencies (Lekstrom-Himes and Gallin, 2000). One of the primary functions of neutrophils is to eliminate invading microorganisms. Neutrophil recognition of invading microorganisms initiates a complex cascade of signal transduction pathways that ultimately direct ingestion and killing of the pathogen. Notwithstanding, the complexity of neutrophil biology extends far beyond microbicidal activity and host defense. Over the past decade, the rapid advance in genomic technologies has provided a diversity of tools that have been instrumental in dissecting the molecular mechanisms of PMN host and pathogen interactions. Moreover, the ability to query the neutrophil transcriptome has provided detailed insight into essential processes such as PMN development (Theilgaard-Monch et al., 2005), and post-phagocytosis sequelae including apoptosis and resolution of inflammation (Kobayashi and DeLeo, 2009), and the role in tissue repair (Theilgaard-Monch et al., 2006). Perhaps more importantly, transcription studies

helped pave the way toward increased understanding of the pathophysiology of several neutrophil disorders (Holland et al., 2007; Kobayashi et al., 2004).

Neutrophils are a primary reason that the human host is highly adept at defending itself against invading bacteria. As such, the vast majority of bacterial infections are minor, may go unnoticed, and are resolved readily. On the other hand, some pathogenic microorganisms have evolved means to circumvent the innate and/or adaptive immune responses and thereby cause disease. Although much of the focus in our efforts to understand host–pathogen interactions has been placed on responses of the host or host cells, advances in genomics and related technologies over the past 10 years now permits comprehensive molecular analyses of bacterial pathogens during this process. We are currently in an era that allows interrogation of pathogens at the whole-genome, transcriptome, and proteome levels to better understand infection processes and disease transmission. These approaches have been made possible by bacterial genome sequencing (Blattner et al., 1997; Fleischmann et al., 1995; Koonin et al., 1996; Medini et al., 2008; Pallen and Wren, 2007; Perna et al., 2001; Tettelin and Feldblyum, 2009; Wren, 2000).

Here, we review our understanding of the neutrophil transcriptome in the context of the inflammatory process and interaction with bacteria. In the second half of the chapter, we provide a brief overview of bacterial responses to the host or host cells. The text is not meant to provide a comprehensive

Genomic and Personalized Medicine, 2nd edition
by Ginsburg & Willard. DOI: http://dx.doi.org/10.1016/B978-0-12-382227-7.00094-X

2013 Published by Elsevier Inc.

review of the literature; rather, we have used selected examples to indicate advances in the area of host–pathogen interactions.

HOST RESPONSE TO BACTERIAL PATHOGENS

Neutrophil Phagocytosis and Host Defense

Neutrophils bind and ingest invading microorganisms at sites of infection through a process known as phagocytosis (Figure 94.1). Bacteria contain a number of conserved molecules such as lipopolysaccharide (LPS), lipoprotein, lipoteichoic acid (LTA), and flagellin that interact with receptors on the surface of neutrophil membranes. Neutrophils express Toll-like receptors 1, 2, and 4–10 (Hayashi et al., 2003) and other pattern-recognition receptors such as the formyl peptide receptor. Ligation of these neutrophil pattern-recognition receptors activates signal transduction pathways that contribute to innate host defense. Neutrophil phagocytosis is enhanced significantly by opsonization of microbes with host antibody and serum complement. Specific antibodies recognize epitopes on microbial surfaces and facilitate the deposition of complement components. Ligation of membrane-bound opsonin receptors initiates cytoskeletal changes that carry out the physical process of phagocytosis. Ingested microbes are sequestered within membrane-bound vacuoles known as phagosomes (Figure 94.1).

PMN phagocytosis initiates a battery of neutrophil microbicidal mechanisms generated from two primary sources: (1) production of superoxide radicals and other secondarily derived reactive oxygen species (ROS), and (2) granules containing antimicrobial peptides, proteins, and degradative enzymes (Figure 94.1). The generation of ROS is mediated by a multi-component, membrane-bound complex called the NADPH-dependent oxidase (Quinn et al., 2006). Neutrophil phagocytosis also triggers degranulation, which involves fusion of cytoplasmic granules with the plasma and/or phagosome membrane. The secretory vesicles, gelatinase granules, and specific granules serve as a reservoir of functionally important membrane proteins such as CR3, formyl peptide receptor, flavocytochrome b_{558}, and β2-integrins. Fusion of primary/azurophilic granules (peroxidase-positive granules) with phagosomes enriches the vacuole lumen with antimicrobial agents including α-defensins, cathepsins, bactericidal-permeability-increasing protein, azurocidin, lysozyme, and elastase. The cumulative antimicrobial activity of neutrophils is thus comprised of ROS and a broad range of antimicrobial peptides and enzymes.

The Neutrophil Transcriptome

Cells of the immune system originate from common hematopoietic progenitor cells in bone marrow. Granulocyte differentiation and development are influenced by the concerted activities of myeloid colony-stimulating factors such as granulocyte colony-stimulating factor (G-CSF) and

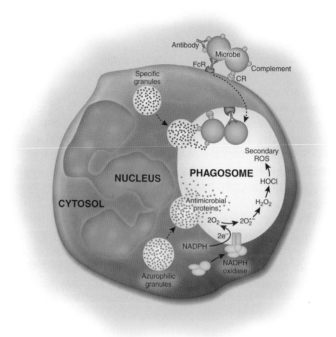

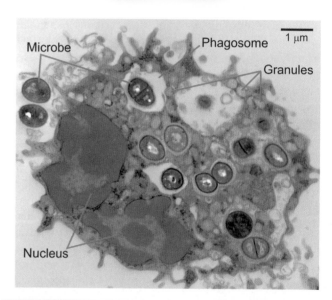

Figure 94.1 Neutrophil phagocytosis and microbicidal processes. *Top panel* is a model of neutrophil phagocytosis and activation of microbicidal processes. See text for details. *Bottom panel* is a scanning electron micrograph of a human neutrophil that has phagocytosed *Staphylococcus aureus*. FcR, Fc receptor; CR, complement receptor; ROS, reactive oxygen species. *Reproduced with permission from Kobayashi and DeLeo, 2009.*

granulocyte–macrophage colony-stimulating factor (GM-CSF) (Dale et al., 1998). During maturation toward mature neutrophils, myeloid precursors sequentially acquire features necessary for microbicidal activity, including receptors for phagocytosis and signaling, granule components, and NADPH oxidase proteins. Transcriptional processes occurring during granulopoiesis have been investigated intensely (Ferrari et al., 2007; Martinelli et al., 2004; Theilgaard-Monch et al., 2004). The transcriptional regulation of neutrophil development is mediated in part by the coordinated activities of several transcription factors and repressors. Transcription factors such as CCAAT/enhancer-binding protein alpha (C/EBPα), spleen focus forming virus proviral integration oncogene (SPI1 or PU.1), retinoic acid receptor, alpha (RARA), CCAAT-box-binding transcription factor (CBF), and v-myb myeloblastosis viral oncogene homolog (MYB) are involved in early granulopoiesis; C/EBPε, PI1 or PU.1, specificity protein 1 (SP1), cut-like homeobox 1 (CUX1 or CDP), homeobox A10 (HOXA10), signal transducer and activator of transcription (STAT)1, STAT3, STAT5, and growth factor independent 1 transcription repressor (GFI-1) are involved in terminal neutrophil differentiation (Friedman, 2002; McDonald, 2004). On the other hand, mature neutrophils are able to initiate the process of phagocytosis and killing of microorganisms without new synthesis of proteins (Kasprisin and Harris, 1977, 1978). Thus, it is provocative that the initial genome-wide PMN transcriptome studies revealed that mature neutrophils also have tremendous capacity for gene transcription (Kobayashi et al., 2002).

Several of these early studies used microarray analysis to gain insight into molecular processes that accompany PMN phagocytosis, and these findings have recently been reviewed (Kobayashi and DeLeo, 2009). Of note, it was discovered that the process of PMN phagocytosis *per se* induces numerous transcriptional changes that are consistent with increased production of pro-inflammatory mediators (Kobayashi et al., 2002, 2003b). For example, PMN phagocytosis results in increased expression of transcripts that encode important inflammatory mediators such as chemokine (C-C motif) ligand (CCL)3, CCL4, CCL20, oncostatin M, chemokine (C-X-C motif) ligand (CXCL)2, CXCL3, vascular endothelial growth factor A (VEGFA or VEGF), interleukin (IL)-6, and tumor necrosis factor (TNF)-α (Kobayashi et al., 2003b). In addition, PMN apoptosis (or programmed cell death) is a notable sequela of phagocytosis. The importance of neutrophil spontaneous apoptosis in maintenance of cellular homeostasis is well recognized, and considerable effort has been expended on determining the molecular mechanisms governing this essential process (Savill, 1997; Serhan and Savill, 2005). However, the process of phagocytosis-induced cell death (PICD) differs from that of spontaneous neutrophil apoptosis (Kobayashi and DeLeo, 2009). Neutrophil transcriptome studies have been instrumental in reshaping our view of the role of PMNs in the resolution of the inflammatory response (Figure 94.2). Notably, genome-wide studies have provided insight into key molecular determinants that regulate neutrophil PICD and have revealed mechanisms by which bacteria exploit this important process.

Bacteria and Neutrophil Survival

PICD is a mechanism to clear tissues of effete neutrophils containing killed microbes, thereby facilitating the resolution of

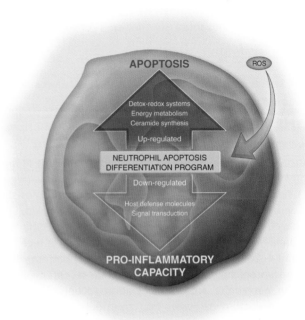

Figure 94.2 Remodeling of the neutrophil transcriptome during phagocytosis. *Reproduced with permission from Kobayashi and DeLeo, 2009.*

infection. The timely removal of spent neutrophils prevents the release of cytotoxic molecules into surrounding host tissues, a phenomenon caused by necrotic lysis. Apoptotic PMNs are recognized and cleared by macrophages, thereby preventing excessive host tissue damage and limiting inflammation at sites of infection (Savill et al., 1989). The process of phagocytosis significantly accelerates the rate of apoptosis in human PMNs (Coxon et al., 1996; Gamberale et al., 1998; Kobayashi et al., 2002; Watson et al., 1996; Zhang et al., 2003). Similarly, neutrophil ingestion of *Escherichia coli*, *Neisseria gonorrhoeae*, *Streptococcus pneumoniae*, *Streptococcus pyogenes*, *Candida albicans*, *Staphylococcus aureus*, *Mycobacterium tuberculosis*, *Burkholderia cepacia*, *Borrelia hermsii*, and *Listeria monocytogenes* significantly accelerates the rate of PMN apoptosis (DeLeo, 2004). PMN apoptosis plays a critical role in regulation of cell turnover and termination of the inflammatory process (Savill et al., 1989). Inasmuch as the process of PMN apoptosis is essential for maintenance of host health, it is not surprising that several bacterial pathogens are capable of exploiting apoptosis/PICD as a potential mechanism of pathogenesis.

Recognition and ingestion of microorganisms results in a complex series of neutrophil signal transduction events including Toll-like receptor activation and the production of additional inflammatory mediators that are potentially pathogen-specific. Bacteria are also capable of secreting molecules that alter function of host cells, including those of the innate immune system. Thus, the neutrophil response to bacterial pathogens involves complex signals induced by ligation of multiple receptors in addition to those participating in Fc- and complement-mediated phagocytosis. The PMN transcriptome has been queried following phagocytosis of many different microorganisms, including *Escherichia coli* (Subrahmanyam et al., 2001; Zhang et al., 2004), attenuated *Yersinia pestis* (Subrahmanyam et al., 2001), *Staphylococcus aureus* (Borjesson et al., 2005; Kobayashi et al., 2003a, 2010), *Streptococcus pyogenes* (or group A *Streptococcus*, GAS) (Kobayashi et al., 2003a), *Burkholderia cepacia* (Kobayashi et al., 2003a), *Listeria monocytogenes* (Kobayashi et al., 2003a), *Borrelia hermsii* (Kobayashi et al., 2003a), *Mycobacterium bovis* (Suttmann et al., 2003), *Candida albicans* (Fradin et al., 2007; Mullick et al., 2004), and *Anaplasma phagocytophilum* (Borjesson et al., 2005; Lee and Goodman, 2006; Sukumaran et al., 2005). Inasmuch as differences in pathogen-induced PMN transcript levels occur in those encoding signal transduction mediators and prominent transcription factors, it is not surprising that the PMN response is not entirely conserved.

Bacterial strategies that promote pathogenesis involve mechanisms that delay neutrophil apoptosis or ultimately cause neutrophil lysis. Although the relatively short lifespan of neutrophils is generally not amenable to long-term survival strategies employed by most intracellular pathogens, some pathogens have adapted their niche to include PMNs as a host cell. *Anaplasma phagocytophilum*, the causative agent of human granulocytic anaplasmosis, was the first bacterial pathogen reported to delay PMN apoptosis, which ultimately promotes replication within an endosomal compartment (Yoshiie et al., 2000). Oligonucleotide microarrays have been used to investigate the molecular basis of *A. phagocytophilum* survival within neutrophils (Borjesson et al., 2005). Functional analysis of infected PMNs confirmed previous reports that

A. phagocytophilum fails to trigger neutrophil production of ROS (Carlyon et al., 2004; Mott and Rikihisa, 2000), and the pathogen delays PMN spontaneous apoptosis (Yoshiie et al., 2000). In addition, PMN uptake of *A. phagocytophilum* occurs at a slow rate compared to that observed with other bacteria (Borjesson et al., 2005). Infection of human PMNs with *A. phagocytophilum* delays up-regulation of transcripts involved in the acute inflammatory response, such as *TNF*, *IL1B*, *CXCL1*, *CXCL2*, *CXCL3*, *CCL3*, *CCL4*, and *CD54* (Borjesson et al., 2005). Also, *A. phagocytophilum*-infected PMNs increase expression levels of several anti-apoptosis genes, including *BIRC2*, *BIRC3*, *CFLAR*, *TNFAIP8*, and *TNIP2*, and decrease expression of numerous apoptosis-inducing factors. Conversely, bacterial pathogens such as *Staphylococcus aureus* and GAS cause direct PMN lysis and/or accelerate bacteria-induced apoptosis to the point of secondary necrosis (Kobayashi et al., 2003a, 2010; Voyich et al., 2005; Wang et al., 2007). It is likely that accelerated PMN lysis contributes to the overall levels of tissue necrosis associated with these infections. The ability of GAS to cause rapid PICD and ultimately PMN lysis is reflected by and/or results from the changes in neutrophil gene expression (Kobayashi et al., 2003a). Interaction of PMNs with GAS elicits transcriptional changes in cell death pathways that include up-regulation of activator protein-1 (AP-1)-complex-related transcripts such as *FOS*, *FOSL1*, *FOSB*, *JUNB*, and *CD40*, and down-regulation of transcripts encoding members of the NF-κB signal transduction pathway (Kobayashi et al., 2003a). These findings indicate that the ability of GAS to exploit PMN fate pathways is an important component of streptococcal pathogenesis and a potential contributor to tissue destruction observed in invasive disease (Musser and DeLeo, 2005).

Blocking apoptosis to facilitate intracellular pathogen survival and promoting rapid lysis to eliminate neutrophils are

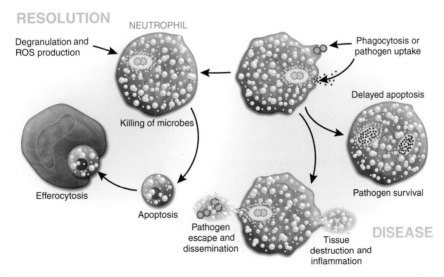

Figure 94.3 Two possible outcomes of bacteria–neutrophil interactions. *Reproduced with permission from Kobayashi and DeLeo, 2009.*

pathogen abilities that are plausible mechanisms of virulence. Genome-wide analyses with human neutrophils and bacteria, coupled with the results of *in vitro* and *in vivo* studies, led to the hypothesis that there are two fundamental outcomes of PMN–bacteria interactions (Figure 94.3). The first possibility is that phagocytosis activates PMNs and ingested microorganisms are killed, leading to PICD and removal of effete neutrophils by macrophages. This process leads to the resolution of infection and is beneficial for the host. Alternatively, pathogens are ingested but not killed, and PMNs either lyse or survive to promote pathogen replication and subsequent disease.

BACTERIAL RESPONSE TO THE HOST
The Genomics Era

More than 1000 bacterial genomes have been sequenced to completion (Liolios et al., 2010). This wealth of genome sequence information has revolutionized our understanding of bacterial physiology and pathogenesis, and how microorganisms adapt to the host. It is now known that bacterial adaptation in general occurs by gain and/or loss of genes and changes in gene sequence or order (Pallen and Wren, 2007). Also, as new genes are routinely identified, gene function or topology can be predicted, and novel vaccine approaches based upon this information can be implemented (Serruto et al., 2009). Microarrays based upon complete bacterial genomes can be used to compare strains and/or track the evolution of outbreaks (Behr et al., 1999; Fitzgerald et al., 2001; Salama et al., 2000; Tenover et al., 2006). Furthermore, genome-scale approaches such as DNA or oligonucleotide microarrays are routinely used to evaluate bacterial gene expression on a global scale and under a variety of conditions, including during interaction with the host or host cells (Table 94.1) (Musser and DeLeo, 2005; Sturdevant et al., 2010). Such studies are directed to better understand commensal interaction with the mammalian host, virulence, and pathogenesis, and to elucidate potential targets for therapeutic maneuvers directed to prevent or treat infections.

As noted above, proteomics has also been used to study pathogen responses to the host on a genome-wide scale. For example, bacterial proteomics has been used successfully to better understand the interaction of *Staphylococcus aureus* with components of innate host defense (Attia et al., 2010) in growth conditions that mimic specific host environments (Friedman et al., 2006), or to assess immune responses during infection (Burlak et al., 2007). Performing bacterial proteomics studies in the context of host–pathogen interaction is technically challenging, therefore the vast majority of bacterial proteomics studies have been directed to identify proteins produced using specific growth conditions *in vitro*. Studies with *S. aureus* have evaluated metabolism, post-translational modifications, stress and starvation responses, and virulence factors such as secreted molecules (Hecker et al., 2010). Proteomics approaches are essential for a comprehensive understanding of

host–pathogen interactions and, as such, merit extensive discussion. However, there is insufficient space in this chapter for such a discussion, and we refer the reader to recent reviews on the topic (Becher et al., 2009; Gotz et al., 2010; Hecker et al., 2010; Volker and Hecker, 2005). Rather, the remainder of this section will review our understanding of host–pathogen interactions based upon knowledge gained from bacterial transcriptome studies, with an emphasis on selected pathogens.

Microarray-based Comparative Genomics

Microarray-based approaches are now widely used to compare genome content among bacterial strains or to determine bacterial transcriptomes during specific culture conditions *in vitro*. Such investigations have provided important insight into the emergence of epidemic strains, bacterial evolution, and virulence and pathogenesis. To these ends, numerous pathogens have been investigated, and here we will discuss a few representative studies that utilized DNA microarrays for comparative genome studies.

Tuberculosis is a significant cause of morbidity and mortality in non-industrialized countries. The disease is caused by *Mycobacterium tuberculosis*, but the live attenuated vaccine strains, known collectively as bacille Calmette–Guérin (BCG), were developed from the related subspecies *Mycobacterium bovis*. BCG was passaged dozens of times to attenuate the strains for development of the vaccine – and many times since – and there is significant variance in vaccine efficacy in human trials (Behr, 2002). To better understand the evolution of the attenuation of BCG, and ultimately the basis for variance in vaccine efficacy, Behr and colleagues compared BCG obtained from multiple sources worldwide using a DNA microarray comprising 99.4% of the *M. tuberculosis* (strain H37Rv) open reading frames (Behr et al., 1999). The studies revealed sequential deletions of regions of DNA in BCG strains relative to the highly virulent H37Rv reference strain, and pinpointed the timing of the loss of DNA. Such an approach to understanding bacterial vaccine efficacy was unprecedented and heralded a new era in which comparative genomics would aid in the development of therapeutics and bacterial vaccines.

Salama and colleagues were among the first to use microarrays for comparative genome hybridization to identify potential bacterial virulence determinants (Salama et al., 2000). The researchers generated a *Helicobacter pylori* DNA microarray based upon the two genome sequences that were available at that time, and then used it to compare genome content and diversity among 15 *H. pylori* clinical isolates. Ultimately, the work identified new candidate virulence genes whose role in disease could then be tested.

The groundbreaking work of Fitzgerald and colleagues (2001) is another early example of how genomics approaches were used to solve an important problem in the area of host–pathogen interactions. Two major findings from those studies have exemplified the utility of genome-wide approaches in the area of infectious diseases. First, using *Staphylococcus aureus* DNA microarrays, the researchers discovered that the

TABLE 94.1	Selected studies of the bacterial transcriptome during host interaction	
Pathogen	**Host interaction**	**References**
Actinobacillus pleuropneumoniae	St Jude porcine lung cell line	Auger et al., 2009
Aeromonas hydrophila	RAW 264.7 murine macrophage-like cell line	Galindo et al., 2003
Anaplasma phagocytophilum	HL60 human promyelocytic cell line and HMEC-1 human endothelial cell line, ISE6 tick cell line	Nelson et al., 2008
Bacillus anthracis	RAW 264.7 murine macrophage-like cell line	Bergman et al., 2007
Borrelia burgdorferi	H4 human neuroglial cell line	Livengood et al., 2008
Brucella melitensis	HeLa human epithelial cell line	Rossetti et al., 2009
Chlamydia trachomatis	HeLa human epithelial cell line	Belland et al., 2003a, b
Clostridium difficile	Caco-2 human colon epithelial cell line	Janvilisri et al., 2010
Escherichia coli	Human primarily neutrophils, human urinary tract	Jandu et al., 2009; Kim et al., 2009; Roos and Klemm 2006; Staudinger et al., 2002
Francisella tularensis	Primary murine bone-marrow-derived macrophages	Wehrly et al., 2009
Lactobacillus plantarum	Mouse caeca	Marco et al., 2009
Leptospira interrogans	J774A.1 murine macrophage-like cell line, THP-1 human macrophage-like cell line	Xue et al., 2010
Listeria monocytogenes	Mouse spleen, P388D1 murine macrophage-like cell line, Caco-2 human colon epithelial cell line	Camejo et al., 2009; Chatterjee et al., 2006; Hain et al., 2006; Joseph et al., 2006
Mycobacterium avium subsp. *paratuberculosis*	Primary bovine monocyte-derived macrophages, bovine gastrointestinal tract	Janagama et al., 2010; Zhu et al., 2008
Mycobacterium tuberculosis complex	Primary human monocyte-derived macrophages, mouse lung, primary murine bone-marrow-derived macrophages, THP-1 human macrophage-like cell line	Cappelli et al., 2006; Dubnau et al., 2002; Fontan et al., 2008; Homolka et al., 2010; Rengarajan et al., 2005; Talaat et al., 2004
Neisseria gonorrhoeae	A431 human epithelial cell line	Du et al., 2005
Neisseria meningitidis	HeLa human epithelial cell line, human brain microvascular endothelial cell line, 16HBE14 human epithelial cell line	Dietrich et al., 2003; Grifantini et al., 2002
Pseudomonas aeruginosa	Primary human lung epithelial cells	Chugani and Greenberg, 2007
Rickettsia prowazekii	HMEC-1 human endothelial cell line	Bechah et al., 2010
Rickettsia typhi	HeLa human epithelial cell line	Ammerman et al., 2008
Streptococcus pneumoniae	THP-1 human macrophage-like cell line, A549 human lung epithelial cell line	Song et al., 2008a, b, 2009
Streptococcus pyogenes	Human primary neutrophils, human blood, mouse skin infection, monkey pharyngitis, Detroit 562 human pharyngeal cells	Graham et al., 2005, 2006; Ryan et al., 2007; Shea et al., 2010; Virtaneva et al., 2005; Voyich et al., 2003
Salmonella enterica serovar Typhimurium	HeLa human epithelial cell line	Hautefort et al., 2008; Sirsat et al., 2011; Wright et al., 2009
Salmonella enterica serovar Typhi	THP-1 human macrophage-like cell line	Faucher et al., 2005, 2006
Shigella flexneri	U937 human macrophage-like cell line, HeLa human epithelial cell line	Lucchini et al., 2005

(*continued*)

TABLE 94.1	(Continued)	
Pathogen	**Host interaction**	**References**
Staphylococcus aureus	Human primary neutrophils, A549 human lung epithelial cells	Garzoni et al., 2007; Voyich et al., 2005
Vibrio cholera	Human vomitus and feces	Larocque et al., 2005
Yersinia pestis	Flea, rat bubo	Vadyvaloo et al., 2010

mecA gene, which confers methicillin resistance, has been transferred into *S. aureus* multiple times. This finding indicates that methicillin-resistant *S. aureus* (MRSA) has evolved independently many times and is not simply a descendant of a single ancestor, as had been hypothesized earlier. Second, the study revealed that the outbreak of toxic-shock syndrome in the 1970s and 1980s was caused by a change in the host environment – use of a superabsorbent tampon – rather than the emergence of a new hypervirulent strain (Fitzgerald et al., 2001).

More recent genomics-based studies of *Staphylococcus aureus* have been conducted to understand recent outbreaks, especially that caused by community-associated (CA)-MRSA (DeLeo et al., 2010). For example, Tenover and colleagues used a microarray-based comparative genome hybridization to identify a set of genes specific to the epidemic CA-MRSA strain USA300 (Tenover et al., 2006). The findings suggested that the unique set of genes may contribute to the success of the USA300 strain. Others used a similar approach to show that CA-MRSA strains were created by multiple horizontal gene-transfer events into specific *S. aureus* lineages (Koessler et al., 2006). Lindsay and colleagues reported that at least 10 dominant lineages exist, and that there is significant gene diversity among and within each lineage (Lindsay et al., 2006). An important implication from these findings is that community-associated invasive disease results in part from host susceptibility, rather than the presence of a unique *S. aureus* virulence molecule or set of molecules. This hypothesis is consistent with our current knowledge of the epidemiology of CA-MRSA infections in the United States. That is, the epidemic USA300 strain has the capacity to cause severe or fatal invasive disease in otherwise healthy individuals, but the vast majority of CA-MRSA infections are relatively mild skin and soft tissue infections (DeLeo et al., 2010).

Although there are caveats to the DNA or oligonucleotide microarray-based comparative genome analyses, such as inability to detect genes or mobile DNA elements in query species not present on the microarray, the approach has dramatically increased our understanding of bacterial diversity, evolution, and virulence (Fitzgerald and Musser, 2001; Joyce et al., 2002).

Bacterial Transcriptome Studies In Vitro

There have been many applications of bacterial transcriptome analyses *in vitro* (in the absence of the host or host cells). For example, studies have been conducted to aid in understanding the role of gene regulatory molecules and/or the molecules controlled by global gene regulators, adaptation to specific growth conditions or nutritional requirements (e.g., temperature, availability of iron or other metals), stress responses, biological processes such as biofilm formation or cell wall biosynthesis, impact of gene deletions, and responses to antimicrobial agents/antibiotics. Here we focus our discussion on a few selected transcriptome studies performed with *Staphylococcus aureus* and GAS, pathogens that are leading causes of human bacterial infections worldwide.

Some of the earliest transcriptome studies with GAS revealed that changes in culture temperature (those that are physiologically relevant) can have a profound impact on GAS gene expression *in vitro* (Smoot et al., 2001). The studies suggested that GAS alters gene expression depending on the site of infection (e.g., skin versus blood). Subsequent work identified virulence molecules or those important for infection that are controlled by GAS two-component gene regulatory systems (Graham et al., 2002; Shelburne et al., 2005; Sitkiewicz and Musser, 2006; Voyich et al., 2004). For instance, an expression microarray study by Shelburne and colleagues identified a novel two-component gene regulatory system, *Spt*R/*Spt*S, that is important for GAS survival in human saliva (Shelburne et al., 2005). Iterative microarray analyses using wild-type and *spt*R/*spt*S deletion mutants revealed the specific molecules controlled by *Spt*R/*Spt*S (Shelburne et al., 2005). A similar approach identified a GAS survival response following exposure to neutrophil microbicides (Voyich et al., 2004). The GAS survival response includes differential regulation of molecules encoding cell wall metabolism, oxidative stress, and virulence, and is controlled in part by a two-component gene regulatory system known as Ihk/Irr. These types of iterative microarray approaches are now widely used to identify molecules important for infection of the host.

Expression microarray work with *Staphylococcus aureus* has identified regulons for multiple global gene regulatory systems, such as accessory gene regulator (Agr) and staphylococcal accessory regulator (Sar) (Dunman et al., 2001), vancomycin resistance-associated sensor/regulator (VraSR) (Kuroda et al., 2003), alternative sigma factor (*sigB* or σ^B) (Bischoff et al., 2004; Pane-Farre et al., 2006), TcaR (McCallum et al., 2004), ArlRS (Liang et al., 2005), MgrA (Luong et al., 2006), CidR (Yang et al., 2006), Aps (Li et al., 2007), and SaeRS (Liang et al., 2006; Nygaard et al., 2010; Rogasch et al., 2006; Voyich et al., 2009). Collectively, these studies have provided information vital to our understanding of how *S. aureus* regulates molecules important for interaction with the host.

The ability of *Staphylococcus aureus* to avoid or resist killing by phagocytic leukocytes – especially neutrophils – is the underlying basis of its ability to cause disease in humans, and it is well known that individuals with defects in phagocyte function are highly susceptible to *S. aureus* infection. Therefore, understanding the response of *S. aureus* to microbicidal components of host phagocytes should provide insight into mechanisms used by the pathogen to survive in the human host. To that end, Richardson and colleagues used *S. aureus* oligonucleotide microarrays to identify genes differently regulated following exposure to nitric oxide, an innate immune effector molecule produced by host cells (Richardson et al., 2006). Notably, the *S. aureus* response to nitric oxide includes up-regulation of molecules involved in iron homeostasis (Richardson et al., 2006), control of which is critical for bacterial survival in the host (Pishchany et al., 2010; Skaar, 2010). Consistent with this finding, subsequent work by Palazzolo-Ballance and colleagues showed that exposure of strain MW2, the prototype CA-MRSA strain, to hydrogen peroxide caused up-regulation of transcripts involved in heme/iron uptake (Palazzolo-Ballance et al., 2008). In addition, exposure of *S. aureus* to neutrophil granule proteins, which occurs in the phagocytic vacuole, caused up-regulation of genes involved in stress response, including those that moderate ROS, and multiple toxins and hemolysins (Palazzolo-Ballance et al., 2008). Subsequent DNA microarray studies indicated that *vraSR* is up-regulated by exposure of *S. aureus* to cationic antimicrobial peptides (Pietiainen et al., 2009), findings consistent with Li and colleagues (2007) and Palazzolo-Ballance and colleagues, and the general notion that VraSR is important for cell wall biosynthesis (Kuroda et al., 2003). Such information has provided an enhanced view of how *Staphylococcus aureus* circumvents destruction by the innate immune system.

Antibiotic resistance in *Staphylococcus aureus* is a major problem in developed countries worldwide (Chambers and DeLeo, 2009). Although the mechanisms for some types of antibiotic resistance in *S. aureus* are known and well characterized, such as resistance to β-lactam antibiotics, the mechanism for glycopeptide intermediate resistance is less well defined. Glycopeptide intermediate resistance is acquired gradually during prolonged treatment, and involves multiple mutations. Several microarray or comparative whole-genome sequence-based studies have investigated the molecular genetic basis for glycopeptide resistance in *S. aureus* and identified genes linked to resistance, many with no previously characterized function (Cui et al., 2005; McCallum et al., 2006a, b; Mwangi et al., 2007; Renzoni et al., 2006). In aggregate, the findings indicate that glycopeptide intermediate resistance is due to altered cell wall metabolism and biosynthesis, which is at least partly controlled by the VraSR regulon and may include post-translational protein processing.

Multiple gene expression profiling studies have provided exciting new insights into biofilm formation by staphylococci, such as the involvement of surfactant-like peptides known as phenol-soluble modulins (Yao et al., 2005) and extensive changes in metabolism (Resch et al., 2005). This topic

is important and merits more discussion, and we refer the reader to a comprehensive review on the subject (Otto, 2009).

Bacterial Transcriptome Analyses during Host Interaction

Elucidation of bacterial transcriptomes *in vitro* as described above is clearly a step forward in our efforts to generate a comprehensive view of bacterial physiology and adaptation in specific environments. Therefore, it is not unfounded to state that the ability to analyze global gene expression in bacterial pathogens during interaction with the host or during infection of the host *in vivo* is a major breakthrough for infectious disease research. Microarray-based studies of global gene expression during host or host–cell interaction have been conducted for numerous bacterial pathogens (Table 94.1). Here, we discuss a few representative studies and highlight the major findings and their implications.

Staudinger and colleagues were the first to evaluate the bacterial transcriptome during interaction with host cells (Staudinger et al., 2002). The researchers investigated global gene expression in a urinary isolate of *Escherichia coli* following phagocytosis by human neutrophils from healthy individuals or those with chronic granulomatous disease (CGD). Phagocytes in individuals with CGD are incapable of generating reactive oxygen species, and patients thus acquire recurrent and severe bacterial and fungal infections. By comparing changes in transcripts following phagocytosis by healthy or CGD neutrophils, Staudinger and colleagues identified molecules used by *Escherichia coli* to moderate reactive oxygen species (e.g., the OxyR regulon), and the authors proposed that the bacterium responds to intraphagosomal hydrogen peroxide rather than superoxide (Staudinger et al., 2002).

Work by Voyich and colleagues immediately followed the Staudinger studies and evaluated the GAS transcriptome during phagocytic interaction with human neutrophils (Voyich et al., 2003). In addition to the identification of novel secreted molecules, one of the key findings was the discovery that Ihk/Irr is up-regulated following neutrophil phagocytosis, and is essential for survival of the pathogen after uptake (Voyich et al., 2003). Iterative GAS microarray studies *in vitro* identified the Ihk/Irr regulon (described above), which engages a survival response in the pathogen (Voyich et al., 2004). Similar studies performed with hospital-associated MRSA and CA-MRSA strains identified SaeRS as potentially important for survival of *Staphylococcus aureus* after phagocytosis by human neutrophils (Voyich et al., 2005). Subsequent studies using animal infection models and an iterative microarray approach verified the importance of SaeRS for infection, and identified molecules that contribute to *S. aureus* survival after phagocytosis (Nygaard et al., 2010; Voyich et al., 2009).

Although technically challenging, some studies have evaluated bacterial transcriptomes during infection *in vivo*. For example, Talaat and colleagues reported global gene expression in *Mycobacterium tuberculosis* during intranasal infection of mice over the course of 28 days (Talaat et al., 2004). Notably, these researchers discovered genes whose expression is dictated by immune responses of the host, and found that transcriptome data with macrophages *in vitro* aligned with their findings in the mouse *in vivo* after 21 days. These results provide an estimate of the immune response at this stage of infection.

Differential gene expression of *Listeria monocytogenes* recovered from the spleens of infected mice indicates that a significant component of the *Listeria* response to host infection is dedicated to virulence and subversion of the host immune system (Camejo et al., 2009). These *in vivo* transcriptome data demonstrated that molecules implicated in virulence were indeed up-regulated during infection. Further, the work identified PrfA and VirR as the two major regulators of *L. monocytogenes* virulence during infection *in vivo*.

Studies by Vadyvaloo and colleagues compared transcriptomes of *Yersinia pestis* during transit through the flea vector and those of the rat bubo (Vadyvaloo et al., 2010). Transit through the flea causes changes in gene expression that prepare the pathogen for enhanced survival following transmission to mammals. Some of this phenotype is related to induction of genes that confer resistance to phagocytosis by host leukocytes.

In a study of unprecedented magnitude, Shea and colleagues investigated the transcriptome of GAS during pharyngeal infection of monkeys, and concomitantly identified host genes whose patterns of expression correlated with those of the pathogen (Shea et al., 2010). The correlation of host and pathogen patterns of gene expression, dubbed the "interactome,"

revealed correlations between genes encoding GAS hyaluronic acid production and host cell endocytic vesicle formation, GAS mevalonic acid synthesis and host $\gamma\delta$ T cells, and GAS responses to host phagocytes and host neutrophils. Notably, changes in the GAS transcriptome *in vivo* during the course of infection were associated with specific phases of pharyngitis and the host response. The interactome approach is at the cutting edge of transcriptome analysis and serves as a springboard for future work in many areas of infectious disease research.

FUTURE PERSPECTIVE

Antibiotic resistance is arguably the single greatest problem for treatment of bacterial infections today. This is a long-term problem, and it will be imperative to develop new diagnostics, therapeutics, and vaccines to treat and/or prevent severe bacterial infections. Genomics approaches are absolutely critical for these endeavors, which have been stymied in the past by a lack of information about host and bacterial pathogen responses during infection. Alternatively, information gained from host genomics and transcriptomics may ultimately be used to augment host defense and/or inform about appropriate prophylaxis to prevent infections. This is a reasonable goal, as such approaches have been successfully implemented for diagnosis of respiratory viral infections (Zaas et al., 2009).

REFERENCES

Ammerman, N.C., Rahman, M.S., Azad, A.F., 2008. Characterization of Sec-translocon-dependent extracytoplasmic proteins of *Rickettsia typhi*. J Bacteriol 190, 6234–6242.

Attia, A.S., Benson, M.A., Stauff, D.L., Torres, V.J., Skaar, E.P., 2010. Membrane damage elicits an immunomodulatory program in *Staphylococcus aureus*. PLoS Pathog 6, e1000802.

Auger, E., Deslandes, V., Ramjeet, M., et al., 2009. Host-pathogen interactions of *Actinobacillus pleuropneumoniae* with porcine lung and tracheal epithelial cells. Infect Immun 77, 1426–1441.

Bechah, Y., El, K.K., Mediannikov, O., et al., 2010. Genomic, proteomic, and transcriptomic analysis of virulent and avirulent *Rickettsia prowazekii* reveals its adaptive mutation capabilities. Genome Res 20, 655–663.

Becher, D., Hempel, K., Sievers, S., et al., 2009. A proteomic view of an important human pathogen – towards the quantification of the entire *Staphylococcus aureus* proteome. PLoS ONE 4, e8176.

Behr, M.A., 2002. BCG – different strains, different vaccines? Lancet Infect Dis 2, 86–92.

Behr, M.A., Wilson, M.A., Gill, W.P., et al., 1999. Comparative genomics of BCG vaccines by whole-genome DNA microarray. Science 284, 1520–1523.

Belland, R.J., Nelson, D.E., Virok, D., et al., 2003a. Transcriptome analysis of chlamydial growth during IFN-gamma-mediated persistence and reactivation. Proc Natl Acad Sci USA 100, 15,971–15,976.

Belland, R.J., Zhong, G., Crane, D.D., et al., 2003b. Genomic transcriptional profiling of the developmental cycle of *Chlamydia trachomatis*. Proc Natl Acad Sci USA 100, 8478–8483.

Bergman, N.H., Anderson, E.C., Swenson, E.E., et al., 2007. Transcriptional profiling of *Bacillus anthracis* during infection of host macrophages. Infect Immun 75, 3434–3444.

Bischoff, M., Dunman, P., Kormanec, J., et al., 2004. Microarray-based analysis of the *Staphylococcus aureus* sigmaB regulon. J Bacteriol 186, 4085–4099.

Blattner, F.R., Plunkett III, G., Bloch, C.A., et al., 1997. The complete genome sequence of *Escherichia coli* K-12. Science 277, 1453–1462.

Borjesson, D.L., Kobayashi, S.D., Whitney, A.R., et al., 2005. Insights into pathogen immune evasion mechanisms: *Anaplasma phagocytophilum* fails to induce an apoptosis differentiation program in human neutrophils. J Immunol 174, 6364–6372.

Burlak, C., Hammer, C.H., Robinson, M.A., et al., 2007. Global analysis of community-associated methicillin-resistant *Staphylococcus aureus* exoproteins reveals molecules produced in vitro and during infection. Cell Microbiol 9, 1172–1190.

Camejo, A., Buchrieser, C., Couve, E., et al., 2009. In vivo transcriptional profiling of *Listeria monocytogenes* and mutagenesis identify new virulence factors involved in infection. PLoS Pathog 5, e1000449.

Cappelli, G., Volpe, E., Grassi, M., et al., 2006. Profiling of *Mycobacterium tuberculosis* gene expression during human macrophage infection: Upregulation of the alternative sigma factor G, a group of transcriptional regulators, and proteins with unknown function. Res Microbiol 157, 445–455.

Carlyon, J.A., Latif, D.A., Pypaert, M., Lacy, P., Fikrig, E., 2004. *Anaplasma phagocytophilum* utilizes multiple host evasion mechanisms to thwart NADPH oxidase-mediated killing during neutrophil infection. Infect Immun 72, 4772–4783.

Chambers, H.F., DeLeo, F.R., 2009. Waves of resistance: *Staphylococcus aureus* in the antibiotic era. Nat Rev Microbiol 7, 2464–2474.

Chatterjee, S.S., Hossain, H., Otten, S., et al., 2006. Intracellular gene expression profile of *Listeria monocytogenes*. Infect Immun 74, 1323–1338.

Chugani, S., Greenberg, E.P., 2007. The influence of human respiratory epithelia on *Pseudomonas aeruginosa* gene expression. Microb Pathog 42, 29–35.

Coxon, A., Rieu, P., Barkalow, F.J., et al., 1996. A novel role for the β2 integrin CD11b/CD18 in neutrophil apoptosis: A homeostatic mechanism in inflammation. Immunity 5, 653–666.

Cui, L., Lian, J.Q., Neoh, H.M., Reyes, E., Hiramatsu, K., 2005. DNA microarray-based identification of genes associated with glycopeptide resistance in *Staphylococcus aureus*. Antimicrob Agents Chemother 49, 3404–3413.

Dale, D.C., Liles, W.C., Llewellyn, C., Price, T.H., 1998. Effects of granulocyte-macrophage colony-stimulating factor (GM-CSF) on neutrophil kinetics and function in normal human volunteers. Am J Hematol 57, 7–15.

DeLeo, F.R., 2004. Modulation of phagocyte apoptosis by bacterial pathogens. Apoptosis 9, 399–413.

DeLeo, F.R., Otto, M., Kreiswirth, B.N., Chambers, H.F., 2010. Community-associated methicillin-resistant *Staphylococcus aureus*. Lancet 375, 1557–1568.

Dietrich, G., Kurz, S., Hubner, C., et al., 2003. Transcriptome analysis of *Neisseria meningitidis* during infection. J Bacteriol 185, 155–164.

Du, Y., Lenz, J., Arvidson, C.G., 2005. Global gene expression and the role of sigma factors in *Neisseria gonorrhoeae* in interactions with epithelial cells. Infect Immun 73, 4834–4845.

Dubnau, E., Fontan, P., Manganelli, R., Soares-Appel, S., Smith, I., 2002. *Mycobacterium tuberculosis* genes induced during infection of human macrophages. Infect Immun 70, 2787–2795.

Dunman, P.M., Murphy, E., Haney, S., et al., 2001. Transcription profiling-based identification of *Staphylococcus aureus* genes regulated by the agr and/or sarA loci. J Bacteriol 183, 7341–7353.

Faucher, S.P., Curtiss III, R., Daigle, F., 2005. Selective capture of *Salmonella enterica* serovar typhi genes expressed in macrophages that are absent from the *Salmonella enterica* serovar typhimurium genome. Infect Immun 73, 5217–5221.

Faucher, S.P., Porwollik, S., Dozois, C.M., McClelland, M., Daigle, F., 2006. Transcriptome of *Salmonella enterica* serovar typhi within macrophages revealed through the selective capture of transcribed sequences. Proc Natl Acad Sci USA 103, 1906–1911.

Ferrari, F., Bortoluzzi, S., Coppe, A., et al., 2007. Genomic expression during human myelopoiesis. BMC Genomics 8, 264.

Fitzgerald, J.R., Musser, J.M., 2001. Evolutionary genomics of pathogenic bacteria. Trends Microbiol 9, 547–553.

Fitzgerald, J.R., Sturdevant, D.E., Mackie, S.M., Gill, S.R., Musser, J.M., 2001. Evolutionary genomics of *Staphylococcus aureus*: Insights into the origin of methicillin-resistant strains and the toxic shock syndrome epidemic. Proc Natl Acad Sci USA 98, 8821–8826.

Fleischmann, R.D., Adams, M.D., White, O., et al., 1995. Whole-genome random sequencing and assembly of *Haemophilus influenzae* Rd. Science 269, 496–512.

Fontan, P., Aris, V., Ghanny, S., Soteropoulos, P., Smith, I., 2008. Global transcriptional profile of *Mycobacterium tuberculosis* during THP-1 human macrophage infection. Infect Immun 76, 717–725.

Fradin, C., Mavor, A.L., Weindl, G., et al., 2007. The early transcriptional response of human granulocytes to infection with *Candida albicans* is not essential for killing but reflects cellular communications. Infect Immun 75, 1493–1501.

Friedman, A.D., 2002. Transcriptional regulation of granulocyte and monocyte development. Oncogene 21, 3377–3390.

Friedman, D.B., Stauff, D.L., Pishchany, G., et al., 2006. *Staphylococcus aureus* redirects central metabolism to increase iron availability. PLoS Pathog 2, e87.

Galindo, C.L., Sha, J., Ribardo, D.A., et al., 2003. Identification of *Aeromonas hydrophila* cytotoxic enterotoxin-induced genes in macrophages using microarrays. J Biol Chem 278, 40,198–40,212.

Gamberale, R., Giordano, M., Trevani, A.S., Andonegui, G., Geffner, J.R., 1998. Modulation of human neutrophil apoptosis by immune complexes. J Immunol 161, 3666–3674.

Garzoni, C., Francois, P., Huyghe, A., et al., 2007. A global view of *Staphylococcus aureus* whole genome expression upon internalization in human epithelial cells. BMC Genomics 8, 171.

Gotz, F., Hacker, J., Hecker, M., 2010. Pathophysiology of staphylococci in the post-genomic era. Int J Med Microbiol 300, 75.

Graham, M.R., Smoot, L.M., Migliaccio, C.A., et al., 2002. Virulence control in group A *Streptococcus* by a two-component gene regulatory system: Global expression profiling and in vivo infection modeling. Proc Natl Acad Sci USA 99, 13,855–13,860.

Graham, M.R., Virtaneva, K., Porcella, S.F., et al., 2005. Group A *Streptococcus* transcriptome dynamics during growth in human blood reveals bacterial adaptive and survival strategies. Am J Pathol 166, 455–465.

Graham, M.R., Virtaneva, K., Porcella, S.F., et al., 2006. Analysis of the transcriptome of group A *Streptococcus* in mouse soft tissue infection. Am J Pathol 169, 927–942.

Grifantini, R., Bartolini, E., Muzzi, A., et al., 2002. Gene expression profile in *Neisseria meningitidis* and *Neisseria lactamica* upon host-cell contact: From basic research to vaccine development. Ann NY Acad Sci 975, 202–216.

Hain, T., Steinweg, C., Chakraborty, T., 2006. Comparative and functional genomics of *Listeria* spp. J Biotechnol 126, 37–51.

Hautefort, I., Thompson, A., Eriksson-Ygberg, S., et al., 2008. During infection of epithelial cells *Salmonella enterica* serovar typhimurium undergoes a time-dependent transcriptional adaptation that results in simultaneous expression of three type 3 secretion systems. Cell Microbiol 10, 958–984.

Hayashi, F., Means, T.K., Luster, A.D., 2003. Toll-like receptors stimulate human neutrophil function. Blood 102, 2660–2669.

Hecker, M., Becher, D., Fuchs, S., Engelmann, S., 2010. A proteomic view of cell physiology and virulence of *Staphylococcus aureus*. Int J Med Microbiol 300, 76–87.

Holland, S.M., DeLeo, F.R., Elloumi, H.Z., et al., 2007. *STAT3* mutations in the hyper-IgE syndrome. N Engl J Med 357, 1608–1619.

Homolka, S., Niemann, S., Russell, D.G., Rohde, K.H., 2010. Functional genetic diversity among *Mycobacterium tuberculosis* complex clinical isolates: Delineation of conserved core and lineage-specific transcriptomes during intracellular survival. PLoS Pathog 6, e1000988.

Janagama, H.K., Lamont, E.A., George, S., et al., 2010. Primary transcriptomes of *Mycobacterium avium* subsp. *paratuberculosis* reveal proprietary pathways in tissue and macrophages. BMC Genomics 11, 561.

Jandu, N., Ho, N.K., Donato, K.A., et al., 2009. Enterohemorrhagic *Escherichia coli* O157:H7 gene expression profiling in response to growth in the presence of host epithelia. PLoS ONE 4, e4889.

Janvilisri, T., Scaria, J., Chang, Y.F., 2010. Transcriptional profiling of *Clostridium difficile* and Caco-2 cells during infection. J Infect Dis 202, 282–290.

Joseph, B., Przybilla, K., Stuhler, C., et al., 2006. Identification of *Listeria monocytogenes* genes contributing to intracellular replication by expression profiling and mutant screening. J Bacteriol 188, 556–568.

Joyce, E.A., Chan, K., Salama, N.R., Falkow, S., 2002. Redefining bacterial populations: A post-genomic reformation. Nat Rev Genet 3, 462–473.

Kasprisin, D.O., Harris, M.B., 1977. The role of RNA metabolism in polymorphonuclear leukocyte phagocytosis. J Lab Clin Med 90, 118–124.

Kasprisin, D.O., Harris, M.B., 1978. The role of protein synthesis in polymorphonuclear leukocyte phagocytosis II. Exp Hematol 6, 585–589.

Kim, Y., Oh, S., Park, S., Kim, S.H., 2009. Interactive transcriptome analysis of enterohemorrhagic *Escherichia coli* (EHEC) O157:H7 and intestinal epithelial HT-29 cells after bacterial attachment. Int J Food Microbiol 131, 224–232.

Kobayashi, S.D., Braughton, K.R., Palazzolo-Ballance, A.M., et al., 2010. Rapid neutrophil destruction following phagocytosis of *Staphylococcus aureus*. J Innate Immun 2, 560–575.

Kobayashi, S.D., Braughton, K.R., Whitney, A.R., et al., 2003a. Bacterial pathogens modulate an apoptosis differentiation program in human neutrophils. Proc Natl Acad Sci USA 100, 10,948–10,953.

Kobayashi, S.D., DeLeo, F.R., 2009. Role of neutrophils in innate immunity: A systems biology-level approach. WIREs Syst Biol Med 1, 309–333.

Kobayashi, S.D., Voyich, J.M., Braughton, K.R., DeLeo, F.R., 2003b. Down-regulation of proinflammatory capacity during apoptosis in human polymorphonuclear leukocytes. J Immunol 170, 3357–3368.

Kobayashi, S.D., Voyich, J.M., Braughton, K.R., et al., 2004. Gene expression profiling provides insight into the pathophysiology of chronic granulomatous disease. J Immunol 172, 636–643.

Kobayashi, S.D., Voyich, J.M., Buhl, C.L., Stahl, R.M., DeLeo, F.R., 2002. Global changes in gene expression by human polymorphonuclear leukocytes during receptor-mediated phagocytosis: Cell fate is regulated at the level of gene expression. Proc Natl Acad Sci USA 99, 6901–6906.

Koessler, T., Francois, P., Charbonnier, Y., et al., 2006. Use of oligoarrays for characterization of community-onset methicillin-resistant *Staphylococcus aureus*. J Clin Microbiol 44, 1040–1048.

Koonin, E.V., Mushegian, A.R., Rudd, K.E., 1996. Sequencing and analysis of bacterial genomes. Curr Biol 6, 404–416.

Kuroda, M., Kuroda, H., Oshima, T., et al., 2003. Two-component system VraSR positively modulates the regulation of cell-wall biosynthesis pathway in *Staphylococcus aureus*. Mol Microbiol 49, 807–821.

Larocque, R.C., Harris, J.B., Dziejman, M., et al., 2005. Transcriptional profiling of *Vibrio cholerae* recovered directly from patient specimens during early and late stages of human infection. Infect Immun 73, 4488–4493.

Lee, H.C., Goodman, J.L., 2006. *Anaplasma phagocytophilum* causes global induction of antiapoptosis in human neutrophils. Genomics 88, 496–503.

Lekstrom-Himes, J.A., Gallin, J.I., 2000. Immunodeficiency diseases caused by defects in phagocytes. N Engl J Med 343, 1703–1714.

Li, M., Cha, D.J., Lai, Y., et al., 2007. The antimicrobial peptide-sensing system *aps* of *Staphylococcus aureus*. Mol Microbiol 66, 1136–1147.

Liang, X., Yu, C., Sun, J., et al., 2006. Inactivation of a two-component signal transduction system, SaeRS, eliminates adherence and attenuates virulence of *Staphylococcus aureus*. Infect Immun 74, 4655–4665.

Liang, X., Zheng, L., Landwehr, C., et al., 2005. Global regulation of gene expression by ArlRS, a two-component signal transduction regulatory system of *Staphylococcus aureus*. J Bacteriol 187, 5486–5492.

Lindsay, J.A., Moore, C.E., Day, N.P., et al., 2006. Microarrays reveal that each of the ten dominant lineages of *Staphylococcus aureus* has a unique combination of surface-associated and regulatory genes. J Bacteriol 188, 669–676.

Liolios, K., Chen, I.M., Mavromatis, K., et al., 2010. The Genomes On Line Database (GOLD) in 2009: Status of genomic and metagenomic projects and their associated metadata. Nucleic Acids Res 38 (Database issue), D346–D354.

Livengood, J.A., Schmit, V.L., Gilmore Jr., R.D., 2008. Global transcriptome analysis of *Borrelia burgdorferi* during association with human neuroglial cells. Infect Immun 76, 298–307.

Lucchini, S., Liu, H., Jin, Q., Hinton, J.C., Yu, J., 2005. Transcriptional adaptation of *Shigella flexneri* during infection of macrophages and epithelial cells: Insights into the strategies of a cytosolic bacterial pathogen. Infect Immun 73, 88–102.

Luong, T.T., Dunman, P.M., Murphy, E., Projan, S.J., Lee, C.Y., 2006. Transcription profiling of the mgrA regulon in *Staphylococcus aureus*. J Bacteriol 188, 1899–1910.

Marco, M.L., Peters, T.H., Bongers, R.S., et al., 2009. Lifestyle of *Lactobacillus plantarum* in the mouse caecum. Environ Microbiol 11, 2747–2757.

Martinelli, S., Urosevic, M., Daryadel, A., et al., 2004. Induction of genes mediating interferon-dependent extracellular trap formation during neutrophil differentiation. J Biol Chem 279, 44,123–44,132.

McCallum, N., Bischoff, M., Maki, H., Wada, A., Berger-Bachi, B., 2004. TcaR, a putative MarR-like regulator of *sarS* expression. J Bacteriol 186, 2966–2972.

McCallum, N., Karauzum, H., Getzmann, R., et al., 2006a. In vivo survival of teicoplanin-resistant *Staphylococcus aureus* and fitness cost of teicoplanin resistance. Antimicrob Agents Chemother 50, 2352–2360.

McCallum, N., Spehar, G., Bischoff, M., Berger-Bachi, B., 2006b. Strain dependence of the cell wall-damage induced stimulon in *Staphylococcus aureus*. Biochim Biophys Acta 1760, 1475–1481.

McDonald, P.P., 2004. Transcriptional regulation in neutrophils: Teaching old cells new tricks. Adv Immunol 82, 1–48.

Medini, D., Serruto, D., Parkhill, J., et al., 2008. Microbiology in the post-genomic era. Nat Rev Microbiol 6, 419–430.

Mott, J., Rikihisa, Y., 2000. Human granulocytic ehrlichiosis agent inhibits superoxide anion generation by human neutrophils. Infect Immun 68, 6697–6703.

Mullick, A., Elias, M., Harakidas, P., et al., 2004. Gene expression in HL60 granulocytoids and human polymorphonuclear leukocytes exposed to *Candida albicans*. Infect Immun 72, 414–429.

Musser, J.M., DeLeo, F.R., 2005. Toward a genome-wide systems biology analysis of host-pathogen interactions in group A *Streptococcus*. Am J Pathol 167, 1461–1472.

Mwangi, M.M., Wu, S.W., Zhou, Y., et al., 2007. Tracking the in vivo evolution of multidrug resistance in *Staphylococcus aureus* by whole-genome sequencing. Proc Natl Acad Sci USA 104, 9451–9456.

Nelson, C.M., Herron, M.J., Felsheim, R.F., et al., 2008. Whole genome transcription profiling of *Anaplasma phagocytophilum* in human and tick host cells by tiling array analysis. BMC Genomics 9, 364.

Nygaard, T.K., Pallister, K.B., Ruzevich, P., et al., 2010. SaeR binds a consensus sequence within virulence gene promoters to advance USA300 pathogenesis. J Infect Dis 201, 241–254.

Otto, M., 2009. *Staphylococcus epidermidis* – the "accidental" pathogen. Nat Rev Microbiol 7, 555–567.

Palazzolo-Ballance, A.M., Reniere, M.L., Braughton, K.R., et al., 2008. Neutrophil microbicides induce a pathogen survival response in community-associated methicillin-resistant *Staphylococcus aureus*. J Immunol 180, 500–509.

Pallen, M.J., Wren, B.W., 2007. Bacterial pathogenomics. Nature 449, 835–842.

Pane-Farre, J., Jonas, B., Forstner, K., Engelmann, S., Hecker, M., 2006. The sigmaB regulon in *Staphylococcus aureus* and its regulation. Int J Med Microbiol 296, 237–258.

Perna, N.T., Plunkett III, G., Burland, V., et al., 2001. Genome sequence of enterohaemorrhagic *Escherichia coli* O157:H7. Nature 409, 529–533.

Pietiainen, M., Francois, P., Hyyrylainen, H.L., et al., 2009. Transcriptome analysis of the responses of *Staphylococcus aureus* to antimicrobial peptides and characterization of the roles of vraDE and vraSR in antimicrobial resistance. BMC Genomics 10, 429.

Pishchany, G., McCoy, A.L., Torres, V.J., et al., 2010. Specificity for human hemoglobin enhances *Staphylococcus aureus* infection. Cell Host Microbe 8, 544–550.

Quinn, M.T., Ammons, M.C., DeLeo, F.R., 2006. The expanding role of NADPH oxidases in health and disease: No longer just agents of death and destruction. Clin Sci (Lond) 111, 1–20.

Rengarajan, J., Bloom, B.R., Rubin, E.J., 2005. Genome-wide requirements for *Mycobacterium tuberculosis* adaptation and survival in macrophages. Proc Natl Acad Sci USA 102, 8327–8332.

Renzoni, A., Barras, C., Francois, P., et al., 2006. Transcriptomic and functional analysis of an autolysis-deficient, teicoplanin-resistant derivative of methicillin-resistant *Staphylococcus aureus*. Antimicrob Agents Chemother 50, 3048–3061.

Resch, A., Rosenstein, R., Nerz, C., Gotz, F., 2005. Differential gene expression profiling of *Staphylococcus aureus* cultivated under biofilm and planktonic conditions. Appl Environ Microbiol 71, 2663–2676.

Richardson, A.R., Dunman, P.M., Fang, F.C., 2006. The nitrosative stress response of *Staphylococcus aureus* is required for resistance to innate immunity. Mol Microbiol 61, 927–939.

Rogasch, K., Ruhmling, V., Pane-Farre, J., et al., 2006. Influence of the two-component system SaeRS on global gene expression in two different *Staphylococcus aureus* strains. J Bacteriol 188, 7742–7758.

Roos, V., Klemm, P., 2006. Global gene expression profiling of the asymptomatic bacteriuria *Escherichia coli* strain 83972 in the human urinary tract. Infect Immun 74, 3565–3575.

Rossetti, C.A., Galindo, C.L., Lawhon, S.D., Garner, H.R., Adams, L.G., 2009. *Brucella melitensis* global gene expression study provides novel information on growth phase-specific gene regulation with potential insights for understanding *Brucella*:host initial interactions. BMC Microbiol 9, 81.

Ryan, P.A., Kirk, B.W., Euler, C.W., Schuch, R., Fischetti, V.A., 2007. Novel algorithms reveal streptococcal transcriptomes and clues about undefined genes. PLoS Comput Biol 3, e132.

Salama, N., Guillemin, K., McDaniel, T.K., et al., 2000. A whole-genome microarray reveals genetic diversity among *Helicobacter pylori* strains. Proc Natl Acad Sci USA 97, 14,668–14,673.

Savill, J., 1997. Apoptosis in resolution of inflammation. J Leukoc Biol 61, 375–380.

Savill, J.S., Wyllie, A.H., Henson, J.E., et al., 1989. Macrophage phagocytosis of aging neutrophils in inflammation: Programmed cell death in the neutrophil leads to its recognition by macrophages. J Clin Invest 83, 865–875.

Serhan, C.N., Savill, J., 2005. Resolution of inflammation: The beginning programs the end. Nat Immunol 6, 1191–1197.

Serruto, D., Serino, L., Masignani, V., Pizza, M., 2009. Genome-based approaches to develop vaccines against bacterial pathogens. Vaccine 27, 3245–3250.

Shea, P.R., Virtaneva, K., Kupko III, J.J., et al., 2010. Interactome analysis of longitudinal pharyngeal infection of cynomolgus macaques by group A *Streptococcus*. Proc Natl Acad Sci USA 107, 4693–4698.

Shelburne III, S.A., Sumby, P., Sitkiewicz, I., et al., 2005. Central role of a bacterial two-component gene regulatory system of previously unknown function in pathogen persistence in human saliva. Proc Natl Acad Sci USA 102, 16,037–16,042.

Sirsat, S.A., Burkholder, K.M., Muthaiyan, A., et al., 2011. Effect of sublethal heat stress on *Salmonella typhimurium* virulence. J Appl Microbiol 110, 813–822.

Sitkiewicz, I., Musser, J.M., 2006. Expression microarray and mouse virulence analysis of four conserved two-component gene regulatory systems in group A *Streptococcus*. Infect Immun 74, 1339–1351.

Skaar, E.P., 2010. The battle for iron between bacterial pathogens and their vertebrate hosts. PLoS Pathog 6, e1000949.

Smoot, L.M., Smoot, J.C., Graham, M.R., et al., 2001. Global differential gene expression in response to growth temperature alteration in group A *Streptococcus*. Proc Natl Acad Sci USA 98, 10,416–10,421.

Song, X.M., Connor, W., Hokamp, K., Babiuk, L.A., Potter, A.A., 2008a. *Streptococcus pneumoniae* early response genes to human lung epithelial cells. BMC Res Notes 1, 64.

Song, X.M., Connor, W., Hokamp, K., Babiuk, L.A., Potter, A.A., 2009. Transcriptome studies on *Streptococcus pneumoniae*, illustration of early response genes to THP-1 human macrophages. Genomics 93, 72–82.

Song, X.M., Connor, W., Jalal, S., Hokamp, K., Potter, A.A., 2008b. Microarray analysis of *Streptococcus pneumoniae* gene expression changes to human lung epithelial cells. Can J Microbiol 54, 189–200.

Staudinger, B.J., Oberdoerster, M.A., Lewis, P.J., Rosen, H., 2002. mRNA expression profiles for *Escherichia coli* ingested by normal and phagocyte oxidase-deficient human neutrophils. J Clin Invest 110, 1151–1163.

Sturdevant, D.E., Virtaneva, K., Martens, C., et al., 2010. Host-microbe interaction systems biology: Lifecycle transcriptomics and comparative genomics. Future Microbiol 5, 205–219.

Subrahmanyam, Y.V., Yamaga, S., Prashar, Y., et al., 2001. RNA expression patterns change dramatically in human neutrophils exposed to bacteria. Blood 97, 2457–2468.

Sukumaran, B., Carlyon, J.A., Cai, J.L., Berliner, N., Fikrig, E., 2005. Early transcriptional response of human neutrophils to *Anaplasma phagocytophilum* infection. Infect Immun 73, 8089–8099.

Suttmann, H., Lehan, N., Bohle, A., Brandau, S., 2003. Stimulation of neutrophil granulocytes with *Mycobacterium bovis* bacillus Calmette-Guerin induces changes in phenotype and gene expression and inhibits spontaneous apoptosis. Infect Immun 71, 4647–4656.

Talaat, A.M., Lyons, R., Howard, S.T., Johnston, S.A., 2004. The temporal expression profile of *Mycobacterium tuberculosis* infection in mice. Proc Natl Acad Sci USA 101, 4602–4607.

Tenover, F.C., McDougal, L.K., Goering, R.V., et al., 2006. Characterization of a strain of community-associated methicillin-resistant *Staphylococcus aureus* widely disseminated in the United States. J Clin Microbiol 44, 108–118.

Tettelin, H., Feldblyum, T., 2009. Bacterial genome sequencing. Methods Mol Biol 551, 231–247.

Theilgaard-Monch, K., Jacobsen, L.C., Borup, R., et al., 2005. The transcriptional program of terminal granulocytic differentiation. Blood 105, 1785–1796.

Theilgaard-Monch, K., Knudsen, S., Follin, P., Borregaard, N., 2004. The transcriptional activation program of human neutrophils in skin lesions supports their important role in wound healing. J Immunol 172, 7684–7693.

Theilgaard-Monch, K., Porse, B.T., Borregaard, N., 2006. Systems biology of neutrophil differentiation and immune response. Curr Opin Immunol 18, 54–60.

Vadyvaloo, V., Jarrett, C., Sturdevant, D.E., Sebbane, F., Hinnebusch, B.J., 2010. Transit through the flea vector induces a pretransmission innate immunity resistance phenotype in *Yersinia pestis*. PLoS Pathog 6, e1000783.

Virtaneva, K., Porcella, S.F., Graham, M.R., et al., 2005. Longitudinal analysis of the group A *Streptococcus* transcriptome in experimental pharyngitis in cynomolgus macaques. Proc Natl Acad Sci USA 102, 9014–9019.

Volker, U., Hecker, M., 2005. From genomics via proteomics to cellular physiology of the gram-positive model organism *Bacillus subtilis*. Cell Microbiol 7, 1077–1085.

Voyich, J.M., Braughton, K.R., Sturdevant, D.E., et al., 2004. Engagement of the pathogen survival response used by group A *Streptococcus* to avert destruction by innate host defense. J Immunol 173, 1194–1201.

Voyich, J.M., Braughton, K.R., Sturdevant, D.E., et al., 2005. Insights into mechanisms used by *Staphylococcus aureus* to avoid destruction by human neutrophils. J Immunol 175, 3907–3919.

Voyich, J.M., Sturdevant, D.E., Braughton, K.R., et al., 2003. Genome-wide protective response used by group A *Streptococcus* to evade destruction by human polymorphonuclear leukocytes. Proc Natl Acad Sci USA 100, 1996–2001.

Voyich, J.M., Vuong, C., DeWald, M., et al., 2009. The SaeR/S gene regulatory system is essential for immune evasion by *Staphylococcus aureus*. J Infect Dis 199, 1698–1706.

Wang, R., Braughton, K.R., Kretschmer, D., et al., 2007. Identification of novel cytolytic peptides as key virulence determinants for community-associated MRSA. Nat Med 13, 1510–1514.

Watson, R.W., Redmond, H.P., Wang, J.H., Condron, C., Bouchier-Hayes, D., 1996. Neutrophils undergo apoptosis following ingestion of *Escherichia coli*. J Immunol 156, 3986–3992.

Wehrly, T.D., Chong, A., Virtaneva, K., et al., 2009. Intracellular biology and virulence determinants of *Francisella tularensis* revealed by transcriptional profiling inside macrophages. Cell Microbiol 11, 1128–1150.

Wren, B.W., 2000. Microbial genome analysis: Insights into virulence, host adaptation and evolution. Nat Rev Genet 1, 30–39.

Wright, J.A., Totemeyer, S.S., Hautefort, I., et al., 2009. Multiple redundant stress resistance mechanisms are induced in *Salmonella enterica* serovar typhimurium in response to alteration of the intracellular environment via TLR4 signalling. Microbiology 155, 2919–2929.

Xue, F., Dong, H., Wu, J., et al., 2010. Transcriptional responses of *Leptospira interrogans* to host innate immunity: Significant changes in metabolism, oxygen tolerance, and outer membrane. PLoS Negl Trop Dis 4, e857.

Yang, S.J., Dunman, P.M., Projan, S.J., Bayles, K.W., 2006. Characterization of the *Staphylococcus aureus* CidR regulon: Elucidation of a novel role for acetoin metabolism in cell death and lysis. Mol Microbiol 60, 458–468.

Yao, Y., Sturdevant, D.E., Otto, M., 2005. Genomewide analysis of gene expression in *Staphylococcus epidermidis* biofilms: Insights into the pathophysiology of *S. epidermidis* biofilms and the role of phenol-soluble modulins in formation of biofilms. J Infect Dis 191, 289–298.

Yoshiie, K., Kim, H.Y., Mott, J., Rikihisa, Y., 2000. Intracellular infection by the human granulocytic ehrlichiosis agent inhibits human neutrophil apoptosis. Infect Immun 68, 1125–1133.

Zaas, A., Chen, M., Varkey, J., et al., 2009. Gene expression signatures diagnose influenza and other symptomatic respiratory viral infections in humans. Cell Host Microbe 6, 207–217.

Zhang, B., Hirahashi, J., Cullere, X., Mayadas, T.N., 2003. Elucidation of molecular events leading to neutrophil apoptosis following phagocytosis: Cross-talk between caspase 8, reactive oxygen species, and MAPK/ERK activation. J Biol Chem 278, 28,443–28,454.

Zhang, X.Q., Kluger, Y., Nakayama, Y., et al., 2004. Gene expression in mature neutrophils: Early responses to inflammatory stimuli. J Leukoc Biol 75, 358–372.

Zhu, X., Tu, Z.J., Coussens, P.M., et al., 2008. Transcriptional analysis of diverse strains *Mycobacterium avium* subspecies *paratuberculosis* in primary bovine monocyte-derived macrophages. Microbes Infect 10, 1274–1282.

Microbial Vaccine Development

M. Anthony Moody

INTRODUCTION

Vaccination is one of the most cost-effective and efficient ways to combat infectious diseases. The worldwide use of vaccines has eradicated one infectious disease (smallpox) from circulation in humans and has the potential to eradicate others such as measles (Katz and Hinman, 2004) and polio (Katz, 2006) if concerted efforts and global cooperation permit. During the nineteenth and much of the 20th century, the development of vaccines was largely empirical, and even for vaccines where efficacy was clear, our understanding of the immunological correlates of protection was often incomplete (Plotkin, 2001, 2008). Advances in vaccinology have paralleled and sometimes preceded development of scientific concepts in immunology that explain vaccine effects. For example, variation against smallpox was discovered before the development of the germ theory of disease (Baker and Katz, 2004), and early vaccines such as Pasteur's rabies vaccine were developed before the advent of modern culture techniques and genetic theory. Even today, licensure of vaccines is usually not based on immunological correlates but rather on demonstration of safety and protection from disease.

Vaccination has traditionally been the domain of public health. Public health interventions are designed to benefit the general population while personalized medicine is intended to benefit individual patients. These definitions suggest that such interventions are in opposition; however, individual people benefit from public health measures and personalized medicine benefits the public in ways both tangible (preservation of health during productive years, reduction in negative side effects that incur additional costs to the healthcare system) and intangible (improved wellbeing of caregivers and relatives). Figure 95.1 displays how vaccines fall along the continuum and shows how some current vaccines differ in their protection of individuals or the population at large.

The toolbox of genomics and personalized medicine has great promise in the development of new vaccines. Our ability to harness this toolbox is, at present, limited by an incomplete understanding of genetic factors contributing to immune response. Genetic contributions to immunity have long been recognized (e.g., the journal *Immunogenetics* was founded in 1974), and, as with other areas of genomic medicine, this field is rapidly evolving. I will not attempt in this chapter to give a comprehensive review of more than a century of research but rather will use examples from the literature to outline current understanding of relevant concepts.

THE IDEAL VACCINE AND MEASUREMENT OF RESPONSE

The ideal vaccine is an immune system stimulus that induces a completely protective response without producing any negative or unwanted side effects at the time of administration, when subsequently challenged by native infection, or later in life during immune senescence. Such a vaccine would require minimal intervention to administer and would not be subject to waning of the immune response over time. To date,

Genomic and Personalized Medicine, 2nd edition
by Ginsburg & Willard. DOI: http://dx.doi.org/10.1016/B978-0-12-382227-7.00095-1

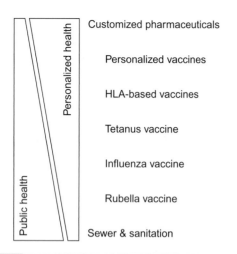

Figure 95.1 Continuum of public and personalized health interventions. Public health interventions such as sewer and sanitation services reduce disease transmission by reducing hazards to all members of a community, while personalized medicine strategies like custom pharmaceuticals are designed to benefit patients at the individual level. Vaccines reduce the transmission of infectious diseases and disease consequences for vaccinated persons. Current vaccine strategies are intended for deployment to large segments of the population while personalized and HLA-based vaccines would target individuals or small segments of the population. Examples of three vaccines currently in use show different models of protection. Tetanus vaccine provides personal protection by preventing disease caused by the soil organism *Clostridium tetani* but does change the likelihood of contracting the organism in a dirty wound. Influenza vaccine reduces disease burden in the immunized and reduces transmission of influenza from vaccinated to susceptible persons. Rubella vaccine prevents the relatively mild disease German measles but is chiefly used to prevent congenital rubella syndrome by reducing transmission to susceptible pregnant women.

TABLE 95.1 Examples of direct and indirect vaccine assessments

Vaccine against	Direct assessment	Indirect assessment	Reference
Influenza	Protection from disease	Mucosal and serum antibody (neutralizing and hemagglutination inhibition)	Plotkin, 2008
Polio	Protection from disease	Mucosal and serum antibody (neutralizing)	Faden et al., 1990
Smallpox	Protection from disease	Serum antibody titer (neutralizing)	Mack et al., 1972; Sarkar et al., 1975
Tetanus	Protection from disease	Animal protection after passive infusion	Edsall, 1959
Tuberculosis	Protection from disease	IFNg release assays	Fletcher, 2007

few vaccines have achieved most of these goals and none have achieved all. When considering the potential impact of genomics on vaccine response, we are faced with large hurdles. As described in other chapters in this text, gene expression is not uniform across populations or across time during the life of a given person. Vaccine doses that are well tolerated in children can have more significant side effects in adults [e.g., diphtheria vaccine (Galazka and Robertson, 1996)] or can be less immunogenic [e.g., varicella-zoster vaccine (Oxman et al., 2005)]. The resulting variability in side effects, efficacy, and duration of protection afforded by vaccines in genetically diverse populations of all ages means that vaccines must be constantly re-evaluated as longitudinal data accumulate in vaccinated populations.

Vaccine success is typically judged on two related criteria: protection from disease and correlates of protection (Plotkin, 2001, 2008). Protection from disease is a simple concept – rates of disease over a period of time are assessed in vaccinated and unvaccinated groups, from which a degree of protection can be calculated. This is easier to assess in animal models where experimental infections can be used, but in humans experimental infection models are difficult, expensive, and sometimes unethical. Studies designed to demonstrate protection in humans are also subject to many confounders, such as participant attrition, poor control of pathogen exposure prior to and after administration of a test vaccine, and self-reporting biases regarding risk and exposure. While protection from disease is considered the *gold standard*, indirect methods of assessment are usually cheaper and more practical (Table 95.1).

Correlates of protection are immunological parameters (e.g., antibody titer, T cell function) that have a plausible mechanism of action against the targeted pathogen, that could be expected to protect from disease challenge, and that are proportional to the observed protective effect (Plotkin, 2001, 2008). For many pathogens, the correlate of protection induced by vaccine matches the immune response found following recovery from disease (e.g., antibodies against influenza), while for other vaccines the protective correlate does not occur in disease survivors (e.g., antibodies against tetanus toxin). To evaluate new vaccines, comparison with the immune response found in patients who recover from a severe infection such as Ebola hemorrhagic fever is usually considered appropriate (Sullivan et al., 2003), although similar if weaker responses can sometimes be found in patients who die (Baize et al., 1999). Some pathogens, such as herpes simplex and varicella-zoster virus, establish latency or persistent infection even in the face of an appropriate immune response, and the correlates of protection are usually considered to be immune responses associated with lack of progression or with suppression of active disease. In all cases, correlates of protection are measurable responses that can either be compared to protection trials or used as endpoints themselves.

The use of correlates of protection is not without problems. For some pathogens, the demarcation between protective and possibly detrimental immune responses is not clear. Dengue virus infection is an example, where the development of anti-virus antibodies can be associated with protection from subsequent challenge (Sabin, 1952) but can also be associated with development of severe dengue hemorrhagic fever upon re-infection (Halstead and O'Rourke, 1977; Halstead et al., 1970). For other pathogens such as HIV-1, no correlate of protection is known so proof of vaccine efficacy must rely upon infection rate comparisons between vaccine and placebo arms (Rerks-Ngarm et al., 2009; Sekaly, 2008).

THE BIOLOGY OF IMMUNE RESPONSE TO VACCINES

Protective vaccines induce responses at the time of vaccination and elicit immunological memory that is both durable and epitope specific (Janeway, 2005). For example, tetanus toxoid and live attenuated measles vaccine induce high titers of protective antibodies that mediate long-term protection after priming and boosting immunizations (Amanna et al., 2007). Vaccine-induced memory is mediated by B lymphocytes (responsible for antibody production) and T lymphocytes (responsible for helper functions/regulation of B cell responses and for antigen-specific cell-mediated killing), both of which can be recruited upon subsequent vaccine or pathogen challenge. Differences in genes expressed in B and T lymphocytes that are directly involved in their effector function can alter vaccine performance (see below for examples); many such genes are known and more are being identified. Additionally, B and T cell development requires the participation of non-lymphocyte cells and complex structures (Janeway, 2005), and induction of immunological memory is dependent on cell trafficking and physical cues from other cell types (Allen et al., 2007). For example, the germinal center reaction requires interaction of multiple T and B lymphocytes with dendritic cells, stromal cells, monocytes, and intracellular matrix. Evidence suggests that induction of high-affinity antibodies requires germinal center formation (Allen et al., 2007). Due to this complex set of interactions, genes not expressed in B and T cells can affect vaccine response.

The innate arm of the immune system performs critical functions for survival from infection through mechanisms such as phagocytosis by neutrophils and monocytes, and killing of infected targets mediated by natural killer (NK) cells. Some of these functions are assisted by adaptive immunity [e.g., antibody-dependent cell-mediated cytotoxicity (ADCC) by NK cells (Tyler et al., 1989), opsonization of bacterial (Reed, 1975) or viral (Van Strijp et al., 1989) targets mediated by preformed antibodies], but these effector mechanisms also participate during primary infection. In fact, intact adaptive immunity is less critical than intact innate immunity for early survival during overwhelming infection (Janeway, 2005). Given these facts, it is not unexpected that vaccination with live-attenuated vectors would involve the innate immune system, and recent work has shown that vaccine responses to killed and subunit vaccines are also dependent on innate immunity. Bacterial and viral components contained within vaccines can trigger pathogen-associated molecular pattern (PAMP) receptors (Tukhvatulin et al., 2010) that result in cytokine production that modifies B and T cell responses. Adjuvants used with some vaccines (e.g., squalene-based emulsions containing bacterial products) can trigger PAMPs; other adjuvants can use alternative pathways, such as alum-adjuvanted hepatitis B vaccine that triggers the inflammasome via nucleotide-binding domain leucine-rich repeat-containing protein 3 (NLRP3) (Kool et al., 2008, Li et al., 2008).

Thus the genes and pathways that can modify vaccine response include those expressed in adaptive immune cells that directly determine specificity, and genes expressed in other cell types that have indirect effects. Recent work has suggested that people may be genetically predisposed to become symptomatic with infectious diseases (Alcaïs et al., 2009), and given the homology between response to infections and vaccines, these genetic differences will likely modify vaccine response. Genetics can result in profound differences in immune response [e.g., Janus kinase 3 (Jak3) deficiency leading to severe combined immunodeficiency (Cacalano et al., 1999)] or subtle differences [e.g., inability to use immunoglobulin (Ig) gene *IGHV1-69* to make antibodies (Sasso et al., 1996)], and these effects are probably not uniform for all vaccines. For now, the field continues to identify pathways and gene candidates that contribute to differences in vaccine response (Table 95.2).

SMALL ANIMAL MODELS

At the start of the 20th century, studies of tumor rejection suggested a role for genetic differences in immune response. This work paralleled the development of well-characterized inbred strains of mice – the number of papers increased through the 1930s (Bittner, 1935) as many factors (with a dizzying array of nomenclatures) were associated with rejection of transplanted tumors. In 1936, Gorer published two landmark papers (Gorer, 1936a, b) that provided the link between tumor rejection and blood group antigens, identifying what came to be known as the histocompatibility-2 (*H2*) gene complex (Klein, 1986). Since then, the contributions of this complex to immune responses have been extensively studied (Table 95.3).

TABLE 95.2	Examples of human genes involved in vaccine response		
Vaccine against	**Associated genes**	**Observed phenomena**	**Reference(s)**
Hepatitis B (subunit)	*IL10*	Antibody titer increased (ACC haplotype)	Hohler et al., 2005
	HLA-DRA, FOXP1	Antibody titer (responder vs non-responder)	Davila et al., 2010
HIV-1 (adenovirus vector)	*HLA-B* (various alleles)	Gag-specific CD8+ T cell response increased	Fellay et al., 2011
Influenza	*HLA-DQB1*	Hemagglutination inhibition titer increased	Gelder et al., 2002
	IL6	Hemagglutination inhibition titer increased	Poland et al., 2008
Measles	*HLA-B*	Seroconversion after one immunization (high for *07, low for *08)	Jacobson et al., 2003
Mumps	*HLA-DQB1*0303*	Antibody titer decreased	Ovsyannikova et al., 2008
	HLA-DRB1, HLA-DQA1, HLA-DQB2, IL10RA, IL12RB1, IL12RB2	Lymphoproliferation (various effects)	Ovsyannikova et al., 2008
Pertussis (acellular)	*TLR4* (and downstream genes)	Antibody titer increased	Kimman et al., 2008
Rubella	*IL12B, IFNGR1*	IFN-γ, IL-10 secretion (various effects)	Jacobson et al., 2009
	DDX58, RARB, TRIM5, TRIM22	Antibody titer (various effects)	Ovsyannikova et al., 2010b
	DDX58, MAVS, RARB, TLR3	IFN-γ, IL-6, TNF-α secretion (various effects)	Ovsyannikova et al., 2010a

IFN-γ, interferon-gamma; IL-6, interleukin-6; IL-10, interleukin-10; TNF-α, tumor necrosis factor alpha.

TABLE 95.3	Examples of genetic variation in vaccine response in mouse studies			
Vaccine against	**Mouse strains**	**Genes/loci**	**Phenomena**	**Reference(s)**
Malaria (subunit)	C57BL/10 and congenics	*H2*-associated loci	Lack of protection/enhancement after vaccination	Tian et al., 1996
Tuberculosis (Ad35-based)	BALB/c, C57BL/6	*H2*	BALB/c less responsive than C57BL/6, differential protection observed	Radošević et al., 2007
Tuberculosis (BCG)	C57BL/10 and A congenics	*H2*	Delayed-type hypersensitivity response under *H2* control, protection correlated	Apt et al., 1993
Influenza (DNA)	BALB/c, C3H, C57BL/10	*H2*	Neuraminidase vaccine protected all strains, hemagglutinin vaccine only protected BALB/c	Chen et al., 1999
Rabies (rhesus diploid)	C57BL/6, C3H	*H2*, Ig, additional loci	C3H more protected, protection inherited dominantly	Templeton et al., 1986
Yersinia pestis (DNA)	BALB/c, outbred	*H2*	Outbred mice not responsive to vaccine	Brandler et al., 1998
Group A streptococcal carbohydrate	A/J, BSVS	*H2*, Ig, additional locus	A/J strain produced higher titer antibodies vs BSVS	Briles et al., 1977
Streptococcus pneumoniae 23F polysaccharide	BALB/c, CBA	*H2*	BALB/c strain produced higher titer antibodies	McCool et al., 2003

H2, histocompatibility-2 complex; Ig, immunoglobulin complex.

Because of low cost, ease of care, and small size, mice are common small animal models in immunology research. Most studies in mice use inbred strains, and thereby enforce genetic restriction based on the particular strain used. Those restrictions are sometimes inadvertent or based on convenience, while others are intentional comparisons. These studies have identified a number of genes associated with variations in vaccine response (examples in Table 95.3). The *H2* gene complex has been frequently implicated (Apt et al., 1993; Brandler et al., 1998; Briles et al., 1977; McCool et al., 2003; Radošević et al., 2007; Templeton et al., 1986; Tian et al., 1996), and this is especially so for vaccines thought to be primarily T-cell-based or for studies that primarily measure T cell response. For vaccines with neutralizing or blocking antibody titer as the primary endpoint, the immunoglobulin (Ig) gene complex has been identified (Briles et al., 1977; Templeton et al., 1986). However, in many cases additional genetic loci have been found to associate with vaccine response, either in genes linked to *H2* (Tian et al., 1996) or in distinct loci (Briles et al., 1977; Templeton et al., 1986), and genes that contribute to vaccine responses in mice continue to be identified.

Mice are not the only small animal models, and in some cases there are specific reasons to avoid them. For example, some HIV-1 assays have poor performance in the presence of mouse serum; this discovery led to retraction of one vaccine candidate despite promising early results (Lacasse et al., 1999; Nunberg, 2002). Other small animal models that have been used in vaccine trials include guinea pigs, hamsters, ferrets, rabbits, and rats. At present, there are fewer reagents and highly characterized strains for each of those species and less detailed genetic information compared with mice. While these other small animal models have not been as extensively characterized, genetic contributions to vaccine response have been described in guinea pigs (Cohen et al., 1987), rats (Stankus and Leslie, 1975), and recently even in transgenic rabbits (Hu et al., 2010).

STUDIES IN NON-HUMAN PRIMATES AND HUMANS

Genomics applied to vaccine development in humans and non-human primates (NHPs) has proceeded along three main tracks: (1) the use of known genetic variations to design vaccines that target a desired response [e.g., vaccines targeted to specific human leukocyte antigen (HLA) types], (2) the use of genetics to dissect the variation in vaccine response among recipients, and (3) screening for genetic variants known to contribute to vaccine or pathogen response to improve study design. The first two have been applied to both NHP and human studies, while the last has not yet become routine in humans.

The goal of vaccinology is to design vaccines that target specific immune responses. With each advance in understanding the basis of adaptive immune response, there has been a corresponding wave of new vaccine approaches. Once pathogen-specific antibodies were discovered to mediate protection from some infections, antisera and antitoxin products were developed (Dale, 1954; Macnalty, 1954), ultimately leading to the current tetanus and diphtheria vaccines. Given the diversity of the human antibody repertoire, targeting specific Ig genes has not been required, although some families of Ig genes appear to be favored (Corti et al., 2010; Sui et al., 2009) or disfavored (Berberian et al., 1993; Ditzel et al., 1996) in some immune responses. Once the mechanism of T cell function was determined, vaccine design began to target T cells in hopes of developing highly specific anti-virus vaccines. Given the lack of known correlates of protection, this approach has been tried to the greatest extent against HIV-1. Initial efforts linked specific T cell and B cell epitopes to target variable regions of the virus (Bartlett et al., 1998). As T cell immunity became better understood, vaccine approaches became more sophisticated, leading to a number of human trials. Based on an adenoviral vector, the Step trial [HVTN 502/Merck 023 (Buchbinder et al., 2008)] was stopped after an interim analysis showed no benefit; further analysis suggested that pre-existing immunity to the vector may have led to undesirable responses in some subjects (Buchbinder et al., 2008; Sekaly, 2008). In contrast, the ALVAC-prime/AIDSVAX-boost Thailand trial was designed to elicit both B and T cell responses. The trial showed a small degree of non-durable protection (Rerks-Ngarm et al., 2009) and at present, work is ongoing to determine if genetically restricted responses may be responsible. Recently, computational approaches have been used to design mosaic immunogens that can optimally cover the range of desired T cell responses, and initial work in NHP trials has been promising (Barouch et al., 2010; Kong et al., 2009; Santra et al., 2010).

The second track is the use of genetic approaches to understand differential response to vaccines. This area of research has been increasingly fruitful. A review in 2007 cited 12 studies of specific alleles and four twin-based studies examining vaccine response (Kimman et al., 2007); some studies cited in that review and more recent work are summarized in Table 95.2. Much of the work has focused on non-responders. For example, specific HLA alleles have been found to associate with lack of response to vaccines against influenza (Gelder et al., 2002), tuberculosis (Newport et al., 2004), hepatitis B (Kruskall et al., 1992), and other diseases. Importantly, this research has shown that lack of response is not universal. In particular, non-responders to standard hepatitis B vaccine had good response to an alternative vaccine (Zuckerman et al., 1997). This is encouraging from a public health perspective, since it appears that the use of customized vaccines could result in greater coverage of the total population, although such approaches will be more costly and difficult to implement globally.

The third track is the use of genetic information to select vaccine trial cohorts. This has been especially important in NHP trials of HIV-1 candidate vaccines due to the lack of correlates of protection and the need to use protection from infection or disease progression as the marker of vaccine efficacy. In rhesus macaques, the HLA allele *Mamu-A*01* was found to protect against disease progression in simian immunodeficiency virus (SIV) (Zhang et al., 2002). For this reason, it has become standard practice to perform vaccine trials in animals with known *Mamu-A*01* alleles, so that progression can be used as a surrogate endpoint (Liang et al., 2005; Subbramanian et al., 2006). As more markers associated with progression of infection emerge, such as the recently described protective *TRIM5* polymorphism (Lim et al., 2010), additional screenings will be applied with the goal of reducing confounders and improving trial efficiency. To date, this approach has not been widely applied to human trials; ultimately, such screenings may be economically and politically difficult to implement.

For specific pathogens, infectious challenge trials in humans have shed light on correlates of protection and specific pathways involved in immune response. Live-attenuated influenza vaccine trials in children demonstrated that the vaccine could induce robust mucosal and systemic antibody responses, and a subsequent infectious challenge trial demonstrated that protection from infection correlated with antibody titer (Belshe et al., 2000). More recently, challenge models of common respiratory viruses (influenza, respiratory syncytial virus, and rhinovirus) have characterized gene expression patterns induced that were highly specific for each infecting pathogen (Zaas et al., 2009). Such patterns may be useful in the development of rapid diagnostics for outbreak settings. Those patterns many also provide clues for immunogen and adjuvant design, and targets for measures of vaccine response (e.g., vaccine induction of gene profiles correlated with protection from infection).

PRACTICAL CONSIDERATIONS

The number of genes known to be involved in response to vaccines is growing steadily (Tables 95.2 and 95.3). At present, no test of a single gene or combination of such tests is likely to predict response to a single vaccine, much less to all vaccines. All parts of the innate and adaptive immune systems contribute to host response to pathogens and vaccines, and while the effect of single genes can be isolated by careful experiment, we do not have a complete list of contributing genes. Even for genes already known to contribute, we do not understand all of the relevant interactions. As such, designing a practical panel for predicting vaccine response in the general population is not yet feasible.

In one sense, the goals of vaccination and those of personalized/genomic medicine are at cross purposes (Figure 95.1). Vaccination is a public health intervention for inducing herd immunity (protection of the unvaccinated through reduced disease transmission, occurs when a sufficient fraction of group members have protective immunity), while personalized medicine is targeted at the individual subject to improve therapy and prevent complications. Mass vaccination efforts, whether for the purpose of eradicating an infectious disease such as smallpox (Fenner, 1982) or providing updated annual protection such as with influenza (Fiore et al., 2009), rely on the rapid distribution of vaccine to as large a population as possible. Current genetic testing methods are not yet robust enough to implement in such a broad manner.

The economics of vaccine development and manufacture may not favor genomic approaches in the general population. Vaccines are administered in one or at most a few immunizations – once the primary sequence is given, most vaccines are infrequently or never given again (Committee on Infectious Diseases 2010; Advisory Committee on Immunization Practices 2010). Unlike medicines for hypertension, diabetes, or other chronic illnesses, vaccines are not highly profitable, and some vaccine manufacturers have abandoned the market (Baker and Katz, 2004). Development of a vaccine effective in only a subset of the population reduces potential market share, and given the attacks leveled at some well-established vaccines [e.g., measles/MMR (measles, mumps, rubella) (Murch et al., 2004; Wakefield et al., 1998), hepatitis B (Ascherio et al., 2001; Herroelen et al., 1991; Nadler, 1993, Zipp et al., 1999), *Haemophilus influenzae* type B (Destefano et al., 2001; Karvonen et al., 1999)], companies may be reluctant to pursue strategies with thin profit margins.

Perhaps more importantly, there are public health considerations that may not be initially apparent. Vaccines developed in industrialized countries are provided to poorer nations through direct purchase, via non-governmental organizations (NGOs), and through the efforts of the World Health Organization (WHO). In many cases they have dramatically reduced disease burden (Centers for Disease Control and Prevention CDC, 2006). In 1999, a rhesus-human reassortant rotavirus vaccine (RotaShield®) was withdrawn from the United States market after report of a small but statistically significant increased risk of intussusception following the first dose (Murphy et al., 2001). Following that withdrawal, there was active debate about the decision and whether the small risk was acceptable in the US or in any other country (Bines, 2006). Analysis suggested that despite the apparent increased risk, overall rates of intussusception (a condition where a segment of the intestine is pulled into the adjoining segment, resulting in partial or complete bowel obstruction) remained constant (Simonsen et al., 2001), and that providing the vaccine to the developing world could have saved hundreds of thousands of lives annually (Weijer, 2000). Ultimately, deployment of a vaccine deemed "not good enough" for the US market was unacceptable for the manufacturer, NGOs, WHO, and world governments (Glass et al., 2004). It is likely that a vaccine used in conjunction with genomic testing in the US or other industrialized markets would have to be deployed using the same testing globally, even if a clear population-level benefit for use of the vaccine without testing could be shown.

As noted above, genetic testing is already being used to detect possible confounders of HIV-1 vaccine candidates in NHPs, and at least one *post-hoc* analysis of a human trial has been performed (Forthal et al., 2007). Testing that is cost-effective and provided in "real time" to use at enrollment could be used to improve the quality of vaccine trials, although ethical and political considerations may make implementation difficult. Regardless, none of these confounders are absolute contraindications to the use of genomics in vaccine development or deployment. Rather, with the application of appropriate scientific and technological advances, each of these hurdles can be overcome and the profound benefits of genomic approaches realized.

CONCLUSIONS

Advances in genomics have great potential to improve vaccine development in the near term, and may lead to better and personalized vaccines in the future. At present, economic factors combined with biological complexity limit our ability to use genomics to individualize vaccines in a systematic way, but the rapid evolution of the field and changes in our ability to identify relevant genes may soon open this path. Regardless, the use of genomics, in both animal models and humans, to improve vaccine trials by quantifying confounders has the potential for real, immediate benefits.

ACKNOWLEDGMENTS

The author would like to thank the following individuals for helpful discussions in preparing this work: Barton F. Haynes and Garnett Kelsoe. Accuracies and insights are due to them; errors are entirely the fault of the author.

The author reports no conflicting financial interests related to this work.

REFERENCES

Advisory Committee on Immunization Practices, 2010. Recommended adult immunization schedule: United States. Ann Intern Med 152, 36–39.

Alcaïs, A., Abel, L., Casanova, J.L., 2009. Human genetics of infectious diseases: Between proof of principle and paradigm. J Clin Invest 119, 2506–2514.

Allen, C.D., Okada, T., Cyster, J.G., 2007. Germinal-center organization and cellular dynamics. Immunity 27, 190–202.

Amanna, I.J., Carlson, N.E., Slifka, M.K., 2007. Duration of humoral immunity to common viral and vaccine antigens. N Engl J Med 357, 1903–1915.

Apt, A.S., Avdienko, V.G., Nikonenko, B.V., Kramnik, I.B., Moroz, A.M., Skamene, E., 1993. Distinct H-2 complex control of mortality, and immune responses to tuberculosis infection in virgin and BCG-vaccinated mice. Clin Exp Immunol 94, 322–329.

Ascherio, A., Zhang, S.M., Hernan, M.A., et al., 2001. Hepatitis B vaccination and the risk of multiple sclerosis. N Engl J Med 344, 327–332.

Baize, S., Leroy, E.M., Georges-Courbot, M.C., et al., 1999. Defective humoral responses and extensive intravascular apoptosis are associated with fatal outcome in Ebola virus-infected patients. Nat Med 5, 423–426.

Baker, J.P., Katz, S.L., 2004. Childhood vaccine development: An overview. Pediatr Res 55, 347–356.

Barouch, D.H., O'Brien, K.L., Simmons, N.L., et al., 2010. Mosaic HIV-1 vaccines expand the breadth and depth of cellular immune responses in rhesus monkeys. Nat Med 16, 319–323.

Bartlett, J.A., Wasserman, S.S., Hicks, C.B., et al., 1998. Safety and immunogenicity of an HLA-based HIV envelope polyvalent synthetic peptide immunogen. DATRI 010 Study Group. Division of AIDS Treatment Research Initiative. AIDS 12, 1291–1300.

Belshe, R.B., Gruber, W.C., Mendelman, P.M., et al., 2000. Correlates of immune protection induced by live, attenuated, cold-adapted, trivalent, intranasal influenza virus vaccine. J Infect Dis 181, 1133–1137.

Berberian, L., Goodglick, L., Kipps, T.J., Braun, J., 1993. Immunoglobulin V_H3 gene products: Matural ligands for HIV gp120. Science 261, 1588–1591.

Bines, J., 2006. Intussusception and rotavirus vaccines. Vaccine 24, 3772–3776.

Bittner, J., 1935. A review of genetic studies on the transplantation of tumours. J Genet 31, 471–487.

Brandler, P., Saikh, K.U., Heath, D., Friedlander, A., Ulrich, R.G., 1998. Weak anamnestic responses of inbred mice to *Yersinia* F1 genetic vaccine are overcome by boosting with F1 polypeptide while outbred mice remain nonresponsive. J Immunol 161, 4195–4200.

Briles, D.E., Krause, R.M., Davie, J.M., 1977. Immune response deficiency of BSVS mice. Immunogenetics 4, 381–392.

Buchbinder, S.P., Mehrotra, D.V., Duerr, A., et al., 2008. Efficacy assessment of a cell-mediated immunity HIV-1 vaccine (the Step Study): A double-blind, randomised, placebo-controlled, test-of-concept trial. Lancet 372, 1881–1893.

Cacalano, N.A., Migone, T.S., Bazan, F., et al., 1999. Autosomal SCID caused by a point mutation in the N-terminus of Jak3: Mapping of the Jak3-receptor interaction domain. EMBO J 18, 1549–1558.

Centers for Disease Control and Prevention CDC, 2006. Vaccine preventable deaths and the Global Immunization Vision and

Strategy, 2006–2015. MMWR Morb Mortal Wkly Rep 55, 511–515.

Chen, Z., Yoshikawa, T., Kadowaki, S., et al., 1999. Protection and antibody responses in different strains of mouse immunized with plasmid DNAs encoding influenza virus haemagglutinin, neuraminidase and nucleoprotein. J Gen Virol 80, 2559–2564.

Cohen, M.K., Bartow, R.A., Mintzer, C.L., Mcmurray, D.N., 1987. Effects of diet and genetics on Mycobacterium bovis BCG vaccine efficacy in inbred guinea pigs. Infect Immun 55, 314–319.

Committee on Infectious Diseases, 2010. Recommended childhood and adolescent immunization schedules–United States. Pediatrics 125, 195–196.

Corti, D., Suguitan Jr., A.L., Pinna, D., et al., 2010. Heterosubtypic neutralizing antibodies are produced by individuals immunized with a seasonal influenza vaccine. J Clin Invest 120, 1663–1673.

Dale, H., 1954. Paul Ehrlich, born March 14, 1854. Br Med J 1, 659–663.

Davila, S., Froeling, F.E., Tan, A., et al., 2010. New genetic associations detected in a host response study to hepatitis B vaccine. Genes and immunity 11, 232–238.

Destefano, F., Mullooly, J.P., Okoro, C.A., et al., 2001. Childhood vaccinations, vaccination timing, and risk of type 1 diabetes mellitus. Pediatrics 108, E112.

Ditzel, H.J., Itoh, K., Burton, D.R., 1996. Determinants of polyreactivity in a large panel of recombinant human antibodies from HIV-1 infection. J Immunol 157, 739–749.

Edsall, G., 1959. Specific prophylaxis of tetanus. J Am Med Assoc 171, 417–427.

Faden, H., Modlin, J.F., Thoms, M.L., McBean, A.M., Ferdon, M.B., Ogra, P.L., 1990. Comparative evaluation of immunization with live attenuated and enhanced-potency inactivated trivalent poliovirus vaccines in childhood: Systemic and local immune responses. J Infect Dis 162, 1291–1297.

Fellay, J., Frahm, N., Shianna, K.V., et al., 2011. Host genetic determinants of T cell responses to the MRKAd5 HIV-1 gag/pol/nef vaccine in the step trial. J Infect Dis 203, 773–779.

Fenner, F., 1982. A successful eradication campaign. Global eradication of smallpox. Rev Infect Dis 4, 916–930.

Fiore, A.E., Shay, D.K., Broder, K., et al., 2009. Prevention and control of seasonal influenza with vaccines: Recommendations of the Advisory Committee on Immunization Practices (ACIP), 2009. MMWR Recomm Rep 58, 1–52.

Fletcher, H.A., 2007. Correlates of immune protection from tuberculosis. Curr Mol Med 7, 319–325.

Forthal, D.N., Landucci, G., Bream, J., Jacobson, L.P., Phan, T.B., Montoya, B., 2007. FcγRIIa genotype predicts progression of HIV infection. J Immunol 179, 7916–7923.

Galazka, A.M., Robertson, S.E., 1996. Immunization against diphtheria with special emphasis on immunization of adults. Vaccine 14, 845–857.

Gelder, C.M., Lambkin, R., Hart, K.W., et al., 2002. Associations between human leukocyte antigens and nonresponsiveness to influenza vaccine. J Infect Dis 185, 114–117.

Glass, R.I., Bresee, J.S., Parashar, U.D., Jiang, B., Gentsch, J., 2004. The future of rotavirus vaccines: A major setback leads to new opportunities. Lancet 363, 1547–1550.

Gorer, P., 1936a. The detection of a hereditary antigenic difference in the blood of mice by means of human group A serum. J Genet 32, 17–31.

Gorer, P., 1936b. The detection of antigenic differences in mouse erythrocytes by the employment of immune sera. Br J Exp Pathol 17, 42–50.

Halstead, S.B., Nimmannitya, S., Cohen, S.N., 1970. Observations related to pathogenesis of dengue hemorrhagic fever. IV. Relation of disease severity to antibody response and virus recovered. Yale J Biol Med 42, 311–328.

Halstead, S.B., O'Rourke, E.J., 1977. Dengue viruses and mononuclear phagocytes. I. Infection enhancement by non-neutralizing antibody. J Exp Med 146, 201–217.

Herroelen, L., De Keyser, J., Ebinger, G., 1991. Central-nervous-system demyelination after immunisation with recombinant hepatitis B vaccine. Lancet 338, 1174–1175.

Hohler, T., Reuss, E., Freitag, C.M., Schneider, P.M., 2005. A functional polymorphism in the IL-10 promoter influences the response after vaccination with HBsAg and hepatitis A. Hepatology 42, 72–76.

Hu, J., Schell, T.D., Peng, X., Cladel, N.M., Balogh, K.K., Christensen, N.D., 2010. Using HLA-A2.1 transgenic rabbit model to screen and characterize new HLA-A2.1 restricted epitope DNA vaccines. J Vaccin Vaccinat 1, 1–10.

Jacobson, R.M., Ovsyannikova, I.G., Poland, G.A., 2009. Genetic basis for variation of vaccine response: Our studies with rubella vaccine. Paediatr Child Health 19, S156–S159.

Jacobson, R.M., Poland, G.A., Vierkant, R.A., et al., 2003. The association of class I HLA alleles and antibody levels after a single dose of measles vaccine. Hum Immunol 64, 103–109.

Janeway, C., 2005. Immunobiology: The immune system in health and disease, 6th ed. Garland Science, New York (Chapter 10).

Karvonen, M., Cepaitis, Z., Tuomilehto, J., 1999. Association between type 1 diabetes and Haemophilus influenzae type b vaccination: Birth cohort study. BMJ 318, 1169–1172.

Katz, S.L., 2006. Polio – new challenges in 2006. J Clin Virol 36, 163–165.

Katz, S.L., Hinman, A.R., 2004. Summary and conclusions: Measles elimination meeting, 16–17 March 2000. J Infect Dis 189 (Suppl 1), S43–S47.

Kimman, T.G., Banus, S., Reijmerink, N., et al., 2008. Association of interacting genes in the toll-like receptor signaling pathway and the antibody response to pertussis vaccination. PLoS One 3, e3665.

Kimman, T.G., Vandebriel, R.J., Hoebee, B., 2007. Genetic variation in the response to vaccination. Community Genet 10, 201–217.

Klein, J., 1986. Seeds of time: Fifty years ago Peter A. Gorer discovered the H-2 complex. Immunogenetics 24, 331–338.

Kong, W.P., Wu, L., Wallstrom, T.C., et al., 2009. Expanded breadth of the T-cell response to mosaic human immunodeficiency virus type 1 envelope DNA vaccination. J Virol 83, 2201–2215.

Kool, M., Petrilli, V., De Smedt, T., et al., 2008. Cutting edge: Alum adjuvant stimulates inflammatory dendritic cells through activation of the NALP3 inflammasome. J Immunol 181, 3755–3759.

Kruskall, M.S., Alper, C.A., Awdeh, Z., Yunis, E.J., Marcus-Bagley, D., 1992. The immune response to hepatitis B vaccine in humans: Inheritance patterns in families. J Exp Med 175, 495–502.

Lacasse, R.A., Follis, K.E., Trahey, M., Scarborough, J.D., Littman, D.R., Nunberg, J.H., 1999. Fusion-competent vaccines: Broad neutralization of primary isolates of HIV. Science 283, 357–362.

Li, H., Willingham, S.B., Ting, J.P., Re, F., 2008. Cutting edge: Inflammasome activation by alum and alum's adjuvant effect are mediated by NLRP3. J Immunol 181, 17–21.

Liang, X., Casimiro, D.R., Schleif, W.A., et al., 2005. Vectored Gag and Env but not Tat show efficacy against simian-human immunodeficiency virus 89.6P challenge in Mamu-A*01-negative rhesus monkeys. J Virol 79, 12,321–12,331.

Lim, S.Y., Rogers, T., Chan, T., et al., 2010. TRIM5α modulates immunodeficiency virus control in rhesus monkeys. PLoS Pathog 6, e1000738.

Mack, T.M., Noble Jr., J., Thomas, D.B., 1972. A prospective study of serum antibody and protection against smallpox. Am J Trop Med Hyg 21, 214–218.

Macnalty, A.S., 1954. Emil von Behring, born March 15, 1854. Br Med J 1, 668–670.

McCool, T.L., Schreiber, J.R., Greenspan, N.S., 2003. Genetic variation influences the B-cell response to immunization with a pneumococcal polysaccharide conjugate vaccine. Infect Immun 71, 5402–5406.

Murch, S.H., Anthony, A., Casson, D.H., et al., 2004. Retraction of an interpretation. Lancet 363, 750.

Murphy, T.V., Gargiullo, P.M., Massoudi, M.S., et al., 2001. Intussusception among infants given an oral rotavirus vaccine. N Engl J Med 344, 564–572.

Nadler, J.P., 1993. Multiple sclerosis and hepatitis B vaccination. Clin Infect Dis 17, 928–929.

Newport, M.J., Goetghebuer, T., Weiss, H.A., Whittle, H., Siegrist, C.A., Marchant, A., 2004. Genetic regulation of immune responses to vaccines in early life. Genes Immun 5, 122–129.

Nunberg, J.H., 2002. Retraction. Science 296, 1025.

Ovsyannikova, I.G., Dhiman, N., Haralambieva, I.H., et al., 2010a. Rubella vaccine-induced cellular immunity: Evidence of associations with polymorphisms in the Toll-like, vitamin A and D receptors, and innate immune response genes. Hum Genet 127, 207–221.

Ovsyannikova, I.G., Haralambieva, I.H., Dhiman, N., et al., 2010b. Polymorphisms in the vitamin A receptor and innate immunity genes influence the antibody response to rubella vaccination. J Infect Dis 201, 207–213.

Ovsyannikova, I.G., Jacobson, R.M., Dhiman, N., Vierkant, R.A., Pankratz, V.S., Poland, G.A., 2008. Human leukocyte antigen and cytokine receptor gene polymorphisms associated with heterogeneous immune responses to mumps viral vaccine. Pediatrics 121, e1091–e1099.

Oxman, M.N., Levin, M.J., Johnson, G.R., et al., 2005. A vaccine to prevent herpes zoster and postherpetic neuralgia in older adults. N Engl J Med 352, 2271–2284.

Plotkin, S.A., 2001. Immunologic correlates of protection induced by vaccination. Pediatr Infect Dis J 20, 63–75.

Plotkin, S.A., 2008. Vaccines: Correlates of vaccine-induced immunity. Clin Infect Dis 47, 401–409.

Poland, G.A., Ovsyannikova, I.G., Jacobson, R.M., 2008. Immunogenetics of seasonal influenza vaccine response. Vaccine 26 (Suppl 4), D35–D40.

Radošević, K., Wieland, C.W., Rodriguez, A., et al., 2007. Protective immune responses to a recombinant adenovirus type 35 tuberculosis vaccine in two mouse strains: CD4 and CD8 T-cell epitope mapping and role of gamma interferon. Infect Immun 75, 4105–4115.

Reed, W.P., 1975. Serum factors capable of opsonizing Shigella for phagocytosis by polymorphonuclear neutrophils. Immunology 28, 1051–1059.

Rerks-Ngarm, S., Pitisuttithum, P., Nitayaphan, S., et al., 2009. Vaccination with ALVAC and AIDSVAX to prevent HIV-1 infection in Thailand. N Engl J Med 361, 2209–2220.

Sabin, A.B., 1952. Research on dengue during World War II. Am J Trop Med Hyg 1, 30–50.

Santra, S., Liao, H.X., Zhang, R., et al., 2010. Mosaic vaccines elicit CD8+ T lymphocyte responses that confer enhanced immune coverage of diverse HIV strains in monkeys. Nat Med 16, 324–328.

Sarkar, J.K., Mitra, A.C., Mukherjee, M.K., 1975. The minimum protective level of antibodies in smallpox. Bull World Health Organ 52, 307–311.

Sasso, E.H., Johnson, T., Kipps, T.J., 1996. Expression of the immunoglobulin VH gene 51p1 is proportional to its germline gene copy number. J Clin Invest 97, 2074–2080.

Sekaly, R.-P., 2008. The failed HIV Merck vaccine study: A step back or a launching point for future vaccine development? J Exp Med 205, 7–12.

Simonsen, L., Morens, D., Elixhauser, A., Gerber, M., Van Raden, M., Blackwelder, W., 2001. Effect of rotavirus vaccination programme on trends in admission of infants to hospital for intussusception. Lancet 358, 1224–1229.

Stankus, R.P., Leslie, G.A., 1975. Genetic influences on the immune response of rats to streptococcal A carbohydrate. Immunogenetics 2, 29–38.

Subbramanian, R.A., Charini, W.A., Kuroda, M.J., et al., 2006. Expansion after epitope peptide exposure in vitro predicts cytotoxic T lymphocyte epitope dominance hierarchy in lymphocytes of vaccinated mamu-a*01+ rhesus monkeys. AIDS Res Hum Retroviruses 22, 445–452.

Sui, J., Hwang, W.C., Perez, S., et al., 2009. Structural and functional bases for broad-spectrum neutralization of avian and human influenza A viruses. Nat Struct Mol Biol 16, 265–273.

Sullivan, N., Yang, Z.Y., Nabel, G.J., 2003. Ebola virus pathogenesis: Implications for vaccines and therapies. J Virol 77, 9733–9737.

Templeton, J.W., Holmberg, C., Garber, T., Sharp, R.M., 1986. Genetic control of serum neutralizing-antibody response to rabies vaccination and survival after a rabies challenge infection in mice. J Virol 59, 98–102.

Tian, J.H., Miller, L.H., Kaslow, D.C., et al., 1996. Genetic regulation of protective immune response in congenic strains of mice vaccinated with a subunit malaria vaccine. J Immunol 157, 1176–1183.

Tukhvatulin, A., Logunov, D., Shcherbinin, D., et al., 2010. Toll-like receptors and their adapter molecules. Biochemistry (Moscow) 75, 1098–1114.

Tyler, D.S., Lyerly, H.K., Weinhold, K.J., 1989. Anti-HIV-1 ADCC. AIDS Res Hum Retroviruses 5, 557–563.

Van Strijp, J.A., Van Kessel, K.P., Van Der Tol, M.E., Verhoef, J., 1989. Complement-mediated phagocytosis of herpes simplex virus by granulocytes. Binding or ingestion. J Clin Invest 84, 107–112.

Wakefield, A.J., Murch, S.H., Anthony, A., et al., 1998. Ileal-lymphoid-nodular hyperplasia, non-specific colitis, and pervasive developmental disorder in children. Lancet 351, 637–641.

Weijer, C., 2000. The future of research into rotavirus vaccine. BMJ 321, 525–526.

Zaas, A.K., Chen, M., Varkey, J., et al., 2009. Gene expression signatures diagnose influenza and other symptomatic respiratory viral infections in humans. Cell Host Microbe 6, 207–217.

Zhang, Z.Q., Fu, T.M., Casimiro, D.R., et al., 2002. *Mamu-A*01* allele-mediated attenuation of disease progression in simian-human immunodeficiency virus infection. J Virol 76, 12,845–12,854.

Zipp, F., Weil, J.G., Einhaupl, K.M., 1999. No increase in demyelinating diseases after hepatitis B vaccination. Nat Med 5, 964–965.

Zuckerman, J.N., Sabin, C., Craig, F.M., Williams, A., Zuckerman, A.J., 1997. Immune response to a new hepatitis B vaccine in healthcare workers who had not responded to standard vaccine: Randomised double blind dose-response study. BMJ 314, 329–333.

CHAPTER

96

Bacterial Infections

Yurong Zhang, Sun Hee Ahn, and Vance G. Fowler, Jr

INTRODUCTION

Microarray analysis of genome-wide transcriptional profiling is increasingly employed in the study of host–pathogen interactions in bacteremia. To date, these techniques have had three primary applications in the area of bacterial pathogenesis: class comparison (contrasting expression profiles of various classes of specimen), prognostic prediction (using expression profiles combined with other factors to predict clinical outcomes), and class discovery (in which important subtypes of specimens are distinguished by class (Simon et al., 2002). Initially, microarray technologies were used to identify the unknown function or regulation of target genes in microorganisms (pathogens). Extending this to a host-based approach, a number of studies have been conducted to identify new target genes or function of genes both *in vitro* and *in vivo*. In this chapter, we describe *in vitro* and *in vivo* microarray studies targeting host responses to bacterial pathogens at the whole-genome transcriptome level.

IN VITRO STUDIES: HOST CELL RESPONSES

To date, there have been numerous expression profiling analyses of the host cell response during pathogen infection. Target cell lines or tissues have been chosen on the basis of study purpose, such as host infection, site of pathogen, or disease-developing area. Most of these studies have been conducted in *in vitro* infection models, and provided the first insights into the complexity of acute host–bacterial interactions (see Table 96.1).

Macrophages

Macrophages have key roles in the host defense to infection, as they mediate microbe phagocytosis and initiate, maintain, and resolve host inflammatory responses by releasing cytokines and chemokines. An innate pattern of macrophage response is triggered by various bacterial pathogen-associated molecular patterns (PAMPs), such as lipopolysaccharide (LPS), lipoteichoic acid (LTA), fimbriae, lipoproteins, glycoproteins, and peptidoglycan (Henderson et al., 1996), through Toll-like receptors (TLR) on the cell surface (Aderem and Ulevitch, 2000). A number of *in vitro* studies using macrophages have investigated the host response to a variety of pathogens, including bacteria (Detweiler et al., 2001; Ehrt et al., 2001; Nau et al., 2002, 2003; Ragno et al., 2001; Rodriguez et al., 2004; Rosenberger et al., 2000; Sauvonnet et al., 2002), fungi (Lorenz et al., 2004), parasites (Andersson et al., 2006), and viruses (Moreno-Altamirano et al., 2004).

A pivotal study of host cell–pathogen interactions demonstrated that macrophage responses to a broad range of bacteria were robust and shared a pattern of gene expression, with common genes encoding for receptors, signal transduction molecules, and transcription factors (Nau et al., 2002). This shared pattern is associated with macrophage activation to mount an immune response. While macrophages may respond to all bacteria in a standard fashion, such as the TLR signaling pathway (Akira et al., 2001), the diversity of bacteria

Genomic and Personalized Medicine, 2nd edition
by Ginsburg & Willard. DOI: http://dx.doi.org/10.1016/B978-0-12-382227-7.00096-3

TABLE 96.1	*In vitro* microarray analysis of host responses to pathogen	
Cell type	**Stimulus**	**Reference**
Macrophage	*Escherichia coli*, EHEC, *Salmonella enterica* serovars Typhi and Typhimurium, *Staphylococcus aureus*, *Listeria monocytogenes*, *Mycobacterium tuberculosis*, *M. bovis* BCG, LPS, LTA, MDP, TB Hsp70, BCG Hsp65, MPA, fMLP, protein A, mannose	Nau et al., 2002
Macrophage	*Leishmania chagasi*	Rodriguez et al., 2004
Macrophage	*Salmonella* Typhimurium, *Salmonella* Typhimurium *phoP* mutant	Detweiler et al., 2001; Rosenberger et al., 2000
Macrophage	*Francisella tularensis*	Andersson et al., 2006
Macrophage	*Yersinia enterocolitica*	Sauvonnet et al., 2002
Macrophage	*Mycobacterium tuberculosis*	Ehrt et al., 2001; Ragno et al., 2001
Macrophage	IFNα, IFNβ, IFNγ, IL-10, IL-12	Nau et al., 2003
Macrophage	*Mycobacterium tuberculosis*, *Escherichia coli*, *Salmonella* Typhi	Nau et al., 2002
Macrophage	*Candida albicans*	Lorenz et al., 2004
Macrophage	Dengue virus serotype 2	Moreno-Altamirano et al., 2004
Epithelial cell	*Bordetella pertussis*	Belcher et al., 2000
Epithelial cell	*Helicobacter pylori*	Guillemin et al., 2002
Epithelial cell	*Neisseria gonorrhoeae*	Binnicker et al., 2003
Epithelial cell	Respiratory syncytial virus	Tian et al., 2002; Zhang et al., 2001, 2003
Epithelial cell	Rhesus rotavirus	Cuadras et al., 2002
Epithelial cell	Human papilloma virus	Wells et al., 2003
PBMC	LPS, *Bordetella pertussis*, *Escherichia coli*, *Staphylococcus aureus*	Boldrick et al., 2002
PBMC	Human T-lymphotropic virus 1	Pise-Masison et al., 2002
Endothelial cell	Dengue virus	Warke et al., 2003
Endothelial cell	Kaposi's sarcoma-associated herpesvirus	Hong et al., 2004; Wang et al., 2004
Human endothelial cell	*Staphylococcus aureus*	Matussek et al., 2005
Dendritic cell	*Escherichia coli*, influenza, *Candida albicans*, LPS, poly I:C, mannan	Huang et al., 2001
Dendritic cell	Human Immunodeficiency Virus (HIV-1), transcriptional activator Tat	Izmailova et al., 2003
T cell	HIV-1	Corbeil et al., 2001; Van't Wout et al., 2003
T cell/fibroblast/skin	Varicella–zoster virus	Jones and Arvin, 2003
Astrocyte (human)	John Cunningham virus	Radhakrishnan et al., 2003
B cell	Epstein–Barr virus, Kaposi's sarcoma-associated herpesvirus	Jenner et al., 2003
Fibroblast	Human cytomegalovirus	Simmen et al., 2001
HeLa	*Chlamydophila pneumoniae*, *C. trachomatis*, *Salmonella* Typhimurium	Hess et al., 2003

(continued)

Cell type	**Stimulus**	**Reference**
Human monocyte	*Candida albicans*	Kim et al., 2005
Keratinocyte	Human papilloma virus-21	Chang and Laimins, 2000
Leukocyte	*Staphylococcus aureus*, LPS	Feezor et al., 2003

TABLE 96.1 (Continued)

BCG, bacille Calmette–Guérin; EHEC, enterohemorrhagic *E. coli*; fMLP, formyl-methionine-leucine-phenylalanine; Hsp65, heat-shock protein 65; IFN, interferon; IL, interleukin; LPS, lipopolysaccharide; LTA, lipoteichoic acid; MDP, muramyl dipeptide; MPA, monophosphoryl lipid A; PBMC, peripheral blood mononuclear cell; poly I:C, polyinosinic-polycytidylic acid; TB Hsp70, *M. tuberculosis* heat-shock protein 70.

and the differences in their pathogenesis may also lead to pathogen-specific responses (Nau et al., 2002). For example, the fitness of *Mycobacterium tuberculosis* within macrophages (Pieters, 2001) may be due to *M. tuberculosis*-specific host responses such as inhibited interleukin-12 production, suggesting one means by which this organism survives host defenses (Nau et al., 2002). As another example of pathogen-specific responses, gene expression changes in macrophages exposed to gram-negative bacteria encompass both those induced by gram-positive bacteria as well as a distinct TLR4 response. This distinct TLR4 response could provide the basis to diagnose clinical gram-negative infections.

Moreover, gene expression profiling during *in vitro* macrophage infection has been used to evaluate specific bacterial virulence factors. For example, Detweiler studied *phoP*, a transcription factor in *Salmonella enterica* serovar Typhimurium required for virulence, by comparing the expression profiles of human monocytic tissue culture cells infected with either the wild-type bacteria or a phoPTn10 mutant strain (Detweiler et al., 2001). Sauvonnet and colleagues investigated the role of YopP, YopM, and the other pYV-encoded factors on host cell gene expression, characterizing the transcriptome alterations in mouse macrophages infected with either *Yersinia enterocolitica* wild strain or mutant strain (Sauvonnet et al., 2002). Finally, Rosenberger and colleagues studied the simultaneous expression of hundreds of genes during *Salmonella* Typhimurium infection of mouse macrophages, and assessed the contribution of bacterial virulence factor LPS in initiating host responses to *Salmonella* (Rosenberger et al., 2000). The identification and evaluation of discrete virulence factors, such as gram-negative LPS or strain-specific encoded factors, could be the basis for more targeted pharmacological treatments in the future.

Epithelial and Endothelial Cells

The gene expression profiles of epithelial and endothelial cells have also been studied to gain insight into the specifics of host–pathogen interaction or the host response to pathogens. For example, microarray technology was used to examine the interaction of *Bordetella pertussis* with human bronchial epithelial cells (Belcher and Drenkow, 2000), and the interaction of *Neisseria gonorrhoeae* with human urethral epithelium (Binnicker et al., 2003). Hess and colleagues

conducted an expression comparison of gene regulation in human epithelial cells during *Chlamydophila pneumoniae*, *Chlamydia trachomatis*, and *Salmonella* Typhimurium infections. The transcriptional responses induced by *C. pneumoniae* and *C. trachomatis* were similar, but the expression pattern induced by *Salmonella* differed substantially. These genus- or group-specific transcriptional response patterns elicited by viable intracellular pathogens may contribute to the different clinical manifestations associated with these different pathogens (Hess et al., 2003).

Endothelial cells (EC) play an important role in host defense against bacteria. For example, *Staphylococcus aureus* infection of human EC induces expression of cytokines and cell surface receptors involved in activating the innate immune response (Matussek et al., 2005). Matussek and colleagues found that many of the up-regulated genes code for proteins involved in innate immunity, such as cytokines, chemokines, and cell-adhesion proteins. Other up-regulated genes encode proteins involved in antigen presentation, cell signaling, and metabolism. Furthermore, intracellular bacteria that survived for days without inducing EC death were permissive to *in vitro* dengue virus infection (Matussek et al., 2005), demonstrating the viability of this bacterium.

Other Cell Types

Other cell types involved in the immune response, such as dendritic cells, leukocytes, monocytes, peripheral blood mononuclear cells (PBMCs), and T and B cells, have also been used to investigate host response to bacteria *in vitro* (Table 96.1). Huang and colleagues used the gene expression profiles to systematically explore how dendritic cells modulate the immune system in response to *Escherichia coli*, *Candida albicans*, and influenza virus, as well as to their associated PAMPs (Huang et al., 2001).

Peripheral blood as a surrogate tissue has also been used in microarray studies of human diseases, including Huntington's disease (Borovecki et al., 2005), cancer (Martin et al., 2001; Osman et al., 2006), and bacterial sepsis (Ardura et al., 2009; Ramilo et al., 2007; Wong et al., 2010). Boldrick and colleagues evaluated human PBMC response to bacteria, and found a remarkably consistent gene expression pattern induced by *in vitro* stimulation of bacterial LPS and diverse killed bacteria (Boldrick et al., 2002). Similarly, Feezor and colleagues investigated differential gene regulation in leukocytes stimulated *in vitro* with *Escherichia coli* LPS or heat-killed *Staphylococcus aureus* (Feezor et al., 2003). These distinct host gene response patterns elicited by gram-positive versus gram-negative bacterial infection raise the exciting possibility that gene expression signatures could serve as a basis for novel diagnostic and prognostic approaches to life-threatening bacterial sepsis.

IN VIVO STUDIES: GENOMIC TECHNOLOGIES

In bacterial microarray studies, the capability to monitor dynamic disease progression is of fundamental interest and relevance. Improving that capability could result in faster and more accurate diagnostic tools of particular use in the clinical setting where, for instance, the rapid onset of bacteria-induced septic shock remains a major cause of mortality in the United States (Vincent and Abraham, 2006). Currently, pathogen-based diagnostic approaches such as blood culture, antigen response, or polymerase chain reaction (PCR) are time-consuming (an expensive commodity in the clinical setting) and also lack sensitivity (Jonathan, 2006; Landry et al., 2008; Rahman et al., 2008). Host response-based diagnostic tools would enhance sensitivity because they could monitor the dynamic host response (as compared to static pathogen activity) and thus lead to treatment that would be more specific to momentary infection progress (Garey et al., 2006). Building upon *in vitro* studies of bacterial infections, *in vivo* studies have been conducted in a range of genomic technologies to better understand bacterial pathogenesis and to identify potential diagnostic and prognostic applications of these technologies.

Transcriptional Profiling

Transcriptional profiling has already changed diagnostic and prognostic approaches in cancer (Alizadeh et al., 2000; Golub et al., 1999; Van de Vijver et al., 2002) and the analysis of expression signatures in bacterial infections may lead to a better understanding of disease pathogenesis and clinical treatments. Peripheral blood mononuclear cell (PBMC) analysis is an accessible source of clinically relevant information, and these cells are the target of expression analysis for many host-based microarray studies. Blood is both a pool and a dynamic migration compartment of leukocytes, containing components of both the innate immune system and the adaptive immune system (such as neutrophils, CD4$^+$ and CD8$^+$ T lymphocytes, B lymphocytes, natural killer cells, and dendritic cells), each of which expresses a unique combination of genes. Using gene expression microarrays, a comprehensive molecular phenotype of immune response cells can be obtained in order to gain insight into the host response to bacterial infection.

Gram-negative Bacteria

Gram-negative bacteria contain LPS molecules in their cell walls that are recognized by host cell-membrane receptors

TLR4, LBP, and CD14. Some species of gram-negative bacteria also contain cytoplasmic membrane proteins that are recognized by TLR1 and TLR2. Upon host cell recognition, gram-negative bacteria trigger an inflammatory response involving leukocytes as well as cytokines such as tumor necrosis factor alpha (TNF-α), interleukin-1 (IL-1), and IL-6.

Transcriptional profiling of *in vivo* infections demonstrates potential diagnostic and/or prognostic roles for microarray studies in bacterial infections. For example, murine-based *in vivo* studies have used microarray analysis to characterize the stages of *Helicobacter pylori* infection and track disease progression (Mueller et al., 2003), as well as to elucidate infection pathogenesis and etiology (Mills et al., 2001; Syder et al., 2003). In humans with bacterial sepsis, Ramilo and colleagues demonstrated the utility of *in vivo* blood leukocyte gene expression signatures in distinguishing acute infections of different etiologies – viral (influenza A), gram-negative bacterial (*E. coli*), and gram-positive bacterial (*S. aureus* and *Streptococcus pneumoniae*) (see Figure 96.1) (Ramilo et al., 2007). Similarly, a recent study by Zaas and colleagues demonstrated the use of host peripheral blood transcriptional profiling as an early classifier for infection etiology (fungal versus bacterial) (Zaas et al., 2010). Using a murine model, they identified host blood gene expression signatures which were specific to candidemia versus healthy controls, and discriminatory for candidemia versus *S. aureus* bacteremia (see Figure 96.2) This signature was also determinant of illness progression related to candidemia (Zaas et al., 2010). Improving the ability to monitor disease progression and efficiently determine infection etiologies is critical to improved clinical care.

Transcriptional profiling data from animal versus human models of the same disease can provide insight into disease pathogenesis specific to animal versus human physiology. In a murine model of melioidosis, a potentially life-threatening form of sepsis caused by the gram-negative bacillus *Burkholderia pseudomallei*, genome-wide transcription profiling of the liver and spleen found increased levels of pro-inflammatory response genes, as well as an increase in transcripts associated with apoptosis, caspase activation, and peptidoglysis (Chin et al., 2010). In patients with melioidosis, an elevated major histocompatibility complex (MHC) class II expression distinguished patients with septicemic melioidosis from patients with other pathogenic sepsis – gram-positive, gram-negative, and fungal (candidemia) (Pankla et al., 2009). The most strongly identified genes in this set include those involved in antigen processing, proteasome complex in the ubiquitin-proteosome system, proteolysis, the inflammatory immune response, apoptosis, and cell death (Pankla et al., 2009). In septic patients compared to healthy controls, a majority of changes observed were common to both septicemic melioidosis and sepsis caused by other infections, including genes related to inflammation, interferon-related genes, and genes expressed by neutrophils, cytotoxic cells, and T cells (Pankla et al., 2009).

Gene expression analysis has also been employed to examine dynamic host response over an extended period

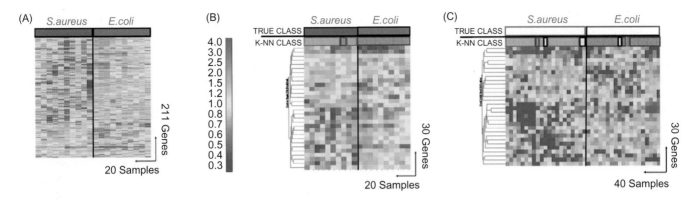

Figure 96.1 Discriminating patients with *Staphylococcus aureus* infections from patients with *E. coli* infections. (A) Hierarchical clustering of 211 genes obtained from Mann–Whitney rank test comparison (*P*<0.01) between two groups: *S. aureus* (10 samples, red rectangle) and *E. coli* (10 samples, blue rectangle) infections. Transformed expression levels are indicated by color scale, with red representing relative high expression and blue indicating relative low expression compared to the median expression for each gene across all donors. (B) A supervised learning algorithm was used to identify 30 genes presenting the highest capacity to discriminate the two classes. Leave-one-out cross-validation of the training set with 30 classifier genes grouped the samples with 95% accuracy. (C) The 30 classifier genes thus identified were tested on an independent set of patients (open rectangles), including 21 new patients with *S. aureus* and 19 with *E. coli* infections. The 40 samples in this test set were predicted with 85% accuracy (predicted class is indicated by light-colored rectangles). *This research was originally published from Ramilio et al., 2007.*

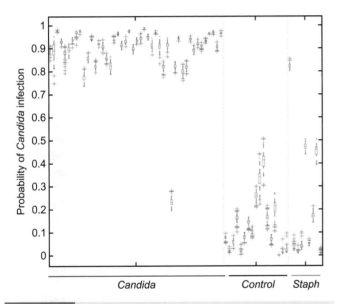

Figure 96.2 Murine blood gene expression signatures distinguish candidemia from *Staphylococcus aureus* bacteremia and from uninfected controls. Each mouse is represented by a red cross. This analysis results in a range of possible predictions for each sample (shown as a blue box and whiskers) depending on which of the remaining samples were used to build the predictive model. Probability of a sample representing a mouse with candidemia is shown on the y-axis. *Reproduced with permission from Zaas et al., 2010.*

of time. A study of 29 individuals with *Salmonella enterica*-induced typhoid fever found that most infected individuals continued to exhibit distinctive peripheral blood transcription profiles, persisting up to nine months post-infection, suggesting that bacterial impact on host immune response may be more profound and prolonged than previously appreciated (Thompson et al., 2009). Such *Salmonella*-induced suppression of T cell and lymphocyte responses had been previously identified in a murine model (Eisenstein et al., 1984; Schwacha et al., 1998; Van der Velden et al., 2005). This recent, extended examination of disease progression proposes that a similar immune repression in humans may provide insight as to why some patients are more susceptible to re-infection, relapse, or becoming persistent carriers for latent infections (Thompson et al., 2009).

Gram-positive Bacteria

Unlike gram-negative bacteria, gram-positive bacteria lack LPS. Instead, LTA and peptidoglycans are major components of the cell wall. The host response to gram-positive bacteria involves TLR2 recognition, as well as a subsequent TLR-independent response with relatively low levels of TNF-α, IL-1, and IL-6, and increased levels of IL-8 (Van Amersfoort et al., 2003). As with transcriptional profiling of gram-negative bacteria, *in vivo* studies of gram-positive bacteria also center on identifying distinct gene expression signatures that may be of diagnostic value.

Although PBMCs are a frequent target of host immunological response studies, their mixed composition of leukocytes (neutrophils, monocytes, macrophages, etc.) limits the ability to identify the contribution of a specific cell subpopulation toward the expression value of an individual gene (Tang et al., 2009). This limitation can largely be overcome with cell-fractioning flow cytometry to separate leukocyte subtypes, followed by gene expression array of the cell subpopulation of interest (Ardura et al., 2009; Wong et al., 2010). Subtype separation may not always be necessary for a diagnostic platform; a recent comparison of the transcript profiles of whole-blood versus fractionated leukocyte subpopulations in septic patients found that the two analyses were congruent in their expression of metabolic and signaling gene pathways (Wong et al., 2010). Network analysis of leukocyte subpopulation profiles showed gene networks involved in antigen presentation, humoral-mediated immunity, and cell-mediated immunity (Wong et al., 2010). In the clinical setting, a study of pediatric *Staphylococcus aureus* patients identified a significant over-expression of innate immunity genes and under-expression of adaptive immunity genes (Ardura et al., 2009). Subsequent flow cytometry analysis into the relative abundance of the immune cell populations found decreased central memory CD4$^+$ and CD8$^+$ T cells and increased numbers of monocytes, the latter of which correlated with higher expression levels of genes related to innate immune response. Furthermore, subpopulation analysis of T cells in *S. aureus*-infected patients found decreased numbers of both central memory CD4+ T cells and CD8+ T cells, but no changes in other T cell subsets (Ardura et al., 2009).

To date, few genome-wide analyses of microbial infection have addressed the interaction between the pathogen or host transcriptomes through the entire course of infection (Shea et al., 2010; Thompson et al., 2009). A recent analysis of group A *Streptococcus* (GAP) infection in cynomolgus macaques monitored gene expression levels in host epithelial cells over a 32-day period. Based on paired host and pathogen transcriptome analysis, Shea and colleagues identified interactions between host and pathogen genes and pathways based on the correlation of gene expression profiles (Shea et al., 2010). Host genes experiencing the first expression reduction in the 32 days included those involved in cytokine production, vesicle formation, metabolism, and signal transduction. Additionally, a signal indicative of an interaction between host T cells and a pathogen metabolic pathway that produces a T-cell-stimulating phospholipid was identified. Still, *in vivo* studies of infectious disease – compared to autoimmune or cancer studies – are few in number (Chaussabel et al., 2010) and reveal a lack of understanding of the comprehensive global and dynamic host response to microbial infection.

Differentiating Gram-negative and Gram-positive Bacteria
In clinical diagnosis, a successful determination of infection etiology is the first step in treatment. It has been shown that host-based gene expression signatures can distinguish among microbial etiologies (Ramilo et al., 2007) and even among viruses (Zaas et al., 2009). Although studies evaluating host gene response to bacterial sepsis have focused on the clinically crucial question of differentiating gram-positive and gram-negative etiologies (Ahn et al., 2010; Boldrick et al., 2002; Feezor et al., 2003; Nau et al., 2002; Pankla et al., 2009; Yu et al., 2004), results have been mixed. Host TLR2 or TLR4 are differentially triggered by gram-positive and gram-negative PAMPs, respectively. Although the resulting signaling pathways

share common components, they retain unique characteristics that account for the specificity of corresponding transcriptional responses (Schmitz et al., 2004). While some early *in vitro* studies have shown quantitative differences in the host response profile to gram-positive versus gram-negative bacteria, others have shown a common host response profile (Boldrick et al., 2002; Feezor et al., 2003; Geiser et al., 1993; Nau et al., 2002; Yu et al., 2004, Yurchenko et al., 2001). An analysis of PBMC expression profiles in *S. aureus* infections versus those in *E. coli* infections yielded 211 genes with significantly different expression levels (p < 0.01) between the two bacterial infections (Ramilo et al., 2007). PBMC transcript profiling identified 30 classifier genes that could discriminate *S. aureus* from *E. coli* infections in an independent patient cohort with 85% accuracy (Ramilo et al., 2007), demonstrating a distinction between the gram-positive versus gram-negative host response.

In contrast, Tang and colleagues demonstrated distinctive gene expression patterns in sepsis versus healthy controls (Tang et al., 2008) and in sepsis versus systemic inflammatory response syndrome (SIRS) (Tang et al., 2009), but were unable to demonstrate a similar pattern among patients with gram-positive versus gram-negative sepsis (Tang et al., 2008, 2009). In conjunction with earlier studies showing a common host response to sepsis (Deutschman et al., 1987), this lack of significant differential gene expression between gram-positive and gram-negative sepsis is further evidence of a host-dependent rather than a pathogen-dependent response. Differing from some previous studies (Feezor et al., 2003; Nau et al., 2002), Tang and colleagues examined clinical patients of various gram-positive/negative infections, and used significantly larger gene sets to investigate the transcript expression profiles of circulating leukocytes (18,664 genes) and PBMCs (54,675 genes). PBMC transcript expression analysis identified 138 discriminatory genes for sepsis versus SIRS that had a predictive accuracy of 86% [95% confidence interval (CI), 0.75–0.93] in external validation. These genes were determined to be involved in suppressing the inflammatory immune response to bacterial-induced sepsis, in line with previous findings (Hotchkiss and Karl, 2003); for example, most genes in the T-cell-mediated inflammatory response were down-regulated.

Most recently, Ahn and colleagues evaluated host gene expression signatures for gram-positive (*S. aureus*) and gram-negative (*E. coli*) sepsis in mice and humans. The investigators used sparse latent factor regression to identify gene "factors," sets of co-expressed genes thought to represent biological sub-pathways, followed by linear regression of these factors. In the murine model, host gene expression PBMC analysis produced a predictive signature for *S. aureus* infection/sepsis. This signature accurately differentiated *S. aureus*-infected from *E. coli*-infected mice or uninfected controls. The signature was consistent and robust across variations of multiple factors (different *S. aureus* and inbred mice strains and post-infection time points) and was validated in outbred mice (Ahn et al., 2010). A corresponding analysis of host gene expression signatures in patients with the same conditions identified an expression signature specific to *S. aureus* that accurately differentiated *S. aureus*-infected from *E. coli*-infected patients or uninfected controls. Testing the discriminative ability of the murine-derived *S. aureus* predictive signature in the human cohort demonstrated that the murine-derived *S. aureus* predictive signature successfully distinguished between *S. aureus*-infected patients versus uninfected controls, but not between *S. aureus*-infected and *E. coli*-infected patients (Ahn et al., 2010).

Mycobacterium

Past reports provide supportive phenotypic evidence of genetic susceptibility to tuberculosis (TB) (Dubos and Dubos, 1952). Even today, only 5–10% of approximately two billion individuals infected with *Mycobacterium tuberculosis* develop active disease. With the advent of the genomic era, *in vitro* and animal models of infection began to elucidate the role of innate immunity pathways in tuberculosis susceptibility. Most recently, *in vivo* studies increasingly utilize blood transcriptional profiling to investigate the nuances of the human immune response in infected patients, the results of which have diagnostic and prognostic potential.

Genome-wide whole-blood transcriptional profiles produced a transcript signature that correlated with tuberculosis progression and state of activity or latency. Berry and colleagues defined a distinct 393-transcript signature in patients with active TB (versus latent TB or healthy controls) which was then validated in two independent cohorts containing all three patient types (Berry et al., 2010). Not only did the signature discriminate for active TB, it also reflected the dynamics of disease progression, diminishing after two months and completely resolving by twelve months post-treatment (Berry et al., 2010). Additional analysis identified an 86-gene signature specific to TB infection versus bacterial and other inflammatory diseases. In patients with active TB, there were significantly reduced percentages of B cells and T cells carrying the CD4 and CD8 antigens, as well as total and central memory T cells carrying the CD4 antigen. Pathway analysis identified significant over-expression of genes downstream involved in both interferon-γ and type I IFNαβ receptor signaling (Berry et al., 2010). Earlier microarray studies of TB had identified a small number of genes and were limited by small sample size (Jacobsen et al., 2007; Mistry et al., 2007), but with genome-wide whole-blood transcription, Berry and colleagues have provided the first human blood transcriptional signature of TB. An *ex vivo* study of PBMC expression profiles in response to TB stimulation by interferon-α, β, ω, γ, IL-12, and TNF-α found that while responses to type I interferons were highly stereotyped, responses to IFNγ and IL12 were mostly limited to a subset of type I interferon-inducible genes (Waddell et al., 2010).

Despite the approximately 600,000 cases of *Mycobacterium leprae* infection worldwide (Pannikar, 2009), research in leprosy pathogenesis has been impeded by the lack of a relevant animal model and by the fact that the bacterium is not culturable *in vitro* (Fine, 1983). Compared to other major bacterial diseases, leprosy remains poorly understood (Scollard et al., 2006) and has not been a frequent target of host-based transcriptional profiling studies (Table 96.1). A 2003 study analyzed skin lesions from 11 patients with two different clinical manifestations of leprosy (tuberculoid leprosy and the more symptomatically severe lepromatous leprosy) and identified differentially expressed genes between the two populations (Bleharski et al., 2003). In lepromatous-form lesions, genes belonging to the leukocyte immunoglobulin-like receptor (LIR) family were significantly up-regulated, as were genes for type 2 cytokines, transforming growth factor-β, and IL-5 (Bleharski et al., 2003). Functionally speaking, increased expression of leukocyte immunoglobulin-like receptor 7 (LIR-7) shifts the balance of IL-12 and IL-10 production towards the latter, resulting in the suppression of cell-mediated immunity and immune responsiveness (Sieling et al., 1993).

Most recently, a seminal genome-wide association study (GWAS) of more than ten thousand leprosy patients and healthy controls (the largest GWAS ever of an infectious disease) identified seven single nucleotide polymorphisms (SNPs) associated with increased susceptibility to *M. leprae* (Zhang et al., 2009).

Genome-wide Association Studies

Though GWAS is becoming an increasingly utilized approach towards investigating disease, the focus has been on complex common disorders, such as cancer and diabetes, and less so on infectious disease (Newport and Finan, 2011). As of late 2010, there were 19 published GWAS reports for infectious disease, representing only six diseases (Newport and Finan, 2011). Findings for *Neisseria meningitidis* (Davila et al., 2010) and leprosy (Zhang et al., 2009) provide evidence that genetic variation in the innate immune system can influence susceptibility to infectious disease. However, the approach in infectious disease GWAS has been almost innately paradoxical – the SNP chips used for GWAS are developed from genetic data of Caucasian populations, but infectious disease in the 21st century disproportionately affects populations in Africa and the Asian subcontinent, the former of which also experiences wider diversity and lower linkage equilibrium (LD) than Caucasians (De Bakker and Telenti, 2010; Newport and Finan, 2011). The limitations imposed by this incongruity are evident in study results – a recent GWAS of TB patients identified SNPs in Gambian and Ghanaian patients but encountered poor reproducibility when tested in a different group of Ghanaian and Malawian patients (Thye et al., 2010). For GWAS to be more relevant to future population studies, there is a need for more diverse and population-specific data (Teo et al., 2010). Another challenge in infectious disease GWAS is the dual-faced nature of pathogenesis, as both the host and pathogen are factors in disease manifestation. And although microbial genomics is an active field of study, with over a thousand sequenced genomes (Fournier et al., 2007), the presence of antibiotic-resistant bacterial strains in even the most developed countries is firsthand evidence that the constant evolutionary capacity of the pathogen itself is the consistently confounding factor in the fight against infectious disease.

Proteomics

Proteomic profiling is another promising tool for investigating host–bacteria interactions, because host–pathogen interactions occur primarily via proteins but mRNA transcripts are not always indicative of cellular protein levels. Pathogen exposure alters host proteins on multiple functional levels including signaling pathways, protein degradation, cytokine and growth factor production, phagocytosis, apoptosis, and cytoskeletal rearrangement (Coiras et al., 2008).

Clinical proteomic profiling is growing and although it is currently not as prevalent as transcriptional profiling, recent studies are establishing its applicability. *In vivo* analyses have identified distinctive protein signatures for diseases such as tuberculosis (Agranoff et al., 2006), *Bartonella henselae*-induced infective endocarditis (Saisongkorh et al., 2010), and infective endocarditis due to 13 different bacteria (see Figure 96.3) (Fenollar et al., 2006). Because cases of blood-culture-negative endocarditis (BCNE) are a considerable diagnostic and

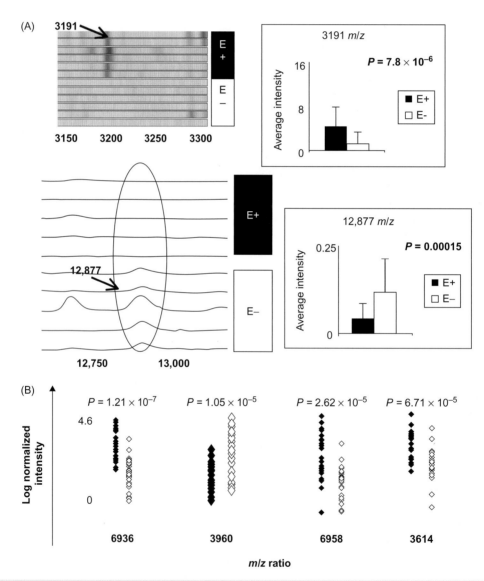

Figure 96.3 Serum proteins that were differentially expressed between endocarditis-positive (E+) and endocarditis-negative (E−) patients. A protein with a mass/charge (m/z) ratio of 3191 (gel view) was up-regulated in endocarditis-positive patients relative to endocarditis-negative patients, whereas a protein with a m/z ratio of 12,877 was down-regulated in endocarditis-positive patients relative to endocarditis-negative patients (spectra view). (A) Average intensities of the 3191 and 12,877 peaks in the endocarditis-positive and the endocarditis-negative patients as well as standard deviations (SDs) and P values are shown in the histograms. (B) Other proteomic markers with differential expression between the endocarditis-positive (black symbols) and the endocarditis-negative (white symbols) patients, plotted as a function of their normalized, log-transformed intensities. *Reproduced with permission from Fenollar et al., 2006.*

therapeutic challenge (Werner et al., 2003), the latter study is particularly encouraging for the clinical value of serodiagnosis. Additionally, this study demonstrated a proteomic signature for a disease with multiple etiologic agents, which bodes well for similar circumstances such as sepsis (Fenollar et al., 2006).

In vivo proteomic assays have also been used to identify novel bacterial virulence factors. In a murine study combining proteomic comparison and metabolic disruption, Alteri and colleagues demonstrated that uropathogenic *E. coli* (UPEC) survival in the host urinary tract relies on peptide import, a metabolic dependency that distinguishes this pathogenic strain from commensal *E. coli* residing in the nutritionally rich intestinal milieu (Alteri et al., 2009). They propose that certain metabolic activity may serve as or be considered virulence factors for pathogenic bacteria. Similarly, proteomic comparisons have identified an antigenic signature specific to a clinically relevant *M. tuberculosis* strain versus laboratory strains (Bahk et al., 2004), as well as a correlation between clinical *Helicobacter pylori* strain polymorphism and *H. pylori*-associated iron deficiency anemia (IDA) (Park et al., 2006). Such pathogenic-versus-nonpathogenic comparisons reveal virulence factors found only in the pathogenic strains, from which novel antibiotics targeting those pathogenic proteins can be developed. This would not only streamline the possible protein candidates for a screen but (considering the abundance of bacteria in healthy human physiology) may also reduce possible detrimental effects to commensal flora composition (Rasko et al., 2008).

Metabolomics

Metabolomic analysis has been explored for parasitic (Olszewski et al., 2009; Wang et al., 2008) and viral infections (Munger et al., 2008; Rodgers et al., 2009; Wikoff et al., 2008), but comparable studies in bacterial infections are still relatively limited (Vaidyanathan, 2005). Two recent studies have examined bacteria and host metabolomes to elucidate disease pathogenesis and host response. In clinical *E. coli* isolates, siderophore production may be a virulence factor in UPEC pathogenesis, rendering that biosynthesis pathway a viable target for anti-virulence drugs (Henderson et al., 2009). Furthermore, siderophores served as a metabolic signature among *E. coli* isolates from patients with recurrent infections (Henderson et al., 2009). In the first analysis of host metabolome during bacterial infection (*Salmonella enterica* Typhimurium), Antunes and colleagues assessed metabolite perturbations and found hormone signaling pathways to be particularly impacted (Antunes et al., 2011). In both studies, genomic data were not predictive of either pathogen (Henderson et al., 2009) or host (Antunes et al., 2011) metabolite abundances, highlighting the need for multidimensional "-omics" approaches in future host–pathogen research.

Ribonomics

RNA interference (RNAi)-based genomic screens have been utilized *in vitro* to investigate the pathogenesis and virulence mechanisms of *S. enterica* Typhimurium, *M. tuberculosis*, and *Listeria monocytogenes* (Hong-Geller and Micheva-Viteva, 2010). Because bacteria mainly replicate outside of host cells and therefore are less susceptible to RNAi silencing than viruses (Lieberman et al., 2003), translation inhibitors discovered via ribonomic analysis may serve as the basis of host-based inhibitor therapeutics that silence those host genes involved in mediating bacterial invasion (Leung and Whittaker, 2005).

FUTURE DIRECTIONS

As with other aspects of host-based immunology studies, the future of bacterial pathogenesis research is likely to see an increased emphasis on *in vivo* human studies (Davis, 2008). In cancer immunotherapy (Ostrand-Rosenberg, 2004) and autoimmunity (Von Herrath and Nepom, 2005) studies, attempts to extend tried-and-true murine protocols to humans have met with very little success. Given the many identified differences between the mouse and human immune systems (Mestas and Hughes, 2004), we should be conscious of the possible limitations in using a murine infection model to explore the human immune response.

Not only should human studies increase in quantity, they should also change in quality – emphasis should be placed on co-institutional and multinational studies with larger patient sample sizes (Davis, 2008; Ramilo and Thomas, 2010). Indeed, biomarkers identified through blood transcriptome studies need validation in large multicenter trials before serving as the basis of new diagnostic tools (Pascual et al., 2008). Finally, although continuing advances in gene and protein profiling technology may allow ever-expanding probe sets and arrays, they should be adopted with rigorous statistical methodology and bioinformatics practices, such as text mining and data mining, in order to produce valid and significant findings (Chaussabel et al., 2008; Müller et al., 2004).

CONCLUSION

In summary, genomic studies of bacterial pathogenesis have made significant progress in identifying the potential diagnostic and prognostic applications of host response signatures. Transcriptional profiling, particularly of blood leukocytes, has successfully identified gene expression signatures in a wide range of bacterial infections. Proteomic profiling of sera has shown similarly promising findings. Successful transfer of these exciting but still investigational techniques to robust platforms would ideally involve the ability to provide accurate information in real time. This would have particular clinical applicability in regard to bacterial sepsis, where diagnostic delay can have catastrophic consequences. Advances in technology and continued refinement of gene expression and proteomic signatures should provide further insights into pathogenesis, facilitating the development of more efficient and cost-effective diagnostic and therapeutic tools.

ACKNOWLEDGMENTS

This work was supported by K24-AI093969 and R01-AI068804.

REFERENCES

Aderem, A., Ulevitch, R.J., 2000. Toll-like receptors in the induction of the innate immune response. Nature 406, 782–787.

Agranoff, D., Fernandez-Reyes, D., Papadopoulos, M.C., et al., 2006. Identification of diagnostic markers for tuberculosis by proteomic fingerprinting of serum. Lancet 368, 1012–1021.

Ahn, S.H., Deshmukh, H., Johnson, N., et al., 2010. Two genes on A/J chromosome 18 are associated with susceptibility to *Staphylococcus aureus* infection by combined microarray and QTL analyses. PLoS Pathog 6, e1001088.

Akira, S., Takeda, K., Kaisho, T., 2001. Toll-like receptors: Critical proteins linking innate and acquired immunity. Nat Immunol 2, 675–680.

Alizadeh, A.A., Eisen, M.B., Davis, R.E., et al., 2000. Distinct types of diffuse large B-cell lymphoma identified by gene expression profiling. Nature 403, 503–511.

Alteri, C.J., Smith, S.N., Mobley, H.L.T., 2009. Fitness of *Escherichia coli* during urinary tract infection requires gluconeogenesis and the TCA cycle. PLoS Pathog 5, e1000448.

Andersson, H., Hartmanova, B., Ryden, P., Noppa, L., Naslund, L., Sjostedt, A., 2006. A microarray analysis of the murine macrophage response to infection with *Francisella tularensis* LVS. J Med Microbiol 55, 1023–1033.

Antunes, L.C., Arena, E.T., Menendez, A., et al., 2011. Impact of *Salmonella* infection on host hormone metabolism revealed by metabolomics. Infect Immun 79 (4), 1759–1769.

Ardura, M.I., Banchereau, R., Mejias, A., et al., 2009. Enhanced monocyte response and decreased central memory T cells in children with invasive *Staphylococcus aureus* infections. PLoS ONE 4, e5446.

Bahk, Y.Y., Kim, S.A., Kim, J.-S., et al., 2004. Antigens secreted from *Mycobacterium tuberculosis*: Identification by proteomics approach and test for diagnostic marker. Proteomics 4, 3299–3307.

Belcher, C.E., Drenkow, J., Kehoe, B., et al., 2000. The transcriptional responses of respiratory epithelial cells to *Bordetella pertussis* reveal host defensive and pathogen counter-defensive strategies. Proc Natl Acad Sci USA 97, 13,847–13,852.

Berry, M.P.R., Graham, C.M., McNab, F.W., et al., 2010. An interferon-inducible neutrophil-driven blood transcriptional signature in human tuberculosis. Nature 466, 973–977.

Binnicker, M.J., Williams, R.D., Apicella, M.A., 2003. Infection of human urethral epithelium with *Neisseria gonorrhoeae* elicits an upregulation of host anti-apoptotic factors and protects cells from staurosporine-induced apoptosis. Cell Microbiol 5, 549–560.

Bleharski, J.R., Li, H., Meinken, C., et al., 2003. Use of genetic profiling in leprosy to discriminate clinical forms of the disease. Science 301, 1527–1530.

Boldrick, J.C., Alizadeh, A.A., Diehn, M., et al., 2002. Stereotyped and specific gene expression programs in human innate immune responses to bacteria. Proc Natl Acad Sci USA 99, 972–977.

Borovecki, F., Lovrecic, L., Zhou, J., et al., 2005. Genome-wide expression profiling of human blood reveals biomarkers for Huntington's disease. Proc Natl Acad Sci USA 102, 11,023–11,028.

Chang, Y.E., Laimins, L.A., 2000. Microarray analysis identifies interferon-inducible genes and Stat-1 as major transcriptional targets of human papillomavirus type 31. J Virol 74, 4174–4182.

Chaussabel, D., Pascual, V., Banchereau, J., 2010. Assessing the human immune system through blood transcriptomics. BMC Biology 8, 84.

Chaussabel, D., Quinn, C., Shen, J., et al., 2008. A modular analysis framework for blood genomics studies: Application to systemic lupus erythematosus. Immunity 29, 150–164.

Chin, C.-Y., Monack, D., Nathan, S., 2010. Genome-wide transcriptome profiling of a murine acute melioidosis model reveals new insights into how *Burkholderia pseudomallei* overcomes host innate immunity. BMC Genomics 11, 672.

Coiras, M., Camafeita, E., López-Huertas, M.R., Calvo, E., López, J.A., Alcamí, J., 2008. Application of proteomics technology for analyzing the interactions between host cells and intracellular infectious agents. Proteomics 8, 852–873.

Corbeil, J., Sheeter, D., Genini, D., et al., 2001. Temporal gene regulation during HIV-1 infection of human CD4+ T cells. Genome Res 11, 1198–1204.

Cuadras, M.A., Feigelstock, D.A., An, S., Greenberg, H.B., 2002. Gene expression pattern in Caco-2 cells following rotavirus infection. J Virol 76, 4467–4482.

Davila, S., Wright, V.J., Khor, C.C., et al., 2010. Genome-wide association study identifies variants in the CFH region associated with host susceptibility to meningococcal disease. Nat Genet 42, 772–776.

Davis, M.M., 2008. A prescription for human immunology. Immunity 29, 835–838.

De Bakker, P.I.W., Telenti, A., 2010. Infectious diseases not immune to genome-wide association. Nat Genet 42, 731–732.

Detweiler, C.S., Cunanan, D.B., Falkow, S., 2001. Host microarray analysis reveals a role for the *Salmonella* response regulator phoP in human macrophage cell death. Proc Natl Acad Sci USA 98, 5850–5855.

Deutschman, C.S., Konstantinides, F.N., Tsai, M., Simmons, R.L., Cerra, F.B., 1987. Physiology and metabolism in isolated viral septicemia: Further evidence of an organism-independent, host-dependent response. Arch Surg 122, 21–25.

Dubos, R., Dubos, J., 1952. The White Plague: Tuberculosis, Man, and Society. Rutgers University Press, New Brunswick, NJ.

Ehrt, S., Schnappinger, D., Bekiranov, S., et al., 2001. Reprogramming of the macrophage transcriptome in response to interferon-gamma and *Mycobacterium tuberculosis*: Signaling roles of nitric oxide synthase-2 and phagocyte oxidase. J Exp Med 194, 1123–1140.

Eisenstein, T.K., Killar, L.M., Stocker, B.A., Sultzer, B.M., 1984. Cellular immunity induced by avirulent *Salmonella* in LPS-defective C3H/HeJ mice. J Immunol 133, 958–961.

Feezor, R.J., Oberholzer, C., Baker, H.V., et al., 2003. Molecular characterization of the acute inflammatory response to infections with gram-negative versus gram-positive bacteria. Infect Immun 71, 5803–5813.

Fenollar, F., Gonçalves, A., Esterni, B., et al., 2006. A serum protein signature with high diagnostic value in bacterial endocarditis: Results from a study based on surface-enhanced laser desorption/ionization time-of-flight mass spectrometry. J Infect Dis 194, 1356–1366.

Fine, P.E., 1983. Natural history of leprosy – aspects relevant to a leprosy vaccine. Int J Lepr Other Mycobact Dis 51, 553–555.

Fournier, P.-E., Drancourt, M., Raoult, D., 2007. Bacterial genome sequencing and its use in infectious diseases. Lancet Infect Dis 7, 711–723.

Garey, K.W., Rege, M., Pai, M.P., et al., 2006. Time to initiation of fluconazole therapy impacts mortality in patients with candidemia: A multi-institutional study. Clin Infect Dis 43, 25–31.

Geiser, T., Dewald, B., Ehrengruber, M.U., Clark-Lewis, I., Baggiolini, M., 1993. The interleukin-8-related chemotactic cytokines GRO alpha, GRO beta, and GRO gamma activate human neutrophil and basophil leukocytes. J Biol Chem 268, 15,419–15,424.

Golub, T.R., Slonim, D.K., Tamayo, P., et al., 1999. Molecular classification of cancer: Class discovery and class prediction by gene expression monitoring. Science 286, 531–537.

Guillemin, K., Salama, N.R., Tompkins, L.S., Falkow, S., 2002. Cag pathogenicity island-specific responses of gastric epithelial cells to Helicobacter pylori infection. Proc Natl Acad Sci USA 99, 15,136–15,141.

Henderson, B., Poole, S., Wilson, M., 1996. Bacterial modulins: A novel class of virulence factors which cause host tissue pathology by inducing cytokine synthesis. Microbiol Rev 60, 316–341.

Henderson, J.P., Crowley, J.R., Pinkner, J.S., et al., 2009. Quantitative metabolomics reveals an epigenetic blueprint for iron acquisition in uropathogenic Escherichia coli. PLoS Pathog 5, e1000305.

Hess, S., Peters, J., Bartling, G., et al., 2003. More than just innate immunity: Comparative analysis of Chlamydophila pneumoniae and Chlamydia trachomatis effects on host-cell gene regulation. Cell Microbiol 5, 785–795.

Hong-Geller, E., Micheva-Viteva, S.N., 2010. Functional gene discovery using RNA interference-based genomic screens to combat pathogen infection. Curr Drug Discov Technol 7, 86–94.

Hotchkiss, R.S., Karl, I.E., 2003. The pathophysiology and treatment of sepsis. N Engl J Med 348, 138–150.

Huang, Q., Liu, D., Majewski, P., et al., 2001. The plasticity of dendritic cell responses to pathogens and their components. Science 294, 870–875.

Izmailova, E., Bertley, F.M., Huang, Q., et al., 2003. HIV-1 Tat reprograms immature dendritic cells to express chemoattractants for activated T cells and macrophages. Nat Med 9, 191–197.

Jacobsen, M., Repsilber, D., Gutschmidt, A., et al., 2007. Candidate biomarkers for discrimination between infection and disease caused by Mycobacterium tuberculosis. J Mol Med 85, 613–621.

Jenner, R.G., Maillard, K., Cattini, N., et al., 2003. Kaposi's sarcoma-associated herpesvirus-infected primary effusion lymphoma has a plasma cell gene expression profile. Proc Natl Acad Sci USA 100, 10,399–10,404.

Jonathan, N., 2006. Diagnostic utility of BINAX NOW RSV – an evaluation of the diagnostic performance of BINAX NOW RSV in comparison with cell culture and direct immunofluorescence. Ann Clin Microbiol Antimicrob 5, 13.

Jones, J.O., Arvin, A.M., 2003. Microarray analysis of host cell gene transcription in response to varicella-zoster virus infection of human T cells and fibroblasts in vitro and SCIDhu skin xenografts in vivo. J Virol 77, 1268–1280.

Kim, H.S., Choi, E.H., Khan, J., et al., 2005. Expression of genes encoding innate host defense molecules in normal human monocytes in response to Candida albicans. Infect Immun 73, 3714–3724.

Landry, M.L., Cohen, S., Ferguson, D., 2008. Real-time PCR compared to Binax NOW and cytospin-immunofluorescence for detection of influenza in hospitalized patients. J Clin Virol 43, 148–151.

Leung, R.K.M., Whittaker, P.A., 2005. RNA interference: From gene silencing to gene-specific therapeutics. Pharmacol Ther 107, 222–239.

Lieberman, J., Song, E., Lee, S.-K., et al., 2003. RNA interference targeting Fas protects mice from fulminant hepatitis. Nat Med 9, 347–351.

Lorenz, M.C., Bender, J.A., Fink, G.R., 2004. Transcriptional response of Candida albicans upon internalization by macrophages. Eukaryot Cell 3, 1076–1087.

Martin, K.J., Graner, E., Li, Y., et al., 2001. High-sensitivity array analysis of gene expression for the early detection of disseminated breast tumor cells in peripheral blood. Proc Natl Acad Sci USA 98, 2646–2651.

Matussek, A., Strindhall, J., Stark, L., et al., 2005. Infection of human endothelial cells with Staphylococcus aureus induces transcription of genes encoding an innate immunity response. Scand J Immunol 61, 536–544.

Mestas, J., Hughes, C.C.W., 2004. Of mice and not men: Differences between mouse and human immunology. J Immunol 172, 2731–2738.

Mills, J.C., Syder, A.J., Hong, C.V., Guruge, J.L., Raaii, F., Gordon, J.I., 2001. A molecular profile of the mouse gastric parietal cell with and without exposure to Helicobacter pylori. Proc Natl Acad Sci USA 98, 13,687–13,692.

Mistry, R., Cliff, J.M., Clayton, C.L., et al., 2007. Gene-expression patterns in whole blood identify subjects at risk for recurrent tuberculosis. J Infect Dis 195, 357–365.

Moreno-Altamirano, M.M., Romano, M., Legorreta-Herrera, M., Sanchez-Garcia, F.J., Colston, M.J., 2004. Gene expression in human macrophages infected with dengue virus serotype-2. Scand J Immunol 60, 631–638.

Mueller, A., O'Rourke, J., Grimm, J., et al., 2003. Distinct gene expression profiles characterize the histopathological stages of disease in Helicobacter-induced mucosa-associated lymphoid tissue lymphoma. Proc Natl Acad Sci USA 100, 1292–1297.

Müller, H.-M., Kenny, E.E., Sternberg, P.W., 2004. Textpresso: An ontology-based information retrieval and extraction system for biological literature. PLoS Biol 2, e309.

Munger, J., Bennett, B.D., Parikh, A., et al., 2008. Systems-level metabolic flux profiling identifies fatty acid synthesis as a target for antiviral therapy. Nat Biotech 26, 1179–1186.

Nau, G.J., Richmond, J.F., Schlesinger, A., et al., 2002. Human macrophage activation programs induced by bacterial pathogens. Proc Natl Acad Sci USA 99, 1503–1508.

Nau, G.J., Schlesinger, A., Richmond, J.F., Young, R.A., 2003. Cumulative Toll-like receptor activation in human macrophages treated with whole bacteria. J Immunol 170, 5203–5209.

Newport, M.J., Finan, C., 2011. Genome-wide association studies and susceptibility to infectious diseases. Brief Funct Genomics 10, 98–107.

Olszewski, K.L., Morrisey, J.M., Wilinski, D., et al., 2009. Host–parasite interactions revealed by Plasmodium falciparum metabolomics. Cell Host Microbe 5, 191–199.

Osman, I., Bajorin, D.F., Sun, T.T., et al., 2006. Novel blood biomarkers of human urinary bladder cancer. Clin Cancer Res 12, 3374–3380.

Ostrand-Rosenberg, S., 2004. Animal models of tumor immunity, immunotherapy and cancer vaccines. Curr Opin Immunol 16, 143–150.

Pankla, R., Buddhisa, S., Berry, M., et al., 2009. Genomic transcriptional profiling identifies a candidate blood biomarker signature for the diagnosis of septicemic melioidosis. Genome Biol 10, R127.

Pannikar, V., 2009. Enhanced global strategy for further reducing the disease burden due to leprosy: 2011–2015. Lepr Rev 80, 353–354.

Park, S.A., Lee, H.W., Hong, M.H., et al., 2006. Comparative proteomic analysis of *Helicobacter pylori* strains associated with iron deficiency anemia. Proteomics 6, 1319–1328.

Pascual, V., Chaussabel, D., Banchereau, J., 2008. A genomic approach to human autoimmune diseases. Ann Rev Immunol 28, 535–571.

Pieters, J., 2001. Entry and survival of pathogenic mycobacteria in macrophages. Microbes Infect 3, 249–255.

Pise-Masison, C.A., Radonovich, M., Mahieux, R., et al., 2002. Transcription profile of cells infected with human T-cell leukemia virus type I compared with activated lymphocytes. Cancer Res 62, 3562–3571.

Radhakrishnan, S., Otte, J., Enam, S., Del Valle, L., Khalili, K., Gordon, J., 2003. JC virus-induced changes in cellular gene expression in primary human astrocytes. J Virol 77, 10,638–10,644.

Ragno, S., Romano, M., Howell, S., Pappin, D.J., Jenner, P.J., Colston, M.J., 2001. Changes in gene expression in macrophages infected with *Mycobacterium tuberculosis*: A combined transcriptomic and proteomic approach. Immunology 104, 99–108.

Rahman, M., Vandermause, M.F., Kieke, B.A., Belongia, E.A., 2008. Performance of Binax NOW Flu A and B and direct fluorescent assay in comparison with a composite of viral culture or reverse transcription polymerase chain reaction for detection of influenza infection during the 2006 to 2007 season. Diagn Microbiol Infect Dis 62, 162–166.

Ramilo, O., Allman, W., Chung, W., et al., 2007. Gene expression patterns in blood leukocytes discriminate patients with acute infections. Blood 109, 2066–2077.

Ramilo, O., Thomas, J., 2010. Farewell to innocence: Untangling septic shock in the postgenomic era. Pediatr Crit Care Med 11, 426–427.

Rasko, D.A., Rosovitz, M.J., Myers, G.S.A., et al., 2008. The pangenome structure of *Escherichia coli*: Comparative genomic analysis of *E. coli* commensal and pathogenic isolates. J Bacteriol 190, 6881–6893.

Rodgers, M.A., Saghatelian, A., Yang, P.L., 2009. Identification of an overabundant cholesterol precursor in hepatitis B virus replicating cells by untargeted lipid metabolite profiling. J Am Chem Soc 131, 5030–5031.

Rodriguez, N.E., Chang, H.K., Wilson, M.E., 2004. Novel program of macrophage gene expression induced by phagocytosis of *Leishmania chagasi*. Infect Immun 72, 2111–2122.

Rosenberger, C.M., Scott, M.G., Gold, M.R., Hancock, R.E., Finlay, B.B., 2000. *Salmonella* Typhimurium infection and lipopolysaccharide stimulation induce similar changes in macrophage gene expression. J Immunol 164, 5894–5904.

Saisongkorh, W., Kowalczewska, M., Azza, S., Decloquement, P., Rolain, J.-M., Raoult, D., 2010. Identification of candidate proteins for the diagnosis of *Bartonella henselae* infections using an immunoproteomic approach. FEMS Microbiol Lett 310, 158–167.

Sauvonnet, N., Pradet-Balade, B., Garcia-Sanz, J.A., Cornelis, G.R., 2002. Regulation of mRNA expression in macrophages after *Yersinia enterocolitica* infection. Role of different Yop effectors. J Biol Chem 277, 25,133–25,142.

Schmitz, F., Mages, J., Heit, A., Lang, R., Wagner, H., 2004. Transcriptional activation induced in macrophages by Toll-like receptor (TLR) ligands: From expression profiling to a model of TLR signaling. Eur J Immunol 34, 2863–2873.

Schwacha, M.G., Meissler Jr., J.J., Eisenstein, T.K., 1998. *Salmonella* Typhimurium infection in mice induces nitric oxide-mediated immunosuppression through a natural killer cell-dependent pathway. Infect Immun 66, 5862–5866.

Scollard, D.M., Adams, L.B., Gillis, T.P., et al., 2006. The continuing challenges of leprosy. Clin Microbiol Rev 19, 338–381.

Shea, P.R., Virtaneva, K., Kupko, J.J., et al., 2010. Interactome analysis of longitudinal pharyngeal infection of cynomolgus macaques by group *A Streptococcus*. Proc Natl Acad Sci USA 107, 4693–4698.

Sieling, P.A., Abrams, J.S., Yamamura, M., et al., 1993. Immunosuppressive roles for IL-10 and IL-4 in human infection. In vitro modulation of T cell responses in leprosy. J Immunol 150, 5501–5510.

Simmen, K.A., Singh, J., Luukkonen, B.G., et al., 2001. Global modulation of cellular transcription by human cytomegalovirus is initiated by viral glycoprotein B. Proc Natl Acad Sci USA 98, 7140–7145.

Simon, R., Radmacher, M.D., Dobbin, K., 2002. Design of studies using DNA microarrays. Genet Epidemiol 23, 21–36.

Syder, A.J., Oh, J.D., Guruge, J.L., et al., 2003. The impact of parietal cells on *Helicobacter pylori* tropism and host pathology: An analysis using gnotobiotic normal and transgenic mice. Proc Natl Acad Sci USA 100, 3467–3472.

Tang, B.M.P., McLean, A.S., Dawes, I.W., Huang, S.J., Cowley, M.J., Lin, R.C.Y., 2008. Gene-expression profiling of Gram-positive and Gram-negative sepsis in critically ill patients. Crit Care Med 36, 1125–1128. doi: 10.1097/CCM.0b013e3181692c0b.

Tang, B.M.P., McLean, A.S., Dawes, I.W., Huang, S.J., Lin, R.C.Y., 2009. Gene-expression profiling of peripheral blood mononuclear cells in sepsis. Crit Care Med 37, 882–888. doi: 10.1097/CCM.0b013e31819b52fd.

Teo, Y.-Y., Small, K.S., Kwiatkowski, D.P., 2010. Methodological challenges of genome-wide association analysis in Africa. Nat Rev Genet 11, 149–160.

Thompson, L.J., Dunstan, S.J., Dolecek, C., et al., 2009. Transcriptional response in the peripheral blood of patients infected with *Salmonella enterica* serovar Typhi. Proc Natl Acad Sci USA 106, 22,433–22,438.

Thye, T., Vannberg, F.O., Wong, S.H., et al., 2010. Genome-wide association analysis identifies a susceptibility locus for tuberculosis on chromosome 18q11.2. Nat Genet 42, 739–741.

Tian, B., Zhang, Y., Luxon, B.A., et al., 2002. Identification of NF-kappaB-dependent gene networks in respiratory syncytial virus-infected cells. J Virol 76, 6800–6814.

Vaidyanathan, S., 2005. Profiling microbial metabolomes: What do we stand to gain? Metabolomics 1, 17–28.

Van Amersfoort, E.S., Van Berkel, T.J.C., Kuiper, J., 2003. Receptors, mediators, and mechanisms involved in bacterial sepsis and septic shock. Clin Microbiol Rev 16, 379–414.

Van de Vijver, M.J., He, Y.D., Van't Veer, L.J., et al., 2002. A gene-expression signature as a predictor of survival in breast cancer. N Engl J Med 347, 1999–2009.

Van der Velden, A.W.M., Copass, M.K., Starnbach, M.N., 2005. *Salmonella* inhibit T-cell proliferation by a direct, contact-dependent immunosuppressive effect. Proc Natl Acad Sci USA 102, 17,769–17,774.

Van't Wout, A.B., Lehrman, G.K., Mikheeva, S.A., et al., 2003. Cellular gene expression upon human immunodeficiency virus type 1 infection of CD4(+)-T-cell lines. J Virol 77, 1392–1402.

Vincent, J.-L., Abraham, E., 2006. The last 100 years of sepsis. Am J Respir Crit Care Med 173, 256–263.

Von Herrath, M.G., Nepom, G.T., 2005. Lost in translation. J Exp Med 202, 1159–1162.

Waddell, S.J., Popper, S.J., Rubins, K.H., et al., 2010. Dissecting interferon-induced transcriptional programs in human peripheral blood cells. PLoS ONE 5, e9753.

Wang, H.W., Trotter, M.W., Lagos, D., et al., 2004. Kaposi sarcoma herpesvirus-induced cellular reprogramming contributes to the lymphatic endothelial gene expression in Kaposi sarcoma. Nat Genet 36, 687–693.

Wang, Y., Utzinger, J., Saric, J., et al., 2008. Global metabolic responses of mice to *Trypanosoma brucei brucei* infection. Proc Natl Acad Sci USA 105, 6127–6132.

Warke, R.V., Xhaja, K., Martin, K.J., et al., 2003. Dengue virus induces novel changes in gene expression of human umbilical vein endothelial cells. J Virol 77, 11,822–11,832.

Wells, S.I., Aronow, B.J., Wise, T.M., et al., 2003. Transcriptome signature of irreversible senescence in human papillomavirus-positive cervical cancer cells. Proc Natl Acad Sci USA 100, 7093–7098.

Werner, M., Andersson, R., Olaison, L., Hogevik, H., 2003. A clinical study of culture-negative endocarditis. Medicine (Baltimore) 82, 263–273.

Wikoff, W.R., Pendyala, G., Siuzdak, G., Fox, H.S., 2008. Metabolomic analysis of the cerebrospinal fluid reveals changes in phospholipase expression in the CNS of SIV-infected macaques. J Clin Invest 118, 2661–2669.

Wong, H.R., Freishtat, R.J., Monaco, M., Odoms, K., Shanley, T.P., 2010. Leukocyte subset-derived genomewide expression profiles in pediatric septic shock. Pediatr Crit Care Med 11, 349–355. doi: 10.1097/PCC.0b013e3181c519b4.

Yu, S.L., Chen, H.W., Yang, P.C., et al., 2004. Differential gene expression in gram-negative and gram-positive sepsis. Am J Respir Crit Care Med 169, 1135–1143.

Yurchenko, V., O'Connor, M., Dai, W.W., et al., 2001. CD147 is a signaling receptor for cyclophilin B. Biochem Biophys Res Commun 288, 786–788.

Zaas, A.K., Aziz, H., Lucas, J., Perfect, J.R., Ginsburg, G.S., 2010. Blood gene expression signatures predict invasive candidiasis. Sci Transl Med 2, 21ra17.

Zaas, A.K., Chen, M., Varkey, J., et al., 2009. Gene expression signatures diagnose influenza and other symptomatic respiratory viral infections in humans. Cell Host Microbe 6, 207–217.

Zhang, F.R., Huang, W., Chen, S.M., et al., 2009. Genomewide association study of leprosy. N Engl J Med 362, 1446–1448.

Zhang, Y., Jamaluddin, M., Wang, S., et al., 2003. Ribavirin treatment up-regulates antiviral gene expression via the interferon-stimulated response element in respiratory syncytial virus-infected epithelial cells. J Virol 77, 5933–5947.

Zhang, Y., Luxon, B.A., Casola, A., Garofalo, R.P., Jamaluddin, M., Brasier, A.R., 2001. Expression of respiratory syncytial virus-induced chemokine gene networks in lower airway epithelial cells revealed by cDNA microarrays. J Virol 75, 9044–9058.

Emerging Viral Infections

Albert D.M.E. Osterhaus and Saskia L. Smits

INTRODUCTION

Infectious viral diseases, both emerging and re-emerging, pose a continuous health threat and disease burden to humans. Since the 1980s, an increased frequency of infectious outbreaks in both humans and animals has been observed (Figure 97.1). Human mortality from recently emerged diseases ranges from about 300 people due to infection with H5N1 avian influenza A virus to tens of millions of people due to acquired immunodeficiency syndrome (AIDS). Emerging viral infections have had a large impact on livestock production as well, by causing direct mortality or because depopulation policies had to be implemented to protect safety of international trade and to control virus spread. Nearly all of the most important human pathogens are either zoonotic or originated as zoonoses before adapting to humans (Kuiken et al., 2005; Smith et al., 2009; Taylor et al., 2001; Woolhouse and Gowtage-Sequeria, 2005), and we are continuously bombarded by novel animal pathogens. Two of the most devastating pandemics in human history, HIV/AIDS and Spanish influenza, started by interspecies transmission of the causative agents (De Wit et al., 2008; Gao et al., 1999; Hirsch et al., 1989; Osterhaus, 2001). Recent outbreaks of infectious diseases, among which severe acute respiratory syndrome (SARS) coronavirus, and swine influenza A virus H1N1 2009 in humans, were also initiated by transmission from the animal host to humans, and have further highlighted this problem (Haagmans et al., 2009; Kuiken et al., 2003; Smith et al., 2009; Song et al., 2005; Tang et al., 2006). The apparently increased transmission

of pathogens from animals to humans is related to a plethora of accelerating environmental and anthropogenic changes, such as increased mobility, and demographic changes, which alter the rate and nature of contacts between animals and humans. Environmental changes are thought to have played a role in the recent increased distribution of the arthropod vector *Aedes aegypti*, which led to large outbreaks of dengue fever in South America and Southeast Asia (Weaver and Reisen, 2010). Intense pig farming in areas where frugivorous bats are common is the probable cause of the introduction of Nipah virus from bats into pig populations in Malaysia, and subsequent transmission to humans (Chua, 2003). HIV-1 and HIV-2 originate from non-human primates (Chen et al., 1996; Keele et al., 2006; Marx et al., 1991), and central African bush hunters have been infected with simian foamy virus (Wolfe et al., 2004). Consequently, there is increased awareness that an ongoing systematic global effort to monitor for emerging and re-emerging pathogens in both animals and humans is needed (Haagmans et al., 2009; Kuiken et al., 2005; Osterhaus, 2001; Wolfe et al., 2007).

Surveillance of viral pathogens in animals should focus both on domestic animals and on wildlife, with key reservoir species that have previously been shown to represent an imminent health threat to humans. In humans, populations with either high exposure to wild or domestic animals, such as hunters, butchers, veterinarians, and zoo workers, or populations with high susceptibility, such as immunocompromized patients, would be key targets for monitoring novel viral pathogens. Collectively, these efforts would result in the identification

Genomic and Personalized Medicine, 2nd edition
by Ginsburg & Willard. DOI: http://dx.doi.org/10.1016/B978-0-12-382227-7.00097-5

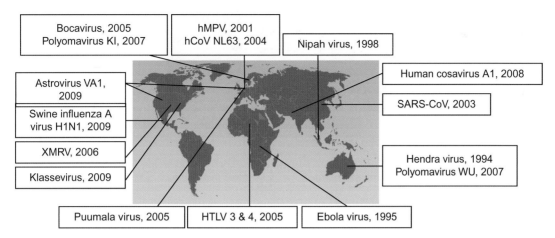

Figure 97.1 Virus threats. This picture shows some of the virus threats worldwide from 1994–2010. Bocavirus, 2005 (Allander et al., 2005); polyomavirus KI, 2007 (Allander et al., 2007); human metapneumovirus (hMPV), 2001 (Van den Hoogen et al., 2001); human coronavirus (hCoV) NL63, 2004 (Fouchier et al., 2004; Van der Hoek et al., 2004); Nipah virus, 1998 (Centers for Disease Control and Prevention, 2009); human common stool associated picornavirus (cosavirus) A genotype 1, 2008 (Kapoor et al., 2008a); SARS coronavirus, 2003 (Drosten et al., 2003); Hendra virus, 1994 (Murray et al., 1995); polyomavirus WU, 2007 (Gaynor et al., 2007); astrovirus VA1, 2009 (Finkbeiner et al., 2009; Smits et al., 2010c); swine influenza A virus H1N1, 2009 (Garten et al., 2009); xenotropic MuLV-related virus (XMRV), 2006 (Urisman et al., 2006); klassevirus, 2009 (Greninger et al., 2009); Ebola virus, 1995 (Hall and Hall, 2007); Puumala virus, 2005 (Tersago et al., 2010); and human T-cell lymphotropic virus 3 and 4 (HTLV 3 & 4), 2005 (Calattini et al., 2005; Wolfe et al., 2005).

of the diversity of viruses to which humans are exposed, the identification of animal pathogens that may threaten us in the future, and, hopefully, their early detection in humans and subsequently their control (Haagmans et al., 2009; Kuiken et al., 2005; Wolfe et al., 2007). It may even be possible to predict which viruses are most likely to cross the species barrier and what changes may be required for the evolution of an animal pathogen into a human-specific pathogen. Taken together, a more effective and rational approach to the prevention of new viral epidemics in humans and animals is essential. Research efforts to mitigate the effects of infectious threats, focusing on improved surveillance and diagnostic capabilities, and also on the development of intervention strategies such as vaccines and antiviral agents, are crucial.

Genomics-based tools are a potential candidate to respond to these challenges. Recent advances in genomics-based tools have allowed more sophisticated understanding of interactions among genes and genetic pathways, the environment, and the host and its pathogens. Already, it is apparent that genomics tools have started to change the practice of medicine. For example, they have been used to assess prognosis and guide therapy in several forms of cancer, to stratify patients according to risk of long-QT syndrome, and to shed light on the response to certain drugs, such as antiepileptic agents (Priori et al., 2003; Siddiqui et al., 2003; Van de Vijver et al., 2002). Our understanding of how living organisms and systems within organisms interact with each other and respond to the environment all changes dramatically in this genomic era. Full benefit from genomics tools in combating viral diseases requires understanding the dynamics of infection with regard

to viral diversity, evolution, and epidemiology, in combination with a better understanding of the pathogenesis of the infection and the molecular basis of the host response to infection. In this chapter, we describe how genomics may aid in mitigating viral threats, by highlighting potential roles and limitations of genomics approaches in surveillance and diagnostics, and development of vaccines and antivirals (Figure 97.2). As the clinical outcome of an infectious disease depends on properties of both the pathogen and the host, a distinction is made between viral genomics and host genomics.

VIRAL GENOMICS

Viral genomics can be defined as the study of viral genomes, functions of their individual genes, and interactions of all of the genes present in a viral genome. Most studies involving viral genomics focus on virus discovery and diversity analyses, which in turn positively influence the development and efficacy of surveillance tools, diagnostics, vaccines, and antivirals (Figure 97.2).

Virus Discovery

Virus discovery often starts with the identification of a viral disease entity, which subsequently requires the identification of the viral pathogen in clinical samples, and association between the disease and the presence of the virus. To increase the sensitivity of detection and reduce the time needed for diagnosis, laboratories today increasingly perform virus species-specific

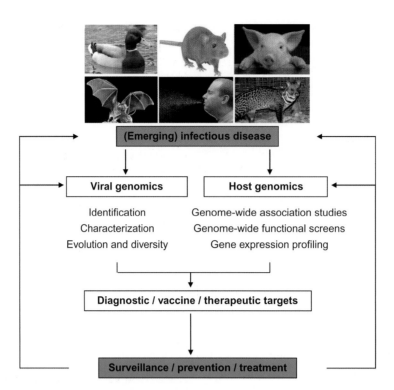

Figure 97.2 Genomics-based approaches used in the control of (emerging) virus diseases from disease to the development of diagnostics, vaccines, and therapeutics [inspired by Seib et al., 2009]. Any (emerging) infectious disease should be studied from a viral and host genomics perspective. Viral genomics aims at identification and characterization of viruses and virus diversity, whereas host genomics focuses on factors that influence how hosts respond to the viral infection. Both types of studies will reveal targets for diagnostic, vaccine, and therapeutic strategies, which in turn will be used in surveillance, prevention, and treatment of the disease.

molecular assays for virus identification in clinical samples. Classical methods used for the identification of known viruses include cell culture-based assays, immunological assays, and molecular detection methods such as polymerase chain reaction (PCR). Although these diagnostic assays are successful, failure rates in determining the etiological cause of the disease can be significant (Granerod and Crowcroft, 2007; Hajjeh et al., 2002), at least in part owing to the limited detection of divergent viruses due to the high specificity of these assays. During the outbreak of pandemic H1N1 influenza virus in 2009, rapid influenza diagnostic tests (RIDTs) were widely used for their ability to quickly detect viral antigen in respiratory clinical specimens. In a comparative study by the Centers for Disease Control and Prevention (CDC), 65 clinical respiratory specimens, which previously tested positive for pandemic H1N1, and seasonal H1N1 and H3N2 influenza virus by reverse transcriptase (RT)-PCR, were tested using three different RIDTs (Table 97.1). The RT-PCR assay is more sensitive than antigen detection assays. Moreover, the sensitivity of the RIDTs also seemed to vary with different influenza virus strains (Centers for Disease Control and Prevention, 2009). The failure rates in determining the etiological cause for example in encephalitis, acute flaccid paralysis, and non A-E hepatitis were reportedly 30–85% (Granerod and Crowcroft, 2007), 12% (Saeed

et al., 2007), and 18–62% (Chu et al., 2001), respectively. In a relatively large proportion of patients suffering from diarrhea (Finkbeiner et al., 2008) and acute respiratory illnesses (Juven et al., 2000), no pathogens could be detected, despite the use of a wide range of sensitive diagnostic assays. There are many clinical syndromes in which viruses are suspected to play a role, but for which current techniques fail to uncover an etiological agent, underlining the limitations of even the best current methodologies for diagnosis of viral infections. Overall, the fact that many acute and chronic diseases with a suspected infectious cause are still of unknown etiology may point to the presence of as-yet-unidentified viruses in the human population. Apparently, the virus-specific assays currently used in routine diagnostic settings are not equipped for the identification of new and previously unrecognized pathogens. Modern genomics-based laboratory techniques, such as generic PCRs, pan-viral microarrays, and sequence-independent amplification combined with high-throughput sequencing, will play an increasingly important role in virus identification and discovery.

With the development of PCR and sequencing techniques in the 1970s and 80s, we have seen an unprecedented increase in the identification and characterization of viral genomes. This led to the development of new species-specific diagnostic assays for viruses infecting animals and humans, which are

TABLE 97.1	Performance characteristics of RT-PCR versus rapid influenza diagnostic tests (RIDTs), point-of-care tests that detect influenza viral nucleoprotein antigen, in patients infected with pandemic H1N1 influenza virus 2009 (pH1N1), seasonal H1N1 virus (sH1N1), or seasonal H3N2 (sH3N2)			
RIDT	**Detection**	**Total no. of specimens positive by RIDT/ Total no. positive by RT-PCR**	**Sensitivity (%)**	
BinaxNow	pH1N1	18/45	40	
	sH1N1	3/5	60	
	sH3N2	12/15	80	
Directigen EZ Flu	pH1N1	21/43	49	
	sH1N1	3/4	75	
	sH3N2	10/12	83	
QuickVue	pH1N1	31/45	69	
	sH1N1	4/5	80	
	sH3N2	12/15	80	

Table adapted from Centers for Disease Control and Prevention (2009).

still being used in routine diagnostic settings, and also allowed the development of generic PCR assays – PCR assays specific for a broader taxonomic range than one virus species (e.g., a whole virus genus or family). These assays decrease time and effort in demonstrating the presence of a known pathogen in clinical samples, although sometimes at the expense of losing some sensitivity compared to PCR-based assays targeting single virus species. In addition, these assays allow the identification of new virus species within already-known viral families, as was demonstrated, for example, for simian homologs of human herpesviruses (Lacoste et al., 2000), picorna-like viruses in seawater (Culley and Steward, 2007), a new begomovirus infecting *Corchorus capsularis* plants (Ha et al., 2006), and astroviruses from roe deer (Smits et al., 2010b). Generic PCR assays thus provide a useful new, modern, genomics-based tool for virus identification, surveillance and diagnostics.

However, the limitations of generic PCR assays become readily apparent in more complex biological situations with diseases caused by co-infections of different viruses or diseases for which no etiological agent has been identified. Also, viruses are constantly evolving, the maximum number of detectable viruses in a single assay is relatively low, and in a diagnostic setting discrimination between subtypes or genera of viruses requires additional labor-intensive procedures for final diagnosis. The limitations of existing viral detection methodologies

prompted DeRisi and coworkers (Wang et al., 2002) to devise a method for comprehensive and unbiased analysis of viral prevalence in a given biological setting. Using a genomics-based approach for virus identification, they have developed a long oligonucleotide (70-mer) DNA microarray with the potential to simultaneously detect hundreds of viruses. It was shown that viruses represented on the microarray were readily detected and identified, as were distinct viruses that were not explicitly represented on the microarray. Distinct hybridization patterns were observed when different subtypes of viruses within a certain family were analyzed, demonstrating the ability to discriminate subtypes. The pan-viral microarray could also be used for virus discovery provided that the new viruses are sufficiently related to those already represented on the microarray to permit specific hybridization. For example, in the diagnostic setting, the use of the pan-viral microarray was shown by identification of human parainfluenza virus 4 in an immunocompetent adult with a life-threatening acute respiratory illness, where an extensive panel of routine clinical diagnostic assays failed to yield a diagnosis (Chiu et al., 2006). During the outbreak of SARS in 2003, the presence of a previously uncharacterized coronavirus was revealed using the pan-viral DNA microarray (Wang et al., 2003), and a novel gammaretrovirus, XMRV, was identified in ~40% of patients with prostate tumors who are homozygous for the R462Q variant of RNAse L (Urisman et al., 2006). Although pan-viral microarrays have considerable advantages over generic PCR assays, especially in terms of broader-reaching and less biased detection of viruses and the possibility of subtyping viruses, highly divergent viruses may still elude identification. Also, implementation as a standard diagnostic assay in clinical settings would require the establishment of a reference library of hybridization signatures for hundreds of individual viral subtypes, training of skilled laboratory workers to read these "viral barcodes," and the ability to create a platform in which the method is cost-effective and reproducible. These barriers may hamper large-scale clinical use of this method.

Knowing that the speed, accuracy, efficiency, and cost-effectiveness of DNA sequencing have improved continuously since its introduction, it is no longer unrealistic to sequence thousands of genes of a single individual with a suspected viral infection. Therefore, the next generations of sequencing techniques have certain important advantages over pan-viral microarrays when used in combination with sequence-independent amplification of nucleic acids. Sequence-independent amplification methods are relatively simple and fast, lack bias towards any virus group, and do not require specific reagents to detect new viruses, even if they are highly divergent from those that are already known. The major limitation of these methods is their limited suitability for samples in which host nucleic acids and viral nucleic acids cannot easily be separated, such as solid tissues or tissue biopsies. The relative contribution of host nucleic acids compared to viral nucleic acids is extremely large, decreasing the chance of identification of viral sequences even when using next-generation sequencing platforms, which are capable of generating more than 400,000 sequencing reads per instrument run (Kuroda et al., 2010). The key to virus identification by sequence-independent amplification methods is increasing the levels of viral nucleic acids while reducing background nucleic acids. This is generally done via filtration, centrifugation, and enzymatic removal of non-particle protected nucleic acids (Delwart, 2007). Already, many new viruses have been identified using sequence-independent nucleic acid amplification techniques (Allander et al., 2005; Finkbeiner et al., 2009; Fouchier et al., 2004; Gaynor et al., 2007; Jones et al., 2005; Kapoor et al., 2008b, 2009; Palacios et al., 2008; Smits et al., 2010b; Van der Hoek et al., 2004; Van Leeuwen et al., 2010; Victoria et al., 2008). Viral genomic analyses of environmentally collected samples or unmanipulated biological samples, such as fecal material, showed a high diversity of viral communities (Breitbart et al., 2002, 2003; Finkbeiner et al., 2008; Venter et al., 2004). Before wide-scale random amplification of nucleic acids and sequencing becomes an accepted medical practice, however, a number of challenges need to be overcome, including development of high-throughput, cost-effective, and reproducible methods, bioinformatics to analyze the generated data, training of skilled laboratory workers to interpret the data, and linking the presence of virus in samples to disease (Delwart, 2007; Service, 2006; Ten Bosch and Grody, 2008). In case new viruses are identified, epidemiological studies and ultimately experimental infection studies in appropriate animal models are required, to prove that the identified virus is truly the causal agent of infection.

Modern genomics-based laboratory techniques will play an increasingly important role in the surveillance and identification of new and previously unrecognized pathogens in both animals and humans. This will provide new targets for development of virus discovery tools, specific diagnostic and surveillance assays, and vaccines and antivirals that can be used in the clinical setting. Although ongoing innovations in genomics-based tools would aid perfection of these techniques, allowing them to become standard in medical practice, implementation into routine diagnostic settings awaits several major technical breakthroughs and improvements in cost-effectiveness and analysis of results.

Virus Diversity

Genomics-based tools are not only useful for virus identification and characterization; they also allow investigation of patterns of virus evolution by in-depth sequencing of viral genomes in individual hosts. There is growing evidence that epidemiological inferences that are made on consensus sequences of populations of viruses within an individual hide information on generic and phenotypic diversity. Moreover, viral lineages identified in individual hosts are often highly divergent, suggesting that they did not evolve in the host via mutations and antigenic drift. Thus, it is clear that single individuals can be superinfected with multiple virus variants, which may have major consequences for development of vaccines and intervention strategies. At present, full genome sequences

of each major human pathogen are available, and over 3000 viral genomes have been completed in recent years. With the advances in sequencing technology and bioinformatics, exponential growth is expected in sequence information of viruses and viral quasispecies in individual hosts. This increasing availability of viral genomic sequences allows a previously unattainable route to investigate phenomena such as viral pathogenesis, virus evolution, development of resistance to treatment, and vaccine development, as illustrated below.

Both hepatitis C virus (HCV) and human immunodeficiency virus (HIV) continue to exert a substantial disease burden on millions of chronically infected patients worldwide. These viruses are highly genetically diverse, both within individuals and globally (Rambaut et al., 2004; Simmonds, 2004; Simmonds et al., 2005). This has hampered and prohibited vaccine development and development of efficient antivirals. High levels of replication, and an error-prone RNA-dependent RNA polymerase and a reverse transcriptase enzyme in HCV and HIV, respectively, result in a cloud of individual quasispecies sequences in any infected person. Minor quasispecies variants that are resistant to antiviral actions of small molecule compounds are readily selected, also because they are often already present in the quasispecies swarm, and become dominant during therapy. A recent study examined treatment-naïve HCV-infected patients for pre-existing resistance mutations for 27 developmental antiviral compounds directed against nonstructural proteins NS3 (protease) and NS5B (RNA-dependent RNA polymerase). The results showed that 44.4% of genotype 1b patients had at least one mutation associated with resistance and 2.7% had mutations in both NS3 and NS5B (Gaudieri et al., 2009). Shimakami and coworkers (2009) commented that an even greater frequency of such mutations would undoubtedly be discovered with "deeper" sequencing methods. The risk of resistance can be diminished by using combinations of small antiviral compounds, although in vitro studies have already showed that double-resistant mutants can be selected, albeit often at an associated fitness cost (Mo et al., 2005). Similar observations have been made in HIV infections. Primary drug resistance mutations in treatment-naïve patients were observed in 5–10% of the patients (Gupta et al., 2008; Paredes and Clotet, 2010). Several primary resistance surveillance studies using more sensitive genomics-based resistance assays have demonstrated increases in detection of HIV-1 variants with higher primary resistance relative to standard population-based sequencing (Metzner et al., 2005; Paredes et al., 2007). If antiretroviral therapy is to remain successful, resistance-associated failure of therapy must be prevented by (1) identifying primary resistance, (2) tailoring antiretroviral therapy to ensure that all different compounds retain full antiviral activity, (3) ensuring long-term adherence by patients, (4) detecting failure of therapy early and identifying emerging resistance mutations, and (5) managing failure of therapy by switching to new antiretroviral regimens (Hirsch, 2008). Here, genomics-based tools aimed at resistance detection could help design more effective personalized antiretroviral therapies.

Another example of how genomics-based approaches influence infectious viral disease prevention methods comes from influenza A virus evolution studies. Computational analyses of sequence data have changed our views on influenza A virus evolution. Genomics data have revealed that genome-wide interactions are a critical aspect of influenza virus evolution. This was shown by a global rise in resistance of influenza A viruses to adamantanes, one of the first-generation antivirals (Bright et al., 2005, 2006). Surprisingly, this rise in resistance was also observed in regions where adamantanes are rarely used. Apparently, the resistance mutation was linked to another beneficial mutation located elsewhere in the genome (Simonsen et al., 2007), resulting in much broader spread of influenza A viruses carrying the resistance mutation. Genomics-based studies also confirmed the importance of reassortment in influenza virus epidemiology and evolution, which sometimes results in generation of viruses with altered antigenic properties and in vaccine failure (Holmes, 2009; Holmes et al., 2005). The raw material for reassortment is provided by co-circulation of multiple and often diverse viral lineages from the same subtype in a given population (Holmes, 2009; Holmes et al., 2005; Nelson et al., 2008), and a global source population of human influenza viruses, located in Southeast Asia, ignites the seasonal influenza epidemics in the rest of the world (Rambaut et al., 2008; Russell et al., 2008). A global influenza surveillance network routinely characterizes genetic and antigenic properties of influenza A viruses, the latter mainly using the classical hemagglutination inhibition assay. This is a binding assay based on the ability of influenza viruses to agglutinate red blood cells with one of their surface proteins, hemagglutinin (HA), and the ability of antisera raised against the same or related strains to block this agglutination. In combination with sequence information on the immunogenic HA1 domain and epidemiological data, antigenic data are used to select strains for the vaccine against seasonal epidemic influenza (Smith et al., 2004). The fact that the source population of human influenza A viruses is known is crucial to vaccine development, as vaccines can fail due to a periodically more accelerated antigenic evolution than expected from genetic data (Smith et al., 2004). Knowing where these variants are generated and intensively surveying that geographic area will aid in formulating vaccine strains. Future genomics-based studies on influenza A virus variability, in combination with comparative pathology studies, may unravel, for example, whether there is a genetic basis for the observed differences in clinical presentation of influenza in humans.

Thus, virus diversity studies can have an impact on vaccine efficacy, antiviral efficacy, and resistance. The challenges for future virus diversity and evolution studies are to integrate viral genomics with data on clinical symptoms and signs, pathology, vaccine development, and antiviral resistance studies. Advances in genomics-based tools, in combination with increasing amounts and more in-depth analysis of individual samples from patients with mild to more severe illness, may allow *a priori* risk assessments regarding severity of disease, and may guide vaccine and therapy options. As mentioned previously, implementation of these methods into routine diagnostic settings will make it necessary to take large hurdles in the future.

HOST GENOMICS

It is common knowledge that immune status, age, nutritional status, and many other factors play an important role in the clinical outcome of a viral infection, underlining the importance of studying both host and pathogen parameters with genomics tools in the battle against viral infections. Following completion of the Human Genome Project, the investigation of underlying heritable mutations capable of inducing or increasing susceptibility to disease has become possible. But these mutations do not explain all patient-to-patient differences in pathology. In contrast to analyzing a patient's DNA, gene expression profiling examines the expression of many different RNAs and/or proteins, and integrates the impact of DNA-based effects and a variety of other factors, including viral infections, that modulate cellular responses. Both analyses lead to a more complete understanding of the mechanisms contributing to pathology, and can drive development of new diagnostic and therapeutic strategies (Figure 97.2).

Genome-wide Association Studies and Functional Screens

The underlying mechanisms of many diseases and efficient treatments remain obscure. It also remains largely unknown how host factors influence the viral life cycle. Genomic research offers new opportunities to determine how diseases occur by characterizing molecular abnormalities underlying disease processes, and to identify host genes required for viral replication. The latter is done by performing genome-wide functional screens using small interfering RNAs (siRNAs), which interfere with the expression of a specific gene, in an infectious *in vitro* culture system. Genome-wide association studies (GWAS) are used to identify genes associated with disease. Surveys of genetic variation, with typically hundreds of thousands of single nucleotide polymorphisms (SNPs) in patients with and without a given disease, are performed to identify variant SNPs that are found significantly more commonly in patients with disease compared with the disease-free controls.

Many SNPs, associated with a wide range of diseases and characteristics, have been described (Cooper et al., 2008; Gudbjartsson et al., 2009; Liu et al., 2008; Miyagawa et al., 2008). For example, some 450 different heritable mutations involved in development of primary cardiomyopathies have been identified, and it was estimated that SNPs are responsible for up to 70% of hypertrophic cardiomyopathy (Margulies et al., 2009). Assays to screen for cardiomyopathy-causing mutations in affected patients are available, with 30–60% likelihood of detecting the genetic basis in a patient with hypertrophic and dilated cardiomyopathies (Margulies et al., 2009), which is likely to increase as more mutations are identified. Ultimately, these types of assays will be accessible to clinicians, who can identify patients and family members at increased risk for disease, and possibly prevent or modulate disease progression with disease-modifying therapies.

Despite the enormous impact of viral infections on human health, relatively few GWAS have been performed to successfully identify disease susceptibility variants in viral infections. Perhaps not surprisingly, most studies that have been done concern chronic, and often lifelong, infections, such as HIV and HCV. Recently, it was shown that several SNPs located near the *IL28B* gene were significantly more common in HCV patients who responded to interferon therapy than in non-responders (Ge et al., 2009; Suppiah et al., 2009; Tanaka et al., 2009), suggesting that there is a genetic basis for efficiency of anti-HCV therapy. Although diagnostic testing to identify likely responders to interferon therapy may be a future possibility, clinical decision-making will be difficult as the effect of the advantageous *IL28B* SNPs is not absolute, and no alternative to interferon therapy exists. Fellay and coworkers (2007) used a GWAS to determine why viral loads prior to onset of AIDS can differ up to five logs between individual patients. They identified polymorphisms associated with *HLA* loci B and C, and concluded that they were major determinants in the difference of circulating viral loads (Fellay et al., 2007). Moreover, their data underline the importance of carrying out similar studies for other chronic, and perhaps also acute, infectious diseases. After this initial study, several others implicated other loci that influence viral load and progression of HIV-1 infection to AIDS (Loeuillet et al., 2008; Pelak et al., 2010). Remarkably, *CCR5*, a gene critical for HIV infection, was not detected at all. This shows the difficulties imposed by strict statistical tests that, although they allow discovery of new genes, also guarantee that many true associations are missed in the process of avoiding false-positive associations (Winkler, 2008).

Another method developed to find HIV and HCV dependency factors is a siRNA genome-wide functional screen (Brass et al., 2008; Li et al., 2009). In this approach, siRNAs were used to knock out some 20,000 genes one by one in a cell line that supported HIV or HCV replication. In these assays, many HIV and HCV dependency factors that influence viral replication were identified. As these factors were not required for cell viability, they are excellent, genetically stable candidates for antiviral drugs. It is of note that 10 of the host genes needed for HIV replication are also important for HCV replication, which could be exploited in cases of co-infection (Li et al., 2009). In addition, several identified HCV dependency factors were found to be required for replication of two other members of the *Flaviviridae* family, West Nile virus and

dengue virus (Krishnan et al., 2008), and for particle formation of multiple viruses, including Marburg, Ebola, and measles viruses (Murray et al., 2005; Shisheva, 2008). The siRNA screens are, however, limited to the identification of cellular genes. Moreover, the effects on disease of the identified gene products in the infected population remain unknown. Similar functional genomics experiments were performed to identify host cell factors that are temporarily dispensable for the host, but crucial for virus replication of a broad range of influenza A viruses, including highly pathogenic avian H5N1 and pandemic H1N1 viruses (Prusty et al., 2011). These host factors provide promising new antiviral targets that can block or alleviate the downstream effects of pathogen infection.

Overall, GWAS and genome-wide functional screens have been shown to elucidate biomarkers of disease risk, provide new biological insights into pathophysiological processes, clarify how known risk factors predispose to disease development, and discover host genes required for replication of viruses. To fully interpret the results of future studies, data should be incorporated from functional and expression experiments and virus-mediated changes in gene expression (Ge et al., 2008). The insights obtained from these types of studies may be used in development of intervention strategies, and hold promise for the personalization of medicine by linking individual genetic information to disease prevention and therapy response. It is highly interesting that different viruses share the need for certain host proteins for replication *in vitro*. Defining host factors and pathways upon which multiple viruses depend could aid in development of common intervention strategies applicable to a large number of viral infections.

Gene Expression Profiling

As the outcome of viral infections depends as much on viral as on host properties, scientists have used a wide array of methods to study how viruses interact with the host and the mechanisms by which the host protects itself against viruses. With the development of expression microarrays and other high-throughput genomics and proteomics techniques, our ability to characterize the key interactions between host and pathogen has greatly increased, and has unraveled possibilities for development of diagnostic assays and multi-targeted therapies designed to limit virus replication and mitigate immunopathology (Andeweg et al., 2008). Treatments may even be tailored to target host response pathways common to many virus infections, which could be used early in infection when an accurate diagnosis is not yet available. Although expression profiling experiments have been performed for many different viruses, including HIV, HCV, SARS coronavirus, influenza A virus, dengue, and West Nile, only a few examples will be discussed here to show the principal applications of genomics/proteomics-based tools to study infections with different pathogenic viruses.

Dengue virus infection is a global public health problem that causes 50–100 million cases of dengue fever (DF) yearly, resulting in 500,000 clinical cases of the life-threatening dengue hemorrhagic fever syndrome (DHF). DF and DHF patients display a similar clinical picture early after infection. In endemic areas with outbreaks of dengue cases, the capacity of medical assistance is limited, and an early discrimination between DF patients and patients at risk of developing DHF would allow a more effective allocation of medical attention. A recent study using genomics-based tools on a well-characterized cohort of dengue patients aimed at obtaining insight into early immune mechanisms associated with dengue severity and identifying biomarkers to predict infection outcome. The results suggested that DF and DHF are extremes of a continuum of the same disease, and not two separate diseases (Nascimento et al., 2009; Sierra et al., 2007). A set of genes were identified that, upon quantification by PCR assays on the first medical visit of a dengue virus-infected patient, could be used to predict whether the patient will develop DHF symptoms a couple of days later (Nascimento et al., 2009).

Peripheral blood gene expression signatures have been identified that accurately distinguish individuals with symptomatic acute respiratory infection from uninfected individuals. In addition, these gene expression signatures distinguish individuals with respiratory viral infection from bacterial infection, and even distinguish between individuals with respiratory viral infections caused by human rhinovirus, influenza A virus, and respiratory syncytial virus (Zaas et al., 2009). Thus, gene expression profiling may serve as a useful diagnostic test for tailoring treatment in acute respiratory infections and could enhance traditional diagnostic assays.

To individualize treatment in chronic hepatitis C virus-infected patients, unbiased proteomic profiling has been used to identify host predictors of virologic response to interferon-based therapy. A serum-based protein signature was identified that could accurately predict whether chronic HCV-infected patients would respond to standard-of-care interferon-based therapy prior to treatment (Patel et al., 2011). As interferon-based therapy is associated with significant side effects and only about half of treated subjects respond to the relatively expensive therapy, accurate prediction of antiviral efficacy is an important tool in the proper tailoring of individual treatment processes.

Gene expression profiling studies in simian immunodeficiency virus (SIV)-infected natural and non-natural hosts have provided a unique view of how host responses influence lentiviral infection outcome. Many non-human primate species in sub-Saharan Africa are infected with SIV. These natural reservoir hosts do not develop AIDS as a result of infection, and appear to live a normal lifespan. In contrast, non-natural hosts such as Asian pig-tailed macaques develop AIDS in a manner similar to HIV-infected humans (Sodora et al., 2009). During acute infection of natural and non-natural hosts, virus replication peaks within a few weeks after infection, and high viral loads remain during the chronic phase of infection. It long remained unclear how natural hosts could sustain high viral loads, without developing disease. Gene expression profiling in

pathogenic and non-pathogenic SIV infection models revealed that a strong type I interferon response is initially induced in both models, but is resolved after peak viral load in non-pathogenic SIV infection models only (Lederer et al., 2009). The natural host also exhibited better preservation of overall cell homeostasis, lower levels of immune activation, no progressive depletion of CD4+ T cells, and preservation of mucosal Th17 cells (Bosinger et al., 2009; Favre et al., 2009; Jacquelin et al., 2009; Lederer et al., 2009; Sodora et al., 2009). These results suggest that active immune down-regulatory mechanisms, rather than intrinsically attenuated immune responses (Mandl et al., 2008), underlie the low level of immune activation and coexistence with SIV infection without progression to AIDS. Human long-term asymptomatic HIV-infected patients with high viral loads and low immune activation status have been described that seem to underscore the studies in natural reservoir non-human primate hosts (Choudhary et al., 2007). These data show that the supposedly longstanding evolutionary coadaptation between primate lentiviruses and natural reservoir hosts, which allowed coexistence of virus and host, may eventually also occur in humans (Choudhary et al., 2007; Sodora et al., 2009). Most importantly, the genomics-based studies in SIV-infected natural and non-natural hosts showed that future research should focus on identifying the mechanisms responsible for chronic immune activation, and should devise immunomodulatory strategies that prevent it.

The zoonotic transmission of SARS coronavirus caused an outbreak in 2003 of severe acute respiratory syndrome, a pneumonic disease in humans, with an overall mortality rate of ~10%. The clinical course and outcome of SARS coronavirus-associated disease were more favorable in children compared to adolescents and adults (Hon et al., 2003; Leung and Chiu, 2004; Wong et al., 2003); elderly patients had a poor prognosis, with mortality rates of up to 50% (Peiris et al., 2003, 2004). This predisposition to disease is also shown when aged and young cynomolgus macaques are experimentally infected with SARS coronavirus, with aged macaques showing significantly more lung pathology upon infection than young adult macaques (Smits et al., 2010a). Gene expression profiling studies revealed that aged macaques display a more prominent host response to infection than young adult macaques, with an increase in differential expression of genes associated with inflammation, whereas type I interferon expression is reduced. This observation triggered another experiment, in which aged macaques were therapeutically treated with type I interferon after infection, which reduced pathology and proinflammatory gene expression without affecting virus replication levels. As the clinical manifestations of SARS coronavirus-associated disease are highly similar to acute lung injury caused by multiple other pathogenic conditions, including sepsis, gastric acid aspiration, and pulmonary infections such as H5N1 avian influenza virus, in multiple host species, a conserved injury pathway was proposed (Imai et al., 2008). Assuming that there is such a conserved injury pathway that can be induced by multiple triggers, including pandemic influenza A viruses, modulation of the host response by type I interferon provides a promising outlook for novel intervention strategies (Smits et al., 2010a). More recent studies in African green monkeys, however, indicate that distinct SARS coronavirus-induced acute lung injury pathways may exist in different host species, which will need to be taken into account when analyzing intervention strategies in these species (Smits et al., 2011).

Microarray expression profiling studies have also been used to determine why newly emerging zoonotic influenza virus infections often cause more severe disease than the seasonal epidemic influenza A viruses. By using genetically engineered influenza A viruses, it was shown that the NS1 protein derived from the 1918 Spanish H1N1 pandemic influenza virus blocked expression of the host antiviral response more efficiently than the NS1 protein from an established seasonal influenza virus (Geiss et al., 2002). Other studies in mice and macaques with genetically engineered influenza viruses containing gene segments from the 1918 H1N1 virus or the pathogenic avian H5N1 influenza A virus suggest that highly pathogenic influenza viruses induce severe disease through aberrant and persistent activation of proinflammatory cytokine and chemokine responses (Baskin et al., 2009; Kash et al., 2004, 2006; Kobasa et al., 2007). In ferrets, it was shown that infection with H5N1 influenza A virus induced severe disease, with strong expression of *CXCL10*, among others. Treatment with an antagonist of the *CXCL10* receptor reduced the severity of symptoms and viral titers compared to control animals (Cameron et al., 2008). These experiments show that comparison of different viruses and strains, with respect to their impact on host gene expression, could lead to development of therapeutic interventions, but also could aid in predicting the pathogenicity of new or past virus strains by using high-throughput profiling techniques (Geiss et al., 2002).

Overall, gene expression profiling studies of viral infections provide insight into pathophysiological processes, elucidate biomarkers of disease risk, and allow knowledge-guided development of host response-modulating therapeutic treatments. Future efforts should focus on comparing and contrasting the outcomes of infection with closely related viruses in different host species, to delineate the various host pathways and viral traits that cause the differences in disease outcome. Integration of genomics/proteomics data obtained for different viruses and different hosts could potentially elucidate common host pathways involved in development of pathology, and allow rational design of diagnostic assays and host-modulating antipathology strategies applicable to a large number of viral infections.

are predominantly applied to identification and diagnosis of viral infections, virus diversity studies, GWAS, genome-wide functional screens, and host gene expression profiling. These studies are delivering new targets of both viral and host origin for the development of diagnostic and surveillance assays, vaccines, and therapeutics (Figure 97.2). From a corporate perspective, large populations of patients could benefit from any vaccine or therapeutic agent that becomes financially of interest to produce. True personalized treatment for single individuals will be more difficult to accomplish. Thus, identifying biomarkers, assays, and host-modifying treatments, that work for and/or against a broad spectrum of different viruses, are of utmost importance. Comparative studies into dynamics of infection with regard to viral diversity, evolution, and epidemiology, in combination with the molecular basis of the host response to infection and pathogenesis, may enable the rational design of multipotent biological response modifiers that will mitigate the effects of viral diseases. By focusing on broad-acting intervention strategies and surveillance for new viral pathogens, we may be better prepared for any outbreak of newly emerging infectious disease.

To realize the full potential of genomics-based tools and data in a clinical setting, several major obstacles need to be overcome. For one, methods need to be cost-effective and in a high-throughput format. Although it has been contemplated that within two decades it will be possible to sequence anyone's entire genome for less than $1000, this may be complicated by ethical, legal, social, and societal issues. Existing DNA sequence and/or method intellectual property claims may have an impact on development of multi-gene diagnostic tests (Chandrasekharan and Cook-Deegan, 2009). In addition, although seemingly far-fetched at present, privacy measures regarding genetic information need to be enacted and signed into law, to prevent abuse of such data by, for example, insurers or employers, which may one day become a realistic situation. Secondly, a large investment in bioinformatics tools, databases and data management is required, to make optimal use of the generated data. Thirdly, clinical diagnosis of disease will largely change when use of genomics-based tools becomes standard operating procedure, and molecular diagnosticians will need to gain experience in interpreting the data generated by these radically different and rapidly evolving techniques. However, as with so many other molecular biology tools that have found their way into the clinical setting, there is no reason why genomics-based tools could not transition from the research setting into the clinic in the future, despite the hurdles that need to be overcome (Ten Bosch and Grody, 2008).

FUTURE CHALLENGES

It is clear that genomics-based tools have obtained a firm foothold in many scientific disciplines. In combating viral infections, either established or newly emerging, genomics-based tools

ACKNOWLEDGMENTS

We thank M. Van Leeuwen and Dr E.G.M. Berkhoff for critical reading of this book chapter.

REFERENCES

Allander, T., Andreasson, K., Gupta, S., et al., 2007. Identification of a third human polyomavirus. J Virol 81, 4130–4136.

Allander, T., Tammi, M.T., Eriksson, M., Bjerkner, A., Tiveljung-Lindell, A., Andersson, B., 2005. Cloning of a human parvovirus by molecular screening of respiratory tract samples. Proc Natl Acad Sci USA 102, 12,891–12,896.

Andeweg, A.C., Haagmans, B.L., Osterhaus, A.D., 2008. Virogenomics: The virus–host interaction revisited. Curr Opin Microbiol 11, 461–466.

Baskin, C.R., Bielefeldt-Ohmann, H., Tumpey, T.M., et al., 2009. Early and sustained innate immune response defines pathology and death in nonhuman primates infected by highly pathogenic influenza virus. Proc Natl Acad Sci USA 106, 3455–3460.

Bosinger, S.E., Li, Q., Gordon, S.N., et al., 2009. Global genomic analysis reveals rapid control of a robust innate response in SIV-infected sooty mangabeys. J Clin Invest 119, 3556–3572.

Brass, A.L., Dykxhoorn, D.M., Benita, Y., et al., 2008. Identification of host proteins required for HIV infection through a functional genomic screen. Science 319, 921–926.

Breitbart, M., Hewson, I., Felts, B., et al., 2003. Metagenomic analyses of an uncultured viral community from human feces. J Bacteriol 185, 6220–6223.

Breitbart, M., Salamon, P., Andresen, B., et al., 2002. Genomic analysis of uncultured marine viral communities. Proc Natl Acad Sci USA 99, 14,250–14,255.

Bright, R.A., Medina, M.J., Xu, X., et al., 2005. Incidence of adamantane resistance among influenza A (H3N2) viruses isolated worldwide from 1994 to 2005: A cause for concern. Lancet 366, 1175–1181.

Bright, R.A., Shay, D.K., Shu, B., Cox, N.J., Klimov, A.I., 2006. Adamantane resistance among influenza A viruses isolated early during the 2005–2006 influenza season in the United States. JAMA 295, 891–894.

Calattini, S., Chevalier, S.A., Duprez, R., et al., 2005. Discovery of a new human T-cell lymphotropic virus (HTLV-3) in Central Africa. Retrovirology 2, 30.

Cameron, C.M., Cameron, M.J., Bermejo-Martin, J.F., et al., 2008. Gene expression analysis of host innate immune responses during lethal H5N1 infection in ferrets. J Virol 82, 11,308–11,317.

Centers for Disease Control and Prevention, 2009. Evaluation of rapid influenza diagnostic tests for detection of novel influenza A (H1N1) virus – United States, 2009. MMWR Morb Mortal Wkly Rep 58, 826–829.

Chandrasekharan, S., Cook-Deegan, R., 2009. Gene patents and personalized medicine – what lies ahead? Genome Med 1, 92.

Chen, Z., Telfier, P., Gettie, A., et al., 1996. Genetic characterization of new West African simian immunodeficiency virus SIVsm: Geographic clustering of household-derived SIV strains with human immunodeficiency virus type 2 subtypes and genetically diverse viruses from a single feral sooty mangabey troop. J Virol 70, 3617–3627.

Chiu, C.Y., Rouskin, S., Koshy, A., et al., 2006. Microarray detection of human parainfluenzavirus 4 infection associated with respiratory failure in an immunocompetent adult. Clin Infect Dis 43, e71–e76.

Choudhary, S.K., Vrisekoop, N., Jansen, C.A., et al., 2007. Low immune activation despite high levels of pathogenic human immunodeficiency virus type 1 results in long-term asymptomatic disease. J Virol 81, 8838–8842.

Chu, C.M., Lin, D.Y., Yeh, C.T., Sheen, I.S., Liaw, Y.F., 2001. Epidemiological characteristics, risk factors, and clinical manifestations of acute non-A-E hepatitis. J Med Virol 65, 296–300.

Chua, K.B., 2003. Nipah virus outbreak in Malaysia. J Clin Virol 26, 265–275.

Cooper, M.L., Adami, H.O., Gronberg, H., Wiklund, F., Green, F.R., Rayman, M.P., 2008. Interaction between single nucleotide polymorphisms in selenoprotein P and mitochondrial superoxide dismutase determines prostate cancer risk. Cancer Res 68, 10,171–10,177.

Culley, A.I., Steward, G.F., 2007. New genera of RNA viruses in subtropical seawater, inferred from polymerase gene sequences. Appl Environ Microbiol 73, 5937–5944.

Delwart, E.L., 2007. Viral metagenomics. Rev Med Virol 17, 115–131.

De Wit, E., Kawaoka, Y., De Jong, M.D., Fouchier, R.A., 2008. Pathogenicity of highly pathogenic avian influenza virus in mammals. Vaccine 26, D54–D58.

Drosten, C., Gunther, S., Preiser, W., et al., 2003. Identification of a novel coronavirus in patients with severe acute respiratory syndrome. N Engl J Med 348, 1967–1976.

Favre, D., Lederer, S., Kanwar, B., et al., 2009. Critical loss of the balance between Th17 and T regulatory cell populations in pathogenic SIV infection. PLoS Pathog 5, e1000295.

Fellay, J., Shianna, K.V., Ge, D., et al., 2007. A whole-genome association study of major determinants for host control of HIV-1. Science 317, 944–947.

Finkbeiner, S.R., Kirkwood, C.D., Wang, D., 2008. Complete genome sequence of a highly divergent astrovirus isolated from a child with acute diarrhea. Virol J 5, 117.

Finkbeiner, S.R., Li, Y., Ruone, S., et al., 2009. Identification of a novel astrovirus (astrovirus VA1) associated with an outbreak of acute gastroenteritis. J Virol 83, 10,836–10,839.

Fouchier, R.A., Hartwig, N.G., Bestebroer, T.M., et al., 2004. A previously undescribed coronavirus associated with respiratory disease in humans. Proc Natl Acad Sci USA 101, 6212–6216.

Gao, F., Bailes, E., Robertson, D.L., et al., 1999. Origin of HIV-1 in the chimpanzee *Pan troglodytes troglodytes*. Nature 397, 436–441.

Garten, R.J., Davis, C.T., Russell, C.A., et al., 2009. Antigenic and genetic characteristics of swine-origin 2009 A(H1N1) influenza viruses circulating in humans. Science 325, 197–201.

Gaudieri, S., Rauch, A., Pfafferott, K., et al., 2009. Hepatitis C virus drug resistance and immune-driven adaptations: Relevance to new antiviral therapy. Hepatology 49, 1069–1082.

Gaynor, A.M., Nissen, M.D., Whiley, D.M., et al., 2007. Identification of a novel polyomavirus from patients with acute respiratory tract infections. PLoS Pathog 3, e64.

Ge, D., Fellay, J., Thompson, A.J., et al., 2009. Genetic variation in IL28B predicts hepatitis C treatment-induced viral clearance. Nature 461, 399–401.

Ge, D., Zhang, K., Need, A.C., et al., 2008. WGAViewer: Software for genomic annotation of whole genome association studies. Genome Res 18, 640–643.

Geiss, G.K., Salvatore, M., Tumpey, T.M., et al., 2002. Cellular transcriptional profiling in influenza A virus-infected lung epithelial cells: The role of the nonstructural NS1 protein in the evasion of the host innate defense and its potential contribution to pandemic influenza. Proc Natl Acad Sci USA 99, 10,736–10,741.

Granerod, J., Crowcroft, N.S., 2007. The epidemiology of acute encephalitis. Neuropsychol Rehabil 17, 406–428.

Greninger, A.L., Runckel, C., Chiu, C.Y., et al., 2009. The complete genome of klassevirus – a novel picornavirus in pediatric stool. Virol J 6, 82.

Gudbjartsson, D.F., Bjornsdottir, U.S., Halapi, E., et al., 2009. Sequence variants affecting eosinophil numbers associate with asthma and myocardial infarction. Nat Genet 41, 342–347.

Gupta, R., Hill, A., Sawyer, A.W., Pillay, D., 2008. Emergence of drug resistance in HIV type 1-infected patients after receipt of first-line highly active antiretroviral therapy: A systematic review of clinical trials. Clin Infect Dis 47, 712–722.

Ha, C., Coombs, S., Revill, P., Harding, R., Vu, M., Dale, J., 2006. Corchorus yellow vein virus, a New World geminivirus from the Old World. J Gen Virol 87, 997–1003.

Haagmans, B.L., Andeweg, A.C., Osterhaus, A.D., 2009. The application of genomics to emerging zoonotic viral diseases. PLoS Pathog 5, e1000557.

Hajjeh, R.A., Relman, D., Cieslak, P.R., et al., 2002. Surveillance for unexplained deaths and critical illnesses due to possibly infectious causes, United States, 1995–1998. Emerg Infect Dis 8, 145–153.

Hall, R.C., Hall, R.C., 2007. The 1995 Kikwit Ebola outbreak – model of virus properties on system capacity and function: A lesson for future viral epidemics. Am J Disaster Med 2, 270–276.

Hirsch, M.S., 2008. Initiating therapy: When to start, what to use. J Infect Dis 197, S252–S260.

Hirsch, V.M., Olmsted, R.A., Murphey-Corb, M., Purcell, R.H., Johnson, P.R., 1989. An African primate lentivirus (SIVsm) closely related to HIV-2. Nature 339, 389–392.

Holmes, E.C., 2009. RNA virus genomics: A world of possibilities. J Clin Invest 119, 2488–2495.

Holmes, E.C., Ghedin, E., Miller, N., et al., 2005. Whole-genome analysis of human influenza A virus reveals multiple persistent lineages and reassortment among recent H3N2 viruses. PLoS Biol 3, e300.

Hon, K.L., Leung, C.W., Cheng, W.T., et al., 2003. Clinical presentations and outcome of severe acute respiratory syndrome in children. Lancet 361, 1701–1703.

Imai, Y., Kuba, K., Neely, G.G., et al., 2008. Identification of oxidative stress and Toll-like receptor 4 signaling as a key pathway of acute lung injury. Cell 133, 235–249.

Jacquelin, B., Mayau, V., Targat, B., et al., 2009. Nonpathogenic SIV infection of African green monkeys induces a strong but rapidly controlled type I IFN response. J Clin Invest 119, 3544–3555.

Jones, M.S., Kapoor, A., Lukashov, V.V., Simmonds, P., Hecht, F., Delwart, E., 2005. New DNA viruses identified in patients with acute viral infection syndrome. J Virol 79, 8230–8236.

Juven, T., Mertsola, J., Waris, M., et al., 2000. Etiology of community-acquired pneumonia in 254 hospitalized children. Pediatr Infect Dis J 19, 293–298.

Kapoor, A., Li, L., Victoria, J., et al., 2009. Multiple novel astrovirus species in human stool. J Gen Virol 90, 2965–2972.

Kapoor, A., Victoria, J., Simmonds, P., et al., 2008a. A highly prevalent and genetically diversified Picornaviridae genus in South Asian children. Proc Natl Acad Sci USA 105, 20,482–20,487.

Kapoor, A., Victoria, J., Simmonds, P., et al., 2008b. A highly divergent picornavirus in a marine mammal. J Virol 82, 311–320.

Kash, J.C., Basler, C.F., Garcia-Sastre, A., et al., 2004. Global host immune response: Pathogenesis and transcriptional profiling of type a influenza viruses expressing the hemagglutinin and neuraminidase genes from the 1918 pandemic virus. J Virol 78, 9499–9511.

Kash, J.C., Tumpey, T.M., Proll, S.C., et al., 2006. Genomic analysis of increased host immune and cell death responses induced by 1918 influenza virus. Nature 443, 578–581.

Keele, B.F., Van Heuverswyn, F., Li, Y., et al., 2006. Chimpanzee reservoirs of pandemic and nonpandemic HIV-1. Science 313, 523–526.

Kobasa, D., Jones, S.M., Shinya, K., et al., 2007. Aberrant innate immune response in lethal infection of macaques with the 1918 influenza virus. Nature 445, 319–323.

Krishnan, M.N., Ng, A., Sukumaran, B., et al., 2008. RNA interference screen for human genes associated with West Nile virus infection. Nature 455, 242–245.

Kuiken, T., Fouchier, R., Rimmelzwaan, G., Osterhaus, A., 2003. Emerging viral infections in a rapidly changing world. Curr Opin Biotechnol 14, 641–646.

Kuiken, T., Leighton, F.A., Fouchier, R.A., et al., 2005. Public health. Pathogen surveillance in animals. Science 309, 1680–1681.

Kuroda, M., Katano, H., Nakajima, N., et al., 2010. Characterization of quasispecies of pandemic 2009 influenza A virus (A/H1N1/2009) by de novo sequencing using a next-generation DNA sequencer. PLoS One 5, e10256.

Lacoste, V., Mauclere, P., Dubreuil, G., et al., 2000. Simian homologues of human gamma-2 and betaherpesviruses in mandrill and drill monkeys. J Virol 74, 11,993–11,999.

Lederer, S., Favre, D., Walters, K.A., et al., 2009. Transcriptional profiling in pathogenic and non-pathogenic SIV infections reveals significant distinctions in kinetics and tissue compartmentalization. PLoS Pathog 5, e1000296.

Leung, C.W., Chiu, W.K., 2004. Clinical picture, diagnosis, treatment and outcome of severe acute respiratory syndrome (SARS) in children. Paediatr Respir Rev 5, 275–288.

Li, Q., Brass, A.L., Ng, A., et al., 2009. A genome-wide genetic screen for host factors required for hepatitis C virus propagation. Proc Natl Acad Sci USA 106, 16,410–16,415.

Liu, Y.Z., Wilson, S.G., Wang, L., et al., 2008. Identification of *PLCL1* gene for hip bone size variation in females in a genome-wide association study. PLoS One 3, e3160.

Loeuillet, C., Deutsch, S., Ciuffi, A., et al., 2008. *In vitro* whole-genome analysis identifies a susceptibility locus for HIV-1. PLoS Biol 6, e32.

Mandl, J.N., Barry, A.P., Vanderford, T.H., et al., 2008. Divergent TLR7 and TLR9 signaling and type I interferon production distinguish pathogenic and nonpathogenic AIDS virus infections. Nat Med 14, 1077–1087.

Margulies, K.B., Bednarik, D.P., Dries, D.L., 2009. Genomics, transcriptional profiling, and heart failure. J Am Coll Cardiol 53, 1752–1759.

Marx, P.A., Li, Y., Lerche, N.W., et al., 1991. Isolation of a simian immunodeficiency virus related to human immunodeficiency virus type 2 from a West African pet sooty mangabey. J Virol 65, 4480–4485.

Metzner, K.J., Rauch, P., Walter, H., et al., 2005. Detection of minor populations of drug-resistant HIV-1 in acute seroconverters. Aids 19, 1819–1825.

Miyagawa, T., Kawashima, M., Nishida, N., et al., 2008. Variant between CPT1B and CHKB associated with susceptibility to narcolepsy. Nat Genet 40, 1324–1328.

Mo, H., Lu, L., Pilot-Matias, T., et al., 2005. Mutations conferring resistance to a hepatitis C virus (HCV) RNA-dependent RNA polymerase inhibitor alone or in combination with an HCV serine protease inhibitor in vitro. Antimicrob Agents Chemother 49, 4305–4314.

Murray, J.L., Mavrakis, M., McDonald, N.J., et al., 2005. Rab9 GTPase is required for replication of human immunodeficiency virus type 1, filoviruses, and measles virus. J Virol 79, 11,742–11,751.

Murray, K., Selleck, P., Hooper, P., et al., 1995. A morbillivirus that caused fatal disease in horses and humans. Science 268, 94–97.

Nascimento, E.J., Braga-Neto, U., Calzavara-Silva, C.E., et al., 2009. Gene expression profiling during early acute febrile stage of dengue infection can predict the disease outcome. PLoS One 4, e7892.

Nelson, M.I., Edelman, L., Spiro, D.J., et al., 2008. Molecular epidemiology of A/H3N2 and A/H1N1 influenza virus during a single epidemic season in the United States. PLoS Pathog 4, e1000133.

Osterhaus, A., 2001. Catastrophes after crossing species barriers. Philos Trans R Soc Lond B 356, 791–793.

Palacios, G., Druce, J., Du, L., et al., 2008. A new arenavirus in a cluster of fatal transplant-associated diseases. N Engl J Med 358, 991–998.

Paredes, R., Clotet, B., 2010. Clinical management of HIV-1 resistance. Antiviral Res 85, 245–265.

Paredes, R., Marconi, V.C., Campbell, T.B., Kuritzkes, D.R., 2007. Systematic evaluation of allele-specific real-time PCR for the detection of minor HIV-1 variants with pol and env resistance mutations. J Virol Methods 146, 136–146.

Patel, K., Lucas, J.E., Thompson, J.W., et al., 2011. High predictive accuracy of an unbiased proteomic profile for sustained virologic response in chronic hepatitis C patients. Hepatology 53, 1809–1818.

Peiris, J.S., Chu, C.M., Cheng, V.C., et al., 2003. Clinical progression and viral load in a community outbreak of coronavirus-associated SARS pneumonia: A prospective study. Lancet 361, 1767–1772.

Peiris, J.S., Guan, Y., Yuen, K.Y., 2004. Severe acute respiratory syndrome. Nat Med 10, S88–S97.

Pelak, K., Goldstein, D.B., Walley, N.M., et al., 2010. Host determinants of HIV-1 control in African Americans. J Infect Dis 201, 1141–1149.

Priori, S.G., Schwartz, P.J., Napolitano, C., et al., 2003. Risk stratification in the long-QT syndrome. N Engl J Med 348, 1866–1874.

Prusty, B.K., Karlas, A., Meyer, T.F., Rudel, T., 2011. Genome-wide RNAi screen for viral replication in mammalian cell culture. Methods Mol Biol 721, 383–395.

Rambaut, A., Posada, D., Crandall, K.A., Holmes, E.C., 2004. The causes and consequences of HIV evolution. Nat Rev Genet 5, 52–61.

Rambaut, A., Pybus, O.G., Nelson, M.I., Viboud, C., Taubenberger, J.K., Holmes, E.C., 2008. The genomic and epidemiological dynamics of human influenza A virus. Nature 453, 615–619.

Russell, C.A., Jones, T.C., Barr, I.G., et al., 2008. The global circulation of seasonal influenza A (H3N2) viruses. Science 320, 340–346.

Saeed, M., Zaidi, S.Z., Naeem, A., et al., 2007. Epidemiology and clinical findings associated with enteroviral acute flaccid paralysis in Pakistan. BMC Infect Dis 7, 6.

Seib, K.L., Dougan, G., Rappuoli, R., 2009. The key role of genomics in modern vaccine and drug design for emerging infectious diseases. PLoS Genet 5, e1000612.

Service, R.F., 2006. Gene sequencing. The race for the $1000 genome. Science 311, 1544–1546.

Shimakami, T., Lanford, R.E., Lemon, S.M., 2009. Hepatitis C: Recent successes and continuing challenges in the development of improved treatment modalities. Curr Opin Pharmacol 9, 537–544.

Shisheva, A., 2008. PIKfyve: Partners, significance, debates and paradoxes. Cell Biol Int 32, 591–604.

Siddiqui, A., Kerb, R., Weale, M.E., et al., 2003. Association of multidrug resistance in epilepsy with a polymorphism in the drug-transporter gene ABCB1. N Engl J Med 348, 1442–1448.

Sierra, B., Alegre, R., Perez, A.B., et al., 2007. HLA-A, -B, -C, and -DRB1 allele frequencies in Cuban individuals with antecedents of dengue 2 disease: Advantages of the Cuban population for HLA studies of dengue virus infection. Hum Immunol 68, 531–540.

Simmonds, P., 2004. Genetic diversity and evolution of hepatitis C virus – 15 years on. J Gen Virol 85, 3173–3188.

Simmonds, P., Bukh, J., Combet, C., et al., 2005. Consensus proposals for a unified system of nomenclature of hepatitis C virus genotypes. Hepatology 42, 962–973.

Simonsen, L., Viboud, C., Grenfell, B.T., et al., 2007. The genesis and spread of reassortment human influenza A/H3N2 viruses conferring adamantane resistance. Mol Biol Evol 24, 1811–1820.

Smith, D.J., Lapedes, A.S., De Jong, J.C., et al., 2004. Mapping the antigenic and genetic evolution of influenza virus. Science 305, 371–376.

Smith, G.J., Vijaykrishna, D., Bahl, J., et al., 2009. Origins and evolutionary genomics of the 2009 swine-origin H1N1 influenza A epidemic. Nature 459, 1122–1125.

Smits, S.L., De Lang, A., Van den Brand, J.M., et al., 2010a. Exacerbated innate host response to SARS-CoV in aged nonhuman primates. PLoS Pathog 6, e1000756.

Smits, S.L., Van den Brand, J.M., De Lang, A., et al., 2011. Distinct severe acute respiratory syndrome coronavirus-induced acute lung injury pathways in two different nonhuman primate species. J Virol 85, 4234–4245.

Smits, S.L., Van Leeuwen, M., Kuiken, T., Hammer, A.S., Simon, J.H., Osterhaus, A.D., 2010b. Identification and characterization of deer astroviruses. J Gen Virol 91, 2719–2722.

Smits, S.L., Van Leeuwen, M., Van der Eijk, A.A., et al., 2010c. Human astrovirus infection in a patient with new onset coeliac disease. J Clin Microbiol 48, 3416–3418.

Sodora, D.L., Allan, J.S., Apetrei, C., et al., 2009. Toward an AIDS vaccine: Lessons from natural simian immunodeficiency virus infections of African nonhuman primate hosts. Nat Med 15, 861–865.

Song, H.D., Tu, C.C., Zhang, G.W., et al., 2005. Cross-host evolution of severe acute respiratory syndrome coronavirus in palm civet and human. Proc Natl Acad Sci USA 102, 2430–2435.

Suppiah, V., Moldovan, M., Ahlenstiel, G., et al., 2009. IL28B is associated with response to chronic hepatitis C interferon-alpha and ribavirin therapy. Nat Genet 41, 1100–1104.

Tanaka, Y., Nishida, N., Sugiyama, M., et al., 2009. Genome-wide association of IL28B with response to pegylated interferon-alpha and ribavirin therapy for chronic hepatitis C. Nat Genet 41, 1105–1109.

Tang, X.C., Zhang, J.X., Zhang, S.Y., et al., 2006. Prevalence and genetic diversity of coronaviruses in bats from China. J Virol 80, 7481–7490.

Taylor, L.H., Latham, S.M., Woolhouse, M.E., 2001. Risk factors for human disease emergence. Philos Trans R Soc Lond B 356, 983–989.

Ten Bosch, J.R., Grody, W.W., 2008. Keeping up with the next generation: Massively parallel sequencing in clinical diagnostics. J Mol Diagn 10, 484–492.

Tersago, K., Verhagen, R., Vapalahti, O., Heyman, P., Ducoffre, G., Leirs, H., 2010. Hantavirus outbreak in Western Europe: Reservoir host infection dynamics related to human disease patterns. Epidemiol Infect 139 (3), 1–10.

Urisman, A., Molinaro, R.J., Fischer, N., et al., 2006. Identification of a novel gammaretrovirus in prostate tumors of patients homozygous for R462Q RNASEL variant. PLoS Pathog 2, e25.

Van den Hoogen, B.G., De Jong, J.C., Groen, J., et al., 2001. A newly discovered human pneumovirus isolated from young children with respiratory tract disease. Nat Med 7, 719–724.

Van der Hoek, L., Pyrc, K., Jebbink, M.F., et al., 2004. Identification of a new human coronavirus. Nat Med 10, 368–373.

Van de Vijver, M.J., He, Y.D., Van't Veer, L.J., et al., 2002. A gene-expression signature as a predictor of survival in breast cancer. N Engl J Med 347, 1999–2009.

Van Leeuwen, M., Williams, M.M., Koraka, P., Simon, J.H., Smits, S.L., Osterhaus, A.D., 2010. Human picobirnaviruses identified by molecular screening of diarrhea samples. J Clin Microbiol 48, 1787–1794.

Venter, J.C., Remington, K., Heidelberg, J.F., et al., 2004. Environmental genome shotgun sequencing of the Sargasso Sea. Science 304, 66–74.

Victoria, J.G., Kapoor, A., Dupuis, K., Schnurr, D.P., Delwart, E.L., 2008. Rapid identification of known and new RNA viruses from animal tissues. PLoS Pathog 4, e1000163.

Wang, D., Coscoy, L., Zylberberg, M., et al., 2002. Microarray-based detection and genotyping of viral pathogens. Proc Natl Acad Sci USA 99, 15,687–15,692.

Wang, D., Urisman, A., Liu, Y.T., et al., 2003. Viral discovery and sequence recovery using DNA microarrays. PLoS Biol 1, e2.

Weaver, S.C., Reisen, W.K., 2010. Present and future arboviral threats. Antiviral Res 85, 328–345.

Winkler, C.A., 2008. Identifying host targets for drug development with knowledge from genome-wide studies: Lessons from HIV-AIDS. Cell Host Microbe 3, 203–205.

Wolfe, N.D., Dunavan, C.P., Diamond, J., 2007. Origins of major human infectious diseases. Nature 447, 279–283.

Wolfe, N.D., Heneine, W., Carr, J.K., et al., 2005. Emergence of unique primate T-lymphotropic viruses among central African bushmeat hunters. Proc Natl Acad Sci USA 102, 7994–7999.

Wolfe, N.D., Switzer, W.M., Carr, J.K., et al., 2004. Naturally acquired simian retrovirus infections in central African hunters. Lancet 363, 932–937.

Wong, G.W., Li, A.M., Ng, P.C., Fok, T.F., 2003. Severe acute respiratory syndrome in children. Pediatr Pulmonol 36, 261–266.

Woolhouse, M.E., Gowtage-Sequeria, S., 2005. Host range and emerging and reemerging pathogens. Emerg Infect Dis 11, 1842–1847.

Zaas, A.K., Chen, M., Varkey, J., et al., 2009. Gene expression signatures diagnose influenza and other symptomatic respiratory viral infections in humans. Cell Host Microbe 6, 207–217.

Sepsis

Stephen F. Kingsmore, Christopher W. Woods, and Carol J. Saunders

INTRODUCTION

An 18-year-old female college freshman and dormitory resident presents with 12 hours of fever and headache to an acute care center, where she is referred to the emergency department. In transit, she develops a petechial rash, and becomes hypotensive and lethargic in the emergency department. She is admitted to the intensive care unit. Blood and cerebrospinal fluid cultures grow *Neisseria meningitidis*, serogroup C. She dies two days later with multiorgan failure, despite aggressive attempts at resuscitation and support. Her physicians considered drotrecogin-α, but chose not to give it in the setting of meningitis. Prophylaxis is provided to the other students in her dormitory, and no one else becomes ill. Her devastated parents wish to know if her siblings are at risk of poor outcome with a similar infection. Molecular genetic testing reveals she is heterozygous for a rare missense mutation in the *TLR4* gene, which neither sibling carries.

A 40-year-old male is seen as an outpatient three days following a hemilaminectomy for nerve root impingement due to an acutely herniated lumbar disc. He has severe lower back pain, fever, tachycardia, and leukocytosis. Physical examination reveals wound infection, with two wound cultures positive for coagulase-negative staphylococci. He is treated as an outpatient with wound debridement and oral antibiotics. However, local infection persists, requiring surgical debridement a month later, followed by six weeks of intravenous antibiotics. He loses 30 lb, and full recovery takes a year.

A 62-year-old woman with a several-month history of fever with negative evaluation, including blood cultures, presents with acute renal failure and severe aortic regurgitation. A transthoracic echocardiogram reveals vegetations on her aortic valve leaflet. Immunohistochemistry at the time of valve replacement confirms *Bartonella henselae*. The patient raises feral cats. Her primary care physician is frustrated that he was unable to diagnose the infection earlier.

These cases highlight the clinical heterogeneity of sepsis and illustrate areas of unmet medical need, such as predicting those at risk for complicated sepsis, developing earlier diagnostic tests for sepsis syndrome, defining the etiologic agent, and determining prognosis. Sepsis severity is influenced by many factors, including the infectious agent, primary site of infection, pathogenicity of the invading microorganism, infective dose, the genetic background of the patient, comorbid conditions, time to clinical intervention, and the type and intensity of healthcare provided. In the past decade, a variety of new techniques have been developed to investigate host molecular mechanisms in complex diseases. In this chapter, we will review the sepsis literature, focusing on host genomic determinants of sepsis outcomes.

DEFINITIONS, INCIDENCE, AND TEMPORAL DYNAMICS

Sepsis is a highly heterogeneous, common clinical syndrome. In adults, it is defined as evidence of infection plus the presence of two or more allostatic changes, known collectively as the systemic inflammatory response syndrome (SIRS) achieving

Genomic and Personalized Medicine, 2nd edition
by Ginsburg & Willard. DOI: http://dx.doi.org/10.1016/B978-0-12-382227-7.00098-7

TABLE 98.1	Sepsis and systemic inflammatory response syndrome definitions
1 Temperature	<36°C or >39°C
2 White blood cell count	>12,000 or <4000 or >10% bands
3 Heart rate	>90 bpm*
4 Respiratory rate	>24 bpm† or PaCO₂ <32 mmHg

Sepsis = SIRS + infection

Severe sepsis = SIRS + infection + end organ damage

Septic shock = severe sepsis + refractory hypotension (<90 mmHg or 40% below baseline)

*Beats per minute.
†Breaths per minute.
SIRS, systemic inflammatory response syndrome.

homeostasis through physiological change (Bone et al., 1989, 1992; Jaimes et al., 2003; Sands et al., 1997) (Table 98.1). The etiologic agent may be bacterial, viral, fungal, or parasitic, with more than 80% of infections originating from a pulmonary, genitourinary, or abdominal source (Bone et al., 1992). Despite rigorous diagnostic work-up, however, the etiologic agent remains uncertain in about half of the patients who are adjudged clinically to have sepsis (Glickman et al., 2010; Heffner et al., 2010). Sepsis with coincident acute organ dysfunction is considered *severe sepsis*. Failure to maintain blood pressure despite adequate hydration is termed *septic shock*. Severe sepsis is a major cause of morbidity and mortality, with an annual incidence of 50–100 cases per 100,000 persons in several industrialized nations (Bone et al., 1992). In the United States, there are approximately 750,000 new cases of severe sepsis annually, with an economic impact approaching $17 billion (Angus et al., 2001; Bone et al., 1992). Despite enormous investments in critical care resources, 10–50% of patients with sepsis die, making it the third-leading cause of infectious death and tenth-leading cause of death overall (Alberti et al., 2002; Angus et al., 2001; McCaig and Nawar, 2006).

Comorbidity is an important determinant of sepsis outcome that has been codified in the chronic illness components of the Acute Physiology and Chronic Health Evaluation II (APACHE II) score (Knaus et al., 1985). Comorbidities include age, immunodeficiency, immunosuppression, chronic renal failure, recent surgery, liver insufficiency, heart failure, and respiratory insufficiency. Another important determinant of sepsis outcome is the clinical severity at time of presentation. For example, temperature of less than 37°C at time of presentation is associated with higher mortality than patients with euthermia or pyrexia (Glickman et al., 2010). The prognostic impact of clinical severity at time of presentation has been also been codified in tools such as APACHE II and the Sequential Organ Failure Assessment (SOFA) scores (Vincent et al., 1998). The latter includes objective criteria for assessment of acute dysfunction of the respiratory, cardiovascular, hepatic, coagulation, renal, and neurologic systems. Concomitant acute dysfunction of more than one system is common, and is termed multiple organ dysfunction syndrome (MODS). The presence of septic shock at presentation or within

24 hours is associated with particularly high mortality (Annane et al., 2003; Glickman et al., 2010), as is elevated blood lactate, which is indicative of tissue hypoxia (Green et al., 2011).

The incidence of severe sepsis has been increasing by approximately 9% annually (CDC, 1991). Between 1979 and 2000, the incidence increased from 82.7/100,000 to 240.4/100,000 (Angus et al., 2001). This increase is multifactorial, resulting from increased awareness and documentation of sepsis; "graying" of the population; greater use of invasive procedures for the diagnosis, monitoring, and management of ill patients; emergence of antibiotic-resistant organisms; and increasing prevalence of immunocompromised patients (e.g., malignancy, AIDS, transplant recipients, diabetes mellitus, alcoholism, or malnutrition) (Parrillo et al., 1990). Furthermore, the relative contribution of etiologic organisms has changed substantially over time: in adult sepsis, gram-positive organisms superseded gram-negative organisms in predominance in 1987 and remain more common (Angus et al., 2001). Fungal sepsis has increased by more than 200% since 1980. Community-acquired and nosocomial sepsis are distinguished by differences in comorbidity as well as etiologic spectra and virulence. For example, the incidence of community-acquired sepsis exhibits seasonal fluctuation, with the peak coinciding with influenza season (Apostolidou et al., 1994; Douglas et al., 1991). Finally, sepsis in teenagers and adults differs from that occurring in neonates and young children in several ways, including predominant etiologic agent, clinical presentation, temporal dynamics, complications (Kingsmore et al., 2008a) and host molecular responses.

Genomic medicine appears well situated to facilitate the individualized diagnosis and treatment of sepsis by identifying those patients at substantial risk for certain infections, rapidly classifying the specific etiologic organism, and stratifying patients according to whether they are likely to have adverse outcomes versus benefits from therapeutic intervention. In patients with suspected sepsis, early, accurate identification of those at higher risk of organ dysfunction or death is critical for effective management and positive outcome. Availability of such predictive tools would enable individualized treatment strategies based on risk, thereby reducing mortality and increasing effectiveness of resource allocation. For example, low-risk patients could be treated in an outpatient setting, versus those requiring hospital admission, intensive care, and more costly or complex medical therapies such as early goal-directed therapy (EGDT) or drotrecogin-α (activated protein C, APC). Early identification of a high-risk genotype may even help guide the introduction of prophylactic therapy. However, the complex pathophysiology and diverse molecular pathology and epidemiology of sepsis, together with the variety of physicians confronted by this syndrome (and of scientific disciplines studying sepsis), have kept progress slow. Much of the literature focuses on the tremendous achievements in elucidating the genome sequences of various bacterial, viral, and fungal pathogens. This chapter, however, focuses on the human genome and its response to the infectious perturbations that result in sepsis.

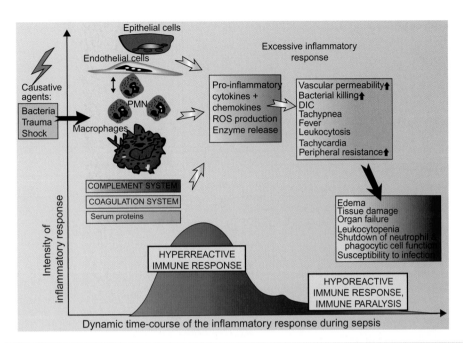

Figure 98.1 Time course and intensity of host immune inflammatory and coagulation responses during SIRS and sepsis. Microbiologic factors cause activation of the innate and cognate immune system, inflammatory response, and coagulation system in sepsis, leading to production of cytokines, activation of the alternate pathway of complement and up-regulation of adhesion and signaling molecules on white blood cells and blood vessel endothelium. ROS, reactive oxygen species; DIC, disseminated intravascular coagulation. Metabolic and neurohormonal systems, which are also important in sepsis progression, are not shown. Recent studies have determined that the pattern of molecular change in SIRS survivors is monophasic, with rapid deviation from normal ranges and gradual return to the latter. In individuals who die, however, there is early divergence from the survivor profile that increases as death approaches. This contrasts with the traditional depiction of a hyperimmune and then hypoimmune host response, as shown here. *Reproduced with permission from Riedemann et al., 2003.*

GENETIC VARIATIONS ASSOCIATED WITH SEPSIS

Microbiologic infections and their complications, such as sepsis and severe sepsis, occur at the interface of host genomes and epigenomes, microbial genomes, and the environment. While exposure to a microbial agent is necessary, it is typically not sufficient to cause sepsis in the host. Regardless of etiology, sepsis and SIRS profoundly perturb the host's immune, inflammatory, coagulation, neurohormonal, and metabolic systems (Figure 98.1). In fact, many of the host molecular perturbations in sepsis-associated SIRS are stereotyped, also occurring in burns, trauma (Calvano et al., 2005; Cobb et al., 2005; McDunn et al., 2008; Qian et al., 2010; Rajicic et al., 2010; Zhou et al., 2010), and other acute conditions such as pancreatitis. The patterns of molecular perturbation in SIRS survivors follow approximately Gaussian deviations from normal ranges, with return to the latter after a variable period (above references; Kingsmore et al., submitted with; Langley et al., 2010). There are notable inter-individual differences in the magnitude and longevity of molecular perturbations in surviving versus deceased SIRS

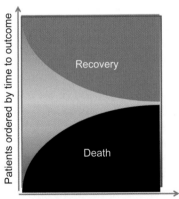

Days to recovery or death

Figure 98.2 Time course of severity of disease in 1700 patients with SIRS due to burns or trauma. Disease severity was measured daily by the Multiple Organ Dysfunction (MOD) score (Marshall, 1997), where red indicates severe illness, and blue, mild illness. The more severe the initial MOD score, the shorter the duration to death or the longer the time to recover. There was no evidence of a biphasic SIRS response.

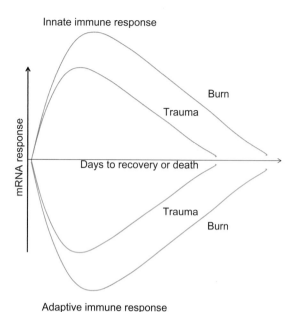

Figure 98.3   Host peripheral leukocyte mRNA responses in patients with SIRS due to burns or trauma. The pattern of molecular change is a rapid deviation from normal ranges with gradual return to the latter. Innate immune response genes are up-regulated, whereas adaptive immune response genes are down-regulated. The traditional host response is biphasic, with a hyperimmune and then hypoimmune component, as shown here. Many of the host molecular perturbations in sepsis-associated SIRS are stereotyped, also occurring in burns, trauma, and other acute conditions such as pancreatitis. The patterns of molecular perturbation in SIRS survivors follow approximately Gaussian deviations from normal ranges, with return to the latter after a variable period. The time to recovery or death is generally longer in burn injuries than in trauma, and is shorter in patients with extremes of illness severity.

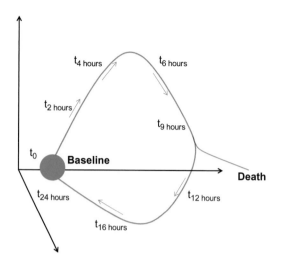

Figure 98.4   Three-dimensional principal component analysis of the time course of the host peripheral leukocyte mRNA responses in patients with SIRS due to burns or trauma. The mRNA variability at baseline is shown by the green circle. Following sepsis-, burn-, or trauma-induced SIRS, the ribonucleogram demonstrates a homeostatic adaptation with time (blue line) that eventually returns to the baseline in survivors. It is hypothesized that there is divergence of the host response in patients who will die (red line), reflecting allostatic overload (maladaptation), that becomes more divergent as time of death nears (Langley et al., 2010). *Adapted from Polpitiya et al., 2009.*

patients that may have important clinical implications (Bellamy and Hill, 1998; Burgner and Levin, 2003; Choi et al., 2001). For example, the time to recovery or death is generally longer in burn injuries than trauma, and is shorter in patients with extremes of illness severity (Figures 98.2 and 98.3) (Langley et al., 2010). The trajectory of molecular perturbations diverges early in the clinical course of surviving patients versus those who die, and becomes more divergent as time of death nears (Figure 98.4) (Kingsmore et al., submitted with; Langley et al., 2010). Thus, the initial host molecular perturbations in SIRS are dichotomous, reflecting return to homeostasis (adaptation) or allostatic overload (maladaptation) (Brame and Singer, 2010; Mongardon et al., 2009). Accordingly, it is useful to evaluate how genomic variation in patients affects individual susceptibility and response to infection.

Death from sepsis shows stronger heritability than death from any other cause, including cardiovascular disease and neoplasia (Bone et al., 1992; Sorensen et al., 1988). Family and twin studies have demonstrated familial or twin aggregation of

some sepsis outcomes and clinical presentations on a common genetic basis (Hartel et al., 2007; Jepson et al., 1995). Many recessive Mendelian disorders, for example, are associated with specific opportunistic infections, frequent episodes of sepsis, or adverse outcomes (Table 98.2). In addition, there are rare variants in many host response genes that do not cause constitutional diseases *per se*, but are nevertheless associated with unusual susceptibility to infectious agents. For example, rare variants in the gene encoding complement factor properdin (*CFP*) or the Toll receptor 4 gene are associated with autosomal dominant susceptibility to meningococcal meningitis (Densen et al., 1987; Smirnova et al., 2003). Other examples of sepsis susceptibility alleles are variants in the major histocompatibility complex, chemokine (C-C motif) receptor 5 (*CCR5*) in human immunodeficiency virus infection, and Toll receptor 2 and solute carrier family 11 (proton-coupled divalent metal ion transporters) member 1 (*SLC11A1*, also known as *NRAMP1*) in mycobacterial infection (Drennan et al., 2004; Liu et al., 1996; Samson et al., 1996; Skamene et al., 1998). New technologies such as exome and whole-genome analysis by next-generation sequencing are enabling assessment of the contribution of rare variant alleles in host genes to individual sepsis susceptibility and adverse outcomes for the first time (Kingsmore and Saunders, 2011). However, since the functional consequences of the vast majority of individual heterozygous missense and nonsense alleles and the cumulative effect of several such

TABLE 98.2	Mendelian recessive illnesses characterized by sepsis	
OMIM no.	**Disease**	**Gene**
102700	SCID, autosomal recessive, T-cell-negative	ADA
186740	Immunodeficiency due to defect in CD3-gamma	CD3G
186780	Immunodeficiency due to defect in CD3-zeta	CD3Z
186790	SCID, autosomal recessive, T-cell-negative, B cell+, NK cell+	CD3D
186830	Immunodeficiency due to defect in CD3-epsilon	CD3E
203750	Alpha-methylacetoacetic aciduria	ACAT1
208900	Ataxia-telangiectasia	ATM
209950	Atypical mycobacteriosis, familial	IKBKG
209950	Atypical mycobacteriosis, familial	IL12B
209950	Atypical mycobacteriosis, familial	IL12RB1
209950	Atypical mycobacteriosis, familial	IFNGR1
209950	Atypical mycobacteriosis, familial	STAT1
209950	Atypical mycobacteriosis, familial	TYK2
209950	Atypical mycobacteriosis, familial	IFNGR2
210900	Bloom syndrome	BLM
212720	Martsolf syndrome	RAB3GAP2
214500	Chediak Higashi syndrome	LYST
217090	Plasminogen deficiency type I	PLG
219100	Cutis laxa, autosomal recessive, type I	EFEMP2
219100	Cutis laxa, autosomal recessive, type I	FBLN5
219700	Cystic fibrosis	CFTR
230400	Galactosemia	GALT
231050	Geleophysic dysplasia	ADAMTSL2
235550	Hepatic veno-occlusive disease with immunodeficiency	SP110
240300	Autoimmune polyendocrine syndrome type I	AIRE
242860	Immunodeficiency–centromeric instability–facial anomalies syndrome	DNMT3B
243700	Hyper-IgE recurrent infection syndrome, autosomal recessive	DOCK8
248190	Hypomagnesemia, renal, with ocular involvement	CLDN19

TABLE 98.2	Continued	
OMIM no.	**Disease**	**Gene**
249100	Familial Mediterranean fever	MEFV
250250	Cartilage–hair hypoplasia	RMRP
251000	Methylmalonic aciduria due to methylmalonyl-CoA mutase deficiency	MUT
251100	Methylmalonic aciduria, cb1A type	MMAA
251110	Methylmalonic aciduria, cb1B type	MMAB
251260	Nijmegen breakage syndrome	NBN
252900	Mucopolysaccharidosis type IIIA (Sanfilippo type A)	SGSH
252930	Mucopolysaccharidosis type IIIC (Sanfilippo type C)	HGSNAT
253260	Biotinidase deficiency	BTD
277300	Spondylocostal dysostosis, autosomal recessive 1	WNT7A
300209	Simpson–Golabi–Behmel syndrome, type 2	OFD1
300240	Hoyeraal–Hreidarsson syndrome	DKC1
300291	Ectodermal dysplasia, hypohidrotic, with immune deficiency	IKBKG
300301	Ectodermal dysplasia, anhidrotic, with immunodeficiency, osteopetrosis, and lymphedema	IKBKG
300400	SCID, X-linked	IL2RG
300635	Lymphoproliferative syndrome, X-linked, 2	XIAP
300755	Agammaglobulinemia, X-linked (XLA)	BTK
301000	Wiskott–Aldrich syndrome	WAS
301835	Arts syndrome	PRPS1
302060	Barth syndrome	TAZ
304790	Immunodysregulation, polyendocrinopathy, and enteropathy, X-linked	FOXP3
308205	Ichthyosis follicularis, atrichia, and photophobia syndrome	MBTPS2
308230	Immunodeficiency with hyper-IgM, type 1	CD40LG
308240	Lymphoproliferative syndrome, X-linked, 1	SH2D1A

OMIM, Online Mendelian Inheritance in Man; SCID, severe combined immunodeficiency.

alleles *in trans* are unknown, it will be some time before we have a sense of the contribution of rare variants to the burden of sepsis susceptibility and outcomes. However, individual alleles of known large effect are discussed briefly in this chapter and in more detail in adjoining chapters in this volume.

A Mendelian pattern of inheritance does not apply to sepsis development or outcomes in most individuals. Like other common, complex disorders, in most sepsis cases, the host's contributing genetic components are likely to be combinations of rare heterozygous alleles (as discussed above), the net effects of multiple common, weak risk and protective factors (common variant hypothesis), or a combination of both. Common risk alleles have traditionally been identified by linkage analysis in multiple large pedigrees or candidate gene-based association studies in ethnically matched cohorts of cases and controls. While such approaches are being extensively applied to microbial infection outcomes and sepsis, to date they have met with limited success in the identification of risk alleles for host-gene/pathogen interaction (Botstein and Risch, 2003; Freimer and Sabatti, 2004; Hill, 2006). Most studies have investigated candidate genes involved in pathogen detection [e.g., toll-like receptors (TLRs)], the inflammatory response [e.g., tumor necrosis factor (*TNF*)-α], or coagulation [e.g., plasminogen activator inhibitor (*PAI*)]. Several reviews of the associations between candidate gene polymorphisms and the risk and outcome of sepsis have been published (Arcaroli et al., 2005; Lin and Albertson, 2004; Majetschak et al., 2002; Mira et al., 1999). More recently, GWAS have shown broad utility in identifying common single nucleotide polymorphisms (SNPs) contributing to human disease risk (Kingsmore et al., 2008b). Published GWAS in infectious diseases have hitherto been limited to malaria, tuberculosis, and HIV (Kingsmore et al., 2008b).

Unraveling the genetic variation in sepsis is complicated, requiring close attention to study design issues such as the selection of an appropriate study population, sample size, and acknowledgement of potential gene–environment interactions. A lack of replication among studies provides considerable concern about the interpretation of results. For example, an initial study may identify an allele with large estimated genetic effects, but subsequent studies fail to corroborate the results (Goring et al., 2001; Hirschhorn et al., 2002; Ioannidis et al., 2001; Lander and Kruglyak, 1995). Biologic explanations of inconsistent results include unacknowledged confounding heterogeneity (such as poorly defined phenotypes), heterogeneous genetic sources for the phenotype (genocopies), population diversity (ethnic ancestry), population-specific linkage disequilibrium (LD), heterogeneous genetic and epigenetic backgrounds, or heterogeneous environmental influences (phenocopies). Analytic reasons for problems with reproducibility include failure to control the rate of false discoveries, lack of power, model misspecification, and heterogeneous bias in estimated effects among studies (Cardon and Bell, 2001; Cardon and Palmer, 2003; Redden and Allison, 2003; Sillanpaa and Auranen, 2004). Among these, the most frequent source of

non-replication has been lack of power due to insufficient sample sizes (Lohmueller et al., 2003; Risch, 2000).

PATHOGEN RECOGNITION/SIGNALING
Toll-like Receptors

There has been substantial progress in defining variants of several genes and pathways implicated in infectious disease susceptibility, but none have caused more excitement than new discoveries about the genes affecting innate immunity. In particular, newly discovered pattern recognition receptors and their associated signaling pathways, namely the toll-like receptors (TLRs), have attracted considerable recent attention (Figure 98.5).

Innate immunity recognition of invading microorganisms is mediated by a set of soluble and membrane receptors that recognize conserved pathogen-associated molecular patterns (PAMPs) shared among each class of infectious agents but absent in higher eukaryotes. TLRs have been conserved by diverse life-forms including plants, insects, and mammals, and have a cytoplasmic domain that bears homology to the interleukin (IL)-1 receptor. Thus far, 10 human TLRs with pathogen-sensing roles have been identified. Activation of TLRs stimulates macrophages, resulting in the elaboration of proinflammatory cytokines as well as antimicrobial molecules such as nitric oxide and defensins. Concurrently, stimulated dendritic cells migrate to lymph nodes and overexpress antigen major histocompatibility complex (MHC) and co-stimulatory molecules (CD80/CD86). Thus, TLRs are an essential link between innate and adaptive immunity.

The discovery and characterization of TLRs as key pattern recognition molecules for pathogens (or PAMPs) (Beutler and Rietschel, 2003) and initiators of innate immune responses (Akira et al., 2001; Creagh and O'Neill, 2006) stimulated many investigators to further characterize the role of polymorphisms in the receptors in susceptibility to infectious disease. Sepsis was an early target for characterization of the relevance of polymorphisms in TLRs. The TLR4 protein was found to activate nuclear factor (NF)-kappa B, resulting in up-regulation of expression of the pro-inflammatory cytokines IL-1, IL-6, and IL-8 in cultured human cells. Moreover, *TLR*-deficient mice are resistant to systemic endotoxin exposure, but remain susceptible to gram-negative infections (Qureshi et al., 1999). Two common missense mutations in the extracellular domain of *TLR4*, Asp229Gly and Thr399Ile, have been associated with varied human responses to inhaled endotoxin (Arbour et al., 2000). Several, but not all, studies have corroborated association of the *TLR4* Asp299Gly variant with increased susceptibility to either gram-negative infections, SIRS, or response to lipopolysaccharide (LPS), a component of the cell wall of gram-negative bacteria (Agnese et al., 2002; Calvano et al., 2006; Child et al., 2003; Lorenz et al., 2002). In addition, Smirnova and colleagues found an excess of rare *TLR4* coding variants in meningococcal cases compared with controls (Smirnova et al., 2003).

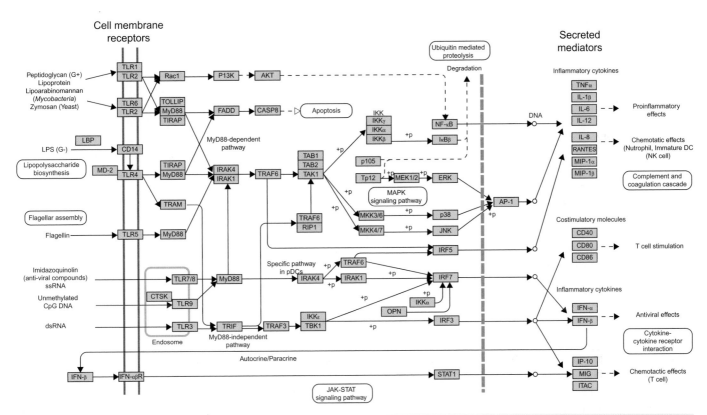

Figure 98.5 Components of the human toll-like receptor signaling pathway. There are two groups of TLR signaling pathways: a MyD88-dependent pathway that leads to the production of proinflammatory cytokines with quick activation of nuclear factor kappa B (NF-κB) and mitogen-activated protein kinase (MAPK), and a MyD88-independent pathway associated with the induction of interferon (IFN)-β and IFN-inducible genes, and maturation of dendritic cells with slow activation of NF-κB and MAPK. *Reproduced from Kanehisa Laboratories, 2010. Copyright Kanehisa Laboratories.*

However, this study, as well as another, much larger, case-control study (Read et al., 2001), did not find an association with the Asp299Gly or Thr399Ile polymorphisms in *TLR4*. Together, these studies suggest that individuals with the 299/399 polymorphisms in TLR4 may have an aberrant response to certain, but not all, gram-negative bacterial diseases, resulting in an increased susceptibility to infection and severity of disease.

TLR2 and *TLR5* have also been targets of interest. *TLR2* is most notable for detecting a wide repertoire of pathogens, including gram-positive and gram-negative bacteria, mycobacteria, fungi, viruses, and parasites. This largely results from an ability to recognize ligands as a heterodimer with *TLR1* or *TLR6* (Lorenz, 2006). Of particular importance, *TLR2* has been linked to the recognition of gram-positive bacteria, which are now the leading cause of sepsis (Martin et al., 2003). TLR2 accomplishes this via response to peptidoglycan, lipoteichoic acid, and a variety of macromolecules in gram-positive bacteria such as *Staphylococcus aureus*. The relationship between TLR2 polymorphisms (primarily Arg753Gln) and *S. aureus* infection has been examined in several studies, with contradictory results (Lorenz et al., 2000; Moore et al., 2004). Therefore, additional investigations are warranted to confirm this association.

TLR5, which recognizes bacterial flagellin from both gram-positive and gram-negative bacteria, activates NF-kappa B and the release of pro-inflammatory cytokines in response to this bacterial antigen (Hayashi et al., 2001). A stop codon polymorphism (Arg392TER) has been identified in the *TLR5* gene and is associated with an increased susceptibility to *Legionella pneumophila* (Hawn et al., 2003), suggesting that this allele may increase susceptibility to pneumonia associated with flagellated organisms.

NOD-like and RIG-like Receptors

TLRs are involved in the recognition of all types of pathogens, regardless of location. In contrast, NOD-like receptors (NLRs; NOD = nucleotide oligomerization domain) primarily recognize cytoplasmic bacterial pathogens (Inohara and Nunez, 2003). Many NLRs have caspase recruitment domains (CARDs), which are involved in the assembly of pro-apoptotic protein complexes and may also participate in NF-kappa B signaling pathways. Hill has studied a truncation variant of the human *CARD8* gene and found that African children homozygous for this inactivating mutation are susceptible to non-typhoidal *Salmonella* bacteremia (Hill, 2006).

A third family of recently described pattern recognition receptors is the DExD/H Box RNA helicase family of RIG-I-like receptor (RLR; RIG = retinoic acid-inducible gene) genes, which include MDA5 as well as RIG-1 (Creagh and O'Neill, 2006). These receptors appear to differentially recognize double-stranded RNA from various viruses (Kato et al., 2006). An amino acid change in MDA5 has been associated with susceptibility to type 1 diabetes, lending support to a viral etiology for this disease (Smyth et al., 2006).

Mannose-binding Lectin

Mannose-binding lectin (MBL) is an acute-phase protein that can opsonize many bacterial and fungal pathogens and activate complement (Kuhlman et al., 1989). Approximately one-third of most human populations are heterozygous for one of several mutations in exon 1, resulting in lower MBL concentrations. Individuals with two low-MBL haplotypes have a higher risk of pneumococcal sepsis (Kronborg and Garred, 2002; Roy et al., 2002). There is less well-replicated evidence for susceptibility to other bacterial pathogens as well as candidiasis.

Although multiple studies indicate that MBL plays a role in mitigating certain pathogens, the high prevalence of low-MBL haplotypes in certain ethnic groups suggests that a relative lack of MBL might be beneficial to the host under other circumstances. In particular, studies of the MBL-MBL Serine Peptidase (MASP) pathway suggest a role in reducing reperfusion injury in the heart, resulting in a protective effect against the complications of myocardial infarction (Walsh et al., 2005).

CD14

CD14 functions as an anchor protein and is a ubiquitous pattern recognition receptor specific for ligands such as LPS. Performing as a co-receptor for TLR4, it is shed by monocytes to facilitate LPS signaling for all other cells. Two polymorphisms at the promoter region of the *CD14* gene have been widely explored. The -159T allele has been associated with increased prevalence of gram-negative infections and sepsis, but not with shock or survival (Gibot et al., 2002; Sutherland et al., 2005). However, other studies have yielded contradictory results, including a recent report that association with -260T allele increased survival in a Brazilian sepsis population (D'Avila et al., 2006).

Intracellular Signaling Molecules

The IL-1 receptor-associated kinase-4 (IRAK4) is an intracellular kinase that transduces signals conveyed by the TLRs and IL-1 receptors. Individuals homozygous for either of two known variant alleles appear to have completely absent expression of IRAK4, thereby inhibiting activation of NF-kappa B and NF-kappa B-dependent pro-inflammatory mediators (Von Bernuth et al., 2005), resulting in increased risk of invasive bacterial disease (Lasker et al., 2005).

CYTOKINE POLYMORPHISMS

Genetic variation in the pro-inflammatory cytokines TNF-α, TNF-β, IL-6, IL-8, the macrophage migration inhibitory factor

(MIF), and the anti-inflammatory cytokines IL-10 and IL-1 RA are the most extensively studied cytokines in relation to sepsis. Patients predisposed to a balanced anti-/pro-inflammatory response appear to have better chances for survival (Figure 98.1). Defects in regulation occur when the balance is shifted, allowing unmitigated inflammation, resulting in tissue and organ damage or the inability to extirpate invading pathogens.

Pro-inflammatory Cytokines

TNF-α is a primary mediator of sepsis and an initial trigger of the immune response. Most studies have investigated the significance of the -308A allele in the promoter region of TNF-α, yielding mixed results for outcome of sepsis and no association with TNF-α expression levels (Mira et al., 1999; Stuber et al., 1996). Potential associations with polymorphisms at positions −376 and −238 are also unconvincing. Bayley and colleagues reviewed the functional impact of genomic variations within the TNF-α locus, and showed that most variations had no impact on TNF-α expression or had contradictory results (Bayley et al., 2004).

A number of polymorphisms have also been described in TNF-β, which binds to the same receptor as TNF-α. In particular, the TNF-β 250 SNP has been the target of multiple investigations. However, linkage disequilibrium with heat shock protein 70 alleles confounds interpretation, and more advanced models will be necessary before inferences of the role of TNF-β variation in sepsis can be made.

IL-1α and IL-1β engage the same receptor and are potent pro-inflammatory cytokines released by macrophages involved in the systemic inflammatory response. IL-1α and IL-1β are capable of inducing the symptoms of septic shock and organ failure in animal models (Leon et al., 1992). Despite the finding that a homozygous TaqI genotype correlates with expression of IL-1β, no association with incidence or outcome of sepsis has been determined (Ma et al., 2002; Pociot et al., 1992).

IL-6 plays a role in lymphocyte stimulation, and its levels are consistently associated with severity and mortality in sepsis (Van der Poll and Van Deventer, 1999). A German study demonstrated improved survival but not incidence of sepsis in association with a promoter variant, -174 G/C (Schluter et al., 2002). It is possible that variation at the *IL*-6 gene may contribute to the outcome of sepsis, but variable association of specific SNPs with IL-6 levels and the potential for linkage disequilibrium with several alleles obscure the clinical significance of these results.

Similar studies to identify polymorphisms in IL-8 and MIF promoter regions and genes are ongoing.

Anti-inflammatory Cytokines

IL-10 is an integral part of the body's anti-inflammatory processes and is involved in the suppression of innate and adaptive immune responses. Its expression is stimulated by inflammation, and it is a potent inhibitor of non-specific inflammatory response. In sepsis, high IL-10 levels have consistently been associated with disease severity (Wunder et al., 2004). Of the several promoter variants reported in this cytokine gene, a SNP at -1082 has been found at a higher frequency in patients with sepsis (particularly pneumococcal) and community-acquired pneumonia; however, more studies are needed.

The IL-1 receptor antagonist (IL1RN) inhibits the pro-inflammatory actions of IL-1 by binding to its receptor. Polymorphisms in IL1RN may be promising candidates for genetic associations, but should be explored in relation to the rest of IL-1 peptides.

COAGULATION

Activation of the coagulation cascade is a key event in the pathogenesis of sepsis. As a result, abnormalities in the antithrombin III, protein C, and tissue-factor inhibitor pathways have been implicated in the pathogenesis of sepsis. Activated protein C (APC) is the only pharmacotherapy shown to be effective in the treatment of a subset of patients with severe sepsis (Bernard et al., 2001). Protein C -1641 AA genotype is associated with decreased survival, more organ dysfunction, and more systemic inflammation in patients having severe sepsis (Walley and Russell, 2007). APC inhibits not only factors Va, Xa, and plasminogen activator inhibitor-1 (PAI-1), but also neutrophil adherence, chemotaxis, and cytokine release. In animal models of septic shock, genetic deficiency of tissue factor reduces mortality (Texereau et al., 2004). In contrast, genetic deficiencies of the anticoagulants thrombomodulin, antithrombin II, and protein C increase mortality. Genetic polymorphisms have been described in genes of hemostatic factors, including thrombin fibrinogen, factor V, PAI-1, protein C, and endothelial protein C receptor, but have not yet been demonstrated to be associated with sepsis risk or outcome.

Factor V and the Protein C Pathway

Factor V influences protein C activation by promoting thrombin generation. Three independent SNPs of factor V have been described, all of which cause partial resistance to inactivation of factor Va by APC, resulting in a pro-thrombotic state (Nicolaes and Dahlback, 2003). The frequency of factor V Leiden (1691G>A) in certain populations suggests that it confers some evolutionary advantage. In a large study of children with meningococcal disease, factor V Leiden heterozygosity was associated with increased incidence of purpura fulminans and a trend toward reduced mortality (Kondaveeti et al., 1999). Factor V Leiden carrier status has been associated with lower 28-day mortality and less vasopressor use (Bernard et al., 2001), but did not determine responsiveness to APC treatment. However, biomarker studies have demonstrated that plasma protein C levels predict usefulness of APC infusion (Shorr et al., 2006).

PAI-1 and the Fibrinolytic System

PAI-1 promotes clot stability, extension, and resistance to lysis, and also acts as an acute-phase reactant (Hoekstra

et al., 2004). Increased PAI-1 is a risk factor for cardiovascular disease, but may confer a survival benefit in meningococcal sepsis (Kornelisse et al., 1996). A common PAI-1 promoter SNP is characterized by 4G and 5G alleles. 4G homozygosity is associated with an increase in PAI-1 transcription and higher rates of myocardial infarction, as well as a higher risk of mortality and vascular complications among patients with sepsis (Haralambous et al., 2003; Hermans and Hazelzet, 2005; Zhan et al., 2005). Genetic polymorphisms of other constituents of the fibrin formation and degradation pathways also deserve further study.

METABOLISM

Energy for intracellular homeostasis is primarily generated by mitochondrial respiration and oxidative phosphorylation (Raha and Robinson, 2000). Mitochondrial genomes replicate independently of the nuclear genome in the process of mitochondrial biogenesis. Mitochondrial proteins are encoded in both the nuclear and mitochondrial genomes. Several lines of evidence support the importance of metabolism as a mechanism of sepsis survival or death: an early indicator of sepsis survival or death is the presence or absence of mitochondrial biogenesis and increased aerobic metabolism, respectively (Brealey et al., 2002, 2004; Carre et al., 2010; Haden et al., 2007; Kingsmore et al., 2011; Suliman et al., 2005). Structural studies have shown mitochondrial derangements, decreased mitochondrial number, and reduced substrate utilization in sepsis death, and progressive fall in total body oxygen consumption with increasing severity of sepsis (Brealey et al., 2002; Fredriksson et al., 2008; Giovannini et al., 1983; Kreymann et al., 1993; Long, 1977; Pyle et al., 2010). Rare inherited mutations in the mitochondrial genome are associated with severe sepsis and death (Akman et al., 2004; Chae et al., 2010), as are inherited mutations in mitochondrial enzymes encoded by the nuclear genome, such as α-methylacetoacetic aciduria, galactosemia, methylmalonic aciduria, some mucopolysaccharidoses and biotinidase deficiency. Specific mitochondrial haplotypes have been associated with sepsis survival and death (Baudouin et al., 2005; Yang et al., 2008). Reactive oxygen species, by-products of respiration that are overproduced during sepsis, damage the mitochondrial genome (Anson et al., 2000; Kantrow et al., 1997; Raha and Robinson, 2000; Shigenaga et al., 1994; Taylor et al., 1995, 1998) by converting guanine nucleotides to 8-hydroxyguanine (Lindahl, 1993). These residues, in turn, create somatic G:C to T:A transversion mutations during mitochondrial biogenesis (Anson et al., 2000; Dianov et al., 2001; Kantrow et al., 1997; Kasai and Nishimura, 1984; Osiewacz, 2002; Raha and Robinson, 2000; Shigenaga et al., 1994; Taylor et al., 1995, 1998), which can further impair mitochondrial respiration and oxidative phosphorylation. However, the extent to which somatic mutations in mitochondrial genomes affect sepsis outcomes has not yet been determined.

NEUROHORMONAL SYSTEM

As discussed above, cytokines are key mediators of host inflammatory responses during infection. While local cytokine production at sites of infection is important, more than 90% of the TNF-α released following LPS comes from the spleen (Huston et al., 2006). Splenic release of inflammatory cytokines is largely controlled by the vagus nerve, a mechanism called the cholinergic anti-inflammatory pathway (Borovikova et al., 2000; Rosas-Ballina et al., 2011). Vagal afferents signal via nicotinic acetylcholine α7 receptors via the celiac ganglion and thence the splenic nerve (Wang et al., 2003). Interestingly, splenic nerve neurons appear to terminate on a subpopulation of memory T lymphocytes through synapse-like endings (Rosas-Ballina et al., 2008). As yet there is no genetic link between sepsis risk or outcome and the cholinergic anti-inflammatory pathway.

Gender is known to affect sepsis susceptibility and outcome (Marriott and Huet-Hudson, 2006). Glucocorticoids and androgens are counter-regulatory anti-inflammatory mediators that normally suppress pro-inflammatory cytokine production, whereas estrogens enhance inflammatory responses (Rettew et al., 2008, 2009). Transient changes in the glucocorticoid sensitivity of leukocytes occur during sepsis (Van Rossum and Van den Akker, 2011), and in the relative levels of glucocorticoids and androgens (Kingsmore et al., unpublished). There is also wide inter-individual variation in glucocorticoid sensitivity, due in part to variation of the glucocorticoid receptor gene. Further work is required to determine the significance of genetic variation in steroid synthesis and metabolism in sepsis.

FUTURE INVESTIGATIONS

To address the complexity of the sepsis response and prediction of outcome, multiple surrogate markers that reflect the nature and severity of the inflammatory response, the status of the coagulation and fibrinolysis systems, and the magnitude of organ injury will likely be useful in stratifying patients at risk for adverse outcome and those who may benefit from interventional therapies. Such strategies will inherently require the use of multiplex approaches that have sufficiently high throughput to be cost-effective (Kingsmore, 2006). A combination of measurements of five serum or plasma cytokines showed high sensitivity and specificity in distinguishing between acutely ill septic and non-septic patients, whereas individual cytokine measurements lacked specificity (Kingsmore et al., 2011). Therefore, isolated candidate gene-by-gene or protein-by-protein approaches may be rendered obsolete in the near future.

INTEGRATIVE GENOMICS, MOLECULAR SIGNATURES, AND SEPSIS

The recognition and treatment of sepsis has posed a daunting challenge for clinicians despite technological advances and improvements in critical care medicine. In addition to a broad array of infectious etiologies, many of the clinical signs of overwhelming sepsis, such as fever, tachycardia, and leukocytosis, are non-specific and can also be seen with systemic inflammation induced by non-infectious causes such as mesenteric ischemia-reperfusion, pulmonary embolus, pancreatitis, or trauma. Regardless, empirical broad-spectrum antibiotic use is the rule as a component of early goal-directed therapy (Rivers et al., 2001).

The characterization of the human genome has provided a mechanism to study infection at the molecular level. Where former techniques have focused on obtaining fluid or tissue from a body compartment and incubating the sample to determine the cause of infection (if any), genomics provides another means. The discovery of surface receptors that recognize microbial invasion and transmit signals to the immune system has led scientists to attempt characterization of the molecular and genetic changes in the host organism as it is invaded by exogenous infections. From as little as 100 ng of total cellular RNA isolated from whole blood or any tissue, the relative expression of all known genes, expressed sequence tags, and open-reading frames can be evaluated simultaneously. More importantly, the global interaction among these thousands of genes can be studied without any specific selection bias. By studying global expression, one can devise methods of class prediction based on global gene expression activity, which can be done without prior knowledge of the function of any of the genes.

Functional genomics relates these technologies to clinical medicine. Although most investigations have focused on the changes in one or relatively few genes in response to a disease or treatment, these newer technologies are exploring the changes in gene expression of the entire genome. This has resulted in functional genomics offering two unique perspectives on sepsis biology – the ability to use "patterns" of gene, protein, or metabolite expression to class-predict or classify tissue responses (i.e., to develop a "signature" or "fingerprint" for a specific tissue, pathogen, or outcome), and the ability to explore the underlying biological changes that occur in health and disease while not being limited to a subset of selected genes, proteins, or metabolites.

Challenges Associated with Applying Genomic Science to Sepsis

Although the technical aspects of data acquisition from microarray analyses are not problematic, quality control, data analysis, and information extraction remain particularly challenging. Traditional statistical and bioinformatic approaches are not appropriate for the simultaneous analyses of the large number of analytes on most microarrays. In addition to using probabilities to define confidence intervals among groups, false discovery rates are often applied (Steinmetz and Davis, 2004; Storey and Tibshirani, 2003). The issues of how such vast quantities of data are analyzed or presented in a manner that can be readily understood are equally problematic. Heat maps, dendrograms, hierarchical clustering, principal component analyses, network analyses, Bayesian factor analysis, molecular cartography, self-organizing maps, correlation analyses, and topographic comparisons are being used to extract meaningful information (McDunn et al., 2008; Polpitiya et al., 2009).

As methods for accruing and analyzing data are improved, investigators can turn to other important issues to maximize the utility of genomic methodologies. For example, transcript abundance in peripheral blood leukocytes may not reflect that in leukocytes at local sites of infection. Furthermore, human samples other than blood, cells, or tissue from the primary site of infection might prove to be useful for genomic analysis.

Pathogen Signatures

Recognition of microbial invasion is one of the hallmarks of innate immunity. Antigen-presenting cells, as well as cells of epithelial and endothelial origin, express surface receptors that can recognize PAMPs. Following the description of TLRs, considerable effort was directed at examining the genome-wide expression response to different microbial pathogens. Nau and colleagues found that human macrophages have a profound genetic perturbation demonstrated following infection by any of a number of exogenous bacteria (Nau et al., 2002). The shared responses alter the baseline expression of genes that encode for other cell-surface receptors, signal transduction proteins, and transcription factors. Other investigators have examined genomic responses to exogenous microbial stimuli using varying populations of cells. Calvano and colleagues examined the genome-wide response of leukocytes from healthy volunteers inoculated with endotoxin. Of 44,000 genes studied, there were 3714 whose expression signal changed significantly with *in vivo* stimulation (Calvano et al., 2005). This proved, on a genomic level, that systemic perturbations of infection were more vast than was clinically evident.

Despite such studies showing vast alterations in gene expression induced by infection, genomic technology is still useful to discriminate among types of infections. Huang and colleagues not only confirmed a large shared response among dendritic cells exposed to *Escherichia coli*, *Candida albicans*, and influenza virus, they also showed a differential response based on the pathogen (Huang et al., 2001). Feezor and colleagues stimulated whole blood from human volunteers with LPS or heat-killed *S. aureus*, and found 758 distinct genes with different expression levels according to gram-negative versus gram-positive exposure (Feezor et al., 2003). The investigators demonstrated that the families of genes affected by different pathogenic stimuli were strikingly unique. For instance, gram-negative infection tended to alter the expression of genes involved in global immune response, signal transduction, and plasma membrane function, while gram-positive infection altered the expression of genes believed to control ribosomal proteins and cell cycle regulation.

Chung and colleagues sought to characterize a genomic profile capable of distinguishing sepsis from presumed sterile systemic inflammation (Chung et al., 2006). Those two cohorts were compared in a parallel analysis of human and murine models. Genomic data from spleens of septic and injured patients were used to create a septic profile that showed an accuracy of 67.1%; the accuracy of the genomic predictor profile in the murine model was as high as 96%.

Ramilo and colleagues have demonstrated distinctive gene expression patterns in peripheral blood leukocytes from patients with confirmed systemic bacterial (*E. coli*, *S. aureus*, *Streptococcus pneumoniae*) and viral (influenza A) infections (Ramilo et al., 2007). Furthermore, distinctive gene expression patterns were observed in patients with respiratory infections of different etiologies, demonstrating the utility of such analyses in blood in the evaluation of infections in different organ systems. Similar results were documented in a cohort of military trainees with febrile respiratory illness (Thach et al., 2005).

Recently, Zaas and colleagues demonstrated that peripheral blood gene expression signatures distinguish individuals with symptomatic acute respiratory infections (ARI) from uninfected individuals with greater than 95% accuracy (Figure 98.6). This signature was observed in human viral challenge studies with live rhinovirus, respiratory syncytial virus, and influenza A, and was able to differentiate viral from bacterial ARI. They validated the accuracy of the signature for classification of infected individuals and healthy controls in an independent influenza A dataset (Zaas et al., 2009).

These studies are the first to show that molecular characterization can be performed to distinguish between the infected and non-infected states even in the face of very similar clinical presentations. These data also serve to highlight the grossly insensitive criteria used in clinical medicine to determine infection or sepsis, and the power of genomic science to provide a molecular characterization of a clinical dilemma. On a macro-system level, the whole body response to either stimulus has been well established: fever, leukocytosis, and tachycardia, with the potential of devolving into multisystem organ failure. Functional genomics is a better test of the cause of the patient's global dysfunction, and may one day be used to guide initial therapy or to tailor ongoing therapy.

Progression Signatures

The need for rapid, accurate identification of disease progression in sepsis has increased dramatically with the upcoming availability of several novel treatment regimens. While novel sepsis therapies are improving sepsis outcomes, they are creating new patient-management and diagnostic challenges for physicians. In 2001, the Food and Drug Administration (FDA) approved APC (Xigris®, Eli Lilly and Company, Indianapolis, IN) for treatment of patients with severe sepsis and an APACHE II score of ≥25 – one of the first examples of the inclusion of a biomarker in the indication for a therapy. In the pivotal phase III trial of APC in severe sepsis (SS) (PROWESS), 28-day mortality was decreased by 6% (Bernard et al., 2001). The greatest reduction in mortality (13%) and cost-effectiveness was observed in the most seriously ill patients (those with APACHE II score ≥25). APC benefit was also observed in patients with pneumonia. In contrast, APC exhibited modest survival benefit and cost-ineffectiveness in patients with

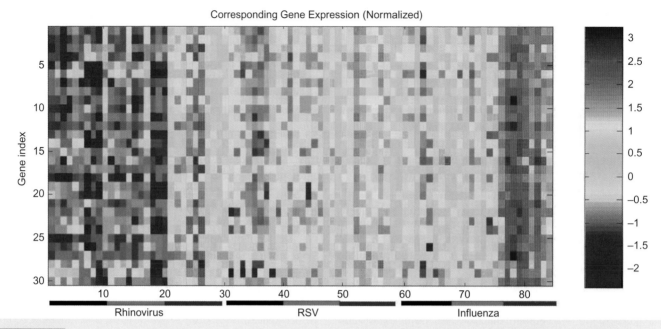

Corresponding Gene Expression (Normalized)

Rhinovirus   RSV   Influenza

Figure 98.6 Heat map showing that infected adult subjects with symptomatic human rhinovirus (HRV), respiratory syncytial virus (RSV), or influenza A infection were distinguished from uninfected individuals by a distinct group of genes demonstrating differential expression among symptomatic individuals as compared to asymptomatic individuals. Whole-blood gene expression was determined pre-inoculation (baseline) and at the time of peak symptoms for each symptomatic individual, and at a matched time-point for each asymptomatic individual. Columns represent subjects, with the first 10 representing baseline gene expression of asymptomatic individuals in the HRV challenge, the next 10 representing time-points matched to peak symptoms for the asymptomatic subjects in the HRV cohort, and the following 10 columns representing time of peak symptoms for the 10 subjects who developed symptomatic HRV infection. A similar layout continues for the RSV and influenza cohorts. Blue and red represent extremes of gene expression, with visually apparent differences between baseline and matched time-points in the asymptomatic individuals versus time of peak symptoms in symptomatic individuals. Bars underneath represent individual groups (black, baseline; red, asymptomatic; blue, symptomatic). *Reproduced with permission from Zaas et al., 2009.*

APACHE II score <25. However, APC therapy is associated with a 1–2% incidence of major bleeding. For these reasons, appropriate use of APC is most likely to occur following development of an objective, accurate, rapid diagnostic test for progression to severe sepsis. Diminution of protein C activity in citrated plasma was recently shown to be a surrogate marker for APC efficacy in patients with severe sepsis (Shorr et al., 2006).

Prompt and accurate diagnosis is especially important for effective sepsis management and contributes significantly to positive outcomes (Rivers et al., 2001). Rate of progression of severe sepsis to organ failure, septic shock, and death is heterogeneous and largely independent of the specific underlying infectious disease process. For example, case fatality rates in those with culture-negative severe sepsis are similar to those with positive cultures (Rangel-Frausto et al., 1995). Case fatality rates in sepsis however, are critically dependent upon disease staging. Current differentiation of SIRS, sepsis, severe sepsis, and septic shock relies exclusively on clinical assessment (Balk, 2000). Thus, rapid, quantitative, objective determination of the stage of sepsis development and likelihood of progression to severe is a key unmet diagnostic need.

Several currently available clinical indices do provide quantitative assessment of staging and severity of sepsis, but with limitations. As noted above, they include SOFA and APACHE II scores, and capillary lactate level (Pacelli et al., 1996). Although it is quantitative, APACHE II score is largely subjective, complex, cumbersome, and has a relatively narrow dynamic range. Blood lactate levels are quantitative, but are limited by false negative normal values in elderly patients and by confounding comorbidities such as liver failure, diabetes mellitus, and the use of certain medications. The clinical picture in sepsis patients is highly dynamic, and assessment of such indices tends to be subjective and to occur with insufficient frequency. Furthermore, sepsis is very heterogeneous in terms of pathogen, source of infection, associated comorbidity, course, and complications. Delayed or errant diagnosis of severe sepsis may result in failure of timely treatment. Thus, an early, rapid, objective diagnostic of severe sepsis would be a significant adjunct to these clinical indices and would significantly advance patient management.

Elevations of IL-6, IL-8, IGFBP1, IL-2sR, and MIF in sepsis have been widely documented. Furthermore, levels of IL-6, IL-8, and IL-2sR have variously shown correlation with sepsis severity, progression to septic shock, organ failure, and mortality. However, these blood analytes are also elevated in many other conditions, and diagnostic sensitivity and specificity of individual analyte levels has been insufficient for diagnosis of sepsis or severe sepsis (Delogu et al., 1995; Martin et al., 1994; Selberg et al., 2000).

Using an antibody microarray on specimens from patients from the PROWESS study (Bernard et al., 2001; Perlee et al., 2004), including sepsis non-survivors, 133 analytes were measured in 139 matched serum and plasma samples (between day 0 and 28) from 12 severe sepsis patients (seven survivors and five who died) and eight normal individuals. A total of 63 candidate sepsis biomarkers were identified (35 in serum and 48 in plasma); each exhibited a greater than two-fold change between severe sepsis patients and controls on day 0. The top biomarker was IL-6 (26-fold increase). This study also identified six analytes with greater than five-fold difference in SS that were not identified in Study 1 (ST2, FGF21, leptin, MCP1, MIP3β, and TNFR1) (Kingsmore et al., 2011). Multivariable models were more effective than single analytes in prediction of sepsis. A combination of measurements of five serum or plasma cytokines showed high sensitivity and specificity in distinguishing between acutely ill, septic, and non-septic patients, whereas individual cytokine measurements lacked specificity.

In an additional study, biomarkers predictive of sepsis death exhibited at least a two-fold difference in average level between sepsis survivors (three patients) and non-survivors (three patients) in the days immediately preceding death (days 0–4). A total of 10 candidate sepsis mortality biomarkers were identified, the most dramatic of which was follistatin, which exhibited a 15-fold difference.

DNA microarray-based gene transcript profiling of the responses of primates to infection has begun to yield new insights into host–pathogen interactions, but this approach remains plagued by challenges and complexities that have yet to be adequately addressed. The rapidly changing nature of acute infectious diseases over time in a host, coupled with the genetic diversity of microbial pathogens, presents unique problems for the design and interpretation of functional genomic studies in this field. Also, there are the more common problems related to heterogeneity within clinical samples, the complex non-standardized confounding variables associated with human subjects, and the complexities posed by the analysis and validation of highly parallel data. Various approaches have been developed to address each of those issues, but there are still significant limitations to be overcome. The resolution of these problems should lead to a better understanding of the dialogue between the host and pathogen.

THERAPEUTICS

SIRS may represent either an exaggeration of a normal physiological response to the infectious process or an imbalance between the pro- and anti-inflammatory processes of the host.

To that end, multiple immunotherapies have been investigated in an effort to modulate the inflammatory response and alter the clinical course of the patient with sepsis. These approaches have been relatively successful in animal models, but, with the exception of APC, have largely failed in clinical trials (Bernard et al., 2001).

In general, clinical trials have failed because of the heterogeneity of the clinical population, the inability to recognize early clinical signs, and an incomplete understanding of the underlying pathophysiology of the septic response. Eichacker and colleagues have even speculated that the failure of the trials was inevitable given the severity of the illness in the patient populations. Their analysis of the clinical trials showed that anti-inflammatory agents were more effective in patients with a higher complication risk, and were potentially harmful in less severely ill patients (Eichacker et al., 2002).

Knowledge of the entire human genomic sequence is also the basis of the fields of pharmacogenetics and pharmacogenomics. Pharmacogenomics seeks an understanding of how genes influence drug response and toxicity, and the discovery of new disease pathways that can be targeted with tailor-made drugs. Pharmacogenetics is the study of the genetic factors involved in the differential response between patients to the same medicine. Polymorphisms account for our differential susceptibility to disease and the variable outcome of treatments. The study of these variants in the human genome should enable pharmacogenetics to define optimal treatment regimens for subsets of the population, allowing a wider range of patients to be treated and more effective outcomes to be produced with any given drug.

Host response to severe sepsis is multifactorial and its pathogenesis is complex, making it unlikely to be successfully addressed by a single drug or intervention. Rather, the trend in therapeutic interventions has been to focus on multimodal therapies targeting specific pathological components of the host response. It is probably naïve to conclude that molecular monitoring of patients with sepsis will rely on the concentration or expression of any one marker, protein, or gene. In fact, prognostic studies conducted over the past 20 years have clearly shown that measurement of single plasma analytes generally lacks the sensitivity or specificity to predict outcome of severe trauma and sepsis. Although numerous individual mediators have plasma concentrations as a group that differ between patients with and without sepsis and between those who survive or die from sepsis (e.g., TNF-α, IL-1β, IL-6, IL-10, and procalcitonin), individual measurements have not proven effective in predicting whether a patient will survive or respond to therapy (Tsalik et al., 2012).

CONCLUSIONS

Sepsis is a highly heterogeneous, genetically complex syndrome initiated by infection and characterized by SIRS with variable progression and outcome. Nucleotide variants in the immune response to infection have been shown to be associated with sepsis outcomes. Genetic, genomic, and functional studies have provided important insights into the mechanisms involved in the pathogenesis of sepsis-induced organ dysfunction and death. Recent advances in whole-genome and exome sequencing studies and integrative genomics will provide valuable information on the interaction of multiple allelic variants and clinical outcome. More precise categorization of patients based on genetic background is likely to lead to individualized treatment.

In summary, the prevalence, disparate presentation, rapid evolution, unpredictable course, high mortality, and availability of complex or costly new therapies for sepsis combine to create a significant and growing need for accurate, rapid, early identification of severe sepsis. Availability of such diagnostics will enable physicians to select patients rapidly and objectively for early, intensive sepsis therapy, thereby decreasing morbidity, mortality, and associated patient care costs.

REFERENCES

Agnese, D.M., Calvano, J.E., Hahm, S.J., et al., 2002. Human toll-like receptor 4 mutations but not CD14 polymorphisms are associated with an increased risk of gram-negative infections. J Infect Dis 186 (10), 1522–1525.

Akira, S., Takeda, K., Kaisho, T., 2001. Toll-like receptors: Critical proteins linking innate and acquired immunity. Nat Immunol 2 (8), 675–680.

Akman, C.I., Sue, C.M., Shanske, S., et al., 2004. Mitochondrial DNA deletion in a child with megaloblastic anemia and recurrent encephalopathy. J Child Neurol 19 (4), 258–261.

Alberti, C., Brun-Buisson, C., Burchardi, H., et al., 2002. Epidemiology of sepsis and infection in ICU patients from an international multicentre cohort study. Intensive Care Med 28 (2), 108–121.

Angus, D.C., Linde-Zwirble, W.T., Lidicker, J., Clermont, G., Carcillo, J., Pinsky, M.R., 2001. Epidemiology of severe sepsis in the United States: Analysis of incidence, outcome, and associated costs of care. Crit Care Med 29 (7), 1303–1310.

Annane, D., Aegerter, P., Jars-Guincestre, M.C., Guidet, B., 2003. Current epidemiology of septic shock: The CUB-Rea network. Am J Respir Crit Care Med 168 (2), 165–172.

Anson, R.M., Hudson, E., Bohr, V.A., 2000. Mitochondrial endogenous oxidative damage has been overestimated. FASEB J 14 (2), 355–360.

Apostolidou, I., Katsouyanni, K., Touloumi, G., Kalpoyannis, N., Constantopoulos, A., Trichopoulos, D., 1994. Seasonal variation of neonatal and infant deaths by cause in Greece. Scand J Soc Med 22 (1), 74–80.

Arbour, N.C., Lorenz, E., Schutte, B.C., et al., 2000. TLR4 mutations are associated with endotoxin hyporesponsiveness in humans. Nat Genet 25 (2), 187–191.

Arcaroli, J., Fessler, M.B., Abraham, E., 2005. Genetic polymorphisms and sepsis. Shock 24 (4), 300–312.

Balk, R.A., 2000. Severe sepsis and septic shock: Definitions, epidemiology, and clinical manifestations. Crit Care Clin 16 (2), 179–192.

Baudouin, S.V., Saunders, D., Tiangyou, W., et al., 2005. Mitochondrial DNA and survival after sepsis: A prospective study. Lancet 366 (9503), 2118–2121.

Bayley, J.P., Ottenhoff, T.H., Verweij, C.L., 2004. Is there a future for TNF promoter polymorphisms? Genes Immun 5 (5), 315–329.

Bellamy, R., Hill, A.V., 1998. Genetic susceptibility to mycobacteria and other infectious pathogens in humans. Curr Opin Immunol 10 (4), 483–487.

Bernard, G.R., Vincent, J.L., Laterre, P.F., et al., 2001. Efficacy and safety of recombinant human activated protein C for severe sepsis. N Engl J Med 344 (10), 699–709.

Beutler, B., Rietschel, E.T., 2003. Innate immune sensing and its roots: The story of endotoxin. Nat Rev Immunol 3 (2), 169–176.

Bone, R.C., Balk, R.A., Cerra, F.B., et al., 1992. Definitions for sepsis and organ failure and guidelines for the use of innovative therapies in sepsis. The ACCP/SCCM Consensus Conference Committee. American College of Chest Physicians/Society of Critical Care Medicine. Chest 101 (6), 1644–1655.

Bone, R.C., Fisher Jr., C.J., Clemmer, T.P., Slotman, G.J., Metz, C.A., Balk, R.A., 1989. Sepsis syndrome: A valid clinical entity. Methylprednisolone Severe Sepsis Study Group. Crit Care Med 17 (5), 389–393.

Borovikova, L.V., Ivanova, S., Zhang, M., et al., 2000. Vagus nerve stimulation attenuates the systemic inflammatory response to endotoxin. Nature 405 (6785), 458–462.

Botstein, D., Risch, N., 2003. Discovering genotypes underlying human phenotypes: Past successes for Mendelian disease, future approaches for complex disease. Nat Genet (33 Suppl.), 228–237.

Brame, A.L., Singer, M., 2010. Stressing the obvious? An allostatic look at critical illness. Crit Care Med 38 (10 Suppl.), S600–S607.

Brealey, D., Brand, M., Hargreaves, I., et al., 2002. Association between mitochondrial dysfunction and severity and outcome of septic shock. Lancet 360 (9328), 219–223.

Brealey, D., Karyampudi, S., Jacques, T.S., et al., 2004. Mitochondrial dysfunction in a long-term rodent model of sepsis and organ failure. Am J Physiol Regul Integr Comp Physiol 286 (3), R491–R497.

Burgner, D., Levin, M., 2003. Genetic susceptibility to infectious diseases. Pediatr Infect Dis J 22 (1), 1–6.

Calvano, J.E., Bowers, D.J., Coyle, S.M., et al., 2006. Response to systemic endotoxemia among humans bearing polymorphisms of the Toll-like receptor 4 (hTLR4). Clin Immunol 121 (2), 186–190.

Calvano, S.E., Xiao, W., Richards, D.R., et al., 2005. A network-based analysis of systemic inflammation in humans. Nature 437 (7061), 1032–1037.

Cardon, L.R., Bell, J.I., 2001. Association study designs for complex diseases. Nat Rev Genet 2 (2), 91–99.

Cardon, L.R., Palmer, L.J., 2003. Population stratification and spurious allelic association. Lancet 361 (9357), 598–604.

Carre, J.E., Orban, J.C., Re, L., et al., 2010. Survival in critical illness is associated with early activation of mitochondrial biogenesis. Am J Respir Crit Care Med 182 (6), 745–751.

Center for Disease Control and Prevention, 1991. Current trends increase in national hospital discharge survey rates for septicemia – United States, 1979–1987. Centers for Disease Control and Prevention, Atlanta, GA, Morbidity and Mortality Weekly Report 39, pp. 31–34.

Chae, J.H., Lim, B.C., Cheong, H.I., Hwang, Y.S., Kim, K.J., Hwang, H., 2010. A single large-scale deletion of mtDNA in a child with recurrent encephalopathy and tubulopathy. J Neurol Sci 292 (1–2), 104–106.

Child, N.J., Yang, I.A., Pulletz, M.C., et al., 2003. Polymorphisms in Toll-like receptor 4 and the systemic inflammatory response syndrome. Biochem Soc Trans 31 (3), 652–653.

Choi, E.H., Zimmerman, P.A., Foster, C.B., et al., 2001. Genetic polymorphisms in molecules of innate immunity and susceptibility to infection with *Wuchereria bancrofti* in south India. Genes Immun 2 (5), 248–253.

Chung, T.P., Laramie, J.M., Meyer, D.J., et al., 2006. Molecular diagnostics in sepsis: From bedside to bench. J Am Coll Surg 203 (5), 585–598.

Cobb, J.P., Mindrinos, M.N., Miller-Graziano, C., et al., 2005. Application of genome-wide expression analysis to human health and disease. Proc Natl Acad Sci USA 102 (13), 4801–4806.

Creagh, E.M., O'Neill, L.A., 2006. TLRs, NLRs and RLRs: A trinity of pathogen sensors that cooperate in innate immunity. Trends Immunol 27 (8), 352–357.

D'Avila, L.C., Albarus, M.H., Franco, C.R., et al., 2006. Effect of CD14 -260C>T polymorphism on the mortality of critically ill patients. Immunol Cell Biol 84 (4), 342–348.

Delogu, G., Casula, M.A., Mancini, P., Tellan, G., Signore, L., 1995. Serum neopterin and soluble interleukin-2 receptor for prediction of a shock state in gram-negative sepsis. J Crit Care 10 (2), 64–71.

Densen, P., Weiler, J.M., Griffiss, J.M., Hoffmann, L.G., 1987. Familial properdin deficiency and fatal meningococcemia: Correction of the bactericidal defect by vaccination. N Engl J Med 316 (15), 922–926.

Dianov, G.L., Souza-Pinto, N., Nyaga, S.G., Thybo, T., Stevnsner, T., Bohr, V.A., 2001. Base excision repair in nuclear and mitochondrial DNA. Prog Nucleic Acid Res Mol Biol 68, 285–297.

Douglas, A.S., Al-Sayer, H., Rawles, J.M., Allan, T.M., 1991. Seasonality of disease in Kuwait. Lancet 337 (8754), 1393–1397.

Drennan, M.B., Nicolle, D., Quesniaux, V.J., et al., 2004. Toll-like receptor 2-deficient mice succumb to *Mycobacterium tuberculosis* infection. Am J Pathol 164 (1), 49–57.

Eichacker, P.Q., Parent, C., Kalil, A., et al., 2002. Risk and the efficacy of anti-inflammatory agents: Retrospective and confirmatory studies of sepsis. Am J Respir Crit Care Med 166 (9), 1197–1205.

Feezor, R.J., Oberholzer, C., Baker, H.V., et al., 2003. Molecular characterization of the acute inflammatory response to infections with gram-negative versus gram-positive bacteria. Infect Immun 71 (10), 5803–5813.

Fredriksson, K., Tjader, I., Keller, P., et al., 2008. Dysregulation of mitochondrial dynamics and the muscle transcriptome in ICU patients suffering from sepsis-induced multiple organ failure. PLoS ONE 3 (11), e3686.

Freimer, N., Sabatti, C., 2004. The use of pedigree, sib-pair and association studies of common diseases for genetic mapping and epidemiology. Nat Genet 36 (10), 1045–1051.

Gibot, S., Cariou, A., Drouet, L., Rossignol, M., Ripoll, L., 2002. Association between a genomic polymorphism within the CD14 locus and septic shock susceptibility and mortality rate. Crit Care Med 30 (5), 969–973.

Giovannini, I., Boldrini, G., Castagneto, M., et al., 1983. Respiratory quotient and patterns of substrate utilization in human sepsis and trauma. J Parenter Enter Nutr 7 (3), 226–230.

Glickman, S.W., Cairns, C.B., Otero, R.M., et al., 2010. Disease progression in hemodynamically stable patients presenting to the emergency department with sepsis. Acad Emerg Med 17 (4), 383–390.

Goring, H.H., Terwilliger, J.D., Blangero, J., 2001. Large upward bias in estimation of locus-specific effects from genome-wide scans. Am J Hum Genet 69 (6), 1357–1369.

Green, J.P., Berger, T., Garg, N., Shapiro, N.I., 2011. Serum lactate is a better predictor of short-term mortality when stratified by C-reactive protein in adult emergency department patients hospitalized for a suspected infection. Ann Emerg Med 57 (3), 291–295.

Haden, D.W., Suliman, H.B., Carraway, M.S., et al., 2007. Mitochondrial biogenesis restores oxidative metabolism during *Staphylococcus aureus* sepsis. Am J Respir Crit Care Med 176 (8), 768–777.

Haralambous, E., Hibberd, M.L., Hermans, P.W., Ninis, N., Nadel, S., Levin, M., 2003. Role of functional plasminogen-activator-inhibitor-1 4G/5G promoter polymorphism in susceptibility, severity, and outcome of meningococcal disease in Caucasian children. Crit Care Med 31 (12), 2788–2793.

Hartel, C., Schultz, C., Herting, E., Gopel, W., 2007. Genetic association studies in VLBW infants exemplifying susceptibility to sepsis – recent findings and implications for future research. Acta Paediatr 96 (2), 158–165.

Hawn, T.R., Verbon, A., Lettinga, K.D., et al., 2003. A common dominant TLR5 stop codon polymorphism abolishes flagellin signaling and is associated with susceptibility to legionnaires' disease. J Exp Med 198 (10), 1563–1572.

Hayashi, F., Smith, K.D., Ozinsky, A., et al., 2001. The innate immune response to bacterial flagellin is mediated by Toll-like receptor 5. Nature 410 (6832), 1099–1103.

Heffner, A.C., Horton, J.M., Marchick, M.R., Jones, A.E., 2010. Etiology of illness in patients with severe sepsis admitted to the hospital from the emergency department. Clin Infect Dis 50 (6), 814–820.

Hermans, P.W., Hazelzet, J.A., 2005. Plasminogen activator inhibitor type 1 gene polymorphism and sepsis. Clin Infect Dis 41 (Suppl. 7), S453–S458.

Hill, A.V., 2006. Aspects of genetic susceptibility to human infectious diseases. Annu Rev Genet 40, 469–486.

Hirschhorn, J.N., Lohmueller, K., Byrne, E., Hirschhorn, K., 2002. A comprehensive review of genetic association studies. Genet Med 4 (2), 45–61.

Hoekstra, T., Geleijnse, J.M., Schouten, E.G., Kluft, C., 2004. Plasminogen activator inhibitor-type 1: Its plasma determinants and relation with cardiovascular risk. Thromb Haemost 91 (5), 861–872.

Huang, Q., Liu, D., Majewski, P., et al., 2001. The plasticity of dendritic cell responses to pathogens and their components. Science 294 (5543), 870–875.

Huston, J.M., Ochani, M., Rosas-Ballina, M., et al., 2006. Splenectomy inactivates the cholinergic anti-inflammatory pathway during lethal endotoxemia and polymicrobial sepsis. J Exp Med 203 (7), 1623–1628.

Inohara, N., Nunez, G., 2003. NODs: Intracellular proteins involved in inflammation and apoptosis. Nat Rev Immunol 3 (5), 371–382.

Ioannidis, J.P., Ntzani, E.E., Trikalinos, T.A., Contopoulos-Ioannidis, D.G., 2001. Replication validity of genetic association studies. Nat Genet 29 (3), 306–309.

Jaimes, F., Garces, J., Cuervo, J., et al., 2003. The systemic inflammatory response syndrome (SIRS) to identify infected patients in the emergency room. Intensive Care Med 29 (8), 1368–1371.

Jepson, A.P., Banya, W.A., Sisay-Joof, F., Hassan-King, M., Bennett, S., Whittle, H.C., 1995. Genetic regulation of fever in *Plasmodium falciparum* malaria in Gambian twin children. J Infect Dis 172 (1), 316–319.

Kanehisa Laboratories, 2010. Toll-like receptor signaling pathway – Homo sapiens (human). KEGG: Kyoto Encyclopedia of Genes and Genomes. <http://www.genome.jp/kegg/pathway/hsa/hsa04620.html>.

Kantrow, S.P., Taylor, D.E., Carraway, M.S., Piantadosi, C.A., 1997. Oxidative metabolism in rat hepatocytes and mitochondria during sepsis. Arch Biochem Biophys 345 (2), 278–288.

Kasai, H., Nishimura, S., 1984. Hydroxylation of deoxyguanosine at the C-8 position by ascorbic acid and other reducing agents. Nucleic Acids Res 12 (4), 2137–2145.

Kato, H., Takeuchi, O., Sato, S., et al., 2006. Differential roles of MDA5 and RIG-I helicases in the recognition of RNA viruses. Nature 441 (7089), 101–105.

Kingsmore, S.F., 2006. Multiplexed protein measurement: Technologies and applications of protein and antibody arrays. Nat Rev Drug Discov 5 (4), 310–320.

Kingsmore, S.F., Kennedy, N., Halliday, H.L., et al., 2008a. Identification of diagnostic biomarkers for infection in premature neonates. Mol Cell Proteomics 7 (10), 1863–1875.

Kingsmore S.F., Lejnine S.J., Driscoll M., Tchernev V.T., 2011. Biomarkers for sepsis: US Patent 8,029,982.

Kingsmore, S.F., Lindquist, I.E., Mudge, J., Gessler, D.D., Beavis, W.D., 2008b. Genome-wide association studies: Progress and potential for drug discovery and development. Nat Rev Drug Discov 7 (3), 221–230.

Kingsmore, S.F., Saunders, C.J., 2011. Deep sequencing of patient genomes for disease diagnosis: When will it become routine? Sci Transl Med 3 (87) 87ps23.

Knaus, W.A., Draper, E.A., Wagner, D.P., Zimmerman, J.E., 1985. APACHE II: A severity of disease classification system. Crit Care Med 13 (10), 818–829.

Kondaveeti, S., Hibberd, M.L., Booy, R., Nadel, S., Levin, M., 1999. Effect of the factor V Leiden mutation on the severity of meningococcal disease. Pediatr Infect Dis J 18 (10), 893–896.

Kornelisse, R.F., Hazelzet, J.A., Savelkoul, H.F., et al., 1996. The relationship between plasminogen activator inhibitor-1 and proinflammatory and counterinflammatory mediators in children with meningococcal septic shock. J Infect Dis 173 (5), 1148–1156.

Kreymann, G., Grosser, S., Buggisch, P., Gottschall, C., Matthaei, S., Greten, H., 1993. Oxygen consumption and resting metabolic rate in sepsis, sepsis syndrome, and septic shock. Crit Care Med 21 (7), 1012–1019.

Kronborg, G., Garred, P., 2002. Mannose-binding lectin genotype as a risk factor for invasive pneumococcal infection. Lancet 360 (9340), 1176.

Kuhlman, M., Joiner, K., Ezekowitz, R.A., 1989. The human mannose-binding protein functions as an opsonin. J Exp Med 169 (5), 1733–1745.

Lander, E., Kruglyak, L., 1995. Genetic dissection of complex traits: Guidelines for interpreting and reporting linkage results. Nat Genet 11 (3), 241–247.

Langley, R., Kingsmore, S., Chen, B., Carin, L., 2010. Methods for diagnosis of sepsis and risk of death. US Patent Application # 20100273207.

Lasker, M.V., Gajjar, M.M., Nair, S.K., 2005. Cutting edge: Molecular structure of the IL-1R-associated kinase-4 death domain and its implications for TLR signaling. J Immunol 175 (7), 4175–4179.

Leon, P., Redmond, H.P., Shou, J., Daly, J.M., 1992. Interleukin 1 and its relationship to endotoxin tolerance. Arch Surg 127 (2), 146–151.

Lin, M.T., Albertson, T.E., 2004. Genomic polymorphisms in sepsis. Crit Care Med 32 (2), 569–579.

Lindahl, T., 1993. Instability and decay of the primary structure of DNA. Nature 362 (6422), 709–715.

Liu, R., Paxton, W.A., Choe, S., et al., 1996. Homozygous defect in HIV-1 coreceptor accounts for resistance of some multiply exposed individuals to HIV-1 infection. Cell 86 (3), 367–377.

Lohmueller, K.E., Pearce, C.L., Pike, M., Lander, E.S., Hirschhorn, J.N., 2003. Meta-analysis of genetic association studies supports a contribution of common variants to susceptibility to common disease. Nat Genet 33 (2), 177–182.

Long, C.L., 1977. Energy balance and carbohydrate metabolism in infection and sepsis. Am J Clin Nutr 30 (8), 1301–1310.

Lorenz, E., 2006. TLR2 and TLR4 expression during bacterial infections. Curr Pharm Des 12 (32), 4185–4193.

Lorenz, E., Mira, J.P., Cornish, K.L., Arbour, N.C., Schwartz, D.A., 2000. A novel polymorphism in the toll-like receptor 2 gene and its potential association with staphylococcal infection. Infect Immun 68 (11), 6398–6401.

Lorenz, E., Mira, J.P., Frees, K.L., Schwartz, D.A., 2002. Relevance of mutations in the TLR4 receptor in patients with gram-negative septic shock. Arch Intern Med 162 (9), 1028–1032.

Ma, P., Chen, D., Pan, J., Du, B., 2002. Genomic polymorphism within interleukin-1 family cytokines influences the outcome of septic patients. Crit Care Med 30 (5), 1046–1050.

Majetschak, M., Obertacke, U., Schade, F.U., et al., 2002. Tumor necrosis factor gene polymorphisms, leukocyte function, and sepsis susceptibility in blunt trauma patients. Clin Diagn Lab Immunol 9 (6), 1205–1211.

Marriott, I., Huet-Hudson, Y.M., 2006. Sexual dimorphism in innate immune responses to infectious organisms. Immunol Res 34 (3), 177–192.

Marshall, J.C., 1997. The Multiple Organ Dysfunction (MOD) score. Sepsis 1 (1), 49–52.

Martin, C., Saux, P., Mege, J.L., Perrin, G., Papazian, L., Gouin, F., 1994. Prognostic values of serum cytokines in septic shock. Intensive Care Med 20 (4), 272–277.

Martin, G.S., Mannino, D.M., Eaton, S., Moss, M., 2003. The epidemiology of sepsis in the United States from 1979 through 2000. N Engl J Med 348 (16), 1546–1554.

McCaig, L.F., Nawar, E.W., 2006. National Hospital Ambulatory Medical Care Survey: 2004 emergency department summary. Adv Data 372, 1–29.

McDunn, J.E., Husain, K.D., Polpitiya, A.D., et al., 2008. Plasticity of the systemic inflammatory response to acute infection during critical illness: Development of the riboleukogram. PLoS One 3 (2), e1564.

Mira, J.P., Cariou, A., Grall, F., et al., 1999. Association of TNF2, a TNF-alpha promoter polymorphism, with septic shock susceptibility and mortality: A multicenter study. JAMA 282 (6), 561–568.

Mongardon, N., Dyson, A., Singer, M., 2009. Is MOF an outcome parameter or a transient, adaptive state in critical illness? Curr Opin Crit Care 15 (5), 431–436.

Moore, C.E., Segal, S., Berendt, A.R., Hill, A.V., Day, N.P., 2004. Lack of association between Toll-like receptor 2 polymorphisms and susceptibility to severe disease caused by *Staphylococcus aureus*. Clin Diagn Lab Immunol 11 (6), 1194–1197.

Nau, G.J., Richmond, J.F., Schlesinger, A., Jennings, E.G., Lander, E.S., Young, R.A., 2002. Human macrophage activation programs induced by bacterial pathogens. Proc Natl Acad Sci USA 99 (3), 1503–1508.

Nicolaes, G.A., Dahlback, B., 2003. Activated protein C resistance (FV(Leiden)) and thrombosis: Factor V mutations causing hyper-coagulable states. Hematol Oncol Clin North Am 17 (1), 37–61. vi.

Osiewacz, H.D., 2002. Mitochondrial functions and aging. Gene 286 (1), 65–71.

Pacelli, F., Doglietto, G.B., Alfieri, S., et al., 1996. Prognosis in intra-abdominal infections: Multivariate analysis on 604 patients. Arch Surg 131 (6), 641–645.

Parrillo, J.E., Parker, M.M., Natanson, C., et al., 1990. Septic shock in humans: Advances in the understanding of pathogenesis, cardiovascular dysfunction, and therapy. Ann Intern Med 113 (3), 227–242.

Perlee, L., Christiansen, J., Dondero, R., et al., 2004. Development and standardization of multiplexed antibody microarrays for use in quantitative proteomics. Proteome Sci 2 (1), 9.

Pociot, F., Molvig, J., Wogensen, L., Worsaae, H., Nerup, J., 1992. A TaqI polymorphism in the human interleukin-1 beta (IL-1 beta) gene correlates with IL-1 beta secretion in vitro. Eur J Clin Invest 22 (6), 396–402.

Polpitiya, A.D., McDunn, J.E., Burykin, A., Ghosh, B.K., Cobb, J.P., 2009. Using systems biology to simplify complex disease: Immune cartography. Crit Care Med 37 (1 Suppl.), S16–S21.

Pyle, A., Burn, D.J., Gordon, C., Swan, C., Chinnery, P.F., Baudouin, S.V., 2010. Fall in circulating mononuclear cell mitochondrial DNA content in human sepsis. Intensive Care Med 36 (6), 956–962.

Qian, W.J., Petritis, B.O., Kaushal, A., et al., 2010. Plasma proteome response to severe burn injury revealed by ^{18}O-labeled "universal" reference-based quantitative proteomics. J Proteome Res 9 (9), 4779–4789.

Qureshi, S.T., Lariviere, L., Leveque, G., et al., 1999. Endotoxin-tolerant mice have mutations in Toll-like receptor 4 (Tlr4). J Exp Med 189 (4), 615–625.

Raha, S., Robinson, B.H., 2000. Mitochondria, oxygen free radicals, disease and ageing. Trends Biochem Sci 25 (10), 502–508.

Rajicic, N., Cuschieri, J., Finkelstein, D.M., et al., 2010. Identification and interpretation of longitudinal gene expression changes in trauma. PLoS ONE 5 (12), e14380.

Ramilo, O., Allman, W., Chung, W., et al., 2007. Gene expression patterns in blood leukocytes discriminate patients with acute infections. Blood 109 (5), 2066–2077.

Rangel-Frausto, M.S., Pittet, D., Costigan, M., Hwang, T., Davis, C.S., Wenzel, R.P., 1995. The natural history of the systemic inflammatory response syndrome (SIRS): A prospective study. JAMA 273 (2), 117–123.

Read, R.C., Pullin, J., Gregory, S., et al., 2001. A functional polymorphism of toll-like receptor 4 is not associated with likelihood or severity of meningococcal disease. J Infect Dis 184 (5), 640–642.

Redden, D.T., Allison, D.B., 2003. Nonreplication in genetic association studies of obesity and diabetes research. J Nutr 133 (11), 3323–3326.

Rettew, J.A., Huet-Hudson, Y.M., Marriott, I., 2008. Testosterone reduces macrophage expression in the mouse of toll-like receptor 4, a trigger for inflammation and innate immunity. Biol Reprod 78 (3), 432–437.

Rettew, J.A., Huet, Y.M., Marriott, I., 2009. Estrogens augment cell surface TLR4 expression on murine macrophages and regulate sepsis susceptibility in vivo. Endocrinology 150 (8), 3877–3884.

Riedemann, N.C., Guo, P.F., Ward, P.A., 2003. Novel strategies for the treatment of sepsis. Nat Med 9 (5), 517–524.

Risch, N.J., 2000. Searching for genetic determinants in the new millennium. Nature 405 (6788), 847–856.

Rivers, E., Nguyen, B., Havstad, S., et al., 2001. Early goal-directed therapy in the treatment of severe sepsis and septic shock. N Engl J Med 345 (19), 1368–1377.

Rosas-Ballina, M., Ochani, M., Parrish, W.R., et al., 2008. Splenic nerve is required for cholinergic anti-inflammatory pathway control of TNF in endotoxemia. Proc Natl Acad Sci USA 105 (31), 11,008–11,013.

Rosas-Ballina, M., Olofsson, P.S., Ochani, M., et al., 2011. Acetylcholine-synthesizing T cells neural signals in a vagus nerve circuit. Science 334 (6052), 98–101.

Roy, S., Knox, K., Segal, S., et al., 2002. MBL genotype and risk of invasive pneumococcal disease: A case-control study. Lancet 359 (9317), 1569–1573.

Samson, M., Libert, F., Doranz, B.J., et al., 1996. Resistance to HIV-1 infection in Caucasian individuals bearing mutant alleles of the CCR-5 chemokine receptor gene. Nature 382 (6593), 722–725.

Sands, K.E., Bates, D.W., Lanken, P.N., et al., 1997. Epidemiology of sepsis syndrome in 8 academic medical centers. JAMA 278 (3), 234–240.

Schluter, B., Raufhake, C., Erren, M., et al., 2002. Effect of the interleukin-6 promoter polymorphism (-174 G/C) on the incidence and outcome of sepsis. Crit Care Med 30 (1), 32–37.

Selberg, O., Hecker, H., Martin, M., Klos, A., Bautsch, W., Kohl, J., 2000. Discrimination of sepsis and systemic inflammatory response syndrome by determination of circulating plasma concentrations of procalcitonin, protein complement 3a, and interleukin-6. Crit Care Med 28 (8), 2793–2798.

Shigenaga, M.K., Hagen, T.M., Ames, B.N., 1994. Oxidative damage and mitochondrial decay in aging. Proc Natl Acad Sci USA 91 (23), 10,771–10,778.

Shorr, A.F., Bernard, G.R., Dhainaut, J.F., et al., 2006. Protein C concentrations in severe sepsis: An early directional change in plasma levels predicts outcome. Crit Care 10 (3), R92.

Sillanpaa, M.J., Auranen, K., 2004. Replication in genetic studies of complex traits. Ann Hum Genet 68 (6), 646–657.

Skamene, E., Schurr, E., Gros, P., 1998. Infection genomics: Nramp1 as a major determinant of natural resistance to intracellular infections. Annu Rev Med 49, 275–287.

Smirnova, I., Mann, N., Dols, A., et al., 2003. Assay of locus-specific genetic load implicates rare Toll-like receptor 4 mutations in meningococcal susceptibility. Proc Natl Acad Sci USA 100 (10), 6075–6080.

Smyth, D.J., Cooper, J.D., Bailey, R., et al., 2006. A genome-wide association study of nonsynonymous SNPs identifies a type 1 diabetes locus in the interferon-induced helicase (IFIH1) region. Nat Genet 38 (6), 617–619.

Sorensen, T.I., Nielsen, G.G., Andersen, P.K., Teasdale, T.W., 1988. Genetic and environmental influences on premature death in adult adoptees. N Engl J Med 318 (12), 727–732.

Steinmetz, L.M., Davis, R.W., 2004. Maximizing the potential of functional genomics. Nat Rev Genet 5 (3), 190–201.

Storey, J.D., Tibshirani, R., 2003. Statistical significance for genome-wide studies. Proc Natl Acad Sci USA 100 (16), 9440–9445.

Stuber, F., Petersen, M., Bokelmann, F., Schade, U., 1996. A genomic polymorphism within the tumor necrosis factor locus influences plasma tumor necrosis factor-alpha concentrations and outcome of patients with severe sepsis. Crit Care Med 24 (3), 381–384.

Suliman, H.B., Welty-Wolf, K.E., Carraway, M.S., Schwartz, D.A., Hollingsworth, J.W., Piantadosi, C.A., 2005. Toll-like receptor 4 mediates mitochondrial DNA damage and biogenic responses after heat-inactivated *E. coli*. FASEB J 19 (11), 1531–1533.

Sutherland, A.M., Walley, K.R., Russell, J.A., 2005. Polymorphisms in CD14, mannose-binding lectin, and Toll-like receptor-2 are associated with increased prevalence of infection in critically ill adults. Crit Care Med 33 (3), 638–644.

Taylor, D.E., Ghio, A.J., Piantadosi, C.A., 1995. Reactive oxygen species produced by liver mitochondria of rats in sepsis. Arch Biochem Biophys 316 (1), 70–76.

Taylor, D.E., Kantrow, S.P., Piantadosi, C.A., 1998. Mitochondrial respiration after sepsis and prolonged hypoxia. Am J Physiol 275 (1), L139–L144.

Texereau, J., Pene, F., Chiche, J.D., Rousseau, C., Mira, J.P., 2004. Importance of hemostatic gene polymorphisms for susceptibility to and outcome of severe sepsis. Crit Care Med 32 (5 Suppl.), S313–S319.

Tsalik, E.L., Jaggers, L.B., Glickman, S.W., et al., 2012. Discriminative value of inflammatory biomarkers for suspected sepsis. J Emerg Med. 43 (1), 97–106.

Thach, D.C., Agan, B.K., Olsen, C., et al., 2005. Surveillance of transcriptomes in basic military trainees with normal, febrile respiratory illness, and convalescent phenotypes. Genes Immun 6 (7), 588–595.

Van der Poll, T., Van Deventer, S.J., 1999. Cytokines and anticytokines in the pathogenesis of sepsis. Infect Dis Clin North Am 13 (2), 413–426. ix.

Van Rossum, E.F., Van den Akker, E.L., 2011. Glucocorticoid resistance. Endocr Dev 20, 127–136.

Vincent, J.L., De Mendonca, A., Cantraine, F., et al., 1998. Use of the SOFA score to assess the incidence of organ dysfunction/failure in intensive care units: Results of a multicenter, prospective study. Working group on "sepsis-related problems" of the European Society of Intensive Care Medicine. Crit Care Med 26 (11), 1793–1800.

Von Bernuth, H., Puel, A., Ku, C.L., et al., 2005. Septicemia without sepsis: Inherited disorders of nuclear factor-kappa B-mediated inflammation. Clin Infect Dis 41 (Suppl. 7), S436–S439.

Walley, K.R., Russell, J.A., 2007. Protein C -1641 AA is associated with decreased survival and more organ dysfunction in severe sepsis. Crit Care Med 35 (1), 12–17.

Walsh, M.C., Bourcier, T., Takahashi, K., et al., 2005. Mannose-binding lectin is a regulator of inflammation that accompanies myocardial ischemia and reperfusion injury. J Immunol 175 (1), 541–546.

Wang, H., Yu, M., Ochani, M., et al., 2003. Nicotinic acetylcholine receptor alpha7 subunit is an essential regulator of inflammation. Nature 421 (6921), 384–388.

Wunder, C., Eichelbronner, O., Roewer, N., 2004. Are IL-6, IL-10 and PCT plasma concentrations reliable for outcome prediction in severe sepsis? A comparison with APACHE III and SAPS II. Inflamm Res 53 (4), 158–163.

Yang, Y., Shou, Z., Zhang, P., et al., 2008. Mitochondrial DNA haplogroup R predicts survival advantage in severe sepsis in the Han population. Genet Med 10 (3), 187–192.

Zaas, A.K., Chen, M., Varkey, J., et al., 2009. Gene expression signatures diagnose influenza and other symptomatic respiratory viral infections in humans. Cell Host Microbe 6 (3), 207–217.

Zhan, Z.Y., Wang, H.W., Chen, D.F., Cheng, B.L., Wang, H.H., Fang, X.M., 2005. Relationship between sepsis and 4G/5G polymorphism within the promoter region of plasminogen activator inhibitor-1 gene. Zhonghua Yi Xue Za Zhi 85 (34), 2404–2407.

Zhou, B., Xu, W., Herndon, D., et al., 2010. Analysis of factorial timecourse microarrays with application to a clinical study of burn injury. Proc Natl Acad Sci USA 107 (22), 9923–9928.

Viral Hepatitis

Nicholas A. Shackel, Keyur Patel, and John McHutchison

INTRODUCTION

Viral hepatitis is a significant global health problem. Hepatitis B virus (HBV) and hepatitis C virus (HCV) infect more than 300 million people. HBV and HCV are complicated by chronic, persistent infection, characterized in a proportion of patients by progressive hepatic injury leading to complications of end-stage liver disease, including hepatocellular carcinoma (HCC). HCC is the fifth most prevalent human malignancy, and the majority of cases can be directly attributed to liver injury secondary to chronic HBV and/or HCV infection. Although both hepatitis A and E are significant health problems, they are typically characterized by a self-limiting course and are not complicated by significant clinical sequelae in the majority of cases. Therefore, research into infectious hepatitis has focused mainly on HBV and HCV pathogenesis, including the development of liver fibrosis, the immune response in acute infection, mechanisms of viral persistence, and the development of HCC. The use of functional genomics approaches has significantly advanced our understanding of viral hepatitis pathogenesis, as well as our understanding of therapeutic strategies.

VIROLOGY OF HEPATITIS VIRUSES

The hepatitis viruses are characterized by specificity for the liver and in particular the hepatocyte (Figure 99.1). The mechanism by which these viruses are specific for the liver is largely unknown, but it is thought to involve hepatocyte receptor and co-receptor interactions and possible involvement of liver-specific pathways such as lipoprotein trafficking and synthesis.

Hepatitis A virus (HAV) is an RNA virus of the *Hepatovirus* genus belonging to the Picornaviridae family (Flehmig, 1990; Lee, 2003; Martin and Lemon, 2006). HAV is a positive-strand RNA virus with a 7.5 kb genome that is translated into a 2225 to 2227 polyprotein, that gives rise to a number of structural and non-structural proteins. The viral particles are 27–32 nm in diameter, and there are distinct genotypes and sub-genotypes of HAV. In contrast to another hepatotrophic RNA virus, hepatitis C, HAV has a high degree of genomic and resultant antigenic conservation. The spontaneous mutation rate of HAV is low, and antibody-mediated immunity from previous exposure or vaccination is effective in preventing HAV infection. HAV is characterized by slow replication, is rarely cytopathic, and is stable in the environment for at least a month.

Genomic and Personalized Medicine, 2nd edition
by Ginsburg & Willard. DOI: http://dx.doi.org/10.1016/B978-0-12-382227-7.00099-9

(A)

Virus	Family	Length	RNA/DNA	Transmission	Chronic	Comments
HAV	Hepatovirus	2.2 kb	RNA+	Feco-oral	No	Can cause fulminant liver failure
HBV	Hepadnaviruses	3.2 kb	DNA+	Blood-borne Sexual Vertical	Yes	Can cause fulminant liver failure
HCV	Flaviviridae	3.0 kb	RNA+	Blood-borne Vertical	Yes	Fulminant liver failure v. rare
HDV	Deltavirus	1.6 kb	RNA–	Blood-borne Sexual	Yes	Requires HBV surface antigen Can cause fulminant liver failure
HEV	HEV-like viruses	7.5 kb	RNA+	Feco-oral	No	Can cause fulminant liver failure

(B)

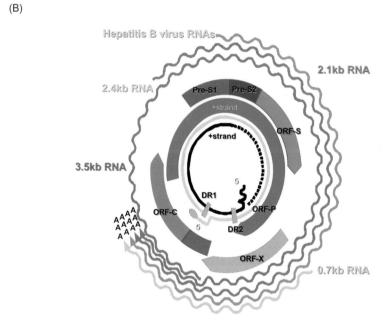

(C)

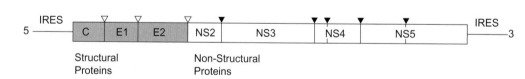

Figure 99.1   Hepatitis virus virology. (A) Summary of the virology of viral hepatitis (HAV to HEV). (B) The HBV circular DNA and the multiple overlapping reading frames. (C) In contrast to hepatitis B, the hepatitis C virus has a linear RNA genome encoding viral proteins. The hepatitis C virus 5′ and 3′ internal ribosome re-entry sites (IRES) are important viral regulatory elements.

Hepatitis B is a circular DNA virus. It belongs to the hepadnaviruses, which include woodchuck hepatitis virus and ground squirrel hepatitis virus (Chisari, 1992; Lee, 1997; Lok, 2000). The partially double-stranded HBV genome is 3.2 kb in length and is organized with multiple overlapping open reading frames (ORFs) (Chisari, 1992; Lee, 1997; Lok, 2000) (Figure 99.1). More than half of the genome is translated in more than one ORF, and this limits the viral mutations that will be tolerated. HBV DNA incorporates into the host genome (Lau and Wright, 1993; Moyer and Mast, 1994). Therefore, recurrence of active infection after the initial exposure is always possible. The viral proteins include the surface (envelope) protein, core, polymerase, and X-proteins. Variations of the nucleotide sequence and corresponding amino acids that constitute the HBV surface protein give rise to eight common HBV genotypes (A–H).

Hepatitis C is a positive-strand RNA virus. It is a member of the Flaviviridae family, which includes flaviviruses and pestiviruses (Shimotohno, 2000). There is considerable diversity in the HCV genome, with at least six distinct genotypes, more than 50 subtypes, and a propensity of the virus to mutate, giving rise to *quasispecies* (Forns and Bukh, 1999; Forns et al., 1999). The virus genome is 9.5 kb, and encodes a large single polyprotein of 3010 to 3033 amino acids (Forns and Bukh, 1999) (Figure 99.1). The 5′ and 3′ untranslated regions are important for the replication of the virus and translation of the polyprotein. The HCV proteins comprise the structural proteins (core protein and the envelope glycoproteins E1, E2, and p7) followed by the non-structural proteins [proteases, helicase (NS2 and NS3), protease cofactor (NS4A), NS4B, replication-associated phosphoprotein (NS5A), and the RNA-dependent polymerase (NS5B)] (Shimotohno, 2000; Simmonds, 1996). Possible receptors for HCV have been identified, including CD-81, SRBP-1, and claudin-1.

Hepatitis D virus (HDV; also known as delta hepatitis) is unlike any other transmissible agent in animals (Taylor, 2006a, b). This minus-strand circular RNA virus requires the hepatitis B surface antigen (HBsAg) for encapsulation and entry into hepatocytes. HDV only coinfects up to 5% of individuals with hepatitis B infection. Importantly, the prognosis of HDV coinfection is worse than with HBV alone.

Hepatitis E virus (HEV) is related to the Calciviradie, but now has been classified into a separate genus of *hepatitis E-like viruses* (Krawczynski et al., 2000; Worm et al., 2002). The HEV genome consists of a single positive RNA strand of 7.5 kb that is translated into three polyproteins, as there are three overlapping open reading frames. Consequently, HEV has a stable genome sequence and exists as a single serotype and at least four unique genotypes. HEV is stable in the environment and is a common waterborne pathogen, especially in developing countries.

ACQUISITION AND PREDISPOSITION TO VIRAL HEPATITIS

Hepatitis A is characterized by an enteric route of infection and classically is transmitted via personal contact, illicit drug use, or ingestion of contaminated food or water (Martin and Lemon, 2006). The incubation period following exposure can be up to 90 days. HAV infection typically occurs in sporadic outbreaks as well as isolated cases, and can be endemic in poorly developed countries with poor sanitation. Although molecular phenotyping of the virus helps trace outbreaks, there is very little known about individual genetic factors that may predispose to HAV infection. Host genetic factors may not be that important, as indicated by epidemiologic studies showing that HAV infection is often transmitted to persons with similar risk factors. Hepatitis E is transmitted similarly to HAV and is characterized by a great propensity for waterborne transmission. Like HAV, HEV is endemic in many poorly

developed countries. HEV has propensity to be associated with miscarriage and increased mortality in pregnant women, especially during the third trimester. The basis for this is unclear, but a number of host factors are clearly important in determining susceptibility to infection.

HBV and HCV enter the host through blood contact by direct inoculation (i.e., needle or transfusion) or via a disrupted percutaneous barrier (i.e., sexual or perinatal transmission) (Shimotohno, 2000; Simmonds, 1996). Entry of HBV or HCV into liver hepatocytes is not understood, and there are thought to be extra-hepatic reservoirs of infection in peripheral blood leukocytes (Cabrerizo et al., 2000; Yoffe et al., 1990). Receptor-based cell entry has been implicated in both HBV and HCV pathogenesis (Cocquerel et al., 2006; Pileri et al., 1998; Yerly et al., 2006). The immunopathogenesis of both HBV and HCV infection is characterized by both innate and adaptive immune response to the virus, resulting in predominantly non-specific inflammation. Typically, neither HBV nor HCV is directly cytopathic, as viral load does not correlate with organ damage and the level of antigen expression in hepatocytes does not correlate with hepatocyte injury (Chisari and Ferrari, 1995). Importantly, both HBV and HCV *in vivo* infectivity is limited to higher primates (chimpanzees and humans).

Predisposing factors to viral hepatitis include a number of non-genetic and genetic factors (Thomas, 2000; Wasley and Alter, 2000; Yee, 2004) (see Table 99.1). Hepatitis B and C are more prevalent in communities with increased rates of intravenous drug use, unsafe therapeutic injections, and use of unscreened blood products (Maddrey, 2000; Merican et al., 2000; Wasley and Alter, 2000). Fortunately, the recognition of routes of transmission, the use of safe injecting practices, and the adoption of volunteer donor blood transfusion services are public health measures that have been widely initiated globally, to control the spread of viral hepatitis. Vertical transmission of HBV is dependent on the stage of maternal HBV infection and the viral antigen expression. Approximately 10–20% of surface-antigen-positive mothers transmit the virus to their offspring. However, both surface antigen and hepatitis B e antigen (HBeAg) expression is associated with a 90% rate of transmission. Maternal acute HBV infection results in a 10% neonatal rate of infection in the first trimester, which increases to 80–90% if acute infection occurs in the third trimester. HCV infection is characterized by lower rates of vertical transmission, with only 2.7–8.4% of offspring being infected. The presence of HIV coinfection increases the rate of vertical transmission of both HBV and HCV. Transmission via sexual activity with mucosal disruption is more prevalent in HBV than in HCV infection.

Genetic factors predispose to persistence of HBV infection, given that the rate of concordance for surface antigen expression is greater in monozygotic compared to dizygotic twins (Table 99.1). The twin concordance data in HCV infection are not as convincing. A number of human leukocyte antigen (HLA) alleles are associated with both the persistence and clearance

TABLE 99.1	Genetic loci associated with susceptibility to viral hepatitis		
Allele/polymorphism	**Hepatitis C**	**Hepatitis B**	**Comment**
HLA-DRB1*1302 and HLA-DRB1*0301		Spontaneous elimination of infection	
HLA-DQA1*0501, DQB1*301, HLA-DRB1*1102 and HLA-DRB1*0301		Persistence of infection	
HLA-DRB1*0101, HLA-DRB1*0401, HLA-DRB1*15, HLA-DRB1*1101, HLA-DRB1*0301, HLA-A*2301, HLA-A*1101, HLA-A*03, HLA-B*57 and HLA-Cw*0102	Spontaneous elimination of infection		
HLA-DRB1*0701, HLA-A*01-B*08-Cw*07-DRB1*0301-DBQ1*0201, HLA-Cw*04, HLA-Cw*04-B*53	Persistence of infection		
TNF promoter	Viral replication and clearance	Viral replication and clearance	Polymorphisms at -308 and -238 best characterized
Interleukin-10		Spontaneous elimination of infection	
Vitamin D receptor		Control of viral replication	Expressed on monocytes and lymphocytes
GDNF family receptor alpha 1	At risk of HCC in HCV		
Chemokine (C-X-C motif) ligand 14	At risk of HCC in HCV		Previously known as SCYB14
IFN-λ3/IL-28B	Spontaneous clearance Interferon treatment response		To date the strongest genetic association demonstrated with either spontaneous clearance or treatment responses to HCV genotype-1 infection

HLA = human leukocyte antigen, TNF = tumor necrosis factor, IFN-λ3 = interferon-λ3, IL-28B = interleukin 28B.

of HBV and HCV (Wang, 2003; Yee, 2004). Interestingly, in contrast to HIV progression, homozygosity for HLA class II locus increases the risk of HBV persistence. The HLA class II locus DRB1*1302 is associated with HBV clearance, and the DQB1*0301 locus is associated with a self-limiting course of HCV infection. Non-HLA immunogenic loci are also implicated in viral hepatitis, with tumor necrosis factor (TNF) promoter polymorphisms resulting in higher TNF secretion being associated with HBV clearance. The killer-cell immunoglobulin-like receptor (KIR) genes interact with HLA class I molecules, and specific KIR heliotypes are associated with HCV clearance (Martin and Carrington, 2005; Williams et al., 2005). Importantly, the majority of genetic predispositions identified in viral hepatitis are linked to viral persistence or clearance, and can be directly implicated in the adaptive immune response. The currently documented genetic disease associations with viral hepatitis are of limited use in clinical practice. Presently, genome-wide association studies are being undertaken to characterize hepatitis genetic susceptibility. Therefore, future clinical practice is likely to see panels of genetic susceptibility markers being screened

to determine prognosis as well as the likelihood of a treatment response.

SCREENING AND DIAGNOSIS OF VIRAL HEPATITIS

Hepatitis A and E are characterized by jaundice, fever, and diarrhea illness. Given the propensity for epidemic outbreaks and known endemic regions of the globe (mainly developing countries), establishing the diagnosis is not difficult. HAV and HEV are readily diagnosed using serology, with acute infection distinguished by the presence of antiviral IgM (Acharya and Panda, 2006; Panda et al., 2006). The diagnosis of HDV is also made by serology, and anti-HDV IgM is detectable 30 days following infection (Fiedler and Roggendorf, 2006; Weston and Martin, 2001).

HBV and HCV infections result in a comparatively non-specific cluster of symptoms, ranging from asymptomatic viral infection (most common in HCV) or jaundice and fever

TABLE 99.2	HBV serology and DNA quantitation
Serologic markers	**Definition/clinical use**
HBsAg – hepatitis B surface antigen	• General marker of infection • First serologic marker to appear after infection • Persistence for >6 months = chronic infection
HBeAg – hepatitis B e antigen	• Indicates active replication of virus
Anti-HBs – antibody to hepatitis B s antigen	• Documents recovery and/or immunity to HBV • Detectable after immunity conferred by hepatitis B vaccination
Anti-HBe – antibody to hepatitis B e antigen	• Marker of reduced level of replication
Anti-HBc (IgM) – antibody to hepatitis B core antigen	• Marker of acute hepatitis B
Anti-HBc (IgG) – antibody to hepatitis B core antigen	• Marker of current or past infection
HBV DNA – hepatitis B virus genomic DNA	• Marker of HBV replication • Used for monitoring response to therapy

HBV, hepatitis B virus.

(30–50% of HBV infections) to fulminant hepatic failure and death (<1% of HBV and exceedingly rare in HCV). Importantly, both HBV and HCV viral replication can be associated with a normal liver panel. Although routine liver tests such as transaminases are used to monitor disease effects on the liver, they do not provide prognostic information and cannot be used for diagnosis of viral hepatitis. The cornerstone of screening and diagnosis is serology for viral-specific antibodies and antigens. The commonly available serology of HBV is complex, with the pattern of antibody and antigen expression determining the nature and timing of infection (Table 99.2).

HCV antibody production is a simpler test of exposure, which becomes positive within four weeks of exposure (Bhandari and Wright, 1995). The commonly used HCV antibody screen is performed with an enzyme immunoassay (EIA) and is prone to false positive readings. Specificity of the EIA has improved with successive generations of the test, but confirmation of a positive result with supplemental methodology is considered mandatory. A positive HCV EIA is routinely confirmed using a recombinant immunoblot assay, in which serum reactivity to a number of HCV recombinant viral proteins is assessed.

Detection of HBV DNA and HCV RNA is now recommended to confirm the diagnosis of chronic hepatitis and to distinguish active viremia from previous exposure (Acharya and Panda, 2006; Mondelli et al., 2005; Patel et al., 2006; Servoss and Friedman, 2004). As viral replication can occur with normal biochemistry and antibody titers do not demonstrate viral replication, quantitative assessment of viral nucleotide presence is used in many centers as the confirmatory diagnostic test for HBV or HCV infection. Quantitation of HBV DNA or HCV RNA is used in planning and following treatment of the hepatitis. HBV DNA can be quantitated using an approach of capturing HBV DNA on a full-length HBV RNA transcript and using antibodies to detect the RNA:DNA hybrid (known as capture assays or Digene™). An alternative approach uses quantitation polymerase chain reaction (PCR)

methodologies to amplify the HBV DNA (COBAS Amplior™) and a variation of the capture approach uses branched DNA (Versant™). Additionally, HBV and HCV viral titers are typically determined once the infection has been confirmed, as the titers carry prognostic significance as well as being used to monitor treatment responses. HBV genotyping is not routinely performed in the clinical setting unless the patient is being considered for interferon-alpha therapy.

An important aspect of viral hepatitis screening and diagnosis involves an assessment of the extent of liver injury attributable to the virus. In particular, the extent of hepatic inflammation and fibrosis has implications on progression and likely response to treatment. Routine biochemistry gives only limited information about the extent of inflammation or fibrosis, and most centers will use imaging of the liver with ultrasound or other modalities combined with a liver biopsy to assess virus-induced liver injury. However, biopsy is an invasive procedure prone to significant sampling and observer error (Siegel et al., 2005). This has led to the development of a number of new noninvasive screening modalities. They include the use of noninvasive ultrasound assessment of liver elastography, which correlates liver "stiffness" with the degree of fibrosis, as well as the use of panels of serum markers to assess the development of liver injury. The most promising approaches use multiple variables to assess for liver injury, and these methods are becoming commonplace in clinical management. Unfortunately, to date both elastography and panels of serum markers are efficient only in distinguishing minimal from advanced fibrosis and cannot efficiently assess the extent of inflammation.

VIRAL HEPATITIS PROGNOSIS AND NATURAL HISTORY

The clinical course and progression of HBV and HCV are highly variable (Chen and Morgan, 2006; Chu and Liaw, 2006;

Fattovich, 2003; Ghany and Seeff, 2006; Thomas and Seeff, 2005). The importance of individually tailored management strategies is paramount, as there is a marked variation in the spectrum of disease even when all other host and viral factors appear equal. HBV infection results in acute self-limiting infection in 90% of cases. The rate of fulminant hepatic failure is 0.1–0.5%. Only 10% of cases go on to develop chronic infection, with a spectrum of disease that varies from the relatively benign chronic surface antigenemia to more aggressive forms of the disease in which individuals have active viral replication characterized by surface antigen expression, e antigen (HBeAg) expression, and high viral DNA titers. However, the propensity of HBV to mutate is 10-fold greater than other DNA viruses, and mutations at positions 1762 and 1764 of the precore protein promoter can give rise to precore mutants in which active HBV infection is not associated with HBeAg expression. Chronic hepatitis can spontaneously clear, with active HBeAg expression being associated with both more rapid disease progression and a spontaneous seroconversion rate [switching from HBeAg to hepatitis B E antigen antibody (HBeAb) expression] of 8–15% per year. However, precore mutants, although characterized by active replication with high titers of HBV DNA, have a spontaneous clearance rate of only 0.5% per year. The presence of normal liver biochemistry is associated with a spontaneous clearance of only 2–5% per year. Surface antigen (HBsAg) carrier state is lowest in individuals with vertical transmission-acquired HBV, and is associated with spontaneous clearance of only 1–2% per year. Active HBV viral replication is associated with intrahepatic inflammation, the progression to cirrhosis, and sequelae including liver failure and the development of hepatocellular carcinoma. In HBeAg antigen-positive individuals, cirrhosis develops at a rate of 2–5.4% per year, with a five-year cumulative risk of 8–20%. The rate of progression to cirrhosis is even higher in HBeAg antigen-negative individuals with active viral replication. The risk of hepatic decompensation with HBV-associated cirrhosis is 16% at five years. Compared to compensated cirrhosis, the development of decompensated cirrhosis results in a decrease in survival from 70% to 55% at one year and 28% to 14% at five years. In the absence of cirrhosis, the incidence of liver-rated deaths is low, in the range of 0–1.06 per 100 person-years. The development of hepatocellular carcinoma (HCC) is uncommon in the absence of cirrhosis, but develops at a rate of 2.8% per year with cirrhosis. The risk of HCC increases with age >45, male gender, HBeAg antigen expression, detectable HBV DNA, and first-degree relatives with HCC.

The prognosis of HCV infection has been the topic of considerable controversy, as the natural history is highly variable (Chen and Morgan, 2006; Ghany and Seeff, 2006; Thomas and Seeff, 2005). Estimates of chronic infectivity range from 55–80%, with most studies between 75% and 80%. Age of infection affects progression and the rate of spontaneous HCV clearance. Infected children have a spontaneous rate of clearance of 40–45%, and develop cirrhosis in only 2–4% of individuals after 20 years of infection. In contrast, adult infection is characterized by <20% rate of spontaneous clearance and cirrhosis in 20–30% of individuals after 20 years. HCV cirrhosis is associated with a 3–4% annual rate risk of liver decompensation and an annual incidence of 1.4–6.9% of developing HCC.

It is clear from the natural history and prognosis that the aim of chronic HBV and/or HCV treatment is to stop the development of cirrhosis and the sequelae of hepatic decompensation and HCC. Genomic medicine is likely to have a significant impact in future determination of viral hepatitis outcomes and identification of individuals at risk for development of cirrhosis and sequelae such as HCC. Genomic studies have made significant contributions to our understanding of HBV and HCV pathogenesis and treatment responses (see Table 99.3).

Pathogenesis of Viral Hepatitis

The pathogenesis of viral hepatitis is unique for HBV and for HCV (Chisari and Ferrari, 1995; Lee and Locarnini, 2004; Tanikawa, 2004). However, there are similarities in the mechanisms of HBV and HCV clearance and persistence. The initial innate immune response is characterized by interferon (IFN) and IFN-stimulated gene (ISG) expression. Chronic infection with either virus is characterized by both antigen-specific and non-specific CD4+ and CD8+ T-cell responses that are pivotal in virus clearance or in chronic infection-responsible ongoing inflammation resulting in liver injury. Furthermore, both HBV and HCV employ the strategy of mutational escape to evade the adaptive immune response. The mode of entry of HBV or HCV into hepatocytes is unknown. Receptor-mediated viral entry is thought to be likely, given the restricted cell population infected (Cocquerel et al., 2006; Pileri et al., 1998; Yerly et al., 2006). In HBV, carboxypeptidase D (gp180) has been shown to interact with the preS portion of the large viral surface protein (Yerly et al., 2006). In HCV, CD-81, scavenger receptor class B type 1, and claudin-1 have all been implicated in viral entry (Cocquerel et al., 2006; Pileri et al., 1998). Currently, no conclusive evidence of a single receptor or the presence of a receptor complex has been demonstrated for either HBV or HCV (Cocquerel et al., 2006; Yerly et al., 2006).

There are marked differences in the pathogenesis of HBV and HCV. Innate immune responses are blunted and do not appear to play an important part in HBV clearance. In contrast, a strong innate immune response is though to be important in HCV clearance. Finally, antibody-derived immunity is present in HBV infection, while it appears to be of no importance in HCV infection.

Studies have started to address the role of non-coding sequences, or microRNAs (miRNAs), in the pathogenesis of viral hepatitis. Bioinformatic approaches have identified putative miRNA sequences in HBV and HCV (Chen, 2009). However, unlike in other DNA viruses such as cytomegalovirus (CMV), these putative miRNA genes in hepatitis B and C have not yet been demonstrated to encode functional miRNA, in part due to the overlapping reading frames of hepatitis B and the high degree of viral diversity generated through mutation seen in hepatitis C. However, miRNAs do play a key role in the

TABLE 99.3	Key findings arising from genomic studies of viral hepatitis		
	Hepatitis C	**Hepatitis B**	**Comments**
Acute infection	IL-28B polymorphism associated with acute HCV clearance	"Stealth virus"	Acute HBV or HCV infection has been partially characterized by genomic studies
	Interferon-stimulated gene expression (Mx1, ISG15) correlated with viral clearance	Innate immune response is abrogated	
		Immune evasion	
Chronic infection	Th1 immuno-phenotype perpetuating chronic injury	T-cell effector function activated and latter B-cell-related gene expression observed	
Treatment responses	IL-28B polymorphism associated with acute HCV response to interferon		HBV treatment responses have been poorly characterized by genomic studies
	Interferon-stimulated gene induction correlated with treatment responses		
Hepatocellular carcinoma	HCV core protein oncogenic	HBV x (HBX) protein oncogenic	
Biomarkers	A number of non-specific inflammatory markers identified	A number of non-specific inflammatory markers identified	
	Potential tumor markers identified		
Future studies	Predicting HCV IFN treatment responses (especially in non-genotype-1 HCV)	Predicting HBV IFN and nucleotide/nucleoside treatment responses	Determining the pathogenesis of and predicting chronicity, disease progression, and development of sequelae such as HCC

Th1 = T helper cell 1, IL-28B = interleukin 28B, IFN = interferon, HBV = hepatitis B virus, HCV = hepatitis C virus.

development of injury associated with viral hepatitis, including fibrosis (see Chapter 78) and hepatocellular carcinoma (Chen, 2009). In the case of hepatocellular carcinomas, many of the genomic DNA rearrangements, deletions, or insertions are associated with genomic regions known to encode miRNAs, regulating key pathogenic pathways, cellular proliferation, and oncogenic potential, including p53, related by miRNA-34a-c, as well as cyclin-D and E (Chen, 2009; He et al., 2007). Further, in liver injury, miRNAs often have in excess of 100 predicted targets, and the functional association between observed changes in miRNA expression and predicted targets has not been demonstrated (Chen, 2009). In hepatitis C, miR-122, the liver-specific miRNA, has been shown to facilitate HCV replication, and a miR-122 binding site has been identified in the HCV genome. Further, cells expressing miR-122 are permissive for HCV infection in vitro, while cells lacking miR-122 are not permissive for viral infection. Additional work has demonstrated 30 cellular miRNAs influenced by both exogenous and endogenous interferons, with 8 miRNAs having near-perfect complementarity to HCV RNA sequence (Chen, 2009). Therefore, as in other diseases, miRNAs appear to have key regulatory roles in

the development of injury, sequelae such as cirrhosis and HCC, as well as determining treatment responses in viral hepatitis (Chen, 2009).

Hepatitis B Virus Pathogenesis

Hepatitis B virus infection results in the formation of a double-stranded HBV genome in the nucleus, which is converted into a covalently closed circular double-stranded DNA (cccDNA). Further, HBV DNA integrates into the host genome, and this has significant implications in the long-term management of HBV, as the complete elimination of the virus is not possible and viral reactivation characterized by active replication is possible in the future. The cccDNA acts as a template for the formation of an RNA replicative intermediate that is prone to a high rate of mutation of 1 in 10^5 bases. However, HBV replication after infection is characterized by low-level replication, reaching levels of 10^2–10^4 genome equivalents per ml for up to six weeks after infection. Once established, HBV is associated with extremely high viral titers of 10^8–10^{13} genome equivalents per ml. This has led to the assertion that HBV initially evades the immune response before becoming established. Importantly, once established, active HBV replication can result in infection of 100% of the intrahepatic hepatocytes.

The innate immune activation in HBV infection is abrogated. However, HBV clearance can occur prior to the induction of an adaptive immune response. This is thought to be mediated by IFN α and β (type I IFN) in a non-classical manner that is proteosome-dependent. Further, antigen-independent natural killer (NK) cell activation is thought to be responsible for IFN-γ induction and induction of an adaptive immune response.

Adaptive immunity in HBV is characterized by CD8+ T-cell response to surface antigen epitopes with secretion of IFN-γ and TNF, which have direct antiviral effects, principally by controlling HBV replication at the stage of formation of the RNA replicative intermediate. In contrast to HCV, HBV antibody-mediated humoral immunity effectively neutralizes the virus. However, antibody production is often absent in chronic HBV infection, by mechanisms that are poorly understood. Viral proteins such as the X protein inhibit proteosome-dependent control of virus replication. Also, both surface antigen and precore protein act as tolerogens, abrogating the T-cell response to the virus.

Functional Genomics Studies of Hepatitis B Virus Pathogenesis

Acute HBV infection in the chimpanzee has been analyzed using microarrays (Wieland et al., 2004a). Importantly, there was no differential gene expression during the initial phase of HBV infection and the first phase of HBV replication. This is in direct contrast to HCV infection, and suggests that HBV infection acts in the initial phase as a "stealth virus," failing to induce a significant innate immune response (Wieland and Chisari, 2005; Wieland et al., 2004a). Intrahepatic gene induction is first seen during the viral clearance phase (Wieland et al., 2004a). Gene expression during the early phase of infection was associated with T-cell receptor and antigen presentation. Following this, T-cell effector function (granzymes), T-cell recruitment (chemokines), and monocyte-activation-associated gene expression was observed (Wieland et al., 2004a, b). A later phase of clearance was associated with the expression of B-cell-related genes. This gene expression profile is consistent with initial innate immune response evasion and a subsequent induction of cell-mediated and humoral immune responses.

In vitro cell models have been utilized to study HBV pathogenesis. However, this approach is impeded by the absence of a suitable model of *in vitro* viral replication. To determine the oncogenic role of hepatitis B virus X protein (HBx) in the development of hepatocellular carcinoma (HCC), gene expression profiles in primary adult human hepatocytes and an HCC cell line (SK-Hep-1) ectopically expressing HBx via an adenoviral system have been studied (Wu et al., 2001). Many genes, including a subset of oncogenes (such as c-myc and c-myb) and tumor-suppressor genes (such as *APC, p53, WAF1*, and *WT1*), were differentially expressed, and cluster analysis showed distinctive gene expression profiles in the two cell types. Therefore, HBx protein altered gene expression as an early event, and favors hepatocyte proliferation that may contribute to liver carcinogenesis (Wu et al., 2001).

Proteomics has also been used to examine different stages of chronic HBV infection. Serum proteomic profiles in chronic HBV infection identified increased haptoglobin beta and alpha 2 chain, apolipoprotein (Apo) A-1 and A-1V, alpha-1 antitrypsin, transthyretin, and DNA topoisomerase 11 beta expression (He et al., 2003). Some of these proteins are among the most abundant serum proteins secreted by the liver, and are generally associated with acute-phase inflammatory responses. It is apparent from this study that different isoforms of some of these proteins showed distinct changes in HBV infection, which differed at times between patients with low inflammatory scores versus high inflammatory scores. Some examples include a decrease in cleaved haptoglobin beta peptides and ApoA-1 fragments in patients with higher inflammatory scores. In contrast, alpha-1 antitrypsin fragments were increased in patients with higher inflammatory scores. An alternate approach studied serum protein profiles and correlated them with disease severity using a surface-enhanced laser desorption/ionization (SELDI) protein chip analysis and artificial neural network models (Poon et al., 2005). They found six fragments with a positive prediction and 24 with a negative prediction of fibrosis stage, and subsequently developed a fibrosis index with excellent predictive values for significant fibrosis and cirrhosis, based on the Ishak fibrosis score. The inclusion of clinical biochemical parameters such as alanine transaminase (ALT), bilirubin, total protein, hemoglobin, and international normalized ratio (INR) strengthened the study's accuracy. In general, however, potential markers of disease severity identified to date are untested in large clinical cohorts of individuals.

Hepatitis C Virus Pathogenesis

Hepatitis C replication involves formation of a negative-sense replicative RNA strand and the subsequent formation of double-stranded RNA (dsRNA). This has several important implications for HCV pathogenesis. First, the formation of RNA intermediates means that there is no stable genomic replicative form of the virus, and as a result HCV must produce new viral RNA and proteins to maintain persistence. Second, the formation of dsRNA associated with HCV replication is a target for endogenous RNA interference and elicits the endogenous IFN response (Yu et al., 2000). Further, the NS5B RNA polymerase of HCV lacks proofreading activity, and as a result virus replication is highly error-prone (1 in 10^3 bases), resulting in remarkable genetic diversity. HCV is currently divided into six major genotypes, with many subtypes that differ by up to 35% in their nucleotide sequence. HCV infection is characterized by a rapid increase in circulating levels to 10^5–10^7 equivalents per ml. The rapid induction of an immune response means that not all hepatocytes are infected, although the true proportion of hepatocytes infected is unknown.

The innate immune response is characterized by induction of type I IFN, interferon-stimulated genes (ISGs), and a natural killer response. The IFN gene expression results from the induction of endogenous RNA interference pathways, the formation of dsRNA that binds to the RNA helicases RIG-1 and MDA5, and binding of the phagocytosed infected cell fragments to toll-like receptor (TLR)-3 (Honda et al., 2006; Kato et al., 2006; Yoneyama et al., 2005). These upstream events then signal through interferon regulatory factor (IRF)-3 phosphorylation, resulting in IFN gene transcription. The IFN gene expression then signals via cognate receptors and activates the JAK/Stat (Janus kinase/signal transducer and activator of transcription) pathway, resulting in the induction of ISGs, including protein kinase R, RNA-specific adenosine deaminase-1 (ADAR-1), P56, and 2'-5'-oligoadenylate synthetase. Most of these ISGs act on the formation of the negative replicative strand of HCV. The cellular innate immune response is characterized by induction of a natural killer cell response. NK cells destroy infected cells in an antigen-independent manner via cytotoxic cell lysis. These activated NK cells secrete large amounts of IFN-γ, which activates and maintains a cellular adaptive immune response.

Adaptive immune responses in HCV infection are characterized by the virus-specific CD4+ and CD8+ T-cell response to multiple HCV epitopes, including many highly promiscuous epitopes formed due to the high spontaneous nucleotide mutation rate associated with HCV replication (Bowen and Walker, 2005a, b). This T-cell response is accompanied and maintained by induction of IFN-γ and TNF, both of which can directly inhibit viral replication without killing an infected cell. Although HCV antibody production is universal in immunocompetent individuals, it does not prevent infection or correlate with outcome. Further, the virus-specific T-cell response is maintained for decades after HCV clearance, in contrast to the antibody responses, which can become undetectable.

Functional Genomics Studies of Hepatitis C Virus Pathogenesis

In the chimpanzee, microarray technology has been used to study acute HCV infection (Bigger et al., 2001; Su et al., 2002). Viral clearance is associated with a marked early induction of interferon-gamma-induced genes such as 2'-5'-oligoadenylate synthetase, Mx1, ISG15, and ISG16, with the later induction of immune T helper cell 1 (Th1)-response-associated transcripts such as MIG (CXCL9) and IP10 (CXCL10) (Shackel et al., 2002). However, in chronic human HCV infection there is a persistent intrahepatic IFN alpha antiviral response, but the virus itself escapes the response via an inhibition of the effector arm (Bowen and Walker, 2005a, b). HCV evades the immune response in many ways. The protein products of viral replication have a number of inhibitory effects: core protein inhibits Fas-mediated apoptosis, E2 inhibits NK cell activation, E2 and NS5a inhibit protein kinase R (PKR), and NS3 inhibits IRF-3 (Karayiannis, 2005). The evasion from the adaptive response is less well defined, but chronically infected individuals have only weak, oligo-/mono-specific or no virus-specific CD4+ or CD8+ T-cell responses. The mechanism of evasion of this

adaptive response is unclear. However, the induction of a persistent, non-specific inflammatory response in chronic HCV infection results in the induction of genes associated with a Th1 immune response. Therefore, in chronic HCV the immune response is characterized by persistent but non-specific immune activation that damages the liver while being insufficient to clear the virus.

The intrahepatic IFN-alpha-induced gene response is variable among individual patients. This response has been identified by microarrays to be higher in patients not responding to pegylated IFN and ribavirin therapy, suggesting an increased resistance to the effector arm that can't be amplified by exogenous therapy (Chen et al., 2005). In contrast, patients who had a sustained viral response (SVR) to pegylated IFN therapy had a lower expression of IFN genes that, by inference, can be amplified by exogenous therapy, resulting in viral clearance.

Comparison of chimpanzees that cleared acute HCV infection with an animal that had virus persistence has provided further insight into the balance between viral clearance and persistence (Su et al., 2002). In these experiments, Su and colleagues observed upregulation of genes associated with the early response (which correlated with viral load), including many IFN-alpha-induced genes: STAT1, 2'-5'-oligoadenylate synthetase, Mx1, ISP15, and p27 (Su et al., 2002). Interestingly, there was the induction of lipid pathway genes such as fatty acid synthetase, sterol response element binding protein (SREBP), and downregulation of PPARα as well as hepatic lipase C and flotillin 2. Importantly, the lipid pathway genes are associated with viral replication, and studies using *in vitro* replicon experiments have demonstrated altered viral replication with geranylgeranylation (Kapadia and Chisari, 2005). Further, the reduction in PPARα would be expected to be associated with insulin resistance, a feature of chronic HCV, but prior to this it was not an expected aspect of acute HCV infection. As noted previously (Bigger et al., 2001), clearance of HCV was associated with the late induction of Th1 transcripts such as CXCL9 and CXCL10, major histocompatibility complex (MHC) expression, and T-cell molecules such as CD8 and granzyme A. Although the induction of IFN-alpha-induced genes early in infection has been observed by others, the timing did not correlate with clearance, as high levels of these transcripts continued in the animal with viral persistence (Bigger et al., 2001; Lanford et al., 2001). Further, functional studies in HCV replicon systems have shown that NS3/4a was able to inhibit interferon alpha antiviral effector function by blocking the phosphorylation of IRF-3, a key protein in the antiviral response (Karayiannis, 2005). Therefore, chronic HCV infection induces a persistent intrahepatic IFN alpha antiviral response, but the virus itself escapes this response via inhibition of the effector arm. However, microarray studies of the intrahepatic IFN-alpha-induced gene response show that this is variable and observed to be higher in patients not responding to pegylated IFN and ribavirin therapy, consistent with resistance of the effector arm of the immune response to amplification by exogenous therapy (Chen et al., 2005). In contrast, patients who had a sustained viral response (SVR) to pegylated IFN therapy had a lower expression of IFN genes, consistent with amplification of the effector arm of the immune response by exogenous therapy, resulting in viral clearance.

Chronic HCV infection has been studied in a number of ways using genomic analysis. Gene expression in liver biopsy material from individuals with chronic HBV or chronic HCV has been compared to a non-diseased control (Honda et al., 2001). Chronic HCV infection was associated with a predominantly anti-inflammatory, pro-proliferative, anti-apoptotic intrahepatic gene profile (Honda et al., 2001). However, the results demonstrated widespread upregulation of pro-inflammatory genes such as IL-2 receptor, CD69, CD44, IFN-gamma-inducible protein, MHC class 1 genes, and monokine induced by gamma IFN. These findings were similar to another study of HCV cirrhosis in which a pro-inflammatory Th1-associated transcript expression predominated (Shackel et al., 2002). Therefore, a Th1 immune response is thought to be responsible for the accelerated fibrogenesis of HCV liver injury, with fibrosis-associated upregulation of a wide variety of genes, including platelet-derived growth factor (PDGF) and transforming growth factor (TGF)-beta 3 (Shackel et al., 2002).

The premalignant potential of intrahepatic HCV infection has been studied by using gene array analysis to compare HCV cirrhosis with and without HCC (McCaughan et al., 2003). The upregulation of many oncogenes (e.g., TEL oncogene), immune genes (IFN-gamma-associated), fibrosis genes (integrins), as well as cell signaling (G-protein-coupled receptor kinase) and proliferation-associated genes (cyclin K), was demonstrated in cirrhosis complicated by HCC. This is consistent with a premalignant cirrhotic response in HCV infection. The data also suggest that there is more cellular proliferation, immune activation, and fibrosis in the liver of patients with HCC than those with cirrhosis alone. A key area of future research will be to ascertain whether such a profile can be recognized before HCC develops. This approach has a direct clinical application in identifying and screening high-risk patients.

Gene array studies of HCV infection have revealed new insights into the development of HCC in HCV, have allowed structural analysis of the HCV RNA genome, and have identified novel markers of HCV intrahepatic injury and HCV-associated HCC. A Smith and colleagues study utilized 13,600 gene microarrays to profile patients with HCV cirrhosis, HCV and HCC, and normal liver (Smith et al., 2003). The results identified 87 upregulated and 45 downregulated genes, that appear to be markers of HCV liver injury (Smith et al., 2003). Importantly, the analysis aimed to exclude genes expressed in normal liver, other forms of cirrhosis, or HCC. Genes such as ILxR (IL-13 receptor a2), CCR4, and cartilage glycoprotein 39 (GP-39) were identified (Smith et al., 2003). However, the study highlights the problems with the interpretation of these large data sets using small numbers of patient samples – does the identified gene expression represent unique disease, or phenotype-associated gene expression, or the stochastic probability of identifying a small cohort of genes from the many thousands being analyzed? Clearly studies such as these, as powerful as they are, need to be validated by alternative methodologies in large patient groups. One approach to validation has been to confirm important gene expression identified in these studies by reverse transcriptase (RT)-PCR in a larger cohort of patients (Shackel et al., 2002; Shackel et al., 2001).

Hepatocellular carcinoma proliferation in HCV-associated liver injury has been studied by array analysis. This has resulted in a plethora of potentially novel tumor markers being identified. They include the serine/threonine kinase 15 (STK15) and phospholipase A2 (PLA2G13 and PLA2G7), which were shown to be increased in more than half of the tumors identified (Smith et al., 2003). However, another study implicated different gene groups in HCV-associated HCC: cytoplasmic dynein light chain, hepatoma-derived growth factor, ribosomal protein L6, TR3 orphan receptor, and c-myc (Shirota et al., 2001). The clustering analysis in this study showed that the expression of 22 genes in HCC related to differentiation of the malignancy, with more than half of the genes being transcription factors or related to cell development or differentiation (Shirota et al., 2001). Although many of these genes can be implicated in HCC development, they were often identified in large gene sets in end-stage disease. Therefore, whether the genes represent cause or effect is unknown. Additionally, the number of differing gene sets being examined by the gene arrays being utilized is almost as great as the number of studies using them. As these gene sets still only represent a fraction of the transcriptome being examined, they selectively identify differentially expressed genes.

Gene array analysis of HCV recurrence in liver transplant allografts has provided novel insights into the molecular mechanisms of viral recurrence (Mansfield and Sarwal, 2004; McCaughan and Zekry, 2004). HCV recurrence in the liver graft is associated with expression of IFN-γ-associated genes, such as CXCL10 (IP-10), CXCL9 (HuMIG), and RANTES (McCaughan and Zekry, 2004). Further, antiviral IFN-α-associated gene expression is seen in chronic HCV recurrence and during acute rejection associated with HCV recurrence (McCaughan and Zekry, 2004). Additionally, upregulation of the nuclear factor (NF)-kappa β pathway during acute rejection in association with HCV recurrence appears to alter cellular apoptosis via changes in the expression of TNF-related apoptosis-inducing ligand (TRIAL)-associated genes (McCaughan and Zekry, 2004). Importantly, chronic HCV recurrence in grafts is associated with Th1-associated gene expression, similar to that seen in chronically HCV-infected individuals that have not been transplanted (McCaughan and Zekry, 2004). In contrast, cholestatic HCV recurrence [affecting bile ducts], which follows an aggressive course, is associated with a Th2 cytokine profile (McCaughan and Zekry, 2004). This suggests that the Th1 immune response suppresses viral replication while being profibrogenic (McCaughan and Zekry, 2000, 2004; Shackel et al., 2002). In cholestatic HCV recurrence, the unchecked viral replication is directly fibrogenic (McCaughan and Zekry, 2000, 2004).

A particular challenge in the study of the effect of viruses on liver cells is the difficulty in infecting liver cells with virus. The studies described below have involved models in which cultured cells are infected with viral proteins or viral genome. Progress in this field has been rapid, and, most recently, a cellular model of HCV infection has been reported that is likely to stimulate further study (Heller et al., 2005; Lindenbach et al., 2005).

Proteomic methodologies have been applied to a number of aspects of HCV-related liver injury. However, to date most proteomic studies have focused on the identification of a number of biomarkers of disease rather than trying to unravel aspects of HCV pathobiology. Proteomic studies have defined potential protein therapeutic targets that interact with HCV in detail. Large-scale proteome analysis of a full-length HCV replicon revealed prominent expression of proteins involved in lipid metabolism (Jacobs et al., 2005). Several *in vitro* proteomic studies have identified proteins that interact with specific HCV proteins. Heat shock protein 27 (Hsp27) was shown to specifically interact with NS5A via the N-terminal regions (Choi et al., 2004). A total of 14 cellular proteins binding to the core protein were identified by proteomics (Kang et al., 2005). These proteins include DEAD-box polypeptide 5 (DDX5) and intermediate microfilament proteins, including cytokeratins (cytokeratin 8, 19, and 18) and vimentin. Interestingly, DDX5 gene polymorphisms are associated with accelerated fibrosis development in HCV-infected individuals (see Chapter 78) (Huang et al., 2006).

The development of HCC and IFN treatment responses are two further aspects of HCV infection studied using proteomics. The study of HCV-related HCC development has shown an overexpression of alpha enolase and is poorly correlated with differentiated HCC (Kuramitsu and Nakamura, 2005; Takashima et al., 2005). The response of hepatocyte cell lines to IFN gamma treatment has uncovered more than 54 IFN response genes, including many novel targets – an approach that may pave the way for novel therapies. Examination of protein extracts that bind to the HCV internal ribosome re-entry sites (IRES) has identified a number of novel protein targets such as Ewing sarcoma breakpoint 1 region protein (EWS) and TRAF-3. The final aspect of HCV liver injury receiving attention is the study of potential biomarkers, such as heat shock protein HSP-70, associated with HCV infection progression to HCC (Takashima et al., 2003).

THERAPEUTICS AND PHARMACOGENOMICS

The principle treatment goal in viral hepatitis is clearance of the virus, with a secondary goal of averting or delaying the onset of cirrhosis, hepatic decomposition, and hepatocellular carcinoma. Immune modulators in the form of IFN treatment have been the mainstay of treatment for years (see Table 99.4). Antiviral therapy has now become an effective treatment option in HBV, but is useful in HCV only when combined with IFN treatment. Finally, HAV and HBV infection are reliably protected against by immunization. There is no prospect in the foreseeable future of a vaccine for HCV. Treatment options for viral hepatitis are summarized in Table 99.4. Predicting individuals' treatment responses based on gene expression is likely to be an area in which genomic medicine will enable highly directed individual therapy in treatment of viral hepatitis.

Treatment of Hepatitis B Virus Infection

Interferon is the only treatment shown to clear HBV infection in chronically infected individuals without the development of drug resistance. In both HBeAg-positive and -negative individuals with chronic HBV infection, IFN treatment for four to six months has been shown to normalize liver function abnormalities, result in clearance of HBsAg and HBeAg, and result in a sustained loss of HBV DNA. In a meta-analysis of 15 studies,

TABLE 99.4	Treatment of viral hepatitis				
Virus	**Chronicity**	**Treatment**		**Vaccine**	**Comments**
HAV	No	Immune globulin		Yes	
HBV	Yes	Immune globulin		Yes	Viral resistance to nucleoside analogs common
		Nucleoside analogs (target DNA polymerase)			
		Immune mediators, i.e., α-interferon			
HCV	Yes	Immune mediators, i.e., α-interferon and ribavirin		No	Small molecular inhibitors in clinical trials
HDV	Yes	Treatment of HBV		No	
HEV	No	Nil		No	Nil specific treatment
					Avoidance

HAV, hepatitis A virus; HBV, hepatitis B virus; HCV, hepatitis C virus; HDV, hepatitis D virus; HEV, hepatitis E virus.

suppression of HBV DNA was seen in 37% of patients, loss of HBeAg was seen in 33% of subjects, and HBeAg seroconversion was seen in 18% of patients. Subjects responded to treatment if they had lower pretreatment HBV DNA levels and higher pretreatment liver transaminase levels. The advent of long-lasting IFN preparations using pegylated IFN has been shown to have an additional benefit over conventional IFN in treating HBV infection, resulting in improved HBsAg and HBeAg seroconversion. However, the current meta-analysis of the treatment outcomes with IFN in HBV does not support its use in preventing HCC. Compared to nucleoside analogs (discussed below), one principle advantage of IFN therapy for HBV is the durability of the treatment response, with <10% of individuals having a relapse in HBeAg expression up to eight years later.

Nucleoside analogs are now being increasingly used to treat HBV infection, and these agents target the HBV DNA polymerase. Lamivudine is the most widely used nucleoside analog. It effectively suppresses HBV replication, as is evident by a >2-log drop in HBV viral DNA titers. Lamivudine results in 16–18% HBeAg seroconversion in one year and 50% after five years. The durability of the response is 77% at three years. However, the sustained viral response in other studies has been reported as 39% at four years, and sustained response to lamivudine following cessation of therapy is significantly less than IFN. Continued treatment with lamivudine results in sustained suppression of HBV viral replication, but is limited by the appearance of mutant forms of the HBV polymerase, typically in a conserved YMDD motif at methionine 204 of the enzyme. After one year, resistance develops in 14–32% of cases, and this increases to 50% at two years and 74% at five years.

Other nucleoside analogs include adefovir, tenofovir, and entecavir. All have significant activity against HBV replication, although both tenofovir and entecavir would appear to have greater activity against HBV. Importantly, resistance with all of these newer agents is uncommon, with entecavir resistance being less than 5% at two years. Unfortunately, resistance to these newer agents appears inevitable. In a situation analogous to highly active antiretroviral therapy (HAART) treatment of HIV-1 infection, combination therapy is now being studied in HBV infection. The conclusive outcome of these studies is not yet available, but the initial results are promising, especially in cases of lamivudine resistance.

Treatment of Hepatitis C Virus Infection

Interferon treatment is the only effective antiviral therapy available for the treatment of hepatitis C. The current recommendations are for a 24–48 week course of treatment using pegylated IFN combined with the antiviral ribavirin. Ribavirin is a guanosine analog able to inhibit the replication of viruses, but in the absence of IFN it has no significant effect on HCV RNA levels. The overall chance of a sustained viral response (SVR) varies according to HCV genotype. In genotype 1 infection, SVR can be achieved in 42–46% of patients, with better response rates of 76–88% for those with genotype 2 or 3

infection. Virologic response to treatment can be predicted from the decline in HCV RNA at 12 weeks. The absence of a 2-log drop or undetectable HCV RNA at week 12 has a highly negative predictive value for the absence of SVR in genotype 1 patients with continued therapy. In genotypes 2 and 3, the drop in viral load at 4 weeks may be predictive of achieving SVR with only 24 weeks of therapy. The treatment of patients with cirrhosis is controversial, but there appears to be a benefit in avoiding progression of disease, decompensation, and the development of HCC. However, in the presence of hepatic decompensation, IFN-based treatment is contraindicated, and patients should be referred for transplant evaluation.

Functional Genomics Studies Related to the Treatment of Viral Hepatitis

IFN alpha is currently part of the standard of care for HCV infection. Several studies have used microarray analysis to identify the mechanisms by which IFN alpha acts on hepatocytes and the hepatitis C virus. IFN alpha activated the multiple signal transducer and activator of transcription factors (STAT) 1, 2, 3, 5 in cultured hepatocytes (Radaeva et al., 2002). Other upregulated genes include a variety of antiviral and tumor-suppressor/pro-apoptotic genes. Downregulated genes include c-myc and c-Met, and the hepatocyte growth factor (HGF) receptor (Radaeva et al., 2002). In a second and comparable study, IFN alpha antiviral efficacy was associated with 6-16 (G1P3) expression. Involvement of STAT3 in IFN-alpha signaling was confirmed (Zhu et al., 2003). Resistance to IFN alpha antiviral activity may be mediated the hepatitis C viral protein, NS5A. To identify the mechanisms through which NS5A blocks interferon activity, the gene expression profile was studied in IFN-treated Huh7 cells expressing NS5A. The strongest effect of NS5A on interferon response was observed for the OAS-p69 gene (Girard et al., 2002). Another key response of hepatocytes to the HCV virus is cellular proliferation. Gene array studies identified upregulation of growth-related genes, in particular wnt-1 and its downstream target gene, WISP (Fukutomi et al., 2005). In another study, cyclin dependent kinases (CDK) activity, hyperphosphorylation of retinoblastoma tumor suppressor protein (Rb), and E2F activation were shown to be associated with hepatocyte proliferation induced by a full-length HCV clone (Tsukiyama-Kohara et al., 2004).

Global quantitative proteomic analysis was performed in a human hepatoma cell line (Huh7) in the presence and absence of IFN, to examine liver-specific responses to IFN and the mechanisms of IFN inhibition of virus replication (Yan et al., 2004). A total of 54 proteins were induced by IFN and 24 were repressed, representing several novel and liver-specific key regulatory components of the IFN response.

Molecular markers used on an individual basis might improve prediction of treatment responses prior to commencement of therapy. Previously, Chen and colleagues examined liver biopsies prior to α-IFN and ribavirin therapy from 16 responders, 15 non-responders, and 20 normal individuals by gene array analysis, and determined that 18 genes were predictive of an SVR (Chen et al., 2005). These investigators identified a gene expression signature of eight genes that could predict the likelihood of achieving an SVR in 30 of 31 individuals (*GIP2/IFI15/ISG15, ATF5, IFIT1, MX1, USP18/UBP43, DUSP1, CEB1*, and *RPS28*) (Chen et al., 2005). The striking outcome of this study was that these genes, known to be involved in interferon responsiveness, were over-expressed in non-responders and formed part of the predictive gene signature profile. Additionally, two genes (*ISG15/IFI15* and *USP18/UBP43*) were identified as part of a previously unrecognized, novel IFN regulatory pathway (Chen et al., 2005; Randall et al., 2006). In a further study of peripheral blood mononuclear cells (PBMCs), a relative lack of interferon-stimulated gene (ISG) expression was associated with a poor response to antiviral therapy with pegylated IFN (Taylor et al., 2007). Further, Feld and colleagues demonstrated unique patterns of liver gene expression that correlated with interferon and ribavirin treatment responses in HCV genotype 1 infection (Feld et al., 2007). The authors used gene arrays to study individuals who had an early virologic response at four weeks of treatment (Feld et al., 2007). These individuals were compared to those who had a slow virologic response, and to individuals with chronic HCV genotype 1 infection who were treatment-naïve (Feld et al., 2007). Their results are intriguing. Individuals with a relapse response (RR) were more likely to have a significant induction of interferon-stimulated genes following 72–96 hours of pegylated IFN compared to sustained responders (SR) or treatment-naïve individuals. The expression of genes known to inhibit the interferon response was increased in SR compared to individuals with an RR. Ribavirin appeared to diminish gene expression that could inhibit interferon responsiveness. To achieve suppression of viral replication and achieve a treatment response following interferon administration, there needs to be a significant induction of interferon-associated gene expression. This means that exogenous interferon appears to only be effective in enhancing the innate immune response if there is a significant increase in interferon-associated gene expression, and that this requires an immune response that is not already near-maximally stimulated (Shackel and McCaughan, 2007).

Pharmacogenomics of Viral Hepatitis

Pharmacogenomic studies of viral hepatitis are currently evolving as treatment options improve and understanding of the disease pathobiology improves. Prior to the use of genomics approaches enabling the simultaneous examination of many thousands of gene polymorphisms in a single experiment, a candidate gene approach was adopted. However, genome-wide association studies (GWAS) that examine genetic variation across whole genomes have recently been used successfully to identify genetic associations that correlate with outcomes in viral hepatitis C treatment. Indeed, the genetic association identified in hepatitis C treatment is one of the strongest known pathogenic genetic variations identified in any human disease to date. These studies in hepatitis C define approaches that have yet to be applied to hepatitis B treatment, so a similar strong genetic association with hepatitis B treatment has not yet been identified.

Recent hepatitis C GWAS studies have identified a number of single nucleotide polymorphisms (SNPs) in the region of the IL-28B gene, which enables personalized tailored therapy on the basis of individual genetic variability. A number of studies have identified that SNPs do not all occur at random and frequently associate, by linkage, with other SNPs in the IL-28B loci. Four landmark papers have now identified a SNP near the IL-28B gene that is associated with a marked difference in treatment response to pegylated interferon in individuals with hepatitis C infection (Ge et al., 2009; Rauch et al., 2010; Suppiah et al., 2009; Tanaka et al., 2009). A fifth manuscript published at the same time describes an association of the same SNP with spontaneous clearance of hepatitis C (Thomas et al., 2009). The IL-28B gene encodes the protein interferon-λ3, which acts through the well-defined JAK-Stat pathway, leading to the induction of hundreds of ISGs that have already been widely implicated in both HCV treatment outcomes and spontaneous viral clearance (see Figure 99.2). The paper by Ge and colleagues analyzed the Individualized Dosing versus Flat Dosing to Assess Pegylated Interferon Therapy (IDEAL) study cohort of 1137 patients, and demonstrated that the IL-28B SNP with CC vs a TT or TC genotype was a stronger predictor of HCV treatment outcome [odds ratio (OR) 7.1] than either baseline viral load (OR 4.2) or fibrosis grade (OR 3.0) (Ge et al., 2009). At the same time, a manuscript by Suppiah and colleagues identified an IL-28B polymorphism in an Australian cohort that was associated with a weaker effect on IFN-α responsiveness (OR 2.0) (Suppiah et al., 2009). A further supporting manuscript by Tanaka and colleagues identified another IL-28B SNP in a Japanese population, which associates by linkage disequilibrium with the SNPs identified in the other two manuscripts (Tanaka et al., 2009). In this case, the effect on interferon responsiveness was even more profound (OR 17.7) (Tanaka et al., 2009). The fourth manuscript described seven SNPs in the IL-28B locus (Rauch et al., 2010). They showed that the SNPs from the Ge and Suppiah studies were in strong linkage disequilibrium, and only 4378 bases apart. However, there were clear racial differences evident in these GWAS studies. The IL-28B SNPs in the Duke study (Ge et al., 2009) were strongly linked to treatment outcomes in Caucasians ($r^2 = 0.52$), but very low linkage was seen in African Americans ($r^2 = 0.07$) (Ge et al., 2009). Further, the alleles identified could not explain the low response rates to therapy and the high rates of viral persistence in African Americans (Ge et al., 2009; Thio and Thomas, 2010). Clearly, the association of IL-28B gene polymorphisms with HCV treatment response is strong, although there are many cofactors, including population genetics, influencing this association.

A final pivotal paper from the Goldstein group (Center for Human Genome Variation) group at Duke University demonstrates that this IL-28B polymorphism influences HCV spontaneous clearance (Thomas et al., 2009). Individuals with the favorable genotype were three times more likely to clear HCV (Thomas et al., 2009). This manuscript demonstrated marked variation in the frequency of this genetic variant in different

TABLE 99.5	Implications raised by identification of IL-28B polymorphism in hepatitis C

Current implications of identification of a favorable IL-28B polymorphism

- Increased spontaneous clearance of genotype-1 HCV infection
- Increased response to interferon treatment in genotype-1 HCV liver disease

Future questions raised by the IL-28B polymorphisms

- What is the effect of the polymorphism on IL-28B?
- Does this IL-28B polymorphism influence a non-coding regulatory element or the coding region of IL-28B?
- Which cells express IL-28B?
- What is the associated beneficial interferon-stimulated gene profile?
- Do the associated treatment outcomes apply to non-genotype-1 disease?

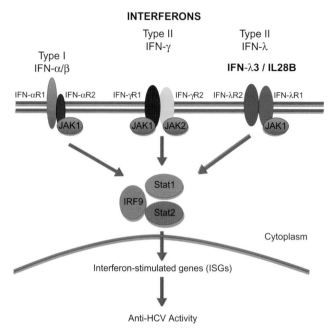

Figure 99.2 Interferon responses in hepatitis C infection. The interferon-signaling pathway has three different receptor complexes (types I–III), which mediate distinct differences in JAK (Janus kinase) and Stat (signal transducer and activator of transcription) signaling within the cell cytoplasm. Subsequent interferon-stimulated gene (ISG) expression is differentially expressed, and this in turn determines the relative anti-hepatitis C virus activity of the pathway. The identified IL-28B polymorphisms are near the gene that encodes interferon-λ3, which signals through the type III interferon receptor complex.

populations. Therefore, HCV treatment responses and spontaneous viral clearance are influenced by variation in multiple genetic loci, one of the strongest of which is the demonstrated IL-28B association. This polymorphism clearly has major implications for current HCV treatment. However, it is unclear what the effect of the polymorphism is on IL-28B. It remains to be determined if this polymorphism influences a non-coding regulatory element or, through a haplotype association, the coding region of the IL-28B gene. Further, it is not known in which cells IL-28B is expressed, or how this influences clearance of HCV from infected hepatocytes. The current implications of this pivotal finding in HCV are summarized in Table 99.5.

Finally, what does this tell us about the role of endogenous and exogenous interferons in HCV eradication? In general, the ISG effect on HCV replication is determined by variably induced ligand (endogenous IFN-α/-β, IFN-γ, and IFN-λ1–3, and exogenous IFN-α2a/b) and receptor (IFNαR1/IFNαR2, IFNγR1/IFNγR2, or IFNλR1/IL-10R2) expression and interactions (see Figure 99.2) (Asselah et al., 2009). Additionally, it is clear that there is an apparently paradoxical relationship between the level of endogenous ISG induction and the subsequent responsiveness of an HCV-infected individual to interferon therapy (Shackel and McCaughan, 2007). Further, there are known differential effects of the many different ISGs on HCV replication (Asselah et al., 2009; Marcello et al., 2006). This can lead us to assume that selective targeting of the interferon pathways, leading to a favorable ISG profile and resulting in HCV eradication, is possible. This is the basis behind new candidate therapeutics such as the clinical trial of IL-29 (IFN-λ1) in HCV (Marcello et al., 2006; Pagliaccetti et al., 2008).

These five pivotal studies have simultaneously identified a significant predictor of HCV treatment response, and at the same time have provided new insight into HCV pathobiology. While gaps in our knowledge of disease pathogenesis remain, a better understanding of host genomics that influence

treatment outcomes means that pharmacogenomic measures have matured into a clinical reality in hepatitis C treatment. The findings will lead to personalized medicine, as IFN-α2a/b treatment response can be determined prior to treatment. Even greater promise is likely to be realized by the development of newer therapeutic agents that more selectively target IFN-λ-associated antiviral responses in HCV infection.

Complementary candidate gene approaches in HCV infection have examined the non-structural protein NS5a, which is known to bind viral RNA and alter HCV replication (Goyal et al., 2006; Kohashi et al., 2006). Amino acid substitutions within the interferon sensitivity-determining region (ISDR) of NS5A correlate with IFN-α treatment responses (Goyal et al., 2006; Kohashi et al., 2006; Watanabe et al., 2005). A meta-analysis of ISDR NS5a mutations demonstrates a relative risk of 4.66–5.73 of IFN-α treatment response compared to non-mutant ISDR HCV (Schinkel et al., 2004). However, this effect is more pronounced in Japanese patients than in European patients (Pascu et al., 2004; Schinkel et al., 2004).

HBV genetic variability has also been studied, to a lesser extent, to identify treatment outcomes. In HBV infection, a group of 82 chronic active carriers received standard IFN-α treatment for six months (King et al., 2002). These patients were concurrently studied for SNPs in genes involved in the JAK/Stat signaling of IFN genes leading to the expression of ISGs (King et al., 2002). Two SNPs were identified that appeared to predict response; one in the promoter region of the ISG MxA and the other in the IFN-regulated eIF-2α gene. This study is significant in demonstrating that host polymorphisms may correlate with treatment response in HBV infection (King et al., 2002).

FUTURE IMPACT OF GENOMICS

Although genomics studies have already made significant contributions to our understanding of viral hepatitis pathogenesis, there are still many unanswered questions. In particular, the molecular pathways mediating acute HBV and HCV infection and the development of chronicity are poorly understood. However, in hepatitis C, the identification of the role ISGs and the IL-28B (IFN-λ3) pathway play in particular infections has led to one of the first examples of personalized therapy based on individual genomic variability. Therefore, in the next five to ten years, genomics studies will help to predict viral hepatitis treatment responses to IFN and will help to predict the development of sequelae such as cirrhosis and HCC. Long-term proteomics promises to develop non-invasive markers of liver injury, and will help to screen individuals for HCC. The identification of genomic susceptibility markers will help to further individualize risk assessment. Individualized patient assessment and tailored therapy will be possible in viral hepatitis due to genomics approaches.

CONCLUSIONS

Clearly, viral hepatitis represents a major global health burden resulting in significant morbidity and mortality as well as accounting for the majority of primary liver cancers. Our understanding of the pathogenesis of chronic viral hepatitis B and C shows that each of these viruses has developed ways of evading the immune system, and that injury is characterized by chronic intrahepatic immune activation that fails to eliminate the virus. The diagnosis of viral hepatitis was previously based on serology, but there is now an increasing reliance on nucleic acid quantification in both hepatitis B and C. The treatment of viral hepatitis is limited, and while eradication is possible for HCV with immune mediators such as interferon, we clearly have a long way to go in attempting to eradicate these viruses and prevent disease progression. One goal in the future is the development of newer therapeutic compounds based on our understanding of the molecular events involved in the immunopathogenesis of these chronic viral infections.

REFERENCES

Acharya, S.K., Panda, S.K., 2006. Hepatitis E virus: Epidemiology, diagnosis, pathology and prevention. Trop Gastroenterol 27, 63–68.

Asselah, T., Bieche, I., Sabbagh, A., et al., 2009. Gene expression and hepatitis C virus infection. Gut 58, 846–858.

Bhandari, B.N., Wright, T.L., 1995. Hepatitis C: An overview. Ann Rev Med 46, 309–317.

Bigger, C.B., Brasky, K.M., Lanford, R.E., 2001. DNA microarray analysis of chimpanzee liver during acute resolving hepatitis C virus infection. J Virol 75, 7059–7066.

Bowen, D.G., Walker, C.M., 2005a. Adaptive immune responses in acute and chronic hepatitis C virus infection. Nature 436, 946–952.

Bowen, D.G., Walker, C.M., 2005b. Mutational escape from CD8+ T cell immunity: HCV evolution, from chimpanzees to man. J Exp Med 201, 1709–1714.

Cabrerizo, M., Bartolome, J., Caramelo, C., Barril, G., Carreno, V., 2000. Molecular analysis of hepatitis B virus DNA in serum and peripheral blood mononuclear cells from hepatitis B surface antigen-negative cases. Hepatology 32, 116–123.

Chen, L., Borozan, I., Feld, J., et al., 2005. Hepatic gene expression discriminates responders and nonresponders in treatment of chronic hepatitis C viral infection. Gastroenterology 128, 1437–1444.

Chen, S.L., Morgan, T.R., 2006. The natural history of hepatitis C virus (HCV) infection. Int J Med Sci 3, 47–52.

Chen, X., 2009. MicroRNA signatures in liver diseases. World J Gastroenterol 15, 1665.

Chisari, F.V., 1992. Hepatitis B virus biology and pathogenesis. Mol Genet Med 2, 67–104.

Chisari, F.V., Ferrari, C., 1995. Hepatitis B virus immunopathogenesis. Ann Rev Immunol 13, 29–60.

Choi, Y.W., Tan, Y.J., Lim, S.G., Hong, W., Goh, P.Y., 2004. Proteomic approach identifies HSP27 as an interacting partner of the hepatitis C virus NS5A protein. Biochem Biophys Res Commun 318, 514–519.

Chu, C.M., Liaw, Y.F., 2006. Hepatitis B virus-related cirrhosis: Natural history and treatment. Semin Liver Dis 26, 142–152.

Cocquerel, L., Voisset, C., Dubuisson, J., 2006. Hepatitis C virus entry: Potential receptors and their biological functions. J Gen Virol 87, 1075–1084.

Fattovich, G., 2003. Natural history and prognosis of hepatitis B. Semin Liver Dis 23, 47–58.

Feld J., Nanda S., Huang Y., et al., 2007. Hepatic gene expression during treatment with peginterferon and ribavirin: Identifying molecular pathways for treatment responses. Hepatology 46(5), 1548-1563.

Fiedler, M., Roggendorf, M., 2006. Immunology of HDV infection. Curr Top Microbiol Immunol 307, 187–209.

Flehmig, B., 1990. Hepatitis A. Baillieres Clin Gastroenterol 4, 707–720.

Forns, X., Bukh, J., 1999. The molecular biology of hepatitis C virus: Genotypes and quasispecies. Clin Liver Dis 3, 693–716. vii.

Forns, X., Purcell, R.H., Bukh, J., 1999. Quasispecies in viral persistence and pathogenesis of hepatitis C virus. Trends Microbiol 7, 402–410.

Fukutomi, T., Zhou, Y., Kawai, S., Eguchi, H., Wands, J.R., Li, J., 2005. Hepatitis C virus core protein stimulates hepatocyte growth:

Correlation with upregulation of wnt-1 expression. Hepatology 41, 1096–1105.

Ge, D., Fellay, J., Thompson, A.J., et al., 2009. Genetic variation in IL28B predicts hepatitis C treatment-induced viral clearance. Nature 461, 399–401.

Ghany, M.G., Seeff, L.B., 2006. Efforts to define the natural history of chronic hepatitis C continue. Clin Gastroenterol Hepatol 4, 1190–1192.

Girard, S., Shalhoub, P., Lescure, P., et al., 2002. An altered cellular response to interferon and up-regulation of interleukin-8 induced by the hepatitis C viral protein NS5A uncovered by microarray analysis. Virology 295, 272–283.

Goyal, A., Hofmann, W.P., Hermann, E., et al., 2006. The hepatitis C virus NS5A protein and response to interferon alpha: Mutational analyses in patients with chronic HCV genotype 3a infection from India. Med Microbiol Immunol (Berl) 196 (1), 11–21.

He, L., He, X., Lim, L., et al., 2007. A microRNA component of the p53 tumour suppressor network. Nature 447, 1130–1134.

He, Q.Y., Lau, G.K., Zhou, Y., et al., 2003. Serum biomarkers of hepatitis B virus infected liver inflammation: A proteomic study. Proteomics 3, 666–674.

Heller, T., Saito, S., Auerbach, J., et al., 2005. An in vitro model of hepatitis C virion production. Proc Natl Acad Sci USA 102, 2579–2583.

Honda, K., Takaoka, A., Taniguchi, T., Type, I., 2006. [interferon] gene induction by the interferon regulatory factor family of transcription factors. Immunity 25, 349–360.

Honda, M., Kaneko, S., Kawai, H., Shirota, Y., Kobayashi, K., 2001. Differential gene expression between chronic hepatitis B and C hepatic lesions. Gastroenterology 120, 955–966.

Huang, H., Shiffman, M.L., Cheung, R.C., et al., 2006. Identification of two gene variants associated with risk of advanced fibrosis in patients with chronic hepatitis C. Gastroenterology 130, 1679–1687.

Jacobs, J.M., Diamond, D.L., Chan, E.Y., et al., 2005. Proteome analysis of liver cells expressing a full-length hepatitis C virus (HCV) replicon and biopsy specimens of post-transplantation liver from HCV-infected patients. J Virol 79, 7558–7569.

Kang, S.M., Shin, M.J., Kim, J.H., Oh, J.W., 2005. Proteomic profiling of cellular proteins interacting with the hepatitis C virus core protein. Proteomics 5, 2227–2237.

Kapadia, S.B., Chisari, F.V., 2005. Hepatitis C virus RNA replication is regulated by host geranylgeranylation and fatty acids. Proc Natl Acad Sci USA 102, 2561–2566.

Karayiannis, P., 2005. The hepatitis C virus NS3/4A protease complex interferes with pathways of the innate immune response. J Hepatol 43, 743–745.

Kato, H., Takeuchi, O., Sato, S., et al., 2006. Differential roles of MDA5 and RIG-I helicases in the recognition of RNA viruses. Nature 441, 101–105.

King, J.K., Yeh, S.H., Lin, M.W., et al., 2002. Genetic polymorphisms in interferon pathway and response to interferon treatment in hepatitis B patients: A pilot study. Hepatology 36, 1416–1424.

Kohashi, T., Maekawa, S., Sakamoto, N., et al., 2006. Site-specific mutation of the interferon sensitivity-determining region (ISDR) modulates hepatitis C virus replication. J Viral Hepat 13, 582–590.

Krawczynski, K., Aggarwal, R., Kamili, S., 2000. Hepatitis E. Infect Dis Clin North Am 14, 669–687.

Kuramitsu, Y., Nakamura, K., 2005. Current progress in proteomic study of hepatitis C virus-related human hepatocellular carcinoma. Expert Rev Proteomics 2, 589–601.

Lanford, R.E., Bigger, C., Bassett, S., Klimpel, G., 2001. The chimpanzee model of hepatitis C virus infections. ILAR J 42, 117–126.

Lau, J.Y., Wright, T.L., 1993. Molecular virology and pathogenesis of hepatitis B. Lancet 342, 1335–1340.

Lee, J.Y., Locarnini, S., 2004. Hepatitis B virus: Pathogenesis, viral intermediates, and viral replication. Clin Liver Dis 8, 301–320.

Lee, S.D., 2003. Recent advance on viral hepatitis A. J Chin Med Assoc 66, 318–322.

Lee, W.M., 1997. Hepatitis B virus infection. N Engl J Med 337, 1733–1745.

Lindenbach, B.D., Evans, M.J., Syder, A.J., et al., 2005. Complete replication of hepatitis C virus in cell culture. Science 309, 623–626.

Lok, A.S., 2000. Hepatitis B infection: Pathogenesis and management. J Hepatol 32, 89–97.

Maddrey, W.C., 2000. Hepatitis B: An important public health issue. J Med Virol 61, 362–366.

Mansfield, E.S., Sarwal, M.M., 2004. Arraying the orchestration of allograft pathology. Am J Transplant 4, 853–862.

Marcello, T., Grakoui, A., Barba-Spaeth, G., et al., 2006. Interferons alpha and lambda inhibit hepatitis C virus replication with distinct signal transduction and gene regulation kinetics. Gastroenterology 131, 1887–1898.

Martin, A., Lemon, S.M., 2006. Hepatitis A virus: From discovery to vaccines. Hepatology 43, S164–S172.

Martin, M.P., Carrington, M., 2005. Immunogenetics of viral infections. Curr Opin Immunol 17, 510–516.

McCaughan G.W., Xiao X.H., Zekry A., Beard M.R., Williams R., Gorrell M.D., 2003. The intrahepatic response to HCV infection. Proceedings of the 11th International Symposium on Viral Hepatitis and Liver Disease, April 2003, Sydney, Australia, pp. 100–106.

McCaughan, G.W., Zekry, A., 2000. Effects of immunosuppression and organ transplantation on the natural history and immunopathogenesis of hepatitis C virus infection. Transpl Infect Dis 2, 166–185.

McCaughan, G.W., Zekry, A., 2004. Mechanisms of HCV reinfection and allograft damage after liver transplantation. J Hepatol 40, 368–374.

Merican, I., Guan, R., Amarapuka, D., et al., 2000. Chronic hepatitis B virus infection in Asian countries. J Gastroenterol Hepatol 15, 1356–1361.

Mondelli, M.U., Cerino, A., Cividini, A., 2005. Acute hepatitis C: Diagnosis and management. J Hepatol 42 (Suppl.), S108–S114.

Moyer, L.A., Mast, E.E., 1994. Hepatitis B: Virology, epidemiology, disease, and prevention, and an overview of viral hepatitis. Am J Prev Med 10 (Suppl.), 45–55.

Pagliaccetti, N.E., Eduardo, R., Kleinstein, S.H., Mu, X.J., Bandi, P., Robek, M.D., 2008. Interleukin-29 functions cooperatively with interferon to induce antiviral gene expression and inhibit hepatitis C virus replication. J Biol Chem 283, 30,079–30,089.

Panda S.K., Thakral D., Rehman S., 2006. Hepatitis E virus. Rev Med Virol 17(3), 151–180.

Pascu, M., Martus, P., Hohne, M., et al., 2004. Sustained virological response in hepatitis C virus type 1b infected patients is predicted by the number of mutations within the NS5A-ISDR: A meta-analysis focused on geographical differences. Gut 53, 1345–1351.

Patel, K., Muir, A.J., McHutchison, J.G., 2006. Diagnosis and treatment of chronic hepatitis C infection. BMJ 332, 1013–1017.

Pileri, P., Uematsu, Y., Campagnoli, S., et al., 1998. Binding of hepatitis C virus to CD81. Science 282, 938–941.

Poon, T.C., Hui, A.Y., Chan, H.L., et al., 2005. Prediction of liver fibrosis and cirrhosis in chronic hepatitis B infection by serum proteomic fingerprinting: A pilot study. Clin Chem 51, 328–335.

Radaeva, S., Jaruga, B., Hong, F., et al., 2002. Interferon-alpha activates multiple STAT signals and down-regulates c-Met in primary human hepatocytes. Gastroenterology 122, 1020–1034.

Randall, G., Chen, L., Panis, M., et al., 2006. Silencing of USP18 potentiates the antiviral activity of interferon against hepatitis C virus infection. Gastroenterology 131, 1584–1591.

Rauch, A., Kutalik, Z., Descombes, P., et al., 2010. Genetic variation in IL28B is associated with chronic hepatitis C and treatment failure: A genome-wide association study. Gastroenterology 138, 1338–1345. 1345 e1331–e1337.

Schinkel, J., Spoon, W.J., Kroes, A.C., 2004. Meta-analysis of mutations in the NS5A gene and hepatitis C virus resistance to interferon therapy: Uniting discordant conclusions. Antivir Ther 9, 275–286.

Servoss, J.C., Friedman, L.S., 2004. Serologic and molecular diagnosis of hepatitis B virus. Clin Liver Dis 8, 267–281.

Shackel, N.A., McCaughan, G.W., 2007. Intrahepatic interferon-stimulated gene responses: Can they predict treatment responses in chronic hepatitis C infection? Hepatology 46, 1326–1328.

Shackel, N.A., McGuinness, P.H., Abbott, C.A., Gorrell, M.D., McCaughan, G.W., 2001. Identification of novel molecules and pathogenic pathways in primary biliary cirrhosis: cDNA array analysis of intrahepatic differential gene expression. Gut 49, 565–576.

Shackel, N.A., McGuinness, P.H., Abbott, C.A., Gorrell, M.D., McCaughan, G.W., 2002. Insights into the pathobiology of hepatitis C virus-associated cirrhosis: Analysis of intrahepatic differential gene expression. Am J Pathol 160, 641–654.

Shimotohno, K., 2000. Hepatitis C virus and its pathogenesis. Semin Cancer Biol 10, 233–240.

Shirota, Y., Kaneko, S., Honda, M., Kawai, H.F., Kobayashi, K., 2001. Identification of differentially expressed genes in hepatocellular carcinoma with cDNA microarrays. Hepatology 33, 832–840.

Siegel, C.A., Silas, A.M., Suriawinata, A.A., Van Leeuwen, D.J., 2005. Liver biopsy 2005: When and how? Clevel Clin J Med 72, 199–201. 206, 208.

Simmonds, P., 1996. Virology of hepatitis C virus. Clin Ther 18 (Suppl. B), 9–36.

Smith, M.W., Yue, Z.N., Korth, M.J., et al., 2003. Hepatitis C virus and liver disease: Global transcriptional profiling and identification of potential markers. Hepatology 38, 1458–1467.

Su, A.I., Pezacki, J.P., Wodicka, L., et al., 2002. Genomic analysis of the host response to hepatitis C virus infection. Proc Natl Acad Sci USA 99, 15,669–15,674.

Suppiah, V., Moldovan, M., Ahlenstiel, G., et al., 2009. IL28B is associated with response to chronic hepatitis C interferon-alpha and ribavirin therapy. Nat Genet 41, 1100–1104.

Takashima, M., Kuramitsu, Y., Yokoyama, Y., et al., 2003. Proteomic profiling of heat shock protein 70 family members as biomarkers for hepatitis C virus-related hepatocellular carcinoma. Proteomics 3, 2487–2493.

Takashima, M., Kuramitsu, Y., Yokoyama, Y., et al., 2005. Overexpression of alpha enolase in hepatitis C virus-related hepatocellular carcinoma: Association with tumor progression as determined by proteomic analysis. Proteomics 5, 1686–1692.

Tanaka, Y., Nishida, N., Sugiyama, M., et al., 2009. Genome-wide association of IL28B with response to pegylated interferon-alpha and ribavirin therapy for chronic hepatitis C. Nat Genet 41, 1105–1109.

Tanikawa, K., 2004. Pathogenesis and treatment of hepatitis C virus-related liver diseases. Hepatobiliary Pancreat Dis Int 3, 17–20.

Taylor, J.M., 2006a. Structure and replication of hepatitis delta virus RNA. Curr Top Microbiol Immunol 307, 1–23.

Taylor, J.M., 2006b. Hepatitis delta virus. Virology 344, 71–76.

Taylor, M.W., Tsukahara, T., Brodsky, L., et al., 2007. Changes in gene expression during pegylated interferon and ribavirin therapy of chronic hepatitis C virus distinguish responders from nonresponders to antiviral therapy. J Virol 81, 3391–3401.

Thio, C.L., Thomas, D.L., 2010. Interleukin-28b: A key piece of the hepatitis C virus recovery puzzle. Gastroenterology 138, 1240–1243.

Thomas, D.L., 2000. Hepatitis C epidemiology. Curr Top Microbiol Immunol 242, 25–41.

Thomas, D.L., Seeff, L.B., 2005. Natural history of hepatitis C. Clin Liver Dis 9, 383–398. vi.

Thomas, D.L., Thio, C.L., Martin, M.P., et al., 2009. Genetic variation in IL28B and spontaneous clearance of hepatitis C virus. Nature 461, 798–801.

Tsukiyama-Kohara, K., Tone, S., Maruyama, I., et al., 2004. Activation of the CKI-CDK-Rb-E2F pathway in full genome hepatitis C virus-expressing cells. J Biol Chem 279, 14,531–14,541.

Wang, F.S., 2003. Current status and prospects of studies on human genetic alleles associated with hepatitis B virus infection. World J Gastroenterol 9, 641–644.

Wasley, A., Alter, M.J., 2000. Epidemiology of hepatitis C: Geographic differences and temporal trends. Semin Liver Dis 20, 1–16.

Watanabe, K., Yoshioka, K., Yano, M., et al., 2005. Mutations in the nonstructural region 5B of hepatitis C virus genotype 1b: Their relation to viral load, response to interferon, and the nonstructural region 5A. J Med Virol 75, 504–512.

Weston, S.R., Martin, P., 2001. Serological and molecular testing in viral hepatitis: An update. Can J Gastroenterol 15, 177–184.

Wieland, S., Thimme, R., Purcell, R.H., Chisari, F.V., 2004a. Genomic analysis of the host response to hepatitis B virus infection. Proc Natl Acad Sci USA 101, 6669–6674.

Wieland, S.F., Chisari, F.V., 2005. Stealth and cunning: Hepatitis B and hepatitis C viruses. J Virol 79, 9369–9380.

Wieland, S.F., Spangenberg, H.C., Thimme, R., Purcell, R.H., Chisari, F.V., 2004b. Expansion and contraction of the hepatitis B virus transcriptional template in infected chimpanzees. Proc Natl Acad Sci USA 101, 2129–2134.

Williams, A.P., Bateman, A.R., Khakoo, S.I., 2005. Hanging in the balance: KIR and their role in disease. Mol Interv 5, 226–240.

Worm, H.C., Van der Poel, W.H., Brandstatter, G., 2002. Hepatitis E: An overview. Microbes Infect 4, 657–666.

Wu, C.G., Salvay, D.M., Forgues, M., et al., 2001. Distinctive gene expression profiles associated with Hepatitis B virus x protein. Oncogene 20, 3674–3682.

Yan, W., Lee, H., Yi, E.C., et al., 2004. System-based proteomic analysis of the interferon response in human liver cells. Genome Biol 5, R54.

Yee, L.J., 2004. Host genetic determinants in hepatitis C virus infection. Genes Immun 5, 237–245.

Yerly, D., Di Giammarino, L., Bihl, F., Cerny, A., 2006. Targets of emerging therapies for viral hepatitis B and C. Expert Opin Ther Targets 10, 833–850.

Yoffe, B., Burns, D.K., Bhatt, H.S., Combes, B., 1990. Extrahepatic hepatitis B virus DNA sequences in patients with acute hepatitis B infection. Hepatology 12, 187–192.

Yoneyama, M., Kikuchi, M., Matsumoto, K., et al., 2005. Shared and unique functions of the DExD/H-box helicases RIG-I, MDA5, and LGP2 in antiviral innate immunity. J Immunol 175, 2851–2858.

Yu, S.H., Nagayama, K., Enomoto, N., Izumi, N., Marumo, F., Sato, C., 2000. Intrahepatic mRNA expression of interferon-inducible antiviral genes in liver diseases: dsRNA-dependent protein kinase overexpression and RNase L inhibitor suppression in chronic hepatitis C. Hepatology 32, 1089–1095.

Zhu, H., Zhao, H., Collins, C.D., et al., 2003. Gene expression associated with interferon alfa antiviral activity in an HCV replicon cell line. Hepatology 37, 1180–1188.

Malaria

Nadia Ponts and Karine G. Le Roch

INTRODUCTION

Malaria is a mosquito-borne disease caused by a eukaryotic protozoan parasite of the genus *Plasmodia*. With 350–500 million cases leading to up to one million deaths per year, malaria remains one of the deadliest infectious diseases in the world. It is widespread in all tropical and subtropical areas. In regions where levels of malaria transmission are consistently high, naturally acquired immunity (NAI) can be observed. In such areas of continuous exposure to malaria, pregnant women, whose immunities are drastically reduced (sometimes even absent), and young children, who have not yet developed their own immunity, are subject to the most severe forms of malaria. Ninety percent of malaria-related deaths occur in children under five years old in sub-Saharan African countries. Malaria is also associated with poverty, and is a major impediment to economic development.

Humans can be infected by four different species of *Plasmodium*: *Plasmodium malariae*, *Plasmodium ovale*, *Plasmodium vivax*, and *Plasmodium falciparum*. Human cases of lethal or febrile malaria have also occurred recently with *Plasmodium knowlesi*, a parasite usually found in macaques in certain forested areas of Southeast Asia, making *P. knowlesi* the fifth species known to cause malaria in humans (Sabbatani et al., 2010). *P. falciparum and P. vivax* are the most common strains infecting humans, *P. falciparum* being responsible for the most severe malignant malaria, leading to coma and death.

Despite the lack of a vaccine, malaria is curable. Traditional medicinal plants have been used for treatment since antiquity. Native Peruvians used the bark of the *Cinchona succirubra* tree for centuries before quinine was isolated from it in 1820. Chloroquine was derived from quinine and became the prophylactic treatment for malaria in 1947. It was the most effective and inexpensive treatment at the time. With cheap access to chloroquine, the World Health Organization (WHO) and the US Centers for Disease Control and Prevention (CDC) launched a worldwide malaria eradication campaign. The combined use of chloroquine and insecticides (such as dichlorodiphenyltrichloroethane, or DDT) in designated malaria-infected areas resulted in the elimination of endemic malaria in developed countries (Europe and North America) and a significant reduction of cases in developing parts of the world. Unfortunately, the emergence of chloroquine-resistant parasites, resistance to "new synthetic" drugs (mefloquine, doxycycline, atovaquone, proguanil hydrochloride) and DDT-resistant mosquitoes led to the reappearance and spread of malaria in most of the developing world. With the lack of concern, funding, or new tools to combat the disease, the eradication campaign failed dramatically in the seventies and eighties. In 1972, artemisinin, a natural product with antimalarial properties used in traditional Chinese

Genomic and Personalized Medicine, 2nd edition
by Ginsburg & Willard. DOI: http://dx.doi.org/10.1016/B978-0-12-382227-7.00100-2

medicine for more than two thousand years, was isolated from *Artemisia annua*. After years of research and clinical trials, artemisinin-based combination therapies (ACTs) were recently adopted, and they are currently our last resort in combating malarial infection. Regrettably, ACT-resistant strains have recently been identified in Southeast Asia. With malaria's huge impact on social and economic issues at both the individual and governmental levels, the absence of an efficient vaccine and the spread of parasite resistance to all antimalarial drugs, there is a dire need for development of new, cost-effective antimalarial strategies.

To identify innovative and long-lasting strategies to eradicate malaria, it is critical to understand both the pathogen and host responses to infection. The availability of the fully sequenced *Plasmodium* and human genomes, the rapid development of genomics, and the recent advent and easy access to next-generation sequencing technologies have started to complete and modernize our understanding of the malaria parasite and its interactions with its hosts. If well-funded and managed, these approaches could revolutionize the way we think about medicine and treatments to eradicate infectious diseases in developing countries. In this chapter, we present the recent genomic approaches that have led to major advances in the understanding of the pathogen and its host as well as their complex interactions. We discuss the impact of these approaches on the development of new therapeutic strategies in the near future, and possible long-term access to personalized medicine in developing countries.

MALARIA AND ITS CAUSAL AGENT
Life Cycle of the Human Malaria Parasite

The malaria parasite has a complex life cycle (Figure 100.1). The parasite is transmitted by the bite of an infected female *Anopheles* mosquito. When a mosquito bites a person, infective sporozoites enter the bloodstream from the mosquito's saliva and quickly invade liver cells. The parasite matures within a hepatocyte into a schizont containing thousands of merozoites. This phase of the disease is generally asymptomatic. After 7–10 days – sometimes years for *P. vivax* and *P. ovale*, which can remain dormant in their hypnozoite form – merozoites are released into the bloodstream, where they invade circulating red blood cells. Parasites mature in ring, trophozoite, and schizont stages in 48–72 hours, depending on the species, to produce 16–32 new daughter cells called merozoites, which are then released to infect fresh red blood cells. This phase of the parasite's life cycle is symptomatic, and can cause fever, headache, vomiting, anemia, hallucinations, coma, and eventually death. The pernicious and exponential parasite growth inside the red blood cells can continue until recovery or death. Following exposure to stress such as fever or anemia, parasites undergo a sexual development to form sexually mature female and male gametocytes. When a mosquito takes a blood meal from an infected individual,

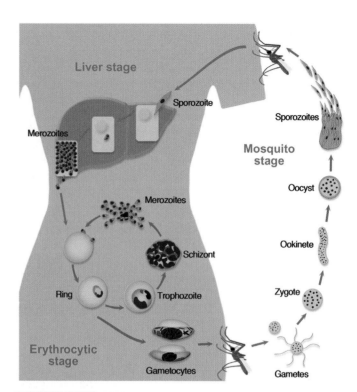

Figure 100.1   The human malaria parasite life cycle. Infective sporozoites enter the bloodstream from the saliva of a female *Anopheles* mosquito and quickly invade liver cells. After ~10 days of differentiation and multiplication, merozoites are released into the bloodstream, where they invade circulating red blood cells. Over the next 48 hours in the case of *P. falciparum*, parasites mature through ring, trophozoite and schizont stages; at this point, 16–32 new daughter cells are released to infect additional red blood cells. The pernicious process continues until recovery or death. Following exposure to stress, parasites in the trophozoite stage follow a different developmental pathway and form sexually mature gametocytes. A mosquito taking a blood meal from an infected person ingests these gametocytes, and sexual reproduction takes place in the mosquito midgut. After maturation and multiplication, sporozoites ultimately pass into the insect's salivary glands and are injected into a new human host when the infected mosquito feeds, beginning a new infectious cycle.

gametocytes are ingested and sexual reproduction starts in the mosquito midgut. After maturation and multiplication, the infected mosquito transmits the disease when biting another human host and the infectious cycle repeats. Understanding the key molecular mechanisms that control the parasite's progression throughout these stages is critical to designing successful antimalarial strategies.

Genomics for Diagnosis

In the absence of a correct diagnosis and prompt treatment, severe malaria leading to death may occur. The first symptoms of malaria are attributed to the erythrocytic stage of the cycle.

They are non-specific and resemble the ones provoked by viral infections such as flu viruses. Thus, diagnosis cannot be solely based on symptoms but is usually pronounced after microscopic examination of blood smears from infected patients (the *Gold Standard*). Malaria rapid diagnostic tests (RDTs) that detect specific malaria-related antigens in infected individuals are also available, and their use is recommended by the WHO to assist in the diagnosis of malaria (see http://www.wpro.who.int/sites/rdt/home.htm for details). However, determination of the exact invading species in a patient, especially in the case of multi-infections, cannot be accomplished by such methods. Nonetheless, available genomic sequences and markers are now easily used to identify parasites in blood samples from patients. Polymerase chain reaction (PCR)-based methods have been extensively developed and optimized to complement standard malaria diagnostic tests and allow better diagnoses (Cnops et al., 2010; Zakeri et al., 2010). Because of their costs and requirements (equipment etc.), however, these methods can only be used as confirmatory tests and in the case of epidemiological surveys, response to drug treatments studies, etc. (Parija, 2010). Considering the constant diminution of diagnostic costs, the increasing robustness of the techniques, and the miniaturization of electronics and equipment, the widespread implementation of such methods and their application for field diagnostic should be nonetheless achievable.

Genome-based (Re-)Discovery of Virulence Genes

The malaria parasite can persist in the infected host for prolonged periods by continuously evading the human immune system. *Plasmodium* uses antigenic variation to alter its surface proteins. In addition to allowing immune evasion, this allows re-infection and persistence, since the parasite is poorly recognized by the host immune system. In *P. falciparum*, the *Var* family of ~60 genes codes for the clonally variant surface molecule PfEMP1 (*P. falciparum* erythrocyte membrane protein 1) (Scherf et al., 2008), and the *Var* family genes are mostly found in the sub-telomeric regions of the chromosomes. *Var* genes are expressed one at a time in each individual parasite by allelic exclusion. The host immune system develops an antibody-mediated immune response against the expressed PfEMP1 (immunodominant). Then, sub-populations of the parasite change their antigenic profile and rise in frequency. The antibodies cannot recognize the new variants – the parasites evade the host immune system and continue the infection. *Var* genes are major virulence factors in *P. falciparum*. They are involved in cytoadherence (Su et al., 2009), i.e., the adhesion of infected red blood cells to the endothelium, a major cause of severe forms of malaria. The sequestered parasites escape spleen clearance, and the infection is maintained until sexual maturation and uptake by a mosquito. There is no *Var* homolog in *P. vivax*. Nonetheless, genome sequence analysis revealed the presence of a variant sub-telomeric family of *Vir* genes that could also be virulence factors (Del Portillo et al., 2004).

The genomic organization of *Var* genes and the mechanisms underlying antigenic variations are extensively studied (Scherf et al., 2008). *Var* genes contain one single intron and are present in highly polymorphic regions of the genome. Their organization is conserved among different strains, but they carry many single nucleotide polymorphisms (SNPs). Recently, all PfEMP1 sequences from seven different genomes of *P. falciparum* were compared (Rask et al., 2010). The result is the identification of ordered domains, including conserved cassettes and recombination hotspots with important potential consequences on variations in virulence among genetically distinct strains of *P. falciparum*. The physical tethering of recombinant regions to the nuclear periphery also very likely influences the location of these recombination hotspots (Scherf et al., 2008). Altogether, the analysis of the *P. falciparum* genome has revealed an enormous potential diversity among the *Var* genes that allows continuous antigenic variation and persistence of the infection.

This gene-switching has crucial consequences in terms of PfEMP1-based vaccine development. There is currently not enough knowledge and understanding of the mechanisms underlying antigenic variation to propose an anti-PfEMP1 vaccine (Dahlbäck et al., 2010). A major difficulty in studying virulence genes is the rapid adaptation of the model parasite strains used in the research labs, leading to important discrepancies when *in vitro* results are transposed to the field. The complete genome sequences for both *P. falciparum* and the human host, coupled with the rapid development of affordable deep sequencing methods, are tools to circumvent such limitations, as illustrated in a recent study where the transcription profiles of *Var* genes were directly studied in *P. falciparum* patient isolates (Blomqvist et al., 2010).

PATHOGENESIS IN THE HUMAN HOST AND IMMUNITY
Host Genetic Factors and Susceptibility/Resistance to Malaria

The idea of personalized medicine has been routinely taken into consideration for many years in the practice of medicine. For any type of disease, including infectious diseases, physicians seek information about the patient's environment, family history and ethnicity. The collected information is then used to determine more precisely the possible evolution of the pathology in the light of eventual relevant genetic factors, to optimize the therapeutic care to be provided. Several independent genetic loci of susceptibility or resistance to diseases have emerged in the recent history of elucidation of the human genome. In the case of malaria, evidence has accumulated over the past few decades demonstrating strong genomic variations in susceptibility/resistance to malaria. For example, certain tribes in Africa are more resistant than others to severe malaria (Modiano et al., 1996). Many determinants of malaria contribute to such differences, both non-genetic and genetic.

They include levels of exposure, the existence of co-infections, and the host's and the parasite's genetics. The synergistic or antagonistic effects among these factors are complex. To untangle genetic from non-genetic factors, cases of malaria infections among families or between twins are frequently surveyed. For example, a cohort study of 258 pairs of twins in Gambia showed that infection is largely under the influence of environmental factors, whereas genetic determinants influence the susceptibility to malaria-fever episodes (Jepson et al., 1995). Identifying and quantifying the role of genetic and non-genetic factors in the resistance/susceptibility to the various aspects of the disease using case-study analyses is not trivial, the first pitfall being to determine the quantifiable metrics that can be measured to assess resistance (e.g., parasitemia, intensity of clinical symptoms, etc.). The use of genetic analyses has provided accurate insights with regard to the genetic factors involved in resistance/susceptibility to malaria. For example, a case-control study involving more than 6000 individuals from various countries found that a single mutation in a mediator protein in the Toll-like-mediated signal transduction pathway, involved in initiating the innate immune response to various pathogens, has a protective effect against severe forms of four different infectious diseases, including malaria (Khor et al., 2007). Very recently, a genome-wide association (GWA) study identified 139 SNPs dispersed in 100 different susceptible loci in the genomes of 1060 Gambian children with severe malaria (Jallow et al., 2009; Table 100.1). Among them, the hemoglobin S (*HbS*) locus, which is involved in sickle cell disease (SCD), is the strongest known determinant against severe malaria due to *P. falciparum*. This locus is probably the most- and best-studied case of innate immunity against malaria.

Human host innate immunity refers to the level of the original resistance/susceptibility of an individual, as encoded within the genome. It is constant and generic, as opposed to adaptive immunity, which is activated in a manner specifically designed to target the infecting pathogen and (eventually) leads to acquired immunity. Historically, innate immunity against malaria has been widely illustrated by the association of natural resistance to various blood-related disorders. Indeed, blood-stage parasites are obligate intra-erythrocytic pathogens, and modifications of the structure and/or biochemistry of the red blood cells have drastic consequences on the parasite.

Blood Diseases and Innate Immunity

The first example of a blood disease trait that confers relative resistance to malaria was published in 1954 and involved *HbS*, the allele of the *HbB* gene encoding β-globin that is involved in SCD (Allison, 1954). In that study, young Ugandan children carrying *HbS* had reduced malaria parasite counts. SCD is an autosomal recessive genetic disorder that is widespread among people whose ancestors originated in African malaria-endemic regions. *HbS* homozygotes develop SCD with lethal consequences, where abnormal hemoglobin deforms erythrocytes until they rupture, causing anemia, blood vessel obstruction, and general ischemic damage. *HbS* heterozygotes, however, show enhanced resistance to severe malaria, with a 10-fold reduced risk (Ackerman et al., 2005; Hill et al., 1991). Two hypotheses are proposed to explain this resistance: the moderate hematological symptoms of heterozygous SCD can impede the parasite's growth, or that the parasite may be more efficiently eliminated through the removal of deformed erythrocytes by the human immune system (Akide-Ndunge et al., 2003). In the case of the second hypothesis, the protective effect against malaria involves the reduction of cytoadherence (Cholera et al., 2008), leading to improved clearance of the infected erythrocytes, mainly in the spleen (Hebbel, 2003). Other hemoglobin mutations have been found around the globe (Kwiatkowski, 2000). For example, the most pre-eminent *HbB* variant found in Southeast Asia, where the *HbS* variant is rare, is the *HbE* variant. *HbS* is typical in sub-Saharan African and a few Middle East regions, whereas *HbC* is localized in Central West Africa, at frequencies lower than *HbS*. The protective effect of *HbE* from severe malaria is suspected but not proven. *HbC*, on the contrary, is definitely associated with relative resistance in regions where *HbS* is rare (Agarwal et al., 2000), possibly by reducing the cytoadherence of infected erythrocytes and enhancing their clearance (Arie et al., 2005; Fairhurst et al., 2005; Tokumasu et al., 2009). Recent GWA studies nonetheless suggest that *HbS* is probably the strongest protective trait of them all by far, with an odds ratio down to 0.08 (May et al., 2007).

There are many regional genetic variants that confer different levels of resistance/susceptibility to malaria. A striking one is the case of the Duffy blood group negative phenotype caused by a SNP in the Fy glycoprotein, or FY (Kwiatkowski, 2000), also known as the Duffy antigen/chemokine receptor (DARC). FY is a glycosylated membrane protein expressed at the surface of mature erythrocytes that serves as a receptor for various chemokines including the Duffy antigen. FY is also a receptor for the invasive forms (merozoites) of the malaria parasites *P. vivax* and *P. knowlesi*. The Duffy blood group negative phenotype is caused by the occurrence of an SNP in the *Fy* locus (the *Fyb* allele) that is very frequent in black populations of African origin (88–100% in Black Africans, 68% in African Americans), but extremely rare in white populations (Levinson and Jawetz, 1996). Individuals with the *Fyb* allele are resistant to *P. vivax*, which explains why *P. vivax* infections are rare in most sub-Saharan populations, whereas they are common in Asian and South American populations.

Other cases of increased resistance to malaria are commonly encountered in regions where other mutation-related blood diseases are frequent, including α- and β-thalassemia (Allen et al., 1997; Flint et al., 1986; Williams et al., 1996), caused by mutations in the α- and β-globin genes respectively. Thalassemias represent a major health problem. Clinical manifestations of thalassemias are very diverse, ranging from mild anemia to fatal outcomes. The Haldane Malaria Hypothesis first proposed, in 1949, that malaria was the selective pressure behind the high frequencies of thalassemia (Haldane, 1949). Various population genetics studies indicated a negative association between thalassemia and malaria. The protective effect of α-thalassemia appears to be restricted to malarial

| TABLE 100.1 | Sample of human genomic loci associated with resistance/susceptibility to severe malaria |

Human chr	Position	SNP ID*	Association (*P*-value)	Odds ratio	Within gene	Nearby genes (within 150 kb, listed by shortest to longest distance)
1	46,033,863	rs6429582	6.0×10^{-5}	1.42 (1.20–1.68)	MAST2	IPP
	46,058,254	rs10890361	9.4×10^{-6}	1.46 (1.24–1.73)		
	46,068,168	rs10890363	9.1×10^{-5}	1.40 (1.18–1.66)	MAST2	IPP, PIK3R3
	46,075,900	rs7556436	5.2×10^{-5}	1.41 (1.20–1.67)	MAST2	PIK3R3
	72,957,189	rs10889990	5.1×10^{-5}	1.29 (1.14–1.46)	–	–
	72,957,277	rs10889991	7.7×10^{-5}	1.28 (1.13–1.45)		
	76,147,441	rs12405994	3.8×10^{-7}	0.31 (0.19–0.51)	–	ASB17, MSH4, ST6GALNAC3
	76,149,217	rs1417402	5.0×10^{-7}	0.31 (0.19–0.51)		
2	231,702,824	rs2136477	5.2×10^{-5}	0.70 (0.59–0.83)	SPATA3	PSMD1, GPR55, HTR2B, ITM2C, CR749389
	231,704,001	rs1543061	9.6×10^{-7}	0.65 (0.55–0.78)		
	231,704,245	rs10192428	5.1×10^{-7}	0.65 (0.55–0.77)		
	231,705,617	rs6750230	1.6×10^{-5}	0.68 (0.57–0.81)		
	231,705,813	rs6750277	6.6×10^{-5}	0.69 (0.58–0.83)		
3	43,107,448	rs488069	7.6×10^{-7}	1.42 (1.24–1.64)	C3orf39	AK126791
4	26,180,252	rs2046784	2.5×10^{-5}	0.76 (0.66–0.86)	–	CCKAR
	26,180,426	rs2130250	3.8×10^{-5}	0.76 (0.67–0.87)		
	107,663,169	rs2949632	6.5×10^{-6}	0.74 (0.65–0.84)	–	SCYE1, MGC16169, AK130051, AK098157
	107,668,048	rs2949641	1.3×10^{-5}	0.74 (0.65–0.85)		
5	43,040,037	rs316414	4.5×10^{-7}	0.60 (0.49–0.73)	–	LOC389289, LOC153684, AK128204, AK075204, ZNF131
	43,052,407	rs222060	7.4×10^{-5}	0.68 (0.57–0.83)		
	43,129,786	rs316413	7.9×10^{-5}	0.68 (0.56–0.83)	–	ZNF131, FLJ10246, LOC389289, MGC42105, AK128204
7	50,396,921	rs1451375	9.1×10^{-5}	0.70 (0.58–0.84)	DDC	AK057400, GRB10, FIGNL1, BC027856
	50,397,660	rs10249420	6.8×10^{-5}	0.69 (0.57–0.83)		
	50,400,107	rs7803788	7.9×10^{-5}	0.69 (0.58–0.83)		
	131,522,141	rs10269601	1.1×10^{-5}	0.75 (0.66–0.85)	–	PLXNA4B
	131,526,982	rs4731867	9.0×10^{-5}	1.29 (1.14–1.47)	PLXNA4B	
8	13,601,864	rs2410094	6.0×10^{-5}	0.38 (0.22–0.63)	–	FLJ25402
	13,604,374	rs1384057	2.4×10^{-5}	0.32 (0.18–0.57)		
10	23,134,834	rs946961	6.9×10^{-6}	0.54 (0.40–0.71)	–	PIP5K2A, ARMC3
	23,138,098	rs11013140	5.6×10^{-6}	0.53 (0.40–0.71)		
	101,094,702	rs10786538	2.5×10^{-5}	1.51 (1.24–1.82)	CNNM1	GOT1, HPSE2
	101,094,992	rs11190062	1.3×10^{-5}	1.53 (1.26–1.85)		
	126,328,434	rs7076268	1.6×10^{-5}	0.67 (0.56–0.81)	FAM53B	LHPP, LOC399818
	126,389,556	rs4962686	7.3×10^{-5}	0.70 (0.59–0.84)	FAM53B	LOC399818, FAM175B, LHPP
11	4,672,522	rs7105156	7.1×10^{-6}	1.35 (1.18–1.53)	OR51E2	OR51E1, BC022401, OR51D1, OR51F1, TRIM68, BC075058, AF439153, OR52I1, OR52I2, OR52R1, C11orf40, 37196, OR51F2, OR52M1
	4,976,561	rs16908208	1.4×10^{-5}	0.62 (0.50–0.78)	–	OR51L1, MMP26, OR51A2, OR52J3, OR51A4, OR52E2, OR51G1, OR51G2, OR51A7, OR51T1, OR52A4, OR52A5, OR51S1

(continued)

TABLE 100.1	Continued				
Human chr	Position	SNP ID*	Association (*P*-value)	Odds ratio	Within gene	Nearby genes (within 150 kb, listed by shortest to longest distance)
	5,017,517	rs2034839	4.3×10^{-5}	1.30 (1.14–1.47)	–	*OR52J3, OR52E2, OR51L1, MMP26,*
	5,018,117	rs2499984	7.7×10^{-5}	1.28 (1.13–1.45)		*OR52A4, OR51A2, OR52A5,*
	5,023,934	rs2500012	4.4×10^{-5}	1.29 (1.14–1.45)		*OR51A4, OR52A1, OR51G1,*
	5,023,965	rs2500013	8.3×10^{-5}	1.27 (1.13–1.43)		*OR51G2, OR51A7*
	5,029,411	rs2959184	2.9×10^{-5}	1.30 (1.15–1.46)		
	5,182,211	rs11036238	3.9×10^{-7}	0.61 (0.50–0.74)	–	*OR51V1, HBB, HBD, HBG1, HBG2, OR52A1, HBE1, OR52A5, OR52A4, OR51B4, OR51B2, OR51B5, OR52E2, OR51B6*
	5,265,472	rs11036635	4.8×10^{-6}	0.74 (0.64–0.84)	–	*OR51B4, HBE1, HBG2, OR51B2, HBG1, HBD, OR51B5, HBB, OR51B6, OR51V1, OR51M1, OR51Q1, OR52A1*
	5,329,266	rs4373970	6.7×10^{-6}	0.62 (0.50–0.77)	–	*OR51B6, OR51B5, OR51B2, OR51M1, OR51B4, OR51Q1, HBE1,OR51I1 HBG2, HBG1, OR51I2, HBD, HBB, OR52D1*
	5,362,699	rs4910780	2.3×10^{-5}	0.64 (0.52–0.79)	–	*OR51M1, OR51B6, OR51Q1, OR51B5, OR51I1, OR51B2, OR51I2, OR51B4, OR52D1, HBE1, UBQLN3, UBQLNL, MGC20470, HBG2*
	5,398,802	rs17359438	1.1×10^{-5}	0.62 (0.50–0.77)	–	*OR51Q1, OR51I1, OR51M1, OR51I2, OR52D1, OR51B6, OR51B5, UBQLN3, UBQLNL, MGC20470, OR51B2, OR51B4, OR52H1*
	5,420,001	rs16931041	1.1×10^{-5}	0.66 (0.55–0.80)	–	*OR51I1, OR51I2, OR51Q1, OR52D1, OR51M1, UBQLN3, UBQLNL, OR51B6, OR51B5, OR52H1, OR51B2, OR52B6, OR51B4*
	5,502,398	rs2010794	4.8×10^{-6}	0.71 (0.62–0.83)	–	*UBQLNL, UBQLN3, OR52H1, OR52D1, OR52B6, OR51I2, TRIM6, CR749260, TRIM6–TRIM34, OR51I1, TRIM34, OR51Q1, OR51M1, TRIM5*
13	92,497,204	rs1444227	3.6×10^{-5}	1.54 (1.25–1.89)	–	*GPC6, GPC5*
	92,510,870	rs1158878	4.2×10^{-5}	1.52 (1.25–1.86)		
14	45,436,312	rs17726637	1.2×10^{-5}	1.54 (1.27–1.88)	–	–
	45,504,335	rs6572335	9.3×10^{-7}	1.61 (1.33–1.95)		
	45,537,581	rs17728971	6.5×10^{-7}	1.63 (1.34–1.98)		
16	56,025,909	rs16987051	2.5×10^{-5}	0.66 (0.55–0.80)	*CIAPIN1*	*COQ9, CCL17, POLR2C, DOK4, CX3CL1, CCL22, CCDC102A*
	56,060,831	rs2161647	4.1×10^{-5}	0.69 (0.58–0.82)	*POLR2C*	*DOK4, COQ9, CIAPIN1, CCDC102A, CCL17, GPR114, CX3CL1*
19	6,892,161	rs460375	1.2×10^{-5}	0.66 (0.55–0.79)	–	*EMR1, EMR4, FLJ25758, VAV1, MBD3L2, ZNF557*
	6,892,867	rs465543	6.1×10^{-5}	0.68 (0.57–0.82)		

*Entry ID in data base SNP at http://www.ncbi.nlm.nih.gov/projects/SNP/.
Table adapted from Jallow et al., 2009.

anemia (Wambua et al., 2006). The protective mechanisms of thalassemias are still unclear. Possible mechanisms include reduced parasite growth in erythrocytes with a thalassemia phenotype, increased binding of antibodies from malaria-immune sera, enhanced splenic clearance of infected erythrocytes, and possibly enhanced early exposure to the less virulent *P. vivax* to protect from later infection with the deadly *P. falciparum* (Kwiatkowski, 2000).

Glucose-6-phosphate dehydrogenase (G6PDH) deficiency, caused by polymorphisms of the erythrocyte enzyme gene *G6PD*, is also believed to provide a certain level of protection against severe malaria (Kwiatkowski, 2000). G6PDH is involved in the oxidative phase of the pentose phosphate pathway, generating the NADPH that is essential in red blood cells to prevent the accumulation of peroxides leading to hemolysis and to the oxidation of ferrous hemoglobin (able to carry oxygen) into ferric hemoglobin (unable to carry oxygen). G6PDH deficiency is associated with protection against severe malaria, possibly through reduction of parasite replication and/or an enhanced phagocytosis of malaria-infected *G6PD*‾erythrocytes (López et al., 2010).

Clearly, consequences of malaria on mortality, especially child mortality, in malaria-endemic regions in general and Africa in particular had critical effects on the evolution of the human genome. The previous examples of gene variants that confer protection against malaria (summarized in Table 100.2) reflect that malaria exerts a very strong selective pressure in the host genome – the strongest known so far. For example, the *HbS* variant is present at very high frequencies in African malaria-endemic regions despite lethal consequences for homozygotic individuals. The high evolutionary pressure of malaria is also illustrated by the diversity of protective genetic variants that have arisen around the world at the population and sub-population level, as shown by the distribution of the various *HbB*, *FY*, *G6PD*, etc. variants. Malaria is so tightly linked to human history that genetic analyses of these loci are used to retrace human migration events within and out of Africa. Recent events of positive selection seem to have emerged from the pressure of the highly virulent *P. falciparum*, and there are certainly many new protective traits to be

discovered and explored. Understanding the molecular basis of the protection conferred by these different genetic variants should provide a better understanding of malaria's pathogenesis and guide our discovery of efficient therapeutic strategies that take individual differences into account.

The Innate Immune Response

All individuals are not born equal in terms of immunity from malaria. The human innate immune response controls primary infection. The serum levels of various pro-inflammatory cytokines – including interferon (IFN)-γ, tumor necrosis factor (TNF), and the interleukins IL-8 and IL-12p40 – increase upon initial infection with *P. falciparum*. The mechanisms of innate immune response appear to be triggered above certain levels of parasite density, or peak parasitemia, in the blood. Peak parasitemia in a given individual is highly predictable for a given parasite strain (Molineaux et al., 2002). Variations can be attributed to differences among humans in making a rapid innate immune response. Innate immune mechanisms lead to parasite clearance up to parasitemias below the threshold that can activate the immune response (Stevenson and Riley, 2004). The parasite effector molecules that trigger the innate immune response seem to be the same among the various *Plasmodium* species (Langhorne et al., 2008), which indicates that variations in the immune response are either non-genetic or are encoded in the human host genome.

Several immunity-related genetic traits were identified that alter immune responses to malaria (Kwiatkowski, 2000). For example, IFN-γ and α receptor variants offer protection against the cerebral form of severe malaria (Aucan et al., 2003; Koch et al., 2002). On the contrary, the *TNF2* variation in the *TNF* promoter affects the binding of the OCT1 transcription factor, which leads to increased levels of serum TNF and enhanced susceptibility to cerebral malaria. However, very low levels of plasma TNF characterize severe malarial anemia and are associated with the variant *TNF-238A* (Knight and Kwiatkowski, 1999; McGuire et al., 1994). The role of the innate immune response is to contain the initial infection to prevent host death (limiting the severity of the infection) and give time for adaptive immunity to develop and proceed to complete clearance of the infection.

Gene	Product	Allele	Geographic distribution	Effect
HbB	β-globin	HbC	Central West Africa	Relative resistance
		HbE	Southeast Asia	Suspected protection against severe malaria
		HbS	Sub-Saharan Africa	Strong protection
			Localized Middle-East regions	Suspected protection against malarial anemia
		β-thalassemia SNPs	Mediterranean populations	Mediterranean populations
HbA	α-globin	α-thalassemia SNPs	Western African descents	Protection against malarial anemia
Fy	Duffy-antigen/chemokine receptor	Fyb	Black populations of African origins	Resistance to *P. vivax* infections
G6PD	glucose-6-phosphate dehydrogenase	G6PD A−	West Africa	Protection against severe malaria

TABLE 100.2 Synopsis of major blood disease traits that confer innate immunity to malaria

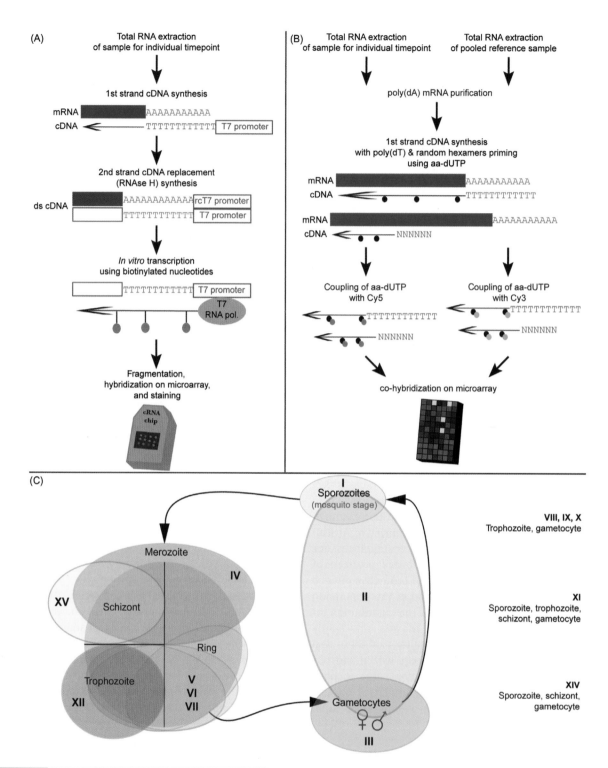

Figure 100.2   Initial transcriptomic approaches for the discovery of gene function. Samples were taken at different stages of the parasite's life cycle. (A) Photolithographic (Affymetrix®) whole-genome complementary RNA (cRNA) microarray approach from Le Roch et al., 2003. The chip consists of spotted 25 bp-long oligos, tiling the *P. falciparum* genome. Biotinylated cRNA probes are hybridized on the chip and revealed with a streptavidin-coupled dye. At each spot, the intensity of the signal (red dots on the chip) is proportional to the number of probes that hybridized, i.e., the amount of cRNA present in the sample. (B) Spotted microarray approach from Bozdech et al., 2003. Short oligos were spotted on the surface of a glass slide. Samples taken at individual time points are labeled with the red dye Cy5 and are hybridized against a Cy3-labeled (green) control (consisting of a pool of all the samples). Expression is quantified according to the Cy5/Cy3 intensity [spot color ranging from green to yellow to red, i.e., from genes down-regulated in the sample vs the control (green), to equally expressed (yellow), and over-expressed (red)]. (C) Representation of the functional mRNA level-based gene clustering from Le Roch et al., 2003. A total of 15 groups (I to XV) were delineated by K-means clustering. Genes within groups I and III are specifically expressed at the mosquito sporozoite and the sexual gametocyte stages, respectively. Group II contains a mix of sporozoite- and gametocyte-expressed genes. Group IV contains genes expressed at the schizont, merozoite, and ring intra-erythrocytic stages; groups V to VII correspond to ring and trophozoite stages; XII is specific to trophozoites, and XV is specific to schizont. Groups VIII to XI and XIV contain genes with more complex patterns of expression involving multiple stages.

The Adaptive Humoral and Cellular Immune Responses

The antibody- or cell-mediated immune responses to malarial infection are no better understood than the innate immune response. Immunoglobulin (Ig)G and IgE seem to be the major effectors of the humoral response, which is directed against blood-stage parasites through inhibition of merozoite invasion, growth inhibition, or enhanced parasite clearance (Perlmann and Troye-Blomberg, 2002). Cell-mediated immune response is targeted against blood-stage parasites and possibly against pre-erythrocytic stages. Since erythrocytes do not express major histocompatibility complex (MHC) antigens, CD8+ "cytotoxic" T cells cannot be involved in the defense against blood-stage parasites. Their role during protection from malaria is actually not clear. CD8+ cells may be involved in regulation, immunosuppression, and inflammation, or they may eventually be involved in protection against pre-erythrocytic parasites. Defense against blood-stage parasites is instead mediated by CD4+ "helper" T cells and various cytokines, such as the interleukin IL-4 and the interferon IFN-γ.

There are many known examples of adaptive immunity-related genetic traits that are associated with variations in susceptibility to malaria. Most of them were discovered by case studies involving individuals from various ethnic groups. The Fulani ethnic group (Burkina Faso, West Africa), for example, is known to be the least susceptible and the most immunologically responsive to malaria among neighboring ethnic groups. Fulani individuals produce very high levels of malaria-specific antibody (IgG and IgM) (Bolad et al., 2005). This exceptional manifestation of an antimalarial humoral response is associated with the interleukin IL-4 variant *IL4-524T* (Luoni et al., 2001). The human leukocyte antigen (HLA) class I allele *Bw53* in the MHC region is also widespread in West Africa, and is associated with reduced risk of severe malaria (Hill et al., 1991). In addition, the HLA class II haplotypes -DRB1 and -DQB1 are suspected to be linked to the resistance to malaria among individuals from the Fulani ethnic group (Lulli et al., 2009). The MHC region on chromosome 6 was shown to be involved in susceptibility to malaria-fever episodes (Jepson et al., 1995) and in the control of blood infection levels (Flori et al., 2003; Garcia et al., 1995; Rihet et al., 1998).

Host resistance to malaria is complex and spreads across both innate resistance and adaptation of the immune response. Innate resistance is crucial for the survival of young children who are still too young to have developed their own immunity and whose immunity inherited from their mothers has faded. On the other hand, the development of adaptive immune responses is crucial in older children and adults living in malaria-endemic areas. Understanding this host resistance is of major importance for the development of future antimalarial strategies. Unfortunately, the mechanisms of the immune response to malaria infection are still not fully understood. Many other genetic traits of resistance are surely to be discovered. The advent of high-throughput sequencing technologies fast-tracked whole genome discoveries of resistance traits. The past couple of years have seen exponential increase in discovery of genetic traits potentially related to a very large panel of phenotypes, and the malaria field is no exception, mainly on account of the coordinated efforts of the MalariaGEN (Malaria Genomic Epidemiology Network) consortium (http://www.malariagen.net; see Table 100.1 and Jallow et al., 2009).

GENOMICS FOR THE DEVELOPMENT OF NEW ANTIMALARIAL STRATEGIES

Ideally, antimalarial strategies are a balanced use of mosquito control measures, anti-*Plasmodium* treatments, and a general improvement of sanitation and awareness. This is how malaria was eradicated from North America and Europe in the early 1900s, and later was efficiently controlled in richer regions of South America and Asia. However, the malaria situation in Africa, especially sub-Saharan Africa, remains dramatic. As a consequence, there is a constant need to develop new antimalarial drugs to keep up with the constant emergence of resistance. In addition, vaccines are currently being developed to target different stages of the parasite's life cycle to prevent infection (pre-erythrocytic vaccines), to reduce the severity of the disease outcome (erythrocytic asexual stages), or to block transmission (gametocyte stages).

The availability of the complete sequences of the *P. falciparum* and *P. vivax* genomes (Carlton et al., 2008; Gardner et al., 2002), coupled with the exponential development of bioinformatics, has revolutionized the malaria research field. Research has been particularly active in the area of transcriptomics. Many approaches have provided an enormous quantity of information at various levels. In a very global manner, large-scale analyses of *P. falciparum*'s transcripts (microarray) allowed the discovery of expressed genes and their functional association to the various stages of the parasite (Figure 100.2, Bozdech et al., 2003; Le Roch et al., 2003). Genome-wide methods have also contributed to the identification and understanding of the various regulatory elements that rule gene expression in the parasite's genome (Elemento et al., 2007; Gonzales et al., 2008; Ponts et al., 2010; Westenberger et al., 2009). In particular, recent analyses suggested that genome-wide changes in chromatin architecture and nucleosome occupancy control gene transcription initiation (Ponts et al., 2010). The more recent RNA-seq technique has shown further promising results for the prediction of splicing events, which reflect the potential downstream variability, and the annotation of new open reading frames and their untranslated flanking regions (Otto et al., 2010). Moreover, transcriptome analysis of *P. falciparum*'s field isolate can identify previously unknown factors involved in pathogenesis (see Daily et al., 2007 and Siau et al., 2007 for an example of virulence factors identified in cerebral malaria). Finally, analyses of transcription profiles of variant surface antigens have identified patterns that are specific to the parasite's sexual stages and could be relevant for new vaccine interventions (Petter et al., 2008; Wang et al., 2010). Furthermore, genomic analyses have opened new leads to understand the importance of both post-transcriptional and post-translational events in the malaria parasite's development (Foth et al., 2003; Ganesan et al., 2008; Ponts et al., 2010) – mechanisms of gene regulation that were often overlooked in the past. Genomic analyses help to identify potential new drug targets by examining transcriptional or post-transcriptional and post-translational variations *in vitro* or *in vivo* (Daily et al., 2005; Hall et al., 2005; Jiang et al., 2008; Le Roch et al., 2008; Llinás et al., 2006; Mok et al., 2008; Natalang et al., 2008), or looking at host–pathogen interactions (e.g., the effect of febrile temperature on the infecting malaria parasite (Oakley et al., 2007).

Target-based Approaches for Vaccine Discovery

In life sciences, the beginning of the 21st century has been marked by the generalization of high-throughput analyses (microarrays, deep sequencing, quantitative mass spectrometry, epigenome mapping, computational modeling, etc.) to mine *Plasmodium*'s genome, looking for genetic elements that may be used to fight malaria. For example, microarray/RNA-seq analyses have been used to identify up-regulated *Plasmodium* genes in infecting forms of the parasite or in patients with severe forms of malaria (Bozdech et al., 2003, 2008; Daily et al., 2004, 2005, 2007; Ganesan et al., 2008; Hall et al., 2005; Le Roch et al., 2003; Le Roch et al., 2004; Llinás et al., 2006; Otto et al., 2010; Siau et al., 2007; Silvestrini et al., 2005; Tarun et al., 2008; Young et al., 2005). Numerous proteomic analyses have surveyed all the surface proteins in *P. falciparum* to identify potential antigens that could be used for rational development of vaccines (Florens et al., 2002, 2004; Lasonder et al., 2002; Le Roch et al., 2004; Sam-Yellowe et al., 2004). For example, the circumsporozoite protein (CSP) was identified as the major surface antigen in sporozoites, the liver stage of the parasite. For an effective vaccine, antigens must be selected according to their role in protective immunity (preventing infection or preventing the disease). Vaccines against malaria could target different aspects of the parasite's life cycle. They could enhance immunity against the pre-erythrocytic phase of the parasite (to block infection) or its blood stages, or prevent transmission.

Liver-stage parasites, or sporozoites, represent a valuable candidate for vaccination, primarily due to the low parasite count at the time of infection and increased chances of parasite clearance. Vaccination with irradiated sporozoites or antigen based on pre-erythrocytic stages, targeting infected hepatocytes, is meant to block infection and has had some success in eliciting partial protection in human volunteers (see Speake and Duffy, 2009 for a review). A whole-parasite *P. falciparum* sporozoite (PfSPZ) vaccine has been recently developed, the Sanaria PfSPZ Vaccine, and entered phase I clinical trials in 2009. The performances of the candidate vaccine are, however, disappointing with only five protected individuals out of the 80 volunteers (Fox, 2010).

Aside from the whole-cell vaccination approach, malaria vaccine development relies on the combination of parasite antigens with adjuvants designed to elicit the cell-mediated human immune response (CMI), which is often too weak to confer protection. The use of viral vectors to deliver and present the parasite's antigens can circumvent the problem. The RTS,S candidate malaria vaccine (or Mosquirix from GlaxoSmithKline) consists of a genetically engineered recombinant subunit vaccine, that combines the hepatitis B surface antigen and epitopes from the sporozoite-specific CSP (Ballou, 2009). It is the most promising malaria vaccine against *P. falciparum*, and has recently entered phase III clinical trials. RTS,S has been developed to protect African infants and young children from clinical and severe disease caused by *P. falciparum*. For a decade, the candidate vaccine, RTS,S, has been evaluated in multiple phase II studies. It was shown to be satisfactorily safe, and to confer significant protection against *P. falciparum*. Nonetheless, the mechanisms regarding how RTS,S vaccination actually protects against malaria are very poorly understood. These important unknown parameters raise serious and legitimate concerns that could compromise the vaccine submission, initially scheduled for as early as 2012. Also, the significance of the risk of encountering increased malaria incidence due to loss of natural immunity because of mass vaccination cannot be assessed. Ultimately, this lack of knowledge represents a major obstacle to the future proliferation of new generations of vaccines.

Other possible vaccines are being investigated, targeting the blood-stage parasites (Goodman and Draper, 2010). The most preponderant variant antigens produced by the malaria parasite are indubitably PfEMP1 encoded by the *Var* genes. However, the constant variant switching of the *Var* genes dramatically complicates the task of vaccination, against pregnancy-associated malaria (PAM) for example. PAM is a severe form of malaria that causes maternal anemia and low birth weight. Numerous studies led to the identification of the receptor chondroitin sulfate A (CSA) on the placenta as the target of infected erythrocytes via the variant antigen PfEMP1, named *VAR2CSA* in the case of PAM. Research for a vaccine against PAM has therefore been focusing on finding other genes that may be involved either in the regulation of *VAR2CSA* expression at the surface of infected erythrocytes or its binding to CSA, rather than *VAR2CSA* itself. A recent microarray-based study of *P. falciparum*'s transcription profiles from 18 infected human placentas at delivery identified about a dozen genes likely to be involved in host–parasite interaction that could be suitable new potential targets for vaccine development (Tuikue Ndam et al., 2008), providing new leads for vaccine discovery.

Merozoite surface proteins (MSP) and apical membrane antigen 1 (AMA1) also possess good antigenic properties.

MSP1 and AMA1 have been particularly investigated as antigens for vaccine candidates (among many others). So far, MSP1 and AMA1 do not seem to induce sufficient immunity to confer protection against clinical malaria, and research is ongoing to improve the adjuvant formulation, to find one that could elicit satisfactory levels of immune response. However, the high levels of genetic variations observed in MSP1 and AMA1 are challenging the viability of these vaccine candidates (Kidgell et al., 2006). More recently, a candidate vaccine based on MSP3 has been developed using the long synthetic peptide approach (the MSP3-LSP vaccine). Phase I trials demonstrated that MSP3-LSP is safe and immunogenic; it can elicit an antibody response in humans and inhibit parasite growth *in vitro*. These LSP approaches combine the abundance of genomic data with the flexibility of peptide synthesis for a faster, rational vaccine design (Olugbile et al., 2010).

Finally, sexual-stage parasite surface proteins are of interest as candidate target antigens for transmission-blocking vaccines. The most prominent variant surface antigen transcribed in both gametocytes and sporozoites of 3D7/NF54 parasite strains is a single variant of the RIFIN protein family. This discovery may lead to the identification of the parasite's binding ligands responsible for adhesion during sexual stages, and potentially to novel vaccine candidates (Wang et al., 2010).

The next step in the fight against malaria is to learn the lessons from the design of a *P. falciparum* vaccine for the rapid development of a *P. vivax* vaccine, especially regarding vaccine evaluation protocols (Mueller et al., 2009). Studies of the interaction between *P. vivax* and red blood cells point out the Duffy antigen as a potential base for a candidate vaccine (Rowe et al., 2009). In the future, the mining of the *P. vivax* genome will certainly identify many other suitable antigens. In *P. falciparum*, several genome-wide proteomic analyses have already surveyed all *P. falciparum* proteins, including surface proteins, to identify candidate antigens to use in the rational development of vaccines (Florens et al., 2002, 2004; Lasonder et al., 2002; Le Roch et al., 2004; Sam-Yellowe et al., 2004). One of the most promising advances brought by the advent of genomics is probably the development of whole-genome protein arrays. Such arrays are used to probe human plasma from individuals before and after malaria season, to identify parasite proteins associated with immunoreactivity (Crompton et al., 2010; Doolan et al., 2008; Sundaresh et al., 2006). Such applications mark the beginning of immuno-genomics, or immunomics.

Target-based Approaches to Drug Discovery, Mechanisms of Action, and Resistances

Current antimalarial drugs originate from traditional medicinal plants used since antiquity (e.g., quinine and artemisinin), or were discovered after intensive screening of compound libraries. Over the past two years, three groups have reported extensive efforts to provide early leads for the identification of new antimalarial compounds: about 0.5% of the four million screened compounds were identified as potential starting points for new antimalarial drug discovery (Gamo et al., 2010; Plouffe et al., 2008). These encouraging discoveries are at the onset of the drug discovery and development process. For example, compounds have to be tested *in vivo*, known mechanisms of resistance must not affect their efficiency, and their toxicity to host cells must be minimal. In a more practical aspect, the compound must be stable under tropical temperatures and must be cheap to produce. In addition, the action on the host must be thoroughly evaluated (including bioavailability, metabolic pathways and speeds, etc.). Finally, before the introduction of a new drug in the field, the absence of known and predictable resistances must be verified, which involves understanding the other resistances and verifying that the new drug will not go through the same pathways. All together, developing drugs that meet all of the requirements to become suitable antimalarials requires many years. The performances of these approaches are therefore limited, and may not allow drug discovery to keep up with the fast emergence of resistances worldwide. In this regard, genomics-based approaches for antimalarial drug discovery can dramatically improve the procedure. A striking example of the power of genomics tools is the discovery of a potent drug by cell proliferation-based compound screening – the spiroindolone NITD609 – followed by the identification of its target using high-density microarrays and sequencing; everything done within a two-year time span (Rottmann et al., 2010).

Many of the current drug therapies are based on chemically engineered variants of already-known antimalarial compounds (e.g., aminoquinolines and/or peroxides) without the development of new chemicals with different mechanisms of action. Intensive exploration of the *Plasmodium* genome could lead to the identification of essential genes or metabolic bottlenecks that could be potential targets for the development of novel antimalarial strategies (rational design). For example, a fosmidomycin-sensitive mevalonate-independent pathway of isoprenoid biosynthesis, absent from higher eukaryotes and located in the plant plastid-like parasite organelle, namely the apicoplast, was identified in *Plasmodium* (Jomaa et al., 1999). Similarly, a type II fatty acid biosynthesis (usually found in plants, fungi, and certain bacteria), also associated with the apicoplast, has been identified as a potential target for drugs such as triclosan (Goodman and McFadden, 2007; Surolia et al., 2002; Waller et al., 2003; Wiesner et al., 2008). However, this pathway was recently found to be essential for the first blood-stage infection only (Vaughan et al., 2009). More recently, the polyamine biosynthetic pathway was identified as a possible target for antimalarial drugs (Becker et al., 2010; Clark et al., 2010; Radivojac et al., 2010). This essential pathway is responsible for the metabolism of important amines, such as spermidine, involved in cellular metabolism. In the long run, identifying drug targets in various metabolic pathways could minimize the risk of resistance emergence.

Typically, the mechanisms by which a drug is toxic for *Plasmodium* are determined by examining the genetic differences between sensitive and resistant strains. Genes carrying a mutation are then potentially associated with the mode of action. Finding these mutations is usually performed using classical genetic tools such as transfection, and can often be a tedious task. In addition, these tools do not permit the discovery of multi-trait resistances. The advent of whole-genome analyses (deep sequencing, microarrays) offered the possibility to rapidly analyze the genetic changes (SNPs, copy number variations or CNVs, insertions/deletions or indels) that occurred in a resistant strain at the whole-genome scale, possibly including organelles' genomes (mitochondria and apicoplasts). Whole-genome scanning using tiling microarrays is often used for this purpose. For example, point mutations in the apicoplast were recently associated with resistance to clindamycin, a drug used in combination with quinine for the treatment of malaria in pregnant women and infants, using tiling microarrays (Dharia et al., 2010). Tiling microarrays also determined that resistance to the spiroindolone NITD609 involves the P-type cation transporter ATPase4 (Rottmann et al., 2010).

Chloroquine resistance (CQR) is probably the best-studied case of drug resistance in *P. falciparum*. CQR is mediated by the CQR transporter gene *Pfcrt* and by the multi-drug resistance gene *Pfmdr1*, which is also involved in resistance to other antimalarials, potentially including artemisinin. As a consequence, the genetic variants of the *Pfcrt* and *Pfmdr1* loci are the subject of many genetic analyses, providing a wide range of information. In terms of resistance emergence, for example, a recent large-scale (460 blood samples) analysis of microsatellite polymorphisms at these two loci proposed that different mechanisms led to the rise of these mutations in Africa and Asia, which may have important consequences for the future evolution of the parasite (Mehlotra et al., 2008). Whole-genome scanning for the presence of evolutionary signatures could indeed provide critical information for understanding the mechanisms of resistance to drugs and identifying new potential targets for drug discovery. These considerations have been briefly reviewed by Gupta and colleagues (2010). For example, a genome-wide SNP analysis of the genomes from almost 200 different *P. falciparum* isolates identified regions of high and low recombination frequencies (hotspots and coldspots), and found that loci associated with known drug resistance, e.g., chloroquine, are likely found in hotspots (Mu et al., 2010). A practical consequence of these observations could be that drug targets should be proteins encoded by genes located in coldspots rather than hotspots. Beyond SNP analyses, tiling microarray analyses found relatively abundant CNVs in *P. falciparum* that may be involved in drug resistance (Kidgell et al., 2006). For example, CNVs have been identified surrounding the *P. falciparum* multi-drug-resistance gene *Pfmdr1* and the first gene involved in the folate pathway (as a

result of widespread use of anti-folate drug). Recently, *in vitro*-derived fosmidomycin resistance was attributed to a CNV in the *pfdxr* gene, that suppresses the drug-induced inhibition of isoprenoid biosynthesis in the parasite (Dharia et al., 2009). Despite their abundance and importance, CNVs have been overlooked in the malaria parasite (Anderson et al., 2009). The fast-increasing performances of deep sequencing technologies should allow the field to develop in the near future. Moreover, novel classes of drugs can arise from genomics drug design, with the development of AT-specific DNA-binding agents that target the AT-rich genome of the malaria parasite (Woynarowski et al., 2007; Yanow et al., 2007).

Ultimately, the use of genome-based strategies to discover drugs should provide a larger and more diverse selection of effective antimalarial drugs (summarized in Figure 100.3), and would make possible the application of pharmacogenomics to the treatment of malaria.

Pharmacogenomics

Pharmacogenomics is the *sine qua non* condition of personalized medicine. It combines biochemistry, chemistry, genomics, genetics, proteomics, etc., to understand how genetics influences the human body's response to drugs (classically correlating gene expression or SNPs with drug efficacy), and to

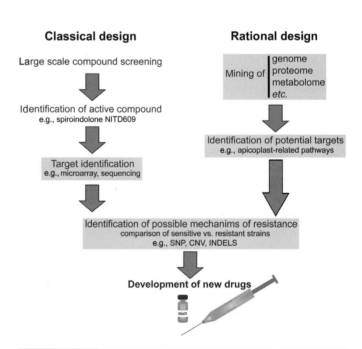

Figure 100.3 Genomics and antimalarial drug development. The shaded boxes indicate steps facilitated by genomics. Classical drug discovery involves compound screening for antimalarial properties followed by target identification of a selected drug candidate. In rational design, "-ome" (genome, proteome, metabolome, etc.) mining identifies potential drug targets (e.g., blocking essential/crucial metabolic pathways). Mechanisms of resistance are then investigated.

optimize drug and drug-combination treatments to individuals. Depending on the patient, the same drug can be toxic or not and effective or not. Antimalarial drugs are no exception (Gil, 2008; Kerb et al., 2009; Mehlotra et al., 2009). Clinical trials have revealed important variations of plasmatic concentrations of drugs, leading to lower effectiveness or increased toxicity, and pharmacogenomics would doubtlessly help to optimize antimalarial chemotherapies and drug management. Polymorphisms in drug-metabolizing enzymes are responsible for such disparities, mainly various cytochrome P450 (CYP) and drug transporters. For example, plasma levels of proguanil and amodiaquine are linked to polymorphisms *CYP2C19* and *CYP2C8* respectively. Polymorphisms in *CYP2C8* may also have important consequences for the use of chloroquine and dapsone, both substrates of the cytochrome (Gil and Gil Berglund, 2007). On the same line, chlorcycloguanil metabolism is modulated by genetic variations at the *CYP2C19/CYP2C9* locus (Janha et al., 2009).

These examples are a sample of the published polymorphisms leading to different efficacies and toxicities of antimalarial drugs. The future should bring more knowledge of this kind, and contribute to a pharmacogenomics paradigm for the treatment of malaria. Incorporated in clinical trials, the advent of pharmacogenomics for malaria would also fast-track the development of new drugs by assisted sampling of patients with a certain genotype.

Vector Control

The malaria parasite's primary hosts and transmission vectors are female mosquitoes of the *Anopheles* genus. There are more than 400 recognized *Anopheles* species, 30 to 40 of which can commonly transmit parasites in humans. *Anopheles gambiae* is so far one of the most studied species because of its transmission of the most deadly form of malaria parasite, *P. falciparum*. So far, efforts to eradicate malaria by eliminating mosquitoes have been the most successful methods. By draining wetlands, improving sanitation, and using pesticides such as DDT, malaria was successfully eradicated from North America and Europe. Unfortunately these standard methods have been more challenging in heavily infected tropical regions, and new methodologies for mosquito control need to be developed.

Novel biological vector control techniques have emerged as new ways to combat vector-borne diseases, including malaria. Progress toward genome-wide analysis of mosquitoes, or genetically modified insects, suggests that wild mosquito populations could be made malaria-resistant. Wild mosquito strains that are completely resistant to malaria parasites have been identified and selected in the mosquito population (Collins et al., 1986; Vernick et al., 1995). Understanding the complex genetic bases of this resistance has been a major concern over the past few years. Large genetic analyses of these strains indicate that several genetic factors in mosquitoes are controlling the level of parasite transmission (Blandin et al., 2009; Riehle et al., 2006). Identifying these complex molecular components and how they are maintained and selected in field populations is critical for the control of malaria transmission.

The parasite must overcome several major barriers in the vector to achieve a successful transmission from one host to another. Once injected in the mosquito midgut, parasites must undergo complex developmental transitions and survive attacks from the mosquito's innate immune system. In the mosquito's midgut, mature gametocytes differentiate into male or female gametes and then fuse in the mosquito gut to produce ookinetes. The ookinetes then travel to the basal lamina to differentiate into oocysts in the gut wall. They further develop into thousands of sporozoites that migrate from the mosquito's body to the salivary glands. Newly infective parasites can then be transmitted to a human when the mosquito takes another blood meal. The majority of the parasite loss throughout this journey can be attributed to luminal and epithelial immune responses. Cellular and humoral factors are known to be the major players in response to microbial agents (Cirimotich et al., 2010). Mosquito pathogens, with no exception to malaria parasites, can either be engulfed by phagocytosis of the mosquito's hemocytes or encapsulated by melanization and antimicrobial effector molecules that are activated by catalytic cascades and intracellular immune signaling pathways (Toll pathway, immune deficient pathway (IMD) and Jak/Stat pathway).

Boosting anti-*Plasmodium* immune responses in mosquitoes can block the insect infection and therefore disease transmission to humans. In laboratories, over-expression of key molecules such as the NF-kappaB-like transcription factor Rel1 in the Toll pathway and Rel2-F in the IMD pathway (Meister et al., 2005), or activation of AKT, or protein kinase B(PKB) signaling pathway signaling (Corby-Harris et al., 2010) can increase parasite clearance. It becomes, therefore, obvious that genetically engineered mosquitoes could help malaria eradication if successfully and safely introduced into the wild population. This replacement should rely upon a mechanism that allows for non-Mendelian inheritance of the locus of interest (James, 2007). This challenging approach is, however, controversial. Mosquitoes are an important link in the food chain. Understanding how genetically modified strains could affect the ecosystem will be critical. Furthermore, the huge genetic diversity existing in all mosquito and parasite species increases the complexity of the challenge. Individual immune responses to the various *Plasmodium* species may not be universal, suggesting that defense mechanisms are both mosquito species- and parasite species-dependent (Garver et al., 2009). These latest results have significant consequences, considering that mosquitoes have evolved quickly. Recently, two studies using whole-genome sequencing and SNP-genotyping array revealed heterogeneous genetic divergences between wild *Anopheles gambiae* species that are morphologically indistinguishable (Lawniczak et al., 2010; Neafsey et al., 2010). This unexpected, widespread genetic divergence has certainly already affected difference in behaviors, physiology, and disease transmission. It also increases the complexity of vector control strategies. Therefore, larger GWA studies to monitor population structure and to better understand parasite infection resistances and susceptibilities are now needed for the development of innovative and sustainable vector control strategies.

PERSONALIZED MEDICINE AND MALARIA: HOW, WHERE, AND WHEN

Genomics has changed the study of malaria pathogenesis. Current and future research is not based on phenotypic changes anymore (such as the presence of sickle cells in a blood sample), but is based on the genotyping of thousands of individuals and the identification of malaria-resistance loci. These approaches, driven by the rapid evolution of high-throughput sequencing technologies, are the basis for genomic epidemiology of malaria (Figure 100.4). This term refers to the use of genomic tools in large-scale studies to detect and analyze genetic variants in the malaria parasite and its hosts in order to improve our understanding of the disease and develop efficient preventive and curative strategies. For example, a direct application of genomic epidemiology is the use of GWA studies to monitor antimalarial

drug resistance spread. Before GWAs, candidate gene analysis was often performed by PCR-based genotyping. The main problem with that approach was that preliminary identification of a candidate gene as well as the suspected polymorphism was required, potentially leading to many false positives. Moreover, such studies are never replicated, leading to false negatives and more false positives, though limited in the case of family studies. With genome-wide techniques, the whole human genome is considered, including genes of unknown function. GWA studies have primarily been permitted by the advent of high-throughput genotyping technologies. In addition, the establishment of large multi-center collaborations such as the Malaria Genomic Epidemiology Network or MalariaGEN (http://www.malariagen. net) allowed the analyses to consider large sample sizes. Finally, the HapMap Project surveyed around four million SNPs while describing the allele combination commonly found in African, Asian, and European individuals, thus providing a large database of genetic variants. A future goal will be to predict the risk for an individual of malaria (and, in the end, other infectious diseases), which involves understanding well the molecular cross-talk between host and pathogen.

On the parasite's side, genomics has led to the thorough analysis of the entire *P. falciparum* genome. The accumulated knowledge can be used at various levels, such as minimizing

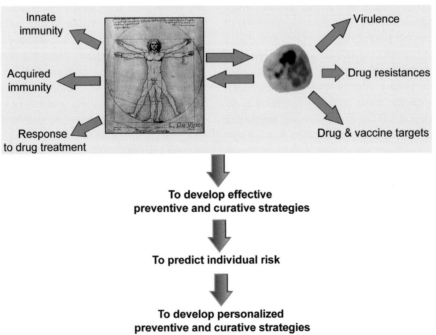

Genomic Epidemiology

To detect and analyze genetic variants at the whole-genome scale

Innate immunity

Acquired immunity

Response to drug treatment

Virulence

Drug resistances

Drug & vaccine targets

To develop effective preventive and curative strategies

To predict individual risk

To develop personalized preventive and curative strategies

Figure 100.4   Whole-genome approaches towards a personalized medicine against malaria. Blue arrows indicate the use of genomics to understand the intrinsic characteristics of the human host and *Plasmodium* involved in clinical malaria (host immunity, parasite virulence factors, etc.) as well as the complex host–parasite interactions that occur. Ultimately, the integrated use of the data generated by genomic analyses will help in developing a personalized medicine against malaria.

the risk of resistance emergence by understanding the mechanisms of action of drugs. For example, a proteomic approach has recently been used in *P. falciparum* to clarify the mode of action of doxycycline, a compound used in chemoprophylactic treatments and in ACTs (Briolant et al., 2010). In terms of resistance, genomic analyses identified CNV as a major form of genetic variation within the *P. falciparum* genome (Cheeseman et al., 2009) that plays a prominent role in the adaptive biology of the parasite. Among other findings, sub-telomeric deletions on chromosomes 2 and 9 have been linked to a loss of infected erythrocyte cytoadherence through deletion of genes encoding adhesion molecules (Biggs et al., 1989; Trenholme et al., 2000). Conversely, mefloquine resistance is conferred through amplification of a multigenic locus on chromosome 5, including the multi-drug transporter *Pfmdr1* gene (Cowman et al., 1994). Moreover, genomics has brought knowledge of the non-visible, i.e., non-coding RNAs (Raabe et al., 2010), small nuclear ribonucleic acid (snRNA) (Bawankar et al., 2009), and small nucleoar RNA (snoRNAs) (Mishra et al., 2009), whose roles during parasite development are currently under investigation. Finally, genomics has shed new light on phylogeny. *Plasmodium* has many genomic features that resemble the ones typically found in plants, including its mechanisms of gene regulation, which are often more related to those of plants than traditional model organisms (Gissot et al., 2009) (e.g., its known family of transcription factors ApiAP2). Such considerations could be very valuable information to aid drug discovery.

Genomics for Malaria Therapies

Once malaria infection is diagnosed, there are many factors to take into account for treatment. The nature of the infecting *Plasmodium* species is of course of prime importance. Then there is the patient, and his or her genome and immune system – all are major factors modulating treatment efficacy. A misunderstanding of the mechanisms behind treatment responsiveness could have terrible consequences for the future, potentially supporting the emergence of more resistant *Plasmodium* strains.

Vaccination represents an attractive solution to eradicate the malaria problem across the world. The Bill and Melinda Gates Foundation states, "we believe a preventive vaccine would provide the best long-term hope to defeat malaria, and would be especially beneficial for those at greatest risk: infants, children, and pregnant women." Aside from undeniable assets, vaccination should be carefully considered and implemented. Innate immunity favors transmission by allowing host survival (reduced acute disease) and higher parasitemia. For example, a very recent study associated genetic variation at the human *HbB* locus with *P. falciparum* transmission rates (Gouagna et al., 2010). As a consequence, vaccines less than 100% effective may exert enough selective pressure on the parasite to promote gametocytogenesis and transmission. Under these circumstances, massive

vaccination campaigns may favor strains with lower transmission rates not included in the vaccine – strains that also could be more virulent. Several studies corroborate this hypothesis and suggest that widespread vaccination could put the unvaccinated at greater risk (Mackinnon et al, 2008). The host's genetic background should also be taken into consideration for optimized vaccination, such as ethnicity-related genetic diversity linked to disease susceptibility (Campbell and Tishkoff, 2008).

The same considerations are true when drug treatments are considered. The use of pharmacogenomics for malaria treatment is a critical step toward optimized chemotherapies. Pharmacogenomics is already applied to the treatment of various diseases, including cancers, HIV, asthma, and diabetes. Such medicine personalization involves the development of a panel of drugs that can be used for various cases, which directly impacts research and development strategies. Old drugs could be re-used for certain cases, or other drugs used in other contexts could be applied to malaria treatment. The bottom line is that several non-scientific problems are pending with pharmacogenomics. The first one is probably to convince pharmaceutical companies to make numerous drug alternatives rather than mass-produce generic "one-size-fits-all" molecules. On the ethical side, advanced screening and ethnicity targeting (drug discrimination) will certainly raise concerns that will need to be addressed.

Future Personalized Medicine for Malaria?

Constantly diminishing costs, over the past few years, of high-throughput genotyping and DNA sequencing technologies have dramatically changed the way science is being done. These changes should soon transform the way physicians assess genetic risk factors and beyond, and give personalized medicine a new meaning. Democratized access to next-generation sequencing and the possibility of having the human genome sequenced for a few dollars will definitely revolutionize the way we think about medicine, treatments, and possible disease eradication, including in developing countries. Diagnosis has already been significantly improving over the past years with the increasing availability of predictive genetic tests such as RDTs. In addition, regional genomic specificities of ethnic groups are extensively collected and analyzed by MalariaGEN. Finally, the understanding of the malaria parasite's biology is growing exponentially.

Francis Collins, former director of the National Human Genome Research Institute and at the time of writing, the Director of the National Institutes of Health in Bethesda, Maryland, identified five lessons from the first decade of the genome era that apply to the future of personalized medicine (Collins, 2010):

1. Free and open access to genome data has had a profoundly positive effect on progress (example of the human genome project).
2. Technology development for sequencing and functional genomics must continue.

3. The success of personalized medicine will depend on continued accurate identification of genetic and environmental factors and the ability to use this information in the real world to influence health behaviors.

4. To achieve the enormous promise of the myriad new drug targets emerging from genomic analyses will require new paradigms of public–private partnership.

5. Good policy decisions, including individual privacy protection, effective education of healthcare providers and the public about genomic medicine, and healthcare system reimbursement for preventive measures will be crucial to reaping the benefits that should flow from the coming revelations about the genome.

These five lessons have important implications in various areas. First, the release of free and open-access human genomic data raises concerns about various ethical principles, such as privacy, anonymity, and security, which concern the avoidance of genetic discrimination. MalariaGEN has developed GWA data release policies and procedures in developing countries (Parker et al., 2009). Far from being finalized, these policies represent a valuable basis for future improvements. The education of healthcare providers will not be straightforward. They will have to agree to provide extra diagnosis that will lead to minimal side effects, trial and errors treatment optimization, etc., which overall will lower cost of treatment. This point is particularly relevant in developing countries, where private providers are the main, if not the only, source of healthcare available. In addition, cost, quality, and availability of malaria treatment is highly variable in low- to middle-income countries and represents a major obstacle to access to antimalarial therapies (Patouillard et al., 2010). Education of public organizations, local communities, and individuals must also be provided to ensure access to medicine and personalized medicine.

Altogether, the ingredients for personalized medicine for malaria are being brought together. The rapid democratization and improvement of genomics technologies will soon overcome obstacles related to cost or equipment. A better understanding of the human genetic factors that influence susceptibility and response to both malaria and the drug/vaccine to treat it, as well as a better understanding of the malaria parasite itself, is certainly necessary to eradicate malaria. Genomics will greatly contribute to this better understanding and will change the future of disease management. In this regard, governmental and non-governmental organizations, academic institutions, and individuals will have to be educated to ensure that the benefits of genomics will be clearly understood and will benefit everyone.

ACKNOWLEDGMENT

We would like to thank Serena Cervantes for her talent in preparing Figure 100.1. We also would like to thank Vance C. Huskins for his help with proofreading the manuscript. KLR is supported by grant from the National Institute of Allergy and Infectious Diseases and the National Institutes of Health (#R01AI085077).

REFERENCES

Ackerman, H., Usen, S., Jallow, M., et al., 2005. A comparison of case-control and family-based association methods: The example of sickle-cell and malaria. Ann Hum Genet 69, 559–565.

Agarwal, A., Guindo, A., Cissoko, Y., et al., 2000. Hemoglobin C associated with protection from severe malaria in the Dogon of Mali, a West African population with a low prevalence of hemoglobin S. Blood 96, 2358–2363.

Akide-Ndunge, O.B., Ayi, K., Arese, P., 2003. The Haldane malaria hypothesis: Facts, artifacts, and a prophecy. Redox Rep 8, 311–316.

Allen, S.J., O'Donnell, A., Alexander, N.D., et al., 1997. Alpha+-thalassemia protects children against disease caused by other infections as well as malaria. Proc Natl Acad Sci USA 94, 14,736–14,741.

Allison, A.C., 1954. Protection afforded by sickle-cell trait against subtertian malarial infection. Br Med J 1, 290–294.

Anderson, T.J.C., Patel, J., Ferdig, M.T., 2009. Gene copy number and malaria biology. Trends Parasitol 25, 336–343.

Arie, T., Fairhurst, R.M., Brittain, N.J., Wellems, T.E., Dvorak, J.A., 2005. Hemoglobin C modulates the surface topography of Plasmodium falciparum-infected erythrocytes. J Struct Biol 150, 163–169.

Aucan, C., Walley, A.J., Hennig, B.J.W., et al., 2003. Interferon-alpha receptor-1 (IFNAR1) variants are associated with protection against cerebral malaria in the Gambia. Genes Immun 4, 275–282.

Ballou, W.R., 2009. The development of the RTS,S malaria vaccine candidate: Challenges and lessons. Parasite Immunol 31, 492–500.

Bawankar, P., Shaw, P.J., Sardana, R., Babar, P.H., Patankar, S., 2009. 5′ and 3′ end modifications of spliceosomal RNAs in Plasmodium falciparum. Mol Biol Rep 37 (4), 2125–2133.

Becker, J.V.W., Mtwisha, L., Crampton, B.G., et al., 2010. Plasmodium falciparum spermidine synthase inhibition results in unique perturbation-specific effects observed on transcript, protein and metabolite levels. BMC Genomics 11, 235.

Biggs, B.A., Culvenor, J.G., Ng, J.S., Kemp, D.J., Brown, G.V., 1989. Plasmodium falciparum: Cytoadherence of a knobless clone. Exp Parasitol 69, 189–197.

Blandin, S.A., Wang-Sattler, R., Lamacchia, M., et al., 2009. Dissecting the genetic basis of resistance to malaria parasites in Anopheles gambiae. Science 326, 147–150.

Blomqvist, K., Normark, J., Nilsson, D., et al., 2010. Var gene transcription dynamics in Plasmodium falciparum patient isolates. Mol Biochem Parasitol 170, 74–83.

Bolad, A., Farouk, S.E., Israelsson, E., et al., 2005. Distinct interethnic differences in immunoglobulin G class/subclass and immunoglobulin M antibody responses to malaria antigens but not in immunoglobulin G responses to nonmalarial antigens in sympatric tribes living in West Africa. Scand J Immunol 61, 380–386.

Bozdech, Z., Llinas, M., Pulliam, B., et al., 2003. The transcriptome of the intraerythrocytic developmental cycle of *Plasmodium falciparum*. PLoS Biol 1, E5.

Bozdech, Z., Mok, S., Hu, G., et al., 2008. The transcriptome of *Plasmodium vivax* reveals divergence and diversity of transcriptional regulation in malaria parasites. Proc Natl Acad Sci USA 105, 16,290–16,295.

Briolant, S., Henry, M., Oeuvray, C., et al., 2010. Absence of association between piperaquine *in vitro* responses and polymorphisms in the *pfcrt, pfmdr1, pfmrp,* and *pfnhe* genes in *Plasmodium falciparum*. Antimicrob Agents Chemother 54, 3537–3544.

Campbell, M.C., Tishkoff, S.A., 2008. African genetic diversity: Implications for human demographic history, modern human origins, and complex disease mapping. Annu Rev Genomics Hum Genet 9, 403–433.

Carlton, J.M., Adams, J.H., Silva, J.C., et al., 2008. Comparative genomics of the neglected human malaria parasite *Plasmodium vivax*. Nature 455, 757–763.

Cheeseman, I.H., Gomez-Escobar, N., Carret, C.K., et al., 2009. Gene copy number variation throughout the *Plasmodium falciparum* genome. BMC Genomics 10, 353.

Cholera, R., Brittain, N.J., Gillrie, M.R., et al., 2008. Impaired cytoadherence of *Plasmodium falciparum*-infected erythrocytes containing sickle hemoglobin. Proc Natl Acad Sci USA 105, 991–996.

Cirimotich, C.M., Dong, Y., Garver, L.S., Sim, S., Dimopoulos, G., 2010. Mosquito immune defenses against *Plasmodium* infection. Dev Comp Immunol 34, 387–395.

Clark, K., Niemand, J., Reeksting, S., et al., 2010. Functional consequences of perturbing polyamine metabolism in the malaria parasite, *Plasmodium falciparum*. Amino Acids 38, 633–644.

Cnops, L., Jacobs, J., Van Esbroeck, M., 2010. Validation of a four-primer real-time PCR as a diagnostic tool for single and mixed *Plasmodium* infections. Clin Microbiol Infect 17 (7), 1101–1107.

Collins, F., 2010. Has the revolution arrived? Nature 464, 674–675.

Collins, F.H., Sakai, R.K., Vernick, K.D., et al., 1986. Genetic selection of a *Plasmodium*-refractory strain of the malaria vector *Anopheles gambiae*. Science 234, 607–610.

Corby-Harris, V., Drexler, A., Watkins de Jong, L., et al., 2010. Activation of Akt signaling reduces the prevalence and intensity of malaria parasite infection and lifespan in *Anopheles stephensi* mosquitoes. PLoS Pathog 6, e1001003.

Cowman, A.F., Galatis, D., Thompson, J.K., 1994. Selection for mefloquine resistance in *Plasmodium falciparum* is linked to amplification of the *pfmdr1* gene and cross-resistance to halofantrine and quinine. Proc Natl Acad Sci USA 91, 1143–1147.

Crompton, P.D., Kayala, M.A., Traore, B., et al., 2010. A prospective analysis of the Ab response to *Plasmodium falciparum* before and after a malaria season by protein microarray. Proc Natl Acad Sci USA 107, 6958–6963.

Dahlbäck, M., Nielsen, M.A., Salanti, A., 2010. Can any lessons be learned from the ambiguous glycan binding of PfEMP1 domains? Trends Parasitol 26, 230–235.

Daily, J.P., Le Roch, K.G., Sarr, O., et al., 2004. *In vivo* transcriptional profiling of *Plasmodium falciparum*. Malar J 3, 30.

Daily, J.P., Le Roch, K.G., Sarr, O., et al., 2005. *In vivo* transcriptome of *Plasmodium falciparum* reveals overexpression of transcripts that encode surface proteins. J Infect Dis 191, 1196–1203.

Daily, J.P., Scanfeld, D., Pochet, N., et al., 2007. Distinct physiological states of *Plasmodium falciparum* in malaria-infected patients. Nature 450, 1091–1095.

Del Portillo, H.A., Lanzer, M., Rodriguez-Malaga, S., Zavala, F., Fernandez-Becerra, C., 2004. Variant genes and the spleen in *Plasmodium vivax* malaria. Int J Parasitol 34, 1547–1554.

Dharia, N.V., Plouffe, D., Bopp, S.E.R., et al., 2010. Genome-scanning of Amazonian *Plasmodium falciparum* shows subtelomeric instability and clindamycin-resistant parasites. Genome Res 20 (11), 1534–1544.

Dharia, N.V., Sidhu, A.B.S., Cassera, M.B., et al., 2009. Use of high-density tiling microarrays to identify mutations globally and elucidate mechanisms of drug resistance in *Plasmodium falciparum*. Genome Biol 10, R21.

Doolan, D.L., Mu, Y., Unal, B., et al., 2008. Profiling humoral immune responses to *P. falciparum* infection with protein microarrays. Proteomics 8, 4680–4694.

Elemento, O., Slonim, N., Tavazoie, S., 2007. A universal framework for regulatory element discovery across all genomes and data types. Mol Cell 28, 337–350.

Fairhurst, R.M., Baruch, D.I., Brittain, N.J., et al., 2005. Abnormal display of PfEMP-1 on erythrocytes carrying haemoglobin C may protect against malaria. Nature 435, 1117–1121.

Flint, J., Hill, A.V., Bowden, D.K., et al., 1986. High frequencies of alpha-thalassaemia are the result of natural selection by malaria. Nature 321, 744–750.

Florens, L., Liu, X., Wang, Y., et al., 2004. Proteomics approach reveals novel proteins on the surface of malaria-infected erythrocytes. Mol Biochem Parasitol 135, 1–11.

Florens, L., Washburn, M.P., Raine, J.D., et al., 2002. A proteomic view of the *Plasmodium falciparum* life cycle. Nature 419, 520–526.

Flori, L., Kumulungui, B., Aucan, C., et al., 2003. Linkage and association between *Plasmodium falciparum* blood infection levels and chromosome 5q31-q33. Genes Immun 4, 265–268.

Foth, B.J., Ralph, S.A., Tonkin, C.J., et al., 2003. Dissecting apicoplast targeting in the malaria parasite *Plasmodium falciparum*. Science 299, 705–708.

Fox, M., 2010. Malaria vaccine trial disappoints. Reuters.

Gamo, F., Sanz, L.M., Vidal, J., et al., 2010. Thousands of chemical starting points for antimalarial lead identification. Nature 465, 305–310.

Ganesan, K., Ponmee, N., Jiang, L., et al., 2008. A genetically hard-wired metabolic transcriptome in *Plasmodium falciparum* fails to mount protective responses to lethal antifolates. PLoS Pathog 4, e1000214.

Garcia, I., Miyazaki, Y., Araki, K., et al., 1995. Transgenic mice expressing high levels of soluble TNF-R1 fusion protein are protected from lethal septic shock and cerebral malaria, and are highly sensitive to *Listeria monocytogenes* and *Leishmania major* infections. Eur J Immunol 25, 2401–2407.

Gardner, M., Hall, N., Fung, E., et al., 2002. Genome sequence of the human malaria parasite *Plasmodium falciparum*. Nature 419, 498–511.

Garver, L.S., Dong, Y., Dimopoulos, G., 2009. Caspar controls resistance to *Plasmodium falciparum* in diverse anopheline species. PLoS Pathog 5, e1000335.

Gil, J.P., 2008. Amodiaquine pharmacogenetics. Pharmacogenomics 9, 1385–1390.

Gil, J.P., Gil Berglund, E., 2007. CYP2C8 and antimalaria drug efficacy. Pharmacogenomics 8, 187–198.

Gissot, M., Kim, K., Schaap, D., Ajioka, J.W., 2009. New eukaryotic systematics: A phylogenetic perspective of developmental gene expression in the Apicomplexa. Int J Parasitol 39, 145–151.

Gonzales, J.M., Patel, J.J., Ponmee, N., et al., 2008. Regulatory hotspots in the malaria parasite genome dictate transcriptional variation. PLoS Biol 6, e238.

Goodman, A.L., Draper, S.J., 2010. Blood-stage malaria vaccines – recent progress and future challenges. Ann Trop Med Parasitol 104, 189–211.

Goodman, C.D., McFadden, G.I., 2007. Fatty acid biosynthesis as a drug target in apicomplexan parasites. Curr Drug Targets 8, 15–30.

Gouagna, L., Bancone, G., Yao, F., et al., 2010. Genetic variation in human HBB is associated with Plasmodium falciparum transmission. Nature Genetics 42, 328–331.

Gupta, B., Awasthi, G., Das, A., 2010. Malaria parasite genome scan: Insights into antimalarial resistance. Parasitol Res 107, 495–499.

Haldane, J.B.S., 1949. Disease and Evolution. Ric Sci Suppl A19, 68–76.

Hall, N., Karras, M., Raine, J.D., et al., 2005. A comprehensive survey of the Plasmodium life cycle by genomic, transcriptomic, and proteomic analyses. Science 307, 82–86.

Hebbel, R.P., 2003. Sickle hemoglobin instability: A mechanism for malarial protection. Redox Rep 8, 238–240.

Hill, A.V., Allsopp, C.E., Kwiatkowski, D., et al., 1991. Common west African HLA antigens are associated with protection from severe malaria. Nature 352, 595–600.

Jallow, M., Teo, Y.Y., Small, K.S., et al., 2009. Genome-wide and fine-resolution association analysis of malaria in West Africa. Nat Genet 41, 657–665.

James, A.A., 2007. Preventing the spread of malaria and dengue fever using genetically modified mosquitoes. J Vis Exp (5), 231.

Janha, R.E., Sisay-Joof, F., Hamid-Adiamoh, M., et al., 2009. Effects of genetic variation at the CYP2C19/CYP2C9 locus on pharmacokinetics of chlorcycloguanil in adult Gambians. Pharmacogenomics 10, 1423–1431.

Jepson, A.P., Banya, W.A., Sisay-Joof, F., et al., 1995. Genetic regulation of fever in Plasmodium falciparum malaria in Gambian twin children. J Infect Dis 172, 316–319.

Jiang, H., Yi, M., Mu, J., et al., 2008. Detection of genome-wide polymorphisms in the AT-rich Plasmodium falciparum genome using a high-density microarray. BMC Genomics 9, 398.

Jomaa, H., Wiesner, J., Sanderbrand, S., et al., 1999. Inhibitors of the nonmevalonate pathway of isoprenoid biosynthesis as antimalarial drugs. Science 285, 1573–1576.

Kerb, R., Fux, R., Mörike, K., et al., 2009. Pharmacogenetics of antimalarial drugs: Effect on metabolism and transport. Lancet Infect Dis 9, 760–774.

Khor, C.C., Chapman, S.J., Vannberg, F.O., et al., 2007. A Mal functional variant is associated with protection against invasive pneumococcal disease, bacteremia, malaria and tuberculosis. Nat Genet 39, 523–528.

Kidgell, C., Volkman, S.K., Daily, J., et al., 2006. A systematic map of genetic variation in Plasmodium falciparum. PLoS Pathog 2, e57.

Knight, J.C., Kwiatkowski, D., 1999. Inherited variability of tumor necrosis factor production and susceptibility to infectious disease. Proc Assoc Am Physicians 111, 290–298.

Koch, O., Awomoyi, A., Usen, S., et al., 2002. IFNGR1 gene promoter polymorphisms and susceptibility to cerebral malaria. J Infect Dis 185, 1684–1687.

Kwiatkowski, D., 2000. Genetic susceptibility to malaria getting complex. Curr Opin Genet Dev 10, 320–324.

Langhorne, J., Ndungu, F.M., Sponaas, A., Marsh, K., 2008. Immunity to malaria: More questions than answers. Nat Immunol 9, 725–732.

Lasonder, E., Ishihama, Y., Andersen, J.S., et al., 2002. Analysis of the Plasmodium falciparum proteome by high-accuracy mass spectrometry. Nature 419, 537–542.

Lawniczak, M.K.N., Emrich, S.J., Holloway, A.K., et al., 2010. Widespread divergence between incipient Anopheles gambiae species revealed by whole genome sequences. Science 330, 512–514.

Le Roch, K., Johnson, J., Ahiboh, H., et al., 2008. A systematic approach to understand the mechanism of action of the bisthiazolium compound T4 on the human malaria parasite, Plasmodium falciparum. BMC Genomics 9, 513.

Le Roch, K., Zhou, Y., Blair, P., et al., 2003. Discovery of gene function by expression profiling of the malaria parasite life cycle. Science 301, 1503–1508.

Le Roch, K.G., Johnson, J.R., Florens, L., et al., 2004. Global analysis of transcript and protein levels across the Plasmodium falciparum life cycle. Genome Res 14, 2308–2318.

Levinson, W.E., Jawetz, E., 1996. Medical Microbiology and Immunology, fourth ed. Appleton & Lange.

Llinás, M., Bozdech, Z., Wong, E.D., Adai, A.T., DeRisi, J.L., 2006. Comparative whole genome transcriptome analysis of three Plasmodium falciparum strains. Nucleic Acids Res 34, 1166–1173.

López, C., Saravia, C., Gomez, A., Hoebeke, J., Patarroyo, M.A., 2010. Mechanisms of genetically-based resistance to malaria. Gene 467, 1–12.

Lulli, P., Mangano, V.D., Onori, A., et al., 2009. HLA-DRB1 and -DQB1 loci in three west African ethnic groups: Genetic relationship with sub-Saharan African and European populations. Hum Immunol 70, 903–909.

Luoni, G., Verra, F., Arcà, B., et al., 2001. Antimalarial antibody levels and IL4 polymorphism in the Fulani of West Africa. Genes Immun 2, 411–414.

Mackinnon, M.J., Gandon, S., Read, A.F., 2008. Virulence evolution in response to vaccination: The case of malaria. Vaccine 26 (Suppl 3), C42–C52.

May, J., Evans, J.A., Timmann, C., et al., 2007. Hemoglobin variants and disease manifestations in severe falciparum malaria. JAMA 297, 2220–2226.

McGuire, W., Hill, A.V., Allsopp, C.E., Greenwood, B.M., Kwiatkowski, D., 1994. Variation in the TNF-alpha promoter region associated with susceptibility to cerebral malaria. Nature 371, 508–510.

Mehlotra, R.K., Henry-Halldin, C.N., Zimmerman, P.A., 2009. Application of pharmacogenomics to malaria: A holistic approach for successful chemotherapy. Pharmacogenomics 10, 435–449.

Mehlotra, R.K., Mattera, G., Bockarie, M.J., et al., 2008. Discordant patterns of genetic variation at two chloroquine resistance loci in worldwide populations of the malaria parasite Plasmodium falciparum. Antimicrob Agents Chemother 52, 2212–2222.

Meister, S., Kanzok, S.M., Zheng, X., et al., 2005. Immune signaling pathways regulating bacterial and malaria parasite infection of the mosquito Anopheles gambiae. Proc Natl Acad Sci USA 102, 11,420–11,425.

Mishra, P.C., Kumar, A., Sharma, A., 2009. Analysis of small nucleolar RNAs reveals unique genetic features in malaria parasites. BMC Genomics 10, 68.

Modiano, D., Petrarca, V., Sirima, B.S., et al., 1996. Different response to Plasmodium falciparum malaria in west African sympatric ethnic groups. Proc Natl Acad Sci USA 93, 13,206–13,211.

Mok, B.W., Ribacke, U., Rasti, N., et al., 2008. Default pathway of var2csa switching and translational repression in Plasmodium falciparum. PLoS ONE 3, e1982.

Molineaux, L., Träuble, M., Collins, W.E., Jeffery, G.M., Dietz, K., 2002. Malaria therapy reinoculation data suggest individual variation of an innate immune response and independent acquisition of antiparasitic and antitoxic immunities. Trans R Soc Trop Med Hyg 96, 205–209.

Mu, J., Myers, R.A., Jiang, H., et al., 2010. *Plasmodium falciparum* genome-wide scans for positive selection, recombination hot spots and resistance to antimalarial drugs. Nat Genet 42, 268–271.

Mueller, I., Moorthy, V.S., Brown, G.V., et al., 2009. Guidance on the evaluation of *Plasmodium vivax* vaccines in populations exposed to natural infection. Vaccine 27, 5633–5643.

Natalang, O., Bischoff, E., Deplaine, G., et al., 2008. Dynamic RNA profiling in *Plasmodium falciparum* synchronized blood stages exposed to lethal doses of artesunate. BMC Genomics 9, 388.

Neafsey, D.E., Lawniczak, M.K.N., Park, D.J., et al., 2010. SNP genotyping defines complex gene-flow boundaries among African malaria vector mosquitoes. Science 330, 514–517.

Oakley, M.S.M., Kumar, S., Anantharaman, V., et al., 2007. Molecular factors and biochemical pathways induced by febrile temperature in intraerythrocytic *Plasmodium falciparum* parasites. Infect Immun 75, 2012–2025.

Olugbile, S., Habel, C., Servis, C., et al., 2010. Malaria vaccines – the long synthetic peptide approach: Technical and conceptual advancements. Curr Opin Mol Ther 12, 64–76.

Otto, T.D., Wilinski, D., Assefa, S., et al., 2010. New insights into the blood-stage transcriptome of *Plasmodium falciparum* using RNA-Seq. Mol Microbiol 76, 12–24.

Parija, S.C., 2010. PCR for diagnosis of malaria. Indian J Med Res 132, 9–10.

Parker, M., Bull, S.J., De Vries, J., et al., 2009. Ethical data release in genome-wide association studies in developing countries. PLoS Med 6, e1000143.

Patouillard, E., Hanson, K., Goodman, C., 2010. Retail sector distribution chains for malaria treatment in the developing world: A review of the literature. Malaria J 9, 50.

Perlmann, P., Troye-Blomberg, M., 2002. Malaria and the immune system in humans. Chem Immunol 80, 229–242.

Petter, M., Bonow, I., Klinkert, M., 2008. Diverse expression patterns of subgroups of the rif multigene family during *Plasmodium falciparum* gametocytogenesis. PLoS One 3, e3779.

Plouffe, D., Brinker, A., McNamara, C., et al., 2008. *In silico* activity profiling reveals the mechanism of action of antimalarials discovered in a high-throughput screen. Proc Natl Acad Sci USA 105, 9059–9064.

Ponts, N., Harris, E.Y., Prudhomme, J., et al., 2010. Nucleosome landscape and control of transcription in the human malaria parasite. Genome Res 20, 228–238.

Raabe, C.A., Sanchez, C.P., Randau, G., et al., 2010. A global view of the nonprotein-coding transcriptome in *Plasmodium falciparum*. Nucleic Acids Res 38, 608–617.

Radivojac, P., Vacic, V., Haynes, C., et al., 2010. Identification, analysis, and prediction of protein ubiquitination sites. Proteins 78, 365–380.

Rask, T.S., Hansen, D.A., Theander, T.G., Gorm Pedersen, A., Lavstsen, T., 2010. *Plasmodium falciparum* erythrocyte membrane protein 1 diversity in seven genomes – divide and conquer. PLoS Comput Biol 6, e1000933.

Riehle, M.M., Markianos, K., Niaré, O., et al., 2006. Natural malaria infection in *Anopheles gambiae* is regulated by a single genomic control region. Science 312, 577–579.

Rihet, P., Traoré, Y., Abel, L., et al., 1998. Malaria in humans: *Plasmodium falciparum* blood infection levels are linked to chromosome 5q31-q33. Am J Hum Genet 63, 498–505.

Rottmann, M., McNamara, C., Yeung, B.K.S., et al., 2010. Spiroindolones, a potent compound class for the treatment of malaria. Science 329, 1175–1180.

Rowe, J.A., Opi, D.H., Williams, T.N., 2009. Blood groups and malaria: Fresh insights into pathogenesis and identification of targets for intervention. Curr Opin Hematol 16, 480–487.

Sabbatani, S., Fiorino, S., Manfredi, R., 2010. The emerging of the fifth malaria parasite (*Plasmodium knowlesi*): A public health concern? Braz J Infect Dis 14, 299–309.

Sam-Yellowe, T.Y., Florens, L., Wang, T., et al., 2004. Proteome analysis of rhoptry-enriched fractions isolated from *Plasmodium* merozoites. J Proteome Res 3, 995–1001.

Scherf, A., Lopez-Rubio, J.J., Riviere, L., 2008. Antigenic variation in *Plasmodium falciparum*. Annu Rev Microbiol 62, 445–470.

Siau, A., Toure, F.S., Ouwe-Missi-Oukem-Boyer, O., et al., 2007. Whole-transcriptome analysis of *Plasmodium falciparum* field isolates: Identification of new pathogenicity factors. J Infect Dis 196, 1603–1612.

Silvestrini, F., Bozdech, Z., Lanfrancotti, A., et al., 2005. Genome-wide identification of genes upregulated at the onset of gametocytogenesis in *Plasmodium falciparum*. Mol Biochem Parasitol 143, 100–110.

Speake, C., Duffy, P.E., 2009. Antigens for pre-erythrocytic malaria vaccines: Building on success. Parasite Immunol 31, 539–546.

Stevenson, M.M., Riley, E.M., 2004. Innate immunity to malaria. Nat Rev Immunol 4, 169–180.

Su, X., Jiang, H., Yi, M., Mu, J., Stephens, R.M., 2009. Large-scale genotyping and genetic mapping in *Plasmodium* parasites. Korean J Parasitol 47, 83–91.

Sundaresh, S., Doolan, D.L., Hirst, S., et al., 2006. Identification of humoral immune responses in protein microarrays using DNA microarray data analysis techniques. Bioinformatics 22, 1760–1766.

Surolia, N., RamachandraRao, S.P., Surolia, A., 2002. Paradigm shifts in malaria parasite biochemistry and antimalarial chemotherapy. Bioessays 24, 192–196.

Tarun, A.S., Peng, X., Dumpit, R.F., et al., 2008. A combined transcriptome and proteome survey of malaria parasite liver stages. Proc Natl Acad Sci USA 105, 305–310.

Tokumasu, F., Nardone, G.A., Ostera, G.R., et al., 2009. Altered membrane structure and surface potential in homozygous hemoglobin C erythrocytes. PLoS ONE 4, e5828.

Trenholme, K.R., Gardiner, D.L., Holt, D.C., et al., 2000. clag9: A cytoadherence gene in *Plasmodium falciparum* essential for binding of parasitized erythrocytes to CD36. Proc Natl Acad Sci USA 97, 4029–4033.

Tuikue Ndam, N., Bischoff, E., Proux, C., et al., 2008. *Plasmodium falciparum* transcriptome analysis reveals pregnancy malaria associated gene expression. PLoS ONE 3, e1855.

Vaughan, A.M., O'Neill, M.T., Tarun, A.S., et al., 2009. Type II fatty acid synthesis is essential only for malaria parasite late liver stage development. Cell Microbiol 11, 506–520.

Vernick, K.D., Fujioka, H., Seeley, D.C., et al., 1995. *Plasmodium gallinaceum*: A refractory mechanism of ookinete killing in the mosquito, *Anopheles gambiae*. Exp Parasitol 80, 583–595.

Waller, R.F., Ralph, S.A., Reed, M.B., et al., 2003. A type II pathway for fatty acid biosynthesis presents drug targets in *Plasmodium falciparum*. Antimicrob Agents Chemother 47, 297–301.

Wambua, S., Mwangi, T.W., Kortok, M., et al., 2006. The effect of alpha+-thalassaemia on the incidence of malaria and other diseases in children living on the coast of Kenya. PLoS Med 3, e158.

Wang, C.W., Mwakalinga, S.B., Sutherland, C.J., et al., 2010. Identification of a major rif transcript common to gametocytes and sporozoites of *Plasmodium falciparum*. Malaria J 9, 147.

Westenberger, S.J., Cui, L., Dharia, N., Winzeler, E., Cui, L., 2009. Genome-wide nucleosome mapping of *Plasmodium falciparum* reveals histone-rich coding and histone-poor intergenic regions and chromatin remodeling of core and subtelomeric genes. BMC Genomics 10, 610.

Wiesner, J., Reichenberg, A., Heinrich, S., Schlitzer, M., Jomaa, H., 2008. The plastid-like organelle of apicomplexan parasites as drug target. Curr Pharm Des 14, 855–871.

Williams, T.N., Maitland, K., Bennett, S., et al., 1996. High incidence of malaria in alpha-thalassaemic children. Nature 383, 522–525.

Woynarowski, J.M., Krugliak, M., Ginsburg, H., 2007. Pharmacogenomic analyses of targeting the AT-rich malaria parasite genome with AT-specific alkylating drugs. Mol Biochem Parasitol 154, 70–81.

Yanow, S.K., Purcell, L.A., Lee, M., Spithill, T.W., 2007. Genomics-based drug design targets the AT-rich malaria parasite: Implications for antiparasite chemotherapy. Pharmacogenomics 8, 1267–1272.

Young, J., Fivelman, Q., Blair, P., et al., 2005. The *Plasmodium falciparum* sexual development transcriptome: A microarray analysis using ontology-based pattern identification. Mol Biochem Parasitol 143, 67–79.

Zakeri, S., Kakar, Q., Ghasemi, F., et al., 2010. Detection of mixed *Plasmodium falciparum* and *P. vivax* infections by nested-PCR in Pakistan, Iran and Afghanistan. Indian J Med Res 132, 31–35.

CHAPTER

HIV Pharmacogenetics and Pharmacogenomics

C. William Wester, Sophie Limou, and Cheryl A. Winkler

INTRODUCTION

The motivation for identifying genes associated with HIV/AIDS phenotypes has gained momentum as effective vaccines and sterilizing (curative) treatments have failed to materialize. Discovery of genes and pathways utilized by human immunodeficiency virus (HIV) in the course of its lifecycle provide new targets for drug development, and knowledge of the genetic correlates of immune regulation is crucial to the design of effective vaccines against HIV. A major challenge is to integrate the information from "-omic" technologies (genomics, proteomics, metabolomics, and epigenomics) to improve clinical care and long-term survival of the estimated 33.3 million persons living with HIV infection, and to translate that knowledge to improve therapeutic options and reduce toxicities associated with combined antiretroviral therapy (cART) (Vidal et al., 2010). Optimally, personalized medicine for HIV-infected individuals would incorporate genetic and biomarker profiling to: (1) guide the optimal timing of initiation of cART treatment; (2) select the best combination of drugs to avoid toxicities and adverse reactions, thus maximizing adherence and efficacy; and (3) predict risk for non-AIDS-defining conditions (specifically cardiovascular, hepatic, and renal diseases, as well as the development of non-AIDS-associated malignancies) related to HIV infection and treatment – this becomes increasingly important in aging populations on cART (Tarr and Telenti, 2010; Rotger et al., 2009). Despite increasing evidence

that genetic variation, particularly in ADME (absorption, distribution, metabolism, and excretion) genes, affects efficacy, tolerability, and adherence, personalized HIV treatment utilizing pharmacogenetic screening is not a part of routine care (Lubomirov et al., 2007; Lubomirov et al., 2010). Cost–benefit analyses and randomized clinical trials are required to bring personalized HIV care to the clinic. Here we review the role of ADME and human leukocyte antigen (*HLA*) genes in drug-related toxicities and their potential utility in personalized medicine (Vidal et al., 2010).

PHARMACOGENETICS OF COMBINED ANTIRETROVIRAL THERAPY TOXICITY

The widespread availability of cART in the developed world has resulted in remarkable reductions in HIV-associated morbidity and mortality over the past two decades (Walensky et al., 2006). Significant progress has been made in this past decade in terms of making these potentially life-saving combination antiretroviral therapies available to persons residing in resource-limited settings of the world. Currently, 6.6 million HIV-infected persons are receiving cART in low- and middle-income countries, the vast majority of whom reside in sub-Saharan Africa (World Health Organization, UNICEF, UNAIDS, 2010). Large numbers of national initiatives offering public non-nucleoside reverse transcriptase inhibitor (NNRTI)-based

Genomic and Personalized Medicine, 2nd edition

by Ginsburg & Willard. DOI: http://dx.doi.org/10.1016/B978-0-12-382227-7.00101-4

2013 Published by Elsevier Inc.

cART have commenced in the region, with preliminary data documenting impressive efficacy outcomes among the vast majority of cART-treated adults (Boulle et al., 2010; Bussmann et al., 2008; Chi et al., 2010; Coetzee et al., 2004; Nglazi et al., 2011).

Antiretroviral (ARV) drugs are also used to prevent HIV transmission or acquisition. Landmark clinical trials presented in Rome in July 2011 (Baeten and Celum, 2011; Cohen, 2011; Thigpen et al., 2011) documented the impact of ARVs in preventing HIV sexual transmission. The HIV Prevention Trials Network HPTN052 study found that HIV-infected persons initiating cART with CD4+ cell counts of 350–550 cells/mm^3 had an unprecedented 96% reduced risk of transmitting HIV to uninfected stable sexual partners (Cohen, 2011). Clinical and prevention synergy was confirmed when early cART also reduced disease progression in infected individuals. Knowledge of one's HIV status is critical for personal decision-making. If one's HIV-infected partner does not take cART, or an HIV-uninfected person is potentially exposed by a non-stable partner, ARV medications can be used orally or topically (microbicide) by uninfected vulnerable persons themselves to prevent infection, i.e., pre-exposure prophylaxis (PrEP). The PartnersPrEP trial demonstrated a greater than 60% reduction in heterosexual HIV acquisition among men and women in Kenya and Uganda, using either daily oral tenofovir plus emtricitabine (co-formulated as Truvada™) or daily oral tenofovir alone (Baeten and Celum, 2011). The TDF2 trial found similar protection with daily oral Truvada™ in Botswana (Thigpen et al., 2011). PrEP efficacy in heterosexuals extends protective evidence from men who have sex with men (Grant et al., 2010) and in women who used topical intermittent 1% tenofovir vaginal gel (pre/post coitus) (Abdool Karim et al., 2010). The use of antiretroviral therapy for mother and/or child to prevent mother-to-child transmission has greatly reduced maternal transmission to infants (Chasela et al., 2010; De Vincenzi, 2011; Marazzi et al., 2011; World Health Organization, 2010). It is therefore likely that the number of ARV-treated individuals who are either HIV seronegative or who do not meet the "when to start" guidelines for antiretroviral therapy (ART) will increase as pre-exposure prophylaxis and ARV treatment is increasingly used to reduce risk of HIV transmission and acquisition.

Effective therapies have promoted immunologic recovery, but adverse metabolic complications and negative health outcomes, including ARV-medication-related toxicities, have emerged as major short- and longer-term health concerns among this rising number of cART-treated persons (Antiretroviral Therapy Cohort Collaboration, 2010; Brady et al., 2010; Deeks and Phillips, 2009; Martinez et al., 2007; Palella et al., 2006). Recent studies showed that ARV treatment is characterized by varying rates of adverse events and responses (Tozzi, 2010). Up to 45% of treatment-naïve patients change or discontinue treatment during their first year of cART, primarily due to poor tolerance or adverse drug reactions rather than

suboptimal virologic response (Elzi et al., 2010; Vo et al., 2008). Genetic variation among human beings accounts for a substantial proportion of this variability (Tozzi, 2010). A number of associations between human genetic variants and predisposition to adverse events have been described, and for some ARV medications, a clear and causal genotype–phenotype correlation has already been established (Tozzi, 2010). Here we highlight and review the most significant associations, listed by major antiretroviral medication drug class, that have been published to date. A list of antiretroviral drugs, their side effects, and associated genetic factors are shown in Table 101.1.

ESTABLISHED PHARMACOGENETIC PREDICTORS IN HIV TREATMENT

Nucleoside (or Nucleotide) Reverse Transcriptase Inhibitors

Abacavir Hypersensitivity Reaction

Abacavir (ABC) is a guanosine nucleoside reverse transcriptase inhibitor utilized as a component in cART for the treatment of HIV-1 (Thompson et al., 2010). The main toxicity associated with abacavir treatment is a potentially life-threatening hypersensitivity reaction, commonly referred to as abacavir hypersensitivity reaction (ABC HSR), which has been documented to occur in approximately 5–8% of ABC-treated patients. Clinically, ABC HSR is characterized by a combination of fever, rash, and the development of constitutional, gastrointestinal, and/or respiratory symptoms, appearing within the first six weeks of treatment, and becoming more severe with continued dosing (Chaponda and Pirmohamed, 2011; Hetherington et al., 2001; Phillips and Mallal, 2009). Upon abacavir discontinuation, signs of ABC HSR rapidly reverse without sequelae. However, subsequent rechallenge in persons with suspected ABC HSR after initial exposure is contraindicated, since it may result in severe morbidity or even mortality (Clay, 2002; Shapiro et al., 2001).

Immunogenetic Basis for Abacavir Hypersensitivity Reactions

The immunological basis of ABC HSRs is not fully understood. Early post-market reports suggested an immunogenetic basis, as the incidence was higher in individuals with European ancestry (Symonds et al., 2002) and familial clustering was observed among family members (Peyrieere et al., 2001). In 2002, two independent groups reported an association between ABC HSR and the *HLA-B*5701* allele (Hetherington et al., 2002; Mallal et al., 2002), which was then replicated by several groups of investigators (Hughes et al., 2004a, b; Martin et al., 2004). *HLA-B*5701* is an allele of the multi-allelic *HLA* class I *HLA-B* locus, coding for cell-surface glycoproteins that present antigens to T-cell receptors to elicit immune responses.

TABLE 101.1 Genetic associations with components of combination antiretroviral regimens

	Generic name	Association	Gene	Function	Reference(s)
Nucleotide reverse transcriptase inhibitors					
3TC	Lamivudine	↑Intracellular exposure	ABCC4 (3724G>A), (4131T>G)	Disposition	Anderson et al., 2006
ZDV	Zidovudine				
ABC	Abacavir	Hypersensitivity reaction	HLA-B*5701†	Toxicity	Mallal et al., 2002
TDF	Tenofovir	Proximal renal tubulopathy	ABCC2 CATC haplotype; 1249G>A	Toxicity	Izzedine et al., 2006
		Peripheral neuropathy	Mitochondrial haplogroups T and L1c	Toxicity	Canter et al., 2010; Hulgan et al., 2005
		Pancreatitis	CFTR 1717-1G>A and SPINK-1-112C>T (also in general population)	Toxicity	Felley et al., 2004
Non-nucleotide reverse transcriptase inhibitors					
EFV	Efavirenz	Increased exposure	CYP2B6*6 (516G>T), CYP2B6*18 (983T>G)	Disposition	Haas et al., 2004; Rotger et al., 2005a, 2009; Tsuchiya et al., 2004
EFV	Efavirenz	CNS side effects	CYP2B6*6 (516G>T), CYP2B6*18 (983T>G)	Toxicity	Haas et al., 2004; Rotger et al., 2005a
NVP	Nevirapine	Rash; hypersensitivity reaction/ rash; hepatotoxicity	HLA-DRB1*0101	Toxicity	Martin et al., 2005a; Vitezica et al., 2008
		Hypersensitivity reaction*	HLA-Cw8-B14 (in some populations)	Toxicity	Littera et al., 2006
EFV	Efavirenz	Reduced plasma exposure; increase in HDL cholesterol	ABCB1 (MDR1) (3435C>T)	Disposition Toxicity	Ribaudo et al., 2010; Rotger et al., 2009
Protease inhibitors					
SQV	Saquinavir	Faster oral clearance	CYP3A5*1, CYP3A5*3	Disposition	Josephson et al., 2007; Mouly et al., 2005
IDV	Indinavir	Faster oral clearance	CYP3A5*3; CYP3A4*1B	Disposition	Anderson et al., 2006; Bertrand et al., 2009
ATV IDV	Atazanavir Indinavir	Unconjugated hyperbilirubinemia and jaundice	UGT1A1*28 (UGTA1-TA7), ABCB1 (MDR1) 3435C>T	Disposition Toxicity	Anderson et al., 2006; Rodriguez-Novoa et al., 2007; Rotger et al., 2005b
LPV	Lopinavir boosted with ritonavir	Clearance rates	SLO1B1*4, SLO1B1*5, CYP3A rs6945984, ABCC2 rs717620	Disposition	Lubomirov et al., 2007; Lubomirov et al., 2010; Lubomirov et al., 2011
	All protease inhibitors (PI's)	Hyperlipidemia	APOE ε2 and ε4 genotypes, APOA5-1131T>C, 64G>C, APOC3 482C>T, 455C>T, 3238C>G, ABCA1 2962A>G	Toxicity Disposition	Rotger et al., 2009; Song et al., 2004; Talmud et al., 2002

CNS, central nervous system; HDL, high density lipoprotein.

†Information/warnings for abacavir and specifically the need for baseline (pre-initiation of abacavir) testing for the genetic variant HLA-B*5701 is contained in the *Guidelines for the use of antiretroviral agents in HIV-1-infected adults and adolescents* (Panel on Antiretroviral Guidelines for Adults and Adolescents, 2011) as well as the Food and Drug Administration (FDA) drug label (effective July 18, 2008; http://www.accessdata.fda.gov/scripts/cder/drugsatfda/index.cfm).

Table adapted from Tozzi, 2010.

False Positive Clinical Diagnosis and Patch Testing

Early studies mistakenly suggested a lower sensitivity of *HLA-B*5701* for ABC HSR in non-Caucasian populations (Hughes et al., 2004a), but it became evident that this observation was driven by false positive clinical diagnoses resulting from nonspecific symptoms associated with ABC HSR. To reduce the overestimation of ABC HSR clinical diagnoses over the true immunological ABC HSR, epicutaneous patch testing was developed. Patch testing can discriminate between false positive clinically diagnosed ABC HSRs and true immunologically mediated ABC HSRs among persons who have previously ingested the medication (Antiretroviral Therapy Cohort Collaboration, 2010; Deeks and Phillips, 2009). Patch testing cannot be used as a predictive screening tool for ABC HSR since prior exposure to abacavir is required for an immune reaction. To date, all clinically diagnosed patients exhibiting a positive result on patch testing carry *HLA-B*5701*, illustrating the robustness of the association between this *HLA* allele and ABC HSR (Mallal et al., 2008; Martin et al., 2004; Phillips et al., 2002; Phillips et al., 2006; Phillips et al., 2007; Saag et al., 2008; Shear et al., 2008).

*Utility of HLA-B*5701 Screening and Translation to Clinical Practice*

Before the development of patch testing, several observational studies had demonstrated that screening for *HLA-B*5701* prior to initiation of abacavir treatment significantly reduced the incidence of ABC HSR and all-cause abacavir discontinuations (Rauch et al., 2006; Reeves et al., 2006; Trottier et al., 2007; Waters et al., 2007; Young et al., 2008; Zucman et al., 2007) (Figure 101.1). A randomized clinical trial using patch testing to accurately diagnose ABC HSR clearly showed that *HLA-B*5701* screening prior to initiation of abacavir treatment fully eliminated immunological ABC HSR. It also significantly reduced clinically diagnosed ABC HSR, suggesting that the confidence in the *HLA-B*5701*-negative screening result impacts the management of patients' symptoms (Phillips et al., 2005). Some *HLA-B*5701*-positive patients tolerate the ABC treatment and do not develop ABC HSR. Therefore, the clinical use of *HLA-B*5701* screening would exclude a subset of *HLA-B*5701*-positive individuals who would not display ABC HSR. However, it would also entirely prevent all ABC HSR (100% negative prediction), and it is therefore an ideal screening test to prevent these potentially fatal hypersensitivity reactions.

As shown in Figure 101.1, *HLA-B*5701* is a sensitive (100%) marker for immunological ABC HSRs in both black and white populations in the United States (Saag et al., 2008), and potentially also in Hispanic and Thai patients (Hughes et al., 2004a; Mosteller et al., 2006). Thus, even if the *HLA-B*5701* frequency varies across populations (Figure 101.2) (Orkin et al., 2010; Poggi et al., 2010; Parczewski et al., 2010; Sanchez-Giron et al., 2011), the test should be generalizable. At a time of increasing migration and admixture of global populations, selective screening on the basis of race or ethnicity is not advisable. Screening for *HLA-B*5701* before the

prescription of abacavir has been shown to have clinical utility – in 2008, the US Food and Drug Administration (FDA) approved changes to the drug's label, and *HLA-B*5701* screening is now recommended in clinical guidelines.

Tenofovir Proximal Tubulopathy

Tenofovir disoproxil fumarate (TDF) is a bioavailable ester prodrug of tenofovir, an acyclic nucleoside reverse transcriptase inhibitor (NRTI) with activity against HIV-1, HIV-2, and hepatitis B virus (Izzedine et al., 2006). The two main toxicities associated with TDF use are nephrotoxicity and bone mineral density (bone porosity) abnormalities. TDF is extensively and rapidly excreted into the urine by the kidneys via glomerular filtration and tubular secretion (Deeks and Phillips, 2009; Izzedine et al., 2006). Current routine practice is to calculate creatinine clearance (using one of the accepted formulas) and only initiate preferred TDF-containing cART if a person's creatinine clearance is greater than 50–60 milliliters per minute, due to the potential of TDF-induced and/or exacerbated nephrotoxicity. Although large prospective trials, including accumulating data among TDF-treated adults residing in resource-limited settings of the world, have shown that TDF is very well tolerated and relatively safe for the kidney, with a very low rate of renal insufficiency, cases of tubular dysfunction, including the development of Fanconi's syndrome, have been reported (Barrios et al., 2004; Coca and Perazella, 2002; Nelson et al., 2007; Rodriguez-Novoa et al., 2009b). Renal abnormalities are relatively common in HIV-infected populations (Kopp and Winkler, 2003; Rodriguez-Novoa et al., 2009b; Yanik et al., 2010). In addition, there is high interindividual variability in the characteristics and severity of renal dysfunction associated with TDF use, which suggests an important underlying role of host genetics (Rodriguez-Novoa et al., 2009a, b). Carriers of two copies of *APOL1* codon-changing variants found only on African-ancestry chromosomes have an increased risk (odds ratio = 30) for HIV-associated nephropathy (Kopp et al., 2011); the potential interaction of *APOL1* variants with TDF or other ART components has not been investigated. Some have postulated that TDF-associated toxicity is mitochondrial in etiology (Rodriguez-Novoa et al., 2009b), although TDF has not been shown to be as mitochondrially toxic as other, earlier NRTIs [e.g., stavudine (d4T), didanosine (ddI)]. The pharmacogenetics of TDF has largely focused on the other possible mechanism of toxicity – interference with the normal function of tubular cells, specifically interference with key transporter proteins (Rodriguez-Novoa et al., 2009a, b). Multidrug resistance (MDR) protein 2, coded by the *ABCC2* gene, is apically located in the cellular membrane of proximal renal tubular cells (Rotger et al., 2005b). Several associations have been reported relating polymorphisms in transporter genes with renal clearance rates and TDF-associated tubulopathy/renal toxicity; the *ABCC2* CATC haplotype [hallmarked by single nucleotide polymorphisms (SNPs) at positions -29, 1249, 3563, and 3972] is a determinant of proximal tubular damage, as is the *ABCC2*-24C allele (Izzedine et al., 2006; Kiser et al., 2008;

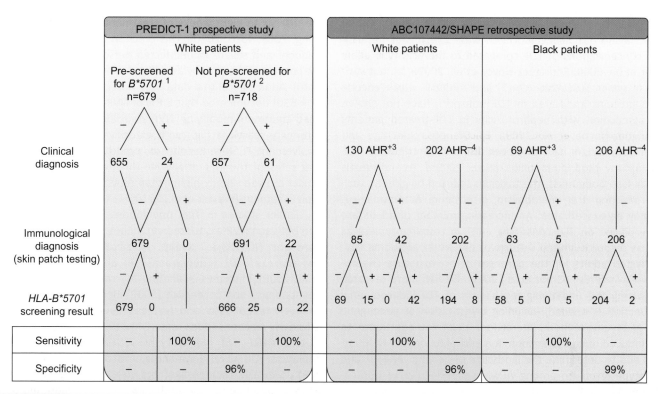

Figure 101.1 Summary of the results obtained in the PREDICT-1 prospective study (Mallal et al., 2008) (left, yellow panel) and the SHAPE retrospective study (Saag et al., 2008) (right, blue panel), focusing on the benefit of *HLA-B*5701* screening before abacavir treatment initiation for prevention of ABC HSR. [1]*HLA-B*5701* screening before abacavir treatment initiation to exclude people suspected of developing AHR; [2]no *HLA-B*5701* screening before abacavir treatment initiation; [3]retrospective identification of individuals who initiated abacavir treatment and who experienced a clinically diagnosed ABC HSR; [4]retrospective identification of individuals who initiated abacavir treatment and who did not experience any ABC HSR. [#]Some individuals could not be tested by skin patch (five in PREDICT-1, three in White-SHAPE, and one in Black-SHAPE).

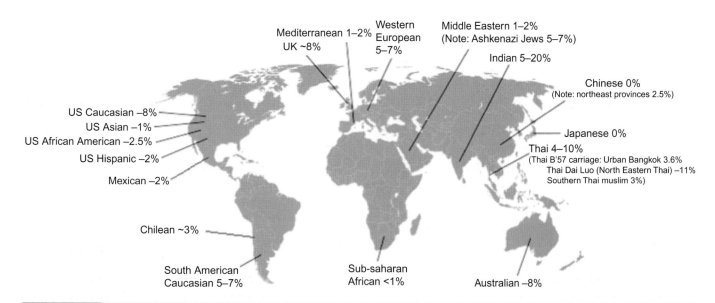

Figure 101.2 *HLA-B*5701* frequency in global populations. *Adapted from a review article by Phillips and Mallal, 2009.*

Rodriguez-Novoa et al., 2009a). In addition, the *ABCC4* (*MRP4*) 3463G allele has been associated with 35% higher intracellular concentrations of TDF compared to the wild-type allele (Kiser et al., 2008; Rodriguez-Novoa et al., 2009a, b), but variants in genes *SLCOA226* (*OAT1*) and *ABCCB1*, which encode other proteins implicated in TDF transport, have not shown any association with nephrotoxicity in TDF-treated patients (Rodriguez-Novoa et al., 2009a, b). Of note, older age and lower body weight have also been found to contribute to TDF tubulopathy (Rodriguez-Novoa et al., 2009b). As individuals age or lose body mass, the toxicities induced by genetic variants affecting drug metabolism, disposition, or plasma levels may be exacerbated. An increasing number of adults are being placed on TDF-containing cART in resource-replete as well as resource-limited settings of the world, including HIV-uninfected adults for the purposes of pre-exposure chemoprophylaxis (PrEP) (Grant et al., 2010). Clearly, more research elucidating the underlying mechanism of TDF-induced renal dysfunction is needed, including investigation of gene–gene and gene–environment interactions that may contribute to TDF-induced renal dysfunction. It is critical to identify new variants and to determine the landscape of genetic variance and haplotype structure among ADME genes in diverse populations to assess the contribution of genetic and environmental factors associated with TDF-induced renal dysfunction.

Non-Nucleoside Reverse Transcriptase Inhibitors
Nevirapine-Associated Toxicities

The NNRTI nevirapine (NVP) is widely prescribed for the treatment of HIV-1, predominantly in resource-limited settings of the world. NVP has been associated with two potentially serious treatment-related toxicities – cutaneous hypersensitivity reactions and hepatotoxicity (Baylor and Johann-Liang, 2004; Tozzi, 2010; Yuan et al., 2011). These toxicities typically occur during the initial weeks of therapy, especially in persons having higher baseline CD4+ cell counts (Yuan et al., 2011). Nevirapine clearance is principally done via the cytochrome P450 pathway. Major metabolic pathways for NVP also involve hydroxylation by CYP2B6 and CYP3A4. Nevirapine induces *CYP2B6* and *CYP3A4* expression over a period of several weeks, increasing its own clearance (Riska et al., 1999; Yuan et al., 2011). Genetic variants in the *ABCB1* gene, which encodes permeability glycoprotein 1 (Pgp), a multidrug resistance protein involved in removing xenobiotic substrates including antiretroviral drugs from tissue and cellular compartments, have been associated with risk of NVP-induced hepatotoxicity (Ciccacci et al., 2010; Haas et al., 2006; Ritchie et al., 2006).

Nevirapine-Associated Cutaneous Hypersensitivity Reactions

Nevirapine-treated persons may develop a cutaneous hypersensitivity reaction similar in scope and severity to ABC HSR, occurring in up to 5% of NVP-treated patients, typically within the first six to eight weeks of treatment (Baylor and Johann-Liang, 2004; Tozzi, 2010). This reaction is also typically more frequent and severe in uninfected persons who receive NVP for post-exposure prophylaxis (Patel et al., 2004; Vidal et al., 2010). Adult clinical trial data from a large cohort in Botswana (n = 650) documented that 5.8% of NVP-treated adults developed treatment-modifying NVP-associated cutaneous hypersensitivity reactions (i.e., with extensive mucous membrane involvement – Stevens–Johnson syndrome), which suggests that rates of this potentially life-threatening toxicity appear similar or even slightly higher than rates reported in resource-replete settings (Wester et al., 2010).

Earlier studies in Thai populations reported an association between NVP-associated cutaneous hypersensitivity reactions and *HLA* class I alleles, specifically with *HLA-B*3505* and *HLA-Cw*0401* (Chantarangsu et al., 2009; Likanonsakul et al., 2009). In studies evaluating 147 cases and 187 controls, Chantarangsu and colleagues (2009) identified an association between *HLA-B*3505* and NVP-associated cutaneous hypersensitivity reactions [odds ratio (OR) ~19]; also predictive were *HLA-Cw*0401* (OR ~5) and the *HLA-B*3501:HLA-Cw*0401* haplotype (OR ~12) (Chantarangsu et al., 2009; Yuan et al., 2011). Similarly, Likanonsakul and colleagues (2009) reported an association between *HLA-Cw*04* and NVP-associated cutaneous hypersensitivity reactions, evaluating a smaller number of cases and controls. Yuan and colleagues (2011) supported these potential associations between *HLA* class I alleles *HLA-B*3505* and *HLA-Cw*0401* and NVP-associated cutaneous hypersensitivity reactions, evaluating a large series of patients having initial CD4+ cell counts of at least 150 cells/mm[3] (Yuan et al., 2011). In their multi-country study (n = 175 cases, with 2:1 matched controls), they found strong associations between *HLA-B*35*, *HLA-Cw*04*, and NVP-associated cutaneous toxicities among Asians, particularly among those of Thai origin, and extended the association to African-ancestry and European-ancestry populations residing outside of Africa. Results showed that among *HLA-Cw*04* carriers, NVP-associated cutaneous toxicities increased in persons concomitantly having the *CYP2B6* 516G→T polymorphism. In contrast, this study did not find any significant associations for *HLA-B*35* in African-ancestry or European-ancestry groups, which the authors suggest is possibly due to the relative infrequency of the *HLA-B*35* allele in these populations (Yuan et al., 2011).

Nevirapine-Associated Hepatotoxicity

NVP-associated hepatotoxicity, which can be severe and potentially life-threatening, typically develops within the first 18 weeks following NVP initiation. It tends to occur with significantly higher frequency in the treatment of cART-naïve females having a baseline CD4+ cell count of greater than 250 cells/mm[3] and in the treatment of cART-naïve males having a baseline CD4+ cell count value of greater than 400 cells/mm[3] (Tozzi, 2010). Patients with chronic hepatitis B and/or C infection also appear to be at higher risk.

Genetic risk for NVP-associated hepatotoxicity, in contrast to NVP-associated cutaneous hypersensitivity reactions, appears to be mediated by *HLA* class II mechanisms, and is not under the influence of *CYP2B6* metabolism variants (Yuan et al., 2011). One study from Australia (Martin et al., 2004; Vidal et al., 2010; Yuan et al., 2011) documented *HLA-DRB1*0101* as a risk for NVP-associated hepatotoxicity, whereas studies from Japan and Sardinia implicated *HLA-Cw*08* (Gatanaga et al., 2007b; Littera et al., 2006). The Australian study (Martin et al., 2005b; Yuan et al., 2011) included 26 NVP-treated cases who experienced an increase in the liver enzyme alanine aminotransferase (ALT), fever and/or rash, and 209 controls. They found that *HLA-DRB1*0101* was significantly associated with hepatic/systemic reactions (OR = 4.8), but not with isolated cutaneous toxicities. This elevated NVP-associated toxicity risk was only seen in persons having a CD4+ cell percentage of 25% or greater (Martin et al., 2005b; Yuan et al., 2011). In their large multi-country NVP genetic association study, Yuan and colleagues confirmed an association between *HLA-DRB1*01* and NVP-associated hepatotoxicity among whites, but they did not find any significant association with *HLA-DRB1*01* in blacks or Asians. They also found a possible association between *HLA-DQB1*05* and NVP-associated hepatotoxicity in whites, which they felt reflected a linkage between *HLA-DRB1*01* and *HLA-DQB1*05*. When evaluating only NVP-treated cases having isolated hepatotoxicity, without concomitant cutaneous and/or systemic involvement, they found that the association with *HLA-DQB1*05* was no longer significant, but that the *HLA-DRB1*01* effect size increased from OR = 3.0 to 3.6 (Yuan et al., 2011).

The risk for NVP-associated hepatotoxicity is also influenced by its metabolism in hepatocytes via the cytochrome P450 system, specifically involving the transporter enzyme P-glycoprotein (Ciccacci et al., 2010; Haas et al., 2006; Ritchie et al., 2006; Tozzi, 2010; Vidal et al., 2010). The 3435C>T variant in the *ABCB1* gene influences the risk of NVP-associated hepatotoxicity. Individuals with the *ABCB1* 3435C allele have a much higher risk of hepatotoxicity, especially among those coinfected with hepatitis B; in such cases, hepatotoxicity occurred 82% of the time (Ritchie et al., 2006). Individuals with the T allele have a correspondingly lower risk of hepatotoxicity (OR = 0.25; p = 0.021) (Ritchie et al., 2006). Similar protective effects of the T allele (*ABCB1* 3435C>T) were also observed by Haas and colleagues (2006) and Ciccacci and coworkers (2010) in studies of HIV-infected adults receiving NVP-based cART in South Africa (Haas et al., 2006) and Mozambique (Ciccacci et al., 2010), respectively. These resource-limited countries are in southern Africa, where HIV infection rates are among the highest in the world, and nevirapine is widely prescribed because of its low cost and efficacy (Table 101.2).

In summary, it appears that fundamentally different pathways are involved with the development of the most significant and potentially severe NVP-associated toxicities – cutaneous hypersensitivity reactions and hepatotoxicity (Yuan

et al., 2011). Both ADME genes and *HLA* class I and class II genes (Tables 101.1 and 101.2) affect risk of NVP-associated toxicity. Cutaneous hypersensitivity reactions seem to be primarily mediated via *HLA* class I mechanisms and influenced by NVP plasma levels regulated by *CYP2B6* 516G>T variants. On the other hand, hepatotoxicity is primarily mediated via *HLA* class II mechanisms (Yuan et al., 2011) with NVP plasma levels or *CYP2B6* genotype showing little influence. However, in persons coinfected with the hepatitis B virus, the *ABCB1* 3435C>T variant is strongly associated with hepatotoxicity. Unfortunately, the presence or absence of these particular ADME and *HLA* variants lack sensitivity, and they therefore have minimal clinical utility independently for persons initiating NVP-containing cART regimens (Yuan et al., 2011), although it is feasible that by combining genetic and environmental risk factors, a more sensitive predictive risk score might be obtained (Lubomirov et al., 2011). Large proportions of NVP-treated persons developing cutaneous and/or hepatic toxicities carried no risk alleles for any of the genes, highlighting the need for in-depth genetics research to identify additional genetic factors associated with NVP toxicities.

Efavirenz Clearance/Drug Levels

The NNRTI efavirenz (EFV) is widely prescribed for the treatment of HIV-1 infection, but central nervous system (CNS) toxicities occur commonly, and population-specific differences in pharmacokinetics and treatment response have been noted (Clifford et al., 2005; Gulick et al., 2006; Phillips et al., 2007; Rauch et al., 2006; Ribaudo et al., 2010; Saag et al., 2008; Shear et al., 2008). The enzyme CYP2B6 metabolizes EFV with minor involvement of CYP3A4 and CYP3A5, and since there is considerable variability in expression levels and function of *CYP2B6*, significant interindividual variability exists in EFV exposure and the development of CNS abnormalities, its main treatment-related toxicity. The CNS toxicity manifests as abnormal and/or vivid dreams (or nightmares), dizziness, depression, feelings of intoxication/lethargy, confusion/abnormal thoughts, euphoria, and aggression (and other forms of unusual behavior), as well as homicidal and suicidal ideation (rarely) (Mukonzo et al., 2009; Tozzi, 2010). As a result, *CYP2B6* polymorphisms and their association with EFV disposition have been extensively evaluated (Tozzi, 2010); the National Center for Biotechnology Informatics (NCBI) database of SNPs (dbSNP) now reports 81 missense or frame-shift variants in the *CYP2B6* gene (Vidal et al., 2010). This number increases up to 119 in the 1000 Genomes Project database (1000 Genomes Project Consortium, 2010).

Haas and colleagues, in their sub-study of HIV-infected patients participating in adult clinical trials group study 5097 (5097s), reported that the T allele of the *CYP2B6* 516G→T polymorphism (Glu172His) was associated with greater EFV plasma exposure (p < 0.0001); the TT genotype most strongly associated with CNS toxicity is nearly seven-fold more frequent in African Americans (20%) than in European Americans (3%).

TABLE 101.2	Nevirapine adverse effects studies		
Genetic risk allele	**Geographic region and continental ancestry**	**Risk group**	**Reference**
Nevirapine hypersensitivity reaction			
HLA-B*3505 and HLA-Cw*04	Thailand	Asian	Peyrieere et al., 2001
HLA-B*3505	Asians, African-descent (blacks) and European-descent (whites) enrolled from Argentina, Australia, Canada, France, Germany, Netherlands, Spain, Taiwan, Thailand, United Kingdom, and USA (Tennessee)	Asian	Yuan et al., 2011
HLA-Cw*04	Asians, African-descent (blacks) and European-descent (whites) enrolled from Argentina, Australia, Canada, France, Germany, Netherlands, Spain, Taiwan, Thailand, United Kingdom, and USA (Tennessee)	All	Yuan et al., 2011
HLA-Cw*04	Thailand	Asian	Likanonsakul et al., 2009
HLA-Cw*08	Japan	Asian	Gatanaga et al., 2007b
HLA-Cw*08	Sardinia	European	Littera et al., 2006
CYP2B6 516T	Asians, African-descent (blacks) and European-descent (whites) enrolled from Argentina, Australia, Canada, France, Germany, Netherlands, Spain, Taiwan, Thailand, United Kingdom, and USA (Tennessee)	All	Yuan et al., 2011
CYP2B6 516TT and HLA-Cw*04	Asians, African-descent (blacks) and European-descent (whites) enrolled from Argentina, Australia, Canada, France, Germany, Netherlands, Spain, Taiwan, Thailand, United Kingdom, and USA (Tennessee)	African-descent	Yuan et al., 2011
Nevirapine hepatotoxicity			
HLA-DRB1*0101	Australia	European	Martin et al., 2005a
HLA-DRB1*01	France	European	Vitezica et al., 2008
HLA-DRB*01	Asians, African-descent (blacks) and European-descent (whites) enrolled from Argentina, Australia, Canada, France, Germany, Netherlands, Spain, Taiwan, Thailand, United Kingdom, and USA (Tennessee)	European-descent	Yuan et al., 2011
ABCB1 3435C>T	South Africa	Black Africans	Haas et al., 2006
ABCB1 3435C>T	(USA) Tennessee	European and African Americans	Haas et al., 2006
ABCB1 3435C>T	Mozambique	Black Africans	Ciccacci et al., 2010

Nevirapine-Associated Adverse Reactions (cutaneous hypersensitivity and/or hepatic toxicities) [reproduced with permission from author David W. Haas, MD (as presented at the *Pharmacogenomics: A Path Towards Personalized HIV Care* meeting; 16–17 June, 2010, Rockville, MD, USA, sponsored by the National Institute of Allergy and Infectious Diseases, National Institutes of Health, Bethesda, MD)].

They reported that EFV exposure was higher for carriers of the T allele; the area under the plasma concentration time curve (AUC) for 0–24 hours, according to G/G, G/T, and T/T genotype, was 44 (n=78), 60 (n=60), and 130 (n=14) μg/ml, respectively (p<0.0001). Carriage of CYP2B6 516T, associated with elevated EFV exposure, was also associated with CNS symptoms at week one (p=0.036) (Haas et al., 2004). Additional polymorphisms have been described that are associated with the loss or reduction in CYP2B6 activity (Vidal et al., 2010). These include 983T→C (allele *CYP2B6*16*), 785A→G (allele *CYP2B6*16*), 593T→C (allele *CYP2B6*27*), and 1132C→T (allele *CYP2B6*28*), which, particularly in homozygous individuals, show evidence for heightened risk for development of high plasma EFV concentrations (Vidal et al., 2010). In addition, although EFV is not a substrate of the permeability glycoprotein (P-glycoprotein) drug transporter encoded by *ABCB1*, Fellay and colleagues (2002) have suggested that a polymorphism in the *ABCB1* gene may be associated with low plasma EFV concentrations.

The utility of EFV pharmacogenetics data in personalized treatment of HIV-1 was shown by a small study where a patient's genotype was used to guide dosage of EFV. Lower doses (400 instead of the usual 600 mg) were administered to HIV-infected adults identified as being carriers of the CYP2B6 516T allele. Gatanaga and colleagues found that the majority (72%) of patients receiving lower EFV dosing had significantly reduced CNS symptoms without compromising virologic efficacy (Gatanaga et al., 2007a). Similar larger-scale randomized trials and cost-effectiveness analyses of targeted CYP2B6 screening need to be performed in order to bring pharmacogenetics to clinical practice. Such studies will greatly inform policy-makers in regions of the world where higher proportions of those initiating cART may be at risk for these potentially debilitating EFV-related CNS toxicities that may compromise drug efficacy due to poor adherence. If unaddressed, poor adherence leading to a loss of virologic control may compromise the long-term success of such regimens and lead to unnecessarily high rates of circulating drug-resistant virus.

Protease Inhibitors

Atazanavir-induced Hyperbilirubinemia

Ritonavir-boosted atazanavir is one of the most commonly prescribed protease inhibitors for first-line cART. Atazanavir (ATV) is typically very well tolerated, but the potential to develop hyperbilirubinemia is a cosmetically unappealing side effect. Early clinical trials suggested that ATV-associated hyperbilirubinemia occurred in 20–48% of patients. This abnormality results from the inhibition of the drug-transporter uridine diphosphate-glucuronosyltransferase 1A1, which is encoded by the gene UGT1A1 (Beutler et al., 1998; Haas et al., 2003; Quirk et al., 2004; Sanne et al., 2003). This enzyme catalyzes the glucuronidation of bilirubin (Ribaudo et al., 2010). An increased number of thymine adenine (TA) repeats has been associated with diminished UGT1A1 activity, and persons homozygous for the seven TA repeats (the 7/7 genotype) have chronic hyperbilirubinemia (Gilbert's syndrome) (Beutler et al., 1998; Quirk et al., 2004). Promoters containing seven TA repeats (allele UGT1A1*28) have been found to be less active than the wild-type six repeats, resulting in lower gene expression and decreased UGT1A1 activity. Levels of UGT1A1 in persons homozygous or even heterozygous for UGT1A1*28 have been shown to be lower than those in persons with the wild-type six repeats (Beutler et al., 1998; Quirk et al., 2004). UGT1A1 is important for the metabolism of numerous medications, and among patients treated for solid tumors, the homozygous UGT1A1*28 genotype has been shown to predict toxicity (including diarrhea and neutropenia) to the anti-neoplastic medication irinotecan, a substrate of UGT1A1 (Quirk et al., 2004). The homozygous UGT1A1*28 genotype has also been associated with hyperbilirubinemia in 353 patients receiving the PI ATV in phase 1 clinical trials (Quirk et al., 2004). Among patients achieving therapeutic serum drug concentrations, the *28/*28 genotype was highly predictive of total serum bilirubin increases to greater than 2.5 mg/dl (Quirk et al., 2004). The UGT1A1*28 genotype clearly appears to be associated with the development of ATV-induced hyperbilirubinemia. However, since ATV-induced hyperbilirubinemia is purely cosmetic and is not associated with any morbidity or mortality, genetic screening and/or testing for UGT1A1*28 in HIV-infected individuals may have a low clinical priority. However, the development of stigmatizing, overt jaundice may lead to poor adherence or discontinuation of atazanavir.

Nephrolithiasis (kidney stones) is a rare but potentially serious toxicity of atazanavir. The stones from patients with ATV-related nephrolithiasis contain the drug, which is insoluble in alkaline fluids. It is well known that adequate absorption of ATV requires an acidic gastric pH, therefore current treatment guidelines recommend avoiding the concomitant administration of atazanavir with proton-pump inhibitors (PPIs). This toxicity does not appear to have a pharmacogenetic association.

Lopinavir/Ritonavir Plasma Concentrations

There is marked interindividual variability in plasma concentrations of most, if not all, antiretroviral medications (Hartkoorn et al., 2010). This is particularly true for PIs, for which therapeutic drug monitoring (TDM) should be considered for optimizing treatment in some circumstances (Poggi et al., 2010). Attaining and maintaining therapeutic plasma PI concentrations are critical for efficacy, as sub-therapeutic levels can lead to the emergence of drug resistance (Hartkoorn et al., 2010). Ritonavir-boosted lopinavir is a commonly prescribed PI, especially in resource-limited settings of the world, where it is typically the only PI available and is reserved for use in second-line cART regimens. In addition to ADME genes, age, gender, and underlying synthetic liver function have all been shown to affect plasma drug concentrations (Hartkoorn et al., 2010).

PIs are primarily metabolized by the CYP3A enzyme system, but they are also substrates for the drug efflux transporters ABCB1, ABCC1, and ABCC2. These proteins affect integral components of PI drug metabolism, specifically absorption, distribution, and clearance (Hartkoorn et al., 2010). In addition, many of the genes encoding these important efflux transporter proteins are highly variant, with functional polymorphisms (Hartkoorn et al., 2010). Organic-anion-transporting polypeptides (OATPs), encoded by the SLCO genes, represent a family of membrane transport proteins involved in the influx of numerous chemical compounds. These OATPs have varied substrate specificity and are often expressed simultaneously (Hartkoorn et al., 2010). Numerous genetic polymorphisms have also been reported in the SLCO1B1, SLCO1A2, and SLCO1B3 genes. SNPs located within the trans-membrane domain of OATP1B1 have been associated with decreased transport function, both in vitro and in vivo (Hartkoorn et al., 2010).

The SLCO1B1 521T→C (Val174Ala) (rs4149056) C allele is frequent in European populations (~15%), but is rare or near-absent in sub-Saharan Africa and Oceania (Hartkoorn et al., 2010; Pasanen et al., 2008). Hartkoorn and colleagues have recently shown that the 521-C allele of *SLCO1B1* was significantly associated with higher plasma lopinavir levels (Hartkoorn et al., 2010). The clinical utility of testing for *SLCO1B1* 521T→C is not apparent. The widespread use of ritonavir-boosted lopinavir is limited in resource-replete countries where the polymorphism is common. In sub-Saharan Africa, where increasing numbers of persons are being treated with this protease inhibitor as a second-line treatment, the polymorphism is rare-to-absent – genetic testing is therefore not required for dosage determinations. No associations with lopinavir plasma levels were observed for functional variants of *SLCO1A2* and *SLCO1B3* (Hartkoorn et al., 2010). Additional studies are ongoing to assess the influence of other SNPs and functional variants within this important *SLCO* gene family (Hartkoorn et al., 2010).

Ritonavir-boosted Protease Inhibitors and Lipodystrophy

Differential metabolic effects of various PIs have been reported (Noor et al., 2004). Metabolic complications of HIV infection and its treatment frequently present as lipodystrophy, an umbrella term encompassing loss of peripheral adipose tissue (lipoatrophy), redistribution of adipose tissue/central fat accumulation (e.g., buffalo hump or visceral fat; lipohypertrophy), and dyslipidemia or abnormalities in plasma lipids. These abnormalities may be associated with insulin resistance (Mulligan et al., 1997; Walli et al., 1998). The biochemical disturbances observed in patients on PI-containing cART regimens resemble metabolic syndrome in the general population, and may increase the risk of cardiovascular disease (Carr et al., 1998a, b; Grundy et al., 2004). Table 101.3 provides a summary of the metabolic effects of HIV PIs and their potential end-organ consequences (Flint et al., 2009). Body composition changes due to HIV infection (i.e., wasting syndrome) or due to PI components of cART (central fat accumulation in varying presentations) may be stigmatizing and affect treatment adherence (Duran et al., 2001).

Metabolic syndrome in the general population is also associated with cardiovascular diseases. Two large prospective studies, comprising 23,437 and 2386 patients respectively, showed an increase in myocardial infarction correlated to PIs, but not to other non-PI cART components (Friis-Moller et al., 2007; Kaplan et al., 2007).

The effects of PIs on dyslipidemia [hypertriglyceridemia, elevated low-density lipoprotein (LDL) cholesterol and decreased high-density lipoprotein (HDL) cholesterol] are acute, manifesting within four weeks following initiation of ritonavir in HIV-infected persons (Purnell et al., 2000). These effects are independent of underlying HIV-induced dyslipidemia, as shown by the observation that ritonavir-boosted liponavir (Lee et al., 2005) administered short-term (four weeks) to HIV-seronegative men also changes lipid profiles unfavorably. Lee and colleagues reported increases in fasting very-low-density lipoprotein (VLDL), free fatty acids and triglycerides, but no changes in fasting LDL, HDL, intermediate-density lipoprotein (IDL), glucose, or insulin-mediated glucose

disposal. Notably, there were no changes in weight, body fat, or abdominal adipose tissue by computed tomography. The effects of PIs on serum lipid levels are independent of their effects on HIV suppression, since not all PIs induce dyslipidemia in HIV-1-infected participants or uninfected patients (Jemsek et al., 2006; Noor et al., 2001).

Ritonavir is associated with acute insulin resistance at high therapeutic doses, and causes body composition changes and insulin resistance. Ritonavir directly inhibits the GLUT4 insulin-regulated transporter, preventing glucose from entering fat and muscle cells, leading to insulin resistance (Rathbun and Rossi, 2002). Ritonavir-boosted tipranavir and lopinavir were shown to increase subcutaneous (limb) fat without evidence of increasing visceral fat or insulin in a 48-week study (Carr et al., 2008). A study of healthy individuals showed differential responses to atazanavir (ATZ) and ritonavir-boosted lopinavir on insulin sensitivity (Noor et al., 2004). After a five-day course, ATZ did not affect insulin sensitivity, in contrast to ritonavir/ lopinavir (LPV/r), which induced insulin resistance. The glycogen storage rate for LPV/r was also significantly lower compared to either ATZ (36%) or placebo (34%) (Noor et al., 2004). However, other studies have reported no difference in insulin sensitivity after four weeks in healthy donors (Dube et al., 2003).

Organ	Effect	Consequence
Adipose tissue	↓Glucose uptake	Lipoatrophy
	↓Triglyceride synthesis	Muscles, intracellular lipid levels, visceral fat disposition
Liver	↑Very low density lipid proteins	↑Triglycerides, pancreatitis, arteriosclerosis
	↑ApoB*	Arteriosclerosis, fatty liver (steatosis)
	↑Hepatic glucose production	Type 1 diabetes
	↑SKEBP-1c†	Insulin resistance, atherosclerosis
Muscle	↓Glucose uptake	Glucose intolerance, type 2 diabetes
	↑Intramyocellular lipid levels	Increased visceral adiposity
Pancreas	↓Insulin secretion	Glucose intolerance, type 2 diabetes

TABLE 101.3 Multiple metabolic effects of protease inhibitors and possible consequences

*Apolipoprotein B is the primary apolipoprotein of low density lipoproteins, responsible for cholesterol transport to tissues.
†SKEBP-1c is a transcription factor sterol regulatory element binding protein-1c, inducing the expression of pathway genes for glucose utilization and fatty acid synthesis.
Adapted from Flint et al., 2009.

The MONARK trial evaluated the potential lipodystrophic effects of ritonavir-boosted lopinavir on ART-naïve HIV-infected patients. Those receiving LPV/r monotherapy had statistically significantly lower limb fat loss (median = −63 g) compared to those on LPV/r+ZDV/3TC combination therapy (−703 g). The proportion of patients having a greater than 20% loss of limb fat was also greater in the group on triple therapy (27.3%) compared to just 3% in the monotherapy arm. In both arms of the study, no changes in trunk fat were noted. These data suggest that LPV/r, and possibly other PIs, may not be the main contributors to lipoatrophy (Rotger et al., 2009).

Because of the enormous health burden of metabolic disorders in the general population, numerous genome-wide association studies (GWAS) have been performed to identify genetic variants associated with various phenotypes associated with metabolic syndrome, type 2 diabetes, serum lipid levels, and body mass in non-HIV infected persons [an online catalog of GWAS genotype–phenotype associations is available at http://www.genome.gov/gwastudies (Hindorff et al., 2009)]. The genetic factors identified to date as contributing to these disorders have had relatively small effect sizes (low penetrance), are multigenic, and interact with environmental factors, including antiretroviral medications. Individually, small-effect-size markers are not good risk predictors, but by combining their additive effects, sensitivity and specificity may be improved, increasing clinical utility.

Using a pharmacogenomics approach, Rotger and colleagues tested 33 GWAS SNPs and nine candidate SNPs previously reported to contribute to serum lipid levels (Rotger et al., 2009). Patients with favorable genetic scores (32%), based on the number of dyslipidemia risk alleles, were found to have lower levels of LDL ("bad" cholesterol) compared to the 53% individuals with unfavorable scores, who had higher levels of LDL. Low levels of HDL ("good" cholesterol) were also associated with unfavorable scores in 42% of patients, compared to just 17% of participants with favorable scores for dyslipidemia risk alleles (Rotger et al., 2009). The study showed that a combination of genetic scores, based on additive scores or weighted for the effect size of the risk alleles, was equally predictive for identifying persons with sustained lipodystrophy. Genes significantly associated with non-HDL cholesterol, HDL cholesterol, and triglyceride levels are indicated in Table 101.1. Individuals receiving the most strongly dyslipidemic ART medications and carrying more risk alleles associated with dyslipidemia (Table 101.4) were much more likely to develop dyslipidemia, as indicated in Figure 101.3. A combination of genetic risk score and antiretroviral treatment information provided the best risk prediction for identifying risk of lipodystrophy at the individual level. Notably, the proportion of the variance in lipid variation explained by genetic factors alone (~7.4%) or ART alone (~6.2%) was similar, indicating that genetic screening for lipid (or other drug-related toxicity) genetic risk factors may have clinical utility in the personalized selection of optimal ART regimens (Tarr and Telenti, 2010).

THE FUTURE OF HIV PHARMACOGENETICS AND PHARMACOGENOMICS

The Application of "Pharmaco-omics" to Personalized Medicine

Although pharmacogenetics studies have increased our knowledge and understanding of antiretroviral therapy pharmacokinetics, efficacy, and toxicity, very few associations will be readily translatable to clinical practice. To date, *HLA-B*5701* screening to prevent ABC HSR is the only example of pharmacogenetics translated to HIV clinical care. Because sustained HIV therapy with multiple drugs is required to avoid viral escape, and because of the complex interactions between genetic and other factors (e.g., age, sex, body weight, drug–drug interactions, and comorbidities), clinical management of HIV-infected patients is challenging. Extensive research efforts are required to discover new host factors and pathways that can be exploited to improve outcomes with antiretroviral treatment.

Hypothesis-free investigations of the genome have not been widely exploited in HIV pharmacogenetics. The major advance of the HapMap project has increased our knowledge of common genetic variations in the human genome, and combined with the development of SNP genotyping arrays, GWAS have provided significant advances in many pathologies, including HIV/AIDS (An and Winkler, 2010; Aouizerat et al., 2011)

TABLE 101.4	Antiretroviral regimens grouped according to their impact on serum lipid levels		
Serum lipid analyzed	**Group 1**	**Group 2**	**Group 3**
HDL cholesterol	NNRTI	PI	No ART NRTI only
Non-HDL cholesterol	No ART NRTI only Atazanavir boosted with ritonavir (ATV/r)	PI (except ATV/r) NNRTI	NA
Triglycerides	No ART	Single PI-containing ART (without ritonavir)	Ritonavir-containing ART (except ATV/r)
	NRTI only Nevirapine-containing ART without PI Atazanavir unboosted	ATV/r Efavirenz	

Because <0.5% of lipid determinations were made during raltegravir, etravirine, or T20 exposure, these agents were not considered in the analysis. NA, not applicable.
Adapted from Rotger et al., 2009.

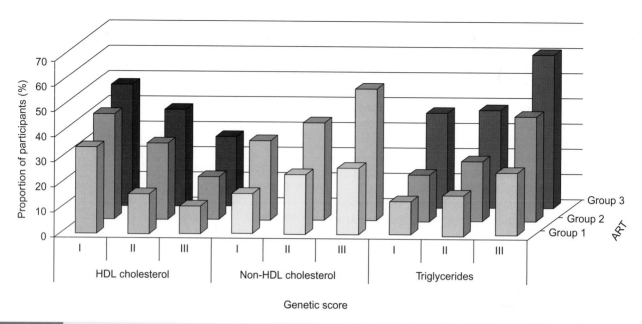

Figure 101.3 Three-dimensional histogram representing the proportion of participants according to genetic score and ART group for each serum lipid level analyzed: HDL cholesterol (blue gradient), non-HDL cholesterol (yellow gradient), and triglycerides (orange gradient). A description of ART groups is provided in Table 101.4. *Adapted from Rotger et al., 2009.*

and AIDS-related complications (dyslipidemia, coronary artery disease, osteoporosis, obesity) (Tarr and Telenti, 2010). A high-throughput pharmacogenetic association study (~1500 SNPs in ADME genes) targeting the pharmacokinetics of lopinavir co-formulated with ritonavir has revealed promising new pharmacokinetic risk alleles (Lubomirov et al., 2010). The next step is a shift from targeted genes (pharmacogenetics) to agnostic GWAS to identify all variants associated with drug-response phenotypes (pharmacogenomics). Pharmacogenomics has provided new insights in the management of commonly prescribed drugs for a number of disorders, including cardiovascular disease, mental illness, cancers, and infectious diseases, as recently reviewed by Wang and colleagues (2011).

Pharmacogenomic research should benefit from the additional advances offered by the 1000 Genomes project (1000 Genomes Project Consortium, 2010), an extension of the HapMap project. This international project is revealing new sequence variants (rare to common) in 1000 human genomes, in populations representing distinct human ancestries. As this variation is revealed, it is being released to public databases and included in new versions of genotyping arrays, thus ensuring better coverage of human genetic diversity. Finally, the exploitation of next-generation sequencing platforms will allow rapid and full coverage of human genomic diversity and the discovery of rare or *de novo* mutations (Metzker, 2010). It is likely that rare variants will be associated with extreme but rare drug toxicities, such as Stevens–Johnson syndrome, a major toxicity of nevirapine (Metry et al., 2001). Rare and *de novo* variants have the potential to have larger effect sizes than common variants, and may account for the missing heritability or unexplained

variance seen with many common diseases and drug-related phenotypes (Hemminki et al., 2011; Manolio et al., 2009).

Beyond the exploration of genetic variants impacting the response to HIV antiretroviral therapy, investigating transcriptomic profiles by *in vitro* and *in vivo* approaches has led to new perspectives in drug-induced toxicity (Cui and Paules, 2010). Since the development of noninvasive biomarkers for clinical applications is one of the most important issues, blood is an excellent biological resource for this kind of investigation. As the study of gene expression patterns focuses only on mRNA, the results do not necessarily reflect the protein levels, and proteomics strategies could thus also participate in unraveling new candidates able to predict therapeutic responses. Interestingly, these technologies have been fruitfully exploited in rheumatoid arthritis response to treatment to identify new putative biomarkers (Bansard et al., 2009).

Epigenetics is the study of heritable and somatic (non-heritable) changes in genes and gene expression that do not involve modifications of primary DNA nucleotide sequence. Epigenetic events include DNA methylation, histone modifications, and miRNA expression — consequences of epigenetic modification are changes in the regulation of gene expression and gene silencing. Recent studies have provided evidence of the important role of pharmacoepigenetics in survival and treatment response in heart disease and cancers (Dejeux et al., 2010; Ji et al., 2009; Mateo Leach et al., 2010).

Metabolomics is the study of small molecules in the blood or urine resulting from natural cellular metabolism. In the case of antiretroviral drugs, biotransformation usually occurs via cytochrome P450 pathways. The study of metabolic profiles of antiretroviral drugs is important for a number of reasons: (1) there is a potential for inhibitory or additive interactions among drug metabolites, (2) metabolites themselves may cause toxicities, (3) the metabolite profile provides an indication of clearance rates of the parent drug, and (4) alterations in metabolite levels may be biomarkers for toxicities. Also, antiretroviral drugs may be metabolized by more than one metabolism enzymatic pathway, as is the case for efavirenz. The major pathway for efavirenz is via CYP2B6, but the minor pathways CYP3A4/3A5 are also utilized. Since the genes encoding these enzymes are extremely polymorphic, the gain or loss of function of any of these enzymes may affect metabolite profiles or clearance patterns, either or both of which may be associated with increased risk of toxicities. The role of metabolic profiles, clearance rates, and underlying genetic variation in key metabolizing genes on drug-induced toxicities has not been established for most antiretroviral drugs, including the widely prescribed efavirenz (Arab-Alameddine et al., 2011.; Lubomirov et al., 2007, 2011; Markwalder et al., 2001; Mutlib et al., 2000).

The application of newer genotyping platforms that are more representative of global populations and contain 2.5–5 million rare-to-common SNPs should lead to the identification of new variant alleles associated with drug transporter genes. Although genetic variation in drug metabolism pathway genes has been well investigated, much less is understood about the extent of variation in drug transporter genes and the potential for interactions with metabolizing variants and transporter genes on drug toxicities. Transporter genes and their proteins are critical because they modulate drug levels in cells and tissues. Thus drug transporters are key players in drug disposition and exposure in tissue compartments harboring HIV (e.g., brain tissue). The *ABCB1* gene encoding the permeability glycoprotein (Pgp) is a member of the MDR subfamily. Pgp is an adenosine triphosphate (ATP)-dependent drug efflux pump that removes drugs from the cell to the blood; Pgp also is a transporter in the blood–brain barrier. The *ABCB1* (*MDR1*) 3435C/T polymorphism predicts both PI nelfinavir plasma concentrations and CD4 T cell recovery (Fellay et al., 2002). In addition, poor-metabolizer alleles and extensive-metabolizing alleles of *CYP2D6*, encoding a cytochrome P450 isozyme, were also correlated with drug plasma concentrations in the same study, although they observed no changes in clinical CD4 T cell recovery. Evidence of additive gene–gene interactions between transporter genes and metabolism genes in bioavailability and disposition of antiretroviral drugs is not extensive. Gene–gene interactions between *CYP2B6* 516G>T and *ABCB1* 3435C>T have been associated with higher nevirapine plasma concentrations and hepatotoxicity in a population from Mozambique (Ciccacci et al., 2010). Other combinations of ultra-fast-metabolizing alleles (*CYP2D6*) and transporter genes (*ABCC2* rs717620) lead to sub-therapeutic levels of the PI ritonavir. Additive interactions between rapid or slow alleles of metabolizing genes and drug transporter genes have also been reported for premature discontinuation of therapy (Lubomirov

et al., 2011). There is a reasonable likelihood that the rates of certain toxicities (e.g., lactic acidosis) that apparently differ among geographic areas or ethnic populations may be due to functional allele frequency differences in transporter and metabolizing genes. It is also likely that more population-specific alleles and rare variants remain to be discovered by GWAS using newer platforms. The challenge will be to identify single-gene effects as well as gene–gene and gene–environment interactions, and translate these findings to the clinic to improve efficacy and reduce toxicities.

Barriers to personalized medicine in HIV-1 clinical care are not insurmountable. Incomplete knowledge of drug metabolism and toxicity pathways limits the identification of specific and sensitive biomarkers that can be used as risk predictors. The integration of "pharmaco-omics" (genomics, proteomics, metabolomics, and epigenomics), using a systems-based approach, should lead to development of less toxic antiretroviral drugs and to an increase in the number of informative biomarkers for risk prediction in personalized HIV-related healthcare. This comprehensive approach is critical as more people initiate treatment to prevent HIV transmission or acquisition, and as the population on ART ages (Arab-Alameddine et al., 2009).

A second barrier is that most common risk alleles have small effect sizes (OR < 1.5) – the corresponding lack of specificity and/or sensitivity makes the use of single-SNP markers for risk prediction problematic. Combining information from multiple small-effect genetic risk factors into an individualized risk score has been shown to predict dyslipidemia (Arnedo et al., 2007; Rotger et al., 2009) and premature discontinuation of ART due to toxicities (Lubomirov et al., 2011). These and other studies provide proof of principle that pre-screening with panels of ADME genes and other genes can be used to predict drug exposure and toxicities to optimize treatment. Determining the utility and cost-effectiveness of genetic screening to inform drug selection requires appropriately controlled, randomized clinical trials (Telenti, 2009). Before the expense of clinical trials can be justified, genetic factors associated with antiretroviral phenotypes must be securely validated by replication in diverse patient groups.

Application for Resource-Limited Settings
Low-Cost Technology Approaches for Abacavir Hypersensitivity Genotyping

High-resolution *HLA* testing is needed to identify the *HLA-B*5701* allele, and to differentiate it from closely related alleles such as *HLA-B*5702*, *HLA-B*5703*, and *HLA-B*5801/5802*, which do not appear to be associated with ABC HSR. A high-resolution method (i.e., DNA sequencing, which is expensive) may not be covered by health insurance, and is not available in some settings. The development of feasible and cost-effective technologies, such as specific amplification or flow cytometry, has greatly helped to implement *HLA-B*5701* screening in clinical practice (Dello Russo et al., 2011; Hammond et al., 2007;

Kostenko et al., 2011; Martin et al., 2005b, 2006). Previous genetic studies suggested absolute linkage disequilibrium (a measure of correlation between alleles of genes on the same chromosome) between the HCP5 rs2395029 and HLA-B*5701. Studies with Europeans from Spain and Mexican Mestizos showed similar correlations between HCP5 and HLA-B*5701, with all HLA-B*5701 carriers also carrying the HCP5 SNP. Since bi-allelic SNPs are more cheaply and easily genotyped than HLA-B*5701, HCP5 genotyping could serve as a simple screening tool for ABC HSR, particularly in settings where sequence-based HLA typing is not available. However, further studies are required to determine if the strong correlation between HLA-B*5701 and HCP5 rs2395029 observed to date is generalizable to other populations.

Development of Pre-Treatment Predictive Scores

Another potential application of personalized genetic medicine in resource-limited settings may be the development of pre-treatment predictive scores. Such pre-treatment predictive scores have previously been proposed in the hepatitis C literature (Abu Dayyeh et al., 2011; Haas et al., 2011). Abu Dayyeh and colleagues were risk stratifying patients for early and sustained virologic responses prior to antiviral therapy based on the presence (or absence) of the IL-28B non-responder genotype (rs12979860), with maximal points (based on a 0-, 3-, and 6-point scale) being given to patients having the most favorable genotype (CC genotype as compared to having the CT or TT genotype) and/or the lowest pre-treatment hepatitis C RNA level (<500,000 IU/ml) (Abu Dayyeh et al., 2011).

For example, if one were to identify specific risk alleles associated with the development of lactic acidosis, a potentially fatal toxicity, one could calculate a pre-treatment predictive score for an individual patient based on genetics (i.e., the presence or absence of a deleterious risk allele) plus a key baseline or laboratory characteristic [i.e., baseline body mass index (BMI) and/or sex, as female sex and high BMI have been shown to be significant risk factors for the development of lactic acidosis (Wester et al., 2007)]. Then, based on an individual's genetics, that is, if he or she had the risk allele, the intermediate (heterozygote) allele state, or did not have the risk allele, a predictive score would be developed. He or she could, for example, be given a score of 0 (no risk allele), 2 (intermediate), or 4 (presence of the risk allele). The same could be done for baseline BMI in a female patient, with overweight females being most at risk. For example, there could be scores of 0 (indicating normal BMI, 18.5–25), 2 (intermediate risk BMI, 25–30), and 4 (highest risk BMI, ≥30). Using statistics, one could then determine an appropriate pre-treatment risk score cutoff based on the negative predictive values (NPVs) of various possible values. If, for example, the NPV was greater than 90% for a score of 4 or higher (out of a possible 8), then such recommendations could be incorporated into regional cART treatment guidelines, assuming that the cost of personalized genetics testing will decrease dramatically over the next few years and therefore be feasible in resource-limited settings. Such an example may not be as relevant now that the majority of adults initiating cART

are receiving far less mitochondrially toxic NRTIs, namely tenofovir (TDF) plus emtricitabine (FTC). A more relevant example may be risk for tenofovir-associated tubulopathy and resultant renal insufficiency, which have been linked to a polymorphism in the ABCC2 transporter (Josephson et al., 2007), if a particular clinical and/or laboratory factor (e.g., being above 40 years of age) was also associated with risk for this potentially untoward complication.

CONCLUSIONS

In summary, there are clinically relevant genetic predictors of cART efficacy and toxicity in addition to HLA-B*5701, which is the only genetic screening test that is currently standard practice in resource-replete settings. Many are of the opinion that the most important predictors of cART efficacy are yet to be discovered. There are some novel, low-cost technological approaches for existing and well-proven HIV pharmacogenetics testing, such as HCP5 rs2395029 SNP genotyping as a proxy for HLA-B*5701 screening, that may have substantial public health relevance and significance, and if implemented on a much larger, global, scale may result in substantial reductions in HIV-treatment-associated morbidity and mortality. At the same time, it will be critically important to enhance pharmacovigilance in resource-limited settings by establishing regional cART outcomes databases, in order to systematically capture and longitudinally monitor these most-at-risk populations while monitoring them for the development of the most clinically relevant toxicities (e.g., lactic acidosis, nevirapine cutaneous hypersensitivity and hepatotoxicity, tenofovir-associated renal toxicity, etc.). Pertinent pharmacogenetic discoveries, however, will only be attainable through the establishment of regional and openly collaborative DNA specimen repositories in resource-limited settings. Such an undertaking could be facilitated by the increasing collective insight and experience of clinicians, ethical review board committee members, and persons with bioinformatics expertise integrally involved in the establishment of such repositories elsewhere. While prioritizing what to study, researchers will need to carefully define their particular phenotype, and the proposed genetics screening recommendation that results from their findings will need to be easy to implement and cost-effective. The clinical benefit of genetic screening will only be determined through randomized clinical trials followed by cost–benefit analysis.

The future of HIV pharmacogenetics and, specifically, personalized medicine is promising, especially as the medical field is moving so rapidly toward individual genome testing. But because of the numerous variants and pathways simultaneously involved and the complexity of HIV disease and its treatment, the challenge is to integrate findings from pharmaco-omics and host genome-wide studies to inform and improve patient care. For both, the development of large, diverse cohorts is essential to confirm the associations previously suggested, to generalize the signals across populations, and to reveal new leads.

REFERENCES

1000 Genomes Project Consortium, 2010. A map of human genome variation from population-scale sequencing. Nature 467, 1061–1073.

Abdool Karim, Q., Abdool Karim, S.S., Frohlich, J.A., et al., 2010. Effectiveness and safety of tenofovir gel, an antiretroviral microbicide, for the prevention of HIV infection in women. Science 329, 1168–1174.

Abu Dayyeh, B.K., Gupta, N., Sherman, K.E., DeBakker, P.I.W., Chung, R.T., (for the AIDS Clinical Trials Group A5178 Study Team), 2011. IL28B alleles exert an additive dose effect when applied to HCV-HIV co-infected persons undergoing peginterferon and ribavirin therapy. PLoS ONE 6, e25753.

An, P., Winkler, C.A., 2010. Host genes associated with HIV/AIDS: Advances in gene discovery. Trends Genet 26, 119–131.

Anderson, P.L., Lamba, J., Aquilante, C.L., Schuetz, E., Fletcher, C.V., 2006. Pharmacogenetic characteristics of indinavir, zidovudine, and lamivudine therapy in HIV-infected adults: A pilot study. J Acquir Immune Defic Syndr 42, 441–449.

Antiretroviral Therapy Cohort Collaboration, 2010. Causes of death in HIV-1-infected patients treated with antiretroviral therapy, 1996–2006: Collaborative analysis of 13 HIV cohort studies. Clin Infect Dis 50, 1387–1396.

Aouizerat, B.E., Pearce, C.L., Miaskowski, C., 2011. The search for host genetic factors of HIV/AIDS pathogenesis in the post-genome era: Progress to date and new avenues for discovery. Curr HIV/AIDS Rep 8, 38–44.

Arab-Alameddine, M., Decosterd, L.A., Buclin, T., Telenti, A., Csajka, C., 2011. Antiretroviral drug toxicity in relation to pharmacokinetics, metabolic profile and pharmacogenetics. Expert Opin Drug Metab Toxicol 7, 609–622.

Arab-Alameddine, M., Di Iulio, J., Buclin, T., et al., 2009. Pharmacogenetics-based population pharmacokinetic analysis of efavirenz in HIV-1-infected individuals. Clin Pharmacol Ther 85, 485–494.

Arnedo, M., Taffe, P., Sahli, R., et al., 2007. Contribution of 20 single nucleotide polymorphisms of 13 genes to dyslipidemia associated with antiretroviral therapy. Pharmacogenet Genomics 17, 755–764.

Bansard, C., Lequerre, T., Daveau, M., et al., 2009. Can rheumatoid arthritis responsiveness to methotrexate and biologics be predicted? Rheumatology (Oxford) 48, 1021–1028.

Barrios, A., Garcia-Benayas, T., Gonzalez-Lahoz, J., Soriano, V., 2004. Tenofovir-related nephrotoxicity in HIV-infected patients. AIDS 18, 960–963.

Baylor, M.S., Johann-Liang, R., 2004. Hepatotoxicity associated with nevirapine use. J Acquir Immune Defic Syndr 35, 538–539.

Bertrand, J., Treluyer, J.M., Panhard, X., et al., 2009. Influence of pharmacogenetics on indinavir disposition and short-term response in HIV patients initiating HAART. Eur J Clin Pharmacol 65, 667–678.

Beutler, E., Gelbart, T., Demina, A., 1998. Racial variability in the UDP-glucuronosyltransferase 1 (UGT1A1) promoter: A balanced polymorphism for regulation of bilirubin metabolism? Proc Natl Acad Sci USA 95, 8170–8174.

Boulle, A., Van Cutsem, G., Hilderbrand, K., et al., 2010. Seven-year experience of a primary care antiretroviral treatment programme in Khayelitsha, South Africa. AIDS 24, 563–572.

Brady, M.T., Oleske, J.M., Williams, P.L., et al., 2010. Declines in mortality rates and changes in causes of death in HIV-1-infected children during the HAART era. J Acquir Immune Defic Syndr 53, 86–94.

Bussmann, H., Wester, C.W., Ndwapi, N., et al., 2008. Five-year outcomes of initial patients treated in Botswana's National Antiretroviral Treatment Program. AIDS 22, 2303–2311.

Canter, J.A., Robbins, G.K., Selph, D., et al., 2010. African mitochondrial DNA subhaplogroups and peripheral neuropathy during antiretroviral therapy. J Infect Dis 201, 1703–1707.

Carr, A., Ritzhaupt, A., Zhang, W., et al., 2008. Effects of boosted tipranavir and lopinavir on body composition, insulin sensitivity and adipocytokines in antiretroviral-naive adults. AIDS 22, 2313–2321.

Carr, A., Samaras, K., Burton, S., et al., 1998a. A syndrome of peripheral lipodystrophy, hyperlipidaemia and insulin resistance in patients receiving HIV protease inhibitors. AIDS 12, F51–F58.

Carr, A., Samaras, K., Chisholm, D.J., Cooper, D.A., 1998b. Pathogenesis of HIV-1-protease inhibitor-associated peripheral lipodystrophy, hyperlipidaemia, and insulin resistance. Lancet 351, 1881–1883.

Celum, C., Baeten, J.M., 2012. Tenofovir-based pre-exposure prophylaxis for HIV prevention: Evolving evidence. Curr Opin Infect Dis. 25, 51–57.

Chantarangsu, S., Mushiroda, T., Mahasirimongkol, S., et al., 2009. HLA-B*3505 allele is a strong predictor for nevirapine-induced skin adverse drug reactions in HIV-infected Thai patients. Pharmacogenet Genomics 19, 139–146.

Chaponda, M., Pirmohamed, M., 2011. Hypersensitivity reactions to HIV therapy. Br J Clin Pharmacol 71, 659–671.

Chasela, C.S., Hudgens, M.G., Jamieson, D.J., et al., 2010. Maternal or infant antiretroviral drugs to reduce HIV-1 transmission. N Engl J Med 362, 2271–2281.

Chi, B.H., Mwango, A., Giganti, M., et al., 2010. Early clinical and programmatic outcomes with tenofovir-based antiretroviral therapy in Zambia. J Acquir Immune Defic Syndr 54, 63–70.

Ciccacci, C., Borgiani, P., Ceffa, S., et al., 2010. Nevirapine-induced hepatotoxicity and pharmacogenetics: A retrospective study in a population from Mozambique. Pharmacogenomics 11, 23–31.

Clay, P.G., 2002. The abacavir hypersensitivity reaction: A review. Clin Ther 24, 1502–1514.

Clifford, D.B., Evans, S., Yang, Y., et al., 2005. Impact of efavirenz on neuropsychological performance and symptoms in HIV-infected individuals. Ann Intern Med 143, 714–721.

Coca, S., Perazella, M.A., 2002. Rapid communication: Acute renal failure associated with tenofovir – evidence of drug-induced nephrotoxicity. Am J Med Sci 324, 342–344.

Coetzee, D., Hildebrand, K., Boulle, A., et al., 2004. Outcomes after two years of providing antiretroviral treatment in Khayelitsha, South Africa. AIDS 18, 887–895.

Cohen, M.S., Chen, Y.Q., McCauley, M., et al., 2011. Prevention of HIV-1 infection with early antiretroviral therapy. NEJM 6, 493–505.

Cui, Y., Paules, R.S., 2010. Use of transcriptomics in understanding mechanisms of drug-induced toxicity. Pharmacogenomics 11, 573–585.

De Vincenzi, I., 2011. Triple antiretroviral compared with zidovudine and single-dose nevirapine prophylaxis during pregnancy and breastfeeding for prevention of mother-to-child transmission of HIV-1 (Kesho Bora study): A randomised controlled trial. Lancet Infect Dis 11, 171–180.

Deeks, S.G., Phillips, A.N., 2009. HIV infection, antiretroviral treatment, ageing, and non-AIDS related morbidity. BMJ 338, a3172.

Dejeux, E., Ronneberg, J.A., Solvang, H., et al., 2010. DNA methylation profiling in doxorubicin treated primary locally advanced breast tumours identifies novel genes associated with survival and treatment response. Mol Cancer 9, 68.

Dello Russo, C., Lisi, L., Lofaro, A., et al., 2011. Novel sensitive, specific and rapid pharmacogenomic test for the prediction of abacavir hypersensitivity reaction: HLA-B*5701 detection by real-time PCR. Pharmacogenomics 12, 567–576.

Dube, M.P., Stein, J.H., Aberg, J.A., et al., 2003. Guidelines for the evaluation and management of dyslipidemia in human immunodeficiency virus (HIV)-infected adults receiving antiretroviral therapy: Recommendations of the HIV Medical Association of the Infectious Disease Society of America and the Adult AIDS Clinical Trials Group. Clin Infect Dis 37, 613–627.

Duran, S., Saves, M., Spire, B., et al., 2001. Failure to maintain long-term adherence to highly active antiretroviral therapy: The role of lipodystrophy. AIDS 15, 2441–2444.

Elzi, L., Marzolini, C., Furrer, H., et al., 2010. Treatment modification in human immunodeficiency virus-infected individuals starting combination antiretroviral therapy between 2005 and 2008. Arch Intern Med 170, 57–65.

Fellay, J., Marzolini, C., Meaden, E.R., et al., 2002. Response to antiretroviral treatment in HIV-1-infected individuals with allelic variants of the multidrug resistance transporter 1: A pharmacogenetics study. Lancet 359, 30–36.

Felley, C., Morris, M.A., Wonkam, A., et al., 2004. The role of CFTR and SPINK-1 mutations in pancreatic disorders in HIV-positive patients: A case-control study. AIDS 18, 1521–1527.

Flint, O.P., Noor, M.A., Hruz, P.W., et al., 2009. The role of protease inhibitors in the pathogenesis of HIV-associated lipodystrophy: Cellular mechanisms and clinical implications. Toxicol Pathol 37, 65–77.

Friis-Moller, N., Reiss, P., Sabin, C.A., et al., 2007. Class of antiretroviral drugs and the risk of myocardial infarction. N Engl J Med 356, 1723–1735.

Gatanaga, H., Hayashida, T., Tsuchiya, K., et al., 2007a. Successful efavirenz dose reduction in HIV type 1-infected individuals with cytochrome P450 2B6 *6 and *26. Clin Infect Dis 45, 1230–1237.

Gatanaga, H., Yazaki, H., Tanuma, J., et al., 2007b. HLA-Cw8 primarily associated with hypersensitivity to nevirapine. AIDS 21, 264–265.

Grant, R.M., Lama, J.R., Anderson, P.L., et al., 2010. Preexposure chemoprophylaxis for HIV prevention in men who have sex with men. N Engl J Med 363, 2587–2599.

Grundy, S.M., Brewer Jr., H.B., Cleeman, J.I., Smith Jr., S.C., Lenfant, C., 2004. Definition of metabolic syndrome: Report of the National Heart, Lung, and Blood Institute/American Heart Association conference on scientific issues related to definition. Circulation 109, 433–438.

Gulick, R.M., Ribaudo, H.J., Shikuma, C.M., et al., 2006. Three- vs four-drug antiretroviral regimens for the initial treatment of HIV-1 infection: A randomized controlled trial. JAMA 296, 769–781.

Haas, D.W., Bartlett, J.A., Andersen, J.W., et al., 2006. Pharmacogenetics of nevirapine-associated hepatotoxicity: An Adult AIDS Clinical Trials Group collaboration. Clin Infect Dis 43, 783–786.

Haas, D.W., Kuritzkes, D.R., Ritchie, M.D., et al., 2011. Pharmacogenomics of HIV therapy: Summary of a workshop sponsored by the National Institute of Allergy and Infectious Diseases. HIV Clin Trials 12, 277–285.

Haas, D.W., Ribaudo, H.J., Kim, R.B., et al., 2004. Pharmacogenetics of efavirenz and central nervous system side effects: An Adult AIDS Clinical Trials Group study. AIDS 18, 2391–2400.

Haas, D.W., Zala, C., Schrader, S., et al., 2003. Therapy with atazanavir plus saquinavir in patients failing highly active antiretroviral therapy: A randomized comparative pilot trial. AIDS 17, 1339–1349.

Hammond, E., Mamotte, C., Nolan, D., Mallal, S., 2007. HLA-B*5701 typing: Evaluation of an allele-specific polymerase chain reaction melting assay. Tissue Antigens 70, 58–61.

Hartkoorn, R.C., Kwan, W.S., Shallcross, V., et al., 2010. HIV protease inhibitors are substrates for OATP1A2, OATP1B1 and OATP1B3 and lopinavir plasma concentrations are influenced by SLCO1B1 polymorphisms. Pharmacogenet Genomics 20, 112–120.

Hemminki, K., Forsti, A., Houlston, R., Bermejo, J.L., 2011. Searching for the missing heritability of complex diseases. Hum Mutat 32, 259–262.

Hetherington, S., Hughes, A.R., Mosteller, M., et al., 2002. Genetic variations in HLA-B region and hypersensitivity reactions to abacavir. Lancet 359, 1121–1122.

Hetherington, S., McGuirk, S., Powell, G., et al., 2001. Hypersensitivity reactions during therapy with the nucleoside reverse transcriptase inhibitor abacavir. Clin Ther 23, 1603–1614.

Hindorff, L.A., Sethupathy, P., Junkins, H.A., et al., 2009. Potential etiologic and functional implications of genome-wide association loci for human diseases and traits. Proc Natl Acad Sci USA 106, 9362–9367.

Hughes, A.R., Mosteller, M., Bansal, A.T., et al., 2004a. Association of genetic variations in HLA-B region with hypersensitivity to abacavir in some, but not all, populations. Pharmacogenomics 5, 203–211.

Hughes, D.A., Vilar, F.J., Ward, C.C., et al., 2004b. Cost-effectiveness analysis of HLA B*5701 genotyping in preventing abacavir hypersensitivity. Pharmacogenetics 14, 335–342.

Hulgan, T., Haas, D.W., Haines, J.L., et al., 2005. Mitochondrial haplogroups and peripheral neuropathy during antiretroviral therapy: An Adult AIDS Clinical Trials Group study. AIDS 19, 1341–1349.

Izzedine, H., Hulot, J.S., Villard, E., et al., 2006. Association between ABCC2 gene haplotypes and tenofovir-induced proximal tubulopathy. J Infect Dis 194, 1481–1491.

Jemsek, J.G., Arathoon, E., Arlotti, M., et al., 2006. Body fat and other metabolic effects of atazanavir and efavirenz, each administered in combination with zidovudine plus lamivudine, in antiretroviral-naive HIV-infected patients. Clin Infect Dis 42, 273–280.

Ji, J., Shi, J., Budhu, A., et al., 2009. MicroRNA expression, survival, and response to interferon in liver cancer. N Engl J Med 361, 1437–1447.

Josephson, F., Allqvist, A., Janabi, M., et al., 2007. CYP3A5 genotype has an impact on the metabolism of the HIV protease inhibitor saquinavir. Clin Pharmacol Ther 81, 708–712.

Kaplan, R.C., Kingsley, L.A., Sharrett, A.R., et al., 2007. Ten-year predicted coronary heart disease risk in HIV-infected men and women. Clin Infect Dis 45, 1074–1081.

Kiser, J.J., Aquilante, C.L., Anderson, P.L., et al., 2008. Clinical and genetic determinants of intracellular tenofovir diphosphate concentrations in HIV-infected patients. J Acquir Immune Defic Syndr 47, 298–303.

Kopp, J., Winkler, C.A., 2003. HIV-associated nephropathy in African Americans. Kidney Int 65, S43–S49.

Kopp, J.B., Nelson, G.W., Sampath, K., et al., 2011. APOL1 variants in focal segmental glomerulosclerosis and HIV-associated nephropathy. J Am Soc Nephrol 22, 2129–2137.

Kostenko, L., Kjer-Nielsen, L., Nicholson, I., et al., 2011. Rapid screening for the detection of HLA-B57 and HLA-B58 in prevention of drug hypersensitivity. Tissue Antigens 78, 11–20.

Lee, G.A., Rao, M.N., Grunfeld, C., 2005. The effects of HIV protease inhibitors on carbohydrate and lipid metabolism. Curr HIV/AIDS Rep 2, 39–50.

Likanonsakul, S., Rattanatham, T., Feangvad, S., et al., 2009. HLA-Cw*04 allele associated with nevirapine-induced rash in HIV-infected Thai patients. AIDS Res Ther 6, 22.

Littera, R., Carcassi, C., Masala, A., et al., 2006. HLA-dependent hypersensitivity to nevirapine in Sardinian HIV patients. AIDS 20, 1621–1626.

Lubomirov, R., Colombo, S., Di Iulio, J., et al., 2011. Association of pharmacogenetic markers with premature discontinuation of first-line anti-HIV therapy: An observational cohort study. J Infect Dis 203, 246–257.

Lubomirov, R., Csajka, C., Telenti, A., 2007. ADME pathway approach for pharmacogenetic studies of anti-HIV therapy. Pharmacogenomics 8, 623–633.

Lubomirov, R., Di Iulio, J., Fayet, A., et al., 2010. ADME pharmacogenetics: Investigation of the pharmacokinetics of the antiretroviral agent lopinavir coformulated with ritonavir. Pharmacogenet Genomics 20, 217–230.

Mallal, S., Nolan, D., Witt, C., et al., 2002. Association between presence of HLA-B*5701, HLA-DR7, and HLA-DQ3 and hypersensitivity to HIV-1 reverse-transcriptase inhibitor abacavir. Lancet 359, 727–732.

Mallal, S., Phillips, E., Carosi, G., et al., 2008. HLA-B*5701 screening for hypersensitivity to abacavir. N Engl J Med 358, 568–579.

Manolio, T.A., Collins, F.S., Cox, N.J., et al., 2009. Finding the missing heritability of complex diseases. Nature 461, 747–753.

Marazzi, M.C., Palombi, L., Nielsen-Saines, K., et al., 2011. Extended antenatal use of triple antiretroviral therapy for prevention of HIV-1 mother-to-child transmission correlates with favourable pregnancy outcomes. AIDS 25, 1611–1618.

Markwalder, J.A., Christ, D.D., Mutlib, A., Cordova, B.C., Klabe, R.M., Seitz, S.P., 2001. Synthesis and biological activities of potential metabolites of the non-nucleoside reverse transcriptase inhibitor efavirenz. Bioorg Med Chem Lett 11, 619–622.

Martin, A.M., Krueger, R., Almeida, C.A., et al., 2006. A sensitive and rapid alternative to HLA typing as a genetic screening test for abacavir hypersensitivity syndrome. Pharmacogenet Genomics 16, 353–357.

Martin, A.M., Nolan, D., Gaudieri, S., et al., 2004. Predisposition to abacavir hypersensitivity conferred by HLA-B*5701 and a haplotypic Hsp70-Hom variant. Proc Natl Acad Sci USA 101, 4180–4185.

Martin, A.M., Nolan, D., James, I., et al., 2005a. Predisposition to nevirapine hypersensitivity associated with HLA-DRB1*0101 and abrogated by low CD4 T-cell counts. AIDS 19, 97–99.

Martin, A.M., Nolan, D., Mallal, S., 2005b. HLA-B*5701 typing by sequence-specific amplification: Validation and comparison with sequence-based typing. Tissue Antigens 65, 571–574.

Martinez, E., Milinkovic, A., Buira, E., et al., 2007. Incidence and causes of death in HIV-infected persons receiving highly active antiretroviral therapy compared with estimates for the general population of similar age and from the same geographical area. HIV Med 8, 251–258.

Mateo Leach, I., Van der Harst, P., De Boer, R.A., 2010. Pharmaco-epigenetics in heart failure. Curr Heart Fail Rep 7, 83–90.

Metry, D.W., Lahart, C.J., Farmer, K.L., Hebert, A.A., 2001. Stevens–Johnson syndrome caused by the antiretroviral drug nevirapine. J Am Acad Dermatol 44, 354–357.

Metzker, M.L., 2010. Sequencing technologies – the next generation. Nat Rev Genet 11, 31–46.

Mosteller, M., Hughes, A., Warren, L., et al., 2006. Pharmacogenetic (PG) investigation of hypersensitivity to Abacavir. In: The 16th International AIDS Conference. International AIDS Society, Toronto, Canada. Abstract 76.

Mouly, S.J., Matheny, C., Paine, M.F., et al., 2005. Variation in oral clearance of saquinavir is predicted by CYP3A5*1 genotype but not by enterocyte content of cytochrome P450 3A5. Clin Pharmacol Ther 78, 605–618.

Mukonzo, J.K., Roshammar, D., Waako, P., et al., 2009. A novel polymorphism in ABCB1 gene, CYP2B6*6 and sex predict single-dose efavirenz population pharmacokinetics in Ugandans. Br J Clin Pharmacol 68, 690–699.

Mulligan, K., Tai, V.W., Schambelan, M., 1997. Cross-sectional and longitudinal evaluation of body composition in men with HIV infection. J Acquir Immune Defic Syndr Hum Retrovirol 15, 43–48.

Mutlib, A.E., Gerson, R.J., Meunier, P.C., et al., 2000. The species-dependent metabolism of efavirenz produces a nephrotoxic glutathione conjugate in rats. Toxicol Appl Pharmacol 169, 102–113.

Nelson, M.R., Katlama, C., Montaner, J.S., et al., 2007. The safety of tenofovir disoproxil fumarate for the treatment of HIV infection in adults: The first 4 years. AIDS 21, 1273–1281.

Nglazi, M.D., Lawn, S.D., Kaplan, R., et al., 2011. Changes in programmatic outcomes during 7 years of scale-up at a community-based antiretroviral treatment service in South Africa. J Acquir Immune Defic Syndr 56, e1–e8.

Noor, M.A., Lo, J.C., Mulligan, K., et al., 2001. Metabolic effects of indinavir in healthy HIV-seronegative men. AIDS 15, F11–F18.

Noor, M.A., Parker, R.A., O'Mara, E., et al., 2004. The effects of HIV protease inhibitors atazanavir and lopinavir/ritonavir on insulin sensitivity in HIV-seronegative healthy adults. AIDS 18, 2137–2144.

Orkin, C., Wang, J., Bergin, C., et al., 2010. An epidemiologic study to determine the prevalence of the HLA-B*5701 allele among HIV-positive patients in Europe. Pharmacogenet Genomics 20, 307–314.

Palella Jr., F.J., Baker, R.K., Moorman, A.C., et al., 2006. Mortality in the highly active antiretroviral therapy era: Changing causes of death and disease in the HIV outpatient study. J Acquir Immune Defic Syndr 43, 27–34.

Panel on Antiretroviral Guidelines for Adults and Adolescents, 2011. Guidelines for the Use of Antiretroviral Agents in HIV-1-Infected Adults and Adolescents. Department of Health and Human Services, <http://www.aidsinfo.nih.gov/ContentFiles/AdultandAdolescentGL.pdf>, pp. 19–20.

Parczewski, M., Leszczyszyn-Pynka, M., Wnuk, A., et al., 2010. Introduction of pharmacogenetic screening for the human leucocyte antigen (HLA) B*5701 variant in Polish HIV-infected patients. HIV Med 11, 345–348.

Pasanen, M.K., Neuvonen, P.J., Niemi, M., 2008. Global analysis of genetic variation in SLCO1B1. Pharmacogenomics 9, 19–33.

Patel, S.M., Johnson, S., Belknap, S.M., et al., 2004. Serious adverse cutaneous and hepatic toxicities associated with nevirapine use by non-HIV-infected individuals. J Acquir Immune Defic Syndr 35, 120–125.

Peyrieere, H., Nicolas, J., Siffert, M., et al., 2001. Hypersensitivity related to abacavir in two members of a family. Ann Pharmacother 35, 1291–1292.

Phillips, E., Mallal, S., 2009. Successful translation of pharmacogenetics into the clinic: The abacavir example. Mol Diagn Ther 13, 1–9.

Phillips, E., Rauch, A., Nolan, D., et al., 2006. Pharmacogenetics and clinical characterization of patch test confirmed patients with abacavir hypersensitivity. Rev Antivir Ther 3, 57.

Phillips, E., Rauch, A., Nolan, D., et al., 2007. Genetic characterization of patients with MHC class I mediated abacavir hypersensitivity reaction. In: The 4th IAS Conference on Pathogenesis, Treatment and Prevention. International AIDS Society, Sydney, Australia. Abstract 49.

Phillips, E.J., Sullivan, J.R., Knowles, S.R., Shear, N.H., 2002. Utility of patch testing in patients with hypersensitivity syndromes associated with abacavir. AIDS 16, 2223–2225.

Phillips, E.J., Wong, G.A., Kaul, R., et al., 2005. Clinical and immunogenetic correlates of abacavir hypersensitivity. AIDS 19, 979–981.

Poggi, H., Vera, A., Lagos, M., et al., 2010. HLA-B*5701 frequency in Chilean HIV-infected patients and in general population. Braz J Infect Dis 14, 510–512.

Purnell, J.Q., Zambon, A., Knopp, R.H., et al., 2000. Effect of ritonavir on lipids and post-heparin lipase activities in normal subjects. AIDS 14, 51–57.

Quirk, E., McLeod, H., Powderly, W., 2004. The pharmacogenetics of antiretroviral therapy: A review of studies to date. Clin Infect Dis 39, 98–106.

Rathbun, R.C., Rossi, D.R., 2002. Low-dose ritonavir for protease inhibitor pharmacokinetic enhancement. Ann Pharmacother 36, 702–706.

Rauch, A., Nolan, D., Martin, A., et al., 2006. Prospective genetic screening decreases the incidence of abacavir hypersensitivity reactions in the Western Australian HIV cohort study. Clin Infect Dis 43, 99–102.

Reeves, I., Churchill, D., Fisher, M., 2006. Screening for HLA-B*5701 reduces the frequency of abacavir hypersensitivity reactions. Antiviral Ther 11, S1–S192.

Ribaudo, H.J., Liu, H., Schwab, M., et al., 2010. Effect of CYP2B6, ABCB1, and CYP3A5 polymorphisms on efavirenz pharmacokinetics and treatment response: An AIDS Clinical Trials Group study. J Infect Dis 202, 717–722.

Riska, P., Lamson, M., MacGregor, T., et al., 1999. Disposition and biotransformation of the antiretroviral drug nevirapine in humans. Drug Metab Dispos. 27, 1488–1495.

Ritchie, M.D., Haas, D.W., Motsinger, A.A., et al., 2006. Drug transporter and metabolizing enzyme gene variants and non-nucleoside reverse-transcriptase inhibitor hepatotoxicity. Clin Infect Dis 43, 779–782.

Rodriguez-Novoa, S., Martin-Carbonero, L., Barreiro, P., et al., 2007. Genetic factors influencing atazanavir plasma concentrations and the risk of severe hyperbilirubinemia. AIDS 21, 41–46.

Rodriguez-Novoa, S., Labarga, P., Soriano, V., et al., 2009a. Predictors of kidney tubular dysfunction in HIV-infected patients treated with tenofovir: A pharmacogenetic study. Clin Infect Dis 48, e108–e116.

Rodriguez-Novoa, S., Labarga, P., Soriano, V., 2009b. Pharmacogenetics of tenofovir treatment. Pharmacogenomics 10, 1675–1685.

Rotger, M., Bayard, C., Taffe, P., et al., 2009. Contribution of genome-wide significant single-nucleotide polymorphisms and

antiretroviral therapy to dyslipidemia in HIV-infected individuals: A longitudinal study. Circ Cardiovasc Genet 2, 621–628.

Rotger, M., Colombo, S., Furrer, H., et al., 2005a. Influence of CYP2B6 polymorphism on plasma and intracellular concentrations and toxicity of efavirenz and nevirapine in HIV-infected patients. Pharmacogenet Genomics 15, 1–5.

Rotger, M., Taffe, P., Bleiber, G., et al., 2005b. Gilbert syndrome and the development of antiretroviral therapy-associated hyperbilirubinemia. J Infect Dis 192, 1381–1386.

Saag, M., Balu, R., Phillips, E., et al., 2008. High sensitivity of human leukocyte antigen-b*5701 as a marker for immunologically confirmed abacavir hypersensitivity in white and black patients. Clin Infect Dis 46, 1111–1118.

Sanchez-Giron, F., Villegas-Torres, B., Jaramillo-Villafuerte, K., et al., 2011. Association of the genetic marker for abacavir hypersensitivity HLA-B*5701 with HCP5 rs2395029 in Mexican Mestizos. Pharmacogenomics 12, 809–814.

Sanne, I., Piliero, P., Squires, K., Thiry, A., Schnittman, S., 2003. Results of a phase 2 clinical trial at 48 weeks (AI424-007): A dose-ranging, safety, and efficacy comparative trial of atazanavir at three doses in combination with didanosine and stavudine in antiretroviral-naive subjects. J Acquir Immune Defic Syndr 32, 18–29.

Shapiro, M., Ward, K.M., Stern, J.J., 2001. A near-fatal hypersensitivity reaction to abacavir: Case report and literature review. AIDS Read 11, 222–226.

Shear, N.H., Milpied, B., Bruynzeel, D.P., Phillips, E.J., 2008. A review of drug patch testing and implications for HIV clinicians. AIDS 22, 999–1007.

Song, Y., Stampfer, M.J., Liu, S., 2004. Meta-analysis: Apolipoprotein E genotypes and risk for coronary heart disease. Ann Intern Med 141, 137–147.

Symonds, W., Cutrell, A., Edwards, M., et al., 2002. Risk factor analysis of hypersensitivity reactions to abacavir. Clin Ther 24, 565–573.

Talmud, P.J., Hawe, E., Martin, S., et al., 2002. Relative contribution of variation within the APOBEC3/A4/A5 gene cluster in determining plasma triglycerides. Hum Mol Genet 11, 3039–3046.

Tarr, P.E., Telenti, A., 2010. Genetic screening for metabolic and age-related complications in HIV-infected persons. F1000 Med Rep 2, 83.

Telenti, A., 2009. Time (again) for a randomized trial of pharmacogenetics of antiretroviral therapy. Pharmacogenomics 10, 515–516.

Thigpen, M.C., Kebaabetswe, P.M., Smith, D.K., et al., 2011. Daily oral antiretroviral use for the prevention of HIV infection in heterosexually active young adults in Botswana: Results from the TDF2 study. In: 6th IAS Conference on HIV Pathogenesis, Treatment and Prevention; 17–20 July 2011, Rome, Italy. International Aids Society.

Thompson, M.A., Aberg, J.A., Cahn, P., et al., 2010. Antiretroviral treatment of adult HIV infection: 2010 recommendations of the International AIDS Society-USA panel. JAMA 304, 321–333.

Tozzi, V., 2010. Pharmacogenetics of antiretrovirals. Antiviral Res 85, 190–200.

Trottier, B., Thomas, R., Nguyen, V.K., et al., 2007. How effectively HLA screening can reduce the early discontinuation of Abacavir in real life. In: The 4th IAS Conference on Pathogenesis, Treatment and Prevention. Sydney, Australia. International AIDS Society. Abstract MOPEB002.

Tsuchiya, K., Gatanaga, H., Tachikawa, N., et al., 2004. Homozygous CYP2B6 *6 (Q172H and K262R) correlates with high plasma efavirenz concentrations in HIV-1 patients treated with standard efavirenz-containing regimens. Biochem Biophys Res Commun 319, 1322–1326.

Vidal, F., Gutierrez, F., Gutierrez, M., et al., 2010. Pharmacogenetics of adverse effects due to antiretroviral drugs. AIDS Rev 12, 15–30.

Vitezica, Z.G., Milpied, B., Lonjou, C., et al., 2008. HLA-DRB1*01 associated with cutaneous hypersensitivity induced by nevirapine and efavirenz. AIDS 22, 540–541.

Vo, T.T., Ledergerber, B., Keiser, O., et al., 2008. Durability and outcome of initial antiretroviral treatments received during 2000–2005 by patients in the Swiss HIV Cohort Study. J Infect Dis 197, 1685–1694.

Walensky, R.P., Paltiel, A.D., Losina, E., et al., 2006. The survival benefits of AIDS treatment in the United States. J Infect Dis 194, 11–19.

Walli, R., Herfort, O., Michl, G.M., et al., 1998. Treatment with protease inhibitors associated with peripheral insulin resistance and impaired oral glucose tolerance in HIV-1-infected patients. AIDS 12, F167–F173.

Wang, L., McLeod, H.L., Weinshilboum, R.M., 2011. Genomics and drug response. N Engl J Med 364, 1144–1153.

Waters, L.J., Mandalia, S., Gazzard, B., Nelson, M., 2007. Prospective HLA-B*5701 screening and abacavir hypersensitivity: A single centre experience. AIDS 21, 2533–2534.

Wester, C.W., Okezie, O.A., Thomas, A.M., et al., 2007. Higher-than-expected rates of lactic acidosis among highly active antiretroviral therapy-treated women in Botswana: Preliminary results from a large randomized clinical trial. J Acquir Immune Defic Syndr 46, 318–322.

Wester, C.W., Thomas, A.M., Bussmann, H., et al., 2010. Non-nucleoside reverse transcriptase inhibitor outcomes among combination antiretroviral therapy-treated adults in Botswana. AIDS 24 (Suppl 1), S27–S36.

World Health Organization, 2010. Recommendations for use of antiretroviral drugs for treating pregnant women and preventing HIV infection in infants: Guidelines on care, treatment and support for women living with HIV/AIDS and their children in resource-constrained settings. <http://www.who.int/hiv/pub/mtct/guidelines/en/>.

World Health Organization, UNICEF, UNAIDS, 2010. Towards Universal Access: Scaling Up Priority HIV/AIDS Interventions in the Health Sector. Progress Report. World Health Organization, Geneva.

Yanik, E.L., Lucas, G.M., Vlahov, D., Kirk, G.D., Mehta, S.H., 2010. HIV and proteinuria in an injection drug user population. Clin J Am Soc Nephrol 5, 1836–1843.

Young, B., Squires, K., Patel, P., et al., 2008. First large, multicenter, open-label study utilizing HLA-B*5701 screening for abacavir hypersensitivity in North America. AIDS 22, 1673–1675.

Yuan, J., Guo, S., Hall, D., et al., 2011. Toxicogenomics of nevirapine-associated cutaneous and hepatic adverse events among populations of African, Asian, and European descent. AIDS 25, 1271–1280.

Zucman, D., Truchis, P., Majerholc, C., Stegman, S., Caillat-Zucman, S., 2007. Prospective screening for human leukocyte antigen-B*5701 avoids abacavir hypersensitivity reaction in the ethnically mixed French HIV population. J Acquir Immune Defic Syndr 45, 1–3.

RECOMMENDED RESOURCES

Food and Drug Administration (FDA)/Center for Drug Evaluation and Research, Office of Communications, Division of Information Services

http://www.accessdata.fda.gov/scripts/cder/drugsatfda/index.cfm (the Drugs@fda website; updated daily) information for abacavir: "ziagen" (NDA # 020977)

Glossary

α-MSH α-melanocyte-stimulating hormone; a peptide produced by post-translational processing of proopiomelanocortin. α-MSH binding to melanocortin receptor 1 (MC1R) triggers melanocytes to switch production of melanin from red/yellow melanin (pheomelanin) to brown/black melanin (eumelanin).

Abl Abelson murine leukemia viral oncogene homolog; a cytoplasmic and nuclear receptor tyrosine kinase that is important in cell differentiation, cell division, cell adhesion, and cellular response to stress. The *ABL1* gene codes for a proto-oncogene, however a common translocation leads to fusion of *ABL1* with another gene (*BCR*). The resulting fusion protein (Abl-Bcr) is linked with development of chronic myeloid leukemia.

Absolute risk The expected frequency of an event in a defined group of people during a specified period of time (incidence). When assigned to an individual person, the probability of experiencing the event within a specified period of time.

ACCE Analytical validity; clinical validity; clinical utility; and ethical, legal, and social issues; a framework for evaluation of family history and genomic tests.

Accountable care organization (ACO) A local healthcare organization and a related set of providers (at a minimum, primary care physicians, specialists, and hospitals) that can be held accountable for the cost and quality of care delivered to a defined population [Dove et al. (2009). *J Am Coll Cardiol* 54, 985–988].

Accurate-mass-and-time-tag alignment Methodology used in LC/MS/MS (the coupling of liquid chromatography with tandem mass spectrometry) analyses to align peptides between different samples.

Action potential An increase in the intracellular potential lasting approximately 300 ms in cardiac myocytes, that results from altered permeability of transmembrane ion channels (K^+, Na^+, Ca^{2+}) and acts as the initiating event for each heart beat.

Activated B cell A lymphocyte, or B cell, that has progressed through the germinal center and is following the pathway of differentiation into a plasma cell. Lymphomas that arise in this post-germinal center area retain features of activated B cells.

Activating *BRAF* mutation The V600E point mutation in the *BRAF* oncogene that may also confer resistance of relapsed colorectal cancer to anti-EGFR antibodies and is also the target of the drug vemurafenib (PL-4032) for the treatment of *BRAF* mutated malignant melanoma and papillary thyroid cancer.

Activating *EGFR* mutation The sequencing of the epidermal growth factor receptor (*EGFR*) gene in patients with non-small-cell lung cancer is designed to detect certain point mutations that are specifically associated with sensitivity to the small molecule anti-EGFR signaling drugs erlotinib and gefitinib. Both the absence of mutations and certain point mutations in the *EGFR* gene are associated with resistance to the drugs.

Activating *KRAS* mutation A point mutation in codon 12 or 13 in the *KRAS* oncogene that is associated with resistance of relapsed/metastatic colorectal cancer (CRC) to anti-EGFR antibody therapeutics (cetuximab, panitumumab).

Acute coronary syndrome Heart attack, usually caused by coronary thrombosis with underlying atherosclerosis.

Adjuvant   A drug or other substance (such as interferon α-2b) that enhances the activity of another drug (such as chemotherapy in the treatment of cancer).

Adjuvant   A compound or formulation that is not itself immunogenic but that modifies the response of the adaptive immune system to vaccine components.

Adjuvant systemic therapy   Treatment that is provided intravenously or orally after surgical excision of a malignancy.

ADME genes   Genes affecting the absorption, distribution, metabolism, and excretion of drugs. These are critical in pharmacogenetics, as all of these factors affect the availability of the drug at the site of its activity.

Adult stem cells   Cells of a postnatal tissue that have the ability to self-renew and to differentiate into all cell types of that particular tissue.

AE   Alternative expression; a process by which many distinct mRNA molecules can be produced from a single gene locus by alternative joining of exons (splicing), or alternative usage of transcription start sites or polyadenylation sites.

AI   Allelic imbalance; production of unequal amounts of transcript from two alleles of the same gene.

Akt protein kinase   A serine/threonine protein kinase that is important in the regulation of cell survival. It is also linked to angiogenesis and tumor formation. Also known as protein kinase B (PKB).

Allele   A genetic term meaning one of two or more alternative forms of a gene at a specific chromosomal location.

Allograft   An organ transplanted from a genetically dissimilar donor of the same species.

Alpha$_1$-antitrypsin deficiency   Genetic disease where homozygotes (PiZZ) have markedly reduced antiprotease activity, resulting in increased neutrophil elastase activity in the lungs and early-onset emphysema. PiZZ is also associated with liver disease characterized by inclusion bodies made up of polymers of Z antitrypsin.

Alzheimer's disease   A progressive, irreversible dementia of unknown etiology associated with older populations. Many genetic factors have been linked to the age of onset of Alzheimer's disease.

Analyte   A substance or chemical constituent that is determined in an analytical procedure [International Society for Biological and Environmental Repositories (2008). *Cell Preserv Technol* 6, 5–58].

Analyte-specific reagents (ASRs)   Antibodies, both polyclonal and monoclonal, specific receptor proteins, ligands, nucleic acid sequences, and similar reagents, which, through specific binding or chemical reactions with substances in a specimen, are intended for use in a diagnostic application for identification and quantification of an individual chemical substance or ligand in biological specimens.

Analytical validation   The analytical procedures (including those involved in specific instruments or devices) that are used to measure a biomarker go through a separate regulatory process from qualification of the biomarkers. This process, called "analytical validation," assesses whether an analytical tool can measure the biomarker accurately and consistently, including whether it has sufficient sensitivity (i.e., can detect the biomarker at sufficiently low concentrations or resolution), and specificity (i.e., detects only that biomarker, as opposed to similar markers).

Angiogenesis   The formation of new blood vessels from pre-existing vessels. Angiogenesis is important in embryonic development and in adults during wound healing. The growth of tumor formations also relies on angiogenesis to provide the tumor with required nutrients and oxygen as well as a means to remove waste. The related term vasculogenesis is used for spontaneous blood-vessel formation.

Annotation   Explanatory information associated with a biospecimen [National Cancer Institute (2007) *NCI Best Practices for Biospecimen Resources*].

Apoptosis   A natural method by which damaged or unwanted cells are eliminated. Also called programmed cell death. Apoptosis is important in cancer prevention, and when apoptotic pathways are altered due to genetic mutations, uncontrolled cell growth can ensue leading to the development of a tumor.

Area under curve   The area under the receiver operator characteristic curve can be interpreted as the discriminative ability of a test. An AUC of 1 represents a perfect test, which unequivocally distinguishes those who will get disease from those who will not. An AUC of 0.5 indicates the test is no better than a random guess.

Asthma   A clinical disorder characterized by reversible airflow obstruction, wheezing and shortness of breath, and airway inflammation.

Atrial fibrillation   An irregularly irregular rhythm originating from the atria of the heart, that causes palpitations and leads to increased risk of stroke and heart failure.

Autophagy   A catabolic process through the lysosomal machinery involving the degradation of a cell's own components, or components of microbes or antigens that the cell has internalized.

Autophosphorylation   Phosphorylation (addition of a phosphate group) of a protein kinase, catalyzed by the kinase's own enzymatic activity.

BabA   Blood group antigen-binding adhesin: an outer membrane protein of *Helicobacter pylori* which binds the Lewis b blood group antigen.

Base pair Two nucleotides – adenine (A), thymine (T), guanine (G) or cytosine (C) – that form a weak bond, and are crucial to the double helix structure of DNA. In the double helix, A and T form a base pair while G and C form another base pair. These nucleotides are attached to the deoxyribose (or sugar group) of the DNA strand.

Bayes information criterion A measure of global model fit of a statistical model based on the likelihood function.

Behavior Actions, an output based on processed sensory inputs and cognition (see cognition).

Benefits In the context of elements of *informed consent* for genomic research, a statement of any benefits that may accrue from research participation, which in most cases will be societal rather than direct health benefit to participants.

Best practice A technique, process, or protocol that has been shown or is otherwise believed to be state-of-the-science in that it provides superior results to those achieved by any other technique, process, or protocol. Best practices may evolve as new evidence emerges. While best practices are consistent with all applicable ethical, legal, and policy statutes, regulations, and guidelines, they differ from guidance, policy, and law in that they are recommendations and are neither enforced nor required [National Cancer Institute (2007) *NCI Best Practices for Biospecimen Resources*].

Biomarker A biomarker is "a characteristic that is objectively measured and evaluated as an indicator of normal biological processes, pathogenic processes, or pharmacologic responses to a therapeutic intervention" [Biomarkers Definitions Working Group (2001). *Clin Pharmacol Ther* 69, 89–95]. Biomarkers come in many forms, including direct physical measurements of disease symptoms (e.g., fever as a biomarker for infection), proteins measured in the blood or spinal fluid, genetic markers, and markers derived through medical imaging technologies.

Biomarker proteomics Unbiased "omic" differential expression analysis of proteins in lysates and/or biofluids to identify changes in protein expression predictive of biomedical conditions such as disease state, disease progression, drug response, etc.

Biomarker qualification The process through which a biomarker becomes accepted by regulatory agencies for use in decision-making regarding medical product or other specific regulatory assessments. Biomarker qualification involves developing sufficient data to justify the use of the biomarker for a specific purpose or in specific circumstances, i.e., a *context of use.*

Biosignature A selected group of genes, derived from class prediction or modular analyses, that is characteristic of a specific condition, and can be used as a diagnostic tool that encompasses our ability to predict the onset, rate of progression, and response to therapy and/or clinical outcome with greater reproducibility and reliability than standard biomarkers.

Biospecimen A quantity of tissue, blood, urine, or other human-derived material. A single biopsy may generate several biospecimens, including multiple paraffin blocks or frozen biospecimens. A biospecimen can comprise subcellular structures, cells, tissue (e.g., bone, muscle, connective tissue, or skin), organs (e.g., liver, bladder, heart, or kidney), blood, gametes (sperm and ova), embryos, fetal tissue, or waste (urine, feces, sweat, hair and nail clippings, shed epithelial cells, or placenta). Portions, or aliquots, of a biospecimen are referred to as samples [National Cancer Institute (2007) *NCI Best Practices for Biospecimen Resources*].

Biospecimen collection In the context of elements of *informed consent* for genomic research, an explanation of how biospecimens will be obtained (e.g., through a blood draw or biopsy specifically for research purposes, or collection of specimens left over following a procedure for a clinical purpose).

Brain connectivity A measure of structural and/or functional connectivity between brain regions. The former uses methods such as diffusion-tensor imaging to infer white-matter tracts between regions of the brain. The latter employs functional MRI to reveal temporally correlated regions of the brain, either during a task or at rest. These metrics can be used as endophenotypes in neuropsychiatric disease.

Brugada syndrome A condition with autosomal dominant inheritance, characterized by a right bundle branch block pattern and abnormal ST segment elevation on the surface electrocardiogram, ventricular tachycardia and fibrillation, fainting, and sudden cardiac death.

Cadherin A family of calcium (Ca^{2+})-dependent transmembrane proteins important in cell adhesion. Classical cadherins include E-cadherin, expressed primarily in epithelial cells, and N-cadherin, primarily expressed in neurons. Other, non-classical cadherins include VE-cadherin, expressed in vascular endothelial cells, protocadherins, expressed primarily in the nervous system, and desmosomal cadherins.

cagPAI The cagA pathogenicity island; part of the genome of *Helicobacter pylori* and one of the major determinants of its virulence.

Calcium-channel blockers A class of antihypertensive drugs, that act as vasodilators and are used in systemic sclerosis patients because of these properties in normotensive subjects.

Calibration Assessment of whether a predicted event rate matches the observed event rate.

Candidate gene   A single gene picked based on known biology that is then related via single nucleotide polymorphism (SNP) variants to a phenotype of interest.

Care transitions   The movements patients make between care settings and healthcare practitioners as their conditions and care needs change during the course of a chronic or acute illness.

Cascade genetic testing   Genetic testing is initiated with a proband and generally involves thorough analysis of target gene(s) and putative mutations; then the identified familial mutation is offered to related individuals. This strategy limits the complexity and cost of subsequent genetic testing in relatives, as single-site analysis is typically sufficient.

CCR5 receptor   The C-C chemokine receptor type 5; a transmembrane protein located on a variety of white blood cells including lymphocytes, macrophages, and microglial cells. Several soluble protein ligands can bind to and activate this receptor as part of inflammatory processes. The human immunodeficiency virus (HIV) uses CCR5 as a receptor for cell entry.

Cdk4   Cyclin-dependent kinase 4; a serine/threonine protein kinase important in the regulation of the cell cycle. Mutations in the *CDK4* gene are common in cancer, and can lead to aberrant proliferation as well as genomic and chromosomal instability.

CDS system   A system that provides "clinical decision support" to its end-users. A CDS system may be a module within a broader clinical software application, such as an electronic health record system or a computerized provider order entry system.

Cell cycle   A series of events that occurs in a cell leading to cell growth and division.

Cellular senescence   A state of growth arrest limiting the replicative life span of a cell.

Centroid   A term used in genomics referring to the average gene expression across a set of samples.

Chemokine   A regulatory protein produced by the cells of the immune system during inflammation, which acts to mobilize or activate white blood cells.

Chemoprevention   The use of drugs, vitamins, or other agents to try to reduce the risk of, or delay the development or recurrence of cancer.

Chemoresistance   Absence of a response to treatment with chemotherapy due to factors present in the cell that render it resistant to the cytotoxic actions. Chemoresistance can be inherent to the cell or acquired in response to treatment with the chemotherapy reagent.

Chemotherapy   A chemical reagent that is used in the treatment of disease, especially cancer, as a means to kill a cell – a cytotoxic chemical therapy.

Chromosome   The organized package of DNA found in the nucleus of the cell. Chromosomes come in pairs, and human cells contain 23 chromosome pairs.

CLIA   Clinical Laboratory Improvement Amendments; the Centers for Medicare and Medicaid Services (CMS) regulate all laboratory testing (except research) performed on humans in the US through CLIA.

Clinical decision support (CDS)   The act of providing clinicians, patients, and other healthcare stakeholders with pertinent knowledge and/or person-specific information, intelligently filtered or presented at appropriate times, to enhance health and healthcare.

Clinical epidemiology   The application of epidemiologic methods to the critical evaluation of diagnostic and therapeutic interventions in clinical practice.

Clinical workflow   An established process describing a series of tasks, how they are accomplished, by whom, in what sequence, and at what priority. It accomplishes a defined step in an activity in the clinical care of patients.

Clonal   Clone; a cell or group of cells that are genetically identical to a common ancestor.

Cluster randomization   A type of randomization method whereby allocation to intervention and control groups is based on groups of individuals rather than individuals.

Coagulation   A cellular and biochemical process which results in fibrin formation.

Codon   In DNA or RNA, the series of three nucleotides that codes for a specific amino acid in a protein.

Cognition   Processing of information by the brain; includes perception, learning, memory, and reasoning (see behavior).

Collisionally induced dissociation   Mass spectrometry technique used with tandem mass spectrometers to generate fragment ions from intact precursor ions to provide qualitative identifications.

Colonization   A process by which disseminated tumor cells grow to form clinically detectable metastatic lesions.

Combined antiretroviral therapy (cART)   Effective antiretroviral therapy generally combines three or four drugs from at least two different classes of antiretroviral drugs to avoid the evolution of escape mutants and to fully suppress HIV replication. These are primarily reverse transcriptase inhibitors (nucleoside and non-nucleoside; NRTIs and NNRTIs) and protease inhibitors, but also entry/fusion inhibitors, co-receptor antagonists and integrase inhibitors. cART is essential for the treatment of HIV/AIDS, since the high mutation rate of HIV allows the virus to accumulate mutations that render a single drug ineffective; however, HIV is generally unable to overcome the multiple mutations needed to escape from multiple drugs simultaneously.

Commercialization In the context of elements of *informed consent* for genomic research, a statement that the research could result in a commercial product such as a test or drug, and whether participants will share in any profits.

Companion diagnostics Biomarker tests designed to enable therapy decisions.

Comparative effectiveness research Collection and analysis of data for evaluating the benefits and harms of alternative healthcare interventions.

Comparative genomic hybridization (CGH) A method to identify DNA copy number alterations, such as gains and losses in a tumor genome. Differentially labeled tumor DNA and reference DNA are hybridized and a map showing DNA copy number alterations as a function of location in the chromosome is generated.

Computerized provider order entry (CPOE) system A clinical software application that allows a clinician to use a computer to directly enter medical orders.

Confidentiality protections In the context of elements of *informed consent* for genomic research, a description of the steps taken to protect participants' privacy and confidentiality, such as coding of specimens/data, limiting access to identifying information, material transfer and data use agreements, physical security measures, and applicable legal protections (e.g., the Genetic Information Non-Discrimination Act, Certificates of Confidentiality).

Congenital disorders of glycosylation Inherited diseases that are caused by mutations in genes primarily affecting the N-glycosylation pathway.

Congenital heart defect (CHD) Structural errors in the heart, formed during embryonic and fetal life. A complex taxonomy is based on anatomic and physiologic consequences.

Connectivity Map (cmap) A collection of genome-wide transcriptional expression data from cultured human cells treated with bioactive small molecules and simple pattern-matching algorithms that together enable the discovery of functional connections between drugs, genes, and diseases through the transitory feature of common gene-expression changes.

Constitutive Produced in relatively constant amounts in a cell without regard for the conditions or environment of the cell, such as a protein or enzyme. Oncogenes, for example, are frequently characterized by being either constitutively expressed or constitutively active.

Contacts In the context of elements of *informed consent*, for genomic research, an explanation of whom participants should contact in the event of questions or concerns.

Copy number The number of copies of a particular gene in an individual.

Copy number variation This term is generally used to define a DNA segment larger than 1 kb in length that is present at variable copy number (deletion or duplication) in comparison with a reference genome.

Correlates of protection Measurable immunological parameters/tests that provide a surrogate of vaccine efficacy. In general, the parameter/test should be a biologically plausible defense against future infection and should be proportional to vaccine efficacy as measured by direct infectious challenge.

Cost-effectiveness analysis A systematic comparison of the relative costs and outcomes of different courses of action; the perspective of the analysis determines which costs and outcomes are relevant.

Costs/payments In the context of elements of *informed consent* for genomic research, a description of any costs to the participant or the participant's insurance, as well as any payments to compensate participants (e.g., for their time, travel, inconvenience).

CPT codes The "Current Procedural Terminology" code set describes the medical, surgical, and diagnostic services for administrative, financial, and analytical purposes.

C-reactive protein A protein that is produced by the liver. Levels of c-reactive protein that are found in the blood increase in response to systemic inflammation and are used as a very general, non-specific test for inflammation.

Cyclin D1 A protein that belongs to the cyclin family of cell cycle proteins that are important in regulating the activity of cyclin-dependent kinases (CDKs) and cell cycle progression.

Cyclin-dependent kinase inhibitor 2A (*CDKN2A*) This gene encodes several proteins, two of which are particularly important in the regulation of cyclin-dependent kinase 4 (CDK4): $p16^{INK4\alpha}$ and $p14^{ARF}$.

Cytochrome P450 superfamily Liver enzyme complexes, which show substantial variability among individuals, involved in drug metabolism.

Cytogenetics Analysis of chromosome structure and rearrangements in disease.

Cytokine A regulatory, cell-signaling protein that is secreted by cells, especially of the immune system and glia, and that acts as an intercellular mediator in the generation of an immune response.

Dacarbazine   DTIC; a chemotherapy reagent used in the systemic treatment of melanoma.

De novo genome assembly   One approach to reconstructing the authentic order of bases on chromosomes after nuclear DNA has been fragmented, cloned, and sequenced. This is a required step since DNA sequencing produces only a fragmentary read of up to 1000 bp (Sanger-based capillary electrophoresis) or up to 600 bp (so-called next-generation sequencing platforms). Assembly can be confounded by the fact that the human genome is comprised (up to 50%) of repetitive sequence.

Decision analysis (modeling)   A discipline comprising the philosophy, theory, methodology, and professional practice necessary to address important decisions in a formal manner. Decision models describe the relationship among all elements of a decision – the known data (including results of predictive models), the decision, and the forecast results of the decision – in order to predict the results of decisions involving many variables. These models can be used in optimization, maximizing certain outcomes while minimizing others.

Deletion   Loss of genetic material; may be as small as a single base or large enough to encompass one or many genes.

Dermis   The inner or lower layer of the two main layers of cells that make up the skin. The dermis contains the blood vessels, lymph vessels, hair follicles, and sweat glands.

Desmin-related myopathy (DRM)   A disease belonging to the group of myofibrillar diseases characterized by muscle weakness and ventricular remodeling, including dilated cardiomyopathy.

Differentiation   The process by which a cell becomes more specialized.

Diffuse parenchyma lung disease (DPLD)   Disease located in the alveolar septa of the lung.

Dilated cardiomyopathy (DCM)   Abnormal dilatation of the (left) cardiac ventricle, usually preceded by mild hypertrophy and usually leading to heart failure.

Dimerization   The formation of one compound from two identical molecules.

Directed differentiation   Generation of somatic cell types from pluripotent stem cells using specific growth conditions and morphogenic signals to promote and induce the changes observed during normal embryonic development toward the cell type of interest.

Direct-to-consumer   Advertising and/or selling laboratory services to consumers instead of, or in addition to, healthcare providers. Direct-to-consumer advertisements are intended to raise consumer awareness and interest in obtaining testing services. Direct-to-consumer marketing, or access, refers to tests that can be obtained by consumers without the need for a healthcare provider intermediary.

Discontinuing participation   In the context of elements of *informed consent* for genomic research, a statement that participants can withdraw at any time without penalty, including any limitations on that right (e.g., that samples/data already distributed for research cannot be called back).

Discrimination   A measure of the ability to rank cases higher than non-cases, which is often measured using the area under the receiver operating characteristic curve (AUC).

DNA repair   An important cellular defense against environmental damage. DNA repair is the process by which replication errors and DNA damage are repaired.

Dormant   A term that applies to cells in a non-dividing state.

Drug disposition   The term refers to all aspects of absorption, distribution, metabolism, and excretion of drugs that affect drug concentrations in the appropriate target tissue(s).

Duplication   The presence of extra genetic material; may result in additional copies of portions of genes, or may be large enough to result in additional copies of many genes.

Duration of storage   In the context of elements of *informed consent* for genomic research, an explanation of the length of time specimens/data will be stored (which may be "unlimited").

Dyspepsia   Non-specific symptoms of heartburn or fullness.

Dystroglycanopathy   A general term used to describe a series of diseases that are caused by mutations in genes that create or modify the O-mannose-based sugar chains of a-dystroglycan that bridges the cell membrane and laminin in the extracellular matrix.

E2F1, E2F3   E2F transcription factors 1 and 3; transcription factors important in the control of the cell cycle and in the action of tumor-suppressor proteins.

EASS (east-Asia-specific sequence)   Genetic sequence in the *cag*PAI pathogenicity island found in east Asian isolates of *Helicobacter pylori*.

Echocardiography (Echo)   Production of graphical images from sonograms of the heart. Used by cardiologists as a noninvasive diagnostic procedure to study structure and function of the heart.

Eco-genetics   The study of the mechanisms of interaction of various classes of environmental exposures and behaviors with genetic variation.

Ectoderm   The outermost of the three germ layers of the embryo, which gives rise to the epidermis, nervous tissue, and mucous membranes.

EGAPP   Evaluation of Genomic Applications in Practice and Prevention; the EGAPP Working Group supports the development of a systematic process for assessing the available evidence regarding the validity and utility of genetic tests for clinical practice.

Electrocardiogram (ECG or EKG)   A graphic tracing of the variations in electrical potential caused by the excitation of the heart muscle and detected on the body surface. The normal electrocardiogram is a scalar representation that shows deflections resulting from cardiac activity as changes in the magnitude of voltage and polarity over time, and comprises the P wave, QRS complex, and T and U waves.

Electronic health record (EHR) system   A system that includes: (1) longitudinal collection of electronic health information for and about persons; (2) immediate electronic access to person- and population-level information by authorized, and only authorized, users; (3) provision of knowledge and decision support that enhance the quality, safety, and efficacy of patient care; and (4) support of efficient processes for healthcare delivery.

Electrospray ionization   Technology used to create gas-phase ions from liquid-phase ions, a technology particularly useful for the analysis of biomolecules.

Embryonic stem (ES) cells   Cells derived from the inner cell mass of blastocysts and grown *in vitro*, which have the ability to self-renew and to differentiate into all cell types of the animal.

EML4-ALK translocation   A translocation of the *EML* and *ALK* (anaplastic lymphoma kinase) genes in patients with non-small-cell lung cancer, whose detection using the fluorescence *in situ* hybridization technique has been associated with a dramatic benefit from treatment with the tyrosine kinase inhibitor crizotinib.

Emphysema   Destruction of alveolar walls as a result of connective tissue (especially elastin) degradation and cellular damage, that results in progressive airway obstruction.

Endomyocardial biopsy (EMB)   A biopsy specimen collected from the left ventricular wall of the heart via access through the jugular vein.

Endothelial vascular barrier   A term that refers to the structural integrity of the endothelial cells of the blood system.

Epidermal growth factor (EGF)   A key extracellular-receptor-binding ligand in normal biology and in disease. Overactivation of its receptor, EGFR, or related tyrosine kinase growth factor receptors, is a key feature of many cancers.

Epidermal growth factor receptor (EGFR)   A protein that drives the growth of some epithelial tumors, such as colon and lung cancers.

Epidermis   The outer or upper layer of the two layers of cells that make up the skin. The epidermis is made up of squamous cells, basal cells, and melanocytes.

Epigenetic change   A heritable change in the regulation of gene expression due to chromatin modification, but not DNA sequence change. DNA methylation and histone modification are thought to be the key mechanisms.

Epistasis   The modification of the phenotypic effect of variation at one gene by variation at one or several additional unlinked genes.

Epithelial-to-mesenchymal transition (EMT)   The conversion of a polarized epithelial cell to a mesenchymal motile cell. EMT is important in normal development and is often inappropriately activated during cancer growth, facilitating tumor invasion and metastasis.

Epitope   The smallest portion of an immunogen or pathogen to which specific immunity can be directed. An epitope can be simple (e.g., a linear peptide 9-mer presented by major histocompatibility complex class I as recognized by a $CD8^+$ T cell) or complex (e.g., a conformational structure composed of multiple protein and carbohydrate molecules as recognized by an antibody).

esiRNA   Endoribonuclease-prepared siRNA; a short interfering (siRNA) molecule synthesized through the digestion of complementary DNA (cDNA) by an RNase III class endoribonuclease.

Evidence-based medicine   Healthcare policies and practice based on systematic review and synthesis of scientific information on the effectiveness of interventions.

Ex nihilo   Latin phrase meaning "out of nothing."

Exome   The exonic portion of the genome; while the exact definition differs, it is often restricted only to "coding" exons.

Exome sequencing   Selective sequencing of the protein coding region, about 1.5%, of the human genome in one or more individuals or a pedigree with a disorder that is believed to be inherited in a Mendelian fashion to identify novel genes associated with causality. Genomics vendors provide targeted enrichment kits to "capture" just the exome for subsequent sequencing.

Exon   The coding sequence in DNA. Newly transcribed messenger RNA consists of both exons (coding sequences) and introns (non-coding sequences). Prior to translation into protein, introns are spliced out leaving only the exons.

Extracellular matrix   The matrix that is laid down by cells, in which they adhere and move.

Extracellular-signal-regulated kinase (ERK)   A widely expressed protein kinase important in several intracellular signaling pathways. ERK is important in proliferation, differentiation, and cell cycle progression.

Extravasation   The exit of tumor cells from the capillary beds into the parenchyma of an organ.

F1 progeny   Progeny that result from the outcross between two genetically distinct individuals.

False negative   Failure of a test result to identify a trait or condition, even though it is present.

False positive   A test result indicates a trait or condition is present when it is not.

Family health history   Collection of health information about blood relatives.

FDA   The Food and Drug Administration is an agency within the US Department of Health and Human Services that is responsible for protecting the public health by assuring the safety and effectiveness of human drugs, vaccines, and other biological products. They also have responsibility for various medical devices, tests, food substances, and cosmetics.

FFPE   Formalin-fixed paraffin-embedded; a method to preserve and embed biopsy tissue in paraffin so that it can be available for histopathological review and other investigative or diagnostics tests.

Field cancerization   The constellation of locoregional changes triggered by long-term or repeated exposure of a field of tissue to a carcinogenic insult (e.g., tobacco or alcohol); field cancerization may induce carcinoma, carcinoma *in situ*, or dysplasia.

Field-effect transistor (FET)   A three-terminal electronic device enabling gate control via the field effect.

G1 phase   The first growth phase of the cell cycle before DNA synthesis begins; when cell cycle progression is interrupted at this phase it is referred to as G1-arrest.

Gain-of-function mutation   A mutation in the DNA sequence that leads to the alteration of the structure of the protein encoded by the DNA, such that the protein has new or enhanced activity.

Gene   A single unit that stores information for a protein or RNA chain that has a functional purpose.

Gene expression profiling (GEP)   Quantitative analysis of gene expression levels of some or all genes expressed in a sample of tissue or cells.

Gene expression signature   A set of genes showing similar expression based on a defined cut-off value at a set significance level.

Gene product   The functional product or protein that is created from the information contained in genes.

Gene signature   A set of genes that are used together to determine prognosis or response to therapy.

Generic PCR assay   PCR assays specific for a broader taxonomic range than one species.

Genetic association study/analysis   A case-control study in which one or more genetic variants within a candidate gene are compared between people with a particular condition and unaffected individuals to seek genotype-phenotype co-occurrence more often than can be readily explained by chance. A method widely used to determine, in families, whether a genetic variant is associated with a disease or trait.

Genetic counseling   The process of helping people understand and adapt to the medical, psychological, and familial implications of genetic contributions to disease. This process integrates interpretation of family and medical histories to assess the chance of disease occurrence or recurrence; education about inheritance, testing, management, prevention, resources and research; and counseling to promote informed choices and adaptation to the risk or condition.

Genetic linkage analysis   A method used to localize genes that are generally inherited as a Mendelian disorder.

Genetic risk factor   A genetic lesion that causes a person or group of people to be particularly susceptible to a specific disease.

Genetic test   A test that examines the genetic information contained inside a person's cells, to determine if that person has or will develop a certain disease. Genetic tests also determine whether or not couples are at a higher risk than the general population for having a child affected with a genetic disorder.

Genetical genomics   An area bringing together traditional genetic analysis and gene expression studies by directly characterizing the genetic influence of gene expression. By combining information on phenotypic traits, pedigree structure, molecular markers, and gene expression, such studies can be used for estimating heritability of mRNA transcript abundances, for mapping expression quantitative trait loci (eQTL), and for inferring regulatory gene networks.

Genome scan   The process of searching for a disease gene using widely spaced markers scattered throughout the genome.

Genome-wide association study (GWAS)   A case-control study in which single nucleotide polymorphism (SNP) haplotypes across the entire genome are compared between people with a particular condition and unaffected individuals to seek haplotype-phenotype co-occurrence more often than can be readily explained by chance.

Genome-wide testing Any genetic testing that spans the genome, including array-based genotyping, exome sequencing, and whole-genome sequencing.

Genomic instability An increased tendency for the genome to acquire mutations and other lesions, which occurs when processes important in maintenance and replication of the genome are dysfunctional.

Genomic medicine The optimization of individuals' health through the use of data on a patient's genome and its downstream products, including messenger RNA, proteins, and metabolites.

Genomic test A test on germline (somatic) genes, mutated genes in a pathological tissue such as a tumor, or expression of RNA in a tissue.

Germinal center The site in the lymph node where B cells proliferate and differentiate into their mature forms.

Germinal center B cell Lymphocyte or B cell that is located in the germinal center and undergoing the process of proliferation and differentiation into a memory B cell. Lymphomas that develop at this stage retain the features of germinal center B cells.

Germline mutation A genetic lesion within the germ or reproductive cells of the body (egg or sperm), which can be passed on to subsequent generations.

Glaucoma A disease that is characterized by the progressive loss of retinal ganglion cells, associated with a distinctive type of optic neuropathy and visual field loss.

Glycomics A general term referring to a comprehensive study of glycomes, which is the aggregate of all glycan (sugar chain) structures in a cell or organism.

Glycosylphosphatidylinositol anchor (GPI-anchor) A glycolipid attached to the C-terminus of selected proteins in the lumen of the endoplasmic reticulum. This often leads to their association in lipid rafts in the plasma membrane.

GMK ganglioside vaccination A cancer vaccination which consists of a melanoma-specific antigen called GM2 ganglioside. The vaccine may stimulate the immune system to produce antibodies against GM2-ganglioside-expressing melanoma cells.

Goldilocks allele A non-synonymous coding variant of relatively important effect on phenotype and of relatively low frequency (usually 0.5 to 5% minor allele frequency) but of adequate frequency to perform a population genotyping study.

Gray (or grey) data or literature The Grey Literature Network Service (http://www.greynet.org) defines grey literature as "information produced on all levels of government, academics, business and industry in electronic and print formats not controlled by commercial publishing i.e. where publishing is not the primary activity of the producing body."

Green fluorescent protein A protein that, when exposed to blue light, emits bright green fluorescence.

Growth factors Signaling molecules that are important in the control of cell growth and differentiation.

Health Level Seven International (HL7) The global authority on standards for interoperability of health information technology.

Health technology assessment Research-based policy analysis that examines the medical, economic, social, and ethical implications of the use of a technology in healthcare.

Healthcare provider A person who delivers healthcare in a systematic and professional way to any individual in need of health services.

Heart failure with normal ejection fraction (HFNEF) A form of heart failure that occurs in the setting of normal chamber size without impaired ventricular filling but abnormal myocardial stiffness during the relaxation phase.

Hemostasis Coagulation in the context of forming fibrin to stem the loss of blood following vascular injury.

Her2-neu A protein that is highly expressed in some breast cancers; such cancers may be treated with an anti-Her2-neu therapeutic antibody such as trastuzumab.

Herd immunity Protection of non-immune members of a population mediated by immunity in the majority of the population. The effect is due to both reduced disease burden overall (fewer contacts to infect the susceptible) and reduction in the number of susceptible members of the population (infected patients transmit to fewer members of the "herd").

Heritability The proportion of variation in the phenotype attributable to genetic factors, including gene–environment interactions.

Heterogeneity Multiple and diverse genetic alterations and mutations that together lead to a single phenotype or trait such as cancer.

Heterozygote and homozygote A region on a chromosome that includes one or many bases that either differ on the two copies of the same chromosome (heterozygote) or are the same but differ from a population of individuals at the same position (homozygote). DNA sequencing and variant calling is one approach to provide unequivocal precision in determining these changes.

Hierarchical clustering Iterative agglomerative clustering method that can be used to generate gene trees and condition trees. Condition tree clusters group samples based on the similarity of their expression profiles across a specified gene list. Clustering is commonly used for the discovery of expression patterns in large datasets.

HIPAA   The Health Insurance Portability and Accountability Act of 1996 constitutes a set of US national standards for the protection of certain health information.

Histone deacetylase (HDAC)   An enzyme that deacetylates acetylated histone residues, thereby shutting off gene transcription. Corticosteroids switch off inflammatory genes by recruiting HDAC2, but this is defective in chronic obstructive pulmonary disease, accounting for the corticosteroid resistance in this disease.

Human homolog of murine Mdm2   HDM2; a nuclear protein that is a target of the transcription factor p53, which can lead to degradation of p53 by an autoregulatory negative feedback loop.

Human leukocyte antigen (HLA) system   The major histocompatibility complex in humans. The super locus contains a large number of genes related to immune system function in humans.

Human papilloma virus (HPV)   A double-stranded DNA virus of the family Papoviridae. It infects only epithelial cells in humans, such as skin and mucous membranes. HPV can be divided into two groups, low and high risk, based on oncogenic potential. The most frequent location for HPV-mediated squamous cell carcinoma in the head and neck is the oropharynx, including the tongue base and tonsils. The most common oncogenic type is HPV-16. There are more than 100 sub-types of HPV.

Hyperplasia   Abnormal increase in the number of otherwise-normal cells in an organ or tissue, leading to enlargement of the tissue.

Hypertrophic cardiomyopathy (HCM)   Unexplained cardiac hypertrophy, usually of, but not limited to, the left ventricle, characterized by variable clinical expression and wide genotypic variability.

Idiopathic   With no known cause.

Immune response   The active response of the immune system to foreign substances or antigens.

Immune senescence   The decline in immune function that occurs in older animals and humans, manifesting as decreased response to both pathogens and vaccines.

Immunogen   An intact organism, portion of an organism, or artificial construct that elicits a response from, and can induce memory in, B cells and/or T cells. An immunogen consists of one or more epitopes.

Immunohistochemistry   Technique for labeling proteins with specific antibodies on microscope slides of human tissues, such as tumors.

Immunotherapy   A treatment that acts by stimulating the immune system to fight the infection or disease.

Implantable cardioverter defibrillator (ICD)   An electrophysiological device implanted under the skin of a patient, with leads in the heart, that monitors the heart rhythm, paces abnormally slow rhythms, and converts dangerous fast rhythms (ventricular tachycardia, ventricular fibrillation) using rapid pacing or an electric shock delivered directly to the heart.

Implementation science   The scientific study of methods to promote the systematic uptake of research findings and other evidence-based practices into routine practice, and hence to improve the quality and effectiveness of health services and care [Eccles and Mittman (2006). *Implement Sci* 1, 1–3].

***In vitro* diagnostic products**   Those reagents, instruments, and systems intended to be used in the diagnosis of disease or other conditions, including a determination of the state of health, with the goal of curing, mitigating, or preventing disease or its sequelae.

***In vitro* diagnostics**   Measurement of soluble or "wet biomarkers," usually in a selected body fluid such as blood, serum, plasma, urine, or synovial fluid, and usually representing modulation of an endogenous substance in these fluids.

***In vivo* diagnostics**   Diagnostics using "dry biomarkers," usually consisting of visual analog scales, questionnaires, performed tasks, or imaging.

Incomplete penetrance   The situation in which only a portion of individuals carrying a mutation express the associated phenotype.

Indel or insertion/deletion polymorphism   A type of DNA variant where at the same locus each autosomal chromosomal copy has a different number of bases. This results in variants of unequal length in the case of a heterozygote.

Induced pluripotent stem (iPS) cells   Cells whose properties for self-renewal and differentiation are similar but not identical to those of embryonic stem cells, that were derived by reprogramming of somatic cells. They are "patient-specific" when derived from a patient's cells for personalized medicine or the study of a disease.

Inflammasome   A protein complex assembled in myeloid cells after activation by molecular pattern stimuli that promotes pro-inflammatory cytokine secretion. The composition varies depending on the stimulus, but can include caspase 1, PYCARD [N-terminal PYRIN-PAAD-DAPIN domain (PYD), C-terminal caspase-recruitment domain], and NLR (NACHT-, LRR-, and PYD-domain-containing protein) family members.

Inflammation A protective response of the body to injury or disease that involves redness, swelling, pain, and a sensation of heat.

Information collection In the context of elements of *informed consent* for genomic research, a description of the information that will be associated with the specimens, including basic information collected from participants (e.g., demographics, family health history), information obtained from medical records, and research data generated by analysis of the specimens.

Integrated delivery system (IDS) A network of healthcare providers and organizations that provides or arranges to provide a coordinated continuum of services to a defined population, and is willing to be held clinically and fiscally accountable for the clinical outcomes and health status of the population served. An IDS may own or could be closely aligned with an insurance product.

Integrative biology Enabling significant discoveries by putting measurement systems and data together.

Interferon α-2b One of the α class of interferons, which are a family of proteins produced by cells of the immune system. Interferons are thought to boost the immune response to cancer, for example, thereby reducing the growth of cancer cells.

Interleukin A cytokine that acts as a chemical messenger for the immune system. Examples include the interleukin 1 family of cytokines (IL-1β and IL-1α) and interleukin 6 (IL-6).

Internal validity A test's result is accurate relative to the actual value in a given biological sample.

Interstitial lung disease (ILD) Disease located in the subepithelial compartment of the alveolar septa of the lung.

Intravasation The entry of tumor cells into the bloodstream.

Intrinsic subtyping A term used in breast cancer genomics specifying a method of tumor classification using a gene set with small variation between biologic replicates of a tumor from the same patient, but large variation compared to tumors from different individuals.

Invasion A process that initiates metastasis and consists of changes in tumor cell adherence to the extracellular matrix, proteolytic degradation of the surrounding tissues, and motility through these tissues.

Isolated limb infusion (ILI) A procedure to deliver high doses of chemotherapy to an extremity (such as the arm or leg), that has been isolated from the rest of the body. Blood flow to and from the extremity is temporarily stopped with a tourniquet. Catheters connected to a pump are used to circulate blood through the isolated extremity. Chemotherapy is infused throughout the isolated extremity using this system of catheters.

Java A set of software products commonly used to build applications that work across computer platforms.

Keratinocytes The most common type of cell in the skin. The keratin produced by these cells provides strength to skin, hair, and nails. Keratinocytes migrate up and out from the deep basal layer of the skin, differentiating as they reach the surface.

Knowledge management A concept that comprises a range of strategies, technologies, and practices that are used in organizations or social structures to identify, create, distribute, and enable adoption of insights and experiences for a specific application.

Lactate dehydrogenase (LDH) An enzyme that is important in cellular energy production, catalyzing the conversion of lactate to pyruvate. Present in many different cell types, it is released into the blood when cells die. LDH, which is elevated in nearly all cancers, is frequently used to monitor treatment.

Large-scale data sharing In the context of elements of *informed consent* for genomic research, an explanation of sharing that may occur outside the control of the researcher/ entity that originally collected the specimens and data [e.g., dbGaP (database of genomes and phenotypes)].

LC/MS/MS The coupling of liquid chromatography with tandem mass spectrometry for quantitative and qualitative analyses of complex mixtures of analytes.

LDT Laboratory-developed tests; in-house developed and validated laboratory tests offered by CLIA-certified laboratories. Many of the currently available genetic tests are LDTs.

Learning healthcare system A healthcare system that has undergone a process of systematic organizational change that has incorporated and synthesized individual and system learning in order to improve and transform care delivery [DeBurca et al. (2000). *Int J Qual Health Care* 12, 457–458].

Leucine-rich repeat sequence A protein structural motif that forms an α/β horseshoe fold.

Linkage disequilibrium The strong correlation (due to non-random inheritance) between variants at two or more polymorphic sites in the genome.

Lipodystrophy A heterogeneous group of disorders characterized by selective or generalized atrophy of anatomical adipose tissue stores.

Locus The physical location on a chromosome of a gene or other DNA sequence.

LOINC®   The "Logical Observation Identifiers Names and Codes" provides a set of universal codes and names to identify laboratory and other clinical observations.

Long-QT syndrome   A condition with autosomal dominant (Romano–Ward) or autosomal recessive (Jervell and Lange-Nielsen) inheritance, characterized by prolongation of the QT interval on the surface electrocardiogram, arrhythmias including ventricular tachycardia with a rotating axis (torsades de pointes), fainting, and sudden cardiac death.

Loss of heterozygosity (LOH)   A deletion or mutation of the wild-type or normal allele at a heterozygous locus (region of the chromosome that harbors both a mutant allele and a normal allele).

Loss-of-function mutation   A mutation that results in loss of or reduced activity of a protein.

Low-penetrance genes   Genes in which the phenotype or trait is expressed only in a small part of the population harboring the gene or genetic lesion.

Lymphoma   A cancer that develops in the lymph nodes and cells of the immune system.

Macro-organization   The organization that represents all micro- and meso-systems. This can range from a single provider practice where the macro-organization and the microsystem are nearly coincident to a national health service (such as the UK's) where the macro-organization encompasses all care delivered for a country.

Malignant transformation   A change in a cancerous growth that renders it capable of invading and destroying neighboring tissue and invading the rest of the body.

Mammary epithelial cells   The epithelial cells originating in both the ductal and lobular structure of the mammary gland.

Mass customization   Production of personalized or custom-tailored goods or services to meet consumers' diverse and changing needs at near mass-production prices. Enabled by technologies such as computerization, the Internet, product modularization, and lean production, it portends the ultimate stage in market segmentation where every customer can have exactly what he or she wants. In the context of healthcare, this refers to individualization of the disease management or care process by customizing evidence-based best practice guidelines to meet the specific needs of a given patient, including incorporation of patient preferences.

Meaningful use   A series of criteria designed to meet requirements set by the US Congress for use of certified electronic health record technology.

Melanin   The pigment produced by melanocytes that imparts color to eyes, hair, and skin. The melanin produced consists of eumelanin (brown/black melanin) and pheomelanin (red/yellow melanin).

Melanocortin-1 receptor (MC1R)   Receptor activated by members of the melanocortin family of hormones such as MSH (melanocyte stimulating hormone). MC1R is a G-protein-coupled transmembrane receptor that controls the production of melanin. Variants in MC1R are a risk factor for melanoma.

Melanocytes   Cells in the skin that produce the pigment melanin.

Melphalan   A chemotherapy drug used to treat multiple myeloma and in-transit melanoma. It is an alkylating agent and acts by introducing crosslinks into the DNA which, in turn, lead to cell death.

Mendelian randomization   A process by which "randomly" assorted genotypic variants associated with an intermediate phenotypic trait (e.g., LDL cholesterol) are assessed for their association with a hard clinical outcome (e.g., myocardial infarction).

Meso-system   The meso-system is a primarily administrative unit that functions to support microsystems and facilitate communication between the microsystem and the macro-organization.

Meta-analysis   A formal statistical analysis that integrates findings from different studies of the same outcome.

Metagenomics   Using 16S rRNA to determine bacterial, viral, and fungal DNA fragments that make up the microbiome.

Metastasis   The spread of cancer from the primary site to distant regions in the body.

Methylation   A reaction in which a methyl group is added to another molecule. Methylation of DNA at promoter regions can alter transcription of a gene.

Methylome   The global methylation state of chromosomal DNA in a cell or tissue sample. Can be measured using microarray chip-based techniques. Methylation of chromosomal DNA is one of the principal epigenetic modifications that can alter gene expression by effects on DNA structure and transcription factor binding.

Microarray analysis   A method to measure the expression of genes across the genome. Expression is measured by hybridization of labeled complementary DNA or RNA (cDNA or cRNA) from a sample to an array of specific DNA sequences covering selected regions or the whole genome.

Microelectromechanical systems (MEMS)   Micro-scale systems integrating mechanical and electronic functionalities.

MicroRNA   Small, approximately 21–25 nucleotide single-stranded RNA molecules encoded in the genome that regulate gene expression mainly through binding complementary sequences in the 3′-untranslated regions of mRNA molecules. There are at least several hundred microRNAs (also known as miRNAs and miRs) in the human genome, and although recently discovered, they play a fundamental role in the biology of the cell.

Microscopic colitis Two medical conditions that cause diarrhea: collagenous colitis and lymphocytic colitis. Both conditions are characterized by a triad of clinicopathological features: chronic watery diarrhea, normal colonoscopy, and characteristic histopathology.

Microsystem The combination of a small group of people who work together on a regular basis, or assemble as needed around the patient to provide care, and the individuals who receive that care (who can also be recognized as members of a discrete subpopulation of patients).

MIP-1α and MIP-1β Macrophage inflammatory proteins; chemokines that are important during acute inflammation for recruitment and activation of granulocytes.

miRNA-seq Massively parallel sequencing of mature miRNAs for expression profiling.

MITF Microphthalmia-associated transcription factor; a transcription factor important in the differentiation and development of melanocytes and in transcription of genes important for melanin synthesis.

Mitochondria The organelles in eukaryotic cells in which respiration and production of energy, in the form of adenosine triphosphate (ATP), occur.

Mitogen-activated and extracellular-signal-regulated protein kinase (MEK) One of a family of dual-specificity protein kinases. These act as a mitogen-activated protein kinase kinase or MAPK kinase.

Mitogen-activated protein kinase (MAPK) A serine threonine kinase that regulates differentiation and proliferation.

Molecular distance to health Molecular score that summarizes the perturbation of the whole genome and represents the molecular distance of a given sample relative to a baseline or healthy controls [Pankla et al. (2009). *Genome Biol* 10, R127].

Morphogen A small molecule or protein that directly acts on cells of a developing embryo to guide their specific differentiation in a concentration-dependent manner.

Multimodal disease classifier A mathematical model that integrates data from several sources (e.g., blood biomarkers, magnetic resonance imaging, cognitive tests) and outputs the probability of disease status or risk in an individual.

Multiple cross mapping (MCM) strategy A technique that exploits shared haplotypes among different inbred strains of mice used in genetic mapping studies to reduce the number of potential candidate genes in a given candidate interval.

Multiple testing problem When many hypotheses are concurrently tested, the likelihood of incorrectly rejecting a null hypothesis is increased.

Multistep carcinogenesis Cancer development is a multistep process, and multiple genetic changes are required before a normal cell becomes fully neoplastic. These genetic changes involve oncogenes, tumor-suppressor genes, and possibly senescence genes.

Muramyl dipeptide (MDP) A peptidoglycan constituent of both Gram-positive and Gram-negative bacteria.

Mutation analysis A diagnostic test that identifies the presence or absence of an abnormal DNA sequence.

Myc v-myc myelocytomatosis viral oncogene homolog; nuclear phosphoprotein and transcription factor important in cell cycle progression, apoptosis, and cellular transformation.

Nanomaterial A material with one dimension less than ~100 nm.

Nanoscale capillary liquid chromatography Use of liquid chromatography columns with small inner diameters (75 μm) and flow rates of ~400 nL per minute. Such columns are used to provide enhanced concentration sensitivity for electrospray ionization mass spectrometry.

Nanowire A material platform for devices with diameter less than ~100 nm.

Necrotic core Destabilizing soft, lipid-rich, arterial plaque component, devoid of supporting collagen.

Neo-adjuvant therapy Treatment that is provided before surgical excision of a malignancy.

Neural crest In development, the region bordering the juncture where the neural tube pinches off from the ectoderm. Cells of the neural crest migrate to various parts of the embryo, ultimately forming melanocytes, bones of the skull, teeth, and adrenal glands, and parts of the nervous system.

Nevus A growth on the skin formed by a cluster of melanocytes. Usually dark and raised. Common and benign nevi are not cancerous; dysplastic nevi can develop into malignant melanoma.

Newborn screening The practice of testing of blood of newborn babies for several genetic, endocrinologic, metabolic, and hematologic diseases that are not apparent at birth.

NF-κB pathway (nuclear factor kappa-light-chain-enhancer of activated B cells) A protein complex that controls the transcription of DNA. NF-κB is found in almost all animal cell types and is involved in cellular responses to stimuli such as stress, cytokines, free radicals, ultraviolet irradiation, oxidized LDL, and bacterial or viral antigens.

N-glycosylation Addition of sugar chains to proteins within the endoplasmic reticulum, linking asparagine and N-acetyl-glucosamine.

Non-nucleoside reverse transcriptase inhibitor (NNRTI) Molecule that blocks the action of HIV reverse transcriptase by binding a pocket of the reverse transcriptase and distorting the shape of the molecule.

Non-synonymous mutation A small nucleotide change in a gene resulting in an amino acid change in its protein.

NT-pro-BNP N-terminal fragment of pro-brain natriuretic peptide, which is a precursor of brain natriuretic peptide, that is secreted by heart muscle cells in response to excessive stretching. It is used as a marker for screening for pulmonary arterial hypertension, heart failure.

Nucleic acid amplification Precise, high multiplication of a DNA or RNA sequence to allow further laboratory analysis.

Nucleolus A distinct region within the nucleus of the cell that is not delimited by a membrane and in which ribosomal RNA (rRNA) is synthesized and assembled into ribosomes.

Nucleoside reverse transcriptase inhibitor (NRTI) Medication based on molecules that are converted in the cell to structures analogous to DNA bases, which are incorporated into the DNA strand transcribed by HIV polymerase (or reverse transcriptase) from HIV RNA, but block the DNA from further elongation.

Nucleotides The units that make up the structure of RNA and DNA.

Nutrigenomics Transient and evolutionary influences on the genome from dietary practices and particular nutrients.

O-6-methylguanine-DNA methyltransferase (MGMT) A DNA repair enzyme. MGMT repairs alkylated guanine in DNA by transferring the alkyl group from the DNA to a residue in the enzyme in a suicide reaction (the enzyme becomes irreversibly inactivated).

Office of the National Coordinator for Health Information Technology (ONC) In the US, the main federal entity in charge of coordinating nationwide efforts to implement and use health information technology and the electronic exchange of health information.

Oncogene A protein that controls the rate of cell growth. Under normal conditions, these genes are called proto-oncogenes and play an important role in cell division. Activating mutations are common in these genes, however, and when mutated they become oncogenes. Oncogenes increase the rate of growth of cells by either accelerating cell division or inhibiting proteins important in slowing cell division (such as tumor-suppressor proteins).

Ontology The formal modeling and representation of concepts of knowledge, the terms and other attributes used to describe these concepts, and the relationships of these concepts.

Open innovation A business model for research and product development first described by John Chesbrough in 2006. In the open innovation approach, companies and institutions source new ideas and inventions from both within and outside their walls, and use both internal and external paths to develop and market them.

Oral tolerance Local and systemic immunological tolerance in response to harmless gut antigens (commensal bacteria, food proteins).

Organ preservation Treatment modalities to achieve locoregional control without surgical resection of important anatomical structures. These modalities aim to maintain physiological functions such as speech, respiration, and swallowing without compromising the locoregional control of cancer in comparison to the more radical surgical treatment modalities. They include refinement in the surgical technique, radiotherapy, and chemoradiotherapy.

Oropharynx One of the three anatomic divisions of the pharynx, that lies posterior to the mouth and is continuous above with the nasopharynx and below with the laryngopharynx. It extends behind the oral cavity from the soft palate above to the level of the hyoid bone below and contains the soft palate, palatine and lingual tonsils, and the base of the tongue.

Oxidative stress Cellular damage caused by the excessive production of reactive oxygen species.

Oxido-reductive stress Increased amounts of reducing equivalents in excess of the normal antioxidant pathways.

p53 A tumor-suppressor protein, encoded by the gene *TP53*, that is important in the regulation of cell growth. Mutations in this gene are common in cancer and lead to unchecked growth of the cell.

PARP [poly (ADP-ribose) polymerase] inhibitors A new class of drugs causing cell death by synthetic lethality selectively only in cancer cells with defective homologous recombination DNA repair mechanism.

Pattern recognition receptor proteins Proteins expressed by cells of the innate immune system to identify pathogen-associated molecular patterns (PAMPs), which are associated with microbial pathogens or cellular stress.

Peptidoglycan A polymer consisting of sugars and amino acids that forms a mesh-like layer outside the plasma membrane of bacteria.

Personal genomic information (PGI) Individual-level information from across the genome, along with interpretations of that information derived from scientific discoveries.

Personal utility Usefulness of information, including, for example, improved comprehension or increased motivation to maintain a healthy lifestyle.

Personalized medicine The optimization of individuals' health through the use of all available patient data, including data on a patient's genome and its downstream products. By this definition, genomic medicine is a subset of personalized medicine.

Pharmacogenetics The study of inter-individual genetic variations in the response to and toxicity of medications – in short, how an individual's genetics influences the favorable or undesirable effects of a particular treatment.

Pharmacogenomics The use of knowledge about a person's genetics, and how their unique genetic make-up influences the choice and/or prescribed dosage of a particular medication or class of medications for that individual. Whereas pharmacogenetics is based on the effects of ADME genes on drug disposition and toxicities, pharmacogenomics is a multigenic approach. One significant benefit of pharmacogenomics is the use of the most efficacious and well-tolerated medication for an individual or group of individuals sharing the same genetic characteristics – thus providing the basis of personalized medicine.

phoP Member of the two-component regulatory system phoQ/phoP, which regulates the expression of genes involved in virulence and resistance to host defense antimicrobial peptides.

Phosphatidylinositol-3 kinase (PI3K) A family of lipid kinases that phosphorylate inositol lipids. They are important in proliferation, survival, cell migration, and vesicular trafficking.

Phosphatidylinositol 3,4,5-trisphosphate (PIP3) The product of phosphatidylinositol-3 kinase (PI3K) phosphorylation of phosphatidylinositol 4,5-bisphosphate. PIP3 is an important intracellular signaling component and activates pathways required for proliferation and survival.

Phosphorylation Addition of a phosphate group to a molecule.

Plaque erosion Term used for thrombosis caused by atherosclerosis without plaque rupture.

Plaque rupture Disruption of the fibrous cap over a necrotic core. It is by far the most common cause of coronary thrombosis.

Pleiotropy The situation in which a single gene affects multiple phenotypes; therefore mutations in the gene can underlie one or many of the phenotypes.

Pluripotency A cell's ability to differentiate into cell derivatives of all three primary embryonic germ layers: ectoderm, mesoderm, and endoderm.

Polymorphism A variation within a gene where two or more alleles exist at a frequency of at least 1% in the general population.

Population stratification Differences in allele frequency and trait predisposition between populations that can lead association studies to spuriously claim an allele is associated with a trait.

Post-translational modification (PTM) Chemical modification of a protein after translation (phosphorylation, glycosylation, acylation, alkylation, s-nitrosylation, proteolytic processing, etc.), which alters the function of the protein.

Predictive value (positive or negative) The possibility that a patient with a positive (or negative) biomarker value has (or does not have) the disease in question.

Preventive medicine Practice focused on protecting, promoting, and maintaining the health of individual persons and defined populations.

Primitive tubular networks Highly patterned and functional vascular channels that can be formed *in vitro* and *in vivo* by certain aggressive human melanoma cells.

Process management Planning and administering the activities necessary to achieve a high level of performance in a process, and identifying opportunities for improving quality, operational performance, and ultimately customer satisfaction [Scott et al. (2009). *Healthc Q* 13 (Spec No), 30–36].

Prognosis The likely or probable outcome of a disease, based on a patient's condition and the normal course of the disease; the chance or prospect of recovery or survival from disease.

Programmatic interface The means by which software components connect and intercommunicate.

Promoter A unique DNA sequence upstream of a gene that is important in regulating transcription of the gene. Binding of transcription factors to the promoter region initiates transcription of the gene.

Protease inhibitor (PI) A drug that inhibits the activity of the HIV protease, which is required by HIV to cleave the precursor HIV proteins produced by the infected cell to the smaller proteins required for the assembly of new HIV virions.

Proteinases Enzymes that degrade proteins and connective tissue, including neutrophil elastase and matrix metalloproteinases. Also called proteases.

Proteomics Qualitative or qualitative/quantitative analysis of complex mixtures of proteins (protein complexes, cell/tissue lysates, biofluids, etc.).

Public-private partnership (PPP) A joint initiative or venture which is funded or operated through a formal partnership of one or more governments or government agencies and one or more private sector companies. Private non-profit organizations may also be involved in PPPs.

PUD   Peptic ulcer disease; ulceration of either the stomach or the duodenum.

Pulmonary arterial hypertension   Increased pulmonary arterial pressure with mean pulmonary arterial pressure (>25 mmHg at rest, >30 mmHg during exercise), or systolic pulmonary pressure (>40 mmHg).

Pulmonary surfactant   A complex mixture of phospholipids and proteins that function to reduce surface tension at the alveolar air interface preventing alveolar collapse and having host defense activity.

Purpose   In the context of elements of *informed consent* for genomic research, a description of the purpose of the research, including an explanation that it involves genetics and a statement concerning whether or not samples/data will be stored for future research.

PZ   Plasticity zone; a zone of genetic hypervariability in *Helicobacter pylori*.

Quantitative trait locus (QTL)   A stretch of genomic DNA closely linked to the genes that underlie the trait in question.

Raf   A family of serine and threonine protein kinases that are important in the regulation of cell division, differentiation, and secretion. In response to growth factor stimulation, Ras proteins are activated, which in turn activate Raf kinases, which leads to activation of the MAP kinase signaling pathway.

Raynaud's phenomenon   A reversible vasospasm of the small arterioles and arteries of the hands, feet, and, less frequently those of other regions – tip of the nose, the earlobes, and the tongue. It may be primary, when no underlying causes for its development can be found and secondary in the context of other disease.

Receiver operator characteristic (ROC) curve   The curve obtained by plotting sensitivity (true positive rate) against specificity (false positive rate) for each possible "cut-off" in a diagnostic test, summarizing the discriminative accuracy for a continuous distribution of test results as the "area under the curve" (AUC). For a genetic profile this may be a simple threshold applied to a count of risk alleles or, usually, a product of both allele count and the disease risk attached to each allele.

Receptor   A protein on the surface of a cell that, when stimulated by binding of a specific factor, leads to distinct intracellular signaling changes. Examples include tyrosine kinase receptors, platelet-derived growth factor receptor (which binds PDGF), and vascular endothelial growth factor receptor (which binds VEGF).

RECIST   Response Evaluation Criteria In Solid Tumors; a set of published rules that define cancer patient response

to treatment: improve (respond), stay the same (stable), or worsen (progress).

Reclassification   A measure of the differences between the prediction methods currently in use and the proposed new method in classifying people into clinically relevant risk categories. A commonly used reclassification measure, the Net Reclassification Improvement, measures whether more cases are moved to higher risk categories than lower risk categories and more controls are moved to lower risk categories than higher risk categories.

Recombinant adenovirus   A replication-deficient virus used as a tool to deliver genes to cells.

Recontact   In the context of elements of *informed consent* for genomic research, when applicable, an explanation that participants might be recontacted, for example to obtain updated information, to collect a new specimen, or for recruitment into additional research.

Relapse-free survival   Survival defined by the absence of signs or symptoms of cancer.

Reprogramming   Inducing a stable change in a somatic cell's identity to the pluripotent stem cell state by initiating an embryonic stem-cell-like gene expression pattern; usually achieved by over-expression of the transcription factors OCT4, SOX2, and KLF4.

Research results   In the context of elements of *informed consent* for genomic research, a statement concerning what kind of individual research results and/or incidental findings (if any) participants will be offered and under what circumstances, as well as a statement concerning access to aggregate results.

Researcher access   In the context of elements of *informed consent* for genomic research, when materials will be stored and/or shared for future research, a description of who may be allowed to study the materials (e.g., researchers from other academic institutions, commercial companies) and the process by which access decisions will be made (e.g., scientific review, ethics review).

RESTful   **Re**presentational **S**tate **T**ransfer-**ful**: refers to a type of software architecture common in the World Wide Web that mediates transactions between web servers.

Retinoblastoma protein (Rb)   An important negative regulator of cell cycle. Acts as a tumor-suppressor by binding to and inhibiting the transcription factor E2F1, which leads to cell cycle arrest.

Risk modifier   A gene that, if mutated, is not causative of cancer by itself but can increase cancer risk when mutated together with another cancer-causing gene.

Risks   In the context of elements of *informed consent* for genomic research, a description of reasonably foreseeable risks, including physical risks associated with biospecimen collection, as well as privacy/confidentiality risks (leading to the possibility of psychosocial harm); group harm may also be possible when the research involves socially defined groups. For research involving storage of materials for future use, it may be appropriate to include a statement that there may be other risks currently unforeseen.

RNA   Ribonucleic acid; a biologic polymer of nucleic acids which serves a variety of cellular functions. Most notably, messenger RNA (mRNA) is copied from DNA as a single strand and serves as the chemical intermediate that is used to convey the amino acid sequence of proteins encoded by DNA.

RNAi   RNA interference; the use of double-stranded RNA molecules that are complementary to a specific gene and introduced into a cell for targeted gene silencing.

RNA-seq   Massively parallel sequencing of small fragments from a complementary DNA (cDNA) library, used for measuring expression, alternative expression, and allelic imbalance, and identifying single nucleotide variants.

S phase   One of the phases of the cell cycle, characterized by DNA synthesis and doubling of the genome.

SabA   Sialic-acid-binding adhesion: an outer membrane protein which binds Lewis blood group antigen x, coded for by *sabA*.

SAGE   Serial Analysis of Gene Expression; a tool for digitally analyzing gene expression patterns.

Selective/selected reaction monitoring   A technique for quantifying specific proteins using a triple quadrupole mass spectrometer, by monitoring ionized peptides and their fragments.

Semiconductor   A material with controllable conductivity between that of metals and insulators.

Sensitivity   The ability of a diagnostic test to detect low levels of an analyte, for example, 1 ng/ml; the ability of a diagnostic test to detect a trait or condition in a population.

Sepsis   Evidence of acute infection plus two or more systemic inflammatory response criteria.

Severe sepsis   Sepsis plus acute lung dysfunction (acute lung injury or acute respiratory distress syndrome), acute encephalopathy, acute liver dysfunction (acute progressive coagulopathy or elevated unconjugated serum bilirubin levels), acute renal insufficiency, or acute cardiovascular dysfunction.

Shotgun proteomics (bottom-up proteomics)
Analysis of complex proteomes via total proteolytic digestion of the proteins, followed by qualitative and quantitative analysis of the proteolytic peptides using nanoscale capillary liquid chromatography coupled to high-resolution, accurate mass tandem mass spectrometers.

shRNA   Short hairpin RNA, which can be cloned into an expression plasmid to be used for post-transcriptional gene silencing.

Sibling risk ratio   The ratio of disease incidence in siblings of an affected individual to the incidence of disease in the general population.

Signal transduction networks/pathways   Proteins that participate in a coordinated manner in transduction of external chemical stimuli to internal changes in the function of cells.

Significance threshold   The statistical significance or the probability that an observation has occurred by chance, which needs to be overcome before rejecting the null hypothesis, and hence deeming the observation an association (e.g., $P < 0.05$).

Single nucleotide polymorphism (SNP)   A DNA sequence variation consisting of a change in a single base pair. Although this change does not necessarily have an adverse effect in and of itself, it occurs frequently in the population and its presence may correlate with disease, drug response, or other phenotypes.

siRNA   Short interfering RNA; short nucleotide sequences that block the expression of specific messenger RNA (mRNA) in terms of protein synthesis.

SNOMED CT   The "Systematized Nomenclature of Medicine – Clinical Terms" provides a comprehensive clinical terminology.

SNP array   A type of DNA microarray that can detect copy number changes as well as individual genotypes due to the inclusion of single nucleotide polymorphism (SNP) probes. SNP arrays can also detect uniparental disomy, absence of heterozygosity, common descent and/or mosaicism within a given individual.

SOAP   **S**imple **O**bject **A**ccess **P**rotocol: refers to a type of software architecture common in the World Wide Web that mediates transactions between web servers.

Social network   A social structure made up of individuals (or organizations) called "nodes," which are connected by one or more specific types of interdependency, such as friendship, common interest, financial exchange, dislike, knowledge, or prestige.

Soluble receptors   Receptors that take on a soluble form and are released into the extracellular space. Soluble receptors often contain the extracellular portion of the membrane-bound receptor and therefore retain the ability to bind ligand.

Somatic mutations   Mutations in the DNA that are acquired after birth and occur in any of the cells of the body with the exception of the germ cells. These mutations are not passed on to subsequent generations.

SPARQL   SPARQL **P**rotocol and **R**DF **Q**uery **L**anguage is a query language used to recover data stored in the Resource Description Framework which underpins the Semantic Web.

Specificity   The ability of a diagnostic test to screen out signals from other than the desired analyte, for example, the test detects mutation X but not mutations Y or Z; the ability of a diagnostic test's result to correlate with only one trait or condition in a population.

Squamous cell carcinoma (SCC)   Carcinoma developing from squamous epithelium and characterized by cuboid cells and keratinization. It is the most frequent malignancy in the oral cavity and oropharynx. Squamous cells line the skin and passages of the respiratory and digestive tracts. The word "squamous" comes from the Latin *squama*, meaning "the scale of a fish or serpent."

Src tyrosine kinase   A tyrosine kinase important in proliferation, survival, and migration. It is an upstream mediator of the MAPK and PI3K signaling pathways.

Statin   A class of drug used to lower cholesterol levels by inhibiting the enzyme HMG-CoA reductase, also referred to as HMG-CoA reductase inhibitors.

Stochastic   Involving probability or chance.

Structural genomic variation   Microscopic and submicroscopic genetic variants that include deletions, duplications, insertions, inversions, and translocations of segments of DNA.

Subphenotype   For a disease with heterogeneous clinical features, cases can be sub-classified on the basis of particular disease manifestation, or phenotype. In lupus, subphenotypes include renal, cutaneous, or hematological manifestations, or the presence of particular immunological features, such as dsDNA antibody production.

Supervised machine learning approaches   Analysis to determine genes or proteins that fit a predetermined pattern.

Supervised methods of analysis   Methods of categorization in which sample labels are given to the algorithm being utilized. Algorithms use data from the labeled samples to predict categorization of unlabeled samples.

Surface plasmon resonance (SPR)   Collective electron resonances in metals that enable field modifications and are highly sensitive to the local dielectric environment.

Surfactant protein A (SP-A)   The most abundant surfactant protein, structurally homologous to a family of innate immune proteins known as collectins.

Surfactant proteins B and C (SP-B and SP-C)   Highly hydrophobic proteins present in pulmonary surfactant that contribute to the surface-tension-reducing activity of surfactant.

Synthetic lethality   Combination of defects in at least two genes or proteins which leads to cell death, while a defect in only one of these is viable.

Systemic sclerosis   A chronic, multisystem connective tissue disease, which is characterized by microangiopathy, fibrosis of the skin and internal organs, and activation of humoral as well as cellular immune responses.

Systems biology   An integrative approach that attempts to understand higher-level operating principles of living organisms, including humans.

T4SS (type IV secretion system)   A multi-protein assembly projecting from the cell wall of bacteria, used to transmit compounds into host cells.

Tag-seq   A massively parallel version of SAGE that forgoes some of the steps that made SAGE more laborious and biased.

Tandem mass spectrometry (MS/MS)   Mass spectrometry using instruments with at least two stages of mass analysis; in proteomic studies typically one stage of mass analysis provides quantitative information on intact peptides (precursor ions) whereas the second stage provides qualitative information on peptide fragments.

Temozolomide (TMZ)   An alkylating chemotherapy reagent used in the treatment of brain tumors and other types of cancer.

Thin-cap fibroatheroma   An atherosclerotic plaque containing a necrotic core separated from the lumen by a thin cap of fibrous tissue.

Thrombosis   Coagulation in the context of pathological events, including myocardial infarction and ischemic stroke.

TLR (Toll-like receptor)   A family of receptors on the cell surface initiating intracellular signaling events.

Toxicogenomics   The effects of chemical, physical, infectious, and behavioral exposures on the genome, reflected in mutations, epigenetic modifications of histones or DNA, and other somatic or transmissible changes in the functional genome.

Transcription factors   Proteins that regulate gene expression by binding to DNA at sites called promoter regions. Binding of transcription factors to these promoter regions leads to transcription or the synthesis of RNA from a unique sequence of the DNA.

Transcriptional modules   Groups of genes coordinately expressed in human diseases that share a similar biologic function. Once modules have been defined for a given biological system, statistical comparisons can be performed between study groups, on a module-by-module basis. Their analysis aids investigating the immunologic mechanisms relevant to human diseases and generates disease-specific transcriptional fingerprints that provide a stable framework for the visualization and functional interpretation of microarray data.

Transient receptor potential cation channel, M-1 (TRPM1)   A calcium-selective ion channel that is part of the transient receptor potential family of ion channel proteins.

Tumor necrosis factor-α (TNF-α) A cytokine made by white blood cells in response to a stimulus. TNF-α can act by boosting the immune response, has a range of pro-inflammatory actions, and may induce necrosis (cell death). It is used in the treatment of some types of cancer.

Tumor-suppressor proteins Proteins that are important in the regulation of cell growth – in particular, these proteins keep a check on cell division. Loss or mutation of these proteins, as often occurs in cancer, can lead to uncontrolled cell growth. Examples of tumor-suppressor proteins include p53, p14^{ARF}, and p16^{INK4a}.

Ubiquitination The addition of ubiquitin (a 76-amino-acid intracellular protein) to other proteins, thereby marking these other proteins for degradation or transport.

***UGT1A1* polymorphisms** Polymorphisms in the *UGT1A1* gene have been strongly linked to toxicity of irinotecan-based treatment for colorectal cancer. One particular polymorphism, homozygosity for the seven-repeat allele (also known as UGT1A1*28) is associated with severe diarrhea when irinotecan is administered.

Ultraviolet B (UVB) signature mutation A C→T or CC→TT transition mutation in DNA that is commonly found in response to exposure to ultraviolet B light.

Unclassified inflammatory bowel disease (IBD-U) Chronic colitis without the clear clinical endoscopic or histological features of Crohn's disease or ulcerative colitis.

Unsupervised hierarchical clustering analysis A method to categorize samples based on patterns of similarly expressed genes without prior knowledge of the underlying biologies of the samples. Hierarchical clustering divides the samples into successively smaller groups of samples.

Unsupervised machine learning approaches Analyses to characterize the components of a dataset, without the *a priori* input or knowledge of a right answer.

Upper aerodigestive tract (UADT) Surgical anatomical term used to denote the lips, oral cavity, sinonasal tract, larynx, pyriform sinus, pharynx, and cervical esophagus.

Uveal tract The middle layer of the wall of the eye.

Validation A process by which the clinical value of a biomarker is established, usually by performing phase I, II, or III studies on retrospectively or prospectively collected samples.

Value of information analysis Procedure for evaluating the potential usefulness of additional information before it has been collected or analyzed.

Variable expressivity The range of severity of phenotypic features that can occur in different individuals with the same genetic condition.

Vasculopathy Disorder of blood vessels.

Viral genomics The study of viral genomes, functions of their individual genes, and interactions of all of the genes present in a viral genome.

Vulnerable plaque Arterial plaque presumed to be dangerous by increasing the risk of thrombosis.

WSS (western-specific sequence) Genetic sequence in the *cag*PAI pathogenicity island found in Western isolates of *Helicobacter pylori*.

Xenograft The transplantation of an organ, tissue, or cells to an individual of another species.

YopP, YopM Effector proteins which function to alter host cell physiology and promote bacterial survival in host tissues.

Index

Page numbers followed by *f* indicate a figure; page numbers followed by *t* indicate tabular material.